CW00971626

GERMAN ENGLISH DICTIONARY

of technical, scientific and general terms

GERMAN ENGLISH DICTIONARY
of technical, scientific and general terms

Including
A list of atomic weights, Specific gravities,
Melting and boiling points of elements,
Abbreviations, Signs and symbols, Botanical
section and an Appendix of new words

A. WEBEL

Routledge
London and New York

First published 1930
by Routledge & Kegan Paul Ltd
Second Edition (Revised) 1937
Reprinted 1945, 1946 and 1949
Third Edition (with an Appendix of new words) 1952
Reprinted 1953, 1958, 1963, 1969, 1974, 1978 and 1982
Reprinted 1989
by Routledge
11 New Fetter Lane, London EC4P 4EE
29 West 35th Street, New York, NY 10001
Printed in Great Britain by
T.J. Press (Padstow) Ltd, Cornwall

ISBN 0 415 040 345

PREFACE

This book was compiled with the object of being purely a private work of reference, but, in the course of time, the collection increased to such an extent that friends persuaded the author to publish the work in order to supply the need for a technical and scientific dictionary and work of reference.

In now presenting this volume to the public, the author entertains the hope that it will prove to be an efficient signpost to the many ways of translation that lie open to one – to advise upon the way to be taken is impossible. This is perhaps most clearly shown by the various renderings of the German word 'viel,' of which the following are a few:

Ich habe *viele* Neuigkeiten: I have *a lot* of news.

Es waren *viele* Leute da: There were *many* people present.

Viele Leute denken: *Many* people think . . . ; *A number* of people think . . . ; *Numerous* people think . . . ; *One section* of the public thinks . . .

Es giebt *viel* Arbeitslosigkeit: There is *much* unemployment; There is *a good deal* of unemployment; There is *a considerable amount* of unemployment.

Ziemlich *viel* Wasser: A fair *quantity* of water . . .

Es giebt *viele* Möglichkeiten: There are *various* possibilities; There are *divers* possibilities; There are *sundry* possibilities.

Er ist *viel* bei uns: He is *often* at our house; He is *frequently* at our house.

For this reason, in the case of Chemical, Botanical and Mineralogical Terms, as many data as possible have been given. These data will, in many cases, help the translator to a clearer understanding of the matter which he is translating and, at the same time, will give the technical man the information he requires without the necessity of having to turn up other works of reference. Considering the many differences in opinion that exist amongst authorities themselves with regard to technical data, the information herein given must only be regarded as approximate, although the author has made every endeavour to select the most favoured view.

A. Webel

CODING TABLE

	A	E	I	O	U	Guide Number.
B	00	01	02	03	04	0
C	05	06	07	08	09	
D	10	11	12	13	14	1
F	15	16	17	18	19	
G	20	21	22	23	24	2
H	25	26	27	28	29	
J	30	31	32	33	34	3
K	35	36	37	38	39	
L	40	41	42	43	44	4
M	45	46	47	48	49	
N	50	51	52	53	54	5
P	55	56	57	58	59	
R	60	61	62	63	64	6
S	65	66	67	68	69	
T	70	71	72	73	74	7
V	75	76	77	78	79	
W	80	81	82	83	84	8
X	85	86	87	88	89	
Y	90	91	92	93	94	9
Z	95	96	97	98	99	

For coding purposes, every word has been provided with a number in two parts. The first part of the number is given, under the designation of "Code Indicator," at the top of each page. It consists of the ten-thousand and thousand series of a five-figure number. The second portion—the hundreds, tens and units—of the five-figure number, is given opposite to each word. As an example will be found, on page 1 on the left-hand side at the bottom, the word "ABBITTEN," which is numbered 00034, i.e., Code Indicator 00 and word number 034.

As words of ten letters can be transmitted in code, two references from the dictionary may be combined by means of the Coding table.

Example: "Angebot Motor heute drahten."

On page 21 will be found the word "ANGEBOT" with a Code Indicator of 01 and a word number 693, i.e., a total number of 01693. The same procedure having been adopted with the other words, the result may be written down as shown below, and this may then be typed out as a telegram confirmation.

01693	ANGEBOT	:	QUOTATION	0169334394
34394	MOTOR	:	MOTOR	
20669	HEUTE	:	TODAY	2066909661
09661	DRAHTEN	:	TELEGRAPH	

The total figures show the two numbers in the form in which they can be coded.

It will be seen from the number 0169334394 that it consists of five groups of two figures. These groups may be converted into letters by means of the Coding table.

It should be noted that the first two lines of figures in the Coding table all commence with 0, the next two lines with 1 and so on. Under the Guide number therefore, may be seen at a glance where the required number is to be found, thus saving time. In order to code words, the resultant groups of letters should be written down commencing with the consonants, for, as will be seen later, the vowel may be used to commence the groups when a number is to be coded. By means of the Coding Table we obtain: 01, BE; 69, SU; 33, JO; 43, LO; 94, YU, and these letters, brought together to form one complete word, read "BESUJOLOYU." The other number, therefore, results in "GASEYAZERE."

The process of decoding is, of course, the inverse of the above.

The rule for coding numbers is:

If the first half of the ten-figure code number represent a number instead of a word, the resultant code word should commence with a vowel. If the second half also represent a number, the last group of two letters of the code word should be written with the vowel first and the consonant last. By this means it is possible to combine a five-figure number and a reference from the dictionary in one code word, irrespective of whether the number is to come first or last. Should, however, the number consist of more than five figures, then a complete code word must be used to transmit it. Thus 314157 would have to be written 0000314157 in order to obtain ten letters for transmission as the code word "ABBAJELEIP." The commencement with a vowel and the ending with a consonant shows, at a glance, that the whole word represents a number.

By taking the above telegram and extending it a little, one might telegraph as follows:

"ANGEBOT ZWEI MOTOREN EINEM ANLASSER HEUTE DRAHTEN."

which would give:

01693	ANGEBOT	:	QUOTATION	0169300002	:	BESUJABAIB
00002	ZWEI	:	TWO			
34394	MOTOR-(EN)	:	MOTOR(S)	3439400001	:	JUKULABAEB
00001	EIN-(EM)	:	ONE			
01841	ANLASSER	:	STARTER	0184120669	:	BEWUDICESU
20669	HEUTE	:	TODAY			
09661	DRAHTEN	:	TELEGRAPH	———	:	DRAHTEN

If it be not possible to add a further word to the telegram, the last word must be transmitted in plain language.

When coding or decoding it must always be borne in mind that, where the context demands it, the singular is also to be understood as the plural, as for example in the above telegram where "Motor" should, in reality, read "Motors" as there are two of them.

After a minimum of practice it should be an easy matter to compile a suitable telegram.

VORWORT

Die Grundlage zu diesem Buche wurde in zwanzigjähriger Arbeit geschaffen. Die Anfänge reichen zurück bis ins Jahr 1907, in dem der Verfasser aus privatem Sammlerinteresse mit der Sammlung begann, ohne zunächst an eine Herausgabe zu denken. Im Laufe der Jahre wuchs die Sammlung dann zu dem jetzt vorliegenden Umfang an, wobei der Verfasser sich bemühte, sie immer wieder auf den neuesten Stand zu bringen, und Freunde rieten ihm dann an, die Sammlung herauszugeben, weil damit einem fühlbaren Mangel abgeholfen werde. So entstand dieses Buch.

Indem der Verfasser nun sein Werk der Öffentlichkeit überreicht, hegt er die Hoffnung, dass es ein deutlicher Wegweiser zu den mannigfachen Möglichkeiten der Übersetzung sein wird. Einen Rat über den bei der Übersetzung einzuschlagenden Weg zu geben, ist unmöglich. Das ist vielleicht am klarsten ersichtlich aus den verschiedenen Übersetzungsmöglichkeiten des deutschen Wortes "viel," wovon einige hier angegeben sind:

Ich habe *viele* Neuigkeiten; I have *a lot* of news.

Es waren *viele* Leute da: There were *many* people present.

Viele Leute denken: *Many* people think . . .
A number of people think . . .
Numerous people think . . .
One section of the public thinks . . .

Es giebt *viel* Arbeitslosigkeit: There is *much* unemployment.
There is *a good deal* of unemployment.
There is *a considerable amount* of unemployment.

Ziemlich *viel* Wasser: A fair *quantity* of water . . .

Es giebt *viele* Möglichkeiten: There are *various* possibilities.
There are *divers* possibilities.
There are *sundry* possibilities.

Er ist *viel* bei uns: He is *often* at our house.
He is *frequently* at our house.

Aus diesem Grunde sind so viele Einzelangaben wie möglich neben die chemischen, botanischen und mineralogischen Ausdrücke gesetzt worden. Diese Angaben werden dem Übersetzer vielfach behilflich sein, einen klaren Begriff über den zu übersetzenden Text zu erhalten, während dem Fachmann die von ihm gewünschte Unterweisung gegeben wird, ohne dass er andere Nachschlagewerke heranzuziehen braucht. Angesichts der vielen Meinungsverschiedenheiten der Autoritäten über technische Tatsachen darf man auch die in diesem Werk angeführten Angaben nur als annähernd betrachten, obschon der Verfasser sich alle Mühe gegeben hat, den bevorzugten Standpunkt auszuwählen.

A. WEBEL.

London, 1929.

KODIERTABELLE

	A	E	I	O	U	Leitnummer.
B	00	01	02	03	04	0
C	05	06	07	08	09	
D	10	11	12	13	14	1
F	15	16	17	18	19	
G	20	21	22	23	24	2
H	25	26	27	28	29	
J	30	31	32	33	34	3
K	35	36	37	38	39	
L	40	41	42	43	44	4
M	45	46	47	48	49	
N	50	51	52	53	54	5
P	55	56	57	58	59	
R	60	61	62	63	64	6
S	65	66	67	68	69	
T	70	71	72	73	74	7
V	75	76	77	78	79	
W	80	81	82	83	84	8
X	85	86	87	88	89	
Y	90	91	92	93	94	9
Z	95	96	97	98	99	

Zu Kodezwecken ist jedes Wort mit einer zweiteiligen Nummer versehen. Der erste Teil der Nummer steht unter dem Namen "Code Indicator" am Kopfe eines jeden Blattes und besteht aus den ersten zwei Ziffern einer fünfstelligen Zahl; der zweite Teil—die drei letzten Ziffern der fünfstelligen Zahl—steht rechts gegenüber jedem Wort. Als Beispiel findet man auf Seite 1, links unten das Wort "ABBITTEN," das mit 00034 nummeriert ist; hier ist also Code Indicator 00 und Wortnummer 034.

Da beim Kodieren Wörter von zehn Buchstaben in Frage kommen, kann man somit zwei Verweisungen aus dem Wörterbuch mittels der Kodiertabelle zusammensetzen. Als Beispiel nehme man: "Angebot Motor heute drahten."

Hier findet man auf Seite 21 das Wort "ANGEBOT" mit einem "Code Indicator" von 01 und einer Wortnummer von 693, also einer Gesamtzahl von 01693. Mit den andern Wörtern verfährt man ebenso und schreibt das Ergebnis—das alsdann als Telegrammbestätigung gebraucht werden kann—wie folgt nieder:

01693	ANGEBOT	:	QUOTATION	0169334394
34394	MOTOR	:	MOTOR	
20669	HEUTE	:	TODAY	2066909661
09661	DRAHTEN	:	TELEGRAPH	

Die Gesamtzahlen zeigen die zwei Zahlen in der Form, wie sie Kodiert werden sollen.

Nun zu der Kodiertabelle. Man ersieht aus der Nummer 0169334394, dass Sie aus fünf zweistelligen Gruppen besteht. Diese Gruppen setzt man jetzt mittels der Tabelle in Buchstaben um. Hierbei bemerkt man, dass die oberen zwei Zifferzeilen der Tabelle alle mit 0, die nächsten zwei mit 1 anfangen u.s.w. Unter der Leitnummer kann man daher sofort ersehen, wo die gebrauchte Zahl zu finden ist und gewinnt dadurch viel Zeit. Um Wörter zu kodieren, fängt man immer mit dem Konsonanten an, da wir den Vokal—wie wir später ersehen werden—als Anfangsbuchstaben benutzen, wenn eine Zahl kodiert werden soll.

Das Kodieren ergibt somit 01, BE ; 69, SU ; 33, JO ; 43, LO ; 94, YU, und diese Buchstaben ergeben als Wort zusammengefasst " BESUJOLOYU." Die andere Zahl ergibt daher " GASEYAZERE."

Das Auskodieren wird naturgemäss umgekehrt behandelt.

Stellt die erste Hälfte der zehnstelligen Kodezahl eine Zahl dar, dann fängt man das Kodieren nach der Tabelle mit einem Vokal an. Wenn die zweite Hälfte eine Zahl darstellt, so setzt man die letzte Gruppe von zwei Buchstaben mit dem Vokal als Anfangs- und dem Konsonanten als Endbuchstaben ein. Nach diesem Verfahren kann man eine fünfstellige Zahl mit einer Verweisung in einem Wort kodieren, in dem die Zahl zu Anfang oder zu Ende gesetzt werden kann. Sollte die Zahl dagegen mehr als fünf Stellen besitzen, so muss man ein Gesamtkodewort benutzen. Z.B. müsste man 314157 kodieren 0000314157, um zehn Buchstaben zu erhalten ; und das Wort würde lauten "ABBAJELEIP." Der Anfang mit einem Vokal und das Ende mit dem Konsonanten bedeutet ohne weiteres, dass das ganze Wort eine Zahl ist.

Nehmen wir jetzt das vorhergehende Telegrammbeispiel und dehnen wir es ein wenig aus, so könnten wir wie folgt telegraphieren: "ANGEBOT ZWEI MOTOREN EINEM ANLASSER HEUTE DRAHTEN," was folgendes ergibt :

01693	ANGEBOT	:	QUOTATION	0169300002	BESUJABAIB
00002	ZWEI	:	TWO		
34394	MOTOR-(EN)		MOTOR(S)	3439400001	JUKULABAEB
00001	EIN-(EM)	:	ONE		
01841	ANLASSER	:	STARTER	0184120669	BEWUDICESU
20669	HEUTE	:	TODAY		
09661	DRAHTEN	:	TELEGRAPH	———	DRAHTEN

Wenn man dem Telegramm nicht noch ein Wort anfügen kann, muss das letzte Wort in der nichtkodierten Form angegeben werden. Beim Kodieren und Auskodieren muss immer berücksichtigt werden, dass da, wo der Sinn es verlangt, unter der Einzahl auch die Mehrzahl zu verstehen ist, wie z.B. bei dem Wort " Motor " (s.o.), das in diesem Falle in Wirklichkeit " Motoren " heissen muss, da nach dem Wort " zwei " die Mehrzahl folgt.

Nach diesen Angaben sollte man mit dem geringsten Mass an Übung ein geeignetes Telegramm leicht zusammenstellen können.

GERMAN TECHNICAL DICTIONARY

CODE INDICATOR 00

A

AALQUAPPE (*f*): (see QUAPPE) 001
AALRAUTE (*f*): (see QUAPPE) 002
AAS (*n.*): carcase; carcass; flesh (hides); scrapings from hides 003
AASEN: to flesh (scrape) hides 004
AASKRÄHE (*f*): carrion crow (*Corvus corone*) 005
AASSCHMIERE (*f*): flesh-side dubbing (leather) 006
AASSEITE (*f*): flesh-side dubbing (leather) 007
ABÄNDERN: to vary; change; alter; modify; amend; qualify; rectify 008
ABÄNDERUNG (*f*): variant; amendment; variation; modification; etc. 009
ABÄNDERUNGSANTRAG (*m*): amendment 010
ABÄNDERUNGSVORSCHLAG (*m*): amendment 011
ABARBEITEN: to rough; rough-hew; work; work off; wear out 012
ABARTEN: to vary; degenerate 013
ABATMEN: to glow; anneal; cupel 014
ABÄTMEN: to glow; anneal; cupel 015
ABÄTZEN: to corrode; eat away 016
ABBAU (*m*): demolition; decomposition; removal; disintegration; destruction; analysis; breaking up; working (Mining) 017
ABBAUEN: to decompose; split up; demolish; disintegrate; remove; break up; analyse; work (Mines) 018
ABBAUORT (*m*): working (Mining) 019
ABBAUSTRECKE (*f*): working (Mining) 020
ABBEIZEN: to taw (skins); dress; dip; corrode; pickle (remove by caustics or corrosives) 021
ABBEIZMITTEL (*n*): corrosive 022
ABBERSTEN: to burst (spring; jump; fly) off 023
ABBESTELLEN: to countermand 024
ABBEZAHLEN: to discharge; pay off; pay in full 025
ABBIEGEN: to bend (break) off; inflect (grammar); deflect; diverge 026
ABBILDEN: to portray; paint; copy; illustrate; picture; describe 027
ABBILDUNG (*f*): portrait; copy; illustration; picture; diagram 028
ABBILDUNGSVERMÖGEN (*n*): resolving power (of a microscope) 029
ABBIMSEN: to rub with pumice stone; fluff; buff (leather) 030
ABBINDEN: to harden; set (cement); detach; unbind 031
ABBINDEZEIT (*f*): setting time (cement) 032
ABBITTE (*f*): apology 033
ABBITTEN: to apologise, beg pardon; beg from; deprecate 034
ABBITTEND: deprecative-(ly), deprecatory, deprecatorily 035
ABBITTE TUN: to apologise; beg pardon 036
ABBLASEHAHN (*m*): blow-off cock 037
ABBLASEVENTIL (*n*): blow-off valve; exhaust valve 038
ABBLÄTTERN: to skin; scale off; peel; exfoliate; flake-off; strip; shed; laminate; desquamate 039
ABBLÄUEN: to blue; dye blue (fabrics); to lose blue colour; wash out (of blue fabrics); beat; thrash 040
ABBLEICHEN: to bleach; whiten; fade; pale; become pale 041
ABBLENDEN: to screen off 042
ABBLICKEN: to brighten; give the blick to (assaying) 043
ABBLITZEN: to misfire; flash 044
ABBORKEN: to bark (of trees); (*n*) barking (of trees) 045
ABBÖSCHEN: to incline, slant, slope 046
ABBOSSIEREN: to emboss 047
ABBRAND (*m*): slag; burnt ore; loss by (combustion) burning; consumption; waste (in metal working) 048
ABBRAUEN: to brew; finish brewing 049
ABBREITEN: to flatten; stretch out 050
ABBREMSEN (*n*): braking, application of the brake 051
ABBRENNBÜRSTE (*m*): sparking contact brush; contact breaking brush 052
ABBRENNEN: to burn off; miss fire; temper (steel); dip; pickle (of metals to remove oxidation); burn away; deflagrate (chem.); corrode; cauterize; sear; singe; fire for the last time; finish firing (of kilns) 053
ABBRENNGLOCKE (*f*): deflagrating jar 054
ABBRENNKONTAKTSTÜCK (*n*): sparking contact piece 055
ABBRENNLÖFFEL (*m.*): deflagrating spoon 056
ABBREVIIEREN: to abbreviate 057
ABBRÖCKELN: to flake-off; break off in little bits; crumble 058
ABBRUCH (*m*): breaking off; damage, demolition; loss; landslide, landslip, declivity 059
ABBRUCH (*m*) ERLEIDEN: to be affected 060
ABBRÜHEN: to parboil; scald; seethe 061
ABDACHEN: to incline; dip; slope; unroof 062
ABDÄCHIG: slanting, sloping, inclined 063
ABDACHUNG (*f*): slope; dip; incline; descent; unroofing 064
ABDAMPF (*m*): exhaust steam 065
ABDAMPFDRUCKREGLER (*m*): exhaust-steam pressure regulator 066

ABDAMPFEN : to evaporate ; exhaust ; volatilize (solids) ; vapourize (see **EINDAMPFEN**) 067
ABDÄMPFEN : to throttle ; damp ; quench (charcoal) ; (see also **ABDAMPFEN**) 068
ABDAMPFENTÖLER (*m*) : exhaust-steam oil separator 069
ABDÄMPFGEFÄSS (*n*) : evaporating pan 070
ABDÄMPFKAPELLE (*f*) : evaporating capsule 071
ABDÄMPFKESSEL (*m*) : evaporating (pan) boiler 072
ABDÄMPFKOLBEN (*m*) : evaporating flask 073
ABDAMPFLEITUNG (*f*) : exhaust (piping) pipe 074
ABDÄMPFRÜCKSTAND (*m*) : residue from evaporation 075
ABDÄMPFSCHALE (*f*) : evaporating dish 076
ABDAMPFTURBINE (*f*) : exhaust steam turbine 077
ABDANKEN : to throw aside ; pay off crew (Naut.); to pension (off) ; retire 078
ABDARREN : to cure ; kiln-dry (malt) ; liquate (metal) 079
ABDARRTEMPERATUR (*f*) : drying (finishing ; curing) ; temperature (of brewing) 080
ABDECKEN : to uncover ; skin ; flay ; cover 081
ABDECKEREIANLAGE (*f*) : flaying plant 082
ABDECKEREIMASCHINE (*f*) : flaying machine 083
ABDECKPLATTE (*f*) : cover plate 084
ABDESTILLIEREN : to distil 085
ABDICHTEN : (see **DICHTEN**) 086
ABDOMEN (*n*) : abdomen, belly (Med.) 087
ABDOMINAL : abdominal 088
ABDÖRREN : to dry (up) ; roast 089
ABDÖRROFEN (*m*) : refining furnace 090
ABDRAHT (*m*) : turnings 091
ABDREHEN : to twist (turn) off ; turn (on a lathe) ; machine ; unscrew 092
ABDREHMASCHINE (*f*) : lathe ; finishing machine 093
ABDREHSPÄNE (*m.pl*) : turnings 094
ABDREHSPINDEL (*f*) : finishing machine (ceramics) 095
ABDREHSTAHL (*m*) : turning tool 096
ABDREHSTAHLVORRICHTUNG (*f*) : turning tool-(s) ; turner 097
ABDRÜCKSCHRAUBE (*f*) : set screw 098
ABDUKTOR (*m*) : abductor (muscle) (Med.) 099
ABDUNSTEN : to evaporate ; vapourize 100
ABDÜNSTEN : to evaporate ; vapourize 101
ABEICHEN : to gauge ; adjust ; measure 102
ABENDGEMÄLDE (*n*) : nocturne (Art) 103
ABENDLAND (*n*) : west ; occident 104
ABENDLÄNDISCH : occidental, western 105
ABERKENNEN : to deprive, dispossess 106
ABERMALIG : renewed ; further ; repeated 107
ABERRATION ( ) : aberration (Photography, etc.) 108
ABERRATION (*f*) **CHROMATISCHE-** : chromatic aberration (Photography) 109
ABFACHEN : to classify ; partition 110
ABFALL (*m*) : falling off ; waste ; by-products ; cuttings ; scrap 111
ABFALLEN : to fall (off) (away), drop (off) (away), decline, decrease, slope, incline (away from), droop, shed ; to contrast adversely 112
ABFALLENERGIE (*f*) : waste energy 113
ABFÄLLEVERWERTUNGSANLAGE (*f*) : plant for the utilization of waste 114
ABFALLGRENZE (*f*) : critical limit 115
ABFALLGUT (*n*) : waste washings ; mud ; fuel recovered from the waste (coal) 116
ABFALLHEFE (*f*) : waste yeast (from Breweries) 117
ABFALLLAUGE (*f*) : spent lye 118
ABFALLMARKE (*f*) : off-shade (dyeing) 119
ABFALLMOMENT (*n*) : breaking down (torque) moment 120
ABFALLPAPIER (*n*) : waste paper 121
ABFALLPRODUKT (*n*) : waste (product) : by-product ; residuary product 122
ABFALLSÄURE (*f*) : residuary acid 123
ABFALLSEIFE (*f*) : waste soap, soap waste 124
ABFALLSTOFF (*m*) : waste material ; sewage 125
ABFALLSTROM (*m*) : waste current (Electricity) 126
ABFALLWARE (*f*) : waste ; culls ; rejected ware ; throw-out 127
ABFALZEN : to remove hair (from skins) (Tanning) 128
ABFÄRBEN : to dye ; colour ; stain ; lose colour ; finish dyeing ; rub off (paper) : streak (Min.) 129
ABFASERN : to unravel ; fray ; lose fibres ; unpick 130
ABFASSEN : to draw up ; catch ; compose ; plane ; bend (iron) 131
ABFEILEN : to file off 132
ABFEILICHT (*n*) : filings 133
ABFERTIGEN : to dispatch ; finish ; complete 134
ABFERTIGUNG (*f*) : readiness, completion 135
ABFERTIGUNGSSCHEIN (*m*) : customs (permit) declaration form 136
ABFIEREN : to ease away, pay out (a rope) 137
ABFINDEN : to settle (with) ; pay (off) ; compound (with) (Commercial) ; compensate, indemnify, satisfy, come to terms 138
ABFINDUNGSGELD (*n*) : indemnity, compensation 139
ABFINDUNGSVERTRAG (*m*) : agreement 140
ABFLACHEN : to straighten, level off, even up, bevel, slope 141
ABFLACHUNG (*f*) : flat, level ; bevel (of crystals) ; rounding off (of a curve) 142
ABFLAMMEN : to grease ; tallow (hides) 143
ABFLAUEN : to wash ; rinse ; scour (ore) ; drop ; go down (of the wind) 144
ABFLECKECHT : fast to rubbing ; non-spotting 145
ABFLECKEN : to stain ; remove spots ; spot (Phot.) 146
ABFLEISCHEN : to flesh ; scrape (tanning) 147
ABFLEISCHMESSER (*n*) : scraper (for fleshing) (Tanning) 148
ABFLIESSEN : to discharge, ebb, flow away 149
ABFLIESSENDES WASSER (*n*) : discharge, waste water 150
ABFLUSS (*m*) : discharge, efflux 151
ABFLUSSKÜHLER (*m*) : efflux condenser 152
ABFLUSSRINNE (*f*) : channel, trench, gully 153
ABFLUSSROHR (*n*) : drain (waste) pipe ; discharge pipe ; gutter 154
ABFLUSS (*m*), **WASSER-** : water outlet 155
ABFLUSSWASSER (*n*) : waste water 156
ABFORMEN : to mould, form 157
ABFRAGEN : to enquire ; ascertain 158
ABFRESSEN : to eat away, corrode, remove by caustics (Metals) 159
ABFRISCHEN : to change ; renew ; steep 160
ABFUHRDÜNGER (*m*) : night soil 161
ABFÜHREN : to discharge ; lead (off) away ; exhaust (steam) ; draw (wire) ; send (goods) ; divert ; purge (Med.) 162

ABFÜHRENDES BRAUSEPULVER (*n*): Seidlitz powder (Pharm.) 163

ABFÜHREND WIRKEN: to have a purgative (effect) action, be (purgative) aperient 164

ABFÜHRGANG (*m*): secretory duct (Med.); efferent duct 165

ABFÜHRMITTEL (*n*): purge, purgative, aperient, laxative (Med.) 166

ABFÜHRMITTEL (*n*), DRASTISCHES: purge, drastic clearing agent (Med.) 167

ABFÜHRUNG (*f*): outlet, discharge, evacuation 168

ABFÜHRUNGSGANG (*m*): secretory duct, efferent duct (Med.) 169

ABFUHRWAGEN (*m*): dust-cart 170

ABFÜLLAPPARAT (*m*): emptying apparatus; racking apparatus; racker (Brewing) 171

ABFÜLLBÜTTE (*f*): racking square (Brewing) 172

ABFÜLLEN: to draw off, decant, empty, rack, skim off, drain (tap) off; fill 173

ABFÜLLKELLER (*m*): racking cellar (brewing) 174

ABFÜLLMASCHINE (*f*): filling machine, bottling machine 175

ABFÜLLRAUM (*m*): racking (decanting, bottling, filling) room (Brewing) 176

ABFÜLLSCHLAUCH (*m*): racking hose (Brewing) 177

ABFURCHEN: to furrow off, divide (separate) by furrows 178

ABFURCHUNG (*f*): segmentation; division 179

ABFÜTTERN: to line, case (Mech.); feed (cattle) 180

ABFUTTERN: to line, case (Mech.); feed (cattle) 181

ABGABE (*f*): delivery; giving off; escape (of gas); tax; duty; output; generation (of steam); draft (commerce) 182

ABGANG (*m*): loss; waste; exit; escape; departure; sale; market; miscarriage 183

ABGANGPAPIER (*n*): waste paper 184

ABGANGSDAMPF (*m*): exhaust steam 185

ABGANGSZEUGNIS (*n*) AUSSTELLEN: to deliver a certificate on leaving school 186

ABGASE (*n.pl*): waste (exit) gases, flue gases, escaping gases 187

ABGASVERWERTUNG (*f*): utilisation of (waste) exhaust gases 188

ABGAUTSCHEN: to couch (paper) 189

ABGEBEN: to give off; dispose of; generate (steam); deliver; pay 190

ABGEGRIFFEN: worn-out 191

ABGELEGEN: remote; distant; matured 192

ABGELEGENHEIT (*f*): isolation, remoteness 193

ABGELEITET: derived 194

ABGEORDNETE (*m*): deputy; delegate; ambassador 195

ABGERBEN: to tan 196

ABGERUNDETER STEIN (*m*): bull-nosed brick 197

ABGESANDTE (*m*): ambassador 198

ABGESCHAÜMTE (*n*): scum; skimmings 199

ABGESCHMECKT: insipid; tasteless; absurd 200

ABGESEHEN VON: apart from; neglecting; not taking into account; leaving out of the question; exclusive (irrespective) of 201

ABGESTANDEN: decayed, brittle, perished, rotten (of wood) 202

ABGESTECKTE MEILE (*f*): measured mile 203

ABGESTUMPFTER KEGEL (*m*): truncated cone 204

ABGIESSEN: to decant; cast (founding); pour off 205

ABGLANZ (*m*): reflection; reflected glory 206

ABGLÄTTEN: to polish; smooth (off) 207

ABGLEICHEN: to equalise; smooth; balance (accounts); level; square; justify (coins) 208

ABGLÜHEN: to anneal; heat; heat to red-heat; cool; cease glowing; mull (wine) 209

ABGRABEN: to dig up; to root up, uproot 210

ABGRATEN: to remove (take off) the burr 211

ABGRATPRESSE (*f*): burr removing press, burring-off press 212

ABGREIFEN: to obtain; determine; read off (from a graph) 213

ABGRENZUNG (*f*): limit, border, boundary, line of demarcation 214

ABGUSS (*m*): casting; cast; pouring off; decanting 215

ABHAAREN: to remove hair (hides); depilate; moult; fall out (of hair) 216

ABHALTEN: to keep away from, protect (from) against 217

ABHANDELN: to transact, bargain, settle, treat; discuss, debate 218

ABHANDLUNG (*f*): treatise; paper; dissertation; transaction 219

ABHANGEN: to depend; slope; hang down 220

ABHÄNGEN: to depend (upon); take (off) down; disconnect (see also ABHANGEN) 221

ABHÄNGIGKEIT (*f*): subjection; dependence; slope 222

ABHÄRTEN: to harden; temper (metal) 223

ABHARZEN: to tap (trees for resin) 224

ABHÄUTEN: to skin, flay, excoriate; cast (skin), peel; scum, free from scum (founding) 225

ABHEBEN: to lift off; uncover; skim off; cut (cards) 226

ABHEBERN: to siphon off 227

ABHELLEN: to clear; clarify 228

ABHETZEN: to harass 229

ABHITZE (*f*): waste heat 230

ABHITZEKESSEL (*m*): waste-heat boiler 231

ABHITZEVERWERTUNG (*f*): utilisation of waste heat 232

ABHOBELN: to plane off 233

ABHOLD: averse; disinclined; unfavourable 234

ABHOLEN: to go to meet, fetch, call for 235

ABHÖREN: to sift (evidence), examine, question, audit, interrogate 236

ABHUB (*m*): scum; skimmings; offal; dross 237

ABIETIN (*n*): Abietine, coniferine, laricine $C_{10}H_{19}O_8$. $2H_2O$ Mp. 185°C. 238

ABIETINSÄURE (*f*): abietic acid; $(C_{44}H_{64}O_5)$; (Mp. 182°C.); (Acidum abietinicum) 239

ABITURIENTENPRÜFUNG (*f*): matriculation (Examination) 240

ABKALKEN: to remove lime (from); unlime; de-lime 241

ABKANTEN: to bevel; edge; remove (selvedge) border; shear off; fold; (cut a square-) corner 242

ABKANTMASCHINE (*f*): folding machine 243

ABKANTPRESSE (*f*): folding press 244

ABKAPPEN: to cut off, decapitate, unhood, lop 245

ABKELTERN: to press (wine) 246

ABKLAPPEN: to lift the flap; swing (aside) out; let down; uncover 247

ABKLÄREN: to clarify; filter; clear; brighten (a colour); de-colour (sugar); decant; boil off (a dye) 248

ABKLÄRFLASCHE (*f*): decanting flask 249

ABKLÄRTOPF (*m*): decanting jar 250

ABKLATSCHEN: to print (colour photography on clay) 251

ABKNEIFEN : to pinch ; nip off 252
ABKNICKEN : to snap off ; break off ; crack-(off) 253
ABKNISTERN : to decrepitate 254
ABKOCHECHT : fast to boiling 255
ABKOCHEN : to decoct, extract ; boil (off) down 256
ABKOCHUNG (*f*) : decoction ; (Decoctum) ; (see **DEKOKT**) 257
ABKOMMEN (*n*) : agreement 258
ABKÖMMLING (*m*) : derivative (Chemical) ; offspring, descendant, child 259
ABKÖPFEN : to top (trees) ; decapitate 260
ABKRATZEISEN (*n*) : scraper 261
ABKRÜCKEN : to rake off 262
ABKÜHLEN : to cool, reduce temperature, quench, refrigerate, anneal 263
ABKÜHLFASS (*n*) : annealing oven ; cooling vat 264
ABKÜHLGESCHWINDIGKEIT (*f*) : cooling (annealing speed (metal) 265
ABKÜHLOFEN (*m*) : annealing oven (glass) 266
ABKÜHLUNGSMITTEL (*n*) : cooling medicine (Med.) 267
ABKUNFT (*f*) : agreement ; breed ; race ; descent 268
ABLADEN : to unload, unship, trans-ship, discharge (cargo) 269
ABLAGERN : to season (wood) ; to store, bond, season, mature 270
ABLASS (*m*) : drain ; escape ; sluice ; remission ; discount 271
ABLASSDRUCK (*m*) : blow-off pressure (of safety valve) 272
ABLASSEN : to draw off ; let run ; anneal (steel) ; fade (colour) ; empty (boilers) 273
ABLASSHAHN (*m*) : blow-off cock, mud (sludge) cock ; discharge (drain, delivery) cock 274
ABLASSLEITUNG (*f*) : delivery (main) pipe, blow off (main) piping 275
ABLASSVENTIL (*n*) : blow-off valve ; escape valve ; blow-down valve 276
ABLATIV (*m*) : ablative 277
ABLATIV (*m*), **UNABHÄNGIGER-** : absolute ablative (ablativus absolutus) 278
ABLAUF (*m*) : lapse, expiration ; gutter ; maturity (of bills) ; issue, result 279
ABLAUFEN LASSEN : to launch (a ship) ; to let run-off, tap 280
ABLAUFGEWICHT (*n*) : launching weight 281
ABLAUFÖL (*n*) : run oil ; expressed oil 282
ABLAUFRINNE (*f*) : sink , gutter, channel 283
ABLAUFROHR (*n*) : discharge (waste) pipe, drain pipe 284
ABLAUFSCHLITTEN (*m*) : launching cradle 285
ABLAUFTRICHTER (*m*) : discharge (discharging) funnel 286
ABLAUGEN : to steep in lye ; wash out lye from 287
ABLÄUTERN : to clear ; clarify ; draw off ; wash (ore) ; refine (sugar) ; filter (see **LÄUTERN**) 288
ABLÄUTERUNGSZEIT (*f*) : clarifying (filtering) period, time required to (filter) clarify 289
ABLEBEN (*n*) : demise, decease, death 290
ABLEGER (*m*) : slip ; scion (of plants) 291
ABLEITEN : to make up (a formula) an equation ; turn aside ; draw off ; let escape ; deal with ; derive ; conduct ; deduct ; determine ; arrive at ; divert 292
ABLEITUNG (*f*) : derivation, etc. 293
ABLEITUNGSMITTEL (*n*) : revulsive ; conductor ; draining medium ; derivative 294
ABLEITUNGSROHR (*n*) : discharge (outlet-) pipe 295
ABLENKEN : to deflect (magnetic needle, etc.) 296
ABLENKEN (*n*) : deflection (of magnetic needle) 297
ABLENKUNG (*f*) : deflection 298
ABLENKUNGSPLATTE (*f*) : deflection plate 299
ABLESBAR : readable 300
ABLESEN : to read off ; pick off 301
ABLESEVORRICHTUNG (*f*) : reading device 302
ABLICHTEN : to reduce ; clear ; shade (Dyeing) ; strip ; lighten 303
ABLÖHNEN : to pay off 304
ABLÖSCHBAR : temperable ; quenchable (metals) ; slakable (lime) 305
ABLÖSEN : to loosen, amputate, sever, take off, to relieve (from duty), to change (guard), to resolve, redeem 306
ABLÖSEND : resolvent (Med.) 307
ABLÜFTEN : to air 308
ABLUFTLUTTE (*f*) : exhaust air duct or exit 309
ABMAISCHEN : to finish mashing (Brewing) 310
ABMALLEN : to mould 311
ABMESSEN : to measure off, dimension ; survey ; gauge 312
ABMESSUNG (*f*) : dimension, measurement ; survey 313
ABMONTIEREN : to dismount, demount, dismantle, take down 314
ABMONTIERUNG (*f*) : demounting, dismantling, taking down 315
ABNAHME (*f*), **KESSEL-** : boiler acceptance 316
ABNAHMEVERSUCH : (*m*) acceptance test (test on taking over plant or machinery) 317
ABNAHMEVORSCHRIFTEN (*f.pl*) : instructions for the permissible reduction (steel tests) 318
ABNARBEN : to buff ; scrape (hides) 319
ABNEHMEN : to take from, amputate, subtract, reduce, accept, skim, remove, deduct decorticate 320
ABNEHMER (*m*) : consumer ; user 321
ABNUTSCHEN : to suck off ; hydro-extract (Dyeing) 322
ABNUTZEN : to wear (out) away 323
ABNUTZUNG (*f*) : wear and tear, wearing 324
ABÖLEN : to de-oil ; remove oil from 325
ABONNENT (*m*) : subscriber 326
ABONNIEREN : to subscribe (to a journal, etc.) 327
ABORTIVMITTEL (*n*) : abortive agent (for producing an abortion) 328
ABORTÖL (*n*) : lavatory oil 329
ABORTUS (*m*) : abortion, miscarriage (Med.) 330
ABPRALLEN : to rebound, reverberate, deflect, reflect, ricochet 331
ABPRALLUNG (*f*) : rebound ; impingement ; reflection (Optics) 332
ABPRALLUNGSWINKEL (*m*) : angle of (deflection) reflection 333
ABPRODUKT (*n*) : waste product ; by-product 334
ABPUTZ (*m*) : plaster 335
ABPUTZEN : to clean, trim, plane, polish, level up ; plaster (walls) 336
ABQUETSCHEN : to squeeze (off) out of or from, crush 337
ABQUICKEN : to purify ; separate (gold from amalgam) ; refine (ore) with mercury 338
ABRAHMEN : to skim ; remove cream from ; dismantle (take off) a frame 339

ABRAUCHEN: to fume; evaporate; vapourize; smoke 340
ABRÄUCHERN: to fumigate; smoke 341
ABRAUM (*m*): rubbish; rubble 342
ABRÄUMEN: to clear-(away): remove 343
ABRAUMLOKOMOTIVE (*f*): engine (locomotive) fitted with rubbish burning furnace, engine for burning rubbish 344
ABRAUMSALZ (*n*): Abraumsalz (a saline deposit from the Stassfurt district of Germany), Stassfurt salt (see **KAINIT, KIESERIT, BORAZIT, KARNALLIT, SYLVIN**) 345
ABRECHEN: to rake off 346
ABRECHNEN: to cast up (figures); settle (accounts); subtract, deduct, allow 347
ABRECHNUNGSKONTOR (*n*): clearing house 348
ABREISSEN: to break off, be interrupted, tear off; choke, go out (of fires) 349
ABREISSER (*m*): designer, tracer; tracing tool 350
ABREISSKALENDER (*m*): tear-off calendar 351
ABRICHTELAUGE (*f*): weak caustic liquor (soap) 352
ABRICHTEN: to teach, train; adjust, proportion, fit out (ships); true, make true, face 353
ABRICHTER (*m*): trainer 354
ABRIEB (*m*): pieces (ground, rubbed) broken off coal, coke, etc.); the (grinding, rubbing) breaking off; dust 355
ABRINDEN: to bark, skin, decorticate 356
ABRISS (*m*): section; sketch; draft; summary; plan 357
ABROLLEN: to unwind, to pay out (a rope or cable) 358
ABRÖSCHEN: to air (paper) 359
ABRUNDEN: to round off; even up; curve 360
ABRUNDUNG (*f*): curvature; rounding off; evening up; rounded join 361
ABRUSSECHT: not crocking; fast to rubbing 362
ABSATZ (*m*): reduction; deposit; sediment; scale; pause; paragraph; market; turnover; heel; precipitate; stage; use; useful purpose 363
ABSATZBASSIN (*n*): settling tank 364
ABSÄTZEN, IN-: alternately, intermittently 365
ABSATZGESTEIN (*n*): sedimentary rock, stratified rock 366
ABSÄTZIG: inferior; intermittent; interrupted; alternately 367
ABSATZWEISE: in steps or stages, intermittently; variable; stage-wise; intermittent 368
ABSATZWEISE BEWEGT: having variable (stroke) movement, actuated intermittently 369
ABSAUGEANLAGE (*f*): suction plant 370
ABSAUGEFLASCHE (*f*): filtering flask 371
ABSAUGEN: to suck off; exhaust; hydro-extract (dyeing); withdraw 372
ABSAUGER (*m*): exhauster; hydro-extractor, (dye); sucker; suction box (paper) 373
ABSCESS (*m.*): abscess (Med.) 374
ABSCHABEN: to grind off, scrape off 375
ABSCHABEWERKZEUG (*n*): scraping (scaling) tool, scraper 376
ABSCHÄLEN: to peel off, exfoliate (see also **SCHÄLEN**) 377
ABSCHALTEN: to cut out; switch off; turn off 378
ABSCHÄTZEN: to appraise, value 379
ABSCHÄTZER (*m*): valuer 380
ABSCHAUM (*m*): scum; refuse; dross 381
ABSCHÄUMEN: to skim 382
ABSCHEIDEN: to separate; precipitate; seclude; refine (metals); divide off 383
ABSCHEIDER (*m*): separator; trap 384
ABSCHEIDUNG (*f*): separation, seclusion, sequestration, precipitation, dividing off, secretion, refining, deposit, sediment 385
ABSCHEIDUNGSMITTEL (*n*): precipitant; separating medium 386
ABSCHEIDUNGSPRODUKT (*n*): secretion; production of separation 387
ABSCHEREN: to shear, cut off; (*n*) shearing 388
ABSCHERUNG (*f*): shear; shearing stress; cutting (shearing) off 389
ABSCHIEFERN: to scale; flake (-off); peel; laminate; exfoliate 390
ABSCHIESSEN: to fade (of colours); fire (shoot) off 391
ABSCHIRMEN: to screen; cover; protect 392
ABSCHLACKEN (*n*): slacking, shifting (of a wall), removal of slag 393
ABSCHLACKEN: to remove slag from; slacken or shift (of walls) 394
ABSCHLAGEN: to chip; take to pieces; cast (metal); sieve (pharmacy); reject; refuse; fall (in price) 395
ABSCHLAGSZAHLUNG (*f*): instalment, prepayment (part), payment on account 396
ABSCHLAGSZAHLUNG (*f*): part payment 397
ABSCHLÄMMEN: to wash (ore); buddle; remove mud; blow off (blow down) boilers 398
ABSCHLEIFEN: to sharpen, rub down (paint, etc.), grind, polish, wear (away) out 399
ABSCHLEUDERN: to centrifuge; throw (fly; shake) off; detrain; hydro-extract 400
ABSCHLICHTEN: to plane, polish, planish, smooth 401
ABSCHLIESSEN: to close, shut- (off), seal, cut-off, conclude, occlude, seclude, terminate, complete, cut-out, bring to a conclusion 402
ABSCHLUSS (*m*): closing, shutting, seal, closing (sealing) arrangement, cut-off, conclusion, occlusion, seclusion, end, termination 403
ABSCHLUSSVORRICHTUNG (*f*): closing-in arrangement, seal 404
ABSCHLUSSVORRICHTUNG (*f*), **HINTERE-**: back-end arrangement (of stokers) 405
ABSCHMELZDRAHT (*m*): safety (cut-out) fuse, fuse wire 406
ABSCHMELZEN: to fuse, melt off 407
ABSCHMELZSICHERUNG (*f*): safety (fuse) cut-out, safety (fusible) plug 408
ABSCHMELZSTREIFEN (*m*): fuse- (strip) 409
ABSCHMELZUNG (*f*): melting; liquation (metal) 410
ABSCHMIRGELN: to rub with emery 411
ABSCHNEIDEAPPARAT (*m*): cutting apparatus; cutter 412
ABSCHNEIDEN: to cut off; clip; isolate; truncate (Maths.); deprive of 413
ABSCHNITT (*m*): segment; section; portion; sub-section; period; part; paragraph 414
ABSCHNITZEL (*n*): chip; clipping; paring 415
ABSCHNÜREN: to mark (measure) off with a cord, to line (lay) out 416
ABSCHÖPFEN: to skim off; ladle out; scum; remove surface of molten metal 417
ABSCHRÄGEN: to bevel off; incline; slope; plane 418
ABSCHRAUBEN: to screw (out) off, to screw apart, unscrew 419
ABSCHRECKEN: to chill; cool; quench (metal); render warm; make lukewarm; take off the chill (of water) 420
ABSCHREIBEN: to write off; depreciate; copy 421

ABSCHREIBUNG (*f*): amount written off; depreciation (accountancy); (act of-) writing off 422
ABSCHRIFT (*f*): copy, summary, transcript 423
ABSCHROTEN: to chop (saw) off, grind roughly 424
ABSCHRUPPEN: to plane off 425
ABSCHUPPEN: to laminate; peel; flake (-off); exfoliate 426
ABSCHWÄCHEN: to soften; weaken; reduce; mellow; water; thin; reduce in strength 427
ABSCHWÄCHER (*m*): reducer, reducing (agent) medium, (Photography) 428
ABSCHWÄCHUNG (*f*): decline, slackening, fall; reduction (Phot.) 429
ABSCHWARTEN: to bark, decorticate (tree-trunks) 430
ABSCHWEFELN: to desulphurize; coke (coal); impregnate with sulphur 431
ABSCHWINGEN: to fan; winnow; centrifuge; hydro-extract 432
ABSCHWIRREN: to whiz; centrifuge 433
ABSCHWITZEN: to sweat off; depilate (leather) 434
ABSCISSE (*f*): abscissa; (see **ABSZISSE**) 435
ABSEHEN: to perceive; intentionally neglect or overlook; look away 436
ABSEHEN VON: to exclude; let alone 437
ABSEIDE (*f*): floss-silk 438
ABSEIGERN: to liquate; separate (by fusion) (Metal); measure the perpendicular (depth) height; plumb 439
ABSEIHBIER (*n*): drawings 440
ABSEIHEN: to draw; filter; strain; percolate; drain 441
ABSENGEN: to singe off 442
ABSENKEN: to lower; descend; (*n*); lowering; descent 443
ABSETZBASSIN (*n*): settling tank 444
ABSETZBOTTICH (*m*): settling vat 445
ABSETZEN: to deposit; reduce; sell; push, (sale); settle; precipitate; put in type 446
ABSETZUNGSTEICH (*m*): settling trough 447
ABSIEDEN: to boil; decoct; extract by boiling 448
ABSINTH (*m*): absinthe; wormwood; (Artemisia absinthium) 449
ABSIPHONIEREN: to siphon off; draw off by means of a syphon 450
ABSITZBÜTTE (*f*): settling tub (Brewing) 451
ABSITZEN LASSEN: to allow to settle 452
ABSITZENLASSEN (*n*): allowing to (stand) settle (of solutions) 453
ABSOLUT: absolute(ly); (in reference to pressure means; pressure above vacuum, *i.e.*, 1 Atm. above gauge pressure. Absolute pressure equals 735.5 m.m. mercury at 0°C., *i.e.*, the abs. press. fluctuates with the column of mercury barometrically) 454
ABSOLUTSCHWARZ (*n*): absolute black (dye) 455
ABSONDERN: to separate; divide; disjoin; segregate; sever; sunder; detach; part; sequester; secrete; seclude; insulate; isolate; exclude; abstract 456
ABSONDERUNG (*f*): separation; secretion; fission; division 457
ABSORBENS (*n*): absorbent 458
ABSORBIEREN: to absorb 459
ABSORPTIONSANLAGE (*f*): absorption (plant); installation 460
ABSORPTIONSFÄHIGKEIT (*f*): absorptive capacity 461
ABSORPTIONSGEFÄSS (*n*): absorption vessel 462
ABSORPTIONSGESCHWINDIGKEIT (*f*): absorption velocity 463
ABSORPTIONSKOEFFIZIENT (*m*): absorption coefficient (of gases) 464
ABSORPTIONSKOLONNE (*f*): absorption column 465
ABSORPTIONSSTREIFEN (*m*): absorption band 466
ABSORPTIONSTURM (*m*): absorption tower 467
ABSORPTIONSVERBINDUNG (*f*): absorption compound 468
ABSORPTIONSVERLUST (*m*): absorption loss 469
ABSORPTIONSVERMÖGEN (*n*): absorptive capacity 470
ABSPANNEN: to slacken: relax; span; measure estrange; alienate; release (pressure); reduce (tension); cut off (steam) 471
ABSPANNUNG (*f*): debility; asthenia (Med.); relaxation; slackening; spanning; measuring 472
ABSPERRBAR: capable of being cut out or cut off as required (referring to pipes, machinery, etc.) 473
ABSPERREN: to cut off (steam etc.); to bar (a road etc.) 474
ABSPERRFLÜSSIGKEIT (*f*): sealing (fluid) liquid 475
ABSPERRHAHN (*m*): stop cock 476
ABSPERRSCHIEBER (*m*): gate valve; slide valve; sluice valve; gate stop valve 477
ABSPERRVENTIL (*n*): cut-off valve; stop valve 478
ABSPERRVENTIL (*n*), **DAMPF-**: steam stop valve 479
ABSPIELEN: to take place; occur; happen 480
ABSPREIZEN: to stay, support, prop, underpin 481
ABSPRENGEN: to burst off; break off; demolish 482
ABSPRENGER (*m*): cutter; glass-cutter 483
ABSPRINGEN: to spring (snap) off; rebound; warp (wood); be red-short (iron) 484
ABSPRITZEN: to wash, cleanse, spray 485
ABSPRUNG (*m*): reflection (Physics.) 486
ABSTAND (*m*): pitch, pitching, space, spacing, clearance, distance-(between), span, interval 487
ABSTÄNDEN, IN GLEICHEN---VON EINANDER: pitched equidistantly 488
ABSTAND (*m*), **IN---HALTEN**: to separate, keep apart 489
ABSTAND NEHMEN: to waive, abandon; object to, resist 490
ABSTANDSFUSS (*m*): distance piece, foot 491
ABSTATTEN: to render (thanks), to pay (a visit) 492
ABSTÄUBECHT: non-crocking (of colours) 493
ABSTECHEN: to tap; run off; butcher; etch (picture) 494
ABSTECHEN (*n*): the running off (tapping) of the liquid iron in blast furnace work 495
ABSTECHER (*m*): tapper (metal); etcher (of pictures) 496
ABSTECKEN: to mark (stake) out; peg out; set out; plot; lay out 497
ABSTEHEN LASSEN: to let stand (of liquids, etc.) 498
ABSTEIFEN: to stiffen 499
ABSTEIGEN: to dismount; fall, descend, decrease 500
ABSTELLEN: to cut off (steam); switch off; anul; put away; turn off; stop; put out of commission 501

ABSTELLHAHN (*m*): cut-off cock; stop cock 502
ABSTELLVENTIL (*n*): cut-off valve; stop valve 503
ABSTERBEN: to die off; fade; wither; mortify; air slack (lime); become opaque 504
ABSTERGIEREN: to cleanse 505
ABSTICH (*m*): the tapping (the liquid iron run-off or tapped from blast furnaces) 506
ABSTICHGENERATOR (*m*): generator from which something is tapped; blast furnace (from which metal is tapped or run off) 507
ABSTICHLOCH (*n*): tap-hole (for molten iron from blast furnace) 508
ABSTIMMBARKEIT (*f*): gradation (Phot.) (the possibility of varying the gradation) 509
ABSTIMMEN: to synchronize, tune (Wireless Tel.) 510
ABSTIMMSCHÄRFE (*f*): clearness (of tuning) (Wireless Tel.) 511
ABSTIMMSPULE (*f*): synchronizing (tuning) coil (Wireless Tel.) 512
ABSTIMMUNG (*f*): synchronizing, tuning (Wireless Tel.) 513
ABSTOSSEN: to repel, reject, knock off, scrape 514
ABSTOSSFETT (*n*): fat scrapings; scraping-fat (from leather) 515
ABSTOSSUNG (*f*): repulsion 516
ABSTOSSUNGSKRAFT (*f*): repelling power 517
ABSTRAHL (*m*): reflected ray 518
ABSTREBEKRAFT (*f*): centrifugal force 519
ABSTREICHMESSER (*n*): scraping-off knife; doctor (Calico) 520
ABSTREIFER (*m*): waterback (of stokers) 521
ABSTREIFRING (*m*): wiping ring, wiper (oil lubrication) 522
ABSTRICH (*m*): scum; dross; skin; litharge 523
ABSTRICHBLEI (*n*): lead scum; crude litharge 524
ABSTUFEN: to grade; graduate 525
ABSTUFUNG (*f*): gradation, graduation; shading 526
ABSTUMPFEN: to blunt; dull; truncate; neutralize (acids) 527
ABSTUMPFUNGSFLÄCHE (*f*): truncating face (Crystalography) 528
ABSTÜTZEN: to prop (ships on the stocks), etc., stay, support, shore 529
ABSUD (*m*): decoction; extract; mordant 530
ABSÜSSEN: to edulcorate; sweeten; purify; wash out (a precipitate) 531
ABSÜSSKESSEL (M): edulcorating vessel 532
ABSÜSSSPINDEL (*f*): hydrometer; sweet water spindle (Sugar); saccharometer 533
ABSYNT (*m*): absinthe (see **ABSINTH**) 534
ABSZISSE (*f*): abscissa (see **ABSCISSE**) 535
ABSZISSE (*f*): abscissa (the horizontal ruling of graph paper) 536
ABTAKELN: to unrig, to clear-(off), dismantle 537
ABTAUEN: to melt, thaw 538
ABTEIL (*n*): compartment (of railway carriage) share 539
ABTEILUNG (*f*): compartment (of a ship, etc) 540
ABTEUFEN: to sink (a shaft) 541
ABTRAG (*m*): excavation; cut; cutting; compensation; payment 542
ABTRAGEN: to settle, carry away, level, clear 543
ABTRÄNKEN: impregnate; saturate; fill up 544
ABTREIBEMITTEL (*n*): abortifacient (Med.); expulsive agent; purge 545
ABTREIBEN: to make leeway (Naut.), drift (of ships); expel; abort (Med.); drive away, repulse, repel; separate, distil, refine, cupel; discharge 546
ABTREIBKAPELLE (*f*): refining cupel (silver) 547
ABTREPPEN: to step; terrace 548
ABTRIFT (*f*): leeway (Naut.) 549
ABTROPFBRETT (*n*): draining board 550
ABTROPFGESTELL (*n*): draining (rack) stand 551
ABWALZEN: to roll off; make smooth or even by rolling; (*n*) smoothing off 552
ABWÄLZEN: (see **ABWALZEN**) 553
ABWÄRME (*f*): waste heat 554
ABWÄRMEVERWERTUNGSANLAGE (*f*): waste heat (utilization) plant 555
ABWARTEN: to await, expect, anticipate 556
ABWASSER (*n*): waste water; sewage (see also **GASWASSER**) 557
ABWÄSSERANLAGE (*f*): sewage (works) disposal) plant 558
ABWASSERKLÄRUNG (*f*): sewage (clarifying) clarification 559
ABWÄSSERN: to free from water; drain 560
ABWASSERORGANISMEN (*pl*): waste water (sewage) organisms 561
ABWECHSELN: to change; vary; alternate; fluctuate 562
ABWECHSELND: alternative-(ly), intermittent-(ly) 563
ABWEHRMITTEL (*n*): preventive; prophylactic 564
ABWEICHEN: to soften, steep, soak (hides); deviate, deflect, diverge, decline, differ, depart, vary; (*n*) diarrhœa, etc. 565
ABWEICHEND: anomalous, varying, dissentient, different, differing 566
ABWEICHUNG (*f*): tolerance, deviation, difference; (amount of)-error, variation, declination (of compass needle); leeway (of ships); divergency, anomaly 567
ABWEICHUNGSKOMPASS (*m*): variation compass 568
ABWELKEN: to wither; shrivel; dry; fade 569
ABWERFEN: to throw off; run off (slag): give off; eject; yield 570
ABWERFOFEN (*m*): refining furnace 571
ABWERFPANNE (*f*): refining pan; list pot (tinning) 572
ABWESEND: absent 573
ABWICKELN: to rectify; develop (of graphs, drawings, etc.); unwind; wind up 574
ABWIEGEMASCHINE (*f*): weighing machine, dosing machine (for chemists, etc). 575
ABWRACKEN: to lay up (a ship) 576
ABWURF (*m*): throw off; offal; rough cast (plaster) 577
ABZAPFEN: to tap, draw off 578
ABZEHREN: to waste away, emaciate, consume 579
ABZIEHBLASE (*f*): retort; alembic 580
ABZIEHEN: to draw (run) -off; decant; rack (liquids); distil; rectify; drain; transfer (designs); strip (dyeing); boil off colour; scrape (leather); clear; escape (gases); deduct; subtract; print; copy 581
ABZIEHHALLE (*f*): racking room (Brewing) 582
ABZIEHKELLER (*m*): racking cellar (Brewing) 583
ABZIEHKOLBEN (*m*): retort; alembic 584
ABZIEHMITTEL (*n*): stripping agent 585
ABZIEHMUSKEL (*m*): abductor (muscle) 586
ABZIEHPAPIER (*n*): transferotype, stripping-paper (Photography) 587
ABZIEHSTEIN (*m*): oil stone, hone, grindstone, whetstone 588
ABZIEHUNG (*f*): abduction (Anatomy) 589

ABZUG (*m*): scum; dross; print; proof sheet; copy; discount; reduction; rebate; flue; outlet; discharge; exit; deduction; hood (for noxious gases); escape 590

ABZUGSGAS (*n*): flue gas; exit gas; waste gas; escaping gas 591

ABZUGSKANAL (*m*): sewer; channel; drain; discharge duct; escape pipe 592

ABZUGSQUELLE (*f*): market, custom(-ers) 593

ABZUGSSCHRANK (*m*): hood (gases) 594

ABZUGSTEMPERATUR (*f*): exit (outlet) temperature; temperature of escaping gases (boilers) 595

ABZUGSVORRICHTUNG (*f*): escape; discharge (arrangement) apparatus; constant level apparatus (for baths) 596

ABZUGSWÄRME (*f*): heat loss; exit (escaping, waste) heat 597

ABZWEIGDRAHT (*m*): shunt wire 598

ABZWEIGROHR (*n*): branch pipe 599

ABZWEIGSTROMKREIS (*m*): derived circuit 600

ABZWEIGUNG (*f*): derivation (of electric current) arm, branch 601

ABZWEIGWIDERSTAND (*m*): shunt resistance 602

ACAJOU (*m*): (see **AKAJOU**) 603

ACAJOUÖL (*n*): (see **AKAJOUÖL**) 604

ACANTHIT (*n*): acanthite (Ag_2S) (87% silver) (see also **SILBERSULFID**); Sg. 7.28 605

ACCELEREN (*pl*): accelerators (agents for increasing the speed of vulcanization of caoutchouc) 606

ACCEPT (*n*): acceptance (Com.) 607

ACCEPTATION (*f*): acceptance (Com.) 608

ACCEPTIEREN: to accept 609

ACCEPT, ZUM-: for acceptance 610

ACCIDENTIEN (*pl*): fine, penalty (in connection with shipments) 611

ACCISE (*f*): excise 612

ACCUMULATORENSÄURE (*f*): battery (accumulator) acid 613

ACENAPHTEN (*n*): acenaphthene; ethylene-naphthene; ethylenenaphthalene (solid hydrocarbon of the auromatic series); sg. 1.0687; Mp. 95°C. Bp. 277°C.[$C_{10}H_6(CH_2)_2$] 614

ACENAPHTHEN (*n*): (see **ACENAPHTEN**) 615

ACENAPHTHENCHINON (*n*): acenaphthene quinone Mp. 261° C. 616

ACENAPHTHENON (*n*): acenaphthenone $C_{12}H_8O$ Mp. 121° C. 617

ACENAPHTHENPIKRAT (*n*): acenaphthene picrate $C_{12}H_{10}.C_6H_2(NO_2)_3OH$; Mp. 161°C 618

ACENAPHTYLEN (*n*): acenaphthylene ($C_{12}H_8$) 619

ACERDESE (*n*): (see **MANGANIT**) 620

ACETAL (*n*): acetal, ethylidene diethyl ester: $CH_3\text{-}CH(O.C_2H_5)_2$; Sg. 0.8314; Bp. 102.2°C (see also **ÄTHYLAL, DIMETHYLACETAL**, and **METHYLAL**) 621

ACETALDEHYD (*n*): acetaldehyde, acetic aldehyde, ethyl aldehyde, ethanol, aldehyde; (C_2H_4O) Sg. 0.8; Bp. 21° C. 622

ACETAMID (*n*): acetamide, acetic acid amine; C_2H_5ON; Sg. 0.9901-1.159; Mp. 83° C.; Bp. 222° C. 623

ACETANILID (*n*): acetanilide, antifebrin, phenylacetamide; $C_6H_5NH\ (COCH_3)$; Sg. 1.21; Mp. 113-116° C.; Bp. 305°C. 624

ACETESSIGÄTHER (*m*): acetoacetic ether (see **ACETESSIGESTER**) 625

ACETESSIGESTER (*m*): acetoacetic ester, ethyl acetoacetate; $C_6H_{10}O_3$; Sg. 1.0282; Bp. 180.6-181.2 626

ACETESSIGSÄURE (*f*): acetoacetic acid 627

ACETIN (*n*): acetin; monoacetin; Sg. 1.2212; [$C_2H_3O_2C_3H_5(OH)_2$] 628

ACETINBLAU (*n*): acetin blue (basic azine dyestuff) 629

ACETINDRUCK (*m*): acetin printing 630

ACETINDULIN (*n*): acetinduline (see **ACETINBLAU**) 631

ACETOCAUSTIN (*n*): acetocaustine (a 50% hydrous solution of trichloroacetic acid) 632

ACETOL (*n*): acetol ($CH_3.CO.CH_2.OH$) 633

ACETON (*n*): acetone, ketopropane, dimethylketone, pyroacetic ether, methylacetyl (C_3H_6O); Sg. 0.79; Mp. -94.3°C.; Bp. 56.5°C. 634

ACETONAL (*n*): acetonal, aluminium-sodium acetate $(CH_3COO)_5Na.Al_2(OH)_2$ 635

ACETONBISULFIT (*n*): acetone bisulphite (Photographic chemical) (acts as a retarder for rapid developers, a preserver for developers, a clarifier for fixing baths, etc.) 636

ACETONCHLOROFORM (*n*): acetone chloroform (see **CHLORETON**) 637

ACETONDICARBONSÄUREESTER (*m*): acetone dicarboxylic ester (ethyl); $C_9H_{14}O_5$; Sg. 1.1107; Bp. 250°C. 638

ACETONITRIL (*n*): acetonitrile, methyl cyanide C_2H_3N; Sg. 0.7906; Mp. -44.9°C; Bp. 81.6°C 639

ACETONNACHLAUF (*m*): second runnings (of) from acetone 640

ACETONÖL (*n*): acetone oil (by-product from distillation of acetone); Bp. 75-130°C. for light sorts; Bp. 130-250°C. for heavy sorts 641

ACETONSÄURE (*f*): acetonic acid; hydroxyisobutyric acid 642

ACETONSULFIT (*n*): acetone sulphite (Photographic chemical) 643

ACETONYLACETON (*n*): acetonyl acetone; $C_6H_{10}O_2$; Sg. 0.9729; Bp. 194°C. 644

ACETOPHENON (*n*): acetophenone, hypnone (C_8H_8O); Sg. 1.033; Mp. 20.5°C.; Bp. 202°C. 645

ACETOPURPURIN (*n*): Acetopurpurin, (a diazo dyestuff) 646

ACETOPYRIN (*n*): (*Antipyrinum acetosalicylicum*); Acetopyrin, (a compound of Aspirin and Phenyldimethylpyrazolon); Mp. 64.5°C. 647

ACETPARAMINOSALOL (*n*): acetyl-para-aminosalol; acetaminosalol; salophen (see **SALOPHEN**) 648

ACETYLACETON (*n*): acetyl acetone; $C_5H_8O_2$; Sg. 0.9745; Bp. 139°C. 649

ACETYLBROMID (*n*): acetyl bromide; C_2H_3OBr.; Bp. 81°C. 650

ACETYLCELLULOSE (*f*): cellulose acetate 651

ACETYLCHLORID (*n*): acetyl chloride; C_2H_3OCl; Sg. 1.1051; Mp. 50.9°C.; Bp. 51-52°C. 652

ACETYLEN (*n*): acetylene; ethine; (C_2H_2); Sg. 0.42-0.91; Mp. -81.5°C.; Bp. 83.6°C. 653

ACETYLENBINDUNG (*f*): acetylene (bond) linkage; triple bond 654

ACETYLENCHLORID (*n*): acetylene chloride (see **ACETYLENHEXACHLORID, ACETYLENTETRACHLORID**) 655

ACETYLENHEXACHLORID (*n*): acetylene hexachloride (see **HEXACHLORÄTHAN**) 656

ACETYLENLAMPE (*f*): acetylene lamp 657

ACETYLENREINIGUNGSMASSE (*f*): acetylene purifying mass, acetylene purifier 658

ACETYLENTETRABROMID (*n*): acetylene tetrabromide, Muthmann's liquid, tetrabromoethane; ($CHBr_2CHBr_2$); Sg. 2.98–3.0; Bp. 239–242°C. (see also **BROMACETYLEN**) 659

ACETYLENTETRACHLORID (*n*): acetylene tetrachloride tetrachloroethane; ($CHCl_2CHCl_2$); Sg. 1.582; Mp. -36°C.; Bp. 147.2°C. 660

ACETYLIEREN: to acetylate 661

ACETYLIERUNG (*f*): acetylation, acetylating 662

ACETYLIERUNGSVERFAHREN (*n*): acetylation process 663

ACETYL-P-KRESOTINSÄURE (*f*): acetyl-para-cresotinic acid, ervasine; $O.COCH_3.CO_2H.CH_3$; Mp. 141.5°C. 664

ACETYLSALICYLSÄURE (*f*): acetylsalicylic acid; aspirin; ($C_2H_3O_2C_6H_9CO_2H$); Mp. 132–135°C. 665

ACETYLSÄURE (*f*): acetic acid; ($HC_2H_3O_2$); Sg. 1.049; Mp. 16.7°C.; Bp. 118.1°C. 666

ACETYLZAHL (*f*): acetyl number 667

ACHAT (*m*): agate (variegated chalcedony); (SiO_2); Sg. 2.59 668

ACHATMÖRSER (*m*): agate mortar 669

ACHATPORZELLAN (*n*): agate ware 670

ACHATSCHELLACK (*m*): shellac substitute 671

ACHIRIT (*n*): (see **DIOPTAS**) 672

ACHMIT (*m*): acmite (Silica, 56%; soda, 12%; red iron oxide, 32%); Sg. 3.54 673

ACHROMATISCH: achromatic 674

ACHROMATISCHE LINSE (*f*): achromatic lens (Photography) 675

ACHROMATISIEREN: to achromatize 676

ACHROMATISMUS (*m*): achromatism (Photography) 677

ACHSBÜCHSE (*f*): axle (box) bearing 678

ACHSDRUCK (*m*): axial pressure; axle-load 679

ACHSE (*f*): axis; axle; shaft; co-ordinate 680

ACHSE (*f*), **GROSSE-**: major axis 681

ACHSE (*f*), **KLEINE-**: minor axis 682

ACHSEL (*f*): axilla (physiology); shoulder 683

ACHSE (*f*), **LÄNGS-**: longitudinal axis 684

ACHSELARTERIE (*f*): axillary artery (Med.) 685

ACHSELHÖHLE (*f*): axilla arm-pit (Med.) 686

ACHSELSTÜCK (*n*): shoulder piece 687

ACHSELTRÄGERISCH: ambidextrous; hypocritical equivocal 688

ACHSEN (*f.pl*): co-ordinates, axes 689

ACHSENBRUCH (*m*): axle fracture 690

ACHSENDRUCK (*m*): axle pressure 691

ACHSE (*f*), **NEUTRALE**: neutral axis 692

ACHSENKREUZ (*n*): intersection (intersecting) point of the axes 693

ACHSENLAGER (*n*): axle bearing; journal (shaft) bearing 694

ACHSENNAGEL (*m*): linch pin 695

ACHSENÖL (*n*): axle oil 696

ACHSENREGLER (*m*): shaft governor 697

ACHSENREIBUNG (*f*): axle friction 698

ACHSENRING (*m*): axle ring 699

ACHSENSCHMIERE (*f*): cart (axle) grease 700

ACHSENWINKEL (*m*): axial angle 701

ACHSE (*f*), **ORDINATEN-**: ordinate axis 702

ACHSE (*f*), **PROPELLER-**: axis of a propeller 703

ACHSE (*f*), **ROTATIONS-**: axis of rotation 704

ACHSE (*f*), **SCHWIMM-**: axis of flotation 705

ACHSE (*f*), **STEUERRAD-**: steering wheel axle 706

ACHSE (*f*), **UMDREHUNGS-**: axis of rotation 707

ACHSLAGER (*n*): journal (shaft) bearing, axle bearing 708

ACHSRECHT: axial- (ly) 709

ACHSSCHENKEL (*m*): axle journal 710

ACHSWECHSELWINDE (*f*): axle-jack 711

ACHSZAPFEN (*m*): axle arm, axle end, journal 712

ACHTBASISCH: octabasic 713

ACHTECK (*n*): octagon 714

ACHTEL (*n*): eighth (as a prefix, octo-, octa-) 715

ACHTELKOHLENEISEN (*n*): octoferric carbide 716

ACTHELSCHWEFELEISEN (*n*): octoferric sulphide 717

ACHTER: aft (Naut.) 718

ACHTERDECK (*n*): quarter deck 719

ACHTERLASTIG: down by the stern (of ships) 720

ACHTERRAUM (*m*): after hold (ships) 721

ACHTERSCHIFF (*n*): after body (ships) 722

ACHTERSPANT (*n*): after frame 723

ACHTERSTEVEN (*m*): stern (rudder) post (Naut.) 724

ACHTERSTEVENHACKE (*f*): stern (rudder) post heel (Naut.) 725

ACHTFLACH: octahedral; (*n*) octahedron 726

ACHTGLIEDERIG: eight-membered 727

ACHTWERTIG: octavalent 728

ACIDIFIKATION (*f*): acidification 729

ACIDIFIZIEREN: to acidify, sour 730

ACIDIFIZIERUNG (*f*): acidification 731

ACIDITÄT (*f*): acidity 732

ACIDOL (*n*): acidole (Betaine hydrochloride), ($CH_2N.(CH_3)_3\ Cl.CO_2H$) (crystals containing 23.8 % HCl.) 733

ACIDOLCHROMFARBSTOFF (*m*): acidole chrome colour 734

ACIDOLCHROMSCHWARZ (*n*): acidole chrome black 735

ACIDYLIEREN: to acylate 736

ACKERBODEN (*m*): arable soil; surface soil 737

ACKERDOPPEN (*f.pl*): valonea 738

ACKEREI (*f*): tillage 739

ACKERERDE (*f*): arable (soil) land; vegetable mould 740

ACKERKAMILLE (*f*): wild camomile (*Matricaria chamomilla*) 741

ACKERKRUME (*f*): arable (soil) land; vegetable mould 742

ACKERMINZE (*f*): corn mint (*Mentha arvensis*) 743

ACNE (*f*): acne (a skin disease) (Med.) 744

ACONITIN (*n*): (see **AKONITIN**) 745

ACONITKNOLLE (*f*): aconite root; monk's-hood; wolf's-bane (*Tubera aconiti; radix aconiti*); (root of *Aconitum napellus*) 746

ACROLEÏN (*n*): propenal, acrylic aldehyde, acrolein, acraldehyde; C_3H_4O; Sg. 0.84; Bp. 52.4°C. 747

ACRYLSÄURE (*f*): acrylic acid; $C_3H_4O_2$; Sg. 1.0621; Mp. 7-8°C.; Bp. 140°C. 748

ACRYLSÄURE-ÄTHYLESTER (*m*): acrylic ester (ethyl); $C_5H_8O_2$; Sg. 0.9393; Bp. 98·5°C. 749

ACTE (*f*): act (see **AKTE**) 750

ACTOL (*n*): actol, silver lactate (*Actolum*) (*Argentum lacticum*); $C_3H_5O_3Ag + H_2O$. 751

ACUT: acute 752

ACYCLISCH: acyclic 753

ACYLIEREN: to acylate 754

ADAMSAPFEL (*m*): Adam's apple 755

ADDIEREN: to add (sum) up, to total 756

ADDIEREN (*n*): addition, totalling 757

ADDIERMASCHINE (*f*): adding machine 758

ADDIERUNG (*f*): addition (com.) 759

ADDITIONELL: additional 760

ADDITIONSREAKTION (*f*): additive reaction 761

ADDITIVE METHODE (*f*): additive method (of three colour photography) 762

ADENIN (*n*): adenine, $N.HC.N.C.C.C.NH_2NH.CH.N$ (a purine derivate) 763

ADEPS LANAE (*n*): adeps lanae, lanum, lanolin, lanalin, laniol (purified wool grease) 764
ADER (*f*): vein; seam; streak 765
ADER (*f*), **BLUT-**: blood-vessel, vein 766
ADERBRUCH (*m*): varicocele (Med.) 767
ADERGESCHWULST (*f*): aneurism, swelling of a vein (Med.) 768
ADERHAUT (*f*): choroid membrane (of eye), chorion (of foetus) (Med.) 769
ADERHÄUTCHEN (*n*): chorion 770
ADERHOLZ (*n*): grain wood 771
ADERIG: veined; streaked; seamed; vascular 772
ADERLASS (*m*): bleeding; venesection 773
ADERN: to vein; seam; streak; marble 774
ADERPRESSE (*f*): tourniquet (Med.) 775
ADER (*f*), **PULS-**: artery 776
ADER (*f*), **SCHLAG-**: artery 777
ADERSYSTEM (*n*): system of blood vessels, arterial system (Med.) 778
ADER (*f*), **THON-**: clay seam 779
ADER (*f*), **WASSER-**: lymphatic vessel (Med.) 780
ADERWASSER (*n*): (blood)-serum (Med.) 781
ADHÄRIEREN: to adhere 782
ADHÄSION (*f*): adhesion 783
ADHÄSIONSFETT (*n*): adhesive grease or fat 784
ADHÄSIV: adhesive 785
ADIABATE (*f*): the adiabatic- (equation, line, or curve) 786
ADIABATISCHE LINIE (*f*): adiabatic (curve) line 787
ADIAKTINISCH: adiactinic (permitting non-actinic rays of light only to pass, and impervious to actinic or chemical rays) 788
ADIAPHANSPAT (*m*), **PYRAMIDALER-**: gehlenite 789
ADIOWANÖL (*n*): (see **AJOWANÖL**) 790
ADIOWANSAMEN (*m*): ajowan, (fruit of *Carum ajowan*) 791
ADIPID: fatty 792
ADIPIDIEREN: to grease 793
ADIPINSÄURE (*f*): adipinic acid, adipic acid (*Acidum adipinicum*); $C_6H_{10}O_4$; Mp. 153°C.; Bp. 216.5°C. 794
ADIPOCIRE (*n*): adipocere 795
ADJUSTIEREN: to set; adjust 796
ADLERSTEIN (*m*): eagle-stone; aetites 797
ADLERVITRIOL (*m*): salzburg vitriol (mixed cupric and ferrous sulphate) 798
ADONIDIN (*n*): adonidine (obtained from *herba Adonis vernalis*) 799
ADONISKRAUT (*n*): pheasant's eye, adonis (*herba Adonis vernalis*) 800
ADONISRÖSCHENKRAUT (*n*): spring pheasant's eye (*herba Adonidis vernalis* of *Adonis vernalis*) 801
ADOUCIEREN: to sweeten; edulcorate; anneal; decarbonize (iron) 802
ADOUCIERGEFÄSS (*n*): edulcorating vessel; annealing pot 803
ADOUCIEROFEN (**M**): annealing furnace 804
ADRENALIN (*n*): adrenaline, epinephrin; $C_9H_{13}NO_3$; Mp. 205-212°C. 805
ADRESSANT (*m*): consignor; sender; writer 806
ADRESSAT (*m*): consignee 807
ADRESSBUCH (*n*): directory 808
ADRESSIERMASCHINE (*f*): addressing machine 809
ADSORBENS (*m*): absorbent 810
ADSORBENS (*n*), **ADSORBENDA** (*pl*): solid solution 811
ADSORPTION (*f*): adsorption (a variation in concentration between the rim layer and the main mass of a disperse system); (the taking up of one substance by the surface of another substance, as of vapour on a hard surface, i.e., absorption without penetration) 812
ADSORPTIONSERSCHEINUNG (*f*): (the displacement of the concentration to the rim of the colloidal solution); adsorption phenomenon 813
ADSTRINGENS (*n*): astringent (Med.) 814
ADSTRINGIEREND: astringent (Med.) 815
ADULAR (*m*): adularia (a white or colourless orthoclase); (see **ORTHOKLAS**) 816
ADUROL (*n*): adurol (photographic developer; either chlorohydroquinone or Bromohydroquine; see **CHLORHYDROCHINON** and **BROMHYDROCHINON**) 817
ADVOKATENBAUM (*m*): avocado pear(tree) (*Laurus persea* or *Persea gratissima*) 818
AEGIRIN (*m*): aegirite (see **AKMIT**) 819
AENIGMATIT (*m*): aenigmatite; $Na_4Fe_9(AlFe)_2(SiTi)_{12}O_{38}$; Sg. 3.75 820
AEROBIONTISCH: aerobic 821
AEROBISCH: aerobic 822
AEROGENGAS (*n*): aerogene gas 823
AEROLITH (*m*): aerolite, meteoric stone 824
AEROMETER (*m*): aerometer (densimeter for liquids) 825
AERONAUT (*m*): aeronaut 826
AERONAUTIK (*f*): aeronautics 827
AERONAUTISCH: aeronautic-(al)-(ly) 828
AEROSTAT (*m*): aerostat 829
AEROSTATIK (*f*): aerostatics 830
AEROSTATISCH: aerostatic-(al-)(ly) 831
AESCHYNIT (*m*): echinite, eschynite $(Ti_2O_5)_4$ Ce (CeO) (Ca Fe); Sg. 5.0 (Min.) 832
AFFE (*m*): monkey; ape; windlass; crane (Mech.) 833
AFFENBROTBAUM (*m*): Baobab tree (*Adansonia digitata*) 834
AFFENBROTBAUMÖL (*n*): (see **BAOBABÖL**) 835
AFFICHE (*f*): poster, placard, bill 836
AFFICHIEREN: to placard, post up bills; advertise 837
AFFIN ÄHNLICH: of a similar affinity; in a similar ratio 838
AFFINATION (*f*): refining (of gold by treating with sulphuric acid) 839
AFFINIEREN: to refine 840
AFFINITÄT (*f*): affinity 841
AFFINITÄTSEINHEIT (*f*): unit of affinity 842
AFFINITÄTSLEHRE (*f*): doctrine of affinity 843
AFFINITÄTSREST (*m*): residual affinity 844
AFFINITÄTSRICHTUNG (*f*): direction of affinity 845
AFFINIVALENZ (*f*): valence; atomicity 846
AFTER-: (as a prefix, pseudo-; false; secondary; after; neo-; back; anal; rectal) 847
AFTER (*m*): anus, rectum, posterior (Med.); (*n*) refuse, offal, garbage, rubbish, residue 848
AFTERBILDUNG (*f*): malformation 849
AFTERDARM (*m*): rectum (Med.) 850
AFTERGEBURT (*f*): afterbirth (Med.) 851
AFTERHOLZ (*n*): dead wood 852
AFTERHORN (*m*): sycamore (*Acer pseudo-platanum*) (see **AHORN, GEMEINER-**) 853
AFTERKEGEL (*m*): conoid (Maths.) 854
AFTERKEGELFÖRMIG: conoid 855
AFTERKIEL (*m*): false-keel (Naut.) 856
AFTERKOHLE (*f*): slack (dust, small) coal 857
AFTERKRISTALL (*m*): pseudo morphous crystal 858

AFTERKUGEL (*f*) : spheroid (Maths.) 859
AFTERKUGELFÖRMIG : spheroid 860
AFTERMEHL (*n*) : pollard, coarse flour 861
AFTERSCHÖRL (*m*) : axinite (mineral) 862
AFTERSILBER (*n*) : silver containing dross 863
AFTERVERMIETEN : to sublet 864
AFTERVERPACHTEN : to sublet 865
AGALMATOLIT (*m*) : (see **PYROPHYLLIT** and **TALK**) 866
AGALMATOLITH (*m*) : agalmatolite (a compact pyrophyllite) (see **PYROPHYLLIT**) 867
AGAR-AGAR (*n*) : agar-agar (from the seaweed, *Gracilaria lichenoides*), Japanese gelatine, or isinglass ; (also obtained from the white seaweeds, *Fucus spinosus* and *Gelidium corneum*) (see **JAPANLEIM**) 868
AGARICINSÄURE (*f*) : agaricic acid (from *Agaric*) (see **AGARIZIN**) 869
AGARIZIN (*n*) : (*Agaricinum*), agaricine (resinic acid from the fruit of *Polyporus officinalis*, *Agaricus albus*) (used as medicine, *Acidum agaricinicum*) 870
AGAT (*m*) : (see **ACHAT**) 871
AGENS (*n*) : agent, principle 872
AGENT (*m*) : agent ; consignee 873
AGENTUR (*f*) : agency 874
AGGLUTINIEREN : to agglutinate 875
AGGREGAT (*n*) : unit ; installation ; plant ; set ; aggregate ; average 876
AGGREGATZUSTAND (*m*) : average ; state of aggregation ; aggregate state 877
AGGREGIEREN : to aggregate 878
AGGRESSIVITÄT (*f*) : aggressiveness 879
AGIO (*n*) : agio ; premium ; change ; exchange 880
AGITAKEL (*m*) : stirrer 881
AGLYKON (*n*) : a cleavage product of gardenia dye 882
AGOMETER (*n*) : agometer 883
AGRIKULTUR (*f*) : agriculture 884
AGRIKULTURCHEMIE (*f*) : agricultural chemistry 885
AGTSTEIN (*m*) : amber (see **BERNSTEIN**) 886
AGURIN (*n*) : agurine, theobromine-sodium-acetate, $C_7H_7N_4O_2Na.CH_3CO_2Na$ (used as a medicine for dropsy) 887
AHLBAUM : (*m*) : honeysuckle (*Lonicera Periclymenum*) 888
AHLBEERE (*f*) : red (black, white) currant ; (*Ribes rubum, ribes nigrum, ribes album*) 889
AHLE (*f*) : awl ; bodkin ; punch ; pricker ; black cherry 890
AHMINGS (*f.pl*) : depth gauge, draught marks (of ships) 891
AHNDEN : to punish 892
AHNE (*f*) : awn ; chaff 893
ÄHNELN (-SICH) : to liken oneself ; be similar ; be like unto 894
ÄHNLICH : similar- (to) ; analogous- (to) 895
ÄHNLICHKEITSSATZ (*m*) : law of similarity 896
AHORN (*m*) : maple tree (*Acer saccharinum*) ; Sg. 0.6-0.8 897
AHORN (*m*), **GEMEINER-** : sycamore tree (*Acer pseudo-platanum*) ; Sg. 0.5-0.8 898
AHORNGEWÄCHSE (*n.pl*) : aceraceae (Bot.) 899
AHORNHOLZ (*n*) : maple wood (see **AHORN**) 900
AHORNLACK (*m*) : maple varnish 901
AHORNSAFT (*m*) : maple sap 902
AHORNSÄURE (*f*) : aceric acid 903
AHORNZUCKER (*m*) : maple sugar 904
ÄHRE (*f*) : ear ; spike ; spicula 905
ÄHRENFRÜCHTE (*f.pl*) : cereals ; grain 906
AIROL (*n*) : airole, Bismuth oxyiodide-gallate $C_6H_2(OH)_3CO_2.Bi(OH)I$ (used as a substitute for iodoform) 907
AIXERÖL (*n*) : (see **OLIVENÖL** and **BAUMÖL**) 908
AJOWAN (*n*) : ajowan 909
AJOWANÖL (*n*) : ajowan oil, ajava oil from seeds of *Carum ajowan*) (about 50% Thymol ; Sg. 0.915 910
AJOWANSAAT (*f*) : ajowan seed (seed of *Carum ajowan*) 911
AKAGIN (*n*) : acagine (mixture of chloride of lime and lead chromate) 912
AKAJOU (*m*) : cashew (see **AKAJOUBAUM**) 913
AKAJOUBALSAM (*m*) : cashew nut oil, cashew oil (*Cardolum vesicans*) (from *Anacardium occidentale*) 914
AKAJOUBAUM (*m*) : cashew tree (*Anacardium occidentale*) 915
AKAJOUNUSS (*f*) : cashew nut 916
AKAJOUÖL (*n*) : cashew nut oil (from *Anacardium occidentale*) 917
AKANTHIT (*m*) : acanthite (see **SILBERGLANZ**) 918
AKAROIDHARZ (*n*) : acaroid resin (from grass-gum tree of the *Xanthorrhoea* species) 919
AKASCHUÖL (*n*) : (see **AKAJOUÖL**) 920
AKAZIE (*f*) : acacia (*Robinia pseudacacia*) (see **AKAZIENHOLZ**) 921
AKAZIENBLÜTE (*f*) : acacia flower (*Flores pruni spinosi*) 922
AKAZIENGUMMI (*n*) : acacia gum ; gum arabic ; Sg. 1.45 923
AKAZIENHOLZ (*n*) : acacia wood, locust wood (*Robinia pseudacacia*) ; Sg. 0.58-0.83 924
AKAZIENÖL (*n*) : acacia oil 925
AKAZIENRINDE (*f*) : acacia bark (*Acaciæ cortex*) 926
AKAZIN (*n*) : gum arabic 927
AKAZINGUMMI (*n*) : gum arabic 928
AKEEÖL (*n*) : akee oil (from seeds of *Blighia sapida*) ; Sg. 0.859. Mp. 25-35°C. 929
AKKOMODAGE (*f*) : (see **AUFMACHUNG**) 930
AKKORD (*m*) : contract 931
AKKORDARBEIT (*f*) : piece work ; contract work 932
AKKORD (*m*) **IM-** : by contract 933
AKKUMULATORENBATTERIE (*f*) : storage battery (Elect.) 934
AKKUMULATORENSÄURE (*f*) : accumulator acid 935
AKKUMULATORENZELLE (*f*) : storage cell 936
AKKUMULATORPLATTE (*f*) : accumulator plate 937
AKMEGELB (*n*) : acme yellow 938
AKMIT (*m*) : acmite, aegirite (a soda-ferric pyroxene) ; $Na.Fe(SiO_3)_2$; Sg. 3.5 939
AKONIT (*n*) : aconite (*Aconitum*) 940
AKONITIN (*n*) : aconitine (*Aconitinum*) (an alkaloid from *Aconitum napellus*) ; $C_{34}H_{47}NO_{11}$; Mp. 195°C. 941
AKONITIN (*n*), **BROMWASSERSTOFFSAURES-** : aconitine hydro-bromide, $C_{34}H_{47}NO_{11}.HBr.2.5H_2O$; Mp. 163°C. 942
AKONITINCHLORHYDRAT (*n*) : aconitine hydrochloride, $C_{23}H_{47}NO_{11}HCl.3H_2O$ 943
AKONITINCHLORID (*n*) : (see **AKONITINCHLORHYDRAT**) 944
AKONITINNITRAT (*n*) : aconitine nitrate ; $C_{34}H_{47}NO_{11}.HNO_3.5.5H_2O$ 945
AKONITINSÄURE (*f*) : aconitic (aconitinic) acid ; Mp. 195°C. ; ($C_{34}H_{47}NO_{11}$) 946
AKONITINSULFAT (*n*) : aconitine sulphate $(C_{34}H_{47}NO_{11})_2.H_2SO_4$ 947

AKONITKNOLLEN (*m*): aconite bulb or tuber (*Tubera* or *Radix aconiti*); aconite root; monkshood, wolfs-bane, friar's cowl (*Aconitum napellus*) 948
AKONITSÄURE (*f*): aconitic acid (see **AKONITINSÄURE**) 949
AKONITWURZEL (*f*): aconite root (see **AKONITKNOLLEN**) (*Radix aconiti*) 950
AKRIDIN (*n*): acridine 951
AKRIDINGELB (*n*): acridine yellow 952
AKRIDINSÄURE (*f*): acridic (acridinic) acid 953
AKROYDMASCHINE (*f*): hot-bulb engine (of the type invented by Akroyd-Stuart) 954
AKRYLSÄURE (*f*): acrylic acid 955
AKTE (*f*): act (law); deed; document; file-(cover); folder 956
AKTENNUMMER (*f*): file number, reference number 957
AKTENPAPIER (*n*): deed (record) paper 958
AKTENSCHRANK (**M**): deed-box, safe; file (filing) cupboard or cabinet 959
AKTENZEICHEN (*n*): reference (on a letter); file (number) reference 960
AKTIE (*f*): share (in a company) 961
AKTIENABSCHNITT (*m*): coupon 962
AKTIENANTEIL (*m*): share 963
AKTIENAUSGABE (*f*): (share)-issue, issue of stock or scrip 964
AKTIENBANK (*f*): joint stock bank 965
AKTIENBESITZER (*m*): shareholder 966
AKTIENBÖRSE (*f*): stock exchange 967
AKTIENDIVIDENDE (*f*): dividend 968
AKTIENFONDS (*m*): share capital, stock 969
AKTIENHANDEL (*m*): stock-jobbing, stock-broking 970
AKTIENINHABER (*m*): shareholder 971
AKTIENKAPITAL (*n*): share capital; stock 972
AKTIENKURS (*m*): market, price 973
AKTIENMAKLER (*m*): stock-broker 974
AKTIENPROMESSE (*f*): scrip 975
AKTIENREITER (*m*): stock-jobber, gambler on the stock exchange 976
AKTIENSCHEIN (*m*): scrip 977
AKTIENZEICHNUNG (*f*): subscription 978
AKTINISCH: actinic 979
AKTINISMUS (*m*): actinism 980
AKTINIUM (*n*): actinium (an element; a radio-active substance in Thorium) 981
AKTINIUMEMANATION (*f*): actinium emanation (a radio-active substance) 982
AKTINOLITH (*m*): actinolite (see **NEPHRIT**) 983
AKTIONÄR (*m*): shareholder 984
AKTIONSTURBINE (*f*): impulse turbine 985
AKTIV: active 986
AKTIVA (*n.pl*): assets (of bankruptcy and commercial) 987
AKTIVEN (*n.pl*): (see **AKTIVA**) 988
AKTIVIEREN: to activate, make active; (*n*) activating, making active 989
AKTIVIERUNG (*f*): activation; making active 990
AKTIVIERUNGSZAHL (*f*): activation number 991
AKTIVITÄT (*f*), **OPTISCHE-**: optical activity, rotatory power 992
AKTIVSCHULDEN (*f.pl*): assets, bills receivable 993
AKTUELL: actual; present; important 994
AKUSTIK (*f*): acoustics 995
AKUT: acute 996
AKZEPTIEREN: to accept 997
ALABANDIN (*m*): alabandite (see **MANGANBLENDE**) 998
ALABANDIT (*m*): alabandite; MnS., Sg. 3.95 (see also **ALABANDIN**) 999
ALABASTER (*m*): alabaster (a finely granular variety of gypsum) ($Ca.SO_4.2H_2O$); Sg. 2.4 000
ALABASTERVERFAHREN (*n*): alabastrine process (Photography) 001
ALANIN (*n*): alanine, d-a-amino- propionic acid, $CH_3.CH(NH_2).CO_2H$ 002
ALANT (*m*): elecampane (*Inula helenium*) 003
ALANTHOL (*n*): Helenin (see **HELENIN**) 004
ALANTKAMPHER (*m*): Elecampane camphor; Helenin (see **HELENIN**) 005
ALANTWURZEL (*f*): scab-wort; elf-wort; horse-heel (root of *Inula Helenium*); Inula; elecampane (*Radix enulæ*; *radix helenii*) 006
ALANYLGLYCIN (*n*): alanylglycine, $NH_2.CH(CH_3).CO.NH.CH_2.CO_2H$ 007
ALAUN (*m*): alum (see **ALUMINIUMAMMONIUMALAUN**; **ALUMINIUMKALIUMALAUN**, **ALUMINIUMNATRIUMALAUN**, **ALUMINIUMRUBIDIUMALAUN**) 008
ALAUNBEIZE (*f*): alum-(inous) mordant 009
ALAUNBRÜHE (*f*): alum steep (leather) 010
ALAUNERDE (*f*): alumina (see **ALUMINIUMOXYD**); (A_2O_3) 011
ALAUNERDEHALTIG: aluminous; aluminiferous; containing alum 012
ALAUNERZ (*n*): alunite (see **ALUNIT**) 013
ALAUN (*m*), **KONZENTRIERTER-**: (*Alumen concentratum*) (see **ALUMINIUMSULFAT, WASSERFREI-**) 014
ALAUNFELS (*m*): alunite; Sg. 2.83; ($K_2O_3.Al_2O_3.4SO_3.6H_2O$) 015
ALAUNFESTIGKEIT (*f*): alum resistance (paper) 016
ALAUNGAR: tawed; steeped; dressed with alum 017
ALAUNGERBEREI (*f*): tawing; tawery (see **WEISSGERBEREI**) 018
ALAUNIG: aluminous; aluminiferous; containing alum 019
ALAUNKIES (*m*): aluminous pyrites 020
ALAUNKUCHEN (*m*): alum cake 021
ALAUNLAUGE (*f*): alum liquor 022
ALAUNLEDER (*n*): alum leather 023
ALAUNMEHL (*n*): powdered alum 024
ALAUNSAUER: aluminate-(of) 025
ALAUNSAURES SALZ (*n*): aluminate 026
ALAUNSCHIEFER (*m*): alunite; alum slate; alum shale (see **ALUNIT**) 027
ALAUNSIEDEREI (*f*): alum works 028
ALAUNSPAT (*m*): (see **ALUNIT** and **ALAUNSTEIN**) 029
ALAUNSTEIN (*m*): alunite; alum stone (see **ALAUNFELS** and **ALUNIT**) 030
ALAUNTON (**M**): alum earth 031
ALAUNWURZEL (*f*): alum root 032
ALBERTYPIE (*f*): albertype, collotype (see **LICHTDRUCK**) (Photography) 033
ALBIDUR: an iron-aluminium alloy; Sg. 2.9-3 034
ALBIT (*m*): albite; soda feldspar; Sg. 2.6; Mp. 1135-1215°C.; ($Al_4Na_4Si_{12}O_{32}$) 035
ALBOLEN (*n*): albolene, paraffin oil, liquid petrolatum; Sg. 0.84-0.94 036
ALBUMIN (*n*): albumin, albumen (from Ox-blood, Casein and egg-white) 037
ALBUMINAT (*n*): albuminate 038
ALBUMIN (*n*), **GERBSAURES-**: albumin tannate 039
ALBUMINPAPIER (*n*): albumin paper (used for photographic purposes) 040
ALBUMINPAPIER (*n*), **BRILLANT-**: double albumenised paper (Photography) 041

ALBUMINSTOFF (*m*): albumen; egg-white; albuminous substance; protein 042
ALBUMINURIE (*f*): albuminuria (Med.) 043
ALBUMINVERFAHREN (*n*): albumen process (Photography) 044
ALBUMOSE (*f*): albumose (cleavage product of egg-white and used for foodstuffs) 045
ALCANNA (*f*): (see ALKANNA) 046
ALCHIMIE (*f*): alchemy 047
ALCHIMIST (*m*): alchemist 048
ALDEHYD (*n*): aldehyde (see ACETALDEHYD) 049
ALDEHYDAMMONIAK (*n*): ammonium aldehyde, $CH_3.CH(OH)NH_2$ 050
ALDOL (*n*): aldol, oxybutyric aldehyde; $C_4H_8O_2$; Sg. 1.1094; Bp. 100°C. 051
ALEBEREITUNG (*f*): ale making (Brewing) 052
ALEMBROTHSALZ (*n*): (salt of) -alembroth; $(2NH_4Cl.HgCl_2.H_2O)$ 053
ALEXANDRIT (*m*): alexandrite (a variety of chrysoberyl) (see CHRYSOBERYLL) 054
ALFA (*f*): esparto 055
ALFENID (*n*): argentan; German silver (see NEUSILBER) 056
ALGAROBILLA (*n*): algarobilla (pods of *Caesalpinia brevifolia*) (contain about 45% tannin) 057
ALGAROTPULVER (*n*): algaroth-(powder); oxychloride of antimony; basic antimony chloride; mercurius vitæ; (Sb.OCl.); (see also ANTIMONOXYCHLORID) 058
ALGE (*f*): alga; sea-weed 059
ALGEBRA (*f*): algebra 060
ALGEBRAISCH: algebraical 061
ALGODONIT (*m*): algodonite (a natural copper and arsenic ore); Cu_3As (about 72% copper) Sg. 7.3 062
ALICYCLISCH: alicyclic 063
ALIPHATISCH: aliphatic 064
ALIQUANT: aliquant (Maths.) 065
ALIQUOT: aliquot (Maths.) 066
ALIT (*n*): alite (a 15% aluminium-iron alloy) 067
ALITIEREN: to alite (annealing of steel in a special powder containing aluminium, in which the aluminium enters the material to a certain depth and forms a chemical combination with it). 068
ALITIERVERFAHREN (*n*): aliting process (see ALITIEREN) 069
ALIZARIN (*n*): alizarin (the colouring principle of madder) (see KRAPPROT) 070
ALIZARINBLAU (*n*): alizarin blue 071
ALIZARINFARBE (*f*): alizarine dye 072
ALIZARINKRAPPLACK (*m*): alizarin madder lake (see also KRAPPROT) 073
ALIZARINROT (*m*): alizarin red (see KRAPPROT) 074
ALKALI (*n*): alkali 075
ALKALIBLAU (*n*): alkali blue 076
ALKALIECHT: fast to alkali 077
ALKALIECHTFARBE (*f*): alkali fast (dye) colour 078
ALKALIECHTROT (*n*): alkali fast red 079
ALKALIHALTIG: containing alkali, alkaline 080
ALKALIHUMAT (*n*): alkali humate (from peat) 081
ALKALILÖSLICH: soluble in alkali 082
ALKALILÖSUNG (*f*): alkali solution 083
ALKALIMETRIE (*f*): alkalimetry 084
ALKALIMETRISCH: alkalimetric 085
ALKALISALZ (*n*): alkali salt 086
ALKALISCH: alkalious 087
ALKALISCHE ENTWICKELUNG (*f*): alkaline development (Photography) 088
ALKALISCHE ERDMETALLE (*n.pl*): alkaline earth metals 089
ALKALISCHE LÖSUNG (*f*): alkaline solution, lye 090
ALKALISCHE LUFT (*f*): alkaline air; (Priestley's name for ammonia) 091
ALKALITÄT (*f*): alkalinity 092
ALKALOID: alkaloid-(al); (*n*) alkaloid 093
ALKALOIDLÖSUNG (*f*): alkaloid solution 094
ALKANNA (*f*): alkanna; alkanet (root of *Anchusa tinctoria*) 095
ALKANNA (*f*), ECHTE-: root of *Lawsonia alba* or *L. inermis* 096
ALKANNAEXTRAKT (*m*): alkannin, alkanna extract (see ALKANNIN) 097
ALKANNA (*f*), FALSCHE-: (see ALKANNA) 098
ALKANNAROT (*n*): alkanna red; anchusin 099
ALKANNAWURZEL (*f*): alkanna root (*Radix alcannae*) (see ALKANNA) 100
ALKANNIN (*n*): alkannin, alkanna extract (from root of *Alkanna* (*Anchusa*) *tinctoria*) 101
ALKEN (*n*): alkene (see OLEFIN) 102
ALKERMES (see KERMES-) 103
ALKOGEL (*n*): alcogel 104
ALKOHOL (*m*): (absolute-) alcohol; ethyl alcohol; (C_2H_5OH); Sg. pure at 60°F. = 0.79; Mp. -112.3°C.; Bp. 78.4°C. 105
ALKOHOL ABSOLUTUS: (see ALKOHOL) 106
ALKOHOLAUSZUG (*m*): alcohol-(ic) extract 107
ALKOHOLFREI: free from alcohol; non-alcoholic 108
ALKOHOLGÄRUNG (*f*): alcoholic fermentation 109
ALKOHOLMESSER (*m*): alcoholometer 110
ALKOHOLOMETER (*n*): alcoholometer 111
ALKOHOLSÄURE (*f*): alcohol acid 112
ALKOSOL (*n*): alcosol 113
ALKYL (*n*): alkyl 114
ALKYLDERIVAT (*n*): alkyl derivate 115
ALKYLEN (*n*): alkylene (see OLEFIN) 116
ALKYLIEREN: to alkylate; introduce alkyl 117
ALLANIT (*m*): allanite (a variety of epidote) (see EPIDOT) 118
ALLANTOISCH: allantoic 119
ALLANTOISSÄURE (*f*): allantoic acid; allantoin 120
ALLANTOSÄURE (*f*): (see ALLANTOISSÄURE) 121
ALLEIN: but; alone 122
ALLEINVERKAUF (*m*): sole agency, monopoly 123
ALLEMONTIT (*m*): (As, Sb.) allemontite; Sg. 6.2 124
ALLERDINGS: undoubtedly, without doubt, no doubt, to be sure, however 125
ALLERLEIGEWÜRZ (*n*): allspice, pimento, pimenta, Jamaica pepper (fruit of *Pimenta officinalis*) 126
ALLFÄLLIG: in all cases 127
ALLGEMEIN: general, universal, common 128
ALLIGATORENÖL (*n*): alligator oil 129
ALLIIEREN: to ally 130
ALLMÄHLICH: gradual-(ly) 131
ALLMONATLICH: once a month, monthly 132
ALLOGAMIE (*f*): allogamy, cross-fertilisation (of ovules of one plant by pollen of another) (Bot.) 133
ALLOISOMERIE (*f*): alloisomerism 134
ALLOMERISMUS (*m*): allomerism 135
ALLOMORPH: allomorphic; allomorphous 136
ALLONGE (*f*): adapter 137
ALLOPALLADIUM (*n*): (Pd.) Allopalladium (a variety of palladium) 138
ALLOPHAN (*m*): allophane, $Al.Si.O_5.5H_2O$; Sg. 1.9 139
ALLOTROP: allotropic 140
ALLOTROPIE (*f*): allotropy; allotropism 141

ALLOXURBASE (*f*) : alloxuric base ; purine base 142

ALLOXURKÖRPER (*m* and *pl*): alloxuric substance(s) ; purine substance(s) 143

ALLOZIMTSÄURE (*f*) : allocinnamic acid 144

ALLUVIUM (*n*) : alluvium (Geology) (a term used to denote formations which are taking place at present) 145

ALLYLACETON (*n*): allyl acetone ; $C_6H_{10}O$; Sg. 0.832 ; Bp. 130°C. 146

ALLYLALKOHOL (*m*): allyl alcohol ; (CH_2CHCH_2OH) ; Sg. 0.8491 ; Mp. -129°C ; Bp. 96.69°C. 147

ALLYLAMIN (*n*): allyl amine; C_3H_7N ; Sg. 0.7631; Bp. 56.5°C. 148

ALLYLÄTHER (*m*): allyl ether ; $C_6H_{10}O$; Sg. 0.8046 ; Bp. 94.3°C. 149

ALLYLBROMID (*n*): allyl bromide ; C_3H_5Br. ; Sg. 1.3980 ; Bp. 71°C. 150

ALLYLCHLORID (*n*): allyl chloride ; C_3H_5Cl. ; Sg. 0.9379 ; Mp. -136.4°C ; Bp. 45°C. 151

ALLYLJODID (*n*): allyliodide: C_3H_5I; Sg. 1.8293; Bp. 102°C. 152

ALLYLSALICYLAT (*n*) : allyl salicylate 153

ALLYLSENFÖL (*n*) : allyl isothiocyanate, mustard gas, allyl mustard oil (a poison gas); C_4H_5NS ; Sg. 1.0057 ; Mp. about -80°C. ; Bp. 148.2°C. 154

ALLYLSULFID (*n*) : allyl sulphide 155

ALLYLSULFOCARBAMID (*n*) : allyl (sulfocarbamide) thiocarbamide 156

ALLYLSULFOCYANID (*n*): allyl (sulfocyanide) thio-cyanide ; allyl (sulfocyanate) thiocyanate 157

ALLYLSULPHOHARNSTOFF (*m*): allylsulphurea (see **THIOSINAMIN**) 158

ALLYL (*n*), **ZIMTSAURES-** : allyl cinnamate 159

ALLZUWEIT : too far 160

ALM (*f*) : mountain pasture 161

ALMANDIN (*m*) : almandite (a variety of garnet) (Mg.Fe.Ca.)$_3$(Al.Fe.)$_2$(SiO_4)$_3$; Sg. 4.2 [162

ALMANDINSPAT (*m*) : **RHOMBOEDRISCHER-** : eudialyte ; Sg. 2.9. 163

ALMERODER TIEGEL (*m*) : Hessian crucible 164

ALOË (*f* and *pl*) : aloe(s) ; (*Aloe vulgaris*) 165

ALOËEXTRAKT (*m*) : aloe extract (from leaves of *Aloe vulgaris*, etc.) 166

ALOË (*f*), **GLÄNZENDE-** : (see **KAPALOË**) 167

ALOEHOLZ (*n*) : aloe wood (wood of *Aloe vulgaris*) 168

ALOE (*f*), **MATTE-** : hepatic aloe, curacao (*Aloe hepatica*) 169

ALOEROT (*n*) : aloe red 170

ALOËSÄURE (*f*) : aloic acid ; aloetic acid 171

ALOETINSÄURE (*f*) : aloetic acid 172

ALOIN (*n*): aloin, barbaloin; $C_{17}H_{18}O_7.\frac{1}{2}H_2O$ [173

ALOXIT (*n*): aloxite (an abrasive) (similar to Alundum, which see) 174

ALPAKA (*n*): same as " **NEUSILBER** ", which see 175

ALPAKKA (*n*): same as " **NEUSILBER** ", which see 176

ALPENROSENGEBÜSCH (*n*): Rhododendron bush 177

ALPHA : alpha (α) 178

ALPHA-DIPHENYLENMETHAN (*n*): (see **FLUOREN**) 179

ALPRANKEN (*m*): bittersweet (*Solanum dulcamara*) 180

ALQUIFOUX (*n*): alquifou ; potter's lead 181

ALRAUN (*m*): alrune; mandrake (*Podophyllum peltatum*) 182

ALRAUNE (*f*) : see (**ALRAUN**) 183

ALRAUNWURZEL (*f*): mandragora (Pharm.); (root of *Podophyllum peltatum*) 184

ALSBALDIG : immediate, direct 185

ALSTONIT (*m*) : (Ba,Ca)CO_3 ; Alstonite ; Sg. 3.7 186

ALTAIT (*m*): altaite (a natural lead telluride); PbTe ; Sg. 8.2 187

ALTEISEN (*n*) : old (scrap) iron 188

ALTERANTIA (*n.pl*) : alteratives 189

ALTERN (*n*) : brittle test (metal) ; tiring ; fatiguing (metal) ; ageing 190

ALTERNIEREND : alternating 191

ALTERSABSCHLEIFUNG (*f*) : Osteo-arthritis, rheumatoid arthritis of joints (Med.) 192

ALTERTUMSKUNDE (*f*) : archeology 193

ALTERUNGSGRENZE (*f*) : maximum brittleness ; limit of brittleness (metal) 194

ALTGESELL (*m*) : bachelor ; foreman 195

ALTHEAWURZEL (*f*) : marsh-mallow (althea) root ; (*Radix althaea officinalis*) 196

ALTHEE (*f*): marsh-mallow ; althaea ; althea (*Althaea officinalis*) 197

ALTHEEWURZEL (*f*) : (see **ALTHEAWURZEL**) 198

ALTMALZ (*n*) : old malt ; stored malt 199

ALTSCHADENWASSER (*n*) : yellow mercurial lotion (Pharm.) 200

ALTWEIBER-KNOTEN (*m*) : false knot, granny 201

ALUCHIHARZ (*n*) : aluchi (acouchi) resin 202

ALUMINA (*n*) : alumina (Al_2O_3) (see **ALUMINIUMOXYD**) 203

ALUMINAT (*n*) : aluminate 204

ALUMINIT (*m*) : aluminite; $Al_2SO_6.9H_2O$; Sg. 1.7. 205

ALUMINIUM (*n*): aluminium (Al) ; Sg. 2.7 ; Mp. 700°C. 206

ALUMINIUMACETAT (*n*) : (see **ALUMINIUMAZETAT**) 207

ALUMINIUMACETOTARTRAT (*n*) : aluminium acetotartrate 208

ALUMINIUMACETOTARTRATLÖSUNG (*f*): (*Liquor aluminii aceticotartarici*) : aluminium acetotartrate solution ; Sg. 1.262 ; (about 45% aluminium acetotartrate) 209

ALUMINIUMALKOHOLAT (*n*) : aluminium alcoholate ($C_4H_9O_3$)Al ; Mp. 140°C. 210

ALUMINIUM (*n*), **AMEISENSAURES-** : aluminium formate (see **TONERDE, AMEISENSAURE-**) 211

ALUMINIUMAMMONIUMALAUN (*m*) (*Alumen ammoniacale*) : aluminium ammonium sulphate, aluminium ammonium alum ; $Al(NH_4)(SO_4)_2.12H_2O$; Sg. 1.6357-1.645 ; Mp. 94.5°C. 212

ALUMINIUMAMMONIUMSULFAT (*n*) : (see **ALAUN**) 213

ALUMINIUMAZETAT (*n*): aluminium acetate (*Aluminicum aceticum*) (see **TONERDE, ESSIGSAURE**) 214

ALUMINIUMAZETAT (*n*), **BASISCHES-** : basic aluminium acetate ; $Al(C_2H_3O_2)_2OH$ 215

ALUMINIUMAZETATLÖSUNG (*f*): (*Liquor aluminii acetici*): aluminium acetate solution; Sg. 1.046 ; about 8% $Al(C_2H_3O_2)_2OH$ [216

ALUMINIUMBLECH (*n*) : sheet aluminium 217

ALUMINIUMBROMID (*n*) (*Aluminium bromatum*) : aluminium bromide ($AlBr_3$), Sg. 2.54 [218

ALUMINIUMBRONZE (*f*) : aluminium bronze (an alloy) ; 90-95% Cu. ; 5-10% Al. ; trace of manganese 219

ALUMINIUMCARBID (*n*): aluminium carbide (Al_4C_3) ; Sg. 2.36 220

ALUMINIUMCHLORAT (*n*): aluminium chlorate (*Aluminium chloricum*) (see TONERDE, CHLORSAURE); $Al(ClO_3)_3$ 221

ALUMINIUMCHLORID (*n*): (*Aluminium chloratum*); aluminium chloride $AlCl_3$; Bp. 183°C.; Mp. 190°C. 222

ALUMINIUM (*n*), **CHLORSAURES-**: aluminium chlorate (see ALUMINIUMCHLORAT) 223

ALUMINIUMDRAHT (*m*): aluminium wire 224

ALUMINIUM (*n*), **ESSIGSAURES-**: aluminium acetate (*Aluminium aceticum*); $Al(C_2H_3O_2)_3$ [225

ALUMINIUM (*n*), **ESSIGWEINSTEINSAURES-**: aluminium acetotartrate 226

ALUMINIUMFABRIKATE (*n.pl*): aluminium manufactures 227

ALUMINIUM (*n*), **FETTSAURES-**: aluminium sebate 228

ALUMINIUMFLUORID (*n*): (*Aluminium fluoratum*); aluminium fluoride; (AlF_3) (Sg. 3.1) [229

ALUMINIUMFLUORWASSERSTOFFSÄURE (*f*): fluo-aluminic acid 230

ALUMINIUMFORMIAT (*n*): aluminium formate (see TONERDE, AMEISENSAURE) 231

ALUMINIUMGIESSEREI (*f*): aluminium foundry 232

ALUMINIUMGUSS (*m*): aluminium casting 233

ALUMINIUM (*n*), **HARZSAURES-**: aluminium resinate; $Al(C_{44}H_{63}O_5)_3$ 234

ALUMINIUM (*n*), **HOLZESSIGSAURES-**: aluminium pyrolignite 235

ALUMINIUMHYDROXYD (*n*): aluminium hydroxide (*Alumina hydrata*) (see ALUMINIUMOXYDHYDRAT) 236

ALUMINIUMJODID (*n*): aluminium iodide; (AlJ_3) (Sg. 2.63); Mp. 185°C; Bp. 360°C. 237

ALUMINIUMKALIUMALAUN (*m*): aluminium-potassium sulphate; aluminium-potassium alum; $AlK(SO_{11})_2+12H_2O$; Sg. 1.751-1.7546; Mp. 105°C. 238

ALUMINIUMKALIUMSULFAT (*n*): (see ALAUN) 239

ALUMINIUMKRYOLITH (*m*): aluminium cryolite; ($AlNa_3F_6$) (Sg. 2.9) 240

ALUMINIUMLAKTAT (*n*): aluminium lactate (see TONERDE, MILCHSAURE) 241

ALUMINIUMMESSING (*n*): aluminium brass (an alloy); 60-68% Cu.; 27-32% Zn.; 1.5-8% Al. 242

ALUMINIUMMETAPHOSPHAT (*n*): aluminium metaphosphate; $Al(PO_3)_3$; Sg. 2.779 [243

ALUMINIUM (*n*), **MILCHSAURES-**: aluminium lactate (see TONERDE, MILCHSAURE-) 244

ALUMINIUMNATRIUMALAUN (*m*): aluminium sodium sulphate; aluminium-sodium alum; $AlNa(SO_4)_2.12H_2O$; Sg. 1.675; Mp. 61°C. 245

ALUMINIUMNATRIUMAZETAT (*n*): aluminium-sodium acetate (see ACETONAL) 246

ALUMINIUMNATRIUMCHLORID (*n*): aluminium-sodium chloride (*Aluminium-natrium chloratum*); $Al_2Cl_6.2NaCl$; Mp. 185°C. [247

ALUMINIUM-NATRIUM-FLUORID: Aluminium-sodium fluoride, Kryolith, Cryolite; $AlNa_3F_6$ (see also KRYOLITH) 248

ALUMINIUMNATRIUMSULFAT (*n*): (see ALAUN) 249

ALUMINIUMNITRAT (*n*): aluminium nitrate (*Aluminium nitricum*); $Al(NO_3)_3+9H_2O$; Mp. 72.8°C. 250

ALUMINIUMNITRID (*n*): aluminium nitride; AlN 251

ALUMINIUM (*n*), **ÖLSAURES-**: aluminium oleate (see TONERDE, ÖLSAURE-) 252

ALUMINIUM-ORTHOPHOSPHAT (*n*): aluminium orthophosphate; $AlPO_4$; Sg. 2.59 253

ALUMINIUMOXALAT (*n*): aluminium oxalate (*Aluminium oxalicum*); $Al_2(C_2O_4)_3$ 254

ALUMINIUM (*n*), **OXALSAURES-**: aluminium oxalate (see ALUMINIUMOXALAT) 255

ALUMINIUMOXYD (*n*) (*Alumina*): aluminium oxide, alumina; Al_2O_3; Sg. 3.85; Mp. 2020°C.; corundum, ruby, sapphire 256

ALUMINIUMOXYDHYDRAT (*n*): (*Aluminia hydrata*), aluminium trihydrate, aluminium hydrate, aluminium hydroxide; $Al(OH)_3$; Sg. 2.423 (see also TONERDEHYDRAT) 257

ALUMINIUM (*n*), **PALMITINSAURES-**: aluminium palmitate; $Al(C_{15}H_{42}O_4)_3$ 258

ALUMINIUMPHOSPHAT (*n*): aluminium phosphate; $AlPO_4$; Sg. 2.55 259

ALUMINIUM (*n*), **PHOSPHORSAURES-**: aluminium phosphate (see ALUMINIUMPHOSPHAT) 260

ALUMINIUMPROFILE (*n.pl*): aluminium sections 261

ALUMINIUMRHODANÜR (*n*): (*Aluminium rhodanatum*): aluminium rhodanide, aluminium sulphocyanate, aluminium thiocyanate, aluminium sulfocyanide; $Al(SCN)_3$ 262

ALUMINIUMROHR (*n*): aluminium pipe 263

ALUMINIUMRUBIDIUM ALAUN (*m*): aluminium rubidium sulphate, aluminium rubidium alum; $AlRb(SO_4)_2.12H_2O$; Sg. 1.8667 [264

ALUMINIUM (*n*), **SALPETERESSIGSAURES-**: aluminium nitroacetate 265

ALUMINIUM (*n*), **SALPETERSAURES-**: aluminium nitrate (see ALUMINIUM-NITRAT) 266

ALUMINIUMSALZ (*n*): aluminium salt 267

ALUMINIUM (*n*), **SCHWEFELESSIGSAURES-**: aluminium sulphacetate 268

ALUMINIUM (*n*), **SCHWEFELSAURES-**: aluminium sulphate (see ALUMINIUMSULFAT) 269

ALUMINIUM (*n*), **SCHWEFLIGSAURES-**: aluminium sulphite 270

ALUMINIUMSILICOFLUORID (*n*): aluminium fluosilicate; $Al_2(SiF_6)_3$ 271

ALUMINIUMSILIKAT (*n*): (*Aluminium silicicum*): aluminium silicate, andalusit, cyanite, disthene; Al_2SiO_5; Sg. 3.022 (see also TONERDE, KIESELSAURE) 272

ALUMINIUMSTANGE (*f*): aluminium bar or rod 273

ALUMINIUM (*n*), **STEARINSAURES-**: aluminium stearate; $Al(C_{18}H_{35}O_2)_3$ 274

ALUMINIUMSULFAT (*n*) (*Aluminium sulfuricum*): aluminium sulphate; $Al_2(SO_4)_3.18H_2O$; Sg. 1.6 275

ALUMINIUMSULFATHYDRAT (*n*): hydrated aluminium sulphate; $Al_2(SO_4)_3.18H_2O$; Sg. 1.62 276

ALUMINIUMSULFAT (*n*), **WASSERFREI**: aluminium sulphate anhydrous; $Al_2(SO_4)_3$; Sg. 2.7; (*Aluminium sulfuricum*) (*Alumen concentratum*) (see also TONERDE, SCHWEFELSAURE) 277

ALUMINIUMSULFID (*n*): aluminium sulphide; Al_2S_3; Sg. 2.37 278

ALUMINIUM SULPHOCYANID (*n*): (see RHODAN-ALUMINIUM) 279

ALUMINIUMWALZWERK (*n*): aluminium rolling mill 280

ALUNDUM (*n*): alundum, aluminium oxide; Al_2O_3; Sg. 4.0; Mp. 2000°C. (an abrasive) 281

ALUNIT (*m*): alunite; $K_2O_3.Al_2O_3.4SO_3.6H_2O$ (11.4% potash); Sg. 2.83 282

AMALGAM (*n*): amalgam; mercury alloy; Ag.Hg.; Sg. 13.9 283
AMALGAMIERSALZ (*n*): amalgam salt (see **QUECKSILBEROXYDSULFAT**) 284
AMARANT (*n*): amaranth (a coal-tar dye, deep red) 285
AMATEURAUFNAHME (*f*): amateur photograph 286
AMAZONENSTEIN (*m*): (see **MIKROKLIN**) 287
AMBER (*m*): amber 288
AMBER, GRAUER- (*m*): ambergris (see **AMBRA**) 289
AMBLYGONIT (*n*): amblygonite (Fluophosphate of Aluminium and Lithium); $PO_4(AlF)Li$; Sg. 3.08 290
AMBOSS (*m*): anvil 291
AMBRA (*m*): ambergris (*Ambra grisea*): Sg. 0.91; Mp. 60°C. 292
AMBRABAUM (*m*): amber tree 293
AMBRAFETT (*n*): ambrein (see **AMBRIN**) 294
AMBRA, FLÜSSIGER- (*m*): liquid amber 295
AMBRA, GELBER- (*m*): yellow (ordinary; common) amber 296
AMBRA, GRAUER- (*m*): ambergris (see **AMBRA**) 297
AMBRAHOLZ (*n*): yellow sandal-wood 298
AMBRAIN (*n*): (see **AMBREIN**) 299
AMBRAÖL (*n*): amber oil; Sg. 0.915-0.975 [300
AMBREIN (*n*): ambrein [a substance similar to cholesterme, from amber, (see **AMBRA**)]; Mp. 55°C. 301
AMBRETTEÖL (*n*): ambrette oil; abelmoschus oil 302
AMBRIN (*n*): ambrein; Mp. 55°C. (see **AMBREIN**) 303
AMEISE (*f*): ant (insect of the order *Formicidæ*) 304
AMEISENÄTHER (*m*): formic ether, ethyl formate; $HCO_2.C_2H_5$; Sg. 0.9231; Mp. -80.5°C.; Bp. 54.05°C. 305
AMEISENEIER (*n.pl*): ant's eggs 306
AMEISENGEIST (*m*): spirit of ants (mixture of formic acid, alcohol and water) (Pharm.) 307
AMEISENSAUER: formate of; combined with formic acid 308
AMEISENSÄURE (*f*): formic acid (*Acidum formicicum*); HCOOH; Sg. 1.2178; Mp. 8.3°C.; Bp. 100.8°C. 309
AMEISENSÄURE-ÄTHYLESTER (*m*): ethyl formate, formic ether, formic ester (ethyl); $C_3H_6O_2$; Sg. 0.948; Mp. -78.9°C; Bp. 54.4°C; orthoethyl formate 310
AMEISENSÄURE-NITRIL (*n*): formic acid nitrile, prussic acid formonitrile; CHN; Sg. 0.6969; Mp. -10-12°C.; Bp. 25.2°C.; hydrogen cyanide 311
AMEISENSAURESSALZ (*n*): formate 312
AMEISENSPIRITUS (*m*): (see **AMEISENGEIST**) 313
AMESIT (*m*): amesite; $H_4(Mg,Fe)_2Al_2SiO_9$; Sg. 2.71 314
AMETALL (*n*): non-metal 315
AMETHYST (*n*): amethyst (a transparent violet variety of quartz) (see **QUARZ**) 316
AMIANT (*m*): amianthus; silky asbestos 317
AMIDGRUPPE (*f*): amido (amino; amide) group 318
AMIDIEREN: to amidate; convert into an amide 319
AMIDO- (prefix): the latest German parctice is to write all these words as amino-) 320
AMIDOACETANILID (*n*), **PARA-**: para-aminoacetanilide; $NH_2C_6H_9NHOCCH_3$; Mp. 162°C. 321
AMIDOANTIPYRIN (*n*): aminoantipyrine 322
AMIDOÄTHYLBENZOESÄURE (*f*): aminoethylbenzoic acid 323
AMIDOAZOBENZOL (*n*): aminoazobenzene; $NH_2C_6N_2C_6H_5$; Mp. 127.4°C; Bp. about 360°C. 324
AMIDOAZOBENZOLBASE (*f*): aminoazobenzene base 325
AMIDOAZOBENZOLCHLORHYDRAT (*n*): amidoazobenzene (aminoazobenzene) hydrochloride; $NH_2C_6H_4N_2C_6H_5HCl$ 326
AMIDOAZOBENZOLDISULFOSÄURE (*f*): aminoazobenzene disulphonic acid 327
AMIDOAZOBENZOL (*n*), **SALZSAURES**: amidoazobenzene hydrochloride (see **AMIDOAZOBENZOLCHLORHYDRAT**) 328
AMIDOAZOBENZOLSULFOSÄURE (*f*): aminoazobenzenesulphonic acid 329
AMIDOAZOTOLUOL (*n*): aminoazotoluene 330
AMIDOAZOTOLUOLCHLORHYDRAT (*n*): amidoazotoluene (aminoazotoluene) hydrochloride; $(CH_3C_6H_4)_2N_2HCl$ 331
AMIDOAZOTOLUOLDISULFOSÄURE (*f*): aminoazo toluenedisulphonic acid 332
AMIDOAZOTOLUOL (*n*), **SALZSAURES-**: amidoazotoluene hydrochloride (see **AMIDOAZOTOLUOLCHLORHYDRAT**) 333
AMIDOAZOTOLUOLSULFOSÄURE (*f*): aminoazotoluenesulphonic acid 334
AMIDOBENZOESÄUREÄTHYLESTER (*m*): aminobenzoic ethylester, ethyl aminobenzoate 335
AMIDOBENZOL (*n*): aminobenzine; amidobenzine; $C_6H_5NH_2$; Sg, 1.0235; Mp. -5.96°C; Bp. 184.4°C. 336
AMIDOBENZOLSULFOSÄURE (*f*): (*Acidum aminobenzolsulfonicum*): aminobenzenesulphonic acid, sulfanilic acid; $C_6H_4NH_2SO_3H$ 337
AMIDOESSIGSÄURE (*f*): aminocetic acid; $COOHCH_2NH_2$); Sg. 1.1607; Mp. 232-236°C. (see also **AMINOESSIGSÄURE**) 338
AMIDOKOHLENSÄURE (*f*): amidocarbonic acid; carbamic acid 339
AMIDOL (*n*): amidol (diaminophenol hydrochloride) (trade name of a photographic developer) (see **DIAMINOPHENOLCHLORHYDRAT**) 340
AMIDOSCHWEFELSÄURE (*f*): amidosulphuric acid; sulfamic acid; (NH_2SO_3H) 341
AMIDOSULFOSÄURE (*f*): amidosulfonic acid; sulfamic acid; NH_2SO_3H 342
AMIDOMETAKRESOLMETHYLÄTHER (*m*): aminometacresolemethylic ether 343
AMIDONAPHTOLÄTHER (*m*): aminonaphthol ether 344
AMIDONAPHTOLDISULFOSÄURE (*f*): amidonaphtholdisulfonic acid, H.-acid; $C_{10}H_4(OH)(NH_2)(SO_3H)_2$ 345
AMIDONAPHTOLMONOSULFOSÄURE (*f*): G.-acid, aminonaphtholmonosulphonic acid (see **GAMMASÄURE**) 346
AMIDONAPHTOLSULFOSÄURE (*f*): aminonaphtholsulphonic acid, Gamma acid, G.-acid (see **GAMMASÄURE**) 347
AMIDOPHENOL (*n*): aminophenol; $C_6H_4(OH)NH_2$; ortho.-Mp. 170°C; meta.-Mp. 122°C.; para.-Mp. 184°C. 348
AMIDOPYRAZOLIN (*n*): amino pyrazoline 349
AMIDOPYRIN (*n*): pyramidon (see **PYRAMIDON**) 350
AMIDOSALICYLSÄURE (*f*): aminosalicylic acid; $C_6H_3COOH(OH)NH_2$ 351
AMIDOSULFONSÄURE (*f*): aminosulphonic acid, aminosulfonic acid, sulfamic acid; $NH_2.SO_3H$ 352

AMIKROSKOPISCH: amicroscopical (an ultra-microscopical particle which can not be made visible) 353

AMINARTIG: amine-like, like amine 354

AMINOBENZOESÄURE (*f*): aminobenzoic acid, benzaminic acid; ortho.-$C_7H_7O_2N$, Mp. 145°C.; meta.-$C_7H_7O_2N$, Sg. 1.5105, Mp. 174°C.; para.-$C_7H_7O_2N$, Mp. 187°C. [355

AMINOBENZOESÄUREISOBUTYLESTER (*m*), **PARA-**: isobutyl para-aminobenzoate (see **CYCLOFORM**) 356

AMINOBENZOESÄUREPROPYLESTER (*m*), **PARA-**: propyl-para-aminobenzoate; $H_2N.C_6H_4.CO.O.C_3H_7$ 357

AMINOBENZOL (*n*): (see **ANILN**) 358

AMINOBERNSTEINSÄURE (*f*): aminosuccinic acid, asparaginic acid; $HCO_2.CH_2CH(NH_2)CO_2H$ 359

AMINO-β-METHYLÄTHYLPROPIONSÄURE (*f*): (see **ISOLENCIN**) 360

AMINOESSIGSÄURE (*f*): amino-acetic acid; $C_2H_5O_2N$; Sg. 1.1607; Mp. 232-236°C.; glycocoll, glycine, aminoethanoic acid (see **GLYCIN**) 361

AMINOFORM (*n*): aminoform (see **HEXAMETHYLENTETRAMIN**) 362

AMINOGLUTARSÄURE (*f*): amino-glutaric acid, glutamic acid (see **GLUTAMINSÄURE**); $HCO_2(CH_2)_2CH(NH_2)CO_2H$ 363

AMINOISOBUTYLESSIGSÄURE (*f*): (see **LEUCIN**) 364

AMINOISOPROPYLBENZOL (*n*): aminoisopropylbenzene, cumidine (see **CUMIDIN**) 365

AMINOISOVALERIANSÄURE (*f*): amino-isovalerianic acid, valine $(CH_3)_2CHCH(NH_2).CO_2H$ 366

AMINOKOHLENSÄURE (*f*): (see **CARBAMINSÄURE**) 367

AMINONAPHTALIN (*n*): amino-naphthalene (see **NAPHTYLAMIN**) 368

AMINONAPHTHOLDISULFOSÄURE (*f*): aminonaphtholdisulphonic acid $(C_{10}H_4(OH)(NH_2)(SO_3H)_2)$ 369

AMINOPHENETOL (*n*), **PARA-**; phenetidin (see **PHENETIDIN**) 370

AMINOPHENOL (*n*): amino phenol; C_6H_7ON; ortho. Mp. 170°C.; meta. Mp. 123°C.; para. Mp. 184°C. decomposes 371

AMINOPROPIONSÄURE (*f*), d-α-; (see **ALANIN**) 372

AMINOQUECKSILBERCHLORID (*n*): (*Hydrargyrum praecipitatum album*): amino-mercuric chloride, mercuric-ammonium chloride, ammoniated mercury chloride, ammoniated mercury, white precipitate; NH_2HgCl [373

AMINOQUECKSILBERCHLORÜR (*n*): amino-mercurous chloride, mercurous-ammonium chloride; NH_2Hg_2Cl 374

AMINOSÄUREHYDRID (*n*): amino-acid hydride, diketopiperazine 375

AMINOXYLOL (*n*): (see **XYLIDIN**) 376

AMINSCHWARZ (*n*): amine black 377

AMMON (*n*): ammonium (see **AMMONIUM**) 378

AMMONAL (*n*): ammonal (trade name of an explosive); (Mixture of 95% ammonium nitrate (NH_4NO_3), aluminium (Al) and charcoal) 379

AMMON-CARBONIT (*n*): (see **AMMON-KARBONIT**) 380

AMMONCHLORID (*n*): ammonium chloride; (NH_4Cl); Sg. 1.52; (see **AMMONIUMCHLORID**) 381

AMMONIAK (*n*): ammonia-(gas); NH_3; Sg. 0.6 382

AMMONIAKALAUN (*m*): ammonia alum; $(NH_4)_2SO_4.Al_2(SO_4)_3+24H_2O$; Sg. 1.645; Mp. 94.5°C. (see also **ALUMINIUM-AMMONIUM-ALAUN**) 383

AMMONIAK (*n*), **AMEISENSAURES-**: (see **AMMONIUM, AMEISENSAURES-**) 384

AMMONIAKBESTIMMUNG (*f*): determination of ammonia 385

AMMONIAKDESTILLATION (*f*): ammonia distillation, ammonia distilling 386

AMMONIAKDESTILLATIONS-APPARAT (*m*): ammonia distilling plant 387

AMMONIAK (*n*) **DICHROMSAURES-**: ammonium bichromate $(NH_4)_2Cr_2O_7$ (see **AMMONIUM, SAURES CHROMSAURES-**) 388

AMMONIAK (*n*) **DOPPELTCHROMSAURES-**: ammonium bichromate (see **AMMONIUM, SAURES CHROMSAURES-**) 389

AMMONIAK (*n*), **ESSIGSAURES-**: (*Ammonium aceticum*) ammonium acetate; $NH_4(C_2H_3O_2)$ Mp. 89°C. 390

AMMONIAK (*n*), **FLÜSSIGES-**: liquid ammonia 391

AMMONIAKFLÜSSIGKEIT (*f*): ammonia water (Pharm.); ammonia (ammoniacal; gas) liquor (see also **GASWASSER**) 392

AMMONIAKGAS (*n*): ammonia gas 393

AMMONIAK (*n*), **GOLDSAURES-**: ammonium aurate fulminating gold 394

AMMONIAKGUMMI (*n*): (*Ammoniacum*): gum ammoniac, ammoniac (from resin of a Persian umbellated plant, *Dorema ammoniacum*) 395

AMMONIAK (*n*), **KOHLENSAURES-**: ammonium carbonate (see **AMMONIUMKARBONAT**) 396

AMMONIAK (*n*), **MOLYBDÄNSAURES-**: ammonium molybdate (see **AMMONIUMMOLYBDAT**) 397

AMMONIAK (*n*), **OXALSAURES-**: ammonium oxalate (see **AMMONIUMOXALAT**) 398

AMMONIAKPFLANZE (*f*)): ammoniac plant (*Dorema ammoniacum*) 399

AMMONIAKPFLASTER (*n*): ammoniac plaster 400

AMMONIAK (*n*), **PHOSPHORSAURES-**: ammonium phosphate (see **AMMONIUM-PHOSPHAT**) 401

AMMONIAK (*n*), **PLATINSAURES-**: ammonium platinate, fulminating platinum 402

AMMONIAKREST (*m*): ammonia residue; amidogen (NH_2) 403

AMMONIAK (*n*), **SALPETERSAURES-**: ammonium nitrate (see **AMMONIUMNITRAT**) 404

AMMONIAKSALZ (*n*): ammonium salt 405

AMMONIAK (*n*), **SALZSAURES-**: ammonium chloride: NH_4Cl; Sg. 1.520 (see also **AMMONIUMCHLORID**) 406

AMMONIAK (*n*), **SAURES OXALSAURES-**: ammonium bioxalate (see **AMMONIUMBIOXALAT**) 407

AMMONIAK (*n*), **SCHWEFELSAURES-**: ammonium sulphate (see **AMMONIUMSULFAT**) 408

AMMONIAKSUPERPHOSPHAT (*n*): ammonia superphosphate 409

AMMONIAK (*n*), **ÜBERCHLORSAURES-**: ammonium perchlorate (see **AMMONIUMPERCHLORAT**) 410

AMMONIAK (*n*), **ÜBERSCHWEFELSAURES-**: ammonium persulphate (see **AMMONIUMPERSULFAT**) 411

AMMONIAK (*n*), **VANADINSAURES-**: ammonium vanadate (see **AMMONIUMVANADAT**) 412

AMMONIAKVERBINDUNG (*f*)): ammonium compound 413

AMMONIAKWASSER (*n*): gas liquor, ammonia liquor (from Gas Works) 414

AMMONIAK (*n*), **WÄSSERIGES-**: aqueous ammonia 415

AMMONIAKWASSERMESSER (*m*): ammonia-water (gas-liquor) meter 416

AMMONIAK (*n*), **WEINSAURES-**: ammonium tartrate (see **AMMONIUMTARTRAT**) 417

AMMONIAKWEINSTEIN (*m*): ammonium potassium tartrate 418

AMMONIUM (*n*): ammonium 419

AMMONIUM (*n*), **AMEISENSAURES-**: (*Ammonium formicicum*): ammonium formate; $HCO_2.NH_4$ 420

AMMONIUM (*n*), **ARSENIGSAURES-**: ammonium arsenite; NH_4AsO_2 421

AMMONIUM (*n*), **ARSENSAURES-**: ammonium arsenate $(NH_4)_3AsO_4.3H_2O$ 422

AMMONIUMAZETAT (*n*): ammonium acetate (see **AMMONIUM, ESSIGSAURES-**) 423

AMMONIUM (*n*), **BALDRIANSAURES-**: ammonium (valerate, valerianate); valeriate; $NH_4C_5H_9O_4$ 424

AMMONIUM (*n*), **BENZOESAURES-**: ammonium benzoate $(NH_4)C_7H_5O_2$ 425

AMMONIUM (*n*), **BERNSTEINSAURES-**: ammonium succinate 426

AMMONIUMBICARBONAT (*n*): (see **AMMONIUM, SAURES KOHLENSAURES-**) 427

AMMONIUMBICHROMAT (*n*): (see **AMMONIUM, SAURES CHROMSAURES-**) 428

AMMONIUMBIFLUORID (*n*): ammonium bifluoride, $NH_4F.HF.$; Sg. 1.211 429

AMMONIUMBIOXALAT (*n*): ammonium bioxalate; $NH_4.HC_2O_4$ 430

AMMONIUM (*n*), **BORSAURES-**: ammonium borate; $(NH_4)_3BO_3$; Sg. 2.4-2.95 431

AMMONIUMBROMID (*n*): (*Ammonium bromatum*: ammonium bromide; $NH_4Br.$; Sg. 2.327 [432

AMMONIUM (*n*), **CASEINSAURES-**: ammonium caseinate 433

AMMONIUMCHLORAT (*n*): ammonium chlorate; $(NH_4)ClO_3$ 434

AMMONIUMCHLORID (*n*): (*Ammonium chloratum*): ammonium chloride, sal ammoniac; NH_4Cl; Sg. 1.522 435

AMMONIUM (*n*), **CHLORSAURES-**: (see **AMMONIUMCHLORAT**) 436

AMMONIUMCHROMAT (*n*): ammonium chromate; $(NH_4)_2CrO_4$; Mp. 185°C.; Sg. 1.886-1.917 437

AMMONIUM (*n*), **CHROMSAURES-**: ammonium chromate (see **AMMONIUMCHROMAT**) 438

AMMONIUMCITRAT (*n*): ammonium citrate, $NH_4.H_2.C_6H_5O$ or $(NH_4)_2.H.C_6H_5O_7$ [439

AMMONIUM (*n*), **CITRONENSAURES-**: ammonium citrate (see **AMMONIUMCITRAT**) 440

AMMONIUMDICHROMAT (*n*): ammonium bichromate, ammonium dichromate; $(NH_4)_2Cr_2O_7$; Sg. 2.151 441

AMMONIUM (*n*), **ESSIGSAURES-**: ammonium acetate (see **AMMONIAK, ESSIGSAURES**) 442

AMMONIUMFERRICYANÜR (*n*): ammonium ferricyanide 443

AMMONIUMFERRISULFAT (*n*): ferric ammonium sulphate (see **EISENALAUN**) 444

AMMONIUMFERROSULFAT (*n*): Mohr's salt, ferrous-ammonium sulphate; $FeSO_4(NH_4)_2SO_4.6H_2O$; Sg. 1.865 445

AMMONIUMFLUORID (*n*): ammonium fluoride (see **FLUORAMMONIUM**) 446

AMMONIUMFLUORID (*n*), **SAURES-**: (see **AMMONIUMBIFLUORID**) 447

AMMONIUM (*n*), **GOLDCHLORWASSERSTOFFSAURES-** ammonium chloraurate; $NH_4AuCl_4.H_2O$ 448

AMMONIUMHYPOSULFIT (*n*): ammonium hyposulphite; $(NH_4)_2S_2O_3.5H_2O$ 449

AMMONIUM (*n*), **ICHTHYOLSULFOSAURES-**: ammonium ichthyolsulfonate (see **ICHTHYOL**) 450

AMMONIUM-IRIDIUMCHLORID (*n*): ammonium-iridium chloride; $(NH_4)_2IrCl_6$; Sg. 2.856. 451

AMMONIUMJODID (*n*): ammonium iodide; (NH_4I): Sg. 2.501; (*Ammonium jodatum*) 452

AMMONIUMKARBONAT (*n*): ammonium carbonate; (*Ammonium carbonicum*); (salt of-) Hartshorn; $NH_4.HCO_3+CO(NH_2).O.NH_4$: Mp. 85°C. 453

AMMONIUM (*n*), **KOHLENSAURES-**: ammonium carbonate (see **AMMONIUMKARBONAT**) 454

AMMONIUMMETAVANADAT (*n*): ammonium metavanadate; HN_4VO_3 455

AMMONIUM (*n*), **MILCHSAURES-**: ammonium lactate 456

AMMONIUM (*n*), **MOLYBDÄNSAURES-**: ammonium molybdate (see **AMMONIUMMOLYBDAT**) 457

AMMONIUMMOLYBDAT (*n*): (*Ammonium molybdaenicum*); ammonium molybdate; $(NH_4)_2MoO_4$; Sg. 2.4-2.95 458

AMMONIUMNATRIUMPHOSPHAT (*n*), **SAURES-**: microcosmic salt, ammonium-sodium diphosphate, $Na(NH_4)HPO_4.4H_2O$ 459

AMMONIUMNITRAT (*n*): (*Ammonium nitricum*): ammonium nitrate; NH_4NO_3; Sg. 1.73; Mp. 160°C. 460

AMMONIUMOXALAT (*n*): (*Ammonium oxalicum*: ammonium oxalate; $(NH_4)_2C_2O_4$; Sg. 1.502 461

AMMONIUM (*n*), **OXALSAURES-**: ammonium oxalate (see **AMMONIUMOXALAT**) 462

AMMONIUMPALLADIUMCHLORID (*n*): ammonium-palladium chloride, ammonium-palladic chloride; $(NH_4)_2PdCl_6$; Sg. 2.418 463

AMMONIUMPERCHLORAT (*n*): (*Ammonium perchloricum*): ammonium perchlorate; NH_4ClO_4 Sg. 1.95 464

AMMONIUMPERSULFAT (*n*): (*Ammonium persulfuricum*): ammonium persulphate; $(NH_4)_2S_2O_8$ 465

AMMONIUMPHOSPHAT (*n*) (*Ammonium phosphoricum*): ammonium phosphate, ammonium diphosphate; $(NH_4)_2HPO_4$; Sg. 1.62 [466

AMMONIUM (*n*), **PHOSPHORSAURES-**: ammonium phosphate (see **AMMONIUMPHOSPHAT**) 467

AMMONIUMPLATINCHLORID (*n*): ammonium chloroplatinate; $(NH_4)_2PtCl_6$ 468

AMMONIUMRHODANÜR (*n*) (*Ammonium rhodanatum*): ammonium thiocyanate, ammonium (sulfocyanate) sulfocyanide; $(NH_4)CNS$; Sg. 1.3; Mp. 159°C. 469

AMMONIUM (*n*), **SALICYLSAURES-**: ammonium salicylate; $NH_4C_7H_5O_3$ 470

AMMONIUM (*n*): **SALPETERSAURES-**: ammonium nitrate (see **AMMONIUMNITRAT**) 471

AMMONIUMSALZ (*n*): ammonium salt 472

AMMONIUM (*n*), **SAURES CHROMSAURES-** *Ammonium bichromicum*): ammonium bichromate; $(NH_4)_2Cr_2O_7$; Sg. 2.15 473

AMMONIUM (*n*), **SAURES KOHLENSAURES-**: ammonium bicarbonate; NH_4HCO_3; Sg. 1.586 474

AMMONIUM (*n*), **SCHWEFELSAURES-**: ammonium sulphate (see **AMMONIUMSULFAT**) 475

AMMONIUM (*n*), **SCHWEFLIGSAURES-**: ammonium sulphite 476

AMMONIUMSILICOFLUORID (*n*): ammonium fluosilicate; $(NH_4)_2SiF_6$ 477

AMMONIUMSULFAT (*n*) (*Ammonium sulfuricum*): ammonium sulphate $(NH_4)_2SO_4$; Sg. 1.77; Mp. 140°C. 478

AMMONIUMSULFHYDRAT (*n*): ammonium hydrosulphide; (NH_4HS) 479

AMMONIUMSULFID (*n*): (*Ammonium sulfuratum*): ammonium sulphide $(NH_4)_2S$ 480

AMMONIUMSULFOCYANAT (*n*): ammonium sulfocyanate; ammonium thiocyanate; NH_4SCN (or NH_4SCy); Sg. 1.308; Mp. 159°C. 481

AMMONIUMSULFOCYANID (*n*): ammonium thiocyanide; (NH_4SCN); Sg. 1.308; Mp. 159°C.; (see also **AMMONIUMSULFOCYANAT**) 482

AMMONIUMTARTRAT (*n*) (*Ammonium tartaricum*): ammonium tartrate (neutral salt); $(NH_4)_2C_4H_4O_6$; Sg. 1.6 483

AMMONIUMTETRACHROMAT (*n*): ammonium tetrachromate; $(NH_4)_2Cr_4O_{13}$; Sg. 2.343 484

AMMONIUMTHALLIUMALAUN (*m*): ammonium-thallium sulphate, ammonium thallium alum; $AlTl(SO_4)_2.12H_2O$; Sg. 2.32 485

AMMONIUMTRICHROMAT (*n*): ammonium trichromate; $(NH_4)_2Cr_3O_{10}$; Sg. 2.329 486

AMMONIUM (*n*), **ÜBERSCHWEFELSAURES-**: ammonium persulphate; $(NH_4)_2S_2O_8$ (see also **AMMONIUMPERSULFAT**) 487

AMMONIUM (*n*), **UNTERPHOSPHORIGSAURES-**: ammonium hypophosphite 488

AMMONIUM (*n*), **UNTERSCHWEFLIGSAURES-**: ammonium (thiosulphate) hyposulphite; $(NH_4)_2S_2O_3$ 489

AMMONIUMURANAT (*n*): ammonium uranate; $(NH_4)_2U_2O_7$ 490

AMMONIUMVANADAT (*n*): ammonium vanadate; NH_4VO_3 491

AMMONIUM (*n*), **VANADINSAURES-**: ammonium vanadate (see **AMMONIUMVANADAT**) 492

AMMONIUMVERBINDUNG (*f*): ammonium compound 493

AMMONIUM (*n*), **WEINSTEINSAURES-**: ammonium tartrate (see **AMMONIUMTARTRAT**) 494

AMMONIUM (*n*), **WOLFRAMSAURES-**: ammonium (wolframate) tungstate; $(NH_4)_2W_2O_{13}.8H_2O$ 495

AMMONIUMZINKCHLORID (*n*): (see **LÖTSALZ**) 496

AMMONIUMZINNCHLORID (*n*): pink salt, ammoniumstannic chloride (a dyeing mordant); $SnCl_4.2NH_4Cl$ (see also **PINKSALZ**) 497

AMMON-KARBONIT (*n*): ammoncarbonite (trade name of an explosive) (90% ammonium nitrate, 3.8% nitroglycerine, and other constituents) 498

AMMONSALPETER (*m*): ammonium nitrate; (NH_4NO_3); Sg. 1.725; Mp. 160°C. 499

AMMONSULFAT (*n*): ammonium sulphate; $(NH_4)_2SO_4$; Sg. 1.7687; Mp. 140°C. (see also **AMMONIUMSULFAT**) 500

AMNIOSFLÜSSIGKEIT (*f*): amniotic fluid 501

AMNIOSSÄURE (*f*): amniotic acid; allantoin 502

AMOORAÖL (*n*): amoora oil (from seeds of *Amoora rohituka*); Sg. 0.939 503

AMORPHISCH: amorphous 504

AMORTISATION (*f*): liquidation, amortization 505

AMORTISATIONSFOND (*m*): sinking fund (Com.) 506

AMORTISIEREN: to wipe (out) off, pay off, liquidate, amortize 507

AMPERE (*n*): ampere (Elect.) 508

AMPEREMETER (*n*): ammeter 509

AMPERESTUNDENZÄHLER (*m*): ampere hour counter 510

AMPEREWINDUNG (*f*): ampere winding 511

AMPFER (*m*): sorrel; dock (*Rumex acetosa*) 512

AMPHIBOL (*m*): amphibole, amphibolite, hornblende (see also **HORNBLENDE**); (a magnesium metasilicate); $MgSiO_3$; Sg. 3.1-3.2 513

AMPHIBOLASBEST (*m*) asbestos (a fibrous kind of amphibole) 514

AMPHIBOLIT (*m*): (see **AMPHIBOL**) 515

AMPHIGENSPAT (*m*), **TRAPEZOIDALER-**: leucite; $KO.SiO_2+AlO_3S_3O_2$; Sg. 2.48 516

AMPHOTER: amphoteric 517

AMPHOTROPIN (*n*): amphotropine; $C_8H_{14}(COOH_2)([CH_2]_6N_4)_2$ (taken internally as a urine antiseptic) 518

AMPLITUDE (*f*): amplitude (Wireless Tel.) 519

AMPULLE (*f*): ampule; phial 520

AMPULLENFÖRMIG: ampulliform 521

AMPULLENFÜLLMASCHINE (*f*): ampule filling machine 522

AMPULLENZUSCHMELZMASCHINE (*f*): ampule sealing machine 523

AMSEL (*f*): blackbird 524

AMSEL, WASSER- (*f*): ousel 525

AMT (*n*): post, position, office (used as a suffix) 526

AMTSBERICHT (*m*): official report 527

AMTSBLATT (*n*): official gazette 528

AMÜSANT: amusing 529

AMYGDALIN (*n*): amygdaline 530

AMYGDALINSÄURE (*f*): amygdalic acid; phenylhydroxyacetic acid; phenylglycolic acid; benzoglycolic acid; para-mandelic acid amgydalinic acid; $C_6H_5CH(OH)COOH$; Mp. 118°C. 531

AMYLACETAT (*n*): (*Amylium aceticum*): amyl acetate, isoamyl acetate, amylacetic ether, banana oil, essence of Jargonelle pears; $C_7H_{14}O_2$; Sg. 0.866; Bp. 148°C. 532

AMYLALKOHOL (*m*): (*Alcohol amylicus*): (active) amyl alcohol, secondary butylcarbinol, amyl hydrate; $C_5H_{12}O$; Sg. 0.8121; Mp. about −134°C; Bp. 131°C.; (see also **FUSELÖL**) 533

AMYLALKOHOL (*m*), **TERTIÄRER-**: tertiary amyl alcohol, amylene hydrate (see **AMYLENHYDRAT**) 534

AMYL (*n*), **AMEISENSAURES-**: amyl formate, isoamyl formate; $CH_2OC_5H_{11}$; Sg. 0.9; Bp. 130.4°C. 535

AMYLAMIN (*n*): amyl amine 536

AMYLAMINCHLORHYDRAT (*n*): amylamine hydrochloride 537

AMYL (*n*), **APFELSAURES-**: amyl malate 538

AMYLAT (*n*): amylate 539

AMYLÄTHER (*m*): diamyl ether, amyl ether, amyl oxide; $C_{10}H_{22}O$; Sg. 0.7807; Bp. 173°C. 540

AMYL (*n*), **BALDRIANSAURES-**: amyl (valerianate) valeriate, apple (essence) oil; $C_4H_9CO_2C_5H_{11}$; Sg. 0.88; Bp. 203.7°C. 541

AMYL (*n*), **BENZOESAURES-**: amyl benzoate 542

AMYL (*n*), **BERNSTEINSAURES-**: amyl succinate 543

AMYLBROMID (*n*): amyl bromide 544

AMYL (*n*), **BUTTERSAURES-**: amyl butyrate, Isoamylbutyric ester; $C_4H_7O_2C_5H_{11}$; Sg. 0.859; Bp. 154°C. 545

AMYLCHLORID (*n*): amyl hydrochloride 546

AMYLEN (*n*) : amylene ; beta-isoamylene ; pental ; pentene ; trimethyl-ethylene ; (C_5H_{10}) ; Sg. 0.666 ; Bp. 37-42°C. 547
AMYLENBROMID (*n*) : amylene bromide 548
AMYLENCHLORAL (*n*) : amylene chloral, dormiol ; $CCl_3.CH(OH)—O—C(CH_3)_2.C_2H_5$; Sg. 1.24 ; Bp. 93°C. 549
AMYLENHYDRAT (*n*) : amylene hydrate, tertiary amyl alcohol ; $(CH_3)_2.C.C_2H_5.OH$; Sg. 0.815 ; Mp. -12°C ; Bp. 99-103°C. 550
AMYL (*n*), **ESSIGSAURES-** : (*Amylium aceticum*) : amyl acetate, banana oil, amylacetic ether, isoamyl acetate ; $CH_3CO_2C_5H_{11}$; Sg. 0.866 ; Bp. 148°C. 551
AMYLJODID (*n*) : amyl iodide ; $C_5H_{11}I$; Sg. 1.4734 ; Bp. 148.2°C 552
AMYLNITRIT (*n*) (*Amylium nitrosum*) : amyl nitrite, isoamyl nitrite ; $C_5H_{11}O_2N$; Sg. 0.88 ; Bp, 99°C. 553
AMYLOID (*n*) : amyloid, amylaceous body or substance, (a substance resembling starch) 554
AMYLOLYSE (*f*) : amylolysis 555
AMYLOXYD (*n*) : amyl oxide (see **AMYLÄTHER**) 556
AMYLOXYDHYDRAT (*n*) : amyl alcohol (see **AMYLALKOHOL**) 557
AMYL (*n*), **PROPIONSAURES-** : amyl propionate 558
AMYLRHODANÜR (*n*) : amyl sulfocyanide 559
AMYL (*n*), **SALICYLSAURES-** : amyl salicylate ; $C_7H_5O_3C_5H_{11}$; Sg. 1.045 ; Bp. about 270°C. 560
AMYL (*n*), **SALPETERSAURES-** : amyl nitrate 561
AMYL (*n*), **SALPETRIGSAURES-** : amyl nitrite (see **AMYLNITRIT**) 562
AMYLSCHWEFELSAUER : amylsulphate of 563
AMYLSULFID (*n*) : amyl sulphide 564
AMYLSULFOCYANAT (*n*) : amyl (sulfocyanate) thiocyanate 565
ANACARDIENÖL (*n*) : anacardium oil (from *Semecarpus anacardium*) 566
ANAEROBISCH : anaërobic 567
ANAESTHETICUM (*n*) : anæsthetic 568
ANALCIM (*m*) : analcime, analcite (a hydrous sodium aluminium metasilicate) ; $NaAl(SiO_3)_2.H_2O$; Sg. 2.27 569
ANALDRÜSE (*f*) : anal gland 570
ANALEPTICUM (*n*) : analeptic, restorative 571
ANALEPTISCH : analeptic, restorative, restoring, invigorating 572
ANALGETICUM (*n*) ; analgetic (an agent for removing pain) 573
ANALGETISCH : analgetic, painless 574
ANALOG : analagous ; similar 575
ANALOGIE (*f*) : analogy ; similarity 576
ANALOGON (*n*) : analogon 577
ANALYSATOR (*m*) : analyzer 578
ANALYSENFORMEL (*f*) : empirical formula 579
ANALYSENREIN : analytically pure 580
ANALYSENTRICHTER (*m*) : analytical funnel 581
ANALYSENWAGE (*f*) : analytical balance 582
ANALYTIKER (*m*) : analyst 583
ANALYTISCH : analytical 584
ANÄMIE (*f*) : anæmia, deficiency of red corpuscles in the blood (Med.) 585
ANÄMIE, PROGRESSIVE PERNICIÖSE- : pernicious anæmia (Med.) 586
ANANAS (*f*)) : pineapple (*Ananas sativa*) 587
ANANASÄTHER (*m*) : pineapple essence ; ethyl butyrate ; ($C_3H_7CO_2C_2H_5$) ; Sg. 0.8788 ; Mp. -93.3°C. ; Bp. 120.6°C. 588
ANANASESSENZ (*f*) : (see **ANANASÄTHER**) 589
ANANASSAFT (*m*) : pineapple juice 590
ANÄSTHESIA (*f*) : anasthesia (loss of feeling) (Med.) 591
ANÄSTHESIE (*f*) : anæsthesia (Med.) 592
ANÄSTHESIE (*f*), **LOKALE-** : local anasthesia 593
ANÄSTHESIN (*n*) : an-(a)-esthesine, ethyl-para-aminobenzoate ; $C_6H_4NH_2CO_2.C_2H_5$; Mp. 90-91°C. 594
ANÄSTHETIK (*f*) : anasthetic 595
ANÄSTHETISCHES MITTEL (*n*) : anasthetic 596
ANÄSTHOL (*n*) : an-(a)-esthol (mixture of methyl chloride and ethyl chloride) 597
ANATAS (*m*) : anatase, octahedrite ; TiO_2 (see **TITANDIOXYD**) ; Sg. 3.84 598
ANATOMIE (*f*) : anatomy 599
ANATTO (*n*) : annatto (see **ORLEAN**) 600
ANÄTZEN : to cauterize ; etch on ; commence to etch. 601
ANBACKEN : to stick to ; bake on ; burn on 602
ANBACKEN (*n*) : baking (caking) on, burning, sticking, adhesion 603
ANBAHNEN : to make a road for ; cut a road 604
ANBAU (*m*) : out (house) buildings ; extension (of buildings) ; wing ; settlement 605
ANBAUEN : to build on to ; fit to 606
ANBEIZEN : to mordant 607
ANBELANGEN : to relate to ; refer to ; concern 608
ANBERAUMEN : to appoint ; fix ; take place 609
ANBETRACHT (*m*) : consideration 610
ANBETRACHT, IN- : in view of, taking into consideration, considering 611
ANBIETEN : to offer, tender 612
ANBLASEN : to start up (blast furnace) ; affix to (by blast) ; (*n*) the starting up of blast furnaces 613
ANBLÄUEN : to blue ; colour blue 614
ANBLUTEN : to bleed ; run (of colours) 615
ANBOHREN : to bore ; tap ; perforate 616
ANBRINGEN : to place ; sell ; dispose of ; lodge (complaint) ; provide ; arrange ; attach ; affix ; fit ; instal ; include ; incorporate 617
ANBRUCH (*m*) : commencement ; rotting ; opening (of anything which is sealed) ; decay ; dawn ; first yield 618
ANBRUCHGEBIET (*n*) : starting point 619
ANBRÜCHIG : decaying ; putrescent ; rotten ; spoiled 620
ANBRÜHEN : to scald ; infuse ; steep (in hot water) 621
ANCHUSA (*f*) : (see **ALKANNA**) 622
ANCHUSASÄURE (*f*) : anchusic acid ; anchusin 623
ANCHUSIN (*n*) : anchusin (see **ALKANNAROT** and **ANCHUSASÄURE**) 624
ANDALUSIT (*m*) : (natural aluminium silicate, andal-(o)-usite, chiastolite ; Al_2SiO_5 ; Sg. 3.15 (see also **ALUMINIUMSILIKAT**) 625
ANDAÖL (*n*) : anda oil 626
ÄNDERN : to alter ; change ; transform ; convert 627
ANDERTHALBFACH : sesqui- ; one and a half times 628
ANDERTHALB-SCHWEFELEISEN (*n*) : iron sesquisulphide (see **FERRISULFID**) 629
ÄNDERUNG (*f*) : departure, change, conversion, transformation 630
ANDESIN (*m*) : andesine, andesite (mixture of anorthite and albite, see **ANORTHIT** and **ALBIT**) ; $(NaAlSi_3O_8)$: $(CaCl_2Si_2O_8)$; Sg. 2.68 631
ANDESIT (*m*) : (see **ANDESIN**) 632

ANDEUTEN: to indicate; point out; illustrate; mean; suggest 633
ANDIROBAÖL (*n*): (see **CARAPAÖL**) 634
ANDISIN (*n*): andisine (a glysoside from *Andira retusa*) 635
ANDORN (*m*): hore-hound: hoar-hound (*Marrubium vulgare*) 636
ANDORNKRAUT (*n*): (see **ANDORN**) 637
ANDRANG (*m*): pressure; congestion; urgency; crowd; rush; throng 638
ANDREHEN: to screw on; commence turning 639
ANDREHUNG (*f*): screwing, threading, turning 640
ANDRERSEITS: on the other hand 641
ANDROPOGONÖL (*n*): (see **CITRONELLAÖL**) 642
ANDRÜCKEN: to press (force) on 643
ANEIGNEN: to assimilate; adapt; acquire; appropriate 644
ANEINANDERSCHLIESSEN: to close (draw) together, lock together, couple, connect, interconnect, join 645
ANEMOMETER (*n*): anemometer, wind gauge (instrument for measuring the velocity of the wind) 646
ANEMONIN (*n*): anemonine, pulsatilla camphor; $C_{10}H_8O_4$ 647
ANERBIETEN (*n*): offer, tender 648
ANERKANNT: recognized, acknowledged 649
ANERKENNEN: to acknowledge; recognize 650
ANERKENNENSWERT: praiseworthy 651
ANEROIDBAROMETER (*n*): aneroid barometer 652
ANETHOL (*n*): anethole, anis camphor, para-allylphenylmethylic ester; $C_6H_4C_3H_5(OCH_3)$; Sg. 0.985; Mp. 22.5°C.; Bp. 233°C. 653
ANFACHEN: to fan; kindle; blow 654
ANFAHREN: to carry near; bring up; attack; approach; arrive; snub; start; speed up; start up; set in motion; fly at 655
ANFAHRVORRICHTUNG (*f*): starting gear 656
ANFALL (*m*): attack; fit; onset; stay; prop 657
ANFALLEN: to accrue; fall; fall on; be given off; go on (dyes) 658
ANFANG (*m*): origin, beginning, start, commencement, outset, onset 659
ANFÄNGLICH: initial, original 660
ANFANGSBOHRER (*m*): short borer (jumper); pitching borer; picker; gad 661
ANFANGSBUCHSTABE (*f*): initial letter; capital 662
ANFANGSDRUCK (*m*): initial pressure 663
ANFANGSGLIED (*n*): initial member 664
ANFANGSPHASE (*f*): initial phase 665
ANFANGSPUNKT (*m*): initial point; spring (of a line); starting (commencing) point 666
ANFANGSWERT (*m*): initial value 667
ANFANGSZUSTAND (*m*): initial state; original condition 668
ANFÄRBEN: to dye; commence dyeing; paint; colour; tint 669
ANFAULEN: to begin to rot; decay; go mouldy; putresce 670
ANFERTIGUNG (*f*): manufacture; construction; composition 671
ANFETTEN: to grease; baste; oil; lubricate; smear with fat 672
ANFEUCHTEN: to damp; moisten; wet 673
ANFEUCHTER (*m*): moistener; sparger (brewing) 674
ANFEUERN: to (light) fire; prime; inflame; incite; excite 675
ANFIRNISSEN: to varnish (over) 676
ANFLIEGEN: to incrust; effloresce; grow (spontaneously); fly on; occur; rush on (dyes) 677
ANFLUG (*m*): incrustation; flight; tinge; onset; copse; small wood 678
ANFLUSS (*m*): swelling; rise; alluvium; afflux 679
ANFORDERUNG (*f*): requirement, demand 680
ANFRAGE (*f*): enquiry 681
ANFRAGEN: to enquire 682
ANFRESSEN: to corrode; gnaw; eat (at) away; pit (of water tubes); (formation of pin-holes by corrosion) 683
ANFRISCHEN: to revive; refresh; freshen; animate; reduce (metal); varnish (paintings) 684
ANFÜHREN: to adduce, lead, guide, quote, direct 685
ANFÜHRUNGSZEICHEN (*n.pl.*): quotation marks; inverted commas 686
ANGABE (*f*): statement; specification; information; declaration (customs) 687
ANGABEN (*f.pl.*): data 688
ANGABEN (*f.pl.*) **NÄHERE-**: further (data) particulars 689
ANGÄNGIG: feasible, possible, applicable 690
ANGEBEN: to give; dictate; show; state; declare 691
ANGEBLICH: alleged; nominal; pretended 692
ANGEBOT (*n*): offer; bid; quotation; estimate; tender 693
ANGEFLOGEN: incrusted 694
ANGEHÄUFE (*n*): accumulation; heap; aggregate; conglomerate 695
ANGEHEN: to begin; solicit; request; grow; be possible; be practicable; be feasible; decay; burn; approach; concern; be absorbed; go on (dyes) 696
ANGEHEN UM: to solicit, request 697
ANGEHÖREN: to belong to; appertain to; attach to 698
ANGEL (*f*): hinge; hook; pivot; pole; axis 699
ANGELEGEN: interesting, important 700
ANGELEGENTLICH: earnest; pressing; urgent 701
ANGELIKA (*f*): angelica (*Angelica officinalis*) 702
ANGELIKAÖL (*n*): angelica (root) oil (from *Angelica officinalis*); Sg. 0.857-0.918; Bp. 65°C. 703
ANGELIKASAMENÖL (*n*): angelica seed oil (from the seeds of *Angelica officinalis*); Sg. 0.856-0.89 704
ANGELIKASÄURE (*f*): angelic acid (from *Archangelica officinalis*); $CH_2CHCH(CH_3)COOH$; Sg. 0.9539; Mp. 45°C.; Bp. 185°C. 705
ANGELIKASPIRITUS (*m*): angelica spirit (*Spiritus angelicae compositus*) (distilled from angelica root) 706
ANGELIKAWURZEL (*f*): angelica root (root of *Angelica officinalis* and *Archangelica officinalis*) 707
ANGELSÄCHSISCH: anglo-saxon 708
ANGEMESSEN: suitable; proper 709
ANGEMESSENER PREIS (*m*): reasonable price 710
ANGER (*m*): green; grass-plot; pasture 711
ANGESCHWEMMT: alluvial 712
ANGESESSEN: resident; settled 713

ANGESICHTS : in view of ; on account of ; due to 714
ANGESTELLTE (*m*) : employee ; official (pl.) staff 715
ANGESTRENGTER BETRIEB (*m*) : overload ; forcing ; forced load 716
ANGEWANDT : practical ; applied 717
ANGEZEIGT : proper, suitable, right, expedient 718
ANGIESSEN : to pour on ; water ; cast on (founding) 719
ANGIOSPERMEN (*f.pl.*) : Angiospermae (Bot.) (flowering plants in which the seeds are enclosed in an ovary) 720
ANGLESIT (*m*) : anglesite (natural lead sulphate) $PbSO_4$; Sg. 6.25 (see also **BLEISULFAT**) 721
ANGLIEDERN : to join on ; link on ; attach to ; connect with 722
ANGLIEDERUNG (*f*) : attachment ; joining ; connection ; union ; linkage ; coalition 723
ANGLÜHEN : to heat ; anneal ; mull (wine) 724
ANGORAHAAR (*n*) : angora (wool) hair, mohair (from the Angora goat ; see **ANGORAZIEGE**) 725
ANGORAWOLLE (*f*) : angora wool, mohair 726
ANGORAZIEGE (*f*) : Angora goat (*Capra hircus angorensis*) 727
ANGOSTURAÖL (*n*) : angostura (bark) oil ; Sg. 0.945 ; Bp. 200-220°C (from bark of *Galipea cusparia* by distillation) 728
ANGREIFBARKEIT (*f*) : assailability (of metal to corrosion) 729
ANGREIFEN : to attack ; affect ; assail ; pit (metal) ; corrode ; fatigue ; grip ; undertake 730
ANGRENZEN : to adjoin ; border on 731
ANGRENZEND : neighbouring, adjoining 732
ANGRIFFSPUNKT (*m*) : point of attack 733
ANHAFTEN : to stick (adhere) to 734
ANHAFTEND : adhesive ; adhering ; inherent 735
ANHAKEN : to hook to 736
ANHALT (*m*) : stop ; pause ; stay ; support ; handle ; hold ; purchase ; clue ; guidance 737
ANHALTEN : to stop ; last, endure 738
ANHALTEND : persistent ; lasting ; continuous ; durable ; adhesive 739
ANHALT GEBEN, GENÜGENDER- : to be sufficient for the purpose, give sufficient (hold, clue) purchase 740
ANHALTPUNKT (*m*) : stopping (point) place ; fulcrum (mechanical) 741
ANHANG (*m*) : postcript ; P.S. ; supplement ; addition ; party ; appendix (books and anatomy) 742
ANHÄNGEN : to connect ; attach ; throw in (machinery) ; append 743
ANHÄNGSEL (*n*) : appendage ; appendix ; amulet 744
ANHEBEN : to raise ; lift ; open (valve) 745
ANHEBEN (*n*) : raising ; lifting 746
ANHEIZEN : to heat up ; start (stoke) up (boilers) 747
ANHOLEN : to (haul tight) tally (the sheets) (Naut.) 748
ANHYDRID (*n*) : anhydride 749
ANHYDRIDBILDUNG (*f*) : anhydride formation 750
ANHYDRIEREN : to eliminate water from 751
ANHYDRISCH : anhydrous 752
ANHYDRIT (*m*) : anhydrite (natural calcium sulphate) ; overburned Gypsum ; anhydrous calcium sulphate ; $CaSO_4$ (about 42% lime) ; Sg. 2.95 (see **CALCIUMSULFAT**) 753
ANHYDROMETHYLENCITRONENSAUER : anhydromethylenecitrate 754
ANHYDROOXYMETHYLENDIPHOSPHORSÄURE (*f*) : Anhydroxymethylenediphosphoric acid 755
ANHYDROSÄURE (*f*) : anhydro-acid 756
ANILIN (*n*) : aminobenzene, phenylamine, aniline ; aniline oil ; C_6H_7N ; Sg. 1.039 ; Mp. -6.2 ; Bp. 184.4°C. 757
ANILINBLAU (*n*) : analine blue 758
ANILINDRUCK (*m*) : aniline printing (Photography) 759
ANILINFARBE (*f*), **FETTLÖSLICHE-** : aniline colour soluble in fat, fat-soluble aniline dye 760
ANILINFARBSTOFF (*m*) : aniline dye 761
ANILINGELB (*n*) : aniline yellow 762
ANILINÖL (*n*) : aniline oil (see **ANILIN**) 763
ANILINSALZ (*n*) : aniline salt (see **CHLORANILIN**) 764
ANILIN (*n*), **SALZSAURES-** : aniline (hydro) chloride (see **CHLORANILIN**) 765
ANILIN (*n*), **SCHWEFELSAURES-** : aniline sulphate 766
ANILINSULFOSÄURE (*f*) : (see **AMIDOBENZOLSULFOSÄURE**) 767
ANION (*n*) : anion (positive "Ion" which see) 768
ANIS (*m*) : anise ; aniseed (*Pimpinella anisum*) 769
ANISÄHNLICH : anise (aniseed) like, like (anise) aniseed 770
ANISALDEHYD (*n*) : anisaldehyde, para-anisic aldehyde, anisic aldehyde ; $C_8H_8O_2$; Sg. 1.134 ; Mp. 0°C. ; Bp. 199.5°C. ; aubepine, para-methoxybenzaldehyde 771
ANISALKOHOL (*m*) : anisic alcohol, anisalcohol ; $C_8H_{10}O_2$; Sg. 1.1202 ; Mp. 45°C. ; Bp. 258.8°C. 772
ANISGEIST (*m*) : spirit of anise 773
ANISIDIN (*n*) : anisidine, para- ; C_7H_9ON ; Sg. 1.0711 ; Mp. 57.2°C. ; Bp. 239.5°C. 774
ANISOL (*n*) : anisol, phenol methyl ether ; C_7H_8O ; Sg. 1.0124 ; Mp. -37.8°C. ; Bp. 153.8°C. 775
ANISÖL (*n*) : (*Oleum anisi viridum*) : aniseed oil, anise oil, Illicium oil, Anise-seed oil (from *Pimpinella anisum*) ; Sg. 0.99 ; Mp. 12.5°C. 776
ANISOTROP : anisotropic 777
ANISSÄURE (*f*) : anisic acid (para-) (from *Illicium anisatum*) ; $C_8H_8O_3$; Sg. 1.385 ; Mp. 184.2°C. ; Bp. 275–280°C. 778
ANKAUFEN : to purchase, buy 779
ANKAUFSRECHNUNG (*f*) : purchase account 780
ANKER (*m*) : anchor ; stay ; grapnel ; armature (electric) 781
ANKERACHSE (*f*) : armature spindle ; armature axis 782
ANKERARM (*m*) : anchor arm 783
ANKER-BLECH (*n*) : core plate (Elect.) 784
ANKERBOHRUNG (*f*) : armature (bore) hole 785
ANKERBOJE (*f*) : (anchor)-buoy (Naut.) 786
ANKERBOLZEN (*m*) : truss bolt 787
ANKER (*m*), **DOPPEL-** (T.) ; H. section armature 788
ANKERDRAHT (*m*) : armature wire 789
ANKEREISEN (*n*) : armature iron 790
ANKERENDSCHEIBE (*f*) : armature end (disc) plate (Elect.) 791

ANKERFLÜGEL (*m*) : palm of an anchor 792
ANKERHALS (*m*) : throat of an anchor 793
ANKERIT (*m*) : ankerite $(Ca,Mn,Mg,Fe)_2(CO_3)_2$; Sg. 3.0 794
ANKER (*m*), **KATT-** : backing anchor 795
ANKERKERN (*m*) : armature core 796
ANKERKETTE (*f*) : chain cable, anchor cable 797
ANKERKLÜSE (*f*) : hawse hole 798
ANKERKREUZ (*n*) : anchor crown 799
ANKER (*m*), **KURZSCHLUSS-** : squirrel cage (Elect.) 800
ANKER (*m*), **LAND-** : land (shore) anchor 801
ANKER (*m*), **DEN-LICHTEN** : to weigh anchor 802
ANKERN : to anchor, cast anchor (nautical) 803
ANKERPOL (*m*) : armature pole 804
ANKER (*m*), **SCHEIBEN-** : disc armature 805
ANKER (*m*), **SCHLEIFRING-** : slip ring (rotor) armature (elect.) 806
ANKERSEGMENT (*n*) : armature segment 807
ANKERSPILL (*n*) : anchor capstan 808
ANKERSPULE (*f*) : armature coil 809
ANKER (*m*), **SPULEN-** : slip ring (rotor) armature (Elect.) 810
ANKERSTICH (*m*) : clinch of a cable 811
ANKERSTROM (*m*) : armature current 812
ANKERTAU (*n*) : cable (for anchors) 813
ANKER, VOR- (GEHEN) LEGEN : to anchor, bring to (anchor) anchorage, to cast anchor 814
ANKER (*m*) **VOR-, TREIBEN** : to drag the anchor 815
ANKERWELLE (*f*) : armature shaft 816
ANKER WERFEN : to anchor, cast anchor, come to an anchorage 817
ANKERWICKELUNG (*f*) : armature winding 818
ANKERWICKLUNG (*f*) : armature winding 819
ANKERWIDERSTAND (*m*) : armature resistance 820
ANKERWINDE (*f*) : capstan, windlass 821
ANKERWINDUNG (*f*): armature winding 822
ANKITTEN : to (fasten with) cement 823
ANKLEBEN : to stick (on) to 824
ANKNÜPFEN : to tie, start ; enter upon, obtain, increase (a business connection) 825
ANKOHLEN : to char (carbonize) partially 826
ANKOMMEN : to come upon ; concern ; commence fermenting (Brewing) 827
ANKÖRNEN : to mark with a centre-punch ; bait, allure, decoy 828
ANKÖRNMASCHINE (*f*) : centring machine 829
ANKREIDEN : to chalk 830
ANKREUTZEN : to place a cross against, mark with a cross, cross 831
ANKÜNDIGUNG (*f*) : announcement, statement, advice, notification 832
ANKUPPELN : to couple (to) 833
ANLAGE (*f*) : plant ; installation ; set 834
ANLAGE (*f*), **AUSGEFÜHRTE-** : installation 835
ANLAGEKOSTEN (*pl.*) : installation costs 836
ANLAGERN : to accumulate ; store ; add (on) ; take up 837
ANLANGEN : to attain, reach, arrive 838
ANLASS (*m*), **BEI-** : on the occasion of, in connection with 839
ANLASSEN : to start (up) ; occasion ; temper ; anneal (glass or metal) 840
ANLASSER (*m*) : starting switch (Mech.), starter 841
ANLASSFARBE (*f*) : temper-(ing) colour (steel) 842
ANLASS GEBEN : to occasion ; cause ; give rise to 843
ANLASSGEFÜGE (*n*) : temper structure (steel) 844
ANLASSGESTÄNGE (*n*) : starting gear 845
ANLASSHEBEL (*m*) : starting lever 846
ANLASSKURVE (*f*) : annealing curve (in connection with properties of steel) 847
ANLÄSSLICH : probably ; apparently 848
ANLASSOFEN (*m*) : tempering (annealing) oven 849
ANLASSTEMPERATUR (*f*) : tempering (annealing) temperature 850
ANLASSVENTIL (*n*) : starting valve 851
ANLASSVORRICHTUNG (*f*) : starting (gear) device 852
ANLASSWIDERSTAND (*m*) : starting resistance 853
ANLASSWIRKUNG (*f*) : tempering (annealing) effect 854
ANLAUF (*m*) : starting up ; run (Mech.) ; onset ; catch ; tappet (Machinery) ; swelling (water) 855
ANLAUFEN : to start up ; set in motion ; become coated (with) ; tarnish ; recolour 856
ANLAUFFARBE (*f*) : tempering colour (Metal) ; (see **BROKATFARBE**) 857
ANLAUFRAMPE (*f*) : ramp with steeper gradient than that ruling 858
ANLAUFSTEIGUNG (*f*) : steeper gradient than that ruling or usual 859
ANLEGEN : to lay on, colour (a drawing) ; to come alongside (Marine) ; (*n*) laying-on (Art), (coating a surface with a flat wash) 860
ANLEIHE (*f*), **EINE . . . AUFNEHMEN** : to take up (raise) a loan, to contract a loan 861
ANLEIMEN : to glue on, stick to 862
ANLEITUNG (*f*) : introduction ; leading ; conduction ; instruction ; guide ; key ; preface 863
ANLIEGEN : to lie near ; enclose ; fit ; concern ; attach ; be interesting ; important 864
ANLIEGEND : attached, enclosed, herewith 865
ANLÖTEN : to solder on ; cause to adhere 866
ANLUVEN : to haul the wind, to luff (naut.) 867
ANMACHEN : to fasten ; join ; affix ; slack (Lime) ; light (Fire) ; mix ; adulterate ; temper (Colours) ; dress (salad) ; raise (steam) 868
ANMELDEBOGEN (*m*) : declaration (registration) form 869
ANMELDEN : to register, declare, report, announce 870
ANMELDUNG (*f*) : announcement ; registration ; provisional (specification) (Patents) 871
ANMELDUNG, PATENT- (*f*) : application (for patent) 872
ANMENGEN : to mix ; blend ; temper ; dilute ; admix 873
ANMERKEN : to note, quote 874
ANMERKUNG (*f*) : remark, note 875
ANNABERGIT (*m*) : annabergite, nickel bloom ; $Ni_3As_2O_8.8H_2O$ (see **ERDKOBALT, GRÜNER-**) Sg. 3.05 876
ANNÄHERND : approximate(ly) 877
ANNAHME (*f*) : assumption ; hypothesis 878
ANNALINE (*f*) : gypsum (see **GYPS**) 879
ANNÄSSEN : to moisten ; damp slightly 880
ANNEKTIEREN : to annex 881
ANNERÖDIT (*m*) : annerödite ; $(Nb_2O_7)_3Y_4$.U, Th,Ce,Pb,Fe,Ca ; Sg. 5.7 882
ANNETZEN : to wet (see **ANNÄSSEN**) 883
ANNIETEN : to rivet on 884
ANNONCE (*f*) : advertisement 885
ANODE (*f*) : anode (positive electrode) 886
ANODENDICHTE (*f*) : anode (anodic) density 887

ANODENSTRAHLEN (*m.pl*) : anode (anodic) rays 888
ANODENSTROM (*m*) : anode current 889
ANÖLEN : to-(coat with)-oil ; grease ; lubricate 890
ANOMAL : anomalous 891
ANOMALISCH : (see ANOMAL) 892
ANOMIT (*m*) : anomite ; $(H,K)_2(Mg,Fe)_2(Al,Fe)_2(SiO_4)_3$; Sg. 3.0 893
ANÖMLEIN (*n*) : bishop's weed (*Aegopodium podagraria*) 894
ANORDNEN : to arrange ; order ; regulate ; classify ; employ ; use ; provide-(for) 895
ANORDNUNG (*f*) : arrangement, classification, disposal, disposition 896
ANORDNUNG, GESAMT- (*f*) : general arrangement 897
ANORGANISCH : inorganic (Chemistry) 898
ANORMAL : abnormal 899
ANORTHIT (*m*) : anorthite (a calcium alum silicate) $CaAl_2Si_2O_8$; Sg. 2.75 900
ANORTHOMER FELDSPAT (*m*) : anorthic feldspar, anorthite, christianite 901
ANOXYDIEREN : to oxidize 902
ANPASSEN : to adapt ; adjust ; fit ; suit 903
ANPASSEND : well-fitting, suitable 904
ANPASSUNGSFÄHIGKEIT (*f*) : adaptability ; fitness ; suitability 905
ANPLÄTZEN : partially to (remove the bark) decorticate 906
ANPRALL (*m*) : impact ; impingement 907
ANQUICKEN : to amalgamate 908
ANRAT (*m*) : advice ; counsel 909
ANRATEN : to advise ; counsel 910
ANRAUCHEN : to smoke 911
ANRÄUCHERN : to smoke ; fumigate 912
ANRECHNEN : to ascribe, count, charge, add-(to) 913
ANRECHT (*n*) : claim ; title 914
ANREGEN : to set going ; incite ; excite ; stimulate ; interest ; suggest ; mention ; cause ; induce ; effect 915
ANREGENDES MITTEL (*n*) : excitant 916
ANREGUNG (*f*), **AUF-** : at the instance of 917
ANREGUNGSMITTEL (*n*) : stimulant (Med.) 918
ANREIBEN : to grind (colour) ; squeegee (Photography) 919
ANREIBER (*m*) : squeegee, roller squeegee (Photography) 920
ANREICHERN : to enrich ; strengthen ; concentrate 921
ANREICHERUNG (*f*) : enrichment, concentration 922
ANREIHEN : to arrange in (attach to) a series 923
ANREISSEN : to trace, tear-(off), to mark out, to sketch down 924
ANREIZEN : to stimulate ; induce 925
ANRICHTEN : to prepare ; dress ; perform ; produce ; cause ; mix (colours) ; assay 926
ANRICHTER (*m*) : assayer ; mixer ; dresser (of ores) 927
ANROSTEN : to corrode, rust 928
ANROSTUNG (*f*) : corrosion, rusting 929
ANRUF (*m*) : call (on the phone, etc.), appeal, hail 930
ANRUFGLOCKE (*f*) : call bell 931
ANRÜHREN : to touch ; stir ; mix ; temper ; refer to 932
ANSAGEN : to announce ; state 933
ANSAMMELN : to collect ; accumulate ; focus 934
ANSATZ (*m*) : deposit ; sediment ; price ; rate ; quotation ; incrustation ; standard (Dyes) ; projection (Mech.) ; item ; lug (Mech.) ; incomplete progression ; one member of a progression (Maths.) ; expression ; attachment ; leprosy ; lepra ; elephantiasis graecorum (Med.) 935
ANSATZBAD (*n*) : initial (first) bath 936
ANSÄTZE (*m.pl*), **GLEITENDE-** : sliding (scales) rates 937
ANSATZPUNKT (*m*) : point of (insertion) attachment ; spring (of an arch) 938
ANSATZSTÜCK (*n*) : connecting (make-up) piece 939
ANSAUGABSCHNITT (*m*) : period of suction 940
ANSAUGEHUB (*m*) : suction stroke (pumps) 941
ANSAUGEN : to suck ; (*n*) suction 942
ANSÄUREN : to acidulate, acidify ; add yeast to, leaven 943
ANSCHAFFUNGSKOSTEN (*f.pl*) : installation costs 944
ANSCHAFFUNGSPREIS (*m*) : installation cost, first cost, prime cost 945
ANSCHÄRFER (*m*), **BLEISTIFT-** : pencil sharpener 946
ANSCHEINEND : apparently 947
ANSCHEINLICH : apparent-(ly) 948
ANSCHICHTEN : to stratify ; place in layers ; pile 949
ANSCHIEBEN : to push, shove 950
ANSCHIEBER (*m*) : lengthening piece 951
ANSCHIESSEN : to crystallize ; shoot ; rush ; be adjacent 952
ANSCHLAG (*m*) : stroke ; impact ; placard ; attempt ; estimate 953
ANSCHLAGEN : to bend (a sail) ; affix, post, placard, value, estimate, nail on-(to) 954
ANSCHLAG (*m*), **FENSTER-** : window rebate or rabbet (joinery) 955
ANSCHLÄGIG : estimated ; ingenious 956
ANSCHLAGSTIFT (*m*) : detent pin 957
ANSCHLÄMMEN : to suspend ; deposit (mud) ; smear (with mud) ; wash ; become muddy 958
ANSCHLEIFEN : to grind (an edge), to edge, sharpen 959
ANSCHLUSS (*m*) : connection ; coupling up ; junction 960
ANSCHLUSSSTÜCK (*n*) : connecting piece 961
ANSCHLUSS (*m*), **TELEPHON-** : telephone -(telephonic) connection 962
ANSCHMELZEN : to fuse to ; weld ; join by (fusion) welding 963
ANSCHMIEDEN : to forge (together) on to, to join by forging 964
ANSCHMIEGEN : to-(cause to-) cling to ; conform to ; press to 965
ANSCHNEIDEN : to cut into ; broach (a question) 966
ANSCHRAUBEN : to fasten with screws, to screw (to) on, bolt on 967
ANSCHUHEN : to shoe (a mast) 968
ANSCHÜREN : to stir up (stoke) (fire) 969
ANSCHUSS (*m*) : crystallization ; crop (of crystals) ; rush ; shooting 970
ANSCHWÄNGERN : to saturate ; impregnate 971
ANSCHWÄNZAPPARAT (*m*) : sprinkler ; sparger 972
ANSCHWEFELN : to fumigate (treat) with sulphur 973
ANSCHWÖDEN : to paint (flesh side of hides) with lime 974
ANSETZEN : to charge (a furnace) ; crystallize ; fit ; attach ; join ; prepare ; mix 975
ANSICHT (*f*) : view ; opinion ; elevation (of drawings) 976

ANSIEDEN: to boil; scorify; blanch; mordant 977
ANSIEDEPROBE (*f*): scorification (assay) test 978
ANSIEDESCHERBEN (*m*): scorifier 979
ANSPANNEN: to put tension on, to place tension upon, to subject to tension (stress, pressure), stress, tighten 980
ANSPITZEN: to point, sharpen 981
ANSPORNEN: to spur on, encourage 982
ANSPRUCH (*m*): application (for patents); demand; consideration 983
ANSPRUCH (*m*), **IN-NEHMEN**: to impose on, make a call upon 984
ANSTALT (*f*): institution; works; establishment; arrangement, preparation 985
ANSTAND (*m*): behaviour, courtesy, manners, politeness, grace, decorum, propriety, demeanour, suitability, scruple, pause, delay, respite, refinement, objection, proposal (of marriage); (*pl*) objections 986
ANSTÄNDE BEIBRINGEN: to raise objections, take exception to 987
ANSTÄNDE HABEN: to receive (offers) proposals (of marriage) 988
ANSTANDSLOS: without (delay, scruple) question 989
ANSTAU (*m*): heaping up; piling up; damming; collection; accumulation 990
ANSTEIGEN: to rise; increase; accumulate 991
ANSTELLBOTTICH (*m*): starting tub (Brewing) 992
ANSTELLIG: fit, practical, suitable, skilled 993
ANSTELLTEMPERATUR (*f*): pitching temperature; initial temperature 994
ANSTELLUNG (*f*): appointment 995
ANSTIEG (*m*): rise; increase 996
ANSTOSS (*m*): impulse; starting point, point of contact 997
ANSTREBEKRAFT (*f*): centripetal force 998
ANSTRENGUNG (*f*): load, stress 999
ANSTRENGUNG, NORMALE- (*f*): normal load (Mech.) 000
ANSTRICHFARBE (*f*): paint, colour (actually the ground coating of paint) 001
ANSTRICHMASSE (*f*): priming (mass), coating compound (for applying by means of painting) 002
ANSTRICHMASSE (*f*), **ROSTSCHÜTZENDE-**: anti-rusting compound 003
ANSUCHEN (*n*): application 004
ANTAGONISMUS (*m*): antagonism 005
ANTAGONISTISCH: antagonistic-(ally) 006
ANTECEDENTIEN (*pl*): antecedents 007
ANTEIGEN: to (make) convert into a paste 008
ANTEIL (*m*): part; portion 009
ANTENNE (*f*): aerial (Wireless Tel.) 010
ANTHERE (*f*): anther (Bot.) 011
ANTHERIDIE (*f*): antheriferous plant, plant having anthers 012
ANTHOPHYLLIT (*m*): (an iron-aluminium metasilicate); anthophyllite (Mg,Fe) (SiO_3) (a variety of amphibole); Sg. 3.15 013
ANTHRACEN (*n*): anthracene; $C_{14}H_{10}$; Sg. 1.147; Mp. 216°C.; Bp. 360°C.; anthracene oil, para-, naphthalene 014
ANTHRACENÖL (*n*): anthracene oil (see **ANTHRACEN**) 015
ANTHRACENPECH (*n*): anthracene pitch 016
ANTHRACHINOLIN (*n*): anthraquinoline 017
ANTHRACHINON (*n*): anthraquinone; $C_6H_4(CO)_2C_6H_4$; Sg. 1.419-1.438; Mp. 284.6°C.; Bp. 380°C. 018
ANTHRAPURPURIN (*n*): anthrapurpurin, purpurine red, trihydroxyanthraquinone; $C_{14}H_5O_2(OH)_3$; Mp. over 330°C.; Bp. 462°C. [019
ANTHRASOL (*n*): anthrasol (purified coal tar) 020
ANTHRASOLSEIFE (*f*): anthrasole soap 021
ANTHRAZEN (*n*): anthracene; $C_6H_4(CH)_2C_6H_4$; Sg. 1.147; Mp. 216°C.; Bp. 360°C. 022
ANTHRAZENSÄURESCHWARZ (*n*): acid anthracene black 023
ANTHRAZIT (*m*): anthracite; Sg. 1.4-1.7; (non-coking form of coal; Ash, 8-10%; Moisture, 10%; Calorific value 7,000 cals.; 90-95% carbon; 3%H.; 3% O., plus N) 024
ANTIARIN (*n*): antiarin; $C_{27}H_{42}O_{10}$ 025
ANTIAROL (*n*): antiarol (a resin and glycoside); $C_9H_{12}O_4$ 026
ANTICHLOR (*n*): sodium thiosulphate, antichlor, hypo. (see **NATRIUMHYPOSULFIT**) 027
ANTIDYSENTERIESERUM (*n*): anti-dysentery serum 028
ANTIDYSENTERIKUM (*n*): anti-dysentery remedy 029
ANTIFEBRIN (*n*): antifebrin (see **ACETANILID**) 030
ANTIFRIKTIONSMETALL (*n*): (85% Zn.; 10% Sb.; 5% Cu.); antifriction metal, bearing metal 031
ANTIGORIT (*m*): antigorite; $H_4(Mg,Fe)_3Si_2O_9$ $(+Al_2O_3)$; Sg. 2.62 032
ANTIKATALYSATOR (*m*): anticatalyser 033
ANTIKATALYTISCH: anticatalytic 034
ANTIKÖRPER (*m*): anti-substance 035
ANTIKORPULENZPRÄPARAT (*m*): anticorpulency preparation 036
ANTIKORPULENZPULVER (*n*): anticorpulency powder 037
ANTIMON (*n*): antimony (*Stibium*); (Sb); Sg. 6.69; Mp. 630°C. (see also **SPIESSGLANZ**) 038
ANTIMON-BARYT (*m*), **PRISMATISCHER-**: valentinite 039
ANTIMONBLEI (*n*): lead-antimony alloy 040
ANTIMONBLENDE (*f*): antimony blende; kermesite; Sb_2S_2O; Sg. 4.562; Mp. 546°C. 041
ANTIMONBLÜTE (*f*): valentinite; antimony bloom (see **ANTIMONTRIOXYD**) 042
ANTIMONBUTTER (*f*): antimony butter; antimony chloride; $(SbCl_3$; Sg. 1.35=38°Be.) 043
ANTIMONCHLORID (*n*): antimony chloride (*Stibium chloratum*); $SbCl_3$; Sg. 3.064; Mp. 73.2°C; Bp. 223.5°C., antimony pentachloride (*Stibium pentachloratum*); $SbCl_5$; Mp. -6°C. 044
ANTIMONCHLORÜR (*n*): antimony trichloride; $SbCl_3$ (*Stibium chloratum*) (see **ANTIMONCHLORID**) 045
ANTIMONDOPPELFLUORID (*n*): antimony bifluoride; SbF_2 046
ANTIMONERZ (*n*): antimony ore 047
ANTIMONFAHLERZ (*n*): freibergite, tetrahedrite; $3(Cu,Ag)_3SbS_3+CuZn_2SbS_4$; Sg. 4.8 [048
ANTIMONFLUORID (*n*): antimony fluoride (see **ANTIMONPENTAFLUORID**; see **ANTIMON TRIFLUORID**) 049
ANTIMONGLANZ (*m*): antimony glance; stibnite; antimonite; Sb_2S_3; Sg. 4.6 050
ANTIMON-GLANZ (*m*), **PRISMATISCHER-**: sylvanite, graphic tellurium (see **SYLVANIT**) 051
ANTIMONGLAS (*n*): antimony glass, fused antimony sulphide (see also **ANTIMONOXYD**) 052
ANTIMONHALOGEN (*n*): antimony halide 053

ANTIMONIGESÄURE (*f*): antimonious acid; antimony trioxide; Sb_2O_3; Sg. 5.2-5.67 [054
ANTIMONIGSAUER: antimonite of; combined with antimonious acid 055
ANTIMONIGSÄUREANHYDRID (*n*): antimony trioxide; Sb_2O_3 (see **ANTIMONTRIOXYD**) 056
ANTIMONIT (*m*): antimonite (see **ANTIMONGLANZ**) 057
ANTIMONKUPFERGLANZ (*m*): antimonial copper glance; wölchite; bournonite; chalcostibite (see **WOLFSBERGIT**) 058
ANTIMONLAKTAT (*n*): antimony lactate, antimonine; $Sb(C_3H_5O_3)_3$ 059
ANTIMONMETALL (*n*): Antimony regulus (*Antimonium regulus*) (see **ANTIMON**) 060
ANTIMON (*n*), **MILCHSAURES-**: antimony lactate (see **ANTIMONLAKTAT**) 061
ANTIMONNATRIUMSULFANTIMONIAT (*n*): antimony-sodium sulfantimoniate; $Na_3SbS_4.9H_2O$; Sg. 1.806-1.864 062
ANTIMONNICKEL (*n*): breithauptite, NiSb (see **BREITHAUPTIT**); Sg. 7.55 063
ANTIMONNICKELGLANZ (*m*): ullmannite (see **ULLMANNIT**) 064
ANTIMONOCKER (*m*): antimony ochre, stibiconite (natural hydrous, antimony oxide; about 74% antimony); $H_2Sb_2O_5$; Sg. 4.5 065
ANTIMONORANGE (*f*): antimony orange (see **ANTIMONPENTASULFID**) 066
ANTIMONOXALAT (*n*): (*Stibium oxalicum*): antimony oxalate; $SbOH.C_2O_4$ 067
ANTIMONOXYCHLORID (*n*): (*Pulvis algarothi*): antimony oxychloride, algarot, algaroth, algarot powder; SbOCl (see also **ALGAROTHPULVER**) 068
ANTIMONOXYCHLORÜR (*n*): antimony oxychloride; SbOCl 069
ANTIMONOXYD (*n*): antimony oxide (any); trioxide, Sb_2O_3; Sg. 5.6; tetroxide, Sb_2O_4; Sg. 4.1; pentoxide, Sb_2O_5; Sg. 3.8 070
ANTIMONOXYSULFURET (*n*): (see **ANTIMONZINNOBER**) 071
ANTIMONPENTACHLORID (*n*): (*Stibium pentachloratum*): antimony pentachloride; $SbCl_5$; Sg. 2.346; Mp. −6°C. 072
ANTIMONPENTAFLUORID (*n*): (*Stibium pentafluoratum*): antimony pentafluoride; SbF_5; Sg. 2.99 073
ANTIMONPENTAJODID (*n*): antimony pentiodide; SbI_5; Mp. 7°C.; Bp. 150°C. 074
ANTIMONPENTASULFID (*n*): antimony pentasulphide (*Stibium sulfuratum aurantiacum*); Sb_2S_5 075
ANTIMONPENTOXYD (*n*): antimony pentoxide; Sb_2O_5; Sg. 3.78 076
ANTIMONSAFFRAN (*m*): antimonial saffron; crocus of antimony 077
ANTIMONSALZ (*n*): antimony salt; a mixture of sodium and antimony fluorides 078
ANTIMONSAUER: antimonate of; combined with antimonic acid 079
ANTIMONSÄURE (*f*): antimonic acid; Sb_2O_5 080
ANTIMONSÄUREHYDRAT (*n*): antimony pentoxide, hydrated; $Sb_2O_5.5H_2O$; Sg. 6.6 081
ANTIMONSILBER (*n*): antimonial silver; dyscrasite; Ag_2Sb; Sg. 9.7 082
ANTIMONSILBERBLENDE (*f*): antimonial silver blende; pyrargyrite; $3Ag_2S.Sb_2S_3$; Sg. 5.8 (contains about 60% Ag.) (see **ROTGILTIGERZ**) 083
ANTIMONSPIEGEL (*m*): antimony mirror 084
ANTIMONSULFID (*n*): antimony trisulphide (see **ANTIMONSULFÜR**); antimony pentasulphide; Sb_2S_5 085
ANTIMONSULFÜR (*n*): antimony trisulphide; Sb_2S_3; Sg. 4.562; Mp. 546°C. 086
ANTIMONTANNAT (*n*): antimony tannate 087
ANTIMONTETROXYD (*n*): antimony tetroxide; Sb_2O_4; Sg. 4.07 088
ANTIMONTRIBROMID (*n*): antimony tribromide; $SbBr_3$; Sg. solid 4.148; Mp. 94.2°C; Bp. 280°C. 089
ANTIMONTRICHLORID (*n*): (*Stibium chloratum*); antimony trichloride; $SbCl_3$; Sg. 2.675-3.064; Mp. 73.2°C; Bp. 223.5°C; butter of antimony 090
ANTIMONTRIFLUORID (*n*): (*Stibium fluoratum*): antimony trifluoride; SbF_3; Sg. 4.379; Mp. about 292°C. 091
ANTIMONTRIJODID (*n*): antimony tri-iodide; SbJ_3; Sg. 4.676 092
ANTIMONTRIOXYD (*n*): (*Stibium oxydatum*): antimony trioxide; Sb_2O_3; Sg. 5.6 093
ANTIMONTRISULFID (*n*): antimony trisulphide; (*Antimonium crudum*; *Stibium sulfuratum nigrum*); Sb_2S_3; Sg. 4.12-4.64; Mp. 546°C. 094
ANTIMONWASSERSTOFF (*m*): antimony hydride; stibine; SbH_3; Sg. 2.3 095
ANTIMONWEISS (*n*): antimony white (see **ANTIMONOXYD**) 096
ANTIMONYLKALIUM (*n*), **WEINSAURES-**: tartar emetic (see **BRECHWEINSTEIN**) 097
ANTIMONZINNOBER (*m*): kermesite; kermes mineral; (*Stibium sulfuratum nigrum*); $Sb_2S_3.Sb_2O_3$; Sg. 4.562; Mp. 546°C. 098
ANTIOXYD (*n*): antioxide (guttapercha dissolved in benzine) 099
ANTIPERIODICUM (*n*): antiperiodic (Med.) 100
ANTIPHLOGISTISCH: antiphlogistic 101
ANTIPODISCH: antipodal 102
ANTIPYRETIKUM (*n*): antipyretic (preventive or remedy for pyrexia "high temperature") (Med.) 103
ANTIPYRIN (*n*) (*Pyrazolonum phenyldimethylicum*): dimethyloxyquinizine, antipyrine, phenazone, phenyldimethyl-pyrazole, anodynine, analgesine, phenylene, pyrazine, sedatine, pyrazoline, oxydimethylquinizine; $C_{11}H_{12}N_2O$; Sg. 1.19; Mp. 113°C.; Bp. 319°C. 104
ANTIQUARISCH: antiquarian; second-hand (books) 105
ANTIQUITÄT (*f*): antiqnity 106
ANTISEPTICUM (*n*): antiseptic 107
ANTISEPTIKUM (*n*): antiseptic 108
ANTISEPTISCH: antiseptic, aseptic 109
ANTISEPTISCHES MITTEL (*n*): antiseptic 110
ANTISTREPTOKOKKENSERUM (*n*): antistreptococcus serum (Med.) 111
ANTISYPHILITICUM (*n*): antisyphilitic, remedy (preventive) for syphilis; syphilis antitoxin 112
ANTITOMER FELDSPAT (*m*): oligoclase 113
ANTIWEINSÄURE (*f*): antitartaric acid 114
ANTONSKRAUT (*n*): willow herb (*Chamaenerion angustifolium*) 115
ANTREFFEN: to meet with; come across 116
ANTREIBEN: to start up; drive; hoop (casks) 117
ANTRETEN: come into, inherit 118
ANTRIEB (*m*): impulse; volition (motive-) power; drive; starter 119
ANTRIEB (*m*), **AUS EIGNEM-**: of his own accord 120

ANTRIEB (*m*), ELEKTRISCHER- : electrical drive 121
ANTRIEBMASCHINE (*f*) : driving engine, prime mover 122
ANTRIEB (*m*), MOTORISCHER- : motordrive 123
ANTRIEBSART (*f*) : type of drive 124
ANTRIEBSSCHEIBE (*f*) : driving pulley 125
ANTRITTSVORLESUNG (*f*) : inaugural lecture 126
ANTWERPENERBLAU (*n*) : Antwerp blue 127
ANTWORTSCHREIBEN : (*n*) reply 128
ANWALT (*m*) : attorney ; deputy ; solicitor 129
ANWÄRMEN : to warm ; dry ; heat (up) 130
ANWÄRMER (*m*) : feed heater ; preheater ; economiser 131
ANWEICHEN : to soak ; soften ; steep 132
ANWEISEN : to ask, direct, instruct ; advise, inform, remit, furnish a cheque for 133
ANWEISSEN : to whiten ; whitewash 134
ANWEISUNG (*f*) : direction(s) ; instruction(s) ; notice ; money order ; cheque ; draft ; bill ; remittance ; note ; assignment ; advice ; method ; designation ; advice note 135
ANWENDBARKEIT (*f*) : applicability, practicability 136
ANWENDEN : to make use of ; apply ; utilise ; employ 137
ANWERFEN : to start (start up, set in motion, put in commission) a motor 138
ANWESEN (*n*) : concern, plant, establishment 139
ANWESEND : present 140
ANWURF (*m*) : deposit ; plastering ; starting 141
ANWURFKURBEL (*f*) : starting crank (of a motor) ; starting handle 142
ANZAHLUNG (*f*) : prepayment, part payment in advance 143
ANZAPFDAMPF (*m*) : bleeder steam (steam bled or drawn off from any intermediate point of a turbine) 144
ANZAPFEN : to tap 145
ANZAPFMASCHINE (*f*) : bleeder ; bleeder turbine ; interstage leak-off turbine 146
ANZAPFSTELLE (*f*) : leak-off (turbines) 147
ANZAPFSTUFE (*f*) : bleeding stage (turbines) 148
ANZAPFUNG (*f*) : tapping ; bleeding (of steam), interstage leak off 149
ANZEIGE (*f*) : report ; information ; advice ; notice ; advertisement 150
ANZEIGENENTWURF (*m*) : advertisement (advertising) design 151
ANZEIGER (*m*) : indicator ; gauge ; informer ; advertiser ; exponent (Maths.) 152
ANZETTELN : to plot ; conspire ; warp (Weaving) 153
ANZIEHEN : to draw, attract, pull up, tighten, draw (together) tight, to pack (stuffing boxes), screw up (a nut) 154
ANZIEHEND : tractive ; attractive ; astringent (Med.) 155
ANZIEHMUSKEL (*m*) : abductor muscle (Med.) 156
ANZIEHUNGSKRAFT (*f*) : attraction, attractive power 157
ANZÜNDEN : to ignite, set on fire, light 158
AORTE (*f*) : aorta (Anat.) 159
AOUARAÖL (*n*) : African palm oil ; Sg. 0.920-0.927 ; Mp. 27-42.5°C. 160
APATIT (*m*) : apatite (natural calcium phosphate plus either fluorine or chlorine ; about 42% phosphorous pentoxide) ; $(PO_4)_3FCa_5$ or $(PO_4)_3ClCa_5$; Sg. 3.2 161
APFELÄTHER (*m*) : malic ether ; ethyl malate ; essence of apple ; $C_4H_9CO_2C_5H_{11}$; Sg. 0.8812 ; Bp. 203.7°C. 162
APFELBAUMHOLZ (*n*) : apple tree wood (*Pirus malus*) ; Sg. 0.65-0.85 163
APFELKERNÖL (*n*) : apple-pip oil (from *Pirus malus*) 164
APFELMOST (*m*) : apple must ; cider 165
APFELSALBE (*f*) : pomatum 166
APFELSAUER : malate of ; combined with malic acid 167
APFELSÄURE (*f*) : (*Acidum malicum*) : apple acid, malic acid, oxysuccinic acid ; $C_4H_6O_5$; Sg. 1.6 ; Mp. 100°C. 168
APFELSÄURE-DIÄTHYLESTER (*m*) : malic diethyl ester ; $C_8H_{14}O_5$; Sg. 1.1280 ; Bp. 149°C. ; diethyl malate 169
APFELSINENSCHALENÖL (*n*) : orange-peel oil ; essence of orange ; Sg. 0.842-0.857 ; (see POMERANZENSCHALENÖL, SÜSSES-) 170
APHLOGISTISCH : aphlogistic 171
APHRODISIACUM (*n*) : aphrodisiac 172
APHRODISIAKUM (*n*) : aphrodisiac (Med.) 173
APLANATISCH : aplanatic (corrected for spherical and chromatic aberration) (of lenses) 174
APLOM (*m*) aplome ; $Ca_3(Fe,Al)_2(SiO_4)_3$; Sg. 3.55 175
APOATROPIN (*n*) : apoatropine, atropamine ; $C_{17}H_{21}O_2N$; Mp. 61°C. 176
APOCHINEN (*n*) : apoquinene 177
APOCHININ (*n*) : (see APOCHINEN) 178
APOCHROMAT (*n*) : apochromat (the name given by Abbe to the objective which gives a clear achromatic aplanatic image) (of microscopes) 179
APOKAMPHERSÄURE (*f*) : apocamphoric acid 180
APOMORPHIN (*n*) : apomorphine ; $C_{17}H_{17}NO_2$ 181
APOMORPHINCHLORIDHYDRAT (*n*) : apomorphine hydrochloride (*Apomorphinæ hydrochloridum*) $C_{17}H_{17}NO_2HCl.H_2O$; Mp. 205°C. 182
APOPHYLLIT (*m*) : apophyllite ; $H_7KCa_4(SiO_3)_8 + 4\frac{1}{2}H_2O$: Sg. 2.35 183
APOPLEXIE (*f*) : apoplexy (Med.) 184
APOTHEKER (*m*) : apothecary ; druggist ; pharmaceutical chemist ; pharmacist 185
APOTHEKERBUCH (*n*) : pharmacopeia 186
APOTHEKERORDNUNG (*f*) : dispensatory (see APOTHEKERBUCH) 187
APOTHEKERWAREN (*f.pl*) : drugs 188
APOTHEKERWESEN (*n*) : pharmacy ; pharmaceutical matters 189
APOTHEKERWISSENSCHAFT (*f*) : pharmacology 190
APPARAT (*m*) : apparatus ; camera (Phot.) 191
APPARATEBAU (*m*) : construction of apparatus 192
APPARATIV : pertaining to apparatus 193
APPARATUR (*f*) : apparatus 194
APPELL (*n*) : appeal 195
APPRETIEREN : to dress ; size ; finish (Cloth) 196
APPRETURECHT : fast to (size) finishing or dressing 197
APPRETURLEIM (*m*) : dressing size 198
APPRETURMASCHINE (*f*) : dressing machine (Textile) ; finishing machine 199
APPRETURMITTEL (*n*) : dressing (finishing) medium 200
APPRETURÖL (*n*) : dressing (finishing) oil 201
APPRETURPRÄPARAT (*n*) : dressing-(preparation) 202
APPRETURPULVER (*n*) : dressing powder 203
APPRETURSEIFE (*f*) : dressing soap 204
APRIKOSENBAUM (*m*) : apricot tree (*Prunus armeniaca*) 205
APRIKOSENKERNÖL (*n*) : apricot kernel oil (from *Prunus armeniaca*) ; Sg. 0.921 206

APRIKOSENPFIRSCH (*m*): nectarine (a variety of peach) 207
APRILBLUME (*f*): (see **WALDRÖSCHEN**) 208
AQUÄDUKT (*n*): aqueduct 209
AQUAMARIN (*n*): aquamarine (a variety of beryl) (see **EISSPAT**); Sg. 2.7 210
AQUA REGIA: aqua regia, nitrohydrochloric acid, chloronitrous acid, chlorazotic acid 211
AQUARELLFARBE (*f*): (transparent)-water-colour 212
AQUARELLMALEREI (*f*): water colour painting 213
ÄQUATORIAL: equatorial 214
AQUAVIT (*m*): aqua vitæ; whisky; spirit(s) 215
ÄQUIDISTANT: equidistant 216
ÄQUIMOLEKULAR: equimolecular 217
ÄQUIVALENT (*n*): equivalent 218
ÄQUIVALENTEINHEIT (*f*): equivalent unit 219
ÄRA (*f*): era 220
ARABIN (*n*): arabin, gummic (arabic) acid 221
ARABINGUMMI (*n*): gum arabic; gum acacia (*Gummi indicum*; *Gummi arabicum*) 222
ARABINSÄURE (*f*): arabic acid; arabin 223
ARABISCHES GUMMI (*n*): (see **ARABINGUMMI**) 224
ARACHINSÄURE (*f*): arachic (arachidic) acid; $CH_3(CH_2)_{18}COOH$; Mp. 77°C. 225
ARACHISÖL (*n*): ground nut oil (see **ERDNUSSÖL**) 226
ARAGONIT (*m*): aragonite (natural calcium carbonate); $CaCO_3$; Sg. 2.95 (see also **CALCIUMCARBONAT**) 227
ARÄOMETER (*n*): areometer; hydrometer 228
ARBEIT (*f*): labour, work, toil, employment, energy 229
ARBEITEN: to work, act, operate, go, run, function; strain, complain (of ships), to labour; (*n*) swelling and contraction (of wood) 230
ARBEITER (*m*): labourer, worker (of any kind) 231
ABREITERKLEIDERSCHRANK (*m*): clothes cupboard or locker (for worker's clothes) 232
ARBEITGEBER (*m*): employer 233
ARBEITNEHMER (*m*): employee 234
ARBEITNEHMERAKTIE (*f*): worker's share 235
ARBEITSBANK (*f*): (work-)bench 236
ARBEITSEINSTELLUNG (*f*): strike (of workers); breakdown (Plant); stoppage; cessation of work 237
ARBEITSERSPARNIS (*f*): saving of labour 238
ARBEITSFÄHIGKEIT (*f*): working capacity, capacity, capability 239
ARBEITSFLÄCHE (*f*): bearing (working) surface 240
ARBEITSFREI: ineffective-(ly), non-effective-(ly) 241
ARBEITSHUB (*m*): working stroke 242
ARBEITSKAMPF (*m*): labour (war) trouble(s) 243
ARBEITSKÖRPER (*m* and *pl.*): (*sing.*) working (fluid) medium; (*pl.*) working media 244
ARBEITSLEISTUNG (*f*): duty; performance; load; output (of work) 245
ARBEITSRÄDER (*n.pl.*): transmitting gears 246
ARBEITSSTROM (*m*): working current 247
ARBEITSTREITIGKEIT (*f*): labour (war) trouble(s) 248
ARBEITSVERMÖGEN (*n*): energy (Physics) 249
ABREITSVERRICHTUNG (*f*): performance (of work) 250
ARBEITWALZE (*f*): working roll 251
ARBITRIUM (*n*): arbitrament 252
ARCHE (*f*): ark; chest; air-box; sound-board 253
ARCHIPEL (*m*): archipelago 254
ARCHITEKT (*m*): architect 255
ARCHIV (*n*): records; archives; registrar's office 256
ARDENNIT (*m*): ardennite; $H_{10}Mn_{10}Al_{10}Si_{10}(V,As)_2O_{55}$; Sg. 3.64 257
ARDOMETER (*n*): ardometer 258
AREAL (*n*): area 259
AREKANUSS (*f*): areca nut (*Semen arecae*) 260
ARFVEDSONIT (*m*): arfvedsonite; $Na_2Fe_3(SiO_3)_4$; Sg. 3.46 261
ARGEMONEÖL (*n*): argemone oil (from seeds of *Argemone mexicana*) 262
ARGENTAN (*m*): argentan: same as "**NEUSILBER**" 263
ARGENTIT (*m*): argentite (see **SILBERGLANZ**) 264
ARGENTOMETER (*n*): argentometer (Photography) 265
ARGENTOPYRIT (*m*): argentopyrite; $AgFe_3S_5$; Sg. 6.47 266
ARGON (*n*): argon (Ar) 267
ARGUMENT (*n*): argument 268
ARGUMENTIEREN: to argue 269
ARGYRODIT (*m*): argyrodite (natural silver and germanium sulphide); Ag_8GeS_6; Sg. 6.26 270
ARGYROPYRIT (*m*): argyropyrite; $Ag_3Fe_7S_{11}$; Sg. 4.2 271
ARISTOPAPIER (*n*): gelatino-chloride emulsion printing-out paper; (sometimes gelatino-argentic printing-out paper); (Photography) 272
ARISTOTYPIE (*f*): (see **ARISTOPAPIER**) 273
ARITHMETIK (*f*): arithmetic 274
ARITHMETISCH: arithmetical 275
ARM (*m*): arm; branch; crank; beam (of Scales); tributary 276
ARM: poor, having a low content, scanty, barren, meagre, weak, thin 277
ARMATUR (*f*): mountings, garniture, armature, fittings 278
ARMATUREN (*f.pl.*): mountings, fittings (valves, gauges, etc.) 279
ARMBLEI (*n*): refined lead; lead free from silver; lead having no silver content 280
ARMENISCHE ERDE (*f*): Armenian earth (see **BOLUS, ROTER-**) 281
ARMFEILE (*f*): arm file 282
ARMIEREN: to arm 283
ARNAUDONGRÜN (*n*): (see **GUIGNETGRÜN**) 284
ARNIKA (*f*): arnica (*Arnica montana*) 285
ARNIKABLÜTEN (*f.pl.*): arnica flowers; arnica; (*Flores arnica montana*) 286
ARNIKABLÜTENÖL (*n*): arnica (flower) oil; Sg. 0.906 287
ARNIKAPAPIER (*n*): arnica paper 288
ARNIKASEIFE (*f*): arnica soap 289
ARNIKAWURZEL (*f*): arnica root (*Rhizoma arnicae*); arnica rhizome 290
AROMA (*n*): aroma; perfume; smell 291
AROMATISCHE VERBINDUNG (*f*): aromatic compound 292
AROMATISIEREN: to scent, perfume 293
ARON (*m*): arum; cuckoo-pint; lords and ladies (*Arum maculatum*) 294
ARONSSTÄRKE (*f*): arrow-root 295
ARONSWURZEL (*f*): arum root 296
ARRAK (*m*): arrack 297
ARRAKÄTHER (*m*): arrack ether 298
ARRAKESSENZ (*f*): arrack essence 299
ARRESTANT (*m*): prisoner 300
ARRETIEREN: to arrest; make prisoner 301
ARRODIEREN: to erode 302

ARROWMEHL (*n*): arrow-root; maranta (Pharm.) 303

ARSANILSAUER: arsanilate, combined with arsanilic acid 304

ARSANYLSÄURE (*f*): atoxyl, atoxylic (arsanilic, para-aminophenylarsinic) acid, arsenic acid anilide; $C_6H_4NH_2.AsO(OH)_2$; Mp. 232°C. 305

ARSAZETIN (*n*): (see **AZETYLARSANILAT**) 306

ARSEN (*n*): arsenic; (As); Sg. 4.7-5.7 307

ARSENAL (*n*): arsenal (military) 308

ARSENBLENDE (*f*): arsenic blende 309

ARSENBLENDE, GELBE- (*f*): orpiment; As_2S_3 310

ARSENBLENDE, ROTE- (*f*): realgar; AsS 311

ARSENBLÜTE (*f*): arsenic bloom; arsenite; As_2O_3; Sg. 3.865; Mp. 200°C. 312

ARSENBROMÜR (*n*): arsenious bromide, arsenic tribromide, arsenic bromide; $AsBr_3$; Sg. 3.66; Mp. 31°C.; Bp. 221°C. 313

ARSENCHLORID (*n*): arsenic chloride; arsenic butter; arsenic trichloride; $AsCl_3$; Sg. 2.2; Mp. 18°C.; Bp. 134°C.; (*Arsenium chloratum*) 314

ARSENDAMPF (*m*): arsenic vapour 315

ARSENDISULFID (*n*): red orpiment, ruby (red) arsenic, red arsenic glass, arsenic disulphide; realgar; As_2S_2; Sg. 3.55; Mp. 307°C.; (*Arsenium sulfuratum rubrum*); (see also **ARSENROT** and **REALGAR**) 316

ARSENEISEN (*n*): arsenic iron; Löllingite; (see **LÖLLINGIT**) 317

ARSENEISENSINTER (*m*): pitticite 318

ARSENFAHLERZ (*n*): tennanite, grey copper ore; $3Cu_3AsS_3.Cu(Cu_4,Fe_2)AsS_4$; Sg. 4.42 319

ARSENFLECK (*m*): arsenic stain 320

ARSENGLAS (*n*): arsenic glass; As_2O_3; (see **ARSENIKGLAS**) 321

ARSENHALOGEN (*n*): arsenic halide 322

ARSENID (*n*): -ic arsenide 323

ARSENIGESÄURE (*f*): arsenious acid; As_2O_3; (*Acidum arsenicosum*); (see **ARSENTRIOXYD**) 324

ARSENIGSAUER: arsenite of; combined with arsenious acid 325

ARSENIK (*m*): (see **ARSEN**) 326

ARSENIKALFAHLERZ (*n*): tennantite; $3Cu_2S.As_2S_3$; Sg. 4.42; (57.5%Cu) 327

ARSENIKALKIES (*m*): leucopyrite; arsenical pyrites; löllingite (see **LÖLLINGIT**) 328

ARSENIKBLEI (*n*): lead arsenide 329

ARSENIKBLÜTE (*f*): arsenite 330

ARSENIKBUTTER (*f*): chloride of arsenic (see **ARSENCHLORID**) 331

ARSENIKERZ (*n*): arsenic ore 332

ARSENIK (*m*), **GEDIEGENER-**: native arsenic; Sg. 4.7-5.7 (see **ARSENIK**) 333

ARSENIK (*m*), **GELBER-**: orpiment (see **AURIPIGMENT** and **ARSENTRISULFID**) 334

ARSENIKGLAS (*n*): arsenic glass; vitreous arsenic trioxide; As_2O_3 (see **ARSENIGESÄURE**) 335

ARSENIKGLAS, GELBES- (*n*): (see **ARSENBLENDE, GELBE-**) 336

ARSENIKGLAS, ROTES- (*n*): (see **ARSENBLENDE, ROTE-**) 337

ARSENIKKALK (*m*): calcium arsenate; $Ca_3(AsO_4)_2$ 338

ARSENIKKIES (*m*): arsenical pyrites; arsenopyrite; mispickel; FeAsS; Sg. 6.1; (46% As.) 339

ARSENIKKOBALT (*m*): skutterudite (see **SKUTTERUDIT**) 340

ARSENIKKOBALTKIES (*m*): skutterudite (see **SKUTTERUDIT**) 341

ARSENIKKÖNIG (*m*): arsenic regulus (see **ARSENIK**) 342

ARSENIKÖL, (ÄTZENDES-) (*n*): caustic oil of arsenic; $AsCl_3$ 343

ARSENIK (*m*), **ROTER-**: realgar (see **REALGAR** and **ARSENDISULFID**) 344

ARSENIKRUBIN (*m*): realgar, ruby arsenic, red arsenic glass, red orpiment, arsenic disulphide; As_2S_2; Sg. 3.5; Mp. 307°C. 345

ARSENIKSALZ (*n*): salt of arsenic; arsenate or arsenite 346

ARSENIKSÄURE (*f*): arsenic acid (see **ARSENSÄURE**) 347

ARSENIKSÄURE (*f*); **OKTAEDRISCHE-**: arsenite 348

ARSENIKSILBER (*n*): arsenical silver 349

ARSENIKSINTER (*m*): Scorodite; $Fe_2O_3.As_2O_5.4H_2O$ 350

ARSENIKVITRIOL (*m*): arsenic sulphate 351

ARSENIKWASSERSTOFF (*m*): arsenic hydride (see **ARSENWASSERSTOFF**) 352

ARSENIK (*m*), **WEISSER-**: white arsenic, arsenic trioxide, arsenious acid (see **ARSENTRIOXYD**) 353

ARSENIT (*m*): arsenite (see **ARSENTRIOXYD**) 354

ARSENJODID (*n*): arsenic iodide; arsenic triiodide; AsI_3; Sg. 4.39; Mp. 140.7°C.; Bp. 394°C. 355

ARSENJODID (*n*): arsenic iodide; AsI_3 (see **ARSENTRIJODID**) 356

ARSENJODÜR (*n*): arsenic diiodide; AsI_2 357

ARSENKIES (*m*): (see **ARSENIKKIES**) 358

ARSENKOBALT (*m*): skutterudite (see **SKUTTERUDIT**) 359

ARSENKUPFER (*n*): arsenical copper, domeykite (natural copper arsenide); Cu_3As; Sg. 7.2 360

ARSENLEGIERUNG (*f*): arsenic alloy 361

ARSENMETALL (*n*): metallic arsenide 362

ARSEN (*n*), **METALLISCHES-**: metallic arsenide 363

ARSENNICKEL (*m*): nickel arsenide; Niccolite (Mineral) (see **NICCOLIT**) 364

ARSENNICKELGLANZ (*m*): Gersdorffite, NiAsS; Sg. 6.2 (see **NICKELGLANZ**) 365

ARSENNICKELKIES (*m*): natural nickel arsenide; Niccolite,; NiAs (44% Ni) 366

ARSENOLITH: arsenolite; As_2O_3; Sg. 3.71 [367

ARSENOPYRIT (*m*): arsenopyrite (see **ARSENIKKIES**) 368

ARSENOXYD (*n*): arsenic oxide; arsenic trioxide; As_2O_3; Sg. 3.865; Mp. 200°C. 369

ARSENPENTACHLORID (*n*): (*Arsenium pentachloratum*): arsenic pentachloride; $AsCl_5$; Mp. -40°C. 370

ARSENPENTAFLUORID (*n*): arsenic pentafluoride; AsF_5; Bp. -53°C. 371

ARSENPENTAJODID (*n*): arsenic pentaiodide; AsJ_5; Sg. 3.93 372

ARSENPENTOXYD (*n*): arsenic pentoxide, arsenic anhydride; As_2O_5; Sg. 4.086 373

ARSENROHR (*n*): arsenic tube 374

ARSENROT (*n*): red arsenic (*Arsenium sulfuratum rubrum*) (see **ARSENDISULFID**) 375

ARSENRUBIN (*m*): realgar; AsS (see **ARSENDISULFID**) 376

ARSENSALZ (*n*): arseniate, arsenate 377

ARSENSAUER: arsenate of; combined with arsenic acid 378

ARSENSÄURE (*f*): arsenic acid; As_2O_5 (*Acidum arsenicicum*) 379

ARSENSÄURE-HYDRAT (*n*) : arsenic acid, hydrated ; $2H_3AsO_4+H_2O$; Mp. 36°C. 380
ARSENSAURES SALZ (*n*) : arseniate, arsenate 381
ARSENSÄURE, UNVOLLKOMMENE- (*f*) : arsenious acid (*Acidum arsenicosum*) (see **ARSENIGESÄURE**) 382
ARSENSILBER (*n*) : arsenical silver ; silver arsenide 383
ARSENSILBERBLENDE (*f*) : arsenical silver glance ; proustite, arsenic-silver blende (natural sulpharsenide of silver) ; Ag_3AsS_3 ; Sg. 5.5 (see also **ROTGILTIGERZ, LICHTES-**) 384
ARSENSULFID (*n*) : (any-) sulphide of arsenic ; (see **ARSENDISULFID** and **ARSENTRISULFID**) 385
ARSENTRIBROMID (*n*) : arsenic tribromide ; arsenious bromide ; $AsBr_3$; Sg. 3.54-3.66 ; Mp. 31°C. ; Bp. 221°C. 386
ARSENTRICHLORID (*n*) : (*Arsenium chloratum*) : arsenic trichloride ; arsenic butter ; $AsCl_3$; Sg. 2.163-2.205 ; Mp.–18°C. ; Bp. 134°C. 387
ARSENTRIFLUORID (*n*) : arsenic trifluoride ; AsF_3 ; Sg. 2.6659-2.734 388
ARSENTRIJODID (*n*) : arsenic triiodide, arsenious iodide ; AsI_3 ; Sg. 4.374 -4.39 ; Mp. 140.7°C. ; Bp. 394°C. 389
ARSENTRIOXYD (*n*) : arsenic trioxide, white arsenic, arsenious acid, arsenous anhydride, arsenous oxide ; As_2O_3 ; Sg. 3.646-4.0 ; Mp. 200°C. 390
ARSENTRISULFID (*n*) ; arsenous sulphide, yellow arsenic sulphide ; arsenic trisulphide ; orpiment (*Arsenium sulfuratum citrinum*) ; As_2S_3 ; Sg. 3.43 ; Mp. 310°C. ; Bp. 700°C. 391
ARSENTRISULFIDHYDRAT (*n*) : arsenic trisulphide hydrated ; $As_2S_3.6H_2O$; Sg. 1.881 392
ARSENÜR (*n*) : -ous arsenide 393
ARSENWASSERSTOFF (*m*) : arsenic hydride ; arseniuretted hydrogen ; arsine ; AsH_3 ; Sg. 2.685 ; Mp. 113.5°C. ; Bp. 54.8°C. [394
ARSENWASSERSTOFFGAS (*n*) : arseniuretted hydrogen gas 395
ARSINIGSÄURE (*f*) : arsinic acid 396
ARSINSÄURE (*f*) : arsonic acid 397
ART (*f*) : description, type, kind, sort 398
ARTERIÖS : arterial 399
ARTERIOSKLEROSE (*f*) : arterio-sclerosis (Med.) 400
ARTESISCH : artesian 401
ARTHRITIS (*f*) : arthritis (Med.) 402
ARTHRITIS DEFORMANS : rheumatic arthritis, rheumatoid arthritis, rheumatic gout, osteoarthritis (*Malum coxæ senile*) (Med.) 403
ARTIKEL (*m*) : article 404
ARTISCHOCKE (*f*) : artichoke (Globe artichoke, *Cynara scolymus* ; Jerusalem artichoke, *Helianthus tuberosus*) 405
ARZNEI (*f*) : medicine ; drugs ; physic 406
ARZNEIAPPLIKATION (*f*) : medication (Med.) 407
ARZNEIESSIG (*m*) : medicated vinegar 408
ARZNEIFORM (*f*) : medicinal form ; medical form 409
ARZNEIFORMEL (*f*) : medical formula 410
ARZNEIGERUCH (*m*) : medicinal smell 411
ARZNEIGESCHMACK (*m*) : medicinal taste 412
ARZNEIGEWICHT (*n*) : medicinal weight 413
ARZNEIHÄNDLER (*m*) : druggist 414
ARZNEIKAPSEL (*f*) : gelatine capsule (for containing nasty tasting medicines) 415
ARZNEIKÖRPER (*m*) : medicinal substance 416
ARZNEIKRÄFTIG : medicinal ; curative ; therapeutic 417
ARZNEIKRAUT (*n*) : medicinal herb ; medicinal plant 418
ARZNEILEHRE (*f*) : pharmacology 419
ARZNEILICH : medical ; medicinal 420
ARZNEIMASS (*n*) : officinal measure 421
ARZNEIMITTEL (*n*) : medicine ; remedy ; drug ; physic 422
ARZNEIMITTELKUNDE (*f*) : pharmacology 423
ARZNEIMITTELTRÄGER (*m*) : dissolvent medicinal tubes or capsules ; (*Cereoli*) ; (for conveying medicines by insertion into the channels of the body) 424
ARZNEIPRÄPARAT (*n*) : medicine, remedy, medicinal preparation 425
ARZNEISIRUP (*m*) : medicated syrup 426
ARZNEISTOFF (*m*) : medicinal substance 427
ARZNEIVERORDNUNG (*f*) : prescription 428
ARZNEIVERSCHREIBUNG (*f*) : prescription 429
ARZNEIWARE (*f*) : drug 430
ARZNEIWARENKUNDE (*f*) : pharmacology 431
ARZNEIWEIN (*m*) : medicinal (medicated) wine (tincture in which wine is used in place of alcohol for extracting the active principle of the drug) (Pharm.) ; (sometimes used in reference to a mixture of a drug in wine) 432
ÄRZTLICH : medical 433
ASAND (*m*) : (see **ASANT**) 434
ASANT, STINKENDER- (*m*) : asafoetida (from roots of *Ferula foetida*) 435
ASANT, WOHLRIECHENDER- (*m*) : gum benzoin (resin from *Styrax benzoin*) 436
ASARUMCAMPHER (*m*) : asarum camphor (from *Asarum* see **HASELWURZEL**) 437
ASBEST (*m*) : asbestos ; Sg. 2.5-2.9 438
ASBESTARTIG : amianthine ; asbestoid ; asbestiform 439
ASBESTFASER (*f*) : asbestos fibre 440
ASBESTGEWEBE (*n*) : asbestos cloth 441
ASBESTISOLIERPLATTE (*f*) : asbestos insulating (board) plate 442
ASBESTPACKUNG (*f*) : asbestos packing 443
ASBESTPAPIER (*n*) : asbestos paper 444
ASBESTPAPPE (*f*) : asbestos board 445
ASBESTPLATTE (*f*) : asbestos plate 446
ASBESTSCHALE (*f*) : asbestos dish 447
ASBESTSCHNUR (*f*) : asbestos twine 448
ASBESTSTRICK (*m*) : asbestos rope 449
ASBESTWAREN (*f.pl*) : asbestos (wares) goods 450
ASBESTWOLLE (*f*) : asbestos wool 451
ASBOLAN (*m*) : asbolite (see **KOBALTMANGANERZ**) 452
ASCHANTINUSSÖL (*n*) : (see **ERDNUSSÖL**) 453
ASCHBLAU (*n*) : zaffre (a crude cobalt oxide) 454
ASCHBLEI (*n*) : bismuth (see **WISMUT**) ; blacklead, graphite (Plumbago) 455
ASCHE (*f*) : ashes 456
ASCHEL (*m*) : sullage (Metal) 457
ÄSCHEL (*m*) : (see **ASCHEL**) 458
ASCHELOCHTÜRE (*f*) : ash-pit door 459
ASCHENABFUHR (*f*) : ash removal 460
ASCHENABTRANSPORT (*m*) : ash-handling plant ; ash-conveying plant ; ash removal 461
ASCHENARM : poor in ash content, having a low ash content (under 12% ash) (fuels) 462
ASCHENBAD (*n*) : ash bath 463
ASCHENBESTANDTEIL (*m*) : ash content ; percentage (%) of ash ; ash constituent 464
ASCHENDÜNGER (*m*) : slag (cinereal) manure 465

ASCHENERMITTELUNG (*f*): determination of ash (-content) 466
ASCHENFALL (*f*): ash (pit; pan) chute 467
ASCHENFALLKLAPPE (*f*): ash-valve, ash-damper, ash-door 468
ASCHENFLECKIG : sullage (speckled) spotted (Metal) 469
ASCHENHALDEN (*f.pl*): ash (slag) heaps 470
ASCHENKASTEN (*m*): ash-box. ash-pan 471
ASCHENKELLER (*m*): ash (pit) basement (of furnaces) 472
ASCHENLAUGE (*f*): lye from ashes 473
ASCHENOFEN (*m*): calcining (furnace) oven 474
ASCHENRAUM (*m*): ash-pit 475
ASCHENREICH: rich in ash content, having a high ash content (over 12% ashes) (fuels) 476
ASCHENSALZ (*n*): potash; K_2CO_3; Sg. 2.3312; Mp. 909°C.; alkali 477
ASCHENTRECKER (*m*): tourmaline 478
ASCHENWÄSCHE (*f*): ash washing plant 479
ASCHENZIEHER (*m*): (see ASCHENTRECKER); ash rake 480
ÄSCHER (*m*): tanner's pit; lime pit; slaked lime (Masonry); ash cistern (Soap); tin ashes (Ceramics) 481
ÄSCHERBRÜHE (*f*): lime liquor (Leather) 482
ÄSCHERFLÜSSIGKEIT (*f*): (see ÄSCHERBRÜHE) 483
ÄSCHERKALK (*m*): lime (Leather) 484
ÄSCHERN : to lime; ash 485
ÄSCHEROFEN (*m*): frit kiln (Ceramics) 486
ASCHLAUCH (*m*): (see WINTERPORREE) 487
ASCHTELLER (*m*): ash plate 488
ASCHWINDE (*f*): ash hoist 489
ÄSCHYNIT (*m*): eschynite (see AESCHYNIT) 490
ÄSCULIN (*n*): esculine; $C_{15}H_{16}O_9$ (from the bark of *Aesculus hippocastanum*) 491
ASEPSIS (*f*): asepsis, antisepsis 492
ASEPTISCH : aseptic, antiseptic 493
ASEPTOL (*n*): aseptol, orthophenolsulphonic acid; $C_6H_4.SO_3H.OH$ 494
ASKANDRUCK (*m*): asphalt process (a pigment process) (Phot.) 495
ASPARAGIN (*n*): asparagine; $CH_2.CO.(NH_2)_2.CH.CO_2H$ 496
ASPARAGINSÄURE (*f*): asparaginic acid, amino-succinic acid, asparagic acid, aspartic acid; $COOHCH_2CH(NH_2)COOH$ 497
ASPE (*f*): aspen tree (*Populus tremula*) 498
ASPHALT (*m*): asphalt (*Asphaltum*); bitumen; mineral pitch; Jew's pitch; Sg. 0.91-1.65 499
ASPHALTBETON (*m*): asphalt concrete 500
ASPHALTBREI (*m*): asphalt pulp 501
ASPHALTFIRNIS (*m*): asphalt varnish (Photography) 502
ASPHALTHALTIG : asphaltic; bitumistic; bituminiferous; containing (bitumen) asphalt 503
ASPHALTIEREN : to asphalt 504
ASPHALTKITT (*m*): asphalt cement; mastic 505
ASPHALTPAPIER (*n*): asphalt paper 506
ASPHALTPECH (*n*): bitumenous (mineral) pitch 507
ASPHALTPFLASTER (*n*): asphalt pavement 508
ASPHALTVERFAHREN (*n*): asphalt process (Photography) 509
ASPIDOSPERMIN (*n*): aspidospermin; $C_{22}H_{30}N_2O_2$; Mp. 205°C. 510
ASPIRIEREN : to aspire; aspirate; suck; draw 511
ASPIRIN (*n*): aspirin; acetylsalicylic acid (see ACETYLSALICYLSÄURE) 512

ASSEKURANZ (*f*): assurance; insurance 513
ASSEKURATEUR (*m*): insurance agent, underwriter 514
ASSEKURIEREN : to assure; insure 515
ASSELINSÄURE (*f*): asselinic acid; $C_{16}H_{31}CO_2H$ 516
ASSIMILATION (*f*): assimilation 517
ASSIMILATIONSVORGANG (*m*): assimilating (assimilation) process, process of assimilation 518
ASSIMILIERBAR : assimilable; capable of being assimilated 519
ASSIMILIEREN : to assimilate 520
ASSISTENTENZEIT (*f*): period of experience as an assistant (generally refers to budding chemists in Germany, and is of at least one year's duration) 521
ASSOCIIEREN : to associate 522
ASSORTIEREN : to sort; assort 523
ASSOUPLIEREN : to render supple or pliable 524
AST (*m*): branch; knot (in wood) 525
ASTATISCHES NADELPAAR (*n*): astatic needles 526
ASTFREI : not having any branches, having all the branches removed 527
ASTHMA (*n*): asthma 528
ASTHMATISCH : asthmatic 529
ASTHMAZIGARETTE (*f*): asthma cigarette, anti-asthmatic cigarette 530
ASTHOLZ (*n*): wood (timber) from branches 531
ÄSTIG : full of branches 532
ASTIGMATISCH : astigmatic 533
ASTIGMATISMUS (*m*): astigmatism 534
ASTREICH : full of branches 535
ASTROPHYLLIT (*m*): astrophyllite; $H_8(K,Na)_4(Fe,Mn)_9Fe_2Ti_4Si_{13}O_{52}$; Sg. 3.35 536
ASUROL (*n*): asurol; $C_{11}H_{14}NO_7HgNa$ 537
ASYL (*n*): asylum 538
ASYMMETRISCH : asymmetric-(al)-(ly) 539
ASYMPTOTE (*f*): asymptote 540
ASYMTOTE (*f*): asymptote 541
ASYMTOTISCH : asymptotic(-al)-(ly) 542
ASYNCHRON : asynchronous 543
ATAKAMIT (*m*): atakamite, atacamite (a basic copper chloride); $Cu_2Cl(OH)_3$; Sg. 3.76 544
ATELIERKAMERA (*f*): studio camera 545
ATEMEINSATZ (*m*): drum (of a German gas mask) 546
ÄTHAL (*n*): ethal; cetyl alcohol; $C_{16}H_{33}OH$; Sg. 0.8176; Mp. 50°C.; Bp. 344°C. 547
ÄTHAN (*n*): ethane, ethyl hydride; C_2H_6; Sg. 0.446; Mp. -171.4°C; Bp. -89.5°C. 548
ÄTHANAL (*n*): (see ACETALDEHYD) 549
ÄTHANSÄURE (*f*): ethanic acid, acetic acid (see ESSIGSÄURE): $CH_3.CO_2H$ 550
ÄTHER (*m*): ether; $(C_2H_5)_2O$; Sg. 0.71994; Mp. 116.2°C.; Bp. 34.97°C (see also ÄTHYLÄTHER); (as a suffix = ether; ester) 551
ÄTHERANLAGE (*f*): ether plant 552
ÄTHERART (*f*): kind of ether 553
ÄTHERARTIG : etheral 554
ÄTHERBILDUNG (*f*): ether formation; etherification 555
ÄTHERIFIZIEREN : to etherify; (*n*) etherification 556
ÄTHERISCH : etheral; essential; volatile (of oils) 557
ATHERMAN : athermanous (incapable of transmitting heat rays; impermeable to radiant heat) 558
ÄTHERPROBER (*m*): ether tester 559
ÄTHERSÄURE (*f*): ether acid 560

ÄTHERSCHWEFELSÄURE (*f*): ethylsulphuric acid; alkylsulphuric acid; $C_2H_5HSO_4$; Sg. 1.316; Bp. 280°C. 561

ÄTHERSCHWINGUNG (*f*): etheral vibration 562

ÄTHERWEINGEIST (*m*): spirit of ether; spirit of wine (solution of ether in alcohol) (see also ÄTHYLALKOHOL) 563

ÄTHERZERSTÄUBER (*m*): ether spray 564

ÄTHIN (*n*): ethine; acetylene (see ACETYLEN) 565

ÄTHOXYKAFFEIN (*n*): ethoxycaffeine; $C_8H_9(O.C_2H_5)N_4O_2$; Mp. 140°C. 566

ÄTHOXYKOFFEIN (*n*): ethoxycaffeine (see ÄTHOXYKAFFEIN) 567

ÄTHOXYL (*n*): ethoxyl 568

ÄTHYL (*n*): ethyl 569

ÄTHYLACETANILID (*n*): ethyl acetanilide 570

ÄTHYLACETAT (*n*): ethyl acetate, acetic ether; vinegar naphtha; $CH_3COOC_2H_5$; Sg. 0.9003-0.9238; Mp. -82.4°C; Bp. 77.15°C. 571

ÄTHYLAL (*n*): ethylal, methylene diethyl ester, $CH_2(O.C_2H_5)_2$; Sg. 0.851; Bp. 89°C. 572

ÄTHYLALKOHOL (*m*): ethyl alcohol, spirit of (ether) wine, grain alcohol, fermentation alcohol; C_2H_6O; Sg. 0.804; Mp. -117.6°C.; Bp. 78.4°C. 573

ÄTHYL-ALPHA-NAPTHYLAMIN (*n*): ethyl-alpha-naphthylamine 574

ÄTHYL, AMEISENSAURES- (*n*): (see AMEISENSÄURE-ÄTHYLESTER) 575

ÄTHYLAMIDOBENZOESÄURE (*f*): ethyl amidobenzoic acid 576

ÄTHYLAMIN (*n*): ethyl amine; C_2H_7N; Sg. 0.7057; Mp.-85.2°C; Bp. 16.6°C. 577

ÄTHYLANILIN (*n*): ethyl aniline, monethylaniline; $C_8H_{11}N$; Sg. 0.9625; Mp. about -80°C.; Bp. 204°C. 578

ÄTHYLAT (*n*): ethylate 579

ÄTHYLÄTHER (*m*): ethyl ether, ether, ethyl oxide, sulphuric ether; $C_4H_{10}O$; Sg. 0.7191; Mp.-117.6°C; Bp. 34.6°C. 580

ÄTHYL (*n*), BALDRIANSAURES- : ethyl (valerate, valerianate) valeriate; $C_5H_9O_2C_2H_5$; Sg. 0.877; Bp. 144.5°C. 581

ÄTHYL (*n*), BENZOESAURES- : ethyl benzoate (see BENZOESÄURE-ÄTHYLESTER) 582

ÄTHYLBENZOL (*n*): ethyl benzene, ethyl benzol; C_8H_{10}; Sg. 0.8759; Mp. -92.8°C.; Bp. 135.7-135.9°C. 583

ÄTHYLBENZYLANILIN (*n*): ethyl benzylaniline; $C_6H_5N(C_2H_5)CH_2C_6H_5$; Sg. 1.034; Bp. 286°C. 584

ÄTHYLBENZYLKETON (*n*): ethyl benzyl ketone; $C_{10}H_{12}O$; Sg. 0.998; Bp. 223-226°C. 585

ÄTHYL (*n*), BERNSTEINSAURES- : ethyl succinate 586

ÄTHYLBLAU (*n*): ethyl blue 587

ÄTHYLBROMID (*n*): (*Aethylium bromatum*): ethyl bromide, bromic ether, hydrobromic ether, monobromoethane; C_2H_5Br.; Sg. 1.5014; Mp. 117.6°C.; Bp. 38.4°C. 588

ÄTHYL (*n*), BUTTERSAURES- : ethyl butyrate, butyric ether; $C_3H_7CO_2C_2H_5$; Sg. 0.879; Mp.-93.3°C.; Bp. 120.6°C. 589

ÄTHYLCARBINOL (*n*): (see PROPYLALKOHOL) 590

ÄTHYLCARBONSÄURECHININESTER (*m*): quinine ethylcarbonate; $CO.OC_2H_5.O.C_{20}H_{23}N_2O$; Mp. 95°C. 591

ÄTHYL (*n*), CHLORAMEISENSAURES- : ethyl chloroformate; $ClCOO.C_2H_5$ 592

ÄTHYL (*n*), CHLORESSIGSAURES : ethyl chloroacetate (see CHLORESIGSÄURE-ÄTHYLESTER) 593

ÄTHYLCHLORID (*n*): monochlorethane, ethyl chloride, Kelene, chelene (*Aethylium chloratum*); C_2H_5Cl; Sg. 0.9176; Mp.-141.6°C.; Bp. 14°C. 594

ÄTHYL (*n*), CHLORKOHLENSAURES- : ethyl chlorocarbonate 595

ÄTHYL (*n*), CHLORPROPIONSAURES- : ethyl chloropropionate 596

ÄTHYL (*n*), CYANESSIGSAURES- : ethyl acetocyanate 597

ÄTHYLDISULFID (*n*): ethyl disulphide; $C_4H_{10}S_2$; Sg. 0.9927; Bp. 152.8-153.4°C. 598

ÄTHYLEN (*n*): ethylene; olefiant gas; C_2H_4; Sg. 0.6095; Mp. 169°C.; Bp. 102.5°C. 599

ÄTHYLENALKOHOL (*m*): ethylene alcohol, ethylene glycol, glycol, glycohol alcohol (see GLYKOL) 600

ÄTHYLENÄTHENYLDIAMIN (*n*): ethylene ethenyldiamine (see LYSIDIN) 601

ÄTHYLENÄTHENYLDIAMINCHLORHYDRAT (*n*): ethylene ethenyldiamine hydrochloride 602

ÄTHYLENBINDUNG (*f*): ethylene linkage; double bond 603

ÄTHYLENBROMID (*n*): ethylene bromide, dibromomethane; $C_2H_4Br_2$; Sg. 2.1823; Mp. 9.95°C. Bp. 129°C. 604

ÄTHYLENCHLORHYDRIN (*n*): ethylene chlorohydrine 605

ÄTHYLENCHLORID (*n*): elayl chloride, dichloroethane; ethylene chloride, vinyl trichloride, Dutch liquid; $C_2H_4Cl_2$; Sg. 1.2656; Mp.-35.3°C.; Bp. 83.7°C. 606

ÄTHYLENDIAMIN (*n*): ethylene diamine; $C_2H_8N_2+H_2O$; Sg. 0.97; Mp. 10°C.; $C_2H_8N_2$; Sg. 0.902; Bp. 117°C. 607

ÄTHYLENDIAMINCHLORHYDRAT (*n*): ethylenediamine hydrochloride 608

ÄTHYLENDIBROMID (*n*): ethylene dibromide (see ÄTHYLENBROMID) 609

ÄTHYLENGLYKOL (*n*): ethylene glycol (see GLYKOL) 610

ÄTHYLENGLYKOLMONOBENZOESÄUREESTER (*m*): ethylene-glycol monobenzoate; $C_9H_{10}O_3$ 611

ÄTHYLENIMIN (*n*): ethylene imine (see PIPERAZIN) 612

ÄTHYLENJODID (*n*): ethylene iodide; $C_2H_4I_2$; Sg. 2.07; Mp. 81-82°C. 613

ÄTHYLENOXYD (*n*): ethylene oxide; C_2H_4O; Sg. 0.8966; Bp. 13.5°C. 614

ÄTHYLENREST (*m*): ethylene residue 615

ÄTHYL (*n*), ESSIGSAURES- : ethyl acetate, vinegar naphtha, acetic ether (see ESSIGÄTHER) 616

ÄTHYLESTER (*m*): ethyl ester 617

ÄTHYLFORMIAT (*n*): ethyl formate; $HCOO.C_2H_5$ (see AMEISENSAURE-ÄTHYLESTER) 618

ÄTHYLGLYKOLSÄUREMENTHOLESTER (*m*): menthol ethylglycolate (see CORYFIN) 619

ÄTHYLGRUPPE (*f*): ethyl group 620

ÄTHYLIDEN (*n*): ethylidene 621

ÄTHYLIDENBROMID (*n*): ethylidene bromide; $C_2H_4Br_2$; Sg. 2.1001; Bp. 112.5°C. 622

ÄTHYLIDENCHLORID (*n*): ethylidene chloride; $C_2H_4Cl_2$; Sg. 1.1895; Mp.-96.7; Bp. 57.5 623

ÄTHYLIDENDIÄTHYLÄTHER (*m*): (see ACETAL) 624

ÄTHYLIDENDIAZETAT (*n*): ethylidene diacetate; $C_2H_4(O_2C_2H_3)_2$ 625

ÄTHYLIDENDIMETHYLÄTHER (*m*): ethylidene dimethylester (see DIMETHYLAZETAL) 626

ÄTHYLIDENJODID (*n*): ethylidene iodide; $C_2H_4I_2$; Sg. 2.84; Bp. 177-179°C. 627

ÄTHYLIDENMILCHSÄURE (*f*): (see MILCHSÄURE, GÄRUNGS-) 628

ÄTHYLIEREN: to ethylate; (*n*) ethylation, ethylating 629

ÄTHYL (*n*), ISOBUTTERSAURES-: ethyl isobutyrate, isobutyric ether; $(CH_3)_2CHCOOC_2H_5$; Sg. 0.889; Bp. 110.5°C. 630

ÄTHYLISOPROPYLÄTHER (*m*): ethyl isopropylether; $C_5H_{12}O$; Sg. 0.7447; Bp. 54°C. 631

ÄTHYLISOPROPYLKETON (*n*): ethyl isopropylketone; $C_6H_{12}O$; Sg. 0.8139; Bp. 114°C. 632

ÄTHYLJODID (*n*): ethyl iodidie, monoiodoethane; C_2H_5I; Sg. 1.98; Mp.–108.5°C; Bp. 72°C. 633

ÄTHYLKARBINOL (*n*): ethyl carbinol, normal (primary) propyl alcohol; $CH_3.CH_2.CH_2.OH$ or C_3H_8O.; Sg. 0.8066-0.8192; Bp. 97.3°C. 634

ÄTHYL (*n*): KOHLENSAURES-: ethyl carbonate, carbonic ether, diethylcarbonic ether; $(C_2H_5)_2CO_3$; Sg. 0.978; Bp. 126°C. 635

ÄTHYLMALONSÄURE (*f*): ethyl-malonic acid; $CH_3CH_2CH(COOH)_2$; Mp. 111.5°C. 636

ÄTHYL (*n*), MALONSAURES-: ethyl malate 637

ÄTHYLMERCAPTAN (*n*): (ethyl-) mercaptan; ethyl hydrosulphide; C_2H_6S; Sg. 0.8391; Bp. 37°C. 638

ÄTHYL (*n*), MILCHSAURES-: ethyl lactate 639

ÄTHYLMORPHIN (*n*): ethyl morphine 640

ÄTHYLMORPHINCHLORHYDRAT (*n*): (see DIONIN) 641

ÄTHYLNARCEINCHLORHYDRAT (*n*): ethylnarcein hydrochloride; $C_{25}H_{31}NO_8.HCl$; Mp. 205.5°C. 642

ÄTHYLNITRAT (*n*): ethyl nitrate, nitric ether; $C_2H_5O_3N$; Sg. 1.1352; Mp.–112°C.; Bp. 86.3°C. 643

ÄTHYLNITRIT (*n*): (*Aether nitrosus*): ethyl nitrite, nitrous ether; $C_2H_5O_2N$; Sg. 0.9; Bp. 17°C. 644

ÄTHYL (*n*), OXALSAURES-: ethyl oxalate 645

ÄTHYL (*n*), PELARGONSAURES-: ethyl pelargonate 646

ÄTHYLPHENYLÄTHER (*m*): ethyl-phenyl-ether; $C_8H_{10}O$; Sg. 0.9792; Mp. –33.5°C; Bp. 155.5°C. 647

ÄTHYL (*n*), PHENYLESSIGSAURES-: ethyl phenylacetate 648

ÄTHYLPHENYLHYDRAZIN (*n*): ethyl-phenyl hydrazine 649

ÄTHYLPHENYLKETON (*n*): ethyl-phenyl-ketone; $C_9H_{10}O$; Sg. 1.0141; Mp. 21°C; Bp. 218°C. 650

ÄTHYL (*n*), PHTALSAURES-: ethyl phthalate 651

ÄTHYLPROPIONAT (*n*): ethyl propionate; $C_2H_5.COO.C_2H_5$ 652

ÄTHYL (*n*), PROPIONSAURES-: ethyl propionate (see ÄTHYLPROPIONAT) 653

ÄTHYLPROPYLÄTHER (*m*): ethyl propylether; $C_5H_{12}O$; Sg. 0.7545; Bp. 63.6°C. 654

ÄTHYLPROPYLKETON (*n*): ethyl propylketone; $C_6H_{12}O$.; Sg. 0.8333; Bp. 122-124°C. [655

ÄTHYLROT (*n*): ethyl red; $C_{23}H_{23}N_2I$ 656

ÄTHYL (*n*), SALICYLSAURES-: ethyl salicylate 657

ÄTHYL (*n*), SALPETERSAURES-: ethyl nitrate (see ÄTHYLNITRAT) 658

ÄTHYL (*n*), SALPETRIGSAURES-: ethyl nitrite (see ÄTHYLNITRIT) 659

ÄTHYLSCHWEFELSAUER: ethylsulphate of; combined with ethylsulphuric acid 660

ÄTHYLSCHWEFELSÄURE (*f*). sulfethylic acid, sulfovinic acid, ethyl-sulphuric acid, monoethyl sulphate; $C_2H_6O_4S$; Sg. 1.316; Bp. 280°C. 661

ÄTHYL (*n*), SEBACINSAURES-: ethyl sebate 662

ÄTHYL (*n*), SEBACYLSAURES-: ethyl sebacylate 663

ÄTHYLSENFÖL (*n*): ethyl mustard oil, ethyl thiocarbimide; C_3H_5NS; Sg. 0.9972; Mp. -5.9°C; Bp. 131-132°C. 664

ÄTHYLSULFHYDRAT (*n*): ethyl hydrosulphide, mercaptan (see MERCAPTAN) 665

ÄTHYLSULFID (*n*): diethyl sulphide; ethyl sulphide; $C_4H_{10}S$; Sg. 0.8368; Mp.–99.5°C.; Bp. 93°C. 666

ÄTHYLSULFOCYANAT (*n*): ethyl sulfocyanate (see ÄTHYL, SULFOCYANSAURES-) 667

ÄTHYL (*n*), SULFOCYANSAURES-: ethyl sulfocyanate, ethyl thiocyanate; $C_2H_5.SCN$ 668

ÄTHYLSULFONSÄURE (*f*): ethylsulphonic acid 669

ÄTHYLTOLUIDIN (*n*): ethyl toluidine, ethyl-ortho-toluidine; $C_6H_4(CH_3)NH_2C_2H_5$; Sg. 0.953; Bp. 214°C. 670

ÄTHYL (*n*), TRICHLORESSIGSAURES-: ethyl trichloracetate 671

ÄTHYLWASSERSTOFF (*m*): ethyl hydride; ethane (see ÄTHAN) 672

ÄTHYLWEINSÄURE (*f*): ethyltartaric acid 673

ÄTHYL (*n*), WEINSAURES-: ethyl tartrate 674

ÄTHYL (*n*), ZIMTSAURES-: cinnamic ether, ethyl cinnamate; $C_6H_5C_2H_2CO_2C_2H_5$; Sg. 1.055; Mp. 12°C.; Bp. 271°C. 675

ATLAS (*m*): atlas; satin 676

ATLASERZ (*n*): false emerald; malachite; $2CuO.CO_2.H_2O$. (about 40% Cu.) (see also MALACHIT) 677

ATLASHOLZ (*n*): satinwood (*Chloroxylon swietenia*); Sg. 0.79-1.04 678

ATLASKIES (*m*): (see ATLASERZ and MALACHIT) 679

ATLASSPAT (*m*): satin spar; gypsum (natural calcium sulphate) (see GIPS) 680

ATLASVITRIOL (*m*): white vitriol; zinc sulphate; $ZnSO_4.7H_2O$; Sg. 1.9661; Mp. 50°C. [681

ATMOSPHÄRE (*f*): atmosphere 682

ATMUNG (*f*): breathing, respiration 683

ATMUNGSAPPARAT (*m*): respirator 684

ATOM (*n*): atom 685

ATOMBINDUNGSKRAFT (*f*): atomic combining power; valence 686

ATOMBINDUNGSVERMÖGEN (*n*): (see ATOMBINDUNGSKRAFT) 687

ATOMGEWICHT (*n*): atomic weight (see table of elements at end of book) 688

ATOMGEWICHTTABELLE (*f*): table of atomic weights (see table of elements at commencement of book) 689

ATOMHALTIG: atomic 690

ATOMHYPOTHESE (*f*): atomic hypothesis 691

ATOMIGKEIT (*f*): atomicity 692

ATOMMODELL (*n*): atomic model 693

ATOMREFRAKTION (*f*): atomic refraction 694

ATOMVERBAND (*m*): atomic union 695

ATOMVERKETTUNG (*f*): atomic linkage 696

ATOMVERSCHIEBUNG (*f*): atomic displacement 697

ATOMVOLUM (*n*): atomic volume 698

ATOMVOLUMEN (*n*): (see ATOMVOLUM) 699

ATOMWÄRME (*f*): atomic heat 700

ATOMZAHL (*f*): atom-(ic) number 701

ATOXYL (*n*): atoxyle (see ARSANYLSÄURE or ATOXYLSÄURE) 702

ATOXYLSÄURE (*f*): atoxylic acid, arsanilic acid, arsenic acid anilide; ($C_6H_4NH_2AsO(OH)_2$); Mp. 232°C. 703

ATRAMENTSTEIN (*m*): inkstone; native copperas; $FeSO_4$ 704

ATROPASÄURE (*f*): atropic acid 705

ATROPHIE (*f*): atrophy (Med.) 706

ATROPIN (*n*): atropine, daturin (an alkaloid from *Datura stramonium*): $C_{17}H_{23}NO_3$; Mp. 115.5°C. 707

ATROPIN (*n*), **SCHWEFELSAURES-**: atropin sulphate (see **ATROPINSULFAT**) 708

ATROPINSULFAT (*n*): atropin sulphate; $(C_{17}H_{23}NO_3)_2.H_2SO_4$; Mp. 187°C. 709

ATTENUIEREN: to attenuate 710

ATTEST (*n*): attestation; certificate 711

ATTICH (*m*): danewort, bloodwort, blood elder, ebulus, dwarf elder, wall-wort (*Sambucus ebulus*) 712

ATTICHBEERE (*f*): danewort berry (see **ATTICH**) 713

ATTICHWURZEL (*f*): dwarf elder root (see **ATTICH**) 714

ÄTZALKALI (*n*): caustic alkali 715

ÄTZAMMONIAK (*n*): caustic ammonia (*Liq. amm.*); ammonium hydroxide; ammonia water; Sg. 0.880; Bp. 38.5°C.; (NH_4OH), (Phot.) 716

ÄTZARTIKEL (*m*): discharge style (Calico); etching article 717

ÄTZBAR: corrodible; capable of being etched; dischargeable (Calico) 718

ÄTZBARYT (*n* and *m*): caustic baryta; barium hydroxide; barium hydrate; $Ba(OH)_2.8H_2O$; Sg. 1.656; Mp. 78°C.; Bp. 103°C. 719

ÄTZBEIZDRUCK (*m*): discharge printing 720

ÄTZBEIZE (*f*): discharge mordant 721

ÄTZDRUCK (*m*): discharge printing; etching 722

ÄTZE (*f*): corrosion; etching; mordant; cauterization; aqua fortis; discharge (Calico) 723

ÄTZEN: to corrode; etch; cauterize; discharge (Calico) 724

ÄTZEND MACHEN: to make (render) caustic; causticize 725

ÄTZFARBE (*f*): discharge (paste) colour (Calico) 726

ÄTZFIGUR (*f*): etching figure; corrosion figure or mark, (mark arising in metal after it has been first stressed and then subjected to a corrosion test) 727

ÄTZGIFT (*n*): caustic poison 728

ÄTZGRUND (*m*): etching ground 729

ÄTZKALI (*n*): caustic potash; potassium hydroxide; KOH; Sg. 2.044; Mp. 360.4°C. 730

ÄTZKALILAUGE (*f*): caustic potash (solution) lye; (solution of KOH in H_2O) 731

ÄTZKALK (*m*): caustic lime; calcium oxide, CaO (becomes calcium hydroxide, $Ca(OH_2)$, on exposure to moist air) 732

ÄTZKALK (*m*), **GEBRANNTER-**: quicklime; calcium-oxide, CaO; Sg. 3.15-3.4; Mp. 2570°C. 733

ÄTZKALK (*m*), **GELÖSCHTER-**: slaked lime; calcium hydroxide; $CaOH_2$; Sg. 2.078; loses water at 580°C. 734

ÄTZKRAFT (*f*): causticity; corrosive power 735

ÄTZLACK (*m*): discharge lake 736

ÄTZLAUGE (*f*): caustic (liquor; lye) solution (see **ÄTZKALI** and **ÄTZNATRON**) 737

ÄTZMAGNESIA (*f*): caustic magnesia 738

ÄTZMITTEL (*m*): escharotic (a substance which destroys tissue) (Med.); chemical discharge; corrosive, caustic 739

ÄTZMITTEL (*n. pl*), **FÜR KATTUNDRUCK** (*m*): discharge agents for calico printing 740

ÄTZNADEL (*f*): etching point (Art) 741

ÄTZNATRON (*n*): caustic soda; NaOH; Sg. 2.13; Mp. 318°C. 742

ÄTZNATRONLAUGE (*f*): caustic soda (solution; lye) liquor 743

ÄTZPAPP (*m*): resist (Calico) 744

ÄTZPASTE (*f*): etching paste 745

ÄTZPULVER (*n*): caustic (corrosive) powder 746

ÄTZSALZ (*n*): caustic salt 747

ÄTZSILBER (*n*): lunar caustic; silver nitrate; $AgNO_3$; Sg. 4.352; Mp. 218 °C. 748

ÄTZSTEIN (*m*): caustic (*Lapis causticus*): caustic potash; fused caustic potash 749

ÄTZSTIFT (*m*): caustic stick 750

ÄTZSTOFF (*m*): corrosive, mordant, caustic 751

ÄTZSTRONTIAN (*m*): caustic strontia, strontium (hydrate) hydroxide (see **STRONTIUMHYDROXYD**) 752

ÄTZSUBLIMAT (*n*): corrosive sublimate; mercuric chloride (*Mercurius sublimatus corrosivus*): $HgCl_2$; Sg. 5.32; Mp. 265°C.; Bp. 303°C. 753

ÄTZTINTE (*f*): caustic ink 754

ÄTZVERFAHREN (*n*): corrosive process (Metal) 755

ÄTZWASSER (*n*): caustic (water) etching liquid; nitric acid; (*Aqua fortis*); mordant; caustic solution 756

ÄTZWEISS (*m*): white discharge (Calico) 757

ÄTZWIRKUNG (*f*): discharge action; corrosive action 758

ÄTZZEICHNUNG (*f*): etching (Art) 759

AUBEPIN (*n*): aubepine (see **ANISALDEHYD**) 760

AUE (*f*): meadow; pasture 761

AUERBRENNER (*m*): Welsbach burner 762

AUERHAHN (*m*): capercaillie; mountain (black)-cock 763

AUERLICHT (*n*): Welsbach light 764

AUERLITH (*m*): auerlite (natural thorium silicophosphate); $SiO_2,ThO_2,P_2O_5,Fe_2O_3,Al_2O_3,CaO,CO_2,H_2O,MgO$; Sg. 4.6 765

AUERMETALL (*n*): Auer's metal (a pyrophorous alloy of cerium and iron, containing about 70% of the rare-earth metals and 30% iron) 766

AUFÄTZEN: to cauterize; open with caustic; corrode (etch) upon 767

AUFBAU (*m*): building up; construction; synthesis; erection; building; composition; state; formation 768

AUFBAUCHEN: to belly (of a sail); bulge; swell 769

AUFBAUMEN: to beam; rear up; wind up; show (metallic) lustre 770

AUFBEREITEN: to dress; prepare 771

AUFBEREITUNGSMASCHINE (*f*): dressing machine 772

AUFBEWAHREN: to store up; save; conserve; stock; preserve; reserve 773

AUFBEWAHRUNG (*f*): care, saving, storage, stocking, preservation, reservation, conservation 774

AUFBLÄHEN: to swell; bulge; inflate 775

AUFBLÄTTERN: to exfoliate (Min. and Med.) open (Plants and Books) 776

AUFBLICK (*m*): looking up; raising of the eyes; fulguration blick (Assaying) 777

AUFBLICKEN: to look up; raise the eyes; brighten; give blick 778
AUFBLÜHEN: to bloom; grow; flourish 779
AUFBRAUSEN: to effervesce; ferment 780
AUFBRECHEN (*n*): puddling (metal), (the turning round constantly of the lumps of iron to obtain uniform decarbonization); breaking up, disrupture 781
AUFBRENNEN: to burn up; refine (Metals); brand; consume (fuels); sulphur (Wine making) 782
AUFBÜRDEN: to saddle-(with), inflict, impose-(upon), burden-(with) 783
AUFDAMPFEN: to evaporate; smoke; steam; vapourize. convert into (steam) vapour 784
AUFDOCKEN: to roll (wind) up; bundle 785
AUFDREHEN: to untwist; unroll; unscrew; unlay (ropes, etc.); throw (Ceramics) 786
AUFDRUCK (*m*): printing; impress; stamp; imprint 787
AUFDUNSTEN: to evaporate; steam; vapourize 788
AUFDÜNSTEN: (see **AUFDUNSTEN**) 789
AUFEINANDERFOLGEND: consecutive; in series 790
AUFEINANDERHÄUFEN: to stack; heap; pile up 791
AUFFÄDELN: to thread on 792
AUFFALLEN: to be noticeable, be striking, be obvious, be easily seen, fall open, fall upon 793
AUFFALLEND: noteworthy; remarkable; striking 794
AUFFALLENDERWEISE: strangely enough, strange to say; noticeably, extraordinarily 795
AUFFANGEDRAHT (*m*): receiving wire, aerial 796
AUFFÄRBEN: to redye; dye (anew) again 797
AUFFASERN: to unravel; separate fibres 798
AUFFASSEN: to comprehend; conceive; perceive; take up; appreciate; understand; view; regard; look at 799
AUFFASSUNG (*f*): comprehension, appreciation, view 800
AUFFLECHTEN: to twist up, plait 801
AUFFRISCHEN: to freshen up; renew; retouch; refresh (Yeast, etc.); change 802
AUFFÜHREN: to enter; represent; present; produce; perform; note 803
AUFFÜHRUNG (*f*): conduct; representation; entering (in an account); presentation; performance 804
AUFGABE (*f*): problem, proposition; statement, note (Com.) 805
AUFGABEBECHERWERK (*n*): bucket elevator 806
AUFGABETRICHTER (*m*): feeding (charging) hopper (of mechanical stokers) 807
AUFGÄREN: to effervesce; ferment 808
AUFGEBEN: to deliver; surrender; give up; charge (Furnaces) 809
AUFGEBOT (*n*): summons; levy; banns; call; time of asking 810
AUFGEDUNSEN: swollen; puffed up; inflated 811
AUFGEHEN: to be absorbed (Dyes); shoot (Plants); leave no remainder (Arith.); rise; open; go up 812
AUFGEIEN: to brail up (sails) 813
AUFGELD (*n*): premium; agio; exchange; change 814
AUFGESCHLOSSEN: decomposed 815
AUFGESOGEN: sucked up; aspirated; drawn up 816
AUFGIESSEN: to pour upon; infuse 817
AUFGUSS (*m*): infusion (Infusum) 818
AUFGUSSAPPARAT (*m*): sparger (Brewing) 819
AUFGUSSGEFÄSS (*n*): infusion vessel; digester 820
AUFGUSSTIERCHEN (*n.pl*): infusoria 821
AUFHAKEN: to unhook, unclasp 822
AUFHALTEN: to stunt, retard, check, hold up, stop, halt 823
AUFHÄNGEN: to suspend 824
AUFHÄNGER (*m*): hanger; rack 825
AUFHÄNGUNG (*f*): suspension 826
AUFHASPELN: to reel, wind, roll up 827
AUFHAUEN: to re-cut (files); cut open 828
AUFHÄUFEN: to amass, heap up, pile, accumulate 829
AUFHEBEN: to lift; raise; be complementary to (colours); reduce (fraction) to lowest terms; cancel; annul; abolish 830
AUFHEFTEN: to bind (books), pin together, fasten 831
AUFHELLBAR: capable of being cleared (clarified, lightened or brightened) 832
AUFHELLEN: to lighten, clear, brighten 833
AUFHELLUNG (*f*): clarifying, lightening, brightening 834
AUFHETZEN: to urge on, incite, stir up 835
AUFHISSEN: to hoist (pull) up 836
AUFHOLEN: to ground a ship 837
AUFHORCHEN: to prick up the ears, listen, harken 838
AUFHÖREN: to cease, stop, discontinue, suspend 839
AUFKATTEN: to fish (an anchor) 840
AUFKEILEN: to wedge on; key on 841
AUFKIMMUNG (*f*): rise of floor line; rise of floors 842
AUFKLAFFEN: to gape, open, burst, rupture, spring, yawn, split 843
AUFKLAPPEN: to open; lift a flap; raise 844
AUFKLÄREN (ÜBER): to inform, enlighten, clear up 845
AUFKLÄRUNG (*f*): information, enlightenment, orientation; (*pl*) data 846
AUFKLAUEN: to make a V-joint (of wood) 847
AUFKLEBEN: to paste on to, stick on to, glue on to 848
AUFKLEISTERN: to paste on to, stick on to, glue on to 849
AUFKLINKEN: to unlatch, unlock 850
AUFKOCHEN: to boil (up); prime (boilers) 851
AUFKOMMEN: to come upon; hit (an idea) 852
AUFKRATZEN: to scratch open; stir; poke (the fire); card (wool); dress; nap (fabrics) 853
AUFKRÄUSEN: to form a head (beer) 854
AUFKREMPELN: to card (Textile) 855
AUFKÜHLEN: to aerate; cool 856
AUFLADEN: to load, charge 857
AUFLAGE (*f*): edition (book); duty; tax; impost; rest; support; buttress; bearing (Mech.) 858
AUFLAGER (*n*): truck, saddle, bearing, bearer, support 859
AUFLAGERDRUCK (*m*): bearing pressure 860
AUFLAGERENTFERNUNG (*f*): span between supports 861
AUFLAGERUNG (*f*): support, setting; buttress, abutment; spring (of an arch); thrust; bearer-bar (of grates) 862
AUFLANGER (*m*): futtock (Naut.) 863
AUFLAUFEN: to wind up; rise; swell; feed (a furnace) 864

AUFLEBEN : to revive ; resuscitate 865

AUFLEGEN : to impose ; print ; publish ; lean upon ; rest on ; be supported on 866

AUFLEHNEN : to lean-(on) ; rebel-(against) ; oppose 867

AUFLEIMEN : to glue on to, stick to, paste upon 868

AUFLESEN : to gather ; pick 869

AUFLIEGEN ; to be supported by 870

AUFLOCKERN : to loosen ; 871

AUFLOCKERUNGSMITTEL (*n*) : loosening agent (for lead converter process and to hasten the roasting) 872

AUFLÖSBAR : soluble 873

AUFLÖSBARKEITSGRENZE (*f*) : disappearing point (of objects under a microscope) 874

AUFLÖSEN : to dissolve, loosen, liquidate, analyze, decompose, resolve, solve 875

AUFLÖSER (*m*) : solvent ; dissolver ; mixer (paper) 876

AUFLÖSUNG (*f*) : solution ; decomposition ; analysis ; dissolution ; liquidation ; mellowness (malt) ; friability ; disappearance ; resolution ; return to original key (music) 877

AUFLÖSUNGSGRENZE (*f*) : limit of resolution (microscopy) 878

AUFLÖSUNGSMITTEL (*n*) : solvent ; dissolvent 879

AUFLÖSUNGSNAPHTA (*f*) : (see SOLVENTNAPHTA) 880

AUFLÖSUNGSPFANNE (*f*) : clarifier (sugar) 881

AUFLÖSUNGSVERMÖGEN (*n*) : solvent power dissolving capacity ; resolving power (microscopy) 882

AUFLÖSUNGSZEICHEN (*n*) : natural, any accidental which cancels a previous accidental (music) 883

AUFLÖTEN : to solder on ; unsolder 884

AUFMACHUNG (*f*) : method of making up (thread for commercial purposes, such as winding on reels, cards, etc.) 885

AUFMAISCHEN : to mash ; (*n*) mashing 886

AUFNAHME (*f*) : absorption ; assimilation ; reception ; exposure (Phot.) ; photograph ; portrait-(ure) 887

AUFNAHMEFÄHIG : receptive, absorptive 888

AUFNAHMEFÄHIGHEIT (*f*)) : admissibility, receptiveness ; capacity ; absorbing power ; absorption (absorptive) capacity ; absorptivity, reception 889

AUFNAHMEKOLBEN (*m*) : absorption flask 890

AUFNAHMEPERSON (*f*) : operator (Photography) 891

AUFNEHMEN : to hold ; absorb ; receive ; take (Phot.) ; take up 892

AUFNEHMERHEIZUNG (*f*) : receiver heating 893

AUFNIETEN : to rivet on 894

AUFPASSEN : to attend, give attention to, watch ; fit on to, try on 895

AUFPLATZEN : to burst open, split, fissure, crack, rupture, break open 896

AUFPRALLEN : to rebound ; impinge (of flames), strike upon (see PRALLEN and PRELLEN) 897

AUFQUELLEN : to swell ; well up (of water) 898

AUFRECHT ERHALTEN : to keep up, keep going, support 899

AUFREIHEN : to place in a row, place in series, string. 900

AUFREISSEN : to lay out ; draw down, arrange (Drawing) ; tear-(open), split 901

AUFRICHTEN : to erect, to right (of ships) 902

AUFRICHTIG : sincere, candid 903

AUFRISS (*m*) : elevation ; sketch ; design ; vertical section 904

AUFROLLEN : to roll up, unroll 905

AUFRUF (*m*) : summons (Law) ; call 906

AUFRÜHREN : to stir-(up), agitate 907

AUFRÜHRER (*m*) : stirrer ; agitator 908

AUFRÜTTELN : to shake up, rouse 909

AUFSAGEN : to recite ; revoke ; give warning ; renounce 910

AUFSÄSSIG : refractory ; hostile 911

AUFSATZ (*m*) : attachment ; head-piece ; top ; set (of china) ; essay ; treatise ; topping (dye) ; cover print (calico) ; ornament ; something formed or made 912

AUFSATZFARBE (*f*) : topping colour (dye) 913

AUFSAUGEN : to suck up ; absorb ; draw up 914

AUFSCHÄUMEN : to froth ; effervesce ; ferment ; foam up 915

AUFSCHICHTEN : to arrange (heap) in layers ; stratify ; pile up 916

AUFSCHIESSEN : to coil (a rope) ; to shoot up 917

AUFSCHLAG (*m*) : impact ; increase ; rise ; advance (in price) ; facing ; bound (of a ball) 918

AUFSCHLAGEBUCH (*n*) : reference book 919

AUFSCHLAGEN : to open (books) ; rise ; advance ; increase (in price) ; unlay (ropes) ; handle (hides) ; unbung (casks) 920

AUFSCHLÄMMEN : to make into a paste ; slime 921

AUFSCHLEIFEN : to grind on 922

AUFSCHLEPPEN : to draw up, haul up (pull up a boat on to the beach, etc) 923

AUFSCHLEPPHELLING (*f*) : hauling slip 924

AUFSCHLIESSBAR : explainable ; capable of being opened ; capable of being decomposed 925

AUFSCHLIESSEN : to decompose ; explain ; open ; disintegrate ; make non-refractory ; disclose ; unlock ; hydrolyse (starch) ; break ; crush (coal) 926

AUFSCHLIESSMASCHINE (*f*) : crusher (ores) ; dissolving machine 927

AUFSCHLUSS (*m*) : disintegration ; explanation, information, elucidation, disclosure, (*pl*) data, information 928

AUFSCHLUSS (*m*) **GEBEN** : to inform, give particulars, give an idea of, explain, clear up 929

AUFSCHMELZEN : to melt on ; fuse on 930

AUFSCHÖPFEN : to scoop up ; dip up 931

AUFSCHRAUBEN : to screw on ; unscrew 932

AUFSCHRIFT (*f*) : address, inscription 933

AUFSCHRUMPFEN : to shrink on 934

AUFSCHUB (*m*) : delay ; adjournment ; putting off 935

AUFSCHÜREN : to stir up, poke, stoke (a fire) 936

AUFSCHÜTTEN : to feed (a fire, etc.) ; pour on 937

AUFSCHÜTTTRICHTER (*m*) : feeding hopper ; hopper (for coal, attached to boiler front) 938

AUFSCHWEISSEN : to weld (on) together 939

AUFSCHWEMMEN : to deposit ; silt ; swell ; float on 940

AUFSEHEN : to look up ; (*n*) sensation 941

AUFSEHER (*m*) : overseer ; inspector ; foreman ; supervisor ; superintendent 942

AUFSETZEN : to draw up ; put down (in writing) ; top (dye) ; put on ; seat ; regain the seat (valves) ; mount ; cover print (calico) 943

AUFSICHT (*f*) : inspection, control, survey, superintendence 944

AUFSICHTSBEHÖRDE (*f*) : board of control 945

AUFSICHTSRAT (*m*) : legal adviser 946

AUFSIEDEN : to boil up ; (*n*) ebullition 947

AUFSPALTUNG ( )*f* : cleavage ; fission ; splitting ; bursting 948
AUFSPANNAPPARAT (*m*) : fixing apparatus 949
AUFSPANNEN : to stress ; put tension on ; strain ; place a stress (strain) upon ; tighten ; stretch 950
AUFSPANNPLATTE (*f*) : fixing (setting-up) plate, (lathes) ; foundation (bed) plate 951
AUFSPEICHERN : to store ; accumulate 952
AUFSPEICHERUNG (*f*) : storeage, accumulation 953
AUFSPERREN : to spread apart, open wide, gape 954
AUFSPRENGEN : to burst-(open) ; explode 955
AUFSPULEN : to spool, wind on to a spool, to roll 956
AUFSPUNDEN : to unbung ; take the bung out of (casks) 957
AUFSTAPELN : to pile ; heap ; stack ; store 958
AUFSTAUCHEN : to shorten (forging) ; dam up ; knock against 959
AUFSTAUEN : to dam, store, hold (heap) up, bank up 960
AUFSTECHEN : to pierce, perforate, cut open ; fix (Lithography) 961
AUFSTECKEN : to hoist (flags, etc.) ; fix, put up 962
AUFSTECKSCHLÜSSEL (*m*) : socket key ; water key 963
AUFSTELLEN : to erect, put (up) together, fit up, compile, make up (a formula), formulate ; mount 964
AUFSTELLUNG (*f*) : curve, graph, schedule, list erection, statement, formula 965
AUFSTICHBOGEN (*m*) : divided sheet (Lithography) 966
AUFSTÖBERN : to ferret out ; discover ; find out 967
AUFSTOSSEN : to chance upon ; discover ; push open ; ferment ; rise ; erupt ; become acid (Wine) 968
AUFSTREBEN : to aspire, have a rising tendency, tend to rise, strive to rise 969
AUFSTREICHEN : to spread on ; stain (paper) ; coat (paint) 970
AUFTAKELN : to rig (Naut.) 971
AUFTALJEN : to bowse (Naut.) 972
AUFTAUCHEN : to occur, crop up, come to light, emerge, rise 973
AUFTOPPEN : to top (a yard) (Naut.) 974
AUFTRAG (*m*) : embankment ; filling ; transfer ; commission ; order ; putting on 975
AUFTRAGEN : to charge (furnace) ; protract (Geometry) ; record ; illustrate graphically ; mark off a point on a graph ; incorporate ; add ; plot in the form of a graph 976
AUFTRAGSERTEILUNG (*f*) : placing of an order 977
AUFTRAGUNG (*f*), RECHNERISCHE- : graphical illustration 978
AUFTREIBEN : to blow up ; sublime ; drive up ; procure ; rise ; round up ; bulge ; bubble (Ceramics) 979
AUFTREIBUNG (*f*) : bulge, bubble (Ceramics) 980
AUFTRETEN : to arise, crop up, occur, happen 981
AUFTRIEB (*m*) : buoyancy 982
AUFTRIEBSKEIL (*m*) : wedge of buoyancy 983
AUFTRITT (*m*) : step ; platform ; appearance 984
AUFWALLEN : to boil up, effervesce, well up 985
AUFWALLUNG (*f*) : ebullition ; effervescence 986
AUFWAND (*m*) : expenditure ; display ; consumption ; expense ; input 987
AUFWÄRMEN : to warm up ; heat ; reheat 988
AUFWÄRTS STREBEND : tending to rise, having a rising tendency 989
AUFWARTUNG (*f*) MACHEN : to call upon, wait upon, visit, make a call upon 990
AUFWEICHEN : to soften, temper, soak 991
AUFWEICHEND : emollient ; softening 992
AUFWEISEN : to exhibit, show, produce 993
AUFWENDEN : to expend ; employ ; devote ; spend ; bestow ; use ; consume 994
AUFWICKELN : to spool, wind on to a spool, roll up 995
AUFWIEGEN : to counterbalance ; outweigh ; balance 996
AUFWINDEN : to haul up, wind (up) on to 997
AUFZÄHLEN : to number ; enumerate ; add ; count 998
AUFZEHRUNG (*f*) : absorption ; consumption ; utilization 999
AUFZEICHNEN : to draw upon ; record ; add ; mark off a point on a graph ; draw down ; incorporate 000
AUFZIEHBRÜCKE (*f*) : draw-bridge ; rouser (Brewing) 001
AUFZIEHEN : to rouse ; agitate (Yeast) ; pull up ; wind up ; elevate ; be absorbed ; mount (Phot.) ; go on (Dyeing) 002
AUFZUG (*m*) : lift ; elevator ; crane ; hoist ; act (of a play) ; elevation ; raising ; cage 003
AUFZUGSEIL (*n*) : hoist (elevator, lift) cable, rope or hawser 004
AUFZUGSMASCHINE (*f*) : hauling engine, winding engine 005
AUGE (*n*) : eye ; retina ; bud (Bot.) ; bubble (in glass, etc.) ; lustre (on fabrics) 006
AUGE, IM . . . BEHALTEN : to remember ; bear in mind ; keep in view 007
AUGENACHAT (*m*) : cat's-eye (Mineral) (see CHRYSOBERYLL) 008
AUGENBAD (*n*) : eye bath 009
AUGENBLICKSBILD (*n*) : snap-shot, instantaneous photograph 010
AUGENBLICKSVERSCHLUSS (*m*) : instantaneous shutter (of cameras) 011
AUGENENTZÜNDUNG (*f*) : ophthalmitis, panopthalmitis, inflammation of the whole of the structure of the eye (Med.) 012
AUGENFÄLLIG : obvious ; clear 013
AUGENFLÜSSIGKEIT, GLASARTIGE- (*f*) : vitreous humor 014
AUGENFLÜSSIGKEIT, WASSERARTIGE- (*f*) : aqueous humor 015
AUGENHEILKUNDE (*f*) : ophthalmics, eye-treatment 016
AUGENHÖHLE (*f*) : eye socket ; orbit (Anat.) 017
AUGENKAMMERWASSER (*n*) : aqueous humor 018
AUGENLINSE (*f*) : crystalline lens ; eye-piece or lens 019
AUGENMARMOR (*m*) : eye-spotted marble 020
AUGENMERK (*n*) : aim ; object (in view) 021
AUGENPINSEL (*m*) : eye brush 022
AUGENPUNKT (*m*) : point of sight 023
AUGENRING (*m*) : iris 024
AUGENSCHEINLICH : apparent-(ly) ; evident-(ly) 025
AUGENSCHWARZ (*n*) : retinal (black) pigment 026
AUGENSPIEGEL (*m*) : opthalmoscope (an instrument for lighting up the interior of the eye) 027
AUGENSTEIN (*m*) : eye stone ; white vitriol ; white copperas ; zinc sulphate (a variety of Chalcedony) ; $ZnSO_4$ 028
AUGENTÄUSCHUNG (*f*) : optical illusion 029

AUGENTROST (*m*): eye-bright (*Euphrasia officinalis*) (Bot.) 030
AUGENWASSER (*n*): collyrium; eye (ophthalmic) water 031
AUGENWURZEL (*f*): dandelion; wood anemone; mountain parsley; Valerian root (*Radix valerianæ*) 032
AUGE (*n*), **STAG-**: stay collar 033
AUGIT (*m*): augite; $Ca(Mg,Fe)Si_2O_6.(Mg,Fe,Ca)Al_2SiO_6$; Sg. 3.25; Mp. 1230-1260°C. 034
AUGITARTIG: augitic 035
AUGITSPAT (*m*), **AXOTOMER-**: babingtonite; Sg. 3.38 036
AUGITSPAT (*m*), **DIATOMER-**: rhodonite 037
AUGITSPAT (*m*), **HEMIPRISMATISCHER-**: amphibole 038
AUGITSPAT (*m*), **PARATOMER-**; augite, pyroxene (see **AUGIT**); acmite (see **ACHMIT**) 039
AUGITSPAT (*m*), **PERITOMER-**: arfvedsonite 040
AUGITSPAT (*m*), **PRISMATISCHER-**: Wollastonite, tabular spar 041
AUGITSPAT (*m*), **PRISMATOIDISCHER-**: epidote 042
AURAMIN (*n*): auramine, (basic yellow dye-stuff), aminotetramethyl-diaminodiphenylmethane $C_{17}H_{24}N_3OCl$ (see **PYOKTANIN, GELBES-**) 043
AURANTIA (*f*): (see **KAISERGELB**) 044
AURANTIAZEENÖL (*n*): (see **NEROLIÖL**) 045
AUREOLIN (*n*): cobalt yellow (see **KOBALTGELB**) 046
AURICHALCIT (*m*): aurichalcite (a basic carbonate of zinc and copper); $2(Zn,Cu)CO_3.3(Zn,Cu)(OH)_2$; Sg. 3.5 047
AURICHLORID (*n*): auric chloride; gold chloride (*Aurum chloratum*); $AuCl_3$ or $Au_3Cl.2H_2O$ 048
AURIPIGMENT (*n*): orpiment (*Auripigmentum*) (a gold-yellow pigment); As_2S_3; Sg. 3.45 (see **ARSENTRISULFID**) 049
AURIVERBINDUNG (*f*): auric compound 050
AUROCHLORID (*n*): aurous chloride; gold chloride; AuCl 051
AUROCHLORWASSERSTOFFSÄURE (*f*): chloraurous acid; $HAuCl_2$ 052
AUROKALIUMCYANID (*n*): potassium aurocyanide; $KAu(CN)_2$ 053
AUROVERBINDUNG (*f*): aurous compound 054
AUSARBEITEN: to calculate, work out, run out, elaborate 055
AUSARBEITUNG (*f*): working out; drawing up; getting out; making up; elaboration 056
AUSARTEN: to degenerate 057
AUSÄTHERN: to extract with ether 058
AUSÄTZEN: to cauterise; destroy by caustics; discharge (colours) 059
AUSBAGGERN: to dredge out 060
AUSBALANCIEREN: to balance; bring into a state of equilibrium; compensate 061
AUSBAUCHEN: to belly; swell; bulge (hollow) out 062
AUSBAUEN: to fit up; equip; provide with or for; exhaust (coal seams); dismantle; take down or out; arrange 063
AUSBEDINGEN: to stipulate; reserve; condition 064
AUSBESSERN: to mend; repair 065
AUSBESSERUNGSARBEIT (*f*): repairs, improvements 066
AUSBESSERWERKSTÄTTEN (*f.pl*): repair shops 067
AUSBEULEN: to swell; round out; bulge 068
AUSBEUTE (*f*): yield; gain; output; crop; profit; production 069
AUSBEUTEERHÖHUNG (*f*): increase in yield, increased yield 070
AUSBEUTEN: to cultivate (Agriculture); work (mine); turn to account; produce; yield; gain; profit 071
AUSBIEGEN: to deflect, bend out 072
AUSBIEGUNG (*f*): deflection 073
AUSBILDUNG (*f*): formation; improvement; development; design; construction; arrangement 074
AUSBLASEDAMPF (*m*): exhaust steam 075
AUSBLASEHAHN (*m*): blow-off cock; drain cock 076
AUSBLASELEITUNG (*f*): escape pipes; blow-off (blow-down) pipe or main 077
AUSBLASEN: to blow out; shut down (a furnace); inflate, pump up; spit out, vomit, erupt; aspirate blow (off) down (boilers or blast furnace); exhaust (steam) 078
AUSBLASEN (*n*): shutting down of a blast furnace (takes place after about 3-6 years of working; used to be 10-15 years) 079
AUSBLEICHEN: to fade; bleach 080
AUSBLEICHVERFAHREN (*n*): bleaching process 081
AUSBLEIEN: to line with lead 082
AUSBLÜHEN: to effloresce; fade; decay; blow; drop; wither 083
AUSBRAND (*m*): combustion 084
AUSBRECHEN: to quarry (stone); work; prune; lop; vomit; break out 085
AUSBREITAPPARAT (*m*): stretching apparatus, stretcher (Textile) 086
AUSBREITEN: to diffuse; spread out; distribute; stretch; extend; propagate; unfold; promulgate; floor (grain, etc) 087
AUSBRENNEN: to burn; cauterize (Med.); bake (bricks) 088
AUSBRINGEN: to yield; produce; extract (gas) 089
AUSBRINGEN (*n*): yield, output, gain, production 090
AUSBRUCH (*m*): eruption; working; explosion; outbreak; first press wine 091
AUSBRUCHQUERSCHNITT (*m*): working face (Mining) 092
AUSBRÜHEN: to parboil; scald; seethe 093
AUSBÜCHSEN: to box 094
AUSBUCHTUNG (*f*): recess; pocket; bulge; corrugation 095
AUSDAMPFEN: to evaporate; steam; vapourize; steam out; cease steaming; fumigate 096
AUSDÄMPFEN: to steam out; smoke out; evaporate; vapourize; clean with steam 097
AUSDEHNBAR: ductile; extensible 098
AUSDEHNEN: to spread; stretch; enlarge; expand; dilate; extend; elongate; distend 099
AUSDEHNUNG (*f*): expansion; extent; spread; distension; dilation; extension; dimension (Maths.) 100
AUSDEHNUNGSKOEFFIZIENT (*m*): co-efficient of expansion 101
AUSDEHNUNGSKUPPELUNG (*f*): expansion joint; flexible coupling 102
AUSDEHNUNGSRAUM (*m*): expansion chamber 103
AUSDEHNUNGSVERMÖGEN (*n*): expansibility; ductility 104
AUSDEHNUNGSZAHL (*f*): coefficient of expansion 105
AUSDEUTEN: to interpret; explain; clear up 106
AUSDOCKEN: to leave the dock (of ships) 107

AUSDREHEN: to turn (out) (of the solid), to hollow out 108
AUSDREHSTAHL (*m*): inside (turning) tool 109
AUSDRUCKEN: to print (in full); finish printing 110
AUSDRÜCKEN: to give vent to (opinion); express (Maths.); press (squeeze) out; wring; express (oils) 111
AUSDUNST (*m*): exhalation; perspiration; evaporation 112
AUSDÜNSTUNGSMESSER (*m*): atmometer; evaporometer; evaporimeter (for determining the quantity of liquid evaporated in a specified period) 113
AUSEINANDERFAHREN: to diverge; break up; part company 114
AUSEINANDERHALTEN: to keep apart; subdivide 115
AUSEINANDERNEHMEN: to dismantle, demount, take to pieces, take down 116
AUSEINANDERSETZEN: explain; make clear 117
AUSERKOREN: chosen; selected 118
AUSERLESEN: select; exquisite; choice; picked; selected 119
AUSERSEHEN: to mark out; choose; select; destine 120
AUSFAHREN: to export; ship; go (carry) round 121
AUSFALL (*m*): outfall (sewage); precipitate; cutting out of the circuit, putting out of commission, breakdown, stoppage (of plant); drop, result, deficit 122
AUSFALLEN: to be lacking, fall out, to turn out 123
AUSFALLWINKEL (*m*): angle of reflection 124
AUSFÄRBEN: to dye; extract colour from; pale 125
AUSFÄRBVORRICHTUNG (*f*): colour extractor 126
AUSFERTIGUNG (*f*), **IN DOPPELTER-**: in duplicate 127
AUSFERTIGUNG (*f*), **IN DREIFACHER-**: in triplicate 128
AUSFETTEN: to extract fat or grease; scour (wool); de-oil 129
AUSFEUERN: to burn out; warm; heat 130
AUSFINDIG MACHEN: to find out; discover 131
AUSFIXIEREN: to fix (fully, thoroughly) completely 132
AUSFLEISCHEN: to flesh (hides) 133
AUSFLIESSEN: to escape, flow out, run out 134
AUSFLOCKEN: to separate in (flakes) flocks 135
AUSFLOCKUNGSGESCHWINDIGKEIT (*f*): speed of separation into flakes or flocks, flaking (flocking) speed 136
AUSFLUCHT (*f*): evasion; subterfuge; flight; exodus 137
AUSFLUSS (*m*): eduction, outflow, effluence, outrush, efflux, discharge, escape, drainage, secretion 138
AUSFLUSSKOEFFIZIENT (*m*): co-efficient of discharge, efflux coefficient 139
AUSFLUSSLOCH (*n*): discharge, opening, outlet 140
AUSFLUSSROHR (*n*): escape pipe 141
AUSFOLGEN: to deliver up 142
AUSFORMEN: to develop; improve; shape; form; mould 143
AUSFORSCHEN: to investigate; enquire into 144
AUSFORSCHUNG (*f*): investigation, research 145
AUSFRÄSEN: to countersink; bead; edge; mill 146
AUSFRIEREN: to concentrate by refrigeration; freeze-(out); congeal; refrigerate 147
AUSFUHR (*f*): export; exportation 148
AUSFÜHRBAR: practicable; feasible; exportable 149
AUSFUHRBEWILLIGUNG (*f*): export permit 150
AUSFÜHREN: to carry out; manufacture; export; arrange; perform; execute; erect; work out; construct (see also **AUSFÜHRUNG**) 151
AUSFUHRGESCHÄFT (*n*): export trade 152
AUSFÜHRLICH: complete; detailed 153
AUSFUHRPRÄMIE (*f*): bounty 154
AUSFÜHRUNG (*f*): outlet; manufacture; construction; arrangement; method of carrying out; type; working out; model; workmanship; design; practice; performance; execution; erection; pattern 155
AUSFÜHRUNGEN (*f.pl*): representations; ideas put forward or worked out, methods; contentions 156
AUSFÜHRUNGSBEISPIEL (*n*): method of (construction) carrying out 157
AUSFÜHRUNGSMITTEL (*n*): purgative 158
AUSFUHRZOLL (*m*): export duty 159
AUSFÜLLEN: to fill; complete; empty; stop up; stuff; pack; grout (brickwork) 160
AUSFÜLLMASSE (*f*): filling (for teeth, etc.); stuffing; grouting; packing 161
AUSFUTTERN: to line 162
AUSFÜTTERN: to line; feed; fatten 163
AUSGABE (*f*): expense; edition; distribution 164
AUSGANGSMATERIAL (*n*): original (raw) material 165
AUSGANGSPRODUKT (*n*): commencing (raw; original) product 166
AUSGANGSPUNKT (*m*): initial (starting) point; point of origin; commencement; spring (of arch) 167
AUSGANGSROHR (*n*): outlet (waste) pipe 168
AUSGANGSSTOFF (*m*): raw material (Minerals) 169
AUSGÄREN: to ferment (thoroughly); weld (steel) 170
AUSGEBEN: to yield; produce; publish; swell; slake (lime); give out 171
AUSGEBOHRTE WELLE (*f*): hollow shaft 172
AUSGEFÜHRT: completed 173
AUSGEHEN: to proceed; start from; emanate; ferment; crop out (minerals); go out 174
AUSGELASSEN: extravagant 175
AUSGEPRÄGT: marked; decided; defined; delineated 176
AUSGERBEN: to tan; weld (steel) 177
AUSGESCHLOSSEN: out of the question, not possible, impossible, excluded, unlikely 178
AUSGESPROCHEN: pronounced, decided-(ly) 179
AUSGESTALTEN: to arrange, design, fit up, equip, model, form, shape, develop 180
AUSGESTALTUNG (*f*): development 181
AUSGEZOGEN: drawn out; solid (not dotted or broken) (of a line); full; shown in full line (see under **LINIE**) 182
AUSGIEBIG: productive; fertile; abundant 183
AUSGIEBIGKEIT (*f*): yield; strength (dye) 184
AUSGLEICH (*m*): settlement (of accounts), compromise, equalization, balance, compensation 185
AUSGLEICHDRAHT (*m*): equalizing wire 186
AUSGLEICHDYNAMO (*f*): balance dynamo 187
AUSGLEICHEN: to level up; settle (accounts); balance; equalize; compensate 188
AUSGLEICHENDE FARBE (*f*): complementary colour 189

AUSGLEICHER (*m*): compensator, equalizer 190
AUSGLEICHFILTER (*m*): equalizing filter (for colour photography) 191
AUSGLEICHSPULE (*f*): equalizing coil 192
AUSGLEICHSTROM (*m*): equalizing current 193
AUSGLÜHER (*m*): annealer; blancher (Minting) 194
AUSGRABEN: to excavate; engrave; dig out 195
AUSGUCK (*m*): crow's nest (Naut.); look-out (Naut.) 196
AUSGUSS (*m*): spout; outlet; sink; drain; gutter; effusion; efflux; infusion; outfall 197
AUSGUSSPFANNE (*f*): mould (for ingots) 198
AUSGUSSROHR (*n*): delivery pipe; drain (waste) pipe 199
AUSGUSSVENTIL (*n*): escape valve 200
AUSGUSSWASSER (*n*): drainage; waste water 201
AUSHAKEN: to unship (rudders); unhook; hook (pick) out 202
AUSHALTEN: to hold out; pick out; sort (Mining) persevere; endure 203
AUSHÄNDIGEN: to hand over, pay over to, hand (round) out; surrender, deliver up 204
AUSHANG (*m*): placard; poster 205
AUSHÄNGEBOGEN (*m*): proof sheet; specimen sheet 206
AUSHÄNGESCHILD (*n*): sign-board 207
AUSHARREN: to persevere; endure; hold out 208
AUSHARZEN: to exude (extract) resin 209
AUSHAUCHEN: to breathe out, aspirate 210
AUSHEBEN: to unship (masts); lift out 211
AUSHECKEN: to hatch; plot 212
AUSHILFE (*f*): stand-by; spare; auxiliary 213
AUSHÖHLER (*m*): excavator 214
AUSHOLEN: to pump out, sound, reach out 215
AUSHOLER (*m*): outhaul, outhauler 216
AUSHÜLSEN: to husk; shell; peel 217
AUSKEHLEN: to channel 218
AUSKEIMEN: to germinate; sprout; stop sprouting 219
AUSKITTEN: to (fill with) cement 220
AUSKLEIDEN: to line; dress; undress 221
AUSKLEIDUNG (*f*): lining 222
AUSKLOPFEN: to scale (boilers), to knock out 223
AUSKOCHEN: to extract; boil out 224
AUSKOCHER (*m*): boiler; extractor; bucking kier 225
AUSKOPIEREN: to copy, print-(out) 226
AUSKOPIERPAPIER (*n*): printing-out paper; P.O.P.; (Phot.) 227
AUSKOPIERVERFAHREN (*n*): printing-out process (Phot.) 228
AUSKÖRNEN: to free from grains; thresh (grain); gin (cotton) 229
AUSKRATZEN: to cross out, scratch out, scrape out 230
AUSKRISTALLISATION (*f*): formation of new crystals (metal smelting) 231
AUSKRISTALLISIEREN: to form new crystals (metal smelting) 232
AUSKRÜCKEN: to rake out; draw 233
AUSKULTATION (*f*): auscultation, sounding of the chest (Med.) 234
AUSKUNFT (*f*): information; resource 235
AUSKUNFT (*f*) **ERTEILEN**: to give information, to inform 236
AUSKUNFTSMITTEL (*n*): expedient; resort 237
AUSKUPPELN: to uncouple, disconnect 238
AUSLADEN: to unload, discharge 239
AUSLADER (*m*): discharger 240
AUSLAGE (*f*): outlay; disbursement; show; display (window); expenditure; expense 241
AUSLAND (*n*): abroad, foreign country 242
AUSLÄNDISCH: foreign; exotic; outlandish 243
AUSLASS (*m*): outlet, delivery, discharge 244
AUSLASSEN: to leave out; exhaust (steam); let escape; discharge; omit; elide; eliminate; delete; cut out 245
AUSLASSHAHN (*m*): discharge cock 246
AUSLASSVENTIL (*n*): exhaust valve 247
AUSLAUF (*m*): outlet; leak; leakage; emission; effusion; efflux; result; drain (opening) 248
AUSLAUFEN: to run (colours); bleed; discharge into; project; run out 249
AUSLAUFSPITZE (*f*): discharge tip (of a burette) 250
AUSLAUGEN: to steam; lixiviate; extract; buck; wash out (steep) in lye 251
AUSLEEREN: to drain; empty; evacuate; clear out 252
AUSLEERENDES MITTEL (*n*): evacuant (such as an emetic, expectorant, purgative, etc.) (Med.) 253
AUSLEGEN: to exhibit; inlay; veneer; interpret; lay out; file (patents) 254
AUSLEGER (*m*): derrick, crane, outrigger, beam, jib (Naut.); interpreter 255
AUSLEGER (*m*), **KRAN-**: crane beam or jib 256
AUSLENKEN: to incline, deflect 257
AUSLENKUNG (*f*): distance from the centre; deflection; inclination 258
AUSLESE (*f*): selection; choice wine 259
AUSLÖSCHEN: to quench, extinguish 260
AUSLÖSEN: to liberate, dissolve out, solve, uncouple, redeem, release 261
AUSLÖSUNG (*f*): redemption; release (of a balance); liberation 262
AUSLOTEN, MIT DER WASSERWAGE-: to level 263
AUSLÜFTEN: to ventilate; air; aerate; extract air 264
AUSMAUERUNG (*f*): underpinning; brickwork setting; masonry 265
AUSMERGELN: to debilitate; exhaust; emaciate 266
AUSMERZEN: to sort (pick, reject, wipe) out, remove anything that is bad, cut out, do away with 267
AUSMESSEN: to span, survey, measure 268
AUSMITTELN: to ascertain; determine 269
AUSMÜNDUNG (*f*): orifice; outlet; mouth; exit 270
AUSNUTZEN: (see **AUSNÜTZEN**) 271
AUSNÜTZEN: to use up; exhaust; make the most of; exploit; consume; make effective 272
AUSNÜTZUNG (*f*): efficiency; exploitation; consumption; fuel 273
AUSNUTZUNG (*f*), **BRENNSTOFF-**: fuel efficiency 274
AUSNUTZUNG (*f*), **WÄRME-**: heat efficiency 275
AUSÖLEN: to oil; de-oil 276
AUSPEILEN: to sound 277
AUSPICHEN: to pitch; tar 278
AUSPRÄGEN: to delineate; define; stamp; impress; coin 279
AUSPRESSEN: to press out; extrude (tubes) 280
AUSPROBIEREN: to test, try, try out, test (thoroughly) exhaustively 281

AUSPUFF (*m*): exhaust; escape (Mach.) 282
AUSPUFFDAMPFMASCHINE (*f*): non-condensing engine, engine exhausting to atmosphere 283
AUSPUFFGAS (*n*): exhaust gas 284
AUSPUFFHAUBE (*f*): exhaust head 285
AUSPUFFHUB (*m*): exhaust stroke 286
AUSPUFFLEITUNG (*f*): exhaust pipe; exhaust main 287
AUSPUFFMASCHINE (*f*): non-condensing engine, engine exhausting to atmosphere 288
AUSPUFFÖFFNUNG (*f*): exhaust port 289
AUSPUFFROHR (*n*): exhaust pipe 290
AUSPUFFSTEUERUNG (*f*): exhaust distribution by means of ports, port exhaust 291
AUSPUFFTOPF (*m*): exhaust head, silencer 292
AUSPUFFVENTIL (*n*): exhaust valve 293
AUSPUMPEN: to pump out; exhaust 294
AUSQUETSCHEN: to squeeze out; crush out; wring out 295
AUSRADIEREN: to erase; scrape out; rub out 296
AUSRÄUCHERN: to fumigate; smoke out 297
AUSRECHNEN: to work out; run out; calculate; estimate 298
AUSREDE (*f*): utterance; evasion; excuse; subterfuge 299
AUSREICHEND: sufficient, adequate, ample, enough; sufficiently 300
AUSREISE (*f*): outward (run, journey) voyage 301
AUSRENKEN: to sprain; dislocate 302
AUSRICHTEN: to do; dress; discover; explore (Mining); straighten out 303
AUSROTTEN: to root up; exterminate 304
AUSRÜCKEN: to throw off, uncouple, throw out of gear, take out the clutch, disconnect, throw out clutch, to disengage 305
AUSRÜCKER (*m*): disengaging (gear) lever 306
AUSRÜCKGABEL (*f*): disengaging fork 307
AUSRÜCKHEBEL (*m*): disengaging lever 308
AUSRÜCKKUPPELUNG (*f*): disengaging (coupling) clutch 309
AUSRÜCKMUFFE (*f*): disengaging clutch 310
AUSRÜCKVORRICHTUNG (*f*): disengaging gear 311
AUSRÜCKWELLE (*f*): disengaging shaft 312
AUSRUNDUNG (*f*): a rounding-off 313
AUSRÜSTEN: to furnish (supply, provide) with, fit out (ships); equip; finish 314
AUSRÜSTUNG (*f*): equipment; supply; finish (of cloth or paper); fitting out 315
AUSRÜSTUNGSSTÜCK (*n*): fitting; mounting 316
AUSSALZBAR: capable of being salted out 317
AUSSALZEN: to separate by adding salt; salt 318
AUSSATZ (*m*): leprosy; scurf; display (of wares) 319
AUSSÄUERN: to de-acidify; detartarize 320
AUSSAUGEN: to suck out; drain; exhaust 321
AUSSCHÄKELN: to unshackle 322
AUSSCHALTEN: to cut out; switch off; isolate; by-pass; disconnect; shut down 323
AUSSCHALTER (*m*): circuit-breaker (Elect.); cut-out; switch 324
AUSSCHALTER (*m*), **SELBST-**: automatic cut-out 325
AUSSCHÄRFEN: to neutralize; deaden; dull 326
AUSSCHEIDEN: to separate; liberate; eliminate (Maths.); precipitate; extract; be (out of the question) of no use 327
AUSSCHEIDUNG (*f*): separated product, by-product (see also **AUSSCHEIDEN**) 328
AUSSCHEIDUNGSMITTEL (*n*): precipitant; separating agent 329
AUSSCHEIDUNGSPRODUKT (*n*): separation product; by-product 330
AUSSCHEREN: to unreeve (ropes) 331
AUSSCHIEBEN: to push out; displace 332
AUSSCHIESSEN (*n*): rake (Naut.); to shoot out 333
AUSSCHIFFEN: to disembark, unload (ships) 334
AUSSCHIFFUNG (*f*): unloading (of ships) 335
AUSSCHIRREN: to unharness; throw out of gear; ungear 336
AUSSCHLACKEN: to slack; clear of (dross) clinker; clinker (furnaces) 337
AUSSCHLAG (*m*): issue; decision; scum; eruption; result; deflection (referring to small displacement); throw (referring to large displacement); run; tapped liquid 338
AUSSCHLAG (*m*): Exanthema, exanthem (a skin eruption produced by typhoid fever, small-pox, measles, etc.) (Med.) 339
AUSSCHLAGBÜTTE (*f*): underback (Brewing) 340
AUSSCHLAGEISEN (*n*): punch; pounding tool (for ores); piercer 341
AUSSCHLAGEN: to punch (knock) out; remove; refuse; express (oils); run off (wort) (Brewing); flatten; planish; line; stretch; bud; effloresce; turn (balance beam); deflect; spew (leather) 342
AUSSCHLAGGEBEND: decisive 343
AUSSCHLAGGEBENDE (*m*): umpire, referee (Com.) 344
AUSSCHLÄMMEN: to blow-off; blow-down (boilers); clean from (mud) slime 345
AUSSCHLEIFEN: to grind out; turn out (Mech.) 346
AUSSCHLEUDERN: to expel by centrifugal force; centrifuge; hydro-extract 347
AUSSCHLIESSLICH: exclusive-(ly) 348
AUSSCHLUSS (*m*): exclusion; exception 349
AUSSCHMELZEN: to liquate (metals); fuse; melt; render 350
AUSSCHMIEREN: to smear; copy; plagiarize; oil; grease 351
AUSSCHNITT (*m*): sector; notch; cut; cutting; vignette; extract (see **ROHAUSSCHNITT**) 352
AUSSCHÖPFEN: to ladle out; scoop out; drain 353
AUSSCHRAUBEN: to screw out, unscrew; rack out (a camera) 354
AUSSCHREIBEN: to advertise for, offer for public tender or competition, invite tenders for 355
AUSSCHREIBUNG (*f*): submission; invitation to tender; specification 356
AUSSCHREITUNG (*f*): excess 357
AUSSCHUBWINKEL (*m*): angle of oscillation (of piston) 358
AUSSCHUSS (*m*): waste; culls; select committee; refuse; dross; garbage; throw-out; reject 359
AUSSCHUSSHUB (*m*): exhaust stroke 360
AUSSCHÜTTELN: to agitate; shake out 361
AUSSCHWEFELN: to fumigate with sulphur; sulphur (casks) 362
AUSSCHWEIFEND: excessive; extravagant 363
AUSSCHWEISSEN: to weld out; clean iron (by welding); bleed 364
AUSSCHWEMMEN: to flush; rinse; flood 365
AUSSCHWENKEN: to whirl; centrifuge; hydro-extract 366
AUSSCHWENKMASCHINE (*f*): centrifugal; hydro-extractor 367
AUSSCHWIRREN: to whiz; centrifuge; hydro-extract 368

AUSSCHWITZEN: to exude; sweat-(out); perspire 369
AUSSEHEN: appearance 370
AUSSEIGERN: to liquate; separate by liquation (metal) 371
AUSSENACHTERSTEVEN (*m*): false post 372
AUSSENAUFNAHME (*f*): outdoor photograph 373
AUSSENBEPLATTUNG (*f*): shell-(plating) 374
AUSSENBORDS: outboard 375
AUSSENDRUCK (*m*): outside pressure; atmospheric pressure 376
AUSSENFALL (*n*): outer halliard 377
AUSSENFRÄSER (*m*): hollow mill 378
AUSSENGANG (*m*): outside strake 379
AUSSENHAUT (*f*): shell-(plating) 380
AUSSENKLÜVER (*m*): flying-jib (Naut.) 381
AUSSENKLÜVERBAUM (*m*): flying jib boom (Naut.) 382
AUSSENKLÜVERFALL (*n*): flying jib halliard 383
AUSSENKLÜVERLEITER (*f*): flying jib stay 384
AUSSENKLÜVERSCHOTE (*f*): flying jib sheet 385
AUSSENKLÜVERSTAMPFSTAG (*n*): martingale stay 386
AUSSENLEITUNG (*f*): outer circuit; outside (main) pipeline 387
AUSSENLUFT (*f*): surrounding (air) atmosphere 388
AUSSENPOL (*m*): external pole 389
AUSSENSCHOTE (*f*): tack (of studding sail) 390
AUSSENWAND (*f*): outer (external) wall 391
AUSSERGEWÖHNLICH: out of the ordinary; extraordinary; exceptional; unusual; exceeding-(ly) 392
AUSSERORDENTLICH: extraordinary, exceptional, unusual, exceedingly 393
ÄUSSERST: very, extremely, exceedingly 394
AUSSETZEN: to set out; line; face; expose; suspend; pause; stop; suspend (interrupt) operation-(s); subject-(to); cease working; lower (a boat) 395
AUSSICHT (*f*), **IN-NEHMEN**: to propose, designate 396
AUSSICHTSREICH: promising; hopeful; having good prospects 397
AUSSICKERN: to trickle (ooze) out; percolate; stop trickling 398
AUSSIEDEN: to boil out; blanch (silver) 399
AUSSOGGEN: to precipitate (in crystals) 400
AUSSÖHNEN: to reconcile; appease 401
AUSSONDERN: to separate; reject; sort; segregate 402
AUSSORTIEREN: to sort out, separate 403
AUSSPANNEN: to stretch; spread; slacken; unharness; carry off 404
AUSSPAREN: to spare out (high lights) (Art); leave out 405
AUSSPARUNG (*f*): sparing out, niche, clearance, play 406
AUSSPERRUNG (*f*): lock-out (of employees by employers) 407
AUSSPRECHEN: to speak out; define; pronounce; express; declare; speak decidedly; discuss 408
AUSSPRENGEN: to blast; sprinkle (water); noise abroad 409
AUSSPRUNGSWINKEL (*m*): angle of reflection 410
AUSSPÜLEN: to wash out; wash away; scour; rinse 411
AUSSTAND (*m*): strike; anything overdue; (*pl*) arrears; debts (outstanding) 412
AUSSTÄNDIGER (*m*): striker (one who refuses to work owing to some grievance) 413
AUSSTANZEN: to stamp out (metal); punch 414
AUSSTATTEN: to provide (fit) with 415
AUSSTÄUBEN: to dust; winnow 416
AUSSTECHEN: to cut out (of the solid) 417
AUSSTEIFUNGSRING (*m*): stiffening ring 418
AUSSTELLEN: to exhibit; find fault; make out (give) (a certificate) 419
AUSSTELLER (*m*): drawer (Com.); exhibitor 420
AUSSTELLUNGSGLAS (*n*): display glass; specimen (jar) tube 421
AUSSTICH (*m*): choicest wine 422
AUSSTÖPFELN: to unplug, unstop 423
AUSSTOPFEN: stop up; stuff; plug 424
AUSSTOSSEN: to expel; eject; eliminate; delete; cut out; elide; oust; flesh (hides); retail (liquor) 425
AUSSTOSSLADUNG (*f*): bursting charge 426
AUSSTOSSPRODUKT (*n*): waste products 427
AUSSTOSSROHR (*n*): expulsion tube 428
AUSSTOSSSYSTEM (*n*): cleansing system 429
AUSSTRAAKEN: to sheer (a line) 430
AUSSTRAHLEN: to emit; be emitted; radiate 431
AUSSTRAHLVERLUST (*m*): radiation loss 432
AUSSTRICH (*m*): stream tin; granular Cassiterite (a natural tin oxide); SnO_2 (about 80% Sn.); Sg. 6.8 (see also **ZINNSTEIN**); blotting (striking) out; erasion; cancellation 433
AUSSTRÖMEN: to flow out; escape; emanate; effuse; be delivered 434
AUSSTRÖMUNG (*f*): eduction, flowing out, effluence, streaming out, effusion, efflux, outrush, escape, emission, outflow, discharge 435
AUSSTRÖMUNGSROHR (*n*): delivery (escape) pipe, discharge pipe 436
AUSSÜSSEN: to edulcorate; wash (precipitates); sparge (Brewing); sweeten; leach 437
AUSSÜSSGLAS (*n*): wash bottle (for precipitates) 438
AUSSÜSSPUMPE (*f*): leaching pump 439
AUSSÜSSROHR (*n*): washing tube 440
AUSSÜSSVORRICHTUNG (*f*): washing arrangement; washing bottle (for precipitates); sparger (Brewing) 441
AUSTAUCHEN: to emerge, rise, appear 442
AUSTAUEN: to melt, thaw-(out) 443
AUSTAUSCH (*m*): exchange; interchange; barter 444
AUSTAUSCHBARKEIT (*f*): interchangeability 445
AUSTENIT (*n*): austenite (metal) (state of steel above the change temperature of 700-900°C.) (microscopical examination shows its form to be that of juxtaposed grains) (properties:- not magnetizable, low conductivity for heat and electricity, low flow limit, high rupturing elongation) 446
AUSTER (*f*): oyster 447
AUSTERNSCHALE (*f*): oyster shell (Sg. 2.09) 448
AUSTRAG (*m*): decision; issue 449
AUSTRETEN: to emerge; project; protrude; extravasate (Med.) 450
AUSTRETENDE KANTE (*f*): trailing edge (of propellers) 451
AUSTRITT (*m*): emergence; outlet; efflux; exit; effusion (blood); outfall 452
AUSTRITTGESCHWINDIGKEIT (*f*): exit (outlet, nozzle) velocity 453
AUSTRITTSSCHLITZ (*m*): outlet (exit) port 454
AUSTRITTSTUTZE (*f*): outlet branch 455
AUSTROCKNEN: to desiccate; drain; dry; season (wood); exsiccate 456
AUSTROCKNEN (*n*): drying, seasoning (of wood); desiccation 457

AUSTROCKNER (*m*): desiccator; drier 458
AUSTRÖPFELN: to drip out; cease dripping 459
AUSÜBEN: to exert (a force); exercise; execute; perpetrate; carry out 460
AUSWACHSEN: to germinate; sprout; grow (up) out of 461
AUSWÄGEN: to calibrate; weigh out 462
AUSWAHL (*f*), **EINE-TREFFEN**: to make a choice or selection, to select (choose) from 463
AUSWALZEN: to roll out 464
AUSWANDERERSCHIFF (*n*): emigrant ship 465
AUSWÄRMEN: to anneal (metal); heat, warm 466
AUSWÄRMOFEN (*m*): annealing furnace 467
AUSWÄRTIG: foreign 468
AUSWASCHEN: to wash out; scour (fabrics); scrub (gases) 469
AUSWÄSSERN: to soak; steep 470
AUSWECHSELBAR: interchangeable, renewable replaceable 471
AUSWECHSELN: to change, replace, renew, exchange, interchange 472
AUSWECHSELUNG (*f*): replacement, renewal 473
AUSWEICHEN: to slip away, give way, yield, soften, evade, shunt on to a siding 474
AUSWEIDEN: to eviscerate, draw 475
AUSWEIS (*m*): return, statement (Com.); notice to quit 476
AUSWEISEN: to prove, record, banish, expel, turn out 477
AUSWEITEN: to distend, widen 478
AUSWERFER (*m*): ejector 479
AUSWERTUNG (*f*): valuation; evaluation 480
AUSWINDEN: to wrench out; unscrew; wring out 481
AUSWIRKEN: to remove; effect; knead (dough) 482
AUSWISCHEN: to wipe out, eradicate, rub out, exterminate, delete, wash out, exclude 483
AUSWITTERN: to effloresce; weather; air; scent; season (wood) 484
AUSWUCHS (*m*): excrescence; growth 485
AUSWUCHTEN: to balance; bring into a state of equilibrium; compensate 486
AUSWUCHTUNG (*f*): balancing 487
AUSWUCHTVORRICHTUNG (*f*): balancing device 488
AUSWURF (*m*): discharge; refuse; trash; throw-out; excrement; sputum (Med.) 489
AUSWURFSBEFÖRDERNDES MITTEL (*n*): expectorant (Med.) 490
AUSZACKEN: to indent; notch; serrate 491
AUSZAHLEN: to pay (out), hand (over), make (over) 492
AUSZEHRUNG (*f*): consumption 493
AUSZIEHBARER TURM (*m*): telescopic (extensible) tower 494
AUSZIEHEN: to extract (drugs); exhaust; fade (colours); ink in; draw out; extend; prolong; undress; rack out (of a camera); extrude (Metal) 495
AUSZIEHTISCH (*m*): extension table 496
AUSZIEHTUSCHE (*f*): drawing ink 497
AUSZIEHUNG (*f*): extraction; undressing; extrusion; extension; racking; drawing out; pulling out; elongation 498
AUSZUG (*m*): extract; essence; tincture; infusion; decoction; extension; bellows; telescopic portion 499
AUSZUG (*m*), **DREIFACHER-**: triple extension (of cameras) 500
AUSZUGGEHALT (*m*): content of extract 501
AUSZUGMEHL (*n*): superfine flour 502
AUTAN (*n*): autan (a mixture of polymerized formaldehyde with alkaline metal oxides) 503
AUTOGAMIE (*f*): autogamy, pairing of like with like; self-fertilization (Bot.) 504
AUTOGEN: autogenous; autogenously; autogene 505
AUTOGEN SCHWEISSVERFAHREN (*n*): autogene welding process (see **SELBSTHITZENDES SCHWEISSEN**) 506
AUTOINTOXIKATION (*f*): auto-intoxication 507
AUTOKATALYSE (*f*): autocatalysis 508
AUTOKATALYTISCH: autocatalytic 509
AUTOKLAV (*m*): autoclave 510
AUTOLYSE (*f*): autolysis 511
AUTOMAT (*m*): automatic machine; automaton 512
AUTOMATISCH: automatic, self-acting 513
AUTOMOBIL (*n*): automobil, motor car 514
AUTOMOBIL-ANLASSER (*m* and *pl.*): automobile (motor-car) starter(s) 515
AUTOMOBIL-ARMATUREN (*f pl.*): automobile (motor-car) fittings 516
AUTOMOBILBELEUCHTUNG (*f*): automobile (motor-car) lighting 517
AUTOMOBILBEREIFUNG (*f*): automobile (motor-car, motor) tyres 518
AUTOMOBILFÜHRER (*m*): chauffeur (of a motor-car) 519
AUTOMOBILIST (*m*): motorist 520
AUTOMOBILMASCHINE (*f*): motor-car engine 521
AUTOMOBILÖL (*n*): motor (automobile) oil 522
AUTOMOBILTEILE (*m pl.*): automobile (motor-car) parts or spares 523
AUTOMOLIT (*m*): (Zn, Fe) Al_2O_4; Automolite; Sg. 4.62 524
AUTOOXYDATION (*f*): auto-oxydation 525
AUTOOXYDATIONSFÄHIG: auto-oxydizable 526
AUTOOXYDATIONSFÄHIGKEIT (*f*): auto-oxydizability 527
AUTORACEMISIERUNG (*f*): autoracemization 528
AUTORENRECHT (*n*): copyright 529
AUTORISIEREN: to authorize 530
AUTOSUGGESTION (*f*): autosuggestion 531
AUTOTYPIE (*f*): autotype, process-block making 532
AUTOXYDABLE: autoxidizable 533
AUTUNIT (*m*): calcouranite, autunite (natural hydrated calcium-uranium phosphate) $Ca(UO_2)_2(PO_4)_2.xH_2O$; Sg. 3.1 534
AVANTURIN (*m*): aventurine (a variety of quartz) (see **QUARZ**) 535
AVANTURINFELDSPAT (*m*): aventurine feldspar (see **ORTHOKLAS** and **OLIGOKLAS**) 536
AVENTURIN (*m*): avanturine, aventurine (see **QUARZ**) 537
AVIS (*m*): advice-(note) 538
AVISIEREN: to adv se 539
AVIVAGE (*f*): reviving; brightening, livening; freshening 540
AVIVIERECHT: not affected by reviving 541
AVIVIEREN: to revive; restore; brighten; liven; enliven; freshen; top; clear; tone 542
AVOCATOFETT (*n*): avocado fat (from *Laurus persea* or *Persea gratissima*) 543
AWARUIT (*m*): awaruite: $FeNi_2$ 544
AXE (*f*): (see **ACHSE**) 545
AXIALDRUCK (*m*): axial pressure 546
AXIALE BEANSPRUCHUNG (*f*): axial thrust 547

AXIALSCHUB (*m*) : axial thrust (of bearings) 548
AXIALSYMMETRISCH : axialsymmetric-(al)-(ly) 549
AXIALTURBINE (*f*) : axial flow turbine 550
AXINIT (*m*) : axinite ($Ca,Fe,Mn,Mg,H_2)Al_4B_2(SiO_4)_3$; Sg. 3.3 551
AXIOMETER (*n*) : axiometer, tell-tale gauge 552
AXOTOMER ANTIMONGLANZ (*m*) : Jamesonite 553
AXT (*f*) : axe, hatchet 554
AZALIE (*f*) : azalea (*Azalea procumbens*) 555
AZELAINSÄURE (*f*) : azelaic acid 556
AZETAT (*n*) : acetate 557
AZETON (*n*) : acetone ; C_3H_6O ; Sg. 0.79 ; Mp. – 94.3°C. ; Bp. 56.48°C. 558
AZETONBISULFIT (*n*) : acetone bisulphite (a retarding agent for rapid developers) (Phot.) 559
AZETONCHLOROFORM (*n*) : acetone chloroform (see **CHLORETONE**) 560
AZETONSULFIT (*n*) : acetone sulphite (see **AZETONBISULFIT**) 561
AZETYLARSANILAT (*n*) : acetyl arsenilate ; sodium acetyl-para-aminophenylarsinate ; acetyl atoxyl ; $CH_3.CO.NH(C_6H_4)AsOHOONa$ 562
AZETYLATOXYL (*n*) : acetyl atoxyl (see **AZETYLARSANILAT**) 563
AZETYLEN (*n*) : acetylene ; C_2H_2 ; Sg. 0.91 ; Mp. – 81.5°C. ; Bp. – 83.6°C. 564
AZETYLENHEXACHLORID (*n*) : acetylene hexachloride (see **HEXACHLORÄTHAN**) 565
AZETYLZAHL (*f*) : acetyl number 566
AZETYLZELLULOSE (*f*) : cellulose acetate, sericose $C_6H_5(CO_2CH_3)_5$ 567
AZIDINECHTGELB (*n*) : azidine fast yellow (see **CURCUMIN**) 685
AZIDINFARBE (*f*) : azidine (dye) colour 569
AZIMUTKOMPASS (*m*) : azimuth compass 570
AZINFARBSTOFF (*m*) : azine (dye) colour 571
AZINGRÜN (*n*) : azine green 572
AZOALIZARINBRAUN (*n*) : azoalizarine brown 573
AZOBENZOL (*n*) : azobenzol, azobenzene, azobenzide, benzeneazobenzene ; $C_{12}H_{10}N_2$; Sg. 1.203 ; Mp. 68°C. ; Bp. 297°C. 574
AZOCYLISCH : azocyclic 575
AZOECHTFARBE (*f*) : fast azo (dye) colour 576
AZOFARBE (*f*) : azo dye ; azo colour 577
AZOFARBSTOFF (*m*) : azo dye 578
AZOKÖRPER (*m*) : azo (body) compound 579
AZOMETHYLEN (*n*) : azomethylene (see **DIAZOMETHAN**) 580
AZOSÄUREFARBSTOFF (*m*) : acid azo dyes or colours 581
AZOT (*n*) : azote ; nitrogen ; N_2 ; Sg. (gas) 0.96737, (liquid) 0.804, (solid) 1.0265 ; Mp. – 210.5°C. ; Bp. – 195.5°C. 582
AZOVERBINDUNG (*f*) : azo compound 583
AZOWALKROT (*n*) : fulling azo red 584
AZOXYBENZOL (*n*) : azoxybenzene ; $C_{12}H_{10}ON_2$; Sg. 1.246 ; Mp. 36°C. 585
AZUR (*m*) : azure, sky-blue 586
AZURBLAU (*n*) : azure-(blue) ; smalt ; cobalt blue (see also **KUPFERCARBONAT**) 587
AZURIT (*m*) : azurite, chessylite, blue copper carbonate (about 46% Cu.) (a basic copper carbonate) (see **KUPFERLASUR**) ; $Cu(CO_3)_2(CuOH)_2$; Sg. 3.8 588
AZURN : azure-(blue) 589
AZUROLBLAU (*n*) : azurol blue 590
AZURSTEIN (*m*) : Lapis lazuli (a translucent azure-blue stone) 591

B

BABBIT-METALL (*n*) : babbit metal (same composition as Britannia metal) (see **BRITANNIAMETALL**) (name generally used in North America for white metal alloys) 592
BABINGTONIT (*m*) : babingtonite $(Ca,Fe,Mn)_2(SiO_3)_2.Fe_2(SiO_3)_3$; Sg. 3.36 593
BACHSTELZE (*f*) : wagtail (*Motacilla alba* ; *M. lugubris* ; *M. Melanope*, etc.) 594
BACHWEIDE (*f*) : osier (*Salix triandra*) 595
BACILLE (*f*) : bacillus 596
BACILLENKRAUT (*n*) : basil (*Calamintha clinopodium* and *C. officinalis*) 597
BACILLUS (*m*) : bacillus 598
BACK (*n*) : forecastle (Naut.) 599
BACKBORD (*m*) : port, larboard (Naut.) 600
BACKBORDANKER (*m*) : port anchor 601
BACKBORDMASCHINE (*f*) : port engine 602
BACKBORDS : port 603
BACKBORDSEITE (*f*) : port side 604
BACKBORDWACHE (*f*) : port watch 605
BACKBRASSEN : to brace back 606
BACKDECK (*n*) : forecastle deck (Naut.) 607
BACKDECKBALKEN (*m*) : forecastle deck beam 608
BACKEN (*f.pl*) : jaws (of a vice) 609
BACKEN (*f.pl* or *m sing.* and *pl.*) : cheek(s) 610
BACKEN : to bake ; fry ; dry ; roast (ores) ; burn (tiles) ; coke (coal) ; cake 611
BACKENDE KOHLE (*f*) : caking (coking) coal ; bitumenous coal 612
BACKENZAHN (*m*) : molar-(tooth) 613
BÄCKEREIEINRICHTUNG (*f*) : bakery equipment 614
BACKKOHLE (*f*) : caking (coking) coal ; bitumenous coal 615
BACKOBST (*n*) : dried fruit 616
BACKOFEN (*m*) : bake oven 617
BACKPFANNE (*f*) : frying (baking) pan 618
BACKPULVER (*n*) : baking powder 619
BACKSPIERE (*f*) : (swing)-boom (Naut.) 620
BACKSTAG (*m*) : backstay, guy 621
BACKSTAGWIND (*m*) : quartering wind 622
BACKSTEIN (*m*) : (burnt-)brick 623
BACKSTEINFABRIK (*f*) : brickworks 624
BACKSTEINOFEN (*m*) : brick kiln 625
BAD (*n*) : bath (either the liquid or the container) ; dip ; steep (Tech.) ; bathing place ; watering place ; spa 626
BADAMÖL (*n*) : (see **CATAPPAÖL**) 627
BADDELEYIT (*m*) : baddeleyite (see **ZIRKONDIOXYD**) ; ZrO_2 ; Sg. 5.75 628
BADEEXTRAKT (*m*) : bath-extract 629
BADESALZ (*n*) : bath salt 630
BADISCHE SÄURE (*f*) : badische acid ; $C_{10}H_6(NH_2)CO_3H$ 631
BADSPANNUNG (*f*) : bath tension ; cell voltage 632
BAGASSE (*f*) : bagasse, megass (the crushed sugar cane after the juice has been squeezed out) (used for fuel, has about 50% moisture and 2200 k.cal. cal. val.) 633

BAGASSEBRÜCKE (*f*): bagasse bridge (a guide piece arranged between the crushing rolls to guide the bagasse coming from one set of rolls into the next set) 634
BAGGER (*m*): dredger; excavator 635
BAGGERBOOT (*n*): dredger, dredging boat 636
BAGGEREIMER (*m*): dredger-bucket 637
BAGGERMASCHINE (*f*): dredger 638
BAGGERN: to dredge, clear out mud 639
BAGGERPRAHM (*m*): dredger, dredging boat or lighter 640
BAGIENBRASSE (*f*): cross-jack brace (Naut.) 641
BAGIENBULIN(E) (*f*): cross-jack bowline 642
BAGIENGEITAU (*n*): cross-jack clew garnet 643
BAGIENHALSE (*f*): cross-jack tack 644
BAGIENRAH (*f*): cross-jack yard 645
BAGIENRAHPEERD (*n*): cross-jack foot-rope 646
BAGIENSCHOTE (*f*): cross-jack sheet 647
BAGIENSEGEL (*n*): cross-jack sail 648
BAHAMAHOLZ (*n*): Brazil wood (*Caesalpinia brasiliensis*) 649
BÄHEN: to foment; toast (bread) 650
BAHIAHOLZ (*n*): (see BAHAMAHOLZ) 651
BAHN (*f*): path; way; road; trajectory; orbit; railway; career; course; breadth (of cloth) 652
BAHN (*f*), AMBOSS-: anvil plate 653
BAHNBRECHEND: pioneer; epoch-making 654
BAHNEN: to smooth a way; make (level) a path; carve out a career 655
BAHNHAMMER (*m*): face hammer 656
BAHNHOFSGEBÄUDE (*n*): station buildings (Railways) 657
BAHNHOFSVORSTEHER (*m*): station master 658
BAHNTELEGRAPH (*m*): railway telegraph 659
BAHNWÄRTER (*m*): watchman, signalman (Railways) 660
BAHNWÄRTERHÄUSCHEN (*n*): watchman's house (at nearly every level crossing on the German State Railways) 661
BAHRE (*f*): barrow; bier 662
BÄHUNG (*f*): fomentation (Med.) 663
BAI (*f*): bay 664
BAISALZ (*n*): bay (sea) salt (from sea-water evaporated) 665
BAISSE (*f*): decline; fall; abatement; diminution in value; drop (in money) 666
BAKE (*f*): beacon, buoy 667
BAKELIT (*m*): bakelite (a synthetic resin); Sg. 1.25 668
BAKTERIE (*f*): bacterium 669
BAKTERIEGIFT (*n*): bactericide 670
BAKTERIEN (*f.pl*): bacteria (see also SPALTPILZ) 671
BAKTERIENKUNDE (*f*): bacteriology 672
BAKTERIENLEHRE (*f*): (see BAKTERIENKUNDE) 673
BAKTERIENSTICKSTOFFDÜNGER (*m*): azotic manure for bacteria 674
BAKTERIENVERNICHTEND: bacteria-destroying, anti-bacteria 675
BAKTERIOLOGIE (*f*): bacteriology 676
BAKTERIOLOGISCH: bacteriological-(ly) 677
BAKTERIOLYSE (*f*): bacteriolysis 678
BALAMTALG (*m*): balam tallow, Siak tallow (from seed of *Palaquium pisang*) 679
BALANCE (*f*): equipoise, balance 680
BALANCERUDER (*n*): balanced rudder 681
BALANCIER (*m*): beam; balancing arrangement 682
BALANCIERDAMPFMASCHINE (*f*): beam engine 683
BALANCIEREN: to balance 684
BALANCIERSPANT (*n*): balance frame 685
BALATA RIEMEN (*m* and *pl*); balata belt(s) 686
BÄLDE, IN-: as soon as possible, at your convenience, in a short time 687
BALDIG: quick; early; prompt; at an early date; speedy 688
BALDRIAN (*m*): Valerian (*Valeriana officinalis*) (see also VALERIAN) 689
BALDRIANÄTHER (*m*): valeric ether (see ÄTHYL, BALDRIANSAURES-) 690
BALDRIANÖL (*n*): Valerian oil; (*Oleum valerianæ*); Sg. 0.940-0.950 691
BALDRIANSAUER: combined with valeric acid, valeriate of 692
BALDRIANSÄURE (*f*): valeric acid; $CH_3(CH_2)_3CO_2H$; Sg. 0.9415; Mp. $-58.5°C.$; Bp. 186.4°C. 693
BALDRIANSÄUREÄTHYLESTER (*m*): valeric ether, ethyl valeriate (see ÄTHYL, BALDRIANSAURES-) 694
BALDRIANSAURES SALZ (*n*): valerate, valeri-(an)-ate 695
BALDRIANWURZEL (*f*): valerian root (*Radix valeriana*) 696
BALG (*m*): skin; bellows; husk; shell; cyst (Med.) 697
BALG AUSZUG (*m*): bellows body, bellows extension (of cameras) 698
BALG KAMERA (*f*): bellows camera 699
BALKEN (*m*): beam; girder; baulk (timber); *Corpus callosum* (Anat.) 700
BALKENENDE (*n*): beam end 701
BALKENKIEL (*m*): bar keel (ships) 702
BALKENKNIE (*n*): beam knee 703
BALKENLAGE (*f*): position (arrangement) of beams 704
BALKENSTRINGER (*m*): beam stringer 705
BALKEN (*m*), WAGE-: balance arm (scales) 706
BALKENWERK (*n*): rafters, beams 707
BALKEN (*m*), ZWISCHENDECK-. between deck beam 708
BALKWEGER (*m*): clamp 709
BALL (*m*): ball; sphere; globe; black ash (soda) 710
BALLAST (*m*): ballast 711
BALLASTBEHÄLTER (*m*): ballast tank 712
BALLAST NEHMEN: to ballast, to take on (ship) ballast 713
BALLASTPFORTE (*f*): ballast port 714
BALLAST (*m*), WASSER-: water ballast 715
BALLEN: to ball; form into balls; conglomerate; bale; pack; bundle; (*m*) bale; pack; bundle; conglomeration 716
BALLEN (*m*): package, parcel, bale 717
BALLENLÄNGE (*f*): length of body (of rolls) 718
BALLENPRESSE (*f*): bale-press, press for bales 719
BALLON (*m*): balloon; carboy; flask; cylinder with plate cover for catching zinc dust from zinc reduction process in muffle furnace (see ZINKSTAUB) 720
BALLONENTLEERER (*m*): carboy emptier 721
BALLONFILTER (*n*): filtering flask 722
BALLONGAS (*n*): balloon gas 723
BALLONKIPPER (*m*): carboy tipper 724
BALSAM (*m*): balsam, balm (see AKAJOUBALSAM, KANADABALSAM, KOPAIVABALSAM, PERUBALSAM, PERUGEN, STORAX, TOLUBALSAM) 725
BALSAMAPFEL (*m*): balsam apple (*Momordica*) 726
BALSAMBAUM (*m*): balsam tree (any of various balsamaceous trees) (see BALSAMPAPPEL and BALSAMTANNE) 727

BALSAMERZEUGEND: balsamiferous; producing balsam 728
BALSAMHARZ (*n*): balsamic resin 729
BALSAMHOLZ (*n*): Balm of Gilead wood (*Xylobalsamum*) 730
BALSAMIEREN: to embalm; perfume 731
BALSAMISCH: balsamic 732
BALSAMPAPPEL (*f*): tacamahac (*Populus balsamifera*) 733
BALSAMTANNE (*f*): balsam fir (*Abies balsamea*); *Abies balsamica*); Norway spruce (*Picea abies*) 734
BAMBUKBUTTER (*f*): (see SHEABUTTER) 735
BAMBUS (*m*): bamboo (varies species of *Bambusa*) Sg. 0.4 736
BAMBUSHOLZ (*n*): bamboo wood (see BAMBUS) 737
BAMBUSROHR (*n*): bamboo cane 738
BAMBUSSTOCK (*m*): bamboo (cane, rod) stick 739
BAMBUSZUCKER (*m*): bamboo sugar; tabasheer; tabashir (siliceous secretion from certain bamboos) 740
BANANE (*f*): banana (*Musa sapientum*) 741
BANCAZINN (*n*): Banca tin (see BANKAZINN) 742
BAND (*n*): band, brace, strap, strip, ribbon, tape, hoop (for casks), tie, string 743
BANDACHAT (*m*): ribbon agate; onyx 744
BANDÄHNLICH: striped; streaked; striated; ligamentous 745
BANDBREMSE (*f*): band brake 746
BÄNDE, DAS SPRICHT-: that speaks volumes 747
BANDEISEN (*n*): hoop iron; iron strap 748
BANDFÖRDERER (*m*): belt conveyor 749
BANDLFETT (*n*): (see DARMFETT) 750
BANDMASS (*n*): tape measure; measuring tape 751
BANDSÄGE (*f*): band saw 752
BÄNDSEL (*n*): lashing 753
BANDSTAHL (*m*): band steel, hoop steel 754
BANDSTRÖMUNG (*f*): laminar flow 755
BANDTRANSPORTEUR (*m*): belt conveyor 756
BANDWEIDE (*f*): osier (*Salix viminalis*) 757
BANDWURM (*m*): tape worm; Cestoda (*Toenia solium*) (Med.) 758
BANDWURMMITTEL (*n*): tape-worm (remedy) remedies, anthelmintic-(s); vermicides (which kill the worm); vermifuges (which expel the worm) 759
BANIANE (*f*): banyan-(tree) (*Ficus bengalensis*) 760
BANIANENBAUM (*m*): banian (banyan) tree (*Ficus bengalensis* or *Ficus religiosa*) 761
BANK (*f*): bench; bed; bank; layer; stratum; reef; set; row 762
BANKANWEISUNG (*f*): bank-note; cheque 763
BANKAUFGELD (*n*): bank agio 764
BANKAUSWEIS (*m*): bank balance 765
BANKAZINN (*n*): Banca tin (silver white in colour) from Banka in East Indies; in refined state has 99.99% tin) 766
BANKAZINN (*n*), **RAFFINIERTES-** refined Banka tin (from Banka, East Indies, 99.9% tin) 767
BANKBEAMTE (*m*): bank-clerk 768
BANKBILANZ (*f*): bank balance 769
BANKBRUCH (*m*): bankruptcy; insolvency; failure 770
BANKEROTT (*m*): bankrupt; insolvent; bankruptcy; insolvency; failure 771
BANKETT (*n*): berm; bench; step; terrace 772
BANKHAMMER (*m*): bench (hand) hammer 773
BANKHOBEL (*m*): joining plane, bench plane 774
BANKMÄSSIGE ZINSEN (*m.pl*): bank (charges) interest 775
BANKNOTENPAPIER (*n*): bank-note paper 776
BANK (*f*), **RETORTEN-**: retort bench 777
BANKROTT: bankrupt, insolvent; (*m*): bankruptcy, insolvency, failure 778
BANKROTT (*m*), **BETRÜGERISCHER-**: fradulent bankruptcy 779
BANKSÄGE (*f*): bench saw 780
BANKSCHRAUBE (*f*): bench (screw) vice 781
BANKSCHRAUBSTOCK (*m*): bench vice 782
BANKULNUSS (*f*): candle-nut (*Aleurites triloba* and *A. moluccana*) 783
BANKULNUSSÖL (*n*): candle-nut oil; lumbang oil (from *Aleurites moluccana*); Sg. 0.923 784
BANKUNTERNEHMEN (*n*): banking (company) undertaking 785
BANKWECHSEL (*m*): bank bill 786
BANKWEGERUNG (*f*): thwart clamp (Naut.) 787
BANKWESEN (*n*): banking 788
BANKZINSEN (*m.pl*): bank (interest) charges, bank rate 789
BANN (*n*): ban; jurisdiction; spell; proscription; excommunication 790
BANNEN: to ban, banish, proscribe, excommunicate 791
BANQUIER (*m*): banker 792
BAOBABÖL (*n*): baobab oil (from seeds of *Adansonia digitata*) 793
BAOBAÖL (*n*): (see BAOBABÖL) 794
BÄR (*m*): bear; debt; ram (Mech.); head; plunger 795
BAR: ready money, in cash; bare, naked 796
BARACKE (*f*): barrack (military) 797
BARBADOSALOE (*f*): Barbados aloe, curacao aloe (see ALOE, MATTE-) 798
BARBADOSASPHALT (*m*): Barbados asphalt (a very pure natural asphalt from Barbados) (see ASPHALT) 799
BARBITURSÄURE (*f*): barbituric acid, malonyl urea (*Acidum barbituricum*); $CO(NHCO)_2CH_2$ 800
BÄRENFENCHEL (*m*): spicknel; spignel meu; bald-money (*Meum athamanticum*) 801
BÄRENKLAU (*f*): bear's breech (*Acanthus mollis*; *A. spinosus*) 802
BÄRENTRAUBE (*f*): bearberry (*Arctostaphylos uva-ursi*) 803
BÄRENTRAUBENBLÄTTER (*n.pl*): bearberry leaves (*folia uvæ ursi* of *Arctostaphylos uva ursi*) 804
BÄRENWURZEL (*f*): saxifrage (*Saxifraga*) 805
BÄRENZUCKER (*m*): (see LAKRITZENSAFT) 806
BARES GELD (*n*): hard cash, ready money, spot cash 807
BARGELD (*n*): ready (money) cash, specie 808
BARGESCHÄFT (*n*): cash business, business on a cash basis 809
BAR, IN-: in cash, spot cash 810
BARIUM (*n*): (see BARYUM) 811
BARK (*f*): bark, barque (ships), barge, lighter 812
BARKASSE (*f*): long boat (Naut.), launch 813
BARKASSE (*f*), **DAMPF-**: steam launch 814
BARKBOOT (*n*): bark, barque (ships) 815
BARKE (*f*): bark; barge; shallow vat; lighter; barque 816
BARKEVIKIT (*m*): barkevikite $(Na_2,Ca,Fe,Mg)_3(Al,Fe)_2Si_3O_{12}(+TiO_2)$; Sg. 3.4 817
BARKSCHIFF (*n*): bark, barque (ships) 818
BÄRLAPP (*m*): club-moss; Lycopodium; (*Lycopodium clavatum*) 819
BÄRLAPPKRAUT (*n*): (see BÄRLAPP) 820

BÄRLAPPMEHL (*n*) : lycopodium (powder), vegetable sulphur (spores of *Lycopodium clavatum*) 821

BÄRLAPPSAMEN (*m*) : vegetable sulphur, lycopodium powder (spores of *Lycopodium clavatum*) 822

BÄRME (*f*) : barm ; yeast (see **HEFE**) 823

BÄRMUTTER (*f*) : womb, uterus (Med.) 824

BARNSTEIN (*m*) : brick 825

BAROCK : baroque, fantastic, bizarre, quaint, odd, decadent, degenerate (of Art) 826

BAROCKPERLEN (*f.pl*) : baroque pearls (ill-shaped pearls) 827

BAROGRAPH (*m*) : barograph, recording barometer 828

BAROMETER (*n*) : barometer 829

BAROMETERSÄULE (*f*) : barometric column 830

BAROMETERSKALA (*f*) : barometer scale 831

BAROMETERSTAND (*m*) : barometric (level) height 832

BARRE (*f*) : bar ; ingot (of precious metal) 833

BARREL (*n*) : barrel (of oil) (mostly 42 American galls. or 42 English galls., but the value may fluctuate between 40-50 American, or 36-42 English galls). 834

BARREN (*m*) : ingot, bar (metals) 835

BARRENFORM (*f*) : ingot mould 836

BARSCH (*m*) : perch (fish) 837

BARSCH : harsh-(ly) ; rough-(ly) 838

BARSCHAFT (*f*) : cash ; property ; ready money ; effects ; liquid assets 839

BART (*m*) : beard ; ward (of a key) ; bur 840

BARTFINNE (*f*) : sycosis, acne mentagra, folliculitis (follicular inflammation) of the hairy parts of the face (Med.) 841

BARTGRAS (*n*) : beard grass (*Polypogon*) 842

BARVORRAT (*m*) : balance in favour (at a bank) 843

BARYSILIT (*m*) : (a natural lead silicate) ; $3PbO.2SiO_2$; Mp. 690°C. 844

BARYT (*m*) : baryta ; BaO or $Ba(OH)_2$; barite ; heavy spar ; $BaSO_4$; Sg. 4.5 845

BARYTERDE (*f*) : barium oxide ; anhydrous baryta ; BaO ; Sg. 4.73-5.46 846

BARYT (*m*), **ESSIGSAURER-** : barium acetate (see **BARYUMAZETAT**) 847

BARYTFLUSSSPAT (*m*) : fluorite with a barium content ; natural calcium-barium fluorite 848

BARYTGELB (*n*) : barium yellow ; barium chromate ; $BaCrO_4$; Sg. 4.498 849

BARYTHYDRAT (*n*) : barium hydrate ; barium hydroxide ; $Ba(OH)_2.8H_2O$; Sg. 1.656 ; Mp. 78°C. ; Bp. 103°C. 850

BARYT (*m*), **KOHLENSAURER-** : barium carbonate (see **BARYUM CARBONAT**) 851

BARYTLAUGE (*f*) : barium hydroxide solution ; baryta water 852

BARYT (*m*), **MANGANSAURER-** : barium manganate (see **BARYUMMANGANAT**) 853

BARYT (*m*), **OXALSAURER-** : barium oxalate (see **BARIUM, OXALSAURES-**) 854

BARYTPAPIER (*n*) : baryta paper (Photography) 855

BARYT (*m*), **SALPETERSAURER-** : barium nitrate (see **BARYUMNITRAT**) 856

BARYT (*m*), **SALPETRIGSAURER-** : barium nitrite (see **BARIUM NITRIT**) 857

BARYT (*m*), **ÜBERMANGANSAURER-** : barium permanganate (see **BARYUMPERMANGANAT**) 858

BARYTWASSER (*n*): solution of barium hydroxide; baryta water 859

BARYTWEISS (*n*) : permanent white ; blanc fixé ; $BaSO_4$; Sg. 4.476 860

BARYUM (*n*) : barium ; Ba (see **BARIUM**) 861

BARYUMALUMINAT (*n*) : barium aluminate 862

BARYUM (*n*), **AMEISENSAURES-** : barium formate 863

BARYUMAMID (*n*) : barium amide ; $Ba(NH_2)_2$; Mp. 280°C. 864

BARYUMARSENID (*n*) : barium arsenide ; Ba_3As_2; Sg. 4.1 865

BARYUMAZETAT (*n*) : barium acetate (*Baryum aceticum*) ; $Ba(C_2H_3O_2)_2.H_2O$; Sg. 2.02 [866

BARYUM (*n*), **BALDRIANSAURES-** : barium (valerianate) valerate 867

BARYUMBIOXYD (*n*) : barium (dioxide) peroxide, barium binoxide (see **BARYUMSUPEROXYD**) 868

BARYUMBORAT (*n*) : (*Baryum boricum*) : barium borate ; $BaO.B_2O_3$; Mp. 1060°C. ; $2BaO.B_2O_3$; Mp. 1002°C. ; $3BaO.B_2O_3$; Mp. 1320°C. 869

BARYUM (*n*), **BORSAURES-** : barium borate (see **BARYUMBORAT**) 870

BARYUMBROMAT (*n*) : barium bromate ; $Ba(BrO_3)_2.H_2O$; Sg. 3.82 871

BARYUMBROMID (*n*) : barium bromide ; $BaBr_2$; Sg. 4.781 ; $BaBr_2.2H_2O$; Sg., 3.827 872

BARYUMCARBID (*n*) : barium carbide ; BaC_2 ; Sg. 3.75 873

BARYUMCARBONAT (*n*) : barium carbonate (*Barium carbonicum*) ; $BaCO_3$; Sg. 4.275 ; Mp. 1360°C. 874

BARYUMCHLORAT (*n*) : barium chlorate (*Baryum chloricum* ; $Ba(ClO_3)_2.H_2O$; Sg. 3.179 ; Mp. 414°C. 875

BARYUMCHLORID (*n*), **KRYST** : (*Baryum chloratum*); barium chloride (cryst) ; $BaCl_2.2H_2O$; Sg. 3.104 ; Mp. 860°C. 876

BARYUMCHLORID (*n*), (**WASSERFREI**) : barium chloride (anhydrous) ; Sg. 3.856 ; $BaCl_2$ [877

BARYUM (*n*), **CHLORSAURES-** : barium chlorate (see **BARIUMCHLORAT**) 878

BARYUMCHROMAT (*n*) : (*Baryum chromicum*) ; barium chromate; $BaCrO_4$; Sg. 4.498 879

BARYUM (*n*), **CHROMSAURES-** : barium chromate (see **BARIUMCHROMAT**) 880

BARYUMCYANID (*n*) : barium cyanide ; $Ba(CN)_2$ 881

BARYUM (*n*), **ESSIGSAURES-** : barium acetate ; $Ba(C_2H_3O_2)_2.H_2O$; Sg. 2.02 882

BARYUMFELDSPAT (*m*) : barium feldspar ; $BaAl_2Si_2O_8$; Sg. 3.35 883

BARYUMFLUORID (*n*) : (*Baryum fluoratum*) ; barium fluoride ; BaF_2 ; Sg. 4.828 ; Mp. 1280°C. 884

BARYUM (*n*), **GOLDCHLORWASSERSTOFFSAURES-** : barium chloraurate ; $Ba(AuCl_4)_2.xH_2O$ [885

BARYUMHYDRAT (*n*) : barium hydrate, barium hydroxide (see **ÄTZBARYT** and **BARYUMHYDROXYD**) 886

BARYUMHYDROXYD (*n*) ; caustic baryta ; barium hydroxide ; barium hydrate (*Bary m hydroxydatum*) ; $Ba(OH)_2.8H_2O$; Sg. 1.656 ; Mp. 78°C. ; Bp. 103°C. 887

BARYUMHYDRÜR (*n*) : barium hydride ; BaH_2 ; Sg. 4.21 888

BARYUMHYPEROXYD (*n*) : barium peroxide ; barium dioxide ; BaO_2 ; Sg. 4.96 889

BARYUMHYPOPHOSPHIT (*n*) : barium hypophosphite ; $Ba(H_2PO_2)_2$; Sg. 2.87 890

BARYUMHYPOSULFAT (*n*) : bariumhyposulphate ; $BaS_2O_6.4H_2O$; Sg. 3.142 891

BARYUMHYPOSULFIT (*n*): barium hyposulphite; $BaS_2O_3.H_2O$; Sg. 3.447 892
BARYUMJODAT (*n*): barium iodate; $Ba(IO_3)_2$ Sg. 5.23 893
BARYUMJODID (*n*): barium iodide; BaI_2; Sg. 5.15; Mp. 740°C. 894
BARYUM (*n*), JODSAURES-: barium iodate (see BARYUMJODAT) 895
BARYUMKARBONAT (*n*): (see BARYUMCARBONAT) 896
BARYUM (*n*), KOHLENSAURES-: barium carbonate (see BARYUMCARBONAT) 897
BARYUMMANGANAT (*n*): (*Baryum manganicum*): barium manganate, Cassel green, manganese green; $BaMnO_4$; Sg. 4.85 898
BARYUMMETASILIKAT (*n*): barium metasilicate; $BaSiO_2$; Sg. 3.77 899
BARYUM (*n*), MILCHSAURES-: barium lactate 900
BARYUMMOLYBDAT (*n*): barium molybdate; $BaMoO_4$; Sg. 4.654 901
BARYUMMONOSULFID (*n*): barium monosulphide, (see BARYUMSULFID) 902
BARYUMNITRAT (*n*): (*Baryum nitricum*): barium nitrate; $Ba(NO_3)_2$; Sg. 3.23-3.244; Mp. 575°C. 903
BARYUMNITRIT (*n*): (*Baryum nitrosum*): barium nitrite; $Ba(NO_2)_2$; Sg. 3.23; $Ba(NO_2)_2.H_2O$; Sg. 3.173 904
BARYUMORTHOSILIKAT (*n*): barium orthosilicate; Ba_2SiO_4 905
BARYUMOXALAT (*n*): barium oxalate (see BARYUM, OXALSAURES-) 906
BARYUM (*n*), OXALSAURES-: (*Baryum oxalicum*): barium oxalate; $BaC_2O_4.H_2O$ 907
BARYUMOXYD (*n*): barium monoxide, barium protoxide, barium oxide, calcined baryta (*Baryum oxydatum*): BaO; Sg. 5.32-5.72 [908
BARYUMOXYDHYDRAT (*n*): barium hydroxide (see BARYUMHYDROXYD) 909
BARYUMPERCHLORAT (*n*): barium perchlorate; $Ba(ClO_4)_2$; Mp. 505°C. 910
BARYUMPERMANGANAT (*n*): (*Baryum permanganicum*): barium permanganate; $BaMn_2O_8$ [911
BARYUMPEROXYD (*n*): (*Baryum hyperoxydatum*): barium peroxide, barium dioxide, barium binoxide; BaO_2; Sg. 4.96 912
BARYUMPEROXYDHYDRAT (*n*): barium peroxide hydrate 913
BARYUM (*n*), PHOSPHORSAURES-: barium phosphate 914
BARYUMPLATINCYANÜR (*n*): barium platinocyanide; $BaPt(CN)_4$ 915
BARYUM (*n*), PROPIONSAURES-; barium propionate 916
BARYUMPYROPHOSPHAT (*n*): barium pyrophosphate; $Ba_2P_2O_7$; Sg. 4.1 917
BARYUMRHODANÜR (*n*): (*Barium rhodanatum*): barium (sulfocyanide) sulfocyanate, barium rhodanide; $Ba(CNS)_2.2H_2O$ 918
BARYUM (*n*), SALPETERSAURES-: barium nitrate (see BARYUMNITRAT) 919
BARYUM (*n*), SCHWEFELSAURES-: barium sulphate (see BARYUMSULFAT) 921
BARYUM (*n*), SCHWEFLIGSAURES-: barium sulphite 922
BARYUMSELENAT (*n*): barium selenate; $BaSeO_4$; Sg. 4.75 923
BARYUMSILICOFLUORID (*n*): barium silicofluoride; $BaSiF_6$; Sg. 4.279 924
BARYUMSULFAT (*n*): blanc fixé, permanent white, barium sulphate; (*Baryum sulfuricum*); $BaSO_4$; Sg. 4.25-4.48 925
BARYUMSULFID (*n*): barium monosulphide, barium sulphide; (*Barium sulfuratum*); BaS; Sg. 4.25 926
BARYUM (*n*), SULFOBENZOESAURES-; barium sulfobenzoate 927
BARYUMSUPEROXYD (*n*): barium peroxide, barium binoxide, barium dioxide; (*Baryum hyperoxydatum*); BaO_2; Sg. 4.958 928
BARYUMSUPEROXYDHYDRAT (*n*): barium peroxide hydrate 929
BARYUMTHIOSULFAT (*n*): barium thiosulphate, (*Barium thiosulfuricum*); $Ba_2Si_4O_3$ 930
BARYUM (*n*), UNTERPHOSPHORIGSAURES-: barium hypophosphite (see BARYUMHYPOPHOSPHIT) 931
BARYUM (*n*), UNTERSCHWEFLIGSAURES-: barium hyposulphite (see BARIUMHYPOSULFIT) 932
BARYUMWOLFRAMAT (*n*): barium tungstate (*Baryum wolframicum*); $BaWO_4$; Sg. 5.023-6.35 933
BARYUM (*n*), WOLFRAMSAURES-: barium tungstate (see BARYUMWOLFRAMAT) 934
BARYUMZYANID (*n*): (see BARYUMCYANID) 935
BARZAHLUNG (*f*): cash payment; spot cash 936
BASALT (*m*): basalt (a basic, volcanic rock) 937
BASALTFELSEN (*m*): basaltic rock 938
BASALTIEREN: to convert (slag) into a material resembling basalt 939
BASE (*f*): base; basis (female-)cousin 940
BASEDOWSCHE KRANKHEIT (*f*); Basedow's disease, exophthalmic goitre (Med.) 941
BASELERBLAU (*n*): basle blue, ferric ferrocyanide; $Fe_4[Fe(CN)_6]_3$ 942
BASELER GRÜN (*n*): basle green; Paris green; cupric acetoarsenite; $3CuOAs_2O_3.Cu(C_2H_3O_2)_2$ 943
BASENBILDER (*m*): base former; basifier 944
BASICITÄT (*f*): basicity 945
BASIEREN: to base; be based 946
BASILICUM (*n*): (sweet)-basil (see BASILIKUM) 947
BASILIENKRAUT (*n*): (sweet)-basil (*herba basilici* of *Ocymum basilicum*) 948
BASILIENÖL (*n*): basil oil; Sg. 0.945-0.987 949
BASILIKUM (*n*): (see BASILIENKRAUT) 950
BASILIKUMKRAUT (*n*): (sweet-) basil 951
BASILIKUMÖL (*n*): basil oil (see BASILIENÖL) 952
BASIS (*f*): base; basis; basement; pedestal; foundation; essence 953
BASISCH: basic-(ally) (see EINBASISCH; ZWEIBASISCH and DREIBASISCH) 954
BASISCHE SÄURE (*f*): basic (monobasic) acid 955
BASISCHE SCHLACKE (*f*): basic slag (Siemens Martin process) (35-55% CaO; 10-20% SiO_2; 10-15% FeO; 5-15% MnO; 2-9% Al_2O_3; 1-9% P_2O_5) (of little value except in case of that with high phosphorous content, which can be used as argicultural manure) 956
BASSFASER (*f*): (see PIASSAVA, AFRIKANISCHE-) 957
BASSIAÖL (*n*): Bassia oil, Mowra oil (from the seeds of *Bassia longifolia* or *B. latifolia*); Sg. 0.896; Mp. 23-29°C. (see also MAHWABUTTER and MOWRAHBUTTER) 958
BASSIN (*n*): tank; cistern; reservoir; basin; bowl; container; receiver 959
BASSIN (*n*), WASSER-: water cistern 960
BAST (*m*): bast (a fibre), inside bark 961
BASTARD (*m*): bastard, mongrel, hybrid cross-(breed) 962
BASTARDFEILE (*f*): bastard file 963
BASTERN (*m*): raw sugar 964

BASTERZUCKER (*m*): raw sugar; bastard sugar 965
BASTFASER (*f*): bast fibre (see also **PIASSAVA, AFRIKANISCHE-**) 966
BASTNÄSIT (*m*): $[(Ce,La,Di)F]CO_3$; bastnäsite; Sg. 5.0 967
BASTPAPIER (*n*): manilla paper 968
BASTSCHWARZ (*n*): bast black 969
BASTSEIDE (*f*): raw silk 970
BASTSEIFE (*f*): degumming soap 971
BÄTING (*f*): (see also **BETING**); bitt (Naut.) 972
BATIST (*m*): cambric; bastiste 973
BATSCHEN: to ret (flax, etc.) 974
BATTERIE (*f*): battery (marine, electric, mechanical, etc.) 975
BATTERIEDECK (*n*): gun deck 976
BATTERIEDRAHT (*m*): battery wire 977
BATTERIEELEKTRODE (*f*): battery electrode 978
BATTERIEGESCHÜTZ (*n*): battery gun 979
BATTERIEKESSEL (*m*): battery boiler-(s) 980
BATTERIEKOHLE (*f*): (battery)-carbon 981
BATTERIELAMPE (*f*): battery lamp 982
BATTERIE (*f*), **PRIMÄR-**: primary battery 983
BATTERIEPRÜFER (*m*): battery (detector) tester 984
BATTERIESCHALTER (*m*): battery switch 985
BATTERIESCHRANK (*m*): battery (box) case 986
BATTERIE (*f*), **SEKUNDÄR-**: secondary battery 987
BATTERIESTROM (*m*): battery current 988
BATTERIE (*f*), **TROCKEN-**: dry battery 989
BATTERIE (*f*): **VERSTÄRKUNGS-**: subsidiary (auxiliary) battery 990
BATTERIEWIDERSTAND (*m*): battery resistance 991
BATTERIEZELLE (*f*): (battery)-cell 992
BAU (*m*): structure; formation; construction; building; make; erection; cultivation 993
BAUART (*f*): type (method of) construction; structure 994
BAUCH (*m*): belly; bulge; bulb (of a retort); bunt (of a sail) 995
BAUCHBRUCH (*m*): hypogastrocele 996
BÄUCHE (*f*): bucking (steeping or boiling in lye or suds) 997
BÄUCHEN: to buck; steep or boil in lye or suds 998
BAUCHEN: to belly, bulge, distend 999
BAUCHFELL (*n*): peritoneum (Med.) 000
BAUCHFELLENTZÜNDUNG (*f*): peritonitis (inflammation of the peritoneum) (Med.) 001
BAUCHFLUSS (*m*): diarrhœa (Med.) 002
BAUCHFÜSSLER (*m*): gastropod 003
BAUCHGORDING (*f*): buntline (Naut.) 004
BAUCHIG: bellied; bulged; convex 005
BAUCHMUSKEL (*m*): abdominal muscle (Med.) 006
BAUCHSPEICHEL (*m*): pancreatic juice (Med.) 007
BAUCHSPEICHELDRÜSE (*f*): pancreas (Med.) 008
BAUCHSPEICHELDRÜSENSEKRET (*n*): pancreatic juice (Med.) 009
BAUCHSTÜCK (*n*): floor timber 010
BAUCHSTÜCKPLATTE (*f*): floor plate 011
BAUCHWASSERSUCHT (*f*): dropsy of the peritoneum, Ascites (Med.) 012
BAUCHWEGERUNG (*f*): ceiling 013
BAUCHWEH (*n*): stomach-ache; gripes; colic; abdominal pain-(s) 014
BAUCHZEISING (*f*): belly (band) gasket, bunt gasket (Naut.) 015
BAUEISEN (*n*): constructional ironwork 016
BAUEN: to make; construct; build; erect; form; grow; produce; cultivate 017
BAUERDE (*f*): soil (suitable for cultivation); arable land 018
BAUERGUT (*n*): farm; homestead 019
BAUERHOF (*m*): farm; homestead 020
BÄUERLICH: rural 021
BAUERNBUTTER (*f*): country butter, farmhouse butter 022
BAUERNSAND (*m*): moulding sand 023
BAUERNWETZEL (*m*): (see **MUMPS**) 024
BAUERSTAND (*m*): peasantry 025
BAUFÄHIG: arable 026
BAUFÄLLIG: dilapidated; decaying; out of repair 027
BAUGERÜST (*n*): building framework, scaffold-(ing) 028
BAUGERÜSTEINSCHALEISEN (*n*): steel framing; structural steelwork (for ferro-concrete buildings) 029
BAUGUSS (*m*): casting for structural (building) purposes 030
BAUHOLZ (*n*): (building)-timber 031
BAU-INGENIEUR (*m*): constructional engineer 032
BAUJAHR (*n*): date when built, year in which constructed 033
BAUKUNST (*f*): architecture 034
BAULÄNGE (*f*): length-(overall) 035
BAULICHE ANORDNUNG (*f*): constructional arrangement 036
BAULOKOMOTIVE (*f*): locomotive for constructional work-(s) 037
BAULOS (*n*): portion under construction; allotment for construction 038
BAUM (*m*): tree; beam (Mech.); bar; pole; arbor; boom (Naut.) 039
BAUMACHAT (*m*): dendritic agate 040
BAUMÄHNLICH: arborescent; tree-like; dendritic 041
BAUMARTIG: (see **BAUMÄHNLICH**) 042
BAUMASCHINEN (*f.pl*): machinery for constructional work-(s) 043
BAUMATERIAL (*n*): building material 044
BAUM (*m*), **BESAN-**: spanker boom (Naut.) 045
BAUMEISTER (*m*): architect; builder; master builder; contractor; constructor 046
BAUMELN; to dangle 047
BAUMFARN (*m*): tree fern (as *Alsophila* and *Cyathea*) 048
BAUMFOCK (*f*): boom fore-sail (Naut.) 049
BAUMFÖRMIG: dendritic; arborescent; tree-like 050
BAUMGARTEN (*m*): orchard; nursery 051
BAUMHARZ (*n*): tree resin 052
BAUMKANTIG: round (rough) edged, (of wood) 053
BAUMKERZE (*f*): Christmas-tree candle 054
BAUMÖL (*n*): a second grade olive oil (*Oleum olivarum*); Sg. 0.914 055
BAUMÖLSEIFE (*f*): olive oil soap; venetian soap 056
BAUMÖRTEL (*m*): (building)-mortar 057
BAUMRINDE (*f*): bark; rind 058
BAUMSCHLAG (*m*): foliage 059
BAUMSCHWAMM (*m*): agaric; very light variety of calcite 060
BAUMSEGEL (*n*): boom sail 061
BAUMSENF (*m*): (common evergreen)-candytuft (*Iberis coronaria*; *I. umbellata*; *I. sempervirens*, etc.) 062
BAUMSTARK: robust; strong 063
BAUMSTEIN (*m*): dendrite (Min.) 064
BAUMTAU (*n*): guest rope (Naut.) 065

BAUMWACHS (*n*): grafting wax; mummy (a mixture of resin, yellow wax, turpentine and other constituents for covering wounds in trees) 066
BAUMWOLLABFALL (*m*): cotton waste 067
BAUMWOLLBLAU (*n*): cotton blue (see **NAPHTYL-ENBLAU**) 068
BAUMWOLLBLÜTEN (*f.pl*): flowers of the cotton plant, (*Gossypium herbaceum*) 069
BAUMWOLLE (*f*): cotton (from various types of *Gossypium*) 070
BAUMWOLLENABFALL (*m*): cotton waste 071
BAUMWOLLENBAUM (*m*): cotton tree (*Bombax*) 072
BAUMWOLLENBLAU (*n*): cotton blue (see **NAPHTYL-ENBLAU**) 073
BAUMWOLLENPFLANZE (*f*): cotton plant (*Gossypium*) 074
BAUMWOLLENSAMEN (*m*): cotton seed 075
BAUMWOLLENSTREICHER (*m*): carder (Textile) 076
BAUMWOLLFÄRBEREI (*f*): wool dyeing or dye-house 077
BAUMWOLLFARBSTOFF (*m*): cotton dye 078
BAUMWOLLPUTZWOLLE (*f*): cotton waste 079
BAUMWOLLSAATÖL (*n*): (see **BAUMWOLLSAMENÖL**) and **COTTONÖL**) 080
BAUMWOLLSAMENÖL (*n*): cotton seed oil (*Oleum gossypii*); Sg. 0.922-0.93 (see **COTTONÖL**) 081
BAUMWOLLSAMENRÜCKSTAND (*m*): cotton-seed residue 082
BAUMWOLLSCHNUR (*f*): cotton yarn 083
BAUMWOLLSTAUDE (*f*): cotton plant (*Gossypium herbaceum*) 084
BAUMWOLLWURZELRINDE (*f*): cotton-root (Gossypium) bark (*Gossypii radicis cortex*) 085
BAUPAPPE (*f*): building board (Paper) 086
BAURISS (*m*): building plan 087
BAUSAND (*m*): building sand; mortar sand 088
BAUSCH (*m*): pad; wad; compress; plug; bundle; lump 089
BAUSCHEN: to bulge; swell out; protrude; refine (tin) 090
BAUSCHUNG (*f*): protrusion; protuberance; bulge; excrescence; lump; swelling; crease (in fabrics) 091
BAUSTEIN (*m*): building stone; brick 092
BAUSTOFF (*m*): constructional (building) material, material 093
BAUUNTERNEHMER (*m*): builder and contractor 094
BAUWEISE (*f*): method of construction, type 095
BAUWERK (*n*): constructional work, building 096
BAUXIT (*m*): bauxite $(Al,Fe)_2O(OH)_4$ or $(Al_2O_3.2H_2O)$ (from Baux in the South of France) (contains 50-60% Al_2O_3); Sg. 2.7 097
BAUXITOFEN (*m*): bauxite kiln 098
BAYBEERENBAUM (*m*): greenheart tree; bebeeru (*Nectandra rodiœi*) 099
BAYERSCHE SÄURE (*f*): Bayer's acid, beta-naphtholsulfonic acid; $C_{10}H_6(OH)(SO_3H)$ 2:8 100
BAYÖL (*n*): bay oil; oil of bay; volatile laurel oil (*Oleum pimentæ acris*) (from leaves of *Pimenta acris*); Sg. 0.924 (see **LORBEERÖL**) 101
BAYRISCH: Bavarian 102
BAYRISCHBLAU (*n*): Bavarian blue 103
BAYRUM (*m*): bay rum (hair wash, by double distillation from leaves and berries of *Pimenta acris*) 104
BAZILLE (*f*): bacillus (see also **BACILLE**) 105
BAZILLOL (*n*): liquor cresoli saponatus (a disinfectant) 106
BEABSICHTIGEN: to intend; have in view 107
BEACHTEN: to consider; notice; observe (bear in) mind 108
BEAMTE (*m*): officer; official 109
BEANSPRUCHEN: to require; claim; stress; bespeak; be expected 110
BEANSPRUCHUNG (*f*): requirement; rating (evaporation per unit of heating surface) (boilers); stressing (of material); load, duty, strain, stress, demand-(s), thrust, working conditions 111
BEANSPRUCHUNG (*f*), **GEMISCHTE-**: mixed stress 112
BEANSTANDEN: to object to; contest 113
BEANTRAGEN: to propose; move 114
BEARBEITEN: to work; make; dress; till; cultivate; edit; machine 115
BEARBEITUNG (*f*): machining 116
BEAUFSCHLAGTE GLEICHDRUCK-TURBINE, PAR-TIELL-: partial admission impulse turbine 117
BEAUFSCHLAGUNG (*f*): admission (of turbines); flow 118
BEAUFSCHLAGUNG (*f*), **ÄUSSERE-**: inward flow 119
BEAUFSCHLAGUNG (*f*), **INNERE-**: outward flow 120
BEAUFSCHLAGUNG (*f*), **INNERE UND ÄUSSERE-**: mixed flow 121
BEAUFSCHLAGUNG (*f*), **PARTIELLE-**: partial admission 122
BEAUFSCHLAGUNGSGRAD (*m*): degree of admission 123
BEAUFSCHLAGUNG (*f*), **VOLLE-**: full admission, full supply 124
BEAUFSICHTIGEN: to superintend; supervise; inspect; survey 125
BEAUFTRAGEN: to charge; authorize; commission 126
BEAUXIT (*m*): (see **BAUXIT**) 127
BEBEERIN (*n*): bebeerine, bebirine (an alkaloid from the bark of *Nectandra Rodiœi* or *Pareira brava*); $C_{18}H_{21}NO_3$; Mp. 214°C. 128
BEBEN: to quake; tremble; vibrate; oscillate 129
BECHER (*m*): cup; bowl; goblet; chalice; calyx (Bot.) 130
BECHERELEVATOR (*m*): bucket elevator 131
BECHERFRÜCHTLER (*m* and *pl*): cupuliferæ (Bot.) 132
BECHERGLAS (*n*): beaker; goblet; chalice 133
BECHERGLASKOLBEN (*m*): Erlenmeyer flask 134
BECHERMENSUR (*f*): measuring cup 135
BECHERRAD (*n*): bucket (blade) wheel 136
BECHERWERK (*n*): bucket conveyor; bucket elevator; bucket transporter 137
BECHERWERKGRUBE (*f*): elevator pit 138
BECKELITH (*m*): bechilite; $Ca_3(Ce,La,Di)_4Si_3O_{15}$; Sg. 4.15 139
BECKEN (*n*): basin; vessel; district; region; coal-field; cymbal; pelvis (Anat.) 140
BECKEN, RUHR- (*n*): the basin of the Ruhr 141
BECQUERELSTRAHLEN (*m.pl*): Becquerel rays (from radio-active substances) 142
BEDACHT (*m*): consideration; care; deliberation; circumspection; prudence; discretion 143
BEDARF (*m*): need; demand; requirement; want; necessity 144
BEDARFSFALL (*m*): requirement 145
BEDAUERN: to regret; pity; deplore 146

BEDECKEN : to cover ; screen ; protect ; shelter ; bottom (sugar) ; case ; shroud 147
BEDENKEN (n) HEGEN : to hesitate 148
BEDENKEN (n) TRAGEN : to hesitate 149
BEDENKLICH : doubtful ; risky ; serious ; questionable 150
BEDEUTEN : to denote ; mean ; import ; imply ; signify 151
BEDEUTEND : considerable ; important 152
BEDEUTUNG (*f*) : importance ; consequence ; signification ; implication 153
BEDIELEN : to cover, board up 154
BEDIENUNG (*f*) : service ; attendance ; attention; attendants ; personnel ; staff 155
BEDINGEN : to stipulate ; limit ; require ; restrict ; be necessary ; need ; necessitate ; want ; occasion 156
BEDINGT : conditional, hypothetical, qualified, dependent on, affected by 157
BEDINGUNG (*f*) : condition, stipulation ; (*pl*) : terms, etc. 158
BEDÜNKEN (*n*): judgement, opinion, discretion 159
BEDÜRFEN : to require, need, want 160
BEDÜRFNIS (*n*) : want ; requirement ; necessity ; need ; demand 161
BEEHREN : to honour ; favour ; have the honour 162
BEEIDIGEN : to declare on oath ; swear 163
BEEIFERN : to strive ; endeavour 164
BEEINFLUSSEN : to influence ; affect 165
BEEINTRÄCHTIGEN : to injure ; encroach upon ; wrong ; prejudice 166
BEEINTRÄCHTIGEND : detrimental 167
BEEINTRÄCHTIGT : affected 168
BEERE (*f*) : berry (*Fructus* or *Baca*) 169
BEERENWANZE (*f*) : currrant (berry) eating lice (*Pentatoma baccarum*) 170
BEERESCHE (*f*): service tree (*Sorbus domestica*) 171
BEERGELB (*n*) : buckthorn yellow 172
BEERSCHWAMM (*m*) : framboesia ; yaws (a negro disease) 173
BEET (*n*) : (garden)- bed 174
BEFÄHIGEN : to enable ; qualify 175
BEFANGEN : embarrassed ; prejudiced ; confused 176
BEFASSEN : to handle ; engage 177
BEFEHL (*m*) : command ; order ; precept 178
BEFEHLEND : imperative 179
BEFESTIGEN : to fasten ; fix ; attach ; strengthen 180
BEFESTIGEN, SICH- : to (get) become firmer (market quotations) 181
BEFESTIGUNG (*f*) : fastening, attachment ; fortification 182
BEFESTIGUNGSKNAGGEN (*f.pl*) : fixing lugs ; lugs for fixing or attachment 183
BEFESTIGUNGSMITTEL (*n*) : fixing agent ; fixing medium 184
BEFEUCHTEN : to moisten ; damp ; wet ; humidify 185
BEFEUCHTIGUNGSAPPARAT (*m*) : damping apparatus 186
BEFINDEN (*n*) : condition ; state ; opinion 187
BEFINDLICH : existing ; situate ; situated 188
BEFLEISSEN : (see BEFLEISSIGEN) 189
BEFLEISSIGEN : to apply oneself 190
BEFLISSEN : studious ; intent 191
BEFÖRDERN : to forward ; promote ; aid ; carry ; transport ; supply ; pass (steam, etc.) ; help 192
BEFÖRDERUNGSMITTEL (*n*) : promoter ; adjuvant (Med.) 193
BEFÖRDERUNGSWEITE (*f*) : lead ; distance to be transported 194
BEFRACHTEN : to load ; freight ; charter (ships) 195
BEFREIEN : to liberate ; release ; relieve ; clear ; free 196
BEFREMDEN : to surprise ; astonish 197
BEFREMDLICH : surprising ; strange 198
BEFREUNDEN : to befriend, be friendly (inclined) disposed towards, regard in a friendly spirit 199
BEFRIEDIGEN : to satisfy ; please 200
BEFRIEDIGUNG (*f*) : satisfaction, settlement 201
BEFRUCHTEN : to fertilize ; fecundate ; fructify 202
BEFUND (*m*) : condition ; finding ; state ; result 203
BEFÜRWORTEN : to speak in favour of ; support ; stand good for 204
BEGABEN : to endow ; bestow-(upon) 205
BEGEBEN : to go ; happen ; give up ; renounce ; desist from ; take place ; repair to 206
BEGEHEN : to commit ; perpetrate ; do 207
BEGEHR (*n*) : desire ; request ; demand ; coveting ; greed 208
BEGICHTEN : to charge (a blast furnace) ; (*n*) : charging 209
BEGIERDE (*f*) : desire ; lust ; eagerness 210
BEGIERIG : desirous, eager 211
BEGIESSEN : to moisten ; irrigate ; water ; sprinkle 212
BEGIPSEN : to plaster 213
BEGLAUBIGEN : to certify ; testify ; attest ; verify ; confirm 214
BEGLEICHUNG (*f*) : settlement, payment, balancing 215
BEGLEITERSCHEINUNG (*f*) : attendant phenomenon 216
BEGLEITFARBE (*f*) : complementary colour 217
BEGLEITKÖRPER (*m*) : accompanying (substance) body 218
BEGLEITSCHREIBEN (*n*) : covering letter 219
BEGLÜCKEN : to favour ; bless 220
BEGNÜGEN : to content ; satisfy 221
BEGREIFEN : to understand ; comprehend ; imply ; comprise 222
BEGREIFLICH : conceivable, comprehensible, intelligible 223
BEGRENZEN : to limit ; terminate ; define ; border 224
BEGRIFF (*m*) : conception ; idea ; notion 225
BEGRÜNDEN : to establish ; base ; prove; found ; float (a company) 226
BEGÜNSTIGEN : to favour ; promote 227
BEGUSS (*m*) : slip (Ceramics) ; engobe 228
BEGUSSMASSE (*f*) : (see BEGUSS) 229
BEGUTACHTEN : to pass judgement on ; arbitrate ; approve ; signify approval 230
BEGUTACHTER (*m*) : arbitrator 231
BEGUTACHTUNG (*f*) : approval, arbitration 232
BEGÜTERT : wealthy ; rich ; endowed with (possessions) land 233
BEGÜTIGEN : to appease 234
BEHAFTET : affected ; afflicted ; attended by 235
BEHAGEN : to please ; (*n*) : pleasure, comfort 236
BEHAGLICH : agreeable ; comfortable ; easy ; pleasing 237
BEHAGLICHKEIT (*f*) : comfort, ease 238
BEHALTEN : to keep ; carry (Maths.) ; retain 239
BEHÄLTER (*m*) : container ; receiver ; receptacle ; tank ; bin ; reservoir, etc. 240

BEHÄLTER (*m*), **WASSER-** : water tank 241
BEHÄLTNIS (*n*): case; container; box; bin; store; magazine; cage 242
BEHANDELN : to treat; handle; manipulate; manage; work; deal with 243
BEHANDLUNG (*f*): handling; manipulation; treatment (metal) 244
BEHANDLUNGSMITTEL (*n*): agent; reagent 245
BEHANDLUNGSVERFAHREN (*n*): (process, method of) treatment 246
BEHARREN : to continue; persevere; persist; stick to 247
BEHARRUNGSKRAFT (*f*): inertia 248
BEHARRUNGSVERMÖGEN (*n*): inertia (*vis inertiæ*) 249
BEHARRUNGSWIRKUNG (*f*): inertia-(effect) 250
BEHARRUNGSZUSTAND (*m*): permanence; resistance; state; state of inertia or equilibrium 251
BEHAUPTUNG (*f*): assertion; proposition; declaration 252
BEHEBEN : to claim; put forward; put an end to; show (set) off (see **HEBEN**) 253
BEHEIZT : heated 254
BEHELF (*m*): shift; device; help; expedient; recourse; resource; contrivance 255
BEHELLIGEN : to molest; annoy; importune 256
BEHEND : nimble; clever; agile; quick 257
BEHENNUSS (*f*): ben nut (seed of *Moringa oleifera*) 258
BEHENÖL (*n*): oil of ben; Sg. 0.921-0.9244; solidifies -5°C.; Mp. 26-32°C. 259
BEHENSÄURE (*f*): behenic acid; $CH_3(CH_2)_{20}.CO_2H$; Mp. 84°C. 260
BEHERBERGEN : to harbour; shelter; give shelter to; contain 261
BEHERRSCHBAR : controllable, governed by, commensurate with 262
BEHERRSCHUNG (*f*): control 263
BEHERZIGEN : to take heart; consider; mind 264
BEHILFLICH : helpful; useful 265
BEHÖRDE (*f*): authority; magistrate; official; court; competence; jurisdiction 266
BEHÖRDLICH : official (state or municipal) 267
BEHUF (*m*): purpose; object; behalf 268
BEHUFE (*m*), **ZU DIESEM-** : for this purpose 269
BEHUFS : on behalf of; in order to 270
BEHÜTEN : to guard; be careful; be cautious; be circumspect 271
BEIBEHALTEN : to retain; continue; keep 272
BEIBLATT (*n*): supplement 273
BEIBOOT (*n*): lifeboat (on board ship): launch 274
BEIBOOT (*n*), **DAMPF-** : steam (launch) lifeboat 275
BEIBRINGEN : to administer; produce; impart; teach; apply; suggest; bring home to 276
BEIBRINGUNGSMITTEL (*n*): vehicle (for drugs) 277
BEIDERSEITIG : joint, mutual, jointly; conjointly, mutually, reciprocal-(ly) 278
BEIDLEBIG : amphibious 279
BEIDREHEN : to bring to (ships), to heave to (ships) 280
BEIFÄLLIG : approving; favourable; assenting 281
BEIFOLGEND : enclosed 282
BEIFÜGEN : to annex 283
BEIFUSS (*m*): wormwood; artemisia (*Artemisia absinthium*); (*Herba artemisiae*); Mugwort (*Artemisia vulgaris*); truss (of a sail) 284
BEIFUSS (*m*), **BITTERER-** : (see **WERMUT**) 285
BEIFUSSKRAUT (*n*): mugwort leaves (from the plant *Artemisia vulgaris*, which is somewhat similar to wormwood); wormwood leaves (from *Artemisia absinthium*) 286
BEIFUSSWURZEL (*f*): wormwood root (*radix Artemisiae*) 287
BEIGEBEN : to give with; add (join) to 288
BEIGESCHMACK (*m*): after-taste; tang; flavour 289
BEIHILFE (*f*): help; assistance 290
BEIKOMMEN : to reach, attain, repair (a loss), nullify (an effect), do away with (a cause) 291
BEIL (*n*): hatchet 292
BEILAGE (*f*): addition; supplement; appendix; enclosure; annex 293
BEILAUFIG : incidental; by the way; besides 294
BEILÄUFIGE TEILE (*m. pl*): accessories, spare parts 295
BEILEGEN : to enclose; settle; add; attribute; attach; ascribe; subjoin; impute 296
BEILEN : to trim timber 297
BEIL (*n*), **GROSSES-** : axe 298
BEILIEGEND : enclosed herewith 299
BEIMENGEN : to admix; add to; mix with 300
BEIMENGUNG (*f*): admixture; impurity 301
BEIMESSEN : to attribute; assume; impute 302
BEIMISCHUNG (*f*): admixture; infusion 303
BEINASCHE (*f*): bone ash (see **KNOCHENASCHE**) 304
BEINHAUT (*f*): periosteum (Med.) 305
BEINHÖHLE (*f*): cotyle, socket (Anatomy) 306
BEINMEHL (*n*): bone meal 307
BEINSCHWARZ (*n*): bone black; animal charcoal; abaiser (see **ELFENBEINSCHWARZ**) 308
BEINWURZ (*f*): comfrey (*Symphytum officinale*) 309
BEIORDNEN : to adjoin; coordinate 310
BEIPFLICHTEN : to assent; agree; approve 311
BEISATZ (*m*): addition; admixture 312
BEISCHIFF (*n*): tender (ship) 313
BEISCHLIESSEN : to annex, enclose 314
BEISCHLUSS (*m*): enclosure 315
BEISEGEL (*n*): light sail 316
BEISETZEN : to set (hoist) sails; put by 317
BEISPIELLOS : unexampled; unparalleled 318
BEISPIELSWEISE : for example; as a case in point; incidentally 319
BEISPRINGEN : to assist; succour 320
BEISSBEERE (*f*): cayenne pepper; capsicum 321
BEISSEND : biting; sharp; pungent; acrid; smarting; burning; stinging 322
BEISSPROBE (*f*): biting test 323
BEISTAND (*m*): assistance; assistant; support; supporter; help; helper; succour 324
BEISTEHEN LASSEN : to spread (let out) sails 325
BEISTEUERN : to contribute 326
BEISTIMMEN : to agree; concur; assent 327
BEITEL (*m*): bore bit 328
BEITRAGEN : to contribute; assist; aid; help; add-(to) 329
BEITREIBEN : to collect; drive in; exact 330
BEITRETEN : to join; accede; take part 331
BEIWERK (*n*): accessory 332
BEIWERT (*m*): coefficient 333
BEIWOHNEN : to be present; cohabit; assist at 334
BEIWORT (*n*): adjective; epithet 335

BEIZÄHLEN : to number with, enumerate 336
BEIZARTIKEL (*m*) : mordant style (Calico) 337
BEIZBAD (*n*) : mordant bath (Calico) 338
BEIZBRÜCHIG : short, brittle (to corrosion) (of iron) 339
BEIZBÜTTE (*f*) : drench pit (Leather) 340
BEIZE (*f*) : corrosive ; corrosion ; mordant (Dye); caustic ; cauterization (Med.) ; drench ; bate (Leather) ; pickle ; etching (Metal) ; staining (wood) ; blanching (Metal) ; saucing (tobacco) 341
BEIZEFARBE (*f*) : mordant (dye) colour 342
BEIZEN : to corrode ; mordant (Dye) ; cauterize (Med.) ; bate, drench (leather) ; stain (wood) ; pickle, etch (Metal) ; blanch (Metal) ; sauce (tobacco) 343
BEIZENDRUCK (*m*) : mordant printing 344
BEIZENFARBSTOFF (*m*) : mordant dye-(stuff) 345
BEIZENGELB (*n*) : mordant yellow 346
BEIZEREI (*f*) : pickling plant 347
BEIZFLÜSSIGKEIT (*f*) : corrosive liquid ; caustic liquor ; mordant liquor ; bate ; drench ; pickle ; stain 348
BEIZKRAFT (*f*) : corrosive (caustic) power or strength 349
BEIZLÖSUNG (*f*) : mordant (staining) solution 350
BEIZMITTEL (*n*) : corrosive ; mordant ; caustic 351
BEJUTEN : to jute, cover with jute 352
BEKALKEN : to lime 353
BEKÄMPFUNGSMITTEL (*n*) : any means of combating pests 354
BEKANNT : known ; usual ; general ; common 355
BEKANNTGEBEN : to make known ; advertise 356
BEKANNTLICH : notoriously ; as is known ; noted; famous ; general 357
BEKLECKEN : to blur ; blotch ; blot ; daub ; stain 358
BEKLECKERN : (see **BEKLECKEN**) 359
BEKLECKSEN : (see **BEKLECKEN**) 360
BEKLEIDEN : to cover ; coat ; line ; face ; box ; clothe ; invest ; occupy ; jacket ; case ; dress 361
BEKLEIDUNG (*f*) : plating, covering, casing, lining, jacket, dress ; bulwark (of ships) 362
BEKLEIDUNG (*f*), **FEUERFESTE-** : fireproof (clothes, covering) lining 363
BEKLEIDUNGSPLANKE (*f*) : garboard ; bulwark (Naut.) 364
BEKLOMMEN : oppressed ; anxious ; uneasy 365
BEKOHLUNGSANLAGE (*f*) : coal conveying plant, coal handling plant ; furnace charging plant ; coaling plant 366
BEKÖSTIGEN : to board ; feed 367
BEKRÄFTIGEN : to confirm ; assert ; strengthen ; emphasize 368
BEKRÄFTIGUNG (*f*) : confirmation ; emphasizing, emphasis 369
BEKRÄNZEN : to wreathe ; crown ; festoon ; encircle 370
BEKRITTELN : to censure ; criticize 371
BEKRUSTEN : to incrust ; crust ; become coated with scale 372
BELAG (*m*) : platform. etc. (see also **BELEG**) 373
BELANG (*m*) : importance ; amount ; consequence 374
BELANGEN : to concern ; bring action against ; accuse 375
BELANGLOS : of no importance ; not affecting the issue ; negligible 376
BELASSEN : to leave 377
BELASTEN : to load ; weight ; burden ; charge ; debit ; strain ; stress ; put tension on 378
BELASTEN, STOSSWEISE- : to shock-load ; load under impact 379
BELÄSTIGEN : to trouble ; molest ; annoy 380
BELASTUNG (*f*) : load ; charge ; debit ; taint (Med.) ; output ; rating ; evaporation (boilers) ; consumption (furnaces) 381
BELASTUNG DURCH EIGENGEWICHT : dead load 382
BELASTUNG (*f*), **RUHENDE-** : dead load 383
BELASTUNGSAUFGABE (*f*) : debit note 384
BELASTUNGSFAKTOR (*m*) : load factor (power plant) 385
BELASTUNGSSCHWANKUNG (*f*) : load fluctuation (plant) 386
BELASTUNG (*f*), **STOSSARTIGE-** : loading under impact, shock-loading 387
BELASTUNGSVENTIL (*n*) : loaded valve 388
BELASTUNGSWECHSEL (*m*) : load fluctuation ; load variation ; interchange of debit notes 389
BELASTUNG (*f*), **ZULÄSSIGE-** : safe load 390
BELAUF (*m*) : amount 391
BELAUFEN : to amount 392
BELAUFEN, SICH- : to amount (come) to 393
BELEBEN : to revive ; quicken ; resuscitate ; brighten ; animate ; freshen (colours) 394
BELEDERN : to cover with leather ; leather 395
BELEG (*m*) : covering ; coating ; lining ; floor ; proof ; document ; voucher ; overlay 396
BELEGE (*m.pl*) : (documentary) evidence 397
BELEGNADEL (*f*) : belaying pin (Naut.) 398
BELEGSCHAFT (*f*) : personnel ; staff ; workers 399
BELEGSTÜCK (*n*) : voucher 400
BELEIDIGUNG (*f*) : offence, injury, insult, wrong, slander, defamation, affront 401
BELEIMEN : to glue ; cover with glue ; gum ; paste 402
BELESEN : well-read ; well versed in literature ; literary 403
BELEUCHTEN : to light ; illumine ; illuminate ; examine ; elucidate ; shed light on 404
BELEUCHTUNG (*f*) : illumination, lighting 405
BELEUCHTUNG (*f*), **KÜNSTLICHE-** : artificial lighting 406
BELEUCHTUNGSARMATUREN (*f.pl*) : lighting fittings (such as electric light or gas fittings) 407
BELEUCHTUNGSARTIKEL (*m.pl*), **BENGALISCHE-** : Bengal lights (fireworks) 408
BELEUCHTUNGSKOHLE (*f*) : (electric light) carbon 409
BELEUCHTUNGSKOHLE (*f*), **ELEKTRISCHE-** : electric light carbon 410
BELEUCHTUNGSLINSE (*f*) : illuminating lens 411
BELEUCHLUNGSMITTEL (*n*) : illuminant 412
BELEUCHTUNGSSCHIRM (*m*) : reflector 413
BELICHTEN : to expose-(to light) 414
BELICHTUNG (*f*) : lighting ; exposure (photos.) 415
BELICHTUNGSDIFFERENZ (*f*) : difference in lighting ; exposure difference (Photography) 416
BELICHTUNGSKARTE (*f*) : exposure card 417
BELICHTUNGSPRODUKT (*n*) : lighting product, product from exposure to light 418
BELICHTUNGSSTÄRKE (*f*) : strength (intensity) of lighting 419
BELICHTUNGSZEIT (*f*) : time (duration) of exposure (Photography) 420
BELIEBIG : optional ; any ; whatever ; given ; desired **421**

BELLADONNA (*f*): belladonna (*Atropa Belladonna*), deadly nightshade, Banewort, Divale, Death's herb 422

BELLADONNABLÄTTER (*n.pl*): (*Folia Belladonnæ*): belladonna leaves (see **BELLADONNA**) 423

BELLADONNAEXTRAKT (*m*): belladonna extract 424

BELLADONNAÖL (*n*): belladonna oil (from seeds of *Atropa belladonna*) 425

BELLADONNAWURZEL (*f*): (*Radix Belladonnæ*): belladonna root (see **BELLADONNA**) 426

BELLADONNIN (*n*): belladonnin (an alkaloid from *Atropa Belladonna*); $C_{17}H_{21}NO_2$ 427

BEMÄCHTIGEN: to master 428

BEMALEN: to paint; colour; decorate 429

BEMÄNTELN: to cover; colour; varnish; palliate; mantle 430

BEMASTEN: to furnish with masts 431

BEMERKEN: to note; remark; observe; perceive 432

BEMERKENSWERT: worthy of note; remarkable; noteworthy 433

BEMESSEN: to measure; proportion; apportion; regulate; dimension 434

BEMESSEN: dimensioned, exact, symmetrical 435

BEMITTELT: well-to-do; wealthy 436

BEMODERT: mouldy; musty 437

BEMOOST: mossy 438

BEMÜHEN: to trouble; endeavour 439

BEMUSTERN: to sample; match; dye to pattern; imitate; ornament; figure 440

BENACHBART: neighbouring; adjacent; adjoining; vicinal 441

BENACHRICHTIGEN: to inform; advise; keep posted 442

BENACHTEILIGEN: to prejudice; hurt; injure; affect adversely 443

BENEBELT: tipsy; befogged; fuddled 444

BENEHMEN: to take away; remove; conduct; behave; (*n*): conduct; behaviour 445

BENEIDEN: to envy; be envious of; be jealous of; grudge 446

BENENNEN: to name; call 447

BENENNUNG (*f*): nomenclature; title; name; appellation; denomination 448

BENETZEN: to wet; moisten; sprinkle; humidify 449

BENETZUNGSWÄRME (*f*): moist heat, heat of humidification 450

BENGALBLAU (*n*): Bengal blue 451

BENGALISCHE BELEUCHTUNGSARTIKEL (*m.pl*): Bengal lights (fireworks) 452

BENGEL (*m*): club; clapper; boor; chap 453

BENÖL (*n*): (see **BEHENÖL**) 454

BENÖTIGEN: to require, be in want of, desire, need, lack 455

BENSÄURE (*f*): (see **BEHENSÄURE**) 456

BENUTZEN: to use; employ 457

BENZALCHLORID (*n*): benzylidene chloride, benzylene chloride, benzal chloride, chlorobenzal, benzyl dichloride; $C_7H_6Cl_2$; Sg. 1.2699; Mp. -16.1°C.; Bp. 203.5°C. 458

BENZALDEHYD (*n*): benzoic aldehyde, benzoyl hydride; benzaldehyde artificial essential oil of almonds; C_7H_6O; Sg. 1.0504; Mp. -26°C.; Bp. 179.1°C. 459

BENZALGRÜN (*n*): benzal green 460

BENZAMID (*n*): benzamide; C_7H_7ON; Sg. 1.341; Mp. 128°C. 461

BENZAMINBLAU (*n*): benzamine blue 462

BENZAMINSÄURE (*f*): benzaminic acid, meta-aminobenzoic acid ($C_6H_4NH_2CO_2H$); Sg. 1.51; Mp. 173-174°C. 463

BENZANILID (*n*): benzoylanilide, phenylbenzamine, benzanilide; $C_{13}H_{11}ON$; Sg. 1.306-1.321; Mp. 161°C. 464

BENZHYDROL (*n*): diphenyl carbinol; $(C_6H_5)_2.CHOH$ Mp. 69°C.; Bp. 298°C. 465

BENZIDIN (*n*): benzidine, para-diaminodiphenol; $C_{12}H_{12}N_2$; Mp. 128°C.; Bp. 401°C. 466

BENZIDIN (*n*), **SCHWEFELSAURES-**: benzidine sulphate, para-diaminodiphenyl sulphate; $C_{12}H_{10}(NH_2)_2SO_4H_2$ 467

BENZIDINSULFOSÄURE (*f*): benzidine sulphonic acid 468

BENZIL (*n*): benzil, dibenzoyl; $C_6H_5CO.COC_6H_5$; Mp. 95°C; Bp. 346°C. 469

BENZIN (*n*): benzine; petroleum ether (*Aether petrolei*): canadol; Sg. 0.635-0.66; Bp. 40-70°C. 470

BENZINANLAGE (*f*): benzin plant 471

BENZINERSATZ (*m*): benzine substitute 472

BENZINGASAPPARAT (*m*): benzine gas apparatus 473

BENZINOFORM (*n*): benzinoform (trade name for Carbon tetrachloride: see **TETRACHLOR-KOHENSTOFF**) 474

BENZINSEIFE (*f*): benzine soap (for cleaning articles of attire) 475

BENZOAT (*n*): benzoate 476

BENZOCHINON (*n*): benzoquinone; $CO(CH.CH)_2CO$; Sg. 1.307; Mp. 115.7°C. 477

BENZOE (*f* and *n*): benzoin-(gum) (*Resina benzoë* from *Styrax benzoin*) 478

BENZOEÄTHER (*m*): benzoic ether; ethyl benzoate; $C_6H_5CO_2C_2H_5$; Sg. 1.0509; Mp. -32.7°C; Bp. 212.9°C. 479

BENZOEBAUM (*m*): benzoin tree (*Styrax benzoin*) 480

BENZOEBLUMEN (*f.pl*): benzoin flowers (flowers of *Styrax benzoin*) 481

BENZOECHTFARBE (*f*): fast benzo (dye) colour 482

BENZOE (*n*), **GEREINIGTES-**: purified benzoin 483

BENZOEHARZ (*n*): benzoin-(resin) (from *Styrax benzoin*) 484

BENZOELORBEER (*m*): spice bush (*Benzoin odoriferum*) 485

BENZOESALZ (*n*): benzoate 486

BENZOESAUER: benzoate of; combined with benzoic acid 487

BENZOESÄURE (*f*): (*Acidum benzoicum*): benzoic acid; phenylformic acid; ($C_6H_5CO_2H$); Sg. 1.266; Mp. 121.25°C.; Bp. 249.2°C. 488

BENZOESÄURE-ANHYDRID (*n*): benzoic anhydride; $C_{14}H_{10}O_3$; Sg. 0.9555; Mp. 42°C.; Bp. 360°C 489

BENZOESÄURE-ÄTHYLESTER (*m*): ethyl benzoate, ethyl benzoic ester; $C_9H_{10}O_2$; Sg. 1.0614; Mp. -34.2°C; Bp. 212.9°C.; benzoic ether 490

BENZOESÄUREBENZYLESTER (*m*): benzyl benzoate; $C_6H_5.CO_2.CH_2.C_6H_5$; (the active principle of Peru balsam) 491

BENZOESÄURECHLORID (*n*): (see **BENZOYLCHLORID**) 492

BENZOESÄUREESTER (*m*): benzoate 493

BENZOESÄUREMETHYLESTER (*m*): methyl benzoate (see **METHYLBENZOAT**) 494

BENZOESÄURESEIFE (*f*): benzoic acid soap 495

BENZOESÄURESULFINID (*n*): bensosulfinide (see **SACCHARIN**) 496

BENZOESCHMALZ (*n*): benzoated (benzoinated) lard 497

BENZOIN (*n*): (gum)-benzoin (see BENZOE); benzoin, phenylbenzoylcarbinol, oxyphenylbenzylketone; $C_6H_5.CH(OH).CO.C_6H_5$; Mp. 135°C. 498

BENZOINIERT: benzoinated; benzoated 499

BENZOL (*n*): benzene, benzol, phenyl hydride, coal naphtha; C_6H_6; Sg. 0.8736; Mp. 5.58°C.; Bp. 80.2°C. 500

BENZOLBINDUNG (*f*): benzene linkage 501

BENZOLDIKARBONSÄURE (*f*), ORTHO-: (see PHTALSÄURE) 502

BENZOLKERN (*m*): benzene nucleus 503

BENZOLKOHLENWASSERSTOFF (*m*): benzene hydrocarbon 504

BENZOLLACK (*m*): benzene varnish 505

BENZOLÖL (*n*): benzol oil (a motor fuel oil, being a mixture of benzene and coal tar oil); Sg. 0 84-0.895 506

BENZOLSULFOCHLORID (*n*): benzene sulfonic chloride; $C_6H_5O_2SCl$; Sg. 1.383; Mp. 14.5°C., Bp. 116.3°C. 507

BENZOLSULFOSAUER: benzene sulphonate, combined with benzenesulphonic acid 508

BENZOLSULFOSÄURE (*f*): benzenesulfonic acid (*Acidum benzolsulfonicum*); phenyl-sulfonic acid; $C_6H_5SO_3H$; Mp. 65–66°C.; Bp. 137°C. 509

BENZOLTINKTUR (*f*): benzol (benzene) tincture 510

BENZOLTRICARBONSÄURE (*f*): Hemimellitic acid (see HEMIMELLITHSÄURE) 511

BENZONAPHTOL (*n*): (*Naphtolum benzoicum*), : benzonaphthol, benzoylnaphthol, Beta-naphthol benzoate; $C_{10}H_7.CO_2.C_6H_5$; Mp. 107°C. 512

BENZONITRIL (*n*): benzonitrile, phenyl cyanide; C_7H_5N; Sg. 1.0228; Mp. -13.1°C.; Bp. 191.3°C. 513

BENZOPHENOL (*n*): (see PHENOL) 514

BENZOPHENON (*n*): diphenylketone, benzophenone; $C_6H_5.CO.C_6H_5$; Sg. 1.098; Mp. 48.5°C.; Bp. 305°C. 515

BENZOSALIN (*n*): benzosaline; $C_6H_4(C_2O_2H_3)(C_7O_2H_5)$; Mp. 84.5°C. 516

BENZOSOL (*n*): benzosol (trade name) (see BENZOYLGUAJAKOL) 517

BENZOTRICHLORID (*n*): benzoic trichloride, phenylchloroform, benzotrichloride, toluene trichloride, bezenyl trichloride; $C_7H_5Cl_3$; Sg. 1.38; Mp. -21.2°C; Bp. 214°C. 518

BENZOYLACETON (*n*): benzoyl acetone; $C_{10}H_{10}O_2$; Sg. 1.0899; Mp. 61°C.; Bp. 261°C. 519

BENZOYL-BETA-NAPHTOL (*n*): beta-naphthol benzoate (see BENZONAPHTOL) 520

BENZOYLCHLORID (*n*): benzyol chloride; C_7H_5OCl; Sg. 1.2291; Mp. -1°C.; Bp. 198.3°C. 521

BENZOYLEKGONIN (*n*): benzoylekgonine; $C_{16}H_{19}NO_4$ 522

BENZOYLGUAJACOL (*n*): benzoyl guaiacol (see BENZOYLGUAJAKOL) 523

BENZOYLGUAJAKOL (*n*): benzosol, guaiacol benzoate; $C_6H_4.O.CH_3.O.CO.C_6H_5$; Sg. 60.5°C. 524

BENZOYLIEREN: to benzoylate; introduce benzoyl into 525

BENZOYLPEROXYD (*n*): benzoyl peroxide (see BENZOYLSUPEROXYD) 526

BENZOYLPSEUDOTROPEIN (*n*): tropacocaine (see TROPACOCAIN) 527

BENZOYLSUPEROXYD (*n*): benzoyl peroxide; $(C_6H_5CO.O)_2$; Mp. 103.5°C. 528

BENZOYLWASSERSTOFF (*m*): benzoyl hydride; benzaldehyde; C_6H_5CHO; Sg. 1.0504; Mp. -13.5°C.; Bp. 179.9°C. 529

BENZYLALKOHOL (*m*): (*Alcohol benzylicus*): benzyl alcohol; C_7H_8O; Sg. 1.0459; Bp. 204.7°C. 530

BENZYL (*n*), AMEISENSAURES-: benzyl formate 531

BENZYLAMIN (*n*): benzyl-amine, benzylphenylamine; C_7H_9N; Sg. 0.9826; Bp. 182°C. [532

BENZYLANILIN (*n*): benzyl aniline; $C_6H_5NHCH_2C_6H_5$; Mp. 33°C.; Bp. 310°C. 533

BENZYLÄTHER (*m*): benzyl ether; $C_{14}H_{14}O$; Sg. 1.0359; Bp. 298°C. 534

BENZYLÄTHYLANILIN (*n*): benzyl ethyl aniline 535

BENZYLAZETAT (*n*): benzyl acetate; $C_9H_9O_2$; Sg. 1.061; Mp. 215.5°C. 536

BENZYL (*n*), BALDRIANSAURES-: benzyl valerianate 537

BENZYLBENZOAT (*n*): benzyl benzoate 538

BENZYL (*n*), BENZOESAURES-: benzyl benzoate 539

BENZYLBENZOL (*n*): benzylbenzene (see DIPHENYL-METHAN) 540

BENZYLBROMID (*n*): benzyl bromide; $C_6H_5CH_2Br$ 541

BENZYL (*n*), BUTTERSAURES-: benzyl butyrate 542

BENZYLCHLORID (*n*): benzyl chloride; C_7H_7Cl; Sg. 1.1135; Mp. -43.2°C; Bp. 179°C. 543

BENZYLCYANID (*n*): benzyl cyanide, phenylacetic acid nitrile; $C_6H_5CH_2CN$; Sg. 1.016; Mp. -24.6°C.; Bp. 233.5°C. 544

BENZYL (*n*), ESSIGSAURES-: benzyl acetate (see BENZYLAZETAT) 545

BENZYLKARBINOL (*n*): benzyl carbinol (see PHENYLÄTHYLALKOHOL) 546

BENZYLMERCAPTAN (*n*): benzyl mercaptan; C_7H_8S; Sg. 1.058; Bp. 195°C. 547

BENZYLMETHYLANILIN (*n*): benzylmethylaniline 548

BENZYL (*n*), PROPIONSAURES-: benzyl propionate 549

BENZYL (*n*), SALICYLSAURES-: benzyl salicylate 550

BENZYLSULFID (*n*): benzyl sulphide; $(CH_2C_6H_5)_2S$; Sg. 1.07; Mp. 49°C. 551

BEOBACHTUNG (*f*): observation; the taking of a reading (thermometer, etc.); consideration; investigation 552

BEOBACHTUNGSROHR (*n*): observation tube or pipe 553

BEOBACHTUNGSTÜR (*f*): inspection door; spy door 554

BEÖLEN: to oil 555

BEPACKEN: to load; charge; pack 556

BEPLANKUNG (*f*): planking 557

BEPLATTUNG (*f*): plating 558

BEPUTZEN: to dress 559

BEQUEM: convenient; comfortable; lazy; conformable; compliable 560

BERAPP (*m*): rough plaster; brown coat; rough cast 561

BERATEN: to advise; consult 562

BERATENDER INGENIEUR (*m*): consulting engineer 563

BERATSCHLAGEN: to consult, deliberate 564

BERATUNG (*f*): consultation; consulting; consultant 565

BERÄUCHERN: to fumigate 566

BERÄUCHERUNG (*f*): fumigation (Med.) 567

BERAUSCHEND: intoxicating 568

BERAUSCHUNGSMITTEL (*n*): intoxicant 569
BERBERIN (*n*): berberine (from root of *Berberis vulgaris*); $C_{20}H_{17}NO_4$; Mp. 145°C. 570
BERBERIN (*n*), **SALZSAURES-**: berberine hydrochloride 571
BERBERITZE (*f*): barberry; berberis (*Berberis aristata*) 572
BERBERITZENBEEREN (*f.pl*): barberries (*Fructus Berberidis*) (from *Berberis vulgaris*) 573
BERBERITZENEXTRAKT (*m*): barberry extract 574
BERBERITZENSAFT (*m*): barberry (berberis) juice 575
BERBERITZENSTRAUCH (*m*): barberry, berberis (*Berberis vulgaris*) 576
BERBERITZENWURZEL (*f*): barberry root (the root of *Berberis vulgaris*) 577
BERECHENBAR: calculable; appreciable 578
BERECHNUNG (*f*): calculation; working (running) out; evaluating; computation (of size; number; duty, etc.) 579
BERECHTIGEN: to entitle; authorize; qualify; justify; warrant; be (justified) justifiable 580
BERECHTIGUNG (*f*): authorization; right; title; qualification; privilege; justification 581
BEREDEN: to talk over; confer; persuade 582
BEREDSAMKEIT (*f*): eloquence 583
BEREGNUNGSANLAGE (*f*): sprinkling (sprinkler) plant (for agricultural purposes) 584
BEREICH (*m*): reach range; domain; district; field; world; scope; compass; province; region 585
BEREICHERN: to enrich 586
BEREICH (*m*), **IN DEM-**: in the region of, within the scope of 587
BEREIFUNG (*f*): tyres, tyre equipment, fitting of tyres 588
BEREITEN: to prepare; make; make ready; manufacture 589
BEREITSTELLUNG (*f*): preparation; provision; production; manufacture 590
BERG (*m*): mountain; heap; rubbish; waste (Mining); rock; berg (ice); breasting (of a rag engine) (paper) 591
BERGAB: downhill 592
BERGADER (*f*): lode; vein; stratum 593
BERGAHORN (*m*): (see **AHORN, GEMEINER-**) 594
BERGALAUN (*m*): rock alum; alumina; Al_2O_3; Sg. 3.73; Mp. 2020° C. 595
BERGAMOTTÖL (*n*): Bergamot oil (*Oleum Bergamottae*) (obtained from the fruit of *Citrus Bergamia Risso*); Sg. 0.884; Bp. 165-190° C. 596
BERGAN: uphill 597
BERGARBEIT (*f*): mining 598
BERGART (*f*): gangue; matrix; metalliferous stratum; mineral ore 599
BERGASCHE (*f*): an inferior kind of mineral blue (see **BERGBLAU**) 600
BERGBALSAM (*m*): petroleum 601
BERGBAU (*m*): mining 602
BERGBLAU (*n*): mineral blue; blue ash; blue basic carbonate of copper; $CuCO_3.Cu(OH)$ 603
BERGBRAUN (*n*): umber 604
BERGBUTTER (*f*): mountain butter; rock butter; impure iron alum; iron or zinc sulphate (see **BERGSEIFE**) 605
BERGE (*m. pl*): waste (Mining) (70-80% ash) 606
BERGE, LESE-: waste pickings 607
BERGEN: to save; shelter; conceal; salvage (shipping); recover; shade (paint); hide; haul in (take in) sails 608
BERGERZ (*n*): crude (raw) ore 609
BERGETURM (*m*): waste (tower) bunker (for storing waste coal) 610
BERGE, WASCH-: waste washings 611
BERGEWASSER (*n*): water for washing waste coal 612
BERGFAHRT (*f*): hill-climb, mountain journey 613
BERGFALL (*m*): land-slide 614
BERGFEIN: native (Min.) 615
BERGFETT (*n*): mountain tallow; ozocerite; Sg. 0.85-0.95; Mp. 55-110° C., average 70° C. 616
BERGFLACHS (*m*): mountain flax; amianthus; asbestos 617
BERGFLEISCH (*n*): mountain flesh (a variety of asbestos) 618
BERGFLUSS (*m*): coloured (opaque) quartz (see **QUARZ**) 619
BERGGANG (*m*): vein; lode (of ore); stratum; gangue 620
BERGGELB (*n*): mountain yellow; yellow ochre 621
BERGGLAS (*n*): rock crystal; transparent quartz; SiO_2; (see **BERGKRISTALL**) 622
BERGGLIMMER (*m*): margarite (a variety of mica) (see **GLIMMER**) 623
BERGGOLD (*n*): rock gold; gold found in rocks 624
BERGGRAT (*m*): ridge, crest (of mountains) 625
BERGGRÜN (*n*): mineral green; malachite green (see **MALACHITGRÜN**); green verditer; green basic copper carbonate; Sg. 3.7-4.0; $Cu_2(OH)_2CO_3$ 626
BERGGUHR (*m*): mountain milk; agaric mineral (earthy and light variety of calcite or calcium carbonate); $CaCO_3$ 627
BERGGUT (*n*): minerals; fossils 628
BERGHAAR (*n*): amianthus; fibrous asbestos 629
BERGHALDE (*f*): refuse dump (Coal) 630
BERGHARZ (*n*): mineral pitch; asphalt 631
BERGHOLZ (*n*): rock wood; woody asbestos 632
BERGIG: mountainous; hilly; undulating 633
BERGKALK (*m*): mountain limestone 634
BERGKIESEL (*m*): rock flint; chert; felsite 635
BERGKOHLE (*f*): coal 636
BERGKORK (*m*): mountain cork (a light form of asbestos) 637
BERGKRANKHEIT (*f*): mountain sickness (experienced by some people when ascending to any height) (Med.) 638
BERGKRISTALL (*m*): rock crystal (transparent and colourless) (see **QUARZ**) 639
BERGKUPFER (*n*): native copper 640
BERGLASUR (*f*): azurite; $CuCO_3.Cu(OH)_2$; (46% Cu) 641
BERGLEDER (*n*): mountain leather (a variety of asbestos) 642
BERGMANN (*m*): miner 643
BERGMEHL (*n*): mountain flour; mountain meal; kieselguhr; fossil meal; infusorial earth; infusorite; (see **KIESELGUHR**) 644
BERGMELISSE (*f*): calamint (*Calamintha*) 645
BERGMILCH (*f*): mountain milk (see **BERGGUHR**) 646
BERGMINZE (*f*): mountain mint (applied to a variety of plants as Calamint, Field basil, etc.) 647

BERGNAPHTA (*f*) : mineral naphtha ; petroleum ; rock oil ; crude oil ; (see **BERGÖL**) 648

BERGÖL (*n*): petroleum ; naphtha ; rock oil ; mineral oil ; crude oil ; Sg. 0.78-0.97 649

BERGPAPIER (*n*): mountain paper (a variety of asbestos) 650

BERGPECH (*n*): mineral pitch ; asphalt (*Asphaltum*) ; bitumen 651

BERGROT (*n*) : realgar ; indian red ; (see **REALGAR**) 652

BERGRÜSTER (*f*) : (see **BERGULME**) 653

BERGRUTSCH (*m*) : land-slip 654

BERGSALZ (*n*): rock salt ; NaCl ; Sg. 2.161 ; Mp. 804° C. ; Bp. 1490° C. 655

BERGSATTEL (*m*) : saddle, dip between two peaks (of mountains) 656

BERGSCHULE (*f*) : School of mines 657

BERGSCHWADEN (*m*): firedamp, chokedamp (Mining) 658

BERGSCHWEFEL (*m*) : natural sulphur 659

BERGSEIFE (*f*): mountain soap (a kind of clay) ; rock butter, rock soap (impure hydrous aluminium silicate); (see **BERGBUTTER**) 660

BERGTALG (*m*) : mountain tallow ; rock butter ; ozocerite (see **BERGWACHS**) 661

BERGTEE (*m*): mountain tea ; wintergreen (leaves of *Gaultheria procumbens*) 662

BERGTEEGEIST (*m*): spirit of wintergreen (from *Gaultheria*) 663

BERGTEEÖL (*n*): oil of wintergreen ; Sg. 1.175-1.185 ; Bp. 218-221° C. 664

BERGTEER (*m*): mineral tar ; pissasphalt ; maltha 665

BERGTORF (*m*) : mountain peat 666

BERGULME (*f*): Wych, Scotch elm, witch elm (*Ulmus montana*) 667

BERGWACHS (*n*): mineral wax ; ozocerite ; Sg. 0.85-0.95 ; Mp. 55-110°C., average 70°C. 668

BERGWERK (*n*) : mine 669

BERGWERKSKUNDE (*f*) : metallurgy 670

BERGWERKSMASCHINEN (*f pl.*) : mining machinery 671

BERGWETTER (*n*) : (see **BERGSCHWADEN**) 672

BERGWISSENSCHAFT (*f*): mining, mineralogy, metallurgy 673

BERGWOLLE (*f*): mineral wool ; asbestos ; amianthus ; (see **ASBEST** and **AMIANT**) 674

BERGZINN (*n*): mined tin ; cassiterite (see **BERGZINNERZ**) 675

BERGZINNERZ (*n*) : mine tin ; cassiterite ; (SnO_2); 79% tin 676

BERGZINNOBER (*m*): native cinnabar (HgS) ; 86% mercury ; (see **ZINNOBER**) 677

BERGZUNDER (*m*): mountain tinder ; asbestos (see **ASBEST** and **AMIANT**) 678

BERICHT (*m*) : report ; notice ; advice ; information ; account 679

BERICHTIGEN : to adjust ; correct ; settle (an account) ; rectify 680

BERICHTIGUNGSTAFELN (*f. pl.*): correction tables 681

BERICHTSWOCHE (*f*): week (under notice) in question 682

BERIESELN : to water ; irrigate ; cause to flow (trickle) over 683

BERIESELUNGSKÜHLER (*m*): Baudelot (surface irrigation) cooler (Brewing) ; water cooling (tower) plant ; spray (drip, irrigation) cooler 684

BERITTEN : mounted 685

BERLINERBLAU (*n*) : Berlin blue ; prussian blue ; ($Fe_4[Fe(CN)_6]_3$) 686

BERLINERBLAUSÄURE (*f*): hydrocyanic acid (HCN) ; Sg. (gas) 0.9483 ; Mp. (liquid anhydrous HCN,)-15°C. ; Bp. 26.1°C. 687

BERLINERROT (*n*) : Berlin red (a red lake colour) 688

BERME (*f*): berm ; bench ; step ; terrace ; counter or reinforcing dam ; ledge 689

BERNSTEIN (*m*): amber ; Sg. 1.08 ; $C_{40}H_{64}O_4$; (*Succinum*) ; (as a prefix = succinic, succinyl-, succino-) 690

BERNSTEINALAUN (*m*) : aluminous amber 691

BERNSTEINLACK (*m*) : amber varnish 692

BERNSTEINÖL (*n*) : oil of amber (*oleum succini*) ; Sg. 0.915-0.975 693

BERNSTEINSAUER : combined with succinic acid ; succinate of 694

BERNSTEINSÄURE (*f*): succinic acid, ethylenesuccinic acid, ethylenedicarboxylic acid $C_4H_6O_4$; Sg. 1.657 ; Mp. 184.5°C. ; Bp. 235°C. 695

BERNSTEINSÄURE-ANHYDRID (*n*) : succinic anhydride ; $C_4H_4O_3$; Sg.1.1036 ; Mp. 119.6°C.; Bp. 261°C. 696

BERNSTEINSÄURE-DIÄTHYLESTER (*m*): diethyl succinate ; succinic ester (ethyl) ; $C_8H_{14}O_4$; Sg. 1.0436 ; Mp. -20.8°C. ; Bp. 216.5°C. 697

BERNSTEINSÄURE-NITRIL (*n*): succinonitrile ; succinic nitrile ; $C_4H_4N_2$; Sg. 0.9848 ; Mp. 54.5°C. ; Bp. 158-160°C. ; ethylene dicyanide ; ethylene cyanide 698

BERNSTEIN (*m*), **SCHWARZER-** : jet 699

BERSTEN : to burst ; explode ; crack ; fracture ; rupture 700

BERTRAM (*m*): pellitory, Spanish chamomile, longwort (*Anacyclus pyrethrum*) 701

BERTRAMSWURZ (*f*): pellitory ; pyrethrum (root of *Anacyclus pyrethrum*) 702

BERTRAMSWURZEL (*f*) : (see **BERTRAMSWURZ**) 703

BERTRAMWURZ (*f*) : (see **BERTRAMSWURZ**) 704

BERTRAMWURZEL (*f*): (see **BERTRAMSWURZ**) 705

BERÜCHTIGT : notorious ; ill-famed 706

BERÜCKEN : to trick, charm, ensnare, trap 707

BERÜCKSICHTIGEN : to recognize (claims), take count of, allow for, make allowance (provision) for, consider, bear in mind, regard, take into account, have regard to 708

BERÜCKSICHTIGUNG (*f*) : consideration 709

BERUF (*m*): calling ; occupation ; profession ; vocation 710

BERUFKRAUT (*n*): any of various plants ; fleabane (*Erigeron acris*) ; horseweed, scabious (*Erigeron canadensis*) 711

BERUFSKLEIDUNG (*f*) : overalls 712

BERUFSKRAUT (*n*) : (see **BERUFKRAUT**) 713

BERUFSLITHOGRAPH (*m*): professional lithographer 714

BERUHEN : to rest upon ; depend (be based) upon 715

BERUHIGEND : quieting ; sedative ; reassuring 716

BERUHIGENDES MITTEL (*n*) : sedative, depressant, anodyne (Med.) 717

BERUHIGUNGSMITTEL (*n*): sedative ; anodyne ; depressant (Med.) 718

BERÜHRUNG (*f*): contact ; touching ; conduction (of heat) 719

BERÜHRUNGSEBENE (*f*) : tangential plane 720

BERÜHRUNGSELEKTRIZITÄT (*f*): galvanism ; contact electricity 721

BERÜHRUNGSFLÄCHE (*f*): contact surface 722
BERÜHRUNGSLINIE (*f*): tangent 723
BERÜHRUNGSSTELLE (*f*): place (point) of contact; mark (made by touching) 724
BERYLL (*m*): beryl; ($Be_3Al_2(SiO_3)_6$); Sg. 2.65-2.75; emerald 725
BERYLLERDE (*f*): beryllia; (BeO); Sg. 3.016 726
BERYLLIUM (*n*): Beryllium (Be); Glucinum; (Gl); (In chemical formulæ either Be or Gl may be used, Be being usually used on the Continent, and Gl in America) 727
BERYLLIUMBROMID (*n*): beryllium bromide, glucinum bromide; $BeBr_2$; Mp. 490-617°C. 728
BERYLLIUMCARBID (*n*): beryllium carbide; Be_2C; Sg. 1.9 729
BERYLLIUMCHLORID (*n*): beryllium chloride, glucinum chloride; $BeCl_2$; Mp. 440-617°C. 730
BERYLLIUMFLUORID (*n*): beryllium fluoride; BeF_2; Sg. 2.1 731
BERYLLIUMJODID (*n*): beryllium iodide; BeJ_2; Sg. 4.2 732
BERYLLIUMMETASILIKAT (*n*): beryllium metasilicate; $BeSiO_3$; Mp. above 2000°C. 733
BERYLLIUMNITRAT (*n*): beryllium nitrate, glucinum nitrate; $Be(NO_3)_2+3H_2O$; Mp. 90°C. 734
BERYLLIUMORTHOSILIKAT (*n*): beryllium orthosilicate; Be_2SiO_4; Sg. 2.46 735
BERYLLIUMOXYD (*n*): beryllium oxide; BeO; Sg. 3.06 736
BERYLLIUMSALZ (*n*): beryllium salt 737
BERYLLIUMSULFAT (*n*): beryllium sulphate; ($BeSO_4$), Sg. 2.443; ($BeSO_4.4H_2O$), Sg. 1.7125; ($BeSO_4.12H_2O$), Sg. 1.713 738
BERYLLONIT (*m*): Beryllonite; $NaBePO_4$; Sg. 2.84 739
BESAGEN: to mean; signify; mention 740
BESAN (*m*): spanker (Naut.); mizzen 741
BESANMARS (*m*): mizzen top 742
BESAN MAST (*m*): mizzen mast (in the case of a barque) 743
BESANRÜSTE (*f*): mizzen channel 744
BESANSEGEL (*n*): spanker (Naut.); mizzen sail 745
BESANSTAG (*n*): mizzen stay 746
BESANSTENGE (*f*): mizzen topmast 747
BESAN (*m*), **STURM-**: storm mizzen or spanker 748
BESATZ (*m*): border; edging; trimming; fettling (Metal); relining of a furnace with refractory material 749
BESATZUNG (*f*): occupation, troops of occupation 750
BESCHÄDIGEN: to damage; hurt; harm; injure 751
BESCHADIGÜNG (*f*): harm, injury, damage 752
BESCHAFFEN: to procure; supply; make; execute; constitute; obtain; condition 753
BESCHAFFENHEIT (*f*): nature; constitution; state; condition; quality; disposition; composition 754
BESCHÄFTIGTE (*m*): worker, employee; (*pl.*) employed, staff, workers, employees 755
BESCHÄFTIGUNG (*f*): occupation; business; employment 756
BESCHAULICH: contemplative 757
BESCHEID (*m*): decision; answer; information 758
BESCHEIDEN: to assign; inform; order; summon; modest, unassuming 759
BESCHEIDEN, SICH-: to be content 760
BESCHEINIGEN: to certify; receipt; attest 761
BESCHENKEN: to present; give; make a present of; make a donation to 762
BESCHEREN: to give; bestow 763
BESCHICKEN: to load; charge; feed; prepare; manage; attend to; convey; handle 764
BESCHICKUNG (*f*): the total charge (of raw material to blast furnaces); conveying plant, conveyor, transporter; alloy (Minting) 765
BESCHICKUNG (*f*), **KOHLEN-**: coal conveying (handling) plant 766
BESCHICKUNGSTRICHTER (*m*): feeding (charging) hopper (of mechanical stokers) 767
BESCHICKUNGSVORRICHTUNG (*f*): feeding (charging) plant or arrangement 768
BESCHIRMEN: to shelter; cover; protect 769
BESCHLAG (*m*): any coating (sweat, moisture on glass, binding on boxes, tarnish on metals, etc.); clout; ironwork; clamp; cleat; fitting; mounting; hooping; shoeing; horseshoe; seizure; fastening; casing; clasp; covering 770
BESCHLAGBÄNDSEL (*n*): furling line 771
BESCHLÄGE (*m.pl*): fittings 772
BESCHLAGEN: skilled, versed, familiar with 773
BESCHLAGEN: to coat, effloresce, sweat, incrust; cover (mount, hoop, fit, bind) with metal etc.; garnish, square (hew) timber, hand (furl) a sail, clamp, case, lute, shoe (horses), be (versed) skilled in, be familiar with 774
BESCHLAG (*m*), **METALL-**: metal (covering; casing), coating 775
BESCHLAGNAHMEN: to confiscate, sequestrate, distrain, seize 776
BESCHLAGTEILE (*m.pl*): fittings (metal bindings, hoop iron, etc., as opposed to boiler fittings) 777
BESCHLEUNIGEN: to accelerate; speed-up; hasten; despatch; quicken 778
BESCHLEUNIGER (*m*): accelerator; also an agent for increasing the speed of vulcanization of caoutchouc 779
BESCHLEUNIGUNG (*f*): acceleration 780
BESCHLEUNIGUNGSREGLER (*m*): speed (acceleration) regulator 781
BESCHLIESSEN: to finish; close; determine; resolve upon 782
BESCHLIESSER (*m*): steward; housekeeper 783
BESCHLUSS (*m*): resolution; conclusion; determination; bond; decree; locking-up 784
BESCHMIEREN: to smear; besmear; grease; oil 785
BESCHMUTZEN: to dirty, soil, stain, foul 786
BESCHNEIDEGLAS (*n*): trimmer, trimming glass, cutting shape (Photography) 787
BESCHNEIDEN: to cut-(off); curtail; trim; pare; dress 788
BESCHNEIDUNG (*f*): cutting down of (wages, etc.); curtailment, circumcision, clipping 789
BESCHÖNIGUNG (*f*): gloss; varnish; palliation; colour 790
BESCHOTTERUNG (*f*): metalling; ballasting (roads) 791
BESCHRÄNKEN: to limit; restrict; confine; curtail; abridge; narrow; constrict 792
BESCHRÄNKTEM, IN-MASSE: in slight measure 793
BESCHREIBEN: to describe; write about; letter (a drawing, etc.) 794

BESCHREIBUNG (*f*): specification, description 795

BESCHULDIGEN: to accuse; charge with 796

BESCHUSSPROBE (*f*): shooting test 797

BESCHÜTZEN: to protect; defend; screen 798

BESCHWÄNGERN: to impregnate, make pregnant 799

BESCHWERDE (*f*): grievance; trouble; annoyance; complaint; malady 800

BESCHWEREN: to load; weight; burden; annoy; complain; importune 801

BESCHWERLICH: burdensome; troublesome; hard-(ly); difficult-(ly); with difficulty 802

BESCHWERUNGSMITTEL (*n*): filler; loading material; weighting (filling) agent 803

BESCHWICHTIGEN: to allay; appease; soothe; hush-(up) 804

BESCHWÖREN: to swear to; conjure; exorcise 805

BESEELEN: to animate: enliven; inspirit 806

BESEGELN: to furnish with sails; sail (the sea); navigate 807

BESEGELUNG (*f*): sails (of a ship); navigation 808

BESEHEN: to view; inspect; examine 809

BESEITIGEN: to do away with; set (put) on one side; remove 810

BESEN (*m*): broom 811

BESENGINSTER (*m*): broom (*Cytisus scoparius*) 812

BESENKRAUT (*n*): (see BESENGINSTER) 813

BESETZEN: to occupy; fill; engage; garrison; set; trim; border; stock; fit; charge (furnace) 814

BESICHTIGEN: to view; inspect; examine; survey; look over 815

BESICHTIGUNG (*f*): survey 816

BESIEGEN: to conquer; defeat; beat 817

BESINNEN: to consider; remember; recollect; think over; ruminate 818

BESINNUNGSLOS: unconscious 819

BESITZ (*m*): possession; property 820

BESITZEN: to occupy; hold; own; possess 821

BESONDER: particular; special; specific; separate; singular; odd; individual 822

BESONNEN: discreet; considerate 823

BESPINNEN: to cover, bind round with; to case 824

BESPONNEN: covered; spun over 825

BESPRECHEN: to discuss; confabulate; arrange; bespeak; confer; talk over 826

BESPRENGEN: to sprinkle 827

BESPRITZEN: to sprinkle; spatter; bespatter; squirt 828

BESSEMERBIRNE (*f*): Bessemer converter 829

BESSEMERN: to bessemerize, subject to the Bessemer process 830

BESSEMERPROZESS (*m*): [acid (sandstone or quartz) brickwork lining to converter. This process very little used on Continent. Used for pig iron with low phosphorous content]; Bessemer process, acid process, converting process 831

BESSEMERSCHLACKE (*f*): acid slag (iron-manganese silicate) (50-60% Si O_2; 20-40% MnO; 10-20% FeO; 1-2% CaO + MgO and 1-5% Al_2O_3) 832

BESSEMERSTAHL (*m*): Bessemer steel 833

BESTAND (*m*): duration; stability; state; amount; stock; balance; residue; consistence; continuance 834

BESTÄNDIG: stable; constant-(ly); continual-(ly); durable; permanent; fast (of colour); steady 835

BESTANDTEIL (*m*): component-(part); constituent-(part); ingredient; (*pl.*) accessories 836

BESTANDTEILE (*m:pl*), FLÜCHTIGE-: volatile matter 837

BESTÄRKEN: to confirm; strengthen; corroborate 838

BESTÄTIGEN: to confirm; corroborate; ratify 839

BESTÄUBEN: to cover with dust or powder; pollinate (Botany) 840

BESTÄUBUNG (*f*): pollination (Bot.); covering with dust, making dusty 841

BESTECHEN: to stitch; bribe; charm; corrupt 842

BESTECHLICH: venal, corruptible, capable of being bribed 843

BESTECK (*n*): case; set (of instruments); cover (at table) 844

BESTECK (*n*), SCHIFFS-: ship's reckoning 845

BESTEHEN: to consist; undergo; subsist; exist; persist; insist; endure; resist; encounter; pass; to be formed; to be composed of; (*n*): composition; existence; persistence 846

BESTEHENBLEIBEND: enduring, still (existing) remaining 847

BESTEHEND: existing 848

BESTELLEN: to order; appoint; engage; arrange; deliver; till; plant; cultivate; convey 849

BESTELLSCHEIN (*m*): order form; requisition; indent 850

BESTELLUNGSFALLE (*m*), IM-: should the order be placed, in case the order is placed, in case of the placing of the order 851

BESTFALL (*m*): best case 852

BESTGEEIGNET: most (appropriate) suitable 853

BESTIMMEN: to determine; fix; design; allot; define; appoint; be certain; intend for; calculate; compute; induce; designate 854

BESTIMMT: fixed; given; certain; positive; appointed; precise 855

BESTIMMTE GLEICHUNG (*f*): determinate equation 856

BESTIMMUNGSAPPARAT (*m*): determining (recording) apparatus, meter 857

BESTRAHLEN: to throw rays on; irradiate; illuminate; light 858

BESTRAHLUNG (*f*): radiation; irradiation, illumination 859

BESTREBEN: to strive; endeavour; make effort 860

BESTREBEN (*n*), DAS- HABEN: to tend, strive, have a tendency 861

BESTREGULIEREND: most easily (regulated) regulable 862

BESTREICHEN: to coat; smear; spread; brush; stroke; sweep over (of gases); touch 863

BESTREITEN: to dispute; defray (costs); contest 864

BESTREUEN: to strew; sprinkle; powder 865

BESTRICKEN: to ensnare; entrap; entangle; snare; trap 866

BESTÜRTZT: amazed; dismayed; surprised; perplexed; confounded 867

BESUDELN: to soil; defile; contaminate; foul 868

BETA : beta 869
BETAGT : aged ; due ; old 870
BETAIN (*n*) : betaine (an animal alkaloid from urine) ; $C_5H_{11}NO_2$; betaine, trimethylglycocoll (sugar-beet alkaloid) ; $C_5H_{11}NO_2.H_2O$ 871
BETAINCHLORHYDRAT (*n*) : betaine hydrochloride; $CH_2N.(CH_3)_3Cl.CO_2H$ 872
BETAINCHLORID (*n*) : betaine hydrochloride (see BETAINCHLORHYDRAT) 873
BETAKELN : to rig (of ships) 874
BETALGEN : to tallow 875
BETA-NAPHTOLÄTHYLÄTHER (*m*) : beta-naphthol-ethyl ether, nerolin ; $C_{10}H_7.O.C_2H_5$ 876
BETA-NAPHTOLKALIUM (*n*) : beta-naphthol potassium ; $C_{10}H_7OK$ 877
BETA-NAPHTOLMETHYLÄTHER (*m*) : beta-naphthol-methyl ether (see YARA-YARA) 878
BETA-NAPHTYLAMINSULFOSÄURE BR (*f*) : (see BRÖNNERSCHE SÄURE) 879
BETÄTIGEN : to prove ; manifest ; operate ; actuate ; set to work 880
BETÄTIGUNG (*f*) : actuation ; valve (spindle) gear 881
BETÄTIGUNGSVORRICHTUNG (*f*) : operating gear 882
BETÄUBUNGSMITTEL (*n*) : narcotic 883
BETE (*f*) : beet ; beetroot (*Beta vulgaris*) 884
BETEEREN : to tar 885
BETEILIGEN : to give a share to ; concern ; participate ; interest 886
BETEILIGUNG (*f*) : participation, interest (in a business) 887
BETELNUSS (*f*) : betel-nut (fruit of *Areca catechu*) 888
BETEUERN : to protest ; swear to 889
BETING (*f*) : bitts (*pl*) 890
BETINGKNIE (*n*) : carrick bitt 891
BETINGSBALKEN (*m*) : bitts cross piece 892
BETINGSBOLZEN (*m*) : bitt bolt 893
BETINGSHOLZ (*n*) : bitt (Naut.) 894
BETITELN : to name ; entitle 895
BETOL (*n*) : betol (see SALIZYLSÄURE-B-NAPHTHYL--ESTER) 896
BETON (*m*) : concrete ; beton 897
BETONBAU (*m*) : concrete construction 898
BETONEN : to accent ; emphasize ; accentuate ; stress 899
BETONIENKRAUT (*n*) : betony (*Betonica officinalis* and *Stachys betonica*) ; (*herba betonicæ*) (Pharm.) 900
BETONIEREN : to concrete ; cover with (lay, form of) concrete 901
BETONIERUNG (*f*) : cementing, concreting, reinforcement 902
BETONSTAMPFER (*m*) : concrete ram-(mer) 903
BETONWAREN (*f.pl*) : concrete (wares) goods 904
BETRACHT (*m*) : consideration ; regard ; account ; view ; examination 905
BETRÄCHTLICH : considerable ; important 906
BETRAG (*m*) : amount ; sum ; total ; figure 907
BETREFF (*m*) : subject (of correspondence) 908
BETREFFEN : to concern ; befall ; catch 909
BETREFFEND : concerned ; in question ; under consideration ; under discussion 910
BETREIBEN : to carry on ; conduct ; pursue ; manage 911
BETRETEN : to tread upon ; enter upon ; start ; struck ; confused ; perplexed ; beaten 912
BETRIEB (*m*) : work ; works ; operation ; management ; trade ; traffic ; action ; machinery ; plant ; commission ; industry 913
BETRIEB (*m*), **ANGESTRENGTER-** : forced working, overload (of plant) 914
BETRIEB (*m*), **ELEKTRISCHER-** : electric traction ; electric drive 915
BETRIEB (*m*), **IM PRAKTISCHEN-** : under practical (actual) working conditions 916
BETRIEB (*m*), **IN- SETZEN** ; to set to work ; start up ; put under load ; put in commission 917
BETRIEBSANLAGE (*f*) : plant ; works ; installation ; machinery 918
BETRIEBSAUSWEIS (*m.pl*) : working record ; proof of work done 919
BETRIEBSBEDINGUNGEN (*f.pl*) : conditions of (working) operation 920
BETRIEBSCHEMIKER (*m*) : industrial (works) chemist 921
BETRIEBSDRUCK (*m*) : working pressure 922
BETRIEBSINGENIEUR (*m*) : work's engineer, chief engineer, engineer-in-charge 923
BETRIEBSKONTROLLE (*f*) : control of -(technical)-operations 924
BETRIEBSKOSTEN (*f.pl*) : working expenses, maintenance, upkeep, running costs 925
BETRIEBSLEITER (*m*) : works manager 926
BETRIEBSORGANISATION (*f*) : works management 927
BETRIEBSRAUM (*m*) : work (room) shop 928
BETRIEBSSICHER : safe-(ly), (as regards working of plant) 929
BETRIEBSSTÖRUNG (*f*) : breakdown 930
BETRIEBSVERHÄLTNISSE (*n.pl*) : working conditions 931
BETRIEBSZEIT (*f*) : working hours ; shift ; season ; campaign ; duration of working 932
BETRIFFT : reference (at heading of letter), refers to 933
BETROFFEN : perplexed, disconcerted, upset, dazed 934
BETRUG (*m*) : fraud ; deceit ; swindle 935
BETRÜGERISCH : fraudulent 936
BETT (*n*) : bed, foundation, seating, base 937
BETTEL (*m*) : trash ; begging 938
BETTEN : to embed ; put to bed ; bed ; seat ; support (rest) on 939
BETTLEINEN (*n*) : bed-linen 940
BETTSTELLE (*f*) : bedstead 941
BETTTUCH (*n*) : sheet 942
BETTUNG (*f*) : bedding (on bed) ; seating ; support ; foundation 943
BETTWACHS (*n*) : grafting wax, mummy 944
BETULIN (*n*) : betulin (a resin from birch bark) ; $C_{30}H_{50}O_2$ or $C_{30}H_{52}O_2$ 945
BETUPFEN : to dab ; dip ; tip ; spot ; dot ; touch 946
BEUCHE (*f*) : bucking ; bucking lye 947
BEUCHEN : to buck ; steep in lye or suds 948
BEUCHFASS (*n*) : bucking (tub) kier 949
BEUCHKESSEL (*m*) : bucking kier 950
BEUCHLAUGE (*f*) : bucking lye 951
BEUCHWASSER (*n*) : buck ; lye ; steep 952
BEUGE (*f*) : bend ; bow 953
BEUGEN : to bend ; bow ; deflect ; diffract (light) ; inflect (words) ; flex ; humble ; (see also BIEGEN) 954
BEUGSAM : flexible ; pliant ; declinable 955
BEUGSAMKEIT : pliancy, flexibility, pliability 956
BEUGUNG (*f*) : flexure, flexion, bending, bend, curvature, curve 957
BEUGUNGSFALL (*m*) : case (Grammar) 958
BEULE (*f*) : boil ; swelling ; bump ; bulge 959
BEULE (*f*) **VON ALEP** : Aleppo evil, Delhi boil, Bagdad boil, oriental sore (a skin ulcer) (Med.) 960

BEURKUNDEN : to authenticate ; attest 961
BEURTEILEN : to judge ; estimate ; value ; criticize 962
BEUTE (*f*) : booty ; spoil ; prize ; gain ; yield ; prey 963
BEUTEL (*m*) : bag ; pouch ; purse ; sac ; cyst (Med.) 964
BEUTELFILTER (*n*) : bag filter 965
BEUTELN : to bolt ; sift ; bag 966
BEUTELSIEB (*n*) : bolting sieve ; bolter 967
BEUTELTUCH (*n*) : bolting cloth 968
BEUTELZEUG (*n*) : bolting (cloth) apparatus 969
BEVOLLMÄCHTIGEN : to empower ; authorize ; give power of attorney 970
BEVORSTEHEN : to approach ; impend ; be imminent 971
BEVORZUGEN : to prefer ; favour 972
BEWACHSEN : foul, rotten (of ships) 973
BEWAHREN : to keep ; preserve ; protect ; guard 974
BEWÄHREN : to verify ; approve ; prove true ; try ; test 975
BEWAHRHEITEN : to prove the truth of ; justify 976
BEWÄHRT : approved ; tried ; tested ; proved excellent ; authentic ; justified 977
BEWAHRUNGSMITTEL (*n*) : preservative ; prophylactic (Med.) 978
BEWÄLTIGEN : to overcome ; master 979
BEWANDERT : versed ; skilled ; experienced 980
BEWANDTNIS (*f*) : case ; state ; circumstance ; condition 981
BEWÄSSERN : to water ; moisten ; sprinkle ; irrigate 982
BEWÄSSERUNG (*f*) : irrigation 983
BEWEGBAR : moveable ; mobile 984
BEWEGEN : to move ; induce ; stir 985
BEWEGENDE KRAFT (*f*) : motive power 986
BEWEGER (*m*) : motor ; mover 987
BEWEGLICH : moveable ; versatile ; mobile ; flexible 988
BEWEGLICHKEIT (*f*) : mobility 989
BEWEGUNG (*f*) : movement, motion, running ; emotion 990
BEWEGUNG (*f*), **FORTSCHREITENDE-** : transfer motion ; continuous (perpetual) motion 991
BEWEGUNGSFREIHEIT (*f*) : freedom of movement 992
BEWEGUNGSFUGE (*f*) : expansion joint (in tramlines, etc.) 993
BEWEGUNGSLEHRE (*f*) : mechanics ; kinematics ; kinetics 994
BEWEGUNGSSITZ (*m*) : medium fit, free fit 995
BEWEHRUNG (*f*) : stiffening ; stiffener ; mounting ; arming ; armament 996
BEWEIS (*m*) : proof ; evidence ; argument (of patents) ; demonstration ; showing 997
BEWEISARTIKEL (*m*) : argument (of patents) 998
BEWEISAUFNAHME (*f*) : evidence 999
BEWEISFÜHRUNG (*f*) : demonstration ; reasoning 000
BEWEISSTÜCK (*n*) : proof, concrete evidence 001
BEWENDEN : to rest ; drop ; submit to ; (*n*) : end ; drop ; rest 002
BEWERBER (*m*) : candidate ; applicant ; aspirant ; wooer ; suitor 003
BEWERBUNG (*f*) : application (for a situation) ; courtship 004
BEWERFEN : to plaster ; pelt ; throw at 005
BEWERKSTELLIGEN : to accomplish ; perform ; effect ; carry out ; cause ; bring about 006
BEWERTEN : to evaluate ; calculate the efficiency (of engine) 007
BEWERTEN (*n*) : evaluation 008
BEWERTUNG (*f*) : (e)-valuation, computation, determination of value of, calculation 009
BEWETTERN : to ventilate 010
BEWETTERUNG (*f*) : ventilation ; weathering (as of a storm) ; making use of suitable means for tiding over a crisis (i.e., accumulating power to tide over a period when insufficient power is being generated for the load) 011
BEWICKELN : to wrap round ; envelop 012
BEWILLIGEN : to grant ; permit ; allow 013
BEWIRKEN : to effect ; cause ; exert ; attain ; bring about 014
BEWIRTSCHAFTEN : to manage ; run ; administer (economically) 015
BEWURF (*m*) : plaster-(ing) 016
BEWUSST : conscious ; aware ; known 017
BEWUSSTLOS : unconscious ; senseless 018
BEWUSSTSEIN (*n*) : consciousness ; knowledge 019
BEYRICHIT (*m*) : Beyrichite (a natural nickel sulphide) ; NiS ; Sg. 4.6 020
BEZAHLUNG (*f*) : payment ; settlement 021
BEZÄHMEN : to tame ; restrain 022
BEZEICHNEN : to mark ; label ; designate ; describe ; represent ; refer to ; illustrate ; symbolize ; indicate ; signify ; characterize ; point out ; note ; sign ; term 023
BEZEICHNEN AUF : to mark off on (such as the point on a graph) 024
BEZEICHNUNG (*f*) : description, term, designation, indication 025
BEZEICHNUNG (*f*), **GESCHÜTZTE-** : registered trade name 026
BEZEICHNUNGSWEISE : method (manner) of notation 027
BEZEUGEN : to testify ; certify ; witness 028
BEZIEHEN : to draw ; obtain ; order ; buy ; enter ; refer ; cover ; advert ; be overcast 029
BEZIEHEN AUF : to base on, refer to 030
BEZIEHEN VON : to obtain-(from) 031
BEZIEHUNG (*f*) : relation-(ship), understanding, communication, business, reference, respect, connection, direction 032
BEZIEHUNG ENTSPRECHEN : to bear the relation, correspond with 033
BEZIFFERN : to figure ; number 034
BEZIRK (*m*) : district ; circuit 035
BEZOARWURZEL (*f*) : contrayerva (Pharmacy) 036
BEZOGEN AUF : corresponding to ; in conformity with ; based on 037
BEZOGENE (*m*) : drawee (Com.) 038
BEZUG (*m*) : relation ; reference ; drawing 039
BEZUG (*m*), **IN-AUF** : with regard to ; in respect of ; as far as concerns ; with reference to 040
BEZÜGLICH : relative ; with reference to ; respecting ; commensurate with 041
BEZUG, MIT- AUF : with regard to ; in respect of ; respecting ; having regard to 042
BEZUGNAHME (*f*) : reference 043
BEZUGSQUELLE (*f*) : source-(of supply) 044
BEZUGSWERT (*m*) : relative value 045
BEZWECKEN : to aim at ; intend ; peg ; have for its object 046
BEZWEIFELN : to doubt, question 047
BEZWINGEN : to subdue ; restrain ; overcome ; compel ; vanquish 048
BHANG (*m*) : (see **HANF, INDISCHER-**) 049

BI- : (see also under **DI-**) 050
BIBER (*m*) : beaver (*Castor fiber*) 051
BIBERGEIL (*n*) : castor ; castoreum (dried preputial follicles and secretions of common beaver) 052
BIBERNELL(E) (*f*) : burnet saxifrage (*Pimpinella saxifraga*) 053
BIBERNELLWURZEL (*f*) : saxifrage root (*radix Pimpinellae*) 054
BIBERSCHWANZ (*m*) : flat roofing-tile ; beaver's tail 055
BICHLORBENZOL (*n*) : dichlorobenzene, dichlorobenzol ; $C_6H_4Cl_2$; Ortho-, Sg. 1.325 ; Mp. 14°C. ; Bp. 179°C. ; meta-, Sg. 1.31 ; Mp. −18°C. ; Bp. 172°C. ; para-, Sg. 1.268 ; Mp. 53°C. ; Bp. 174°C. 056
BICHROMAT (*n*) : bichromate, dichromate 057
BICYCLISCH : bicyclic 058
BIEBERIT (*m*) : (a natural cobalt sulphate) ; $CoSO_4.7H_2O$ 059
BIEDER : upright ; honest ; true ; loyal 060
BIEGBAR : flexible ; pliant ; pliable ; ductile ; capable of being bent 061
BIEGE (*f*) : bend ; bow ; curvature ; curve 062
BIEGEMASCHINE (*f*) : bending machine 063
BIEGEMUSKEL (*m*) : flexor muscle (Med.) 064
BIEGEN : to bend ; curve ; diffract ; refract (light) ; deflect ; inflect (words) ; warp (wood) ; (see aslo **BEUGEN**) 065
BIEGENDES MOMENT (*n*) : bending moment 066
BIEGSAM : flexible ; pliable ; pliant ; ductile 067
BIEGSAMKEIT (*f*) : deflectibility, flexibility, pliancy, springiness, suppleness, capacity for being bent or deflected or sprung 068
BIEGUNG (*f*) : bending ; deflection (of a shaft) ; curve ; curvature ; bend : flexure (Medical) 069
BIEGUNGSBEANSPRUCHUNG (*f*) : bending (stress) strain 070
BIEGUNGSFESTIGKEIT (*f*) : bending strength, strength of flexure 071
BIEGUNGSMOMENT (*n*) : bending moment, moment of flexure 072
BIEGUNGSPFEIL (*m*) : amount of deflection, direction of deflection 073
BIEGUNGSSCHWINGUNG (*f*) : oscillation due to deflection 074
BIEGUNGSSPANNUNG (*f*) : bending stress 075
BIENENHARZ (*n*) : propolis ; bee's varnish ; bee's glue ; hive dross 076
BIENENKORBOFEN (*m*) : bee-hive coke oven 077
BIENENWABE (*f*) : honeycomb 078
BIENENWACHS (*n*) : bee's wax ; Sg. 0.965-0.969 ; Mp. 63.5°C. 079
BIENENWACHS (*n*), **GELBES-** : yellow bee's wax (*Cera flava*) 080
BIENENWACHS (*n*), **WEISSES-** : white bee's wax (*Cera alba*) 081
BIER (*n*) : beer ; ale ; Sg. 1.03 082
BIERBLUME (*f*) : head (froth) on beer 083
BIERBRAUER (*m*) : brewer 084
BIERBRAUEREI (*f*) : brewery 085
BIERESSIG (*m*) : malt vinegar 086
BIERGRAND (*m*) : underback (Brewing) 087
BIERHEFE (*f*) : barm ; yeast (cells and spores of *Saccharomyces cerevisii*) 088
BIERKESSEL (*m*) : boiler ; vat ; boiling copper (for beer) 089
BIERPECH (*n*) : brewer's pitch 090
BIERPROBER (*m*) : beer tester ; beer taster ; hydrometer 091
BIERSCHÖNE (*f*) : fining (for beer) 092
BIERSIEDER (*m*) : beer boiler 093
BIERSTEIN (*m*) : beer scale 094
BIERUNTERSUCHUNG (*f*) : beer (testing, examination) analysis 095
BIERWAGE (*f*) : hydrometer 096
BIERWÜRZE (*f*) : wort 097
BIESTMILCH (*f*) : colostrum 098
BIETEN : to offer ; bid ; show ; tender 099
BIGNONIENGEWÄCHSE (*n.pl.*) : *Bignoniaceae* (Bot.) 100
BILAKTAT (*n*) : bilactate 101
BILANZ (*f*) : balance ; balance-sheet 102
BILANZAUSFERTIGUNG (*f*) : preparation of balance-sheet 103
BILD (*n*) : figure ; picture ; diagram ; idea ; print ; image ; portrait ; description 104
BILDEBENE (*f*) : plane of projection (Geometry) 105
BILDEN : to form ; shape ; fashion ; educate ; train ; civilize ; compose 106
BILDEN (*n*) : formation, development, education, shaping 107
BILDEND : component, constituent ; plastic (Art) ; forming 108
BILDERACHAT (*m*) : figured agate 109
BILDERBESCHREIBUNG (*f*) : iconography 110
BILDERKUNDE (*f*) : iconology 111
BILDERSTÜRMEND : iconoclastic 112
BILDERSTÜRMER (*m*) : iconcolast 113
BILDERSTÜRMEREI (*f*) : iconoclasm 114
BILDFELD (*n*) : (angle of)-view (Photography) 115
BILDFORMAT (*n*) : size of (picture) print (Photography) 116
BILDHAUER (*m*) : sculptor 117
BILDHAUERKITT (*m*) : badigeon 118
BILDHAUERLEIM (*m*) : (see **BILDHAUERKITT**) 119
BILDHAUERMARMOR (*m*) : statuary marble 120
BILDLICH : figurative-(ly) ; typical-(ly) 121
BILDLOS : amorphous 122
BILDMARMOR (*m*) : figured marble ; variegated marble 123
BILDMESSER (*m*) : view meter (Photography) 124
BILDSAM : plastic ; flexible ; ductile 125
BILDSAMKEIT (*f*) : ductility (Metal) 126
BILD (*n*), **SCHIEFES-** : false (wrong) impression 127
BILDSCHÖN : extremely beautiful 128
BILDSTECHER (*m*) : engraver 129
BILDSTEIN (*m*) : figured stone ; agalmatolite ; $(H_2O.Al_2O_3.4SiO_2)$ 130
BILDSUCHER (*m*) : view finder (Photography) 131
BILDUNG (*f*) : formation ; shape ; structure ; organization ; learning ; culture 132
BILDUNGSENERGIE (*f*) : energy of formation 133
BILDUNGSGESCHWINDIGKEIT (*f*) : speed of formation 134
BILDUNGSGLEICHUNG (*f*) : equation of formation ; structural equation 135
BILDUNGSWÄRME (*f*) : heat of formation 136
BILDUNGSWEISE (*f*) : method (mode) of formation 137
BILDUNGSWERK (*n*) : sculpture ; carving 138
BILD (*n*), **VERBRANNTES-** : burnt-out (print) picture, over-exposed (print) picture (Photography) 139
BILD (*n*), **VERSCHWOMMENES-** : fuzzy (blurred) image (Photography) 140

BILDWÖLBUNG (*f*): curvature of the field, aberration of form (Photography) 141
BILGEPUMPE (*f*): bilge pump 142
BILGEROHR (*n*): bilge pipe 143
BILGEWASSER (*n*): bilge, bilgewater 144
BILLARD (*n*): billiards 145
BILLARDBEUTEL (*m*): pocket (of a billiard table) 146
BILLARDKREIDE (*f*): billiard chalk 147
BILLARDKUGEL (*f*): billiard ball 148
BILLARDLOCH (*n*): pocket (of a billiard table) 149
BILLARDSTOCK (*m*): (billiard)-cue 150
BILLET (*n*): ticket; note 151
BILLIG: cheap; just; equitable 152
BILLIGEN: to approve-(of) 153
BILSENKRAUT (*n*): henbane (*herba hyoscyami* of *Hyoscyamus niger*) 154
BILSENKRAUTSAMENÖL (*n*): henbane seed oil (from seeds of *Hyoscyamus niger*) 155
BILSENKRAUT (*n*), **SCHWARZES-**: (black) henbane, hog's bean, poison tobacco, insane root (*Hyoscyamus niger*) 156
BIMSEN: to-(smooth by the aid of)-pumice stone; rub with pumice stone; (*n*): pumice-stoning 157
BIMSSTEIN (*m*): pumice stone (*Pumex*; *Lapis pumicis*); pumice; Sg. 0.92 158
BIMSSTEINARTIG: like (similar to) pumice stone 159
BIMSSTEINSEIFE (*f*): pumice stone soap 160
BINÄR: binary; composed of two parts 161
BINDE (*f*): band; bandage; ligature; tie; necktie 162
BINDEDRAHT (*m*): binding wire 163
BINDEGEWEBE (*n*): connective tissue 164
BINDEGLIED (*n*): connecting (link) member 165
BINDEHAUT (*f*): conjunctiva 166
BINDEKRAFT (*f*): combining power; binding power (of cement) 167
BINDEMITTEL (*n*): cement; binder; agglutinant; medium; matrix; binding agent or material 168
BINDEN: to bind; bond; fix; combine; unite; tie; constrain; harden; cement; link (join) up; set (of cement) 169
BINDEN, LOS-: to untie 170
BINDESTRICH (*m*): hyphen 171
BINDETON (*m*): ball clay; white, plastic -(fat)-clay 172
BINDEVERMÖGEN (*n*): combining power; binding power; plasticity 173
BINDEWORT (*n*): conjunction 174
BINDEZEIT (*f*): setting time (of cement) 175
BINDFADEN (*m*): twine; string; packthread 176
BINDUNG (*f*): combination; union; bond; linkage; binding; ligature 177
BINDUNG (*f*), **DOPPEL-**: double (bond) linkage 178
BINDUNG (*f*), **DREIFACHE-**: triple linkage 179
BINDUNGSISOMERIE (*f*): isomerism of the union of atoms 180
BINDUNGSWÄRME (*f*): heat of combination 181
BINDUNGSWEISE (*f*): method (mode) of union, combination or linkage 182
BINITROBENZOL (*n*): dinitrobenzene (see **DINITROBENZOL**) 183
BINITROCHLORBENZOL (*n*): dinitrochlorobenzol, dinitrochlorobenzene; $C_6H_3(NO_2)_2Cl$; Sg. 1.69; Mp. 37-53°C. 184
BINITROTOLUOL (*n*): dinitrotuluene (see **DINITROTOLUOL**) 185
BINNENACHTERSTEVEN (*m*): inner sternpost (Naut.) 186
BINNENBORDS: inboard 187
BINNENDRUCK (*m*): internal (stress, tension) pressure 188
BINNENGANG (*m*): inside strake 189
BINNENHANDEL (*m*): home trade; domestic trade; internal trade 190
BINNENHINTERSTEVEN (*m*): inner sternpost 191
BINNENKLÜVER (*m*): inner jib (Naut.) 192
BINNENSCHIFFFAHRT (*f*): inland navigation 193
BINNENSPANT (*n*): rider plate 194
BINNENVORDERSTEVEN (*m*): apron (Naut.) 195
BINNENWEGER (*m* and *pl*): ceiling 196
BINSE (*f*): rush; bull-rush; bast; lake scirpus (*Scirpus lacustris*) 197
BINSENFÖRMIG: rush (shaped) like; reed (shaped) like 198
BIOCHEMIE (*f*): biochemistry 199
BIOLOG (*m*): biologist 200
BIOLOGIE (*f*): biology 201
BIOTIT (*m*): biotite (a magnesium-aluminium silicate; one of the forms of Mica); $H_4Mg(Al_2)_3Si_6O_{24}$; Sg. 2.8-3.2; (black, brown or green colour) 202
BIRKE (*f*): birch-(tree) (*Betula alba* or *Betula lenta*); Sg. 0.47-0.77 203
BIRKENBLÄTTER (*n.pl*): birch leaves (*folia betulæ* of *Betula alba*) (see **BIRKE**) 204
BIRKENGEWÄCHSE (*n.pl*): *Betulaceæ* (Bot.) 205
BIRKENHOLZ (*n*): birch wood (see **BIRKE**) 206
BIRKENKNOSPENÖL (*n*): birch bud oil 207
BIRKENÖL (*n*): birch oil (see **BIRKENRINDÖL**) 208
BIRKENRINDE (*f*): birch-bark 209
BIRKENRINDENTEER (*m*): birch-bark tar; Sg. 0.937 210
BIRKENRINDÖL (*n*): oil of betula; oil of sweet birch; Wintergreen oil; Sg. 1.183; Birch oil; Sg. 0.956; (from *Betula alba*) 211
BIRKENTEER (*m*): birch tar; Sg. 1.153 212
BIRKENTEERÖL (*n*): birch tar oil (from birch-tar from *Betula alba*); Sg. 0.956 213
BIRNBAUM (*m*): pear tree (*Pyrus communis*); Sg. 0.66-0.72 214
BIRNBAUMHOLZ (*n*): pear tree wood 215
BIRNE (*f*): pear; any pear shaped object; converter (Met.); bulb (Elect.) 216
BIRNE (*f*), **BESSEMER-**: bessemer converter 217
BIRNE (*f*), **GLAS-**: glass bulb, electric light bulb 218
BIRNENÄTHER (*m*): (see **BIRNENÖL**) 219
BIRNENKONTAKT (*m*): pear push (Elect.) 220
BIRNENÖL (*n*): pear oil (a solution of amyl acetate in alcohol in proportion of 1 : 10) 221
BIRNFÖRMIG: pear shaped; pyriform 222
BIRNKERNÖL (*n*): pear-pip oil (from *Pirus communis*) 223
BIRNMOST (*m*): pear juice; perry 224
BIRNÖL (*n*): pear oil; amyl acetate; $CH_3CO_2C_5H_{11}$; Sg. 0.8659; Bp. 148°C. 225
BIRNWEIN (*m*): perry 226
BIRZSTRAUCH (*m*), **DEUTSCHER-**: German tamarisk (*Myricaria germanica* or *Tamarix germanica*) 227
BISAM (*m*): musk (secretion from *Moschus moschiferus*) 228
BISAMARTIG: musky 229
BISCHOFIT (*m*): bischofite; $MgCl_2.6H_2O$; Sg. 1.59 230
BISKUIT (*m*): biscuit 231
BISKUITPORZELLAN (*n*): biscuit porcelain (the term applied to unglazed porcelain) 232

BISLANG: up to the present 233
BISMARCKBRAUN (*n*): (see **BRAUNSALZ**) 234
BISMUTH (*n*): bismuth (see **WISMUT**) 235
BISMUTIN (*m*): bismuthinite (natural bismuth sulphide); Bi_2S_3; Sg. 6.4 236
BISSEN (*m*): morsel; bit; bolus (Med.) 237
BISTER (*m*): bistre, bister (see **MANGANBRAUN**) 238
BISTERBRAUN (*n*): bistre 239
BISTUM (*n*): episcopate; bishopric 240
BISULFAT (*n*): bisulphate (see **NATRIUMBISULFAT**) 241
BISULFIT (*n*): bisulphite (see **NATRIUMBISULFIT**) 242
BISWEILEN: sometimes; occasionally 243
BITTERERDE (*f*): magnesia; (MgO): Sg. 3.22; Mp. 2800°C. 244
BITTERESCHENGEWÄCHSE (*n.pl*): *Simarubaceæ* (Bot.) 245
BITTERFRÜCHTIG: bitter fruited 246
BITTERKALK (*m*): dolomite; $(Ca.Mg)CO_3$ 247
BITTERKALKSPAT (*m*): (see **BITTERKALK**) 248
BITTERKLEE (*m*): buck bean (*folia trefolii fibrini* of *Menyanthes trifoliata*); trefoil (*Trefolium*) 249
BITTERKLEESALZ (*n*): oxalic acid; $CO_2H.CO_2H.2H_2O$; Sg. 1.653; Mp. 187°C. 250
BITTERLOS: free from bitterness 251
BITTERMANDEL (*f*): bitter almond (*Amygdalus communis*) 252
BITTERMANDELGEIST (*m*): spirit (essence) of bitter almonds 253
BITTERMANDELÖL (*n*): oil of bitter almonds (*Oleum amygdalarum amarum æthereum*); benzaldehyde; C_6H_5CHO; Sg. 1.0504; Mp. -13.5°C.; Bp. 179.9°C. 254
BITTERMANDELÖLKAMPFER (*m*): (see **BENZOIN**) 255
BITTERMANDELÖL (*n*), **KÜNSTLICHES-**: nitrobenzene; Sg. 1.19867; Mp. 8.7 °C.; Bp. 210.85°C. 256
BITTERMANDELWASSER (*n*); bitter almond water (a distillate obtained from *Amygdalus communis*; water containing Epsom salt) 257
BITTERMITTEL (*n*): bitter-(s) 258
BITTERSALZ (*n*): Epsom salts; epsomite; magnesium sulphate; $MgSO_4$ or $MgSO_4.7H_2O$; Sg. 2.65 and 1.6784 respectively 259
BITTERSÄURE (*f*): picric acid; $C_6H_2(NO_2)_3OH$; Sg. 1.767; Mp. 122°C. 260
BITTERSCHMECKEND: bitter-tasting 261
BITTERSPAT (*m*): magnesite; $MgCO_3$; dolomite; $(Ca.Mg)CO_3$; (see also **MAGNESIT** and **DOLOMIT**) 262
BITTERSTEIN (*m*): nephrite (see **NEPHRIT**); saussurite (see **SAUSSURIT**); jade, jadeite (see **JADEIT**) 263
BITTERSTOFF (*m*): bitter principle 264
BITTERSÜSS (*n*): bittersweet; woody nightshade (*Solanum dulcamara*) 265
BITTERSÜSSSTENGEL (*m*): bittersweet stalk (*Stipites dulcamarae*) 266
BITTERWASSER (*n*): bitter water; water containing epsom salts; bitter almond water 267
BITTERWURZEL (*f*): gentian root; bitter root (from *Gentiana lutea*) 268
BITUMEN (*n*): bitumen; (Sg. 0.85) 269
BITUMINÖS: bitumenous; bituminous 270
BIXIN (*n*): bixine (dark-red vegetable dyestuff from seeds of *Bixa orellana*) 271
BIYAKUSHI (*f*): Japanese biyakushi (*Anemone anomala*) 272
BLÄHEN: to inflate; swell; cause flatulency 273
BLÄHSUCHT (*f*): flatulency; wind (Med.) 274
BLÄHUNGSMITTEL (*n*): carminative 275
BLÄHUNGSTREIBEND: carminative 276
BLÄHUNGSTREIBENDES MITTEL (*n*): carminative 277
BLAKEN: to smoke 278
BLAMIEREN: to expose to ridicule; make a fool of; disgrace 279
BLANCHIEREN: to blanch; whiten 280
BLANCHISSURE (*f*): light spot (in dyeing) 281
BLANK: bright; clear; clean; polished; naked; bare; white; blank 282
BLANKKOCHEN: to boil down sugar without graining 283
BLANKOKREDIT (*m*): open credit 284
BLANKPUTZEN: to polish; scour 285
BLANKTRAN (*m*): clear; light-yellow cod liver oil 286
BLASAPPARAT (*m*): blast apparatus 287
BLASBALG (*m*): (see **BLASEBALG**) 288
BLÄSCHEN (*n*): small bubble; bleb; vesicle 289
BLÄSCHENFLECHTE (*f*): Herpes Zoster, shingles (Med.) 290
BLASE (*f*): bubble; blister; bleb; bladder-(shaped vessel); bladder (Med.) 291
BLASEAPPARAT (*m*): blower, blast apparatus 292
BLASEBALG (*m*): bellows 293
BLASEBALGKAMERA (*f*): bellows-body camera 294
BLASELAMPE (*f*): blast-lamp; blow-lamp 295
BLASEN: to blow; smelt in a blast furnace (Metal); blow-off (safety valve); inject (steam) 296
BLASENANTISEPTIKUM (*n*): bladder antiseptic (Med.) 297
BLASENENTSTAUBUNG (*f*): forced draught dust extraction 298
BLASENGÄRUNG (*f*): bubbling fermentation 299
BLASENGRÜN: sap-green 300
BLASENKATARRH (*m*): catarrh of the bladder 301
BLASENKRANKHEIT (*f*): pemphigus, pompholyx (the forming of bullous eruptions on the skin) disease of the bladder (Med.) 302
BLASENLEIDEN (*n*): bladder trouble (Med.) 303
BLASENPFLASTER (*n*): blistering tissue (*tela vesicatoria*); cantharides cerate; Spanish fly (dried beetle, *Cantharis vessicatoria*) 304
BLASENROST (*m*): bladder rust, pine cluster-cup; currant (rust) mildew (from the pine, *Pinus strobus*) (*Peridermium strobi*) 305
BLASENSÄURE (*f*): uric acid; $CO(NH)_2COC_2CO(NH)_2$; Sg. 1.855-1.893 306
BLASENSTEIN (*m*): urinary calculus; stone in the bladder (Med.) 307
BLASENTANG (*m*): bladder (sea) wrack; seaweed (*Fucus vesiculosus*) 308
BLASENZIEHEND: blistering; vesicatory 309
BLASEOFEN (*m*): blast furnace; blowing furnace 310
BLÄSEREI (*f*): blowing; blowing shop (glass) 311
BLASEROHR (*n*): blast pipe; nozzle (of bellows); blow-pipe 312
BLASETISCH (*m*): blow-lamp table 313
BLASIG: blistery; blistered; vesicular 314
BLASIGER KUPFERSTEIN (*m*): pimple metal 315
BLASROHR (*n*): blast pipe, blow pipe, nozzle of bellows 316
BLÄSSE (*f*): pallor (due to lack of blood) (Med.) 317
BLASSGELB: pale yellow 318

BLASVERLUST (*m*): blow-off losses (either from safety valves or blow-off valves) 319
BLASVORGANG (*n*): blowing (blast) process (of the metal in bessemer process) 320
BLATT (*n*): leaf (*folia*); lamina; lamella; plate; sheet (paper): flake; newspaper 321
BLÄTTCHENARTIG: lamelliform; laminiform; lamellar 322
BLÄTTCHENPULVER (*n*): leaf powder 323
BLATTEISEN (*n*): sheet iron 324
BLATTER (*f*): blister; pimple; (*pl*), smallpox 325
BLÄTTERERZ (*n*): nagyagite; foliated tellurium (see NAGYAGIT) 326
BLÄTTERIG: leafy; foliated; laminated; lamellar 327
BLÄTTERKIES (*m*): lamellar pyrites 328
BLÄTTERKOHLE (*f*): slaty coal; foliated coal 329
BLÄTTERMAGNET (*m*): lamellar magnet 330
BLÄTTERN: to flake; exfoliate; to turn over pages 331
BLATTERN (*f.pl*): variola, small pox (Med.) 332
BLÄTTERSTEIN (*m*): variolite 333
BLÄTTERTELLUR (*f*): (see BLÄTTERERZ) 334
BLÄTTERTON (*m*): foliated clay 335
BLÄTTERZEOLITH (*m*): heulandite; flaky (foliated) zeolite 336
BLATTFEDER (*f*): plate spring 337
BLATTGELB (*n*): xanthophyll 338
BLATTGOLD (*n*): leaf gold; gold leaf; gold foil 339
BLATTGRÜN (*n*): chlorophyl 340
BLATTKEIM (*m*): plumule (of malt, etc.); acrospire 341
BLATTKUPFER (*n*): sheet copper 342
BLATTLACK (*m*): shellac 343
BLATTMETALL (*n*): sheet (leaf) metal 344
BLATT (*n*), **RIEMEN-**: blade of an oar 345
BLATTSILBER (*n*): silver (leaf) foil 346
BLATTZINN (*n*): tin foil; sheet tin 347
BLAU: blue; (*n*): Blue 348
BLAUBEERKRAUT (*n*): blueberry, bilberry, blæberry (*herba myrtylli* of *Vaccinium myrtillus*) 349
BLAUBRÜCHIG: blue short (metal) 350
BLAUDRUCK (*m*): ferroprussiate printing 351
BLÄUE (*f*): blue; blueness 352
BLAUE ERDE (*f*): blue earth, earthy vivianite (a natural hydrous ferrous phosphate); $Fe_3P_2O_8+8H_2O$; Sg. 2.65 353
BLAUEISENERDE (*f*): (see BLAUE ERDE) 354
BLAUEISENERZ (*n*): vivianite (see VIVIANIT) 355
BLAUEISENSPAT (*m*): (see BLAUE ERDE) 356
BLAUEISENSPAT (*m*), **FASERIGER-**: crocidolite (see KROKYDOLITH) 357
BLAUEISENSTEIN (*m*): (see BLAUE ERDE) 358
BLÄUEL (*m*): blue starch; beetle 359
BLAUEN: to blue; dye blue; turn blue; wash blue; beat 360
BLÄUEN: (see BLAUEN) 361
BLAUEREI (*f*): indigo dye (house) works 362
BLAUERZ (*n*): (see BLAUE ERDE) 363
BLAUFARBE (*f*): blue; smalt; Sg. 2.44; cobalt blue (cobalt-potassium silicate) 364
BLAUFARBENGLAS (*n*): smalt (potash-cobalt glass, by fusing sand and potash with cobalt oxide) 365
BLAUFÄRBER (*m*): blue dyer 366
BLAUFILTER (*m.* and *n*): blue filter (for three colour printing or photography) 367
BLAUGAS (*n*): Blaugas; oil gas (termed after the inventor) (a liquid lighting gas) 368
BLAUHOLZ (*n*): logwood; campeachy wood (*Haematoxylon campechianum*) 369
BLAUHOLZEXTRAKT (*m*): logwood extract, hematine extract, hematine paste; Sg. 1.25 370
BLAUHOLZSPÄNE (*m.pl*): logwood chips 371
BLAUHOLZTINKTUR (*f*): logwood liquor; Sg. 1.075 372
BLAUKOHL (*m*): red cabbage 373
BLAUKREUZGESCHOSS (*n*): blue-cross-(gas)-shell or projectile (containing diphenylchloroarsine, $(C_6H_5)_2AsCl$) 374
BLAUKÜPE (*f*): blue vat 375
BLAUMÜHLE (*f*): smalt mill 376
BLAUOFEN (*m*): flowing furnace 377
BLAUPAUSE (*f*): blue print; blue-line print 378
BLAUPROBE (*f*): blue test 379
BLAUSALZ (*n*): potassium ferrocyanide; yellow prussiate of potash; $K_4Fe(CN)_6.3H_2O$ 380
BLAUSAND (*m*): the very coarsest smalt 381
BLAUSAUER: cyanide (prussiate) of; combined with hydrocyanic (prussic) acid 382
BLAUSÄURE (*f*): hydrocyanic (prussic) acid (*Acidum hydrocyanicum*); (HCN); Sg. (gas) 0.9483; Mp. (liquid anhydrous HCN) −15°C.; Bp. 21.6°C. 383
BLAUSÄUREGAS (*n*): hydrogen cyanide (see BLAUSÄURE) 384
BLAUSAURES SALZ (*n*): prussiate, cyanate 385
BLAUSÄUREVERBINDUNG (*f*): cyanide 386
BLAUSCHÖRL (*m*): cyanite; $Al_2O_3.SiO_2$ 387
BLAUSCHWARZ: blue-black 388
BLAUSPAT (*m*): lazulite (see LAZULITH) 389
BLAUSTEIN (*m*): blue-stone; blue vitriol; lazulite (see LAZULITH); blue metal; blue matt) (metal) 390
BLAUSTICH (*m*): bluish (tinge) tint 391
BLAUSTIFT (*m*): blue pencil 392
BLAUSTOFF (*m*): cyanogen; (C_2N_2); Sg. 1.8064 compared with air; liquifies at 21°C.; solidifies at −34°C. 393
BLAUSUCHT (*f*): cyanosis (Med.) 394
BLECH (*n*): plate; vane; sheet; sheet-iron; sheet metal; foil 395
BLECHBEARBEITUNG (*f*): sheet-metal working 396
BLECHBIEGEMASCHINE (*f*): plate bending machine 397
BLECHBÜCHSE (*f*): tin (can) box; canister 398
BLECHDRUCK (*m*): tin-plate printing, printing on tin 399
BLECHEN: tin; of tin-(plate) 400
BLECHERN: (see BLECHEN) 401
BLECHFLASCHE (*f*): tin bottle 402
BLECHGEFÄSS (*n*): tin vessel 403
BLECHHAFEN (*m*): tin can 404
BLECHHAMMER (*m*): sheet iron forge 405
BLECHKANNE (*f*): tin can 406
BLECHKASSETTE (*f*): dark-slide, metal sheath (Photo.) 407
BLECHKASTEN (*m*): tin box; tin case 408
BLECH (*n*), **KESSEL-**: boiler plate 409
BLECHKONSTRUKTION (*f*): plate (sheet-metal) construction 410
BLECHLACK (*m*): tin-(plate) varnish 411
BLECHLEHRE (*f*): plate gauge 412
BLECHMANTEL (*m*): metal casing 413
BLECHRÄHMCHEN (*n*): metal sheath (Photo.) 414
BLECHSCHEIBE (*f*): sheet, plate 415
BLECHSCHERE (*f*): plate shears 416
BLECHSCHERMASCHINE (*f*): (plate)-shearing machine 417
BLECH (*n*), **SCHWARZES-**: black (iron) plate 418

BLECHSTOSS (*m*): butt-(of plates), butt joint 419
BLECHTAFEL (*f*): sheet iron 420
BLECHTRICHTER (*m*): tin funnel 421
BLECHTROMMEL (*f*): tin (metal) drum or cylindrical box 422
BLECHWALZE (*f*): plate roll 423
BLECHWARE (*f*): tin ware 424
BLECH (*n*), **WEISSES-**: tin plate 425
BLEI (*n*): lead (Pb.); Sg. 11.34; Mp. 327°C.; Bp. 1525°C. 426
BLEIAKKUMULATOR (*m*): lead accumulator 427
BLEI (*n*), **AMEISENSAURES-**: lead formate 428
BLEIANODE (*f*): lead(en) anode 429
BLEIANTIMONERZ (*n*): zinkenite (natural lead-antimony ore) 430
BLEI (*n*), **ANTIMONSAURES-**: lead antimonate, Naples yellow; $Pb_3(SbO_4)_2$ 431
BLEIARBEIT (*f*): plumbing; lead smelting 432
BLEIARBEITER (*m*): plumber 433
BLEIARMATUREN (*f pl.*): lead fittings 434
BLEI (*n*), **ARSENSAURES-**: lead arsenate; $Pb_3(AsO_4)_2$; Sg. 6.42 435
BLEIASCHE (*f*): lead dross; lead ash (the grey film of oxide on lead exposed to air) 436
BLEIAUSKLEIDUNG (*f*): lead lining 437
BLEIAZETAT (*n*): (*Plumbum aceticum*): lead acetate, sugar of lead; $Pb(C_2H_3O_2)_2.3H_2O$; Sg. 2.5; Bp. 280°C.; loses $3H_2O$ at 75°C 438
BLEIAZETAT (*n*), **BASISCHES-**: (see **BLEISUBAZETAT**) 439
BLEI (*n*), **BASISCH ESSIGSAURES-**: basic lead acetate (see **BLEISUBAZETAT**) 440
BLEIBAUM (*m*): lead tree (*Arbor saturni*) 441
BLEIBEN: to remain; rest; last; stay; be permanent; be durable; be fast; be persistent; be stable 442
BLEIBEND: permanent; durable; stable; fast (of colours); persistent 443
BLEIBLECH (*n*): sheet lead 444
BLEIBLÜTE (*f*): mimetite; $(3Pb_3As_2O_8.PbCl_2)$ 445
BLEIBORAT (*n*): lead borate (see **BLEI, BORSAURES-**) 446
BLEI (*n*), **BORSAURES-**: lead borate; (*Plumbum boricum*); $Pb(BO_2)_2.H_2O$; Sg. 5.6 447
BLEIBROMID (*n*): lead bromide; $PbBr_2$; Sg. 6.572-6.611 448
BLEICARBONAT (*n*): (*Plumbum carbonicum*): lead carbonate; $PbCO_3$; Sg. 6.57; white lead ore, cerrusite (see also **CERUSSIT**) 449
BLEICHANSTALT (*f*): bleachery; bleaching works 450
BLEICHAPPARAT (*m*): bleaching apparatus 451
BLEICHBAD (*n*): bleaching liquid 452
BLEICHBEIZE (*f*): bleaching mordant 453
BLEICHE (*f*): bleaching works; bleach, pallor 454
BLEICHEN: to bleach; whiten; blanch; lose colour; fade; turn white or pale 455
BLEICHER (*m*): bleacher 456
BLEICHERDE (*f*): Fuller's earth (aluminium-magnesium silicate) 457
BLEICHEREI (*f*): bleachery; bleaching works; bleaching 458
BLEICHFASS (*n*): bleaching vat 459
BLEICHFLECK (*m*): bleaching stain 460
BLEICHFLÜSSIGKEIT (*f*): bleaching liquid; chlorine water 461
BLEICHGUT (*n*): material for bleaching 462
BLEICHHOLLÄNDER (*m*): poacher; poaching engine; bleaching engine 463
BLEICHHÜLFSMITTEL (*n*): bleaching agent; bleaching (aid) assistant 464
BLEICHKALK (*m*): chloride of lime; bleaching powder; $Ca(ClO)_2.4H_2O$; calcium hypochlorite (*Calcaria chlorata*); (see **BLEICHPULVER**) 465
BLEICHKASTEN (*m*): bleaching vat 466
BLEICHLAUGE (*f*): bleaching lye 467
BLEICHLORID (*n*): (*Plumbum chloratum*): (see **CHLORBLEI**) 468
BLEI (*n*), **CHLORSAURES-**: lead chlorate 469
BLEICHMITTEL (*n*): bleaching agent; decolourant 470
BLEICHÖL (*n*): bleaching oil 471
BLEICHPRODUKT (*n*): bleaching preparation 472
BLEICHPROZESS (*m*): bleaching process 473
BLEICHPULVER (*n*): bleaching powder; chlorinated lime; bleaching lime; hypochlorite of lime; (see **BLEICHKALK**) 474
BLEICHROMAT (*n*): lead chromate, Leipzig yellow (*Plumbum chromicum*); $PbCrO_4$; Sg. 6.123; (see also **CHROMGELB**) 475
BLEICHROMAT (*n*), **BASISCHES-**: basic lead chromate (see **CHROMROT**) 476
BLEICHROMAT (*n*), **NEUTRALES-**: (see **BLEICHROMAT**) 477
BLEI (*n*), **CHROMSAURES-**: (see **BLEICHROMAT**) 478
BLEICHSALZ (*n*): bleaching salt; chloride of lime; sodium hypochlorite; $NaOCl$ 479
BLEICHSEIFE (*f*): bleaching soap 480
BLEICHSODA (*n*): bleaching soda; sodium hypochlorite; (see **BLEICHSALZ**) 481
BLEICHSUCHT (*f*): chlorosis (Med.) 482
BLEICHWASSER (*n*): bleaching liquid; chlorine water; Eau de Javelle; (see **CHLORKALI-LÖSUNG**) 483
BLEICHWASSERSÜCHTIG: leucophlegmatic (Med.) 484
BLEICHWIRKUNG (*f*): bleaching (action) effect 485
BLEICHZWECK (*m*): bleaching purpose 486
BLEIDICHLORID (*n*): lead dichloride; lead chloride (*Plumbum chloratum*); $PbCl_2$; Sg. 5.8; Mp. 498°C.; Bp. 861-954°C. 487
BLEIDICHTUNG (*f*): lead joint-(ing) 488
BLEIDIOXYD (*n*): lead peroxide (see **BLEIOXYD, BRAUNES-**) 489
BLEIDRAHT (*m*): lead wire 490
BLEIEINFASSUNG (*f*): leading (for glass) 491
BLEIERDE (*f*): earthy (lead spar) cerussite; $PbO.CO_2$; (77.5% lead) 492
BLEIERZ (*n*): lead ore (see **BLEIGLANZ**) 493
BLEIESSIG (*m*): vinegar of lead; basic lead acetate solution in water (*Liquor plumbi subacetici*); Sg. 1.238 494
BLEI (*n*), **ESSIGSAURES-**: (*Plumbum aceticum*): lead acetate, sugar of lead; $Pb(C_2H_3O_2)_2.3H_2O$; Sg. 2.5; Bp. 280°C. 495
BLEIEXTRAKT (*m*): (see **BLEIESSIG**) 496
BLEIFAHLERZ (*n*): bournonite; (see **BOURNONIT**) 497
BLEIFARBIG: lead coloured; leaden hued 498
BLEIFEDER (*f*): lead pencil 499
BLEI (*n*), **FETTSAURES-**: lead sebate 500
BLEIFLASCHE (*f*): lead bottle 501
BLEIFLUORID (*n*): lead fluoride; PbF_2; Sg. 8.241 502
BLEIFOLIE (*f*): lead foil 503
BLEIFREI: lead-free; free from lead 504
BLEIGELB (*n*): lead chromate; $PbCrO_4$; Sg. 6.123; massicot (PbO); (92.8% lead) 505

BLEI (n), GERBSAURES- : lead tannate 506
BLEIGEWICHT (*n*): lead weight, sinker 507
BLEIGIESSER (*m*): plumber 508
BLEIGLANZ (*m*): lead glance; galenite; galena; lead sulphide; (PbS); (86.6% lead); Sg. 7.13-7.7 509
BLEIGLANZ (*m*), **OKTAEDRISCHER- :** Steinmannite; Pb,SbS_4; Sg. 6.8 510
BLEIGLAS (*n*): lead glass; Anglesite (Mineral) (see BLEISULFAT) 511
BLEIGLAS (*n*): lead glass, flint glass (see FLINTGLAS) 512
BLEIGLASFENSTER (*n*): leaded glass window-(s); leaded light-(s) 513
BLEIGLÄTTE (*f*): litharge; (PbO); Sg. 9.375; Mp. 888°C.; (see BLEIOXYD, GELBES-) 514
BLEIGLÄTTEANLAGE (*f*): litharge plant 515
BLEIGLAZUR (*f*): lead (glaze) glazing 516
BLEIGLIMMER (*m*): micaceous cerrusite; $PbO.CO_2$; (see ZERRUSIT) 517
BLEIGRAU: lead-grey 518
BLEIGRIES(S) (*m*): lead gravel, powdered lead 519
BLEIHALTIG: containing lead; plumbiferous 520
BLEI (*n*), **HARZSAURES- :** lead resinate; $Pb(C_{20}H_{29}O_2)_2$ 521
BLEI (*n*), **HOLZESSIGSAURES- :** lead pyrolignite 522
BLEI (*n*), **HOLZÖLSAURES- :** lead lignoleate 523
BLEIHORNERZ (*n*): phosgenite (natural chlorocarbonate of lead); (see PHOSGENIT) 524
BLEIHORNSPAT (*m*): (see BLEIHORNERZ) 525
BLEIHYDROXYD (*n*): hydrated lead oxide, lead hydroxide, lead hydrate (*Plumbum hydroxydatum*); $2PbO.H_2O$; Sg. 7.59 526
BLEIJODID (*n*): lead iodide; PbI_2; Sg. 6.12-6.16; Mp. 358°C.; Bp. 861-954°C. 527
BLEIKAMMER (*f*): lead chamber 528
BLEIKAMMERVERFAHREN (*n*): lead chamber process (of Sulphuric acid) 529
BLEIKARBONAT (*n*): lead carbonate; (see BLEICARBONAT) 530
BLEIKARBONAT (*n*), **BASISCHES- :** basic lead carbonate (grey coating formed on lead when exposed to moist air); (see also BLEIWEISS) 531
BLEIKIEL (*m*): lead keel 532
BLEI (*n*), **KOHLENSAURES- :** lead carbonate (see BLEICARBONAT) 533
BLEIKOLIK (*f*): lead cholic (*colica saturnina*) (Med.) 534
BLEIKÖNIG (*m*): lead regulus 535
BLEILASUR (*f*): linarite; $PbO.CuO.SO_3.H_2O$ 536
BLEILEGIERUNG (*f*): lead alloy 537
BLEI, LEINÖLSAURES- (*n*): lead linoleate; $Pb(C_{18}H_{31}O_2)_2$; lead plaster 538
BLEILOT (*n*): lead solder; plummet 539
BLEILÖTER (*m*): lead (burner) solderer; plumber 540
BLEILÖTEREI (*f*): lead soldering, plumbing 541
BLEIMANGANAT (*n*): (*Plumbum manganicum*): lead manganate; $PbMnO_4$ 542
BLEI (*n*), **MANGANSAURES- :** lead manganate (see BLEIMANGANAT) 543
BLEIMEHL (*n*): lead meal, lead dust, finely pulverized lead 544
BLEIMENNIGE (*f*): minium; red lead; red lead oxide; plumbo-plumbic oxide; Pb_3O_4; (90.6% lead); Sg. 9.07 545
BLEIMETAPHOSPHAT (*n*): lead metaphosphate; $Pb(PO_3)_2$; Mp. 800°C. 546
BLEIMETASILIKAT (*n*): lead metasilicate; $PbSiO_3$; Mp. 766°C. 547
BLEI (*n*), **MILCHSAURES- :** lead lactate 548
BLEIMOLYBDAT (*n*): lead molybdate; $PbMoO_4$; Sg. 6.62 549
BLEIMULDE (*f*): lead-pig 550
BLEIMULM (*m*): earthy galena; (see BLEIGLANZ) 551
BLEINIERE (*f*): bindheimite (natural hydrous antimonate of lead) 552
BLEINITRAT (*n*): lead nitrate (*Plumbum nitricum*); $Pb(NO_3)_2$; Sg. 4.41-4.53 553
BLEIOCHER (*m*): lead ochre 554
BLEI (*n*), **ÖLSAURES- :** lead oleate; $Pb(C_{18}H_{33}O_2)_2$ 555
BLEIORTHOSILIKAT (*n*): lead orthosilicate; Pb_2SiO_4; Mp. 740-746°C. 556
BLEIOXALAT (*n*): (*Plumbum oxalicum*): lead oxalate; PbC_2O_4 557
BLEI (*n*), **OXALSAURES- :** lead oxalate (see BLEIOXALAT) 558
BLEIOXYCHLORID (*n*): lead oxychloride; $PbO.PbCl_2$; $2PbO.PbCl_2$; $4PbO.PbCl_2$; $5PbO.PbCl_2$, Turner's yellow, patent yellow; $7PbO.PbCl_2$, Cassel yellow 559
BLEIOXYD (*n*): lead oxide; (*Plumbum oxydatum*); (see BLEIHYDROXYD; BLEIOXYD, BRAUNES-; BLEIOXYD, GELBES-; BLEIOXYD, ROTES-) 560
BLEIOXYD (*n*), **BORSAURES- :** lead borate (see BLEI, BORSAURES-) 561
BLEIOXYD (*n*), **BRAUNES- :** brown lead oxide, lead peroxide, lead dioxide, anhydrous plumbic acid, lead superoxide; PbO_2; Sg. 8.91; (*Plumbum hyperoxydatum*) 562
BLEIOXYD (*n*), **GELBES- :** yellow lead oxide, plumbous oxide, litharge, lead protoxide, massicot, lead monoxide; (*Lithargyrum*); PbO; Sg. 9.375; Mp. 888°C. 563
BLEIOXYDHYDRAT (*n*): (see BLEIHYDROXYD) 564
BLEIOXYD, ROTES- (*n*): red lead; plumbo-plumbic oxide; (see BLEIMENNIGE) 565
BLEIPEROXYD (*n*): lead dioxide; PbO_2; Sg. 8.91; (see BLEIOXYD, BRAUNES-) 566
BLEIPERSULFAT (*n*): (*Plumbum persulfuricum*); lead persulphate; $PbS_2O_8.2H_2O$ 567
BLEIPFLASTER (*n*): lead plaster; (see BLEI, LEINÖLSAURES-) 568
BLEI (*n*), **PHOSPHORSAURES- :** lead phosphate 569
BLEIRHODANÜR (*n*): lead sulfocyanide 570
BLEIROHR (*n*): lead pipe 571
BLEIRÖHRE (*f*): lead pipe 572
BLEIRÖSTPROZESS (*m*): lead roasting process 573
BLEIROT (*n*): (see BLEIMENNIGE) 574
BLEISALBE (*f*): lead ointment; cerate of lead subacetate 575
BLEISALPETER (*m*): lead nitrate; (*Plumbum nitricum*); $Pb(NO_3)_2$; Sg. 4.53 576
BLEI (*n*), **SALPETERSAURES- :** lead nitrate; (see BLEISALPETER) 577
BLEI (*n*), **SALPETRIGSAURES- :** lead nitrite 578
BLEISALZ (*n*): lead salt; lead acetate 579
BLEI (*n*), **SALZSAURES- :** lead chloride; (see CHLORBLEI) 580
BLEISAMMLER (*m*): lead (accumulator) storage battery 581

BLEISAUER: plumbate of, combined with plumbic acid 582

BLEISÄURE (*f*): plumbic acid; (anhydrous PbO_2 has Sg. 8.91); (see **BLEIOXYD, BRAUNES-**) 583

BLEISAURES SALZ (*n*): plumbate 584

BLEISCHALE (*f*): lead dish 585

BLEISCHAUM (*m*): lead scum 586

BLEISCHNEE (*m*): lead snow (a basic lead sulphate of the formula $PbSO_4.Pb(OH)_2$); Sg. 6.0 587

BLEISCHWAMM (*m*): lead sponge 588

BLEISCHWÄRZE (*f*): graphite, wad; (see **GRAPHIT**) 589

BLEI (*n*), **SCHWEFELSAURES-**: lead sulphate; (see **BLEISULFAT**) 590

BLEI (*n*), **SCHWEFLIGSAURES-**: lead sulphite 591

BLEISELENID (*n*): lead selenide; PbSe; Sg. 8.1 592

BLEISICHERUNG (*f*): lead cut-out, lead fuse 593

BLEISIKKATIV (*n*): lead borate; lead drier (for paints); (*Plumbum boricum*); PbB_2O_6; Sg. 5.598 594

BLEISILIKAT (*n*): lead silicate; $PbSiO_3$ 595

BLEISPAT (*m*): (see **ZERRUSIT**) 596

BLEISPAT, ROTER- (*m*): crocoite; crocoisite (a natural lead chromate, about 70% lead); $PbCrO_4$ 597

BLEISPAT (*m*), **WEISSER-**: cerusite; (see **ZERRUSIT**); $PbOCO_2$; (about 78% lead) 598

BLEISPIEGEL (*m*): specular galena; PbS; 86.6% lead; (see **BLEIGLANZ**) 599

BLEISTEIN (*m*): lead matt 600

BLEISTIFT (*m*): lead pencil 601

BLEISTIFT-(*m*) SCHÄRFER (*m*): pencil sharpener 602

BLEISUBAZETAT (*n*): (*Plumbum subaceticum*): lead subacetate, monobasic lead acetate; $Pb_2O(CH_3COO)_2$ 603

BLEISULFAT (*n*): (*Plumbum sulfuricum*): lead sulphate, anglesite; $PbSO_4$; Sg. 6.34 604

BLEISULFATRÜCKSTAND (*m*): lead-sulphate residue 605

BLEISULFID (*n*): (*Plumbum sulfuratum*): lead sulphide, galenite, galena; PbS; Sg. 7.65 [606

BLEISUPEROXYD (*n*): brown lead oxide, lead peroxide, lead superoxide, lead dioxide, anhydrous plumbic acid; PbO_2; Sg. 8.91; (*Plumbum hyperoxydatum*) 607

BLEITELLURID (*n*): lead telluride; PbTe; Mp. 917°C. 608

BLEITETRACHLORID (*n*): lead tetrachloride; $PbCl_4$; Sg. 3.18; Mp. -15°C. 609

BLEITHIOSULFAT (*n*): (*Plumbum thiosulfuricum*): lead thiosulphate, lead hyposulphite; PbS_2O_3 610

BLEITUTTE (*f*): lead assaying crucible 611

BLEI (*n*), **ÜBERSCHWEFELSAURES-**: lead persulphate; (see **BLEIPERSULF**) 612

BLEI (*n*), **UNTERSCHWEFLIGSAURES-**: lead hyposulphite, lead thiosulphate; PbS_2O_3 613

BLEIVERGIFTUNG (*f*): lead poisoning 614

BLEIVERHÜTTUNG (*f*): lead smelting 615

BLEIVITRIOL (*m*): lead vitriol; lead sulphate; $PbSO_4$; Anglesite (Mineral) 616

BLEIVITRIOLSPAT (*m*): Anglesite (see **BLEISULFAT**) 617

BLEIWASSER (*n*): lead water; Goulard water (a 1% solution of basic lead acetate) 618

BLEIWEISS (*n*): basic lead carbonate, white lead, (colour), ceruse, lead flake; $2PbCO_3.Pb(OH)_2$; lead subcarbonate 619

BLEIWEISSERSATZ (*m*): white lead substitute 620

BLEIWEISSSALBE (*f*): lead carbonate ointment 621

BLEIWOLFRAMAT (*n*): lead tungstate; $PbWO_4$; Sg. 8.18 622

BLEIWURF (*m*): heaving the lead (Naut.) 623

BLEIZINNOBER (*m*): minium, red lead; (see **BLEIMENNIGE**) 624

BLEIZUCKER (*m*): sugar of lead; lead acetate; $Pb(C_2H_3O_2)_2.3H_2O$; Sg. 2.5; Bp. 280°C. 625

BLEIZUCKER (*m*), **BRAUNER-**: (see **BLEI, HOLZESSIGSAURES-**) 626

BLEIZUCKERSALBE (*f*): lead acetate ointment 627

BLENDE (*f*): blend; glance; blind; screen; diaphragm; (Phot.); stop (Phot.); border; circular slide valve; Blende; zinc blende; sphalerite (67% Zn.); (see **ZINKSULFID**) 628

BLENDEN: to blind; dazzle 629

BLENDEÖFFNUNG (*f*): screen (diaphragm) opening (of a camera) 630

BLENDERÖSTOFEN (*m*): blende roasting furnace 631

BLENDUNG (*f*): dazzle, glare 632

BLEUELN: to beat; beetle; blue 633

BLICK (*m*): look; view; glance; flash; fulguration; blick; brightening (of molten silver); shine 634

BLICKFEUER (*n*): signal fire 635

BLICKGOLD (*n*): refined gold still containing silver 636

BLICKSILBER (*n*): refined silver 637

BLIND: blind; dull; tarnished; false 638

BLINDDARM (*m*): caecum 639

BLINDDARMENTZÜNDUNG (*f*): appendicitis (inflammation of the appendix vermiformis) (Med.) 640

BLINDFLANSCH (*m*): blank flange 641

BLINDLINGS: blindly; rashly 642

BLINKEN: to gleam; sparkle; glitter; twinkle 643

BLINKERN: (see **BLINKEN**) 644

BLINZELN: to blink; wink 645

BLITZABLEITER (*m*): lightning conductor 646

BLITZLICHT (*n*): flash light; flare; limelight 647

BLITZLICHTMISCHUNG (*f*): flash-light mixture 648

BLITZLICHTPRÄPARAT (*n*): flash-light preparation 649

BLITZPULVER (*n*): lycopodium (from spores of *Lycopodium clavatum*) 650

BLITZRÖHRE (*f*): fulgurite 651

BLITZSCHLAG (*m*): lightning stroke 652

BLITZSCHLAG (*m*), **KALTER-**: a flash of lightning which does not cause fire when it strikes 653

BLITZSCHUTZ (*m*): lightning (arrester) conductor 654

BLITZSINTER (*m*): fulgurite (Mineralogy) 655

BLITZSTRAHL (*m*): lightning flash 656

BLOCK (*m*): block; pig; ingot; sow (lead); pad (paper) 657

BLOCKABSTREIFKRAN (*m*): stripper (for stripping the chill-mould from the casting) (Metal) 658

BLOCKBLEI (*n*): pig lead 659

BLOCKBOJE (*f*): wooden buoy 660

BLOCK (*m*), **EINSCHEIBIGER-**: single pulley block 661

BLOCKHAHN (*m*): stop cock 662
BLOCKIEREN: to block; fix, jamb 663
BLOCKMÜHLE (*f*): block mill 664
BLOCKNAGEL (*m*): block pin 665
BLOCKSCHEIBE (*f*): block sheave, pulley block 666
BLOCKWALZE (*f*): cogging roll (rolling mills) 667
BLOCK (*m*), **ZWEISCHEIBIGER-**: double pulley block 668
BLÖDE: dim-sighted; imbecile; bashful 669
BLÖDIT (*m*): bloedite; $Na_2Mg(SO_4)_2.4H_2O$; Sg. 2.25 670
BLÖDSINN (*m*): imbecility; idiocy; insanity; feeblemindedness; nonsense; dementia; mental weakness (Med.) 671
BLOMSTRANDIN (*m*): blomstrandite; Nb_2O_5, Ta_2O_5, TiO_2, UO, FeO, CaO, H_2O; Sg. 4.2 672
BLOSS: bare-(ly); naked-(ly); nude; mere-(ly); sole-(ly); only 673
BLÖSSE (*f*): bareness; unhaired hide; pelt; nakedness 674
BLOSSLEGEN: to lay bare; strip 675
BLÜHEN: to bloom, blossom, flower 676
BLUMENBLATT (*n*): petal (Botanical) 677
BLUMENDECKBLATT (*n*): sepal (Botanical) 678
BLUMENDÜNGER (*m*): flower manure 679
BLUMENGÄRTNER (*m*): florist; nurseryman 680
BLUMENKOHL (*m*): cauliflower 681
BLUMENSCHEIDE (*f*): sheath (Botanical) 682
BLUMENSEITE (*f*): hair-side (Leather) 683
BLUMENSTAUB (*m*): pollen 684
BLUMENSTENGEL (*m*): flower-stalk (Botanical) 685
BLUMENSTIEL (*m*): peduncle (Botanical) 686
BLUMENTEE (*m*): imperial tea 687
BLUMENZWIEBEL (*f*): bulb-(of flowers) (Botanical) 688
BLUMIG: flowery; bloomy 689
BLUT (*n*): blood; Sg. 1.05 690
BLUTADER (*f*): vein; blood-vessel 691
BLUTALBUMIN (*n*): blood albumen (from ox-blood) 692
BLUTARM: bloodless; anæmic; very poor 693
BLUTARMUT (*f*): anæmia (Med.), deficiency of red corpuscles in the blood 694
BLUTBAHN (*f*): blood-vessel 695
BLUTBAUM (*m*): campeachy tree, logwood tree (*Haematoxylon Campechianum*) 696
BLUTBLUME (*f*): arnica; blood flower (*Haemanthus*) 697
BLUTBRECHEN (*n*): haematemesis (vomiting of blood from the stomach) (Med.) 698
BLUTDRUCK (*m*): blood pressure 699
BLUTDÜNGER (*m*): blood manure 700
BLUTEGEL (*m*): leech (*Hirudo*) 701
BLUTEN: to bleed; suffer; run; bleed (of colours) 702
BLÜTENBLATT (*n*): petal (Botanical) 703
BLÜTENDECKE (*f*): perigon, calyx (Botanical) 704
BLÜTENGEIST (*m*): blossom spirit 705
BLÜTENHÜLLENBLATT (*n*): sepal (one of the leaves forming the calyx) (Botanical) 706
BLÜTENKÄTZCHEN (*n*): catkin (Botanical) 707
BLÜTENKÖRBCHEN (*n*): anthodium 708
BLÜTENKRONE (*f*): corolla (Bot.) 709
BLÜTENÖL (*n*): flower oil (scent, perfume) 710
BLÜTENPFLANZEN (*f.pl*): *Phanerogamae*, phanerogamous plants, flowering plants (Bot.) 711
BLÜTENRIECHSTOFF (*m*): scent; perfume 712
BLÜTENSPELZE (*f*): glume (Bot.) 713
BLÜTENSPROSS (*m*): (flower)-shoot (Bot.) 714
BLÜTENSTAND (*m*): inflorescence (Bot.) 715
BLÜTENSTAUB (*m*): pollen (Bot.) 716
BLUTERKRANKHEIT (*f*): hæmophilia, haemorrhage diathesis (a condition of uncontrollable haemorrhage following a slight abrasion or wound) (Med.) 717
BLÜTEZEIT (*f*): time (period) for flowering or blossoming (of plants) 718
BLUTFARBE (*f*): blood colour; blood pigment 719
BLUTFARBIG: blood coloured; blood red; crimson 720
BLUTFARBSTOFF (*m*): colouring matter of the blood; blood pigment 721
BLUTFASERSTOFF (*m*): fibrin (see **ALBUMIN**) 722
BLUTFLECK (*m*): bloodstain 723
BLUTFLECKEN (*m*): (see **BLUTFLECK**) 724
BLUTFLECKENKRANKHEIT (*f*): purpura (purple patches or spots on the skin due to subcutaneous hæmorrhages) (Med.) 725
BLUTFLUSS (*m*): hæmorrhage, bleeding, emission of blood from its usual container (Med.) 726
BLUTFLÜSSIGKEIT (*f*): blood plasma 727
BLUTGEFÄSS (*n*): bloodvessel 728
BLUTGESCHWULST (*f*): hæmatocele (swelling due to effusion of blood under the testicle covering) (Med.) 729
BLUTGESCHWÜR (*n*): abscess 730
BLUT (*n*), **GETROCKNETES-**: dried blood, coagulated blood 731
BLUTGIERIG: blood-thirsty; sanguinary 732
BLUTHARNEN (*n*): hæmaturia, blood in the urine (Med.) 733
BLUTHIRSE (*f*): (see **HÜHNERFUSSGRAS**) 734
BLUTHOLZ (*n*): logwood; campeachy wood (see **BLAUHOLZ**) 735
BLUTHUND (*m*): bloodhound 736
BLUTHUSTEN (*m*): hæmoptysis, hæmorrhage from larynx, trachea or lungs (Med.) 737
BLUTKOHLE (*f*): blood charcoal (used as a decolourant) 738
BLUTKÖRPERCHEN (*n*): corpuscle (either red or white) 739
BLUTKRAUT (*n*): any blood red coloured plants or plants with blood stanching properties 740
BLUTKRAUT (*n*), **GELBES-**: goldenseal, turmeric (*Hydrastis canadensis*) 741
BLUTLAUF (*m*): blood circulation 742
BLUTLAUGENSALZ (*n*): potassium ferrocyanide; potassium ferricyanide 743
BLUTLAUGENSALZABSCHWÄCHER (*m*): potassium ferricyanide reducer (Phot.) 744
BLUTLAUGENSALZ (*n*), **GELBES-**: (*Kalium ferrocyanatum*): potassium ferrocyanide, yellow potassium prussiate; $K_4Fe(CN)_6.3H_2O$; Sg. 1.89 745
BLUTLAUGENSALZ (*n*), **ROTES-**: (*Kalium ferricyanatum*): potassium ferricyanide, red potassium prussiate; $K_3Fe(CN)_6$; Sg. 1.8109 746
BLUTLEER: bloodless; anæmic 747
BLUTLOS: (see **BLUTLEER**) 748
BLUTMEHL (*n*): dried blood, blood powder, blood meal (an artificial manure) 749
BLUTMELASSE (*f*): blood molasses 750
BLUTREINIGEND: blood-purifying 751
BLUTREINIGUNGSMITTEL (*n*): blood purifying medicine 752
BLUTREINIGUNGSTEE (*m*): blood purifying tea (usually made from walnut leaves, *Folia juglandis*); (see **HOLZTEE**) 753

BLUTROT: bloodred; crimson 754
BLUTRUHR (*f*): dysentery (Med.) 755
BLUTSCHLAG (*m*): apoplexy (Med.) 756
BLUTSERUM (*n*): blood serum, liquor sanguinis (the colourless fluid in which the corpuscles float) 757
BLUTSEUCHE (*f*): anthrax (both internal and external) (Med.) 758
BLUTSTEIN (*m*): bloodstone, hematite; Fe_2O_3; 70% iron 759
BLUTSTILLEND: blood-stanching; hemastatic; styptic 760
BLUTSTILLSTIFT (*m*): blood stanching (pencil) stick (a styptic) 761
BLUTSTURZ (*m*): hæmorrhage 762
BLUTSUCHT (*f*): hemophilia; hæmophilia (Med.) 763
BLUTUNTERSUCHUNG (*f*): blood examination 764
BLUTVERLUST (*m*): loss of blood 765
BLUTWASSER (*n*): (blood)-serum; lymph (an alkaline albuminous fluid from the blood) 766
BLUTWOLLE (*f*): dead wool 767
BLUTWURZ (*f*): any plant with red sap or blood stanching properties; blood-root (*Radix sanguinariæ* of *Sanguinaria canadensis* or *Rhizoma tormentillæ* of *Potentilla tormentilla*) 768
BLUTWURZEL (*f*): (see **BLUTWURZ**) 769
BÖ (*f*): squall (of weather) 770
BOCK (*m*): buck; he goat; bock beer (Brewing); trestle; jack; horse; ram; block; strut (Technical); blunder; mistake 771
BOCKASCHE (*f*): coal ashes 772
BOCKHOLZ (*n*): guaiacum wood; lignum vitæ; Sg. 0.79-1.61 773
BOCKIG: resistant; refractory (meaning to butt against, like a goat) 774
BÖCKIG: (see **BOCKIG**) 775
BOCKLEDER (*n*): goat leather; kid; buckskin 776
BOCKNUSS (*f*): souari (butter)nut (fruit of *Caryocar butyrosum* or *C. nuciferum*) 777
BOCKSÄURE (*f*): hircic acid (old term) 778
BOCKSEIFE (*f*): mountain soap (a kind of clay-impure iron or zinc sulphate) 779
BOCKSHORN (*n*): fenugreek (*Fœnum græcum*) 780
BOCKSHORNSAMEN (*m*): (*Semen Fœnugræci*): fenugreek, trigonella seeds 781
BODEN (*m*): soil; earth; ground; land; base; floor; bottom; loft; garret; attic; drum-end (boilers); header; head 782
BODENANALYSE (*f*): soil analysis 783
BODENART (*f*): type of soil 784
BODENAUSZUG (*m*): (lens-)extension, base-board (of a camera) 785
BODENBEARBEITUNG (*f*): tilling-(of the soil) 786
BODENFLÄCHE (*f*): floor space 787
BODENFRÄSER (*m*): ground mill (Agric.) 788
BODENGANG (*m*): bottom strake 789
BODENHALS (*m*): neck; tube; outlet branch at the bottom 790
BODENHEFE (*f*): grounds; dregs; lees 791
BODEN (*m*), **KIES-**: gravel soil 792
BODENKOLLOID (*n*): soil colloid 793
BODENKÖRPER (*m*): sediment; precipitate; solid; soil substance; body or substance at the bottom 794
BODENNÄHRSTOFF (*m*): soil nutrient; manure 795
BODENPLANKE (*f*): bottom plank 796
BODEN (*m*), **SAND-**: sandy soil 797
BODENSATZ (*m*): deposit; precipitate; sediment; dregs; bottoms; bottom; grounds; lees 798
BODENSCHÄTZE (*m.pl*): mineral wealth 799
BODENSTEIN (*m*): foundation brickwork for blast furnaces; bottom stone or masonry 800
BODENTEIG (*m*): undercrust; underdough 801
BODEN (*m*), **TON-**: clay soil 802
BODENUNTERSUCHUNG (*f*): soil (investigation) examination 803
BODENVENTIL (*n*): bottom valve; foot valve 804
BODENWRANGE (*f*): (iron)-floor plate, (wooden)-floor timber 805
BODMEREI (*f*): bottomry 806
BOGEN (*m*): bow; arc (Elect.); sheet (paper); bend; curve; arch 807
BOGENBRÜCKE (*f*): arched bridge 808
BOGENFLAMME (*f*): arc flame 809
BOGENGANG (*m*): arcade 810
BOGENLAMPE (*f*): arc lamp 811
BOGENLICHT (*f*): arc light 812
BOGEN (*m*), **PAPIER-**: sheet of paper 813
BOGENSÄGE (*f*): bow saw 814
BOGENSCHWEISSUNG (*f*): arc welding 815
BOGENSTÜCK (*n*): curved piece; bend 816
BOHLE (*f*): thick (heavy) plank; board; bowl 817
BOHNE (*f*): bean 818
BOHNEN: to wax 819
BOHNENÄHNLICH: bean-like 820
BOHNENBAUM (*m*): laburnam (*Cytisus laburnum*) 821
BOHNENERZ (*n*): granular iron ore (see **BOHNERZ**) 822
BOHNENKRAUT (*n*): summer savoury (*Herba satureiæ*); (*Satureia hortensis*) 823
BOHNENMEHL (*n*): bean meal, bean flour (from the seeds of *Phaseolus vulgaris*) 824
BOHNENÖL (*n*), **CHINESISCHES-**: (see **SOJABOHNENÖL**) 825
BOHNERMASSE (*f*): polishing wax 826
BOHNERWACHS (*n*): floor-(polishing) wax 827
BOHNERZ (*n*): pea ore; oölitic limonite (see **BRAUNEISENERZ**) 828
BOHNWACHS (*n*): polishing wax 829
BOHRBANK (*f*): boring bench; drilling bench 830
BOHRDECKEL (*m*): hole cover 831
BOHREISEN (*n*): bit, drill 832
BOHREN: to bore; drill 833
BOHRER (*m*): borer; drill; gimlet; auger; perforator; bit; boring tool 834
BOHRER (*m*), **HOLZ-**: auger 835
BOHRER (*m*), **LOCH-**: auger 836
BOHRER (*m*), **SCHLAG-**: percussion borer or drill 837
BOHRER (*m*), **STOSS-**: percussion borer or drill 838
BOHRER (*m*), **ZIMMERMANNS-**: auger 839
BOHRFAUSTEL (*n*): miner's (borer's) hammer or mallet 840
BOHRFUTTER (*n*): boring socket, drill chuck 841
BOHRGESTELL (*n*): boring frame 842
BOHRGUT (*n*): material gained by boring 843
BOHRKNARRE (*f*): ratchet drill, rack brace 844
BOHRKOPF (*m*): boring head, drill head; cutter-head (turbine tube cleaners) 845
BOHRKURBEL (*f*): crank brace 846
BOHRLOCH (*n*): bore hole; hole drilled 847
BOHRMASCHINE (*f*): boring (drilling) machine 848
BOHRMASCHINE (*f*), **HORIZONTAL-**: horizontal boring machine 849
BOHRMASCHINE (*f*), **VERTIKAL-**: vertical boring machine 850

BOHRMASCHINE (*f*), **ZYLINDER-**: cylinder boring machine 851
BOHRMEHL (*n*): borings; dust from boring 852
BOHRÖL (*n*): oil obtained by boring 853
BOHRSCHLAMM (*m*): lime; sludge 854
BOHRSCHMAND (*m*): slime, sludge (from boring) 855
BOHRSPÄNE (*m.pl*): drillings 856
BOHRSTANGE (*f*): boring rod or bar 857
BOHRSTOCK (*m*): boring stock; borer 858
BORUNG (*f*): boring; bore; hole; opening; drilling 859
BOHRUNGSTOLERANZ (*f*): boring tolerance 860
BOHRVORRICHTUNG (*f*): boring apparatus 861
BOHRWELLE (*f*): cutter bar, boring spindle 862
BOHRWERK (*n*): boring machine 863
BOHRWERKZEUG (*n*): boring (cutter, cutters, tool, tools, implement, implements) apparatus, bit, borer, drill 864
BOHRWURM (*n*): ship's borer (*Teredo navalis*) 865
BÖIG: squally 866
BOJE (*f*): buoy (Naut.) 867
BOL (*m*): (see **BOLUS**) 868
BOLLE (*f*): bulb; onion (see **SOMMERZWIEBEL**) 869
BOLLWERK (*n*): bulwark (Naut.) 870
BOLOGNESERFLASCHE (*f*): Bologna flask 871
BOLUS (*m*): bole (mixture of Al_2O_3,Fe_2O_3 and SiO_2); Sg. 2.35 872
BOLUS (*m*), **ROTER-**: reddle, red bole (an ochrous red clay), (bole containing oxide of iron) 873
BOLUS (*m*), **WEISSER-**: kaolin, china clay, white bole; $Al_2O_3.2SiO_2.2H_2O$; (see **BOLUS**) 874
BOLZEN (*m*): bolt; peg; pin; prop; stay-(bolt); rivet 875
BOLZENDREHBANK (*f*): bolt lathe 876
BOLZENKEIL (*m*): bolt key 877
BOLZENKOPF (*m*): bolt head 878
BOLZENMUTTER (*f*): nut for a bolt 879
BOLZENSCHEIBE (*f*): washer for a bolt 880
BOLZEN (*m*), **VERBINDUNGS-**: tie (binding) bolt 881
BOMBE (*f*): bomb; shell; bomb-(shell) 882
BOMBE, KALORIMETRISCHE- (*f*): bomb calorimeter 883
BOMBENROHR (*n*): bomb tube 884
BOMSE (*f*): support (Ceramics) 885
BONDUCNUSSÖL (*n*): bonducella nut oil (from seeds of *Caesalpinia bonducella*) 886
BONIFIKATION (*f*): indemnity, bounty, bonus, compensation 887
BONIFIZIEREN: to indemnify, compensate, make good 888
BONITIERUNG (*f*): valuation; appraisement 889
BOOT (*n*): boat 890
BOOT (*n*), **GROSSES-**: long boat 891
BOOTSBAU (*m*): boat building 892
BOOTSBAUER (*m*): boat builder 893
BOOTSCHWERT (*n*): lee-board (Naut.) 894
BOOTSDAVIT (*m*): boat davit 895
BOOTSDECK (*n*): boat deck 896
BOOTSHAKEN (*m*): boathook 897
BOOTSHAUS (*n*): boat house 898
BOOTSKLAMPE (*f*): cleat 899
BOOTSKRAN (*m*): boat crane 900
BOOTSLACK (*m*): boat varnish 901
BOOTSMASCHINE (*f*): engine for a boat 902
BOOTSRIEMEN (*m*): oar 903
BOOTSSCHWERT (*n*): leeboard (Naut.) 904
BOOTSZELT (*n*): awning of a boat 905
BOR (*n*): boron (B); Sg. 2.45 906
BORACIT (*n*): boracite (see **BORAZIT**) 907
BORAT (*n*): borate 908
BORÄTHAN (*n*): borethane; boroethane 909
BORAX (*m*): (*Natrium Boracicum*): borax (see also **NATRIUMTETRABORAT**); $Na_2B_4O_7.10H_2O$; Sg. 1.692-1.757 910
BORAXANLAGE (*f*): borax plant 911
BORAX (*m*), **GEWÖHNLICHER PRISMATISCHER-**: ordinary prismatic borax (see **BORAX**) 912
BORAXGLAS (*n*): borax glass 913
BORAXHONIG (*m*): borax honey 914
BORAXKALK (*m*): calcium borate (see **CALCIUM-BORAT**) 915
BORAX (*m*), **OKTAEDRISCHER-**: octahedral borax (see **JUWELIERBORAX**) 916
BORAXPERLE (*f*): borax bead 917
BORAXSAUER: (see **BORSAUER**) 918
BORAXSÄURE (*f*): boric acid (see **BORSÄURE**) 919
BORAXSÄURE (*f*), **PRISMATISCHE-**: sassoline, native boracic acid (Boracic acid about 56%, water 44%); BoO_4H_3; Sg. 1.48 920
BORAXSEIFE (*f*): borax soap 921
BORAXSPAT (*m*): boracite (found at Stassfurt, Germany) (see **BORAZIT**) 922
BORAXWEINSTEIN (*m*): boryl potassium tartrate (*Tartarus boratus*) 923
BORAZIT (*m*): boracite (a natural mineral con--taining borax, found at Stassfurt, Germany); $Mg_7Cl_2B_{16}O_{30}$; Sg. 2.6-2.7 924
BORBUTAN (*n*): borobutane 925
BORCARBID (*n*): boron carbide; B_6C or B_2C_2; Sg. 2.51-2.7 926
BORCHLORID (*n*): boron chloride 927
BORD (*m*): board (ship); border; edge; rim 928
BORD, AN- GEHEN: to go (aboard) on board ship 929
BORDANLAGE (*f*): marine (installation) plant 930
BORDEAUX: bordeaux 931
BÖRDELEISEN (*n*): bordering tool, hatchet stake 932
BÖRDELMASCHINE (*f*): socketing machine 933
BÖRDELN: to border, angle, flange, socket 934
BORDFREI: free on board, f.o.b. 935
BORDSPRACHE (*f*): nautical language 936
BORDWAND (*f*): ship's wall 937
BORETSCH (*m*): borage (*Borago officinalis*) 938
BORETSCHBLÜTEN (*f.pl*): borage flowers (flowers of *Borago officinalis*) 939
BORFLUORWASSERSTOFF (*m*): fluoboric acid; hydrofluoboric acid; HBF_4 940
BORFLUORWASSERSTOFFSÄURE (*f*): (see **BOR-FLUORWASSERSTOFF**) 941
BORFLUSSSÄURE (*f*): fluoboric acid, hydrofluoboric acid; HBF_4 942
BORGEN: to borrow; lend; obtain on credit 943
BORGPARDUNE (*f*): preventer backstay (Naut.) 944
BORGSTENGE (*f*): spare topmast 945
BORIUM (*n*): boron (see **BOR**) 946
BORKALK (*m*): calcium borate (see **CALCIUM-BORAT**) 947
BORKE (*f*): bark (*Cortex*); crust; scab; rind 948
BORKEN: to bark, remove bark, decorticate 949
BORN (*m*): spring; well; salt pit 950
BORNEOKAMPFER (*m*): Borneo camphor (see **BORNEOL**) 951
BORNEOL (*n*): borneol, bornyl alcohol, borneo camphor, isoborneol; $C_{10}H_{18}O$; Sg. 0.8083; Mp. 204°C.; Bp. 212°C. 952
BORNEOLISOVALERIANSÄUREESTER (*m*): borneol isovalerianate; $C_{10}H_{17}.O.C_5H_9O$; (Bp. 255°C.) 953

BORNEOL (*n*), **RECHTS-** : dextroborneol ; $C_{10}H_{18}O$; Sg. 0.8083 ; Mp. 204°C ; Bp. 212°C. 954
BORNIT (*m*) : horseflesh ore, peacock ore, erubescite, variegated copper ore, bornite, peacock copper ore, purple copper ore (natural copper-iron sulphide ; 62%Cu.) ; Cu_5FeS_4 ; Sg. 5.15 ; (see also **BUNTKUPFERERZ**) 955
BORNYLAZETAT (*n*) : (see **BORNYL, ESSIGSAURES-**) 956
BORNYL, (*n*), **BALDRIANSAURES-** : bornyl valer--ianate 957
BORNYL (*n*), **ESSIGSAURES-** : bornyl acetate, artificial pine oil ; Mp. 29°C. ; Bp. 98°C. [958
BORNYLVALERIANAT (*n*) : bornyl valerate 959
BORNYVAL (*n*) : (see **BORNEOLISOVALERIAN--SÄUREESTER**) 960
BOROCALCIT (*m*) : (see **BOROKALZIT**) 961
BOROCARBID (*n*) : boron carbide ; B_6C ; Sg. 2.7 962
BOROCITRONSAUER : borocitrate 963
BOROGLYCERID (*n*) : boroglyceride 964
BOROGLYCERIN (*n*) : boroglycerine ; $C_3H_5.BO_3$ 965
BOROKALZIT (*m*) : borocalcite (a natural calcium borate) ; $CaB_4O_7+4H_2O$ 966
BOROL (*n*) : borol (mixture of boric acid and sodium bisulphate) 967
BORONATROKALZIT (*n*) : boronatrocalcite (a natural hydrous calcium-sodium borate) ; $CaNaB_5O_9+6H_2O$; Sg. 1.73 968
BOROVERTIN (*n*) : borovertine, hexamethylenetetramine triborate ; $(CH_2)_6N_4.3HBO_2$ [969
BOROWOLFRAMSÄURE (*f*) : borotungstic acid, borowolframic acid ; $B_2O_3(WO_3)_9.24H_2O$; Sg. 3.0 970
BORREE (*m*) : (see **WINTERPORREE**) 971
BORSALBE (*f*) : boric acid ; boracic acid ; boric (boracic) ointment 972
BORSAUER : borate of ; combined with boric (boracic) acid 973
BORSÄURE (*f*) : boric (boracic) acid ; H_3BO_3 ; Sg. 1.4347 ; Mp. 184°C. ; (*Acidum boricum*) 974
BORSÄUREANHYDRID (*n*) : boric anhydride ; B_2O_3 975
BORSÄURESEIFE (*f*) : boric (boracic) acid soap 976
BORSAURES NATRON (*n*) : borate of soda, borax (see **BORAX**) 977
BORSÄUREWEINSTEIN (*m*) : (see **BORAXWEINSTEIN**) 978
BÖRSE (*f*) : purse ; exchange 979
BÖRSENGESCHÄFT (*n*) : stock (money) market 980
BÖRSENPAPIERE (*n.pl*) : stocks, shares, securities 981
BORSILICID (*n*) : boron silicide (see **SILICIUM--BORID**) 982
BORST (*m*) : burst ; crack ; fissure ; rupture 983
BORSTAHL (*m*) : boron steel 984
BORSTE (*f*) : bristle 985
BORSTICKSTOFF (*m*) : boron nitride; (BN) 986
BORSTIG : cracked ; bristly ; surly 987
BORTE (*f*) : border ; edging ; lace 988
BORTRIBROMID (*n*) : boron tribromide ; BBr_3 ; Sg. 2.65 989
BORTRICHLORID (*n*) : boron trichloride ; BCl_3 ; Sg. 1.434 990
BORTRIFLUORID (*n*) : boron trifluoride ; BF_3 ; Mp. –127°C. ; Bp. –101°C. 991
BORTRIJODID (*n*) : boron triiodide ; BJ_3 ; Sg. 3.3 992
BORTRIOXYD (*n*) : boron trioxide ; B_2O_3 ; Sg. 1.79 993
BORVERBINDUNG (*f*) : boron compound 994
BORWASSERSTOFF (*m*) : boron hydride 995
BORWOLFRAMSÄURE (*f*) : borotungstic acid ; borowolframic acid ; $B_2O_3(WO_3)_9.24H_2O$; Sg. 3.0 996
BÖSARTIG : malignant ; wicked ; ill-natured 997
BÖSCHEN : to slope ; bank ; slant ; bevel 998
BÖSCHUNG (*f*) : slope ; sloping ; embankment ; bank ; slant ; bevel 999
BÖSE : bad ; evil ; harmful ; wicked ; noxious sore ; aching 000
BÖSER VORSATZ (*m*) : malice (prepense) aforethought) 001
BÖSES WESEN (*n*) : epilepsy (Med.) ; evil spirit 002
BÖSES WETTER (*n*) : choke damp (Mining) ; bad weather 003
BOSHAFT: malicious ; wicked ; malignant 004
BOSHEIT (*f*) : wickedness ; malice ; anger ; malignity 005
BOSSIRWACHS (*n*) : moulding wax 006
BOSTWICKGITTER (*n*) : sliding lattice-work door (as of a lift) 007
BOTANIK (*f*) : botany 008
BOTANIKER (*m*) : botanist 009
BOTANISCH : botanical 010
BOTE (*m*) : messenger ; ambassador 011
BOTSCHAFT (*f*) : message ; tidings ; news ; embassy 012
BÖTTCHER (*m*) : cooper 013
BOTTICH (*m*) : coop ; vat ; tub ; tun ; tank ; vessel ; back 014
BOTTICHGÄRUNG (*f*) : tub (vat) fermentation 015
BOTTICHKÜHLER (*m*) : attemperator (Brewing) 016
BOTTLEREI (*f*) : pantry 017
BOUANDJOBUTTER (*f*) : bouandjo butter (from the seeds of *Allanblackia floribunda*) 018
BOUGIE (*f*) : candle ; catheter (Med.) 019
BOULANGERIT (*m*) : boulangerite ; $Pb_6Sb_4S_{11}$; Sg. 6.0 020
BOURNONIT (*m*) : cog-wheel ore, bournonite (a lead-copper sulphantimonide) ; $PbCuSbS_3$; Sg. 5.8 021
BOUTEILLE (*f*) : bottle 022
BOYKOTTIERUNG (*f*) : boycotting 023
BRACH : fallow ; unploughed 024
BRACH LIEGEN : to be dull (of the market) 025
BRACHYTYPES MANGANERZ (*n*) : braunite ; Mn_2O_3 ; Sg. 4.85 026
BRACK (*m*) : rubbish ; refuse 027
BRACKGUT (*n*) : rubbish ; refuse 028
BRACKHOLZ (*n*) : decayed (brittle, perished, rotten) wood 029
BRACKIG : brackish 030
BRAMBRASSE (*f*) : top-gallant brace (Naut.) 031
BRAMSALING (*f*) : top-gallant cross-tree 032
BRAMSEGEL (*n*) : top-gallant sail 033
BRAMMENKOKILLE (*f*) : chill mould (for ingots) (Metal) (of rectangular shape) 034
BRAMME (*f*) : bloom ; ingot (Metal) ; slab (of iron) 035
BRAND (*m*) : burning ; fire ; charge (of furnace) ; blight ; smut (Botanical) ; gangrene (death of tissue) (Med.) ; slough ; brand ; combustion ; burn 036
BRANDERZ (*n*) : bitmenous shale ; idrialite 037
BRANDFEST : fireproof 038
BRANDFLECK (*m*) : (see **BRANDFLECKEN**) 039
BRANDFLECKEN (*m*) : burn ; mark of burning ; scald 040

BRANDGESCHOSS (*n*): incendiary (bomb)-shell 041
BRANDGOLD (*n*): refined gold 042
BRANDGRANATE (*f*): (see **BRANDGESCHOSS**) 043
BRANDHARZ (*n*): empyreumatic resin 044
BRANDIG: burnt; blasted; gangrenous 045
BRAND (*m*), **KALTER-**: gangrene 046
BRANDMARK (*n*): brand; mark 047
BRANDMARKEN: to brand; mark 048
BRANDÖL (*n*): oil which is obtained by the process of destructive distillation 049
BRANDPROBE (*f*): fire (assay) test 050
BRANDRISS (*m*): (fire)-crack (due to fire) 051
BRANDSCHADEN (*m*): damage caused by fire 052
BRANDSCHIEFER (*n*): bitumenous shale; (see **BRANDERZ**) 053
BRANDSILBER (*n*): refined silver 054
BRANDSTEIN (*m*): brick 055
BRANDSTIFTUNG (*f*): arson; incendiarism 056
BRANDUNG (*f*): surge; surf; seething; breakers 057
BRANDWUNDE (*f*): burn, wound caused by a burn 058
BRANDZIEGEL (*m*): (fire)-brick 059
BRÄNKE (*f*): yeast tub (Brewing) 060
BRANNTWEIN (*m*): spirits (brandy, etc.) 061
BRANNTWEINBLASE (*f*): still (for spirits) 062
BRANNTWEINBRENNER (*m*): distiller 063
BRANNTWEINBRENNEREI (*f*): distillery, distillation 064
BRANNTWEINHEFE (*f*): alcohol ferment (*Saccharomyces cerevisiae*) (see also **OBERGÄRIGE HEFE**) 065
BRANNTWEINMIXTUR (*f*): brandy mixture (Pharmacy) 066
BRANNTWEINPROBER (*m*): alcoholometer 067
BRANNTWEINSPÜLICHT (*m*): wash (produced by fermentation of the wort) 068
BRANNTWEINWAGE (*f*): alcoholometer 069
BRANSCHE (*f*): branch (of trade); line (of goods); department 070
BRANSTIG: having a burnt smell or taste 071
BRASCHEN-WÄSCHE (*f*): combustible matter remaining in clinker (a sort of waste washings; obtained by breaking and washing the clinker) 072
BRASILEIN (*n*): brasilein (the colouring principle of redwood); $C_{16}H_{12}O_5$ 073
BRASILHOLZ (*n*): Brazil wood (of *Caesalpinia crista*); (see also **BRASILIENHOLZ**) 074
BRASILIENHOLZ (*n*): brasil wood (see also **BRASIL-HOLZ**); (see **FERNAMBUKHOLZ**) 075
BRASILINSÄURE (*f*): brasilinic acid 076
BRASILNUSSÖL (*n*): (see **SAPUCAJAÖL**) 077
BRASILSÄURE (*f*): brasilic acid 078
BRASSBAUM (*m*): outrigger 079
BRASSE (*f*): brace 080
BRASSEN: to brace; (*m*) bream (the fish) (*Abramis brama*, *A. blicca* and *A. vimba*) 081
BRASSENBLOCK (*m*): brace block 082
BRASSIDINSÄURE (*f*): brassidic acid; $C_{22}H_{42}O_2$; Mp. 65°C. 083
BRASSINSÄURE (*f*): brassic acid; erucic acid 084
BRATEN (*m*): roast; grill; broil; fry; burn; scorch; (also the verb to roast etc.) 085
BRATENFETT (*n*): dripping 086
BRATFRISCHARBEIT (*f*): roasting and refining 087
BRATSPILL (*n*): windlass 088
BRAU (*m*): brew; total quantity brewed; amount brewed; malt liquor 089
BRÄU (*m*): (see **BRAU**) 090
BRAUBOTTICH (*m*): brewing vat 091
BRAUBURSCH (*m*): brewery journeyman 092
BRAUCH (*m*): custom; usage 093
BRAUCHBAR: useable; serviceable; fit; suitable 094
BRAUCHEN: to use; want; need; make use of; employ; require; necessitate 095
BRAUE (*f*): brow (of eye) 096
BRAUEN: to brew 097
BRAUER (*m*): brewer 098
BRAUEREI (*f*): brewery; brewing 099
BRAUEREIEINRICHTUNG (*f*): brewery (installation) plant 100
BRAUEREIGLASUR (*f*): brewery glaze 101
BRAUEREIMASCHINEN (*f.pl*): brewery machinery 102
BRAUERLACK (*m*): brewer's varnish 103
BRAUERPECH (*n*): brewer's pitch 104
BRAUHAUS (*n*): brewery 105
BRAUKESSEL (*m*): kettle; copper; boiler; pan (Brewing) 106
BRAUKUFE (*f*): brewer's tub 107
BRAUMEISTER (*m*): master brewer 108
BRAUN: brown; bay (horse) 109
BRAUNBLEIERZ (*n*): pyromorphite (green, brown or variegated lead ore; natural lead chlorophosphate); $3Pb_3P_2O_8.PbCl_2$; Sg. 6.9 to 7.0; (see also **PYROMORPHIT**) 110
BRAUNBLEIOXYD (*n*): plattnerite; PbO_2; Sg. 9.4 111
BRÄUNE (*f*): angina; cynanche; quinsy (Med.); brownness 112
BRAUNEISENERZ (*n*): brown iron ore; limonite; (hydrated iron oxide); $2Fe_2O_3.3H_2O$; 30-60% iron; Sg. 3.6-4.2 113
BRAUNEISENERZ, OOLITHISCHES- (*n*): (variety of brown iron ore); oolithic brown iron ore; (see **BRAUNEISENERZ**) 114
BRAUNEISENSTEIN (*m*): (see **BRAUNEISENERZ**) 115
BRAUNELLE (*f*): self-heal (*Prunella vulgaris*) 116
BRÄUNEN: to burnish (see **BRÜNIEREN**) 117
BRAUNER BLEIZUCKER (*m*): (see **BLEI, HOLZESSIG-SAURES-**) 118
BRAUNER GLASKOPF (*m*): limonite (see **BRAUN-EISENERZ**) 119
BRAUNERZ (*n*): brown ore; vivianite (see **BLAUE ERDE**) 120
BRAUNFISCH (*m*): porpoise (*Delphinus phocaena*) 121
BRAUNFISCHTRAN (*m*): (see **MEERSCHWEINTRAN**) 122
BRAUNHEIL (*n*): (see **BRAUNELLE**) 123
BRAUNHOLZ (*n*): logwood (heartwood of *Hæmatoxylon campechianum*) 124
BRAUNHOLZPAPIER (*n*): browns; nature brown (paper) 125
BRAUNIT (*m*): braunite ($3Mn_2O_3MnO.SiO_2$); manganese sesquioxide; manganic oxide (Mn_2O_3); Sg. 4.5-4.8 126
BRAUNKALK (*m*): dolomite; $CaMg(CO_3)_2$ or $(CaMgFe)CO_3$ 127
BRAUNKOHLE (*f*): brown coal; lignite (calorific value about 2500 to 5000 calories) (C equals 55-75%) (German brown coal soft and earthy; lignite is hard) 128
BRAUNKOHLENBITUMEN (*n*): brown coal (lignite) bitumen; montan wax (up to 5% ash; 25-50% resin; Mp. 80-90°C.) (see also **MONTANWACHS** and **ERDHARZ**) 129
BRAUNKOHLENFEUERUNG (*f*): lignite furnace; lignite (brown-coal) firing 130
BRAUNKOHLENFORMATION (*f*): tertiary system 131

BRAUNKOHLENKOKS (*m* and *pl*): brown coal (lignite) coke 132
BRAUNKOHLENLÖSCHE (*f*): lignite (brown-coal) dust 133
BRAUNKOHLENPECH (*n*): lignite (bitumenous) pitch 134
BRAUNKOHLENPULVER (*n*): Cologne earth, Van Dyke brown, Cassel brown 135
BRAUNKOHLENTEER (*m*): lignite (bitumenous) tar (distilled from bitumenous brown coals) (60% paraffin; 33% crude oil; 2% distillation gas; 2% coke; moisture, etc.); Sg. 0.82-0.95 136
BRAUNKOHLENTEERPECH (*n*): bitumenous (lignite) pitch 137
BRAUNKOHLENWACHS (*n*): brown coal (lignite) wax 138
BRAUNKRÄUSEN (*f.pl*): fuzzy heads (Brewing) 139
BRAUNLACK (*m*): brown lake (colour) 140
BRAUNROT (*n*): colcothar; Indian red 141
BRAUNSALZ (*n*): Bismark brown (Mark G); chrysoidine (Mark R); (diazobenzene hydrochloride plus meta-phenylenediamine; a basic azo-dyestuff) 142
BRAUNSCHLIFF (*m*): steamed mechanical wood (Paper) 143
BRAUNSCHWEIGERGRÜN (*n*): Brunswick green (see **BREMERGRÜN**; also **KUPFERCARBONAT**) 144
BRAUNSPÄNE (*m.pl*): logwood (shavings) chips (from the heartwood of *Hœmatoxylon campechianum*) 145
BRAUNSPAT (*m*): dolomite (see **BRAUNKALK**) (a variety of magnesite, ankerite or dolomite which turns brown on exposure); Sg. 2.9 [146
BRAUNSTEIN (*m*): manganese dioxide; black manganese oxide; pyrolusite; MnO_2; 63% Mn; Sg. 5.026 147
BRAUNSTEINELEMENT (*n*): Leclanchè battery 148
BRAUNSTEINRAHM (*m*): bog manganese; wad; $MnO.MnO_2.H_2O$; Sg. 3.5 149
BRAUNSTEIN, ROTER- (*m*): rhodochrosite (natural manganese carbonate); $MnCO_3$ 150
BRAUNSTEINSCHAUM (*m*): (see **BRAUNSTEINRAHM**) 151
BRAUNSTEIN, SCHWARZER- (*m*): Hausmannite (a black manganese oxide); 2MnO plus MnO_2 (see also **HAUSMANNIT**) 152
BRAUNTRAN (*m*): blubber; thick cod oil 153
BRAUNWURZGEWÄCHSE (*n.pl*): *Scrophulariaceæ* (Bot.) 154
BRAUPFANNE (*f*): kettle pan; copper (Brewing) 155
BRAUSCH: decayed, brittle, perished, rotten (of wood) 156
BRAUSE (*f*): effervescence; sprinkler; rose; douche; fermentation 157
BRAUSEBAD (*n*): shower bath 158
BRAUSELITHIUMZITRAT (*n*): effervescent lithium citrate 159
BRAUSEMISCHUNG (*f*): effervescent mixture 160
BRAUSEN: to effervesce; froth; roar; buzz; ferment 161
BRAUSEND: effervescent 162
BRAUSEPULVER (*n*): effervescent powder; Seidlitz powder 163
BRAUSESALZ (*n*): effervescent (saline) salt 164
BRAUSEWEIN (*m*): sparkling wine 165
BRAUT IN GRÜNEN: (see **KAPUZINERKRAUT**) 166
BRAUT IN HAAREN: (see **KAPUZINERKRAUT**) 167
BRAUWASSER (*n*): brewing water; water for brewing purposes; liquor (Brewing) 168
BRAUWESEN (*n*): brewing business 169
BRECCIE (*f*): breccia (angular pieces of rock joined by a matrix) (Geology) 170
BRECCIENARTIG: brecciated; like breccia 171
BRECHARZNEI (*f*): emetic 172
BRECHBACKEN (*f.pl*): crusher jaws 173
BRECHBAR: breakable; brittle; refrangible 174
BRECHBOHNE (*f*): kidney bean (*Phaseolus vulgaris*) 175
BRECHEISEN (*n*): crowbar 176
BRECHEN: to break; burst; refract; mine; quarry; crush; pulverize; retch; vomit; pluck; fold; bend; blend (colours); boil-off (silk); crease; crinkle; change (dye-baths); decompose; fractionize; crook; wear-out; ruin; bevel; chamfer; (see also **FLOCKEN**) 177
BRECHENERREGEND: emetic 178
BRECHER (*m*): breaker; crusher; grinding mill (for coal, ores etc.) 179
BRECHGEBIRGE (*m*): rock which is capable of being worked by means of sledge hammer and wedges 180
BRECHHAUFEN (*m*): broken piece (Brewing) 181
BRECHKÖRNER (*n.pl*): castor beans (seeds of *Ricinus communis*) 182
BRECHMITTEL (*n*): emetic; vomiting agent (Med.) 183
BRECHNUSS (*f*): nux vomica (dried ripe seed of *Strychnos nux vomica*) 184
BRECHPULVER (*n*): emetic powder 185
BRECHPUNKT (*m*): point of change; change point; point of break; breaking point; crank point 186
BRECHREIZ (*m*): vomiting 187
BRECHSTANGE (*f*): crow-bar; pinch-bar 188
BRECHTSCHES DOPPELSALZ (*n*): Brecht's double salt, potassium-bimagnesium sulphate 189
BRECHUNG (*f*): breaking, etc. (see **BRECHEN**); refraction (Physics) 190
BRECHUNGSEBENE (*f*): plane of refraction (Optics) 191
BRECHUNGSEXPONENT (*m*): refractive index 192
BRECHUNGSGESETZ (*n*): law of refraction 193
BRECHUNGSINDEX (*m*): refractive index 194
BRECHUNGSKOEFFIZIENT (*m*): coefficient of refraction 195
BRECHUNG (*f*), **STRAHLEN-**: refraction 196
BRECHUNGSVERHÄLTNIS (*n*): (see **BRECHUNGS-EXPONENT**) 197
BRECHUNGSVERMÖGEN (*n*): refractive power; refraction 198
BRECHUNGSWINKEL (*m*): angle of refraction 199
BRECHWALZWERK (*n*): crushing (grinding) mills; crushing rolls or rollers 200
BRECHWEIN (*m*): emetic wine; wine of antimony; antimonial wine; solution of tartar emetic in sherry 201
BRECHWEINSTEIN (*m*): tartar emetic; antimony potassium tartrate (*Tartarus stibiatus*); $K(SbO)C_4H_4O_6$; Sg. 2.6 202
BRECHWEINSTEINERSATZ (*m*): tartar emetic substitute 203
BRECHWERK (*n*): crusher; grinder 204
BRECHWURZ (*f*): ipecacuanha-(root); (*Radix ipecacuanhæ*) (dried root of *Cephælis ipecacuanha* or *Uragoga ipecacuanha*) 205
BREI (*m*): pulp; puree; mash; broth; paste; magma; porridge; pap; slip (Ceramics) 206
BREIARTIG: pulpy; pappy; pasty; thickly fluid; viscous 207
BREIIG: (see **BREIARTIG**) 208

BREISLAUCH (*m*): (see SCHNITTLAUCH) 209
BREIT: broad; flat; wide; large 210
BREITBEIL (*n*): broad axe; broad bill 211
BREITE (*f*): breadth; width; latitude 212
BREITEISEN (*n*): broad tool; flat billet (Metal) 213
BREITEN: to spread; extend; flatten 214
BREITENMETAZENTRUM (*n*): transverse meta-centre 215
BREITENTRÄGHEITSMOMENT (*n*): lateral moment of inertia (shipbuilding) 216
BREITFLANSCHIG: broad flanged 217
BREITFÜSSIGE ZAHNSTANGE (*f*): cogged rail; flat-bottomed rack rail 218
BREITHACKE (*f*): mattock 219
BREITHAMMER (*m*): sledge hammer 220
BREITHAUE (*f*): mattock 221
BREITHAUPTIT (*m*): breithauptite (natural nickel antimonide); NiSb; Sg. 7.55 222
BREITLEDER (*n*): sole leather 223
BREITSCHNITTHOLZ (*n*): planks, boles, boards 224
BREITSEITE (*f*): broadside 225
BREITSEITGESCHÜTZ (*n*): broadside gun 226
BREITSPUR (*f*): broad-gauge (Railways) 227
BREITSPURIG: broad-gauge (Railways); having a broad gauge 228
BREITUMSCHLAG (*m*): poultice; cataplasm (Med.) 229
BREMER BLAU (*n*): Bremen blue (a blue-green colour); $Cu(OH)_2$; (see also KUPFER--CARBONAT) 230
BREMER GRÜN (*n*): (see BREMER BLAU and KUPFERCARBONAT) 231
BREMSANORDUNG (*f*): braking arrangement; arrangement of brakes 232
BREMSANTRIEB (*m*): brake control 233
BREMSARBEIT (*f*): braking energy; braking effect 234
BREMSAUSRÜSTUNG (*f*): brake equipment 235
BREMSBACKE (*f*): brake block or shoe 236
BREMSBAND (*n*): brake band 237
BREMSBANDHALTER (*m*): brake-band holder 238
BREMSBANDKUPPLUNG (*f*): brake-band coupling 239
BREMSBANDSPANNUNG (*f*): brake-band tension 240
BREMSBANDSTÜTZE (*f*): brake-band holder; brake-band support 241
BREMSBANDUNTERSTÜTZUNG (*f*): brake-band support or supporting-(arrangement) 242
BREMSBAR: brakeable; capable of being braked 243
BREMSBARE ACHSE (*f*): braked wheel 244
BREMSBARER WAGEN (*m*): braked carriage, car or waggon 245
BREMSBERG (*m*): braking incline 246
BREMSBERGFÖRDERUNG (*f*): haulage over a braking incline 247
BREMSBRUTTO (*n*): gross weight to be braked 248
BREMSDRAHTSEILBAHN (*f*): brake ropeway 249
BREMSDREIECK (*n*): brake truss-bar or beam 250
BREMSDRUCK (*m*): brake pressure 251
BREMSDRUCKREGLER (*m*): brake (braking) pressure regulator 252
BREMSDYNAMOMETER (*n*): brake dynamometer 253
BREMSE (*f*): brake; gadfly; horsefly (*Gastrophilus equi* and *Hippobosca equina*) 254
BREMSE (*f*), **BAND-**: band brake 255
BREMSEISEN (*m*): brake hanger suspension link 256
BREMSE (*f*), **LUFTDRUCK-**: atmospheric (pneumatic) brake, air brake 257
BREMSE (*f*), **MAGNET-**: magneto-electric brake 258
BREMSE (*f*), **MAGNETISCHE-**: magnetic brake 259
BREMSEN: to brake; apply the brake; damp (see under WELLE); (*n*) braking 260
BREMSEND: braking; as (like) a brake 261
BREMSENFUTTER (*n*): brake lining 262
BREMSENMADE (*f*): gadfly (maggot) bot (that which attacks the ox is *Oestrus bovis*, and that which attacks human beings is *Oestrus homonis*) 263
BREMSENÖL (*n*): brake oil; horsefly (gadfly) oil 264
BREMSER (*m*): brakesman 265
BREMSERHAUS (*n*): brakesman's (cabin, hut) box 266
BREMSERHÜTTE (*f*): (see BREMSERHAUS) 267
BREMSERSITZ (*m*): brakesman's seat 268
BREMSERSTAND (*m*): brakesman's platform; place where brakesman stands; brakesman's seat 269
BREMSE (*f*), **VAKUUM-**: vacuum brake 270
BREMSE, ZWEIKLOTZIGE- (*f*): two (double) block brake 271
BREMSFEDER (*f*): brake spring 272
BREMSFLÄCHE (*f*): brake surface; braking surface 273
BREMSFLÜSSIGKEIT (*f*): braking (fluid) liquid 274
BREMSGEHÄNGE (*n*): brake-block suspension arrangement or hanger 275
BREMSGEHÄUSE (*n*): brake casing; brake housing 276
BREMSGENERATOR (*m*): generator for braking; braking generator 277
BREMSGESTÄNGE (*n*): brake rigging; brake rods; brake levers 278
BREMSGEWICHT (*n*): brake counter-(balance)-weight 279
BREMSGRENZNEIGUNG (*f*): maximum braking gradient ; steepest incline which can be descended with the aid of brakes 280
BREMSHAHN (*m*): brake valve; brake cock 281
BREMSHANDRAD (*n*): brake handwheel 282
BREMSHASPEL (*m.* and *f*): brake (reel, winding drum) pulley 283
BREMSHEBEL (*m*): brake valve handle; brake lever 284
BREMSHEBELFÜHRUNG (*f*): brake lever guide 285
BREMSHEBELVERBINDUNG (*f*): brake lever coupling 286
BREMSHUNDERTSTEL (*n*): percentage of brake power 287
BREMSKEGEL (*m*): brake cone 288
BREMSKETTE (*f*): brake chain 289
BREMSKETTENROLLE (*f*): brake chain sheave 290
BREMSKLOTZ (*m*): brake block or shoe 291
BREMSKLOTZGEHÄNGE (*n*): brake-block suspension arrangement or hanger 292
BREMSKLOTZHALTER (*m*): brake block or shoe holder 293
BREMSKLOTZSTELLVORRICHTUNG (*f*): brake block adjusting gear 294
BREMSKNÜPPEL (*m*): brake scotch 295
BREMSKOLBEN (*m*): brake piston 296
BREMSKONTROLLE (*f*): brake control 297
BREMSKONUS (*m*): brake cone 298
BREMSKRAFT (*f*): brake power 299
BREMSKÜHLWASSER (*n*): brake cooling water 300
BREMSKUPPLUNG (*f*): brake coupling 301
BREMSKURBEL (*f*): brake screw handle 302
BREMSLAMELLE (*f*): friction disc 303
BREMSLÄNGE (*f*): braking length or distance 304

BREMSLEINE (*f*): brake cord 305
BREMSLEITUNG (*f*): brake pipe 306
BREMSLEITUNGSWAGEN (*m*): waggon fitted with brake pipes 307
BREMSLUFT (*f*): air for brake; braking air 308
BREMSLUFTKLAPPE (*f*): flap (clappet) valve for brake 309
BREMSLÜFTMAGNET (*m*): brake magnet 310
BREMSLÜFTUNGSMAGNET (*m*): (see BREMSLÜFT-MAGNET) 311
BREMSLÜFTWEG (*m*): brake (play) lift or travel 312
BREMSMAGNET (*m*): brake magnet 313
BREMSMITTEL (*n*): braking device or gear 314
BREMSMOMENT (*n*): braking moment 315
BREMSMOTOR (*m*): brake motor 316
BREMSMUTTER (*f*): brake screw nut 317
BREMSPERIODE (*f*): braking period 318
BREMSPLATTFORM (*f*): brake platform 319
BREMSPROZENT (*n*): percentage of brake power 320
BREMSPRÜGEL (*m*): brake scotch 321
BREMSREGLUNG (*f*): brake regulation 322
BREMSREGULIERUNG (*f*): brake regulation 323
BREMSRICHTUNG (*f*): brake direction 324
BREMSSCHALTER (*m*): brake control valve or switch 325
BREMSSCHALTUNG (*f*): braking (brake) connection 326
BREMSSCHEIBE (*f*): brake (disc) pulley 327
BREMSSCHLAUCH (*m*): brake (hose)pipe 328
BREMSSCHLITTEN (*m*): brakeshoe; drag 329
BREMSSCHLUSS (*m*): brake application 330
BREMSSCHUH (*m*): brake block or shoe 331
BREMSSCHUHHALTER (*m*): brake shoe holder 332
BREMSSCHUTZWIDERSTAND (*m*): brake safety resistance 333
BREMSSCHWENGEL (*m*): brake handle 334
BREMSSEIL (*n*): brake cord 335
BREMSSPERKLINKE (*f*): brake pawl 336
BREMSSPERRAD (*n*): brake ratchet wheel 337
BREMSSPINDEL (*f*): brake spindle 338
BREMSSPINDELLAGER (*n*): brake spindle bracket or bearing 339
BREMSSTANGENSICHERUNG (*f*): brake rod safety device 340
BREMSSTÄRKE (*f*): brake power; braking power or force 341
BREMSSTELLUNG (*f*): braking position 342
BREMSSTEUERUNG (*f*): brake control 343
BREMSSTOSS (*m*): brake pull 344
BREMSSTROM (*m*): braking current 345
BREMSSTROMSTÄRKE (*f*): strength of braking current 346
BREMSSTUFE (*f*): brake contact (braking) stage 347
BREMSSYSTEM (*n*): braking system 348
BREMSTOPF (*m*): brake cylinder 349
BREMSTRÄGER (*m*): brake support 350
BREMSTRAVERSE (*f*): brake cross tie; brake beam 351
BREMSTROMMEL (*f*): brake drum 352
BREMSÜBERSCHUSS (*m*): excess of braking (force) power 353
BREMSUNG (*f*): braking 354
BREMSVENTIL (*n*): brake valve 355
BREMSVENTILHANDGRIFF (*m*): brake-valve handle 356
BREMSVENTILLUFTBEHÄLTER (*m*): brake valve air reservoir 357
BREMSVERSUCH (*m*): brake test 358
BREMSVORRICHTUNG (*f*): braking device or gear; braking arrangement 359
BREMSWAGEN (*m*): braked carriage or waggon; brake-van 360
BREMSWALZE (*f*): brake drum 361
BREMSWEG (*m*): brake path, length or distance 362
BREMSWELLE (*f*): brake shaft 363
BREMSWELLENBOCK (*m*): brake shaft bracket, support or carrier 364
BREMSWELLENHALTER (*m*): brake shaft bracket, support, holder or carrier 365
BREMSWELLENHEBEL (*m*): brake shaft lever 366
BREMSWERK (*n*): braking (arrangement, gear) device 367
BREMSWERT (*m*): brake (braking) value or efficiency; retardation-(efficiency) 368
BREMSWIDERSTAND (*m*): braking or brake resistance 369
BREMSWIDERSTANDSCHALTUNG (*f*): braking (resistance) connection; armature short-circuit connection 370
BREMSWINKEL (*m*): bell crank lever 371
BREMSWIRKUNG (*f*): braking action or effect 372
BREMSZAHNRAD (*n*): brake gear-wheel or cog-wheel 373
BREMSZANGE (*f*): brake clip 374
BREMSZAUM (*m*): prony brake 375
BREMSZUGBAND (*n*): brake rod 376
BREMSZUGLASCHE (*f*): brake rod 377
BREMSZUGSCHERE (*f*): brake rod 378
BREMSZUGSTANGE (*f*): brake rod 379
BREMSZYLINDER (*m*): brake cylinder 380
BREMSZYLINDERHEBEL (*m*): brake cylinder lever 381
BREMSZYLINDERTRÄGER (*m*): brake cylinder bracket or support 382
BRENKAS (*n*): East Indian tin 383
BRENKE (*f*): yeast tub (Brewing) 384
BRENNAPPARAT (*m*): distilling apparatus; still; branding machine 385
BRENNARBEIT (*f*): burning; firetesting; pickling (see GELBBRENNE) 386
BRENNBAR: combustible; burnable 387
BRENNBARE BESTANDTEILE (*m.pl*): combustible matter; combustible (constituent) content 388
BRENNBARE LUFT (*f*): (old name for)-hydrogen; (H) 389
BRENNBARKEIT (*f*): combustibility 390
BRENNBAR, LEICHT-: easily combustible 391
BRENNBAR, SCHWER-: difficultly combustible, difficult to burn 392
BRENNBLASE (*f*): alembic; still 393
BRENNCYLINDER (*m*): moxa (Pharmacy) 394
BRENNDAUER (*f*): duration of combustion or burning 395
BRENNE (*f*): dip, pickle (see GELBBRENNE) 396
BRENNEN: to burn; roast; calcine; fire; distil; bake; anneal; char; scale; brand; scorch; cauterize; smart; curl (hair) 397
BRENNEND: pungent, etc., ardent 398
BRENNER (*m*): burner; distiller; brickmaker; lime-burner 399
BRENNERDÜSE (*f*): burner nozzle 400
BRENNEREI (*f*): distillery (spirits) 401
BRENNEREIANLAGE (*f*): distilling (distillery) plant 402
BRENNEREIEINRICHTUNG (*f*): distillery plant 403
BRENNEREIMAISCHE (*f*): distillery mash 404
BRENNERKAPSEL (*f*): burner tip, cover or cap 405
BRENNERSCHIRM (*m*): burner shade 406
BRENNERVORRICHTUNG (*f*): burner device or arrangement 407

BRENNERZANGE (*f*) : burner pliers 408
BRENNGAS (*n*) : heating, combustible or fuel gas 409
BRENNGESCHWINDIGKEIT (*f*) : rate of combustion 410
BRENNGLAS (*n*) : burning glass ; magnifying glass 411
BRENNGUT (*n*) : fuel, combustible-(material) 412
BRENNHELM (*m*) : still-head 413
BRENNHOLZ (*n*) : firewood 414
BRENNHOLZMASCHINEN (*f.pl*) : firewood machinery 415
BRENNKALENDER (*m*) : table of lighting-up times 416
BRENNKAPSEL (*f*) : sagger 417
BRENNKEGEL (*m*) : melting or fusible cones ; pyrometric cones (see **SEGERKEGEL**) 418
BRENNKOLBEN (*m*) : alembic ; distilling flask ; still 419
BRENNKÖPFE (*m.pl*) : channels for supplying separate gas and air to the furnace (Siemens Martin process). (They last 300-350 heatings) 420
BRENNLUFT (*f*) : air for combustion 421
BRENNMATERIAL (*n*) : fuel ; combustible 422
BRENNNESSEL (*f*) : stinging nettle (*Urtica*) 423
BRENNNESSELKRAUT (*n*) : nettle-(herb) ; (*Herba urticæ*) 424
BRENNOFEN (*m*) : burning oven ; bake-oven ; kiln ; furnace ; roasting furnace ; calcining furnace 425
BRENNÖL (*n*) : burning, lamp or fuel oil 426
BRENNÖLRÜCKSTAND (*m*) : fuel oil residue 427
BRENNPALME (*f*) : wine palm ; caryota palm (*Caryota urens*) 428
BRENNPUNKT (*m*) : focus ; focal point ; spontaneous ignition point (of oil as opposed to **FLAMM-PUNKT**, flashpoint, i.e., 20-60°C. above temperature of flash point) 429
BRENNPUNKT (*m*), **CHEMISCHER-** : actinic focus (Photography) 430
BRENNPUNKT (*m*), **OPTISCHER-** : visual focus 431
BRENNPUNKTSABSTAND (*m*) : focal distance 432
BRENNROHRMUNDSTÜCK (*n*) : mouth ; burner ; nozzle 433
BRENNSÄURE (*f*) : pickling acid 434
BRENNSCHICHT (*f*) : fuel bed or layer 435
BRENNSCHICHTHÖHE (*f*) : thickness of (fire) fuel bed (of furnaces) 436
BRENNSCHWINDUNG (*f*) : shrinkage in firing 437
BRENNSILBER (*n*) : amalgam for silvering 438
BRENNSPIEGEL (*m*) : burning reflector 439
BRENNSTAHL (*m*) : blister steel 440
BRENNSTOFF (*m*) : fuel ; combustible ; phlogiston (Stahl's theory ; old chemistry) 441
BRENNSTOFFAUSNUTZUNG (*f*) : fuel efficiency 442
BRENNSTOFFBETT (*n*) : fuel bed 443
BRENNSTOFFECONOMISER (*m*) : fuel economizer 444
BRENNSTOFFKOSTEN (*f.pl*) : fuel costs 445
BRENNSTOFFMASCHINEN (*f.pl*) : fuel machinery 446
BRENNSTOFFNUTZUNGSGRAD (*m*) : fuel efficiency 447
BRENNSTOFFSCHICHT (*f*) : fuel bed or layer 448
BRENNSTOFFSTRAHLUNG (*f*) : radiation of heat from fuel 449
BRENNSTOFFTRICHTER (*m*) : fuel hopper (of furnaces) 450
BRENNSTOFFVERBRAUCH (*m*) : fuel consumption 451
BRENNSTOFFZIEGEL (*m*) : briquette 452
BRENNSTOFFZUFUHR (*f*) : fuel supply ; transport of fuel 453
BRENNSTUNDE (*f*) : lamp hour (Elect.) 454
BRENNTEMPERATUR (*f*) : fusible temperature ; temperature of (burning) combustion ; temperature of fusion 455
BRENNWEITE (*f*) : focal (length) distance 456
BRENNWERT (*m*) : fuel value ; calorific value 457
BRENNWERTBESTIMMUNG (*f*) : determination of (heat) calorific value ; calorimetry 458
BRENNZIEGEL (*m*) : firebrick ; fuel briquette 459
BRENNZYLINDER (*m*) : moxa (Pharmacy) 460
BRENZ (*n*) : empyreuma ; combustible ; (as a prefix, pyro-) (see also **PYRO-**) 461
BRENZAPFELSÄURE (*f*) : maleic acid ; $COOH(CH)_2COOH$; Sg. 1.59 ; Mp. 136-137°C. ; Bp. 160°C. 462
BRENZCAIN (*n*) : pyrocain (Pharm.) 463
BRENZCATECHIN (*n*) : pyrocatechinic acid ; catechol ; oxyphenic acid ; ortho-dioxybenzene; pyrocatechin ; pyrocatechol ; $C_6H_6O_2$; Sg. 1.34-1.371 ; Mp. 104°C. ; Bp. 245°C. 464
BRENZCATECHINDIÄTHYLESTER (*m*) : pyrocatechin diethylester ; Mp. 44°C. 465
BRENZCATECHINDIMETHYLÄTHER (*m*) : pyrocatechin dimethylester, veratrol ; Sg. 1.1 ; Mp. 22.5°C. ; Bp. 205°C. 466
BRENZCATECHINMONOÄTHYLÄTHER (*m*) : pyrocatechin monoethylester ; Mp. 21°C. ; Bp. 216.5°C. 467
BRENZCATECHINMONOMETHYLÄTHER (*m*) : pyrocatechin monomethylester (see **GUAJACOL**) 468
BRENZELN : to have a burning flavour or taste 469
BRENZESSIGÄTHER (*m*) : acetone ; C_3H_6O ; Sg. 0.79 ; Mp. – 94.3°C. ; Bp. 56.48°C. 470
BRENZESSIGGEIST (*m*) : (see **BRENZESSIGÄTHER**) 471
BRENZGALLUSSÄURE (*f*) : pyrogallic acid ; pyrogallol ; pyro ; (*Acidum pyrogallicum*) ; $C_6H_3(OH)_3$; Sg. 1.463 ; Mp. 132.5°C. ; Bp. 210 – 293°C. ; (by heating gallic acid to 210°C.) 472
BRENZHARZ (*n*) : empyreumatic resin 473
BRENZKATECHIN (*n*) : (see **BRENZCATECHIN**) 474
BRENZLICH : empyreumatic ; tarry 475
BRENZLICHE SÄURE (*f*) : pyro acid 476
BRENZLIG : empyreumatic ; tarry 477
BRENZLIGE SÄURE (*f*) : pyro acid 478
BRENZÖL (*n*) : empyreumatic oil 479
BRENZSÄURE (*f*) : pyro acid 480
BRENZSCHLEIMSÄURE (*f*) : pyromucic acid ; $C_5H_4O_3$; Mp. 134°C. 481
BRENZTEREBINSÄURE (*f*) : pyroterebic acid 482
BRENZTRAUBENALKOHOL (*m*) : pyroracemic alcohol 483
BRENZTRAUBENSÄURE (*f*) : pyroracemic acid ; $C_3H_4O_3$; Sg. 1.2649 ; Mp. 13.6°C ; Bp. 65°C. 484
BRENZWEINSÄURE (*f*) : pyrotartaric acid ; $C_5H_8O_4$; Mp. 97°C. ; Bp. 302 to 304°C. 485
BRETT (*n*) : board ; plank ; shelf ; office 486
BRETTERSTOSS (*m*) : stack of (wooden)-boards 487
BRETTMÜHLE (*f*) : saw-mill ; board-mill 488
BREWSTERIT (*m*) : brewsterite ; Sg. 2.16 489
BREZEL (*f*) : cracknel 490
BRIEF (*m*) : letter ; paper ; packet ; epistle ; charter ; brief 491
BRIEFCHEN (*n*) : note ; short letter 492
BRIEFGELD (*n*) : postage 493
BRIEFLICH : by letter ; in writing ; written 494
BRIEFMARKE (*f*) : postage stamp 495

BRIEFPAPIER (*n*): letter paper; writing paper 496
BRIEFSCHAFTEN (*f.pl*): papers; documents; writings 497
BRIEFSTEMPEL (*m*): postmark 498
BRIEFTISCH (*m*): writing desk 499
BRIEFUMSCHLAG (*m*): envelope; wrapper 500
BRIEFUMSCHLAGMASCHINE (*f*): envelope machine (Paper) 501
BRIEFWECHSEL (*m*): correspondence 502
BRIESE (*f*): breeze, light wind 503
BRIGANTINE (*f*): brigantine 504
BRIGG (*f*): brig (ship) 505
BRIGGSCHONER (*m*): brigantine 506
BRIGGSEGEL (*n*): spanker (Naut.) 507
BRIGHTSCHE KRANKHEIT (*f*): Bright's disease 508
BRIKETT (*n*): briquette 509
BRIKETTFABRIK (*f*): briquetting (briquette) works 510
BRIKETTIERPRESSE (*f*): briquetting press 511
BRIKETTIERUNGSANLAGE (*f*): briquetting plant 512
BRIKETTKOHLEN (*f.pl*): briquette coals (suitable for briquetting) (size from 0—4 m.m.) 513
BRIKETTPECH (*m*): briquetting pitch 514
BRIKETTPRESSE (*f*): briquet-(ting) press 515
BRILLANTALIZARINBLAU (*n*): brilliant alizarine blue (a thiazine dyestuff) 516
BRILLANTGELB (*n*): brilliant yellow (see AZIDIN-ECHTGELB and CURCUMIN) 517
BRILLANTKARMIN (*n*): brilliant carmine 518
BRILLE (*f*): spectacles, eyeglasses; (stuffing)-gland 519
BRILLE (*f*), **STOPFBÜCHSEN-**: stuffing box gland 520
BRILLIANTGRÜN (*n*): brilliant green 521
BRILLIANTINE (*f*): brilliantine 522
BRINELLHÄRTE (*f*): hardness (of metal), according to Brinell's method; ball hardness test; (this scale used for steel hardened by quenching; carried out with 10 mm. ball and 3000 kg. load) 523
BRINELLPRESSE (*f*): Brindell press (for hardness test of metal) 524
BRINGEN, MIT SICH-: to involve 525
BRISE (*f*): breeze; light wind 526
BRITANNIAMETALL (*n*): Britannia metal (70-94% Sn; 4-15% Sb; 0-5% Cu; 0-5% Zn; 0-9% Pb. for rolling, stamping and pressing. For casting plates, 90.7 % Sn; 7.8%Sb; 1.5% Cu; for casting spoons, 85.5%Sn; 14.4%Sb; 0.1% Cu.) 527
BRITISCHGUMMI (*n*): British gum, dextrin (see DEXTRIN) 528
BRIXENERGRÜN (*n*): (see SCHWEINFURTGRÜN) 529
BROCHANTIT (*m*): brochantite (a basic copper sulphate); $CuSO_4.3Cu(OH)_2$; Sg. 3.9 530
BRÖCKELIG: brittle; fragile; friable; crumbling 531
BRÖCKELN: to crumble; break up 532
BROCKEN (*m*): lump; fragment; crumb; morsel 533
BROCKENGLAS (*n*): cullet; broken glass 534
BRODEL (*m*): vapour, steam 535
BRODELN: to bubble 536
BRODEM (*m*): vapour; steam; foul air (Mining) 537
BROENNERSALZ (*n*): Broenner salt 538
BRÖGGERIT (*m*): broeggerite (see URANPECH-BLENDE) 539
BROKAT (*m*): brocade 540
BROKATFARBE (*f*): bronze colour (powdered metals or metal alloys) (light shades = about 83% Cu and 17% Zn; red shades = about 92%Cu. and 8% Zn); brocade dye 541
BROM (*n*): bromine (Br); Sg. 3.1883; Mp. 7.3°C.; Bp. 58.7°C. 542
BROMACETYLEN (*n*): acetylene bromide, acetylene tetrabromide, tetrabromoethane; Muthmann's liquid; $C_2H_2Br_4$; Sg. 2.943; Bp. 239-242°C. 543
BROMALHYDRAT (*n*): bromal hydrate; $CBr_3.CHO.H_2O$; Mp. 53.5°C. 544
BROMALIN (*n*): bromaline; $[(CH_2)_6N_4].C_2H_5Br$; Mp. 200°C. 545
BROMAMMON (*n*): ammonium bromide (*Ammonium bromatum*); NH_4Br; Sg. 2.327-3.327 546
BROMAMMONIUM (*n*): (see BROMAMMON) 547
BROMANTIMON (*n*): antimony bromide 548
BROMARGYRIT (*m*): bromargyrite (natural silver bromide); AgBr; Sg. 5.9 549
BROMARSEN (*n*): arsenic bromide; $AsBr_3$; Sg. 3.66; Mp. 31°C.; Bp. 221°C. 550
BROMÄTHER (*m*): ethyl bromide 551
BROMÄTHYL (*n*): ethyl bromide (see ÄTHYL-BROMID) 552
BROMÄTHYLEN (*n*): ethylene bromide; $CH_2Br CH_2Br$; Sg. 2.189; Bp. 129-131°C. 553
BROMBARIUM (*n*): barium bromide, $BaBr_2.2H_2O$ 554
BROMBEERBLÄTTER (*n.pl*): blackberry leaves 555
BROMBEERE (*f*): blackberry; bramble (*Rubus fruticosus*) 556
BROMBEERGEBÜSCH (*n*): blackberry (bramble, brambleberry) bush (*Rubus fruticosus*) 557
BROMBEERKRAUT (*n*): blackberry, bramble (*herba rubi fruticosi* of *Rubus fruticosus*) 558
BROMBENZOL (*n*): benzene bromide; C_6H_5Br; Sg. 1.5219; Mp. -30.5°C.; Bp. 156.2°C. 559
BROMBERNSTEINSÄURE (*f*): bromosuccinic acid; monobromosuccinic acid; $C_2H_3Br(COOH)_2$; Mp. 159-160°C. 560
BROMCALCIUM (*n*): calcium bromide (see also BROMKALZIUM) 561
BROMCYAN (*n*): cyanogen bromide; bromocyanogen; CNBr; Mp. 52°C.; Bp. 61°C. 562
BROMDÄMPFE (*m.pl*): bromine vapours 563
BROMDERIVAT (*n*): bromine derivative 564
BROMEIN (*n*): bromeine, codeine hydrobromide 565
BROMEISEN (*n*): iron (ferrous) bromide; $FeBr_2$ or Fe_3Br_8: Sg. 4.636; Mp. 27°C. 566
BROMELIN (*n*): bromelin (an enzyme from pineapple *Ananas sativa*) 567
BROMEOSINSILBER (*n*): eosinated bromide of silver (Photography) 568
BROMESSIGSÄURE (*f*): bromoacetic acid, monobromoacetic acid; $C_2H_3O_2Br$; Mp. 50°C.; Bp. 118°C. 569
BROMETON (*n*): brometon, tertiary tribromobutylalcohol (see TRIBROMBUTYLALKOHOL, TERTIÄRER-) 570
BROMFABRIKEINRICHTUNG (*f*): bromine plant 571
BROMFLASCHE (*f*): bromine bottle 572
BROMFLUOR (*n*): (see BROMFLUORID) 573
BROMFLUORID (*n*): bromine fluoride; BrF_3; Mp. 5°C.; Bp. 130 to 140°C. 574
BROM (*n*), **FLÜSSIGES-**: liquid bromine 575
BROMGEHALT (*m*): bromine content 576

BROMGOLDKALIUM (*n*): potassium (auribromide) bromaurate 577
BROMHYDRAT (*n*): hydrobromide; bromine hydrate 578
BROMHYDROCHINON (*n*): bromohydroquinone, adurol; $C_6H_3(OH)_2Br$; Mp. 109.5°C. 579
BROMID (*n*): bromide (an -ic bromide as opposed to BROMÜR, an -ous bromide) 580
BROMIDPAPIER (*n*): bromide paper (Photography) 581
BROMIEREN: to brominate; (*n*) bromination 582
BROMIERUNG (*f*) bromination 583
BROMIPIN (*n*): bromipin (a bromine addition product of Sesame oil; 10% has Sg. 0.996 and 33⅓% has Sg. 1.301) 584
BROMIPINKAPSEL (*f*): bromipin capsule (see BROMIPIN) 585
BROMISMUS (*m*): brominism 586
BROMISOVALERIANSÄUREFORMYLESTER (*m*), α-: formyl (alpha-bromoisovalerianate) alpha-bromoisovalerate; $(CH_3)_2HC.CHBr.CO_2.C_{10}H_{17}$ 587
BROMIT (*m*): bromite, silver bromide (see BROMARGYRIT) 588
BROMKADMIUM (*n*): cadmium bromide (see CADMIUM BROMID) 589
BROMKALIUM (*n*): potassium bromide (*Kalium bromatum*); KBr; Sg. 2.69 to 2.749; Mp. 730-750°C.; Bp. 1435°C. 590
BROMKALZIUM (*n*): calcium bromide; $CaBr_2$ or $CaBr_2.6H_2O$; Sg. (a) 3.353; Mp. (a) 680-760°C.; (b) 35°C.; Bp. (a) 806-812°C.; (b) 149-150°C. 591
BROMKAMPHER (*m*): bromocamphor; monobromated camphor (*Camphora monobromata*) $C_{10}H_{15}BrO$; Mp. 76°C.; Bp. 274°C. 592
BROMKUPFER (*n*): copper bromide 593
BROMLAUGE (*f*): bromide lye (solution of sodium hypobromide and bromine) 594
BROMLITHIUM (*n*): lithium bromide; LiBr; Sg. 3.466; Mp. 442-547°C. 595
BROMLÖSUNG (*f*): bromine solution 596
BROMMAGNESIUM (*n*): magnesium bromide; $MgBr_2.6H_2O$ 597
BROMMETALL (*n*): metallic bromide 598
BROMNAPHTHALIN (*n*): (mono)-bromonaphthalene; $C_{10}H_7Br$; Sg. 1.49; Mp. 6.2°C.; Bp. 279°C. 599
BROMNATRIUM (*n*); sodium bromide; NaBr; Sg. 3.014-3.203; Mp. 760°C.; Bp. 1455°C.; $NaBr.2H_2O$; Sg. 2.176; Mp. 757.7°C. 600
BROMNATRIUMHYDRAT (*n*): sodium hydrobromide, hydrated sodium bromide, $NaBr.2H_2O$; Sg. 2.176; Mp. 757.7°C. 601
BROMOCHINAL (*n*): bromoquinal (the quinine salt of dibromosalicylic acid); $C_{20}H_{24}N_2O_2.2(C_6H_2Br_2OH.COOH)$; Mp. 197.5°C. 602
BROMOCOLL (*n*): bromocoll; (20%Br; 30% tannin; 30% glue; 10% H_2O) 603
BROMOCOLLLÖSUNG (*f*): bromocoll solution (bromocoll and distilled water in proportion of 1:3 parts, plus hot solution of borax in distilled water in proportion 6:54) 604
BROMOCOLLSALBE (*f*): bromocoll ointment 605
BROMOCOLLSEIFE (*f*): bromocoll soap 606
BROMOFORM (*m*): bromoform, formyl tribromide, tribromomethane, methenyl tribromide; $CHBr_3$; Sg. 2.9045; Mp. 9°C.; Bp. 148.5-150.5°C. 607
BROMOFORMIUM (*n*): (*Bromoformum*): bromoform, officinal bromoform; Sg. 2.83; Bp. 149.5°C. 608
BROMOISOVALERYL-PHENOKOLL (*n*): (see BRO-PHENIN) 609
BROMOL (*n*): bromol, tribromophenol; $C_6H_3OBr_3$; Mp. 95-96°C. 610
BROMOPLATINAT (*n*): (potassium) bromoplatinate, K_2PtBr_6; Sg. 4.658; (ammonium) bromoplatinate, $(NH_4)_2PtBr_6$; Sg. 4.265 611
BROMOPYRIN (*n*): bromopyrine, monobromoantipyrine; Mp. 114°C. 612
BROMPHENOL (*n*): bromophenol (C_6H_5OBr); ortho.- Bp. 195°C.; Meta.- Mp. 33°C.; Bp. 236.5°C.; para.- Mp. 64°C.; Bp. 238°C. 613
BROMPHOSPHOR (*m*): phosphorus bromide 614
BROMRADIUM (*n*): radium bromide; $RaBr_2$ 615
BROMSALZ (*n*): bromine salt 616
BROMSALZGEMISCH (*n*): bromine salt mixture (mixture of sodium bromide and sodium bromate in a molecular ratio of 2:1); $2NaBr.NaBrO_3$ 617
BROMSAUER: bromate of; combined with bromic acid 618
BROMSÄURE (*f*): bromic acid (*Acidum bromicum*) ($HBrO_3$); Sg. 3.188; Bp. 100°C. 619
BROMSCHWEFEL (*m*): sulphur bromide 620
BROMSILBER (*n*): silver bromide; AgBr; Sg. 6.473; Mp. 427°C. 621
BROMSILBERGELATINE (*f*): gelatino-bromide of silver (Photography) 622
BROMSILBERGELATINEPAPIER (*n*): bromide paper (Photography) 623
BROMSILBERKOLLODIUM (*n*): collodio-bromide of silver (Photography) 624
BROMSILBERPAPIER (*n*): bromide paper (Photography) 625
BROMSILIZIUM (*n*): silicon bromide 626
BROMSPAT (*m*): bromyrite; AgBr; about 60% silver 627
BROMSTYROL (*n*): bromstyrol (see HYACINTHIN) 628
BROMTOLUOL (*n*): bromotoluene (C_7H_7Br); ortho.- Sg. 1.4222; Mp. −25.9°C.; Bp. 180.3°C.; meta.- Sg. 1.4099; Mp. −39.8°C.; Bp. 183.7°C.; para.- Sg. 1.3898; Mp. 28.5°C.; Bp. 183.6C. 629
BROMÜBERTRÄGER (*m*): bromine carrier 630
BROMÜR (*n*): bromide (an -ous bromide as opposed to BROMID, an -ic bromide) 631
BROMURAL (*n*): bromural, alpha-monobromoisovalerylurea; $(CH_3)_2CH.CHBr.CO.HN.CO.NH_2$; Mp. 147-154°C. 632
BROMWASSER (*n*): bromine water 633
BROMWASSERSTOFF (*m*): hydrogen bromide; hydrobromic acid; HBr; Sg. (gas) 2.71; Mp. 86.13°C.; Bp. 68.7°C. 634
BROMWASSERSTOFFSAUER: hydrobromide of; combined with hydrobromic acid 635
BROMWASSERSTOFFSÄURE (*f*): hydrobromic acid (*Acidum hydrobromicum*); (HBr in aqueous solution); Sg. 1.78 636
BROMWASSERSTOFFSAURES SALZ (*n*): hydro-bromate, (of a non-metallic base); bromide 637
BROMWISMUT (*m.* and *n*): bismuth bromide 638
BROMYRIT (*m*): bromyrite (Mineral) (natural silver bromide); AgBr; about 60% Ag.; (see BROMARGYRIT) 639
BROMZAHL (*f*): bromine number 640
BROMZINK (*n*): zinc bromide; $ZnBr_2$; Sg. 4.219; Mp. 394°C.; Bp. 650°C. 641
BROMZINN (*n*): tin bromide 642
BRONCE (*f*): bronze (copper-tin alloy); brass 643

BRONCHIALDRÜSEN (*f.pl*): bronchial glands 644
BRONCHIALKATARRH (*m*): bronchial catarrh (Med.) 645
BRONCHIEN-KRANKHEIT (*f*): disease of the bronchi (Med.) 646
BRONCHOPNEUMONIE (*f*): broncho (catarrhal, lobular) pneumonia (Med.) 647
BRONCIEREN: (see BRONZIEREN) 648
BRÖNNERSCHE SÄURE (*f*): Broenner's acid, beta-naphthylamine-beta-sulfonic acid; $C_{10}H_6(NH_2)SO_3H$, 2 : 6 649
BRONZE (*f*): (see BRONCE) 650
BRONZEBÜCHSE (*f*): brass bush 651
BRONZEDRAHT (*m*): bronze(brass) wire 652
BRONZEFARBE (*f*): bronze colour 653
BRONZEFARBENE SCHATTEN (*m.pl*): bronzed shadows (Photography) 654
BRONZEFUTTER (*n*): brass bush, brass packing 655
BRONZEGIESSEREI (*f*): brass foundry 656
BRONZEGRÜN (*n*): bronze green (a mixture of lead chromate and Berlin blue) (see BLEICHROMAT and BERLINERBLAU) 657
BRONZEGUSS (*m*): bronze casting 658
BRONZELACK (*m*): bronze varnish 659
BRONZEMESSER (*m*): bronze-plates (paper) 660
BRONZEROT (*n*): bronze red 661
BRONZETINKTUR (*f*): bronzing liquid 662
BRONZIEREN: to bronze 663
BRONZIERTE SCHATTEN (*m.pl*): bronzed shadows. (Photography) 664
BRONZIT (*m*): bronzite; $(Mg.Fe)_2(SiO_3)_2$; Sg. 3.3 665
BROOKIT (*m*): brookite (see RUTIL); TiO_2; Sg. 4.17 666
BROPHENIN (*n*): brophenine, bromoisovaleryl-phenocoll; $C_2H_5O.NH.CO.CH_2.NH.CO.CHBr.CH(CH_3)_2$; Mp. 157°C. 667
BROSAME (*f*): crumb; scrap 668
BRÖSCHEN (*n*): sweetbread 669
BROSCHIEREN: to stitch; sew; work; bind in paper cover; make up in pamphlet form 670
BROSCHÜRE (*f*): stitched book; pamphlet; brochure 671
BROTKORB (*m*), DEN-ZU HOCH HÄNGEN: to take unfair advantage 672
BROTKORN (*n*): breadstuff 673
BROTLAST (*f*): bread room (on ships) 674
BROTLOS: without bread; unemployed; without sustenance; unprofitable 675
BROTMASSE (*f*): breadstuff 676
BROTRAFFINADE (*f*): loaf sugar 677
BROTZUCKER (*m*): loaf sugar 678
BROWNSCHE BEWEGUNG (*f*): Brownian movement (the rapid movement of finely divided particles suspended in water) 679
BROYEUSE (*f*): (see WALZENQUETSCHE) 680
BRUCH (*m*): fracture; rupture; fraction (Maths.); break; breakage; shortness (Metal); breach; crack; cleft; fold; crease; wrinkle; crimp; quarry; flaw; fragments; scrap; hernia (Med.); (*pl.*) marsh; moor 681
BRUCHBAND (*n*): suspensor-(y bandage), truss (Med.) 682
BRUCHBLEI (*n*): scrap lead 683
BRUCHDEHNUNG (*f*): breaking tension; stretch (paper); rupturing elongation (Metal) 684
BRUCHEISEN (*n*): scrap iron; old iron 685
BRUCHFEST: resisting breakage 686
BRUCHFESTIGKEIT (*f*): tensile strength; breaking (strain) stress; transverse strength 687
BRUCHFLÄCHE (*f*): surface of fracture 688
BRUCHGEWICHT (*n*): fractional weight 689
BRUCHGLAS (*n*): cullet; broken glass 690
BRUCHGRAMM (*n*): fraction of a gram 691
BRUCHHEIDE (*f*): heath (*Erica*) (Bot.) 692
BRÜCHIG: brittle; short (Metal); friable; fragile; full of (breaks) cracks; clastic; fragmental (Geol.) 693
BRÜCHIG WERDEN: to become brittle etc. (see BRÜCHIG); become tender (fabrics) 694
BRUCH (*m*), KNOCHEN-: fracture (Med.) 695
BRUCHKRAUT (*n*): rupture-wort (*herba herniariae* of *Herniaria glabra*) 696
BRUCHMETALL (*n*): scrap (broken) metal 697
BRUCHMOMENT (*n*): moment of rupture; breaking moment 698
BRUCHPROBE (*f*): breaking test 699
BRUCHSCHIENE (*f*): splint (Med.) 700
BRUCHSCHLAGZAHL (*f*): number of blows to rupture, number of impacts to cause fracture (metal testing) 701
BRUCHSICHERHEIT (*f*), EINE 4 FACHE-: a factor of safety of 4 702
BRUCHSICHERHEIT (*f*): factor of safety 703
BRUCHSPANNUNG (*f*): ultimate stress, breaking (strain) stress 704
BRUCHSTEIN (*m*): quarry stone; ashlar; rubble 705
BRUCHSTRICH (*m*): fractional line 706
BRUCHSTÜCK (*n*): fragment; shred; (*pl.*) scraps; debris; shreds; fragments 707
BRUCHTEIL (*m*): fraction; portion 708
BRUCHTRAGBAND (*n*): suspensor-(y bandage), truss (Med.) 709
BRUCH (*m*), UNTERLEIBS-: hernia (Med.) 710
BRUCH (*m*), ZU-GEHEN: to fracture, break, rupture, collapse 711
BRUCIN (*n*): brucine (*Brucinum*); $C_{23}H_{26}N_2O_4.4H_2O$; Mp. 100°C. (crystallized); Mp. 178°C. (anhydrous) 712
BRUCINCHLORHYDRAT (*n*): brucine hydrochloride $C_{23}H_{26}N_2O_4HCl$ 713
BRUCINLÖSUNG (*f*): brucine solution 714
BRUCINNITRAT (*n*): brucine nitrate; $C_{23}H_{26}N_2O_4HNO_3.2H_2O$; Mp. 230°C. 715
BRUCINSULFAT (*n*): brucine sulphate; $(C_{23}H_{26}N_2O_4)_2.H_2SO_4.7H_2O$ 716
BRUCIT (*m*): brucite (natural hydrated magnesia); $Mg(OH)_2$; Sg. 2.39 717
BRÜCKE (*f*): pons varolii, bridge of Varolius (a portion of the brain) (Med.); bridge 718
BRÜCKE (*f*), KOMMANDO-: bridge (Naut.) 719
BRÜCKENBAHN (*f*): roadway over bridge 720
BRÜCKENBAU (*m*): bridge-building 721
BRÜCKENDECK (*m*): bridge deck 722
BRÜCKENDRAHT (*m*): bridge wire 723
BRÜCKENGLIED (*n*): floating pier 724
BRÜCKENHAUS (*n*): bridge house 725
BRÜCKENKOPF (*m*): bridge-head (Military) 726
BRÜCKENPROBE (*f*): loading test of bridge 727
BRÜCKENSAUERSTOFF (*m*): connecting oxygen; bridging oxygen 728
BRÜCKENSCHALTUNG (*f*), WHEATSTONESCHE-: Wheatstone bridge, Wheatstone's bridge (for measuring electrical resistance) 729
BRÜCKENSCHANZE (*f*): bridge-head (Military) 730
BRÜCKENSCHLÜSSEL (*m*): bridge key 731
BRÜCKENWAGE (*f*): platform-(weighing machine) balance; weighbridge 732
BRÜCKENZOLL (*m*): (bridge)-toll 733

BRÜDEN (*m*): any liquid used in evaporators (as milk; sugar juice; lye, etc.) or steam from them 734

BRÜDENDAMPF (*m*): the vapours generated by "BRÜDEN" 735

BRÜDENKOMPRESSION (*f*): compression of "BRÜDENDAMPF" by means of their own steam 736

BRÜHE (*f*): soup; liquor; liquid; broth; gravy; dye-bath; tan-juice; tan-liquor; ooze; drench; sauce (Tobacco) 737

BRÜHEN: to scald; scour; treat with hot water 738

BRÜHHEISS: scalding hot 739

BRÜHMESSER (*m*): barkometer (Leather) 740

BRUMMER (*m*): blue-bottle-(fly), (*Misca vomitoria*) 741

BRUNELLE (*f*): self-heal (*Prunella vulgaris*) 742

BRUNIEREN: to brown; burnish; polish 743

BRÜNIEREN: to burnish (metal is covered with acid, antimony trichloride $SbCl_3$, and left to action of air. Basic iron salt formed and after multiple repetition of process the heated piece is rubbed with wax. This acts as anti-corrosion rust preventative) 744

BRUNIERER (*m*): burnisher 745

BRUNIERSTEIN (*m*): burnishing stone; bloodstone 746

BRUNNEN (*m*): well; spring; fountain; waters (Mineral); aperture 747

BRUNNENFLASCHE (*f*): syphon; mineral-water bottle 748

BRUNNENKRESSE (*f*): water cress 749

BRUNNENSALZ (*n*): brine salt; well salt 750

BRUNNEN (*m*), **SCHRAUBEN-**: screw aperture, screw trunk 751

BRUNNENWASSER (*n*): well water 752

BRÜNSTIG: ardent; fervent; hot 753

BRUST (*f*): breast; chest; bosom; brisket; hearth (Blast furnaces); (as a prefix, thoracic, of the chest, chest, pectoral) 754

BRUSTALANT (*m*): elecampane (see **ALANT**) 755

BRUSTANGST (*f*): angina pectoris (Med.) 756

BRUSTARZNEI (*f*): chest medicine; pectoral medicine 757

BRUSTBEERE (*f*): jujube (*Fructus jujubae*) 758

BRUSTBEIN (*n*): sternum, breast bone (Med.) 759

BRUSTBILD (*n*): head-and-shoulders; bust portrait (Photography) 760

BRUSTBOHRER (*m*): breast drill 761

BRUSTBRÄUNE (*f*): angina pectoris (Med.) 762

BRUSTDRÜSE (*f*): mammary glands; breast; nipple; sweetbread 763

BRUSTDRÜSENKRANKHEIT (*f*): disease of the breasts, disease of the breast 764

BRUSTELIXIR (*n*): chest elixir (solution of liquorice extract and aniseed oil in fennel water and ammonia water) 765

BRÜSTEN: to plume oneself 766

BRUSTFELL (*n*): pleura (Med.) 767

BRUSTFELLENTZÜNDUNG (*f*): pleurisy (Med.) 768

BRUSTFIEBER (*n*): bronchitis (Med.) 769

BRUSTGANG (*m*): thoracic duct (Med.) 770

BRUSTHOLZ (*n*): cutwater (Naut.) 771

BRUSTKASTEN (*m*): thorax (Med.) 772

BRUSTKORB (*m*): thorax (Med.) 773

BRUSTKRAMPF (*m*): asthma (Med.) 774

BRUSTKRANKHEIT (*f*): phthisis; consumption (Med.) 775

BRUSTLEHNE (*f*): breastwork, parapet 776

BRUSTLEIER (*f*): brace 777

BRUSTMESSER (*m*): stethometer (Med.) 778

BRUSTMILCH (*f*): human milk 779

BRUST (*f*), **OFFENE-**: open hearth (open at the bottom) (blast furnaces) 780

BRUSTPULVER (*n*): pectoral powder; compound liquorice powder (Pharm.) 781

BRUSTREINIGEND: expectorant 782

BRUSTREINIGUNGSMITTEL (*n*): expectorant (Med.) 783

BRUSTTEE (*m*): (*Species pectorales*): pectoral tea 784

BRÜSTUNG (*f*): parapet; breastwork 785

BRUSTWALZE (*f*): breast-roll (Paper) 786

BRUSTWARZE (*f*): papilla, nipple (Med.) 787

BRUSTWASSERSUCHT (*f*): hydrothorax (dropsy of the pleura) (Med.) 788

BRUSTWEHR (*f*): breastwork, parapet 789

BRUSTWERK (*n*): breastwork, parapet 790

BRUSTWURZEL (*f*): (see **ANGELIKA** and **ENGELWURZ**) 791

BRUT (*f*): gross; brood; hatch; spawn 792

BRUTBIENE (*f*): drone 793

BRÜTEN: to brood; hatch; incubate; spawn 794

BRUTHENNE (*f*): broody (sitting) hen 795

BRUTKASTEN (*m*): incubator 796

BRUTOFEN (*n*): incubator-(oven) 797

BRUTSCHRANK (*m*): incubator 798

BRUTTO: gross (of weights) 799

BRUTTOFORMEL (*f*): a gross molecular formula 800

BRUTTOGEWICHT (*n*): gross weight 801

BRUTTOTONNENGEHALT (*m*): gross tonnage 802

BRUTTOVERKAUFSPREIS (*m*): gross sales price 803

BRYONIE (*f*): bryony; bryonia (*Bryonia*) 804

BRYOPHYTEN (*pl*): *Bryophyta* (Bot.) (mosses, including the liverworts and true mosses) 805

BUBONENPEST (*f*): bubonic plague 806

BUCCOBLÄTTER (*n.pl*): buchu-(leaves); bucco; buku; bucku (Pharm.); (*Folia bucco*); (dried leaves of *Barosma betulina*; *B. serratifolia*: *B. crenulata* etc.) 807

BUCCUBLÄTTER (*n.pl*): buchu leaves (see **BUCCOBLÄTTER**) 808

BUCCUÖL (*n*): buchu oil 809

BUCH (*n*): book; quire 810

BUCHBINDER (*m*): book-binder 811

BUCHBINDEREIMASCHINEN (*f.pl*): book-binding machinery 812

BUCHBINDERGOLD (*n*): leaf gold 813

BUCHBINDERLACK (*m*): book-binder's varnish 814

BUCHBINDERLEIM (*m*): book-binding glue 815

BUCHBINDERPAPPE (*f*): binder's board (Paper) 816

BUCHDRUCK (*m*): book printing, typography (see **TYPOGRAPHIE**) 817

BUCHDRUCKER (*m*): printer; typographer 818

BUCHDRUCKEREI (*f*): printing works; press; printing; typography 819

BUCHDRUCKPRESSE (*f*): printing press 820

BUCHDRUCKSCHWÄRZE (*f*): printer's ink 821

BUCHE (*f*): beech (*Fagus sylvatica*) 822

BUCHECKER (*f*): beech nut 823

BUCHECKERNÖL (*n*): beechnut oil (from *Fagus sylvatica*); Sg. 0.921 824

BUCHEICHEL (*f*): beech nut 825

BUCHEL (*f*): beech nut 826

BUCHEN: beech; beechen; of beech; to book; place to an account 827

BÜCHEN: beech; beechen; of beech 828
BUCHENHOLZ (*n*): beech wood; Sg. 0.66-0.83; (see **BUCHE**) 829
BUCHENHOLZKOHLE (*f*): beech wood charcoal 830
BUCHENHOLZTEER (*m*): beech tar 831
BUCHENHOLZTEERÖL (*n*): beech tar oil 832
BUCHENKERNÖL (*n*): (*Oleum fagi silvaticae*); beech kernel oil 833
BUCHENRINDE (*f*): beech bark 834
BUCHFÜHREN (*n*): book-keeping 835
BUCHGOLD (*n*): leaf-gold 836
BUCHHALTER (*m*): book-keeper; accountant 837
BUCHHALTUNG (*f*), **DOPPELT-**: double entry book-keeping 838
BUCHHALTUNG (*f*), **EINFACH-**: single entry book-keeping 839
BUCHHÄNDLER (*m*): book-seller; publisher 840
BUCHKAMERA (*f*): book camera 841
BUCHNUSS (*f*): beech nut (fruit of *Fagus sylvatica*) 842
BUCHSBAUM (*m*): box tree (*Buxus sempervirens*); Sg. 0.79-1.19 843
BUCHSBAUM (*m*), **DEUTSCHER-**: spindle tree (*Euonymus europaeus*) 844
BUCHSBAUMGEWÄCHSE (*n.pl*): *Buxaceae* (Bot.) 845
BUCHSBAUMHOLZ (*n*): boxwood (see **BUCHS-BAUM**) 846
BUCHSE (*f*): (see **BÜCHSE**) 847
BÜCHSE (*f*): box; chest; case; can; tin; pot; rifle; bush; box; socket (Machinery) 848
BÜCHSE (*f*), **LEERLAUF-**: no-load bush 849
BÜCHSENFLEISCH (*n*): tinned meat 850
BÜCHSENFÖRMIG: box-shaped 851
BÜCHSENMACHER (*m*): gunsmith 852
BÜCHSENMETALL (*n*): bush (bearing) metal 853
BÜCHSENMÜHLE (*f*): barrel mill 854
BÜCHSENPULVER (*n*): gunpowder 855
BÜCHSENSCHMIED (*m*): gun-smith 856
BÜCHSENSCHUSS (*m*): gunshot 857
BÜCHSE (*f*), **RAD-**: wheel bush 858
BUCHSTABE (*f*): letter; character; type 859
BUCHSTABENRECHNUNG (*f*): algebra 860
BUCHSTABIEREN: to spell 861
BUCHSTÄBLICH: literal-(ly); verbal-(ly) 862
BUCHT (*f*): bay; inlet; sinus; round (of a deck 863
BUCHWEIZEN (*m*): buckwheat (*Fagopyrum esculentum*) 864
BUCKEL (*m*): hump; knob; knot 865
BÜCKLING (*m*): red herring; bloater; bow 866
BUCKOBLÄTTER (*n.pl*): buchu (Pharm.); (see **BUCCOBLÄTTER**) 867
BUCKUBLÄTTER (*n.pl*): buchu (Pharm.); (see **BUCCOBLÄTTER**) 868
BUDGET (*n*): budget 869
BÜFFEL (*m*): buffalo 870
BUFFER (*m*): buffer (see under **PUFFER**) 871
BUG (*m*): bend; bow (and of a boat) 872
BUGANKER (*m*): bow anchor 873
BUGBAND (*n*): breast hook 874
BÜGEL (*m*): bow stirrup, trap, hoop, guard, frame; handle (of pail); gimbol (Naut.); crown, bridge, stay, strap, head, end, grip, yoke, dog, bracket, crossbar; any curved thing of stirrup shape 875
BÜGELECHT: fast to ironing 876
BÜGELEISEN (*n*): flat-(iron) 877
BÜGELN: to iron 878
BUGFLAGGE (*f*): jack 879
BUGFLAGGENSTOCK (*m*): jack staff 880
BUGGESCHÜTZ (*n*): bow-chaser-(gun) 881
BUGHOLZ (*n*): hawse piece 882
BUGRIEMEN (*m*): bow oar 883
BUG (*m*), **SCHARFER-**: sharp bow, pointy bow 884
BUGSIERSCHIFF (*n*): tugboat 885
BUGSIERTAU (*n*): tow rope 886
BUGSPRIET (*n*): bowsprit (Naut.) 887
BUGSPRIETGATT (*n*): bowsprit (bed) opening 888
BUGSPRIETWANTEN (*n.pl*): bowsprit shrouds 889
BUGSPRIETWULING (*f*): (bowsprit)- gammoning (Naut.) 890
BUG (*m*), **VÖLLIGER-**: bluff bow, blunt bow 891
BUGWELLE (*f*): bow wave 892
BÜHNE (*f*): platform; scaffold; stage 893
BUHNLOCH (*n*): hole for nog (Mining) 894
BÜKEN: to buck (steep, boil) in lye or suds 895
BUKETT (*n*): bouquet 896
BULBARPARALYSE (*f*): **PROGRESSIVE-**: bulbar palsy, chronic labio-glosso-laryngeal paralysis 897
BULBOCAPNIN (*n*): bulbocapnine (an alkaloid) (see **CORYDALISALKALOID**); $C_{19}H_{19}NO_4$ 898
BULIN (*f*): bowline 899
BULINE (*f*): bowline 900
BULLAUGE (*n*): bullseye 901
BUMMELN: to crawl; travel slowly 902
BUMMELZUG (*m*): crawling (slow) train 903
BUND (*n*): bundle; bunch; truss; (*m*) band, bond, tie, alliance, confederacy, league; collar (used with loose flanges; on pipe lines) 904
BÜNDEL (*n*): bundle; bunch; parcel; packet 905
BUNDESGENOSS (*m*): confederate; ally 906
BUNDESGENOSSENSCHAFT (*f*): alliance, confederacy, league 907
BUNDESRAT (*m*): federal council 908
BUNDESTAG (*m*): federal diet 909
BÜNDIG: flush; valid; concise; bound; binding 910
BÜNDNIS (*n*): alliance; confederation; covenant; agreement 911
BUNDSTAHL (*m*): faggot steel 912
BUNKER (*m* and *pl*): bunker-(s) (Coal, etc.) 913
BUNKERBESCHICKUNGSANLAGE (*f*): bunker charging (filling) plant 914
BUNKERDECKEL (*m*): bunker cover 915
BUNSENBRENNER (*m*): Bunsen burner 916
BUNSENELEMENT (*n*): bunsen's cell 917
BUNSENIT (*m*): bunsenite (natural nickelous oxide); (see **NICKELOXYDUL**) 918
BUNSENKOHLE (*f*): bunsen-coal 919
BUNT: coloured; variegated; motley; mottled; speckled; gay; fancy; gaudy 920
BUNTBLEIERZ (*n*): variegated pyromorphite, green lead ore (a natural chlorophosphate of lead); (see **BRAUNBLEIERZ** and **PYRO-MORPHIT**) 921
BUNTDRUCK (*m*): colour print-(ing) 922
BUNTFÄRBEN: to stain; colour; dye; tint; mottle; marble (Paper); (*n*) dyeing; staining etc. 923
BUNTFARBIG: variegated; gaudy coloured 924
BUNTKUPFERERZ (*n*): bornite (natural copper-iron sulphide); Cu_3FeS_3; about 60% copper; Sg. 4.9-5.1; (see also **BORNIT**) 925

BUNTKUPFERKIES (*m*): (see BUNTKUPFERERZ) 926
BUNTPAPIER (*n*): marbled paper; stained paper 927
BÜRDE (*f*): load; charge; burden 928
BUREAU (*n*): bureau; office; (see also BÜRO-) 929
BUREAUBEAMTE (*m*): clerk 930
BUREAUCHEF (*m*): chief clerk 931
BUREAUMÖBEL (*n*): office furniture (either as a whole or as a single piece) 932
BUREAUSCHRANK (*m*): office cabinet 933
BÜRETTE (*f*): burette 934
BÜRETTENBÜRSTE (*f*): burette brush 935
BÜRETTENGESTELL (*n*): burette stand 936
BÜRETTENHALTER (*m*): burette holder 937
BÜRETTENKLEMME (*f*): burette clamp or holder 938
BÜRETTENSCHWIMMER (*m*): burette float 939
BÜRGE (*m*): security, warrant, warranty, bail, surety, referee, voucher 940
BÜRGE (*m*), **ALS-BESTELLEN**: to give a person's name as a reference or security 941
BÜRGEN: to vouch for; give bail; go bail for; give security for; stand security for; stand surety for; warrant; stand good for; guarantee 942
BÜRGER (*m*): citizen; townsman; civilian 943
BÜRGERKRIEG (*m*): civil war 944
BÜRGERLICH: civil; common; civilian 945
BÜRGERMEISTER (*m*): mayor; burgomaster 946
BÜRGERSTAND (*m*): middle class; civilian 947
BÜRGERSTEIG (*m*): sidewalk; pavement 948
BÜRGSCHAFT (*f*): security; surety; bail; warranty; guarantee 949
BURGUNDER (*m*): burgundy (wine); Sg. 0.99 950
BURGUNDERHARZ (*n*): Burgundy (resin) pitch; pine resin (*Resini pini; Pix Burgundica*); (see FICHTENHARZ) 951
BURNETTISIEREN: to impregnate (impregnation of) wood with zinc chloride; (as a noun (*n*)) 952
BÜRO (*n*): office, bureau (see also BUREAU-) 953
BÜROLEIM (*m*): mucilage 954
BÜROMÖBEL (*n*): (see BUREAUMÖBEL) 955
BÜROORGANISATION (*f*): office management 956
BÜRSTE (*f*): brush 957
BÜRSTENFASER (*f*): bristle fibre (from the cocoa-nut) 958
BÜRSTENHALTER (*m*): brush holder (Elect.) 959
BÜRSTENWALZE (*f*): brush roll 960
BUSCH (*m*): bush; copse; tuft; thicket 961
BUSCHBAU (*m*): facine (construction) work 962
BÜSCHEL (*m*): tuft; bunch; cluster; brush (Electrical) 963
BÜSCHELFÖRMIG: clustered; tuft-like 964
BUSCHHOLZ (*n*): undergrowth, underwood 965
BUSCHIG: bushy 966
BUSEN (*m*): bosom; breast; bay; gulf 967
BUSSE (*f*): pennance; fine; penalty; atonement; penitence 968
BÜSSEN: to atone for; make good; expiate; pay for; suffer; repent 969
BUSSFERTIG: penitent; repentant 970
BUSSOLE (*f*), **SINUS-**: sine compass, sine galvanometer 971
BUSSOLE (*f*), **TANGENTEN-**: tangent compass or galvanometer 972
BÜSSUNG (*f*): expiation; atonement, penance 973
BÜSTE (*f*): bust; bust portrait; head and shoulders (Photography) 974
BUTAN (*n*): butane 975
BUTAN (*n*) **NORMAL**: butane, normal; C_4H_{10}; Sg. 0.6; Mp. −135°C.; Bp. 6°C. 976
BUTAN (*n*), **ISO-**: isobutane; C_4H_{10}; Sg. 0.6029; Mp. −145°C.; Bp. −10.2°C. 977
BUTENBORDS: outboard 978
BUTENKLÜVER (*m*): flying jib (Naut.) 979
BUTENKLÜVERBAUM (*m*): flying jib boom (Naut.) 980
BUTENKLÜVERFALL (*n*): flying jib halliard 981
BUTENKLÜVERLEITER (*f*): flying jib stay 982
BUTENKLÜVERSCHOTE (*f*): flying jib sheet 983
BUTTE (*f*): (see BÜTTE) 984
BÜTTE (*f*): tub; vat; back; tank; chest; coop; barrel; flounder-(fish) (*Pleuronectes flesus*) 985
BÜTTEL (*m*): bailiff; beadle; jailor 986
BÜTTENFÄRBUNG (*f*): pulp colouring, vat colouring (Paper) 987
BÜTTENGEFÄRBT: unbleached (Paper); vat-coloured (Paper) 988
BÜTTENLEIMUNG (*f*): vat sizing, pulp sizing (Paper) 989
BÜTTENPAPIER (*n*): vat (hand-made) paper 990
BÜTTENPAPIERFABRIK (*f*): vat-mill (Paper) 991
BUTTER (*f*): butter; Sg. 0.94 992
BUTTERÄHNLICH: buttery; like butter; butyraceous 993
BUTTERAROMA (*f*): butter aroma 994
BUTTERARTIG: (see BUTTERÄHNLICH) 995
BUTTERÄTHER (*m*): butyric ether, ethyl butyrate; (see BUTTERSÄURE-ÄTHYLESTER) 996
BUTTERBAUM (*m*): butter-tree (*Bassia* species) 997
BUTTERBLUME (*f*): buttercup (*Ranunculus* species) 998
BUTTERBOHNE (*f*): butter bean (roasted seed of *Vateria indica*) 999
BUTTERFARBE (*f*): butter colour-(ing) 000
BUTTERFASS (*n*): churn; butter-tub 001
BUTTERFETT (*n*): butter fat; butyrin 002
BUTTERGÄRUNG (*f*): butyric fermentation 003
BUTTERGELB (*n*): butter yellow 004
BUTTERIG: buttery; like butter; butyraceous 005
BUTTER (*f*), **INDISCHE-**: (see GHEABUTTER) 006
BUTTERMILCH (*f*): butter milk 007
BUTTERN: to churn; butter; make butter 008
BUTTERNUSS (*f*): butternut (*Juglans cinerea*) 009
BUTTERSALZ (*n*): butter salt 010
BUTTERSAUER: butyrate of; combined with butyric acid 011
BUTTERSÄURE (*f*): (*Acidum Butyricum*): butyric acid, propylformic acid; $C_4H_8O_2$; Sg. 0.9534; Mp. −7.9 to −6.5°C.; Bp. 155-162°C. 012
BUTTERSÄUREANHYDRID (*n*): butyric anhydride; $C_8H_{14}O_3$; Sg. 0.978; Bp. 191-193°C. 013
BUTTERSÄUREÄTHER (*m*): butyric ether (see BUTTERSÄURE-ÄTHYLESTER) 014
BUTTERSÄURE (*f*)-**ÄTHYLESTER** (*m*): ethyl-butyrate; butyric ester (ethyl); $C_6H_{12}O_2$; Sg. 0.8807; Mp. about −80°C.; Bp. 119.9°C.; butyric ether 015
BUTTERSÄUREESTER (*m*): butyric acid ester 016
BUTTERSÄURE (*f*), **ISO-**; isobutyric acid; $C_4H_8O_2$; Sg. 0.9487; Mp. −47°C.; Bp. 154.4°C.-155.5°C. 017

BUTTERSÄURE-NITRIL (*n*): butyric nitrile; C_4H_7N; Sg. 0.796; Bp. 117°C. 018
BÜTTNER (*m*): cooper 019
BUTYLALDEHYD (*n*): butyl-(ic) aldehyde; C_4H_8O; Sg. 0.817; Bp. 73-74°C. 020
BUTYLALKOHOL (*m*): butyl alcohol; (*Alcohol butylicus*); $C_4H_{10}O$; Sg. 0.8094; Bp. 116°C. 021
BUTYL (*n*), **AMEISENSAURES-**: butyl formate 022
BUTYLAMIN (*n*): butylamine; $C_4H_{11}N$; Sg. 0.7401; Bp. 77°C. 023
BUTYLÄTHER (*m*): butyl ether; $C_8H_{18}O$; Sg. 0.7685; Bp. 140.5°C. 024
BUTYL (*n*), **BALDRIANSAURES-**: butyl valerate 025
BUTYL (*n*), **BENZOESAURES-**: butyl benzoate 026
BUTYLBROMID (*n*): butyl bromide; C_4H_9Br; Sg. 1.299; Bp. 100.4°C. 027
BUTYL (*n*), **BUTTERSAURES-**: butyl butyrate 028
BUTYLCHLORAL (*n*): butyl chloral 029
BUTYLCHLORAL-ANTIPYRIN (*n*): butylchloral-antipyrine, butylhypnal; $C_{11}H_{12}N_2O.C_4H_5Cl_3O.H_2O$ 030
BUTYLCHLORALHYDRAT (*n*): butyl chloral-hydrate, butyl chloral hydras, trichloro-butylidene glycol; $C_4H_7Cl_3O_3$; Mp. 78°C. 031
BUTYLCHLORID (*n*): butyl chloride; C_4H_9Cl; Sg. 0.8874; Bp. 77.6°C. 032
BUTYL (*n*), **ESSIGSAURES-**: butyl acetate 033
BUTYLESSIGSÄURE (*f*): butylacetic acid; isohexoic acid; isocaproic acid; $(CH_3)_2CH(CH_2)_2COOH$; Sg. 0.925; Bp. 199°C. 034
BUTYLHYPNAL (*n*): (see **BUTYLCHLORAL-ANTI-PYRIN**) 035
BUTYLISOSULFOCYANAT (*n*): butyl isosulfocyanate; C_4H_9CNS 036.
BUTYLJODID (*n*): butyl iodide; C_4H_9I; Sg 1.6166; Bp. 131.4°C. 037
BUTYLVERBINDUNG (*f*): butyl compound 038
BUTYRON (*n*): (see **DIPROPYLKETON**) 039
BUTYRONITRIL (*n*): (see **ISOBUTTERSÄURE-NITRIL**) 040
BUTYRYLCHLORID (*n*): butyryl chloride; C_4H_7OCl; Sg. 1.0277; Bp. 101.5°C. 041
BUTYRYLCHLORID (*n*) **ISO-**: isobutyryl chloride; C_4H_7OCl; Sg. 1.0174; Bp. 92°C. 042
BYROLIN (*n*): byrolin (mixture of boric acid, wool grease, glycerine and water) 043
BYROLINGELEE (*n*): byrolin jelly 044
BYROLINSEIFE (*f*): byrolin soap 045
BYSSUS (*m*): (see **MUSCHELSEIDE**) 046
BYTOWNIT (*m*): bytownite (mixture of albite and anorthite) (see **ALBIT** and **ANORTHIT**); Sg. 2.72 047

C

[Refer also to K and Z for any words not given under letter C.]

CACAO (*m*): cocoa (from plant *Theobroma cacao*) (see also **KAKAO**) 048
CACAOBOHNE (*f*): cocoa bean (from *Theobroma cacao*) 049
CACAOBUTTER (*f*): cocoa butter (*Butyrum Cacao; Oleum cacao*): Sg. 0.9702; Mp. 33°C.; (from cocoa beans of *Theobroma cacao*) 050
CACAOFETT (*n*): cocoa butter (see **CACAOBUTTER**) 051
CACHOU (*n*): catechu; cutch (from *Acacia catechu* and *Areca catechu*) 052
CACHOUBRAUN (*n*): cutch brown (a basic azo dyestuff) 053
CACHOUTIEREN: to dye with catechu 054
CADAVERIN (*n*): cadaverine (a ptomaine from human corpses); $C_5H_{16}N_2$ 055
CADEÖL (*n*): cade oil; Sg. 0.98-1.06; (from wood of *Juniperus oxycedrus*) (see also **KADDICH, KADDIGÖL, KADINÖL** and **KADIÖL**) 056
CADETSCHE FLÜSSIGKEIT (*f*): Cadet's fuming liquid, alcarsin (a spontaneously inflammable liquid from distilling dry potassium acetate with arsenious anhydride) 057
CADMIUMACETAT (*n*) (see **CADMIUMAZETAT**) 058
CADMIUMAMALGAM (*n*): cadmium amalgam 059
CADMIUMAZETAT (*n*): (*Cadmium aceticum*); cadmium acetate; $Cd(C_2H_3O_2)_2.3H_2O$ 060
CADMIUMBROMID (*n*): (*Cadmium bromatum*): cadmium bromide; $CdBr_2$; Sg. 5.192-5.196; Mp. 568°C-580°C.; Bp. 806-863°C. 061
CADMIUMCARBONAT (*n*): cadmium carbonate; $CdCO_3$; Sg. 4.258 062
CADMIUMCEMENT (*m* and *n*.): cadmium cement 063
CADMIUMCHLORID (*n*): $CdCl_2$; cadmium chloride; Sg. 4.049; Mp. 568°C; Bp. 861-964°C; $CdCl_2.2H_2O$; Sg. 3.327 064
CADMIUMDIHYDROARSENAT (*n*): cadmium dihydroarsenate; $H_4Cd(AsO_4)_2.2H_2O$; Sg. 3.241 065
CADMIUMDIHYDROPHOSPHAT (*n*): cadmium dihydrophosphate; $H_4Cd(PO_4)_2.2H_2O$; Sg. 2.741 066
CADMIUMELEKTRODE (*f*): cadmium electrode 067
CADMIUM (*n*), **ESSIGSAURES-**: cadmium acetate (see **CADMIUMAZETAT**) 068
CADMIUMFARBE (*f*): cadmium colour 069
CADMIUMFLUORID (*n*): cadmium fluoride; CdF_2; Sg. 5.994-6.64 070
CADMIUMGELB (*n*): cadmium yellow, cadmium sulphide (see **CADMIUMSULFID**) 071
CADMIUMHALTIG: cadmiferous, containing cadmium 072
CADMIUMHYDROXYD (*n*): cadmium hydrate; cadmium hydroxide, (*Cadmium hydroxydatum*); $Cd(OH)_2$; Sg. 4.79 073
CADMIUMJODID (*n*): (*Cadmium jodatum*); Cadmium iodide; CdI_2; Sg. 5.64-5.974; Mp. 385°C.-404°C.; Bp. 708-719°C. 074
CADMIUM (*n*), **KOHLENSAURES-**: cadmium carbonate 075
CADMIUMLEGIERUNG (*f*): cadmium alloy 076
CADMIUMLÖSUNG (*f*): cadmium solution 077
CADMIUMMETALL (*n*): cadmium metal 078
CADMIUMMETASILIKAT (*n*): cadmium metasilicate; $CdSiO_3$ 079
CADMIUMNITRAT (*n*): cadmium nitrate; $Cd(NO_3)_2.4H_2O$; Sg. 2.455; Mp. 59.5°C.; Bp. 132°C. 080
CADMIUMORANGE (*f*): cadmium orange 081

CADMIUMOXYD (*n*): (*Cadmium oxydatum*): cadmium oxide; CdO; Sg. 6.95-8.1 082

CADMIUMOXYDHYDRAT (*n*): cadmium hydroxide (see CADMIUMHYDROXYD) 083

CADMIUMPYROARSENAT (*n*): cadmium pyro-arsenate; $Cd_2As_2O_7$; Sg. 5.474 084

CADMIUMQUADRANTOXYD (*n*): cadmium quadrant oxide; Cd_2O; Sg. 8.192 085

CADMIUMROT (*n*): cadmium red 086

CADMIUM (*n*), **SALPETERSAURES-**: cadmium nitrate (see CADMIUMNITRAT) 087

CADMIUMSALZ (*n*): cadmium salt 088

CADMIUM (*n*), **SCHWEFELSAURES-**: cadmium sulphate (see CADMIUMSULFAT) 089

CADMIUMSELENID (*n*): cadmium selenide; CdSe; Sg. 5.81 090

CADMIUMSUBOXYD (*n*): cadmium suboxide; Cd_4O or Cd_2O 091

CADMIUMSULFAT (*n*): (*Cadmium sulfuricum*): cadmium sulphate; $CdSO_4$; Sg. 4.72; Mp. 1000°C; $3CdSO_4.8H_2O$; Sg. 3.09 [092

CADMIUMSULFID (*n*): (*Cadmium sulfuratum*): cadmium sulphide; cadmium yellow; CdS· Sg. 3.906-4.5; Bp. 980°C. 093

CADMIUMSULFURET (*n*): cadmium sulphide (see CADMIUMSULFID) 094

CADMIUMTELLURID (*n*): cadmium telluride; $CdTe_2$ 095

CADMIUMVERBINDUNG (*f*): cadmium compound 096

CADMIUMZEMENT (*m* and *n*): cadmium cement 097

CADOGEL (*n*): cadogel (a colloidal tar preparation from the fractional distillation of Cade oil) 098

CAESIUM (*n*): caesium (see CÄSIUM) 099

CAFFEINSALZ (*n*): caffein salt (see under KAFFEIN) 100

CAINCASÄURE (*f*): cahincic acid 101

CAINCAWURZEL (*f*): cahinca-(root) 102

CAINCIN (*n*): cahincin 103

CAJEPUTGEIST (*m*): spirit of cajeput 104

CAJEPUTÖL (*n*): (see KAJEPUT-): 105

CAJÜTE (*f*): cabin (Naut.) 106

CALABARBOHNE (*f*): calabar bean (see KALA-BARBOHNE) 107

CALAMEN (*n*): calamene (a sesqui-terpene); $C_{15}H_{24}$; Sg. 0.9323; Bp. 255-258°C. 108

CALAMEON (*n*): Calmus (Calamus) camphor (see KALMUSKAMPHER) 109

CALAMIN (*m*): calamine (see KIESELZINKERZ); methylamine (see METHYLAMIN) (from *Acorus calamus*) 110

CALAVERIT (*m*): calaverite; $(Au,Ag)Te_2$; Sg. 9.25; (see also GOLDTELLURID) (about 39%Au) 111

CALCIDUM (*n*): calcidum (a saturated aqueous solution of calcium chloride); (see CALCIUM-CHLORID) 112

CALCINIEREN: to calcine (see also KALSINIER- and KALZINIER-) 113

CALCINIEROFEN (*m*): calcining furnace 114

CALCIT (*m*): calcite; $CaCO_3$; Sg. 2.7; (see CAL-CIUMCARBONAT) (56% CaO) 115

CALCIUM (*n*): calcium (see also KALCIUM and KALZIUM) 116

CALCIUMACETO-SALICYLICUM (*n*): calcium aceto-salicylicum 117

CALCIUMALUMINAT (*n*): calcium aluminate; $CaAl_2O_4$; Sg. 3.671 118

CALCIUM (*n*), **AMEISENSAURES-**: calcium formate 119

CALCIUM (*n*), **APFELSAURES-**: calcium malate 120

CALCIUMARSENIAT (*n*): (see CALCIUM, ARSEN-SAURES-) 121

CALCIUMARSENID (*n*): calcium arsenide; Ca_3As_2; Sg. 2.5 122

CALCIUMARSENIT (*n*), calcium arsenite; $Ca_2As_2O_5$ 123

CALCIUMARSENIT (*n*), **NORMAL-**: normal calcium arsenite; $Ca_3(AsO_3)_2$ 124

CALCIUM (*n*), **ARSENSAURES-**: calcium arsenate; $Ca_3(AsO_4)_2.3H_2O$ 125

CALCIUMAZETAT (*n*): calcium acetate; (*Calcium aceticum*); $Ca(C_2H_3O_2)_2.2H_2O$ 126

CALCIUM (*n*), **BALDRIANSAURES-**: calcium valerate 127

CALCIUM (*n*), **BASISCH GERBSAURES-**: basic calcium tannate; $CaOH.C_{14}H_9O_9$ 128

CALCIUM (*n*), **BENZOESAURES-**: calcium benzoate; $Ca(C_7H_5O_2)_2.3H_2O$ 129

CALCIUMBICHROMAT (*n*): calcium bichromate; $CaCr_2O_7$ 130

CALCIUMBISULFIT (*n*): (*Calcium bisulfurosum*): calcium bisulphite; $CaH_2(SO_3)_2$ 131

CALCIUM (*n*), **BLEISAURES-**: calcium plumbate (see CALCIUMPLUMBAT) 132

CALCIUMBORAT (*n*): calcium borate; $CaO.B_2O_3$; Mp. 1095°C.; $2CaO.B_2O_3$; Mp. 1225°C. (see also BORONATROCALCIT and COLEMANIT) 133

CALCIUM (*n*), **BORSAURES-**: calcium borate (see CALCIUMBORAT) 134

CALCIUMBROMID (*n*): calcium bromide; $CaBr_2$; Sg. 3.32; Mp. 680-760°C.; Bp. 806-812°C. 135

CALCIUMCARBID (*n*): calcium carbide; CaC_2; Sg. 2.22 136

CALCIUMCARBONAT (*n*): chalk, limestone, marble, calcium carbonate; $CaCO_3$; Sg. 2.712-2.934; calcite, Iceland spar; (*Calcium carbonicum*) 137

CALCIUMCHLORAT (*n*): calcium chlorate (see CALCIUM, CHLORSAURES-) 138

CALCIUMCHLORID (*n*): (*Calcium chloratum*): calcium chloride; $CaCl_2$; Sg. 2.22; $CaCl_2.6H_2O$; Sg. 1.654 139

CALCIUMCHLORID (*n*), **BASISCHES-**: basic calcium chloride; $3CaO.CaCl_2.15H_2O$ 140

CALCIUM (*n*), **CHLORSAURES-**: calcium chlorate; $Ca(ClO_3)_2.2H_2O$; Mp. 100°C. 141

CALCIUMCHROMAT (*n*); (*Calcium chromicum*): Calcium chromate; $CaCrO_4$ 142

CALCIUM (*n*), **CHROMSAURES-**: calcium chromate (see CALCIUMCHROMAT) 143

CALCIUM (*n*). **CITRONENSAURES-**: calcium citrate (see CALCIUMZITRAT) 144

CALCIUMCYANAMID (*n*): calcium cyanamide; $CaCN_2$ 145

CALCIUM (*n*), **ESSIGSAURES-**: calcium acetate (see CALCIUMAZETAT) 146

CALCIUMFERRICYANID (*n*): calcium ferricyanide; $Ca_3[Fe(CN)_6]_2.10H_2O$ 147

CALCIUMFERROCYANID (*n*): calcium ferrocyanide; $Ca_2Fe(CN)_6.12H_2O$ 148

CALCIUMFERROCYANÜR (*n*): calcium ferrocyanide, $Ca_2Fe(CN)_6.12H_2O$ 149

CALCIUMFLUORID (*n*): calcium fluoride; CaF_2; (*Calcium fluoratum*); Sg. 3.15-3.19; Mp. 1300°C; fluorspar 150

CALCIUM (*n*), **GERBSAURES-**: (see CALCIUM, BASISCH GERBSAURES-) 151

CALCIUM (*n*), **GOLDCHLORWASSERSTOFFSAURES-**: calcium chloraurate; $Ca(AuCl_4)_2.6H_2O$ 152

CALCIUM (*n*), HARZSAURES- : calcium resinate ; $Ca(C_{44}H_{62}O_4)_2$ 153
CALCIUM (*n*), HIPPURSAURES- : calcium hippurate 154
CALCIUMHYDRID (*n*) : calcium hydride ; CaH_2 ; Sg. 1.7 155
CALCIUMHYDROSULFID (*n*) : (see CALCIUM-SULFHYDRAT) 156
CALCIUMHYDROSULFIT (*n*) : calcium hydrosulphite ; CaS_2O_4 or $CaS_2O_4.1\frac{1}{2}H_2O$ [157
CALCIUMHYDROSULFÜR (*n*) : (see CALCIUMSULF-HYDRAT) 158
CALCIUMHYDROXYD (*n*) : (*Calcium hydroxydatum*) : calcium hydroxide ; calcium hydrate ; slaked lime ; $Ca(OH)_2$; lime hydrate (*Calcaria hydrica*) ; Sg. 2.078 159
CALCIUMHYDRÜR (*n*) : calcium hydride ; CaH_2 ; Sg. 1.7 160
CALCIUMHYPOCHLORIT (*n*) : calcium hypochlorite (see CHLORKALK) 161
CALCIUMHYPOPHOSPHIT (*n*) : (see CALCIUM, UNTERPHOSPHORIGSAURES-) 162
CALCIUMHYPOSULFIT (*n*) : (see CALCIUMTHIO-SULFAT) 163
CALCIUM (*n*), ISOBUTTERSAURES- : calcium isobutyrate 164
CALCIUM (*n*), JODBEHENSAURES- : calcium iodobehenate 165
CALCIUMJODID (*n*) : calcium iodide ; CaI_2 ; Sg. 3.956 ; Mp. 631-740°C. ; Bp. 708-719°C. 166
CALCIUM (*n*), KARBOLSAURES- : calcium (carbolate) phenolate 167
CALCIUM (*n*), KOHLENSAURES- : calcium carbonate (see CALCIUMCARBONAT) 168
CALCIUMLAKTAT (*n*) : (*Calcium lacticum*) : calcium lactate, calcinol ; $Ca(C_3H_5O_3)_2.5H_2O$ 169
CALCIUMLEGIERUNG (*f*) : calcium alloy 170
CALCIUM (*n*), LEINÖLSAURES- : calcium linoleate ; $Ca(C_{18}H_{31}O_2)_2$ 171
CALCIUMMETALL (*n*) : calcium metal 172
CALCIUMMETAPHOSPHAT (*n*) : calcium metaphosphate ; $Ca_2P_4O_3$ or $Ca(PO_3)_2$ 173
CALCIUMMETASILIKAT (*n*) : calcium metasilicate ; $CaSiO_3$; Sg. 2.92 ; wollastonite 174
CALCIUM (*n*), MILCHSAURES- : calcium lactate (see CALCIUMLAKTAT) 175
CALCIUMNITRAT (*n*) : (*Calcium nitricum*) : calcium nitrate ; $Ca(NO_3)_2$; Sg. 2.36 ; $Ca(NO_3)_2.4H_2O$; Sg. 1.82 ; Mp. 42°C. ; Bp. 132°C. 176
CALCIUMNITRID (*n*) : calcium nitride ; Ca_3N_2 ; Sg. 2.63 177
CALCIUMNITRIT (*n*) : calcium nitrite ; $Ca(NO_2)_2$; Sg. 2.294 ; $Ca(NO_2)_2.H_2O$; Sg. 2.231 178
CALCIUM (*n*), ÖLSAURES- : calcium oleate 179
CALCIUMORTHOPLUMBAT (*n*) : calcium orthoplumbate (see CALCIUMPLUMBAT) 180
CALCIUMORTHOSILIKAT (*n*) : calcium orthosilicate ; Ca_2SiO_4 ; Mp. 2130°C. 181
CALCIUMOXALAT (*n*) : (*Calcium oxalicum*) : calcium oxalate ; CaC_2O_4 182
CALCIUM (*n*), OXALSAURES- : calcium oxalate (see CALCIUMOXALAT) 183
CALCIUMOXYD (*n*) : (*Calcium oxydatum*) : calcium oxide ; burnt lime ; (*Calcaria usta*) ; CaO ; Sg. 3.15-3.4 ; Mp. 2570°C. 184
CALCIUMPENTASULFID (*n*) : Calcium pentasulphide ; CaS_5 185
CALCIUMPERBORAT (*n*) : (*Calcium perboricum*) : calcium perborate ; $CaBO_3$ or CaB_4O_8 186
CALCIUMPERMANGANAT (*n*) : (*Calcium permanganicum*) : calcium permanganate ; $CaMn_2O_8$; Sg. 1.8 or $Ca(MnO_4)_2.5H_2O$ 187
CALCIUMPEROXYD (*n*) : (*Calcium hyperoxydatum*) : calcium peroxide, calcium superoxide ; CaO_2 188
CALCIUMPHOSPHAT (*n*) : (*Calcium phosphoricum*) : calcium phosphate (see DICALCIUMPHOSPHAT ; MONOCALCIUMPHOSPHAT ; TRICALCIUM-PHOSPHAT) 189
CALCIUMPHOSPHAT (*n*), EINFACH SAURES- : (see DICALCIUMPHOSPHAT) 190
CALCIUMPHOSPHAT (*n*), PRIMÄRES- : primary calcium phosphate (see MONOCALCIUM-PHOSPHAT) 191
CALCIUMPHOSPHAT (*n*), SEKUNDÄRES- : secondary calcium phosphate (see DICALCIUM-PHOSPHAT) 192
CALCIUMPHOSPHAT (*n*), TERTIÄRES- : tertiary calcium phosphate (see TRICALCIUM-PHOSPHAT) 193
CALCIUMPHOSPHID (*n*) : calcium phosphide ; Ca_3P_2 or Ca_2P_2 or $Ca_2P_2O_7$; Sg. 2.51 194
CALCIUM (*n*), PHOSPHORMILCHSAURES- : calcium lactophosphate (*Calcium phospho-lacticum*) (mixture of calcium lactate, calcium acid latcate and calcium acid phosphate) 195
CALCIUM (*n*), PHOSPHORSAURES- : calcium phosphate (monobasic), superphosphate, acid calcium phosphate, calcium biphosphate ; $CaH_4(PO_4)_2.2H_2O$; Sg. 2.22 196
CALCIUMPLUMBAT (*n*) : calcium plumbate (*Calcium plumbicum*) ; Ca_2PbO_4 197
CALCIUM (*n*), PROPIONSAURES- : calcium propionate 198
CALCIUMPYROPHOSPHAT (*n*) : calcium pyrophosphate ; $Ca_2P_2O_7$ 199
CALCIUM (*n*), PYROPHOSPHORSAURES- : calcium pyrophosphate (see CALCIUMPYROPHOSPHAT) 200
CALCIUMRHODANÜR (*n*) : calcium (sulfocyanide) thiocyanate (see RHODANKALZIUM) 201
CALCIUMSACCHARAT (*n*) : calcium saccharate 202
CALCIUM (*n*), SALPETERSAURES- : calcium nitrate (see CALCIUMNITRAT) 203
CALCIUM (*n*), SAURES KOHLENSAURES- : calcium bicarbonate 204
CALCIUM (*n*), SCHWEFELSAURES- : calcium sulphate (see CALCIUMSULFAT) 205
CALCIUM (*n*) : SCHWEFLIGSAURES- : calcium sulphite (see CALCIUMSULFIT, NEUTRALES-) 206
CALCIUMSELENAT (*n*) : calcium selenate ; $CaSeO_4$; Sg. 2.93 207
CALCIUMSILICID (*n*) : calcium silicide ; $CaSi_2$; Sg. 2.5 208
CALCIUMSILICOFLUORID (*n*) : calcium silicofluoride ; $CaSiF_6$; Sg. 2.662 209
CALCIUMSILICOTITANAT (*n*) : titanite, calcium silico-titanate ; $CaTiSiO_5$; (see TITANIT) 210
CALCIUMSILIKAT (*n*) : calcium silicate (see DICALCIUMSILIKAT and MONOCALCIUM-SILIKAT) 211
CALCIUMSUBCHLORID (*n*) : calcium subchloride ; Ca_2Cl_2 ; Sg. 2.08 212
CALCIUMSULFAT (*n*) : (*Calcium sulfuricum*) : calcium sulphate, gypsum, plaster of Paris ; anhydrous $= CaSO_4$; Sg. 2.96 ; Mp. 1360°C. ; hydrous $= CaSO_4 + 2H_2O$; Sg. 2.32 213
CALCIUMSULFHYDRAT (*n*) : calcium hydrosulphide ; $Ca(SH)_2$ 214

CALCIUMSULFID (*n*): (*Calcium sulfuratum*): calcium sulphide; calcic liver of sulphur (*Hepar calcis*); CaS; Sg. 2.8 215

CALCIUMSULFIT (*n*): (*Calcium sulfurosum*): calcium sulphite; $CaSO_3.2H_2O$ 216

CALCIUMSULFIT (*n*), **NEUTRALES-**: neutral calcium sulphite; $CaSO_3$ (see **CALCIUMSULFIT**) 217

CALCIUMSUPEROXYD (*n*): calcium peroxide (see **CALCIUMPEROXYD**) 218

CALCIUMTEILCHEN (*n*): calcium particle 219

CALCIUMTHIOSULFAT (*n*): (*Calcium thiosulfuricum*): calcium thiosulphate; $CaS_2O_3.H_2O$ 220

CALCIUM (*n*), **ÜBERMANGANSAURES-**: calcium permanganate, acerdol; $Ca(MnO_4)_2.4H_2O$ 221

CALCIUM (*n*), **UNTERPHOSPHORIGSAURES-**: calcium hypophosphite; $Ca(H_2PO_2)_2$ 222

CALCIUM (*n*), **UNTERSCHWEFLIGSAURES-**: calcium hyposulphite (see **CALCIUMTHIOSULFAT**) 223

CALCIUM (*n*), **WEINSTEINSAURES-**: calcium tartrate; $CaC_4H_4O_6.H_2O$ 224

CALCIUMWOLFRAMAT (*n*): calcium tungstate; calcium wolframate; $CaWO_4$; Scheelite; Sg. 6.062 225

CALCIUMZITRAT (*n*): calcium citrate (*Calcium citricum*) 226

CALCIUMZYANAMID (*n*): calcium cyanamide, lime nitrogen, nitrolim; $CaCN_2$ 227

CALCIUMZYANID (*n*): calcium cyanide (*Calcium cyanatum*); $Ca(CN)_2$ 228

CALCUL (*m*): calculation, reckoning 229

CALCULIEREN: to calculate, reckon 230

CALEFACTOR (*m*): heater, calefactor, fireman, stoker 231

CALESZENZ (*f*): heating 232

CALIBER (*m*): gauge; pass (rolling mills); calibre, bore, internal diameter 233

CALIBERMÄSSIG: true to (gauge) scale 234

CALIBERZIRKEL (*m*): calipers 235

CALINLEGIERUNG (*f*): calin alloy (a metal foil, consisting of 86.5% Pb; 12.5% Zn; 1%Cu) 236

CALISAYARINDE (*f*): calisaya bark 237

CALMUSÖL (*n*): (see **KALMUSÖL**) 238

CALOMEL (*n*): calomel; mercurous chloride; HgCl; Sg. 6.993 239

CALOMELOL (*n*): colloidal calomel (contains about 80% HgCl and 20% albuminous substance) 240

CALORIMETER (*n*): calorimeter (an apparatus for measuring the heat generated by various physical or chemical processes) 241

CALORIMETRIE (*f*): calorimetry 242

CALORIMETRISCH: calorimetric-(al)-(ly) 243

CALORIMETRISCHE BOMBE (*f*): bomb calorimeter 244

CAMBAHOLZ (*n*): camwood (see **GABANHOLZ**) 245

CAMÉE-BILD (*n*): cameo print (Photography) 246

CAMELIA (*f*): camellia (see **KAMELIA**) 247

CÄMENTIEREN: to cement 248

CAMERAL: financial 249

CAMERALIST (*m*): financier 250

CAMERALWESEN (*n*): finance, financial matters 251

CAMERALWISSENSCHAFT (*f*): finance (the science) 252

CAMILLE (*f*): camomile (*Anthemis nobilis*) 253

CAMPANE (*f*): bell jar 254

CAMPECHEHOLZ (*n*): logwood; campeachy wood (see **BLAUHOLZ**) 255

CAMPHANSÄURE (*f*): camphanic acid 256

CAMPHEN (*n*): camphene (a terpene hydrocarbon); $C_{10}H_{16}$; Sg. 0.842; Mp. 49°C.; Bp. 159.5°C. 257

CAMPHER (*m*): gum camphor, camphor, laurel (Japan, Formosa) camphor (from the trees *Laurus camphora* and *cinnamomum camphora*); $C_{10}H_{16}O$; Sg. 0.811; Mp. 176.4°C.; Bp. 209.1°C. (see also **KAMPHER**) 258

CAMPHERDESTILLATIONSANLAGE (*f*): camphor distillation plant 259

CAMPHERERSATZMITTEL (*n*): camphor substitute 260

CAMPHER (*m*), **RECHTS-**: (see also **KAMPHER**); dextrocamphor; $C_{10}H_{16}O$; Sg. 0.811; Mp. 176.4°C.; Bp. 209.1°C. 261

CAMPHERSÄURE (*f*), **GEWÖHNLICHE-**: (ordinary)-camphoric acid, $C_{10}H_{16}O_4$: Mp. 187°C. [262

CAMPHERSÄUREMETHYLESTER (*m*), **NEUTRALER-**: neutral methyl camphorate, $C_{12}H_{20}O_4$; Bp. 259°C. 263

CAMPHERSÄURE, R-(*f*): (see **CAMPHERSÄURE, GEWÖHNLICHE**) 264

CAMPHOR (*m*): camphor (see **CAMPHER**) 265

CAMPHOR (*m*), **ALANT-**: Elecampane camphor; Helenin; (see **HELENIN**) 266

CAMPHORBLUMEN (*f.pl*): flowers of camphor 267

CAMPHOR (*m*), **BORNEO-**: Borneo camphor (see **BORNEOL**) 268

CAMPHORERSATZ (*m*): camphor substitute 269

CAMPHORESSIG (*m*): camphorated vinegar 270

CAMPHORGEIST (*m*): spirit of camphor 271

CAMPHORGERUCH (*m*): camphor (smell) odour 272

CAMPHORHALTIG: camphorous, camphorated, containing camphor, camphoric 273

CAMPHORHOLZ (*n*): camphor wood (Japanese = *Cinnamomum camphora*) (see also **KAMPHER-LORBEERBAUM**); camphorwood (Borneo = *Dryobalanops aromatica*); Sg. 0.9-1.06 [274

CAMPHORHOLZÖL (*n*): camphor-wood oil (from *Dryobalanops aromatica*); Sg. 1.155 275

CAMPHOR (*m*), **JAPAN-**: Japan camphor (from *Cinnamomum camphora*) (see **CAMPHER** and **CAMPHER, RECHTS-**) 276

CAMPHORKRAUT (*n*): (see **KAMPHERKRAUT**) 277

CAMPHOR (*m*), **KÜNSTLICHER-**: artificial camphor 278

CAMPHOR (*m*), **LAURINEEN-**: laurel camphor (from *Cinnamomum camphora*) (see **CAMPHER**) 279

CAMPHOR (*m*), **LINKS-**: levocamphor (see **CAMPHER**) 280

CAMPHORMILCH (*f*): camphorated emulsion 281

CAMPHOR (*m*), **MIT- SÄTTIGEN**: to camphorate, saturate with camphor 282

CAMPHOR (*m*), **MIT-WASCHEN**: to camphorate, wash with camphor 283

CAMPHORNAPHTOL (*n*): camphornaphthol (a condensation product from camphor and beta-naphthol; employed medically) 284

CAMPHORÖL (*n*): camphor oil (from *Cinnamomum camphora*); Sg. 0.87-1.04 285

CAMPHORÖL (*n*), **LEICHTES-**: light camphor oil (Sg. up to 0.89) (see **CAMPHORÖL**) 286

CAMPHORÖL (*n*), **SCHWARZES-**: dark camphor oil; Sg. 0.9-1.0; (see **CAMPHORÖL**) 287

CAMPHORÖL (*n*), **SCHWERES-**: heavy camphor oil; Sg. 0.9-1.0; (see **CAMPHORÖL**) 288

CAMPHORÖL (*n*), **WEISSES-**: light camphor oil; Sg. up to 0.89; (see **CAMPHORÖL**) 289

CAMPHORPHORON (*n*): camphorphorone 290

CAMPHOR (*m*), **RECHTS-**: (see **CAMPHER, RECHTS-**) 291

CAMPHOR (*m*), **ROH-**: crude camphor 292

CAMPHORSAUER: camphorated, camphorate of, combined with camphoric acid 293

CAMPHORSÄURE (*f*): camphoric acid (*Acidum camphoricum*); $C_8H_{14}(CO_2H)_2$; Sg. 1.228; Mp. 208°C. 294

CAMPHORSÄUREMETHYLESTER (*m*): camphoric methyl ester; methyl camphorate; Sg. 0.993 295

CAMPHORSEIFE (*f*): camphor soap 296

CAMPHORSPIRITUS (*m*): spirit of camphor 297

CAMPHOR (*m*), **SYNTHETISCHER-**: synthetic camphor 298

CANADABALSAM (*m*): canada balsam (see **KANADABALSAM**) 299

CANADOL (*n*): (see **KANADOL**) 300

CANARIENBAUM (*m*), **GEMEINER-**: common canary tree (*Canarium commune*) 301

CANARIÖL (*n*): canary oil (from seeds of *Canarium commune*) 302

CANCRINIT (*m*): cancrinite; $H_6Na_6Ca(NaCO_3)_2Al_8(SiO_4)_9$; Sg. 2.45 303

CANDELILLAPFLANZE (*f*): candelilla plant 304

CANDELILLAWACHS (*n*): candelilla wax; Sg. 0.983; Mp. 67.5°C. 305

CANDLENUSSÖL (*n*): candle-nut oil, lumbang oil (from the tree *Aleurites moluccana*); Sg. 0.926 306

CANELLARINDE (*f*): canella bark (bark of *Canella alba*) (see **CANELLE**) 307

CANELLE (*f*): canella, white cinnamon, Bahama white wood, false winter's bark, wild canilla (*Canella alba*) 308

CANELLENÖL (*n*): canella oil (from *Canella alba*); Sg. 0.93 309

CANNABIN (*n*): cannabin (the active resin of Indian hemp, *Cannabis indica*) 310

CANNABINOL (*n*): cannabinol (the active principle of Hashish); $C_{21}H_{30}O_2$; Bp. 215°C. 311

CANNABINON (*n*): cannabinon (the balsam resin of Indian hemp, *Cannabis indica*) 312

CANNELKOHLE (*f*): cannel coal; Sg. 1.24-1.34 [313

CANTHARIDEN (*pl*): cantharides, Spanish fly, blistering (beetle) fly (*Cantharis Vesicatoria*) 314

CANTHARIDENÄTHER (*m*): cantharides ether 315

CANTHARIDEN-KAMPFER (*m*): cantharides camphor (see **CANTHARIDIN**) 316

CANTHARIDENVERGIFTUNG (*f*): cantharides poisoning, Spanish fly poisoning 317

CANTHARIDIN (*n*): cantharidin, cantharides camphor; lactone of cantharidic acid (the active principle of Spanish flies); $C_{10}H_{12}O_4$; Mp. 210°C. 318

CANTHARINSÄURE (*f*): cantharidic acid 319

CANTHARSÄURE (*f*): cantharic acid 320

CAOUTSCHUK (*m*): caoutchouc; Sg. about 0.9 (see **KAUTSCHUK**) 321

CAPILLAR (*n*): capillary (see also **KAPILLAR**) 322

CAPILLARITÄT (*f*): capillarity, capillary attraction 323

CAPPELINIT (*m*): cappelenite; $3BaSiO_3.2Y_2(SiO_3)_3.5YBO_3$; Sg. 4.4 324

CAPRIBLAU (*n*): capri blue 325

CAPRINALDEHYD (*n*): (see **DECYLALDEHYD**) 326

CAPRINSÄURE (*f*): capric acid; $CH_3.(CH_2)_8.COOH$; (see **DECYLSÄURE**); Sg. 0.8858; Mp. 31.3°C.; Bp. 268.4°C. 327

CAPRINSÄUREÄTHYLESTER (*m*): ethyl (caprate) caprinate; Sg. 0.862; Bp. 244°C. 328

CAPRINSÄUREISOAMYLESTER (*m*): isoampl (caprate) caprinate; Bp. 275-290°C. 329

CAPRINSÄUREMETHYLESTER (*m*): methyl (caprate) caprinate; Mp. 18°C.; Bp. 223.5°C. 330

CAPRONSÄURE (*f*): capronic (caproic, hexoic, hexylic, pentylformic) acid; $C_5H_{11}COOH$; Sg. 0.922; Mp. −5.2°C.; Bp. 205°C. 331

CAPRONSÄUREÄTHYLESTER (*m*): ethyl capronate; Sg. 0.873; Bp. 167°C. 332

CAPRONSÄURE (*f*), **D-α-AMINO-N-**: (see **NORLEUCIN**) 333

CAPRONSÄURE (*f*), **D-α, ε-DIAMINO-N-**: (see **LYSIN**) 334

CAPRONSÄURE (*f*), **ISO-**: caproic acid, iso-; $(CH_3)_2.CH(CH_2)_2COOH$; Mp. −9.5°C.; Bp. 202.5°C. 335

CAPRONSÄUREMETHYLESTER (*m*): methyl capronate; Sg. 0.904; Bp. 149.6°C. 336

CAPRYLALKOHOL (*m*): caprylic alcohol; capryl alcohol; octoic alcohol; secondary octylic alcohol; $CH_3(CH_2)_5CHOHCH_3$; Sg. 0.832; Bp. 179°C. 337

CAPRYLALKOHOL (*m*), **PRIMÄR-**: octylic alcohol, primary caprylic alcohol, $CH_3.(CH_2)_6CH_2OH$; Sg. 0.83; Mp. −17.9°C.; Bp. 198.5°C. [338

CAPRYLALKOHOL (*m*), **SEKUNDÄR-**: secondary caprylic alcohol, metnyl-hexylcarbinol; $CH_3.CH(OH).(CH_2)_5.CH_3$; Sg. 0.819; Bp. 179°C. 339

CAPRYLEN (*n*): caprylene; C_8H_{16}; Sg. 0.72; Bp. 123°C. 340

CAPRYLSÄURE (*f*): caprilic acid, caprylic acid; octic (octylic) acid; $(C_7H_{15}COOH)$; Sg. 0.919; Mp. 17°C.; Bp. 235-237°C. 341

CAPRYLSÄUREÄTHYLESTER (*m*): ethyl caprylate; Bp. 205.8°C 342

CAPRYLSÄUREMETHYLESTER (*m*): methyl caprylate; Sg. 0.887; Bp. 192.9°C 343

CAPSICUMEXTRAKT (*m*): capsicum extract (from pepper plant, *Capsicum fastigiatum*) 344

CAPSICUMPFLASTER (*n*): capsicum plaster 345

CAPUT MORTUUM (*n*): caput mortuum, colcothar; (see **ENGLISCHROT**) 346

CARAGHEENMOOS (*n*): caragheen moss, Irish moss; (*Chondrus crispus*) 347

CARAPABAUM (*m*): carapa-tree (*Carapa guianensis*) 348

CARAPAÖL (*n*): carapa oil (from seeds of *Carapa guianensis*) 349

CARBAMID (*n*): Carbamide, urea; $CO(NH_2)_2$; Sg. 1.323; Mp. 132°C. 350

CARBAMID (*n*), **OXALSAURES-**: carbamide oxalate 351

CARBAMID (*n*), **SALPETERSAURES-**: carbamide nitrate 352

CARBAMINSAUER: carbamate of; combined with carbamic acid 353

CARBAMINSÄURE (*f*): carbamic acid; $CONH_2OH$ 354

CARBANILID (*n*): carbanilide, diphenylurea; $(NHC_6H_5)CO(NHC_6H_5)$; Mp. 235°C.; Bp. 260°C. 355

CARBANILSÄURE (*f*): carbanilic acid 356

CARBAZOL (*n*): carbazole, iminophenyl, diphenylimide; $C_{12}H_9N$; Mp. 238°C.; Bp. 335-338°C. 357

CARBAZOLBLAU (*n*): carbazole blue (by melting carbazole wich oxalic acid) 358

CARBAZOLGELB (*n*): carbazole yellow (a yellow azo dyestuff) 359

CARBAZOLKALIUM (*n*): potassium carbazolate; $C_{13}H_8NK$ 360

CARBID (*n*) : carbide ; (see CALCIUMCARBID and KARBID) 361
CARBIDKOHLE (*f*) : carbide carbon 362
CARBIDÖFENANLAGE (*f*) : carbide furnace plant 363
CARBINOL (*n*) : (see METHYLALKOHOL) 364
CARBOCYCLISCH : carbocyclic 365
CARBOLFUCHSIN (*n*) : carbol-fuchsin (1 gr. of basic fuchsin dissolved in 15 cc. absolute alcohol and made up to 100 cc. with a 5% solution of carbolic acid) 366
CARBOLGLYZERINSEIFE (*f*) : carbolic glycerine soap 367
CARBOLIGNUM (*n*) : carbolignum, wood charcoal 368
CARBOLINEUM (*n*) : carbolineum (a mixture of heavy coal tar oils, used for wood preserving) Sg. 1.128 ; Bp. 230-295°C. 369
CARBOLKALK (*m*) : carbolated lime 370
CARBOLPULVER (*n*) : carbolic powder 371
CARBOLSALBE (*f*) : phenol ointment 372
CARBOLSAUER : carbolate, combined with carbolic acid, phenylate, phenolate 373
CARBOLSÄURE (*f*) : carbolic acid, phenol, phenic (phenylic) acid, hydroxybenzene, phenyl hydrate ; C_6H_5OH ; Sg. 1.068 ; Mp. 43°C. ; Bp. 182.6°C. ; (see also PHENOL) 374
CARBOLSÄUREFLASCHE (*f*) : carbolic acid (bottle) carboy 375
CARBOLSCHWEFELSAUER : sulphocarbolate of ; combined with sulphocarbolic acid ; sulfocarbolate of 376
CARBOLSCHWEFELSÄURE (*f*) : sulphocarbolic (sulfocarbolic ; sulphophenic ; phenolsulphonic) acid ; $C_6H_5SO_3H$ 377
CARBOLSEIFE (*f*) : carbolic soap 378
CARBOLSULFOSÄURE (*f*) : (see CARBOLSCHWEFEL-SÄURE) 379
CARBOLWASSER (*n*) : aqueous solution of phenol 380
CARBONAT (*n*) : carbonate 381
CARBONATHALTIG : containing carbonate-(s) 382
CARBONATHÄRTE (*f*) : carbonate hardness (of water) (temporary hardness caused by bicarbonates, which is removable by boiling) 383
CARBONATION (*f*) : carbonation ; coking ; (*n*) carbonate ion 384
CARBONATWASSER (*n*) : carbonate water (water containing carbonates) 385
CARBON BLACK : carbon black (see GASRUSS) 386
CARBONIEREN : to carbonate 387
CARBONISATION (*f*) : carbonization ; carbonation (see also KARBONISATION) 388
CARBONISIEREN : to carbonize ; carbonate ; coke (see also KARBONISIERUNG) 389
CARBONISIERFLÜSSIGKEIT (*f*) : carbonizing liquid 390
CARBONIT (*n*) : carbonite (trade name of a German safety explosive) 391
CARBONPAPIER (*n*) : carbon paper 392
CARBONSÄURE (*f*) : carboxylic acid ; carbonic acid 393
CARBONYL (*n*) : carbonyl 394
CARBONYLCHLORID (*n*) : carbonyl chloride (see CHLORKOHLENOXYD) 395
CARBONYLSAUERSTOFF (*m*) : carbonyl oxygen 396
CARBORUND (*n*) : carborundum (a silicon carbide) SiC ; Sg. about 3.15 397
CARBORUNDUM (*n*) : carborundum (see CARBORUND) 398
CARBOSTYRIL (*n*) : (see OXYCHINOLIN, ALPHA-) 399
CARBOXYTARTRONSÄURE (*f*) : carboxytartronic acid (see DIOXYWEINSÄURE) 400
CARBOZYKLISCH : carbocyclic 401
CARBURATEUR (*m*) : carburettor 402
CARBURIEREN : to carburet ; carburize 403
CARBURIERUNG (*f*) : carburetting, carburization 404
CARCEL (*f*) : carcel (a French lighting unit ; the light emitted by a lamp consuming 42 g. refined rape oil in 1 hour, with a flame 40mm. high) 405
CARDAMOME (*f*) : cardamom (Bot.) (from plant, *Elettaria cardamomum*) 406
CARDIOIDE (*f*) : cardioid 407
CARDOBENEDIKTENKRAUT (*n*) : blessed thistle herb (*Herba cardui benedicti*) 408
CARDOL (*n*) : cardol (obtained from cashew oil from the fruit of *Anacardium occidentale*) ; $C_{32}H_{50}O_3.H_2O$ 409
CARMIN (*n*) : Cochineal (of which the colouring principle is carminic acid, $C_{17}H_{18}O_{10}$; from dried bodies of insects *Coccus cacti*) 410
CARMOISIN (*n*) : crimson (see KARMOISIN-) 411
CARNALLIT (*m*) : carnallite (from Stassfurt district of Germany) ; $KMgCl_3.6H_2O$; Sg. 1.6 412
CARNAUBAERSATZ (*m*) : carnauba substitute 413
CARNAUBAPALME (*f*) : carnauba palm (*Copernica cerifera*) (see also WACHSPALME) 414
CARNAUBASÄURE (*f*) : carnaubic acid 415
CARNAUBAWACHS (*n*) : carnauba wax (*Cera carnaubæ*) ; Sg. 0.995 ; Mp. 83-91°C. 416
CARNIN (*n*) : carnine (an animal alkaloid from fresh meat) ; $C_7H_8N_4O_3$ 417
CARNOTIT (*m*) : carnotite (a radio-active mineral) $K_2O.V_2O_5.2U_2O_2.3H_2O$ 418
CARO'SCHE SÄURE (*f*) : (see OXYSCHWEFELSÄURE or SULFOMONOPERSÄURE) : permonosulphuric acid, Caro's acid, oxysulphuric acid ; H_2SO_5 ; Mp. 45°C. 419
CARRAGHEENMOOS (*n*) : carragheen moss, chondrus, Irish moss, killeen, rock-salt moss, pearl-moss ; pig-wrack (*Chondrus crispus* ; *Fucus crispus*) 420
CARREAU (*n*) : check ; square 421
CARREAUGRAPHIE (*f*) : carreaugraphy (a reproduction process) 422
CARRIÈRE MACHEN : to better oneself ; to carve out one's career 423
CARTHAMIN (*n*) : carthamine, carthaminic acid (an insoluble red substance from Safflower, *Carthamus tinctorius*) ; $C_{14}H_{16}O_7$ 424
CARTHAMINSÄURE (*f*) : carthaminic acid (see CARTHAMIN) 425
CARVACROL (*n*) : carvacrol ; $(CH_3)_3.CH.OH$; Sg. 0.981 ; Mp. 0°C. ; Bp. 237.7°C. 426
CARVOL (*n*) : carvol (see CARVON) 427
CARVON (*n*) : carvone, carvol ; $C_{10}H_{14}O$; Sg. 0.96 ; Bp. 230°C. 428
CASCARILLA (*f*) : cascarilla, sweet-wood bark, eleuthera bark (bark of *Croton eluteria*) 429
CASCARILLAÖL (*n*) : cascarilla oil (from *Croton eluteria*) 430
CASEIN (*n*) : casein (see also KASEIN) 431
CASEINSALBE (*f*) : casein ointment 432
CASEINSAUER : caseinate 433
CASEMATTE (*f*) : central battery ; casemate 434
CÄSIUM (*n*) : cesium ; coesium ; Cs ; Sg. 1.87 ; Mp. 28.5°C. ; Bp. 670°C. 435

CÄSIUMALAUN (*m*): cesium alum; caesium aluminium sulphate; $Cs_2SO_4.Al_2(SO_4)_3.24H_2O$; Sg. 2.02; Mp. 117°C. 436

CÄSIUMAZID (*n*): caesium azide; CsN_3; Mp. 310-318°C. 437

CÄSIUMBINITRAT (*n*): caesium binitrate; $CsNO_3.HNO_3$; Mp. 100°C. 438

CÄSIUMBROMID (*n*): caesium bromide; CsBr; Sg. 4.377-4.455 439

CÄSIUMCARBONAT (*n*): caesium carbonate; Cs_2CO_3 440

CÄSIUMCHLORID (*n*): caesium chloride; CsCl; Sg. 3.969-3.982; Mp. 631-646°C. 441

CÄSIUM (*n*), **GOLDCHLORWASSERSTOFFSAURES-**: caesium chloraurate; $CsAuCl_4$ or $CsAuCl_4.\frac{1}{2}H_2O$ 442

CÄSIUMHYDRID (*n*): caesium hydride; CsH; Sg. 2.7 443

CÄSIUMHYDROXYD (*n*): caesium hydroxide; CsOH; Sg. 3.675 444

CÄSIUMJODAT (*n*): caesium iodate; $CsJO_3$; Sg. 4.831 445

CÄSIUMJODID (*n*): caesium iodide; CsJ; Sg. 4.508-4.523; Mp. 621°C. 446

CÄSIUMNITRAT (*n*): caesium nitrate; $CsNO_3$; Sg. 3.6415-3.687; Mp. 414°C. 447

CÄSIUMOXYD (*n*): caesium oxide; Cs_2O; Sg. 4.36 448

CÄSIUMPENTASULFID (*n*): caesium pentasulphide; Cs_2S_5; Sg. 2.806 449

CÄSIUMPERJODAT (*n*): caesium periodate; $CsJO_4$; Sg. 4.259 450

CÄSIUMPERMANGANAT (*n*): caesium permanganate; $CsMnO_4$; Sg. 3.597 451

CÄSIUMPEROXYD (*n*): caesium peroxide; Cs_2O_4; Sg. 3.77; Mp. 515°C. 452

CÄSIUMSELENAT (*n*): caesium selenate; Cs_2SeO_4; Sg. 4.4528 453

CÄSIUMSILICOFLUORID (*n*): caesium silicofluoride; Cs_2SiF_6; Sg. 3.372 454

CÄSIUMSULFAT (*n*): caesium sulphate; Cs_2SO_4; Sg. 4.2434; Mp. 1019°C. 455

CÄSIUMSULFID (*n*): casium sulphide: Cs_2S_2; Cs_2S_3; Cs_2S_4; Cs_2S_5; Cs_2S_6 456

CÄSIUMSUPEROXYD (*n*): caesium peroxide; Cs_2O_2; Sg. 4.47; Cs_2O_3; Sg. 4.25; Cs_2O_4; Sg. 3.68 457

CÄSIUMTRINITRAT (*n*): caesium trinitrate; $CsNO_3.2HNO_3$; Mp. 32-36°C. 458

CÄSIUMVERBINDUNG (*f*): cesium compound 459

CASSAWASTÄRKE (*f*): tapioca 460

CASSELERGELB (*n*): cassell yellow (lead oxychloride); $7PbO.PbCl_2$ 461

CASSELMANNS GRÜN (*n*): Casselmann's green, basic copper sulphate (see **KUPEFRSULFAT, BASISCHES-**) 462

CASSETTE (*f*): dark-slide (Photography); holder, case, portfolio (see **KASSETTE**) 463

CASSETTENSCHNÄPPER (*m*): dark-slide-(spring)-catch 464

CASSIA (*f*): cassia, cinnamon (bark of *Cinnamomum cassia*) (see also **KASSIA**) 465

CASSIAÖL (*n*): cassia oil; Chinese cinnamon oil (about 70-90% cinnamic aldehyde) (*Oleum Cassiæ*; *Oleum cinnamomi cassiæ*); Sg. 1.25-1.45; Bp. about 250°C. 466

CASSIENRINDE (*f*): cassia bark (bark of *Cinnamomum cassia*) 467

CASSIOPEIUM (*n*): (see **LUTETIUM**: on list of elements) (an element) 468

CASSITERIT (*m*): cassiterite, tinstone, black tin (see **AUSSTRICH** and **ZINNSTEIN**) 469

CASSIUSPURPUR (*m*): purple of cassius (see **GOLDPURPUR**) 470

CASTILLIANERSEIFE (*f*): castile soap 471

CASTORÖL (*n*): (see **RICINUSÖL**) 472

CATAPPAÖL (*n*): catappa oil, wild almond oil (from seeds of *Terminalia catappa*) 473

CATAPPENBAUM (*m*): catappa tree, wild almond (*Terminalia catappa*) 474

CATHARTINSÄURE (*f*): cathartic acid 475

CAUTEL (*f*): precaution 476

CAY-CAYBUTTER (*f*): (see **COCHINCHINAWACHS**) 477

CEARAWACHS (*n*): carnauba wax (from the leaves of the wax palm *Corypha cerifera*) (see **CARN--AUBAWACHS**) 478

CEDER (*f*): cedar (see **CEDERNHOLZ**) 479

CEDERNGUMMI (*n*): cedar gum (from *Lallitris arborea*) 480

CEDERNHARZ (*n*): cedar resin (white resin exuded from cedar trees) 481

CEDERNHOLZ (*n*): cedar, red cedar (*Juniperus virginiana*); Sg. 0.4-1.32 482

CEDERNHOLZÖL (*n*): cedar wood oil (from the wood of *Juniperus virginiana*); Sg. 0.955 483

CEDERNÖL (*n*): cedar oil (see **CEDERNHOLZÖL**); Sg. 0.88 to 0.96; (from the leaves of *Juniperus virginiana*) 484

CEDERNUSSÖL (*n*): cedar nut oil (from seeds of *Pinus cembra*); Sg. 0.93 485

CEDER (*f*), **SIBIRISCHE-**: Siberian cedar-tree (*Pinus cembra*) 486

CEDERTANNE (*f*): moulmein cedar, toon (toona) tree (*Cedrela toona*) 487

CEDIEREN: to cede, cancel 488

CEDRATÖL (*n*): citron oil (*Oleum citri*) (from fruit-skins of *Citrus limomum Risso*); Sg. 0.854-0.861 489

CEDRELAHOLZ (*n*): cedar (toon, toona) wood (see **CEDERTANNE**) 490

CEDROÖL (*n*): (see **CEDRATÖL**) 491

CEKAS (*m*): a nickel-iron-chromium alloy (electrical resistance material) 492

CELASTERÖL (*n*): celastrus oil (from seeds of *Celastrus paniculatus*) 493

CELLOIDIN (*n*): celloidin, celluidine, (pure nitrocellulose) (a form of pyroxylin) 494

CELLOIDINPAPIER (*n*): collodio-chloride (celloidin) paper (Photography) 495

CELLOTROPIN (*n*): cellotropin; $C_{19}H_{20}O_8$; Mp. 184.5°C. 496

CELLULOIDSCHALE (*f*): celluloid dish 497

CELLULOSE (*f*): (see **ZELLULOSE** and **ZELLSTOFF**) 498

CELLULOSEAZETAT (*n*): cellulose acetate 499

CELOSIAÖL (*n*): celosia oil (from seeds of *Celosia cristata*) 500

CELTIUM (*n*): celtium (an element) 501

CENSIEREN: to criticize; censor; censure; review 502

CENSUR (*f*): censorship; report; review; censure; criticism 503

CENTRIFUGE (*f*): (see **ZENTRIFUGE**) 504

CENTRUMBOHRER (*m*): centre bit 505

CER (*n*): cerium; Ce; Sg. 6.92; Mp. 645°C. (see also **ZER**) 506

CERARGYRIT (*m*): cerargyrite (natural silver chloride); AgCl; Sg. 5.36 507

CERAT (*n*): cerate; encaustic paste; cerate paste (used to give a brilliant surface to a

photographic print without the use of collodion or hot rollers; composed of white wax, benzol, gum elemi, spike oil and lavender essence) 508
CERATPAPIER (*n*): cerate paper 509
CERCARBID (*n*): cerium carbide; CeC_2; Sg. 5.23 510
CERCHLORID (*n*): cerium chloride; $CeCl_3$; Sg. 3.92; Mp. 848°C. 511
CERDIOXYD (*n*): cerium dioxide, ceric oxide; CeO_2; Sg. 6.74-7.65 512
CEREALIEN (*pl*): cereals 513
CEREAWACHS (*n*): carnauba wax (see KARNAUBA-WACHS) 514
CERERERZ (*n*): cerite, cererite 515
CERESIN (*n*): ceresin, ozocerite, cerin, cerosin, earth (mineral) wax; Sg. 0.918-0.94 516
CERESIT (*n*): ceresit (a waterproofing material for cement) 517
CERIN (*n*): cerin (see CERESIN) 518
CERIOXYD (*n*): ceric oxide; CeO_2; Sg. 7.65 [519
CERISULFAT (*n*): cerium (ceric) sulphate; $Ce(SO_4)_2.4H_2O$ 520
CERIT (*m*): cerite (a mineral, basic cerium silicate containing lanthanum, didymium, calcium, iron, and aluminium); Sg. 4.9-5.0; $H_3(Ca,Fe)(La,Di,Ce)_3Si_3O_{13}$ 521
CERITMETALL (*n*): cerite metal 522
CERITSALZ (*n*): cerite salt 523
CERIUMFLUORID (*n*): cerium (ceric) fluoride; $CeF_4.H_2O$ 524
CERIUM (*n*), **GOLDCHLORWASSERSTOFFSAURES-**: cerium chloraurate; $CeCl_3,AuCl_3+10H_2O$ 525
CERIUMMETALL (*n*): cerium metal 526
CERIUMNITRAT (*n*): cerium (cerous) nitrate; $Ce(NO_3)_3.6H_2O$ 527
CERIUM (*n*), **OXALSAURES-**: cerium oxalate, cerous oxalate; $Ce_2(C_2O_4)_3.9H_2O$ 528
CERIUM (*n*), **SALICYLSAURES-**: cerium salicylate 529
CERIUM (*n*), **SALPETERSAURES-**: cerium nitrate (see CERIUMNITRAT) 530
CERIUMSALZ (*n*): cerium salt 531
CERIUM (*n*), **SCHWEFELSAURES-**: cerium sulphate (see CERISULFAT) 532
CERIUMVERBINDUNG (*f*): cerium compound (see CERI- and CERO-) 533
CERIVERBINDUNG (*f*): ceric compound 534
CERMETALL (*n*): cerium metal 535
CEROTIN (*n*): cerotine (an encaustic paste made from vegetable wax) (Photography); (see also CERAT) 536
CEROTINSÄURE (*f*): cerotic acid (from *Corypha cerifera*) 537
CEROVERBINDUNG (*f*): cerous compound 538
CERSILICID (*n*): cerium silicide; $CeSi_2$; Sg. 5.67 539
CERSULFAT (*n*): cerium sulphate; $Ce_2(SO_4)_3$; Sg. 3.912; $Ce_2(SO_4)_3.8H_2O$; Sg. 2.886 [540
CERSULFID (*n*): cerium sulphide; Ce_2S_3; Sg. 5.02 541
CERTAPARTIE (*f*): charter party (Com.) 542
CERUSSIT (*m*): cerussite, white lead ore (natural lead carbonate) (see BLEISPAT and WEISS-BLEIERZ); $PbCO_3$; Sg. 6.55 543
CESSION (*f*): cession, cancellation 544
CETYLALKOHOL (*m*): cetyl (cetylic) alcohol; palmityl alcohol; hecdecatylic alcohol; ethal; ethol; normal primary hexadecyl alcohol; $C_{16}H_{33}OH$; Sg. 0.82; Mp. 50°C.; Bp. 344°C. 545
CETYLJODID (*n*): cetyl iodide 546
CH-: (see also under SCH-) 547
CHABASIT (*m*): chabazite; $(CaNa_2K_2)Al_2(SiO_3)_4.6H_2O$; Sg. 2.1 548
CHAGRIN (*m*): shagreen 549
CHALCANTHIT (*m*): blue vitriol, copper vitriol chalcanthite (natural hydrous copper sulphate) (see KUPFERVITRIOL and KUPFERSULFAT); $CuSO_4.5H_2O$; Sg. 2.21 550
CHALCEDON (*m*): chalcedony, calcedony (Mineralogy); SiO_2; Sg. 2.62 (see ACHAT) 551
CHALCOPYRIT (*m*): chalcopyrite (see KUPFERKIES) 552
CHALKOLITH (*m*): chalcolite (natural copper-uranium phosphate); $Cu(UO_2)_2(PO_4)_2.xH_2O$ 553
CHALKOPYRIT (*m*): yellow copper ore, chalcopyrite (copper pyrites) (a copper-iron sulphide), $CuFeS_2$; Sg. 4.25 (see KUPFERKIES) 554
CHALKOSIN (*m*): chalcocite (a cuprous sulphide); Cu_2S; Sg. 5.65 (see KUPFERGLANZ) 555
CHALKOSTIBIT (*m*): chalcostibite (see WOLFS-BERGIT) 556
CHALKOTRICHIT (*m*): chalcotrichite, plush copper (see ROTKUPFERERZ) 557
CHALZEDON (*m*): (see CHALCEDON) 558
CHAMÄLEON (*n*): chameleon; chameleon mineral (a natural potassium manganate) 559
CHAMÄLEONLÖSUNG (*f*): potassium permanganate solution 560
CHAMOSIT (*m*): chamoisite; $(Fe,Mg)_3Al_2Si_2O_{10}.3H_2O$; Sg. 3.2 561
CHAMOTTE (*f*): chamotte (Ceramics); grog; fireclay 562
CHAMOTTETIEGEL (*m*): fireclay crucible 563
CHAMOTTETON (*m*): fireclay 564
CHAMOTTETON, GEMAHLENER- (*m*): ground fireclay 565
CHAMOTTEZIEGEL (*m*): firebrick 566
CHAMPACAFETT (*n*): champaca fat (from seeds of *Media champaca*) 567
CHANGIEREND: changeable; shot (of fabrics) 568
CHAPPESEIDE (*f*): spun silk; Schappe silk 569
CHARAKTER (*m*): character 570
CHARAKTERISIEREN: to characterize 571
CHARAKTERISTISCH: characteristic-(al)-(ly) 572
CHARGE (*f*): charge (ore and flux) 573
CHARGIEREN: to charge; load 574
CHARGIERKRAN (*m*): loading crane 575
CHARGIERMASCHINE (*f*): charging machine (Metal) 576
CHARGIERWAGEN (*m*): charging (waggon) truck 577
CHARLTONWEISS (*n*): charlton white (see LITHO-PON) 578
CHARNIER (*n*): joint, hinge 579
CHARPIE (*f*): lint 580
CHARTEPARTIE (*f*): charter party (Com.) 581
CHARTOTYPIE (*f*): talbotype (Photography) 582
CHASSIS (*m*): chassis, carriage (see also FARBTROG) 583
CHAUFFEUR (*m*): chauffeur (of a motor car) 584
CHAULMOOGRAÖL (*n*): gynocardia oil, chaulmoogra oil (*Oleum Chaulmoograe*); Sg. 0.94; Mp. 26°C. 585
CHAULMOOGRASÄURE (*f*): chaulmoogra acid; $C_{17}H_{31}CO_2H$; Mp. 68°C. 586
CHAUSSEE (*f*): main road; highway; highroad 587
CHECK (*m*): cheque 588
CHEF (*m*): chef; chief; head; principal 589
CHEMIE (*f*): chemistry 590

CHEMIGRAPHIE (*f*): chemigraphy, zinc etching (Photography) (relief etching on metals; mostly on zinc) 591
CHEMIKALIEN (*n.pl*): chemicals 592
CHEMIKER (*m*): chemist 593
CHEMISCH: chemical-(ly) 594
CHEMISCHBLAU (*n*): chemic blue (indigo extract; sodium coerulin sulphate); $C_{16}H_8N_2O_2(SO_3Na)_2$ 595
CHEMISCHE EINRICHTUNGEN (*f.pl*): chemical (apparatus) fittings 596
CHEMISCHE ENTWICKELUNG (*f*): chemical development, alkaline development (Photography) 597
CHEMISCHER APPARAT (*m*): chemical apparatus 598
CHEMISCHER BRENNPUNKT (*m*): actinic focus (Photography) 599
CHEMISCHER SCHLEIER (*m*): chemical fog (Photography) 600
CHEMISCHES GEFÄSS (*n*): chemical vessel 601
CHEMISCHE WAGE (*f*): chemical (analytical) balance or scales 602
CHEMISCHGELB (*n*): cassel yellow (lead oxychloride) 603
CHEMISCHGRÜN (*n*): sap-green 604
CHEMISCHROT (*n*): Venetian red 605
CHEMISCHREIN: chemically pure 606
CHEMISCH-TECHNISCH: techno-chemical 607
CHEMISMUS (*m*): chemism 608
CHESSYLITH (*m*): chessylite (see AZURIT) 609
CHESTERKÄSE (*m*): Cheshire cheese 610
CHEVILLE (*f*): glosser, polisher, burnisher (a polished rod for glossing and softening silk) 611
CHEVILLIEREN (*n*): glossing, polishing or softening process for silk 612
CHEVILLIERMASCHINE (*f*): glossing machine (for silk) 613
CHIASTOLITH (*m*): chiastolite (see ANDALUSIT) 614
CHICAGOBLAU (*n*): chicago blue 615
CHICA-(ROT) (*n*): chica red (a Brazilian native dye, obtained from the roots of *Bigonia chica*); $C_8H_8O_3$ 616
CHIFFER (*f*): cipher 617
CHIFFRE (*f*): cipher 618
CHILEKUPFER (*n*) IN BARREN (*f.pl*): chili bars (copper-) 619
CHILENISCH: Chilian 620
CHILESALPETER (*m*): Chile saltpeter, sodium nitrate (see NATRIUM, SALPETERSAURES-) 621
CHILISALPETER (*m*): chili saltpeter, sodium nitrate (see NATRIUMNITRAT) 622
CHINA (*f*): cinchona; Peruvian bark (see CHINA-RINDE) 623
CHINAALKALOID (*n*): cinchona alkaloid (from cinchona bark of various species of *Cinchona*) 624
CHINABASE (*f*): cinchona base; quinine base 625
CHINAEISENWEIN (*m*): bitter iron wine; quinine and iron wine (compounded with ferric ammonium citrate) 626
CHINAEXTRAKT (*m*): cinchona extract 627
CHINAGERBSÄURE (*f*): quinotannic acid (from *Cinchona succirubra*) 628
CHINAGRAS (*n*): Chinese nettle (*Urtica nivea* or *Urtica boehmeria*) (see RAMIE) 629
CHINALDIN (*n*): quinaldine; $C_9H_6N.CH_3$; Sg. 1.1013; Bp. 246-247°C. 630
CHINALDINCYANIN (*n*): quinaldine-cyanine, ethyl red (see ÄTHYLROT) 631
CHINARINDE (*f*): (*Cortex Chinæ*): cinchona bark (bark of various species of cinchona); china bark, pale bark, crown bark, loxa bark (from *Cinchona officinalis*); Belivian bark, royal bark, yellow cinchona bark (from *Cinchona calisaya*); red cinchona bark (from *Cinchona succirubra*); grey cinchona bark, Jesuit's bark, Peruvian bark (from *Cinchona Peruviana*) 632
CHINAROT (*n*): cinchona red 633
CHINASAUER: quinate of; combined with quinic acid 634
CHINASÄURE (*f*): quinic acid; kinic acid; chinic acid (*Acidum chinicum*); $C_6H_7(OH)_4COOHH_2O$; (from *Vaccinium myrtillus*); Sg. 1.637; Mp. 160°C. 635
CHINASILBER (*n*): China silver (a nickel alloy) (NEUSILBER plus a percentage of Ag) 636
CHINATINKTUR (*f*): tincture of cinchona 637
CHINATOXIN (*n*): cinchona toxin 638
CHINAWEIN (*m*): quinine wine (extraction of cinchona bark with alcohol, sherry and hydrochloric acid) 639
CHINAWURZEL (*f*): cinchona root (*Rhizoma chinae* of *Smilax china*) 640
CHINAZOLIN (*n*): quinazoline 641
CHINDOLIN (*n*): quindoline 642
CHINEN (*n*): quinene; quinine; $C_{20}H_{24}N_2O_2.3H_2O$; Mp. 57°C. 643
CHINEONAL (*n*): chineonal, quinine diethylbarbiturate (two parts quinine and one part veronal) 644
CHINESISCHER TALG (*m*): Chinese tallow, vegetable wax (*Oleum Stillingiae*) (from the seeds of *Stillingia sebifera*, the Chinese tallow tree) 645
CHINESISCHGRÜN (*n*): Chinese green; lokao 646
CHINESISCHROT (*n*): Chinese red; red mercuric sulphide; HgS; Sg. 8.06-8.12; sublimes at 446°C. (see also ZINNOBER) 647
CHINETUM (*n*): quinetum 648
CHINICIN (*n*): quinicine 649
CHINID (*n*): quinide 650
CHINIDIN (*n*): quinidine; $C_{20}H_{24}N_2O_2$; Mp. 168°C. 651
CHINIEREN: to print chiné; cloud 652
CHININ (*n*): (see CHINEN) 653
CHININALKALOID (*n*): quinine alkaloid (see CHININALKALOID, WASSERFREIES- and CHININALKALOID, WASSERHALTIGES-) 654
CHININALKALOID (*n*), WASSERFREIES-: anhydrous quinine alkaloid; $C_{20}H_{24}N_2O_2$; Mp. 174.9°C. 655
CHININALKALOID (*n*), WASSERHALTIGES-: hydrous quinine alkaloid; $C_{20}H_{24}N_2O_2.3H_2O$; Mp. 57°C. 656
CHININ (*n*), BALDRIANSAURES-: quinine valerate 657
CHININBISULPHAT (*n*): quinine bisulphate; $C_{20}H_{24}N_2O_2.H_2SO_4.7H_2O$; Mp. 160°C. 658
CHININBROMHYDRAT (*n*): quinine hydrobromide; $C_{20}H_{24}N_2O_2.HBr.H_2O$; Mp. 152-200°C. [659
CHININCHLORHYDRAT (*n*): quinine hydrochloride (see CHININCHLORID) 660
CHININCHLORID (*n*): quinine hydrochloride; $C_{20}H_{24}N_2O_2.HCl.2H_2O$; Mp. 156-200°C. 661
CHININCITRAT (*n*): quinine citrate; $(C_{20}H_{24}N_2O_2)_2.C_6H_8O_7.7H_2O$ 662
CHININ (*n*), CITRONSAURES-: quinine citrate (see CHININCITRAT) 663
CHININEISEN (*n*): citrate of iron and quinine 664
CHININEISENCITRAT (*n*): (see CHININEISEN) 665

CHININ (*n*), GERBSAURES- : quinine tannate 666
CHININ (*n*), MILCHSAURES- : quinine lactate 667
CHININNITRAT (*n*) : quinine nitrate ; $C_{20}H_{24}N_2O_2.HNO_3.H_2O$ 668
CHININPERLE (*f*) : quinine pearl 669
CHININ (*n*), PHOSPHORSAURES- : quinine phosphate 670
CHININPILLE (*f*) : quinine pill 671
CHININ (*n*), SALICYLSAURES- : quinine salicylate ; $C_{20}H_{24}N_2O_2.C_7H_6O_3.H_2O$; Mp. 183-187°C. 672
CHININSALZ (*n*) : quinine salt 673
CHININ (*n*), SALZSAURES- : quinine hydrochloride ; $C_{20}H_{24}N_2O_2.HCl.2H_2O$; Mp. 156-200.°C. 674
CHININSÄURE (*f*) : quininic acid ; $CH_3OC_9H_5NCO_2H$ 675
CHININ (*n*), SCHWEFELSAURES- : quinine sulphate (see CHININSULFAT) 676
CHININSULFAT (*n*) : quinine sulphate ; $(C_{20}H_{24}N_2O_2)_2.H_2SO_4.8H_2O$ 677
CHININWASSER (*n*) : quinine water 678
CHININWEIN (*m*) : (see CHINAWEIN) 679
CHININ (*n*), WEINSAURES- : quinine tartrate 680
CHINIT (*m*) : quinite 681
CHINIZARIN (*n*) : quinizarin 682
CHINOIDIN (*n*) : quinoidine 683
CHINOL (*n*) : quinol ; hydroquinone ; $C_6H_4(OH)_2$; Sg. 1.33 ; Mp. 169°C. ; Bp. 285°C. 684
CHINOLIN (*n*) : quinoline, leucoline ; chinoline ; C_9H_7N ; Sg. 1.0944 ; Mp.– 22.6°C ; Bp. 238°C. 685
CHINOLINBASE (*f*) : quinoline base (of the general formula, $C_nH_{2n}-11N$) 686
CHINOLINBLAU (*n*) : quinoline blue ; cyanine (see CYANIN) 687
CHINOLIN-4-CARBONSÄURE (*f*) : (see CINCHONIN-SÄURE) 688
CHINOLINGELB (*n*) : quinoline yellow 689
CHINOLINROT (*n*) : quinoline red 690
CHINOLINSÄURE (*f*) : quinolinic acid 691
CHINOLIN (*n*), WEINSAURES- : quinoline tartrate 692
CHINOLIN-WISMUTRHODANAT (*n*) : (see CRURIN) 693
CHINON (*n*) : chinone ; quinone ; benzoquinone ; $CO(CH.CH)_2CO$; Sg. 1.307 ; Mp. 115.7°C. 694
CHINOPYRIDIN (*n*) : quinopyridine ; quinopyridene 695
CHINOSOL (*n*) : quinosol ; $(C_9H_7NO)_2H_2SO_4$ (see KALI, OXYCHINOLINSULFOSAURES-) 696
CHINOTROPIN (*n*) : quinotropin (compound of urotropin and quinic acid) ; Mp. about 122°C. 697
CHINOVABITTER (*n*) : quinova bitter ; quinovin 698
CHINOVASÄURE (*f*) : quinovic acid (from *Potentilla tormentilla*) 699
CHINOVIN (*n*) : (see CHINOVABITTER) 700
CHINOVOSE (*f*) : quinovose 701
CHINOXALIN (*n*) : quinoxaline 702
CHINOXALINFARBSTOFFE (*m.pl*) : quinoxaline dyestuffs of the quinone imide group 703
CHINOYL (*n*) : quinoyl 704
CHINUCLIDIN (*n*) : quinuclidine 705
CHIRALKOL (*n*) : chiralkol (a solid alcohol paste, 86% absolute alcohol and 14% curd soap) 706
CHIRETTA (*f*) : chiretta ; chirata ; bitterstick ; (dried plant *Swertia chirayita*) 707
CHIRURG (*m*) : surgeon 708
CHIRURGIE (*f*) : surgery 709
CHIRURGISCHE ARTIKEL (*m.pl*) : surgical appliances 710
CHIRURGISCHE OPERATION (*f*) : surgical operation 711
CHLOANTHIT (*m*) : chloanthite (a nickel arsenide) (see ARSENNICKELKIES) ; $NiAs_2$; Sg. 6.5 712
CHLOR (*n*) : chlorine ; Cl 713
CHLORACETON (*n*) : chloroacetone, monochloroacetone ; C_3H_5OCl ; Sg. 1.162 ; Bp. 119°C ; monochlorinated acetone 714
CHLORACETYL (*n*) : acetyl chloride ; (CH_3COCl) ; Sg. 1.105 ; Mp. 50.9°C. 715
CHLORAL (*n*) : (*Chloralum anhydricum*) : chloral, trichloroacetic aldehyde ; $CCl_3.CHO$; Sg. 1.512 ; Mp. – 57.5°C. Bp. 97.7°C. 716
CHLORALAMID (*n*) : (*Chloralum formamidatum*) : chloral formamide ; $(C_3H_4Cl_3NO_2)$; Mp. 114-115°C. 717
CHLORALHYDRAT (*n*) : (*Chloralum hydratum*) : chloral hydrate ; $C_2H_3O_2Cl_3$; Sg. 1.5745 ; Mp. 57°C. ; Bp. 97.5°C. 718
CHLORALUM (*n*) : (see CHLORALUMINIUM) 719
CHLORALUMINIUM (*n*) : aluminium chloride ; Al_2Cl_6 ; or $Al_2Cl_6.12H_2O$; Mp. 190°C. 720
CHLORAMINBRAUN (*n*) : chloramine brown 721
CHLORAMMONIUM (*n*) : ammonium chloride ; NH_4Cl ; Sg. 1.52 (see also SALMIAK) 722
CHLORAMYL (*n*) : amyl chloride 723
CHLORANIL (*n*) : chloranil, tetrachloroquinone ; $C_6Cl_4O_2$; Mp. 285°C. 724
CHLORANILIN (*n*) : aniline salt, aniline (hydro)-chloride ; (C_6H_8NCl) ; ortho.- Sg. 1.2125 ; Bp. 208.8°C ; Meta.- Sg. 1.2149 ; Bp. 230°C.; para.- Sg. 1.1704 ; Mp. 71°C .; Bp. 232.3°C. 725
CHLORANOL (*n*) : chloranol (a photographic chemical ; consists of monomethylpara-aminophenol and chlorohydroquinone in proportion 2 : 1) ; $2(C_6H_4OH.NH.CH_3)$. $C_6H_4(OH)_2Cl$ 726
CHLORARGYRIT (*m*) : cerargyrite (see HORN-SILBER) 727
CHLORARSENIK (*m*) : arsenic chloride 728
CHLORARSENIKLÖSUNG (*f*) : arsenic chloride solution ; solution of arsenious acid ; hydrochloric solution of arsenic 729
CHLORAT (*n*) : chlorate 730
CHLORÄTHYL (*n*) : ethyl chloride (see ÄTHYL-CHLORID) 731
CHLORÄTHYLEN (*n*) : ethylene chloride (see ÄTHYLENCHLORID) 732
CHLORÄTHYLENCHLORID (*n*) : chloroethylene chloride ; monochloroethylene chloride ; trichloroethane ; vinyl trichloride ; monochlorinated Dutch liquid ; $(CH_2)_2Cl_3$; Sg. 1.458 ; Bp. 114°C. 733
CHLORÄTHYLIDEN (*n*) : ethylidene chloride 734
CHLORATION (*f*) : chlorination ; (*n*) chlorate ion 735
CHLORBARIUM (*n*) : barium chloride ; $BaCl_2.2H_2O$; Sg. 3.097 ; Mp. 860°C. 736
CHLORBENZOESÄURE (*f*) : (*Acidum chlorbenzoicum*) ; Chlorbenzoic acid ; $C_6H_4Cl.CO_2H$ 737
CHLORBENZOL (*n*) : chlorobenzene ; monochlorbenzol ; C_6H_5Cl ; Sg. 1.1064 ; Mp. – 44.9°C.; Bp. 132°C. 738
CHLORBENZOYL (*n*) : benzoyl chloride (see BENZOYLCHLORID) 739
CHLORBENZYL (*n*) : benzyl chloride (see BENZYL-CHLORID) 740

CHLORBLEI (n): lead chloride; $PbCl_2$; Sg. 5.88; Mp. 498°C.; Bp. 861-954°C. (see BLEI-CHLORID) 741
CHLORBLEICHE (f): chlorine bleach-(ing) 742
CHLORBLEISPAT (m): phosgenite 743
CHLORBROM (n): bromine chloride 744
CHLORBROMSILBER (n): silver chlorobromide (Phot.); embolite (natural silver chlorobromide); Ag(ClBr) 745
CHLORBROMSILBERPAPIER (n): silver chlorobromide paper (Photography) 746
CHLORCALCIUM (n): calcium chloride (*Calcium chloratum*); $CaCl_2$; Sg. 2.22 or $CaCl_2.6H_2O$; Sg. 1.654 747
CHLORCHROM (n): chromium chloride; chromic chloride; $CrCl_3$ or $CrCl_3.6H_2O$; Sg. 2.757; Bp. 1200-1500°C. 748
CHLORCHROMSAUER: chlorochromate, combined with chlorochromic acid 749
CHLORCHROMSÄURE (f): chlorochromic acid 750
CHLORCROTONSÄURE (f): chlorocrotonic acid 751
CHLORCYAN (n): cyanogen chloride; CNCl; Mp. –5°C.; Bp. 13°C. 752
CHLORECHT: fast to chlorine 753
CHLOREISEN (n): iron chloride; ferric chloride; $FeCl_3$; Sg. 2.804; Mp. 301°C.; $FeCl_3.6H_2O$; Mp. 370°C. 754
CHLOREN: to chlorinate; gas (Bleaching) 755
CHLORENTWICKLER (m): chlorine generator 756
CHLORESSIGSAUER: chloroacetate, combined with chloroacetic acid 757
CHLORESSIGSÄURE (f): (*Acidum monochloraceticum*): chloroacetic acid; $C_2H_3O_2Cl$; Sg. 1.3978; Mp. about 61-63°C.; Bp. 185-189°C. 758
CHLORESSIGSÄURE-ÄTHYLESTER (m): chloroacetic ethylester, ethyl chloroacetate; $C_4H_7O_2Cl$; Sg. 1.1585; Bp. 144.9°C. 759
CHLORESSIGSÄURE (f): DI-: (*Acidum dichloraceticum*): dichloracetic acid, $CHCl_2.CO_2H$; Sg. 1.5216; Bp. 190°C. 760
CHLORESSIGSÄURE (f), MONO-: (*Acidum monochloraceticum*): monochloracetic acid; $CH_2Cl.CO_2H$; Sg. 1.3978; Mp. 63°C.; Bp. 186°C. 761
CHLORESSIGSÄURE (f), TRI-: (*Acidum trichloraceticum*): trichloracetic acid, $CHCl_3.CO_2H$; Sg. 1.62; Mp. 52.3°C.; Bp. 195-200°C. 762
CHLORETON (n): chloretone, acetone chloroform, tertiary trichlorobutyl alcohol; $CCl_3(CH_3)_2COH.\frac{1}{2}H_2O$; Mp. 80-81°C. 763
CHLOR (n), FLÜSSIGES-: chlorine water, bleaching liquor; liquid chlorine 764
CHLORGAS (n): chlorine gas; Cl_2; Sg. 2.491; Mp. –102°C.; Bp. –33.6°C. 765
CHLORGOLD (n): gold chloride; auric chloride; (*Aurum chloratum*); $AuCl_3$ or $AuCl_3.2H_2O$ [766
CHLORGOLDKALIUM (n): gold and potassium chloride, potassium chloraurate, potassium aurichloride; $AuCl_3.KCl.2H_2O$ 767
CHLORGOLDNATRIUM (n): sodium chloraurate; $NaAuCl_4.2H_2O$ 768
CHLORHALTIG: containing chlorine 769
CHLORHEPTOXYD (n): chlorine heptoxide, perchloric anhydride; Cl_2O_7; Bp. 82°C. 770
CHLORHYDRAT (n): hydrochloride; chlorine hydrate; $Cl_2.8H_2O$; Sg. 1.23 771
CHLORHYDRIN (n): glycerine dichlorohydrin, dichloroisopropyl alcohol, alpha-propenyl-dichlorohydrin, chlorohydrin, dichlorohydrin; $ClH_2C.CH(OH).CH_2Cl$; Sg. 1.396; Bp. 176.5°C. 772
CHLORHYDRIN (n), DI-: (see CHLORHYDRIN) 773
CHLORHYDRIN (n), MONO-: monochlorohydrin, chlorophenylene-glycol, alpha-propylene-chlorohydrin; $CH_2(OH).CH(OH).CH_2Cl$; Sg. 1.325 774
CHLORHYDROCHINON (n): chlorohydroquinone, adurol; $C_6H_3(OH)_2Cl$; Mp. 103.5°C. 775
CHLORID (n): chloride (an -ic chloride as opposed to -ous chloride; see CHLORÜR) 776
CHLORIEREN: to chlorinate; gas; chemic-(k), (bleach with calcium hypochlorite) 777
CHLORIERER (m): chlorinator 778
CHLORIERUNG (f): chlorination 779
CHLORIG: chlorous 780
CHLORIGE SÄURE (f): chlorous acid 781
CHLORIGSAUER: chlorite of; combined with chlorous acid 782
CHLORIGSÄURE (f): (see CHLORIGE SÄURE) 783
CHLORIN (n): chlorine (see CHLOR) 784
CHLORIRIDIUM (n): iridium chloride 785
CHLORION (n): chlorine ion 786
CHLORISIEREN: to chemic-(k) (bleach with calcium hypochlorite) 787
CHLORIT (m): chlorite (natural magnesium-aluminium hydrosilicate); $6Mg(Fe)O,Al_2O_3,3SiO_2$ plus $4H_2O$; Sg. 2.7 788
CHLORITOID (m): chloritoid; $H_2FeAl_2SiO_7$; Sg. 3.5 (see also CHLORITSPAT) 789
CHLORITSPAT (m): masonite, spathic chlorite, foliated chlorite (see CHLORIT and CHLORITOID); Sg. 3.5 790
CHLORJOD (n): iodine chloride; iodine monochloride; ICl; Mp. 25°C.; Bp. 101°C. [791
CHLORKADMIUM (n): cadmium chloride (see CADMIUMCHLORID) 792
CHLORKALI (n): chloride of potash; potassium chloride; potassium hypochlorite; KCl; Sg. 1.987; Mp. 772°C.; Bp. 1500°C. [793
CHLORKALILÖSUNG (f): Javelle water; solution of chlorinated potassa (used as bleaching and disinfecting agent) 794
CHLORKALIUM (n): (see CHLORKALI) 795
CHLORKALK (m): chloride of lime (*Calcaria chlorata*); $Ca(ClO)_2.4H_2O$ (see BLEICH-PULVER) 796
CHLORKALKBAD (n): chemic-(k) vat; bleaching vat 797
CHLORKALKFLÜSSIGKEIT (f): solution of chlorinated lime; bleaching (powder) solution 798
CHLORKALKKUFE (f): (see CHLORKALKBAD) 799
CHLORKALKLÖSUNG (f): (see CHLORKALKFLÜSSIGKEIT) 800
CHLORKNALLGAS (n): chlorine detonating gas (explosive mixture of chlorine and hydrogen) 801
CHLORKOBALT (n): (see KOBALTCHLORÜR) 802
CHLORKOHLENOXYD (n): carbonyl chloride; phosgene; $COCl_2$; Sg. 1.421; Mp. –75°C.; Bp. 8.2°C. 803
CHLORKOHLENOXYDÄTHER (m): ethyl chlorocarbonate (see CHLORKOHLENSÄUREESTER) 804
CHLORKOHLENSAUER: chlorocarbonate, combined with chlorocarbonic acid 805
CHLORKOHLENSÄURE (f): chlorocarbonic acid 806
CHLORKOHLENSÄUREÄTHYLESTER (m): chlorocarbonic ether, ethyl chlorocarbonate (see CHLORKOHLENSÄUREESTER) 807
CHLORKOHLENSÄUREESTER (m): chlorocarbonic ester, ethyl chlorocarbonate; CHO_2Cl; Sg. 1.144; Bp. 91.3°C 808

CHLORKOHLENSTOFF (*m*): carbon chloride; carbon tetrachloride; CCl_4; Sg. 1.5835; Mp. – 22.95°C.; Bp. 76.74°C. 809

CHLORKOHLENSTOFF, VIERFACH- (*m*): (see **CHLORKOHLENSTOFF**) 810

CHLORKUPFER (*n*): copper chloride (see **KUPFER-CHLORID**) 811

CHLORLITHIUM (*n*): lithium chloride; LiCl; Sg. 1.998-2.074; Mp. 602°C. 812

CHLORMAGNESIUM (*n*): magnesium chloride (see **MAGNESIUMCHLORID**) 813

CHLORMANGAN (*n*): manganese chloride; $MnCl_2$ or $MnCl_2.4H_2O$ 814

CHLORMESSER (*m*): chlorometer (an instrument for testing the bleaching power of chloride of lime) 815

CHLORMETALL (*n*): metallic chloride 816

CHLORMETALLOID (*n*): a metalloid chloride 817

CHLORMETHYL (*n*): methyl chloride; chloromethane (*Methylum chloratum*); CH_3Cl; Sg. 0.9197; Mp. – 91.5°C.; Bp. 23.73°C. 818

CHLORMETHYLMENTHYLÄTHER (*m*): chloromethylmenthyl ether (see **FORMAN**) 819

CHLORNAPHTALIN (*n*): chloronaphthalin: chloronaphthalene 820

CHLORNATRIUM (*n*): sodium chloride (see **NATRI-UMCHLORID**) 821

CHLORNATRON (*n*): sodium hypochlorite; NaOCl 822

CHLORNATRONLÖSUNG (*f*): Labarraque's solution; solution of chlorinated soda; Eau de Labarraque 823

CHLORNICKEL (*n*): nickel chloride (see **NICKELCHLORID**) 824

CHLORNICKELAMMONIAK (*n*): nickel-ammonium chloride; $Ni(NH_3)_2Cl_2.4NH_3$ 825

CHLORNITROBENZOL (*n*): chloronitrobenzol; chloronitrobenzene; $C_6H_4Cl(NO_2)$; meta.- Sg. 1.534; Mp. 44°C.; Bp. 236°C.; ortho.- Sg. 1.368; Bp. 245.5°C.; para.- Sg. 1.52; Mp. 83°C.; Bp. 242°C. 826

CHLORNITROBENZOLSULFOSÄURE (*f*): chloronitrobenzenesulfonic acid 827

CHLOROBENZAL (*n*): benzal chloride; $C_6H_5CHCl_2$; Sg. 1.295; Mp. 16.1°C.; Bp. 212.4°C. [828

CHLOROBENZIL (*n*): dichlorobenzil; $C_6H_5COCCl_2C_6H_5$ 829

CHLOROFORM (*n*): chloroform, methenyl trichloride, trichloromethane; $CHCl_3$; Sg. 1.5039; Mp. – 63.2°C.; Bp. 61.2°C.; (from the action of bleaching powder—chloride of lime—and water on alcohol) 830

CHLOROFORMCOLCHICIN (*n*): chloroform colchicine; $C_{22}H_{25}NO_6.2CHCl_3$ 831

CHLOROPAL (*m*): chloropal; SiO_2 and Al_2O_3 832

CHLOROPHYLL (*n*): chlorophyll, leaf green 833

CHLOROPHYLLFARBSTOFF (*m*): chlorophyll colouring matter 834

CHLOROPHYLLFÜHREND: chlorophyll-bearing 835

CHLOROPLATINAT (*n*): chloroplatinate; of potassium, K_2PtCl_6; Sg. 3.499; of Sodium, Na_2PtCl_6; Sg. 2.5; of Ammonium, $(NH_4)_2PtCl_6$; Sg. 3.034 836

CHLOROSE (*f*): chlorosis (Med.) 837

CHLOROSPINELL (*m*): chlorospinel (a magnesia iron spinel); $Mg(Al_2,Fe_2)O_4$; Sg. 3.6 838

CHLORPHENOL (*n*): chlorophenol; (C_6H_5OCl); ortho.- Mp. 7°C.; Bp. 176°C.; meta.- Mp. 28.5°C.; Bp. 214°C.; para.- Sg. about 1.3; Mp. 37°C.; Bp. 217°C. 839

CHLORPHOSPHOR (*m*): phosphoric chloride, phosphoric perchloride; phosphorous pentachloride; PCl_5; Sg. 3.6; Bp. 162°C. 840

CHLORPHTALSÄURE (*f*): chlorophthalic acid; $C_6H_3Cl(COOH)_2$ 841

CHLORPIKRIN (*n*): chloropicrin, trinitromethane, nitrotrichloromethane, nitrochloroform; (CO_2NCl_3); Sg. 1.666; Mp. – 69.2°C.; Bp. 112°C. 842

CHLORPLATIN (*n*): platinic (platinum) chloride; $PtCl_4$ or $PtCl_4.5H_2O$ 843

CHLORPLATINSÄURE (*f*): chloroplatinic acid; platinic chloride; $H_2PtCl_6.H_2O$; Sg. 2.431 844

CHLORPROPIONSAUER: chloropropionate 845

CHLORQUECKSILBER (*n*): mercury chloride (see **QUECKSILBERCHLORÜR** and **QUECKSILBER-CHLORID**); Calomel (see **KALOMEL**) 846

CHLORSAUER: chlorate of; combined with chloric acid 847

CHLORSÄURE (*f*): chloric acid (*Acidum chloricum*); $HClO_3$; Sg. 1.12 848

CHLORSÄUREANHYDRID (*n*): chloric anhydride 849

CHLORSAURES KALI (*n*): potassium chlorate (see **KALIUMCHLORAT**) 850

CHLORSAURES SALZ (*n*): chlorate 851

CHLORSCHWEFEL (*m*): sulphur chloride (see **SCHWEFELCHLORÜR** and **SCHWEFELCHLORID**) 852

CHLORSCHWEFEL (*m*), **EINFACH-**: sulphur monochloride (see **SCHWEFELCHLORÜR**) 853

CHLORSCHWEFEL (*m*), **ZWEIFACH-**: sulphur dichloride (see **SCHWEFELCHLORID**) 854

CHLORSILBER (*n*): silver chloride; AgCl; Sg. 5.561; Mp. 451°C.; cerargyrite (see **CER-ARGYRIT**) 855

CHLORSILBERGELATINE (*f*): gelatino-chloride of silver 856

CHLORSILBERGELATINEPAPIER (*n*): chloride paper (Photography) 857

CHLORSILBERKOLLODIUM (*n*): collodio-chloride of silver 858

CHLORSILIZIUM (*n*): silicon tetrachloride; $SiCl_4$; Sg. 1.524; Mp. – 89°C.; Bp. 56.9°C. 859

CHLORSODA (*f*): (see **CHLORNATRON**) 860

CHLORSTICKSTOFF (*m*): nitrogen chloride; NCl_3; Sg. 1.653 861

CHLORSTROM (*m*): current (stream) of chlorine 862

CHLORSTRONTIUM (*n*): strontium chloride; $SrCl_2$ or $SrCl_2.6H_2O$ 863

CHLORSULFONSÄURE (*f*): (*Acidum chlorsulfonicum*); chlorosulfonic acid; $SO_2Cl(OH)$; Sg. 1.785; Bp. 82°C. 864

CHLORTITAN (*n*): titanium chloride (see **TITAN-CHLORID**) 865

CHLORTOLUOL (*n*): (C_7H_7Cl); benzyl chloride, chlorotoluene; ortho.- Sg. 1.0807; Mp. – 34°C.; Bp. 159.4°C.; meta.- Sg. 1,0722; Mp. – 47.8°C.; Bp. 162.2°C.; para.- Sg. 1.0697; Mp. 7.4°C.; Bp. 162.3°C. 866

CHLORÜBERTRÄGER (*m*): chlorine carrier 867

CHLORÜR (*n*): chloride; protochloride; (an -ous chloride as opposed to **CHLORID**, an -ic chloride) 868

CHLORVANADIUM (*n*): vanadium chloride; divanadyl tetrachloride; hypovanadic hydrochloride; $(VO)_2Cl_4.5H_2O$; Sg. 3.23 [869

CHLORVERBINDUNG (*f*): chlorine compound 870

CHLORVERFLÜSSIGUNG (*f*): chlorine liquefaction 871

CHLORVERFLÜSSIGUNGSANLAGE (*f*): chlorine liquifying plant 872
CHLORWASSER (*n*): chlorine water 873
CHLORWASSERPUMPE (*f*): chlorine water pump 874
CHLORWASSERSTOFF (*m*): hydrogen chloride; hydrochloric acid; HCl; Sg. 1.269 875
CHLORWASSERSTOFFÄTHER (*m*): ethyl chloride; C_2H_5Cl; Sg. 0.9214; Mp. – 140.85°C.; Bp. 12.5°C. 876
CHLORWASSERSTOFF HYDRAT (*n*): hydrochloric (muriatic) acid, hydrated (see SALZSÄURE-HYDRAT) 877
CHLORWASSERSTOFFSAUER: chloride (hydrochloride) of; combined with hydrochloric acid 878
CHLORWASSERSTOFFSÄURE (*f*): hydrochloric acid; HCl; Sg. 1.269; muriatic acid; spirits of salts 879
CHLORWISMUT (*n*): bismuth (tri)-chloride; $BiCl_3$; Sg. 4.56; Mp. 227°C. 880
CHLORZINK (*n*): zinc chloride; $ZnCl_2$; Sg. 2.91; Mp. 262°C.; Bp. 730°C. 881
CHLORZINKANLAGE (*f*): zinc chloride plant 882
CHLORZINKLAUGE (*f*): zinc chloride (lye) solution 883
CHLORZINN (*n*): tin chloride (see ZINNCHLORÜR and ZINNCHLORID) 884
CHLORZYAN (*n*): (see CHLORCYAN) 885
CHOCOLADE (*f*): chocolate (see SCHOKOLADE) 886
CHOLALSÄURE (*f*): cholalic (cholic) acid (*Acidum cholalicum*); $C_{26}H_{43}NO_6$; Mp. 133°C.; Bp. 140°C. 887
CHOLEINSÄURE (*f*): choleinic (choleic) acid; $C_{26}H_{45}NSO_7$ 888
CHOLERA (*f*): cholera (Med.) 889
CHOLERA (*f*): ASIATISCHE-: Asiatic cholera (Med.) 890
CHOLERA (*f*), SPORADISCHE-: sporadic cholera, cholerine, choleraic diarrhœa (Med.) 891
CHOLERATROPFEN (*m* and *pl.*): cholera drops 892
CHOLESTERIN (*n*): cholesterin, cholesterol; $C_{27}H_{45}.OH.H_2O$; Sg. 1.067; Mp. 148.5°C.; Bp. 360°C. 893
CHOLIN (*n*): choline (a ptomaine from cadaveric putrefraction); $C_5H_{15}NO_2$ 894
CHOLSÄURE (*f*): (see CHOLALSÄURE) 895
CHONDRODIT (*m*): chondrodite, humite; $Mg_5[Mg(F.OH)]_2[SiO_4]_3$; Sg. 3.15 896
CHORDE (*f*): chord 897
CHRISTDORN (*m*): holly (*Ilex aquifolium*) 898
CHRISTOPHSKRAUT (*n*): baneberry (*Actæa*) 899
CHRISTOPHSWURZ (*f*): (see CHRISTOPHSKRAUT) 900
CHRISTPALMÖL (*n*): castor oil; Sg. 0.96-0.97 [901
CHRISTWURZ (*f*): (see ADONISRÖSCHENKRAUT) 902
CHROM (*n*): chromium; Cr; Sg. 6.92; Mp. 1520°C.; Bp. 2200°C.; Chrome 903
CHROMACETAT (*n*): chromous acetate (see CHROMOACETAT) 904
CHROMALAUN (*m*): chrome alum; chromium-potassium sulphate; $K_2SO_4.Cr_2(SO_4)_3.24H_2O$; Sg. 1.813 905
CHROMALAUNLAUGE (*f*): chrome alum (lye) solution 906
CHROM (*n*), AMEISENSAURES-: chromium formate 907
CHROMAT (*n*): chromate 908
CHROMATISCH: chromatic 909
CHROMATISCHE ABERRATION (*f*): chromatic aberration (Photography) 910
CHROMAT, SAURES- (*n*): dichromate 911
CHROMATSCHWARZ (*n*): chromate black 912
CHROMAZETAT (*n*): (*Chromium aceticum*): chromium (chromous) acetate; $Cr(C_2H_3O_2)_3.H_2O$ 913
CHROMBEIZE (*f*): chrome mordant 914
CHROMBEIZEN: to chrome-mordant 915
CHROMBLAU (*n*): chrome blue (see KALIUMBICHROMAT) 916
CHROMBLEI (*n*): crocoite; $PbO.CrO_3$; 68.9% lead and 31.1% CrO_3 917
CHROMBLEISPAT (*m*): (see CHROMBLEI) 918
CHROMCHLORAT (*n*): chromium chlorate; $Cr(ClO_3)_3$ 919
CHROMCHLORID (*n*): chromic chloride; chromium sesquichloride; Cr_2Cl_6; Sg. 2.757; Bp. 1200-1500°C. (*Chromium chloratum*) 920
CHROM (*n*), CHLORSAURES-: chromium chlorate (see CHROMCHLORAT) 921
CHROMCHLORÜR (*n*): chromous chloride 922
CHROMCHROMAT (*n*): chromium (chromic) chromate (*Chromium chromatum*) (a chrome mordant) 923
CHROMDICHLORID (*n*): chromium dichloride; $CrCl_2$; Sg. 2.751 924
CHROMDIFLUORID (*n*): chromium difluoride; CrF_2; Sg. 4.11 925
CHROMECHT: fast to (chrome) potassium dichromate 926
CHROMEISEN (*n*): chromite (see CHROMIT) 927
CHROMEISENERZ (*n*): (see CHROMIT) 928
CHROMEISENSTEIN (*m*): chromic iron ore; chromite (see CHROMIT) 929
CHROMERZ (*n*): chromium ore 930
CHROMERZ (*n*), OKTAEDRISCHES-: chromite (see CHROMIT) 931
CHROM (*n*), ESSIGSAURES-: chromium acetate, chromous acetate (see CHROMAZETAT) 932
CHROMFARBE (*f*): chrome (dye) colour 933
CHROMFLUORID (*n*): chromic fluoride; (*Chromium fluoratum*); $CrF_3.4H_2O$; Sg. 3.78 [934
CHROMGAR: chrome tanned 935
CHROMGELATINE (*f*): chrome gelatine; bichromated gelatine 936
CHROMGELB (*n*): chrome yellow; Leipzig yellow; lead chromate; $PbCrO_4$; Sg. 6.123; Cologne yellow (lead chromate and sulphate) (see KÖLNERGELB) 937
CHROMGERBEEXTRAKT (*m*): chrome tanning extract 938
CHROMGERBEREI (*f*): chrome (tanning) tannery 939
CHROMGERBUNG (*f*): chrome tanning 940
CHROMGLIMMER (*m*): chrome mica; fuchsite (see also FUCHSIT) 941
CHROMGRÜN (*n*): chrome green (chromic oxide); Cr_2O_3 942
CHROMHALTIG: chromiferous; containing chromium 943
CHROM (*n*), HARZSAURES-: chromium resinate 944
CHROMHYDROXYD (*n*): chromic hydroxide (*Chromium oxydatum hydratum*); $Cr_2(OH)_6.4H_2O$ or $Cr(OH)_3$ 945
CHROMHYDROXYDUL (*n*): chromous hydroxide 946
CHROMIACETAT (*n*): chromic acetate; $Cr(C_2H_3O_2)_3.5H_2O$ 947
CHROMICHLORID (*n*): chromic chloride (*Chromium chloratum*): $CrCl_3$ or $CrCl_3.6H_2O$; Sg. 2.757; Bp. 1200-1500°C. 948
CHROMIERARTIKEL (*m*): chrome style (Calico) 949
CHROMIEREN: to chrome 950

CHROMIHYDROXYD (*n*): chromic hydroxide (see CHROMHYDROXYD) 951
CHROMISALZ (*n*): chromic salt 952
CHROMISULFAT (*n*): chromic sulphate (*Chromium sulfuricum*); $Cr_2(SO_4)_3$; Sg. 3.012 [953
CHROMIT (*m*): chromite (natural iron chromate; about 68% chromic oxide); $FeCr_2O_4$; Sg. 4.6 954
CHROMITSTEIN (*m*): chromite brick; Sg. 4.0 [955
CHROMIVERBINDUNG (*f*): chromic compound 956
CHROMKALI (*n*): potassium dichromate; potassium bichromate; red potassium chromate (*Kalium bichromicum*); $K_2Cr_2O_7$; Sg. 2.692 Mp. 396°C. 957
CHROMKARBID (*n*): chromium carbide 958
CHROMLAUGE (*f*): chrome lye 959
CHROMLEDER (*n*): chrome leather 960
CHROMLEDERSCHWARZ (*n*): chrome leather black 961
CHROMLEIM (*m*): (see CHROMGELATINE) 962
CHROMMETALL (*n*): chromium metal 963
CHROMMETAPHOSPHAT (*n*): chromium metaphosphate; $Cr(PO_3)_3$; Sg. 2.974 964
CHROM (*n*), MILCHSAURES-: chromium lactate 965
CHROMNATRON (*n*): sodium chromate; $Na_2CrO_4.10H_2O$; Sg. 2.71; Mp. 19.92 966
CHROMNICKELSTAHL (*m*): chrome-nickel steel 967
CHROMNITRAT (*n*): chromium (chromic) nitrate (*Chromium nitricum*) 968
CHROMOACETAT (*n*): chromous acetate; $Cr(C_2H_3O_2)_2.H_2O$ 969
CHROMO-ARISTOPAPIER (*n*): baryta paper (Photography) 970
CHROMOCHLORID (*n*): chromous chloride 971
CHROMODIESSIGSÄURE (*f*): dichromoacetic acid $[Cr(OH)_2(C_2H_3O_2)](C_2H_4O_2)_2$ 972
CHROMOESSIGSÄURE (*f*): chromoacetic acid; $[Cr_2O(C_2H_3O_2)_4](C_2H_4O_2)_2$ 973
CHROMOGEN (*n*): chromogen (see CHROMOTROPSÄURE) 974
CHROMOHYDROXYD (*n*): chromous hydroxide 975
CHROMOION (*n*): chromous ion 976
CHROMO-MONO-ESSIGSÄURE (*f*): mono-chromoacetic acid $[Cr(OH)(C_2H_3O_2)_2](C_2H_4O_2)$ [977
CHROMOPHOR: chromophore; chromophoric; chromophorous 978
CHROMOPHOTOGRAPHIE (*f*): colour photography (the term is also often erroneously applied to crystoleum painting) 979
CHPOMOPHOTOLITHOGRAPHIE (*f*): chromo-photolithography (a method of printing in process work) 980
CHROMOPHOTOTYPIE (*f*): chromo-phototype (a method of printing in process work) 981
CHROMOPHOTOXYLOGRAPHIE (*f*): chromo-photoxylography (a method of printing in process work with the aid of wood) 982
CHROMOPHOTOZINKOGRAPHIE (*f*): chromo-photo-zincography (a method of printing in process work, with the aid of zinc) 983
CHROMOPHOTOZINKOTYPIE (*f*): chromo-photo-zincotype (a method of printing in process work) 984
CHROMORANGE (*f*): chrome orange (basic lead chromate); $PbCrO_5$; (also mixture of CHROMGELB and CHROMROT, which see) 985
CHROMOTROPSÄURE (*f*): chromotropic acid, dioxynaphthalene-disulfonic acid; OH:OH: SO_3H:SO_3H = 1:8:3:6 986
CHROMOTYPOGRAPHIE (*f*): chromo-typography (a method of printing in process work) 987
CHROMOVERBINDUNG (*f*): chromous compound 988
CHROMOXYCHLORID (*n*): chromium oxychloride; chromyl chloride; (*Chromium oxychloratum*); CrO_2Cl_2; Sg. 1.915; Bp. 117°C. 989
CHROMOXYD (*n*): chromic oxide; Cr_2O_3; Sg. 5.04; Mp. 1990°C.; chromium sesquioxide; chrome green: (*Chromium oxydatum*) 990
CHROMOXYD (*n*), CHROMSAURES-: chromic chromate (see CHROMCHROMAT) 991
CHROMOXYDGRÜN (*n*), DECKENDES-: chrome oxide; Cr_2O_3 992
CHROMOXYDGRÜN (*n*): FEUERIGES-: (see GUIGNETGRÜN) 993
CHROMOXYDGRÜN (*n*), LASIERENDES-: viridian (see GUIGNETGRÜN) 994
CHROMOXYD (*n*), HARZSAURES-: chromium resinate (produced by melting resin with chromium sulphate) 995
CHROMOXYDHYDRAT (*n*): chromic (hydrate) hydroxide; (*Chromium oxydatum hydratum*); $Cr(OH)_3$ 996
CHROMOXYDNATRON (*n*): sodium chromite 997
CHROMOXYD (*n*). SALPETERSAURES-: chromium (chromic) nitrate (see CHROMNITRAT) 998
CHROMOXYDSALZ (*n*): chromic salt 999
CHROMOXYDSALZLÖSUNG (*f*): solution of a chromic salt 000
CHROMOXYD (*n*), SCHWEFELSAURES-: (see CHROM-SULFAT) 001
CHROMOXYD (*n*), SCHWEFLIGSAURES-: (see CHROM-SULFIT) 002
CHROMOXYDUL (*n*): chromous oxide 003
CHROMOXYDULVERBINDUNG (*f*): chromous compound 004
CHROMOXYD (*n*), WOLFRAMSAURES-: (see CHROM-WOLFRAMAT) 005
CHROMOXYFLUORID (*n*): chromium oxyfluoride; CrO_2F_2 006
CHROMOXYLOGRAPHIE (*f*): chromoxylography (a method of printing in process work) 007
CHROMPHOSPHID (*n*): chromium phosphide; CrP; Sg. 5.71 008
CHROM (*n*), PHOSPHORSAURES-: chromium (chromic) phosphate; $Cr(PO_4)$ 009
CHROMRESINAT (*n*): chromium resinate 010
CHROMROT (*n*): chrome red; Derby red; basic lead chromate; $PbCrO_4.Pb(OH)_2$ 011
CHROMSALPETERSÄURE (*f*): chromonitric acid 012
CHROM (*n*), SALPETERSAURES-: chromium nitrate (see CHROMNITRAT) 013
CHROMSALZ (*n*): chromium salt; chromate 014
CHROMSAUER: chromate of; combined with chromic acid 015
CHROMSAUER, EINFACH-: chromate of 016
CHROMSAUER, ZWEIFACH-: bichromate of 017
CHROMSÄURE (*f*): chromic acid; chromium trioxide; CrO_3; Sg. 2.67-2.82; Mp. 196°C.; (*Acidum chromicum*) 018
CHROMSÄUREANHYDRID (*n*): chromic anhydride (see CHROMSÄURE) 019
CHROMSÄUREELEMENT (*n*): bichromate (pile) battery 020
CHROMSÄURESALZ (*n*): chromic acid salt; chromate 021
CHROMSAURES KALI (*n*): potassium chromate (see KALI, CHROMSAURES-) 022
CHROMSAURES SALZ (*n*): chromate 023
CHROMSCHWARZ: chrome black 024

CHROMSCHWEFELSÄURE (*f*): chromosulphuric acid 025
CHROM (*n*), **SCHWEFELSAURES-**: chromium sulphate (see **CHROMSULFAT**) 026
CHROM (*n*), **SCHWEFLIGSAURES-**: chromium sulphite (see **CHROMSULFIT**) 027
CHROMSESQUICHLORID (*n*): (see **CHROMCHLORID**) 028
CHROMSESQUIOXYD (*n*): chromium sesquioxide (see **CHROMOXYD**) 029
CHROMSILICID (*n*): chromium silicide; Cr_3Si; Sg. 6.52 030
CHROMSILICOFLUORID (*n*): chromium fluosilicate 031
CHROMSILIZIUM (*n*): (see **CHROMSILICID**) 032
CHROMSTAHL (*m*): chrome steel; chromium steel 033
CHROMSULFAT (*n*): (*Chromium sulfuricum*): chromium sulphate; $Cr_2(SO_4)_3$; Sg. 3.012 034
CHROMSULFID (*n*): chromium sulphide; CrS.; Sg. 4.08 035
CHROMSULFIT (*n*): chromium sulphite (*Chromium sulfurosum*) 036
CHROMSULFÜR (*n*): chromium sulphide (see **CHROMSULFID**) 037
CHROMTRICHLORID (*n*): chromic chloride, chromium sesquichloride, chromium chloride; (*Chromium chloratum*); $CrCl_3$; Sg. 2.361-2.757; Bp. 1200-1500°C. 038
CHROMTRIFLUORID (*n*): chromium (chromic) fluoride, chromium trifluoride; CrF_3; (*Chromium fluoratum*): Sg. 3.78 039
CHROMTRIOXYD (*n*): chromium trioxide (see also **CHROMSÄURE**); CrO_3; Sg. 2.67 040
CHROMVANADIUMSTAHL (*m*): chrome-vanadium steel 041
CHROMVERBINDUNG (*f*): chromium compound 042
CHROM (*n*), **WEINSAURES-**: chromium tartrate 043
CHROMWOLFRAMAT (*n*): (*Chromium wolframicum*): chromium (wolframate) tungstate; $Cr_2(WO_4)_3$ 044
CHROM (*n*), **WOLFRAMSAURES-**: chromium tungstate (see **CHROMWOLFRAMAT**) 045
CHROMYLCHLORID (*n*): chromyl chloride (see **CHROMOXYCHLORID**) 046
CHROMZINNOBER (*m*): chrome red, Austrian cinnabar (see **CHROMROT**) 047
CHRONIK (*f*): chronicle 048
CHRONISCH: chronic-(ally) 049
CHRONOGRAPH (*m*): chronograph, time recorder 050
CHRONOLOGIE (*f*): chronology 051
CHRONOMETER (*n*): chronometer 052
CHRONOSKOP (*n*): chronoscope, time meter 053
CHRYSALIDENÖL (*n*): chrysalis oil (from the chrysalis of the silk worm); Sg. 0.9105 [054
CHRYSAMIN (*n*): chrysamine (a yellow coal-tar dyestuff) 055
CHRYSANILIN (*n*): chrysaniline (see **PHOSPHIN**) 056
CHRYSAROBIN (*n*): chrysophanic acid; chrysarobin; araroba; chrysarobinum (Pharm.); (from the tree, *Andira araroba aguiar*); $C_{30}H_{26}O_7$; Sg. 0.92; Mp. 157°C. 057
CHRYSAROBINSEIFE (*f*): chrysarobin soap 058
CHRYSOBERYLL (*m*): chrysoberyl (beryllium aluminate) (Mineralogy); $BeAl_2O_4$; Sg. 3.7 059
CHRYSOIDIN (*n*): chrysoidine (an orange dyestuff for leather, wool, silk, cotton and for colouring liqueurs, etc.) (see **BRAUNSALZ**) 060
CHRYSOKOLL (*m*): chrysocolla (a hydrous copper silicate); $CuSiO_3.2H_2O$; Sg. 2.2; (see **KIESELKUPFER**) 061
CHRYSOLITH (*m*): chrysolite (Mineralogy) (see **OLIVIN**); Sg. 2.78 062
CHRYSOLITH (*m*), **HEMIPRISMATISCHER-**: humite, chondrodite 063
CHRYSOPHAN (*m*): chrysophane, xanthophyllite (see **XANTHOPHYLLIT**) 064
CHRYSOPHANSÄURE (*f*): chrysophanic acid (from *Cassia angustifolia*) (see **CHRYSAROBIN**) 065
CHRYSOPHENIN (*n*): chrysophenine (yellow synthetic dyestuff) 066
CHRYSOPRAS (*m*): chrysopras (Mineralogy); Sg. 2.65; (see **CHALCEDON** and **ACHAT**) 067
CHRYSOTIL (*m*): chrysotile (see **SERPENTIN**) 068
CHRYSOTILASBEST (*m*): chrysotile, amianthus (see **SERPENTIN**) 069
CHYLUS (*m*): chyle (a milky fluid) (Med.) 070
CIBAROT (*n*): ciba red 071
CICHORIE (*f*): chicory; endive; succory (*Cichorium intybus*) 072
CICHORIENWURZEL (*f*): chicory root (*Radix cichorii* of *Cichorium intybus*) 073
C.I.F.DOKUMENTE (*n.pl*): c.i.f. documents, shipping documents 074
CINCHONIDIN (*n*): cinchonidine (an alkaloid); $C_{19}H_{22}N_2O$; Mp. 200.5°C. 075
CINCHONIDINBISULFAT (*n*): cinchonidine bisulphate; $C_{19}H_{22}N_2O.H_2SO_4.5H_2O$ 076
CINCHONIDINCHLORHYDRAT (*n*): cinchonidine hydrochloride; $C_{19}H_{22}N_2O.HCl.H_2O$ 077
CINCHONIDINCHLORID (*n*): cinchonidine hydrochloride (see **CINCHONIDINCHLORHYDRAT**) 078
CINCHONIDINSULFAT (*n*): cinchonidine sulphate; $(C_{19}H_{22}N_2O)_2.H_2SO_4.3H_2O$; Mp. 205.3°C. 079
CINCHONIN (*n*): cinchonine (an alkaloid); $C_{19}H_{22}N_2O$; Mp. 240 to 264°C. 080
CINCHONINBISULFAT (*n*): cinchonine bisulphate; $C_{19}H_{22}N_2O.H_2SO_4.4H_2O$ 081
CINCHONINCHLORHYDRAT (*n*): cinchonine hydrochloride; $C_{19}H_{22}N_2O.HCl.2H_2O$ 082
CINCHONINCHLORID (*n*): (see **CINCHONINCHLORHYDRAT**) 083
CINCHONINNITRAT (*n*): cinchonine nitrate; $C_{19}H_{22}N_2O.HNO_3.\frac{1}{2}H_2O$ 084
CINCHONINSÄURE (*f*): cinchoninic acid 085
CINCHONIN (*n*), **SCHWEFELSAURES-**: cinchonine sulphate (see **CINCHONINSULFAT**) 086
CINCHONINSULFAT (*n*): cinchonine sulphate; $(C_{19}H_{22}N_2O)_2.H_2SO_4.2H_2O$; Mp. 198.5°C. 087
CINCHONISMUS (*m*): quinism (caused by quinine in the blood) (Med.) 088
CINCHONSÄURE (*f*): cinchonic acid 089
CINNABAR (*m*): (see **ZINNOBER**) 090
CINNAMALDEHYD (*n*): cinnamaldehyde; cinnamic aldehyde; cinnamyl aldehyde; $C_6H_5CH.CH.CHO$; Sg. 1.13; Mp. −8°C.; Bp. 248°C. 091
CINNAMOL (*n*) cinnamol, cinnamene (see **STYROL**) 092
CINNAMYLALDEHYD (*n*): cinnamyl aldehyde, cinnamic aldehyde (see **CINNAMALDEHYD**) 093
CINNAMYLALKOHOL (*m*): cinnamyl alcohol, styryl alcohol, styrone; $C_6H_5CH:CH:CH_2OH$; Sg. 1.04; Mp. 33°C.; Bp. 257°C. 094
CINNAMYLÄTHER (*m*): (see **ZIMTSÄUREÄTHYLESTER**) 095

CINNAMYLCOCAIN (*n*) : cinnamyl cocain (a coca alkaloid ; see COCAALKALOID) ; $C_{19}H_{23}NO_4$ 096

CINNAMYLEKGONIN (*n*) : cinnamylekgonine ; $C_{18}H_{21}NO_4$ 097

CINNAMYLSÄURE (*f*) : cinnamylic acid, cinnamic acid (see also ZIMTSÄURE) ; beta-phenylacrylic acid ; $C_6H_5CHCHCOOH$; Sg. 1.248 ; Mp. 133°C. ; Bp. 300°C. 098

CIRCA : about ; approximately 099

CIRKA : (see CIRCA) 100

CIRKULIEREN : to circulate 101

CIRRHOSE (*f*) : cirrhosis (usually of the liver) (Med.) 102

CISSOIDE (*f*) : cissoid 103

CISSTELLUNG (*f*) : cis position 104

CISTACEEN (*f.pl*) : (see CISTROSENGEWÄCHSE) 105

CISTERNE (*f*) : cistern, tank 106

CISTROSENGEWÄCHSE (*n.pl*) : *Cistaceae* (Bot.) 107

CITAT (*n*) : citation ; quotation 108

CITIEREN : to cite ; quote ; summon 109

CITRACONSÄURE (*f*) : citraconic acid ; $C_5H_6O_4$; Sg. 1.617 110

CITRACONSÄUREANHYDRID (*n*) : citraconic anhydride ; $C_5H_4O_3$; Sg. 1.2504 ; Mp. 7°C. ; Bp. 214°C. 111

CITRACONSÄURE-DIÄTHYLESTER (*m*) : citraconic ; diethylester ; $C_9H_{14}O_4$; Sg. 1.0468 ; Bp. 230.3°C. 112

CITRAL (*n*) : citral ; $C_9H_{15}.COH$; Sg. 0.894 ; Bp. 228°C. 113

CITRIN (*n*) : false topaz, yellow cairngorm, citrine (Mineralogy) (see QUARZ) 114

CITRONAT (*m*) : candied (lemon) peel 115

CITRONE (*f*) : citron, lemon 116

CITRONELLAL (*n*) : citronellal (a terpene aldehyde) 117

CITRONELLAÖL (*n*) : citronella oil (*Oleum citronellæ* ; *Oleum andropogonis nardi*, distilled from leaves of *Andropogon schoenanthus*) ; Sg. 0.886-0.920 118

CITRONELLOL (*n*) : citronellol ; $C_{10}H_{19}.OH$; Sg. 0.864 ; Bp. 225°C. 119

CITRONENBAUM (*m*) : lemon tree (*Citrus medica*) 120

CITRONENBEIFUSS (*m*) : southern-wood (*Artemisia abrotanum*) 121

CITRONENGELB : lemon-yellow 122

CITRONENHOLZ (*n*) : candle-wood 123

CITRONENKERNÖL (*n*) : citron-(seed) oil (from *Citrus limonum*) 124

CITRONENKRAUT (*n*) : balm mint (*herba melissa officinalis*) 125

CITRONENMELISSE (*f*) : balm mint (see CITRONENKRAUT) 126

CITRONENÖL (*n*) : lemon oil (*Oleum citri*) (from the skins of the fruit of *Citrus limonum risso*) ; Sg. 0.854-0.861 ; Bp. 177°C. 127

CITRONENPRESSE (*f*) : lemon squeezers 128

CITRONENSAFT (*m*) : lemon juice 129

CITRONENSAUER : citrate of ; combined with citric acid 130

CITRONENSÄURE (*f*) : (*Acidum citricum*) : citric acid ; oxytricarballylic acid ; $(CO_2HCH_2)_2C(OH)CO_2H$; Sg. 1.54 ; Mp. 153°C. 131

CITRONENSÄURE-(*f*) HYDRAT (*n*) : citric acid, hydrated ; $C_6H_8O_7+H_2O$; Sg. 1.542 ; Mp. 153°C. 132

CITRONENSÄURETRIÄTHYLESTER (*m*) : citric triethylester ; $C_{12}H_{20}O_7$; Sg. 1.1369 ; Bp. 185°C. 133

CITROTHYMÖL (*n*) : lemon-thyme oil 134

CIVIL : civil ; moderate 135

CLAIRCIEREN : to clarify ; clay (Sugar) 136

CLAUDETIT (*m*) : claudetite (see ARSENTRIOXYD) 137

CLAUSIUS-RANKINESCHER KREISPROZESS (*m*) : adiabatic process (the comparative process by which expansion of the steam takes place without heat exchange with the surroundings. A cyclic process (cycle) by Clausius-Rankine) 138

CLEVES' SÄURE (*f*) : Cleves' acid, alpha naphthol-sulfonic acid ; $C_{10}H_6(OH)(SO_3H)$ 1 : 5 139

CLUPANODONSÄURE (*f*) : clupanodonic acid ; $C_{13}H_{19}CO_2H$ 140

COAKS (*m* and *f.pl*) : coke 141

COBALT (*m*) : cobalt (see KOBALT) 142

COCAALKALOID (*n*) : coca alkaloid (base obtained from the leaves of various species of *Coca*) 143

COCABLÄTTER (*n.pl*) : coca leaves (*folia cocae*) (leaves of *Erythroxylon coca*) 144

COCAIN (*n*) : cocaine (*Cocainum*) ; $C_{17}H_{21}NO_4$; (from the leaves of *Erythroxylon coca*) ; Mp. 98°C. (see also KOKAIN) 145

COCAMIN (*n*) : cocamine (see ISATROPYLCOCAIN) 146

COCAWEIN (*m*) : coca wine (*Vinum cocae*) 147

COCCERIN (*n*) : coccerine (a wax from cochineal ; (see COCHENILLE) ; $C_{30}H_{60}(C_{31}H_{61}O_3)_2$; Mp. 106°C. 148

COCCINSÄURE (*f*) : coccinic acid 149

COCCIONELLE (*f*) : (see COCHENILLE) 150

COCHENILLE (*f*) : cochineal (dried bodies of female insects *Coccus cacti* ; the colouring principle is $C_{17}H_{18}O_{10}$) 151

COCHENILLEROT (*n*) : cochineal red, crimson lake, scarlet (purple) lake (see KARMIN) 152

COCHENILLESÄURE (*f*) : cochinillic acid 153

COCHINCHINAWACHS (*n*) : cochin-china wax (from various species of *Irvingia*) 154

COCHINEAL (*n*) : (see COCHENILLE) 155

COCOLITH (*m*) : cocolite 156

COCOSPFLAUME (*f*) : (see ICOCA, WESTINDISCHE-) 157

COCUSBAUM (*m*) : cocoa-nut palm (*Cocos nucifera*) 158

COCUSNUSSÖL (*n*) : cocoa-nut oil (see COCUSÖL) 159

COCUSÖL (*n*) : cocoanut-oil (*Oleum Cocos*) ; Mp. 24.5°C. ; Sg. 0.912 160

CODEIN (*n*) : codeine (an opium alkaloid) ; $C_{18}H_{21}NO_3.H_2O$; Mp. 154.9°C. 161

CODEINCHLORHYDRAT (*n*) : codeine hydrochloride ; $C_{18}H_{21}NO_3.HCl.H_2O$; Mp. 264°C. 162

CODEINHYDROBROMID (*n*) : codeine hydrobromide, bromeine 163

CODEINHYDROCHLORID (*n*) : codein hydrochloride (see CODEINCHLORHYDRAT) 164

CODEINPHOSPHAT (*n*) : codeine phosphate ; $C_{18}H_{21}NO_3PO_4.2H_2O$; Mp. 235°C. 165

CODEIN (*n*), PHOSPHORSAURES- : codeine phosphate (see CODEINPHOSPHAT) 166

CODEIN (*n*) : SALZSAURES- : codein hydrochloride (see CODEINCHLORHYDRAT) 167

CODEINSULFAT (*n*) : codeine sulphate ; $(C_{18}H_{21}NO_3)_2.H_2SO_4.5H_2O$; Mp. 278°C. 168

CODEONAL (*n*) : codeonal (a mixture of 2 parts codeine diethylbarbiturate and 15 parts sodium diethylbarbiturate) 169

CODÖL (*n*) : cod oil ; rosin oil ; retinol ; rosinol ; (product of dry distillation of resin) ; Sg. 0.98-1.11 170

COELESTIN (*m*): celestite (see **CÖLESTIN**) 171
COERULEUM (*n*): coeruleum (see **HIMMELBLAU**, **CÖLINBLAU**, and **CÖRULEUM**) 172
COFFEIN (*n*): caffein (see **KAFFEIN**) 173
COGNAC (*m*): cognac (see **KOGNAK**) 174
COGNAK (*m*): cognac (see **KOGNAK**) 175
COGNAKÖL (*n*): cognac oil, oenanthic ether; oenanthylic ether; $CH_3(CH_2)_5CO.C_2H_5$ 176
COHOBIEREN: to cohobate 177
COLATUR (*f*): filtrate (see **KOLATUR**) 178
COLCHICEIN (*n*): colchiceine; $C_{15}H_9(OCH_3)_3(NH.CO.CH_3)COOH$ 179
COLCHICIN (*n*): colchicine (from *Colchicum autumnale*); (*Colchicinum*); $C_{22}H_{25}NO_6$ or $C_{15}H_9(OCH_3)_3(NH.CO.CH_3)CO.OCH_3$; average Mp. 145°C., (commences to soften at 120°C. and is completely melted at 160°C.) 180
COLCHICINSÄURE (*f*): colchicinic acid; $C_{15}H_9(OH)_3(NH_2)COOH$ 181
COLCHICUMEXTRAKT (*m*): colchicum extract (*Extractum colchici*) 182
COLCHICUMPRÄPARAT (*n*): colchicum preparation 183
COLCHICUMTINKTUR (*f*): colchicum tincture (*Tinctura colchici*) 184
COLCHICUMWEIN (*m*): colchicum wine (*Vinum colchici*) 185
COLCOTHAR (*n*): colcothar, caput mortuum (see **ENGELROT** and **EISENOXYD**) 186
COLEMANIT (*m*): colemanite (a natural borate of calcium); $Ca_2B_6O_{11}.5H_2O$; Sg. 2.4 187
CÖLESTIN (*m*): celestite (a natural strontium sulphate); $SrSO_4$; Sg. 4.0; (see also **STRONTIUMSULFAT**) 188
CÖLESTINBLAU (*n*): celestine blue, sky-blue, azure 189
CÖLINBLAU (*n*): ceruleum (see **CÖRULEUM**) (composed of stannic anhydride, silicon dioxide, and cobaltous oxide in proportion of 5; 3; 2) 190
COLLARGOL (*n*): (*Argentum Colloidale*): collargol, colloidal silver, *Argentum credé* 191
COLLARGOLSALBE (*f*): (*Unguentum credé*): Collargol ointment 192
COLLAURIN (*n*): collaurin, colloidal gold (*Aurum colloidale*) 193
COLLODIN (*n*): collodine (a ptomaine from putrefying horse-flesh or mackerel); $C_8H_{11}N$; also a vegetable glue) gluten 194
COLLODIUM (*n*): collodium; collodion; pyroxylin, flexible collodion (see also **COLLODIUMLÖSUNG**) 195
COLLODIUMLÖSUNG (*f*): collodion, collodion solution, flexible collodion, pyroxylin (solution of a mixture of trinitrocellulose and tetranitrocellulose in some solvent such as alcohol) 196
COLLODIUMPAPIER (*n*): collodium paper (Photography) 197
COLLODIUMWOLLE (*f*): collodion wool, nitrocellulose, gun cotton, colloxylin; $C_6H_7O_5(NO_2)_3$; Sg. 0.1-0.3 198
COLLOID (*n*): colloid (see **KOLLOID**) (as a prefix = colloidal) 199
COLLOXYLIN (*n*): (see **KOLLODIUMWOLLE**) 200
COLOMBOSÄURE (*f*): calumbic acid (from *Jateorrhiza palmata*) 201
COLOMBOWURZEL (*f*): calumba root (root of *Jateorrhiza palmata*); (*Radix colombo*) 202
COLONNE (*f*): column 203
COLOPHONIUM (*n*): colophony, rosin, resin (see **KOLOPHONIUM**) 204
COLORIMETRISCH: colorimetric 205
COLUMBIT (*m*): columbite (a natural iron columbate and tantalate); ($FeNb_2O_6$,$FeTa_2O_6$) or $(Fe,Mn)[(Nb,Ta)O_3]_2$; (see also **NIOBIT**); Sg. 5.87 206
COLZAÖL (*n*): (see **RAPSÖL**) 207
COMBINIEREN: to combine 208
COMMIS (*m*): clerk; merchant's clerk; confidential clerk 209
COMMISSTELLE (*f*): clerkship 210
COMPASS-NADEL (*f*): compass needle 211
COMPASSSCHEIBE (*f*): compass card 212
COMPOSITEN (*pl*): *Compositæ* (Bot.) 213
COMPOSITION (*f*): composition; tin-composition (tin solution in *aqua regia*) 214
COMPOUNDDRAHT (*m*): compound wire 215
COMPOUNDDYNAMO (*f*): compound dynamo 216
COMPOUNDWICKELUNG (*f*): compound winding 217
COMPRESSIONSMASCHINE (*f*): compressing engine; condensing engine 218
COMPRIMIEREN: to compress 219
COMPTOIR (*n*): counting house 220
CONCREMENT (*n*): concrement, concretion (Med.) 221
CONDENSTOPF (*m*): steam trap 222
CONDITIONIEREN: to condition 223
CONDURANGIN (*n*): condurangine (a glycoside from *Marsdenia condurango*) 224
CONDURANGORINDE (*f*): condurango bark, condor vine, eagle vine; Mata-perro (bark of *Gonolobus condurango*) (see also **GEIER-RINDE**) 225
CONEPHRIN (*n*): conephrin (a solution of cocaine and paranephrine, ready for use as a local anæsthetic) 226
CONFEKT (*n*): confection 227
CONGLOBIEREN: to heap up; conglomerate 228
CONGLOMERAT (*n*): conglomerate 229
CONGOBLAU (*n*): congo blue 230
CONGOGELB (*n*): congo yellow 231
CONGOROT (*n*): congo red 232
CONGOVIOLETT (*n*): congo violet 233
CONHYDRIN (*n*): conhydrine, oxyconine, oxyconiine; $C_8H_{17}NO$; Mp. 120.6°C.; Bp. 220-225°C. 234
CONICEIN (*n*): coniceine; $C_8H_{15}N$ 235
CONIFERE (*f*): conifer (see **KONIFERE**) 236
CONIFERIN (*n*): coniferine (see **ABIETIN**) 237
CONIIN (*n*): conine; coniine (*Coniinum*) (alkaloid of *Conium maculatum*); $C_8H_{17}N$; Sg. 0.855; Mp. −2.5°C.; Bp. 166-170 °C. 238
CONIINCHLORHYDRAT (*n*): coniine hydrochloride, $C_8H_{17}N.HCl$; Mp. 209°C. 239
CONIINEXTRAKT (*m*): (*Extractum conii*); coniine extract 240
CONIINSALBE (*f*): (*Unguentum conii*): coniine ointment 241
CONIUMSÄURE (*f*): coniic acid (from *Conium maculatum*) 242
CONNOSSEMENT (*n*): bill of lading (Com.) 243
CONSTATIEREN: to ascertain; verify 244
CONSTRUCTEUR (*m*): designer, constructor 245
CONSTRUKTIONSEINZELHEITEN (*f.pl*): constructional details 246
CONSUM-VEREIN (*m*): co-operative (store) company, society or association 247
CONTO (*n*): account 248

CONTOCORRENT (*m.* and *n*): account current (Banking) 249
CONTOR (*n*): office 250
CONUS (*m*): cone 251
CONVEYORANLAGE (*f*): conveyor plant, transporter plant 252
COPAÏVABALSAM (*m*): copaiba (see KOPAIVA-BALSAM) 253
COPAIVÄOL (*n*): copaiba oil; Sg. 0.895-0.905; Bp. 250-275°C. 254
COPALDRYER (*m*): copal dryer (see also KOPAL-) 255
COPRA (*f*): copra (dried kernel of coco-nut or cocoa-nut) (see under KOKOS-) 256
COQUIMBIT (*m*): coquimbite (a natural ferric sulphate); $Fe_2(SO_4)_3.9H_2O$ 257
CORALLIN (*n*), GELBES-: aurine, yellow corraline, rosolic acid (see also ROSOLSÄURE) 258
CORDIERIT (*m*): cordierite, dichroite, water sapphire, iolite (natural aluminium-iron-magnesium silicate); $(Mg,Fe)_2Al_4Si_5O_{18}$; Sg. 2.63 259
CORDIT (*n*): cordite (an explosive; about 58% nitroglycerin, 37% gun cotton, 5% mineral jelly); Sg. 1.56 260
CORIANDER (*m*): coriander-(seed) (of *Coriandrum sativum*) 261
CORIANDERÖL (*n*): coriander oil; Sg. 0.86-0.878; (distilled from fruit of *Coriandrum sativum*) 262
CORIUM (*n*): hide (for leather making) (see also LEDERHAUT) 263
CORNWALLKESSEL (*m*): Cornish boiler 264
COROZONUSS (*f*): ivory nut; corozo nut (seeds of *Phytelephas macrocarpa*) 265
COROZONUSSÖL (*n*): corozo-nut oil, cohune-(nut)-oil, cahune-(nut)- oil, cahoun-(nut)-oil; Mp. 19°C. 266
CORRODIEREN: to corrode 267
CÖRULEIN (*n*): cerulein 268
CÖRULEUM (*n*): ceruleum; sky-blue (a cobalt colour) (see ZÖRULEUM and CÖLINBLAU) 269
CORUNDUM (*n*): corundum (see KORUND) 270
CORYBULBIN (*n*): corybulbine (an alkaloid) (see CORYDALISALKOID); $C_{21}H_{25}NO_4$ 271
CORYDALIN (*n*): corydaline (an alkaloid) (see CORYDALISALKALOID); $C_{22}H_{27}NO_4$ 272
CORYDALINGRUPPE (*f*): corydaline group 273
CORYDALISALKALOID (*n*): corydalis alkaloid (any alkaloid from *Corydalis cava*) 274
CORYDIN (*n*): corydine (an alkaloid) (see CORYDALISALKALOID; $C_{21}H_{23}NO_4$ 275
CORYCAVAMIN (*n*): corycavamine (an alkaloid) (see CORYDALISALKALOID); $C_{21}H_{21}NO_5$ [276
CORYCAVIDIN (*n*): corycavidine (an alkaloid) (see CORYDALISALKALOID); $C_{22}H_{25}NO_4$ [277
CORYCAVIN (*n*): corycavine (an alkaloid) (see CORYDALISALKALOID); $C_{23}H_{23}NO_6$ 278
CORYCAVINGRUPPE (*f*): corycavine group 279
CORYFIN (*n*): coryfine, menthol ethylglycolate (by action of ethoxyacetyl chloride on Menthol); $C_{10}H_{19}O.CO.CH_2OC_2H_5$; Bp. 155°C. 280
CORYTUBERIN (*n*): corytuberine (an alkaloid) (see CORYDALISALKALOID); $C_{19}H_{23}NO_4$ [281
CORYTUBERINGRUPPE (*f*): corytuberine group 282
COTARNIN (*n*): cotarnine (*Cotarninum*); $C_{12}H_{13}NO_3.H_2O$ 283
COTARNINHYDROCHLORID (*n*): (see COTARNIN, SALZSAURES- and STIPTIZIN) 284
COTARNIN (*n*), PHTALSAURES-: (*Cotarninum phtalicum*): cotarnine phthalate, styptol (a styptic similar to STYPTIZIN, which see) 285
COTARNIN (*n*), SALZSAURES-: (*Cotarninum hydrochloricum*): cotarnine hydrochloride, stypticin (as the name implies a styptic or blood stanching agent); $C_{12}H_{14}NO_3Cl$; Mp. 143°C. 286
COTOIN (*n*): cotoine, methyl trioxybenzophenone (from coto bark; see COTORINDE); $C_2H_6O_3$; Mp. 130.5°C. 287
COTONISIEREN: to degum (Silk) 288
COTORINDE (*f*): coto bark (believed to be obtained from *Palicourea densifolia*) 289
COTTA (*f*): bloom; lump; cake (of metal) 290
COTTONÖL (*n*): cotton-seed oil (*Oleum gossypii* from seeds of *Gossypium herbaceum*); Sg. 0.922-0.93 291
COTUNNIT (*m*): cotunnite (a natural lead chloride); $PbCl_2$; Sg. 5.22 292
COULANUSSÖL (*n*): coula-nut oil (from seeds of *Coula edulis*) 293
COULEUR (*f*): colour; caramel; burnt sugar; medium fine smalt (see BLAUFARBE); (*pl*) fancy (coloured) goods 294
COULISSE (*f*): groove, channel; wings, scenes (Theatre); connecting link (Locomotives), coulisse (that which slides); a slide; slot-hole (see also KULISSE) 295
COULISSENFIEBER (*n*): stage fright 296
COULISSENLADEN (*m*): venetian blind 297
COULISSENSTEUERUNG (*f*): link motion 298
COUPÉ (*n*): compartment (of railway carriage) 299
COUPON (*n*): coupon 300
COURANT (*n*): currency 301
COUVERT (*n*): envelope; cover; wrapper 302
COVELLIN (*m*): covelline, covellite, indigo copper (natural indigo-blue copper sulphide, with about 66% copper and 34% sulphur); CuS; Sg. 4.6 303
C-PROZESS (*m*): C process (the name applied to the process for extracting toluene from coal gas, by washing with thin tar) 304
CRACKDESTILLATION (*f*): destructive distillation; cracking process 305
CRACKPROZESS (*m*): cracking process (oils) (a distillation process of crude oil under pressure of 10-20 atms., yielding approx. double the quantity of benzene to that obtained by the ordinary method of distillation) 306
CREATININ (*n*): creatinine (an animal alkaloid from urine); $C_4H_7N_3O$ 307
CREDÉSCHES SILBER (*n*): (*Argentum credé*): colloidal silver, collargol 308
CREME (*f*): cream; polishing cream; polish; ointment; salve; unction; unguent (*Unguentum*) 309
CREMOR TARTARI (*m*): cream of tartar; potassium bitartrate (see WEINSTEIN) 310
CREOLIN (*n*): creoline (a trade name for a disinfectant and wood-preserving agent; contains 25-30% creosote oil) 311
CRESYLIT (*n*): cresylite (an explosive consisting of molten trinitrocresol) 312
CRETINISMUS (*m*): cretinism (a kind of idiocy) (Med.) 313
CRIGHTON ÖFFNER (*m*): Crighton opener (Textile) 314
CRISTOBALIT (*n*): cristobalite (see SILICIUM-DIOXYD); Sg. 2.412 315
CROCETIN (*n*): crocetin (a dye from saffron); $C_{10}H_{14}O_2$ 316

CROCUS (*m*): oxide of iron, crocus martis, ferric hydroxide (see **EISENHYDROXYD**); saffron, crocus (Bot.) (*Crocus sativus*) 317
CROOKESIT (*m*): crookesite; (Cu, Ag, $Tl)_2Se$; Sg. 6.9 318
CROTONALDEHYD (*n*): crotonic aldehyde; C_4H_6O; Sg. 0.8557; Bp. 103.5°C. 319
CROTONCHLORALHYDRAT (*n*): (see **BUTYL-CHLORALHYDRAT**) 320
CROTONÖL (*n*): croton oil (from seeds of *Croton tiglium*); Sg. 0.94-0.96 321
CROTONOLSÄURE (*f*): crotonolic acid, tiglic acid; methylcrotonic acid; (from *Croton tiglium*) $CH_3CHC(CH_3)CO_2H$; Sg. 0.964; Mp. 65°C.; Bp. 198.5°C. 322
CROTONSAAT (*f*): croton seeds (seeds of *Croton tiglium*) 323
CROTONSÄURE (*f*): crotonic acid; $CH_3CHCHCO_2H$; Sg. 0.973; Mp. 72°C.; Bp. 185°C. 324
CROTONSÄUREÄTHYLESTER (*m*): crotonic ethylester; $C_6H_{10}O_2$; Sg. 0.9208; Bp. 138°C. 325
CROUP (*m*): croup, hoarseness (Med.) 326
CROUPÖS: croupous (Med.) 327
CRURIN (*n*): crurine, quinoline-bismuth thiocyanate; $C_9H_7N.HSCN.Bi(SCN)_3$ 328
CRUSOCREATININ (*n*): crusocreatinine (an animal alkaloid from fresh meat); $C_5H_8N_4O$ 329
CUBEBE (*f*): cubeb (see **KUBEBE**) 330
CUBEBINSÄURE (*f*): cubebinic acid (from *Piper cubeba*) 331
CUCURBITACEENÖL (*n*): cucurbitaceæ oil (see **KÜRBISKERNÖL, GURKENKERNÖL, MELONEN-KERNÖL, SCHWAMM-KÜRBISKERNÖL** and **WASSERMELONENÖL**) 332
CUITESEIDE (*f*): boiled-off silk 333
CUMALINSÄURE (*f*): coumalic (cumalic) acid 334
CUMARILSÄURE (*f*): coumarilic (cumarilic) acid 335
CUMARIN (*n*): cumarin, coumarin, cumaric anhydride, tonka bean camphor; $C_9H_6O_2$; Sg. 0.9348; Mp. 67°C; Bp. 290.5°C. 336
CUMARINSÄURE (*f*): coumarinic (cumarinic) acid 337
CUMARON (*n*): cumar, cumarone resin, para-coumarone-(resin), benzofurane resin, cumaron; C_8H_6O; Sg. 1.0767-1.096; Bp. 172°C. 338
CUMARONHARZ (*n*): cumarone resin, benzofurane resin (see **CUMARON**) 339
CUMARONNITROSIT (*n*): cumarone nitrosite; Mp. 115°C. 340
CUMARONPIKRAT (*n*): coumarone picrate; Mp. 102.5°C 341
CUMARSÄURE (*f*): $(C_9H_8O_3)$; cumaric acid, coumaric acid; ortho.- Mp. 201°C.; meta.- Mp. 191°C.; para.- Mp. 206.°C. 342
CUMARSÄUREANHYDRID (*n*): (see **KUMAR-SÄUREANHYDRID**) 343
CUMIDIN (*n*): cumidine; $C_9H_{13}N$; Sg. 0.9526 (liquid); Bp. 225°C.; pseudocumidine, trimethylaminobenzene; Mp. 62°C.; Bp. 236°C. 344
CUMINÖL (*n*): cumin oil; Sg. 0.915; (from fruit of *Cuminum cyminum*) (see **KÜMMELÖL**) 345
CUMINSÄURE (*f*): cumic (cuminic) acid 346
CUMOL (*n*): cumene, cumol, isopropylbenzene; C_9H_{12}; Sg. 0.8798; Bp. 152.9°C.; pseudocumol, trimethylbenzene; Sg. 0.8643; Bp. 169.8°C. 347
CUPFERRON (*n*): cupferron, ammonium-nitroso-beta-phenylhydroxylamine; $C_6H_9O_2N_3$ [348
CUPOL (*m*): cupola 349
CUPOLSTEIN (*m*): cupola brick 350
CUPRICARBONAT (*n*): cupric carbonate (*Cuprum carbonicum*); $Cu_2(OH)_2CO_3$; Sg. 3.7-4.0; (see also under **KUPRI-**) 351
CUPRICHLORID (*n*): cupric chloride (*Cuprum bichloratum*); $CuCl_2$; Sg. 3.054; Mp. 498°C. 352
CUPRIOXYD (*n*): cupric oxide (*Cuprum oxydatum*); CuO; Sg. 6.32; Mp. 1064°C. 353
CUPRISULFAT (*n*): cupric sulphate (*Cuprum sulfuricum*); $CuSO_4.5H_2O$; Sg. 2.284 [354
CUPRIT (*m*): cuprite, ruby copper, red copper ore (see **ROTKUPFERERZ**) 355
CUPRIVERBINDUNG (*f*): cupric compound 356
CUPROCHLORID (*n*): cuprous chloride (*Cuprum chloratum*); CuCl; (see also under **KUPRO-**) 357
CUPROION (*n*): copper ion 358
CUPROMANGAN (*n*): cupro-manganese 359
CUPROMERCURIJODID (*n*): cupromercuric iodide; Cu_2HgI_4 360
CUPROOXYD (*n*): cuprous oxide (*Cuprum oxydulatum*); Sg. 5.75-6.09; Mp. 1210°C.; Bp. 1800°C. 361
CUPROVERBINDUNG (*f*): cuprous compound 362
CURAÇAOSCHALENÖL (*n*): curaçao peel oil 363
CURARE (*f*): curare (an arrow poison from various types of *Strychnos* growing in South America) 364
CURCASÖL (*n*): curcas oil (from *Latropha curcas* or *Curcas purgans*); Sg. 0.919 365
CURCUMA (*f*): curcuma (see **KURKUMA**) 366
CURCUMIN (*n*): curcumin, brilliant yellow, azidine fast yellow (a dyestuff from **CURCUMA**, which see) 367
CURIN (*n*): curine (an alkaloid from **CURARE**, which see); $C_{18}H_{19}O_3N$ 368
CUTOLINE (*f*): cutoline (a blood stanching agent) 369
CUVETTE (*f*): bulb; trough; bath (see **KUVETTE**) 370
CYAN (*n*): cyanogen (C_2N_2) or (CN) (given in German formulæ sometimes as Cy); Sg. 0.866 compared with air; liquefies at 21°C.; solidifies at −34°C.; (as a prefix means, cyanide; prussiate) 371
CYANALKALI (*n*): alkali cyanide 372
CYANALKYL (*n*): alkyl cyanide 373
CYANAMID (*n*): (see **CALCIUMCYANAMID**) 374
CYANAMIDCALCIUM (*n*): calcium cyanamide; $CaCN_2$ 375
CYANAMIDNATRIUM (*n*): sodium cyanamide; Na_2CN_2 376
CYANAMIDODIKOHLENSÄURE (*f*): cyanamidodicarboxylic acid; cyanocarbamic acid 377
CYANAMMONIUM (*n*): ammonium cyanide (*Ammonium cyanatum*); $NH_4.CN$ 378
CYANAT (*n*): cyanate 379
CYANÄTHER (*m*): cyanic ether 380
CYANÄTHYL (*n*): ethyl cyanide; C_2H_5CN; Sg. 0.7799; Mp. −103.5°C.; Bp. 97.08°C. 381
CYANBARYUM (*n*): barium cyanide; $Ba(CN)_2$ 382
CYANBENZOL (*n*): cyanobenzene 383
CYANBROMID (*n*): cyanogen bromide 384
CYANCALCIUM (*n*): calcium cyanide (*Calcium cyanatum*); $(CaCN)_2$ 385

CYANCHLORID (*n*): cyanogen chloride; CNCl; Mp. –5°C.; Bp. 13°C. 386
CYANDOPPELSALZ (*n*): double cyanide 387
CYANEISEN (*n*): iron cyanide; $Fe_4[Fe(CN)_6]_3$ 388
CYANEISENKALIUM (*n*): potassium ferro-cyanide; $K_4Fe(CN)_6.3H_2O$ 389
CYANESSIGSAUER: cyanoacetate, combined with cyanoacetic acid 390
CYANESSIGSÄURE (*f*): cyanoacetic acid; $CH_2CNCOOH$; Mp. 66.25°C 391
CYANESSIGSÄURECHLORID (*n*): cyanoacetic chloride 392
CYANGOLD (*n*): gold cyanide 393
CYANGOLDKALIUM (*n*): potassium auri- (auro-) cyanide; $KAu(CN)_2$ 394
CYANHALTIG: cyanous 395
CYANID (*n*): cyanide (an -ic cyanide as opposed to CYANÜR an -ous cyanide) 396
CYANIG: cyanous 397
CYANIGE SÄURE (*f*): cyanous acid 398
CYANIN (*n*): cyanine, chinoline blue, quinoline blue, cyanine iodide (from action of amyl iodide on quinoline; $C_{28}H_{35}N_2I$ 399
CYANISIEREN: to cyanize 400
CYANISIERUNG (*f*): cyanidation 401
CYANIT (*m*): cyanite, disthene (an aluminium silicate) (see ALUMINIUMSILIKAT); Al_2SiO_5; Sg. 3.61 402
CYANJODID (*n*): cyanogen iodide; iodine cyanide; ICN; Mp. 146.5°C. 403
CYANKALI (*n*): (see CYANKALIUM) 404
CYANKALISALZ (*n*): potassium cyanide, cyanide powder (see CYANKALIUM) 405
CYANKALIUM (*n*): potassium cyanide (*Kalium cyanatum*); KCN; Sg. 1.52 406
CYANKOBALT (*n*): cobalt cyanide 407
CYANKUPFER (*n*): copper cyanide (see KUPFER-CYANID) 408
CYANMETALL (*n*): metallic cyanide 409
CYANMETHYL (*n*): methyl cyanide; CH_3CN; Sg. 0.7897; Mp. –41°C.; Bp. 83°C. 410
CYANNATRIUM (*n*): sodium cyanide (*Natrium cyanatum*); NaCN; 411
CYANOGEN (*n*): cyanogen (see CYAN) 412
CYANOMETER (*n*): cyanometer (an instrument for measuring degrees of blueness) 413
CYANOTYP (*n*): cyanotype 414
CYANOTYPVERFAHREN (*n*): cyanotype-(process), (Photography) 415
CYANPLATIN (*n*): platinum cyanide 416
CYANPROPIONSÄURE (*f*): cyanopropionic acid 417
CYANQUECKSILBER (*n*): mercury cyanide; $Hg(CN)_2$; Sg. 4.018 (see QUECKSILBERCYANID) 418
CYANSAUER: cyanate of; combined with cyanic acid 419
CYANSÄURE (*f*): cyanic acid; CNOH; Sg. 1.15 420
CYANSAURES SALZ (*n*): cyanate 421
CYANSEIFE (*f*): cyanogen soap (a soap containing potassium cyanide, used in photography for removing silver stains from the hands) 422
CYANSILBER (*n*): silver cyanide; AgCN; Sg. 3.95 423
CYANTOLUOL (*n*): cyanotoluene 424
CYANÜR (*n*): cyanide (an -ous cyanide as opposed to an ic- cyanide, CYANID) 425
CYANURSÄURE (*f*): (iso)-cyanuric acid, tricyanic acid, tricarbimide; $CO(NHCO)_2NH.2H_2O$ Sg. 1.768 426
CYANVERBINDUNG (*f*): cyanogen compound; any cyanide 427
CYANWASSERSTOFF (*m*): hydrogen cyanide; prussic acid; hydrocyanic acid (*Acidum hydrocyanicum*); HCN; Sg. 0.9483 428
CYANWASSERSTOFFSAUER: cyanide of; combined with hydrocyanic (prussic) acid 429
CYANWASSERSTOFFSÄURE (*f*): hydrocyanic (prussic) acid; HCN; Sg. 0.9483; Mp. –15°C.; Bp. 26.1°C. 430
CYANWASSERSTOFFSÄUREVERGIFTUNG (*f*): prussic acid poisoning (Med.) 431
CYANWASSERSTOFFSEIFE (*f*): potassium cyanide soap (Photography); (see CYANSEIFE) 432
CYANZINK (*n*): zinc cyanide; $Zn(CN)_2$ 433
CYCLAMEN (*m*): cyclamen; (see also ERDBROT) 434
CYCLISCH: cyclic 435
CYCLOBUTAN (*n*): cyclobutane; C_4H_8; Sg. 0.7038; Bp. 11-12°C. 436
CYCLOFORM (*n*): cycloform, Isobutyl paramino-benzoate; $C_5H_{11}NO_2$; Mp. 65°C. 437
CYCLOGERANIUMSÄURE (*f*): cyclogeranic acid 438
CYCLOHEPTAN (*n*): cycloheptane; C_7H_{14}; Sg. 0.8252; Bp. 117.5°C. 439
CYCLOHEXAN (*n*): cyclohexane; C_6H_{12}; Sg. 0.7764; Mp. 6.4°C.; Bp. 80.9°C. 440
CYCLOHEXANOL (*n*): cyclohexanol; (see HEXALIN) 441
CYCLOIDE (*f*): cycloid 442
CYCLOOLEFIN (*n*): cyclo-olefine 443
CYCLOPARAFFIN (*n*): cyclo-paraffin 444
CYCLOPENTADIEN (*n*): cyclopentadien, pyropentylene; C_5H_6; Sg. 0.815; Bp. 41°C. 445
CYCLOPENTAN (*n*): cyclopentane; C_5H_{10}; Sg. 0.7506; Bp. 50.8°C. 446
CYCLOPROPAN (*n*): cyclopropane; C_3H_6; Mp. –126°C.; Bp. about –35°C. 447
CYLINDERN (*m.pl*): casts (of blood etc.) (Med.) 448
CYMARIN (*n*): cymarine (active principle of the roots of *Apocynum cannabinum*); Mp. 130-140°C. 449
CYMOL (*n*): cymene, cymol, isopropyltoluene; $C_{10}H_{14}$; ortho-, Sg. 0.875; Mp. –182°C.; Bp. 181°C.; meta-, Sg. 0.862; Mp. –25°C.; Bp. 176°C.; para-, Sg. 0.855; Mp. –74°C; Bp. 177°C. 450
CYMOPHAN (*m*): cymophane, cat's-eye; (see CHRYSOBERYLL) 451
CYPERBLAU (*n*): cyprus blue 452
CYPERN (*n*): cyprus 453
CYPERVITRIOL (*m*): old term for copper sulphate; blue vitriol; $CuSO_4.5H_2O$; Sg. 2.284 454
CYPRESSE (*f*): cypress (*Cupressus*) 455
CYPRESSENHOLZ (*n*): cypress wood; (see CYPRESSE); Sg. 0.65 456
CYPRESSENNUSS (*f*): cypress cone 457
CYPRESSENÖL (*n*): cypress oil (distilled from leaves of *Cupressus sempervirens*); Bp. 160-250°C. 458
CYPRISCHE UMBRA (*f*): raw umber; (see MANGANBRAUN) 459
CYSTENFLÜSSIGKEIT (*f*): cystic fluid (Med.) 460
CYSTINURIE (*f*): cystinuria 461
CYSTINURIKER (*m*): a sufferer from cystinuria 462
CYSTOPURIN (*n*): cystopurine (a double compound composed of 1 mol. hexamethylenetetramine plus 2 mols. sodium acetate); $C_6H_{12}N_4+2(CH_3COONa.3H_2O)$ 463
CYSTOSIN (*n*): cystosine; HN.OC.N.C.CH.CH.NH₂ (a pyrimidine base) 464

D

DACH (*n*): dome; roof; house; crown; shelter 465
DACHBAU (*m*): roofing 466
DACHBEDECKUNGSMATERIAL (*n*): roofing (composition) material 467
DACHBEDECKUNGSSTOFF (*m*): roofing (composition) material 468
DACHBEKLEIDUNG (*f*): roofing 469
DACHBLECH (*n*): roofing plate of sheet metal or corrugated iron 470
DACHBODEN (*m*): loft 471
DACHEL (*m*): lump; bloom (Metal) 472
DÄCHEL (*m*): (see DACHEL) 473
DACHKAMMER (*f*): attic; garret; loft 474
DACHKITT (*m*): roof cement 475
DACHKOHLE (*f*): upper coal 476
DACHLACK (*m*): roof-(ing) varnish 477
DACHLAUB (*n*): common house leek (*Sempervivum tectorum*) 478
DACHPAPPE (*f*): roofing board; tarred felt; composition roofing 479
DACHPFANNE (*f*): pantile 480
DACHRINNE (*f*): gutter 481
DACHSCHIEFER (*m*): roofing slate 482
DACHSCHINDEL (*f*): shingle 483
DÄCHSEL (*m*): adze 484
DACHSPARREN (*m*): rafter 485
DACHSTEIN (*m*): tile; roofing slate; roof rock (Mining); bitumenous shale 486
DACHSTUHL (*m*): rafters, roof supports 487
DACHTRAUSE (*f*): eaves 488
DACHUNG (*f*): roof-(ing) 489
DACHWIPPE (*f*): lever (Rolling mills) 490
DACHWURZ (*f*): common house leek (*Sempervivum tectorum*) 491
DACHZIEGEL (*m*): roofing tile 492
DACHZIEGELEI (*f*): tile kiln; tile works 493
DAFERN: provided, if, in case 494
DAFÜRHALTEN (*n*): opinion; notion; idea 495
DAFÜRHALTENS, MEINES-: to my mind; in my opinion; according to my idea 496
DAFÜRSPRECHEN: to speak for, show, prove 497
DAGUERREOTYPIEPROZESS (*m*): daguerreotype process (named after the inventor; a positive copper plate process of photography) 498
DAHERUM: thereabouts 499
DAHIER: in this place 500
DAHINSTEHEN: to be (undecided) uncertain 501
DAHLIA (*f*): dahlia (the flower and plant); a general term for fuchsine mixtures (used for dyeing silk) 502
DAHLSCHE SÄURE (*f*): Dahls acid, beta-naphthylaminesulfonic acid; $C_{10}H_6(NH_2)SO_3H$ 2 : 5 503
DAHMENIT (*n*): dahmenite (about 91.5% ammonium nitrate; 2% Potassium bichromate and 6½% naphthalin: an explosive) 504
DAMASCIN (*n*): damascene 505
DAMAST (*m*): damask 506
DAMASTSTAHL (*m*): damask steel; damascined steel 507
DAMASZENER: damascus; damascene; damask 508
DAMASZENER PFLAUME (*f*): (see KRIECHE and SYRISCHE PFLAUME) 509
DAMASZIEREN: to damescene (damask) (of steel) 510
DAMASZIERUNG (*f*): damascening (of steel, etc.) 511
DAME (*f*): lady; queen (Cards); queen (Chess); king (Draughts) 512
DAMENBRETT (*n*): draught board (for the game of draughts) 513
DAMENSPIEL (*n*): (the game of)-draughts 514
DAMENSTEIN (*m*): man or piece in the game of draughts 515
DAMM (*m*): dam; dike; bank; pier; causeway; embankment; perineum (Med.) 516
DAMMAR (*n*): (*Resina Dammar*): gum (damar) dammar, resin damar (obtained from *Shorea Wiesneri*); Sg. 1.04-1.12; Mp. 120°C.; up to 1% moisture and 1% ash contents 517
DAMMARHARZ (*n*): dammar gum (see DAMMAR) (also obtained from *Dammar orientalis*) 518
DÄMMEN: to dam; stop up; curb; bank up 519
DAMMERDE (*f*): mould; humus; pit sand (Founding) 520
DAMMRAMPE (*f*): ramp; raised road 521
DAMMTÜRE (*f*): damming door (Mining) 522
DAMPF (*m*): vapour; steam; fume; smoke 523
DAMPFABSCHEIDER (*m*): steam separator 524
DAMPFABSPERRVENTIL (*n*): steam stop valve 525
DAMPFARMATUREN (*f.pl*): steam fittings (mountings, valves, gauges etc.) 526
DAMPFARTIKEL (*m*): steam style (Calico) 527
DAMPFAUSLASSROHR (*n*): escape pipe, steam outlet 528
DAMPFAUSPUFF (*m*): steam exhaust 529
DAMPFAUSSTRÖMUNG (*f*): steam outlet 530
DAMPFBAD (*n*): steam (vapour) bath 531
DAMPFBEDARF (*m*): steam demand; steam consumption; steam required 532
DAMPFBILDUNG (*f*) steam-formation, formation of steam 533
DAMPFBLASE (*f*): steam bubble; a still heated by steam 534
DAMPFBOOT (*n*): steam (launch) boat 535
DAMPFCHLOR (*n*): steam chemicking 536
DAMPFDARRE (*f*): steam (drying) kiln 537
DAMPFDICHT: steam tight 538
DAMPFDICHTE (*f*): steam density (weight of vapour in grams, divided by the volume it occupies in cu. cm.) 539
DAMPF (*m*): DIREKTER-: live steam 540
DAMPFDOM (*m*): steam dome 541
DAMPFDRUCK (*m*): steam pressure; steam printing 542
DAMPFDRUCKMESSER (*m*): manometer; (steam)-pressure gauge 543
DAMPFDURCHTRITT (*m*): steam flow, steam flow opening 544
DAMPFECHT: fast to steam 545
DAMPFEINSTRÖMUNG (*f*): steam inlet 546
DAMPFEN: to (give) steam; smoke; fume; evaporate; vapourize 547
DÄMPFEN: to damp (see also WELLE); put out (fire); smother; subdue (colour or sound); quench; suppress; steam; stew 548
DAMPFENTNAHME (*f*): withdrawal of steam; steam outlet; load (of boilers); leak-off (of turbines) 549
DAMPFENTWÄSSERUNGSAPPARAT (*m*): steam dryer, steam drying apparatus 550
DAMPFENTWICKLER (*m*): steam generator 551
DAMPFER (*m*): steamer; steamboat; steamship; 552

DÄMPFER (*m*): damper; steamer; steam cooker 553
DAMPFERZEUGER (*m*): steam generator 554
DAMPFERZEUGUNG (*f*): steam (production) generation; evaporative capacity 555
DAMPFESSE (*f*): funnel, steam-pipe 556
DAMPFFASS (*n*): digestor; autoclave; steam sterilizer (any vessel having an internal pressure over 1 atm., due to vapour or gas received from other apparatus) 557
DAMPFFÖRMIG: in the form of steam; vaporous 558
DAMPF (*m*), GEMISCHTER-: mixed steam 559
DAMPFGRÜN (*n*): steam green (a nitroso dye-stuff) 560
DAMPFGUMMI (*n*): dextrin (see DEXTRIN) 561
DAMPFHAHN (*n*): steam cock 562
DAMPFHAMMER (*m*): steam hammer 563
DAMPFHEIZUNG (*f*): steam heating; radiator 564
DAMPFHOLZSCHLIFF (*m*): steamed mechanical wood pulp 565
DAMPFHÜLLE (*f*): steam jacket; vaporous envelope 566
DAMPF (*m*), INDIREKTER-: exhaust steam 567
DAMPFKANAL (*m*): steam pipe (port) 568
DAMPFKESSEL (*m*): steam boiler; steam generator; boiler; steamer (Cooking) 569
DAMPFKESSELBLECH (*n*): boiler plate 570
DAMPFKOCHAPPARAT (*m*): steam boiling apparatus, boiling (steaming) pan, copper. cooker 571
DAMPFKOLBEN (*m*): steam piston 572
DAMPFKRAFTMASCHINE (*f*): (either) steam engine (or) steam turbine 573
DAMPFLASTWAGEN (*m*): steam (tractor) lorry 574
DAMPFLEISTUNG (*f*): (see VERDAMPFLEISTUNG) 575
DAMPFLOKOMOTIVE (*f*): steam locomotive, engine 576
DAMPF MACHEN: to get up steam; form steam, vaporize 577
DAMPFMANTEL (*m*): steam jacket 578
DAMPFMASCHINE (*f*): steam engine; locomotive 579
DAMPFMESSER (*m*): steam meter; pressure gauge; steam gauge; manometer 580
DAMPFMÜHLE (*f*): steam mill 581
DAMPFPACKUNG (*f*): steam packing 582
DÄMPFPFANNE (*f*): stew (pot) pan 583
DAMPFPUMPE (*f*): steam pump 584
DAMPFREDUZIERVENTIL (*n*): (steam)-reducing valve 585
DAMPFREIBUNG (*f*): steam friction 586
DAMPFROHR (*n*): steam pipe; steam tube (of water-tube boilers) 587
DAMPFSAMMLER (*m*): steam collector, steam drum (Boilers), receiver 588
DAMPFSCHIEBER (*m*): steam slide valve 589
DAMPFSCHIFF (*n*): steamer, steamship, ss 590
DAMPFSCHLANGE (*f*): steam coil 591
DAMPFSCHREIBER (*m*): recording steam meter 592
DAMPFSCHWARZ (*n*): steam black-(dye) 593
DAMPFSCHWELUNG (*f*): steam distillation 594
DAMPFSEITIG: on the steam side 595
DAMPFSICHERHEITSVENTIL (*n*): steam safety valve 596
DAMPFSPANNUNG (*f*): steam (tension) pressure 597
DAMPF (*m*), STICKENDER-: choke damp (Mining) 598
DAMPFSTRAHL (*m*): steam jet 599
DAMPFSTRAHLSAUGER (*m*): steam jet aspirator 600
DAMPFSTROM (*m*): steam current; steam flow; path of the steam 601
DAMPFSTROMMESSER (*m*): steam flow meter 602
DAMPFTABELLE (*f*): steam table 603
DAMPFTEMPERATUR (*f*): steam temperature; total temperature (as opposed to superheat) 604
DAMPFTOPF (*m*): digester; autoclave; steam sterilizer (see DAMPFFASS) 605
DAMPFTRICHTER (*m*): steam funnel; uptake 606
DAMPFTROCKNER (*m*): steam drier 607
DAMPFTURBINE (*f*): steam turbine 608
DAMPFÜBERDRUCK (*m*): absolute steam pressure (see ABSOLUT) 609
DAMPFÜBERHITZER (*m*): (steam)-superheater 610
DAMPF (*m*), ÜBERHITZTER-: superheated steam 611
DAMPFUHR (*f*): steam (dial) meter, steam gauge 612
DAMPF (*m*), UNGEMISCHTER-: unmixed steam 613
DÄMPFUNGSSTROMKREIS (*m*): damping curcuit 614
DÄMPFUNGSWIDERSTAND (*m*): damping resistance 615
DAMPFVENTIL (*n*): steam valve 616
DAMPFVERTEILER (*m*): steam distributor 617
DAMPFWAGEN (*m*): steam car; loco.; steam waggon 618
DAMPFWÄRME (*f*): heat of steam; vaporizing temperature 619
DAMPFWASSER (*n*): condensate 620
DAMPFWASSERSAMMLER (*m*): collector, receiver, header, steam drum (of boilers) 621
DAMPFWASSERSCHLAG (*m*): (steam: water)-hammer or hammering-(in pipe lines) 622
DAMPFWASSERTOPF (*m*): steam trap 623
DAMPFWEG (*m*): steam (port; passage) pipe; path of the steam; steam-flow 624
DAMPFZÄHLER (*m*): recording steam meter; counter; teller 625
DAMPFZEIGER (*m*): steam dial; (indicating-) steam meter; steam gauge 626
DAMPFZUSTAND (*m*): steam (condition) state 627
DAMPFZYLINDERENTWÄSSERUNGSAPPARAT (*m*): steam cylinder drainage apparatus 628
DANAIT (*m*): danaite, cobalt pyrites; $(Fe,Co)AsS$ with $(Fe,Co)(AsS)_2$ 629
DANBURIT (*m*): danburite (a calcium-boron silicate); $CaB_2(SiO_4)_2$; Sg. 2.99 630
DANKESBEZEUGUNGEN (*f.pl*): thanks 631
DANZIGERBLAU (*n*): Danzig blue 632
DAPHNEÖL (*n*): daphne oil (from the seeds of *Daphne gnidium*) 633
DAPHNIN (*n*): daphnin (a glycoside from the plant *Daphne mezereum*) 634
DARAUFLASSEN (*n*): doubling (Brewing) 635
DARBEN: to starve; be in need; want 636
DARBIETEN: to offer; present 637
DARLEGEN: to show; display; demonstrate; exhibit; explain; lay down; prove 638
639
DARLEHN (*n*): loan 640
DARM (*m*): intestine; gut
DARMADSTRINGENS (*n*): intestinal astringent (Med.) 641
DARMBANDWURM (*m*): (see BANDWURM) 642
DARMBAUCHBRUCH (*m*): gastrocele (Med.) 643
DARMBEIN (*n*): haunch bone; ilium. (Med.) 644

DARMBEINMUSKEL (*m*): iliac muscle 645
DARMBEWEGUNG (*f*): peristalsis, peristaltic contraction, worm-like motion of the intestine (Med.) 646
DARM (*m*), **BLINDER-**: caecum (Med.) 647
DARMBRUCH (*m*): enterocele (Med.) 648
DARM (*m*), **DICKER-**: colon (Med.) 649
DARMDRÜSE (*f*): intestinal gland, Peyer's patch (formed by a number of intestinal glands) (Med.) 650
DARM (*m*), **DÜNNER-**: small intestine (Med.) 651
DARMENTZÜNDUNG (*f*): enteritis, inflammation of the intestine (Med.) 652
DARMFÄULE (*f*): dysentery (Med.) 653
DARMFELL (*n*): peritoneum (Med.) 654
DARMFETT (*n*): a low grade fat from the intestines of pigs 655
DARM (*m*), **GERADER-**: rectum (Med.) 656
DARMGICHT (*f*): iliac passion (Med.) 657
DARMGICHTIG: iliac (Med.) 658
DARMHAUT (*f*): peritoneum (Med.) 659
DARMKRANKHEIT (*f*): intestinal disease (Med.) 660
DARM (*m*), **LANGER-**: ileum (Med.), (lowest part of small intestine, is preceded by jejunum and followed by caecum) 661
DARM (*m*), **LEERER-**: jejunum (Med.) 662
DARMPERISTALTIK (*f*): intestinal peristalsis (the worm-like movement of the intestine) 663
DARMSAFT (*m*): intestinal juice; gastric juice 664
DARMSAITE (*f*): catgut 665
DARMSCHWANZ (*f*): vermiform appendix (blind process from the caecum) (Med.) 666
DARMSTEIN (*m*): enterolith (an intestinal concretion) (Med.) 667
DARMTIERE (*f.pl*): matazoa (organisms of the intestines) (Med.) 668
DARMVERSCHLIESSUNG (*f*): constriction (stoppage) of the bowel, constipation, intestinal obstruction (Med.) 669
DARMWAND (*f*): wall of the intestine 670
DARRARBEIT (*f*): kiln drying; liquation (Metal) 671
DARRBODEN (*m*): drying floor; kiln floor 672
DARRBRETT (*n*): drying board 673
DARRE (*f*): drying; kiln-drying; drying room; malt kiln (Brewing); liquation hearth (Metal); consumption; phthisis (Med.) 674
DARREN: to dry; kiln-dry; kiln; torrefy; smelt; liquate (Metal) 675
DARREN (*n*): torrefaction, liquation (Metal); (kiln)-drying 676
DARRFAX (*m*): kilnman (Brewing) 677
DARRFLÄCHE (*f*): drying surface 678
DARRGEKRÄTZ (*n*): slag (Metal) 679
DARRKAMMER (*f*): drying (room) chamber 680
DARRKUPFER (*n*): liquated (smelted) copper 681
DARRMALZ (*n*): cured malt; kiln dried malt 682
DARROFEN (*m*): drying (kiln) oven; liquation hearth (Metal) 683
DARRRAUM (*m*): drying (curing) chamber of a malt kiln 684
DARRSCHRANK (*m*): drying oven; drying cupboard 685
DARRSTAUB (*m*): (malt)-dust 686
DARSTELLEN: to prepare; produce; amount to; make; manufacture; show; exhibit; display; illustrate; describe; represent; construct; depict; illustrate 687
DARSTELLEND: descriptive, illustrative representative 688
DARSTELLUNG (*f*): statement; preparation, etc. (see **DARSTELLEN**) 689
DARSTELLUNGSWEISE (*f*)): method of (preparation; representation; construction) production, etc. 690
DARTUN: to prove; verify; show 691
DASEIN (*n*): presence; being; existence 692
DASYMETER (*n*): gas meter (see **GASWAGE**) 693
DATEN (*pl.* of **DATUM**): dates (Chronology); data, particulars 694
DATIEREN: to date 695
DATOLITH (*m*): datolite (a basic calcium-boron orthosilicate); $Ca(B.OH)SiO_4$; Sg. 2.9 [696
DATTEL (*f*): date (fruit) 697
DATTELBAUM (*m*): date palm (*Phœnix dactylifera*) 698
DATTELPALMENHOLZ (*n*): date-palm wood (see **DATTLEBAUM**) 699
DATTELPFLAUME (*f*): date plum (*Diospyros kaki* and *D. lotus*); persimmon (*Diospyros virginiana* and *D. texana*) 700
DATUM (*n*): date; (*pl.* **DATEN**) 701
DATURAÖL (*n*): datura oil (from the seeds of *Datura strammonium*) 702
DATURIN (*n*): atropine, daturin (see **ATROPIN**) 703
DATURINSÄURE (*f*): daturic acid; $C_{17}H_{34}O_2$ [704
DAUER (*f*): duration; permanence; durability; continuance; lasting; keeping; enduring; length 705
DAUERBEANSPRUCHUNG (*f*): continuous load; constant stress-(ing) 706
DAUERBETRIEB (*m*): continuous (load) working 707
DAUERBIER (*n*): lager beer 708
DAUERBRUCH (*m*): lasting (complete)-rupture (due to weakening or tiring of the material, caused by repeated local stresses in steel overstepping the elastic limit) 709
DAUERENTLADUNG (*f*): continuous (evacuation) discharge 710
DAUERFARBE (*f*): permanent colour 711
DAUERGEWÄCHS (*n*): perennial-(plant) 712
DAUERHAFT: durable; solid; permanent, etc.; fast (of colours); tough (of leather) 713
DAUERHAFTIGKEIT (*f*): durability, solidity, etc. (see **DAUERN**) 714
DAUERLEISTUNG (*f*): continuous (load) output; duty (Machinery) 715
DAUERMAGNET (*m*): permanent magnet 716
DAUERND: permanent-(ly) 717
DAUERPAPIER (*n*): ready sensitized paper (Photography) 718
DAUERPRÄPARAT (*n*): permanent preparation [719
DAUERSCHLAGPROBE (*f*): continuous impact (shock) test; continuous impact (shock) test pieces (Steel) 720
DAUERSCHLAGWERK (*n*): repeated impact testing machine 721
DAUERSPORE (*f*): resting spore; inactive spore 722
DAUERWÄSCHE (*f*): any linen articles of attire made washable by an insoluble coating 723
DAUM (*m*): (see **DAUMEN**) 724
DAUMEN (*m*): thumb; cam (Mechanical); inch 725
DAUMEN, AUSLASS- (*m*): exhaust cam 726
DAUMENRAD (*m*): cam wheel 727
DAUMENSCHEIBE (*f*): cam disc 728
DAUMENSCHRAUBE (*f*): thumb screw 729
DAUMENWELLE (*f*): cam shaft 730
DÄUMLING (*m*): small cam, knob; stall (for the thumb) 731

DAVID (*m*): davit (Naut.) 732
DAVIT (*m*): davit (Naut.) 733
DAVIT (*m*), **ANKER-**: anchor davit 734
DAVITHALTER (*m*): davit clamp 735
DAVITSPUR (*f*): davit step 736
DAZU: further, furthermore, moreover, in addition, over and above this 737
DAZUMAL: then; at that time 738
DEBATTE (*f*): debate 739
DECANAL (*n*): (see **DECYLALDEHYD**) 740
DECARBONISIEREN: to decarbonize 741
DECHLORISIEREN: to dechlorize; (*n*) dechlorizing (process of rendering inactive the free chlorine from bleaching processes, by the employment of reducing agents) 742
DECHSEL (*m*): adze 743
DECHSELN: to adze 744
DECK (*n*): deck 745
DECKBALKEN (*m*): deck beam 746
DECKBALKENUNTERSCHLAG (*m*): central (deck-beam) stringer 747
DECKBALKENWEGER (*m*): deck-beam clamp 748
DECKBAND (*n*): deck hook 749
DECKBELASTUNG (*f*): deck (load) weight 750
DECKBLATT (*n*): wrapper (of cigars); bract (Bot.) 751
DECKE (*f*): cover; cap; integument; coat; roof; top; ceiling; blanket; coverlet; cleansing; purging (Sugar) (also liquor) 752
DECKEL (*m*): cover; lid; top; cap 753
DECKELABFALL (*m*): carding waste (Textile) 754
DECKELDOHLE (*f*): box culvert 755
DECKELN: to (provide with) lid or cover; cover; top (Paper); cleanse; wash; purge; protect; cover (Colours) 756
DECKEL (*m*), **PUMPEN-**: pump cover 757
DECKELRIEMEN (*m*): deckle (Paper) 758
DECKEN: to cover; screen; provide for; cleanse; wash; (*n*): covering; cleansing; washing; covering power; opacity (of paint) 759
DECKENSTRICH (*m*): covering coat; top coat; daub (Leather) 760
DECKENVENTILATOR (*m*): ceiling fan 761
DECKER (*m*): coverer; wrapper 762
DECKFÄHIG: having a good covering quality; opaque 763
DECKFARBE (*f*): body colour; second coat, covering colour (which allows the ground to show through); opaque colour 764
DECKFENSTER (*n*): skylight 765
DECKFIRNIS (*m*): covering varnish (Etching); protecting varnish 766
DECKGLAS (*n*): glass cover 767
DECKGLÄSCHEN (*n*): glass-cover-plate; covering glass plate (Microscopes) 768
DECKGRÜN (*n*): opaque (covering) green, chrome gr en (see **CHROMOXYD**) 769
DECKHAUS (*n*): deck house 770
DECKHAUT (*f*): covering skin, integument, exterior tegument 771
DECKKLÄRE (*f*): clairce (liquor for cleaning sugar crystals) 772
DECKKLÄRSEL (*n*): (see **DECKKLÄRE**) 773
DECKKRAFT (*f*): covering power; protective power; body (of colours) 774
DECKLACK (*m*): deck varnish; covering varnish 775
DECKLADUNG (*f*): deck cargo, deck load 776
DECK (*n*), **LOSES-**: preventer deck 777
DECKMITTEL (*n*): covering material 778
DECKOFFIZIER (*m*): warrant officer (Naut.) 779
DECKPANZER (*m*): deck armour 780
DECKPAPP (*m*): (see **DECKPAPPE**) 781
DECKPAPPE (*f*): resist paste 782
DECKPLATTE (*f*): covering (roofing) slabs, slates, tiles or flags; rider plate, deck plate 783
DECKSAUFBAU (*m*): erection (structure) on deck, deck erection 784
DECKSBEPLANKUNG (*f*): deck planking 785
DECKSBEPLATTUNG (*f*): deck plating 786
DECKSBOLZEN (*m*): deck bolt 787
DECKSEL (*n*): (see **DECKKLÄRE**) 788
DECKSHÖHE (*f*): height between decks 789
DECKSIRUP (*m*): syrup (obtained by cleansing sugar crystals with clairce) (see **DECKKLÄRE**) 790
DECKSLAST (*f*): deck cargo, deck load 791
DECKSNAGEL (*m*): deck nail 792
DECKSNAHT (*f*): deck seam 793
DECKSPLANKE (*f*): deck plank 794
DECKSPLATTE (*f*): deck plate 795
DECKSSCHRAUBE (*f*): deck screw 796
DECKSTOPFEN (*m*): flanged stopper 797
DECKSTRAAK (*m*): sheer of a deck 798
DECKSTRINGER (*m*): deck stringer 809
DECKSTÜTZE (*f*): deck pillar 800
DECKUNG (*f*): covering; cover; remittance; security; congruence (Geom.); density (Phot.); body (Colours) 801
DECKWEISS (*n*): (see **LITHOPON**) 802
DECKZIEGEL (*m*): cover (covering; roofing) tile; coping tile 803
DECOCT (*n*): decoction 804
DECOLORIMETER (*n*): decolorimeter (an apparatus) for measuring the decolourizing power of bone ash) (Sugar) 805
DECREPITIEREN: to decrepitate 806
DECREUSAGE (*f*): determination of gum (Silk) 807
DECROLIN (*n*): decrolin 808
DECYLALDEHYD (*n*): decylic (caprinic, decoic) aldehyde, decanal; $CH_3.(CH_2)_8.CHO$; Sg. 0.828; Bp. 208°C. 809
DECYL-BETA-NAPHTHOCINCHONINSÄURE (*f*): decyl-beta-naphthocinchoninic acid (from beta-naphthylamine plus pyroracemic acid); Mp. 237°C. (see **NAPHTYLAMIN** and **BRENZ-TRAUBENSÄURE**) 810
DECYLSÄURE (*f*): decylic acid; caprinic acid; decoic acid (see **CAPRINSÄURE**) 811
DEDUKTIV: deductive 812
DEFEKT (*m*): defect; (as an adjective, defective) 813
DEFIBRINIERT: defibrinated 814
DEFINIEREN: to define 815
DEFINIERT: definite, defined 816
DEFIZIT (*n*): deficit 817
DEFLAGRIEREN: to deflagrate 818
DEFORMATION (*f*): deformation (Metal) 819
DEFORMIEREN: to deform 820
DEGALLIEREN: to degall 821
DEGENERATION (*f*): degeneration (Med.) 822
DEGOMMIEREN: to de-gum 823
DEGORGIEREN: to wash out; cleanse (from slime) (Wines) 824
DEGRAISSIEREN: to de-oil; remove fat, oil or grease; scour 825
DEGRAS (*n*): degras, sod oil, moellon, [the (waste) fat obtained by tanning chamois leather, and which consists of an oxydated oil] (Leather-dressing) 826
DEGRASSIEREN: (see **DEGRAISSIEREN**) 827
DEGRAS (*n*), **WEISSGERBER-**: sod oil, degras, Sg. 0.92 – 0.95; Mp. 18 – 30°C. 828

DEHNBAR: extensible; malleable; ductile; stretchable; expansible (of gases); elastic; distendable; dilatable; tensile; flexible 829
DEHNBARKEIT (*f*): ductility 830
DEHNEN: to draw out, elongate 831
DEHNUNG (*f*): elongation (Metals); expansion (Gases) 832
DEHYDRACETSÄURE (*f*): dehydracetic acid; $C_8H_8O_4$ 833
DEHYDRATION (*f*): dehydration 834
DEHYDROTHIO-(M)-XYLIDIN (*n*): dehydrothio-(meta)-xylidine, aminotoluenyl-ortho-amino-thioxylenol; $C_6H_2(CH_3)_2NSCC_6H_3(CH_3)NH_2$; Mp. 107°C.; Bp. 283°C. 835
DEHYDROTHIO-P-TOLUIDIN (*n*): dehydrothio-para-toluidine, amino-benzenyl-ortho-aminothiocresol; $C_{14}H_{12}N_2S.3H_2S$; Mp. 191°C.; Bp. 434°C. 836
DEHYDROTHIO-P-TOLUIDINSULFOSÄURE (*f*): dehydrothio-para-toluidinesulfonic acid; $C_{14}H_9NS(NH_2)SO_3H$ 837
DEICH (*m*): dike; bank; dam 838
DEICHEN: to dam; make a dam or dike 839
DEICHSEL (*f*): pole; shaft 840
DEIL (*m*): bloom (Metal); lump; cake 841
DEISSEL (*m*): adze 842
DEISSELN: to adze 843
DEKADISCH: decadic 844
DEKADISCHE LOGARITHMEN: common (logs) logarithms 845
DEKAHYDRAT (*n*): decahydrate ($10H_2O$) 846
DEKALIN (*n*): fully hydrogenated decahydronaphthalin (used as a substitute for turpentine oil) 847
DEKANTIERAPPARAT (*m*): decanting (clarifying, settling) apparatus 848
DEKANTIEREN: to decant; (*n*) decanting 849
DEKANTIERGEFÄSS (*n*): decanting (clearing) vessel 850
DEKANTIERGLAS (*n*): decanting (jar) glass 851
DEKANTIERTOPF (*m*): decanting jar (of earthenware) 852
DEKANTIERZYLINDER (*m*): decanting cylinder 853
DEKAPIEREN (*n*): dipping, pickling (treatment before electrolytic treatment (Metal), to remove oxydation, by dipping in diluted hydrochloric acid; consists of corroding, washing, scouring and de-greasing) 854
DEKARBONISIEREN: to decarbonize 855
DEKATIEREN: to hot-press (steam) cloth; shrink (Textile) 856
DEKATIERMASCHINE (*f*): hot-pressing (shrinking) machine (Textile) 857
DEKAUSTIZIEREN: to decausticize 858
DEKLARIEREN: to declare, give out, let it be understood-(that) 859
DEKLINATIONSKOMPASS (*m*): declining compass 860
DEKLINATIONSNADEL (*f*): declining needle 861
DEKLINIEREN: to decline; deviate 862
DEKOKT (*n*): decoction (*Decoctum*) (aqueous extract of vegetable substances) 863
DEKOKTPRESSE (*f*): decoction press 864
DEKOR (*n*): decoration (Ceram.) (see **MUFFELDEKOR** and **SCHARFFEUERDEKOR**) 865
DEKORATION (*f*): decoration 866
DEKORATIONSMALEREI (*f*): decorative painting 867
DEKORIEREN: to decorate; paint (Ceramics) 868
DEKREPITIEREN: to decrepitate 869
DEKRETIEREN: to decree 870
DEKROLIN (*n*): decrolin (a cleavage medium) 871
DELCREDERE (*n*): commission of guarantee, delcredere, security (Com.) 872
DELPHIN (*m*): dolphin (*Delphinus delphis*) 873
DELPHINBLAU (*n*): dolphin blue 874
DELPHININ (*n*): delphinine (from seeds of *Delphinium staphisagria*); $C_{22}H_{35}NO_6$; Mp. 119°C. 875
DELPHINTRAN (*m*): dolphin oil; porpoise oil (from *Delphinus globiceps*); Sg. 0.927 876
DELTAMETALL (*n*): delta metal; (a copper alloy) (when cast has 55.94% Cu; 41.61% Zn; 0.72% Pb; 0.87% Fe; 0.81% Mn; 0.013% P.) 877
DELTAPAPIER (*n*): gelatino-chloride paper 878
DELTA SÄURE (*f*): (see **F SÄURE**) 879
DEMANT (*m*): diamond; adamant; Sg. 3.5 880
DEMANTBLENDE (*f*): eulytite; eulytine 881
DEMANTBLENDE (*f*), **DODEKAEDRISCHE-**: eulytine, bismuth blende; Sg. 5.97 882
DEMANTSPAT (*m*): adamantine spar; corundum; Al_2O_3 883
DEMARGARINIERUNGSPROZESS (*m*): demargarinizing process 884
DEMENTSPRECHEND: correspondingly; accordingly; in conformity with; commensurate with 885
DEMGEGENÜBER: as an off-set to (this) that, on the contrary 886
DEMGEMÄSS: accordingly; therefore 887
DEMNACH: accordingly; therefore 888
DEMNÄCHST: shortly; soon-(after); thereafter; next to that 889
DEMONSTRATION (*f*): demonstration 890
DEMONSTRIEREN: to demonstrate 891
DEMONTIEREN: to dismantle, demount, take to pieces, take down 892
DEMPGORDING (*f*), **BESAN-**: spanker brail (Naut.) 893
DEMZUFOLGE: consequently; in consequence; therefore; on that account; due to that 894
DENATURIEREN: to denature; denaturize (see **VERGÄLLEN**) 895
DENATURIERUNG (*f*): denaturing, denaturizing, denaturization 896
DENATURISIEREN: (see **DENATURIEREN**) 897
DENDRACHAT (*m*): dendritic agate 898
DENDRITENACHAT (*m*): dendritic agate 899
DENDRITENSTRUKTUR (*f*): dendritic structure (Metal) (fern-like arrangement of Perlite in steel) 900
DENGUE (*n*): dengue, three-day fever, break-bone fever, dandy fever (a West Indian contagious disease) (Med.) 901
DENITRIEREN: to denitrate 902
DENITRIFIZIEREN: to denitrify 903
DENKART (*f*): way (method) of thinking; mind 904
DENKEN: to think, conceive, imagine, believe, suppose, remember, reason, muse, concentrate 905
DENKKRAFT (*f*): thinking (intellectual) power, concentration 906
DENKLEHRE (*f*): logic 907
DENSIMETER (*n*): densimeter, hydrometer (an apparatus for measuring specific gravities) 908
DENSIMETRISCH: densimetric 909
DENSOGRAPH (*m*): densograph (Photography) 910
DEPESCHE (*f*): dispatch; message 911
DEPHLEGMATION (*f*): dephlegmation (Rectification of Spirits) 912
DEPHLEGMATOR (*m*): dephlegmator (Rectification of Spirits) 913

DEPLACEMENT (*n*): displacement 914
DEPLACEMENTSBERECHNUNG (*f*): computation of displacement 915
DEPLACEMENTSKURVE (*f*): displacement curve 916
DEPLACEMENTSSCHWERPUNKT (*m*): centre of (buoyancy) displacement 917
DEPLACEMENTSSCHWERPUNKTSKURVE (*f*): centre of buoyancy curve 918
DEPLACEMENTSSKALA (*f*): scale of displacement 919
DEPLACIEREN: to displace 920
DEPLACIERUNG (*f*): displacement 921
DEPOLARISATION (*f*): depolarization 922
DEPOLARISATOR (*m*): depolarizer 923
DEPOLARISIEREN: to depolarize 924
DEPOLYMERISIERUNG (*f*): de-polymerization 925
DEPONIEREN: to dispose 926
DEPRESSION (*f*): (see GEFRIERPUNKTSERNIE-DRIGUNG); depression 927
DEPRIMIEREN: to depress 928
DEPUTAT (*n*): compensation, allowance 929
DERB: solid; compact; firm; strong; dense; hard; stout; rough; rude, massive (Mineral); coarse; sturdy; blunt 930
DERBGEHALT (*m*): solid (cubic) contents 931
DERBHEIT (*f*): solidity, etc. (see DERB) 932
DERBHOLZ (*n*): close-grained wood 933
DERBYROT (*n*): Derby red (see CHROMROT) 934
DERICIN (*n*): dericine (trade name of a product obtained from Castor oil) 935
DERICINÖL (*n*): dericine oil 936
DERIVAT (*n*): derivative 937
DERIVIEREN: to derive, be derived 938
DERMATISCH: cutaneous 939
DERMATOL (*n*): dermatol, bismuth subgallate, bassic bismuth gallate (*Bismuthum subgallicum*); $C_6H_2(OH)_3.CO_2Bi(OH)_2$ 940
DERMATOLOGIE (*f*): dermatology 941
DERRICKKRAN (*m*): derrick 942
DERZEITIG: present; of the time being 943
DESAGGREGATOR (*m*): disintegrator 944
DESAGREGATION (*f*): disaggregation; disintegration 945
DESAMIDIEREN: to deaminize; deamidize; remove amidogen 946
DESCLOIZIT (*m*): descloizite; $(PbOH)VO_4$; Sg. 6.0 947
DESINFEKTION (*f*): disinfection; sterilization 948
DESINFEKTIONSAPPARAT (*m*): disinfecting apparatus 949
DESINFEKTIONSKRAFT (*f*): disinfecting (disinfective, disinfectant) power 950
DESINFEKTIONSLAUGE (*f*): disinfecting liquid 951
DESINFEKTIONSMITTEL (*n*): disinfectant 952
DESINFEKTIONSPULVER (*n*): disinfecting powder, carbolic powder 953
DESINFEKTIONSSEIFE (*f*): disinfectant soap 954
DESINFEKTIONSWIRKUNG (*f*): disinfectant action, disinfective action 955
DESINFIZIENS (*n*): disinfectant 956
DESINFIZIEREN: to disinfect; deodorize; sterilize 957
DESINFIZIEREND: disinfecting, sterilizing 958
DESINTEGRATOR (*m*): disintegrator 959
DESINTEGRIEREN: to disintegrate 960
DESMIN (*m*): desmine, stilbite (see also STILBIT); $(Na_2,Ca)Al_2Si_6O_{16}.6H_2O$); Sg. 2.15 961
DESMOTROPIE (*f*): desmotropy; desmotropism 962
DESODORATIONSMITTEL (*n*): deodorizer 963
DESODORIEREN: to deodorize 964
DESODORISIEREN: to deodorize 965
DESOXYDATION: de-oxydation, extraction of oxygen 966
DESOXYDATIONSMITTEL (*n*): deoxydizing agent; deoxydating agent 967
DESOXYDIEREN: to deoxidize; deoxydate 968
DESSENUNGEACHTET: nevertheless 969
DESSIN (*m* and *n*): design; pattern 970
DESTILLAT (*n*): distillate 971
DESTILLATION, FRAKTIONIERTE- (*f*): fractional distillation 972
DESTILLATIONSANLAGE (*f*): distilling (distillation) plant 973
DESTILLATIONSGAS (*n*): distillation gas (from the distillation of brown coal tar: 28% CH_4; 32% C_2H_4; 5% H; 2% CO; 3% CO_2; 3% H_2S; 7% heavy hydrocarbons; 3.5% O, and other constituents; calorific value 7,000–9,000 calories) 974
DESTILLATIONSGEFÄSS (*n*): distilling vessel; still; alembic; distilling retort 975
DESTILLATIONSPRODUKT (*n*): distillation product, product of distillation 976
DESTILLATIONSPROZESS (*m*): reduction process of zinc in muffle furnace 977
DESTILLATIONSROHR (*n*): distilling tube 978
DESTILLATMAISCHE (*f*): distillery mash 979
DESTILLERIE (*f*): distillery 980
DESTILLIERAPPARAT (*m*): distilling apparatus; still; alembic 981
DESTILLIERBLASE (*f*): distilling (vessel) retort; body of still; still; alembic 982
DESTILLIEREN: to distill 983
DESTILLIERGEFÄSS (*n*): (see DESTILLATIONS-GEFÄSS) 984
DESTILLIERHELM (*m*): distilling head (of a still) 985
DESTILLIERHOLZKOHLE (*f*): distillery charcoal 986
DESTILLIERKOLBEN (*m*): distilling (flask) retort 987
DESTILLIERKOLONNE (*f*): distilling column (a series of stills) 988
DESTILLIERT: distilled 989
DESTILLIERTOPF (*m*): distilling pot 990
DETACHEUR (*m*): spot-cleaner (Dyeing) 991
DETACHIEREN: to remove (clean out) spots (Textile) 992
DETACHIERMITTEL (*n*): spot-cleaner (Dye); cleaning agent (for removing spots) 993
DETACHUR (*f*): spot-cleaning (Textile) 994
DETAIL-GESCHÄFT (*n*): retail business 995
DETAILKUNDSCHAFT (*f*): retail trade 996
DETAILREICH: full of detail, detailed; bringing out (showing up) the details 997
DETAILZEICHNUNG (*f*): detail drawing 998
DETEKTOR (*m*): detector 999
DETEKTOR (*m*), WELLEN-: wave detector 000
DETERMINANTE (*f*): determinant (Maths.) 001
DETRITUS (*m*): detritus, débris, rubbish, waste (matter), disintegrated matter 002
DEUL (*m*): bloom (Metal); lump; cake 003
DEUTEN: to point; explain; show; illustrate; make (clear; distinct; plain); evident; interpret 004
DEUTSCHE FOLIE (*f*): German foil (from a tin alloy) 005
DEUTSCHES GESCHIRR (*n*): stampers (Paper) 006
DEVALVATION (*f*): devaluation; deflation (stabilization of a deflated currency at a fraction of its previous value) 007

DEVALVATION (*f*), **EFFEKTIVE-** : issue of new gold coinage at a multiple of its previous value, for the purpose of stabilizing the deflation in currency 008
DEVALVATION (*f*), **IDEELE-** : the same as " **EFFEKTIVE DEVALVATION** " except that no coinage is issued. It however becomes a unit of calculation, thus operating and being accepted as if it were in circulation. The actual coinage is assumed as being issued later 009
DEVISE (*f*) : device ; motto ; bill (of exchange) 010
DEWARSCHE FLASCHE (*f*) : Dewar flask 011
DEWARSCHES GEFÄSS (*n*) : Dewar flask 012
DEXEL (*m*) : adze 013
DEXTRIN (*n*) : dextrin, British gum, starch gum, artificial (vegetable) gum, gommeline (obtained by roasting or heating starch) ; $C_{12}H_{20}O_{10}$ or $(C_6H_{10}O_5)_n$ 014
DEXTRONSÄURE (*f*) : dextronic (dextrogluconic) acid 015
DEXTROSE (*f*) : fructose, dextrose, glucose, dextroglucose, grape (corn) sugar ; (*Saccharum amylaceum*) : $C_6H_{12}O_6$; Sg. 1.555 ; Mp. 95°C. 016
DEZENTRALISATION (*f*) : de-centralization 017
DEZIGRAMM (*n*) : decigram 018
DEZIMAL : decimal 019
DEZIMALBRUCH (*m*) : decimal fraction 020
DEZIMALWAGE (*f*) : decimal balance 021
D-G-DIAGRAMM (*n*) (**DAMPFVERBRAUCH-GEGEN-DRUCK-DIAGRAMM**) : steam-consumption-back-pressure diagram 022
DI- : (see also under **BI-**) 023
DIABAS (*m*) : diabase (Mineralogy) ; (composition :—plagioclase feldspar, augite, magnetite and apatite with or without olivine) 024
DIABETESMITTEL (*n*) : remedy for diabetes 025
DIABETIKER (*m*) : one who suffers from diabetes 026
DIACETBERNSTEINSÄUREÄTHYLESTER (*m*) : diacetosuccinic ethylester, ethyl diacetossuccinate 027
DIACETYL (*n*) : diacetyl ; $C_4H_6O_2$; Sg. 0.9793 ; Bp. 88°C. 028
DIACETYLMORPHIN (*n*) : diacetylmorphine, heroin $C_{21}H_{23}O_5N$; Mp. 173°C. 029
DIACETYLTANNIN (*n*) : diacetyltannin, tannigen, acetyltannin ; $C_{18}H_{14}O_{11}$ 030
DIACHYLONSALBE (*f*) : diachylon ointment 031
DIAGNOSE (*f*) : diagnosis (Med.) 032
DIAGONAL : diagonal 033
DIAGONALBAND (*n*) : diagonal tie 034
DIAGONALBOOT (*n*) : diagonal-built boat 035
DIAGONALE (*f*) : diagonal 036
DIAGONALSCHIENE (*f*) : diagonal rider, diagonal tie plate 037
DIAGONALSCHNEIDER (*m*) : diagonal cutter 038
DIAGRAMM (*n*) : diagram 039
DIAKTINISCH : diactinic (permitting actinic or chemical rays of light only to pass) 040
DIAL (*n*) : (see **DIALLYLBARBITURSÄURE**) 041
DIALKYLAMIN (*n*) : dialkylamine 042
DIALLAG (*m*) : diallage ; (Ca, Mg, Fe)O,SiO_2 ; Sg. 3.25 043
DIALLYLBARBITURSÄURE (*f*) : diallylbarbituric acid ; $(C_3H_5)_2C(CO.HN)_2CO$; Mp. 170.5°C. 044
DIALYSABEL : dialysable 045
DIALYSATOR (*m*) : dialyser 046
DIALYSE (*f*) : dialysis (the separation of colloids from crystalloids, or the diffusion of matter in solution through membranes) 047
DIAMAGNETISCH : diamagnetic 048
DIAMAGNETISCHER KÖRPER (*m*) : diamagnetic (non-magnetic) body or substance (a substance which is rejected by a magnet) 049
DIAMAGNETISMUS (*m*) : diamagnetism 050
DIAMANT (*m*) : diamond (pure carbon) ; (C) ; Sg. 3.5 051
DIAMANTARTIG : adamantine ; like diamond 052
DIAMANTBLAU (*n*) : diamond blue 053
DIAMANTGLANZ (*m*) : adamantine lustre 054
DIAMANTIN (*n*) : (see **ALUNDUM**) 055
DIAMANTITE (*f*) : (see **ALUNDUM**) 056
DIAMANTMÖRSER (*m*) : diamond mortar ; steel mortar 057
DIAMANTPLATTE (*f*) : diamond plate 058
DIAMANTSCHWARZ (*n*) : diamond black 059
DIAMANTSPAT (*m*) : adamantine spar ; corundum ; Al_2O_3 ; Sg. 4.0 ; Mp. 1750-1800°C. 060
DIAMETRAL ENTGEGENGESETZT : diametrically opposed 061
DIAMIDOBENZOL (*n*) : diaminobenzene ; phenylene-diamine ; $C_6H_4(NH_2)_2$; (see **PHENYLENDIAMIN**) 062
DIAMIDOPHENOL (*n*) : diaminophenol ; $[C_6H_3OH(NH_2)_2]$; ortho., Mp. 49°C. ; Bp. 299°C. ; para., Mp. 53°C. ; Bp. 302°C. 063
DIAMIDOTOLUOL (*n*) : diaminotoluene ; $C_6H_3(CH_3)(NH_2)_2$; Mp. 99°C. ; Bp. 280°C. 064
DIAMINOOXYDIPHENYLPHENOLAT (*n*) : diamino-oxydiphenyl phenolate ; $NaO.NH_2.C_6H_3.C_6H_4.NH_2$ 065
DIAMINOPHENOLCHLORHYDRAT (*n*) : diaminophenol hydrochloride, amidol (a photographic [developer] chemical) ; $C_6H_3(NH_2)_2OH.2HCl$; Sg. 1.223 066
DIAMINOPHENOLHYDROCHLORID (*n*) : (see **DIAMINOPHENOLCHLORHYDRAT**) 067
DIAMINORESORZIN (*n*) : diaminoresorcinol ; $C_6H_2.OH.OH.NH_2.HCl.NH_2.HCl$ 068
DIAMINOVALERIANSÄURE (*f*) : (see **ORNITHIN**) 069
DIAMINSCHWARZ (*n*) : diamine black 070
DIAMOL (*n*) : diamol, diaminophenol hydrochloride (a photographic developer) ; (ee **DIAMINOPHENOLCHLORHYDRAT**) 071
DIAMYLAMIN (*n*) : diamyl amine 072
DIANENBAUM (*m*) : silver tree (*Arbor dianae*) 073
DIANILBLAU (*n*) : dianile blue 074
DIANISIDIN (*n*) : dianisidine ; $[C_6H_3(OCH_3)NH_2]_2$; 1 ; 3 ; 4 ; Mp. 137°C. 075
DIANISIDINCHLORHYDRAT (*n*) : dianisidine hydrochloride 076
DIANISIDINSULFAT (*n*) : dianisidine sulphate 077
DIANOL (*n*) : dianol (glycerine lactate) ; $C_3H_5OH_2-O.C_3H_6O_3$ 078
DIANYLROT (*n*) : dianile red 079
DIAPHAN : diaphanous ; transparent 080
DIAPHANASKOP (*n*) : diaphanoscope (lanternoscope, alethoscope, pantoscope, are similar instruments for looking at a photograph through a single lens) 081
DIAPHRAGMA (*n*) : diaphragm ; throttle diaphragm (of steam meter) 082
DIAPHRAGMEN : plural of **DIAPHRAGMA** (which see) 083
DIAPOSITIV (*n*) : diapositive, transparency (Photography) 084

DIAPOSITIVPLATTE (*f*): diapositive plate (Photography) 085
DIARRHÖE (*f*): diarrhœa (Med.) 086
DIASPIRIN (*n*): diaspirin, succinylsalicylic acid; $C_{18}H_{14}O_8$; Mp. 179°C. 087
DIASPOR (*m*): diaspore; $Al(OH)_3.Al_2O_3$ or $AlO(OH)$; Sg. 3.37 088
DIASTAFOR (*n*): (trade name) (see DIASTASE) 089
DIASTASE (*f*): diastase (a ferment which converts starch into sugar; is extracted from malt) 090
DIASTATISCH: diastatic, separative (having properties of a diastase or pertaining to a diastase) 091
DIÄT (*f*): diet (food and also legislative assembly) 092
DIÄTETICUM (*n*): dietetic 093
DIÄTETISCH: dietary, diatetic-(ally) 094
DIÄTETISCHES MITTEL (*n*): dietetic 095
DIATHERMAN: diathermanous, transcalent, capable of transmitting heat, permeable to heat 096
DIÄTHYLAMIN (*n*): diethylamine; $C_4H_{11}N$; Sg. 0.7116; Mp. about −40°C.; Bp. 55.5°C. 097
DIÄTHYLAMINCHLORHYDRAT (*n*): diethylamine hydrochloride 098
DIÄTHYLANILIN (*n*): diethylaniline; $C_{10}H_{15}N$; Sg. 0.9344; Mp. −38.8°C.; Bp. 215.5°C. 099
DIÄTHYLBARBITURSÄURE (*f*): diethylbarbituric acid, veronal, barbital, malonurea, diethylmalonylurea (*Acidum diaethylbarbituricum*); $(C_2H_5)_2C(CONH)_2CO$; Mp. 182-191°C (see DIÄTHYLMALONYLHARNSTOFF) 100
DIÄTHYLDIPHENYLHARNSTOFF (*m*): diethyldiphenyl urea 101
DIÄTHYLENDIAMIN (*n*): diethylenediamine (see PIPERAZIN) 102
DIÄTHYLGLYKOKOLL-GUAJAKOL (*n*), SALZSAURES-(see GUJASANOL) 103
DIÄTHYLKETON (*n*): diethylketone; $C_5H_{10}O$; Sg. 0.8175; Bp. 102.7°C. 104
DIÄTHYLMALONYLHARNSTOFF (*m*): (*Urea diaethylmalonylica*): diethylmalonylurea (see DIÄTHYLBARBITURSÄURE) 105
DIÄTHYLOXAMID (*n*): diethyloxamide 106
DIÄTHYLSULFAT (*n*): diethylsulphate; $(OC_2H_5)_2SO_2$; Sg. 1.185; Bp. 208°C.; or $(C_2H_5)_2SO_4$; solidifies at −24.5°C.; (an oil distilled from sulfethylic acid, see ÄTHYLSCHWEFELSÄURE) 107
DIÄTHYLSULFONMETHYLÄTHYLMETHAN (*n*) (see METHYLSULFONAL) 108
DIÄTHYLTOLUIDIN (*n*): diethyltoluidine 109
DIATOMACEEN (*f.pl*): *Diatomaceæ* (a group of small aquatic plants of the single-celled *Algæ* species) 110
DIATOMIT (*n*): diatomite (a porous material made from KIESELGUR); Sg. 0.3 111
DIATONISCH: diatonic 112
DIAZANILSCHWARZ (*n*): diazanile black 113
DIAZETYLAMINOAZOTOLUOL (*n*): diacetylaminoazotoluene, pellidol; $(CH_3)C_6H_4.N:N.C_6H_4(CH_3).N(COCH_3)_2$ 114
DIAZETYLDERIVAT (*n*): diacetyl derivate 115
DIAZETYLMORPHINHYDROCHLORID (*n*): diacetylmorphine hydrochloride (see HEROIN) 116
DIAZINBLAU (*n*): diazine blue 117
DIAZINSCHWARZ (*n*): diazine black 118
DIAZOBENZOL (*n*): diazobenzene 119
DIAZOBENZOLCHLORHYDRAT (*n*): diazobenzene hydrochloride 120
DIAZOBENZOLSULFOSÄURE (*f*): diazobenzenesulphonic acid 121
DIAZODRUCK (*m*): diazotype printing 122
DIAZOECHTFARBE (*f*): fast diazo (dye) colour 123
DIAZOESSIGESTER (*m*): diazoacetic ester; $C_4H_6O_2N_2$; Sg. 1.083; Mp. −22°C.; Bp. 84°C. 124
DIAZOESSIGSÄURE (*f*): diazoacetic acid 125
DIAZOFARBSTOFF (*m*): diazo dye 126
DIAZOIMID (*n*): diazoimide 127
DIAZOMETALLVERBINDUNG (*f*): diazo metal compound, diazotate 128
DIAZOMETHAN (*n*): diazomethane, azomethylene, CH_2N_2; Bp. −23.5°C.; solidifies at −145°C. 129
DIAZOSCHWARZ (*n*): diazo black 130
DIAZOTAT (*n*): diazotate 131
DIAZOTIERBAR: diazotisable 132
DIAZOTIEREN: to diazotize 133
DIAZOTIERUNG (*f*): diazotization 134
DIAZOVERBINDUNG (*f*): diazo compound 135
DIAZURIN (*n*): diazurine 136
DIBENZOYLIEREN: to have two benzol groups added, forming dibenzoyl 137
DIBENZYL (*n*): dibenzyl; $C_{14}H_{14}$; Sg. 0.9416; Mp. 52°C.; Bp. 284°C. 138
DIBROMANTHRACEN (*n*): dibromoanthracene; $C_6H_4C_2Br_2C_6H_4$; Mp. 221°C. 139
DIBROMBENZOL (*n*): dibromobenzene; $C_6H_4Br_2$; ortho.—Sg. 1.977; Mp. −1°C.; Bp. 223.8°C.; meta.—Sg. 1.955; Mp. 1.5°C.; Bp. 219.4°C.; para.—Sg. 1.8408; Mp. 89.3°C.; Bp. 219°C. 140
DIBROMBERNSTEINSÄURE (*f*): dibromosuccinic acid 141
DIBROMPROPYLDIÄTHYLBARBITURSÄURE (*f*): dibromopropyldiethylbarbituric acid (contains 41.6% bromine); $(C_2H_5)_2C.CO.NH.CO.N.CO.CH_2.CHBr.CH_2Br$; Mp. 125°C. 142
DIBROMSALICYLSÄURE (*f*): dibromosalicylic acid 143
DICALCIUMMETAPHOSPHAT (*n*): dicalcium metaphosphate; $Ca_2(PO_3)_4.4H_2O$ 144
DICALCIUMPHOSPHAT (*n*): dibasic calcium phosphate; dicalcium orthophosphate; secondary calcium phosphate; bicalcic phosphate; $CaHPO_4.2H_2O$; Sg. 2.306 145
DICALCIUMSILIKAT (*n*): dicalcium silicate; $2CaO.SiO_2$ 146
DICARBONSÄURE (*f*): dicarboxylic acid 147
DICHININ (*n*): dichinine 148
DICHININ, KOHLENSAURES- (*n*): dichinine carbonate 149
DICHINOL (*n*): diquinol 150
DICHINOLIN (*n*): diquinoline 151
DICHINOYL (*n*): dichinoyl 152
DICHLORACETON (*n*): dichloroacetone; $C_3H_4OCl_2$; Sg. 1.3843; Mp. 43°C.; Bp. 173°C. 153
DICHLORANILIN (*n*): dichloroaniline 154
DICHLORÄTHAN (*n*): dichloroethane (see ÄTHYLENCHLORID) 155
DICHLORÄTHER (*m*): dichloroether, dichloroethyl oxide; $CH_2ClCHClOC_2H_5$; Sg. 1.175; Bp. 143°C. 156
DICHLORÄTHYLÄTHER (*m*): dichloroethyl ester, $CH_2.CHCl.O.CHCl.CHCH_3$ 157
DICHLORÄTHYLEN (*n*): ethylene dichloride (see ÄTHYLENCHLORID) 158
DICHLORBENZIDIN (*n*): dichlorobenzidine; $C_6H_3ClNH_2:C_6H_3Cl.NH_2$; Mp. 133°C. 159

DICHLORBENZOL (*n*): dichlorobenzene; $C_6H_4Cl_2$; ortho.—Sg. 1.3278; Bp. 179°C.; Mp. 14°C.; meta.—Sg. 1.307; Bp. 172°C.; Mp. −18°C.; para.—Sg. 1.2499; Mp. 52.7°C.; Bp. 174°C. 160

DICHLORESSIGSÄURE (*f*): dichloroacetic acid; bichloroacetic acid; $CHCl_2COOH$; Sg. 1.573; Mp. −4°C.; Bp. 189-194°C. 161

DICHLORHYDRIN (*n*): dichlorohydrin, glycerin dichlorohydrin, dichloroisoprophyl alcohol; $C_3H_6OCl_2$; Sg. 1.396; Bp. 174°C. 162

DICHLORMETHAN (*n*): dichloromethane; methylene (chloride) bichloride; CH_2Cl_2; Sg. 1.2615; Mp. −97°C.; Bp. 42°C. 163

DICHLORPHTALSÄURE (*f*): dichlorophthalic acid; $C_6H_2Cl_2(CO_2)_2$ 164

DICHROISMUS (*m*): dichroism (the variation in the colours of crystals, due to the light passing through them in different directions) 165

DICHROITISCH: dichroic 166

DICHROMSAUER: dichromate of; combined with dichromic acid 167

DICHROMSÄURE (*f*): dichromic acid 168

DICHT: tight; impervious; firm; dense; thick; compact; massive; solid 169

DICHTBRENNEN: to vitrify (Ceramics) 170

DICHTE (*f*): density; solidity; bulking quality (of paper); firmness; tightness; thickness; compactness; massiveness 171

DICHTE (*f*), **KRITISCHE-**: critical density 172

DICHTEMESSER (*m*): densimeter; hydrometer (of water) (see **DENSIMETER**) 173

DICHTEN: to tighten; make tight; pack; caulk; densify; condense; lute, seal; compress 174

DICHTESCHREIBER (*m*): recording densimeter 175

DICHTHALTEN (*n*): tightness 176

DICHTHAMMER (*m*): caulking hammer 177

DICHTHEIT (*f*): tightness; density (see **DICHT**) 178

DICHTIGKEIT (*f*): density, tightness (see **DICHT**) 179

DICHTIGKEITSMESSER (*m*): (see **DENSIMETER**) 180

DICHTMASCHINE (*f*): caulking machine 181

DICHTPOLEN: to toughen by poling (Metal) 182

DICHTUNG (*f*): packing; stuffing; joint; jointing; washer; poem; poetry; poesy; composition; invention; fable; fiction 183

DICHTUNGSMITTEL (*n*): packing; jointing; luting (caulking) material; packing material 184

DICHTUNGSPAPPE (*f*): packing cardboard 185

DICHTUNGSRING (*m*): packing ring, grummet, joint ring 186

DICHTUNGSSCHEIBE (*f*): packing disc; washer 187

DICHTUNGSSTOFF (*m*): packing material 188

DICHTWERG (*n*): oakum 189

DICK: thick; dense; inspissated; viscous; syrupy 190

DICKAUSZUG (*m*): inspissated extract 191

DICKDARM (*m*): colon; large intestine (Med.) 192

DICKE (*f*): thickness; bigness; bulkiness; fatness; swollenness; consistency; viscosity; dregs; mother 193

DICKEN: to thicken; inspissate; concentrate 194

DICKENMESSER (*m*): thickness gauge; calipers 195

DICKFARBE (*f*): thick colour; body-colour 196

DICKFLÜSSIG: viscous; thickly liquid; syrupy 197

DICKMAISCHE (*f*): thick mash (Brewing); decoction 198

DICKMILCH (*f*): curdled milk; curds 199

DICKÖL (*n*): thick (heavy) oil; oxydated (air resinified) turpentine oil (for porcelain painting); linseed oil or spike oil (Paint and Varnish manufacturing) 200

DICKPFANNE (*f*): concentration pan 201

DICKSAFT (*m*): thick juice; syrup (Sugar); latex 202

DICKSCHALIG: thick-shelled; husky (Grain) 203

DICKWANDIG: thick-walled 204

DICKZIRKEL (*m*): thickness gauge; calipers 205

DICYAN (*n*): dicyanogen; cyanogen gas; C_2N_2; Sg. 1.8064 compared with air; liquefies 21°C.; solidifies −34°C. 206

DICYCLOPENTADIEN (*n*): dicyclopentadien; $C_{10}H_{12}$; Mp. 32.5°C.; Bp. 170°C. 207

DIDYM (*n*): didymium: (Di); (no longer regarded as an element but divided up into neodymium and praseodymium) 208

DIDYM (*n*), **GOLDCHLORWASSERSTOFFSAURES-**: didymium chloraurate; $DiCl_3,AuCl_3+10H_2O$ 209

DIDYM (*n*), **SALIZYLSAURES-**: didymium salicylate; $Di_2(C_6H_4OH.CO_2)_6$; (a by-product from the manufacture of incandescent mantles) 210

DIDYMSALZ (*n*): didymium salt 211

DIELE (*f*): board; plank; (paving)-slab; batten; floor-board 212

DIELEKTRIKUM (*n*): dielectric (the medium separating the plates of a condenser) (Electric) 213

DIELEKTRISCH: dielectric 214

DIELEKTRIZITÄT (*f*): dielectricity 215

DIELEKTRIZITÄTSKONSTANTE (*f*): dielectric constant 216

DIELEN: to board, plank 217

DIENEN: to serve; act 218

DIENLICH: serviceable; fit; expedient 219

DIENST (*m*): service; duty; employment 220

DIENST (*m*), **TELEGRAPHEN-**: telegraphic service 221

DIESBEZÜGLICH: in this connection; relative (relating, referring, adverting) to this 222

DIETRICH (*m*): skeleton key 223

DIFFERENTIAL (*n*): differential 224

DIFFERENTIALBOGENLAMPE (*f*): differential arc lamp 225

DIFFERENTIALELEKTROMETER (*n*): differential electrometer 226

DIFFERENTIALFORMEL (*f*): differential formula 227

DIFFERENTIALGALVANOMETER (*n*): differential galvanometer 228

DIFFERENTIALGLEICHUNG (*f*): differential equation 229

DIFFERENTIAL-KOLBEN (*m*): differential piston 230

DIFFERENTIALRECHNUNG (*f*): differential calculus 231

DIFFERENTIALSCHALTUNG (*f*): differential system 232

DIFFERENTIALSPULE (*f*): differential coil 233

DIFFERENTIALWAGE (*f*): differential balance 234

DIFFERENTIIEREN: to differentiate 235

DIFFERENZ (*f*): difference 236

DIFFERENZIEREN: to differentiate 237

DIFFERENZZUGMESSER (*m*): differential draught gauge 238

DIFFERIEREND: differing 239

DIFFUNDIEREN: to diffuse 240

DIFFUS: diffuse 241
DIFFUSEUR (*m*): diffuser (Sugar, etc.) 242
DIFFUSION (*f*): diffusion 243
DIFFUSIONSABFALLWASSER (*n*): waste water from diffusion 244
DIFFUSIONSABWASSER (*n*): waste water from diffusion 245
DIFFUSIONSBATTERIE (*f*): diffusion (diffuser) battery (Sugar) 246
DIFFUSIONSBETRIEBSWASSER (*n*): working fluid for diffusion 247
DIFFUSIONSFÄHIG: diffusible 248
DIFFUSIONSGESCHWINDIGKEIT (*f*): speed of diffusion 249
DIFFUSIONSKOEFFIZIENT (*m*): coefficient of diffusion 250
DIFFUSIONSPRODUKT (*n*): diffusion product 251
DIFFUSIONSVERFAHREN (*n*): diffusion process 252
DIFFUSIONSVERMÖGEN (*n*): diffusibility 253
DIFFUSITÄT (*f*): diffusivity 254
DIFFUSOR (*m*): diffuser 255
DIFLUORDIPHENYL (*n*): difluorodiphenyl; $FC_6H_4.C_6H_4F$; Sg. 1.04; Mp. 87-89°C.; Bp. 254.5°C. 256
DIGALEN (*n*): digalene (*Digitoxinum solubile*), [a white amorphous product from digitalis leaves (*folia digitalis*)] 257
DIGALLUSSÄURE (*f*): digallic acid; tannic acid; gallotannic acid; $C_{10}H_{14}O_9$ 258
DIGERIEREN: to digest 259
DIGERIERFLASCHE (*f*): digestion (bottle) flask; digesting (bottle) flask 260
DIGESTIONSKOLBEN (*m*): digestion (pressure) flask 261
DIGESTIVSALZ (*n*): digestive salt; potassium chloride (see **KALIUMCHLORID**) 262
DIGESTOR (*m*): digestor, autoclave 263
DIGESTORIUM (*n*): hood 264
DIGIPAN (*n*): digipan (a digitalis preparation) 265
DIGITALIGENIN (*n*): digitaligenin; $C_{22}H_{30}O_3$ 266
DIGITALIN (*n*): digitalin (the active principle of *Digitalis purpurea*); $C_{35}H_{56}O_{14}$; Mp. 217°C. 267
DIGITALISBLÄTTER (*n.pl*): digitalis leaves (*folia digitalis*) (the leaves of *Digitalis purpurea*) (see also under **FINGERHUT**) 268
DIGITALISGLYKOSID (*n*): digitalis (glycoside) glucoside 269
DIGITALISSAMEN (*m* and *pl*): digitalis seed (seed of *Digitalis purpurea*) 270
DIGITALOSE (*f*): digitalose; $C_7H_{14}O_5$ 271
DIGITONIN (*n*): digitonine; $C_{54}H_{92}O_{28}$ 272
DIGITOXIGENIN (*n*): digitoxigenin; $C_{22}H_{32}O_4$; Mg. 230°C. 273
DIGITOXIN (*n*): digitoxin (a glycoside extracted from the leaves of *Digitalis purpurea*); $C_{34}H_{54}O_{11}$ 274
DIGITOXINSÄURE (*f*): digitoxic acid; $C_{34}H_{56}O_{12}$ 275
DIGITOXOSE (*f*): digitoxose; $C_6H_{12}O_4$ 276
DIHARNSTOFF (*m*): diurea; urazine 277
DIHETEROATOMIG: diheteratomic 278
DIHYDRAT (*n*): dihydrate ($2H_2O$) 279
DIHYDRATISCH: dihydric 280
DIISOBUTYLAMIN (*n*): diisobutylamine 281
DIISOPROPYLKETON (*n*): diisopropylketone; $C_7H_{14}O$; Sg. 0.8062; Bp. 123.7°C. 282
DIJODCARBAZOL (*n*): diiodocarbazole; $C_{12}H_7I_2N$ 283
DIJODOFORM (*n*): diiodoform, ethylene tetraiodide, iodoethylene; CI_2 (95% I); Mp. 187°C.; or $I_2CH.CHI_2$ 284
DIJODPARAPHENOLSULFAT (*n*): diiodoparaphenol sulphate 285
DIJODSALICYLSÄURE (*f*): diiodosalicylic acid; $C_6H_2I_2OHCOOH$; Mp. 220-230°C. 286
DIKALIUMPHOSPHAT (*n*): potassium hydrogen phosphate, dibasic potassium phosphate, dipotassium orthophosphate, potassium monophosphate; K_2HPO_4 287
DIKETOPIPERAZIN (*n*): (see **AMINOSÄUREHYDRID**) 288
DIKTIEREN: to dictate, cause 289
DILATOMETER (*n*): dilatometer (an apparatus, after the style of a thermometer, for measuring small changes of volume) 290
DILL (*m*): anethum, garden dill (*Anethum graveolens*) 291
DILLE (*f*): socket; nozzle 292
DILLÖL (*n*): dill oil; oil of dill; Sg. 0.905-0.915 293
DILLÖLSAMEN (*m*): dill oil seed (*Semen anethi*) (fruit of *Anethum graveolens*) 294
DILLSAMEN (*m*): (see **DILLÖLSAMEN**) 295
DILUVIUM (*n*): diluvium, drift (quarternary clay, gravel or sand of the glacial, interglacial or post-glacial formation) (Geol.) 296
DIMENSION (*f*): dimension 297
DIMENSIONIEREN: to dimension 298
DIMER: dimeric 299
DIMETHYLACETAL (*n*): (see **DIMETHYLAZETAL**) 300
DIMETHYL-ALPHA-NAPHTYLAMIN (*n*): dimethyl-alpha-naphthylamine; $C_{12}H_{13}N$; Bp. 273°C. 301
DIMETHYLAMIDOANTIPYRIN (*n*): dimethylamido antipyrin (see **PYRAMIDON**) 302
DIMETHYLAMIN (*n*): dimethylamine; C_2H_7N; Sg. 0.6865; Bp. 7.3°C. 303
DIMETHYLAMINCHLORHYDRAT (*n*): dimethylamine hydrochloride 304
DIMETHYLAMINOANTIPYRIN (*n*): dimethylaminoantipyrine (see **PYRAMIDON**) 305
DIMETHYLAMINOBENZALDEHYD (*n*): (para)-dimethylaminobenzaldehyde; $C_6H_4[N(CH_3)_2]CHO$ 1 : 4; Mp. 73°C. 306
DIMETHYLAMINOBENZOL (*n*): benzeneazodimethylaniline, dimethylaminoazobenzene, dimethylaminobenzol, butter yellow; $C_{14}H_{15}N_3$; Mp. 116°C. 307
DIMETHYLANILIN (*n*): dimethylaniline; $C_8H_{11}N$; Sg. 0.9555; Mp. 2.5°C.; Bp. 193.1°C. 308
DIMETHYLANTHRACEN (*n*): dimethylanthracene (from coal tar); $C_{16}H_{14}$; Mp. 224.5°C. 309
DIMETHYLARSENSÄURE (*f*): dimethylarsenic acid (see **KAKODYLSÄURE**) 310
DIMETHYLAZETAL (*n*): dimethylacetal, ethylidene-dimethylester; $CH_3CH(OCH_3)_2$; Sg. 0.859-0.879; Bp. 62°C. 311
DIMETHYLBENZOL (*n*): dimethylbenzene, xylol, xylene; $C_6H_4(CH_3)_2$; meta.—Sg. 0.867; Mp. −54°C.; Bp. 139°C.; ortho.—Sg. 0.876; Mp. −28°C.; Bp. 142°C.; para.—Sg. 0.862; Mp. 15°C.; Bp. 138°C. 312
DIMETHYLCARBINOL (*n*): (see **ISOPROPYLALKOHOL**) 313
DIMETHYLDIOXIM (*n*): (see **DIMETHYLGLYOXIM**) 314
DIMETHYLDIPHENYLHARNSTOFF (*m*): dimethyldiphenyl urea (see **ZENTRALIN**) 315

DIMETHYLGLYOXIM (*n*): (alpha-) dimethylglyoxime, butane dioxime; $(CH_3CNOH)_2$; Mp. 232-234.5°C. 316
DIMETHYLHYDROCHINON (*n*): dimethyl hydroquinone 317
DIMETHYLIERT: dimethylated; dimethyl- 318
DIMETHYLKARBINOL (*n*): dimethyl carbinol (see **PROPYLALKOHOL, ISO-**) 319
DIMETHYLKETON (*n*): dimethyl ketone, acetone (see **ACETON**) 320
DIMETHYLNAPHTHALIN (*n*): dimethyl naphthalene (from coal tar); $C_{12}H_{12}$; Sg. 1.008; Mp. -20°C.; Bp. 264°C.; alpha.—Sg. 1.0176; Bp. 263°C.; beta—Bp. 265°C. [321
DIMETHYLNAPHTHALINPIKRAT (*n*): dimethylnaphthalene picrate; Mp. 180°C.; alpha.—Mp. 141°C.; beta.—Mp. 118°C. 322
DIMETHYL-O-TOLUIDIN (*n*): dimethyl orthotoluidine; $C_6H_4.CH_3.N.(CH_3)_2$ 323
DIMETHYLOXAMID (*n*): dimethyl oxamide 324
DIMETHYLOXYCHINIZIN (*n*): dimethyloxyquinizine (see **ANTIPYRIN**) 325
DIMETHYLPHENOL (*n*): dimethylphenol. xylenol (from coal tar); $(CH_3)_2C_6H_3.OH$ (see **XYLENOL**) 326
DIMETHYLPHENYLPYRAZOLON (*n*): phenyl- dimethylpyrazole (see **ANTIPYRIN**) 327
DIMETHYLPHOSPHIN (*n*): dimethyl phosphine, secondary methyl phosphine, $PH(CH_3)_2$; Bp. 25°C. 328
DIMETHYLPHOSPHINSÄURE (*f*): dimethylphosphinic acid; $(CH_3)_2PO_2H$ 329
DIMETHYLPIPERAZIN (*n*), **WEINSAURES-**: (see **LYZETOL**) 330
DIMETHYLPYRIDIN (*n*): dimethylpyridine (see **LUTIDIN**) 331
DIMETHYLRESORCIN (*n*): dimethylresorcine 332
DIMETHYLSULFAT (*n*): dimethylsulphate; $C_2H_6O_4S$; Sg. 1.3276-1.3516; Mp. -10°C. Bp. 188-188.6°C. 333
DIMETHYLTOLUIDIN (*n*): dimethyltoluidine 334
DIMETHYLXANTHIN (*n*): dimethylxanthine (see **THEOBROMIN**) 335
DIMORPH: dimorphous 336
DIMORPHIE (*f*): dimorphism 337
DIMORPHIN (*m*): dimorphite; As_4S_3; Sg. 2.6 338
DIMORPHISCH: dimorphous 339
DIMORPHISMUS (*m*): dimorphism (existence in two different crystal forms) 340
DINAPHTYL (*n*): dinaphthyl 341
DINAPHTYLEN (*n*): dinaphthylene 342
DINAPHTYLIN (*n*): dinaphthyline 343
DINAS (*m*): dinas 344
DINASSTEIN (*m*): dinas brick (refractory brick made from quartz) 345
DINASTON (*m*): dinas clay 346
DINASZIEGEL (*m*): dinas brick 347
DINATON (*m*): dinas clay 348
DINATRIUMHYDROPHOSPHAT (*n*): sodium phosphate, dibasic; disodium phosphate, hydrodisodic phosphate; $Na_2HPO_4+12H_2O$; Sg. 1.52-1.545; Mp. 35°C. 349
DINATRIUMPHOSPHAT (*n*): dibasic sodium phosphate; disodium phosphate; disodium orthophosphate; hydrodisodic phosphate; $Na_2HPO_4.12H_2O$; Sg. 1.524; Mp. 35°C. 350
DINGEN: to hire; bargain; engage 351
DINICOTINSÄURE (*f*): dinicotinic acid 352
DINITRANILIN (*n*): dinitroaniline ; $C_6H_3NH_2(NO_2)_2$; Sg. 1.62; Mp. 187.7°C. 353
DINITRIEREN: to introduce two nitro groups, forming dinitro- 354
DINITROANISOL (*n*): dinitroanisol 355
DINITROBENZOL (*n*): dinitrobenzene; $(C_6H_4O_4N_2)$; ortho.—Sg. 1.59; Mp. 116.5°C.; Bp. 319°C.; meta.—Sg. 1.369; Mp. 89.7°C.; Bp. 302.8°C.; para.—Sg. 1.625; Mp. 172°C.; Bp. 299°C. 356
DINITROCHLORBENZOL (*n*): dinitrochlorobenzene (see **BINITROCHLORBENZOL**) 357
DINITRODIPHENYLAMIN (*n*): dinitrodiphenylamine 358
DINITROGLYCERINSPRENGSTOFF (*m*): dinitroglycerine explosive 359
DINITROGLYZERIN (*n*): dinitroglycerine (an explosive); $C_3H_5(OH)(ONO_2)_2$ 360
DINITROGLYZERINSPRENGSTOFF (*m*): dinitroglycerine (explosive) blasting charge 361
DINITRONAPHTALIN (*n*): dinitronaphthalene; $C_{10}H_6(NO_2)_2$; 1:5; Mp. 217°C. 362
DINITROPHENOL (*n*): dinitrophenol; $C_6H_3OH(NO_2)_2$; 2:4; Sg. 1.68; Mp. 114.5°C. [363
DINITROTOLUOL (*n*): dinitrotoluene; $C_6H_3CH_3(NO_2)_2$; Sg. 1.3208; Mp. 70.5°C. 364
DINKEL (*m*): spelt; German wheat 365
DIOGEN (*n*): diogen, sodium aminonaphthol-disulfonate (a photographic developer) 366
DIOGENAL (*n*): diogenal (see **DIBROMPROPYLDIÄTHYLBARBITURSÄURE**) 367
DIONIN (*n*): dionin, ethylmorphine hydrochloride; $C_{19}H_{23}NO_3HCl.2H_2O$; Mp. 124-125°C. 368
DIOPHANTISCHE GLEICHUNG (*f*): indeterminate equation 369
DIOPSID (*m*): diopside (natural calcium-magnesium silicate; $CaMgSi_2O_6$, with also a certain amount of calcium iron silicate; $CaFeSi_2O_6$); Sg. 3.3. 370
DIOPTAS (*m*): dioptas; H_2CuSiO_4; Sg. 3.3 371
DIORIT (*m*): diorite (granitoid rock) (mainly hornblende with a feldspar) (Mineralogy) 372
DIORSELLINSÄURE (*f*): diorsellinic acid 373
DIOXYANTHRACHINON (*n*): dioxyanthraquinone, chrysazine; $C_{14}H_6O_2(OH)_2$1:8; Mp. 191°C.; (see **KRAPPROT**); 1:2 (see **ALIZARIN**); 1:4 = quinazarin, chinizarin, quinizarin, dihydroxyanthraquinone; $C_{13}H_8O_5$; Mp. 194.5°C.; 1:5 = anthrarufine; $C_{14}H_6O_2(OH)_2$; Mp. 280°C. 374
DIOXYBENZOESÄURE (*f*): dioxybenzoic acid 375
DIOXYBENZOL (*n*): dioxybenzene; dihydroxybenzene (see **BRENZCATECHIN**; **RESORCIN** and **HYDROCHINON**) 376
DIOXYBENZOLHEXAMETHYLENTETRAMIN (*n*): dioxybenzenehexamethylenetetramine; $C_8H_{12}N_4.C_6H_6O_2$ 377
DIOXYBERNSTEINSÄURE (*f*): tartaric acid (see **WEINSTEINSÄURE**) 378
DIOXYCHINOLIN (*n*): dioxyquinoline; dihydroxyquinoline 379
DIOXYCHINON (*n*): dioxyquinone; dihydroxyquinone 380
DIOXYD (*n*): dioxide 381
DIOXYDIAMINOARSENOBENZOL (*n*): (see **SALVARSAN**) 382
DIOXYDIAMINOARSENOBENZOLDICHLORHYDRAT (*n*): (see **SALVARSAN**) 383
DIOXYNAPHTALIN (*n*): dioxynaphthalene; dihydroxynaphthalene; $C_{10}H_6(OH)_2$; Mp. 60°C. 384
DIOXYNAPHTALINDISULFOSÄURE (*f*): (see **CHROMOTROPSÄURE**) 385

DIOXYPURIN (*n*): (see XANTHIN) 386

DIOXYSTEARINSÄURE (*f*): dioxystearinic acid, dihydroxystearic acid; $C_{17}H_{33}(OH)_2COOH$; Mp. 135°C.-142°C. 387

DIOXYTOLUOL (*n*): dioxytoluene; dihydroxytoluene; orcin; orcinol; methylresorcinol; $CH_3C_6H_3(OH_2).H_2O$; Sg. 1.29; Mp. 56°C.; Bp. 288°C. 388

DIOXYWEINSÄURE (*f*): dioxytartaric acid; $HCO_2.C(OH)_2.C(OH)_2.CO_2H$ 389

DIPEPTID (*n*): dipeptide 390

DIPHENAL (*n*): diphenal (see DIAMINO-OXYDI-PHENYLPHENOLAT) 391

DIPHENSÄURE (*f*): diphenic acid 392

DIPHENYL (*n*): diphenyl; $C_{12}H_{10}$; Sg. 0.9919; Mp. 70.5°C.; Bp. 254.9°C. 393

DIPHENYLAMIN (*n*): diphenylamine, phenylaniline; $C_{12}H_{11}N$. Sg. 1.159; Mp. 54°C; Bp. 302-310°C. 394

DIPHENYLARSENCHLORÜR (*n*): diphenylarsenious chloride 395

DIPHENYLBORCHLORID (*n*): diphenylboron chloride 396

DIPHENYLENDANILODIHYDROTRIAZOL (*n*): (see NITRON) 397

DIPHENYLFARBSTOFF (*m*): diphenyl dye-(stuff) (for dyeing cotton goods) 398

DIPHENYLIMID (*n*): (see KARBAZOL and CARBAZOL) 399

DIPHENYLMETHAN (*n*): diphenylmethane, benzylbenzene; $C_{13}H_{12}$; Sg. 1.0008-1.0056; Mp. 23-27°C.; Bp. 261-264.7°C. 400

DIPHENYLMETHANFARBSTOFF (*m*): diphenylmethane dye-stuff 401

DIPHENYLOXYD (*n*): (see PHENYLÄTHER) 402

DIPHOSPHORSÄURE (*f*): diphosphoric (pyrophosphoric) acid 403

DIPLOGENER KUPHONSPAT (*m*): epistilbite (see EPISTILBIT) 404

DIPLOSAL (*n*): diplosal, salicylosalicylic acid; $C_6H_4CO.O.C_6H_4.COOH.OH$; Mp. 147°C. 405

DIPOL: dipolar 406

DIPROPÄSIN (*n*): dipropaesin; $CO(HN.C_6H_4.COOC_3H_7)_2$; Mp. 171.5°C. 407

DIPROPYLAMIN (*n*): dipropylamine 408

DIPROPYLBARBITURSÄURE (*f*): dipropyl-barbituric acid; proponal; Mp. 145°C. (see PROPONAL) 409

DIPROPYLKETON (*n*): dipropylketone, butyron; $C_3H_7.CO.C_3H_7$; Sg. 0.82; Bp. 143.5°C. 410

DIPROPYLMALONYLHARNSTOFF (*m*): dipropylmalonylurea, proponal; Mp. 145°C. (see PROPONAL) 411

DIPSOMANIE (*f*): dipsomania, alcoholism, drunkenness, (a craving for intoxicants) 412

DIPTERITISÖL (*n*): dipteritis oil 413

DIPTHERIE (*f*): diphtheria (Med.) 414

DIPTHERITIS (*m*): diphtheria (Med.) 415

DIREKT: direct-(ly) 416

DIREKTFARBSTOFF (*m*): direct dye 417

DIREKTIONSRAT (*m*): board of directors 418

DIREKTKOPIEREN (*n*): direct (copying) printing 419

DIREKTOR (*m*): director; manager; superintendent 420

DIREKTSCHWARZ (*n*): direct black 421

DIRIGENT (*m*): director; manager; head; chief; superintendent 422

DIRIGIEREN: to direct; manage; rule; superintend 423

DISACCHARID (*n*): disaccharide (one of the cane sugar group of sugars); $C_{12}H_{22}O_{11}$ (see ROHRZUCKER and MALTOSE) 424

DISÄURE (*f*): diacid; dibasic acid; pyro acid 425

DISCHWEFELSÄURE (*f*): disulphuric acid; pyrosulphuric acid (see SCHWEFELSÄURE, RAUCHENDE-) 426

DISCRASIT (*m*): dyscrasite (see ANTIMONSILBER) 427

DISKONTINUIERLICH: discontinuous-(ly); periodic-(al)-(ly); intermittent-(ly) 428

DISKONTINUITÄT (*f*): discontinuity; intermittence 429

DISKONTO (*m*): discount 430

DISKREPANZ (*f*): discrepancy 431

DISKRETIONS-TAGE (*m.pl*): days of grace (Com.) 432

DISKUSSION (*f*): discussion 433

DISKUTIEREN: to discuss 434

DISMEMBRATOR (*m*): grinding (breaking down) mill 435

DISPACHE (*f*): statement of average 436

DISPACHEUR (*m*): accountant, surveyor of averages 437

DISPENSIEREN: to dispense; exempt; excuse 438

DISPERGIEREN: to disperse: 439

DISPERS: disperse; dispersed 440

DISPERSE PHASE (*f*): disperse phase; sol (Colloidal chemistry) 441

DISPERSES SYSTEM (*n*): disperse system 442

DISPERSION (*f*): dispersion (see LICHTZERSTREUUNGSVERMÖGEN) 443

DISPERSION (*f*), **ANOMALE-**: anomalous dispersion 444

DISPERSIONSMETHODE (*f*): dispersion method (method of colloidal synthesis in which the degree of dispersion of a coarsely disperse system is so long increased until it reaches the region of colloidal dimensions) 445

DISPERSIONSMITTEL (*n*): dispersion medium; dispersive agent 446

DISPERSIONSVERMÖGEN (*n*): dispersive power 447

DISPERSITÄTSGRAD (*m*): degree of dispersiveness 448

DISPERSOID (*n*): dispersoid, disperse system 449

DISPONENT (*m*): manager; director; agent; superintendent; accountant 450

DISPONIBEL: available; at disposal 451

DISPONIEREN: to dispose; manage 452

DISPOSITION (*f*): disposition, disposal, arrangement 453

DISSIMILATION (*f*): dissimilation 454

DISSIPATOR (*m*): dissipator (see also GITTER-SCHORNSTEIN) 455

DISSOCIIERBAR: dissociable (see under DISSOZIATION) 456

DISSOCIIEREN: to dissociate 457

DISSOUSGAS (*n*): (see AZETYLEN) 458

DISSOZIATION (*f*): dissociation 459

DISSOZIATIONSGRAD (*m*): dissociation degree 460

DISSOZIATIONSKONSTANTE (*f*): dissociation constant 461

DISSOZIATIONSPRODUKT (*n*): dissociation product 462

DISSOZIATIONSSPANNUNG (*f*): dissociation tension 463

DISSOZIATIONSTEMPERATUR (*f*): temperature of dissociation 464

DISSOZIATIONSTHEORIE (*f*): dissociation theory 465
DISSOZIATIONSVERMÖGEN (*n*): dissociation (dissociative) power 466
DISSOZIATIONSWÄRME (*f*): dissociation heat 467
DISSOZIATIONSZUSTAND (*m*): dissociation condition 468
DISSOZIIERBAR: dissociable 469
DISSOZIIEREN: to dissociate 470
DISTEL (*f*): thistle (*Onopordon acanthium*) 471
DISTELÖL (*n*): nettle oil (from seeds of *Onopordon acanthium*) 472
DISTHEN (*n*): disthene, cyanite; Al_2SiO_5; Sg. 3.6 (see **ALUMINIUMSILIKAT**) 473
DISTHENSPAT (*m*), **EUTOMER-**: diaspore 474
DISTHENSPAT (*m*), **PRISMATISCHER-**: kyanite 475
DISTILLIEREN: to distil 476
DISTRIKT (*m*): district 477
DISULFAMINSÄURE (*f*): disulfamic acid 478
DISULFOSÄURE (*f*): disulfonic acid 479
DITARINDE (*f*): dita bark (bark of *Alstonia scholaris*) 480
DITHIOKOHLENSÄURE (*f*): dithiocarbonic acid; dithiolcarbonic acid; (HSCOSH) 481
DITHIONIG: hyposulphurous 482
DITHIONIGSAUER: hyposulphite of; combined with hyposulphurous acid 483
DITHIONSÄURE (*f*): dithionic acid; hyposulphuric acid 484
DITHIOPHOSPHORSÄURE (*f*): dithiophosphoric acid 485
DITHIOSALIZYLSÄURE (*f*): dithiosalicylic acid 486
DITMARISIEREN: to treat by the Ditmar process, to ditmarize (named after the inventor, Dr. R. Ditmar), to latex (i.e., dip the filling material for upholstery in latex) (see **LATEX**) 487
DIÜBERJODSÄURE (*f*): diperiodic acid 488
DIURETICUM (*n*): diuretic (an agent for increasing the secretion of the kidneys) 489
DIURETIN (*n*): diuretine, theobromin-sodium salicylate 490
DIVARICATINSÄURE (*f*): divaricatinic acid 491
DIVARICATSÄURE (*f*): divaricatic acid 492
DIVERGIEREND: diverging; divergent 493
DIVIDEND (*m*): dividend (Maths.) 494
DIVIDENDE (*f*): dividend (Com.) 495
DIVIDIEREN: to divide 496
DIVIDIVIEXTRAKT (*m*): divi-divi extract (from fruit of *Caesalpinia copiaria*) 497
DIVISION (*f*): division 498
DIVISOR (*m*): divisor (Maths.) 499
DIWOLFRAMCARBID (*n*): diwolfram (ditungsten) carbide; W_2C; Sg. 16.06 500
DIWOLFRAMSÄURE (*f*): ditungstic acid 501
DIZIMTSÄURE (*f*): dicinnamic acid 502
DIZYAN (*n*): dicyanogen; cyanogen gas; (see **DICYAN**) 503
DIZYANDIAMID (*n*): dicyanodiamide; $(CN_2H_2)_2$; Mp. 204.5°C. 504
DOCHT (*m*): wick 505
DOCHTKOHLE (*f*): cored carbon 506
DOCHTSCHMIERUNG (*f*): wick lubrication 507
DOCK (*n*): dock 508
DOCKE (*f*): bundle; skein; mandrel; plug; baluster; dock; doll; rail 509
DOCKEN: to wind (yarn, etc.); dock (a ship) 510
DOCKPUMPE (*f*): dock pump 511
DOCKSTÜTZE (*f*): dock shore 512
DOCKTOR (*n*): dock-gate 513
DOCK (*n*), **TROCKEN-**: dry dock 514
DODECYLSÄURE (*f*): (see **LAURINSÄURE**) 515
DODEKAEDER (*n*): dodecahedron 516
DOGMATIK (*f*): dogmatics 517
DOHLE (*f*): jackdaw (*Corvus monedula*): sewer 518
DOHLEN (*m*): sewer; small culvert; sink; gutter-bridge; drain 519
DOKTOR (*m*): Doctor (title); doctor (see **AB-STREICHMESSER**) 520
DOKTORARBEIT (*f*): thesis 521
DOKUMENT (*n*): document 522
DOLDE (*f*): umbel; strobile (Botany); cone (of hops) 523
DOLDENBLUMEN (*f.pl*): *Umbelliferæ* (Botany) 524
DOLDENBLUMIG: umbelliferous (Botany) 525
DOLDENBLÜTIG: umbelliferous (Botany) 526
DOLDENPFLANZE (*f*): umbellated plant (as *Dorema ammoniacum*) 527
DOLDENTRAUBE (*f*): corymb (an erect cluster) (Botany) 528
DOLDIG: umbelliferous, umbellate, umbellated 529
DOLIÖL (*n*): (see **BASSIAÖL**) 530
DOLLE (*f*), **RIEMEN-**: rowlock 531
DOLLE (*f*), **RUDER-**: rowlock 532
DOLMETSCHER (*m*): interpreter 533
DOLOMIT (*m*): dolomite (a calcium-magnesium carbonate): $CaMg(CO_3)_2$; Sg. 2.9 534
DOLOMITMÜHLE (*f*): dolomite mill 535
DOM (*m*): dome; cupola; cover; receiver; cathedral 536
DOMA (*n*): dome (Cryst.) 537
DOMBAÖL (*n*): (see **TACAMAHACFETT**) 538
DOMEYKIT (*m*): domeykite (a natural arsenic-copper ore); Cu_3As (about 82% copper); Sg. 7.3 539
DOMHAUBE (*f*): dome cover; domed (hood) cap 540
DOMIZILIERT: payable, domi iliated (Com.) 541
DONARIT (*n*): donarite (an explosive; 80% Ammonium nitrate, 12% trinitrotoluene, 3.8% nitroglycerine, 4% flour and 0.2% collodion cotton) 542
DONLÄGIG: inclined; slanting; sloping 543
DONNERKEIL (*m*): thunderbolt; belemnite 544
DONNERSTRAHL (*m*): flash of lightning 545
DOPPEL: (as a prefix equals double, duplex, di-, bi-, twin, two) 546
DOPPELBILD (*n*): double image 547
DOPPELBINDUNG (*f*): double bond; double union 548
DOPPELBODEN (*m*): double bottom; false bottom; jacket 549
DOPPELBODEN (*m*), **ZELLENFÖRMIGER-**: cellular double bottom 550
DOPPELBRECHUNG (*f*): double refraction 551
DOPPELBRUCH (*m*): compound fraction (Maths.); compound fracture (Med.) 552
DOPPELCASSETTE (*f*): double dark-slide (Photography) 553
DOPPELCHROMSAUER: bichromate (dichromate) of 554
DOPPELDEUTIG: ambiguous, equivocal 555
DOPPELENDER (*m*): double ended boiler 556
DOPPELENDIG: double ended 557
DOPPELFARBIG: dichromatic 558
DOPPELFARBIGKEIT (*f*): dichroism 559
DOPPELFLUORID (*n*): difluoride, bifluoride 560
DOPPELGÄNGER (*m*): double; person gifted with second sight 561

DOPPELGÄNGERBILD (*n*): double photograph, two-position photograph (two positions of a sitter on one negative) 562
DOPPELGESTALTUNG (*f*): dimorphism 563
DOPPELGEWINDE (*n*): double thread 564
DOPPELGLEIS (*n*): double track, double way, double set of lines (Railways. etc.) 565
DOPPELGRIFF (*m*): double stopping (on a stringed instrument) 566
DOPPELHAUE (*f*): mattock 567
DOPPELHOBEL (*m*): double plane 568
DOPPELJODID (*n*): double iodide (silver iodide and potassium iodide solution) 569
DOPPELJODQUECKSILBER (*n*): biniodide of mercury; mercury iodide; mercuric iodide; HgI_2 570
DOPPELKESSEL (*m*): double boiler; double-ended boiler; battery boiler 571
DOPPELKLEMME (*f*): double connector, double terminal, double clamp 572
DOPPELKOHLENSAUER: bicarbonate of 573
DOPPELKOKOS (*m*): (see **SEEKOKOS**) 574
DOPPELKREUZ (*n*): double sharp (Music) 575
DOPPELLAUT (*m*): diphthong 576
DOPPELLEBIG: amphibious 577
DOPPELLEITUNG (*f*): duplicate main, duplicate pipe line, duplicate wire 578
DOPPELOBJEKTIV (*n*): double lens (Photography) 579
DOPPELPAPIER (*n*): duplex paper (Paper) 580
DOPPELPÖLLER (*m*): double bitton, double bollard 581
DOPPELPUNKT (*m*): colon (:) 582
DOPPELSALZ (*n*): double salt 583
DOPPELSCHRAUBENDAMPFER (*m*): twin-screw steamer 584
DOPPELSCHWEFELSAUER: bisulphate of (see **DOPPELTSCHWEFELSAUER**) 585
DOPPELSCHWEFLIGSAUER: bisulphite of (see **DOPPELTSCHWEFLIGSAUER**) 586
DOPPELSEITIG: bilateral 587
DOPPELSINN (*m*): ambiguity 588
DOPPELSINNIG: ambiguous; equivocal 589
DOPPELSPAT (*m*): Iceland spar; ($CaCO_3$); (56%CaO) 590
DOPPELSTARK: double strength 591
DOPPELSTÜCK (*n*): duplicate 592
DOPPELSUPERPHOSPHAT (*n*): double superphosphate 593
DOPPELT: double; duplex; twofold; twice; bi-; di-; bin- 594
DOPPELTBORSAUER: biborate of 595
DOPPELTBRECHEND: double refracting 596
DOPPELTCHLORZINN (*n*): stannic chloride; $SnCl_4$; Sg. 2.2788; Mp. – 33°C.; Bp. 114°C. 597
DOPPELTCHROMSAUER: bichromate (dichromate) of 598
DOPPELTCHROMSAURES KALI (*n*): bichromate of potash; $K_2Cr_2O_7$; Sg. 2.692; Mp. 396°C. 599
DOPPELTCHROMSAURES NATRON (*n*): bichromate of soda; $Na_2Cr_2O_7.2H_2O$; Sg. 2.52 600
DOPPELTHOCHRUND: convexo-convex 601
DOPPELTHOHL: concavo-concave 602
DOPPELTKOHLENSAUER: bicarbonate of 603
DOPPELTKOHLENSAURES NATRON (*n*): bicarbonate of soda; $NaHCO_3$; Sg. 2.2 604
DOPPELTRANSPORTVERFAHREN (*n*): double-transfer carbon process (Photography) 605
DOPPELTSCHWEFELEISEN (*n*): iron bisulphuret 606
DOPPELTSCHWEFELSAUER: bisulphate of; -acid sulphate 607
DOPPELTSCHWEFLIGSAUER: bisulphite of; -acid sulphite 608
DOPPELTSCHWEFLIGSAURER KALK (*m*): (see **CALCIUMBISULFIT**) 609
DOPPEL-T-TRÄGER (*m*): I beam; joist, girder, etc. of I section 610
DOPPELTWEINSAUER: bitartrate of; -acid tartrate 611
DOPPELVERBAND (*m*): double bandage; double bond 612
DOPPELWANDIG: double-walled 613
DOPPELWASSERGLAS (*n*): silicate of potash and soda (see **NATRONWASSERGLAS** and **KALI-WASSERGLAS**) 614
DOPPELWIRKEND: double-acting 615
DOPPELZENTNER (*m*): one-tenth of a metric ton, about 2 cwts. 616
DOPPELZERSETZUNG (*f*): double decomposition 617
DORMIOL (*n*): dormiol (see **AMYLENCHLORAL**) 618
DORMIOLLÖSUNG (*f*): dormiol solution (a 50% aqueous solution of dormiol; see **DORMIOL**); Sg. 1.125 619
DORMONAL (*n*): dormonal (see **NATRIUM, DI-ÄTHYLBARBITURSAURES-**) 620
DORN (*m*): thorn; spine; pin; punch; spike; drift; bolt; tongue; mandrel; tap cinder; puddling slag; slag (of copper) 621
DÖRNERSCHLACKE (*f*): tap cinder; puddling slag; bulldog (Metal) 622
DORNFORTSATZ (*m*): spinal process (either spinous or transverse process) (Med.) 623
DORNSTEIN (*m*): thornstone (a gypsum incrustation; a residue from the passage of brine through the graduating apparatus) (see **GRADIERWERK**) 624
DORREN: to dry; bake; wither; kiln-dry 625
DÖRREN: to dry; desiccate; cure; scorch 626
DORSCH (*m*): torsk (*Gadus callarius*) (Icht.) 627
DORSCHLEBERÖL (*n*): (see **DORSCHLEBERTRAN**) 628
DORSCHLEBERTRAN (*m*): cod liver oil (from *Gadus callarius*); Sg. 0.922-0.93 629
DOSE (*f*): box 630
DOSENBAROMETER (*n*): aneroid barometer 631
DOSENBAUM (*m*): mountain pine (*Pinus montana*) 632
DOSENFÜLLMASCHINE (*f*): box filling machine 633
DOSENLIBELLE (*f*): box-level (Photography) 634
DOSENTELEPHON (*n*): box telephone 635
DOSIEREN: to dose; determine 636
DOSIERMASCHINE (*f*): dosing machine, machine for making up exact doses of medicine, etc. 637
DOSIERUNG (*f*): dosing; dose; determination 638
DOSIS (*f*): dose 639
DOSSIERUNG (*f*): slope 640
DOST (*m*): marjoram (*Origanum vulgare*); thyme (*Thymus vulgaris*) 641
DOSTEN (*m*): (see **DOST**) 642
DOSTENKRAUT (*n*): origan (origanum) herb, marjoram (*herba origani*) (*Origanum vulgare*); thyme (*Thymus vulgaris*) 643
DOSTENÖL (*n*): origan (origanum, thyme), oil Sg. 0.925 644
DOTTER (*m*): yolk (of eggs); vitellus 645
DOTTERÖL (*n*): (see **LEINDOTTERÖL**) 646
DOUBLIEREN: to double; concentrate 647

DOUMPALME (*f*): Egyptian doom-palm (*Hyphæne thebaica*) 648
DRACHENBLUT (*m*): dragon's blood (*Sanguis draconis*) (resin from fruit of *Daemonorops draco*; *Calamus draco*; and resin from the trees *Dracoena draco* and *Eucalyptus resinifera*) 649
DRACHENFLIEGE (*f*): dragon fly (*Libellulidæ* family of insects) 650
DRACHENKRAUT (*n*): dragon wort (*Dracunculus vulgaris*) 651
DRACHENWURZ (*f*): dragon wort (*Dracunculus vulgaris*) 652
DRAGIEREN: to coat (pills, sweets, etc.) 653
DRAHT (*m*): wire; thread; twist (Spinning); grain (Wood) 654
DRAHT (*m*), **ABSCHMELZ-**: fuse wire 655
DRAHTARBEIT (*f*): wire (filigree) work 656
DRAHTBANK (*f*): wire drawing (bench) machine 657
DRAHT (*m*), **BESPONNENER-**: covered (cased) wire 658
DRAHT (*m*), **BLANKER-**: bare (naked) wire 659
DRAHTBÜRSTE (*f*): wire brush 660
DRAHTEN: of wire; to wire; cable; telegraph 661
DRÄHTERN: of wire; wiry; to wire 662
DRAHTFASER (*f*): wire mould (Paper) 663
DRAHTGAZE (*f*): wire gauze 664
DRAHTGEFLECHT (*n*): wire-work; wire netting 665
DRAHTGEWEBE (*n*): wire gauze; wire netting 666
DRAHTGITTER (*n*): wire (trellis) grating or screen 667
DRAHTGLAS (*n*): wire (wired) glass (glass with wire netting embedded in it) 668
DRAHTKLEMME (*f*): wire clamp; terminal (Elect.) 669
DRAHT, KNÜPPEL FÜR-: wire billets 670
DRAHTLEHRE (*f*): wire gauge 671
DRAHTLOS: wireless 672
DRAHTLOSE TELEGRAPHIE (*f*): wireless telegraphy 673
DRAHTMASKE (*f*): wire mask 674
DRAHTNETZ (*n*): wire (net) gauze; wire (netting) network; network (system) of wires 675
DRAHTPUPPE (*f*): puppet 676
DRAHTSCHERE (*f*): wire shears 677
DRAHT (*m*), **SEIDENUMSPONNENER-**: silk-covered wire 678
DRAHTSEIL (*n*): wire (cable; cord) rope 679
DRAHTSEILBAHN (*f*): rope tramway, ropeway, cableway; aerial ropeway, or cableway 680
DRAHTSEILSCHMIERE (*f*): wire rope (grease) lubricant 681
DRAHTSICHERUNG (*f*): wire fuse 682
DRAHTSIEB (*n*): wire (screen) sieve 683
DRAHTSPANNER (*m*): wire stretcher 684
DRAHTSPIRALE (*f*): solenoid 685
DRAHTSTIFT (*m*): wire tack or nail 686
DRAHTTROSSE (*f*): wire hawser 687
DRAHTZANGE (*f*): wire-cutters; pliers 688
DRAHTZIEHBANK (*f*): wire drawing (bench) machine 689
DRAHTZIEHEN (*n*): wire drawing 690
DRAHTZIEHER (*m*): wire drawer 691
DRAHTZIEHEREI (*f*): wire drawing; wire mill 692
DRAHTZIEHMASCHINE (*f*): wire-drawing machine 693
DRAHTZUG (*m*): wire mill; wire drawing 694
DRAINIERBAR: drainable 695
DRAINIEREN: to drain; irrigate 696
DRAINROHR (*n*): drain-pipe 697
DRAINSTRANG (*m*): drain-(age) pipe-(line) 698
DRAINWASSER (*n*): drainage water 699
DRALL (*m*): twist; rifling (of guns); torsion; torque; tight; robust; strong 700
DRALLVEKTOR (*m*): torsion vector 701
DRANG (*m*): throng; pressure; impulse; craving; urgency 702
DRÄNGEN: to press; crowd; force; urge; constrict; squeeze; impel; drive (Gases) 703
DRASS (*m*): dregs (of oil) 704
DRASTISCH: drastic-(ally) 705
DRAUFSICHT (*f*): plan (of drawings) 706
DRECHSELBANK (*f*): turning-lathe, lathe 707
DRECHSELN: to turn (on a lathe); elaborate 708
DRECHSLERLACK (*m*): Russian varnish 709
DRECK (*m*): filth; dirt; dung; rubbish; offal; excrement; fæces 710
DREGANKER (*m*): grapnel, grappling-(iron), drag-(anchor) 711
DREGIEREN: to drag, dredge 712
DREGTAU (*n*): drag-rope; painter (Naut.) 713
DREH-: (as a prefix), rotary; revolving; turning; rolling; twisting; torsional 714
DREHACHSE (*f*): axis of rotation; pin; axle; spindle; knife-edge (of balance) 715
DREHARM (*m*) **DER KRAFT** (*f*): lever (arm) of the force 716
DREHBANK (*f*): lathe 717
DREHBANKFUTTER (*n*): (lathe)-chuck 718
DREHBANKSPINDEL (*f*): mandrel, mandril 719
DREHBANKSUPPORT (*m*): slide rest 720
DREHBAR: rotary; turnable,; twistable 721
DREHBEANSPRUCHUNG (*f*): torsional stress; torque 722
DREHBEWEGUNG (*f*): rotary motion; rotation; revolution; turning movement 723
DREHBLENDE (*f*): rotating (wheel); revolving diaphragm (Photography) 724
DREHBOHRER (*m*): rotary drill; twist-drill 725
DREHBRÜCKE (*f*): swing bridge 726
DREHDORNPRESSE (*f*): arbor press 727
DREHEISEN (*n*): turning tool; chisel, gouge 728
DREHEISENMESSWERK (*n*), **MIT-**: of the moving iron type (Elect.) 729
DREHEN: to turn, rotate, revolve, twist, roll 730
DREHENDES ELEMENT (*n*): revolving (rotating) part 731
DREHER (*m*): turner; rotator; crank; handle; winch 732
DREHFELD (*n*): rotary field 733
DREHFENSTER (*n*): pivoted window 734
DREHFESTIGKEIT (*f*): torsional strength, torque 735
DREHGESTELL (*n*): bogie frame (Engines) 736
DREHHERZ (*n*): lathe dog 737
DREHKALZINIEROFEN (*m*): rotating calcining oven 738
DREHKNAUF (*m*): handle 739
DREHKRAFT (*f*): torsion-(al force); torque 740
DREHKRAN (*m*): turning (slewing) crane 741
DREHKRANKHEIT (*f*): staggers (of sheep) 742
DREHKREUZ (*n*): turnstile; sparger (Brewing) 743
DREHKUPPEL (*f*): revolving dome 744
DREHMOMENT (*n*): twisting (torsional; static; rotation) moment; torque 745
DREHOFEN (*m*): revolving furnace; rotary furnace; converter 746
DREHPAAR (*n*): couple (of forces); force-couple 747
DREHPUNKT (*m*): centre of motion; turning point; pivot; axis (centre) of rotation; pole (of moment) 748

DREHPUNKT, AUGENBLICKLICHER- : instantaneous (momentary) centre (of rotary motion) 749
DREHREEP (*n*) : tie 750
DREHRICHTUNG (*f*) : direction of rotation 751
DREHSCHEIBE (*f*) : disc-chuck ; potter's wheel ; turntable ; turning (revolving) platform 752
DREHSCHEMEL (*m*) : bogie 753
DREHSCHNELLE (*f*) : rotary velocity 754
DREHSCHWINGUNG (*f*) : torsional oscillation 755
DREHSINN (*m*) : direction of rotation 756
DREHSPÄNE (*m.pl*) : turnings ; shavings 757
DREHSPANNUNG (*f*) : torsional (strain ; force) stress ; torque 758
DREHSPIEGEL (*m*) : rotary mirror 759
DREHSPINDEL (*f*) : revolving spindle 760
DREHSPULE (*f*) : revolving coil 761
DREHSTAHL (*m*) : turning tool ; chisel, gouge 762
DREHSTROM (*m*) : rotary current ; alternating current ; A.C. ; multiphase current (Elect.) 763
DREHSTROMMOTOR (*m*) : A.C. (alternating current) motor 764
DREHTURM (*m*) : revolving turret 765
DREHUNG (*f*) : rotation ; revolution ; turning ; torsion ; twisting 766
DREHUNGSACHSE (*f*) : axis of rotation ; turning axis 767
DREHUNGSEBENE (*f*) : plane of rotation 768
DREHUNGSGRAD (*m*) : degree of rotation 769
DREHUNGSKRAFT (*f*) : rotatory force 770
DREHUNGSMOMENT (*n*) : moment of rotation ; moment of torsion ; torque 771
DREHUNG (*f*), **SPEZIFISCHE-** : specific rotation (the angle by which the plane of polarization is rotated) 772
DREHUNGSTRÄGHEIT (*f*) : inertia due to rotation 773
DREHUNGSVERMÖGEN (*n*) : rotation ; rotatory power 774
DREHUNGSVERMÖGEN (*n*), **MOLEKULARES-** : molecular (rotating power) rotation (the product of specific rotation and molecular weight of a compound, or the hundredth part of such product) 775
DREHUNGSWINKEL (*m*) : angle of rotation (see **DREHUNG, SPEZIFISCHE-**) 776
DREHWAGE (*f*) : torsion balance 777
DREHWERK (*n*) : turning machine 778
DREHWUCHS (*m*) : twisted growth 779
DREHZAHL (*f*) : number of revolutions ; speed ; velocity 780
DREHZYLINDER (*m*) : revolving cylinder 781
DREIACHSIG : triaxial 782
DREIATOMIG : triatomic 783
DREIBASISCH : tribasic 784
DREIBASISCHE SÄURE (*f*) : tribasic acid 785
DREIBEIN (*n*) : tripod 786
DREIBEINSTATIV (*n*) : tripod stand (Photography, etc.) 787
DREIBLATT (*n*) : trefoil (*Trefolium* species of plant) 788
DREIBLÄTTERIG : three-leaved 789
DREIBRENNER (*m*) : triple burner 790
DREIBUND (*m*) : triple alliance 791
DREIDIMENSIONAL : tridimensional 792
DREIECK (*n*) : triangle 793
DREIECK (*n*), **GLEICHSCHENKLIGES-** : isosceles triangle 794
DREIECK (*n*), **GLEICHSEITIGES-** : equilateral triangle 795
DREIECKIGER MUSKEL (*m*) : deltoid (muscle) (uniting shoulder blade and arm) 796
DREIECK (*n*), **RECKTWINKELIGES-** : right-angled triangle 797
DREIECKSCHALTUNG (*f*) : triangle connection, delta connection 798
DREIECKSLEHRE (*f*) : trigonometry 799
DREIECK (*n*), **SPHÄRISCHES-** : spherical triangle 800
DREIECK (*n*), **SPITZWINKELIGES-** : acute-angled triangle 801
DREIECK (*n*), **STUMPFWINKELIGES-** : obtuse-angled triangle 802
DREIECK (*n*), **UNGLEICHSEITIGES-** : scalene triangle 803
DREIFACH : triple ; threefold ; treble ; tri- ; three- ; thrice ; three times 804
DREIFACHBASISCH : tribasic 805
DREIFACHCHLORANTIMON (*n*) : antimony trichloride (see **ANTIMONTRICHLORID**) 806
DREIFACHCHLORJOD (*n*) : iodine trichloride (see **JODTRICHLORID**) 807
DREIFACHE AUSFERTIGUNG (*f*) : triplicate 808
DREIFACHSCHWEFELARSEN (*n*) : arsenic trisulphide (see **ARSENTRISULFID**) 809
DREIFARBENDRUCK (*m*) : three-colour print-(ing) 810
DREIFARBENPHOTOGRAPHIE (*f*) : three-colour photograph-(y) 811
DREIFARBENRASTER (*m*) : three-colour screen (Photography) 812
DREIFLÄCHIG : three-faced 813
DREIFLAMMIG : three-flamed 814
DREIFUSS (*m*) : tripod 815
DREIFÜSSIG : three-footed ; tripedal 816
DREIFUSSSTATIV (*n*) : tripod stand 817
DREIGLIEDRIG : three-membered ; trinomial (Maths.) 818
DREIHALSIG : three-necked 819
DREIHEIT (*f*) : triad 820
DREIJÄHRIG : three years old ; triennial 821
DREIKANTFEILE (*f*) : triangular file 822
DREIKANTIG : triangular ; three-cornered 823
DREIKANTSCHABER (*m*) : three-square scraper 824
DREIKLANG (*m*) : triad (Music) 825
DREILEITERSYSTEM (*n*) : three wire system (Elect.) 826
DREIMASTGAFFELSCHONER (*m*) : three-masted fore and aft schooner 827
DREIMASTSCHONER (*m*) : three-masted schooner (Naut.) 828
DREINAMIG : trinomial 829
DREIPHASENSTROM (*m*) : three phase current (Elect.) 830
DREIPHASIG : three-phase 831
DREIRAD (*n*) : tricycle 832
DREISÄURIG : triacid-(ular) 833
DREISCHENKELIG : three-legged 834
DREISEITIG : three-sided ; trilateral; triangular; three-faced 835
DREISILBIG : trisyllabic, having three syllables 836
DREIST : bold ; confident ; courageous 837
DREISTEINWURZEL (*f*) : feverroot ; feverwort ; horse gentian (*Triosteum*, species of plant) 838
DREISTELLIG : to three places (of decimals) 839
DREITEILIG : three-part ; tripartite ; in three parts 840
DREI-UND-EINACHSIG : hexagonal (Cryst.) 841
DREIWANDIG : three-walled ; triple walled 842
DREIWEGHAHN (*m*) : three-way cock 843
DREIWEGSTÜCK (*n*) : three-way piece (such as a T or a Y piece, of pipes) 844

DREIWEGVERBINDUNG (*f*): three-way connection 845
DREIWERTIG: trivalent 846
DREIWERTIGE SÄURE (*f*): tribasic acid 847
DREIWINKEL (*m*): triangle 848
DREIWINKELIG: triangular; three-cornered 849
DREIZACK (*m*): trident 850
DREIZACKIG: three-pronged 851
DREIZÄHLIG: triple; threefold; ternary 852
DREIZWEITELTAKT (*m*): $\frac{3}{2}$-time, three minims in a bar (Music) 853
DRELL (*m*): tick-(ing) 854
DRESCHEN: to thresh 855
DRESCHER (*m*): thresher (Agric.) 856
DRESCHFLEGEL (*m*): flail (for threshing cereals) 857
DRESCHMASCHINE (*f*): threshing machine, thrashing machine, thresher 858
DRESSIEREN: to dress; train; break-(in); drill 859
DRILLBOHRER (*m*): drill; borer 860
DRILLEN: to drill; bore; turn 861
DRILLICH (*m*): tick-(ing) 862
DRILLING (*m*): triplet; lantern; spring-wheel 863
DRILLINGMASCHINE (*f*): three-cylinder (triple-cylinder) engine 864
DRILLINGSALZ (*n*): triple salt 865
DRILLINGSCASSETTE (*f*): triple (multiple) dark-slide (Photography) 866
DRILLINGSDAMPFMASCHINE (*f*): three-cylinder (triple-cylinder) engine 867
DRILLINGSWIRKUNG (*f*): triple (action) effect; three-throw 868
DRINGEN: to press; rush; urge; force; penetrate 869
DRINGEND: urgent; pressing 870
DRINGLICH: pressing; urgent 871
DRITTELSAUER: tribasic 872
DROGE (*f*): drug 873
DROGENGESCHÄFT (*n*): drug trade 874
DROGENHÄNDLER (*m*): druggist 875
DROGENKUNDE (*f*): pharmacology 876
DROGERIEGESCHÄFT (*n*): drug trade; drug shop; druggist's 877
DROGET-(T) (*m*): drugget 878
DROGIST (*m*): druggist 879
DROHEN: to threaten; menace 880
DRÖHNEN: to rumble; boom; roar 881
DROSSEL (*f*): thrush (Bird); throttle 882
DROSSELADER (*f*): jugular vein 883
DROSSELBEIN (*n*): collar bone; clavicle 884
DROSSELDIAGRAM (*n*): throttle diagram 885
DROSSELFLANSCH (*m*): throttle flange (of steam meters, etc.) (a flange fitted with a throttling diaphragm or some similar arrangement, such as tapered bore) 886
DROSSELGARN (*n*): water twist (Textile) 887
DROSSELKLAPPE (*f*): admission (inlet, throttle) valve, governor valve 888
DROSSELN: to throttle 889
DROSSELREGULATOR (*m*): throttle governor 890
DROSSELSCHEIBE (*f*): throttle diaphragm (steam meters, etc.) (a diaphragm in a pipe, with an opening less than that of the internal diameter of the pipe itself) 891
DROSSELSPULE (*f*): choking coil (Elect.) 892
DROSSELUNG (*f*): throttling, constriction, taper, swage, restriction, choking 893
DROSSELVENTIL (*n*): throttle valve 894
DRUCK (*m*): pressure; thrust; force; compression; oppression; weight; impression; load; head (of water); print-(ing) 895
DRUCKANSTIEG (*m*): pressure rise 896
DRUCKAPPARAT (*m*): printing (pressure) apparatus 897
DRUCKAUTOMAT (*m*): automatic pressure apparatus (for supplying fluids by means of compressed air; only operates when fluid and compressed air are simultaneously available) 898
DRUCKBEANSPRUCHUNG (*f*): pressure (stress) strain, compressive stress, compression 899
DRUCKBIRNE (*f*): pressure apparatus (for supplying fluids by means of compressed air) 900
DRUCKBUCHSTABE (*m*): type 901
DRUCKEN: to print; stamp; impress; (*n*): printing (Photography, etc.); stamping; impressing 902
DRÜCKEN: to force, discharge, deliver (of pumps, etc..); press, squeeze, oppress; efface oneself 903
DRUCKENTWICKELUNG (*f*): pressure (development) increase; generation of pressure 904
DRUCKER (*m*): printer 905
DRÜCKER (*m*): trigger, handle 906
DRUCKEREI (*f*): printing works; printing; print works (Calico) 907
DRUCKEREIMASCHINEN (*f.pl*): printing machinery 908
DRUCKERSCHWÄRZE (*f*): printer's ink 909
DRUCKFARBE (*f*): printing (colour) ink 910
DRUCKFASS (*n*): autoclave 911
DRUCKFEDER (*f*): pressure spring 912
DRUCKFEHLER (*m*): printer's error; printing error; misprint; typographical error 913
DRUCKFESTIGKEIT (*f*): pressure (thrust) resistance; compression (resistance) elasticity; compressive strength 914
DRUCKFILTER (*m*): pressure filter 915
DRUCKFIRNIS (*m*): printer's varnish 916
DRUCKFLASCHE (*f*): pressure (bottle) flask; digester 917
DRUCKGASFEUERUNGSANLAGE (*f*): pressure system gas firing plant 918
DRUCKGEFÄLLE (*n*): pressure drop 919
DRUCKGEFÄSS (*n*): pressure vessel, vessel under pressure 920
DRUCK (*m*), **GLEICHBLEIBENDER-**: constant pressure 921
DRUCK (*m*), **GLEICHER-**: constant pressure, equal pressure 922
DRUCKHÖHE (*f*): head (of water) 923
DRUCKKNOPF (*m*): push-button (Elect.); press-button, bell-push 924
DRUCKKOLBEN (*m*): piston; pressure flask 925
DRUCK (*m*), **KRITISCHER-**: critical pressure 926
DRUCKLAGER (*n*): thrust bearing, thrust block 927
DRUCKLEGUNG (*f*): publishing 928
DRUCKLEITUNG (*f*): pressure (piping; pipe; main); pipeline; discharge (delivery) piping (of pumps) 929
DRUCKLUFT (*f*): compressed air (see also **LUFTDRUCK-**) 930
DRUCKLUFTANLAGE (*f*): compressed air (pneumatic) plant 931
DRUCKLUFTARMATUREN (*f.pl*): compressed air (pneumatic) fittings 932
DRUCKLUFTAUSRÜSTUNG (*f*): compressed air (pneumatic) equipment 933
DRUCKLUFTENTÖLUNG (*f*): compressed air de-oiling 934
DRUCKLUFTHAMMER (*m*): pneumatic hammer 935
DRUCKLUFTMEISSEL (*m*): pneumatic chisel 936

DRUCKLUFTPEGEL (*m*): compressed air water gauge or depth gauge 937
DRUCKLUFTPUMPE (*f*): compressed air pump 938
DRUCKLUFTSCHLAUCH (*m*): compressed air (pneumatic) flexible hose or tubing 939
DRUCKLUFTSPRITZAPPARAT (*m*): aerograph, compressed air (sprayer) spraying apparatus 940
DRUCKLUFTTURBINE (*f*): compressed air turbine 941
DRUCKLUFTWERKZEUG (*n*): pneumatic tool (s) 942
DRUCKMESSER (*m*): pressure (meter) gauge; manometer 943
DRUCKMESSUNG (*f*): pressure measurement 944
DRUCKMINDERUNGSVENTIL (*n*): pressure reduction valve, throttle valve, pressure reducing valve 945
DRUCKÖL (*n*): oil under pressure 946
DRUCKÖLFEUERUNGSANLAGE (*f*): pressure system oil firing plant 947
DRUCK (*m*), **OSMOTISCHER-**: osmotic pressure 948
DRUCKPAPIER (*n*): ; printing paper; plate paper (Paper) 949
DRUCKPLATTE (*f*): printing (engraving) plate; gelatine plate (Photography) 950
DRUCKPROBE (*f*): pressure test; hydraulic (compression) test; proof (Printing); squeezing test (Grain) 951
DRUCKPUMPE (*f*): pressure (force, compression) pump 952
DRUCKRAUM (*m*): pressure chamber 953
DRUCK-RAUM-DIAGRAMM (*n*): pressure-volume diagram, P-V diagram 954
DRUCKREGLER (*m*): pressure regulator 955
DRUCKREGULATOR (*m*): pressure regulator 956
DRUCKRING (*m*): thrust collar 957
DRUCKROHR (*n*): pressure (tube) pipe; force pipe; discharge (delivery) pipe (of pumps) 958
DRUCKSACHEN (*f.pl*): printed matter (such as catalogues, pamphlets, leaflets, etc.); printed papers (on envelope or wrapper for postal purposes) 959
DRUCKSCHLAUCH (*m*): (flexible-) pressure pipe or tube; force (pipe) hose 960
DRUCKSCHRAUBE (*f*): pressure (adjusting) screw, thumb-screw 961
DRUCKSCHRIFT (*f*): publication; type 962
DRUCKSCHWARZ (*n*): printing black 963
DRUCKSEITE (*f*): page; pressure side; discharge side (of pumps); side on which pressure or thrust is exerted 964
DRUCKSPANFASS (*n*): pressure chip cask (Brewing) 965
DRUCKSPANNUNG (*f*): compression strain; compressive stress; crushing load 966
DRUCKSTEIGERUNG (*f*): pressure rise; pressure increase 967
DRUCKSTEMPEL (*m*): piston; ram 968
DRUCKSTUFE (*f*): pressure stage (Turbines) 969
DRUCKSTUTZE (*f*): delivery (discharge) branch 970
DRUCKSTUTZEN (*m*): discharge (delivery) branch 971
DRUCKTINTE (*f*): printing ink 972
DRUCKTOPF (*m*): autoclave 973
DRUCKUNTERSCHIED (*m*): pressure (variation; fluctuation) difference 974
DRUCKUNTERSCHIEDMESSER (*m*): pressure difference meter 975
DRUCKVENTIL (*n*): pressure (discharge) valve 976
DRUCK (*m*), **VERMINDERTER-**: reduced pressure 977
DRUCKVERMINDERUNG (*f*): pressure (diminuition; reduction) decrease; pressure (fall) drop 978
DRUCKVOLUMENDIAGRAMM (*n*): pressure-volume diagram, P-V diagram 979
DRUCKWALZE (*f*): pressure (roller; roll) cylinder; printing roller 980
DRUCKWASSER (*n*): water under pressure; (as a prefix hydraulic) 981
DRUCKWASSERSAMMLER (*m*): hydraulic accumulator 982
DRUCKWELLE (*f*): pressure wave; thrust shaft 983
DRUCKZEUG (*n*): printing material 984
DRUCKZUGMESSER (*m*): draught (gauge) meter 985
DRUCKZUNAHME (*f*): pressure (rise) increase 986
DRÜSCHEN (*n*): small gland 987
DRUSE (*f*): druse; hollow (of ore); glanders 988
DRÜSE (*f*): gland 989
DRUSEN (*f.pl*): dregs; lees; husks; glanders 990
DRUSENASCHE (*f*): calcined (wine) lees 991
DRÜSENBEULE (*f*): (see **DRÜSENGESCHWULST**) 992
DRUSENBRANTWEIN (*m*): spirits distilled from lees 993
DRUSENFÖRMIG: drusy; hollowed 994
DRÜSENGESCHWULST (*f*): glandular swelling (such as Bubo, any inflammatory swelling of a lymphatic gland; scrofula or struma, a tubercular swelling of a lymphatic gland) (Med.) 995
DRÜSENKRANKHEIT (*f*): scrofula; struma; tuberculous glands 996
DRÜSENKROPF (*m*): goitre; bronchocele (swelling of the thyroid gland) (Med.) 997
DRUSENÖL (*n*): grape-seed oil (see **COGNAKÖL**) 998
DRUSENSCHWARZ (*n*): Frankfort black (from wine lees and vine wood) 999
DRÜSENSCHWELLUNG (*f*): swelling of the glands (Med.) 000
DRUSICHT: drusy, hollowed 001
DRUSIG: drusy; hollowed 002
DRÜSIG: glandular 003
DUALISMUS (*m*): dualism 004
DUALISTISCH: dualistic 005
DUALITÄT (*f*): duality 006
DÜBEL (*m*): dowel, plug, peg, pin 007
DÜBEL (*m*), **HOLZ-**: wooden (plug, dowel) peg 008
DUBLETTEN (*n*): doubling (a thin flake of real precious stone cemented on to glass) 009
DUBOISIN (*n*), **SCHWEFELSAURES-**: duboisin sulphate 010
DUCHT (*f*): thwart 011
DUCKEN: to humble; duck; give in; submit; stoop 012
DUCKER (*m*): siphon 013
DUCKSTEIN (*m*): calcareous tufa; trass; sometimes also dolomite (see **DOLOMIT**) 014
DUCTUS THORACICUS (*m*): thoracic duct (Med.) 015
DUFRENIT (*m*): dufrenite (see **GRÜNEISENERZ**) 016
DUFTEN: to smell fragrant; be (odorous; perfumed; fragrant; vapourous) misty; give off (vapour) perfume 017
DUFTSTOFF (*m*): perfume, scent 018
DUGONGÖL (*n*): dugong (sea-cow) oil, (from *Halicore australis* and *Halicore indicus*) 019
DÜKER (*m*): siphon 020
DUKTILITÄT (*f*): ductility (Metal) 021
DUMPERBLOCK (*m*): pitch block 022
DUMPERKLAPPLÄUFER (*m*): pitch whip 023
DUMPF: dull; hollow; muffled; damp; musty; close; stifling 024

DUMPFHEIT (*f*): torpor; dulness; stupor; insensibility 025
DUMPING (*n*): dumping (of goods) (a word introduced since the war) 026
DUNG (*m*): dung; manure; fertilizer 027
DÜNGEGIPS (*m*): gypsum for manuring purposes 028
DÜNGEKALK (*m*): lime for manuring purposes 029
DÜNGEMITTEL (*n*): manure; fertilizer 030
DÜNGEN: to manure, fertilize 031
DÜNGERFABRIKATION (*f*): fertilizer manufacture 032
DÜNGERSALZ (*n*): saline manure, dung salt 033
DÜNGERWERT (*m*): fertilizing (manurial) value 034
DÜNGESALZ (*n*): dung salt 035
DÜNGJAUCHE (*f*): liquid manure 036
DÜNGPULVER (*n*): powdered manure 037
DUNGSALZ (*n*): saline manure, dung salt 038
DUNGZWECK (*m*): fertilizing (manurial) purposes 039
DÜNKEL (*m*): conceit; arrogance 040
DUNKEL: dark; dim; obscure; dull; cloudy; deep (of colours); (*n*): darkness; obscurity 041
DUNKELBLAU (*n*): dark (deep) blue 042
DUNKELBRAUN: dark brown, deep brown 043
DUNKELFARBIG: dark (deep) coloured 044
DUNKELKAMMER (*f*): dark-room; camera obscura 045
DUNKELN: to darken; deepen; sadden (Colours); dim; obscure; dull; cloud 046
DUNKELROTGILTIGERZ (*n*): dark red silver ore (see **ROTGILTIGERZ**, **DUNKLES-**) 047
DUNKELROTGLUT (*f*): dull-red heat 048
DUNKELROTGÜLTIGERZ (*n*): dark-red silver ore; pyrargyrite; $3Ag_2S.Sb_2S_3$ (60% silver) 049
DUNKELZELT (*n*): dark tent (Photography) 050
DUNKELZIMMER (*n*): dark-room (Photography) 051
DUNKELZIMMERLAMPE (*f*): dark-room lamp (Photography) 052
DÜNKEN: to seem; appear 053
DÜNN: thin; dilute; slender; rare; small; narrow (small) bore (of pipes) 054
DÜNNBIER (*n*): small beer 055
DÜNNDARM (*m*): small intestine (Med.) 056
DÜNNFLÜSSIG: thinly liquid; watery 057
DÜNNSCHLAGEN: to beat out-(thin) 058
DÜNNSTEIN (*m*): thin matte (Copper); table diamond (Jewellery) 059
DÜNNWALZEN: to roll out-(thin) 060
DÜNNWANDIG: thin-walled 061
DUNST (*m*): vapour; steam; fume; damp; fine (small) shot 062
DUNSTABZUG (*m*): hood (for fumes) 063
DUNSTABZUGROHR (*n*): vent pipe 064
DÜNSTEN: to steam; vapour; smoke; fume; stew 065
DUNSTFANG (*m*): hood 066
DUNSTHÜLLE (*f*): vapourous envelope; atmosphere 067
DUNSTKREIS (*m*): atmosphere 068
DUNSTLOCH (*n*): vent 069
DUOTAL (*n*): guaiacol carbonate, duotal; $(C_6H_4OCH_3)_2CO_3$; Mp. 78-87°C. 070
DUOWALZE (*f*): twin rolls (Rolling mills) 071
DUOWALZWERK (*n*): twin rolling mill 072
DUPLEXSYSTEM (*n*): duplex system 073
DUPLEX-VERFAHREN (*n*): combined Bessemer and Martin process (Metal) 074
DUPLIKAT (*n*): duplicate 075
DUPLIKATFRACHTBRIEF (*m*): duplicate bill of lading; B/L. 076
DUPLIKATOR (*m*): duplicator 077
DUPLO, IN-: in duplicate 078
DURALUMIN (*n*): duralumin (see **DURALUMINIUM**) 079
DURALUMINIUM (*n*): duralumin (aluminium alloy) (93.2-95.5% Al; 3.5-5.5% Cu; 0.5 Mg.; 0.5-0.8% Mn; 2.8 specific gravity; resists weather, sea water and some acids but not hydrochloric acid and lyes) 080
DURANAMETALL (*n*): durana metal (copper alloy) 64.78% Cu; 2.22% Sn; 29.5% Zn; 1.7% Al; 1.71% Fe) 081
DURATOL (*n*): duratol, benzyl-para-aminophenol hydrobromide (a photographic developer) 082
DURCH: through; via; by means of 083
DURCHARBEITEN: to work through; knead (Dough); elaborate; pole (Dye) 084
DURCHÄTZEN: to corrode; eat through 085
DURCHAUS NICHT: in no wise; not at all 086
DURCHBEIZEN: to corrode; eat through 087
DURCHBEUTELN: to bolt (Flour) 088
DURCHBIEGEN: to sag; deflect 089
DURCHBIEGUNG (*f*): sag; deflection 090
DURCHBILDEN: to design, construct, develop, perfect, form, improve 091
DURCHBILDUNG (*f*): formation, development, perfection, improvement, construction, design 092
DURCHBOHREN: to bore through; punch; perforate; penetrate through 093
DURCHBRECHEN: to break through; open; pierce; cut; punch; perforate; break out; appear; collapse; fall in 094
DURCHBROCHEN: openwork; perforated 095
DURCHBRUCH (*m*): collapse; escape (Gas); eruption; aperture; opening; breach; crevasse; crevice 096
DURCHBRUCHSGESTEIN (*n*): (see **ERUPTIVGESTEIN**) 097
DURCHDRINGBAR: permeable; penetrable; pervious 098
DURCHDRINGEN: to penetrate, pervade, permeate, press (force a way) through 099
DURCHDRINGUNG (*f*): penetration 100
DURCHDRÜCKEN: to press (force; squeeze; strain; filter) through 101
DURCHFAHRT (*f*): passage (gateway) through; thoroughfare 102
DURCHFALL (*m*): diarrhoea (Med.); amount of riddlings falling through a grate; riddlings 103
DURCHFALLEN: to fall through; be transmitted (Light); collapse (Blast furnace charge) 104
DURCHFÄRBBAR: capable of being dyed through and through 105
DURCHFÄRBEN: to dye (stain) through and through 106
DURCHFEDERUNG (*f*): sagging, springing, deflection 107
DURCHFEUCHTEN: to soak; saturate 108
DURCHFLUSS (*m*): flow; flowing through 109
DURCHFLUSSMESSER (*m*): flow meter 110
DURCHFLUSSQUERSCHNITT (*m*): flow area, section of waterway 111
DURCHFRESSEN: to corrode; eat through; pit (of metals) 112
DURCHFRESSUNG (*f*): corrosion, pitting 113
DURCHFRIEREN: to chill-(to the bone): freeze through 114

DURCHFÜHRBAR : practicable ; feasible ; capable of being carried out 115

DURCHFÜHREN : to carry. out ; lead (convey) through ; execute ; accomplish ; conduct ; wash (Tinning process) 116

DURCHGANG (*m*) : passage ; transit ; gangway 117

DURCHGÄNGIG : permeable ; penetrable ; pervious ; general ; usual 118

DURCHGANGSFRACHT (*f*) : through (freight) rate 119

DURCHGANGSHAHN (*m*) : two-way (straight-through) cock 120

DURCHGANGSQUERSCHNITT (*m*) : cross sectional area 121

DURCHGANGSQUERSCHNITT (*m*), VOLLER- : full bore, full flow area 122

DURCHGANGSWIDERSTAND (*m*) : flow resistance 123

DURCHGANG (*m*), WÄRME- : heat transfer 124

DURCHGEBEN : to pass through ; filter ; strain 125

DURCHGEHEN : to go through ; run away ; be transmitted (Light) ; examine ; pierce ; pervade ; pass through ; race (of engines) 126

DURCHGEHEN (*n*) DER MASCHINE (*f*) : racing of an engine 127

DURCHGEHENDS : generally ; usually ; throughout ; completely 128

DURCHGERBEN : to tan thoroughly 129

DURCHGERBUNGSZAHL (*f*) : tanning number ; degree of tanning 130

DURCHGESEIHTES (*n*) : filtrate 131

DURCHGIESSEN : to pour through ; filter ; strain 132

DURCHGLÜHEN : to heat to red heat 133

DURCHGRIFF (*m*) : attractive power (exerted by an anode, through the grid, upon the electrons) (Wireless Tel.) 134

DURCHGUSS (*m*) : filter ; strainer ; gutter ; sink ; filtrate ; filtration ; pouring through 135

DURCHHEIZEN : to heat thoroughly 136

DURCHKOCHEN : to boil thoroughly 137

DURCHKOMMEN : to come through ; recover 138

DURCHKREUZEN : to cross ; traverse ; intersect 139

DURCHLASS (*m*) : filter ; sieve ; passage ; opening ; port ; strainer ; transmitter (light) ; culvert 140

DURCHLASSFARBE (*f*) : transmitted colour 141

DURCHLÄSSIG : pervious ; permeable ; penetrable 142

DURCHLAUF (*m*) : sieve ; collander ; passage ; diarrhoea (Med.) 143

DURCHLAUFEN : to run through ; cover (a distance) 144

DURCHLAUFENDE WELLE (*f*) : continuous shaft 145

DURCHLEITEN : to conduct ; pass-(through) 146

DURCHLEUCHTEN : to illuminate 147

DURCHLOCHEN : to punch, perforate,, pierce, bore 148

DURCHLÖCHERN : to perforate ; pierce ; punch ; bore ; hole ; make holes in 149

DURCHLOCHT : perforated 150

DURCHLÜFTEN : to air ; ventilate ; aerate (see LÜFTEN) 151

DURCHLÜFTUNG (*f*) : airing, aeration ; ventilation 152

DURCHMACHEN : to experience ; go through 153

DURCHMESSEN : to measure ; traverse ; pass through 154

DURCHMESSER (*m*) : diameter 155

DURCHMESSER (*m.pl*), KONJUGIERTE- : conjugate diameters 156

DURCHMESSER (*m*), LICHTER- : internal diameter 157

DURCHNÄSSEN : to soak ; steep ; wet 158

DURCHNEHMEN : to go over ; criticize ; analyse 159

DURCHNETZEN : to soak ; steep ; wet 160

DURCHPAUSEN : to trace 161

DURCHPRESSEN : to press (force ; squeeze ; strain ; filter) through 162

DURCHRÄUCHERN : to fumigate ; smoke 163

DURCHRECHNEN : to calculate ; evaluate ; run out ; check 164

DURCHREIBEN : to rub through ; strain ; chafe ; gall ; rub sore 165

DURCHREISSEN : to tear asunder ; break ; tear up 166

DURCHRÜHREN : to stir 167

DURCHRÜTTELN : to riddle, shake 168

DURCHSÄTTIGEN : to saturate 169

DURCHSAUGEN : to suck through 170

DURCHSÄUREN : to acidify ; make sour ; leaven 171

DURCHSCHEINEN : to shine through ; be translucent ; be (semi-) transparent 172

DURCHSCHEINEND : transparent ; translucent 173

DURCHSCHLAG (*m*) : filter ; strainer ; punch ; piercer ; opening ; breach ; carbon copy ; drift ; meeting of headings (Mining) 174

DURCHSCHLAGBODEN (*m*) : perforated bottom 175

DURCHSCHLAGEN : to filter ; strain ; sieve ; punch ; perforate ; beat (break) through ; shine through ; penetrate ; blot (paper) ; act ; operate 176

DURCHSCHNEIDEN : to intersect ; cross ; cut (through) across ; pass (through) over ; traverse 177

DURCHSCHNITT (*m*) : cut ; section ; intersection ; average ; mean ; profile 178

DURCHSCHNITT (*m*) IM- : on an average 179

DURCHSCHNITTLICH : average ; mean ; on an average 180

DURCHSCHNITTSPROBE (*f*) : average (sample) test 181

DURCHSCHNITTSWERT (*m*) : average (mean) value 182

DURCHSCHNITTSZAHL (*f*) : average ; mean 183

DURCHSCHNITTSZIFFER (*f*) : average ; mean 184

DURCHSCHUSS (*m*) : woof ; weft ; interleaf ; lead (Printing) ; space lines 185

DURCHSCHÜTTELN : to shake ; agitate ; riddle 186

DURCHSEIHEN : to strain ; filter ; percolate 187

DURCHSETZEN : to permeate ; infiltrate ; intermingle (Mining) ; sieve (Mining) ; deal with ; sift ; use ; intersect ; carry (put) through ; pass through ; mix ; add 188

DURCHSETZT : mixed, streaked 189

DURCHSETZT, STARK- : having a high content-(of) 190

DURCHSETZZEIT (*f*) : time taken by the raw material to pass from the throat to the hearth of blast furnaces 191

DURCHSICHT (*f*) : view ; perusal ; examination ; inspection ; revision 192

DURCHSICHTIG : transparent, clear 193

DURCHSICKERN : to trickle through ; percolate ; strain ; filter 194

DURCHSIEBEN : to sift ; sieve ; screen ; bolt ; riddle 195

DURCHSTECHEN: to pierce; cut; stab; smelt (Metal); perforate; pass; feed to rolling mills 196
DURCHSTOSSEN: to push (punch) through; perforate 197
DURCHSTRAHLEN: to radiate; shine through; irradiate; penetrate (Rays) 198
DURCHSTREICHEN: to cross out; run (go) through; sift; screen 199
DURCHSTREICHENDE LINIE (*f*): trajectory 200
DURCHSTRÖMEN: to flow (rush) through 201
DURCHTEUFEN: to sink (a shaft through a particular layer of subsoil) 202
DURCHTRÄNKEN: to saturate; impregnate; infiltrate 203
DURCHTRÄNKUNG (*f*): impregnation (see DURCHTRÄNKEN) 204
DURCHTREIBEN: to drive (force) through; distil 205
DURCHTRETEN: to pass through 206
DURCHTRIEBEN: cunning; clever 207
DURCHTRITT (*m*): passage, entrance, thoroughfare 208
DURCHWACHSEN: to grow through; interpenetrate; intermingle 209
DURCHWACHSENE KOHLEN (*f.pl*): lumps of coal with foreign matter embedded (size 0-10mm.) 210
DURCHWÄSSERN: to soak; drench 211
DURCHWEG (*m*): thoroughfare; passage 212
DURCHWEG: throughout 213
DURCHWERFEN: to throw through; sift; screen; bolt; traject 214
DURCHWÜHLEN: to rake up; ransack; dig through; grub; rummage 215
DURCHWURF (*m*): sieve; screen; riddle; riddlings; screenings 216
DURCHZEICHNEN: to trace (of drawings) 217
DURCHZIEHEN: to draw through; intermix; interweave; traverse 218
DURCHZIEHGLAS (*n*): slide (of microscope) 219
DURCHZUG (*m*): draught; passage; girder; circulation; stripping; stripper (Plates) 220
DURFTIG: needy; scanty; poor 221
DUROCHINON (*n*): duroquinone 222
DUROL (*n*): durene; $C_6H_2(CH_3)_4$; Mp. 79-81°C.; Bp. 189-191°C. 223
DÜRR: dry; arid; withered; dead; barren; sterile; plain; lean 224
DÜRSCHENÖL (*n*): (see ICHTHYOLÖL, ROHES-) 225
DÜRSTENBLUT (*n*): (see ICHTHYOLÖL, ROHES-) 226
DURSTIG: thirsty; eager 227
DURSTSTILLEND: thirst-quenching 228
DÜSE (*f*): nozzle; blast-pipe; tuyère 229
DÜSENBODEN (*m*): bottom of melting pan, fitted with nozzles made of burnt clay for Bessemer and magnesite for Thomas, processes, for letting in the air 230
DÜSENMESSER (*m*): nozzle meter (Steam) 231
DÜSENREGLER (*m*): nozzle regulator 232
DÜSENSCHREIBER (*m*): recording steam meter 233
DÜSENSTOCK (*m*): nozzle pipe, blast connection (of blast furnaces) 234
DÜSENWÄCHTER (*m*): (see DÜSENREGLER) 235
DÜTE (*f*): paper bag; assay crucible; glass cylinder 236
DÜTENMASCHINE (*f*): paper-bag machine 237
DWARSBALKEN (*m*): cross beam (Naut.) 238
DYMAL (*n*): (see DIDYM, SALIZYLSAURES-) 239
DYNAKTINOMETER (*n*): actinometer (an instrument for measuring the actinism of the sun's rays, usually by means of sensitive paper) (Photography) 240
DYNAMIK (*f*): dynamics 241
DYNAMISCH: dynamic 242
DYNAMIT (*n*): dynamite (the name given to all explosives the active principle of which is nitroglycerine) 243
DYNAMO (*f*): dynamo, generator (Elect.) 244
DYNAMOANKER (*m*): armature 245
DYNAMO (*f*), **AUSGLEICH-**: balance dynamo 246
DYNAMO (*f*), **AUSSENPOL-**: external pole dynamo 247
DYNAMOBÜRSTE (*f*): dynamo brush (Elect.) 248
DYNAMO (*f*), **ERREGER-**: exciting dynamo 249
DYNAMO (*f*), **INNENPOL-**: internal pole dynamo 250
DYNAMOMASCHINE (*f*): dynamo 251
DYNAMO (*f*), **MEHRPOLIGE**: multipolar dynamo 252
DYNAMOMETER (*n*): dynamometer 253
DYNAMOÖL (*n*): dynamo oil 254
DYNAMORAUM (*m*): dynamo (generator) room 255
DYNAMO (*f*), **VIELPOL-**: multipolar dynamo 256
DYNAMO (*f*), **ZUSATZ-**: supplementary dynamo 257
DYNE (*f*): dyne (C.G.S. unit of force) 258
DYSMENORRHOË (*f*): dysmenorrhœa (form of menstruation) (Med.) 259
DYSPEPSIE (*f*): dyspepsia (Med.) 260
DYSPEPTISCH: dyspeptic 261
DYSPROSIUMCHLORID (*n*): dysprosium chloride; $DyCl_3$; Sg. 3.67 262
DYSTEKTISCH: dystectic 263
DYSTOMGLANZ (*m*), **DIPRISMATISCHER-**: bournonite 264
DYSTOMGLANZ (*m*), **DODEKAEDRISCHER-**: tennantite 265
DYSTOMGLANZ (*m*), **HEMIPRISMATISCHER-**: plagionite 266
DYSTOMGLANZ (*m*), **HEXAEDRISCHER-**: stannine 267
DYSTOMGLANZ (*m*), **PRISMATOIDISCHER-**: wolchite; Sg. 5.75 268
DYSTOMGLANZ (*m*), **RHOMBOEDRISCHER-**: zinckenite 269
DYSTOMGLANZ (*m*), **TETRAEDRISCHER-**: tetrahedrite, fahlore 270
DYSTOMSPAT (*m*), **PRISMATISCHER-**: datholite, datolite 271

E

EBBANKER (*m*): ebb anchor 272
EBBE (*f*): ebb-(tide) 273
EBBEN: to ebb; decline 274
EBEN: even; level; flat; smooth; exact; just; plane; plain; precise-(ly) 275
EBENBAUM (*m*): ebony tree (*Diospyros ebenum*) 276
EBENBILD (*n*): image; likeness 277
EBENBÜRTIG SEIN: to be as good (satisfactory; efficient) as; be equally good; be of equal birth 278
EBENDA: at that same place; (also the name of a German publication) 279
EBENE (*f*): plane; plain; level; flat; smooth surface 280

EBENEN: to level; even; plane; flatten; smooth 281
EBENHEIT (*f*): flatness (Photography, etc.) 282
EBENHOLZ (*n*): ebony-(wood) (from *Diospyros ebenum*); Sg. 1.0-1.5 283
EBENHOLZLACK (*m*): ebony varnish 284
EBENIEREN: to ebonize 285
EBENIST (*m*): ebonist; cabinet maker 286
EBENMASS (*n*): symmetry; harmony 287
EBENSO: similarly 288
EBER (*m*): (wild)-boar (*Sus scrofa*) 289
EBERESCHE (*f*): mountain ash; quicken tree (*Pyrus Aucuparia*); Sg. 0.69-1.13 290
EBERRAUTE (*f*): abrotanum (*Artemisia abrotanum*) (an aromatic plant) usually termed Southernwood 291
EBERREIS (*n*): (see EBERRAUTE) 292
EBERWURZ (*f*): carline thistle (*Carlina vulgaris*) 293
EBONIT (*n*): ebonite, vulcanized rubber, vulcanite (rubber which has been hardened by treatment with sulphur) (see also KAUTSCHUK) 294
EBULLIEREN: to boil up; break out; bubble 295
EBURNEUMVERFAHREN (*n*): Eburneum process (Photography) 296
ECGONIN (*n*): ecgonine; $C_9H_{15}NO_3.H_2O$; Mp. 198°C. 297
ECGONINCHLORHYDRAT (*n*): ecgonine hydrochloride; $C_9H_{15}NO_3HCl$; Mp. 246°C. 298
ECHAPPEÖL (*n*): recovered oil 299
ECHINOPSÖL (*n*): echinops oil (from seeds of *Echinops ritro*) 300
ECHT: real; genuine; true; proper; pure; unadulterated; fast (of colours); ingrained 301
ECHTBEIZENFARBE (*f*): fast mordant (dye) colour 302
ECHTBLAU (*n*): fast blue (see NAPHTYLENBLAU) 303
ECHTDAMPFFARBE (*f*): fast steam (dye) colour 304
ECHTDRUCKFARBE (*f*): fast printing (dye) colour 305
ECHTFARBE (*f*): ingrained colour, fast colour 306
ECHTGELB (*n*): fast (ingrained) yellow; aniline yellow 307
ECHTHEITSGRAD (*m*): degree of fastness (of dyes) 308
ECHTHEITSPRÜFUNG (*f*): fastness (testing) test (of colours) 309
ECHTROT (*n*): fast red (see BLEIMENNIGE) 310
ECHTSCHWARZ (*n*): fast black 311
ECHTWOLLFARBE (*f*): fast wool (dye) colour 312
ECK (*n*): angle; corner; edge; summit; facet (of a crystal) 313
ECKBESCHLAG (*m*): corner (cleat, clamp) binding 314
ECKE (*f*): angle; corner; edge; plane (solid) angle; summit; facet (of a crystal); quoin; coin (Arch.) 315
ECKEISEN (*n*): angle iron 316
ECKEN: to corner, jamb, cant, tilt, turn slightly on edge, stick, get (run) out of truth, skew, run out of line, get out of alignment 317
ECKER (*f*): acorn (nut of *Quercus robur*, etc.); beechnut 318
ECKERDOPPE (*f*): acorn cup 319
ECKFEILE (*f*): angular file 320
ECKHOLZ (*n*): squared timber 321
ECKIG: angular; cornered; angled; awkward; edged 322
ECKPFEILER (*m*): corner (pillar) column 323
ECKSÄULE (*f*): corner (pillar) column; prism 324
ECKSÄULIG: prismatic 325
ECKSTEIN (*m*): corner (curb) stone 326
ECKVENTIL (*n*): angle valve 327
ECZEM (*m*): eczema (skin disease) (Med.) 328
EDEL: noble; refined (Metals); inert (Gases); precious (Stones); vital (Organs); rich (Mining); generous (Wine) 329
EDELERDE (*f*): rare (rich) earth 330
EDELERZ (*n*): rich ore 331
EDELESCHE (*f*): ash tree (*Fraxinus excelsior*) (see ESCHENHOLZ) 332
EDELFICHTE (*f*): silver fir (see EDELTANNE) 333
EDELGALMEI (*m*): Smithsonite (a natural zinc carbonate; $ZnCO_3$; 52% Zn) 334
EDELGAS (*n*): rare gas; inert gas (either Helium, Neon, Xenon, Krypton or Argon) 335
EDELHIRSCH (*m*): stag (*Corvus elaphus*) 336
EDELMETALL (*n*): precious metal 337
EDELMÜTIG: noble; generous; magnanimous 338
EDELPASSUNG (*f*): wringing fit (Mech.) 339
EDELROST (*m*): basic copper carbonate (green coating formed on copper exposed to moist air) (see KUPFERCARBONAT) 340
EDELSALZ (*n*): refined salt (see NATRIUMCHLORID) 341
EDELSTAHL (*m*): refined steel; steels which resist oxydation (the best sorts of steel such as crucible steel, electro steel and special steels, including nickel steel, chromium steel, manganese steel, tungsten steel, vanadium steel, molybdenum steel) 342
EDELSTEIN (*m*): precious stone; jewel; gem 343
EDELSTEIN (*m*), **KÜNSTLICHER-**: artificial precious stone (such as artificial rubies and sapphires) 344
EDELTANNE (*f*): silver fir (*Abies picea*) (see also HARZKIEFER and FICHTE) 345
EDELTANNENHOLZ (*n*): pitchpine wood (see HARZKIEFER) 346
EDELTANNENÖL (*n*): pine oil, pine needle oil; Sg. 0.85-0.9; Bp. 150-185°C. (see FICHTEN-NADELÖL) 347
EDELWEISS (*n*): edelweiss (*Leontopodium alpinum*) 348
EDELWILD (*n*): deer 349
EDINGTONIT (*m*): Edingtonite; Sg. 2.71 350
EDINOL (*n*): edinol, aminooxybenzyl alcohol (a photographic developer) (para-amino-saligenin, potassium hydroxide and potassium bromide); hydrochloride of meta-amino-ortho-oxy-benzoyl alcohol, $C_6H_3.OH.NH_2HCl.CH_2OH$; Mp. 135-142°C. 351
EDINOLENTWICKLER (*m*): edinol developer 352
EDINOLHYDROCHINON (*n*): edinol-hydroquinone (a photographic developer) 353
EDISONFASSUNG (*f*): Edison socket 354
EDISONIT (*m*): (see TITANDIOXYD); Sg. 4.26 355
EDITIEREN: to edit; publish 356
EDULKORIEREN: to edulcorate; wash 357
EFEU (*m*): ivy (*Hedera helix*) 358
EFEUGEWÄCHSE (*n.pl*): *Hederaceae* (Bot.) 359
EFFEKT (*m*): effect 360
EFFEKTEN (*m.pl*): effects; bills, paper (Banking) 361
EFFEKTENBÖRSE (*f*): Stock Exchange 362
EFFEKTENHANDEL (*m*): Stock Exchange business 363
EFFEKTENHÄNDLER (*m*): stock jobber 364
EFFERVESZIEREN: to effervesce; froth; ferment 365

EFFLORESCIEREN : to effloresce ; bloom 366
EFFLORESZIEREN : to effloresce ; bloom 367
EFFUSIVGESTEIN (*n*) : volcanic rock, effusive rock 368
EGAL : equal ; even 369
EGALISIEREN : to equalize ; even ; level ; flatten ; plane ; smooth 370
EGALISIERER (*m*) : equalizer ; levelling agent 371
EGALISIERUNGSVERMÖGEN (*n*) : equalizing power 372
EGALITÄT (*f*) : equality 373
EGEL (*m*) : leech 374
EGELKRAUT (*n*) : spearwort ; lesser spearwort (*Ranunculus flammula*) ; greater spearwort (*Ranunculus lingua*) 375
EGGE (*f*) : harrow ; selvedge ; selvage 376
EGGEN : to harrow 377
EGOUTTEUR (*m*) : dandy roll (Paper) 378
EGRENIERMASCHINE (*f*) : gin (Cotton) 379
EHELICHEN : to marry 380
EHERN : brazen ; brass ; bronze 381
EHESCHEIDUNG (*f*) : divorce 382
EHEST : soonest ; next ; premier ; first 383
EHRBEGIERDE (*f*) : ambition 384
EHRE (*f*) : honour ; reputation ; character ; reverence ; credit ; respect ; esteem ; deference 385
EHREN- : (as a prefix), honorary (of a title, degree, or post) 386
EHRENAMT (*n*) : post of honour ; honorary post 387
EHRENBEZEIGUNG (*f*) : mark of honour or esteem 388
EHRENERKLÄRUNG (*f*) : apology, reparation 389
EHRENMITGLIED (*n*) : honorary member 390
EHRENPREIS (*m*) : speed-well, veronica-(herb) (*Veronica officinalis*) 391
EHRENPUNKT (*m*) : point of honour 392
EHRENRETTUNG (*f*) : vindication ; apology 393
EHRENRÜHRIG : libellous ; defamatory ; injurious 394
EHRENSTELLE (*f*) : place (post) of honour 395
EHRENSTUFE (*f*) : degree (of honour) 396
EHRERBIETIG : respectful ; reverent ; deferential 397
EHRFURCHT (*f*) : awe, veneration, respect, reverence 398
EHRGEFÜHL (*n*) : sense of honour 399
EHRGEIZ (*m*) : ambition 400
EHRLICH : honourable, honest, fair, just 401
EHRLICH-HATA 606 : (see SALVARSAN) 402
EHRWÜRDIG : reverend ; venerable 403
EI (*n*) : egg ; ovum 404
EIALBUMIN (*n*) : egg (albumen) albumin, ovalbumin 405
EIBE (*f*) : yew-(tree) (see EIBENBAUM) 406
EIBENBAUM (*m*) : yew tree (*Taxus baccata*) ; Sg. 0.74-0.94 407
EIBENGEWÄCHSE (*n.pl*) : *Taxaceæ* (Bot.) 408
EIBISCH (*m*) : marshmallow (dried root, flowers and leaves of *Althæa officinalis*) 409
EIBISCHBLÄTTER (*n.pl*) : althea (marshmallow) leaves (*Folia Althæae*) 410
EIBISCHWURZEL (*f*) : Althea (marshmallow) root (*Radix Althaeæ*) 411
EICHAMT (*n*) : testing (gauging) office 412
EICHAPFEL (*m*) : oak apple ; gall-(nut) 413
EICHBAUM (*m*) : (see EICHE) 414
EICHE (*f*) : oak (British = *Quercus robur*) (see also SOMMEREICHE and WINTEREICHE) ; gauge, standard, calibrater 415
EICHEL (*f*) : acorn (nut of *Quercus robur*, etc.) ; gland 416
EICHELDOPPE (*f*) : acorn cup 417
EICHELFÖRMIG : acorn shaped ; glandiform 418
EICHELKAFFEE (*m*) : acorn coffee (made from the seeds of *Quercus robur*) 419
EICHELKAKAO (*m*) : acorn cocoa (made from the seeds of *Quercus robur*) 420
EICHELÖL (*n*) : acorn oil (from *Quercus agrifolia*) 421
EICHELZUCKER (*m*) : quercite ; acorn sugar ; quercitol ; $C_6H_7(OH)_5$; Sg. 1.5845 ; Mp. 234°C. 422
EICHEN : to gauge ; measure ; standardize ; test ; calibrate ; stamp or adjust (Weights and Measures) ; graduate 423
EICHEN (*adj.* and *adv.*) : oaken ; of oak ; (*n*) : ovule, little egg 424
EICHENGERBSÄURE (*f*) : quercitannic acid ; oak tannin 425
EICHENGERBSTOFF (*m*) : (see EICHENGERBSÄURE) 426
EICHENHOLZ (*n*): oak wood (see SOMMEREICHE and WINTEREICHE) 427
EICHENHOLZEXTRAKT (*m*) : oak (oak wood) extract 428
EICHENHOLZLACK (*m*) : oak varnish 429
EICHENMISTEL (*f*) : oak mistletoe (*Viscum quercinum*) ; (mistletoe which only grows on oak ; this is very rare) (see MISTEL) 430
EICHENMOOS (*n*) : oak moss (*Lichen quercinum*) 431
EICHENRINDE (*f*) : oak bark (from *Quercus robur*, *Q. pendunculata* or *Q. sessiliflora*) ; quercitron (bark of *Quercus nigra* or *Q. tinctoria*) 432
EICHENROT (*n*) : oak red 433
EICHENSAMEN (*m*) : acorn (nut of *Quercus robur*, etc.) 434
EICHENWERFTKÄFER (*m*) : wood-eater, wood-beetle (*Lymexylon navale*) 435
EICHER (*m*) : gauge-(r) ; tester ; calibrater ; inspector of weights and measures 436
EICHFÄHIG : capable of being (adjusted) calibrated ; adjustable ; standardizable 437
EICHHOLZ (*n*) : (see EICHENHOLZ) 438
EICHMASS (*n*) : standard measure ; gauge 439
EICHMEISTER (*m*) : calibrater ; gauger ; tester ; adjuster 440
EICHMETALL (*n*) : similar to Delta metal (see DELTAMETALL) 441
EICHNAGEL (*m*) : gauge mark 442
EICHSCHEIN (*m*) : certificate oı calibration 443
EICHUNG (*f*) : calibration (see EICHEN) 444
EID (*m*) : oath ; affidavit 445
EIDBRUCH (*m*) : perjury 446
EIDECHSE (*f*) : lizard (*Lacerta* genus) 447
EIDECHSENARTIG : lacertine 448
EIDERENTE (*f*) : eider duck (*Somateria mollissima*) 449
EIDGENOSS (*m*) : confederate 450
EIDLICH : sworn ; on oath 451
EIDOTTER (*m*) : yolk of egg 452
EIDOTTERFETT (*n*) : lecithin (*Lecithinum*) ; ovalecithin 453
EIERALBUMIN (*n*) : egg albumin ; ovalbumin 454
EIERKONSERVE (*f*) : egg preservative 455
EIERÖL (*n*): egg-yolk oil (from hen, duck and goose eggs) 456
EIERPULVER (*n*) : egg powder ; custard powder 457
EIERSCHALE (*f*) : egg-shell 458

EIERSTOCK (*m*): ovary 459
EIERWEISS (*n*): egg-white 460
EIFER (*m*): zeal; eagerness; fervour; passion; ardour 461
EIFERN: to be zealous; be jealous; be angry; declaim 462
EIFERSUCHT (*f*): jealousy; envy 463
EIFÖRMIG: ovoid; egg-shaped; oval 464
EIGELB (*n*): egg-yolk 465
EIGEN: own; individual; proper; specific; special; odd; exact; self-contained; peculiar; dead– 466
EIGENARBEIT (*f*): no-load work 467
EIGENARBEITSVERMÖGEN (*n*): specific (intrinsic, internal) energy 468
EIGENARTIG: peculiar; singular; individual; original 469
EIGENBELASTUNG (*f*): dead-load 470
EIGENBEWEGUNG (*f*): individual (proper; specific) motion 471
EIGENDREHUNG (*f*): individual (proper; specific) rotation 472
EIGENENERGIE (*f*): specific (intrinsic; internal) energy 473
EIGENFREQUENZ (*f*): individual (specific) frequency 474
EIGENGEWICHT (*n*): specific gravity; dead-weight; own (specific; proper) weight; tare (Railways) 475
EIGENHÄNDIG: single-handed; with one's own hand 476
EIGENLAST (*f*): dead-load 477
EIGENMÄCHTIG: despotic; arbitrary 478
EIGENMITTEL (*n*): specific-(remedy) 479
EIGENNAME (*m*): proper name 480
EIGENNÜTZIG: selfish 481
EIGENREIBUNG (*f*): internal friction 482
EIGENS: expressly; particularly 483
EIGENSCHAFT (*f*): property; quality; peculiarity; nature; condition; attribute 484
EIGENSCHAFTSWORT (*n*): adjective; epithet 485
EIGENSCHAFTSZEICHEN (*n*): distinctive mark 486
EIGENSCHWINGUNG (*f*): individual (oscillation) vibration; natural (oscillation) vibration 487
EIGENSINN (*m*): stubbornness; wilfulness 488
EIGENTLICH: proper; true; real; intrinsic; actual-(ly) 489
EIGENTUM (*n*): property; ownership; possession-(s) 490
EIGENTÜMLICH: peculiar; characteristic; specific 491
EIGENVERGRÖSSERUNG (*f*): actual (real) magnification or enlargement 492
EIGENWÄRME (*f*): specific heat; total heat (of a body) (see also KÖRPERWÄRME) 493
EIGENWIDERSTAND (*m*): internal resistance 494
EIGNEN: to suit; qualify; appropriate; be adapted; be fit; be proper; be capable of 495
EIGON (*m*): eigon (compound of egg-white, iodine and bromine) 496
EIKONOGEN (*n*): eikonogen, sodium salt of amino-β-naphthol-β-mono-sulfonic acid (a photographic developer); $NH_2C_{10}H_5(OH)SO_3Na.2\frac{1}{2}H_2O$ 497
EILGUT (*n*): express goods; goods sent by express goods or mail trains 498
EILGUT (*n*), ALS-: by passenger train 499
EILZUG (*m*): express train 500
EIMER (*m*): bucket; pail 501
EINACHSIG: uniaxial; having one axle or axis 502
EINANKERUMFORMER (*m*): rotary converter (Elect.) 503
EINÄSCHERN: to incinerate; calcine; convert (burn) to ashes 504
EINATMEN: to inhale; breathe in 505
EINATOMIG: monatomic 506
EINAXIG: uniaxial 507
EINBADSCHWARZ (*n*): one dip black (Dye) 508
EINBADVERFAHREN (*n*): one (single) dip or bath process 509
EINBALSAMIEREN: to embalm 510
EINBALSAMIERUNG (*f*): embalming 511
EINBAND (*m*): cover; binding 512
EINBASISCH: monobasic 513
EINBASISCHE SÄURE (*f*): monobasic acid 514
EINBAU (*m*): housing accessory (Rolling mills) 515
EINBAUEN: to instal; incorporate; couple up; build in-(to); fit in-(to); arrange; equip; brick up; set (Boilers) 516
EINBAUEN (*n*): brickwork, setting, etc. 517
EINBAUSTÜCK (*n*): housing accessory 518
EINBAUTEN (*pl*): installations 519
EINBEHALTEN: to keep back; retain 520
EINBEIZEN: to etch in 521
EINBETTEN: to embed; bed (down) in 522
EINBEULEN: to bend in; deflect inwards; hump inwards; swell inwards; buckle; cockle 523
EINBIEGEN: to bend in; turn down; sag; deflect; indent 524
EINBIEGUNG (*f*): curvature; deflection 525
EINBILDEN: to imagine; think; pride oneself 526
EINBILDUNG (*f*): imagination; fancy; conceit 527
EINBINDEN: to bind-(in); fasten 528
EINBLASEDRUCK (*m*): injection pressure (of Diesel motor) 529
EINBLASEN: to blow-(in); inject; insufflate 530
EINBLICK (*m*): insight 531
EINBRECHEN: to break in; begin; force an entry-(into) 532
EINBRENNEN: to anneal; burn in; cauterize; brand 533
EINBRENNFARBE (*f*): colour (paint) for porcelain painting 534
EINBRINGEN: to bring in; yield; insert 535
EINBRUCH (*m*): breaking-(in or down); forcible entry 536
EINBRUCH MACHEN: to cut; hole; trench; break ground; force (effect) an entrance 537
EINBRÜHEN: to scald; steep in boiling water 538
EINBUCHTUNG (*f*): inward bulge or curvature; concavity 539
EINBÜRGERN: to adopt; naturalize; be adopted (naturalized); come (make a way) to the fore 540
EINBUSSE (*f*): loss; damage 541
EINDAMPFAPPARAT (*m*): vapourizing apparatus; steam drier 542
EINDAMPFEN: to vapourize; concentrate; inspissate; boil down; steam; smoke (see ABDAMPFEN) 543
EINDAMPFKESSEL (*m*): evaporator; vaporizer 544
EINDAMPFSCHALE (*f*): evaporator, drier, drying cup 545
EINDAMPFUNG (*f*): vapourization 546
EINDECKEN: to cover-(in or over); roof 547
EINDEUTIG: plain; clear; unequivocal 548
EINDICKEN: to thicken; inspissate; concentrate (Solutions) 549

EINDICKEN (*n*): concentration 550
EINDICKUNG (*f*): concentration 551
EINDIMENSIONAL: unidimensional 552
EINDOCKEN: to dock (a ship) 553
EINDORREN: to dry up; shrink 554
EINDRINGEN: to press in; force in; penetrate; infiltrate; (*n*): entry; infiltration; penetration 555
EINDRINGLICH: impressive; forcible; express-(ly) 556
EIN DRITTEL GESÄTTIGTES CALCIUMPHOSPHAT (*n*): (see MONOCALCIUMPHOSPHAT) 557
EINDRUCK (*m*): impression; indentation; stamp 558
EINDRUCKEN: to imprint; impress; stamp; flatten; press; compress; indent; crush in 559
EINDRÜCKEN: (see EINDRUCKEN) 560
EINEN: to agree; unite 561
EINENDIG: single ended 562
EINENGEN: to concentrate; confine; compress; narrow; contract; swage; constrict 563
EINER (*m*): unit; digit 564
EINERLEI OB: irrespective of (no matter) whether 565
EINERSEITS: on the one hand 566
EINFACH: simple; single; elementary; once; primary; plain; proto-; mono-; -ous (Chemical); purely 567
EINFACHBASISCH: monobasic 568
EINFACHBROMJOD (*n*): iodine monobromide; IBr; Mp. 36°C. 569
EINFACHCHLORJOD (*n*): iodine monochloride; ICl; Mp. 25°C.; Bp. 101°C. 570
EINFACHCHLORSCHWEFEL (*m*): sulphur monochloride; sulphur protochloride; S_2Cl_2; Sg. 1.7094; Mp. – 80°C.; Bp. 138°C. 571
EINFACHCHLORZINN (*n*): tin protochloride; stannous chloride; $SnCl_2$; Mp. 249.3°C.; Bp. 603-628°C. 572
EINFACH CHROMSAUER: chromate of 573
EINFACHER KÖRPER (*m*): element; simple substance 574
EINFACH GEWÄSSERT: monohydrate 575
EINFACHKOHLENSAUER: neutral carbonate of 576
EINFACH SAURES KALZIUMPHOSPHAT (*n*): dibasic calcium phosphate (see DICALCIUMPHOSPHAT) 577
EINFACHSCHWEFELEISEN (*n*): ferrous sulphide; iron protosulphide; FeS; Sg. 3.4-4.75; Mp. 1179°C. 578
EINFACHSCHWEFELMETALL (*n*): -ous sulphide; monosulphide; protosulphide 579
EINFACHSCHWEFELZINN (*n*): stannous sulphide; tin protosulphide (see ZINNSULFÜR) 580
EINFACHTRANSPORTVERFAHREN (*n*): single-transfer carbon process (Photography) 581
EINFACHWIRKEND: single acting 582
EINFALL (*m*): falling in; decay; idea; inroad; incidence (Physics); inclination; occurrence 583
EINFALLEN: to fall in; occur-(to); strike one; happen; interrupt; fall upon; dip (Min.); be incident; incline-(towards) 584
EINFALLEND: incident (Physics) 585
EINFALLSLOT (*n*): perpendicular; ordinate 586
EINFALLSWINKEL (*m*): angle of incidence 587
EINFALZEN: to rabbet (of wood) 588
EINFALZUNG (*f*): rabetting 589
EINFÄRBEN: to dye thoroughly 590
EINFARBIG: monochromatic 591
EINFASSEN: to include; enclose; fill; case; barrel; bind 592
EINFETTEN: to grease; oil 593
EINFINDEN: to appear; arrive 594
EINFLAMMIG: single-flamed 595
EINFLECHTEN: to weave in; interweave 596
EINFLIESSEN: to flow in 597
EINFLÖSSEN: to instil; infuse 598
EINFLUSS (*m*): influx; influence 599
EINFLUSSROHR (*n*): influx (inlet) pipe; suction pipe 600
EINFORMEN: to mould 601
EINFÖRMIG: uniform 602
EINFRESSEN: to eat into; corrode; devour 603
EINFRIEREN: to freeze-(up); congeal 604
EINFÜGEN: to insert; fit in; splice; rabbet; dovetail 605
EINFUHR (*f*): importation; supply; bringing in; import 606
EINFÜHREN: to introduce; inaugurate; import; install 607
EINFÜHRUNG (*f*): inlet; feeding; introduction; inauguration; importation; installation; feed (branch) opening or inlet 608
EINFÜLLTRICHTER (*m*): charging (filling) funnel 609
EINGABE (*f*): presentation; petition 610
EINGANG (*m*): entrance; importation; introduction; receipt; mouth; inlet 611
EINGÄNGIG: single-threaded (of screws) 612
EINGEBEN: to give-(in); administer; present; deliver; surrender 613
EINGEBRANNTE PHOTOGRAPHIE (*f*): Ceramic (burnt-in) photograph 614
EINGEDENK: mindful-(of) 615
EINGEHEN: to go in-(to); enter; arrive; come to hand; shrink; delve into; agree to 616
EINGEHEN AUF: to delve into, agree to 617
EINGEHEND: exhaustive; close; re-entrant (Geom.); comprehensive; intricate; thorough-(ly); circumstantial 618
EINGESCHOBEN: intercostal 619
EINGESETZTES SCHWEISSEISEN (*n*): blister steel 620
EINGESTEHEN: to confess; admit; allow 621
EINGETAUCHT: immersed 622
EINGEWEIDE (*n*): entrails; bowels; viscera 623
EINGIESSEN: to pour in; infuse; transfuse; cast-(in); fill-(in) 624
EINGLEISIG: having a single line of metals (Railways); one-way 625
EINGRABEN: to engrave; dig in; trench 626
EINGREIFEN: to catch-(in); lock; interlock; infringe; grip; gear; engage; (*n*): gear-(ing); intervention; interference; catch; grip 627
EINGREIFEND: radical 628
EINGUSS (*m*): pouring in; mould; filling; infusion; grout-(ing); cast-(ing) 629
EINGUSSLOCH (*n*): grouting hole; aperture for filling 630
EINHALT (*m*): check; restraint; interruption; stop; impediment; suspension 631
EINHALTEN: to check; restrain; interrupt; prevent; keep-(in); take in; cease; stop; retain; discharge; impede; prohibit; suspend 632
EINHÄNDIGEN: to hand in or over; deliver 633
EINHÄNGEN: to ship a rudder 634
EINHAUEN: to cut into; cut up; cut open 635
EINHEIMISCH: native; indigenous; interior; domestic; home-made 636

EINHEIT (*f*) : unit-(y) 637
EINHEIT (*f*), **ABSOLUTE-** : absolute unit 638
EINHEIT (*f*), **ARBEITS-** : unit of work 639
EINHEIT (*f*), **KRAFT-** : unit of (power) force 640
EINHEIT (*f*), **LÄNGEN-** : unit of length 641
EINHEITLICH : unitary ; uniform ; homogenous ; centralized 642
EINHEITLICH ALS GANZES : as a self-contained unit 643
EINHEITLICHKEIT (*f*) : uniformity ; homogenity ; centralization 644
EINHEIT (*f*), **STROM-** : unit of current 645
EINHEIT (*f*), **WÄRME-** : unit of heat, heat (thermal) unit ; 1 W.E. = 3.968 B.T.U. ; 1 W.E. per Kg. = $\frac{9}{5}$ B.T.U. per lb. 646
EINHEIT (*f*), **ZEIT-** : unit of time 647
EINHELLIG : unanimous ; united 648
EINHOLEN : to overtake ; obtain ; haul in 649
EINHOLER (*m*) : inhaul (Naut.) 650
EINHÜLLEN : to wrap ; envelop ; embed ; involve ; (en)-case ; house 651
EINHÜLLENDES MITTEL (*n*) : dulcent 652
EINHYDRATIG : monohydrate 653
EINIG : united ; agreed ; agreeing 654
EINIGE : some ; a few 655
EINIGEN : to unite ; agree 656
EINIGERMASSEN : somewhat ; in some degree ; partly 657
EINJÄHRIG : annual ; of (for) one year 658
EINKALKEN : to lime ; treat (soak) with lime-(water) 659
EINKÄLKEN : (see **EINKALKEN**) 660
EINKASSIEREN : to cash ; pay in 661
EINKAUF (*m*) : purchase 662
EINKAUFSPREIS (*m*) : purchase (buying ; cost) price 663
EINKELLERN : to cellar ; store ; lay in 664
EINKERBEN : to notch ; indent 665
EINKERKERN : to immure, incarcerate, imprison 666
EINKERNIG : uninuclear ; mononuclear 667
EINKLAMMERN : to enclose in brackets or parenthesis ; cramp 668
EINKLANG (*m*) : harmony ; sympathy ; unison ; tune ; reconciliation 669
EINKLANG, IN- BRINGEN : to reconcile 670
EINKOCHEN : to boil down ; evaporate ; thicken ; inspissate 671
EINKOMMEN (*n*) : income ; revenue ; receipts ; profit ; (*verb*) : to solicit ; apply for 672
EINKRYSTALL (*m*) : monocrystal 673
EINKUPPELN : to engage, couple, throw into (mesh) gear 674
EINLADEN : to invite ; load into 675
EINLAGE (*f*) : enclosure ; insertion ; investment ; inlaying ; filler (Cigars) ; filling ; printing-frame pad (Photo.) ; darkslide carrier (Phot.) 676
EINLAGE FÜR PAPPE (*f*) : middles (Paper) 677
EINLAGERN : to infiltrate ; deposit ; store ; stratify 678
EINLASS (*m*) : inlet ; admission ; insertion ; injection ; feed ; suction 679
EINLASSEN, SICH- : to dabble in, embark in 680
EINLASSÖFFNUNG (*f*) : inlet (opening), feed inlet 681
EINLASSVENTIL (*n*) : feed (inlet) valve 682
EINLAUF (*m*) : inlet 683
EINLAUF (*m*), **DOPPELSEITIGER-** : double inlet 684
EINLAUFECHT : unshrinkable 685
EINLAUF (*m*), **EINSEITIGER-** : single inlet 686
EINLAUFEN : to enter ; arrive ; run in ; shrink ; contract ; call ; put in (at a port) 687
EINLAUFSCHACHT (*f*) : gully ; sump 688
EINLAUGEN : to buck ; soak (steep) in lye 689
EINLEBEN IN : to throw oneself heart and soul into ; live in (a part) 690
EINLEGEN : to put (place ; lay) in ; soak ; steep ; pickle ; preserve ; embed ; insert, inlay ; enclose 691
EINLEITEN : to introduce ; conduct ; pass ; insert ; convey in ; inject ; feed ; preface ; initiate ; start ; prepare ; cause 692
EINLEITUNGSROHR (*n*) : inlet (feed) pipe 693
EINLEUCHTEN : to be (clear ; plain ; obvious) evident 694
EINLEUCHTEND : plain, evident, obvious, clear, comprehensible 695
EINLÖSEN : to (cash) have cashed 696
EINLÖTEN : to solder in 697
EINMACHEN : to preserve ; conserve ; pickle ; temper (Lime) ; knead ; wrap up 698
EINMAISCHEN : to dough in ; (*n*) : doughing in (Brewing) 699
EINMALIG : single ; solitary 700
EINMAUERN : to wall up, immure, to brick in, brick up, to set (of boilers) 701
EINMAUERUNG (*f*) : brickwork-(setting), masonry, bricking-in, walling-up, setting 702
EINMISCHEN : to mix in ; intermix ; blend ; interfere 703
EINMÜNDEN : to empty ; discharge ; flow-into (of a river) ; inosculate 704
EINMÜTIG : united ; of one mind ; unanimous ; concordant 705
EINNAHME (*f*) : takings ; receipts ; revenue ; income ; turn-over 706
EINNEHMEN : to take in ; deceive ; receive ; occupy ; captivate ; overcome ; prejudice ; gather ; accept ; capture ; collect ; include ; infatuate ; be dizzy 707
EINÖDE (*f*) : solitude ; desert ; waste ; wild 708
EINÖLEN : to oil ; grease 709
EINORDNEN : to arrange ; classify 710
EINPACKEN : to pack up ; pack (stuff) in ; embed ; wrap ; lag ; bale 711
EINPHASENSTROM : single-phase current 712
EINPHASEN-WECHSELSTROM (*m*) : single-phase (A.C.) alternating current 713
EINPHASIG : single-phase ; monophase 714
EINPÖKELN : to pickle ; corn ; salt ; embalm 715
EINPÖKLER (*m*) : embalmer 716
EINPOLDYNAMO (*f*) : unipolar dynamo 717
EINPOLIG : unipolar 718
EINPRÄGEN : to imprint ; impress ; inculcate 719
EINPRESSEN : to press (force) in ; compress ; indent 720
EINPUDERN : to powder ; grind to powder 721
EINPUMPEN : to pump in 722
EINQUANTIG : single 723
EINQUELLEN : to steep ; soak 724
EINRAHMEN : to frame 725
EINRÄUCHERN : to fumigate ; smoke 726
EINRÄUMEN : to clear up ; stow (put) away ; furnish ; yield ; concede ; rake in 727
EINRECHNEN : to add in ; comprise ; take into account ; allow for ; reckon upon 728
EINREDE (*f*) : objection ; protest ; opposition ; remonstrance 729
EINREDEN : to talk into ; persuade ; interrupt ; object 730
EINREFFEN : to reef (Naut.) 731
EINREIBEN : to rub in-(to) ; smear ; grate 732

EINREIBUNG (*f*): rubbing in; smearing; liniment; embrocation 733
EINREICHEN: to hand in; present; send in; forward; tender; deliver 734
EINREIHIG: in a single row; single series 735
EINREISSEN: to rend; demolish; tear-(down or up); spread 736
EINRENNEN: to melt (run) down 737
EINRICHTEN: to arrange; adjust; organize; establish; set right; contrive; fit up; manage; adapt 738
EINRICHTUNG (*f*): contrivance; arrangement; furniture; setting (Bones); fittings; installation; equipment; plant; apparatus; device; machinery; fitting up; establishment 739
EINRISS (*m*): fissure; rent; crack; scratch 740
EINRITZEN: to scratch, etch 741
EINROSTEN: to rust in; be covered with rust 742
EINRÜCKEN: to engage (Machinery); throw in (gear or clutch); mesh; put in clutch 743
EINRÜCKHEBEL (*m*): engaging lever 744
EINRÜCKKUPPELUNG (*f*): engaging coupling 745
EINRÜCKVORRICHTUNG (*f*): engaging gear 746
EINRÜHREN: to stir up; beat; mix 747
EINSÄEN: to sow in 748
EINSÄGEN: to saw (cut) into 749
EINSALBEN: to grease; anoint; embalm 750
EINSALZEN: to salt; corn; pickle 751
EINSAM: alone; solitary; retired 752
EINSAMMELN: to collect; gather; garner; store; accumulate 753
EINSATZ (*m*): insertion; putting in; charge (Furnace); liner (Mech.); segment; deposit; set (of weights or turbine nozzles, etc.); case; internal casing (Turbines); (see EINGESETZT) 754
EINSATZ-AUFSTREUPULVER (*n*): case-hardening powder 755
EINSATZGEWICHTE (*n.pl*): nest (set) of weights 756
EINSATZHÄRTEMITTEL (*n*): case-hardening (compound) agent 757
EINSATZHÄRTEPULVER (*n*): case-hardening powder 758
EINSATZHÄRTUNG (*f*): case hardening 759
EINSATZKASTEN (*m*): hardening trough, annealing box 760
EINSATZKESSEL (*m*): set of (kettles) boilers 761
EINSATZOFEN (*m*): hardening (tempering, annealing, case-hardening) furnace 762
EINSATZÖFFNUNG (*f*): charging opening 763
EINSATZRINGE (*m.pl*): fitting rings 764
EINSATZSTAHL (*m*): case-hardening steel 765
EINSATZSTÜCK (*n*): distance (packing) piece, inserted piece 766
EINSATZTÜR (*f*): charging door (Metal); firing door (Fuels) 767
EINSATZZYLINDER (*m*): cylinder liner 768
EINSÄUERN: to acidify; sour; pickle; vinegar; leaven 769
EINSAUGEMITTEL (*n*): absorbent 770
EINSAUGEN: to absorb; suck (soak) up or in; imbibe 771
EINSAUGUNG (*f*): suction; absorption 772
EINSAUGUNGSMITTEL (*n*): absorbent 773
EINSÄURIG: monacid 774
EINSCHABEN: to grind in, machine to fit 775
EINSCHABUNG (*f*): machining (to fit) 776
EINSCHALEISEN (*n*): framing, steel framework, casing iron; structural steelwork (for ferro-concrete buildings) 777
EINSCHALEN: to encase, case, enclose 778
EINSCHALTEN: to let in; put in; insert; couple up; switch in; fit; introduce; cut in; intercalate 779
EINSCHALTER (*m*): circuit closer, switch 780
EINSCHALTUNG (*f*): switching in, joining in the circuit, insertion, intercalation, interpolation, introduction 781
EINSCHALUNG (*f*): casing, enclosure, mould, form; falsework (for ferro-concrete building construction); (see also SCHALUNG) 782
EINSCHÄRFEN: to enjoin; impress; inculcate 783
EINSCHÄTZEN: to estimate; evaluate; calculate; compute; reckon 784
EINSCHENKEN: to pour in; fill 785
EINSCHEREN: to reeve (Nautical) 786
EINSCHICHTEN: to embed; stratify; arrange in layers; pack 787
EINSCHICHTIG: single-layered 788
EINSCHIEBEN: to shove in; put in; insert; introduce; push in 789
EINSCHIFFEN: to ship, embark, go aboard, load 790
EINSCHLÄFERENDES MITTEL (*n*): soporific, narcotic (agent for promoting sleep) (Med.) 791
EINSCHLÄFERN: to narcotize; make drowsy; lull (into security) to sleep 792
EINSCHLÄFERND: narcotic; soporific 793
EINSCHLAG (*m*): wrapper; envelope; plait; fold; woof; weft; striking in 794
EINSCHLAGEN: to strike in; break; punch; wrap up; cover; follow; adopt; dip (Sheet Metal); sulphur (Wine); sink in; succeed; include 795
EINSCHLÄGIG: pertinent; appropriate; belonging (relating) to; connected with; inclusive of 796
EINSCHLEIFEN: to grind in; engrave; machine 797
EINSCHLEIFPASTE (*f*): grinding-(in) paste 798
EINSCHLIESSEN: to enclose; lock in; include; surround; form; comprise 799
EINSCHLIESSLICH: inclusive; including; comprising 800
EINSCHLUCKEN: to absorb; gulp down; swallow 801
EINSCHLÜPFEN: to slip in; soak in 802
EINSCHLUSS (*m*): inclusion; enclosure; seal; constituent; content; parenthesis 803
EINSCHLUSSROHR (*n*): sealed tube 804
EINSCHLUSSTHERMOMETER (*n*): enclosed-scale thermometer 805
EINSCHMALZEN: to oil; grease 806
EINSCHMELZEN: to melt down; fuse; cast; recast 807
EINSCHMELZGLAS (*n*): fusible glass 808
EINSCHMELZKOLBEN (*m*): melting flask 809
EINSCHMIEREN: to grease; smear; oil; lubricate 810
EINSCHNEIDEND: cutting, incisive-(ly), considerable, extensive 811
EINSCHNITT (*m*): cut; incision; notch; excavation; indentation; segment; cutting (Railways) 812
EINSCHNITTSTOLLEN (*m*): advance heading; through cutting (Mining) 813
EINSCHNÜREN: to bind (lace) up; constrict; throttle; restrict; narrow; swage 814
EINSCHRÄNKEN: to confine; limit; restrain; reduce; restrict; circumscribe; curtail; abrogate; abbreviate; qualify (something said) 815

EINSCHRAUBEN : to screw in 816
EINSCHREIBEN : to inscribe ; register ; note ; book ; enter ; write in 817
EINSCHREIBEN (*n*) : enclosure (at the bottom of correspondence) 818
EINSCHRUMPFEN : to shrink ; shrivel 819
EINSCHÜTTEN : to pour (out) in 820
EINSCHWEFELN : to sulphurize 821
EINSCHWINGEN : to oscillate (a shaft) 822
EINSEHEN : to perceive ; understand ; look into ; (*n*) : insight ; judgement 823
EINSEIFEN : to soap ; lather 824
EINSEITIG : one sided ; unilateral ; partial ; single-sided 825
EINSENKEN : to sink in ; be depressed ; deflect 826
EINSETZEN : to set (put) in ; insert ; plant ; preserve (Fruit) ; establish ; instal ; place ; fix ; step (a mast) ; caseharden (Metal) ; be set up ; occur ; take place ; incorporate ; charge (to a furnace) 827
EINSETZER (*m*) : combustion boat 828
EINSETZGEWICHTE (*n.pl*) : nest (set) of weights 829
EINSETZTÜR (*f*) : charging door (Metal) ; firing door (Fuels) 830
EINSICHT (*f*) : insight ; examination ; intelligence ; discernment 831
EINSICHTIG : discerning ; intelligent ; prudent ; judicious 832
EINSICKERN : to soak in ; infiltrate 833
EINSIEDLER (*m*) : hermit 834
EINSPANNEN : to grip ; stretch ; (*n*) : gripping ; stretching 835
EINSPIELEN : to balance ; practice (Music) ; become perfect by playing 836
EINSPRACHE (*f*) : exception ; objection ; protest 837
EINSPRENGEN : to sprinkle ; intersperse ; disseminate ; interstratify ; burst open 838
EINSPRENGLING (*m*) : component, constituent (usually referring to the crystals " sprinkled " throughout a rock) 839
EINSPRENGUNG (*f*) : dissemination 840
EINSPRINGEN : to spring in ; shrink (Fabric) ; re-enter (of an angle) ; catch (Locks) ; burst inwards 841
EINSPRITZEN : to squirt in ; inject ; syringe 842
EINSPRITZER (*m*) : injector ; syringe 843
EINSPRITZHAHN (*m*) : injection cock 844
EINSPRITZKONDENSATOR (*m*) : jet (injection) condenser 845
EINSPRITZSTRAHL (*m*) : (condensing)-jet 846
EINSPRITZUNG (*f*) : injection ; syringing 847
EINSPRUCH (*m*) : objection ; opposition ; exception 848
EINSTAMPFEN : to compress ; ram down ; stamp in-(to) 849
EINSTANDSPREIS (*m*) : prime cost ; cost price 850
EINSTÄUBEN : to dust ; powder ; grind to dust or powder ; pulverize 851
EINSTAUBVERFAHREN (*n*) : powder process (a process for producing transparencies on glass or prints on paper, by means of an organic sticky substance, sensitized with potassium or ammonium bichromate, exposed to the action of light, when certain sticky portions will dry, powder being then dusted on the remaining sticky parts) 852
EINSTAUCHMASCHINE (*f*) : expanding tool ; expander ; expanding machine (for tubes) 853
EINSTECHEN : to stick in ; pierce ; puncture ; punch 854
EINSTEHEN : to answer ; guarantee ; become possessed of 855
EINSTEIGETÜRE (*f*) : access door 856
EINSTELLBAR : adjustable 857
EINSTELLBAR, HAND- : hand operated 858
EINSTELLEN : to put(set) in ; adjust ; stop, standardize (a solution) ; set at ; incorporate ; include ; suspend ; regulate ; put (lay) up ; engage (Help) ; focus ; true ; align ; cease ; appear ; set (Magnet) ; turn up ; set to blow-off (Safety valve) 859
EINSTELLEND : aligning 860
EINSTELLER (*m*) : regulator ; thermostat 861
EINSTELL-LOUPE (*f*) : focussing magnifier (Photography) 862
EINSTELLSCHRAUBE (*f*) : focussing screw (Photography) 863
EINSTELLTUCH (*n*) : focussing cloth (Photography) 864
EINSTELLUNG (*f*) : alignment ; adjustment ; setting 865
EINSTICH (*m*) : puncture 866
EINSTIMMEN : to agree ; accord ; chime in ; consent 867
EINSTIMMIG : unanimous-(ly) 868
EINSTOSSEN : to thrust (ram ; break ; push ; stave ; kick ; force) in 869
EINSTREICHEN : to nick ; slit ; run ; take (rake ; fill) in 870
EINSTREUEN : to strew ; sprinkle ; disseminate 871
EINSTRÖMEN : to flow in ; stream in 872
EINSTROM-TURBINE (*f*) : single-flow turbine ; uniflow turbine 873
EINSTRÖMUNGSROHR (*n*) : inlet pipe 874
EINSTRÖMVENTIL (*n*) : inlet valve 875
EINSTUFIG : single-stage 876
EINSTÜRZEN : to fall in ; collapse ; demolish 877
EINSUMPFEN : to soak ; wet 878
EINTAUCHEN : to dip ; plunge ; immerse ; steep ; sink ; dive ; duck ; submerge 879
EINTAUCHER (*m*) : dipper (Photography) ; plunger ; sinker 880
EINTAUSCH (*m*) : exchange ; barter 881
EINTEIGEN : to make into a paste or dough 882
EINTEILEN : to divide-(up) ; distribute ; separate ; graduate ; part ; calibrate ; subdivide ; cut (saw) up ; classify 883
EINTEILUNG (*f*) : division ; subdivision ; graduation ; separation ; classification ; scale ; distribution 884
EINTEILUNGSBOGEN (*m*) : divided sheet (Lithography) 885
EINTÖNIG : monotonous 886
EINTÖNIGKEIT (*f*) : monotony 887
EINTRACHT (*f*) : harmony ; unity ; unanimity 888
EINTRAG (*m*) : entry ; woof ; weft ; damage ; prejudice 889
EINTRAGEN : to carry in ; introduce ; enter ; record ; yield ; plot (a line on a graph) ; book ; post ; incorporate ; register 890
EINTRÄGLICH : profitable ; lucrative 891
EINTRÄNKEN : to soak ; steep ; impregnate 892
EINTRÄUFELN : to drop in ; instil 893
EINTREFFEN : to arrive ; happen ; coincide ; agree 894
EINTREIBEN : to drive (rub) in ; collect ; exact (Payment) 895

EINTRETEN: to enter; set in; commence; occur; take place; happen; begin; supervene 896
EINTRETENDE FLÜGELKANTE (*f*): leading (forward) edge (of propeller) 897
EINTRETENDE KANTE (*f*): forward (leading) edge 898
EINTRICHTERN: to pour in by means of a funnel; inculcate 899
EINTRITT (*m*): entry; entrance; admission; appearance; commencement; incidence (Optics); inlet 900
EINTRITTSROHR (*n*), **DAMPF-**: steam inlet pipe, supply steam pipe 901
EINTRITTSSCHLITZ (*m*): inlet port 902
EINTROCKNEN: to dry (up) in; shrink; wither; wilt 903
EINTROCKNUNG (*f*): drying; shrinking 904
EINTROCKNUNGSPROZESS (*m*): drying process 905
EINTRÖPFELN: (see **EINTROPFEN**) 906
EINTROPFEN: to drop in; instal 907
EIN-UND-EINACHSIG: orthorhombic (Crystallography) 908
EINVERLEIBEN: to incorporate; embody 909
EINVERLEIBUNG (*f*): incorporation; embodiment 910
EINVERNEHMUNG (*f*): understanding 911
EINVERSTANDEN: agreed 912
EINVERSTÄNDNIS (*n*): agreement; understanding; intelligence 913
EINWÄGELÖFFEL (*m*): weighing-in spoon 914
EINWÄGEN: to weigh in 915
EINWALKEN: to full-close (Texile); force (oil, etc.) by fulling; oil; shrink by fulling 916
EINWALZAPPARAT (*m*): tube beader, tube expander 917
EINWALZEN: to expand (bead) tubes (into a boiler drum) 918
EINWALZSTELLE (*f*): tube hole (hole in the tube plate of boiler drums) 919
EINWAND (*m*): objection; exception; pretext 920
EINWANDERN: to immigrate 921
EINWANDFREI: unobjectionable; unexceptionable; without exception; satisfactory; flawless; objectionless; unbiassed 922
EINWÄRTS: inward-(s) 923
EINWÄSSERIG: one-bath (Dye) 924
EINWÄSSERN: to steep; soak; put in water 925
EINWEBEN: to weave in; shrink; intersperse; interweave 926
EINWEIBIG: monogynous; monogynian 927
EINWEICHEN: to steep; digest; soak; soften (by soaking); infuse; macerate; ret (Flax) 928
EINWENDEN: to object; oppose; take exception 929
EINWERFEN: to throw in; object 930
EINWERTIG: univalent 931
EINWERTIGE SÄURE (*f*): monobasic acid 932
EINWICKELMASCHINE (*f*): wrapping (packing) machine 933
EINWICKELN: to wrap up-(in); enclose; envelop 934
EINWICKELPAPIERE (*n.pl*): wrapping papers (Paper) 935
EINWIEGEN: to weigh in 936
EINWILLIGEN: to consent; approve; agree; acquiesce 937
EINWILLIGUNG (*f*): consent, assent, subscription, permission, approval, agreement, acquiescence 938
EINWINDEN: to wind in; entwine 939
EINWIRKEN: to act; influence; work (weave) in; affect 940
EINWIRKUNG (*f*): action; influence; effect; interweaving 941
EINWOHNER (*m*): inhabitant; resident; dweller 942
EINWURF (*m*): objection; reply; exception; pretext; slot (of post-box) 943
EINZAHLEN: to pay in 944
EINZAHLUNG (*f*): payment; instalment; share; deposit 945
EINZAHNEN: to indent; tooth; dove-tail; cog 946
EINZEL: single; individual; separate; sole; solo; various; unit- 947
EINZELARBEIT (*f*): output per unit 948
EINZELBLECH (*n*): single(tin)-plate 949
EINZELDÜSE (*f*): single nozzle 950
EINZELFALL (*m*): particular (individual; single) case 951
EINZELHEIT (*f*): detail; item; details; particulars 952
EINZELKRAFT (*f*): concentrated (single) force 953
EINZELLAST (*f*): concentrated (single) load 954
EINZELLIG: unicellular; single-celled 955
EINZELMÜNDUNG (*f*): single (port) opening 956
EINZELN: single; separate; solitary; sole; individual 957
EINZELNEN, IM-: in detail, minutely 958
EINZELÖLER (*m*): single lubricator 959
EINZELPOTENTIAL (*n*): single potential 960
EINZELREGELUNG (*f*): single regulation; governing by cutting out single nozzles 961
EINZELSCHIEBER (*m*): single (plain) valve 962
EINZELTEILE (*m.pl*): separate (individual, various) -parts or portions 963
EINZELVERBAND (*m*): single bandage; single bond 964
EINZELVERKAUF (*m*): retail 965
EINZELVORGANG (*m*): single (separate) process or reaction 966
EINZELWIRKUNGSGRAD (*m*): individual efficiency; efficiency per unit 967
EINZIEHEN: to draw (pull; soak) in; absorb; reduce; lessen; cash; contract; collect; confiscate; infiltrate; enter; shrink; retire; inhale; imbibe; taper; arrest; suppress; call in; withdraw 968
EINZIEHMESSER (*n*): reed-hook (Weaving) 969
EINZIEHUNGSMITTEL (*n*): absorbent 970
EINZIG: single; sole; only; unique 971
EINZUCKERN: to sugar; preserve 972
EINZUG (*m*): entry; entrance 973
EINZWÄNGEN: to force in; wedge in; pinch; confine; constrain 974
EINZWINGEN: to force (upon) into 975
EINZYLINDRIG: single-cylinder 976
EIRUND: oval 977
EIS (*n*): ice, (H_2O); Sg. 0.917 978
EISÄHNLICH: icy; like (resembling) ice 979
EISAMMER (*f*): snow bunting (*Plectrophanes nivalis*) 980
EISAPPARAT (*m*): refrigerating apparatus 981
EISARTIG: (see **EISÄHNLICH**) 982
EISBEERE (*f*): snowberry (an American shrub; *Symphoricarpus racemosus*) 983
EISBEIN (*n*): hip-bone (Med.) 984
EISBLUME (*f*): ice plant (see **EISKRAUT**) 985
EISBRECHER (*m*): ice breaker 986
EISEN: to break ice; ice, cover with ice 987

EISEN (*n*): iron (*Ferrum*); Fe; Sg. 7.85-7.88; Mp. 1530°C.; Bp. 2450°C. (see also **FERRI-** and **FERRO-**) 988
EISENABBRAND (*m*): iron waste 989
EISENABFALL (*m*): iron scrap; scrap iron; iron filings 990
EISENABFALLHÄNDLER (*m.pl*): scrap-iron merchants, scrap-metal merchants 991
EISENABGANG (*m*): iron waste or scrap 992
EISENABSTICH (*m*): tapping 993
EISENACETAT (*n*): iron acetate (see **EISENAZETAT**) 994
EISENALAUN (*m*): iron alum; either (ferric potassium sulphate, *Alumen ferratum*, $K_2Fe_2(SO_4)_4.24H_2O$; Sg. 1.806) or (ferric ammonium sulphate: $(NH_4)_2Fe_2(SO_4)_4.24H_2O$ 995
EISENALBUMINAT (*n*): iron (ferric) albuminate (*Ferrum albuminatum*) 996
EISENALBUMINATLÖSUNG (*f*): *Liquor ferri albuminati*): iron-albuminate solution (contains 0.4% iron) 997
EISENAMMONIAKALAUN (*m*): ferric ammonium sulphate (see **EISENALAUN**) 998
EISENANTIMONERZ (*n*): berthierite 999
EISENANTIMONGLANZ (*m*): (see **EISENANTIMONERZ**) 000
EISENARM: poor in iron; having low iron content 001
EISENARSENIK (*n*): iron arsenide; $FeAs_2$ 002
EISENARTIG: ferruginous; ferrous; iron-like; chalybeate; like iron 003
EISENASBEST (*m*): fibrous silica 004
EISENAUSSCHEIDER (*m*): iron separator 005
EISENAZETAT (*n*): ferric acetate (see **EISEN, ESSIGSAURES-**) 006
EISENAZETATFLÜSSIGKEIT (*f*): (*Liquor ferri acetici*): iron acetate (liquid) solution (solution of semibasic iron acetate; contains about 4.9% iron) 007
EISENBAHN (*f*): railway; railroad; train 008
EISENBAHNBEDARF (*m*): railway supplies 009
EISENBAHNGESETZ (*n*): railway (act) law 010
EISENBAHNGLEIS (*m* and *n*): railway track 011
EISENBAHNGRÜN (*n*): (see **CHROMGRÜN**) 012
EISENBAHNKREUZUNG (*f*): level (railway) crossing 013
EISENBAHNSCHIENE (*f*): iron rail, railway rail 014
EISENBAHNWAGEN (*m*): railway (carriage, coach, car) waggon 015
EISEN (*n*), **BALDRIANSAURES-**: iron valerate 016
EISENBALLEN (*m*): iron (puddle) ball 017
EISENBAND (*n*): iron hoop 018
EISEN (*n*), **BASISCHES-**: basic (steel) iron, mild steel (from Thomas process) 019
EISENBAUM (*m*): ironwood tree (*Ostrya virginica*) 020
EISENBAUTEN (*m.pl*): constructions of structural iron-work 021
EISENBEIZE (*f*): iron (liquor) mordant; iron acetate; iron nitrate; (see **EISENACETAT** and **EISENNITRAT**) 022
EISENBEIZE (*f*), **ESSIGSAURE-**: acetate iron mordant 023
EISENBEIZE (*f*), **SALPETERSAURE-**: nitrate iron mordant 024
EISENBEIZE (*f*), **SCHWEFELSAURE-**: sulphate iron mordant 025
EISEN (*n*), **BENZOESAURES-**: iron benzoate, ferric benzoate; $Fe_2(C_7H_5O_2)_6$ 026
EISENBERGWERK (*n*): iron mine 027
EISENBETON (*m*): ferroconcrete; reinforced concrete 028
EISENBETONBAU (*m*): ferro-concrete (reinforced concrete) construction 029
EISENBETONSCHIFF (*n*): ferro-concrete (reinforced concrete) ship 030
EISENBLAU (*n*): vivianite (see **BLAUE ERDE**) 031
EISENBLAUDRUCK (*m*): cyanotype process, ferroprussiate (blue) process, negative cyanotype (method of reproducing engineer's drawings) 032
EISENBLAUERDE (*f*): earthy vivianite (see **BLAUE ERDE**) 033
EISENBLAUSAUER: ferrocyanide of; combined with ferrocyanic acid 034
EISENBLAUSÄURE (*f*): ferrocyanic acid 035
EISENBLAUSPAT (*m*): vivianite (see **BLAUE ERDE**) 036
EISENBLAUSPAT (*m*), **FASERIGER-**: crocidolite (see **KROKIDOLITH**) 037
EISENBLECH (*n*): iron plate; sheet iron 038
EISENBLENDE (*f*): pitchblende; uraninite (see **PECHBLENDE**) 039
EISENBLOCK (*f*): ingot; bloom, block of iron 040
EISENBLUMEN (*f.pl*): iron flowers; ferric chloride $FeCl_3$; Sg. 2.804; Mp. 301°C. 041
EISENBLÜTE (*f*); aragonite (see **ARAGONIT**) 042
EISENBOR (*n*): ferro-boron 043
EISENBORID (*n*): iron boride; FeB; Sg. 7.15 044
EISENBRAUNKALK (*m*): dolomite (see **DOLOMIT**) 045
EISENBROMID (*n*): iron bromide; ferric bromide; $FeBr_3$ 046
EISENBROMÜR (*n*): ferrous bromide; $FeBr_2.6H_2O$, Sg. 4.636; Mp. 27°C. 047
EISENBROMÜRBROMID (*n*): ferrosoferric bromide 048
EISEN (*n*), **BRÜCHIGES-**: short iron 049
EISENBRÜHE (*f*): iron mordant (see **HOLZESSIG-SAURES EISEN** and **EISENBEIZE**) 050
EISENCARBID (*n*): iron carbide; Fe_3C; Sg. 7.07-7.396 051
EISENCARBONAT (*n*): iron carbonate, ferrous carbonate; $FeCO_3$ and $FeCO_3.H_2O$; Sg. 3.7-3.87; Mp. 470°C. 052
EISENCARBONYL (*n*): iron carbonyl; $Fe(CO)_4$; Sg. 1.996; $Fe(CO)_5$; Sg. 1.4664; $Fe_2(CO)_9$; Sg. 2.085 053
EISENCHINAWEIN (*m*): quinine and iron wine 054
EISENCHININ (*n*): quinine and iron 055
EISENCHININ (*n*), **CITRONSAURES-**: iron-quinine citrate, ferrous quinine citrate (*Ferri et quininæ citras*) 056
EISENCHLORID (*n*): iron (ferric) chloride; iron sesquichloride (*Ferrum sesquichloratum*); Flores martis; perchloride of iron; Fe_2Cl_6; (see **FERRICHLORID**) 057
EISENCHLORIDLÖSUNG (*f*): ferric chloride solution (*Liquor Ferri sesquichlorati*); Sg. 1.281; about 10% iron 058
EISENCHLORÜR (*n*): iron protochloride; ferrous chloride (*Ferrum chloratum*); $FeCl_2$; Sg. 2.988 059
EISENCHLORÜRCHLORID (*n*): ferrosoferric chloride 060
EISENCHLORWASSERSTOFF (*m*): ferrichloric acid 061
EISENCHROM (*n*): chromic iron; chromite; ferro-chromium; $FeCr_2O_4$; (68% chromic oxide); Sg. 4.4 062

EISENCHROMAT (*n*): (*Ferrum chromatum*): iron (ferric) chromate; $Fe_2(CrO_4)_3$ 063
EISEN (*n*), **CHROMSAURES-**: iron chromate (see EISENCHROMAT) 064
EISENCHRYSOLITH (*m*): hyalosiderite; fayalite (see HYALOSIDERIT and FAYALIT) 065
EISEN (*n*), **CITRONSAURES-**: iron citrate; $Fe_2(C_6H_5O_7)_2.6H_2O$ 066
EISENCYANFARBE (*f*): iron-cyanogen pigment 067
EISENCYANID (*n*): iron cyanide; ferric (ferri-) cyanide; $Fe(CN)_3$ 068
EISENCYANKALIUM (*n*): potassium ferrocyanide (see FERROCYANKALIUM) 069
EISENCYANÜR (*n*): ferrous (ferro-) cyanide; $Fe(CN)_2$ 070
EISENCYANÜRZINKOXYD (*n*): zinc ferrocyanide; $Zn_2Fe(CN)_6.3H_2O$ 071
EISENCYANVERBINDUNG (*f*): iron-cyanogen (ferro- or ferri-cyanogen) compound 072
EISENDIALYSAT (*n*): dialysed iron 073
EISENDIBROMID (*n*): ferrous bromide (see EISENBROMÜR) 074
EISENDICHLORID (*n*): iron (ferrous) chloride (see EISENCHLORÜR) 075
EISENDIFLUORID (*n*): ferrous fluoride; FeF_2; Sg. 4.09 076
EISENDIJODID (*n*): ferrous iodide (see EISEN-JODÜR); Mp. 177°C. 077
EISENDISULFID (*n*): iron (ferric) sulphide; FeS_2; Sg. 4.86; marcasite; iron pyrites 078
EISENDRAHT (*m*): iron wire 079
EISENDREHSPÄNE (*m.pl*): iron (chips, turnings, filings) shavings 080
EISENEINLAGE (*f*): core (of iron) 081
EISENERDE (*f*): ferruginous earth; iron earth; sort of hard stoneware (Ceramics) 082
EISENERDE (*f*), **BLAUE-**: earthy vivianite (see BLAUE ERDE) 083
EISENERZ (*n*): iron ore 084
EISENERZ (*n*), **AXOTOMES-**: ilmenite (see ILMENIT) 085
EISENERZ (*n*), **DODEKAEDRISCHES-**: franklinite 086
EISENERZ (*n*), **HEXAEDRISCHES-**: iserine 087
EISENERZ (*n*), **OKTAEDRISCHES-**: magnetite 088
EISENERZ (*n*), **RHOMBOEDRISCHES-**: hematite, red clay ironstone (see HEMATIT) 089
EISEN (*n*), **ESSIGSAURES-**: iron (ferric) acetate; $Fe_2(C_2H_3O_2)_6$ 090
EISENEXTRAKT (*m*): iron extract 091
EISENEXTRAKT, APFELSAURER-: iron extract malate 092
EISENFARBE (*f*): iron (grey) colour 093
EISENFARBIG: iron (grey) coloured 094
EISENFEILE (*f*): iron filings 095
EISENFEILICHT (*n*): (see EISENFEILE) 096
EISENFEILSPÄNE (*m.pl*): (see EISENFEILE) 097
EISEN (*n*), **FEINKÖRNIGES-**: fine grained iron 098
EISENFLECK (*m*): iron spot, rust mark, iron (mould) stain 099
EISENFLUORID (*n*): iron fluoride, ferrous fluoride; $FeF_2.8H_2O$; Sg. 4.1 100
EISENFREI: free from iron; not having an iron content 101
EISENGANG (*m*): iron lode; iron ore vein 102
EISENGANS (*f*): (iron)-pig 103
EISEN (*n*), **GESCHMIEDETES-**: wrought iron 104
EISENGESTELL (*n*): iron (stand) frame 105
EISENGEWERBE (*n*): iron (trade) industry 106
EISENGEWINNUNG (*f*): iron (production; gain) yield 107
EISENGIESSER (*m*): iron founder 108
EISENGIESSEREI (*f*): iron (foundry) founding 109
EISENGILBE (*f*): yellow orchre 110
EISENGITTER (*n*): iron grating 111
EISENGLANZ (*m*): iron glance; specular iron ore (type of hematite; Fe_2O_3 (70% Fe) (see EISENOXYD) 112
EISENGLAZUR (*f*): iron (glaze) glazing 113
EISENGLIMMER (*m*): micaceous iron ore; type of hematite; Fe_2O_3 (see EISENOXYD) 114
EISENGRANAT (*m*): iron garnet; almandite (see ALMANDIN) 115
EISENGRAU: iron grey 116
EISENGRAUPEN (*f.pl*): granular bog iron ore (approx. $2Fe_2O_3.3H_2O$) 117
EISEN (*n*), **GROBKÖRNIGES-**: coarse grained iron 118
EISENGRUND (*m*): iron liquor (by action of pyroligneous acid on iron filings) 119
EISENGUSS (*m*): iron casting; cast iron 120
EISENGUSSWARE (*f*): cast iron ware, hardware 121
EISENHALTIG: ferruginous; ferrous; ferriferous; containing iron; chalybeate (of water) 122
EISENHAMMER (*m*): forge hammer, iron-works 123
EISENHAMMERSCHLAG (*m*): iron hammer scale 124
EISENHANDEL (*m*): iron trade 125
EISENHÄNDLER (*m*): ironmonger 126
EISENHART (*m*): ferriferous gold sand 127
EISEN (*n*), **HARZSAURES-**: iron resinate 128
EISENHOLZ (*n*): iron wood (*Ostrya virginica*) 129
EISEN (*n*), **HOLZESSIGSAURES-**: iron pyrolignite; iron (acetate) liquor (see HOLZESSIGSAURES EISEN and EISENACETAT) 130
EISENHUT (*m*): aconite; monkshood (*Aconitum napellus*) (*Herba aconiti*) 131
EISENHÜTCHEN (*n*): (see EISENHUT) 132
EISENHUTKNOLLEN (*m*): aconite root, monkshood, wolfsbane (*Tubera* or *Radix aconiti* of *Aconitum napellus*) 133
EISENHÜTTE (*f*): ironworks; forge 134
EISENHÜTTENKUNDE (*f*): iron metallurgy 135
EISENHUTVERGIFTUNG (*f*): aconite poisoning (by *Aconitum napellus*) (Med.) 136
EISENHUTWURZEL (*f*): aconite root (root of *Aconitum napellus*) 137
EISENHYDROXYD (*n*): (*Ferrum hydroxydatum*): iron hydrate, ferric hydroxide, ferric hydrate, rust; $Fe(OH)_3$; Sg. 3.4-3.9 138
EISENHYDROXYDUL (*n*): ferrous hydroxide 139
EISENJODID (*n*): iron (ferric) iodide 140
EISENJODÜR (*n*): ferrous iodide; $FeI_2.4H_2O$; Sg. 2.873; Mp. 177°C. 141
EISENJODÜRJODID (*n*): ferrosoferric iodide; Fe_3I_8 142
EISENKALIUM (*n*): potassium ferrate 143
EISENKALIUMALAUN (*m*): iron potassium alum; ferric potassium sulphate; $Fe_2(SO_4)_3K_2SO_4.24H_2O$; Sg. 1.806 144
EISENKALIUM (*n*), **OXALSAURES-**: iron-potassium oxalate 145
EISENKALIUM (*n*), **SCHWEFELSAURES-**: ferric-potassium sulphate (see EISENKALIUMALAUN) 146
EISENKALIUM (*n*), **WEINSAURES-**: iron-potassium tartrate, ferrous-potassium tartrate 147
EISENKALK (*m*): calcined iron 148

EISENKARBID (*n*): iron carbide; Fe_3C; Sg. 7.07-7.396 149
EISENKARBURET (*n*): iron carbide 150
EISENKEIL (*m*): iron wedge 151
EISENKERN (*m*): core (of iron) 152
EISENKIEL (*m*): iron keel 153
EISENKIES (*m*): pyrite, iron pyrites, iron sulphide FeS_2 (about 47% iron) 154
EISENKIESEL (*m*): ferruginous (flint) quartz (see QUARZ) 155
EISENKIES (*m*), **HEXAGONALER-**: magnetic pyrites, pyrrhotite (see MAGNETKIES) 156
EISENKIES (*m*), **RHOMBISCHER-**: white iron pyrites, marcasite (see EISENDISULFID) 157
EISENKITT (*m*): iron cement; iron rust cement 158
EISENKLUMPEN (*m*): iron pig 159
EISENKOBALTERZ (*n*): cobaltite; cobalt glance; $CoAsS$; 35.5% cobalt; safflorite 160
EISENKOBALTKIES (*m*): (see EISENKOBALTERZ) 161
EISENKOHLENOXYD (*n*): iron carboxide 162
EISEN (*n*), **KOHLENSAURES-**: iron carbonate; $FeCO_3$ or $FeCO_3.H_2O$; Sg. 3.7-3.87; Mp. 470°C. 163
EISENKOHLENSTOFF (*m*): iron carbide (see EISEN-KARBID) 164
EISENKONSTRUKTION (*f*): iron (construction) building; structural (constructional) iron-work 165
EISENKORN (*n*): grain (particle) of iron 166
EISENKRAM (*m*): ironmongery 167
EISENKRAUT (*n*): verbena herb; vervain (*Herba verbenæ*) 168
EISENKRAUTGEWÄCHSE (*n.pl*): *Verbenaceæ* (Bot.) 169
EISENKRISTALL (*m*): iron crystal 170
EISENLACK (*m*): iron varnish 171
EISENLAKTAT (*n*): (*Ferrum lacticum*): iron lactate, ferrous lactate; $Fe(C_3H_5O_3)_2.3H_2O$ 172
EISENLEBERERZ (*n*): hepatic iron ore 173
EISENLIKÖR (*m*): iron (liqueur) liquor; iron acetate liquor 174
EISENMANGAN (*n*): ferromanganese; iron manganate (see SPIEGELEISEN) 175
EISENMANGANERZ (*n*): manganiferous iron ore 176
EISENMANGANPEPTONAT (*n*): peptonated (iron manganate) ferromanganese (*Ferrum-manganopeptonatum*) 177
EISENMANGANSACCHARAT (*n*): saccharated (iron manganate) ferromanganese (*Ferrum Manganosaccharatum*) 178
EISENMANGANSILIKATSCHLACKE (*f*): acid slag (see BESSEMERSCHLACKE) 179
EISENMANN (*m*): foliated sort of hematite 180
EISENMASSE (*f*): bloom (Iron) 181
EISEN (*n*), **MATTES-**: syrupy molten iron, viscous liquid (molten) iron 182
EISENMEHL (*n*): iron meal 183
EISENMENNIGE (*f*): iron oxide, red; iron minium; red-(iron)-ochre; Fe_2O_3; Sg. 5.12-5.24; Mp. 1548°C. 184
EISENMETASILIKAT (*n*): iron metasilicate; $FeSiO_3$; Sg. 3.44 185
EISEN (*n*), **MILCHSAURES-**: (see EISENLAKTAT) 186
EISENMOHR (*m*): black iron oxide; ethiops martialis; Fe_3O_4 (as a black powder); earthy magnetite (see MAGNETIT) 187
EISENMONOSULFID (*n*): ferrous sulphide (see EISENSULFÜR) 188
EISENMONOXYD (*n*): Iron monoxide (see EISENOXYDUL) 189
EISENMULM (*m*): earthy iron ore 190
EISENNICKEL (*n*): ferro-nickel; Ni_3Fe; Sg. 7.8 191
EISENNIERE (*f*): eagle stone 192
EISENNITRAT (*n*): (*Ferrum nitricum*): iron (ferric) nitrate; $Fe_2(NO_3)_3.9H_2O$; Sg. 1.684; Mp. 47.2°C. (used as a mordant) 193
EISENNITRID (*n*): iron nitride; Fe_2N; Sg. 6.35 194
EISENNITRIT (*n*): iron nitrite 195
EISEN (*n*), **NUKLEINSAURES-**: iron nucleinate 196
EISENÖL (*n*): Oleum martis (Pharm.) 197
EISENOLIVIN (*n*): fayalite (see FAYALIT) 198
EISENOXALAT (*n*): (*Ferrum oxalicum*): iron (ferrous) oxalate; $FeC_2O_4.2H_2O$ 199
EISEN (*n*), **OXALSAURES-**: iron oxalate (see EISEN-OXALAT) 200
EISENOXYCHLORID (*n*): iron oxychloride 201
EISENOXYCHLORIDLÖSUNG (*f*), **DIALYSIERTE-**: dialysed iron oxychloride solution (*Liquor Ferri oxychlorati dialysati*); Sg. 1.045; about 3.5% iron) 202
EISENOXYD (*n*): (*Ferrum oxydatum*) (*Caput mortuum*): iron (ferric) oxide, iron ore, red iron oxide, iron trioxide, Indian (Venetian) red, hematite, colcothar, crocus martis, iron sesquioxide; Fe_2O_3; Sg. 5.12-5.24; Mp. 1548°C. 203
EISENOXYDACETAT (*n*): (see FERRIACETAT) 204
EISENOXYDAMMONIUM (*n*), **CITRONENSAURES-**: ammonio-citrate of iron, ammonioferric citrate, ferric-ammonium citrate, iron-ammonium citrate 205
EISENOXYDAMMONIUM (*n*), **OXALSAURES-**: ferric ammonium oxalate; $(NH_4)_3Fe(C_2O_4)_3$ 206
EISENOXYDARM: poor in ferric oxide, having a low ferric oxide content, containing very little ferric oxide 207
EISENOXYDCHLORID (*n*): ferric chloride (see EISENCHLORID) 208
EISENOXYD (*n*), **CHROMSAURES-**: iron (ferric) chromate (see EISENCHROMAT) 209
EISENOXYD (*n*), **CITRONENSAURES-**: citrate of iron, ferric citrate, $Fe_2(C_6H_5O_7)_2.6H_2O$ 210
EISENOXYDFARBE (*f*): iron oxide; Venetian red; Fe_2O_3 211
EISENOXYDHALTIG: containing ferric oxide 212
EISENOXYDHYDRAT (*n*): ferric hydroxide; $Fe(OH)_3$; Sg. 3.4-3.9 (see EISENHYDROXYD) 213
EISENOXYDLAKTAT (*n*): ferric lactate 214
EISENOXYDMETAPHOSPHAT (*n*): ferric metaphosphate; $Fe(PO_3)_3$; Sg. 3.02 215
EISENOXYDNATRON (*n*), **OXALSAURES-**: sodium ferric oxalate 216
EISENOXYD (*n*), **OXALSAURES-**: ferric oxalate; $Fe_2(C_2O_4)_3$ 217
EISENOXYDOXYDUL (*n*): ferrosoferric oxide; iron ethiops; Sg. 5.16; Fe_3O_4 (see MAGNE-TIT) 218
EISENOXYD (*n*), **SALPETERSAURES-**: ferric nitrate (see EISENNITRAT) 219
EISENOXYDSALZ (*n*): ferric salt 220
EISENOXYDSCHWARZ (*n*): iron oxide black; Fe_3O_4 (see EISENOXYDULOXYD) 221
EISENOXYD (*n*), **SCHWEFELSAURES-**: ferric sulphate (see EISENOXYDSULFAT) 222
EISENOXYDSULFAT (*n*): (*Ferrum sulfuricum oxydatum*): ferric sulphate, iron sulphate; $Fe_2(SO_4)_3$; Sg. 3.097 223
EISENOXYDUL (*n*): iron (ferrous) oxide; FeO 224

EISENOXYDULACETAT (*n*): ferrous aectate; $Fe(C_2H_3O_2)_2.4H_2O$ 225
EISENOXYDULAMMONIAK (*n*), SCHWEFELSAURES- : ferrous-ammonium sulphate (see AMMONIUMFERROSULFAT) 226
EISENOXYDULAMMONIUM (*n*), SCHWEFELSAURES- : ferrous ammonium sulphate (a double salt of iron and ammonium); $Fe(NH_4)_2.2SO_4.6H_2O$ 227
EISENOXYDULCARBONAT (*n*): spathic iron; siderite; $FeCO_3$; Sg. 3.8 228
EISENOXYDUL (*n*), ESSIGSAURES- : ferrous acetate 229
EISENOXYDULHYDRAT (*n*): ferrous hydroxide 230
EISENOXYDULLAKTAT (*n*): ferrous lactate (see EISENLAKTAT) 231
EISENOXYDULNIOBAT (*n*): ferrous niobate; $FeNb_2O_6$ 232
EISENOXYDUL (*n*), OXALSAURES- : ferrous oxalate (see EISENOXALAT) 233
EISENOXYDULOXYD (*n*): black ferric oxide, iron ethiops, magnetic iron oxide, ferrosoferric oxide; Fe_3O_4 (see MAGNETIT) 234
EISENOXYDUL (*n*), SALPETERSAURES- : ferrous nitrate 235
EISENOXYDULSALZ (*n*): ferrous salt 236
EISENOXYDUL (*n*), SCHWEFELSAURES- : ferrous sulphate (see EISENOXYDULSULFAT) 237
EISENOXYDULSULFAT (*n*): ferrous sulphate, copperas, (*Ferrum sulfuricum*); $FeSO_4$; Sg. 2.99; $FeSO_4.7H_2O$; Sg. 1.875-1.899; Mp. 64°C.; green vitriol; green copperas, iron vitriol, hydrous ferrous sulphate, iron protosulphate 238
EISENOXYDULTANTALAT (*n*): ferrous tantalate; $FeTa_2O_6$ 239
EISENOXYDULVERBINDUNG (*f*): ferrous compound 240
EISENOXYDVERBINDUNG (*f*): ferric compound 241
EISENOXYD (*n*), WASSERHALTIGES- : hydrated iron oxide, limonite, brown iron ore (see BRAUNEISENSTEIN) 242
EISENPASTILLE (*f*): iron (lozenge) pastille 243
EISENPECHERZ (*n*): limonite; $2Fe_2O_3.3H_2O$ [244
EISENPEPTONAT (*n*): iron (ferric) peptonate (*Ferrum peptonatum*) 245
EISENPHOSPHAT (*n*): iron (ferric) phosphate; $FePO_4.4H_2O$; Sg. 2.87 246
EISENPHOSPHID (*n*): iron phosphide; FeP, Sg. 5.76; Fe_2P, Sg. 6.56; Fe_2P_3, Sg. 4.5; Fe_3P, Sg. 6.74 247
EISEN (*n*), PHOSPHORSAURES- : iron phosphate (see EISENPHOSPHAT) 248
EISENPLASMON (*n*): iron plasmon 249
EISENPLATTE (*f*): iron plate 250
EISENPLATTENBEKLEIDUNG (*f*): steel (casing) plating 251
EISENPORTLANDCEMENT (*m*): iron (slag) Portland cement 252
EISENPRÄPARAT (*n*): iron preparation 253
EISENPULVER (*n*): iron powder 254
EISEN (*n*), PYROPHOSPHORSAURES- : iron pyrophosphate; $Fe_4(P_2O_7)_3$ 255
EISENQUARZ (*m*): iron (ferriferous) quartz 256
EISENRAHM (*m*): a porous variety of hematite (see HÄMATIT) 257
EISENREGAL (*n*): iron shelf 258
EISENREICH: rich in iron, having a high iron content 259
EISENRHODANID (*n*): ferric (thiocyanate; sulfocyanate) sulfocyanide 260
EISENRHODANÜR (*n*): ferrous (thiocyanate; sulfocyanate) sulfocyanide 261
EISENRÖHRE (*f*): iron pipe or tube 262
EISENROST (*m*): iron rust (see EISENHYDROXYD) 263
EISENROT (*n*): colcothar (a red iron oxide from calcining copperas) 264
EISENSACCHARAT (*n*): iron saccharate (*Ferrum saccharatum*) 265
EISENSACCHARAT, KOHLENSAURES- (*n*): carbonated iron saccharate 266
EISENSÄGE (*f*): iron cutting saw, metal saw 267
EISEN (*n*), SALICYLSAURES- : iron salicylate 268
EISENSALMIAK (*m*): ammoniated iron (Pharm.) 269
EISEN (*n*), SALPETERSAURES- : iron nitrate (see EISENNITRAT) 270
EISENSALZ (*n*): iron salt 271
EISEN (*n*), SALZSAURES- : iron (ferric) chloride (see EISENCHLORID) 272
EISENSAU (*f*): pig (block, sow) of iron 273
EISENSAUER: ferrate of; combined with ferric acid 274
EISENSÄUERLING (*m*): chalybeate (iron) water 275
EISENSÄURE (*f*): ferric acid (not known in a free state) 276
EISEN (*n*), SAURES- : acid (steel) iron (usually by Bessemer process) 277
EISENSCHAUM (*m*): kish (Metal); a porous variety of hematite (Min.) (see HÄMATIT) 278
EISENSCHEIDER (*m*): iron separator 279
EISENSCHIFFBAU (*m*): iron shipbuilding 280
EISENSCHLACKE (*f*): iron (dross) slag 281
EISENSCHMIED (*m*): blacksmith 282
EISEN (*n*), SCHMIEDBARES- : malleable iron 283
EISENSCHMIEDE (*f*): smithy, forge 284
EISENSCHÖRL (*m*): iron schorl 285
EISENSCHROT (*m* and *n*): iron scrap 286
EISENSCHÜSSIG: ferriferous; ferruginous; ferrous; containing iron 287
EISENSCHUTZ (*m*): protection of iron 288
EISENSCHUTZFARBE (*f*): antirusting paint; iron protecting paint 289
EISENSCHUTZLACK (*m*): antirusting paint or varnish 290
EISENSCHUTZMITTEL (*n*): antirusting composition; anti-incrustant 291
EISENSCHWAMM (*m*): iron sponge; porous (spongy) iron 292
EISENSCHWARZ (*n*): graphite; lampblack; iron black (precipitated antimony); iron oxide black (see EISENOXYDULOXYD); copperas black (Dye) 293
EISENSCHWÄRZE (*f*): graphite; ground (blacklead) graphite; currier's ink (Leather); earthy magnetite (Min.) (see MAGNETIT); iron liquor (Dye) 294
EISEN (*n*), SCHWEFELSAURES- : iron sulphate (see also EISENVITRIOL, EISENOXYDSULFAT and EISENOXYDULSULFAT) 295
EISENSELENÜR (*n*): ferrous selenide 296
EISENSEPARATOR (*m*): iron separator 297
EISENSESQUISULFID (*n*): iron sesquisulphide (see FERRISULFID) 298
EISENSILBERGLANZ (*m*): sternbergite (see STERNBERGIT) 299
EISENSILICID (*n*): iron silicide; FeSi, Sg. 6.17; $FeSi_2$, Sg. 5.4; Fe_2Si, Sg. 7.0; Fe_3Si_2, Sg. 6.36 300
EISENSILICIUM (*n*): iron silicide (see EISENSILICID) 301

EISENSINTER (*m*): iron (scale) dross 302
EISENSOMATOSE (*f*): iron somatose 303
EISENSPÄNE (*m.pl*): iron (turnings; borings; shavings) filings; scrap-(iron) 304
EISENSPAT (*m*): siderite; spathic iron ore; (natural iron carbonate); $FeCO_3$; (48% Fe); Sg. 3.8 305
EISENSPAT (*m*), **REIFER-**: (see **BLAUE ERDE**) 306
EISENSPIEGEL (*m*): hematite (see **HÄMATIT**); specular iron; Fe_2O_3; 70% Fe 307
EISEN (*n*), **SPRÖDES-**: brittle iron 308
EISEN (*n*), **STEARINSAURES-**: iron (ferric) stearate; $Fe(C_{18}H_{35}O_2)_3$ 309
EISENSTEIN (*m*): ironstone; iron ore 310
EISENSTEINMARK (*n*): bole (see **BOL**); lithomarge with an iron content 311
EISENSTEIN (*n*), **SPATIGER-**: (see **EISENSPAT**) 312
EISENSUBLIMAT (*n*): iron sublimate; ferric chloride; $FeCl_3$; Sg. 2.804; Mp. 301°C. 313
EISENSULFAT (*n*): iron sulphate (see **EISENOXYDSULFAT** and **EISENOXYDULSULFAT**) 314
EISENSULFID (*n*): iron (ferric) sulphide; Fe_2S_3 (see **EISENDISULFID**) 315
EISENSULFÜR (*n*): iron (protosulphide) sulphuret; ferrous sulphide (*Ferrum sulfuratum*); FeS; Sg. 4.75-5.4; Mp. 1179°C. 316
EISENSUMPFERZ (*n*): bog iron ore; brown hematite; $2Fe_2O_3$ plus $3H_2O$; (about 60% Fe) 317
EISENTINKTUR (*f*): tincture of iron 318
EISENTINKTUR (*f*), **AROMATISCHE-**: aromatic tincture of iron (see **EISENZUCKERLÖSUNG**) 319
EISENTITAN (*n*): ilmenite; menaccanite; $FeTiO_3$ 36.8% Fe 320
EISENTON (*m*): clay ironstone 321
EISENTONGRANAT (*m*): almandite (see **ALMANDIN**) 322
EISENTRICHLORID (*n*): iron (ferric) chloride; iron trichloride; $FeCl_3$ (see **EISENCHLORID**) 323
EISENTRIFLUORID (*n*): ferric fluoride; FeF_3; Sg. 3.18 324
EISENTUTTE (*f*): iron assaying crucible 325
EISEN (*n*), **UNTERPHOSPHORIGSAURES-**: iron hypophosphite 326
EISENUNTERSUCHUNG (*f*): iron (examination; analysis) testing 327
EISENVITRIOL (*m*): iron vitriol (see **EISEN-OXYDULSULFAT**); $FeSO_4.7H_2O$; Sg. 1.85 328
EISENWARE (*f*): iron-ware, hardware 329
EISENWASSER (*n*): chalybeate (iron) water 330
EISEN (*n*), **WEICHES-**: soft iron 331
EISENWEIN (*m*): iron wine 332
EISEN (*n*), **WEINSAURES-**: iron tartrate 333
EISENWEINSTEIN (*m*): iron and potassium tartrate; ferrous potassium tartrate; tartrated iron 334
EISENWERK (*n*): iron works 335
EISENWOLLE (*f*): iron wool 336
EISEN (*n*), **WULST-**: bulb iron 337
EISEN (*n*), **ZÄHES-**: fibrous iron 338
EISENZEMENT (*m*): iron cement 339
EISENZINKBLENDE (*f*): marmatite (Min.) 340
EISENZINKSPAT (*m*): monkheimite; ferruginous calamine $(Fe.Zn)CO_3$; Sg. 4.17 341
EISENZINNERZ (*n*): ferriferous cassiterite (see **AUSSTRICH** and **ZINNSTEIN**) 342
EISENZUCKER (*m*): ferric (iron) saccharate (*Ferrum oxydat. sacch.*) 343
EISENZUCKERLÖSUNG (*f*): ferric saccharate solution 344
EISERN: of iron; iron; strong 345
EISERNE KÖRBE FÜR SÄUREBALLONS: iron baskets for (glass) acid carboys 346
EISESSIG (*m*): (*Acetum glaciale*): glacial acetic acid, methanecarboxylic acid; (CH_3COOH); Sg. 1.055; Bp. 117.5°C. 347
EISGLAS (*n*): frosted glass 348
EISHAUT (*f*): ice doubling 349
EISJACHTSEGELN (*n*): ice-yachting 350
EISJACHTSEGLER (*m*): ice-yachtsman 351
EISKELLER (*m*): ice cellar 352
EISKRAUT (*n*): ice plant (*Mesembryanthemum crystallinum*) 353
EISMASCHINE (*f*): ice (refrigerating) machine; (*pl*): refrigerating plant 354
EISMASCHINENRINGE (*m. pl*): ice machine rings 355
EISPFLANZE (*f*): ice plant (see **EISKRAUT**) 356
EISPUNKT (*m*): freezing point 357
EISSCHRANK (*m*): refrigerator 358
EISSPAT (*m*): crystalline feldspar; sanidine; aquamarine (see also **ORTHOKLAS**); $K_2O.Al_2O_3.6SiO_2$ 359
EISSTEIN (*m*): cryolite; cryolith (see **KRYOLITH**) 360
EISTRICHTER (*m*): ice funnel 361
EISVERSTÄRKUNG (*f*): ice doubling 362
EISYACHT (*f*): ice yacht 363
EISZACKEN (*m*): icicle 364
EISZAPFEN (*m*): icicle 365
EISZONE (*f*): ice (frozen) zone 366
EITER (*m*): pus; matter 367
EITERABFLUSS (*m*): discharge 368
EITERÄHNLICH: purulent 369
EITERARTIG: (see **EITERÄHNLICH**) 370
EITERBEFÖRDERND: suppurative 371
EITERBEULE (*f*): abscess 372
EITERBLÄSCHEN (*n*): pustule (Med.) 373
EITERBRUST (*f*): empyema (pus outside the pleura) 374
EITERERZEUGEND: suppurative 375
EITERFLUSS (*m*): discharge; discharging (of pus) 376
EITERGANG (*m*): fistula 377
EITERGESCHWULST (*f*): abscess (Med.) 378
EITERICHT: purulent; mattery 379
EITERIG: purulent; mattery 380
EITERN: to suppurate; fester; discharge (of pus) 381
EITERND: suppurative 382
EITERSTOFF (*m*): pus; purulent matter 383
EITERTRIEFEN (*n*): blennorhœa (Med.) 384
EITERUNG (*f*): suppuration; festering; discharging (of pus) 385
EITERZIEHEND: suppurative; suppurating 386
EIWEISS (*n*): albumin; albumen; egg-white; protein; (as a prefix, albuminous: albuminoid) 387
EIWEISSÄHNLICH: albuminoid 388
EIWEISSARM: poor in (albumen) protein; having low albumen content 389
EIWEISSART (*f*): kind (variety; sort) of (albumen) protein 390
EIWEISSARTIG: albuminous; albuminoid 391
EIWEISSDRÜSE (*f*): salivary gland 392
EIWEISSFÖRMIG: albuminous 393
EIWEISSHALTIG: albuminous; containing (albumen) protein 394
EIWEISSHARNEN (*n*): albuminuria; albuminous urine (Med.) 395
EIWEISSKÖRPER (*m*): protein; albumenous substance 396

EIWEISSKRYSTALL (*m*): albumen crystal 397
EIWEISSLEIM (*m*): gluten protein 398
EIWEISS (*n*), MUSKEL-: muscle fibrin (see SYNTONIN) 399
EIWEISSNAHRUNG (*f*): albuminous food 400
EIWEISSPAPIER (*n*): albumen paper; albuminized paper 401
EIWEISSREICH: rich in albumen or protein; having a high (albumen) protein content 402
EIWEISSSALZ (*n*): albuminoid 403
EIWEISSSPALTEND: proteolytic 404
EIWEISSSPALTPRODUKT (*n*): albumin (protein) cleavage product, product of proteolysis 405
EIWEISSSPALTUNG (*f*): proteolysis; albumen or protein cleavage 406
EIWEISSSTOFF (*m*): albuminous substance; protein 407
EIWEISSTRÜBUNG (*f*): albumen (coagulation) clouding (Brewing) 408
EIWEISSVERBINDUNG (*f*): albuminous compound; protein; albuminate 409
EIZELLE (*f*): egg cell; ovum 410
EJEKTOR (*m*): ejector 411
EKASILIZIUM (*n*): ekasilicon 412
EKGONIN (*n*): Ecgonine; $C_9H_{15}NO_3.H_2O$; Mp. 198°C. 413
EKLAMPSIE (*f*): eclampsia (Med.) 414
EKLAMPSIE IM WOCHENBETT (*n*): puerperal convulsions, puerperal eclampsia (convulsions in childbed) (*Eclampsia gravidarum, parturientuum, vel puerperarum*) (Med.) 415
EKLAMPSIE IN DER SCHWANGERSCHAFT: puerperal convulsions, puerperal eclampsia (Med.) (see also EKLAMPSIE IM WOCHENBETT) 416
EKLEKTISCH: eclectic 417
EKLEPSIS (*f*): desquamation, peeling 418
EKLIPSE (*f*): eclipse 419
EKONOMISER (*m*): economiser 420
EKRASIT (*n*): ecrasite (an explosive, consisting of the ammonium salt of trinitrocresol) 421
EKSTATISCH: ecstatic 422
EKZEM (*m*): eczema (Med.) 423
ELAEOSTEARINSÄURE (*f*): eleostearic acid, elaeostearic acid (see ELÄOSTEARINSÄURE) 424
ELAIDINSÄURE (*f*): elaidic acid; $C_{18}H_{34}O_2$; Mp. 51.5°C. 425
ELAINSÄURE (*f*): oleic acid; $C_8H_{17}CHCH(CH_2)_7CO_2H$; Sg. 0.8908; Mp. –10.5°C.; Bp. 222.4°C. 426
ELAINSPAT (*m*), PYRAMIDALER-: scapolite, wernerite, meionite 427
ELAINSPAT (*m*), RHOMBOEDRISCHER-: nepheline 428
ELÄOKOKKAÖL (*n*): elaeococca oil, Tung chou oil, Japanese (Chinese) wood oil (from the tree *Elaeococca vernicia*); Sg. 0.93-0.94 429
ELÄOLITH (*m*): elaeolite (nephelite of massive structure; see NEPHELIN) 430
ELÄOMARGARINSÄURE (*f*): eleomargaric acid; $C_{17}H_{31}CO_2H$; Mp. 43.5-48°C. 431
ELÄOSTEARIN (*n*), β-: beta (β) eleostearin; Mp. 61°C. 432
ELÄOSTEARINSÄURE (*f*): eleostearic acid; $C_{17}H_{31}CO_2H$; Mp. 71°C. 433
ELÄOSTEARINSÄURE (*f*), α-: alpha-eleostearic acid (see ELÄOMARGARINSÄURE) 434
ELÄOSTEARINSÄURE (*f*), β-: beta-eleostearic acid (see ELÄOSTEARINSÄURE) 435
ELASTICITÄT (*f*): elasticity; flexibility 436
ELASTICITÄTSTHEORIE (*f*): theory of elasticity 437
ELASTISCH: elastic; flexible 438
ELASTISCHE LINIE (*f*): elastic (curve) line 439
ELASTIZITÄT (*f*): elasticity 440
ELASTIZITÄTSGRENZE (*f*): elastic limit 441
ELASTIZITÄTSGRUNDGESETZ (*n*): fundamental law of elasticity 442
ELASTIZITÄTSKOEFFIZIENT (*m*): coefficient of elasticity 443
ELASTIZITÄTSMODUL (*m*): modulus of elasticity 444
ELASTIZITÄTSMODULUS (*m*): modulus of elasticity 445
ELATERIN (*n*): elaterin; $C_{20}H_{28}O_5$; Mp. 216°C.; $C_{28}H_{38}O_7$ (the active principle of *Echallium elaterium* or *Memordica elaterium*; used as purgative) 446
ELATERIT (*m*): elaterite (Mineralogy); C_nH_{2n}; Sg. 1.1. 447
ELAYLCHLORID (*n*): (see ÄTHYLENCHLORID) 448
ELEFANTENLAUS (*f*): cashew nut (*Anacardium occidentale*) 449
ELEFANTENLÄUSEÖL (*n*), OSTINDISCHES-: (see ANACARDIENÖL) 450
ELEFANTENLÄUSEÖL (*n*), WESTINDISCHES-: (see AKAJOUÖL) 451
ELEFANTENNUSS (*f*): ivory nut, corozo nut (*Phytelephas macrocarpa*) 452
ELEKTAWOLLE (*f*): first-class wool; A.1. wool 453
ELEKTRIKER (*m*): electrician 454
ELEKTRISCH: electric-(al)-(ly) 455
ELEKTRISCHE BELEUCHTUNG (*f*): electric lighting 456
ELEKTRISCHE EINHEIT (*f*): electrical unit 457
ELEKTRISCHE KOHLE (*f*): carbon 458
ELEKTRISCHE LEITFÄHIGKEIT (*f*): electrical conductivity 459
ELEKTRISCHE LOKOMOTIVE (*f*): electric locomotive 460
ELEKTRISCHER KOCHAPPARAT (*m*): electrical cooking utensil 461
ELEKTRISCHER SCHLAG (*m*): electric shock 462
ELEKTRISCHES FERNTHERMOMETER (*n*): electrical distance thermometer 463
ELEKTRISCHES THERMOMETER (*n*): electrical thermometer 464
ELEKTRISIERBAR: electrifiable 465
ELEKTRISIEREN: to electrify; (*n*): electrification 466
ELEKTRISIERMASCHINE (*f*): electrical machine, galvanic battery 467
ELEKTRISIERUNG (*f*): electrification 468
ELEKTRIZITÄT (*f*): electricity 469
ELEKTRIZITÄT (*f*), DYNAMISCHE-: dynamic electricity 470
ELEKTRIZITÄT (*f*), FREIE-: free electricity 471
ELEKTRIZITÄT (*f*), GALVANISCHE-: galvanic electricity 472
ELEKTRIZITÄT (*f*), GEBUNDENE-: bound electricity 473
ELEKTRIZITÄTSLEHRE (*f*): electrology, electricity 474
ELEKTRIZITÄTSLEITER (*m*): conductor (of electricity) 475
ELEKTRIZITÄTSMESSER (*m*): electrometer 476
ELEKTRIZITÄTSSTROM (*m*): electric current 477
ELEKTRIZITÄT (*f*), STATISCHE-: static electricity 478
ELEKTRIZITÄTSWAGE (*f*): electrometer 479
ELEKTRIZITÄTSWERK (*n*): electricity works, electric generating station 480
ELEKTRIZITÄTSZÄHLER (*m*): electricity (electric) meter 481

ELEKTRIZITÄTSZEIGER (*m*) : electroscope 482
ELEKTROCHEMIE (*f*) : electrochemistry 483
ELEKTROCHEMIKER (*m*) : electro-chemist 484
ELEKTROCHEMISCH : electrochemical 485
ELEKTRODE (*f*) : electrode 486
ELEKTRODENKESSEL (*m*) : electro-boiler (in which the water is heated by electrodes) 487
ELEKTRODENSCHWEISSUNG (*f*) : electro-welding 488
ELEKTRODYNAMIK (*f*) : electro-dynamics 489
ELEKTRODYNAMISCH : electro-dynamic 490
ELEKTROHÄNGEBAHN (*f*) : suspended electric railway 491
ELEKTRO-INGENIEUR (*m*) : electrical engineer 492
ELEKTROLYSE (*f*) : electrolysis (the decomposition of electrolytes by means of electric current) 493
ELEKTROLYSIEREN : to electrolyse ; galvanize 494
ELEKTROLYT (*m*) : electrolyte 495
ELEKTROLYTEISEN (*n*) : electrolytic iron 496
ELEKTROLYTISCH : electrolytic ; galvanic 497
ELEKTROLYTISCHER APPARAT (*m*) : electrolytic apparatus 498
ELEKTROLYT KUPFER (*n*) : electrolyte copper (99.95% Cu ; 0.035% Si ; 0.0025% S ; 0.003% Ag.) 499
ELEKTROLYTLÖSUNG (*f*) : electrolyte solution 500
ELEKTROLYTSILBER (*n*) : electrolytic silver 501
ELEKTROMAGNET (*m*) : electro-magnet 502
ELEKTROMAGNETISCH : electro-magnetic 503
ELEKTROMAGNETISMUS (*m*) : electro-magnetism 504
ELEKTROMECHANIK (*f*) : electro-mechanics 505
ELEKTROMECHANISCH : electro-mechanical 506
ELEKTROMEDIZINISCH : electro-medical 507
ELEKTROMETER (*n*) : electrometer 508
ELEKTROMETRISCH : electrometric 509
ELEKTROMOBIL (*n*) : electromobile ; electrically driven automobile 510
ELEKTROMOTOR (*m*) : electromotor ; electric motor 511
ELEKTROMOTORISCH : electromotive 512
ELEKTROMOTORISCHE KRAFT (*f*) : electro-motive force, EMF 513
ELEKTRON (*n*) : electron (diminutive negative-electricity carriers which compose cathode rays) ; electron (a light metal, magnesium alloy : Sg. 1.75 to 2.0 ; over 80% Mg ; some Zn. ; and up to 10% Al.) 514
ELEKTRONEGATIV : electro-negative 515
ELEKTRONENAUSSTRAHLUNG (*f*) : emission of electrons (from a cathode) 516
ELEKTRONENFLUSS (*m*) : flow (stream) of electrons 517
ELEKTRONENRELAIS (*n*) : electron relay 518
ELEKTRON-LEICHTMETALL (*n*) : electron-(light metal) (80-99% Mg. and 0.5-20% other metals ; Sg. 1.7-1.8 ; Mp. 630-650°C.) (see also ELEKTRON) 519
ELEKTROOFEN (*m*) : electric (oven) furnace 520
ELEKTROPHOR (*m*) : electrophorous 521
ELEKTROPLATTIEREN : to electro-plate ; (*n*) : electro-plating 522
ELEKTROPLATTIERUNG (*f*) : electro-plating 523
ELEKTROPOSITIV : electro-positive 524
ELEKTROSCHMELZOFEN (*m*) : electric smelting furnace 525
ELEKTROSCHNELLFÖRDERER (*m*) : electric high-speed (transporter, conveyor) conveying plant 526
ELEKTROSKOP (*n*) : electroscope 527
ELEKTROSPEICHER (*m*) : electro-accumulator 528
ELEKTROSPEICHERUNG (*f*) : electro-accumulator installation 529
ELEKTROSTAHL (*m*) : electro-steel 530
ELEKTROSTATISCH : electrostatic 531
ELEKTROSYNTHESE (*f*) : electro-synthesis 532
ELEKTROTECHNIK (*f*) : electro-technics 533
ELEKTROTECHNIKER (*m*) : electrician ; electrical engineer 534
ELEKTROTHERMIE (*f*) : thermo-electricity 535
ELEKTROTHERMISCH : electro-thermal 536
ELEKTROTYP (*n*) : electro-type 537
ELEKTRUM (*m*) : electrum (natural gold and silver alloy ; about 40% silver) ; AuAg ; Sg. 15.4 ; (*n*) : (an ancient alloy of gold and silver in proportion of 3 : 1 (see also NEUSILBER) 538
ELEMENT (*n*) : cell (Elect.), battery ; element ; part 539
ELEMENTAR : elementary ; rudimentary 540
ELEMENTARISCH : elementary, elemental, preparatory, rudimentary 541
ELEMENTARSTEIN (*m*) : pyrite (see PYRIT) 542
ELEMENTE (*n.pl*) : elements, parts 543
ELEMENTENMESSUNG (*f*) : stoichiometry 544
ELEMENT (*n*), GALVANISCHES- : galvanic (battery) cell 545
ELEMIHARZ (*n*) : gum elemi ; manila copal (*Resina elemi*) (a resin from tree as *Carnarium commune*) 546
ELEMIÖL (*n*) : elemi oil ; Sg. 0.87-0.91 547
ELENTIER (*n*) : elk, moose (*Alces machlis*) 548
ELEOLITH (*m*) : eleolite ; $(Na_2,K,Ca)Al_2Si_2O_8$; Sg. 2.61 549
ELEPHANTIASIS (*f*) : elephantiasis arabum (Med.) 550
ELEVATION (*f*) : elevation 551
ELEVATIONSWINKEL (*m*) : angle of elevation 552
ELEVATOR (*m*) : elevator ; hoist ; lift 553
ELEVATOREIMER (*m*) : elevator bucket 554
ELEVENZEIT (*f*) : apprenticeship (period of practical experience) (generally refers to budding chemists in Germany and is of three years' duration) 555
ELFENBEIN (*n*) : ivory ; Sg. 1.83 556
ELFENBEINDRECHSLER (*m*) : ivory turner 557
ELFENBEINERN : of ivory, ivory 558
ELFENBEIN (*n*), GEBRANNTES- : ivory black ; burnt ivory 559
ELFENBEIN (*n*), KÜNSTLICHES- : celluloid ; artificial ivory 560
ELFENBEINNUSS (*f*) : ivory nut (see ELEFANTENNUSS) 561
ELFENBEINPAPPE (*f*) : ivory (paper) board 562
ELFENBEINSCHWARZ (*n*) : ivory black ; bone black, ebony black (*Ebur ustum nigrum*) (about 10% carbon, remainder mineral, mostly calcium phosphate) (by calcining ivory in closed crucibles ; used as a pigment and and as an ingredient of black varnish) 563
ELFENBEINSUBSTANZ (*f*) : dentine 564
ELIMINIEREN : to eliminate 565
ELIXIR (*n*) : elixir (a kind of tincture prepared by the aid of extracts) (Pharm.) 566
ELLAGENGERBSÄURE (*f*) : ellagic acid ; $C_2H_4O_8$ 567
ELLAGENSÄURE (*f*) : ellagic acid ; $C_{14}H_6O_8$ 568
ELLAGSÄURE (*f*) : ellagic acid (from *Potentilla tormentilla*) 569
ELLBOGEN (*m*) : elbow 570
ELLE (*f*) : ell ; yard ; ulna (Anat.) 571
ELLENBOGEN (*m*) : elbow 572

ELLENHANDEL (*m*): drapery-(trade) 573
ELLENWAREN (*f.pl*): dry goods; draper's goods; drapery 574
ELLENWARENHÄNDLER (*m*): draper 575
ELLER (*f*): alder (see **ERLE**) 576
ELLERN: alder, of alder 577
ELLIPSE (*f*): ellipse 578
ELLIPSENZIRKEL (*m*): trammel, eliptic compasses 579
ELLIPSOID (*n*): ellipsoid 580
ELLIPTISCH: elliptic-(al)-(ly) 581
ELSE (*f*): alder (see **ERLE**); sailmaker's awl 582
ELSEBAUM (*m*): (see **ERLE**) 583
ELSEBEERBAUM (*m*): service tree (*Pyrus domestica*) 584
ELSEBEERE (*f*): service berry (of *Pyrus domestica*) 585
ELUTION (*f*): steam, extract 586
ELUTOR (*m*): extractor (Molasses) 587
ELUTRIEREN: to wash; elutriate 588
EMAIL (*n*): enamel 589
EMAILLESPRITZAPPARAT (*m*): enamel (aerograph) sprayer or spraying apparatus 590
EMAILLEWEISS (*n*): (see **LITHOPON**) 591
EMAILFARBE (*f*): enamel (paint) colour 592
EMAILGLAS (*n*): enamel (fusible) glass 593
EMAILLACK (*m*): enamel varnish 594
EMAILLE (*n*): enamel 595
EMAILLEMÜHLE (*f*): enamel mill 596
EMAILLIEREN: to enamel 597
EMAILLIEROFEN (*m*): enamel (stove) furnace 598
EMAILLIERSODA (*f*): enamelling soda 599
EMAILLIERTES GEFÄSS (*n*): enamelled vessel 600
EMAILMÜHLE (*f*): enamel mill 601
EMAILSCHILD (*n*): enamel (sign; badge) shield 602
EMBALLAGE (*f*): bale, packing, box, case 603
EMBALLIEREN: to bale, pack, box, case 604
EMBOLIE (*f*): embolism (lodgement of some foreign matter in a blood vessel) (Med.) 605
EMBOLIT (*m*): embolite (natural silver chlorobromide); Ag(Cl.Br); Sg. 6.0 606
EMETIN (*n*): (alkaloid from root of *Cephaelis ipecacuanha*); emetine, ($C_{30}H_{40}N_2O_5$); Mp. 68°C. 607
EMINENT: eminent-(ly) 608
EMISSIONSVERMÖGEN (*n*): emissive power 609
EMOLLIENS (*n*): emollient 610
EMPFANG (*m*): receipt; reception 611
EMPFANGANTENNE (*f*): receiving aerial (Wireless Tel.) 612
EMPFANGEN: to receive; accept; conceive; welcome 613
EMPFÄNGER (*m*): receiver 614
EMPFANGFERNHÖRER (*m*): (telephone)-receiver (Wireless Tel.) 615
EMPFANGHÖRER (*m*): (telephone)-receiver 616
EMPFÄNGLICH: receptive; susceptible; sensible 617
EMPFÄNGLICHKEIT (*f*): susceptibility 618
EMPFANGNAHME (*f*): receipt, reception 619
EMPFÄNGNIS (*n*): conception 620
EMPFANGSANZEIGE (*f*): acknowledgment of receipt 621
EMPFANGSAPPARAT (*m*): receiving apparatus 622
EMPFANGSBESCHEINIGUNG (*f*): acknowledgment of receipt; receipt form; receipt 623
EMPFANGSLUFTDRAHT (*m*): receiving (aerial) wire (Wireless Tel.) 624
EMPFANGSSCHALTUNG (*f*): receiver connection (Wireless Tel.) 625
EMPFANGSSCHEIN (*m*): receipt, receipt form 626
EMPFANGSSTATION (*f*): receiving station 627
EMPFANGSSTELLE (*f*): receiving station 628
EMPFANGVERMITTLER (*m*): detector (Wireless Tel.) 629
EMPFEHL (*m*): recommendation, compliment 630
EMPFEHLEN: to recommend; commend; present one's respects 631
EMPFEHLENSWERT: worthy of recommendation; to be recommended; advisable; eligible; preferential 632
EMPFEHLUNGSBRIEF (*m*): letter of (introduction) recommendation 633
EMPFINDBAR: susceptible, sensible, sensitive, perceptible 634
EMPFINDEN: to feel; experience; perceive; be sensible of 635
EMPFINDLICH: irritable; sensible; sensitive; severe; susceptible; delicate; perceptible 636
EMPFINDUNGSVERMÖGEN (*n*): sensitiveness, susceptibility 637
EMPHYSEM (*m*): (subcutaneous)-Emphysema (distention of the areolar tissue with gas or air) (Med.) 638
EMPIRIE (*f*): empiricism 639
EMPIRIK (*f*): empiricism 640
EMPIRIKER (*m*): empiric 641
EMPIRISCH: empiric-(al)-(ly) 642
EMPLEKTIT (*m*): emplektite; Cu_2S,Bi_2S_3 or $CuBiS_2$; Sg. 5.2-6.3 643
EMPORARBEITEN: to struggle upwards, work one's way up 644
EMPORBRINGEN: to raise, bring to a higher level 645
EMPORDRINGEN: to struggle upwards, force a way up 646
EMPORHEBEN: to raise (lift) up, elevate, exalt 647
EMPORKOMMEN: to rise; emerge; mount; ascend; prosper; grow; soar; elevate 648
EMPORRAGEN: to tower -(up) 649
EMPORSTREBEN: to strive upwards, have an upward tendency, aspire 650
EMPORTREIBEN: to drive (force) upwards, sublimate 651
EMPYROFORM (*n*): empyroform (condensation product of birch tar and formaldehyde) 652
EMPYROFORMPASTE (*f*): empyroform paste 653
EMPYROFORMPULVER (*n*): empyroform powder 654
EMPYROFORMSALBE (*f*): empyroform (salve) ointment 655
EMPYROFORMSEIFE (*f*): empyroform soap 656
EMPYROFORMTINKTUR (*f*): empyroform tincture 657
EMSER SALZ (*n*), **KÜNSTLICHES-**: sodium bicarbonate, Ems salt (see **NATRIUMBIKARBONAT**) 658
EMULGIEREN: to emulsify 659
EMULGIERMASCHINE (*f*): emulsifier; emulsifying machine 660
EMULGIERUNG (*f*): emulsification 661
EMULSION (*f*): emulsion 662
EMULSIONIEREN: to emulsify 663
EMULSIONSGELATINE (*f*): emulsion gelatine 664
EMULSIONSMÖRSER (*m* and *pl*): emulsion mortar-(s) 665
EMULSIONSÖL (*n*): emulsion oil 666
EMULSIONSPAPIER (*n*): emulsion paper 667
EMULSOID (*n*): colloidal emulsion 668
ENANTIOTROPIE (*f*): enantiotropy; enantiotropism 669

ENARGIT (*m*) : enargite (natural sulph-arsenide of copper) ; Cu_3AsS_4 ; Sg. 4.36-4.44 670
ENCYKLOPÄDIE (*f*) : encyclopædia, encyclopedia 671
ENDAUSBEUTE (*f*) : final (end) yield 672
ENDBAR : terminable 673
ENDDRUCK (*m*) : final (end ; terminal) pressure 674
ENDE (*n*) : end ; finish ; limit ; purpose ; termination ; terminus ; conclusion ; close ; issue 675
ENDE (*n*), **ÄUSSERSTES-** : extremity 676
ENDE (*n*), **BALKEN-** : beam end 677
ENDECK (*n*) : summit ; terminal angle 678
ENDECKE (*f*) : (see **ENDECK**) 679
ENDEN : to end ; terminate ; finish ; put a stop to ; conclude ; close ; bring to an issue 680
ENDESGENANNTE (*m*) : undersigned, subscriber 681
ENDFLÄCHE (*f*) : end face 682
ENDGESCHWINDIGKEIT (*f*) : terminal (end, final) speed or velocity 683
ENDGLEID (*n*) : terminal (number) member 684
ENDGÜLTIG : final ; conclusive ; finally ; eventual 685
ENDIGEN : (see **ENDEN**) 686
ENDLAGE (*f*) : end (final) position 687
ENDLAUGEN (*f.pl*) ; foots 688
ENDLICH : finite ; final ; late ; at last ; finally ; after all ; ultimate ; concluding 689
ENDLICHIT (*m*) : endlichite (an intermediate variety between mimetite and vanadinite (see **MIMETIT** and **VANADINIT**) 690
ENDLOS : infinite ; endless ; boundless 691
ENDNÜANCE (*f*) : final (end) shade or nuance 692
ENDOCARDITIS (*f*) : endocarditis, inflammation of the lining membrane of the heart (Med.) 693
ENDÖFFNER (*m*) **NACH GARNETT** : Garnett's end opener (Textile) 694
ENDOSMOTISCH : endosmotic 695
ENDOSPERM (*n*) : endosperm 696
ENDOTHERMISCH : endothermic ; endothermal ; consuming heat 697
ENDOTHERMISCHE REAKTION (*f*) : [heat (used) absorbed], endothermal reaction 698
ENDPRODUKT (*n*) : final product 699
ENDPUNKT (*m*) : end point ; terminus ; extremity 700
ENDSILBE (*f*) : end syllable ; termination 701
ENDSTÄNDIG : end ; terminal ; standing at the end 702
ENDSTATION (*f*) : terminus, block station 703
ENDSTÜCK (*n*) : nozzle (Brick) ; end piece 704
ENDTEMPERATUR (*f*) : final temperature 705
ENDURTEIL (*n*) : decision, final decision 706
ENDWERT (*m*) : final (end) value 707
ENDZIEL (*n*) : aim, object, goal, end, eventual objective 708
ENDZOTTE (*f*) : tail-end, streamer (of water fungi) 709
ENDZUSTAND (*m*) : final (state) condition 710
ENDZWECK (*m*) : aim, end, object, goal, purpose, design, eventual objective 711
ENERGIE (*f*) : energy 712
ENERGIEABNAHME (*f*) : energy (reduction) decrease 713
ENERGIEÄNDERUNG (*f*) : change (fluctuation ; alteration) of energy 714
ENERGIEAUFWAND (*m*) : energy (expenditure ; used) consumption 715
ENERGIEGLEICHUNG (*f*) : energy equation 716
ENERGIEINHALT (*m*) : energy content 717
ENERGIEVERLUST (*m*) : energy loss 718
ENERGIEZUFUHR (*f*) : energy (addition) supply 719
ENERGISCH : energetic-(ally) 720
ENFLEURAGE (*f*) : perfuming (the saturation of fat with the scent from the leaves of flowers) 721
ENG : close ; narrow ; tight ; confined ; constricted 722
ENGBRÜSTIG : asthmatic-(al), narrow-chested 723
ENGE (*f*) : corner ; narrowness ; constriction ; swaging, etc. 724
ENGELROT (*n*) : colcothar (a red iron oxide from calcining copperas) 725
ENGELSÜSS (*n*) : polypodium (type of fern) (*Rhizoma Polypodii*) 726
ENGELSÜSSWURZEL (*f*) : polypody root (*Radix Polypodii*) 727
ENGELWURZ (*f*) : angelica (root of *Archangelica officinalis*) (*Radix angelicæ*) 728
ENGELWURZEL (*f*) : (see **ENGELWURZ**) 729
ENGHALSFLASCHE (*f*) : narrow-necked bottle 730
ENGLÄNDER (*m*) : universal screw spanner, universal screw wrench, universal screw hammer ; monkey wrench 731
ENGLISCH : english 732
ENGLISCHBLAU (*n*) : royal blue (see **KUPFER-CARBONAT**) 733
ENGLISCHE ERDE (*f*) : rotten stone ; tripoli (see **TRIPEL**) 734
ENGLISCHES GEWÜRZ (*n*) : pimenta ; allspice (fruit of *Pimenta officinalis*) 735
ENGLISCHE HAUT (*f*) : gold beater's skin 736
ENGLISCHE KRANKHEIT (*f*) : rickets, rachitis (Med.) 737
ENGLISCHES PULVER (*n*) : algaroth powder ; antimony oxychloride ; SbOCl 738
ENGLISCHES ROT (*n*) : English red ; colcothar ; red oxide of iron ; Fe_2O_3 (see **ALGAROTH-PULVER**) 739
ENGLISCHER SCHWEISS (*m*) : sweating sickness, miliary fever 740
ENGLISCHES PFLASTER (*n*) : court plaster 741
ENGLISCHES SALZ (*n*) : Epsom salt (see **MAGNESIUMSULFAT**) 742
ENGLISCHGELB (*n*) : patent yellow ; Turner's yellow ; lead oxychloride ; $5PbO.PbCl_2$ 743
ENGLISCHGRÜN (*n*) : patent green, Paris green **KUPFERACETATARSENIAT**) 744
ENGLISCHPFLASTER (*n*) : court plaster 745
ENGLISCHROT (*n*) : colcothar ; red oxide of iron ; English red ; Fe_2O_3 ; trip ; jeweller's red (see **EISENOXYD**) ; Sg. 5.1 746
ENGLISCHSALZ (*n*) : Epsom salt ; $MgSO_4$; Sg. 2.65 (see **MAGNESIUMSULFAT**) 747
ENGLOCHIG : with small holes, pores, or meshes 748
ENGMASCHIG : close-meshed ; fine-meshed 749
ENGOBE (*m*) : engobe ; slip 750
ENGOBEFARBEN (*f.pl*) ; engobe (slip) colours 751
ENGOBIEREN : to coat with engobe 752
ENGRINGIG : narrow-ringed, close-ringed 753
ENGROS : wholesale 754
ENHARMONISCH : enharmonic (Music) 755
ENKAUSTISCH : encaustic (having colours burnt in) 756
ENLEVAGE (*f*) : discharge (Dye) 757
ENORM : enormous-(ly) 758
ENORMITÄT (*f*) : enormity 759

ENSTATIT (*m*): enstatite (an orthorhombic member of the pyroxene group; magnesium metasilicate; $MgSiO_3$; Sg. 3.2 (see also AMPHIBOL) 760
ENTALKOHOLISIEREN: to dealcoholize; extract alcohol from 761
ENTAMIDIEREN: to deaminize; deamidize 762
ENTARTEN: to deteriorate; degenerate 763
ENTARTUNG (*f*): degeneration (of living cells in the body) (Med.) 764
ENTASCHUNGSANLAGE (*f*): ash removal (ash handling) plant 765
ENTBASTEN: to degum-(silk) (see DEGUMMIEREN) 766
ENTBEHREN: to be without; lack; dispense with; do without; want 767
ENTBEHRLICH: unnecessary; dispensable; not wanted 768
ENTBENZOLIERUNG (*f*): extraction of (benzol) benzene 769
ENTBINDEN: to untie; disengage; set free; loosen; liberate; release; evolve (Gases); deliver (Med.); exonerate; absolve 770
ENTBINDUNGSANSTALT (*f*): lying-in hospital (Med.) 771
ENTBINDUNGSFLASCHE (*f*): generating flask 772
ENTBINDUNGSROHR (*n*): delivery tube 773
ENTBINDUNGSZANGE (*f*): forceps (Med.) 774
ENTBITTERN: to deprive of bitterness; extract bitterness from 775
ENTBLÄTTERN: to strip (deprive) of leaves 776
ENTBLEIEN: to extract lead from 777
ENTBLÖSSEN; to uncover; strip; deprive; denude; bare; expose 778
ENTBLÖSST: bare, deprived of; destitute; denuded 779
ENTBRECHEN: to abstain from, forbear 780
ENTCARBONISIEREN: to decarbonize (water, etc.) 781
ENTCHLORUNG (*f*): dechlorination 782
ENTDECKEN: to discover; disclose; detect 783
ENTDECKER (*m*): discoverer; detector; detective; explorer 784
ENTDENATURIEREN: to free from denaturants 785
ENTDUFTEN: to deodorize; (*n*): deodorization 786
ENTDÜFTUNGSANLAGE (*f*): deodorizing plant 787
ENTE (*f*): crucible (a vessel used in coal analysis or testing, in place of a platinum crucible); duck (*Anatidæ* family of birds) 788
ENTEISENEN: to extract iron from; purify (soften) water 789
ENTEISENUNGSANLAGE (*f*): water-softening plant; plant for the extraction of iron 790
ENTENFETT (*n*): duck fat; Sg. 0.93; Mp. 25-34°C 791
ENTENGRÜTZE (*f*): duckweed (see WASSER-LINSE) 792
ENTERN: to board, grapple (Naut.) 793
ENTEROLITH (*m*): bezoar (an intestinal concretion of animals); enterolith (an intestinal concretion of human beings) (Med.) 794
ENTFAHREN: to escape-(from) 795
ENTFALLEN: to drop; slip; fall; be precipitated; escape 796
ENTFALTEN: to unfold; display; evolve; develop 797
ENTFÄRBEN: to bleach; extract colour from; make (grow) pale; fade; change colour; clear (Phot.); decolour 798
ENTFÄRBUNGSKRAFT (*f*): bleaching power 799
ENTFÄRBUNGSMITTEL (*n*): decolourizing (bleaching) agent; decolourant 800
ENTFÄRBUNGSPULVER (*n*): bleaching powder; calcium hypochlorite; $Ca(ClO)_2.4H_2O$ 801
ENTFÄRBUNGSPULVER (*n*), **AMERIKANISCHES-**: bleaching powder (magnesium hydrosilicate; for bleaching paraffin) 802
ENTFASERN: to free from fibres; string; shred (Beans) 803
ENTFERNEN: to withdraw; remove; depart; extract; absent; eliminate; separate; alienate; discharge; retire 804
ENTFERNT: distant, remote 805
ENTFERNUNGSKRAFT (*f*): centrifugal force 806
ENTFESSELN: to unbind, release, set free 807
ENTFETTEN: to extract (remove) fat, oil or grease; de-oil; scour; degrease 808
ENTFETTUNGSANLAGE (*f*): grease (oil) extracting plant; degreasing plant; fat (extracting) removing plant; deoiling plant 809
ENTFETTUNGSAPPARAT (*m*): degreasing apparatus, grease (removing) extracting apparatus 810
ENFILZEN: to remove felt from 811
ENTFLAMMBAR: inflammable 812
ENTFLAMMEN: to inflame; kindle; flash; fire; be inflamed 813
ENTFLAMMUNGSPUNKT (*m*): flash point 814
ENTFLAMMUNGSTEMPERATUR (*f*): ignition temperature; flash point 815
ENTFLEISCHEN: to flesh; strip of flesh 816
ENTFLIESSEN: to flow; emanate 817
ENTFREMDEN: to estrange; abandon; conceal; alienate 818
ENTFRITTEN: to decohere 819
ENTFRITTER (*m*): decoherer 820
ENTFRITTUNG (*f*): decoherence 821
ENTFÜHREN: to carry off; lead in wrong direction; abduct; mislead 822
ENTFUSELN: to rectify; remove fusel oil from 823
ENTGASEN: to degas; extract gas from; coke 824
ENTGASUNG (*f*): extraction of gas, de-gasification 825
ENTGEGEN: against; towards; opposite; counter 826
ENTGEGENARBEITEN: to counteract, work against, work towards 827
ENTGEGENBLICKEN: to look forward to 828
ENTGEGEN GEHEN: to go to meet; go towards; go against; face; encounter; oppose; go counter to 829
ENTGEGENGESETZT: opposite; opposed; contra; contrary; inverse; reverse; counter 830
ENTGEGENHALTEN: to contrast, hold against, oppose 831
ENTGEGENKOMMEN: to meet, aid, further; to prevent, hinder, obviate, to influence for better or worse 832
ENTGEGENLAUFEN: to run counter to 833
ENTGEGENWIRKEN: to counteract, act counter to 834
ENTGEGENSEHEN: to look forward to; expect 835
ENTGEGNEN: to reply; respond; retort 836
ENTGEHEN: to escape; avoid; elude 837
ENTGEISTEN: to dealcoholize 838
ENTGELTEN: to pay for; atone; suffer-(for); requite; recompense 839
ENTGERBERN: to wash (Wool) 840
ENTGIFTEN: to remove (extract) poison 841
ENTGLASEN: to devitrify 842

ENTGLEISEN : to run off (the rails), to be derailed 843
ENTGLIEDERN : to dismember, remove (deprive of) members 844
ENTGOLDEN : to remove (extract) gold 845
ENTGRANNEN : to awn 846
ENTGRÄTEN : to bone-(fish) 847
ENTGRATEN : to remove the burr (see ABGRATEN) 848
ENTGÜLTIGEN : to make invalid, invalidate 849
ENTGUMMIEREN : to degum ; scour ; wash 850
ENTHAAREN : to depilate ; scrape off (remove) hair 851
ENTHAARUNGSMITTEL (*n*) : depilatory 852
ENTHALOGENISIEREN : to dehalinate ; dehalogenate 853
ENTHALTEN : to contain ; include ; abstain ; be contained in ; forbear ; be shown on ; comprise 854
ENTHALTSAM : temperate ; abstemious ; forbearing 855
ENTHÄRTEN : to soften ; anneal 856
ENTHARZEN : to remove (extract) resin 857
ENTHÄUTEN : to skin ; flay 858
ENTHEBEN : to exonerate ; exempt ; dismiss ; dispense 859
ENTHÜLLEN : to unfold ; unveil ; disclose ; reveal 860
ENTHÜLSEN : to husk ; shell ; peel ; skin 861
ENTHYDRATISIEREN : to dehydrate 862
ENTKALKEN : to decalcify ; delime ; extract (remove) lime 863
ENTKÄLKEN : (see ENTKALKEN) 864
ENTKALKUNGSFELD (*n*) : decalcifying bed, bed for removing (extracting) the chalk from water 865
ENTKALKUNGSMITTEL (*n*) : de-liming (decalcifying) agent 866
ENTKARBOXYLIEREN : to decarboxylate 867
ENTKARBOXYLIERUNG (*f*) : decarboxylation 868
ENTKEIMEN : to degerminate ; germinate ; sprout 869
ENTKERNEN : to stone (Fruit) 870
ENTKIESELN : to desilicify ; extract (remove) silica from 871
ENTKLEIDEN : to undress ; strip ; remove ; divest 872
ENTKOHLEN : to decarbonize 873
ENTKOHLENSTOFFT : decarbonized 874
ENTKOHLUNG (*f*) : decarbonization 875
ENTKOMMEN : to escape 876
ENTKÖRNEN : to free from grains ; gin (Cotton) ; thresh (Grain) 877
ENTKÖRPERN : to disembody 878
ENTKRÄFTEN : to exhaust ; weaken ; relax ; debilitate ; enervate ; fatigue ; invalidate 879
ENTKUPFERN : to extract copper from 880
ENTLADEN : to unload ; discharge ; explode ; go off ; evacuate ; draw off ; exonerate ; relieve 881
ENTLADER (*m*) : discharger (Elect.) 882
ENTLADESTROM (*m*) : discharging current 883
ENTLADUNG, ELEKTRISCHE- (*f*) : electric discharge 884
ENTLANG : along 885
ENTLASSEN : to dismiss ; discharge 886
ENTLASTEN : to unload ; reduce pressure ; balance ; discharge ; ease ; free-(from) 887
ENTLASTETES VENTIL (*n*) : equilibrum (balanced) valve 888
ENTLASTUNGSBOGEN (*m*) : saving arch (Brickwork) 889
ENTLASTUNGSSCHIEBER (*m*) : balanced slide valve 890
ENTLASTUNGSVENTIL (*n*) : easing valve, pressure reducing valve, by-pass valve 891
ENTLEEREN : to empty ; discharge ; evacuate 892
ENTLEERUNG (*f*) : depletion (Med.) ; evacuation, discharge, emptying 893
ENTLEERUNGSAPPARAT (*m*) : emptying (discharging, evacuating) apparatus 894
ENTLEGEN : distant ; remote 895
ENTLEHNEN : to borrow ; take ; derive 896
ENTLEIMEN : to degum ; scour ; wash ; degelatine 897
ENTLEIMT : degummed, degelatined 898
ENTLEUCHTEN : to deprive of light ; make non-luminous ; shine ; radiate 899
ENTLÜFTEN : to de-aerate ; extract air from 900
ENTLÜFTUNG (*f*) : air extraction, de-aeration 901
ENTLÜFTUNGSHAHN (*m*) : air cock 902
ENTLÜFTUNGSVORRICHTUNG (*f*) : de-aerating plant 903
ENTMAGNETISIEREN : to demagnetize 904
ENTMAGNETISIERUNG (*f*) : demagnetization 905
ENTMASTEN : to dismast (Naut.) 906
ENTMETHYLIEREN : to demethylate 907
ENTMISCHEN : to disintegrate, separate-(a mixture into its component parts) 908
ENTMISCHUNG (*f*) : disintegration ; separation into component parts 909
ENTMUTIGEN : to discourage ; dishearten 910
ENTNAHME (*f*) : taking ; drawing ; draft ; getting ; borrowing ; depriving ; extracting ; outlet ; exit ; withdrawal ; discharge ; removal ; leak-off (Steam) 911
ENTNAHME AUF LONDON : draft on London 912
ENTNAHMEROHR (*n*) : outlet (discharge) pipe 913
ENTNAHMETURBINE (*f*) : bleeder (leak-off) turbine 914
ENTNÄSSEN : to dry ; extract moisture 915
ENTNEBELN : to dissipate (mist) fog ; extract moisture from gases 916
ENTNEBELUNGSANLAGE (*f*) : fog removing plant ; mist condensing plant ; moisture extraction plant 917
ENTNEHMEN : to take ; draw ; get ; borrow ; deprive ; remove from ; see ; extract from ; observe ; note ; be obvious from ; withdraw 918
ENTNERVEN : to unnerve, enervate 919
ENTÖLEN : to deoil 920
ENTÖLER (*m*) : oil separator 921
ENTÖLUNGSAPPARAT (*m*) : deoiling plant 922
ENTOMOLOG (*m*) : entomologist 923
ENTOMOLOGISCH : entomological-(ly) 924
ENTPHOSPHOREN : to dephosphorize 925
ENTPOLYMERISIEREN : to depolymerize 926
ENTPRESSEN : to press ; squeeze out of ; extort 927
ENTPÜLPT : de-pulped, have the pulp (extracted) removed 928
ENTQUALMEN : to go up in steam, vapour or smoke 929
ENTQUELLEN : to spring, flow from 930
ENTRAHMEN : to skim ; remove cream ; separate (Milk) 931
ENTRICHTEN : to pay (discharge) debts 932
ENTRIESELN : to trickle away 933
ENTRINDEN : to bark ; decorticate ; remove bark from (see also SCHÄLEN) 934
ENTRINNEN : to escape ; run away 935

ENTROPIE (*f*) : entropy 936
ENTROPIEDIAGRAMM (*n*) : entropy diagram 937
ENTROPIETAFEL (*f*) : entropy diagram 938
ENTROSTEN : to remove (clean) rust from 939
ENTRÜSTEN : to provoke anger 940
ENTSAGEN : to give up ; renounce ; relinquish ; resign 941
ENTSANDEN : to (extract) remove sand from 942
ENTSATZ (*m*) : relief ; rescue ; removal ; raising of a blockade 943
ENTSÄUERN : to deacidify ; deoxidize ; free from acid ; remove (extract) acid ; detartarize 944
ENTSÄUERUNG (*f*) : extraction of acid, detartarization 945
ENTSÄUERUNGSAPPARAT (*m*) : acid extractor, detartarizer 946
ENTSCHÄDIGEN : to indemnify ; compensate 947
ENTSCHALEN : to shell ; peel ; scour ; degum ; wash (Silk) 948
ENTSCHÄLEN : to bark, decorticate (see also **SCHÄLEN**) 949
ENTSCHEIDEN : to decide ; determine ; be final ; decree ; arbitrate ; referee 950
ENTSCHEIDUNG (*f*) : decision ; crisis ; finality ; determination ; arbitration 951
ENTSCHEIDUNGSPUNKT (*m*) : crisis 952
ENTSCHEINEN : to de-bloom (oils, etc.) ; remove (lustre, light, shine)- polish 953
ENTSCHEINUNG (*f*) : de-blooming, removal of fluorescence from mineral oils 954
ENTSCHEINUNGSMITTEL (*n*) : de-blooming agents 955
ENTSCHIEDEN, GANZ- : decidedly, certainly, assuredly 956
ENTSCHLACKEN : to clinker, remove slag or clinker 957
ENTSCHLACKUNGSANLAGE (*f*) : slag removal plant 958
ENTSCHLICHTEN : to undress (Linen) ; free from size 959
ENTSCHLUSS (*m*) : resolution ; decision ; resolve ; determination ; purpose 960
ENTSCHWEFELN : to de-sulphurize 961
ENTSCHWEFELUNG (*f*) : de-sulphurization 962
ENTSCHWEFELUNGSSCHLACKE (*f*) : de-sulphurizing slag (for electro steel) ; (lime, fluor spar, and sand) (is snow white in colour) 963
ENTSCHWEISSEN : to scour (Wool) ; de-grease 964
ENTSCHWELEN : to distil (Fuel) ; to be distilled 965
ENTSEIFEN : to rinse ; free from soap 966
ENTSILBERN : to de-silver ; remove silver from 967
ENTSPANNEN : to reduce tension ; reduce in pressure ; release (Spring) ; relax ; cut off (Steam) ; expand (Gases) 968
ENTSPANNUNG (*f*) : pressure drop ; decrease in (tension) pressure ; expansion 969
ENTSPANNUNGSKÜHLUNG (*f*) : cooling by (reduction of pressure) expansion 970
ENTSPANNUNGSVENTIL (*n*) : pressure-reducing valve, expansion valve 971
ENTSPRECHEN : to accord with ; correspond ; represent ; answer ; agree with ; conform ; concord ; comply with ; suit ; satisfy (Maths.) 972
ENTSPRECHEND : suitably (see also **ENTSPRECHEN**) ; according to ; corresponding to ; commensurate with ; consequent upon ; in conformity with 973
ENTSPRINGEN : to spring ; rise ; escape ; originate 974
ENTSPRÜHEN : to spit (gush) forth (of fire) 975
ENTSTÄNKERN : to deodorize ; remove smell from 976
ENTSTÄNKERUNGSPATRONE (*f*) : deodorising cartridge 977
ENTSTAUBEN : to dust off ; extract dust from ; remove dust from 978
ENTSTAUBUNGSANLAGE (*f*) : dust extraction (extracting ; removal) plant 979
ENTSTEHEN : to originate ; begin ; arise ; occur ; be formed ; be caused ; be induced by ; be occasioned ; be generated ; be produced ; be set up ; result 980
ENTSTEHEND, EBEN- : nascent ; incipient 981
ENTSTEHUNG (*f*) : origin ; genesis ; nascence ; formation ; generation ; production 982
ENTSTEHUNGSART (*f*) : mode (nature) of origin 983
ENTSTEHUNGSGRUND (*m*) : reason, cause 984
ENTSTEHUNGSWEISE (*f*) : (see **ENTSTEHUNGSART**) 985
ENTSTEHUNGSZUSTAND (*m*) : nascent state 986
ENTSTELLEN : to distort ; disfigure ; deface ; deform 987
ENTSTRÖMEN : to flow ; stream ; issue ; gush ; escape from 988
ENTWÄSSERN : to dehydrate ; extract water from ; concentrate ; rectify ; dephlegmate ; drain 989
ENTWÄSSERT : anhydrous (see also **ENTWÄSSERN**) 990
ENTWÄSSERUNG (*f*) : extraction of water, drainage, drain-(pipe) 991
ENTWÄSSERUNGSHAHN (*m*) : drain cock 992
ENTWÄSSERUNGSMASCHINE (*f*) : wet machine (Paper) 993
ENTWÄSSERUNGSMITTEL (*n*) : dehydrating agent 994
ENTWÄSSERUNGSROHR (*n*) : drain pipe 995
ENTWEICHEN : to leak ; escape ; abscond ; dissipate 996
ENTWEICHUNGSHAHN (*m*) : discharge (delivery, escape) cock 997
ENTWEICHUNGSVENTIL (*n*) : delivery (escape) valve 998
ENTWENDEN : to steal ; embezzle ; purloin ; pilfer 999
ENTWERFEN : to design ; plan ; draw down ; project ; plot (Curves) ; sketch 000
ENTWERFER (*m*) : designer ; draughtsman 001
ENTWERTEN : to depreciate ; cancel (Stamps) ; reduce (decrease) in value 002
ENTWICKELN : to develop ; evolve ; expand ; disengage ; generate (Gases) ; unfold ; set up ; display 003
ENTWICKELUNG (*f*) : development ; evolution ; generation ; disengagement ; production ; unfolding ; explanation 004
ENTWICKELUNGSBAD (*n*) : developing bath ; developer 005
ENTWICKELUNGSDRUCK (*m*) : generating pressure ; bromide-(print) (Photography) 006
ENTWICKELUNGSGEFÄSS (*n*) : generating vessel ; generator (Gas) 007
ENTWICKELUNGSLEHRE (*f*) : theory of evolution 008
ENTWICKELUNGSMITTEL (*n*) : developer (Photography) 009
ENTWICKELUNGSROHR (*n*) : delivery tube 010
ENTWICKELUNGSTEMPO (*n*) : rate (speed) of development, evolution or formation 011
ENTWICKLER (*m*) : generator, developer 012

ENTWICKLERLÖSUNG (*f*): developing solution, developing bath, developer (Photography) 013
ENTWICKLUNGSBAD (*n*): developing bath 014
ENTWICKLUNGSFÄHIGKEIT (*f*): developing capacity 015
ENTWICKLUNGSPAPIER (*n*): developing paper (Photography) 016
ENTWICKLUNGSSUBSTANZ (*f*): developing (agent) principle 017
ENTWICKLUNGSWEISE (*f*): metamorphosis 018
ENTWINDEN: to extricate; wrest from 019
ENTWURF (*m*): sketch; outline; plan; draft; project; proposal; drawing; design(-ing) 020
ENTWURFSSKIZZE (*f*): rough sketch; lay-out sketch 021
ENTWURZELN: to uproot, root out, eradicate 022
ENTZIEHEN: to extract; take away; withdraw; deprive 023
ENTZIFFERN: to decipher; explain 024
ENTZUCKERN: to extract sugar from 025
ENTZUCKERUNG (*f*): extraction of sugar 026
ENTZÜNDBAR: inflammable 027
ENTZÜNDBAR, LEICHT-: easily inflammable 028
ENTZÜNDEN: to ignite; kindle; inflame; take fire; prime 029
ENTZÜNDLICH: inflammable; inflammatory 030
ENTZÜNDLICHKEIT (*f*): inflammability 031
ENTZÜNDUNG (*f*): ignition; igniting; inflammation (Med.); (as a medical suffix equals, -itis) 032
ENTZÜNDUNGSGESCHWULST (*f*): phlegmon (inflammation of an area of connective tissue) (Med.) 033
ENTZÜNDUNGSPROBE (*f*): ignition test 034
ENTZÜNDUNGSPUNKT (*m*): ignition (kindling; burning) point; flash point 035
ENTZÜNDUNGSWIDRIG: antiphlogistic, having inflammation healing qualities 036
ENTZWEI: asunder; apart; in two 037
ENZIAN (*m*): gentian (*Gentiana lutea*); yellow gentian 038
ENZIANEXTRAKT (*m*): gentian extract 039
ENZIANWURZEL (*f*): bitter root; gentian root (*Radix gentianae*); (root of *Gentiana lutea*) 040
ENZYM (*n*): enzyme (the active principle of a ferment) 041
ENZYMATISCH: enzymatic; enzymic 042
EOSIN (*n*): eosine (alkali salt of tetrabromofluorescein); $C_{20}H_8Br_4O_5$ 043
EOSINBLAUSTICH (*m*): erythrosin (Photography) 044
EOSINSÄURE (*f*): eosinic (eosic) acid 045
EPHEDRIN (*n*): ephedrine (alkaloid from the plant *Ephedra vulgaris*); $C_{10}H_{15}NO$; Bp. 225°C. 046
EPHEU (*m*): ivy (*Hedera helix*) 047
EPICHLORHYDRIN (*n*): epichlorohydrine; $ClH_2C.CH.CH_2.O$; Sg. 1.191; Bp. 117°C. 048
EPIDEMIE (*f*): epidemic, wide-spread disease (Med.) 049
EPIDEMISCH: epidemic 050
EPIDERMISZELLE (*f*): epidermis cell 051
EPIDOT (*m*): epidote; $HCa_2(Al)_3,(Fe)_2Si_3O_{13}$ or $Ca_2(Al.OH)(Al,Fe)_2(SiO_4)_3$; Sg. 3.4 052
EPIHYDRINSÄURE (*f*): epihydrinic acid 053
EPILEPSIE (*f*): epilepsy (Med.) 054
EPILEPTISCHES IRRESEIN (*n*): epileptic insanity (Med.) 055
EPILOG: (*m*): epilogue 056
EPINEPHRIN (*n*): epinephrin (see **ADRENALIN**) 057
EPIPHYSE (*f*): epiphysis (*epiphysis cerebri*; *glandula pinealis*); pineal gland of the brain) 058
EPIRENAN (*n*): (see **ADRENALIN**) 059
EPISODE (*f*): episode 060
EPISTILBIT (*m*): epistilbite; $CaO.Si_3O_2+AlO_3.Si_3O_2+H_2O$; Sg. 2.245 061
EPITHELIALKREBS (*m*): rodent ulcer (Med.) 062
EPITHELKREBS (*m*): epithelioma, cancroid, epithelial cancer (Med.) 063
EPITHELZELLE (*f*): epithelial cell 064
EPOCHEMACHEND: epoch-making 065
EPPICH (*m*): celery 066
EPSOMIT (*n*): epsomite (natural hydrous magnesium sulphate) (see **MAGNESIUMSULFAT** and **BITTERSALZ**); $MgSO_4+7H_2O$ 067
EPULIS (*f*): epulis (a tumour projecting from the gum) (Med.) 068
EQUIPIEREN: to equip; fit out 069
EQUISETSÄURE (*f*): equisetic (aconitic) acid 070
ERACHTEN (*n*): opinion; judgment; (as a verb= to opine; judge; consider; regard; think) 071
ERACHTENS, MEINES-: in my opinion; to my mind 072
ERBAUEN: to build; construct 073
ERBAUUNG (*f*): construction, building 074
ERBEBEN: to shake; tremble; shudder 075
ERBFOLGE (*f*): line of succession; heritage 076
ERBGRIND (*m*): dermato-mycosis achorina, tinea favosa (a contagious skin affection caused by a fungus *Achorion Schönleinii*) (Med.) 077
ERBIETEN: to offer; proffer 078
ERBINERDE (*f*): erbia; erbium oxide 079
ERBITTEN: to request; persuade; beg 080
ERBIUM (*n*): erbium (Er.) 081
ERBIUMOXYD (*n*): erbium oxide; Er_2O_3; Sg. 8.64 082
ERBIUMPRÄPARAT (*n*): erbium preparation 083
ERBIUMSULFAT (*n*): erbium sulphate; $Er_2(SO_4)_3$; Sg. 3.678; $Er_2(SO_4)_3.8H_2O$; Sg. 3.18 084
ERBIUMVERBINDUNG (*f*): erbium compound 085
ERBLASEN: to subject to (obtain by; produce by) blast (Blast furnace work) 086
ERBLASENES EISEN (*n*): the iron produced (Blast furnace work) 087
ERBLASSEN: to fade; die; pale; expire 088
ERBLEICHEN: (see **ERBLASSEN**) 089
ERBLICH: hereditary; inheritable 090
ERBLOS: disinherited; without heirs 091
ERBRECHEN: to break open; vomit 092
ERBSE (*f*): pea 093
ERBSENBAUM (*m*): Siberian acacia 094
ERBSENERZ (*n*): pea ore; kind of limonite (see **LIMONIT**) 095
ERBSENFÖRMIG: pea shaped; pisiform; pea-like 096
ERBSENGROSS: of the size of a pea 097
ERBSENGRÜN: pea-green 098
ERBSENSTEIN (*m*): pisolite; granular limestone; arragonite (see **ARRAGONIT**) 099
ERDACHSE (*f*): axis of the earth 100
ERDALKALIMETALL (*n*): alkaline earth metal 101
ERDALKALISCH: alkaline-earthy 102
ERDAPFEL (*m*): potato (see also **ERDBROT**); truffle (see also **ERDARTISCHOKE**) 103
ERDART (*f*): type (kind; sort) of earth or clay 104
ERDARTIG: earthy 105

ERDARTISCHOKE (*f*): Jerusalem artichoke (*Helianthus tuberosus*) (a potato-shaped edible tuber) 106
ERDASPHALT (*m*): earthy asphalt 107
ERDBAHN (*f*): earth's orbit 108
ERDBALL (*m*): terrestrial globe 109
ERDBANK (*f*): terrace, bank of earth, embankment 110
ERDBEBEN (*n*): earthquake 111
ERDBEBENMESSER (*m*): (see **SEISMOGRAPH**) 112
ERDBEERBLÄTTER (*n.pl*): strawberry leaves (*Herba fragariae*) (leaves of *Fragaria vesca*) 113
ERDBEERE (*f*): strawberry (*Fragaria vesca*) 114
ERDBEERKRAUT (*n*): (see **ERDBEERBLÄTTER**) 115
ERDBEERSAMENÖL (*n*): strawberry seed oil (from *Fragaria*); Sg. 0.9345 116
ERDBESCHREIBEND: geographical 117
ERDBESCHREIBUNG (*f*): geography 118
ERDBILDUNGSKUNDE (*f*): geology 119
ERDBIRNE (*f*): Jerusalem artichoke; truffle; potato (see **ERDARTISCHOKE**) 120
ERDBODEN (*m*): ground; soil; earth; floor 121
ERDBOHRER (*m*): earth (ground, soil) borer 122
ERDBROT (*n*): sow-bread (*Cyclamen europæum*) 123
ERDE (*f*): earth; soil; ground; Sg. 1.5-2.2 124
ERDE (*f*), **GELBE-**: yellow ochre (an earthy iron oxide) (clay plus Fe_2O_3); (see also **OCKER**) 125
ERDEICHEL (*f*): earth-nut; pea-nut (*Arachis hypogæa*); heath nut (*Lathyrus tuberosus*) 126
ERDEICHELÖL (*n*): (see **ERDNUSSÖL**) 127
ERDEN: to earth, lead to earth (Electricity); of earth, earthen 128
ERDENGE (*f*): isthmus 129
ERDENKEN: to devise; invent; conceive; think of 130
ERDE (*f*), **ROTE-**: red earth 131
ERDE (*f*), **SELTENE-**: rare earth 132
ERDFAHL: earth (clay) coloured 133
ERDFARBE (*f*): earth colour; mineral colour 134
ERDFARBIG: earth (clay) coloured 135
ERDFLÄCHE (*f*): earth (area) surface 136
ERDFLACHS (*m*): amianthus 137
ERDFORSCHER (*m*): geologist 138
ERDFORSCHUNG (*f*): geology 139
ERDGANG (*m*): vein; tunnel; gallery; working; passage; adit 140
ERDGAS (*n*): natural gas (the main constituent is methane; see **METHAN**) 141
ERDGELB (*n*): yellow ochre (clay plus Fe_2O_3) (see also **OCKER**) 142
ERDGERUCH (*m*): earthy smell; musty smell 143
ERDGESCHMACK (*m*): earthy taste 144
ERDGESCHOSS (*n*): ground floor 145
ERDGLAS (*n*): selenite (transparent gypsum); $CaSO_4.2H_2O$; Sg. 2.3; (see also **SELENIT**) 146
ERDGRÜN (*n*): mineral (sap) green; green verditer (green basic copper carbonate); $Cu_2(OH)_2CO_3$; Sg. 3.7-4.0 147
ERDGÜRTEL (*m*): zone 148
ERDHAKEN (*m*): parapet (of culvert) 149
ERDHARZ (*n*): asphalt; fossil resin; bitumen; (see **MONTANWACHS** and **BRAUNKOHLEN-BITUMEN**) 150
ERDHARZ, GELBES- (*n*): amber 151
ERDHARZIG: asphaltic; bitumenous 152
ERDHARZ (*n*), **SCHWARZES-**: asphalt, bitumen, mineral pitch, petroleum asphalt 153
ERDHAUE (*f*): pickaxe 154
ERDIG: earthy; terrestrial 155
ERDINNERE (*n*): interior of the earth 156
ERDKABEL (*n*): underground cable; subterranean cable 157
ERDKALK (*m*): limestone 158
ERDKOBALT (*m*): asbolite (see **KOBALTMANGAN-ERZ**) 159
ERDKOBALT, GRÜNER- (*m*): annabergite; $Ni_3As_2O_8.8H_2O$; Sg. 3.0 160
ERDKOBALT, ROTER- (*m*): earthy erythrite (see **ERYTHRIN**) 161
ERDKOBALT, ROTER STRAHLIGER- (*m*): cobalt bloom; erythrite; $CO_3As_2O_8.8H_2O$; Sg. 2.95 162
ERDKOHLE (*f*): earthy coal; lignitic earth; brown coal; peat 163
ERDKÖRPER (*m*): terrestrial body 164
ERDKREIS (*m*): sphere; globe; orb; ball (of the earth) 165
ERDKREISLINIE (*f*): horizon 166
ERDKRUSTE (*f*): earth's crust 167
ERDKUGEL (*f*): terrestrial globe 168
ERDKUNDE (*f*): geography; geology; geognosy; knowledge of the earth 169
ERDLEHRE (*f*): geology 170
ERDLEITUNG (*f*): underground (line) main or pipe-line, earth (circuit, conduit) wire 171
ERDLINIE (*f*): ground (floor) line 172
ERDMAGNETISCH: earth magnetic 173
ERDMAGNETISMUS (*m*): earth (terrestrial) magnetism 174
ERDMANDEL (*f*): chufa (*Cyperus esculentus*); ground nut (also see **ERDNUSS**) 175
ERDMANDELÖL (*n*): chufa oil (from *Cyperus esculentus*) 176
ERDMEHL (*n*): diatomaceous earth; kieselguhr; siliceous earth; infusorial earth 177
ERDMESSER (*m*): geometer, geometrician 178
ERDMESSKUNST (*f*): geometry 179
ERDMESSUNG (*f*): geodesy 180
ERDMETALL (*n*): earth metal 181
ERDMETALL, ALKALISCHES- (*n*): alkaline earth metal 182
ERDMOOS (*n*): club-moss, lycopodium, vegetable sulphur 183
ERDMOOS (*n*), **PURGIERENDES-**: Iceland moss, Cetraria 184
ERDNUSS (*f*): earth (pig) nut; ground (pea) nut (*Arachis hypogæa*) 185
ERDNUSSÖL (*n*): ground nut oil; earth (pea) nut oil; Sg. 0.916-0.922 (*Oleum arachidis* or *Oleum arachis*) 186
ERDOBERFLÄCHE (*f*): earth's (surface) area 187
ERDÖL (*n*): petroleum; mineral (rock; crude) oil 188
ERDÖLERZEUGNIS (*f*): petroleum (rock oil) product 189
ERDPECH (*n*): mineral pitch; asphalt (*Asphaltum*); bitumen 190
ERDPECHHALTIG: asphaltic; bitumenous 191
ERDPOL (*m*): earth pole 192
ERDPOTENTIAL (*n*): earth potential 193
ERDPROBE (*f*): soil (test) sample 194
ERDRAUCH (*m*): fumitory (*Fumaria officinalis*) 195
ERDREICH (*n*): ground; earth; land; soil; empire; kingdom of the earth 196
ERDRINDE (*f*): earth's crust 197
ERDRÖHRE (*f*): drainage pipe 198
ERDROSSELN: to throttle; strangle 199

ERDRÜCKEN : to stifle ; repress ; smother ; oppress ; (*m*) : ridge ; range of hills ; high land 200
ERDSACK (*m*) : sandbag 201
ERDSAFT (*m*) : (see **ERDÖL**) 202
ERDSALZ (*n*) : rock-salt ; saltpetre ; saline efflorescence on soil 203
ERDSCHATZ (*m*) : mineral wealth 204
ERDSCHICHT (*f*) : layer of earth ; stratum ; subsoil 205
ERDSCHIERLING (*m*) : (see **SCHIERLING**) 206
ERDSCHLACKE (*f*) : earthy slag 207
ERDSCHLUSS (*m*) : earth leakage (Electric), earth connection (Electric) 208
ERDSCHOLLE (*f*) : clod of earth 209
ERDSCHWAMM (*m*) : mushroom ; fungus ; toadstool 210
ERDSCHWEFEL (*m*) : lycopodium ; club moss ; vegetable sulphur (spores of *Lycopodium clavatum*) 211
ERDSCHWEFELSAMEN (*m*) : (see **ERDSCHWEFEL**) 212
ERDSEIFE (*f*) : a kind of clay 213
ERDSTEIN (*m*) : eagle stone ; ætites 214
ERDSTRICH (*m*) : zone 215
ERDSTROM (*m*) : earth current 216
ERDSTROMKREIS (*m*) : earth circuit 217
ERDSTUFE (*f*) : terrace 218
ERDSTURZ (*m*) : landslip ; landslide 219
ERDTALG (*m*) : mineral tallow ; ozocerite ; hatchetin ; hatchettite ; Sg. 0.85-0.95 ; Mp. 55-110°C. 220
ERDTALK (*m*) : earthy talc 221
ERDTEER (*m*) : mineral tar ; maltha ; pissasphalt ; bitumen 222
ERDTEIL (*m*) : continent ; part of the world 223
ERDUNG (*f*) : earthing (Elect.) 224
ERDWACHS (*n*) : soft paraffin ; mineral wax ; ozocerite ; Sg. 0.85-0.95 ; Mp. 55-110°C. (see **OZOKERIT** and **PARAFFIN**) 225
ERDWÄRME (*f*) : heat of the earth 226
EREIGNEN : to happen ; occur ; chance 227
EREIGNIS (*n*) : event ; occurrence ; happening 228
EREILEN : to overtake ; hurry (hasten) up to 229
EREMAKAUSIE (*f*) : eremacausis (slow combustion) 230
ERFAHREN : to learn ; discover ; hear ; endure ; undergo ; gather ; experience ; understand ; be expert ; be versed ; be experienced 231
ERFAHRUNG (*f*) : experience ; knowledge ; practice ; skill 232
ERFAHRUNGSBEWEIS (*m*) : proof by experience, practical proof 233
ERFAHRUNGSGEMÄSS : empirical ; in accordance with experience ; usual 234
ERFAHRUNGSGEMÄSS, WEIL- : because it has been [proved that] shown by experience that 235
ERFAHRUNGSKREIS (*m*) : sphere of knowledge, range of experience 236
ERFAHRUNGSLOS : inexperienced, non-practical 237
ERFAHRUNGSMÄSSIG : (see **ERFAHRUNGSGEMÄSS**) 238
ERFAHRUNGSREICH : experienced, skilled, versed, practical 239
ERFAHRUNGSSATZ (*m*) : empirical formula 240
ERFASSEN : to seize ; grasp ; comprehend 241
ERFINDEN : to invent ; discover ; devise ; find out ; design ; fabricate 242
ERFINDER (*m*) : inventor ; designer 243
ERFINDERISCH : ingenious ; inventive 244
ERFINDUNG (*f*) : invention ; discovery ; device ; ingenuity 245
ERFINDUNGSGABE (*f*) : (see **ERFINDUNGSKRAFT**) 246
ERFINDUNGSKRAFT (*f*) : inventive faculty 247
ERFINDUNGSREICH : ingenious, inventive 248
ERFOLG (*m*) : result ; success ; consequence ; issue 249
ERFOLGEN : to follow ; ensue ; result ; take place ; occur ; be effected ; succeed ; happen 250
ERFOLGLOS : unsuccessful 251
ERFOLGREICH : successful ; having (with) excellent results 252
ERFORDERN : to require ; demand ; afford ; necessitate ; want 253
ERFORDERNIS (*n*) : requisite 254
ERFORSCHEN : to investigate ; explore ; discover ; fathom ; enquire into 255
ERFORSCHER (*m*) : investigator, discoverer 256
ERFRIEREN : to freeze 257
ERFRISCHEN : to freshen ; refresh ; refrigerate ; cool 258
ERFRISCHENDES MITTEL (*n*) : refrigerant ; cooling drink 259
ERFRISCHUNG (*f*) : refreshment 260
ERFÜLLEN : to fulfil ; fill-(up) ; impregnate ; satisfy (Maths.) ; accomplish ; realize 261
ERG (*n*) : erg (C.G.S. unit of work and energy) 262
ERGÄNZEN : to supply ; complete ; make up ; restore ; perfect ; replenish ; renew ; supplement 263
ERGÄNZEND : complementary ; supplementary 264
ERGÄNZUNG (*f*) : completion ; restoration ; supplement ; complement ; supply ; replenishment 265
ERGÄNZUNGSBAND (*m*) : supplement ; supplementary volume 266
ERGÄNZUNGSFARBE (*f*) : complementary colour 267
ERGÄNZUNGSSTÜCK (*n*) : make-up piece 268
ERGÄNZUNGSWINKEL (*m*) : supplementary (complementary) angle 269
ERGEBEN : to yield ; show ; appear ; result ; submit ; devote oneself ; give ; amount to ; resign ; be devoted ; be obedient ; be resigned ; surrender ; be addicted ; (be)-humble 270
ERGEBEN, SICH OHNE WEITERES- AUS : to be obvious from 271
ERGEBNIS (*n*) : result ; consequence ; product ; yield ; sequel 272
ERGIEBIG : productive ; rich ; plentiful ; fertile 273
ERGIESSEN : to pour forth ; discharge ; empty 274
ERGIESSUNG (*f*) : efflux, discharge, effusion 275
ERGLÜHEN : to glow ; burn out (of boiler tubes) 276
ERGMETER (*n*) : ergmeter 277
ERGOTIN (*n*) : Ergotin ; ergot extract 278
ERGOTININ (*n*) : ergotinine ; $C_{35}H_{40}N_4O_6$; Mp. 205°C. 279
ERGOTISMUS (*m*) : ergotism (*Morbus cerealis*) (Med.) 280
ERGREIFEN : to grasp ; seize ; catch ; take (hold) up ; resort to ; affect ; strike ; assume ; apprehend 281
ERGRÜBELN : to think (worry) out, search out, meditate upon 282
ERGRÜNDEN : to investigate ; sound , fathom ; penetrate 283

ERGUSS (*m*) : discharge ; effusion ; overflow 284
ERGUSSGESTEIN (*n*) : effusive rock ; volcanic rock 285
ERHABEN : high ; raised ; relief ; convex ; prominent ; sublime ; exalted ; lofty ; illustrious ; grand 286
ERHABENE ARBEIT (*f*) : relief ; relievo ; bas relief 287
ERHABENHEIT (*f*) : elevation ; nobility ; grandeur ; loftiness ; eminence ; sublimity ; convexity ; prominence ; protuberance 288
ERHALTEN : to keep ; preserve ; save ; maintain ; receive ; support ; obtain ; gain ; acquire ; be accorded ; conserve 289
ERHALTEN BLEIBEN : to remain constant 290
ERHALTUNGSMITTEL (*n*) : preservative ; antiseptic ; subsistence 291
ERHÄRTEN : to harden, set, confirm, declare 292
ERHÄRTUNG (*f*) : hardening ; decoration ; confirmation 293
ERHEBEN : to raise ; rise ; elevate ; lift ; levy ; obtain ; collect 294
ERHEBLICH : considerable ; important ; decided ; weighty ; cogent 295
ERHEBUNG (*f*) : collection ; elevation ; raising ; promotion ; levy 296
ERHEBUNGSWINKEL (*m*) : angle of elevation 297
ERHELLEN : to light ; illuminate ; brighten ; lighten ; expose (Photography) ; become clear ; appear ; clarify ; enlighten 298
ERHELLUNGSKESSEL (*m*) : clarifying pan 299
ERHITZEN : to heat ; become hot ; get angry ; stimulate ; inflame 300
ERHITZUNG (*f*) : heating 301
ERHÖHBAR : capable of being raised, heightened, increased, susceptible of increase, increasable 302
ERHÖHEN : to raise ; elevate ; erect ; heighten ; advance ; increase ; rise 303
ERHÖHUNG (*f*) : rise, elevation, advance, increase 304
ERHOLEN : to recover ; take breath ; amuse oneself ; come to ; recreate ; recuperate ; get better ; become convalescent 305
ERHOLUNG (*f*) : recovery ; recreation ; relaxation ; recuperation ; convalescence 306
ERHOLUNGSSTUNDE (*f*) : leisure hour 307
ERHOLUNGSTEMPERATUR (*f*) : temperature of recuperation (Metal) 308
ERINNERN : to remind ; state ; mention ; remember ; admonish ; recollect 309
ERINNERUNG (*f*) : memory ; recollection ; remembrance ; reminder ; admonition 310
ERINNERUNGSKRAFT (*f*) : memory 311
ERIODICTYONSÄURE (*f*) : eriodictyonic acid (from *Eriodictyon glutinosum*) 312
ERKALTEN : to cool ; grow cold ; chill (Metal) 313
ERKÄLTEN : to cool ; catch cold ; grow (become) cold ; chill 314
ERKALTUNG (*f*) : cooling ; chilling 315
ERKÄLTUNG (*f*) : cold ; catarrh (Med.) ; chill 316
ERKAUFEN : to obtain ; get ; buy ; purchase ; bribe 317
ERKENNBAR : evident, obvious, recognizable, realizable 318
ERKENNEN : to detect ; realize ; recognize ; know ; distinguish ; acknowledge ; discern ; perceive ; understand ; credit (Com.) ; diagnose (Med.) ; decide ; pass sentence 319
ERKENNEN LASSEN : to show ; illustrate ; (not so obvious as **ZEIGEN**) 320
ERKENNTLICH : grateful ; thankful 321
ERKENNTNIS (*f* and *n*) : perception ; realization ; knowledge ; cognisance 322
ERKENNUNG (*f*) : detection ; recognition ; acknowledgment ; diagnosis (see also **ERKENNEN**) 323
ERKENNUNGSZEICHEN (*n*) : distinctive mark ; characteristic ; countersign 324
ERKER (*m*) : projection ; balcony 325
ERKIESEN : to choose, elect, select 326
ERKLÄREN : to explain ; illustrate ; declare ; announce ; make clear ; define ; expound 327
ERKLÄRLICH : explicable ; obvious ; evident 328
ERKUNDEN : to explore 329
ERKUNDIGEN : to enquire ; find out 330
ERKUNDIGUNG (*f*) : enquiry ; information ; research 331
ERLANGBAR : attainable 322
ERLANGEN : to attain ; reach ; obtain ; acquire 333
ERLANGERBLAU (*n*) : (see **BERLINERBLAU**) 334
ERLANGER LEDER (*n*) : glove kid 335
ERLASS (*m*) : order ; decree ; edict ; reduction ; pardon ; allowance ; remission 336
ERLASSEN : to issue ; proclaim ; remit ; exempt ; let off ; dispense with 337
ERLAUBEN : to allow ; permit ; grant ; license 338
ERLAUBNIS (*f*) : permit ; permission ; leave ; licence ; license 339
ERLAUBNISBRIEF (*m*) : letters patent, license 340
ERLAUBNISSCHEIN (*m*) : license, permit 341
ERLÄUTERN : to explain ; illustrate ; elucidate 342
ERLE (*f*) : alder (*Alnus glutinosa*) ; Sg. 0.42-0.64 343
ERLEDIGEN : to dispose of ; settle ; arrange ; vacate ; discharge ; empty 344
ERLEDIGUNG (*f*) : reply, response, settlement, disposal, arrangement, vacation, vacancy 345
ERLEICHTERN : to facilitate ; make lighter ; make easier ; relieve ; alleviate ; simplify ; mitigate 346
ERLE (*f*), **KLEBRIGE-** : alder (see **ERLE**) 347
ERLEN : of alder wood 348
ERLENBAUM (*m*) : alder tree (see **ERLE**) 349
ERLENHOLZ (*n*) : alder wood 350
ERLENMEYER-KÖLBCHEN (*n*) : Erlenmeyer flask 351
ERLESEN : to choose ; select ; pick out 352
ERLEUCHTEN : to illuminate ; light ; enlighten 353
ERLISTEN : to obtain by fraud 354
ERLÖS (*m*) : proceeds 355
ERLÖSCHEN : to put out ; go out ; extinguish ; expire ; efface 356
ERLÖSEN : to redeem ; save ; realize (Money) ; rescue ; release 357
ERLÜGEN : to fabricate, lie, invent, falsify 358
ERMÄCHTIGEN : to empower ; authorize 359
ERMÄCHTIGUNG (*f*) : warrant, authorization 360
ERMAHNEN : to exhort ; remind ; warn ; admonish 361
ERMANGELUNG (*f*) : default ; want ; lack ; failing ; deficiency 362
ERMÄSSIGEN : to moderate ; reduce ; decrease, limit ; abate 363
ERMÄSSIGUNG (*f*) : reduction, fall, drop, decline, decrease, limitation, moderation 364
ERMESSEN : to judge ; conceive ; measure ; consider ; weigh 365
ERMITTELN : to ascertain ; arrive at ; determine ; find out 366

ERMÖGLICHEN: to make (possible) feasible, to render practicable, to enable to be carried out 367
ERMÜDBARKEIT (*f*): weakening; fatiguing; (Metal) 368
ERMÜDEN: to tire; weaken; fatigue; exhaust; weary 369
ERMÜDUNGSBRUCH (*m*): fatiguing (weakening) rupture (due to repeated local stress overstepping elastic limit of metal) 370
ERNÄHREN: to nourish; feed; support; maintain 371
ERNÄHREND: nutritive; nourishing 372
ERNÄHRUNG (*f*): nutrition; alimentation; support; subsistence; maintenance 373
ERNÄHRUNGSKANAL (*m*): alimentary canal 374
ERNÄHRUNGSKUNDE (*f*): diatetics 375
ERNÄHRUNGSSTOFF (*m*): nutritive (alimentary; food) substance 376
ERNÄHRUNGSSUBSTANZ (*f*): (see **ERNÄHRUNGS-STOFF**) 377
ERNÄHRUNGSWERT (*m*): nutritive value 378
ERNANNTE (*m*): nominee 379
ERNENNEN: to nominate; appoint; name; designate 380
ERNENNUNG (*f*): nomination, appointment 381
ERNEUEN: to renew; renovate; revive; refresh (Colours); regenerate (Steam); repair; recommence 382
ERNEUERN: (see **ERNEUEN**) 383
ERNIEDRIGEN: to debase; lower; humble; decrease; depress; degrade 384
ERNTE (*f*): harvest; crop; gathering; campaign (Sugar) 385
ERNTEMASCHINE (*f*): harvesting machine; harvester; reaper 386
ERODIEREN: to erode 387
ERÖFFNEN: to open; discover; inaugurate; disclose; start; publish 388
ERÖFFNEND: aperient (Med.) 389
ERPICHT: bent upon, intent upon 390
ERPRESSEN: to press out of, exact, extort 391
ERPROBEN: to test; try; prove 392
ERQUICKEN: to quicken; refresh; revive; resuscitate 393
ERRECHNEN: to work out; calculate; compute 394
ERREGEN: to excite; agitate; stimulate; stir up; irritate; create; provoke; incite 395
ERREGEND: exciting 396
ERREGER (*m*): exciter; exciting cause; agitator; irritant; stimulant 397
ERREGERDYNAMO (*f*): exciting dynamo 398
ERREGER LAGER (*n*): exciter bearing 399
ERREGERSALZ (*n*): exciting salt 400
ERREGERSTROM (*m*): exciting current 401
ERREGERTURBINE (*f*): exciter turbine 402
ERREGERVERLUST (*m*): exciter loss (Elect.) 403
ERREGUNG (*f*): excitation, agitation, excitement 404
ERREGUNGSFLÜSSIGKEIT (*f*): exciting fluid (in a voltaic cell) 405
ERREGUNGSMITTEL (*n*): stimulant 406
ERREICHEN: to reach; attain; achieve; get; obtain 407
ERRICHTEN: to erect; set up; establish; draw (upright lines on a graph) 408
ERRINGEN: to wrest from 409
ERRUNGENSCHAFT (*f*): acquisition; conquest; achievement; improvement 410
ERSÄTTIGEN: to fill, satiate, satisfy 411
ERSATZ (*m*): replacement; substitution; equivalent; reparation; compensation; spare; substitute; duplicate; extra 412
ERSATZBRENNE (*f*): substitute pickle or dip (Metal) 413
ERSATZDECKEL (*m*): spare lid; replace (spare; extra) cover 414
ERSATZFASERSTOFF (*m*): fibre substitute 415
ERSATZGEWICHT (*n*): equivalent weight 416
ERSATZKLEBSTOFF (*m*): (glue, gum) adhesive substitute 417
ERSATZLEBENSMITTEL (*n* and *pl*): substitute foodstuff-(s) 418
ERSATZLIEFERUNG (*f*): substitution, duplication 419
ERSATZMITTEL (*n*): substitute; surrogate 420
ERSATZPFLICHT (*f*): compensation; reparation 421
ERSATZSTÜCK (*n*): spare part; replace part 422
ERSATZTEIL (*m*): spare (replace) part 423
ERSAUFEN: to be drowned 424
ERSCHAFFEN: to create; produce 425
ERSCHALLEN: to sound; resound; ring; echo 426
ERSCHEINEN: to appear; be clear; be published 427
ERSCHEINUNG (*f*): phenomenon; appearance; apparition 428
ERSCHEINUNGSFORM (*f*): phase; manifestation 429
ERSCHLEICHEN: to get (obtain) by (fraud) fraudulent means 430
ERSCHLIESSEN: to open-(up); disclose; infer; tap 431
ERSCHÖPFEN: to exhaust; drain 432
ERSCHÖPFEND: exhaustive; thorough; comprehensive 433
ERSCHÜTTERN: to shake; thrill; tremble 434
ERSCHÜTTERND: concussive; thrilling 435
ERSCHÜTTERUNG (*f*): concussion; percussion; shock; shaking; commotion; vibration 436
ERSCHWEREN: to weight; reduce (Dyes); weighten; make difficult; aggravate 437
ERSETZEN: to replace; substitute; displace; compensate; recover; supersede; indemnify 438
ERSICHTLICH: visible; evident; obvious 439
ERSPARNIS (*f*): saving-(s); economy 440
ERSTARREN: to solidify; congeal; freeze; harden; set; coagulate; stiffen 441
ERSTARRUNGSDIAGRAMM (*n*): solidification diagram (of metal smelting) 442
ERSTARRUNGSFORM (*f*): system of solidification, form of solidification (Metal) 443
ERSTARRUNGSGESTEIN (*n*): (see **ERUPTIVGESTEIN**) 444
ERSTARRUNGSKURVE (*f*): solidification curve (Metal) 445
ERSTARRUNGSPUNKT (*m*): freezing point; solidification (coagulation) point 446
ERSTARRUNGSSCHAUBILD (*n*): solidification diagram (of metal smelting) 447
ERSTARRUNGSSYSTEM (*n*): system of solidification, form of solidification (Metal) 448
ERSTARRUNGSVORGANG (*m*): solidification process (Metal smelting) 449
ERSTARRUNGSWÄRME (*f*): heat of fusion; heat evolved on solidification 450
ERSTATTEN: to return; render; repay; restore; reimburse; compensate 451
ERSTATTUNG (*f*): restitution, compensation 452

ERSTEHEN: to buy (at an auction); arise; originate 453
ERSTELLEN: to put up; erect; instal; fit 454
ERSTICKEN: to suffocate; choke; suppress; smother 455
ERSTICKUNG (*f*): asphyxia (Med.), suffocation 456
ERSTMILCH (*f*): colostrum 457
ERSTREBEN: to strive, tend, have a tendency, endeavour 458
ERSTRECKEN: to stretch; extend; amount; reach 459
ERSUCHEN: to request; solicit; beseech; entreat; petition; implore; (*n*): supplication; request 460
ERTAPPEN: to catch; seize; detect 461
ERTEILEN: to give; bestow; grant (Patents); make over; place (Orders) 462
ERTÖNEN: to (re)-sound 463
ERTRAG (*m*): revenue; amount; sum; yield; produce; proceeds; profit 464
ERTRAGEN: to bear; endure; tolerate; suffer; support 465
ERTRÄGLICH: bearable, endurable; profitable; suitable; supportable 466
ERTRÄGLICHKEIT (*f*): mediocrity, tolerableness 467
ERUCASÄURE (*f*): erucic acid (from *Sinapis alba*); $CH_3(CH_2)_7CH:CH(CH_2)_{11}CO_2H$; Sg. 0.8602; Mp. 34°C.; Bp. 254.5°C. 468
ERUPTIVGESTEIN (*n*): eruptive rock, igneous rock, unstratified rock 469
ERVASIN (*n*): (see **ACETYL-P-KRESOTINSÄURE**) 470
ERWACHSEN: grown; adult; to grow; to accrue; spring up; arise 471
ERWÄGEN: to consider; weigh; ponder; discuss 472
ERWÄHLEN: to choose, elect 473
ERWÄRMEN: to heat; warm 474
ERWÄRMUNG (*f*): heating, warming 475
ERWÄRMUNGSKRAFT (*f*): calorific (heating) power 476
ERWECKEN: to waken; rouse; animate; resuscitate; incite 477
ERWEICHEN: to soften; soak; mellow; move; mollify 478
ERWEICHEND: softening; emollient; demulcent 479
ERWEICHENDES MITTEL (*n*): emollient (used for softening and relaxing the place of application) (Med.) 480
ERWEICHUNGSMITTEL (*n*): emollient; demulcent 481
ERWEISEN: to prove; demonstrate; show; be found; evince 482
ERWEITERN: to expand; distend; widen; extend; enlarge; dilate; stretch; amplify 483
ERWERBSAM: industrious 484
ERWERBSFLEISS (*m*): industry 485
ERWIRKEN: to effect, bring about, cause 486
ERWÜRGUNG (*f*): strangulation (Med.) 487
ERYSIPEL (*f*): erysipelas (a skin disease) (Med.) 488
ERYSIPELAS (*m*): erysipelas (a skin disease) (Med.) 489
ERYTHREN (*n*): erythren (synthetic caoutchouc) 490
ERYTHRIN (*m*): erythrite, cobalt bloom, $Co_3As_2O_8.8H_2O$; Sg. 3.0 (see **ERDKOBALT, ROTER STRAHLIGER-** and **KOBALTBLÜTE**); (*n*): erythrin (an alcohol from mosses and seaweed; see **ERYTHRIT**); (*n*): erythrin (a dye) 491
ERYTHRIT (*n*): erythrit (saponification product from erythrin (*n*); $C_4H_{10}O_4$; (see **SALPETERSÄUREESTER**) 492
ERYTHRIT (*n*), **PENTA-**: pentaerythrit; $C_5H_{12}O_4$ 493
ERYTHROL-(TETRA)-NITRAT (*n*): erythrol tetranitrate (see **SALPETERSÄUREESTER**) 494
ERYTHROSIN (*n*): erythrosine (a dyestuff from iodization of fluorescein) 495
ERZ (*n*): ore; metal; bronze; brass 496
ERZABFÄLLE (*pl*): tailings 497
ERZADER (*f*): ore vein 498
ERZÄHNLICH: metallic 499
ERZANBRUCH (*m*): ore (in its native state) 500
ERZARBEITER (*m*): metal (bronze) worker 501
ERZART (*f*): type of ore 502
ERZARTIG: metallic 503
ERZAUFBEREITUNG (*f*): ore (cleansing) dressing 504
ERZAUFBEREITUNGSANLAGE (*f*): ore dressing plant 505
ERZBRECHMASCHINE (*f*): ore crusher 506
ERZBRUCH (*m*): mine. 507
ERZDRUSE (*f*): hollow, crystalline (crystallized) ore 508
ERZEIGEN: to render; display; confer; manifest; show; prove 509
ERZEISEN (*n*): mine iron (obtained from ore without fluxes) 510
ERZEN: brazen 511
ERZEUGEN: to produce; generate; beget; engender; breed 512
ERZEUGER (*m*): producer; generator; grower 513
ERZEUGNIS (*n*): product; production; produce 514
ERZEUGUNGSKOSTEN (*f.pl*): costs of production 515
ERZEUGUNGSMENGE (*f*): amount (quantity) produced 516
ERZEUGUNGSWÄRME (*f*), **GLEICHE-**: constant total heat (of steam) 517
ERZFARBE (*f*): bronze (brass) colour 518
ERZFÖRDERUNG (*f*): mining, ore raising 519
ERZFRISCHEN (*n*): refining 520
ERZFÜHREND: ore bearing 521
ERZGEBIRGE (*n*): ore bearing mountain 522
ERZGICHT (*f*): ore and limestone charge (to blast furnaces) 523
ERZGIESSER (*m*): brass (bronze) founder 524
ERZGIESSEREI (*f*): bronze (brass) foundry 525
ERZGRÄBER (*m*): miner 526
ERZGRUBE (*f*): mine; pit 527
ERZHALTIG: ore bearing; containing ore 528
ERZHÜTTE (*f*): smelting (house) works 529
ERZIEHERISCH: educational-(ly), leading 530
ERZIELEN: to obtain; attain; strive after; produce; raise; realize; aim at 531
ERZKIES (*m*): metalliferous pyrites 532
ERZ, KLAUB- (*n*): picked ore 533
ERZKUNDE (*f*): metallurgy 534
ERZLAGER (*n*): ore bed; deposit of (metal) ore; seam; lode; vein 535
ERZ, LINSEN- (*n*): oolitic ore 536
ERZMESSER (*m*): mine surveyor 537
ERZMETALL (*n*): heavy (ore) metal 538
ERZMUTTER (*f*): matrix 539
ERZNIERE (*f*): kidney-(shaped) ore 540
ERZPOST (*f*): charge of ore and limestone 541
ERZPROBE (*f*): ore (test, assay) sample 542

ERZ (*n*), **PURPUR-** : (see **KIESABBRAND**) 543

ERZQUETSCHE (*f*) : crusher 544

ERZREICH : rich in ore ; having a high ore content 545

ERZSATZ (*m*) : charge of ore and limestone 546

ERZSCHAUM (*m*) : scoria, cinder, slag 547

ERZ (*n*), **SCHEID-** : picked (screened) ore 548

ERZSCHEIDER (*m*) : ore screener or separator 549

ERZSCHLICH (*m*) : dressed ore ; concentrate (60% tin) (obtained by reducing ore in shaft or reverberatory furnace) 550

ERZSTAHL (*m*) : ore (direct mine) steel (obtained from ore without using fluxes) 551

ERZSTAUB (*m*) : ore dust 552

ERZSTOCK (*m*) : solid (massive) deposit of ore 553

ERZWALZWERK (*n*) : ore crushing mill 554

ERZWÄSCHE (*f*) : ore (washing) washery 555

ERZZERREIBER (*m*) : ore grinder 556

ERZZIEGEL (*m*) : ore briquette 557

ESCHE (*f*) : ash-(tree) (*Fraxinus excelsior*) 558

ESCHE (*f*), **CHINESISCHE-** : Chinese ash (*Fraxinus chinensis*) 559

ESCHEL (*m*) : zaffer ; zaffre ; cobalt blue (finest grade of smalt) ; black spot (Founding) 560

ESCHEN : of ash ; ash 561

ESCHENBAUM (*m*) : ash tree (*Fraxinus excelsior*) 562

ESCHENBLÄTTER (*n.pl*) : ash leaves (*Herba fraxini*) 563

ESCHENHOLZ (*n*) : ash wood (*Fraxinus excelsior*) ; Sg. 0.52-0.95 564

ESCHLAUCH (*m*) : shalot (see **SCHALOTTE**) 565

ESCOBEIZE (*f*) : esco mordant 566

ESDRAGON (*m*) : tarragon (see **ESTRAGON**) 567

ESEL (*m*) : donkey ; (wooden)-horse ; scoop, (Brewing) ; stand ; frame 568

ESELSBRÜCKE (*f*) : Pons asinorum 569

ESELSHAUPT (*n*) : cap 570

ESELSHAUPT (*n*), **MARSSTENGE-** : topmast cap 571

ESELSHAUPTSTÜTZE (*f*) : cap stanchion 572

ESELSHUF (*m*) : coltsfoot 573

ESERIN (*n*) : eserine ; physostigmine (see **PHYSOSTIGMIN**) 574

ESERINÖL (*n*) : eserine oil (solution of 0.2g. physostigmine salicylate, dried at 100°C., in 40 g. olive oil) 575

ESKOBEIZE (*f*) : esco mordant 576

ESPARSETTE (*f*) : sainfoin (*Onobrychis sativa*) 577

ESPARTOGRAS (*n*) : esparto grass (*Stipa tenacissima*) 578

ESPE (*f*) : aspen (*Populus tremula*) 579

ESPENSTOFF (*m*) : aspen pulp (Paper) 580

ESSE (*f*) : forge ; stack ; chimney ; hearth ; flue ; funnel ; uptake 581

ESSEISEN (*n*) : tuyere (Metal) ; pipe 582

ESSEN : to eat, feed, dine 583

ESSENFEGER (*m*) : chimney-sweep 584

ESSENKEHRER (*m*) : chimney-sweep 585

ESSENKLAPPE (*f*) : damper 586

ESSENVERLUST (*m*) : chimney loss 587

ESSENZ (*f*) : essence 588

ESSENZUG (*m*) : chimney draught 589

ESSE (*f*), **SCHMIEDE-** : smith's forge 590

ESSIG (*m*) : vinegar (*Acetum*) 591

ESSIGAAL (*m*) : vinegar eel (a vinegar nematode) (*Anguillula aceti*) 592

ESSIGÄLCHEN (*n*) : (see **ESSIGAAL**) 593

ESSIG (*m*), **AROMATISCHER-** : aromatic vinegar (solution of a series of etheral oils in alcohol, water and diluted acetic acid), (Pharm.) 594

ESSIGARTIG : acetic ; acetous ; vinegar-like 595

ESSIGÄTHER (*m*) : (*Aether aceticus*) : acetic ether, ethyl acetate, vinegar naphtha ; ($CH_3CO_2C_2H_5$) ; Sg. 0.9 ; Mp. – 82.4°C. ; Bp. 77.2°C. 596

ESSIGBILDUNG (*f*) : acetification 597

ESSIGESSENZ (*f*) : vinegar essence (see **ESSIGSÄURE**) 598

ESSIGESTER (*m*) : acetic ether, ethyl acetate (see **ESSIGÄTHER**) 599

ESSIGFABRIK (*f*) : vinegar factory 600

ESSIGGÄRUNG (*f*) : acetic fermentation 601

ESSIGGEIST (*m*) : acetone ; CH_3COCH_3 ; Sg. 0.79 ; Mp. – 94.3°C. ; Bp. 56.48°C. 602

ESSIGGURKE (*f*) : pickled (cucumber) gherkin 603

ESSIGGUT (*n*) : alcoholic liquid for quick vinegar process 604

ESSIGHEFEN (*f.pl*) : vinegar dregs or lees 605

ESSIGHONIG (*m*) : oxymel (mixture of honey and vinegar) 606

ESSIGMESSER (*m*) : acetometer ; oxymeter 607

ESSIGMUTTER (*f*) : mother of vinegar 608

ESSIGNAPTHA (*f*) : ethyl acetate ; $CH_3CO_2C_2H_5$ 609

ESSIGPILZ (*m*) : vinegar plant ; mother of vinegar 610

ESSIGPROBER (*m*) : (see **ESSIGPRÜFER**) 611

ESSIGPRÜFER (*m*) : vinegar tester ; acetimeter 612

ESSIGROSE (*f*) : damask rose ; French (red) rose 613

ESSIGSALZ (*n*) : acetate 614

ESSIGSAUER : acetate of ; combined with acetic acid ; acetic ; acetous 615

ESSIGSÄURE (*f*) : (*Acidum aceticum*) : acetic acid, vinegar acid, methanecarboxylic acid ; ($HC_2H_3O_2$) ; Sg. 1.05 ; Mp. 16.7°C. ; Bp. 118°C. 616

ESSIGSÄUREAMYLESTER (*m*) : amyl acetate (see **AMYLACETAT**) 617

ESSIGSÄUREANHYDRID (*n*) : acetic-(acid) anhydride ; acetyl (acetic) oxide ; $C_4H_6O_3$; Sg. 1.082 ; Bp. 139.5 °C. 618

ESSIGSÄUREANLAGE (*f*) : acetic acid plant 619

ESSIGSÄUREÄTHER (*m*) : acetic ether ; ethyl acetate ; $CH_3CO_2C_2H_5$ (see **ESSIGÄTHER**) 620

ESSIGSÄURE-ÄTHYLÄTHER (*m*) : acetic ether, ethyl acetate ; $C_4H_8O_2$; Sg. 0.9048 ; Mp. -82.4°C ; Bp. 77.2°C. (see also **ESSIGÄTHER**) 621

ESSIGSÄUREÄTHYLESTER (*m*) : (see **ESSIGÄTHER**) 622

ESSIGSÄURECHLORID (*n*) : (see **ACETYLCHLORID**) 623

ESSIGSÄUREGÄRUNG (*f*) : acetic fermentation 624

ESSIGSÄUREISOAMYLESTER (*m*) : (see **AMYLACETAT**) 625

ESSIGSÄURE (*f*)-**KRYSTALLE** (*n.pl.*) : acetic acid crystals 626

ESSIGSÄUREMESSER (*m*) : acetometer 627

ESSIGSAURES SALZ (*n*) : acetate 628

ESSIGSCHAUM (*m*) : flower of vinegar 629

ESSIGSIRUP (*m*) : oxymel (mixture of vinegar and honey) 630

ESSIGSPRIT (*m*) : vinegar essence 631

ESSIGSTÄNDER (*m*) : vinegar tun ; graduator (Quick vinegar process) 632

ESSIGSTUBE (*f*) : vinegar room ; warm room (Quick vinegar process) 633

ESSIGWEINSTEINSAUER : acetotartrate 634

ESSIGWOLFRAMSAUER : acetotungstate 635

ESSIGZUCKER (*m*) : oxysaccharum 636
ESSLÖFFEL (*m*) : table-spoon 637
ESSWAREN (*f.pl*) : eatables, food, victuals 638
ESSZEIT (*f*) : meal time 639
ESTER (*m*) : ester (such as ethyl-ester ; see ÄTHYLESTER) 640
ESTERBILDUNG (*f*) : esterification ; ester formation 641
ESTERGUMMI (*n*) : ester gum 642
ESTERIFIKATION (*f*) : esterification 643
ESTERIFIZIEREN : to esterify 644
ESTERZAHL (*f*) : ester number 645
ESTRAGON (*n*) : tarragon (*Artemisia dracunculus*) 646
ESTRAGONKRAUT (*n*) : tarragon herb (*Herba dracunculi*) 647
ESTRAGONÖL (*n*) : tarragon oil ; Sg. 0.9-0.949 648
ESTRICH (*m*) : layer of mortar ; plaster floor 649
ESTRICHGIPS (*m*) : estrich gypsum (see CALCIUM-SULFAT) 650
ESTRICHSTEIN (*m*) : type of paving stone 651
ESTRIFICATION (*f*) : esterification 652
ESTRIFIZIEREN : to esterify 653
ETABLIEREN : to establish, found, form, go into (business) 654
ETABLISSEMENT (*n*) : establishment 655
ETAGE (*f*) : story, floor (of building-s) 656
ETAGENROST (*m*) : step (story) grate 657
ETHIK (*f*) : ethics 658
ETHIKISCH : ethical 659
ETHNOGRAPHISCH : ethnographical-(ly) 660
ETHNOLOGIE (*f*) : ethnology, the science of mankind 661
ETIKETT (*n*) : tag ; label ; ticket 662
ETIKETTE (*f*) : (see ETIKETT) 663
ETIKETTENLACK (*m*) : label varnish 664
ETIKETTENLEIM (*m*) : label (gum) glue 665
ETIKETTIERMASCHINE (*f*) : labelling machine 666
ETUI (*n*) : case ; box 667
ETWA : about, perhaps, indeed 668
ETYMOLOGIE (*f*) : etymology, the derivation of words 669
EUCAIN (*n*) : (see EUKAIN) 670
EUCALYPTUSBLÄTTER (*n.pl*) : (see EUKALYPTUS-BLÄTTER) 671
EUCHLOR (*n*) : euchlorine (mixture of chlorine and chlorine peroxide ; a yellow gas) 672
EUCHROMATISCHER OPALIN-ALLOPHAN (*m*) : Chrysocolla 673
EUCODIN (*n*) : (see KODEINBROMMETHYLAT) 674
EUDIOMETER (*n*) : eudiometer (apparatus for gas analysis) 675
EUGENGLANZ (*m*) : polybasite (see POLYBASIT) 676
EUGENOL (*n*) : eugenol, eugenic acid, caryophyllic acid ; $C_3H_5C_6H_3(OH)OCH_3$; Sg. 1.07 ; Bp. 252°C. 677
EUGENSÄURE (*f*) : eugenic acid ; eugenol (see EUGENOL) 678
EUKAIN (*n*) : eucaine, beta-eucaine, Benzoylvinyldiacetonealkamine, trimethylbenzoxypiperidine hydrochloride (*Eucainum hydrochloricum B.*) ; $C_{15}H_{21}O_2N$; Mp. 78°C. 679
EUKALYPTOL (*n*) : eucalpytol, cajeputol, cineole ; $C_{10}H_{18}O$; Sg. 0.927 ; Mp. −1 to −3°C. ; Bp. 176°C. 680
EUKALYPTUSBAUM (*m*) : Eucalyptus, blue gum, Australian fever tree (*Eucalyptus globulus*) 681
EUKALYPTUSBLÄTTER (*n.pl*) : eucalyptus leaves (*Folia eucalypti* of *Eucalpytus globulus*) 682
EUKALYPTUSÖL (*n*) : eucalyptus oil (*Oleum eucalypti*) ; Sg. 0.85-0.94 ; Bp. 170-233 °C. 683
EUKASIN (*n*) : eucasin, casein ammonia 684
EUKLAS (*m*) : euclase ; $HBeAlSiO_5$; Sg. 3.0 [685
EUKODIN (*n*) : (see EUCODIN) 686
EUKOLIT (*m*) : eucolite ; $(Na,K,H)_{13}(Ca,Fe)_6(Si,Zr)_{20}O_{52}Cl$ $(+Nb_2O_5)$; Sg. 3.0 687
EULACHONÖL (*n*) : eulachon oil (from *Thaleicthys pacificus*) ; Sg. 0.87 688
EULE (*f*) : owl 689
EUPHONISCH : euphonic 690
EURYSKOP (*n*) : euryscope, symmetrical doublet, rectilinear doublet (of lenses) 691
EUSTACHISCH : eustachian 692
EUTEKTIKUM (*n*) : eutectic alloy ; easily flowing, well liquified alloy (Metal) ; (an intimate mixture of the crystals of two substances) 693
EUTEKTISCH : eutectic 694
EUTEKTISCHE LÖSUNG (*f*) : (see EUTEKTIKUM) : eutectic solution, eutectic alloy (Metal) 695
EUTEKTISCHES GEMENGE (*n*) : eutectic mixture (see EUTEKTIKUM) 696
EUTEKTOIDER KOHLENSTOFFSTAHL (*m*) : eutectic (eutectoid) carbon steel 697
EUTOMER DISTHENSPAT (*m*) : diaspore 698
EUTOMER KOBALTKIES (*m*) : gersdorffite 699
EUTOMGLANZ (*m*), DIRHOMBOEDRISCHER- : molybdenite 700
EUTOM-GLANZ (*m*), ELASTISCHER- : tellurwismuth 701
EUTOMGLANZ (*m*), PRISMATISCHER- : sternbergite 702
EUTOM-GLANZ (*m*), RHOMBOEDRISCHER- : tetradymite 703
EUXANTHINSÄURE (*f*) : euxanthinic acid ; $C_{19}H_{16}O_{11}$ 704
EVAKUIEREN : to evacuate ; empty ; exhaust ; (*n*) : evacuation 705
EVAKUIERUNGSKESSEL (*m*) : vacuum (pan) boiler 706
EVAPORATOR (*m*) : evaporator 707
EVAPORIEREN : to evaporate 708
EVENTUALITÄT (*f*) : eventuality 709
EVENTUELL : eventual-(ly), perhaps, in the event of 710
EVERNINSÄURE (*f*) : everninic acid 711
EVERNSÄURE (*f*) : evernic acid 712
EVOLUTE (*f*) : evolute 713
EVOLVENTE (*f*) : involute 714
EVOLVIEREN : to evolve 715
EVOMIEREN : to vomit 716
EWER (*m*) : lighter, wherry (Boat) 717
EWERSEGEL (*n*) : lug sail 718
EWIGBLUMEN (*f.pl*) : everlastings, immortelles (*Helichrysum bracteatum*, etc.) 719
EWIGLICH : eternally, perpetually 720
EXAKT : exact, correct-(ly) 721
EXALGIN (*n*) : exalgin, methyl acetanilide (see METHYLACETANILID) 722
EXAMEN (*n*) : examination 723
EXAMINIEREN : to examine 724
EXCELSIORMÜHLE (*f*) : excelsior mill 725
EXCENTER (*m*) : eccentric 726
EXCENTRICITÄT (*f*) : eccentricity 727
EXCENTRISCH : eccentric 728
EXCITANS (*n*) : excitant 729
EXEMPEL (*n*) : example 730
EXEMPEL, ZUM- : for example (*exempli gratia*, e.g.) 731
EXEMPLAR (*n*) : copy, sample, specimen, pattern, model 732

EXEMPLARISCH : exemplary 733
EXERZIEREN: to drill, practice, exercise 734
EXERZIERTORPEDO (*m*) : dummy torpedo 735
EXHAUSTDAMPF (*m*) : exhaust steam 736
EXHAUSTERGEBLÄSE (*n*) : exhauster 737
EXHAUSTOR (*m*) : exhauster ; ventilator ; dust-catcher 738
EXISTENZ (*f*) : existence ; subsistence 739
EXISTENZFÄHIG : capable of existing 740
EXISTIEREN : to exist ; subsist 741
EXKRETION (*f*) : excretion 742
EXOSMOTISCH : exosmotic 743
EXOTHERM : exothermic ; exothermal ; giving heat 744
EXOTHERMISCH : exothermal, exothermic, giving heat 745
EXOTHERMISCHE REAKTION (*f*) : (heat externally appreciable ; heat given off) exothermal reaction 746
EXPANDIEREN : to expand 747
EXPANDIERRAUM (*m*) : expansion chamber 748
EXPANSION (*f*) : expansion piece ; expansion 749
EXPANSIONSADIABATE (*f*) : adiabatic curve of expansion 750
EXPANSIONSGRAD (*m*) : degree of expansion 751
EXPANSIONSKAMMER (*f*) : expansion chamber 752
EXPECTORANS (*n*) : expectorant (Med.) 753
EXPEDIENT (*m*) : dispatcher, consignor 754
EXPEDIEREN : to dispatch ; forward ; send ; consign ; expedite 755
EXPEDITION (*f*) : office ; forwarding; expedition ; dispatch ; consignment 756
EXPERIMENT (*n*) : experiment 757
EXPERIMENTATOR (*m*) : experimentor 758
EXPERIMENTIEREN : to experiment 759
EXPERIMENTIERTISCH (*m*) : experimenting table 760
EXPLIZIEREN : to explain 761
EXPLODIERBAR : explosive 762
EXPLODIEREN : to explode 763
EXPLOSIBEL : explosive 764
EXPLOSIONSARTIG : explosive 765
EXPLOSIONSDRUCK (*m*) : explosion pressure 766
EXPLOSIONSFÄHIG : explosive 767
EXPLOSIONSICHERES GEFÄSS (*n*) : explosion proof vessel 768
EXPLOSIONSSICHER : explosion-proof 769
EXPLOSIONSSTOSS (*m*) : explosive (explosion) impact, impulse or shock 770
EXPLOSIONSTURBINE (*f*) : explosion turbine 771
EXPLOSIONSVORGANG (*m*) : explosive process 772
EXPLOSIONSWÄRME (*f*) : heat of explosion 773
EXPLOSIONSWELLE (*f*) : explosion wave 774
EXPLOSIV (*n*) : explosive 775
EXPLOSIVSTOFF (*m*) : explosive-(substance) ; bursting charge ; gunpowder 776
EXPLOSIVSTOFFCHEMIE (*f*) : explosive chemistry 777
EXPONENT (*m*) : exponent (Maths.) 778
EXPONIEREN : to expose (Photography) 779
EXPORT (*m*) : export 780
EXPORTFIRMA (*f*) : export firm, exporters 781
EXPORTIEREN : to export 782
EXPOSITION (*f*) : exposure (Photography) 783
EXPOSITIONMESSER (*m*) : exposure meter (Photography) (see also DYNAKTINOMETER) 784
EXPOSITIONSFAKTOR (*m*) : exposure factor (Photography) 785
EXPOSITIONSZEIT (*f*) : (duration of)-exposure (Photography) 786
EXPRESS : express-(ly), for the purpose 787
EXSICCATOR (*m*): desiccator; exsiccator; drier 788
EXSUDAT (*n*) : exudation 789
EXSUDIEREN : to exude 790
EXTEMPORIEREN : to extemporize 791
EXTENSITÄTSFAKTOR (*m*) : (see QUANTITÄTS-FAKTOR) 792
EXTINKTEUR (*m*) : extinguisher (Fire) 793
EXTRA : extra, additional, special 794
EXTRAAUSBEUTE (*f*) : extra (additional) yield 795
EXTRABLATT (*n*): special edition (of newspaper) 796
EXTRAHIEREN : to extract 797
EXTRAHIERUNG (*f*) : extraction 798
EXTRAKT (*m*) : extract (by inspissation of vegetable substances and the sap of plants) 799
EXTRAKTBRÜHE (*f*) : extracted liquor 800
EXTRAKT (*m*), DICKER- : thick extract (of a consistency which cannot be poured when cold) 801
EXTRAKT (*m*), DÜNNER- : thin extract (of the consistency of fresh honey) 802
EXTRAKTEUR (*m*) : extractor 803
EXTRAKT (*m*), FLUID- : fluid extract (see FLUID-EXTRAKT) 804
EXTRAKT (*m*), FLÜSSIGER- : liquid extract (see FLUIDEXTRAKT) 805
EXTRAKTGEHALT (*m*) : content of extract 806
EXTRAKTION (*f*) : extraction 807
EXTRAKTIONSANLAGE (*f*) : extracting (extraction) plant 808
EXTRAKTIONSAPPARAT (*m*): extracting (apparatus) plant 809
EXTRAKTIONSFETT (*n*) : bone (fat) grease (fat or grease obtained from bones by extraction) 810
EXTRAKTIONSFLÜSSIGKEIT (*f*) : extraction (liquid; liquor) fluid ; liquid extract 811
EXTRAKTIONSHÜLSE (*f*) : extraction (shell) thimble 812
EXTRAKTIONSKOLONNE (*f*) : extraction plant 813
EXTRAKTIVSTOFF (*m*) : extractive material or substance 814
EXTRAKT (*m*), TROCKNER- : dry extract (of a consistency which may be pulverized) 815
EXTRAKTWOLLE (*f*) : extract wool 816
EXTRAMOLEKULAR : extramolecular 817
EXTRAPOLATION (*f*) : extrapolation (assuming a curve to continue in a fixed manner outside actual existing limits) 818
EXTRAPOLIEREN : to extrapolate 819
EXTRAVASAT (*n*) : extravasation (the escape of any fluid of the body, from the vessel in which it is normally contained, into the surrounding tissue (Med.) 820
EXTREM : extreme-(ly) 821
EXTREM (*n*) : extreme, extremity 822
EXTREMFALL (*m*) : extreme case 823
EXTREMITÄT (*f*) : extremity 824
EXZELSIORMÜHLE (*f*) : excelsior mill 825
EXZENTER (*m*) : eccentric 826
EXZENTERBOLZEN (*m*) : eccentric bolt 827
EXZENTERBÜGEL (*m*) : eccentric hoop, eccentric strap 828
EXZENTERGABEL (*f*) : eccentric fork 829
EXZENTERREGULATOR (*m*) : eccentric governor 830
EXZENTERRING (*m*) : eccentric ring 831
EXZENTERSCHEIBE (*f*) : eccentric disc, eccentric sheave 832
EXZENTERSTANGE (*f*) : eccentric rod 833
EXZENTRISCH : eccentric 834
EXZENTRIZITÄT (*f*) : eccentricity 835
EXZERPT (*n*) : excerpt 836
EXZESSIV : excessive 837

F

FABEL (*f*): fable, legend, saga, fiction, tale, story 838
FABELLEHRE (*f*): mythology 839
FABELZEIT (*f*) mythical period 840
FABRIK (*f*): factory; works; plant; mill; manufactory 841
FABRIKANT (*m*): manfuacturer 842
FABRIKARBEIT (*f*): labour; manufactured article 843
FABRIKARBEITER (*m*): operative; factory hand; mill hand; factory worker 844
FABRIKAT (*n*): manufacture; fabric; make 845
FABRIKATION (*f*): manufacture; production; works 846
FABRIKATIONSANLAGE (*f*): manufacturing plant 847
FABRIKATORISCH: industrial 848
FABRIKATUR (*f*): manufacture; production 849
FABRIKBAU (*m*): factory construction 850
FABRIKBETRIEB (*m*): factory (management; practice) operation 851
FABRIKDIREKTOR (*m*): works manager 852
FABRIKGOLD (*n*): gold leaf 853
FABRIKKOMMISSAR (*m*): factory inspector 854
FABRIKMÄSSIG: industrially; by machinery 855
FABRIKÖL (*n*): oil for use in factories 856
FABRIKPRAXIS (*f*): factory practice 857
FABRIKPREIS (*m*): price ex. works or factory; cost price; works cost price; wholesale price 858
FABRIKSTADT (*f*): manufacturing town 859
FABRIKWARE (*f*): manufactured article 860
FABRIKWESEN (*n*): manufacturing industry 861
FABRIKZEICHEN (*n*): trade mark; maker's mark 862
FABRIZIEREN: to make; fabricate; manufacture; produce 863
FÄCES (*pl*): feces; fæces; excrement 864
FACETTE (*f*): facet 865
FACETTIERT: faceted 866
FACH (*n*): branch; division; department; compartment; line; profession; cell; subject; shelf; row; specialty 867
FACHARTIG: cellular 868
FACHAUSDRUCK (*m*): technical (term) expression; trade term 869
FACHBILDUNG (*f*): technical education; professional education 870
FÄCHELN: to fan 871
FACHEN: to classify, file, index, arrange in order, partition; fan, blow 872
FÄCHER (*m*): fan; flapper 873
FÄCHERBRENNER (*m*): fan-tail burner 874
FÄCHER (*m*), **ELEKTRISCH ANGETRIEBENE-**: electrically driven fan, electric fan 875
FÄCHERFÖRMIG: fan shaped 876
FÄCHERGERSTE (*f*): bearded barley 877
FÄCHERIG: cellular; divided into compartments 878
FÄCHERPALME (*f*): Palmyra palm (*Borassus flabelliformis*) 879
FÄCHERPLATTE (*f*): gusset plate 880
FACHGENOSSE (*m*): colleague (in the same profession) 881
FACHKENNTNIS (*f*): business (special) knowledge, trade (technical) knowledge 882
FACHLICH: professional; technical 883
FACHLITERATUR (*f*): technical (trade) literature 884
FACHMANN (*m*): technical man; expert; specialist; professional 885
FACHMÄNNIG: technical; professional; expert; specialist 886
FACHMÄSSIG: professional-(ly) 887
FACHORDNUNG (*f*): classification; arranging in order 888
FACHSCHULE (*f*): technical (school) college 889
FACHSPRACHE (*f*): technical (professional; trade) terms, terminology or language 890
FACHSTUDIUM (*n*): technical (professional) studies 891
FACHWERK (*n*): framework; panelling; latticework 892
FACHWISSENSCHAFT (*f*): special branch of science; specialty 893
FACHZEITSCHRIFT (*f*): trade (technical) journal or periodical 894
FACIT (*n*): sum; result; answer; total; product; amount 895
FACKEL (*f*): torch 896
FACKELBAUM (*m*): marsh elder (*Sambucus*) 897
FACKELFÖHRE (*f*): Scotch (pine) fir (*Pinus sylvestris*) (see **KIEFER**) 898
FACKELGLANZ (*m*): perfect clearness (of wines) 899
FACKELKOHLE (*f*): cannel coal; Sg.1.24-1.34 900
FACKELKRAUT (*n*): Great Mullein (*Verbascum Thapsus*) 901
FACKELN: to blaze, flare, trifle, flash, mislead 902
FACKELPALME (*f*): sago palm (*Sagus rumphii* or *Sagus lævis*) 903
FACON (*f*): fashion; style; cut; manner; mode; method; form 904
FACONNIEREN: to fashion; figure; make; cut; form 905
FACTIS (*pl*): india-rubber substitutes 906
FÄDCHEN (*n*): thread; filament 907
FADE: insipid; flat; dull; stale; tasteless 908
FADEN (*m*): thread; cord; fibre (Wood); filament; string; twine; grain; fathom; hypha (see **HYPHE**) 909
FADENÄHNLICH: threadlike; fibrous; filamentous 910
FADENALGE (*f*): filamentous alga 911
FADENFÖRMIG: filiform (see also **FADENÄHNLICH**) 912
FADENGERADE: as straight as a die 913
FADENGLAS (*n*): spun (filigree) glass 914
FADENGOLD (*n*): gold thread 915
FADENHOLZ (*n*): cord wood 916
FADENKÄFER (*m*): linear bug (destroys the sugar cane) (*Phœnacantha australica Kirkaldy*) 917
FADENLÄNGE (*f*): length of (fibre) thread 918
FADENNASS: wringing wet 919
FADENNUDELN (*f.pl*): vermicelli 920
FADENPILZ (*m*): mould fungus, mould 921
FADENRECHT: straight 922
FADENSCHEINIG: threadbare 923
FADENWERG (*n*): oakum 924
FADENWÜRMER (*m.pl*): Nematoda, round-worms (Tapeworms) (Med.) 925
FADENZÄHLER (*m*): thread counter (Textile) 926
FADENZELLE (*f*): hypha (see **HYPHE**) 927
FADENZELLGEWEBE (*n*) (see **HYPHENGEWEBE**) 928
FADENZIEHEND: ropy 929
FÄDIG: thready; fibrous; filamentous 930
FAECES (*pl*): feces, fæces, excreta 931

FAHAMBLÄTTER (*n.pl*) : faham leaves (*Angræcum fragrans*) 932
FÄHIGKEIT (*f*) : capacity ; capability ; ability ; aptness ; faculty ; talent ; qualification 933
FAHL : fallow ; fawn ; dun ; earth-coloured ; pale ; ashy ; faded ; sallow 934
FAHLERZ (*n*) : tetrahedrite ; fahl-ore ; grey copper ore ; $3Cu_2S. Sb_2S_3$; about 52% Cu ; Sg. about 5.0 935
FAHLSTEIN (*m*) : light grey slate 936
FAHNE (*f*) : lug (of an accumulator plate) ; flag, standard ; vane ; beard (of a quill) 937
FAHNE (*f*) **LUFT-** : air vane 938
FAHRBAR : portable ; transportable ; movable ; passable ; practicable ; navigable 939
FAHRBARES SCHWEISSFEUER (*n*) : flying fire (for tubes), portable fire 940
FÄHRBOOT (*n*) : ferry boat 941
FÄHRE (*f*) : ferry-(boat) 942
FAHREN : to go ; travel ; ride ; drive ; move ; journey ; pass ; proceed ; convey ; transport ; row 943
FAHRGELD (*n*) : fare ; carriage ; toll 944
FAHRGESTELL (*n*) : bogie waggon, trolley, car frame 945
FAHRKARTE (*f*) : ticket 946
FAHRLÄSSIGKEIT (*f*) : negligence, carelessness, want of dilgence or application 947
FAHRPLAN (*m*) : time-table 948
FAHRRAD (*n*) : bicycle ; cycle 949
FAHRRADLACK (*m*) : bicycle (cycle) varnish or lacquer 950
FAHRRADLENKSTANGE (*f*) : handle bar (of a cycle) 951
FAHRRADÖL (*n*) : cycle oil 952
FAHRRADRAHMEN (*m*) : frame (of a cycle) 953
FAHRSCHALTER (*m*) : controller 954
FAHRSTUHL (*m*) : elevator ; hoist ; lift 955
FAHRT (*f*) : passage ; journey ; run ; trip ; drive ; voyage ; ride ; course ; movement 956
FÄHRTE (*f*) : track ; trail ; trace 957
FAHRVORRICHTUNG (*f*) : driving arrangement or gear 958
FAHRWASSER (*n*) : navigable water, channel (also used figuratively as diverted into other channels) 959
FAHRWEG (*m*) : carriage (road) way 960
FAHRZEUG (*n*) : vehicle ; vessel 961
FÄKAL : fecal 962
FÄKALIEN (*n.pl*) : feces ; fæces ; fecal matter 963
FÄKALSTOFF (*m*) : fecal matter or substance 964
FAKSIMILEDRUCK (*m*) : facsimile printing 965
FAKTIS (*m* and *pl*): caoutchouc substitute (actually not a substitute, but a thinning material for caoutchouc mixtures) (chemical compound of vegetable and animal oils with sulphur) 966
FAKTIS (*m*), **BRAUNER-** : brown vulcanized oil (a compound of unsaturated glycerides with sulphur, having a sulphur content up to 20%) 967
FAKTISMASSE (*f*) : mass of caoutchouc substitute (see **FAKTIS**) 968
FAKTISSORTE (*f*) : kind (sort) of caoutchouc substitute (see **FAKTIS**) 969
FAKTIS (*m*), **WEISSER-** : white vulcanized oil (an addition product of the unsaturated acids, from rape oil with sulphur chloride) 970
FAKTOR (*m*) : factor ; manager ; agent ; foreman 971
FAKTORENTABELLE (*f*) : table of factors 972
FAKTUR (*f*) : invoice (Com.) 973
FAKTURA (*f*) : invoice (Com.) 974
FAKTURIEREN : to invoice (Com.) 975
FÄKULENZ (*f*) : feculence ; sediment ; dregs 976
FAKULTÄT (*f*) : faculty (of Universities) 977
FALB pale yellow ; fallow ; pale ; cream 978
FALL (*m*) : fall ; decay ; descent ; case ; ruin ; event ; yield ; downfall ; decline ; waterfall ; halliard (Naut.) 979
FÄLLBAR : precipitable ; ready to fell 980
FÄLLBARKEIT (*f*) : precipitability ; readiness for felling 981
FALLBLOCK (*m*) : halliard block 982
FÄLLBOTTICH (*m*) : precipitating vat 983
FALLE (*f*) : trap ; bolt ; snap ; latch ; valve 984
FALLEISEN (*n*) : latch 985
FALLEN : to fall ; separate; drop ; yield ; dip (Mining) ; incline ; decline ; produce ; decay ; decrease ; diminish ; be deposited ; be precipitated 986
FÄLLEN : to precipitate ; drop ; fell ; pull down ; pass (Judgement) 987
FALLHAMMER (*m*) : drop (steam) hammer ; pile driver 988
FALLHÖHE (*f*) : height or length of drop or fall 989
FALLHOLZ (*n*) : fallen wood 990
FALLIEREN : to fail ; go bankrupt ; go into liquidation ; become insolvent 991
FÄLLIG : due ; payable 992
FÄLLIGKEITSTERMIN (*m*) : date of falling due (of payments, etc.) 993
FALLIMENT (*n*) : failure (in business), bankruptcy 994
FALLIT : bankrupt 995
FALLITENGESETZ (*n*) : bankruptcy law 996
FÄLLKESSEL (*m*) : precipitating vessel 997
FALLKLAPPE (*f*) : flap 998
FALLLEITBLOCK (*m*) : halliard (guide) leading block 999
FÄLLMITTEL (*n*) : precipitant 000
FALLMOMENT (*n*) : falling moment 001
FALLREEP (*n*) : manrope 002
FALLREEPSKNOTEN (*m*) : diamond knot 003
FALLREEPSLEITER (*f*) : accommodation ladder (Marine) 004
FALLROHR (*n*) : downpipe ; waste pipe ; downcomer (Boilers) 005
FALLSCHACHT (*m*) : chute (for ashes, etc.) 006
FALLSCHIRM (*m*) : parachute 007
FALLSUCHT (*f*) : epilepsy (a disturbance of the cerebral functions) (Med.) 008
FALLSÜCHTIG : epileptic 009
FALLTÜR (*f*) : trap-door 010
FÄLLUNG (*f*) : precipitation ; felling 011
FÄLLUNGSMITTEL (*n*) : precipitant 012
FÄLLUNGSPRODUKT (*n*): precipitate 013
FÄLLUNGSWERT (*m*) : precipitation value 014
FALLVERSCHLUSS (*m*) : drop-shutter (Photography) 015
FALLWERK (*n*) : stamp ; pile driver 016
FALLWINKEL (*m*) : inclination ; angle of (declination) inclination 017
FALL-WIND (*m*) : eddy (gust of) wind 018
FALSCH : false base ; counterfeit ; spurious ; pseudo ; artificial ; adulterated ; wrong ; treacherous ; deceitful ; forged 019
FALSCHBLAU : navy-blue 020
FÄLSCHEN : to falsify ; adulterate ; counterfeit ; forge 021
FÄLSCHLICH : falsely 022
FÄLSCHUNGSMITTEL (*n*) : adulterant 023
FALTENFILTER (*n*) : folded (plaited) filter 024

FALTENREICH : folded, creased, wrinkled 025
FALTENWERFEN (*n*) : creasing ; crimping ; puckering ; draping 026
FALTENWURF (*m*) : drapery 027
FALTER (*m*) : folder ; creaser ; butterfly ; third stomach of ruminants 028
FALTSCHACHTEL (*f*) : folding (cardboard) box 029
FALTUNGSPUNKT (*m*) : plait point ; point of folding 030
FALZ (*m*) : fold ; groove ; notch ; flute ; rabbet ; rebate (Joinery) 031
FALZEN : to fold ; groove ; notch flute ; rabbet ; rebate ; pare ; join ; trim ; solder 032
FALZHOBEL (*m*) : rabbet (rebating) plane 033
FALZZANGE (*f*) : pliers 034
FAMILIENVERHÄLTNISSE (*n.pl*) : family (home) affairs 035
FAMOS : famous ; fine ; excellent 036
FANATISCH : fanatic-(al)-(ly) 037
FANG (*m*) : capture ; catch ; trap ; fang ; talon ; stab ; snare 038
FANGDRAHT (*m*) : guard wire 039
FANGEN : to capture ; catch ; trap ; secure ; snare ; seize 040
FANGGARN (*n*) : snare 041
FANGSTOFF (*m*) : stuff ; pulp (Paper) 042
FANGVORRICHTUNG (*f*) : trap ; trapping (catching device) 043
FANTASIE (*f*) : fantasy, fantasia, extemporization, imagination, fancy 044
FANTASIEREN : to extemporize, imagine, improvise 045
FARAD (*n*) : farad ; (electrical unit of capacity) 046
FARBBAD (*n*) : dye-bath 047
FARBBAND (*n*) : typewriter ribbon ; any inked ribbon 048
FÄRBBAR : capable of being dyed, coloured or stained 049
FARBBASE (*f*) : colour base 050
FARBBEIMENGUNG (*f*) : colour admixture 051
FARBBEIZE (*f*) : mordant 052
FARBBRÜHE (*f*) : dye liquor 053
FARBE (*f*) : colour ; dye ; pigment ; paint ; hue ; tint ; complexion ; ink ; stain ; medium fine smalt (see **BLAUFARBE**) 054
FÄRBE (*f*) : dyeing 055
FÄRBEARTIKEL (*m* and *pl*) : dyed style (Calico) 056
FÄRBEBAD (*n*) : dye bath 057
FÄRBEBRÜHE (*f*) : dyeing liquor 058
FÄRBEFASS (*n*) : dye tub 059
FÄRBEFLÜSSIGKEIT (*f*) : dyeing (colouring) liquid ; staining fluid 060
FARBEGANG (*m*) : sheer strake 061
FÄRBEHOLZ (*n*) : dye wood 062
FÄRBEKUFE (*f*) : dyeing vat 063
FÄRBELACK (*m*) : lac dye 064
FÄRBEMITTEL (*n*) : colouring (agent ; medium) matter ; pigment ; dye 065
FÄRBEN : to colour ; tinge ; dye ; stain 066
FARBENABBEIZMITTEL (*n*) : paint removing agents ; colour (dye) remover 067
FARBENABSTREICHMESSER (*n*) : knife for removing excess dye or colour 068
FARBENAUFTRAG (*m*) : colouring ; painting ; application of (colour) paint 069
FARBENBILD (*n*) : coloured image ; spectrum 070
FARBENBINDEMITTEL (*n*) : colour agglutinants 071
FARBENBLIND : colour-blind 072
FARBENBRECHUNG (*f*) : colour (blending) refraction 073
FARBENBRETT (*n*) : palette (Art.) 074
FARBENBRÜHE (*f*) : dyeing liquor 075
FARBENCHEMIE (*f*) : colour chemistry 076
FÄRBEND : colouring, dyeing, staining 077
FARBENDIAPOSITIV (*n*) : coloured (diapositive) transparency (Photography) 078
FARBENDISTEL (*f*) : (see **FÄRBERDISTEL**) 079
FARBENDRUCK (*m*): colour print-(ing) 080
FARBENEMPFINDLICH : sensitive to colours, orthochromatic 081
FARBENERDE (*f*) : colour-(ed) earth 082
FARBENERZEUGEND : colour producing ; chromogenic 083
FARBENFABRIK (*f*) : colour (dye ; paint) works or factory 084
FARBEN, FEUERSICHERE- (*f.pl*) : fireproof paints 085
FARBENFILTER (*m.* and *n.*) : colour filter, colour screen (Photography) 086
FARBEN FÜR LACKE (*pl*) : varnish colours 087
FARBEN FÜR SEIFEN (*pl.*) : soap colours 088
FARBENGEBUNG (*f*) : coloration 089
FARBEN, GIFTFREIE- (*pl*) harmless (poisonless) colours 090
FARBENGLANZ (*m*) : colour brilliancy 091
FARBENGLAS (*n*) : stained (coloured) glass 092
FARBENHOLZ (*n*) : logwood, dyewood 093
FARBENKASTEN (*m*) : paint-box 094
FARBENKÖRPER (*m*) : colouring matter ; colour ; pigment 095
FARBENLACK (*m*) : lake ; colour varnish 096
FARBENLAGE (*f*) : coat-(ing) (of colour) 097
FARBENLEHRE (*f*) : chromatics ; science of colour 098
FARBENLEITER (*f*) : colour scale 099
FARBENLOS : colourless ; achromatic ; pallid ; pale 100
FARBENLOSIGKEIT (*f*) : achromatism 101
FARBENLÖSUNG (*f*) : dye, colour, staining (solution) 102
FARBENMALZ (*n*) : colouring malt 103
FARBENMASS (*n*) : colorimeter 104
FARBENMESSER (*m*) : colorimeter ; (*n*) : colour (palette) knife 105
FARBENMESSUNG (*f*) : colorimetry 106
FARBENMÜHLE (*f*) : paint mill 107
FARBENPHOTOGRAPHIE (*f*) : colour photography 108
FARBENPROBE (*f*) : colour (dye) test 109
FARBENRAND (*m*) : iris 110
FARBENREIBEN (*n*) : colour (paint) grinding 111
FARBENREIBER (*m*) : colour (paint) grinder 112
FARBENREIBMASCHINE (*f*) : paint grinder ; colour (paint) grinding machine 113
FARBENREICH : many-coloured, of many colours 114
FARBEN, SÄUREFESTE- (*pl*) : acid proof paints 115
FARBENSCHILLER (*m*) : colour play ; surface colour 116
FARBENSCHMELZOFEN (*m*) : enamelling furnace 117
FARBENSKALA (*f*) : colour scale 118
FARBENSPEKTRUM (*n*) : coloured spectrum 119
FARBENSPIEL (*n*) : colour play ; opalescence 120
FARBEN, SPRITLÖSLICHE- (*pl*): alcohol-soluble colours 121
FARBENSTEINDRUCK (*m*) : colour lithography 122
FARBENSTIFT (*m*) : coloured (pencil ; crayon) pastel 123

FARBENSTOFF (*m*): dye; dye-stuff; colouring matter; pigment 124
FARBENSTRAHL (*m*): coloured ray 125
FARBENSTUFE (*f*): colour shade; gradation; graduation; tinge; tint 126
FARBENSTUFENMESSER (*m*): tintometer 127
FARBENTAFEL (*f*): cake of colour or paint 128
FARBENTIEFE (*f*): depth of colour 129
FARBENTON (*m*): hue (colour-)tone 130
FARBENTUBE (*f*): tube of colour 131
FARBENUMSCHLAG (*m*): colour change 132
FARBENVERÄNDERUNG (*f*): colour change; discolouration 133
FARBENWAREN (*f.pl*): dyes; paints; pigments; colours; dyestuffs 134
FARBENWECHSEL (*m*): colour (play) change 135
FARBENWERT (*m*): colour value 136
FARBEN, WETTERFESTE- (*pl*): weather-proof (paints) colours 137
FARBENZERSTREUUNG (*f*): chromatism (Optics) 138
FÄRBEPROZESS (*m*): dyeing (colouring) process 139
FÄRBER (*m*): dyer 140
FÄRBERBAUM (*m*): Venetian sumac (*Cotinus cotinus*); Shumac (*Rhus glabra*) 141
FÄRBERBEERE (*f*): buckthorn-(berry); (*Rhamnus catharticus*) 142
FÄRBERDE (*f*): colour earth 143
FÄRBERDISTEL (*f*): Carthamus, dyer's weed, safflower, thistle (false, American, African, bastard, Dejer's, Dyer's) saffron (dried florets of *Carthamus tinctorius*) 144
FÄRBEREI (*f*): dyeing; dye (house) works; colouring; staining; painting; tinting 145
FÄRBEREIARTIKEL (*m* and *pl*): dyeing article-(s) 146
FÄRBEREICHE (*f*): dyer's oak, quercitron (*Quercus coccinea tinctoria*) 147
FÄRBEREIMASCHINEN (*f.pl*): dyeing machinery 148
FÄRBERFLECHTE (*f*): dyer's moss; archil; orchil; orseille 149
FÄRBERGINSTER (*m*): dyer's broom (*Genista tinctoria*) 150
FÄRBERKNÖTERICH (*m*): a species of Indigofera (*Polygonum tinctorium*) 151
FÄRBERKREUZDORN (*m*): buckthorn (*Rhamnus catharticus*) 152
FÄRBERLACK (*m*): lack dye (see **LACK-LACK**) 153
FÄRBERMAULBEERBAUM (*m*): dyer's mulberry, old fustic (*Morus tinctoria*) (see **FUSTIK, ALTER-**) 154
FÄRBEROCHSENZUNGE (*f*): alkanet; alkanna (*Anchusa tinctoria*) 155
FÄRBERÖL (*n*): dyer's oil 156
FÄRBERRINDE (*f*): quercitron bark (Bark of *Quercus coccinea tinctoria*) 157
FÄRBERRÖTE (*f*): madder (*Rubia tinctorum*) (see **KRAPP**) 158
FÄRBERWAID (*m*): woad (*Isatis tinctoria*) 159
FÄRBERWAU (*m*): yellow weed; dyer's weed; weld (*Reseda luteola*) 160
FÄRBERWURZEL (*f*): madder (root of *Rubia tinctorum*) 161
FÄRBESIEB (*n*): colour (strainer) sieve 162
FÄRBESTOFF (*m*): dye-(stuff); pigment; colouring matter 163
FARBETABLETT (*n*): dye tablet 164
FÄRBEVERMOGEN (*n*): colouring (dyeing; tinctorial) power, property or value 165
FARBE (*f*), **WASSER-**: water colour 166
FÄRBEWURZEL (*f*): (see **FÄRBERWURZEL**) 167
FARBEXTRAKT (*m*): colour extract 168
FARBFERTIG: ready for dyeing 169
FARBFILTER (*n*): colour filter 170
FARBFLOTTE (*f*): dye bath 171
FARBGEHALT (*m*): dye (colour) content 172
FARBGELATINE (*f*): coloured gelatine (Photography) 173
FARBGLAS (*n*): stained (coloured) glass 174
FARBGLASKOMPOSITIONEN (*f.pl*): stained glass compounds 175
FARBGLASUR (*f*): colouring (coloured) glaze, (Ceramics) 176
FARBGUT (*n*): goods for dyeing 177
FARBHOLZ (*n*): dye wood 178
FARBHOLZEXTRAKT (*m*): dye wood extract 179
FARBIG: coloured; stained; dyed; tinted (Paper) 180
FARBKOCHAPPARAT (*m*): colour boiling apparatus: colour boiler 181
FARBKOKE (*m.pl*): colour cokes 182
FARBKÖRPER (*m*): colouring (material) substance: colouring matter 183
FARBKRAFT (*f*): colouring (tinctorial; dyeing) power or capacity 184
FARBLACK (*m*): lake 185
FARBLOS: colourless; achromatic; pallid; pale 186
FARBLÖSUNG (*f*): colour-(ed) solution; colouring solution; dye-(ing) solution 187
FARBMALZ (*m*): colouring malt (by roasting moistened kiln-dried malt in rotating drums at 150-200°C., for dark beers; or by allowing green malt to saccharify and then vatting it at low temperature, for light beers) 188
FARBMALZTROMMEL (*f*): colouring-malt drum (Brewing) 189
FARBMATERIAL (*n*): dyestuff; colouring (matter) material; dye 190
FARBMESSEND: colorimetric 191
FARBMISCHER (*m*): colour mixer 192
FARBMÜHLE (*f*): paint mill 193
FARBÖL (*n*): colour oil 194
FARBRASTER (*m*): colour screen (Three-colour photography) 195
FARBRASTERPLATTE (*f*): colour-screen plate (for colour photography) 196
FARBSÄURE (*f*): colour acid 197
FARBSCHICHT (*f*): layer of colour 198
FARBSCHLEIER (*m*): green fog, iridescent stain, chemical fog (green stains on photographic plates, usually due to ammonia developer) 199
FARBSTIFT (*m*): coloured pencil; crayon; pastel 200
FARBSTOFF (*m*): dyestuff; dye; colouring (matter) material; pigment 201
FARBSTOFF (*m*): **BASISCHER-**: basic colour-(ing matter) 202
FARBSTOFFERZEUGEND: colour producing; chromogenic 203
FARBSTOFFLÖSUNG (*f*): dye (colour; staining) solution 204
FARBSTUFE (*f*): colour (gradation; tinge; tint) shade 205
FARBTIEFE (*f*): depth of colour; depth of dyeing 206
FARBTON (*m*): tone; tint; shade 207
FARBTROG (*m*): colour (stain; dye) trough 208
FÄRBUNG (*f*): colouring; dyeing; staining; painting; tinting; hue; colouration; tinge; dye 209

.BWECHSEL (*m*): colour (change; variation) play 210
ARBWERK (*n*): colour (paint; dye) works 211
FARBWERT (*m*): colour value 212
FARBZERSTÄUBER (*m*): aerograph, sprayer for paint 213
FARIN-(ZUCKER) (*m*): brown sugar; moist sugar 214
FARMER'SCHER ABSCHWÄCHER (*m*): Farmer's reducer, potassium ferricyanide reducer (Photography) 215
FARNKRAUT (*n*): fern 216
FARNKRAUTÄHNLICH: fern-like 217
FARNKRAUTEXTRAKT (*m*): (male)- fern extract (*Extractum filicis maris*) 218
FARNKRAUTÖL (*n*): male fern oil (see FARNÖL) 219
FARNÖL (*n*): fern oil (from the rhizome of *Dryopteris filix-mas marginalis*); Sg. 0.85; Bp. 140-250°C. 220
FARNSAMEN (*m*): fern seed 221
FARNWURZEL (*f*): fern root (*Rhizoma filicis*); Filixmas; male fern; shield fern; Aspidium (root of *Aspidium filix mas*) 222
FARRE (*m*): bullock 223
FARREN (*m*): (see FARN) 224
FASCH (*m*): Aphthæ, a form of catarrh (Med.) 225
FASCHINE (*f*): fascine, hurdle 226
FASELN: to breed, prosper, unravel 227
FASEN (*m*): thread, fibre, filament 228
FASER (*f*): fibre; filament; thread; string; grain (Wood); vascular tissue 229
FASERÄHNLICH: fibrous; filamentous; fibre-like 230
FASERALAUN (*m*): halotrichite; $FeSO_4.Al_2(SO_4)_3.24H_2O$ (see also HAARSALZ) 231
FASERARTIG: fibrous; filamentous; fibre-like 232
FASERFÖRMIG: fibrous; fibre-like; filamentous 233
FASERGEHALT (*m*): fibre content 234
FASERGEWEBE (*n*): fibrous tissue 235
FASERGEWINNUNG (*f*): production of fibre 236
FASERGEWINNUNGSMASCHINE (*f*): thread making machine 237
FASERGIPS (*m*): fibrous gypsum 238
FASERHAUT (*f*): fibrous membrane 239
FASERIG: fibrous; stringy; filamentous 240
FASERKALK (*m*): fibrous calcite; arragonite; (natural calcium carbonate); Sg. 2.95 (see CALCIUMCARBONAT) 241
FASERKIESEL (*m*): sillimanite (see SILLIMANIT) 242
FASERKNORPEL (*m*): fibrous cartilage 243
FASERLOS: fibreless 244
FASERN: to tease; fuzz; mottle (Paper); become (mottled) variegated; unravel 245
FASER (*f*), NEUTRALE-: neutral fibre 246
FASERPULVER (*n*): powdered fibre 247
FASERSTOFF (*m*): fibrin; fibrous substance; fibrous material; gluten; cellulose 248
FASERSTOFFARTIG: fibrinous 249
FASERSTOFFINDUSTRIE (*f*): textile industry 250
FASERUNG (*f*): teasing; fuzzing 251
FASERZEOLITH (*m*): natrolite (see NATROLITH) 252
FASRIG: fibrous; stringy; filamentous 253
FASS (*n*): barrel; cask; keg; vat; tank; tube; drum; tun; pipe; hogshead 254
FASSBAND (*n*): hoop-iron 255
FASSBINDER (*m*): cooper 256
FASSBODEN (*m*): head (of casks) 257
FASSDAUBE (*f*): stave (of a cask) 258
FASS, EISERNES- (*n*): iron (barrel; cask) drum 259
FASSEN: to grasp; seize; take; set; put; mount; comprehend; comprise; understand; include; conceive; contain; hold; cask; tun; barrel; compose oneself; express oneself; calm; prepare; apprehend 260
FASSFÜLLMASCHINE (*f*): barrel (cask) filling machine 261
FASSGÄRUNGSSYSTEM (*n*): tub (vat, barrel -(cleaning)) system (Brewing) 262
FASSGELÄGER (*n*): bottoms; cask (sediment; precipitate) deposit 263
FASS, GEPICHTES- (*n*): pitched (cask) barrel 264
FASSGUT (*n*): coal which passes with the water through the bottom sieve of the settling bed (Coal washing) 265
FASSHAHN (*m*): spigot; faucet; cock; tap 266
FASS, HÖLZERNES- (*n*): wooden (cask) barrel 267
FASSLICH: comprehensible; conceivable 268
FASSMESSUNG (*f*): gauging 269
FASSONEISEN (*n*): section iron 270
FASSPACKMASCHINE (*f*): barrel-packing machine 271
FASS, PAPIERENES- (*n*): paper barrel 272
FASSPECH (*n*): cooper's pitch; pitch in barrels or casks 273
FASSREINIGUNGSANLAGE (*f*): barrel (cask) cleaning or purifying plant 274
FASSSPUND (*m*): bung (of casks) 275
FASS, STÄHLERNES- (*n*): steel (barrel) drum 276
FASSTALG (*m*): tallow in casks 277
FASSUNG (*f*): seizing, etc. (see FASSEN); wording; frame; composure; setting; countenance; socket; holder; rim 278
FASSUNG (*f*), LAMPEN-: lamp holder 279
FASSUNGSGABE (*f*): gift of perception 280
FASSUNGSKRAFT (*f*): power of comprehension 281
FASSUNGSLOS: disconcerted, confused 282
FASSUNGSRAUM (*m*): capacity 283
FASSUNGSVERMÖGEN (*n*): (holding)-capacity; power of comprehension 284
FASSWAREN (*f.pl*): goods in barrels or casks 285
FASSWEISE: by (in) casks, barrels or drums 286
FASSZWICKEL (*m*): try-cock 287
FASTAGE (*f*): casks; cooperage; barrels 288
FASTENZEIT (*f*): lent, time of fasting 289
FASTNACHT (*f*): carnival, Shrove tide 290
FATAL: calamitous; fatal; disagreeable 291
FATALISMUS (*m*): fatalism 292
FATALITÄT (*f*): fatality, misfortune, calamity 293
FAUL: rotten; foul; bad; dirty; putrid; decaying; lazy; idle; slow; sluggish; short (Metal); brittle 294
FAULBAUM (*m*): alder buckthorn; water elder; black alder (*Rhamnus frangula*); bird cherry (*Prunus padus*) 295
FAULBAUMRINDE (*f*): black alder-tree bark (*Cortex frangulæ*) 296
FAULBRUCH (*m*): shortness; brittleness 297
FAULBRÜCHIG: short; brittle 298
FAULBUTTE (*f*): fermenting trough 299
FAULBÜTTE (*f*): (see FAULBUTTE) 300
FÄULE (*f*): (dry)-rot; blight; rottenness; putrefaction; decomposition 301
FAULEN: to rot; putrefy; ferment; decompose; become septic; decay 302
FAULEND: putrescent; septic; rotting; decomposing; decaying 303
FAULENZEN: to be lazy; idle 304
FAULENZER (*m*): idler (also Mechanical); ready-reckoner; Y-connection piece (Brewing) 305
FAULER HEINZ (*m*): athanor (Old chemistry) 306

FAULFIEBER (*n*): putrid fever, Typhus fever (Med.) 307
FÄULNIS (*f*): rot; sepsis; putrefaction; corruption; decomposition; rottenness; rotting; decay 308
FÄULNISALKALOID (*n*): ptomaine 309
FÄULNISBAKTERIEN (*pl*): putrefactive bacteria 310
FÄULNISBASE (*f*): ptomaine 311
FÄULNISBEFÖRDERND: putrefactive; septic 312
FÄULNISBEWIRKEND: putrefactive; septic 313
FÄULNISERREGEND: putrefactive; septic 314
FÄULNISFÄHIG: putrefiable 315
FÄULNISGIFT (*n*): septic poison 316
FÄULNISHEMMEND: antiseptic; aseptic 317
FÄULNISHINDERND: antiseptic; aseptic 318
FÄULNISPRODUKT (*n*): decomposition product, product from (rotting) putrefaction 319
FÄULNISUNFÄHIG: unputrefiable 320
FÄULNISWIDRIG: antiseptic; aseptic 321
FAULTIER (*n*): sloth (animal) (*Bradypodidæ tridactylus*) 322
FAUM (*m*): foam; froth 323
FÄUSTEL (*n*): miner's hammer 324
FAUSTGELENK (*n*): wrist 325
FAUSTGERECHT: dextrous 326
FAUSTKAMPF (*m*): pugilistic match, boxing 327
FAUSTREGEL (*f*): rule of thumb (rule correct enough for general purposes) 328
FAUSTSCHLAG (*m*): punch (with the clenched fist) 329
FAVORISIEREN: to favour 330
FAVORIT (*m*): favourite 331
FAYALIT (*m*): fayalite (an iron silicate); Fe_2SiO_4; Sg 4.2. 332
FAYENCE (*f*): fine pottery, delf 333
FAYENCE (*f*), **FEINE-**: (see **STEINGUT, ENGLISCHES-**) 334
FAYENCE (*f*), **GEMEINE-**: majolica (see **MAJOLIKA**) 335
FAYENCEMASSE (*f*): fine pottery clay; Sg. 2.6 [336
FEBRICULA (*n*): febricula, slight fever (Med.) 337
FECHTER (*m*): fighter, boxer, gladiator, fencer 338
FEDER (*f*): feather; plume; pen; spring (Mech.); bristle; taper (of bricks) 339
FEDERÄHNLICH: feather-like; spring-like 340
FEDERALAUN (*m*): feather alum; holotrichite; $FeSO_4.Al_2(SO_4)_3.24H_2O$; epsomite; $MgSO_4.7H_2O$; amianthus (see **AMIANT**) 341
FEDERANGEL (*f*): spring hinge 342
FEDERANSCHUSS (*m*): feathery crystallization 343
FEDERARTIG: feathery; feather-like; springy; spring-like 344
FEDERBAROMETER (*m*): aneroid barometer 345
FEDERBART (*m*): web of a feather 346
FEDERBELASTUNG (*f*): valve spring, spring loading 347
FEDERBESEN (*m*): feather (dusting) brush or duster 348
FEDER (*f*), **BLATT-**: plate spring 349
FEDERBLECH (*n*): spring steel 350
FEDERBÜCHSE (*f*): pen(case) box 351
FEDERBUSCH (*m*): plume; crest; tuft 352
FEDERCHEN: little (feather; pen) spring; plumule (Bot.) 353
FEDEREINSTELLUNG (*f*): setting-(of a spring) 354
FEDERERZ (*n*): feather ore; Jamesonite (see **JAMESONIT**); plumosite; Pb_2SbS_4: Sg. 5.8 [355
FEDERFÖRMIG: feathery; plumiform 356
FEDERGEHÄUSE (*n*): spring case 357
FEDERGIPS (*m*): fibrous gypsum 358
FEDERHART: spring-hard; elastic 359
FEDERHÄRTE (*f*): elasticity 360
FEDERHARZ (*n*): india-rubber; caoutchouc; elaterite (*Gummi elasticum*); Sg. 0.9 361
FEDERHEBER (*m*): spring lever 362
FEDERICHT: feathery; feathered; feather-like; fleecy 363
FEDERIG: (see **FEDERICHT**) 364
FEDERKIEL (*m*): quill 365
FEDERKLINKE (*f*): spring latch 366
FEDERKRAFT (*f*): elasticity; springiness 367
FEDER (*f*), **LITHOGRAPHISCHE-**: lithographic pen 368
FEDERMESSER (*n*): pen-knife 369
FEDERMISTELGEWÄCHSE (*n.pl*): *Myzodendraceae* (Bot.) 370
FEDERN: to be elastic; lose feathers; moult; be flexible; be springy; taper; feather; spring 371
FEDERNELKE (*f*): sweet-william (*Dianthus barbatus*) 372
FEDERREGULATOR (*m*): spring regulator 373
FEDERRIEGEL (*m*): spring bolt 374
FEDERSALZ (*n*): feather salt (such as natural alum) 375
FEDERSCHLOSS (*n*): spring lock 376
FEDERSPANNUNG (*f*): spring tension 377
FEDER (*f*), **SPIRAL-**: spiral (helical) spring 378
FEDERSPULE (*f*): quill 379
FEDERSTAHL (*m*): spring steel 380
FEDERSTAUB (*m*): down 381
FEDERSTEIN (*m*): tongued brick; tapered brick 382
FEDERTELLER (*m*): spring (cap, collar) washer 383
FEDER UND NUT (**MIT-**): tongue and groove (tongued and grooved) 384
FEDERUNG (*f*): spring, deflection; springing; taper; feather 385
FEDERVENTIL (*n*): spring valve 386
FEDERVIEH (*n*): feathered world; poultry 387
FEDERWAGE (*f*): spring balance 388
FEDERWEISS (*n*): French chalk; fibrous gypsum; asbestos; amianthus; magnesium silicate; $Mg_3Si_4O_{11}$; Sg. 2.7 389
FEDERWELLE (*f*): flexible shaft 390
FEDERWISCHER (*m*): pen-wiper 391
FEDERZEICHNUNG (*f*): pen-(and ink) drawing 392
FEDERZIRKEL (*m*): spring-bows, spring dividers (Drawing instruments) 393
FEGE (*f*): sieve; ridde; screen; sweeping-(s) 394
FEGEFEUER (*n*): hell-fire; purgatory 395
FEGEN: to sweep; clean; winnow; wipe; screen; (*n*): cleansing, etc. 396
FEGESAND (*m*): scouring sand 397
FEGESCHOBER (*m*): scum-pan (Salt) 398
FEGSEL (*n*): sweepings 399
FEHDE (*f*): quarrel; feud 400
FEHL: in vain, wrong 401
FEHLBAR: fallible 402
FEHLBARKEIT (*f*): fallibility 403
FEHLBETRAG (*m*): deficit; deficiency 404
FEHLDRUCK (*m*): misprint, printer's error 405
FEHLEN: to fail; err; lack; be missing; miss; be absent; ail; be defective (deficient); be lacking 406
FEHLER (*m*): error; mistake; fault; defect; want; flaw; failing 407
FEHLERFREI: faultless; sound; clear 408
FEHLERGRENZE (*f*): tolerance, limit of error 409
FEHLERHAFTIGKEIT (*f*): defectiveness; incorrectness; faultiness 410

FEHLERLOS : faultless, sound, clear, flawless 411
FEHLERLOSIGKEIT (*f*) : faultlessness, flawlessness 412
FEHLERMÖGLICHKEIT (*f*) : possibility of error 413
FEHLERQUELLE (*f*) : source of error 414
FEHLERVOLL : faulty, full of mistakes 415
FEHLGEBURT (*f*) : miscarriage ; abortion (Med.) 416
FEHLGEHEN : to go (wrong) astray 417
FEHLGRIFF (*m*) : mistake ; blunder 418
FEHLGUSS (*m*) : waste (Founding) 419
FEHLINGSCHE LÖSUNG (*f*) : Fehling's solution 420
FEHLSCHLAGEN : to be disappointed ; go wrong ; fail ; miscarry 421
FEHLSCHLUSS (*m*) : wrong conclusion ; false inference 422
FEHLSCHRITT (*m*) : wrong (move) step ; error ; mistake ; blunder 423
FEHLTRETEN : to stumble, make a false step 424
FEHLTRITT (*m*) : fault, false step 425
FEHN (*n*) : fen, moor, marsh 426
FEIER (*f*) : holiday, recess, celebration, solemnity, festival 427
FEIERLICH : solemn, awful, festive 428
FEIERN : to celebrate ; observe ; rest from work ; make holiday ; be idle ; strike ; solemnize ; honour ; take leisure 429
FEIERZEIT (*f*) : public holidays, feast days 430
FEIGBOHNE (*f*) : lupine 431
FEIGE : rotten ; cowardly ; (*f*) : fig 432
FEIGENBAUM (*m*) : fig-tree (*Ficus carica*) 433
FEIGENBAUM (*m*), ÄGYPTISCHER- : Egyptian fig tree (*Ficus sycomorus*) 434
FEIGENBLATT (*n*) : fig-leaf 435
FEIGHEIT (*f*) : cowardice 436
FEIGLING (*m*) : coward 437
FEIGWARZE (*f*) : Condyloma (a warty growth) (Med.) 438
FEIGWARZE (*f*), BREITE- : syphilitic condyloma, mucous (patch) tubercle ; (an eruption on the mucous membrane) (Med.) 439
FEIL : for sale ; mercenary ; venal 440
FEILBAR : capable of being filed 441
FEILE (*f*) : file ; phial ; polish ; rasp 442
FEILE (*f*), BASTARD- : bastard file 443
FEILE (*f*), FLACHSPITZE- : taper-flat file 444
FEILE (*f*), HALBRUND- : half round file 445
FEILEN : to file, polish, rasp 446
FEILENHAUER (*m*) : file cutter 447
FEILENHAUMASCHINE (*f*) : file cutting machine 448
FEILENHEFT (*n*) : file handle 449
FEILENHIEB (*m*) : file cut 450
FEILENZAHN (*m*) : file tooth 451
FEILE (*f*), VOR- : bastard file 452
FEILHEIT (*f*) : venality 453
FEILICHT (*n*) : filings 454
FEILKLOBEN (*m*) : filing vice 455
FEILSCHEN : to bargain ; haggle 456
FEILSEL (*n*) : (see FEILICHT) 457
FEILSPÄNE (*m.pl*) : (see FEILICHT) 458
FEILSTAUB (*m*) : (see FEILICHT) 459
FEILSTRICH (*m*) : stroke of a file 460
FEIN : fine-(ly) ; polite ; nice ; refined ; delicate ; acute ; small ; minute ; miniature 461
FEINBLECH (*n*) : sheet metal 462
FEINBRENNEN : to refine (Metals) 463
FEINBRENNER (*m*) : refiner 464
FEIND : hostile, enemy, inimical ; (*m*) enemy, foe 465
FEINDLICH : hostile, inimical, enemy 466
FEINDLICHKEIT (*f*) : hostility 467
FEINDSCHAFT (*f*) : enmity, hatred, hostility 468
FEINDSCHAFTLICH : hostile 469
FEINDSELIG : hostile, hateful 470
FEINE (*f*) : fineness ; delicacy ; refinement ; purity 471
FEINEISEN (*n*) : refined iron 472
FEINEN : to refine 473
FEINERN : to refine 474
FEINFASERIG : fine-fibred 475
FEINFEUER (*n*) : refinery ; refining furnace 476
FEINFILTER (*n*) : fine filter (see also ULTRAFILTER) 477
FEINGEFÜHL (*n*) : delicacy 478
FEINGEHALT (*m*) : fineness (Metals) ; standard (Coinage) 479
FEINGEHALTSTEMPEL (*m*) : hall-mark 480
FEINGEMAHLEN : finely ground, triturated 481
FEINGEPULVERT : finely (powdered) crushed 482
FEINGESPITZT : sharp, pointed 483
FEINGOLD (*n*) : pure (refined) gold (24-carat gold) 484
FEINGUSS (*m*) : first (A1, best) quality casting 485
FEINHEIT (*f*) : fineness ; purity ; elegance ; delicacy 486
FEINHEITSGRAD (m) : degree of fineness 487
FEINJÄHRIG : narrow-ringed (of timber) 488
FEINKIES (*m*) : fine ore ; fines 489
FEINKÖRNIG : fine-grained ; finely granular 490
FEINKRYSTALLINISCH : fine-crystalled ; finely crystalline 491
FEINLOCHIG : small (fine) holed ; fine pored ; finely porous 492
FEINMACHEN : to refine 493
FEINMACHER (*m*) : refiner 494
FEINMAHLEN : to grind (to powder; to dust ; small) fine ; triturate 495
FEINMAHLMASCHINE (*f*) : fine grinder 496
FEINMALER (*m*) : miniature painter 497
FEINMASCHIG : fine-meshed 498
FEINMECHANIK (*f*) : fine mechanics 499
FEINOFEN (*m*) : refining furnace 500
FEINPASSUNG (*f*) : wringing fit, tight fit 501
FEINPLATIN (*n*) : refined platinum 502
FEINPORIG : fine-pored ; finely pored 503
FEINPROZESS (*m*) : refining process 504
FEINPULVERISIERT : finely powdered 505
FEINSCHLACKE (*f*) : refinery slag 506
FEINSCHMECKER (*m*) : epicure 507
FEINSCHROT (*n* and *m*) : fine groats ; finely crushed malt 508
FEINSEIFE (*f*) : fancy (toilet) soap (75-82% fatty acid content) 509
FEINSICHTIG : sharp-sighted 510
FEINSILBER (*n*) : fine (refined) silver 511
FEINSINN (*m*) : delicacy 512
FEINSODA (*f*) : finely pulverised soda 513
FEINSPRIT (*m*) : (*Spiritus rectificatissimus*) : rectified alcohol 514
FEINSTEIN (*m*) : refined garnierite for gaining copper (see GARNIERIT) 515
FEINSTEINZEUG (*n*) : first quality stoneware or hardware 516
FEINTALG (*m*) : refined tallow ; Sg. 0.943-0.95 ; Mp. 31-50°C. 517
FEINVERTEILT : finely (divided) distributed 518
FEINZELLIG : fine-celled ; finely cellular 519
FEINZERTEILT : finely (distributed) divided 520
FEINZEUG (*n*) : pulp (Paper) 521
FEINZEUGHOLLÄNDER (*m*) : beater ; beating (engine) machine ; hollander (Paper) 522
FEINZINK (*n*) : refined zinc (99.8-99.9% zinc) 523
FEINZINN (*n*) : grain tin 524

FEINZUCKER (*m*): refined sugar 525
FELBEL (*n*): velveteen 526
FELD (*n*): field; land; ground; pane; panel; compartment; sphere; department 527
FELDAPOTHEKE (*f*): field dispensary 528
FELDARBEITER (*m*): farm (field) labourer or hand; agricultural labourer 529
FELDARZT (*m*): Army doctor 530
FELDBAHN (*f*): portable (field; light; small gauge; narrow gauge); railway or tramway 531
FELDBAU (*m*): farming; agriculture; husbandry; tillage 532
FELDBLUME (*f*): field (wild) flower 533
FELDBRAND (*m*): clamp burning (Ceramics) 534
FELDCHIRURGUS (*m*): Army surgeon 535
FELDDIENST (*m*): active sevice; service in the field (Army) 536
FELD (*n*), ELEKTRISCHES-: electric field 537
FELD (*n*), ELEKTROMAGNETISCHES-: electromagnetic field 538
FELDERBAU (*m*): panel work (Mining) 539
FELDFLASCHE (*f*): water-bottle (Army) 540
FELDFRUCHT (*f*): farm (product) produce 541
FELDGEHÄGE (*n*): covert; warren 542
FELDGERÄT (*n*): agricultural (farming) implements; field equipment 543
FELDGESCHIRR (*n*): (see FELDGERÄT) 544
FELDGRAU (*n*): field-grey; greenish grey 545
FELDHOSPITAL (*n*): field hospital 546
FELDHUHN (*n*): partridge 547
FELDHUNDSKAMILLE (*f*): dog's fennel (*Anthemis cotula*) 548
FELD, INS- FÜHREN: to give publicity to 549
FELDKÜMMEL (*m*): wild carraway; wild thyme (see FELDTHYMIAN) 550
FELDKÜMMELÖL (*n*): wild caraway oil (from fruit of *Carum carvi*) 551
FELDLAGER (*n*): camp 552
FELDMAGNET (*m*): field magnet (Elect.) 553
FELDMESSEN (*n*): land surveying 554
FELDMESSKUNST (*f*): surveying 555
FELDMOHN (*m*): field poppy, common poppy (see MOHN) 556
FELDOFEN (*m*): clamp kiln (Ceramics); field oven (Army) 557
FELDRAUCH (*m*): fumitory (*Fumaria officinalis*) 558
FELDRAUTE (*f*): (see FELDRAUCH) 559
FELDRÜSTER (*f*): (see ULME) 560
FELDSCHMIEDE (*f*): travelling (portable, field) forge 561
FELDSCHWAMM (*m*): mushroom 562
FELDSENF (*m*): wild mustard, charlock (*Sinapis arvensis*) (see also HEDERICH) 563
FELDSPAT (*m*): feldspar (see BARYUMFELDSPAT; KALIFELDSPAT; ORTHOKLAS; PLAGIOKLAS) 564
FELDSPATARTIG: feldspathic 565
FELDSPATGESTEIN (*n*): feldspathic rock 566
FELDSPATHALTIG: feldspathic; containing feldspar 567
FELDSPATIG: (see FELDSPATHALTIG) 568
FELDSPATPORZELLAN (*n*): feldspar porcelain (by heating china clay with feldspar as flux, plus chalk, gypsum and quartz) (up to 60% clay, 40% quartz and 30% feldspar) 569
FELDSPULE (*f*): field coil (Elect.) 570
FELDSTÄRKE (*f*): field (density) strength (Elect.) 571
FELDSTECHER (*m*): field glass 572
FELDSTEIN (*m*): compact feldspar; boulder; landmark 573
FELDTELEGRAPH (*m*): field telegraph 574
FELDTHYMIAN (*m*): wild thyme (*Thymus serpyllum*) (see also THYMIAN) 575
FELDTHYMIANÖL (*n*): wild thyme oil (from *Thymus serpyllum*); Sg. 0.89-0.92 576
FELDULME (*f*): (see ULME) 577
FELDWEG (*m*): field path 578
FELDWICKLUNG (*f*): field winding 579
FELDWINDE (*f*): wild convolvulus, field bindweed (*Convolvulus arvensis*) 580
FELDWIRTSCHAFT (*f*): agriculture 581
FELDZEUG (*n*): ordnance; munitions 582
FELDZIEGEL (*m*): clamp brick (brick burnt in a clamp kiln) 583
FELDZUG (*m*): campaign (Army) 584
FELGE (*f*): fallow; felly; felloe (rim of a wheel) 585
FELGEN: to fit rims to wheels 586
FELGENBREMSE (*f*): rim brake 587
FELGENHAUER (*m*): wheelwright 588
FELGENKEIL (*m*): rim wedge 589
FELGE (*f*), RAD-: rim (of a wheel) 590
FELGE (*f*), STAHL-: steel rim 591
FELL (*n*): skin; hide; pelt; fur 592
FELLHÄNDLER (*m*): dealer in hides and skins 593
FELLLEIM (*m*): hide glue 594
FELLSCHMITZER (*m*): dyer (of skins) 595
FELLSPÄNE (*m.pl*): hide parings 596
FELLZURICHTEREI (*f*): skin (hide) dressing works 597
FELPEL (*m*): (see FELBEL) 598
FELS (*m*): rock-(s); cliff 599
FELSABHANG (*m*): precipice 600
FELSALAUN (*m*): rock alum; alumina; Al_2O_3; Sg. 3.73; Mp. 2020°C. 601
FELSART (*f*): kind of rock 602
FELSARTIG: rocky; rock-like 603
FELSBLOCK (*m*): boulder (of rock) 604
FELSBRUCH (*m*): quarry 605
FELSEN (*m*): rock-(s); cliff 606
FELSENABHANG (*m*): (see FELSABHANG) 607
FELSENADER (*f*): rock (vein) seam; dike 608
FELSENARTIG: rock-like, rocky 609
FELSENFEST: firm as a rock 610
FELSENGEBIRGE (*n*): rocky mountains 611
FELSENGERÖLLE (*n*): broken rocks 612
FELSENGRAS (*n*): Iceland moss, Cetraria (*Cetraria islandica*) 613
FELSENHART: rocky; hard as rock 614
FELSENKLIPPE (*f*): cliff 615
FELSENKLUFT (*f*): chasm (between rocks) 616
FELSENRIFF (*n*): reef of rocks 617
FELSENRITZE (*f*): crevice (in rocks) 618
FELSENRÜCKEN (*m*): ridge (of rock) 619
FELSENSCHICHT (*f*): layer (stratum) of rock 620
FELSENSCHICHTUNG (*f*): stratification of rock 621
FELSENSPITZE (*f*): peak (crag) of a rock 622
FELSENWAND (*f*): wall of rock, precipice, face of a precipice 623
FELSGLIMMER (*m*): mica (as Muscovite; Biotite; Phlogopite; Lepidolite) 624
FELSIG: rocky 625
FELSÖL (*n*): petroleum; mineral oil; rock oil; Sg. 0.78-0.97 626
FENCHEL (*m*): fennel (*Fœniculum capillaceum* and *Fœniculum officinale*) 627
FENCHEL (*m*), GEMEINER-: fennel (*Fœniculum officinale*) 628
FENCHELHOLZ (*n*): sassafras wood (*Sassafras officinale*) 629

FENCHELHONIG (*m*) : fennel honey 630
FENCHELÖL (*n*) : fennel oil (*Oleum fœniculi*) (from fruit of *Fœniculum capillaceum*) ; Sg. 0.965-0.975 ; Bp. 160-220°C. 631
FENCHELWASSER (*n*) : fennel water 632
FENCHELWURZEL (*f*) : (*Radix fœniculi*) : fennel root 633
FENDER (*m*) : fender (Naut.) 634
FENSTER (*n*) : window 635
FENSTERFLÜGEL (*m*) : casement 636
FENSTERGITTER (*n*) : window grating ; lattice 637
FENSTERGLAS (*n*) : window glass 638
FENSTERGLIMMER (*m*) : muscovite (see **MUSKOVIT**) 639
FENSTERLADEN (*m*) : shutter (of windows) 640
FENSTERRAHMEN (*m*) : window frame 641
FENSTERRIEGEL (*m*) : sashbolt 642
FENSTERSCHEIBE (*f*) : windowpane 643
FERCH (*m*) : mine damp 644
FERGUSONIT (*m*) : Fergusonite (natural tantalate and columbate of Yttrium, with Uranium, etc.) ; Y(Nb,Ta)O_4 ; Sg. 5.8 645
FERIEN (*f.pl*) : vacation ; holidays 646
FERMENT (*n*) : ferment (the microbes which cause fermentation) ; enzyme (the active principles of a ferment) 647
FERMENTATION (*f*) : fermentation (see also under **GÄREN**) 648
FERMENTDIAGNOSTIKUM (*n*) : ferment diagnostic 649
FERMENTIEREN : to ferment 650
FERMENTINTOXIKATION (*f*) : an auto-intoxication due to ferments in the blood 651
FERMENTWIRKUNG (*f*) : fermentation 652
FERN : distant ; far ; remote ; (as a prefix= distance) 653
FERNAMBUKHOLZ (*n*) : Brazil wood (*Cœsalpinia brasiliensis* or *C. bijuga*) 654
FERNGLAS (*n*) : telescope 655
FERNHÖRER (*m*) : (telephone)-receiver 656
FERNKRAFTWERK (*n*) : power station at a distance ; distant power station 657
FERNLEITUNG (*f*) : main (steam or electric, to or from a distance) ; distance (main ; lead ; line) conduit 658
FERNOBJEKTIV (*n*) : telephotographic lens (Photography) 659
FERNROHR (*n*) : telescope 660
FERNRUF (*m*) : telephone 661
FERNSCHALTER (*m*) : distance switch 662
FERNSCHREIBER (*m*) : telegraph ; distance recorder 663
FERNSEHEN (*n*), **ELEKTRISCHES-** : electric (telephotoscopy) telescopy, (photographing or seeing at a distance) 664
FERNSPRECHAMT (*n*) : telephone exchange 665
FERNSPRECHER (*m*) : telephone ; 'phone 666
FERNSPRECHSTELLE (*f*) : call station, call box ('phone) 667
FERNTHERMOMETER (*n*) : distance thermometer 668
FERNWIRKUNG (*f*) : carrying distance (Wireless Tel.) 669
FERNZEICHNUNG (*f*) : perspective drawing 670
FEROLEHOLZ (*n*) : Satinwood (see **ATLASHOLZ**) 671
FERRIACETAT (*n*) : ferric acetate ; $Fe(C_2H_3O_2)_3.2H_2O$ 672
FERRIAMMONIUMCITRAT (*n*) : ferric ammonium citrate 673
FERRIBROMID (*n*) : ferric (iron) bromide ; ferric (tribromide) sesquibromide ; $FeBr_3$ 674
FERRICHLORID (*n*) : ferric chloride ; perchloride of iron ; $FeCl_3$; Sg. 2.804 ; Mp. 301°C. (see **EISENCHLORID**) 675
FERRICHLORWASSERSTOFF (*m*) : ferrichloric acid 676
FERRICHLORWASSERSTOFFSÄURE (*f*) : ferrichloric acid 677
FERRICHROMAT (*n*) : iron (ferric) chromate (see **EISENCHROMAT**) 678
FERRICYAN (*n*) : ferricyanogen 679
FERRICYANCALCIUM (*n*) : (see **CALCIUMFERRI-CYANID**) 680
FERRICYANEISEN (*n*) : ferrous ferricyanide ; Turnbull's blue ; $Fe_3[Fe_4(CN)_6]_2$ 681
FERRICYANID (*n*) : ferricyanide ; ferric cyanide (see **EISENCYANID**) 682
FERRICYANKALIUM (*n*) : potassium ferricyanide ; red potassium prussiate ; $K_3Fe(CN)_6$; Sg. 1.8109 683
FERRICYANNATRIUM (*n*) : sodium ferricyanide ; red sodium prussiate (*Natrium ferricyanatum*) $Na_3Fe(CN)_6.H_2O$ 684
FERRICYANVERBINDUNG (*f*) : ferricyanide ; ferric cyanide 685
FERRICYANWASSERSTOFF (*m*) : ferricyanic acid ; hydrogen ferricyanide ; $H_3Fe(CN)_6$ 686
FERRICYANWASSERSTOFFSÄURE (*f*) : (see **FERRI-CYANWASSERSTOFF**) 687
FERRIDOXALAT (*n*) : ferric oxalate (see **EISENOXYD, OXALSAURES-**) 688
FERRIEISENCYANÜR (*n*) : ferric ferro-iron cyanide ; prussian blue ; $Fe_4[Fe(CN)_6]_3$ 689
FERRIFERROCYANID (*n*) : (see **FERRIEISEN-CYANÜR**) 690
FERRIFERROJODID (*n*) : ferroso-ferric iodide 691
FERRIFERROOXYD (*n*) : ferroso (magnetic iron) oxide ; black ferric oxide ; iron ethiops ; $FeO.Fe_2O_3$ 692
FERRIHYDROXYD (*n*) : ferric hydroxide, rust, ferric hydrate ; ferric sesquioxide, hydrated ; hydrous iron peroxide, hydrated iron oxide, iron hydrate ; $Fe(OH)_3$; Sg. 3.4–3.9 693
FERRIION (*n*) : ferric ion 694
FERRIJODID (*n*) : ferric iodide 695
FERRIKALIUMSULFAT (*n*) : ferric potassium sulphate (see **EISENKALIUMALAUN**) 696
FERRILAKTAT (*n*) : ferric lactate 697
FERRINITRAT (*n*) : ferric nitrate (*Ferrum nitricum*) (used as a mordant) ; $Fe(NO_3)_3.9H_2O$; Sg. 1.6835 ; Mp. 47.2°C. 698
FERRIOXALAT (*n*) : ferric oxalate 699
FERRIOXYD (*n*) : ferric oxide ; Fe_2O_3 ; (see **EISENOXYD**) 700
FERRIPEPTONAT (*n*) : (see **EISENPEPTONAT**) 701
FERRIPHOSPHAT (*n*) : iron (ferric) phosphate ; $FePO_4.4H_2O$ 702
FERRIPYRIN (*n*) : (see **FERROPYRIN**) 703
FERRIRHODANID (*n*) : iron (ferric) thiocyanate ; iron (ferric) sulfocyanate 704
FERRISALZ (*n*) : ferric salt 705
FERRISUBLAKTAT (*n*) : ferric sublactate 706
FERRISULFAT (*n*) : ferric sulphate ; $Fe_2(SO_4)_3$; Sg. 3.097 ; (see **EISENOXYDSULFAT**) 707
FERRISULFID (*n*) : ferric sulphide ; FeS_2 ; (see **EISENDISULFID**) 708
FERRIT (*n*) : ferrite (a soft, pure iron) 709
FERRIZYAN (*n*) : ferricyanogen ; (see **FERRICYAN**) 710
FERROACETAT (*n*) : ferrous acetate ; (see **EISEN-OXYDULACETAT**) 711
FERROALUMINIUM (*n*) : (10–20% Al ; no C) ; ferroaluminium (an iron-aluminium alloy) 712

FERROBOR (*n*) : ferroboron (an iron-boron alloy) 713

FERROBROMID (*n*) : ferrous (iron) bromide; $FeBr_2.6H_2O$; Sg. 4.636 ; Mp. 27°C. 714

FERROCHLORID (*n*) : ferrous (iron) chloride; $FeCl_2$; Sg. 2.988 ; (see EISENCHLORÜR) 715

FERROCHROM (*n*) : ferrochrome ; ferro chromium ; (iron and chromium alloy) ; (25–75% Cr. ; 2–8% Carbon) 716

FERROCYAN (*n*) : ferrocyanogen 717

FERROCYANBLAU (*n*) : (see FERRIFERROCYANID) 718

FERROCYANCALCIUM (*n*) : (see CALCIUMFERROCYANID) 719

FERROCYANEISEN (*n*) : ferric ferrocyanide ; prussian blue ; (see FERRIEISENCYANÜR) 720

FERROCYANID (*n*) : ferrous (ferro) cyanide ; (see EISENCYANÜR) 721

FERROCYANKALIUM (*n*) : potassium ferrocyanide ; yellow prussiate of potash ; $K_4Fe(CN)_6.3H_2O$; Sg. 1.89 722

FERROCYANNATRIUM (*n*) : sodium ferrocyanide ; yellow prussiate of soda ; $Na_4Fe(CN)_6.12H_2O$; Sg. 1.458 723

FERROCYANVERBINDUNG (*f*) : ferrocyanide 724

FERROCYANWASSERSTOFF (*m*) : ferro cyanic acid ; hydrogen ferrocyanide ; $H_4Fe(CN)_6$ 725

FERROCYANWASSERSTOFFSÄURE (*f*) : (see FERRO-CYANWASSERSTOFF) 726

FERROCYANZINK (*n*) : zinc ferrocyanide ; $Zn_2Fe(CN)_6.3H_2O$ 727

FERROFERRICYANID (*n*) : (see FERRIFERROCYANID) 728

FERROFERRIOXYD (*n*) : (see FERRIFERROOXYD) 729

FERROHYDROXYD (*n*) : ferrous hydroxide 730

FERROION (*n*) : ferrous ion 731

FERROJODID (*n*) : ferrous iodide ; $FeI_2.4H_2O$; Sg. 2.873 ; Mp. 177°C. 732

FERROKALIUMSULFAT (*n*) : ferrous potassium sulphate 733

FERROKARBONAT (*n*) : ferrous carbonate ; $FeCO_3$; Sg. 3.7–3.87 ; Mp. 470°C. 734

FERROLAKTAT (*n*) : ferrous lactate ; (see EISEN-LAKTAT) 735

FERROLEGIERUNG (*f*) : ferro (iron) alloy 736

FERROMAGNESIUM (*n*) : ferromagnesium (an iron-magnesium alloy) 737

FERROMANGAN (*n*) : ferro-manganese ; Spiegeleisen (an iron-manganese alloy ; 20–80% Mn ; 5–7% C.) 738

FERROMANGANSILIZIUM (*n*) : (see SILIKOSPIEGEL) 739

FERROMOLYBDÄN (*n*) : ferro-molybdenum (normal 52%Mo ; 1.8%C. ; refined 0.3%C. only ; high percent. 75–85%Mo ; 3–6%C.) ; (a molybdenum-iron alloy) 740

FERRONICKEL (*n*) : ferro-nickel (50% Ni ; 0.6% C.) ; (An iron-nickel alloy) 741

FERRONITRAT (*n*) : ferrous nitrate 742

FERROOXALAT (*n*) : ferrous oxalate ; (see EISEN-OXALAT) 743

FERROOXYD (*n*) : ferrous oxide ; FeO 744

FERROPHOSPHAT (*n*) : ferrous phosphate 745

FERROPHOSPHOR (*n*) : ferro-phosphorous ; (15-25% P.) ; (alloy of iron and phosphorous) 746

FERROPYRIN (*n*) : ferropyrine, ferripyrine, iron (ferric) chloride-antipyrine ; $(C_{11}H_{12}N_2O)_3.Fe_2Cl_6$; Mp. 220–225°C. 747

FERROSALZ (*n*) : ferrous salt 748

FERROSILIZIUM (*n*) : ferro-silicon ; (50% Si ; 49.3% Fe ; 0.3% C ; 0.35% Mn ; under 0.04% P & S) ; (an iron-silicon alloy) 749

FERROSTYPTIN (*n*) : ferrostyptin ; $(CH_2)_6N_4.HCl.FeCl_3$; Mp. 111°C. 750

FERROSULFAT (*n*) : ferrous sulphate ; $FeSO_4.7H_2O$; Sg. 1.8987 ; Mp. 64°C. ; (see EISENOXYDULSULFAT) 751

FERROSULFID (*n*) : ferrous sulphide ; FeS ; Sg. 4.75–5.4 ; Mp. 1179°C. ; (see EISEN-SULFÜR) 752

FERROTITAN (*n*) : ferro-titanium ; (32%Ti ; 3.2%C ; sometimes 56%Ti ; 0.5%C.) ; (alloy of iron and titanium) 753

FERROTYPIE (*f*) : ferrotype (wet-process positive on iron plate which is coated with a black or brown varnish) (Photography) 754

FERROTYPLACK (*m*) : ferrotype varnish, crystal varnish, benzol varnish (Photography) 755

FERROVANADIN (*n*) : ferrovanadium ; (see FERRO-VANADIUM) 756

FERROVANADIUM (*n*) : ferro-vanadium ; (30–50%V ; 1–3%C) ; (an iron-vanadium alloy) 757

FERROWOLFRAM (*n*) : ferro-tungsten ; [65–84% W ; 0.4–3.3%C(carbon content fluctuates)] ; (an iron-tungsten alloy) 758

FERROZIRKON (*n*) : ferrozirconium (an iron-zirconium alloy) 759

FERROZYAN (*n*) : (see FERROCYAN) 760

FERTIG : ready ; finished ; complete ; done ; skilful ; insolvent ; prepared ; practised ; accomplished ; fluent ; dexterous 761

FERTIGEN : to prepare ; make ; do ; dispatch ; finish 762

FERTIGER (*m*) : finisher, maker, consigner 763

FERTIGGUSS (*m*) : finished casting 764

FERTIGKALIBER (*n*) : finishing pass (Rolling Mills) 765

FERTIGMACHEN : to finish ; adjust ; (*n*) finishing ; adjustment 766

FERTIGMACHER (*m*) : finisher ; foreman 767

FERTIGROLLGANG (*m*) : finishing (rolls) train 768

FERTIGUNG (*f*) : finish-(ing); performing, making 769

FERTIGWALZE (*f*) : finishing roll 770

FERULASÄURE (*f*) : ferulic acid 771

FESSEL (*f*) : fetters ; chain ; shackles ; handcuffs 772

FESSELBEIN (*n*) : pastern 773

FESSELN : to fetter ; chain ; shackle ; handcuff ; fasten ; attach ; captivate 774

FEST : solid ; firm ; proof ; steadfast ; fast ; compact ; strong ; fixed ; steady ; tight ; close ; impervious to ; immovable ; (*n*) festival ; feast ; holiday 775

FESTBRENNEN DES LAGERS (*n*) : seizing of the bearing 776

FESTE (*f*) : fortress ; firmament ; solidity ; density ; firmness 777

FESTER KÖRPER (*m*) : solid, solid body, solid substance 778

FESTER SPIRITUS (*m*) : solid spirit 779

FESTES LAND (*n*) : terra firma ; continent ; (see FESTLAND) 780

FESTFRESSEN : to seize ; jamb ; stick 781

FESTHALTEN : to hold fast ; retain ; fix ; adhere ; cling ; be tenacious ; be retentive ; keep in position 782

FESTHALTEN (*n*) : retention, etc. 783

FESTIGEN : to consolidate ; make (fast) firm ; establish ; confirm 784

FESTIGKEIT (*f*): solidity; tenacity (Metal); strength (Mech.); stress; durability; resistance to stress 785
FESTIGKEIT (*f*), **ABSOLUTE-**: ultimate tensile strength, absolute tensile strength (the tractive force necessary to cause tearing of a body when extended above the elastic limit) 786
FESTIGKEIT (*f*), **RÜCKWIRKENDE-**: limit of resistance to compression; resistance to crushing strain or stress 787
FESTIGKEITSEIGENSCHAFT (*f*): stress property, tensile property (Metal) 788
FESTIGKEITSMODUL (*m*): breaking modulus 789
FESTIGKEITSPRÜFER (*m*): testing machine (Paper) 790
FESTKLEBEN: to stick fast; adhere 791
FESTLAND (*n*): main land; (see **FESTES LAND**) 792
FESTLEGEN: to fix; place; determine; bed; moor 793
FESTMACHEN: to make fast; fasten; settle; fix; attach; affix; furl; hand (Sails); moor 794
FESTMETER (*n*): theoretical cubic meter (of wood) 795
FESTSCHEIBE (*f*): fast pulley 796
FESTSCHRIFT (*f*): publication for a festival or anniversary 797
FESTSETZEN: to establish; settle; fix; determine; stipulate 798
FESTSITZ (*m*): tight fit 799
FESTSTEHEN: to stand fast; be settled or fixed; remain (be) stationary; be stable; be established 800
FESTSTEHENDE DAMPFMASCHINE (*f*): stationary (fixed) engine 801
FESTSTELLEN: to establish; ascertain; determine; confirm; settle; fix 802
FESTUNG (*f*): fortification, fortress 803
FESTUNGSACHAT (*m*): fortification agate 804
FESTWEICH: semi-solid 805
FETT: fatty; oily; rich; fertile; profitable; strong; adipose; greasy; lucrative 806
FETT (*n*): fat (Sg. 0.94); grease, oil; see also **TALG**): adipose tissue (Med.); dressing 807
FETTABSCHEIDER (*m*): grease separator; grease extractor; de-oiler 808
FETTALKOHOL (*m*): fat alcohol 809
FETTARM: having a low fat content; poor in fat 810
FETTAROMATISCH: aliphatic-aromatic 811
FETTAUFFANGGEFÄSS (*n*): fat catching vessel; grease (trap) catcher 812
FETTAUGE (*f*): exophthalmia, protrusion of the eyeballs in exophthalmic goitre 813
FETTAUSSCHEIDER (*m*): grease extractor, de-oiler 814
FETTDICHTES PAPIER (*n*): grease-proof paper 815
FETTDRÜSE (*f*): sebaceous gland 816
FETTE (*f*): fatness; greasiness; richness 817
FETTEN: to oil; grease; paste; lubricate; fatten; stuff (Leather) 818
FETTES PUDDELN (*n*): pig-boiling (Iron) 819
FETTFABRIKATIONSMASCHINE (*f*): fat manufacturing plant 820
FETTFÄNGER (*m*): grease (trap) catcher 821
FETTFARBE (*f*): fat-soluble (colour) dye, oil-soluble dye 822
FETTFLECK (*m*): grease spot 823
FETTGAR: oil-tanned (Leather) 824
FETTGAS (*n*): oil gas 825
FETTGEHALT (*m*): fat content 826
FETTGEWEBE (*n*): adipose tissue 827
FETTGEWINNUNG (*f*): fat extraction 828
FETTGLANZ (*m*): greasy lustre 829
FETTHALTIG: containing fat; fatty 830
FETTHÄRTUNG (*f*): fat hardening, hardening of liquid fats 831
FETTHÄRTUNGSANLAGE (*f*): fat hardening plant 832
FETTHARZ (*n*): oleoresin (mixture of resin and essential oils of plants from which they exude) 833
FETTHAUT (*f*): adipose tissue 834
FETTHEFE (*f*): fatty yeast (containing about 17% fat) 835
FETTHEIT (*f*): fatness; greasiness; richness 836
FETTIG: fatty; fat; greasy; oily; sebaceous; adipose 837
FETTIGE INFILTRATION (*f*) **DES HERZENS**: fatty accumulation on the heart (Med.) 838
FETTIGE METAMORPHOSE (*f*) **DES HERZENS**: fatty degeneration of the heart (Med.) 839
FETTKALK (*m*): fat lime (see **WEISSKALK**) 840
FETTKOHLE (*f*): rich coal [(will coke) ash 10-14%; moisture 3-5%; calorific value 5500]; fat coal (rich in volatile matter) 841
FETTKÖRPER (*m*): fatty (greasy) substance, matter or compound; aliphatic compound 842
FETTKRAM (*m*): chandlery 843
FETTKRAUT (*n*): butterwort (*Pinguicula vulgaris*) (a carnivorous plant) 844
FETTKÜGELCHEN: fat globule 845
FETTLEIBIG: obese 846
FETTLEIBIGKEIT (*f*): corpulence, obesity, polysarcia (Med.) 847
FETTLÖSLICH: fat-soluble; soluble in fat 848
FETTNOPPE (*f*): grease-spot (in fabrics) 849
FETTPRÄPARAT (*n*): fat preparation; fatty preparation; grease preparation 850
FETTPUDDELN (*n*): pig-boiling (Iron) 851
FETTQUARZ (*m*): greasy quartz 852
FETTREICH: rich in fat, having a high fat content 853
FETTREIHE (*f*): fatty series; aliphatic series 854
FETTRESERVEMITTEL (*n*): fat (wax) resist 855
FETTSAUER: sebacic, sebate of; combined with sebacic acid 856
FETTSÄURE (*f*): sebacic (fatty) acid (see **STEARIN** and **ELAINSÄURE**) 857
FETTSÄUREDESTILLATIONSANLAGE (*f*): sebacic acid distilling plant 858
FETTSAURES SALZ (*n*): sebate, sebacic salt 859
FETTSCHMIERBÜCHSE (*f*): grease box 860
FETTSCHMIERE (*f*): fat liquor; stuffing (Leather); fat; lubricant (Mech.) 861
FETTSCHWEISS (*m*): fatty sweat; yolk (of wool) 862
FETTSEIFE (*f*): soap from fats 863
FETTSEIN (*n*): ropiness (of wine) 864
FETTSPALTEND: fat-cleaving; lipolytic 865
FETTSPALTER (*m*): fat cleavage agent; lipolytic 866
FETTSPALTUNG (*f*): fat cleavage; lipolysis 867
FETTSPALTUNGSANLAGE (*f*): fat cleavage plant 868
FETTSPALTUNGSMITTEL (*n*): fat cleavage agent; lipolytic 869
FETTSTEIN (*m*): eleolite (see **ELEOLITH**) 870
FETTSTIFT (*m*): wax (grease) pencil or crayon 871
FETTSUCHT (*f*): corpulence, obesity, polysarcia (Med.) 872

FETTTON (m): Fuller's earth 873
FETTUNG (f): oiling, etc. (see FETTEN) 874
FETTVERBINDUNG (f): fatty (aliphatic) compound 875
FETTWACHS (n): adipocere 876
FETTWACHS (m): adipocere (Med.) 877
FETTWOLLE (f): grease wool; yolk wool 878
FETTZELLE (f): fat (adipose) cell, sebaceous cell 879
FETZEN (m): shread; tatter; scrap; trifle; rag; shoddy 880
FEUCHT: moist; humid; damp; wet; slack 881
FEUCHTEN: to wet, damp, moisten, humidify 882
FEUCHTER (m): moistener; damper 883
FEUCHTHEIT (f): dampness; moisture; humidity; wetness 884
FEUCHTIGKEIT (f): moisture. etc. (see FEUCHT-HEIT); humor; phlegm 885
FEUCHTIGKEITSGEHALT (m): moisture content 886
FEUCHTIGKEITSMESSER (m): hygrometer (instrument for measuring moisture content) 887
FEUCHTIGKEITSMESSINSTRUMENT (n): hygrometer, instrument for measuring moisture content 888
FEUCHTIGKEITSZEIGER (m): hygroscope (instrument for indication of moisture content) 889
FEUER (n): fire; ardour; spirit 890
FEUERANSTALT (f): fire-regulations 891
FEUERARTIG: igneous 892
FEUERASSEKURANZ (f): fire-insurance-(office) 893
FEUERBALL (m): fireball, ball of fire 894
FEUERBERG (m): volcano 895
FEUERBESTÄNDIG: fire-resisting; fire-proof; refractory; heat-resisting 896
FEUERBESTATTUNG (f): cremation 897
FEUERBLENDE (f): pyrostilpnite; Ag_3SbS_3; Sg. 4.2 (see PYROSTILPNIT) 898
FEUERBOCK (m): andiron 899
FEUERBOHNE (f): scarlet runner-(bean) (*Phaseolus multiflorus*) 900
FEUERBRAND (m): firebrand 901
FEUERBRÜCKE (f): fire-bridge; bridge-wall (Boilers) 902
FEUERBÜCHSE (f): firebox, furnace, combustion chamber 903
FEUERBÜCHSENDECKE (f): crown plate, crown of furnace, firebox or combustion chamber 904
FEUEREIFER (m): zeal, ardour 905
FEUEREIMER (m): firebucket 906
FEUERFANGEND: inflammable 907
FEUERFARBE (f): the colour of fire, fiery red 908
FEUERFEST: fireproof; refractory (from Seger cone Nos. 26-30) (see SEGERKEGEL) 909
FEUERFESTER KITT (m): fireproof cement; fireclay 910
FEUERFESTER MÖRTEL (m): fireproof mortar, fireclay 911
FEUERFESTER STEIN (m): firebrick (Sg. 2.2-2.7) 912
FEUERFESTER TON (m): fireclay 913
FEUERFESTER ZIEGEL (m): firebrick; firetile; Sg. 2.2-2.7 914
FEUERFESTES PRODUKT (n): refractory product 915
FEUERFÜHRUNG (f): firing; furnace control 916
FEUERGAS (n): furnace gas 917
FEUERGATTER (n): fireguard 918
FEUERGEFÄHRLICH: inflammable: combustible 919
FEUERGEWÖLBE (n): fire (furnace) arch, coking arch 920
FEUERGLOCKE (f): fire (alarm) bell 921
FEUERGRADMESSER (m): pyrometer 922
FEUERHAKEN (m): poker; rake 923
FEUERHERD (m): hearth, fireplace 924
FEUERHOLZ (n): firewood 925
FEUERIG: fiery; burning; hot; igneous; ardent; spirited; generous (Wine); bright; loud (Colours) 926
FEUERKASSE (f): fire-insurance office 927
FEUERKITT (m): fireproof cement; fireclay 928
FEUERKRAUT (n): scarlet lichen 929
FEUERKRÜCKE (f): rake 930
FEUERKUGEL (f): fireball 931
FEUERKUNST (f): pyrotechnics 932
FEUERLACK (m): fireproof varnish 933
FEUERLÄRM (m): fire alarm 934
FEUERLEITUNG (f): priming; train; fuse 935
FEUERLOS: fireless; lustreless (of gems); without fire; without spirit 936
FEUERLÖSCHAPPARAT (m): fire-extinguishing apparatus, fire extinguisher 937
FEUERLÖSCHEINRICHTUNG (f): (see FEUERLÖ-SCHAPPARAT) 938
FEUERLÖSCHMITTEL (n): fire-extinguishing agent or medium; fire extinguisher 939
FEUERLUFT (f): furnace gas (Scheele's name for Oxygen); air for combustion 940
FEUERMANN (m): fireman, stoker 941
FEUERMATERIAL (n): fuel 942
FEUERMELDEANLAGE (f): fire alarm-(installation) 943
FEUERMELDER (m): fire alarm 944
FEUERMESSER (m): pyrometer 945
FEUERMESSUNG (f): pyrometry 946
FEUERN: to fire; heat; burn; kindle; stoke; animate; spark; rouse 947
FEUEROFEN (m): stove; furnace 948
FEUEROPAL (m): fire opal (see OPAL) 949
FEUERPORZELLAN (n): refractory (fireproof) porcelain 950
FEUERPROBE (f): fire-test 951
FEUERPUNKT (m): focus (Optics); hearth (Mining) 952
FEUERRAD (n): Catherine wheel (Fireworks) 953
FEUERRAUM (m): combustion chamber; firebox; furnace; fireplace; hearth 954
FEUERRAUMWANDUNG (f): combustion chamber (furnace) wall-(s) 955
FEUERREGEN (m): rain of fire 956
FEUERROHR (n): flue; fire-tube (Boilers); fire-arm 957
FEUERROHRKESSEL (m): fire-tube boiler 958
FEUERROST (m): grate (of furnaces or domestic grates) 959
FEUERROT: fiery red 960
FEUERSAFT (m): slag-bath (Iron) 961
FEUERSBRUNST (f): fire, conflagration 962
FEUERSCHADEN (m): damage by fire 963
FEUERSCHIFF (n): fire-float; lightship 964
FEUERSCHUTZMITTEL (n): fire protecting agent (for impregnating inflammable articles) 965
FEUERSCHWADEN (m): fire-damp 966
FEUERSCHWAMM (m): amadou; German tinder (a fungus; *Polyporus fomentarius*) 967
FEUERSGEFAHR (f): danger of fire 968
FEUERSICHER: fireproof 969
FEUERSNOT (f): distress caused by fire 970
FEUERSPEIEND: volcanic, spitting fire 971
FEUERSPEIENDER BERG (m): volcano 972

FEUERSPRITZE (*f*): fire-engine (on land); fire-float (on water) 973
FEUERSTEIN (*m*): flint; Sg. 2.5-2.7 974
FEUERSTOFF (*m*): caloric 975
FEUERTON (*m*): fireclay 976
FEUERUNG (*f*): firing; fire; heating; fuel; furnace; fireplace; hearth; grate; stoker; stoking 977
FEUERUNG (*f*), **FLÜSSIGKEITS-**: liquid-fuel furnace 978
FEUERUNG (*f*), **MECHANISCHE-**: mechanical grate or stoker, mechanical firing or stoking 979
FEUERUNGSANLAGEN (*f.pl*): port blocks (of mill heating furnace); furnace installations 980
FEUERUNGSAUSKLEIDUNG (*f*): furnace (fire-brick) lining 981
FEUERUNGSDECKE (*f*): crown (of a furnace) 982
FEUERUNGSGEWÖLBE (*n*): furnace arch 983
FEUERUNGSKONTROLLE (*f*): furnace control 984
FEUERVERSICHERUNG (*f*): fire insurance 985
FEUERVERSICHERUNGSGESELLSCHAFT (*f*): fire insurance company 986
FEUERWEHR (*f*): fire-brigade 987
FEUERWERK (*n*): firework-(s) 988
FEUERWERKEREI (*f*): pyrotechnics 989
FEUERWERKPULVER (*n*): gunpowder in fireworks 990
FEUERZANGE (*f*): fire tongs 991
FEUERZEUG (*n*): lighter; tinder-box; match 992
FEUERZUG (*m*): flue; fire-tube; tinder; train (of gunpowder); pass (of a boiler) 993
FEURIG: fiery; burning; hot; igneous; ardent; spirited; generous (Wine); bright; loud (Colour) 994
FEURIGER SCHWADEN (*m*): firedamp 995
FFEIN: (a technical method of expressing "**SEHR FEIN**," very fine; see **FFS**, under abbreviations) 996
FIBER (*f*): fibre, string 997
FIBERMASCHINE (*f*): fibre machine (any kind of machine for treating fibre) 998
FIBRIN (*n*): fibrin 999
FIBROLITH (*m*): fibrolite, sillimanite (see **SILLI-MANIT**) 000
FIBRÖS: fibrous 001
FICELLIEREN: to wire (Bottles) 002
FICHTE (*f*): spruce (*Picea excelsa*), white fir, white deal, pine, fir; Sg. 0.35-0.6 003
FICHTEN: of pine, pine 004
FICHTENAPFEL (*m*): pine cone 005
FICHTENHARZ (*n*): pine resin; resin; common resin (*Resina pini*); colophony (see **BUR-GUNDERHARZ** and **FICHTENPECH**); Sg. 1.08; Mp. 100-140°C. 006
FICHTENHOLZ (*n*): pine (fir) wood; spruce wood (see **FICHTE**) 007
FICHTENLOHE (*f*): pine tan-(bark) 008
FICHTENLOHEEXTRAKT (*m*): pine tan extract 009
FICHTENNADEL (*f*): pine needle 010
FICHTENNADELEXTRAKT (*m*): pine needle extract 011
FICHTENNADELÖL (*n*): pine needle oil; Sg. 0.853-0.905; Bp. 150-185°C. (see also **EDELTAN-NENÖL**) 012
FICHTENNADELÖL (*n*), **KÜNSTLICHES-**: (see **BORNYL, ESSIGSAURES-**); artificial pine oil 013
FICHTENNADELPRÄPARAT (*n*): pine needle preparation 014
FICHTENPECH (*n*): spruce pitch; pine resin; rosin; common resin (*Pix burgundica*) (see **FICHTENHARZ**) 015
FICHTENRINDE (*f*): pine bark 016
FICHTENRINDEXTRAKT (*m*): pine bark extract 017
FICHTENSAMENÖL (*n*): pine seed oil (from seeds of *Pinus abies*) 018
FICHTENSPAN (*m*): pine (shaving) chip 019
FICHTENSPROSSEN (*m* or *f.pl*): pine (sprouts) shoots (*Turiones pini*) 020
FICHTENWOLLE (*f*): pine (fir) wool 021
FICHTENZAPFEN (*m*): pine cone 022
FICHTENZUCKER (*m*): pinite, pine sugar (from *Pinus lambertiana*); $C_7H_{14}O_6$; Mp. 186°C. 023
FIDEIKOMMISS (*m*): entail 024
FIEBER (*n*): fever; ague 025
FIEBERANFALL (*m*): attack of (fever) ague 026
FIEBERARTIG: febrile; feverish 027
FIEBERARZNEI (*f*): antipyretic; agent for lowering temperature 028
FIEBER, AUSSETZENDES-: (*n*): intermittent fever 029
FIEBERBUSCHSAMENÖL (*n*): (see **GEWÜRZBUSCH-SAMENÖL**) 030
FIEBERFEST: immune from fever 031
FIEBERFROST (*m*): paroxysm of shivering (with the ague); shivering fit, rigor (Med.) 032
FIEBERHAFT: feverish 033
FIEBERHITZE (*f*): (feverish)-temperature, febrile heat, high temperature, pyrexia 034
FIEBERISCH: feverish-(ly) 035
FIEBERKÄLTE (*f*): (see **FIEBERFROST**) 036
FIEBER (*n*), **KALTES-**: ague 037
FIEBERKLEE (*m*): buckbean (*Menyanthes trifoliata*); bogbean; marsh trefoil 038
FIEBERKRANK: feverish 039
FIEBERKRAUT (*n*): feverfew (*Matricaria parthenium* and *Pyrethrum parthenium*) 040
FIEBERMITTEL (*n*): antipyretic; febrifuge 041
FIEBERN: to rave; have fever; be in a fever 042
FIEBERPULVER (*n*): fever (ague) powder; (antimonial powder) 043
FIEBERRINDE (*f*): Cinchona bark, Calisaya bark, Peruvian bark (from *Cinchona calisaya*) 044
FIEBERSCHAUDER (*m*): shivering fit, fit of ague 045
FIEBERWURZEL (*f*): feverwort, horse gentian (*Triosteum perfoliatum*) 046
FIEDELBOGEN (*m*): (fiddle-)bow; bow drill (Mech.) 047
FIEDELBOHRER (*m*): drill 048
FIEDER (*f*): leaflet; pinnule (Bot.) 049
FIEDERIG: feathered; pinnate 050
FIEREN: to (ease away) veer (of ropes) 051
FIGUR (*f*): figure; illustration, diagram 052
FIGURBASS (*m*): figured bass (Music) 053
FIGURENDRUCK (*m*): figured printing (Calico) 054
FIGURENDRUCKARTIKEL (*m*): figured style (Calico) 055
FIGURENPORZELLAN (*n*): figure porcelain (porcelain for decorative purposes) 056
FIGURIEREN: to figure 057
FIGÜRLICH: figurative-(ly) 058
FIKTIV: imaginary, fictitious-(ly) 059
FILET (*m*): network, netting 060
FILIAL (*n*): branch (of an office, works, establishment, depot, etc.); affiliated company 061
FILIALANSTALT (*f*): branch establishment; affiliated establishment 062
FILIALLEITER (*m*): branch manager 063
FILIASÄURE (*f*): filicic acid (see **FILIXSÄURE**) 064
FILIBUSTIER (*m*): filibuster 065
FILICINSÄURE (*f*): filicinic acid (see **FILIXSÄURE**) 066
FILIGRAN (*n*): filigree 067

FILIGRANARBEIT (*f*) : filigree work 068
FILIXSÄURE (*f*) : filicic acid (from *Aspidium filix mas*) 069
FILM (*f*) : film ; skin 070
FILTER (*n*) : filter 071
FILTERANLAGE (*f*) : filter (filtering, filtration) plant 072
FILTERGESTELL (*n*) : filter stand 073
FILTERKEGEL (*m*) : filter cone 074
FILTERKIES (*m*) : filter-(ing) gravel 075
FILTERKOHLE (*f*) : filter-(ing) charcoal 076
FILTERMASSE (*f*) : filter (pulp) mass 077
FILTERPAPIER (*n*) : filter-(ing) paper 078
FILTERPLATTE (*f*) : filter plate 079
FILTERPRESSE (*f*) : filter press 080
FILTERPRESSENPLATTE (*f*) : filter press plate 081
FILTERPRESSENRAHMEN (*m*) : filter press frame 082
FILTERPRESSPLATTE (*f*) : filter-press plate 083
FILTERPRESSRING (*m*) : filter press ring 084
FILTERRAHMEN (*m*) : filter frame 085
FILTERSAND (*m*) : filter-(ing) sand 086
FILTERSCHICHT (*f*) : filter-bed, filtering layer 087
FILTERSCHLAUCH (*m*) : filtering tube (for dry process dust extraction from coal) 088
FILTERSCHONER (*m*) : filter cone 089
FILTERSTATIV (*n*) : filter stand 090
FILTERSTEIN (*m*) : porous stone 091
FILTERSTOFF (*m*) : filter cloth ; filtering material 092
FILTERTRÄGER (*m*) : filter (ring) holder 093
FILTERTUCH (*n*) : filter cloth 094
FILTERTURM (*m*) : filter tower 095
FILTERWÄGER (*m*) : filter weigher ; weighing tube 096
FILTERWAND (*f*) : filter-bed ; filter-plate 097
FILTRAT (*n*) : filtrate 098
FILTRATIONSFÄHIG : filterable ; capable of (filtration) being filtered 099
FILTRATIONSFÄHIGKEIT (*f*) : filtrative capacity, permeability 100
FILTRIERAPPARAT (*m*) : filtering apparatus ; filter 101
FILTRIERBAR : filterable ; capable of being filtered 102
FILTRIERBEUTEL (*m*) : filtering bag ; percolator 103
FILTRIEREN : to strain ; filter 104
FILTRIERER (*m*) : strainer ; filter-(er) 105
FILTRIERFLÄCHE (*f*) : filtering (surface) area 106
FILTRIERFLASCHE (*f*) : filtering flask 107
FILTRIERGESTELL (*n*) : filter stand 108
FILTRIERKANNE (*f*) : percolator 109
FILTRIERKONUS (*m*) : filter cone 110
FILTRIERKORB (*m*) : filtering basket 111
FILTRIERMATERIAL (*n*) : filtering material 112
FILTRIERPAPIER (*n*) : filter paper (Paper) 113
FILTRIERPLATTE (*f*) : filter plate ; drainer plate (Paper) 114
FILTRIERPUMPE (*f*) : filter pump 115
FILTRIERRING (*m*) : filter ring 116
FILTRIERROHR (*n*) : filter tube 117
FILTRIERSACK (*m*) : filter bag ; percolator 118
FILTRIERSCHALE (*f*) : filter dish 119
FILTRIERSIEB (*n*) : filter-(ing) sieve 120
FILTRIERSTATIV (*n*) : filter stand 121
FILTRIERSTUTZEN (*m*) : filter-(ing) jar 122
FILTRIERTRICHTER (*m*) : filtering funnel ; colander 123
FILTRIERTUCH (*n*) : filter-(ing) cloth ; percolator ; straining cloth 124
FILTRIERUNG (*f*) : filtration ; filtering 125
FILTRUM (*n*) : filter 126
FILZ (*m*) : felt ; felt-hat ; tomentum ; slime ore (Metal) ; blanket ; snub 127
FILZDICHTUNG (*f*) : felt washer 128
FILZEN : to felt ; to be miserly, snub ; (as an adjective) of felt 129
FILZIG : felt-like ; tomentous ; stingy ; fluffy 130
FILZISOLIERPLATTE (*f*) : felt insulating (board) plate 131
FILZKRAUT (*n*) : cotton rose, downweed (*Filago*) 132
FILZSCHLAUCH (*m*) : jacket (Paper) 133
FILZUNTERLAGE (*f*) : felt blanket 134
FILZWOLLE (*f*) : felted wool 135
FIMMEL (*m*) : sledge hammer ; wedge ; fimble (male) hemp (*Cannabis sativa*) 136
FINANZ (*f*) : finance 137
FINANZAUSSCHUSS (*m*) : finance committee 138
FINANZIELL : financial-(ly) 139
FINANZKUNDE (*f*) : finance 140
FINANZLAGE (*f*) : financial position 141
FINANZWESEN (*n*) : finance 142
FINDEN : to find ; discover ; think ; recognize ; understand ; exist ; be found 143
FINDLING (*m*) : foundling ; erratic block (Geol.) 144
FINGER (*m*) : finger 145
FINGERBEINE (*n.pl*) : phalanges (Anat.) 146
FINGERBRETT (*n*) : fingerboard (of a musical instrument) 147
FINGERFUTTER (*n*) : finger stall 148
FINGERHUT (*m*) : thimble ; foxglove ; digitalis (*Digitalis purpurea*) (see also under DIGITALIS) 149
FINGERHUTBLÄTTER (*n.pl*) : digitalis (foxglove) leaves (*folia digitalis*) 150
FINGERHUT (*m*), ROTER- : purple fox-glove, digitalis, fairy glove, fox-glove (*Digitalis purpurea*) 151
FINGERKRAUT (*n*) : cinquefoil (*Potentilla*) 152
FINGERLING (*m*), RUDER- : rudder pintle (Naut.) 153
FINGERN : to finger 154
FINGERPLATTE (*f*) : finger plate (of a door) 155
FINGERPROBE (*f*) : rule of thumb 156
FINGERSATZ (*m*) : fingering 157
FINGERSPRACHE (*f*) : dumb language, the language of signs 158
FINGERSTEIN (*m*) : belemnite 159
FINGERZEIG (*m*) : hint ; cue ; pointing with the finger ; indication ; sign ; pointer 160
FINGIEREN : to simulate ; forge ; invent ; pretend 161
FINGIEREND : (see FINGIERT) 162
FINGIERT : fictitious, imaginary, pretended, assumed 163
FINNE (*f*) : pimple ; fin ; postule ; stud ; knot ; claw ; pane (of a hammer) 164
FINNWAL (*m*) : fin-whale, fin-back-(whale) (*Balaenoptera musculus*) 165
FINSTER : dark ; gloomy ; obscure ; dim ; morose 166
FINSTERNIS (*f*) : darkness ; dimness ; obscurity ; gloom ; eclipse (Astron.) 167
FIPS (*m*) : fillip 168
FIRMA (*f*) : firm ; company ; name ; title ; heading ; style ; sign 169
FIRMAMENT (*n*) : firmament 170
FIRMAMENTSTEIN (*m*) : opal (see OPAL) 171
FIRMA (*f*), UNTER DER- : by the name of, under the style of, under the heading of 172

FIRMAZIT (*n*): quartz glass as clear as water; Sg. about 2.2; Mp. about 2000°C. 173
FIRMELUNG (*f*): confirmation (Religion) 174
FIRMENSCHILD (*n*): name-plate 175
FIRN: old; of last year (*m*): glacier snow; névé; firn 176
FIRNBLAU (*n*): glacier blue 177
FIRNER (*m*): (see FIRN) 178
FIRNIS (*m*): varnish; colourless oil coating (mostly linseed oil, which gradually hardens) 179
FIRNISEN: to varnish; glaze; gloss 180
FIRNISERSATZ (*m*): varnish substitute 181
FIRNISFABRIKANLAGE (*f*): plant at varnish making works 182
FIRNISPAPIER (*n*): glazed paper 183
FIRNISPRÄPARAT (*n*): varnish preparation 184
FIRSTE (*f*): ridge; roof; crown; coping; top 185
FIRSTENBALKEN (*m*): ridge piece 186
FIRSTENSTEMPEL (*m*): roof prop (Mining) 187
FIRSTENZIEGEL (*m*): ridge-tile 188
FIRSTSTOLLEN (*m*): top heading (Mining) 189
FISCH (*m*): fish 190
FISCHABDRUCK (*m*): ichthyolite; petrified fish in rocks 191
FISCHANGEL (*f*): fish-hook 192
FISCHARTIG: fish-like 193
FISCHAUGENSTEIN (*m*): apophyllite (see APOPHYL-LIT) 194
FISCHBEIN (*n*): whalebone; fishbone 195
FISCHBESCHREIBUNG (*f*): Ichthyography, Ichthyology, science of fishes 196
FISCHBLASE (*f*): fish bladder; sounds; isinglass 197
FISCHDAMPFER (*m*): trawler 198
FISCHDAVIT (*m*): fish davit 199
FISCHDÜNGER (*m*): fish manure 200
FISCHEIDECHSE (*f*): ichthyosaurus 201
FISCHEN: to fish 202
FISCHER (*m*): fisherman 203
FISCHERBOOT (*n*): fishing boat 204
FISCHEREI (*f*): fishing; fishery 205
FISCHEREIGESETZ (*n*): fishing (fishery) law 206
FISCHESSEND: ichthyophagous 207
FISCHESSER (*m*): ichthyophagist 208
FISCHGUANO (*m*): fish (guano) manure 209
FISCHIG: fishy 210
FISCHKÖRNER (*pl*): Indian berries (*Cocculus indicus*) 211
FISCHKUNDIGE (*m*): ichthyologist 212
FISCHKUNDIGER (*m*): Ichthyologist 213
FISCHLEHRE (*f*): ichthyology (the science of fishes) 214
FISCHLEIM (*m*): isinglass; fish glue; ichthyocolla; Sg. 1.11 215
FISCHLEIMGUMMI (*n*): sarcocolla 216
FISCHMEHL (*n*): fish meal 217
FISCHÖL (*n*): fish oil; ichthyol (thick blackish-browr fluid) 218
FISCHÖL (*n*), AMERIKANISCHES-: American fish oil (see MENHADENÖL) 219
FISCHÖL (*n*), JAPANISCHES-: Japanese fish oil (see SARDINENÖL) 220
FISCHRECHT (*n*): fishing rights 221
FISCHSCHUPPE (*f*): scale (of fish) 222
FISCHSCHUPPENAUSSCHLAG (*m*): ichthyosis, fish skin disease (a skin disease) (Med.) 223
FISCHSCHWANZ (*m*): fish tail 224
FISCHSPECK (*m*): blubber 225
FISCHTAKEL (*m*): fishing tackle 226
FISCHTORPEDO (*m*): fish torpedo 227
FISCHTRAN (*m*): fish (train) oil 228
FISCHUNG (*f*), MAST-: mast partner 229
FISCHVERWERTUNG (*f*): fish utilization 230
FISCHVERWERTUNGSANLAGE (*f*): fish utilization plant 231
FISCHZEIT (*f*): fishing season 232
FISETHOLZ (*n*): young fustic (*Rhus cotinus*) 233
FISETTHOLZ (*n*): (see FISETHOLZ) 234
FISTEL (*f*): fistula; falsetto; reed; pipe 235
FISTELHOLZ (*n*): young fustic (*Rhus cotinus*) 236
FISTELKASSIE (*f*): purging cassia (*Cassia fistula*) 237
FISTULÖS: fistular; fistulous 238
FITZEN: to fold, wrinkle, whip, tease out, make up into skeins 239
FIX: fixed; quick; smart 240
FIXAGE (*f*): fixing 241
FIXATEUR (*m*): fixer (Perfumery) 242
FIXATIV (*n*): fixing solution, fixer 243
FIXATORLÖSUNG (*f*): fixing solution 244
FIXBLEICHE (*f*): chloride of lime bleaching 245
FIXE BESTANDTEILE (*m.pl*): fixed constituents 246
FIXER KOHLENSTOFF (*m*): fixed carbon (coal analysis) 247
FIXFÄRBEREI (*f*): fixed dyeing 248
FIXIERBAD (*n*): fixing bath 249
FIXIERBAD (*n*), NEUTRALES-: neutral fixing bath (sodium thiosulphate and water in proportion of 1:4) (Photography) 250
FIXIERBAD (*n*), SAURES-: acid fixing bath (the neutral fixing bath plus sodium bisulphite solution in proportion 100: 5 by volume) (Photography) (see also FIXIERBAD, NEUT-RALES-) 251
FIXIEREN: to fix; harden; set; settle; establish; mordant (Silk); stare at; determine; prove 252
FIXIERENDE WIRKUNG (*f*): fixing action 253
FIXIERFLÜSSIGKEIT (*f*): fixing (liquor) liquid; fixing bath 254
FIXIERLÖSUNG (*f*): fixing solution, fixer (Photography) 255
FIXIERMITTEL (*n*): fixing agent; fixative; fixer; hardening agent 256
FIXIERNATRON (*n*): hyposulphite of soda; hyposodium thiosulphate (see NATRIUM-HYPOSULFIT) 257
FIXIERNATRONLÖSUNG (*f*): sodium thiosulphate solution (Photography) 258
FIXIERNATRONREST (*m*): sodium thiosulphate residue 259
FIXIERSALZ (*n*): fixing salt (hyposodium-thiosulphate): $Na_2S_2O_3.5H_2O$; Sg. 1.729; Mp. 48°C. 260
FIXIERSALZ (*n*), SAURES-: acid fixing salt 261
FIXIERSALZZERSTÖRER (*m*): fixing salt destroyer (for extracting the last traces of sodium thiosulphate from plates and prints) (Photography). 262
FIXIERTON (*m*): fixing clay (a clay with high silicic acid content for fixing tar dyes) 263
FIXIERUNG (*f*): fixing; fixation; hardening 264
FIXIERUNGSMITTEL (*n*): fixing agent; fixer; fixative; hardening agent 265
FIXPREIS (*m*): fixed (firm) price 266
FIXPUNKT (*m*): fixed point 267
FIXSTERN (*m*): fixed star 268
FIXUM (*n*): fixed (stated) sum or salary 269
FLACH: flat; level; plain; shallow; superficial; insipid 270
FLACHBRUNNENWASSER (*n*): shallow-well water 271

FLÄCHE (*f*): surface; face; flat; plain; plane; sheet; area; superficies 272
FLACHEISEN (*n*): flat iron, flats (Metal section) 273
FLACHEN: to flatten, level 274
FLÄCHEN: to level, plane 275
FLÄCHENANZIEHUNG (*f*): surface (adhesion) attraction 276
FLÄCHENGRÖSSE (*f*): area 277
FLÄCHENINHALT (*m*): area 278
FLÄCHENMASS (*n*): area; square (surface) measure 279
FLÄCHENMESSUNG (*f*): surface measurement; planimetry 280
FLÄCHENRAUM (*m*): area 281
FLÄCHENWINKEL (*m*): plane angle 282
FLÄCHENZAHL (*f*): square number; number of faces 283
FLACHFEILE (*f*): flat file 284
FLACHGÄNGIG: flat threaded (of screws) 285
FLACHGEWUNDEN: planispiral 286
FLÄCHHEIT (*f*): flatness, etc. (see FLACH) 287
FLACHHOHLEISEN (*n*): gouge 288
FLACHKEIL (*m*): flat key or wedge 289
FLACHKIEL (*m*): flat keel (Naut.) 290
FLACHMEISSEL (*m*): flat chisel 291
FLACHRELIEF (*n*): bas-relief 292
FLACHS (*m*): flax (*Linum usitatissimum*) 293
FLACHSARTIG: flaxen 294
FLACHSBAU (*m*): flax cultivation 295
FLACHSCHABER (*m*): flat scraper 296
FLACHSCHIEBER (*m*): slide valve 297
FLACHSCHIENE (*f*): plate rail 298
FLACHSCHNITT (*m*): horizontal section 299
FLACHSDARRE (*f*): flax drying-house 300
FLACHSDOTTERPFLANZE (*f*): (see LEINDOTTER-PFLANZE) 301
FLÄCHSEN: flaxen 302
FLACHSERNTE (*f*): flax (harvest) crop 303
FLACHSFASER (*f*): flax (bast) fibre 304
FLACHSRAUFE (*f*): ripple, flax comb 305
FLACHSRÖSTE (*f*): (see RÖSTE) 306
FLACHSSAMEN (*m*): flax seed; linseed (seed of *Linum usitatissimum*) 307
FLACHSSCHEBEN (*pl*): flax residue (92% combustible matter; 6.5% moisture; 1.5% ashes; calorific value, dry, 4100 cals.) 308
FLACHSSTEIN (*m*): amianthus; asbestos 309
FLACHSWERG (*m* and *n*): flax tow (short flax fibre for making tow) 310
FLACHWULSTEISENEN (*n*): flat bulb iron 311
FLACHZANGE (*f*): pliers (having flat jaws) 312
FLACKERN: to flicker; flare 313
FLACKERND: flickering; fitful 314
FLACON (*m*): flagon; small bottle; phial; scent bottle 315
FLADEN (*m*): pancake, cake of dung (sewage disposal) 316
FLADER (*f*): flaw; curl; streak; speckle; knot; vein 317
FLADERIG: streaked; veined 318
FLADERSCHNITT (*m*): (see TANGENTIALSCHNITT): knotty or irregular vein (grain) section 319
FLAGGEN: to dress (a ship); to beflag; signal with flags; (*n*): flag-signalling 320
FLAGGENFALL (*n*): ensign halliard 321
FLAGGENGAFFEL (*m*): monkey gaff 322
FLAGGENKNOPF (*m*): truck (Naut.) 323
FLAGGENLEINE (*f*): ensign halliard 324
FLAGGENSTOCK (*m*): flagstaff 325
FLAGGENTOPP (*m*): flag pole 326
FLAGGENTUCH (*n*): bunting 327
FLAGGSCHIFF (*n*): flag ship 328

FLAMME (*f*): flame; light; flash 329
FLAMMEN: to flame; burn; glow; blaze; sear; singe; water; cloud (Fabrics, etc.) 330
FLAMMENBACH (*m*): stream of fire 331
FLAMMENBOGEN (*m*): electric arc; flaming arc 332
FLAMMENERSTICKEND: flame-killing, flame-extinguishing (by depriving a fire of the air necessary for combustion, or by introducing a superabundance of some gas, as nitrogen, etc.) 333
FLAMMENFLÄCHE (*f*): flame surface 334
FLAMMENLOSE FEUERUNG (*f*): flameless furnace (see note to OBERFLÄCHENVERBRENNUNG) 335
FLAMMENOFEN (*m*): reverbatory (flame) furnace; converter 336
FLAMMENOFENFRISCHEN (*n*): puddling-(process) (Metal) 337
FLAMMENOPAL (*m*): fire opal 338
FLAMMENSÄULE (*f*): column of fire 339
FLAMMENSTULPE (*f*): upturned (inverted, upsidedown) flame duct 340
FLAMMENWERFER (*m*): flame thrower 341
FLAMMFARBIG: flame (rainbow) coloured 342
FLAMMICHT: watered (Fabrics); veined; grained (Wood); flame-like 343
FLAMMIEREN: to flame; burn; glow; blaze; sear; singe; water; cloud; streak (with colour) 344
FLAMMKOHLE (*f*): flaming (bitumenous) coal (with high volatile; its coke too small for steelworks) 345
FLAMMOFEN (*m*): (see FLAMMENOFEN) 346
FLAMMOFENFRISCHEN (*n*): puddling (Metal) 347
FLAMMPUNKT (*m*): flash point (i.e., lowest temperature at which oil vapours form a combustible mixture with air) 348
FLAMMPUNKTPRÜFER (*m*): flash-point tester 349
FLAMMROHR (*n*): fire-tube; flue 350
FLAMMROHRKESSEL (*m*): fire-tube boiler 351
FLAMMROHRKESSEL, ZWEI- (*m*): two-flued Cornish boiler 352
FLANKIEREN: to flank, make a flank movement, protect the flanks 353
FLANSCH (*m*): flange; gland 354
FLANSCHBEFESTIGUNG (*f*): flange-coupling 355
FLANSCHE (*f*): (see FLANSCH) 356
FLANSCHEN: to flange 357
FLANSCHENAUFWALZMASCHINE (*f*): flanging roll, flanging machine 358
FLANSCHENDREHBANK (*f*): flange lathe 359
FLANSCHENFRÄSMASCHINE (*f*): flange milling machine 360
FLANSCHENKUPPELUNG (*f*): flanged coupling 361
FLANSCHENROHR (*n*): flanged pipe 362
FLANSCHENWALZE (*f*): flanging roll 363
FLANSCHRING (*m*): flange ring 364
FLANSCHWULSTEISEN (*n*): flanged bulb iron, bulb rail 365
FLANTSCH (*m*): (see FLANSCH) 366
FLANTSCHE (*f*): (see FLANSCH) 367
FLASCHE (*f*): bottle; flask; phial; jar; (gas)-cylinder; casting box; block (Mech.) 368
FLASCHENABTEILUNG (*f*): bottle department 369
FLASCHEN, BLEI- (*f.pl*): lead bottles 370
FLASCHENELEMENT (*n*): bottle battery; bottle cell 371
FLASCHENFÜLLAPPARAT (*m*): bottle filling apparatus, bottling apparatus 372
FLASCHENFUTTER (*n*): bottle (rack) case 373
FLASCHENGESTELL (*n*): bottle rack 374

FLASCHENGLAS (*n*): bottle glass 375
FLASCHENGRÜN (*n*): bottle green 376
FLASCHENKAPPE (*f*): bottle cap 377
FLASCHENKAPSEL (*f*): bottle capsule 378
FLASCHENKÜRBIS (*m*): bottle gourd; calabash 379
FLASCHENPAPIER (*n*): bottle wrapping (Paper) 380
FLASCHENVERSCHLUSS (*m*): bottle stopper; sealing arrangement for bottles 381
FLASCHENZUG (*m*): set of pulleys; block and tackle; pulley block 382
FLASCHE (*f*), **SCHMIEDEEISERNE-**: wrought iron bottle or cylinder 383
FLASCHE (*f*), **WOULFFSCHE-**: Woulff (bottle) flask 384
FLASER (*f*): curl; streak; speckle; flaw; knot 385
FLATTERHAFT: flighty; fickle; volatile; inconstant 386
FLATTERIG: (see **FLATTERHAFT**) 387
FLATTERN: to flutter; wave; flicker; be flighty; flirt 388
FLATTERRUSS (*m*): lampblack 389
FLATTIERFEUER (*n*): choked fire 390
FLATULENZ (*f*): flatulence, flatulency 391
FLAU: faint; weak; dull; slack; easy; low; indifferent; languid; stagnant (of the market) (Comm.) 392
FLAUEN: to buddle; become (weak; faint; dull; languid) slack 393
FLAUMFETT (*n*): neutral lard (see **SCHMALZ, BESTE SORTE-**) 394
FLECHSE (*f*): tendon; sinew 395
FLECHTE (*f*): lichen; plait; twist; skin eruption (shingles; herpes); tress; hurdle; scab 396
FLECHTEN: to plait; wreathe; twist; interweave; braid; intertwine; bind; fasten 397
FLECHTENAUSSCHLAG (*m*): Herpes (Med.) (a skin eruption) 398
FLECHTENROT (*n*): orcein 399
FLECHTENSALBE (*f*): herpes ointment 400
FLECHTENSÄURE (*f*): fumaric acid; COOHCH-CHCOOH 401
FLECHTENSEIFE (*f*): herpes soap 402
FLECHTENSTÄRKEMEHL (*n*): lichenin; moss starch 403
FLECHTENSTOFF (*m*): lichen substance 404
FLECHTING (*f*): upper shroud (Naut.) 405
FLECHTSTRÖMUNG (*f*): turbulent flow 406
FLECHTWEIDE (*f*): osier (*Salix triandra*) (a type of willow used for basket making) 407
FLECHTWERK (*n*): wickerwork 408
FLECK (*m*): spot; speck; stain; flaw; patch; blot; blemish; place; (*pl.*): maculae (locally spotted skin) (Med.) 409
FLECKEN: to spot; speckle; blemish; mottle; patch; blot; stain; (*m*): taint; village; hamlet; (see also **FLECK**) 410
FLECKENLOS: spotless 411
FLECKENMAL (*n*): pigmentary mole (*Naevus pigmentosus*) (Med.) 412
FLECKENPORPHYR (*m*): tufaceous porphyry (see **PORPHYRTUFF**) 413
FLECKENPULVER (*n*): scouring powder (for removing spots) 414
FLECKENPUTZEREI (*f*): spot-cleaning (Textile) (see **DETACHUR**) 415
FLECKENTFERNUNGSKRAFT (*f*): spot removing power, cleaning power 416
FLECKENWASSER (*n*): scouring (cleaning) water; scouring (cleaning) drops; liquid for removing stains 417
FLECKFIEBER (*n*): spotted fever; petechial fever; typhus; purples (Med.) 418
FLECKIG: spotted; stained; speckled 419
FLECKKUGEL (*f*): scouring (ball) soap 420
FLECKSCHIERLING (*m*): spotted (poison) hemlock (*Conium maculatum*) 421
FLECKSEIFE (*f*): scouring (cleaning) soap 422
FLECKSTEIN (*m*): scouring stone 423
FLECKSTIFT (*m*): scouring stick 424
FLECKSTORCHSCHNABELWURZEL (*f*): geranium; cranesbill; stork's-bill (dried rhizome of *Geranium maculatum*) 425
FLECKTYPHUS (*m*): (see **KRIEGSPEST**) 426
FLECKWASSER (*n*): (see **FLECKENWASSER**) 427
FLECTIEREN: to inflect 428
FLEDERFISCH (*m*): flying fish (*Exocoetus* family of fish) 429
FLEDERMAUS (*f*): bat (*Chiroptera* family of mammals) 430
FLEDERMAUSBRENNER (*m*): batswing burner 431
FLEGEL (*m*): flail; boor; clown; churl; rude (insolent; coarse) person 432
FLEICHERVOGEL (*m*): butcher bird (shrike species: *Lanius*) 433
FLEISCH (*n*): flesh; meat; (fruit)-pulp; cellular tissue (of plants); sarcolemma (of muscles) 434
FLEISCHAUSWUCHS (*m*): fleshy excrescence 435
FLEISCHBRUCH (*m*): sarcocele (Med.) 436
FLEISCHBRÜHE (*f*): bouillon; gravy; meat broth 437
FLEISCHDÜNGEMEHL (*m*): flesh meal manure 438
FLEISCHEN: to flesh 439
FLEISCHER (*m*): butcher 440
FLEISCHERN: of flesh; fleshy; carnal 441
FLEISCHERTALG (*m*): unmelted tallow 442
FLEISCHEXTRAKT (*m*): extract of meat, meat extract 443
FLEISCHFARBE (*f*): flesh colour, carnation 444
FLEISCHFARBIG: flesh-coloured 445
FLEISCHFASER (*f*): fibre (of flesh) 446
FLEISCHFRESSEND: carnivorous 447
FLEISCHGEWÄCHS (*n*): sarcoma (a fleshy tumour) (Med.) 448
FLEISCHGIFT (*n*): ptomaine; meat toxin 449
FLEISCHGUMMI (*n*): sarcocol; sarcocolla 450
FLEISCHIG: meaty; fleshy; pulpy 451
FLEISCHKEGEL (*m*): wattle (of a turkey) 452
FLEISCHKLEE (*m*): trefoil (*Trifolium*) 453
FLEISCHKOHLE (*f*): animal charcoal; bone black; abaiser 454
FLEISCHKONSERVE (*f*): preserved meat 455
FLEISCHKOST (*f*): meat diet 456
FLEISCHLAUCH (*m*): (see **WINTERPORREE**) 457
FLEISCHLEIM (*m*): sarcocol; sarcocolla 458
FLEISCHLICH: carnal 459
FLEISCHLOS: fleshless 460
FLEISCHMADE (*f*): maggot 461
FLEISCHMEHL (*n*): flesh meal 462
FLEISCHMILCHSÄURE (*f*): sarcolactic acid 463
FLEISCHSAFT (*m*): extract of meat; meat juice 464
FLEISCHTON (*m*): flesh colour, carnation 465
FLEISCH (*n*), **WILDES-**: hypersarcosis 466
FLEISCHZUCKER (*m*): inosite; $C_6H_6(OH)_6$ 467
FLEMINGIN (*n*): flemingin (colouring principle from *Flemingia congesta*); $C_{12}H_{12}O_3$ 468
FLETE (*m*): (see **GLATTROCHE**) 469

FLIEDER (*m*): elder (*Sambucus nigra* and *Sambucus ebulus*) 470
FLIEDERBEERE (*f*): elderberry (*Fructus sambuci*) 471
FLIEDERBLAU (*n*): lilac-(colour) 472
FLEIDERBLÜTEN (*f.pl*): elder flowers (*Flores sambuci*) 473
FLIEDER (*m*), SPANISCHER-: lilac (*Syringa*) 474
FLIEGE (*f*): fly; sight (Gun) 475
FLIEGEN: to fly 476
FLIEGENBEKÄMPFUNGSMITTEL (*n*): preparations for combating the fly nuisance 477
FLIEGEND: overhung; flying 478
FLIEGEND ANORDNEN: to overhang (a wheel on a shaft) 479
FLIEGENDE ANORDNUNG (*f*): overhung (type) arrangement (of turbine wheels) 480
FLIEGENDES RAD (*n*): overhung wheel (Turbines) 481
FLIEGEND SITZEN: to be overhung (of a wheel on a shaft) 482
FLIEGENFÄNGER (*m*): fly-catcher 483
FLIEGENGIFT (*f*): fly (killer, exterminator) poison 484
FLIEGENHOLZ (*n*): bitterwood; quassia wood (wood of *Quassia amara*) 485
FLIEGENKOBALT (*m*): fly stone (native cobalt arsenide) 486
FLIEGENLEIM (*m*): fly glue 487
FLIEGENPAPIER (*n*): fly-paper 488
FLIEGENPILZ (*m*): fly agaric, fly fungus, fly amanita (*Amanita muscaria*); toadstool (*Agaricus muscarius*) 489
FLIEGENSCHIMMELPILZ (*m*): fly mould fungus (*Empusa muscae*) 490
FLIEGENSCHWAMM (*m*): fly agaric; fly fungus (*Amanita muscaria*) 491
FLIEGEN (*f.pl*), SPANISCHE-: Spanish flies, blistering flies, cantharides, blistering beetles (*Cantharis vesicatoria*) 492
FLIEGENSTEIN (*m*): fly stone (native cobalt arsenide); arsenic (*Arsenicum metallicum*) (see ARSEN) 493
FLIEGENVERTILGUNGSMITTEL (*n*): fly (killer) destroyer 494
FLIEGER (*m*): middle staysail (Naut.) 495
FLIEHKRAFT (*f*): centrifugal force 496
FLIESE (*f*): flag-(stone); floor tile; wall tile; paving brick; Dutch tile 497
FLIESS (*n*): fleece (of wool); (*m*): flow-(ing) (of water) 498
FLIESSEN: to flow; melt; run; elapse; blot; bleed (colours); (*n*): flow; strain; stress (Metal) 499
FLIESSEND: running, flowing, fluid, fluent, liquid 500
FLIESSFEDER (*f*): fountain pen 501
FLIESSGLÄTTE (*f*): litharge 502
FLIESSGOLD (*n*): stream gold, gold found in streams 503
FLIESSGRENZE (*f*): yield point 504
FLIESSHARZ (*n*): (see FLUSSHARZ) 505
FLIESSOFEN (*m*): pyrites kiln 506
FLIESSPAPIER (*n*): blotting (absorbent) paper 507
FLIESSWASSER (*n*): flowing water 508
FLIMMER (*m*): mica (see also GLIMMER); tinsel; spangle; glitter; glimmer 509
FLIMMERN: to glitter; glisten; sparkle; scintillate; vibrate; twinkle 510
FLINT (*m*): flint; (SiO_2); Sg. 2.5-2.6 511
FLINTE (*f*): gun; musket 512
FLINTENKOLBEN (*m*): butt (of a gun) 513
FLINTENLAUF (*m*): gun barrel 514
FLINTENSCHAFT (*m*): stock (of a gun) 515
FLINTENSCHROT (*n*): gun (small) shot 516
FLINTENSCHUSS (*m*): gunshot 517
FLINTGLAS (*n*): flint glass (composed of potassium and lead silicates) 518
FLINTMEHL (*n*): flint meal, crushed flint 519
FLINTSTEIN (*m*): flint (see FLINT) 520
FLINZ (*m*): siderite; $FeCO_3$ (about 48% Fe) (see also EISENSPAT) 521
FLITTER (*m*): spangle; tinsel; Dutch gold (see FLITTERGOLD) 522
FLITTERERZ (*n*): glittering (flaky) ore 523
FLITTERGLANZ (*m*): tinsel 524
FLITTERGLAS (*n*): crushed glass (used for frosting) 525
FLITTERGOLD (*n*): Dutch metal; tombac (a sort of brass; a copper-zinc alloy) 526
FLITTERIG: showy; tinsel; tawdry 527
FLITTERN: to glitter; sparkle; glisten 528
FLITTERSAND (*m*): micaceous (sparkling) sand 529
FLOCKE (*f*): flock; flake; flaky-stone; hemp 530
FLOCKEN: to flake; crack (Oil); (*n*): cracking (the removal of slime from oil) 531
FLOCKENARTIG: flaky, lamellar 532
FLOCKENERZ (*n*): mimetite; $3Pb_3As_2O_8.PbCl_2$; Sg. 7.22 (natural lead chloroarsenate) 533
FLOCKENGRAPHIT (*n*): flaky graphite 534
FLOCKENPAPIER (*n*): flock paper 535
FLOCKENSALPETER (*m*): efflorescent saltpeter 536
FLOCKENSEIDE (*f*): floss silk 537
FLOCKIG: flaky; filamentous; exfoliated; laminated; efflorescent 538
FLOCKSEIDE (*f*): floss silk; silk waste 539
FLOCKWOLLE (*f*): flock wool; short wool; wool waste 540
FLOHFARBE (*f*): puce colour 541
FLOHKRAUT (*n*): fleabane (*Erigeron canadensis*; *Erigeron acris*, etc.) 542
FLOHKRAUTÖL (*n*): flea bane oil; Sg.0.85; Bp. 175-180°C. 543
FLOHSAMEN (*m*): fleawort seed (*Semen psyllii*); fleabane seed; flea seed (*Plantago psyllium*) 544
FLOR (*m*): bloom; blooming; gauze; nap; pile; crape 545
FLORA (*f*): flora 546
FLORENTINERFLASCHE (*f*): Florentine receiver (for essential oils) 547
FLORENTINERLACK (*m*): Florentine lake, carmine lake (see KARMIN) 548
FLORICINOL (*n*): trade name for a varnish obtained from castor oil 549
FLORIDABLEICHERDE (*f*): Florida bleaching earth (an Aluminium-magnesium-hydrosilicate; for bleaching fats); $4MgO,3Al_2O_3,25SiO_2$ 550
FLORIDAPHOSPHAT (*n*): Florida phosphate (about 39% P_2O_5) 551
FLORIDAWASSER (*n*): Florida water 552
FLORIDIN (*n*): (see FLORIDABLEICHERDE) 553
FLORIEREN: to flourish 554
FLOSS (*n*): pig-(iron) (Metal); raft; float; buoy 555
FLOSSBETT (*n*): pig mould (Metal) 556
FLOSSE (*f*): fin; float; raft 557
FLÖSSEN: to float; infuse; skim; rinse 558
FLOSSHOLZ (*n*): floated (timber) wood 559
FLOSSIG: finny; having floats 560
FLOSSOFEN (*m*): flowing furnace 561
FLOTATION (*f*): flotation 562

FLOTATIONSVERFAHREN (*n*): flotation (process) method 563
FLOTTE (*f*): dye (liquor) bath; fleet; navy 564
FLOTTENFLÜSSIGKEIT (*f*): dye liquor 565
FLOTTENGEFÄSS (*n*): dye (colour) bath 566
FLOTTENSALZ (*n*): borax (see **BORAX**) 567
FLOTTENSTAND (*m*): dye (liquor) level 568
FLOTTGRAS (*n*): duckweed (see **WASSERLINSE**) 569
FLOTTILLE (*f*): flottilla (Marine) 570
FLOTTMACHEN: to float (a ship) 571
FLOTTSEIDE (*f*): untwisted silk; thrown silk 572
FLOTTSTAHL (*m*): ingot steel; run steel 573
FLÖTZ (*n*): layer; stratum; seam; vein; bed; lode 574
FLÖTZASCHE (*f*): clay marl 575
FLÖTZERZ (*n*): ore in seams, veins, strata or beds 576
FLÖTZGEBIRGE (*n*): secondary rocks, stratified mountains 577
FLÖZ (*n*): (see **FLÖTZ**) 578
FLUATIEREN: to waterproof, weatherproof (to cover stones, bricks or cement with soluble fluosilicates of metals, as a protection against the weather); (*n*): waterproofing, weatherproofing 579
FLUATLÖSUNG (*f*): waterproofing (weatherproofing) solution (a solution of magnesium fluosilicate) 580
FLUCHT (*f*): flight; line; row; play; swing; escape; disappearance; vanishing 581
FLUCHTEBENE (*f*): vanishing plane 582
FLÜCHTEN: to flee, fly, escape 583
FLUCHTHOLZ (*n*): level; rule 584
FLÜCHTIG: fugitive; fleeting; fragile; cursory; fleet; fickle; hasty; transient; superficial; volatile 585
FLÜCHTIGE BESTANDTEILE (*m.pl*): volatile matter (of coal, etc.) 586
FLÜCHTIGE SALBE (*f*): ammonia liniment; volatile liniment 587
FLÜCHTIGES LAUGENSALZ (*n*): ammonium carbonate (see **AMMONIUMKARBONAT**) 588
FLÜCHTIGES ÖL (*n*): volatile (essential) oil 589
FLÜCHTIGKEIT (*f*): volatility, etc. (see **FLÜCHTIG**) 590
FLÜCHTIG MACHEN: to volatilize 591
FLÜCHTIGMACHUNG (*f*): volatilization 592
FLUCHTLINIE (*f*): vanishing line, sight-line 593
FLUCHTPUNKT (*m*): vanishing (disappearing; visual) point 594
FLUCHTRECHT: flush 595
FLUDER (*n*): weir; channel 596
FLUG (*m*): flight; flying; flock; swarm; soaring; covey 597
FLUGASCHE (*f*): light (flying) ashes; flue ashes; flue dust 598
FLUGASCHENABLAGERUNG (*f*): flue ash (dust) deposit 599
FLUGASCHENFÄNGER (*m*): grit catcher; (flue-)dust catcher 600
FLUGBAHN (*f*): trajectory; line of flight 601
FLUGBEULEN (*f.pl*): shingles; herpes (a skin eruption) (Med.) 602
FLUGBIENE (*f*): worker (of bees) 603
FLUGBLATT (*n*): flysheet; handbill; pamphlet 604
FLÜGEL (*m*): grand piano; wing; flap; lobe; vane; branch; casement-(window); blade (Propeller); sail; aisle 605
FLÜGELFENSTER (*n*): casement-(window) 606
FLÜGELFLÄCHE (*f*): blade area (Propellers) 607
FLÜGELFÖRMIG: wing shaped 608
FLÜGELGEBLÄSE (*n*): fan blower 609
FLÜGELHAHN (*m*): butterfly cock 610
FLÜGEL (*m*), **HYDROMETRISCHER-**: water measuring (hydrometric) vane or propeller, current meter 611
FLÜGELKANTE (*f*), **AUSTRETENDE-**: following edge, trailing edge (of propeller blade) 612
FLÜGELKANTE (*f*), **EINTRETENDE-**: forward edge (of propeller blade) 613
FLÜGELKÜHLER (*m*): fan cooler 614
FLÜGELMUTTER (*f*): wing nut 615
FLÜGELN: to wing 616
FLÜGEL (*m*), **PROPELLER-**: blade of a screw propeller 617
FLÜGELRAD (*n*): screw propeller; worm (screw) wheel 618
FLÜGELSCHRAUBE (*f*): thumb screw, wing screw 619
FLÜGELSTANGE (*f*): vane (propeller) rod or spindle 620
FLÜGELTÜR (*f*): folding door 621
FLÜGELWELLE (*f*): propeller shaft 622
FLUGFISCH (*m*): (see **FLEDERFISCH**) 623
FLUGKRAFT (*f*): power (force) of flight 624
FLÜGLIG-, DREI-: three bladed 625
FLUGLINIE ( )*f*: trajectory; line of flight 626
FLUGMASCHINE (*f*): aeroplane 627
FLUGPLATZ (*m*): aerodrome 628
FLUGS: quickly; immediately; speedily; hastily 629
FLUGSAND (*m*): quicksand 630
FLUGSCHRIFT (*f*): pamphlet 631
FLUGSTAUB (*m*): flue dust; smoke; fume (Metal); soot from chimney; flying dust 632
FLUGSTAUBKAMMER (*f*): condensing chamber; dust catcher 633
FLUGSTAUBKONDENSATOR (*m*): smoke condenser 634
FLUGWASSER (*n*): spray 635
FLUGWESEN (*n*): aviation 636
FLUGZEUGHALLE (*f*): aeroplane hangar 637
FLUGZEUGTEILE (*m.pl*): aeroplane parts 638
FLUID (*n*): fluid, liquid 639
FLUIDEXTRAXT (*m*): fluid extract; liquid extract (in fluid extract of plants, the fluid extract is taken to be equal to the quantity of the air dried part of the plant used) (Pharm.) 640
FLUIDUM (*n*): liquid, fluid 641
FLUKTUIEREN: to fluctuate 642
FLUKTUÖS: fluctuating 643
FLUOCERIT (*m*): flucerine, fluocerine, fluocerite; $(Ce,La,Di)_2OF_4$; Sg. 5.8 644
FLUOR (*n*): fluorine; F_2; Sg. compared with air 1.14; Mp. −223°C.; Bp. −187°C. 645
FLUORALUMINIUM (*n*): aluminium fluoride (see **ALUMINIUMFLUORID**) 646
FLUORALUMINIUMNATRIUM (*n*): aluminium sodium fluoride 647
FLUORAMMONIUM (*n*): ammonium fluoride; NH_4F; (*Ammonium fluoratum*) 648
FLUORANTIMON (*n*): antimony fluoride; trifluoride, SbF_3; pentafluoride, SbF_5 649
FLUORANTIMONIUM (*n*): antimony fluoride (see **ANTIMONPENTAFLUORID** and **ANTIMON-TRIFLUORID**) 650
FLUORARSEN (*n*): arsenic fluoride 651
FLUORBARYUM (*n*): barium fluoride; BaF_2; Sg. 4.828; Mp. 1280°C. 652
FLUORBENZOL (*n*): fluoro (fluo) benzene 653
FLUORBLEI (*n*): lead fluoride (see **BLEIFLUORID**) 654

FLUORBOR (*n*): boron fluoride 655
FLUORBROM (*n*): bromine fluoride (see BROM-FLUORID) 656
FLUORCALCIUM (*n*): calcium fluoride, CaF_2; Sg. 3.15-3.19; Mp. 1300°C. (*Calcium fluoratum*) 657
FLUORCERIUM (*n*): cerium (ceric) fluoride; $CeF_4.H_2O$ 658
FLUORCHROM (*n*): chromium (chromic) fluoride; $CrF_3.4H_2O$; Sg. 3.78 659
FLUOREISEN (*n*): iron (ferrous) fluoride; $FeF_2.8H_2O$; Sg. 4.1 660
FLUOREN (*n*): fluorene, alpha-diphenylene-methane; $(C_6H_4)_2CH$; Mp. 116°C.; Bp. 295°C. 661
FLUORESCEIN (*n*): diresorcinolphthalein, tetra-oxyphthalophenon-anhydride, fluoresceïn, resorcinolphthalein; $O(C_6H_3OH)_2C.C_6H_4COO.H_2O$ 662
FLUORESZEIN (*n*): fluorescein; $C_{20}H_{12}O_5$ (see FLUORESCEIN) 663
FLUORESZEINCHLORID (*n*): fluorescein chloride 664
FLUORESZENZ (*f*): fluorescence 665
FLUORESZENZSCHIRM (*m*): fluorescent screen 666
FLUORESZIEREN: to fluoresce 667
FLUORESZIEREND: fluorescent 668
FLUORID (*n*): fluoride 669
FLUORIT (*m*): fluorite, chlorophane, fluor spar (a calcium fluoride); CaF_2: (about 51% calcium and 49% fluorine); Sg. 3.18 670
FLUORJOD (*n*): iodine fluoride 671
FLUORKALIUM (*n*): potassium fluoride; KF; Sg. 2.454; Mp. about 800°C. (*Kalium fluoratum*) 672
FLUORKALZIUM (*n*): calcium fluoride; CaF_2; Sg. 3.15-3.18; Mp. 1300°C. 673
FLUORKOHLENSTOFF (*m*): carbon fluoride 674
FLUORLITHIUM (*n*): lithium fluoride; LiF; Sg. 2.601; Mp. 801°C. 675
FLUORMAGNESUM (*n*): magnesium fluoride; MgF_2; Sg. 2.472; Mp. 1396°C. (see MAG-NESIUMFLUORID) 676
FLUORMETALL (*n*): metallic fluoride 677
FLUORNATRIUM (*n*): sodium fluoride; fluorol; NaF; Sg. 2.766; Mp. 980°C. 678
FLUORPHOSPHOR (*n*): phosphorous fluoride 679
FLUORPRÄPARATE, ORGANISCHE- (*n.pl*): organic fluorine preparations 680
FLUORSALZ (*n*): fluoride 681
FLUORSCHWEFEL (*m*): sulphur fluoride 682
FLUORSELEN (*n*): selenium fluoride 683
FLUORSILBER (*n*): silver fluoride (see SILBER-FLUORID) 684
FLUORSILIZIUM (*n*): silicon fluoride 685
FLUORSILIZIUMVERBINDUNG (*f*): fluosilicate 686
FLUORSTAHL (*m*): ingot steel; hard steel; mild (homogenous) steel (in which the wrought iron is produced in a liquid state) 687
FLUORTELLUR (*n*): tellurium fluoride 688
FLUORTITAN (*n*): titanium fluoride 689
FLUORTOLUOL (*n*): fluoro (fluo) toluene 690
FLUORÜR (*n*): fluoride 691
FLUORVERBINDUNG (*f*): fluorine compound 692
FLUORWASSERSTOFF (*m*): hydrogen fluoride; HF; hydrofluoric acid (*Acidum hydrofluoricum*); H_2F_2; Sg. 0.98; Sg. gas 0.7126; anyhydrous liquid Mp. −92.3°C; Bp. 19.44°C. 693
FLUORWASSERSTOFFSAUER: hydrofluoride of; combined with hydrofluoric acid 694
FLUORWASSERSTOFFSÄURE (*f*): hydrofluoric acid (*Acidum hydrofluoricum*); HF; (see FLUOR-WASSERSTOFF) 695
FLUORZINK (*n*): zinc fluoride; ZnF_2; Sg. 4.612; Mp. 734°C. 696
FLUORZINN (*n*): tin fluoride 697
FLUR (*f*): field; meadow; floor; vestibule; plain; hall; landing 698
FLURGANG (*m*): corridor 699
FLUSS (*m*): flux; fluorspar; enamel; paste (Gem); fusion; flow; river; stream; catarrh; issue; discharge; rheumatism; melt (Metal); running; flush (Cards); salin (see SALIN) 700
FLUSSARTIG: river-like; rheumatic; catarrhal 701
FLUSSÄTHER (*m*): fluoric ether 702
FLUSSBETT (*n*): river bed 703
FLUSSDAMPFER (*m*): river steamer 704
FLUSSEISEN (*n*): ingot iron; ingot metal; mild (cast) steel; structural steel; homogenous iron; wrought iron produced in the liquid state 705
FLUSSEISENBLECH (*n*): mild steel plate 706
FLUSSERDE (*f*): earthy fluorite (see FLUORIT) 707
FLUSSFIEBER (*m*): rheumatic fever; influenza 708
FLUSSGOLD (*n*): stream (river) gold 709
FLUSSHARZ (*n*); (gum-)animé; copal; gum copal; cowrie (see KOPAL) 710
FLÜSSIG: liquid; fluid; rheumatic; melted 711
FLÜSSIGER EXTRAKT (*m*): fluid extract; liquid extract 712
FLÜSSIGES CHLORZINK (*n*): zinc chloride solution 713
FLÜSSIGKEIT (*f*): liquid; fluid; liquor; humor (Med.) 714
FLÜSSIGKEIT (*f*), **ENTSPANNTE-**: reduced pressure fluid 715
FLÜSSIGKEIT (*f*), **IDEALE-**: perfect (ideal) liquid (a liquid free of friction) 716
FLÜSSIGKEIT (*f*), **REIBUNGSFREIE-**: ideal (perfect) liquid (a liquid free of friction) 717
FLÜSSIGKEITSDRUCK (*m*): fluid (hydraulic) pressure 718
FLÜSSIGKEITSMASS (*n*): liquid (fluid) measure 719
FLÜSSIGKEITSMENGE (*f*): amount (quantity) of liquid or fluid 720
FLÜSSIGKEITSMESSER (*m*): liquid meter 721
FLÜSSIGKEITSPRISMA (*n*): liquid prism 722
FLÜSSIGKEITSREIBUNG (*f*): fluid friction 723
FLÜSSIGKEITSSÄULE (*f*): column of liquid 724
FLÜSSIGKEITSSPIEGEL (*m*): surface of a liquid 725
FLÜSSIGKEITSVERTEILER (*m*): distributor 726
FLÜSSIGKEITSWÄRME (*f*): heat of liquid; liquid (fluid; visible; sensible) heat 727
FLÜSSIGKEITSZERSTÄUBER (*m*): fluid (liquid) atomizer 728
FLÜSSIGKEIT (*f*), **VOLKOMMENE-**: ideal (perfect) liquid (a liquid free from friction) 729
FLÜSSIGMACHEN: to liquefy; melt; (*n*): liquefaction; melting 730
FLÜSSIGMACHUNG (*f*): liquefaction; melting 731
FLÜSSIGWERDEND: liquescent 732
FLUSSKIESELSAUER: fluosilicate of; combined with fluosilicic acid 733
FLUSSKIESELSÄURE (*f*): fluosilicic acid; hydrofluosilicic (silicofluoric; hydrosilicifluoric; hydrofluorsilicic; sand) acid; H_2SiF_6 734

FLUSSKOMMISSION (*f*): river-(s pollution) commission 735
FLUSSMITTEL (*n*): fluxing medium; flux; antirheumatic 736
FLUSSOFEN (*m*): flowing furnace 737
FLUSSPULVER (*n*): flux powder 738
FLUSSRADDAMPFER (*m*): river paddle steamer 739
FLUSSREINIGUNGSKOMMISSION (*f*): rivers pollution commission 740
FLUSSRÖSTE (*f*): stream retting process of flax (to obtain bast fibre) (largely in Belgium) (Textile) (see RÖSTE) 741
FLUSSSAND (*m*): river sand 742
FLUSSSAUER: hydrofluoride of; combined with hydrofluoric acid 743
FLUSSSÄURE (*f*): hydrofluoric acid; HF (see FLUORWASSERSTOFF) 744
FLUSSSCHIFF (*n*): river (steamer) boat 745
FLUSSSCHMIEDEEISEN (*n*): ingot iron; wrought iron produced in the liquid state 746
FLUSS (*m*), SCHWARZER-: black flux (a reducing flux; mainly potassium carbonate and carbon) 747
FLUSSSPAT (*m*): fluorspar; Derbyshire spar; fluorite; calcium fluoride; CaF_2; Sg. 3.15 (see also FLUORIT) 748
FLUSSSPATERDE (*f*): earthy fluorite (see FLUSSSPAT) 749
FLUSSSPATSÄURE (*f*): hydrofluoric acid; HF; (see FLUORWASSERSTOFF) 750
FLUSSSTOFF (*m*): humour (Med.) 751
FLUSSTEIN (*m*): fluorite (see FLUORIT) 752
FLUSSTINKTUR (*f*): anti-rheumatic tincture 753
FLUSSWASSER (*n*): river water 754
FLUSSWASSERSTOFFSÄURE (*f*): (see FLUOR-WASSERSTOFF) 755
FLUSS (*m*), WEISSER-: white flux (an oxydating flux of potassium carbonate, nitrate and nitrite); leucorrhœa (*Fluor albus*), white discharge-(of females) (Med.) 756
FLUT (*f*): flood; deluge (high-)tide; waves; water; flow; flood tide; torrent; inundation; stream; crowd 757
FLUTANKER (*m*): flood anchor 758
FLUTTOR (*n*): flood gate 759
FLUTWELLE (*f*): tidal wave 760
FLUTZEICHEN (*n*): high-water mark 761
FLUTZEIT (*f*): high water-(time) 762
FOCKBAUCHGORDING (*f*): fore buntline (Naut.) 763
FOCKBRASSE (*f*): fore brace (Naut.) 764
FOCKBULIN-(E) (*f*): fore bowline (Naut.) 765
FOCKE (*f*): fore-sail (Naut.) 766
FOCKHALS (*m*): fore tack (Naut.) 767
FOCKLEESEGEL (*n*): fore studding sail 768
FOCKMARS (*m*): fore top 769
FOCKMAST (*m*): fore mast 770
FOCKMASTSALING (*f*): fore mast cross tree 771
FOCKRAH (*f*): fore yard 772
FOCKRAHPEERD (*n*): fore yard foot-rope 773
FOCKSEGEL (*n*): foresail 774
FOCKSPIERE (*f*): fore yard boom 775
FOCKSTAG (*n*): fore stay 776
FOCK (*f*), STAG-: stay foresail 777
FOCKSTAGSEGEL (*n*): fore stay sail 778
FOCK (*f*), STURM-: storm jib 779
FOCKWANTEN (*n.pl*): fore rigging 780
FOCUS (*m*): focus 781
FOCUSDIFFERENZ (*f*): focal difference (non-coincidence of foci) 782
FOCUSLÄNGE (*f*): focal length 783
FOCUSMESSER (*m*): focimeter (an instrument for determining the difference between the visual focus of a lens and the chemical focus) 784
FOCUSTIEFE (*f*): depth of focus 785
FOCUSWEITE (*f*): focal length, focus 786
FÖHRE (*f*): fir-(tree); Scotch pine (see KIEFER) 787
FÖHREN: fir, of fir 788
FÖHRENMISTEL (*f*): fir mistletoe (mistletoe which only grows on fir trees) (see MISTEL) 789
FÖHRENSAMENÖL (*n*): fir-seed oil (from seeds of *Pinus sylvestris*) 790
FOLGE (*f*): sequence; consequence; series; set; future; result; succession; order; issue; conclusion; compliance 791
FOLGEN: to follow; succeed; obey; attend (see also FOLGE) 792
FOLGEND: subsequent; next; following; consecutive 793
FOLGENDE (*n*): following 794
FOLGENDERMASSEN: as follows; in the following manner 795
FOLGENDERWEISE: (see FOLGENDERMASSEN) 796
FOLGENLOS: without (effect) consequences 797
FOLGENSCHWER: portentous; important 798
FOLGEPUNKTE (*m.pl*): consequent (points) poles (Magnetism); points of (inference) conclusion (in a discourse, etc.) 799
FOLGERECHT: consequent; conclusive; consistent; logical-(ly) 800
FOLGEREIHE (*f*): order of succession; sequence; series 801
FOLGERICHTIGKEIT (*f*): logical (result) sequence 802
FOLGERN: to conclude; infer 803
FOLGERUNG (*f*): conclusion; inference; deduction 804
FOLGESATZ (*m*): conclusion; inference; deduction 805
FOLGEWIDRIG: inconsequent 806
FOLGEWIRKUNG (*f*): resultant; consequence 807
FOLGEZEIT (*f*): future, futurity, posterity 808
FOLGLICH: consequently; in consequence; therefore; of course; accordingly 809
FOLGSAM: obedient; docile; tractable 810
FOLIANT (*m*): folio (Volume) 811
FOLIE (*f*): foil; leaf metal; film (Photography) 812
FOLIIEREN: to silver; cover with foil; page; folio (a book) 813
FOLIKEL (*m*): follicule 814
FOLIO (*n*): folio, page 815
FOLIOFORMAT (*n*): folio size 816
FOLIUM (*n*): foil, leaf 817
FOLTERKAMMER (*f*): torture chamber 818
FOMITIN (*n*): a fluid extract from two kinds of fungi (*Fomes cinnamomeus* and *F. igniarius*) 819
FOND (*m*): foundation; base; basis; capital; funds; ground (Dye); bottom 820
FONDS (*m*): funds, capital, stock-(s) 821
FONDSBESITZER (*m*): stockholder; shareholder 822
FONZIERMASCHINE (*f*): staining machine (Paper) 823
FORCIEREN: to force; overtax; overload (Plant) 824
FORCIERKRANKHEIT (*f*): strain disease (Metal) 825
FORCIERUNGSVERSUCH (*m*): overload-(ing) test, forcing test 826

FÖRDERANLAGE (*f*) : conveying plant, transporter 827

FÖRDERBAND (*n*) : belt conveyor ; conveyor belt 828

FÖRDERER (*m*): furtherer, promoter; patron; supplier ; conveyor ; conductor, transporter 829

FÖRDERGERÜST (*n*) : hoist (frame, framework) framing 830

FÖRDERGESTÄNGE (*n*) : mine track or roadway 831

FÖRDERGURT (*m*) : belt conveyor ; conveyor belt 832

FÖRDERHÖHE (*f*) : supply height ; head (Pump) 833

FÖRDERKOHLE (*f*) : pit coal (coal as it leaves the pit) 834

FÖRDERLEISTUNG (*f*) : quantity delivered, delivery, duty, output (of pumps, etc.) 835

FÖRDERLICH : serviceable ; speedy ; useful ; conducive ; beneficial 836

FÖRDERMASCHINE (*f*) : driving machine, driving (hauling) engine 837

FORDERN : to demand ; ask ; require ; postulate ; summon ; challenge ; claim ; exact 838

FÖRDERN : to further ; forward ; advance ; supply ; dispatch ; convey ; raise ; expedite ; draw out ; promote ; transport 839

FÖRDERNIS (*f*) : help ; furtherance 840

FÖRDERQUANTUM (*n*) : output 841

FÖRDERSCHACHT (*m*) : hauling shaft (Mining) ; mine-shaft 842

FÖRDERSCHNECKE (*f*) : screw conveyor 843

FÖRDERSEIL (*n*) : haulage (cable) rope 844

FÖRDERSTEINSALZ (*n*) : crude rock salt (see **HALIT**) 845

FÖRDERSTOLLEN (*m*) : supply gallery (Mining) ; adit 846

FORDERUNG (*f*) : demand, etc. (see **FORDERN**) 847

FÖRDERUNG (*f*) : furthering ; hauling ; raising ; conveying ; supply-(ing) ; transport-(ing) ; output (see also **FÖRDERN**) 848

FÖRDERUNGSMITTEL (*n*) : adjuvant (Pharm.) 849

FÖRDERWAGEN (*m*) : coal (wagon) trolley, tram, supply waggon 850

FÖRDERWIDERSTAND (*m*) : supply resistance ; supply head ; discharge head (against which a pump works) 851

FORELLENEISEN (*n*) : mottled pig iron 852

FORENSISCH : forensic, learned in the law, for law purposes 853

FORENSISCHE UNTERSUCHUNG (*f*) : examination (testing) in compliance with law requirements 854

FORM (*f*) : form ; figure ; shape ; cut ; size ; mould ; pattern ; tuyére ; lines (of a ship) ; frame (Soap) ; conformation 855

FORMALDEHYD (*n*) : formalin, formic aldehyde, paraform, formaldehyde, oxymethylene, paraformaldehyde ; CH_2O ; Sg. 0.8153 (Gas) ; Bp. -21°C. 856

FORMALDEHYDAZETAMID (*n*) : formaldehyde-acetamide (see **FORMICIN**) 857

FORMALDEHYD-GELATINE (*f*) : formaldehyde gelatine (see **GLUTOL**) 858

FORMALDEHYDHARZ (*n*): formaldehyde resin (see **BAKELIT** and **CUMARON**) 859

FORMALDEHYDSEIFE (*f*) : formaldehyde soap 860

FORMALDEHYDSULFOXYLAT (*n*) : formaldehyde sulfoxylate 861

FORMALIEN (*pl*) : formalities 862

FORMALIN (*n*) : formalin (see **FORMALDEHYD**) 863

FORMALINSEIFE (*f*) : formalin soap 864

FORMALITÄT (*f*) : formality 865

FORMAMID (*n*) : formamide, methanamine ; CH_3ON ; Sg. 1.1337 ; Mp. about -1°C. ; Bp. 85-95°C. 866

FORMAMINT (*n*) : formamint (compound of formaldehyde and lactose) (a disinfectant) 867

FORMAN (*n*) : forman, chloromethylmenthyl ether ; $C_{10}H_{19}OCH_2Cl$ 868

FORMÄNDERUNG (*f*) : change of form ; strain ; deformation 869

FORMÄNDERUNGSARBEIT (*f*) : strain, energy, work (energy) of deformation 870

FORMANILID (*n*) : formanilide, phenylformamide ; $C_6H_5NH.CHO$; Mp. 46°C. 871

FORMARBEIT (*f*) : casting ; moulding ; mould (pattern) making 872

FORMART (*f*) : type of (form ; mould) pattern 873

FORMAT (*n*) : size ; form (usually of books) ; formate (Chem.) 874

FORMBAR : plastic ; shapeable ; capable of being formed or moulded 875

FORMBEKLEIDUNG (*f*) : mould case 876

FORMEL (*f*) : formula (Maths., Chem., etc.) 877

FORMELBILD (*n*) : structural formula 878

FORMELL : formal-(ly) 879

FORMELZEICHEN (*n* and *pl*) : formula sign-(s) 880

FORMEN : to form ; shape ; fashion ; mould ; cast ; frame (Soap) ; make 881

FORMENLEHRE (*f*) : etymology (the science of the derivation of words) 882

FORMENMACHER (*m*) : pattern maker ; moulder 883

FORMENÖL (*n*) : oil for forms or moulds 884

FORMENSCHNEIDER (*m*) : (see **FORMENMACHER**) 885

FORMER (*m*) : moulder, sand-moulder 886

FORMERDE (*f*) : moulding (modelling) clay 887

FORMEREI (*f*) : moulding ; forming ; modelling ; moulding house 888

FORMERZ (*n*) : rich silver ore 889

FORMGEBUNG (*f*) : shaping ; fashioning ; moulding ; modelling 890

FORMIAT (*n*) : formate (Chem.) 891

FORMICIN (*n*) : formicin, formaldehyde-acetamide ; $CH_3CO.NHCH_2OH$; Sg. 1.25 892

FORMIEREN : to form ; shape ; fashion ; mould ; cast ; pattern ; make ; frame (Soap) 893

FÖRMIG : formed ; shaped ; moulded (as a suffix = like) 894

FORMIN (*n*) : formin (see **HEXAMETHYLENTETRA-MIN**) 895

FORMIZIN (*n*) : (see **FORMICIN**) 896

FORMKASTEN (*m*) : moulding box 897

FÖRMLICH : formal-(ly) ; ceremonious ; plain ; downright 898

FÖRMLICHKEIT (*f*) : formality ; form ; ceremony 899

FÖRMLING (*m*) : moulded article 900

FORMLOS : without form, shapeless ; informal 901

FORMMASCHINE (*f*) : moulding machine 902

FORMOL (*n*) : formol (see **FORMALDEHYD**) 903

FORMONUKLEINSAUER : formonucleinate 904

FORMOTANNINSTREUPULVER (*n*) : formotannine powder for strewing 905

FORMSAND (*m*) : moulding sand 906

FORMSCHEIBE (*f*) : small (tiny) pane of glass ; turning top of a potter's wheel 907

FORM (*f*), **SCHLANKE-** : slim (lean) form 908

FORMSTEIN (*m*) : stone pipe ; stone for moulding ; shaped brick 909

FORMSTÜCK (*n*): section iron; make-up piece; adaptor; template pipe; adapting pipe; (any)-shaped part 910
FORMTISCH (*m*): moulding frame (Candlemaker's) 911
FORMULAR (*n*): (printed)-form 912
FORMULIEREN: to formulate 913
FORMVERÄNDERUNG (*f*): change of form, deflection, deformation 914
FORM (*f*), **VÖLLIGE-**: full form (of body of ship) 915
FORMWECHSEL (*m*): change of form; accomodation bill (Comm.) 916
FORMYLSÄURE (*f*): formic acid; HCOOH; Sg. 1.2178; Mp. 8.3°C.; Bp. 100.8°C. 917
FORMYLTRICHLORID (*n*): (see **CHLOROFORM**) 918
FORMYLZELLULOSE (*f*): cellulose ester 919
FORSCHEN: to investigate; search; enquire into; be inquisitive 920
FORSCHER (*m*): investigator; enquirer; searcher 921
FORSCHSINN (*m*): spirit of enquiry; enquiring turn of mind 922
FORSCHUNG (*f*): investigation; enquiry; research 923
FORSCHUNGSGEBIET (*n*): field of investigation; domain of research 924
FORSCHUNGSGEIST (*m*): spirit of (enquiry; research) investigation 925
FORSCHUNGSREISE (*f*): voyage of discovery 926
FORSTERIT (*m*): fosterite (a magnesium olivine) Mg_2SiO_4; Sg. 3.3 927
FORSTGERÄUME (*n*): forest clearing 928
FORSTKULTUR (*f*): forest culture, afforestation 929
FORSTKUNDE (*f*): forrestry; woodcraft 930
FORSTREVIER (*n*): forest (district) region 931
FORSTWESEN (*n*): forest matters, forestry, afforestation, any affairs relating to a forest 932
FORSTWIRTSCHAFT (*f*): forest economy 933
FORSTWISSENSCHAFT (*f*): forestry-(science), wood-craft 934
FORT: on; along; forth; away; off; (*n*): fort; fortress; (as a prefix=to continue; persist; keep on) 935
FORTAN: henceforth; henceforward; onward; further 936
FORTBAU (*n*): extension (of building) 937
FORTBESTEHEN: to continue; exist; persist subsist; endure; last 938
FORTBEWEGUNG (*f*): locomotion; progression; propulsion 939
FORTBILDUNG (*f*): continuation of study 940
FORTDAUER (*f*): permanence; duration; lasting; continuance 941
FORTDAUERND: incessant; continuous; permanent 942
FORTEXISTENZ (*f*): continued existence, life after death 943
FORTFAHREN: to continue; set off; depart; go on; drive off; proceed 944
FORTFLIEGEN: to fly away; be wafted away 945
FORTFLIESSEN: to flow away; run off 946
FORTFÜHREN: to lead away; carry away; continue; convey; prosecute 947
FORTGANG (*m*): progress; advance; departure 948
FORTHELFEN: to help on 949
FORTKOMMEN: to get (away) on; progress; escape; prosper; succeed 950
FORTLASSEN: to let (allow to) go; leave out; omit; dispense with; permit to depart; elide 951
FORTLAUFEN: to run away; flow away; escape 952
FORTLAUFEND: continual; continuous; successive; incessant; constant 953
FORTLEITEN: to remove; draw off; lead away; conduct; (*n*): leading away; removal 954
FORTPFLANZEN: to propagate; transmit; communicate; transplant 955
FORTREISSEN: to sweep away; break (tear) down 956
FORTSATZ (*m*): continuation; process (Med.); extension piece (Mech.) 957
FORTSCHAFFEN: to take away; remove; dismiss; get rid of; transport; discharge; discard 958
FORTSCHREITEN: to progress, continue, proceed, advance, improve; (*n*): progress (see **FORTSCHRITT**) 959
FORTSCHREITEND: progressive; onward; continued 960
FORTSCHRITT (*m*): advance; progress; continuance; improvement; development 961
FORTSCHWEMMEN: to wash away 962
FORTSETZEN: to continue; carry on; put away; proceed 963
FORTSETZUNG (*f*): continuation; prosecution; pursuit 964
FORTSTRÖMEN: to stream (flow) away 965
FORTTRAGEN: to transport, carry away 966
FORTTREIBEN (*n*): propulsion; to propel, drive away, prosecute 967
FORTWÄHREND: continual; continuous; permanent; constant; incessant; perpetual; lasting 968
FORTZUCHT (*f*): propagation 969
FOSSIL: fossil, fossilized; (*n*): fossil 970
FOSSILBILDUNG (*f*): fossilization, formation of fossils; petrifaction 971
FOSSILHARZ (*n*): fossil resin, Amber, Asphalt, Ozocerite (see **BERNSTEIN, ASPHALT, OZOKERIT**) 972
FOSSILIENHALTIG: fossiliferous 973
FOURIERREIHE (*f*): Fourier series (a harmonic series) 974
FOURNIER (*n*): veneer 975
FOURNIEREN: to veneer; inlay 976
FOURNIERHOLZ (*n*): veneering (inlaying) wood 977
FOURNIERPRESSE (*f*): veneer-(ing) press 978
FOURNIERSÄGE (*f*): veneer saw 979
FOWLERSCHE LÖSUNG (*f*): Fowler's solution (*Liquor Kalii arsenicosi*) (1% As_2O_3) 980
FRACHT (*f*): freight (by sea); cargo; load; shipment; carriage (by land); transport charges 981
FRACHTBRIEF (*m*): way bill; bill of lading, B/L 982
FRACHTDAMPFER (*m*): cargo steamer, tramp steamer 983
FRACHTEN: to freight, ship, load, transport, carry 984
FRACHTFREI: carriage (paid) free; including freight charges 985
FRACHTGELD (*n*): freight; cartage; transport charges; carriage 986
FRACHTGUT (*n*): freight; lading; cargo 987
FRACHTGUT (*n*), **ALS-**: by goods train 988
FRACHTKOSTEN (*f.pl*): freight-(charges) 989
FRACHTMÄKLER (*m*): shipping agent 990

FRACHTSCHIFF (*n*): cargo boat, merchantman, merchant ship 991
FRACHTSPESEN (*f.pl*): (see FRACHTGELD) 992
FRACHTUNFREI: carriage forward, excluding (exclusive of) carriage or freight 993
FRACHTVERKEHR (*m*): goods traffic 994
FRAGEBOGEN (*m*): enquiry sheet 995
FRAGEN: to ask; question; enquire; interrogate; demand 996
FRAGEWEISE (*f*): method of interrogation 997
FRAGEZEICHEN (*n*): note of interrogation; question mark 998
FRAGLICH: questionable; doubtful; aforementioned; in question 999
FRAGMENT (*n*): fragment 000
FRAGMENTARISCH: fragmentary 001
FRAKTION (*f*): fraction 002
FRAKTIONIERAPPARAT (*m*): fractionizing (fractionating) apparatus 003
FRAKTIONIEREN: to fractionate; fractionize 004
FRAKTIONIERKOLBEN (*m*): fractionating (fractionizing) flask 005
FRAKTIONIERKOLONNE (*f*): fractionating (fractionizing) column (a series of fractionizing apparatus) 006
FRAKTIONIERROHR (*n*): fractionating (fractionizing) tube 007
FRAKTIONIERTE DESTILLATION (*f*): fractional distillation (vapourizing by stages and then re-condensing) 008
FRAKTIONIERUNG (*f*): fractionation, fractionization 009
FRAKTIONSKOLBEN (*m*): (see FRAKTIONIER-KOLBEN) 010
FRAKTUR (*f*): fracture 011
FRANCO: post-free, post-paid, carriage paid, free 012
FRANGULA (*f*): frangula, arrow wood, Persian berries, (alder-) buckthorn (from *Rhamnus frangula*) 013
FRANGULAWEIN (*m*): frangula wine 014
FRANKFURTERSCHWARZ (*n*): Frankfurt black (see RABENSCHWARZ) 015
FRANKIEREN: to pre-pay, stamp (Letters) 016
FRANKIERMASCHINE (*f*): stamping machine (for postage stamps) 017
FRANKLINIT (*m*): franklinite; (Fe,Mn,Zn)O. $(Fe,Mn)_2O_3$; Sg. 5.15-5.9 018
FRANKLINSCHE TAFEL (*f*): Franklin's plate 019
FRANSEN: to fringe; put a valence round 020
FRANZBAND (*m*): calf binding (Book) 021
FRANZBOHNE (*f*): French bean 022
FRANZBRANNTWEIN (*m*): cognac; brandy 023
FRANZBRANNTWEINESSENZ (*f*): cognac (brandy) essence 024
FRANZGOLD (*n*): French leaf gold 025
FRANZOBST (*n*): wall fruit; dwarf tree fruit 026
FRANZOSE (*m*): Frenchman; monkey wrench (Mech.) 027
FRANZOSENHARZ (*n*): guaiacum; gum guaiac (from *Guaiacum sanctum* and *G. officinale*) 028
FRANZOSENHOLZ (*n*): guaiacum wood; lignum vitæ (Sg. 0.79); Balsam wood 029
FRANZOSENHOLZÖL (*n*): guaiacum wood oil; Sg. 0.965-0.975 030
FRANZÖSISCHE BEEREN (*f.pl*): Avignon berries (see GELBBEEREN) 031
FRANZTOPAS (*m*): smoky topaz (see TOPAS) 032
FRAPPANT: striking, surprising 033
FRAPPIEREN: to strike, surprise, astonish 034
FRÄSEN: to mill 035
FRÄSER (*m*): milling (cutter) machine 036
FRÄSMASCHINE (*f*): milling (edging) machine 037
FRÄSWERK (*n*): milling (edging) machine 038
FRAUENARZT (*m*): gynecologist (a doctor specially versed in female complaints) 039
FRAUENBEERE (*f*): hawthorn berry 040
FRAUENBIRKE (*f*): weeping birch (*Betula alba pendula*) 041
FRAUENBISS (*m*): Germander (*Teucrium*) 042
FRAUENBLUME (*f*): Scarlet pimpernel (*Anagallis arvensis*) 043
FRAUENDISTEL (*f*): Scotch thistle (*Onopordum acanthium*) 044
FRAUENEIS (*n*): selenite (Min.); $CaSO_4.2H_2O$; Moonstone 045
FRAUENFEIND (*m*): woman hater; misogynist 046
FRAUENFLACHS (*m*): flax (see FLACHS) 047
FRAUENGLAS (*n*): selenite (Min.); $CaSO_4.2H_2O$ (see ERGDLAS) 048
FRAUENGLAS, RUSSISCHES- (*n*): muscovite; $H_2KAl_3(SiO_4)_3$ (see MUSKOVIT) 049
FRAUENHAAR (*n*): maidenhair (Bot.) (*Adiantum capillus-veneris*) 050
FRAUENHASS (*m*): hatred of women; misogyny 051
FRAUENHEILKUNDE (*f*): gynecology (that portion of medicial science dealing with female complaints) 052
FRAUENKÄFER (*m*): ladybird (*Coccinellidæ* family of beetles) 053
FRAUENMILCH, LIEB- (*f*): Liebfrauenmilch (a Rhenish wine) 054
FRAUENSPAT (*m*): selenite (Min.) (see FRAUEN-GLAS) 055
FRAUENSPIEGEL (*m*): Venus's looking-glass (*Specularia hybrida*) (*Campanulaceae* order of plants) 056
FREGATTE (*f*): frigate (Marine) 057
FREI: free; uncombined; liberal; bold; open; post-paid; pre-paid; frank; exempt; unconfined; unchained; vacant; independent; disengaged 058
FREIBERGIT (*m*): freibergite (from Freiberg in Saxony), tetrahedrite (see ANTIMONFAHLERZ) (about 32% silver) (similar to polytelite, Great Britain) 059
FREIBLEIBEND: not binding (of prices) 060
FREIBORD: free on board; f.o.b. 061
FREIBORD (*m*): free board 062
FREIBRIEF (*m*): permit; patent; licence; charter 063
FREICOUVERT (*n*): stamped envelope 064
FREIDENKER (*m*): free-thinker; agnostic 065
FREIDREHEN: to throw (Ceramics); unscrew (of screws, nuts, etc.) 066
FREIEIGEN: freehold 067
FREIE, INS- (*n*): to the atmosphere; into the open-(air) 068
FREIESLEBENIT (*m*): freieslebenite (natural lead-silver sulfantimonide $(Pb,Ag_2)_3Sb_4S_{11}$ (about 25% silver) Sg. 6.3 069
FREIGEBIG: generous; prodigal; liberal; munificent 070
FREIGEBUNG (*f*): release; freeing 071
FREIGEIST (*m*): freethinker; agnostic 072
FREIGEPÄCK (*n*): free luggage 073
FREIGESINNT: (see FREIGEBIG) 074
FREIGUT (*n*): freehold property, goods free of duty 075
FREIGUTSBESITZER (*m*): freeholder (of property) 076

FREIHAFEN (*m*): free port 077
FREIHALTEN: to pay one's way; to pay -(for someone); treat 078
FREIHANDEL (*m*): free trade 079
FREIHANDZEICHNUNG (*f*): free hand-(drawing) 080
FREI HÄNGEND: freely suspended 081
FREIHEIT (*f*): freedom; liberty; franchise; privilege; license 082
FREIHEITSBRIEF (*m*): Charter-(patent) 083
FREIHEITSGRAD (*m*): degree of freedom 084
FREIHEITSKRIEG (*m*): war of independence 085
FREIHEITSURKUNDE (*f*): (see **FREIHEITSBRIEF**) 086
FREIHERZIG: frank; generous; open-hearted 087
FREILASSUNG (*f*): freeing, emancipation 088
FREILAUF (*m*): by-pass opening 089
FREILICH: certainly; to be sure; of course; frankly 090
FREILICHTAUFNAHME (*f*): outdoor photograph-(y) 091
FREIMACHEN: to set free; liberate; disengage; prepay; emancipate 092
FREIMARKE (*f*): postage stamp 093
FREIMAURER (*m*): free mason 094
FREIMÜTIG: frank; sincere; candid 095
FREIPASS (*m*): passport 096
FREISINN (*m*): free-mindedness, liberality, freethought 097
FREISPRECHEN: to acquit; dispense 098
FREISPRECHUNG (*f*): acquittal, absolution 099
FREISTAAT (*m*): republic; free state 100
FREISTUNDE (*f*): leisure hour 101
FREIUNG (*f*): emancipation 102
FREIWACHE (*f*): watch below (Naut.) 103
FREIWERDEN: to become free; be set free; be liberated; be disengaged; make (be) so bold; (*n*): liberation; disengagement 104
FREIWERDEND: nascent; liberated; disengaged 105
FREI WERK (*n*): ex works 106
FREIWILLIG: voluntary; free; spontaneous-(ly) 107
FREMD: foreign; strange; exotic; extraneous; peculiar; unusual 108
FREMDARTIG: odd; heterogeneous; extraneous; uncommon 109
FREMDENAMT (*n*): alien office 110
FREMDENBUCH (*n*): visitor's book 111
FREMDKÖRPER (*m*): foreign (matter; substance) body) 112
FREMDMETALL (*n*): foreign metal (in ores, etc.) 113
FREQUENTIEREN: to frequent 114
FREQUENZ (*f*): frequency; crowd; traffic; attendance 115
FRESKOFARBE (*f*): fresco colour 116
FRESKOGEMÄLDE (*n*): fresco-(painting) 117
FRESSEND: corrosive 118
FRESSZELLE (*f*): phagocyte; eating cell; leucocyte; white blood corpuscle 119
FREUDE (*f*): joy; comfort; enjoyment; delight; satisfaction; pleasure; gladness; heartiness; cheerfulness 120
FREUEN: to be glad; rejoice; gladden; enjoy oneself; be happy; be cheerful 121
FREUNDLICH: friendly; cheerful; kind; pleasant; affable 122
FREVEL (*m*): misdeed; crime; misdemeanour; outrage 123
FREVELHAFT: wicked; malicious; outrageous; criminal 124
FREVELN: to outrage, commit a crime 125
FRIEDE (*f*): peace; tranquility 126
FRIEDEN (*m*): (see **FRIEDE**) 127
FRIEDENSBRUCH (*m*): breach of the peace 128
FRIEDENSEINLEITUNGEN (*f.pl*): peace preliminaries 129
FRIEDENSFUSS (*m*): peace (footing) establishment 130
FRIEDENSPFEIFE (*f*): pipe of peace 131
FRIEDENSRICHTER (*m*): justice of the peace 132
FRIEDENSSCHLUSS (*m*): conclusion of peace 133
FRIEDENSUNTERHANDLUNGEN (*f.pl*): peace negotiations 134
FRIEDENSVERTRAG (*m*): peace treaty 135
FRIEDLICH: peaceable; peaceful; pacific 136
FRIEREN: to freeze; congeal; feel cold; (*n*); freezing; shivering; ague 137
FRIERPUNKT (*m*): freezing point 138
FRIESEL (*m*): miliaria (skin eruption usually due to perspiration) (Med.) 139
FRIESEL (*m*), **-FIEBER** (*n*): miliary fever, purples (Med.) 140
FRIKTION (*f*): friction 141
FRIKTIONSHAMMER (*m*): friction hammer 142
FRIKTIONSKALANDER (*m*): friction calender (Paper) 143
FRIKTIONSKUPPELUNG (*f*): friction clutch 144
FRIKTIONSMESSER (*m*): friction meter 145
FRIKTIONSPRESSE (*f*): friction press 146
FRIKTIONSRAD (*n*): friction wheel 147
FRIKTIONSSCHEIBE (*f*): friction (plate) disc 148
FRISCH: fresh; green (Hides or Wood); cheerful; vigorous; gay; cool; new; recent; crude; raw 149
FRISCHARBEIT (*f*): (re)-fining process 150
FRISCHBLEI (*n*): refined lead 151
FRISCHDAMPF (*m*): live steam 152
FRISCHDAMPFLEITUNG (*f*): main (live) steam pipe 153
FRISCHEISEN (*n*): refined iron 154
FRISCHEN: to refine; fine; revive; reduce (Lead); puddle; refresh; polish 155
FRISCHEN (*n*): refining (of metal); reducing 156
FRISCHER (*m*): finer; refiner 157
FRISCHEREI (*f*): refinery; finery 158
FRISCHEREIROHEISEN (*n*): forge pig (Iron) 159
FRISCHESSE (*f*): refining furnace; refinery; puddling furnace 160
FRISCHFEUER (*n*): refining (fire) furnace 161
FRISCHFEUEREISEN (*n*): charcoal iron, refined (fined) iron 162
FRISCHFEUERSCHLACKEN (*f.pl*): refinery slag (55-60% Fe) (Steel works process by-product) 163
FRISCHGEFÄLLT: freshly felled; freshly precipitated; newly felled (Trees) 164
FRISCHGESTEIN (*n*): solid rock 165
FRISCHHERD (*m*): fining (refining) furnace or forge, puddling furnace 166
FRISCHLING (*m*): scoria (Metal); young boar 167
FRISCHOFEN (*m*): refining (fining) furnace 168
FRISCHPFANNE (*f*): (s)-melting pan 169
FRISCHPROZESS (*m*): refining (fining) process 170
FRISCHSCHLACKE (*f*): refinery (ash; cinders) slag; dross (see **FRISCHFEUERSCHLACKEN**) 171
FRISCHSTAHL (*m*): refinery (refined) steel; German steel; fined steel 172
FRISCHSTÜCK (*n*): liquation cake (Copper) 173
FRISCHUNG (*f*): refining; reviving; refreshing; reducing (Lead); fining; puddling; polishing 174
FRISCHVERFAHREN (*n*): refining process 175

FRISCHWIRKUNG (*f*): refining 176
FRIST (*f*): time; space of time; respite; days of grace; notice 177
FRISTEN: to prolong; delay; postpone 178
FRISTENWEISE: at stated intervals 179
FRISTVERLÄNGERUNG (*f*): extension of time 180
FRITOFEN (*m*): frit kiln; calcar (Glass) 181
FRITTE (*f*): frit 182
FRITTEN: to frit; sinter; cohere 183
FRITTER (*m*): coherer; fritter (as apple fritters) 184
FRITTERKLEMME (*f*): coherer terminal 185
FRITTERPRÜFER (*m*): coherer tester 186
FRITTERRÖHRE (*f*): coherer tube 187
FRITTER (*m*), VAKUUM-: vacuum coherer 188
FRITTERWIDERSTAND (*m*): coherer resistance 189
FRITTUNG (*f*): fritting; sintering; coherence 190
FROH: glad; joyous; joyful; cheerful; happy; gay; merry; blithe 191
FROHLOCKEN: to exult, triumph 192
FROMM: gentle; pious; brave; religious; devout 193
FRÖMMELN: to affect piety, be a hypocrite 194
FROMMEN: to avail; benefit 195
FRÖNEN: to indulge; be a slave; drudge; toil 196
FRONLEICHNAM (*m*): Corpus Christi 197
FRONLEICHNAMSFEST (*n*): Corpus Christi day 198
FRONT-(E) (*f*): front, fore part, face 199
FRONTSCHOTT (*n*): front bulkhead 200
FROSCH (*m*): frog; cam; arm (Mach.); heel (of violin bow) 201
FROSCHADER (*f*): ranular vein 202
FROSCHARTEN (*f.pl*): batrachia; amphibia; anura 203
FROSCHKLEMME (*f*): draw tongs 204
FROSCHLAICH (*m*): frog's spawn 205
FROSCHLAICHPFLASTER (*n*): lead plaster; Pb $(C_{18}H_{31}O_2)_2$ 206
FROSCHLEINGESCHWULST (*f*): ranula (a mouth disease; a cystic swelling of the fraenum linguae) (Med.) 207
FROSCHLÖFFEL (*m*): water plantain (*Alisma natans*) 208
FROSCHMÄUSEKRIEG (*m*): batrachomyomachia, the battle of the frogs and mice (an old poem) 209
FROST (*m*): frost; cold; coldness; chill; frozen (soil) rock 210
FROSTBEULE (*f*): chilblain (Med.) 211
FROSTFIEBER (*n*): ague 212
FROSTIG: frosty, chilly, frigid 213
FROSTMISCHUNG (*f*): freezing mixture 214
FROSTMITTEL (*n*): chilblain remedy 215
FROSTPUNKT (*m*): freezing point 216
FROSTSALBE (*f*): chilblain ointment 217
FROSTSCHNITT (*m*): frozen section 218
FROSTSCHUTZPRÄPARAT (*n*): protector against frost 219
FROTTIERBÜRSTE (*f*): flesh brush 220
FROTTIEREN: to rub; brush 221
FRUCHT (*f*): fruit (*Fructus*); grain; crop; fœtus; embryo; profit; produce; product; result; harvest; effect 222
FRUCHTABTREIBUNGSMITTEL (*n*): abortefacient 223
FRUCHTACKER (*m*): cornfield 224
FRUCHTAROMA (*n*): fruit aroma 225
FRUCHTÄTHER (*m*): fruit (ether) essence (see ÄTHYLNITRIT; AMYLACETAT and AMYLNITRIT) 226
FRUCHTAUGE (*f*): (fruit)-bud 227
FRUCHTBALG (*m*): follicle 228
FRUCHTBAR: fruitful; fertile; plentiful; prolific 229
FRUCHTBARKEIT (*f*): fecundity; fruitfulness; fertility; plentifulness 230
FRUCHTBARMACHUNG (*f*): fertilization 231
FRUCHTBAUM (*m*): fruit tree 232
FRUCHTBEHÄLTER (*m*): pericarp (Bot.) 233
FRUCHTBILDUNG (*f*): fructification, formation of fruit 234
FRUCHTBLATT (*n*): carpellary leaf, carpophyll 235
FRUCHTBODEN (*m*): ovary (Bot.); fruitful soil; granary 236
FRUCHTBRAND (*m*): ergot 237
FRUCHTBRINGEND: fruitful, productive 238
FRUCHTEN: to be of use; bear fruit; have effect; avail 239
FRUCHTESSENZ (*f*): fruit essence (see FRUCHT-ÄTHER) 240
FRUCHTFELD (*n*): cornfield 241
FRUCHTGARTEN (*m*): orchard 242
FRUCHTGELEE (*n*): jam (fruit)-jelly 243
FRUCHTHALTER (*m*): uterus 244
FRUCHTHANDEL (*m*): fruit (corn) trade 245
FRUCHTHÄUTCHEN (*n*): epicarp (the outer of the three coverings of fruits) 246
FRUCHTHÜLLE (*f*): (seed)-capsule; pericarp (the covering of fruits) 247
FRUCHTHÜLSE (*f*): husk, shell 248
FRUCHTKEIM (*m*): embryo, germ 249
FRUCHTKERN (*m*): kernel 250
FRUCHTKNOTEN (*m*): ovary; germ; seed bud 251
FRUCHTKÖRPER (*m*): fruit receptacle 252
FRUCHTLOS: barren, fruitless 253
FRUCHTMARK (*n*): pulp, mesocarp (the middle of the three fruit coverings) 254
FRUCHTMUS (*n*): sauce, jam, stewed fruit 255
FRUCHTSAFT (*m*): fruit juice 256
FRUCHTSÄURE (*f*): fruit acid 257
FRUCHTSTAND (*m*): syncarpy, arrangement of fruit on its stem 258
FRUCHTSTIEL (*m*): peduncle (Bot.) 259
FRUCHTTRAGEND: fruit bearing 260
FRUCHTWASSER (*n*): amniotic fluid 261
FRUCHTWECHSEL (*m*): rotation of crops 262
FRUCHTWEIN (*m*): fruit wine; cider 263
FRUCHTZUCKER (*m*): fruit sugar; levulose; fructose; $C_6H_{12}O_6$; Sg. 1.555; Mp. 95°C. 264
FRUCTOSE (*f*): fructose (one of the grape sugar group of sugars) (see FRUCHTZUCKER and DEXTROSE) 265
FRÜH: early; premature; soon; forward; speedy 266
FRÜHE (*f*): early morning; earliness; prime 267
FRÜHEICHE (*f*): (see WINTEREICHE) 268
FRÜHER: earlier; formerly; sooner; former; prior 269
FRÜHGEBURT (*f*): premature birth 270
FRÜHJAHR (*n*): Spring 271
FRÜHLING (*m*): spring; youth; premature (birth) child 272
FRÜHREIF: precocious; early ripe 273
FRÜHSOMMERKATARRH (*m*): hay-fever (*Catarrhus æstivus*) (Med.) 274
FRÜHSTÜCK (*n*): breakfast 275
FRÜHSTÜCKEN: to breakfast 276
FRÜHZEITIG: early; premature; precocious; soon; forward; untimely 277
FRÜHZÜNDUNG (*f*): premature ignition 278

F SÄURE (*f*): F acid, beta-naphthylamine-delta-sulfonic acid; $C_{10}H_6(NH_2)SO_3H$ 2 : 7 279
FUCHS (*m*): fox; reynard; trestle; jack; flue; freshman; fluke; sorrel (chestnut; bay) horse; unmeltable piece of iron (Metal); uptake (of boilers) 280
FUCHSBRÜCKE (*f*): rear seal wall (bridge wall at the entrance to the flue) 281
FUCHSEN: to fluke 282
FUCHSGAS (*n*): flue gas 283
FUCHSIG: foxy; fox coloured; carroty; chestnut (reddish) brown 284
FUCHSIN (*n*): fuchsine (Dye); magenta, aniline red, roseine, harmaline, fuchsiacine, ruby; triamidotolyl-diphenyl carbinol; HOC $(C_6H_3CH_3NH_2)(C_6H_4NH_2)_2$ 285
FUCHSIT (*m*): fuchsite, chrome mica Sg.2·75 [286
FUCHSROT (*n*): reddish brown; ginger; fox colour 287
FUCHSSCHWANZ (*m*): pad saw, hand saw 288
FUCHSRÄUDE (*f*): Alopecia areata, disease of the hair (Med.) 289
FUDER (*n*): cartload; fuder; measure (Wine, etc.) 290
FUG (*m*): right; authority 291
FUGATO (*n*): fugue, fugal passage (Music) 292
FUGE (*f*): joint; juncture; junction; slit; groove; suture; seam; rabbet; rebate; fugue (Music) 293
FUGEBANK (*f*): joiner's bench; rabbeting plane 294
FUGEN: to join; unite; add; rabbet; rebate; groove 295
FÜGEN: to unite, add, join, rabbet, rebate, dispose, accommodate oneself to 296
FUGENGELENK (*n*): articulation 297
FUGEWORT (*n*): conjunction 298
FÜGLICH: reasonably; well; easily; suitable; proper; fit; pertinent; convenient 299
FÜGSAM: tractable; pliant; adaptive; yielding; supple; agreeable 300
FÜGUNG (*f*): joining; resignation; decree; joint; disposition; articulation 301
FÜHLBAR: sensible; perceptible; appreciable; palpable 302
FÜHLBARE WÄRME (*f*): sensible heat 303
FÜHLEN: to feel; (*n*): feeling; perception; sensation 304
FÜHLLOS: unfeeling, imperceptible, insensible 305
FÜHLORGAN (*n*): sensitive (part) device (of a pressure regulator, etc.) 306
FÜHLUNG (*f*): feeling; sensation; touch; contact; perception 307
FÜHRBAR: manageable, transportable 308
FUHRE (*f*): cart; load; carriage; carrying; conveyance; vehicle 309
FÜHREN: to guide; lead; carry; convey; bear; wear; construct; carry on; cause; point to (the fact); bring; conduct; keep (Books); run (Plant); deal; harbour; drive; wage 310
FÜHREND: leading, etc. (see **FÜHREN**) (largely used as a suffix meaning, bearing, containing) 311
FÜHRER (*m*): driver; guide; leader, conductor 312
FUHRGELD (*n*): carriage 313
FUHRSTRASSE (*f*): highway, carriage road 314
FÜHRUNG (*f*): conduction; guide; guidance; leading, etc. (see **FÜHREN**) 315
FÜHRUNGSSCHLITTEN (*m*): guide bar 316
FÜHRUNGSSTANGE (*f*): guide bar; guide rod 317
FUHRWERK (*n*): vehicle; carriage; cart (see also **WAGEN**) 318
FUHRWERKSWAGE (*f*): weigh-bridge 319
FÜLLAPPARAT (*m*): filling (bottling) apparatus 320
FÜLLBAR: capable of being filled 321
FÜLLBRETT (*n*): panel 322
FÜLLDRUCK (*m*): filling pressure (of gas cylinders, etc.) 323
FÜLLE (*f*): filling; fullness; plenty; sufficiency; amplitude; abundance; body (Colour); depth; packing; contents; stuffing 324
FÜLLEN: to fill-(up); pour; put-(in); load; stuff; pack; replenish; bottle; plump (Leather); (*n*): filling; foal; colt; filly 325
FÜLLER (*m*): filler; loader; charger 326
FULLERERDE (*f*): Fuller's earth (an aluminium-magnesium hydro-silicate) 327
FÜLLHAUS (*n*): filling (room) house 328
FÜLLHOLZ (*n*): filling timber 329
FÜLLHORN (*n*): cornucopia 330
FÜLLKEIL (*m*): wedge 331
FÜLLKELLE (*f*): filling ladle 332
FÜLLKORB (*m*): basket (Mining) 333
FÜLLKÖRPER (*m*): filter bed (which is alternately filled and emptied with the waste water of sewage works; (see also **TROPFKÖRPER**); filling meterial (see **FÜLLMATERIAL** and also **RASCHIGRINGE**) 334
FÜLLLÖFFEL (*m*): filling ladle 335
FÜLL (*n*), **LUKEN-**: hatchway coaming 336
FÜLLMASCHINE (*f*): filling machine 337
FÜLLMASSE (*f*): massecuit (mixture of syrup and cane sugar crystals); filling (material) mass packing; stuffing 338
FÜLLMATERIAL (*n*): filling (material) medium; stuffing; loading; packing; filler 339
FÜLLMITTEL (*n*): (see **FÜLLMATERIAL**) 340
FÜLLÖFFNUNG (*f*): charging opening; filling opening 341
FÜLLPLATTE (*f*): liner 342
FÜLLROHR (*n*): filling (feed) pipe; spout 343
FÜLLSCHACHT (*m*): charging hopper 344
FÜLLSEL (*n*): stuffing; forcemeat 345
FÜLLSPANT (*n*): filling timber 346
FÜLLSTOFF (*m*): (see **FÜLLMATERIAL**) 347
FÜLLSTRICH (*m*): filling level 348
FÜLLSTÜCK (*n*): filling piece, liner 349
FÜLLTRICHTER (*m*): (filling)-funnel; (charging)-hopper 350
FÜLLUNG (*f*): filling; stuffing; charge; charging; packing; panel; saturation; admission; cut-off (Engine) 351
FÜLLUNGSGRAD (*m*): degree of admission, degree of charge 352
FÜLLUNGSMATERIAL FÜR ZAHNÄRZTE: dental filling 353
FULMENIT (*n*): fulmenite (an explosive) 354
FULMINAT (*n*): fulminate (salt of fulminic acid, CNOH) 355
FULMINIEREN: to fulminate 356
FULWABUTTER (*f*): (see **GHEABUTTER**) 357
FUMARSÄURE (*f*): fumaric acid (from plant *Fumaria officinalis*); COOHCHCHCOOH; Sg. 1.625; Mp. 287°C. 358
FUMARSÄURE-DIÄTHYLESTER (*m*): fumaric diethylester; $C_8H_{12}O_4$; Sg. 1.0508; Bp. 218.5°C 359
FUMMEL (*m*): welt (of shoes) 360
FUND (*m*): discovery, finding 361

FUNDAMENT (*n*): foundation; base; basis; baseplate; basement; bedplate 362
FUNDAMENTALSATZ (*m*): basic (rule) principle 363
FUNDAMENTANKER (*m*): foundation stay, foundation anchor 364
FUNDAMENTBOLZEN (*m*): foundation bolt 365
FUNDAMENTIEREN: to lay a foundation 366
FUNDAMENTPLATTE (*f*): foundation plate, base plate, bed plate 367
FUNDAMENTPLATTE (*f*), GEMEINSAME-: common baseplate, combined baseplate 368
FUNDAMENTSCHRAUBE (*f*): foundation screw, straining screw 369
FUNDAMENTSTEIN (*m*): foundation stone 370
FUNDGRUBE (*f*): mine-shaft 371
FUNDIEREN: to found 372
FÜNDIG: ore-bearing 373
FUNDORT (*m*): place; locality; (where something is found); habitat 374
FUNDSTÄTTE (*f*): (see FUNDORT) 375
FÜNFATOMIG: pentatomic 376
FÜNFBLATT (*n*): cinquefoil 377
FÜNFECK (*n*): pentagon 378
FÜNFECKIG: pentagonal 379
FÜNFFACH: fivefold; quintuple 380
FÜNFFACHCHLORPHOSPHOR (*m*): phosphorous pentachloride; phosphoric chloride; PCl_5; Sg. 3.6; Bp. 160-165°C. 381
FÜNFFACHSCHWEFELANTIMON (*n*): antimony (penta)-sulphide; Sb_2S_3; Sg. 4.562; Mp. 546°C. 382
FÜNFFÄLTIG: five-fold, quintuple 383
FÜNFGESANG (*m*): voice quintet 384
FÜNFGLIEDERIG: five-membered 385
FÜNFKANTIG: pentagonal 386
FÜNFKLANG (*m*): fifth (Music) 387
FÜNFMASTBARK (*f*): five masted barque 388
FÜNFMASTVOLLSCHIFF (*n*): five masted ship 389
FÜNFSEITIG: five-sided; pentahedral 390
FÜNFSTIMMIG: quintet, set for five voices (Music) 391
FÜNFWERTIG: pentavalent; quinquivalent 392
FÜNFZIGSTELLÖSUNG (*f*): solution of 1 in 50 [393
FÜNFZIGSTELPERMANGANAT (*n*): 1 : 50 permananate, a 1 in 50 solution of permanganate of potash 394
FUNGICID: fungicidal 395
FUNGIEREN: to function; act; work; officiate; operate; discharge 396
FUNGÖS: spongy, fungous 397
FUNKE (*m*): spark; sparkle; glimpse 398
FUNKELN: to sparkle; scintillate; glitter; glisten; twinkle 399
FUNKEN: to sparkle; glitter; spark; (*m*): (see FUNKE) 400
FUNKENBILDUNG (*f*): sparking; formation of sparks 401
FUNKENDÄMPFUNG (*f*): spark damping (Wireless Tel.) 402
FUNKENENTLADUNG (*f*): electric spark discharge 403
FUNKENFÄNGER (*m*): spark catcher 404
FUNKENGEBEND: scintillating, glistening, twinkling, producing sparks 405
FUNKENGEBER (*m*): sparking (device; coil) plug; spark transmitter 406
FUNKENHOLZ (*n*): touchwood 407
FUNKENINDUKTOR (*m*): induction coil (Elect.) 408
FUNKENKARREN (*m*): spark car (Wireless Tel.) 409
FUNKENLÄNGE (*f*): sparking distance, length of spark 410
FUNKENSAMMLER (*m*): spark condenser 411
FUNKENSPEKTRUM (*n*): spark spectrum 412
FUNKENSPIEL (*n*): play of sparks 413
FUNKENSPRÜHEN (*n*): emission of sparks; twinkling; scintillation; glittering 414
FUNKENSTRECKE (*f*): spark gap (Elect.) 415
FUNKENTELEGRAPHIE (*f*): radio-telegraphy, wireless telegraphy 416
FUNKENWURM (*m*): glow-worm 417
FUNKGERÄT (*n*): wireless apparatus 418
FUNKSTELLE (*f*): wireless station 419
FUNKTION (*f*): function 420
FURAN (*n*): furane 421
FURANALDEHYD (*n*): (see FURFUROL) 422
FURANKERN (*m*): furane nucleus 423
FURANRING (*m*): furane (ring) cycle 424
FÜRBASS: further; onward 425
FÜRBITTE (*f*): intercession; mediation 426
FURCHE (*f*): furrow; groove; channel; wrinkle 427
FURCHEN: to furrow; crease; wrinkle; knit; groove; channel 428
FURCHENSPATEL (*f*): grooved spatula 429
FURCHT (*f*): fear; fright; horror; terror; apprehension; dread 430
FURCHTBAR: fearful; frightful; horrible; terrible; dreadful; formidable 431
FÜRCHTEN: to fear; be afraid; dread; apprehend 432
FURCHTLOS: fearless; intrepid 433
FURCHTSAM: timid; faint hearted; timorous; fearful 434
FURCHUNG (*f*): furrowing; grooving; segmentation; creasing 435
FÜRDER: further-(more); henceforward 436
FURFURAMID (*n*): furfuramide (a crystalline mass from the action of ammonia on furfurol) 437
FURFURAN (*n*): furfurane; C_4H_4O; Sg. 0.9086; Bp. 31.6°C. 438
FURFUROL (*n*): furfurol, furfural, furol, pyromucic aldehyde, furfuraldehyde, artificial oils of ants, furfurane-carboxylic aldehyde, furfurane aldehyde; $C_5H_4O_2$; Sg. 1.1594; Bp. 160.7°C. 439
FURNIER (*n*): veneer 440
FURNIEREN: to veneer; inlay 441
FUROL (*n*): (see FURFUROL) 442
FÜRSORGE (*f*): care; provision; solicitude; providence; precaution; attention 443
FÜRSORGER (*m*): guardian; trustee; provider 444
FÜRSPRACHE (*f*): intercession; mediation; defence 445
FÜRSPRECHER (*m*): advocate; intercessor 446
FÜRSTENBIRN (*f*): Bergamot (fruit of *Citrus Bergamia Risso*) (see also BERGAMOTT) 447
FURT (*f*): ford 448
FURUNKEL (*m*): boil (Med.) 449
FÜRWORT (*n*): pronoun; intercession; mediation; good word 450
FUSEL (*m*): fusel oil (see FUSELÖL); bad spirits or liquor 451
FUSELGERUCH (*m*): odour of fusel oil 452
FUSELHALTIG: containing fusel oil 453
FUSELIG: containing fusel oil; intoxicated 454
FUSELÖL (*n*): fusel oil, grain oil, potato spirit, fermentation amyl alcohol; $C_5H_{10}O$; Sg. 0.81; Bp. 130°C. 455

FUSIONSPUNKT (*m*): fusing (fusion; melting point) 456
FUSS (*m*): foot; base; establishment; standard; footing; pedal; low; bottom; inferior; pedestal; stem; leg; stand 457
FUSSBAD (*n*): foot bath 458
FUSSBEUGE (*f*): instep 459
FUSSBLATT (*n*): sole of foot; May apple (*Podophyllum peltatum*) 460
FUSSBODEN (*m*): floor; flooring; ground; bottom 461
FUSSBODENBELAG (*m*): floor covering 462
FUSSBODENLACK (*m*): flooring varnish; floor polish 463
FUSSBODENÖL (*n*): flooring oil 464
FUSSBODENPAPIER (*n*): flooring (carpet) paper 465
FUSSBODENPLATTE (*f*): flooring tile; paving tile 466
FUSSBODENZIEGEL (*m*): paving tile or brick 467
FUSSBREMSE (*f*): foot brake 468
FUSSBRETT (*n*): pedal; footboard; foot-rest 469
FUSSDECKE (*f*): floor covering, carpet, linoleum, etc. 470
FUSSDREHBANK (*f*): foot lathe 471
FUSSEN AUF: to base on; rest on; perch on; depend (rely) on; gain a footing on 472
FUSSFALL (*m*): prostration 473
FUSSFÄLLIG: prostrate 474
FUSSGANG (*m*): footpath; walk 475
FUSSGÄNGER (*m*): pedestrian; walker 476
FUSSGEFÄSS (*n*): vessel with feet 477
FUSSGELENK (*n*): ankle joint 478
FUSSGESTELL (*n*): pedestal; base; basis; foot 479
FUSSGLÄTTE (*f*): black impure litharge 480
FUSSKONTAKT (*m*): floor contact 481
FUSSLAGER (*n*): foot bearing, step bearing 482
FUSSLIEK (*n*): foot rope 483
FUSSMEHL (*n*): the most inferior grade of flour 484
FUSSNOTE (*f*): foot note 485
FUSSPFAD (*m*): footpath; walk 486
FUSSPFUND (*n*): foot-pound 487
FUSSPLATTE (*f*): sole (base) plate, foot, footplate 488
FUSSPUNKT (*m*): nadir (Astron.); foot 489
FUSSREGISTER (*n*): pedal stop (Music) 490
FUSSREISE (*f*): walking tour 491
FUSSSCHIENE (*f*): foot rail 492
FUSSSICHER: sure-footed 493
FUSSSOHLE (*f*): sole of the foot 494
FUSSSPUR (*f*): track, trail 495
FUSSSTEIG (*m*): foot-path 496
FUSSSTOCK (*m*): foot-rule 497
FUSSTASTE (*f*): organ pedal 498
FUSSTEPPICH (*m*): carpet; rug 499
FUSSTRITT (*m*): treadle; footboard, footstep, kick 500
FUSSUNTERLAGE (*f*): plinth, pedestal, base 501
FUSSVENTIL (*n*) UND SAUGKORB (*m*): footvalve and strainer 502
FUSSWEG (*m*): foot-way, foot-path 503
FUSSWURZEL (*f*): tarsus 504
FUSSZAPFEN (*m*): heel tenon (of a mast) 505
FUSTAGE (*f*): barrels; casks; package 506
FUSTEL (*m*): (see FISETHOLZ) 507
FUSTIK (*n*): fustic (see FISETHOLZ) 508
FUSTIK (*m*), ALTER-: old fustic (*Morus tinctoria*, or *Broussonetia tinctoria*) 509
FUSTIKHOLZ (*n*): fustic wood (Cuba wood); (heartwood of *Chlorophora tinctoria* or *Manchura tinctoria*) 510
FUSTIK (*m*), JUNGER-: young fustic (see FISETHOLZ) 511
FUTTER (*n*): food; feed; fodder; forage; provender; lining; packing; casing; case; coating; covering; sheath; stuffing; chuck (Lathe) 512
FUTTERAL (*n*): case; casing; box; covering; sheath 513
FUTTER (*n*), BACKEN-: jaw chuck 514
FUTTERBOHNE (*f*): horse bean (*Faba vulgaris*) 515
FUTTER (*n*), BOHR-: drill chuck 516
FUTTER (*n*), DREHBANK-: lathe chuck 517
FUTTER (*n*), EINSCHLAG-: socket chuck 518
FUTTERERBSE (*f*): field pea (*Pisum arvense*) 519
FUTTERGERSTE (*f*): (winter)-barley; oats (for feeding purposes) 520
FUTTERGETREIDE (*n*): grain for feeding purposes; fodder 521
FUTTERGEWÄCHS (*n*): forage; fodder-(plants) 522
FUTTERGRAS (*n*): grazing grass; green fodder 523
FUTTERHEFE (*f*): yeast for fodder (the waste yeast from breweries dried and used with cattle fodder) 524
FUTTERKALK (*m*): food lime (dicalcium phosphate plus some tricalcium phosphate) 525
FUTTERKLEE (*m*): clover; red clover; purple clover (*Trifolium pratense*) 526
FUTTER (*n*), KLEMM-: clamping chuck 527
FUTTERKORN (*n*): grain for feeding; fodder 528
FUTTERKRAUT (*n*): forage; fodder (plants) 529
FUTTER (*n*), KREUZ-: socket chuck-(split into four parts) 530
FUTTER (*n*), KREUZSPUND-: socket chuck-(split into four parts) 531
FUTTERMITTEL (*n*): fodder; forage; feeding material; lining 532
FUTTERMITTELBEREITUNG (*f*): preparation of fodder 533
FUTTERN: (see FÜTTERN) 534
FÜTTERN: to feed; line; case; coat; cover; sheathe; pack; stuff 535
FUTTERRÜBE (*f*): (common)-turnip (*Brassica rapa*) 536
FUTTERSCHWINGE (*f*): winnowing fan 537
FUTTER (*n*), SPANN-: clamping chuck 538
FUTTER (*n*), SPUND-: socket chuck 539
FUTTER (*n*), STACHEL-: holdfast chuck, chuck with pins or holdfasts 540
FUTTERSTEIN (*m*): lining brick 541
FUTTERSTOFF (*m*): feeding stuff; lining; coating 542
FUTTERSTROH (*n*): forage (feeding) straw 543
FUTTER (*n*), TEILSPUND-: socket chuck- (split into halves) 544
FÜTTERUNG (*f*): feeding; feed; food; forage; provender; fodder; lining; casing; sheathing; coating 545
FUTTER (*n*), UNIVERSAL-: universal chuck 546
FUTTERWICKE (*f*): (common)-vetch (*Vicia sativa*) 547
FUTTERWURZEL (*f*): forage root 548
FUTTERZEUG (*n*): lining material 549
FUTURISCH: futurist; future 550
FUTURUM (*n*): future tense 551

G

GABANHOLZ (*n*): barwood; camwood (from *Baphia nitida*) 552
GABE (*f*): gift; dose; alms; donation; offering; talent 553
GABEL (*f*): fork; shaft (of a cart); tendril; crutch 554
GABELARTIG: bifurcated; forked 555
GABELFÖRMIG: (see GABELARTIG) 556
GABELGELENK (*n*): fork (link) joint 557
GABELIG: forked, bifurcated 558
GABELKLAMMER (*f*): forked clamp 559
GABELKRAUT (*n*): common agrimony (*Agrimonia eupatoria*) 560
GABELN: to pitchfork, to stick a fork into, to feed (coke on to a fire with a fork) 561
GABELRUTSCHE (*f*): bifurcated chute 562
GABELSCHLÜSSEL (*m*): fork spanner 563
GABELTEILUNG (*f*): forking; bifurcation 564
GABELZAPFEN (*m*): cross head pin 565
GABELZINKE (*f*): prong 566
GADININ (*n*): gadinine (a ptomaine from putrid cod-fish); $C_7H_{16}NO_2$ 567
GADOLEINSÄURE (*f*): gadoleic acid; $C_{19}H_{37}CO_2H$; Mp. 24.5°C. 568
GADOLINERDE (*f*): gadolinia; gadolinium oxide (see GADOLINIUMOXYD) 569
GADOLINIT (*m*): gadolinite (a silicate of rare earth metals) (one of the datolite group); $Be_3FeSi_2O_{10}Y_2$; Sg. 4.3 570
GADOLINIUM (*n*): gadolinium (Gd) 571
GADOLINIUMBROMID (*n*): gadolinium bromide; $GdBr_3.6H_2O$; Sg. 2.844 572
GADOLINIUMCHLORID (*n*): gadolinium chloride; $GdCl_3$; Sg. 4.52; $GdCl_3.6H_2O$; Sg. 2.424 [573
GADOLINIUM (*n*), GOLDCHLORWASSERSTOFF-SAURES-: gadolinium chloraurate; $GdCl_3.AuCl_3+10H_2O$ 574
GADOLINIUMNITRAT (*n*): gadolinium nitrate; $Gd(NO_3)_3.6{\cdot}5H_2O$; Sg. 2.332; $Gd(NO_3)_3.5H_2O$; Sg. 2.406 575
GADOLINIUMOXYD (*n*): gadolinium oxide; Gd_2O_3; Sg. 7.407 576
GADOLINIUMSULFAT (*n*): gadolinium sulphate; $Gd_2(SO_4)_3$; Sg. 4.139; $Gd_2(SO_4)_3.8H_2O$; Sg. 3.010 577
GAFFEL (*f*): fork, gaff 578
GAFFELBAUM (*m*): gaff 579
GAFFELFALL (*m*): throat halliard (Naut.) 580
GAFFELKLAUE (*f*): gaff hook 581
GAFFELPIEK (*f*): gaff peak 582
GAFFELSCHONER (*m*): fore and aft schooner 583
GAFFELSEGEL (*n*): try-sail, gaff sail 584
GAFFEL (*f*), VORTREISEGEL-: fore trysail gaff 585
GAGAT (*m*): jet. (a form of lignite) 586
GAGATEN: jet, of jet 587
GAGATKOHLE (*f*): (see GAGAT) 588
GAGE (*f*): wages, pay, salary, earnings 589
GAGEL (*f*): wax myrtle (*Myrica cerifera*) 590
GAGELGEWÄCHSE (*n.pl*): *Myricaceae* (Bot.) 591
GAHNIT (*m*): gahnite (a zinc spinel); ZnO,Al_2O_3; Sg. 4.34 592
GÄHRBOTTICH (*m*): fermenting vat 593
GÄHRDAUER (*f*): duration (length) of fermentation 594
GÄHRE (*f*): fermentation, ferment, fermenting, yeast, leaven 595
GÄHREN: to ferment; effervesce; work; froth; bubble 596
GÄHRMITTEL (*n*): leaven, yeast, ferment, barm 597
GÄHRSTOFF (*m*): zyme (Med.); leaven, yeast, ferment, barm 598
GÄHRUNG (*f*): zymosis (Med.); fermentation, effervescence, bubbling, working 599
GÄHRUNGSFÄHIG: zymotic (Med.); capable of fermenting; fermentable 600
GÄHRUNGSLEHRE (*f*): zymology (Med.) 601
GÄHRUNGSPROZESS (*m*): process of fermentation 602
GÄHRUNGSVORGANG (*m*): process of fermentation 603
GAIACOL (*n*): guaiacol (see GUAJAKOL) 604
GAIDINSÄURE (*f*): gaidinic acid; $C_{16}H_{30}O_2$; Mp. 39°C. 605
GAIS (*f*): (she)-goat; doe 606
GALAKTOSE (*f*): galactose (one of the grape sugar group of sugars) (from hydrolytic cleavage of milk sugar) (see DEXTROSE) 607
GALAMBUTTER (*f*): vegetable butter (see SHEA-BUTTER) 608
GALAMGUMMI (*n*): gum arabic 609
GALANGAWURZEL (*f*): galangal; colic root; Chinese ginger (*Rhizoma galangæ*) (Rhizome of *Alpinia officinarum*) 610
GALANGAWURZELÖL (*n*): galangal oil; Sg. 0.921 611
GALANTERIE (*f*): gallantry, politeness, millinery 612
GALANTERIEARBEIT (*f*): fancy goods, jewelry, trinkets 613
GALBANHARZ (*n*): galban (gum) resin; galbanum (from *Ferula galbaniflua*) 614
GALBANUM (*n*): galbanum (see GALBANHARZ) 615
GALEASSE (*f*): galease (Naut.) 616
GALEGOL (*n*): galega (Goat's rue) extract (from *Galega officinalis*) 617
GALENISCHESPRÄPARAT (*n*): (Galenica) galenical-(preparation) (a general term for all medicinal preparations) 618
GALGANT (*m*): galangal (see GALANGAWURZEL) 619
GALGANTWURZEL (*f*): (see GALANGAWURZEL) 620
GALGANTWURZELÖL (*n*): (see GALANGAWURZELÖL) 621
GALGENHOLZ (*n*): touchwood 622
GALITZENSTEIN (*m*): white (zinc) vitriol; zinc sulphate; $ZnSO_4.7H_2O$; Sg. 1.966; Mp. 50°C. 623
GALITZENSTEIN, BLAUER- (*m*): blue vitriol; copper sulphate; bluestone; $CuSO_4.5H_2O$; Sg. 2.284 624
GALIZENSTEIN (*m*): (see GALITZENSTEIN) 625
GALLAPFEL (*m*): (nut)-gall; gallnut; galla; oak apple (containing up to 60% tannic acid) 626
GALLÄPFELAUSGUSS (*m*): infusion of nut galls 627
GALLÄPFELEXTRAKT (*m*): infusion (extract) of nutgalls; gallnut extract 628
GALLÄPFELGERBSÄURE (*f*): gallotannic acid; tannic acid; tannin; $C_{10}H_{14}O_9$ 629
GALLÄPFELSÄURE (*f*): gallic acid, trihydroxybenzoic (trioxybenzoic) acid; $C_6H_2(OH)_3CO_2H.H_2O$; Sg. 1.694; Mp. 220/240°C. 630
GALLAT (*n*): gallate 631
GALLE (*f*): gall-(nut); bile; protuberance; nodule; kidney ore (Mineral); flaw; defect; sandiver; rancour; spite; choler; quagmire 632

GALLEICHE (*f*): gall oak 633
GALLEN: to gall; treat with gall nut; remove gall from 634
GÄLLEN: (see GALLEN) 635
GALLENABFÜHREND: cholagogic; expelling bile 636
GALLENABFÜHRUNGSMITTEL (*n*): cholagog (any drug for expelling bile) 637
GALLENABTREIBUNGSMITTEL (*n*): cholagog (any drug for expelling bile) 638
GALLENARTIG: billiary 639
GALLENBEHÄLTER (*m*): gall bladder 640
GALLENBITTER: bitter as gall; (*n*): picromel 641
GALLENBLASE (*f*): gall bladder 642
GALLENBLASENSTEIN (*m*): gall stone (see GALLEN-STEIN) 643
GALLENBRAUN (*n*): bilirubin; biliphæin; cholepyrrhin; hematoidin; bilifulvin; $C_{16}H_{18}N_2O_3$; Mp. 192°C. 644
GALLENDARM (*m*): duodenum 645
GALLENFARBSTOFF (*m*): bile pigment 646
GALLENFETT (*n*): cholesterol; cholestrin; $C_{27}H_{45}OH.H_2O$; Sg. 1.067; Mp. 148.5°C.; Bp. 360°C. 647
GALLENFETTSÄURE (*f*): bile acid 648
GALLENFIEBER (*n*): bilious fever 649
GALLENFISTEL (*f*): biliary fistula 650
GALLENFLUSS (*m*): discharge (flow) of bile 651
GALLENGANG (*m*): bile duct 652
GALLENGELB (*n*): billirubin (see GALLENBRAUN) 653
GALLENGRÜN (*n*): billiverdin 654
GALLENSÄURE (*f*): bile acid 655
GALLENSEIFE (*f*): (ox)-gall soap 656
GALLENSTEIN (*m*): gall stone; biliary calculus; hepatic calculus; cholelithiasis (Med.) 657
GALLENSTEINFETT (*n*): cholesterol (see GALLEN-FETT) 658
GALLENSTEINKOLIK (*f*): Hepatalgia (see HEPAT-ALGIE) 659
GALLENSTOFF (*m*): bile (substance) constituent; bilin 660
GALLENSUCHT (*f*): jaundice 661
GALLENSÜCHTIG: bilious 662
GALLENSÜSS (*n*): picromel 663
GALLENTALG (*m*): (see GALLENFETT) 664
GALLENTREIBEND: cholagogic, expelling bile 665
GALLENZUCKER (*m*): picromel 666
GALLERIE (*f*): gallery, lobby 667
GALLERT (*n*): jelly; gelatine; glue; gluten (see also GEL) 668
GALLERTÄHNLICH: gelatinous; jelly-like; colloidal 669
GALLERTARTIG: (see GALLERTÄHNLICH) 660
GALLERTE (*f*): (see GALLERT) 671
GALLERTFILTER (*m*): colloid filter (for filtering colloidal solutions) 672
GALLERTIG: (see GALLERTÄHNLICH) 673
GALLERTMOOS (*n*): iceland moss; carrageen; Chondrus (dried plant *Chondrus crispus*) 674
GALLERTSÄURE (*f*): pectic acid 675
GALLERTSUBSTANZ (*f*): jelly-like substance; colloid-(al substance) 676
GALLETSEIDE (*f*): silk waste 677
GALLICHLORID (*n*): gallic chloride (see GALLIUM-CHLORID) 678
GALLIEREN: to gall; treat with gall-nut extract 679
GALLIG: billiary; bilious; bitter (like gall); choleric; rancourous 680
GALLIHYDROXYD (*n*): gallic hydroxide 681
GALLION (*n*): head of a ship, nose of a ship, figurehead 682
GALLIPOTHARZ (*n*): gallipot (by distillation from turpentine) (see FICHTENHARZ) 683
GALLISALZ (*n*): gallic salt 684
GALLISIEREN: to gallise (Wine) 685
GALLITZENSTEIN (*m*): (see GALITZENSTEIN) 686
GALLIUM (*n*): gallium; (Ga) 687
GALLIUMCHLORID (*n*): gallium chloride; $GaCl_3$; Sg. 2.29 688
GALLOGEN (*n*): (see ELLAGSÄURE) 689
GALLONE (*f*): gallon (English equals 4.5459631 litres) (American equals 3.78543 lits. in commercial practice often taken as 27 American galls. equals 100 litres) 690
GALLSCHE KETTE (*f*): Gall's chain 691
GALLSEIFE (*f*): (ox)-gall soap 692
GALLUS (*m*): nut gall; gall nut 693
GALLUSGERBSÄURE (*f*): (see GALLÄPFELGERB-SÄURE) 694
GALLUSSAUER: gallate of; combined with gallic acid 695
GALLUSSÄURE (*f*): gallic acid (*Acidum gallicum*); trioxybenzoic (trihydroxybenzoic) acid; Sg. 1.694; Mp. 222-240°C. (see GALLÄPFEL-SÄURE) 696
GALLUSSÄUREGÄRUNG (*f*): gallic fermentation 697
GALMEI (*m*): calamine; $2ZnO.SiO_2.H_2O$: (about 54% zinc) (see ZINCSPAT) 698
GALMEI (*m*), EDLER-: calamine (52.2% Zn) (see ZINKSPAT) 699
GALOPPIEREN: to gallop 700
GALVANISCH: galvanic; voltaic 701
GALVANISIERANLAGE (*f*): galvanizing plant 702
GALVANISIERAPPARAT (*m*): galvanizing apparatus, galvaniser 703
GALVANISIEREN: to galvanize 704
GALVANISIERUNG (*f*): galvanization 705
GALVANISMUS (*m*): galvanism; voltaism 706
GALVANOGRAPHISCH: electro-type 707
GALVANOMETER (*n*): galvanometer 708
GALVANOMETER (*n*), VERTIKAL-: balance galvanometer 709
GALVANOPLASTIK (*f*): galvanoplastics; electrometallurgy; electro-typing process 710
GALVANOPLASTISCH: galvanoplastic 711
GALVANOPLATTIERUNG (*f*): electro-plating, electro-deposition 712
GALVANOSKOP (*n*): galvanoscope 713
GALVANOSTEGIE (*f*): electro-plating 714
GALVANOSTEGISCH: electro-depositive, electro-plating 715
GAMANDER (*m*): germander (*Teucrium scorodonia*) 716
GAMBE (*f*): jamb (Ceramics) 717
GAMBIER (*n*): (see GAMBIR) 718
GAMBIR (*n*): Gambir, gambier (extracted from stems and leaves of *Nauclea* (*Uncaria*) *Gambir* (tanning material) (also Pharm.) (see also under KATECHU) 719
GAMBIR-KATECHU (*m*): (see GAMBIR) 720
GAMMASÄURE (*f*): Gamma acid; G. acid; aminonaphtholsulphonic acid; $C_{10}H_5(OH)(NH_2)(SO_3H)$ 721
GANG (*m*): motion; action; operation; (state of-) working; running; course; path; passage; turn (of a screw); seam; gallery; vein (Mining); thread; gear (Mach.); duct; canal (Med.); carriage; movement; alley; errand; gangway; gait; going; pace; walk; round; shift; functioning; strake (of plates or planks) 722

GÄNG: customary, current, usual, general, fashionable 723
GANGART (*f*): matrix ; gangue ; gait 724
GANGARTIG: gangue, in veins (seams, layers) 725
GANGBAR: passable ; customary ; saleable ; current ; marketable ; workable ; pervious ; practicable ; feasible ; officinal (Pharm.) 726
GANGBARKEIT (*f*): practicability ; currency 727
GANGBORD (*m*): gangway (Naut.) 728
GANGERZ (*n*): vein ore 729
GANGGEBIRGE (*n*): mountains containing vein ore 730
GANGGESTEIN (*n*): gangue (Mining) 731
GANGHAFT: gangue, veined, streaked, seamed 732
GANGHÖHE (*f*): pitch (of screw) 733
GÄNGIG: customary, usual, general, current, fashionable; veined, streaked, seamed, in layers 734
GÄNGIG, LINKS-: left handed (of screws), left hand threaded 735
GÄNGIG, RECHTS-: right handed (of screws) right hand threaded 736
GANG (*m*), **IN- SETZEN**: to engage, couple, start-(up), set in motion, put into commission, set on foot 737
GANGLIE (*f*): ganglion, (*pl*): ganglia (see **NERVEN-KNOTEN**) 738
GANGLIENZELLE (*f*): ganglion cell (see **NERVEN-KNOTEN**) 739
GANGMASSE (*f*): gangue (Mining) 740
GANGRÄN (*f*): gangrene (Med.) 741
GANGRÄN (*f*), **SYMMETRISCHE-**: symmetrical gangrene, Raynaud's disease (Med.) 742
GANGSPILL (*n*): capstan 743
GANGSPILLKOPF (*m*): capstan (drum)-head 744
GANGSPILLSPAKE (*f*): capstan bar 745
GANGSPILLWELLE (*f*): capstan barrel, capstan drum, capstan spindle 746
GANGSTEIN (*m*): (see **GANGMASSE**) 747
GANGTIEFE (*f*): depth of a screwed thread 748
GANG (*m*), **TOTER-**: end play (Mech.) 749
GANG (*m*), **WASSER-**: water-way 750
GANGWEG (*m*): gangway (around boilers) 751
GANGWEISE: in veins (seams, layers); (*f*): method of operation, manner of working (Mech.) 752
GANISTER (*m*): ganister 753
GANISTERSTEIN (*m*): ganister brick 754
GANS (*f*): goose ; pig (Metal); lump ; hard rock (Mining); barnacle 755
GÄNSEAUGEN (*f.pl*): inverted commas, quotation marks 756
GÄNSEBAUM (*m*): (see **PLATANE** and **AHORN, GEMEINER-**) 757
GÄNSEBLUME (*f*): daisy 758
GÄNSEFEDER (*f*): goose (feather) quill 759
GÄNSEFETT (*n*): goose fat ; Sg. 0.93 ; Mp 25-34°C. 760
GÄNSEFÜSSCHEN (*n.pl*): inverted commas ; quotation marks (" ") 761
GÄNSEGARBE (*f*): wild tansy (*Tanacetum vulgare*) 762
GÄNSEHAUT (*f*): gooseflesh ; creeps 763
GÄNSEKIEL (*m*): goose-quill 764
GÄNSEKRAUT (*n*): wild tansy (*Tanacetum vulgare*) 765
GÄNSELEBERPASTETE (*f*): pâte de foie gras 766
GANSER (*m*): gander 767
GÄNSERICH (*m*): gander ; wild tansy (*Tanacetum vulgare*) 768
GANTER (*m*): support (for casks); gantry ; frame 769
GANZ: whole ; complete ; full ; all ; very ; perfect ; quite ; entire-(ly) ; total-(ly) ; (*f*): pig (Metal) 770
GANZE (*n*): whole-(number) ; total-(ity) ; integer 771
GANZFORM (*f*): pig-mould (Metal) 772
GANZHOLLÄNDER (*m*): beater ; beating engine (Paper) 773
GANZHOLZ (*n*): round (unhewn) timber 774
GÄNZLICH: wholly ; absolutely ; full-(y) ; complete-(ly) ; whole ; utter-(ly) ; total-(ly) ; entire-(ly) 775
GANZMAHLEN (*n*): beating (Paper) 776
GANZSTOFF (*m*): pulp (Paper) 777
GANZSTOFFHOLLÄNDER (*m*): (see **GANZHOLLÄNDER**) 778
GANZZAHLIG: integral 779
GANZZEUG (*n*): (see **GANZSTOFF**) 780
GANZZEUGHOLLÄNDER (*m*): refining engine (Paper) (see **GANZSTOFFHOLLÄNDER**) 781
GANZZEUGKASTEN (*m*): stuff chest (Paper) 782
GAR: done ; purified ; cooked; finished ; complete ; refined (Metal) ; tanned ; dressed; (Leather) ; quite ; very ; fully ; at all ; even ; prepared ; entirely ready 783
GARAGE (*f*): garage (for motor cars, etc.) 784
GARANTIE (*f*): guarantee ; security 785
GARANTIEREN: to guarantee, secure 786
GARARBEIT (*f*): refining (Metal) 787
GARAUS (*m*): ruin ; finishing stroke ; end ; collapse 788
GARBE (*f*): sheaf ; faggot ; pile (Iron) ; carraway ; yarrow ; milfoil (*Achillea millefolium*) 789
GÄRBEISEN (*n*): (see **GERBEISEN**) 790
GARBENBINDER (*m*): (sheaf)-binder or binding machine (Agric.) 791
GÄRBOTTICH (*m*): fermenter ; fermenting (vat) tub 792
GARBRAND (*m*): finishing burn (Ceramics) 793
GARBRENNEN: to fire (burn) to maturity or thoroughly 794
GÄRBSTAHL (*m*): tilted (shear) steel ; polishing steel ; burnisher 795
GÄRBÜTTE (*f*): fermenter ; fermenting (vat) tub 796
GÄRCHEMIE (*f*): zymurgy ; fermentation chemistry 797
GÄRDAUER (*f*): duration of fermentation 798
GARDENIAÖL (*n*): gardenia oil 799
GARDENIAÖL (*n*) **KÜNSTLICHES-**: synthetic gardenia oil 800
GARE (*f*): finished (refined) state ; batch ; dressed state ; dressing (Leather) ; tawing paste ; mellowness 801
GÄRE (*f*): fermentation ; yeast ; bouquet (Wine) 802
GAREISEN (*n*): refined iron ; trial (test) rod or probe (to test melted copper) 803
GÄREN: to ferment 804
GAREN: to dress ; refine ; finish ; cook sufficiently 805
GARERZ (*n*): roasted ore 806
GÄRFASS (*n*): fermenting cask 807
GARFRISCHEN: to refine thoroughly 808
GÄRFÜHRUNG (*f*): method of fermentation 809
GARGANG (*m*): good working order (Metal) ; thorough refining ; complete reduction and good operation (Blast furnace) 810
GÄRGEFÄSS (*n*): fermenting (fermentation) tub or vessel 811
GARGEKRÄTZ (*n*): refinery slag 812

GARHERD (*m*): refining hearth 813
GÄRKAMMER (*f*): fermenting (fermentation) room, chamber or cellar 814
GÄRKELLER (*m*): fermenting (fermentation) cellar 815
GÄRKÖLBCHEN (*n*): fermentation saccharimeter (Urinalysis) 816
GÄRKRAFT (*f*): fermenting power 817
GÄRKRÄFTIG: strongly fermenting 818
GÄRKRÄTZE (*f*): refinery slag 819
GARKUPFER (*n*): refined copper 820
GARLEDER (*n*): dressed (tanned) leather 821
GARMACHEN: to refine; dress; finish; tan 822
GÄRMITTEL (*n*): ferment 823
GARN (*n*): yarn; thread; twine; worsted; net; snare; decoy 824
GARNEN: thread, of thread 825
GARNET (*m*): (see **GRANAT**) 826
GARNHANDEL (*m*): cotton trade 827
GARNIEREN: to garnish; trim 828
GARNIERIT (*m*): garnierite, numeaite ($NiO.MgO.SiO_2.H_2O$) (7-8% nickel); (natural nickel and magnesium silicate); Sg. 2.55 (see also **NUMEAIT**) 829
GARNIERUNG (*f*): garnishing; trimming 830
GARNISON (*f*): garrison 831
GARNITUR (*f*): garniture; trimming; fittings; mounting-(s); armature 832
GARN (*n*), **MERCERISIERTES-**: mercerized yarn (from cold treatment in soda lye) 833
GARNPRESSPAPPE (*f*): boards for cotton yarn (Paper) 834
GAROFEN (*m*): refining furnace 835
GARPROBE (*f*): refining (assay) test 836
GARSCHAUM (*m*): kish (Iron); refining (fining) form or scum (deposited graphite) 837
GÄRSCHAUM (*m*): scum -(of ferments) 838
GARSCHEIBE (*f*): disc of refined copper 839
GARSCHLACKE (*f*): refining slag 840
GARSPAN (*m*): coating of copper on test rod 841
GARSTIG: mean; bad; dirty; ugly; foul; filthy; indecent; obscene; vile 842
GÄRSTOFF (*m*): ferment 843
GARSTÜCK (*n*): lump of purified salt 844
GÄRTANK (*m*): fermenting (fermentation) tank 845
GÄRTÄTIGKEIT (*f*): fermenting activity 846
GARTEN (*m*): garden 847
GARTENANEMONE (*f*): garden anemone (*Anemone coronaria*) 848
GARTENARBEIT (*f*): gardening 849
GARTENBAU (*m*): horticulture 850
GARTENBAUKUNST (*f*): (see **GARTENBAU**) 851
GARTENBAUVEREIN (*m*): horticultural society 852
GARTENBLUME (*f*): garden flower 853
GARTENBUTTERBLUME (*f*): (*Ranunculus bulbosus*) bulbous buttercup 854
GARTENDISTEL (*f*): artichoke 855
GARTENGERÄT (*n*): gardening tools 856
GARTENKRAUT (*n*): Southern wood (*Artemisia abrotanum*) (see **EBERRAUTE**) 857
GARTENKRESSENSAMENÖL (*n*): (see **KRESSEN-SAMENÖL**) 858
GARTENKUNST (*f*): gardening; horticulture 859
GARTENLAUBE (*f*): bower; arbour; name of a German periodical 860
GARTENMESSER (*n*): pruning knife 861
GARTENSALAT (*m*): cabbage lettuce (*Lactuca scariola sativa*) 862
GARTENSCHWAMM (*m*): mushroom 863
GARTENWALZE (*f*): garden roller 864
GARTENWESEN (*n*): gardening, horticulture 865
GARTENWURZ (*f*): (see **EBERRAUTE**) 866
GARTENZAUN (*m*): garden hedge 867
GÄRTNER (*m*): gardener 868
GÄRTNEREI (*f*): horticulture; gardening; nursery-(garden) 869
GÄRTNERISCH: horticultural 870
GÄRTÜCHIG: vital (of yeast) 871
GÄRUNG (*f*): fermentation 872
GÄRUNG (*f*), **ALKOHOLISCHE-**: alcoholic fermentation 873
GÄRUNGERREGEND: fermentative; zymotic; zymogenic 874
GÄRUNGERZEUGEND: (see **GÄRUNGERREGEND**) 875
GÄRUNGHEMMEND: anti-fermentative; anti-zymotic 876
GÄRUNGSALKOHOL (*m*): ethyl alcohol (see **ÄTHYLALKOHOL**) 877
GÄRUNG (*f*), **SAURE-**: oxydation 878
GÄRUNGSBUTTERSÄURE (*f*): butyric acid (see **BUTTERSÄURE**) 879
GÄRUNGSBUTYLALKOHOL (*m*): isobutylalkohol (see **ISOBUTYLALKOHOL**) 880
GÄRUNGSCAPRONSÄURE (*f*): (see **CAPRONSÄURE**) 881
GÄRUNGSCHEMIE (*f*): fermentation chemistry; zymurgy 882
GÄRUNGSERREGER (*m*): ferment 883
GÄRUNGSFÄHIG: fermentable; capable of (fermenting) being fermented 884
GÄRUNGSFÄHIGKEIT (*f*): fermentability; fermenting capacity; fermentative capacity 885
GÄRUNGSKRAFT (*f*): fermentative power 886
GÄRUNGSLEHRE (*f*): zymology 887
GÄRUNGSMESSER (*m*): zymometer 888
GÄRUNGSMITTEL (*n*): ferment 889
GÄRUNGSPILZ (*m*): fermentation fungus 890
GÄRUNGSVORGANG (*m*): fermentation process 891
GÄRUNGSWIDRIG: anti-fermentative; anti-zymotic 892
GÄRUNGSZEIT (*f*): duration of fermentation; time to complete (finish) any technical product 893
GÄRUNG (*f*), **UNTERDRUCK-**: vacuum fermentation 894
GÄRUNGVERHINDERND: anti-fermentative; anti-zymotic 895
GÄRVERMÖGEN (*n*): fermenting (fermentative) capacity; fermentability 896
GARWAGE (*f*): brine gauge (Salt) 897
GÄRWÄRME (*f*): heat of fermentation 898
GAS (*n*): gas 899
GASABLEITUNGSROHR (*n*): gas (delivery-)pipe 900
GASABSORPTION (*f*): gas absorption 901
GASABSPERRBLASE (*f*): gas receiver 902
GASÄHNLICH: (see **GASARTIG**) 903
GASANALYSE (*f*): gas analysis 904
GASANALYSENAPPARAT (*m*): gas (testing) analysing apparatus 905
GASANALYTISCH: gas analytical 906
GASANGRIFF (*m*): gas attack 907
GASANLAGE (*f*), **SAUG-**: suction gas plant 908
GASANSTALT (*f*): gas works 909
GASANZÜNDER (*m*): gas lighter 910
GASAPPARAT (*m*): gas apparatus 911
GASARMATUREN (*f.pl*): gas fittings 912
GASART (*f*): type (kind; sort; variety) of gas 913

GASARTIG: gasiform; gaseous; like gas 914
GASAUSBRINGEN (*n*): extraction of gas, gas extraction 915
GASBATTERIE (*f*): gas battery 916
GASBEHÄLTER (*m*): gas holder; gasometer 917
GASBEHEIZUNG (*f*): gas heating; gas fire 918
GASBELEUCHTUNG (*f*): gas lighting 919
GASBELEUCHTUNGSANSTALT (*f*): gas works 920
GASBELEUCHTUNGSGESELLSCHAFT (*f*): gas company 921
GASBEREITUNG (*f*): gas (preparation) making 922
GASBESCHIESSUNG (*f*): gas bombardment 923
GASBESICHTIGUNG (*f*): gas inspection 924
GASBEUTEL (*m*): gasbag, gas holder 925
GASBILDEND: gas-forming; gas producing; gasifying 926
GASBILDNER (*m*): gas (former) producer 927
GASBILDUNG (*f*): gas (formation) production; gasification 928
GASBLASE (*f*): gas bubble; gas retort 929
GASBLEICHE (*f*): gas bleaching 930
GASBRENNER (*m*): gas burner 931
GÄSCHEN: to foam, froth, ferment, effervesce, bubble 932
GÄSCHEND: yeasty; foaming, fermenting, bubbling, effervescing 933
GÄSCHT (*m*): yeast, froth, foam 934
GASDICHT: gas tight 935
GASDICHTE (*f*): gas density 936
GASDRUCK (*m*): gas pressure 937
GASDRUCKREGLER (*m*): gas pressure regulator 938
GASDRUCKREGULATOR (*m*): gas pressure regulator 939
GASDYNAMO (*f*): gas dynamo 940
GAS (*n*), **EDEL-**: rare gas, inert gas (either Helium, Neon, Xenon, Krypton or Argon) 941
GASEINSTELLER (*m*): gas regulator 942
GASEN: to gasify 943
GASENTBINDUNG (*f*): generation (production) of gas, gasification 944
GASENTBINDUNGSFLASCHE (*f*): gas (generator; bottle; retort) flask 945
GASENTBINDUNGSROHR (*n*): gas delivery pipe 946
GASENTWICKELUNG (*f*): gas (evolution) generation 947
GASENTWICKELUNGSAPPARAT (*m*): gas producer 948
GASENTWICKLER (*m*): gas generator 949
GASERZEUGER (*m*): gas (generator) producer 950
GASERZEUGUNG (*f*): gas production; gasification, gas generation 951
GASERZEUGUNGSAPPARAT (*m*): gas producer 952
GASFABRIK (*f*): gas works 953
GASFANG (*m*): gas (catcher, take) trap or outlet (for blast furnaces at the top) 954
GASFEUERUNG (*f*): gas furnace; (furnace for)-gas firing 955
GASFEUERUNGSANLAGE (*f*): gas firing plant or furnace 956
GAS (*n*), **FLÜSSIGES-**: liquid gas 957
GASFÖRMIG: (see **GASARTIG**) 958
GASGEBLÄSE (*n*): gas (blast) apparatus 959
GASGEMENGE (*n*): gas-(eous) mixture 960
GASGEMISCH (*n*): (see **GASGEMENGE**) 961
GASGENERATOR (*m*): gas generator; gas producer 962
GASGESCHOSS (*n*): gas (projectile) shell 963
GASGESETZ (*n*): gas law 964
GASGEWINDE (*n*): gas thread 965
GASGLEICHUNG (*f*): gas equation 966
GASGLOCKE (*f*): gasometer, gas bell 967
GASGLÜHLICHT (*n*): incandescent (gas)-light 968
GASGRANATE (*f*): gas (shell; bomb) grenade 969
GASGRANATENANGRIFF (*m*): gas-(shell) attack 970
GASHALTER (*m*): gasometer: gasholder 971
GASHÄLTER (*m*): (see **GASHALTER**) 972
GASHALTIG: gas-containing; having a gas content 973
GASIEREN: to gas; gasify; singe (Textile) 974
GASIERTES GARN (*n*): singed yarn 975
GASINTERFEROMETER (*n*): optical apparatus for the determination of the composition of a gas mixture 976
GASKALK (*m*): gas lime 977
GASKALORIMETER (*n*): gas calorimeter 978
GASKAMPFFLASCHE (*f*): gas cylinder (for cloud gas attacks) 979
GASKANDELABER (*m*): gas chandelier 980
GASKETTE (*f*): gas cell (Elect.) 981
GASKOCHER (*m*): gas (cooker; burner; ring) stove 982
GASKOHLE (*f*): gas coal (about 85% carbon) 983
GASKOKS (*m*): gas coke (10-20% ash content, and about 20% moisture) 984
GAS (*n*), **KOMPRIMIERTES-**: compressed gas 985
GASKONSTANTE (*f*): gas constant 986
GASKRAFTMASCHINE (*f*): gas engine 987
GASKÜHLER (*m*): gas (refrigerator) cooler 988
GASLAMPE (*f*): gas lamp 989
GASLEITUNG (*f*): gas (conduction; flue; main; conduit) pipe 990
GASLEITUNGSNETZ (*n*): system (network) of gas mains 991
GASLEITUNGSROHR (*n*): gas (main) pipe 992
GASLICHT (*n*): gas light 993
GASLICHTPAPIER (*n*): gaslight paper (Photo.) 994
GASMASCHINE (*f*): gas engine 995
GASMESSER (*m*): gas meter 996
GASMESSROHR (*n*): eudiometer; gas measuring tube 997
GASMESSUNG (*f*): gas measurement 998
GASMESSUNGSROHR (*n*): (see **GASMESSROHR**) 999
GASMINE (*f*): gas shell 000
GASMOTOR (*m*): gas engine 001
GASOFEN (*m*): gas (furnace) stove; gas oven; gas-fired kiln 002
GASÖL (*n*): gas oil; Sg. 1.306-1.408; Bp. 315-343°C. 003
GASOLIN (*n*): gasoline, petrol, motor-spirit; Sg. 0.64-0.7; Bp. 70-80°C. 004
GASOLINGAS (*n*): gasoline gas 005
GASOMETER (*n*): gas holder, gasometer 006
GASÖS: gaseous 007
GAS (*n*), **PERMANENTES-**: permanent gas 008
GASPFEIFE (*f*): gas pipe 009
GASPRESSUNG (*f*): gas pressure 010
GASPROBE (*f*): gas test-(ing) 011
GASPRÜFER (*m*): gas tester 012
GASPUMPE (*f*): gas pump 013
GASQUELLE (*f*): gas well 014
GASRAUM (*m*): gas volume; gas chamber 015
GASREINIGER (*m*): gas (purifier) scrubber 016
GASREINIGUNG (*f*): gas cleaning, gas scrubbing, gas (purification) purifying 017
GASREINIGUNGSANLAGE (*f*): gas (scrubbing, cleaning) purifying plant 018
GASREINIGUNGSMASSE (*f*): gas purifying mass 019
GASRETORTE (*f*): gas retort 020
GASROHR (*n*): gas pipe 021

GASRÖHRE (*f*): gas (pipe) jet 022
GASRUSS (*m*): gas black, lampblack, carbon black 023
GASSACK (*m*): gas bag 024
GASSAMMLER (*m*): gas (collector) tank 025
GASSAUGER (*m*): gas suction fan 026
GASSCHLAUCH (*m*): flexible gas pipe or tube (piping or tubing) 027
GASSE (*f*): lane; alley; channel (Founding) 028
GASSELBSTZÜNDER (*m*): automatic gas lighter 029
GASSENKEHRER (*m*): scavenger 030
GASSENKOT (*m*): street (offal) mud 031
GASSENRINNE (*f*): sewer 032
GASSTRAHL (*m*): gas jet 033
GASSTROM (*m*): gas (current) stream 034
GASSUMPF (*m*): a low, dense gas cloud 035
GAST (*m*): guest; visitor; stranger; customer; star (Actor) 036
GÄST (*f*): yeast 037
GASTEER (*m*): gas tar 038
GASTIEREN: to entertain, be a guest 039
GASTLICH: hospitable 040
GASTLICHKEIT (*f*): hospitality 041
GASTRALGIE (*f*): gastralgia, neuroses of the stomach (Med.) 042
GASTRICH: gastric 043
GASTRITIS (*f*): gastritis (Med.) 044
GASTRITIS (*f*), **AKUTE-**: acute gastritis (*Gastritis glandularis acuta*) (Med.) 045
GASTROCKNER (*m*): gas drier 046
GASTROENTEROSTOMIE (*f*): gastro-enterostomy (the forming of a communication between the stomach and the intestine at some distance from the pylorus) (Surg.) 047
GASTROLLE (*f*): star (part) rôle 048
GASTURBINE (*f*): gas turbine 049
GASUHR (*f*): gas meter (both wet or dry types) 050
GASVENTIL (*n*): gas valve 051
GASVERFLÜSSIGUNG (*f*): gas (liquefaction) liquefying 052
GASVERFLÜSSIGUNGSANLAGE (*f*): gas liquefying plant 053
GASVOLUMETRISCH: gasometric 054
GASWAGE (*f*): gas balance (for automatically determining the specific gravity of gases) 055
GASWANNE (*f*): gas (pneumatic) trough 056
GASWASCHAPPARAT (*m*): gas washing apparatus; gas scrubber 057
GASWASCHAUFSATZ (*m*): gas washing arrangement 058
GASWASCHER (*m*): gas (washer) scrubber 059
GASWASCHTURM (*m*): gas (scrubber) washing tower 060
GASWASSER (*n*): gas liquor; ammonia liquor (from gas works) 061
GASWECHSEL (*m*): gas exchange 062
GASWERFER (*m*): gas (projector) thrower; gas cylinder 063
GASWERFERANGRIFF (*m*): gas attack 064
GASTWIRT (*m*): landlord; innkeeper 065
GASTWIRTSCHAFT (*f*): inn-(keeping) 066
GASWOLKE (*f*): gas (wave) cloud 067
GASWOLKENANGRIFF (*m*): cloud gas attack; gas wave attack 068
GASZÄHLER (*m*): gas meter 069
GASZENTRIFUGE (*f*): gas centrifuge (an apparatus for cooling and sometimes heating, also purifying gases) 070
GASZERLEGUNG (*f*): gas separation 071
GASZÜNDER (*m*): gas lighter 072
GASZUSTAND (*m*): gaseous (state) condition 073
GATSCH (*m*): crude paraffin 074
GAT(T) (*n*): orifice, opening, hole, mouth 075
GATTER (*n*): grating; lattice; railing; helm; rudder 076
GATTER (*n*), **DAMPF-**: steam saw-mill 077
GATTERN: to refine (Tin) 078
GATTERWERK (*n*): lattice-work 079
GATTIEREN: to mix; classify; sort 080
GATTUNG (*f*): genus; kind; sort; class; race; species; family; gender; description; classification; heading 081
GATTUNGSFALL (*m*): genitive case (Grammar) 082
GATTUNGSNAME (*m*): generic name; common noun 083
GAU (*m*): district; county; province; region 084
GAUCHBROT (*n*): wood sorrel (*Oxalis acetosella*) 085
GAUCHHEIL (*n*): scarlet pimpernel (*Anagallis arvensis*); self-heal (*Prunella vulgaris*) 086
GAUDE (*f*): (see **FÄRBERWAU**) 087
GAUFRIEREN: to emboss; goffer; print 088
GAUGERICHT (*n*): petty sessions (Law) 089
GAUKELHAFT: deceptive; delusive; illusive; juggling 090
GAUKELN: to deceive; delude; dazzle; juggle; flit about 091
GAUKLERBLUME (*f*): maidenwort 092
GAULTHERIAÖL (*n*): gaultheria oil, Wintergreen oil (*Oleum gaultheriæ*) (obtained from *Gaultheria procumbens*); Sg. 1.18; Bp. about 219°C. 093
GAUMEN (*m*): palate 094
GAUNER (*m*): cheat; swindler; sharper; rogue; rook; impostor; scoundrel 095
GAUTSCHEN: to couch (Paper) 096
GAUTSCHROLLE (*f*): couch-(ing) roll (Paper) 097
GAYERDE (*f*): native saltpeter earth; saltpeter sweepings 098
GAYLUSSACTURM (*m*): Gay-Lussac tower 099
GAYLÜSSIT (*m*): gaylussite (a hydrous calcium-sodium carbonate); $Na_2O,CaO,2CO_2,5H_2O$ or $Na_2CO_3.CaCO_3.5H_2O$; Sg. 1.94 100
GAZE (*f*): gauze; canvas; net 101
GAZEFILTER (*n*): gauze filter 102
GEÄDER (*n*): veins; grain; veined structure; marbling; system of blood vessels; veinous system 103
GEBÄCK (*n*): pastry; baker's wares; baking; batch 104
GEBÄLK (*n*): timber-work; beams; baulks; rafters. 105
GEBÄRANSTALT (*f*): nursing home, lying-in hospital 106
GEBÄRDE (*f*): gesture; demeanour; mien 107
GEBÄREN: to bear; bring forth; produce; (*n*): child-birth; delivery; parturition 108
GEBAREN: to appear, seem, behave 109
GEBÄREND: procreative 110
GEBÄRMUTTER (*f*): uterus; womb; matrix 111
GEBÄRMUTTERBLUTUNG ()*f*: hæmorrhage from the womb (Med.) 112
GEBÄRMUTTERBRUCH (*m*): hysterocele 113
GEBÄRMUTTERSCHNITT (*m*): Cæsarian section 114
GEBÄUDE (*n*): building; structure; mine; system; edifice; fabric; burrow 115
GEBEFALL (*m*): Dative case (grammar) 116
GEBEIN (*n*): skeleton; remains; bones 117
GEBEN: to give; yield; show; express; sell; abate; relent; prove; bestow; stretch; render; present; confer; exist; deal 118

GEBER (*m*): giver; donor; transmitter (Telegraphy); sender 119
GEBERDE (*f*): gesture, demeanour, mien 120
GEBETT (*n*): bedding 121
GEBIET (*n*): territory; region; district; province; department; line; sphere; jurisdiction; domain; realm; field; branch 122
GEBIETEN: to command; order; rule 123
GEBIETERISCH: imperious; imperative; commanding; dictatorial 124
GEBILD (*n*): structure; system; organisation; form; image; formation; conformation (Geology); composition; creation 125
GEBILDE (*n*): (see **GEBILD**) 126
GEBILDET: educated; accomplished 127
GEBINDE (*n*): bundle; skein; hank; barrel cask; truss; joint; bond; row (of tiles) 128
GEBIRG (*n*): gangue (Mining); chain (of mountains); highlands 129
GEBIRGE (*n*): (see **GEBIRG**) 130
GEBIRGIG: mountainous 131
GEBIRGSART (*f*): kind of stone or rock 132
GEBIRGSBILDUNG (*f*): mountainous (rock) formation 133
GEBIRGSEBENE (*f*): table-land 134
GEBIRGSFORMATION (*f*): (see **GEBIRGSBILDUNG**) 135
GEBIRGSKAMM (*m*): ridge (of mountains) 136
GEBIRGSKUNDE (*f*): orology; geognosy 137
GEBIRGSLAND (*n*): hills, highlands 138
GEBIRGSLEHRE (*f*): (see **GEBIRGSKUNDE**) 139
GEBIRGSLOKOMOTIVE (*f*): mountain locomotive, locomotive for mountain work 140
GEBIRGSRÜCKEN (*m*): ridge (of mountains) 141
GEBIRGSZUG (*m*): chain of mountains 142
GEBISS (*n*): bit; set of teeth 143
GEBISS (*n*), **KÜNSTLICHES-**: set of false teeth 144
GEBLÄSE (*n*): blast; bellows; blower; fan; forced draught (fan) blower; ventilator 145
GEBLÄSELAMPE (*f*): blast (blow) lamp 146
GEBLÄSELUFT (*f*): blast air, forced draught 147
GEBLÄSEMASCHINE (*f*): blast (blowing) engine; blower; fan engine 148
GEBLÄSEMESSER (*m*): blast meter; draught meter 149
GEBLÄSEOFEN (*m*): blast furnace 150
GEBLÄSERÖHRE (*f*): blast pipe; tuyere 151
GEBLÄSEWERK (*n*): blower, fan, blast apparatus 152
GEBLÄSEWIND (*m*): blast; draught 153
GEBLÄTTERT: foliated; laminated 154
GEBLÜMT: figured, flowered 155
GEBOGEN: bent, hooked; vaulted 156
GEBOT (*n*): command-(ment); bid-(ding); precept; offer 157
GEBRÄU (*n*): brewing; brew; amount brewed 158
GEBRAUCH (*m*): use; custom; fashion; usage; employment; vogue; consumption 159
GEBRAUCHEN: to use; employ; make use of; consume; need; necessitate 160
GEBRÄUCHLICH: usual; customary; ordinary; in vogue; general 161
GEBRÄUCHLICHKEIT (*f*): custom; vogue 162
GEBRAUCHSANWEISUNG (*f*): directions for use; instructions for use 163
GEBRAUCHSFÄHIG: useable; serviceable; capable of being used 164
GEBRAUCHSFERTIG: ready for use 165
GEBRAUCHSGEGENSTAND (*m*): commodity 166
GEBRAUCHSMUSTER (*n*): protection. (There are two kinds of patent in Germany; one is the protection of a definite arrangement which is "**GEBRAUCHSMUSTER**," and the other is the patenting of an idea which is "**PATENT**.") 167
GEBRAUCHSWASSER (*n*): useable water, drinking water, fresh (clean) water 168
GEBRAUCHT: used, second-hand 169
GEBRÄUDE (*f*): brewing; brew 170
GEBRÄUNT: (see **BRÜNIEREN**) 171
GEBRAUSE (*f*): roaring; rushing 172
GEBRECH: brittle; soft; fragile 173
GEBRECHEN: to be lacking; (*n*) want; defect; infirmity; fault 174
GEBRECHLICH: frail; infirm; crippled; breakable; fragile; brittle 175
GEBREMSTE WELLE (*f*): damped wave 176
GEBROCHENE KANTEN (*f.pl.*): rounded edges 177
GEBROTEL (*n*): boiling; bubbling, foaming 178
GEBÜHR (*f*): due; duty; charge; fee; decency 179
GEBÜHREN: to deserve 180
GEBÜHREND: duly; befitting; due 181
GEBUND (*n*): bunch; skein; bundle; truss; hank 182
GEBUNDEN: obliged, bound; latent (of heat) 183
GEBUNDENE WÄRME (*f*): latent heat 184
GEBURT (*f*): birth; labour; offspring; descent; race; origin; delivery; production; generation; creation; nativity 185
GEBURTSARBEIT (*f*): labour (of birth) 186
GEBURTSHILFE (*f*): midwife-(ry); obstetrics; accouchement 187
GEBURTSHILFLICH: obstetric 188
GEBURTSKUNDE (*f*): obstetrics 189
GEBURTSLEHRE (*f*): obstetrics 190
GEBURTSNOT (*f*): labour (of birth) 191
GEBURTSORT (*m*): place of birth, birthplace 192
GEBURTSSCHEIN (*m*): birth certificate 193
GEBURTSTAG (*m*): birthday 194
GEBÜSCH (*n*): thicket; wood; bushes; undergrowth; underwood; coppice; copse 195
GECRACKT: cracked (of oil) 196
GEDÄCHTNIS (*n*): memory; remembrance; commemoration; memorial; recollection; monument 197
GEDÄCHTNISBUCH (*n*): memorandum book 198
GEDÄCHTNISFEIER (*f*): commemoration, anniversary 199
GEDÄCHTNISKUNST (*f*): mnemonics 200
GEDÄCHTNISMÜNZE (*f*): commemorative medal 201
GEDÄCHTNISSCHWÄCHE (*f*): forgetfulness, Amnesia (Med.) 202
GEDÄCHTNISTAG (*m*): anniversary 203
GEDÄCHTNISÜBUNG (*f*): memory training 204
GEDÄCHTNISZEICHEN (*n*): souvenir 205
GEDÄMPFT: steamed (see **DÄMPFEN**) 206
GEDÄMPFTE WELLE (*f*): damped wave 207
GEDANKE (*m*): thought; opinion; purpose; aim; object; design; meaning 208
GEDANKENFOLGE (*f*): train of thought 209
GEDANKENLEER: thoughtless, poverty of thought 210
GEDANKENREICH: thoughtful, fertile in ideas 211
GEDANKENSTRICH (*m*): hyphen, dash 212
GEDÄRM (*n*): entrails; intestines; bowels 213
GEDÄRMVORFALL (*m*): prolapsus (Med.) 214

GEDECK (*n*): covering; roof-(ing); cover (at table) 215
GEDEIHEN: to thrive; grow; prosper; proceed; swell (Slaking lime); increase 216
GEDENKBUCH (*n*): memorandum book 217
GEDENKEN: to think of; be mindful; remember; intend; propose; purpose; (*n*): memory; remembrance 218
GEDIEGEN: pure; solid; massive; unmixed; genuine; superior; native (Minerals) 219
GEDIEGENHEIT (*f*): native state (Minerals); purity; intrinsic worth; solidity 220
GEDRÄNGE (*n*): thronging; crowd; dilemma; distress; need; press 221
GEDRÄNGT: constricted; crowded; squeezed; driven; forced; urged; crushed; pressed; thrust 222
GEDRIT (*m*): (similar to Amphibol) (see **AMPHI-BOL**) 223
GEDRUNGEN: compact; compelled; thickset; solid 224
GEDRUNGENE BAUART (*f*): compact (construction) type 225
GEDÜFT (*n*): smell, odour, perfume 226
GEDUNSEN: bloated, puffed up, turgid 227
GEECKT: angled, cornered 228
GEEIGNET: suitable; appropriate; capable of; fit; qualified 229
GEFAHR (*f*): danger; risk; peril 230
GEFÄHRDEN: to endanger; risk; imperil; compromise 231
GEFÄHRLICH: dangerous; risky 232
GEFAHRLOS: free from (without) danger, peril or risk; not dangerous, perilous or risky 233
GEFÄHRTE (*m*): companion; associate; colleague; comrade 234
GEFÄLL (*n*): fall; gradient; incline; drop; decline; income; revenue; taxes; head (Water) 235
GEFÄLLE (*n*): (see **GEFÄLL**) 236
GEFÄLLE (*n*), **MIT-ZUFLIESSEN**: to run by gravity 237
GEFALLEN: to please; (*n*): pleasure; favour 238
GEFALLEN TUN: to oblige; please; give pleasure; render a service; do a favour 239
GEFÄLLIG: agreeable; pleasing; obliging; complaisant 240
GEFÄLLIGKEIT (*f*): favour; kindness; complaisance 241
GEFÄLLIGST: please; pray; I beg 242
GEFÄLLVERGÜTUNG (*f*): drawback (Com.) 243
GEFÄLTEL (*n*): fold, pleat, gather 244
GEFANGEN: caught, imprisoned, captured, captivated 245
GEFÄSS (*m*); vessel; receptacle; blood-vessel; hilt (of sword) 246
GEFÄSSBAROMETER (*n*): cistern barometer 247
GEFÄSSBAU (*m*): vascular structure (Med.) 248
GEFÄSSEINMÜNDUNG (*f*): anastomosis; communication of blood vessels (Med.) 249
GEFÄSSHAUT (*f*): vascular membrane 250
GEFÄSSIG: vascular 251
GEFÄSSKRYPTOGAMEN (*pl*): Phænogamous plants, Vasculares, vascular plants, flowering plants 252
GEFÄSSKUNDE (*f*): ceramic art 253
GEFÄSSOFEN (*m*): closed furnace 254
GEFÄSSSYSTEM (*n*): vascular system 255
GEFASST: collected, prepared, ready, steady, calm, cool; written 256
GEFECHT (*n*): fight-(ing); fray; combat; skirmish; battle 257
GEFERTIGTE (*m*): subscriber (used in Austria) 258
GEFESSELT: chained, enchained, fettered 259
GEFIEDER (*n*): plumage 260
GEFIEDERT: pinnate; feathered 261
GEFILDE (*n*): open country, plain, fields 262
GEFLAMMT: watered, clouded (Fabrics, etc.) 263
GEFLECHT (*n*): network; plaited (woven; hurdle) work; texture; scrofula; plexus (Med.); plaiting 264
GEFLECKT: spotted, stained, speckled 265
GEFLIMMER (*n*): glittering, shimmering, glistening, scintillation 266
GEFLISSEN: studious; industrious; diligent; assiduous 267
GEFLISSENTLICH: wilful; intentional-(ly) 268
GEFLÜGEL (*n*): fowl-(s); poultry; birds 269
GEFLÜGELT: winged 270
GEFLÜSTER (*n*): whispering 271
GEFOLGE (*n*): train; attendants; retinue; followers; suite; consequences 272
GEFRÄSSKRYPTOGAMEN (*pl*): *Pteridophyteæ* (Bot.) 273
GEFRÄSTE STÄBE (*m.pl*): bars in dead lengths 274
GEFRIERAPPARAT (*m*): freezing apparatus; refrigerator 275
GEFRIERBAR: freezable; congealable 276
GEFRIEREN: to freeze; congeal; refrigerate 277
GEFRIERPUNKT (*m*): freezing point 278
GEFRIERPUNKTSERNIEDRIGUNG (*f*): lowering (reduction) of freezing point 279
GEFRIERSALZ (*n*): freezing salt 280
GEFRIERSCHUTZMITTEL (*n*): preventative against freezing, lagging, covering 281
GEFRORENE (*n*): ice-(cream); something frozen 282
GEFÜGE (*n*): structure; texture; bed; stratum; joining; joint; layer; groove 283
GEFÜGIG: pliable; pliant; flexible; accommodating 284
GEFÜHL (*n*): feeling; touch; sensation; sensitiveness; consciousness; sentiment 285
GEFÜHLLOS: heartless, unfeeling, apathetic, insensible 286
GEFÜHLVOLL: sensitive, tender, feeling 287
GEFÜTTERTES PAPIER (*n*): lined linen paper (Paper) 288
GEGABELT: bifurcated; forked 289
GEGABELTER HEBEL (*m*): forked lever 290
GEGABELTE RUTSCHE (*f*): bifurcated chute 291
GEGEBENENFALLS: if necessary; in an emergency; in a given case; under certain circumstances 292
GEGEN: toward-(s); against; to; about; compared with; opposite; counter; contra; back-(stay); near; inverted; reverse; anti- 293
GEGENABSICHT (*f*): opposite intention or purpose 294
GEGENANTWORT (*f*): answer, rejoinder, reply 295
GEGENARBEIT (*f*): negative work (due to back pressure) 296
GEGENARZNEI (*f*): antidote (see **GEGENMITTEL**) 297
GEGENBILD (*n*): counterpart; contrast 298
GEGENBOGEN (*m*): inverted arch 299
GEGENBÖSCHUNG (*f*): counterscarp 300
GEGEND (*f*): region; quarter; district; neighbourhood 301

GEGENDIAGONALE (*f*): counterbrace; diagonal 302
GEGENDIENST (*m*): return (reciprocal) service 303
GEGENDIENSTE (*m.pl*) LEISTEN: to reciprocate, do a return service, requite services 304
GEGENDRUCK (*m*): back-pressure; counter pressure; resistance; reaction 305
GEGENDRUCKKOLBEN (*m*): dummy piston (Turbine); counter piston 306
GEGENDRUCKMASCHINE (*f*): back-pressure engine 307
GEGENDRUCKTURBINE (*f*): back-pressure turbine 308
GEGENEINANDERSTELLUNG (*f*): comparison 309
GEGENEINANDERSTOSSEN: to collide; butt (of plates) 310
GEGENFARBE (*f*): complementary colour 311
GEGENFÄRBUNG (*f*): contrast (staining) colouring 312
GEGENFEUER (*n*): passing of flames to the front of the furnace 313
GEGENFLANSCH (*m*): counter flange 314
GEGENFORDERUNG (*f*): counter-claim 315
GEGENGEWALT (*f*): reprisal; retaliation 316
GEGENGEWICHT (*n*): counterpoise; counter-(balance) weight 317
GEGENGIFT (*n*): antidote; antitoxin (see GEGENMITTEL) 318
GEGENHALL (*m*): echo, resonance 319
GEGENHALTER (*m*): dolly; holder-up (for riveting) 320
GEGENKLAGE (*f*): counter-plea; recrimination 321
GEGENKOLBEN (*m*): counter (dummy) piston 322
GEGENKRAFT (*f*): counter (force) stress; opposing force; reaction 323
GEGENKRAFT, ELASTISCHE- (*f*): elastic reaction 324
GEGENLAUF (*m*): anti-clockwise motion, counter (reverse) running or rotation 325
GEGENLAUFEN: to run (in opposite directions) anti-clockwise; run counter to 326
GEGENMAUER (*f*): counter-wall, buttress, counterscarp 327
GEGENMINE (*f*): counter-mine 328
GEGENMITTEL (*n*): antidote; remedy; prophylactic; antitoxin; preventative; emetic; cure 329
GEGENMUTTER (*f*): check nut, locknut 330
GEGENPFEILER (*m*): buttress 331
GEGENPOL (*m*): reciprocal pole; antipole 332
GEGENPOLARE (*f*): (see GEGENPOL) 333
GEGENREDE (*f*): contradiction, objection, response, reply 334
GEGENREIZMITTEL (*n*): counter-irritant (Med.) 335
GEGENRING (*m*): packing ring 336
GEGENSATZ (*m*): opposition; opposite; contrast; return; antithesis 337
GEGENSCHALTUNG (*f*): opposite connection, counter connection 338
GEGENSCHEIN (*n*): reflection; opposition (Astronomy) 339
GEGENSCHRÄGE (*f*): counterbrace; diagonal 340
GEGENSEITE (*f*): reverse; opposite side 341
GEGENSEITIG: reciprocal-(ly); mutual-(ly); common-(ly); opposite 342
GEGENSEITIGKEIT (*f*): reciprocity 343
GEGENSONNE (*f*): parhelion; mock sun 344
GEGENSPANT (*n*): reverse frame, reversed frame 345
GEGENSTAND (*m*): object; matter; subject; substance; article 346
GEGENSTEIGERUNG (*f*): anti-climax 347
GEGENSTRAHL (*m*): reflection; reflected ray 348
GEGENSTREBE (*f*): counter-(strut) brace; diagonal 349
GEGENSTROM (*m*): counter (inverse) current or flow 350
GEGENSTROMKÜHLER (*m*): counter-current cooler 351
GEGENSTROMVERFAHREN (*n*): counter-current (regenerative) process 352
GEGENSTROMVORWÄRMER (*m*): counter-current (preheater) feed-heater 353
GEGENSTÜCK (*n*): counterpart; match; antithesis; companion 354
GEGENSTÜTZE (*f*): buttress 355
GEGENTEIL (*n*): contrary; opposite; converse; (*m*): adversary 356
GEGENTEIL, IM-: on the (other hand) contrary 357
GEGENÜBER: over; against; opposite; in place of; instead of; compared with 358
GEGENÜBER LIEGEND: opposite 359
GEGENÜBER STELLEN: to compare; arrange side by side (applied to comparative tables); confront; oppose 360
GEGENUMDRUCK (*m*): reversed reprint (Lithography) 361
GEGENVERHÖR (*n*): cross-examination 362
GEGENWALL (*m*): counterscarp 363
GEGENWART (*f*): present; presence; present day 364
GEGENWÄRTIG: present; at present 365
GEGENWELLE (*f*): countershaft 366
GEGENWINKEL (*m*): alternate (opposite) angle 367
GEGENWIRKEND: counteracting; re-acting 368
GEGENWIRKENDES MITTEL (*n*): re-agent 369
GEGENWIRKUNG (*f*): counter-action; re-action 370
GEGENZEUGE (*f*): witness for the (opposition) defence 371
GEGLIEDERT: articulate-(d) 372
GEGNER (*m*): opponent; enemy; adversary; antagonist 373
GEGNERISCH: antagonistic 374
GEHÄGE (*n*): fence, enclosure, hedge, stockade 375
GEHALT (*m*): content-(s); capacity; constituent-(s); extent; yield; standard; strength; salary; pay; wages; value; proportion; worth 376
GEHALTEN: sober, steady, obliged, bound 377
GEHALTLOS: superficial; valueless; worthless; not having any capacity 378
GEHALTREICH: solid, valuable 379
GEHALTSANSPRUCH (*m*): salary (expected) required 380
GEHANDHABT: dealt with, enforced 381
GEHÄNGE (*n*): slope; hanging; pendant; festoon; declivity; suspension 382
GEHÄUSE (*n*): case; box; shell; housing; capsule; casing (turbines, etc.) 383
GEHBAR: practicable, workable, passable 384
GEHEGE (*n*): enclosure; hedge; fence; stockade 385
GEHEIM: secret; private; hidden; concealed; clandestine 386
GEHEIME TINTE (*f*): invisible ink 387
GEHEIMKAMERA (*f*): detective camera 388
GEHEIMKUNST (*f*): magic 389

GEHEIMMITTEL (*n*): secret remedy; patent medicine; nostrum; arcanum 390
GEHEIMNIS (*n*): secret; secrecy; mystery; arcanum 391
GEHEIMNISVOLL: mystical; mysterious 392
GEHEIMRAT (*m*): privy councillor 393
GEHEIMSCHREIBEN (*n*): cryptograph 394
GEHEIMVERFAHREN (*n*): secret process 395
GEHEISS (*n*): command; order; bidding; injunction 396
GEHEN: to go; walk; travel; move; pass; act; fare; proceed; operate; function; work; run 397
GEHEN, IN SEE-: to put to sea (Naut.) 398
GEHEUL (*n*): howl-(ing) 399
GEHILFE (*m*): assistant; helper; mate; adjunct; aid; colleague; junior clerk 400
GEHIRN (*n*): brain-(s); cerebrum (Med.) 401
GEHIRNANHANG (*m*): pituitary gland (of the brain) (see HYPOPHYSE) 402
GEHIRNENTZÜNDUNG (*f*): inflammation of the brain, cerebral inflammation (Med.) 403
GEHIRNERSCHÜTTERUNG (*f*): concussion of the brain 404
GEHIRNFETT (*n*): cerebrin 405
GEHIRNGESCHWULST (*f*): brain tumour, cerebral tumour (Med.) 406
GEHIRNHAUT (*f*): membrane covering the brain 407
GEHIRNHÄUTCHEN (*n*), **OBERES-**: dura mater 408
GEHIRNHÄUTCHEN (*n*), **UNTERES-**: pia mater 409
GEHIRN (*n*), **KLEINES-**: cerebellum 410
GEHIRNKRANKHEIT (*f*): cerebral affection 411
GEHIRNLEHRE (*f*): craniology 412
GEHIRNLOS: brainless 413
GEHIRNSYPHILIS (*m*): brain syphilis (Med.) 414
GEHIRNWASSERSUCHT (*f*): hydrocephalus, hydrops capitis, water on the brain (Med.) 415
GEHIRNWUT (*f*): mania, madness 416
GEHÖLZ (*n*): coppice; copse; wood; thicket 417
GEHÖR (*n*): ear; hearing; audience 418
GEHORCHEN: to obey 419
GEHÖREN: to belong; be (proper; fit) becoming; appertain to 420
GEHÖRIG: belonging; requisite; fit; due; proper; appropriate; enormous; suitable 421
GEHÖRLEHRE (*f*): acoustics 422
GEHÖRNERV (*m*): auditory nerve 423
GEHÖRNT: horned 424
GEHORSAM: obedient; dutiful; (*m*): obedience 425
GEHORSAMKEIT (*f*): obedience 426
GEHÖRTROMMEL (*f*): tympanum, ear-drum 427
GEHRE (*f*): bevel; mitre; wedge; oblique (diagonal) direction 428
GEHREN (*m*): gore, gusset, fold; to bevel, mitre 429
GEHRUNGSHOBEL (*m*): mitre-plane, bevel plane 430
GEHRUNGSMASS (*n*): bevel rule 431
GEHRUNGSSCHNEIDER (*m*): bevelling machine; mitre cutter 432
GEHRUNGSWINKEL (*m*): bevel angle, mitre angle 433
GEHT, ES-UM: it is a question of 434
GEHÜLFE (*f*): (see GEHILFE) 435
GEI (*f*): guy-(rope), stay 436
GEIERRINDE (*f*): Condurango bark (bark of *Marsdenia condurango*) (see CONDURAN-GORINDE) 437
GEIFERN: to splutter; drivel, slaver 438
GEIGE (*f*): violin; fiddle 439
GEIGEN: to fiddle; play (on) the violin; bow 440
GEIGENHARZ (*n*): resin; colophony; Sg. 1.08; Mp. 100-140°C. 441
GEIGENLACK (*m*): violin varnish 442
GEIL: luxuriant; fat; rich; lewd; lascivious; rank; proud (of flesh) 443
GEILEN: to manure; prune; castrate 444
GEIN (*n*): (see HUMUS) 445
GEINSAUER: (see HUMUSSAUER) 446
GEISS (*f*): (she)-goat; mountain goat; roe 447
GEISSBART (*m*): goat's beard (usually yellow goat's beard, *Tragopogon pratensis*) 448
GEISSBLATT (*n*): honeysuckle; woodbine (*Lonicera periclymenum*) 449
GEISSBLATTGEWÄCHSE (*n.pl*): *Caprifoliaceæ* (Bot.) 450
GEISSBOCK (*m*): (he)-goat 451
GEISSEL (*f*): flagellum; sarcasm; whip; lash; scourge; (*m*): hostage 452
GEISSFUSS (*m*): goatsfoot (Bot.); corner chisel; parting tool; punch (Tech.) 453
GEISSLERSCHES ROHR (*n*): Geissler tube 454
GEISSRAUTE (*f*): goat's rue (*Galega officinalis*) 455
GEIST (*m*): spirit; mind; ghost; soul; alcohol 456
GEISTESANLAGE (*f*): culture, talent 457
GEISTESGEGENWART (*f*): presence of mind 458
GEISTESKRANKHEIT (*f*): insanity (Med.) 459
GEISTESSTÖRUNG (*f*): insanity (Med.) 460
GEISTIG: alcoholic; spirituous; strong; generous (of liquors); volatile; spiritual; intellectual; mental 461
GEISTLICH: spiritual; clerical; religious; ecclesiastical 462
GEISTLICHE (*m*): clergyman; priest; divine; ecclesiastic 463
GEISTLOS: vacant; dull; spiritless; unintellectual 464
GEISTREICH: cultured, intellectual, talented, intelligent, ingenious 465
GEITAU (*n*): clue garnet, clue line, brail (for sails) 466
GEITAUBLOCK (*m*): clue-line block 467
GEITAUKLAPPLÄUFER (*m*): clue-line whip 468
GEIZ (*m*): avarice; stinginess; greed; greediness; covetousness 469
GEIZEN: to be covetous; be stingy; covet; be avaricious; be greedy 470
GEKENNZEICHNET DADURCH: characterized by the fact that 471
GEKNICKT: cracked, snapped (see KNICKEN); angular 472
GEKNISTER (*n*): crunching; crackling; rustling; (de)-crepitation 473
GEKÖRNT: granulated; granular; in grains 474
GEKRÄTZ (*n*): waste; refuse; dross; slag 475
GEKREUZT: crossbred; crossed 476
GEKRÖPFTE WELLE (*f*): crank-(ed) shaft 477
GEKRÖS: mesenteric (Med.) 478
GEKRÖSE (*n*): mesentery (Med.); giblets; frill 479
GEKRÜMMT: hooked, bent 480
GEKÜNSTELT: artificial, affected 481
GEL (*n*): gel (the solid or jelly-like mass produced by the coagulation of a colloidal solution) 482
GELÄGE (*n*): deposit-(s); dregs; foots; bottoms 483
GELÄNDE (*n*): open country, arable land 484
GELÄNDER (*n*): railing; balustrade; trellis 485
GELÄNDERDOCKE (*f*): baluster 486

GELANGEN: to arrive; attain; reach 487
GELASS (*m*): room; space; heritage 488
GELASSEN: patient; calm; composed; cool; collected; deliberate; self-possessed 489
GELATINE (*f*): gelatine (*Gelatinum*) 490
GELATINEFABRIKANLAGE (*f*): gelatine manufacturing plant 491
GELATINEFOLIE (*f*): sheet gelatine 492
GELATINE FÜR LICHTDRUCK: heliographic gelatine (for photographic printing) 493
GELATINE FÜR PHOTOGRAPHIE: gelatine for photographic purposes 494
GELATINEKAPSEL (*f*): gelatine capsule; medical capsule (*Capsulæ gelatinosæ*) 495
GELATINELEIM (*m*): gelatine glue; gelatine 496
GELATINEPAPIER (*n*): gelatine paper 497
GELATINEPLATTE (*f*): gelatine plate (Photo.) (see DRUCKPLATTE) 498
GELATINIEREN: to gelatinize; convert into a gel (Colloidal chemistry) (see GEL) 499
GELATINISIEREN: (see GELATINIEREN) 500
GELATINÖS: gelatinous 501
GELÄUFIG: fluent; ready; familiar; easy; skilful; current; voluble 502
GELÄUFIGKEIT (*f*): facility; fluency; familiarity; skill; volubility 503
GELÄUT (*n*): chimes; ringing; pealing; tinkling 504
GELÄUTE (*n*): (see GELÄUT) 505
GELB (*n*): yellow 506
GELBAMMER (*f*): yellow-hammer 507
GELBBEEREN (*f.pl*): Avignon berries; buckthorn berries; Persian berries (fruit of *Rhamnus frangula*) (see also KREUZBEEREN) 508
GELBBEEREN (*f.pl*), **CHINESISCHE-**: yellow berries, Hoai-hoa, waifa (the buds of *Sophora japonica*) 509
GELBBEERENEXTRAKT (*m*): Persian berry extract 510
GELBBEIZEN (*n*): yellowing (Metal) 511
GELBBLAUSAURES KALI (*n*): potassium ferrocyanide (see FERROCYANKALIUM) 512
GELBBLEIERZ (*n*): wulfenite; $PbMoO_4$; (natural lead molybdate); Sg. 6.85 (see WULFENIT) 513
GELBBRAUN: yellowish brown 514
GELBBRENNARBEIT (*f*): pickling 515
GELBBRENNE (*f*): dip, pickle (Electroplating metal) (for polishing copper and its alloys) (see VORBRENNE and GLANZBRENNE and MATTBRENNE) 516
GELBBRENNEN: to dip; pickle (Metal) 517
GELBBRENNER (*m*): pickler 518
GELBBRENNPROZESS (*m*): pickling process 519
GELBBRENNSÄURE (*f*): pickling acid 520
GELBE (*n*): yellow; yolk (of egg); jaundice 521
GELBE ARSENBLENDE (*f*): orpiment (see ARSEN-TRISULFID) 522
GELBEISENERZ (*n*): yellow (ironstone) iron ore (see also GELBEISENSTEIN) 523
GELBEISENERZ (*n*), **OKRIGES-**: yellow ochre; copiapite (an earthy iron oxide) 524
GELBEISENKIES (*m*): pyrite; fool's gold; FeS_2; about 47% iron 525
GELBEISENSTEIN (*m*): yellow ironstone; limonite (see LIMONIT; XANTHOSIDERIT) 526
GELB (*n*), **ENGLISCHES-**: English yellow (see MONTPELLIERGELB) 527
GELBERDE (*f*): yellow ochre (see ERDE, GELBE-; and OCKER) 528
GELBES FIEBER (*n*): yellow fever (Med.) 529
GELBFÄRBEN: to yellow, colour yellow; fade (of photographs) 530
GELBFÄRBUNG (*f*): yellowing; yellow colouring; yellow discolouration; fading; yellow stain (Photo.) 531
GELBFILTER (*m* and *n*): yellow filter (Photo.) 532
GELBFINK (*m*): yellow hammer 533
GELBGIESSER (*m*): brass founder 534
GELBGIESSEREI (*f*): brass foundry 535
GELBGLUT (*f*): yellow incandescence 536
GELBGRÜN: yellowish green 537
GELBGUSS (*m*): yellow brass (a copper-zinc alloy) (57-70% Cu; 43-30% Zn; with often a trace of tin and lead); Sg. 8.1 538
GELBHEIT (*f*): yellowness 539
GELBHOLZ (*n*): fustic wood; Cuba wood (*Chlorophora* or *Manchura tinctoria*); old fustic; yellow Brazil wood (*Morus tinctoria*); prickly ash; yellow wood; angelica tree (*Xanthoxylum americanum* or *Fagara clavaherculis*) 540
GELBHOLZEXTRAKT (*m*): yellow wood extract; fustic extract 541
GELBHOLZ (*n*), **UNGARISCHES-**: young fustic (*Rhus cotinus*) 542
GELBILDUNG (*f*): gel formation (coagulation of a colloidal solution or sol) 543
GELBKALI (*n*): potassium ferrocyanide (see FERROCYANKALIUM) 544
GELBKRAUT (*n*): yellow weed; weld; dyer's weed (*Reseda luteola*) 545
GELBKREUZGESCHOSS (*n*): "yellow cross" gas shell (see KNOBLAUCHGAS) 546
GELBKUPFER (*n*): brass; yellow copper; bronze (see GELBGUSS) 547
GELBKUPFERERZ (*n*): chalcopyrite; copper pyrites; yellow copper ore; $CuFeS_2$; about 35% Cu 548
GELBLICH: yellowish 549
GELBLING (*m*): yellow hammer 550
GELBMÜNZE (*f*): coin; piece of money 551
GELBNATRON (*n*): sodium ferrocyanide (see FERROCYANNATRIUM) 552
GELBROT: yellowish red 553
GELBSCHEIBE (*f*): yellow screen (Photo.) 554
GELBSCHLEIER (*m*): yellow stain, pyro stain (a non-actinic yellow colour imparted to a photographic plate when developed with pyrogallol and soda) 555
GELBSCHOTE (*f*), **CHINESISCHE-**: Wongshy, Wongsky (fruit of *Gardenia florida*) 556
GELBSTICHIG: yellow tinged; yellowish 557
GELBSUCHT (*f*): jaundice; yellow jaundice (Med.) 558
GELBSÜCHTIG: icteric 559
GELBWURZ (*f*): curcuma (rhizome of *Curcuma tinctoria*) (see KURKUMA) 560
GELBWURZEL (*f*): turmeric root (of *Hydrastis canadensis*) (see KURKUMA) 561
GELD (*n*): money; coin; cash 562
GELDANGELEGENHEITEN (*f.pl*): money matters 563
GELDANLAGE (*f*): investment 564
GELDANLEIHE (*f*): loan 565
GELDANWEISUNG (*f*): money order, postal order 566
GELDAUSLEIHER (*m*): money lender 567
GELD (*n*), **BARES-**: ready money, spot cash 568
GELDERSATZ (*m*): reimbursement 569
GELDHÜLFE (*f*): pecuniary assistance, subsidy 570
GELDKLEMME (*f*): tightness of money 571

GELDKURS (*m*): exchange 572
GELDLEIHWERT (*m*): discount, rate of interest 573
GELDMANGEL (*m*): tightness (shortage) of money 574
GELDPRÄGEN (*n*): minting, coining 575
GELDPREIS (*m*): rate of exchange 576
GELDSCHRANK (*m*): safe 577
GELDSENDUNG (*f*): remittance 578
GELDSTAND (*m*): state of the money market 579
GELDSTRAFE (*f*): fine 580
GELDSTÜCK (*n*): coin; piece of money 581
GELDUMLAUF (*m*): circulation (of money) 582
GELDVERLEGENHEIT (*f*): monetary (financial) difficulties or embarrassment 583
GELDVERSCHREIBUNG (*f*): promissory note 584
GELDVORSCHUSS (*m*): advance (of money) 585
GELDWECHSEL (*m*): exchange (of money) 586
GELDWUCHER (*m*): usury 587
GELEE (*n*): jelly 588
GELEGEN: opportune; convenient; situated 589
GELEGENHEIT (*f*): opportunity; occasion; convenience; locality 590
GELEGENTLICH: occasional; opportune; accidental; incidental; in due course 591
GELEHRIG: docile; teachable; tractable; intelligent 592
GELEHRSAM: (see GELEHRIG) 593
GELEHRSAMKEIT (*f*): learning; erudition 594
GELEHRTE (*m*): savant; scholar; learned man; man of letters 595
GELEHRTER (*m*): (see GELEHRTE) 596
GELEIS (*n*): track (Railway); line; rut 597
GELEISE (*n*): (see GELEIS) 598
GELEITEN: to accompany; conduct; escort; attend 599
GELEIT (*n*), SICHERES-: safe-conduct 600
GELEITSSCHIFF (*n*): convoy 601
GELENK (*n*): joint; link; hinge; join; connection 602
GELENKBAND (*n*): ligament (Med.) 603
GELENKBASALT (*m*): flexible basalt 604
GELENKFLÜSSIGKEIT (*f*): synovia; synovial fluid 605
GELENKGESCHWULST (*f*): white swelling 606
GELENKIG: flexible; pliable; supple; nimble; pliant; articulate-(d) 607
GELENKKNOPF (*m*): condyle 608
GELENKKUPPELUNG (*f*): flexible coupling 609
GELENKNEUROSE (*f*): neuralgia of joints (Med.) 610
GELENK (*n*), PUMPEN-: pump coupling 611
GELENKQUARZ (*m*): flexible sandstone; itacolumite 612
GELENKRHEUMATISMUS (*m*): rheumatism of the joints (Med.) 613
GELENKRHEUMATISMUS (*m*), AKUTER-: acute rheumatism, rheumatic fever (Med.) 614
GELENKSAFT (*m*): synovia 615
GELENKSCHMIERE (*f*): (see GELENKSAFT) 616
GELENKSCHWAMM (*m*): white swelling 617
GELENKSTEIFHEIT (*f*): ankylosis; stiffness in the joints 618
GELENKVERBINDUNG (*f*): articulation 619
GELENKVERRENKUNG (*f*): dislocation 620
GELENKVERWACHSUNG (*f*): ankylosis, immobility of joints (Med.) 621
GELENKWASSER (*n*): synovia; synovial fluid 622
GELENKWASSERSUCHT (*f*): hydrarthrosis, chronic synovitis or effusion of serum into joints (Med.) 623
GELICHTER (*n*): gang; set; lot; kind; sort; cast; stamp 624
GELIEFERN: to coagulate; set; curdle 625
GELIND: gentle; soft; mild; smooth; tender 626
GELINGEN: to succeed; be achieved; be successful; prosper; become possible 627
GELOBEN: to promise; pledge; vow 628
GELOCHT: pierced, punched, perforated 629
GELODURATKAPSEL (*f*): gelodurate capsule; gelatine capsule (Med.) 630
GELÖSTER KÖRPER (*m*): solute; body in solution 631
GELSEMIEN (*n*): gelsemium; wild woodbine; yellow jasmine (*Gelsemium sempervirens*) 632
GELSEMIENSÄURE (*f*): gelsemic acid (from *Gelsemium sempervirens*) 633
GELSEMIENWURZEL (*f*): gelsemium root (see GELSEMIUMWURZEL) 634
GELSEMIN (*n*): gelsemin (an alkaloid from *Gelsemium sempervirens*) 635
GELSEMININ (*n*): gelseminine (an alkaloid from *Gelsemium sempervirens*); $C_{22}H_{26}N_2O_3$; Mp. 172°C. 636
GELSEMININCHLORHYDRAT (*n*): gelseminine hydrochloride; $C_{22}H_{26}N_2O_3.HCl$; Mp. 330°C. 637
GELSEMINSÄURE (*f*): gelsemic acid (see GELSEMIENSÄURE) 638
GELSEMIUM (*n*): gelsemium (see GELSEMIEN) 639
GELSEMIUMWURZEL (*f*): gelsemium root (of *Gelsemium sempervirens*) (*Radix gelsemii*) 640
GELTEN: to be worth; be applicable; be apparent; be noticeable; operate; have weight; be current; be valid; pass; prevail; be operative; come into (be in) operation; take effect; be considered; apply; concern; hold good; obtain; serve as; act as 641
GELTUNG (*f*): validity; worth; value; currency; duration; importance 642
GELTUNG, ZUR- KOMMEN: to have the value; exercise the (take) effect; apply; have the importance; come into operation 643
GELÜBDE (*n*): vow 644
GEMACH (*n*): chamber; room; closet; apartment; comfortable; softly; by degrees; convenient 645
GEMÄCHLICH: slow; easy; comfortable 646
GEMÄCHT (*n*): genitals; work, handiwork, making, workmanship 647
GEMAHNEN: to remind 648
GEMÄLDE (*n*): picture; painting; drawing; description 649
GEMÄLDEAUSSTELLUNG (*f*): exhibition of pictures 650
GEMÄLDEFIRNIS (*m*): picture varnish 651
GEMÄLDELACK (*m*): picture varnish 652
GEMÄSS: according to; in conformity (accordance) with; agreeable; commensurable; commensurate with; conformable 653
GEMÄSSIGT: temperate; moderate 654
GEMÄUER (*n*): walls, masonry, brickwork 655
GEMEIN: common; general; ordinary; vulgar; simple; public; profane 656
GEMEINDE (*f*): community; municipality; congregation; parish 657
GEMEINGUT (*n*): joint (common) property 658
GEMEINHEIT (*f*): commonness; vulgarity; meanness; coarseness; disgrace; shame 659
GEMEINNÜTZIG: of (general) public utility 660

GEMEINSAM: mutual; common; joint; combined; general 661
GEMEINSCHAFT (*f*): community; society; partnership; intercourse 662
GEMEINSCHAFTLICH: common; mutual; joint 663
GEMENGANTEIL (*m*): constituent part of a mixture; proportion of an ingredient in a mixture 664
GEMENGE (*n*): mixture; frit (Glass); mélée; medley; fray 665
GEMENGSTOFF (*m*): constituents-(s) [ingredient-(s)] of a mixture 666
GEMESSEN: strict; formal; precise 667
GEMISCH (*n*): mixture; mixing; medley 668
GEMISCHTWARENFÄRBEREI (*f*): dyeing of textiles of mixed composition such as a mixture of wool and cotton 669
GEMME (*f*): gem; cameo 670
GEMS (*f*): chamois 671
GEMSE (*f*): (see **GEMS**) 672
GEMSENLEDER (*n*): chamois leather 673
GEMSLEDER (*n*): (see **GEMSENLEDER**) 674
GEMÜLL (*n*): rubbish; waste; dust; dry mould 675
GEMÜLM (*n*): (see **GEMÜLL**) 676
GEMÜNDE (*f*): opening, mouth, discharge (of rivers) 677
GEMÜSE (*n*): vegetables; greens 678
GEMÜSEGARTEN (*m*): kitchen garden 679
GEMÜSEPFLANZEN (*f.pl*): culinary plants; vegetables 680
GEMÜSEPULVER (*n*): vegetable powder, dried vegetables 681
GEMUSTERT: figured (fancy) fabrics 682
GEMÜT (*n*): mind; heart; feeling; soul; nature; temper; disposition; temperament 683
GEMÜTLICH: good (hearted) natured; genial; pleasant; comfortable; agreeable 684
GEMÜTLOS: morose; unfeeling 685
GEMÜTSWAHNSINN (*m*): moral insanity, emotional (impulsive, affective) insanity (Med.) 686
GENAU: accurate; exact; true; precise; close; tight; intimate; minute; parsimonious 687
GENAUIGKEIT (*f*): accuracy; precision (see **GENAU**) 688
GENEALOGIE (*f*): genealogy 689
GENEALOGISCH: genealogical 690
GENEHM: agreeable; acceptable 691
GENEHMIGEN: to ratify, accept, grant, give, approve, permit, sanction, admit 692
GENEIGT: disposed; favourable (also inclined, etc., see **NEIGEN**) 693
GENERAL (*m*): general 694
GENERALISIEREN: to generalize 695
GENERALNENNER (*m*): common denominator 696
GENERALSCHIEBER (*m*): common damper (usually placed in the main flue instead of having a separate damper at the outlet of each individual boiler) 697
GENERATION (*f*): generation, procreation 698
GENERATOR (*m*): generator; producer 699
GENERATORGAS (*n*): producer gas 700
GENERATORGASANLAGE (*f*): producer gas plant 701
GENERATORWASSERGAS (*n*): semi-water gas, Dowson gas, mixed gas, Mond gas 702
GENERELL: general-(ly), universal-(ly) 703
GENERISCH: generic 704
GENESEN: to recover; convalesce; be delivered; get better; grow (become) well 705
GENETISCH: genetic 706
GENEVER (*m*): Geneva; Holland gin 707
GENEVERESSENZ (*f*): gin essence 708
GENF (*m*): Geneva (the town) 709
GENIAL: gifted; having genius 710
GENICK (*n*): back of the neck; nape 711
GENICKDRÜSE (*f*): cervical gland 712
GENIE (*n*): genius; talent; capacity; ingenuity 713
GENIECORPS (*n*): engineer's corps 714
GENIEREN: to incommode; trouble; be (shy) backward; be afraid; embarrass 715
GENIESSBAR: edible; palatable; relishable 716
GENIESSEN: to enjoy; take; taste; partake; eat or drink; benefit 717
GENIST (*n*): genista, broom 718
GENOSS (*m*): companion; comrade; associate; partner; mate; accomplice; colleague; (*pl*): company 719
GENOSSE (*m*): (see **GENOSS**) 720
GENOSSENSCHAFT (*f*): fellowship; partnership; company; association; co-operative society 721
GENOSSENSCHAFTSGESETZ (*n*): company law 722
GENOSSENSCHAFTSMEIEREI (*f*): co-operative dairy 723
GENOSSENSCHAFTSWEGE, AUF-: by means of co-operation 724
GENRE (*n*): kind; sort; style (Calico); genre 725
GENTELES GRÜN (*n*): tin green, copper stannate 726
GENTHIT (*m*): genthite, nickelgymnite (natural hydrous magnesium-nickel silicate) (see **NICKELGYMNIT**) 727
GENTIANASÄURE (*f*): gentianin, gentianic acid (from *Gentiana lutea*) 728
GENTIANIN (*n*): gentianin, gentianic acid (see **GENTIANASÄURE**) 729
GENUG: enough; sufficient; adequate 730
GENÜGEN: to be enough; suffice; satisfy; be sufficient; be adequate 731
GENUGSAM: sufficient; enough; adequate 732
GENÜGSAM: temperate; easily pleased; contented; frugal; moderate 733
GENUGTUUNG (*f*): satisfaction; compensation 734
GENUSS (*m*): enjoyment; pleasure; benefit; gratification; usufruct 735
GENUSSMENSCH (*m*): epicure 736
GENUSSMITTEL (*n*): appetizer (of any kind) 737
GENUSSSÜCHTIG: pleasure-seeking; epicurean 738
GENUSSZWECK (*m*): table purpose-(s) 739
GEODÄSIE (*f*): geodesy, land surveying 740
GEODÄT (*m*): land surveyor 741
GEODÄTISCH: geodetic-(al) 742
GEOKRONIT (*m*): geokronite (see **SCHULZITE**) 743
GEOLOGIE (*f*): geology 744
GEOLOGISCH: geological-(ly) 745
GEOMETRIE (*f*): geometry 746
GEOMETRIE (*f*), **ANALYTISCHE-**: analytic geometry 747
GEOMETRIE (*f*), **DARSTELLENDE-**: descriptive geometry 748
GEOMETRISCH: geometric-(al)-(ly) 749
GEORGINE (*f*): dahlia 750
GEPAART: paired, etc. (see **PAAREN**); conjugate; germinate (Bot.) 751
GEPÄCK (*n*): luggage; baggage; package 752

GEPÄCKSCHEIN (*m*): luggage ticket 753
GEPANZERT: protected, cased 754
GEPRÄGE (*n*): impression; stamp; coinage; character 755
GERADE: straight; even; upright; direct; plain; erect; honest; (*f*): straight (line) curve (on a graph, etc.); upright (vertical; inclined) straight line 756
GERADEZU: straight on; immediately; absolutely; directly 757
GERADFÜHRUNG (*f*): guide, slide 758
GERADFÜHRUNGSSTANGE (*f*): guide (slide) bar or rod 759
GERADLINIG: rectilinear-(ly) 760
GERANIAL (*n*): (see CITRAL) 761
GERANIOL (*n*): geraniol; $C_{10}H_{17}OH$; Sg. 0.882; Mp. – 15°C.; Bp. 229-230°C. 762
GERANIUM (*n*): geranium, Crane's-bill, stork's bill, alum root (from *Geranium maculatum*) 763
GERANIUMLACK (*m*): (see ROTHOLZLACK) 764
GERANIUMÖL (*n*): (*Oleum andropogonis*): geranium oil, palmarosa oil, gingergrass oil; Sg. 0.886-0.906; Mp. – 16°C.; Bp. 210-230°C. 765
GERANIUMSÄURE (*f*): geranic acid 766
GERANYL (*n*): geranyl 767
GERANYLACETAT (*n*): geranyl acetate; $C_{10}H_{17}.C_2H_3O_2$; Sg. 0.91-0.917 768
GERANYL, ESSIGSAURES- (*n*): geranyl acetate (see GERANYLACETAT) 769
GERANYLMETHYLÄTHER (*m*): geranyl-methyl ether 770
GERÄT (*n*): apparatus; implement; tools; utensils; plant; furniture; goods and chattels; effects; appliances 771
GERÄTE (*n*): (see GERÄT) 772
GERÄTE (*n.pl*), **FORSTKULTUR-**: implements (appliances) for forest culture or afforestation 773
GERÄTEGLAS (*n*): apparatus glass 774
GERÄTE (*n.pl*), **LANDWIRTSCHAFTLICHE-**: agricultural implements or appliances 775
GERATEN: to come; call; light; get; turn out; prosper; prosperous; advantageous; successful; advisable 776
GERÄTESCHRANK (*m*): locker 777
GERATEWOHL, AUFS- (*n*): as a speculation; at a venture 778
GERÄTSCHAFTEN (*f.pl*): apparatus; implements; tools; utensils; appliances 779
GERÄT (*n*), **STEUER-**: steering apparatus 780
GERAUM: ample; long; spacious; roomy; large 781
GERÄUMIG: (see GERAUM) 782
GERÄUSCH (*n*): noise; murmur; clatter; bustle; stir; rustling 783
GERÄUSCHLOS: noiseless-(ly); silent-(ly) 784
GERÄUSCHVOLL: noisy 785
GERBANLAGE (*f*): tanning plant 786
GERBAUSZUG (*m*): tanning extract 787
GERBEBRÜHE (*f*): tan-(ner's) liquor; ooze; dyeing liquor (see also LOH-) 788
GERBEFLÜSSIGKEIT (*f*): (see GERBEBRÜHE) 789
GERBEISEN (*n*): weld iron, wrought iron (Metal) 790
GERBEMITTEL (*n*): tanning medium; tan 791
GERBEN: to tan; dress; curry; taw; hull (Grain); polish (Metal); tilt; refine (Steel); harden (Photo.) 792
GERBEN, SÄMISCH-: to chamois 793
GERBEN, WEISS-: to taw 794
GERBER (*m*): tanner; currier; tawer 795
GERBERBAUM (*m*): young fustic (*Rhus cotinus*) 796
GERBEREI (*f*): tanning; tannery; tawing 797
GERBEREIBEIZE (*f*): tanner's mordant; tanning mordant 798
GERBERFETT (*n*): dégras (Leather); stuff-(ing); mœllon dégras (see DEGRAS) 799
GERBERGRUBE (*f*): tan pit 800
GERBERHOF (*m*): tanner's yard; tan yard 801
GERBERKALK (*m*): slaked (gas) lime 802
GERBERLOHE (*f*): tan bark (waste from tanneries) (high moisture up to 80%. If 60% after passing through tan presses, calorific value= 1100-1400 cals. Air-dried, has moisture 30-35% and can be mechanically dried to 10-12%); tan liquor; ooze 803
GERBERMESSER (*n*): fleshing knife 804
GERBERSTRAUCH (*m*): tanner's sumac (*Rhus glabra* and *R. coriaria*); ink plant (*Coriaria*) 805
GERBERTALG (*m*): tanner's tallow 806
GERBERWEIDE (*f*): willow 807
GERBERWOLLE (*f*): skin wool 808
GERBEVERMÖGEN (*n*): tanning power 809
GERBEVERSUCH (*m*): tanning (trial; experiment) test 810
GERBEXTRAKT (*m* and *n*): tanning extract 811
GERBLEIM (*m*): size (Paper) 812
GERBMATERIAL (*n*): tanning (material) principle 813
GERBMEHL (*n*): tanning material 814
GERBSAUER: tannate of; combined with tannic acid 815
GERBSÄURE (*f*): tannic acid; tannin; gallotannic (digallic) acid; $C_{10}H_{14}O_9$ 816
GERBSTAHL (*m*): tilted (refined) steel; shear steel; burnisher; polishing steel 817
GERBSTOFF (*m*): tanning (material) matter; tan; tannin (see GERBSÄURE) 818
GERBSTOFFAUSZUG (*m*): tanning extract; tannin 819
GERBSTOFFBEIZE (*f*): tannin mordant 820
GERBSTOFFBESTIMMUNG (*f*): determination of (tannin) tanning matter 821
GERBSTOFFEXTRAKT (*m*): (see GERBSTOFFAUSZUG) 822
GERBUNG (*f*): tanning, etc. (see GERBEN) 823
GERECHT: righteous; right; just; legitimate; fit; skilled; lawful; suitable 824
GERECHTSAME (*f*): privilege; right; title 825
GEREIBE (*n*): rubbing; friction 826
GEREICHEN: to redound; turn out; be; cause; conduce (tend) to 827
GERICHT (*n*): court; judgement; jurisdiction; tribunal; dish 828
GERICHTLICH: judicial; legal; forensic; by law 829
GERICHTSAKTEN (*f.pl*): records 830
GERICHTSAMT (*n*): court; tribunal 831
GERICHTSAMTMANN (*m*): judge 832
GERICHTSANWALT (*m*): attorney-at-law 833
GERICHTSBARKEIT (*f*): jurisdiction 834
GERICHTSBEFEHL (*m*): writ; summons 835
GERICHTSHAUS (*n*): law court, court house 836
GERICHTSHOF (*m*): court; tribunal; court of law; law-court 837
GERICHTSKOSTEN (*pl*): law (expenses) costs 838
GERICHTSRAT (*m*): counsellor; counsel; justice 839
GERIESEL (*n*): trickling; ripping 840
GERIFFELT: corrugated 841

GERILLT : grooved 842
GERING : small ; slight ; deficient ; inferior ; limited ; low ; base ; little ; humble ; petty ; trifling ; unimportant 843
GERINGACHTUNG (*f*) : disdain ; contempt 844
GERINGFÜGIG : petty ; trifling ; insignificant ; trivial 845
GERINGHALTIG : below standard ; worthless ; of little value ; futile ; having small content 846
GERINGSCHÄTZIG : deprecatory ; disrespectful derogatory ; despicable ; contemptible 847
GERINNBAR : coagulable ; congealable 848
GERINNE (*n*) : channel ; gutter ; running ; flowing ; stream ; conduit 849
GERINNEN : to curdle ; coagulate ; congeal 850
GERINNUNG (*f*) : coagulation ; coagulum 851
GERINNUNG (*f*) IM HERZEN : thrombosis of the heart, heart clotting (Med.) 852
GERINNUNGSMASSE (*f*) : coagulum 853
GERIPPE (*n*) : skeleton ; framework ; framing ; carcase 854
GERIPPE (*n*), SCHIFFS- : framing (ribs), (of a ship) 855
GERIPPTES PAPIER (*n*) : laid paper (Paper) 856
GERIPPTES PAPIER (*n*), AZURBLAU- : azure laid paper 857
GERMANIUM (*n*) : germanium (Ge) 858
GERMANIUMCHLORID (*n*) : germanium chloride ; $GeCl_4$; Sg. 1.887 859
GERMANIUMCHLOROFORM (*n*) : germanium chloroform ; $GeHCl_3$; Bp. 72°C. 860
GERMANIUMDIOXYD : germanium dioxide ; GeO_2 ; Sg. 4.703 861
GERMANIUMFLUORWASSERSTOFFSÄURE (*f*) : fluogermanic acid 862
GERMANIUMOXYCHLORID (*n*) : germanium oxychloride ; $GeOCl_2$ 863
GERMANIUMOXYD (*n*) : germanium (germanic) oxide 864
GERMANIUMOXYDUL (*n*) : germanous oxide 865
GERMANIUMSÄURE (*f*) : germanic acid 866
GERMANIUMSULFID (*n*) : germanium (germanic) sulphide 867
GERMANIUMWASSERSTOFF (*m*) : germanium hydride 868
GERMER (*m*) : (American)-white hellebore (*Veratrum viride*) (see also JERVIN and NIESWURZ) 869
GERN : gladly ; willingly ; with pleasure 870
GERNE : (see GERN) 871
GERÖLLE (*n*) : boulders, water worn rocks 872
GERSDORFFIT (*m*) : gersdorffite (a sulfarsenide of nickel) ; NiAsS. ; Sg. 6.0 (see NICKEL--GLANZ) 873
GERSTE (*f*) : barley 874
GERSTENBIER (*n*) : beer brewed from barley 875
GERSTENBRÜHE (*f*) : barley water 876
GERSTENGRAUPEN (*f.pl*) : pearl barley 877
GERSTENGRÜTZE (*f*) : pearl barley 878
GERSTENKORN (*n*) : barleycorn ; sty (*Hodeolum*) (a boil on the edge of the eyelid) (Med.) 879
GERSTENMALZ (*n*) : barley malt 880
GERSTENÖL (*n*) : barley oil (from *Hordeum vulgare*) 881
GERSTENSCHLEIM (*m*) : barley water 882
GERSTENTRANK (*m*) : barley water 883
GERSTENWOLF (*m*) : (see GRAUPENHOLLÄNDER) 884
GERSTENZUCKER (*m*) : barley sugar ; maltose (see MALTOSE) 885
GERTE (*f*) : whip ; switch ; rod 886
GERTENKRAUT (*n*) : Southernwood (see EBER--RAUTE) 887
GERTENWEIDE (*f*) : osier (a form of willow) (*Salix triandra*) 888
GERTENWURZ (*f*) : Southernwood (see EBER--RAUTE) 889
GERUCH (*m*) : smell ; odour; scent ; savour ; fragrance 890
GERUCHLOS : savourless ; odourless ; unable to smell ; without scent 891
GERUCHSNERV (*m*) : olfactory nerve 892
GERÜCHT (*n*) : report ; rumour 893
GERUHEN : to condescend, be pleased to, deign 894
GERÜLL (*n*) : rubble 895
GERÜMPEL (*n*) : lumber ; trash ; rubbish 896
GERÜST (*n*) : frame-(work) ; scaffold-(ing) ; crate ; framing ; trestle ; stage ; rack ; staging 897
GERÜSTBOCK (*m*) : trestle 898
GERÜSTSTANGE (*f*) : scaffold pole 899
GESÄGT : serrated ; serrate 900
GESÄME (*n*) : seeds 901
GESAMT : total ; entire ; whole ; joint ; common ; united ; collective 902
GESAMTALKALINITÄT (*f*) : total alkalinity 903
GESAMTANORDNUNG (*f*) : general arrangement 904
GESAMTARBEIT (*f*) : total work, gross work 905
GESAMTAUSGABE (*f*) : total expenditure ; complete edition 906
GESAMTBETRAG (*m*) : total amount ; sum total 907
GESAMTERTRAG (*m*) : total (profit, return, income, revenue, amount) yield 908
GESAMTHÄRTE (*f*) : total hardness (Water analysis) 909
GESAMTHEIT (*f*) : totality 910
GESAMTLÖSLICHES (*n*) : total soluble matter (Water analysis) 911
GESAMTMENGE (*f*) : total quantity (Water or Coal analysis, etc.) 912
GESAMTPREIS (*m*) : total price 913
GESAMTPUBLIKUM (*n*) : general public 914
GESAMTRÜCKSTAND (*m*) : total scale ; total residue (Water or Coal analysis, etc.) 915
GESAMTSÄURE (*f*) : total acid 916
GESAMTSCHAFT (*f*) : corporation, body 917
GESAMTSCHWEFEL (*m*) : total sulphur 918
GESAMTSTICKSTOFF (*m*) : total nitrogen 919
GESAMTVERBRAUCH (*m*) : total consumption 920
GESAMTVERHALTEN (*n*) : general behaviour 921
GESAMTVOLUM (*n*) : total volume 922
GESAMTWÄRME (*f*) : total heat 923
GESAMTWÄRMEGEFÄLLE (*n*) : total heat drop 924
GESAMTWERT (*m*) : total value 925
GESAMTWIRKUNGSGRAD (*m*) : total (overall) efficiency ; (η) 926
GESAMTZAHL (*f*) : total (figure) number 927
GESANDTE (*m*) : ambassador ; consul ; messenger 928
GESANDTE, PÄPSTLICHER- (*m*) : Papal nuncio 929
GESANDTSCHAFT (*f*) : embassy, legation, consulate 930
GESANG (*m*) : song ; singing ; canto ; poem ; hymn ; chorale ; chant 931
GESANGSTIMME (*f*) : voice, part (Music) 932
GESANGVEREIN (*m*) : choral society 933
GESÄSS (*n*) : seat ; floor ; bottom ; anus (Anat.) 934
GESÄTTIGT : saturated 935
GESCHABSEL (*n*) : scrapings 936

GESCHÄFT (*n*): business; affair; calling; trade; house; bargain; dealings; occupation; employment 937
GESCHÄFTIG: busy; fussy; active; officious; industrious 938
GESCHÄFTLICH: business; business-like; professional; commercial 939
GESCHÄFTLICHE BEZIEHUNGEN (*f.pl*): business relations 940
GESCHÄFTSABSCHLUSS (*m*): transaction 941
GESCHÄFTSABSCHNITT (*m*): business (season) period 942
GESCHÄFTSANTEIL (*m*): share in a business, debenture 943
GESCHÄFTSBETRIEB (*m*): transaction of business, method (way) of (carrying on) doing business, line of business, business, routine, working of the business 944
GESCHÄFTSBUCHPAPIER (*n*): account book paper 945
GESCHÄFTSFÜHRER (*m*): agent; manager (of a business) 946
GESCHÄFTSHAUS (*n*): business house, firm 947
GESCHÄFTSKARTE (*f*): business (trade) card 948
GESCHÄFTSKENNTNIS (*f*): knowledge of business 949
GESCHÄFTSLEITER (*m*): manager (of a business) 950
GESCHÄFTSSTOCKUNG (*f*): trade stagnation, depression 951
GESCHÄFTSTÄTIGKEIT (*f*): briskness (Com.) 952
GESCHÄFTSVERBINDUNG (*f*): connection (in business), business relation-(ship) 953
GESCHÄFTSZWEIG (*m*): line of business, branch 954
GESCHEHEN: to happen; occur; take place; be done 955
GESCHEIT: sensible; clever; prudent; judicious 956
GESCHENK (*n*): gift; present; donation 957
GESCHICHTE (*f*): history; story; tale 958
GESCHICHTLICH: historical 959
GESCHICK (*n*): skill; aptitude; order; fate; destiny; address; dexterity; ability 960
GESCHICKLICHKEIT (*f*): skill; cleverness; art; dexterity; adroitness 961
GESCHICKT: skilled; clever; qualified; fit; suitable (see also SCHICKEN) 962
GESCHIEBE (*n*): rubble; boulder; shoving; pushing 963
GESCHILF (*n*): place which is over (grown) run with weeds or reeds 964
GESCHIRR (*n*): vessel-(s); ware; utensils; tools; apparatus; gear; harness; plant; mill; vat (Leather); crockery; implements 965
GESCHLECHT (*n*): sex; genus; kind; race family; stock; generation; gender 966
GESCHLECHTLICH: sexual; generic 967
GESCHLECHTSART (*f*): generic character; genus 968
GESCHLECHTSBAUM (*m*): genealogical tree 969
GESCHLECHTSDRÜSE (*f*): genital gland; ovary; testicle 970
GESCHLECHTSFALL (*m*): genitive case (Grammar) 971
GESCHLECHTSFOLGE (*f*): line of descent, lineage 972
GESCHLECHTSKUNDE (*f*): genealogy 973
GESCHLECHTSPFLANZE (*f*): phanerogam; spermophyte; flowering plant 974
GESCHLECHTSREIFE (*f*): puberty 975
GESCHLECHTSREIZEND: aphrodisiac 976
GESCHLECHTSTEILE (*m.pl*): genitals 977
GESCHLECHTSTRIEB (*m*): sexual instinct 978
GESCHLECHTSWORT (*n*): article (Grammar) 979
GESCHLINGE (*n*): frame, bed 980
GESCHLOSSENE KETTE (*f*): closed (endless) chain 981
GESCHLOSSEN KREISEND: circulating in a closed cycle 982
GESCHMACK (*m*): taste; flavour; liking; savour; relish; fancy 983
GESCHMACKLOS: tasteless; insipid 984
GESCHMACKSKORRIGENS (*n*): taste (palate) corrective 985
GESCHMACKSRICHTUNG (*f*): taste, direction of one's taste 986
GESCHMACKVOLL: tasteful; elegant 987
GESCHMEIDE (*n*): jewels, trinkets, jewellery, wrought metal 988
GESCHMEIDIG: pliable; flexible; soft; ductile; supple; malleable; versatile; plastic; pliant; yielding 989
GESCHMEIDIGKEIT (*f*): ductility; suppleness; flexibility; malleability 990
GESCHNITTENE SCHÄRFE (*f*): clear-cut outline (of photographs, etc.) 991
GESCHOSS (*n*): projectile; shell; missile; story; floor (of building) 992
GESCHOSSAUFZUG (*m*): shell crane, shell davit (Naut.) 993
GESCHOSSFÜLLUNG (*f*): projectile (charge) filling 994
GESCHOSSTREIBEND: ballistic 995
GESCHOSSTREIBMITTEL (*n*): gunpowder (black powder; smokeless powder) 996
GESCHRÄNK (*n*), **FEUER-**: furnace front plate, lower half front 997
GESCHREI (*n*): cry; outcry; rumour; disrepute; scream-(s); shout; clamour 998
GESCHÜR (*n*): dross 999
GESCHÜTZ (*n*): gun; cannon; ordnance; artillery; arms; armament 000
GESCHÜTZBRONZE (*f*): (89-91% Cu; rest Sn.); gunmetal 001
GESCHÜTZPFORTE (*f*): gun port (Naut.) 002
GESCHÜTZPULVER (*n*): gun powder 003
GESCHÜTZROHR (*n*): (gun)-barrel 004
GESCHÜTZWEITE (*f*): bore, calibre 005
GESCHÜTZZUG (*m*): artillery train 006
GESCHWADER (*n*): squadron (Naut.) 007
GESCHWÄNGERT: pregnant with; full of; replete with 008
GESCHWEIGEN: not to mention; to say nothing about; neglect 009
GESCHWIND: fast; quick; immediate; swift; prompt 010
GESCHWINDE (*n*): speed, velocity, quickness, dispatch, haste 011
GESCHWINDIGKEIT (*f*): speed; velocity; quickness; dispatch; haste 012
GESCHWINDIGKEIT (*f*): **ANFANGS-**: initial speed 013
GESCHWINDIGKEIT (*f*), **END-**: terminal speed 014
GESCHWINDIGKEIT (*f*), **KOLBEN-**: piston speed 015
GESCHWINDIGKEITSÄNDERUNG (*f*): alteration (change, fluctuation) in speed 016
GESCHWINDIGKEITSMESSER (*m*): speed indicator; trochometer; speedometer 017
GESCHWINDIGKEITSREGLER (*m*): speed regulator or governor 018

GESCHWINDIGKEITSSTUFE (*f*): velocity stage (Turbines) 019
GESCHWINDIGKEITSÜBERSCHREITUNG (*f*): exceeding the speed limit 020
GESCHWINDIGKEITSVERHÄLTNIS (*n*): speed (velocity) ratio 021
GESCHWINDIGKEIT (*f*), **UMDREHUNGS-**: speed of rotation 022
GESCHWINDIGKEIT (*f*), **UMFANGS-**: peripheral speed 023
GESCHWORENE (*m*): juryman; one who has been sworn in 024
GESCHWORENENGERICHT (*n*): jury 025
GESCHWULST (*f*): swelling; tumour (Med.) 026
GESCHWÜR (*n*): ulcer; abscess; swelling 027
GESCHWÜRBILDUNG (*f*): ulceration 028
GESELL (*m*): mate; comrade; fellow; partner; journeyman 029
GESELLE (*m*): (see **GESELL**) 030
GESELLEN: to accompany; ally; associate; join 031
GESELLIG: social; sociable; convivial 032
GESELLSCHAFT (*f*): society; company; partnership; club; community; party; association 033
GESELLSCHAFTSVERTRAG (*m*): articles of association; deed of partnership 034
GESENK (*n*): hollow; slot; punch; stamp; recess; die; swaging 035
GESENKGUSSSTÜCK (*n*): die-casting 036
GESENK, IM-SCHMIEDEN: to swage 037
GESENKSCHMIEDESTÜCK (*n*): swaged forging 038
GESENKTIEFE (*f*): depth of (slot) recess 039
GESETZ (*n*): law; rule; precept; statute; decree; commandment 040
GESETZBUCH (*n*): code (of laws) 041
GESETZENTWURF (*m*): (legislative)-bill draft 042
GESETZGEBEND: legislative 043
GESETZGEBUNG (*f*): legislation 044
GESETZLICH: lawful; legal-(ly); statutory; legitimate-(ly) 045
GESETZLICH GESCHÜTZT: patented 046
GESETZLOS: lawless; illegal 047
GESETZMÄSSIG: regular; lawful; legal; legitimate; conformable to law 048
GESETZMÄSSIGKEIT (*f*): regularity; lawfulness; legitimacy; conformity to law; legality 049
GESETZT: supposing; assuming; serious; settled; steady; given; granted (see also **SETZEN**) 050
GESETZVORSCHLAG (*m*): motion; bill 051
GESETZWIDRIG: illegal; unlawful 052
GESICHT (*n*): face; countenance; look; vision; sight; visage; mien 053
GESICHT (*n*), **KUPFERIGES-**: Acne Rosacea, facial inflammation (Med.) 054
GESICHTSACHSE (*f*): visual axis 055
GESICHTSBILDUNG (*f*): physiognomy 056
GESICHTSFARBE (*f*): colour, complexion 057
GESICHTSFELD (*n*): field of (view) vision; horizon; angle of view (of a lens) 058
GESICHTSKREIS (*m*): (see **GESICHTSFELD**) 059
GESICHTSLINIE (*f*): facial (visual) line; eye-line 060
GESICHTSMALER (*m*): portrait painter 061
GESICHTSPUNKT (*m*): visual (view) point; point of (sight) view; fact; view; opinion 062
GESICHTSSTRAHL (*m*): visual ray 063
GESICHTSTÄUCHUNG (*f*): optical illusion 064
GESICHTSWEITE (*f*): range of vision 065
GESICHTSWINKEL (*m*): visual (optic; facial) angle; angle of view (of a lens) 066
GESICHTSZUG (*m*): feature 067
GESIMS (*n*): moulding; cornice; shelf; entablature 068
GESIMSHOBEL (*m*): moulding plane 069
GESINNT: disposed; minded (see **SINNEN**) 070
GESINNUNG (*f*): mind; disposition; opinion; conviction; sentiment; intention 071
GESINNUNGSLOS: unprincipled; characterless 072
GESITTET: mannered; bred; polite; moral; civilized; polished 073
GESONNEN SEIN: to be inclined; intend; be minded; be disposed (see **SINNEN**) 074
GESPANN (*n*): team (of horses); group, battery 075
GESPANNT: intent, intense, eager (see also **SPANNEN**) 076
GESPENST (*n*): spectre; ghost; apparition; vision 077
GESPINST (*n*): spun yarn- (goods); (textile)-fabric; web 078
GESPINSTFASER (*f*): fibre for textile fabrics 079
GESPRÄCH (*n*): talk; conversation; discourse 080
GESPRÄCHIG: communicative; sociable; talkative; affable; loquacious 081
GESTADE (*n*): shore; coast; bank; land; beach 082
GESTALT (*f*): form; shape; figure; aspect; manner; stature 083
GESTALTEN: to form; shape; fashion; appear 084
GESTALTENLOS: amorphous; formless; shapeless 085
GESTALTEN, SICH GÜNSTIGER-: to look up, brighten, improve 086
GESTALTUNG (*f*): form; type; formation; size; configuration; condition; state; appearance; construction; shape; figure; modelling 087
GESTÄNDNIS (*n*): confession; avowal; admission; acknowledgement 088
GESTÄNGE (*n*): rods; poles; bars 089
GESTANK (*m*): stench; evil smell; stink 090
GESTATTBAR: permissible; allowable; warrantable 091
GESTATTEN: to permit; consent to; warrant; allow; grant 092
GESTEHEN: to coagulate; curdle; clot; congeal; confess; admit; own; grant; acknowledge 093
GESTEIN (*n*): rock; stones; geological formation-(s); minerals 094
GESTEINBOHRER (*m*): rock-drill 095
GESTEINKUNDE (*f*): petrology; petrography; mineralogy; geognosy; geology 096
GESTEINKUNDLICH: petrographic 097
GESTEINSLEHRE (*f*): (see **GESTEINKUNDE**) 098
GESTEINSREST (*m*): rock (residue) remains 099
GESTELL (*n*): frame; jack; stand; trestle; support; base; foot; scaffold; shelf; hearth (Blast furnaces); chassis (Automobile); housing; framing; carriage 100
GESTEUERTES REGELVENTIL (*n*): automatic regulating valve (actuated by valve gear) 101
GESTIKULIEREN: to gesticulate, make signs 102
GESTIRN (*n*): constellation; star 103
GESTIRNT: starred; starry 104
GESTÖBER (*n*): shower; snow-storm; dust-storm; drift; drizzle 105
GESTRÄUCH (*n*): bushes; shrubbery; shrubs 106
GESTRECKTER WINKEL (*m*): angle of 180°. 107

GESTRECKTES EISEN (*n*): wrought iron 108
GESTRICHELT: dotted-(line); chain-dotted 109
GESTRICHENES PAPIER (*n*): coated paper 110
GESTRÜPP (*n*): underwood; thicket; undergrowth; bushes; briars 111
GESTÜBBE (*m*): (see GESTÜBE) 112
GESTÜBE (*m*): brasque (Metal); (coal)-dust; cement; clay 113
GESUCH (*n*): request; demand; want advertisement; application; petition 114
GESUCH (*n*), PATENT-: application (for patent) 115
GESUCHTHEIT (*f*): affectation; choiceness 116
GESUMPF (*n*): bog; marsh; quagmire 117
GESUND: sound; healthy; wholesome; well; salubrious 118
GESUNDEN: to recover; convalesce; become convalescent. 119
GESUNDHEIT (*f*): health; salubrity; soundness 120
GESUNDHEITLICH: sanitary; hygienic 121
GESUNDHEITSAMT (*n*): Board of Health; Sanitary Board; Ministry of Health; Health Ministry. 122
GESUNDHEITSLEHRE (*f*): hygiene 123
GESUNDHEITSPROBE (*f*): quarantine 124
GESUNDHEITSWESEN (*n*): sanitary (affairs) matters; sanitation 125
GESUNDHEITSZUSTAND (*m*): state of health 126
GETÄFEL (*n*): wainscot-(ting); inlaying 127
GETAST (*n*): touching, sense of touch 128
GETIEGERT: striped, marked (like a tiger) 129
GETIER (*n*): animals 130
GETÖSE (*f*): noise, din, bustle 131
GETRÄNK (*n*): beverage; drink; potion 132
GETRAUEN: to dare; venture; trust 133
GETREIDE (*n*): grain; cereals; corn; crops 134
GETREIDEART (*f*): kind of grain; cereals; crops 135
GETREIDEÄTHER (*m*): grain alcohol (see ÄTHYL-ALKOHOL) 136
GETREIDEBAU (*m*): grain growing 137
GETREIDEBODEN (*m*): corn loft, granary 138
GETREIDEBRANNTWEIN (*m*): whisky from grain 139
GETREIDEHANDEL (*m*): corn trade 140
GETREIDEKORN (*n*): grain 141
GETREIDEMARKT (*m*): corn market 142
GETREIDEMÜHLE (*f*): flour-mill 143
GETREIDEPRÜFER (*m*): grain tester 144
GETREIDEREINIGUNG (*f*): grain cleaning 145
GETREIDEREINIGUNGSMASCHINE (*f*): grain cleaning machine 146
GETREIDEROST (*m*): rust (of cereals) (Bot.) 147
GETREIDESPEICHER (*m*): grain elevator; granary 148
GETREIDESTEIN (*m*): beer scale 149
GETREIDEWURM (*m*): corn weevil (*Sitophilus granarius*) 150
GETRENNT: separate-(ly); independent-(ly) 151
GETRIEBE (*n*): driving; working; driving gear; gear; gearing; machinery; pinion; reduction gearing; drive; mechanism 152
GETRIEBEGEHÄUSE (*n*): gear (casing, housing) case 153
GETRIEBENES BLECH (*n*): worked (hammered) plate 154
GETRIEBENES EISEN (*n*): wrought iron 155
GETRIEBERAD (*n*): gear wheel 156
GETROST: confident; courageous; of good cheer; trusting 157
GETRÜBT: troubled; gloomy; turbid; cloudy; clouded; opaque; dull; dim; thick; muddy 158
GETRÜMMER (*n*): (see TRÜMMER) 159
GETÜMMEL (*n*): bustle; tumult; turmoil 160
GEÜBT: expert; skilled; versed 161
GEVIERT (*n*): square; quadrature 162
GEVIERTE (*n*): (see GEVIERT) 163
GEVIERTSCHEIN (*m*): quartile, quadrate, tetragon (Astrol.) 164
GEWÄCHS (*n*): plant; growth; growing; vintage; excrescence 165
GEWÄCHSEN SEIN: to be suitable for; be applicable to 166
GEWÄCHSHAUS (*n*): greenhouse; conservatory; hothouse 167
GEWÄCHSKUNDE (*f*): botany 168
GEWÄCHSLEHRE (*f*): botany 169
GEWÄCHSREICH (*n*): vegetable kingdom 170
GEWACHSTES PAUSPAPIER (*n*): waxed tracing paper (Paper) 171
GEWAHR: aware; perceiving 172
GEWÄHR (*f*): surety; security; guarantee; warrant; assurance; (exchange)-value (of currency) 173
GEWAHREN: to perceive; observe; notice; become sensible of 174
GEWÄHREN: to give; furnish; grant; be a surety for; impart; afford; assure 175
GEWÄHRLEISTEN: to give; furnish; grant; impart; ensure; guarantee; warrant; go bail for 176
GEWÄHRLEISTUNG (*f*): guarantee, security 177
GEWÄHRLEISTUNGSZAHLEN (*f.pl*): guaranteed-(load) figures 178
GEWAHRSAM (*m*): custody; prison; proviso; surety; safety; possession; wary; keeping 179
GEWAHR WERDEN: to become aware of; perceive; notice; observe 180
GEWALT (*f*): power; force; violence; might; supremacy 181
GEWALTBRIEF (*m*): warrant 182
GEWALTIG: powerful; forcible; huge; vast; violent; enormous; gigantic; potent; intense 183
GEWALTRAUB (*m*): usurpation 184
GEWALTSAM: violent; forcible 185
GEWALTSCHEIN (*m*): warrant 186
GEWALTTAT (*f*): outrage, violence 187
GEWALTTRÄGER (*m*): one in whom certain powers are vested; attorney 188
GEWAND (*n*): garment; clothing; dress; drapery; raiment; cloth 189
GEWANDHAUS (*n*): cloth-hall 190
GEWANDT: clever; skilled; versed; quick; adroit; dexterous; versatile; experienced 191
GEWÄRTIG: expectant; attentive; waiting 192
GEWÄRTIGEN: to expect; attend; await 193
GEWÄSSER (*n*): waters; stream; flood; running water 194
GEWÄSSERKUNDE (*f*): hydrology 195
GEWÄSSERT: watered, sprinkled with water; hydrated, hydrous 196
GEWÄSSERT, EINFACH-: monohydrate 197
GEWEBE (*n*): tissue; texture; (textile)-fabric; web; weft; gauze 198
GEWEBEDRAHT (*m*): gauze wire 199
GEWEBEFARBSTOFF (*m*): histohematin (for staining tissue before examination) 200
GEWEBELEHRE (*f*): histology (the science of tissue) 201
GEWEHR (*n*): gun; rifle; weapon; firearm 202
GEWEHRFEUER (*n*): rifle fire 203

GEWEHRHÄNDLER (*m*): gunsmith, armourer 204
GEWEHRLAUF (*m*): barrel (of a gun) 205
GEWEHRÖL (*n*): rifle oil 206
GEWEHRPULVER (*n*): gunpowder 207
GEWEHRSCHAFT (*m*): (gun or rifle)-stock 208
GEWEIH (*n*): horns; antlers 209
GEWELLT: wavy; undulated; corrugated 210
GEWERBE (*n*): trade; occupation; industry; profession; craft 211
GEWERBEAUSSTELLUNG (*f*): trade (industrial) exhibition 212
GEWERBESALZ (*n*): common salt (for industrial purposes); industrial salt 213
GEWERBESCHULE (*f*): technical (industrial) school 214
GEWERBFLEISS (*m*): industry 215
GEWERBKUNDE (*f*): technology 216
GEWERBKUNDLICH: technological 217
GEWERBLICH: industrial; professional 218
GEWERBSTÄTIGKEIT (*f*): (manufacturing)-industry; trade 219
GEWERBSZWEIG (*m*): branch of (trade) industry 220
GEWERBTÄTIG: industrial 221
GEWERBTÄTIGKEIT (*f*): industry, trade 222
GEWERK (*n*): trade; factory; works; work; corporation; guild; mine; manufacture; machine-(ry) 223
GEWERKSCHAFT (*f*): mining company 224
GEWERKSMANN (*m*): manufacturer 225
GEWICHT (*n*): weight; gravity; importance; stress 226
GEWICHTIG: weighty; heavy; important 227
GEWICHTLEHRE (*f*): statics 228
GEWICHTLOS: unimportant, without weight 229
GEWICHTSABGANG (*m*): loss (deficiency) in weight 230
GEWICHTSABNAHME (*f*): decrease in weight 231
GEWICHTSANALYSE (*f*): gravimetric analysis 232
GEWICHTSANALYTISCH: gravimetric-(al)-(ly) 233
GEWICHTSBELASTUNG (*f*): weighting, loading; weight loading; lever loading; dead-weight (of valves) 234
GEWICHTSBESTIMMUNG (*f*): determination (computation) of weight 235
GEWICHTSEINHEIT (*f*): unit of weight 236
GEWICHTSMENGE (*f*): (amount of)-weight 237
GEWICHT (*n*), SPEZIFISCHES-: specific (weight) gravity 238
GEWICHTSPROZENT (*n*): percentage of weight 239
GEWICHTSSATZ (*m*): set of weights 240
GEWICHTSSTÜCK (*n*): weight 241
GEWICHTSTEIL (*m*): part of (by) weight 242
GEWICHTSVERHÄLTNIS (*n*): proportion (ratio) of weight 243
GEWICHTSVERLUST (*m*): loss (decrease) in weight 244
GEWICHTSZUNAHME (*f*): increase (gain) in weight 245
GEWICHTSZUSAMMENSETZUNG (*f*): composition of (by) weight 246
GEWIEGT: experienced; skilled 247
GEWILLT: willing; inclined; disposed 248
GEWINDE (*n*): winding; coil; skein; thread (turn) of a screw 249
GEWINDEANSCHLUSS (*m*): screwed connection 250
GEWINDEBOHRER (*m*): screw tap 251
GEWINDE (*n*), DOPPELTES-: double thread 252
GEWINDE (*n*), EINFACHES-: single thread 253
GEWINDEEISEN (*n*): screw plate 254
GEWINDE (*n*), FLACHES-: square thread 255
GEWINDEGANG (*m*): thread of a screw 256
GEWINDEGLAS (*n*): screwed glass bottle or tube 257
GEWINDEKLUPPE (*f*): screw stock 258
GEWINDELEHRE (*f*): (screw)-thread gauge 259
GEWINDE (*n*), MUTTER-: female thread 260
GEWINDESCHNEIDBACKEN (*f.pl*): screw dies 261
GEWINDESTAHL (*m*): screw tool 262
GEWINDESTEIGUNG (*f*): pitch of thread, screw pitch 263
GEWINN (*m*): yield; winning; gain; profit; proceeds; production; acquisition; advantage 264
GEWINNANTEIL (*m*): share of profits 265
GEWINNBAR: obtainable 266
GEWINNBETEILIGUNG (*f*): share of profits 267
GEWINNBRINGEND: profitable, lucrative 268
GEWINNEN: to obtain; earn; get; profit; win; gain; extract; produce; secure; acquire 269
GEWINN- UND VERLUST-RECHNUNG (*f*): Profit and Loss account 270
GEWINNUNGSANLAGE (*f*): extraction plant 271
GEWIRBELT: vertebrate (see WIRBELN) 272
GEWIRK (*n*): texture, weaving 273
GEWIRR (*n*): complication, confusion, entanglement 274
GEWISS: sure; certain; fixed; indeed; certainly; steady 275
GEWISSEN (*n*): conscience; consciousness 276
GEWISSENHAFT: conscientious-(ly); careful-(ly) 277
GEWISSENLOS: unconscientious; unscrupulous; unprincipled 278
GEWISSENSBISS (*m*): sting of conscience, remorse 279
GEWISSENSFALL (*m*): question of conscience 280
GEWISSENSPFLICHT (*f*): conscientious duty, bounden duty 281
GEWISSENSZWEIFEL (*m*): scruple, doubt 282
GEWISSERMASSEN: in(to) a certain degree; as it were; so to speak 283
GEWISSHEIT (*f*): certainty; proof 284
GEWITTER (*n*): storm; thunderstorm; tempest 285
GEWITTERLUFT (*f*): oppression, depression, heaviness (due to a thunderstorm) 286
GEWITTERSCHWÜLE (*f*): oppressiveness, sultriness (due to a thunderstorm) 287
GEWITTERSTURM (*m*): thunder-storm 288
GEWITTERVOGEL (*m*): storm petrel (*Procellaria pelagica*) 289
GEWITTERWOLKE (*f*): thundercloud 290
GEWOGEN: well-disposed; favourable; weighed; kind; friendly 291
GEWÖHNEN: to accustom; habituate; inure; use; train; domesticate 292
GEWOHNHEIT (*f*): custom; usage; use; habit; fashion; practice 293
GEWÖHNLICH: usual; customary; vulgar; common; ordinary; trivial 294
GEWÖHNLICHER STEIN (*m*): common brick, ordinary brick 295
GEWOHNT: usual, accustomed 296
GEWÖLBE (*n*): vault; arch; crown (of a furnace); store-(house); cellar; shop 297
GEWÖLBEBOGEN (*m*): arch (of a vault) 298
GEWÖLBEDRUCK (*m*): thrust-(of arch) 299
GEWÖLBE (*n*), FEUERFESTES-: fire-brick (refractory) arch 300
GEWÖLBEPFEILER (*m*): buttress, column 301
GEWÖLBERING (*m*): arch ring 302

GEWÖLBESCHLUSSSTEIN (*m*) : key-stone 303
GEWÖLBESTEIN (*m*) : voussoir, arch-brick 304
GEWÖLBESTÜTZE (*f*) : arch support, column, flying buttress 305
GEWÖLK (*n*) : (banks of) clouds 306
GEWOLLT : required, desired, wished for 307
GEWUNDEN : coiled ; wound ; spiral ; twisted ; tortuous ; sinuous 308
GEWÜRM (*n*) : worms ; vermin ; reptiles ; creeping things 309
GEWÜRZ (*n*) : spice-(s) ; seasoning ; aromatics ; groceries 310
GEWÜRZARTIG : spicy ; aromatic 311
GEWÜRZBÜCHSE (*f*) : spice box 312
GEWÜRZBUSCHSAMENÖL (*n*) : spice bush seed oil, Benjamin bush seed oil (from seeds of *Lindera benzoin*) ; Mp. 46°C. 313
GEWÜRZESSIG (*m*) : aromatic vinegar 314
GEWÜRZEXTRAKT (*m*) : spice extract 315
GEWÜRZHAFT : spicy ; aromatic 316
GEWÜRZHANDEL (*m*) : spice (grocer's) trade ; grocery trade 317
GEWÜRZIG : spicy, aromatic 318
GEWÜRZNELKE (*f*) : clove ; caryophyllus (dried buds of *Caryophyllus aromaticus* or *Eugenia aromatica*) 319
GEWÜRZNELKENÖL (*n*) : clove oil ; Sg. 1.05-1.07 ; Bp. 250-260°C. (see also NELKENÖL) 320
GEWÜRZPULVER (*n*) : aromatic powder 321
GEWÜRZSALZ (*n*) : spice salt (mixture of spice extracts with salt) 322
GEWÜRZSTRAUCH (*m*) : spice bush (*Calycanthus*) (Bot.) 323
GEWÜRZSTRÄUCHER (*m*) : *Calycanthaceæ* (Bot.) 324
GEWÜRZT : spiced ; aromatic 325
GEWÜRZWAREN (*f.pl*) : spices ; groceries 326
GEWÜRZWEIN (*m*) : spiced wine 327
GEZAHNT : toothed ; indented ; dentate ; dentated ; serrate-(d) 328
GEZÄNGT : (hammered into square blocks and welded), shingled (Metal) 329
GEZÄNK (*n*) : dispute, disputing, wrangling 330
GEZEIT (*f*) : (flood)-tide ; term 331
GEZELT (*n*) : pavilion ; tent ; sensorium 332
GEZEUG (*n*) : utensils ; tools ; implements 333
GEZIEFER (*n*) : vermin, insects 334
GEZIEFERKENNER (*m*) : entomologist 335
GEZIEFERLEHRE (*f*) : entomology 336
GEZIEMEN : to become ; befit ; beseem ; be (decent ; convenient) proper 337
GEZWEITEILT : bipartite ; bifurcated ; binary 338
GHEABUTTER (*f*) : ghea butter, indian butter, Fulwa (Phulwara) butter (from seeds of *Bassia butyracea*) ; Sg. 0.897 ; Mp. 39°C. 339
GHEDDAWACHS (*n*) : East Indian wax 340
GIBBSIT (*m*) : gibbsite, hydrargillite (an aluminium hydroxide) ; $Al(OH)_3$; Sg. 2.35 (see also HYDRARGILLIT) 341
GICHT (*f*) : top ; mouth ; throat (of blast furnace or crucible) ; charge (total charge of ore and limestone or coke) (Blast furnaces) ; gout ; arthritis (Med.) 342
GICHTARTIG : gouty ; arthritic 343
GICHTBRUCH (*m*) : palsy (Med.) 344
GICHTBRÜCHIG : paralytic ; palsied ; gouty ; arthritic 345
GICHTBRÜCHIGKEIT (*f*) : palsy (Med.) 346
GICHTBÜHNE (*f*) : blast furnace platform 347
GICHTDECKEL (*m*) : top cover (Blast furnaces) 348
GICHTGAS (*n*) : blast furnace gas (8-10% CO_2 ; 25-30% CO ; 1-2% H : 55-60% N ; cal. val. 800-900 WE/cu.m. ; gas temperature 120-200° C.) 349
GICHTIG : gouty, arthritic 350
GICHTÖL (*n*) : antiarthritic oil 351
GICHTPAPIER (*n*) : antiarthritic paper 352
GICHTROSE (*f*) : peony 353
GICHTRÜBE (*f*) : white bryony (*Bryonia dioica*) 354
GICHTSCHMERZ (*m*) : gouty pain 355
GICHTUNG (*f*) : charging (of furnace) (see GICHT) 356
GICHTVERSCHLUSS (*m*) : throat (seal) cover ; gas (catcher) trap, (for blast furnaces at the top) 357
GICHTWASSER (*n*) : antiarthritic water 358
GICHTWATTE (*f*) : antiarthritic (wadding) wool (as " Thermogene " wool) 359
GICHTWIDRIG : antiarthritic 360
GIEBEL (*m*) : gable 361
GIEBELDACH (*n*) : gabled roof 362
GIEBELFELD (*n*) : tympanum (the recessed space of the pediment, or sometimes applied to the pediment itself) (Architecture) 363
GIEBELFENSTER (*n*) : dormer window 364
GIEBELIG : gabled 365
GIEBELN : to gable 366
GIEKBAUM (*m*) : main boom (Naut.) 367
GIEKSEGEL (*n*) : mainsail (Naut.) 368
GIESELGALMEI (*m*) : Siliceous calamine, smithsonite (Zn_2SiO_4 plus H_2O) (see ZINKSPAT) 369
GIESSBACH (*m*) : torrent 370
GIESSBECKEN (*n*) : basin ; ewer 371
GIESSEL (*m*) : spout (for molten metal) 372
GIESSEN : to pour ; cast ; found ; mould ; water ; coat (Phot.) 373
GIESSEN, HOHL- : to cast hollow 374
GIESSEN, KALT- : to cast cold 375
GIESSEN, MASSIV- : to cast solid 376
GIESSER (*m*) : founder ; caster ; moulder ; melter ; smelter ; pourer ; pouring vessel ; ewer ; rose ; watering can 377
GIESSERDE (*f*) : earth for moulds 378
GIESSEREI (*f*) : foundry ; casting ; founding ; watering 379
GIESSEREIEISEN (*n*) : foundry pig-(iron) 380
GIESSEREIKOKS (*m.* and *pl*) : foundry coke 381
GIESSEREIKRAN (*m*) : foundry crane 382
GIESSEREIROHEISEN (*n*) : foundry pig-(iron) 383
GIESSEREISCHWÄRZE (*f*) : moulder's black-(ing) 384
GIESSERZ (*n*) : bronze 385
GIESSFÄHIG : capable of being poured or cast 386
GIESSFORM (*f*) : (casting)-mould 387
GIESSGESCHWINDIGKEIT (*f*) : casting speed 388
GIESSHAUS (*n*) : foundry 389
GIESSHÜTTE (*f*) : foundry 390
GIESSKANNE (*f*) : watering-can 391
GIESSKANNENAUFSATZ (*m*) : rose (of a watering-can) 392
GIESSKASTEN (*m*) : casting (box) mould 393
GIESSKUNST (*f*) : founding 394
GIESSLÖFFEL (*m*) : casting ladle 395
GIESSMODELL (*n*) : casting (model) pattern 396
GIESSMUTTER (*f*) : matrix ; mould 397
GIESSOFEN (*m*) : founding (smelting) furnace 398
GIESSPFANNE (*f*) : foundry ladle, casting ladle 399
GIESSPFANNE (*f*), STAHL- : steel ladle 400
GIESSRINNE (*f*) : sink, gutter 401

GIESSSAND (*m*): moulding sand; sand for moulds 402
GIESSTEMPERATUR (*f*): casting temperature 403
GIESSTIEGEL (*m*): melting-pot 404
GIESSTOPF (*m*): pot or jar with a lip for pouring 405
GIESSWAGEN (*m*): casting trolley (Steel-works) 406
GIFT (*n*): poison; venom; virus; malice; toxin; Realgar (Tanning) (see REALGAR) 407
GIFTABTREIBEND: antitoxic; antidotal 408
GIFTARZNEI (*f*): antidote; antitoxin; poison-(ous) drug 409
GIFTFEST: immune -(to poison) 410
GIFTFREI: free from poison; harmless; non-poisonous; non-injurious 411
GIFTGETREIDE (*n*): poisoned grain (see GIFT-WEIZEN) 412
GIFTHAFER (*m*): poisoned oats (see GIFTWEIZEN) 413
GIFTHALTIG: venomous, poisonous, containing poison 414
GIFTHÜTTE (*f*): arsenic works 415
GIFTIG: poisonous; toxic; virulent; malicious; venomous 416
GIFTJASMIN (*m*): Carolina jasmine (*Gelsemium sempervirens*) 417
GIFTKIES (*m*): arsenopyrite; mispickel; FeAsS; Sg. 6.1 (about 58% As_2O_3) 418
GIFTKRAUT (*n*): monk's hood, wolf's bane (see EISENHUT) 419
GIFTKUNDE (*f*): toxicology (the science of poisons) 420
GIFTLATTICH (*m*): acrid lettuce (*herba lactucæ virosæ*) (*Lactuca virosa*) 421
GIFTLATTICHSAFT (*m*): lactucarium (from *Lactuca virosa* and *Lactuca sativa*) (Pharmacy) 422
GIFTLEHRE (*f*): toxicology (the science of poisons) 423
GIFTLOS: non-poisonous; non-injurious; harmless; free from poison 424
GIFTMEHL (*n*): white arsenic powder; arsenic trioxide; flowers of arsenic; As_2O_3; (see ARSENTRIOXYD and ARSENIGESÄURE) 425
GIFTMITTEL (*n*): antidote; antitoxin 426
GIFTPFLANZE (*f*): (any)-poisonous plant 427
GIFTPULVER (*n*): poisonous powder; crude powdered arsenic trioxide (see GIFTMEHL) 428
GIFTREICH: highly poisonous; toxic 429
GIFTROSE (*f*): oleander (a highly poisonous shrub) 430
GIFTSCHWAMM (*m*): poisonous fungus of the mushroom variety 431
GIFTSTEIN (*m*): white arsenic; arsenic trioxide (see ARSENTRIOXYD and ARSENIGE SÄURE) 432
GIFTSTOFF (*m*): poison-(ous matter) 433
GIFTVOLL: toxic, highly poisonous 434
GIFTWEIZEN (*m*): (*Fructus frumentis venenatus*): grain which has been poisoned with arsenic or strychnine; used for exterminating field pests 435
GIFTWENDE (*f*): swallow-wort (*Cynanchum vincetoxicum*); white swallow-wort (*Vincetoxicum officinale*) 436
GIFTWIDRIG: antitoxic; antidotal 437
GIFTWIRKUNG (*f*): poisonous (action) effect 438
GIFTWURZEL (*f*): contrayerva; swallow-wort (*Vincetoxicum officinale*) 439
GIG (*f*): gig (Naut.) 440
GIGANTISCH: gigantic, enormous 441
GILBE (*f*): yellow-(ish) colour; yellow (ochre) substance; dyer's weed (*Reseda luteola*) 442
GILBEN: to turn yellow; colour yellow 443
GILBWURZ (*f*): turmeric; curry; curcuma (from *Curcuma longa*) 444
GILBWURZEL (*f*): (see GILBWURZ) 445
GILDE (*f*): guild; company; corporation 446
GILEADISCHER BALSAM (*m*): Balm of Gilead (see MEKKABALSAM) 447
GILLUNG (*f*): counter 448
GILLUNGSHOLZ (*n*): counter wood 449
GILLUNGSPLATTE (*f*): counter plate 450
GILSONIT (*m*): gilsonite (a natural asphalt); Sg. 1.07 451
GILTIG: valid; binding; current; good; auriferous (Mining) (see also GÜLTIG) 452
GINGELLIKRAUT (*n*): (see NIGERPFLANZE) 453
GINGERGRAS (*n*): ginger-grass (*Andropogon*) (see INGWERGRAS) 454
GINGERGRASÖL (*n*): ginger-grass oil (distilled from *Andropogon*) (see GERANIUMÖL) 455
GINSENGWURZEL (*f*): ginseng root (*Panax*); (*Radix ginseng* or *Radix panacis quinquefolii* of *Panax quinquefolium*) 456
GINSTER (*m*): broom; green-broom; hogweed; bannal (*Cytisus scoparius*) 457
GINSTERBLÜTEN (*f.pl*): broom flowers (*Flores genistæ*) 458
GINSTERPFLANZE (*f*): broom (*Planta genista*) 459
GIPFEL (*m*): top; summit; climax; pinnacle 460
GIPFELN: to rise to a point, reach a climax 461
GIPS (*m*): gypsum; stucco; plaster of Paris; gypsite (hydrous calcium sulphate); $CaSO_4.2H_2O$ (about 33% lime); Sg. 2.17 (see CALCIUMSULFAT) 462
GIPSABDRUCK (*m*): plaster cast 463
GIPSABGUSS (*m*): plaster cast 464
GIPSANWURF (*m*): plastering 465
GIPSARBEIT (*f*): stucco work 466
GIPSARBEITER (*m*): plasterer 467
GIPSARTIG: gypseous 468
GIPSBINDE (*f*): plaster of Paris bandage 469
GIPSBREI (*m*): plaster of Paris paste 470
GIPSBRENNEREI (*f*): gypsum calcination; plaster (gypsum) kiln 471
GIPSDRUSE (*f*): crystallized gypsum, gypsite (see GIPS and CALCIUMSULFAT) 472
GIPSEN: to plaster 473
GIPSER (*m*): plasterer 474
GIPSERDE (*f*): earthy gypsum 475
GIPSGUHR (*m*): earthy gypsum 476
GIPSGUSS (*m*): plaster cast-(ing) 477
GIPSHALTIG: calcareous, containing gypsum 478
GIPSHÄRTE (*f*): (see NICHTCARBONATHÄRTE) 479
GIPSKALK (*m*): plaster lime 480
GIPSMALEREI (*f*): fresco painting 481
GIPSMEHL (*n*): powdered (gypsum) plaster 482
GIPSMERGEL (*m*): gypseous marl 483
GIPSMÖRTEL (*m*): stucco; plaster 484
GIPSOFEN (*m*): plaster (gypsum) kiln 485
GIPSSPAT (*m*): selenite; gypsum spar; gypsite (see GIPS) 486
GIPSSTEIN (*m*): gypseous stone; plaster stone; hard calcium sulphate deposit 487
GIPSSTEINARTIG: gypseous 488
GIPSTEER (*m*): mixture of plaster of Paris and tar 489
GIRIEREN: to endorse (Bills), circulate 490
GIRO (*n*): endorsement 491
GISCHEN: to foam; froth; ferment; bubble; effervesce 492

GISMONDIN (*m*) : gismondite ; $CaAl_2Si_2O_8.4H_2O$; Sg. 2.3 493
GITTER (*n*) : grating ; lattice ; latticed screen ; screen ; railing ; fence ; fencing ; paling ; grid (Elect.) 494
GITTERARTIG : latticed ; lattice-like ; grated 495
GITTERAUFBAU (*m*) : lattice construction (Crystallography) 496
GITTERBATTERIE (*f*) : grid battery (Wireless Tel.) 497
GITTERFARBE (*f*) : grating colour 498
GITTERFENSTER (*n*) : lattice-window 499
GITTERFÖRMIG : (see **GITTERARTIG**) 500
GITTERHORDE (*f*) : grating, hurdle, latticed screen 501
GITTERKONDENSATOR (*m*) : grid condenser (Wireless Tel.) 502
GITTERKRAFT (*f*) : lattice force (Crystallography) 503
GITTERLAUBE (*f*) : trellis (arbour) work 504
GITTERMAST (*m*) : lattice-work mast 505
GITTERN : to lattice, chequer 506
GITTERSCHORNSTEIN (*m*) : dissipator (chequered openings at the side of the top of a chimney, for distributing the flue gases in the atmosphere) 507
GITTERSPANNUNG (*f*) : grid tension (Wireless Tel.) 508
GITTERVERLAGERUNG (*f*) : lattice rearrangement (Crystallography) 509
GITTERWERK (*n*) : chequer work, chequered portion, trellis work ; grating, railing 510
GLACÉ-KARTON (*m*) : glazed board-(s) (Paper) 511
GLACÉLEDER (*n*)- **FABRIK** (*f*) : glacé (patent) leather factory 512
GLACIEREN : to gloss ; glaze ; freeze 513
GLANDERN : to calender 514
GLANZ (*m*) : lustre ; glitter ; gleam ; polish ; gloss ; glazing ; glaze ; glance (Mineral) ; splendour ; brightness ; brilliance ; water (Precious stones) 515
GLANZBLENDE (*f*) : alabandite ; alabandine (see **MANGANBLENDE**) 516
GLANZBRAUNSTEIN (*m*) : hausmannite (see **MANGANOXYDOXYDUL**) 517
GLANZBRENNE (*f*) : polishing pickle (a mixture of nitric acid of Sg. 1.84, nitric acid of Sg. 1.38 and common salt in proportion of 100 : 75 : 1 parts by weight) (Electroplating) 518
GLANZDRUCKFARBE (*f*) : glossy and brilliant printing colours 519
GLÄNZE (*f*) : glaze ; size ; polishing material ; glazing ; gloss ; polish-(er) 520
GLANZEFFEKT (*m*) : lustre ; gloss ; brilliance 521
GLANZEISENERZ (*n*) : specular iron ore ; hematite ; Fe_2O_3 ; 70% Fe (see **HÄMATIT**) 522
GLANZEISENSTEIN (*m*) : (see **GLANZEISENERZ**) 523
GLÄNZEN : to shine ; glisten ; gleam ; glitter ; gloss ; brighten ; glaze ; polish ; planish ; lustre ; sparkle 524
GLÄNZEND : bright, glossy, brilliant, sparkling (see **GLÄNZEN**) 525
GLÄNZER (*m*) : burnisher ; finisher 526
GLANZERZ (*n*) : plumbago, graphite (see **GRAPHIT**) 527
GLANZFEIN : brilliant 528
GLANZFIRNIS (*m*) : glazing varnish 529
GLANZGOLD (*n*) : brilliant gold (for porcelain painting) ; burnished gold ; gold foil 530
GLANZGOLDFEUER (*n*) : brilliant-gold fire (Ceram.) 531
GLÄNZHAMMER (*m*) : planishing hammer 532
GLANZKARTON (*m*) : glazed cardboard (Paper) 533
GLANZKOBALT (*m*) : cobalt glance ; cobaltite, CoAsS, 36% cobalt, Sg. 6.3 ; smaltite, $CoAs_2$, 28% cobalt (see **SMALTIT**) 534
GLANZKOHLE (*f*) : anthracite ; carbon with metallic lustre ; 85-95% carbon (see **ANTHRAZIT**) 535
GLANZLACK (*m*) : brilliant (glossy) varnish 536
GLANZLEDER (*n*) : patent leather 537
GLANZLEDERLACK (*m*) : patent leather (varnish) enamel 538
GLANZLEINWAND (*f*) : glazed linen 539
GLANZLOS : lustreless ; dull ; dim 540
GLANZMANGANERZ (*n*) : manganite ; $Mn_2O_3.H_2O$ (see also **MANGANIT**) 541
GLANZMESSING (*n*) : polished brass 542
GLANZMETALL (*n*) : speculum metal (2 parts Cu. 1 part Zn) 543
GLANZÖL (*n*) : flooring oil ; brilliant (glossy) oil 544
GLANZPAPIER (*n*) : glazed paper 545
GLANZPLATIN (*n*) : brilliant (burnished) platinum (for painting porcelain) ; platinum glance 546
GLANZRUSS (*m*) : a lustrous form of soot ; lampblack 547
GLANZSCHLEIFEN : to burnish ; polish 548
GLANZSEITE (*f*) : bright side 549
GLANZSILBER (*n*) : silver glance ; brilliant (burnished) silver (for painting porcelain) ; argentite (see **GLANZERZ**) 550
GLANZSPAT (*m*) : (see **FELDSPAT**) 551
GLANZSTÄRKE (*f*) : gloss starch 552
GLANZSTOFF (*m*) : artificial silk (made by cuprammonium process) ; art silk (Trade name) 553
GLANZWEISS (*n*) : talc ; $H_2O.3MgO.4SiO_2$ (mixture of basic aluminium sulphate and potassium sulphate) (Paper) 554
GLAS (*n*) : glass ; Sg. 2.5-3.45 ; tumbler 555
GLASABFALL (*m*) : cullet ; broken glass 556
GLASACHAT (*m*) : obsidian-(ite) ; volcanic glass (a hydrous silica) 557
GLASÄHNLICH : glass-like ; glassy ; vitreous 558
GLASAPPARAT (*m*) : glass apparatus 559
GLASARTIG : (see **GLASÄHNLICH**) 560
GLASASCHE (*f*) : alkali 561
GLASBEARBEITUNG (*f*) : glass working 562
GLASBILDUNG (*f*) : vitrification ; formation of glass 563
GLASBIRNE (*f*) : glass (bulb) globe ; electric light bulb 564
GLASBLASE (*f*) : bubble in glass 565
GLASBLASEN (*n*) : glass-blowing 566
GLASBLASER (*m*) : glass-blower 567
GLASBLASEREI (*f*) : glass blowing 568
GLASBLASERÖHRE (*f*) : blow-pipe for glass 569
GLASBRENNEN (*n*) : glass annealing 570
GLASBROCKEN (*m.pl*) : cullet, broken glass 571
GLASBÜCHSE (*f*) : glass box 572
GLASBÜRSTE (*f*) : glass brush 573
GLASCYLINDER (*m*) : lamp glass or chimney ; glass cylinder 574
GLASDOSE (*f*) : glass box 575
GLASELEKTRICITÄT (*f*) : vitreous (positive) electricity 576
GLASEMBALLAGE (*f*) : glass packing 577
GLASEN : to vitrify 578
GLASER (*m*) : glazier 579
GLASERDIAMANT (*m*) : glazier's diamond 580
GLASERKITT (*m*) : (glazier's)-putty 581

GLÄSERN : glassy, vitreous, of glass 582
GLASERSATZ (*m*) : glass substitute 583
GLASERZ (*n*) : argentine ; argentite ; silver glance ; Ag_2S ; 87% silver (see SILBERGLANZ) 584
GLASFABRIK (*f*) : glass (works) factory 585
GLASFADEN (*m*) : glass thread (defect in glass) 586
GLASFARBE (*f*) : colours for painting on glass or staining glass 587
GLASFEUCHTIGKEIT (*f*) : vitreous humour 588
GLASFLASCHE (*f*) : glass (bottle) flask ; decanter 589
GLASFLUSS (*m*) : glass flux ; paste for imitation gems (see MAINZERFLUSS) 590
GLASFLÜSSIGKEIT (*f*) : vitreous humour 591
GLASFRITTE (*f*) : frit ; batch ; glass composition 592
GLASGALLE (*f*) : glass gall ; sandiver 593
GLASGERÄTE (*n.pl*) : glass (utensils ; apparatus) vessels 594
GLASGERÄTSCHAFTEN (*f.pl*) : (see GLASGERÄTE) 595
GLASGESPINST (*n*) : spun glass ; glass cloth 596
GLASGLANZ (*m*) : vitreous lustre ; frost ; pounded glass 597
GLASGLÄNZEND : glassy-lustrous ; glittering like glass 598
GLASGLOCKE (*f*) : bell jar ; glass bell 599
GLASGRÜN : bottle-green 600
GLASHAFEN (*m*) : glass pot ; crucible 601
GLASHAFENTON (*m*) : glass-(pot) crucible clay 602
GLASHAHN (*m*) : glass cock 603
GLASHART : brittle ; hard as glass ; casehardened ; chilled 604
GLASHÄRTE (*f*) : glass hardness ; chilling (Steel) 605
GLASHÄRTEN (*n*) : tempering of glass 606
GLASHAUS (*n*) : glass house ; conservatory ; hot-house ; greenhouse ; studio (Photo.) 607
GLASHELL : clear as glass ; diaphanous 608
GLASHERSTELLUNG (*f*) : glass manufacture 609
GLASHÜTTE (*f*) : glass (factory) works ; glass house 610
GLASICHT : glassy ; vitreous 611
GLASIEREN : to glaze ; varnish ; ice 612
GLASIG : glassy ; vitreous 613
GLASINSTRUMENT (*n*) : glass instrument 614
GLASKALK (*m*) : glass gall ; sandiver 615
GLASKITT (*m*) : putty ; glass cement 616
GLASKOLBEN (*m*) : glass flask 617
GLASKOPF (*m*), BRAUNER- : limonite ; $2Fe_2O_3.3H_2O$ (see BRAUNEISENERZ) 618
GLASKOPF (*m*), EIGENTLICHER- : hematite (variety of specular iron ore) ; Fe_2O_3 ; 10% Fe (see HÄMATIT) 619
GLASKOPF (*m*), GELBER- : (see GLASKOPF, BRAUNER-) 620
GLASKOPF (*m*), ROTER- : red iron ore, Hematite (70% Fe) (see ROTEISENSTEIN and EISENOXYD) (variety of specular iron ore) 621
GLASKOPF, SCHWARZER- (*m*) : psilomelane ; a kind of brick (a Manganese ore) (see PSILO-MELAN) 622
GLASKORB (*m*) : crate 623
GLASKÖRPER (*m*) : vitreous humour 624
GLASKRAUT (*n*) : wall pelitory (*Parietaria officinalis*) ; (jointed)-glasswort (*Salicornia herbacea*) 625
GLASKUGEL (*f*) : glass ball, glass globe 626
GLASLACK (*m*) : glass varnish 627
GLASLAVA (*f*) : volcanic glass ; obsidian ; obsidianite ; hyalite 628
GLASLEINEN (*n*) : spun glass, glass cloth 629
GLASLINSE (*f*) : glass lens 630
GLASMACHERSEIFE (*f*) : glass maker's soap ; manganese dioxide (see MANGANPEROXYD) 631
GLASMALEREI (*f*) : glass staining or painting 632
GLASMALZ (*n*) : brittle malt 633
GLASMEHL (*n*) : glass meal ; crushed (powdered) glass ; glass powder 634
GLASMESSER (*m*) : vitrometer 635
GLASOFEN (*m*) : glass furnace 636
GLASOPAL (*m*) : hyalite ; volcanic glass (hydrous silica) ; opal (see OPAL) 637
GLASPAPIER (*n*) : glass paper 638
GLASPASTE (*f*) : glass paste (for imitation gems) (see MAINZER FLUSS) 639
GLASPECH (*n*) : hard (stone) pitch 640
GLASPFEIFE (*f*) : blow-pipe 641
GLASPLATTE (*f*) : glass plate 642
GLASPORZELLAN (*n*) : vitreous (transparent) porcelain 643
GLASQUARZ (*m*) : hyalite ; transparent quartz (a hydrous silica) 644
GLASRÄUMER (*m*) : bottle brush 645
GLASRETORTE (*f*) : glass retort 646
GLASROHR (*n*) : glass (tube) tubing 647
GLASRÖHRE (*f*) : (see GLASROHR) 648
GLASSALZ (*n*) : glass gall ; sandiver 649
GLASSATZ (*m*) : glass composition ; batch ; frit 650
GLASSCHAUM (*m*) : (see GLASSALZ) 651
GLASSCHEIBE (*f*) : glass pane ; pane of glass ; glass plate ; plate of glass 652
GLASSCHERBEN (*f.pl*) : cullet ; broken glass 653
GLASSCHLACKE (*f*) : (see GLASSALZ) 654
GLASSCHLEIFEN (*n*) : glass (polishing) grinding 655
GLASSCHLEIFER (*m*) : glass (polisher) grinder 656
GLASSCHMELZ (*m*) : enamel 657
GLASSCHNEIDER (*m*) : glass cutter 658
GLASSCHÖRL (*m*) : axinite (see AXINIT) 659
GLASSCHRAUBE (*f*) : glass screw (such as a screwed bottle-neck) 660
GLASSEIFE (*f*) : glass maker's soap (Manganese dioxide) (see MANGANPEROXYD) 661
GLASSPLITTER (*m*) : shiver of glass, piece of broken glass 662
GLASSPRITZE (*f*) : glass syringe 663
GLASSTAB (*m*) : glass rod ; glass stirrer 664
GLASSTEIN (*m*) : axinite (see AXINIT) ; paste (for imitation jewellery) ; glass brick (used in place of window) 665
GLASSTOPFEN (*n*) : glass stopper 666
GLASTAFEL (*f*) : plate (pane) of glass 667
GLASTIEGEL (*m*) : glass (crucible) pot ; melting pot 668
GLASTOPF (*m*) : glass crucible (pot), melting pot 669
GLASTRÄNE (*f*) : glass tear ; Prince Rupert's drop 670
GLASTRICHTER (*m*) : glass funnel 671
GLASTROPFEN (*m*) : (see GLASTRÄNE) 672
GLASÜBERZUG (*m*) : flashing (a thin layer of coloured glass on ordinary glass) 673
GLASUR (*f*) : glaze ; glazing ; varnish ; frosting ; icing ; enamel ; gloss ; silicate mixture (of fusible cones) (see SEGERKEGEL) 674
GLASURASCHE (*f*) : glaze ash 675
GLASURBLAU (*n*) : zaffre (a cobalt oxide) 676
GLASURENFABRIK (*f*) : glaze factory 677

GLASURERZ (*n*): alquifou; potter's ore; galena (natural lead sulphide); PbS (see **BLEIGLANZ**) 678
GLASURFARBE (*f*): glaze colour (Ceram.) 679
GLASUR (*f*), **FARBLOSE-**: colourless glaze (Ceram.) 680
GLASURMASSE (*f*): glazing mass (a creamy mass for glazing porcelain; see **PORZELLANGLASUR**) 681
GLASURMÜHLE (*f*): glazing (glaze) mill 682
GLASURSAND (*m*): glaze sand 683
GLASURSCHICHT (*f*): (layer of)-glaze 684
GLASWANNE (*f*): glass pot, crucible 685
GLASWAREN (*f.pl*): glass ware 686
GLASWOLLE (*f*): spun glass; glass wool 687
GLASZYLINDER (*m*): lamp glass or chimney; glass cylinder 688
GLATT: smooth; plain; flat; glossy; polished; glazed; even; slippery; sleek 689
GLÄTTAHLE (*f*): broach 690
GLATTBRAND (*m*): glaze burn (Ceramics) 691
GLATTBRENNEN: to subject to the glost burn (Ceramics); (*n*): firing on the glaze 692
GLATTBRENNOFEN (*m*): glaze (glost) kiln 693
GLATTDECKSSCHIFF (*n*): flush-decked ship 694
GLÄTTE (*f*): smoothness; polish; litharge, PbO, Sg. 9.375, Mp. 888°C.; massicot (Mineral), PbO, about 93% lead 695
GLÄTTEISEN (*n*): smoothing iron 696
GLÄTTEN: to smoothe; polish; burnish; plane; planish; glaze; satin (Paper); calender (Cloth) 697
GLÄTTER (*m*): polisher, smoother, burnisher 698
GLATTE WELLE (*f*): free (unloaded) shaft 699
GLATTFÄRBEREI (*f*): plain dyeing 700
GLATTFEILE (*f*): smoothing file 701
GLÄTTFRISCHEN (*n*): reduction of litharge 702
GLÄTTMASCHINE (*f*): glazing machine 703
GLATTOFEN (*m*): glost (finishing) kiln (Ceramics) 704
GLÄTTPRESSE (*f*): smoothing press; calender 705
GLATTROCHE (*m*): ray, roach (*Raja batis*) 706
GLATTSCHERBE (*f*): potsherd 707
GLATTSCHLEIFER (*m*): polisher; burnisher 708
GLÄTTWERKZEUG (*n*): polishing (burnishing) tool 709
GLÄTTZAHN (*m*): polishing (smoothing) tool 710
GLATTZÜNGIG: smooth-tongued 711
GLAUBE (*f*): faith; belief; credit; creed 712
GLAUBEN: to believe; think; trust; suppose; imagine; fancy 713
GLAUBENSABFALL (*m*): apostasy 714
GLAUBENSLEHRE (*f*): dogma, doctrine 715
GLAUBENSWISSENSCHAFT (*f*): theology 716
GLAUBENSWÜRDIG: credible 717
GLAUBENSZWEIFEL (*m*): scepticism, doubt 718
GLAUBERIT (*m*): glauberite (natural sodium-calcium sulphate); $Na_2SO_4.CaSO_4$; Sg. 2.75 719
GLAUBERSALZ (*n*): Glauber's salt; sodium sulphate (see **NATRIUMSULFAT**): $Na_2SO_4.10H_2O$; Sg. 1.45 720
GLAUBERSALZ (*n*), **CALCINIERTES-**: calcinated Glauber's salt; salt cake (92-99% sodium sulphate) 721
GLAUBERSALZ (*n*), **KRISTALLISIERTES-**: crystallized Glauber's salt; Glauber's salt crystals 722
GLAUBHAFT: authentic, credible 723
GLÄUBIGE (*m* and *f*): believer 724
GLÄUBIGER (*m*): creditor (Com.) 725
GLAUBLICH: likely; probable; credible 726
GLAUCIUMÖL (*n*): glaucium oil, horn poppy oil (from seeds of *Chelidonium glaucium* or *Glaucium glaucium*) 727
GLAUKODOT (*m*): glaucodot; (Fe,CO) $(As,S)_2$; Sg. 6.0 728
GLAUKONIT (*m*): greensand, glauconite (hydrous iron-potassium silicate); $K_2O,FeO,Fe_2O_3,SiO_2,Al_2O_3,H_2O$; Sg. 2.3-3.0 729
GLAUKOPHAN (*m*): Glaucophane (rare monoclinic form of amphibole); $Na_2(Al,Fe)_2(SiO_3)_4.Ca(Mg,Fe)_3(SiO_3)_4$; Sg. 3.1 730
GLAZIEREN: to glaze (see also under **GLAS-**) 731
GLEICH: equal; like; alike; similar; equally; directly; constant; uniform; level; even; immediately 732
GLEICHACHSIG: co-axial-(ly), concentric-(ally) 733
GLEICHARTIG: homogeneous; of the same kind; analogous 734
GLEICHBAR: comparable 735
GLEICHBAUEN: to construct on similar lines (Maths. etc.) 736
GLEICHBEDEUTEND: synonymous; equivalent 737
GLEICHBELASTUNG (*f*): uniformly distributed load 738
GLEICHBLEIBEND: constant; uniform; invariable 739
GLEICHDEUTIG: (see **GLEICHBEDEUTEND**) 740
GLEICHDRUCK (*m*), **IN-ARBEITEN**: to work on the impulse principle (Turbines) 741
GLEICHDRUCKRAD (*n*): Impulse wheel (turbines) 742
GLEICHDRUCKTURBINE (*f*): impulse turbine 743
GLEICHE (*f*): level course of brickwork, uniformity, equality, evenness (see **GLEICH**) 744
GLEICHEMPFINDUNG (*f*): sympathy 745
GLEICHEN: to be (like) equal; make alike; equalize; smoothe; level; liken; adjust; size; resemble 746
GLEICHEN DRUCKES, LINIE-: isobar, constant pressure line 747
GLEICHENTFERNT: equidistant 748
GLEICHER (*m*): equalizer; equator 749
GLEICHER DRUCK (*m*): constant (uniform) pressure 750
GLEICHERWEISE: similarly; likewise 751
GLEICHEWINKELEISEN (*n.pl*): equal angles 752
GLEICHFALLS: likewise; also 753
GLEICHFARBIG: isochromatic; of the same colour 754
GLEICHFÖRMIG: uniform; even; monotonous; homogeneous 755
GLEICHGELTEND: equivalent 756
GLEICHGERICHTET: acting in the same direction; similarly directed; direct induced 757
GLEICHGESINNT: of the same mind 758
GLEICHGESTALTET: isomorphous; of the same shape 759
GLEICHGESTIMMT: congenial; tuned alike 760
GLEICHGEWICHT (*n*): equilibrium, balance, equipoise 761
GLEICHGEWICHT (*n*), **INDIFFERENTES-**: indifferent equilibrium 762
GLEICHGEWICHT (*n*), **LABILES-**: unstable equilibrium 763
GLEICHGEWICHTSBEDINGUNG (*f*): condition of equilibrium 764
GLEICHGEWICHTSLAGE (*f*): condition of equilibrium 765
GLEICHGEWICHTSLEHRE (*f*): statics 766

GLEICHGEWICHTSPUNKT (*m*): centre of gravity 767

GLEICHGEWICHTSSTÖRUNG (*f*): unbalancing; displacement (disturbance) of equilibrium 768

GLEICHGEWICHT (*n*), **STABILES-**: stable equilibrium 769

GLEICHGEWICHTSZUSTAND (*m*): state (condition) of equilibrium 770

GLEICHGROSS: equal-(size) 771

GLEICHGÜLTIG: indifferent; immaterial; equivalent 772

GLEICHGÜLTIGKEIT (*f*): (matter of)-indifference; equivalence 773

GLEICHHEIT (*f*): equality; sameness; likeness; similarity; uniformity; conformity; identity 774

GLEICHIONIG: having a common ion 775

GLEICHKLANG (*m*): consonance; unison 776

GLEICHLASTIG: on an even keel (of ships) 777

GLEICHLAUF (*m*): clockwise (parallel) motion, forward (running) rotation 778

GLEICHLAUFEND: parallel; rotating clockwise 779

GLEICHLAUTEND: consonant 780

GLEICHMÄSSIG: proportionate; symmetrical; uniform; similar; homogeneous; equal; even-(ly) 781

GLEICHMÄSSIGE BELASTUNG (*f*): equally distributed (uniform) load 782

GLEICHMUT (*m*): equanimity 783

GLEICHNAMIG: like; of the same kind or name; homologous; homonymous 784

GLEICHNIS (*n*): simile; parable; similitude; image; comparison 785

GLEICHRICHTEN: to rectify (Electricity) 786

GLEICHRICHTER (*m*): rectifier (Elect.) 787

GLEICHRICHTER (*m*), **STROM-**: current rectifier 788

GLEICHSAM: so to speak; almost; as (it were) if 789

GLEICHSCHENKELIG: isosceles; having equal (flanges; legs) sides 790

GLEICHSEITIG: equilateral; double-faced (Fabrics); reciprocal 791

GLEICHSETZEN: to balance, equalize, put in the form of an equation 792

GLEICHSETZUNG (*f*): equivalent (term); expression (Maths.) 793

GLEICHSINNIG: synonymous 794

GLEICHSTELLEN: to equalize, compare 795

GLEICHSTROM (*m*): direct current; continuous current; parallel (current) flow; uniflow; unaflow 796

GLEICHSTROMANLAGE (*f*): continuous current or direct current plant 797

GLEICHSTROMBOGENLAMPE (*f*): continuous current arc lamp; direct current arc lamp 798

GLEICHSTROMDAMPFMASCHINE (*f*): continuous current steam engine; uniflow (unaflow) steam engine 799

GLEICHSTROMDYNAMO (*f*): continuous (direct) current dynamo 800

GLEICHSTROMLEITUNG (*f*): continuous (direct) current line or main 801

GLEICHSTROMMASCHINE (*f*): direct current dynamo (Elect.) 802

GLEICHSTROMUMFORMER (*m*): continuous (direct) current converter or transformer 803

GLEICHTEILIG: homogeneous; of equal parts 804

GLEICHUNG (*f*): equation; equalization; adjustment; sizing 805

GLEICHUNGSLEHRE (*f*): algebra 806

GLEICHVERHALTEND: of similar (behaviour) properties 807

GLEICHVIEL: just as much; equally much; all the same; equally 808

GLEICHWEIT: equidistant; of uniform width; parallel 809

GLEICHWERTIG: equivalent; of equal value 810

GLEICHWERTIGKEIT (*f*): equivalence 811

GLEICHWINKLIG: having equal angles; equiangular 812

GLEICHWOHL: yet; however; notwithstanding; nevertheless 813

GLEICHZEITIG: simultaneous-(ly); contemporary; at the same time; synchronous; isochronous; contemporaneous-(ly) 814

GLEICHZEITIGKEIT (*f*): synchronism 815

GLEIS (*m* and *n*): track; metals; road; permanent way; line (Railways) 816

GLEISE (*f*): (see **GLEIS**) 817

GLEISSEILBAHN (*f*): cable railway 818

GLEISSEN: to shine, glisten, glitter; to be hypocritical, dissemble 819

GLEITBACKEN (*m*): sliding (guide) block 820

GLEITBAHN (*f*): chute; slide; slips (Shipbuilding); slide-bar; guide 821

GLEITEN: to glide; slide; slip; chute; (*n*): sliding, etc. 822

GLEITEND: sliding 823

GLEITFLÄCHE (*f*): gliding plane; chute; slipping (sliding) surface 824

GLEITKLOTZ (*m*): slide block 825

GLEITKONTAKT (*m*): sliding contact 826

GLEITLAGER (*n*): plain bearing 827

GLEITMODUL (*m*): transverse modulus of elasticity 828

GLEITSCHIENE (*f*): slide bar, guide 829

GLEITSITZ (*m*): sliding fit 830

GLEITSPANNUNG (*f*): sliding (displacement) pressure 831

GLEITSTANGE (*f*): slide bar, guide 832

GLEITVENTIL (*n*): slide valve 833

GLETSCHER (*m*): glacier 834

GLETSCHERSPALTE (*f*): crevasse 835

GLIDIN (*n*): glidin (a preparation of wheat gluten, used as a food) 836

GLIED (*n*): member; term (Maths.); limb; joint; link (Chain); degree; section; part; portion; factor; file; rank 837

GLIEDERBAND (*m*): ligament 838

GLIEDERBAU (*m*): structure, articulation 839

GLIEDERFLUSS (*m*): rheumatism of the joints 840

GLIEDERFUGE (*f*): joint, articulation 841

GLIEDERGICHT (*f*): gout in the joints, arthritis 842

GLIEDERKESSEL (*m*): sectional boiler 843

GLIEDERKRANKHEIT (*f*): disease of the joints 844

GLIEDERN: to join; organize; articulate 845

GLIEDERSCHWAMM (*m*): white swelling 846

GLIEDERUNG (*f*): organization; jointing; articulation; formation; configuration; splitting up into (sections) members 847

GLIEDLOSIGKEIT (*f*): Anarthria, affection of the speech (Med.) 848

GLIEDMASS (*n*): limb 849

GLIEDWASSER (*n*): synovial fluid 850

GLIEDWEISE: by sections; in ranks; member by member 851

GLIMMEN: to glimmer; glow 852

GLIMMER (*m*): glimmer, glow; Mica (for chemical composition, see MUSKOVIT, LEPIDOLITH, BIOTIT, and PARAGONIT) 853
GLIMMERARTIG: micaceous; glowing 854
GLIMMERHALTIG: micaceous; containing mica 855
GLIMMERIG: micaceous; glowing; glimmering 856
GLIMMERPLÄTTCHEN (*n*): (thin or small) mica plate 857
GLIMMERPLATTE (*f*): mica plate 858
GLIMMERSAND (*m*): micaceous sand 859
GLIMMERSCHIEFER (*m*): micaceous (shale) schist; mica slate 860
GLIMMERTON (*m*): micaceous clay 861
GLIMMLAMPE (*f*): glow-lamp 862
GLIMPF (*m*): mildness, forbearance, indulgence 863
GLIMPFLICH: moderate; gentle; light-(ly); easy; easily; forbearing 864
GLIMPFLICH DAVONKOMMEN: to escape practically unscathed 865
GLITSCHEN: to slide 866
GLITSCHPULVER (*n*): talcum powder (see MAGNES--IUMSILIKAT) 867
GLITZERN: to glisten; glitter; gleam 868
GLOBULIN (*n*): globulin (a class of proteid) 869
GLOBULOID (*n*): globule 870
GLOBUS (*m*): globe 871
GLÖCKCHEN (*n*): bell-jar; hand (small) bell 872
GLOCKE (*f*): bell; bell-jar; receiver (of air pump); clock; dome; reservoir; container; glass-shade 873
GLOCKE (*f*), FORTSCHELL-: constant (continuously) active or ringing bell 874
GLOCKE (*f*), ISOLIER-: bell insulator 875
GLOCKENBLUME (*f*): bell-shaped flower, bluebell 876
GLOCKENBRONZE (*f*): bell metal (77-80% Cu; 23-20% Sn.) 877
GLOCKENERZ (*m*): bell metal (see GLOCKEN--BRONZE) 878
GLOCKENFORM (*f*): bell shape 879
GLOCKENFÖRMIG: bell-shaped 880
GLOCKENGIESSER (*m*): bell founder 881
GLOCKENGUT (*n*): bell metal (see GLOCKEN--BRONZE) 882
GLOCKENKLÖPPEL (*m*): bell hammer 883
GLOCKENMETALL (*n*): bell metal (see GLOCKEN--BRONZE) 884
GLOCKENSPEICHER (*m*), DAMPF-: steam (dome) receiver 885
GLOCKENSPEISE (*f*): bell metal (see GLOCKEN--BRONZE) 886
GLOCKENTELEGRAPH (*m*): bell-telegraph 887
GLOCKENTRICHTER (*m*): bell funnel 888
GLOCKENTURM (*m*): bell tower; steeple 889
GLOCKENVERFAHREN (*n*): bell (process) method 890
GLOSSIEREN: to gloss; comment-(upon): annotate 891
GLOTTISÖDEM (*n*): oedema of the glottis 892
GLOTZAUGENKROPF (*m*): Basedow's disease, exophthalmic goitre (Med.) 893
GLOVERTURM (*m*): Glover tower 894
GLUCINERDE (*f*): glucina; beryllia (see BERYLL--ERDE) 895
GLÜCK (*n*): (good)-luck; fortune; success; happiness; chance 896
GLÜCKEN: to succeed; prosper 897
GLÜCKLICH: lucky; fortunate; happy 898
GLÜCKSELIG: happy; blissful; lucky 899
GLÜHAPPARAT (*m*): heating (glowing, annealing) apparatus (see GLÜHSCHÄLCHEN and GLÜH-OFEN) 900
GLÜHASCHE (*f*): embers; glowing ashes; hot ashes 901
GLÜHBEHANDLUNG (*f*): annealing 902
GLÜHBIRNE (*f*): incandescent bulb; valve (Wireless) 903
GLÜHE (*f*): glow-(ing); incandescence 904
GLÜHEISEN (*n*): glowing iron; red-hot iron 905
GLÜHEN: to glow; ignite; calcine; anneal; heat; mull (Wine) 906
GLÜHEND: incandescent; ardent; glowing 907
GLÜHEN, ROT-: to heat red hot; red-heat; heat to red heat (Metal) 908
GLÜHFADEN (*m*): incandescent filament 909
GLÜHFARBE (*f*): glowing red colour 910
GLÜHFEUER (*n*): glowing fire; annealing (heating) furnace; hardening-on kiln (Ceram.) 911
GLÜHFRISCHEN (*n*): cementation (tempering, annealing) process (Metal) 912
GLÜHGEFÄSS (*n*): annealing box (Metal) 913
GLÜHHAUBE (*f*): hot bulb (of semi-diesel engines) 914
GLÜHHITZE (*f*): glowing (red) heat 915
GLÜHKATHODE (*f*): incandescent cathode; valve (Wireless Tel.) 916
GLÜHKATHODEAPPARAT (*m*): Röntgen apparatus 917
GLÜHKOPF (*m*): hot bulb (of semi-diesel engines) 918
GLÜHKOPFMASCHINE (*f*): Diesel-type engine, hot-bulb engine (invented by Akroyd-Stuart and sometimes called AKROYD--MASCHINE) 919
GLÜHKÖRPER (*m*): incandescent (substance) body; incandescent mantle; glowing body 920
GLÜHLAMPE (*f*): incandescent (glow) lamp 921
GLÜHLAMPENÄTZPASTE (*f*): etching paste for incandescent lamps 922
GLÜHLAMPENFADEN (*m*): incandescent (lamp) filament 923
GLÜHLAMPENLACK (*m*): varnish for incandescent lamps 924
GLÜHLICHT (*n*): incandescent light 925
GLÜHLICHTBRENNER (*m*): incandescent burner 926
GLÜHLICHTCHEMIKALIEN (*n.pl*): incandescent light chemicals 927
GLÜHLICHTKÖRPER (*m*): incandescent mantle 928
GLÜHLICHTSTRUMPF (*m*): incandescent mantle 929
GLÜHOFEN (*m*): glowing (annealing; heating) furnace; hardening-on kiln (Ceramics) 930
GLÜHROHR (*n*): glow pipe 931
GLÜHROHRZÜNDUNG (*f*): tube (glow-pipe) ignition 932
GLÜHRÜCKSTAND (*m*): hot scale 933
GLÜHSAND (*m*): refractory sand 934
GLÜHSCHÄLCHEN (*n*): ignition capsule; cupel 935
GLÜHSCHALE (*f*): cupel 936
GLÜHSCHIFFCHEN (*n*): combustion boat 937
GLÜHSPAN (*m*): iron scale; hammer (mill) scale; iron oxide 938
GLÜHSTOFF (*m*): incandescent material 939
GLÜHSTRUMPF (*m*): incandescent mantle 940
GLÜHTIEGEL (*m*): crucible 941
GLÜHUNG (*f*): glowing; ignition; calcination; annealing; mulling (wine); incandescence 942

GLÜHVERLUST (*m*): heat (radiation; ignition) loss 943
GLÜHWACHS (*n*): gilder's wax 944
GLÜHWEIN (*m*): mulled wine 945
GLÜHWIND (*m*): Sirocco 946
GLÜHWURM (*m*): glow-worm 947
GLUKONSÄURE (*f*): gluconic acid 948
GLUKOSE (*f*): glucose; grape sugar corn sugar; $C_6H_{12}O_6$; (see **DEXTROSE**) 949
GLUKOSID (*n*): glucoside; glycoside 950
GLUKOSURIE (*f*): glucosuria; glycosuria; sugar in the urine 951
GLUT (*f*): glow; incandescence; heat; fire; ardour; passion 952
GLUTAMINSÄURE (*f*): glutamic acid; (see **AMINOGLUTARSÄURE**) 953
GLUTARSÄURE (*f*): glutaric acid; pyro-tartaric acid; $COOH(CH_2)_3COOH$; Sg. 1.192; Mp. 97°C.; Bp. 200-300°C. 954
GLUTARSÄURE-DIÄTHYLESTER (*m*): glutaric diethylester; $C_9H_{16}O_4$; Sg. 1.0284; Bp. 237°C. 955
GLUTASCHE (*f*): embers; hot ashes 956
GLUTEN: to glow; (*n*): gluten 957
GLUTENFIBRIN (*n*): (see **PFLANZENFIBRIN**) 958
GLUTESSE (*f*): blast forge 959
GLUTINÖS: glutinous 960
GLUTMESSER (*m*): pyrometer 961
GLUTOFORM (*n*): (see **GLUTOL**) 962
GLUTOL (*n*): glutol, formaldehyde gelatine (obtained by placing gelatine flakes in formaldehyde solution) 963
GLUTROT: glowing (fiery) red 964
GLYCERIDISOMERIE (*f*): glyceride isomerism 965
GLYCERIN (*n*): (*Glycerinum*), glycerine, glycerol; $C_3H_8O_3$; Sg. 1.2604; Mp. 20°C.; Bp. 290°C.; (see also **GLYZERIN**) 966
GLYCERINPHOSPHORSÄURE (*f*): (*Acidum glycerinophosphoricum*); glycerophosphoric acid, glycerinophosphoric acid; $C_3H_5(OH)_2.O.PO(OH)_2$; Sg. 1.125 967
GLYCERINSÄURE (*f*): glyceric acid 968
GLYCID (*n*), **SALZSAURES-**: (see **EPICHLOR--HYDRIN**) 969
GLYCIN (*n*): glycine; glycocoll; aminoacetic acid; $COOHCH_2NH_2$; Sg. 1.16; Mp. 232-236°C. 970
GLYCIUM (*n*): beryllium; glucinum; (Be) or (Gl) 971
GLYCOL (*n*): glycol, ethylene (alcohol) glycol, glycohol alcohol; (see **GLYKOL**) 972
GLYCOLSÄURE (*f*): glycolic acid, oxyacetic (hydroxyacetic) acid; $CH_2OHCOOH$; Mp. 78°C. 973
GLYCO-METALL (*n*): glyco-metal (85.5% Zn; 5% Sn; 2.4% Cu; 2% Al; 4.7% Pb) 974
GLYCOSID (*n*): glucoside; glycoside 975
GLYCYLALANIN (*n*): glycylalanine; $NH_2.CH_2.CO.NH.CH(CH_3).CO_2H$ 976
GLYCYLGLYCIN (*n*): glycylglycine; $NH_2.CH_2.CO.NH.CH_2.CO_2H$ 977
GLYCYRRHIZIN (*n*): glycyrrhizin 978
GLYKOCHOLSÄURE (*f*): glycocholic acid; cholic acid; $C_{26}H_{43}NO_6$; Mp. 133°C.; Bp. 140°C. 979
GLYKOKOLL (*n*): glycocoll; (see **AMINOESSIG--SÄURE** and **GLYCIN**) 980
GLYKOL (*n*): glycol, ethylene (alcohol) glycol, glycohol alcohol; $C_2H_6O_2$; Sg. 1.1098; Mp. -17.4°C; Bp. 197.4°C 981
GLYKOLSÄURE (*f*): glycolic acid; $CH_2OHCOOH$; Mp. 78°C; (see **GLYCOLSÄURE**) 982
GLYKOLSCHWEFELSÄURE (*f*): glycolsulphuric acid 983
GLYKOLURSÄURE (*f*): glycoluric acid 984
GLYKOLYTISCH: glycolytic 985
GLYKOSAMIN (*n*): glycosamine; $C_6H_{13}O_5N$ 986
GLYKOSE (*f*): glucose; (see **GLUKOSE**) 987
GLYKOSID (*n*): glucoside; glycoside 988
GLYKOSURIE (*f*): (see **GLUKOSURIE**) 989
GLYKURONSÄURE (*f*): glucuronic (glycuronic) acid; $C_6H_{10}O_7[COH-(CH.OH)_4-COOH]$ 990
GLYOXALSÄURE (*f*): glyoxalic acid 991
GLYOXYLSÄURE (*f*): glyoxylic acid; glyoxalic acid 992
GLYPHINSTEATIT (*m*), **PSEUDOMORPHER-**: soapstone (see **SEIFENSTEIN**) 993
GLYPHINSTEATIT (*m*), **UNTEILBARER-**: agalmatolite Sg. 2.8 994
GLYZERIN (*n*): glycerol; glycerin; $C_3H_8O_3$; Sg. 1.26; Mp. 20°C.; Bp. 290°C. (see also **GLYCERIN**) 995
GLYZERINDESTILLATIONSANLAGE (*f*): glycerine distilling plant 996
GLYZERINERSATZ (*m*): glycerine substitute 997
GLYZERINGEWINNUNGSANLAGE (*f*): glycerine extracting plant 998
GLYZERINPECH (*n*): glycerine pitch 999
GLYZERINPHOSPHORSAUER: glycerophosphate 000
GLYZERINPHOSPHORSÄURE (*f*): glycero-phosphoric acid (see **GLYCERINPHOSPHORSÄURE**) 001
GLYZERINSEIFE (*f*): glycerine soap 002
GLYZIN (*n*): glycine (see **AMINOESSIGSÄURE**) 003
GNADE (*f*): grace; clemency; mercy; pardon; favour; goodwill 004
GNADENKRAUT (*n*): hyssop (*Hyssopus officinalis*) 005
GNADENKRAUTÖL (*n*): hyssop oil (from *Hyssopus officinalis*); Sg. 0.93 006
GNADENREICH: gracious 007
GNADENZEICHEN (*n*): mark of favour 008
GNÄDIG: gracious; merciful; favourable; kind; propitious 009
GNEIS (*m*): gneiss; crystalline rock 010
GNEISGLIMMERSCHIEFER (*m*): (see **GNEIS**) 011
GNEISGRANULIT (*m*): (see **GNEIS**) 012
GNEIST (*m*): scud (Leather); scurf 013
GNOSKOPIN (*n*): gnoscopine (an opium alkaloid); $C_{22}H_{23}NO_7$ 014
GOABUTTER (*f*): Goa butter (from seeds of *Guttifere garcinia*); Sg. 0.895; Mp. 41°C. 015
GOAPULVER (*n*): goa powder; chrysarobin (see **CHRYSAROBIN**) 016
GOETHIT (*m*): (see **GÖTHIT**) 017
GOLD (*n*): Gold (Au); Sg. 19.2; Mp. 1062°C.; Bp. 2530°C. (*Aurum*) (see also **AURO-** and **AURI-**) 018
GOLDADER (*f*): golden vein (of ores) 019
GOLDADLER (*m*): golden eagle 020
GOLDAMALGAM (*n*): gold amalgam; Au,Hg; Sg. 15.47 021
GOLDAMMER (*f*): yellow-hammer, yellow-bunting (*Emberiza citrinella*) 022
GOLDANSTRICH (*m*): gilding 023
GOLDARBEIT (*f*): gold working, goldsmith's work 024
GOLDARBEITER (*m*): goldsmith 025
GOLDBAD (*n*): gold bath, toning bath (Photography) 026

GOLDBARREN (*m*): gold ingot 027
GOLDBERGWERK (*n*): gold mine 028
GOLDBERYLL (*m*): chrysoberyl (see **CHRYSOBERYLL**) 029
GOLDBLATT (*n*): gold (leaf) foil 030
GOLDBLECH (*n*): gold plate 031
GOLDBLUME (*f*): gold bloom, marigold (*Calendula officinalis*) 032
GOLDBORTE (*f*): gold (lace, border) fringe 033
GOLDBRAUN: auburn 034
GOLDBROMID (*n*): auric bromide, gold bromide; $AuBr_3$ 035
GOLDBROMÜR (*n*): gold bromide, aurous bromide; AuBr 036
GOLDBRONZE (*f*): (see **MUSCHELGOLD** and **GOLDKUPFER**) 037
GOLDBRONZE (*f*), **ECHTE-**: (see **MUSCHELGOLD**) 038
GOLDCHLORID (*n*): gold (auric) chloride; $AuCl_3$ plus $2H_2O$ or $AuCl_3$; (*Aurum chloratum*) 039
GOLDCHLORIDKALIUM (*n*): gold-potassium chloride (see **CHLORGOLDKALIUM**) 040
GOLDCHLORÜR (*n*): aurous chloride; AuCl 041
GOLDCHLORWASSERSTOFF (*m*): chlorauric acid 042
GOLDCHLORWASSERSTOFFSAUER: chloraurate, combined with chlorauric acid 043
GOLDCHLORWASSERSTOFFSÄURE (*f*): chlorauric acid 044
GOLDCYANID (*n*): auric cyanide; $Au(CN)_3.6H_2O$ 045
GOLDCYANÜR (*n*): aurous cyanide; AuCN 046
GOLDDRAHT (*m*): gold wire 047
GOLDEN: golden; of gold; to gild; tone (Phot.); (*n*): toning (Phot.) 048
GOLDERDE (*f*): auriferous earth 049
GOLDERDWURZEL (*f*): ipecacuanha (root of *Cephaelis ipecacuanha*) 050
GOLDERZ (*n*): gold ore 051
GOLDFADEN (*m*): gold thread 052
GOLDFÄLLUNGSMITTEL (*n*): gold precipitant 053
GOLDFARBE (*f*): orpiment, gold colour 054
GOLDFARBIG: gold-coloured 055
GOLDFISCH (*m*): gold fish 056
GOLDFLUORID (*n*): auric fluoride; AuF_3 057
GOLDGEHALT (*m*): gold content, proportion of gold (in an alloy), number of carats 058
GOLDGELB: golden-yellow 059
GOLDGELBER STOFF (*m*): canary medium (Photography) 060
GOLDGESCHIRR (*n*): gold-plate 061
GOLDGESPINST (*n*): spun gold 062
GOLDGEWICHT (*n*): gold (troy) weight 063
GOLDGLÄTTE (*f*): gold litharge (litharge after slow cooling; red coloured) (see **BLEIGLÄTTE**) 064
GOLDGLIMMER (*m*): yellow mica 065
GOLDGRIES (*m*): gold-dust 066
GOLDGRUBE (*f*): gold-mine 067
GOLDGRUNDFIRNIS (*m*): gold-size-(varnish); gold priming (ground) varnish 068
GOLDHAARIG: golden-haired 069
GOLDHÄHNCHEN (*n*): golden-crested wren (*Regulus cristatus*) 070
GOLDHALTIG: auriferous; containing gold 071
GOLDHYDROXYD (*n*): gold (auric) hydroxide; auric hydrate; $Au(OH)_3$ 072
GOLDIG: golden 073
GOLDJODID (*n*): gold (auric) iodide; AuI_3 074
GOLDJODÜR (*n*): aurous iodide; AuI 075
GOLDKÄFER (*m*): gold beetle 076
GOLDKÄFERLACK (*m*): gold-beetle varnish 077
GOLDKALIUMBROMÜR (*n*): potassium aurobromide 078
GOLDKALIUMCYANÜR (*n*): potassium (cyanaurate) aurocyanide; $KAu(CN)_2$ 079
GOLDKARBID (*n*): gold carbide; Au_2C_2 080
GOLDKIES (*m*): auriferous (pyrites; sand) gravel 081
GOLDKLEE (*m*): yellow trefoil; golden trefoil (*Anemone hepatica*) 082
GOLDKLUMPEN (*m*): ingot (lump, nugget) of gold 083
GOLDKOCHKÖLBCHEN (*n*): gold boiling flask 084
GOLD (*n*), **KOLLOIDALES-**: colloidal gold, collaurin (see **COLLAURIN**) 085
GOLDKÖNIG (*m*): gold regulus 086
GOLDKORN (*n*): grain of gold 087
GOLDKRAUT (*n*): groundsel (*Senecio vulgaris*) 088
GOLDKUPFER (*n*): Mannheim gold (a type of brass); pinchbeck (copper-zinc alloy) 089
GOLDLACK (*m*): gold varnish; aventurine; wall-flower (*Cherianthus cheiri*) 090
GOLDLAHN (*m*): gold tinsel, plate gold, flattened gold-wire 091
GOLDLEGIERUNG (*f*): gold alloy 092
GOLDLEIM (*m*): gold size 093
GOLDLEISTE (*f*): gilt cornice 094
GOLDLEISTENLACK (*m*): gilt-cornice varnish 095
GOLDLOT (*n*): gold solder (alloys of gold, silver, copper and sometimes zinc; or pure gold for soldering platinum) 096
GOLDMACHER (*m*): alchemist 097
GOLDMACHERKUNST (*f*): alchemy 098
GOLDMÜNZE (*f*): gold (medal) medallion; gold coin 099
GOLDNIEDERSCHLAG (*m*): gold precipitate 100
GOLDNITROSYLCHLORID (*n*): gold nitrosyl chloride; $AuCl_3,NOCl$ 101
GOLDOCHER (*m*): finest yellow ochre 102
GOLDOCKER (*m*): (see **GOLDOCHER**) 103
GOLDOXYD (*n*): gold (auric) oxide; Au_2O_3; (see also **GOLDHYDROXYD**) 104
GOLDOXYDUL (*n*): aurous oxide; Au_2O 105
GOLDOXYDULNATRON (*n*), **UNTERSCHWEFLIGSAURES-**: sel d'or (Photography), gold hyposulphite; $Na_2S_2O_3.Au_2S_2O_3.4H_2O$ 106
GOLDOXYDULVERBINDUNG (*f*): aurous compound 107
GOLDOXYDVERBINDUNG (*f*): auric compound 108
GOLDPAPIER (*n*): gold (gilt) paper 109
GOLDPFLAUME (*f*): (see **ICOCA, WESTINDISCHE-**) 110
GOLDPHOSPHID (*n*): gold phosphide; Au_2P_3; Sg. 6.67 111
GOLDPLATTIEREN: to gold-plate 112
GOLDPLATTIERT: gold-plated 113
GOLDPLATTIERUNG (*f*): gold plating 114
GOLDPRASEODYMBROMID (*n*): auric praseodymium bromide; $PrBr_3,AuBr_3,+10H_2O$ 115
GOLDPROBE (*f*): gold (test) assay 116
GOLDPROBENGLÜHAPPARAT (*m*): gold assaying heating apparatus 117
GOLDPROBENKOCHAPPARAT (*m*): gold assaying boiling apparatus 118
GOLDPROBENTAFEL (*f*): gold assaying table 119
GOLDPURPUR (*m*): purple of Cassius; gold tin (purple) precipitate, (mixture of Au and SnO_2) 120
GOLDQUARZ (*m*): auriferous quartz 121
GOLDREGEN (*m*): laburnum; golden rain (see **BOHNENBAUM**) 122
GOLDREICH: rich in gold, having a high gold content 123
GOLDRUBINGLAS (*n*): ruby glass (coloured by purple of Cassius) 124

GOLDRUTE (*f*): goldenrod (*Solidago virgaurea*) 125
GOLDSALPETER (*m*): gold nitrate 126
GOLDSALZ (*n*): gold salt; gold-sodium chloride (*Auro-Natrium chloratum*); $NaAuCl_4.2H_2O$ 127
GOLDSAUER: aurate of; combined with auric acid 128
GOLDSÄURE (*f*): auric acid 129
GOLDSCHALE (*f*): gold (cup) dish; cupel 130
GOLDSCHAUM (*m*): Dutch metal; imitation gold; tinsel (see METALLGOLD) 131
GOLDSCHEIDER (*m*): gold refiner 132
GOLDSCHEIDEWASSER (*n*): aqua regia; nitro-muriatic (chlorazotic) acid 133
GOLDSCHEIDUNG (*f*): gold refining 134
GOLDSCHLAG (*m*): gold leaf; gold foil 135
GOLDSCHLÄGER (*m*): gold beater 136
GOLDSCHLÄGERHAUT (*f*): gold beater's skin 137
GOLDSCHMELZHITZE (*f*): gold melting-point, degree of heat (temperature) at which gold melts (i.e., about 1062°C.) 138
GOLDSCHMIED (*m*): goldsmith 139
GOLDSCHMIEDEEMAILLE (*n*): goldsmith's enamel 140
GOLDSCHNITT (*m*): gold edge; gilt edge 141
GOLDSCHWEFEL (*m*): antimony pentasulphide (see ANTIMONPENTASULFID) 142
GOLDSELENID (*n*): gold selenide; auric selenide Au_2Se_3; Sg. 4.65 143
GOLDSIEGELLACK (*m*): gold sealing wax; aventurine 144
GOLDSILBER (*n*): electrum (natural gold-silver alloy) (40% silver) 145
GOLDSILBERSULFID (*n*): gold-silver sulphide; Ag_3AuS_2 146
GOLDSILBERTELLURID (*n*): gold-silver telluride; (Mineral) Sylvanite, Au_2Te_2,Ag_2Te; (Mineral), Petzite, $3Ag_2Te,Au_2Te$ 147
GOLDSPINNER (*m*): gold spinner 148
GOLDSTAUB (*m*): gold-dust 149
GOLDSTEIN (*m*): chrysolite, olivine (see OLIVIN) 150
GOLDSTICKEREI (*f*): gold embroidery 151
GOLDSTOFF (*m*): gold-brocade, gold material 152
GOLDSTÜCK (*n*): gold coin. gold piece 153
GOLDSTUFE (*f*): gold-ore 154
GOLDSULFID (*n*): auric sulphide; Au_2S_3 155
GOLDSULFÜR (*n*): aurous sulphide; Au_2S 156
GOLDTELLURID (*n*): gold telluride; Au_2Te_4; Mp. 472°C. 157
GOLDTHIOSCHWEFELSÄURE (*f*): aurothiosulphuric acid 158
GOLDTROPFEN (*m.pl*): ethereal ferric chloride tincture 159
GOLDÜBERZUG (*m*): gold coating 160
GOLDVERBINDUNG (*f*): gold compound 161
GOLDWAGE (*f*): gold (scales) balance 162
GOLDWIESENBLUME (*f*): marsh marigold (*Caltha palustris*) 163
GOLDWÄSCHE (*f*): gold washing, washing the gold sand 164
GOLDWÄSCHER (*m*): gold washer 165
GOLDWIRKER (*m*): gold weaver 166
GOLDWURZ (*f*): (greater)-Celandine, chelandine, tetterwort, swallow-wort (*Chelidonium majus*) 167
GOLDZIEHER (*m*): gold-wire drawer 168
GOLDZYANID (*n*): auric cyanide; $Au(CN)_3.6H_2O$ 169
GOLDZYANÜR (*n*): aurous cyanide; AuCN 170
GOMMELINE (*f*): gommeline (see DEXTRIN) 171
GONDEL (*f*): gondola; basket (car) of a balloon 172
GÖNNEN: to wish; grant; favour; permit; not to grudge 173
GÖNNER (*m*): well-wisher; patron; protector 174
GONORRHÖE (*f*): gonorrhœa (Med.) 175
GONOSAN (*n*): gonosan (solution of Kawa gum in East Indian sandalwood oil, ratio 1 : 4; Sg. 1.0 (a gonorrhœa remedy) 176
GÖSCH (GÖSCHE) (*f*): jack (small flag) 177
GÖSCHSTOCK (*m*): jack staff 178
GOSE (*f*): pale ale (two of the most famous are Berliner Gose and Leiziger Gose) 179
GOSLARIT (*m*): goslarite (a hydrated zinc sulphate); $ZnSO_4+7H_2O$ 180
GOSSE (*f*): gutter; sewer; drain; kennel 181
GOSSENSTEIN (*m*): sink 182
GOTHISCH: gothic 183
GÖTHIT (*m*): göthite; $Fe_2O_3.H_2O$ or $FeO(OH)$; Sg. 4.4-5.25 (see LEPIDOKROKIT) 184
GOTT BEWAHRE: God forbid 185
GÖTTERBAUM (*m*): tree of heaven; ailanthus; tree of the Gods (*Ailanthus glandulosas*) 186
GÖTTERDÄMMERUNG (*f*): twilight of the Gods 187
GÖTTERFABEL (*f*): mythology 188
GÖTTERLEHRE (*f*): mythology 189
GÖTTERSAGE (*f*): legend; myth 190
GÖTTERSITZ (*m*): seat of the Gods, Olympus 191
GÖTTERSPEISE (*f*): ambrosia; the food of the Gods 192
GÖTTERTRANK (*m*): nectar; the drink of the Gods 193
GÖTTERZEIT (*f*): the age of mythology 194
GOTTESACKER (*m*): God's acre; churchyard; cemetery; burial ground 195
GOTTESDIENST (*m*): Divine service 196
GOTTESGELEHRSAMKEIT (*f*): theology, divinity 197
GOTTESGELEHRTHEIT (*f*): theology; divinity 198
GOTTESKÄFER (*m*): lady-bird (*Coccinella bipunctata*) 199
GOTTESLÄSTERLICH: blasphemous 200
GOTTESLEHRE (*f*): theology; divinity 201
GOTTESLEUGNERISCH: atheistical 202
GOTTESLEUGNUNG (*f*): atheism 203
GOTTHEIL (*n*): (see BRUNELLE) 204
GOTTLOS: godless, impious, wicked, ungodly 205
GOTTSELIG: godly; pious; blessed 206
GÖTZE (*f*): idol, false God 207
GÖTZENBILD (*n*): idol 208
GÖTZENDIENER (*m*): idolator 209
GÖTZENDIENST (*m*): idolatry 210
GOUDRON (*m*): tar; bitumen; asphalt 211
GOUVERNEMENT (*n*): government 212
GOUVERNEUR (*m*): governor 213
GRAB (*n*): grave; tomb; sepulchre 214
GRABARTIG: sepulchral 215
GRABBELN: to scratch; itch; grope 216
GRABEISEN (*n*): graving tool 217
GRABEN: to dig; engrave; cut; trench; (*n*): digging; engraving; cutting; entrenching; (*m*): ditch; drain; trench 218
GRABENBÖSCHUNG (*f*): counterscarp 219
GRÄBER (*m*): digger; spade 220
GRABESGERÜST (*n*): catafalque 221
GRABGEWÖLBE (*n*): vault 222
GRABHÜGEL (*m*): cairn, mound, barrow, tumulus 223
GRABLEGUNG (*f*): burying; interment 224

GRABMAL (*n*): tomb, monument, grave (head) stone 225
GRABMEISSEL (*m*): graving tool, graver, burin (see **GRABSTICHEL**) 226
GRABSCHEIT (*n*): spade; shovel 227
GRABSCHRIFT (*f*): epitaph; inscription 228
GRABSTICHEL (*m*): graver; (en)-graving tool; chisel (see **GRABMEISSEL**) 229
GRAD (*m*): degree; grade; rank; step; stage 230
GRADABTEILUNG (*f*): graduation; scale 231
GRADBOGEN (*m*): graduated arc; sextant 232
GRADEINTEILUNG (*f*): graduation; scale 233
GRADIEREISEN (*n*): graving tool, chisel 234
GRADIEREN: to graduate; grade; refine; test with hydrometer 235
GRADIERHAUS (*n*): graduation house 236
GRADIERUNG (*f*): graduation; hydrometer test; grading 237
GRADIERWAGE (*f*): brine gauge; water balance 238
GRADIERWERK (*n*): graduation works; graduation apparatus; cooler; cooling arrangement (for concentrating brine or any other solution by exposing a large surface to air) 239
GRADIG: having degrees; graded; of —— degrees 240
GRADLEITER (*f*): scale of degrees 241
GRADMESSER (*m*): graduator 242
GRADMESSUNG (*f*): graduation, measuring of degrees 243
GRADTEILUNG (*f*): graduation; scale 244
GRADUELL: gradual-(ly) 245
GRADUIEREN: to graduate 246
GRADUIERUNG (*f*): graduation 247
GRAFIT (*n*): graphite; wad; plumbago; black-lead; mineral (native) carbon, (opaque crystalline carbon); (C); Sg. 2.3 248
GRAM (*m*): grief; sorrow; affliction; aversion; anger 249
GRÄMELN: to fret; be (morose) irritable 250
GRÄMEN: to grieve; worry; fret 251
GRÄMLICH: surly; sullen; peevish; morose; irritable 252
GRAMM (*n*): gramme 253
GRAMMATIK (*f*): grammar 254
GRAMMATIKER (*m*): grammarian 255
GRAMMATISCH: gramatical-(ly) 256
GRAMMATIT (*m*): grammatite (see **AKTINOLITH**) 257
GRAMMENFLASCHE (*f*): pycnometer (for measuring specific gravities of liquids) 258
GRAMMESCHER RING (*m*): gramme ring 259
GRAMMKALORIE (*f*): small calorie; gramme calorie (in German this calorie is always written with a small "c" as opposed to "C" for kilogram calorie) 260
GRAN (*m*): grain 261
GRANADILHOLZ (*n*): granadilla wood; red ebony (see **GRANATILL-**) 262
GRANALIEN (*pl*): granulated metal 263
GRANAT (*m*): garnet (an iron-alum silicate); $Fe_3Al_2Si_3O_{12}$; Sg. 3.75-4.2; pomegranate (*Punica granatum*); (see **PYROP** and **MELANIT**) 264
GRANATAPFEL (*m*): pomegranate (fruit of *Punica granatum*) 265
GRANATÄPFELSCHALE (*f*): pomegranate skin or peel 266
GRANATBAUM (*m*): pomegranate tree (*Punica granatum*) 267
GRANATBAUMHOLZ (*n*): pomegranate wood (Sg. 1.36) 268
GRANATBLÜTE (*f*): pomegranate blossom (of *Punica granatum*) 269
GRANATDODEKAEDER (*n*): rhombic dodecahedron 270
GRANATE (*f*): grenade; bomb; shell; garnet (see **GRANAT**); pomegranate (see **GRANATAPFEL**) 271
GRANATENHAGEL (*m*): hail (rain) of shell-(s) (Military) 272
GRANATENSTÜCK (*n*): shell-splinter 273
GRANATFLUSS (*m*): artificial garnet 274
GRANATILLHOLZ (*n*): granadilla wood; Physic nut wood (see **GRANADIL**) 275
GRANATILLÖL (*n*): physic nut oil (from nut of *Curcas purgans*) (see also **CROTONÖL**) 276
GRANATILLSAMEN (*m*): physic nut (nut of *Curcas purgans*) 277
GRANATKARTÄTSCHE (*f*): shrapnel shell 278
GRANATKERN (*m*): pomegranate seed 279
GRANATLEHRE (*f*): shell gauge 280
GRANATOEDER (*n*): rhombic dodecahedron 281
GRANAT (*m*), **PRISMATOIDISCHER-**: Staurolite 282
GRANAT (*m*), **PYRAMIDALER-**: idocrase 283
GRANATRINDE (*f*): pomegranate bark (from *Punica granatum*) 284
GRANATROT: garnet-(red) 285
GRANATSTEIN (*m*): garnet (the gem) 286
GRANATWURZEL (*f*): pomegranate root (of *Punica granatum*) 287
GRANATWURZELRINDE (*f*): pomegranate root bark (see **GRANATRINDE**) 288
GRAND (*m*): coarse sand; gravel; underback (Brewing) 289
GRANDIG: gravelly; containing gravel 290
GRANDMEHL (*n*): coarse meal 291
GRANDSTEIN (*m*): granite (see **GRANIT**) 292
GRANIEREN: to granulate; grain 293
GRANIT (*m*): granite; Sg. 2.5-2.96 (a rock composed of quartz, mica and feldspar) 294
GRANITARTIG: granite-like; granitic; granitiform 295
GRANITEN: granitic 296
GRANITFELS (*m*): granitic rock; granite rock 297
GRANITFELSEN (*m*): granitic rock; granite rock 298
GRANITFÖRMIG: granitic; granite-like; granitiform 299
GRANITISCH: granitic 300
GRANNE (*f*): awn; beard; bristle; needle 301
GRANULATION (*f*): granulation 302
GRANULATIONSBEFÖRDEREND: promoting granulation 303
GRANULIERAPPARAT (*m*): granulating apparatus 304
GRANULIEREN: to granulate; grain 305
GRANULIERMASCHINE (*f*): granulating machine 306
GRANULIERUNG (*f*): granulation 307
GRANULIERWALZE (*f*): granulating roll (Printing) 308
GRANULÖS: granular 309
GRÄNZE (*f*): (see **GRENZE**) 310
GRAPEN (*m*): mixing pot on three legs; iron pot 311
GRAPHISCH: graphic 312
GRAPHISCHE ANSTALT (*f*): art printers (see also under **KUNSTANSTALT**) 313
GRAPHISCHE DARSTELLUNG (*f*): graphical representation 314
GRAPHIT (*n*): graphite (see **GRAFIT**) 315

GRAPHITÄHNLICH : (see **GRAPHITARTIG**) 316
GRAPHITARTIG : graphite like ; graphitoidal 317
GRAPHITIEREN: to blacklead ; coat with graphite 318
GRAPHITISCH : graphitic 319
GRAPHITÖLSCHMIERAPPARAT (*m*): graphite oil lubricating apparatus 320
GRAPHITSÄURE (*f*) : graphitic acid 321
GRAPHITSCHMELZTIEGEL (*m*): graphite crucible (**PASSAUER TIEGEL** and **IPSER TIEGEL**) 322
GRAPHITSCHMIERE (*f*) : graphite lubricant 323
GRAPHITSPITZE (*f*) : carbon (of an arc lamp) 324
GRAPHITSTIFT (*m*): lead pencil; blacklead-(pencil) 325
GRAPHITTIEGEL (*m*): graphite crucible 326
GRAS (*n*): grass 327
GRASANGER (*m*) : green, plot of grass 328
GRASARTIG : gramineous 329
GRASBUTTER (*p*): grass butter (from cattle fed on grass) 330
GRASEBENE (*f*) : Savanna, pampa, prairie, steppe, grassy plain 331
GRASEN : to graze ; cut grass 332
GRASFASER (*f*): grass-(y) fibre 333
GRASFINK (*m*) : chaffinch (*Fringilla caelebs*) 334
GRASFLECK (*m*): grass stain; patch of grass, plot of grass, green 335
GRASFRESSEND : graminiverous ; herbiverous 336
GRASFROSCH (*m*): green frog, grass frog (*Rana temporaria*) 337
GRASGRÜN : grass-green 338
GRASHALM (*m*): blade of grass 339
GRASHÜPFER (*m*): grass-hopper 340
GRASIG : grassy ; grassgrown 341
GRASLAND (*n*): meadow land 342
GRASLAUCH (*m*): (see **SCHNITTLAUCH**) 343
GRASLEINEN (*n*): grass cloth 344
GRASLILIE (*f*): Asphodel (*Narthecium ossifragum*) 345
GRASÖL (*n*): Singapore (verbena) citronella oil (from fruit of *Tetranthera citrada* ; Sg. 0.9); grass oil, Melissa oil, East Indian Verbena oil, Lemongrass oil (from the grass *Andropogon citratus* or *A. nardus* ; Sg. 0.9) 346
GRASPFERD (*n*): grasshopper 347
GRASPILZ (*m*): toadstool 348
GRASPLATZ (*m*): grass-plot, green 349
GRASREICH : grassy 350
•**GRASROST** (*m*): grass rust, mildew (*Uromyces* or *Trichobasis*) 351
GRASSCHNECKE (*f*) : slug 352
GRASSIEREN : to rage ; spread ; prevail 353
GRÄSSLICH : horrible ; frightful ; hideous ; shocking ; ghastly 354
GRASUNG (*f*) : grazing 355
GRASWEIDE (*f*) : pasture 356
GRASWURZEL (*f*) : couch-grass (*Agropyrum repens* or *Triticum repens*); dog-grass ; Graminis ; Triticum ; grass root 357
GRAT (*m*): edge ; ridge ; burr ; groin (of roof) ; rabbet 358
GRÄTE (*f*) : fishbone ; edge ; ridge ; spine 359
GRATHOBEL (*m*) : rabbet plane 360
GRATIAL (*n*) : donation, gift 361
GRATIFIKATION (*f*): supplement, bonus ; gratification 362
GRÄTIG : bony ; full of (fish)-bones 363
GRÄTING (*f*) : grating 364
GRÄTING (*f*), **PUMPEN-** : pump grating 365
GRATISBEILAGE (*f*) : free supplement 366
GRATULIEREN : to congratulate 367

GRAU : gray ; grey ; grizzled ; sombre 368
GRAUBRAUNSTEIN (*m*): manganite ; grey manganese ore ; $Mn_2O_3.H_2O$; Sg. 4.3 (see **MANGANIT**) 369
GRAUBRAUNSTEINERZ (*n*): pyrolusite ; MnO_2 ; 63% Mn (see **PYROLUSIT**) 370
GRAUE ERDE (*f*) : grey earth 371
GRAUEISENERZ (*n*): marcasite ; white iron pyrite ; FeS_2 ; about 47% Fe (see **MARKASIT**) 372
GRÄUEL (*m*) : horror, outrage, aversion, detestation, abomination 373
GRAUEN : to grow grey ; dawn ; be in dread of ; have a horror of ; (*n*): horror ; dread ; fear ; abhorrence 374
GRAUENHAFT : horrible ; dreadful ; uncanny 375
GRAUE QUECKSILBERSALBE (*f*) : mercurial ointment 376
GRAUFAHLERZ (*n*) : (see **GRAUKUPFERERZ**) ; grey copper ore, chalcocite, tennantite 377
GRAUGOLDERZ (*n*): nagyagite (11% silver) (see **NAGYAGIT**) 378
GRAUGÜLTIGERZ (*n*): tetrahedrite ; grey copper ore ; $3Cu_2S.Sb_2S_3$; Sg. about 5.0 (about 52% Cu) 379
GRAUGUSS (*m*): (grey)-pig-iron casting ; grey cast (pig) iron ; Sg. 7.03-7.13 380
GRAUHEIT (*f*): greyness 381
GRAUKALK (*m*): grey lime ; grey chalk ; pyrolignite of lime ; crude calcium acetate (see **CALCIUMAZETAT**) 382
GRAUKEHLCHEN (*n*): white-throat, warbler (member of the genus *Sylvia*) 383
GRAUKOBALTERZ (*n*) : jaipurite ; grey cobalt ore 384
GRAUKUPFERERZ (*n*) : chalcocite ; copper glance ; Cu_2S ; (79.8% Cu) ; tennantite ; grey copper ore ; $3Cu_2S.As_2S_3$ (58% Cu.) 385
GRAULICH : greyish ; grizzly ; shocking ; horrible ; atrocious 386
GRÄULICH : (see **GRAULICH**) 387
GRAUMANGANERZ (*n*): manganite ; grey manganese ore (see **GRAUBRAUNSTEIN**) 388
GRAUMANGANERZ, LICHTES- (*n*): polianite (see **POLIANIT**) 389
GRAUMONTSAMEN (*m*): pumpkin seed ; pepo (seed of *Cucurbita pepo*) 390
GRAUPAPPEL (*f*): white poplar (*Populus alba*) (see **PAPPEL**) 391
GRAUPE (*f*): peeled (barley) grain ; groat ; knot (in cotton) ; large grain-(s) of ore 392
GRAUPELN : to sleet ; drizzle 393
GRAUPEN (*f.pl*) : groats ; pearl barley 394
GRAUPENERZ (*n*): granular ore 395
GRAUPENGRÜTZE (*f*): barley groats 396
GRAUPENHOLLÄNDER (*m*) : grain (hollander) shelling machine (for shelling spelt) 397
GRAUPENKOBALT (*m*): smaltite ; $CoAs_2$ (28% cobalt) (see **SPEISKOBALT**) 398
GRAUPENSCHLEIM (*m*) : barley water 399
GRAUPENSCHÖRL (*m*): granular schorl ; Sg. 3.17 400
GRAUPIG : granular 401
GRAUS (*m*): gravel ; smalls ; horror ; shudder ; dread 402
GRAUSAM : horrible ; inhuman ; cruel ; barbarous ; terrible 403
GRAUSEN : to shudder ; awe ; fill with dread ; be awed 404
GRAUSIG : gruesome ; horrible ; dreadful 405
GRAUSPIEGEL (*m*) : grey spiegel iron 406

GRAUSPIESSGLANZ (*m*): stibnite; grey antimony ore; antimonite; antimony glance; antimony sulphuret; Sb_2S_3; Sg. 4.6 (71% antimony) 407
GRAUSPIESSGLANZERZ (*n*): (see GRAUSPIESS-GLANZ) 408
GRAUWACKE (*f*): graywacke; transitional rock 409
GRAUWACKENSANDSTEIN (*m*): trap sandstone 410
GRAVEUR (*m*): engraver, graver 411
GRAVIEREN: to engrave; aggravate; grave in 412
GRAVIERER (*m*): engraver, graver 413
GRAVIERKUNST (*f*): engraver's art 414
GRAVIERMASCHINE (*f*): engraving machine 415
GRAVIERMEISSEL (*m*): graver, graving tool 416
GRAVIERUNG (*f*): engraving 417
GRAVIMETRISCH: gravimetric-(al)-(ly) 418
GRAVITÄT (*f*): solemnity, gravity, seriousness 419
GRAVITATION (*f*): gravitation, gravity 420
GRAVITATIONSGESETZ (*n*): law of gravitation 421
GRAVITATIONSKRAFT (*f*): (force of)-gravitation 422
GRAVITIEREN: to gravitate 423
GRAVÜRE (*f*): engraving 424
GRAZIÖS: graceful-(ly) 425
GREENOCKIT (*m*): greenockite (natural cadmium sulphide); CdS; Sg. 5.0 426
GREGARIEN (*f.pl*): Gregarindæ (parasites of the alimentary canal) (Med.) 427
GREIFBAGGER (*m*): grab dredger 428
GREIFBAR: tangible; palpable; seizable 429
GREIFEN: to grasp; seize; grab; grip; catch; snatch; excavate; cake; take; take root; comprehend 430
GREIFEN, HERAUS-: to take (usually one case as an example) 431
GREIFER (*m*): grab (Mech.); catch 432
GREIFKLAUE (*f*): talon, claw 433
GREIFZIRKEL (*m*): calipers 434
GREISENALTER (*n*): senility, senile marasmus (Med.) 435
GRELL: bright; dazzling; glaring; shrill; hard; sharp 436
GRELLES ROHEISEN (*n*): white pig iron; Sg. 7.58-7.73; Mp. 1075°C. 437
GRENADILLHOLZ (*n*): granadilla wood (see GRANA-DILHOLZ and GRANATILLHOLZ) 438
GRENADIN (*n*): grenadine 439
GRENAT (*m*): garnet; Sg. 3.75-4.2 440
GRENZALKOHOL (*m*): limit (saturated) alcohol 441
GRENZE (*f*): limit; boundary; frontier; bound; end; limitation; border 442
GRENZEN: to border, bound, limit 443
GRENZENLOS: unlimited; infinite; unbounded; boundless 444
GRENZFALL (*m*): limit-(ing) case; very (best) worst case 445
GRENZFLÄCHE (*f*): boundary surface; surface of contact 446
GRENZGESETZ (*n*): limit law 447
GRENZKOHLENWASSERSTOFF (*m*): limit hydro-carbon (as paraffin) 448
GRENZKURVE (*f*): limit-(ing) curve 449
GRENZLEHRE (*f*): limit gauge 450
GRENZLEISTUNGSTURBINE (*f*): high duty turbine 451
GRENZLINIE (*f*): boundary line, line of demarcation 452
GRENZMAL (*n*): boundary stone, landmark 453
GRENZRING (*m*): limit ring 454
GRENZSCHEIDE (*f*): boundary 455
GRENZVERBINDUNG (*f*): terminal (compound) member; saturated compound 456
GRENZWERT (*m*): limit-(ing) value; end (terminal value 457
GRENZWINKEL (*m*): critical angle 458
GRENZZOLLBEHÖRDE (*f*): frontier customs 459
GRETCHEN IM BUSCH: (see KAPUZINERKRAUT) 460
GREUEL (*m*): horror; detestation; outrage; abomination; aversion 461
GREULICH: abominable; detestable; horrible; shocking; atrocious 462
GREZE (*f*): raw silk 463
GREZSEIDE (*f*): raw silk 464
GRIES (*m*): grit; gravel; coarse sand; groats; small coal (see GRUS); urinary calculus (Med.) 465
GRIESELIG: (see GRIESIG) 466
GRIESELN: to crumble, grind, break in pieces 467
GRIESFLECHTE (*f*): herpes (a skin eruption) (Med.) 468
GRIESIG: gritty; gravelly; calculus 469
GRIESKOHLE (*f*): dust coal, small coal, smalls 470
GRIESMEHL (*n*): pollard 471
GRIESS (*m*): (see GRIES) 472
GRIESSMITTEL (*n*): remedy for (gravel) urinary calculus 473
GRIESSSTEIN (*m*): urinary calculus; gravel (Med.); jade; jadeite; nephrite (see JADEIT) 474
GRIFF (*m*): handle; grasp; grip; knob; touch; feel; shaft; haft 475
GRIFFBRETT (*n*): fingerboard (of a stringed instrument) 476
GRIFFEL (*m*): style; stylus; slate pencil; pencil; style; pistil (Bot.) 477
GRIFFELARTIG: styloid, styliform 478
GRIFFELFÖRMIG: styliform; styloid 479
GRIFFSTOPFEN (*m*): stopper with thumb piece 480
GRILLE (*f*): cricket; whim; crotchet; vagary; caprice; freak 481
GRILLENHAFT: capricious 482
GRILLENKRANKHEIT (*f*): hypochondriasis (a mental disorder akin to melancholia) (Med.) 483
GRIMASSE (*f*): grimace 484
GRIMM (*m*): rage; fury; wrath 485
GRIMMDARM (*m*): colon (Med.) 486
GRIMMEN: to gripe; fume; rage; (*n*): colic (Med.) 487
GRIMMIG: grim; furious; fierce 488
GRIND (*m*): scab; scurf; crust; mange 489
GRINDELIENKRAUT (*n*): grindelia leaves (from *Grindelia robusta* and *G. squamosa*) 490
GRINDHOLZ (*n*): (see FAULBAUM) 491
GRINDIG: scabby; scurfy 492
GRINDSTEIN (*m*): (see GRANIT) 493
GRINDTRAN (*m*): (see DELPHINTRAN) 494
GRINSEN: to grin; weep; stare (of colours); sneer; begin to melt (of metal) 495
GRIPPE (*f*): influenza (Med.) 496
GROB: coarse; thick; heavy; rough; large; gross; rude; ill-bred; clumsy; uncouth 497
GROBBERGE (*m.pl*): rough (pit) coal, (in the rough as mined, containing waste matter, before breaking and washing) (Mining) 498
GROBBRECHEN: to break up into pieces, to grind coarsely 499
GROBDRÄHTIG: coarse-threaded 500
GROBFÄDIG: coarse-threaded 501
GROBFASERIG: coarse-fibred 502

GROBFEILE (*f*) : coarse (rough) file 503
GROBGEPULVERT : coarsely (powdered) crushed 504
GROBGEWICHT (*n*) : gross weight 505
GROBJÄHRIG : wide-ringed (of timber) 506
GROBKORN (*n*) : coarse (large) grain 507
GROBKÖRNIG : coarsely (grained) granular ; large grained 508
GROBKRYSTALLINISCH : coarsely crystalline 509
GRÖBLICH : rather (gross) coarse ; coarsely ; grossly 510
GROBMÖRTEL (*m*) : coarse mortar ; concrete 511
GROBPORIG : large (coarse) pored 512
GROBSCHMIED (*m*) : blacksmith ; farrier 513
GROBSCHROT (*n* and *m*) : coarse groats ; coarsely ground malt 514
GROBSTRASSE (*f*) : breaking-down mill (Rolling mills) 515
GRODEN (*m*) : alluvium, regained land 516
GROG (*m*) : grog, rum and water 517
GROLL (*m*) : rancour, grudge, hatred, resentment 518
GROLLEN : to bear ill-will 519
GROSS : great ; large ; tall ; huge ; high ; vast ; capital (Letter) 520
GROSSARTIG : grand ; sublime ; magnificent 521
GROSSBAUM (*m*), EISJACHT- : ice yacht boom 522
GROSSBETRIEB (*m*) : wholesale trade ; operation on a large scale 523
GROSSBLÄTTERIG : coarsely foliated or laminous ; large-leaved 524
GROSSBRASSE (*f*) : main brace (Naut.) 525
GRÖSSE (*f*) : quantity ; size ; magnitude ; amount ; bulk ; largeness ; greatness ; power ; grandeur ; enormity ; degree 526
GRÖSSENLEHRE (*f*) : geometry ; mathematics 527
GRÖSSENORDNUNG (*f*) : order of magnitude 528
GRÖSSENREIHE (*f*) : order of magnitude 529
GRÖSSERN : to increase ; enlarge 530
GROSSFASERIG : large (coarse) fibred 531
GROSSGAFFEL (*f*) : main gaff (Naut.) 532
GROSSGEFÄSS (*n*) : large size (container, receiver) vessel ; vat 533
GROSSGESTALT (*f*) : colossus 534
GROSSGEWERBE (*n*) : manufacture on a large scale 535
GROSSGEWICHT (*n*) : gross weight 536
GROSSHALS (*m*) : main tack (Naut.) 537
GROSSHANDEL (*m*) : wholesale (business) trade 538
GROSSHÄNDLER (*m*) : wholesale (supplier ; dealer) merchant 539
GROSSHERZIG : magnanimous ; generous 540
GROSSHIRN (*n*) : cerebrum 541
GROSSINDUSTRIE (*f*) : mass (production) manufacture 542
GROSSJÄHRIG : of age 543
GROSSKLÜVER (*m*) : main jib (Naut.) 544
GROSSKOKS (*m*) : large (lump) coke 545
GROSSKRAFTWERK (*n*) : super-power station 546
GROSSMACHT (*f*) : great power 547
GROSSMARS (*m*) : main top (Naut.) 548
GROSSMARSSTENGE (*f*) : main topmast (Naut.) 549
GROSSMASCHIG : large meshed 550
GROSSMAST (*m*) : main mast (Naut.) 551
GROSSMÜTIG : (see GROSSHERZIG) 552
GROSSPIEKFALL (*n*) : main peak halliard (Naut.) 553
GROSSRAH (*f*) : main yard (Naut.) 554
GROSSROIL (*n*) : main royal (Naut.) 555
GROSSROILSEGEL (*n*) : main royal 556
GROSSSEGEL (*n*) : main sail 557
GROSSSTADT (*f*) : large town, metropolis 558
GROSSSTAG (*n*) : main stay (Naut.) 559
GRÖSSTENTEILS : for the (most part) major portion ; mainly ; chiefly 560
GROSSTOPPNANT (*f*) : main lift (Naut.) 561
GROSSULAR (*m*) : grossularite, essonite, green garnet, cinnamon stone (a calcium-aluminium garnet) ; $Ca_3Al_2(SiO_4)_3$; Sg. 3.5 562
GROSSUNTERBRAMSEGEL (*n*) : lower main topgallant sail 563
GROSSUNTERMARSSEGEL (*n*) : lower main topsail 564
GROSSWANTEN (*n.pl*) : main (sheets) rigging 565
GROTTE (*f*) : grotto 566
GROTTENARBEIT (*f*) : rock-work 567
GROTTENWERK (*n*) : rock-work, rockery 568
GRÜBCHEN (*n*) : little (hole) opening ; dimple ; lacuna (Biology) 569
GRUBE (*f*) : mine ; quarry ; hole ; cavity ; pit ; ditch ; trench ; grave 570
GRÜBEL (*m*) : refinement, subtilty, hypercriticism 571
GRÜBELKRANKHEIT (*f*) : spleen 572
GRÜBELN : to ponder ; brood ; meditate ; be hypercritical ; rummage ; rake 573
GRUBENARBEITER (*m*) : miner 574
GRUBENAXT (*f*) : miner's pick 575
GRUBENBAHN (*f*) : mine railway 576
GRUBENBAU (*m*) : mining ; underground working 577
GRUBENBETRIEB (*m*) : mining 578
GRUBENBLENDE (*f*) : miner's lamp 579
GRUBENGAS (*n*) : marsh gas ; methane ; fire-damp (explosive mixture of methane and air) (see METHAN) 580
GRUBENGUT (*n*) : minerals (Mining) 581
GRUBENKLEIN (*n*) : rubbish ; slack (Mining) ; smalls 582
GRUBENLAMPE (*f*) : miner's lamp ; safety lamp 583
GRUBENPULVER (*n*) : blasting (charge) powder 584
GRUBENRÖSTE (*f*) : trench rotting process (for flax ; to obtain bast fibre) (see RÖSTE) 585
GRUBENSAND (*m*) : pit sand 586
GRUBENSCHACHT (*m*) : mine shaft 587
GRUBENSCHLACKE (*f*) : slag 588
GRUBENSCHMELZ (*m*) : a goldsmith's enamel, Email champ-levé 589
GRUBENVENTILATOR (*m*) : mine (colliery) fan 590
GRUBENVERKOHLUNG (*f*) : pit charcoal-burning 591
GRUBENWETTER (*n*) : mine (fire) damp (see GRUBENGAS) 592
GRUBIG : pock-marked, pitted 593
GRÜBLER (*m*) : hypercritic, refiner 594
GRUDEKOKS (*m*) : semi-coke, coke breeze : [a solid residue from brown coal-(tar)-distillation or paraffin manufacture] (15-25% ash ; 15-25% moisture ; calorific value, dry, 6000-7000 calories) 595
GRUDEOFEN (*m*) : coke-breeze stove (suitable for burning coke-breeze in dwellings) 596
GRUFT (*f*) : grave, vault, cavity, cave, sepulchre 597
GRUMMET (*n*) : second-crop, aftermath 598
GRÜN : green ; fresh ; verdant ; unripe ; raw ; (*n*) : green 599
GRÜNBAUM (*m*) : privet (*Ligustrum vulgare*) 600
GRÜNBLEI (*n*) : pyromorphite (see BRAUNBLEIERZ) 601
GRÜNBLEIERZ (*n*) : green lead ore ; green pyromorphite ; green mimetite (see BRAUN-BLEIERZ) 602

GRUND (*m*): ground; reason; grounds; sediment; bottom; motive; argument; estate; rudiments; dregs; elements; foundation; base; basis; priming; soil; territory; cause; earth; lees 603
GRUNDANSTRICH (*m*): first coat (of paint, etc.) ground 604
GRUND, AUF-: based on; on account of; in accordance with; by virtue of 605
GRUND (*m*), **AUF- DESSEN**: on the strength of it 606
GRUNDBASS (*m*): fundamental (figured, thorough) bass (Music) 607
GRUNDBAU (*m*): foundation, substructure 608
GRUNDBEDEUTUNG (*f*): primary meaning 609
GRUNDBEGRIFF (*m*): fundamental principle, basic idea; (*pl*): fundaments, rudiments 610
GRUNDBELASTUNG (*f*): essential (main, general) load (of power station) 611
GRUNDBESITZ (*m*): real estate; landed property 612
GRUNDBESTANDTEIL (*m*): element-(ary constituent); primary component 613
GRUNDBÜCHSE (*f*): slide box 614
GRUNDEIGENTUM (*n*): real estate, landed property 615
GRUNDEINHEIT (*f*): fundamental unit 616
GRUNDEINKOMMEN (*n*): ground rent 617
GRUNDEISEN (*n*): probe 618
GRÜNDEN: to ground; found; establish; groove; float (a company); fathom; (see **GRUNDIEREN**) 619
GRUNDENTWURF (*m*): project, lay-out, ground plan, rough sketch 620
GRÜNDER (*m*): founder; promoter; establisher 621
GRUNDE, ZU-GEHEN: to be wrecked, to perish, to go to the dogs 622
GRUNDFALSCH: radically (wrong) false; fundamentally (wrong) false 623
GRUNDFARBE (*f*): ground (primary) colour; priming colour; bottom colour (Dye) (see **GRUNDIERFARBE**) 624
GRUNDFEHLER (*m*): radical error 625
GRUNDFEUCHTIGKEIT (*f*): soil moisture 626
GRUNDFIRNIS (*m*): priming (ground) varnish 627
GRUNDFLÄCHE (*f*): surface; area; base; basis; floor space 628
GRUNDFLÜSSIGKEIT (*f*): suspending liquid (Colloids) 629
GRUNDFORM (*f*): primary (fundamental) form or type 630
GRUNDGEBIRGE (*n*): primitive rock 631
GRUNDGELEHRT: learned, erudite 632
GRUNDGESETZ (*n*): fundamental law 633
GRUNDHAKEN (*m*): drag hook 634
GRUNDHEFE (*f*): sediment, lees 635
GRUNDIERBAD (*n*): bottoming bath 636
GRUNDIEREISEN (*n*): grounding tool (Etching) 637
GRUNDIEREN: to ground; prime; size; stain; bottom; prepare; (see **GRÜNDEN**) 638
GRUNDIERFARBE (*f*): primary colour, priming colour, ground colour, first coat; bottom colour (Dyeing) 639
GRUNDIERLACK (*m*): ground (priming) varnish 640
GRUNDIRRTUM (*m*): radical error 641
GRUNDKAPITAL (*n*): stock 642
GRUNDKÖRPER (*m*): fundamental (parent) substance; basic substance 643
GRUNDKRAFT (*f*): primary (primitive) force 644
GRUNDLAGE (*f*): foundation; groundwork; base; basis; consideration; principle; matrix; fundament; rudiments; basement 645
GRUNDLAGER (*n*): main bearing 646
GRUNDLAST (*f*): main load (of power plants) 647
GRUNDLEGEND: fundamental; basic 648
GRUNDLEHRE (*f*): first principles, basis 649
GRÜNDLICH: fundamental; profound; thorough -(ly); solid; radically; elementary 650
GRUNDLINIE (*f*): ground line; base line; basis; outline 651
GRUNDLOS: groundless, without foundation, bottomless, without reason, causeless 652
GRUNDMASSE (*f*): ground mass; stroma (Anat.) 653
GRUNDMÖRTEL (*m*): concrete 654
GRUNDPLATTE (*f*): base-plate; bed-plate; foundation plate 655
GRUNDPREIS (*m*): lowest (minimum) price, rock bottom price, regulation price, list price 656
GRUNDPROBLEM (*n*): fundamental (basic) problem 657
GRUNDREGEL (*f*): axiom; principle; fundamental (elementary) rule 658
GRUNDRENTE (*f*): ground rent 659
GRUNDRISS (*m*): ground (plan) sketch; ground outline; horizontal section 660
GRUNDSATZ (*m*): axiom; principle; law; maxim 661
GRUNDSCHICHT (*f*): ground course; fundamental (primary) layer 662
GRUNDSEE (*f*): ground swell 663
GRUNDSETZLICH: fundamentally; basically 664
GRUNDSETZLICH VERSCHIEDEN: diametrically opposed; fundamentally (radically) different 665
GRUNDSTEIN (*m*): foundation stone; corner stone; pyrite; (see **PYRIT**) 666
GRUNDSTEUER (*f*): ground rent; land tax 667
GRUNDSTOCK (*m*): matrix 668
GRUNDSTOFF (*m*): element(-s); raw material; base; basic substance; radical foundation 669
GRUNDSTRICH (*m*): first coat; priming; ground coat; down-stroke 670
GRUNDSTÜCK (*n*): piece of ground or land; main piece or part 671
GRUNDSTÜTZE (*f*): foundation (support) pier 672
GRUNDSUBSTANZ (*f*): (see **GRUNDSTOFF**) 673
GRUNDSUPPE (*f*): grounds, dregs, sediment 674
GRUNDTEILCHEN (*n*): atom 675
GRUNDTEXT (*m*): original (subject matter) text 676
GRUNDTON (*m*): ground (foundation) tone key note 677
GRÜNDUNG (*f*): foundation; first coat; priming; establishment 678
GRÜNDUNGSSOHLE (*f*): ground line, base line, floor line 679
GRUNDURSACHE (*f*): principal (prime) reason or cause 680
GRUNDVENTIL (*n*): foot valve 681
GRUNDVERSUCH (*m*): fundamental (experiment) test 682
GRUNDWAHRHEIT (*f*): fundamental truth 683
GRUNDWASSER (*n*): ground (underground) water 684

GRUNDWERK (*n*): ground-work 685
GRUNDZAHL (*f*): unit; base; base number; cardinal number 686
GRUNDZINS (*m*): ground rent 687
GRÜNE (*n*): green -(ness); verdure; greens; green food 688
GRÜNEISENERZ (*n*): green iron ore; dufrenite; $Fe_2PO_4(OH)_3$; Sg. 3.4 689
GRÜNEISENSTEIN (*m*): (see **GRÜNEISENERZ**) 690
GRÜNEN: to become (grow) green or verdant, thrive 691
GRÜNERDE (*f*): green earth; glauconite; celadonite; terra verde (colouring principle of which is ferrous silicate); (see **GLAUKONIT**) 692
GRÜNFEUER (*n*): green fire 693
GRÜNFILTER (*m and n*): green filter (for three colour printing or photography) 694
GRÜNFINK (*m*): green finch; green linnet; (*Ligurinus chloris*) 695
GRÜNGELB: greenish-yellow 696
GRÜNHOLZ (*n*): dwarf pine, mountain pine; (*Pinus montana*) 697
GRÜNKALK (*m*): gas lime 698
GRÜNKREUZGAS (*n*): green cross gas 699
GRÜNLICH: greenish 700
GRÜNMALZ (*n*): green malt (germinated moist malt) (Brewing) 701
GRÜNÖL (*n*): green oil; anthracene oil (see **ANTHRACEN**) 702
GRÜNSAND (*m*): Greensand, Glauconite (hydrous iron-potassium silicate) (see **GLAUKONIT**); blue clay, marl (earthy calcium carbonate; Sg. about 1.75) 703
GRÜNSÄURE (*f*): verdic acid 704
GRÜNSCHLEIER (*m*): green fog (see **FARBSCHLEIER**) 705
GRÜNSPAN (*m*): verdigris; Sg. 1.9 (see **KUPFER-ACETAT** and **KUPFERACETAT, BASISCHES-**) 706
GRÜNSPAN (*m*), **BASISCHER-**: basic verdigris, blue verdigris, basic copper acetate; $Cu(C_2H_3O_2)_2.Cu(OH)_2+5H_2O$ (see also **KUPFERACETAT, BASISCHES-**) 707
GRÜNSPAN (*m*), **BLAUER-**: blue verdigris (see **GRÜNSPAN, BASISCHER-**) 708
GRÜNSPANBLUMEN (*f.pl*): acetous salt 709
GRÜNSPANGEIST (*m*): spirit of verdigris (see **ESSIGSÄURE**) 710
GRÜNSPAN (*m*), **NEUTRALER-**: neutral verdigris (see **GRÜNSPAN**) 711
GRÜNSPAT (*m*): a variety of diopside (see **DIOPSID**) 712
GRÜNSTAR (*m*): glaucoma (a disease of the eye) (Med.) 713
GRÜNSTEIN (*m*): green stone; syenite; green porphyry; diorite 714
GRÜNSUCHT (*f*): chlorosis (a form of anæmia) (Med.) 715
GRÜNWURZEL (*f*): fumitory (*Fumaria officinalis*) 716
GRUPPE (*f*): group; cluster; set; syndicate 717
GRUPPENWEISE: in (sets; clusters) groups 718
GRUPPIEREN: to group; arrange in (sets) groups; classify 719
GRUPPIERUNG: grouping 720
GRUS (*m* and *n*): smalls; slack; small coal; gravel (see **GRIES**) 721
GRUSEL (*m*): fright; shivering; shuddering 722
GRUSELIG: shuddering; awful 723
GRUSHALTIG: containing dust or small matter 724
GRUSKOHLE (*f*): small coal; smalls (Coal) 725
GRUSS (*m*): greeting; salutation; salute 726
GRÜSSEN: to greet; salute; send one's respects (compliments) to 727
GRÜTZBREI (*m*): porridge; gruel 728
GRÜTZE (*f*): groats; peeled grain; grit; brains 729
GRÜTZSCHLEIM (*m*): gruel 730
G-SALZ (*n*): G-salt (2, 6, 8-Naphthol-disulfonic acid) 731
G-SÄURE (*f*): amidonaphtholmonosulphonic acid, G. acid (see **GAMMASÄURE**) 732
GUAJABENBAUM (*m*): guava (*Psidium guayava*) 733
GUAJACOL (*n*): guaiacol, pyrocatechin monomethylester, protocatechin methylester, monomethylcatechol; $C_7H_8O_2$; Sg. 1.1385; Mp. 28.3°C; Bp. 205°C. 734
GUAJAK (*m*): guaiacum (see **GUAJAKBAUM** and **GUAJAKHARZ**) 735
GUAJAKBAUM (*m*): guaiacum tree (*Guajacum officinale*) 736
GUAJAKHARZ (*n*): guaiacum resin (*Resina guajaci*) (resin from *Guaiacum officinale*); Sg. 1.2; Bp. 85°C. 737
GUAJAKHOLZ (*n*): lignum vitæ; guaiacum wood; Sg. 0.79-1.61 (see **GUAJAKBAUM**) 738
GUAJAKHOLZÖL (*n*): guaiacum wood oil (see **GUAJAKÖL**) 739
GUAJAKINSÄURE (*f*): guaiacinic acid 740
GUAJAKÖL (*n*): guaiacum oil; guaiac (wood) oil; Sg. 0.97 741
GUAJAKOL (*n*): guaiacol (see **GUAJACOL**) 742
GUAJAKOL (*n*), **BALDRIANSAURES-**: guaiacol valerate 743
GUAJAKOLDERIVAT (*n*): guaiacol derivate 744
GUAJAKOLKARBONAT (*n*): guaiacol carbonate (see **DUOTAL**) 745
GUAJAKOL, KOHLENSAURES- (*n*): guaiacol carbonate (see **DUOTAL**) 746
GUAJAKOLSULFONSAUER: guaiacolsulfonate 747
GUAJAKSÄURE (*f*): guaiacic acid 748
GUANIDIN (*n*): guanidine, iminourea; $NHC(NH_2)_2$ 749
GUANIDIN (*n*), **KOHLENSAURES-**: guanidine carbonate 750
GUANIDIN (*n*), **SALPETERSAURES-**: guanidine nitrate 751
GUANIDINSULFOCYANAT (*n*): guanidine (sulfocyanate) thiocyanate 752
GUANIN (*n*): guanine (an animal alkaloid from flesh and guano) (a purine derivate); $C_5H_5N_5O$ 753
GUANO (*m*): guano, fertilizer (decomposed excrement of seabirds; contains 10-15% phosphoric acid) 754
GUANOFABRIKATION (*f*): guano manufacture 755
GUANOFABRIKATIONSANLAGE (*f*): guano manufacturing plant 756
GUANOSUPERPHOSPHAT (*m*): guano superphosphate 757
GUARANIN (*n*): guaranine (see **KAFFEIN**) 758
GUARANTIEREN: to guarantee, secure 759
GUÄTHOL (*n*): (see **BRENZCATECHINMONOÄTHYL-ÄTHER**) 760
GUCKEN: to look; peep; spy 761
GUCKFENSTER (*n*): grille 762
GUCKKASTEN (*m*): camera-obscura 763
GUCKLOCH (*n*): peep-hole; spy-hole 764
GUHR (*f*): guhr; kieselguhr (see **KIESELGUHR**) 765
GUIGNET(S)GRÜN (*n*): emerald green, viridian, Guignet's green; $Cr_2O(OH)_4$ (see **SMARAGD-GRÜN**) 766

GUILLOTINEVERSCHLUSS (*m*) drop-shutter (Photography) 767
GUILLOTINIEREN : to guillotine 768
GUIZOTIE (*f*), **ABESSINISCHE-** : (see **NIGER--PFLANZE**) 769
GUJASANOL (*n*): gujasanol, diethylglycocoll-guaiacol hydrochloride; $OCH_3C_6H_4O.COCH_2N(C_2H_5)_2.HCl$; Mp. 184°C. 770
GÜLLE (*f*) : liquid manure 771
GÜLTIG : valid; legal; binding; good; authentic; current; auriferous (Mineral); applicable (see **GILTIG**) 772
GÜLTIGEN : to make (legal) valid 773
GÜLTIGKEIT (*f*) : validity; availability; legality; applicability 774
GUMMI (*n*): gum; india-rubber; caoutchouc (see also **KAUTSCHUK**) 775
GUMMIARABICUM (*n*): gum arabic; acacia gum; Sg. 1.45 776
GUMMIARABIKUM (*n*): gum arabic (see **GUMMI--ARABICUM**) 777
GUMMI (*n*), **ARABISCHES-** : (see **GUMMIARABI--CUM**) 778
GUMMIART (*f*) : variety (kind; sort) of gum 779
GUMMIARTIG : gum-like; gummy 780
GUMMIARTIKEL (*m*): (india)-rubber (articles, wares) goods 781
GUMMIBAND (*n*) : rubber band; elastic band 782
GUMMI (*n*), **BASSORINHALTIGES-** : (see **TRAGANT**) 783
GUMMIBAUM (*m*): gum (rubber) tree (*Ficus elastica*) 784
GUMMICHT : gummy 785
GUMMIELASTICUM (*n*): (india)-rubber; elastic (see **KAUTSCHUK**) 786
GUMMIEREN : to gum; size 787
GUMMIERMASCHINE (*f*): gumming machine (Paper) 788
GUMMIERSATZ (*m*) : gum substitute 789
GUMMIERZ (*n*) : (see **URANPECHBLENDE**) 790
GUMMIFABRIK (*f*) : rubber factory 791
GUMMIFADEN (*m*): thread of elastic, elastic thread 792
GUMMIFAHNE (*f*) : rubber flag (for precipitates) 793
GUMMIFICHTE (*f*): balsam fir (see **KANADA--BALSAM**) 794
GUMMIFLASCHE (*f*): rubber bottle; hot-water bottle 795
GUMMIG : gummy 796
GUMMIGESCHWULST (*f*) : syphilis (Med.) 797
GUMMIGUT (*n*): gamboge (*Gambogium*) (gum resin of *Garcinia morella*); Sg. 1.22 798
GUMMIGUTT (*n*) : (see **GUMMIGUT**) 799
GUMMIHALTIG : gummy; containing gum 800
GUMMIHARZ (*n*): gum resin (*Gummi resinæ*) (mixture of gum and resin) (see **GUMMIGUT**, **KAUTSCHUK** and **GUTTAPERCHA**) 801
GUMMIHOLZ (*n*): eucalyptus, blue-gum tree, Australian fever tree (*Eucalyptus globulus*) 802
GUMMIKITT (*m*) : india rubber cement 803
GUMMIKNOTEN (*m*): syphilis (Med.) 804
GUMMI (*n*), **KORDOFAN-** : Cordofan gum (see **GUMMIARABICUM**) 805
GUMMILACK (*m*): gum lac; lac; shellac; lacca (resin from the insect, *Coccus lacca*) 806
GUMMILÖSUNG (*f*): gum solution, mucilage (made from gum arabic and distilled water in the proportion of 1 : 2 by weight) (Photo.); india-rubber solution (rubber dissolved in benzol or chloroform) 807
GUMMIPASTA (*f*): Marshmallow paste (see **ALTHEAWURZEL** and **LEDERZUCKER**) 808
GUMMIPASTE (*f*) : (see **GUMMIPASTA**) 809
GUMMIPFLANZE (*f*) : Grindelia (see **GRINDELIEN-KRAUT**) 810
GUMMIPFLASTER (*n*): lead plaster (see **BLEI--PFLASTER**) 811
GUMMIPFROPF (*m*) : rubber stopper 812
GUMMIPFROPFEN (*m*) : (see **GUMMIPFROPF**) 813
GUMMIQUETSCHWALZE (*f*) : (india-rubber)-roller squeegee 814
GUMMIROHR (*n*) : rubber tube 815
GUMMISÄURE (*f*) : gummic (arabic) acid 816
GUMMISCHLAUCH (*m*): rubber (tube; hose) tubing 817
GUMMISCHLEIM (*m*): (gum)-mucilage; acacia mucilage (see also **GUMMIARABICUM**) 818
GUMMISCHUH (*m*) : rubber shoe, galosh 819
GUMMISCHWAMM (*m*) : rubber sponge 820
GUMMI (*n*), **SENEGAL-** : Senegal gum (see **GUMMI--ARABICUM**) 821
GUMMISIRUP (*m*): syrup of acacia (see also **DEXTRIN**) 822
GUMMISTEIN (*m*): hyalite (a hydrous silica); (see **OPAL**) 823
GUMMISTEMPEL (*m*) : rubber stamp 824
GUMMISTOFF (*m*): rubber (substance) material 825
GUMMISTOPFEN (*m*) : rubber stopper 826
GUMMISTRUMPF (*m*): rubber stocking (for varicose veins) 827
GUMMITRAGANT (*m*): gum tragacanth (see **TRAGANT**) 828
GUMMIVERBINDUNG (*f*) : rubber connection 829
GUMMIWAREN (*f.pl*) : rubber goods 830
GUMMIWÄSCHE (*f*): india-rubber collars, fronts and cuffs (made of celluloid) 831
GUMMIWASSER (*n*): gum water (a thin gum used for fixing, in art) 832
GUMMI (*n*), **ZERASINHALTIGES-** : cerasin bearing gum (such as Cherry gum) 833
GUMMÖS : gummy 834
GUNDEL (*m*): wild thyme, creeping thyme (*Thymus vulgaris*) (see also **THYMIAN**) 835
GUNDELBEERE (*f*) : (see **GUNDELREBE**) 836
GUNDELKRAUT (*n*) : (see **GUNDEL**) 837
GUNDELREBE (*f*) : ground ivy (*Glechoma hederacea*) 838
GUNDERMANN (*m*) : (see **GUNDELREBE**) 839
GUNDERMANNKRAUT (*n*): (*herba hederae terrestris*): ground ivy 840
GUNST (*f*) : favour; kindness; goodwill 841
GUNSTBEWERBUNG (*f*): striving to gain favour 842
GUNSTBEZEUGUNG (*f*) : (showing of) favour 843
GUNSTBRIEF (*m*): letter asking (permission) a favour; letter granting (permission) a favour; license, permit, grant 844
GÜNSTIG : favourable; propitious; kind; beneficial; efficient-(ly); advantageous; suitable; appropriate; expedient; friendly 845
GUNSTSCHEIN (*m*) : (see **GUNSTBRIEF**) 846
GUR (*f*) : (see **KIESELGUHR** and **GUHR**) 847
GURDYNAMIT (*n*): guhr (fossil-meal) dynamite 848
GURGEL (*f*): throat; gullet 849
GURGELADER (*f*) : jugular vein 850
GURGELMITTEL (*n*) : gargle (Med.) 851
GURGELN : to gargle; gurgle 852
GURGELWASSER (*n*): gargle 853

GURGIFETT (*n*) : Gurgi fat (from *Garcinia morella*) Mp. 33.5°C. 854

GURJUN (*n*) : wood oil (see **HOLZÖL, CHINESISCHES-**) 855

GURKE (*f*) : cucumber (*Cucumis sativus*); gherkin (*Cucumis dipsaceus*) 856

GURKENGLYZERINSEIFE (*f*) : glycerine and cucumber soap 857

GURKENKERNÖL (*n*) : gurkin-(seed) oil (from *Cucumis sativus*) 858

GURKENKRAUT (*n*) : borage ; bee bread (*Borago officinalis*) 859

GURT (*m*) : girdle ; strap ; belt ; girth ; flange ; boom , chord (Building construction) 860

GURTEISEN (*n*) : boom or flange plate (Building construction) 861

GÜRTEL (*m*) : girdle ; belt ; girth ; sash 862

GÜRTELBAND (*n*) : Zoster, Zona (Med.) ; belt ; zone ; waist-band, girth (see also **HERPES**) 863

GÜRTELROSE (*f*) : Zoster, Zona (Med.) (see **HERPES**) 864

GÜRTEN : to gird ; girdle (see also **GURT**) 865

GURTTRANSPORTEUR (*m*) : belt conveyor 866

GUSS (*m*) : cast ; casting ; founding ; cast iron ; gate (of a mould) ; gutter ; jet (of liquid) ; font (of type) ; mash-liquor (Brewing) ; downpour (Rain) ; spout 867

GUSSBLASE (*f*) : flaw (in a casting) ; air bubble ; curl 868

GUSSBLEI (*n*) : cast lead (see also **BLEI**) 869

GUSSEISEN (*n*) : cast iron (C.I.) ; pig iron ; Sg. 7.03-7.73 ; Mp. 1075-1275°C. 870

GUSSEISENSPÄNE (*m.pl*) : cast iron (chips ; borings ; filings) turnings 871

GUSSEISERN : cast-iron ; of cast-iron 872

GUSSFLASCHE (*f*) : moulding flask 873

GUSSFORM (*f*) : (casting)-mould (see **KOKILLE**) ; (soap)-frame 874

GUSSHAUT (*f*) : casting skin 875

GUSSKASTEN (*m*) : casting (moulding) box 876

GUSSKERN (*m*) : core (of a casting) 877

GUSSLOCH (*n*) : casting hole 878

GUSSMESSING (*n*) : cast brass (" **GELBGUSS** ", which see, but higher zinc content) 879

GUSSMETALL (*n*) : cast metal 880

GUSSMODELL (*n*) : casting (foundry) pattern 881

GUSSMÖRTEL (*m*) : concrete ; cement 882

GUSSMUTTER (*f*) : matrix 883

GUSSNAHT (*f*) : casting (seam) burr 884

GUSSNARBE (*f*) : fault, flaw, curl (in casting) 885

GUSSPFANNE (*f*) : (casting)-ladle (Metal) 886

GUSSPUTZMASCHINE (*f*) : casting cleaning machine 887

GUSSREGEN (*m*) : downpour-(of rain) 888

GUSSATZ (*m*) : set of castings 889

GUSSCHADEN (*m*) : fault, flaw, defect (in casting) 890

GUSSCHALE (*f*) : (see **KOKILLE**) 891

GUSSCHLICKER (*m*) : (casting)-slip (Ceramics) 892

GUSS (*m*), **SCHMIEDBARER-** : malleable casting 893

GUSSCHROT (*m*) : casting scrap 894

GUSSPÄNE (*m.pl*) : cast iron (chips ; borings ; filings) turnings 895

GUSSPANNUNG (*f*) : casting stress, stress set up during the casting of metal 896

GUSSTAHL (*m*) : cast steel (C.S.) (see **STAHL**) 897

GUSSTEIN (*m*) : sink ; drain 898

GUSSTRUKTUR (*f*) : casting structure (characterized by coarse crystals, state of steel after it has solidified and cooled without afterwards being worked or heat-treated) 899

GUSSTÜCK (*n*) : cast-(ing) 900

GUSSTÜCK (*n*), **LEICHTES-** : light casting 901

GUSSTÜCK (*n*), **SCHWERES-** : heavy casting 902

GUSSTIEGEL (*m*) : casting ladle 903

GUSSTRICHTER (*m*) : casting funnel 904

GUSSWAREN (*f.pl*) : castings 905

GUSSWERK (*n*) : cast, casting, cast metal-work 906

GUT : good ; well ; efficient ; excellent ; kind ; generous ; beneficent ; (*n*) : good (valuable ; useful)-material ; possession ; property ; estate ; goods ; ware ; goods (materials) to be treated 907

GUTACHTEN (*n*) : opinion ; judgement ; decision ; verdict ; advice 908

GUTARTIG : good-natured ; mild ; good-hearted 909

GUTBERATEN : well-advised 910

GUTBRINGEN : to credit, place to the credit of 911

GÜTE (*f*) : goodness ; worth ; quality ; efficiency; excellence ; kindness ; favour ; beneficence ; generosity ; value 912

GÜTEGRAD (*m*) : efficiency (usually total) (η) 913

GÜTER (*n.pl*) : goods ; wares ; commodities 914

GÜTERABSCHÄTZUNG (*f*) : valuation of goods 915

GÜTERADRESSE (*f*) : goods address 916

GÜTERANSCHLAG (*m*) : valuation (of goods) 917

GÜTERBAHNHOF (*m*) : goods station 918

GÜTEREXPEDITION (*f*) : forwarding (receiving) office or agency 919

GÜTERHANDEL (*m*) : Estate Agency-(business) 920

GÜTERVERSICHERUNG (*f*) : insurance -(of goods) 921

GÜTERVERTRETER (*m*) : trustee 922

GÜTERWAGEN (*m*) : goods (truck) waggon (Railways) ; lorry (Road transport) 923

GÜTERWAGEN (*m*), **GEDECKTER-** : covered goods waggon, baggage waggon, luggage van 924

GÜTERWAGEN (*m*), **OFFENER-** : open truck (Railways, etc.) 925

GÜTERZUG (*m*) : goods (freight) train 926

GÜTERZUGLOKOMOTIVE (*f*) : goods locomotive, goods engine 927

GÜTEZIFFER (*f*) : (see **QUALITÄTSKOEFFIZIENT**) 928

GUTGEWICHT (*n*) : good (fair) weight ; allowance 929

GUTHABEN (*n*) : balance ; credit ; account ; (outstanding)-claim 930

GUTHEISSEN : to endorse, approve, consent ; (*n*) : endorsement, approbation, consent 931

GÜTIG : good ; kind ; gracious ; indulgent ; charitable 932

GÜTLICH : friendly ; amicable 933

GUTMÜTIG : good-natured ; good-hearted ; kind 934

GUTSAGEN : to stand security, go bail for ; (*n*) : security, bail, surety 935

GUTSBESITZER (*m*) : land (estate) owner, landed proprietor 936

GUTSCHEIN (*m*) : Treasury receipt (in connection with post-war export tax from Germany) 937

GUTSCHMECKER (*m*) : epicure 938

GUTSCHREIBEN : to credit ; place to the credit of 939

GUTSCHRIFT (*f*) : credit-(note) 940

GUT (*n*), **STENGE-** : topmast rigging (Naut.) 941

GUTSTEUER (*f*) : property tax 942

GUTTAPERCHA (*n*) : guttapercha (latex or milk-juice from trees of the family, *Sapotaceæ*) ; Sg. 0.98 ; Mp. 120°C. 943

GUTTAPERCHAFLASCHE (*f*) : guttapercha bottle 944

GUTTAPERCHA (*n*), GEREINIGTES- : refined guttapercha 945
GUTTAPERCHAHARZ (*n*): guttapercha resin (obtained by extraction from guttapercha) 946
GUTTAPERCHA (*n*), ROHES-: crude guttapercha 947
GUTTAPERCHAWAREN (*f.pl*): guttapercha (wares) goods 948
GUTTAPERCHAZAHNKITT (*m*): guttapercha (dental filling ; dental stopping) tooth cement 949
GUTTI (*n*): (see GUMMIGUT) 950
GUTWILLIG : willing ; obliging ; voluntary 951
GUTZIEHEND : well-drawing ; having a good draught (Furnaces) 952
GYMNASIUM (*n*): college ; gymnasium in Germany ; grammar school ; high school ; preparatory school for the University 953
GYMNEMASÄURE (*f*): gymnemic acid (from *Gymnema silvestris*) 954
GYMNIT (*m*): gymnite ; $Mg_4Si_3O_{10}.6H_2O$; Sg. 2.15 955
GYMNOSPERMEN (*pl*): *Gymnospermæ* (Bot.) 956
GYNOKARDIAÖL (*n*): gynocardia oil (believed to be the same as Chaulmoogra oil, but not certain) (see CHAULMOOGRAÖL) 957
GYNOVAL (*n*): gynoval (a valerian preparation) 958
GYPS (*m*): gypsum ; plaster of Paris (see GIPS) 959
GYROMETER (*n*): gyrometer, speed meter for liquids 960

H

HAAR (*n*): hair ; pile ; nap ; filament ; wool 961
HAARADER (*f*): capillary vein 962
HAARALAUN (*m*): capillary alum 963
HAARARTIG : capillary ; hair-like 964
HAARBALG (*m*): hair follicle 965
HAARBEIZE (*f*): depilatory 966
HAARBÜSCHEL (*m*): tuft-(of hair) 967
HAARDRAHT (*m*): hair wire, very fine wire 968
HAAREISEN (*n*): (see PÄLLISEN) 969
HAAREN : to scrape off (remove) hair, fall out (lose) hair, depilate 970
HAARERZ (*n*): capillary ore 971
HAARFARBE (*f*): hair-dye ; colour of hair 972
HAARFÄRBEMITTEL (*n*): hair dye 973
HAARFARN (*m*): Maiden-hair fern (*Adiantum capillus Veneris*) 974
HAARFASER (*f*): filament 975
HAARFASERIG : filamentous ; capillary 976
HAARFEDER (*f*): hair spring 977
HAARFILZ (*m*): hair felt 978
HAARFLECHTE (*f*): braid, plaited hair 979
HAARFLUG (*m*): lichen 980
HAARFÖRMIG : capillary ; hair-shaped ; hair-like ; capilliform 981
HAARGEFÄSS (*n*): capillary vessel 982
HAARICHT : hair-like 983
HAARIG : hairy ; of hair ; pilous ; pilose 984
HAARKANAL (*n*): capillary duct 985
HAARKIES (*m*): capillary pyrites ; millerite (see MILLERIT) 986
HAARKUPFER (*n*): capillary copper 987
HAARLEMER ÖL (*n*): Haarlem oil ; Dutch drops (sulphurated and terebinthinated linseed oil) 988
HAARLOS : hairless ; bald ; napless (Fabrics) 989
HAARNADEL (*f*): hairpin 990
HAARÖL (*n*): hair-oil 991
HAARPFLANZE (*f*): capillary plant 992
HAARPFLEGE (*f*): care of the hair 993
HAARPFLEGEMITTEL (*n*): hair restorer, hair preserver 994
HAARPOMADE (*f*): hair pomatum 995
HAARRISS (*m*): (hair)-crack ; craze 996
HAARRISSIG : crazed 997
HAARROHR (*n*): capillary tube 998
HAARRÖHRCHEN (*n*): capillary tube 999
HAARRÖHRE (*f*): (see HAARROHR) 000
HAARRÖHRENANZIEHUNG (*f*): capillary attraction 001
HAARSÄCKCHEN (*n*): hair follicle 002
HAARSALZ (*n*): hair salt (natural hydrous sulphate of iron and aluminium), halotrichite, fibrous alunogen ; $FeSO_4.Al_2(SO_4)_3.24H_2O$; Sg. 1.65 003
HAARSCHABER (*m*): scraping knife (Tanning) 004
HAARSCHARF : extremely (sharp) exact 005
HAARSCHUPPEN (*f.pl*): scurf-(of the head) 006
HAARSCHWEFEL (*m*): fibrous native sulphur 007
HAARSCHWEIF (*m*): coma ; tail (of a comet) 008
HAARSEIL (*n*): seton (a thread of silk or other material to keep a wound open) 009
HAARSIEB (*n*): hair sieve 010
HAARSILBER (*n*): capillary silver ; filamentous natural silver 011
HAARSTRANG (*m*): sulphur wort (*Peucedanum officinale*) ; Haarstrang (Geology) ; seton (Surgery) 012
HAARSTRANGWURZEL (*f*): peucedanum root (root of *Peucedanum officinale*) 013
HAARTUCH (*n*): hair-cloth 014
HAARVERTILGUNGSMITTEL (*n*): depilatory 015
HAARVITRIOL (*m*): capillary epsomite (see MAGNESIUMSULFAT and BITTERSALZ) 016
HAARWACHS (*n*): pomatum 017
HAARWASSER (*n*): hair-wash ; shampoo 018
HAARWEIDE (*f*): osier, rod (used for making baskets) (*Salix triandra*) 019
HAARWURZEL (*f*): hair root 020
HAARZANGE (*f*): tweezers 021
HABE (*f*): property ; goods ; chattels ; possessions ; effects ; wealth 022
HABEN : to have ; hold ; possess ; be ; (*n*): credit (Com.) 023
HABGIER (*f*): covetousness ; greed ; avarice 024
HABHAFT : possessing 025
HABHAFT WERDEN : to become possessed of ; obtain 026
HABICHT (*m*): hawk 027
HABICHTS INSELN (*f.pl*): the Azores (a group of islands in the Atlantic, off Portugal) 028
HABICHTSKRAUT (*n*): hawkweed (*Hieracium*) 029
HABILITATION (*f*): habilitation ; admission to a faculty 030
HABILITATIONSSCHRIFT (*f*): inaugural lecture 031
HABILITIEREN : to habilitate 032
HABITUS (*m*): habit 033
HABRONEMERZ (*n*), PRISMATOIDISCHES- : Göthite 034
HABRONEMERZ (*n*), UNTEILBARES- : brown hematite, limonite, limnite 035

HABSCHAFT (*f*): (see **HABE**) 036
HABSELIGKEIT (*f*): (see **HABE**) 037
HABSUCHT (*f*): avidity; avarice; covetousness; greed-(iness) 038
HACK (*m*): hack; stroke 039
HACKAXT (*f*): chopper, chopping axe 040
HACKBANK (*f*): (chopping)-block 041
HACKBEIL (*n*): chopper 042
HACKBLOCK (*m*): (chopping)-block 043
HACKBRETT (*n*): chopping board, mincing board 044
HACKE (*f*): hoe; mattock; pick-(axe); hatchet; heel 045
HACKEN: to hack; hoe; chop; mince; (*m*): heel 046
HACKENFÖRMIG: barked, uncinate (Bot.) 047
HACKENLEDER (*n*): heel leather 048
HACKER (*m*): cutter, chopper 049
HÄCKERLING (*m*): chopped straw 050
HÄCKERLINGSBANK (*f*): chaff-cutter 051
HÄCKERLINGSMASCHINE (*f*): chaff-cutter 052
HACKE (*f*), **RUDER-**: rudder heel 053
HACKMASCHINE (*f*): hacking (cutting, chopping) machine 054
HACKSÄGE (*f*): hack-saw 055
HÄCKSELMASCHINE (*f*): chaff (cutting machine) cutter 056
HADER (*m*): rag (see also **LUMPEN**); strife; dispute; quarrel; brawl 057
HADEREI (*f*): wrangling; quarrelling; brawling 058
HADERER (*m*): brawler, wrangler 059
HADERIG: short (of iron) 060
HADERLUMP (*m*): rag-man 061
HADERLUMPEN (*m.pl*): rags 062
HADERN: to quarrel, wrangle, dispute, brawl 063
HADERNDRESCHER (*m*): rag beater (Paper); rag-cleaning machine 064
HADERNKOCHER (*m*): rag boiler (Paper) 065
HADERNLADE (*f*): rag chest 066
HADERNLAGER (*n*): rag (stock) dump 067
HADERNPAPIER (*n*): rag paper (Paper) 068
HADERNREINIGUNG (*f*): rag-cleaning (by wet or dry process) (Paper) 069
HADERNSCHNEIDER (*m*): rag-cutter 070
HADERNSTANZER (*m*): rag-cutter (with vertical knives) (Paper) 071
HADERNSTÄUBER (*m*): rag-cleaner (for cleaning the cut rags) (Paper) 072
HADERNSURROGAT (*n*): rag substitute (such as straw, esparto grass, wood, etc.) (Paper) 073
HADERSCHNEIDER (*m*): rag-cutter 074
HADERSÜCHTIG: brawling, quarrelsome 075
HADRIG: (see **HADERIG**) 076
HAEMORRHOIDEN (*pl*): hæmorrhoids, piles (Med.) 077
HAFEN (*m*): harbour; port; haven; pot 078
HAFENABGABE (*f*): harbour dues, port dues 079
HAFENANKER (*m*): moorings 080
HAFENBAUM (*m*): bar 081
HAFENDAMM (*m*): pier, jetty; breakwater 082
HAFENGAT-(T) (*m*): harbour-mouth 083
HAFENGELD (*n*): anchorage, harbour dues 084
HAFENKAI (*n*): quay 085
HAFENRÄUMER (*m*): dredger 086
HAFER (*m*): oats (*Avena sativa*) 087
HAFERBAU (*m*): cultivation of oats 088
HAFERBREI (*m*): gruel, oatmeal (gruel) porridge 089
HAFERDISTEL (*f*): common (field) thistle 090
HAFERGRÜTZE (*f*): groats 091
HAFERKAKAO (*m*): oat cocoa 092
HAFERMEHL (*n*): oatmeal 093
HAFERÖL (*n*): oat oil (from *Avena sativa*) 094
HAFERSCHLEHE (*f*): (see **KRIECHE**) 095
HAFERSCHLEIM (*m*): gruel; oatmeal gruel; porridge 096
HAFERSEIM (*m*): (see **HAFERSCHLEIM**) 097
HAFERSPELZ (*m*): oat spelt 098
HAFERSPELZTEER (*m*): oat-spelt tar (extracted from oat spelt by dry distillation) 099
HAFERSPREU (*f*): oat-chaff 100
HAFF (*n*): gulf, bay 101
HAFNER (HÄFNER) (*m*): potter 102
HAFNERERZ (*n*): potter's ore; galena; alquifou (see **BLEIGLANZ**) 103
HAFNIUM (*n*): Hafnium (an element) 104
HAFT (*f*): custody; arrest; confinement; prison; imprisonment; (*m*): clasp; hold; brace 105
HAFTBAR: liable; responsible; bound 106
HAFTBEFEHL (*m*): warrant (of arrest) 107
HAFTBRIEF (*m*): warrant-(of arrest) 108
HAFTDAUER (*f*): term of imprisonment 109
HAFTEN: to adhere; cling; stick; be liable; grip; hold; answer for; cleave to; (*n*): adhering; adhesion 110
HAFTFESTIGKEIT (*f*): adhesive resistance 111
HAFTPFLICHT (*f*): liability, responsibility 112
HAFTSITZ (*m*): wringing fit 113
HAFTSPANNUNG (*f*): adhesive force 114
HAFTUNG (*f*): liability; security; surety; bail 115
HAG (*m*): hedge; fence; pile (of tiles); bush; grove; meadow; coppice 116
HAGEBUCHE (*f*): hornbeam (see **HORNBAUM**) 117
HAGEBUTTE (*f*): hip, haw 118
HAGEBUTTENSCHALEN (*f.pl*): hip (haw) peels (*Fructus cynosbati*) 119
HAGEDORN (*m*): hawthorn (*Crataegus oxyacantha*) 120
HAGEL (*m*): hail; small shot; grape shot 121
HAGELN: to hail 122
HAGELSTURM (*m*): hail storm 123
HAGELWETTER (*n*): hail storm 124
HAGER: haggard; thin; meagre; slender; lean; gaunt; lank 125
HAGEROSE (*f*): dog-rose 126
HAHN (*m*): cock; stop-cock; tap; faucet; rooster 127
HAHNENBALKEN (*m*): roost; perch 128
HAHNENFUSS (*m*): crowfoot, ranunculus (Bot.); cock's-foot; scribble (of writing) 129
HAHNENKAMM (*m*): cock's comb 130
HAHNSCHLÜSSEL (*m*): cock key 131
HAHNSITZ (*m*): cock seating 132
HAHNSTOPFEN (*m*): cock stopper 133
HAI (*m*): shark (*Hypoprian brevirostris*); dog-fish (*Squalus acanthius*) 134
HAIFISCH (*m*): (see **HAI**) 135
HAIFISCHLEBERTRAN (*m*): shark-liver oil; Sg. 0.93 136
HAIFISCHTRAN (*m*): shark oil; dog-fish oil; Sg. 0.92 137
HAIN (*m*): grove; wood; thicket 138
HAINAMPFER (*m*): wood sorrel (*Oxalis acetosella*) 139
HAINBUCHE (*f*): hornbeam (see **HORNBAUM**) 140
HÄKCHEN (*n*): little hook; clasp; crochet 141
HÄKEL (*n*): crochet-hook 142
HÄKELARBEIT (*f*): crochet-work 143
HÄKELGARN (*n*): crochet thread 144
HAKELKRAUT (*n*): Pasque flower, wind flower, pulsatilla (*Anemone pulsatilla*) 145

HÄKELN: to crochet; hook 146
HAKEN (*m*): hook; clasp; catch; clamp; claw; clutch; rabble (Puddling); grappling iron; difficulty; (as a verb=to hook; grapple) 147
HAKENFÖRMIG: hooked; hooklike 148
HAKENHAUE (*f*): mattock 149
HAKENKEIL (*m*): nose (key) wedge 150
HAKENKETTE (*f*): hooked-link chain (chain composed of hooked links) 151
HAKENPROBE (*f*): hook test (Sugar) 152
HAKENSCHLÜSSEL (*m*): hooked key, pick-lock, hooked spanner 153
HAKIG: hooked 154
HALB: half 155
HALBALAUN (*m*): impure alum 156
HALBALDEHYD (*n*): semi-aldehyde 157
HALBART (*f*): sub-species 158
HALBATLAS (*m*): satinet 159
HALBBALKEN (*m*): half beam 160
HALBBASISCH: semi-basic 161
HALBBILD (*n*): half-length portrait 162
HALBBILDUNG (*f*): semi-(civilization, culture) education 163
HALBBLEICHE (*f*): half bleach 164
HALBBÜRTIG: half-blood 165
HALBCHLORSCHWEFEL (*m*): sulphur (subchloride) monochloride; (see **SCHWEFELCHLORÜR**) 166
HALBCYLINDRISCH: semi-cylindrical 167
HALBDECK (*n*): half deck (Naut.) 168
HALBDECKFLÜGLER (*m*): (see **HALBFLÜGLER**) 169
HALBDUNKEL (*n*): twilight; gloaming; dusk; dawn; gloom; semi-darkness 170
HALBDURCHLÄSSIG: semi-permeable 171
HALBDURCHMESSER (*m*): radius 172
HALBDURCHSICHTIG: semi-transparent 173
HALBE (*n*): half 174
HALBEDELSTEIN (*m*): semi-precious stone 175
HALBEIRUND: semi-oval; semi-ovoid 176
HALBEN: on account of; on behalf of; by (reason) virtue of 177
HALBEN: to halve, bisect 178
HALBER: (see **HALBEN**) 179
HALBER STICH (*m*): half hitch 180
HALBFABRIKAT (*n*): intermediate product; semi-manufacture 181
HALBFLÄCHNER (*m*): hemihedron 182
HALBFLÜGLER (*m.pl*): hemiptera; Rhynchota; bug 183
HALBFLÜSSIG: semi- (liquid) fluid 184
HALBFRANZBAND (*m*): half -(calf) leather binding for books 185
HALBGAR: half-cooked, underdone 186
HALBGAS (*n*): semi-gas 187
HALBGASFEUERUNG (*f*): semi-gas furnace 188
HALBGESICHT (*n*): side-face, profile 189
HALBGOTT (*m*): demi-god 190
HALBGUT (*n*): tin with high lead content; tin-lead alloy 191
HALBHARZ (*n*): crude resin 192
HALBHEIT (*f*): imperfection; indecision; incompleteness; superficiality 193
HALBHOLLÄNDER (*m*): washing engine (Paper) 194
HALBIEREN: to halve; bisect 195
HALBIERTES ROHEISEN (*n*): mottled pig iron (2.5% C and 1% Si); (mean between white and grey fracture pig iron) 196
HALBIERUNGSPERIODE (*f*): (see **RADIOAKTIVE PERIODE**) 197
HALBIERUNGSPUNKT (*m*): point of bisection 198
HALBINSEL (*f*): peninsula 199
HALBJAHR (*n*): half year 200
HALBJÄHRLICH: half-yearly; every six months 201
HALBKOKS (*m*): coke breeze, semi-coke 202
HALBKOLLOID (*n*): semi-colloid 203
HALBKREIS (*m*): semi-circle 204
HALBKREISFÖRMIG: semi-circular 205
HALBKUGEL (*f*): hemisphere 206
HALBKUGELIG: hemispherical 207
HALBLEDERBAND (*m*): half leather (binding) bound book 208
HALBMAHLEN: to break up roughly; half-crush; half-grind 209
HALBMAST: half-mast 210
HALBMATTGLASUR (*f*): semi-matt glaze 211
HALBMECHANISCH: semi-mechanical, semi-automatic 212
HALBMESSER (*m*): radius 213
HALBMETALL (*n*): semi-metal 214
HALBMONATLICH: fortnightly; every two weeks 215
HALBMOND (*m*): half moon 216
HALBNASS: semi-wet 217
HALBNITRIL (*n*): semi-nitrile 218
HALBNORMAL: semi-normal 219
HALBOPAL (*m*): semi-opal; (see **OPAL**) 220
HALBORTHOOXALSÄURE (*f*): semi-ortho-oxalic acid 221
HALBPART (*m*): halves 222
HALBPORZELLAN (*n*): semi-porcelain 223
HALBRACEMISCH: semi-racemic 224
HALBRUND: half-round (Metal); semi-circular 225
HALBRUNDEISEN (*n*): half-round (iron) beading 226
HALBRUNDFEILE (*f*): half-round file 227
HALBSÄURE (*f*): oxide 228
HALBSCHATTEN (*m*): half-shadow, half-shade; mezzotint 229
HALBSCHATTENAPPARAT (*m*): half- (shade) shadow apparatus (Polarization) 230
HALBSCHATTENSYSTEM (*n*): half- (shade) shadow system (Polarization) 231
HALBSCHEID (*f*): moiety, half 232
HALBSCHLAG (*m*): mixed breed; mongrel 233
HALBSEIDE (*f*): half-silk (a mixture of silk and wool) 234
HALBSEITE (*f*): half (page, side) sheet; column 235
HALBSOPRAN (*m*): mezzosoprano 236
HALBSTOFF (*m*): half stuff (Paper) 237
HALBSTOFFHOLLÄNDER (*m*): breaker (Paper) 238
HALBTIEF: shallow; half the depth 239
HALBTINTE (*f*): mezzotint 240
HALBTON (*m*): half-tone (Art) 241
HALBTONVERFAHREN (*n*): half-tone process 242
HALBTROCKEN: half dry; semi-dry 243
HALBTUCH (*n*): (see **KIRSEI**) 244
HALBVERDECK (*n*): quarter-deck 245
HALBWALZE (*f*): granulating roll (Printing) 246
HALBWASSERGAS (*n*): semi-water-gas, Dowson gas, mixed gas, Mond gas 247
HALBWATTLAMPE (*f*): half-watt lamp 248
HALBWEGS: half-way 249
HALBWEICH: half (semi-; partly) soft 250
HALBWERTSDICKE (*f*): the thickness of a substance, through which the rays of a radioactive substance are reduced to half 251
HALBWERTSPERIODE (*f*): (see **RADIOAKTIVE PERIODE**) 252
HALBWERTSZEIT (*f*): (see **RADIOAKTIVE PERIODE**) 253

HALBWOLLE (*f*): half-wool (a mixture of wool and cotton) 254
HALBWOLLFARBSTOFF (*m*): colouring matter for half wool 255
HALBZEUG (*n*): half stuff (Paper); semi-finished material 256
HALBZINN (*n*): tin with high lead content; tin-lead alloy 257
HALBZIRKEL (*m*): semi-circle 258
HALDE (*f*): heap; dump (of refuse, coal or ash); slope; declivity 259
HALDENSCHLACKE (*f*): waste slag; slag from a slag heap 260
HALFA (*f*): esparto grass 261
HÄLFTE (*f*): half; moiety 262
HALFTER (*f*): halter 263
HALIT (*m*): halite, rock salt, common salt (sodium chloride); NaCl; Sg. 2.15-2.35 264
HALL (*m*): sound; peal; resonance; clang 265
HALLE (*f*): hall; porch; vestibule; portico 266
HALLEN: to sound; resound; reverberate 267
HALLOYSIT (*m*): Halloysite (natural hydrous aluminium silicate of a clayey kind); $Al_2O_3SiO_2.xH_2O$; Sg. about 2.0 268
HALLUZINATION (*f*): hallucination 269
HALM (*m*): blade; stalk; straw; haulm 270
HALMFRÜCHTE (*f.pl*): cereals 271
HALMKNOTEN (*m*): joint in a straw 272
HALMLESE (*f*): gleanings 273
HALOCHEMIE (*f*): chemistry of salts 274
HALOGEN: halogen; halide of; (*n*): halogen 275
HALOGENALKYL (*n*): alkyl halide 276
HALOGENATOM (*n*): halogen atom 277
HALOGENBENZOESÄURE (*f*): halobenzoic acid 278
HALOGENID (*n*): halide 279
HALOGENIEREN: to halogenate; halinate 280
HALOGENIERUNG (*f*): halogenation; halination 281
HALOGENINDIGO (*m*): halogen indigo 282
HALOGENKOHLENSTOFF (*m*): carbon halide 283
HALOGENMETALL (*n*): metallic halide 284
HALOGENPYRIDIN (*n*): halopyridine 285
HALOGENQUECKSILBER (*n*): mercury halide 286
HALOGENSCHWEFEL (*m*): sulphur halide 287
HALOGENSILBER (*n*): silver halide 288
HALOGENSUBSTITUIERTES CHINON (*n*): haloquinone 289
HALOGENÜR (*n*): halide 290
HALOGENVERBINDUNG (*f*): halogen compound 291
HALOGENWASSERSTOFF (*m*): hydrogen halide 292
HALOGENWASSERSTOFFSÄURE (*f*): hydrohalic acid 293
HALOID (*n*): halide; haloid salt; (metallic salt of a halogen) 294
HALOIDSALZ (*n*): (see **HALOID**) 295
HALOTRICHIT (*m*): halotrichite (see **HAARSALZ**) 296
HALOXYLEN (*n*): haloxylene 297
HALS (*m*): neck; throat; stem; collar; tack (Naut.) 298
HALSADER (*f*): jugular vein (Anat.) 299
HALSBAND (*n*): necklace; collar; (neck)-tie; cravat 300
HALSBEIN (*n*): collarbone; clavicle (Anat.) 301
HALS (*m*), **BESAN-**: spanker tack (Naut.) 302
HALSBINDE (*f*): (neck)-tie; cravat; neckerchief 303
HALSBLOCK (*m*): tack block (Naut.) 304
HALSBOLZEN (*m*): throat bolt (Naut.) 305
HALSBRÄUNE (*f*): quinsy; tonsilitis; Angina pectoris (a throat affection) (Med.) 306
HALSBRECHEND: neck-breaking; dangerous 307
HALSE (*f*): tack (of sails) (Naut.) 308
HALSE (*f*), **BESAN-**: spanker tack (Naut.) 309
HALSEN: to wear; embrace 310
HALSENTZÜNDUNG (*f*): sore-throat, inflammation of the throat (Med.) 311
HALSGESCHWÜR (*n*): abscess of the neck or throat 312
HALSLAGER (*n*): neck bearing, journal 313
HALSMANDEL (*f*): tonsil (Anat.) 314
HALSPULSADER (*f*): carotid artery (Anat.) 315
HALSRÖHRE (*f*): wind-pipe; trachea (Anat.) 316
HALSSCHLINGE (*f*): noose 317
HALSSTARRIG: headstrong, stiff-necked, obstinate, stubborn 318
HALSSTARRKRAMPF (*m*): stiff neck 319
HALSSUCHT (*f*): bronchitis (inflammation of the bronchi); bronchiectasis (morbid dilation of the bronchi) (Med.) 320
HALSTUCH (*n*): cravat; neckerchief 321
HALSWEH (*n*): pain in the throat 322
HALSWIRBEL (*m*): cervical vertebra (Anat.) 323
HALSZÄPFLEIN (*n*): uvula (Anat.) 324
HALT (*m*): hold; holding; halt; stop; firmness; yield; support; purchase 325
HALTBAR: stable; strong; durable; fast; permanent; tenable; valid 326
HALTBARKEIT (*f*): keeping quality; stability; durability; permanence; steadfastness; tenacity; life 327
HALTEN: to hold; keep; deliver (an address); take; hold on; remain; last; stop; include; retain; celebrate; maintain; resist, detain; endure; contain; value; support; deem; constrain 328
HALTEN, DEN KURS-: to hold (keep) the course (Naut.) 329
HALTEN, SCHADLOS-: to indemnify 330
HALTEPUNKT (*m*): clue; stopping place 331
HALTER (*m*): holder; support; handle; hold; keeper; receptacle 332
HÄLTER (*m*): holder; reservoir; receptacle 333
HALTERARM (*m*): support-(ing arm) 334
HALTESTELLE (*f*): stopping place, halt 335
HALTEZEIT (*f*): (duration of)-halt; pause 336
HALTIG: rich (Ore); containing; holding; yielding; bearing 337
HALTIG, GOLD-: auriferous; containing gold 338
HALTLOS: unstable; unsteady; loose; infirm; untenable; tender (Paper) 339
HALTUNG (*f*): attitude; bearing; state; condition; harmony; support; principle; holding; maintenance; demeanour; mien 340
HAMAMELIS (*f*): witch hazel (bark and leaves) (*Cortex et folia hamamelidis virginianæ*) (see **ZAUBERHASEL**) 341
HÄMATIN (*n*): haematin; $C_{32}H_{30}N_4FeO_3$ 342
HÄMATIT (*m*): hematite; haematite; bloodstone; specular iron ore; red iron ore (natural iron oxide; Fe_2O_3; about 70% iron) (see also **ROTEISENSTEIN** and **EISENOXYD**); hematite (foundry pig iron, having low phosphorous content) 343
HÄMATITROHEISEN (*n*): hematite pig iron 344
HÄMATOGEN (*n*): Haematogen (a food preparation) 345
HÄMATOPAN (*n*): haematopan (a hæmoglobin food preparation) 346

HÄMATOXYLIN (*n*): haematoxylin, $C_{16}H_{14}O_6.3H_2O$; Mp. 140°C.; logwood (see BLAU-HOLZ) 347
HAMBERGIT (*m*): hambergite; $Be_2(OH)BO_3$; Sg. 2.35 348
HAMBURGERBLAU (*n*): Hamburg blue (see BERLINERBLAU and KUPFERCARBONAT) 349
HÄMISCH: malicious; spiteful 350
HAMMATSCH (*m*): sodium bicarbonate (Brew.) (see NATRIUM, DOPPELKOHLENSAURES)- 351
HAMMEL (*m*): wether; mutton 352
HAMMELFETT (*n*): mutton fat; Sg. 0.92 353
HAMMELFLEISCH (*n*): mutton 354
HAMMELPELZ (*m*): sheepskin 355
HAMMELTALG (*m*): mutton (tallow) suet; Sg. 0.937-0.961; Mp. 44-51°C. 356
HAMMER (*m*): hammer; forge 357
HAMMERBAHN (*f*): hammer face 358
HÄMMERBAR: malleable 359
HAMMERBÄR (*m*): hammer head 360
HÄMMERBARKEIT (*f*): malleability 361
HAMMEREISEN (*n*): wrought iron (W.I.) 362
HAMMERFINNE (*f*): pane of a hammer 363
HAMMERGAR: tough (Copper); tough-pitch 364
HAMMERGARES KUPFER (*n*): tough (tough-pitch) copper; refined copper (99.5-99.8% copper) 365
HAMMER (*m*), HOLZ-: mallet 366
HAMMERINDUKTOR (*m*): hammer inductor 367
HÄMMERN: to hammer; forge 368
HAMMERSCHLAG (*m*): hammer scale, iron scale; (60-75% Fe) (steelworks process byproduct); stroke with a hammer 369
HAMMERSCHMIED (*m*): hammer smith; forge (foundry) smith 370
HAMMERSCHMIEDESTÜCK (*n*): drop-forging 371
HAMMERSTIEL (*m*): haft (handle) of a hammer 372
HAMMERUNTERBRECHER (*m*): hammer interrupter 373
HAMMER (*m*), WAGNERSCHER-: Wagner's interruptor 374
HAMMERWERK (*n*): iron foundry, ironworks, forge 375
HÄMOCHININ (*n*): hemoquinine 376
HÄMOGALLOL (*n*): haemogallol (a blood preparation given in cases of chlorosis) 377
HÄMOGLOBIN (*n*): hemoglobin (main constituent of red corpuscles of blood) 378
HÄMOL (*n*): haemol (a medicine prescribed for chlorosis) 379
HÄMORRHOIDAL-(ISCH): haemorrhoidal 380
HÄMORRHOIDALMITTEL (*n*): haemorrhoid (pile) cure or remedy 381
HÄMORRHOIDEN (*n.pl*): hemorrhoids, piles 382
HAMSTER (*m*): German marmot 383
HAMSTERVERTILGUNGSMITTEL (*n*): German marmot (destroyer; destroying agent) killer 384
HAND (*f*): hand; side 385
HAND (*f*), AN-: by means of; due to; on account of; judging (from) by 386
HANDARBEIT (*f*): hand work; manual labour; handiwork; handicraft; needlework 387
HANDARBEITER (*m*): manual labourer; mechanic 388
HAND (*f*), AUS FREIER-: freehand 389
HANDAUSGABE (*f*): pocket edition 390
HANDBEIL (*n*): hatchet 391
HANDBESCHICKT: hand-fired (of furnaces); hand-fed, supplied by hand 392
HANDBETRIEB (*m*): hand (drive) power, hand operation 393
HANDBOHRER (*m*): gimlet, awl 394
HANDBREMSE (*f*): hand brake 395
HANDBUCH (*n*): manual; handbook; compendium 396
HANDDRUCK (*m*): hand printing (see KLOTZ-DRUCK); hand pressure, squeeze of the hand, hand grip 397
HANDEIMER (*m*): pail 398
HANDEL (*m*): commerce; trade; business; affair; transaction; action (Law); quarrel; traffic 399
HANDEL (*m*), IN DEN- BRINGEN: to put on the market, bring on to the market 400
HANDELN: to act; trade; treat; deal; bargain 401
HANDELSABGABE (*f*): duty 402
HANDELSAMT (*n*): Board of Trade 403
HANDELSANGELEGENHEIT (*f*): commercial (business; trade) matter or affair 404
HANDELSANTIMON (*n*): commercial antimony [98-99% Sb. silverwhite and (shining) glittering] 405
HANDELSARTIKEL (*m*): commercial article; commodity 406
HANDELSAUSSCHUSS (*m*): trade committee 407
HANDELSBANK (*f*): merchant bank 408
HANDELSBEFLISSEN: studying (trade) business 409
HANDELSBEFLISSENER: (*m*) apprentice (to a trade); junior clerk 410
HANDELSBENZOL (*n*): commercial benzene (see BENZOL) 411
HANDELSBERICHT (*m*): trade (commercial) report 412
HANDELSBETRIEB (*m*): traffic; commerce; business 413
HANDELSBLEI (*n*): commercial lead (99.99% lead) (bluish white in colour) 414
HANDELSBRIEF (*m*): commercial (business) letter 415
HANDELSBÜNDNIS (*n*): commercial treaty 416
HANDELSCHEMIKER (*m*): commercial (analytical) chemist 417
HANDELSFIRMA (*f*): trading company 418
HANDELSFLAGGE (*f*): merchant flag 419
HANDELSFLOTTE (*f*): trading fleet, merchant fleet, merchant marine 420
HANDELSFREIHEIT (*f*): free trade 421
HANDELSGEBRAUCH (*m*): commercial practice or usage 422
HANDELSGENOSSENSCHAFT (*f*): partnership, firm 423
HANDELSGERICHT (*n*): commercial court 424
HANDELSGESCHÄFT (*n*): commercial business 425
HANDELSGESELLSCHAFT (*f*): partnership, firm, trading company, commercial society 426
HANDELSGESETZ (*n*): commercial law 427
HANDELSGEWICHT (*n*): commercial weight; avoirdupois 428
HANDELSKAMMER (*f*): Chamber of Commerce 429
HANDELSKAPITAL (*n*): trading capital 430
HANDELSLEUTE (*pl*): tradesmen; tradespeople; merchants 431
HANDELSMANN (*m*): tradesman; merchant 432
HANDELSMARINE (*f*): merchant marine 433
HANDELSPHTALSÄURE (*f*): commercial phthalic acid, phthalic anhydride (see PHTALSÄURE-ANHYDRID) 434
HANDELSPLATZ (*m*): mart 435
HANDELSREISENDE (*m*): commercial traveller 436

HANDELSSCHIFF (*n*): trading vessel, merchant ship, merchantman 437
HANDELSSILBER (*n*): commercial silver 438
HANDELSSORTE (*f*): commercial kind (type; form; sort; variety) 439
HANDELSSPERRE (*f*): prohibition; protection 440
HANDELSSTAND (*m*): business community 441
HÄNDEL (*m.pl*) SUCHEN: to pick a quarrel, to look for trouble 442
HANDELSVERBINDUNG (*f*): commercial (association) union; business connection 443
HANDELSVEREIN (*m*): commercial (association) union 444
HANDELSVERKEHR (*m*): business intercourse 445
HANDELSVERTRAG (*m*): commercial treaty 446
HANDELSVORRAT (*m*): stock-in-trade 447
HANDELSWESEN (*n*): business matters, commercial affairs 448
HANDELSWISMUT (*m* and *n*): commercial bismuth (99.3-99.7 % bismuth) 449
HANDELSZEICHEN (*n*): trade mark 450
HANDELSZINK (*n*): spelter; commercial zinc 451
HANDELSZINN (*n*): commercial tin (silverwhite in colour) 452
HANDELSZWEIG (*m*): branch of trade 453
HANDELT, ES-SICH UM: it is a question of 454
HANDELTREIBEND: commercial, trading 455
HANDFASS (*n*): keg, small cask 456
HANDFESSEL (*f*): manacle, hand-cuff 457
HANDFEST: strong, sturdy, stout, firm 458
HANDGEBRAUCH (*m*): hand (daily) use 459
HANDGELENK (*n*): wrist-(joint) 460
HANDGRANATE (*f*): hand grenade 461
HANDGREIFLICH: obvious; palpable; downright; evident 462
HANDGRIFF (*m*): handle; grip; grasp; knack; manipulation; hand-rail; handshake 463
HANDHABE (*f*): handle; haft; shaft; tiller (Naut.) 464
HANDHABEN: to handle; manipulate; manage; maintain; administer; deal with 465
HANDHABUNG (*f*): handling; manipulation; administration; method of dealing with 466
HANDHAMMER (*m*): hand hammer 467
HANDHEISS: lukewarm; of the temperature of the hand 468
HANDKARREN (*m*): hand-cart, hand-barrow 469
HANDKURBEL (*f*): cranked handle 470
HANDLANGER (*m*): hodman, bricklayer's mate, labourer 471
HÄNDLER (*m*): dealer; retailer; merchant 472
HANDLICH: handy; manageable; tractable 473
HANDLICHKEIT (*f*): handiness; tractability 474
HANDLUNG (*f*): trade; commerce; commercial establishment; action; plot; deed; act; transaction; shop; business 475
HANDLUNGSDIENER (*m*): merchant's clerk 476
HANDLUNGSGESETZ (*n*): commercial law 477
HANDLUNGSINHABER (*m*): merchant 478
HANDLUNGSREISENDE (*m*): commercial traveller 479
HANDLUNGSSCHULE (*f*): commercial (college) school 480
HANDLUNGSSPESEN (*pl*): business (expenses) disbursements, outgoings 481
HANDLUNGSUNKOSTEN (*pl*): business (expenses) disbursements; outgoings 482
HANDLUNGSWEISE (*f*): method of (acting) dealing 483
HANDMÜHLE (*f*): hand mill 484
HANDPAPIER (*n*): (see BÜTTENPAPIER) 485
HANDPFLEGE (*f*): manicuring 486
HANDPUMPE (*f*): hand pump 487
HANDRAD (*n*): hand-wheel 488
HANDRAMME (*f*): paving ram; hand ram 489
HANDRÜCKEN (*m*): back of the hand 490
HANDSÄGE (*f*): hand saw 491
HANDSCHRIFT (*f*): handwriting; manuscript; signature 492
HANDSCHRIFTLICH: in (manuscript) writing; written; in black and white 493
HANDSCHUH (*m*): glove; gauntlet 494
HANDSCHULDSCHEIN (*m*): note of hand; I.O.U.; promissory note 495
HANDSPAKE (*f*): hand spike 496
HANDSPEICHE (*f*): hand spike 497
HANDSTEUERUNG (*f*): hand gear 498
HANDSTRICH (*m*): hand moulding 499
HANDSTÜCK (*n*): a piece of anything about the size of a fist (usually applied to coal) 500
HANDTELLER (*m*): flat (palm) of the hand 501
HANDTUCH (*n*): towel 502
HANDTUCHDRELL (*m*): towelling 503
HAND (*f*), UNTER DER-: by private treaty; by private contract 504
HANDVERKAUF (*m*): retail 505
HANDVOLL (*f*): handful 506
HAND, VON-: by hand 507
HANDWAFFEN (*f.pl*): small-arms 508
HANDWAGE (*f*): hand (scales) balance 509
HANDWAGEN (*m*): hand-cart, hand-barrow 510
HANDWÄRME (*f*): heat (warmth; temperature) of the hand 511
HANDWERK (*n*): handicraft; handiwork; trade; guild 512
HANDWERKER (*m*): mechanic; artisan; workman; craftsman 513
HANDWERKERVEREIN (*m*): trade-union 514
HANDWERKSBURSCHE (*m*): journeyman 515
HANDWERKSINNUNG (*f*): guild 516
HANDWERKSJUNGE (*m*): apprentice 517
HANDWERKSLEUTE (*pl*): mechanics, artisans, workmen, craftsmen 518
HANDWERKSMÄSSIG: (in a)-workmanlike-(manner); mechanical-(ly) 519
HANDWERKSZEUG (*n*): tools; instruments 520
HANDWÖRTERBUCH (*n*): pocket (abridged) dictionary 521
HANDZEICHNUNG (*f*): hand (freehand) drawing 522
HANF (*m*): hemp (bast fibre of *Cannabis sativa*) (see RÖSTE) 523
HANFARTIG: hempen; hemp-like 524
HANFBAU (*m*): cultivation of hemp 525
HANFBRECHE (*f*): hemp-brake 526
HANFDARRE (*f*): hemp kiln 527
HÄNFEN: hempen, of hemp 528
HANFHECHEL (*f*): hatchel, hemp-comb 529
HANF (*n*), INDISCHER-: Indian hemp, chang (*herba Cannabis indicæ* of *Cannabis sativa var. indica*) 530
HANFKAMM (*m*): (see HANFHECHEL) 531
HANFKRAUT (*n*): hemp agrimony (*Agrimonia*) 532
HANFLIDERUNG (*f*): hemp packing 533
HÄNFLING (*m*): limpet 534
HANFNESSEL (*f*): galeopsis herb (*Herba galeopsidis*); hemp nettle (*Galeopsis tetrahit*) 535
HANFÖL (*n*): hemp-(seed)-oil; Sg. 0.926 (from seed of *Cannabis sativa*) 536
HANFPACKUNG (*f*): hemp packing 537
HANFPFLANZE (*f*): hemp plant (*Cannabis sativa*; *Cannabis indica*) (yields the drug Hashish) 538

HANFSAAT (*f*): hemp seed (seed of *Cannabis sativa*) 539
HANFSAMEN (*m*): hemp seed (seed of *Cannabis sativa*) 540
HANFSEIL (*n*): hemp (cord) rope 541
HANFSEILSCHMIERE (*f*): hemp rope (grease) lubricant 542
HANFSTENGEL (*m*): hemp stalk 543
HANFTAU (*n*): hemp rope 544
HANG (*m*): slope; declivity; inclination; acclivity; propensity; bent 545
HÄNGE (*f*): drying (house; loft) place; ager (Dye); hinge 546
HÄNGEBAHN (*f*): suspended railway; aerial (ropeway) cableway 547
HÄNGEBOCK (*m*): hanging (suspended) bearing frame, hanger 548
HÄNGEBODEN (*m*): drying house (loft, place) 549
HÄNGEBRÜCKE (*f*): suspension bridge 550
HÄNGEEISEN (*n*): suspension (iron) rod 551
HÄNGELAGER (*n*): hanging (suspended) bearing, hanger bearing 552
HÄNGEMATTE (*f*): hammock 553
HÄNGEMUSKEL (*m*): suspensory muscle 554
HÄNGEN: to hang; cling; adhere; fasten; depend upon; suspend; attach; dangle; depend from 555
HANGEN: (see HÄNGEN) 556
HÄNGENBLEIBEN (*n*): seizing, sticking, jambing; (Verb=to seize, stick, hang, hang fire, jamb) 557
HÄNGEND: hanging; pendent; suspended 558
HÄNGEN (*n*) **DER GICHTEN**: arching of the charge during operation (which suddenly collapses in an explosive manner and interrupts the operation of the blast furnace) 559
HANGER (*m*): hanger; pendant; filling futtock 560
HANGERBAND (*n*): sling 561
HÄNGESCHE (*f*): weeping ash (see ESCHENBAUM) 562
HÄNGESCHLOSS (*n*): padlock 563
HÄNGESTANGE (*f*): suspension rod 564
HÄNGESTREBE (*f*): suspension (strut) stay 565
HÄNGETRÄGER (*m*): suspension girder 566
HÄNGEWEIDE (*f*): weeping willow (see WEIDEN--BAUM) 567
HÄNGEWERK (*n*): truss (Building Construction) 568
HANKSIT (*m*): hanksite; $9Na_2SO_4.3Na_2CO_3.KCl$; Sg. 2.56 569
HANSA (*f*): hanseatic union or league 570
HANSEL (*m*): return wort (Brewing) 571
HÄNSELN: to play tricks (up)-on; make a fool of 572
HANTEL (*f*): dumb-bell 573
HANTIEREN: to work; handle; manage; manipulate; trade; be (busy) occupied 574
HANTIERUNG (*f*): manipulation; occupation; management 575
HAPERN: to hamper, stick; stop, falter 576
HÄREN: hairy, of hair, to lose hair, fall out (of hair) 577
HÄRING (*m*): herring 578
HÄRINGSFANG (*m*): herring-fishery 579
HÄRINGSÖL (*n*): herring oil; Sg. 0.925 (obtained from *Clupea harengus*) 580
HÄRINGSTRAN (*m*): (see HÄRINGSÖL) 581
HARKE (*f*): rake 582
HARKEN: to rake 583
HARM (*m*): sorrow; grief; harm; trouble; affliction 584
HÄRMEN: to grieve; worry; afflict; trouble 585
HARMLOS: harmless 586
HARMONIE (*f*): harmony, concord 587
HARMONIEGESETZ (*n*): harmonic law 588
HARMONIEREN: to harmonize, agree 589
HARMONIK (*f*): harmonic 590
HARMONISCH: harmonic (Maths., etc.); harmonious 591
HARMONISCHE LINIE (*f*): harmonic (curve) line 592
HARMOPHANER KUPHONSPAT (*m*): (see SKOLEZIT) 593
HARMOTOM (*m*): harmotome (a barium zeolite) (see also STAUROLITH); $H_2(Ba,K_2)Al_2[SiO_3]_5+4H_2O$; Sg. 2.47 594
HARN (*m*): urine 595
HARNABSATZ (*m*): urinary sediment 596
HARNABSONDERUNG (*f*): diuresis 597
HARNANALYSE (*f*): urine (analysis) test; analysis (test) of urine 598
HARNANTISEPTIKUM (*n*): urine antiseptic 599
HARNARTIG: urinous, like urine 600
HARNBENZOESÄURE (*f*): benzolaminoacetic acid, benzaminoacetic acid, hippuric acid, benzoylglycin, benzoylglycocoll; $CH_2NHCOC_6H_5COOH$; Sg. 1.37; Mp. 190°C. 601
HARNBESCHAUUNG (*f*): uroscopy 602
HARNBESCHWERDEN (*f.pl*): difficulty in urinating 603
HARNBLASE (*f*): bladder 604
HARNBLASENBLUTFLUSS (*m*): haematruria (the passing of blood with the urine) (Med.) 605
HARNBLASENBRUCH (*m*): rupture 606
HARNBLASENGRIES (*m*): gravel, urinary calculus (Med.) 607
HARNBLASENSTEINSCHNITT (*m*): lithotomy (removal of urinary calculus from the bladder by means of cutting) (Med.) 608
HARNBRENNEN (*n*): burning in the bladder 609
HARNDESINFIZIENS (*f*): urine disinfectant 610
HARNDRANG (*m*): micturition 611
HARNEN: to urinate 612
HARNFARBSTOFF (*m*): urinary pigment 613
HARNFLUSS (*m*): discharge of urine 614
HARNGANG (*m*): urethra 615
HARNGRIES (*m*): arena (Med.) 616
HARNISCH (*m*): harness; armour 617
HARNKOLLOID (*n*): urinary colloid 618
HARNKRAUT (*n*): yellow seed (*Reseda luteola*) 619
HARNLEHRE (*f*): urinology 620
HARNLEITER (*m*): catheter; urethra 621
HARNMESSER (*m*): urinometer 622
HARNRÖHRE (*f*): urethra 623
HARNRÖHRENSCHNITT (*m*): urethrotomy 624
HARNRÖHRENVERENGERUNG (*f*): stricture 625
HARNRUHR (*f*): diabetes 626
HARNRUHR (*f*), **ZUCKERLOSE-**: polyuria, diabetes insipidus (Med.) 627
HARNSALZ (*n*): microcosmic salt (sodium-ammonium phosphate); $NaNH_4HPO_4.4H_2O$ 628
HARNSAND (*m*): gravel; urinary calculus 629
HARNSAUER: combined with uric acid; urate of 630
HARNSÄURE (*f*): uric (lithic) acid; uric oxide (*Acidum uricum*); $C_5H_4N_4O_3$; Sg. 1.87] 631
HARNSTEIN (*m*): stone in the bladder; urinary calculus 632
HARNSTOFF (*m*): urea; carbamide; $CO(NH_2)_2$; Sg. 1.32; Mp. 132.5°C. 633
HARNSTOFFBESTIMMUNGSAPPARAT (*m*): urea determining apparatus 634

HARNSTOFF (*m*), SALPETERSAURER- : urea nitrate 635
HARNSTRENGE (*f*) : strangury, dysury (*Stillicidium urinæ*) 636
HARNSYSTEM (*n*) : urinary system (Med.) 637
HARNTREIBEND : diuretic 638
HARNTREIBENDES MITTEL (*n*) : diuretic 639
HARNUNTERSUCHUNG (*f*) : urinalysis ; investigation (analysis ; testing) of urine 640
HARNUNTERSUCHUNGSAPPARAT (*m*) : urine testing apparatus 641
HARNWAGE (*f*) : urinometer 642
HARNWEG (*m*) : urinary duct 643
HARNZAPFER (*m*) : catheter 644
HARNZUCKER (*m*) : sugar in urine 645
HARNZWANG (*m*) : strangury (Med.) 646
HARREN : to wait ; stay ; delay ; tarry ; expect ; anticipate 647
HART : hard ; hardy ; stern ; refractory (Minerals) ; hardly ; harsh ; rough ; uneasy ; stiff ; rigorous ; austere ; solid 648
HARTBAUM (*m*) : wild (Cornelian) cherry, cornel, dogwood (a hardy shrub "*Cornus sanguinea*") 649
HARTBLEI (*n*) : hard lead (a lead-antimony alloy ; 80-98% Pb. ; 20-2% Sb.) 650
HARTBORST (*m*) : crack formed during hardening (Steel) 651
HARTBORSTE (*f*) : (see HARTBORST) 652
HARTBRANDSTEIN (*m*) : hard burned (stock) brick ; clinker 653
HARTBRAUNSTEIN (*m*) : Braunite ; Mn_2O_3 ; Sg. 4.85 654
HÄRTE (*f*) : hardness ; hardiness ; rigour ; severity ; temper 655
HÄRTEBESTIMMUNG (*f*) : determination of hardness 656
HÄRTE (*f*), BLEIBENDE- : permanent hardness (of water) 657
HÄRTEGEFÄSS (*n*) : hardening trough (Metal) 658
HÄRTEGRAD (*m*) : degree of hardness ; temper (of steel) 659
HÄRTEKOHLE (*f*) : hardening (temper) carbon 660
HÄRTEMITTEL (*n*) : hardening (compound) agent ; hardening mixture 661
HÄRTEN : to harden ; temper (Steel) 662
HÄRTENRISS (*m*) : (see HARTBORST) 663
HÄRTENSKALA (*f*) : scale of hardnesses 664
HÄRTEOFEN (*m*) : hardening (annealing, tempering) furnace (see GLÜHOFEN) 665
HÄRTEPRÄPARAT (*n*) : hardening (tempering) preparation 666
HÄRTEPRÜFAPPARAT (*m*) : hardness testing apparatus 667
HÄRTEPULVER (*n*) : hardening (cementing, tempering, case-hardening) powder (Metal) 668
HÄRTEREIHE (*f*) : scale of hardness 669
HÄRTERN (*m*) : Cornel, Cornelian (wild) cherry (*Cornus sanguinea*) 670
HÄRTESALZ (*n*) : hardening salt (Steel) 671
HÄRTE (*f*), VERÄNDERLICHE- : temporary (variable) hardness (of water) 672
HARTFLÜGLER (*m.pl*) : *Coleoptera* (an order of insects) 673
HARTGELÖTET : brazed ; hard-soldered 674
HARTGESINNT : hard-hearted 675
HARTGLAS (*n*) : hard-(ened) glass 676
HARTGUMMI (*n*) : hard rubber ; vulcanite ; vulcanized rubber ; ebonite (see EBONIT) 677
HARTGUMMIWAREN (*f.pl*) : hard rubber (vulcanite) goods 678
HARTGUSS (*m*) : chill casting ; chilled casting 679
HARTGUSSFORM (*f*) : chill mould 680
HARTGUSSTEIL (*m*) : chill casting 681
HARTGUSSTIEGEL (*m*) : case-hardening (cementing) crucible or pot 682
HARTGUSSWALZE (*f*) : chilled iron roll 683
HARTHARZ (*n*) : hard (solid) resin 684
HARTHERZIG : hard-hearted 685
HARTHEU (*n*) : St. John's Wort (*Hypericum*) 686
HARTHOLZ (*n*) : hard wood (see also HORNBAUM) 687
HARTKOBALTERZ (*n*) : skutterudite 688
HARTKÖPFIG : stubborn, headstrong 689
HARTLEIBIG : costive 690
HÄRTLINGE (*m.pl*) : difficultly meltable tin-iron, tin-copper alloys from liquation of raw tin 691
HARTLOT (*n*) : hard solder (see SCHLAGLOT) 692
HARTLÖTEN : to hard-solder ; braze 693
HARTLOTWASSER (*n*) : hard soldering fluid (a solution of phosphoric acid in alcohol) 694
HARTMANGAN (*m*) : rhodonite, managnese spar (see MANGANSPAT) 695
HARTMANGANERZ (*n*) : psilomelane (se PSILO-MELAN) 696
HARTMEISSEL (*m*) : cold chisel 697
HARTNÄCKIG : stiff-necked ; stubborn ; obstinate ; pertinacious 698
HARTNAGEL (*m*) : clout 699
HARTPAPIER (*n*) : hard paper (made by winding various layers of cellulose fibre with a binding substance) 700
HARTPARAFFIN (*n*) : hard paraffin (see PARAFFIN) 701
HARTPORZELLLAN (*n*) : hard porcelain (see FELDSPATPORZELLAN) 702
HARTRIEGEL (*m*) : privet (*Ligustrum vulgare*) ; cornel ; dogwood (see HARTBAUM) 703
HARTRIEGELÖL (*n*) : privet oil (from *Cornus sanguinea*) 704
HARTSEHNIG : sinewy 705
HARTSPIRITUS (*m*) : solid (methylated) spirit (denatured alcohol with some substance of solid form ; for burning purposes) 706
HARTSTAHL (*m*) : hard steel (12% manganese) 707
HARTSTEINGUT (*n*) : hard white-ware 708
HARTTROCKEN-GLANZÖL (*n*) : hard-drying brilliant oil 709
HARTTROCKENÖL (*n*) : hard-drying oil 710
HÄRTUNG (*f*) : hardening tempering 711
HÄRTUNGSKOHLE (*f*) : temper carbon (in which iron carbide remains in solution when the metal is suddenly quenched. It considerably affects the hardness). 712
HARTWAREN (*f.pl*) : hard-ware 713
HARTZINK (*n*) : hard zinc (an iron-zinc alloy ; $FeZn_{10}$) 714
HARZ (*n*) : resin ; gum (*Resina*) ; colophony ; Sg. 1.09 ; Mp. 100-140°C. 715
HARZARTIG : resinous ; resin-like 716
HARZBAUM (*m*) : pine (pitch) tree, (any conifer or tree yielding resin or gum) 717
HARZCERAT (*n*) : resin cerate 718
HARZDESTILLATION (*f*) : resin distillation (see HARZÖL) 719
HARZDESTILLATIONSANLAGE (*f*) : resin distilling plant 720
HARZ (*n*), ECHTES- : common resin, colophony (see HARZ) 721
HARZ (*n*), EIGENTLICHES- : common resin, colophony (see HARZ) 722
HARZELEKTRIZITÄT (*f*) : resinous (negative) electricity 723

HARZEN : to be (resinous) sticky ; gather resin ; extract resin from ; exude resin ; resin 724
HARZESSENZ (*f*) : resin essence (vacuum distillation product from colophony) 725
HARZFACKEL (*f*) : resin torch 726
HARZFARBE (*f*) : resin colour (mixture of resin with pigments) 727
HARZFICHTE (*f*) : pitch-pine (see HARZKIEFER) 728
HARZFIRNIS (*m*) : resin varnish 729
HARZFLECK (*m*) : resin spot (Paper) 730
HARZFLUSS (*m*) : flow (exudation) of resin 731
HARZGALLE (*f*) : resinous pore -(in wood) 732
HARZGANG (*m*) : resin (duct) channel 733
HARZGAS (*n*) : resin gas 734
HARZGEBEND : resiniferous ; yielding resin 735
HARZGEIST (*m*) : (see HARZSPIRITUS) 736
HARZGERUCH (*m*) : resin-(ous) odour 737
HARZGLANZ (*m*) : resinous lustre 738
HARZHALTIG : resinous, containing resin 739
HARZIG : resinous ; resiny 740
HARZKANAL (*m*) : resin (duct) channel 741
HARZKARBOLLÖSUNG (*f*) : carbolic acid solution of resin (Paper) 742
HARZKESSEL (*m*) : resin (boiler) melting pan 743
HARZKIEFER (*f*) : pitch pine (*Pinus palustris*) ; Sg. 0.5-0.92 744
HARZKITT (*m*) : resinous cement ; resin (mastic, putty) cement 745
HARZKOHLE (*f*) : bitumenous coal 746
HARZ (*n*), KUNST- : synthetic resin 747
HARZLACK (*m*) : resin (lake) varnish 748
HARZLEIM (*m*) : resin (size) glue (Paper) (colophony and sodium carbonate boiled together) 749
HARZLÖSUNGSMITTEL (*n*) : resin solvent 750
HARZMÜHLE (*f*) : varnish (gum ; resin) mill or crusher 751
HARZÖL (*n*) : resin oil, retinol, rosinol, codoil ; Sg. 0.98-1.11 (product of dry distillation of resin) 752
HARZPRODUKT (*n*) : resin product (see HARZÖL) 753
HARZREICH : resinous, rich in resin, having a high resin content 754
HARZREISSEN (*n*) : tapping trees for resin 755
HARZSALBE (*f*) : resin cerate 756
HARZSAUER : resinate of, combined with resin (resinic) acid 757
HARZSÄURE (*f*) : resin-(ic) acid (see also AGARIZIN) 758
HARZSCHARREN (*n*) : tapping trees for resin 759
HARZSCHMELZANLAGE (*f*) : resin melting plant 760
HARZSCHMELZEN (*n*) : resin melting 761
HARZSEIFE (*f*) : resin soap 762
HARZSPIRITUS (*m*) : resin spirit (distillation product from colophony) 763
HARZTANNE (*f*) : pitch-fir (see HARZKIEFER) 764
HARZTEER (*m*) : resinous tar 765
HARZTRAGEND : resin-bearing, resiniferous 766
HASARD (*m*) : hazard, risk, chance 767
HASCHEN : to catch ; seize ; snatch ; aspire to ; aim at 768
HASCHIEREN : to hatch ; cross-shade (of drawings) 769
HASCHISCH (*n*) : hashish (a drug from Indian hemp, *Cannabis indica*) 770
HASCHISH (*n*) : (see HASCHISCH) 771
HASCHÜR (*f*) : graving, hatching (on a printing roll) 772
HASE (*m*) : hare (*Lepus europaeus*) ; coward 773
HASEL (*f*) : hazel (the bush "*Corylus avellana*" and the nut) 774
HASELHUHN (*n*) : heath-cock 775
HASELKÄTZCHEN (*n*) : hazel catkin 776
HASELMAUS (*f*) : dormouse (*Mus musculus*) 777
HASELNUSS (*f*) : hazel (filbert) nut (see HASEL) 778
HASELNUSSHOLZ (*n*) : filbert wood, hazel wood ; Sg. 0.56-0.71 779
HASELNUSSÖL (*n*) : hazel-nut oil (from seeds of *Corylus avellana*) ; Sg. 0.916 780
HASELRÜSTER (*f*) : (see BERGULME) 781
HASELRUTE (*f*) : hazel rod (used as a divining rod) 782
HASELSTRAUCH (*m*) : hazel bush (see HASEL) 783
HASELWURZ (*f*) : hazel (cabaric) root ; asarum (*Asarum canadense*) ; Asarabacca (*Asarum europœum*) (*Radix asari*) 784
HASELWURZEL (*f*) : cabaric root (see HASELWURZ) 785
HASENAMPFER (*m*) : wood sorrel (*Oxalis acetosella*) 786
HASENKLEE (*m*) : clover ; trefoil ; hare's-foot trefoil (*Trefolium arvense*) 787
HASENKOHL (*m*) : wood sorrel (*Oxalis acetosella*) 788
HASENLATTICH (*m*) : wall-lettuce (*Prenantes muralis*) 789
HASENLEIM (*m*) : hare (rabbit) glue 790
HASENSCHARTE (*f*) : harelip (a cleft upper lip) (Med.) 791
HASENSCHWANZ (*m*) : hare's-tail grass (*Lagurus ovatus*) 792
HASPE (*f*) : hasp ; staple ; hinge ; clamp 793
HASPEL (*m*) : reel ; windlass ; winch ; capstan ; staple 794
HASPEL (*m*), DREH- : winch, capstan, windlass 795
HASPELN : to reel ; wind-(up) 796
HASPELRAD (*n*) : chain (sheave, wheel) pulley 797
HASPEL (*m*), SCHEIBEN- : disc winch 798
HASPEL (*m*), TROMMEL- : drum winch 799
HASS (*m*) : hate ; hatred ; spite ; enmity 800
HASSEN : to hate ; abhor ; detest ; loathe 801
HASSENSWERT : odious ; detestable ; hateful 802
HÄSSLICH : ugly ; bad ; gross ; vicious ; nasty ; wicked ; loathsome 803
HAST (*f*) : haste ; hurry ; speed 804
HASTEN : to hasten ; hurry ; speed 805
HASTIG : hasty ; precipitate ; passionate ; irritable ; precipitous 806
HAU (*m*) : hewing ; cut ; cutting ; stroke ; felling 807
HAUBAR : ready for felling (of wood) 808
HAUBE (*f*) : hood ; cap ; cupola ; dome ; crown ; cover ; top ; crest (of birds) ; tuft ; coiffure 809
HAUBEN : to cap, dome, cover, crown, hood, top 810
HAUBENKOPF (*m*) : milliner's block 811
HAUBENLERCHE (*f*) : crested lark (*Alauda cristata*) ; cone protector (Ceramics) 812
HAUBENSCHACHTEL (*f*) : band-box 813
HAUBITZE (*f*) : dividing head (Machine part) ; howitzer-(gun) 814
HAUBITZGRANATE (*f*) : (howitzer)-shell 815
HAUBITZZÜNDER (*m*) : howitzer fuse 816
HAUBLOCK (*m*) : chopping-block 817
HAUBUCHE (*f*) : (see HORNBAUM) 818
HAUCH (*m*) : breath ; blast ; breathing ; tinge ; breeze ; whiff ; aspiration ; exhalation 819
HAUCHEN : to breathe ; aspirate ; exhale ; blow 820
HAUE (*f*) : hoe ; mattock ; pick-(axe) 821
HAUEN : to hew ; chop ; cut ; strike ; fell ; chisel ; beat ; thrash 822

HAUER (*m*): hewer; cutter; pick; miner; boar; tusk 823
HAUERIT (*m*): hauerite; MnS_2; Sg. 3.45 824
HAUFE (*m*): heap; pile; amount; number; mass; crowd; swarm; hoard; multitude; batch; couch (Brewing) 825
HÄUFELN: to heap-(one on top of another); accumulate in small piles 826
HÄUFEN: to heap; pile; accumulate; amass; increase; hoard; augment 827
HAUFEN (*m*): pile (for copper roasting process); piling (in puddling process of metal) (see also HAUFE) 828
HAUFENFÜHREN (*n*): couching; flooring (Brewing) 829
HAUFENVERKOHLUNG (*f*): charcoal burning in piles 830
HAUFENWEISE: in (heaps) piles 831
HAUFENWOLKE (*f*): cumulus (of clouds) 832
HÄUFIG: frequent-(ly); abundant-(ly); often; copious-(ly) 833
HÄUFIGKEIT (*f*): frequency; frequence 834
HÄUFUNG (*f*): heaping, piling, accumulation, crowding, swarming, hoarding 835
HAUHECHELWURZEL (*f*): rest harrow root (*Radix ononidis* of *Ononis spinosa*) 836
HAUMEISSEL (*m*): gouge; chisel 837
HAUPT (*n*): main; chief; head; principal; leader; top; beginning 838
HAUPTABSICHT (*f*): main (intention) purpose 839
HAUPTACHSE (*f*): main axis 840
HAUPTADER (*f*): cephalic vein 841
HAUPTAGENTUR (*f*): chief agency 842
HAUPTANTEIL (*m*): chief or main (portion; part) constituent 843
HAUPTARTIKEL (*m*): leading article (in a newspaper); chief (principal) article 844
HAUPTBALKEN (*m*): architrave; roof principals 845
HAUPTBEGRIFF (*m*): main (idea) conception 846
HAUPTBESTANDTEIL (*m*): main (chief; principal) constituent 847
HAUPTBETRAG (*m*): total amount; sum total 848
HAUPTBEWEIS (*m*): principal (main) proof 849
HAUPTBINDUNG (*f*): principal (union; bond;) linkage; union (bond; linkage) of the first order 850
HAUPTBRENNPUNKT (*m*): principal (main) focus or focal point 851
HAUPTBRENNWEITE (*f*): principal focal (length) distance 852
HAUPTBUCH (*n*): ledger (usually main or private ledger) 853
HAUPTBUCHSTABE (*m*): capital letter 854
HAUPTDECK (*n*): main deck (Naut.) 855
HAUPTEINFAHRT (*f*): main entrance (for traffic) 856
HAUPTEINGANG (*m*): main entrance (for pedestrians) 857
HAUPTFARBE (*f*): primary (main; principal) colour 858
HAUPTFEDER (*f*): main spring 859
HAUPTFEHLER (*m*): main (chief, principal) error, mistake, or fault 860
HAUPTFIGUR (*f*): chief (principal) figure 861
HAUPTFORM (*f*): chief form; master mould 862
HAUPTFRAGE (*f*): main (principal) question 863
HAUPTFUNKSTELLE (*f*): main wireless station 864
HAUPTGÄRUNG (*f*): principal (main) fermentation 865
HAUPTGEBÄUDE (*n*): main (principal) building 866
HAUPTGESCHOSS (*n*): first floor (of buildings) 867
HAUPTGESETZ (*n*): fundamental law 868
HAUPTGESTELL (*n*): main (principal) housing, framing, carriage or support 869
HAUPTGRUND (*m*): principal (fundamental) reason 870
HAUPTGRUNDSATZ (*m*): fundamental (principle) law 871
HAUPTKETTE (*f*): main (principal) chain 872
HAUPTKIEL (*m*): main keel (Naut.) 873
HAUPTKISSEN (*n*): pillow 874
HAUPTKLASSE (*f*): chief (main; principal) class 875
HAUPTKLAUSEL (*f*): principal (main) clause 876
HAUPTLABORATORIUM (*n*): chief laboratory 877
HAUPTLAGER (*n*): main bearing; chief camp, headquarters 878
HAUPTLEHRE (*f*): principal (chief) teaching or doctrine 879
HAUPTLEHRER (*m*): headmaster (of a school) 880
HAUPTLICHTMASCHINE (*f*): main dynamo 881
HAUPTLINIE (*f*): main line 882
HAUPTMANN (*m*): captain 883
HAUPTMANNSCHAFT (*f*): captaincy 884
HAUPTMAST (*m*): main-mast 885
HAUPTNENNER (*m*): common denominator 886
HAUPTNIEDERLAGE (*f*): main (stores; warehouse) depot 887
HAUPTÖLBAD (*n*): white steep (Dye) 888
HAUPTPFEILER (*m*): main (pillar) column; principal support 889
HAUPTPFLASTER (*n*): opium plaster 890
HAUPTPHASE (*f*): main phase 891
HAUPTPOSTAMT (*n*): General (head) Post Office 892
HAUPTPRÄMIE (*f*): first prize 893
HAUPTPUNKT (*m*): chief (main; principal) point 894
HAUPTQUARTIER (*n*): head-quarters 895
HAUPTQUELLE (*f*): main (principal) source; fountain-head 896
HAUPTQUERBALKEN (*m*): main cross-beam; architrave 897
HAUPTRAUM (*m*): main hold (on board ship) 898
HAUPTREDAKTEUR (*m*): chief editor 899
HAUPTREGISTER (*n*): main index or register 900
HAUPTREPARATUR (*f*): main repair-(s) 901
HAUPTROHR (*n*): main; main (pipe) tube 902
HAUPTROLLE (*f*): chief part; leading rôle; principal character 903
HAUPTRUDER (*n*): stroke-oar 904
HAUPTSACHE (*f*): main thing; main point; chief matter 905
HAUPTSACHE, IN DER-: mainly; chiefly; principally; essentially 906
HAUPTSÄCHLICH: mainly; chiefly; principally; essentially 907
HAUPTSATZ (*m*): fundamental (principle; rule) law; main point; axiom; main proposition (Maths.) 908
HAUPTSCHALTER (*m*): main (commutator) switch (Elect.) 909
HAUPTSCHERGANG (*m*): main sheer strake (Naut.) 910
HAUPTSCHIFF (*n*): flagship (Naut.) 911
HAUPTSCHLUSSBOGENLAMPE (*f*): series arc lamp 912
HAUPTSCHLÜSSEL (*m*): master key 913
HAUPTSEGEL (*n*): main-sail (Naut.) 914
HAUPTSERIE (*f*): chief (main) series 915
HAUPTSICHERUNG (*f*): main (cut-out) fuse (Elect.) 916

HAUPTSITZ (*m*): principal residence 917
HAUPTSPANT (*n*): midship frame or section (Shipbuilding) 918
HAUPTSPIRALE (*f*): primary (main) coil 919
HAUPTSPULE (*f*): (see **HAUPTSPIRALE**) 920
HAUPTSTADT (*f*): metropolis; capital; chief town 921
HAUPTSTAGSEGEL (*n*): main staysail (Naut.) 922
HAUPTSTAMM (*m*): main (stem) trunk 923
HAUPTSTAMMAKTIEN (*f.pl*): capital stock 924
HAUPTSTIMME (*f*): principal (solo) voice; leading part 925
HAUPTSTRASSE (*f*): main-road, highway, main thoroughfare, arterial road 926
HAUPTSTROM (*m*): primary current (Elect.); main stream (of rivers); main flow 927
HAUPTSTROMDYNAMO (*f*): series dynamo 928
HAUPTSTROMERREGUNG (*f*): series excitation 929
HAUPTSTROMKREIS (*m*): principal circuit, main circuit 930
HAUPTSTROMMOTOR (*m*): series motor 931
HAUPTSTÜCK (*n*): main portion, headpiece, chapter, chief item 932
HAUPTSTÜTZE (*f*): mainstay; principal (main) support 933
HAUPTSUMME (*f*): sum total; principal (sum) amount 934
HAUPTTEILHABER (*m*): senior partner 935
HAUPTTYPUS (*m*): main (principal; chief) type 936
HAUPTURSACHE (*f*): main (principal) cause 937
HAUPTWELLE (*f*): main shaft, driving shaft 938
HAUPTWINDLEITUNG (*f*): main air duct 939
HAUPTWORT (*n*): principal word; noun, substantive 940
HAUPTWÜRZE (*f*): first wort (Brewing) 941
HAUPTZAHL (*f*): cardinal number 942
HAUPTZEUGE (*f*): chief (principal) witness 943
HAUPTZUG (*m*): principal (trait, train, draught); main pass (of boilers); principal direction (of stresses or loads) 944
HAUPTZUGLOKOMOTIVE (*f*): main-line engine 945
HAUPTZWECK (*m*): main (aim, object) purpose 946
HAUS (*n*): house; home; family; firm; company; casing; housing; framing; dwelling; residence 947
HAUSARBEIT (*f*): domestic work 948
HAUSARZNEI (*f*): domestic remedy 949
HAUSARZT (*m*): family doctor; house surgeon 950
HAUSBESITZER (*m*): landlord; house-owner; proprietor 951
HAUSBRAND (*m*), **FÜR-**: domestic (household) coal or fuel 952
HAUSEN: to reside (keep)-house; live; dwell; manage; behave badly; (*m*): sturgeon 953
HAUSENBLASE (*f*): isinglass; Sg. 1.11 (see **FISCHLEIM**) 954
HAUSENROGGEN (*m*): caviare 955
HAUSFARBE (*f*), **GELBE-**: yellow ochre 956
HAUSFLUR (*f* and *m*): (entrance)-hall; floor 957
HAUSGERÄT (*n*): furniture; utensils 958
HAUSGESINDE (*n*): domestic servants 959
HAUSHALT (*m*): housekeeping; household 960
HAUSHALTEN (*n*): house-keeping, domestic economy, management 961
HAUSHÄLTER (*m*): housekeeper; steward; economist; house-holder 962
HAUSHALTMASCHINE (*f*): household machine (such as knife cleaners, etc.) 963
HAUSHALTUNGSGERÄT (*n*): household utensils 964
HAUSHALTUNGSKUNST (*f*): domestic economy 965
HAUSHALTUNGSSEIFE (*f*): household soap 966
HAUSHUND (*m*): housedog 967
HAUSIEREN: to hawk; go from door to door 968
HAUSIERER (*m*): hawker; pedlar 969
HAUSKÄFER (*m*): death-watch (*Anobium*) 970
HAUSKEHRICHT (*m* and *n*): household (domestic) dust, rubbish or waste 971
HAUSLAUCH (*m*): houseleek (*Sempervivum tectorum*) 972
HAUSLAUF (*m*): (see **HAUSLAUCH**) 973
HAUSLEINWAND (*f*): home-spun linen 974
HÄUSLICH: domestic; economical; homely; thrifty 975
HAUSMANNIT (*m*): hausmannite (a rare manganese mineral); (Sg. 4.856) (see **MANGAN-OXYDOXYDUL**); Mn_3O_4 or $(Mn,Zn)\ Mn_2O_4$ 976
HAUSMASCHINE (*f*): household machine (such as knife cleaners, etc.) 977
HAUSMIETE (*f*): rent 978
HAUSMITTEL (*n*): household (domestic) remedy 979
HAUSMÜLL (*n*): house (domestic) rubbish or refuse (up to 50% ash; up to 30% moisture; calorific value about 1200 W.E/kg.) 980
HAUSSCHABE (*f*): cockroach, blackbeetle 981
HAUSSCHLÜSSEL (*m*): door-key; key of the house 982
HAUSSCHUH (*m*): slipper 983
HAUSSCHWAMM (*m*): dry-rot 984
HAUSSE (*f*): rise in prices 985
HAUSSEIFE (*f*): household (common) soap; curd-soap 986
HAUSSORGE (*f*): domestic (worry) care 987
HAUSSTEIN (*m*): ashlar 988
HAUSSUCHUNGSBEFEHL (*m*): warrant to search a house 989
HAUSTELEPHON (*n*): house telephone, intercommunicating telephone 990
HAUSTIER (*n*): domestic animal 991
HAUSTUCH (*n*): homespun-(cloth) 992
HAUSTÜR (*f*): house-door, street-door, front-door 993
HAUSVERWALTER (*m*): steward 994
HAUSWANZE (*f*): bug 995
HAUSWESEN (*n*): domestic affairs; domestic economy; household 996
HAUSWIRTSCHAFT (*f*): domestic economy, housekeeping 997
HAUSWURZ (*f*): houseleek (*Sempervivum tectorum*); stonecrop (*Sedum*); (*Crassulacea*) 998
HAUSWURZEL (*f*): (see **HAUSWURZ**) 999
HAUSZINS (*m*): house rent 000
HAUT (*f*): skin; hide; membrane; film; coat; cuticle; integument; epidermis 001
HAUTABSONDERUNG (*f*): secretion from the sebaceous or sudoriparous glands of the skin 002
HAUTALGE (*f*): dulse, seaweed (*Rhodymenia palmata*, or *Iridæa edulis*) 003
HAUTARTIG: skin-like 004
HAUTAUSSCHLAG (*m*): skin eruption (Med.) 005
HAUTBOE (*f*): hautboy; oboe (a musical instrument) 006
HAUTBRÄUNE (*f*): croup; laryngitis (inflammation of the larynx) (Med.) 007
HÄUTCHEN (*n*): thin skin; film; pellicle; membrane 008
HAUTDRÜSENKRANKHEIT (*f*): scrofula, struma (tubercular affection of the lymphatic glands) (Med.) 009

HAUT (*f*), **DURCHSICHTIGE-**: transparent (film) skin; cornea (of the eye) 010
HÄUTEN: to (cast)-skin; peel; desquamate; exfoliate 011
HAUTENTZÜNDUNG (*f*): inflammation of the skin 012
HAUTFARBE (*f*): complexion 013
HAUTGESCHWULST (*f*): œdema, swelling (Med.) 014
HAUT (*f*), **HARTE-**: hard skin (see **HAUTHORN**); sclera, sclerotic-(coat) (of the eye) (Med.) 015
HAUTHORN (*n*): horn (*Cornu cutaneum*) (Med.) 016
HÄUTIG: cutaneous; membranous; skinny; cuticular 017
HAUTJUCKEN (*n*): skin iritation 018
HAUTKLEIE (*f*): pityriasis rubra, dermatitis exfoliativa, eczema foliaceum (a skin affection) (see also **KLEIENGRIND**) (Med.) 019
HAUTKRANKHEIT (*f*): skin disease (Med.) 020
HAUTLEHRE (*f*): dermatology 021
HAUTLEIM (*m*): glue or size from hides 022
HAUTMOOS (*n*): nettlerash (Med.) 023
HAUTPILZE (*m.pl*): *Hymenomycetes* 024
HAUTPULVER (*n*): hide (skin) powder 025
HAUTREINIGEND: cosmetic 026
HAUTREINIGENDES MITTEL (*n*): cosmetic (for the complexion) 027
HAUTRÖTE (*f*): Erythema (flush, redness of the skin) (Med.) 028
HAUTSUBSTANZ (*f*): hide substance 029
HAUTTALG (*m*): sebaceous matter 030
HÄUTUNG (*f*): skinning; shedding (casting) of skin; desquamation; exfoliation 031
HAUTWASSERSUCHT (*f*): Anasarca, dropsy (Med.) 032
HAUTWURM (*m*): Glanders (Med.) 033
HAUYN (*m*): hauyn, nosean, haüyne, haüynite (a mineral of the sodalite group); $3NaAlSiO_4.(Ca,Na_2)SO_4$; Sg. 2.4 034
HAVARIE (*f*): (see **HAVERIE**) 035
HAVARIE (*f*), **GROSSE-**: gross average 036
HAVEREI (*f*): damage (by sea); average (see **HAVERIE**) 037
HAVERIE (*f*): damage by sea; average; (also sometimes for a "breakdown" of machinery involving damage), damage, havoc 038
HAVERIE (*f*), **KLEINE-**: particular average 039
HAZELNUSSHOLZ (*n*); hazel wood (see **HASELNUSS-HOLZ**) 040
HEBAMME (*f*): midwife 041
HEBARZNEIKUNST (*f*): obstetrics (medical science dealing with childbirth) 042
HEBÄRZTLICH: obstetric 043
HEBE (*f*): pulley; lever 044
HEBEARM (*m*): lever; lifting arm 045
HEBEBALKEN (*m*): lever; beam 046
HEBEBAND (*n*): truss 047
HEBEBAUM (*m*): spike, lever, pole 048
HEBEBOCK (*m*): (lifting)-jack 049
HEBEDAUMEN (*m*): cam; lifter; tappet 050
HEBEEISEN (*n*): crowbar 051
HEBEKOPF (*m*): cam, lifter, tappet 052
HEBEKRAFT (*f*): lifting (power) force 053
HEBEKUNDE (*f*): (see **HEBARZNEIKUNST**) 054
HEBEL (*m*): lever 055
HEBELADE (*f*): crane 056
HEBELANORDNUNG (*f*): arrangement of levers; system of levers 057
HEBELBREMSE (*f*): lever brake 058
HEBELKRAFT (*f*): leverage 059
HEBELN: to lever; raise; lift 060
HEBELPRESSE (*f*): lever press 061
HEBELPRESSWERK (*n*): lever press 062
HEBELSCHALTER (*m*): lever switch 063
HEBELUMSCHALTER (*m*): lever commutator 064
HEBELWAGE (*f*): beam (lever) scale or balance 065
HEBELWERK (*n*): system of levers; lever arrangement 066
HEBELWIRKUNG (*f*): leverage 067
HEBEMAGNET (*m*): lifting magnet 068
HEBEMASCHINE (*f*): winding engine, hoisting engine, (*pl*): hoisting (engines) machinery 069
HEBEMUSKEL (*m*): lěvator (Anat.) 070
HEBEN: to lift; raise; elevate; go; further; remove; hoist; lever; rise; cancel (Maths.); heave; reduce; increase; exalt; give prominence to; settle; relieve; fetch; receive; wind up; syphon; haul up 071
HEBEPRAHM (*m*): lighter, pontoon 072
HEBEPUMPE (*f*): lift-(ing) pump 073
HEBEPUNKT (*m*): fulcrum 074
HEBER (*m*): syphon; lifter; lever; levator (Anat.); hoist; (grain)-elevator 075
HEBERBAROMETER (*m*): syphon barometer 076
HEBERLEITUNG (*f*): syphon pipe 077
HEBERPUMPE (*f*): syphon pump 078
HEBESTANGE (*f*): crowbar; handspike 079
HEBETISCH (*m*): lifting table (Rolling mills) 080
HEBEVORRICHTUNG (*f*): lifting (gear) apparatus; lever 081
HEBEWERK (*n*): lifting (gear; device) arrangement; lift; gin; lever; hoist; winch; (screw)-jack; elevator 082
HEBEWINDE (*f*): jack (for raising waggons), lifting winch; windlass 083
HEBEZEUG (*n*): (see **HEBEWERK**) 084
HEBEZWINGE (*f*): lifting (dolly) bar 085
HECHELN: to comb; hatchel; hackle (Flax); criticise; satirize 086
HECHT (*m*): pike (Fish) 087
HECHTGRAU: pike (light) grey 088
HECK (*n*): stern (Naut.); fence, latticework, brushwood, enclosure, hedge 089
HECKBALKEN (*m*): transom 090
HECKE (*f*): hedge; hatch; brood; brushwood 091
HECKEN: to hatch; breed 092
HECKENBAUM (*m*): cornel, dog-wood (see **HARTBAUM**) 093
HECKENBUCHE (*f*): (see **HORNBAUM**) 094
HECKENHOPFEN (*m.pl*): wild hops (*Humulus lupulus*) 095
HECKENROSE (*f*): wild rose, dog rose 096
HECKENSCHLEHE (*f*): sloe, blackthorn (see **SCHLEHE**) 097
HECKGESCHÜTZ (*n*): stern chaser (Naut.) 098
HECKHOLZ (*n*): privet (see **HARTRIEGEL**) 099
HECKLATERNE (*f*): poop lantern (Naut.) 100
HECKPFORTE (*f*): stern-chase port (Naut.) 101
HECKRADDAMPFER (*m*): stern-wheel steamer 102
HECKRELING (*f*): taff-rail (Naut.) 103
HECKSPANT (*n*): stern frame 104
HEDE (*f*): tow; oakum (see **FLACHSWERG**) 105
HEDENBERGIT (*m*): hedenbergite (a variety of pyroxene); $CaFe(SiO_3)_2$; Sg. 3.5 106
HEDERICH (*m*): wild (hedge) mustard, charlock (*Sinapis arvensis*); jointed charlock, white charlock, wild radish (*Raphanus raphanistrum*) 107
HEDERICHÖL (*n*): charlock oil (from *Raphanus*); Sg. 0.918 108
HEDERICHVERNICHTUNGSPULVER (*n*): charlock (killer) destroying powder 109

HEDONAL (*n*): hedonal, methylpropylcarbinol-urethan; Mp. 76°C. 110
HEER (*n*): host; army; force; multitude; troops 111
HEERESFLUCHT (*f*): desertion 112
HEERSCHAR (*f*): host; army; force; troops 113
HEERSCHAU (*f*): review 114
HEERSTRASSE (*f*): highway; military road 115
HEERWESEN (*n*): military (army) affairs or matters 116
HEFE (*f*): yeast; barm (Brewing); lees; dregs; sediment; ferment 117
HEFEAPPARAT (*m*): yeast propagating apparatus 118
HEFEART (*f*): type of (ferment) yeast; race (breed; sort; kind) of yeast 119
HEFEERNTE (*f*): crop of yeast 120
HEFEFETT (*n*): yeast fat (see **FETTHEFE**) (extracted from yeast by means of alcohol) 121
HEFEGABE (*f*): amount (quantity) of yeast for a pitching (Brewing) 122
HEFEGEBEN (*n*): pitching; the addition of yeast to the wort (Brewing) 123
HEFEGUT (*n*): yeast (especially suitable for spirit distilling) 124
HEFENAHRUNG (*f*): yeast food 125
HEFENART (*f*): (see **HEFEART**) 126
HEFENBROT (*n*): leavened bread; bread baked with yeast 127
HEFENEXTRAKT (*m*): yeast extract (extraction of cell contents of yeast by means of boiling) 128
HEFENPFLANZE (*f*): yeast plant (see **HEFENPILZ** 129
HEFENPILZ (*m*): yeast fungus; yeast plant; *Saccharomyces cerevisiæ* for beer; *S. ellipsoideus* and *S. pastorianus*, for wines) 130
HEFENPULVER (*n*): baking powder 131
HEFENTEIG (*m*): leavened dough 132
HEFENVERMEHRUNG (*f*): yeast increase 133
HEFE (*f*), **OBERGÄRIGE-**: top yeast (very rapid and stormy ferment) 134
HEFEPFROPFEN (*m*): yeast (plug) stopper 135
HEFEPILZ (*m*): (see **HEFENPILZ**) 136
HEFEPROPAGIERUNG (*f*): yeast (propagation) culture 137
HEFEPROPAGIERUNGSAPPARAT (*m*): apparatus for yeast culture, yeast propagator 138
HEFERASSE (*f*): race of yeast (see **HEFEART**) 139
HEFEREINZUCHT (*f*): pure yeast culture 140
HEFEREINZUCHTAPPARAT (*m*): apparatus for pure yeast culture, pure yeast propagator 141
HEFESCHAUM (*m*): yeast scum 142
HEFESUBSTANZ (*f*): yeast substance; ferment 143
HEFETABLETT (*n*): yeast tablet 144
HEFETÄTIGKEIT (*f*): activity (action) of yeast; yeast activity 145
HEFETRIEB (*m*): head-(of yeast) 146
HEFE (*f*), **UNTERGÄRIGE-**: bottom yeast (slow and quiet ferment) 147
HEFEVERFUTTERUNG (*f*): the use of fodder yeast for feeding cattle (see also **FUTTERHEFE**) 148
HEFEWANNE (*f*): yeast trough 149
HEFEZELLE (*f*): yeast cell (of *Saccharomyces*) (see **HEFENPILZ**) 150
HEFEZUCHT (*f*): yeast culture 151
HEFE ZUSETZEN: to add yeast; leaven (see also **HEFEGEBEN**) 152
HEFIG: yeasty; yeast-like; having (lees; dregs) sediment 153
HEFNERKERZE (*f*): Hefner candle (Photometric unit) 154
HEFNERLAMPE (*f*): Hefner lamp, amylacetate lamp 155
HEFT (*n*): part; number (of a periodical); volume; copybook; handle; hilt; halt 156
HEFTEL (*m*): clasp, fastening, pin, hook 157
HEFTEN: to fasten; hook; pin; stitch; sew; attach; fix (rivet) the eye upon 158
HEFTFADEN (*m*): stitching (basting) thread 159
HEFTIG: violent; severe; fervent; earnest; intense; vehement; impetuous 160
HEFTIGKEIT (*f*): violence; intensity; ardour; vehemence 161
HEFTNADEL (*f*): stitching (bookbinder's) needle; fibula (Anatomy) 162
HEFTPFLASTER (*n*): sticking plaster, adhesive plaster, gummed tape, gummed paper 163
HEFTSCHRAUBE (*f*): fastening screw 164
HEFTWEISE: serially; in parts (volumes; numbers) 165
HEFTZWIRN (*m*): stitching (basting) thread 166
HEGEN: to enclose; foster; cherish; fence; keep; entertain; expect; anticipate; preserve 167
HEGEZEIT (*f*): close season 168
HEGONON (*n*): hegonon (a silver-albumen preparation, with 7% silver) 169
HEHL (*n*): secrecy; concealment 170
HEHLEN: to secrete; conceal; hide 171
HEHR: sublime; grand; majestic; holy; august; exalted 172
HEIDE (*f*): common; heath; thicket; heather (see **HEIDEKRAUT**); (*m*): heathen; pagan 173
HEIDEBESEN (*m*): heath broom (*Cytisus scoparius*) 174
HEIDEBODEN (*m*): moorland, heath 175
HEIDEBUSCH (*m*): gorse; furze (*Ulex europæus*) 176
HEIDEFLECHTE (*f*): Iceland moss, Cetraria (*Cetraria islandica*) 177
HEIDEFUTTER (*n*): heath fodder 178
HEIDEGEWÄCHSE (*n.pl*): *Ericaceæ* (Bot.) 179
HEIDEGINSTER (*m*): (see **HEIDEBUSCH**) 180
HEIDEGRÜTZE (*f*): buckwheat groats (see **BUCHWEIZEN**) 181
HEIDEKORN (*n*): buckwheat (*Fagopyrum esculentum*) 182
HEIDEKRAUT (*n*): heather (*Calluna vulgaris*) 183
HEIDELBEERE (*f*): whortleberry (*Vaccinium*); bilberry (*Myrtillus*) 184
HEIDELBEERSTRAUCH (*m*): bilberry bush 185
HEIDELERCHE (*f*): wood-lark (*Alauda arborea*) 186
HEIDERAUCH (*m*): peat 187
HEIDETORF (*m*): (heath)-heat 188
HEIKEL: critical; dainty; delicate; difficult; captious 189
HEIL: whole; sound; unhurt; safe; well; healed; (*n*): welfare; happiness; safety; prosperity; healer; cure 190
HEILANSTALT (*f*): hospital; sanatorium 191
HEILART (*f*): curative method 192
HEILBAD (*n*): medicinal (mineral) bath 193
HEILBAR: curable 194
HEILBRUNNEN (*m*): mineral (well) spring 195
HEILBUTT (*m*): halibut (*Hippòglossus vulgaris*) 196
HEILDECKERWURZEL (*f*): (see **TORMENTILLWURZEL**) 197
HEILEN: to cure; heal; become (make) well 198
HEILGYMNASTIK (*f*): kinesitherapeutics (see **KINESITHERAPIE**) 199

HEILIG: holy; solemn; sacred; sanctified; inviolable 200
HEILIGBUTT (*m*): halibut (see **HEILBUTT**) 201
HEILIGEN: to sanctify; hallow; make (keep) holy 202
HEILIGENHARZ (*n*): guaiacum- (resin) (see **GUAJAK**) 203
HEILIGENHOLZ (*n*): holy wood; lignum vitæ; lignum sanctum (a type of lignum vitæ) (see **GUAJAKBAUM** and **GUAJAKHOLZ**) 204
HEILKRAFT (*f*): curative power 205
HEILKRÄFTIG: curative; therapeutic; healing; medicinal 206
HEILKRAUT (*n*): officinal (medicinal) herb or plant 207
HEILKUNDE (*f*): medicine; therapeutics; medical science; the art of healing 208
HEILKUNST (*f*): (see **HEILKUNDE**) 209
HEILLOS: profligate, godless, wicked 210
HEILMITTEL (*n*): remedy; cure; medicament; medicine 211
HEILMITTELLEHRE (*f*): pharmacology; pharmacy (the science of medicines) (*Materia medica*) 212
HEILPFLASTER (*n*): healing (sticking) plaster 213
HEILPFLASTER, ENGLISCHES- (*n*): court plaster 214
HEILQUELL (*m*): mineral (medicinal) spring or well 215
HEILQUELLE (*f*): (see **HEILQUELL**) 216
HEILSAM: wholesome; healing; curative; remedial; salutary; beneficial; salubrious 217
HEILSERUM (*n*): healing serum; antitoxin 218
HEILSTÄTTE (*f*): sanatorium 219
HEILSTOFF (*m*): curative; remedy 220
HEILSTOFFLEHRE (*f*): (see **HEILMITTELLEHRE**) 221
HEILTRANK (*m*): medicine, potion, draught 222
HEILUNG (*f*): cure; curing; healing 223
HEILWASSER (*n*): curative (mineral; medicinal) water 224
HEIL WERDEN: to be healed; heal; become (healed) well 225
HEILWERT (*m*): therapeutic (healing; curative) value 226
HEILWISSENSCHAFT (*f*): medical science 227
HEILWURZ (*f*): medicinal herb 228
HEIM: home-(ward); (*n*): home; dwelling; native-(town; country) place 229
HEIMAT (*f*): fatherland; native country; habitat 230
HEIMATLICH: native 231
HEIMATLOS: homeless; abandoned 232
HEIMCHEN (*n*): cricket (*Gryllus domesticus*) 233
HEIMFALLEN: to devolve upon, revert to 234
HEIMISCH: native; domestic; home; indigenous 235
HEIMLICH: secret-(ly); private-(ly); stealthy; stealthily; secluded 236
HEIMREISE (*f*): homeward voyage (of ships, etc.) homeward (journey) run 237
HEIMSUCHEN: to afflict; haunt; visit; requite 238
HEIMTÜCKISCH: malicious; mischievous 239
HEIMWEH (*n*): homesickness; nostalgia (Med.) 240
HEINTZIT (*m*): heintzite; $KMg_2B_9O_{16}.8H_2O$; Sg. 2.13 241
HEIRAT (*f*): wedding, marriage, nuptials 242
HEIRATEN: to marry, take or give in marriage 243
HEIRATSAUFGEBOT (*n*): banns of marriage 244
HEIRATSSCHEIN (*m*): marriage certificate 245
HEISCHEN: to demand; require; desire; ask; postulate 246
HEISER: hoarse; husky 247
HEISS: hot; ardent; torrid; fervid; passionate 248
HEISSBLASEN (*n*): heating to white heat (Water gas production process); hot-blast 249
HEISSBLÜTIG: hot-blooded; warm-blooded 250
HEISSBRÜCHIG: hot-short 251
HEISSDAMPF (*m*): hot steam 252
HEISSDAMPFENTNAHME (*f*): hot steam outlet 253
HEISSDAMPFVENTIL (*n*): hot steam valve 254
HEISSEN: to call; name; bid; denominate; order; be (said) called; mean; be; enjoin; direct; command; signify; nominate; hoist (Flags, etc.) 255
HEISSGAR: kishy 256
HEISSGRÄDIG: difficult to fuse; difficultly fusible (of ores) 257
HEISSLUFT (*f*): hot air 258
HEISSLUFTMASCHINE (*f*): hot-air engine 259
HEISSLUFTMOTOR (*m*): hot-air motor 260
HEISSMASCHINE (*f*): hoisting (machine) engine 261
HEISS-SATINIERMASCHINE (*f*): (hot)-burnisher (Photography) 262
HEISSWASSERRÜCKLAUF (*m*): return water-(pipe) to boiler 263
HEISSWASSERTRICHTER (*m*): hot-water funnel 264
HEISSWASSERVORLAUF (*m*): discharge (pipe; piping; main) from boiler 265
HEITER: clear; serene; cheerful; merry; bright; glad 266
HEITERN: to cheer-(up); become cheerful 267
HEIZAPPARAT (*m*): heating (apparatus) appliance 268
HEIZBAR: capable of being heated 269
HEIZDAMPF (*m*): heating steam; steam for heating purposes 270
HEIZE (*f*): charge (of pig) (Steel) 271
HEIZEFFEKT (*m*): heating effect; pyrometric effect; fuel efficiency 272
HEIZEFFEKTMESSER (*m*): calorimeter 273
HEIZEN: to heat; fire (Boilers) 274
HEIZER (*m*): fireman; stoker; heater; heating (apparatus) appliance 275
HEIZERSTAND (*m*): firing floor; stoking (level) platform; place where fireman stands; footplate (of Locos.) 276
HEIZFLÄCHE (*f*): heating surface 277
HEIZFLÄCHENBELASTUNG (*f*): rating, evaporative rating, load on the heating surface, evaporation per unit of heating surface per unit of time (Boilers) 278
HEIZGAS (*n*): fuel gas; gas for heating; hot gases 279
HEIZGASVORWÄRMER (*m*): economiser, feed water heater 280
HEIZKESSEL (*m*): heating boiler (vessel; pan; copper) 281
HEIZKÖRPER (*m*): heating (body) element; radiator 282
HEIZKÖRPERLACK (*m*): varnish for radiators 283
HEIZKÖRPERVERKLEIDUNG (*f*): radiator lining 284
HEIZKRAFT (*f*): heating power; calorific (power) value 285
HEIZKRANZ (*m*): ring burner 286
HEIZLOCH (*n*): stoke-hole 287
HEIZMANTEL (*m*): heating jacket 288
HEIZMATERIAL (*n*): fuel; heating material 289
HEIZMITTEL (*n*): fuel; heating medium 290

HEIZOBERFLÄCHE (*f*): heating surface 291
HEIZÖL (*n*): heating oil 292
HEIZPLATTE (*f*): heating plate; hot-plate 293
HEIZRAUM (*m*): heating chamber; firebox, combustion chamber, furnace; stoke-hole 294
HEIZROHR (*n*): heating tube; fire-tube 295
HEIZROHRBÜNDEL (*n*): nest of (heating)-tubes 296
HEIZRÖHRE (*f*): (see **HEIZROHR**) 297
HEIZSCHLANGE (*f*): heating coil 298
HEIZSCHLANGENSYSTEM (*n*): system of heating coils 299
HEIZSTOFF (*m*): fuel 300
HEIZTÜR (*f*): fire-door 301
HEIZUNG (*f*): heating; firing; fuel; furnace; heating (plant) installation 302
HEIZUNGSANLAGE (*f*): heating (plant) installation; furnace installation; stoking plant 303
HEIZUNGSARMATUREN (*f.pl*): fittings (mountings) for heating plants 304
HEIZVERLUST (*m*): heat loss-(es) 305
HEIZVORRICHTUNG (*f*): heating (apparatus; appliance; plant; installation) arrangement) 306
HEIZWERT (*m*): heat value; calorific value (formula for finding calorific value = 81 carbon plus 290 $\left(\text{hydrogen minus } \frac{\text{oxygen}}{8}\right)$ plus 25 sulphur minus 6 water = calories per kilogram) 307
HEIZWERT BESTIMMUNG (*f*): determination of calorific (heat)-value or pyrometic-(effect) value 308
HEIZWERT, UNTERER- (*m*): lower calorific value 309
HEIZWERTUNTERSUCHUNG (*f*): calorimeter (calorimetric) test 310
HEIZWERT (*m*) **WIE VERFEUERT**: calorific value, as fired (Coal analysis) 311
HEIZWIRKUNG (*f*): heating (pyrometric) effect 312
HEIZZWECK (*m*): heating purpose 313
HEKTISCH: hectic 314
HEKTOGRAPH (*m*): hectograph 315
HEKTOGRAPHENMASSE (*f*): copying (hectograph) paste, jelly or gelatine 316
HEKTOGRAPHENTINTE (*f*): hectograph ink 317
HELD (*m*): hero, champion 318
HELDENALTER (*n*): heroic age 319
HELDENGEDICHT (*n*): epic-(poem) 320
HELDENMÜTIG: heroic 321
HELDENTAT (*f*): deed of (heroism) valour, brave deed 322
HELENIN (*n*): Helenin (from the root of *Inula helenium*); $C_{15}H_{20}O_2$; Bp. 192°C; Mp. 74°C.; Elecampane camphor, Alant camphor, alant acid anhydride, alantolactone, Inula camphor 323
HELFEN: to help; assist; be of use; avail; save; remedy; aid; succour 324
HELFER (*m*): helper; assistant 325
HELFERSHELFER (*m*): aider, abettor, accomplice 326
HELINDONFARBSTOFF (*m*): helindon colouring matter 327
HELIOCHROMIE (*f*): photochromatism (Natural colour photography) 328
HELIOCHROMOSKOP (*n*): heliochromoscope (for three colour photography) 329
HELIOFARBSTOFF (*m*): helio colouring matter 330
HELIOGRAPH (*m*): heliograph (an instrument for producing sun photographs, or for signalling by flashing the sun's rays from a mirror) 331
HELIOGRAPHIE (*f*): heliography; photography; sun (drawing, painting) photography 332
HELIOGRAVÜRE (*f*): photogravure, photo-engraving (from half-tone copper plate process) 333
HELIOLATRIE (*f*): sun-worship 334
HELIOMESSER (*m*): heliometer (originally an astronomical instrument for measuring the sun's diameter, but now used for measuring small angles) 335
HELIOMETER (*n*): heliometer (astronomical instrument for measuring small angles) (instrument for determining heat absorption in the atmosphere) 336
HELIOSKOP (*n*): helioscope (arrangement for weakening the sun's rays for telescopic observation of the sun) 337
HELIOTHERAPIE (*f*): sun-cure, treatment of disease by means of the sun (Med.) 338
HELIOTHERMOMETER (*n*): heliometer; sunshine recorder (an instrument for determining heat absorption in the atmosphere) (now superseded by optical pyrometers, thermocouples, etc.) 339
HELIOTROP (*m*): heliotrope, turnsole, cherry pie, (*Heliotropium peruvianum*); heliotrope (a variety of chalcedony) (see **CHALCEDON**); heliotrope (the colour) 340
HELIOTROPESSENZ (*f*): heliotrope (extract) essence 341
HELIOTROPIN (*n*): heliotropin, piperonal, piperonyl aldehyde; $C_6H_3(CH_2O_2)COH$; Mp. 37°C.; Bp. 263°C. 342
HELIOTROPISMUS (*m*): heliotropism, the turning of plants towards the light 343
HELIOTROPÖL (*n*): heliotrope oil 344
HELIOZENTRISCH: heliocentric, using the sun as the centre point 345
HELIOZOEN (*f.pl*): heliozoa, Rhizopoda, a division protozoa, sun animalcule (*Actinophrys sol*) 346
HELIUM (*n*): helium (He) 347
HELL: bright; brilliant; light; pale; clear; transparent; pellucid; plain; limpid; loud; shrill; luminous; distinct 348
HELLBLAU: light-blue 349
HELLBRAUN: light-brown 350
HELLDUNKEL (*n*): twilight, dusk, clare obscure, chiaroscuro 351
HELLE (*f*): brightness; luminosity; clearness; clarity; transparency; distinctness (see **HELL**) 352
HELLEGAT (*n*): storeroom 353
HELLEN: to brighten (make)-clear, clarify, lighten, make distinct 354
HELLERKRAUT (*n*): penny cress (*Thlaspi arvense*) 355
HELLFARBIG: light-coloured 356
HELLIGKEIT (*f*): (see **HELLE**) 357
HELLING (*f*): stocks, building slip (Boats) 358
HELLING (*f*), **GEDECKTE-**: roofed (covered) building slip 359
HELLINGKRAN (*m*): slipway crane 360
HELLNUSSBRAUN (*n*): light nut-brown 361
HELLROT: light-red 362
HELLSEHEN (*n*): second sight, clairvoyance 363
HELLSICHTIG: clear-sighted 364
HELLTRAN (*m*): clear, light-yellow cod-oil (see **CODÖL**) 365
HELLWEISS: clear (bright) white 366
HELM (*m*): helmet; head (alembic) of a still; cupola; dome 367

HELM (*n* and *m*): helm, rudder; handle, haft, hilt 368
HELMFÖRMIG: helmet-like; helmet-shaped; galeate-(d) (Bot.) 369
HELMITOL (*n*): helmitol $C_7H_6O_7.(CH_2)_6N_4$ 370
HELMKRAUT (*n*): helmet flower, Scutellaria, skullcap (*Scutellaria lateriflora*) 371
HELMSCHNABEL (*m*): beak (nose) of a still 372
HELMSTOCK (*m*): tiller (Naut.) 373
HEMATEIN (*n*): hematine (extracted from Logwood *Haematoxylin campechianum*) 374
HEMD (*n*): shirt; smock; mould; shell (of blast furnace) 375
HEMELLITHSÄURE (*f*): hemellitic acid 376
HEMERALOPIE (*f*): night-blindness, nyctalopia (inability to see in a poor light) (Med.) 377
HEMIALBUMOSE (*f*): semi-albumose (intermediate stage between the egg-whites and the peptones) 378
HEMIËDER (*n*): hemihedron; of (having) hemihedral form 379
HEMIËDRIE (*f*): hemihedrism 380
HEMIËDRISCH: hemihedral 381
HEMIMELLITHSÄURE(*f*): hemimellitic acid 382
HEMIMORPHIT (*m*): hemimorphite, electric calamine (see **KIESELZINKERZ**) 383
HEMISPHÄRE (*f*): hemisphere 384
HEMITROP: hemitrope; twinned (Cryst.) 385
HEMMEN: to stop; curb; arrest; check; stem; brake; clog; retard; hinder; restrain; dam; restrict; obstruct; constrict; stanch (Blood) 386
HEMMKETTE (*f*): drag chain 387
HEMMNIS (*n*): obstruction; check; curb; restriction; hindrance (see **HEMMEN**) 388
HEMMSCHUH (*m*): drag-(shoe) 389
HEMMUNG (*f*): stanching (Blood); stopping; obstruction; stoppage; check; stemming; restriction; catch; escapement; constriction 390
HEMMUNGSEBENE (*f*): checking (arresting) plane 391
HENGST (*m*): stallion; jack 392
HENKEL (*m*): handle; lug; ear; hook; ring 393
HENNABLÄTTER (*n.pl*): Henna leaves (leaves of *Lawsonia alba*) 394
HENNE (*f*): hen; orpine; live-long; tuberous stonecrop (*Sedum telephium*) 395
HENNEGATT (*n*): helm port (Naut.) 396
HEPATALGIE (*f*): liver complaint, hepatic colic, hepatalgia (Med.) 397
HEPATICA (*f*): liverwort, liverleaf (*Anemone hepatica*) 398
HEPATISCH: hepatic 399
HEPATISCHE LUFT (*f*): hydrogen sulphide (see **SCHWEFELWASSERSTOFF**) 400
HEPATISCHER GERUCH (*m*): an odour similar to that of hydrogen sulphide 401
HEPATITIS (*f*): inflammation of the liver, hepatitis (Med.) 402
HEPATOLOGIE (*f*): the science of the liver (Med.) 403
HEPTACARBOCYCLISCH: heptacarbocyclic 404
HEPTAHYDRAT (*n*): heptahydrate ($7H_2O$) 405
HEPTAN (*n*): Heptane, dipropylmethane, methyl hexane, heptyl hydride; C_7H_{16}; Sg. 0.694; Bp. 95-100°C. 406
HEPTANAPHTEN (*n*): heptanaphthene 407
HEPTINSÄURE (*f*): heptinic acid 408
HEPTYLALDEHYD (*n*): heptylic aldehyde 409
HEPTYLALKOHOL (*m*): heptylic alcohol; $CH_3(CH_2)_6OH$; Sg. 0.83; Mp. −36.5°C.; Bp. 175°C. 410
HEPTYLSÄURE (*f*): heptylic acid, oenanthylic (oenanthic) acid, hepotic acid; $CH_3(CH_2)_5COOH$; Sg. 0.92; Mp. −10.5°C.; Bp. 223°C. 411
HER: here; hither; since; ago 412
HERAB: down; downwards; down here 413
HERABDRÜCKEN: to depress; press down; force down 414
HERABFALLEN: to fall down, sink, descend 415
HERABFLIESSEN: to flow down; run down 416
HERABGEHEN: to go down, descend, sink 417
HERABHÄNGEN: to hang down, depend 418
HERABLASSEN: to lower; let down; condescend; stoop; be (courteous; sociable) affable; deign 419
HERABMINDERN: to diminish; decrease; lessen; reduce; modify 420
HERABSETZEN: to degrade; decrease; disparage; reduce; lower; lessen; depress; depreciate; abate; throttle 421
HERABSETZUNG (*f*): degradation; lowering; decrease; reduction; throttling; fall; drop; undervaluation; abasement 422
HERABSINKEN: to fall, sink, become (debased) degraded; submerge 423
HERABSTIMMEN: to lower, tune down, depress, deject 424
HERABSTÜRZEN: to throw down, precipitate, plunge into the depths 425
HERABWÜRDIGEN: to degrade, debase, abase, depreciate, disparage 426
HERALDIK (*f*): heraldry 427
HERALDISCH: heraldic 428
HERAN: near; up; on; towards; along; here; onwards 429
HERANBILDEN: to bring up; educate 430
HERANKOMMEN: to approach; come towards 431
HERANNAHEN: to approach, draw near 432
HERANRÜCKEN: to advance, push (forward) towards, draw near 433
HERANWACHSEN: to grow up 434
HERANZIEHEN: to attract, draw towards 435
HERAUF: up; up here; upwards 436
HERAUFGEHEN: to ascend, go up 437
HERAUFKOMMEN: to rise, advance, progress, come up 438
HERAUFSTEIGEN: to ascend, climb, rise 439
HERAUFWACHSEN: to spring (grow) up 440
HERAUFZIEHEN: to (draw, pull, haul) up 441
HERAUS: out; out here; outwards; forth 442
HERAUSARBEITEN: to (work) struggle out 443
HERAUSBEKOMMEN: to get back; get out; elicit; solve; obtain; receive 444
HERAUSBRINGEN: to get (force) out; turn (bring) out; draw (find) out 445
HERAUSFINDEN: to find out, discover 446
HERAUSFORDERN: to challenge; provoke; defy 447
HERAUSGABE (*f*): editing; edition; publication; issue; giving up; surrender 448
HERAUSGEBEN: to edit; publish; issue; give (out) up; give back; deliver up; surrender 449
HERAUSGEBER (*m*): editor; publisher 450
HERAUSGREIFEN: to single out; choose; pick out; extract 451
HERAUSHEBEN: to (lift, pick, take) out 452

HERAUSKOMMEN: to come (get; work) out; be (edited) published; result; amount; appear; become known; issue 453
HERAUSLÖSEN: to dissolve out 454
HERAUSNEHMBAR: removable from 455
HERAUSNEHMEN: to take (pick) out of; choose; remove from; extract 456
HERAUSPRESSEN: to press (force) out, extort, squeeze out 457
HERAUSRAGEN: to project; jut out 458
HERAUSREDEN: to speak out, speak one's mind 459
HERAUSREISSEN: to tear (pull)out, extricate 460
HERAUSSCHLAGEN: to beat out; strike from; make-(out)-of; obtain 461
HERAUSSETZEN: to set (put) out, eject 462
HERAUSSTEHEN: to stand out, project 463
HERAUSSTELLEN: to turn (put) out; prove 464
HERAUSTRETEN: to protrude; emerge; step out of; withdraw; retire; (*n*): extravasation; protuberance (Med.) 465
HERAUSWACHSEN: to grow out, develop 466
HERAUSZIEHEN: to draw (pull) out, extract 467
HERB: harsh; sour; sharp; acid; tart; rough; raw; bitter; astringent; severe; austere 468
HERBE: (see **HERB**) 469
HERBEI: hither; here; near; on; forward 470
HERBEIFÜHREN: to bring-(about); cause; produce; occasion; induce 471
HERBEISCHAFFEN: to bring; collect; procure; furnish; produce 472
HERBERGE (*f*): inn; shelter; publichouse; harbour 473
HERBERGEN: to shelter, harbour 474
HERBIGKEIT (*f*): harshness, acerbity, bitterness 475
HERBORIST (*m*): herbalist 476
HERBST (*m*): autumn 477
HERBSTFIEBER (*n*): (see **HEUFIEBER**) 478
HERBSTLICH: autumnal 479
HERBSTROSE (*f*): hollyhock (*Althœa rosea*) 480
HERBSTZEITLOSE (*f*): meadow (wild) saffron, Colchicum; meadow (autumn) crocus; (*Colchicum autumnale*) 481
HERBSTZEITLOSESAMEN (*m*): meadow (wild) saffron seed; colchicum seed (*Semen colchici*) 482
HERBSTZEITLOSEWURZEL (*f*): meadow (wild) saffron root; colchicum root (*Radix colchici*) 483
HERCYNIT (*m*): Hercynite; $FeAl_2O_4$; Sg. 3.93 484
HERD (*m*): hearth; centre; home; focus; house; seat; fireplace; source 485
HERDE (*f*): herd; flock; drove; crowd; multitude 486
HERDEISEN (*n*): hearth plate; poker 487
HERDFORM (*f*): hearth mould 488
HERDFRISCHARBEIT (*f*): refinery process; hearth process (Metal) 489
HERDFRISCHEISEN (*n*): hearth refined iron, bloomery iron 490
HERDFRISCHEN (*n*): open hearth refining, open hearth process (Iron); litharge reduction (Lead) 491
HERDFRISCHPROZESS (*m*): refinery (open-hearth) process (Iron); Siemens-Martin process 492
HERDFRISCHROHEISEN (*n*): refinery pig iron 493
HERDFRISCHSCHLACKE (*f*): refinery (slag) cinders 494
HERDFRISCHSTAHL (*m*): refined (fined) steel; hearth refined steel; open-hearth steel; Siemens-Martin open-hearth steel 495
HERDFRISCHVERFAHREN (*n*): (see **HERDFRISCH-PROZESS**) 496
HERDGEWÖLBE (*n*): furnace crown, hearth arch 497
HERDGUSS (*m*): open sand casting 498
HERDLÖFFEL (*m*): ladle 499
HERDPROBE (*f*): assay 500
HERDRAUM (*m*): heating chamber 501
HERDSCHMELZVERFAHREN (*n*): (see **HERDFRISCH-PROZESS**) 502
HERDSOLE (*f*): sole; hearth (bed) bottom 503
HEREIN: in; come in!; enter!; in here; inwards 504
HEREINBEZIEHEN (*n*): incorporation 505
HEREINBRECHEN: to break in, force a way in 506
HEREINLASSEN: to let in, admit 507
HEREINZUBEZIEHEN: to incorporate 508
HERFLIESSEN: to flow (from) on, issue, be derived, originate 509
HERFÜHREN: to lead (hither) on; usher in 510
HERGANG (*m*): affair; circumstances; course of events 511
HERGEBEN: to deliver; yield; hand over; give up; surrender 512
HERGEBRACHT: established; conventional; customary; usual 513
HERGEHÖRIG: pertinent, apposite 514
HERHALTEN: to hold (forth) out; suffer; pay; offer; tender; submit 515
HERING (*m*): herring 516
HERINGSLAKE (*f*): herring pickle 517
HERINGSMILCH (*f*): soft-roe (of herrings) 518
HERINGSÖL (*n*): (see **HÄRINGSÖL**) 519
HERINGSTRAN (*m*): (see **HÄRINGSÖL**) 520
HERKOMERTYPIE (*f*): art printing process after Herkomer 521
HERKOMMEN: to come (descend) from; advance; originate; be derived; arise; approach; proceed; (*n*): origin-(ation); descent; derivation; extraction; usage; custom 522
HERKÖMMLICH: customary; usual; traditional 523
HERKOTYPIE (*f*): (see **HERKOMERTYPIE**) 524
HERKUNFT (*f*): origin-(ation); arrival; derivation; extraction; descent 525
HERLEITEN: to derive; deduce; conduct; lead from 526
HERLITZE (*f*): (see **HARTBAUM**) 527
HERMAPHRODIT (*m*): hermaphrodite (an organism in which the two sexes are combined in one) 528
HERMELIN (*n*): ermine 529
HERMETISCH: hermetic-(ally) 530
HERMETISCHER VERSCHLUSS (*m*): hermetical seal 531
HERMETISCH VERSCHLOSSEN: hermetically sealed 532
HERNACH: hereafter; afterwards 533
HERNEHMEN: to deduce, derive, obtain (get) from 534
HEROIN (*n*): Heroin, diacetylmorphine hydrochloride (*Diacetylmorphinum hydrochloricum*); $C_{17}H_{17}NO(O.CO.CH_3)_2.Cl$; Mp. 173°C. 535
HEROINE (*f*): heroine 536
HEROISCH: heroic 537
HEROISMUS (*m*): heroism 538
HERPES (*f*): Herpes Zoster, shingles (Med.) 539
HERRECHNEN: to enumerate; specify 540

HERRENWÄSCHEGESCHÄFT (*n*): shirt-making trade 541
HERRICHTEN: to prepare; arrange; adjust; season (Wood); fit up; assemble; erect; put together 542
HERRLICH: magnificent; grand; lordly; excellent; glorious; splendid 543
HERRLICHKEIT (*f*): grandeur; splendour; glory; excellence; magnificence 544
HERRSCHAFT (*f*): command, dominion, control, mastery 545
HERRSCHBEGIERIG: ambitious, desirous of power 546
HERRSCHEN: to rule; reign; govern; prevail; sway; obtain; exist; be prevalent; predominate 547
HERRSCHSÜCHTIG: imperious, thirsting for power 548
HERRÜHREN: to proceed from; be caused by; be due to; be on account of; originate from; be attributable to 549
HERSAGEN: to recite; repeat; rehearse 550
HERSTAMMEN: to descend, originate, be derived, be extracted 551
HERSTELLEN: to produce; establish; construct; manufacture; make; prepare; restore; repair; renovate; recondition; reconstitute 552
HERSTELLER (*m*): producer; maker; constructor; manufacturer; restorer; repairer 553
HERSTELLUNG (*f*): restoration, recovery, re-establishment, production, construction, manufacture, preparation, repair 554
HERSTELLUNGSMASCHINEN (*f.pl*): manufacturing machinery, machinery for (making, producing, constructing) manufacturing 555
HERSTELLUNGSMITTEL (*n*): restorative 556
HERSTELLUNGSWEISE (*f*): manner of (preparation) production, method of (construction) manufacture 557
HERTZSCHE WELLE (*f*): Hertzian wave (electromagnetic wave of low frequency) 558
HERÜBER: over-(here); across; hither 559
HERUM: round; about 560
HERUMDREHEN: to misconstrue, turn round 561
HERUMSCHIFFEN: to sail round, circumnavigate, double, sail about 562
HERUMSCHWENKEN: to wave about, flourish, swing, rinse, brandish 563
HERUNTER: down (here); downwards 564
HERUNTERBRINGEN: to bring down, reduce 565
HERUNTERKOMMEN: to come down; decay; decline; descend (fall) in the social scale; be reduced 566
HERUNTERLASSEN: to let down; lower; shorten (Sails) 567
HERUNTERSETZEN: to put down; reduce; lower; undervalue; degrade; depreciate; deprecate 568
HERVOR: forward; out; forth 569
HERVORBRINGEN: to produce; bring (about) forth; cause; beget; elicit; procreate; utter 570
HERVORGEHEN: to arise; result; go forth; be (shown) illustrated by; proceed 571
HERVORHEBEN: to make (render) prominent; bring out; emphasize; raise; make conspicuous; lay stress on; set off 572
HERVORRAGEN: to project; stand out; be (prominent) noteworthy; excel 573
HERVORRAGEND: projecting; prominent; outstanding; signal; salient; conspicuous 574
HERVORRUFEN: to call forth; bring (into play) about; produce; develop; evoke; cause 575
HERVORRUFUNG (*f*): development (Photography) 576
HERVORSTEHEND: prominent; projecting; outstanding; conspicuous 577
HERVORTAUCHEN: to appear, emerge, come forth 578
HERVORTRETEN: to stand out; step forward; appear; be conspicuous 579
HERVORTUN: to distinguish oneself; push (put) oneself forward 580
HERZ (*n*): heart; core; courage; centre; sympathy; marrow 581
HERZADER (*f*): aorta (Med.) 582
HERZAHLEN: to pay (down) up, pay out 583
HERZÄHLEN: to enumerate, count out, reckon up 584
HERZARTERIE (*f*): aorta (Med.) 585
HERZBEKLOMMEN: anxious, oppressed 586
HERZBEUTEL (*m*): pericardium (Med.) (see HERZHAUT) 587
HERZBLATT (*n*): diaphragm (Med.); darling; dear 588
HERZBRÄUNE (*f*): angina pectoris (Med.) 589
HERZBRECHEND: heart-breaking 590
HERZEN: to embrace; hug; caress 591
HERZERWEITERUNG (*f*): dilatation of the heart 592
HERZFELL (*n*): pericardium (see HERZHAUT) 593
HERZFELLENTZÜNDUNG (*f*): pericarditis, inflammation of the pericardium (Med.) 594
HERZFÖRMIG: cordate; heart-shaped; cordiform 595
HERZGEGEND (*f*): cardiac region (Med.) 596
HERZGEKRÖSE (*n*): mesocardium (the connecting membrane of the heart) (Med.) 597
HERZGESPANN (*n*): motherwort (*Leonurus cardiaca* (Bot.); heart-burn; cardialgia (Med.) 598
HERZGRUBE (*f*): procardium, cardiac region (Med.) 599
HERZHAFT: hearty; brave; strong (of taste); courageous; stout-hearted 600
HERZHAUT (*f*): the fibro-serous sac of the heart (the inner part of which is called the endocardium and the outer, the pericardium) (Med.) 601
HERZHAUT (*f*), ÄUSSERE-: pericardium, the outer layer of the sac of the heart (Med.) 602
HERZHAUT (*f*), INNERE-: endocardium, the inner layer of the sac of the heart (Med.) 603
HERZIG: hearty; dear; lovely; charming; beloved; sweet 604
HERZKAMMER (*f*): chamber of the heart; auricle (the receiving chamber); ventricle (the discharging chamber) (Med.) 605
HERZKLAPPE (*f*): cardiac valve; valve of the heart (Med.) 606
HERZKLOPFEN (*n*): palpitation (of the heart) (see also PALPITATION) (Med.) 607
HERZKRANKHEIT (*f*): heart disease (Med.) 608
HERZKRANKHEIT (*f*), ANGEBORENE-: congenital heart disease (Med.) 609
HERZLICH: hearty; heartfelt; loving; affectionate; extremely; cordial-(ly); sincere 610
HERZLOS: heartless; unfeeling; unsympathetic; faint-hearted 611
HERZMITTEL (*n*): heart remedy 612

HERZPOLYPEN (*m.pl*): thrombosis of the heart, heart clotting (Med.) 613
HERZSENTE (*f*): main breadth line (of ships) 614
HERZSTÄRKEND: cordial, cardiac 615
HERZSTÜCK (*n*): cross-over, crossing, frog, (catch)-points (Railways) 616
HERZSTÜCKSPITZE (*f*): crossing-nose (Railways) 617
HERZTONICUM (*n*): heart (stimulant) tonic 618
HERZU: here; hither; near 619
HERZVORKAMMER (*f*): auricle (see **HERZKAMMER**) 620
HERZWASSER (*n*): pericardial fluid (Med.) 621
HERZWEH (*n*): heartache; grief; heart-burn; cardialgia (Med.) 622
HERZWUNDE (*f*): wound of the heart (Med.) 623
HERZZERREISSEND: heart-rending 624
HESPERIDINSÄURE (*f*): hesperidic acid 625
HESSENFLIEGE (*f*): Hessian fly; wheat fly; (*Cecidomyia destructor*) 626
HESSISCH: Hessian 627
HESSISCH-FARBSTOFF (*m*): Hessian colouring matter 628
HESSIT (*m*): hessite (a rare silver telluride); Ag_2Te; Sg. 8.38 629
HETEROATOMIG: heteroatomic 630
HETEROCYCLISCH: heterocyclic 631
HETERODOX: heterodox 632
HETERODOXIE (*f*): heterodoxy 633
HETEROGEN: heterogeneous-(ly) 634
HETEROGENITÄT (*f*): heterogeneity 635
HETEROLOG: heterologous 636
HETEROMORPHIT (*m*): heteromorphite (see **JAMESONIT**) 637
HETEROTOMER FELDSPAT (*m*): albite 638
HETEROZYKLISCH: heterocyclic 639
HETOKRESOL (*n*): hetocresol; meta-cresol cinnamate 640
HETOL (*n*): hetol (see **NATRIUM, ZIMTSAURES-**) 641
HETRALIN (*n*): (see **DIOXYBENZOLHEXAMETHYL-ENTETRAMIN**) 642
HETZEN: to hunt; set on; bait; incite; chase; pursue; provoke; harass 643
HEU (*n*): hay 644
HEUASTHMA (*f*): (see **HEUFIEBER**) 645
HEUBINDER (*m*): trusser of hay 646
HEUBLUMEN (*f.pl*): graminis (triticum, couch-grass, dog-grass) flowers, (*flores graminis*) (flowers of *Agropyron repens*) 647
HEUBODEN (*m*): hay-loft 648
HEUBUND (*n*): truss of hay 649
HEUCHELEI (*f*): hypocrisy; (dis)-simulation; cant 650
HEUCHELN: to dissemble; pretend; feign; (dis)-simulate; be hypocritical 651
HEUCHLER (*m*): hypocrite 652
HEUCHLERISCH: hypocritical 653
HEUER: this (year) season; (*m*): haymaker; (*f*): hire; rent; lease 654
HEUERN: to hire; rent; lease 655
HEUERNTE (*f*): hay harvest; hay crop 656
HEUFIEBER (*n*): hay fever, *Catarrhus æstivus* (Med.) 657
HEUGABEL (*f*): pitch (hay) fork 658
HEULANDIT (*m*): heulandite; $CaAl_2Si_6O_{16}.5H_2O$; Sg. 2.15 659
HEULEN: to howl; yell; roar; scream 660
HEUPFERD (*n*): (green)-grasshopper; locust; (member of the family of *Locustidæ*) 661
HEURECHEN (*m*): hay-rake 662
HEUSAME (*m*): hay-seed 663
HEUSCHOBER (*m*): hay-rick, hay-stack 664
HEUSCHREKE (*f*): (see **HEUPFERD**) 665
HEUSCHREKENBAUM (*m*): locust tree (*Robinia pseudacacia*) 666
HEUSCHREKENBEKÄMPFUNGSMITTEL (*n*): locust-fuge; locust (destroyer) killer 667
HEUSSLERSCHE LEGIERUNG (*f*): Heussler's alloy (copper-aluminium-manganese alloy of varying composition, attracted by a magnet) 668
HEUTE: today 669
HEUTIG: modern; of today; present-day; up-to-date 670
HEUTZUTAGE: now-a-days; at the present time; today 671
HEXACHLORÄTHAN (*n*): hexachloroethane, carbon trichloride, perchloroethane, carbon hexachloride, tetrachloroethylene dichloride; C_2Cl_6; Sg. 1.999; Bp. 185°C. sublimes 672
HEXACHLORBENZOL (*n*): hexachlorobenzene; C_6Cl_6; Sg. 1.569; Mp. 227°C.; Bp. 326°C. 673
HEXAEDER (*n*): hexahedron 674
HEXAEDRON (*n*): hexahedron 675
HEXAHYDRAT (*n*): hexahydrate ($6H_2O$) 676
HEXAHYDROBENZOL (*n*): cyclohexane; C_6H_{12} (see **CYCLOHEXAN**) 677
HEXAHYDROPHENOL (*n*): (see **HEXALIN**) 678
HEXAHYDROPYRAZIN (*n*): (see **PIPERAZIN**) 679
HEXAHYDROPYRIDIN (*n*): (see **PIPERIDIN**) 680
HEXAL (*n*): (see **HEXAMETHYLENTETRAMIN, SULFOSALIZYLSAURES-**) 681
HEXALIN (*n*): hexalin, cyclohexanol, hexahydrophenol, naphthenol; $C_6H_{11}OH$; Sg. 0.944; Mp. 16.5°C.; Bp. about 160°C. 682
HEXAMETHYLENTETRAMIN (*n*): hexamethylenetetramine, cystamin, formin, cystogen, aminoform, urotropin, hexamine; $N_4(CH_2)_6$ 683
HEXAMETHYLENTETRAMIN-ÄTHYLBROMID (*n*): see **BROMALIN**) 684
HEXAMETHYLENTETRAMIN (*n*), **NUKLEIN-SAURES-**: hexamethylenetetramine nucleinate 685
HEXAMETHYLENTETRAMIN (*n*), **SULFOSALIZYL-SAURES-**: hexamethylenetetramine sulfo-(thio-) salicylate; $(CH_2)_6N_4.SO_3H.C_6H_2(OH)COOH+H_2O$ 686
HEXAMETRISCH: hexametric-(al) 687
HEXAN (*n*): hexane, caproyl hydride, hexyl hydride; C_6H_{14}; Sg. 0.66; 69°C. 688
HEXANAPHTEN (*n*): hexanaphthene; cyclohexane; C_6H_{12} 689
HEXANDISÄURE (*f*): (see **ADIPINSÄURE**) 690
HEXANITROÄTHAN (*n*): hexanitroethane; $C_2(NO_2)_6$; Mp. 142°C. 691
HEXANITRODIPHENYL (*n*): hexanitrodiphenyl (an explosive); $(NO_2)_3H_2C.CH_2(NO_2)_3$; Mp. 230°C. 692
HEXAVANADINSÄURE (*f*): hexavanadic acid 693
HEXE (*f*): witch; hag; sorceress 694
HEXEN: to practice (sorcery) witchcraft, to bewitch 695
HEXENKRAUT (*n*): mandragora; devil's apple; mandrake; enchanter's nightshade (*Circæa lutetiana* and *Mandragora vernalis*) 696
HEXENKREIS (*m*): magic circle 697
HEXENKUNST (*f*): witchcraft 698
HEXENMEHL (*n*): witch meal; lycopodium; vegetable sulphur (*Lycopodium clavatum*) 699
HEXENMEISTER (*m*): sorcerer; wizard 700
HEXENSCHUSS (*m*): lumbago; rheumatic twinge; shooting pain 701

HEXEREI (*f*): sorcery, witchcraft; bewitching 702
HEXINSÄURE (*f*): hexinic acid 703
HEXOPHAN (*n*): (see **OXYPHENYLCHINOLINDIKARBONSÄURE**) 704
HEXORAN (*n*): hexoran (carbon tetrachloride in solution; 90% CCl_4) 705
HEXOSE (*f*): hexose (one of the grape sugar group of sugars) (see **DEXTROSE**) 706
HIBRIDISCH: hybrid, mongrel 707
HICKORYÖL (*n*): Hickory oil (from seeds of *Carya alba*) 708
HIEB (*m*): stroke; cut; slash; blow 709
HIENFONGESSENZ (*f*): hienfong essence 710
HIER: here; present; at this juncture 711
HIERAN: hereon; hereat; at this; on this; hereupon; by this 712
HIERAUF: hereupon; upon this; up here; thereafter; then; afterwards 713
HIERAUS: hence; from (this) here 714
HIERAUSSEN: out here 715
HIERBEI: hereby; herewith, enclosed, with this 716
HIERDURCH: through here; by this means, due to this 717
HIERFÜR: for this, in place (instead) of this, herefor 718
HIERGEGEN: (as)-against this, counter to this, on the other hand 719
HIERHER: hither; here; this way 720
HIERHIN: hither; this way; in this direction; to this place 721
HIERIN: herein, in this 722
HIERMIT: herewith; with this; hereupon; enclosed 723
HIERNACH: hereafter; according to this; from this; herefrom; after this 724
HIEROGLYPHIK (*f*): hieroglyphics 725
HIERSEIN (*n*): presence-(here) 726
HIERÜBER: over here; regarding this; on this account; hereat 727
HIERUM: hereabout; concerning this; on this account 728
HIERUNTER: under this, among these 729
HIERVON: of (from) this; hereof; herefrom 730
HIERZU: hereto; to this; moreover 731
HIESIG: of (in) this place, city or country; this; the present 732
HIESIGE (*pl*): people belonging to this place 733
HILFE (*f*): help; aid; relief; succour; remedy; assistance; assistant; auxiliary; accessory; stand-by; subsidiary (see **HÜLFE**) 734
HILFELEISTUNG (*f*): help, aid, assistance, relief, succour 735
HILFLICH: helpful; adjuvant, adjutory (Med.) 736
HILFLOS: helpless 737
HILFREICH: helpful 738
HILFSARBEITER (*m*): assistant; helper; mate; colleague 739
HILFSBEDÜRFTIG: needy, in want of (aid, help) assistance, indigent 740
HILFSBEIZE (*f*): auxiliary mordant 741
HILFSDÜNGER (*m*): auxiliary fertilizer 742
HILFSEINRICHTUNG (*f*): auxiliary (Mechanical) 743
HILFSGELD (*n*): subsidy 744
HILFSGENOSS (*m*): confederate, assistant, ally, colleague 745
HILFSKESSEL (*m*): donkey boiler, spare boiler, stand-by boiler 746
HILFSKREUZER (*m*): merchant cruiser 747
HILFSLEHRER (*m*): assistant (master) teacher 748
HILFSMASCHINE (*f*): donkey engine, spare engine; stand-by engine, auxiliary engine, pilot engine; (*pl*): auxiliary (standby) plant or machinery; auxiliaries (of machinery) 749
HILFSMITTEL (*n*): help; aid; expedient; remedy; adjuvant (Med.) 750
HILFSQUELLE (*f*): resource; expedient 751
HILFSRELAIS (*n*): auxiliary relay 752
HILFSSCHIEBER (*m*): auxiliary (slide)-valve 753
HILFSSTAG (*n*): spring stay (Naut.) 754
HILFSSTEUER (*f*): subsidy (of taxation) 755
HILFSSTROM (*m*): auxiliary current 756
HILFSTEILUNG (*f*): auxiliary (graduation) division 757
HILFSTOFF (*m*): auxiliary (accessory) product, stuff or medium 758
HILFSTURBINE (*f*): auxiliary (stand-by) turbine 759
HILFSVENTIL (*n*): auxiliary valve 760
HILFSVORRICHTUNG (*f*): auxiliary (stand-by) device, contrivance, appliance or arrangement 761
HILFSZEITWORT (*n*): auxiliary verb 762
HIMBEERÄTHER (*m*): raspberry (ether) essence 763
HIMBEERE (*f*): raspberry (*Rubus idæus*) 764
HIMBEERÖL (*n*): raspberry oil 765
HIMBEERSAFT (*m*): raspberry juice 766
HIMBEERSPAT (*m*): rhodocrosite (a natural manganese carbonate); Sg. 3.45 (see also **MANGANCARBONAT**) 767
HIMBEERSTAUDE (*f*): raspberry-bush 768
HIMBEERSTRAUCH (*m*): raspberry-bush 769
HIMMEL (*m*): sky; heaven; climate; canopy; firmament 770
HIMMELBLAU: sky-blue; azure; cobalt; (*n*): coeruleum (a cobalt colour) (see **ZÖRULEUM**) 771
HIMMELBROT (*n*): manna (a saccharine exudation from tamarisk) (*Tamarix mannifera*) (see also **MANNA**) 772
HIMMELFAHRTSFEST (*n*): the feast of the Ascension 773
HIMMELSBESCHREIBUNG (*f*): astronomy (science of the stars); uranography (description of the stars visible to the naked eye) 774
HIMMELSBOGEN (*m*): the vault of heaven 775
HIMMELSERSCHEINUNG (*f*): celestial (phenomenon) apparition, heavenly portent 776
HIMMELSFESTE (*f*): firmament 777
HIMMELSGEGEND (*f*): point of the compass; quarter of the heavens 778
HIMMELSGERSTE (*f*): naked barley 779
HIMMELSGEWÖLBE (*n*): the vault of heaven, firmament 780
HIMMELSGÜRTEL (*m*): zone 781
HIMMELSKARTE (*f*): map (chart) of the heavens 782
HIMMELSKORN (*n*): naked barley 783
HIMMELSKÖRPER (*m*): celestial (heavenly) body 784
HIMMELSKOST (*f*): ambrosia (the food or elixir of the Gods) 785
HIMMELSKUNDE (*f*): astronomy; science of the stars 786
HIMMELSLÄNGE (*f*): longitude 787
HIMMELSLEHRE (*f*): uranology 788
HIMMELSLUFT (*f*): ether 789
HIMMELSMEHL (*n*): earthy gypsum (see **GYPS**) 790

HIMMELSMESSKUNST (*f*): uranometry, uranometria (a catalogue and charts of the stars visible to the naked eye) 791
HIMMELSPFERD (*n*): (see HEUPFERD) 792
HIMMELSPUNKT (*m*): point of the compass; quarter of the heavens 793
HIMMELSRAND (*m*): horizon 794
HIMMELSRICHTUNG (*f*): point of the compass; quarter of the heavens 795
HIMMELSSCHLÜSSEL (*m*): primrose; primula (*Primula vulgaris*) 796
HIMMELSSCHLÜSSELBLÜTEN (*f.pl*): primrose-(flowers) (*flores primulæ*) 797
HIMMELSSTRICH (*m*): zone; latitude; climate 798
HIMMELSTEIN (*m*): sapphire (see KORUND, EDLER-) 799
HIMMELSTRANK (*m*): nectar (the drink of the Gods) 800
HIMMELSZEICHEN (*n*): zodiacal sign (one of the twelve signs of the Zodiac) 801
HIMMELSZELT (*n*): canopy of heaven 802
HIMMLICH: celestial; heavenly 803
HIN: there; thither; away; out; along; gone; hence 804
HINAB: down-(there); downwards 805
HINABGEHEN: to go down, descend, fall, sink 806
HINABLASSEN: to let down, lower 807
HINABSTÜRZEN: to precipitate, cast (throw) down, crash (fall) down 808
HINAN: up-(there); upwards; towards 809
HINANSTEIGEN: to ascend, climb, mount, go up, rise 810
HINANZIEHEN: to draw towards, attract, march (towards) upwards 811
HINAUF: up-(there); upwards 812
HINAUFKOMMEN: to come up, struggle up-(wards), fight one's way to the top 813
HINAUFSTEIGEN: to mount, climb, ascend, step up, rise 814
HINAUS: out; beyond; forth; outwards 815
HINAUSGEHEN: to go out, surpass, transcend 816
HINAUSSCHIEBEN: to expel; push out; defer; postpone; prolong 817
HINAUSTREIBEN: to drive out, expel, eject 818
HINAUSWASCHEN: to wash out 819
HINAUSWEISEN: to turn (throw) out, expel, eject, banish 820
HINAUSWEISUNG (*f*): ejection, expulsion, extradition 821
HINAUSZIEHEN: to stretch-(out), prolong, protract, pull out, set out 822
HINBEGEBEN: to repair (resort) to 823
HINBLICK (*m*): look; view; regard 824
HINBRINGEN: to carry; pass (Time); squander 825
HINDERN: to hinder, prevent, impede, check 826
HINDERNIS (*n*): hindrance; obstacle; impediment 827
HINDURCH: through; out 828
HINEIN: in-(side); into 829
HINEINDRINGEN: to penetrate; force a way in-(to) 830
HINEINFINDEN: to become familiar with, become used to, understand 831
HINEINRAGEN: to be keyed into (Brickwork), be let into, project into 832
HINFAHREN: to go; drive; pass away; die; depart; convey (carry) to 833
HINFALLEN: to fall down, decay 834
HINFÄLLIG: decaying; frail; perishable; weak 835
HINFÄLLIG SEIN: not to hold good any longer, not to be reliable any longer 836
HINFORT: henceforth; future 837
HINGEBEN: to give up; surrender; resign (oneself) to; indulge in 838
HINGEGEN: on the contrary; in contradistinction to; on the other hand 839
HINGEHEN: to go; pass; elapse 840
HINGESTRECKT: stretched out, prostrate 841
HINHALTEN: to hold (out) off; proffer; delay; defer; put off 842
HINKEN: to limp; be lame; be imperfect 843
HINLANGEN: to reach to, suffice 844
HINLÄNGLICH: sufficient; adequate; enough 845
HINLENKEN: to turn to, deflect to, guide to, steer to 846
HINNEHMEN: to take away from; carry away; ravish; transport; take (with) along; bear; put up with; suffer 847
HINNEIGEN: to incline (bend) towards 848
HINPASSEN: to fit, suit, be fit, be suitable, match 849
HINRAFFEN: to snatch (sweep, take) away 850
HINREICHEN: to suffice; do; hand; reach over 851
HINREICHEND: sufficient; adequate; enough 852
HINREISSEN: to tear (carry) away or along; snatch away; transport; charm; delight 853
HINREISSEND: charming, enchanting, delightful 854
HINRICHTEN: to turn; direct; execute; spoil; ruin 855
HINSCHAFFEN: to convey, transport 856
HINSCHIFFEN: to ship (convey, transport) to, sail to-(wards) 857
HINSCHLÄNGELN: to wind, twist, meander 858
HINSCHREIBEN: to write down 859
HINSCHWINDEN: to pass away, disappear, vanish 860
HINSICHT (*f*): respect; regard; view; consideration 861
HINSICHTLICH: with regard to; regarding; in respect of; touching; anent 862
HINSINKEN: to sink (fall) down; descend; faint; swoon 863
HINSTRECKEN: to stretch (out) forth, knock down 864
HINSTRICH (*m*): departure, exit, passing (flowing) away 865
HINSTURZ (*m*): downfall; forward dash 866
HINTAN: behind; after 867
HINTEN: behind; aft; in the rear; at the back-(end) 868
HINTER: behind; back; after; hind-(er); at the rear; less-(than) 869
HINTERBACKE (*f*): buttock 870
HINTERBLEIBEN: to remain-(behind), survive 871
HINTERBLIEBENE (*m.* and *f*): survivor 872
HINTERBRINGEN: to give information of; inform of 873
HINTERBRINGUNG (*f*): information 874
HINTERDARM (*m*): rectum (Anat.) 875
HINTERDECK (*n*): poop (Naut.) 876
HINTER DEM- (or DER): outlet, exit (Mech.) 877
HINTERDREHBANK (*f*): backing-off lathe 878
HINTERDREHEN: to turn back 879
HINTEREINANDER: one after another; in succession; in series (Elect.) 880

HINTEREINANDER LIEGEND: arranged (set) one behind the other, arranged in series, arranged in tandem (referring to two only), set (back-to-back) in battery 881

HINTEREINANDER SCHALTUNG (*f*): series connection (Elect.); tanden connection 882

HINTERGEBÄUDE (*n*): outhouse; building at the rear; rear (back) part of a building 883

HINTERGEHEN: to go behind; deceive; impose upon 884

HINTERGRUND (*m*): background (Phot., etc.); rear 885

HINTERHALB: behind, at the rear of 886

HINTERHALT (*m*): reserve; ambush 887

HINTERHALTEN: to hide, hold back, conceal, reserve 888

HINTERHAUPT (*n*): occiput; back part of head (Anat.) 889

HINTERHER: behind; after-(wards) 890

HINTERKLEIDEN (*n*): backing (Photography) 891

HINTERKOPF (*m*): (see HINTERHAUPT) 892

HINTERLASS (*m*): estate, bequest 893

HINTERLASSEN: to leave-(behind); bequeath 894

HINTERLEGEN: to deposit; consign 895

HINTERLEGUNG (*f*): offer, lodgment, deposit-ing (of a security) 896

HINTERLEGUNGSDATUM (*n*): date of expiry; date for lodging a patent 897

HINTERLEIB (*m*): hind quarters; back; dorsum 898

HINTERLIST (*f*): cunning; deceit; fraud; wile; artifice 899

HINTERMAST (*m*): mizzenmast (Naut.) 900

HINTERRAD (*n*): trailing wheel; rear wheel; stern wheel 901

HINTERRAUM (*m*): after hold (Ships) 902

HINTERSCHIFF (*n*): afterbody (Ships) 903

HINTERSITZ (*m*): back-seat, rear-seat 904

HINTERST: hindmost; hindermost; last 905

HINTERSTEVEN (*m*): stern post 906

HINTERSTEVENHACKE (*f*): heel knee 907

HINTERSTEVENKNIE (*n*): sternson 908

HINTERTEIL (*m*): back (rear) part or portion, hind (hinder) part, stern 909

HINTERTREFFEN (*n*): rear-guard 910

HINTERTREIBEN: to hinder, frustrate, thwart, baffle, prevent 911

HINTERVERDECK (*n*): quarter-deck (Naut.) 912

HINTERWÄRTS: backward-(s) 913

HINTRITT (*m*): decease; death; departure; demise 914

HINÜBER: over; across; beyond 915

HIN UND HER: hither and thither; to and fro 916

HIN UND HER GEHEN: to move backwards and forwards; reciprocate (Mech.); (*n*): reciprocation (Mech.) 917

HINUNDHERGEHEND: alternate, reciprocating, moving (to and fro) backward and forward 918

HIN UND WIEDER: now and again; here and there 919

HIN UND ZURÜCK: there and back 920

HINUNTER: down-(there); downstairs; downwards 921

HINWEG: off; away; over; forth; (*m*): the way (there) thither 922

HINWEGNEHMEN: to take away, remove, deduct 923

HINWEIS (*m*): hint; reference 924

HINWEISEN: to show, point the way, direct; allude, hint 925

HINWEISUNG (*f*): (see HINWEIS) 926

HINWERFEN: to throw (jot) down; drop (remarks) 927

HINWIEDERUM: once more, once again, in return, again, on the other hand 928

HINWURF (*m*): hurried (rough) sketch; throwing (flinging, jotting) down 929

HINZEICHNEN: to draw down; plot 930

HINZU: besides; to it; towards; in addition 931

HINZUFÜGEN: to add; subjoin; annex; join 932

HINZUKOMMEN: to be added; come to-(ward); approach 933

HINZUKOMMEND: additional; adventitious 934

HINZUSETZEN: to add to 935

HINZUTRITT (*m*): approach; appearance; accession 936

HINZUZIEHEN: to draw to-(wards) 937

HIPPOSIN (*n*): horse-dung meal 938

HIPPURSAUER: hippurate of 939

HIPPURSÄURE (*f*): hippuric acid; benzaminoacetic (benzolaminoacetic) acid; benzoylglycin; benzoylglycocoll; $CH_2NHCOC_6H_5COOH$; Sg. 1.37; Mp. 190°C. 940

HIRN (*n*): brain; cerebrum 941

HIRNANHANG (*m*): (see HYPOPHYSE) 942

HIRNBLUTUNG (*f*): hæmorrhage into the brain, cerebral apoplexy (Med.) 943

HIRNBRÜTEN (*n*): madness (see SCHWERMUT) 944

HIRNDECKEL (*m*): skull, cranium 945

HIRNENTZÜNDUNG (*f*): inflammation of the brain, cerebritis (Med.) 946

HIRNERSCHÜTTERUNG (*f*): concussion of the brain 947

HIRNFETT (*n*): cerebrin 948

HIRNFLÄCHE (*f*): cross-section, cross-cut (of wood) 949

HIRNHAUT (*f*): Meninges (Med.) 950

HIRNHAUTBLUTUNG (*f*): hæmorrhage (extravasation of blood) into cerebral meninges (Med.) 951

HIRNHAUTENTZÜNDUNG (*f*): meningitis (Med.) 952

HIRNHAUTENTZÜNDUNG (*f*), **AKUTE-**: simple idiopathic (cerebral lepto)-meningitis (Med.) 953

HIRNHAUTENTZÜNDUNG (*f*), **TUBERKULÖSE-**: tubercular (granular) meningitis, acute hydrocephalus (Med.) 954

HIRNHOLZ (*n*): cross-cut timber, wood-cut across the grain; the wood exposed by a section through a tree trunk vertical to its axis 955

HIRNKRANK: weak-minded, crazy 956

HIRNKRANKHEIT (*f*): disease of the brain, madness, insanity (Med.) 957

HIRNLEIN (*n*): cerebellum 958

HIRNLEHRE (*f*): craniology 959

HIRNLOS: brainless, silly 960

HIRNROTZ (*m*): glanders (Med.) 961

HIRNSCHÄDEL (*m*): skull; cranium 962

HIRNSCHÄDELBRUCH (*m*): fracture of the skull 963

HIRNSCHÄDELHAUT (*f*): pericranium (Anatomy) 964

HIRNSCHÄDELMOOS (*n*): parmelia (*Muscus cranii humani*) (a lichen *Parmelia perlata*) 965

HIRNSCHALE (*f*): skull; cranium 966

HIRNSCHALENHAUT (*f*): scalp (Anatomy) 967

HIRNSCHLAG (*m*): cerebral apoplexy, hæmorrhage into the brain (Med.) 968

HIRNSCHNITT (*m*): a section through a tree trunk vertical to its axis; cross cut-(section) 969

HIRNWASSERSUCHT (*f*): hydrocephalus, hydrops capitis, water on the brain (Med.) 970
HIRNWUT (*f*): frenzy, madness 971
HIRSCH (*m*): stag. deer; hart; buck (*Cervus elaphus*) 972
HIRSCHARTIG: cervine 973
HIRSCHBEERDORN (*m*): buckthorn (see HIRSCH-DORN) 974
HIRSCHBRUNST (*f*): boletus (*Boletus cervinus*) (a type of mushroom) 975
HIRSCHDORN (*m*): buckthorn (*Rhamnus catharticus*) 976
HIRSCHGEWEIH (*n*): antlers (of a stag) 977
HIRSCHHORN (*n*): hartshorn (see AMMONIUM-KARBONAT); stag's horn 978
HIRSCHHORNFLECHTE (*f*): Iceland moss (see ISLÄNDISCH-(ES) MOOS) 979
HIRSCHHORNGEIST (*m*): spirit of hartshorn (solution of ammonia) (*Aqua ammoniæ*) 980
HIRSCHHORNSALZ (*n*): (salt of)-hartshorn; ammonium carbonate (see AMMONIUM-KARBONAT) 981
HIRSCHHORNSCHWARZ (*n*): hartshorn black; bone black 982
HIRSCHHORNSPIRITUS (*m*): (see HIRSCHHORN-GEIST) 983
HIRSCHKÄFER (*m*): stag-beetle (*Lucanus cervus*) 984
HIRSCHKALB (*n*): fawn 985
HIRSCHKUH (*f*): hind 986
HIRSCHLATTICH (*m*): colt's-foot 987
HIRSCHLEDER (*n*): buckskin 988
HIRSCHTALG (*m*): deer (suet) tallow; Sg. 0.907; Mp. 49-53°C. 989
HIRSCHZUNGE (*f*): Hart's tongue fern (*Scolopendrium vulgare*) 990
HIRSCHZUNGENBLÄTTER (*n.pl*): Hart's tongue leaves (leaves of hart's tongue fern or *Scolopendrium vulgare*) 991
HIRSE (*f*): millet (seed of the grass, *Panicum miliaceum*) 992
HIRSEBREI (*m*): millet (pap) porridge 993
HIRSEDRÜSE (*f*): sebaceous gland 994
HIRSEFIEBER (*n*): miliaria; miliary fever (see FRIESEL) 995
HIRSENEISENSTEIN (*m*): oolitic hematite (see HÄMATIT) 996
HIRSENERZ (*n*): (see HIRSENEISENSTEIN) 997
HIRSENFÖRMIG: miliary 998
HIRSEÖL (*n*): millet oil (from *Panicum miliaceum*) 999
HIRSEÖLSÄURE (*f*): millet oleic acid; $C_{17}H_{31}CO_2H$ 000
HIRT (*m*): shepherd; herdsman; pastor 001
HIRTE (*m*): (see HIRT) 002
HIRTENHUND (*m*): shepherd's dog 003
HIRTENMÄSSIG: pastoral 004
HIRTENTASCHE (*f*): shepherd's purse (*Bursa bursa-pastoris*) 005
HIRTENTÄSCHELKRAUT (*n*): shepherd's purse (*Herba bursæ pastoris*) 006
HIRTENTÄSCHLEIN (*n*): (see HIRTENTASCHE) 007
HISINGERIT: hisingerite, SiO_2 and Al_2O_3; Sg. 2.75 008
HISSE (*f*): hoist, block and tackle 009
HISSEN: to pull up, hoist, raise, elevate 010
HISSETAU (*n*): rope, halliard 011
HISTER (*m*): dung-beetle, dor-beetle (*Geotrupes stercorarius*) 012
HISTIDIN (*n*): histidine, d-α-amino-β-imidazolylpropionic acid, $C_6H_9O_2N_3$ 013
HISTOCHEMIE (*f*): histochemistry; chemistry of the tissues 014
HISTORIE (*f*): history, tale, narrative, story 015
HISTORIKER (*m*): historian 016
HISTORISCH: historic(al)-(ly) 017
HITZBLASE (*f*): blister, pimple, pustule 018
HITZBLATTER (*f*): (see HITZBLASE) 019
HITZBLÜTIG: hot-blooded, choleric 020
HITZE (*f*): heat; hotness,; ardour; passion; batch; warmth; heating (Metal) 021
HITZEBESTÄNDIG: stable on heating; fast to ironing or hot-pressing (Colour); heat-resisting 022
HITZEDENATURIERUNG (*f*): heat (denaturing) denaturizing 023
HITZEEINWIRKUNG (*f*): action (influence) of heat 024
HITZEGRAD (*m*): degree of heat 025
HITZEGRADMESSER (*m*): pyrometer (see PYRO-METER) 026
HITZEGRADMESSUNG (*f*): pyrometry 027
HITZEMESSER (*m*): pyrometer (see PYROMETER) 028
HITZEMESSUNG (*f*): pyrometry; heat measurement 029
HITZEN: to heat; warm 030
HITZIG: hot; heating; inflammatory; acute; passionate; ardent; burning; fervent; hasty; spirited 031
HITZKÖPFIG: hot-headed, rash, precipitate 032
HJELMIN (*m*): Hjelmite; $(Nb,Ta)_2O_7$, WO_3, UO_2, $(Ca,Mg,Zn)O$; Sg. 5.8 033
HOBEL (*m*): plane 034
HOBELBANK (*f*): carpenter's (planing) bench, joiner's bench, lathe 035
HOBELEISEN (*n*): plane iron 036
HOBELMASCHINE (*f*): planing machine 037
HOBELN: to plane; smooth 038
HOBELNASE (*f*): plane horn 039
HOBELSPÄNE (*m.pl*): shavings 040
HOBOE (*f*): hautboy; oboe 041
HOCH: high; tall; deep; great; sublime; light; lofty; noble; eminent; dear; expensive; brilliant; intense 042
HOCHACHTEN: to esteem, respect, value highly, prize, have great regard for 043
HOCHACHTUNG (*f*): esteem, respect, regard 044
HOCHAMT (*n*): High-mass 045
HOCHBAUTEN (*m.pl*): works (constructions) above ground-(level) 046
HOCHBEANSPRUCHT: high-duty (Mech.) 047
HOCHBEHÄLTER (*m*): elevated tank 048
HOCHBERÜMT: very (highly) renowned or famous 049
HOCHBLAU: light-blue; azure 050
HOCHDRUCK (*m*): high pressure (usually 15-60 atms.); relief printing 051
HOCHDRUCKARMATUR (*f*): high-pressure (fitting; fittings; mounting; mountings; valve) valves 052
HOCHDRUCKENDE (*n*): high pressure end (turbines) 053
HOCHDRUCKGEBLÄSE (*n*): high-pressure blower-(s) 054
HOCHDRUCKKESSEL (*m*): high pressure boiler 055
HOCHDRUCKLEITUNG (*f*): high pressure main 056
HOCHDRUCKSCHAUFEL (*f*): high pressure (vane) blade (Steam Turbines) 057
HOCHDRUCKSYSTEM (*n*): high pressure system 058
HOCHDRUCKTEIL (*m*): high pressure part 059

HOCHDRUCKZYLINDER (*m*) : high pressure cylinder, H.P. cylinder 060
HOCHEBENE (*f*) : tableland 061
HOCHEHRWÜRDIG : right reverend 062
HOCHEMAIL (*n*) : embossed enamel 063
HOCHENTZÜCKUNG (*f*) : ecstasy 064
HOCHERHABEN : in high relief ; sublime 065
HOCHFAHREND : high-flown, imperious, haughty 066
HOCHFARBIG : highly coloured 067
HOCHFEIN : superfine 068
HOCHFEUERFEST : highly refractory, highly fire-proof ; of high fire-resisting quality (from Seger-kegel 31-36 (see also SEGERKEGEL) 069
HOCHFREQUENT : high-frequency 070
HOCHFREQUENZ (*f*) : high frequency (Elect.) 071
HOCHFREQUENZSTROM (*m*) : high frequency current 072
HOCHGEBIRGE (*n*) : highlands, mountain range, peaks 073
HOCHGEEHRT : highly (esteemed) honoured 074
HOCHGERICHT (*n*) : gallows, place of execution ; supreme penal court, high court of Justice 075
HOCHGESINNT : high-minded noble(of character), magnanimous 076
HOCHGESPANNT : at high tension ; highly (stretched ; strained) superheated, (the latter of steam) ; spanned ; cocked ; pinched ; excited ; tight ; intense ; highly wrought ; exaggerated 077
HOCHGESTIMMT : high-pitched (see also HOCH-GESINNT) 078
HOCHGLANZ (*m*) : high (polish ; lustre) brilliancy 079
HOCHGRADIG : in high degree ; intense 080
HOCHGRÜN : bright (light) green 081
HOCHHERZIG : noble-minded ; high-spirited ; high-minded ; magnanimous ; proud 082
HOCHHUBSICHERHEITSVENTIL (*n*) : high lift safety valve 083
HOCHKANTE (*f*) : edge 084
HOCHKANTIG : edgewise ; on edge 085
HOCHKONZENTRIERT : highly concentrated 086
HOCHKRAUT (*n*) : dill (see DILL) 087
HOCHLAND (*n*) : high-lands, high country, uplands 088
HOCHLEISTUNG (*f*) : high duty (Mech.) 089
HOCHMEISTER (*m*) : Grand master 090
HOCHMESSE (*f*) : High Mass 091
HOCHMOLECULAR : high-molecule, highly-molecular 092
HOCHMOLEKULAR : of high molecular (value) content 093
HOCHMOOR (*n*) : bog, fen, marsh 094
HOCHMOORARTIG : boggy, marshy 095
HOCHMOORBODEN (*m*) : marshy (soil) land 096
HOCHMUT (*m*) : haughtiness ; pride ; arrogance 097
HOCHOFEN (*m*) : blast furnace 098
HOCHOFENARMATUREN (*f.pl*) : blast-furnace fittings 099
HOCHOFENAUFZUG (*m*) : blast-furnace (elevator, lift) hoist 100
HOCHOFENBETRIEB (*m*) : blast furnace (working) operation 101
HOCHOFENGANG (*m*) : blast furnace (working) operation 102
HOCHOFENGAS (*n*) : blast furnace gas ; (see GICHTGAS) 103
HOCHOFENGASFEUERUNG (*f*) : blast furnace gas firing or furnace, Harrison furnace 104
HOCHOFENGERÜST (*n*) : blast-furnace (framework) framing 105
HOCHOFENSCHLACKE (*f*) : blast furnace slag 106
HOCHOZONISIERT : highly ozonized 107
HOCHPROZENTIG : of high percent-(age) 108
HOCHRENTABEL : highly-profitable 109
HOCHROT : bright red ; brick-red- (vermilion) ; deep-red- (vermilion) ; crimson ; brilliant red ; scarlet 110
HOCHRUND : convex 111
HOCHSCHARLACH (*m*) : cochineal scarlet ; crimson (see HOCHROT) 112
HOCHSCHÄTZEN : to esteem, value highly, prize 113
HOCHSCHMELZEND : high-melting ; having a high melting-point 114
HOCHSCHULE (*f*) : High School ; College ; Institution ; Academy ; University 115
HOCHSEE (*f*) : ocean, high-sea, main 116
HOCHSEETORPEDOBOOT (*n*) : ocean going torpedo boat 117
HOCHSELIG : late ; deceased 118
HOCHSENSIBEL : highly (sensitive) sensible 119
HOCHSIEDEND : high-boiling ; having a high boiling-point 120
HOCHSINNIG : high-minded ; (see also HOCH-GESINNT) 121
HOCHSPANNUNG (*f*) : high (tension) pressure 122
HOCHSPANNUNGSSTROM (*m*) : high tension current 123
HOCHSPANT (*n*) : high frame 124
HOCHSPANTSYSTEM (*n*) : high frame system 125
HÖCHST : highest ; utmost ; maximum ; peak ; extremely 126
HOCHSTÄMMIG : tall, lofty, having a tall trunk (of trees) 127
HÖCHSTBELASTUNG (*f*) : peak-load ; maximum duty (Mech.) 128
HÖCHSTDRUCK (*m*) : maximum-pressure ; super-high pressure, super-pressure (usually over 60 Atms.) 129
HÖCHSTDRUCKKESSEL (*m*) : super-high pressure boiler 130
HÖCHSTDRUCKTURBINE (*f*) : super-high pressure turbine 131
HÖCHSTENS : at-(the)-most or best 132
HÖCHSTFALL (*m*) : maximum case 133
HÖCHSTLEISTUNG (*f*) : peak load ; maximum output 134
HOCHSTRASSE (*f*) : highway, main road 135
HOCHSTREBEND : ambitious, aspiring 136
HÖCHSTZULÄSSIG : maximum (permissible) admissible 137
HOCHTOURIG : high-speed, high-velocity, having a high (number of revolutions) velocity 138
HOCHTRABEND : pompous ; bombastic ; high-stepping (of horses) 139
HOCHVERBRECHEN (*n*) : capital crime 140
HOCHVERDIENT : well-deserved, meritorious, well-earned 141
HOCHVERGÄREND : highly (attenuating) fermenting 142
HOCHVERRAT (*m*) : high treason 143
HOCHWALD (*m*) : forest 144
HOCHWASSER (*n*) : high-water, high-tide 145
HOCHWASSERSTAND (*m*) : high-water mark or level 146
HOCHWEIDE (*f*) : mountain (pasturage) pasture 147
HOCHWERTIG : of high valence, of first-class quality, high-value, high grade ; high-duty (Mech.) 148

HOCHWICHTIG: highly important 149
HOCHWÖLBER (*m*): high arch, arch with a high rise 150
HOCHZEIT (*f*): wedding; marriage; nuptials 151
HOCHZEITSFEST (*n*): wedding-feast 152
HOCHZEITSGEDICHT (*n*): epithalamium (a song or poem in honour of a newly wedded pair) 153
HOCKEN: to put in heaps, crouch, cower, perch, get on another's back, take on one's back 154
HÖCKEN: to retail, hawk 155
HÖCKER (*m*): hump; lump; bump; bunch; protuberance; knob; hawker; huckster; cam (Mech.) 156
HÖCKERIG: knotty; knobby; rough; hunch-backed; humped; lumpy; bumpy; rugged; uneven; tuberous 157
HODE (*f*): testicle (Med.) 158
HODEN (*m*): (see **HODE**) 159
HODENBRUCH (*m*): scrotocele, scrotal hernia (Med.) 160
HODENFÖRMIG: testiculate 161
HODENGESCHWULST (*f*): tumour of the scrotum (Med.) 162
HODENSACK (*m*): scrotum (Med.) 163
HOF (*m*): areola (of lichens); aureole (a white or coloured corona round the sun or moon); halo, corona; yard; court; residence, farm, manor, household 164
HOFARZT (*m*): court physician 165
HOFDICHTER (*m*): poet-laureate 166
HOFFART (*f*): pride; insolence; vanity; haughtiness; arrogance 167
HOFFEN: to hope-(for); except; anticipate; look for 168
HOFFENTLICH: it is to be hoped 169
HOFFNUNG (*f*): hope; expectation; anticipation 170
HOFFNUNGSLOS: hopeless, unpromising 171
HOFFNUNGSVOLL: hopeful, promising 172
HOFGERICHT (*n*): High Court of Justice 173
HOFHUND (*m*): watch-dog, house-dog 174
HÖFLICH: civil; courteous; polite; obliging 175
HOFLIEFERANT (*m*): purveyor to the (court) royal household 176
HOFMEISTERN: to find fault with, censure, criticize 177
HÖHE (*f*): height; altitude; head (of water); amount; summit; depth; loftiness; elevation; hill; mound (see also **HOCH**) 178
HOHEIT (*f*): highness; greatness; loftiness; elevation; grandeur; majesty; sovereignty; sublimity 179
HÖHENKREIS (*m*): circle of altitude (Astronomy) 180
HÖHENMESSER (*m*): astrolabe (Astronomy); altimeter (height meter) 181
HÖHENMESSKUNST (*f*): altimetry 182
HÖHENRAUCH (*m*): peat; mist 183
HÖHENRICHTUNG (*f*): height, altitude, elevation 184
HÖHENSCHNITT (*m*): sectional elevation 185
HÖHENZIRKEL (*m*): circle of altitude (Astronomy) 186
HÖHEPUNKT (*m*): zenith, peak, climax 187
HÖHER: higher; superior; greater; grander; loftier (see also **HOCH**; **HÖHE**; and **HOHEIT**) 188
HÖHERWERTIG: of higher (valence) value; polyvalent 189
HOHE SEE (*f*): high seas, the main, ocean 190
HOHES LICHT (*n*): high-light 191
HOHL: hollow; concave; empty; shallow 192
HOHLAÜGIG: hollow-eyed 193
HOHLBEIL (*n*): (see **HOHLDECHSEL**) 194
HOHLBOHRER (*m*): auger 195
HOHLBOHRSTAHL (*m*): hollow boring tool 196
HOHLDECHSEL (*m*): hollow (grooving) adze 197
HOHLDEICHSEL (*m*): (see **HOLHDECHSEL**) 198
HOHLDEISSEL (*m*): (see **HOHLDECHSEL**) 199
HOHLDIELE (*f*): hollow slab 200
HOHLDOCKE (*f*): mandrel 201
HÖHLE (*f*): cave; grotto; cavern; hole; cavity; excavation; opening; hollow; den; burrow 202
HOHLEISEN (*n*): gouge 203
HÖHLEN: to hollow; excavate; dig out; socket 204
HÖHLENBEWOHNER (*m*): cave-dweller, troglodyte 205
HOHLERHABEN: concavo-convex 206
HOHLE WELLE (*f*): hollow shaft 207
HOHLFEILE (*f*): round file 208
HOHLGEGOSSEN: hollow-cast 209
HOHLGESCHLIFFEN: hollow ground; concave 210
HOHLGESCHWÜR (*n*): fistula 211
HOHLGEWINDE (*n*): female thread 212
HOHLGLAS (*n*): hollow glass-ware; concave glass 213
HOHLGLASWAREN (*f.pl*): hollow glassware 214
HOHLGUSS (*m*): hollow casting 215
HOHLHAND (*f*): palm (hollow) of the hand 216
HOHLHOBEL (*m*): hollow (chamfering) plane 217
HOHLIG: cavernous 218
HOHLKANT (*n*): hollow (moulding) chamfer 219
HOHLKEHLE (*f*): hollow (well rounded) throat piece, hollow, groove, rabbet, recess, furrow, gutter, fluting, channel 220
HOHLKEHLHOBEL (*m*): rabbet plane 221
HOHLKEIL (*m*): hollow key 222
HOHLKIEL(*m*): hollow keel 223
HOHLKIRSCHE (*f*): black-alder (see **FAULBAUM**) 224
HOHLKUGEL (*f*): hollow (ball) sphere 225
HOHLLAUCH (*m*): (see **SCHNITTLAUCH**) 226
HOHLLINSE (*f*): concave lens 227
HOHLMASS (*n*): dry measure 228
HOHLMAUER (*f*): hollow wall 229
HOHLMEISSEL (*m*): gouge 230
HOHLMUSCHEL (*f*): (see **HOHLMEISSEL**) 231
HOHLPRISMA (*n*): hollow prism 232
HOHLRAUM (*m*): hollow space; empty (space) chamber; blow-hole (Metal) 233
HOHLROST (*m*): hollow grate (see also **MULDEN-ROST**) 234
HOHLROSTSTAB (*m*). hollow (fire-bar) grate-bar (to provide for internal cooling of the bar) 235
HOHLRUND: concave 236
HOHLRUNDUNG (*f*): concavity, hollow 237
HOHLSÄULE (*f*): hollow column 238
HOHLSCHABER (*m*): grooved (fluted) scraper 239
HOHLSCHICHT (*f*): air space 240
HOHLSCHNITT (*m*): hollow section 241
HOHLSPAT (*m*): chiastolite (see **ANDALUSIT**) 242
HOHLSPIEGEL (*m*): concave mirror 243
HOHLSTAB (*m*): catheter (Med.) 244
HOHLSTEIN (*m*): hollow (brick) stone, gutter stone 245
HÖHLUNG (*f*): cavity; excavation; hollow; concavity; socket 246
HOHLWALZE (*f*): hollow (cylinder) roll 247
HOHLWAREN (*f.pl*): hollow ware 248
HOHLWEG (*m*): narrow (passage) pass, defile 249

HOHLWELLE (*f*): hollow shaft 250
HOHLZAHN (*m*): hemp-nettle (*Galeopsis*) (Botany) 251
HOHLZIEGEL (*m*): hollow (brick) tile; gutter tile 252
HOHLZIEGELSTEIN (*m*): hollow brick 253
HOHLZIRKEL (*m*): inside calipers 254
HOHLZYLINDER (*m*): hollow cylinder 255
HOHN (*m*): scorn; sarcasm; mockery; disdain; sneer; derision; defiance; scoff 256
HÖHNEN: to scorn, sneer, jeer, scoff, deride, disdain, defy, hold in contumely, laugh (to scorn) at 257
HÖHNISCH: sneering, jeering, scornful, disdainful, scoffing, derisive, defiant 258
HÖKEN: (see **HÖCKEN**) 259
HOLD: kind; friendly; lovely; charming; propitious; affectionate; favourable; gracious; sweet; pleasant 260
HOLDER (*m*): elder (see **HOLUNDER**) 261
HOLDSELIGKEIT (*f*): graciousness, gracefulness, sweetness; loveliness 262
HOLEN: to fetch; get; go for; draw (Breath); haul 263
HOLK (*m*): hulk (Naut.) 264
HOLLÄNDER (*m*): beating engine; hollander (Paper); Dutchman 265
HOLLÄNDERBLAU (*n*): Holland blue (see **WASCHBLAU**) 266
HOLLÄNDERKASTEN (*m*): vat 267
HOLLÄNDERN: to pulp (beat) rags in a hollander (Paper) 268
HOLLÄNDERWEISS (*n*): Dutch white 269
HOLLÄNDISCH: Dutch 270
HOLLÄNDISCHE FLÜSSIGKEIT (*f*): Dutch liquid; ethylene chloride (see **ÄTHYLENCHLORID**) 271
HOLLÄNDISCHEN CHEMIKER, ÖL DER-: (see **ÄTHYLENCHLORID**) 272
HOLLÄNDISCHES GESCHIRR (*n*): hollander (Paper) (see **HOLLÄNDER**) 273
HOLLÄNDISCHES ÖL (*n*): Dutch oil; Haarlem oil (see **HAARLEMERÖL**) 274
HÖLLE (*f*): hell 275
HÖLLENANGST (*f*): mortal (terror) fright 276
HÖLLENBRAND (*m*): hellfire 277
HÖLLENMASCHINE (*f*): infernal machine 278
HÖLLENÖL (*n*): curcas oil (from *Jatropha curcas*); castor oil; lowest grade of olive oil (see **TOURNANTÖL**) 279
HÖLLENSTEIN (*m*): *Lapis infernalis*; lunar caustic; silver nitrate (see **SILBERNITRAT**) 280
HÖLLENSTEINHALTER (*m*): silver nitrate (container) holder 281
HÖLLISCH: hellish, infernal 282
HOLLUNDER (*m*): elder (see **HOLUNDER**) 283
HOLM (*m*): holme, island, islet; cross-beam, rail 284
HOLOCAIN (*n*): holocain; $C_{18}H_{22}O_2N_2.HCl$ 285
HOLOEDER (*n*): holohedral form 286
HOLOEDRIE (*f*): holohedrism 287
HOLOEDRISCH: holohedral 288
HOLPER (*m*): hillock, roughness, unevenness 289
HOLPERIG: uneven; rough; rugged 290
HOLPERN: to jolt 291
HOLPRIG: (see **HOLPERIG**) 292
HOLUNDER (*m*): elder (*Sambucus canadensis*) (see **HOLDER** and **HOLLUNDER**) 293
HOLUNDERBEERE (*f*): elderberry (*Fructus sambuci*) 294
HOLUNDERBEERENÖL (*n*): elderberry oil (from the seeds of *Sambucus racemosa*); Sg. 0.917 295
HOLUNDERBLÜTEN (*f.pl*): elder flowers (*flores sambuci*) 296
HOLUNDERBLÜTENÖL (*n*): elder-flower oil 297
HOLUNDERHOLZ (*n*): elder wood; Sg. 0.5—0.8 (see also **HOLUNDER**) 298
HOLUNDERÖL (*n*): elder oil 299
HOLUNDER (*m*), **ROTER-**: red elder (*Sambucus racemosa*) 300
HOLUNDERSCHWAMM (*m*): elder fungus (*Fungus sambuci*) 301
HOLUNDER (*m*), **SPANISCHER-**: lilac (*Syringa vulgaris*) 302
HOLUNDER (*m*), **TÜRKISCHER-**: lilac (*Syringa persica*) 303
HOLZ (*n*): wood (*Lignum*); timber; forest; thicket; grove; copse; bush; stick; pole; stock 304
HOLZABFÄLLE (*m.pl*): wood (waste) refuse 305
HOLZÄHNLICH: wood-like; ligneous 306
HOLZALKOHOL (*m*): wood alcohol (see **METHYLALKOHOL**) 307
HOLZANBAU (*m*): cultivation of forests, afforestation 308
HOLZAPFEL (*m*): crab-apple (fruit of *Pyrus malus*) 309
HOLZAPFELBAUM (*m*): crab (wild) apple tree (*Pyrus malus*) 310
HOLZARBEIT (*f*): wood-work, carpentry, joinery, wood-carving 311
HOLZARBEITER (*m*): wood-worker, carpenter, joiner, carver 312
HOLZARM: short of wood 313
HOLZARTIG: woody; ligneous; ligniform 314
HOLZASBEST (*m*): ligneous (woody; fibrous) asbestos 315
HOLZASCHE (*f*): wood ashes 316
HOLZASSEL (*f*): wood louse (*Limnoria lignorum*) 317
HOLZÄTHER (*m*): methyl ether (see **METHYLÄTHER**) 318
HOLZÄTHER, ESSIGSAURER- (*m*): methyl acetate (see **METHYLACETAT**) 319
HOLZAXT (*f*): (wood)-felling axe 320
HOLZBAST (*m*): wood bast, inside bark, wood fibre 321
HOLZBAU (*m*): cultivation of timber, afforestation 322
HOLZBEARBEITUNG (*f*): wood working 323
HOLZBEARBEITUNGSMASCHINEN (*f.pl*): woodworking machinery 324
HOLZBEDECKUNG (*f*): wood-(en) cover-(ing) 325
HOLZBEIZE (*f*): wood (stain) mordant 326
HOLZBEIZEN (*n*): wood staining; to stain wood 327
HOLZBILD (*n*): carving-(in wood) 328
HOLZBILDHAUER (*m*): wood carver 329
HOLZBILDNER (*m*): wood carver 330
HOLZBIRNE (*f*): wild pear (*Pyrus communis*) 331
HOLZBOCK (*m*): sawing trestle 332
HOLZBODEN (*m*): wood-loft 333
HOLZBOHRER (*m*): wood borer, wood beetle (*Lymexylon navale*); goat-moth (*Cossus ligniperda*); *Xylophaga gerst.*; goat-moth caterpillar; type of butterfly (*Xylotropha*); auger-(tool); boring machine (for wood); borer, drill, gimlet 334
HOLZBOHRMUSCHEL (*f*): ship-worm (*Teredo navalis*) 335
HOLZBRAME (*f*): brambles, underwood, undergrowth 336
HOLZBRAND (*m*): gangrene 337

HOLZBRANNTWEIN (*m*): wood (alcohol) spirit (see METHYLALKOHOL) 338
HOLZBÜCHSE (*f*): wooden box 339
HOLZBÜNDEL (*m* and *n*): bundle of wood, fagot 340
HÖLZCHEN (*n*): little stick; splinter; small piece of wood 341
HOLZDESTILLATION (*f*): wood (distilling) distillation 342
HOLZDESTILLATIONSANLAGE (*f*): wood distilling plant 343
HOLZDESTILLIEREN (*n*): wood distilling 344
HOLZDRAHT (*m*): wood wire 345
HOLZDREHEREI (*f*): wood (turnery) turning 346
HOLZDÜBEL (*m*): dowel-(joint) 347
HOLZEN: to cut (gather) wood 348
HÖLZERN: wooden; of wood; ligneous; stiff; awkward 349
HOLZERSATZ (*m*): wood substitute (see HOLZ-MASSE) 350
HOLZESSIG (*m*): wood vinegar; pyroligneous (vinegar) acid (*Acetum pyrolignosum*); $C_2H_5O_2$; Sg. 1.025 351
HOLZESSIG (*m*), **DESTILLIERTER-**: distilled wood vinegar, crude acetic acid (see ESSIGSÄURE) 352
HOLZESSIG (*m*), **REINER-**: (see HOLZESSIG, DESTILLIERTER) 353
HOLZESSIGSAUER: pyrolignite of; combined with pyroligneous acid 354
HOLZESSIGSÄURE (*f*): (see HOLZESSIG) 355
HOLZESSIGSAURES EISEN (*n*): iron liquor; iron acetate liquor (from the action of pyroligneous acid on iron fillings) 356
HOLZFARBE (*f*): wood colour; wood (mordant) stain 357
HOLZFARREN (*m*): tree-fern 358
HOLZFASER (*f*): wood (fibre) pulp; ligneous fibre; grain (of wood) 359
HOLZFASERSTOFF (*m*): lignin; cellulose; lignocellulose; chemical fibre; wood (fibre) pulp (see ZELLULOSE) 360
HOLZFASS (*n*): wooden barrel or cask 361
HOLZFÄULE (*f*): dry-rot 362
HOLZFÄULNIS (*f*): dry-rot 363
HOLZFEILE (*f*): rasp 364
HOLZFEUERUNG (*f*): wood-(burning)-furnace, chip furnace, sawdust furnace; wood firing 365
HOLZFLOSS (*n*): raft 366
HOLZFORM (*f*): wooden mould 367
HOLZFÖRMIG: woody, ligniform, ligneous 368
HOLZ (*n*), **FOSSILES-**: fossil (fossilized) wood 369
HOLZFREI: free from wood (Paper) 370
HOLZFRESSER (*m*): *Xylophaga Gerst.*; type of butterfly (*Xylotropha*); wood-eater, wood-beetle (*Lymexylon navale*); goat-moth (*Cossus ligniperda*) 371
HOLZFUHRE (*f*): carrying (transporting, conveying) of wood; cartload of wood 372
HOLZFUSS (*m*): wooden (foot)base 373
HOLZFUSSBODEN (*m*): wood floor-(ing) 374
HOLZFUSSBODEN, KOMPRIMIERTER- (*m*): compressed wood flooring 375
HOLZ (*n*), **GANZ-**: whole log-(of wood) 376
HOLZGAS (*n*): wood gas (an illuminating gas from resinous wood, by dry distillation; calorific value 3000/4000 calories) 377
HOLZGASDARSTELLUNG (*f*): production of wood gas 378
HOLZGEIST (*m*): wood (pyroxylic) spirit, wood alcohol, wood naphtha, methyl alcohol, methyl (hydrate) hydroxide, acetone alcohol, columbian spirit (see also METHYLALKOHOL); CH_3OH; Sg. 0.8; Mp. −97.8°C.; Bp. 65.75°C. 379
HOLZGEISTÖL (*n*): wood spirit oil; wood alcohol oil 380
HOLZGERÜST (*n*): wooden (scaffolding, framing) staging 381
HOLZGESTELL (*n*): wooden (frame) stand 382
HOLZGRIFF (*m*): wooden (handle; shaft) haft 383
HOLZGRÜNDUNG (*f*): grounding, priming (first-coating)-(of wood) 384
HOLZGUMMI (*n*): wood (gum) resin; xylose (see KIRSCHGUMMI) 385
HOLZHACKER (*m*): wood-cutter; wood-pecker 386
HOLZ (*n*), **HALB-**: half section, log of wood split down the centre 387
HOLZHAMMER (*m*): mallet, wooden hammer 388
HOLZHANDEL (*m*): timber (wood) trade 389
HOLZHÄNDLER (*m*): timber (wood) merchant 390
HOLZHAU (*m*): clearing, place where trees have been felled 391
HOLZ HAUEN: to cut wood, fell trees 392
HOLZHAUER (*m*): wood-cutter 393
HOLZHAUFEN (*m*): pile (stack) of wood 394
HOLZHOF (*m*): timber (wood) yard 395
HOLZHÜLSE (*f*): wooden (cover) case 396
HOLZICHT: woody; ligneous 397
HOLZIG: (see HOLZICHT) 398
HOLZIMPRÄGNIERUNG (*f*): wood impregnation 399
HOLZKÄFER (*m*): (see HOLZFRESSER) 400
HOLZKALK (*m*): pyrolignite of lime; lime acetate; $Ca(C_2H_3O_2)_2.H_2O$ 401
HOLZKASTEN (*m*): wooden (box; case) vat 402
HOLZKIRSCHE (*f*): wild cherry (*Cerasus campron-iana* and *Cerasus avium*) 403
HOLZKISTE (*f*): wooden box or chest 404
HOLZKLOTZ (*m*): wood-(en) block 405
HOLZKOHLE (*f*): wood charcoal; Sg. 0.44 (for pine charcoal) 406
HOLZKOHLENBRIKETT (*n*): wood charcoal briquette 407
HOLZKOHLENEISEN (*n*): charcoal iron 408
HOLZKOHLENKLEIN (*n*): wood charcoal dust 409
HOLZKOHLENOFEN (*m*): wood charcoal (oven) kiln 410
HOLZKOHLENPULVER (*n*): charcoal powder 411
HOLZKOHLENROHEISEN (*n*): charcoal pig iron 412
HOLZKOHLENTEER (*m*): wood (charcoal) tar 413
HOLZKOHLESTAUB (*m*): wood charcoal dust 414
HOLZKONSERVIERUNGSMITTEL (*n* and *pl*): wood preserver-(s); wood preservative-(s); wood preserving agent-(s) 415
HOLZKONSERVIERUNGSÖL (*n*): wood preserving oil 416
HOLZ (*n*), **KREUZ-**: quarter section, log of wood split into four 417
HOLZKUGEL (*f*): wooden ball 418
HOLZ (*n*), **KÜNSTLICHES-**: artificial wood, xylolite 419
HOLZKUPFER (*n*): wood copper; fibrous olivenite (see OLIVENIT) 420
HOLZKUPFERERZ (*n*): (see HOLZKUPFER) 421
HOLZLACK (*m*): stick lack; wood varnish 422
HOLZLADE (*f*): wood-lifting (timber) jack 423
HOLZLAGER (*n*): wood (timber) yard; pile (stack) of wood 424
HOLZMANGEL (*m*): deficiency of wood, shortage of wood 425

HOLZMANTEL (*m*): wooden case or covering 426
HOLZMASSE (*f*): wood-like mass; xylolite (an artificial wood); lignolite (a magnesium compound with sawdust) 427
HOLZMEHL (*n*): wood (flour) powder; sawdust 428
HOLZMEISSEL (*m*): wood (mortise) chisel 429
HOLZMELASSE (*f*): wood molasses 430
HOLZMOSAIK (*f*): wood-inlaying, wood mosaic-(work) 431
HOLZNAGEL (*m*): (wooden)-peg; tree-nail (a nail made of hard wood) 432
HOLZOBST (*n*): wild fruit-(of any kind) 433
HOLZOFEN (*m*): wood-burning oven or stove 434
HOLZÖL (*n*): wood oil (see ELÄOKOKKAÖL) 435
HOLZÖL (*n*), CHINESISCHES-: Gurjun balsam (from *Dipterocarpus trinervis*): Chinese (china) wood oil, tung oil (Sg. 0.94) (from the nuts of *Elaëcocca dryandra*) (see ELÄOKOK-KAÖL) 436
HOLZÖLFIRNIS (*m*): wood oil varnish; Gurjun balsam (see HOLZÖL, CHINESISCHES-) 437
HOLZÖL (*n*), JAPANISCHES-: Japanese wood oil (see ELÄOKOKKAÖL) 438
HOLZÖLSATZ (*m*): wood oil (residue) sediment 439
HOLZÖLSAUER: lignoleate 440
HOLZOPAL (*m*): wood opal 441
HOLZPAPPE (*f*): (wood)-pulp board (Paper) 442
HOLZPASTA (*f*): (see HOLZMASSE) 443
HOLZPECH (*n*): wood pitch 444
HOLZPFAHL (*m*): wooden (pile) pier 445
HOLZPFLASTER (*n*): wood-paving 446
HOLZRASPEL (*f*): (wood)-rasp 447
HOLZRECHEN (*m*): wooden rake; wooden sluice (strainer, grating) 448
HOLZREICH: rich (abounding) in wood 449
HOLZRIEMENSCHEIBE (*f*): wooden belt pulley 450
HOLZROT (*n*): red-wood extract 451
HOLZSAUER: pyrolignite of; combined with pyroligneous acid 452
HOLZSÄURE (*f*): pyroligneous acid (see HOLZESSIG) 453
HOLZSAURES EISEN (*n*): iron liquor (see HOLZES-SIGSAURES EISEN) 454
HOLZSCHACHTEL (*f*): wooden box 455
HOLZSCHEIT (*n*): log (billet, lath) of wood 456
HOLZSCHLAG (*m*): felling (of trees); wood-cutting 457
HOLZSCHLÄGEL (*m*): mallet 458
HOLZSCHLÄGER (*m*): wood-cutter, woodman 459
HOLZSCHLEIFAPPARAT (*m*): wood grinder (Paper) 460
HOLZSCHLEIFEN (*n*): planing (grinding, sand-papering, rubbing) of wood 461
HOLZSCHLEIFER (*m*): wood-(pulp) grinder (Paper) 462
HOLZSCHLEIFEREI (*f*): mechanical wood-pulp mill (Paper) 463
HOLZSCHLIFF (*m*): mechanical wood-pulp (Paper) (see ZELLULOSE) 464
HOLZSCHNEIDER (*m*): wood-carver; sawyer 465
HOLZSCHNITT (*m*): wood-cut; engraving 466
HOLZ (*n*), SCHNITT-: converted (cut, sawn) wood 467
HOLZSCHNITTDRUCK (*m*): wood-cut (the process or the resulting print) 468
HOLZSCHNITZER (*m*): wood-carver 469
HOLZSCHNITZEREI (*f*): (wood)-carving, xylography 470
HOLZSCHOBER (*m*): stack-(of wood) 471
HOLZSCHOPPEN (*m*): wood-shed 472
HOLZSCHUH (*m*): wooden shoe, clog 473
HOLZSCHWAMM (*m*): fungus (on wood); dry rot (of wood) 474
HOLZSPAN (*m*): wood (shaving) chip 475
HOLZSPIRITUS (*m*): wood (spirit) alcohol (see METHYLALKOHOL) 476
HOLZSPLITTER (*m*): splinter-(of wood) 477
HOLZSTAMM (*m*): trunk (of wood) 478
HOLZSTEIN (*m*): fossil wood; variety of quartz (Mineral) (see QUARZ) 479
HOLZSTICH (*m*): woodcut, engraving 480
HOLZSTÖCKEL (*m.pl*): wood-blocks; wood-paving 481
HOLZSTOFF (*m*): wood pulp; wood cellulose (see ZELLULOSE) 482
HOLZSTOFFKARTON (*m*): wood-pulp board (Paper) 483
HOLZSTOSS (*m*): pile-(of wood) 484
HOLZSTUCK (*m*): (see HOLZMASSE) 485
HOLZSUBSTANZ (*f*): cellulose (see ZELLULOSE); wood substance 486
HOLZTAFEL (*f*): board 487
HOLZTAFELDRUCK (*m*): wood-cut (the process or the resulting print) 488
HOLZTAUBE (*f*): wood-pigeon 489
HOLZTEE (*m*): (*Species lignorum*): blood purifying tea 490
HOLZTEER (*m*): wood tar; *pix liquida* 491
HOLZTEERÖL (*n*): wood-tar oil; Sg. 0.867 (distilled from wood tar from *Pinus palustris*) 492
HOLZTRÄNKUNG (*f*): wood (pickling) impregnation 493
HOLZUNG (*f*): wood; forest 494
HOLZVERBINDUNG (*f*): wood-(en) joint 495
HOLZVERKLEIDUNG (*f*): wood lining 496
HOLZVERKOHLUNG (*f*): charcoal burning; carbonization of wood 497
HOLZVERKOHLUNGSANLAGE (*f*): charcoal burning plant 498
HOLZVERKOHLUNGSVERFAHREN (*n*): charcoal burning (wood carbonizing) process 499
HOLZVERZUCKERUNG (*f*): wood saccharification 500
HOLZWAND (*f*): wooden partition 501
HOLZWAREN (*f.pl*): wooden (articles) wares 502
HOLZWERK (*n*): wood-work, wainscoting, timber-work, timbers 503
HOLZ (*n*), WINDSCHIEFES-: wood which has been bent or twisted by the wind, warped wood 504
HOLZWOLLE (*f*): wood wool; fine wood 505
HOLZWURM (*m*): caterpillar of the goat-moth (*Cossus ligniperda*) (see also HOLZFRESSER) 506
HOLZZELLSTOFF (*m*): lignocellulose (see HOLZ-FASERSTOFF) 507
HOLZZEMENT (*n* and *m*): wood cement 508
HOLZZEUG (*n*): wood pulp; wood cellulose (see ZELLULOSE) 509
HOLZZINN (*n*): wood tin; fibrous cassiterite (see ZINNSTEIN; ZINNOXYD and CASSITERIT) 510
HOLZZINNERZ (*n*): (see HOLZZINN) 511
HOLZZUCKER (*m*): wood sugar; xylose (see KIRSCHGUMMI) 512
HOLZZUNDER (*m*): tinder, touchwood; amadou, German tinder (partly decayed wood) 513
HOMATROPIN (*n*): homatropine; $C_{16}H_{21}NO_3$; Mp. 95.5°C. 514
HOMILETIK (*f*): homiletics, homiletic art 515
HOMILETISCH: homiletic-(al)-(ly) 516
HOMILIE (*f*): homily 517

HOMOBRENZCATECHIN (n): homopyrocatechol 518
HOMOCHROM: homochromous; of uniform colour; monogenetic (Dye) 519
HOMOGEN: homogeneous-(ly) 520
HOMOGENEITÄT (f): homogeneity 521
HOMOGENISIEREN: to homogenize, make homogeneous 522
HOMOGENITÄT (f): homogeneity 523
HOMOGUAJAKOL (n): Homoguaiacol, creosol, homopyrocatecholmonomethyl ester (see KREOSOL) 524
HOMOKAFFEESÄURE (f): homocaffeic acid 525
HOMOKAMPHERSÄURE (f): homocamphoric acid 526
HOMOLOG: homologous; similar; agreeing; (m): homologue 527
HOMOLOGIE (f): homology 528
HOMONYM: homonymous; (n): homonym 529
HOMÖOPATHISCH: homœopathic 530
HOMÖOPATIE (f): homœopathy 531
HOMÖOPATISCHES ARZNEIMITTEL (n): homœopathic (medicine) remedy 532
HOMOPHTALSÄURE (f): homophthalic acid 533
HONETT: honest, genteel, decent, respectable 534
HONIG (m): honey; Sg. 1.45 (also applied to any galenical preparation having honey as a base) (Pharm.) 535
HONIGAPFEL (m): honey apple 536
HONIGAROMA (n): honey-perfume (used in making artificial honey) 537
HONIGARTIG: honey-like; melleous 538
HONIGBAU (m): honey culture 539
HONIGBEREITUNG (f): preparation of honey; production of honey, mellification 540
HONIGBIENE (f): honey-bee, worker (*Apis mellifica*) 541
HONIG (m) BIENEN-: bee's honey 542
HONIGBLUME (f): nectar bearing flower; honey-flower (*Melianthus*) 543
HONIGBUSSARD (m): honey-buzzard (*Pernis apivorus*) 544
HONIGERZEUGEND: honey-producing 545
HONIGESSIG (m): oxymel; mixture of vinegar and honey 546
HONIGFARBE (f): honey colour 547
HONIGFARBIG: honey coloured 548
HONIGFLIEGE (f): honey-gnat (*Mellio*) 549
HONIGFRESSEND: mellivorous 550
HONIGGELB: honey yellow 551
HONIG, GEREINIGTER- (m): purified (refined) honey 552
HONIGHARNRUHR (f): diabetes mellitus; sweet diabetes; sugar diabetes (sugar in the urine) (Med.) 553
HONIG (m), JUNGFERN-: virgin honey (see HONIGSEIM) 554
HONIGKELCH (m): nectary (the organ of a flower which produces nectar) 555
HONIGKLEE (m): yellow melilot, common melilot (*herba meliloti* of *Melilotus officinalis*) 556
HONIGKUCHEN (m): honey cakes; honey-comb; gingerbread 557
HONIGKUCKUCK (m): honey-guide (Ornithology) (*Indicator Sparrmani*) 558
HONIG (m), KUNST-: artificial honey (from beet sugar) 559
HONIGMAGEN (m): crop, honey-bag (of bees) 560
HONIGMUND (m): mellifluence (of speech) 561
HONIGORGAN (n): nectary (see HONIGKELCH) 562
HONIGSAFT (m): nectar (sweet secretion from flowers) 563
HONIGSÄURE (f): oxymel (mixture of vinegar and honey) 564
HONIGSCHEIBE (f): honey-comb 565
HONIGSEIFE (f): honey soap 566
HONIGSEIM (m): liquid honey; virgin honey (best quality, drained from the comb) 567
HONIGSTEIN (m): honey stone; mellite; $Al_2C_{12}O_{12}.18H_2O$; Sg. 1.6 568
HONIGSTEINSÄURE (f): mellitic acid; $C_{12}H_6O_{12}$ 569
HONIGSÜSS: as sweet as honey 570
HONIGTAU (m): honey-dew (a sweet secretion from insects, such as Aphides) 571
HONIGTRANK (m): mead 572
HONIGTUSCHFARBE (f): (see HONIGFARBE) 573
HONIGWABE (f): honey-comb 574
HONIGWASSER (n): hydromel; honey water 575
HONIGZELLE (f): alveolus; honey cell 576
HONIGZUCKER (m): (see LÄVULOSE) 577
HONNEUR (m): honour 578
HONORANT (m): acceptor (of a bill) 579
HONORAR (n): fee, payment 580
HONORIEREN: to honour (a cheque), pay (Fees) 581
HONORIERUNG (f): honouring (of cheques), fee, payment; acceptance (of bills) 582
HOPFEN: to hop; skip; jump; add hops to; (m): hop-(s) (*Humulus lupulus*) 583
HOPFENBALLEN (m): hop pocket (Brewing) 584
HOPFENBAU (m): hop culture, hop growing 585
HOPFENBAUER (m): hop grower 586
HOPFENBAUM (m): hop tree (*Ptelea trifoliata*) 587
HOPFENBITTER (n): lupamaric acid; lupulin (see (see HOPFENSÄURE); hop dust (see HOPFENDRÜSE) 588
HOPFENBITTERSÄURE (f): (see HOPFENBITTER) 589
HOPFENBITTERSTOFF (m): (see HOPFENBITTER) 590
HOPFENDARRE (f): hop -(drying) kiln 591
HOPFENDRÜSE (f): lupulin; bitter principle of hops; hop dust (glandular trichomes of hops) 592
HOPFENERSATZ (m): hop substitute (bitter principle such as quassin) 593
HOPFENEXTRAKT (m): hop extract 594
HOPFENGARTEN (m): hop-garden 595
HOPFENHARZ (n): hop resin 596
HOPFENKALTRAUM (m): hop cooling chamber or room (Brewing) 597
HOPFENKEIM (m): hop bud, seed or shoot 598
HOPFENKONSERVIERUNG (f): hop (conservation, preservation) storage 599
HOPFENKÜHLRAUM (m): hop cooling chamber or room 600
HOPFENLAGER (n): hop store 601
HOPFENMEHL (n): hop dust; lupulin (see HOPFENDRÜSE) 602
HOPFENMEHLTAU (m): hop blight (see HOPFENSCHIMMEL) 603
HOPFENÖL (n): hop oil; Sg. 0.87 604
HOPFENPFLANZE (f): hop lant (see HOPFEN) 605
HOPFENSACK (m): hop pocket (Brewing) 606
HOPFENSÄURE (f): lupamaric acid (the bitter principle of the hop, *Humulus lupulus*) 607
HOPFENSCHIMMEL (m): hop mildew (*Sphaerocheca castagnei*); hop blight (*Aphis* and other insects) 608
HOPFENSTANGE (f): hop-pole 609
HOPFENSTAUB (m): hop dust; lupulin (see HOPFENDRÜSE) 610
HOPFENSTOPFEN (n): dry hopping (Brewing) 611

HOPFENSURROGAT (*n*): hop substitute (see QUASSIAHOLZ) 612
HOPFENTREBER (*pl*): spent hops 613
HOPFENTRIEB (*m*): frothy head; first stage of fermentation (Brewing) 614
HOPFENZAPFEN (*m*): hop cone; hop catkin 615
HOPSEN: to hop; jump; skip 616
HÖRBAR: audible 617
HÖRBARKEIT (*f*): audibility 618
HÖRBARKEITSGRENZE (*f*): limit of audibility 619
HORBEL (*f*): water-fowl, coot (*Fulica atra*) 620
HORCHEN: to listen; hearken; heed; obey; be attentive 621
HORCHPOSTEN (*m*): listening post (Military) 622
HORCHROHR (*n*): speaking tube, hearing trumpet; stethoscope (Med.) 623
HORCHSAM: heedful, attentive, obedient 624
HORDE (*f*): hurdle; hurdle-work wall; grating; latticed screen; horde; gang; pen; fold 625
HÖREN: to hear; listen; hearken; heed; obey; be attentive; (*n*): auscultation; sounding of the chest (Med.); hearing 626
HORIZONT (*m*): horizon 627
HORIZONTAL: horizontal 628
HORIZONTALE (*f*): horizontal line 629
HORN (*n*): horn; Sg. 1.69; corn; feeler; peak; hoof; point; crescent; headland; hard skin; projection; tongue (of land) 630
HORNABFÄLLE (*m.pl*): horn chips or waste 631
HORN (*n*), **AMBOSS-**: anvil beak 632
HORNAMBOSS (*m*): horn (beak) anvil 633
HORNARBEIT (*f*): horn-work 634
HORNARTIG: hornlike, horny, like horn 635
HORNBAUM (*m*): hornbeam (*Carpinus betulus*); Sg. 0.62-0.82 636
HORNBLEI (*n*): phosgenite (see PHOSGENIT) 637
HORNBLEIERZ (*n*): (see HORNBLEI) 638
HORNBLENDE (*f*): hornblende (a complex natural silicate); Sg. 3.25; Mp. 1180-1220°C. (see also AMPHIBOL and AUGIT) 639
HORNBLENDEANDESIT (*m*): andesite (volcanic rock of porphyritic structure) (see ANDESIN) 640
HORNBLENDEASBEST (*m*): asbestos (a fibrous variety of hornblende) (see AMPHIBOL and HORNBLENDE) 641
HORNBLENDEBASALT (*m*): basalt (a basic, volcanic rock) 642
HORNBLENDEFELS (*m*): amphibolite (see AMPHI-BOL) 643
HORNBLENDEGNEIS (*m*): gneiss (see GNEIS) 644
HORNBLENDEGRANIT (*m*): granite (see GRANIT) 645
HORNBLENDE (*f*), **LABRADORISCHE-**: Labrador hornblende 646
HORNBLENDESCHIEFER (*m*): amphibolite (see AMPHIBOL) 647
HORNDRECHSLER (*m*): horn-turner 648
HORNEN: to become horny 649
HÖRNERBAUM (*m*): dog-wood, Cornelian cherry, wild cherry, cornel (see HARTBAUM) 650
HÖRNERN: of horn; horny 651
HÖRNERSICHERUNG (*f*): horn fuse 652
HÖRNERV (*m*): auditory nerve (Med.) 653
HORNERZ (*n*): horn silver; cerargyrite (see CERARGYRIT) 654
HORNFEILE (*f*): horn (hoof) file or rasp 655
HORNFÖRMIG: corniform 656
HORNFÜSSIG: hoofed 657
HORNGEWEBE (*n*): horny tissue 658
HORNHAUT (*f*): cornea; horny-(layer of the)-skin; callosity 659
HORNHÄUTIGKEIT (*f*): callosity 660
HORNHAUTSCHNITT (*m*): couching for a cataract (Med.) 661
HORNIG: horny; horn-like; callous 662
HÖRNIG: horned 663
HORNIS (*f*): hornet (*Vespa crabro*) 664
HORNISSE (*f*): (see HORNIS) 665
HORNIST (*m*): horn-player, bugler 666
HORNKIRSCHE (*f*): Cornelian cherry (see HART-BAUM) 667
HORNKLÜFTIG: cloven-hoofed 668
HORNKOBALT (*n*): asbolite (see KOBALTMANGAN-ERZ) 669
HORNKRAUT (*n*): mouse-ear, chickweed (*Cerastium*): rupture-wort (*Herniaria glabra*) (*herba herniariæ glabræ*) 670
HORN (*n*), **KUNST-**: artificial horn 671
HORN (*n*), **KÜNSTLICHES-**: artificial horn 672
HORNLÖFFEL (*m*): horn spoon 673
HORNMEHL (*n*): horn meal; crushed horn (an artificial manure) 674
HORNMEHLFABRIKATION (*f*): manufacture of horn meal 675
HORNMOHN (*m*): horn poppy (*Glaucium glaucium*) 676
HORNMOHNÖL (*n*): (see GLAUCIUMÖL) 677
HORN (*n*), **NATUR-**: natural horn (see HORN) 678
HORNPISANG (*m*): horned viper (*Cerastes cornutus*) 679
HORNPRODUKT (*n*): horn product 680
HORNQUECKSILBER (*n*): horn quicksilver (native mercurous chloride) (see QUECKSILBER-CHLORÜR) 681
HORNSCHICHT (*f*): horny layer of the skin; epidermis (*Stratum corneum*) 682
HORNSILBER (*n*): horn silver; silver chloride; cerargyrite; AgCl; Sg. 5.9 (see CERARGYRIT) 683
HORNSPÄNE (*m.pl*): horn (chips; splinters) shavings 684
HORNSTEIN (*m*): horn stone; chalcedony (a variety of quartz): SiO_2; Sg. 2.65 (see QUARZ and CHALCEDON) 685
HORNSTOFF (*m*): keratin; horn principle 686
HORNSTRAUCH (*m*): dog-wood, cornel (see HART-BAUM) 687
HORNSTRAUCHGEWÄCHSE (*n.pl*): *Cornaceæ* (Bot.) 688
HORNSUBSTANZ (*f*): horn principle, keratin 689
HORNVIEH (*n*): horned cattle 690
HORNWAREN (*f.pl*): horn ware-(s); horn goods 691
HOROGRAPHIE (*f*): horography 692
HOROSKOP (*n*): horoscope 693
HÖRROHR (*n*): ear-trumpet; stethoscope 694
HÖRSAAL (*m*): lecture room; auditorium 695
HORST (*m*): eyrie; thicket 696
HORT (*m*): treasure; safety, safe retreat, shield 697
HORTENSIA (*f*): hydrangea (*Hydrangea hortensis*) (Botanical) 698
HÖRWEITE (*f*): hearing-distance 699
HÖRZEUGE (*m*): oral witness 700
HOSE (*f*): breeches, trousers; spout 701
HOSENROHR (*n*): syphon pipe (of hot-blast oven) 702
HOSE (*f*), **WASSER-**: water-spout 703
HOSPITAL (*n*): hospital 704
HOSPITALITÄT (*f*): hospitality 705
HOSPIZ (*n*): hospice 706
HOSTIE (*f*): host, consecrated wafer 707
HOTBODEN (*m*): crown-(of a hat) 708

HOURDISTEIN (*m*): hollow flue-cover, hollow arched brick 709
HOUSUNG (*f*), MAST- : mast housing 710
H-SÄURE (*f*): H-acid (see AMIDONAPHTOLDI-SULFOSÄURE) 711
HUB (*m*): lift (of a valve); lifting; stroke (Machinery); throw (of a crank); travel (of a piston) 712
HUBBEWEGUNG (*f*): lifting (movement) motion; reciprocating motion (to and fro motion); linear motion (motion in one direction only) 713
HUBBRÜCKE (*f*): lifting bridge 714
HÜBEL (*m*): hillock; mound; tumulus 715
HUBERS REAGENS (*n*): Hubers reagent (aqueous solution of ammonium molybdate and potassium ferrocyanide mixed; for free mineral acids) 716
HUBHÖHE (*f*): height of lift; length of stroke; lift (of pumps); throw (of a crank) 717
HUBLÄNGE (*f*): length of stroke 718
HÜBNERIT (*m*): Hubnerite (natural manganese tungstate, about 75% tungsten trioxide); $MnWO_4$; Sg. 7.18-7.35 719
HUBPUMPE (*f*): lift pump 720
HÜBSCH : handsome, pretty, nice, good 721
HUBVENTIL (*n*): lift valve 722
HUB, VOLLER- (*m*): full (lift; stroke) bore 723
HUBZAHL (*f*): number of strokes 724
HUBZÄHLER (*m*): indicator (of engines); (stroke)-counter 725
HUDELEI (*f*): bungling-(work); vexation 726
HUF (*m*): hoof 727
HUFBESCHLAG (*m*): horse-shoes; shoeing-(of horses) 728
HUFEISEN (*n*): horseshoe 729
HUFEISENFÖRMIG : horse-shoe-(shaped) 730
HUFEISENKRAUT (*n*): horse-shoe vetch (*Hippocrepis*) 731
HUFEISENMAGNET (*m*): horse-shoe magnet 732
HUFFETT (*n*): hoof (ointment; grease) fat 733
HUFLATTICH (*m*): coltsfoot (*Folia farfaræ*) (*Tussilago farfara*) 734
HUFLATTICHBLÄTTER (*n.pl*): coltsfoot leaves (*Folia farfaræ*) 735
HUFLATTICHBLÜTEN (*f.pl*): coltsfoot flowers (*Flores farfaræ*) 736
HUFLATTIG (*m*): coltsfoot (*Folia farfaræ*) 737
HUFNADEL (*m*): hobnail; horseshoe nail 738
HUFSCHLAG (*m*): horseshoeing; hoof-mark; hoof-beat 739
HUFSCHMIED (*m*): farrier 740
HÜFTADER (*f*): sciatic vein 741
HÜFTBEIN (*n*): hip-bone; haunch-bone; pelvis (Anat.) (see LENDE-) 742
HÜFTBRUCH (*m*): hip fracture 743
HÜFTE (*f*): hip; haunch 744
HÜFTGELENK (*n*): hip joint 745
HUFTIERE (*n.pl*): hoofed mammals, ungulata 746
HÜFTLOCH (*n*): obturator notch (the angle between the ischium and the pubis. When this angle is closed by a bone it is termed *Obturator foramen*) (Ornithology) 747
HÜFTNERV (*m*): sciatic nerve (Med.) 748
HÜFTPFANNE (*f*): socket in the pelvis or hip-bone 749
HÜFTSCHMERZ (*m*): sciatica, pain in the sciatic nerve (Med.) 750
HÜFTSTÜCK (*n*): haunch 751
HÜFTVERRENKUNG (*f*): dislocation of the hip 752
HÜFTWEH (*n*): sciatica (neuralgia of the sciatic nerve) (Med.) 753
HÜGEL (*m*): hill; hillock; mound; knob; knodule; knoll; tumulus 754
HÜGELAMEISE (*f*): red ant (*Formica rufa*) 755
HÜGELIG : hilly; hill-like 756
HUHN (*n*): hen; fowl; any female bird 757
HUHNÄPFELCHEN (*n*): hip, haw (hawthorn fruit) (see HAGEDORN) 758
HÜHNCHEN (*n*): chicken; pullet 759
HÜHNERARTIG : gallinaceous 760
HÜHNERAUGE (*n*): corn (Med.) 761
HÜHNERAUGENMITTEL (*n*): corn cure 762
HÜHNERAUGENOPERATEUR (*m*): corn-cutter 763
HÜHNERAUGENPFLASTER (*n*): corn plaster 764
HÜHNERAUGENRING (*m*): corn ring 765
HÜHNERAUGENSCHNEIDER (*m*): corn-cutter 766
HÜHNERBISS (*m*): chickweed (*Stellaria media*) 767
HÜNHERBLINDHEIT (*f*): night-blindness (see NACHT-BLINDHEIT); day-sight, hemeralopia 768
HÜHNERBRUST (*f*): pigeon-breast (*Pectus gallinaceum*) (Med). 769
HÜHNERDARM (*m*): chickweed (*Stellaria media*) 770
HÜHNEREI (*n*): hen's egg 771
HÜHNEREIWEISS (*n*): white of hen's egg; albumen; albumin; protein; egg-white 772
HÜHNEREIWEISSARTIG : albuminous; like egg-white 773
HÜHNERFETT (*n*): chicken fat; Sg. 0.93; Mp. 25-34°C. 774
HÜHNERFUSSGRAS (*n*): cock's-foot grass (*Andropogon ischæmum*) 775
HÜHNERHÄNDLER (*m*): poulterer 776
HÜHNERHAUS (*n*): hen-house 777
HÜHNERHUND (*m*): pointer, spaniel, setter 778
HÜHNERKORB (*m*): hen-coop 779
HÜHNERKORN (*n*): maize, Indian corn (see MAIS) 780
HÜHNERLEITER (*f*): hen-roost 781
HÜHNERSTANGE (*f*): hen-roost 782
HÜHNERSTEIGE (*f*): hen-roost 783
HÜHNERTOD (*m*): henbane (see BILSENKRAUT) 784
HÜHNERWICKE (*f*): vetch, tare (see WICKE) 785
HÜHNERZUCHT (*f*): poultry (rearing) farming 786
HULD (*f*): grace; kindness; favour; charm 787
HULDIGEN : to devote oneself; pay homage 788
HULDIGUNG (*f*): homage, admiration, favour 789
HULDIGUNGSEID (*m*): oath of allegiance 790
HULDREICH : gracious, favourable 791
HÜLFE (*f*): (see HILFE) 792
HÜLFREICH : helpful 793
HÜLFSMASCHINEN (*f.pl*): auxiliary machinery, auxiliary engines 794
HÜLFSMITTEL (*n*): resource, aid 795
HÜLFSMITTEL (*n.pl*), FINANZIELLE- : funds, financial resources 796
HÜLFSPERSONAL (*n*): staff, assistants 797
HULK (*m*): hulk (Naut.) 798
HÜLLBLATT (*n*): involucral leaf, involucrum (Botany) 799
HÜLLE (*f*): cover-(ing); case; casing; wrapper; wrapping; envelope; sheath; veil; integument; raiment; mask; hood 800
HÜLLEN : to cover; case; wrap; envelop; sheathe; muffle; hide; mask; veil 801
HÜLLFRÜCHTLER (*pl*): *Cupuliferæ* (Bot.) 802
HÜLLROHR (*n*): jacket; (en)-casing tube 803
HÜLSE (*f*): hull; husk; pod; shell; case; socket; collar; brass; bush (of bearings) 804

HÜLSEN: to shell, husk; case, cover, encase; bush-(Bearings) 805
HÜLSENARTIG: leguminous, pod-like 806
HÜLSENBAUM (*m*): locust tree (*Robinia pseudacacia*); carob tree (*Ceratonia siliqua*) 807
HÜLSENFRUCHT (*f*): legume; pulse 808
HULSENFRUCHTARTIG: leguminous; pod-like 809
HÜLSENGEWÄCHS (*n*): leguminous plant 810
HÜLSENPALME (*f*): holly tree (see **STECHPALME**) 811
HÜLSENPFLANZE (*f*): leguminous plant 812
HÜLSIG: leguminous, pod-like; husky 813
HÜLSSTRAUCH (*m*): common holly (*Ilex aquifolium*) 814
HÜLSSTRAUCHGEWÄCHSE (*n.pl*): *Aquifoliaceæ* (Bot.) 815
HUMAN: humane 816
HUMANISIEREN: to humanize 817
HUMANITÄT (*f*): humanity 818
HUMAT (*n*): humate-(of) 819
HUMIFIZIERUNG (*f*): humification, formation of (mould) humus 820
HUMIN (*n*): humin (the soluble constituent matter of brown coal) 821
HUMINSAUER: huminate, combined with humic acid 822
HUMINSÄURE (*f*): humic acid; $C_{18}H_{16}O_7$ 823
HUMIT (*m*): humite, ; $Mg_3[Mg(F,OH)]_2[SiO_4]_2$ (see also **CHONDRODIT**) 824
HUMMEL (*f*): bumblebee; drone 825
HUMMEN: to hum, drone 826
HUMMER (*m*): lobster 827
HUMMERSCHERE (*f*): lobster's claw 828
HUMOR (*m*): humour 829
HUMORIST (*m*): humourist 830
HUMPELN: to hobble; limp; bungle 831
HUMPEN (*m*): bumper; tankard 832
HUMUS (*m*): vegetable (mould) earth; humus (ferric oxide and compounds) (the black or brown mass into which plants decompose, and of which brown coal or peat is largely formed) 833
HUMUSARTIG: mouldy, humus (mould)-like 834
HUMUSLÖSUNG (*f*): humus solution 835
HUMUSSAUER: humate of, combined with humus acid 836
HUMUSSÄURE (*f*): mould acid, humus acid $C_{28}H_{26}O_7$ 837
HUMUSSTOFF (*m*): humus, mould (see **HUMUS**) 838
HUND (*m*): dog; hound; cur 839
HUNDEARTIG: canine 840
HUNDEBLUME (*f*): dandelion, lion's tooth (*Leontodon taraxacum*) 841
HUNDEBLUMENBLÄTTER (*n.pl*): dandelion leaves (leaves of *Leontodon taraxacum*; used as a vegetable) 842
HUNDEBLUMENWURZEL (*f*): dandelion root (root of *Taraxacum officinale*; used for medicinal purposes) 843
HUNDEHÜTTE (*f*): dog's kennel 844
HUNDEKOT (*m*): canine (fæces) feces; (*Album graecum*) 845
HUNDEKRANKHEIT (*f*): distemper-(of dogs) 846
HUNDELEBEN (*n*): dog's life 847
HUNDEMÜDE: dead-tired, tired out 848
HUNDERT: hundred 849
HUNDERTBLÄTTERING: centifolious 850
HUNDERTFÄLTIG: hundred-fold; centuple 851
HUNDERTFUSS (*m*): centipede (*Chilopoda*) 852
HUNDERTGRADIG: centigrade 853
HUNDERTJÄHRIG: centennial 854
HUNDERTSTE (*m*): centesimal 855
HUNDERTSTEL (*n*): hundredth-(part) 856
HUNDERTTEILIG: centesimal; centigrade 857
HUNDESEUCHE (*f*): distemper-(of dogs) 858
HUNDEWACHE (*f*): dog-watch (Naut.) 859
HUNDEZAHNSPAT (*m*): dog-tooth spar; calcite (see **CALCIUMCARBONAT**) 860
HUNDEZUCHT (*f*): dog (training) breeding 861
HÜNDIN (*f*): bitch 862
HUNDSAPFEL (*m*): mandrake, mandragora (*Mandragora vernalis* and *M. autumnalis*) 863
HUNDSBAUM (*m*): spindle tree, strawberry tree (Indian)-arrow wood, bitter ash, wahoo (*Euonymus atropurpureus*) 864
HUNDSBEERBAUM (*m*): dogwood, cornel (see **HARTBAUM**); flowering dogwood (*Cornus florida*) 865
HUNDSBLUME (*f*): (see **HUNDEBLUME**) 866
HUNDSDORN (*m*): hawthorn (see **HAGEDORN**) 867
HUNDSHAI (*m*): spotted dog-fish (two species; larger, *Scyllium canicula* and lesser, *S. catulus*) 868
HUNDSKAMILLE (*f*): dog's (stinking) camomile; mayweed (*Anthemis cotula*) 869
HUNDSKIRSCHE (*f*): (white)-bryony, bryonia (*Bryonia alba* and *Byronia dioica*) 870
HUNDSKRAMPF (*m*): cynic spasm 871
HUNDSPETERSILIE (*f*): dog's poison, fool's parsley (*Aethusa cynapium*) 872
HUNDSPINTE (*f*): rope's point, artificial swaging of a rope's end (Naut.) 873
HUNDSPÜNTE (*f*): rope's point (see **HUNDSPINTE**) 874
HUNDSROSE (*f*): wild briar rose, dog-rose (*Rosa canina*) 875
HUNDSRÜBE (*f*): bryony (*Bryonia*) (see also **ZAUNRÜBE** and **HUNDSKIRSCHE**) 876
HUNDSSCHIERLING (*m*): dog's poison, fool's parsley (*Aethusa cynapium*) 877
HUNDSSTERN (*m*): Sirius, dog-star 878
HUNDSTAGE (*m.pl*): dog-days (*Dies caniculares*) 879
HUNDSTOD (*m*): dog's bane (*Apocynum androsæmifolium*) 880
HUNDSVEILCHEN (*n*): dog's violet (*Viola canina*) 881
HUNDSWURZEL (*f*): yam (*Dioscorea batatas* or *D. decaisneana*) 882
HUNDSWUT (*f*): hydrophobia, rabies, canine madness (Med.) 883
HUNDSZUNGE (*f*): hound's tongue (*Cynoglossum officinale*) 884
HUNDSZUNGENKRAUT (*n*): hound's tongue herb (from *Cynoglossum officinale*) 885
HUNDSZUNGENWURZEL (*f*): hound's tongue root (from *Cynoglossum officinale*) 886
HÜNE (*m*): giant (see **RIESE**) 887
HUNGER (*m*): hunger; appetite; desire; longing 888
HUNGERKORN (*n*): ergot (of rye) 889
HUNGERKUR (*f*): starvation (fasting) cure or diet 890
HUNGERN: to be hungry, starve, fast, desire, long for 891
HUNGERPEST (*f*): famine fever, relapsing fever (infectious disease due to parasite, *Spirillium obermeieri*) (Med.) 892
HUNGERSNOT (*f*): famine 893
HUNGERSTEIN (*m*): salt pan scale 894
HUNGERTOD (*m*): death due to starvation 895
HUNGRIG: hungry; starving 896
HÜPFEN: to hop; skip; jump; leap 897

HÜRDE (*f*): hurdle; hurdle-work wall; pen: screen; fold; (com)-pound 898
HÜRDENGEFLECHT (*n*): hurdle (basket) work 899
HURTIG: quick; nimble; speedy; swift; prompt; agile 900
HUSTEN: to cough; have a cough; (*m*): cough; tussis (Med.) 901
HUSTENANFALL (*m*): fit of coughing 902
HUSTENFIEBER (*n*): catarrhal fever 903
HUSTENLÖSEND: easing a cough, cough-loosening 904
HUSTENLÖSENDES MITTEL (*n*): cough remedy, cough-loosening medicine 905
HUSTENMITTEL (*n*): cough (mixture; medicine) remedy 906
HUSTENPRÄPARAT (*n*): cough (preparation) medicine 907
HUSTENSTILLEND: pectoral 908
HUSTENTEE (*m*): cough tea (such as linseed tea) 909
HUSTENTROPFEN (*m.pl*): cough (pectoral) drops 910
HUT (*f*): guard-(ing); keeping; watch; care; pasture; (*m*): hat; cap; bonnet; cover; lid; top; loaf (Sugar); scum (Wine) 911
HÜTEN: to watch; guard; keep; take care; tend; protect; beware 912
HUTFARBE (*f*): hat dye 913
HUTFARBSTOFF (*m*): hat (colouring matter) dye 914
HUTFILZ (*m*): hat felt 915
HUTFORM (*f*): hat-block 916
HUTFUTTER (*n*): hat-lining 917
HUTFUTTERAL (*n*): hat-box 918
HUTKREMPE (*f*): hat-brim 919
HUTLACK (*m*): hat varnish 920
HUTMACHER (*m*): hatter 921
HUTPILZ (*m*): pileate (agaricaceous) fungus; mushroom 922
HUTSTEIFE (*f*): hat stiffening 923
HÜTTE (*f*): (metallurgical or glass)-works; mill; smelting house; forge; foundry; hut; cabin; cottage; shed 924
HÜTTE (*f*), **DECKS-**: poop 925
HÜTTENAFTER (*m*): residue (refuse) from a foundry; (foundry)-slag 926
HÜTTENARBEIT (*f*): smelting; founding; foundry work 927
HÜTTENBEWOHNER (*m*): cottager 928
HÜTTENDECK (*n*): poop deck 929
HÜTTENGERICHT (*n*): court of mines 930
HÜTTENGLAS (*n*): pot metal (Glass) 931
HÜTTENHERR (*m*): owner (proprietor) of a foundry 932
HÜTTENKATZE (*f*): lead colic; foundry travelling crab (Mech.) 933
HÜTTENKOKS (*m.* and *pl*): smeltery (foundry) coke, smelting works' coke 934
HÜTTENKUNDE (*f*): metallurgy 935
HÜTTENKUNDIGER (*m*): metallurgist 936
HÜTTENMÄNNISCH: metallurgical 937
HÜTTENNICHTS (*n*): *Nihilum album*; white tutty (impure zinc oxide) 938
HÜTTENPROBIERKUNST (*f*): metallurgical assaying 939
HÜTTENPRODUKT (*n*): metallurgical product 940
HÜTTENPROZESS (*m*): metallurgical process 941
HÜTTENPRÜFER (*m*): metallurgical assayer 942
HÜTTENRAUCH (*m*): smelter smoke; arsenical fumes; white arsenic; flowers of arsenic (see ARSENOXYD) 943
HÜTTENREISE (*f*): campaign (Metal) 944
HÜTTENSALZ (*n*): purified (refined) salt, table salt 945
HÜTTENTRICHTER (*m*): conical funnel 946
HÜTTENWERK (*n*): smelting works; foundry; forge; mill (plant); rolling mill 947
HÜTTENWESEN (*n*): smelting; metallurgy 948
HÜTTENZINN (*n*): grain tin 949
HUTZUCKER (*m*): loaf sugar 950
HYACINTH (*m*): hyacinth, jacinth (a zirconium silicate) (see ZIRKON) 951
HYACINTHE (*f*): hyacinth (flower) 952
HYACINTHGRANAT (*m*): essonite 953
HYACINTHIN (*n*): hyacinthin (obtained by bromination of Cinnamic acid or Styrol; used for perfumes); Sg. 1.395 954
HYALIT (*m*): hyalite (a variety of opal) (a hydrous silica) (see OPAL) 955
HYALOPHAN (*m*): hyalophane (rare barium feldspar); $(K_2, Ba) Al_2Si_4O_{12}$; Sg. 2.78 956
HYALOSIDERIT (*m*): hyalosiderite (see OLIVIN) 957
HYÄNASÄURE (*f*): hyenic acid, hyaenic acid (from anal glands of *Hyaena striata*); $C_{24}H_{49}CO_2H$; Mp. 77.5°C. 958
HYÄNE (*f*): hyena; hyaena (*Hyaena striata*) 959
HYDANTOINSÄURE (*f*): hydantoic acid 960
HYDNOCARPUSSÄURE (*f*): hydnocarpus acid; $C_{17}H_{31}CO_2H$; Mp. 60°C. 961
HYDRANT (*m*): hydrant 962
HYDRANTROHR (*n*): hydrant stand-pipe 963
HYDRARGILLIT (*m*): hydrargillite (natural hydrous aluminium oxide); $Al(OH)_3$; Sg. 2.35 (see also GIBBSIT) 964
HYDRARGYLLIT (*n*), **KÜNSTLICHES-**: artificial hydrargillite (see ALUMINIUMOXYDHYDRAT) 965
HYDRASTIN (*n*): hydrastine (alkaloid from the root of *Hydrastis canadensis*); $C_{21}H_{21}NO_6$; Mp. 131°C. 966
HYDRASTINCHLORHYDRAT (*n*): hydrastine hydrochloride; $C_{21}H_{21}NO_6.HCl$ 967
HYDRASTININ (*n*): hydrastinine; $C_{11}H_{13}NO_3$; Mp. 116.5°C. 968
HYDRASTIS (*f*): hydrastis, golden seal, yellow (orange) root; (Indian)-turmeric (root of *Hydrastis canadensis*) 969
HYDRASTISWURZEL (*f*): golden seal root (*Rhizoma hydrastidis*) (see HYDRASTIS) 970
HYDRAT (*n*): hydrate; hydroxide (salts or compound containing water) 971
HYDRATATION (*f*): hydration 972
HYDRATHALTIG: hydrated 973
HYDRATISCH: hydrated 974
HYDRATISIEREN: to hydrate; be-(come) hydrated 975
HYDRATROPASÄURE (*f*): hydrotropic acid 976
HYDRATWASSER (*n*): hydrate water; water of hydration 977
HYDRAULIK (*f*): hydraulics 978
HYDRAULISCH: hydraulic 979
HYDRAULISCHE PRESSE (*f*): hydraulic press 980
HYDRAULISCHER MÖRTEL (*m*): hydraulic cement 981
HYDRAZIN (*n*): hydrazine; $(NH_2)_2$; Sg. 1.013; Mp. 1.4°C.; Bp. 113.5°C. 982
HYDRAZIN-(DIAMID) (*n*): hydrazine-(diamide); N_2H_4; Sg. 1.003-1.0114; Mp. 1.4°C.; Bp. 113.5°C. 983
HYDRAZINHYDRAT (*n*): hydrazine hydrate; $N_2H_4+H_2O$ Sg. 1.0305 984
HYDRAZINHYDRATLÖSUNG (*f*): solution of hydrazine hydrate 985

HYDRAZINSULFAT (*n*): hydrazine sulphate; $N_2H_4.H_2SO_4$ 986
HYDRAZOBENZOL (*n*): hydrazobenzene (see AZOBENZOL) 987
HYDRAZOVERBINDUNG (*f*): hydrazo compound 988
HYDRID (*n*): hydride (compound of hydrogen and one element) 989
HYDRIEREN: to hydrogenize; hydrogenate 990
HYDRIERUNG (*f*): hydrogenation 991
HYDROAROMATISCH: hydroaromatic 992
HYDROBORAZIT (*m*): hydroboracite; $CaMgB_6O_{11}.6H_2O$; Sg. 1.95 993
HYDROBROMSÄURE (*f*): hydrobromic acid; hydrogen bromide; HBr 994
HYDROCHININ (*n*): hydroquinine, dihydroquinine; $C_{20}H_{26}N_2O_2.HCl.2H_2O$ 995
HYDROCHINON (*n*): hydroquinone; quinol; hydrochinone; para-dioxybenzene; dihydroxylbenzene; $C_6H_4(OH)_2$; Sg. 1.33; Mp. 169°C.; Bp. 285°C. 996
HYDROCHINONENTWICKLER (*m*): hydroquinone developer (composed of potash, sodium sulphite crystals and hydroquinone (paradioxybenzol), in the proportion 65 : 45 : 5 by weight) (Photo.) 997
HYDROCHINONLÖSUNG (*f*): hydroquinone solution (hydroquinone developer plus about 5 parts water; see also HYDROCHINONENTWICKLER) (Photo.) 998
HYDROCHINOXALIN (*n*): hydroquinoxaline 999
HYDROCHLORSÄURE (*f*): hydrochloric acid; HCl (see CHLORWASSERSTOFF) 000
HYDROCOLLODIN (*n*): hydro-collodine (a ptomaine from putrefying fibrin of bullock's blood); $C_8H_{13}N$ 001
HYDROCRESOL (*n*): hydrocresol 002
HYDROCUMARON (*n*): hydrocoumarone; hydrocumarone 003
HYDROCUMARSÄURE (*f*): hyrdocoumaric (hydrocumaric) acid 004
HYDROCYANSÄURE (*f*): hydrocyanic acid; formonitrile (see CYANWASSERSTOFF) 005
HYDRODYNAMIK (*f*): hydrodynamics 006
HYDRODYNAMISCH: hydrodynamic 007
HYDROELEKTRISCH: hydroelectric 008
HYDROGEL (*n*): hydrogel (the coagulated watery solution of a colloid) 009
HYDROGENIERT: hydrogenated 010
HYDROGENISIEREN: to hydrogenate, harden (Oils) 011
HYDROGENISIERUNG (*f*): hydrogenation, hardening (of oils) 012
HYDROGENLICHT (*n*): Drummond's light, oxycalcium light (by heating lime with a blowpipe flame fed with oxygen) 013
HYDROGENPOL (*m*): negative pole 014
HYDROGRAPHIE (*f*): hydrography 015
HYDROGRAPHISCH: hydrographical 016
HYDROIDPOLYP (*m*): medusoid, hydrozoa 017
HYDROJODSÄURE (*f*): hydriodic acid; hydrogen iodide (see JODWASSERSTOFF) 018
HYDROKAFFEESÄURE (*f*): hydrocaffeic acid 019
HYDROKETTE (*f*): hydro-cell; hydro-element (Elec.) 020
HYDROKOTARNIN (*n*): hydrocotarnine (an opium alkaloid); $C_{11}H_{12}NO_2(OCH_3)$; Mp. 50-55°C. 021
HYDROL (*n*): hydrol, tetramethyldiaminobenzhydrol (see TETRAMETHYLDIAMINO-BENZHYDROL) 022
HYDROLITH (*n*): calcium hydride; CaH_2 023
HYDROLOGIE (*f*): hydrology (the science of liquids at rest) 024
HYDROLOGISCH: hydrological-(ly) 025
HYDROLYSE (*f*): hydrolysis (decomposition by the action of water) 026
HYDROLYSEGESCHWINDIGKEIT (*f*): hydrolysis velocity 027
HYDROLYSIEREN: to hydrolyse 028
HYDROMAGNESIT (*m*): hydromagnesite; $3MgCO_3,Mg(OH)_2.3H_2O$; Sg. 2.17 029
HYDROMECHANIK (*f*): hydromechanics 030
HYDROMECHANISCH: hydromechanical 031
HYDROMEDUSEN (*f.pl*): hydromedusæ, hydrozoa (such as *Physophora hydrostatica*) 032
HYDROMETER (*n*): hydrometer; aerometer (apparatus for comparing the density of liquids) 033
HYDROMETRIE (*f*): hydrometry 034
HYDROMETRISCH: hydrometric 035
HYDROMETRISCHE WAGE (*f*): hydrometer, aerometer (apparatus for comparing the density of liquids) 036
HYDROOXYGENGAS (*n*): oxyhydrogen gas, detonating gas (mixture of hydrogen and oxygen in proportion 2 : 1. combustion product H_2O) 037
HYDROPATHIE (*f*): hydropathy, hydrotherapy (the treatment of disease by water) (Med.) 038
HYDROPATHISCH: hydropathic 039
HYDROPEROXYD (*n*): hydrogen peroxide (see WASSERSTOFFSUPEROXYD) 040
HYDROPHAN (*m*): hydrophane, white opal (see OPAL) 041
HYDROPHOBIE (*f*): hydrophobia, rabies, canine madness (Med.) 042
HYDROPHTALSÄURE (*f*): hydrophthalic acid 043
HYDROSCHWEFLIG: hydrosulphurous 044
HYDROSCHWEFLIGE SÄURE (*f*): hydrosulphurous acid; $H_2S_2O_4$ 045
HYDROSOL (*n*): hydrosol (a colloid dissolved in water) 046
HYDROSTATIK (*f*): hydrostatics 047
HYDROSTATISCH: hydrostatic-(ally) 048
HYDROSULFIT (*n*): hyposulphite, hydrosulphite 049
HYDROSULFITELEKTROLYSE (*f*): hyposulphite electrolysis 050
HYDROTECHNIK (*f*): hydrotechnics 051
HYDROTECHNISCH: hydrotechnical 052
HYDROTHIONSÄURE (*f*): hydrosulphuric acid; hydrogen sulphide (see SCHWEFELWASSER-STOFF 053
HYDROVERBINDUNG (*f*): hydro compound (by adding hydrogen) 054
HYDROXYD (*n*): hydroxide (an -ic hydroxide) 055
HYDROXYDUL (*n*): hydroxide (an -ous hydroxide) 056
HYDROXYGENLICHT (*n*): oxy-hydrogen light 057
HYDROXYLAMIN (*n*): hydroxylamine, oxammonium; $NH_2.OH$; Sg. 1.204-1.227; Mp. 33°C.; Bp. 70°C. 058
HYDROXYLAMINCHLORHYDRAT (*n*): hydroxylamine hydrochloride; oxammonium hydrochloride; NH_2OHHCl; Mp. 151°C. 059
HYDROXYLAMIN (*n*), **SALZSAURES-**: hydroxylamine (hydrochlorate) hydrochloride; $NH_2.OH.HCl$ (see HYDROXYLAMINCHLORHYDRAT) 060
HYDROXYLAMIN (*n*), **SCHWEFELSAURES-**: hydroxylamine sulphate (see HYDROXYLAMIN-SULFAT) 061

HYDROXYLAMINSULFAT (*n*): hydroxylamine sulphate, oxammonium sulphate; $(NH_2OH)_2.H_2SO_4$; Mp. 140°C. 062
HYDROXYLHALTIG: containing hydroxyl 063
HYDROXYLIEREN: to hydroxylate 064
HYDROXYLION (*n*): hydroxyl ion 065
HYDROXYL-IONEN-KONZENTRATION (*f*): hydroxyl-ion concentration, concentration of hydroxyl ions 066
HYDROXYLJON (*n*): hydroxyl ion 067
HYDROZIMTSÄURE (*f*): hydrocinnamic acid; $C_9H_{10}O_2$; Sg. 1.071; Mp. 48.7°C.; Bp. 279.8°C. 068
HYDROZINKIT (*m*): hydrozincite, zinc bloom; $ZnCO_3.2Zn(OH)_2$; Sg. 3.69 (see ZINK-BLÜTE) 069
HYDROZOEN (*pl*): hydrozoa, medusoids (see MEERNESSEL and HYDROMEDUSEN) 070
HYDRÜR (*n*): hydride; sub-hydride 071
HYETOMETER (*n*): rain-gauge 072
HYGIAMATABLETTEN (*n.pl*): hygiama tablets 073
HYGIENE (*f*): hygiene 074
HYGIENISCH: hygienic-(ally) 075
HYGRIN (*n*): Hygrine (a coca alkaloid; see COCAALKALOID); $C_8H_{15}NO$ 076
HYGROGRAPH (*m*): recording hygrometer (see HYGROMETER) 077
HYGROMETER (*n*): hygrometer (instrument for measuring humidity of atmosphere) 078
HYGROMETRISCH: hygrometric 079
HYGROSKOPISCH: hygroscopic 080
HYGROSKOPISCHES WASSER (*n*): hygroscopic water (moisture absorbed from the atmosphere) 081
HYGROSKOPIZITÄT (*f*): hygroscopicity; hygroscopic (power) capacity 082
HYMENOMYCETE (*f*): *Hymenomycetes* (Fungi) 083
HYMENOPTEREN (*pl*): hymenoptera 084
HYMNE (*f*): hymn 085
HYOSCIN (*n*): hyoscine; scopolamine; $C_{17}H_{21}NO_4$; Mp. 50-59°C.; (*Hyoscinum*) (see SKOPOLAMIN) 086
HYOSCINBROMHYDRAT (*n*): hyoscine (scopolamine) hydrobromide; $C_{17}H_{21}NO_4HBr.3H_2O$; Mp. 191°C. 087
HYOSCINSULFAT (*n*): hyoscine sulphate; $(C_{17}H_{21}NO_4)_2.H_2SO_4.2H_2O$ 088
HYOSCYAMIN (*n*): (*Hyoscyaminum*): hyoscyamine (an alkaloid from the seeds of *Hyoscyamus niger*); $C_{17}H_{23}NO_3$; Mp. 108.5°C. 089
HYOSCYAMINBROMHYDRAT (*n*): hyoscyamine hydrobromide; $C_{17}H_{23}NO_3.HBr$; Mp. 191.5°C. 090
HYOSCYAMINCHLORHYDRAT (*n*): hyoscyamine hydrochloride; $C_{17}H_{23}NO_3.HCl$. 091
HYOSCYAMINSULFAT (*n*): hyoscyamine sulphate; $(C_{17}H_{23}NO_3)_2.H_2SO_4$; Mp. 198.9°C. 092
HYOSZIN (*n*): (see HYOSCIN) 093
HYOSZYAMIN (*n*): (see HYOSCYAMIN) 094
HYPERAZIDITÄT (*n*): hyper-acidity; super-acidity 095
HYPERBEL (*f*): hyperbola 096
HYPERBEL (*f*), GLEICHSEITIGE-: rectangular hyperbola 097
HYPERBOLISCH: hyperbolic 098
HYPERBOLOID (*n*): hyperboloid 099
HYPERBOLOIDE (*f*): hyperboloid 100
HYPERBOLOIDISCHES SCHEIBENPROFIL (*n*): hyperbolic disc section 101
HYPERBORÄISCH: hyperborean, northern 102
HYPERCHLORAT (*n*): perchlorate 103
HYPERGLUKÄMIE (*f*): hyperglycemia 104
HYPERION (*m*): hyperion 105
HYPERJODAT (*n*): periodate 106
HYPERKRITIK (*f*): hypercriticism 107
HYPEROXYD (*n*): hyperoxide; peroxide 108
HYPERSTHEN (*m*): hypersthene (iron-magnesium metasilicate); $(Fe,Mg)SiO_3$; Sg. 3.4 109
HYPERTROPHIE (*f*) DES HERZENS: hypertrophy of the heart (increase in weight and size) (Med.) 110
HYPHE (*f*): hypha (a filamentous cell of fungi) 111
HYPHENGEWEBE (*n*): hyphal tissue 112
HYPHENMASSE (*f*): hyphal mass (of fungi) 113
HYPNON (*n*): hypnone, acetophenone (see ACETOPHENON) 114
HYPNOTICUM (*n*): hypnotic, soporific (Med.) 115
HYPNOTISCH: hypnotic-(ally); soporific-(ally) 116
HYPOCHLORIT (*n*): hypochlorite 117
HYPOCHLORITLÖSUNG (*f*) ELEKTROLYTISCHE-: electrolytic hypochlorite solution 118
HYPOCHLORITMASSE (*f*), FESTE-: solid hypochlorite 119
HYPOCHONDER (*m*): hypochondriac (one who suffers with hypochondriasis) (see HYPOCHONDERIE) 120
HYPOCHONDERIE (*f*): Hypochondriasis, mental disorder, a type of melancholia (Med) 121
HYPOCHONDRISCH: hypochondrical, splenetic 122
HYPOCYCLOIDE (*f*): hypocycloid 123
HYPODERMATISCH: hypodermic-(ally) 124
HYPODERMATISCHE INJEKTION (*f*): hypodermic injection 125
HYPODERMATISCHE INJEKTION DER ARZNEIMITTEL: hypodermic medication 126
HYPOGÄASÄURE (*f*): hypogeic acid; $C_{16}H_{30}O_2$; Mp. 33°C. 127
HYPOGEN: hypogene 128
HYPOKRIT (*m*): hypocrite 129
HYPOKRITISCH: hypocritical 130
HYPOPHOSPHITSIRUP (*f*): sirup of hypophosphites (containing hypophosphite of manganese, calcium, iron, and sodium plus quinine) 131
HYPOPHYSE (*f*): hypophysis, pituitary gland (of the brain) (*Hypophysis cerebri; Glandula pituitaria*) (Med.); hypophysis, apophysis (Bot.) 132
HYPOPHYSENEXTRAKT (*m*): hypophysene extract, extract of the pituitary gland 133
HYPOSTASE (*f*): hypostasis 135
HYPOSULFIT (*n*): hyposulphite, thiosulphate 136
HYPOTENUSE (*f*): hypotenuse 137
HYPOTHEK (*f*): mortgage; security 138
HYPOTHEKARANLEIHEN (*n*): mortgage loan 139
HYPOTHEKARISCH: by mortgage 140
HYPOTHEKARISCH VERPFÄNDEN: to mortgage, to take up a loan on something as security 141
HYPOTHEK (*f*) AUFNEHMEN AUF-: to mortgage, take up a loan on something as security 142
HYPOTHEKIEREN: to hypothecate, pawn, mortgage 143
HYPOTHESE (*f*): hypothesis 144
HYPOTHESIS (*f*): hypothesis 145
HYPOTHETISCH: hypothetical-(ly) 146

HYPTISÖL (*n*): Hyptis oil (from seeds of *Hyptis spicigera*) 147
HYRGOL (*n*): hyrgol (colloidal mercury which is soluble in water) 148
HYSTERESIS (*f*): hysteresis (Magnetism) 149
HYSTERIE (*f*): hysteria (Med.) 150
HYSTERISCH: hysterical-(ly) 151
HYSTERISCHE KÜGEL (*pl*): *Globus hystericus* (the feeling as of a lump rising in the throat with hysterical people) (Med.) 152
HYSTERO-EPILEPSIE (*f*): hystero-epilepsy 153
HYSTEROLOGIE (*f*): hysterology 154

I

IATROCHEMIE (*f*): iatrochemistry 155
IATROL (*n*): iatrol; $NH.C_6H_5O_2.C_2H_5OI_2$ 156
IBISCH (*m*): marsh-mallow (see ALTHEE) 157
IBIT (*n*): ibit, bismuth oxyiodotannate 158
ICH: I 159
ICHNOGRAPHIE (*f*): ichnography 160
ICHTHALBIN (*n*): ichthalbin, ichthyol albuminate 161
ICHTHYOL (*n*): Ichthyol (Chemistry); ammonium-ichthyol sulfonate, ammonium sulfoichthyolate (*Ammonium sulfoichthyolicum*) (Med.) 162
ICHTHYOL-EIWEISS (*n*): ichthalbin, ichthyol albuminate 163
ICHTHYOLERSATZ (*m*): ichthyol substitute 164
ICHTHYOLOGIE (*f*): ichthyology (see also FISCH-LEHRE) 165
ICHTHYOLÖL (*n*), ROHES-: (see ICHTHYOLROHÖL) 166
ICHTHYOLROHÖL (*n*): ichthyol (sulfonated hydrocarbons from dry distillation of "Stinkstein") (see STINKSTEIN) 167
ICHTHYOLSEIFE (*f*): ichthyol soap 168
ICHTHYOLSULFOSAUER: sulfoichthyolate, ichthyolsulfonate 169
ICHTHYOLSULFOSÄURE (*f*): ichthyol, ammonium-ichthyol sulfonate, ammonium sulfoichthyolate (Pharm.) 170
ICHTHYOPHTHALM (*m*): apophyllite (see APOPHYL-LIT) 171
ICHTOFORM (*n*): ichthoform, ichthyolformaldehyde 172
ICOCAÖL (*n*): Icoca oil (from *Chrysobalanus icoca*) 173
ICOCA (*f*), WESTINDISCHE-: West Indian icoca, gold plum (*Chrysobalanus icoca*) 174
IDEAL: ideal, perfect; (*n*): ideal 175
IDEALISCH: ideal-(ly) 176
IDEALISMUS (*m*): idealism 177
IDEALISTISCH: idealistic-(ally) 178
IDEE (*f*): idea; notion; conception 179
IDENTIFIZIEREN: to identify, recognize 180
IDENTIFIZIERUNG (*f*): identification 181
IDENTISCH: identical-(ly) 182
IDENTITÄT (*f*): identity 183
IDIOMATISCH: idiomatic-(ally) 184
IDIOT (*m*): idiot, imbecile, feeble-minded person 185
IDIOTIE (*f*): idiocy, feeble-mindedness, imbecility, insanity (Med.) 186
IDIOTISCH: idiotic-(ally) 187
IDIOTISMUS (*m*): idiocy, feeble-mindedness, imbecility, insanity (Med.); idiom (of a language) 188
IDOKRAS (*m*): vesuvianite; californite; idocrase (natural complex calcium-aluminium silicate, with traces of various other constituents); Sg. 3.4; $(H,F)(Ca,Fe,Mg,Mn)_2(Al,Fe,B)(SiO_4)_2$ 189
IDOZUCKERSÄURE (*f*): idosaccharic acid 190
IDRIALIT (*m*): idrialite; $C_{42}H_{28}O$; Sg. 1.5 191
IDYLLE (*f*): idyl-(l) 192
IDYLLISCH: idyllic, pastoral 193
I-EISEN (*n*): I-beam, joist, girder; H-beam, I or H section iron 194
IGASURSÄURE (*f*): strychnic acid (from the seeds of *Strychnos ignatii*) 195
IGEL (*m*): hedgehog; urchin (*Erinaceus europæus*) 196
IGELARTIG: prickly; echinate 197
IGELFISCH (*m*): globe (hedgehog) fish, (*Diodon*); sea-urchin (*Echinus esculentus*) 198
IGELSTEIN (*m*): echinite (see AESCHYNIT) 199
IGNATIUSBOHNE (*f*): St. Ignatius bean (seed of *Strychnos ignatii*) (1.5% strychnine content) 200
IGNORANT (*m*): ignoramus 201
IGNORIEREN: to ignore, overlook, neglect 202
IKONOKLASTISCH: iconoclastic (see also BILDER-STÜRMEND) 203
IKONOMETER (*n*): view meter (Phot.) 204
IKOSITETRAEDER (*n*): icositetrahedron 205
ILIISCH: iliac (Med.) 206
ILLEGITIM: illegitimate 207
ILLIPEFETT (*n*): ellipi fat (see MAHWAHBUTTER) 208
ILLIPEÖL (*n*): ellipi oil (see BASSIAÖL and MAH-WAHBUTTER) 209
ILLIPETALG (*m*): (see BASSIAÖL) 210
ILLIUM (*n*): illium (nickel-alloy, about 60% Ni; 21% Cr; 6% Cu; 5% Mo; 1% Si; 2% W.; 0.98% Mn; 1% Al; 0.76% Fe); Mp. 1300°C. 211
ILLUMINATION (*f*): illumination 212
ILLUMINIEREN: to illuminate 213
ILLUSORISCH: illusory 214
ILLUSTRATION (*f*): illustration 215
ILLUSTRIEREN: to illustrate 216
ILMENIT (*m*): ilmenite; $FeTiO_3$; (51% TiO_2); menaccanite; titaniferous iron (see also TITANEISEN) 217
ILTIS (*m*): polecat; fitchet (*Mustela putorius*) 218
ILVAIT (*m*): ilvaite; $HCaFe_2FeSi_2O_9$; Sg. 4.0 219
IMAGINÄR: imaginary 220
IMBIBIEREN: to imbibe 221
IMBIBITION (*f*): the addition of water to the crushed sugar cane in order to extract the residue of the cane juice, when again pressed through rolls 222
IMBISS (*m*): lunch; luncheon; light meal; breakfast 223
IMIDAZOLYLPROPIONSÄURE (*f*), D-α-AMINO-β-: (see HISTIDIN) 224
IMID (*n*), DIPHENYL-: (see KARBAZOL and CAR-BAZOL) 225
IMIDOÄTHER (*m*): imido ester 226
IMIDOSULFONSÄURE (*f*): imidosulfonic acid; disulfamic acid 227
IMIDOTHIOÄTHER (*m*): imido thio ester 228
IMINOBIPHENYL (*n*): (see CARBAZOL) 229

IMITATION (*f*): imitation; (as a prefix=imitation; artificial) 230
IMMEDIALFARBE (*f*): immedial dye 231
IMMEDIALSCHWARZ (*n*): immedial black 232
IMMER: always; ever; continually; continuously; perpetually; constantly; eternally; evermore 233
IMMERFORT: always; constantly; continually; continuously; perpetually 234
IMMERGRÜN: evergreen; (*n*): evergreen; periwinkle ("*Vinca*" genus of plants) 235
IMMERHIN: still; after all; always; however; nevertheless; yet 236
IMMERSCHÖN (*n*): everlasting; immortelle (*Helichrysum arenarium*) 237
IMMERWÄHREND: everlasting; perpetual; endless 238
IMMOBIL: immovable; fixed 239
IMMOBILIEN (*pl*): property, house property, real estate 240
IMMORALISCH: immoral-(ly) 241
IMMORTELLE (*f*): (see IMMERSCHÖN) 242
IMMUNISIEREN: to-(make)-immune 243
IMMUNISIERUNG (*f*): immunization 244
IMMUNITÄT (*f*): immunity 245
IMMUNKÖRPER (*m*): immune body 246
IMOGEN (*n*), **SCHWEFLIGSAURES-**: imogene sulphite (see IMOGENSULFIT) 247
IMOGENSULFIT (*n*): imogene sulphite (a photographic developer) 248
IMPEDANZ (*f*): inductive resistance, apparent resistance (Elect.); impedance 249
IMPERATIV (*m*): imperative mood 250
IMPERATORISCH: imperious-(ly) 251
IMPERIALBLAU (*n*): imperial blue 252
IMPERMEABEL: impermeable 253
IMPERMEABILITÄT (*f*): impermeability 254
IMPFARZT (*m*): inoculator, vaccinator 255
IMPFBAR: inoculable 256
IMPFEN: to inoculate; vaccinate; graft (Plants) 257
IMPFFLÜSSIGKEIT (*f*): inoculating fluid (see IMPFSTOFF) 258
IMPFSTIFT (*m*): inoculating pencil 259
IMPFSTOFF (*m*): lymph; serum; vaccine 260
IMPFUNG (*f*): inoculating; vaccination 261
IMPONIEREN: to impose; impress 262
IMPORT (*m*): import 263
IMPORTIEREN: to import 264
IMPOST (*m*): impost, customs, duty, tax, charge, levy 265
IMPOTENZ (*f*): impotency (Med.) 266
IMPRÄGNIEREN: to impregnate 267
IMPRÄGNIERFLÜSSIGKEIT (*f*): impregnating (wood preserving) fluid or liquid 268
IMPRÄGNIERKESSEL (*m*): impregnating (impregnation) pan 269
IMPRÄGNIERMITTEL (*n*): impregnating (impregnation) preparation, substance, medium or agent 270
IMPRÄGNIERÖL (*n*): impregnating oil (a coal-tar oil for impregnating railway sleepers) 271
IMPRÄGNIERÖLAPPARAT (*n*): impregnating oil apparatus 272
IMPRÄGNIERPFANNE (*f*): impregnating (impregnation) pan 273
IMPRÄGNIERTROG (*m*): impregnating trough 274
IMPRÄGNIERUNG (*f*): impregnation 275
IMPRÄGNIERUNGSFARBE (*f*): impregnating colour 276
IMPRÄGNIERUNGSMITTEL (*n*): impregnating (agent, preparation) substance 277
IMPROL (*n*): improl (an impregnating agent for waterproofing fibres) 278
IMPROVISIEREN: to improvise 279
IMPULS (*m*): impulse 280
IMPULSFREQUENZ (*f*): impulse frequency 281
IMPULSIV: impulsive-(ly) 282
IMSTANDE: able to; capable of; in a position to 283
INAKTIVITÄT (*f*): inactivity 284
INANSPRUCHNAHME (*f*): claim; duty; demand (load; strain; stress) upon; stressing; the taking up of 285
INBEGRIFF (*m*): total; sum; inclusion; essence; epitome; summary; contents; purport; tenor; abstract 286
INBEGRIFFEN: inclusively 287
INBETRIEBNAHME (*f*): starting up; putting into commission (Machinery) 288
INBETRIEBSETZUNG (*f*): starting up, putting in commission, putting under (steam) load (of machinery or other plant) 289
INBRUNST (*f*): ardour; fervour 290
INBRÜNSTIG: ardent-(ly), fervent-(ly) 291
INCIDENT: casual, incidental, secondary 292
INCIDENZ (*f*): incidence 293
INCIDENZWINKEL (*m*): angle of incidence 294
INCLUDIEREN: to include 295
INDALIZARIN (*n*): indalizarine 296
INDAMIN (*n*): indamine (a dye for imparting a violet colour to cotton); $NH_2.C_6H_4.N:C_6H_4:NH$ 297
INDANTHREN (*n*): indanthrene 298
INDANTHRENFARBE (*f*): indanthrene dye 299
INDANTHRENFARBSTOFF (*m*): indanthrene colouring matter (blue tar-dyes used as vat-dyes) 300
INDEM: while; whilst; when; since; because 301
INDEMNIFIZIEREN: to indemnify 302
INDEMNITÄT (*f*): indemnity, reparations 303
INDEN (*n*): indene; C_9H_8; Sg. 1.0059; Bp. 181.3°C. 304
INDES: meanwhile; however; in the meantime 305
INDESSEN: (see INDES) 306
INDEX (*m*): index 307
INDEX (*m*), **BRECHUNGS-**: refractive index 308
INDEXZIFFER (*f*): index figure 309
INDIANISCH: Indian (see INDISCH) 310
INDIANISCHROT (*n*): Indian red 311
INDIEN (*n*): India 312
INDIENNE (*f*): indienne, a kind of cotton fabric 313
INDIFFERENT: indifferent-(ly) 314
INDIG (*m*): indigo; Sg. 1.01; natural indigo blue 315
INDIGEN: indigenous; native 316
INDIGKARMIN (*m*): indigo carmine, indigo extract, soluble indigo, sodium (indigotindisulfonate) coerulinsulfate; $C_{16}H_8N_2O_2(SO_3Na)_2$ 317
INDIGNITÄT (*f*): indignity, disgrace, affront 318
INDIGO (*m*): (see INDIG) 319
INDIGOBLAU (*n*): indigo blue; synthetic indigo blue; indigotin; $C_{16}H_{10}N_2O_2$; Sg. 1.35; Bp. 390°C. 320
INDIGOBLAU (*n*), **LÖSLICHES-**: soluble indigo (see INDIGKARMIN) 321
INDIGOBLAUSCHWEFELSÄURE (*f*): indigosulphuric acid; indigosulfonic acid; indigo disulfonic acid; sulfindigotic acid 322
INDIGODERIVAT (*n*): indigo derivate 323

INDIGOERSATZ (*m*): indigo substitute 324

INDIGOEXTRAKT (*m*): indigo extract (see INDIG--KARMIN) 325

INDIGOFARBE (*f*): indigo dye 326

INDIGOFARBSTOFF (*m*): indigotin (see INDIGO--BLAU) 327

INDIGOFERA-ARTEN (*f.pl*): *Indigofera* species (Botany) 328

INDIGOIDER FARBSTOFF (*m*): indigoide colouring matter 329

INDIGO (*m*), KOLLOIDALER-: colloidal indigo 330

INDIGOKÜPE (*f*): indigo (blue) vat, (solution of indigo white in alkaline fluids) 331

INDIGOLEIM (*m*): indigo (gelatine) gluten 332

INDIGOLITH (*m*): indicolite, blue tourmaline; $(H,Li,Na,K)_9Al_3[B.OH]_2Si_4O_{19}$ $(+Fe_2O_3,MnO,MgO,FeO)$; Sg. 3.09 333

INDIGOLÖSUNG (*f*): indigo solution 334

INDIGO, NATÜRLICHER- (*m*): natural indigo blue (see INDIG) 335

INDIGOPAPIER (*n*): indigo paper 336

INDIGOPRÄPARAT (*n*): indigo (derivative) preparation 337

INDIGOROT (*n*): indigo red; indirubin 338

INDIGOSALZ (*n*): indigo salt 339

INDIGOSÄURE (*f*): indigotic acid 340

INDIGOSCHWEFELSAUER: indigosulphate, combined with indigosulphuric acid 341

INDIGOSCHWEFELSÄURE (*f*): indigosulphuric acid (see INDIGOBLAUSCHWEFELSÄURE) 342

INDIGOSTOFF (*m*): indigotin; indigo blue (see INDIGOBLAU) 343

INDIGOSTRAUCH (*m*): indigo plant (*Indigofera tinctoria*) 344

INDIGOSUBSTITUTIONSPRODUKT (*m*): indigo substitution product 345

INDIGOSULFOSÄURE (*f*): indigosulfonic acid (see INDIGOBLAUSCHWEFELSÄURE) 346

INDIGOSUSPENSION (*f*): indigo suspension 347

INDIGOSYNTHESE (*f*): indigo synthesis 348

INDIGO, SYNTHETISCHER- (*m*): synthetic indigo (see INDIGOBLAU) 349

INDIGOTIN (*n*): indigotin, indigo blue (see INDI--GOBLAU) 350

INDIGOTINFÄRBUNG (*f*): indigotin colouring 351

INDIGOTINSÄURE (*f*): indigotic acid 352

INDIGOWEISS (*n*): indigo white 353

INDIGSTEIN (*m*): indicolite; blue tourmaline (see INDIGOLITH) 354

INDIGWEISS (*n*): indigo white 355

INDIKAN (*n*): indican (a glycoside substance from indigo plants) 356

INDIKATIV (*m*): indicative 357

INDIKATOR (*m*): indicator 358

INDIKATORDIAGRAMM (*n*): indicator diagram 359

INDIKATORFEDER (*f*): indicator spring 360

INDIKATORHAHN (*m*): indicator cock 361

INDIKATORKOLBEN (*m*): indicator piston 362

INDIKATORPAPIER (*n*): indicator paper 363

INDIKATORROHR (*n*): indicator pipe 364

INDIKATORSCHNUR (*f*): indicator cord 365

INDIKATORSTIFT (*m*): indicator pencil 366

INDIKATORZYLINDER (*m*): indicator cylinder 367

INDIREKT: indirect 368

INDIREKTENDRUCK (*m*): off-set print 369

INDISCH: Indian (see INDIANISCH) 370

INDISCHE FEIGE (*f*): prickly pear; Indian fig; (*Opuntia ficus-indica*); Banyan tree (*Ficus bengalensis*) 371

INDISCHER FLACHS (*m*): jute (*Corchorus capsularis* and *Corchorus olitorius*) 372

INDISCHER HANF (*m*): Indian hemp, (*Cannabis indica*; *Cannabis sativa*) 373

INDISCHGELB (*n*): Indian yellow; jaune indien; cobalt yellow (potassium-cobalt nitrite) (see also KALIUMKOBALTNITRIT and PURREÉ) 374

INDISCHHANFTINKTUR (*f*): tincture of Indian hemp 375

INDISCHROT (*n*): Indian red (a natural red powder) 376

INDISPONIBEL: not available; not to be disposed of 377

INDIUM (*n*): indium (In) 378

INDIUMCHLORID (*n*): indium chloride 379

INDIUMJODID (*n*): indium iodide; mono.-InI; Mp. 350°C.; di.-InI_2; Mp. 212°C.; tri.-InI_3; Mp. 200°C. 380

INDIUMOXYD (*n*): indium oxide; In_2O_3; Sg. 7.179 381

INDIUMSULFAT (*n*): indium sulphate; $In_2(SO_4)_3$; Sg. 3.438 382

INDIUMVERBINDUNG (*f*): indium compound 383

INDIVIDUALITÄT (*f*): individuality 384

INDIVIDUELL: individual-(ly) 385

INDIVIDUEN (*n.pl*): individuals (plural of INDIVIDUUM) 386

INDIVIDUUM (*n*): individual 387

INDIZES (*m.pl*): indices (plural of index) 388

INDIZIEREN: to indicate 389

INDOFORM (*n*): indoform, salicylic acid; methylene acetate; $C_{10}H_{10}O_5$; Mp. 108.5°C. 390

INDOL (*n*): indole, ketole; C_8H_7N; Mp. 52°C.; Bp. 254°C. 391

INDOLBLAU (*n*): indole blue 392

INDOLCARBONSÄURE (*f*): indole-carbonic acid 393

INDOLPROPIONSÄURE (*f*), L-α-AMINO-β-: (see TRYPTOPHAN) 394

INDOSSAMENT (*n*): endorsement 395

INDOSSANT (*m*): endorser 396

INDOSSIEREN: to endorse 397

INDOXYLSÄURE (*f*): indoxylic acid; $C_9H_7NO_3$ 398

INDOXYLSCHWEFELSÄURE (*f*): indoxylsulphuric acid 399

INDUKTION (*f*): induction 400

INDUKTIONOFEN (*m*): induction (furnace) oven (for electro steels) 401

INDUKTIONSAPPARAT (*m*): inductor 402

INDUKTIONSELEKTRIZITÄT (*f*): induction electricity 403

INDUKTIONSKOEFFIZIENT (*m*): co-efficient of induction 404

INDUKTIONSOFEN (*m*): induction furnace (see INDUKTIONOFEN) 405

INDUKTIONSROLLE (*f*): induction coil (Elect.) 406

INDUKTIONSSPULE (*f*): (see INDUKTIONSROLLE) 407

INDUKTIONSSTROM (*m*): induction current (Elect.) 408

INDUKTIONSVERMÖGEN (*n*): inductive capacity 409

INDUKTIV: inductive 410

INDUKTIVITÄT (*f*): inductivity 411

INDUKTOR (*m*): inductor 412

INDUKTORSPULE (*f*): induction coil 413

INDULIN (*n*): indulin (a class of dyes) 414

INDURIEREN: to indurate; make durable; harden 415

INDUSTRIE (*f*): industry, trade 416

INDUSTRIEAUSSTELLUNG (*f*): industrial exhibition 417

INDUSTRIEBAHN (*f*): portable railway 418

INDUSTRIEBANK (*f*): industrial bank 419

INDUSTRIECHEMIKALIEN (*n.pl*): industrial chemicals 420
INDUSTRIEGUMMI (*n*): industrial (commercial) rubber 421
INDUSTRIELAND (*n*): industrial country 422
INDUSTRIELL : industrial-(ly) 423
INDUSTRIESTAAT (*m*): industrial state 424
INDUSTRIESTADT (*f*): industrial (manufacturing town 425
INDUSTRIEZWEIG (*m*): branch of industry 426
INDUSTRIÖS : industrious 427
INDUZIEREN : to induce 428
INDUZIERENDER STROM (*m*): inducing current 429
INEGAL : unequal 430
INEINANDER : into (each other) one another 431
INEINANDERFLECHTEN : to braid, plait, interlace, intertwine, interweave 432
INEINANDERFÜGEN : to couple, join 433
INEINANDERGREIFEN : to gear, engage, interlock, grip 434
INEINANDERGREIFEN (*n*): gearing, interlocking, gripping, engaging 435
INEMPFANGSNAHME (*f*): reception 436
INFAM : infamous, atrocious 437
INFAMIE (*f*): infamy 438
INFEKTIEREN : to infect 439
INFEKTION (*f*): infection 440
INFEKTIONSFÄHIG : infective, capable of infecting 441
INFEKTIONSKRANKHEIT (*f*): infectious disease (Med.) 442
INFEKTIÖS : infectious (Med.) 443
INFINITÄT (*f*): infinity 444
INFINITIV (*m*): infinitive 445
INFIZIEREN : to infect 446
INFLATION (*f*): inflation (of money) 447
INFLEKTIEREN : to inflect, decline, conjugate (Grammar) 448
INFLORESZENZ (*f*): infloresence (Bot.) 449
INFLUENZ (*f*): influence 450
INFLUENZA (*f*): influenza (Med.) 451
INFLUENZMASCHINE (*f*): inductive machine 452
INFÖKUND : sterile 453
INFOLGE : on account of ; owing (due) to 454
INFORMATION (*f*): report, information ; enquiry 455
INFORMATIONEN (*f.pl*): data, particulars 456
INFORMIEREN : to inform, acquaint with ; enquire 457
INFUNDIERAPPARAT (*m*): infusion apparatus ; sparger 458
INFUNDIERBÜCHSE (*f*): infusion (box) vessel ; digester 459
INFUNDIEREN : to infuse 460
INFUNDIERGEFÄSS (*n*): infusion vessel ; digester 461
INFUSION (*f*): infusion (*Infusum*) 462
INFUSIONSMETHODE (*f*): infusion, maceration (Perfumery) 463
INFUSORIENERDE (*f*): infusorial earth (*Terra infusoria*) (see SILICIUMDIOXYD and KIESEL-GUHR) ; kieselguhr, siliceous earth, tripolite, fossil flour, diatomaceous earth ; Sg. 2.2 [464
INGENIEUR (*m*): engineer 465
INGENIEURKUNST (*f*): Engineering 466
INGENIEURWESEN (*n*): (civil) engineering 467
INGENIEURWISSENSCHAFTEN (*f.pl*): (the science of) engineering 468
INGREDIENS (*n*): ingredient 469
INGREDIENSPIRIT (*m*): alcoholic (essence) ingredient 470
INGREDIENZ (*f*): ingredient 471
INGRIMM (*m*): wrath (violent)-anger, rage 472
INGRIMMIG : wrathful ; fur ous ; fiercely 473
INGUSS (*m*): ingot-(mould) 474
INGWER (*m*): ginger (*Zingiber officinale*) 475
INGWERBIER (*n*): ginger-beer 476
INGWER (*m*), **GELBER-** : (see GELBWURZ) 477
INGWERGRAS (*n*): ginger grass (*Andropogon*) (see also GINGERGRAS) 478
INGWERÖL (*n*): ginger oil (from *Zingiber officinale*); Sg. 0.88-0.885 479
INGWERWURZEL (*f*): ginger root (*Rhizoma zingiberis* of *Zingiber officinale*) 480
INHABEN : to possess, hold, own 481
INHABER (*m*): holder ; owner ; proprietor ; possessor ; occupant ; bearer 482
INHALATION (*f*): inhalation, breathing in 483
INHALIERAPPARAT (*m*): breathing (inhaling) apparatus 484
INHALIEREN : to inhale ; take breath 485
INHALT (*m*): content-(s) ; area (of surface) ; volume (of solids) ; tenor ; capacity (of vessel) ; substance ; subject matter 486
INHALTLOS : unmeaning ; empty ; having no contents 487
INHALTREICH : significant ; having a rich content 488
INHALTSMASS (*n*): cubic measure, measure of capacity 489
INHALTSÜBERSICHT (*f*): table of contents ; index ; summary 490
INHALTSVERZEICHNIS (*n*): table of contents ; index 491
INHALTSVOLL : significant ; full of meaning 492
INHIBITORIUM (*n*): inhibition 493
INHOLZ (*n*): square timber 494
INHOMOGEN : inhomogeneous ; not homogeneous 495
INHOMOGENITÄT (*f*): inhomogeneity 496
INITIALSPRENGSTOFF (*m*): priming (of explosives) 497
INITIALZÜNDER (*m*): priming (Explosives) 498
INITIIEREN : to initiate ; explode (Explosives) 499
INJEKTION (*f*): injection 500
INJEKTION DER ARZNEIMITTEL : medication 501
INJEKTOR (*m*): injector 502
INJEKTORARTIG : like (similar to, in a similar manner to) an injector 503
INJIZIEREN : to inject 504
INJURIE (*f*): injury, insult, offence, slander, defamation, affront 505
INJURIENKLAGE (*f*): libel action, action for defamation of character 506
INKARNAT : incarnate 507
INKASSOSPESEN (*f.pl*): charges for collecting or recovering sums of money 508
INKASSO (*n*), **ZUM-** : for cashing 509
INKASTEIN (*m*): pyrite (see PYRIT) 510
INKLINATORIUM (*n*): inclination compass 511
INKLUDIEREN : to include 512
INKLUSIV : inclusive, including 513
INKONSEQUENZ (*f*): inconsequence ; inconsistency 514
INKRUSTIEREN : to incrust ; become (caked) incrusted ; cake ; become coated with scale 515
INKULPIEREN : to inculpate, accuse, charge, incriminate, arraign, impeach, indict, denounce 516
INLAGE (*f*): enclosure 517
INLAND (*n*): inland, interior, native country 518
INLÄNDISCH : inland ; native ; indigenous ; domestic ; internal 519

INLIEGEND : enclosed 520

INNE : in ; within 521

INNEBEHALTEN : to keep back, retain, detain, reserve 522

INNEHABEN : to hold ; occupy ; keep ; possess ; own 523

INNEHALTEN : to confine ; restrict ; stay ; dam ; observe ; keep ; pull up ; stop ; cease ; pause ; discontinue 524

INNEN : within ; in ; at home 525

INNENAUFNAHME (*f*) : interior-(photograph-(y)) 526

INNENBEPLANKUNG (*f*) : inside planking 527

INNENBORDS : inboard (Naut.) 528

INNENDRUCK (*m*) : internal pressure 529

INNENFLÄCHE (*f*) : internal surface 530

INNENHAUT (*f*) : internal (membrane) skin ; endocarp (the inner layer of the fruit wall covering the seeds of fruit) 531

INNENLEITUNG (*f*) : internal (inner) circuit 532

INNENMASS (*n*) : inside measure-(ment) 533

INNENPOL (*m*) : internal pole 534

INNENRAUM (*m*) : interior-(space or chamber) ; inside (of buildings) ; (as a prefix=indoor) 535

INNENSTEVEN (*m*) : inner post 536

INNENVERBRENNUNGSMASCHINE (*f*) : internal combustion engine 537

INNENVERZAHNUNG (*f*) : internal or inside gear or gearing 538

INNER : internal ; interior ; inward ; inner ; inside ; intrinsic 539

INNERE (*n*) : interior, inside ; soul, heart 540

INNERHALB : within ; inside 541

INNERLICH : inward ; internal ; interior ; profound ; cordial ; sincere ; intrinsic ; mental-(ly) 542

INNEWOHNEND : inherent 543

INNIG : intimate ; cordial ; earnest ; sincere ; heartfelt ; hearty ; profound ; deep ; fervent ; ardent 544

INNUNG (*f*) : corporation, guild, society 545

INOKULIEREN : to inoculate 546

INOSINSÄURE (*f*) : inosic acid 547

INOXYDIEREN (*n*) : conversion of the surface of cast iron or wrought iron into ferroso-ferric oxide, Fe_3O_4, to act as anti-rust agent 548

INOXYDOFEN (*m*) : furnace for conversion of surface of cast iron or wrought iron into ferroso-ferric oxide (Fe_3O_4) ; pieces heated to 800-900°C. and alternately exposed to action of oxydating and reducing furnace gases 549

INOYÖL (*n*) : (see POGAÖL) 550

IN PRAXI : in practice 551

INSASS (*m*) : (see INSASSE) 552

INSASSE (*m*) : inhabitant ; inmate 553

INSBESONDERE : especially ; in particular ; particularly 554

INSCHRIFT (*f*) : inscription ; legend 555

INSEKT (*m*) : insect 556

INSEKTENBESTÄUBUNG (*f*) : insect pollination (Bot.) 557

INSEKTENBLÜTEN (*f.pl*) : (*Flores pyrethri*) : Persian (pellitory) insect flowers ; insect powder (flowers of *Pyrethrum roseum*) 558

INSEKTENFRESSEND : insectivorous 559

INSEKTENKENNER (*m*) : entomologist 560

INSEKTENKUNDE (*f*) : entomology 561

INSEKTENLEHRE (*f*) : (see INSEKTENKUNDE) 562

INSEKTENPLAGE (*f*) : plague of insects, blight 563

INSEKTENPULVER (*n*) : insecticide ; insect powder 564

INSEKTENSCHUTZMITTEL (*n*) : insect (preventive) killer 565

INSEKTICID : insecticidal 566

INSEL (*f*) : island ; isle 567

INSELMEER (*n*) : archipelago 568

INSELT (*n*) : suet, tallow 569

INSELTLICHT (*n*) : tallow candle 570

INSELVOLK (*n*) : insular people, islanders 571

INSERANT (*m*) : advertiser 572

INSERAT (*n*) : insertion, advertisement 573

INSERIEREN : to insert (Advertisements) 574

INSGEHEIM : privately, secretly 575

INSGESAMT : altogether 576

INSIEGEL (*n*) : seal 577

INSIGNIEN (*f.pl*) : insignia 578

INSOFERN : as far as ; inasmuch as ; according as ; in so far(-as) 579

INSOLVENT : insolvent 580

INSOLVENZ (*f*) : insolvency 581

INSPIZIEREN : to inspect, examine, superintend 582

INSTABIL : unstable 583

INSTALLATION (*f*) : establishment, installation 584

INSTANDHALTUNG (*f*) : care ; maintenance ; keeping in (trim ; first class order) repair 585

INSTÄNDIG : earnest ; urgent ; instant ; particular ; immediate ; important 586

INSTANDSETZUNG (*f*) : putting in (order, trim) repair ; restoration ; reinstatement 587

INSTINKT (*m*) : instinct 588

INSTINKTLICH : instinctive-(ly) 589

INSTITUT (*n*) : institute, establishment, academy, institution 590

INSTRUIEREN : to instruct 591

INSTRUMENT (*n*) : instrument 592

INSULANER (*m*) : islander 593

INSULARISCH : insular 594

INSULIN (*n*) : insulin 595

INTAKT : intact 596

INTEGRABEL : capable of being integrated 597

INTEGRAL (*n*) : integral 598

INTEGRAL (*n*), BESTIMMTES- : definite integral ; ($\int_1^2$) (i.e., with limits) 599

INTEGRALFORMEL (*f*) : integral formula 600

INTEGRALRECHNUNG (*f*) : integral calculus 601

INTEGRAL (*n*), UNBESTIMMTES- : indefinite integral ($\int$) (i.e., with no limits) 602

INTEGRALZAHL (*f*) : integer 603

INTEGRALZEICHEN (*n*) : integration sign 604

INTEGRATION (*f*) : integration 605

INTEGRATION (*f*), PARTIELLE- : partial integration ; integration by parts 606

INTEGRATIONSKONSTANTE (*f*) : integration constant 607

INTEGRATOR (*m*) : integrator 608

INTEGRIEREN : to integrate 609

INTEGRIEREND : integrating ; integrant ; integral 610

INTEGRIERWERK (*n*) : integrating gear 611

INTELLEKT (*m*) : intellect 612

INTELLEKTUALISTISCH : intellectual-(ly) 613

INTELLIGENT : intelligent 614

INTELLIGENZBLATT (*n*) : advertiser, newspaper 615

INTELLIGENZBÜRO (*n*) : enquiry office, advertising office 616

INTELLIGENZKONTOR (*n*) : enquiry office, advertising office 617

INTENDANT (*m*): manager, director, steward, superintendent, intendant 618
INTENDANZ (*f*): commissariat, superintendence, management 619
INTENSITÄT (*f*): intensity; strength 620
INTENSIV: intensive-(ly) 621
INTERALLIIERTE KOMMISSION (*f*): interallied commission 622
INTERESSANT: interesting 623
INTERESSE (*n*): interest; usury; profit; advantage 624
INTERESSE (*n*), **EIN-HEGEN**: to take an interest in 625
INTERESSENT (*m*): interested party, partaker, sharer, shareholder, party concerned 626
INTERESSIEREN: to-(take an) interest, concern 627
INTERFERENZBILD (*n*): interference figure 628
INTERFEROMETER (*n*): interferometer (apparatus for the optical determination of the composition of a gas mixture) 629
INTERIM (*n*): interim 630
INTERKOSTAL: intercostal (Marine and Anatomy) 631
INTERKOSTALPLATTE (*f*): intercostal plate 632
INTERMEDIÄR: intermediary; intermediate 633
INTERMITTIEREND: intermittent-(ly); interrupted-(ly) 634
INTERN: internal-(ly) 635
INTERNIEREN: to intern 636
INTERPOLATION (*f*): interpolation 637
INTERPOLIEREN: to interpolate (to plot values between certain limits actually available); (*n*): interpolating, interpolation 638
INTERPUNKTION (*f*): punctuation 639
INTERSTITIELL: interstitial 640
INTERZELLULAR: intercellular 641
INTIM: intimate-(ly) 642
INTIMITÄT (*f*): intimacy 643
INTOXIKATION (*f*): intoxication, drunkenness 644
INTRACUTANE ARZNEIAPPLIKATION (*f*): endermic medication (rubbing remedies into the skin) (Med.) 645
INTRAMOLEKULAR: intramolecular 646
INTRAMUSKULÄR: intra-muscular 647
INTRANSITIV: intransitive 648
INTRANSITIVUM (*n*): intransitive verb 649
INTRIGANT: intriguing; (*m*): intriguer 650
INTRUSIVGESTEIN (*n*): (see **PLUTONIT**) 651
INTUITION (*f*): intuition 652
INULIN (*n*): inulin, alant starch (obtained from the chicory root, *Cichorium intybus* and other plants) 653
INVALID: invalid; weak 654
INVALIDE (*f*): invalid 655
INVALIDIEREN: to make invalid, invalidate 656
INVAR (*n*): invar, ferro-nickel (36% Ni; 64% steel) 657
INVENTAR (*n*): inventory; stock-taking 658
INVENTARIUM (*n*): inventory 659
INVENTARSTÜCKE (*n.pl*): fixtures 660
INVENTIEREN: to make an inventory, list, schedule 661
INVENTUR (*f*): inventory 662
INVERSION (*f*): inversion [Sugar; change of rotation); etc.] 663
INVERTIEREN: to invert 664
INVERTZUCKER (*m*): invert sugar (mixture of equal molecules of Dextrose and Levulose); $C_6H_{12}O_6$ 665
INVESTIEREN: to invest (Money, etc.) 666
INWÄRTIG: internal-(ly), inward-(ly) 667
INWÄRTS: inwards; inwardly; internally 668
INWENDIG: inward; inside; interior; internal-(ly) 669
INWIEFERN: in what respect; how; to what extent; in which way 670
INWOHNEND: inherent 671
INZWISCHEN: in the meantime; however; meanwhile; in the interim 672
IOD (*n*): iodine (see **JOD**) 673
IOLITH (*m*): iolite (see **CORDIERIT**) 674
ION (*n*): ion (components of chemical compounds freed by electrolysis) 675
IONENBEWEGLICHKEIT (*f*): ionic mobility 676
IONENBEWEGUNG (*f*): ionic movement 677
IONENFORM (*f*): ionic form 678
IONENGESCHWINDIGKEIT (*f*): ionic velocity 679
IONENGLEICHUNG (*f*): ionic equation 680
IONENREAKTION (*f*): ionic reaction 681
IONENREIBUNG (*f*): ionic friction 682
IONENSPALTUNG (*f*): ionic cleavage; ionization 683
IONENWANDERUNG (*f*): ionic migration 684
IONENZUSTAND (*m*): ionic state 685
IONISATION (*f*): ionization (see **IONISIERUNG**) 686
IONISATIONSWÄRME (*f*): heat of ionization 687
IONISCH: ionic 688
IONISIERBAR: ionizable 689
IONISIEREN: to ionize (set free ions) 690
IONISIERUNG (*f*): ionization, ionic cleavage (setting free of ions) 691
IONISIERUNGSSPANNUNG (*f*): pressure of ionization 692
IPECACUANHA (*f*): ipecacuanha -(root) (see also **BRECHWURZEL**) 693
IPECACUANHASÄURE (*f*): ipecacuanhic acid (from *Uragoga ipecacuanha*) (see also under **BRECHWURZ**) 694
IPSER TIEGEL (*m*): Ipser (graphite)-crucible 695
IRDEN: earthen; of (earth) clay; clay; fictile 696
IRDENWARE (*f*): earthenware; crockery 697
IRDISCH: earthly; terrestrial; earthy; perishable 698
IRGEND: any; at all; some 699
IRGENDWIE: anyhow; somehow 700
IRGENDWO: anywhere; somewhere 701
IRIDISIEREN: to irisate; iridise; iridesce 702
IRIDIUM (*n*): iridium (Ir); (Sg. crystalline, 22.42; spongy, 15.86) 703
IRIDIUMCHLORID (*n*): iridium chloride 704
IRIDIUMLAMPE (*f*): iridium lamp (electric glow-lamp the filament of which is of pure iridium) 705
IRIDIUMOXYD (*n*): iridium oxide 706
IRIDIUMSALMIAK (*m*): ammonium-iridium chloride 707
IRIDIUMSCHWAMM (*m*): white spongy iridium; Sg. 15.86 708
IRIDOSMIUM (*m*): iridosmine (natural iridium-osmium alloy; up to 77% iridium and up to 49% osmium); (Ir.Os) plus (Rh.Fe.Pd); Sg. 19.0—21.12 709
IRISBLENDE (*f*): iris diaphragm 710
IRISCHES MOOS (*n*): Irish moss, pearl moss, carragheen moss, killeen, pig-wrack, rock-salt moss (*Chrondrus crispus*) 711
IRISIEREN: to irisate; iridise; iridize; iridesce; (*n*): iridescence 712
IRISIERENDES GLAS (*n*): iridescent (lustrous) glass 713
IRISIERSALZ (*n*): irisating (irising, iridising) salt 714

IRISIERUNG (*f*): iridescence 715
IRISÖL (*n*): iris (orris) oil (*Oleum iridis*) (see also **VEILCHENWURZ**); Mp. 44-50°C. 716
IRISVIOLETT (*n*): iris violet 717
IRLÄNDISCH: Irish 718
IRON (*n*): irone; $C_{13}H_{20}O$ (the scented component of orris root) 719
IRONIE (*f*): irony 720
IRONISCH: ironical-(ly) 721
IRRATIONAL: irrational 722
IRRBLOCK (*m*): erratic block 723
IRRE: doubtful; disconcerted; astray; wrong; puzzled; perplexed; confused; wandering; (*f*): wandering; confusion; labyrinth; wrong (road) course; (*m* and *f*): lunatic; mad person 724
IRREFÜHREN: to mislead; lead astray; alienate 725
IRREFÜHREND: misleading 726
IRREN: to err; be wrong; go (lead) astray; be insane; be mad; be delirious; be (deceived) mistaken; mislead; puzzle; disturb; wander; deceive 727
IRRENANSTALT (*f*): lunatic asylum; mad-house; bedlam 728
IRRENARZT (*m*): alienist; doctor for mental diseases 729
IRREPARABEL: irreparable 730
IRRESEIN (*n*): insanity; madness 731
IRREVERSIBEL: irreversible (immune to the action of water in colloidal chemistry) 732
IRRFAHRT (*f*): wandering; vagary 733
IRRGANG (*m*): labyrinth; maze 734
IRRGÄNGIG: labyrinthine 735
IRRGARTEN (*m*): (see **IRRGANG**) 736
IRRGLAUBE (*f*): heterodoxy, heresy 737
IRRGLÄUBIG: heterodox, heretical 738
IRRIG: erroneous; false; mistaken; wrong 739
IRRITIEREN: to irritate 740
IRRLEHRE (*f*): false (doctrine) teaching; heresy 741
IRRLEHRIG: heterodox, heretical 742
IRRLICHT (*n*): will-o'-the-wisp; *Ignis fatuus* 743
IRRSINN (*m*): insanity; delirium; madness 744
IRRTUM (*m*): error; mistake; fault; deception 745
IRRTÜMLICH: erroneous-(ly) 746
IRRTÜMLICHERWEISE: erroneously; by mistake 747
IRRUNG (*f*): error, mistake, difference 748
IRRWISCH (*m*): will-o'-the-wisp, *Ingnis fatuus* 749
ISABELLENFARBIG: cream-coloured; light (bay) buff 750
ISABELLFARBE (*f*): dun, buff, cream-colour 751
ISANSÄURE (*f*): isanic acid; $C_{13}H_{19}CO_2H$; Mp. 41°C. 752
ISÄTHIONSÄURE (*f*): isethionic acid 753
ISATIN (*n*): isatic acid anhydride, isatic acid lactime, isatin, ortho-aminobenzoylformic acid; $C_6H_4CO:C(OH).N$; Mp. 200°C. 754
ISATROPASÄURE (*f*): isatropic acid 755
ISATROPYLCOCAIN (*n*): isatropylcocaine (an alkaloid from *Truxillococa*); $C_{38}H_{46}N_2O_8$ 756
I-SÄURE (*f*): I-acid, amidonaphtholsulfonic acid (see **AMIDONAPHTOLSULFOSÄURE**) 757
I.S-DIAGRAMM (*n*): i.s-diagram (total heat entropy diagram) 758
ISENTROPE (*f*): adiabatic (line) curve 759
ISENTROPISCH: adiabatic 760
ISERIN (*m*): (see **ILMENIT** and **TITANEISEN**) 761
ISLÄNDISCH: Iceland: icelandic 762
ISLÄNDISCH-(ES) MOOS (*n*): Iceland moss, cetraria (*Cetraria islandica*) 763
ISOBARE (*f*): isobar; line (curve) of constant pressure 764
ISOBARENSCHAR (*f*): group of isobars, group of lines of constant pressure 765
ISOBORNEOL (*n*): isoborneol (see **BORNEOL**) 766
ISOBORNYL (*n*), **AMEISENSAURES-**: isobornyl formate 767
ISOBORNYL (*n*), **ESSIGSAURES-**: isobornyl acetate (see **BORNYL, ESSIGSAURES-**) 768
ISOBUTTERSAUER: isobutyrate, combined with isobutyric acid 769
ISOBUTTERSÄURE (*f*): isobutyric acid (see **BUTTERSÄURE, ISO-**) 770
ISOBUTTERSÄUREANHYDRID (*n*): isobutyric anhydride; $C_8H_{14}O_3$; Sg. 0.9574; Bp. 181.5°C. 771
ISOBUTTERSÄURE-ÄTHYLESTER (*m*): Isobutyric ester (ethyl); ethyl isobutyrate; $C_6H_{12}O_2$; Sg. 0.8710; Mp. – 93.3°C.; Bp. 110.1°C. 772
ISOBUTTERSÄURE-NITRIL (*n*): isobutyric nitrile; C_4H_7N; Bp. 107-108°C. 773
ISOBUTYL (*n*): isobutyl 774
ISOBUTYLALDEHYD (*n*): isobutyl aldehyde, isobutyryl aldehyde; C_4H_8O; Sg. 0.7938; Bp. 63°C. 775
ISOBUTYLALKOHOL (*m*): isobutyl alcohol, isopropylcarbinol, methylpropanol; $C_4H_{10}O$; Sg. 0.798-0.8057; Bp. 106.4°C.-107.2°C. (see **METHYLPROPANOL**) 776
ISOBUTYL (*n*), **AMEISENSAURES-**: isobutyl formate 777
ISOBUTYLAMIN (*n*): isobutyl amine 778
ISOBUTYL (*n*), **BALDRIANSAURES-**: isobutyl valerate 779
ISOBUTYL (*n*), **BENZOESAURES-**: isobutyl benzoate 780
ISOBUTYLBROMID (*n*): isobutyl bromide 781
ISOBUTYL (*n*), **BUTTERSAURES-**: isobutyl butyrate 782
ISOBUTYLCHLORID (*n*): isobutyl chloride 783
ISOBUTYLENBROMID (*n*): isobutylene bromide 784
ISOBUTYL (*n*), **ESSIGSAURES-**: isobutyl acetate 785
ISOBUTYL (*n*), **ISOBUTTERSAURES-**: isobutyl isobutyrate 786
ISOBUTYLJODID (*n*): isobutyl iodide 787
ISOBUTYLPRÄPARAT (*n*): isobutyl preparation 788
ISOBUTYLRHODANÜR (*n*): isobutyl sulfocyanide 789
ISOBUTYL (*n*), **SALPETERSAURES-**: isobutyl nitrate 790
ISOBUTYL (*n*), **SALPETRIGSAURES-**: isobutyl nitrite 791
ISOBUTYRYLCHLORID (*n*): (see **BUTYRYLCHLORID, ISO-**) 792
ISOCHINOLIN (*n*): isoquinoline; C_9H_7N; Sg. 1.1025; Mp. 24.6°C; Bp. 240.5°C 793
ISOCHRON: isochronous (equal in point of time) 794
ISOCORYBULBIN (*n*): isocorybulbin (an alkaloid; see **CORYDALISALKALOID**); $C_{21}H_{25}NO_4$ 795
ISOCROTONSÄURE (*f*): isocrotonic acid 796
ISOCUMARIN (*n*): isocumarin; isocourmarin 797
ISOCYANSÄURE (*f*): isocyanic acid 798
ISOCYANURSÄURE (*f*): isocyanuric acid (see **CYANURSÄURE**) 799
ISOCYCLISCH: isocyclic 800
ISODEHYDROTHIO-M-XYLIDIN (*n*): isodehydrothio-meta-xylidine; Mp. 121°C. 801

ISOELEKTRISCH : isoelectric 802
ISOERUCASÄURE (*f*) : isoerucic acid ; $C_{21}H_{41}CO_2H$; Mp. 55°C. 803
ISOEUGENOL (*n*) : (see EUGENOL) 804
ISOFERULASÄURE (*f*) : isoferulic acid 805
ISOGON (*n*) : a polygon having equal sides and equal angles 806
ISOHYDRISCH : isohydric 807
ISOKAMPHERSÄURE (*f*) : isocamphoric acid 808
ISOLATION (*f*) : isolation ; insulation (Elect.) 809
ISOLATIONSLACK (*m*) : insulation varnish 810
ISOLATIONSPRÜFER (*m*) : insulation detector 811
ISOLATIONSSTOFF (*m*) : insulating material ; lagging ; non-conducting (material) composition 812
ISOLATOR (*m*) : insulator (Elect.) ; non-conductor 813
ISOLEUCIN (*n*) : isoleucin, amino-β-methylethylpropionic acid ; $CH_3.CH(C_2H_5).CH.CH(NH_2).CO_2H$ 814
ISOLIERASPHALT (*m*) : insulating asphalt 815
ISOLIERBAND (*n*) : insulating tape 816
ISOLIEREN : to insulate (Elect.) ; isolate 817
ISOLIEREND : isolating, insulating 818
ISOLIERFIRNIS (*m*) : insulating varnish 819
ISOLIERGLOCKE (*f*) : insulating bell, bell shaped insulator 820
ISOLIERHÜLLE (*f*) : insulating (envelope) covering 821
ISOLIERLACK (*m*) : insulating varnish 822
ISOLIERMASSE (*f*) : insulating (mass) material, non-conducting composition 823
ISOLIERMITTEL (*n*) : (see ISOLATIONSSTOFF) 824
ISOLIERPAPPE (*f*) : insulating cardboard 825
ISOLIERPECH (*n*) : insulating pitch 826
ISOLIERPLATTE (*f*) : insulating (board) plate 827
ISOLIERSCHEMEL (*m*) : insulating chair or stool 828
ISOLIERSCHLAUCH (*m*) : insulating hose 829
ISOLIERSCHNUR (*f*) : insulating rope or cord 830
ISOLIERT : insulated 831
ISOLIERUNG (*f*) : insulation ; isolation 832
ISOLINOLENSÄURE (*f*) : (see LINOLENSÄURE) 833
ISOLOG : isologous 834
ISOMER : isomeric 835
ISOMERIE (*f*) : isomerism 836
ISOMERIEFALL (*m*) : case of isomerism 837
ISOMERIEMÖGLICHKEIT (*f*) : possibility of isomerism 838
ISOMERISATION (*f*) : isomerization 839
ISOMERISCH : isomeric 840
ISOMERISIEREN : to isomerize 841
ISOMERISIERUNG (*f*) : isomerization, isomerizing 842
ISOMETRISCHE DARSTELLUNG (*f*) : isometric (representation) illustration 843
ISOMORPH : isomorphous 844
ISOMORPHIE (*f*) : isomorphism 845
ISOMORPHISMUS (*m*) : isomorphism 846
ISOÖLSÄURE (*f*) : isooleic acid ; $C_{18}H_{34}O_2$; Mp. 44.5°C. 847
ISOP (*m*) : hyssop (*Hyssopus officinalis*) 848
ISOPHTALSÄURE (*f*) : isophtalic acid 849
ISOPIKRINSÄURE (*f*) : isopicric acid 850
ISOPÖL (*n*) : hyssop oil ; Sg. 0.932 (distilled from *Hyssopus officinalis*) 851
ISOPREN (*n*) : isopren (a product of the dry distillation of caoutchouc) ; C_5H_8 ; Sg. 0.6793 ; Bp. 33.5°C. 852
ISOPROPYLALKOHOL (*m*) : isopropyl alcohol ; $CH_3CHOHCH_3$ or $(CH_3)_2CH.OH$; Sg. 0.8 ; Bp. 82.8°C. 853
ISOPROPYLBENZOL (*n*) : isopropylbenzene (see CUMOL) 854
ISOPROPYLBROMID (*n*) : isopropyl bromide 855
ISOPROPYLCARBINOL (*n*) : isopropylcarbinol (see ISOBUTYLALKOHOL) 856
ISOPROPYLJODID (*n*) : isopropyl iodide 857
ISOPURPURINSÄURE (*f*) : isopurpuric acid 858
ISORICINOLSÄURE (*f*) : isoricinolic acid (isomer of ricinolic acid) (see RICINOLSÄURE) 859
ISOSAFROL (*n*) : isosafrol ; $C_{10}H_{10}O_2$; bp. 247°C. 860
ISOSMOTISCH : isosmotic 861
ISOTHERME (*f*) : isothermal (line) curve ; line (curve) of constant temperature 862
ISOTHERMISCH : isothermal-(ly) 863
ISOTONIE (*f*) : isotonicity 864
ISOTONISCH : isotonic 865
ISOTROP : isotropic 866
ISOVALERIANSÄURE (*f*) : isovaler-(ian-)ic acid ; $C_5H_{10}O_2$; Sg. 0.9298 ; Mp. −51°C. ; Bp. 173.7°C. 867
ISOVALERIANSÄURE -ÄTHYLESTER (*m*) : isovaler-(ian)-ic ethylester ; ethyl isovaler-(ian)-ate $C_7H_{14}O_2$; Sg. 0.8851 ; Bp. 134.3°C. 868
ISOXYLOL (*n*) : isoxylene ; isoxylol (see XYLOL, META-) 869
ISOXYLSÄURE (*f*) : isoxylic acid 870
ISOZIMMTSÄURE (*f*) : isocinnamic acid 871
ISTICIN (*n*) : isticin ; 1,8-dioxyanthraquinone ; Mp. 191°C. 872
ITACONSÄURE (*f*) : itaconic acid ; $C_5H_6O_4$; Sg. 1.573-1.632 ; Mp. 161 °C. 873
ITACONSÄURE-DIÄTHYLESTER (*m*) : itaconic diethylester ; $C_9H_{14}O_4$; Sg. 1.0461 ; Bp. 227.9°C. 874
ITALIENISCH : Italian 875
ITALIENISCHE ERDE (*f*) : terra di Siena, mahogany brown 876
ITALIENISCHROT (*n*) : Italian red 877
ITROL (*n*) : itrol, silver citrate ; $C_6H_5O_7Ag_3$ (*Argentum citricum*) 878
IVAKRAUT (*n*) : *Achillea moschata* (*Herba achilleæ moschatæ* ; *Herba ivæ moschatæ*) 879
IVAÖL (*n*) : iva oil ; Sg. 0.934 (from *Achillea moschata*) 880
IVARANCHUSAÖL (*n*) : vetiver oil ; cuscus oil ; Sg. 1.015 (from root of *Andropogon muricatus*) 881
IVARANCHUSAWURZEL (*f*) : vetiver root (*Radix ivaranchusæ* ; *Radix vetiveris*) (from *Andropogon ivaranchusa*) 882
IXIOLITH (*m*) : tantalite plus SnO_2 (see also TANTALIT) 883

J

JABORANDIBLÄTTER (*n.pl*) : Jaborandi leaves (*Folia pilocarpi* ; *Folia jaborandi*) (leaves of *Pilocarpus microphyllus* or *Pilocarpus jaborandi*) 884
JABORANDIÖL (*n*) : Jaborandi oil (Sg. 0.88) (from the leaves of *Pilocarpus pennatifolius*) 885
JACARANDABRAUN (*n*) : jacaranda brown 886
JACHT (*f*) : yacht 887
JACKE (*f*) : jacket ; jerkin 888
JACKSTAG (*n*) : jack stay (Naut.) 889
JADE (*m*) : jade (see JADEIT and NEPHRIT) 890
JADEIT (*m*) : jade, jadite (natural metasilicate of sodium and aluminium) ; $NaAl(SiO_3)_2$ Sg. 3.33 891

JAGARAZUCKER (*m*): jaggery-(sugar) (saccharine juice from the Indian date palm, *Phœnix sylvestris* and others) 892
JAGD (*f*): chase; hunti-(ng); shoot-(ing); huntsmen; game; hubbub; pursuit 893
JAGDFLINTE (*f*): fowling piece 894
JAGDFREVEL (*m*): poaching 895
JAGDGERECHTIGKEIT (*f*): shooting (hunting) license 896
JAGDGEWEHR (*n*): sporting gun 897
JAGDPATRONE (*f*): sporting cartridge 898
JAGDRECHT (*n*): shooting (rights) license 899
JAGDREVIER (*n*): hunting-ground 900
JAGDWESEN (*n*): sporting (hunting) matters; venery; woodcraft, huntsmanship 901
JAGEA (*m*): jib topsail (Naut.) 902
JAGEN: to hunt; chase; drive; rush; dash; race; pursue; gallop: (*n*): hunting; chasing 903
JÄGER (*m*): huntsman; hunter; sportsman; chasseur; rifleman; ranger; gamekeeper 904
JÄGEREI (*f*): venery, huntsmanship, woodcraft 905
JAGERMAST (*m*): jigger mast 906
JAGGERY (*m*): (see JAGARAZUCKER) 907
JAGRE (*m*): (see JAGARAZUCKER) 908
JÄH: steep; abrupt; hasty; rapid; sudden; precipitous; precipitate 909
JÄHE (*f*): steepness; abruptness; precipitation; precipice 910
JÄHLINGS: headlong; abruptly; precipitously 911
JAHN (*m*): swath 912
JAHNS-GASERZEUGER (*m*): Jahn's gas producer (for extracting gas from waste washings and also from rough waste) (Coal mining) 913
JAHR (*n*): year 914
JAHRANLEIHE (*f*): annuity 915
JAHRBUCH (*n*): year-book; annual; almanac; annals; chronicle 916
JAHRE (*n.pl*), DIE ACHTZIGER-: the eighties 917
JAHRESBERICHT (*m*): annual report 918
JAHRESFEIER (*f*): anniversary 919
JAHRESFRIST (*f*): period (space) of a year 920
JAHRESHÄLFTE (*f*): half-year 921
JAHRESRING (*m*): annual ring (of trees) 922
JAHRESRINGBILDUNG (*f*): the formation of annual rings (Trees) 923
JAHRESSCHRIFT (*f*): annual publication 924
JAHRESTAG (*m*): anniversary; birthday 925
JAHRESVERSAMMLUNG (*f*): annual meeting 926
JAHRESVIERTEL (*n*): quarter-(of a year) 927
JAHRESWECHSEL (*m*): New Year 928
JAHRESZEIT (*f*): season 929
JAHRGANG (*m*): year; year's (annual) course or set; year's (annual) crop; vintage 930
JAHRGEHALT (*m*): yearly (annual) wages (salary, stipend, pension) 931
JAHRGELD (*n*): annuity, yearly (annual) allowance, pension 932
JAHRGEWÄCHS (*n*): annual-(plant) 933
JAHRHUNDERT (*n*): century; age; a hundred years 934
JÄHRIG: a year old 935
JÄHRLICH: annual-(ly); yearly; per annum 936
JAHRLOHN (*m*): annual (wages, stipend) salary 937
JAHRMARKT (*m*): yearly (annual) fair 938
JAHRRECHNUNG (*f*): annual account; era, calculation of the year 939
JAHRRENTE (*f*): annual income 940
JAHRTAUSEND (*n*): millenium; a thousand years 941
JAHRWEISE: annually; in (by) years 942
JAHRWUCHS (*m*): year's (annual) growth 943
JAHRZAHL (*f*): date, year 944
JAHRZEHEND (*n*): decennium 945
JAHRZEHNT (*n*): decade; ten years 946
JÄHZORN (*m*): violent passion; sudden anger; choler 947
JÄHZORNIG: choleric, passionate 948
JAKARANDAHOLZ (*n*): rosewood, South America (see PALISANDERHOLZ) 949
JAKOBSKRAUT (*n*): groundsel (*Senecio vulgaris*) 950
JAKOBSKREUZKRAUT (*n*): staggerwort, ragwort (*Senecio jacobæa*) 951
JAKOBSLAUCH (*m*): chives (*Allium schoenoprasum*) (also *Allium fistulosum*) 952
JAKOBSLEITER (*f*): Jacob's ladder (*Polemonium cœruleum*) (Bot.) 953
JAKONNET (*n*): jaconette (Med.) 954
JALAPA (*f*): jalap (root of *Exogonium purga*); scammony (*Ipomœa purga*) 955
JALAPAHARZ (*n*): jalap resin; jalapin (the active principle of the drug) (from *Exogonium purga*) 956
JALAPAWURZEL (*f*): jalap root (root of *Exogonium purga*) 957
JALAPE (*f*): jalap (see JALAPA) 958
JALAPENHARZ (*n*): (see JALAPAHARZ) 959
JALAPENKNOLLEN (*f.pl*): jalap root (*Tubera jalapæ* of *Exogonium purga*) 960
JALAPINSÄURE (*f*): jalapic acid 961
JALAPPE (*f*): (see JALAPA) 962
JALLAPPIN (*n*): jalapin (see JALAPAHARZ) 963
JALON (*n*): jalone (collargol preparation); stake, picket 964
JALONNIEREN: to stake out-(ground) 965
JALOUSIEAPPARAT (*m*): venetian blind (apparatus, device) arrangement, (over which water trickles, for wet process dust extraction from coal) 966
JALOUSIEKASSETTE (*f*): blind dark-slide (Phot.) 967
JALOUSIEN (*pl*): Venetian blinds 968
JALOUSIERAUCHSCHIEBER (*m*): venetial blind (flue) damper 969
JALOUSIEVERSCHLUSS (*m*): blind shutter (Phot.) 970
JAMAIKAPFEFFER (*m*): Jamaica pepper, allspice, pimenta (fruit of *Pimenta officinalis*) 971
JAMBAÖL (*n*): jamba oil (obtained from a species of *Brassica*) 972
JAMBISCH: iambic 973
JAMBULFRUCHT (*f*): jambul seed 974
JAMESONIT (*m*): jamesonite, feather ore (natural lead-antimony sulphide); $PbFeSb_6S_{14}$ or $Pb_3Sb_2S_6$; Sg. 5.6 975
JAMMER (*m*): lamentation; misery; woe; sorrow; compassion; pity; distress; trouble; wretchedness 976
JAMMERGESICHT (*n*): rueful (countenance) expression 977
JÄMMERLICH: woeful; miserable; piteous; pitiable; deplorable; wretched 978
JAMMERLIED (*n*): dirge, lament 979
JAMMERN: to lament; move to (inspire with) pity; mourn; moan; wail; grieve 980
JAMMERVOLL: miserable, woeful, pitiable, lamentable 981
JÄNNER (*m*): January (in Austria only) 982
JANUSGRÜN (*n*): janus green 983

JANUSSCHWARZ (*n*): janus black 984
JAPANIEREN: to Japan 985
JAPANISCH: Japan; Japanese 986
JAPANISCHES WACHS (*n*): Japan wax (from the kernels of *Rhus succedanea*) (*Cera japonica*); Sg. 0.975 987
JAPANKAMPFER (*m*): Japan camphor (see CAM-PHOR, JAPAN-) 988
JAPANLACK (*m*): Japan-(lacquer); (concentrated sap of the plant *Rhus vernicifera*); (varnish prepared by heating linseed oil with litharge and thinning with naphtha, or mixing asphaltum with linseed oil, turpentine and gum) 989
JAPANLEIM (*m*): Japanese vegetable glue, Agar-agar, Japanese gelatine, Japanese isinglass (from *Gelidium, Eucheuma, Rhodophyceæ* and *Sphæro coccus*) 990
JAPANSÄURE (*f*): Japan acid; $C_{20}H_{40}(CO_2H)_2$; Mp. 117.5°C. 991
JAPANTALG (*m*): (see JAPANWACHS); Sg. about 1; Mp. 52°C. 992
JAPANTRAN (*m*): (see SARDINENÖL and also HERINGSÖL) 993
JAPANWACHS (*n*): Japan vegetable wax (obtained from *Rhus* species of plant); Sg. 0.975; Mp. 53°C. 994
JASMAL (*n*): jasmal, methylenphenylglycol ether; $C_9H_{10}O_2$; Sg. 1.1334; Bp. 218°C. 995
JASMIN (*m*): jasmine; jessamine (white = *Jasmium officinale*; winter, yellow = *Jasmium nudiflorum*) (see also GELSEMIEN) 996
JASMIN (*m*), ARABISCHER-: white-flowering, evergreen (Jessamine) Jasmine (*Jasminum sambac*) 997
JASMINBLÜTENÖL (*n*): jasmine-flower oil) see JASMINÖL) 998
JASMIN, GEMEINER- (*m*): syringa; lilac (*Syringa vulgaris*) 999
JASMINHOLZ (*n*): jasmine wood; Sg. 0.77 (see also JASMIN) 000
JASMINÖL (*n*): jasmine oil; Sg. 1.013; (from *Jasmium grandiflorum*) 001
JASMIN, WILDER- (*m*): (see JASMIN, GEMEINER-) 002
JASPACHAT (*m*): jasper agate (see JASPIS) 003
JASPIS (*m*): jasper (see also QUARZ); Sg. 2.65; $SiO_2(Fe_2O_3)$; 004
JASPISGUT (*n*): jasper ware 005
JASPOPAL (*m*): jasper opal (see OPAL) 006
JÄTARBEIT (*f*): weeding 007
JÄTEN: to weed; (*n*): weeding 008
JÄTER (*m*): weeder 009
JÄTHAUE (*f*): weeding hoe 010
JAUCHE (*f*): dung water; suds; liquid manure; any filthy liquid; ichlor (the ethereal fluid in the veins of the gods); serum (watery part of the blood); matter; discharge; pus (Med.) 011
JAUCHZEN: to shout; exult; rejoice 012
JAUNE BRILLANT: jaune brilliant, cadmium sulphide (see CADMIUMSULFID) 013
JAVAOLIVENÖL (*n*): java olive oil (from seeds of *Sterculia fœtida*) 014
JAVELLESCHE LAUGE (*f*): javelle water; eau de javelle (Sodium hypochlorite solution) (see CHLORKALILÖSUNG) 015
JE: always; ever; every; each 016
JECOLEINSÄURE (*f*): jecoleic acid; $C_{18}H_{35}CO_2H$ 017
JECORINSÄURE (*f*): jecorinic acid; $C_{17}H_{29}CO_2H$ 018
JEDENFALLS: in any case; by all means; probably; at all events; however 019
JEDER: each; every; any; everyone; either 020
JEDERMANN: every one; one; mankind; everybody 021
JEDERZEIT: always, at any time, ever 022
JEDESMAL: always; every time 023
JEDOCH: however; yet; nevertheless; notwithstanding 024
JEDWEDER: each, every-(one) 025
JEFFERSONIT (*m*): jeffersonite; Ca(Fe,Mn,Mg, Zn)O,SiO_2; Sg. 3.5 026
JEGLICH: every; any; everything; each; everyone 027
JELÄNGERJELIEBER (*n*): honeysuckle (*Lonicera periclymenum*) (Bot.) 028
JEMALS: ever; at any time 029
JEMAND: somebody; anybody; someone; anyone 030
JE NACH: according to 031
JENENSER GLAS (*n*): jena glass 032
JENER: that one; that person; the other; the former; yonder 033
JENSEIT: beyond; on the other side of; opposite 034
JENSEITIG: yonder, opposite 035
JEREMEIEWIT (*m*): jeremeiewite; Al_2O_3,B_2O_3; Sg. 3.28 036
JERICHOROSE (*f*): Rose of Jerico (*Anastatica hierochuntica*) 037
JERUSALEMARTISCHOCKE (*f*): Jerusalem artichoke (*Helianthus tuberosus*) 038
JERUSALEMER BALSAM (*m*): Jerusalem balsam (see KANADABALSAM) 039
JERUSALEMSBLUME (*f*): Cross of Jerusalem, Cross of Malta, Flower of Constantinople; scarlet campion (*Lychnis coronaria*) 040
JERVASÄURE (*f*): jervic acid (from *Veratrum album*) 041
JERVIN (*n*): jervine (the active principle of the rhizome of Hellebore) (see NIESWURZ) (*Rhizoma veratri*) 042
JESUIT (*m*): jesuit (Member of the Society of Jesus) 043
JESUITENORDEN (*m*): Jesuit's order, the Society of Jesus 044
JESUITENPULVER (*n*): calisaya, cinchona bark, Jesuits' bark, Peruvian bark (bark of *Cinchona calisaya*) 045
JESUITENRINDE (*f*): Jesuit's bark (see JESUITEN-PULVER) 046
JESUITENTEE (*m*): Mexican tea, American wormseed, goose-foot [*Herba Chenopodii-ambrosioides* of *Chenopodium ambrosioides*] 047
JETSCHWARZ (*n*): jet black 048
JETZIG: present; actual; modern; current 049
JETZT: now; at present 050
JEWEILIG: : for the time being; at (for) a given moment or time; existing; obtaining 051
JEWEILIGE VERHÄLTNISSE (*n.pl*): conditions (obtaining) existing; present conditions 052
JEWEILS MASSGEBEND: ruling, obtaining, in force 053
JIGGERMAST (*m*): jiggermast 054
JOCH (*n*): yoke; pillar; pile; trestle; standard; support; column; upright; arch; crossbeam; transom 055
JOCHBAUM (*m*): hornbeam (see HORNBAUM) 056
JOCHBOLZEN (*m*): yoke bolt 057
JOCHEN: to yoke; couple 058

JOCHHANDELBAUM (*m*): juniper (*Juniperus communis*) 059
JOCHHOLZ (*n*): yoke; cross-beam, transom, cross-bar 060
JOD (*n*): iodine (I); (J is used in German chemical formulæ) (obtained from seaweed or Chili saltpeter) (as a prefix=iodide of) 061
JODADDITIONSPRODUKT (*n*): iodine addition product 062
JODALLYL (*n*): allyl iodide (see ALLYLJODID) 063
JODAMMONIUM (*n*): ammonium iodide (see AMMONIUMJODID) 064
JODAMPULLE (*f*): iodine ampule 065
JODANLAGE (*f*): iodine plant 066
JODANTIMON (*n*): antimony iodide (see ANTI-MONTRIJODID and ANTIMONPENTAJODID) 067
JODARGYRIT (*m*): iodyrite (natural silver iodide) AgI; Sg. 5.65 (about 45% Ag.) 068
JODARSEN (*n*): arsenic iodide (see ARSENTRIJODID) 069
JODAT (*n*): iodate 070
JODÄTHYL (*n*): ethyl iodide; iodoethane (see ÄTHYLJODID) 071
JODAZETON (*n*): acetone iodide; $CH_3.CO.CH_2I$ 072
JODAZID (*n*): iodine azide 073
JODBARYUM (*n*): barium iodide (see BARYUM-JODID) 074
JODBEHENSAUER: iodobehenate 075
JODBENZOL (*n*): iodobenzene; C_6H_5I; Sg. 1.8606; Mp. –28.5°C.; Bp.188.45°C.; any iodine derivative of benzene 076
JODBLEI (*n*): lead iodide (see BLEIJODID) 077
JODBROMID (*n*): iodine bromide (mono.=IBr.; Mp. 36°C.; penta.=IBr_5; Tri.=IBr_3) 078
JODBROMSILBERCOLLODIUMEMULSION (*f*): iodo-bromide collodion emulsion (Phot.) 079
JODBROMSILBERGELATINEEMULSION (*f*): iodo-bromide gelatine emulsion (Phot.) 080
JODCADMIUM (*n*): cadmium iodide (see CADMIUM-JODID) 081
JODCALCIUM (*n*): calcium iodide (see CALCIUM-JODID) 082
JODCHINOLIN (*n*): iodoquinoline 083
JODCHLOR (*n*): iodine chloride (either mono.— or Tri.—) (see JODCHLORID and JODCHLORÜR) 084
JODCHLORID (*n*): iodine chloride; iodine tri-chloride; ICl_3; Mp. 33°C. 085
JODCHLORÜR (*n*): iodine chloride; iodine mono-chloride; ICl; Sg. 3.18-3.28; Mp. 25°C.; Bp. 101°C. 086
JODCOLLODIUM (*n*): iodised collodion 087
JODCYAN (*n*): cyanogen iodide; iodine cyanide; ICN; Mp. 146.5°C. 088
JODDAMPF (*m*): iodine vapour 089
JODDIOXYD (*n*): iodine dioxide; IO_2; Sg. 4.2 090
JODEISEN (*n*): iron iodide (see EISENJODID and EISENJODÜR) 091
JODEOSIN (*n*): erythrosin, potassium salt of tetraiodofluorescine; $C_{20}H_6I_4O_5K_2$ (used in orthochromatic photography) 092
JODEX (*n*): iodex (a salve of mineral fat with 5% iodine) 093
JODFERRATOSE (*f*): iodine ferratose 094
JODGEHALT (*m*): iodine content 095
JODGOLD (*n*): gold iodide (see GOLDJODID and GOLDJODÜR) 096
JODGRÜN (*n*): iodine green 097
JODHALTIG: iodiferous; containing iodine 098
JODHYDRAT (*n*): hydriodide 099
JODID (*n*): -ic iodide 100
JODIDCHLORID (*n*): iodochloride 101
JODID (*n*), DOPPEL-: double iodide (silver iodide and potassium iodide solution) 102
JODIEREN: to iodise; iodize; iodate 103
JODIJODAT (*n*): iodine iodate 104
JODIMETRISCH: iodimetric 105
JODINE (*f*): iodine (see JOD) 106
JODINROT (*n*): scarlet-(red) (see QUECKSILBER-JODID) 107
JODISMUS (*m*): iodism 108
JODIT (*n*): iodite; iodyrite (natural silver iodide; 45% Ag.); AgI; Sg. 5.65 109
JODKADMIUM (*n*): cadmium iodide (see CAD-MIUMJODID) 110
JODKALI (*n*): patassium iodide (see KALIUM-JODID) 111
JODKALIUM (*n*): (see JODKALI) 112
JODKALZIUM (*n*): calcium iodide (see CALCIUM-JODID) 113
JODKUPFER (*n*): copper iodide (see KUPFER-JODID and KUPFERJODÜR) 114
JODLITHIUM (*n*): lithium iodide; LiI; Sg. 4.06; Mp. 330-446°C.; or $LiI.3H_2O$; Mp. 72°C. 115
JODLÖSUNG (*f*): iodine solution 116
JODMAGNESIUM (*n*): magnesium iodide (see MAGNESIUMJODID) 117
JODMETALL (*n*): metallic iodide 118
JODMETHYL (*n*): methyl iodide; iodomethane (see METHYLJODID) 119
JODMITTEL (*n*): iodiferous remedy; remedy containing iodine 120
JODMONOCHLORID (*n*): iodine monochloride; ICl; Sg. 3.18-3.28; Mp. 25°C.; Bp. 101°C. 121
JODNAPHTALIN (*n*): iodoxynaphthalene 122
JODNATRIUM (*n*): sodium iodide (see NATRIUM-JODID) 123
JODNATRON (*n*): (see JODNATRIUM) 124
JODOBENZOËSÄURE (*f*): iodoxybenzoic acid 125
JODOBENZOL (*n*): iodoxybenzene; $C_6H_5IO_2$; 126
JODOFORM (*n*): iodoform; triiodomethane; formyl triiodide; methenyl triiodide; CHI_3; Sg. 4.0; Mp. 119°C. 127
JODOFORMERSATZ (*m*): iodoform substitute 128
JODOFORMSEIFE (*f*): iodoform soap 129
JODOL (*n*): iodol 130
JODOLSEIFE (*f*): iodol soap 131
JODOMETRIE (*f*): iodometry 132
JODOMETRISCH: iodometric 133
JODONIUM (*n*): iodonium 134
JODOSOBENZOL (*n*): iodosobenzene; C_6H_5IO 135
JODOSOL (*n*): iosol, iodosol, thymol iodide, thymotol, thymiodol, thymodin, thymiode, $C_6H_2(CH_3)(OH)(C_3H_7)I$; Mp. 69°C. 136
JODOTHYRIN (*n*): iodothyrine (an iodine compound from the thyroid gland) 137
JODOXYNAPHTOCHINON (*n*): iodohydroxynaph-thoquinone; $C_{10}H_4O_2(OH)I$ 138
JODPAPIER (*n*): iodized paper (Phot.) 139
JODPENTOXYD (*n*): iodine pentoxide; I_2O_5; Sg. 4.799 140
JODPHOSPHONIUM (*n*): phosphonium iodide 141
JODPHOSPHOR (*m*): phosphorous iodide (see PHOSPHORJODID and PHOSPHORJODÜR) 142
JODPHTALSÄURE (*f*): iodophthalic acid 143
JODPRÄPARAT (*n*): iodine preparation 144
JODPROBE (*f*): iodine test 145

JODQUECKSILBER (*n*): mercury (mercuric) iodide (see QUECKSILBERJODID) 146
JODRADIUM (*n*): radium iodide 147
JODSALBE (*f*): iodine (ointment) salve; ointment containing iodine, as iodex (see JODEX) 148
JODSALIZYLSAUER: iodosalicylate 149
JODSALZ (*n*): iodide; iodine salt 150
JODSAUER: iodate of 151
JODSÄURE (*f*): iodic acid; HIO_3; Sg. 4.629; Mp. 110°C. 152
JODSCHALE (*f*): iodine cup 153
JODSCHWEFEL (*m*): sulphur iodide; sulphur moniodide (see SCHWEFELJODÜR) 154
JODSEIFE (*f*): iodine soap 155
JODSILBER (*n*): silver iodide (see SILBERJODID and SILBERJODÜR) 156
JODSILIZIUM (*n*): silicon iodide (see SILIZIUM--JODID) 157
JODSODA (*f*): sodium iodide (see NATRIUMJODID) 158
JODSTÄRKE (*f*): starch iodide; iodized starch 159
JODSTICKSTOFF (*m*): nitrogen iodide; $N_2H_3I_3$; Sg. 3.5 160
JODTHYMOL (*n*): thymoliodide (see JODOSOL) 161
JODTINKTUR (*f*): tincture of iodide (iodine dissolved in alcohol) 162
JODTOLUOL (*n*): iodotoluene 163
JODTRICHLORID (*n*): iodine trichloride; ICl_3; Mp. 33°C; Sg. 3.11 164
JODÜR (*n*): -ous iodide 165
JODVERBINDUNG (*f*): iodine compound 166
JODWASSER (*n*): iodine water 167
JODWASSERSTOFF (*m*): hydrogen iodide; hydriodic acid (*Acidum hydrojodicum*); HI; Sg. 4.374; Mp. –51.3°C.; Bp. –35.6°C. 168
JODWASSERSTOFFÄTHER (*m*): ethyl iodide (see ÄTHYLJODID) 169
JODWASSERSTOFFSAUER: hydriodate of, combined with hydriodic acid 170
JODWASSERSTOFFSÄURE (*f*): hydriodic acid; hydrogen iodide (see JODWASSERSTOFF) 171
JODWASSERSTOFFSAURES SALZ (*n*): hydriodate (of a non-metallic base); iodide 172
JODWISMUT (*m* and *n*): bismuth iodide (see WISMUTJODID) 173
JODYRIT (*m*): iodyrite (see JODIT and JODARGYRIT) 174
JODZAHL (*f*): iodine number 175
JODZIMMTSÄURE (*f*): iodocinnamic acid 176
JODZINK (*n*): zinc iodide (see ZINKJODID) 177
JODZINN (*n*): tin iodide (see ZINNJODID and ZINNJODÜR) 178
JODZINNOBER (*m*): red mercuric iodide (see JODINROT) 179
JOGHURTPRÄPARAT (*n*): yoghurt preparation (see YOGHURT) 180
JOHANNESKRAUT (*n*): (see JOHANNISKRAUT) 181
JOHANNISBEERBLATTBRÄUNE (*f*): a disease of the leaves of the red currant bush, caused by a fungus, *Gloiosporium ribis* 182
JOHANNISBEERBLÄTTER (*n.pl*), SCHWARZE-: black currant leaves (*Herba ribium nigrorum*) 183
JOHANNISBEERE (*f*): currant (see JOHANNIS--BEERE, ROTE-; J. WEISSE- and J. SCHWARZE) 184
JOHANNISBEERE (*f*), ROTE-: red currant (*Ribes rubrum*) 185
JOHANNISBEERE (*f*), SCHWARZE-: black currant (*Ribes nigrum*) 186
JOHANNISBEERE (*f*), WEISSE-: white currant (*Ribes album*) 187
JOHANNISBEERSAFT (*m*): currant juice 188
JOHANNISBEERSTRAUCH (*m*): currant bush 189
JOHANNISBEERWEIN (*m*): (red)-currant wine 190
JOHANNISBLUME (*f*): arnica (*Arnica montana*); daisy 191
JOHANNISBLUT (*n*): St. John's Wort (*Hypericum perforatum*); cochineal (see COCHENILLE) 192
JOHANNISBLUTPFLANZE (*f*): St. John's Wort (*Hypericum perforatum*) 193
JOHANNISBROT (*n*): St. John's bread; carob bean; locust bean (fruit of *Ceratonia siliqua*) 194
JOHANNISBROTBAUM (*m*): St. John's bread (*Ceratonia siliqua*) 195
JOHANNISFEST (*n*): St. John's day, Midsummer day 196
JOHANNISGÜRTEL (*m*): club moss, lycopodium (see *Lycopodium clavatum*) (see also BÄRLAPP) 197
JOHANNISKÄFER (*m*): glow-worm (*Lampyris noctiluca*) 198
JOHANNISKRAUT (*n*): St. John's Wort (*Hypericum perforatum*) 199
JOHANNISLAUCH (*m*): (see JAKOBSLAUCH) 200
JOHANNISPFLAUME (*f*): (see KRIECHE) 201
JOHANNISTAG (*m*): St. John's day, Midsummer day 202
JOHANNISWURM (*m*): glow-worm (*Lampyris noctiluca*) 203
JOHANNISWÜRMCHEN (*n*): glow worm (*Lampyris noctiluca*) 204
JOHANNISWURZ (*f*): male fern, aspidium; anacyclus 205
JOHANNISWURZEL (*f*): male fern (*Dryopteris filix-mas marginalis*); aspidium; shield fern; filix mas 206
JOHANNIT (*m*): (see URANVITRIOL) 207
JOHIMBEHEBAUM (*m*): (see YUMBEHOABAUM) 208
JOHIMBIN (*n*): Yohimbine (see YOHIMBIN) 209
JOHIMBOASÄUREMETHYLESTER (*m*): (see YOHIM--BIN) 210
JOHNIT (*m*): (see TÜRKIS) 211
JOLLE (*f*): yawl, jolly boat, wherry (Naut.) 212
JON (*n*): ion (see ION) 213
JONIUM (*n*): ionium 214
JONKE (*f*): Chinese junk 215
JONON (*n*): ionone; $C_{13}H_{20}O$ (a synthetic perperfume) 216
JONQUILLE (*f*): jonquil (*Narcissus jonquilla*) (Bot.) 217
JOTHION (*n*): iothion, diiodohydroxypropane about 80% iodine); $C_3H_5I_2(OH)$; Sg. 2.45 218
JOULE (*n*): joule (a unit of work) 219
JOURNAL (*n*): Journal; day book (Com.); newspaper 220
JOURNALIST (*m*): journalist 221
JOURNALISTIK (*f*): journalism 222
JOURNALISTISCH: journalistic 223
JOVIALITÄT (*f*): joviality 224
JUBEL (*m*): jubilation; exultation; rejoicing; shout of joy (as a prefix=jubilee) 225
JUBELFEIER (*f*): jubilee 226
JUBELFEST (*n*): jubilee 227
JUBELJAHR (*n*): jubilee-(year) 228
JUBELN: to rejoice, be jubilant, shout for joy, exult 229
JUBILÄUM (*n*): jubilee 230
JUBILIEREN: (see JUBELN) 231
JUCHT (*n*): Russian leather; Muscovy (hides) leather 232
JUCHTEN (*m*): (see JUCHT) 233

JUCHTENLEDER (*n*): (see JUCHT) 234
JUCHTENÖL (*n*): birch-tar oil (see BIRKENTEERÖL) 235
JUCHTENROT (*n*): Janus red; Juft's red; Russian (Muscovy) leather red 236
JUCKEN: to itch; scratch; rub; (*n*): itching; pruritus; irritation (Med.) 237
JUDAISMUS (*m*): judaism 238
JUDASBAUM (*m*): Judas tree (*Cercis siliquastrum*) 239
JUDASOHR (*n*): Jew's ear, Judas's ear (a fungus on elder trees) (*Hirneola Auricula-Judas*) 240
JUDASSCHWAMM (*m*): (see JUDASOHR) 241
JUDE (*m*) Jew; miser; usurer 242
JUDEN (*m.pl*): (see JUDE); refinery scraps (Metal) 243
JUDENBAUM (*m*): Christ's thorn (*Paliurus aculeatus*); Jew's thorn; garland thorn 244
JUDENDORN (*m*): zizyphus 245
JUDENDORNBEERE (*f*): jujube; zizyphus 246
JUDENHARZ (*n*): Jew's pitch (see ASPHALT) 247
JUDENKIRSCHE (*f*): winter cherry; ground cherry; alkekengi (*Fructus physalis alkekengi*) 248
JUDENKRAUT (*n*): milfoil, yarrow (*Achillea*) 249
JUDENLEIM (*m*): Jew's pitch (see ASPHALT) 250
JUDENPAPPEL (*f*): Jew's mallow (*Corchorus olitorius*) 251
JUDENPECH (*n*): Jew's pitch (see ASPHALT) 252
JUDENPILZ (*m*): Jew's ear (see JUDASOHR) 253
JUDENSCHWAMM (*m*): (see JUDENPILZ) 254
JUDENTUM (*n*): judaism 255
JUDENZOPF (*m*): Polish ringworm, *Plica Polonica* (disease of the scalp due to lice and filth) (Med.) 256
JÜDISCH: Jewish 257
JUFFER (*f*): dead-eye 258
JUFTEN (*m*): (see JUCHT) 259
JUGELBEERE (*f*): bilberry, whortleberry, blaeberry, blueberry (*Vaccinium myrtillus*) 260
JUGEND (*f*): youth; adolescence; young people 261
JUGENDLICH: youthful; juvenile 262
JUKKA (*f*): (see YUKKA) 263
JUNG: young; youthful; juvenile; fresh; recent; early; green; new (Wine); overpoled (Copper) 264
JUNGE (*m*): boy; lad; youth; apprentice; (*n*): cub; offspring; young (of animals) 265
JÜNGER: newer; later; latest; younger; junior; (*m*): disciple; follower; votary 266
JUNGFER (*f*): virgin; miss; maid; spinster; chambermaid; dead-eye 267
JUNGFERBLÜTE (*f*): sundew (*Drosera rotundifolia*) (see also SONNENTAU and SONNENKRAUT-) 268
JUNGFER IN GRÜNEN: (see KAPUZINERKRAUT) 269
JUNGFER IN HAAREN: (see KAPUZINERKRAUT) 270
JUNGFERLICH: maiden; virginal 271
JUNGFERNGLAS (*n*): selenite (see ERDGLAS) 272
JUNGFERNGOLD (*n*): native gold 273
JUNGFERNGRAS (*n*): stellaria, stitchgrass, starwort, stitchwort (*Stellaria holostea*, *S. media*, and *S. graminea*) 274
JUNGFERNHERZ (*n*): dicentra, dielytra; bleeding heart (*Dicentra spectabilis*) 275
JUNGFERNHONIG (*m*): virgin honey (see HONIG-SEIM) 276
JUNGFERNINSELN (*f.pl*): virgin islands 277
JUNGFERNKRANKHEIT (*f*): anæmia, chlorosis (see ANÄMIE and BLEICHSUCHT) (Med.) 278
JUNGFERNMETALL (*n*): native metal 279
JUNGFERNMILCH (*f*): virgin's milk, benzoin (Chem.) (1 part benzol tincture to 30 parts rosewater) (see BENZOIN) 280
JUNGFERNNELKE (*f*): maiden pink (*Dianthus virgineus*) 281
JUNGFERNÖL (*n*): virgin oil; olive oil (a highly refined olive oil) (the oil obtained by the pressure of the olives on themselves) (see OLIVENÖL) 282
JUNGFERNPERGAMENT (*n*): a fine, thin parchment 283
JUNGFERNRAUB (*m*): rape; ravishment 284
JUNGFERNREBE (*f*): Virginia Creeper (*Ampelopsis quinquefolia*) 285
JUNGFERNSCHWEFEL (*m*): virgin (native) sulphur 286
JUNGFERNWACHS (*n*): virgin wax 287
JUNGFERNZEUGUNG (*f*): parthenogenesis; asexual reproduction 288
JUNGFRAU (*f*): virgin; maid 289
JUNGFRÄULICH: virgin, maidenly, modest 290
JUNGGESELL (*m*): bachelor 291
JÜNGLING (*m*): young man; youth 292
JÜNGST: youngest; last; latest; newest; lately; recently; date; recent 293
JUNKE (*f*): Chinese junk 294
JUNKER (*m*): young nobleman; squire; youngster 295
JURIDISCH: juridical, legal 296
JURISPRUDENZ (*f*): jurisprudence 297
JURIST (*m*): jurist, lawyer 298
JURISTISCH: juridical, legal 299
JURY (*f*): jury 300
JUST: just, even now, exactly, but just, just now 301
JUSTIEREN: to adjust; justify 302
JUSTIERGEWICHT (*n*): standard weight 303
JUSTIERTISCH (*m*): adjusting (table) bench 304
JUSTIERWAGE (*f*): adjusting balance 305
JUSTIZ (*f*): justice 306
JUSTIZAMT (*n*): court of law 307
JUSTIZIAR (*m*): justiciary 308
JUSTIZKAMMER (*f*): Court of Justice 309
JUSTIZKOLLEGIUM (*n*): Court of Justice 310
JUSTIZKOMMISSÄR (*m*): attorney at law 311
JUSTIZPFLEGE (*f*): administration of justice 312
JUSTIZWESEN (*n*): the law; legal (matters) affairs; judicature 313
JUTE (*f*): jute (bast fibre of *Corchorus* species of plant) 314
JUTEFABRIKAT (*n*): jute (goods; manufactures) wares 315
JUTEFÄRBEREI (*f*): jute dyeing 316
JUTEFARBSTOFF (*m*): jute colouring matter 317
JUTEKOHLSCHWARZ (*n*): jute carbon black 318
JUTESCHWARZ (*n*): jute black 319
JUTIEREN: to jute, cover with jute 320
JUTIERMASCHINE (*f*): jute winding machine 321
JUWEL (*n*): jewel 322
JUWELENHANDEL (*m*): jeweller's trade 323
JUWELENKÄSTCHEN (*n*): jewel (case) casket 324
JUWELIER (*m*): jeweller; lapidary 325
JUWELIERARBEIT (*f*): jeweller's work; jewelry 326
JUWELIERBORAX (*m*): borax (for soldering and melting purposes); $Na_2B_4O_7.5H_2O$ 327
JUX (*m*): filth; trash; spree; lark; frolic; joke 328
JUXEN: to lark; joke; make game of; frolic 329

K

Refer also to C for words not found under K.

KABBELIG : turbulent (of water, etc.) 331
KABEL (*f* and *n*) : cable ; cable's length ; lot ; share 332
KABELBAHN (*f*) : cable-way, cable-tramway 333
KABELFABRIK (*f*) : cable works 334
KABELGARN (*n*) : rope yarn 335
KABELGATT (*n*) : cable stage 336
KABELISOLIERÖL (*n*) : cable insulating oil 337
KABELISOLIERWACHS (*n*) : cable insulating wax 338
KABELJAU (*m*) : cod (fish) (*Gadus callarias* or *Gadus morrhua*) 339
KABELJAULEBERÖL (*n*) : (see LEBERTRAN) 340
KABELJAULEBERTRAN (*m*) : cod liver oil ; Banks oil (see LEBERTRAN) 341
KABELKASTEN (*m*) : cable box 342
KABELKRAN (*m*) : cable crane 343
KABELLÄNGE : cable's length (120 fathoms) 344
KABELLEGUNG (*f*) : cable laying 345
KABELMUFFE (*f*) : cable (sleeve) socket 346
KABELN : to draw lots, allot 347
KABEL (*n*), OBERIRDISCHES- : aerial cable 348
KABELPRÜFUNG (*f*) : cable test 349
KABELSCHIFF (*n*) : cable (laying) ship 350
KABELSEELE (*f*) : core of a cable 351
KABELSEIL (*n*) : cable 352
KABELSPLEISSUNG (*f*) : cable splice 353
KABELTAU (*n*) : cable 354
KABEL (*n*), TRANSATLANTISCHES- : transatlantic cable 355
KABELUNG (*f*) : allotment, share, lot 356
KABEL (*n*), UNTERIRDISCHES- : underground cable 357
KABEL (*n*), UNTERSEEISCHES- : submarine cable 358
KABELVERBINDUNG (*f*) : cable joint 359
KABELWACHS (*n*) : cable wax 360
KABELWERK (*n*) : cable works 361
KABESTAN (*m*) : capstan 362
KABINE (*f*) : cabin (Naut.) 363
KABINETT (*n*) : cabinet 364
KABLIAU (*m*) : cod-(fish) (see KABELJAU) 365
KABLIAULEBERTRAN (*m*) : (see KABELJAULEBER--TRAN) 366
KABUSKRAUT (*n*) : cabbage (*Brassica oleracea*) 367
KACHEKTISCH : cachectic (impoverished in body or mind) (Med.) 368
KACHEL (*f*) : Dutch (stove) tile ; hollow earthenware vessel 369
KACHELFORM (*f*) : tile mould (for Dutch tiles) 370
KACHELMASCHINE (*f*) : tile-(making) machine 371
KACHELOFEN (*m*) : tile (earthenware) oven or stove 372
KADAVER (*n*) : cadaver, carcase 373
KADAVERVERWERTUNGSANLAGE (*f*) : cadaver (carcase) utilization plant 374
KADDICH (*m*) : cade (*Juniperus oxycedrus*) ; juniper (*Juniperus communis*) 375
KADDIG (*m*) : (see KADDICH) 376
KADDIGÖL (*n*) : oil of cade (see CADEÖL) 377
KADEÖL (*n*) : (see CADEÖL) 378
KADINÖL (*n*) : (see CADEÖL) 379
KADIÖL (*n*) : (see CADEÖL) 380
KADMIUM (*n*) : Cadmium (see also CADMIUM) 381
KÄFER (*m*) : beetle ; chafer ; coleopter 382
KÄFERARTIG : coleopterous 383
KÄFERSCHRÖDER (*m*) : stag-beetle (*Lucanus cervus*) 384
KAFFEE (*m*) : coffee 385
KAFFEEAUSZUG (*m*) : coffee extract 386
KAFFEEBAUM (*m*) : coffee (tree) shrub (*Coffea arabica* or *Coffea liberica*) 387
KAFFEEBOHNE (*f*) : coffee bean 388
KAFFEEBOHNENÖL (*n*) : coffee bean oil (from beans of *Coffea arabica*) 389
KAFFEEBRENNEN (*n*) : coffee roasting 390
KAFFEEBRENNER (*m*) : coffee roaster 391
KAFFEEDÜNGER (*m*) : coffee manure 392
KAFFEEERSATZMITTEL (*n*) : coffee substitute 393
KAFFEEESSENZ (*f*) : coffee essence 394
KAFFEEEXTRAKT (*m*) : coffee extract 395
KAFFEEGERBSÄURE (*f*) : caffetannic acid 396
KAFFEEGLASUR (*f*) : coffee glaze 397
KAFFEEKANNE (*f*) : coffee pot 398
KAFFEELÖFFEL (*m*) : tea-spoon 399
KAFFEEMASCHINE (*f*) : coffee (making) machine 400
KAFFEEMÜHLE (*f*) : coffee (mill) grinder 401
KAFFEEÖL (*n*) : coffee oil 402
KAFFEEPAUKE (*f*) : coffee roaster 403
KAFFEERÖSTEN (*n*) : coffee roasting 404
KAFFEESATZ (*m*) : coffee grounds 405
KAFFEESÄURE (*f*) : caffeic acid 406
KAFFEESEIHE (*f*) : coffee strainer 407
KAFFEETASSE (*f*) : coffee cup 408
KAFFEETRICHTER (*m*) : coffee strainer 409
KAFFEETROMMEL (*f*) : coffee roaster 410
KAFFEEZEUG (*n*) : coffee service 411
KAFFEIN (*n*) : caffeine (*Coffeinum*) ; theine ; guaranine ; trimethylxanthine ; methyltheobromine ; $C_8H_{10}N_4O_2.H_2O$; Mp. 230.5°C. 412
KAFFEINBENZOAT (*n*) : caffeine benzoate ; $C_8H_{10}N_4O_2.C_7H_6O_2$ 413
KAFFEINBROMHYDRAT (*n*) : caffeine hydrobromide ; $C_8H_{10}N_4O_2.HBr.2H_2O$ 414
KAFFEIN (*n*), BROMWASSERSTOFFSAURES- : (see KAFFEINBROMHYDRAT) 415
KAFFEINCHLORHYDRAT (*n*) : caffeine hydrochloride ; $C_8H_{10}N_4O_2.HCl.2H_2O$ 416
KAFFEIN (*n*), CITRONENSAURES- : (see KAFFEIN--ZITRAT) 417
KAFFEIN (*n*), SALICYLSAURES- : caffeine salicylate 418
KAFFEINZITRAT (*n*) : caffeine citrate ; $C_8H_{10}N_4O_2.C_6H_8O_7$ 419
KÄFIG (*m*) : (bird)-cage ; prison 420
KAFILLER (*m*) : knacker (one who deals in dead horses) 421
KAFILLEREI (*f*) : knacker's trade ; knacker's yard 422
KAGNEBUTTER (*f*) : Kagné butter (from the seeds of *Allanblackia sacleuxii*) 423
KAHL : bald ; naked ; bare ; barren ; threadbare ; napless (Fabric) ; mere ; poor ; callow ; waste ; leafless ; empty ; sterile 424
KAHLFRAS (*m*) : the stripping of all the leaves from trees by insects 425
KAHLHEIT (*f*) : Alopecia, baldness (Med.) ; bareness, poverty, barrenness, bleakness, sterility (see also KAHL) 426
KAHLHOLZ (*n*) : branchless wood 427

KAHLKÖPFIGKEIT (*f*): baldness, Alopecia (Med.) 428

KAHLWILD (*n*): game (deer) without antlers or horns 429

KAHM (*m*): mould (on wine, etc.) 430

KAHMEN: to mould-(er); become mouldy 431

KAHMHAUT (*f*): pellicle, -(of mould) 432

KAHMIG: mouldy; stale; ropy (of wine) 433

KAHN (*m*): boat; skiff 434

KAHNBAR: navigable-(for small craft) 435

KAHNBEIN (*n*): scaphoid bone (either of the wrist or ankle) (Anat.) 436

KAHNBRÜCKE (*f*): boat (pontoon) bridge 437

KAHNFAHREN (*n*): boating 438

KAHNFÖRMIG: boat-shaped; scaphoid (Med.); carinate-(d) (Bot.); cimbiform 439

KAHN (*m*), **KLEINER-**: skiff 440

KAI (*m*): quay; pier; wharf; jetty 441

KAIGEBÜHR (*f*): pier toll; wharfage-(dues); wharfage (quay) charges 442

KAIMAUER (*f*): quay wall, jetty wall 443

KAINIT (*m*): Kainite (Mineralogy); $MgSO_4.KCl.3H_2O$; Sg. 2.15 (from the Stassfurt district of Germany) 444

KAINKAWURZEL (*f*): cahinca root 445

KAINOSIT (*m*): kainosite, cenosite; $H_4Ca_2(Y_2CO_3)(Si_2O_7)$; Sg. 3.4 446

KAISER (*m*): emperor 447

KAISERBLAU (*n*): smalt (see **SMALTE** and **BLAU-FARBE**) 448

KAISERGELB (*n*): Paris yellow, mineral yellow (see **BLEICHROMAT**); Imperial yellow (an orange colour; ammonium salt of hexanitrodiphenylamine); $(NO_2)_3C_6H_2.NH.C_6H_2(NO_2)_3$ 449

KAISERGRÜN (*n*): imperial green, Paris green, Schweinfurt green (copper acetoarsenite); $(C_6H_3O_2)_2Cu+3(CuAs_2O_4)$ (see **KUPFER-ACETATARSENIAT**) 450

KAISERLICH: imperial; cæsarean 451

KAISERREICH (*n*): empire 452

KAISERROT (*n*): imperial red-(dye); colcothar (see **EISENOXYD** and **ENGELROT**) 453

KAISERSCHLANGE (*f*): boa constrictor 454

KAISERSCHNITT (*m*): hysterotomy, laparotomy, Cæsarean (section) operation, abdominal section (An incision through the anterior abdominal wall) (Surgical) 455

KAISERWURZ (*f*): masterwort; felon (grass) wort (such as *Imperatoria ostruthium*; see also **KAISERWURZEL**) 456

KAISERWURZEL (*f*): sulphurwort (*Peucedanum officinale*) (see also **KAISERWURZ**) 457

KAJAPUTÖL (*n*): cajuput oil (from leaves of *Melaleuca leucadendron*) (see also **CAJEPUT-**) 458

KAJEPUTGEIST (*m*): spirit of cajuput 459

KAJEPUTÖL (*n*): (see **KAJAPUTÖL**) 460

KAJÜTE (*f*): cabin (of boats) 461

KAKADU (*m*): cockatoo (*Cacatua galerita* and *Cacatua leadbeateri*) 462

KAKAO (*m*): cacao; cocoa (see also **CACAO**) 463

KAKAOBAUM (*m*): cacao (cocoa) tree (*Theobroma cacao*) 464

KAKAOBOHNE (*f*): cacao (cocoa) bean or nib (bean or seed of *Theobroma cacao*) 465

KAKAOBUTTER (*f*): cacao (cocoa) butter (see **CACAOBUTTER**) (*Oleum cacao*) 466

KAKAO (*m*), **ENTÖLTER-**: defatted (deoiled) cocoa (the removal of a quantity of the natural fat) 467

KAKAO (*m*), **LÖSLICHER-**: soluble cocoa (cocoa which has been treated to make it more soluble with water) 468

KAKAOMASSE (*f*): cacao (cocoa) paste (contains up to 55% fat) 469

KAKAOMÜHLE (*f*): cocoa (grinding) mill 470

KAKAOPULVER (*n*): cocoa powder (contains up to 35% fat) 471

KAKAORÖSTEN (*n*): cocoa roasting (roasting the cocoa nibs) 472

KAKAORÖSTMASCHINE (*f*): cocoa roasting (drum) machine 473

KAKAOROTTEN (*n*): cocoa fermentation 474

KAKAOSAME (*f*): cacao (cocoa) bean, or nib (bean or seed of *Theobroma cacao*) 475

KAKAO (*m*), **UNGEROTTETER-**: unfermented cocoa (sun-dried cocoa) 476

KAKERLAK (*m*): cockroach (*Blatta orientalis*) 477

KAKODYL (*n*): cacodyl; tetramethyl diarsine; $As_2(CH_3)_4$; Mp. $-6°C.$; Bp. 170°C. 478

KAKODYLOXYD (*n*): cacodyl oxide; $As_2(CH_3)_4O$ 479

KAKODYLPRÄPARAT (*n*): cacodyl preparation 480

KAKODYLSAUER: cacodylate, combined with cacodylic acid 481

KAKODYLSÄURE (*f*): (*Acidum kakodylicum*): cacodylic acid; dimethylarsenic acid; $(CH_3)_2AsOOH$; Mp. 200°C. 482

KAKOPHONIE (*f*): cacophony 483

KAKOXEN (*m*): kakoxen (a hydrous iron phosphate); $2Fe_2O_3,P_2O_5,12H_2O$; Sg. 2.3 484

KAKTUSPFLANZE (*f*): cactus plant (*Opuntia coccinellifera*) 485

KAKTUSSCHILDLAUS (*f*): cochineal insect (*Coccus cacti*) (the colouring principle of which is carminic acid; $C_{17}H_{18}O_{10}$) 486

KALABARBOHNE (*f*): calabar bean; ordeal bean (seed of *Physostigma venenosum*) 487

KALABASSE (*f*): calabash (shell of fruit of calabash tree, *Crescentia cujete*); bottle-gourd 488

KALAIT (*m*): (see **TÜRKIS** and **KALLAIT**) 489

KALAMIN (*m*): calamine (see **ZINKSPAT** and **KIE-SELZINKERZ**) 490

KALAMUS (*m*): calamus (see **KALMUS**) 491

KALANDER (*m*): calender (a machine with polished, heated rollers, for polishing parchment, etc). 492

KALANDER (*m*), **FRIKTIONS-**: friction calender 493

KALANDERN: to calender 494

KALB (*n*): calf; fawn 495

KALBEN (*n.pl*): chocks (Naut.); (verb), to calve 496

KÄLBERDRÜSE (*f*): calf's sweet-bread 497

KÄLBERKROPF (*m*): bastard parsley (*Anthriscus*) 498

KÄLBERLAB (*n*): rennet 499

KÄLBERMAGEN (*m*): rennet 500

KÄLBERN: to calve 501

KALBFELL (*n*): calfskin 502

KALBFLEISCH (*n*): veal 503

KALBLEDER (*n*): calf leather 504

KALBSKARBONADE (*f*): veal cutlet 505

KALBSLAB (*n*): rennet 506

KALBSMILCH (*f*): calf's sweet-bread 507

KALBSPERGAMENT (*n*): vellum 508

KALCINIEREN: to calcine (see also **CALCINIER-**) 509

KALCINIERUNG (*f*): calcination 510

KALCIT (*m*): calcite (see **CALCIUMCARBONAT** and **CALCIT**) 511

KALCIUM (*n*): calcium (see also **CALCIUM**) 512
KALCIUMCYANAMID (*n*): calcium cyanamide; lime nitrogen; cyanamide; nitrolim (see **CALCIUMCYANAMID**) 513
KALDAUNE (*f*): tripe; (*pl*): bowels; intestines; entrails 514
KALENDER (*m*): calender; almanac 515
KALFATDOCKEN (*n.pl*): graving docks 516
KALFATEISEN (*n*): caulking (iron) tool 517
KALFATERER (*m*): caulker 518
KALFATERN: to caulk 519
KALFATHAMMER (*m*): caulking (hammer) mallet 520
KALFATWERG (*n*): oakum 521
KALI (*n*): (caustic)-potash (see **ÄTZKALI**); potassium hydroxide; potassa (see **KALIUM-HYDROXYD**); potash; K_2O (see **KALIUM-OXYD**); potash (see **KALIUMCARBONAT**); potassium (K) (see **KALIUM**); saltwort (*Salsola kali*) 522
KALIALAUN (*m*): potash alum (see **ALAUN**) 523
KALIALAUN (*m*), **GEBRANNTER-**: calcined aluminium-potassium sulphate (see **ALAUN**) 524
KALIAMMONIAKSUPERPHOSPHAT (*n*): potash-ammonia superphosphate 525
KALIAMMONSALPETER (*m*): potassium-ammonium nitrate 526
KALIAPPARAT (*m*): potash apparatus 527
KALI (*n*), **ARSENSAURES-**: potassium arsenate (see **KALIUMARSENIAT**) 528
KALIATURHOLZ (*n*): red sandalwood (from *Pterocarpus santalinus*) 529
KALIBER (*n*): groove; gauge; roll (opening) pass (Rolling mills); calibre; bore 530
KALIBERLEHRE (*f*): calibre (ring) gauge 531
KALIBERRING (*m*): ring (calibre) gauge 532
KALIBICARBONAT (*n*): potash bicarbonate (see **KALIUMBICARBONAT**) 533
KALIBLAU (*n*): Prussian blue (see **FERRIFERRO-CYANID**) 534
KALI (*n*), **BLAUSAURES-**: potassium cyanide (see **CYANKALIUM**) (also see **KALI, GELBES BLAUSAURES-** and **KALI, ROTES BLAUSAURES-**) 535
KALIBRIEREN: to calibrate; gauge; test; groove (a roll) 536
KALI (*n*), **CHLORSAURES-**: potassium chlorate (see **KALIUMCHLORAT**) 537
KALI (*n*), **CHROMSAURES-**: potassium chromate; (K_2CrO_4); Sg. 2.7319; Mp. 971°C. 538
KALI (*n*), **DICHROMSAURES-**: (see **CHROMKALI**) 539
KALI (*n*), **DOPPELTCHROMSAURES-**: potassium bichromate (see **CHROMKALI**) 540
KALI (*n*), **DOPPELTKOHLENSAURES-**: (see **KALIUM-BICARBONAT**) 541
KALIDÜNGER (*m*): potash (fertilizer) manure (see also **ABRAUMSALZ**) 542
KALIDÜNGESALZ (*n*): potassium salt for (manuring) fertilizing purposes; potash (manure) fertilizer (see also **ABRAUMSALZ**) 543
KALIEISENALAUN (*m*): ferric potassium sulphate (see **EISENALAUN**) 544
KALIEISENCYANÜR (*n*): potassium ferrocyanide (see **FERROCYANKALIUM**) 545
KALI (*n*), **EISENSAURES-**: potassium ferrate; K_2FeO_4 546
KALI (*n*), **ESSIGSAURES-**: potassium acetate (see **KALIUMACETAT**) 547
KALIFELDSPAT (*m*): potash feldspar (natural potassium-aluminium silicate) (see **ORTHO-KLAS** and **MIKROKLIN**) 548
KALIFORM (*f*): potash mould 549
KALI (*n*), **GELBBLAUSAURES-**: potassium ferrocyanide (see **FERROCYANKALIUM**) 550
KALI (*n*), **GELBES BLAUSAURES-**: potassium ferrocyanide, yellow prussiate of potash (see **BLAUSALZ**) 551
KALI (*n*), **GELBES CHROMSAURES-**: (see **KALIUM-CHROMAT**) 552
KALIGLAS (*n*): potash glass 553
KALIGLIMMER (*m*): potash mica, muscovite (see **MUSKOVIT**); margarite (see **MARGARIT**) 554
KALIHALTIG: containing potash 555
KALIHYDRAT (*n*): potassium hydroxide (see **KALIUMHYDROXYD**) 556
KALIKALK (*m*): potash lime 557
KALI (*n*), **KARBOLSAURES-**: potassium carbolate, potassium phenolate; C_6H_5KO 558
KALI (*n*), **KIESELSAURES-**: potassium silicate 559
KALIKODRUCK (*m*): calico printing 560
KALI (*n*), **KOHLENSAURES-**: potassium carbonate (see **KALIUMCARBONAT**) 561
KALIKUGEL (*f*): potash bulb 562
KALILAUGE (*f*): caustic potash solution; potash lye (solution of KOH in H_2O) (see **ÄTZKALI-**) (*Liquor kali caustici*); Sg. 1.1385 (15% KOH) 563
KALIMAGNESIA (*f*): potash magnesia 564
KALIMAGNESIAANLAGE (*f*): potash-magnesia plant 565
KALIMAGNESIA (*f*), **SCHWEFELSAURE-**: potash-magnesia sulphate; potassium-magnesium sulphate 566
KALIMETALL (*n*): (metallic)-potassium; potassium metal) 567
KALI (*n*), **MILCHSAURES-**: (see **KALIUMLAKTAT**) 568
KALI (*n*), **MOLYBDÄNSAURES-**: potassium molybdate (see **KALIUMMOLYBDAT**) 569
KALI (*n*), **NEUTRALES OXALSAURES-**: neutral potassium oxalate 570
KALINIT (*m*): kalinite, potash alum (see **ALAUN**) 571
KALI (*n*), **OXYCHINOLINSULFOSAURES-**: potassium oxyquinoline sulfonate, quinosol, chinosol; $C_9H_6.NO.SO_3K+H_2O$ 572
KALIPFLANZE (*f*): glasswort; marsh samphire; alkaline plant (*Salicornia*) 573
KALI (*n*), **PHOSPHORSAURES-**: potassium phosphate (see **KALIUMPHOSPHAT**) 574
KALI (*n*), **ROTES BLAUSAURES-**: potassium ferricyanide, red prussiate of potash; $K_3Fe(CN)_6$; Sg. 1.8109 575
KALI (*n*), **ROTES CHROMSAURES-**: red potassium chromate (see **CHROMKALI**) 576
KALISALPETER (*m*): potassium nitrate; nitre; saltpetre; Sg. 2.12 (see **SALPETER** and **KALIUMNITRAT**) 577
KALISALPETER (*m*), **ROHER-**: crude saltpetre 578
KALISALPETERSUPERPHOSPHAT (*n*): potassium nitrate superphosphate 579
KALI (*n*), **SALPETRIGSAURES-**: potassium nitrite (see **KALIUMNITRIT**) 580
KALISALZ (*n*): potash (potassium) salt (see also **ABRAUMSALZ**) 581
KALISALZLAGER (*n*): bed (deposit) of potash salt (see also **ABRAUMSALZ**) 582
KALI (*n*), **SAURES CHROMSAURES-**: potassium bichromate (see **CHROMKALI**) 583
KALI (*n*), **SAURES OXALSAURES-**: acid potassium oxalate (see **KALIUMBIOXALAT**) 584
KALI (*n*), **SAURES SCHWEFELSAURES-**: potassium bisulphate (see **KALIUMBISULFAT**) 585

KALI (*n*), **SAURES SCHWEFLIGSAURES-**: potassium bisulphite (see **KALIUMBISULFIT**) 586

KALI (*n*), **SAURES WEINSAURES-**: (see **CREMOR TARTARI**) 587

KALISCHMELZE (*f*): potassium melt 588

KALISCHWEFELLEBER (*f*): liver of sulphur; hepar; sulphurated potassa (see **SCHWEFEL-LEBER**) 589

KALI (*n*), **SCHWEFELSAURES-**: potassium sulphate (see **KALIUMSULFAT, NEUTRALES-**) 590

KALI (*n*), **SCHWEFLIGSAURES-**: potassium sulphite (see **KALIUMSULFIT**) 591

KALISEIFE (*f*): potash soap 592

KALISPRENGPULVER (*n*): potash (gunpowder) bursting charge (for mining) 593

KALISUPERPHOSPHAT (*n*): potash superphosphate 594

KALI (*n*), **ÜBERMANGANSAURES-**: potassium permanganate (see **KALIUMPERMANGANAT**) 595

KALI (*n*), **ÜBERSCHWEFELSAURES-**: potassium persulphate (see **KALIUMPERSULFAT**) 596

KALIUM (*n*): potassium (K) 597

KALIUMACETAT (*n*): potassium acetate (*Kalium aceticum*); $KC_2H_3O_2$; Mp. 292°C. 598

KALIUMACETAT (*n*), **EINFACHSAURES-**: monobasic potassium acetate; $KC_2H_3O_2.C_2H_4O_2$; Mp. 148°C. 599

KALIUMACETATLÖSUNG (*f*): potassium acetate solution (*Liquor Kalii acetici*); Sg. 1.178; about 33% $CH_3.CO_2K$ 600

KALIUMACETAT (*n*), **NEUTRALES-**: neutral potassium acetate; $KC_2H_3O_2$; Mp. 292°C. 601

KALIUMACETAT (*n*), **ZWEIFACHSAURES-**: dibasic potassium acetate; $KC_2H_3O_2.2C_2H_2O_2$; Sg. 1.47; Mp. 112°C. 602

KALIUMALAUN (*m*): potassium alum; $K_2SO_4.Al_2(SO_4)_3$ plus $24H_2O$ (see also **ALAUN**) 603

KALIUMALUMINAT (*n*): potassium aluminate 604

KALIUM (*n*), **AMEISENSAURES-**: potassium formate 605

KALIUM (*n*), **AMYLSCHWEFELSAURES-**: potassium amylsulphate 606

KALIUM (*n*), **ANTIMONSAURES-**: potassium antimonate 607

KALIUMANTIMONYLTARTRAT (*n*): tartar emetic, potassium antimonyl tartrate (see **BRECH-WEINSTEIN**) 608

KALIUMARSENIAT (*n*): (*Kalium arsenicicum*): potassium arsenate; KH_2AsO_4 609

KALIUM (*n*), **ARSENIGSAURES-**: potassium arsenite; $KAsO_2$ 610

KALIUM (*n*), **ARSENSAURES-**: potassium arsenate (see **KALIUMARSENIAT**) 611

KALIUM (*n*), **ÄTHYLSCHWEFELSAURES-**: potassium ethylsulphate 612

KALIUM (*n*), **BALDRIANSAURES-**: potassium valerate 613

KALIUM (*n*), **BENZOESAURES-**: potassium benzoate 614

KALIUMBICARBONAT (*n*): potassium bicarbonate, potassium acid carbonate, baking soda (*Kalium bicarbonicum*); $KHCO_3$; Sg. 2.17 615

KALIUMBICHROMAT (*n*): potassium bichromate (see **CHROMKALI**) 616

KALIUMBIOXALAT (*n*): (*Kalium bioxalicum*): potassium binoxalate, acid potassium oxalate, salt of sorrel, *Sal acetosella*, essential salt of lemon; $KHC_2O_4.H_2O$ (Sg. anhydrous, 2.09) 617

KALIUMBISULFAT (*n*): (*Kalium bisulfuricum*): acid potassium sulphate, potassium bisulphate; $KHSO_4$; Sg. 2.355 618

KALIUMBISULFIT (*n*): (*Kalium bisulfurosum*): potassium acid sulphite, potassium bisulphite; $KHSO_3$ 619

KALIUMBITARTRAT (*n*): potassium bitartrate (see **KALIUM, SAURES WEINSTEINSAURES-**) 620

KALIUMBORFLUORID (*n*): potassium-boron fluoride; KBF_4; Sg. 2.498 621

KALIUMBRECHWEINSTEIN (*m*): antimony-potassium tartrate (see **BRECHWEINSTEIN**) 622

KALIUMBROMAT (*n*): potassium bromate; $KBrO_3$; Sg. 2.34-3.24; Mp. 424—434°C. 623

KALIUMBROMID (*n*): (*Kalium bromatum*); potassium bromide; KBr; Mp. 730°C.; Bp. 1435°C.; Sg. 2.73-2.756 624

KALIUM (*n*), **BROMSAURES-**: potassium bromate (see **KALIUMBROMAT**) 625

KALIUMCARBONAT (*n*): (*Kalium carbonicum*): potassium carbonate, potash, salts of tartar, pearl-ash; K_2CO_3; Sg. 2.29; Mp. 909°C.; $K_2CO_3+2H_2O$; Sg. 2.043 626

KALIUMCARBONATLÖSUNG (*f*): (*Liquor Kalii carbonici*): potassium carbonate solution; Sg. 1.336; about 33% K_2CO_3 (see **KALIUM-CARBONAT**) 627

KALIUMCARBONAT (*n*), **NEUTRALES-**: (see **KALIUM-CARBONAT**) 628

KALIUMCHLORAT (*n*): (*Kalium chloricum*; potassium chlorate, potassium oxymuriate; $KClO_3$; Sg. 2.337; Mp. 357°C. 629

KALIUM (*n*), **CHLORCHROMSAURES-**: potassium chlorochromate 630

KALIUMCHLORID (*n*): (*Kalium chloratum*): potassium chloride; KCl; Mp. 772°C.; Bp. 1500°C.; Sg. 1.977-1.989 631

KALIUM (*n*), **CHLORSAURES-**: potassium chlorate; $KClO_3$; (see **KALIUMCHLORAT**) 632

KALIUMCHROMALAUN (*m*): chrome alum (see **CHROMALAUN**) 633

KALIUMCHROMAT (*n*): (*Kalium chromicum*); potassium chromate; K_2CrO_4; Mp. 971°C.; Sg. 2.731 634

KALIUMCHROMAT (*n*), **NEUTRALES-**: (see **KALIUM-CHROMAT**) 635

KALIUM (*n*), **CHROMSAURES-**: potassium chromate; K_2CrO_4 (see **KALIUMCHROMAT**) 636

KALIUM (*n*), **CITRONENSAURES-**: potassium citrate; $K_3C_6H_5O_7.H_2O$; Sg. 1.98 637

KALIUMCYANAT (*n*): potassium cyanate; KO(CN) or KOCy; Sg. 2.048 638

KALIUMCYANID (*n*): potassium cyanide (see **CYANKALIUM**) 639

KALIUM (*n*), **CYANSAURES-**: potassium cyanate 640

KALIUMDICHROMAT (*n*): (see also **CHROMKALI**): bichromate of potassium; red potassium chromate; $K_2Cr_2O_7$; Sg. 2.7; Mp. 396°C. 641

KALIUMDIHYDROARSENAT (*n*): potassium dihydroarsenate; KH_2AsO_4; Sg. 2.85-2.87 642

KALIUMDIHYDROPHOSPHAT (*n*): potassium phosphate, monobasic; KH_2PO_4; Sg. 2.338; Mp. 96°C. 643

KALIUMDITHIONAT (*n*): potassium dithionate; $K_2S_2O_6$; Sg. 2.278 644

KALIUMDOPPELSALZ (*n*): potassium double salt (see **KAINIT** and **KARNALLIT**) 645

KALIUM, DOPPELTKOHLENSAURES- (*n*): potassium bicarbonate (see **KALIUMBICARBONAT**) 646

KALIUMEISENCYANID (*n*): potassium ferricyanide; $K_6Fe_2(CN)_{12}$ (see **FERRICYAN-KALIUM**) 647

KALIUMEISENCYANÜR (*n*): potassium ferrocyanide; $K_4Fe(CN)_6$ plus $3H_2O$ (see **FERRO-CYANKALIUM**) 648

KALIUMEISENSULFAT (*n*): (see **EINSENALAUN**) 649

KALIUM (*n*), **ESSIGSAURES-**: (*Kalium aceticum*): potassium acetate; $KC_2H_3O_2$; Mp. 292°C. 650

KALIUM (*n*), **ESSIGWOLFRAMSAURES-**: potassium acetotungstate 651

KALIUMFERRAT (*n*): potassium ferrate (see **KALI, EISENSAURES-**) 652

KALIUMFERRICYANID (*n*): potassium ferricyanide (see **FERRICYANKALIUM**) 653

KALIUMFERRICYANÜR (*n*): potassium ferricyanide (see **BLUTLAUGENSALZ, ROTES-**) 654

KALIUMFERRIOXALAT (*n*): ferric-potassium oxalate, iron potassium oxalate 655

KALIUMFERRISULFAT (*n*): ferric potassium sulphate (see **EISENALAUN**) 656

KALIUMFERROCYANID (*n*): potassium ferrocyanide (see **FERROCYANKALIUM**) 657

KALIUMFERROCYANÜR (*n*): potassium ferrocyanide (see **BLUTLAUGENSALZ** and **BLUTLAUGEN-SALZ, GELBES-**) 658

KALIUM (*n*), **FETTSAURES-**: potassium sebate 659

KALIUMFLUORID (*n*): (*Kalium fluoratum*): potassium fluoride; KF; Mp. about 800°C.; Sg. 2.454-2.481 660

KALIUMGOLDBROMÜR (*n*): potassium aurobromide 661

KALIUMGOLDCYANID (*n*): potassium auricyanide 662

KALIUMGOLDCYANÜR (*n*): potassium (cyanaurate) aurocyanide; $KAu(CN)_2$ 663

KALIUM (*n*), **GUAJAKOLSULFONSAURES-**: potassium sulfoguaiacolate 664

KALIUMHALOGEN (*n*): potassium halide 665

KALIUMHYDRAT (*n*): potassium hydroxide; KOH (*Kalium hydratum*) (see **KALIUMHYDR-OXYD**) 666

KALIUMHYDRID (*n*): potassium hydride; KH; Sg. 0.8 667

KALIUMHYDROCARBONAT (*n*): potassium hydrocarbonate (see **KALIUMBICARBONAT**); $KHCO_3$; Sg. 2.17 668

KALIUMHYDROSULFAT (*n*): potassium hydrosulphate (see **KALIUMBISULPHAT**); $KHSO_4$; Sg. 2·355 669

KALIUMHYDROXYD (*n*): (*Kalium causticum*): caustic potash, potassium hydrate (*Kalium hydratum*); potassium hydroxide, potassa; KOH; Mp. 360.4°C.; Sg. 2.044; KOH+H_2O; Sg. 1.987 670

KALIUMHYDRÜR (*n*): potassium subhydride; K_2H 671

KALIUMHYPOCHLORIT (*n*): potassium hypochlorite; KOCl 672

KALIUMIRIDICHLORID (*n*): potassium (iridichloride) chloriridate 673

KALIUMIRIDIUMCHLORID (*n*): potassium iridium chloride; K_2IrCl_6; Sg. 3.546 674

KALIUM (*n*), **ISOBUTTERSAURES-**: potassium isobutyrate 675

KALIUM (*n*), **ISOBUTYLSCHWEFELSAURES-**: potassium isobutylsulphate 676

KALIUMJODAT (*n*): potassium iodate; KIO_3; Sg. 3.89 677

KALIUMJODID (*n*): potassium iodide (*Kalium jodatum*) (KI); Sg. 3.1; Mp. 680°C.; Bp. 1420°C. 678

KALIUMKARBONAT (*n*): (see **KALIUMCARBONAT**) 679

KALIUM (*n*), **KIESELFLUOR-**: potassium fluosilicate 680

KALIUMKOBALTNITRIT (*n*): (*Cobalti-Kalium nitrosum*): cobalt-potassium nitrite, Indian yellow, cobalt yellow, potassium cobalt nitrite (double compound of cobalt nitrite and potassium nitrite); $2CoK_3(NO_2)_6.3H_2O$ or $Co_2(NO_2).6KNO_2.3H_2O$ or $Co(NO_2)_3.3KNO_2.3H_2O$ 681

KALIUMKOHLENOXYD (*n*): potassium (carboxide) hexacarbonyl 682

KALIUM (*n*), **KRESOTSULFONSAURES-**: potassium sulfocresotate 683

KALIUMKUPFERCYANÜR (*n*): potassium cuprocyanide 684

KALIUMLAKTAT (*n*): potassium lactate (*Kalium lacticum*); $C_3H_5O_3K$ 685

KALIUMMAGNESIUMCARBONAT (*n*): potassium-magnesium carbonate; $K_2CO_3.MgCO_3$ 686

KALIUMMAGNESIUMSULFAT (*n*): (*Kalio-magnesium sulfuricum*): potassium-magnesium sulphate; $K_2SO_4.MgSO_4$ 687

KALIUM (*n*), **MANGANSAURES-**: potassium manganate; K_2MnO_4 688

KALIUMMETABISULFIT (*n*): potassium (pyrosulphite) metabisulphite; $K_2S_2O_5$ 689

KALIUMMETALL (*n*): potassium metal 690

KALIUMMETAPHOSPHAT (*n*): potassium metaphosphate; KPO_3; Sg. 2.258 691

KALIUMMETASULFIT (*n*): potassium metasulphite 602

KALIUMMETAWOLFRAMAT (*n*): potassium metatungstate, potassium metawolframate; $K_2W_4O_{13}$ 693

KALIUM (*n*), **MILCHSAURES-**: potassium lactate (see **KALIUMLAKTAT**) 694

KALIUM (*n*): **MOLYBDÄNSAURES-**: potassium molybdate (see **KALIUMMOLYBDAT**) 695

KALIUMMOLYBDAT (*n*): (*Kalium molybdænicum*): potassium molybdate (usually potassium trimolybdate, see **KALIUMTRIMOLYBDAT**) 696

KALIUMMONOSULFID (*n*): potassium monosulphide; potassium (sulphide) sulfuret; K_2S; Sg. 2.13 697

KALIUM (*n*), **MYRONSAURES-**: potassium myronate; sinigrin; $C_{10}H_{18}KNS_2O_{10}$ 698

KALIUMNATRIUMCARBONAT (*n*): potassium-sodium carbonate; $K_2CO_3.Na_2CO_3$ 699

KALIUMNATRIUMKARBONAT (*n*): potassium-sodium carbonate; $KNaCO_3.6H_2O$ 700

KALIUMNATRIUMTARTRAT (*n*): (*Kalio-Natrium tartaricum*): potassium-sodium tartrate (*Tartarus natronatus*), Seignette salt, Rochelle salt; $KNaC_4H_4O_6.4H_2O$; Sg. 1.77; Mp. about 75°C. 701

KALIUMNITRAT (*n*): (*Kalium nitricum*); potassium nitrate, nitre, saltpetre, KNO_3; Sg. 2.092-2.109; Mp. 337°C. 702

KALIUMNITRIT (*n*): (*Kalium nitrosum*): potassium nitrite; KNO_2; Sg. 1.915 703

KALIUMORTHOPHOSPHAT (*n*), **NEUTRALES-**: neutral potassium orthophosphate; K_3PO_4 (see also **KALIUMPHOSPHAT**) 704

KALIUM (*n*), **OSMIUMSAURES-**: potassium osmate 705

KALIUMOXALAT (*n*): potassium oxalate (see KALIUMOXALAT, NEUTRALES-; KALIUM-BIOXALAT; KALIUMTETRAOXALAT) 706

KALIUMOXALAT (*n*), NEUTRALES-: (*Kalium oxalicum*): potassium oxalate, neutral potassium oxalate; $K_2C_2O_4.H_2O$; Sg. 2.08 707

KALIUMOXALAT (*n*), SAURES-: (see KALIUM-BIOXALAT) 708

KALIUM (*n*), OXALSAURES-: potassium oxalate (see KALIUMOXALAT) 709

KALIUMOXYD (*n*): (*Kalium oxydatum*); potassium oxide; K_2O; Sg. 2.32-2.656 710

KALIUMOXYDHYDRAT (*n*): potassium hydroxide (see KALIUMHYDROXYD) 711

KALIUMPALLADIUMCHLORID (*n*): potassium-palladium chloride; potassium-palladic chloride; K_2PdCl_6; Sg. 2.77 712

KALIUMPALLADIUMCHLORÜR (*n*): potassium-palladious chloride; PdK_2Cl_4 713

KALIUMPARAWOLFRAMAT (*n*) potassium paratungstate, potassium parawolframate; $K_{10}W_{12}O_{41}$ 714

KALIUMPENTATHIONAT (*n*): potassium pentathionate; $K_2S_5O_6$; Sg. 2.1123 715

KALIUMPERCARBONAT (*n*): potassium percarbonate (see KALIUM, ÜBERKOHLENSAURES-) 716

KALIUMPERCHLORAT (*n*): potassium perchlorate; $KClO_4$; Sg. 2.52-2.524 717

KALIUMPERJODAT (*n*): potassium periodate; KIO_4; Sg. 3.618 718

KALIUMPERMANGANAT (*n*): (*Kalium hypermanganicum*); potassium permanganate; $KMnO_4$; Sg. 2.7032 719

KALIUMPERSULFAT (*n*): (*Kalium persulfuricum*): potassium persulphate (from electrolysis of potassium sulphate); $K_2S_2O_8$ 720

KALIUMPHOSPHAT (*n*): (*Kalium phosphoricum*): potassium phosphate (see KALIUMORTHO-PHOSPHAT, NEUTRALES-) 721

KALIUMPHOSPHAT (*n*), EINFACHSAURES-: dibasic potassium phosphate (see DIKALIUMPHOS-PHAT) 722

KALIUMPHOSPHAT (*n*), ZWEIFACHSAURES-: monobasic potassium phosphate (see MONO-KALIUMPHOSPHAT) 723

KALIUM (*n*), PHOSPHORSAURES-: potassium phosphate (see KALIUMPHOSPHAT) 724

KALIUMPLATINCHLORID (*n*): potassium (platinichloride) chloroplatinate; platinum-potassium chloride; K_2PtCl_6 725

KALIUMPLATINCHLORÜR (*n*): potassium chloroplatinite; K_2PtCl_4; Sg. 3.29 726

KALIUMPLATINCYANÜR (*n*): potassium platinocyanide 727

KALIUMPLATINOCHLORID (*n*): potassium (platinochloride) chloroplatinite; K_2PtCl_4; Sg. 3.29 728

KALIUM (*n*), PROPIONSAURES-: potassium propionate 729

KALIUM (*n*), PROPYLSCHWEFELSAURES-: potassium propylsulphate 730

KALIUMPYROSULFIT (*n*): potassium pyrosulphite (see KALIUMMETABISULFIT) 731

KALIUMQUECKSILBERCYANID (*n*): potassium mercuricyanide 732

KALIUMQUECKSILBERCYANÜR (*n*): potassium mercurocyanide 733

KALIUMQUECKSILBERJODID (*n*): potassium mercuriiodide 734

KALIUMREGULUS (*m*): potassium regulus, regulus of potassium (see KALIUM) 735

KALIUMRHODANAT (*n*): potassium (thiocyanate) sulfocyanate 736

KALIUMRHODANID (*n*): potassium (thiocyanate) sulfocyanate 737

KALIUMRHODANÜR (*n*): (*Kalium rhodanatum*): potassium sulfocyanide, potassium sulfocyanate, potassium rhodanide; potassium thiocyanate; KCNS; Sg. 1.886-1.906; Mp. 161.2-172.3°C. 738

KALIUM (*n*), SALICYLSAURES-: potassium salicylate 739

KALIUM (*n*), SALPETERSAURES-: potassium nitrate, saltpetre (see KALIUMNITRAT) 740

KALIUM (*n*), SALPETRIGSAURES-: potassium nitrite (see KALIUMNITRIT) 741

KALIUMSALZ (*n*): potassium salt 742

KALIUMSALZ (*n*), SAURES-: potassium acid-salt, acid potassium-salt 743

KALIUM (*n*), SAURES CHROMSAURES-: potassium bichromate (see CHROMKALI) 744

KALIUM (*n*), SAURES KOHLENSAURES-: potassium bicarbonate (see KALIUMBICARBONAT) 745

KALIUM (*n*), SAURES OXALSAURES-: potassium binoxalate, salt of sorrel, acid potassium oxalate (see KALIUMBIOXALAT) 746

KALIUM (*n*), SAURES SCHWEFLIGSAURES-: potassium bisulphite (see KALIUMBISULFIT) 747

KALIUM (*n*), SAURES WEINSTEINSAURES-: bitartrate of potash, potassium acid tartrate, argol, tartar (*Tartarus*), cream of tartar (*Cremor tartari*), tartar acid potassium tartrate, potassium bitartrate; $KC_4H_5O_6$; Sg. 1.956 748

KALIUM (*n*), SCHWEFELSAURES-: potassium sulphate; K_2SO_4 (see KALIUMSULFAT, NEUTRALES-) 749

KALIUM (*n*), SCHWEFLIGSAURES-: potassium sulphite (see KALIUMSULFIT) 750

KALIUMSELENAT (*n*): potassium selenate; K_2SeO_4; Sg. 3.066 751

KALIUMSILBERCYANÜR (*n*): potassium argentocyanide 752

KALIUMSILICOFLUORID (*n*): potassium (fluosilicate) silicofluoride; K_2SiF_6; Sg. 2.665 753

KALIUMSILIKAT (*n*): potassium silicate 754

KALIUMSULFAT (*n*): (see KALIUMSULFAT, NEUTRALES- and KALIUMBISULFAT) 755

KALIUMSULFAT (*n*), NEUTRALES-: (*Kalium sulfuricum*): potassium sulphate, *tartarus vitriolatus*, *arcanum duplicatum*; K_2SO_4; Sg. 2.652-2.67; Mp. 1072°C. 756

KALIUMSULFID (*f*): (*Kalium sulfuratum*): potassium sulphide, potassium sulfuret K_2S; Sg. 1.805 757

KALIUMSULFIT (*n*): (*Kalium sulfurosum*): potassium sulphite; K_2SO_3 758

KALIUMSULFIT (*n*), SAURES-: potassium bisulphite (see KALIUMBISULFIT) 759

KALIUMSULFOCREOSOT (*n*): potassium sulfocreosote 760

KALIUMSULFOCYANAT (*n*): potassium sulfocyanide; potassium thiocyanate; potassium sulfocyanate; potassium rhodanide; KS(CN) or KSCy or KCNS (see KALIUM-RHODANÜR) 761

KALIUMSULFOGUAJAKOL (*n*): potassium sulfoguaiacol 762

KALIUMSULFURET (*n*): potassium sulfuret (see KALIUMSULFID) 763

KALIUMTARTRAT (*n*): potassium tartrate (see KALIUM, SAURES WEINSTEINSAURES-) 764
KALIUMTETRACHROMAT (*n*): potassium tetrachromate; $K_2Cr_4O_{13}$; Sg. 2.649 765
KALIUMTETRAOXALAT (*n*): (*Kalium tetroxalicum*) potassium tetroxalate, commercial salt of sorrel; $KH_3(C_2O_4)_2.2H_2O$ 766
KALIUMTETRATHIONAT (*n*): potassium tetrathionate; $K_2S_4O_6$; Sg. 2.2963 767
KALIUMTRICHROMAT (*n*): trichromate of potassium; $K_2Cr_3O_{10}$; Sg. 2.648 768
KALIUMTRIJODID (*n*): potassium triiodide; KI_3; Sg. 3.498 769
KALIUMTRIMOLYBDAT (*n*): potassium trimolybdate; $K_2Mo_3O_{10}$ (see also KALIUMMOLYBDAT) 770
KALIUMTRITHIONAT (*n*): potassium trithionate; $K_2S_3O_6$; Sg. 2.304 771
KALIUM (*n*), **ÜBERCHLORSAURES-**: potassium perchlorate (see KALIUMPERCHLORAT) 772
KALIUM (*n*), **ÜBERKOHLENSAURES-**: (*Kalium percarbonicum*): potassium percarbonate; $K_2C_2O_6$ 773
KALIUM (*n*), **ÜBERJODSAURES-**: potassium periodate (see KALIUMPERJODAT) 774
KALIUM (*n*), **ÜBERMANGANSAURES-**: potassium permanganate (see KALIUMPERMANGANAT) 775
KALIUM (*n*), **ÜBERSCHWEFELSAURES-**: potassium persulphate (see KALIUMPERSULFAT) 776
KALIUM (*n*), **UNTERCHLORIGSAURES-**: potassium hypochlorite (see KALIUMHYPOCHLORIT) 777
KALIUM (*n*), **UNTERPHOSPHORIGSAURES-**: potassium hypophosphite 778
KALIUM (*n*), **UNTERSCHWEFLIGSAURES-**: potassium hyposulphite 779
KALIUMVERBINDUNG (*f*): potassium compound 780
KALIUMWASSERGLAS (*n*): potassium silicate 781
KALIUMWASSERSTOFF (*m*): potassium hydride (see KALIUMHYDRID) 782
KALIUM (*n*), **WEINSAURES-**; potassium tartrate (see KALIUMTARTRAT) 783
KALIUMWOLFRAMAT (*n*): (*Kalium wolframicum*): potassium tungstate, potassium wolframate; K_2WO_4 784
KALIUM (*n*), **WOLFRAMSAURES-**: potassium tungstate (see KALIUMWOLFRAMAT) 785
KALIUMXANTHOGENAT (*n*): potassium xanthogenate; $CS(OC_2H_5)SK$ 786
KALIUM (*n*), **XANTHOGENSAURES-**: potassium xanthogenate (see KALIUMXANTHOGENAT) 787
KALIUMZINKCYANÜR (*n*): cyanide of zinc and potassium 788
KALIUMZYANID (*n*): potassium cyanide (see CYANKALIUM) 789
KALIVERBINDUNG (*f*): potash (potassium) compound 790
KALI (*n*), **VIERFACH OXALSAURES-**: potassium tetroxalate (see KALIUMTETRAOXALAT) 791
KALIWASSERGLAS (*n*): potassium silicate (45 parts quartzsand, 30 parts potash and 25 parts carbon) 792
KALI (*n*), **WOLFRAMSAURES-**: potassium tungstate (see KALIUMWOLFRAMAT) 793
KALI (*n*), **XANTHOGENSAURES-**: (see KALIUMXANTHOGENAT) 794
KALK (*m*): lime (*Calcaria*); calcium; limestone; chalk; calx (old chemical term) (see CALCIUMCARBONAT and CALCIUMOXYD) 795
KALKABSATZTEICH (*m*): chalk (lime) settling trough 796
KALKAMMONSALPETER (*m*): calcium-ammonium nitrate 797
KALKANSTRICH (*m*): whitewash 798
KALKANWURF (*m*): plaster; parget 799
KALKARM: deficient in lime; poor in lime content 800
KALK (*m*), **ARSENSAURER-**: calcium arsenate; $Ca_3(ASO_4)_2$ 801
KALKARTIG: calcareous 802
KALKARTIGKEIT (*f*): chalkiness; calcareousness 803
KALKÄSCHER (*m*): lixiviated ashes (Soap); soap waste; lime pit (Leather) 804
KALKBACK (*m*): lime back (Sugar) 805
KALKBEUCHE (*f*): lime (boil; buck; steep) bucking lye 806
KALK BEWERFEN, MIT-: to plaster, roughcast, coat with plaster 807
KALKBEWURF (*m*): coat of plaster; plastering; roughcast 808
KALKBLAU (*n*): blue verditer; Neuwied blue (mixture of $Cu(OH)_2$ and $CaSO_4$) (see KUPFERBLAU) 809
KALKBODEN (*m*): lime (chalky) soil; chalky ground; calcareous soil 810
KALKBORAT (*n*): calcium borate (see CALCIUMBORAT) 811
KALKBREI (*m*): lime (paste) cream (from lime slaking) 812
KALKBRENNEN: to burn (calcine) lime; (*n*): lime (calcining) burning 813
KALKBRENNER (*m*): lime burner 814
KALKBRENNEREI (*f*): lime kiln 815
KALKBRUCH (*m*): limestone quarry; chalk pit 816
KALKBRÜHE (*f*): milk of lime (see KALKMILCH); whitewash; lime water (see KALKWASSER) 817
KALK (*m*), **CARBOLSAURER-**: lime (calcium) phenylate 818
KALK (*m*), **CHROMSAURER-**: calcium chromate (see CALCIUMCHROMAT) 819
KALK (*m*), **CITRONENSAURER-**: lime (calcium) citrate (see CALCIUMZITRAT) 820
KALK (*m*), **DOPPELTSCHWEFLIGSAURER-**: lime (calcium) bisulphite (see CALCIUMBISULFIT) 821
KÄLKE (*f*): lime (Leather) 822
KALKECHT: fast to lime 823
KALKEISENSTEIN (*m*): ferruginous limestone 824
KALKEN: to lime; mix with lime; steep (soak) in lime water; cover (dress) with lime 825
KÄLKEN: (see KALKEN) 826
KALKENTFERNUNG (*f*): extraction (removal) of lime (see KALK) 827
KALKERDE (*f*): lime; calcium oxide; calcareous earth (see CALCIUMOXYD) 828
KALK (*m*), **ESSIGSAURER-**: lime (calcium) acetate (see CALCIUMAZETAT) 829
KALKFÄLLUNG (*f*): lime precipitation 830
KALKFARBE (*f*): lime colour (lime and colour mixed) 831
KALKFELDSPAT (*m*): Anorthite (see ANORTHIT) 832
KALKFELS (*m*): calcareous rock 833
KALKFLIESS (*m*): schorl; Sg. 3.17 834
KALKFREI: free from lime; lime-free 835
KALK (*m*), **GARER-**: slaked lime 836
KALKGEBIRGE (*n*): calcareous (chalky) formation or mountain chain 837
KALK (*m*), **GEBRANNTER-**: quicklime; Sg. 0.8 (see CALCIUMOXYD) 838

KALK, GELÖSCHTER- (*m*) : slaked lime ; $Ca(OH)_2$ (see CALCIUMHYDROXYD) 839
KALKGLAS (*n*) : lime glass 840
KALKGLIMMER (*m*) : lime mica, margarite ; CaO,Al_2O_3,SiO_2,H_2O ; Sg. 2.95 841
KALKGRANAT (*m*) : lime garnet ; grossularite (see GROSSULAR) 842
KALKGUSS (*m*) : grout-(ing) ; watery lime mortar 843
KALKHALTIG : containing lime ; calcareous 844
KALK (*n*), **HOLZESSIGSAURER-** : pyrolignite of lime ; calcium pyrolignite (crude Calcium acetate) 845
KALKHÜTTE (*f*) : lime kiln 846
KALKHYDRAT (*n*) : hydrate of lime ; calcium hydroxide (see CALCIUMHYDROXYD) 847
KALK (*m*), **HYDRAULISCHER-** : hydraulic (cement) lime ; Sg. 2.75 848
KALKICHT : like lime 849
KALKIG : calcareous ; chalky ; containing lime 850
KALKKASTEN (*m*) : lime chest 851
KALKKELLE (*f*) : trowel 852
KALK (*m*), **KOHLENSAURER-** : calcium carbonate, chalk (see CALCIUMCARBONAT) 853
KALK (*m*), **KOHLENSAURER PRÄCIPITIERTER-** : precipitated chalk (*Calcium carbonicum præcipitatum*) 854
KALKLICHT (*n*) : lime-light ; calcium light ; Drummond light 855
KALKLICHT (*n*), **DRUMMONDS-** : Drummond's light 856
KALKLÖSCHAPPARAT (*m*) : lime slaking apparatus 857
KALK (*m*), **LÖSCHBARER-** : slakable lime (see WEISSKALK) 858
KALKLÖSCHEN : to slake lime ; (*n*) : lime slaking 859
KALKLÖSCHTROMMEL (*f*) : lime slaking drum 860
KALKLÖSUNG (*f*) : solution of lime ; calcium hydroxide (see CALCIUMHYDROXYD) 861
KALK (*m*), **MAGERER-** : lime which does not slake easily 862
KALKMALEREI (*f*) : fresco painting 863
KALKMEHL (*n*) : lime powder ; air slaked lime 864
KALKMERGEL (*m*) : lime (calcareous) marl 865
KALKMESSER (*m*) : calcimeter (for determination of CO_2 in limestone etc.) 866
KALKMILCH (*f*) : milk of lime ; $Ca(OH)_2$ (from lime slaking) 867
KALK (*m*), **MILCHSAURER-** : calcium lactate (see CALCIUMLAKTAT) 868
KALKMÖRTEL (*m*) : lime mortar (a mixture of slaked lime and sand) 869
KALKMULDE (*f*) : trough (hod) for mortar 870
KALKOFEN (*m*) : lime kiln 871
KALK (*m*), **OXALSAURER-** : calcium oxalate (see CALCIUMOXALAT) 872
KALK (*m*), **PHOSPHORMILCHSAURER-** : calcium lactophosphate (*Calcium phospho-lacticum*) (mixture of calcium lactate, calcium acid lactate and calcium acid phosphate) 873
KALK (*m*), **PHOSPHORSAURER-** : calcium phosphate (see CALCIUMPHOSPHAT) 874
KALKSALPETER (*m*) : calcium nitrate (see CALCIUM-NITRAT) 875
KALK (*m*), **SALPETERSAURER-** : calcium nitrate (see CALCIUMNITRAT) 876
KALKSALZ (*n*) : lime salt ; calcium salt 877
KALKSAND (*m*) : calcareous sand 878
KALKSANDSTEIN (*m*) : calcareous (calciferous) sandstone (Geol.) ; sand-lime brick 879
KALKSANDZIEGEL (*m*) : sand-lime brick (Ceramics) 880
KALKSAUERBAD (*n*) : lime sour bath (Bleaching) 881
KALK (*m*), **SAURER SCHWEFLIGSAURER-** : calcium bisulphite (see CALCIUMBISULFIT) 882
KALKSCHAUM(*m*) : lime(froth) scum 883
KALKSCHIEFER (*m*) : calcareous (shale) slate 884
KALKSCHLAMM (*m*) : lime (sludge) mud 885
KALKSCHLEIER (*m*) : lime-coating (a white powdery deposit of lime) 886
KALKSCHWEFELLEBER (*f*) : sulphurated lime ; calcic liver of sulphur ; lime hepar (see CALCIUMSULFID) 887
KALK (*m*), **SCHWEFELSAURER-** : calcium sulphate (see CALCIUMSULFAT) 888
KALK, SCHWEFLIGSAURER- (*m*) : lime (calcium) sulphite (see CALCIUMSULFIT) 889
KALKSEIFE (*f*) : lime (calcium) soap 890
KALKSINTER (*m*) : calcareous (calc) sinter ; stalactite ; calcareous tufa (loose porous carbonate of lime deposit) 891
KALKSPAT (*m*) : calcspar, calcite, calcium carbonate, calcareous spar ; $CaCO_3$ (about 56% CaO) ; Sg. 2.712-2.715 (see also CALCIUMCARBONAT) 892
KALKSPAT (*m*), **DICHTER-** : (see KALKSTEIN) 893
KALKSPAT (*m*), **ERDIGER-** : (see KREIDE) 894
KALKSPAT (*m*), **FASERIGER-** : fibrous calcite (see FASERKALK) 895
KALKSPAT (*m*), **KRISTALLINISCHER-** : marble (see MARMOR) 896
KALKSPAT (*m*), **KRISTALLISIERTER-** : (see KALKSPAT) 897
KALKSTEIN (*m*) : limestone (Sg. green = 3.18 ; white, = 3.16) 898
KALKSTEINGUT (*n*) ; limestone whiteware 899
KALKSTICKSTOFF (*m*) : crude calcium cyanamide ; lime nitrogen ; nitrolim (see CALCIUM-ZYANAMID) 900
KALK (*m*), **SULFOCARBOLSAURER-** : calcium sulfocarbolate, calcium sulphophenate, calcium sulfophenate, calcium sulfophenylate, calcium phenolsulfonate ; $Ca(C_6H_4OHSO_3)_2.5H_2O$ 901
KALKTIEGEL (*m*) : lime crucible 902
KALKTONGRANAT (*m*) : grossularite (see GROS-SULAR) 903
KALK (*m*), **ÜBERBORSAURER-** : (see CALCIUMPER-BORAT) 904
KALK (*m*), **ÜBERMANGANSAURER-** : (see CALCIUM-PERMANGANAT) 905
KALKULATION (*f*) : calculation 906
KALKULATOR (*m*) : calculator 907
KALKULATOR (*m*), **NACH-** : checker (one who checks a calculation) 908
KALKULATOR (*m*), **VOR-** : calculator (one who gets out a calculation) 909
KALK (*m*), **UNGELÖSCHTER-** : quicklime ; Sg. 0.8 (see CALCIUMOXYD) 910
KALK (*m*), **UNLÖSCHBARER-** : unslakable lime (see MAGERER KALK) 911
KALK, UNTERPHOSPHORIGSAURER- (*m*) : lime hypophosphite ; $Ca(H_2PO_2)_2$ 912
KALK (*m*), **UNTERSCHWEFLIGSAURER-** : (see CALCIUMTHIOSULFAT) 913
KALKURANIT (*m*) : calcouranite, autunite ; CaO, $2U_2O_3,P_2O_5.8H_2O$; Sg. 3.1 (see also AUTUNIT) 914

KALK, VANADINSAURER- (*m*): calcium vanadate 915
KALKVERBINDUNG (*f*): lime (calcium) compound (see CALCIUM) 916
KALK (*m*), **VERWITTERTER-**: air-slaked lime 917
KALKWAND (*f*): plaster wall 918
KALKWASSER (*n*): lime water (calcium hydroxide) ($Ca(OH)_2$ solution, from agitation of slaked lime with water) 919
KALK, WEINSAURER- (*m*): lime (calcium) tartrate (see CALCIUM, WEINSTEINSAURES-) 920
KALKWEISSE (*f*): lime-wash; lime-water 921
KALLAIT (*m*): callaite; $H(Al_2,Ca_3,Cu_3,Fe_3)(OH)_4PO_4$; Sg. 2.75 (see also TÜRKIS) 922
KALLIGRAPH (*m*): calligraphist 923
KALLIGRAPHIE (*f*): calligraphy 924
KALLILITH (*m*): Kallilite (a bismuth, nickel sulphide) (Min.) 925
KALLIT (*m*): similar to turquoise; Sg. 2.8 926
KALM: calm, quiet, still 927
KALMER (*m*): squid, cuttle-fish (*Sepia officinalis*) 928
KALMIE (*f*): mountain laurel; American laurel; calico bush (*Kalmia latifolia*) 929
KALMUS (*m*): calamus; calmus; sweet (flag; grass) cane (Rhizome of *Acorus calamus*) 930
KALMUS (*m*), **JAPANISCHER-**: Japanese Calamus (*Acorus spuriosus*) 931
KALMUSKAMPHER (*m*): calamus (calmus) camphor; $C_{15}H_{26}O_2$ (from the distillation of calamus oil); Mp. 168°C. 932
KALMUSÖL (*n*): calamus oil; sweet flag oil; Sg. 0.97; Bp. 170-300°C. (from *Acorus calamus*) 933
KALMUSWURZEL (*f*): calamus root; flag root (root of *Acorus calamus*) (*Rhizoma calami*) 934
KALMUSWURZEL (*f*), **JAPANISCHE-**: Japanese calamus root (root of *Acorus spuriosus*) 935
KALMUSWURZELÖL (*n*), **JAPANISCHES-**: Japanese calamus root oil (from root of *Acorus spuriosus*); Sg. 0.992; Bp. 210-290°C. 936
KALOCHROM (*m*): red lead ore, crocoite (see ROTBLEIERZ) 937
KALOMEL (*n*): calomel (see QUECKSILBER-CHLORÜR) 938
KALOMEL (*n*), **KOLLOIDALES-**: colloidal calomel (see KALOMELOL) 939
KALOMELOL (*n*): calomelol, colloidal calomel (75% Hg_2Cl_2 and 25% albuminoids) 940
KALORIE (*f*): calorie 941
KALORIE (*f*), **GROSSE-**: large calorie, kilogram calorie 942
KALORIE (*f*), **KLEINE-**: small calorie, gram calorie 943
KALORIMETER (*n*): calorimeter (apparatus for determining the specific heat of a substance) 944
KALORIMETRIE (*f*): calorimetry 945
KALORIMETRISCH: calorimetric 946
KALORIMETRISCHE BOMBE (*f*): bomb calorimeter 947
KALORISCHE BOMBE (*f*): (bomb)-calorimeter 948
KALORISIEREN: to calorize 949
KALORISIERUNG (*f*): calorizing 950
KALOTTE (*f*), **KUGEL-**: spherical calotte, calotte (a flattened dome) 951
KALT: cold; cool; frigid; insensible; indifferent; apathetic; reserved; passionless; calm 952
KALTBEARBEITUNG (*f*): cold (hammering, hardening) working (of metal) 953
KALTBLASEN: to cool by blast 954
KALTBLASEN (*n*): cooling (Water gas); blast-cooling 955
KALTBLÄSIG: refractory (Metal) 956
KALTBLEIBEN: to remain (keep) cool, keep one's temper 957
KALTBLÜTER (*m*): cold blooded animal 958
KALTBLÜTIG: cold-blooded, indifferent, calm, deliberate 959
KALTBLÜTIGKEIT (*f*): cold-bloodedness, coolness, calmness, indifference 960
KALTBRUCH (*m*): cold shortness (Metal) 961
KALTBRÜCHIG: cold short (Metal) 962
KÄLTE (*f*): cold-(ness); chill-(iness); indifference; frigidity; insensibility; apathy; reserve 963
KÄLTEANÄSTHETIK (*f*): freezing anæsthetic 964
KÄLTECHEMIE (*f*): cryochemistry 965
KÄLTEEMPFINDLICH: sensitive to cold 966
KÄLTEERZEUGEND: frigorific; cold producing; producing cold 967
KÄLTEERZEUGUNG (*f*): refrigeration; production of cold 968
KÄLTEERZEUGUNGSANLAGE (*f*): refrigerating (ice) plant 969
KÄLTEERZEUGUNGSMASCHINE (*f*): refrigerating (ice) machine 970
KÄLTEGRAD (*m*): degree below zero; degree of (cold) frost 971
KÄLTELEISTUNG (*f*): refrigerating (capacity) duty 972
KÄLTEMASCHINE (*f*): refrigerating (ice) machine 973
KÄLTEMISCHUNG (*f*): freezing mixture 974
KÄLTEN: to chill; cool; ice; become (grow) cold; refrigerate 975
KALTER BLITZSCHLAG (*m*): a flash of lightning which does not cause fire when it strikes 976
KALTER BRAND (*m*): gangrene (Med.) 977
KALTER SCHLAG (*m*): a flash of lightning which does not cause fire when it strikes 978
KALTER SCHWEISS (*m*): cold sweat 979
KÄLTESCHUTZPRÄPARAT (*n*): protector against cold 980
KALTES FIEBER (*n*): ague (Med.) 981
KÄLTETRÄGER (*m*): cooling medium; cold conductor; conductor of cold 982
KÄLTETRÜBUNG (*f*): clouding (albumen coagulation due to over-cooling) 983
KALTE ZONE (*f*): frigid zone 984
KALTGESÄTTIGT: cold-saturated (Phot.) 985
KALTGESCHLAGEN: cold pressed (of oil) 986
KALTGUSS (*m*): cold casting 987
KALTHÄRTEN: to cold-harden (Metal) 988
KALTHÄRTEN (*n*): cold hardening (Steel) 989
KALTKEILEN (*n*): quarrying without blasting 990
KALTLAGERUNG (*f*): cold storage 991
KALTLEIM (*m*): cold glue (a term for all adhesive media which do not require heating before use) 992
KÄLTLICH: chilly, frigid 993
KALTMEISSEL (*m*): cold chisel 994
KALTPOLIERTINTE (*f*): cold polishing ink 995
KALTRECKEN: to cold-draw; (*n*): cold-working; cold-drawing (Metal) 996
KALTSÄGE (*f*): cold saw 997
KALTSÄGEMASCHINE (*f*): cold sawing machine 998
KALTSCHMIED (*m*): brazier 999
KALTSCHUTZPRÄPARAT (*n*): preservative from cold or frost; preventative against cold or frost 000

KALTSINNIG : cold, indifferent, frigid, insensible 001
KALTWALZWERK (*n*) : cold rolling mill 002
KALTWASSERFARBE (*f*) : cold water (paint) colour 003
KALTWASSERKUR (*f*) : hydropathy (treatment of disease by means of cold water) 004
KALTWASSERPROBE (*f*) : cold water test ; water pressure test ; hydraulic test 005
KALTWASSERPUMPE (*f*) : cold water pump 006
KALTZIEHEN : to cold-draw ; (*n*) : cold-drawing (Metal) 007
KALUSZIT (*m*) : caluscite, syngenite ; K_2O,SO_3 plus CaO,SO_3 plus H_2O ; Sg. 2.6 008
KALZINIEREN : to calcine (see also CALCINIER-) 009
KALZINIEROFEN (*m*) : calcining furnace or oven 010
KALZIT (*n*) : calcite (see CALCIT ; CALCIUM-CARBONAT and KALKSPAT) 011
KALZIUM (*n*) : calcium (see under CALCIUM) 012
KAMALA (*f*) : kamala (an Indian drug from the the fruit of *Mallotus philippinensis*) 013
KAMBAHOLZ (*n*) : cam wood ; logwood (see BLAUHOLZ) 014
KAMEL (*n*) : camel 015
KAMELGARN (*n*) : mohair 016
KAMELGRAS (*n*) : camel grass (*Andropogon laniger*) 017
KAMELHAAR (*n*) : camel-hair (the hair of the dromedary, *Camelus dromedarius*) 018
KAMELIA (*f*) : camellia (*Camellia japonica* ; *C. reticulata* and *C. oleifera*) 019
KAMELIENÖL (*n*) : camellia oil 020
KAMELLIE (*f*) : (see KAMELIA) 021
KAMELWOLLE (*f*) : (see KAMELHAAR) 022
KÄMELWOLLE (*f*) : (see ANGORAHAAR) 023
KAMEO (*m*) : cameo (a gem carved in relief) 024
KAMERA (*f*) : camera (Photography) 025
KAMERAD (*m*) : comrade ; fellow ; mate ; companion ; associate 026
KAMERADLICH : sociable, companionable 027
KAMERADSCHAFT (*f*) : fellowship, society, comradeship 028
KAMERADSCHAFTLICH : sociable ; companionable 029
KAMFER (*m*) : camphor (see CAMPHER) 030
KAMILLE (*f*) : camomile (*Flores chamomillæ*) (see HUNDSKAMILLE and FELDHUNDS-KAMILLE) 031
KAMILLENÖL (*n*) : camomile oil ; chamomile oil ; Sg. 0.9-0.94 (from flowers of *Anthemis nobilis*) 032
KAMIN (*m*) : chimney-(stack) ; fireplace ; fireside 033
KAMINFEGER (*m*) : chimney-sweep 034
KAMINGESTELL (*n*) : (fire)-dogs 035
KAMINKAPPE (*f*) : cowl, chimney top 036
KAMINKLAPPE (*f*) : damper ; register 037
KAMINSCHIRM (*m*) : fire-screen 038
KAMINSIMS (*m*) : mantle (chimney) piece 039
KAMM (*m*) : comb ; cog-tooth (of wheel) ; ridge ; crest ; tuft ; carder ; ward (of a key) 040
KAMMARTIG : pectinal, pectinate, comb-like 041
KÄMMELN : to card-(wool) 042
KÄMMELUNG (*f*) : wool-carding 043
KÄMMEN : to comb ; card ; notch ; slice ; dovetail 044
KAMMER (*f*) : chamber ; room ; ventricle (of heart) ; caisson (Engineering) ; cabin ; cavity ; board ; office ; cabinet ; header 045
KÄMMER (*m*) : carder ; comber 046
KAMMERDIENER (*m*) : valet 047
KAMMER (*f*), **DUNKEL-** : dark-room (Photography) 048
KAMMER (*f*), **DUNKELE-** : camera obscura 049
KÄMMEREI (*f*) : exchequer ; finances ; Board of Finance ; wool combing 050
KÄMMERER (*m*) : chamberlain ; treasurer 051
KÄMMERERIT (*m*) : Kammererite, Ripidolite (see RIPIDOLITH) 052
KAMMERGUT (*n*) : crown (domains) lands 053
KAMMERKOLLEGIUM (*n*) : exchequer ; Board of Finance 054
KAMMERKRISTALL (*m*) : chamber crystal 055
KÄMMERLING (*m*) : valet, chamberlain ; waste-(wool) 056
KAMMERMUSIK (*f*) : chamber music 057
KAMMEROFEN : chamber coke oven 058
KAMMERPROZESS (*m*) : chamber process 059
KAMMERSÄULE (*f*) : cellular voltaic pile 060
KAMMERSÄURE (*f*) : chamber acid (Sulphuric acid process) 061
KAMMERSCHLAMM (*m*) : chamber sludge (see SELENSCHLAMM) 062
KAMMERSOHLE (*f*) : chamber floor 063
KAMMERTUCH (*n*) : cambric 064
KAMMERWESEN (*n*) : finances 065
KAMMERWISSENSCHAFT (*f*) : science of finance, finance 066
KAMMFETT (*n*) : melted horse grease 067
KAMMFÖRMIG : comb-shaped ; pectinal ; pectinate 068
KAMMGARN (*n*) : worsted 069
KAMMGRAS (*n*) : dog's tail-(grass) (*Cynosurus cristatus*) ; couch grass, triticum, graminis, dog-grass (*Agropyrum repens*) ; knot-grass, twitch (wheat) grass, dog-wheat (*Triticum repens*) 070
KAMMKIES (*m*) : cockscomb pyrites (a variety of marcasite) (see MARKASIT) 071
KAMMLAGER (*n*) : thrust bearing, thrust block, pinion bearing 072
KAMMMACHER (*m*) : comb-maker 073
KAMMMACHEREI (*f*) : comb-making 074
KAMMMUSCHEL (*f*) : scallop (*Pecten*) 075
KAMMRAD (*n*) : cog-wheel ; spur-wheel 076
KAMMSETZER (*m*) : carder 077
KAMM (*m*), **STAUB-** : tooth-comb 078
KAMMWALZE (*f*) : pinion ; gear (wheel) roll ; helical gear ; doffer 079
KAMMWELLE (*f*) : pinion shaft 080
KAMMWOLLE (*f*) : carded wool ; worsted 081
KAMMZAHN (*m*) : tooth of a comb 082
KAMMZAPFEN (*m*) : thrust journal, pinion journal 083
KAMPAGNE (*f*) : campaign ; season (Beet and Cane sugar industry) 084
KAMPANE (*f*) : bell-jar 085
KAMPESCHEHOLZ (*n*) : campeachy wood ; logwood (see BLAUHOLZ) 086
KAMPF (*m*) : combat ; engagement ; conflict ; brush ; battle ; fight ; pugilism ; contest ; trouble ; wrestling ; struggle ; strife 087
KÄMPFEN : to battle ; contend ; fight ; struggle ; strive ; endeavour ; contest 088
KÄMPFER (*m*) : combatant ; fighter ; pugilist ; abutment ; buttress ; transom (Wood) ; impost (Stone or Concrete) 089
KAMPFER (*m*) : camphor (see CAMPHER and CAMPHOR) 090
KAMPFERCHINON (*n*) : camphor quinone ; $C_8H_{14}(CO)_2$ 091

KAMPFGAS (*n*): poison gas (for use in war) 092
KAMPFHAHN (*m*): gamecock 093
KAMPFPLATZ (*m*): field of battle; arena; cock-pit 094
KAMPHEN (*n*): camphene (see CAMPHEN) 095
KAMPHER (*m*): camphor (see CAMPHER and CAMPHOR) 096
KAMPHERBLUMEN (*f.pl*): flowers of camphor 097
KAMPHERDARSTELLUNG (*f*): production of camphor, camphor production 098
KAMPHERERZEUGUNG (*f*): production of camphor, camphor production 099
KAMPHERGEWINNUNG (*f*): production of camphor, camphor production 100
KAMPHERKRAUT (*n*): Southern-wood (*Artemisia abrotanum*) (see EBERRAUTE) 101
KAMPHERLORBEERBAUM (*m*): camphor laurel tree (*Cinnamomum Camphora*; *Laurus camphora*) 102
KAMPHERN: to camphorate 103
KAMPHERÖL (*n*): camphor oil (see CAMPHORÖL) 104
KAMPHERÖL (*n*), **BLAUES DICKFLÜSSIGES-**: blue viscous camphor oil (Sg. 0.955); (from the wood of the Camphor laurel tree *Laurus camphora*) 105
KAMPHERÖL (*n*), **ROHES-**: crude camphor oil 106
KAMPHERÖL (*n*), **WEISSES-**: light camphor oil (see CAMPHORÖL) 107
KAMPHERSÄURE (*f*): camphoric acid (see CAMPHORSÄURE) 108
KAMPHER (*m*), **SUMATRA-**: Sumatra (Borneo) camphor (see BORNEOL) 109
KAMPHERTEILEN (*m.pl*): camphor particles 110
KAMPHOL (*n*): (see BORNEOL) 111
KAMPHOR (*m*): camphor (see CAMPHER and CAMPHOR) 112
KAMPHORSÄURE (*f*): (see CAMPHORSÄURE) 113
KAMPRATUBE (*f*): a tube made of a prepared parchment (for packing powders and ointments) 114
KAMPYLIT (*m*): campylite; $Pb_5Cl[(As,P)O_4]_3$; Sg. 7.2 115
KANADABALSAM (*m*): (*Balsamum canadense*): Canada balsam, balsam of fir, Canada turpentine (from *Abies balsamica*); Sg. 0.985-0.999 116
KANADISCHER TERPENTIN: Canada turpentine (see KANADABALSAM) 117
KANADISCHGELB (*n*): Canadian yellow 118
KANADOL (*n*): canadol, petroleum ether, gasoline; Sg. 0.65-0.7; Bp. 40-80°C. 119
KANAL (*m*): canal; channel; conduit; sewer; drain; pipe 120
KANALBRÜCKE (*f*): aqueduct 121
KANALDAMPFER (*m*): channel steamer 122
KANALGAS (*n*): sewer gas 123
KANALISATIONSANLAGE (*f*): sewage plant 124
KANALISATIONSROHR (*n*): sewer (water; gas; electric) pipe or conduit; drain pipe 125
KANALISATION (*f*), **STRASSEN-**: street (drainage) sewer-(s) 126
KANALJAUCHE (*f*): liquid sewage 127
KANAL (*m*), **LUFT-**: air duct 128
KANALOFEN (*m*): tunnel kiln 129
KANALSCHLEUSE (*f*): canal lock 130
KANALSTRAHL (*m*): canal ray 131
KANANGAÖL (*n*): cananga oil; Sg. 0.89-0.94 (from flowers of *Cananga odorata*) 132
KANANGAWASSER (*n*): cananga water 133
KANAPEE (*n*): couch; sofa 134
KANARIENGRAS (*n*): canary grass (*Phalaris canariensis*) 135
KANARIENSAMEN (*m.pl*): canary seed (from canary grass; see KANARIENGRAS) 136
KANARIENSEKT (*m*): canary wine (a dry, white wine from the Canary islands) 137
KANARIENVOGEL (*m*): canary; wild canary (*Serinus canarius*) 138
KANASTER (*m*): canaster tobacco 139
KANDELABER (*m*): chandelier; candelabra 140
KANDIDAT (*m*): candidate, applicant 141
KANDIEREN: to candy 142
KANDIOLIN (*n*): candiolin; $C_6H_{10}O_4(CaPO_4)_2$ 143
KANDIS (*m*): sugar candy 144
KANDISZUCKER (*m*): sugar candy 145
KANEEL (*m*): (Chinese)-cinnamon; cassia (bark of *Cinnamomum cassia*) 146
KANEEL, BRAUNER- (*m*): Ceylon cinnamon (bark of *Cinnamomum zeylanicum*) 147
KANEELGRANAT (*m*): cinnamon stone; essonite (a chalk-clay granite) (see GRANAT) 148
KANEELSTEIN (*m*): (see KANEELGRANAT) 149
KANEL (*m*): (see KANEEL) 150
KANELLARINDE (*f*): canella bark (see CANELLE) 151
KANINCHEN (*n*): rabbit; coney (*Lepus cuniculus*) 152
KANINCHENHAAR (*n*): rabbit's hair (Textile) 153
KANINCHENSEPTICHÄMIE (*f*): an infectious (contagious) disease or pestilence of the group, *Septicæmia hæmorrhagica*, peculiar to rabbits 154
KANINLEIM (*m*): rabbit glue 155
KANISTER (*m*): canister (a tin or wooden box) 156
KANKER (*m*): canker (a disease of fruit trees) 157
KANNABIN (*n*): (see CANNABIN) 158
KANNE (*f*): can; jug; pot; litre; tankard; quart (Measure) 159
KANNELIEREN: to channel; flute; groove; chamfer 160
KANNELIERUNG (*f*): channeling, fluting, chamfering, grooving 161
KANNELKOHLE (*f*): cannel coal; Sg. 1.24-1.34 162
KANNENGIESSER (*m*): pewterer 163
KANNENMASS (*n*): quart-(measure) (of liquids) 164
KANNENSTAUDE (*f*): pitcher-plant (insectivorous plant having pitcher-like receptacles) 165
KANNENTRÄGER (*m*): pitcher plant (see KANNENSTAUDE) 166
KANNENZINN (*n*): pewter (a lead-tin alloy) (80% Sn and 20% Pb) 167
KANNEVAS (*m*): canvas (a coarse kind of cloth) 168
KANONE (*f*): cannon; gun; piece (of ordnance) 169
KANONENBOOT (*n*): gunboat 170
KANONENBRONZE (*f*): gun metal; 89-91% Cu; rest Sn. 171
KANONENDONNER (*m*): thunder (boom) of a gun, cannonade 172
KANONENFABRIK (*f*): ordnance (works) factory 173
KANONENGUT (*n*): gun-metal (see KANONENBRONZE) 174
KANONENLAUF (*m*): gun barrel 175
KANONENMETALL (*n*): (see KANONENGUT) 176
KANONENPARK (*m*): gun (artillery) park 177
KANONENPULVER (*n*): gun powder 178

KANONENSCHUSSWEITE (*f*): gun (cannon) range, range of a (cannon) gun 179
KANONIEREN: to shoot, fire, cannonade 180
KANTE (*f*): edge; border; brim; lace; margin; corner; ledge; selvedge 181
KANTE (*f*), **EINTRETENDE-**: forward edge 182
KÄNTELN: to cant, tilt, turn on edge, tip up 183
KANTEN: to edge; border; square (Stones); cant; tilt; tip-(up); turn on edge; throw; up-end 184
KANTENSCNHEIDERWERKZEUG (*n*): edging tool 185
KANTENWEISE: edgewise; on edge 186
KANTE (*f*), **SCHARFE-**: sharp edge 187
KANTE (*f*), **SCHRÄGE-**: bevel edge 188
KANTHAKEN (*m*): cant hook 189
KANTHARIDE (*f*): cantharis, cantharides, Spanish-fly (*Cantharis vesicatoria*) (see also under **CANTHARIDEN**) 190
KANTHOLZ (*n*): beams, posts, square timber, scantling (Wood) 191
KANTIG: edged; angular; cornered 192
KANTSPANT (*n*): cant frame 193
KANYABUTTER (*f*): West African tallow-tree butter (from *Pentadesma butyracea*) 194
KANZEL (*f*): pulpit 195
KANZLEI (*f*): chancery; government (lawyer's) office 196
KANZLEIARCHIV (*n*): records, archives, rolls 197
KANZLEIPAPIER (*n*): official (writing; note; letter) paper; foolscap paper 198
KANZLEITINTE (*f*): record ink 199
KANZLER (*m*): chancellor 200
KAOLIN (*n*): kaolin; terra alba; bolus alba; argilla; white bole; porcelain clay; China clay (a hydrous aluminium silicate); $2SiO_2.Al_2O_3.2H_2O$; Sg. 2.5 201
KAOLINERDE (*f*): (see **KAOLIN**) 202
KAOLINISIEREN: to kaolinize 203
KAOLINISIERUNG (*f*): kaolinization 204
KAOLINIT (*m*): kaolinite (see **KAOLIN**) 205
KAOUTSCHUK (*m*): caoutchouc; rubber (see **KAUTSCHUK**) 206
KAP (*m*): cape (Geographical) 207
KAPALOE (*f*): cape aloe (*Aloe lucida*) 208
KAPANN (*m*): capon (a young castrated cock) 209
KAPAZITÄT (*f*): capacity 210
KAPAZITÄTSFAKTOR (*m*): (see **QUANTITÄTSFAKTOR**) 211
KAPELLE (*f*): cupel; sand-bath; subliming dish; capsule; priming cap; chapel; band; choir 212
KAPELLENASCHE (*f*): bone ash 213
KAPELLENFORM (*f*): cupel mould 214
KAPELLENGOLD (*n*): fine gold 215
KAPELLENOFEN (*m*): cupelling (cupellation; assay; sublimation) furnace 216
KAPELLENPROBE (*f*): cupel test; cupellation; test by refining 217
KAPELLENRAUB (*m*): loss in cupellation 218
KAPELLENSILBER (*n*): fine silver 219
KAPELLENSTATIV (*n*): cupel (support) stand 220
KAPELLENTRÄGER (*m*): cupel (holder) stand 221
KAPELLENZUG (*m*): loss in cupellation 222
KAPELLIEREN: to cupel 223
KAPELLMEISTER (*m*): musical director, bandmaster, conductor 224
KAPER (*m*): captor; privateer; pirate; (*f*): caper (the shrub and also the pickled bud of *Capparis spinosa*) 225
KAPERN: to capture; catch 226
KAPERNBRÜHE (*f*): caper sauce (see **KAPER**) 227
KAPERNSTRAUCH (*m*): caper-(shrub) (see **KAPER**) 228
KAPILLAR: capillary; (*n*): capillary 229
KAPILLARAFFINITÄT (*f*): capillary (affinity) attraction 230
KAPILLARAKTIVITÄT (*f*): capillary activity 231
KAPILLARANALYSE (*f*): capillary analysis 232
KAPILLARANZIEHUNG (*f*): capillary attraction 233
KAPILLARERHEBUNG (*f*): capillary rise 234
KAPILLARFLASCHE (*f*): capillary (bottle) flask 235
KAPILLARGEFÄSS (*n*): capillary vessel 236
KAPILLARITÄT (*f*): capillarity; capillary attraction 237
KAPILLARKRAFT (*f*): capillary force 238
KAPILLARROHR (*n*): capillary tube 239
KAPILLARSENKUNG (*f*): capillary fall 240
KAPILLARSPANNUNG (*f*): capillary (tension) stress; capillary pressure 241
KAPITAL (*n*): capital (Money, etc.); principal 242
KAPITALERHÖHUNG (*f*): increase of capital 243
KAPITALERNIEDRIGUNG (*f*): decrease of capital 244
KAPITALISIEREN: to convert into capital 245
KAPITALIST (*m*): capitalist 246
KAPITALKONTO (*n*): capital account 247
KAPITALRECHNUNG (*f*): capital account 248
KAPITÄN (*m*): captain 249
KAPITEL (*n*): chapter 250
KAPLAKEN (*m*): primage, hat-money 251
KAPNOGRAPH (*m*): recording dust meter; "Kapnograph" 252
KAPOKBAUM (*m*): kapok tree, kapoc tree (*Eriodendron anfractuosum*; or *Bombax pentandrum*) 253
KAPOKFASER (*f*): kapok fibre (from the fruit of the kapok tree; see **KAPOKBAUM**) 254
KAPOKÖL (*n*): kapok (kapoc) oil; Sg. 0.922 (from *Bombax ceila*, *Bombax pentandrum* and *Eriodendron anfractuosum*) 255
KAPPE (*f*): cap; hood; top; dome; crown; cowl; cape; coping; ridge; cover 256
KAPPELN: to top 257
KAPPEN: to top (lop) trees; cut; chop; caponize; castrate; cap; cover; hood 258
KAPPENFLASCHE (*f*): bottle with cap 259
KAPPENFÖRMIG: hood-shaped, hooded 260
KAPPENROHR (*n*): lens hood, flange (of cameras) 261
KAPPHAHN (*m*): capon (a young castrated cock) 262
KAPPLAKEN (*m*): primage, hat-money 263
KAPPZIEGEL (*m*): gutter-tile 264
KAPRINSÄURE (*f*): capric acid 265
KAPRIOLE (*f*): caper; trick 266
KAPSEL (*f*): capsule; case; box; (priming)-cap; chill; mould (Founding); housing; casing; cover; sagger (Ceramics) 267
KAPSELARTIG: capsular; case-like; boxlike 268
KAPSELFÖRMIG: capsular; case-(box)-shaped 269
KAPSELGUSS (*m*): casting in chills; chill casting; chilled work 270
KAPSELIG: capsular 271
KAPSELMASCHINE (*f*): capsuling machine, machine for making capsules 272
KAPSELMOTOR (*m*): box motor 273
KAPSELN: to capsule, enclose in a capsule (case or housing), cover, house, encase 274
KAPSELTON (*m*): sagger clay (Ceramics) 275
KAPSELUNG (*f*): casing, enclosing; case 276

KAPUZE (*f*) : cape ; cowl ; hood 277
KAPUZINER (*m*) : capuchin-(monk) 278
KAPUZINERKRAUT (*n*) : Love-in-a-Mist (*Nigella damascena*) 279
KAPUZINERKRESSE (*f*) : Indian cress (small)-nasturtium (*Tropæolum minus*) 280
KAPUZINERKRESSENÖL (*n*) : Indian cress oil (from seeds of *Tropæolum majus*) 281
KAPWEIN (*n*) : Cape wine 282
KARAIBISCH : Caribbean 283
KARAMEL (*m*) : caramel ; $C_{12}H_{18}O_9$ (by heating cane sugar to 400-420°F., when the sugar molecule loses its water) ; colour ; sugar colour-(ing) ; burnt sugar 284
KARAMELISIEREN : to form caramel (Sugar) 285
KARAT (*n*) : carat (of gold) ; pure gold containing 24 carats 286
KARATGEWICHT (*n*) : troy weight 287
KARATIEREN : to alloy (gold) 288
KARATIERUNG (*f*) : alloying (of gold) 289
KARATIERUNG (*f*), **GEMISCHTE-** : mixed alloying (alloying of gold with silver and copper mixed) 290
KARATIERUNG (*f*), **ROTE-** : alloying of gold with copper 291
KARATIERUNG (*f*), **WEISSE-** : alloying of gold with silver 292
KARATIG : — carat ; containing — carats-(of gold) 293
KARÄTIG : (see **KARATIG**) 294
KARBAMID (*n*) : carbamide (see **HARNSTOFF** and **CARBAMID**) 295
KARBANIL (*n*) : carbanile, phenyl cyanate, phenyl isocyanate (see **PHENYLISOCYANAT**) 296
KARBATSCHEN : to scourge, whip, castigate, flagellate 297
KARBAZOL (*n*) : (see **CARBAZOL**) 298
KARBE (*f*) : caraway (*Carum carvi*) 299
KARBID (*n*) : carbide (see also **CALCIUMCARBID**) 300
KARBIDENZERLEGUNG (*f*) : decomposition of carbide (generation of acetylene gas) 301
KARBIDGAS (*n*) : acetylene (see **ACETYLEN**) 302
KARBIDKOHLE (*f*) : iron carbide, separation of iron carbide in solid form from the metal when cooled slowly (Blast furnace work) 303
KARBIDOFEN (*m*) : carbide furnace 304
KARBIDÖFENANLAGE (*f*) : carbide furnace plant 305
KARBODYNAMIT (*n*) : carbodynamite (an explosive consisting of 90% nitro-glycerin and 10% burnt cork) 306
KARBOLINEUM (*n*) : (see **CARBOLINEUM**) 307
KARBOLISMUS (*m*) : (see **PHENOLVERGIFTUNG**) 308
KARBOLKALK (*m*) : disinfectant powder, carbolic powder 309
KARBOLÖL (*n*) : carbolic oil (from fractional distillation of coal tar) 310
KARBOLSÄURE (*f*) : (see **PHENOL** and **CARBOLSÄURE**) 311
KARBOLSÄURE (*f*), **ROHE-** : cresylic acid (see **KRESYLSÄURE**) ; crude carbolic acid ; Mp. 40°C. ; Bp. 180.5°C. 312
KARBONADE (*f*) : chop, cutlet 313
KARBONAT (*n*) : carbonate 314
KARBONISATION (*f*) : carbonation, carbonization ; removal of the vegetable matter mixed with wool by damping it with hydrochloric acid, centrifuging and drying in a special oven (Wool) 315
KARBONISIEREN : to carbonate, carbonize ; remove the vegetable matter mixed with wool (see also **CARBONISIEREN**) 316
KARBONIT (*n*) : carbonit (a safety explosive) 317
KARBORUNDUM (*n*) : carborundum (see **SILIZIUMCARBID** and **CARBORUND**) 318
KARBORUNDUMSTEIN (*m*) : carborundum brick ; Sg. 3.2 319
KARBOZINK (*n*) : zinc carbonate ; $ZnCO_3$; Sg. 4.44 (anti-corrosion agent ; rust preventative) 320
KARBOZYKLISCH : carbocyclic 321
KARBUNKEL (*m*) : carbuncle (Med.) 322
KARBYLOXIM (*n*) : (see **KNALLSÄURE**) 323
KARDAMOM (*m*) : cardamom (fruit of *Elettaria repens*) 324
KARDAMOMENÖL (*n*) : cardamom oil ; Sg. 0.9 (from seeds of *Elettaria repens*) 325
KARDANGELENK (*n*) : cardan joint 326
KARDANISCHE AUFHÄNGUNG (*f*) : cardan-(ic) suspension 327
KARDAN (*n*), **KUGEL-** : universal (ball) joint 328
KARDANWELLE (*f*) : cardan shaft 329
KARDE (*f*) : carding machine, carder (Textile) 330
KARDEEL (*n*) : strand (of ropes), cable, hawser, rope ; (*pl*) : jears (Naut.) 331
KARDEEL (*n*), **TAU-** : rope strand 332
KARDEN : to card ; comb (Wool) 333
KARDENDISTEL (*f*) : Fuller's thistle, teasel (*Dipsacus fullonum*) 334
KARDINAL (*m*) : cardinal 335
KARDMASCHINE (*f*) : carding machine (Textile) 336
KARDOBENEDIKTENFLUIDEXTRAKT (*m*) : (*Extractum cardui benedicti fluidum*) ; blessed thistle fluid extract 337
KARDOBENEDIKTENKRAUT (*n*) : holy thistle, blessed thistle (*Herba cardui benedicti* of *Centaurea benedicta*, or *Cnicus benedictus*) 338
KARDOL (*n*) : (*Cardolum*) (see **AKAJOUBALSAM**) 339
KARENZ (*f*) : (period of)-grace ; period of rest ; waiting period ; respite ; (see **KARENZZEIT**, **KARENZVERPFLICHTUNG** and also **MORATORIUM**) ; absence of assets, insolvency 340
KARENZVERPFLICHTUNG (*f*) : in Germany, an obligation in a contract of employment not to accept another similar situation within a pre-determined radius during a pre-determined period 341
KARENZZEIT (*f*) : extension of time, respite ; period of waiting ; period of grace (as for payments) ; period of probation (as of a situation) ; duration of obligation (see **KARENZVERPFLICHTUNG**) 342
KARFREITAG (*m*) : Good Friday 343
KARFUNKEL (*m*) : carbuncle (either the gem or the medical term) 344
KARG : scanty ; stingy ; niggardly ; close-(fisted) ; poor (Soil, etc.) ; sparing ; miserly ; parsimonious ; penurious 345
KARGEN : to be sparing ; be penurious ; be niggardly ; be stingy 346
KÄRGLICH : sparing ; spare ; scanty ; penurious ; stingy ; poor 347
KARIKATUR (*f*) : caricature 348
KARIKIEREN : to caricature 349
KARITEBUTTER (*f*) : (see **SHEABUTTER**) 350
KARLSBADER-SALZ (*n*) : Carlsbad salt (45% sodium sulphate, 35% sodium bicarbonate, 18% sodium chloride and 2% potassium sulphate) 351
KARMESIN : crimson 352

KARMIN (*m*): carmine (a red dye) (Carminic acid with Al and Ca, obtained from cochineal) 353
KARMIN (*m*), BLAUER-: (see WOLFRAMBLAU) 354
KARMIN (*m*), BRILLANT-: brilliant carmine 355
KARMINLACK (*m*): carmine lake (see KARMIN) 356
KARMINROT (*n*): carmine red; crimson lake; cochineal 357
KARMINSÄURE (*f*): carminic acid; $C_{22}H_{22}O_{13}$ or $C_{17}H_{18}O_{10}$ (*Acidum carminicum*) (from the cochineal insect) (see COCHENILLE) 358
KARMOISIN (*n*): crimson 359
KARMOISINROT (*n*): crimson 360
KARNALLIT (*m*): carnallite (a potassium double salt, obtained from the Stassfurt mines, Germany); $MgCl_2.KCl.6H_2O$; Sg. 1.6 [361
KARNAUBAWACHS (*n*): carnauba wax; Brazil wax (from *Copernica cerifera*); Sg. 0.995; Mp. 85°C. 362
KARNEOL (*m*): carnelian, cornelian (bright red variety of chalcedony) (Fe_2O_3): Sg. 2.25 (see CHALCEDON) 363
KARNIES (*m*): cornice 364
KARNIESS (*m*): (see KARNIES) 365
KARNOTIT (*m*): (see CARNOTIT) 366
KÄRNTHNERISCH: Carinthian 367
KÄRNTISCH: Carinthian 368
KÄRNTNERISCH: Carinthian 369
KARO (*n*): check; chequer; square; diamonds (Cards) 370
KAROGRAPHIE (*f*): (see CARREAUGRAPHIE) 371
KAROTTE (*f*): carrot 372
KARPFEN (*m*): carp (Fish) (*Cyprinus carpio*) 373
KARPHOLITH (*m*): carpholite (Mineralogy); $H_4MnAl_2Si_2O_{10}$; Sg. 2.94 374
KARPOLITH (*m*): carpolite (a fossilized fruit) 375
KARPOLOGIE (*f*): carpology (science of plant fruits) 376
KARPOLOGISCHE AUSSTELLUNG (*f*): fruit exhibition 377
KARPOLOGISCHES SYSTEM (*n*): carpological system (a plant system based on the composition of the fruit) 378
KARRAGHEENMOOS (*n*): carragheen moss (see CARRAGHEENMOOS) 379
KARRE (*f*): (wheel)-barrow; cart 380
KARREE (*n*): square (Mil.) 381
KARREN: to cart; remove (work) with a cart, truck or barrow; (*m*): (wheel)-barrow; cart; truck 382
KÄRRNER (*m*): carter; drayman 383
KARST (*m*): mattock; hoe 384
KARSTEN: to hoe 385
KARTÄTSCHE (*f*): grape-shot, cannister 386
KARTÄTSCHENGRANATE (*f*): shrapnel shell 387
KARTÄTSCHENKUGEL (*f*): shrapnel ball 388
KARTÄUSER (*m*): Carthusian friar 389
KARTÄUSERPULVER (*n*): Carthusian powder; kermes mineral (see ANTIMONZINNOBER and ANTIMONSULFÜR) 390
KARTE (*f*): card; map; chart; menu; charter; list; ticket 391
KARTEN: to play at cards; contrive, concert 392
KARTENBLATT (*n*): card; one card of a series 393
KARTENFABRIK (*f*): card making works 394
KARTENHAUS (*n*): chart house (Naut.) 395
KARTENMACHER (*m*): card (maker) manufacturer 396
KARTENPAPIER (*n*): cardboard; cartridge paper 397
KARTENSPIEL (*n*): game of cards, playing cards, pack of cards 398
KARTENSTECHER (*m*): map (chart) engraver 399
KARTHARTINSÄURE (*f*): carthartinic acid (from *Cassia angustifolia*) 400
KARTOFFEL (*f*): potato 401
KARTOFFELBRANNTWEIN (*m*): potato spirit (see also FUSELÖL) 402
KARTOFFELBREI (*m*): mashed potatoes 403
KARTOFFELFUSELÖL (*n*): fusel oil (obtained as a by-product from the distillaton of potato spirit) (see FUSELÖL) 404
KARTOFFELKNOLLEN (*m*): tuber 405
KARTOFFELKONSERVE (*f*): potato-jam 406
KARTOFFELKRANKHEIT (*f*): potato rot (see KAR-TOFFELPEST) 407
KARTOFFELKRAUT (*n*): potato tops 408
KARTOFFELMEHL (*n*): potato flour (*Farina*) 409
KARTOFFELMUS (*n*): mashed potatoes 410
KARTOFFELPEST (*f*): potato (disease) rot, (often due to the potato fly, *Anthomyia tuberosa*) 411
KARTOFFELPFLANZE (*f*): potato-(plant) 412
KARTOFFELPFLANZMASCHINE (*f*): potato planting machine 413
KARTOFFELPÜLPE (*f*): potato pulp (from the extraction of starch from potatoes) 414
KARTOFFELSTÄRKE (*f*): potato starch 415
KARTOFFELZELLE (*f*): potato cell 416
KARTOFFELZUCKER (*m*): potato sugar (glucose from potato starch) (see DEXTROSE) 417
KARTON (*m*): carton; pasteboard-(box); cardboard-(box) 418
KARTONIEREN: to bind in boards 419
KARTONMASCHINE (*f*): cardboard machine (Paper) 420
KARTONNAGE (*f*): cardboard box 421
KARTONNAGENLEIM (*m*): box-maker's glue or paste 422
KARVOL (*n*): (see CARVON) 423
KARVON (*n*): carvone, carvol (see CARVON) 424
KARUSELL (*n*): round-about 425
KARWEELBAU (*m*): carvel (work) building 426
KARWEELBOOT (*n*): carvel-built boat 427
KARWOCHE (*f*): Passion week 428
KARYOCERIT (*m*): caryocerite (Mineralogy); Sg. 4.3 429
KASCHMIR (*m*): Cashmere 430
KASCHMIRSCHWARZ (*n*): cashmere black 431
KASCHMIRWOLLE (*f*): cashmere wool 432
KASCHUTIEREN: to dye with (catechu) cutch 433
KÄSE (*m*): cheese; any cheese shaped object 434
KÄSEARTIG: cheesy; cheese-like; curdled; caseous 435
KÄSEBUTTER (*f*): curds; cream cheese 436
KÄSEFARBE (*f*): cheese colour-(ing) (see also KASEINFARBE) 437
KÄSEFORM (*f*): cheese-mould 438
KASEIN (*n*): casein (1-5% ash) (see also CASEIN) 439
KASEINAMMONIAK (*n*): casein ammonia, eucasin 440
KASEINFARBE (*f*): casein colour (made with curds and used for outside painting on account of resistance to weather conditions) 441
KASEINGEWINNUNG (*f*): production of casein 442
KASEINLEIM (*m*): casein (paste, glue) adhesive 443
KASEINLÖSUNG (*f*): casein solution 444
KASEINNATRIUM (*n*): sodium caseinate (see PLASMON) 445
KASEINPRÄPARAT (*n*): casein preparation 446
KASEINSALBE (*f*): casein ointment 447
KASEINSAUER: caseinate 448

KÄSELAB (*n*): rennet (an enzyme) 449
KÄSEMADE (*f*): cheese (maggot) mite (*Tyroglyphus siro*) 450
KASEMATTE (*f*): casemate 451
KASEMATTSCHIFF (*n*): ironclad with (central battery) casemate 452
KÄSEMILBE (*f*): cheese mite (*Tyroglyphus siro*) 453
KÄSEN: to curd; curdle; turn milk into cheese 454
KÄSEPAPPEL (*f*): wild (malva) mallow (*Malva sylvestris*) 455
KÄSEREI (*f*): cheese factory 456
KASERNE (*f*): barracks 457
KÄSESÄURE (*f*): lactic acid; ethylidenelactic acid; alpha-hydroxypropionic acid; $CH_3CHOHCOOH$; Sg. 1.25 458
KÄSESTOFF (*m*): casein (see under **KASEIN**) 459
KÄSEWASSER (*n*): whey 460
KÄSIG: (see **KÄSEARTIG**) 461
KASKADE (*f*): cascade, waterfall 462
KASKADENSCHALTUNG (*f*): multi-valve connection (Wireless Tel.) 463
KASKARILLENRINDE (*f*): cascarilla (eluthera; sweetwood) bark (bark of *Croton eleuteria*) 464
KASPISCH: Caspian 465
KASSA (*f*): cash (see **KASSE**) 466
KASSE (*f*): money-box; chest; till; cash-box; safe; ready money; cash; pay office; ticket-office; cashier's office 467
KASSELERBRAUN (*n*): Cassel brown, Cologne earth (see **BRAUNKOHLENPULVER**) 468
KASSELERERDE (*f*): Cassel earth 469
KASSELERGELB (*n*): (see **CASSELERGELB**) 470
KASSENFÜHRER (*m*): cashier 471
KASSENSCHRANK (*m*): safe 472
KASSEROLLE (*f*): casserole 473
KASSETTE (*f*): sagger (Ceramics); case; casket; holder; coffer; portfolio; dark-slide (Phot.) (see **CASSETTE**) 474
KASSIA (*f*): cassia (see also **CASSIA**) 475
KASSIABLÜTENÖL (*n*): cassia bud oil 476
KASSIE (*f*): (see **CASSIA**) 477
KASSIENBLÜTEN (*f.pl*): cassia buds (buds of *Cinnamomum cassia*) 478
KASSIENRINDE (*f*): cassia bark (bark of *Cinnamomum cassia*) 479
KASSIER (*m*): cashier (Com.); treasurer; ticket agent; booking clerk 480
KASSIERER (*m*): cashier; treasurer; ticket-agent; booking clerk 481
KASSITERIT (*m*): cassiterite (see **AUSSTRICH** and **ZINNSTEIN**) 482
KASSIUSPURPUR (*m*): purple of Cassius (see **GOLD-PURPUR**) 483
KASTANIE (*f*): chestnut (sweet or Spanish) (*Castanea vesca*) 484
KASTANIENBAUM (*m*): chestnut tree (*Castanea vesca*) 485
KASTANIENBAUM (*m*), **WILDER-**: horse-chestnut-(tree) (*Aesculus hippocastanum*) 486
KASTANIENBLÄTTER (*n.pl*): chestnut leaves (*Folia castaneœ vescœ* of *Castanea vesca*) 487
KASTANIENBRAUN: chestnut-brown; chestnut-(coloured) (see **MANGANBRAUN**) 488
KASTANIENGEHÖLZ (*n*): grove of chestnut trees 489
KASTANIENHOLZ (*n*): chestnut wood 490
KASTANIENHOLZEXTRAKT (*m*): chestnut-(wood)-extract (from *Castanea sativa*) 491
KÄSTCHEN: casket; alveolus; cavity; socket; alveole 492
KASTE (*f*): corporation; caste 493
KASTEIEN: to chastise, castigate, chasten, flagellate, mortify, scourge 494
KASTEIUNG (*f*): chastisement, castigation, flagellation, mortification, scourging 495
KASTEN (*m*): hutch; setting (of jewels); chest; box; case; trunk; frame; body; vat; back; coffer; press; crucible (of furnaces); flask (Founding); sagger (Ceramics); casket; collet; alveolus; alveole; cavity; socket 496
KASTENBLAU (*n*): pencil blue; indigo blue (Calico printing) 497
KASTENFORM (*f*): flask mould 498
KASTENGUSS (*m*): box casting 499
KASTENKIELSCHWEIN (*n*): box keelson 500
KASTEN (*m*), **LUFT-**: wind-chest 501
KASTENMACHER (*m*): box-maker 502
KASTENSCHLOSS (*n*): case (box) lock 503
KASTENSTRINGER (*m*): box stringer 504
KASTEN (*m*), **WASSER-**: water tank 505
KASTORÖL (*n*): castor oil (see under **RICINUS-**) 506
KAT (*n*): (see **KATT**) 507
KATALEPSIE (*f*): catalepsy, trance (Med.) 508
KATALOG (*m*): catalogue, pamphlet, brochure 509
KATALONISCH: Catalonian 510
KATALYSATOR (*m*): catalyser (an agent which, while itself remaining unaltered, accelerates a reaction) 511
KATALYSE (*f*): catalysis 512
KATALYSIEREN: to catalyse 513
KATALYSIEREND: catalysing (aiding combustion) 514
KATALYSIS (*f*): catalysis 515
KATALYTISCH: catalytic 516
KATAPHORESE (*f*): cataphoresis 517
KATAPLËIT (*m*): catapleiite; $Na_2ZrSi_3O_9.2H_2O$; Sg. 2.8 518
KATARAKT (*m*): cataract 519
KATARAKTSTEUERUNG (*f*): cataract-(valve)-control gear 520
KATARRH (*m*): catarrh (inflammation or congestion of mucous membrane) (Med.) 521
KATARRHALFIEBER (*n*): influenza (Med.) 522
KATARRHALISCH: catarrhal 523
KATASTER (*m*): land register 524
KATASTROPHE (*f*): catastrophy, calamity, disaster, misfortune, mishap 525
KATATYPIE (*f*): a photographic copying process (a catalytic process in which the positive is not obtained from the negative by the action of light, but by contact) 526
KATECHIN (*n*): catechin, catechu substitute (a dye) 527
KATECHINSÄURE (*f*): catechuic acid; catechin (see **KATECHIN**) 528
KATECHISIEREN: to catechise 529
KATECHISMUS (*m*): catechism 530
KATECHU (*m*): catechu, cutch, terra japonica (see **KATECHU, BRAUNER-** and **KATECHU, GELBER-**) 531
KATECHUBRAUN (*n*): cutch brown 532
KATECHU (*m*), **BRAUNER-**: black catechu, Pegu-catechu, catechu, cutch, terra japonica (extracted from wood of *Acacia catechu*) (ash up to 4%, vegetable residue up to 15%) (Dyeing) 533
KATECHU (*m*), **GELBER-**: gambier (see **GAMBIR**) (ash up to 5%, vegetable residue up to 15%) 534

KATECHUGERBSÄURE (*f*) : catechutannic acid 535
KATECHUPALME (*f*) : pinang, catechu palm (*Areca catechu*) 536
KATECHUSÄURE (*f*) : catechuic acid ; catechin (see KATECHIN) 537
KATEGORIE (*f*) : category, division, class, style, rank, order, head 538
KATEGORISCH : categorical 539
KATER (*m*) : tom-cat 540
KATHARINE (*f*) : dried plum (of *Prunus domestica*) 541
KATHARTISCH : cathartic, evacuant, purgative, abstergent, cleansing 542
KATHARTISCHES MITTEL (*n*) : cathartic, purgative, purge, physic, medicine 543
KATHEDER (*m*) : lecture (chair) desk ; pulpit ; professor's chair 544
KATHEDRALE (*f*) : cathedral 545
KATHODE (*f*) : cathode (negative electrode) 546
KATHODENDICHTE (*f*) : cathode density 547
KATHODENRAUM (*m*) : cathode space ; space around the cathode 548
KATHODENRÖHRE (*f*) : valve (Wireless Tel.) 549
KATHODENSTRAHL (*m*) : cathode ray 550
KATHODENSTROM (*m*) : cathode current 551
KATHODOCHEMISCH : cathodochemical 552
KATHOLICISMUS (*m*) : catholicism 553
KATHOLIK (*m*) : catholic 554
KATHOLISCH : catholic 555
KATION (*n*) : cation ; kation (negative ion) (see ION) 556
KATIONENAUSTAUSCH (*m*) : cation (kation) exchange or interchange 557
KATJANGÖL (*n*) : ground nut oil (see ERDNUSSÖL) 558
KATJANÖL (*n*) : (see ERDNUSSÖL) 559
KATT (*f*) : cat, cat of nine tails ; cat-ship (Naut.) 560
KATTANKER (*m*) : backing anchor (Naut.) 561
KATTBLOCK (*m*) : cat block (Naut.) 562
KATTDAVIT (*m*) : cat davit (Naut.) 563
KATTFALL (*n*) : cat fall (Naut.) 564
KATTHAKEN (*m*) : cat hook 565
KATTLÄUFER (*m*) : cat rope 566
KATTSPUREN (*f.pl*) : riders 567
KATTSTEERT (*m*) : night pennant, riding light 568
KATTTAKEL (*n*) : cat tackle 569
KATTUN (*m*) : calico ; cotton 570
KATTUNDRUCK (*m*) : calico printing 571
KATTUNDRUCKER (*m*) : calico printer 572
KATTUNDRUCKEREI (*f*) : calico printing -(works) 573
KATTUNEN : of (cotton) calico 574
KATTUNFABRIK (*f*) : calico factory ; cotton mill 575
KATTUN (*m*), **GEDRUCKTER-** : print 576
KATTUNLEINWAND (*f*) : union, linen with a cotton weft 577
KATTUNPORPHYR (*m*) : tufaceous porphyry (see PORPHYRTUFF) 578
KATTUNPRESSE (*f*) : calico press 579
KATTUNWEBER (*m*) : calico weaver 580
KÄTZCHEN (*n*) : kitten ; catkin ; amentum (Bot.) 581
KÄTZCHENBLÜTLER (*m*) : amentaceous plant (Bot.) 582
KATZE (*f*) : cat (domestic=*Felis domestica* ; wild=*Felis catus*) ; thread in the stuff (Paper) ; catkin ; amentum (Bot.) ; travelling crab (Mech.) 583
KATZENARTIG : feline ; cat-like 584
KATZENAUGE (*f*) : cat's eye, sunstone (a variety of quartz ; see QUARZ ; KORUND and CHRYSO-BERYLL) 585
KATZENAUGENHARZ (*n*) : Dammar resin (see DAMMAR) 586
KATZENBLEI (*n*) : (a form of) mica 587
KATZENDARM (*m*) : cat-gut 588
KATZENGLAS (*n*) : cat gold ; yellow mica 589
KATZENGLIMMER (*m*) : cat gold ; yellow mica ; glimmer 590
KATZENGOLD (*n*) : (see KATZENGLIMMER) 591
KATZENKLAR (*n*) : (a form of) mica 592
KATZENKOPF (*m*) : hemicephalus ; (*pl*) : cat's paws (Naut.) 593
KATZENKRAUT (*n*) : cat-thyme (*Teucrium marum*) 594
KATZENLUCHS (*m*) : (see LUCHS) 595
KATZENMINZE (*f*) : catnip ; catmint ; field balm (a herb, *Nepeta cataria*) 596
KATZENMÜNZE (*f*) : (see KATZENMINZE) 597
KATZENPETER (*m*) : (see OHRSPEICHELDRÜSEN-ENTZÜNDUNG) 598
KATZENPFÖTCHEN (*n*) : everlasting (*Gnaphalium diaecum*) (Bot.) ; (*pl*) : crow's feet 599
KATZENPFÖTCHEN, ROTE- (*n.pl*) : cat's foot flowers (*Flores gnaphalii rosei*) 600
KATZENSAPHIR (*m*) : cordierite (see CORDIERIT) 601
KATZENSILBER (*n*) : cat silver, mica, white mica 602
KATZENSTERZ (*m*) : horse-tail-(grass), equisetum, bottlerush, field horsetail (*Equisetum hyemale* and *E. arvense*) 603
KATZENWEDEL (*m*) : horse-tail-(grass), equisetum (see KATZENSTERZ) 604
KATZENWURZ (*f*) : valerian root (see BALDRIAN-) 605
KATZENZINN (*n*) : wolfram (see WOLFRAM) 606
KAUBAR : masticable ; capable of being chewed 607
KAUDERN : to gobble ; gibber ; jabber ; higgle ; haggle ; bargain ; negotiate 608
KAUDERWÄLSCH : gibberish, jargon ; (*n*) : gibberish, jargon, rot, twaddle 609
KAUEN : to chew ; masticate ; gnaw 610
KAUERN : to cower, squat, fawn, cringe, crouch, stoop, shrink 611
KAUF (*m*) : purchase ; bargain ; buying 612
KAUFANSCHLAG (*m*) : estimate, bid, quotation, offer, tender ; sale-ticket, sale-placard 613
KAUFBAR : purchaseable, obtainable, procurable 614
KAUFBLEI (*n*) : commercial lead 615
KAUFBRIEF (*m*) : contract of (sale) purchase ; bill of sale 616
KAUFBUCH (*n*) : purchase-book, journal (Com.) 617
KAUFEN : to buy ; purchase ; procure ; obtain 618
KÄUFER (*m*) : buyer ; purchaser ; customer ; client ; merchant ; trader 619
KAUFFAHRER (*m*) : merchantman (Naut.) 620
KAUFFAHRTEISCHIFF (*n*) : merchantman, merchant ship, merchant vessel (Naut.) 621
KAUFGELD (*n*) : purchase money 622
KAUFGUT (*n*) : merchandise 623
KAUFHANDEL (*m*) : trade ; commerce 424
KAUFHAUS (*n*) : warehouse 625
KAUFHERR (*m*) : (wholesale)-merchant 626
KAUFKONTRAKT (*m*) : contract to purchase ; bill of sale 627
KAUFKRAFT (*f*) : purchasing power 628

KAUFLADEN (*m*): store; shop 629
KAUFLEUTE (*pl*): buyers, merchants, tradesmen 630
KÄUFLICH: purchaseable; saleable; commercial; marketable; procurable; obtainable; corruptible; venal 631
KÄUFLICHKEIT (*f*): corruptibility 632
KAUFLUST (*f*): active buying, keen competition 633
KAUFMANN (*m*): merchant; salesman; tradesman; retailer; shopkeeper; trader; commercial traveller 634
KAUFMÄNNISCH: mercantile; commercial; business-like 635
KAUFMANNSCHAFT (*f*): mercantile community, commerce 636
KAUFMANNSGUT (*n*): goods, merchandise, wares, commodities 637
KAUFMANNSINNUNG (*f*): trading company 638
KAUFPLATZ (*m*): market-(place) 639
KAUFPREIS (*m*): first-cost, prime-cost; purchase price, buying-price 640
KAUFSCHOSS (*m*): stamp duty (on conveyances) 641
KAUFSTEUER (*f*): stamp-duty (on conveyance) 642
KAUFSUMME (*f*): purchase (money) price 643
KAUFVERTRAG (*m*): contract-(to purchase); bill of sale 644
KAUFWEISE: by purchase 645
KAUFWERT (*m*): purchase price; (marketable)-value 646
KAUFZINK (*n*): commercial zinc 647
KAULBARS (*m*): (see KAULBARSCH) 648
KAULBARSCH (*m*): Ruffe, Pope (a fresh water perch, *Acerina cernua*) 649
KAULBÖRS (*m*): (see KAULBARSCH) 650
KAULFROSCH (*m*): tadpole 651
KÄULIG: ball-like, spherical-(ly), globular, globose, globate, globe-shaped, round, globulous 652
KAULQUAPPE (*f*): tadpole 653
KAUM: scarce-(ly); hardly; difficultly; with difficulty 654
KAUMITTEL (*n*): masticatory; chewing gum (anything for chewing, as Betel nut, from *Areca catechu*) 655
KAUPFEFFER (*m*): Betel pepper; betel leaf (*Piper betle*) 656
KAURI (*m*): Cowrie, Kauri, animé, gum copal, resin copal, copal (a resin from various trees) (see also DAMMAR and KOPAL) (Sg. 1.018-1.067) 657
KAURIHOLZ (*n*): Cowrie pine, Kauri (*Agathis australis*); Sg. 0.5-0.6 658
KAURIKOPAL (*m*): kauri copal (from *Dammara australis*) (see also KOPAL) 659
KAUSALITÄT (*f*): causality 660
KAUSALITÄTSGESETZ (*n*): law of causality 661
KAUSCH (*f*): bullseye, thimble (of ropes) 662
KAUSCHE (*f*): bullseye, thimble (of ropes) 663
KAUSTIK (*f*): caustic, mordant, corrosive, etching fluid; etching (the art) 664
KAUSTIKA (*pl*): (see KAUSTIK) 665
KAUSTISCH: caustic; mordant; erosive; corrosive; eating; biting 666
KAUSTIZIEREN: to causticize; eat; bite; mordant; corrode; erode; cauterize 667
KAUSTIZIERUNG (*f*): causticization; cauterization 668
KAUSTIZITÄT (*f*): causticity 669
KAUTABAK (*m*): chewing tobacco; plug; twist 670
KAUTEL (*f*): precaution 671
KAUTERISIEREN: to cauterize; cauterise; sear; burn 672
KAUTION (*f*): bail, security, surety, bond 673
KAUTION STELLEN: to (give) go bail, stand bail, act as surety, give security 674
KAUTSCHEN: to couch; press (Paper) 675
KAUTSCHUK (*m*): caoutchouc; elaterite; rubber (*Gummi elasticum*) (empirical formula = $C_{10}H_{16}$); Sg. 0.9 676
KAUTSCHUKABFÄLLE (*m.pl*): caoutchouc (rubber) waste 677
KAUTSCHUKÄHNLICH: rubber-like 678
KAUTSCHUKERSATZSTOFF (*m*): caoutchouc (rubber) substitute 679
KAUTSCHUKFABRIKAT (*n*): caoutchouc (rubber) wares, manufactures or article 680
KAUTSCHUKFIRNIS (*m*): rubber varnish 681
KAUTSCHUKGEWEBE (*n*): covered elastic 682
KAUTSCHUKHARZ (*n*): caoutchouc (rubber) resin (the organic impurities in crude caoutchouc) 683
KAUTSCHUKIN (*n*): (see KAUTSCHUKÖL) 684
KAUTSCHUKKITT (*m*): caoutchouc cement, rubber solution (caoutchouc dissolved in organic solvents) 685
KAUTSCHUKLATEX (*m*): rubber latex (see LATEX) 686
KAUTSCHUKLIDERUNG (*f*): caoutchouc packing 687
KAUTSCHUKMILCH (*f*): latex (see LATEX) 688
KAUTSCHUKÖL (*n*): caoutchouc dissolved in turpentine; caoutchouc oil (obtained from the dry distillation of caoutchouc; Sg. 0.64-0.87) 689
KAUTSCHUKPARAGRAPH (*m*): elastic (rule) paragraph (a rule which is capable of being made to fit one's wishes) 690
KAUTSCHUKPFLASTER (*n*): (india)-rubber (tape) plaster (*Collemplastrum adhæsivum extensum*; *Collemplastrum americanum elasticum*) (for bandaging) 691
KAUTSCHUK (*m*), **ROH-**: crude caoutchouc 692
KAUTSCHUKROHR (*n*): rubber tube 693
KAUTSCHUKRÖHRE (*f*): rubber tube 694
KAUTSCHUKSCHWAMM (*m*): rubber sponge 695
KAUTSCHUKSTOFF (*m*): rubber cloth; caoutchouc (see KAUTSCHUK) 696
KAUTSCHUKSUBSTANZ (*f*): caoutchouc (see KAUTSCHUK) 697
KAUTSCHUKSURROGAT (*n*): rubber substitute 698
KAUTSCHUKSYNTHESE (*f*): caoutchouc synthesis 699
KAUTSCHUKVERDICHTUNGSMATERIAL (*n*): rubber packing material 700
KAUTSCHUKWAREN (*f.pl*): rubber goods 701
KAUZ (*m*): (odd)-fellow; chap; screech (barn) owl (*Strix flammea*); tawny (brown) owl (*Syrnium aluco*) 702
KAUZAHN (*m*): molar, grinder, grinding (molar) tooth, double tooth 703
KAVAWURZEL (*f*): kava root (*Radix kava-kava* of *Macropiper methysticum*) 704
KAVERNE (*f*): cavity; cavern 705
KAWAHARZ (*n*): kawa gum 706
KECK: bold; hardy; daring; stout; rash; quick; lively; alert; pert; impudent; saucy; forward; fearless 707
KECKHEIT (*f*): boldness, rashness, hardiness, daring, audacity, impudence (see KECK) 708
KECKLICH: boldly, rashly, daringly (see KECK) 709

KEEP (*f*): notch, channel (Naut.) 710
KEFIR (*n*): kephir (a Caucasian beverage, from the action of the fungus, *Kephir grains* on cow's milk) 711
KEFIRKÖRNER (*n.pl*): kephir (fungus) grains 712
KEFIRPILZ (*m*): kefir (fungus) grains 713
KEGEL (*m*): cone; pin; skittle; nine-pin (see also SEGERKEGEL) 714
KEGEL (*m*), **ABGESTUMPFTER-**: truncated cone 715
KEGELÄHNLICH: conical, cone-like, conoidal 716
KEGELBAHN (*f*): skittle-alley 717
KEGELBRECHER (*m*): conical (breaker, crusher) grinder (mill for very hard materials) 718
KEGELFLÄCHE (*f*): conical surface 719
KEGELFLASCHE (*f*): conical flask 720
KEGELFÖRMIG: cone-shaped; conical; coniform 721
KEGEL (*m*), **GERADER-**: upright cone 722
KEGELGESTALT (*f*): conic form 723
KEGELGLAS (*n*): conical glass 724
KEGEL (*m*), **HAHN-**: cone (plug) of a cock 725
KEGELIG: cone-shaped, conical, coniform 726
KEGELIGE SPITZEN (*f.pl*): cones, conical shaped recesses-(in a settling tank for coal and waste washing. The mud settles at the bottom point of the cone and the water runs slowly away at the top) 727
KEGELLINIE (*f*): parabola 728
KEGELLINIG: parabolical 729
KEGELN: to bowl; roll; throw; play at skittles; form a cone; make of cone shape 730
KEGELRAD (*n*): cone (conical) wheel; bevel wheel 731
KEGELRADMASCHINE (*f*): bevelling machine 732
KEGELSCHNITT (*m*): conic section 733
KEGELSCHNITTLINIE (*f*): ellipse; trajectory; parabola; hyperbola 734
KEHLADER (*f*): jugular vein (Med.) 735
KEHLBRÄUNE (*f*): quinsy (Med.) (see BRÄUNE) 736
KEHLBUCHSTABE (*m*): guttural-(letter) 737
KEHLDECKE (*f*): arched (floor) ceiling 738
KEHLDECKEL (*m*): epiglottis (Med.) 739
KEHLE (*f*): throat; larynx (Med.); channel; throttle; flute; gutter; gorge; neck-piece (Mech.); throat-piece; channeling; groove 740
KEHLEN: to channel, flute, chamfer, groove 741
KEHLGESCHWULST (*f*): goitre, bronchocele, Derbyshire neck (an enlargement of the thyroid gland) (Med.) 742
KEHLHOBEL (*m*): fluting (chamfering, cornice, moulding, hollowing) plane 743
KEHLIG: grooved, fluted, channelled 744
KEHLKNOCHEN (*m*): larynx (see KEHLKOPF) 745
KEHLKNORPEL (*m*): larynx (see KEHLKOPF) 746
KEHLKOPF (*m*): Adam's apple; larynx; (as a prefix—laryngeal) (Med.) 747
KEHLKOPFENTZÜNDUNG (*f*): (acute)-catarrhal laryngitis (*Laryngitis catarrhalis*) (Med.) 748
KEHLKOPFKATARRH (*m*): laryngeal catarrh 749
KEHLKOPFÖDEM (*n*): œdema of the (glottis) larynx; œdema glottidis; phlegmonous laryngitis (Med.) 750
KEHLKOPFPFEIFEN (*n*): wheezing 751
KEHLKOPFPOLYPEN (*m.pl*): polypi (growths, tumours) of the larynx (Med.) 752
KEHLKOPFSPIEGEL (*m*): (see KEHLSPIEGEL) 753
KEHLKOPFTUBERKULOSE (*f*): tuberculosis of the larynx, laryngeal tuberculosis 754
KEHLLAUT (*m*): guttural sound 755
KEHLRINNE (*f*): gutter 756
KEHLSCHNITT (*m*): laryngotomy, tracheotomy (Surgical) (opening of the windpipe and a metal tube inserted) 757
KEHLSCHWEISSUNG (*f*): corner (angle, valley, channel) weld 758
KEHLSCHWINDSUCHT (*f*): laryngeal phthisis (Med.) 759
KEHLSPIEGEL (*m*): laryngoscope (Med.) (a mirror for exploring the larynx and naso-pharynx) 760
KEHLTON (*m*): guttural (tone, accent) sound 761
KEHLUNG (*f*): fluting, grooving, chamfering, hollowing, moulding 762
KEHLZÄPFLEIN (*n*): uvula (a conical prolongation hanging from the soft palate) (Med.) 763
KEHLZIEGEL (*m*): corner-tile, gutter-tile 764
KEHR (*f*): turn-(ing), return 765
KEHRE (*f*): (see KEHR) 766
KEHREN: to turn; return; sweep; brush 767
KEHRER (*m*): sweeper 768
KEHRICHT (*n*): sweepings; rubbish; refuse; muck; dust 769
KEHRSALPETER (*m*): saltpetre sweepings 770
KEHRSALZ (*n*): salt sweepings 771
KEHRSEITE (*f*): reverse-(side) 772
KEICH (*m*): asthma, shortness of breath 773
KEICHEN: to pant, gasp, puff, blow, cough, labour for breath; (*n*): puffing, panting, convulsive breathing 774
KEICHHUSTEN (*m*): (see KEUCHHUSTEN) 775
KEIFEN: to scold; chide; squabble; upbraid; yelp (of dogs) 776
KEIL (*m*): wedge; cuneus; key; cotter; key-stone; (thunder)-bolt; pin 777
KEILÄHNLICH: wedge-shaped; cuneate; cuneiform; sphenoid 778
KEILARTIG: (see KEILÄHNLICH) 779
KEILBEIN (*n*): sphenoid (cuneiform) bone 780
KEILBOLZEN (*m*): wedge bolt 781
KEILEN: to key; wedge; split; cleave; thrash 782
KEILER (*m*): wild-boar (*Sus scrofa*) 783
KEILFÖRMIG: wedge-shaped; cuneate; cuneiform; sphenoid 784
KEILHACKE (*f*): pick-axe 785
KEILHAUE (*f*): pick-axe 786
KEILINSCHRIFT (*f*): cuneiform inscription 787
KEILKOMPENSATION (*f*): wedge (compensation) compensating arrangement or device 788
KEILLOCHHAMMER (*m*): stone splitting hammer, stone breaking hammer (a mechanical tool for splitting stone, concrete etc.) 789
KEILNUT (*f*): key-way 790
KEILNUTE (*f*): key-way 791
KEILPAAR (*n*): pair of wedges; double wedge 792
KEILPRESSE (*f*): wedge press 793
KEILRAD (*n*): mortice wheel 794
KEILRING (*m*): wedge-shaped ring, key-ring, conical ring 795
KEILSCHRIFT (*f*): cuneiform characters 796
KEILSTEIN (*m*): wedge brick; tapered (feather) brick 797
KEILSTÜCK (*n*): wedge 798
KEILTREIBER (*m*): key (wedge) driver 799
KEILVERBINDUNG (*f*): keying 800
KEIM (*m*): germ; origin; embryo; shoot; sprig; bud; nucleus (of crystallization) 801
KEIMAPPARAT (*m*): germinating apparatus 802
KEIMARRETIERUNG (*f*): germ arresting (by disinfecting wounds, as by painting with iodine) 803

KEIMBLATT (*n*) : cotyledon ; the seed leaf (Bot.) 804
KEIMBLÄTTERIG : cotyledonous 805
KEIMBLATTLOS : acotyledonous 806
KEIMEN : to germinate ; shoot ; bud ; begin ; sprout ; spring up 807
KEIMFÄHIG : germinable ; capable of germinating 808
KEIMFÄHIGKEIT (*f*) : germinability, germinating (capability) capacity 809
KEIMHAUT (*f*) : blastoderm (primitive layer of cells from subdivision of germ cell after fertilization) 810
KEIMHÜLLE (*f*) : perisperm (remains of nucleus of an ovule which is not all absorbed during development of the embryo) 811
KEIMKNOSPE (*f*) : prolific germ 812
KEIMKRAFT (*f*) : germinating (power) capacity ; vitality 813
KEIMLING (*m*) : germ ; embryo (Bot.) 814
KEIMSTOFF (*m*) : blastema ; germinal matter 815
KEIMSUBSTANZ (*f*) : (see **KEIMSTOFF**) 816
KEIMTÖTEND : germicidal 817
KEIMUNG (*f*) : germination 818
KEIMWIRKUNG (*f*) : germ action ; action of nuclei 819
KEIMWÜRZELCHEN (*n*) : radicle, plant embryo 820
KEIMZELLE (*f*) : germ (embryonic) cell 821
KEIN : no ; not (one) a ; none ; no one ; not any 822
KEINERLEI : not of any (kind) sort 823
KEINERSEITS : not on either side 824
KEINESFALLS : not in any case ; not on any account ; in no wise 825
KEINESWEGS : not by any means ; not at all 826
KEINMAL : not once ; never ; not at all 827
KEKUNAÖL (*n*) : candlenut oil (from *Aleurites triloba*) 828
KELCH (*m*) : cup ; chalice ; goblet ; calyx ; infundibulum 829
KELCHARTIG : cup-like 830
KELCHBLATT (*n*) : sepal (the individual leaf which goes to make up the calyx of a flower) 831
KELCHFÖRMIG : cup-shaped 832
KELCHGLAS (*n*) : cup-shaped test glass ; chalice glass ; tumbler 833
KELCHMENSUR (*f*) : measuring cup 834
KELEN (*n*) : kelene (see **ÄTHYLCHLORID**) 835
KELLE (*f*) : ladle ; scoop ; trowel ; fish-slice 836
KELLEN : to ladle ; scoop 837
KELLER (*m*) : cellar 838
KELLEREI (*f*) : cellarage 839
KELLERGESCHOSS (*n*) : basement 840
KELLERGRUNDFLÄCHE (*f*) : cellar floor space 841
KELLERHALS (*m*) : mezereon (*Daphne mezereum*) ; spurge laurel (*Daphne laureola*) ; cellar entrance 842
KELLERSCHNECKE (*f*) : slug 843
KELLERZINS (*m*) : cellarage 844
KELLNER (*m*) : waiter ; butler ; bar-tender ; barman 845
KELOID (*n*) : cheloid, cheloma, keloid (Med.) 846
KELP (*m*) : kelp, barilla, sodium carbonate (the ashes from seaweed burning) (see **NATRIUMKARBONAT**) ; kelp, sea-weed, tangle (*Laminaria digitata* and *L. stenophylla* ; contains sodium carbonate, iodine and potash) 847
KELTER (*f*) : wine-press ; oil-press 848
KELTEREIGERÄTE (*n.pl*) : grape pressing (appliances) apparatus 849
KELTERER (*m*) : wine-presser 850
KELTERFASS (*n*) : grape-tub 851
KELTERN : to press (wine) grapes 852
KELTERZUBER (*m*) : grape-tub 853
KENNBAR : knowable ; recognizable ; distinguishable ; remarkable 854
KENNEN : to know ; be acquainted with 855
KENNER (*m*) : connoisseur ; judge ; professional ; knower 856
KENNLINIE (*f*) : graph, curve, diagram, characteristic line 857
KENNTLICH : cognizable, knowable, remarkable, discernible, distinguishable 858
KENNTNIS (*f*) : knowledge ; skill ; information ; notice ; cognizance ; science 859
KENNTNIS, IN-SETZEN : to advise ; apprise ; inform 860
KENNWORT (*n*) : codeword (Com.) 861
KENNZAHL (*f*) : index (of logarithm) 862
KENNZEICHEN (*n*) : mark ; stamp ; sign ; token ; indication ; symbol ; characteristic ; symptom ; criterion ; distinguishing mark 863
KENNZEICHNEN : to characterize ; mark ; denote ; distinguish 864
KENNZEICHNEND FÜR : characteristic of 865
KENNZIFFER (*f*) : index (of logarithms) 866
KENNZUG (*m*) : characteristic- (feature) 867
KENTERN : to capsize ; turn turtle ; overturn ; cant ; list ; heel over ; tip (up) over) 868
KERAMIK (*f*) : ceramics ; science (art) of pottery 869
KERAMIKLACK (*m*) : ceramic varnish 870
KERAMISCH : ceramic 871
KERAMISCHE FARBE (*f*) : potter's colours ; ceramic colours 872
KERAMISCHE MASSE (*f*) : clay 873
KERAMISCHER BETRIEB (*m*) : pottery work-(s) 874
KERAMOHALIT (*m*) : (see **HAARSALZ**) 875
KERARGYRIT (*m*) : cerargyrite (see **CERARGYRIT**) 876
KERB (*m*) : notch, nick, groove, indent-(ation) 877
KERBBIEGEPROBE (*f*) : notch bending test-(piece) 878
KERBE (*f*) : notch ; nick ; groove ; indent-(ation) 879
KERBEL (*m*) : chervil (*Anthriscus cerefolium*) 880
KERBEN : to notch, nick, groove, indent, incise, cut, mill (Coins) 881
KERBGRUND (*m*) : bottom of the notch 882
KERBIG : notched, serrated, jagged, milled 883
KERBSÄGE (*f*) : cross-cut saw 884
KERBSCHLAGPROBE (*f*) : notch impact (test) test-piece 885
KERBWIRKUNG (*f*) : notch effect 886
KERBZÄHIGKEIT (*f*) : notch (strength) tenacity 887
KERBZÄHNIG : notched, indented ; crenate-(d) (Botany) 888
KERKER (*m*) : jail ; prison ; dungeon ; gaol 889
KERKERN : to incarcerate, imprison, commit (send) to prison, confine, immure 890
KERMESBEEREN (*f.pl*) : (see **KERMESKÖRNER**) 891
KERMESIN (*n*) : crimson (see **KARMOISIN**) 892
KERMESIT (*m*) : Kermesite (see **ANTIMONBLENDE**) 893
KERMESKÖRNER (*n.pl*) : kermes grains (dried female of kermes insect) 894
KERMESSCHILDLAUS (*f*) : kermes insect (*Coccus ilicis*) 895
KERN (*m*) : nucleus ; kernel ; pip ; stone ; pith ; core ; heart ; gist ; curd (Soap) ; grain ; marrow ; essence ; substance 896

KERNBOHNEN (*f.pl*): haricot beans 897
KERNEN: to take the (pips; core; stone) kernel out of; granulate; pick; seed 898
KERNFARBE (*f*): nuclear stain 899
KERNFAUL: rotten to the core 900
KERNFEST: solid as a rock 901
KERNFLEISCH (*n*): pulp, pith 902
KERNFRUCHT (*f*): stone fruit 903
KERNGEBUNDEN: united to the nucleus 904
KERNGEHÄUSE (*n*): core 905
KERNGESUND: sound (healthy) to the core; thoroughly (sound) healthy 906
KERNGUSS (*m*): (founding)-cored work 907
KERNHAFT: solid; strong; pithy; robust; vigorous 908
KERNHEFE (*f*): seed yeast 909
KERNHOLZ (*n*): heart-wood; doura; duramen; (the inner dark rings of tree trunks) 910
KERNIG: strong; substantial; stout; solid; robust; vigorous; full (Liquor); compact (Leather); pithy; full of (pips) seeds; pregnant 911
KERNIGKEIT (*f*): fullness (Leather); body; pithiness (see also KERNIG) 912
KERNISOMERIE (*f*): nuclear isomerism 913
KERNKAMMER (*f*): cell 914
KERNMASS (*n*): calibre, bore 915
KERNMEHL (*n*): firsts; best grade; first quality -(flour) 916
KERNOBST (*n*): stone fruit 917
KERNÖL (*n*): rape-oil, rapeseed oil (see RÜBÖL) 918
KERNRISS (*m*): cracks in tree trunks, extending from the centre of the trunk towards the periphery 919
KERNSALZ (*n*): rock salt 920
KERNSAND (*m*): core sand 921
KERNSCHUSS (*m*): point-blank shot 922
KERNSEIFE (*f*): curd (grain) soap; superfine soap 923
KERNSPRUCH (*m*): pithy saying, apophthegm 924
KERNSTEIN (*m*): coccolith (see KOKKOLITH) 925
KERNSUBSTANZ (*f*): nucleus, nuclear substance 926
KERNSYNTHESE (*f*): nuclear synthesis 927
KERNTALG (*m*): (see ROHKERN) 928
KERNTEIL (*m*): core 929
KERNTHEORIE (*f*): nuclear theory 930
KERNTROCKEN: thoroughly (bone) dry 931
KERNTRUPPEN (*pl*): picked troops 932
KERNWAREN (*f.pl*): first quality goods or wares 933
KERNWOLLE (*f*): prime wool; best kind of wool 934
KERNZONE (*f*): core (of steel or metal) 935
KERZE (*f*): candle; taper; wax-light; burner 936
KERZE (*f*), KOMPOSITIONS-: composition candle (candle formed from a variety of materials) 937
KERZE (*f*), MISCH-: composition candle (made from a mixture of paraffin and stearin) 938
KERZENBAUM (*m*): candle tree (*Parmentiera cerifera*) 939
KERZENBEERBAUM (*m*): candle-nut, candleberry tree (*Aleurites triloba* and *A. moluccana*) 940
KERZENBEERSTRAUCH (*m*): candleberry-myrtle, wax myrtle (*Myrica cerifera*) 941
KERZENDOCHT (*m*): candle wick 942
KERZENFABRIKATION (*f*): candle manufacture 943
KERZENFISCH (*m*): candle fish, enlachon, onlachan (*Thaleicthys pacificus*) 944
KERZENGERADE: straight as a die 945
KERZENGIESSMASCHINE (*f*): candle-making machine 946
KERZENMACHER (*m*): candle-maker 947
KERZENMASSE (*f*): candle-grease, material from which the candle is made 948
KERZENNUSSÖL (*n*): (see BANKULNUSSÖL) 949
KERZENSCHEIN (*m*): candle-light, the light of (tapers, a candle) candles 950
KERZENSTÄRKE (*f*): candle power 951
KERZENSTRAUCH (*m*): candleberry, bayberry, tallow-shrub, wax (myrtle) berry, candleberry myrtle (*Myrica cerifera*) 952
KERZE (*f*), SPIRITUSHARTWACHS-: solid spirit wax candle (containing solid spirit, see HARTSPIRITUS) 953
KERZE (*f*), STEARIN-: stearin candle 954
KERZE (*f*), TALG-: tallow candle 955
KERZE (*f*), WACHS-: wax candle 956
KESSEL (*m*): kettle; cauldron; boiler; copper; basin; reservoir; kennel; vessel; kier; tank (of drum shape); drum; shell 957
KESSELABDECKUNG (*f*): boiler covering, boiler cover, lagging 958
KESSELABDECKUNGSMATERIAL (*n*): boiler lagging, non-conducting composition, non-conducting material 959
KESSELABLASSHAHN (*m*): boiler blow-off cock, mud (sludge) cock 960
KESSELABMESSUNGEN (*f.pl*): boiler dimensions, size of boiler 961
KESSELAGGREGAT (*n*): boiler (installation) unit 962
KESSELANKER (*m*): truss bolt, boiler stay 963
KESSELANLAGE (*f*): boiler (installation) plant 964
KESSELANZUG (*m*): boiler covering 965
KESSELARMATUREN (*f.pl*): boiler mountings 966
KESSELART (*f*): type of boiler 967
KESSELASCHE (*f*): potash (see POTTASCHE) 968
KESSELAUFHÄNGUNG (*f*): boiler suspension, method of suspension of a boiler 969
KESSEL AUS PORZELLAN: porcelain vessel 970
KESSELBATTERIE (*f*): battery of boilers; pair of boilers; mortar-battery 971
KESSELBAUART (*f*): type of boiler 972
KESSELBAUER (*m*): boiler maker, boiler engineering firm, boiler constructor 973
KESSELBEDIENUNG (*f*): boiler (attendants) attendance 974
KESSELBEKLEIDUNG (*f*): boiler casing 975
KESSELBETRIEB (*m*): boiler operation 976
KESSELBLECH (*n*): boiler-plate 977
KESSELBOCK (*m*): drum-stool, saddle 978
KESSELBODEN (*m*): drum (head) end, (either manhole end or plain end) 979
KESSELBOLZEN (*m*): boiler stay 980
KESSELDECKE (*f*): boiler top 981
KESSELDRUCK (*m*): boiler pressure, working-pressure 982
KESSELEINHEIT (*f*): boiler unit 983
KESSELEINMAUERUNG (*f*): boiler-brickwork, boiler-setting, setting, brickwork-setting, brickwork 984
KESSELELEMENT (*n*): boiler (section) element, section (element) of tubes 985
KESSELENTWURF (*m*): boiler (proposal, project) design 986
KESSELEXPLOSION (*f*): boiler explosion 987
KESSELFABRIK (*f*): boiler works 988
KESSELFABRIKANT (*m*): boiler-maker, boiler (constructor) manufacturer 989

KESSELFEUERUNG (*f*): firebox, furnace, combustion chamber 990
KESSELFÖRMIG: kettle (drum) shaped; cauldron (basin) shaped; hollow-(ed); concave 991
KESSELFLICKER (*m*): tinker 992
KESSELFUNDAMENT (*n*): boiler seating or foundation 993
KESSELGARNITUR (*f*): boiler fittings 994
KESSELGRÖSSE (*f*): size of boiler 995
KESSELHAUS (*n*): boiler-house 996
KESSELHAUSEINRICHTUNG (*f*): boiler-house equipment 997
KESSELHAUSFLUR (*f*): boiler-house floor 998
KESSELHEIZER (*m*): fireman, stoker 999
KESSELHEIZFLÄCHE (*f*): boiler heating surface 000
KESSELHÖHE (*f*): height of boiler 001
KESSELINHALT (*m*): boiler (volume) capacity 002
KESSELISOLIERUNG (*f*): boiler-covering, lagging, non-conducting (composition) material 003
KESSELKOHLE (*f*) steam (boiler) coal 004
KESSELKONTROLLE (*f*): boiler inspection, boiler (supervision, superintendence) management 005
KESSELKÖRPER (*m*): boiler shell 006
KESSELLAGERUNG (*f*): boiler(setting)seating 007
KESSELLEISTUNG (*f*): boiler (duty, rating, load, capacity) output 008
KESSELLEISTUNGSVERSUCH (*m*): load test, evaporative test, steaming test (of boilers) 009
KESSELMACHER (*m*): boiler-maker; coppersmith 010
KESSELMANTEL (*m*): boiler-shell; boiler-casing 011
KESSELMAUERWERK (*n*): boiler brickwork, brickwork-setting, boiler-setting, setting, brickwork, seating 012
KESSELNUMMER (*f*): boiler-number (usually to be found on the name plate) 013
KESSEL, OBER- (*m*): upper drum; top drum; steam drum 014
KESSELPAUKE (*f*): kettle-drum 015
KESSELPLATTE (*f*): boiler plate 016
KESSELPROBE (*f*): boiler test, boiler trial 017
KESSELPRÜFER (*m*): boiler inspector 018
KESSELPRÜFUNG (*f*): boiler (inspection) testing 019
KESSELRAUM (*m*): boiler (room) house 020
KESSELRAUMSCHOTT (*n*): boiler room bulkhead (Naut.) 021
KESSELREIHE (*f*): battery (line) of boilers 022
KESSELREINIGUNG (*f*): (any)-boiler cleaning (such as sooting, scaling, etc.) 023
KESSELREINIGUNGSVORRICHTUNG (*f*): boiler cleaning (arrangement) device 024
KESSELREVISION (*f*): boiler inspection 025
KESSELREVISIONSGESELLSCHAFT (*f*): Boiler Inspection Company 026
KESSELROHR (*n*): boiler tube 027
KESSELRÖHRE (*f*): boiler tube 028
KESSELSCHILD (*n*): name-plate 029
KESSELSCHLAMMABLASS (*m*): boiler blow-off -(valve) 030
KESSELSCHMIED (*m*): brazier; boiler-smith; boiler-maker; copper-smith 031
KESSELSCHMIEDE (*f*): boiler-shop, boiler-(maker's) works 032
KESSELSCHMIEDEARBEIT (*f*): boiler (forging, smithing) smith's work 033
KESSELSCHUSS (*m*): shell-ring, course of plates 034
KESSEL (*m*), SEIFEN-: soap boiler 035
KESSELSPANNUNG (*f*): boiler pressure 036
KESSELSPEISEPUMPE (*f*): boiler feed pump 037
KESSELSPEISEWASSER (*n*): boiler feed-(water) 038
KESSELSPEISEWASSERREINIGUNG (*f*): feed water purification 039
KESSELSPEISUNG (*f*): boiler feeding 040
KESSELSTAHL (*m*): boiler steel 041
KESSELSTEIN (*m*): boiler scale, incrustation, fur (deposit) in boilers or kettles; compass-brick 042
KESSELSTEINABKLOPFER (*m*): boiler scaling tool 043
KESSELSTEINGEGENMITTEL (*n*): anti-scale-forming (material, substance, composition, remedy)-agent; boiler-compound; anti-incrustant; disincrustant 044
KESSELSTEINMITTEL (*n*): anti-incrustant, boiler compound (see KESSELSTEINGEGENMITTEL) 045
KESSELSTEINPULVER (*n*): anti-incrustant (disincrustant) powder 046
KESSELSTUHL (*m*): drum-stool 047
KESSELSTÜTZE (*f*): boiler (drum) stool, boiler support-(s), boiler seating, boiler framing (of water tube boilers) 048
KESSELSTUTZEN (*m*): stand-pipe 049
KESSELTAL (*n*): cauldron-shaped valley 050
KESSELTEIL (*m*): boiler-part; (*pl*): spares, spare-parts 051
KESSELTRÄGER (*m*): boiler (drum) stool, boiler support, boiler seating, boiler framing (of water tube boilers) 052
KESSELTYP (*m*): type of boiler 053
KESSEL, UNTER- (*m*): lower (mud) drum 054
KESSELUNTERSUCHUNG (*f*): boiler (inspection) testing 055
KESSELUNTERSUCHUNGSBEHÖRDE (*f*): (official)-boiler (inspection) testing authorities 056
KESSELVERKLEIDUNG (*f*): boiler casing, or lining 057
KESSELVERLUSTE (*m.pl*): boiler-losses, total losses through the boiler 058
KESSELVERNIETUNG (*f*): riveting of (boiler-plates, drum-plates) shell-plates 059
KESSELVERSTÄRKUNG (*f*): (boiler)-stay-bolt, (or any kind of boiler strengthening or stiffening arrangement) 060
KESSELWAGEN (*m*): railway tank waggon; tank (car) truck 061
KESSELWAND (*f*): boiler shell, boiler wall 062
KESSELWANDUNG (*f*): boiler shell; drum-plate (either shell-plate or tube plate); boiler walls 063
KESSELWÄRTER (*m*): boiler attendant 064
KESSELWASSERPRÜFER (*m*): boiler-water tester (for determining the hardness of boiler feed water) 065
KESSELZUBEHÖR (*n*): boiler accessories 066
KESSELZUG (*m*): boiler (flue) pass 067
KESSLER (*m*): brazier; tinker 068
KESSLERSCHE FLUATE (*n.pl*): soluble metal fluosilicates for protecting stone and brick against weather influences 069
KETOFORM (*f*): keto form 070
KETON (*n*): ketone (class of organic compounds related to aldehydes) (Chem.) 071
KETONIMID (*n*): ketone imide (such keton in which the oxgyen of the carbonyl group is replaced by the group NH) 072
KETONÖL (*n*): ketone oil 073
KETONSÄURE (*f*): keto acid; ketonic acid 074
KETONSPALTUNG (*f*): ketonic cleavage 075
KETSCH (*f*): ketch (a two masted vessel) (Naut.) 076

KETTE (*f*): catena; chain; succession; train; sequence; series; cordon; warp (of fabric); slavery; bondage; bond; cell (Elect.); element; circuit; concatenation; (*pl*): fetters 077
KETTE (*f*), ANKER-: anchoring (tightening, straining) chain 078
KETTE (*f*), BLOCK-: block chain 079
KETTE (*f*), ENDLOSE-: endless chain 080
KETTE (*f*), GEGEN-: back-stay chain 081
KETTE (*f*), GELENK-: sprocket chain (chain made up of flat link plates) 082
KETTE (*f*), GLIEDER-: open link chain 083
KETTEL (*f*): small chain 084
KETTE (*f*), LASCHEN-: sprocket chain (chain made up of flat link plates) 085
KETTELHAKEN (*m*): cabin hook 086
KETTELN: to fasten (bind) with a small chain 087
KETTEN: to chain; bind; fetter; tie; link; connect 088
KETTENANKER (*m*): moorings (Naut.) 089
KETTENARTIG: chain-like; catenarian; catenary 090
KETTENBLOCK (*m*): chain-block 091
KETTENBOGEN (*m*): catenary 092
KETTENBOLZEN (*m*): link-pin-(of a chain) 093
KETTENBRUCH (*m*): fractional progression, continued fraction (Maths.) 094
KETTENBRÜCKE (*f*): chain (suspension) bridge 095
KETTENDRUCK (*m*): warp-printing 096
KETTENFLASCHENZUG (*m*): chain pulley block 097
KETTENFÖRMIG: (see KETTENARTIG) 098
KETTENFRÄSE (*f*): chain cutter moulding machine 099
KETTENFRÄSMASCHINE (*f*): chain cutter moulding machine 100
KETTENGARN (*n*): warp, water (Textile) 101
KETTENGEBIRGE (*n*): chain of mountains, mountain-chain 102
KETTENGLIED (*n*): link-(of a chain) 103
KETTENHAKEN (*m*): chain hook 104
KETTENHUND (*m*): watch-dog 105
KETTENISOMERIE (*f*): chain isomerism 106
KETTENKASTEN (*m*): chain locker 107
KETTENKLÜSE (*f*): chain pipe 108
KETTENKUPPELUNG (*f*): chain coupling 109
KETTENLASCHE (*f*): link plate 110
KETTENLINIE (*f*): catenary, chain, catenarian (chain) curve or line 111
KETTENLOS: chainless; unchained 112
KETTENNIETUNG (*f*): parallel (chain) riveting 113
KETTENNUSS (*f*): chain wheel 114
KETTENRAD (*n*): sprocket wheel; chain wheel; chain pulley 115
KETTENRECHNUNG (*f*): (see KETTENREGEL) 116
KETTENREGEL (*f*): double rule of three, chain-rule (Arithmetic) 117
KETTENRING (*m*): chain-ring 118
KETTENROHR (*n*): chain pipe 119
KETTENROLLE (*f*): chain (wheel, sheave) pulley 120
KETTENROST (*m*): chain grate-(stoker) 121
KETTENROSTSTAB (*m*): chain grate stoker link 122
KETTENSCHAKE (*f*): chain link 123
KETTENSCHÄKEL (*m*): chain shackle 124
KETTENSCHLUSS (*m*): sorites, a logical formula 125
KETTENSCHUTZ (*m*): chain guard 126
KETTENSTAB (*m*): flat link 127
KETTENSTICH (*m*): chain stitch (Embroidery); chain-knot 128
KETTENSTOPPER (*m*): chain stopper 129
KETTENTRIEB (*m*): chain drive 130
KETTENTRILLER (*m*): sustained (shake) trill (Music) 131
KETTENTROMMEL (*f*): chain drum 132
KETTENVERANKERUNG (*f*): chain anchorage 133
KETTENWERK (*n*): chains; chain-work 134
KETTENWINDE (*f*): chain jack 135
KETTENWIRBEL (*m*): chain wheel 136
KETTENZUG (*m*): chain pulley block 137
KETTE (*f*), ROLLE-: roller chain 138
KETTE (*f*), RÜCKHALT-: back-stay chain 139
KETTE (*f*), SCHIFFS-: ship's chain 140
KETTE (*f*), SPANN-: tightening (straining) chain 141
KETTE (*f*), TRAG-: suspension chain 142
KETTE (*f*), ZAHN-: sprocket chain 143
KETTLER (*m*): chain-maker 144
KETUNÖL (*n*): (see BANKULNUSSÖL) 145
KETZER (*m*): heretic 146
KETZEREI (*f*): heresy 147
KETZERHAFT: heretical 148
KETZERISCH: heretical 149
KETZERLICH: heretical 150
KEUBEL (*m*): screen, sieve 151
KEUCHEN: to gasp; pant; puff; blow (see also KEICHEN) 152
KEUCHHUSTEN (*m*): whooping-cough 153
KEUCHHUSTENMITTEL (*n*): remedy for whooping cough 154
KEULE (*f*): pestle; club; leg; joint (of meat) 155
KEULEN: to club, strike with a club 156
KEULENFÖRMIG: club-shaped; clavate; claviform (Zoology and Botany) 157
KEULENKÜRBIS (*m*): bottle-gourd, calabash (see KALABASSE) 158
KEUSCH: chaste; pure; modest 159
KEUSCHHEIT (*f*): chastity; purity; modesty; continence 160
KIBDELOPHAN (*m*): (see TITANEISEN) 161
KICHERERBSE (*f*): chick pea (*Cicer arietinum*) (a leguminous plant) 162
KIDLEDER (*n*): kid-(leather) 163
KIEFE (*f*): jawbone, jaw, maxilla; gill-(of a fish) 164
KIEFEN: (as a prefix; maxillary) 165
KIEFENKNOCHEN (*m*): maxilla (Anatomy) 166
KIEFENMUSKEL (*m*): muscle of the jaw 167
KIEFER (*m*): jaw-bone; jaw; maxilla; gill (of a fish) 168
KIEFER (*f*): Scotch (fir) pine (*Pinus sylvestris*); pine, fir, deal, red or yellow fir, yellow deal (see also FICHTE) 169
KIEFERGEHÖLZ (*n*): pine grove 170
KIEFERIG: jawed, having jaws 171
KIEFERN: fir, of fir 172
KIEFERNADEL (*f*): pine needle 173
KIEFERNADELÖL (*n*): pine needle oil (*Oleum pini*) (see FICHTENNADELÖL) 174
KIEFERNEULE (*f*): pine moth (*Panolis piniperda*) 175
KIEFERNHARZ (*n*): pine resin; colophony; Sg. 1.08; Mp. 100-140°C. 176
KIEFERNHOLZ (*n*): pine wood 177
KIEFERNNADEL (*f*): pine needle 178
KIEFERNSAMENÖL (*n*): (see FÖHRENSAMENÖL) 179
KIEFERNSTOFF (*m*): fir pulp (Paper) 180
KIEFERTEER (*m*): pine tar 181
KIEFERTEERÖL (*n*): pine-tar oil (distilled from pine tar) (Sg. 0.97) 182
KIEFERZAPFEN (*m*): pine-cone, fir-cone 183

KIEL (*m*): quill; keel (of ships); straw; halm; feather; carina (keel on the breast-bone of birds) 184
KIELBANK (*f*): gridiron; careening wharf (Naut.) 185
KIELBOLZEN (*m*): keel bolt 186
KIELEN: to feather, furnish with feathers, become fledged; to fit with a keel 187
KIELFEDER (*f*): quill, feather 188
KIELFÖRMIG: carinate, keel-shaped 189
KIELGANG (*m*): garboard strake (Naut.) 190
KIELGANGPLANKE (*f*): garboard plank 191
KIELGANGPLATTE (*f*): garboard plate 192
KIELHOLEN: to heel over, lay over on its side, to careen, keel-haul (of ships) 193
KIELKLOTZ (*m*): keel block 194
KIELLASCHUNG (*f*): keel scarf 195
KIELLINIE (*f*): middle line 196
KIELPLATTE (*f*): keel plate 197
KIELRAUM (*m*): hold (Naut.) 198
KIELSCHWEIN (*n*): keelson 199
KIELSCHWEIN (*n*), **EINGESCHOBENES-**: intercostal keelson 200
KIELSCHWERT (*n*): sliding keel 201
KIELSTÜCK (*n*): keel piece 202
KIELSTÜTZE (*f*): keel shore 203
KIELÜBERZUG (*m*): false keel 204
KIELWASSER (*n*): wake (Naut.) 205
KIEME (*f*): gill; branchia (of fish) 206
KIEN (*m*): resinous pine-(wood) 207
KIENAPFEL (*m*): pine cone 208
KIENBAUM (*m*): pine-(tree) (see **KIEFER**) 209
KIENEN: pine, fir, of (pine) fir 210
KIENFACKEL (*f*): pine-torch 211
KIENHARZ (*n*): (see **KIEFERNHARZ**) 212
KIENHOLZ (*n*): resinous root-wood of pine tree 213
KIENIG: resinous 214
KIENÖL (*n*): pine oil; oil of turpentine; Sg. 0.855-0.876; Bp. 156-161°C. 215
KIENRUSS (*m*): pine soot; soot (lamp) black 216
KIENTEER (*m*): pine tar (see **HOLZTEER**) 217
KIENTEERPECH (*n*): pine pitch 218
KIENZOPF (*m*): bladder-rust (of pine trees) (*Peridermium pini*) 219
KIES (*m*): pyrites; gravel; Sg. 1.75 220
KIESABBRAND (*m*): calcined (roasted) pyrites (see also **SCHWEFELKIES**) (60% Fe, and 0.3-5.0% S) 221
KIESARTIG: pyritic, gravelly; gritty; pyritous (see also **KIESIG**) 222
KIESBODEN (*m*): gritty (gravelly) soil 223
KIESEL (*m*): flint; pebble; silica (Si); (see **SILIZIUM** and **SILICIUM**) (as a prefix=siliceous; silicic; siliciferous; silica; flint; gravel) 224
KIESELARTIG: flinty; siliceous 225
KIESELERDE (*f*): silica; siliceous earth; silicic acid (see **KIESELSÄURE**) 226
KIESELFLUORALUMINIUM (*n*): aluminium fluosilicate 227
KIESELFLUORAMMONIUM (*n*): ammonium fluosilicate 228
KIESELFLUORBARYUM (*n*): barium fluosilicate (see **BARYUMSILICOFLUORID**) 229
KIESELFLUORBLEI (*n*): lead fluosilicate 230
KIESELFLUORCALCIUM (*n*): calcium fluosilicate (see **CALCIUMSILICOFLUORID**) 231
KIESELFLUOREISEN (*n*): iron fluosilicate 232
KIESELFLUORKALIUM (*n*): potassium fluosilicate (see **KALIUMSILICOFLUORID**) 233
KIESELFLUORKUPFER (*n*): copper fluosilicate (see **KUPFERSILICOFLUORID**) 234
KIESELFLUORMAGNESIUM (*n*): magnesium fluosilicate (see **MAGNESIUMSILICOFLUORID**) 235
KIESELFLUORMANGAN (*n*): manganese (fluosilicate) silicofluoride (see **MANGANSILICO-FLUORID**) 236
KIESELFLUORNATRIUM (*n*): sodium fluosilicate (see **NATRIUMSILICOFLUORID**) 237
KIESELFLUORSALZ (*n*): fluosilicate; silicofluoride 238
KIESELFLUORSTRONTIUM (*n*): strontium (fluosilicate) silicofluoride 239
KIESELFLUORVERBINDUNG (*f*): fluosilicate 240
KIESELFLUORWASSERSTOFF (*m*): sand acid; hydrofluosilicic (hydrosilicofluoric) acid; silicofluoric acid; fluosilicic acid; H_2SiF_6 (*Acidum hydrosilicofluoricum*) 241
KIESELFLUORWASSERSTOFFSAUER: fluosilicate of, combined with fluosilicic acid 242
KIESELFLUORWASSERSTOFFSÄURE (*f*): (see **KIESEL-FLUORWASSERSTOFF**) 243
KIESELFLUORZINK (*n*): zinc fluosilicate 244
KIESELFLUSSSÄURE (*f*): fluosilicic acid (see **KIESELFLUORWASSERSTOFF**) 245
KIESELGALMEI (*m*): siliceous calamine; hydrous zinc silicate (see **KIESELZINKERZ**) 246
KIESELGESTEIN (*n*): quartz rocks 247
KIESELGIPS (*m*): vulpinite (form of anhydrite) (see **ANHYDRIT**) 248
KIESELGLAS (*n*): flint glass; siliceous calamine (see **KIESELZINKERZ**) 249
KIESELGRUND (*m*): pebbly ground 250
KIESELGUHR (*f*): diatomaceous earth; earthy opal (see **OPAL**); kieselguhr; fossil meal; infusorial earth; siliceous earth; tripolite (see also **SILICIUMDIOXYD** and **INFUSORIEN-ERDE**) 251
KIESELGUHR (*m*), **GESCHLÄMMTER-**: elutriated infusorial earth 252
KIESELGUHRSTEINE (*m.pl*): fossil meal bricks 253
KIESELGUR (*f*): (see **KIESELGUHR**) 254
KIESELHALTIG: siliceous; siliciferous; containing silica 255
KIESELHART: flint-hard, hard as flint 256
KIESELIG: flinty, siliceous 257
KIESELKALK (*m*): siliceous limestone 258
KIESELKALKEISEN (*n*): ilvaite (see **ILVAIT**) 259
KIESELKALKSCHIEFER (*m*): siliceous (schist; slate) shale (see **QUARZ**) 260
KIESELKALKSPAT (*m*): wollastonite (see **WOLLA-STONIT**) 261
KIESELKALKSTEIN (*m*): siliceous limestone 262
KIESELKLUMPEN (*m*): pudding-stone, consolidated gravel, conglomerate (a rock formed from water-worn debris of the size of pebbles and upwards, consolidated by a clay matrix) 263
KIESELKREIDE (*f*): siliceous chalk 264
KIESELKRYSTALL (*m*): silica crystal; rock crystal (see **BERGKRYSTALL**) 265
KIESELKUPFER (*n*): chrysocolla; $CuO,SiO_2,2H_2O$; Sg. 2.15 266
KIESELMALACHIT (*m*): chrysocolla 267
KIESELMANGAN (*n*): rhodonite (see also **RHODONIT** and **MANGANSPAT**) 268
KIESELMETALL (*n*): silicon (see **SILICIUM**) 269
KIESELPANZER (*m*): siliceous skeleton (of dead plants) 270
KIESELSAND (*m*): gravel 271
KIESELSANDSTEIN (*m*): siliceous sandstone 272
KIESELSAUER: silicate of; combined with silicic acid 273

KIESELSAUERES SALZ (*n*): silicate; $M_m(SiO_3)_n$ 274

KIESELSÄURE (*f*): silicic acid; SiO_2 (see SILIZIUMDIOXYD) 275

KIESELSÄUREANHYDRID (*n*): silicic anhydride; silicon dioxide; SiO_2 (see SILIZIUMDIOXYD) 276

KIESELSÄURE (*f*), **GESCHMOLZENE-**: quartz glass (see QUARZGLAS) 277

KIESELSÄUREHALTIG: siliceous; containing silicic acid 278

KIESELSÄUREHYDRAT (*n*): silicic acid; H_2SiO_3 279

KIESELSÄURESALZ (*n*): (see KIESELSAUERES SALZ) 280

KIESELSÄURESEIFE (*f*): silicated soap 281

KIESELSAURES SALZ (*n*): silicate; $M_m(SiO_3)_n$ 282

KIESELSCHIEFER (*m*): siliceous (slate, shale) schist (see QUARZ) 283

KIESELSINTER (*m*): siliceous sinter; opal (see OPAL) 284

KIESELSPAT (*m*): albite (see ALBIT) 285

KIESELSTEIN (*m*): gravel stone; stone in gravel; flint (pebble) stone 286

KIESELSTEIN (*m*): (see KIESEL, and KIESELKLUMPEN) 287

KIESELSTOFF (*m*): silicon (see SILICIUM) 288

KIESELTON (*m*): argillaceous schist 289

KIESELWASSERSTOFF (*m*): hydrogen silicide 290

KIESELWASSERSTOFFSÄURE (*f*): hydrosilicic acid 291

KIESELWEISS (*n*): silex white; siliceous (whiting) whitening 292

KIESELWISMUT (*n* and *m*): eulytite, bismuth blende (see WISMUTBLENDE) 293

KIESELWOLFRAMSÄURE (*f*): silicotungstic (silicowolframic) acid; $4H_2SiO_2.12WO_3.22H_2O$ 294

KIESELZINK (*n*): calamine (see GALMEI) 295

KIESELZINKERZ (*n*): calamine, siliceous calamine, electric calamine, hemimorphite, Smithsonite, Willemite; (natural hydrous zinc silicate; about 55% zinc); $Zn_2(OH)_2SiO_3$; Sg. 3.45 (see also ZINKSPAT) 296

KIESEN: to choose, select, pick, elect, cull 297

KIESERIT (*m*): Kieserite (natural magnesium sulphate) from Stassfurt district of Germany); $MgSO_4.H_2O$; (Sg. 2.54) 298

KIESERITWÄSCHEREI (*f*): kieserite (washery) washing 299

KIESFILTER (*n*): gravel filter 300

KIESGANG (*m*): gravel-walk 301

KIESHALTIG: pyritiferous; gravelly; pyritous; containing gravel 302

KIESIG: gravelly; flinty; gritty; siliceous; silicic; siliciferous; resembling (containing) gravel or flints 303

KIESSAND (*m*): gravel, coarse sand 304

KIESWEG (*m*): gravel-walk 305

KIESZINK (*n*): (see KIESELZINKERZ) 306

KILOAMPERE (*n*): kiloampere 307

KILOGRAMM (*n*): kilogram (2.205 lbs.) 308

KILOGRAMMETER (*n*): kilogram-meter 309

KILOGRAMMKALORIE (*f*): kilogramcalorie (always designated by a large C, as opposed to a small c for the gram calorie) 310

KILOMETER (*n*): kilometer (about $\frac{5}{8}$ mile) 311

KILOWATT (*n*): kilowatt 312

KIMM (*f*): bilge; (*m*): horizon 313

KIMMBEPLANKUNG (*f*): bilge planking 314

KIMMBEPLATTUNG (*f*): bilge plating 315

KIMME (*f*): notch, edge, border, brim; horizon 316

KIMMEN: to notch; fit with a brim 317

KIMMGANG (*m*): bilge strake 318

KIMMKIEL (*m*): bilge keel 319

KIMMKIELSCHWEIN (*n*): bilge keelson 320

KIMMPLANKE (*f*): bilge plank 321

KIMMPLATTE (*f*): bilge plate 322

KIMMSENTE (*f*): bilge harpin 323

KIMMSTRINGER (*m*): bilge stringer 324

KIMMSTÜCK (*n*): second futtock (Naut.) 325

KIMMSTÜTZPLATTE (*f*): bilge bracket 326

KIMMUNG (*f*): bilge; floor-heads (Naut.) 327

KIMMWASSER (*n*): bilge-water 328

KIMMWEGER (*m* and *pl*): ceiling of the floor heads 329

KIMMWEIDEN (*f.pl*): basket-frame 330

KIND (*n*): child; infant; baby; babe; embryo; foetus 331

KINDBETT (*n*): childbed (Med.) 332

KINDBETTFIEBER (*n*): puerperal (childbed) fever (see also PUERPERALFIEBER); miliary fever (Med.) 333

KINDERBALSAM (*m*): soothing syrup 334

KINDERBLATTERN (*pl*): small-pox (Med.) 335

KINDERGICHTER (*f.pl*): convulsions-(of children) (Med.) 336

KINDERJAHRE (*n.pl*): infancy, childhood 337

KINDERLÄHMUNG (*f*): infantile-(spinal) paralysis (Med.) 338

KINDERMORD (*m*): infanticide (the action) 339

KINDERMÖRDER (*m*): infanticide (the doer) 340

KINDERNÄHRMITTEL (*n*): infant food 341

KINDERNÖTE (*f.pl*): labour-(of childbirth) 342

KINDERPECH (*n*): meconium (Med.) 343

KINDERPOCKEN (*pl*): small-pox (Med.) 344

KINDERPUDER (*m*): infant powder 345

KINDERRAUB (*m*): kidnapping 346

KINDERSCHLEIM (*m*): vernix caseosa 347

KINDERSCHMIERE (*f*): (see KINDERSCHLEIM) 348

KINDERSPIEL (*n*): child's play, simple (easy) matter 349

KINDERZAHN (*m*): milk-tooth 350

KINDESNOT (*f*): labour-(of childbirth) 351

KINDESPECH (*n*): meconium (Med.) 352

KINDESWASSER (*n*): amniotic fluid (a fluid round the embryo) 353

KINDHAFT: child-like, childish 354

KINDHEIT (*f*): childhood; infancy 355

KINDISCH: childish-(ly) 356

KINDLICH: child-like, filial 357

KINDSPECH (*n*): meconium (Med.) 358

KINEMA (*f*): cinema 359

KINEMATIK (*f*): kinematics 360

KINEMATOGRAPH (*m*): cinematograph 361

KINEMATOGRAPHISCH: cinematographic 362

KINEMOGRAPH (*m*): kinemograph, recording kinemometer 363

KINEMOMETER (*n*): kinemometer 364

KINESIATRIK (*f*): kinesitherapeutics (see KINESITHERAPIE) 365

KINESITHERAPIE (*f*): kinesitherapeutics, cure by movement (Med.) 366

KINETIK (*f*): kinetics 367

KINETISCH: kinetic 368

KINETISCHE ENERGIE (*f*): kinetic energy 369

KINN (*n*): chin 370

KINNBACKEN (*m*): jaw; jawbone; maxilla (Med.); mandible (of animals) 371

KINNBACKENDRÜSE (*f*): maxilliary gland (Med.) 372

KINNBACKENKRAMPF (*m*): lock-jaw, tetanus (Med.) 373
KINNBACKENÖL (*n*): dolphin (porpoise) oil (from jawbone of *Delphinus globiceps* and *Delphinus phocœna*) 374
KINNBACKENZWANG (*m*): lock-jaw, tetanus (Med.) 375
KINNGRÜBCHEN (*n*): (see **KINNGRUBE**) 376
KINNGRUBE (*f*): dimple-(in the chin) 377
KINNGRÜBLEIN (*n*): (see **KINNGRUBE**) 378
KINNLADE (*f*): jaw-(bone); maxilla (Med.) 379
KINO (*n*): resin (gum) kino, kino (bark extract of *Pterocarpus marsupium*) (a dyeing and tanning material) 380
KINOBAUM (*m*): Dhak tree, Palas tree (*Butea frondosa*) 381
KINOBAUMÖL (*n*): Dhak tree (palas tree) oil (from *Butea frondosa*) 382
KINOFILM (*f*): cinematograph film 383
KINOGERBSÄURE (*f*): kinotannic acid 384
KINO (*n*), **MALABAR-**: malabar kino 385
KINOMASCHINE (*f*): cinematograph machine 386
KINO (*n*), **PTEROCARPUS-**: (see **KINO**) 387
KINOSORTE (*f*): variety (sort, kind) of kino 388
KIPPAPPARAT (*m*): tilting (tipping; dumping) apparatus or device 389
KIPPBAR: tippable, tiltable, capable of being (tipped, tilted, up-ended) canted 390
KIPPBECHERWERK (*n*): bucket-tipping arrangement; bucket elevator 391
KIPPBELASTUNG (*f*): tipping (stress) load 392
KIPPBÜHNE (*f*): tipping platform 393
KIPPE (*f*): edge; brink; ledge; assay balance; see-saw; tipping (device) arrangement 394
KIPPEN: to tilt; tip; tip over; cant; upend; overturn; overbalance; capsize 395
KIPPER (*m*): tipping (device) arrangement; tipper; tipping (tray) waggon 396
KIPPGEFÄSS (*n*): tipping (bucket) vessel 397
KIPPKARRE (*f*): tip-cart 398
KIPPKARREN (*m*): tip-cart 399
KIPPKASTEN (*m*): tipping skip 400
KIPPKÜBEL (*m*): tipping skip 401
KIPPLAGER (*n*): swing (tipping, pivotal) bearing 402
KIPPMISCHER (*m*): tipping mixer (Metal) 403
KIPPMOMENT (*n*): tipping moment 404
KIPPOFEN (*m*): tilting furnace; tipping furnace 405
KIPPROST (*m*): tipping (dumping) grate 406
KIPPSCHALE (*f*): well dish (Phot.) 407
KIPPSCHER APPARAT (*m*): Kipp apparatus 408
KIPPSICHER: stable 409
KIPPSICHERHEIT (*f*): stability 410
KIPPVORRICHTUNG (*f*): tipping (device) arrangement 411
KIPPWAGEN (*m*): tipping (truck) waggon; dumping (car; waggon; cart) truck 412
KIPPWERK (*n*): tipping (device) arrangement 413
KIPPWINDE (*f*): tipping winch 414
KIRCHE (*f*): church (either the building or the abstract term "the Church") 415
KIRCHENBANN (*m*): excommunication 416
KIRCHENBAU (*m*): building (erection) of a church 417
KIRCHENBUCH (*n*): parish register 418
KIRCHENDIENER (*m*): sacristan; sexton 419
KIRCHENFEST (*n*): church (religious) feast or festival 420
KIRCHENFLUCH (*m*): anathema 421
KIRCHENGEBIET (*n*): diocese 422
KIRCHENGEMEINDE (*f*): parish 423
KIRCHENGESCHICHTE (*f*): ecclesiastical history 424
KIRCHENGESETZ (*n*): canon 425
KIRCHENGESETZLICH: canonical 426
KIRCHENGESETZMÄSSIG: canonical 427
KIRCHENJAHR (*n*): ecclesiastical year 428
KIRCHENRECHT (*n*): canon-law 429
KIRCHENSCHIFF (*n*): nave 430
KIRCHENSPALTUNG (*f*): schism 431
KIRCHFALK (*m*): kestrel, wind-hover (*Falco tinnunculus*) 432
KIRCHHOF (*m*): churchyard; cemetery; graveyard 433
KIRCHLICH: ecclesiastic-(al) 434
KIRCHSCHWALBE (*f*): black (house) martin (*Chelidon urbica*) 435
KIRCHSPRENGEL (*m*): diocese 436
KIRCHTURM (*m*): steeple, spire 437
KIRCHWEIHE (*f*): consecration-(of a church) 438
KIRMES (*f*): fair, wake 439
KIRNEN (*n*): the agglomeration of the fats and oils with the milk in the manufacture of margarine from Olein 440
KIRR: tame; tractable 441
KIRRE: (see **KIRR**) 442
KIRSCHÄTHER (*m*): cherry (ether) essence 443
KIRSCHBAUM (*m*): cherry tree (*Prunus avium*) 444
KIRSCHBAUMHOLZ (*n*): cherry wood (of *Prunus avium*); Sg. 0.51-0.71 445
KIRSCHBEISSER (*m*): haw-finch (*Coccothraustes vulgaris*) 446
KIRSCHBRANNTWEIN (*m*): cherry brandy 447
KIRSCHE (*f*): cherry 448
KIRSCHENSTIEL (*m*): cherry stalk 449
KIRSCHGUMMI (*n*): cherry gum (*Gummi nostras*); xylose (a gum with a cerasin content, from *Prunus avium*) 450
KIRSCHHARZ (*n*): (see **KIRSCHGUMMI**) 451
KIRSCHHOLZ (*n*): cherry wood (see **KIRSCHBAUMHOLZ**) 452
KIRSCHKERN (*m*): cherry-stone 453
KIRSCHKERNÖL (*n*): cherry kernel oil (from *Prunus cerasus*); Sg. 0.92 454
KIRSCHLORBEER (*m*): cherry laurel (*Prunus lauro-cerasus*) 455
KIRSCHLORBEERBLÄTTER (*n.pl*): cherry-laurel leaves (*folia lauro-cerasi* of *Prunus lauro-cerasus*) 456
KIRSCHLORBEERÖL (*n*): cherry laurel oil; Sg. 1.054-1.066 (from *Prunus lauro-cerasus*) 457
KIRSCHLORBEERWASSER (*n*): cherry-laurel water (a distillate from the leaves of *Prunus lauro-cerasus*) 458
KIRSCHPFLAUME (*f*): (see **PFLAUME, TÜRKISCHE-**) 459
KIRSCHROT: cherry-red (of colours and glowing coals) 460
KIRSCHROTGLÜHEND: cherry-red (of fires) 461
KIRSCHROTGLUT (*f*): cherry-red heat 462
KIRSCHSAFT (*m*): cherry juice 463
KIRSCHWASSER (*n*): cherry water; kirsch; kirschwasser (distilled from fermented cherry juice) 464
KIRSEI (*m*): kersey (a light wollen cloth, named after the village of Kersey in England) 465
KISSEN (*n*): cushion; pillow; bolster; pad 466
KISSENÜBERZUG (*m*): cushion-cover; pillow-case 467
KISSENZIEHE (*f*): cushion-cover; pillow-case 468
KISTCHEN (*n*): small (box; chest) case; casket 469

KISTE (*f*) : chest ; case ; box ; trunk ; coffer ; crate ; packing case 470
KISTENPLOMBEN (*f.pl*) : case (chest ; box) seals ; leaden seals 471
KISTENZUCKER (*m*) : muscovado ; unrefined sugar 472
KITON (*n*) : kiton (a water-soluble tar-preparation used for making macadam roads) 473
KITT (*m*) : cement ; lute ; mastic ; putty 474
KITTE (*f*) : bevy ; covey ; flock ; swarm ; throng ; party 475
KITTEL (*m*) : smock ; frock ; gown 476
KITTEN : to cement ; lute ; putty ; glue 477
KITT (*m*), **FEUERFESTER-** : fireproof cement 478
KITTÖL (*n*) : putty oil 479
KITZEL (*m*) : tickling ; itching ; longing ; desire ; titillation 480
KITZELN : to tickle ; itch ; gratify ; please ; titillate 481
KLADDE (*f*) : first (rough) draft ; rough sketch ; day-book ; waste-book 482
KLADSTEIN (*m*) : place brick (brick not fully burned) 483
KLAFF (*m*) : bark, yelp ; ejaculation, shout 484
KLAFFEN : to gape ; yawn ; be ajar ; split ; open ; burst ; rupture ; spring ; bark ; clamour ; yelp 485
KLÄFFEN : to bark ; clamour ; yelp ; brawl 486
KLAFFSTELLE (*f*) : rupture, hole, gap, burst, opening, yawning hole, sprung place 487
KLAFTER (*f*) : cord (of wood) ; fathom (6 ft.) 488
KLAFTERHOLZ (*n*) : cord (stack) wood 489
KLAFTERIG : of a (cord) fathom, containing a cord 490
KLAFTERN : to (stack) pile up (wood) ; fathom 491
KLAFTERWEISE : by the (fathom) cord 492
KLAGBAR : deplorable, lamentable ; actionable 493
KLAGE (*f*) : complaint ; indictment ; grievance ; (law)-suit ; action ; accusation ; impeachment ; lament ; elegy ; moan ; wail ; dirge ; moaning ; plaint ; lamentation ; sorrow 494
KLAGEFALL (*m*) : accusative case (Grammar) 495
KLAGELIED (*n*) : (see KLAGGESANG) 496
KLAGEN : to lament ; grieve ; mourn ; complain ; sue ; accuse ; take action (at law) ; bewail ; deplore ; bemoan ; sorrow ; grumble ; litigate 497
KLAGEND : plaintive 498
KLAGENSWERT : deplorable, lamentable 499
KLÄGER (*m*) : plaintiff, accuser, litigant 500
KLAGEULE (*f*) : screech owl, barn owl (*Strix flammea*) 501
KLAGGESANG (*m*) : dirge, mournful song, threnody ; elegy, lament-(ation) 502
KLAGGESUCH (*n*) : complaint 503
KLÄGLICH : lamentable ; mournful ; wretched ; pitiful ; pitiable 504
KLAGLOS : uncomplaining 505
KLAGPUNKT (*m*) : point at issue ; charge, indictment 506
KLAGSACHE (*f*) : action, suit-(at law) 507
KLAGSCHRIFT (*f*) : plaint, accusation in writing ; actionable (paragraph) communication 508
KLAGTON (*m*) : plaintive (tone) voice 509
KLAMEIEISEN (*n*) : horsing iron 510
KLAMEIEN : to horse up 511
KLAMM : tight ; stiff ; close ; compact ; narrow ; clammy 512
KLAMMER (*f*) : clamp ; brace ; clasp ; holdfast ; cramp (iron) ; parenthesis ; curved bracket ( ) ; clip ; dowel (Mech.) 513
KLAMMER, ECKIGE- (*f*) : (square)-bracket [ ] 514
KLAMMERN : to cramp ; clamp ; rivet ; fasten ; clasp ; cling to ; brace ; bracket 515
KLAMMER (*f*), **RUNDE-** : round (curved) bracket ( ) 516
KLAMMERSATZ (*m*) : bracketed sentence, sentence in parenthesis 517
KLAMPE (*f*) : cleat, clamp, holdfast, brace, cramp-iron 518
KLANG (*m*) : sound ; tone ; ring ; clang ; timbre 519
KLANGBODEN (*m*) : sound-(ing) board 520
KLANGLEHRE (*f*) : acoustics 521
KLANGLOS : without sound, soundless, noiseless, mute, toneless, unmusical 522
KLANGMESSER (*m*) : phonometer (for measuring sound) 523
KLANGVOLL : full-sounding, robust, sonorous ; rich (of the voice) 524
KLAPPBANK (*f*) : folding bench 525
KLAPPBRÜCKE (*f*) : bascule bridge 526
KLAPPE (*f*) : flap ; cuff ; trap ; lid ; damper ; flap (clack) valve ; key ; stop ; leaf ; bascule ; drop (trap) door 527
KLAPPE (*f*), **FLIEGEN-** : fly-flap 528
KLAPPE (*f*), **KESSELRAUCH-** : boiler damper 529
KLAPPEN : to clap ; flap ; fit ; suit ; agree ; coincide ; strike ; clatter ; (as a prefix = valvular) 530
KLAPPENKOLBEN (*m*) : bucket, piston with flaps ; perforated piston 531
KLAPPENSITZ (*m*) : (valve)-seating 532
KLAPPENSITZFLÄCHE (*f*) : valve seating 533
KLAPPENVENTIL (*n*) : clack (flap) valve 534
KLAPPENVERSCHLUSS (*m*) : flap shutter (Phot.) 535
KLAPPER (*f*) : rattle, clapper ; flier (Weaving) 536
KLAPPERESPE (*f*) : (see ZITTERESPE) 537
KLAPPERN : to clatter, rattle ; chatter (of valves on their seats) 538
KLAPPERNUSS (*f*) : bladder-nut (*Staphylea*) 539
KLAPPERROSE (*f*) : corn (common) poppy (*Papaver rhoeas*) 540
KLAPPERSCHLANGE (*f*) : rattle snake (*Crotalus adamanteus*) 541
KLAPPERSCHLANGENWURZEL (*f*) : snakeroot, bugbane (*Actœa racemosa*) 542
KLAPPERSTEIN (*m*) : eagle stone 543
KLAPPFENSTER (*n*) : sky-light 544
KLAPPIG : valvular 545
KLAPPLÄUFER (*m*) : whip 546
KLAPPROST (*m*) : tipping (dumping) grate 547
KLAPPROTHIN (*m*) : lazulite (see LAZULIT) 548
KLAP-(P)-S (*m*) : clap, smack, slap 549
KLAPPSTUHL (*m*) : folding (stool) chair, camp-stool) 550
KLAPPTISCH (*m*) : folding table 551
KLAPPTÜR (*f*) : drop (trap) door 552
KLAPSEN : to clap ; smack ; slap ; flap 553
KLAR : clear ; distinct ; obvious ; comprehensible ; bright ; transparent ; evident ; plain ; pure ; rare (of air) ; pellucid ; limpid ; light (of colours) 554
KLÄRANLAGE (*f*) : clarifying plant 555
KLÄRAPPARAT (*m*) : clarifying (settling) apparatus 556
KLÄRBAD (*n*) : clearing bath (a solution used to clear negatives from stains due to development) (Photography) 557

KLÄRBASSIN (*n*): settling (tank) reservoir 558
KLÄRBOTTICH (*m*): clearing (clarifying; settling) tub or vat 559
KLÄRE (*f*): clarifier; clear (clarified) liquid; thin paste; clarity; clearness; delicacy; purity; coal-dust; bone-ash 560
KLÄREN: to clear; clarify; purify; fine; settle; defecate; clear up; cleanse; brighten; polish; become clear; rectify; debloom (see ENTSCHEINEN) 561
KLÄRFASS (*n*): clearing (clarifying) cask or tub; clarifier; settling (cask) tub; settler; fining or cleansing (cask) tub; tun 562
KLÄRFILTER (*n*): (clarifying)-filter 563
KLÄRFLASCHE (*f*): decanting (bottle) flask 564
KLÄRGEFÄSS (*n*): clarifier; clarifying (settling) apparatus, vessel or tank 565
KLARHEIT (*f*): clearness; brightness; clarity; transparency; plainness; evidence; rarity (of atmosphere); purity; lightness; fineness 566
KLARIEREN: to clear (Customs and Commercial) 567
KLARIFIZIEREN: to clarify, clear, purify, settle, fine, defecate, cleanse 568
KLÄRKASTEN (*m*): settler (Paper); clarifying tank 569
KLÄRKESSEL (*m*): clarifier 570
KLARKOCHEN: to boil till clear 571
KLÄRLICH: clearly, plainly, evidently, obviously, undoubtedly 572
KLÄRMITTEL (*n*): clarifying (clearing) agent; clarifier; fining 573
KLÄRPFANNE (*f*): clarifier; clearing (defecating) pan (Sugar) 574
KLÄRSEL (*n*): clarcé; clarified (refined) sugar; fine (filtered) liquor or syrup 575
KLÄRSTAUB (*m*): bone ash 576
KLÄRSUMPF (*m*): clarifying sump, settling (tank) sump 577
KLÄRUNG (*f*): clarification; clarifying; clearing, etc. (see KLÄREN) 578
KLÄRUNGSMITTEL (*n*): (see KLÄRMITTEL) 579
KLÄRZENTRIFUGE (*f*): clarifying centrifuge 580
KLASSE (*f*): class; rank; rate; rating- (of a ship) 581
KLASSENEINTEILUNG (*f*): classification 582
KLASSIERSIEB (*n*): classifying (sorting) sieve (Coal) 583
KLASSIFIKATION (*f*): classification 584
KLASSIFIZIEREN: to classify; arrange; class; group; dispose; distribute 585
KLASSIFIZIERUNG (*f*): classification 586
KLASTISCH: clastic (produced by disintegration of rock) (Geology) 587
KLATSCHE (*f*): gossip (the action); fly-flap 588
KLATSCHEN: to clap; smack; pop; splash; slap; crack; flap; clash; chat; gossip; babble; prate; tell tales; clatter; applaud 589
KLATSCHEREI (*f*): idle talk, gossip (the action) 590
KLATSCHGESCHICHTE (*f*): scandal; gossip 591
KLATSCHHAFT: gossiping, talkative, tell-tale 592
KLATSCHROSE (*f*): corn (common) poppy (*Papaver rhoeas*) (see MOHN) 593
KLATSCHWEIB (*n*): gossip (the person) 594
KLAU (*f*): jaw, claw; throat (Naut.) 595
KLAUBEN: to pick; sort; cull; sift 596
KLAUE (*f*): claw; paw; hoof; clutch; grasp; hook; jaw; fang; talon; (anchor)-fluke; ungula (Bot.) 597
KLAUEN: to claw, grasp, grip, clutch 598
KLAUENFETT (*n*): neat's-foot oil (*Oleum tauri pedum; Oleum bubulum*); Sg. 0.916 599
KLAUENFÖRMIG: ungulate 600
KLAUENHAMMER (*m*): claw-hammer 601
KLAUENKUPPELUNG (*f*): claw coupling, claw clutch 602
KLAUENMEHL (*n*): hoof meal 603
KLAUENÖL (*n*): (see KLAUENFETT) 604
KLAUENSEUCHE (*f*): foot-and-mouth disease (*Aptha epizootica*) (a cattle disease) 605
KLAUFALL (*n*): throat halliard (Naut.) 606
KLAUFALL (*n*), **BESAN-**: spanker throat-halliard 607
KLAUIG: provided with (having; fitted with) claws 608
KLAUSE (*f*): cell; hole; closet; pit; sink; hermitage; defile 609
KLAUSEL (*f*): clause; stipulation; condition; appendix; musical (period) phrase 610
KLAUSNER (*m*): hermit, recluse 611
KLAUSUR (*f*): confinement, isolation, seclusion; (book)-clasp 612
KLAVIER (*n*): clavier; clavichord; harpsichord; piano-(forte); key-board (having keys similar to a piano); console 613
KLAVIERAUSZUG (*m*): arrangement for the piano 614
KLAVIERDRAHT (*m*): piano wire 615
KLAVIERKASTEN (*m*): piano-case 616
KLAVIERSCHLÜSSEL (*m*): C clef; tuning-key 617
KLAVIERSTIMMER (*m*): piano-tuner 618
KLAVIERTASTEN (*f.pl*): piano keys (often erroneously referred to as "notes") 619
KLEBÄTHER (*m*): collodion (see COLLODIUM) 620
KLEBE (*f*): dodder (*Cuscuta*) (genus of parasitic plants) 621
KLEBEBAND (*n*): adhesive tape 622
KLEBELAUS (*f*): crab louse 623
KLEBEMASSE (*f*): adhesive substance, cement 624
KLEBEMITTEL (*n*): adhesive; agglutinant; gum; paste; any adhesive substance 625
KLEBEN: to stick; adhere; glue; gum; paste; cling; lute; cement; cleave; mount (Phot.) 626
KLEBEND: adhesive; agglutinant; glutinative; adherent; clinging; sticking 627
KLEBEPFLASTER (*n*): adhesive plaster; sticking plaster 628
KLEBEPFLASTER (*n*), **ENGLISCHES-**: court plaster 629
KLEBER (*m*): gluten; adhesive; gluer; sticker; gum; moutant (the substance used to stick a photograph to its mount) 630
KLEBERBROT (*n*): gluten bread 631
KLEBERIG: sticky; adhesive; viscid; glutinous; viscous; clammy 632
KLEBERIGKEIT (*f*): stickiness; viscosity; viscidity 633
KLEBERLEIM (*m*): gluten glue 634
KLEBERPROTEIN (*n*): gluten protein 635
KLEBESTOFF (*m*): adhesive substance; gluten; cement; mastic (see KLEBSTOFF) 636
KLEBKRAFT (*f*): adhesive power 637
KLEBKRAUT (*n*): cleavers; goose-grass (*Galium aparine*) 638
KLEBRICHT: (see KLEBRIG) 639
KLEBRIG: sticky; adhesive; viscid; glutinous; viscous; clammy 640
KLEBSAMENGEWÄCHSE (*n.pl*): *Pittosporaceæ* (Bot.) 641
KLEBSAND (*m*): luting sand (Ceramics) 642

KLEBSCHÄBEN (*f.pl*): awn, chaff (pieces of the broken flax stem) 643
KLEBSTOFF (*m*): adhesive substance; matrix; cement; mastic; paste; size; glue; gum; mucilage; any preparation for sticking purposes 644
KLECKEN: to blot; blur; daub; spot (see under KLECKS-) 645
KLECKS (*m*): blot; splotch; blur; spot 646
KLECKSEN: (see KLECKEN) 647
KLECKSER (*m*): dauber 648
KLECKSIG: blotted, covered with (spots) blots, spotted, blurred 649
KLECKSPAPIER (*n*): waste (scribbling) paper 650
KLEE (*m*): clover; trefoil (genus *Trifolium*) 651
KLEEBAUM (*m*): hop tree (*Ptelea trifoliata*) 652
KLEEBLATT (*n*): clover (trefoil) leaf 653
KLEEBLÜTEN (*f.pl*): trefoil-(flowers) (*Flores trifolii*) 654
KLEEBUSCH (*m*): common holly (*Ilex aquifolium*) 655
KLEE (*m*), **DREIBLÄTTRIGER-**: three-leafed clover 656
KLEE (*m*), **EWIGER-**: (purple)-medic, alfalfa, lucerne, Spanish (Brazilian, Chilean, French) clover (*Medicago sativa*) 657
KLEE (*m*), **LUZERNER-**: lucerne, alfalfa (see KLEE, EWIGER-) 658
KLEESALZ (*n*): salt of sorrel; acid potassium oxalate; potassium binoxalate (see KALIUM-BIOXALAT) 659
KLEESALZKRAUT (*n*): wood sorrel (*Oxalis acetosella*) 660
KLEESAUER: oxalate of; combined with oxalic acid 661
KLEESÄURE (*f*): oxalic acid (see OXALSÄURE) 662
KLEESEIDE (*f*): (flax)-dodder (see KLEBE) (*Cuscuta epilinum*) 663
KLEE (*m*), **SPANISCHER-**: lucerne, alfalfa (see KLEE, EWIGER-): sainfoin (see KLEE, TÜRKISCHER-) 664
KLEESTRAUCH (*m*): hop tree (*Ptelea trifoliata*) 665
KLEE (*m*), **TÜRKISCHER-**: sainfoin (*Onobrychis sativa*) (a leguminous fodder) 666
KLEI (*m*): clay; loam; marl 667
KLEIABSUDBAD (*n*): bran decoction (Dye) 668
KLEIACKER (*m*): clay (soil) land 669
KLEIBEN: to paste, glue, stick, adhere (see KLEBEN) 670
KLEIBODEN (*m*): clay soil 671
KLEID (*n*): garment; dress; clothing; garb; gown; habit; suit; raiment 672
KLEIDEN: to dress; clothe; fit; suit; become; adorn; cover; deck 673
KLEIDERBEHÄLTER (*m*): wardrobe 674
KLEIDERBÜRSTE (*f*): clothes-brush 675
KLEIDERHÄNDLER (*m*): (second-hand)-clothes-dealer 676
KLEIDERHANDLUNG (*f*): clothes-dealer's; second-hand clothes shop 677
KLEIDERMACHER (*m*): tailor 678
KLEIDERMACHERIN (*f*): dress-maker 679
KLEIDERSCHRANK (*m*): wardrobe 680
KLEIDERSTOFF (*n*): dress material, cloth, etc. 681
KLEIDERTRACHT (*f*): fashion, mode, costume 682
KLEIDSAM: becoming 683
KLEIDUNG (*f*): clothing; clothes; dressing; dress; costume; garment-(s); raiment; garb; suit; habit; drapery 684
KLEIDUNGSSTÜCK (*n*): garment 685
KLEIE (*f*): bran; pollard; clay; loam; marl 686
KLEIENARTIG: branny; clayey; furfuraceous 687
KLEIENBAD (*n*): bran (drench; steep) bath 688
KLEIENBEIZE (*f*): bran (drench; steep) liquid 689
KLEIENBROT (*n*): bran bread 690
KLEIENGRIND (*m*): pityriasis, furfur, porrigo (a generic term indicating branny exfoliation of the skin) (see also HAUTKLEIE) (Med.) 691
KLEIENMEHL (*n*): pollard 692
KLEIIG: branny, furfuraceous, clayey 693
KLEIN: minute; small; petty; little; trifling; short; mean; paltry; insignificant; (*n*): smalls (Mining); giblets 694
KLEINACHTEL (*n*): small octavo (Bookbinding) 695
KLEINASIEN (*n*): Asia Minor 696
KLEINBAHN (*f*): narrow gauge railway 697
KLEINBAUER (*m*): small farmer 698
KLEINBETRIEB (*m*): small business; work on a small scale 699
KLEINBILD (*n*): miniature 700
KLEINBOGENFORM (*f*): small folio 701
KLEINEN: to crush, break up into small pieces 702
KLEINERN: to reduce-(fractions) (Maths.) 703
KLEINES GUT (*n*): small (farm) estate 704
KLEINFÄRBER (*m*): clothes dyer 705
KLEINFÜGIG: insignificant, mean 706
KLEINGEFÜGE (*n*): micro structure (Metal) 707
KLEINGEISTIG: frivolous; mean-spirited; narrow-minded 708
KLEINGELD (*n*): small change 709
KLEINGUT (*n*): small (wares) articles 710
KLEINHANDEL (*m*): retail (trade) business 711
KLEINHÄNDLER (*m*): retailer 712
KLEINHEIT (*f*): smallness; minuteness; littleness; meanness; pettiness; insignificance 713
KLEINHERZIG: faint-hearted 714
KLEINHIRN (*n*): cerebellum (Med.) 715
KLEINIGKEIT (*f*): trifle; small matter; bagatelle; (*pl*): toys; trinkets; little things 716
KLEINKOKS (*m*): small coke, coke dust, coke breeze (size 0-10mm.; ash about 12%; moisture 10%; calorific value about 5200 cals.) 717
KLEINKÖRNIG: small-grained; minutely granular 718
KLEINKRUSTER (*m*): small crustacean 719
KLEINLAUT: discouraged; dejected, low-spirited, despondent, quiet 720
KLEINLEBEWESEN (*n*): micro-organism 721
KLEINLICH: petty; paltry; mean; frivolous; narrow (small) minded 722
KLEIN MACHEN: to find fault with, belittle, reduce, diminish, deprecate, depreciate, decry, disparage, abase, debase, traduce, underrate, under-value, malign, degrade, censure 723
KLEINMALER (*m*): miniature-painter 724
KLEINMASCHIG: fine-meshed, close-meshed 725
KLEINMESSER (*m*): micrometer (see MIKROMETER) 726
KLEINMÜHLE (*f*): crushing (pulverizing) mill 727
KLEINMÜTIG: faint-hearted, despondent, pusillanimous, discouraged, dejected 728
KLEINOD (*n*): jewel; gem; trinket; ornament; treasure; (*pl*): insignia 729
KLEINSCHMIED (*m*): lock-smith 730
KLEINSCHMIEDSWAREN (*f.pl*): hardware 731
KLEINSINNIG: narrow-minded 732

KLEINSTÄDTISCH: provincial 733
KLEINSTSPIEL (*n*): minimum play (Mechanical) 734
KLEINSTSPIEL (*n*), **NEGATIVES-**: medium force fit, tight fit 735
KLEINSÜCHTIG: mean, paltry, petty, base 736
KLEINVERKEHR (*m*): retail trade 737
KLEINWASSERRAUMKESSEL (*m*): boiler with small water-space; (often used for a boiler of the flash type) 738
KLEISTER (*m*): (thin)-paste; size; starch paste; moutant (see **KLEBER**) 739
KLEISTERN: to paste; size; stick 740
KLEISTERTIEGEL (*m*): paste-pot 741
KLEISTRIG: pasty; sticky 742
KLEMMBACKEN (*m* and *f.pl*): (clamp)-jaw-(s) (Mech.) 743
KLEMME (*f*): clamp; nippers; terminal; tongs; dilemma; difficulty; tight corner; embarrassment; defile; very narrow passage 744
KLEMMEN: to pinch; squeeze; bind; cramp; stick; jamb; lock; seize (Mech.) 745
KLEMMENSPANNUNG (*f*): terminal voltage 746
KLEMMER (*m*): pinchcock; pince-nez; miser 747
KLEMME (*f*), **VERBINDUNGS-**: binding clamp 748
KLEMMFALL (*m*): dilemma, difficulty, embarrassment, tight corner 749
KLEMMFUTTER (*n*): clamping (elastic) chuck 750
KLEMMHAKEN (*m*): cramp-iron, grappling iron, holdfast 751
KLEMMPLATTE (*f*): clamp, cleat, clip 752
KLEMMSCHRAUBE (*f*): set-screw; binding (clamp) screw; screw-clamp; terminal (Elect.) 753
KLEMMUNG (*f*): sticking, seizing, jambing 754
KLEMPERN: to work in tin 755
KLEMPNER (*m*): tinker; brazier 756
KLEMPNERARBEIT (*f*): tin-work, tin-ware 757
KLEMPNERWARE (*f*): tin-ware 758
KLETTE (*f*): bur (flower-head of burdock); burdock; bardana; clotbur (*Arctium lappa*) 759
KLETTEN: to stick to, adhere, cling to; card-(wool) 760
KLETTENKERBEL (*m*): bastard parsley (*Anthriscus*) 761
KLETTENKOPF (*m*): bur (the flower head of Burdock, *Arctium lappa*) 762
KLETTENÖL (*n*): (see **KLETTENSAMENÖL**) 763
KLETTENSAMENÖL (*n*): burdock-(seed)-oil (from seeds of *Arctium lappa*) 764
KLETTENWURZEL (*f*): lappa; burdock root (*Radix bardanæ*) (from *Lappa officinalis*) 765
KLETTENWURZELÖL (*n*): burdock root oil 766
KLETTERN: to climb; clamber; scramble 767
KLICKE (*f*): clique; set; coterie 768
KLIMA (*n*): climate; clime 769
KLIMAXZUCKER (*m*): climax sugar (a starch sugar) (Brewing) 770
KLIMMEN: to climb; clamber 771
KLIMPERN: to strum (on a musical instrument); jingle; tinkle 772
KLINGE (*f*): blade (of a knife); sword 773
KLINGEL (*f*): bell; hand-bell 774
KLINGELDRAHT (*m*): bell-wire 775
KLINGELN: to ring; tinkle; jingle 776
KLINGELTRANSFORMATOR (*m*): bell transformer 777
KLINGELZIEHER (*m*): bell-pull 778
KLINGEN: to sound; ring; clink; tinkle; jingle; chime; resound; reverberate 779
KLINGEN (*n*): ringing; sounding; chiming; head-noises, noises in the ears (Med.) 780
KLINGEND: sounding, resonant, sonorous, resounding 781
KLINGENPROBE (*f*): sample blade; blade test 782
KLINGENSTAHL (*m*): steel for knife or sword blades 783
KLINGGLAS (*n*): flint glass (see **KIESELGLAS**) 784
KLINGSTEIN (*m*): clinkstone; phonolite (a volcanic rock consisting of nepheline, sanidine feldspar and other constituents) (see **PHONOLITH**) 785
KLINIK (*f*): clinic, hospital 786
KLINISCH: clinical 787
KLINKE (*f*): latch; handle; (door)-catch; jack 788
KLINKEN: to latch or unlatch (**AUFKLINKEN**, to open; **ZUKLINKEN**, to shut) 789
KLINKENKANTE (*f*): tripping edge 790
KLINKER (*m*): clinker; Dutch brick; clinker (from fuel; Sg. 1.7-2.0) 791
KLINKERBAU (*m*): clinker (clincher) work 792
KLINKERBOOT (*n*): clinker (clincher)-built boat 793
KLINKE (*f*), **SPERR-**: pawl 794
KLINKE (*f*), **TÜR-**: latch of a door, door latch 795
KLINKHAKEN (*m*): (latch)-hook or staple 796
KLINOCHLOR (*m*): clinochlore, chlorite (Mineral) (see **CHLORIT**) 797
KLINOHUMIT (*m*): clinohumite; $Mg_7[Mg(F.OH)]_2[SiO_4]_4$; Sg. 3.15 798
KLIPPE (*f*): cliff; rock; crag 799
KLIPPER (*m*): clipper; sculptor's mallet 800
KLIPPIG: rocky, craggy 801
KLIRREN: to clink; clatter; clash; clank; rattle; jangle 802
KLISCHE (*n*): block (Printing) 803
KLISCHIERMETALL (*n*): a complex alloy having a low melting point 804
KLOAKE (*f*): cloaca; sewer 805
KLOAKENWASSER (*n*): sewage 806
KLOBEN (*m*): pulley; block; vice; pincers; log; balk (of timber); clamp; cramp; holdfast; staple 807
KLOBEN (*m*), **FEIL-**: filing vice 808
KLOBENHOLZ (*n*): log-wood, wood for logs 809
KLOBENSÄGE (*f*): frame-saw 810
KLOBENSEIL (*n*): pulley-rope 811
KLOPF (*m*): knock, rap, blow, bang 812
KLOPFEN: to beat; knock; tap; rap; throb 813
KLOPFER (*m*): clapper; beater; knocker; (morse)-tapper, [(Wireless)-telegraphy] 814
KLOPFERMAGNET (*m*): tapper magnet 815
KLOPFERSPULE (*f*): tapper coil 816
KLOPFHOLZ (*n*): mallet; beater 817
KLOPFSIEB (*n*): shaking (tapping) sieve or screen 818
KLOPFWERKZEUG (*n*): scaling tool 819
KLÖPPEL (*m*): beater; knocker; clapper; drum-stick; (lace)-bobbin; hammer-(of a bell, etc.) 820
KLÖPPELGARN (*n*): lace-making thread or yarn 821
KLÖPPELHOLZ (*n*): bobbin 822
KLÖPPELMASCHINE (*f*): bobbin-maschine 823
KLÖPPELN: to make lace 824
KLÖPPELZWIRN (*m*): lace-making thread or yarn 825
KLOSETTPAPIER (*n*): toilet paper (Paper) 826
KLOSET (*n*), **WASSER-**: water closet 827
KLOSS (*m*): clod; dumpling; lump 828

KLOSSIG : doughy ; lumpy 829
KLOSTER (*n*) : cloister ; convent ; abbey ; monastery 830
KLOTZ (*m*) : block ; log ; stump ; stock ; lump ; bed ; foundation ; slab 831
KLOTZ (*m*), AMBOSS- : (anvil)-block 832
KLOTZBREMSE (*f*) : lever brake 833
KLOTZDRUCK (*m*) : slop-pad printing, hand printing 834
KLOTZEN : to stare ; look fixedly at ; slop-pad (Calico) 835
KLOTZHOLZ (*n*) : wood in blocks 836
KLOTZMODELL (*n*) : block model 837
KLUB (*m*) : club 838
KLUB (*m*), (EIS)-JACHT- : (ice)-yacht-(ing) club 839
KLUFT (*f*) : cleft ; chasm ; gap ; gulf ; tongs ; ravine ; cavern ; abyss ; grotto 840
KLÜFTEN : to cleave ; split 841
KLUFTHOLZ (*n*) : log-wood, wood for logs 842
KLÜFTIG : cleft, cracked, full of chasms ; split 843
KLÜFTUNG (*f*) : cleaving ; splitting ; segmentation 844
KLUG : clever ; intelligent ; sharp ; knowing ; wise ; smart ; skilful ; shrewd ; prudent ; sagacious ; sensible ; subtle ; judicious 845
KLÜGELEI (*f*) : sophistry, cavilling 846
KLÜGELN : to criticize, be (wise) clever, be sophisticated 847
KLUGHEIT (*f*) : sense, understanding, judgement, sagacity, good sense, foresight, policy, discretion, wisdom, prudence, prescience, insight, judiciousness, discernment, knowledge, learning, enlightenment, erudition, reason, shrewdness, wit, intelligence, cleverness, genius 848
KLÜGLER (*m*) : wise-acre, sophist, critic 849
KLÜGLICH : prudently ; judiciously, wisely 850
KLUMPEN (*m*) : lump ; ball ; heap ; mass ; ingot ; bloom ; nugget ; agglomeration ; clot ; clod ; knob ; cluster ; clump 851
KLUMPENGOLD (*n*) : ingot-gold 852
KLUMPFUSS (*m*) : club-foot 853
KLUMPFÜSSIG : club-footed 854
KLUMPIG : clodded, full of clods, lumpy, knobby, clotted 855
KLUNKER (*m*) : clot, drop ; clod (of dirt) ; pendant 856
KLUNKERIG : clotted ; bedraggled 857
KLUNKERWOLLE (*f*) : clottings 858
KLUPPE (*f*) : pincers ; tongs ; diestock ; screw-stock ; trap 859
KLUPPENBACKEN (*f.pl*) : screw dies 860
KLUPPZANGE (*f*) : forceps 861
KLÜSE (*f*) : hawse hole 862
KLÜSE (*f*), ANKERKETTEN- : hawse (pipe) hole (Naut.) 863
KLÜSE (*f*), HECK- : stern hawse-hole 864
KLÜSENROHR (*n*) : hawse pipe 865
KLÜSENZAPFEN (*m*) : hawse plug 866
KLÜSHOLZ (*n*) : hawse piece, hawse timber 867
KLÜVER (*m*) : (flying)-jib (Naut.) 868
KLÜVERBAUM (*m*) : jib-boom (Naut.) 869
KLÜVERBAUMBACKSTAG (*n*) : jib-boom guy 870
KLÜVERBAUMNETZ (*n*) : jib-boom net 871
KLÜVERBAUMPEERD (*n*) : jib-boom foot-rope 872
KLÜVERFALL (*n*) : jib halliard 873
KLÜVERFOCK (*n*) : fore-top stay-sail 874
KLÜVERHALS (*m*) : jib tack 875
KLÜVERKRAN (*m*) : jib-crane 876
KLÜVERLEITER (*f*) : jib stay 877
KLÜVERSCHOTE (*f*) : jib sheet 878
KLÜVERSTAG (*n*) : jib stay 879
KLÜVERSTAMPFSTAG (*n*) : martingale stay 880
KLYSTIER (*n*) : clyster ; enema (a rectal injection of liquid) (Med.) 881
KLYSTIEREN : to administer an enema (Med.) 882
KLYSTIERRÖHRE (*f*) : catheter, enema tube (Med.) 883
KLYSTIERSPRITZE (*f*) : enema syringe (Med.) 884
KNABBERN : to gnaw ; nibble 885
KNABE (*m*) : boy, lad, youth 886
KNABENMÄSSIG : boyish 887
KNABENSCHANDEREI (*f*) : sodomy 888
KNACKEN : to crack ; break 889
KNACKERN : to crackle, crepitate 890
KNACKER (*m*), NUSS- : nut-cracker 891
KNACKMANDEL (*f*) : unshelled almond 892
KNACKWURST (*f*) : smoked sausage 893
KNAGGE (*f*) : cam ; catch ; pig ; stop ; lug ; bracket ; knot ; snag ; hinge (see KNAGGEN) 894
KNAGGEN (*m*) : cam, catch, lug, projection, tappet, lifter, brace, stay (see also KNAGGE) 895
KNAGGENSTEUERUNG (*f*) : cam gear (Mech.) 896
KNALL (*m*) : detonation ; explosion ; report ; crack ; pop ; clap ; fulmination 897
KNALLBLEI (*n*) : fulminating lead 898
KNALLBONBON (*m*) : cracker 899
KNALLBÜCHSE (*f*) : pop-gun 900
KNALLEN : to detonate ; fulminate ; report ; crack ; pop ; clap ; go off ; burst ; explode ; knock (of valves) 901
KNALLERBSE (*f*) : detonating ball 902
KNALLFLAMME (*f*) : oxyhydrogen flame 903
KNALLGAS (*n*) : detonating (fulminating) gas ; oxyhydrogen gas (see HYDROOXYGENGAS) 904
KNALLGASFLAMME (*f*) : oxyhydrogen flame 905
KNALLGASGEBLÄSE (*n*) : oxyhydrogen blow-pipe 906
KNALLGASLICHT (*n*) : oxyhydrogen light 907
KNALLGASKETTE (*f*) : chain of oxyhydrogen gas bubbles 908
KNALLGLAS (*n*) : Prince Rupert's drop ; anaclastic glass 909
KNALLGOLD (*n*) : fulminating gold 910
KNALLLUFT (*f*) : detonating gas (see KNALLGAS) 911
KNALLPULVER (*n*) : detonating (fulminating powder) 912
KNALLPYROMETER (*n*) : explosion pyrometer 913
KNALLQUECKSILBER (*n*) : fulminate of mercury ; mercuric fulminate ; $(CNO)_2Hg$ 914
KNALLSATZ (*m*) : detonating composition 915
KNALLSAUER : fulminate of ; combined with fulminic acid 916
KNALLSÄURE (*f*) : fulminic acid ; nitro-acetonitrile ; CNOH 917
KNALLSILBER (*n*) : fulminating silver ; $(CNO)_2Ag_2$ 918
KNALLZÜNDMITTEL (*n*) : detonating priming 919
KNAPP : close ; tight ; narrow ; near ; scarce ; scanty ; exact ; barely-(sufficient) ; mean ; poor ; hard ; concise ; short ; a tight fit ; a narrow shave, etc. 920
KNAPPE (*m*) : esquire ; miner ; workman 921
KNAPPERN : to crunch, gnaw, grind 922
KNAPPHEIT (*f*) : shortage, scarcity, tightness (see also KNAPP) 923
KNAPPSACK (*m*) : knapsack ; wallet 924
KNAPPSCHE FLÜSSIGKEIT (*f*) : Knapp's reagent (for the determination of grape sugar) 925

KNARRE (*f*): rattle, crackle 926
KNARRE (*f*), BOHR-: ratchet drill 927
KNARREN: to crackle; rattle; creak; squeak; drill (Mech.) 928
KNAST (*m*): knot (in wood); log 929
KNASTER (*m*): canaster-(tobacco) 930
KNATTERN: to gnash, crackle, rattle; chatter (Mech.) 931
KNAUEL (*m*): (see KNÄUEL) 932
KNÄUEL (*m*): ball; skein; coil (of string, etc.); convolution; hank 933
KNÄUELN: to coil, wind-(into a ball) 934
KNÄUELDRÜSE (*f*): sweat gland 935
KNÄUELFÖRMIG: convolute-(d) 936
KNAUF (*m*): knob; head; top; stud; button; capital 937
KNAUSERIG: niggardly, mean, close (tight) fisted, stingy, sordid 938
KNEBEL (*m*): stick; gag; clog; knuckle; branch; club; bell-clapper; cross-bar 939
KNEBELGEBISS (*n*): snaffle-bit 940
KNEBELN: to gag; bind; garotte 941
KNECHT (*m*): (man)-servant; trestle; horse; jear-bits (Naut.) 942
KNECHTSCHAFT (*f*): servitude, slavery 943
KNEIF (*m*): clasp-knife; hedge-bill 944
KNEIFEN: to nip; squeeze; pinch; grip 945
KNEIFER (*m*): pince-nez 946
KNEIFZANGE (*f*): nippers; pincers; pliers; cutting nippers 947
KNEIPE (*f*): pincers; nippers; pliers; tavern; public-house; drinking party 948
KNEIPEN: to nip; pinch; squeeze; grip; tipple; frequent a public-house; drink beer 949
KNETBAR: plastic, pliable, easily moulded 950
KNETEN: to knead; mould; pug (Ceramics) 951
KNETER (*m*), VAKUUM-: vacuum kneader 952
KNETMASCHINE (*f*): kneading machine 953
KNETPRESSE (*f*): kneading press (Soap) 954
KNETTROG (*m*): kneading trough 955
KNICK (*m*): break; crack; flaw; angle; sharp bend 956
KNICKBEANSPRUCHUNG (*f*): breaking (stress, strain) load 957
KNICKEN: to bend; crack; break; buckle; collapse; snap, (due to axial compressive stress) 958
KNICKFESTIGKEIT (*f*): resistance to (snapping, breaking strain), collapse; breaking-(stress) resistance 959
KNICKHOLZ (*n*): brush-wood 960
KNICKPUNKT (*m*): point of rupture; breaking point; break 961
KNICKSTAG (*n*): preventer stay (Naut.) 962
KNICKUNG (*f*): cracking, buckling, breaking, collapsing; collapse; angle 963
KNICKUNGSWINKEL (*m*): angle formed by two (sides) legs of an angle; angle of a bend 964
KNIE (*n*): knee; angle; bend; elbow (Mech.); angular joint; knee-bend 965
KNIEBEUGE (*f*): (see KNIEKEHLE) 966
KNIEBEUGUNG (*f*): genuflexion 967
KNIEBUG (*m*): (see KNIEKEHLE) 968
KNIEEN: to kneel; bend 969
KNIEFLECHSE (*f*): hamstring 970
KNIEFÖRMIG: knee-shaped; geniculate 971
KNIEGELENK (*n*): knee (elbow) joint 972
KNIEHEBEL (*m*): toggle (lever) joint; bell-crank lever; bent (knee) lever 973
KNIEKEHLE (*f*): bend of the knee; popliteal space (Med.) 974
KNIEKISSEN (*n*): hassock 975
KNIEPLATTE (*f*): knee plate, knee 976
KNIEPOLSTER (*n*): hassock 977
KNIEROHR (*n*): bent (tube) pipe; elbow; (long)-bend (of piping); knee (piece) pipe 978
KNIERÖHRE (*f*): (see KNIEROHR) 979
KNIESCHEIBE (*f*): kneecap; patella (Anat.) 980
KNIESCHWAMM (*m*): white swelling (a scrofulous disease of a joint) (Med.) 981
KNIESTÜCK (*n*): elbow; bend; knee-piece; half-length (three-quarter length) portrait 982
KNIE (*n*), VORSTEVEN-: stemson (Naut.) 983
KNIFF (*m*): pinch; crease; trick; dodge; device; artifice; twitch; fold; stratagem 984
KNIFFLEHRE (*f*): trickery, casuistry, deception (the science of conscience) 985
KNIFFMASCHINE (*f*): crimping machine 986
KNIRREN: to creak 987
KNIRSCHEN: to crackle; grate; crunch; rustle; gnash; grind 988
KNIRSCHPULVER (*n*): (a sort of)-coarse gunpowder 989
KNISTERGOLD (*n*): Dutch gold; tinsel 990
KNISTERN: to crackle; crepitate; rustle (Silk); crumple 991
KNITTER (*m*): crease; wrinkle; fold 992
KNITTERGOLD (*n*): Dutch gold; tinsel 993
KNITTERIG: crumpled; creased; rumpled; wrinkled; irritable 994
KNITTERN: to crackle; crumple; crease; rumple; crepitate; rustle (of silk); be (vexed) irritable 995
KNOBLAUCH (*m*): garlic (*Allium sativum*) 996
KNOBLAUCHARTIG: garlic-like 997
KNOBLAUCHGAS (*n*): mustard gas; yellow cross gas; Yperite; dichlorodiethyl sulphide; $(C_2H_4Cl)_2S$ 998
KNOBLAUCHÖL (*n*): garlic oil (Sg. 1.053) (from *Allium sativum*) 999
KNOBLAUCHWURZEL (*f*): garlic root (*Radix allii sativi of Allium sativum*) 000
KNÖCHEL (*m*): knuckle; joint; ankle; (*pl*): dice; vertebra 001
KNÖCHELGELENK (*n*): ankle-joint 002
KNÖCHELN: to raffle; play at dice 003
KNÖCHELSPIEL (*n*): dice (the game); a game with dice 004
KNOCHEN (*m*): bone 005
KNOCHENÄHNLICH: bone-like; osseous 006
KNOCHENANSATZ (*m*): epiphysis (one bone growing directly on to another, instead of being separated by cartilage) (Anat.) 007
KNOCHENARBEITER (*m*): bone (turner) worker 008
KNOCHENARTIG: bony, osseous 009
KNOCHENASCHE (*f*): bone ash (*Ebur ustum album*) (68-85% basic calcium phosphate; 3-10% calcium carbonate; up to 3% magnesium phosphate; about 4% calcium fluoride and some caustic lime) 010
KNOCHENAUSWUCHS (*m*): growth on a bone, protuberance of a bone 011
KNOCHENBESCHREIBUNG (*f*): osteography (the description of bones) 012
KNOCHENBILDEND: bone-forming 013
KNOCHENBRAND (*m*): necrosis (injury to, or inflammation of bony matter, resulting in its decay and death) 014
KNOCHENBRECHER (*m*): bone (crusher) breaker; Bog Asphodel (*Narthecium ossifragum*) 015
KNOCHENBRUCH (*m*): fracture 016

KNOCHENDÄMPFAPPARAT (*m*): bone steaming apparatus; bone steamer 017
KNOCHENDÜNGER (*m*): bone manure 018
KNOCHENENTFETTUNGSANLAGE (*f*): bone (fat) grease extraction plant, bone de-greasing plant 019
KNOCHENERDE (*f*): bone earth; phosphate of lime; calcium phosphate; bone ash (see KNOCHENASCHE and TRICALCIUMPHOSPHAT) 020
KNOCHENERWEICHUNG (*f*): (*Mollities ossium*), osteomalacia, malacosteon (softening of the bones) (Med.) 021
KNOCHENEXTRAKT (*m*): bone extract (a substitute for meat-extract used in Germany during the late war) 022
KNOCHENFÄULE (*f*): caries (ulceration or mortification of osseous tissue) (Med.) 023
KNOCHENFÄULNIS (*f*): (see KNOCHENFÄULE) 024
KNOCHENFETT (*n*): bone (fat) grease; Mp. 20-30°C. 025
KNOCHENFETT-EXTRAKTIONSAPPARAT (*m*): (see KNOCHENENTFETTUNGSANLAGE) 026
KNOCHENFORTSATZ (*m*): apophysis (protuberance or process of a bone) 027
KNOCHENFRAS (*m*): caries (see KNOCHENFÄULE) (Med.) 028
KNOCHENFÜGUNG (*f*): articulation-(of bones) 029
KNOCHENGEBÄUDE (*n*): bony structure, skeleton 030
KNOCHEN (*m.pl*), **GEKÖRNTE-**: ground (granulated) bones 031
KNOCHENGELENK (*n*): joint (of bones) 032
KNOCHENGERIPPE (*n*): skeleton 033
KNOCKENGERÜST (*n*): bony framework, skeleton 034
KNOCHENGUANO (*m*): bone guano, bone manure, bone fertilizer 035
KNOCHENHAUER (*m*): butcher 036
KNOCHENHAUS (*n*): charnel-house 037
KNOCHENHAUT (*f*): periosteum (an adherent membrane covering the bones) (Anat.) 038
KNOCHENKOHLE (*f*): bone black; animal charcoal; abaiser [about 90% mineral substance (principally calcium phosphate) and 10% carbon] 039
KNOCHENKOHLENSUPERPHOSPHAT (*n*): animal charcoal superphosphate 040
KNOCHENKOPF (*m*): condyl 041
KNOCHENKRANKHEIT (*f*): disease of the bones (Med.) 042
KNOCHENLEHRE (*f*): osteology (the science of bones) 043
KNOCHENLEIM (*m*): bone glue; osteocolla 044
KNOCHENLOS: boneless 045
KNOCHENMANN (*m*): skeleton; Death 046
KNOCHENMARK (*n*): marrow (soft matter in bones) 047
KNOCHENMEHL (*n*): bone meal; bone dust (crushed or ground bones) 048
KNOCHENMEHLAMMONSALPETER (*m*): bone-meal-ammonium nitrate 049
KNOCHENMEHLSUPERPHOSPHAT (*n*): bone meal superphosphate 050
KNOCHENMÜHLE (*f*): bone mill 051
KNOCHENNAHT (*f*): suture (the junction of bones by their jagged ends) 052
KNOCHENÖL (*n*): bone oil; Dippel's oil (from refined bone fat by pressing); Neat's-foot oil (see KLAUENÖL) 053
KNOCHENPFANNE (*f*): socket (of a joint) (Anat.) 054
KNOCHENPHOSPHAT (*n*): bone superphosphate (see KNOCHENASCHE, KNOCHENKOHLE and KNOCHENMEHL) 055
KNOCHENPORZELLAN (*n*): bone (porcelain) china (consists of kaolin, bone-ash, orthoclase and quartz) 056
KNOCHENPRÄZIPITAT (*n*): bone precipitate 057
KNOCHENSÄURE (*f*): (ortho)-phosphoric acid; H_3PO_4; Sg. 1.884; Mp. 38.6°C. 058
KNOCHENSCHROT (*m*): crushed bones 059
KNOCHENSCHWARZ (*n*): bone black (see KNOCHEN-KOHLE) 060
KNOCHENSCHWIND (*m*): atrophy-(of bones) (wasting away of osseous matter) 061
KNOCHENSTEIN (*m*): osteocolla 062
KNOCHENVERBINDUNG (*f*): articulation-(of bones) 063
KNOCHENVERKOHLUNGSOFEN (*m*): bone carbonizing oven 064
KNOCHENVERWERTUNGSANLAGE (*f*): bone utilization plant 065
KNOCHENWAREN (*f.pl*): bone wares, bone articles, bone goods 066
KNOCHENWUCHS (*m*): ossification, bone-formation, the formation of bone 067
KNOCHEN (*m.pl*), **ZERBROCHENE-**: broken (crushed) bones 068
KNOCHENZERLEGUNG (*f*): osteotomy, the decomposition of bone 069
KNÖCHERN: of bone; bony; osseous 070
KNOCHIG: bony 071
KNOLLBEIN (*n*): Elephantiasis (an unnatural enlargement of any part of the body) (Med.) 072
KNOLLE (*f*): lump; clod; knob; nodule; bulb; tuber (*Tubera* or *Radix*); tubercle; knot; protuberance; knuckle 073
KNOLLEN (*m* and *pl*): tumour, tumulus, knoll (see also KNOLLE) 074
KNOLLENARTIG: tuberous, bulbous, globular, spherical, globulous, globate, globose, globe-shaped, ball-shaped 075
KNOLLENGEWÄCHS (*n*): tuberous (bulbous) plant 076
KNOLLENKRAUT (*n*): licorice, sweet-root (*Glycyrrhiza glabra* or *G. glandulifera*) 077
KNOLLENKREBS (*m*): cheloid, cheloma, keloid (Med.) 078
KNOLLGRAS (*n*): oat-grass (*Arrhenatherum bulbosum* or *A. nodosum*) 079
KNOLLHAFER (*m*): (see KNOLLGRAS) 080
KNOLLIG: knobby; knotty; tuberous; bulbous; immense 081
KNOLLSELLERIE (*f*): heart-celery, celery grown specially for the root (*Apium graveolens*) 082
KNOLLWICKE (*f*): earth-nut (see ERDNUSS) 083
KNOPF (*m*): button; stud; head; knob; knot; top; ball; boss; condyle (Anat.) 084
KNOPFDECKEL (*m*): cover with knob 085
KNÖPFEN: to button 086
KNOPFFORM (*f*): button-mould 087
KNOPFFORTSATZ (*m*): condyloid apophysis (see KNOCHENFORTSATZ and KNOCHENKOPF) 088
KNOPFKRAUT (*n*): (field)-scabious (*Knautia arvensis*) 089
KNOPFLACK (*m*): button varnish 090
KNOPFLOCH (*n*): button-hole 091
KNOPFNAHT (*f*): suture (seam) (Surgical) 092
KNOPPER (*f*): nut-gall; gallnut (see GALLAPFEL); valonia (acorn cups of *Quercus ægilops*) (up to 30% tannin) 093
KNOPPEREISEN (*n*): (see ZAINEISEN) 094

KNORPEL (*m*): cartilage; gristle 095
KNORPELARTIG: gristly, cartilaginous 096
KNORPELBAND (*n*): cartilage, cartilaginous tissue, gristle-(between bones) 097
KNORPELIG: cartilaginous; gristly 098
KNORPELKOHLE (*f*): screened brown coal or lignite 099
KNORPELKRANKHEIT (*f*): disease of a cartilage (Med.) 100
KNORPELKRAUT (*n*): white stonecrop (*Sedum album*) 101
KNORPELLEIM (*m*): chondrin 102
KNORPELTANG (*m*): carragheen; chondrus (see CARRAGHEENMOOS) 103
KNORPLIG: (see KNORPELIG) 104
KNORREN (*m*): protuberance; excrescence; knob; knot; gnarl; snag; knuckle 105
KNORRIG: knobbed; knotty (of wood); gnarled 106
KNOSPE (*f*) bud 107
KNOSPEN: to bud; burst (put) forth 108
KNOSPENBILDUNG (*f*): gemmation, reproduction by buds (Zool.); germination, budding (Bot.) 109
KNOSPENHÄUTCHEN (*n*): hymen (Bot.) 110
KNOSPENTRAGEND: gemmiferous, having (bearing) buds 111
KNOSPENTREIBEN (*n*): gemmation, budding (Bot.) 112
KNOSPIG: bud-like, full of buds, having buds 113
KNOTE (*m*): knot; snob 114
KNOTEN: to tie in knots; knot 115
KNOTEN (*m*): hitch-(in business matters); knot; hitch (Naut.); knot, one sea mile, 1/60th of a degree of latitude (Marine); knot, node, nodule, bend, tubercle, knob, plot, joint, capsule; bow, tie; tangle; growth; ganglion; difficulty; point of (rest, intersection) support 116
KNOTENABSTAND (*m*): distance between points of (rest) support 117
KNOTENADER (*f*): sciatic vein 118
KNOTENBLECH (*n*): gusset-(plate) 119
KNOTENFÄNGER (*m*): strainer 120
KNOTENFANGPLATTE (*f*): strainer-plate (Paper) 121
KNOTENGELENK (*n*): multiple (tie-bar) joint 122
KNOTENHOLZ (*n*): gnarled (knotty, knotted) wood 123
KNOTENPUNKT (*m*): system (nodal; centre) point; knot; point of (support; rest; juncture) intersection; spring (of a number of arches); assemblage point; point of origin (end point) of a number of lines 124
KNOTENPUNKTBEWEGUNG (*f*): motion of point of intersection, etc. (see KNOTENPUNKT) 125
KNOTENPUNKTGESCHWINDIGKEIT (*f*): velocity of point of intersection, etc. (see KNOTEN--PUNKT) 126
KNOTENPUNKTWEG (*m*): path of point of interesction 127
KNOTENVERBINDUNG (*f*): joint connection; connection of (joists; beams) members, etc. 128
KNOTENWURZ (*f*): knotted figwort (*Scrophularia nodosa*) 129
KNÖTERICH (*m*): knot grass (*Herba polygoni avicul*) (*Polygonum aviculare*) 130
KNÖTERICHARTIGE GEWÄCHSE (*n.pl*): *Polygonaceæ* (a subdivision of the *Monochlamydeæ* order of plants, and comprises *Fagopyrum*, *Oxyria*, *Polygonum*, and *Rumex*) 131
KNOTERICHGEWÄCHSE (*n.pl*): *Polygonaceæ* (Bot.) 132
KNOTIG: knotty; tuberous; nodular; jointed; tubercular; knobby; caddish; rude 133
KNUFFEN: to cuff; knock; pommel; thump; beat 134
KNÜPFEN: to join; unite; attach; knit; tie; fasten; knot 135
KNÜPPEL (*m*): billet (Metal); cudgel; club; stick 136
KNÜPPELDAMM (*m*): corduroy road (road made up with wood logs or sticks) 137
KNÜPPEL (*m.pl*) **FÜR DRAHT** (*m*): wire billets 138
KNÜPPELHOLZ (*n*): billet wood; faggot; sticks 139
KNÜPPELSCHERE (*f*): billet shears 140
KNURREN: to snarl; growl; rumble; purr; grumble; grunt 141
KNURRIG: snarling, growling, rumbling 142
KNUSPERIG: crisp; crackling 143
KNUTE (*f*): knout (a type of whip common in Russia) 144
KNÜTTELN: to club, cudgel 145
KOAGULATION (*f*): coagulation 146
KOAGULATIONSWÄRME (*f*): heat of coagulation 147
KOAGULIERBAR: coagulable 148
KOAGULIEREN: to coagulate 149
KOAGULIERUNGSMITTEL (*n*): coagulant; coagulating agent 150
KOAKS (*m*): coke (see KOKS) 151
KOALISIEREN: to coalesce, combine, unite 152
KOAXIAL: co-axial 153
KOBALT (*m*): cobalt (Co) 154
KOBALTANTIMONID (*n*): cobalt antimonide; $CoSb$; Sg. 8.12; $CoSb_2$; Sg. 7.76 155
KOBALTARSENID (*n*): cobalt arsenide; Co_3As_2; Sg. 7.82; Co_2As_3; Sg. 7.35; $CoAs$; Sg. 7.62; $CoAs_2$; Sg. 6.97 156
KOBALTARSENKIES: (*m*) cobalt pyrites, danaite (see DANAIT) 157
KOBALT (*n*), **ARSENSAURES-**: cobalt arsenate, cobaltic arsenate, natural erythrine; $Co_3(AsO_4)_2.8H_2O$; Sg. 2.95 158
KOBALTBESCHLAG (*m*): cobalt efflorescence; earthy erythrite (see ERDKOBALT, ROTER STRAHLIGER-; KOBALTBLÜTE; and ERY--THRIN) 159
KOBALTBLAU (*n*): cobalt blue, Thenard's blue, Vienna blue, Leyden blue (a cobaltiferous clay, by heating cobalt compounds with clay) 160
KOBALTBLUME (*f*): cobalt bloom; erythrite (a natural cobalt arsenate); $Co_3(AsO_4)_2.8H_2O$; Sg. 3.0 161
KOBALTBLÜTE (*f*): (see KOBALTBLUME and ERYTHRIN) 162
KOBALTBORID (*n*): cobalt boride; Co_2B; Sg. 7.9 163
KOBALTBROMID (*n*): cobalt bromide, cobaltous bromide; $CoBr_2$; Sg. 4.909 164
KOBALTCARBONAT (*n*): (see KOBALT, KOHLEN--SAURES-) 165
KOBALTCARBONYL (*n*): cobalt carbonyl; $Co(CO)_4$; Sg. 1.73 166
KOBALTCHLORID (*n*): cobaltic chloride, cobalt chloride; $CoCl_3$; Sg. 2.94 167
KOBALTCHLORÜR (*n*): cobaltous chloride; $CoCl_2$; Sg. 3.348; $CoCl_2.6H_2O$; Sg. 1.84; Mp. 86.75°C. 168
KOBALTCHROMSTAHL (*m*): cobalt-chromium steel; K.S. magnet steel (see K.S.-STAHL) 169

KOBALTERZ (*n*): cobalt ore 170
KOBALT (*n*), ESSIGSAURES-: cobalt (cobaltous) acetate; $Co(C_2H_3O_2)_2.4H_2O$; Sg. 1.7 171
KOBALTFARBSTOFF (*m*): cobalt colouring matter 172
KOBALTFLUORID (*n*): cobalt fluoride; CoF_2; Sg. 4.43 173
KOBALTGELB (*n*): cobalt yellow (see KALIUM--KOBALTNITRIT) 174
KOBALTGLANZ (*m*): cobalt glance; cobaltite; (Co,Fe)As; Sg. 6.2 (about 35% cobalt) 175
KOBALTGLAS (*n*): (see BLAUFARBENGLAS) 176
KOBALT (*n*), GOLDCHLORWASSERSTOFFSAURES-: cobalt chloraurate; $Co(AuCl_4)_2.8H_2O$ 177
KOBALTGRAUPEN (*f.pl*): amorphous grey cobalt 178
KOBALTGRÜN (*n*): cobalt green, Rinmann's green, cobalt zincate 179
KOBALTHALTIG: cobaltiferous 180
KOBALT (*n*), HARZSAURES-: cobalt (cobaltous) resinate; $Co(C_{44}H_{62}O_4)_2$ 181
KOBALTHYDROXYD (*n*): cobalt (cobaltous) hydroxide, cobalt hydrate; $Co(OH)_2$ Sg. 3.597 182
KOBALTIAK (*n*): cobaltiac; cobaltammine 183
KOBALTIAKSALZ (*n*): cobaltiac salt; cobaltammine salt 184
KOBALTICYANKALIUM (*n*): potassium cobalticyanide 185
KOBALTICYANWASSERSTOFF (*m*): cobalticyanic acid 186
KOBALTIKALIUMNITRIT (*n*): potassium cobaltinitrite (see KALIUMKOBALTNITRIT) 187
KOBALTIN (*m*): cobaltite, cobalt glance; $CoAs_2,CoS_2$; Sg. 6.15 188
KOBALTIVERBINDUNG (*f*): cobaltic compound 189
KOBALTJODÜR (*n*): cobaltous (cobalt) iodide; $CoI_2.6H_2O$ 190
KOBALTKALIUMNITRIT (*n*): cobalt-potassium nitrite (see KALIUMKOBALTNITRIT) 191
KOBALTKARBONAT (*n*): (see KOBALT, KOHLEN--SAURES-) 192
KOBALTKIES (*m*): cobalt pyrites; linnæite (see KOBALTNICKELKIES) 193
KOBALT (*n*), KOHLENSAURES-: cobalt (cobaltous) carbonate; $CoCO_3$ 194
KOBALTLEGIERUNG (*f*): cobalt alloy 195
KOBALT (*n*), LEINÖLSAURES-: cobalt (cobaltous) linoleate; $Co(C_{18}H_{31}O_2)_2$ 196
KOBALTMANGANERZ (*n*): earth cobalt; asbolite; $(Co,Cu)O.2MnO_2.4H_2O$; Sg. 2.2 (3-6% Co.) 197
KOBALTMONOXYD (*n*): (see KOBALTOXYDUL) 198
KOBALT-NICKEL-ARSENIDEN (*n.pl*): cobalt-nickel arsenides 199
KOBALTNICKELKIES (*m*): cobalt pyrites; linnæite; $(Ni,Co,Fe)_3S_4$; Sg. 4.9 200
KOBALTNITRAT (*n*): cobalt-(ous) nitrate; $Co(NO_3)_2.6H_2O$; Sg. 1.83; Mp. 56°C. 201
KOBALTNITRIT (*n*): cobalt nitrite; $Co(NO_2)_3$ 202
KOBALTOCHLORID (*n*): cobaltous chloride (see KOBALTCHLORÜR) 203
KOBALTOSALZ (*n*): cobaltous salt 204
KOBALTOVERBINDUNG (*f*): cobaltous compound 205
KOBALT (*n*), OXALSAURES-: cobalt oxalate 206
KOBALTOXYD (*n*): cobalt oxide, RKO; cobaltic oxide (*Cobaltum oxydatum*); Co_2O_3; Sg. 5.18 207
KOBALTOXYDHYDRAT (*n*): (*Cobaltum hydroxydatum*): cobalt-(ic) hydroxide, cobalt hydrate; $Co(OH)_3$ 208
KOBALTOXYDKALI (*n*), SALPETRIGSAURES-: cobalt-potassium nitrite (see KALIUMKOBALTNITRIT) 209
KOBALTOXYDOXYDUL (*n*): cobaltocobaltic oxide; Co_3O_4; Sg. 6.073 210
KOBALTOXYDUL (*n*): cobaltous (cobalt) oxide; cobalt oxide, PO; cobalt protoxide; CoO; Sg. 5.68; Mp. 2860°C. (*Cobaltum oxydulatum*) 211
KOBALTOXYDULACETAT (*n*): cobaltous acetate; $Co(C_2H_2O_2)_2.4H_2O$; Sg. 1.7031 212
KOBALTOXYDUL (*n*), ARSENSAURES-: cobalt arsenate; cobalt oxide, A.K.O.; cobalt rose; $Co_3(AsO_4)_2.8H_2O$; Sg. 2.95 213
KOBALTOXYDULHYDRAT (*n*): (*Cobaltum hydroxydulatum*): cobalt-(ous) hydroxide; cobalt hydrate; $Co(OH)_2$; Sg. 3.597 214
KOBALTOXYDUL (*n*), KOHLENSAURES-: cobalt-(ous) carbonate; cobalt oxide, KOH; $CoCO_3$ 215
KOBALTOXYDULOXYD (*n*): (see KOBALTOXYD--OXYDUL) 216
KOBALTOXYDULPHOSPHAT (*n*): (see KOBALT--OXYDUL, PHOSPHORSAURES-) 217
KOBALTOXYDUL (*n*), PHOSPHORSAURES-: cobalt-(ous) phosphate, cobalt red; cobalt oxide, PKO; $Co_3(PO_4)_2.2H_2O$ 218
KOBALTOXYDULSALZ (*n*): cobaltous salt 219
KOBALTOXYDULVERBINDUNG (*f*): cobaltous compound 220
KOBALTOXYDUL-ZINKOXYD (*n*): cobalt zincate, cobalt green, Rinmann's green 221
KOBALTOXYDUL-ZINNOXYD (*n*): cobalt stannate, ceruleum 222
KOBALTOXYDVERBINDUNG (*f*): cobaltic compound 223
KOBALTPHOSPHID (*n*): cobalt phosphide; Co_2P; Sg. 6.4 224
KOBALT (*n*), PHOSPHORSAURES-: cobalt phosphate (see KOBALTOXYDUL, PHOSPHORSAURES-) 225
KOBALTROSA (*n*): cobalt rose (see KOBALT--OXYDUL, ARSENSAURES-) 226
KOBALTROT (*n*): cobalt red (see KOBALTOXYDUL, PHOSPHORSAURES-) 227
KOBALT (*n*), SALPETERSAURES-: cobalt nitrate (see KOBALTNITRAT) 228
KOBALTSALZ (*n*): cobalt salt 229
KOBALTSCHWÄRZE (*f*): asbolane 230
KOBALT (*n*), SCHWEFELSAURES-: cobalt (cobaltous) sulphate; $CoSO_4$; Sg. 3.5; Mp. 989°C.; $CoSO_4.7H_2O$; Sg. 1.92; Mp. 96.8°C.; Bp. 420°C. 231
KOBALTSELENAT (*n*): cobalt selenate; $CoSeO_4+6H_2O$; Sg. 2.32 232
KOBALTSESQUIOXYD (*n*): cobalt sesquioxide; Co_2O_3 233
KOBALTSILICID (*n*): cobalt silicide; $CoSi_2$; Sg. 5.3; CoSi; Sg. 6.3; Co_2Si; Sg. 7.1 [234
KOBALTSILIKAT (*n*): (*Cobaltum silicicum*): cobalt silicate (see BLAUFARBE) 235
KOBALTSPEISE (*f*): cobalt speiss (see SPEISKOBALT) 236
KOBALTSPIEGEL (*m*): specular cobalt, transparent cobalt ore 237
KOBALTSTUFE (*f*): piece of cobalt ore 238
KOBALTSULFAT (*n*): cobalt sulphate (see KOBALT, SCHWEFELSAURES-) 239
KOBALTSULFID (*n*): cobalt sulphide; CoS; Sg. 5.45 240
KOBALTULTRAMARIN (*n*): cobalt ultramarine (see KOBALTBLAU) 241
KOBALTVERBINDUNG (*f*): cobalt compound 242

KOBALTVIOLETT (*n*) : cobalt violet (see KOBALT-OXYDULPHOSPHAT) 243
KOBALTVITRIOL (*m*) : cobalt vitriol ; cobaltous sulphate (see KOBALT, SCHWEFELSAURES-) 244
KOBEL (*m*) : dove-cot ; small cabin 245
KOBEN (*m*) : pig-sty ; (small)-cabin, hut 246
KOBER (*m*) : basket 247
KOBOLD (*m*) : goblin ; gnome ; hobgoblin ; sprite ; elf ; kobold 248
KOCH (*m*) : cook 249
KOCHAPFEL (*m*) : cooking-apple 250
KOCHAPPARAT (*m*) : cooking (boiling) apparatus ; industrial process apparatus ; digester 251
KOCHBECHER (*m*) : beaker 252
KOCHBIRNE (*f*) : cooking pear 253
KOCHBUCH (*n*) : cookery-book 254
KOCHDAUER (*f*) : duration of (boiling) cooking 255
KOCHECHT : fast to boiling 256
KOCHEMULSION (*f*) : boiled emulsion 257
KOCHEN : to boil ; seethe ; cook ; digest ; (*n*) : boiling ; cooking ; digesting ; ebullition ; seething ; cookery 258
KOCHENILLE (*f*) : (see COCHENILLE) 259
KOCHER (*m*) : cooker ; boiler ; hot-plate ; digester (Paper) 260
KÖCHER (*m*) : quiver ; case 261
KÖCHERBAUM (*m*) : coral tree, coral flower (*Erythrina corallodendron*) ; cockscomb, coral tree (*Erythrina crista-galli*) 262
KÖCHERFLIEGE (*f*) : May-fly (*Phryganea*) 263
KÖCHERJUNGFER (*f*) : May-fly (*Phryganea*) 264
KÖCHERWURM (*m*) : ship worm (*Teredo navalis*) (see also HOLZBOHRER and HOLZFRESSER) 265
KOCHFLASCHE (*f*) : boiling flask 266
KOCHFLOTT (*n*) : stove (Naut.) 267
KOCHGEFÄSS (*n*) : boiler ; boiling (cooking) vessel, pan or utensil ; bucking kier (Paper) 268
KOCHGERÄT (*n*) : boiler, boiling (cooking) pan-(s) vessel or utensil-(s) ; bucking kier (Paper) 269
KOCHGESCHIRR (*n*) : (see KOCHGEFÄSS) 270
KOCHHEISS : boiling hot 271
KOCHHITZE (*f*) : boiling heat 272
KOCHHOLZ (*n*) : firewood 273
KÖCHIN (*f*) : cook 274
KOCHKELLE (*f*) : pot ladle, basting ladle 275
KOCHKESSEL (*m*) : boiling (pot) pan ; kettle ; boiler ; digester (Paper) ; copper ; cauldron 276
KOCHKLÄRE (*f*) : filtered (liquor) syrup (Sugar) 277
KOCHKÖCHIN (*f*) : cordon-bleu 278
KOCHKÖLBCHEN (*n*) : small boiling flask ; fractional distillation flask 279
KOCHKOLBEN (*m*) : boiling flask 280
KOCHKRAUT (*n*) : pot herbs 281
KOCHKUNST (*f*) : (science of)-cookery ; culinary art 282
KOCHLAUGE (*f*) : boiling lye 283
KOCHLÖFFEL (*m*) : pot ladle ; basting spoon 284
KOCHMASCHINE (*f*) : (see KOCHAPPARAT) 285
KOCHOFEN (*m*) : oven, cooking stove 286
KOCHPFANNE (*f*) : stewpan, saucepan 287
KOCHPUDDELN (*n*) : pig boiling 288
KOCHPUNKT (*m*) : boiling point 289
KOCHROHR (*n*) : boiling (distilling) tube 290
KOCHRÖHRE (*f*) : (see KOCHROHR) 291
KOCHSALZ (*n*) : cooking salt ; common (kitchen) salt ; sodium chloride ; Sg. 2.05-2.17 (see NATURIUMCHLORID) 292
KOCHSALZGEIST (*m*) : muriatic acid (see SALZ-SÄURE) 293
KOCHSALZHALTIG : containing cooking salt 294
KOCHSALZSÄURE (*f*) : muriatic acid (see SALZ-SÄURE) 295
KOCHSTÜCK (*n*) : piece-(of meat)-suitable for boiling 296
KOCHTOPF (*m*) : boiling-pot 297
KOCHWASSER (*n*) : boiling water ; water suitable for boiling 298
KOCHZEUG (*n*) : kitchen (cooking) utensils 299
KOCHZUCKER (*m*) : brown (powdered ; cooking) sugar 300
KOCKPIT (*n*) : cockpit (Naut.) 301
KODAMIN (*n*) : codamine (an opium alkaloid) ; $C_{18}H_{18}NO(OH)(OCH_3)_2$ 302
KODEIN (*n*) : codeine (see CODEIN) 303
KODEINBROMMETHYLAT (*n*) : codeine bromomethylate (*Codeinum methylobromatum*) ; $C_{18}H_{21}NO_3(CH_3Br)$; Mp. 261 C. 304
KÖDER (*m*) : bait ; lure ; decoy 305
KÖDERN : to bait ; lure ; allure ; decoy ; entice ; inveigle ; tempt 306
KODIFIZIEREN : to codify, code, precis, systematize, make a digest of 307
KODIZILL (*n*) : codicil, postscript to a will 308
KOEFFIZIENT (*m*) : coefficient 309
KOEMEÖL (*n*) : (see TELFAIRIAÖL) 310
KOERZIBEL : coercible 311
KOERZIT (*n*) : same as K.S.-Stahl but only 18% cobalt (see K.S.-STAHL) 312
KOERZITIVKRAFT (*f*) : coercive force (Magnetism) 313
KOFFEIN (*n*) : (see KAFFEIN) 314
KOFFEINAGEL (*m*) : belaying pin (Naut.) 315
KOFFEIN (*n*), BROMWASSERSTOFFSAURES- : caffeine hydrobromide (see KAFFEINBROMHYDRAT) 316
KOFFEIN (*n*): CITRONENSAURES- : coffeine citrate (see KAFFEINZITRAT) 317
KOFFEIN (*n*), SALICYLSAURES- : coffeine salicylate 318
KOFFER (*m*) : coffer ; trunk ; box ; chest 319
KOFFERDECKDAMPFER (*m*) : trunk steamer 320
KOFFERKESSEL (*m*) : waggon boiler, rectangular boiler 321
KOGNAK (*m*) : cognac (see also COGNAK) 322
KOGNAKÄTHER (*m*) : cognac ether, ethyl cocoinate cocoinic (cocoic) ether ; $C_{15}H_{30}O_2$; Sg. 0.855 ; (see also COGNAKÖL) 323
KOGNAKESSENZ (*f*) : cognac essence (see COGNAKÖL) 324
KOGNAKÖL (*n*) : cognac oil (see COGNAKÖL) 325
KOHÄRENZ (*f*) : coherence, coherency, connection 326
KOHÄRER (*m*) : coherer 327
KOHÄRERRÖHRE (*f*) : coherer tube 328
KOHÄSION (*f*) : cohesion ; connection ; coherence ; coherency 329
KOHL (*m*) : cabbage ; cole (*Brassica oleracea*) 330
KOHLBAUER (*m*) : cabbage-grower 331
KOHLBECKEN (*m*) : (see KOHLENBECKEN) 332
KOHLE (*f*) : coal ; charcoal ; carbon ; (C) ; Sg. 3.5 333
KOHLE (*f*), ABGESCHWEFELTE- : coke (see under KOKS) 334
KOHLECYLINDER (*m*) : carbon cylinder 335
KOHLEDRUCK (*m*) : carbon print-(ing) (Phot.) 336
KOHLEELEKTRODE (*f*) : carbon electrode 337
KOHLEFARBE (*f*) : carbon (black) colour (see BEINSCHWARZ, ELFENBEINSCHWARZ, RUSS, FRANKFURTER SCHWARZ) 338
KOHLEFILTER (*n*) : charcoal (boneblack) filter 339

KOHLE (*f*), GESCHWEFELTE- : coke (see under KOKS) 340

KOHLE (*f*), GLÜHENDE- : live coal 341

KOHLEGREIFER (*m*) : coal grab 342

KOHLEGRUS (*m*) : small coal 343

KOHLEHYDRAT (*n*) : carbo-hydrate (see MONOSACCHARID ; DISACCHARID ; POLYSACCHARID) 344

KOHLE (*f*), KÜNSTLICHE- : artificial carbon 345

KOHLELAMPE (*f*) : carbon lamp 346

KOHLEN : to char ; carbonize ; carburize ; chatter ; be verbose 347

KOHLENANALYSE (*f*) : coal analysis 348

KOHLENANZÜNDER (*m* and *pl*) : coal igniter, fire-lighter 349

KOHLENARBEITER (*m*) : collier ; miner 350

KOHLENART (*f*) : type (kind, sort) of coal 351

KOHLENARTIG : coaly ; carbonic ; like (coal ; carbon) charcoal 352

KOHLENAUFZUG (*m*) : coal (lift ; elevator) hoist 353

KOHLENAUSNÜTZUNG (*f*) : utilization of coal ; coal (fuel) efficiency 354

KOHLENAUSSCHALTER (*m*) : carbon switch ; carbon cut-out 355

KOHLENBANSEN (*m*) . coal stack 356

KOHLENBATTERIE (*f*) : carbon battery 357

KOHLENBAUER (*m*) : charcoal burner 358

KOHLENBECKEN (*n*) : charcoal (coal) pan ; brazier ; chafing dish ; warming pan 359

KOHLENBEFÖRDERER (*m*) : (coal)-transporter, conveyor 360

KOHLENBEFÖRDERUNGSANLAGE (*f*) : coal-conveying plant, coal-handling plant 361

KOHLENBEHÄLTER (*m*) : coal bunker ; silo 362

KOHLENBENZIN (*n*) : benzene ; benzol (see BENZOL) 363

KOHLENBERGBAU (*m*) : coal mining 364

KOHLENBERGWERK (*n*) : coal mine, colliery 365

KOHLENBESCHICKUNG (*f*) : coal (conveying) handling plant, coal conveyor 366

KOHLENBLEI (*n*) : carburet of lead 367

KOHLENBLEISPAT (*m*) : cerussite ; white lead ore ; (see BLEISPAT and WEISSBLEIERZ) 368

KOHLENBLEIVITRIOLSPAT (*m*) : lanarkite (see LANARKIT) 369

KOHLENBLENDE (*f*) : anthracite (see ANTHRAZIT) (4-14% volatile ; 70-85% carbon) 370

KOHLENBLUME (*f*) : bitumenous clay 371

KOHLENBOHRER (*m*) : charcoal borer 372

KOHLENBRECHER (*m*) : coal (breaker) crusher 373

KOHLENBRENNEN (*n*) : charcoal (coal) burning 374

KOHLENBRENNER (*m*) : charcoal burner 375

KOHLENBRENNEREI (*f*) : charcoal (works) kiln ; charcoal burning 376

KOHLENBÜHNE (*f*) : coaling (quay ; stage) platform 377

KOHLENBUNKER (*m*) : coal bunker ; coal hopper ; silo 378

KOHLENBÜRSTE (*f*) : carbon brush (Elect.) 379

KOHLENDAMPF (*m*) : gases from burning coal ; coal gases ; smoke ; fumes 380

KOHLENDÄMPFER (*m*) : coal extinguisher 381

KOHLENDESTILLATION (*f*) : coal distillation, coking process 382

KOHLENDIOXYD (*n*) : carbon dioxide ; CO_2 (see KOHLENSÄURE) 383

KOHLENDISULFID (*n*) : carbon (disulphide) bisulphide (see SCHWEFELKOHLENSTOFF) 384

KOHLENDITHIOLSÄURE (*f*) : dithiolcarbonic acid (see DITHIOKOHLENSÄURE) 385

KOHLENDUNST (*m*) : gases (vapour) from coals ; smoke-(from burning coals) ; fumes ; coal gases 386

KOHLENEISEN (*n*) : iron carbide (see EISENKARBID) 387

KOHLENEISENSTEIN (*m*) : blackband (SPHÄROSIDERIT, which see, mixed with coal ; a carbonaceous iron carbonate) [clay (ironstone) band mixed with coal or carbon] 388

KOHLENELEKTRODE (*f*) : carbon electrode 389

KOHLENELEMENT (*n*) : carbon pile 390

KOHLENELEVATOR (*m*) : coal (lift ; hoist) elevator 391

KOHLENERSPARNIS (*f*) : coal-saving 392

KOHLENFADEN (*m*) : carbon filament 393

KOHLENFELD (*n*) : coal-field, coal-mining district 394

KOHLENFEUER (*n*) : coal (charcoal) fire 395

KOHLENFEUERUNG (*f*) : coal-firing ; coal fired (grate) furnace 396

KOHLENFILTER (*n*) : charcoal filter 397

KOHLENFLÖZ (*n*) : coal (measure) seam 398

KOHLENFÖRDERANLAGE (*f*) : coal conveying (handling) plant 399

KOHLENFORM (*f*) : charcoal mould 400

KOHLENFORSCHUNG (*f*) : coal research 401

KOHLENFUTTER (*n*) : carbonaceous lining 402

KOHLENGALMEI (*m*) : calamine, smithsonite (see GALMEI) 403

KOHLENGAS (*n*) : coal gas 404

KOHLENGESTÜBE (*n*) : coal (dust) slack 405

KOHLENGEWINNUNG (*f*) : coal winning, coal getting, coal hewing 406

KOHLENGLUT (*f*) : live coals 407

KOHLENGRAU : dark-grey 408

KOHLENGRIESS (*m*) : coal dust (see also KRYPTOL) 409

KOHLENGRUBE (*f*) : coal mine 410

KOHLENGRUPPE (*f*) : carboniferous group 411

KOHLENGRUS (*m*) : smalls, slack -(coal) 412

KOHLENHACKE (*f*) : coal pick ; miner's pick 413

KOHLENHALDE (*f*) : coal-dump 414

KOHLENHALTER (*m*) : (carbon)-brush-holder ; carbon (clip) holder 415

KOHLENHALTERBRÜCKE (*f*) : carbon brush (yoke) holder 416

KOHLENHALTIG : carboniferous ; carbonaceous ; containing (carbon) coal 417

KOHLENHANDEL (*m*) : coal-trade 418

KOHLENHÄNDLER (*m*) : coal merchant 419

KOHLENHOF (*m*) : coal (depôt) yard 420

KOHLENHOLZ (*n*) : wood for charcoal 421

KOHLENHORNBLENDE (*f*) : anthracite 422

KOHLENHÜTTE (*f*) : charcoal kiln 423

KOHLENHYDRAT (*n*) : carbohydrate (see KOHLEHYDRAT) 424

KOHLENKALK (*m*) : carboniferous limestone 425

KOHLENKALKSPAT (*m*) : anthraconite 426

KOHLENKALKSTEIN (*m*) : carboniferous limestone 427

KOHLENKARBONIT (*n*) : kohlencarbonite (trade name of an explosive) 428

KOHLENKARRE (*f*) : coal-barrow, coal-skip, coal-truck, coal-trolley 429

KOHLENKARREN (*m*) : (see KOHLENKARRE) 430

KOHLENKASTEN (*m*) : coal (grab ; bucket) skip ; coal box 431

KOHLENKIPPE (*f*) : coal tipper ; coal tipping device 432

KOHLENKIPPER (*m*) : (see KOHLENKIPPE) 433

KOHLENKIPPWAGEN (*m*) : coal tip-(ping) waggon 434

KOHLENKLÄRE (*f*) : coal dust 435

KOHLENKLEIN (*n*) : small coal ; slack 436

KOHLENKNAPPHEIT (*f*) : coal shortage 437

KOHLENKORB (*m*): coal bunker; coal basket; coal hopper 438
KOHLENLAGER (*n*): coal (depot) store; colliery; coal (field; seam) bed 439
KOHLENLICHT (*n*): carbon light 440
KOHLENLÖSCHE (*f*): coal (dust) slack; coal pit refuse 441
KOHLENLUFT (*f*): carbonic (fixed) air 442
KOHLENMEHL (*n*): (see KOHLENSTAUB) 443
KOHLENMEILER (*m*): charcoal pile 444
KOHLENMESSFAHRT (*f*): trial run to ascertain the coal consumption 445
KOHLENMIKROPHON (*n*): carbon microphone, carbon transmitter 446
KOHLENMONOXYD (*n*): carbon monoxide; CO (see KOHLENOXYD) 447
KOHLENNÄSSHAHN (*m*): coal spraying cock 448
KOHLENOFEN (*m*): charcoal kiln; coal stove or furnace 449
KOHLENÖL (*n*): (see STEINKOHLENTEERÖL) 450
KOHLENÖLSÄURE (*f*): carbolic acid (see PHENOL) 451
KOHLENOXYCHLORID (*n*): carbon oxychloride; carbonyl chloride (see KOHLENSTOFFOXY--CHLORID) 452
KOHLENOXYD (*n*): carbon monoxide; CO; Sg. 0.9674; Mp. – 207°C.; Bp. – 190°C. 453
KOHLENOXYDAPPARAT (*n*): carbon monoxide apparatus 454
KOHLENOXYDEISEN (*n*): iron carbonyl 455
KOHLENOXYDHÄMOGLOBIN (*n*): carbohemoglobin 456
KOHLENOXYDKALIUM (*n*): potassium (carboxide) hexacarbonyl 457
KOHLENOXYDKNALLGAS (*n*): explosive gas (carbon monoxide and oxygen) 458
KOHLENOXYDNICKEL (*n*): nickel (carbonyl) tetracarbonyl; $Ni(CO)_4$; Sg. 1.32; Mp. – 25°C.; Bp. 43°C. 459
KOHLENOXYDREICH: rich in (CO) carbon monoxide; with (having) high CO content 460
KOHLENPLATZ (*m*): coal-hole; place where charcoal is made 461
KOHLENPROBE (*f*): coal sample 462
KOHLENPULVER (*n*): charcoal powder (for case-hardening steel), carbon powder 463
KOHLENRAUM (*m*): coal bunker (on ships) 464
KOHLENRISS (*m*): charcoal sketch 465
KOHLENRUTSCHE (*f*): coal chute 466
KOHLENSACK (*m*): belly (widest part of blast furnace chamber, between the two truncated cone shaped parts) 467
KOHLENSÄGE (*f*): charcoal saw 468
KOHLENSANDSTEIN (*m*): carboniferous sandstone 469
KOHLENSAUER: carbonate of; combined with carbonic acid; carbonic 470
KOHLENSÄURE (*f*): carbonic acid, carbon dioxide; CO_2 (*Acidum carbonicum*); Sg. gas. 1.53; liquid 1.057; solid 1.56; Mp. – 65°C.; Bp. – 78.2°C. 471
KOHLENSÄUREANHYDRID (*n*): carbonic anhydride; carbon dioxide; CO_2 472
KOHLENSÄUREANLAGE (*f*): carbon dioxide plant 473
KOHLENSÄUREANTRIEB (*m*): carbonic acid gas (electro-gas) drive 474
KOHLENSÄUREANZEIGER (*m*): CO_2 indicator 475
KOHLENSÄUREARMATUREN (*f.pl*): carbonic acid fittings 476
KOHLENSÄUREASSIMILATION (*f*): assimilation of carbonic acid (by plants) 477
KOHLENSÄUREBESTIMMUNGSAPPARAT (*m*): apparatus for determining the carbonic acid content; CO_2 (meter) recorder 478
KOHLENSÄUREDIÄTHYLESTER (*m*): carbonic diethylester, diethyl carbonate; $C_5H_{10}O_3$; Sg. 0.9762; Bp. 126.4°C. 479
KOHLENSÄUREFABRIKATION (*f*): carbon dioxide manufacture 480
KOHLENSÄUREFLASCHE (*f*): carbonic acid cylinder 481
KOHLENSÄUREGAS (*n*): carbonic acid gas; carbon dioxide; CO_2 482
KOHLENSÄUREGEHALT (*m*): CO_2 content; carbonic acid (carbon dioxide) content 483
KOHLENSÄUREHALTIG: carbonated; containing carbonic acid 484
KOHLENSÄUREHERSTELLUNGSANLAGE (*f*): carbonic acid manufacturing plant 485
KOHLENSÄURELÄUTWERK (*n*): bell actuated by carbonic acid (Railways) 486
KOHLENSÄUREMESSER (*m*): anthracometer; instrument for measuring CO_2; CO_2 meter 487
KOHLENSÄUREREGISTRATOR (*m*): CO_2 (indicator) recorder 488
KOHLENSAURES SALZ (*n*): carbonate 489
KOHLENSAURES WASSER (*n*): carbonated water, sodawater, mineral water 490
KOHLENSÄUREVERFLÜSSIGUNGSANLAGE (*f*): carbonic acid liquefying plant 491
KOHLENSÄUREWASSER (*n*): carbonated water; soda water; mineral water 492
KOHLENSCHACHT (*m*): (coal)-(mine)-shaft 493
KOHLENSCHÄLCHEN (*n*): charcoal dish, or capsule 494
KOHLENSCHALTER (*m*): carbon switch 495
KOHLENSCHAUFEL (*f*): coal shovel 496
KOHLENSCHEIDER (*m*): coal (separator) separating plant 497
KOHLENSCHICHT (*f*): thickness of fire (on stokers); coal (layer, seam) stratum 498
KOHLENSCHIEFER (*m*): bitumenous (coal-bearing) shale 499
KOHLENSCHIFF (*n*): coaling (vessel) ship; collier 500
KOHLENSCHIPPE (*f*): coal shovel 501
KOHLENSCHLACKE (*f*): (coal)-cinders; clinker 502
KOHLENSCHLAMM (*m*): waste washings, sludge, smudge (mud from coal washing); wet coal dust (size 0-½ or ¾ mm.) 503
KOHLENSCHLICHTE (*f*): black wash (Founding) 504
KOHLENSCHOPPEN (*m*): coal shed 505
KOHLENSCHUPPEN (*m*): coal shed; coal (cellar; depot) store 506
KOHLENSCHÜTTE (*f*): coal scuttle 507
KOHLENSCHWARZ (*n*): charcoal black; coal black 508
KOHLENSCHWEFEL (*m*): carbo-sulphuret 509
KOHLENSIEB (*n*): screen, riddle, sieve (for coal) 510
KOHLENSORTE (*f*): (see KOHLENART) 511
KOHLENSPAT (*m*): anthraconite 512
KOHLENSPEICHER (*m*): coal (depôt) store; coal shed; coal cellar; bunker; silo 513
KOHLENSPITZE (*f*): carbon point 514
KOHLENSPRITZHAHN (*m*): coal spraying cock 515
KOHLENSTAUB (*m*): coal dust; charcoal dust; powdered (fuel) coal; pulverized (fuel) coal 516
KOHLENSTAUBFEUERUNG (*f*): powdered (pulverized) coal or fuel furnace or firing, coal-dust furnace or firing 517

KOHLENSTAUBMÜHLE (*f*): coal-dust mill 518
KOHLENSTICKSTOFF (*m*): cyanogen (see CYAN) 519
KOHLENSTICKSTOFFSÄURE (*f*): picric acid (see PIKRINSÄURE) 520
KOHLENSTIFT (*m*): charcoal (pencil) stick; carbon pencil; carbon (for arc lamps) 521
KOHLENSTOFF (*m*): carbon; C; (*Carboneum*) 522
KOHLENSTOFFBESTIMMUNGSAPPARAT (*m*): apparatus for the determination of carbon content 523
KOHLENSTOFFBINDUNG (*f*): carbon linkage 524
KOHLENSTOFFCHLORID (*n*): carbon tetrachloride (see TETRACHLORKOHLENSTOFF or KOHLENSTOFFTETRACHLORID) 525
KOHLENSTOFFDICHLORID (*n*): carbon (bichloride) dichloride; tetrachloroethylene, tetrachloroethene (see TETRACHLORÄTHYLEN) 526
KOHLENSTOFFEISEN (*n*): iron carbide (see EISENCARBID) 527
KOHLENSTOFFENTZIEHUNG (*f*): decarbonization; carbon (removal) extraction 528
KOHLENSTOFF (*m*), FESTER-: fixed carbon (Coal analysis) 529
KOHLENSTOFFGEHALT (*m*): carbon content 530
KOHLENSTOFFHALTIG: carboniferous; carbonaceous; containing carbon 531
KOHLENSTOFFKALZIUM (*n*): calcium carbide (see CALCIUMCARBID) 532
KOHLENSTOFFKERN (*m*): carbon nucleus 533
KOHLENSTOFFKETTE (*f*): carbon chain 534
KOHLENSTOFFMETALL (*n*): (metal)-carbide 535
KOHLENSTOFFOXYBROMID (*n*): carbon oxybromide; $COBr_2$; Sg. 2.48 536
KOHLENSTOFFOXYCHLORID (*n*): carbon oxychloride, phosgene, carbonyl chloride; $COCl_2$; Sg. 1.392-1.432; Mp. – 75°C.; Bp. 8.2°C. 537
KOHLENSTOFFREICH: rich in carbon; with (having) high carbon content 538
KOHLENSTOFFSILIZIUM (*n*): carbon silicide (see SILIZIUMKARBID) 539
KOHLENSTOFFSTAHL (*m*): carbon steel 540
KOHLENSTOFFSTEIN (*m*): carbon brick (Sg. 3.0) 541
KOHLENSTOFFSUBOXYD (*n*): carbon suboxide; C_3O_2; Sg. 1.1137 542
KOHLENSTOFFSULFID (*n*): carbon disulphide (see SCHWEFELKOHLENSTOFF) 543
KOHLENSTOFFTETRABROMID (*n*): carbon tetrabromide; CBr_4; Sg. 3.42 544
KOHLENSTOFFTETRACHLORID (*n*): carbon tetrachloride, tetrachloromethane; CCl_4; perchloromethane; Sg. 1.5883-1.5947; Mp. – 22.95°C; Bp. 76.74°C 545
KOHLENSTOFFTETRAJODID (*n*) carbon tetriodide; CJ_4; Sg. 4.32 546
KOHLENSTOFFTRICHLORID (*n*): carbon trichloride (see HEXACHLORÄTHAN) 547
KOHLENSTOFFVERBINDUNG (*f*): carbon compound; carburet 548
KOHLENSTOFFWERKZEUGSTAHL (*m*): carbon tool steel 549
KOHLENSTURZBAHN (*f*): coal chute 550
KOHLENSUBOXYD (*n*): carbon suboxide (see KOHLENSTOFFSUBOXYD) 551
KOHLENSULFID (*n*): carbon disulphide (see SCHWEFELKOHLENSTOFF) 552
KOHLENSULFIDSALZ (*n*): thiocarbonate; sulphocarbonate 553
KOHLENTEER (*m*): coal tar (see STEINKOHLENTEER) 554
KOHLENTEERÖL (*n*): coal-tar oil (see also STEINKOHLENTEERÖL) 555
KOHLENTEERPECH (*n*): coal tar pitch 556
KOHLENTEERSEIFE (*f*): coal tar soap 557
KOHLENTIEGEL (*m*): carbon (charcoal) crucible; crucible with carboniferous lining 558
KOHLENTOPF (*m*): charcoal pan 559
KOHLENTRÄGER (*m*): coal-man, coal-porter 560
KOHLENTRANSPORTANLAGE (*f*): coal-conveying plant, coal-handling plant 561
KOHLENTRICHTER (*m*): coal-hopper 562
KOHLENTRIMMER (*m*): coal trimmer 563
KOHLENVERBRAUCH (*m*): coal consumption 564
KOHLENVERGASUNG (*f*): coal distillation, coking process; extraction of gas from coal 565
KOHLENVERLADEKRAN (*m*): coaling (loading) crane 566
KOHLENVORRAT (*m*): coal supply; stock of coal 567
KOHLENVORSCHUB (*m*): coal feeding (device) arrangement 568
KOHLENWAGE (*f*): coal weigher, coal weighing machine 569
KOHLENWAGEN (*m*): coal (waggon; cart) truck tender (of an engine) 570
KOHLENWÄSCHE (*f*): coal washing plant 571
KOHLENWASSER (*n*): water for washing coal, washing water 572
KOHLENWASSERSTOFF (*m*): hydrocarbon 573
KOHLENWASSERSTOFFABLASSSCHRAUBE (*f*): screw plug of hydrocarbon outlet valve 574
KOHLENWASSERSTOFFABLASSVENTIL (*n*): hydrocarbon outlet valve 575
KOHLENWASSERSTOFFFÄNGER (*m*): hydrocarbon collector 576
KOHLENWASSERSTOFFGAS (*n*): hydrocarbon gas; carburetted hydrogen 577
KOHLENWASSERSTOFFGAS, LEICHTES- (*n*): methane (see METHAN) 578
KOHLENWASSERSTOFFGAS, SCHWERES- (*n*): ethylene (see ÄTHYLEN) 579
KOHLENWASSERSTOFFHALTIG: hydrocarboniferous; hydrocarbonaceous; containing hydrocarbon-(s) 580
KOHLENWASSERSTOFF (*m*), LEICHTER-: light hydrocarbon 581
KOHLENWASSERSTOFFSAMMLER (*m*): hydrocarbon collector 582
KOHLENWASSERSTOFF (*m*), SCHWERER-: heavy hydrocarbon 583
KOHLENWASSERSTOFFVERBINDUNG (*f*): hydrocarbon (compound of carbon and hydrogen) 584
KOHLENWERK (*n*): colliery, coal-mine 585
KOHLENWIDERSTAND (*m*): carbon resistance 586
KOHLENZECHE (*f*): (coal)-mine 587
KOHLENZEICHNUNG (*f*): charcoal drawing 588
KOHLENZIEGEL (*m*): coal briquette 589
KOHLEPAPIER (*n*): carbon paper 590
KÖHLER (*m*): charcoal burner; miner; collier 591
KÖHLEREI (*f*): charcoal burning; charcoal (works) kiln 592
KÖHLERKRAUT (*n*): speedwell (*Veronica*) (Bot.) 593
KOHLESTAUB (*m*): charcoal dust 594
KOHLESTIFT (*m*): carbon-(rod), (for arc lamps); charcoal pencil, carbon pencil 595
KOHLE-TEER (*m*): coal tar 596
KOHLEZUSAMMENSETZUNG (*f*): composition of coal, analysis of coal 597
KOHLFEUER (*n*): coal (charcoal) fire 598

KOHLHOLZ (*n*): (common)-privet (*Ligustrum vulgare*): wood for charcoal 599
KOHLIG: like coal; coal-bearing 600
KOHLKEIMCHEN (*n*): sprout 601
KOHLKOPF (*m*): head of cabbage 602
KOHLPALME (*f*): cabbage (palm) palmetto (*Sabal palmetto*) 603
KOHLPFANNE (*f*): (see KOHLENBECKEN) 604
KOHLRABI (*m*): kohl-rabi, cabbage-turnip (*Brassica caulorapa*) 605
KOHLRÜBE (*f*): swede (*Brassica rutabaga*); rape, colza, coleseed (colza, summer rape, *Brassica campestris*; coleseed, winter rape, *Brassica napus*) 606
KOHLSAAT (*f*): colza; rapeseed; coleseed (see KOHLRÜBE) 607
KOHLSAATÖL (*n*): colza oil; rape oil (see RAPSÖL and RÜBÖL) 608
KOHLSAME (*m*): coleseed (see KOHLRÜBE) 609
KOHLSCHWARZ: coal-black 610
KOHLUNG (*f*): carbonization (also of metal in blast furnace work); charring; carburization 611
KOHLUNGSZONE (*f*): carbonization zone (in which metal is carbonized in blast furnace work) 612
KOHLWEISSLING (*m*): cabbage butterfly (*Pieris brassicæ*) 613
KOHOBIEREN: to cohobate; distil repeatedly 614
KOJE (*f*): cabin; stateroom; berth 615
KOK (*m*): coke (Sg. 1.0) (see under KOKS) 616
KOKA (*f*): coca (*Erythroxylon coca*) 617
KOKABLÄTTER (*n.pl*): coca leaves 618
KOKAÏN (*n*): cocaine (see COCAIN) 619
KOKAIN (*n*), **BENZOESAURES-**: cocaine benzoate 620
KOKAIN (*n*), **BROMWASSERSTOFFSAURES-**: cocaine hydrobromide 621
KOKAINCHLORHYDRAT (*n*): cocaine hydrochloride; $C_{17}H_{21}NO_4.HCl$; Mp. 183-191°C. 622
KOKAINCHLORHYDRATLÖSUNG (*f*): cocaine hydrochloride solution 623
KOKAINOL (*n*): (see ANÄSTHESIN) 624
KOKAIN (*n*), **SALICYLSAURES-**: cocaine salicylate 625
KOKAIN (*n*), **SALZSAURES-**: cocaine hydrochloride (see KOKAINCHLORHYDRAT) 626
KOKE (*m.* and *f*): coke (Sg. 1.0) (see under KOKS) 627
KOKER (*m*): sagger; trunk 628
KOKEREI (*f*): coking, distillation of coal; coking plant 629
KOKEREIBETRIEB (*m*): coke oven plant 630
KOKEREIMASCHINEN (*f.pl*): coking machinery, coke-oven (work's) machinery 631
KOKER (*m*), **MAST-**: mast trunk 632
KOKER (*m*), **RUDER-**: rudder case, rudder trunk 633
KOKETTIEREN: to coquet; flirt 634
KOKILLE (*f*): (chill-mould) (casting mould for metal, of truncated cone-shape mostly of square section, of hematite pig iron, lasts 100-200 castings, widest at bottom, but in Gathmann process widest at top) 635
KOKILLENGESPANN (*n*): group of chill-moulds (a number of chill-moulds on an iron base plate, grouped around a funnel; usually 4-6 chill-moulds) 636
KOKKELSKORN (*n*): Indian berry (*Cocculus indicus*) seed of *Anamirta cocculus*) 637
KOKKOLITH (*m*): coccolith (a calcareous spicule) (Zool.) 638
KOKOMFETT (*n*): (see GOABUTTER) 639
KOKON (*m*): cocoon (of silkworms) 640
KOKONFADEN (*m*): cocoon fibre 641
KOKOS (*m*): coco; cocoa-(nut) palm; cocoanut tree (*Cocos nucifera*) (see also COCOS) 642
KOKOSBAUM (*m*): (see KOKOS) 643
KOKOSBUTTER (*f*): cocoanut (oil) butter (see KOKOSNUSSÖL) 644
KOKOSFASER (*f*): coir; cocoanut fibre (from the fruit shells of the cocoanut palm) 645
KOKOSFETT (*n*): cocoanut oil (see KOKOSNUSSÖL) 646
KOKOSHOLZ (*n*): cocus wood (*Brya ebenus*); Sg. 1.48-1.61 647
KOKOSNUSS (*f*): cocoanut (fruit of *Cocos nucifera*) 648
KOKOSNUSSÖL (*n*): cocoanut oil; coco-nut-(palm) -oil (*Oleum cocos*) (from Copra); Sg. 0.912; Mp. 20-28°C. 649
KOKOSNUSSROT (*n*): cocoanut red 650
KOKOSÖL (*n*): (see KOKOSNUSSÖL) 651
KOKOSPALME (*f*): cocoanut palm (*Cocos nucifera*) 652
KOKOSSEIFE (*f*): cocoanut oil soap 653
KOKOSTALG (*m*): cocoanut oil (see KOKOSNUSSÖL) 654
KOKOSTALGSÄURE (*f*): cocinic acid 655
KOKS (*m* and *pl*): coke; Sg. 1.0 (residue of dry distillation of coal, but also known in a natural state) 656
KOKSABFALL (*m*): coke (refuse) waste; refuse (waste) coke 657
KOKSASCHE (*f*): small coke, coke dust, coke breeze (preferably called "Kleinkoks" or "Koksgries") (see KLEINKOKS and KOKSGRIES) 658
KOKSBLECH (*n*): coke sheet iron 659
KOKSBRASCHEN (*f.pl*): combustible matter remaining in clinker, a sort of "waste washings" (obtained by breaking and washing clinker) (see BRASCHENWÄSCHE) 660
KOKSBRECHER (*m*): coke (breaker) crusher 661
KOKSBRENNEN (*n*): coking; coke-burning 662
KOKSFILTER (*n*): coke filter 663
KOKSGICHT (*f*): coke charge to blast furnaces 664
KOKSGRIES (*m*): small coke; coke (dust) breeze; (0.10 mm. size; broken coke; ash about 12%; moisture about 10%; calorific value about 5200) 665
KOKSGRUS (*m*): dross (see KOKSGRIES) 666
KOKSKLEIN (*n*): (see KOKSGRIES) 667
KOKSKOHLE (*f*): coking coal (0-10 mm. in size) 668
KOKSLÖSCHE (*f*): cinder-(s); clinker 669
KOKSOFEN (*m*): coke oven 670
KOKSOFENBESCHICKUNGSANLAGE (*f*): coke-oven (feeding) charging plant 671
KOKSOFENGAS (*n*): coke-oven gas 672
KOKSOFENGASREINIGUNG (*f*): coke-oven gas (cleaning) scrubbing 673
KOKSOFENGASREINIGUNGSANLAGE (*f*): coke-oven gas (cleaning) scrubbing plant 674
KOKSPLATZ (*m*): place where coke is unloaded 675
KOKSROHEISEN (*n*): charcoal pig iron 676
KOKSTEER (*m*): coke tar 677
KOKUMBUTTER (*f*): (see GOABUTTER) 678
KOLALIKÖR (*m*): cola liqueur 679
KOLANUSS (*f*): cola nut; Guru; Soudan coffee (from *Cola acuminata*) 680
KOLAROT (*n*): cola red 681
KOLATORIUM (*n*): colatorium; strainer (a woollen or linen cloth) 682

KOLATUR (*f*) : strained liquid (see KOLIEREN) 683
KOLAWEIN (*m*) : cola wine 684
KOLAWIRKUNG (*f*) : action (effect) of cola 685
KOLBE (*f*) : (see KOLBEN) 686
KOLBEN (*m*) : flask ; alembic ; still ; butt-(end) ; club ; piston ; soldering iron ; feeler ; horn ; bloom (Metal) ; bolthead ; matrass ; cucurbit (old Chemistry) ; head ; spadix ; spike (Bot.) ; bulb ; nodule ; protuberance ; excrescence ; plunger ; mace ; knob ; burnisher ; demi-john 687
KOLBENANSATZ (*m*) : piston extension 688
KOLBENAUFGANG (*m*) : piston (rise) ascent ; up-stroke (of a piston) 689
KOLBENBÄRLAPP (*n*) : (see BÄRLAPP) 690
KOLBENBESCHLEUNIGUNG (*f*): piston acceleration 691
KOLBENBODEN (*m*) : piston head 692
KOLBENBOLZEN (*m*) : piston pin 693
KOLBENBRUCH (*m*) : piston fracture 694
KOLBENBÜRSTE (*f*) : bottle brush 695
KOLBENDAMPFMASCHINE (*f*) : reciprocating (piston) steam engine 696
KOLBENDECKE (*f*) : junk ring ; follower-(plate) 697
KOLBENDECKEL (*m*) : (see KOLBENDECKE) 698
KOLBENDECKELSCHRAUBE (*f*) : junk-ring pin 699
KOLBENDIAGRAMM (*n*) : piston diagram 700
KOLBENDICHTUNG (*f*) : piston packing 701
KOLBENDRUCK (*m*) : piston pressure 702
KOLBENDURCHMESSER (*m*) : piston diameter 703
KOLBENFEDER (*f*) : snap piston ring ; elastic ring ; piston spring 704
KOLBENFLÜGE (*m.pl*) : piston up-strokes 705
KOLBENFORM (*f*) : piston shape 706
KOLBENFÖRMIG : club-shaped ; piston-shaped ; nodular 707
KOLBENFÜHRUNG (*f*) : piston guide-(s) 708
KOLBENGEFÄSS (*n*): alembic, still 709
KOLBENGESCHWINDIGKEIT (*f*) : piston speed 710
KOLBENGEWICHT (*n*) : piston weight 711
KOLBENGLAS (*n*) : alembic 712
KOLBENGLEITFLÄCHE (*f*) : piston sliding surface 713
KOLBENGRAS (*n*) : fox-tail grass (*Alopecurus bulbosus*) 714
KOLBENHEIZUNG (*f*) : piston heating 715
KOLBENHINGANG (*m*) : forward stroke (of a piston) 716
KOLBENHÖHE (*f*) : piston (length) depth 717
KOLBENHUB (*m*) : piston (travel) stroke 718
KOLBENHUBRAUM (*m*) : piston stroke volume 719
KOLBENINDIKATOR (*m*) : piston indicator 720
KOLBENKÖRPER (*m*) : piston body ; piston head 721
KOLBENKRAFTDIAGRAMM (*n*) : piston pressure diagram 722
KOLBENKÜHLUNG (*f*) : piston cooling 723
KOLBENLAUF (*m*) : piston (motion ; stroke) travel 724
KOLBENLAUFBAHN (*f*) : working surface of piston 725
KOLBENLAUFFLÄCHE (*f*) : (see KOLBENLAUFBAHN) 726
KOLBENLIDERUNG (*f*) : piston packing 727
KOLBENMASCHINE (*f*) : piston engine 728
KOLBENMASSE (*f*) : piston mass 729
KOLBENMOOS (*n*) : (see BÄRLAPP) 730
KOLBENMOTOR (*m*) : piston (engine) motor 731
KOLBENNABE (*f*) : piston boss 732
KOLBENNIEDERGANG (*m*) : piston descent ; down-stroke 733
KOLBENNUTE (*f*) : piston ring ; groove 734
KOLBEN, OFFENER- (*m*) : truck piston 735
KOLBENPUMPE (*f*) : plunger pump 736
KOLBENREIBUNG (*f*) : piston friction 737
KOLBENRING (*m*) : piston ring ; packing ring 738
KOLBENRINGFUGE (*f*) : piston ring joint 739
KOLBENRINGSCHLOSS (*n*) : piston ring lock 740
KOLBENRÖHRE (*f*) : piston chamber ; pump barrel 741
KOLBENRÜCKGANG (*m*) : return piston-stroke 742
KOLBENSCHEIBE (*f*) : piston disc 743
KOLBENSCHIEBER (*m*) : piston valve 744
KOLBENSCHIEBERBÜCHSE (*f*) : piston valve liner 745
KOLBENSCHIEBEREINSATZ (*m*) : piston valve liner 746
KOLBENSCHIEBERGEHÄUSE (*n*) : piston valve chest 747
KOLBENSCHIEBERKASTEN (*m*) : piston valve chest 748
KOLBENSCHIEBERSTEUERUNG (*f*) : piston valve control 749
KOLBENSCHILF (*n*) : (see ROHRKOLBENSCHILF) 750
KOLBENSCHIMMEL (*m*) : club mould (*Aspergillus* species) 751
KOLBENSCHLOSS (*n*) : piston (ring) lock 752
KOLBENSCHUB (*m*) : piston (thrust) motion 753
KOLBENSPIEL (*n*) : piston (play ; motion ; travel) stroke 754
KOLBENSPIELRAUM (*m*) : piston clearance 755
KOLBENSTANGE (*f*) : piston rod 756
KOLBENSTANGENBUND (*m*) : collar ; shoulder (of a piston rod) 757
KOLBENSTANGENDICHTUNG (*f*) : piston rod packing 758
KOLBENSTANGENENDE (*n*) : piston rod (marine) end 759
KOLBENSTANGENFLANSCH (*m*) : piston rod flange 760
KOLBENSTANGENFÜHRUNG (*f*) : piston rod guide 761
KOLBENSTANGENGEWINDE (*n*) : piston rod thread 762
KOLBENSTANGENKEGEL (*m*) : piston rod taper 763
KOLBENSTANGENKONUS (*m*) : piston rod taper 764
KOLBENSTANGENKRAFT (*f*) : piston pressure 765
KOLBENSTANGENKUPPLUNG (*f*) : piston rod coupling 766
KOLBENSTANGENLIDERUNG (*f*) : piston rod packing 767
KOLBENSTANGENMUTTER (*f*) : piston rod nut 768
KOLBENSTANGENPACKUNG (*f*) : piston rod packing 769
KOLBENSTANGENSCHAFT (*m*) : piston rod 770
KOLBENSTANGENTRAVERSE (*f*) : crosshead-(beam) 771
KOLBENSTREICH (*m*) : blow with a club 772
KOLBENTRAGEND : Spadiceous (Bot.) 773
KOLBENTRÄGER (*m*) : flask (support) stand ; mace-bearer 774
KOLBENTRAGFLÄCHE (*f*) : piston bearing surface 775
KOLBENÜBERLAUF (*m*) : over-run (of a piston) 776
KOLBENVENTIL (*n*) : piston (slide)-valve 777
KOLBENVORLAUF (*m*) : forward piston-stroke 778
KOLBENWASSERMESSER (*m*) : piston (positive) water meter 779
KOLBENWEG (*m*) : piston (travel ; path ; course ; stroke ; play) motion 780
KOLBENWEGDIAGRAMM (*n*) : piston diagram ; stroke diagram 781

KOLBENWEGLINIE (*f*): line indicating piston (course; travel) path 782
KOLBENWEGSINOIDE (*f*): piston travel sinusoid 783
KOLBIG: knobbed, knotty, knotted, knobby, club-like, nodular 784
KOLBWURZ (*f*): white water lily (*Nymphœa alba*) 785
KOLCHIZIN (*n*): (see COLCHICIN) 786
KOLDERKRAUT (*n*): scarlet pimpernel (*Anagallis arvensis*) 787
KOLDERSTOCK (*m*): whip-staff (Naut.) 788
KOLEMANIT (*m*): (see COLEMANIT) 789
KOLIERAPPARAT (*m*): straining (filtering) apparatus; strainer, filter 790
KOLIEREN: to strain, filter, percolate (separate a liquid from the suspended solid matter contained therein); (*n*): straining 791
KOLIERRAHMEN (*m*): filter-(ing) frame 792
KOLIERTUCH (*n*): filter-(ing) cloth; (woollen or linen cloth) 793
KOLIK (*f*): colic; enteralgia (Med.) 794
KOLKOTHAR (*n*): (see COLCOTHAR) 795
KOLLARGOL (*n*): (see COLLARGOL) 796
KOLLAURIN (*n*): (see COLLAURIN) 797
KOLLEG (*n*): college; course of lectures; hall (see also KOLLEGIUM) 798
KOLLEGE (*m*): colleague 799
KOLLEGIAL: collegiate 800
KOLLEGIUM (*n*): board, council, commission, college, lecture, hall, society, administrative board 801
KOLLEKTIEREN: to collect 802
KOLLEKTOR (*m*): commutator (Elect.); collector 803
KOLLEKTORBÜRSTE (*f*): collector brush, commutator brush (Elect.) 804
KOLLEKTORMOTOR (*m*): commutator motor 805
KOLLER (*n*): collar, jerkin; (*pl*): rage, madness 806
KOLLERFARBE (*f*): yellow ochre 807
KOLLERGANG (*m*): edge mill; vertical (pug) mill; grinding (crushing) mill; rolling and sorting mill (Paper) 808
KOLLERIG: mad, choleric 809
KOLLERMÜHLE (*f*): (see KOLLERGANG) 810
KOLLERN: to rumble; gobble; rave; roll; sort (Paper) 811
KOLLERSTOFF (*m*): material to be rolled and sorted (Paper) 812
KOLLI (*pl* of KOLLO): consignment; bales, bundles 813
KOLLIDIN (*n*): collidine (a homologue of pyridine); $C_8H_{11}N$ 814
KOLLISION (*f*): collision 815
KOLLISIONSSCHOTT (*n*): collision bulkhead (Naut.) 816
KOLLO (*n* and *m*): consignment, bale, bundle 817
KOLLODIN (*n*): (see COLLODIN) 818
KOLLODION (*n*): collodion (see COLLODIUM) 819
KOLLODIONIEREN: to collodionize; (*n*): collodionizing 820
KOLLODIUM (*n*): collodion (see also COLLODIUM) 821
KOLLODIUMLÖSUNG (*f*): collodion solution (see COLLODIUMLÖSUNG) 822
KOLLODIUMVERFAHREN (*n*), NASSES-: wet collodion process (Photography) 823
KOLLODIUMVERFAHREN (*n*), TROCKENES-: collodiobromide process (Photography) 824
KOLLODIUMWOLLE (*f*): collodion cotton; pyroxylin; soluble guncotton (see COLLODIUMWOLLE) 825
KOLLOID (*n*): colloid (a body which is difficult to crystallize and to diffuse) (see SOL and GEL) (as a prefix=colloidal) 826
KOLLOIDAL: colloidal 827
KOLLOIDALE FORM (*f*): colloidal form 828
KOLLOIDALER KÖRPER (*m*): colloid, colloidal substance (see KOLLOID) 829
KOLLOIDALES GOLD (*n*): colloidal gold (see COLLAURIN) 830
KOLLOIDALES PRÄPARAT (*n*): colloidal preparation 831
KOLLOIDALES SILBER (*n*): colloidal silver (see COLLARGOL) 832
KOLLOIDCHEMIE (*f*): colloid-(al) chemistry 833
KOLLOIDCHEMISCH: colloidochemical 834
KOLLOID (*n*), GELATINEARTIGES-: (see GEL, HYDROGEL and ORGANOGEL) 835
KOLLOID (*n*), GELÖSTES-: (see SOL, HYDROSOL and ORGANOSOL) 836
KOLLOID-KONZENTRATION (*f*): colloid concentration 837
KOLLOIDÖL (*n*): colloidal oil 838
KOLLOID (*n*), SCHUTZ-: protective colloid (for stabilizing an otherwise instable colloid, i.e., preventing it from coagulating) 839
KOLLOIDSTOFF (*m*): colloid-(substance); colloidal substance 840
KOLLOIDSUBSTANZ (*f*): (see KOLLOIDSTOFF) 841
KOLLOIDZUSTAND (*m*): colloidal state 842
KOLLYRIT (*m*): allophan (see ALLOPHAN) 843
KÖLNERGELB (*n*): Cologne yellow (a mixture of 60% gypsum, 25% chrome yellow, and 15% lead sulphate) 844
KÖLNERWASSER (*n*): Eau de Cologne 845
KÖLNISCHE ERDE (*f*): Cologne earth (see BRAUNKOHLENPULVER) 846
KÖLNISCHES WASSER (*n*): (see KÖLNERWASSER) 847
KOLOMBOSÄURE (*f*): (see COLOMBOSÄURE) 848
KOLOMBOWURZEL (*f*): calumba-(root); columbo (root of *Jateorrhiza palmata*) 849
KOLON (*m*): colon (Med.) 850
KOLONIALZUCKER (*m*): muscovado; (unrefined)-cane sugar 851
KOLONNE (*f*): column 852
KOLONNENAPPARAT (*n*): column apparatus 853
KOLONNENMÜHLE (*f*): column mill (in which the various drums are arranged in series) 854
KOLOPHONEISENERZ (*n*): pitticite (Min.) 855
KOLOPHONIUM (*n*): resin; rosin; colophony; Sg. 1.08; Mp. 100-140°C. (*Colophonium*) 856
KOLOPHONSÄURE (*f*): colophonic acid 857
KOLOQUINTE (*f*): colocynth; bitter gourd; bitter cucumber; bitter apple (fruit of *Citrullus colocynthis*) 858
KOLORIMETER (*n*): colorimeter (an instrument for measuring strength or quality of a substance by a comparison of its colour with a standard) 859
KOLORIMETER (*n*), SPEKTRAL-: spectral colorimeter 860
KOLORIMETRIE (*f*): colorimetry 861
KOLORIMETRISCH: colorimetric 862
KOLOSS (*m*): colossus 863
KOLOSSAL: colossal 864
KOLUMBIT (*m*): columbite (see COLUMBIT and NIOBIT) 865
KOLUMBOWURZEL (*f*): (see KOLOMBOWURZEL) 866
KOLUMNE (*f*): column (of a page) 867

KOLUMNENMASS (*n*): rule, scale 868
KOLUMNENWEISE: in columns 869
KOMANSÄURE (*f*): comanic acid 870
KOMBINATION (*f*): combination 871
KOMBINATIONSDRUCK (*m*): combination printing (Photography); mixed printing process 872
KOMBINIEREN: to combine 873
KOMBOBUTTER (*f*): Kombo butter (from the fruit of *Pycnanthus kombo*) 874
KOMBÜSE (*f*): galley, cook house (Naut.) 875
KOMENSÄURE (*f*): comenic acid 876
KOMET (*m*): comet (Astron.) 877
KOMETENBAHN (*f*): orbit of a comet 878
KOMETENSCHWEIF (*m*): tail of a comet 879
KOMISCH: comic-(al); funny; strange-(ly) 880
KOMITE (*n*): committee 881
KOMMA (*n*): comma 882
KOMMANDANT (*m*): commander 883
KOMMANDITÄR (*m*): sleeping-partner (Com.) 884
KOMMANDITE (*f*): sleeping-partnership; branch (Com.) 885
KOMMANDITGESELLSCHAFT (*f*): Joint Stock Company 886
KOMMANDOBRÜCKE (*f*): (conning)-bridge (Marine) 887
KOMMANDOTURM (*m*): conning tower 888
KOMMEN: to come; happen; be; come (about) out; arrive (at); approach; arise 889
KOMMEN, HINTER-: to discover, find out 890
KOMMEN LASSEN: to send for 891
KOMMERS (*m*): drinking party 892
KOMMERZ (*m*): commerce 893
KOMMIS (*m*): clerk 894
KOMMISSAR (*m*): commissioner 895
KOMMISSION (*f*): commission, committee 896
KOMMISSIONÄR (*m*): commission agent 897
KOMMISSIONSGESCHÄFT (*n*): commission agency 898
KOMMUNALGARDE (*f*): civil guard 899
KOMMUNALSTEUER (*f*): municipal (local) tax 900
KOMMUNISMUS (*m*): communism 901
KOMMUNIZIEREN: to communicate 902
KOMMUTATOR (*m*): commutator, current reverser (Elect.) 903
KOMMUTIEREN: to commute 904
KOMÖDIANT (*m*): comedian, actor 905
KOMÖDIE (*f*): comedy; play 906
KOMPAKT: compact 907
KOMPARTIMENT (*n*): compartment 908
KOMPASS (*m*): compass 909
KOMPASSDOSE (*f*): compass box 910
KOMPASSHAUS (*n*): binnacle 911
KOMPASSHÄUSCHEN (*n*): binnacle 912
KOMPASS (*m*), KREISEL-: gyroscopic compass 913
KOMPASSROSE (*f*): compass card, rhumb card (Naut.) 914
KOMPASSSTRICH (*m*): point of the compass 915
KOMPATIBILITÄT (*f*): compatibility 916
KOMPATIBILITÄTSGLEICHUNG (*f*): compatibility equation 917
KOMPENSATION (*f*): compensation 918
KOMPENSATIONSEINRICHTUNG (*f*): compensating device 919
KOMPENSATIONSLEITUNG (*f*): compensation lead 920
KOMPENSATOR (*m*): compensator 921
KOMPENSIEREN: to compensate 922
KOMPETENZ (*f*): competence; competition 923
KOMPILATOR (*m*): compiler 924
KOMPLEMENTÄR: complementary 925
KOMPLEMENTWINKEL (*m*): complement of an angle 926
KOMPLET: complete 927
KOMPLEX: complex 928
KOMPLEXION (*n*): complexion 929
KOMPLEXSALZ (*n*): complex salt 930
KOMPLIKATION (*f*): complication 931
KOMPLIZIEREN: to complicate 932
KOMPLIZIERT: complicated 933
KOMPONENT: component 934
KOMPONENTE (*f*): component (of force) 935
KOMPONIEREN: to compose (Music) 936
KOMPOSITBAU (*m*): composite building 937
KOMPOSITBILD (*n*): composite (portrait) picture 938
KOMPOSITION (*f*): composition; tin composition (Dye) 939
KOMPOSITIONSRAKEL (*f*): composition doctor (of bronze) (Textile) 940
KOMPOST (*m*): compost (a not-easily decomposable mixture containing plant food); mixed fertilizer; humus (see also HUMUS) 941
KOMPOTT (*n*): compote; preserved (stewed) fruit 942
KOMPRESSIBILITÄT (*f*): compressibility 943
KOMPRESSION (*f*): compression 944
KOMPRESSION (*f*), ADIABATISCHE-: adiabatic compression 945
KOMPRESSIONSGRAD (*m*): degree of compression 946
KOMPRESSIONSHAHN (*m*): compression (tap) cock 947
KOMPRESSIONSPUMPE (*f*): compression pump 948
KOMPRESSIONSVOLUMEN (*n*): compression volume 949
KOMPRESSIONSWELLE (*f*): compression wave 950
KOMPRESSOR (*m*): compresser 951
KOMPRIMIEREN: to compress 952
KOMPRIMIERMASCHINE (*f*): compresser, compressing machine, press 953
KOMPROMISSVORSCHLAG (*m*): proposal of compromise 954
KOMPROMITTIEREN: to compromise, commit oneself 955
KOMPTOIR (*n*): office, bureau 956
KONCHOIDE (*f*): conchoid 957
KONCHYLIEN (*f.pl*): shells; shellfish; conches 958
KONCHYLIOLOGIE (*f*): conchology (the science of shells) 959
KONDENSABSCHEIDER (*m*): condensate separator or trap 960
KONDENSAT (*n*): condensate 961
KONDENSATION (*f*): condensation; condensing plant 962
KONDENSATION (*f*), NASSE-: jet condensation 963
KONDENSATION (*f*), OBERFLÄCHEN-: surface condensation, condensation by contact 964
KONDENSATIONSANLAGE (*f*): condensing plant 965
KONDENSATIONSAPPARAT (*m*): condenser; condensing apparatus 966
KONDENSATIONSDAMPFMASCHINE (*f*): condensing steam engine 967
KONDENSATIONSGEFÄSS (*n*): condensing vessel 968
KONDENSATIONSMASCHINE (*f*): condensing engine 969
KONDENSATIONSMETHODE (*f*): condensation method (method of colloidal synthesis in which the size of the molecules are so long increased by suitable means until they reach the region of colloidal dimensions) 970

KONDENSATIONSROHR (*n*): condensing tube; adapter 971
KONDENSATIONSRÖHRE (*f*): (see KONDENSATIONSROHR) 972
KONDENSATIONSTURM (*m*): condensing (cooling) tower, water cooling tower 973
KONDENSATION (*f*), **TROCKENE-**: surface condensation 974
KONDENSATOR (*m*): condenser 975
KONDENSATOR (*m*), **OBERFLÄCHEN-**: surface condenser 976
KONDENSATORROHR (*n*): condenser tube 977
KONDENSATOR (*m*), **RÖHREN-**: tubular condenser 978
KONDENSATPUMPE (*f*): condensate pump 979
KONDENSIERBAR: condensable 980
KONDENSIEREN: to condense; compress; liquify (of gases or vapours) 981
KONDENSTOPF (*m*): trap, steam trap, drain trap, steam (drier) separator 982
KONDENSWASSER (*n*): condensate 983
KONDENSWASSERABLEITER (*m*): condensate drain 984
KONDENSWASSERENTÖLER (*m*): condensate (deoiler) deoiling plant 985
KONDENSWASSERMESSER (*m*): condensate meter 986
KONDENSWASSERRÜCKLEITER (*m*): condensate return pipe 987
KONDITIONIERAPPARAT (*m*): conditioning apparatus (Silk) 988
KONDITIONIEREN: to condition 989
KONDITOR (*m*): confectioner 990
KONDITORWAREN (*f.pl*): confectionery 991
KONEPHRIN (*n*): (see CONEPHRIN) 992
KONFEKT (*n*): confectionery; confection (*Confectio* or *Conditum*) (raw plant substances or drugs coated with sugar) 993
KONFEKTION (*f*): confection (see KONFEKT) 994
KONFISZIEREN: to confiscate 995
KONFÖDERIEREN: to league, unite 996
KONGELIEREN: to congeal 997
KONGLOMERAT (*n*): conglomerate 998
KONGOBLAU (*n*): Congo blue (see also CONGO-) 999
KONGRUENT: congruent 000
KONGRUIEREN: to coincide, agree, be congruous, be consistent, be compatible, conform, accord, be consonant 001
KONIFERE (*f*): conifer (Bot.) 002
KONIFERENÖL (*n*): conifer oil 003
KÖNIG (*m*): king; regulus 004
KÖNIGBLAU (*n*): king's blue, cobalt blue, smalt (see BLAUFARBE) 005
KÖNIGIN (*f*): queen 006
KÖNIGINMETALL (*n*): queen's metal (tin alloy) 007
KÖNIGLICH: kingly; regal; royal; king-like; imperial 008
KÖNIGLICH GESINNTER (*m*): royalist 009
KÖNIGREICH (*n*): kingdom; realm 010
KÖNIGSALBE (*f*): resin cerate; basilicon ointment (Pharmacy) 011
KÖNIGSBLAU (*n*): king's blue; royal blue; cobalt blue; smalt (see BLAUFARBE) 012
KÖNIGSBLUME (*f*): peony, paeony 013
KÖNIGSCHINARINDE (*f*): calisaya bark; Jesuit's bark; yellow cinchona bark; Peruvian bark 014
KÖNIGSGELB (*n*): king's yellow (As_2S_3); chrome yellow (see CHROMGELB); massicot; yellow arsenic sulphide (see ARSENTRISULFID) 015
KÖNIGSGRÜN (*n*): Paris green (see KAISERGRÜN) 016
KÖNIGSHOLZ (*n*), **ROTES-**: Myall, violet wood (see VIOLETTHOLZ) 017
KÖNIGSKERZE (*f*): (common)-mullein (*Flores verbasci*) (*Verbascum*) 018
KÖNIGSKRANKHEIT (*f*): scrofula (Med.) 019
KÖNIGSKRAUT (*n*): field basil (*Clinopodium vulgare*) 020
KÖNIGSMORD (*m*): regicide (the action) 021
KÖNIGSMÖRDER (*m*): regicide (the doer) 022
KÖNIGSPALME (*f*): king palm (*Oreodoxa regia*) 023
KÖNIGSPISANG (*m*): plantain (*Plantago*) 024
KÖNIGSROSE (*f*): peony, paeony 025
KÖNIGSSÄURE (*f*): aqua regia; nitrohydrochloric (nitromuriatic; chlorazotic; chloronitrous) acid (*Acidum nitrohydrochloricum*) (mixture of HNO_3 and HCl in proportion 1-2 to 1-4) 026
KÖNIGSSCHLANGE (*f*): boa constrictor 027
KÖNIGSWASSER (*n*): (see KÖNIGSSÄURE) 028
KÖNIGSWÜRDE (*f*): kingly mien, regal dignity, royalty 029
KÖNIGTUM (*n*): kingdom; kingship 030
KONIIN (*n*): conine; coniine (see CONIIN) 031
KONISCH: conic-(al); coniform 032
KONIZITÄT (*f*): conicity; conical taper 033
KONJUGIERT: conjugate; coordinate 034
KONJUNKTURTAFEL (*f*): price curve (graphically illustrating the rise and fall of prices for various commodities and currencies) 035
KONKAV: concave 036
KONKREMENT (*n*): concrement; concretion (Med.) 037
KONKRET: concrete (as opposed to abstract); real, actual 038
KONKURRENZ (*f*): competition; competitors 039
KONKURRENZFÄHIG: capable of competing 040
KONKURRIEREN: to compete 041
KONKURS (*m*): bankruptcy; failure; assignment 042
KÖNNEN: to be able; be permitted; know; have power; be skilled 043
KÖNNEN, UMHIN-: to be able to (avoid, forbear) help 044
KONNOSSEMENT (*n*): bill of lading 045
KONSEQUENT: consequent-(ly), consequential 046
KONSERVE (*f*): conserve; preserve; preservative; confection; electuary (Pharmacy) 047
KONSERVENFABRIK (*f*): jam (conserve, preserve) factory 048
KONSERVIEREN: to preserve, conserve 049
KONSERVIERUNG (*f*): conservation; protection; preservation 050
KONSERVIERUNGSFIRNIS (*m*): preserving (protecting) varnish 051
KONSERVIERUNGSLACK (*m*): protecting (preserving) varnish 052
KONSERVIERUNGSMITTEL (*n*): preservative 053
KONSERVIERUNGSPRÄPARAT (*n*): preservative 054
KONSIGNATÄR (*m*): consignee 055
KONSIGNATION (*f*): consignment 056
KONSIGNIEREN: to consign, send, forward 057
KONSISTENT: consistent-(ly), dense, viscous 058
KONSISTENZ (*f*): consistency; viscosity; viscidity; density 059
KONSISTENZMESSER (*m*): viscosimeter; densimeter 060
KONSKRIPIEREN: to conscript, levy 061

KONSOLE (*f*): console, bracket 062
KONSOLIDIEREN: to consolidate 063
KONSOLLAGER (*n*): console (bracket) support or bearing 064
KONSORTIUM (*n*): syndicate 065
KONSTANT: constant-(ly) 066
KONSTANTAN (*n*): (60% Cu; 40% Ni): constantan (nickel alloy) (used for thermo elements) 067
KONSTANTANDRAHT (*m*): constantan wire (Elect.) 068
KONSTANTE (*f*): constant 069
KONSTANZ (*f*): constancy; constant 070
KONSTATIEREN: to ascertain; establish; verify; prove 071
KONSTITUTIONSFORMEL (*f*): constitution-(al) formula 072
KONSTITUTIONSWASSER (*n*): water of constitution 073
KONSTRUIEREN: to construct; build 074
KONSTRUKTEUR (*m*): constructor, builder, manufacturer, producer 075
KONSTRUKTION (*f*): design, construction 076
KONSTRUKTIONSELEMENT (*n*): element (method) of construction 077
KONSTRUKTIONSFEHLER (*m*): fault (error) in construction, constructional fault 078
KONSTRUKTIONSINGENIEUR (*m*): constructional engineer, constructor, builder 079
KONSTRUKTIONSRISS (*m*): construction (working) drawing or plan 080
KONSTRUKTIONSSTAHL (*m*): constructional steel 081
KONSTRUKTIONSTEILE (*m. pl.*): constructional parts 082
KONSTRUKTIONSZEICHNUNG (*f*): construction (working) drawing 083
KONSULENT (*m*): adviser 084
KONSUM (*m*): consumption, utilization, use 085
KONSUMENT (*m*): consumer 086
KONSUMTIV: consumptive-(ly) 087
KONTAKT (*m*): contact 088
KONTAKTABDRUCK (*m*): contact print (Phot.) 089
KONTAKTDRUCK (*m*): contact print, contact printing (Photography.) 090
KONTAKT (*m*) **ELEKTRISCHER-**: electric-(al) contact 091
KONTAKTFEDER (*f*): contact spring; connecting spring 092
KONTAKTFLÄCHE (*f*): contact surface 093
KONTAKTGEBER (*m*): contact transmitter 094
KONTAKTGIFT (*n*): contact poison 095
KONTAKTHEBEL (*m*): contact lever 096
KONTAKTKNOPF (*m*): contact button 097
KONTAKTREAKTION (*f*): contact reaction, catalysis 098
KONTAKTRING (*m*): contact ring 099
KONTAKTSCHRAUBE *f*): contact screw 100
KONTAKTSPALTER (*m*): fat cleavage agent, lipolytic 101
KONTAKTSTIFT (*m*): contact pin 102
KONTAKTSTÖPSEL (*m*): connecting plug 103
KONTAKTSTÜCK (*n*): contact-(piece) 104
KONTAKTSUBSTANZ (*f*): contact substance; catalyser 105
KONTAKTVERFAHREN (*n*): contact process 106
KONTAKTWIRKUNG (*f*): contact action; catalysis 107
KONTERBRASSE (*f*): preventer brace 108
KONTERMUTTER (*f*): back-nut, check-nut, lock nut 109
KONTERRAKEL (*f*): counter-doctor (Textile) 110
KONTINUIERLICH: continuous-(ly) 111
KONTINUITÄT (*f*): continuity 112
KONTO (*n*): account 113
KONTOAUSZUG (*m*): extract of account (Com.) 114
KONTOKORRENT (*n*): account current, current account 115
KONTOR (*n*): office 116
KONTORCHEF (*m*): head (chief) clerk 117
KONTORUTENSILIEN (*n.pl*): office (utensils, implements) requirements 118
KONTRAKT (*m*): contract 119
KONTRAKTILITÄT (*f*): contractibility 120
KONTRAKTION (*f*): contraction (of metal, etc.) 121
KONTRÄR: contrary, adverse, counter 122
KONTRAST (*m*): contrast 123
KONTRASTFÄRBUNG (*f*): contrast (colouring) staining 124
KONTRAUMDRUCK (*m*): reversed reprint (Lithography) 125
KONTROLLAMT (*n*): board of control 126
KONTROLLANLAGE (*f*): (any)-controlling apparatus or plant 127
KONTROLLE (*f*): control; inspection, governing, regulation, supervision, superintendence, management, register 128
KONTROLLER (*m*): controller 129
KONTROLLERKURBEL (*f*): controller (reversing) handle 130
KONTROLLERWIDERSTAND (*m*): controller resistance 131
KONTROLLIERAMT (*n*): railway clearing office 132
KONTROLLIERBLATT (*n*): counterfoil 133
KONTROLLIEREN: to control; exercise a check upon 134
KONTROLLMANOMETER (*n*): controlling pressure gauge 135
KONTROLLUHR (*f*): control clock (for checking employees' time); indicator, register (Photography) 136
KONTROLLVERSAMMLUNG (*f*): roll-call 137
KONTROLLVERSUCH (*m*): control (experiment) test 138
KONTROLLVORRICHTUNG (*f*): control gear, controlling device 139
KONTUMAZ (*f*): contumacy (Law); segregation; quarantine (see **QUARANTÄNE**) 140
KONUS (*m*): cone 141
KONVENTIONELL: conventional-(ly) 142
KONVERGENT: convergent 143
KONVERGENZ (*f*): convergence 144
KONVERGIEREN: to converge 145
KONVERGIEREND: converging (Maths.) (etc.), convergent 146
KONVERSATION (*f*): conversation 147
KONVERSATIONSLEXIKON (*n*): encyclopædia 148
KONVERSIONSSALPETER (*m*): converted saltpeter (sodium nitrate to potassium nitrate by double decomposition with pot. chlor.) 149
KONVERTER (*m*): (converter for the production of mild steel by the air refining process) 150
KONVERTERAUSWÜRFE (*m.pl*): converter waste (30% Fe) (steelworks process byproduct) 151
KONVERTERFUTTER (*n*): converter lining 152
KONVERTERPROZESS (*m*): converter process 153
KONZENTRAT (*n*): concentrate 154
KONZENTRATION (*f*): concentration 155
KONZENTRATIONSANLAGE (*f*): concentrating (concentration) plant 156
KONZENTRATIONSSTEIN (*m*): refined garnierit (for gaining copper) (70-75% Cu.) (result of re-roasting) 157

KONZENTRATIONSVERHÄLTNIS (*f*): concentration ratio 158
KONZENTRIEREN: to concentrate; saturate (a solution) see EINDAMPFEN) 159
KONZENTRIERTE LÖSUNG (*f*): concentrated solution, saturated solution 160
KONZENTRIERUNG (*f*): concentration; saturation 161
KONZENTRIERUNGSAPPARAT (*m*): concentrating apparatus 162
KONZENTRISCH: concentric (ally) 163
KONZEPTPAPIER (*n*): foolscap (Paper) 164
KONZERN (*n*): concern, firm, company 165
KONZESSION (*f*): concession, permission, licence, authorization 166
KONZESSIONIEREN: to concede; be constructed (made suitable) for; be allowed; be permitted; be passed by 167
KONZESSIONSDRUCK (*m*): pressure permitted-(by the boiler inspection authorities) 168
KONZESSIONSPFLICHT (*f*): obligation (duty) to take out a licence or to obtain (consent) permission 169
KONZESSIONSPFLICHT UNTERLIEGEN: to require a licence, need (authorization, consent) permission 170
KONZIS: concise 171
KOORDINATEN (*f.pl*): coordinates 172
KOORDINATENACHSE (*f*): coordinate axis 173
KOORDINATENSYSTEM (*n*): coordinate system 174
KOPAIVABALSAM (*m*): Jesuit's balsam, Copaiba, Balsam capivi, capivi balsam, copaiba balsam (*Balsamum copaivæ*); Sg. 0.935-0.998 175
KOPAIVAÖL (*n*): copaiba oil (distilled from *Balsamum copaivæ*); Sg. 0.9; Bp. 250-275°C. 176
KOPAIVASÄURE (*f*): copaivic acid 177
KOPAL (*m*): resin copal, gum copai, copal, cowrie, Kaurie, Animé; Sg. 1.018-1.067; Mp. 180/340°C. (a resin from various trees; see DAMMAR and KAURI) 178
KOPALDRYER (*m*): copal dryer 179
KOPALERSATZ (*m*): copal substitute 180
KOPALFIRNIS (*m*): copal varnish 181
KOPALHARZ (*n*): copal-(resin) 182
KOPALLACK (*m*): copal varnish 183
KOPALÖL (*n*): copal oil (Sg. 0.868/0.907, according to derivation) 184
KÖPER (*m*): twill; marsella; huckaback 185
KÖPERN: to twill 186
KOPF (*m*): head; top; crown (of hat); bowl (of pipe) (enlarged)-end; knob; summit; front; lead; sconce; mind; genius 187
KOPFARBEIT (*f*): head (brain) work; study 188
KOPFBANK (*f*): surface lathe 189
KOPFBEDECKUNG (*f*): head-covering, hat, head-dress, head-gear 190
KOPFBEERE (*f*): ipecacuanha (*Cephælis ipecacuanha*) 191
KOPFBESTEUERUNG (*f*): poll-tax 192
KOPFBOHRER (*m*): trephine, trepan (a surgical instrument for removing a circular piece of bone from the cranium) 193
KOPFBOLZEN (*m*): set bolt 194
KOPFDECKE (*f*): scalp; hood 195
KOPFDRUCKMASCHINE (*f*): machine for printing headings on business papers 196
KOPFDRÜSE (*f*): cephalic gland 197
KÖPFEN: to top, behead; lop; cup; decapitate 198
KOPFFIEBER (*n*): brain fever (Med.) 199
KOPFFRÄSER (*m*): (see STIRNFRÄSER) 200
KOPFFÜSSLER (*m*): cephalopod (Mollusc) 201
KOPFGELD (*n*): poll-tax 202
KOPFGICHT (*f*): megrim, megraine (a neuralgic head pain, usually of the temporal nerve) (Med.) 203
KOPFGRIND (*m*): dandruff, scurf; scald head (an eruptive affection of the scalp) (Med.) 204
KOPFGRINDKRAUT (*n*): field scabious (*Knautia arvensis*) 205
KOPFHÄLTER (*m*): head-rest (Photography, etc.) 206
KOPFHAUT (*f*): scalp 207
KOPFKEILBEIN (*n*): sphenoid bone (Anat.) 208
KOPFKISSEN (*n*): pillow 209
KOPFKLEE (*m*): common clover (*Trifolium*) 210
KOPFLÄNGS: headlong 211
KOPFLASTIG: top-heavy 212
KOPFLATTICH (*m*): cabbage lettuce (*Lactuca scariola sativa*) 213
KOPFLAUGE (*f*): head-wash 214
KÖPFLINGS: headlong 215
KOPFLOS: headless; acephalous; silly; stupid 216
KOPFPUTZ (*m*): headdress; coiffure; head-gear 217
KOPFRECHNEN (*n*): mental arithmetic 218
KOPFRING (*m*): nozzle block (Zinc) 219
KOPFROSE (*f*): erysipelas (Med.) 220
KOPF (*m*), RUDER-: rudder head 221
KOPFSALAT (*m*): cabbage lettuce (*Lactuca scariola sativa*) 222
KOPFSCHEU: skittish; timid; shy 223
KOPFSCHIMMEL (*m*): Mucor mould (*Mucor mucedo*) 224
KOPFSCHIRM (*m*): reflector (Phot.) 225
KOPFSCHMERZ (*m*): headache (Med.) 226
KOPFSCHRAUBE (*f*): round headed screw 227
KOPFSCHUPPEN (*f.pl*): dandruff, scurf (Med.) 228
KOPFSCHUPPENMITTEL (*n*): scurf remedy 229
KOPFSTEIN (*m*): paving stone; head stone 230
KOPFSTEUER (*f*): poll tax; capitation tax 231
KOPFSTIMME (*f*): head-voice; falsetto 232
KOPFSTÜCK (*n*): boss; head-piece; mouth-piece; header (of pipes) 233
KOPFÜBER: head over heels 234
KOPFÜBERSCHRIFT (*f*): main (principal) heading 235
KOPF, VERLORENER- (*m*): dead head; discard; shrinkage (so-called funnel which has been used in casting steel ingots, for filling out pipe) (Metal) 236
KOPFWASCHPULVER (*n*): hair washing powder 237
KOPFWASSER (*n*): hair-wash; shampoo 238
KOPFWASSERSUCHT (*f*): hydrocephalus (Med.) 239
KOPFWEH (*n*): headache (Med.) 240
KOPFWEIDE (*f*): white willow (*Salix alba*) 241
KOPFWUT (*f*): brain fever (Med.) 242
KOPFZAHL (*f*): number-(of people); head-(of cattle) 243
KOPIE (*f*): copy; duplicate; print (Photography) 244
KOPIERAUTOMAT (*n*): automatic printing machine (Photography) 245
KOPIEREN: to copy; print (Phot.) 246
KOPIERKAMERA (*f*): copying camera 247
KOPIERPAPIER (*n*): duplicating (copying) paper 248
KOPIERRAHMEN (*m*): printing frame (Photography) 249

KOPIERTINTE (*f*): copying ink 250
KOPIERVERFAHREN (*n*): copying (printing) process (Phot., etc.) 251
KOPPE (*f*): top, tuft, head, peak, summit 252
KOPPEL (*f*): couple; tie; enclosure; chain, strap, belt; leash-(of dogs) 253
KOPPELN: to couple; tie; leash (dogs); unite; join; enclose; fence 254
KÖPPELN: to froth (in after-fermentation) 255
KOPPELUNG (*f*): coupling 256
KOPPELWIRTSCHAFT (*f*): rotation of crops 257
KOPPEN: to lop; eruct, hiccough, belch, eructate; top, behead, cup, decapitate 258
KOPROLITH (*m*): coprolite 259
KOPROSTERIN (*n*): coprosterine (from the reduction of cholesterol in the intestines); $C_{27}H_{47}OH$ 260
KORALLE (*f*): coral (Sg. red, 2.7; white, 2.5) 261
KORALLENARTIG: coralloid; coralline; coral-like 262
KORALLENBLUME (*f*): coral flower (*Erythrina corallodendron*) 263
KORALLENFISCHEREI (*f*): coral-fishing; coral fishery 264
KORALLENMOOS (*n*): coralline (*Corallina officinalis*) 265
KORALLENRIFF (*n*): coral reef 266
KORALLENRINDE (*f*): hornwrack (*Corallina officinalis*) 267
KORALLENROT (*n*): coral red (see **CHROMROT**) 268
KORALLENWURZEL (*f*): common polypody (*Polypodium vulgare*) 269
KORALLIN (*n*): A solution of rosolic acid in alcohol 270
KORALLINE (*f*): coralline 271
KORB (*m*): basket; hamper; pannier; crate; canister; corf; cage (Mining); sack; rejection; cold-shoulder; dismissal 272
KORB (*m*), **EISERNER-**: iron basket (used for acid carboys) 273
KORBFLASCHE (*f*): demijohn; carboy; bottle in a wicker case 274
KORBGITTER (*n*): hurdle-work 275
KORBLACK (*m*): basket varnish 276
KORBMACHER (*m*): basket maker 277
KORBROST (*m*): basket shaped grate, trough grate 278
KORBWAREN (*f.pl*): basket wares 279
KORBWEIDE (*f*): osier, common oiser (*Salix viminalis*) 280
KORDEL (*f*): cord 281
KORDIT (*n*): cordite (an explosive) 282
KORDUAN (*m*): cordovan, goatskin, cordwain, Spanish leather, Cordovan leather 283
KORIANDER (*m*), **RÖMISCHER-**: (see **NARDENSAME**) 284
KORIANDER (*m*), **SCHWARZER-**: (see **NARDENSAME**) 285
KORINTHE (*f*): currant 286
KORK (*m*): cork-(stopper); cork (bark of cork tree, *Quercus suber*); Sg. 0.24 287
KORKABFÄLLE (*m.pl*): cork waste 288
KORKBAUM (*m*): cork tree (*Quercus suber*) 289
KORKBETON (*m*): cork (concrete) composition 290
KORKBOHRER (*m*): cork borer 291
KORKBOJE (*f*): cork buoy 292
KORKBRECHER (*m*): cork (crusher) breaker 293
KORKEICHE (*f*): cork tree (see **KORKBAUM**) 294
KORKEN: to cork; of cork 295
KORKERSATZMITTEL (*n*): cork substitute 296
KORKFLOSSEN (*pl*): cork floats 297
KORKHOLZ (*n*): cork wood (see **KORKBAUM**) 298
KORKKOHLE (*f*): burnt cork 299
KORKKOMPOSITION (*f*): cork composition 300
KORK (*m*), **KUNST-**: artificial cork 301
KORKMAHLGANG (*m*): cork grinder 302
KORKMASCHINE (*f*): corking machine 303
KORKMASSE (*f*): cork mass 304
KORKMEHL (*n*): cork (powder) meal; ground cork 305
KORKMETALL (*n*): cork metal (trade name for a very light metal which is practically pure magnesium and has a Sg. 1.76) 306
KORKMÜHLE (*f*): cork mill 307
KORKMULLEREI (*f*): cork milling 308
KORKPFROPF (*m*): cork-(stopper) 309
KORKPFROPFEN (*m*): cork-(stopper) 310
KORKPLATTE (*f*): cork plate 311
KORKSAUER: suberic 312
KORKSÄURE (*f*): suberic acid 313
KORKSCHEIBE (*f*): sheet cork; cork disc; cork pulley 314
KORKSCHROT (*m*): cork meal, ground cork 315
KORKSCHWARZ (*n*): cork black 316
KORKSEGMENT (*n*): cork segment 317
KORKSOLE (*f*): cork sole 318
KORKSPUND (*m*): cork-(stopper) 319
KORKSTEIN (*m*): cork brick; cork (board) plate; Sg. 0.25-0.4 320
KORKSTOFF (*m*): suberin 321
KORKSTOPFEN (*m*): cork-(stopper) 322
KORKSTÖPSEL (*m*): cork stopper, cork 323
KORKSURROGAT (*n*): cork substitute 324
KORKTEILCHEN (*n*): cork particle 325
KORKTEPPICH (*m*): linoleum; cork-lino 326
KORKULME (*f*): (see **ULME**) 327
KORKZIEHER (*m*): corkscrew 328
KORN: (*n*): grain; granulation; proud flesh (Med.); fineness; standard (of coinage); corn; rye; boon (of flax); button (Assaying); pea (Coal); small pill; granule 329
KORNABBRUCH (*m*): corn refuse 330
KORNÄHRE (*f*): ear of (grain, rye) corn; spica (Astron.) 331
KORNALKOHOL (*m*): grain alcohol (see **ÄTHYLALKOHOL**) 332
KORNÄTHER (*m*): grain ether, brandy-(ether) (see **ÄTHYLALKOHOL**) 333
KORNBAU (*m*): cultivation of cereals 334
KORNBILDUNG (*f*): granulation; crystallization (Metal) 335
KORNBILDUNGSPUNKT (*m*): granulating (crystallizing) point (Metal) 336
KORNBLAU (*n*): corn-flower blue 337
KORNBLEI (*n*): grain (finely granulated) lead 338
KORNBLUMEN (*f.pl*): corn-flowers (*Flores cyani*) 339
KORNBODEN (*m*): corn-loft, granary 340
KORNBÖRSE (*f*): corn exchange 341
KORNBRAND (*m*): corn-blight 342
KORNBRANNTWEIN (*m*): whisky; grain spirits 343
KORNBÜRSTE (*f*): button brush (Assaying) 344
KÖRNCHEN (*n*): granule; (little)-grain; very small pill (*Granula*) (Pharmacy) 345
KORNEA (*f*): cornea 346
KORNEISEN (*n*): granular-fracture iron 347
KORNELBAUM (*m*): cornel tree (see **HARTRIEGEL**) 348
KORNELEVATOR (*m*): grain-elevator 349
KORNELKIRSCHE (*f*): cornelian cherry (see **HARTBAUM**) 350

KORNELLE (*f*): cornel, dog-wood (see HARTRIEGEL) 351
KÖRNELN: to granulate 352
KÖRNELUNG (*f*): granulation 353
KÖRNEN: to granulate; grain; corn (Gunpowder); obtain grain; run to seed; (*n*): granulation 354
KÖRNER (*m*): centre punch; (as a prefix = grained, granular) 355
KÖRNERASANT (*m*): granular asafoetida 356
KÖRNERFRESSEND: gramniverous 357
KÖRNERFRITTER (*m*): granular coherer 358
KÖRNERFRUCHT (*f*): cereal; grain 359
KÖRNERGUMMI (*n*): granular gum, gum in grains 360
KÖRNERLACK (*m*): seed-lac 361
KÖRNERLEDER (*n*): grained leather; shagreen 362
KÖRNERMARKE (*f*): centre (mark) punch 363
KÖRNERSCHILDLAUS (*f*): cochineal insect (see COCHENILLE) 364
KÖRNERSTEIN (*m*): granite (see GRANIT) 365
KÖRNERZINN (*n*): grain (granular) tin 366
KORNESSENZ (*f*): grain essence, brandy-(essence) 367
KORNFEGE ()*f*: winnowing machine 368
KORNFELD (*n*): corn-field 369
KORNFÖRMIG: granular; in grain form; granulated 370
KORNFRAS-(S) (*m*): corn-blight 371
KORNFUSELÖL (*n*): fusel oil from grain (see FUSELÖL) 372
KORNGARBE (*f*): sheaf-(of corn) 373
KORNGÄRTE (*f*): privet (see HARTRIEGEL) 374
KORNGERTE (*f*): (see KORNGÄRTE) 375
KORNGRÖSSENVERHÄLTNIS (*n*): comparison (ratio) of the size of grains 376
KORNHANDEL (*m*): corn trade 377
KORNHÄNDLER (*m*): corn (chandler) merchant 378
KÖRNHAUS (*n*): granulating house (Explosives) 379
KÖRNIG: granular; in grains; granulated; gritty; grained; having (showing) a grain; brittle (Malt); seedy; pithy; nervous 380
KÖRNICKRISTALLINISCH: granular-crystalline 381
KORNKÄFER (*m*): corn (grain) weevil (*Sitophilus granarius*) 382
KORNKAMMER (*f*): granary 383
KORNKASTEN (*m*): corn hopper; corn bin 384
KORNKLUFT (*f*): assayer's tongs 385
KORNKOCHEN (*n*): boiling to grain (Sugar) 386
KORNKUPFER (*n*): granulated copper 387
KORNLADE (*f*): corn-bin 388
KORNLEDER (*n*): grained leather 389
KORNMARKT (*m*): corn market 390
KORNMASS (*n*): corn measure 391
KORNMUTTER (*f*): ergot 392
KORNPRÜFER (*m*): grain tester 393
KORNPUDDELN (*n*): puddling of granular iron 394
KORNPULVER (*n*): granulated (gun)-powder 395
KORNRADE (*f*): corn-cockle (*Lychnis githago*) 396
KORNREICH: productive (of grain); granular, gritty 397
KORNROSE (*f*): corn-rose, burnet rose; common red poppy (*Papaver rhoeas*); corn cockle (see KORNRADE) 398
KORNSIEB (*n*): granulating (winnowing; grain) sieve 399
KORNSTAHL (*m*): granulated steel 400
KORNSTECHER (*m*): corn (grain) weevil (*Sitophilus granarius*) 401
KORNSTEUER (*f*): grain (corn) tax or duty 402
KÖRNUNG (*f*): sizing, breaking into small pieces (Coal); granulation, graining, grain 403
KORNWAGE (*f*): button (assay) balance; grain scales 404
KORNWAGEN (*m*): corn (grain) cart or wagon 405
KORNWICKE (*f*): tares; wild vetch (*Vicia*) 406
KORNWURM (*m*): corn (grain) weevil (see KORNKÄFER) 407
KORNZANGE (*f*): assayer's tongs 408
KORNZINN (*n*): grain tin 409
KORNZUCKER (*m*): granulated sugar 410
KÖRPER (*m*): body; substance; compound; carcass; carcase; corpse 411
KÖRPERANLAGE (*f*): temperament, constitution (Med.) 412
KÖRPERBAU (*m*): build; structure of a body; frame 413
KÖRPERBESCHAFFENHEIT (*f*): constitution; composition; structure-(of a body) 414
KÖRPERBILDUNG (*f*): formation (structure) of a body 415
KÖRPERCHEN (*n*): small body; particle; corpuscle 416
KÖRPERFARBE (*f*): body colour 417
KÖRPER (*m*), FESTER-: solid (body) substance 418
KÖRPERFETT (*n*): body fat (from the bodies of animals or fishes) 419
KÖRPER (*m*), FLÜSSIGER-: fluid (body) substance 420
KÖRPERFÜLLE (*f*): corpulence 421
KÖRPERGRÖSSE (*f*): stature 422
KÖRPERGRUPPE (*f*): group of (substances; compounds) bodies 423
KÖRPERHAFT: corporeal 424
KÖRPERKLASSE (*f*): class of (substances; compounds) bodies 425
KÖRPERKRAFT (*f*): bodily (physical) strength 426
KÖRPERLEHRE (*f*): somatology (the science of material substances) 427
KÖRPERLICH: bodily; corporeal; corporal; material; solid; corpuscular 428
KÖRPERLOS: immaterial; without (form) substance; intangible 429
KÖRPERMASS (*n*): cubature, cubic measure-(ment), measure of capacity 430
KÖRPERMESSUNG (*f*): stereometry 431
KÖRPERREICH (*n*): material world, realm of material things 432
KÖRPERSCHAFT (*f*): corporation 433
KÖRPERSCHWÄCHE (*f*): bodily weakness, debility 434
KÖRPERSTOFF (*m*): (organic)-matter 435
KÖRPERTEILCHEN (*n*): particle 436
KÖRPER (*m*), VENTIL-: valve body 437
KÖRPERWÄRME (*f*): body-temperature, temperature of body 438
KÖRPERWELT (*f*): material world, world of material things 439
KORPS (*m*): corps 440
KORPULENZ (*f*): corpulence, corpulency 441
KORREKTIVMITTEL (*n*): corrective; corrigent 442
KORREKTOR (*m*): corrector; proof-reader 443
KORREKTUR (*f*): correction; proof; revision; checking; proof-reading 444
KORREKTURABZUG (*m*): (proof-reader's)-proof 445
KORRELAT (*n*): correlative 446
KORRESPONDIEREN: to correspond 447
KORRIGIEREN: to correct 448
KORRODIEREN: to corrode 449
KORROSION (*f*): corrosion 450

KORUBIN (*n*): (Trade name) ($Fe_2+Al_2O_3$); an artificial corundum (see KORUND and ALUMINIUMOXYD) 451

KORUND (*m*): corundum (Al_2O_3); Sg. 3.95; Mp. 1750-1800°C. (see also ALUMINIUM-OXYD) 452

KORUND (*m*), **DODEKAEDRISCHER-**: spinelle 453

KORUND (*m*), **EDLER-**: ruby (red); sapphire (blue to colourless); emery (blue-grey to blue); Al_2O_3 454

KORUND (*m*), **OKTAEDRISCHER-**: gahnite, automalite 455

KORUND (*m*), **PRISMATISCHER—**: chrysoberyl 456

KORVETTE (*f*): corvette (Naut.) 457

KOSCHENILLE (*f*): (see COCHENILLE) 458

KOSCHENILLELACK (*m*): crimson lake (see KARMIN) 459

KOSEKANTE (*f*): cosecant 460

KOSINUS (*m*): cosine 461

KOSMETIK (*f*): cosmetic 462

KOSMETISCH: cosmetic 463

KOSMETISCHE PRÄPARATE (*n.pl*): toilet preparations 464

KOSMISCH: cosmic 465

KOSOBLÜTEN (*f.pl*): cusso; koussa; brayera (from *Hagenia abyssinica*) 466

KOST (*f*): food; fare; board; diet; victuals 467

KOSTBAR: costly; expensive; valuable; precious; choice; good; excellent; tasty 468

KOSTEN: to cost; taste; try; (*f.pl*): costs; expenses; charges 469

KOSTENANSCHLAG (*m*): estimation, price, quotation, estimate. offer, bid, tender 470

KOSTENANSCHLAG (*m*) **MACHEN**: to estimate, get out an estimate or quotation, put forward a price or quotation, quote, tender, etc. 471

KOSTENAUFWAND (*m*): expenditure, disbursement, out-goings 472

KOSTENBERECHNUNG (*f*): bill of charges 473

KOSTENERSATZ (*m*): compensation 474

KOSTENFREI: free of charge; free of cost; gratis; cost free; all expenses defrayed; carriage (post) paid 475

KOSTFREI: including (inclusive of) board, board-free 476

KOSTGÄNGER (*m*): boarder 477

KOSTGELD (*n*): board; alimony 478

KOSTHAUS (*n*): boarding-house 479

KÖSTLICH: costly; dainty; excellent; choice; precious; delicious; exquisite 480

KOSTSCHULE (*f*): boarding-school 481

KOSTSPIELIG: expensive; costly; dear 482

KOSTÜMIEREN: to dress, drape 483

KOT (*m*): feces; fæces; excreta; excrement; dung; mud; dirt; mire; filth; (*n*): cot, shed 484

KOTABGANG (*m*): defecation 485

KOTABZUCHT (*f*): sink, sewer 486

KOTANGENTE (*f*): cotangent 487

KOTARNIN (*n*): (see COTARNIN) 488

KOTARTIG: fecal; feculent 489

KOTAUSFÜHREND: purgative; cathartic 490

KOTAUSLEEREND: purgative; cathartic 491

KÖTE (*f*): box, cupboard; cot-(tage); fetlock-joint 492

KÖTENGELENK (*n*): fetlock-joint; pastern 493

KÖTER (*m*): cur; dog 494

KOTGANG (*m*): drain 495

KOTGERUCH (*m*): fecal odour 496

KOTGRUBE (*f*): sewer; cesspool 497

KOTIG: fecal; stercoraceous; dirty; filthy; foul; muddy 498

KOTLACHE (*f*): slough, puddle 499

KOTONISIEREN: to degum (Silk) 500

KOTSCHLEUSE (*f*): sewer 501

KOTSCHLINGE (*f*): mealy guelder-rose, wayfaring tree (*Viburnum lantana*) 502

KOTSTEIN (*m*): fecal concretion 503

KOTTONÖL (*n*): cotton seed oil (see BAUM-WOLLSAMENÖL) 504

KOUSSOBLÜTEN (*f.pl*): kousso flowers (see KOSOBLÜTEN) 505

KOVOLUM (*n*): covolume 506

KOVOLUMEN (*n*): (see KOVOLUM) 507

KRABBE (*f*): crab; little (small) child 508

KRABBECHT: fast to crabbing 509

KRABBELN: to tickle; itch; crawl; grope 510

KRACH (*m*): crack; crash 511

KRACHEN: to crack-(le); crash; rustle (of silk); fail; become (go) bankrupt; be ruined 512

KRÄCHZEN: to croak; groan; caw 513

KRACKE (*f*): jade, poor (sorry) horse 514

KRACKEN (*n*): cracking (of oils) (see CRACK-PROZESS) 515

KRAFT (*f*): power; force; capacity; strength; vigour; stress; energy; validity; density; intensity (Phot.); (as a preposition and adverb=on the strength of; by virtue of; due to; in consequence of) 516

KRAFT (*f*), **ABSOLUTE-**: absolute force 517

KRAFTANGRIFF (*m*): force, stress, attack of a (force) stress 518

KRAFTANSTRENGUNG (*f*): effort, energy, strain, exertion of (power) force 519

KRAFT (*f*), **ANZIEHUNGS-**: attractive force, force of attraction 520

KRAFTARZNEI (*f*): tonic (Med.) 521

KRAFT (*f*), **AUFTRIEBS-**: (force of)-buoyancy 522

KRAFTAUFWAND (*m*): energy (power) consumption, effort, force (required) necessary 523

KRAFTAUSDRUCK (*m*): forceful (pithy) expression 524

KRAFTBEDARF (*m*): power (strength, force) required 525

KRAFT (*f*), **BESCHLEUNIGENDE-**: accelerating force 526

KRAFT (*f*), **BEWEGENDE-**: moving force, motive power 527

KRAFTBRÜHE (*f*): strong broth 528

KRAFT (*f*), **DREH-**: twisting (torsional) force 529

KRAFTECK (*n*): force parallelogram 530

KRAFT (*f*), **EINE FRISCHE-**: a new hand, new blood 531

KRAFTEINHEIT (*f*): unit of force 532

KRÄFTE (*f.pl*), **JÜNGERE-**: younger (shoulders) blood 533

KRÄFTEKOMPONENTE (*f*): force component 534

KRAFT (*f*), **ELEKTROMAGNETISCHE-**: electro-magnetic force 535

KRAFT (*f*), **ELEKTROMOTORISCHE-**: electromotive force (EMF). 536

KRÄFTEMASSSTAB (*m*): scale of (power) forces; force scale 537

KRÄFTEPAAR (*n*) force couple 538

KRÄFTEPLAN (*m*): diagram of forces 539

KRÄFTEPOLYGON (*n*): force polygon 540

KRAFTERSPARNIS (*f*): saving in power, power (saving) economy 541

KRAFT (*f*), **EXPANSIONS-**: expansive force 542

KRAFTFAHRZEUG (*n*): power driven vehicle 543

KRAFT (*f*), **FEDER-**: elastic force 544

KRAFTGAS (*n*): power gas; producer gas 545

KRAFTGASANLAGE (*f*): power (producer) gas plant 546
KRAFTGASMASCHINE (*f*): gas engine 547
KRAFTHAMMER (*m*): power hammer 548
KRÄFTIG: strong; vigorous; powerful; robust; forcible; rugged; nourishing; thick; valid; efficacious; plump (of hides) 549
KRÄFTIGEN: to strengthen, refresh, invigorate, energize; intensify, (Phot.); amplify (Wireless Tel.) 550
KRÄFTIGLICH: strongly, forcefully, vigorously, powerfully 551
KRÄFTIGUNG (*f*): intensification (Photography, etc.) 552
KRAFT (*f*), KONSTANTE-: constant force 553
KRAFT (*f*), LEBENDIGE-: live (load) force (*vis viva*) 554
KRAFTLEHRE (*f*): dynamics 555
KRAFTLEITUNG (*f*): power circuit (Elect.) 556
KRAFTLINIE (*f*): line of force 557
KRAFTLOS: impotent, weak, without strength, powerless; invalid, ineffectual 558
KRAFT (*f*), MAGNETISCHE-: magnetic force 559
KRAFTMASCHINE (*f*): power engine, motor, prime mover 560
KRAFTMEHL (*n*): starch; amylum 561
KRAFTMESSER (*m*): dynamometer 562
KRAFTMITTEL (*n*): forceful methods, energetic means; cordial 563
KRAFTPAPIER (*n*): Kraft brown (Paper) 564
KRAFT (*f*), RESULTIERENDE-: resulting force 565
KRAFTSAFT (*m*): essence 566
KRAFTSAMMLER (*m*): accumulator 567
KRAFT (*f*), SCHWER-: force of gravity 568
KRAFT (*f*), SPANN-: elastic force 569
KRAFTSTECKDOSE (*f*): power-(wall)-plug (Elect.) 570
KRAFT (*f*), STOSS-: percussive force, impact force, impulse 571
KRAFT (*f*), TORSIONS-: twisting (torsional) force 572
KRAFT (*f*), TRAG-: carrying (capacity) power, portative force 573
KRAFT (*f*), TRIEB-: impulsive force, driving power, propelling force, propulsive force 574
KRAFT (*f*), VERÄNDERLICHE-: variable force, fluctuating power 575
KRAFTVERBRAUCH (*m*): power used, power consumption, in-put 576
KRAFTVOLL: forcible, powerful, strong 577
KRAFTWAGEN (*m*): autocar, lorry, power propelled vehicle 578
KRAFTWERK (*n*): power station 579
KRAFT (*f*), WIDERSTANDS-: resisting force, resistance 580
KRAFTWIRKUNG (*f*): effect (action) of a force or stress 581
KRAFTWORT (*n*): strong (forceful) word or expression; word of authority 582
KRAFT (*f*), ZENTRIFUGAL-: centrifugal force 583
KRAFT (*f*), ZENTRIPETAL-: centripetal force 584
KRAFT (*f*), ZUG-: tractive force 585
KRAGEN (*m*): collar; cape; flange; cravat (Building); neck (of bottles) 586
KRAGENBLUME (*f*): Stellaria, starwort, stichgrass, stichwort (*Stellaria holostea*) 587
KRAGEN (*m*), SCHORNSTEIN-: cravat 588
KRAGEN (*m*), STEUERRUDER-: rudder coat 589
KRAGSTEIN (*m*): corbel, console, bracket 590
KRÄHE (*f*): crow (*Corvus*) 591
KRÄHEN: to crow; (*n*): crowing 592
KRÄHENAUGEN (*n.pl*): nux vomica 593
KRÄHENFUSS (*m*): scrawl-(ing writing); crow's foot; crow-bar 594
KRÄHENNEST (*n*): crow's nest 595
KRÄHENRABE (*f*): carrion crow (*Corvus corone*) 596
KRÄHENZEHE (*f*): water plantain (*Alisma*) 597
KRAHN (*m*): (see KRAN) 598
KRALLE (*f*): claw; clutch; talon 599
KRALLEN: to claw; clutch; scratch 600
KRALLENARTIG: claw-like 601
KRALLENFÖRMIG: claw-like, claw-shaped 602
KRALLICHT: claw-like 603
KRALLIG: clawed 604
KRAM (*m*): stuff; lot; retail; lumber; store; shop; retail articles; wares 605
KRAMEN: to rummage; retail 606
KRÄMER (*m*): retailer; storekeeper; tradesman; shopkeeper; trader; mercer 607
KRÄMEREI (*f*): trading; shop-keeping; retailing 608
KRÄMERGEWICHT (*n*): avoirdupois -(weight) 609
KRAMHANDEL (*m*): retail trade 610
KRAMKAMMER (*f*): lumber (store) room 611
KRAMME (*f*): cramp, claw 612
KRAMMEN: to claw, clutch 613
KRAMMETSBEERE (*f*): juniper-berry (fruit of *Juniperus communis*) 614
KRAMMETSVOGEL (*m*): field-fare (*Turdus pilaris*) 615
KRAMPE (*f*): cramp; cramp iron; clasp; staple 616
KRÄMPE: (see KREMPE) 617
KRAMPEN: to cramp, fasten, clasp 618
KRAMPF (*m*): cramp; spasm; convulsion (Med.) 619
KRAMPFADER (*f*): varicose vein 620
KRAMPFARTIG: convulsive; spasmodic 621
KRAMPFARZNEI (*f*): antispasmodic 622
KRAMPFEN: to get the cramp, contract; clasp convulsively; (*m.pl*): convulsions (muscular contractions) (Med.) 623
KRAMPFHAFT: convulsive; spasmodic 624
KRAMPFKRAUT (*n*): meadow-sweet (*Spiræa ulmaria*) 625
KRAMPFLINDERND: antispasmodic 626
KRAMPFMITTEL (*n*): antispasmodic 627
KRAMPFSTILLEND: antispasmodic 628
KRAMPFSTILLENDES MITTEL (*n*): antispasmodic 629
KRAMPFWURZEL (*f*): valerian root (root of *Valeriana officinalis*) 630
KRAN (*m*): cock; faucet; crane; hoist (Mech.) 631
KRANAUSLEGER (*m*): crane beam or jib 632
KRANBALKEN (*m*): crane beam or jib; cathead (Naut.) 633
KRANBAUM (*m*): crane beam or jib; derrick 634
KRANBEERE (*f*): cranberry (fruit of *Vaccinium oxycoccos*); bilberry (fruit of *Vaccinium myrtillus*) 635
KRAN (*m*), CHARGIER-: loading crane 636
KRÄNGEN: to cant, list, tip, heel, incline 637
KRANGESTELL (*n*): crane frame 638
KRÄNGUNG (*f*): cant, list, inclination 639
KRÄNGUNGSHEBEL (*m*): heeling lever 640
KRÄNGUNGSPENDEL (*n*): clinometer pendulum 641
KRÄNGUNGSVERSUCH (*m*): heeling (inclination) experiment or test, clinometer test 642
KRÄNGUNGSWINKEL (*m*): heeling angle, angle of (list) inclination 643
KRANHAKEN (*m*): crane hook 644

KRANICH (*m*) : crane (*Grus communis*) 645
KRANICHBEERE (*f*) : cranberry (*Vaccinium oxycoccos*) 646
KRANICHHALS (*m*) : geranium, meadow crane's-bill (*Geranium pratense*) 647
KRANICHSCHNABEL (*m*) : crane's-bill, pelargonium (of the order *Geraniaceæ*) 648
KRANICHSCHNABELZANGE (*f*) : crane's-bill (Surgical) 649
KRANK : ill ; diseased ; ailing ; sick ; rotten (of wood) 650
KRANKABEL (*n*) : crane cable 651
KRAN (*m*), **KABEL-** : cable crane 652
KRANKE (*m* and *f*) : invalid, patient, sick (ailing) person 653
KRÄNKELN : to be (ailing ; sickly) unwell ; sicken 654
KRÄNKEN : to hurt ; vex ; grieve ; offend ; afflict ; insult ; mortify ; injure ; wrong ; make ill 655
KRANKENANSTALT (*f*) : hospital 656
KRANKENATTEST (*m*) : certificate of illness, doctor's certificate 657
KRANKENBERICHT (*m*) : bulletin 658
KRANKENBETT (*n*) : sick-bed 659
KRANKENHAUS (*n*) : hospital ; infirmary 660
KRANKENHAUSSAAL (*m*) : ward (of a hospital) 661
KRANKENHEILANSTALT (*f*) : hospital ; sanatorium 662
KRANKENKASSE (*f*), **(STAATS)-** : Health Insurance (National)- 663
KRANKENKOST (*f*) : diet-(for sick people) (sick-diet, light diet, full diet, milk-diet, etc.) 664
KRANKENLAGER (*n*) : sick-bed 665
KRANKENPFLEGE (*f*) : (sick)-nursing 666
KRANKENSCHIFF (*n*) : hospital-ship 667
KRANKENSTUBE (*f*) : sick (room) chamber 668
KRANKENWAGEN (*m*) : ambulance waggon 669
KRANKENWÄRTER (*m*) : male nurse, male attendant 670
KRANKENWÄRTERIN (*f*) : female (nurse) attendant 671
KRANKETTE (*f*) : crane (hoist) chain 672
KRANKHAFT : diseased ; morbid 673
KRANKHEIT (*f*) : disease ; illness ; sickness ; malady ; distemper ; complaint ; ailment ; trouble 674
KRANKHEIT (*f*), **ENGLISCHE-** : rickets (Med.) 675
KRANKHEITSENTSCHEIDUNG (*f*) : crisis 676
KRANKHEITSERREGER (*m*) : exciter (exciting cause ; excitant) of disease or illness 677
KRANKHEITSERSCHEINUNG (*f*) : appearance (outbreak) of a disease ; symptom 678
KRANKHEITSERZEUGEND : pathogenic 679
KRANKHEITSHALBER : due to (on account of) ill health 680
KRANKHEITSKUNDE (*f*) : pathology 681
KRANKHEITSLEHRE (*f*) : pathology 682
KRANKHEITSSTOFF (*m*) : morbid matter 683
KRANKHEITSVERLAUF (*m*) : progress of a disease 684
KRANKHEITSZEICHEN (*n*) : symptom (Med.) 685
KRÄNKLICH : sickly ; in poor health ; delicate ; weak ; infirm ; ailing ; invalid 686
KRÄNKLICHKEIT (*f*) : ill health, bad health 687
KRÄNKUNG (*f*) : vexation ; grief ; mortification ; insult 688
KRAN (*m*), **LAUF-** : travelling crane 689
KRAN (*m*), **SCHWIMM-** : floating crane 690
KRANSEIL (*n*) : crane (cable) rope 691
KRANSTÄNDER (*m*) : crane upright ; crane support 692
KRANZ (*m*) : wreath ; crown ; border ; circle ; club ; rim ; brim ; garland ; festoon ; valence 693
KRANZADER (*f*) : coronal vein 694
KRANZARTERIEN (*f.pl*) : Coronary arteries (Med.) 695
KRANZBLUME (*f*) : milkwort (*Polygala vulgaris*) 696
KRÄNZCHEN (*n*) : little wreath, etc. ; (social)-circle 697
KRÄNZEN : to wreath, crown (adorn) with a wreath or garland, encircle ; decorticate, bark (Trees) 698
KRANZFÖRMIG : coronoid, wreath-shaped, wreath-like 699
KRANZ (*m*), **RAD-** : rim (of a wheel) 700
KRAPP (*m*) : madder (either the plant or its root, *Rubia tinctorum*) 701
KRAPPBAU (*m*) : madder cultivation 702
KRAPPEXTRAKT (*m*) : madder extract 703
KRAPPFARBE (*f*) : madder dye 704
KRAPPFÄRBEREI (*f*) : madder dyeing 705
KRAPPFARBSTOFF (*m*) : alizarin (see **KRAPPROT**) 706
KRAPPGELB (*n*) : madder yellow (from root of *Rubia tinctorum*) ; xanthine ; ureous acid ; $[C_5H_2N_4(OH)_2]2:6$ 707
KRAPPGEWÄCHSE (*n.pl*) : *Rubiaceæ* (Bot.) 708
KRAPPKARMIN (*n*) : madder carmine 709
KRAPP (*m*), **KÜNSTLICHER-** : artificial madder (see **KRAPPROT**) 710
KRAPPLACK (*m*) : madder lake (see **KRAPPROT**) 711
KRAPPLACKFARBSTOFF (*m*) : madder (colouring matter) dye-stuff (the colouring principle is alizarin) 712
KRAPPLACK (*m*), **NATÜRLICHER-** : natural madder lake (from the roots of *Rubia tinctorum*) 713
KRAPP (*m*), **OSTINDISCHER-** : East Indian madder (*Rubia munjista*) 714
KRAPPPFLANZE (*f*) : madder (*Rubia tinctorum*) 715
KRAPPPRÄPARAT (*n*) : madder preparation 716
KRAPPROSA (*f*) : madder rose 717
KRAPPROT (*n*) : madder red ; alizarin, dioxyanthraquinone ; $C_6H_4(CO)_2C_6H_2(OH)_2$; Mp. 289°C. ; Bp. 430°C. 718
KRAPPRÜCKSTAND (*m*) : madder residue 719
KRAPPWURZEL (*f*) : madder root (root of *Rubia tinctorum*) (*Radix rubiæ tinctorum*) 720
KRATER (*m*) : crater 721
KRATZ (*m*) : scratch, scrape 722
KRATZBEERE (*f*) : bramble-berry, blackberry ; gooseberry (see **BROMBEERE** and **STACHEL-BEERE**) 723
KRÄTZBLEI (*n*) : slag lead 724
KRATZBÜRSTE (*f*) : stiff (scrubbing) brush ; scraper ; cross (irritable) person ; steel-wire brush (Mech.) 725
KRATZDISTEL (*f*) : Fuller's thistle, teasel (*Dipsacus fullonum*) 726
KRATZE (*f*) : scraper ; card 727
KRÄTZE (*f*) : itch, psora, scabies (inflammation of the skin due to parasite *Acarus scabiei* or *Sarcoptes hominis*) (Med.) ; mange ; waste metal (dross, filings, cuttings, scrapings, sweepings) 728
KRATZEISEN (*n*) : scraping-iron, scraper 729
KRATZEN : to scratch ; claw ; scrape ; abrase ; card (Wool) ; grate ; tickle ; rabble (Metal) 730
KRATZER (*m*) : scraper ; scratcher ; rake-(r) 731
KRÄTZER (*m*) : (see **KRATZER**) 732

KRÄTZFRISCHEN (*n*): refining (melting down) of waste-(metal) 733
KRÄTZIG: itchy; scabious; fibrous, (Glass); scabby; psoric (Med.) 734
KRÄTZKRAUT (*n*): field scabious (*Knautia arvensis*) 735
KRÄTZKUPFER (*n*): copper from (refuse) waste 736
KRATZMASCHINE (*f*): carding-machine (Wool) 737
KRÄTZMESSING (*n*): brass (cuttings) filings 738
KRÄTZMILBE (*f*): itch worm (see **KRÄTZE**) 739
KRÄTZSALBE (*f*): ointment for the itch (see **KRÄTZE**) 740
KRAUS: crisp; frizzled; curly; crinkled; curled; nappy; wavy (of wood); plaited; ruffled 741
KRAUSE (*f*): frill; crispness; ruff-(le) 742
KRAUSEEISEN (*n*): (see **ZAINEISEN**) 743
KRÄUSELBRENNEISEN (*n*): milling tool (Coins) 744
KRÄUSELEISEN (*n*): curling (iron) tongs 745
KRÄUSELN: to curl; crisp; frill; crimp; mill (Coins); nap; be ruffled; (*n*): frilling (the wrinkling of the gelatine on a photographic plate) 746
KRÄUSELWERK (*n*): milling; milling tool (Coins) 747
KRAUSEMINZBLÄTTER (*n.pl*): curled mint leaves (*folia menthæ crispæ of Mentha crispa*) 748
KRAUSEMINZE (*f*): curled mint (*Mentha crispa*) (*folia et herba menthæ crispæ*); spearmint (*Mentha viridis*; *Mentha spicata*) 749
KRAUSEMINZÖL (*n*): spearmint oil; Sg. 0.935 750
KRÄUSEN: to become (curled) curly; ruffle (see also **KRÄUSELN**); (*f.pl*): head (Brewing) 751
KRAUSKOPF (*m*): curly-headed person; countersink 752
KRAUSSALAT (*m*): chicory, succory, endive (*Cichorium intybus*) 753
KRAUSTABAK (*m*): shag 754
KRAUT (*n*): herb (*Herba*); plant; weed; leaves; cabbage; vegetable; sumac. 755
KRAUTACKER (*m*): cabbage (garden, field) plot 756
KRAUTARTIG: herbaceous 757
KRÄUTERARTIG: (see **KRAUTARTIG**) 758
KRÄUTERAUSZUG (*m*): tincture 759
KRÄUTERER (*m*): herbalist 760
KRÄUTERESSIG (*m*): aromatic vinegar 761
KRÄUTERFRESSEND: (see **KRAUTFRESSEND**) 762
KRÄUTERKISSEN (*n*): pillow stuffed with herbs 763
KRÄUTERKUNDE (*f*): botany 764
KRÄUTERLEHRE (*f*): botany 765
KRÄUTERLIKÖR (*m*): herbal liqueur 766
KRÄUTERPRESSE (*f*): herb press 767
KRÄUTERREICH: rich in herbs; (*n*): vegetable kingdom 768
KRÄUTERSEIFE (*f*): aromatic soap 769
KRÄUTERTRANK (*m*): herbal decoction 770
KRÄUTERWEIN (*m*): medicated wine 771
KRÄUTERZUCKER (*m*): conserve; confection (Pharmacy) 772
KRAUTFÖRMIG: dendriform, dendroid, dendritic, herb-like 773
KRAUTFRESSEND: herbivorous 774
KRAUTGARTEN (*m*): kitchen garden 775
KRAUTHACKE (*f*): hoe 776
KRAUTHANDEL (*m*): market-gardening 777
KRAUTHÄNDLER (*m*): market-gardener 778
KRAUTHOLUNDER (*m*): dwarf elder, danewort (*Sambucus ebulus*) 779
KRAUTSALAT (*m*): cabbage lettuce (*Lactuca scariola sativa*) 780
KRAWATTE (*f*): cravat; tie; scarf 781
KRAWEELBOOT (*n*): carvel-built boat 782
KREATIN (*n*): creatine; $C_4H_9N_3O_2$ (obtained from meat) 783
KREATININ (*n*): creatinine (see **CREATININ**) 784
KREBS (*m*): ulcer; cancer (Med.); crab; crawfish; crayfish; grain; hard particle (in clay, etc.); knot (in ore, etc.); canker (Botanical) 785
KREBSARTIG: cancerous; cankerous; cancriform; crab-like; crustaceous 786
KREBSBILDUNG (*f*): cancer formation 787
KREBSKREIS (*m*): Tropic of Cancer 788
KREBSWEIDE (*f*): osier (*Salix viminalis*) 789
KREBSWURZ (*f*): beechdrops (*Leptamnium virginianum*) 790
KREDENZEN: to taste, try 791
KREDIT (*m*): credit 792
KREDIT-AKTIE (*f*): credit share 793
KREDITBRIEF (*m*): letter of credit 794
KREDITGESUCH (*n*): application for credit or for an advance (of money) 795
KREDITIEREN: to credit (Com.) 796
KREIDE (*f*): chalk; calcium carbonate; $CaCO_3$; Sg. 1.52-2.73 (see **KALZIUMCARBONAT**); crayon 797
KREIDEARTIG: chalky; cretaceous; like chalk; chalk-like 798
KREIDEGEBILDE (*n*): cretaceous group, chalk formation 799
KREIDE (*f*), **GESCHLÄMMTE-**: (see **SCHLÄMM-KREIDE**) 800
KREIDEGRUBE (*f*): chalk-pit 801
KREIDEHALTIG: cretaceous; containing chalk 802
KREIDEMEHL (*n*): powdered chalk 803
KREIDEN: to chalk 804
KREIDEPAPIER (*n*): enamelled paper; baryta paper (paper coated with barium sulphate emulsion) (Phot.) 805
KREIDEPULVER (*n*): chalk powder; powdered chalk; pipe-clay 806
KREIDESANDSTEIN (*m*): upper green sandstone 807
KREIDESTEIN (*m*): chalk-stone 808
KREIDESTIFT (*m*): crayon; chalk 809
KREIDEWEISS: chalky-white, deathly pale 810
KREIDEZEICHNUNG (*f*): chalk drawing, crayon drawing (sometimes loosely applied to pastel drawing) 811
KREIDEZEIT (*f*): (upper)-cretaceous period (Geol.) 812
KREIDIG: chalky; cretaceous; covered with chalk 813
KREIDIGES BILD (*n*): chalky print (Phot.) 814
KREIS (*m*): circle; circuit; orbit; district; ring; sphere 815
KREISABSCHNITT (*m*): segment 816
KREISAUSSCHNITT (*m*): sector 817
KREISAUSSCHNITTFÖRMIG: segmental 818
KREISBAHN (*f*): circular path; orbit 819
KREISBEWEGUNG (*f*): circular (rotary) motion or movement; rotation 820
KREISBOGEN (*m*): arc-(of a circle); circular (Roman) arch 821
KREISBRIEF (*m*): circular letter 822
KREISCHEN: to shriek; scream; sizzle (of fat); cry (of tin) 823
KREISDREHUNG (*f*): rotation, revolution 824

KREISEL (*m*): (spinning)-top; gyroscope; staggers (Vet.) 825

KREISELKRAFT (*f*): centrifugal force 826

KREISELMOMENT (*n*): centrifugal moment 827

KREISELN: to spin, whirl round, revolve, eddy 828

KREISELPUMPE (*f*): centrifugal pump 829

KREISELRAD (*n*): turbine 830

KREISELWIND (*m*): whirl-wind 831

KREISELWIPPER (*m*): wagon (tipper, tipping device) tipping arrangement 832

KREISELWIRKUNG (*f*): gyroscopic (centrifugal) effect 833

KREISEN: to circle; revolve; circulate; rotate; whirl round 834

KREISEVOLVENTE (*f*): involute 835

KREISFLÄCHE (*f*): circular (area) surface 836

KREISFÖRMIG: circular; round; rotund 837

KREISGANG (*m*): revolution; rotation; circular motion; labyrinth 838

KREISINHALT (*m*): area (of a circle) 839

KREISLAUF (*m*) cycle; circulation (Liquid or Gas); circular (path) course; circuit; rotation; revolution; orbit; cyclic process 840

KREISLAUFEND: circulatory 841

KREISLINIE (*f*): circumference; circular line 842

KREISMESSUNG (*f*): cyclometry 843

KREISPROZESS (*m*): cyclic process; cycle 844

KREISPUMPE (*f*): circulating pump 845

KREISRUND: circular 846

KREISSÄGE (*f*): circular saw 847

KREISSCHEIBE (*f*): circular disc 848

KREISSCHICHT (*f*): circular layer; annual ring (of a tree) 849

KREISSCHREIBEN (*n*): circular letter 850

KREISSEGMENT (*n*): segment of a circle 851

KREISSEKTOR (*m*): sector of a circle 852

KREISSEN: to labour (in childbirth) 853

KREISSTRÖMUNG (*f*): circulation 854

KREISUMFANG (*m*): circumference; periphery 855

KREISVORGANG (*m*): cyclic process 856

KRELUTION (*f*): crelution (an antiseptic, green fluid, containing 66% cresol and having a Sg. 1.054) 857

KREMPE (*f*): brim (of hat); carder; carding machine (Textile); flange (where two pieces join) 858

KREMPEL (*f*): card 859

KREMPELN: to card (Wool) 860

KREMPLER (*m*): wool-carder 861

KREMSERWEISS (*n*): Kremnitz (Cremnitz) white; silver white (see BLEIWEISS) 862

KREMULSION (*f*): cremulsion (an emulsifying cresol preparation) 863

KREOSOL (*n*): homoguaiacol, creosol, homopyrocatecholmonomethylester; $C_6H_3.CH_3(OCH_3).OH(1:3:4)$; Bp. 220°C. 864

KREOSOT (*n*): creosote (from coal tar by fractional distillation; Sg. 1.07) (from beechwood tar by fractional distillation Sg. 1.08 Bp. 215° C.) 865

KREOSOTAL (*n*): (*Creosotum carbonicum*): creosotal, cresote carbonate (from beechwood creosote) 866

KREOSOT (*n*), BALDRIANSAURES-: creosote (valerianate) valerate 867

KREOSOT (*n*), BENZOESAURES: creosote benzoate 868

KREOSOTKARBONAT (*n*): creosote carbonate (see KREOSOTAL) 869

KREOSOT (*n*), KOHLENSAURES-: creosote carbonate (see KREOSOTAL) 870

KREOSOTÖL (*n*): creosote oil (heavy oil); Sg. 1.05 871

KREPIEREN: to die (of animals) 872

KREPITIEREN: to crepitate; crackle 873

KREPP (*m*): crape; crêpe 874

KREPPEN: to wave, crape, frieze 875

KREPPFLOR (*m*): crape 876

KRESALOL (*n*): cresalol; $C_6H_4(OH).COO.C_6H_4.CH_3$ 877

KRESEPTON (*n*): cresepton (a disinfectant from creosote) 878

KRESOL (*n*): cresol (C_7H_8O); ortho.—Sg. 1.0427; Mp. 30°C.; Bp. 188°C.; meta.—Sg. 1.0350; Mp. about 4°C.; Bp. 200.5°C.; para.—Sg. 1.0340; Mp. 36°C.; Bp. 201.1°C. (see also KRESYLSÄURE); ORTHO-—cresyl alcohol, ortho-oxytoluene, ortho-cresylic acid, ortho-methylphenol; META-—meta-cresylic acid, meta-oxytoluene, cresylic acid, meta-methylphenol; PARA-—para-cresylic acid, para-oxytoluene, para-methylphenol 879

KRESOLNATRON (*n*): soda (sodium) cresolate 880

KRESOLPUDER (*n*): cresol powder (a vermin killer) 881

KRESOLSCHWEFELSÄURE (*f*): cresol-sulphuric acid 882

KRESOLSEIFE (*f*): cresol soap (a disinfectant containing 50% crude cresol; Bp. about 201°C.) 883

KRESOLSEIFENLÖSUNG (*f*): liquor cresoli saponatus; Sg. 1.04; about 50% crude cresol 884

KRESORCIN (*n*): cresorcinol 885

KRESOTINGELB (*n*): cresotine yellow 8[illegible]

KRESOTINSAUER: cresotate of, combined with cresotic acid 887

KRESOTINSÄURE (*f*): cresotic (cresotinic) acid; $C_6H_3COOHOHCH_3$; Mp. 151°C. 888

KRESOTSULFONSAUER: sulfocresotate 889

KRESSE (*f*): cress 890

KRESSENSAMENÖL (*n*): cress oil (from *Lepidium sativum*) 891

KRESYL (*n*): cresyl 892

KRESYLBLAU (*n*): cresyl blue 893

KRESYLOL (*n*): cresylol (see KRESOL, META-) 894

KRESYLSÄURE (*f*): cresylic acid, cresol, meta-cresylic acid; $CH_3C_6H_4OH$; Sg. 1.042; Mp. 10.9°C.; Bp. 202°C. 895

KREUZ (*n*): cross; crucifix; loins; backbone; spine; rump; clubs (Cards); cross-bar; cross-over; point of intersection; sharp (Music); affliction 896

KREUZARM (*m*): cross-bar 897

KREUZBALKEN (*m*): cross (bar) beam 898

KREUZBAND (*n*): newspaper-wrapper 899

KREUZBAND (*n*), UNTER-: by book post 900

KREUZBAUM (*m*): turn-pike 901

KREUZBEEREN (*f.pl*): buckthorn (Persian) berries (see GELBBEEREN) 902

KREUZBEERENEXTRAKT (*m*): buckthorn-berry extract 903

KREUZBEEREN (*f.pl*), PERSISCHE-: Persian berries (see also GELBBEEREN) 904

KREUZBEEREXTRAKT (*m*): buckthorn-berry extract 905

KREUZBEERLACK (*m*): buckthorn-berry lake 906

KREUZBEERSTRAUCH (*m*): buckthorn (*Rhamnus frangula*) 907

KREUZBEIN (*n*): sacrum (*Os sacrum*) (Anat.) 908

KREUZBERG (*m*): Calvary 909

KREUZBLUME (*f*): milkwort (*Polygala vulgaris*) 910

KREUZBLÜTLER (*m.pl*): *Cruciferæ*, cruciferous plants 911

KREUZBOGEN (*m*): ogive, a pointed arch 912

KREUZBRAMRAH (*f*): mizzen top-gallant yard (Naut.) 913

KREUZBRASSE (*f*): mizzen top brace (Naut.) 914

KREUZDORN (*m*): buckthorn (*Rhamnus frangula*) 915

KREUZDORNBEEREN (*f.pl*): (see **KREUZBEEREN**) 916

KREUZDORNGEWÄCHSE (*n.pl*): *Rhamnaceæ* (Bot.) 917

KREUZDORNÖL (*n*): buckthorn oil (from seeds of *Rhamnus cathartica*) 918

KREUZEN: to cross; crucify; thwart; clash; cruise; tack (Naut.) 919

KREUZER (*m*): cruiser 920

KREUZERJACHT (*f*): cruising yacht 921

KREUZER (*m*), **PANZER-**: armour-plated cruiser 922

KREUZFAHRT (*f*): crusade; pilgrimage; cruise 923

KREUZFLÜGEL (*m*): transept 924

KREUZFÖRMIG: cruciform; cross-shaped 925

KREUZHAHN (*m*): four-way cock 926

KREUZHOLZ (*n*): buckthorn (see **KREUZDORN**); cross-piece (of wood); mistletoe (*Viscum album*) 927

KREUZIGEN: to crucify 928

KREUZIGUNG (*f*): crucifixion 929

KREUZKNOTEN (*m*): carrick bend, double knot, sailor's knot 930

KREUZKOPF (*m*): cross-head (Mech.) 931

KREUZKOPFBACKEN (*f.pl*): guide blocks, cross-head guides 932

KREUZKOPFBOLZEN (*m*): cross head pin 933

KREUZKOPFKEIL (*m*): cross head cottar 934

KREUZKOPFLAGER (*n*): cross head bearing 935

KREUZKOPFZAPFEN (*m*): cross head pin 936

KREUZKRAUT (*n*): groundsel (*Senecio vulgaris*) 937

KREUZKRAUT (*n*), **BITTERES-**: milkwort (*herba polygalæ amaræ*) (*Polygala amara, Polygala amarella, Polygala calcarea*) 938

KREUZKÜMMEL (*m*): cumin; cummin; (*Cuminum cyminum*) 939

KREUZKURVEN (*f.pl*): cross curves (Shipbuilding) 940

KREUZMARS (*m*): mizen top (Naut.) 941

KREUZMARSSEGEL (*n*): mizen top sail 942

KREUZMARSSTENGE (*f*): mizen topmast 943

KREUZMASS (*n*): T-square 944

KREUZMAST (*m*): mizen mast (in the case of a full-rigged ship) 945

KREUZMEISSEL (*m*): cross cutting chisel 946

KREUZNAHT (*f*): cross-seam 947

KREUZPUNKT (*m*): point of intersection 948

KREUZRAH (*f*): mizen topyard (Naut.) 949

KREUZROIL (*n*): mizen royal 950

KREUZROILSEGEL (*n*): mizen royal-(sail) 951

KREUZRÜSTE (*f*): mizen channel 952

KREUZSCHEISEGEL (*n*): mizen skysail 953

KREUZSCHMERZ (*m*): lumbago (Med.) 954

KREUZSCHRAFFIEREN: to cross-hatch (Drawing) 955

KREUZSEGEL (*n*): mizen topsail 956

KREUZSKEISEGEL (*n*): mizen sky-sail 957

KREUZSTAG (*n*): mizen stay 958

KREUZSTEIN (*m*): cross stone; chiastolite; harmotome; staurolite (see **ANDALUSIT**; **HARMOTOM** and **STAUROLITH**) 959

KREUZSTENGE (*f*): mizen topmast 960

KREUZUNG (*f*): crossing (of rails); cruising (of ships); cross-breeding 961

KREUZUNTERBRAMSEGEL (*n*): lower mizen top-gallant sail (Naut.) 962

KREUZUNTERMARSSEGEL (*n*): lower mizen top-sail 963

KREUZVERHÖR (*n*): cross-examination 964

KREUZWEISE: crosswise, across 965

KREUZZUCHTWOLLE (*f*): cross-bred wool (Textile) 966

KRIEBELKRANKHEIT (*f*): ergotism (*Morbus cerealis*) (Med.) 967

KRIECHE (*f*): plum tree; wild bullace (*Prunus insititia*) 968

KRIECHEN: to creep; trail; crawl; cringe; be servile 969

KRIECHPFLANZE (*f*): creeper, creeping plant 970

KRIECHPFLAUME (*f*): (see **KRIECHE**) 971

KRIEG (*m*): war; hostility; strife 972

KRIEGEN: to make (wage) war; seize; obtain; get; dispute 973

KRIEGSAKADEMIE (*f*): military college 974

KRIEGSAMT (*n*): War Office 975

KRIEGSAUFGEBOT (*n*): mobilisation 976

KRIEGSAUFRUF (*m*): mobilisation 977

KRIEGSBEDARF (*m*): military stores 978

KRIEGSFLOTTE (*f*): Navy 979

KRIEGSHAFEN (*m*): military or naval port 980

KRIEGSMINISTERIUM (*n*): War Office 981

KRIEGSPEST (*f*): typhus (malignant, jail, petechial) fever 982

KRIEGSSCHIFF (*n*): man-o'-war 983

KRIMINELL: criminal-(ly) 984

KRIMPEN: to shrink-(cloth) (see also **KRÜMPEN**) 985

KRIMPFÄHIGKEIT (*f*): shrinking property; felting property (Wool) 986

KRIMPFREI: non-shrinking 987

KRINGEL (*m*): cracknel 988

KRIPPE (*f*): crib; manger; fence; hurdle; caisson 989

KRISE (*f*): crisis, turn (Med.) 990

KRISEN (*pl* of **KRISIS** or **KRISE**): crises 991

KRISIS (*f*): crisis, depression 992

KRISPELN: to (grain; crisp) pebble (Leather, etc.) 993

KRISTALL (*m*): crystal; crystal glass 994

KRISTALLACHSE (*f*): crystal axis; crystallographic axis 995

KRISTALLARTIG: crystalline; crystal-like 996

KRISTALLAUSSCHEIDUNG (*f*): separation of crystals 997

KRISTALLAUSSCHUSS (*m*): crop of crystals 998

KRISTALLBAU (*m*): crystal structure 999

KRISTALLBENZOL (*n*): benzene of crystallization 000

KRISTALLBILDUNG (*f*): formation of crystals; crystallization; granulation (Sugar) 001

KRISTALLBILDUNGSPUNKT (*m*): granulating point (Sugar) 002

KRISTALLBLAU (*n*): crystal blue 003

KRISTALLCHLOROFORM (*n*): chloroform of crystallization 004

KRISTALLDRÜSE (*f*): cluster of crystals 005

KRISTALLEN: crystalline 006

KRISTALLFABRIK (*f*): glass works 007

KRISTALLFEUCHTIGKEIT (*f*): crystalline (vitreous) humour 008

KRISTALLFLÄCHE (*f*): crystal face; facet 009

KRISTALLFLASCHE (*f*) : carafe ; decanter ; water-bottle 010
KRISTALLFORM (*f*) : crystal form 011
KRISTALLGLAS (*n*) : crystal glass ; artificial crystal 012
KRISTALLGLASUR (*f*) : crystalline glaze 013
KRISTALLHELL : clear as crystal 014
KRISTALLIG : crystalline 015
KRISTALLIN : crystalline 016
KRISTALLINE (*f*) : celluloid varnish (trade name) (see **ZELLULOIDLACK**) 017
KRISTALLINISCH : crystalline ; clear ; transparent 018
KRISTALLISATION (*f*) : crystallization ; crystallizing 019
KRISTALLISATIONSANLAGE (*f*) : crystallizing plant 020
KRISTALLISATIONSAPPARAT (*m*) : crystallizing apparatus, crystallizer 021
KRISTALLISATIONSBASSIN (*n*) : crystallizing (basin ; tank) cistern 022
KRISTALLISATIONSDRUCK (*m*) : pressure of crystallization 023
KRISTALLISATIONSFÄHIG : crystallizable 024
KRISTALLISATIONSGEFÄSS (*n*) : crystallizing vessel ; crystallizer 025
KRISTALLISATIONSWÄRME (*f*) : heat of crystallization (Metal) 026
KRISTALLISATOR (*m*) : crystallizer 027
KRISTALLISIERBAR : crystallizable 028
KRISTALLISIERBEHÄLTER (*m*) : crystallizing (receptacle) tank 029
KRISTALLISIEREN : to crystallize 030
KRISTALLISIERGEFÄSS (*n*) : crystallizing (vessel ; pan) dish ; crystallizer 031
KRISTALLISIERKASTEN (*m*) : crystallizing vessel 032
KRISTALLISIERKESSEL (*m*) : crystallizing vessel 033
KRISTALLISIERSCHALE (*f*) : crystallizing dish 034
KRISTALLISIERUNG (*f*) : crystallization ; crystallizing 035
KRISTALLISIERUNGSPUNKT (*m*) : crystallizing point ; granulating pitch (Sugar) 036
KRISTALLKEIM (*m*) : crystal nucleus ; seed crystal 037
KRISTALLKERN (*m*) : nucleus of crystallization 038
KRISTALLKUNDE (*f*) : crystallography (the science of crystals) 039
KRISTALLLEHRE (*f*) : crystallography (the science of crystals) 040
KRISTALLLINSE (*f*) : crystalline lens 041
KRISTALLMASSE (*f*) : crystal mass 042
KRISTALLOBERFLÄCHE (*f*) : crystal surface 043
KRISTALLOGRAPHIE (*f*) : crystallography (the science of crystals) 044
KRISTALLOGRAPHISCH : crystallographic 045
KRISTALLSÄURE (*f*) : fuming sulphuric acid in crystalline form 046
KRISTALLSKELETT (*n*) : crystal skeleton 047
KRISTALLSODA (*f*) : soda crystals ($Na_2CO_3.10H_2O$) 048
KRISTALLSTEIN (*m*) : rock crystal, transparent quartz (see **QUARZ**) 049
KRISTALLWAREN (*f.pl*) : crystal (glass) ware 050
KRISTALLWASSER (*n*) : crystal water ; water of crystallization 051
KRISTALLWASSERFREI : free from water of crystallization 052
KRISTALLWASSERHALTIG : containing crystal water 053
KRISTALLWIEGE (*f*) : crystallizer 054
KRISTALLZINN (*n*) : grain tin 055
KRISTALLZUCKER (*m*) : crystallized (refined) sugar 056
KRITERIUM (*n*) : criterion 057
KRITIK (*f*) : criticism (see **KRITISIEREN**) 058
KRITISCH : critical 059
KRITISIEREN : to review, criticize ; upbraid, reproach, rebuke, censure 060
KRITTELN : to carp at ; criticize ; find fault 061
KRITZELN : to scribble ; scratch 062
KROKIDOLITH (*m*) : crocidolite ; $Na_2Fe_2(SiO_3)_4$; Sg. 3.4 063
KROKOIT (*m*) : crocoite (see **ROTBLEIERZ**) 064
KROKONSÄURE (*f*) : croconic acid 065
KROKOS (*m*) : crocus (*Crocus aureus* and *C. vernus*) (French, Spanish) saffron (*Crocus sativus*) 066
KROKYDOLITH (*m*) : crocidolite ; $NaFeSiO_3$; Sg. 3.4 067
KROLLEN : to curl 068
KRONBEERE (*f*) : bilberry (*Vaccinium myrtillus*) 069
KRONBLATT (*n*) : petal 070
KRONE (*f*) : crown ; corolla ; halo ; corona ; wreath ; coronet ; top ; crest 071
KRÖNEN : to crown ; surmount, top 072
KRONENARTIG : coronal ; crown-like ; coronary (Anat.) 073
KRONENAUFSATZ (*m*) : column (of a still) 074
KRONENBOHRER (*m*) : crown bit 075
KRONENGOLD (*n*) : crown gold ; 18 carat gold 076
KRONGLAS (*n*) : crown glass (composed of potassium and sodium silicates plus calcium and sometimes aluminium) 077
KRONRAD (*n*) : crown wheel 078
KRONSÄGE (*f*) : circular saw 079
KRONSTEIN (*m*) : crown, square (Bricks) 080
KRONZAHN (*m*) : eye tooth 081
KROPF (*m*) : crop ; bend ; excrescence ; breasting (Paper) ; goitre ; bronchocele (enlargement of the thyroid gland) (Med.) 082
KROPFADER (*f*) : varicose vein 083
KROPFBEIN (*n*) : larynx 084
KROPFIG : strumous, strumose ; having goitre 085
KROPFMITTEL (*n*) : anti-strumatic 086
KROPFROHR (*n*) : bent (tube) pipe 087
KRÖPFUNG (*f*) : right-angled bend ; corner moulding 088
KROPFWURZ (*f*) : figwort (*Scrophularia*) 089
KROPFZYLINDER (*m*) : glass cylinder with widened upper part ; hydrometer jar 090
KROQUIEREN : to sketch, make a plan 091
KRÖSEL (*m*) : glazier's iron 092
KRÖSELN : to crumble (Glass) ; groove 093
KRÖTE (*f*) : toad 094
KRÖTENDISTEL (*f*) : lesser meadow rue (*Thalictrum minus*) 095
KRÖTENKRAUT (*n*) : groundsel (*Senecio vulgaris*) 096
KRÖTENMELDE (*f*) : thorn-apple (*Datura stramonium*) 097
KROTONÖL (*n*) : croton oil ; Sg. 0.95 (from *Croton tiglium*) 098
KROTONSÄURE (*f*) : crotonic acid ; $CH_3CHCHCO_2H$; Sg. 0.97 ; Mp. 72°C. ; Bp. 185°C. 099
KROUP (*m*) : croup (a throat disease) (Med.) 100
KROZEINSULFOSÄURE (*f*) : (see **BAYERSCHE SÄURE**) 101
KRÜCKE (*f*) : crutch ; rake ; rabble ; scraper ; crook ; scoop 102
KRÜCKENFÖRMIG : crooked 103

KRUG (*m*): pitcher; jug; mug; jar; pot; tankard 104
KRUKE (*f*): earthenware (pot; pitcher) jar 105
KRÜLLEN: to crumple; curl 106
KRULLFARN (*m*): maiden-hair fern (*Adiantum Capillus-veneris*) 107
KRUME (*f*): crumb; bit; black earth; vegetable mould 108
KRÜMELIG: crumbly; crumbling 109
KRÜMELN: to crumble 110
KRÜMELZUCKER (*m*): dextrose; (dextro)-glucose; $C_6H_{12}O_6$ (see **DEXTROSE**) 111
KRUMM: crooked; curved; bent; twisted; arched; bowed 112
KRUMMACHSE (*f*): crank-(ed axle) 113
KRUMMÄSTIG: gnarled (of trees) 114
KRUMMDARM (*m*): ileum; (as a prefix = iliac) 115
KRÜMME (*f*): bend, curvature, crookedness 116
KRÜMMEN: to curve; bend; crook; warp; crumple; wind; deflect 117
KRÜMMER (*m*): (quarter)-bend; elbow; bent pipe; swan-neck pipe (Brick) 118
KRUMMHOLZ (*n*): crooked (curved) piece of wood; bent wood; knee timber (Shipbuilding); wale (Naut.) 119
KRUMMHOLZBAUM (*m*): knee-pine, dwarf mountain pine (*Pinus montana pumilio*) 120
KRUMMHOLZFICHTE (*f*): (see **KRUMMHOLZBAUM**) 121
KRUMMHOLZKIEFER (*f*): (see **KRUMMHOLZBAUM**) 122
KRUMMHOLZÖL (*n*): templin oil; knee-pine oil; Sg. 0.9 123
KRUMMLINIG: curvilinear 124
KRÜMMUNG (*f*): curvature; curve; bend; winding; sinuosity; turn; deflection; crookedness 125
KRÜMMUNGSHALBMESSER (*m*): radius of bend 126
KRÜMMUNGSMITTELPUNKT (*m*): centre of curvature 127
KRÜMMUNGSRADIUS (*m*): radius of curvature 128
KRUMMZAPFEN (*m*): crank 129
KRUMMZIRKEL (*m*): bow compasses 130
KRÜMPELN: to crumple; pucker; crinkle 131
KRÜMPEN: (see **KRIMPEN**) 132
KRÜMPFMASCHINE (*f*): shrinking machine (Textile) 133
KRUPP (*m*): croup (a throat disease) (Med.) 134
KRUPPIN (*n*): "kruppin" (30% nickel steel with Cu and Mn, used for electrical resistances) 135
KRUSTACEEN (*pl*): crustacea 136
KRUSTE (*f*): crust; incrustation; fur; scale; scurf; scab 137
KRUSTENARTIG: crustaceous 138
KRUSTENTIER (*n*): crustacean 139
KRYOCHEMIE (*f*): cryochemistry 140
KRYOHYDRATISCH: cryohydric 141
KRYOLITH (*m*): kryolith, cryolite ($AlF_3.3NaF$) (from Ivetut, Greenland only) (used as a solvent for obtaining aluminium from bauxite) (see also **ALUMINIUM-NATRIUM-FLUORID**) (Sg. 2.95-3.0) 142
KRYOLITHIONIT (*m*): cryolithionite; $Li_3Na_3Al_2F_{12}$; Sg. 2.8 143
KRYOPHOR (*n*): cryophorus 144
KRYOSKOPIE (*f*): cryoscopy 145
KRYOSTAT (*m*): cryostat 146
KRYPTOGAMEN (*pl*): *Cryptogamœa* (Bot.) 147
KRYPTOGAMISCH: cryptogamous (Botanical) 148
KRYPTOL (*n*): Kryptol (used as a heating resistance for muffle furnaces; is a mixture of carbon, graphite, carborundum, etc.) 149
KRYPTOLMUFFELOFEN (*m*): muffle furnace fitted with kryptol heating resistance 150
KRYPTOPIN (*n*): cryptopine (an opium alkaloid $C_{19}H_{17}NO_3(OCH_3)_2$; Mp. 217°C. 151
KRYSTALL (*m*): crystal (see **KRISTALL**) 152
K.S.-STAHL (*m*): (**KOBALTCHROMSTAHL**): cobalt-chromium steel (contains chromium, tungsten and 35% cobalt; is called K.S. magnet steel) 153
KUBATUR (*f*): cubature 154
KUBEBE (*f*): cubeb-(berry); cubeb pepper (fruit of *Piper cubeba*) 155
KUBEBENÖL (*n*): cubeb-(berry)-oil (from fruit of *Piper cubeba*) (Sg. 0.91; Bp. 175/280°C.) 156
KÜBEL (*m*): vat; tub; pail; bucket; hod 157
KUBIEREN: to cube; raise to the third power 158
KUBIK: cube, cubic 159
KUBIKBERECHNUNG (*f*): cubature 160
KUBIKINHALT (*m*): cubic contents 161
KUBIKMASS (*n*): cubic contents 162
KUBIKMETER (*m*): cubic metre 163
KUBIKWURZEL (*f*): cubic root; cube root 164
KUBIKZAHL (*f*): cube-(number) 165
KUBIKZENTIMETER (*m* and *n*): cubic centimeter 166
KUBISCH: cubic-(al) 167
KUBISCHER SALPETER (*m*): sodium nitrate, Chili saltpeter (see **NATRIUMNITRAT**) 168
KUBUS (*m*): cube 169
KÜCHE (*f*): kitchen; cooking; cookery; culinary art 170
KUCHEN (*m*): cake; caked lump (of coke, etc.); clot (of blood); press residue 171
KÜCHENABFÄLLE (*m.pl*): kitchen refuse 172
KÜCHENGEWÄCHS (*n*): vegetables; potherbs 173
KÜCHENSALZ (*n*): cooking (common) salt (see **SALZ**) 174
KÜCHENSCHELLE (*f*): pulsatilla (see **HAKELKRAUT**) 175
KÜCHENSCHELLENKRAUT (*n*): pulsatilla (*Herba pulsatillœ*) (see **HAKELKRAUT**) 176
KÜCHENZWIEBEL (*f*): common onion (*Allium cepa*) 177
KÜCHLEIN (*n*): chicken; pullet; little cake; lozenge 178
KUCKUCKSBEIN (*n*): coccyx (Anat.) 179
KUCKUCKSKLEE (*m*): wood sorrel, clover sorrel (*Oxalis acetosella*) 180
KUFE (*f*): vat; reservoir; tun; tub; tank; runner (of sails, sledge, etc.) 181
KÜFER (*m*): cooper; butler 182
KUFE (*f*), **SEGELSCHLITTEN-**: ice boat runner 183
KUFF (*f*), **SCHONER-**: Dutch built schooner 184
KUGEL (*m*): bulb; ball; globule; sphere; bullet; globe; molecule; drop; bead; shot; (as a prefix = spherical) 185
KUGELABSCHNITT (*m*): segment of a sphere 186
KUGELÄHNLICH: spheroidal; spherical 187
KUGELAPPARAT (*m*): bulb apparatus; spherical apparatus 188
KUGELARTIG: spherical, globular 189
KÜGELCHEN (*n*): small (ball) bulb; pearl; bead; globule; pellet; pea; drop 190
KUGELDIORIT (*m*): globular diorite (a granitoid rock) 191
KUGELDREIECK (*n*): spherical triangle 192
KUGELDREIECKLEHRE (*f*): spherical trigonometry) 193
KUGELDRUCK (*m*): indentation pressure, pressure of ball test (method of testing hardness of steel with a ball) 194

KUGELFEST : bullet-proof 195
KUGELFLÄCHE (*f*) : spherical surface 196
KUGELFLASCHE (*f*) : spherical (balloon) flask 197
KUGELFORM (*f*) : spherical mould 198
KUGELFÖRMIG : globular ; spherical ; globose ; globate ; globulous 199
KUGELGELENK (*n*) : ball and socket joint ; enarthrosis (Anat.) 200
KUGELGEWÖLBE (*n*) : cupola 201
KUGELGIESSEN : to cast bullets ; (*n*) : bullet casting 202
KUGELIG : (see **KUGELFÖRMIG**) : 203
KUGELKARTE (*f*) : planisphere, a sphere projected on a plane 204
KUGELKOCHER (*m*) : spherical boiler (for purification of rags for paper-making) 205
KUGELKÜHLER (*m*) : ball (spherical) condenser 206
KUGELLAGER (*n*) : ball-bearing 207
KUGELLAGERSCHALE (*f*) : ball bearing (cup, seating) brass or bush 208
KUGELLEHRE (*f*) : spherics ; ball-gauge 209
KUGELLINSE (*f*) : spherical (globe) lens (Phot.) 210
KUGELMASSLIEBE (*f*) : globularia (plants of the order, *Selaginaceæ*) 211
KUGELMÜHLE (*f*) : ball mill 212
KUGELN : to roll ; bowl 213
KUGELOBJEKTIV (*n*) : spherical lens, globe lens (Phot.) 214
KUGELRANUNKEL (*f*) : globe-flower, trollius (*Trollius europæus*) 215
KUGELROHR (*n*) : bulb tube 216
KUGELRUND : round, globe-shaped (see **KUGEL-FÖRMIG**) 217
KUGELSCHALE (*f*) : spherical (bush) seating or brass 218
KUGELSCHARNIER (*n*) : ball and socket joint 219
KUGELSCHNITT (*m*) : spherical section 220
KUGELSEGMENT (*n*) : segment of a sphere 221
KUGELSEKTOR (*m*) : sector of a sphere 222
KUGELSPIEGEL (*m*) : spherical mirror 223
KUGELSPURLAGER (*n*) : ball thrust bearing 224
KUGELSTOPFEN (*m*) : bulb (ball ; globular) stopper 225
KUGELTEE (*m*) : gunpowder tea 226
KUGELVENTIL (*n*) : ball valve 227
KUGELVORLAGE (*f*) : spherical (receiver) header 228
KUGELWINKEL (*m*) : spherical angle 229
KUGELZANGE (*f*) : gas pliers 230
KUGELZAPFEN (*m*) : ball journal, ball pivot, ball and socket joint 231
KUGELZONE (*f*) : spherical zone 232
KUH (*f*) : cow 233
KUHBLUME (*f*) : marsh marigold (*Caltha palustris*) 234
KUHBUTTER (*f*) : meadow saffron (*Colchicum autumnale*) 235
KUHDILL (*m*) : corn (chamomile) camomile (*Anthemis arvensis*) 236
KUHFLADEN (*m*) : cow dung 237
KUHFUSS (*m*) : crow bar, nail claw, nail drawer 238
KÜHL : cool ; fresh 239
KÜHLANLAGE (*f*) : cooling (condensing, refrigerating) plant 240
KÜHLAPPARAT (*m*) : cooling (refrigerating) apparatus ; cooler ; refrigerator ; condenser 241
KÜHLBALKEN (*m*) : water-back (of stokers) 242
KÜHLBOTTICH (*m*) : cooling tub ; cooler 243
KÜHLBÜCHSE (*f*) : cooling (box) jacket 244
KÜHLEN : to cool ; refrigerate ; ice ; refresh ; temper 245
KÜHLEND : cooling ; refreshing 246
KÜHLER (*m*) : condenser ; cooler ; refrigerator ; attemperator 247
KÜHLER (*m*), **GEGENSTROM-** : counter-current cooler 248
KÜHLERGESTELL (*n*) : condenser (stand) support 249
KÜHLERMANTEL (*m*) : condenser jacket 250
KÜHLERRETORTE (*f*) : condenser retort 251
KÜHLER (*m*), **RÖHREN-** : tubular condenser 252
KÜHLER (*m*), **RÜCKFLUSS-** : return flow (cooler) condenser 253
KÜHLFASS (*n*) : cooling (vessel ; tub) vat ; cooler 254
KÜHLFLÄCHE (*f*) : cooling surface 255
KÜHLGEFÄSS (*n*) : cooling (refrigerating) vessel ; cooler ; refrigerator ; condenser 256
KÜHLGELÄGER (*n*) : wort sediment ; dregs (Brewing) ; sediment from cooling 257
KÜHLKAMMER (*f*) : cooling chamber 258
KÜHLMANTEL (*m*) : water jacket (of motors) 259
KÜHLMASCHINEN (*f.pl*) : refrigerating machinery 260
KÜHLMITTEL (*n*) : refrigerant ; cooling (remedy ; agent) medium 261
KÜHLOFEN (*m*) : annealing (oven) furnace ; cooling (oven) furnace 262
KÜHLPFANNE (*f*) : cooling pan ; cooler 263
KÜHLROHR (*n*) : cooling (coil) tube ; condenser ; refrigerating pipe 264
KÜHLSALZ (*n*) : refrigerating (freezing ; cooling) salt 265
KÜHLSCHIFF (*n*) : cooling (copper, pan)- back, cooler (Brewing) 266
KÜHLSCHIFFTRUB (*m*) : cooling back (cooler) dregs or sediment (Brewing) 267
KÜHLSCHLANGE (*f*) : cooling (refrigerating) coil, condensing coil, coil condenser ; worm (for distilling) 268
KÜHLSPIRALE (*f*) : cooling coil (see **KÜHLSCHLANGE**) 269
KÜHLTE (*f*) : gale, fresh wind, breeze 270
KÜHLTRANK (*m*) : cooling drink ; reducing (cooling) medicine or draught 271
KÜHLTURM (*m*) : cooling tower 272
KÜHLUNG (*f*) : cooling, attemperation, refrigeration 273
KÜHLVORRICHTUNG (*f*) : cooling (apparatus) arrangement 274
KÜHLWAGEN (*m*) : cold-storage (refrigerator) wagon 275
KÜHLWASSER (*n*) : cooling (tempering) water ; lead water (Pharm.) 276
KÜHLWIRKUNG (*f*) : cooling effect 277
KÜHN : bold ; daring ; courageous ; audacious ; intrepid 278
KUHPOCKEN (*f.pl*) : cow-pox (the disease communicated on vaccination as an inoculation against small-pox) 279
KUHPOCKENGIFT (*n*) : vaccine virus 280
KUHPOCKENIMPFUNG (*f*) : vaccination (see **KUHPOCKEN**) 281
KUHPOCKENSTOFF (*m*) : vaccine-(matter) 282
KUHZUNGE (*f*) : sorrel (*Rumex acetosella*) 283
KUKURUÖL (*n*) : (see **MAISÖL**) 284
KUKURUZ (*m*) : corn ; maize 285
KÜLBCHEN (*n*) : ball ; piece ; lump (Glass) 286
KULILAWANÖL (*n*) : culilawan oil 287
KULINARISCH : culinary 288
KULISSE (*f*) : (see **COULISSE**) 289

KULISSENFÜHRUNG (*f*): slot hole 290
KULISSENSTEIN (*m*): link-block 291
KULISSENSTEUERUNG (*f*): link motion 292
KULMINATION (*f*): culmination 293
KULMINIEREN: to culminate, reach the (zenith) highest point 294
KULÖR (*f*): (see ZUCKERCOULEUR) 295
KULT (*m*): cult, worship, ritual, ceremonial 296
KULTIVATOR (*m*): cultivator (Agricultural) 297
KULTUR (*f*): culture; civilization; cultivation 298
KULTURAPPARAT (*m*): bacteriological-(culture)-apparatus 299
KULTURGERÄTE (*n.pl*): apparatus (appliances, implements) for purposes of cultivation; agricultural (apparatus, appliances) implements 300
KULTURHEFE (*f*): culture yeast 301
KULTURLAND (*n*): civilized country 302
KULTURSTAAT (*m*): civilized (nation) state 303
KULTURSTUFE (*f*): civilized stage 304
KULTURVERSUCH (*m*): cultivation (test) experiment 305
KUMARIN (*n*): (CUMARIN): cumarin, coumarin, cumaric anhydride, Tonka bean camphor (from Tonka bean; see TONKABOHNE); C_6H_4.O.CH.CH.CO; Mp. 68°C.; Bp. 205-290°C. 306
KUMARON (*n*): (see CUMARON) 307
KUMARSÄUREANHYDRID (*n*): cumaric anhydride (see KUMARIN) 308
KUMIDIN (*n*): cumidine (see CUMIDIN) 309
KUMINSÄURE (*f*): cumic (cuminic) acid 310
KUMME (*f*): basin, vessel, bowl, trough 311
KÜMMEL (*m*): caraway (*Carum carui*); Kümmel (Liqueur); Coriander (*Coriandrum sativum*) 312
KÜMMELÖL (*n*): caraway-(seed) oil (*Oleum carui*); Sg. 0.91; Bp. 175-230°C.; cumin oil (see CUMINÖL) 313
KÜMMEL (*m*), **RÖMISCHER-**: cummin; cumin (*Cuminum cyminum*) 314
KÜMMELSPREUÖL (*n*): (see KÜMMELÖL) 315
KÜMMELTRAUBE (*f*): muscatel (muscadine) grape 316
KUMMER (*m*): sorrow; trouble; care; grief; sadness; affliction; mourning; vexation 317
KÜMMERLICH: needy; scanty (see also KUMMER-VOLL) 318
KÜMMERN: to trouble; concern; grieve; care 319
KUMMERVOLL: sorrowful, grieved, afflicted, depressed, dejected, mourning, sad, mournful, distressing, melancholy, disconsolate, doleful, rueful, lugubrious, piteous, sorry, dismal 320
KUMOL (*n*): cumene (see CUMOL) 321
KUMPAN (*m*): fellow, companion, comrade 322
KUMPE (*f*): basin; vessel; bowl; trough (Dye) 323
KUMPF (*m*): basin, vessel, bowl, trough (Dye) 324
KUMT (*n*): horse-collar 325
KUMTMACHER (*m*): harness-maker, saddler 326
KUMULATIV: cumulative 327
KUMULIEREND: cumulating, cumulative 328
KUMYS (*m*): kumis-(s) (a beverage from fermented mare's milk) 329
KUND: known; public 330
KUNDBAR: notorious 331
KUNDE (*f*): knowledge; information; news; science; notice; (*m*): client; customer 332
KÜNDEN: to publish 333
KUND GEBEN: to make known; announce; make public 334
KUNDGEBUNG (*f*): publication, demonstration 335
KUNDIG: learned; expert; skilful; intelligent; versed; aware; experienced; familiar 336
KÜNDIGEN: to give notice, give warning, hand in one's resignation, publish, recall 337
KÜNDIGUNG (*f*): notice; warning 338
KÜNDIGUNGSFRIST (*f*): length (amount) of notice 339
KUNDMACHUNG (*f*): declaration, proclamation, publication 340
KUNDSCHAFT (*f*): custom; patronage; notice; intelligence; information; knowledge; goodwill; clientele; debtors; practice; clients; customers 341
KUNDSCHAFTEN: to find out, reconnoitre 342
KÜNFTIG: future; coming; to be; next; to come 343
KÜNFTIGHIN: in (for) the future; hereafter; henceforth; from now onwards 344
KUNST (*f*): art; address; profession; skill; dexterity; work-(of art); trick; (as a prefix= artificial; technical; art) 345
KUNSTAKADEMIE (*f*): Academy of (Fine)-Art 346
KUNSTANSTALT (*f*), **GRAPHISCHE-**: fine art printers (which includes the printing of technical drawings, graphs and photographs, as well as the usual fine-art printing) 347
KUNSTARBEIT (*f*): work of art; art-(work) 348
KUNSTASCHE (*f*) artificial ash 349
KUNSTASPHALT (*m*): pitch (for roofing boards and for tarring purposes) 350
KUNSTAUSDRUCK (*m*): technical (expression) term 351
KUNSTAUSSTELLUNG (*f*): art exhibition 352
KUNSTBRONZE (*f*): art bronze (for statues) (80-86% Cu; 18-10% Zn; 2-4% Sn) 353
KUNSTBUTTER (*f*): (oleo)-margarine; artificial butter 354
KUNSTDRUCK (*m*): (fine)-art print-(ing) 355
KUNSTDRUCKPAPIER (*n*): art paper 356
KUNSTDÜNGER (*m*): fertilizer; artificial manure 357
KUNSTEIS (*n*): artificial ice 358
KÜNSTELN: to elaborate, be artful, be artificial, produce in an artistic manner 359
KUNSTFADEN (*m*): artificial fibre (for art, silk and paper) 360
KUNSTFERTIG: experienced, skilled, technical 361
KUNSTFERTIGKEIT (*f*): technical skill; address; experience; ability; capability; dexterity; craft-(smanship); knack 362
KUNSTFETT (*n*): artificial (fat) grease 363
KUNSTFEUERWERKEI (*f*): pyrotechnics 364
KUNSTFLEISS (*m*): industry 365
KUNSTGÄRTNER (*m*): horticulturalist, florist, nurseryman 366
KUNSTGÄRTNEREI (*f*): nursery-(gardening); horticulture 367
KUNSTGEMÄSS: in accordance with the rules of art, technically correct 368
KUNSTGESCHICHTE (*f*): history of art 369
KUNSTGEWEBE (*n*): artificial web 370
KUNSTGEWERBE (*n*): useful art 371
KUNSTGEWERBESCHULE (*f*): polytechnic-(school) 372
KUNSTGLANZ (*m*): artificial lustre 373
KUNSTGRIFF (*m*): artifice; device; trick; dexterity; craft; knack 374

KUNSTHÄNDLER (*m*): art dealer 375
KUNSTHARZ (*n*): artficial resin (see **CUMARON-HARZ** and **BAKELIT**) 376
KUNSTHEFE (*f*): artificial yeast 377
KUNSTHEILMITTEL (*n*): artificial (synthetic) remedy 378
KUNSTHOLZ (*n*): artificial wood (see **HOLZMASSE**) 379
KUNSTHONIG (*m*): artificial honey 380
KUNSTHORN (*n*): artificial horn, horn substitute 381
KUNSTLEDER (*n*): artificial leather; leather substitute 382
KUNSTLEHRE (*f*): technology; technics 383
KÜNSTLER (*m*): artist; artificer; virtuoso 384
KÜNSTLERDRUCK (*m*): artist's proof 385
KÜNSTLERISCH: artistic 386
KÜNSTLERLITHOGRAPHIE (*f*): artists' lithograph (see **ORIGINALLITHOGRAPHIE**) 387
KÜNSTLICH: artificial; false; artful; ingenious; mechanical; synthetic (see also **KUNST-**) 388
KÜNSTLICHER ZUG (*m*): mechanical draught 389
KÜNSTLICHE SEIDE (*f*): artificial silk; art. silk 390
KUNSTMALEREIFARBE (*f*): artist's (colour) paint 391
KUNSTMARMOR (*m*): artificial marble 392
KUNSTMITTEL (*n*): artificial (means) remedy 393
KUNSTPAPIER (*n*): art paper 394
KUNSTPRODUKT (*n*): artificial (synthetic) product 395
KUNSTREICH: artistic, ingenious 396
KUNSTRICHTER (*m*): art critic 397
KUNSTSACHE (*f*): matter of art, work of art 398
KUNSTSAMMLUNG (*f*): art collection 399
KUNSTSCHREINER (*m*): cabinet maker 400
KUNSTSCHULE (*f*): School of Art; Art Academy 401
KUNSTSEIDE (*f*): artificial (art) silk 402
KUNSTSPRACHE (*f*): technical language; technical terminology 403
KUNSTSTECHER (*m*): art engraver 404
KUNSTSTEIN (*m*): artificial stone (see also **KALKSANDSTEIN**) 405
KUNSTSTICKEREI (*f*): art needlework 406
KUNSTSTÜCK (*n*): feat; trick; work of art 407
KUNSTTISCHLER (*m*): cabinet maker 408
KUNSTVERSTÄNDIGER (*m*): connoisseur, expert 409
KUNSTVOLL: (see **KUNSTREICH**) 410
KUNSTWERK (*n*): work of art 411
KUNSTWIDRIG: contrary to the rules of art 412
KUNSTWOLLE (*f*): shoddy 413
KUNTERBUNT: parti-coloured; topsy-turvey 414
KÜPE (*f*): vat; boiler; copper; tub; kier 415
KÜPE (*f*), **DEUTSCHE-**: German vat, soda vat (see **KÜPE, SODA-**) 416
KÜPE (*f*), **GÄRUNGS-**: fermentation vat (see **KÜPE, WARME-**) 417
KÜPE (*f*), **HYDROSULFIT-**: hyposulphite vat (the reducing agent being sodium hydrosulphite, calcium hydrosulphite, or zinc hydrosulphite, amongst others) 418
KÜPE (*f*), **INDIGO-**: indigo vat 419
KÜPE (*f*), **KALTE-**: cold-(indigo)-vat (used for cotton and silk) (see **KÜPE, VITRIOL-**) 420
KUPELLIEREN: to cupel 421
KÜPEN: to vat-(dye) 422
KÜPENARTIKEL (*m*): vat style (Calico) 423
KÜPENBAD (*n*): dye liquor 424
KÜPENFÄRBER (*m*): vat dyer 425
KÜPENFÄRBEREI (*f*): vat dyeing 426
KÜPENFARBSTOFF (*m*): vat dye 427
KÜPENFÄRBUNG (*f*): vat dyeing 428
KÜPENGERUCH (*m*): vat (odour) smell 429
KÜPENINDIGO (*m*): vat indigo, vat dye 430
KÜPE (*f*), **POTTASCHE-**: potash vat (consisting of indigo, madder and potash) 431
KÜPE (*f*), **SODA-**: German vat, soda vat (consisting of indigo, soda, and slaked lime) 432
KÜPE (*f*), **VITRIOL-**: copperas vat, vitriol vat (consisting of indigo, ferrous sulphate, and slaked lime) 433
KÜPE (*f*), **WAID-**: woad vat (consisting of indigo, woad, madder and slaked lime) 434
KÜPE (*f*), **WARME-**: warm-(indigo)-vat, fermentation vat (at a temperature of 50-60°C., for wool) (see **KÜPE, WAID-**; **KÜPE, SODA-**; and **KÜPE, POTTASCHE-**) 435
KUPFER (*n*): copper (Cu) 436
KUPFERABFÄLLE (*m.pl*): copper waste 437
KUPFERACETAT (*n*): (*Cuprum aceticum*): copper acetate, cupric acetate, crystallized verdigris (see **KUPFER, ESSIGSAURES-**); $Cu(C_2H_3O_2)_2.H_2O$; Sg. 1.9 438
KUPFERACETATARSENIAT (*n*): copper (cupric) acetoarsenite, Imperial green, Paris green (*Cuprum acetico-arsenicicum*); $Cu(C_2H_3O_2)_2.3CuAs_2O_4$ (see **KUPFERARSENITAZETAT**) 439
KUPFERACETAT (*n*), **BASISCHES-**: blue verdigris, green verdigris, basic copper acetate, verdigris, copper subacetate; $CuO\text{-}Cu(C_2H_3O_2)_2.6H_2O$ 440
KUPFERALAUN (*m*): copper aluminate 441
KUPFERAMMONIUM (*n*): cupr-(o)-ammonium; Schweitzer's reagent 442
KUPFERAMMONIUMCHLORID (*n*): copper-ammonium chloride, cupr-(o)-ammonium chloride 443
KUPFERAMMONIUMCHROMATLÖSUNG (*f*): cupr-(o)-ammonium chromate solution (referred to in the dyeing industry as copper chromate) 444
KUPFERAMMONIUMSULFAT (*n*): copper-ammonium sulphate; cuprammonium sulphate; cupric-ammonium sulphate,; copper aminosulphate; amino-cupric sulphate; $CuSO_4.4NH_3.H_2O$ 445
KUPFERANODE (*f*): copper anode 446
KUPFERANTIMONGLANZ (*m*): chalcostibite (see **WOLFSBERGIT**) 447
KUPFERAPPARAT (*m*): copper apparatus 448
KUPFER, ARSENIGSAURES- (*n*): copper arsenite; cupric (copper ortho)-arsenite (*Cuprum arsenicosum*); Scheele's green; $CuHAsO_3$ 449
KUPFERARSENITAZETAT (*n*): Schweinfurt green (double compound of copper arsenite and copper arsenate); $Cu(C_2H_3O_2)_2.3Cu(AsO_2)_2$ (see **KUPFERACETATARSENIAT**) 450
KUPFER (*n*), **ARSENSAURES-**: copper arsenate 451
KUPFERASCHE (*f*): copper (scale) ash 452
KUPFERAUSSCHLAG (*m*): red rash (Med.) 453
KUPFERAUTOTYPIE (*f*): copper autotype process 454
KUPFERBAD (*n*): copper bath 455
KUPFERBEDARF (*m*): demand for copper 456
KUPFERBEIZE (*f*): copper mordant (Dyeing) (such as cupric chloride, cupric acetate, cupric sulphate, cupric nitrate, etc.) (see also **NACHKUPFERN**) 457
KUPFERBERGBAU (*m*): copper mining 458
KUPFERBESCHLAG (*m*): copper sheeting 459

KUPFERBLATT (*n*): copper-plate print; sheet copper 460
KUPFERBLAU (*n*): chrysocolla (see **KIESEL-KUPFER**); copper blue (basic copper carbonate, $2CuCO_3.Cu(OH)_2$ (see **BERGBLAU**); blue (verditer) verdigris (see **KUPFERCARBONAT BASISCHES-**); Azurite (see **AZURIT**) 461
KUPFERBLAUSÄURE (*f*): cuprous prussic acid 462
KUPFERBLECH (*n*): sheet copper; copper plate; copper foil 463
KUPFERBLEIVITRIOL (*m*): linarite (see **LINARIT**) 464
KUPFERBLÜTE (*f*): capillary red copper ore, capillary cuprite (see **ROTKUPFERERZ**); copper bloom 465
KUPFERBOLZEN (*m*): copper bolt 466
KUPFER, BORSAURES- (*n*): copper (cupric) borate; $CuBO_4$ 467
KUPFERBRANDERZ (*n*): cuprous coal 468
KUPFERBRAUN (*n*): copper brown (an azo dyestuff); tile ore; earthy ferruginous cuprite (see **CUPRIT** and **ZIEGELERZ**) 469
KUPFERBROMID (*n*): copper (cupric) bromide 470
KUPFERBROMÜR (*n*): cuprous bromide; CuBr; Sg. 4.72 471
KUPFERCARBONAT (*n*): basic copper carbonate, mineral blue, copper carbonate, malachite, cupric carbonate, Bremen green, Brunswick green, Verditer green; $CuCO_3+Cu(OH)_2$ or $Cu_2(OH)_2CO_3$; Sg. 3.85 (*Cuprum carbonicum*) 472
KUPFERCARBONAT (*n*), **BASISCHES-**: basic copper carbonate (green coating formed on copper exposed to moist air) (see **KUPFERCARBONAT** and **BERGBLAU**) 473
KUPFERCARBONAT (*n*), **NEUTRALES-**: neutral copper carbonate; $CuCO_3$ (is not known) 474
KUPFERCHLORAT (*n*): copper chlorate; $Cu(ClO_3)_2.6H_2O$; Mp. 65°C 475
KUPFERCHLORID (*n*): copper (cupric) chloride (*Cuprum bichloratum*) ($CuCl_2.2H_2O$; Sg. 2.5) or ($CuCl_2$; Sg. 3.06; Mp. 498°C.) 476
KUPFERCHLORÜR (*n*): cuprous chloride (*Cuprum chloratum*); CuCl; Sg. 3.53 477
KUPFERCHROMAT (*n*): copper (cupric) chromate (*Cuprum chromatum*); $CuCrO_4$ 478
KUPFER, CHROMSAURES- (*n*): copper chromate; basic cupric chromate; $CuCrO_4.2CuO.2H_2O$ (see also **KUPFERAMMONIUMCHROMATLÖ-SUNG**) 479
KUPFER (*n*), **CITRONENSAURES-**: copper citrate 480
KUPFERCYANID (*n*): copper (cupric) cyanide; $Cu(CN)_2$ 481
KUPFERCYANÜR (*n*): cuprous cyanide 482
KUPFERDORN (*m*): liquated copper slag 483
KUPFERDRAHT (*m*): copper wire 484
KUPFERDRAHTSPULE (*f*): copper wire-(induction)-coil 485
KUPFERDRUCK (*m*): copperplate; copperplate (print) printing 486
KUPFERDRUCKEREI (*f*): copper-plate printing; copperplate printing-works 487
KUPFERDRUCKERPRESSE (*f*): rolling press 488
KUPFERDRUCKERSCHWÄRZE (*f*): Frankfurt black (see **REBENSCHWARZ**) 489
KUPFERDRUCKPAPIER (*n*): plate paper 490
KUPFERELEKTRODE (*f*): copper electrode; negative electrode (of voltaic cell) 491
KUPFER (*n*), **ELEKTROLYTISCHES-**: electrolytic copper (the purest form of copper, refined by electrolysis) 492
KUPFERERZ (*n*): copper ore 493
KUPFERERZEUGUNG (*f*): production of copper, copper smelting 494
KUPFERERZGEWINNUNG (*f*): copper mining, mining of copper ore 495
KUPFERERZLAGERSTÄTTE (*f*): bed of copper ore 496
KUPFERERZ (*n*), **SCHWARZ-**: melaconite (see **MELAKONIT**) 497
KUPFER (*n*), **ESSIGSAURES-**: copper acetate, cupric acetate; green verdigris (see **KUPFER-ACETAT**); $Cu(C_2H_3O_2)_2.H_2O$; Sg. 1.9 498
KUPFERFAHLERZ (*n*): tetrahedrite; tennantite (see **TENNANTIT**) 499
KUPFERFARBE (*f*): copper colour 500
KUPFERFEDERERZ (*n*): copper bloom; capillary cuprite (see **KUPFERBLÜTE**) 501
KUPFERFEILICHT (*n*): copper filings 502
KUPFERFEILSPÄNE (*m.pl*): copper filings 503
KUPFER (*n*), **FETTSAURES-**: copper sebate 504
KUPFERFOLIE (*f*): copper foil 505
KUPFERFRISCHEN (*n*): copper (fining) refining 506
KUPFERGARMACHEN (*n*): copper refining 507
KUPFER (*n*), **GEDIEGENES-**: native copper 508
KUPFERGEHALT (*m*): copper (content) alloy 509
KUPFERGEIST (*m*): spirit of verdigris (see **ESSIGSÄURE**) 510
KUPFERGEWINNUNG (*f*): extraction (production) of copper (from the ore, etc.); copper mining 511
KUPFERGEWINNUNG (*f*), **ELEKTROLYTISCHE-**: electrolytic extraction of copper 512
KUPFERGEWINNUNG (*f*), **NASSE-**: wet process copper extraction (generally employed for poor ores) 513
KUPFERGEWINNUNG (*f*), **TROCKNE-**: dry process copper extraction (generally employed for rich ores) 514
KUPFERGLANZ (*m*): copper glance; chalcocite; redruthite (natural copper sulphide; Cu_2S; about 79.85% copper; Sg. 5.75) (see also **KUPFERSULFÜR**) 515
KUPFERGLAS (*n*): chalcocite (see **KUPFERGLANZ** and **KUPFERSULFÜR**) 516
KUPFERGLIMMER (*m*): chalcophyllite; micaceous copper; copper mica 517
KUPFERGLÜHSPAN (*m*): copper scale 518
KUPFERGOLD (*n*): Mannheim gold (a sort of brass) 519
KUPFERGRÜN: rust-like, eruginous; (*n*): copper green, verditer, subacetate of copper, verdigris (see **KUPFERACETAT, BASCHISCHES-**); chrysocolla (see **CHRYSOKOLL**) 520
KUPFERGUSS (*m*): copper casting 521
KUPFERHALTIG: cupr-(e)-ous; cupriferous; containing copper 522
KUPFERHAMMERSCHLAG (*m*): copper scale 523
KUPFER, HARZSAURES- (*n*): copper (cupric) resinate; $Cu(C_{20}H_{29}O_2)_2$ 524
KUPFERHAUT (*f*): copper sheath-(ing), copper casing 525
KUPFERHORNERZ (*n*): copper muriate 526
KUPFERHÜTTE (*f*): copper (smeltery) smelting works 527
KUPFERHYDROXYD (*n*): copper (cupric) hydroxide; copper hydrate (*Cuprium hydroxydatum*); $Cu(OH)_2$; Sg. 3.37 528
KUPFERHYDROXYDUL (*n*): cuprous hydroxide 529
KUPFERIG: coppery, like-copper 530
KUPFERIGES GESICHT (*n*): Acne Rosacea, facial inflammation (Med.) 531

KUPFERINDIG (*m*): indigo copper; covellite (natural indigo-blue copper sulphide; about 66% Cu); CuS (see also **COVELLIN**) 532
KUPFERION (*n*): copper ion 533
KUPFERJODID (*n*): copper (cupric) iodide 534
KUPFERJODÜR (*n*): cuprous iodide; CuI; Sg. 5.289-5.653 535
KUPFERKALK (*m*): copper oxide 536
KUPFERKESSEL (*m*): copper (kettle) boiler 537
KUPFERKIES (*m*): copper pyrites, chalcopyrite, yellow copper ore (natural copper-iron sulphide); $CuFeS_2$ (about 35% copper) or $Cu_2S.Fe_2S_3$; Sg. 4.2 538
KUPFERKIES, BUNTER- (*m*): bornite (see **BORNIT**) 539
KUPFERKIESE, NICKELHALTIGE- (*m.pl*): nickeliferous copper pyrites (1-6% nickel, from Sudbury, Canada) 540
KUPFERKIESHALTIG: containing copper pyrites 541
KUPFERKLEBEMEHL (*n*): powder for protecting the grape vine against the ravages of the fungus *Peronospora viticola* 542
KUPFERKNALLSÄURE (*f*): copper fulminate, cuprofulminic acid 543
KUPFER, KOHLENSAURES- (*n*): copper carbonate (see **KUPFERCARBONAT**) 544
KUPFERKÖNIG (*m*): copper regulus 545
KUPFERKONVERTER (*m*): copper converter 546
KUPFERLAZUR (*f*): copper blue, mineral blue; azurite (about 69% copper); chessylite; $3Cu_3C_2O_7.7H_2O$; Sg. 3.8 (see also **AZURIT** and **BERGBLAU**) 547
KUPFERLEGIERUNG (*f*): copper alloy-(ing) 548
KUPFER, LEINÖLSAURES- (*n*): copper linoleate 549
KUPFERLÖSUNG (*f*): copper solution 550
KUPFERMANGAN (*n*): cupromanganese 551
KUPFERMETABROMAT (*n*): copper metabromate; $Cu(BO_2)_2$; Sg. 3.859 552
KUPFERMETALL (*n*): cuprite 553
KUPFER (*n*), **METALLISCHES-**: metallic copper 554
KUPFERMINERAL (*n*): copper mineral 555
KUPFERMÜNZE (*f*): copper coin 556
KUPFERN: to (coat with; treat with) copper 557
KUPFERNAGEL (*m*): copper nail 558
KUPFERNICKEL (*m*): copper nickel; niccolite (see **NICKELIN** and **NICCOLIT**) 559
KUPFERNIEDERSCHLAG (*m*): copper precipitate 560
KUPFERNIET (*n*): copper rivet 561
KUPFERNITRAT (*n*): (*Cuprum nitricum*): copper nitrate, cupric nitrate; $Cu(NO_3)_2+3H_2O$; Sg. 2.17; Mp. 114.5°C. 562
KUPFEROCHER (*m*): (see **KUPFEROCKER**) 563
KUPFEROCKER (*m*): tile ore; earthy ferruginous cuprite (see **CUPRIT** and **ZIEGELERZ**) 564
KUPFEROXYD (*n*): (*Cuprum oxydatum*): copper monoxide, copper (cupric) oxide, black copper oxide; CuO; Sg. 6.4; Mp. 1064°C. (see also **CUPRI-** and **KUPRI-**) 565
KUPFEROXYD (*n*), **CHROMSAURES-**: cupric chromate (see **KUPFERCHROMAT**) 566
KUPFEROXYDHYDRAT (*n*): cupric hydroxide (see **KUPFERHYDROXYD**) 567
KUPFEROXYD (*n*), **KOHLENSAURES-**: cupric carbonate (see **KUPFERCARBONAT**) 568
KUPFEROXYDSALZ (*n*): cupric salt 569
KUPFEROXYD (*n*), **SCHWEFELSAURES-**: (see **KUPFERSULFAT**) 570
KUPFEROXYDUL (*n*): red copper oxide, copper (cuprous) oxide, (*Cuprum oxydulatum*); Cu_2O; Sg. 5.92; Mp. 1210°C.; Bp. 1800°C. (see also **CUPRI-** and **KUPRI-**) 571
KUPFEROXYDULOXYDSULFIT (*n*): cupro-cupric sulphite 572
KUPFEROXYDULSALZ (*n*): cuprous salt 573
KUPFEROXYDULSULFIT (*n*): cuprous sulphite 574
KUPFEROXYDULVERBINDUNG (*f*): cuprous compound 575
KUPFEROXYDVERBINDUNG (*f*): cupric compound 576
KUPFER (*n*), **PALMITINSAURES-**: copper palmitate 577
KUPFERPASTE (*f*): copper paste (containing about 8% Ca., and 16% Cu; used as a grape vine fungus remedy) 578
KUPFERPECHERZ (*n*): tile ore (see **KUPFEROCKER**) 579
KUPFERPHOSPHID (*n*): copper phosphide, cuprous phosphide; Cu_3P_2; Sg. 6.67; Cu_2P; Sg. 6.4 580
KUPFERPLATTE (*f*): copper plate; copper sheeting; sheet copper 581
KUPFERPOL (*m*): copper (negative) pole (Elect.) 582
KUPFERPRESSE (*f*): copper-(rolling)-press 583
KUPFERPROBE (*f*): copper assay 584
KUPFERPRODUKTION (*f*): production (yield) of copper 585
KUPFERRADIERUNG (*f*): copper etching 586
KUPFERRAUCH (*m*): copper (efflorescence; smoke) fumes; green vitriol; copperas (see **EISENOXYDULSULFAT**) 587
KUPFERREGEN (*m*): copper rain (particles thrown from surface of molten metal) 588
KUPFERREICH: rich in copper, cupriferous, having a high copper content 589
KUPFERREICHER EISENKIES (*m*): cupriferous iron pyrites 590
KUPFERRHODANAT (*n*): copper (cupric) thiocyanate or sulphocyanate 591
KUPFERRHODANID (*n*): (see **KUPFERRHODANAT**) 592
KUPFERRHODANÜR (*n*): cuprous (thiocyanate) sulphocyanide (*Cuprum rhodanatum*); $Cu_2(CNS)_2$ 593
KUPFERROHR (*n*): copper (tube) pipe 594
KUPFERRÖHRE (*f*): (see **KUPFERROHR**) 595
KUPFERROHSTEIN (*m*): raw copper matt (from roasting of copper) 596
KUPFERROST (*m*): copper rust; verdigris (see **GRÜNSPAN**) 597
KUPFERROT: coppery, copper coloured; (*n*): red (virgin) copper, cuprite (see **ROTKUPFERERZ**) 598
KUPFERRÖTE (*f*): copper colour; (see also **KUPFERROT** (*n*)) 599
KUPFERSACCHARAT (*n*): copper saccharate; $C_{12}H_{22}O_{11}.CuO$ 600
KUPFER, SALPETERSAURES- (*n*): copper (cupric) nitrate; $Cu(NO_3)_2.3H_2O$; Sg. 2.17; Mp. 114.5°C. 601
KUPFERSALZ (*n*): copper salt 602
KUPFER (*n*), **SALZSAURES-**: copper (cupric) chloride (see **KUPFERCHLORID**) 603
KUPFERSAMMETERZ (*n*): velvet copper ore; cyanotrichite 604
KUPFERSÄURE (*f*): copper sesquioxide 605
KUPFERSCHAUM (*m*): copper scum (Metal); tyrolite (Mineral) 606
KUPFERSCHEIBE (*f*): copper disc 607
KUPFERSCHIEFER (*m*): cupriferous slate, copper slate (obtained from the district around Mansfeld) (a black shale containing copper ore) 608

KUPFERSCHLAG (*m*): dross-(from copper) 609
KUPFERSCHLANGE (*f*): copper coil 610
KUPFERSCHMIED (*m*): brazier, copper smith 611
KUPFER (*n*), **SCHWARZ-**: black (coarse) copper (see **SCHWARZKUPFER**) 612
KUPFERSCHWARZ (*n*): copper-black 613
KUPFERSCHWÄRZE (*f*): black copper- (oxide); melaconite (see **MELAKONIT**); tenorite (see **TENORIT**) 614
KUPFERSCHWEFELKALK (*m*): cuprous calcium sulphate 615
KUPFER (*n*), **SCHWEFELSAURES-**: copper sulphate (see **KUPFERVITRIOL**) 616
KUPFER (*n*), **SCHWEFLIGSAURES-**: copper sulphite 617
KUPFERSILBER (*n*): copper-silver alloy 618
KUPFERSILBERGLANZ (*m*): stromeyerite (see **STROMEYERIT**) 619
KUPFERSILICID (*n*): copper silicide, silicon-copper; Cu_2Si; Sg. 6.9; Cu_4Si; Sg. 7.53 620
KUPFERSILICOFLUORID (*n*): copper silicofluoride, copper fluosilicate, cupric (silicofluoride) fluosilicate; $CuF_2SiF_4.6H_2O$; Sg. 2.182 [621
KUPFERSINTER (*m*): copper scale; copper pyrites (see **KUPFERKIES**) 622
KUPFERSMARAGD (*m*): emerald copper; dioptase (see **DIOPTAS**) 623
KUPFERSPÄNE (*m.pl*): (see **KUPFERFEILSPÄNE**) 624
KUPFERSPIRITUS (*m*): spirit of verdigris (see **ESSIGSÄURE**) 665
KUPFERSTAHL (*m*): copper steel 626
KUPFERSTANNAT (*n*): copper stannate, tin green (see **GENTELES GRÜN**) 627
KUPFER (*n*), **STEARINSAURES-**: copper (cupric) stearate; $Cu(C_{18}H_{35}O_2)_2$ 628
KUPFERSTECHEN (*n*): copper engraving; engraving on (of) copper 629
KUPFERSTECHER (*m*): copper-(plate)-engraver 630
KUPFERSTEIN (*m*): copper matt (from roasting of copper) 631
KUPFERSTEIN, BLASIGER- (*m*): pimple metal 632
KUPFERSTICH (*m*): copper engraving; copper-plate print 633
KUPFERSTICHPLATTE (*f*): plate for copper (etching) engraving 634
KUPFERSULFAT (*n*): copper (cupric) sulphate; blue stone (*Cuprum sulfuricum*); $CuSO_4.5H_2O$; Sg. 2.272 ,(see **KUPFER-VITRIOL**) 635
KUPFERSULFAT (*n*), **BASISCHES-**: basic copper sulphate, Casselmann's green, $CuSO_4.Cu(OH)_2+4H_2O$ 636
KUPFERSULFATLÖSUNG (*f*): copper sulphate solution 637
KUPFERSULFID (*n*): copper (cupric) sulphide; CuS; (*Cuprum sulfuratum*); Sg. 3.98; Mp. 1100°C. 638
KUPFER (*n*), **SULFOCARBOLSAURES-**: copper sulfophenate 639
KUPFERSULFÜR (*n*): cuprous sulphide 640
KUPFERTIEGEL (*m*): copper crucible 641
KUPFERUNG (*f*): treatment with copper; copper (sheeting) layer 642
KUPFERURANIT (*m*): chalcolite (see **CHALKOLITH**) 643
KUPFERVERBINDUNG (*f*): copper compound 644
KUPFERVERGIFTUNG (*f*): copper poisoning 645
KUPFERVERKLEIDUNG (*f*): copper (casing, facing, sheeting) lining 646
KUPFERVITRIOL (*m*): blue vitriol; chalcanthite; copper sulphate (see **KUPFERSULFAT**) 647
KUPFERVOLTAMETER (*n*): copper voltameter 648
KUPFERWALZWERK (*n*): copper rolling-mill 649
KUPFERWAREN (*f.pl*): copper (wares, goods) articles 650
KUPFERWASSER (*n*): copperas (see **EISENOXYDUL-SULFAT**); copper water; cement water (Mining) 651
KUPFERWASSERSTOFF (*m*): copper hydride 652
KUPFER (*n*), **WEINSAURES-**: copper tartrate 653
KUPFERWERK (*n*): copper work-(s) 654
KUPFERWISMUTGLANZ (*m*): Emplektite (see **EMPLEKTIT**) 655
KUPFERWURZEL (*f*): Bog Asphodel, Lancashire asphodel (*Narthecium ossifragum*) 656
KUPFERZEMENT (*m*): copper precipitate 657
KUPFERZIEGELERZ (*n*): tile ore (see **CUPRIT** and **ZIEGELERZ**) 658
KUPFERZYANID (*n*): (see **KUPFERCYANID**) 659
KUPFERZYANÜR (*n*): (see **KUPFERCYANÜR**) 660
KUPFRIG: coppery; copperlike 661
KUPHONGLIMMER (*m*), **RHOMBOEDRISCHER-**: Brucite 662
KUPHONSPAT (*m*), **DIPLOGENER-**: Epistilbite (see **EPISTILBIT**) 663
KUPHONSPAT (*m*), **HARMOPHANER-**: (see **SKOLEZIT**) 664
KUPHONSPAT (*m*), **HEMIPRISMATISCHER-**: Heulandite 665
KUPHONSPAT (*m*), **HETEROMORPHER-**: Hydrolite, Gmelinite; Sg. 2.1 666
KUPHONSPAT (*m*), **HEXAEDRISCHER-**: Analcime; $NaO.SiO_2+AlO_3Si_3O_2+H_2O$; Sg. 2.25 [667
KUPHONSPAT (*m*), **MAKROTYPER-**: (see **LEVYN**) 668
KUPHONSPAT (*m*), **MEGALOGONER-**: (see **BREW-STERIT**) 669
KUPHONSPAT (*m*), **PRISMATISCHER-**: mesotype, soda mesotype; $NaO.SiO_2.AlO_3.SiO+H_2O$; Sg. 2.25 670
KUPHONSPAT (*m*), **PRISMATOIDISCHER-**: Stilbite 671
KUPHONSPAT (*m*), **RHOMBOEDRISCHER-**: Chabacite, chabasie; $CaO.SiO_2+AlO_3Si_3O_2$; Sg. 2.1 672
KUPOLOFEN (*m*): cupola (furnace or kiln) (for melting pig metal) 673
KUPOLSTEINE (*m.pl*): cupola bricks 674
KUPON (*n*): coupon 675
KUPPE (*f*): top; tip; summit; vaulted arch 676
KUPPEL (*f*): cupola; dome; cover; top; cap; spire; (lamp)-shade; peak (see also **KOPPEL**) 677
KUPPELDACH (*n*): vaulted (dome-shaped) roof 678
KUPPELGEWÖLBE (*n*): round (Roman) arch; dome; cupola 679
KUPPELHAKEN (*m*): draw hook, drag hook 680
KUPPELN: to couple; leash; tie (see **KOPPELN**); develop (Dye) 681
KUPPELOFEN (*m*): (see **KUPOLOFEN**) 682
KUPPELSTANGE (*f*): coupling (connecting) rod, drag bar, draw bar 683
KUPPELUNG (*f*): join, joining, coupling, connection joint; clutch (Motors) 684
KUPPELUNG (*f*), **ELASTISCHE-**: flexible coupling 685
KUPPELUNGSBAD (*n*): developing bath (Dyeing) 686
KUPPELUNGSFLANSCH (*m*): coupling (connecting) flange 687
KUPPELUNGSFLOTTE (*f*): developing liquor (Dyeing) 688
KUPPELUNGSFUTTER (*n*): clutch lining 689

KUPPELUNGSHAKEN (*m*): draw hook, coupling hook 690
KUPPELUNGSHEBEL (*m*): coupling (connecting) lever 691
KUPPELUNGSKETTE (*f*): coupling chain 692
KUPPELUNGSMUFFE (*f*): coupling box 693
KUPPLER (*m*): coupler; developer (Dye); match-maker 694
KUPPLUNG (*f*): coupling, clutch (see **KUPPELUNG**) 695
KUPPLUNG (*f*), **ELASTISCHE-**: flexible coupling 696
KUPPLUNG (*f*), **REIBUNGS-**: friction clutch 697
KUPRICHLORID (*n*): cupric chloride (see **CUPRI--CHLORID** and **KUPFERCHLORID**) 698
KUPRICHROMAT (*n*): cupric chromate (see **KUP--FERCHROMAT**) 699
KUPRIKARBONAT (*n*): cupric carbonate (see **KUPFERCARBONAT**) 700
KUPRISALZ (*n*): cupric salt 701
KUPRISALZLÖSUNG (*f*): cupric salt solution, solution of cupric salt 702
KUPRIT (*m*): cuprite (natural red copper oxide; about 89% Cu); Cu_2O (see also **ROTKUP--FERERZ**) 703
KUPRIVERBINDUNG (*f*): cupric compound 704
KUPROCHLORID (*n*): cuprous chloride (see **CUPRO--CHLORID** and **KUPFERCHLORÜR**) 705
KUPRO-KUPRISULFIT (*n*): cupro-cupric sulphite 706
KUPROMANGAN (*n*): cupro-manganese (see **MAN--GANKUPFER**) 707
KUPROSALZ (*n*): cuprous salt 708
KUPROSILICIUM (*n*): cupro-silicon, silicon copper 709
KUPROVERBINDUNG (*f*): cuprous compound 710
KUR (*f*): cure; (medical)-treatment; baths; waters; electorate (in Germany) 711
KURAND (*m*): (male)-ward, minor 712
KURARE (*n*): (*Curare*): Curare, Woorali, Urari (an Indian poison) 713
KURAREPRÄPARAT (*n*): curare preparation 714
KURARIN (*n*): curarine (the active principle of Curare) (see **KURARE**) 715
KURBEL (*f*): crank; winch 716
KURBELABSTAND (*m*): crank distance (from dead centre) 717
KURBELACHSE (*f*): crank-(ed) axle 718
KURBELARM (*m*): crank (lever; arm; web) body; lever arm 719
KURBELBILGE (*f*): crank (race) bilge 720
KURBELBLATT (*n*): crank (cheek) web 721
KURBELDRUCK (*m*): crank pressure; pressure on lever or handle 722
KURBELERHEBUNG (*f*): angle of crank to horizontal 723
KURBELFOLGE (*f*): succession (series) of cranks 724
KURBELGEHÄUSE (*n*): crank case (Motor cars), crank chamber 725
KURBELGETRIEBE (*n*): crank (gear) mechanism 726
KURBELGRIFF (*m*): crank handle; windlass (winch) handle 727
KURBELHALBMESSER (*m*): crank (lever) radius or throw 728
KURBELHEFT (*n*): hand grip 729
KURBELHUB (*m*): crank lifting projection 730
KURBELKAMMER (*f*): crank chamber 731
KURBELKÖRPER (*m*): crank (arm; web) body 732
KURBELKREIS (*m*): crank (lever) circle 733
KURBELKRÖPFUNG (*f*): cranked portion of shaft 734
KURBELKRÖPFUNGSWANGE (*f*): crank (cheek) web 735
KURBELLAGER (*n*): crank shaft (main) bearing; pillow block 736
KURBELLAGERBOCK (*m*): crank (main) bearing pedestal 737
KURBELLAGERSCHALE (*f*): bearing (brasses; steps) bush 738
KURBELN: to turn; work; wind 739
KURBELNABE (*f*): crank boss 740
KURBELRAUM (*m*): crank chamber 741
KURBELSCHALTER (*m*): lever (handle) switch 742
KURBELSCHEIBE (*f*): crank disc 743
KURBELSCHENKEL (*m*): crank (cheek) web 744
KURBELSCHLAG (*m*): knocking-(of the crank) 745
KURBELSCHLEIFE (*f*): crank (slot) guide 746
KURBELSPERRE (*f*): crank lock 747
KURBELSTANGE (*f*): connecting rod 748
KURBELSTELLER (*m*): crank lock 749
KURBELSTELLUNG (*f*): crank position 750
KURBELTOTLAGE (*f*): crank dead-centre 751
KURBELTRIEB (*m*): crank (gear) mechanism 752
KURBELVERSETZUNG (*f*): angle between cranks 753
KURBELWANGE (*f*): crank web 754
KURBELWARZE (*f*): crank pin 755
KURBELWEG (*m*): crank (crank-pin) path 756
KURBELWELLE (*f*): crankshaft 757
KURBELWELLE (*f*), **DOPPELT GEKRÖPFTE-**: two-throw crank-shaft 758
KURBELWELLE (*f*), **DREIFACH GEKRÖPFTE-**: three-throw crank-shaft 759
KURBELWELLENBUND (*m*): shoulder on crank shaft 760
KURBELWELLENLAGER (*n*): crank shaft (main) bearing; pillow block 761
KURBELWELLENSCHAFT (*m*): body of crankshaft 762
KURBELWERK (*n*): crank (mechanism) gear 763
KURBELWINDE (*f*): winch with crank or handle 764
KURBELZAPFEN (*m*): crank pin 765
KURBELZAPFENLAGER (*n*): crank pin (steps; brasses) bush 766
KÜRBIS (*m*): gourd; pumpkin (*Cucurbita pepo*) 767
KÜRBISBAUM (*m*): gourd tree; calabash tree (*Crescentia cujete*) 768
KÜRBISFÖRMIG: gourd-shaped 769
KÜRBISKERNÖL (*n*): pumpkin-(seed) oil (from *Cucurbita pepo*); Sg. 0.919-0.925 770
KÜRBISKORN (*n*): pumpkin seed; pepo (seed of *Cucurbita pepo*) 771
KÜRBISSAMEN (*m*): pumpkin seed; pepo (seed of *Cucurbita pepo*) 772
KÜREN: to choose; elect; select; pick 773
KURHAUS (*n*): pump room-(of a spa); sanatorium 774
KURIOS: curious, strange, singular, rare, unique, unusual, queer, weird, extraordinary 775
KURKUMA (*f*): turmeric; curcuma; curry; Indian saffron (roots of *Curcuma longa*) (see **CURCUMA** and **GELBWURZEL**) 776
KURKUMAGELB (*n*): curcumin (yellow colouring matter in Curcuma) (see **CURCUMIN**) 777
KURKUMAPAPIER (*n*): turmeric paper 778
KURKUMAWURZEL (*f*): turmeric-(root) (*Rhizoma curcumæ*) (see **KURKUMA**) 779
KURKUMIN (*n*): curcumin (see **KURKUMAGELB**) 780
KUROMOJIÖL (*n*): curomoji oil 781

KURORT (*m*): health resort; inland watering place; spa 782
KURRENT: current 783
KURRENTBUCHSTABEN (*m.pl*): italics 784
KURS (*m*): exchange; circulating; course (of a vessel, exchange, etc.) 785
KURSAAL (*m*): kursaal; pump-room-(of a spa) 786
KURSBUCH (*n*): railway guide 787
KÜRSCHNER (*m*): furrier 788
KÜRSCHNERGARE (*f*): skin-dressing 789
KÜRSCHNERWAREN (*f.pl*): skins, furs 790
KURSIEREN: to be current; circulate 791
KURSIV: italic 792
KURSIVSCHRIFT (*f*): italics 793
KURSUS (*m*): course-(of instruction) 794
KURSWERT (*m*): exchange-(value); rate of exchange 795
KURSZETTEL (*m*): stock and share list 796
KURVE (*f*): curve (line of a)-graph 797
KURVE (*f*), **ALGEBRAISCHE-**: algebraic curve 798
KURVE (*f*), **EBENE-**: plane curve 799
KURVE (*f*), **EXPANSIONS-**: cut-off curve, expansion curve 800
KURVE (*f*), **METAZENTRISCHE-**: metacentric curve 801
KURVENAST (*m*): branch of a curve 802
KURVENBILD (*n*): curve; graph 803
KURVENBLATT (*n*): diagram, graph 804
KURVENHEBEL (*m*): curved lever 805
KURVENKNICKUNG (*f*): break-(of a curve), sudden change of direction of a curve 806
KURVENLINEAL (*n*): curve (template) templet 807
KURVENSCHAR (*f*): group (system) of curves 808
KURVE (*f*), **TRANSZENDENTE-**: transcendental curve 809
KUR (*f*), **WASSER-**: hydropathy, water cure 810
KURZ: short; brief; concise; succinct; curt; laconic; compendious; fleeting; transient; temporary; terse; ephemeral; transitory; abrupt; pithy; concentrated; abridged 811
KURZATMIG: asthmatic, short-winded 812
KURZBRÜCHIG: short; brittle 813
KURZDAUERND: transitory, transient, short-lived, ephemeral, fleeting, temporary, brief 814
KÜRZE (*f*): shortness; brevity; briefness; terseness; pithiness 815
KURZE FLOTTE (*f*): concentrated dye liquor 816
KÜRZEN: to shorten, compress, abbreviate, curtail, be brief, etc. (see **KURZ**) 817
KURZFLOSSIG: micropterous 818
KURZFÜSSIG: breviped, short-footed 819
KURZGEBAUT: short built, of short construction 820
KURZGEFASST: short, brief, concise, pithy, succinct, compendious, terse 821
KURZGESCHWÄNZT: brevicaudate, short-tailed 822
KURZHAARIG: short-haired; short (nap) staple (Wool) 823
KURZKÖPFIG: brachycephalic, short-headed 824
KÜRZLICH: late; recent; shortly; soon; lately; recently; briefly 825
KURZSCHLIESSEN: to short-circuit (Elect.) 826
KURZSCHLIESSER (*m*): short-circuiter 827
KURZSCHLUSS (*m*): short-circuit (Elect.) 828
KURZSCHLUSSANKER (*m*): squirrel-cage (Elect.) 829
KURZSCHLUSSFUNKE (*f*): short circuit spark 830
KURZSCHREIBEKUNST (*f*): the art of shorthand, stenography 831
KURZSCHRIEBEN (*n*): shorthand, stenography 832
KURZSCHREIBER (*m*): shorthand-writer, stenographer 833
KURZSCHRIFT (*f*): shorthand 834
KURZSICHTIG: near-sighted; short-sighted; short-dated (of bills) 835
KURZSILBIG: short-syllabic, short-syllabled; curt, concise, laconic, short, brief 836
KURZSTÄBCHEN (*n*): bacillus 837
KURZUM: in a nutshell; to be brief; in short; to put it (shortly) briefly; in a word 838
KÜRZUNG (*f*): shortening, abbreviation, curtailment, etc. (see **KÜRZEN**) 839
KÜRZUNGSZEICHEN (*n*): sign of abbreviation, apostrophe 840
KURZWAREN (*f.pl*): small wares; fancy (goods) wares; hardware 841
KURZWEG: only; simply; plainly (see also **KURZUM**) 842
KURZWEIL (*f*): pastime; amusement; diversion 843
KURZWELLIG: of short wave-length 844
KURZZEITIG: temporary, fleeting, short-lived, transient, transitory, ephemeral 845
KUSSOBLÜTEN (*f.pl*): cusso flowers; Koussa flowers (of *Hagenia abyssinia*) 846
KÜSTE (*f*): coast; shore; beach; strand 847
KÜSTENFAHRER (*m*): coaster (Naut.) 848
KÜSTENFAHRT (*f*): coastal trade (Marine) 849
KÜSTENHANDEL (*m*): coastal trade (Marine) 850
KÜSTENLOTSE (*m*): coastal (coasting) pilot 851
KÜSTENPROVINZ (*f*): maritime province-(of a country) 852
KÜSTENSCHIFFAHRT (*f*): coastal trade (Marine) 853
KÜSTENWACHE (*f*): coast-guard 854
KÜSTER (*m*): sexton; sacristan 855
KUSTOS (*m*): keeper; custodian; warder; curator 856
KUTIKULA (*f* and *pl*): cuticle; (*pl*): epidermis 857
KUTIKULARPLÄTTCHEN (*n*): cuticular scale 858
KUTSCHE (*f*): coach; carriage 859
KUTSCHENBAUER (*m*): carriage (coach) builder 860
KUTSCHENLACK (*m*): carriage varnish 861
KUTSCHER (*m*): coachman; driver 862
KUTSCHERHANDSCHUHE (*m.pl*): driving gloves 863
KUTTE (*f*): cowl 864
KUTTELFISCH (*m*): cuttle fish, common cuttle, squid (*Sepia officinalis*) 865
KUTTER (*m*): cutter (Naut.) 866
KUTTERTAKELUNG (*f*): cutter rig (Naut.) 867
KUVERT (*n*): envelope; cover 868
KUVERTLEIM (*m*): envelope gum 869
KUVERTMASCHINE (*f*): envelope machine 870
KUVETTE (*f*): bulb; trough; bath (see **CUVETTE**) 871
KUZELLAMPE (*f*): Kuzel lamp (an electric glow-lamp invented by Kuzel) 872
KYANÄTHIN (*n*): kyanethine; cyanethine 873
KYANIDIN (*n*): kyanidine; cyanidine 874
KYANISIEREN: to kyanize; cyanize (impregnate with corrosive sublimate) 875
KYNURENSÄURE (*f*): kynurenic acid 876

L

LAB (*n*): rennet (an enzyme) 877

LABDANGUMMI (*n*): labdanum, ladanum (the gum resin of *Cistus creticus*) 878

LABDANHARZ (*n*): labdanum; ladanum (gum resin of *Cistus creticus*) 879

LABDANÖL (*n*): ladanum oil (from distillation of LABDANHARZ, which see); Sg. 1.011 880

LABDANUM (*n*): (see LABDANHARZ) 881

LABDRÜSE (*f*): peptic gland 882

LABEN: to recruit; refresh; revive; comfort; curdle (coagulate) with rennet 883

LABESSENZ (*f*): (essence of)-rennet 884

LABEXTRAKT (*m*): rennet extract 885

LABFERMENT (*n*): rennet ferment 886

LABGEWINNUNG (*f*): rennet extraction 887

LABIALE (*f*): labial (of pronunciation) 888

LABIL: labile; unstable 889

LABILITÄT (*f*). instability 890

LABKRAUT (*n*): galium (*Galium*) (Bot.) 891

LABMAGEN (*m*): rennet bag 892

LABORANT (*m*): chemist; laboratory assistant 893

LABORATIONSTAXE (*f*): chemist's tax 894

LABORATORIENEINRICHTUNG (*f*): laboratory fittings 895

LABORATORIUM (*n*): laboratory 896

LABORATORIUMSAPPARAT (*m*): laboratory apparatus 897

LABORATORIUMSAUSRÜSTUNG (*f*): laboratory fittings 898

LABORATORIUMSGERÄT (*n*): laboratory utensils 899

LABORATORIUMSTISCH (*m*): laboratory (table) bench 900

LABORATORIUMSÜBERHITZER (*m*): laboratory superheater (for superheating gases and vapours) 901

LABORIEREN: to labour; suffer from; practice chemistry; work as a chemist 902

LABPRODUKT (*n*): rennet product 903

LABPULVER (*n*): rennet powder 904

LABRADOR (*m*): labradorite (see LABRADORIT); $(Na_2,Ca)O, Al_2O_3,3SiO_2$; Sg. 2.7 905

LABRADORISCHE HORNBLENDE (*f*): Labrador hornblende 906

LABRADORIT (*m*): Labradorite (isomorphous mixture of albite and anorthite) (see ALBIT, ANORTHIT and LABRADOR); Sg. 2.7; Mp. 1210-1280°C 907

LABRADORSTEIN (*m*): labradorite; Labrador stone (see LABRADOR and LABRADORIT) 908

LABYRINTHAUSFÜHRUNG (*f*): labyrinth type (of packing), labyrinth arrangement 909

LABYRINTHDICHTUNG (*f*): labyrinth packing 910

LACHE (*f*): laugh; laughter; pool; backwater (of rivers); puddle; stagnant water; slough 911

LÄCHELN: to smile 912

LACHEN: to laugh 913

LACHENKNOBLAUCH (*m*): water germander (*Teucrium scordium*) 914

LÄCHERLICH: laughable; ludicrous; ridiculous; absurd 915

LACHGAS (*n*): laughing gas (see STICKSTOFF-OXYDULGAS) 916

LACHIG: marshy, swampy 917

LACHKRAMPF (*m*): hysterical laughter 918

LACHS (*m*): salmon (*Salmo salar*) 919

LACHSFARBIG: salmon-pink; salmon-coloured 920

LACHSÖL (*n*): salmon oil (from *Salmo salar*); Sg. 0.926 921

LACK (*m*): (resin dissolved in turpentine or linseed oil varnish), lacquer, lac, japan, lake, varnish; sealing-wax; wallflower (see GOLDLACK) 922

LACKARBEIT (*f*): lacquer-work, Japan ware 923

LACKBAUM (*m*): candle-nut tree (*Aleurites moluccana*); lacquer tree (*Rhus vernicifera*) 924

LACKBAUM (*m*), **MALABARISCHER-**: Dhak tree, Palas tree (*Butea frondosa*) 925

LACKBAUMÖL (*n*): (see BANKULNUSSÖL) 926

LACKBILDNER (*m*): lake former 927

LACK (*m*), **CHINESISCHER-**: Chinese lacquer (solution of gum mastic, gum sandarac and Chinese wood oil in alcohol) 928

LACKESTER (*m*): lac ester (compound obtained from the action of colophony on alcohol) 929

LACKEXTRAKT (*m*): lac extract 930

LACKFABRIKATION (*f*): varnish manufacture 931

LACKFARBE (*f*): lake; lac dye; varnish colour; wallflower colour; carmine (see KARMIN) 932

LACKFARBIG: lake-coloured 933

LACKFARBSTOFF (*m*): lake colour; lac dye (see LACK-LACK) 934

LACKFIRNIS (*m*): lac varnish; lake varnish 935

LACKFLECHTE (*f*): tartarean moss (*Roccella tinctoria*); cudbear (*Lecanora tartarea*) 936

LACK (*m*), **GUMMI-**: gum lac 937

LACKHARZ (*n*): gum-lack 938

LACKHOLZ (*n*): mountain pine (*Pinus montana*) 939

LACKIERARBEIT (*f*): lacquer work 940

LACKIEREN: to lacquer; japan; varnish 941

LACKINDUSTRIE (*f*): varnish industry 942

LACK (*m*), **JAPANISCHER-**: Japan (lacquer) lac (from the lac tree, or lacquer tree *Rhus vernicifera*) 943

LACK (*m*), **KÖRNER-**: seed (grain) lac 944

LACKLACK (*m*): lac (lake) dye (from the insect *Coccus ficus* or *Coccus lacca*) 945

LACKLEDER (*n*): patent (japanned) leather 946

LACKMUS (*m*): litmus (blue dye from various lichens) 947

LACKMUSBLAU (*n*): litmus blue 948

LACKMUSFARBSTOFF (*m*): litmus (dye) colouring matter (dye from various lichens) 949

LACKMUSFLECHTE (*f*): litmus lichen; archil (*Variolaria rocella* and *V. lecanora*) (see ORSEILLE) 950

LACKMUSPAPIER (*n*): litmus paper 951

LACK (*m*), **ROTER-**: (see ROTHOLZLACK) 952

LACK (*m*), **RUSSISCHER-**: Russian varnish 953

LACKSATZ (*m*): varnish (residue) sediment 954

LACKSCHWARZ (*n*): black lake 955

LACKSCHICHT (*f*): layer of varnish 956

LACK (*m*), **SCHWEFLIGER-**: sulphurous varnish 957

LACK (*m*), **STOCK-**: stick lac 958

LACK (*m*), **VENETIANISCHER-**: Venetian lake (see KARMIN) 959

LACKVIOLE (*f*): wallflower (see GOLDLACK) 960

LACMUS (*m*): (see LACKMUS) 961

LACTARIN (*n*): lactarin (purified casein) 962

LACTON (*n*): lactone 963

LACTONSÄURE (*f*): lactone (lactonic) acid; galactonic acid 964
LACTOPHENIN (*n*): lactophenine, lactylphenetidine; $C_6H_4.NHCOCH(OH)CH_3.OC_2H_5$; Mp. 117.5°C. 965
LACTUCARIUM (*n*): lactucarium, lettuce opium (see **GIFTLATTICHSAFT**) 966
LACTUCASÄURE (*f*): lactucic acid 967
LACTUCERIN (*n*): lactucerin (from *Lactuca Virosa*); $C_{15}H_{24}O$ 968
LACTUCIN (*n*): lactucin (from *Lactuca virosa*); $C_{11}H_{14}C_4$ 969
LACTUCON (*n*): lactucon (see **LACTUCERIN**) 970
LACTUCOPIKRIN (*n*): lactucopicrin (from *Lactuca virosa*); $C_{44}H_{32}O_{21}$ 971
LADANGUMMI (*n*): (see **LABDANHARZ**) 972
LADANHARZ (*n*): (see **LABDANHARZ**) 973
LADANUM (*n*): (see **LABDANHARZ**) 974
LADE (*f*): chest; box; trunk; case; flask (Founding); frame; press; bin 975
LADEBAUM (*m*): (loading)-crane, derrick, boom 976
LADEBORD (*m*): skid 977
LADEBRIEF (*m*): bill of lading 978
LADEDAUER (*f*): duration of charge 979
LADEDICHTE (*f*): charging density 980
LADEFÄHIGKEIT (*f*): loading capacity, charge capacity 981
LADEGELD (*n*): loading charges 982
LADEKRAN (*m*): loading crane or hoist 983
LADEKURVE (*f*): charging curve 984
LADELINIE (*f*): load line 985
LADELUKE (*f*): cargo hatchway, loading hatchway or hatch 986
LADEMASCHINE (*f*): charging machine 987
LADEMASS (*n*): amount of charge, quantity of charge, charge 988
LADEN: to load; freight; charge; summon; invite; (*m*): store; shop; shutter 989
LADENBANK (*f*): counter (of a shop) 990
LADENDIEB (*m*): shop-lifter 991
LADENDIENER (*m*): shop (sales) man 992
LADENFELD (*n*): pane (panel) of a shutter 993
LADENFENSTER (*n*): shop window 994
LADENJUNGE (*f*): errand boy 995
LADENKASSE (*f*): till, cash drawer 996
LADENPREIS (*m*): retail (selling) price 997
LADENRIEGEL (*m*): sash-bolt (of a window) 998
LADENTISCH (*m*): (shop)-counter 999
LADEPFORTE (*f*): cargo port, ballast port, loading port 000
LADEPOTENTIAL (*n*): charging potential 001
LADERAUM (*m*): hold (of a ship); charging space; explosion chamber 002
LADESTELLE (*f*): charging (station) place (for accumulators, etc.); loading station 003
LADESTROM (*m*): charge current, charging current 004
LADETAKEL (*n*): guying tackle; loading tackle 005
LADEWASSERLINIE (*f*): load-(water)-line 006
LADEWIDERSTAND (*m*): charging resistance 007
LADUNG (*f*): charge; freight; load; cargo; charging; loading; invitation 008
LADUNG (*f*) **NEHMEN**: to take in cargo 009
LADUNGSBRIEF (*m*): bill of lading 010
LADUNGSDICHTE (*f*): charging density 011
LADUNGSDICHTIGKEIT (*f*): charge density 012
LADUNGSFÄHIGKEIT (*f*): loading capacity; tonnage (of ships) 013
LADUNGSFLASCHE (*f*): Leyden jar 014
LADUNGSKOEFFIZIENT (*m*): coefficient of charge 015
LADUNGSSÄULE (*f*): secondary battery 016
LADUNGSSINN (*m*): nature of a charge 017
LADUNGSSTROM (*m*): charging current 018
LAFETTE (*f*): gun-carriage 019
LAFETTENSCHWANZ (*m*): trail (of a gun-carriage) 020
LAGE (*f*): layer; stratum; bed; coat; coating; state; position; aspect; bearing; site; condition; circumstances; situation; attitude; quire (Paper); trim (of a ship); lap (Textile) 021
LÄGEL (*n*): barrel; keg; cringle; hank 022
LAGEPLAN (*m*): plan of site 023
LAGER (*n*): bed; layer; stratum; store; stock; storehouse; camp; dump; warehouse; support; supply; sediment; couch; lair; stand; bearing; bush (Mach.); shop; show-room; plummer-block (Mach.) 024
LAGER (*n*), **AUF-HABEN**: to have in stock 025
LAGER (*n*), **AUF-NEHMEN**: to take into (add to) stock 026
LAGERBAUM (*m*): gawn-tree (for casks) 027
LAGERBOCK (*m*): plummer block; pedestal bearing 028
LAGERBOLZEN (*m*): bearing bolt 029
LAGERBUCH (*n*): stock book 030
LAGERBÜCHSE (*f*): bush; bearing; brass; step (Mech.) 031
LAGERDECKEL (*m*): bearing cap, bearing cover 032
LAGERDECKELBOLZEN (*m*): bearing cap bolt, bearing cover bolt 033
LAGERECHT: fast to storing 034
LAGER (*n*), **EINFACHES-**: plain bearing 035
LAGERFASS (*n*): storage (vat; tun) cask 036
LAGERFLÄCHE (*f*): bearing surface; warehouse (store-room) area 037
LAGERFUTTER (*n*): pillow, pillow bush, bearing bush, bearing (brass) step 038
LAGERGEBÜHREN (*f.pl*): warehouse charges (see also **LAGERGELD**) 039
LAGERGELD (*n*): warehouse charges; storage-(charges); warehousing-(charges); dock (dues) charges; stowage 040
LAGERHALS (*m*): axle arm, bearing neck 041
LAGERHAUS (*n*): warehouse; bonded stores 042
LAGERKELLER (*m*): storage cellar 043
LAGERKONTO (*n*): warehouse (stock) account 044
LAGER (*n*), **KURBEL-**: crank-bearing 045
LAGER (*n*), **KURBELZAPFEN-**: crank-pin bearing 046
LAGERMETALL (*n*): (60% Pb; 20% Sb; 20% Sn); bearing metal, bush metal, white metal, antifriction metal 047
LAGERMETALL (*n*), **ENGLISCHES-**: (90% Zn; 1.5% Sb; 7% Cu; 1.5% Sn; also 59.4% Zn; 39.8% Sn; 0.5% Cu; 0.3% Pb; or 53% Sn; 33% Pb; 10.6% Sb; 2.4% Cu; 1% Zn); English bearing metal, white metal 048
LAGERN (*n*): storage, arrangement; stratification; bearing, mounting, supporting 049
LAGERN: to store; be stored; lie down; (en)-camp; lay down; deposit; place; stock; harbour; be (borne) supported; mount; bear; support; rest upon 050
LAGERÖLUNG (*f*): bearing lubrication 051
LAGERPFANNE (*f*): pillow, pillow bush, bearing (bush) brass or step 052
LAGERPFLANZEN (*f.pl*): *Thallophytes* (Bot.) (see **THALLOPHYTEN**) 053

LAGERPLATTE (*f*): bed (base, foundation) plate, bearing-plate 054
LAGERPLATZ (*m*), **STABEISEN-**: bar-iron (depot, yard) store 055
LAGERPROZESS (*m*): storage process 056
LAGERPUNKT (*f*): point of support 057
LAGERRAUM (*m*): store-(room); warehouse; housing space; storage (room) space; show-room 058
LAGERREIF: aged 059
LAGERSCHALE (*f*): (see **LAGERBÜCHSE**) 060
LAGERSCHMIERUNG (*f*): bearing lubrication 061
LAGER (*n*), **SELBSTEINSTELLENDES-**: self-aligning bearing 062
LAGER (*n*), **STARRES-**: rigid bearing 063
LAGERSTÄTTE (*f*): deposit; bed (of ores) 064
LAGERUNG (*f*): arrangement; stratification; bedding; bearing; bed; support; storage; lying down; encampment; grain (of stone) 065
LAGER (*n*), **WELLEN-**: shaft bearing 066
LAGEVERÄNDERUNG (*f*): change of position 067
LAHM: lame; halt; paralysed; crippled 068
LÄHMEN: to paralyse; cripple; lame; maim; enervate; disable 069
LÄHMUNG (*f*): lameness; palsy; paralysis; atrophy; paraplegia (Med.) 070
LAHN (*m*): tinsel; flattened (plate) wire; lawn 071
LAIB (*m*): loaf 072
LAIBUNGSDRUCK (*m*): pressure due to shrinking, specific pressure on the face of a hole 073
LAICH (*m*): spawn 074
LAICHE (*f*): spawning time 075
LAICHEN: to spawn 076
LAICHTEICH (*m*): breeding pool 077
LAICHZEIT (*f*): spawning time 078
LAIE (*m*): layman; novice; (*pl*): laity 079
LAIENKREISEN (*m.pl*): laymen's circles, the lay mind 080
LAKE (*f*): brine; pickle 081
LAKEN (*n*): sheet; cloth; shroud 082
LAKKIERARBEIT (*f*): lacquer work (see **LACKIER-**) 083
LAKMUS (*m*): (see **LACKMUS**) 084
LAKRITZ (*m*): (see **LAKRITZE**) 085
LAKRITZE (*f*): liquorice (root of *Glycyrrhiza glabra*) 086
LAKRITZENHOLZ (*n*): stick liquorice 087
LAKRITZENPRÄPARAT (*n*): liquorice preparation 088
LAKRITZENSAFT (*m*): liquorice juice 089
LAKRIZ (*m*): (see **LAKRITZE**) 090
LAKTARINSÄURE (*f*): lactarinic acid 091
LAKTARSÄURE (*f*): lactaric acid 092
LAKTAT (*n*): lactate 093
LAKTON (*n*): lactone 094
LAKTOSE (*f*): lactose, milk sugar (one of the cane sugar group of sugars) (*Saccharum lactis*) (see **ROHRZUCKER**) 095
LAKTUKARIUM (*n*): lettuce opium; lactucarium 096
LAKTYLPHENETIDIN (*n*): lactylphenetidine (see **LACTOPHENIN**) 097
LALLEMANTIAÖL (*n*): lallemantia oil (from seeds of *Lallemantia iberica*); Sg. 0.934 098
LALLEN: to stammer; babble; lisp 099
LAMBERTSNUSS (*f*): filbert 100
LAMBERTSNUSSBAUM (*m*): (see **HASEL**) 101
LAMBERTNUSSHOLZ (*n*): (see **HAZELNUSSHOLZ**) 102
LAMELLAR: laminated; foliated; lamellar; laminate; laminar; wavy 103
LAMELLE (*f*): gill, rib (of tubes); layer, segment, lamina; (*pl*): laminæ 104
LAMELLENKIEL (*m*): lamellar keel 105
LAMELLENMAGNET (*m*): lamellar magnet 106
LAMELLENROHR (*n*): ribbed tube, gilled tube 107
LAMENTIEREN: to lament 108
LAMINAR: laminated, foliated, lamellar, laminate, laminar, wavy 109
LAMINARIAART (*f*): type of laminaria 110
LAMINARIASÄURE (*f*): laminaric acid (acid from various types of Laminaria) 111
LAMINIEREN: to laminate; draw (Spinning) 112
LAMINIERWERK (*n*): drawing frame 113
LAMM (*n*): lamb 114
LAMMEN: to lamb, ewe, yean 115
LÄMMERBLUME (*f*): pilewort (*Ranunculus ficaria*) 116
LÄMMERGEIER (*m*): Lämmergeier, bearded vulture (*Gypaëtus barbatus*) 117
LÄMMERKRAUT (*n*): Antirrhinum, snapdragon (*Antirrhinum majus*) 118
LÄMMERPELZ (*m*): lambskin 119
LÄMMERSCHWÄNZCHEN (*n*): lamb's tail; yarrow, millefoil, milfoil (*Achillea millefolium*) 120
LÄMMERWOLKE (*f*): curl-cloud, mare's-tail, cat's-tail, cirrus (Meteorological) 121
LÄMMERWOLLE (*f*): lamb's wool 122
LAMMFELL (*n*): lambskin 123
LAMMZEIT (*f*): lambing (yeaning, ewing) time 124
LAMPE (*f*): lamp 125
LAMPE (*f*), **ELEKTRISCHE-**: electric lamp 126
LAMPE (*f*), **GAS-**: gas lamp 127
LAMPE (*f*), **GEBLÄSE-**: blast lamp 128
LAMPE (*f*), **LICHTPAUS-**: printing machine lamp 129
LAMPENANZÜNDER (*m*): lamp lighter 130
LAMPENARBEIT (*f*): blast-lamp work 131
LAMPENBRENNER (*m*): burner 132
LAMPENDOCHT (*m*): lamp wick 133
LAMPE (*f*), **NERNST-**: Nernst-(electric)-lamp 134
LAMPENFASSUNG (*f*): lamp fitting 135
LAMPENFIEBER (*n*): stage fright 136
LAMPENGESTELL (*n*): lamp-stand 137
LAMPENGLOCKE (*f*): (lamp)-globe; lamp-shade 138
LAMPENRUSS (*m*): lampblack 139
LAMPENSCHIRM (*m*): lamp shade 140
LAMPENSCHWARZ (*n*): lampblack 141
LAMPENZYLINDER (*m*): lamp chimney, lamp-glass 142
LAMPE (*f*), **PETROLEUM-**: petroleum lamp 143
LAMPE (*f*), **POLARISATIONS-**: polarization lamp 144
LAMPE (*f*), **QUECKSILBERDAMPF-**: mercury-vapour lamp 145
LAMPE (*f*), **SPIRITUS-**: spirit lamp 146
LAMPROCHROMATISCHER OPALIN-ALLOPHAN (*m*): Allophane (see **ALLOPHAN**) 147
LANAFUCHSIN (*n*): lanafuchsine 148
LANARKIT (*m*): lanarkite; PbO, SO_3, PbO_2; Sg. 6.35 149
LANCIERROHR (*n*), **TORPEDO-**: torpedo launching tube, torpedo tube 150
LAND (*n*): land; estate; country; state; ground; soil; (as a prefix = rural; country; agricultural; provincial) 151
LANDACCISE (*f*): land-tax 152
LAND, AN-GEHEN: to land, go ashore, disembark 153

LANDANKER (*m*): shore anchor 154
LANDANWACHS (*m*): alluvium 155
LANDARBEIT (*f*): agricultural (labour) work, agriculture 156
LANDARBEITER (*m*): agricultural labourer, worker on the land 157
LANDBAU (*m*): agriculture 158
LANDBESITZER (*m*): land-holder, landed proprietor 159
LANDEINWÄRTS: inland, up-country 160
LANDEN: to land, disembark, go ashore 161
LANDENGE (*f*): isthmus 162
LÄNDERLOS: stateless; without lands 163
LANDESBESCHREIBUNG (*f*): topography (description of a country) 164
LANDESBRAUCH (*m*): custom-(of a country) 165
LANDESERZEUGNIS (*n*): home (product) produce 166
LANDESGEBIET (*n*): territory, possession 167
LANDESGEBRAUCH (*m*): custom-(of a country) 168
LANDESGERICHT (*n*): high court 169
LANDESGESCHICHTE (*f*): history-(of a country) 170
LANDESGESETZ (*n*): statute law; law of the land 171
LANDESGEWÄCHS (*n*): native plants; home-produce 172
LANDESGRENZE (*f*): boundary, frontier 173
LANDESKAMMER (*f*): board of finance 174
LANDESKASSE (*f*): (public)-treasury 175
LANDESKOLLEGIUM (*n*): local government board 176
LANDESMÜNZE (*f*): legal (coinage) currency 177
LANDESPRODUKT (*n*): home produce 178
LANDESSCHATZ (*m*): public (national) treasure; treasure-trove 179
LANDESSCHULD (*f*): national debt 180
LANDESSCHULE (*f*): Board school 181
LANDESSITTE (*f*): custom, manners-(of a country) 182
LANDESSPERRE (*f*): protection, prohibition, tariff reform, anti-dumping regulations 183
LANDESSPRACHE (*f*): native language, dialect 184
LANDESTRACHT (*f*): national costume 185
LANDESVALUTA (*f*): lawful currency; rate of exchange 186
LANDESVERFASSUNG (*f*): constitution-(of a country) 187
LANDESVERORDNUNG (*f*): statute law 188
LANDESVERRAT (*m*): high treason 189
LANDESVERSAMMLUNG (*f*): national assembly 190
LANDESVERWALTER (*m*): viceroy 191
LANDESVERWALTUNG (*f*): administration (government) of a country 192
LANDESVERWEISUNG (*f*): banishment, exile, transportation 193
LANDESWEHRUNG (*f*): legal currency 194
LANDESWOHL (*n*): common (good) weal, national welfare 195
LANDFIEBER (*n*): epidemic-(fever); endemic fever 196
LANDFRACHT (*f*): carriage (overland)-freight 197
LANDGERICHT (*n*): county court, petty session, assize 198
LANDGRENZE (*f*): boundary, frontier, land mark 199
LANDGUT (*n*): estate; country seat 200
LANDHANDEL (*m*): inland trade 201
LANDHAUS (*n*): country (seat) house; manor 202
LANDKARTE (*f*): map 203
LANDKRANKHEIT (*f*): endemic disease 204
LANDKUNDIG: notorious; well-known; generally known; commonly known; common knowledge 205
LANDLÄUFIG: current; customary 206
LANDLEBEN (*n*): country life 207
LÄNDLICH: rural; rustic; countrified 208
LANDMACHT (*f*): continental (land) power 209
LANDMANN (*m*): countryman; farmer; rustic; peasant 210
LANDMARK (*f*): boundary; frontier; territory 211
LANDMARKE (*f*): land-mark 212
LANDMESSER (*m*): surveyor; geometer; geometrician 213
LANDMESSKUNST (*f*): surveying, geodesy 214
LANDPARTIE (*f*): picnic; outing 215
LANDPOLIZEI (*f*): rural (police) constabulary 216
LANDRECHT (*n*): provincial (statute, municipal, common) law 217
LANDRECHTLICH: according to common law 218
LANDRÖTE (*f*): madder (see FÄRBERRÖTE) 219
LANDSCHAFT (*f*): landscape; province; state; district 220
LANDSCHAFTLICH: provincial, rural 221
LANDSCHAFTSGÄRTNEREI (*f*): landscape gardening 222
LANDSCHAFTSLINSE (*f*): landscape lens (Phot.) 223
LANDSCHAFTSMALEREI (*f*): landscape painting 224
LANDSEE (*m*): lake 225
LANDSEUCHE (*f*): epidemic; endemic disease 226
LANDSITZ (*m*): country seat 227
LANDSMANN (*m*): countryman 228
LANDSPITZE (*f*): promontory, cape 229
LANDSTADT (*f*): inland (country; provincial) town 230
LANDSTEUER (*f*): land-tax 231
LANDSTRASSE (*f*): highway; high road; main road; arterial road 232
LANDSTRICH (*m*): district; climate; tract-(of country) 233
LANDSTURM (*m*): general levy (troops) 234
LANDSTURZ (*m*): land-slip, land-slide 235
LANDTAG (*m*): diet; legislature; legislative assembly 236
LANDTIER (*n*): land animal 237
LANDTRANSPORT (*m*): road transport, cartage, carriage by land 238
LANDÜBLICH: general, national, common, usual, customary 239
LANDUNG (*f*): landing, debarcation 240
LANDUNGSBOOT (*n*): shore boat 241
LANDUNGSGEBÜHREN (*f.pl*): landing charges 242
LANDUNGSKOSTEN (*pl*): landing charges 243
LANDUNGSPLATZ (*m*): landing place 244
LANDUNGSSPESEN (*f.pl*): landing charges 245
LANDVOGEL (*m*): land bird 246
LANDVOGT (*m*): governor-(of a province) 247
LANDWÄRTS: landward 248
LANDWEHR (*f*): militia; yeomanry; home service troops 249
LANDWIND (*m*): land-wind, breeze blowing off the land 250
LANDWIRT (*m*): farmer; landlord; innkeeper; husbandman 251
LANDWIRTSCHAFT (*f*): agriculture; husbandry; farming 252
LANDWIRTSCHAFTLICH: agricultural 253
LANDZUNGE (*f*): tongue (neck) of land, promontory, point 254

LANG: long; high; tall; protracted; lengthy; prolonged; ropy (of wine) 255
LANGANHALTEND: protracted, lengthy, prolonged, continued 256
LANGBEINIT (*m*): langbeinite; $K_2Mg_2[SO_4]_3$; Sg. 2.83 257
LANGBLÄTTERIG: long-leaved 258
LANGE: long; far; by far 259
LÄNGE (*f*): length; longitude; tallness 260
LÄNGEBRUCH (*m*): longitudinal (fracture) rupture 261
LANGEIFÖRMIG: of extended oval shape 262
LANGEN: to suffice; reach; take; seize; attain 263
LÄNGEN: to lengthen, stretch, extend, elongate 264
LÄNGENAUSDEHNUNG (*f*): elongation; linear expansion 265
LÄNGENBRUCH (*m*): longitudinal (fracture) rupture 266
LÄNGENDURCHSCHNITT (*m*): longitudinal section 267
LÄNGENGRAD (*m*): degree of longitude 268
LÄNGENKREIS (*m*): degree of longitude 269
LÄNGENMASS (*n*): linear (long) measure 270
LÄNGENMETAZENTRUM (*m*): longitudinal metacentre 271
LANGEN'SCHE GLOCKE (*f*): Langen's gas bell (for catching gas from blast furnaces) (fitted at top of furnace) 272
LÄNGENSCHNITT (*m*): longitudinal section 273
LÄNGENSCHWINGUNG (*f*): longitudingal (oscillation) vibration 274
LÄNGENTRÄGHEITSMOMENT (*n*): longitudinal moment of inertia (Shipbuilding) 275
LANGFRISTIGER VERTRAG (*m*): contract (agreement) of long duration 276
LANGGESTIELT: long-handled 277
LANGGESTRECKT: extended; elongated; stretched 278
LANGHAARIG: long-haired; shaggy; flossy (of silk); long-staple (of wool) 279
LANGHALSIG: long-necked 280
LANGHOBEL (*m*): long plane 281
LANGHOLZ (*n*): the wood exposed by a tangential section through a tree trunk; balks of timber; beams; grain (cleft) wood 282
LANGJÄHRIG: of many years standing 283
LANGKREIS (*m*): ellipse, oval 284
LANGLEBEND: long lived 285
LÄNGLICH: oblong; longish; oval; elliptical; longitudinal-(ly) 286
LÄNGLICHRUND: elliptical; oval 287
LANGMUT (*f*): patience; forbearance 288
LANGRUND: oval; elliptical 289
LÄNGS: along; (as a prefix=longitudinal) 290
LÄNGSACHSE (*f*): longitudinal axis 291
LANGSALING (*f*): trestle trees 292
LANGSAM: slow-(ly); dull; tardy; backward; easily; gently 293
LANGSAMBINDER (*m*): slow-setting cement 294
LANGSAMLAUFEND: slow-speed 295
LÄNGSANSICHT (*f*): longitudinal (side) view or elevation 296
LÄNGSBALKEN (*m*): centre timber (of sails); longitudinal beam 297
LÄNGSBAND (*n*): tie plate, longitudinal tie 298
LANGSCHLIFF (*m*): long-fibred mechanical pulp 299
LÄNGSDURCHSCHNITT (*m*): longitudinal section 300
LÄNGSEBENE (*f*): longitudinal plane 301
LANGSEIN (*n*): ropiness (of wine) 302
LÄNGSFASER (*f*): longitudinal fibre; grain-(of wood) 303
LANGSICHTIG: long-sighted 304
LÄNGSNAHT (*f*): longitudinal seam 305
LANGSPLISSUNG (*f*): long splice 306
LÄNGSRICHTUNG (*f*), **IN DER-**: longitudinally 307
LÄNGSRISS (*m*): sheer plan (of ships) 308
LÄNGSSALING (*f*): trestle trees 309
LÄNGSSCHIENE (*f*): tie plate 310
LÄNGSSCHIFFS: fore and aft (Nautical) 311
LÄNGSSCHNEIDER (*m*): slitter (of paper) 312
LÄNGSSCHNITT (*m*): longitudinal section 313
LÄNGSSCHOTT (*n*): longitudinal bulkhead 314
LÄNGSSEIT GEHEN: to go alongside 315
LÄNGSSEIT KOMMEN: to come alongside (Marine) 316
LÄNGSSPANT (*n*): longitudinal frame 317
LÄNGSSTREIFEN (*m*): longitudinal (strip) stripe 318
LÄNGST: longest; long (ago) since 319
LANGSTÄBCHEN (*n*): bacillus 320
LANGSTAPELIG: long-staple (of wool) 321
LÄNGSTENS: at the (most) longest 322
LÄNGSTRÄGER (*m*): longitudinal girder 323
LÄNGSVERBAND (*m*): longitudinal framing, longitudinal strengthening, longitudinal stiffener, longitudinal stiffening, longitudinal binding, longitudinal tie, etc. 324
LANGTRÄGER (*m*): longitudinal grate bar bearer, longitudinal bearer bar, longitudinal (support, bearer, girder) beam 325
LANGWEILE (*f*): tedium; boredom; ennui; irksomeness 326
LANGWEILEN: to tire; bore; be (tedious) irksome 327
LANGWELLIG: long-wave; of long wave-length 328
LANGWIERIG: wearisome; protracted; lingering 329
LANOLIN (*n*): lanolin (see **ADEPS LANAE**) 330
LANOLINSEIFE (*f*): lanolin soap 331
LANTHAN (*n*): lanthanum (La) 332
LANTHANAMMONIUMNITRAT (*n*): lanthanum-ammonium nitrate; $La(NO_3)_3.2NH_4.NO_3.4H_2O$ 333
LANTHANCARBID (*n*): lanthanum carbide; LaC_2; Sg. 5.02 334
LANTHANCHLORID (*n*): lanthanum chloride; $LaCl_3$; Sg. 3.947 335
LANTHAN (*n*), **GOLDCHLORWASSERSTOFFSAURES-**: lanthanum chloraurate; $LaCl_3,AuCl_3+5H_2O$ 336
LANTHANIT (*m*): lanthanite; $La_2[CO_3]_3+9H_2O$; Sg. 2.65 337
LANTHANKALIUMSULFAT (*n*): lanthanum-potassium sulphate; $La_2(SO_4)_3.3K_2SO_4$ 338
LANTHANKARBONAT (*n*): lanthanum carbonate; $La_2(CO_3)_3.3H_2O$ 339
LANTHANNITRAT (*n*): lanthanum nitrate; $La(NO_3)_3.6H_2O$ 340
LANTHANOXYD (*n*): lanthanum oxide (trioxide or sesquioxide); La_2O_3; Sg. 6.41-6.48 341
LANTHAN (*n*), **SALPETERSAURES-**: lanthanum nitrate (see **LANTHANNITRAT**) 342
LANTHANSALZ (*n*): lanthanum salt 343
LANTHANSULFAT (*n*): lanthanum sulphate; $La_2(SO_4)_3$; Sg. 3.6; $La_2(SO_4)_3+9H_2O$; Sg. 2.821-2.853 344
LANTHANSULFID (*n*): lanthanum sulphide; La_2S_3; Sg. 4.911 345

LANTHOPIN (*n*): lanthopine (an opium alkaloid); $C_{23}H_{25}NO_4$ 346
LANZE (*f*): lance; spear; harpoon 347
LANZENFÖRMIG: lanciform, lance-shaped, lanceolate-(d) 348
LANZETTE (*f*): lancet 349
LANZETTENFÖRMIG: lanceolate 350
LAPIS-LAZULI (*m*): Lapis-lazuli (see LASURIT) 351
LÄPPCHEN (*n*): flap; rag; lobe; duster; shred; patch 352
LAPPEN (*m*): lap (see also LÄPPCHEN) 353
LAPPENLOS: acotyledonous (Bot.) 354
LAPPENRING (*m*): flap ring, flanged ring 355
LAPPENSAMMLER (*m*): rag merchant, rag man 356
LAPPIG: lobed; lobate; flabby; flaccid; ragged; tattered 357
LÄRCHE (*f*): larch (*Larix europœa*; *Pinus larix*) 358
LÄRCHENBAUM (*m*): (see LÄRCHE) 359
LÄRCHENEXTRAKT (*m*): larch extract (containing about 25% tannin) (from *Larix europœa* and *Pinus larix*) 360
LÄRCHENHARZ (*n*): larch resin; Venetian (Venice) turpentine (see LÄRCHE) 361
LÄRCHENHOLZ (*n*): larch wood; Sg. 0.44-0.8 (see LÄRCHE) 362
LÄRCHENPILZ (*m*): purging agaric (*Polyphorus officinalis*) 363
LÄRCHENSCHWAMM (*m*): (see LÄRCHENPILZ) 364
LÄRCHENTANNE (*f*): larch (*Pinus larix*) 365
LARICIN (*n*): laricine (see ABIETIN) 366
LÄRM (*m*): noise; alarm; uproar; row 367
LÄRMEN: to create a disturbance; make a (row) noise; give an alarm; (*m*): noise; alarm; uproar; row 368
LÄRMGLOCKE (*f*): tocsin; alarm bell 369
LARVE (*f*): larva; grub; chrysalis; mask 370
LARVENZUSTAND (*m*): chrysalis state 371
LARYNGOSKOP (*n*): laryngoscope (Med.) (see also under KEHL-) 372
LARYNGOTOMIE (*f*): laryngotomy (Surgical) (see also under KEHL) 373
LASCHE (*f*): strap; butt strap; cover plate; lash; clip; fish-plate; latchet; lashing; gusset; tie 374
LASCHEN: to lash, fish-(together), tie, clip, join 375
LASCHENBOLZEN (*m*): fish bolt 376
LASCHENSCHRAUBE (*f*): fish-(plate) screw 377
LASCHENVERBINDUNG (*f*): fishing 378
LASCHEVORGANG (*n*): shock bending test by Lasche's method (Steel) 379
LASCHING (*f*): lashing 380
LASCHUNG (*f*): fishing (of plates) 381
LASCHUNG (*f*), KIEL-: keel scarf 382
LASERKRAUT (*n*): laser-wort (*Laserpitium* or *Silphium*) 383
LASERPFLANZE (*f*): laser-wort (*Laserpitium* or *Silphium*) 384
LASIEREN: to glaze 385
LASIEREND: transparent 386
LASIONIT (*m*): lasionite; $2Al_2(PO_4)_4$, $Al_2Fl_2(OH)_2,7H_2O$; Sg. 2.33 387
LASSBESITZ (*m*): copyhold 388
LASSBESITZER (*m*): copyholder 389
LASSEN: to let; leave; allow; permit; have; desist; look; suit; appear; become 390
LÄSSIG: negligent; indolent; lazy; sluggish; slothful 391
LASSOBAND (*n*): India-rubber tape (for making containers of any kind air-tight) 392
LAST (*f*): load; burden; encumbrance; weight; charge; cargo; freight; tonnage; hold 393
LASTBAR: oppressive; capable of (carrying, supporting) bearing a load 394
LASTEN: to load; charge; weigh; weight; oppress; encumber; freight 395
LASTENAUFZUG (*m*): goods lift 396
LASTENFREI: unencumbered 397
LASTENMASSSTAB (*m*): dead-weight scale, displacement curve (Shipbuilding) 398
LASTER (*n*): vice; crime 399
LÄSTERN: to slander; revile; abuse; calumniate; blaspheme; defame 400
LÄSTERREDE (*f*): slander 401
LÄSTERSCHRIFT (*f*): libel 402
LASTERVOLL: profligate 403
LÄSTERWORT (*n*): blasphemy; invective 404
LASTGELD (*n*): tonnage, freight 405
LÄSTIG: burdensome; irksome; onerous; tedious; troublesome; boring 406
LASTIG: freighted, loaded, charged, weighted 407
LASTRAUM (*m*): stowage, hold (Naut.) 408
LASTSAND (*m*): ballast-(sand) 409
LASTSCHIFF (*n*): lighter, cargo-boat 410
LAST (*f*), SCHIFFS-: freight 411
LASTTIER (*n*): beast of burden 412
LASTTRÄGER (*m*): porter 413
LASTWAGEN (*m*): lorry 414
LAST (*f*), WANDERNDE-: live load 415
LAST (*f*), WASSER-: water hold 416
LASTZUG (*m*): goods train 417
LASTZUGSMASCHINE (*f*): traction engine; goods engine (Railways) 418
LASUR (*f*): varnish; glazing; azure; ultramarine; azurite; lapis lazuli (see LASURIT) 419
LASURBLAU (*n*): sky-blue, azure (see BERGBLAU and KUPFERCARBONAT) 420
LASURFARBE (*f*): azure; transparent colour (Art); glazing colour (covering colour which allows the ground to show through) 421
LASURGRÜN (*n*): green bice 422
LASURIT (*m*): lapis lazuli, lazurite; azurite; $3NaAlSiO_4.Na_2S_3$; Sg. about 3.0 423
LASURLACK (*m*): transparent varnish; clear varnish; glaze 424
LASURQUARZ (*m*): siderite (see EISENSPAT) 425
LASURSPAT (*m*): lazulite (see LAZULIT) 426
LASURSTEIN (*m*): azurite; lapis lazuli (see LASURIT) 427
LATEIN (*n*): latin-(language) 428
LATEINISCH: latin; roman (letters) 429
LATEINRAH (*f*): lateen yard 430
LATEINSEGEL (*n*): lateen sail 431
LATENT: latent, intrinsic 432
LATENTES BILD (*n*): latent image (Phot.) 433
LATENTE VERDAMPFUNGSWÄRME (*f*): latent heat of evaporation 434
LATENTE WÄRME (*f*): latent heat 435
LATENZ (*f*): latency 436
LATERAL: lateral-(ly) 437
LATERALSKLEROSE (*f*): lateral sclerosis 438
LATERNA MAGICA (*f*): magic lantern 439
LATERNBILD (*n*): lantern slide 440
LATERNE (*f*): lantern; lamp 441
LATEX (*m*), LATICES (*pl*): latex (a coarse suspension of caoutchouc globules in water); and [a milky exudation (Botany)] 442
LATEXIEREN: to latex (see DITMARISIEREN) 443

LATEXKITT (*m*): latex cement (caoutchouc globules in watery suspension) 444
LATRINE (*f*): latrine, water-closet 445
LATSCHE (*f*): slut; sloven; dwarf (knee) pine (see LATSCHENKIEFER) 446
LATSCHENKIEFER (*f*): dwarf (knee) pine (*Pinus montana pumilio*) 447
LATSCHENKIEFER-(N)-ÖL (*n*): dwarf pine oil (see FICHTENNADELÖL) 448
LATSCHENÖL (*n*): templin oil (*Oleum pini pumilionis*) (see FICHTENNADELÖL) 449
LATSCHIG: slovenly; damp; sloppy, slushy, muddy, wet, dirty 450
LATTE (*f*): lath; batten; brace 451
LATTEN: to lath, batten 452
LATTENHOLZ (*n*): (wood for)-laths or battens 453
LATTENWERK (*n*): lath-work; lattice-work 454
LATTENZAUN (*m*): pailing 455
LATTE (*f*), **STRAAK-**: batten for drawing curves 456
LATTICH (*m*): lettuce (*Lactuca scariola sativa*) 457
LATTUNG (*f*): (see LATTENWERK) 458
LATWERGE (*f*): electuary; confection; (*Electuaria*) (paste-like medicinal preparations, from solid, liquid or semi-liquid substances, for internal use) (Pharmacy) 459
LATZ (*m*): bib; flap; stomacher 460
LATZMÜTZE (*f*): cap with side (ear) flaps 461
LAU: tepid; lukewarm; mild; indifferent 462
LAUB (*n*): foliage; leaves 463
LAUBANIT (*m*): laubanite (a hydrous aluminium-calcium silicate) 464
LAUBAPFEL (*m*): oak-apple (see GALLAPFEL) 465
LAUBE (*f*): arbour; bower; hall 466
LAUBERDE (*f*): leaf (vegetable) mould 467
LAUBGANG (*m*): avenue; arcade 468
LAUBGITTER (*n*): trellis-work, lattice-work 469
LAUBGRÜN (*n*): leaf green (see CHROMGRÜN) 470
LAUBHOLZ (*n*): wood from deciduous (leafed) trees (as opposed to firs); leafy wood, leaved wood 471
LAUBIG: leaved; leafy; leaf-like; deciduous 472
LAUBKNOSPE (*f*): leaf bud 473
LAUBLOS: leafless 474
LAUBPFLANZE (*f*): (see KRYPTOGAMEN) 475
LAUBROST (*m*): (vine)-mildew 476
LAUBSÄGE (*f*): fret-saw 477
LAUBSÄGEBOGEN (*m*): fret-saw frame 478
LAUBSTIEL (*m*): Petiole, leaf-stalk (Bot.) 479
LAUCH (*m*): leek (*Allium porrum*) 480
LAUCHARTIG: leek-like 481
LAUCH (*m*), **GEMEINER-**: (see WINTERPORREE) 482
LAUCHGRÜN: leek-green 483
LAUCHKNOBLAUCH (*m*): (Spanish)-garlic (*Allium sativum*) 484
LAUCH (*m*), **SCHNITT-**: chives (*Allium schoenoprasum*) 485
LAUCHSCHWAMM (*m*): leek fungus (*Marasmium*) 486
LAUCH (*m*), **SPANISCHER-**: (see WINTERPOREE) 487
LAUDANIDIN (*n*): laudanidine (an opium alkaloid) $C_{17}H_{15}N(OH)(OCH_3)_3$ 488
LAUDANIN (*n*): laudanine (an opium alkaloid); $C_{17}H_{15}N(OH)(OCH_3)_3$; Mp. 166°C. 489
LAUDANOSIN (*n*): laudanosine (an opium alkaloid); $C_{17}H_{13}N(OCH_3)_4$; Mp. 89°C. 490
LAUER (*m*): sour (low; tart) wine; wine of the second press; (*f*): ambush; eavesdropping; spying; listening; lurking 491
LAUERN: to lurk; watch; spy; eavesdrop; ambush; lie in wait; listen; pry 492
LAUF (*m*): course; path; run; race; progress; current; way (of a ship); barrel (of a gun); career; movement; motion; running; track; chute 493
LAUFBAHN (*f*): (race)-course; career; run; race (Mech.); slide; chute; path; track 494
LAUFBLECH (*n*): foot-board -(of an engine) 495
LAUFBOHNE (*f*): scarlet runner (*Phaseolus multiflorus*) 496
LAUFBRETT (*n*): baseboard (Photography) 497
LAUFBURSCHE (*m*): errand boy; messenger; office boy 498
LAUFEN: to run; flow; go; ooze; move; turn; trend; leak; walk; sail; operate; circulate; act; function; continue; work 499
LAUFEND: running; current; consecutive; regularly; continuously 500
LAUFENDE NUMMER (*f*): running number (being the number of the column or line of a schedule or table, the columns or lines of which are numbered); serial number; consecutive number; current number 501
LAUFENDES BAD (*n*): running bath; wash bath (Phot.); standing bath (Dye) 502
LAUFENDES GUT (*n*): running rigging (Naut.) 503
LÄUFER (*m*): racer; runner; messenger; muller (Colours); rammer; slider; stretcher (Brickwork); stair-carpet; rotor; armature (Elect.); (tackle)-fall (Naut.); Bishop (Chess); sucker; shoot; guy-rope 504
LÄUFERMÜHLE (*f*): edge mill 505
LÄUFERPLANKE (*f*): cross timber (of sails), runner 506
LÄUFERSCHUH (*m*): runner shoe 507
LÄUFER (*m*), **SEGELSCHLITTEN-**: ice boat runner 508
LAUFFEUER (*n*): wild-fire; train of gunpowder; running fire 509
LAUFGEWICHTSWAGE (*f*): balance with poise or sliding weight 510
LAUFGLASUR (*f*): flow glaze (Ceramics) 511
LAUFGRABEN (*m*): communication trench; sap (Military) 512
LAUFKARREN (*m*): trolley, skip, truck 513
LAUFKATZE (*f*): crab, trolley, travelling crab (of cranes); run-way, aerial transporter 514
LAUFKATZE (*f*), **FERNSTEUERUNG-**: travelling crab with distance control 515
LAUFKATZE (*f*), **FÜHRERSTAND-**: self-contained travelling crab; travelling crab complete with driver's platform 516
LAUFKRAN (*m*): travelling crane; portable (crane) hoist 517
LAUFKRANZ (*m*): blade rim (of turbine) 518
LAUFPASS (*m*): dismissal; passport 519
LAUFPFANNE (*f*): cooler (Sugar) 520
LAUFPLANKEN (*f.pl*): gangway (Naut.) 521
LAUFRAD (*n*): rotor; runner wheel (Turbines) 522
LAUFRADSCHAUFEL (*f*): runner wheel blade, moving blade (Turbines) 523
LAUFRIEMEN (*m*): strap, belt 524
LAUFRING (*m*): ball race (of bearings) 525
LAUFROLLE (*f*): roller 526
LAUFSCHAUFEL (*f*): moving blade; bucket (Turbines, etc.) 527
LAUFSCHAUFELSEGMENT (*n*): bucket segment 528
LAUFSCHIENE (*f*): rail 529
LAUFSCHRANKEN (*f.pl*): sliding barrier-(s) 530
LAUFSITZ (*m*): medium fit 531

LAUFSITZ (*m*), **LEICHTER-**: free fit 532
LAUFSPINDEL (*f*): (bearing)-spindle 533
LAUFSTAG (*n*): man rope, life line; stay-rope (Naut.) 534
LAUFTAU (*n*): fall (Naut.) 535
LAUFWELLE (*f*): line (intermediate) shaft 536
LAUFWERK (*n*): traversing (running) gear; rollers 537
LAUFWERKGEHÄUSE (*n*): roller casing 538
LAUFWIDERSTAND (*m*): rolling (motion; frictional) resistance 539
LAUFWINDE (*f*): crab; travelling hoist 540
LAUFZAPFEN (*m*): neck journal 541
LAUFZEIT (*f*): period of validity (Patent) 542
LAUFZIRKEL (*m*): caliper compasses 543
LAUGE (*f*): lye (usually an alkaline solution); liquor; brine; buck (Bleaching) 544
LAUGE (*f*), **JAVELLE'SCHE-**: Eau de Javelle (see **CHLORKALILÖSUNG**) 545
LAUGEN: to lye; steep in lye; buck; lixiviate 546
LAUGENARTIG: resembling (like) lye; alkaline; lixivial 547
LAUGENASCHE (*f*): potash (see **POTTASCHE**) 548
LAUGENBAD (*n*): lye-(bath); liquor; steep 549
LAUGENECHT: fast to lye 550
LAUGENFLÜSSIGKEIT (*f*): liquor; alkaline solution; brine; lye (see **LAUGE**) 551
LAUGENHAFT: (see **LAUGENARTIG**) 552
LAUGENKRAUT (*n*): mountain arnica (*Arnica montana*) 553
LAUGENMESSER (*m*): alkalimeter 554
LAUGENPUMPE (*f*): lye pump 555
LAUGENSALZ (*n*): alkaline (lixivial) salt 556
LAUGENSALZ, FLÜCHTIGES- (*n*): sal volatile (ammonium carbonate) (see **AMMONIUM--KARBONAT**) 557
LAUGENSALZIG: alkaline; lixivial; resembling (like) alkaline or lixivial salt 558
LAUGENSALZ, MINERALISCHES- (*n*): sodium carbonate (see **NATRIUMKARBONAT**) 559
LAUGENSALZ, VEGETABILISCHES- (*n*): potassium carbonate (see **KALIUMCARBONAT**) 560
LAUGENSTÄNDER (*m*): lye container 561
LAUGENTUCH (*n*): bucking cloth 562
LAUGENVERDAMPFER (*m*): lye evaporator 563
LAUGENWASSER (*n*): (see **LAUGENFLÜSSIGKEIT**) 564
LAUGEREI (*f*): lixiviation; bucking; steeping 565
LAUGIG: (see **LAUGENARTIG**) 566
LAUHEIT (*f*): tepidity, lukewarmness, mildness, indifference 567
LAUMONTIT (*m*): laumontite; $CaAl_2Si_4O_{12}+4H_2O$; Sg. 2.3 568
LAUNE (*f*): humour; temper; mood; caprice; vein; frame of mind; whim 569
LAUNEN: to be moody, have whims or fancies, be capricious, be (variable, fickle) changeable, be fitful 570
LAUNENHAFT: (see **LAUNISCH**) 571
LAUNIG: humorous; droll; witty; moody 572
LAUNISCH: moody; fitful; splenetic; capricious; variable; fickle; changeable; whimsical 573
LAURANTHAZEEN (*f.pl*): *Loranthaceæ* (Bot.) (see **RIEMENBLUME**) 574
LAURIN (*n*): laurin; $C_{22}H_{30}O_3$; Mp. 44.5°C. 575
LAURINEENKAMPHER (*m*): (see **CAMPHOR, LAUR--INEEN**) 576
LAURINFETT (*n*): laurin (see **LAURIN**) 577
LAURINSÄURE (*f*): lauric acid; $C_{12}H_{24}O_2$; Mp. 43.5°C.; Bp. 102°C. 578
LAURIONIT (*m*): laurionite; PbCl[OH] 579
LAURIT (*m*): laurite; $(Ru,Os)S_2$; Sg. 7.0 580
LAUROCERIN (*n*): (see **LAURIN**) 581
LAUROSTEARIN (*n*): laurin (see **LAURIN**) 582
LAUROSTEARINSÄURE (*f*): laurostearic acid (see **LAURINSÄURE**) 583
LAUROSTEARINSÄUREÄTHYLÄTHER (*m*): lauric ethyl ether; $C_{12}H_{23}O_2.C_2H_5$; Bp. 269°C. 584
LAUS (*f*): louse (a vermin which, according to which part of the human body it infests-head, clothes, or pubic region—is termed *Pediculus capitis, pediculus vestimenti* or *pediculus pubis*) (Med.); fault, defect, blemish, flaw, imperfection, (as a light spot in dyeing, a knot in wool) 585
LAUSCHEN: to listen; eavesdrop; watch; spy; pry 586
LÄUSEBAUM (*m*): black alder tree (*Alnus glutinosa*) 587
LAUSEKÖRNER (*n.pl*): stavesacre seed (*Cocculus indicus*); sabadilla seed 588
LÄUSEKRAUT (*n*): red rattle, marsh louse-wort (*Pedicularis palustris*) 589
LÄUSEMITTEL (*n*): (see **LAUSTOD**) 590
LÄUSEPULVER (*n*): insect powder (see **LAUSTOD**) 591
LÄUSESUCHT (*f*): phthiriasis, pediculosis (bites from any of the three body lice; see **LAUS**) (according to the particular location of the vermin as head, body or pubic region, it is termed *Phthiriasis capitis, Phthiriasis corporis, Phthiriasis pubis* and *Pediculosis vestimenti*) (Med.) 592
LAUSTOD (*m*): lice killer, vermin (insect) killer, vermicide 593
LAUT: loud; audible; public; forte; aloud; according to; as per; in accordance with; (*m*): sound; tone 594
LAUTARIT (*m*): lautarite; $Ca[IO_3]_2$; Sg. 4.6 595
LAUTBAR: audible; public; notorious; known 596
LAUTE (*f*): lute (Musical Instrument); crutch (Dyeing) 597
LAUTEN: to sound; run; read; purport 598
LÄUTEN: to ring; toll; chime; peal 599
LAUTER: pure; clear; genuine; unalloyed; mere; unmixed; undefiled; upright 600
LÄUTER (*m*): bell-ringer; sour (low; tart) wine (see **LAUER**); (as a prefix=refining; purifying; clarifying) 601
LÄUTERBATTERIE (*f*): underlet (Brewing) 602
LAUTERBEIZE (*f*): white liquor (in turkey red dyeing) 603
LÄUTERBODEN (*m*): false bottom; strainer 604
LÄUTERBOTTICH (*m*): straining (clarifying) vat or tub 605
LÄUTERFLASCHE (*f*): washing bottle (for gases) 606
LÄUTERHAUS (*n*): refining house (Sugar) 607
LÄUTERKESSEL (*m*): defecator (Sugar); refining (copper) pan 608
LÄUTERMAISCHE (*f*): mash (liquor) liquid (Brewing) 609
LÄUTERN: to purify; strain; clarify; refine; filter; clear; defecate; rectify; purge (see **ABLÄUTERN**) 610
LÄUTEROFEN (*m*): refining (kiln) furnace 611

LÄUTERPFANNE (*f*): refining pan; defecator (Sugar) 612
LÄUTERTUCH (*n*): filter-(ing) cloth 613
LÄUTERUNG (*f*): purification; refining; clearing; clarifying; clarification; defecating; rectifying; rectification 614
LÄUTERUNGSMITTEL (*n*): purifying (clarifying) agent or medium 615
LAUTHEIT (*f*): loudness, resonance, sonorousness, audibility, physiological strength-(of sound) 616
LAUTIEREN: to spell phonetically 617
LÄUTINDUKTOR (*m*): ringing inductor 618
LAUTLEHRE (*f*): phonetics 619
LAUTLOS: mute, silent 620
LAUTPHYSIOLOGIE (*f*): physiology of sound 621
LAUTSPRECHENDES FUNKTELEFON (*n*): loud speaker (Wireless telephony or telegraphy) 622
LAUTSYSTEM (*n*): system of sounds; phonetic system 623
LÄUTWERK (*n*): bell, ringing apparatus 624
LAUWARM: lukewarm; tepid 625
LAVA (*f*): lava (rock ejected from the earth in a molten state) (see **BIMSSTEIN**) 626
LAVATÖL (*n*): residual olive oil (a very low quality of olive oil) 627
LAVENDEL (*m*): lavender (*Lavandula vera* or *L. officinalis*) 628
LAVENDELBLÜTEN (*f.pl*): lavender flowers (*Flores lavandulæ*) 629
LAVENDELBLÜTENÖL (*n*): (*Oleum lavandulæ florum*): lavender flower oil; Sg. 0.89 630
LAVENDELGEIST (*m*): lavender (spirit) essence (see **LAVENDELSPIRITUS**) 631
LAVENDELHEIDE (*f*): wild rosemary, marsh rosemary (*Rosmarinus officinalis*); marsh andromeda (*Andromeda polifolia*) 632
LAVENDELÖL (*n*): lavender oil (see **SPIKÖL**) (*Oleum lavandulæ*) (from *Lavandula vera*); Sg. 0.88-0.9; Bp. 186-192°C. 633
LAVENDELSPIRITUS (*m*): (*Spiritus lavendulæ*): lavender spirit (distilled from 1 part of lavender flowers to 3 parts of spirit of wine) 634
LAVENDELWASSER (*n*): lavender water 635
LÅVENIT (*m*): låvenite; $Na[ZrO.F](Mn,Ca,Fe)[SiO_3]_2$; Sg. 3.55 636
LAVIEREN: to tack (Naut.) 637
LÄVOGYR: levorotatory; levogyratory; rotating (gyrating) to the left (see also **LINKS-**) 638
LAVOISIT (*n*): lavoisit (trade name for a specially prepared calcium hypochlorite, for generating a very pure oxygen) 639
LÄVULINSÄURE (*f*): levulinic acid; $C_5H_8O_3$; Sg. 1.1395; Mp. 33°C.-37.2°C.; Bp. 250-253°C. 640
LÄVULINSÄURE-ÄTHYLESTER (*m*): levulinic ethyl-ester; $C_7H_{12}O_3$; Sg. 1.0156; Bp. 205.2°C. 641
LÄVULOSE (*f*): levulose (see **DEXTROSE**) (obtained from Inulin) (see **INULIN**) 642
LAWINE (*f*): avalanche 643
LAWSONIE, WEISSE- (*f*): white Lawsonia (*Lawsonia alba*) 644
LAWSONIT (*m*): lawsonite; $H_4CaAl_2Si_2O_{10}$; Sg. 3.085 645
LAX: lax; loose 646
LAXANS (*n*): (see **LAXANZ**) 647
LAXANS (*n*), **MILDES-**: mild laxative 648
LAXANZ (*f*): laxative; aperient; purgative 649
LAXANZ (*f*), **MILDE-**: mild laxative 650
LAXATIV: laxative; (*n*): laxative 651
LAXATIVES TONICUM (*n*): laxative tonic 652
LAXHEIT (*f*): laxity, looseness 653
LAXIEREN: to purge; take a (purgative; aperient) laxative 654
LAXIEREND: purging; aperient; laxative 655
LAXIERKONFEKT (*n*): laxative-(confection) (as confection of senna) 656
LAXIERMITTEL (*n*): laxative; purgative; aperient 657
LAXIERSALZ (*n*): laxative salt (as Glauber salt) 658
LAXIERSALZ, ENGLISCHES- (*n*): Epsom salt (see **MAGNESIUMSULFAT**) 659
LAXITÄT (*f*): laxity 660
LAZARETT (*n*): (military)-hospital 661
LAZARETTFIEBER (*n*): (see **KRIEGSPEST**) 662
LAZARETTGEHÜLFE (*m*): hospital (nurse) orderly 663
LAZULINBLAU (*n*): lazuline blue 664
LAZULIT-(H) (*m*): lazulite; $(Mg,Fe,Ca)[AlOH]_2[PO_4]_2$; Sg. 3.06 665
LAZURSTEIN (*m*): lapis lazuli (see **LAPIS LAZULI**) 666
LEADHILLIT (*m*): leadhillite; $[PbSO_4)_4[CO_3]_2+H_2O$; Sg. 6.4 667
LEBEN: to live; exist; dwell; be alive; (*n*): life; existence 668
LEBENDER KALK (*m*): quicklime (see **KALK, GEBRANNTER-**) 669
LEBENDIG: alive; live; living; lively; vivacious; active 670
LEBENDIGER KALK (*m*): quicklime (see **KALK, GEBRANNTER-**) 671
LEBENSART (*f*): mode of living 672
LEBENSASSEKURANZ (*f*): life insurance 673
LEBENSBAHN (*f*): career 674
LEBENSBAUM (*m*): arbor vitæ; tree-of-life; white cedar (*Thuja occidentalis*) 675
LEBENSBAUM (*m*), **GEWÖHNLICHER-**: arbor vitæ (Canadian)-white cedar (*Thuja occidentalis*) 676
LEBENSBEDÜRFNISSE (*n.pl*): necessaries of life 677
LEBENSBESCHREIBER (*m*): biographer 678
LEBENSBESCHREIBUNG (*f*): biography 679
LEBENSDAUER (*f*): lifetime; existence; life; period of life; durability 680
LEBENSFÄHIG: capable of (existing) living; vital; viable (Med.) 681
LEBENSFÄHIGKEIT (*f*): viability; vitality; capacity for living 682
LEBENSGESCHICHTE (*f*): biography 683
LEBENSGROSS: full-length (of portraits) 684
LEBENSGRÖSSE (*f*): life-size 685
LEBENSHALTUNGSKOSTEN (*f.pl*): cost of living 686
LEBENSHOLZ (*n*): lignum vitæ (from *Guaiacum officinale*) (see **GUAJAK**) 687
LEBENSKRAFT (*f*): vitality; vital force; vigour 688
LEBENSLANG: life-long 689
LEBENSLAUF (*m*): career 690
LEBENSLEHRE (*f*): biology 691
LEBENSLUFT (*f*): vital air; oxygen 692
LEBENSLUFTMESSER (*m*): eudiometer (apparatus for determining the quantity of a single gas in a gaseous mixture) 693
LEBENSMITTEL (*n* and *pl*): provisions; food; victuals; necessaries of life 694
LEBENSORDNUNG (*f*): diet; regimen 695
LEBENSPROZESS (*m*): vital (process) functions 696
LEBENSREGEL (*f*): maxim, rule of conduct 697

LEBENSRENTE (*f*): annuity (Insurance) 698
LEBENSRETTUNG (*f*): life-saving 699
LEBENSRETTUNGSAPPARAT (*m*): life-saving apparatus 700
LEBENSSAFT (*m*): vital fluid; latex (Bot.) (see **LATEX**) 701
LEBENSSTRAFE (*f*): capital punishment 702
LEBENSTEILE (*m.pl*): vital parts; vitals 703
LEBENSUNTERHALT (*m*): subsistence; living; livelihood; sustenance 704
LEBENSUNTERHALTUNGSKOSTEN (*f.pl*): cost of living 705
LEBENSVERRICHTUNG (*f*): vital function 706
LEBENSVERSICHERUNG (*f*): life insurance 707
LEBENSVERSICHERUNGSURKUNDE (*f*): life policy 708
LEBENSVORGANG (*m*): vital (process) functions 709
LEBENSWANDEL (*m*): (course of-)life; conduct 710
LEBENSWÄRME (*f*): vital heat 711
LEBENSWASSER (*n*): aqua-vitæ; spirits; cordial 712
LEBENSWEISE (*f*): mode of life; habit; diet 713
LEBENSWEISHEIT (*f*): worldly wisdom 714
LEBENSZEICHEN (*n*): sign of life 715
LEBENSZEIT (*f*): life-time; age 716
LEBENSZIEL (*n*): goal of life, aim of life, purpose in life 717
LEBER (*f*): liver; hepar; (as a prefix=hepatic) 718
LEBERABSCESS (*m*): abscess of the liver, hepatic abscess (Med.) 719
LEBERALOE (*f*): (see **ALOE, MATTE-**) 720
LEBERANEMONE (*f*): liverwort, liverleaf (see **HEPATICA**) (*Hepatica*) 721
LEBERANSCHWELLUNG (*f*): enlargement of tne liver (Med.) 722
LEBERBESCHWERDE (*f*): liver complaint 723
LEBERBLÜMCHEN (*n*): (see **LEBERBLUME**) 724
LEBERBLUME (*f*): liverwort; liverleaf (*Hepatica*) 725
LEBERBRAUN: liver-coloured 726
LEBERBRUCH (*m*): hepatocele (a hernial affection of the liver in which it obtrudes through the wall of the abdomen) (Med.) 727
LEBERDISTEL (*f*): (see **SKARIOL**) 728
LEBEREISENERZ (*n*): pyrrhotite; magnetic pyrites (see **PYRRHOTIN** and **MAGNETKIES**) 729
LEBERENTZÜNDUNG (*f*): hepatitis, inflammation of the liver (Med.) 730
LEBERENTZÜNDUNG (*f*), **AKUTE-**: acute inflammation of the liver; hepatitis; acute congestion of the liver (Med.) 731
LEBERENTZÜNDUNG (*f*), **CHRONISCHE INTERSTITIELLE-**: interstitial hepatitis, cirrhosis of the liver, gin drinker's liver (Med.) 732
LEBERERZ (*n*): hepatic ore; hepatic (liver-brown) cinnabar 733
LEBERFLECK (*m*): Chloasma (a pigmentary discolouration of the skin) (Med.) 734
LEBERFLECKEN (*m* and *pl*): (see **LEBERFLECK**) 735
LEBERGANG (*m*): hepatic duct 736
LEBERKIES (*m*): pyrrhotite (see **MAGNETKIES** and **PYRRHOTIN**) 737
LEBERKLEE (*m*): (see **HEPATICA** and **KLEE**) 738
LEBERKLETTE (*f*): (hemp)-agrimony (*Agrimonia eupatoria*) 739
LEBERKOLIK (*f*): Hepatic colic (see **HEPATALGIE**) 740
LEBERKRANK: suffering with the liver, having a liver complaint 741
LEBERKRANKHEIT (*f*): liver complaint 742
LEBERKRAUT (*n*): hepatica; liverwort; liverleaf [*Herba hepaticæ -(nobilis)*] (of *Hepatica triloba*) (see **HEPATICA**) 743
LEBERLEIDEN (*n*): liver complaint 744
LEBEROPAL (*m*): menilite (brown opaque opal) (see **OPAL**) 745
LEBERSCHMERZ (*m*): Hepatalgia (see **HEPATALGIE**) 746
LEBERSTÄRKE (*f*): glycogen; animal starch; $(C_6H_{10}O_5)_n$ 747
LEBERSTEIN (*m*): hepatite (variety of barite) (see **BARYT**) (Min.); gallstone; biliary calculus 748
LEBERTRAN (*m*): codliver oil (*Oleum jecoris aselli*); Sg. 0.92 (see also **KABELJAU**) 749
LEBEWESEN (*n*): organism; living being 750
LEBHAFT: lively; brisk; active; strong; vivacious; bright; brilliant; vivid; gay; sprightly; vigorous; smart 751
LEBLANC-PROZESS (*m*): Leblanc-(soda)-process 752
LEBLOS: lifeless; inanimate; impassive 753
LECANORSÄURE (*f*): lecanoric acid (see **LEKANORSÄURE**) 754
LECH (*m*): regulus; matt; metal 755
LECHZEN: to thirst; be parched; long (pine) for; split; gape; crack (of earth, due to drought) 756
LECITHIN (*n*): (*Lecithinum*): lecithin, ova-lecithin; $C_{44}H_{90}NPO_9$ 757
LECITHOL (*n*): lecithol (a chemically pure lecithin from egg-white) 758
LECK (*m* and *n*): leak-(age); (as an adj.=leaky; leaking; dripping) 759
LECKDAMPF (*m*): leakage steam (from stuffing boxes) 760
LECKEN: to leak; drip; trickle; lick; touch; sweep-(over) (as of gases over a heating surface) 761
LECKER: dainty; nice; tasty 762
LECKEREI (*f*): dainty; sweetmeat; daintiness; delicacy 763
LECK SEIN: to have sprung a leak, be leaky 764
LECKSTEIN (*m*): drip stone 765
LECKVERLUST (*m*): leakage loss 766
LECK WERDEN: to spring a leak (Naut., etc), become leaky 767
LEDER (*n*): leather; skin 768
LEDERABFALL (*m*): leather (cuttings) waste 769
LEDERÄHNLICH: leather-like 770
LEDERAPPRETUR (*f*): leather (dressing; finishing) finish 771
LEDERARTIG: tough, leathery, like leather, coriaceous 772
LEDERBAND (*m*): leather binding; leather (thong) strap 773
LEDERBEREITER (*m*): currier, leather dresser 774
LEDERBEREITUNG (*f*): leather (treating, tanning, preparation) dressing 775
LEDERBITUMEN (*n*): leather bitumen (a material consisting of leather waste and bitumen, for making insulating plates, etc.) 776
LEDERBRAUN: tawny 777
LEDER (*n*), **CHAGRIN-**: shagreen 778
LEDERCREME (*f*): leather (dressing; grease; oil) cream 779
LEDERERSATZ (*m*): leather substitute 780
LEDERETUI (*n*): leather (case) wallet 781
LEDERFABRIK (*f*): tannery 782

LEDERFÄRBUNG (*f*): leather dyeing 783
LEDERFETT (*n*): dègras, dubbing (see DEGRAS) 784
LEDERGELB: buff 785
LEDERGERBUNG (*f*): (leather)-tanning 786
LEDER (*n*), GESCHIRR-: harness leather 787
LEDER (*n*), GLACÉ-: glacé leather, patent leather 788
LEDERGUMMI (*n*): (india)-rubber 789
LEDERHANDEL (*m*): leather trade 790
LEDERHARZ (*n*): (india)-rubber 791
LEDERHAUT (*f*): corium, cutis, cutis vera, true skin (innermost layer of skin) 792
LEDERIMITATION (*f*): leather imitation, imitation leather 793
LEDER (*n*), KALBKID-: kid-(leather) 794
LEDERKALK (*m*): quicklime 795
LEDERKOHLE (*f*): leather (coal) charcoal (used for case-hardening, is obtained by heating leather waste) 796
LEDERKOLLODIUM (*n*): enamel collodion (for coating photographic prints to impart a brilliant surface; consists of pyroxylin, methylated alcohol and ether) 797
LEDER (*n*), KUNST-: artificial leather 798
LEDERLACK (*m*): leather varnish 799
LEDERLEIM (*m*): glue from leather waste; size; skin glue; cartilage glue 800
LEDERLEIMFABRIKATION (*f*): leather glue (skin glue) manufacture 801
LEDER (*n*), LOHGARES-: bark-tanned leather 802
LEDERMANSCHETTE (*f*): leather (collar) cuff 803
LEDERMEHL (*n*): ground leather; leather meal 804
LEDERN: (of)-leather; leathern 805
LEDER (*n*), OBER-: leather for uppers (Bootmaking) 806
LEDERÖL (*n*): leather oil 807
LEDERPACKUNG (*f*): leather packing 808
LEDERPAPIER (*n*): leather-paper (Paper) 809
LEDERPAPPE (*f*): leather board (Paper) 810
LEDERRIEMEN (*m*): leather (belt) strap 811
LEDER (*n*), RIEMEN-: belt leather 812
LEDER (*n*), SATTLER-: saddler's leather 813
LEDERSCHMIERE (*f*): degras; dubbing 814
LEDERSCHWÄRZE (*f*): black dye for leather; leather blacking; black leather-dye; currier's black 815
LEDER (*n*), SOHL-: sole leather 816
LEDERSTRAUCH (*m*): hop tree (*Ptelea trifoliata*) 817
LEDERSTREIFEN (*m*): strip of leather, leather thong 818
LEDERSURROGAT (*n*): leather substitute 819
LEDERTUCH (*n*): leather cloth 820
LEDERZUCKER (*m*): marshmallow paste (see ALTHEAWURZEL) 821
LEDERZUCKER, BRAUNER- (*m*): liquorice paste 822
LEDIG: exempt; devoid; unmarried; single; free; vacant; unemployed; unencumbered 823
LEDIGLICH: only; solely; merely; entirely; quite; purely 824
LEE (*n*): lee, lee-side (Naut.) 825
LEEANKER (*m*): lee anchor 826
LEEBULIN-(E) (*f*): lee bowline 827
LEEGIERIGKEIT (*f*): slackness 828
LEELÄUFER (*m*), EISJACHT-: ice yacht lee runner 829
LEER: empty; blank; void; vacant 830
LEERDARM (*m*): jejunum (see DARM, LEERER-) 831
LEERE (*f*): vacuum; emptiness; vacancy; void; gap; blank; vacuity; gauge; pattern (see LEHRE) 832
LEEREN: to empty; evacuate; drain; vacate 833
LEERFASS (*n*): empty (cask) vat; emptying vat (Paper) 834
LEERGEBRANNT: burnt out 835
LEERGEHEN: to run without load, to run light 836
LEERGEHEND: working (running) on no load, running light, no-load 837
LEERGEWICHT (*n*): dead-weight 838
LEERHEIT (*f*): vacancy, vacuity, emptiness, futility, inanity; inanition, exhaustion from hunger (Med.) 839
LEERLAUF (*m*): no-load (Elect., etc.) 840
LEERLAUFBÜCHSE (*f*): no-load bush 841
LEERLAUFDÜSE (*f*): no-load (empty; light-running) nozzle 842
LEERLAUFEN: to run without load, to run light 843
LEERLAUFEND: no-load 844
LEERLAUFSVERLUST (*m*): no-load loss 845
LEERSCHEIBE (*f*): loose pulley 846
LEERVERKAUF (*m*): forward (speculative) sale 847
LEESEGEL (*n*): studding sail 848
LEEWÄRTS: leeward 849
LEEWEG (*m*): leeway 850
LEFZE (*f*): lip 851
LEGALISIEREN: to legalize, validate, make valid 852
LEGAT (*m*): legacy, bequest; legate, consul, ambassador 853
LEGATION (*f*): legation, embassy, consulate 854
LEGEEISEN (*n*): hearth plate 855
LEGEL (*m*), REFF-: reef cringle 856
LEGEN: to lay-(down); put; place; deposit; plant; lie-(down); abate; devote oneself 857
LEGENDE (*f*): myth, allegory, tradition, fiction, fable, saga, legend, text, wording, subject matter, explanatory notes 858
LEGENDENHAFT: legendary, mythical, fabled, fabulous, fictitious, traditional, allegorical 859
LEGEN, EIN RUDER-: to put a helm 860
LEGESTEIN (*m*): coping stone 861
LEGIEREN: to alloy; bequeath 862
LEGIERUNG (*f*): alloy-(ing); alligation 863
LEGION (*f*): legion 864
LEGION (*f*), DIE ZAHL IST-: the number is legion 865
LEGISLATUR (*f*): legislature 866
LEGITIM: legitimate 867
LEGITIMIEREN: to legitimize 868
LEGUMINOSE (*f*): legume; pea meal; leguminous plant 869
LEHDE (*f*): fallow (waste) land 870
LEHEN (*m*): fief; fee; loan 871
LEHENSSYSTEM (*n*): feudal system 872
LEHM (*m*): loam; mud; clay 873
LEHMARTIG: loamy, clayey 874
LEHMBODEN (*m*): clay-(ey) soil 875
LEHMEN: loamy, clayey; to plaster (smear) with clay or mud 876
LEHMFORM (*f*): loam (clay) mould 877
LEHMFORMEREI (*f*): loam (clay) moulding; clay modelling 878
LEHMGRUBE (*f*): loam-pit, clay-pit 879
LEHMGUSS (*m*): loam casting 880
LEHMHÜTTE (*f*): mud hut 881
LEHMIG: loamy; clayey 882
LEHMKALK (*m*): argillocalcite; argillaceous limestone 883
LEHMKERN (*m*): loam (clay) core 884

LEHMKITT (*m*): loam lute 885
LEHMMERGEL (*m*): loamy marl 886
LEHMSTEIN (*m*): unburnt brick; adobe; sun-dried brick 887
LEHMWAND (*f*): clay (mud) wall 888
LEHNE (*f*): incline; slope; prop; support; back (of chair); railing; hand-rail; back-rest; arm (of chair); balustrade 889
LEHNEN: to lean; slope; incline; recline; lend; borrow 890
LEHNSATZ (*m*): lemma, an assumed proposition 891
LEHNWORT (*n*): borrowed word 892
LEHRAMT (*n*): tutorship, professorship 893
LEHRANSTALT (*f*): school, academy, polytechnic, college, university, etc. 894
LEHRBAR: tractable, teachable 895
LEHRBOGEN (*m*): centre, centering, wooden frame for arches (Building) 896
LEHRBRETT (*n*): pattern, template 897
LEHRBRIEF (*m*): apprentice's indenture 898
LEHRBUCH (*n*): text-book; manual 899
LEHRE (*f*): teaching; instruction; tenet; doctrine; dogma; science; lesson; apprenticeship; pattern; model; gauge; balance; centre; centering; inference; conclusion; template; rule; precept; moral; (as a suffix=ology) 900
LEHREN: to teach; instruct; inform; prove; gauge 901
LEHREND: instructive, didactic, preceptive 902
LEHRER (*m*): teacher; instructor; professor; tutor; master 903
LEHRERSEMINAR (*n*): teachers' training college 904
LEHRE (*f*), SCHIEB-: sliding (gauge) calipers 905
LEHRFACH (*n*): teaching; branch of study 906
LEHRGANG (*m*): course-(of instruction or lectures) 907
LEHRGEBÄUDE (*n*): system 908
LEHRGEHÜLFE (*m*): assistant teacher, pupil teacher 909
LEHRGELD (*n*): premium (for an apprentice); fee (for instruction) 910
LEHRJAHRE (*n.pl*): school-days; (years of)-apprenticeship 911
LEHRKURSUS (*m*): course of study 912
LEHRLING (*m*): apprentice; pupil; scholar; student; novice; learner 913
LEHRMEINUNG (*f*): hypothesis; dogma 914
LEHRMEISTER (*m*): master, teacher, instructor 915
LEHRMITTEL (*n*): means of (material for) instruction 916
LEHRREICH: instructive 917
LEHRSAM: tractable, docile, pliant, yielding, teachable 918
LEHRSATZ (*m*): teaching; dogma; aphorism; precept; thesis; proposition; theorem (Maths.) 919
LEHRSPANT (*n*): mould 920
LEHRSPRUCH (*m*): aphorism, maxim (see LEHR-SATZ) 921
LEHRSTUHL (*m*): professor's chair 922
LEHRSTUNDE (*f*): lesson, lecture 923
LEHRVERTRAG (*m*): apprentice's indenture 924
LEHRWIDRIG: heterodox 925
LEHRZEIT (*f*): apprenticeship 926
LEIB (*m*): body; womb; belly; trunk; waist 927
LEIBEN: to exist 928
LEIBESBESCHAFFENHEIT (*f*): Constitution (conformation) of the body, Diathesis (Med.) 929
LEIBESERBE (*m*): descendant, offspring; (*pl*): issue 930
LEIBESFRUCHT (*f*): fœtus; embryo 931
LEIBESKRAFT (*f*): physical strength 932
LEIBESÖFFNUNG (*f*): movement, motion, stool 933
LEIBESSCHWÄCHE (*f*): corporal (weakness) infirmity 934
LEIBESSTÄRKE (*f*): corporal strength, strength of body 935
LEIBESSTRAFE (*f*): corporal punishment 936
LEIBE, ZU- GEHEN: to probe (the matter) 937
LEIBGEDINGE (*n*): jointure, settlement, annuity -(for life) 938
LEIBLICH: bodily, corpor-(e)-al, natural 939
LEIBLOS: incorporeal 940
LEIBRENTE (*f*): annuity (Insurance) 941
LEIBRENTNER (*m*): annuitant 942
LEIBSCHMERZEN (*m.pl*): stomach-ache 943
LEIBSCHNEIDEN (*n*): gripes, flatulency, colic 944
LEIBWASSERSUCHT (*f*): ascites, dropsy of the abdomen (Med.) 945
LEICHDORN (*m*): corn; callosity (Med.) 946
LEICHDORNSALBE (*f*): corn ointment 947
LEICHDORNSCHNEIDER (*m*): chiropodist 948
LEICHE (*f*): corpse; cadaver; funeral 949
LEICHENALKALOID (*n*): putrefactive alkaloid; ptomaine 950
LEICHENARTIG: cadaverous 951
LEICHENBASE (*f*): putrefactive base; ptomaine 952
LEICHENBESCHAUER (*m*): coroner 953
LEICHENBESORGER (*m*): undertaker 954
LEICHENBESTATTER (*m*): undertaker 955
LEICHENBLÄSSE (*f*): deathly pallor 956
LEICHENDIEB (*m*): body-snatcher 957
LEICHENDUFT (*m*): cadaverous odour 958
LEICHENEULE (*f*): screech-owl, barn owl (*Strix flammea*) 959
LEICHENFETT (*n*): adipocere; corpse fat 960
LEICHENGEDICHT (*n*): elegy, funeral poem, song of mourning 961
LEICHENGERUCH (*m*): cadaverous odour 962
LEICHENHAFT: cadaverous; funereal 963
LEICHENHEMD (*n*): shroud, winding sheet 964
LEICHENKAMMER (*f*): mortuary; dissecting chamber 965
LEICHENÖFFNUNG (*f*): autopsy, post-mortem (Med.) 966
LEICHENSCHAU (*f*): Necropsy (a post-mortem examination and inspection of a body) (Med.); inquest 967
LEICHENSCHAUSTÄTTE (*f*): morgue, mortuary 968
LEICHENSTEIN (*m*): tomb-stone 969
LEICHENTUCH (*n*): pall; shroud 970
LEICHENUNTERSUCHUNG (*f*): (see LEICHEN-SCHAU) 971
LEICHENVERBRENNUNG (*f*): cremation 972
LEICHENWACHS (*n*): adipocere 973
LEICHNAM (*m*): corpse; cadaver; (dead)-body 974
LEICHT: light; faint; easy; simple; mild; slight; easily; soft; readily; lightly; softly; slightly 975
LEICHTBEWEGLICH: mobile; easily movable 976
LEICHTENTZÜNDLICH: easily (ignited) inflammable 977
LEICHTER (*m*): lighter (Boat) 978
LEICHTFASSLICH: easily (understood) comprehensible 979

LEICHTFLÜCHTIG: readily volatile 980
LEICHTFLÜGLER (*m.pl*): lepidoptera (an order of insects) (Ent.) 981
LEICHTFLÜSSIG: mobile; easily (dissolved) liquefiable; easily (soluble) fusible; easily flowing; thinly liquid 982
LEICHTFLÜSSIGKEIT (*f*): fluidity 983
LEICHTFRUCHT (*f*): light grain (oats, etc.) 984
LEICHTGEBLEICHT (*n*): easy bleach (Paper) 985
LEICHTGLAUBIG: credulous 986
LEICHTIGKEIT (*f*): simplicity; lightness; facility; ease; readiness 987
LEICHTLEBIG: light-hearted 988
LEICHTLICH: easily, lightly 989
LEICHTLÖSLICH: easily (dissolved) soluble 990
LEICHTMETALL (*n*): light metal (having a specific gravity of less than 3.0; usually applied to **MAGNALIUM** and **ELEKTRON-LEICHTMETALL**, which see) 991
LEICHTÖL (*n*): light oil 992
LEICHTREDUZIERBAR: easily reducible 993
LEICHTSINNIG: frivolous; thoughtless 994
LEICHTSPAT (*m*): light spar; gypsum (see **GYPS**) 995
LEICHTVERDERBLICH: perishable 996
LEID (*n*): sorrow; pain; wrong; harm; mourning; hurt; grief; injury 997
LEIDEN: to bear; tolerate; suffer; ail; undergo; endure; (*n*): suffering; distress; ailment; malady 998
LEIDENER FLASCHE (*f*): Leyden jar 999
LEIDENSCHAFT (*f*): passion; emotion; rage; desire 000
LEIDENSCHAFTLICH: vehement, impassioned, passionate 001
LEIDENSCHAFTSLOS: dispassionate 002
LEIDER: unfortunately; alas 003
LEIDIG: fatal; evil; tiresome; miserable; disagreeable 004
LEIDLICH: tolerable; fair; passable; moderate; mediocre 005
LEIDWESEN (*n*): sorrow; regret; grief; lamentation 006
LEIER (*f*): lyre (Instrument) 007
LEIERN: to play the lyre; harp; drawl 008
LEIHBIBLIOTEK (*f*): lending library, circulating library 009
LEIHEN: to lend; borrow; hire; confer 010
LEIHWEISE: on loan 011
LEIHWEISE ÜBERLASSEN: to loan 012
LEIK (*n*): bolt rope (see **LIEK**); (foot)-leech-(rope) 013
LEIKEN: to rope (a sail) 014
LEIKNADEL (*f*): roping needle 015
LEIM (*m*): glue; size 016
LEIMARTIG: gluey; glutinous; gelatinous (see also **LEIMIG**) 017
LEIMBEREITUNG (*f*): glue (preparation) manufacture, preparation of glue 018
LEIMBLOCKSCHNEIDEMASCHINE (*f*): glue-(block) cutting machine 019
LEIMBRÜHE (*f*): (hot)-liquid glue 020
LEIMDÄMPFER (*m*): glue extractor 021
LEIMEN: to glue; size; lime; distemper; glue (water-colour) paint 022
LEIMEND: adhesive; agglutinative 023
LEIMEN, GANZ-: to hardsize (Paper) 024
LEIMER (*m*): gluer, sizer 025
LEIMFABRIK (*f*): glue factory 026
LEIMFABRIKATION (*f*): glue manufacture 027
LEIMFARBE (*f*): distemper; size colour (Paper) 028
LEIM (*m*), **FARBLOSER-**: gelatine 029
LEIMFASS (*n*): size (glue) tank (Paper) 030
LEIMFESTIGKEIT (*f*): resistance (imperviousness) due to sizing (Paper) 031
LEIM, FLÜSSIGER- (*m*): liquid glue; mucilage 032
LEIMFORM (*f*): glue mould 033
LEIMFUGE (*f*): glued joint 034
LEIMGALLERTE (*f*): glue jelly 035
LEIMGEBEND: yielding glue; gelatinous 036
LEIMGEBENDER STOFF (*m*): glue-yielding (gelatinous) substance 037
LEIMGEWINNUNG (*f*): production (extraction) of glue 038
LEIMGLANZ (*m*): size (Leather) 039
LEIMGRUND (*m*): sizing 040
LEIMGUT (*n*): glue stock; material for making glue 041
LEIMIG: glutinous; gluey; viscous; viscid; sticky; gummy; adhesive 042
LEIMKESSEL (*m*): glue or size (boiler) copper 043
LEIMKITT (*m*): putty; joiner's cement 044
LEIMKRAUT (*n*): catchfly (*Silene quinquevulnera*) 045
LEIMLEDER (*n*): scrap (waste) leather (for glue making) 046
LEIMLÖSUNG (*f*): glue solution 047
LEIMNIEDERSCHLAG (*m*): lye (soap) precipitate (Curd soap) 048
LEIM (*m*), **PFLANZLICHER-**: gluten (see also **COLLODIN**) 049
LEIMPINSEL (*m*): glue brush 050
LEIMPROBER (*m*): glue tester 051
LEIMPULVER (*n*): powdered glue; glue powder 052
LEIMSCHNEIDEMASCHINE (*f*): glue cutting machine 053
LEIMSEIFE (*f*): filled soap (containing glycerine and lye) 054
LEIMSIEDER (*m*): glue boiler 055
LEIMSORTE (*f*): kind of glue 056
LEIMSTOFF (*m*): gluten 057
LEIMSUBSTANZ (*f*): gelatinous susbtance 058
LEIMSÜSS (*n*): glycine (see **AMINOESSIGSÄURE**) 059
LEIMTABLETT (*n*): glue tablet 060
LEIMTAFEL (*f*): glue tablet 061
LEIMTAFELSCHNEIDEMASCHINE (*f*): glue-(tablet) cutting machine 062
LEIMTIEGEL (*m*): glue pot 063
LEIMTOPF (*m*): glue pot 064
LEIMTRÄNKE (*f*): glue-(for bookbinding) 065
LEIMTROCKENANLAGE (*f*): glue drying plant 066
LEIMUNG (*f*): glueing; sizing; distempering 067
LEIMUNG IM BOGEN: sheet sizing (Paper) 068
LEIMUNG IM STOFF: pulp sizing (Paper) 069
LEIMUNG (*f*), **VEGETABILISCHE-**: vegetable sizing (see **BÜTTENLEIMUNG**) 070
LEIMVERGOLDUNG (*f*): water-(size) gilding 071
LEIMWASSER (*n*): glue water; size; lime water 072
LEIM (*m*), **WEISSER-**: gelatine 073
LEIMZUCKER (*m*): glycine (see **AMINOESSIG-SÄURE**) 074
LEIMZWINGE (*f*): glue press, clamping frame, joiner's clamp 075
LEIN (*m*): flax (see **FLACHS**); linseed 076
LEINBAU (*m*): flax cultivation 077
LEINBAUM (*m*): elm-(tree) (*Ulmus campestris*) 078
LEINDOTTER (*m*): cameline (*Camelina sativa*) 079
LEINDOTTERÖL (*n*): cameline oil (from seeds of *Camelina sativa*); Sg. 0.925 080

LEINDOTTERPFLANZE (*f*): cameline, gold of pleasure (*Camelina sativa*) 081
LEINDRUCKEREI (*f*): linen printing 082
LEINE (*f*): line; cord; rope; lash; halliard (Naut.) 083
LEINEN (*n*): linen; (as an *adj.* = linen; of linen) 084
LEINENFÄRBEREI (*f*): linen (printing) dyeing 085
LEINENFÄRBUNG (*f*): linen dyeing 086
LEINENFASER (*f*): linen (fibre) thread 087
LEINENGARN (*n*): linen yarn 088
LEINENPROBE (*f*): linen sample 089
LEINENZEUG (*n*): linen material 090
LEINENZWIRN (*m*): linen thread 091
LEINFELD (*n*): flax field 092
LEINFINK (*m*): linnet (*Linota cannabina*) 093
LEINFIRNIS (*m*): linseed varnish; thick linseed oil 094
LEINGRÜN (*n*): pale green (also *adj.*, pale-green) 095
LEINKNOTEN (*m*): husk of (linseed) flaxseed 096
LEINKRAUT (*n*): toadflax; yellow toadflax; wild flax (*Linaria vulgaris*) 097
LEINKUCHEN (*m*): linseed-(oil)-cake 098
LEINMEHL (*n*): linseed (flaxseed) meal 099
LEINÖL (*n*): linseed (boiled) oil (*Oleum lini*) (from **LEINSAMEN**, which see); Sg. 0.924-0.938 100
LEINÖLBEREITUNG (*f*): linseed oil manufacture 101
LEINÖLERSATZ (*m*): linseed oil substitute 102
LEINÖLFIRNIS (*m*): linseed oil varnish 103
LEINÖLFIRNISERSATZ (*m*): linseed oil varnish substitute; drying substitute 104
LEINÖL (*n*), **KALTGESCHLAGENES-**: cold-pressed linseed oil 105
LEINÖLSAUER: linoleate, combined with linoleic acid 106
LEINÖLSÄURE (*f*): linoleic acid (see **LINOLSÄURE**); $[CH_3(CH_2)_4.CH:CHCH_2CH:CH(CH_2)_7:COOH]$ 107
LEINÖLSAURES SALZ (*n*): linoleate 108
LEINÖLTROCKENPROZESS (*m*): linseed oil drying process 109
LEINPFAD (*m*): towing-path-(along a river, etc.) 110
LEINPFLANZE (*f*): flax (see **FLACHS**) 111
LEINSAAT (*f*): linseed; flaxseed (seed of *Linum usitatissimum*) 112
LEINSAATÖL (*n*): linseed oil (see **LEINÖL**) 113
LEINSAATÖLFIRNIS (*m*): linseed oil varnish 114
LEINSAMEN (*m*): (see **LEINSAAT**) 115
LEINSAMENMEHL (*n*): flaxseed (linseed) meal 116
LEINSAMENÖL (*n*): linseed oil (see **LEINÖL**) 117
LEINTUCH (*n*): sheet-(ing); linen-(stuff); linen material; canvas 118
LEINWAND (*f*): linen; canvas (for Art) 119
LEINWANDBAND (*m*): cloth binding 120
LEINWAND (*f*), **GEBLEICHTE-**: bleached linen 121
LEINWAND (*f*), **GROBE-**: coarse linen, canvas, sackcloth, sacking, bagging 122
LEINWANDHANDEL (*m*): linen trade 123
LEINWANDHÄNDLER (*m*): linen-draper 124
LEINWANDKITTEL (*m*): linen smock 125
LEINWAND (*f*), **UNGEBLEICHTE-**: unbleached linen 126
LEINWANDVERBINDUNG (*f*): linen bandage 127
LEINWAND (*f*), **VERPACKUNGS-**: bagging, sacking, sackcloth 128
LEINWANDWEBER (*m*): linen weaver 129
LEINWANDWEBEREI (*f*): (see **LEINWEBEREI**) 130
LEINWEBER (*m*): linen weaver 131
LEINWEBEREI (*f*): linen (manufacture)weaving, linen (factory) weaving mill 132
LEINWEBERSTUHL (*m*): linen weaver's loom 133
LEINZEUG (*n*): linen-(material) 134
LEIPZIGERGELB (*n*): Leipzig yellow, chrome yellow (see **CHROMGELB** and **BLEICHROMAT**) 135
LEISE: low; soft; slight; gentle; imperceptible; light; gently; softly; imperceptibly 136
L-EISEN (*n*): L irons, angle-irons, angles 137
LEISTBAR: serviceable 138
LEISTE (*f*): band; ledge; fillet; selvedge; border; list; guide-(plate); face; bracket; clamp; groin; facing; shoulder; projection; cornice; beading; moulding 139
LEISTEN: to do; act; perform; confer; accomplish; effect; render; pay; meet; deal with; afford; produce; give; fulfil 140
LEISTEN (*m*): (shoemaker's)-last 141
LEISTENBAND (*n*): inguinal ligament 142
LEISTENBEULE (*f*): Bubo, Apostema inguinis (swelling of inguinal gland) (Med.) 143
LEISTENBRUCH (*m*): bubonocele, inguinal (hernia) rupture (Med.) 144
LEISTENGESCHWULST (*f*): swelling or tumour in the groin 145
LEISTENHOBEL (*m*): rabbet plane, egee plane 146
LEISTENWERK (*n*): moulding 147
LEISTUNG (*f*): load; working; duty; output; evaporation; work; effect; performance; capacity; service; power (Mech.); doing; rendering; activity; payment; production 148
LEISTUNGSBEDARF (*m*): output; duty 149
LEISTUNGSFÄHIG: capable, serviceable, able, powerful, efficient, substantial, first-class 150
LEISTUNGSFÄHIGKEIT (*f*): serviceability; power; ability; capacity; duty; rating; efficiency; load 151
LEISTUNGSFAKTOR (*m*): load (output) factor 152
LEISTUNGSREGELUNG (*f*): load regulation 153
LEISTUNGSREGLER (*m*): load regulator (Mech.) 154
LEISTUNGSSPITZE (*f*): maximum output; peak-(load) 155
LEISTUNGSVERSUCH (*m*): load test, evaporative test (of boilers)- 156
LEISTUNGSVERTEILUNG (*f*): distribution of (output) load (of generating plant) 157
LEITACHSE (*f*): leading axle, front axle, fore axle, sliding axle (see also **LENK**) 158
LEITARM (*m*): crank 159
LEITBAR: manageable; tractable; capable of being guided; docile 160
LEITBLECH (*n*): guide (baffle) plate 161
LEITBLOCK (*m*): leading block; guide block 162
LEITEN: to conduct; lead; guide; pass; govern; preside over; manage; transmit; convey; direct 163
LEITENDER DIREKTOR (*m*): managing director 164
LEITER (*m*): conductor; leader; guide; director; manager; governor; stay (Naut.) 165
LEITER (*f*): ladder; scale, gamut: rack (of a waggon) 166
LEITERBAUM (*m*): ladder (beam) side 167
LEITER, SCHLECHTER- (*m*): bad conductor (of insulation) 168
LEITERSPROSSE (*f*): ladder step or rung 169
LEITFADEN (*m*): clue; guide; manual; key; thread; hint; indication 170

LEITFÄHIG: conducting; conductive 171
LEITFÄHIGKEIT (*f*): conductivity; conductibility; conducting (power) capacity 172
LEITFOSSIL (*n*): *Belemnitella quadrata* 173
LEITKLAMPE (*f*): leading cleat 174
LEITKRANZ (*m*): diaphragm (guide-wheel; guide-blade disc) rim 175
LEITLINIE (*f*): guide line; directrix 176
LEITMITTEL (*n*): vehicle 177
LEITRAD (*n*): diaphragm; guide-wheel; guide-blade disc; fixed wheel 178
LEITRADBÜCHSE (*f*): diaphragm (guide-wheel; guide-blade disc) bush or liner 179
LEITRADNABE (*f*): diaphragm (guide-wheel; guide-blade disc) boss or hub 180
LEITRADPUMPE (*f*): turbine pump, pump with guide wheel 181
LEITRIEMEN (*m*): leash; rein 182
LEITROHR (*n*): conducting (delivery; conduit) pipe, tube, duct or main 183
LEITRÖHRE (*f*): (see **LEITROHR**) 184
LEITROLLE (*f*): guide pulley, guide roller 185
LEITSATZ (*m*): rule; guiding principle 186
LEITSCHAUFEL (*f*): guide blade; bucket; vane 187
LEITSCHEIBE (*f*): diaphragm (guide blade disc); (of turbines) 188
LEITSCHIENE (*f*): guide rail 189
LEITSPINDEL (*f*): leading screw, guide spindle, lathe spindle 190
LEITSPINDELBANK (*f*): slide lathe 191
LEITSPINDELDREHBANK (*f*): leading screw lathe 192
LEITSTANGE (*f*): radius bar or rod, parallel motion guide rod 193
LEITSTERN (*m*): polestar; guiding star 194
LEITTON (*m*): leading note (Music) 195
LEITUNG (*f*): conduction (Steam); conducting; transmission; circuit (Elect.); line; feeder; wire; cable; lead (Elect.); leading; direction; conduit; guide; guidance; management; care; government; (railway)-points; cross-over; switch (Railway); pipe-(line); main; duct; channel; canal 196
LEITUNG (*f*), **NEGATIVE-**: negative feeder (Elect. three-wire system) 197
LEITUNG (*f*), **POSITIVE-**: positive feeder (Elect. three-wire system) 198
LEITUNGSBEHÖRDE (*f*): board (committee) of directors, governors or guardians 199
LEITUNGSDRAHT (*m*): conducting (telegraph; conduction) wire or wiring 200
LEITUNGSFÄHIGKEIT (*f*): (see **LEITFÄHIGKEIT**) 201
LEITUNGSKRAFT (*f*): conducting power; conductivity 202
LEITUNGSROHR (*n*): (see **LEITROHR**) 203
LEITUNGSRÖHRE (*f*): (see **LEITROHR**) 204
LEITUNGSVERLUSTE (*m.pl*): pipe losses; losses due to conduction 205
LEITUNGSVERMÖGEN (*n*): conducting power; conductivity 206
LEITUNGSWÄRME (*f*): (heat of)-conduction 207
LEITUNGSWASSER (*n*): tap (town) water; water in pipelines, conduits or mains 208
LEITUNGSWIDERSTAND (*m*): resistance (Elect.); conduction resistance 209
LEITZAUM (*m*): bridle, rein 210
LEITZELLE (*f*): guide box or bucket 211
LEITZUNGEN (*f.pl*): points, catch-points (Railways) 212
LEKANORSÄURE (*f*): lecanoric acid; orsellic (diorsellinic) acid 213
LEKTION (*f*): lesson; lecture; rebuke 214
LEKTÜRE (*f*): reading 215
LEMNISCHE ERDE (*f*): Lemnian earth (see **BOLUS, WEISSER-**) 216
LEMONGRASÖL (*n*): lemon-grass oil; melissa oil; verbena oil (from *Andropogon citratus*); Sg. about 0.9 217
LENDE (*f*): loin-(s); haunch; thigh; hip (see **HÜFT-**) 218
LENDENGEGEND (*f*): lumbar region 219
LENDENGICHT (*f*): sciatica (Med.) 220
LENDENKNOCHEN (*m*): hip bone, thigh bone 221
LENDENKRANKHEIT (*f*): sciatica (Med.) 222
LENDENMUSKEL (*m*): psoas (Anat.) 223
LENDENSCHMERZ (*m*): lumbago (Med.) 224
LENDENWEH (*n*): lumbago (Med.) 225
LENDENWIRBEL (*m*): lumbar vertebra (Anat.) 226
LENKACHSE (*f*): leading axle, front axle, guiding axle, sliding axle (see also **LEIT-**) 227
LENKBAR: (see **LEITBAR**) 228
LENKEN: to guide; direct; steer; drive; govern; rule; order 229
LENKER (*m*): ruler, etc. (see **LENKEN** and **LEITEN**); steersman; steering-rod 230
LENKRAD (*n*): guide wheel; steering wheel 231
LENKROLLE (*f*): leading pulley, guide pulley 232
LENKSEIL (*n*): guide-rope 233
LENKSTANGE (*f*): connecting rod (for steering), steering rod, handlebar 234
LENKUNG (*f*): ruling; steering, etc. (see **LENKEN** and **LEITEN**) 235
LENKVORRICHTUNG (*f*): guiding (leading) apparatus, steering gear, governing device 236
LENS: free, clear, empty, bare, exhausted 237
LENSEN: to scud under bare poles (Naut.) (see **LENZEN**) 238
LENZ (*m*): spring; prime 239
LENZEJEKTOR (*m*): bilge ejector 240
LENZEN: to scud (run) (of ships before the wind) (see **LENSEN**) 241
LENZIN (*m*): light spar; Gypsum (see **GYPS**) 242
LENZLICH: vernal 243
LENZPUMPE (*f*): bilge pump, emptying (clearing) pump (for dry docks) (for pumping the last remaining water after the main pumps have ceased to suck) 244
LEONIT (*m*): leonite; $K_2Mg[SO_4]_2+H_2O$; Sg. 2.38 245
LEOPOLDIT (*m*): leopoldite; KCl; Sg. 1.95 246
LEPIDOKROKIT (*m*): lepidocrocite; $Fe_2O_3.(Mn)H_2O$; Sg. 3.7 247
LEPIDOLIT-(H) (*m*): lepidolite (a lithium mica; 4-6% Li); Li_2O, $Al_2O_3,(F,OH)_2$, $3SiO_2$; Sg. 3.0-3.7 248
LEPIDOMELAN (*m*): lepidomelane; $K_2O,2H_2O$, $12(Fe,Mg)O,3(Fe,Al)_2O_3,12SiO_2$; Sg. 2.9 (see also **BIOTIT**) 249
LEPTANDRIN (*n*): leptandrine (from roots of *Veronica virginica*) 250
LERCHE (*f*): lark; (woodlark, *Alauda arborea*; skylark, *Alauda arvensis*); larch (*Larix europæa*) (Bot.) (see **LÄRCHE**) 251
LERCHENBLUME (*f*): primula, primrose (*Primula vulgaris*) 252
LERCHENKLAUE (*f*): field larkspur (*Delphinium consolida*) 253
LERNBAR: capable of being learnt 254
LERNEN: to learn; assimilate; acquire; study 255

LERNEN, AUSWENDIG- : to learn by (heart) rote 256
LERNEN, KENNEN- : to become acquainted with ; make the acquaintance of 257
LESART (*f*) : reading ; manner (method ; mode ; style) of reading 258
LESBAR : readable, legible, plain, decipherable 259
LESBARKEIT (*f*) : legibility 260
LESE (*f*) : vintage ; gathering ; gleaning ; harvest ; pickings ; collecting ; collection 261
LESEBAND (*n*) : conveyor (from which boys pick out the clean lump waste and throw it into waggons) (Coal) 262
LESEBERGE (*m.pl*) : waste pickings (Mining) 263
LESEBUCH (*n*) : reader ; reading book 264
LESEHOLZ (*n*) : wind-fallen wood 265
LESEN : to screen ; gather ; glean ; cull ; collect pick ; lecture ; read 266
LESENSWERT : worth reading 267
LESENSWÜRDIG : worth reading 268
LESERLICH : legible 269
LESESAAL (*m*) : reading-room ; lecture-room 270
LESESTEIN (*m*) : bog (brown) iron ore (see **BRAUNEISENERZ**) ; boulder 271
LETAL : lethal ; fatal 272
LETHARGIK (*f*) : lethargy 273
LETTE (*f*) : clay, loam 274
LETTEN (*m*) : loam ; potter's clay 275
LETTENARTIG : clayey, loamy 276
LETTENBODEN (*m*) : clay-soil 277
LETTERDRUCK (*m*) : letter-press ; printing 278
LETTERNGUT (*n*) : type metal 279
LETTERNMETALL (*n*) : type metal 280
LETTERNMETALL (*n*), **FEINES- :** special type metal (60% Pb ; 25% Sb ; 15% Sn) 281
LETTERNMETALL (*n*), **GEWÖHNLICHES- :** ordinary type metal (75% Pb ; 23% Sb ; 2% Sn) 282
LETTIG : clayey ; loamy 283
LETZT : last, lowest, final, ultimate, latest, extreme 284
LETZTE (*m, f* and *n*) : last ; latest ; lowest ; conclusion ; end ; termination ; hindermost ; finish 285
LETZTENS : lastly, finally, recently 286
LETZTERWÄHNT : latter, last (named) mentioned 287
LETZTGEMELDET : last quoted 288
LETZTGENANNT : latter ; last-(named) ; last mentioned 289
LEUCHÄMIE (*f*) : leukemia ; leukæmia ; leucocythæmia (Med.) 290
LEUCHTARM (*m*) : (gas)-bracket 291
LEUCHTBAKE (*f*) : light-(signal)-beacon, beacon 292
LEUCHTBOJE (*f*) : light-buoy 293
LEUCHTBRENNER (*m*) : illuminating burner 294
LEUCHTE (*f*) : light ; lamp ; lantern ; beacon ; mouse-ear (Bot.) 295
LEUCHTEN : to-(give)-light ; shine ; gleam ; illuminate ; glimmer ; glare ; beam ; lighten 296
LEUCHTEND : luminous ; lucid ; bright ; shining ; phosphorescent ; illuminating ; illuminant 297
LEUCHTER (*m*) : candle-stick 298
LEUCHTERDILLE (*f*) : candle-stick socket 299
LEUCHTFARBE (*f*) : luminous paint 300
LEUCHTFEUER (*n*) : beacon-fire 301
LEUCHTGAS (*n*) : coal gas, lighting gas, illuminating gas, town gas (calorific value about 21000 B.T.U.) 302
LEUCHTGASFABRIKATION (*f*) : coal-gas manufacture 303
LEUCHTGASUNTERSUCHUNG (*f*) : coal-gas (testing) analysis 304
LEUCHTGESCHOSS (*n*) : star-shell ; Verey-light 305
LEUCHTHÜLLE (*f*) : luminous envelope ; photosphere 306
LEUCHTKÄFER (*m*) : firefly ; glow-worm 307
LEUCHTKRAFT (*f*) : luminosity ; illuminating power 308
LEUCHTKRAFTBESTIMMUNG (*f*) : photometry 309
LEUCHTKRAFTMESSUNG (*f*) : photometry 310
LEUCHTKUGEL (*f*) : fire ball ; Roman candle (Pyrotechnics) 311
LEUCHTMASSE (*f*) : luminous (phosphorescent) material, substance or mass 312
LEUCHTMATERIAL (*n*) : phosphorescent substance 313
LEUCHTMITTEL (*n*) : illuminant 314
LEUCHTÖL (*n*) : illuminating (lighting) oil 315
LEUCHTPETROLEUM (*n*) : kerosene ; solar oil ; Sg. 0.76-0.86 ; Bp. 150-300 °C. 316
LEUCHTRAKETE (*f*) : rocket ; signal rocket 317
LEUCHTSCHIFF (*n*) : lightship 318
LEUCHTSPIRITUS (*m*) : lighting spirit ; also trade name for a special lighting spirit 319
LEUCHTSTEIN (*m*) : phosphorescent stone ; phosphorous ; Bologna (phosphorous) stone 320
LEUCHTSTOFF (*m*) : luminous substance ; photogene ; solar oil (see **LEUCHTPETROLEUM**) 321
LEUCHTTURM (*m*) : lighthouse 322
LEUCHTWERT (*m*) : illuminating value ; luminosity ; phosphorescence 323
LEUCHTWIRKUNG (*f*) : phosphorescence 324
LEUCHTWURM (*m*) : glow-worm 325
LEUCHTZIFFERBLATT (*n*) : watch-face with phosphorescent figures ; phosphorescent-(watch)-dial 326
LEUCHTZWECK (*m*) : lighting purpose 327
LEUCIN (*n*) : leucine, amidocaproic acid ; $CH_3CH_2CH_2CH_2CHNH_2COOH$; amino-iso-butylacetic acid, $(CH_3)_2CH.CH_2.CH(NH_2).CO_2H$ 328
LEUCIT (*m*) : leucite (an aluminium-potassium metasilicate) $(K,Na)AlSi_2O_6$; Sg. 2.5 329
LEUGBAR : deniable 330
LEUGNEN : to deny ; retract ; disavow ; disclaim ; recant ; revoke ; recall ; abjure ; disown ; gainsay 331
LEUGNUNG (*f*) : recantation, retraction, denial, disavowal, revocation, abjuration 332
LEUKÄMIE (*f*) : Leucocythæmia (see **LEUCHÄMIE**) (Med.) 333
LEUKOBASE (*f*) : leuco base 334
LEUKONIN (*n*) : leuconine (a preparation of antimony with about 98% of Sodium meta-antimoniate) 335
LEUKOPHAN (*m*) : leucophane ; $5(Ca,Be)O, 5SiO_2,2NaF$; Sg. 2.97 (Min.) 336
LEUKOVERBINDUNG (*f*) : leuco compound 337
LEUKOXEN (*m*) : leukoxene (see **TITANIT**) 338
LEUMUND (*m*) : reputation, conduct, report, rumour 339
LEUTE (*pl*) : people ; persons ; men ; the world ; hands ; servants ; crowd ; throng ; guests ; populace 340
LEUTSELIG : humane ; affable ; popular ; kind ; pleasant ; gentle 341
LEUWAGEN (*m*) : carriage of a tiller (Naut.) 342
LEUWAGEN (*m*), **EISJACHT- :** ice yacht horse 343

LEUWASGASVERFAHREN (*n*): Besemfelder's method of gas production 344
LEUZIT (*m*): (see LEUCIT) 345
LEVANTISCH: Levantine 346
LEVIGIEREN: to levigate; pulverize; rub to dust 347
LEVKOJE (*f*): sea-stock (*Matthiola sinuata*) 348
LEVKOJESTOCK (*m*): stock (*Matthiola incana*) 349
LEVYN (*m*): levyne (Mineralogy); Sg. 2.15 [350
LEXIKOGRAPH (*m*): lexicographer, a compiler of a dictionary 351
LEXIKON (*n*): lexicon, dictionary 352
LEXIKON (*n*), KONVERSATIONS-: encyclopædia 353
LEYDENERBLAU (*n*): Leyden blue (see KOBALT--BLAU) 354
LEYDNER FLASCHE (*f*): Leyden jar 355
LEZITHIN (*n*): lecithin (*Lecithinum*) (a fatty acid glyceride) 356
LIANE (*f*): liana (a general term for the climbing, creeping and woody plants of the tropics) 357
LIBELL (*n*): libel, a defammatory writing 358
LIBELLE (*f*): level; water-level; dragon-fly (*Libellula*) 359
LIBERALISMUS (*m*): liberalism 360
LIBETHENIT (*m*): libethenite (natural copper phosphide); $[CuOH]CuPO_4$; Sg. 3.7 361
LICHT: light; pale; bright; luminous; clear; interior; lucid; internal; in the clear (of pipe diameters and widths); (*n*): light,; candle; taper; lamp; any object giving light 362
LICHTABSORPTION (*f*): absorption of light 363
LICHTANLAGE (*f*): lighting plant 364
LICHTART (*f*): type (sort) of light 365
LICHTÄTHER (*m*): luminiferous ether 366
LICHTBAD (*n*): (electric)-light bath 367
LICHTBILD (*n*): photograph 368
LICHTBILDKUNST (*f*): photography 369
LICHTBILDMESSKUNST (*f*): photogrammetry (see PHOTOGRAMMETRIE) 370
LICHTBLAU: light-blue 371
LICHTBOGEN (*m*): luminous (electric) arc; arc 372
LICHTBOGENOFEN (*m*): (electric)-arc (oven) furnace (for electro steels) 373
LICHTBOGENOFEN, DIREKTER-: direct arc furnace (heating by arc passing direct from one electrode through the slag to the metal bath, and then into the other electrode) 374
LICHTBOGENOFEN, INDIREKTER-: indirect arc oven (heating by radiation; arc passes from one electrode to the other, radiating heat to the metal bath) 375
LICHTBOGENSCHWEISSUNG (*f*): arc welding 376
LICHTBRECHEND: refracting; refractive 377
LICHTBRECHUNG (*f*): refraction (of light); optical refraction 378
LICHTBRECHUNGSVERMÖGEN (*n*): refractive power; refractivity 379
LICHTBÜSCHEL (*m*): pencil of rays-(of light) 380
LICHTDRUCK (*m*): photographic (photomechanical) printing; collotype; heliotype; phototype 381
LICHTE BREITE (*f*): inside (internal) width, width in the clear 382
LICHTECHT: fast to light 383
LICHTEFFEKT (*m*): luminous effect 384
LICHTE HÖHE (*f*): inside (internal) height, height in the clear 385
LICHTEINHEIT (*f*): unit of light (Elect., etc.) 386
LICHTE LÄNGE (*f*): inside (internal) length, length in the clear 387
LICHTELEKTRISCH: photoelectric 388
LICHTEMPFINDLICH: sensitive-(to light); optically sensitive; sensitized 389
LICHTEMPFINDLICHKEIT (*f*): sensitiveness-(to light) 390
LICHTEN: to light; expose; clear; thin; lift; raise; hoist; weigh (Anchor); lighten; unload; light up; illuminate; (*n*): weighing (Anchor) 391
LICHTENTWICKLUNG (*f*): evolution of light 392
LICHTER (*m*): lighter, barge 393
LICHTERFABRIK (*f*): candle factory 394
LICHTERFAHRZEUG (*n*): lighter (Naut.) 395
LICHTERLOH: blazing; ablaze 396
LICHTERSCHEINUNG (*f*): luminous phenomenon 397
LICHTERZEUGEND: photogenic; producing light; light-producing 398
LICHTE WEITE (*f*): inside (internal) diameter; (i.d.); width in the clear 399
LICHTFARBEN: light-coloured 400
LICHTFARBENDRUCK (*m*): photomechanical colour printing 401
LICHTFARBIG: light-coloured 402
LICHTFILTER (*m* and *n*): light filter (Phot.) 403
LICHTFLACHDRUCKVERFAHREN (*n*): (see PHOTO--LITHOGRAPHIE) 404
LICHTFLECK (*m*), ZENTRALER-: (central)-flare spot (a fogged spot on a photographic plate) 405
LICHTFORM (*f*): candle mould 406
LICHTFORTPFLANZUNG (*f*): transmission of light 407
LICHTGEBEND: luminous; illuminating; giving light 408
LICHTGELB: light-yellow 409
LICHTGIESSEN (*n*): candle moulding 410
LICHTGIESSER (*m*): chandler 411
LICHTGLANZ (*m*): lustre, brightness 412
LICHTGRÜN: light-green 413
LICHTHELL: very (bright) light 414
LICHTHOF (*m*): halation (Phot.) 415
LICHT (*n*), HOHES-: high-light 416
LICHTISOMERISATION (*f*): light isomerization 417
LICHTKABEL (*n*): electric light cable 418
LICHTKATALYTISCH: light catalyti 419
LICHTKEGEL (*m*): luminous cone; cone of (light)-rays 420
LICHTKÖRPER (*m*): luminous body; luminary 421
LICHTKRAFT (*f*): rapidity (of a lens) 422
LICHTKREIS (*m*): halo; circle of light 423
LICHTKUPFERDRUCK (*m*): photogravure (see HELIOGRAVÜRE) 424
LICHTLEHRE (*f*): optics; photology; the science of light 425
LICHTLEITUNG (*f*): lighting circuit (Elect.) 426
LICHTLOS: dark, unilluminated, obscure, unenlightened, rayless, overcast, black, unintelligible, incomprehensible, ignorant, gloomy, dismal 427
LICHTMAGNET (*m*): light magnet 428
LICHTMASCHINE (*f*): electric light engine; dynamo 429
LICHTMATERIE (*f*): luminous matter 430
LICHTMESSE (*f*): Candlemas 431
LICHTMESSER (*m*): photometer (see PHOTO--METER) 432
LICHTMESSKUNST (*f*): photometry 433
LICHTMESSUNG (*f*): photometry 434

LICHTMYRTE (*f*): wax myrtle (*Myrica cerifera*) 435
LICHTNUSSÖL (*n*): (see **BANKULNUSSÖL**) 436
LICHTPAPIER (*n*): photographic paper, photogenic paper, paper which is sensitive to light, printing paper; waxed paper (Phot.) 437
LICHTPAUSAPPARAT (*m*): printing (machine) apparatus (for blue and white prints) 438
LICHTPAUSE (*f*): photographic tracing; photostat; print (from a tracing); phototype 439
LICHTPAUSEINRICHTUNG (*f*): printing machine (for blue and white prints) 440
LICHTPAUSLAMPE (*f*): printing machine (arc)-lamp 441
LICHTPAUSPAPIER (*n*): printing paper (for blue prints termed ferro-prussiate; for white prints termed ferro-gallic) 442
LICHTPAUSVERFAHREN (*n*): heliographic printing, cyanotype, blue-print process, ferro-prussiate process (for making blue-prints of engineers' drawings) (printing from tracings by means of electric arc lamps, mercury vapour lamps, etc.) 443
LICHTPOLYMERISATION (*f*): light polymerization 444
LICHTPUNKT (*m*): luminous (bright) spot; ray; point of light 445
LICHTQUELLE (*f*): source of light 446
LICHTREFLEX (*m*): reflection (of light) 447
LICHTROT (*n*): light red (see **BLEIMENNIGE**) 448
LICHTSCHEIN (*m*): shine (of a light) 449
LICHTSCHEU: afraid of light, shunning light 450
LICHTSCHIRM (*m*): lamp-shade 451
LICHTSCHLEIER (*m*): light fog (due to light getting at photographic plates) 452
LICHTSCHLUCKEND: absorptive; light-absorbing 453
LICHTSEHEN (*n*): photopsia (Med.) 454
LICHTSEITE (*f*): bright (luminous) side 455
LICHTSPALTER (*m*): prism 456
LICHTSTÄRKE (*f*): strength of light 457
LICHTSTRAHL (*m*): ray (beam) of light; luminous ray 458
LICHTTALG (*m*): candle tallow 459
LICHTUNG (*f*): clearing; glade; thinning 460
LICHTVOLL: luminous; lucid; clear; resplendent; bright 461
LICHTWEITE (*f*): i.d.; inside (internal) diameter; width in the clear 462
LICHTWELLE (*f*): wave of light 463
LICHTWIRKUNG (*f*): action of light; luminous effect 464
LICHTWURM (*m*): (see **JOHANNISKÄFER**) 465
LICHTZEICHEN (*n*): illuminated sign 466
LICHT (*n*), **ZERSTREUTES-**: diffused light 467
LICHTZERSTREUUNGSKRAFT (*f*): light (diffusing) dispersive power 468
LICHTZERSTREUUNGSVERMÖGEN (*n*): light (diffusing) dispersive capacity 469
LICHTZIEHEN (*n*): candle-making 470
LICHTZIEHER (*m*): chandler, candle-maker 471
LICHTZUG (*m*): candle mould 472
LICHTZUTRITT (*m*): admittance of light 473
LID (*n*): lid (of eye, etc.) 474
LIDERN: to pack (Tech.) 475
LIDERUNG (*f*): packing; gasket 476
LIDERUNGSBOLZEN (*m*): gland bolt or stud, packing bolt or stud 477
LIEB: dear; beloved; agreeable; esteemed; valued; favourite 478
LIEBE (*f*): love; favour; passion; affection; charity; beloved; dear 479
LIEBEN: to love; like; fancy; value; cherish 480
LIEBENSWÜRDIG: lovable; amiable; kind 481
LIEBENSWÜRDIGKEIT (*f*): amiability; kindness 482
LIEBER: rather; sooner; better; dear 483
LIEBERSCHES KRAUT (*n*): hemp-nettle (*Herba galeopsidis*) 484
LIEBESAPFEL (*m*): tomato (*Lycopersicum esculentum*) 485
LIEBESBAUM (*m*): (see **JUDASBAUM**) 486
LIEBESBLUME (*f*): agapanthus; African lily (*Agapanthus umbellatus*) 487
LIEBESGRAS (*n*): quaking grass; briza (*Briza*) (Bot.) 488
LIEBESWUT (*f*): Erotomania (general term); (in men, termed) Satyriasis; (in women, termed) Nymphomania; (insanity due to excessive sexual excitement) (Med.) 489
LIEBFRAUENBETTSTROH (*n*): lady's bedstraw; yellow bedstraw (*Galium verum*) 490
LIEBFRAUENMILCH (*f*): Liebfrauenmilch (a particular brand of Rhenish wine) 491
LIEBHABER (*m*): lover; amateur; fancier 492
LIEBHABEREI (*f*): hobby, fondness, favourite occupation 493
LIEBIGIT (*m*): liebigite, uranothallite; $2CaCO_3.U[CO_3]_2+10H_2O$ 494
LIEBIG'SCHER KÜHLER (*m*): Liebig condenser 495
LIEBKRAUT (*n*): (see **LABKRAUT**) 496
LIEBLICH: sweet; charming; lovely; delightful; pleasing; enjoyable 497
LIEBLINGSBESCHÄFTIGUNG (*f*): favourite occupation 498
LIEBLOS: unkind; uncharitable; loveless 499
LIEBREICH: amiable, kind, charitable 500
LIEBREIZ (*m*): charm 501
LIEBSTOCK (*m*): lovage (*Levisticum*) 502
LIEBSTÖCKEL (*m*): lovage; sea-parsley (*Levisticum officinale*) 503
LIEBSTÖCKELWURZEL (*f*): lovage root (*Radix levistici of Levisticum officinale*) 504
LIEBSTOCKÖL (*n*): lovage (levisticum) oil; Sg. 0.93-1.03 505
LIEBSTOCKWURZEL (*f*): lovage root (*Radix levistici*) 506
LIED (*n*): song; air 507
LIEDERLICH: careless; disorderly; loose; dissolute; negligent 508
LIEFERANT (*m*): purveyor; supplier; furnisher; contractor; creditor 509
LIEFERBAR: to be (delivered, furnished) supplied; capable of being supplied, delivered or furnished 510
LIEFERFIRMA (*f*): contractor; supplier 511
LIEFERGRAD (*m*): supply efficiency (ratio of actual suction volume to stroke volume) (Compressors) 512
LIEFERMENGE (*f*): quantity (amount) delivered 513
LIEFERN: to supply; deliver; consign; furnish; yield; produce; afford 514
LIEFERTERMIN (*m*): time (date) of delivery 515
LIEFERUNG (*f*): supply; purveying; delivery; issue; number (of books); consignment (see also **LIEFERN**) 516
LIEFERUNGSBEDINGUNGEN (*f.pl*): terms (conditions) of delivery; conditions of contract 517
LIEFERUNGSPREIS (*m*): price delivered; price including delivery 518

LIEFERUNGSSCHEIN (*m*): delivery note 519
LIEFERUNGSVERTRAG (*m*): contract-(to supply) 520
LIEFERUNGSWEISE: in (numbers) issues (of books) 521
LIEFERZEIT (*f*): time of delivery 522
LIEGEGELD (*n*): demurrage 523
LIEGEHAUS (*n*): quarantine-house 524
LIEGEN: to lie; be situate-(d); matter; signify; recline 525
LIEGEND: horizontal; extended; lying; situate-(d); recumbent; inclined; slanting; reclining 526
LIEGENDE GRÜNDE (*m.pl*): (see **LIEGENDE GÜTER**) 527
LIEGENDE GÜTER (*n.pl*): immovables; landed estate; real estate 528
LIEGENDE SCHRIFT (*f*): italics 529
LIEGENDES GELD (*n*): dead capital 530
LIEGEN, VOR ANKER-: to ride (lie) at anchor (Naut.) 531
LIEGEZEIT (*f*): period of (lying) quarantine 532
LIEGEZEIT (*f*), **EXTRA-**: period of demurrage 533
LIEK (*n*): bolt rope, leech-rope (Naut.) 534
LIEKEN: to rope (a sail) 535
LIEKGUT (*n*): bolt rope 536
LIEKNADEL (*f*): bolt rope needle, roping needle 537
LIEKTAU (*n*): bolt rope 538
LIESCHGRAS (*n*): Timothy-grass; herd's-grass; cat's-tail-(grass) (*Phleum pratense*) 539
LIÉVRIT (*m*): Lievrite (see **ILVAIT**) 540
LIGATUR (*f*): ligature 541
LIGNIN (*n*): (see **ZELLULOSE**) 542
LIGNIT (*m*): lignite (60-70% carbon, hydrogen, oxygen, ashes); Sg. 1.25 543
LIGNOCERINSÄURE (*f*): lignocerinic acid; $CH_3(CH_2)_{22}CO_2H$; Mp. 80.5°C. 544
LIGNOSULFIT (*n*): lignosulphite (the solid constituent of celluoid solution) 545
LIGROIN (*n*): ligroin (see **BENZIN**) 546
LIGUSTER (*m*): privet (*Ligustrum vulgare*) 547
LIKÖR (*m*): liquer; liqueur; cordial 548
LIKÖRBOHNE (*f*): liqueur bean 549
LILA (*m*): lilac (*Syringa vulgaris*) 550
LILACIN (*n*): lilacine, terpineol, terpilenol; $C_{10}H_{17}OH$; Sg. 0.93; Mp. 33.5°C.; Bp. 214°C. 551
LILAK (*m*): lilac (*Syringa vulgaris*) 552
LILIE (*f*): lily (*Lilium*) 553
LILIENARTIG: liliaceous 554
LILIENBAUM (*m*): tulip tree (*Liriodendron tulipifera*) 555
LILIENZWIEBEL (*f*): bulb of a lily 556
LIMAHOLZ (*n*): Lima wood (of *Cæsalpinia echinata*) 557
LIMETTE (*f*): lime (*Citrus limetta*) 558
LIMETTÖL (*n*): lime oil (from *Citrus limetta*); Sg. 0.882 559
LIMITUM (*n*): limit, maximum, highest price 560
LIMONE (*f*): citron (*Citrus medica*) 561
LIMONE (*f*), **AMERIKANISCHE WILDE-**: American wild lemon (*Podophyllum peltatum*) 562
LIMONEN (*n*): limonene (a constituent of peppermint oil) 563
LIMONE, SAURE- (*f*): lemon (*Citrus limonum*) 564
LIMONE, SÜSSE-: (*f*): lime (*Citrus medica var. acida* or *Citrus limetta*) 565
LIMONGRAS (*n*): lemon grass (*Andropogon citratus*) 566
LIMONGRASÖL (*n*): (see **LEMONGRASÖL**) 567
LIMONIT (*m*): limonite, brown iron ore (natural hydrated ferric oxide); $Fe_4O_3(OH)_6$; Sg. 3.8 (see also **BRAUNEISENSTEIN**) 568
LIMONÖL (*n*): (see **CEDRATÖL**) 569
LINALOË (*f*): linaloa, lign-aloe, calambac, agalwood, eagle-wood, agalloch, agallochum, agillochum, aloes-wood 570
LINALOEHOLZ (*n*): linaloe wood (see **LINALOË**) 571
LINALOËÖL (*n*): linaloe oil (from **LINALOË**, which see); Sg. about 0.88 572
LINALOOL (*n*): linalool; $C_{10}H_{17}OH$; Sg. 0.87; Bp. about 197°C. 573
LINALYLACETAT (*n*): linalyl acetate, Bergamiol; $C_{10}H_{17}C_2H_3O_2$; Bp. 109°C. 574
LINARIT (*m*): linarite (natural basic lead and copper sulphate); $[(Pb,Cu)OH]_2SO_4$; Sg. 5.4 575
LINCRUSTA (*f*): lincrusta (pulverized cork and cellulose paper, soaked with oil and resin and compressed. It is then pressed and used as a wall-paper, i.e., raised paper) 576
LIND: gentle; soft; smooth; mild; tender; scoured (Silk) 577
LINDE: (see **LIND**); (*f*): Linden tree; lime tree (*Tilia europæa*; *T. cordata* and *T. ulmifolia*) 578
LINDELUFT (*f*): Linde air; air rich in oxygen (92 volumetric parts of oxygen and 8 volumetric parts of nitrogen) 579
LINDENBLÜTEN (*f.pl*): linden flowers (*Flores tiliæ*) (see also **LINDE**) 580
LINDENBLÜTENÖL (*n*): linden flower oil; lime blossom oil (expressed from *Flores tiliæ*) 581
LINDENGEWÄCHSE (*n.pl*): *Tiliaceæ* (Bot.) 582
LINDENHOLZ (*n*): linden wood; white wood; lime tree wood; bass wood; (see **LINDE**); Sg. 0.32-0.59 583
LINDENKOHLE (*f*): linden wood charcoal 584
LINDENSAMENÖL (*n*): lime seed oil (from seeds of *Tilia parvifolia* and *Tilia ulmifolia*) 585
LINDERN: to alleviate; temper; soften; relieve; mitigate; alloy; palliate; soothe 586
LINDERND: palliative; soothing; lenitive 587
LINDERUNG (*f*): alleviation, mitigation, palliation, comfort (see **LINDERN**) 588
LINDERUNGSMITTEL (*n*): lenitive; palliative; soothing agent 589
LINDIGKEIT (*f*): lenity, mildness, mercy, gentleness 590
LINEAL (*n*): straight-edge; rule; ruler; (as an adj. = lineal) 591
LINEATUR (*f*): screen plate (Phot.) 592
LINIE (*f*): line; lineage; family; race; ancestry; descent; row; equator; stripe; streak 593
LINIE (*f*), **AUSGEZOGENE-**: full line (as opposed to dotted line) 594
LINIENBATTERIE (*f*): line battery 595
LINIENBLATT (*n*): page with guide lines ruled on it (for placing under notepaper); ruled sheet 596
LINIENFÖRMIG: linear 597
LINIEN (*f.pl*) **GLEICHEN DRUCKES** (*m*): lines of constant pressure, isobars 598
LINIEN (*f.pl*) **GLEICHER TEMPERATUR**: lines of constant temperature, isothermal lines, isotherms 599
LINIENPAAR (*n*): pair of lines 600
LINIENPAPIER (*n*): lined (ruled) paper 601
LINIENRISS (*m*): lines, (of a ship); outline 602
LINIENSCHIFF (*n*): ship of the line, warship, battle-ship, man-o'-war 603

LINIENSPEKTRUM (*n*) : line spectrum 604
LINIENSTROM (*m*) : line current 605
LINIENSYSTEM (*n*) : diatonic scale (Mus.) 606
LINIENWÄHLER (*m*) : distributing switch 607
LINIENZIEHER (*m*) : ruling pen 608
LINIENZUG (*m*) : line, plotted line, curve (usually of graphs) 609
LINIE (*f*), **PUNKTIERTE-** : dotted line 610
LINIEREN : to rule ; draw lines ; sketch ; delineate 611
LINIERFARBE (*f*) : ruling ink 612
LINIERFEDER (*f*) : drawing-pen 613
LINIERMASCHINE (*f*) : ruling machine (Printing) 614
LINIE (*f*), **SCHWACH AUSGEZOGENE-** : light (thin) line, lightly or thinly drawn or ruled line 615
LINIE (*f*), **STARK AUSGEZOGENE-** : heavy (thick) line, heavily or thickly drawn or ruled line ; full line (as opposed to dotted line) 616
LINIE (*f*), **STRICHPUNKTIERTE-** : chain dotted line 617
LINIE (*f*), **VOLLE-** : full line (on drawings) 618
LINIIEREN : (see **LINIEREN**) 619
LINIIERMASCHINE (*f*) : ruling machine 620
LINIIERFARBE (*f*) : ruling (colour) ink 621
LINIIERTINTE (*f*) : ruling ink 622
LINIMENT (*n*) : liniment 623
LINK : left ; wrong ; awkward 624
LINKISCH : awkward ; clumsy ; left-handed 625
LINKRUSTAMASSE (*f*) : lincrusta mass (see **LIN-CRUSTA**) 626
LINKRUSTATAPETE (*f*) : lincrusta wall paper (see **LINCRUSTA**) 627
LINKS : to (on) the left ; left ; (as a prefix=levo- ; left-handed) (see also **LÄVO-**) 628
LINKSDREHEND : levorotatory 629
LINKSDREHUNG (*f*) : levorotation ; left-handed polarization 630
LINKSGÄNGIG : left-handed (of screws, etc.) 631
LINKSGEWINDE (*n*) : left-handed thread (of screws) 632
LINKSMILCHSÄURE (*f*) : levolactic acid 633
LINKSPOLARISATION (*f*) : levorotation ; left-handed polarization 634
LINKSPROPELLER (*m*) : left-handed propeller 635
LINKSSÄURE (*f*) : levo acid 636
LINKSSCHRAUBE (*f*) : left-handed (screw) propeller 637
LINKS SEIN : to be left-handed 638
LINKSUMDRUCK (*m*) : reversed reprint (Lithography) 639
LINKSWEINSÄURE (*f*) : levotartaric acid ; $C_4H_6O_6$ 640
LINNEIT (*m*) : linnæite (see **KOBALTNICKELKIES**) 641
LINNEN : linen, of linen ; (*n*) : linen, canvas 642
LINOLENSÄURE (*f*) : linolenic acid ; isolinolenic acid ; $C_{18}H_{30}H_2$ 643
LINOLEUM (*n*) : linoleum ; lino 644
LINOLEUMÄHNLICH : linoleum-like, like linoleum 645
LINOLEUMERSATZSTOFF (*m*) : linoleum substitute 646
LINOLEUMFABRIK (*f*) : linoleum factory 647
LINOLEUMFABRIKATION (*f*) : linoleum manufacture 648
LINOLEUMLACK (*m*) : linoleum varnish 649
LINOLEUMMASSE (*f*) : linoleum mass 650
LINOLEUMSTREIFEN (*m*) : linoleum strip 651
LINOLEUMTAPETE (*f*) : linoleum (wall-paper) wall-covering 652
LINOLEUMWICHSE (*f*) : linoleum (wax) polish 653
LINOLSÄURE (*f*) : linoleic acid ; $C_{18}H_{32}O_2$ (see **LEINÖLSÄURE**) ; Mp. – 18°C. 654
LINOTYPIE (*f*) : enlargements on canvas (Phot.) 655
LINSE (*f*) : lens ; lentil (seeds of *Ervum lens*) ; bob (of pendulum) ; linch-pin ; lens (of the eye) 656
LINSENÄHNLICH : (see **LINSENARTIG**) 657
LINSENARTIG : lenticular ; lens-shaped 658
LINSENBEIN (*n*) : sesamoid bone (small, round, osseous mass where a tendon passes over a bony projection, as in the case of the patella) (Anat.) 659
LINSENDRÜSE (*f*) : lenticular gland 660
LINSENENTZÜNDUNG (*f*) : inflammation of the crystalline lens (of the eye) (Med.) 661
LINSENERZ (*n*) : liroconite, oolitic limonite, pea ore ; $18CuO,4Al_2O_3,5As_2O_5,5SH_2O$; Sg. 2.85 662
LINSENFÖRMIG : lens-shaped ; lenticular ; lentiform 663
LINSENGLAS (*n*) : lens 664
LINSENHAFT : lenticular, lens-shaped 665
LINSENKÜHLER (*m*) : lens cooler 666
LINSENÖFFNUNG (*f*) : lens aperture 667
LINSENSUPPE (*f*) : lentil soup 668
LIPANIN (*n*) : lipanine (pure olive oil with 6% free oleic acid) 669
LIPOJODIN (*n*) : lipojodin (contains 41% Iodine) 670
LIPPE (*f*) : lip ; groove ; notch ; mortice 671
LIPPENBLUMEN (*f.pl*) : *Labiatæ* (Bot.) 672
LIPPENBUCHSTABE (*m*) : labial-(letter) 673
LIPPENLAUT (*m*) : labial-(sound) 674
LIPPENPOMADE (*f*) : lip salve 675
LIPPIG : lipped 676
LIQUESZIEREN : to liquefy ; melt ; liquate ; dissolve ; fuse ; make liquid 677
LIQUEUR (*m*) : (see **LIKÖR**) 678
LIQUID : liquid ; payable 679
LIQUIDIEREN : to fail, go bankrupt, go into liquidation, become insolvent, wind up ; liquidate, settle, clear 680
LIQUOR (*m*) : liquor (a solution of a gas or solid drugs in water or other non-oily solvents) (Pharm.) 681
LIROKONIT (*m*) : liroconite (see **LINSENERZ**) 682
LISPEL (*m*) : lisp 683
LISPELN ; to lisp ; whisper ; murmur ; rustle ; (*n*) : lisp-(ing) ; whispering ; rustling ; murmuring 684
LIST (*f*) : slyness ; cunning ; finesse ; art ; wile ; ruse ; craft ; deceit ; artifice ; trick ; stratagem 685
LISTE (*f*) : list ; register ; catalogue ; roll ; inventory ; panel 686
LISTENPREIS (*m*) : list (catalogue) price 687
LISTIG : artful ; cunning ; wily ; crafty ; deceitful ; sly 688
LISTOFORMSEIFE (*f*) : listoform soap 689
LITER (*m*) : litre (about 1.8 pints) 690
LITERARISCH : literary 691
LITERAT (*m*) : man of letters, literateur, author, writer ; scholar, savant ; (*pl*) : literati 692
LITERATOR (*m*) : man of letters (see **LITERAT**) 693
LITERATUR (*f*) : literature 694
LITERKOLBEN (*m*) : litre flask 695
LITHARGYRUM (*n*) : litharge (see **BLEIGLÄTTE** and **BLEIOXYD, GELBES-**) 696
LITHION (*n*) : lithia (see **LITHIUM**) 697
LITHIONGLIMMER (*m*) : lithia mica ; lithium-bearing mica (see **LEPIDOLITH**) 698

LITHIOPHILIT (*m*): lithiophilite; Li(Mn,Fe) PO_4; Sg. 3.45 699
LITHIOPIPERAZIN (*n*): lithiopiperazine (combination of piperazine and lithium salts, for dissolving uric acid) 700
LITHIUM (*n*): lithia; lithium (Li) 701
LITHIUMACETAT (*n*): lithium acetate 702
LITHIUMAMID (*n*): lithium amide; $LiNH_2$; Sg. 1.178 703
LITHIUMARSENAT (*n*): lithium arsenate; Li_3AsO_4; Sg. 3.07 704
LITHIUMBENZOAT (*n*): lithium benzoate; $LiC_6H_5CO_2$ 705
LITHIUM (*n*), **BENZOESAURES-**: lithium benzoate (see **LITHIUMBENZOAT**) 706
LITHIUMBROMID (*n*): lithium bromide; LiBr; Sg. 3.464; Mp. 442-547°C. 707
LITHIUMCARBONAT (*n*): lithium carbonate; Li_2CO_3; Sg. 2.111; Mp. 618-710°C. 708
LITHIUMCHLORID (*n*): lithium chloride; LiCl; Sg. 2.068; Mp. 602°C. 709
LITHIUMCITRAT (*n*): lithium citrate; $Li_3C_6H_5O_7.4H_2O$ 710
LITHIUM (*n*), **CITRONENSAURES-**: lithium citrate (see **LITHIUMCITRAT**) 711
LITHIUMERZ (*n*): lithium ore (see **LEPIDOLITH**) 712
LITHIUMFLUORID (*n*): lithium fluoride; LiF; Sg. 2.601; Mp. 801°C. 713
LITHIUMGLIMMER (*m*): lithia mica (see **LEPIDO-LITH**) 714
LITHIUM (*n*), **GOLDCHLORWASSERSTOFFSAURES-**: lithium chloraurate; $LiAuCl_4$; $LiAuCl_4+2H_2O$ and $LiAuCl_4+4H_2O$ 715
LITHIUMIMID (*n*): lithium imide; Li_2NH; Sg. 1.303 716
LITHIUMJODID (*n*): lithium iodide; LiI; Sg. 4.061; Mp. 330-446°C. 717
LITHIUMKARBONAT (*n*): (see **LITHIUMCARBONAT**) 718
LITHIUM (*n*), **KOHLENSAURES-**: lithium carbonate (see **LITHIUMKARBONAT**) 719
LITHIUMMETASILIKAT (*n*): lithium metasilicate; Li_2SiO_3; Sg. 2.529-2.61 720
LITHIUMNITRAT (*n*): lithium nitrate; $LiNO_3$; Sg. 2.39 721
LITHIUMNITRIT (*n*): lithium nitrite; $LiNO_2$; Sg. 1.671 722
LITHIUMPERCHLORAT (*n*): lithium perchlorate; $LiClO_4$; Sg. 2.429 723
LITHIUMPHOSPHAT (*n*): lithium phosphate; Li_3PO_4; Sg. 2.41 724
LITHIUM (*n*), **PHOSPHORSAURES-**: lithium phosphate (see **LITHIUMPHOSPHAT**) 725
LITHIUMSALICYLAT (*n*): lithium salicylate 726
LITHIUM (*n*), **SALICYLSAURES-**: lithium salicylate 727
LITHIUM (*n*), **SALPETERSAURES-**: lithium nitrate (see **LITHIUMNITRAT**) 728
LITHIUMSALZ (*n*): lithium salt 729
LITHIUM (*n*), **SCHWEFELSAURES-**: lithium sulphate (see **LITHIUMSULFAT**) 730
LITHIUMSILICID (*n*): lithium silicide; Li_6Si_2; Sg. 1.12 731
LITHIUMSULFAT (*n*): lithium sulphate; Li_2SO_4; Sg. 2.21; $Li_2SO_4+H_2O$; Sg. 2.02-2.054 732
LITHIUMSULFID (*n*): lithium sulphide; Li_2S; Sg. 1.66 733
LITHOGRAPH (*m*): lithographer 734
LITHOGRAPHIE (*f*): lithography; lithograph; (as a prefix=lithographic) 735
LITHOGRAPHIEFARBE (*f*): lithographic colour 736
LITHOGRAPHIEREN: to lithograph (draw, etch or write on stone) 737
LITHOGRAPHIERPAPIER (*n*): lithographic paper, transfer paper 738
LITHOGRAPHIESTEIN (*m*): lithographic stone 739
LITHOGRAPHIETINTE (*f*): lithographic ink 740
LITHOGRAPHISCH: lithographic 741
LITHOGRAPHISCHER STEIN (*m*): lithographic stone 742
LITHOPONE (*f*): lithopone; $BaSO_4.ZnS$; Sg. 4.3 743
LITHOPONEANLAGE (*f*): lithopone plant 744
LITHOPONEFABRIK (*f*): lithopone factory 745
LITHOPONEWEISS (*n*): (see **LITHOPONE**) 746
LITZE (*f*): braid; cord; string; lace; strand; thread 747
LIZENZ (*f*): licence 748
LIZENZGEBÜHR (*f*): (amount of)-royalty 749
LIZENZIEREN: to licence; permit to make or sell under licence or on payment of royalty 750
LIZENZWEISE: on a royalty basis, as agent 751
LOB (*n*): praise; fame; commendation; eulogy 752
LOBELIAKRAUT (*n*): lobelia herb; Indian tobacco (*Herba lobeliæ-(inflatæ)*; of *Lobelia inflata*) 753
LOBELIENKRAUT (*n*): (see **LOBELIAKRAUT**) 754
LOBEN: to praise; laud; commend; approve; value; extol 755
LÖBLICH: commendable, laudable 756
LOBREDE (*f*): eulogy; panegyric 757
LOCH (*n*): hole; gap; eye; pore; breach; cavity; opening; orifice; pocket (Billiards); prison; foramen (Anat.); aperture; perforation 758
LOCHBOHRER (*m*): piercer, auger 759
LOCHEISEN (*n*): punch 760
LOCHEN: to punch; perforate 761
LÖCHERIG: full of holes; perforated; porous 762
LÖCHERN: to perforate, punch (make) holes in 763
LÖCHERPILZ (*m*): boletus (*Boletus polyporus*) 764
LÖCHERSCHWAMM (*m*): (see **LÖCHERPILZ**) 765
LOCHFEILE (*f*): riffler 766
LOCHKAMERA (*f*): pinhole camera 767
LOCHKREIS (*m*): pitch diameter 768
LOCHLEHRE (*f*): hole gauge 769
LOCHMASCHINE (*f*): perforating (punching) machine 770
LOCHSÄGE (*f*): fretsaw, lock-saw, key-hole saw, compass-saw 771
LOCHSCHEIBE (*f*): perforated (dividing) plate 772
LOCHSTANZE (*f*): punching machine, perforating machine 773
LOCHSTANZMASCHINE (*f*): perforating (punching, hole-cutting) machine 774
LOCHSTEMPEL (*m*): punch 775
LOCHTASTER (*m*): inside callipers 776
LOCHTIEFE (*f*): depth of (hole) cavity 777
LOCHWEITE (*f*): width (diameter) of hole; aperture 778
LOCKE (*f*): lock; curl; flock; bait; allurement; temptation; enticement; lure; decoy; attraction 779
LOCKEN: to decoy; lure; call; entice; allure; induce; bait; attract; tempt; curl 780
LOCKER: loose; spongy; slight; not compact; dissolute; disorderly; light 781

LOCKERN: to loosen; make loose; slacken; relax; lighten 782
LOCKERUNG (*f*): loosening 783
LOCKIG: curly, curled 784
LODE (*f*): coarse woollen cloth or stuff 785
LODEN (*n*): (see **LODE**) 786
LODERN: to blaze; flame; flare 787
LOEWEIT (*m*): loeweite; $2Na_2Mg[SO_4]_2 + 5H_2O$; Sg. 2.38 788
LÖFFEL (*m*): spoon; ladle 789
LÖFFELBAUM (*m*): (see **KALMIE**) 790
LÖFFELBOHRER (*m*): centre bit, spoon bit 791
LÖFFELBUG (*m*): spoon bow (of ships) 792
LÖFFELFÖRMIG: spoon-shaped; cochleariform (Bot.), cochleated, cochleate, cochleary, cochlean, cochleous 793
LÖFFELKRAUT (*n*): scurvy grass (*Herba cochleariæ*) (*Cochlearia officinalis*) 794
LÖFFELN: to ladle-(out) 795
LOG(*n* and *m*): log (Naut.) (an automatic apparatus for registering mileage and speed of vessels); log, logarithm (Maths.) 796
LOGARITHM (*m*): logarithm, log 797
LOGARITHMENTAFEL (*f*): logarithmic table, table of logarithms 798
LOGARITHMISCH: logarithmic 799
LOGARITHMUS (*m*): logarithm 800
LOGARITHMUS (*m*), **BRIGGSCHER-**: Brigg's logarithm 801
LOGARITHMUS (*m*), **NATÜRLICHER-**: natural logarithm 802
LOGBRETT (*n*): log board 803
LOGBUCH (*n*): logbook 804
LOGE (*f*): lodge (Freemasonry); box (Theatre) 805
LOGEMENT (*n*): lodgings 806
LOG-(G)-EN: to heave the log (Naut.) 807
LOGGER (*m*). lugger (Naut.); one who heaves the log 808
LOGGERSEGEL (*n*): lug sail 809
LOGGLAS (*n*): log glass 810
LOGIEREN: to lodge, dwell 811
LOGIERHAUS (*n*): (common)-lodging house 812
LOGIK (*f*): logic 813
LOGISCH: logical-(ly) 814
LOGLEINE (*f*): log line 815
LOGROLLE (*f*): log reel 816
LOH: blazing; burning; ablaze; glaring; flaming; (*f*): morass; marsh; bog; (*n*): (see **LOHE**) (see also **GERB-**) 817
LOHASSEL (*f*): wood louse (*Limnoria terebanum*) 818
LOHBALLEN (*m*): tan-ball, tan-cake; peat 819
LOHBEIZE (*f*): tan liquor; tanning; tan-pit 820
LOHBODEN (*m*): peat soil 821
LOHBRÜHE (*f*): tan bark liquor; ooze (Leather); tannin (see **GERBSÄURE**) 822
LOHE (*f*): tan bark (waste from tanneries; high moisture up to 80%; if moisture 60% after passing through tan presses, calorific value is 1100-1400 cals.; if air dried, has moisture 30-35% and can be mechanically dried to 10-12% moisture); tan liquor; ooze; (see **GERBERLOHE**); flame, flare, blaze; mildew 823
LOHEEXTRAKT (*m*): tan-bark extract (see **LOHE**) 824
LOHEICHE (*f*): tan-bark oak 825
LOHEN: to tan; steep-(in tan liquor); blaze; flame; flare 826
LOHEPRESSE (*f*): tan press 827
LÖHER (*m*): tanner 828
LOHFARBE (*f*): tan colour (see also **LOHBRÜHE**) 829
LOHFARBIG: tan coloured; tawny; auburn 830
LOHFASS (*n*): tan vat 831
LOHFEUER (*n*): blazing fire 832
LOHGAR: (bark)-tanned 833
LOHGERBER (*m*): (bark)-tanner 834
LOHGERBEREI (*f*): (bark)-tanning or tannery 835
LOHGRUBE (*f*): tan pit 836
LOHHAUS (*n*): bark-kiln 837
LOHKUCHEN (*m*): tan-cake 838
LOHMESSER (*m*): barkometer 839
LOHMÜHLE (*f*): tanning mill; bark mill (for breaking up the bark) 840
LOHN (*m*): reward; wages; pay; recompense; salary; deserts 841
LOHNARBEITER (*m*): jobber 842
LOHNEN: to pay; reward; recompense; requite; repay; be worth while 843
LOHNEND: remunerative, paying 844
LOHNFÄRBER (*m*): job dyer 845
LOHNTAG (*m*): pay-day 846
LOHNUNG (*f*): pay; payment; reward (see **LOHN**) 847
LÖHNUNG (*f*): (see **LOHNUNG**) 848
LÖHNUNGSTAG (*m*): pay-day 849
LOHNVERHÄLTNISSE (*n.pl*): rates of wages 850
LOHPROBE (*f*): bark test 851
LOHPROBER (*m*): barkometer 852
LOHPRÜFER (*m*): barkometer 853
LOHPULVER (*n*): tan powder 854
LOHRINDE (*f*): tan bark; oak bark (bark of *Quercus suber*, etc.) 855
LOHRÖL (*n*): (see **LORBEERÖL**) 856
LOHSCHNEIDER (*m*): bark cutter 857
LOKAL (*n*): locality; place; tavern; premises; office; show-room; shop (as an adj. = local-(ly); stationary) 858
LOKALATION (*f*): the wasting of an electrode in an open circuit, due to very slight leaks or local currents 859
LOKALBEHÖRDE (*f*): local authority 860
LOKALE VERSTÄRKUNG (*f*): local intensification (Photography, etc.) 861
LOKALFARBE (*f*): natural (local) colour 862
LOKALISIEREN: to localize; restrict-(an effect) 863
LOKALSTROM (*m*): local current (Elect. and Galvanism) 864
LOKAO (*n*): locao, lokao, chinese green (obtained from the bark of *Rhamnus utilis*) 865
LOKOMOBILE (*f*): locomobile (see **LOKOMOTIVE**) 866
LOKOMOTIVE (*f*): locomotive, engine 867
LOKOMOTIVE (*f*), **GEKUPPELTE-**: coupled locomotive 868
LOKOMOTIVE (*f*), **KRAN-**: crane locomotive 869
LOKOMOTIVENFÜHRER (*m*): engine driver 870
LOKOMOTIVE (*f*), **TENDER-**: tender locomotive 871
LOKOMOTIVFÜHRER (*m*): engine driver 872
LOKOMOTIVFÜHRERSTAND (*m*): locomotive cab, foot-plate 873
LOKOMOTIVSCHUPPEN (*m*): locomotive shed, engine (house) shed 874
LOKOMOTIVVERDACHUNG(*f*): locomotive cab 875
LOLCH (*m*): Darnel (*Lolium temulentum* or *Lolium arvense*); rye-grass (*Lolium perenne*) 876
LÖLLINGIT (*m*): löllingite (natural iron arsenide); leucopyrite; $FeAs_2$; Sg. 7.25 877

LOOFAH (*n*): loofah (fibrous part of the fruit of the towel-gourd, *Luffa ægyptiaca*; used as a washing flannel) 878
LORANDIT (*m*): lorandite; $TlAsS_2$; Sg. 5.53 879
LORANSKIT (*m*): loranskite; $Ta_2O_5,Ce_2O_3,Fe_2O_3,Y_2O_3,CaO,ZrO_2,H_2O$; Sg. 4.4 880
LORBEER (*m*): laurel; noble laurel; bay (*Laurus nobilis*) 881
LORBEERBAUM (*m*), **EDLER-**: noble laurel; laurel tree; bay tree; sweet laurel; sweet bay tree (*Laurus nobilis*) 882
LORBEERBAUM (*m*), **INDISCHER-**: Indian laurel (*Laurus indica*) 883
LORBEERBLATT (*n*): laurel leaf; bay leaf (*Folia lauri* of *Laurus nobilis*) 884
LORBEEREN (*f.pl*): laurel berries 885
LORBEERGEWÄCHSE (*n.pl*): *Lauraceæ* (Bot.) 886
LORBEERHOLZ (*n*): bay wood; Sg. 0.82 (*Laurus nobilis*) 887
LORBEERKRANZ (*m*): laurel crown, victor's crown (*Corona triumphalis*) 888
LORBEERKRAUT (*n*): spurge laurel (*Daphne laureola*) 889
LORBEERÖL (*n*): (*Oleum laurinum*): bay leaf oil, laurel oil; Sg. 0.924 (from *Laurus nobilis*) (see **BAYÖL**) 890
LORBEERÖL (*n*), **INDISCHES-**: Indian laurel oil (from *Laurus indica*) 891
LORBEERROSE (*f*): oleander 892
LORBEERTALG (*m*): laurel tallow (from the seeds of *Tetranthera laurifolia*) 893
LORBEERWEIDE (*f*): bay-leaved willow (*Salix pentandra*) 894
LORCHEL (*f*): mushroom (*Gyromitra*) 895
LORETIN (*n*): loretin; $C_9H_4NI(OH)SO_3H$ [896
LOS: loose; free; on; up; porous; acquitted; detached; slack (see **LÖSEN**); (*n*): lot; share; fate; prize; destiny; ticket (in a lottery) 897
LOSARBEITEN: to work loose; become (free, disengaged, uncoupled); loose (see **LÖSEN**) 898
LÖSBAR: soluble; resolvable 899
LÖSBARKEIT (*f*): solubility 900
LOSBINDEN: to untie, uncouple, disconnect, set free 901
LOSBRENNEN: let off, discharge (Pyrotechnics) 902
LOSBRÖCKELN: to crumble off 903
LÖSCHARBEIT (*f*): charcoal process (Metal); quenching; tempering (Steel); unloading; work of discharging (Cargo) 904
LÖSCHE (*f*): charcoal (dust) bed; coal dust; slack; culm; clinker; quenching (trough) tub 905
LÖSCHEIMER (*m*): quenching tub; fire bucket 906
LÖSCHEN: to temper (Steel); discharge, unload (Cargo); disembark; land; to extinguish, quench; slake (Lime); blot, cancel, annul; be cancelled, lapse, (Patents); go out; (*f.pl*): dust, slack (from coal) (see **LÖSCHE**) 907
LÖSCHGELD (*n*): unloading charges; wharfage 908
LÖSCHKALK (*m*): quicklime (see **CALCIUMOXYD**) 909
LÖSCHKOHLE (*f*): quenched charcoal 910
LÖSCHMASCHINEN (*f.pl*): unloading (discharging) machinery 911
LÖSCHPAPIER (*n*): blotting paper 912
LÖSCHPLATZ (*m*): wharf; place of (delivery) unloading 913
LÖSCHUNG (*f*): extinguishing; quenching; slaking (Lime); blotting; tempering (Steel); discharge; discharging; unloading; cancelling 914
LÖSCHUNGSSPESEN (*pl*): unloading (landing) charges 915
LÖSCHWASSER (*n*): quenching (tempering) water 916
LÖSCHWEDEL (*m*): brush, sprinkler 917
LOSDREHEN: to unscrew; twist off 918
LOSDRÜCKEN: to push off; fire (a gun); detach 919
LOSE: loose; porous; detached (see **LOS**) 920
LÖSEGELD (*n*): ransom 921
LÖSEMITTEL (*n*): solvent; expectorant; purgative 922
LÖSEN: to dissolve; solve; loosen; untie; redeem; fire (Gun); relieve; make (Money); get (become) loose; unscrew; unfix; uncouple; disconnect; detach; disengage; unloosen; take (a ticket); absolve; unbind 923
LOSEN: to draw lots 924
LÖSEND: dissolving; solvent; expectorant; purgative 925
LÖSEVERMÖGEN (*n*): dissolving power 926
LOSE WERDEN: to get loose, become loose; loosen 927
LÖSEWIRKUNG (*f*): dissolving (solvent) action or effect 928
LOSGEHEN: to become loose; come (go) off; attack; commence; start 929
LOSKAUFEN: to ransom; redeem; buy out 930
LOSKETTEN: to unchain 931
LOSKIEL (*m*): false keel 932
LOSKITTEN: to detach (disengage) anything which is fastened with cement or an agglutinant; ungum, unstick, unglue 933
LOSKNÜPFEN: to undo, untie, unbutton 934
LOSKOMMEN: to become loose; escape; be acquitted 935
LOSKUPPELN: to disconnect, uncouple, disengage 936
LOSLASSEN: to let go, release 937
LÖSLICHKEIT (*f*): solubility 938
LÖSLICHKEITSPRODUKT (*n*): product of solubility 939
LOSLÖSEN: to set free; liberate; separate; untie; detach; loosen 940
LOSLÖTEN: to unsolder 941
LOSMACHEN: to cast off (a rope), to cast loose; let go; loosen 942
LOSNEHMEN: to dismount, loosen, disconnect, uncouple, dismantle, unfix, remove 943
LOSNIETEN: to unrivet 944
LOSOPHAN (*n*): Losophan; tri-iodo-meta-cresol; $C_6H(I_3)OHCH_3$ 945
LOSPLATZEN: to burst (off) out; explode 946
LOSREISSEN: to tear off, pull off 947
LÖSS (*m*): loess (fine, porous, siliceous and calcareous earth) 948
LOSSAGEN: to renounce; release; acquit; absolve; break away from 949
LOSSCHÄKELN: to unshackle 950
LOSSCHEIBE (*f*): loose pulley 951
LOSSCHLAGEN: to sell; get rid of; beat off; dispose of; knock off 952
LOSSCHNALLEN: to unbuckle 953
LOSSCHNÜREN: to undo, unlace, untie 954

LOSSCHRAUBEN: to screw (out) off, to screw apart, unscrew 955
LOSSCHÜTTELN: to shake (off) loose 956
LOSSPANNEN: to relax; unharness 957
LOSSPRECHEN: to release; acquit; absolve 958
LOSSPRENGEN: to burst (spring) off 959
LOSSPRINGEN: to fly off 960
LOSTRENNEN: to separate; tear apart; rip; unstitch 961
LOSUNG (*f*): casting (drawing) of lots; excrement; watchword; countersign; sign; signal 962
LÖSUNG (*f*): solution; loosening; discharge (of gun) (see LÖSEN) 963
LÖSUNG, FESTE- (*f*): compact (solid) solution 964
LÖSUNG (*f*), **FÜNFZIGSTEL-**: solution of 1 in 50 965
LÖSUNGSBENZOL (*n*): a mixture containing alkylated benzenes 966
LÖSUNGSDRUCK (*m*): solution pressure 967
LÖSUNGSFÄHIGKEIT (*f*): solubility; dissolving capacity 968
LOSUNGSGELD (*n*): price of a lottery ticket 969
LÖSUNGSMITTEL (*n*): solvent 970
LÖSUNGSTENSION (*f*): solution (tension) pressure 971
LÖSUNGSVERMÖGEN (*n*): dissolving (solvent) power 972
LÖSUNGSWÄRME (*f*): heat of solution 973
LÖSUNGSWASSER (*n*): solvent water 974
LOSWEICHEN: to soak off 975
LOSWERDEN: to sell, get rid of 976
LOSWICKELN: to unwind, unravel 977
LOT (*n*): solder; plummet; plumb-bob; plumb; line depending vertically on a graph; upright (perpendicular; vertical)-(line); sounding lead (Naut.) 978
LÖTAPPARAT (*m*): soldering (apparatus) equipment 979
LÖTBAR: solderable; capable of being soldered 980
LOTBLEI (*n*): sounding lead 981
LÖTBRENNER (*m*): soldering burner 982
LÖTE (*f*): solder; soldering 983
LÖTEINRICHTUNG (*f*): (see LÖTAPPARAT) 984
LÖTEISEN (*n*): soldering iron 985
LOTEN: to plumb; sound; take soundings (Naut.) 986
LÖTEN: to solder; braze; (*n*): soldering; brazing 987
LÖTFUGE (*f*): soldered joint 988
LÖTIG: fine; pure (of precious metal) 989
LÖTIGKEIT (*f*): fineness; purity (of precious metal) 990
LÖTKOLBEN (*m*): soldering (bit) iron; copper bit 991
LÖTLAMPE (*f*): soldering lamp, blow-lamp 992
LOTLEINE (*f*): plumb line 993
LÖTMATERIAL (*n*): solder; soldering material; material to be soldered 994
LÖTMETALL (*n*): solder 995
LÖTMITTEL (*n*): solder; soldering (medium) material 996
LÖTNAHT (*f*): soldered (joint) seam; seam to be soldered or brazed 997
LÖTOFEN (*m*): soldering furnace 998
LÖTÖL (*n*): soldering oil 999
LÖTPASTE (*f*): soldering paste 000
LÖTPULVER (*n*): soldering powder 001
LOTRECHT: perpendicular; vertical; upright 002
LÖTROHR (*n*): blow-pipe 003
LÖTROHRANALYSE (*f*): blow-pipe analysis 004
LÖTROHRBESTECK (*n*): complete blowpipe set 005
LÖTRÖHRE (*f*): (see LÖTROHR) 006
LÖTROHRFLAMME (*f*): blow-pipe flame 007
LÖTRÖHRLAMPE (*f*): blow-lamp 008
LÖTROHRPROBE (*f*): blow-pipe (test) test-piece 009
LÖTROHRPROBIERKUNST (*f*): blow-pipe assaying 010
LÖTROHRVERSUCH (*m*): blow-pipe (test) experiment 011
LÖTSALZ (*n*): zinc-ammonium chloride, flux; $ZnCl_2.5NH_3.H_2O$ 012
LOTSCHNUR (*f*): plumb line 013
LOTSE (*m*): pilot 014
LOTSEN: to pilot 015
LOTSENBOOT (*n*): pilot boat 016
LOTSENFISCH (*m*): pilot fish (*Naucrates ductor*) 017
LOTSENFLAGGE (*f*): pilot flag 018
LÖTSPITZE (*f*): soldering bit 019
LÖTSTELLE (*f*): soldered (seam) place, spot (place, seam) to be soldered 020
LOTTER: disorderly, vagrant, slack, loose, negligent, lazy, idle 021
LOTTERIE (*f*): lottery, game of chance 022
LOTTERIELOS (*n*): lottery ticket 023
LOTTERIG: (see LOTTER) 024
LÖTUNG (*f*): soldering; brazing; adhesion; agglutination 025
LOTUS (*m*): Lotus (Bot.) 026
LOTUSBAUM (*m*): Zizyphus (Bot.) (*Zizyphus lotus*) 027
LOTUSBLUME (*f*): Egyptian lotus flower (*Nymphæa lotus*) 028
LOTUSKLEE (*m*): bird's foot trefoil (*Lotus corniculatus* and *Lotus major*) 029
LOTUSPFLAUME (*f*): green ebony; date plum; lotus (*Diospyros lotus*) 030
LOTUSWEGEDORN (*m*): lotus (*Zizyphus lotus*) 031
LÖTWASSER (*n*): soldering (water; liquid) fluid 032
LOTWURZ (*f*): *Onosma* (Bot.), golden drop 033
LÖTZINN (*n*): pewter (tin) for soldering; (pure Sn. with Mp. 230°C.; or alloy of Sn. and Pb. with Mp. between 180-240°C.) 034
LOUPE (*f*): magnifying glass; microscope 035
LÖWE (*m*): lion (*Felis leo*) 036
LÖWEIT (*m*): löweite, loeweite (see LOEWEIT) 037
LÖWENAUSSATZ (*m*): leontiasis (an abnormal enlargement of the face) (Med.) 038
LÖWENKLAU (*f*): (see BÄRENKLAU) 039
LÖWENMAUL (*n*): snapdragon (*Antirrhinum majus*) 040
LÖWENSCHWANZ (*m*): motherwort (*Leonurus cardiaca*) 041
LÖWENZAHN (*m*): dandelion (*Taraxacum officinale*) 042
LÖWENZAHNWURZEL (*f*): dandelion root (*Radix taraxaci*) (of *Taraxacum officinale*) 043
LUCHIG: porous 044
LUCHS (*m*): lynx (*Felis lynx*); sharp (sly; cunning) person 045
LUCHSSAPHIR (*m*): cordierite (see CORDIERIT) 046
LUCHSSTEIN (*m*): belemnite (a fossil) 047
LUCIENHOLZ (*n*): St Lucia (Malaheb) cherry wood (wood of *Prunus malaheb*) 048
LÜCKE (*f*): gap; crevasse; orifice; void; space; slit; deficiency; aperture; loophole; flaw; opening; breach; break; defect; chasm; crack 049

LÜCKENHAFT: having (loopholes) gaps; defective; interrupted; incomplete; deficient; porous (see also LÜCKE) 050
LÜCKENHAFTIGKEIT (*f*): incompleteness; porousness (see also LÜCKE) 051
LÜCKENLOS: flawless; non-porous; uninterrupted 052
LUCKIG: porous; honeycombed (Metal) 053
LÜCKIG: honeycombed (Metal) (see also LÜCKENHAFT) 054
LUDER (*n*): carrion; carcass; lure; bait 055
LUDWIGIT (*m*): ludwigite; $4(Mg,Fe)O,Fe_2O_3, B_2O_3$; Sg. 4.0 056
LUFT (*f*): air; Sg. 0.0012; breeze; atmosphere (see also under WIND) 057
LUFTABSCHLUSS (*m*): exclusion of air; air-seal 058
LUFTANFEUCHTER (*m*): air (damping) moistening apparatus 059
LUFTARTIG: aeriform; gaseous; vaporous; airy; air-like; ethereal 060
LUFT (*f*), ATMOSPHÄRISCHE-: atmospheric air; atmosphere (about 23% oxygen; 75% nitrogen, by weight, plus traces of various other gases; Sg. 0.0012) 061
LUFTAUSDEHNUNGSPYROMETER (*n*): air expansion-pyrometer 062
LUFTBAD (*n*): air bath 063
LUFTBEFEUCHTIGUNGSANLAGE (*f*): air moistening plant 064
LUFTBEFEUCHTIGUNGSAPPARAT (*m*): air moistening apparatus 065
LUFTBEHÄLTER (*m*): air (holder; reservoir; chamber; receiver) container; lungs 066
LUFTBESCHAFFENHEIT (*f*): composition of the air 067
LUFTBESCHREIBUNG (*f*): aerography; meteorology 068
LUFTBESTÄNDIG: stable in air; not affected by air; air-proof; air-resisting 069
LUFTBLASE (*f*): air (bladder) bubble; vesicle 070
LUFTBREMSE (*f*): air brake 071
LUFTBRUST (*f*): pneumothorax (air or gas in the pleural cavity) (Med.) 072
LUFTBUFFER (*m*): air buffer 073
LUFTDICHT: air-tight; hermetical-(ly)-(sealed); air-proof 074
LUFTDICHTIGKEIT (*f*): atmospheric density 075
LUFTDRAHT (*m*): aerial-(wire) (Wireless Tel.) 076
LUFTDRUCK (*m*): air pressure; atmospheric (pneumatic) pressure; (as a prefix=pneumatic) (see also DRUCKLUFT-) 077
LUFTDRUCKBREMSE (*f*): air brake 078
LUFTDRUCKHAMMER (*m*): pneumatic hammer 079
LUFTDRUCKKESSEL (*m*): boiler (drum, cylinder) under atmospheric pressure 080
LUFTDRUCKMASCHINE (*f*): pneumatic engine 081
LUFTDRUCKMESSER (*m*): barometer 082
LUFTDRUCKMESSER (*m*), FEDER-: aneroid barometer 083
LUFTDRUCKMESSER (*m*), HEBER-: syphon barometer 084
LUFTDURCHLÄSSIG: permeable to air 085
LUFTDURCHLÄSSIGKEIT (*f*): air permeability 086
LUFTECHT: fast to air or atmospheric (influence) action; air-proof 087
LUFTEINTRITT (*m*): air (entrance) inlet 088
LUFTELEKTRIZITÄTSMESSER (*m*): electrometer (an instrument for measuring the electricity in the atmosphere) 089
LUFTEMPFINDLICH: sensitive to air 090
LÜFTEN: to air; aerate; ventilate; weather; raise; lift (see also DURCHLÜFTEN) 091
LÜFTER (*m*): aerator; ventilator 092
LUFTERHITZER (*m*): air heater 093
LUFTERSCHEINUNG (*f*): atmospheric phenomenon; meteor 094
LUFTERSCHEINUNGSLEHRE (*f*): meteorology 095
LUFTFAHRER (*m*): aeronaut 096
LUFTFAHRT (*f*): aeroplane trip; balloon ascent; aeronautics; ballooning; airship voyage 097
LUFT (*f*), FALSCHE-: false air (air which mixes with the gases of combustion after combustion of the fuel has taken place but before the point is reached at which the flue gases are tested) 098
LUFTFANG (*m*): ventilator, ventilating duct 099
LUFTFEST: air-proof; air-tight 100
LUFTFEUCHTIGKEIT (*f*): atmospheric moisture; moisture in the air 101
LUFTFEUCHTIGKEITSMESSER (*m*): hygrometer (an instrument for measuring the moisture in the atmosphere) 102
LUFTFILTER (*n*): air filter 103
LUFT (*f*), FIXE-: fixed air (Old name): carbon dioxide; CO_2 (see KOHLENSÄURE) 104
LUFT, FLÜSSIGE- (*f*): liquid air 105
LUFTFÖRMIG: gaseous; aeriform (see LUFTARTIG) 106
LUFT (*f*), FRISCHE -: fresh air 107
LUFTGAS (*n*): air gas 108
LUFTGEFÄSS (*n*): air-vessel 109
LUFTGEKÜHLT: air-cooled 110
LUFTGLEICHGEWICHTSLEHRE (*f*): aerostatics 111
LUFTGRANULATION (*f*): air granulation 112
LUFTGÜTE (*f*): quality of the (atmosphere) air; salubrity 113
LUFTGÜTEMESSER (*m*): eudiometer (an apparatus for determining the amount of a certain gas in a gaseous mixture) 114
LUFTGÜTEMESSUNG (*f*): eudiometry 115
LUFTHAHN (*m*): air cock 116
LUFTHALTIG: containing air; aerated 117
LUFTHAMMER (*m*): pneumatic hammer 118
LUFTHÄRTUNG (*f*): air hardening 119
LUFTHEFE (*f*): (see MINERALHEFE) 120
LUFTHEFEVERFAHREN (*n*): process in which air is passed through the mash during fermentation 121
LUFTHEIZER (*m*): air heater; calorifier 122
LUFTHEIZUNG (*f*): hot air heating apparatus or plant; air-heating 123
LUFTIG: light; voluminous; aerial; airy; windy; flighty 124
LUFTION (*n*): atmospheric ion 125
LUFTKABEL (*n*): aerial cable 126
LUFTKALK (*m*): air-slaked lime; gypsum 127
LUFTKANAL (*m*): air (channel; passage) duct; flue 128
LUFTKASTEN (*m*): air (wind) box 129
LUFTKESSEL (*m*): (air)-receiver 130
LUFTKISSEN (*n*): air cushion 131
LUFTKLAPPE (*f*): air-valve; ventilator 132
LUFTKOMPRESSOR (*m*): air compressor 133
LUFT (*f*), KOMPRIMIERTE-: compressed air 134
LUFTKREIS (*m*): atmosphere; surrounding (air) atmosphere 135
LUFTKÜHLER (*m*): air-cooler 136
LUFTKUNDE (*f*): aerology 137
LUFTLACK (*m*): air-proof varnish 138
LUFTLEER: exhausted; void-(of air) 139
LUFTLEERE (*f*): vacuum; air exhaustion 140

LUFTLEEREMESSER (*m*): vacuometer; vacuum (meter) gauge 141
LUFTLEERER RAUM (*m*): vacuum; void 142
LUFTLEHRE (*f*): aerology 143
LUFTLEITER (*m*): aerial (Wireless Tel.); air conductor 144
LUFTLEITUNG (*f*): overhead wire, aerial, aerial (line, main) wire; air duct 145
LUFTLOCH (*n*): air-hole, vent 146
LUFTMALZ (*n*): air-dried malt 147
LUFTMANTEL (*m*): air-jacket 148
LUFTMEISSEL (*m*): pneumatic chisel 149
LUFTMESSER (*m*): aerometer; air-gauge; barometer 150
LUFTMESSKUNST (*f*): pneumatics 151
LUFTMÖRTEL (*m*): air (lime) mortar (needing air to make it set); air-dried mortar 152
LUFTNITRIT (*n*): nitrite from the atmosphere; atmospheric nitrogen; nitrogen in the air 153
LUFTOFEN (*m*): air furnace 154
LUFTPERSPEKTIV (*n*): aerial perspective 155
LUFTPRÜFER (*m*): air tester 156
LUFTPUFFER (*m*): (air)-dash-pot 157
LUFTPUFFERKOLBEN (*m*): dash-pot piston 158
LUFTPUFFERZYLINDER (*m*): dash-pot cylinder 159
LUFTPUMPE (*f*): air pump; pneumatic pump 160
LUFTPUMPENGLOCKE (*f*): air-pump receiver 161
LUFTPYROMETER (*n*): air pyrometer 162
LUFTRAUM (*m*): air-space; atmosphere 163
LUFTREINIGER (*m*): air purifier 164
LUFTREINIGUNG (*f*): air purification; ventilation 165
LUFTREINIGUNGSANLAGE (*f*): air-purifying plant; ventilating plant 166
LUFTRÖHRE (*f*): air (duct; tube) pipe (Mech.); wind-pipe; trachea (Anat.); air-vessel (Bot.); (as a prefix=tracheal; bronchial; laryngeal) 167
LUFTRÖHRENÄSTE (*m.pl*): bronchiæ (Anat.) 168
LUFTRÖHRENBRÄUNE (*f*): cynanche (affection of the throat, including such troubles as croup, quinsy, etc.) 169
LUFTRÖHRENBRUCH (*m*): bronchocele, Derbyshire neck, goitre (an enlargement of the thyroid gland) (Med.) 170
LUFTRÖHRENDECKEL (*m*): epiglottis (Anat.) 171
LUFTRÖHRENENTZÜNDUNG (*f*): inflammation of the wind-pipe (including such troubles as bronchitis, croup, laryngitis) (Med.) 172
LUFTRÖHRENKOPF (*m*): larynx (Anat.) 173
LUFTRÖHRENSCHNITT (*m*): tracheotomy (Surg.) (see KEHLSCHNITT) 174
LUFTRÖHRENSCHWINDSUCHT (*f*): bronchial phthisis (tuberculosis); consumption of the bronchiæ (Med.) 175
LUFTRÖHRENSPALT (*m*): glottis (Anat.) 176
LUFTSAUERSTOFF (*m*): atmospheric oxygen; oxygen in the air 177
LUFTSAUGAPPARAT (*m*): aspirator 178
LUFTSAUGER (*m*): aspirator 179
LUFTSÄULE (*f*): air-column; column of air; blast 180
LUFTSÄURE (*f*): carbonic acid (see KOHLENSÄURE) 181
LUFTSCHACHT (*f*): air-shaft 182
LUFTSCHEIBE (*f*): ventilator (usually refers to the ventilators in panes of glass) 183
LUFTSCHICHT (*f*): air course in brickwork or fuel bed; stratum (layer) of air 184
LUFTSCHIFF (*n*): airship 185
LUFTSCHIFFER (*m*): aeronaut 186
LUFTSCHIFFHALLE (*f*): air-ship shed or hangar 187
LUFTSCHIFFKUNDE (*f*): aeronautics 188
LUFTSCHLAUCH (*m*): (flexible)-air pipe, tube or hose 189
LUFTSCHLITZ (*m*): air space (of grates) 190
LUFTSCHÖPFEN: to take in air; breathe; aspirate; respirate; (*n*): respiration 191
LUFTSCHWERE (*f*): specific gravity of air 192
LUFTSEGEL (*n*): windmill sail 193
LUFTSPALTE (*f*): air vent; air space 194
LUFTSPIEGELUNG (*f*): mirage 195
LUFTSTAUB (*m*): (atoms of)-dust floating in the air 196
LUFTSTEIN (*m*): aerolite, a meteoric stone; air-dried brick 197
LUFTSTICKSTOFF (*m*): atmospheric nitrogen; nitrogen in the air 198
LUFTSTOSS (*m*): blast 199
LUFTSTRICH (*m*): climate 200
LUFTSTROM (*m*): air current; current of air 201
LUFTSTRÖMUNG (*f*): air current; current of air; flow of air 202
LUFTTEILCHEN (*n*): particle of air 203
LUFTTROCKEN: air-dried; air-seasoned (of wood) 204
LUFTTROCKNEN: to air-dry; air-season (of wood) 205
LUFTÜBERSCHUSS (*m*): excess air (the amount of air actually required above the theoretical amount to burn fuel) 206
LÜFTUNG (*f*): airing; aeration; ventilating; ventilation; weathering; raising; lifting 207
LÜFTUNGSANLAGE (*f*): ventilating plant 208
LÜFTUNGSKANAL (*m*): ventilating (shaft) channel 209
LÜFTUNGSSCHLOT (*m*): ventilating shaft or pipe 210
LÜFTUNGSSCHORNSTEIN (*m*): ventilating (chimney, tower, pipe) shaft 211
LÜFTUNGSZEIT (*f*): period of ventilation 212
LUFTVENTIL (*n*): air-valve 213
LUFTVERDICHTER (*m*): air-compressor 214
LUFTVERDRÄNGUNG (*f*): air displacement 215
LUFTVERDÜNNUNG (*f*): rarefaction of the (air) atmosphere 216
LUFTVERFLÜSSIGER (*m*): air liquefier 217
LUFTVERFLÜSSIGUNGSANLAGE (*f*): air liquefying plant 218
LUFTVERKEHR (*m*): aerial (communication) traffic 219
LUFTVORWÄRMUNG (*f*): air heating 220
LUFTWAGE (*f*): aerometer; air-gauge 221
LUFTWÄGEKUNST (*f*): aerostatics 222
LUFTWEG (*m*): air (way) passage 223
LUFTWIDERSTAND (*m*): air resistance 224
LUFTWIRKUNG (*f*): atmospheric (air) action or effect 225
LUFTWURZEL (*f*): aerial root (Bot.) 226
LUFTZELLE (*f*): air cell 227
LUFTZIEGEL (*m*): air-dried brick 228
LUFTZUFUHR (*f*): admission of air, air-supply 229
LUFTZUFUHRKASTEN (*m*): air-box 230
LUFTZUG (*m*): air-current; draught; air-duct; shaft; windage 231
LUFTZÜNDER (*m*): pyrophorous (ignites spontaneously in air) 232
LUFTZUTRITT (*m*): air (access; supply; inlet) admission 233
LÜGE (*f*): lie; falsehood; untruth 234

LÜGEN : to lie ; tell a lie ; deceive ; be affected ; be false ; (*n*) : lying 235

LUGEN : to look out ; lurk ; spy ; peer ; watch 236

LUGGER (*m*) : lugger (Naut.) 237

LUGGERSEGEL (*n*) : lug sail 238

LÜGNER (*m*) : liar 239

LUGOLSCHE LÖSUNG (*f*) : Lugol solution, potassium iodide 240

LUKE (*f*) : hatchway, hatch (on board ship) ; trap-door ; dormer-(window) 241

LUKE (*f*), GROSSE- : main hatch or hatchway 242

LUKENBALKEN (*m*) : hatchway beam 243

LUKENBÜGEL (*m*) : hatchway bar 244

LUKENDECKEL (*m*) : hatchway cover ; trap-door 245

LUKENFÜLL (*n*) : hatchway coaming 246

LUKENGRÄTING (*f*) : hatchway grating 247

LUKENSTRINGER (*m*) : hatchway stringer 248

LULLEN : to lull ; hush ; hum ; sing ; croon 249

LUMINAL (*n*) : Luminal, phenylethyl-malonylurea ; $C_{12}H_{12}O_3N_2$ 250

LUMINALNATRIUM (*n*) : sodium phenylethylbarbiturate ; $C_{12}H_{11}O_3N_2Na$ 251

LUMINATORVERFAHREN (*n*) : luminator process of water softening (in which the scale is kept in a soft condition by first passing the feed over a corrugated aluminium plate before its entry into the boiler) 252

LUMINESZENZ (*f*) : luminescence 253

LUMINESZENZANALYSE (*f*) : luminescence analysis (a microscopical purity analysis of chemical crystals by means of ultra-violet rays, in which the object holder and condenser system must be of rock crystal) 254

LUMINESZENZMIKROSKOP (*n*) : luminescence microscope 255

LUMINOPHOR (*n*) : phosphorescent stone 256

LUMP (*m*) : ragamuffin ; rascal ; scoundrel ; scamp ; rag 257

LUMPEN (*m*) : rag ; tatter ; titler (coarse sugar loaf) 258

LUMPENBREI (*m*) : first rag pulp (Paper) 259

LUMPENBÜTTE (*f*) : rag tub (Paper) 260

LUMPENGESINDEL (*n*) : rabble ; riffraff 261

LUMPENHANDEL (*m*) : rag trade 262

LUMPENHÄNDLER (*m*) : rag merchant 263

LUMPENKAMMER (*f*) : rag sorting room 264

LUMPENPAPIER (*n*) : rag paper 265

LUMPENREISSMASCHINE (*f*) : rag tearing machine 266

LUMPENSAMMLER (*m*) : rag-picker ; rag-man 267

LUMPENSCHNEIDER (*m*) : rag cutter (Paper) 268

LUMPENWOLLE (*f*) : shoddy 269

LUMPENZUCKER (*m*) : titler (coarse loaf sugar) ; lump (loaf) sugar 270

LUMPEREI (*f*) : trash ; trumpery ; trifle ; rubbish ; meanness 271

LUMPIG : tattered, ragged, stingy, sordid, mean 272

LUNGE (*f*) : lung ; lights ; (as a prefix = pulmonary) 273

LUNGENATROPIE (*f*) : atropy of the lungs, senile emphysema (Med.) 274

LUNGENBLUME (*f*) : marsh gentian (*Gentiana pneumonanthe*) 275

LUNGENBLUTUNG (*f*) : hæmorrhage into the lungs (extravasation of blood into the lungs) (Med.) 276

LUNGENBRAND (*m*) : gangrene of the lungs (dying off of a portion of the lung) (Med.) 277

LUNGENCOLLAPSUS (*m*) : collapse of the lungs, apneumatosis (reduction in size of whole or part of the lung) (Med.) 278

LUNGENECHINOCOCCUS (*m*) : hydatid of the lung (hydatid cysts in the lung) (Med.) 279

LUNGENEMPHYSEM (*n*) : emphysema of the lungs (excess of air in the lungs) (Med.) 280

LUNGENENTZÜNDUNG (*f*) : inflammation of the lungs (phneumonia of various types) (see PNEUMONIE) (Med.) 281

LUNGENFIEBER (*n*) : pulmonary fever (Med.) 282

LUNGENFLECHTE (*f*) : (see LUNGENMOOS) 283

LUNGENFLÜGEL (*m*) : lobe-(of the lungs) 284

LUNGENGESCHWÜR (*n*) : pulmonary cavity ; ulcerated lung ; (Med.) 285

LUNGENKOMPRESSION (*f*) : compression of the lung (reduction in size of whole or part of the lung due to pressure on the pleural surface) (Med.) 286

LUNGENKRANKHEIT (*f*) : pulmonary disease (Med.) 287

LUNGENKRAUT (*n*) : lungwort (*Herba pulmonariæ*) (*Pulmonaria officinalis*) 288

LUNGENMOOS (*n*) : pulmonary moss (*Lichen pulmonariæ*) 289

LUNGENÖDEM (*n*) : œdema of the lungs (infiltration of pulmonary tissue with serous fluid) (Med.) 290

LUNGENSCHÜTZER (*m*) : respirator ; lung-protector 291

LUNGENSCHWINDSUCHT (*f*) : phthisis, consumption, pulmonary tuberculosis (Med.) 292

LUNGENSUCHT (*f*) : pulmonary (consumption) tuberculosis ; phthisis (Med.) 293

LUNGENSÜCHTIG : phthistical, consumptive 294

LUNGENTUBERKULOSE (*f*) : (see LUNGEN--SCHWINDSUCHT) 295

LUNGERN : to pipe (Metal) ; idle ; long (yearn) for 296

LUNKER (*m*) : pipe (Metal) (a hollow left in the casting due to volumetric difference between liquid and solid states of the metal ; usually funnel shaped) 297

LUNKERN : to pipe (Metal) (see LUNKER) 298

LUNNIT (*m*) : lunnite ; $[Cu.OH]_3PO_4$; Sg. 3.9 299

LÜNSE (*f*) : linch-pin 300

LUNTE (*f*) : slow-match ; fuse ; taper ; match 301

LUNTE RIECHEN : to smell a rat 302

LUPE (*f*) : magnifying glass ; magnifier ; microscope 303

LUPINE (*f*) : lupine (*Lupinus polyphyllus ; L. albus ; L. luteus ;* and other species) (Bot.) 304

LUPINE (*f*), ZWERG- : dwarf lupine (*Lupinus nanus*) 305

LUPININ (*n*) : lupinine (extracted from seeds of *Lupinus luteus*) ; $C_{21}H_{40}N_2O_2$; Mp. 69°C ; Bp. 256°C. 306

LUPPE (*f*) : loop ; bloom ; ingot ; billet (Metal) 307

LUPPENEISEN (*n*) : iron blooms, ingot (billet) iron 308

LUPPENFEUER (*n*) : smelting furnace (Metal) 309

LUPPENFRISCHARBEIT (*f*) : bloomery (Metal) 310

LUPPENFRISCHHÜTTE (*f*) : bloomery (Metal) 311

LUPPENMACHEN (*n*) : blooming (the compressing of the lumps of metal into blooms) (Puddling) 312

LUPPENSTAHL (*m*) : bloom steel 313

LUPPENWALZWERK (*n*) : blooming rolls 314

LUPULIN (*n*): lupulin (glandular trichomes from the fruit of the hop plant, *Humulus lupulus*; used for medical and brewing purposes) 315
LUPULINSÄURE (*f*): lupulic (lupulinic) acid 316
LUSSATIT (*n*): lussatite (see SILICIUMDIOXYD); Sg. 2.04 317
LUST (*f*): pleasure; joy; mirth; desire; inclination; lust 318
LUSTBARKEIT (*f*): festivity; amusement; enjoyment; diversion; pleasure 319
LÜSTER (*m*): lustre; gloss; glaze 320
LÜSTERGLASUR (*f*): lustrous glaze (Ceramics) 321
LÜSTERN: desirous; lustful; wanton; lascivious; covetous; longing; to hanker; long (lust) for; desire; answer (of a ship to the helm) 322
LÜSTERPORZELLAN (*n*): lustre porcelain 323
LÜSTER (*m*), **URAN-WISMUTGOLD-**: uranium-bismuth-gold lustre (a mixture of uranium lustre, bismuth lustre and brilliant gold in proportion of 20 : 10 : 1 by weight) (Ceramic) 324
LUSTFAHRT (*f*): excursion; outing; pleasure (cruise) trip 325
LUSTFEUER (*n*): bonfire 326
LUSTFEUERWERK (*n*): fireworks, pyrotechnic display 327
LUSTGARTEN (*m*): pleasure ground 328
LUSTGAS (*n*): laughing gas; nitrous oxide (see STICKSTOFFOXYDULGAS) 329
LUSTHAUS (*n*): summer house 330
LUSTIG: merry; gay; jolly; funny; comical; amusing; droll; happy; humorous 331
LUSTRIEREN: to lustre; give (impart) lustre to; glaze; gloss and stretch (Silk) 332
LUSTSCHIFF (*n*): pleasure boat 333
LUSTSEUCHE (*f*): syphilis (a venereal disease) (Med.) 334
LUSTSPIEL (*n*): comedy 335
LUTECIN (*m*): (see CHALCEDON) 336
LUTEOKOBALTCHLORID (*n*): luteocobaltic chloride 337
LUTEOLIN (*n*): luteolin (a dye contained in a decoction of *Reseda luteola*) 338
LUTETIUM (*n*): cassiopeium, lutecium (Lu) (an element) 339
LUTH (*f*): outrigger, boom (Naut.) 340
LUTIDIN (*n*): lutidine, dimethylpyridine (a homologue of pyridine); C_7H_9N 341
LUTIDINSÄURE (*f*): lutidinic acid 342
LUTIEREN: to lute 343
LUTSCHEN: to suck 344
LUTTER (*m*): low (sour; tart) wine; low brandy; first still (Distilling) 345
LUTTERN: to distil weak (low) wine or spirit 346
LUTTERPROBER (*m*): low wine tester 347
LUV (*f*): weatherside; luff (Naut.) 348
LUVANKER (*m*): weather anchor (Naut.) 349
LUVBAUM (*m*): outrigger, boom (Naut.) 350
LUVBRASSE (*f*): weather-brace (Naut.) 351
LUVEN: to luff, ply to windward (Naut.) 352
LUVGIERIG: carrying a weather helm, luffing, plying to windward 353
LUVLÄUFER (*m*), **EISJACHT-**: ice yacht weather runner 354
LUVSEITE (*f*): weather side, weather beam 355
LUXUS (*m*): luxury; (as a prefix=fancy; ornamental; specially high-class; de luxe) 356
LUXUSARTIKEL (*m*): (article of)-luxury (any article not regarded as a necessity) 357
LUXUSAUSGABE (*f*): (see PRACHTAUSGABE) 358
LUZERNE (*f*): lucern; alfalfa 359
LUZERNERKLEE (*m*): lucern; alfalfa 360
LYCETOL (*n*): lycetol (see LYZETOL) 361
LYCOPODIUM (*n*): lycopodium; vegetable sulphur; club-moss (*Lycopodium clavatum* and the spores thereof) 362
LYCOPODIUMSÄURE (*f*): lycopodium acid; $C_{15}H_{29}CO_2H$ 363
LYDDIT (*n*): lyddite (melted picric acid; an explosive) (see PIKRINSÄURE) 364
LYDISCHER STEIN (*m*): Lydian stone; touch-stone 365
LYGOSIN (*n*): lygosine, diorthodioxydibenzalacetone 366
LYGOSINNATRIUM (*n*): sodium lygosinate (*Natrium lygosinatum*) 367
LYGOSINPRÄPARAT (*n*): lygosine preparation 368
LYKOPODIUM (*n*): lycopodium, club-moss, vegetable sulphur (spores of *Lycopodium clavatum*) 369
LYMPHATISCH: lymphatic 370
LYMPHDRÜSE (*f*): lymphatic gland 371
LYMPHE (*f*): lymph; vaccine 372
LYMPHEBEHÄLTER (*m*): vaccine (lymph) tube or container 373
LYMPHGEFÄSS (*n*): lymphatic vessel 374
LYMPHGLAS (*n*): vaccine (lymph) glass 375
LYSALBINSÄURE (*f*): lysalbinic acid (the cleavage product from alkaline hydrolysis of egg-white or albumin) (used for forming colloidal solutions of metals and indigo) 376
LYSARGIN (*n*): lysargin (a special name for colloidal silver obtained by the use of LYSALBIN-SÄURE or PROTALBINSÄURE, which see) 377
LYSIDIN (*n*): lysidine, ethylene-ethenyldiamine, methylglyoxalidine; $CH_3.CNCH_2CH_2NH$; Mp. 105-106°C.; Bp. 199°C. 378
LYSIN (*n*): lysine, d-α,ϵ-diamino-n-caproic acid; $NH_2.CH_2(CH_2)_3CH(NH_2).CO_2H$ 379
LYSOFORM (*n*): (*Liquor Formaldehydi saponatus*): lysoform (a solution of formaldehyde in alcoholic potash soap solution; is a disinfectant) 380
LYSOL (*n*): lysol (a disinfectant); Sg. 1.04 [381
LYSOL (*n*), **BETA-**: (second quality)-Lysol 382
LYSOLSEIFE (*f*): lysol soap 383
LYTHOL (*n*): lythol 384
LYZETOL (*n*): lycetol; dimethylpiperazine tartrate (*Lycetolum*); $C_6H_{14}N_2.C_4H_6O_6$; Mp. 250°C. 385

M

MAAL (*n*): mole; mark 386
MAAT (*m*): mate (Naut.) 387
MACERATION (*f*): maceration (the addition of water to the crushed sugar cane in order to extract the residue of the cane juice, when again pressed through rolls) 388
MACERIEREN: to macerate 389
MACGILP (*m*): meglip, maglip, megilp (a mixture of boiled linseed oil and mastic varnish; used by artists for thinning oil colours or for glazing) 390
MACHBAR: capable of being made or constructed, feasible, practicable 391
MACHE (*f*): make; workmanship; production; manufacture; making 392
MACHE-EINHEIT (*f*): Mache unit, of radio-active substances, (i.e., the potential drop at the electroscope, converted into

electrostatic units and multiplied by 1000) 393

MACHEN: to make; do; cause; procure; get; deal; trade; construct; build; manufacture; create; produce; fabricate; effect; amount to; total; constitute; perform 394

MACHER (*m*): maker, builder, constructor, manufacturer 395

MACHT (*f*): might; force; power; strength; potency; authority; army 396

MACHTHABERISCH: dictatorial, despotic, lordly 397

MÄCHTIG: mighty; huge; vast; powerful; strong; thick; big; proficient; potent; immense 398

MÄCHTIG, EINER SPRACHE- SEIN: to be conversant with (proficient in) a language 399

MÄCHTIGKEIT (*f*): mightiness; thickness; depth; size; power; potency; vastness; immensity 400

MACHTLOS: impotent, powerless, weak 401

MACHWERK (*n*): poor (bungling) work 402

MACIS (*f*): mace (see **MUSKATBLÜTE**) 403

MACISBLÜTE (*f*): mace flower (see **MUSKATBLÜTE**) 404

MACISÖL (*n*): mace oil (see **MUSKATBLUTENÖL**) 405

MACKINTOSHIT (*m*): mackintoshite; $UO_3.3ThO_2.3SiO_2.6H_2O$ (plus various other constituents); Sg. 4.5 406

MADE (*f*): maggot; mite; worm 407

MADENKRAUT (*n*): soapwort (*Saponaria officinalis*) 408

MADIAÖL (*n*): madia oil (from fruit of *Madia sativa*) 409

MADIG: maggoty, worm-eaten 410

MADRASÖL (*n*): (see **ERDNUSSÖL**) 411

MAFURATALG (*m*): mafura tallow (a vegetable tallow from the kernels of *Trichilia emetica*) 412

MAFURRAH (*f*): mafura (*Trichilia emetica*) 413

MAFURRAHTALG (*m*): mafura tallow (see **MAFUR-ATALG**) 414

MAGADI-SODA (*f*): Magadi soda (soda ash, calcined from the natural soda obtained from the soda lakes in lower Egypt, by the Magadi Soda Co.; "magadi" is the African word for soda) 415

MAGAZIN (*n*): magazine, store-house, repository, warehouse 416

MAGD (*f*): maid; servant; virgin 417

MAGDALAROT (*n*): Magdala red 418

MAGEN (*m*): stomach; gizzard; maw; (as a prefix = gastric-; gastro-) 419

MAGENARZNEI (*f*): stomachic (see **MAGENMITTEL**) 420

MAGENAUSSPÜLUNG (*f*): lavage (a method of washing out the stomach) (Med.) 421

MAGENBESCHWERDEN (*pl*): indigestion 422

MAGENBLÄHUNG (*f*): flatulence (Med.) 423

MAGENBREI (*m*): chyme (condition of food after undergoing action of gastric juice) 424

MAGENBRENNEN (*n*): heart-burn; cardialgia (Med.) 425

MAGENDARMKATARRH (*m*): gastro-intestinal catarrh (Med.) 426

MAGENDARMPROBE (*f*): (see **MAGENDARM-SCHWIMMPROBE**) 427

MAGENDARMSCHNITT (*m*): (see **GASTROENTERO-STOMIE**) 428

MAGENDARMSCHWIMMPROBE (*f*): gastro-intestinal flotation test (to determine whether a still born child has ever breathed) (Med.) 429

MAGENDRÜCKEN (*n*): heartburn, cardialgia (Med.) 430

MAGENDRÜSE (*f*): gastric (peptic) gland; pancreas; sweetbread 431

MAGENENTZÜNDUNG (*f*): gastritis (inflammation of the stomach) (Med.) 432

MAGENFIEBER (*n*): gastric (typhoid; enteric) fever (Med.) 433

MAGENFLÜSSIGKEIT (*f*): gastric (pancreatic) fluid or juice (contains hydrochloric acid and pepsin) 434

MAGENHAUT (*f*): membrane lining the stomach 435

MAGENINHALT (*m*): stomach content-(s) 436

MAGENKATARRH (*m*): catarrh of the stomach (secretion of mucus from the stomach) 437

MAGENKRANKHEIT (*f*): disorder of the stomach 438

MAGENKRAUT (*n*): (see **WERMUT**) 439

MAGENMITTEL (*n*): stomachic (an agent for promoting the functional activity of the stomach) (Med.) 440

MAGENMUND (*m*): pylorus; œsophagus; esophagus (orifice of the stomach) 441

MAGENPFÖRTNER (*m*): (see **MAGENMUND**) 442

MAGENPULVER (*n*): stomachic powder 443

MAGENSAFT (*m*): (see **MAGENFLÜSSIGKEIT**) 444

MAGENSCHLUND (*m*): (see **MAGENMUND**) 445

MAGENSCHMERZ (*m*): stomach-ache; gastralgia, gastrodynia, neuroses of the stomach; praecordial pain, heartburn (Med.) 446

MAGENSCHWÄCHE (*f*): dyspepsia (form of indigestion) (Med.) 447

MAGENSEKT (*m*): stomachic wine (see **MAGEN-MITTEL**) 448

MAGENSTÄRKEND: stomachic 449

MAGENSTÄRKUNG (*f*): stomachic; tonic 450

MAGENSTÄRKUNGSMITTEL (*n*): (see **MAGEN-MITTEL**) 451

MAGENSTEIN (*m*): gastric concretion 452

MAGENTA (*f*): magenta, violet (see **FUCHSIN**) 453

MAGENTRANK (*m*): stomachic draught 454

MAGENTROPFEN (*m.pl*): stomachic drops 455

MAGENWEH (*n*): (see **MAGENSCHMERZ**) 456

MAGENWURZ (*f*): cuckoo-pint (*Acorus maculatum*); sweet flag (*Acorus calamus*) 457

MAGER: lean; slender; thin; meagre; haggard; weak; poor; sterile; gaunt 458

MAGERER KALK (*m*): lime which does not slake easily 459

MAGERES ÖL (*n*): mineral oil 460

MAGERKOHLE (*f*): anthracite; non-caking coal (ash, 8-10%; moisture, 10%; calorific value about 7000 cals.) 461

MAGERMILCH (*f*): skimmed milk 462

MAGERN: to make (become; grow) lean (thin or poor); shorten; diminish the plasticity (Ceramics) 463

MAGERUNGSMITTEL (*n*): plasticity-diminishing substance (Ceramics) 464

MAGISTER (*n*): magistery, subnitrate 465

MAGISTRAL (*m*): roasted copper pyrites (containing copper sulphate) 466

MAGISTRAT (*m*): magistrate 467

MAGISTRATUR (*f*): magistracy 468

MAGNALIUM (*n*): magnalium [(80-90% Al; 20-10% Mg.) an aluminium magnesium alloy; that with 10% Mg. has Sg. 2.5 and Mp. 650-700°C.; that with over 10% Mg. has Sg. 2.25; brittleness increases with increasing Mg. content] 469

MAGNAT (*m*): magnate (a person of influence or high rank) 470

MAGNESIA (*f*): magnesia; magnesium oxide (see MAGNESIUMOXYD) 471

MAGNESIABLEICHFLÜSSIGKEIT (*f*): magnesium hypochlorite solution (see MAGNESIUM-HYPOCHLORIT); (Bleaching) 472

MAGNESIA, CITRONENSAURE- (*f*): magnesia citrate (see MAGNESIUM, CITRONENSAURES-) 473

MAGNESIA (*f*), **DOPPELTKOHLENSAURE-**: magnesium bicarbonate (see MAGNESIUMBIKARBONAT) 474

MAGNESIAEISENGLIMMER (*m*): magnesia mica, biotite (see BIOTIT) 475

MAGNESIA, GEBRANNTE- (*f*): calcined magnesia; MgO; magnesium oxide (see MAGNESIUMOXYD) 476

MAGNESIAGERÄT (*n*): magnesia utensils, utensils made of magnesia 477

MAGNESIAGLIMMER (*m*): magnesium mica; biotite (see BIOTIT) 478

MAGNESIAHÄRTE (*f*): magnesia hardness (of water) 479

MAGNESIAMENGE (*f*): quantity of (magnesia) magnesium 480

MAGNESIAMILCH (*f*): magnesia milk 481

MAGNESIA (*f*), **ÖLSAURE-**: magnesium oleate (see MAGNESIUMOLEAT) 482

MAGNESIATONGRANAT (*m*): Pyrop (see PYROP) 483

MAGNESIA (*f*), **ÜBERBORSAURE-**: magnesium perborate (see MAGNESIUMPERBORAT) 484

MAGNESIAWEISS (*n*): magnesia white (a white body colour used in paper manufacture) 485

MAGNESIT (*m*): magnesite (natural magnesium carbonate); $MgCO_3$; Sg. 3.0 (see MAGNESITSPAT and MAGNESIUMCARBONAT) 486

MAGNESITFARBE (*f*): magnesite colour 487

MAGNESIT, GEBRANNTER- (*m*): (see MAGNESIA, GEBRANNTE-) 488

MAGNESIT (*m*), **KUNST-**: artificial magnesite 489

MAGNESITSPAT (*m*): magnesite (see MAGNESIT) 490

MAGNESITSTEIN (*m*): magnesite brick (Sg. 3.4-3.6) 491

MAGNESITZIEGEL (*m*): magnesite brick; Sg. 3.4-3.6 492

MAGNESIUM (*n*): magnesium; (Mg) 493

MAGNESIUMACETAT (*n*): magnesium acetate (*Magnesium aceticum*); $Mg(C_2H_3O_2)_2$ [494

MAGNESIUMALUMINAT (*n*): magnesium aluminate; $MgO.Al_2O_3$; Sg. 3.57 495

MAGNESIUMBAND (*n*): magnesium tape 496

MAGNESIUM (*n*), **BENZOESAURES-**: magnesium benzoate 497

MAGNESIUMBIKARBONAT (*n*): magnesium bicarbonate; $MgHCO_3$ 498

MAGNESIUMBISULFIT (*n*): magnesium bisulphite; $MgH_2(SO_3)_2$ 499

MAGNESIUMBLITZLICHT (*n*): magnesium flash light (Photography) 500

MAGNESIUMBLITZPULVER (*n*): magnesium flashlight powder (Photography) 501

MAGNESIUMBORAT (*n*): magnesium borate (*Magnesium boricum*); $Mg_3B_2O_6$ 502

MAGNESIUMBOROCITRAT (*n*): magnesium borocitrate; $Mg(BO_2)_2.Mg_3(C_6H_5O_7)_2.H_2O$ [503

MAGNESIUM (*n*), **BORSAURES-**: magnesium borate (see MAGNESIUMBORAT) 504

MAGNESIUMBROMID (*n*): magnesium bromide; $MgBr_2.6H_2O$ 505

MAGNESIUMCARBONAT (*n*): (*Magnesium carbonicum*): (light)-magnesium carbonate, magnesite; $MgCO_3$; Sg. 3.04 506

MAGNESIUMCARBONAT (*n*), **BASISCHES-**: basic magnesium carbonate; heavy magnesium carbonate (*Magnesia alba*); $4MgCO_3.Mg(OH)_2.5H_2O$; Sg. 2.18 507

MAGNESIUMCHLORAT (*n*): magnesium chlorate; $Mg(ClO_3)_2$ 508

MAGNESIUMCHLORID (*n*): (*Magnesium chloratum*): magnesium chloride; $MgCl_2$, Sg. 2.177, Mp. 708°C.; $MgCl_2+6H_2O$, Sg. 1.562 509

MAGNESIUMCHROMAT (*n*): magnesium chromate; $MgCrO_4+7H_2O$; Sg. 1.761 510

MAGNESIUMCITRAT (*n*): magnesium citrate (*Magnesium citricum*); $Mg_3(C_6H_5O_7)_2$ [511

MAGNESIUM (*n*), **CITRONENSAURES-**: magnesium citrate (see MAGNESIUMCITRAT) 512

MAGNESIUMDRAHT (*m*): magnesium wire 513

MAGNESIUM (*n*), **ESSIGSAURES-**: magnesium acetate (see MAGNESIUMACETAT) 514

MAGNESIUMFLUORID (*n*): (*Magnesium fluoratum*): magnesium fluoride; MgF_2; Sg. 2.472; Mp. 1396°C. 515

MAGNESIUM (*n*), **GOLDCHLORWASSERSTOFFSAURES-**: magnesium chloraurate; $Mg(AuCl_4)_2.8H_2O$ 516

MAGNESIUM (*n*), **HARZSAURES-**: magnesium resinate 517

MAGNESIUMHYDRAT (*n*): magnesium (hydrate) hydroxide (see MAGNESIUMHYDROXYD) 518

MAGNESIUMHYDROARSENAT (*n*): magnesium hydroarsenate; $MgHAsO_4+\frac{1}{2}H_2O$; Sg. 3.155 519

MAGNESIUMHYDROPHOSPHAT (*n*): magnesium hydrophosphate, magnesium hydrogen phosphate; $MgHPO_4+H_2O$; Sg. 2.326 [520

MAGNESIUMHYDROXYD (*n*): magnesium hydroxide, magnesium hydrate; $Mg(OH)_2$; Sg. 2.36 521

MAGNESIUMHYPOCHLORIT (*n*): magnesium hypochlorite; $Mg(OCl)_2$ 522

MAGNESIUMJODID (*n*): magnesium iodide; $MgI_2.8H_2O$ 523

MAGNESIUMKARBONAT (*n*): (see MAGNESIUMCARBONAT) 524

MAGNESIUM (*n*), **KIESELSAURES-**: magnesium silicate (see MAGNESIUMSILIKAT) 525

MAGNESIUM (*n*), **KOHLENSAURES-**: magnesium carbonate (see MAGNESIUMCARBONAT) 526

MAGNESIUMLAKTAT (*n*): magnesium lactate; $Mg(C_3H_5O_3)_2.3H_2O$ or $.H_2O$ 527

MAGNESIUMLEGIERUNG (*f*): magnesium alloy 528

MAGNESIUMLICHT (*n*): magnesium light 529

MAGNESIUMMETASILIKAT (*n*): magnesium metasilicate; $MgSiO_3$; Sg. 3.06 530

MAGNESIUM (*n*), **MILCHSAURES-**: magnesium lactate (see MAGNESIUMLAKTAT) 531

MAGNESIUMNITRAT (*n*): (*Magnesium nitricum*): magnesium nitrate; $Mg(NO_3)_2+6H_2O$; Sg. 1.464; Mp. 90°C. 532

MAGNESIUMOLEAT (*n*): magnesium oleate; $Mg(C_{18}H_{33}O_2)_2$ 533

MAGNESIUM (*n*), **ÖLSAURES-**: magnesium oleate (see **MAGNESIUMOLEAT**) 534

MAGNESIUMORTHOSILIKAT (*n*): magnesium orthosilicate; Mg_2SiO_4; Sg. 3.21 535

MAGNESIUM (*n*), **OXALSAURES-**: magnesium oxalate 536

MAGNESIUMOXYD (*n*): (*Magnesium oxydatum*): magnesium oxide, (calcined)-magnesia; (*Magnesia usta*); MgO; Sg. 3.22-3.65; Mp. 2800°C. 537

MAGNESIUMPERBORAT (*n*): magnesium perborate; MgB_4O_7 538

MAGNESIUMPERHYDROL (*n*): magnesium perhydrol (trade name for Magnesium peroxide; see **MAGNESIUMSUPEROXYD**) 539

MAGNESIUMPEROXYD (*n*): magnesium peroxide (see **MAGNESIUMSUPEROXYD**) 540

MAGNESIUMPHOSPHAT (*n*): (*Magnesium phosphoricum*): magnesium phosphate-(dibasic), magnesium hydrogen phosphate, dimagnesium ortho-phosphate; $MgHPO_4.3H_2O$; Sg. 2.123 541

MAGNESIUMPHOSPHAT (*n*), **TERTIÄRES**: tertiary magnesium phosphate; $Mg_3(PO_4)_2.7H_2O$ 542

MAGNESIUM, PHOSPHORSAURES- (*n*): magnesium phosphate; $MgHPO_4.7H_2O$ (see **MAGNESIUMPHOSPHAT**) 543

MAGNESIUMPULVER (*n*): magnesium powder 544

MAGNESIUMPUTZLICHT (*n*): magnesium flashlight 545

MAGNESIUMPYROPHOSPHAT (*n*): magnesium pyrophosphate; $Mg_2P_2O_7$; Sg. 2.4 546

MAGNESIUMRICINAT (*n*): magnesium ricinate 547

MAGNESIUM (*n*), **SALICYLSAURES-**: magnesium salicylate; $Mg(C_7H_5O_3)_2.4H_2O$ 548

MAGNESIUM, SALPETERSAURES- (*n*): magnesium nitrate (see **MAGNESIUMNITRAT**) 549

MAGNESIUMSALZ (*n*): magnesium salt 550

MAGNESIUMSALZLÖSUNG (*f*): magnesium salt solution 551

MAGNESIUM (*n*), **SCHWEFELSAURES-**: magnesium sulphate (see **BITTERSALZ** and **MAGNESIUMSULFAT**) 552

MAGNESIUM (*n*), **SCHWEFLIGSAURES-**: magnesium sulphite (see **MAGNESIUMSULFIT**) 553

MAGNESIUMSILICOFLUORID (*n*): magnesium (fluosilicate) silicofluoride; $MgSiF_6.6H_2O$ 554

MAGNESIUMSILIKAT (*n*): magnesium silicate (*Magnesium silicicum*); $3MgSiO_3.5H_2O$; Sg. 2.7 555

MAGNESIUMSILIKATNIEDERSCHLAG (*m*): magnesium silicate precipitate 556

MAGNESIUMSULFAT (*n*): (*Magnesium sulfuricum*): magnesium sulphate, Epsom salt; $MgSO_4$, Sg. 2.66; $MgSO_4+5H_2O$, Sg. 1.718; $Mg.SO_4+7H_2O$, Sg. 1.68-1.69 557

MAGNESIUMSULFID (*n*): magnesium sulphide; MgS; Sg. 2.82 558

MAGNESIUMSULFIT (*n*): (*Magnesium sulfurosum*): magnesium sulphite; $MgSO_4.6H_2O$ 559

MAGNESIUMTHIOSULFAT (*n*): (*Magnesium thiosulfuricum*): magnesium thiosulphate, magnesium hyposulphite; MgS_2O_3 560

MAGNESIUMSUPEROXYD (*n*): magnesium peroxide (*Magnesium hyperoxydatum*); MgO_2 561

MAGNESIUMTARTRAT (*n*): magnesium tartrate 562

MAGNESIUM (*n*), **UNTERCHLORIGSAURES-**: magnesium hypochlorite (see **MAGNESIUMHYPOCHLORIT**) 563

MAGNESIUM (*n*), **UNTERPHOSPHORIGSAURES-**: magnesium hypophosphite; $[Mg(H_2PO_2)_2.6H_2O]$ 564

MAGNESIUM (*n*), **UNTERSCHWEFLIGSAURES-**: magnesium hyposulphite (see **MAGNESIUMTHIOSULFAT**) 565

MAGNESIUMVERBINDUNG (*f*): magnesium compound 566

MAGNESIUMZITRAT (*n*): magnesium citrate (see **MAGNESIUMCITRAT**) 567

MAGNESIUM (*n*), **ZITRONENSAURES-**: magnesium citrate (see **MAGNESIUMCITRAT**) 568

MAGNET (*m*): magnet; magneto 569

MAGNETANKER (*m*): armature 570

MAGNETAPPARATE (*m.pl*): magnets 571

MAGNET (*m*), **BLEIBENDER-**: permanent magnet 572

MAGNETBÜNDEL (*n*): compound magnet, fagot magnet 573

MAGNETEISEN (*n*): magnetic iron ore; magnetite (see **EISENOXYDULOXYD** and **MAGNETIT**) 574

MAGNETEISENERZ (*n*): (see **MAGNETEISEN**) 575

MAGNETEISENSTEIN (*m*): magnetite (45-70% Fe) (see **MAGNETEISEN**) 576

MAGNETELEKTRISCH: magneto-electric 577

MAGNETELEKTRIZITÄT (*f*): magneto-electricity 578

MAGNETINDUKTION (*f*): magnetic induction 579

MAGNETINDUKTOR (*m*): magneto-electric inductor 580

MAGNETISCH: magnetic 581

MAGNETISCHE KRAFTLINIEN (*f.pl*): magnetic lines 582

MAGNETISCHE LEGIERUNG (*f*): (see **HEUSSLERSCHE LEGIERUNG**) 583

MAGNETISCHER KÖRPER (*m*): magnetic body (a substance which is attracted by a magnet) 584

MAGNETISIERBAR: magnetizable 585

MAGNETISIERBARKEIT (*f*): magnetizability 586

MAGNETISIEREN: to magnetize; mesmerize; (*n*): magnetization; mesmerization; magnetizing; mesmerizing 587

MAGNETISIERUNG (*f*): magnetization; magnetizing 588

MAGNETISMUS (*m*): magnetism; mesmerism 589

MAGNETISMUS (*m*), **FREIER-**: free magnetism 590

MAGNETISMUS (*m*), **GEBUNDENER-**: bound magnetism 591

MAGNETISMUS (*m*), **PERMANENTER-**: permanent magnetism 592

MAGNETISMUS (*m*), **REMANENTER-**: residual magnetism 593

MAGNETIT (*m*): (natural ferrosoferric oxide), magnetite, magnetic iron ore, Lodestone, iron ethiops; Fe_3O_4 (about 73% iron); Sg. 5.16) 594

MAGNETITLAMPE (*f*): magnetite lamp (an arc lamp, the one electrode being composed of magnetite powder and the other of copper) 595

MAGNETITPULVER (*n*): magnetite powder 596

MAGNETKERN (*m*): magnet core 597

MAGNETKIES (*m*): magnetic pyrites; pyrrhotite (a natural iron sulphide), FeS, Sg. 4.59; magnetopyrite, Fe_7S_8, Sg. 4.55 598

MAGNETKIES (*m*), **NICKELHALTIGER-**: nickeliferous magnetic pyrites (1-6% nickel; obtained from Sudbury, Canada) 599
MAGNETNADEL (*f*): compass needle, magnetic needle 600
MAGNETOPYRIT (*m*): magnetopyrite (see MAGNETKIES) 601
MAGNET (*m*), **PERMANENTER-**: permanent magnet 602
MAGNETPOL (*m*): magnet (magnetic) pole 603
MAGNETSCHEIDER (*m*): magnet-(ic) separator (Min.) 604
MAGNETSPIEGEL (*m*): magnet mirror 605
MAGNETSPULE (*f*): magnet coil, magnetic coil 606
MAGNETSTAB (*m*): magnetic bar; magnetized bar, bar magnet 607
MAGNETSTAHL (*m*): magnet (magnetic) steel 608
MAGNETZÜNDUNG (*f*): magneto ignition 609
MAGNET (*m*), **ZUSAMMERGESETZTER-**: compound magnet, fagot magnet 610
MAGNOLIAMETALL (*n*): magnolia metal (approx. 17.8% Sb, 78.3% Pb, 0.04% Cu, 3.86% Sn; or 78% Pb, 16% Sb, 6% Sn) 611
MAGNOLIENGEWÄCHSE (*n pl*): *Magnoliaceæ* (Bot.) 612
MAGSAMEN (*m*): poppy seed 613
MAHAGONI (*n*): mahogany (wood of *Swietenia mahagoni*) 614
MAHAGONIBRAUN (*n*): mahogany brown, terra di Siena 615
MAHAGONIHOLZ (*n*): mahogany wood (Sg. 0.7-1.0) (wood of *Swietenia mahagoni*) 616
MAHAGONILACK (*m*): mahogany varnish 617
MÄHEN: to mow; reap; cut 618
MAHL (*n*): meal; banquet; repast; mole; mark; assembly 619
MAHLEN: to grind; mill; crush; pound 620
MAHLFEINHEIT (*f*): grinding (milling) fineness 621
MAHLGANG (*m*): mill-stones 622
MAHLGUT (*n*): material to be (ground) milled; ground (milled) material; stock 623
MAHLKORN (*n*): grist 624
MAHLSTEIN (*m*): mill-stone 625
MAHLZAHN (*m*): molar-(tooth); grinder 626
MAHLZEIT (*f*): meal; repast; mealtime; an idiomatic greeting at meal times, wishing a good appetite (GUTEN APPETIT) for the forthcoming meal 627
MÄHMASCHINE (*f*): mower, mowing (reaping) machine 628
MAHNBAR: notifiable 629
MAHNBRIEF (*m*): letter of advice, notification, reminder 630
MÄHNE (*f*): mane 631
MAHNEN: to remind; urge; dun; warn; admonish; notify; inform; apprise; advise; make aware; acquaint; make application for-(payment) 632
MAHNSCHREIBEN (*n*): (see MAHNBRIEF) 633
MAHULABUTTER (*f*): (see MAHWABUTTER) 634
MAHWABUTTER (*f*): mahwa butter, ellipi oil, ellipi fat; Mp. 23/30°C. (see also BASSIAÖL) 635
MAHWAFRUCHT (*f*): mahwa nut (seed of *Bassia latifolia*) 636
MAI (*m*): May (the month) 637
MAIBAUM (*m*): birchtree (see BIRKE) 638
MAIBLUME (*f*): lily-of-the-valley (the plant and flower, *Convallaria majalis*) 639
MAIBLUMENKRAUT (*n*): lily-of-the-valley (*Herba convallariæ*) (*Convallaria majalis*) 640
MAIBUTTER (*f*): May butter (butter from cattle fed on grass in May) 641
MAIGLÖCKCHEN (*n*): lily-of-the-valley (see MAI-BLUMENKRAUT) 642
MAIGLÖCKCHENBLÜTENÖL (*n*): lily-of-the-valley-(flower)-oil 643
MAIGLÖCKCHENWURZEL (*f*): lily-of-the-valley root (*Radix convallariæ* of *Convallaria majalis*) 644
MAIKRAUT (*n*): (see WALDMEISTER) 645
MAILLECHORT (*n*): (see NEUSILBER) 646
MAINZER FLUSS (*m*): an easily meltable glass invented by "Strass" (from whom it takes its name), for making spurious precious stones (see also STRASS) 647
MAIS (*m*): maize; Indian corn (*Zea mays*) 648
MAISBRANNTWEIN (*m*): corn (maize) whisky 649
MAISCH (*m*): mash; mashing; grape-juice (see also MEISCH) 650
MAISCHAPPARAT (*m*): mashing (mash) apparatus or machine 651
MAISCHBOTTICH (*m*): mash (mashing) tub or tun; fermenting vat 652
MAISCHE (*f*): mash-(ing); grape-juice 653
MAISCHEN: to mash; mix 654
MAISCHE (*f*), **WEINGARE-**: mash (Distilling) 655
MAISCHE (*f*), **ZWEITE-**: second mash; after-mash 656
MAISCHGITTER (*n*): stirrer; rake (Brewing) 657
MAISCHGUT (*n*): mash; material for mashing 658
MAISCHHOLZ (*n*): scoop 659
MAISCHKESSEL (*m*): mash (copper) boiler 660
MAISCHKUFE (*f*): (see MAISCHBOTTICH) 661
MAISCHPFANNE (*f*): mash (copper) pan 662
MAISCHPROZESS (*m*): mashing process 663
MAISCHPUMPE (*f*): mash pump 664
MAISCHUNG (*f*): mash-(ing) 665
MAISCHVENTIL (*n*): mash (trap) valve 666
MAISCHWASSER (*n*): mash liquor 667
MAISCH (*m*), **WEINGARER-**: (distilling)-mash 668
MAISCHWÜRZE (*f*): mash wort; grain wash 669
MAISCH (*m*), **ZWEITER-**: second mash, after-mash 670
MAISIN (*n*): maize protein 671
MAISKOLBEN (*m*): corncob (see MAIS) 672
MAISKRANKHEIT (*f*): pellagra (a cerebro-spinal disease) (Med.) 673
MAISMEHL (*n*): Indian corn (maize) meal 674
MAISMEHLKLEBER (*m*): zein 675
MAISÖL (*n*): corn (maize) oil (from *Zea mays*); Sg. 0.923-0.96 676
MAISPISTILLE (*n.pl*): cornsilk; zea; stigmata maydis (Pharm.) 677
MAISPROTEIN (*n*): maize protein 678
MAISSPIRITUS (*m*): maize (corn) spirits 679
MAISSTÄRKE (*f*): maize (corn) starch 680
MAITRANK (*m*): a wine flavoured with woodruff, *Asperula odorata* 681
MAIZENA (*f*): A proprietary trade mark for Corn Starch 682
MAIZENASÄURE (*f*): maize acid (from *Zea mays*) 683
MAJESTÄTISCH: majestic-(ally) 684
MAJOLIKA (*n*): majolica (an Italian type of pottery in which SnO_2 is used for glazing) (Ceramics) 685
MAJOLIKAFARBE (*f*): majolica colour 686
MAJOLIKAGLASUR (*f*): majolica (glaze) glazing (opaque due to tin dioxide; SnO_2) 687

MAJOLIKAMASSE (*f*): majolica mass (containing between 25 and 40% $CaCO_3$) 688
MAJORAN (*m*): (sweet)-marjoram (*Origanum vulgare* and *O. majorana*) (*Herba majoranæ*) 689
MAJORANÖL (*n*): majoram oil; Sg. 0.9 690
MAJORENN: of age 691
MAJORENNITÄT (*f*): majority 692
MAJORITÄT (*f*): majority, major (part) portion 693
MAKADAMISIEREN: to macadamize, pave (coat) with macadam 694
MAKADAMSTRASSE (*f*): macadam road, road which has been made up with macadam 695
MAKEL (*m*): spot; blemish; flaw; stain; defect; fault 696
MÄKELEI (*f*): broking; brokerage; criticism 697
MÄKELIG: censorious, critical 698
MAKELLOS: unspotted, unblemished, spotless, immaculate 699
MÄKELN: to criticize; act as broker 700
MAKLER (*m*): broker; agent; jobber; fault-finder; carper; critic 701
MÄKLER (*m*): (see MAKLER) 702
MAKLERGEBÜHR (*f*): brokerage 703
MAKLER (*m*), **SCHIFFS-**: ship broker 704
MAKLIG: easy (of ships); stained, blemished, spotted 705
MAKRELE (*f*): mackerel (*Scomber scombrus*) 706
MAKRONE (*f*): macaroon 707
MAKROPHOTOGRAPHIE (*f*): macrophotography, enlarging; enlargement-(of photographic negatives) 708
MAKROSTRUKTUR (*f*): a structure which can be seen with the naked eye 709
MAKROTYPER KUPHONSPAT (*m*): (see LEVYN) 710
MAKULATUR (*f*): waste paper 711
MAKULATURBOGEN (*m*): waste sheet 712
MAL: time; times; once; just; (*n*): time; mark; spot; mole; sign; token; stigma (Biol.) 713
MALABARTALG (*m*): Malabar tallow (a vegetable tallow from the seeds of *Vateria indica*; Mp. 36.5°C.) (see also PINEYTALG) 714
MALACHIT (*m*): malachite (see also ATLASERZ) (green basic copper carbonate; 40-72% Cu); $Cu_2CO_4.H_2O$; Sg. 3.9 (see also KUPFERCARBONAT) 715
MALACHITGRÜN (*n*): malachite green; benz-aldehyde green (see also KUPFERCARBONAT, BASISCHES- and BERGGRÜN) 716
MALACHITKIESEL (*m*): chrysocolla (see KIESEL--KUPFER) 717
MALAIISCH: Malay-(an) 718
MALAIISCHER ARCHIPEL (*m*): Straits Settlements 719
MALAKKANUSS (*f*): Malacca (marking) nut 720
MALAKOLITH (*m*): malacolite (a calcium magnesium silicate); $Ca_3MgSi_4O_{12}$; Sg. 3.3 (see also DIOPSID) 721
MALAKON (*m*): malacon; $ZrSiO_4.H_2O$; Sg. 3.8 722
MALAMBORINDE (*f*): malambo bark 723
MALARIA (*f*): malaria (Med.) 724
MALARIN (*n*): acetophenonphenitide, malarin; $C_6H_5C(CH_3):N.C_6H_4OC_2H_5$; Mp. 88°C. 725
MALAXIEREN: to malax-(ate); knead; soften; mollify 726
MALDONIT (*m*): maldonite; Au_2Bi; Sg. 9.0 727
MALEINSÄURE (*f*): maleic (malenic; maleinic) acid; $COOH(CH)_2COOH$; Sg. 1.59; Mp. 136.5°C.; Bp. 160°C. 728
MALEINSÄUREANHYDRID (*n*): maleic anhydride; $C_4H_2O_3$; Sg. 0.9339; Mp. 57°C; Bp. 202°C. 729
MALEINSÄURE-DIÄTHYLESTER (*m*): malic-(acid)-diethylester, diethyl malate; $C_8H_{12}O_4$; Sg. 1.0674; Bp. 225°C. 730
MALEN: to paint; colour; depict; picture 731
MALER (*m*): painter; artist 732
MALERDRUCK (*m*): art printing 733
MALEREI (*f*): painting; picture; art of painting 734
MALERFARBE (*f*): painter's colour; art (paint) colour 735
MALERFIRNIS (*m*): painter's varnish; picture varnish 736
MALERGOLD (*n*): painter's gold (see MUSCHEL--GOLD) 737
MALERISCH: picturesque 738
MALERKOLIK (*f*): painter's colic (*Colica pictorum*) (due to intestinal lead poisoning) (Med.) 739
MALERKRANKHEIT (*f*): (see MALERKOLIK) 740
MALERKUNST (*f*): art of painting 741
MALERLACK (*m*): painter's lacquer 742
MALERLEIM (*m*): painter's (glue) size 743
MALERSILBER (*n*): silver powder 744
MALERSTICH (*m*): art printing 745
MALERSTOCK (*m*): mahl-stick, maul-stick (Art) 746
MALL (*f*): mould 747
MALLARDIT (*m*): mallardite (a manganese sulphate); $MnSO_4+7H_2O$ 748
MALLBODEN (*m*): mould loft floor 749
MALLEBENE (*f*): moulding side 750
MALLEBREIN (*n*): mallebrein (a 25% solution of albuminium chlorate) (see TONERDE, CHLOR--SAURE-) 751
MALLEN: to mould 752
MALLKANTE (*f*): moulding edge 753
MALL (*f*), **RUDER-**: rudder mould 754
MALONSAUER: malonate, combined with malonic acid 755
MALONSÄURE (*f*): malonic (methanedicarbonic) acid; $C_3H_4O_4$; Mp. 132 to 135.6°C. 756
MALONSÄURE-DIÄTHYLESTER (*m*): malonic-(acid)-diethylester, Ethyl malonate; $C_7H_{12}O_4$; Sg. 1.0553; Mp. −49.8°C.; Bp. 198.4°C. 757
MALONSÄURE-NITRIL (*n*): malonic nitrile; $C_3H_2N_2$; Sg. 1.063; Mp. 30°C.; Bp. 219°C. 758
MALONYLHARNSTOFF (*m*): malonyl urea, barbituric acid (see BARBITURSÄURE) 759
MALPIGHISCHES NETZ (*n*): (see SCHLEIMSCHICHT) 760
MALSTEIN (*m*): boundary stone; monumental stone 761
MALTOBIOSE (*f*): (see MALTOSE) 762
MALTOSE (*f*): maltose, malt sugar; $C_{12}H_{22}O_{11}.H_2O$; loses H_2O at 100°C. (see also ROHRZUCKER) 763
MALTYL (*n*): maltyl (a malt extract) 764
MALVE (*f*): mallow (*Malva silvestris*) 765
MALVENBLÄTTER (*f.pl*): mallow leaves (*Folia malvæ of Malva silvestris*) 766
MALVENBLÜTEN (*f.pl*): mallow flowers 767
MALVENGEWÄCHSE (*n.pl*): *Malvaceæ* (Bot.) 768
MALWE (*f*): (see MALVE) 769

MALZ (*n*): malt (barley grain, steeped and prepared for use in Brewing) 770
MALZAROMA (*n*): malt aroma 771
MALZAUFGUSS (*m*): infusion of malt; wort 772
MALZAUSZUG (*m*): malt extract; wort 773
MALZBEREITUNG (*f*): malting 774
MALZBESCHAFFENHEIT (*f*): constitution of the malt 775
MALZBODEN (*m*): malt loft 776
MALZBOTTICH (*m*): steeping vat 777
MALZ (*n*), **DARR-**: kiln-dried malt 778
MALZDARRE (*f*): malt kiln 779
MALZDIASTASE (*f*): diastase (see **DIASTASE**) 780
MALZEIWEISS (*n*): diastase (see **DIASTASE**) 781
MALZEN: to malt; (*n*): malting 782
MÄLZEN: (see **MALZEN**) 783
MALZER (*m*): maltster 784
MÄLZEREI (*f*): malting; malt house 785
MÄLZEREIEINRICHTUNG (*f*): malting (installation) plant (Brewing) 786
MÄLZEREIMASCHINEN (*f.pl*): malting machinery (Brewing) 787
MALZESSIG (*m*): malt vinegar 788
MALZEXTRAKT (*m*): malt extract 789
MALZEXTRAKTANLAGE (*f*): malt extract (extracting; extraction) plant 790
MALZEXTRAKTPULVER (*n*): malt-extract powder 791
MALZ (*n*), **FARB-**: colouring malt 792
MALZFABRIK (*f*): malt house 793
MALZGERSTE (*f*): malting barley 794
MALZ (*n*), **GERSTEN-**: barley malt 795
MALZ (*n*), **GRÜN-**: germinated, moist malt 796
MALZHÄUFEN (*n*): couching 797
MALZHAUS (*n*): malt house 798
MALZKRÜCKE (*f*): scoop 799
MALZPRÄPARAT (*n*): malt preparation 800
MALZPROBE (*f*): malt test-(ing); sample of malt 801
MALZPUTZMASCHINE (*f*): malt cleaning machine 802
MALZSCHROT (*n*): crushed malt; malt grist 803
MALZSTAUB (*m*): malt dust 804
MALZSTEUER (*f*): malt tax 805
MALZTENNE (*f*): malt floor (Brewing) 806
MALZTENNENKÜHLUNG (*f*): cooling of the malt floor 807
MALZTREBER (*pl*): spent (malt) grains; brewer's grains 808
MÄLZUNG (*f*): malting 809
MÄLZUNGSSCHWAND (*m*): malting (loss) shrinkage 810
MALZZERKLEINERUNG (*f*): malt crushing 811
MALZZUCKER (*m*): maltose; malt sugar (see **MALTOSE**) 812
MALZZUCKERLÖSUNG (*f*): malt-sugar (maltose) solution 813
MAMAIBAUM (*m*): mammee tree (*Mammea americana*) 814
MAMMON (*m*): Mammon; pelf 815
MAMMUT (*n*): mammoth (*Elephas antiquus*) 816
MAMMUTBAUM (*m*): mammoth tree, giant redwood (*Sequoia gigantea*) 817
MAN: (some-)one; a person; they; somebody; we; you; people; men; only 818
MANCH: many a -(one or thing); some; several; many 819
MANCHERLEI: many; various; several; diverse; varied 820
MANCHESTER (*m*): corduroy, velveteen 821
MANCHESTERBRAUN (*n*): Manchester brown 822

MANCHETTE (*f*): cuff; sleeve (see **MANSCHETTE**) 823
MANCHMAL: often; sometimes 824
MANDARINE (*f*): mandarin-(orange) (*Citrus nobilis*) 825
MANDARINENÖL (*n*): mandarin oil (see **MANDARINENSCHALENÖL**) 826
MANDARINENSCHALENÖL (*n*): mandarin-(peel)-oil; Sg. 0.855; Bp. 177°C. 827
MANDAT (*n*): mandate, authorization 828
MANDEL (*f*): almond (*Amygdalus communis*); tonsil; fifteen; shock 829
MANDELARTIG: almond-like; amygdaline 830
MANDELBAUM (*m*): almond tree (*Amygdalus communis*) 831
MANDELBAUMHOLZ (*n*): almond wood, Indian (*Crickrassia tabularis*); Sg. 0.85-0.88 832
MANDELBLÜTE (*f*): almond blossom (see **MANDELBAUM**) 833
MANDELBRÄUNE (*f*): mumps (or swelling of the salivary or parotid glands) (Med.) (see also **BRÄUNE**) 834
MANDELDRÜSE (*f*): tonsil 835
MANDELEITERGESCHWULST (*f*): abscess of the tonsils; tonsillar abscess (Med.) 836
MANDELENTZÜNDUNG (*f*): tonsilitis; inflammation of the tonsils (Med.) 837
MANDELFÖRMIG: almond-shaped; amygdaloid 838
MANDELHOLZ (*n*): almond wood (from *Amygdalus communis*) (see also **MANDELBAUMHOLZ**) 839
MANDELKERN (*m*): almond-(kernel) 840
MANDELKLEIE (*f*): almond (meal) bran (used for cosmetics) (see **MANDELKUCHEN**) 841
MANDELKUCHEN (*m*): almond cake (the residue after expressing the oil from almonds, which, after being ground, is termed almond meal) 842
MANDELMILCH (*f*): almond milk; milk (emulsion) of almond (Pharm.) 843
MANDELN: to set in sheaves (shocks; heaps) of fifteen; mangle 844
MANDELÖL (*n*): (*Oleum amygdalarum*): almond oil (from seeds of *Amygdalus communis*); Sg. 0.91-1.06; Bp. 180°C. (see **BITTERMANDELÖL** and **SÜSSMANDELÖL**) 845
MANDELÖL (*n*), **ÄTHERISCHES-**: (see **BITTERMANDELÖL**) 846
MANDELÖL (*n*), **BITTER-**: (see **BITTERMANDELÖL**) 847
MANDELÖL (*n*), **KÜNSTLICHES-**: artificial oil of almonds (see **BENZALDEHYD**) 848
MANDELÖL (*n*), **SÜSSES-**: sweet almond oil (from seeds of *Amygdalus communis*); Sg. 0.919 849
MANDELÖL (*n*), **WILDES-**: (see **CATAPPAÖL**) 850
MANDELSÄURE (*f*): mandelic acid 851
MANDELSÄURENITRIL (*n*): mandelic nitrile 852
MANDELSÄURETROPINESTER (*m*): atropine-mandelic acid 853
MANDELSEIFE (*f*): almond soap 854
MANDELSTEIN (*m*): amygdaloid; tonsillar concretion 855
MANDELSTORAX (*m*): amygdaloid storax 856
MANDELSYRUP (*m*): syrup of almonds 857
MANDELWEISE: by fifteens 858
MANDRAGORA (*f*): mandragora, mandrake (*Mandragora autumnalis* and *M. vernalis*; also *Podophyllum Peltatum*) 859
MANGAN (*n*): manganese (Mn) 860
MANGANALAUN (*m*): manganese alum 861

MANGANALUMINIUMBRONZE (*f*): manganese-aluminium bronze (bronze with a manganese content up to 10%; addition of Ni, Pb or Zn increases its mechanical properties) 862

MANGANANTIMONID (*n*): manganese antimonide; MnSb; Sg. 5.6 863

MANGANAZETAT (*n*): (*Manganum aceticum*): manganese acetate, manganous acetate; $Mn(C_2H_3O_2)_2.4H_2O$; Sg. 1.6 864

MANGAN (*n*), **BENZOESAURES-**: manganese benzoate 865

MANGAN (*n*), **BERNSTEINSAURES-**: manganese succinate 866

MANGANBISTER (*m*): manganese bistre-(brown) (see **MANGANOXYDHYDRAT**) 867

MANGANBLENDE (*f*): alabandite (natural manganese sulphide); MnS; Sg. 4.0 (see also **MANGANSULFID**) 868

MANGANBORAT (*n*): (*Manganum boricum*): manganese (manganous) borate; MnB_4O_7 [869

MANGANBORID (*n*): manganese boride; MnB; Sg. 6.2 870

MANGAN, BORSAURES- (*n*): manganese borate (see **MANGANBORAT**) 871

MANGANBRAUN (*n*): raw umber (an aluminium silicate with a high content of ferric hydroxide and manganese peroxide); Manganese brown (see **MANGANOXYDHYDRAT**) 872

MANGANBROMID (*n*): manganese bromide; $MnBr_2$; Sg. 4.385 873

MANGANBRONZE (*f*): (82-85% Cu; 14-17% Sn; 0.25-0.6% Mn; or 65-90% Cu; 0-3% Sn; 2-15% Zn; 5-20% Mn); manganese bronze 874

MANGANCARBID (*n*): manganese carbide; Mn_3C; Sg. 6.888; Mp. 1217°C. 875

MANGANCARBONAT (*n*): (*Manganum carbonicum*): manganese carbonate, manganous carbonate; $MnCO_3$; Sg. 3.125 876

MANGANCHLORID (*n*): manganese (manganic) chloride 877

MANGANCHLORÜR (*n*): manganous chloride (*Manganum chloratum*); $MnCl_2.4H_2O$; Sg. 2.0; Mp. 88°C.; Bp. 106°C. 878

MANGAN (*n*), **CHROMSAURES-**: manganese chromate 879

MANGANDIFLUORID (*n*): manganese (di)-fluoride, manganous fluoride; MnF_2; Sg. 3.98; Mp. 856°C. 880

MANGANDIOXYD (*n*): black manganese oxide; manganese (binoxide; peroxide) dioxide (*Manganum hyperoxydatum*); MnO_2; Sg. 5.026 881

MANGANEISEN (*n*): ferromanganese 882

MANGANEISENSTEIN (*m*): triplite (see **TRIPLIT**) 883

MANGANEPIDOT (*m*): piemontite (see **PIEMONTIT**) 884

MANGANERZ (*n*): manganese ore (50-55% Mn; 1-7% Fe) 885

MANGANERZ (*n*), **BRACHYTYPES-**: (see **HART-BRAUNSTEIN**) 886

MANGANERZ, GRAUES- (*n*): manganite; pyrolusite (see **MANGANIT**; **BRAUNSTEIN** and **PYROLUSIT**) 887

MANGANERZ (*n*), **PRISMATISCHES-**: (see **WEICH-BRAUNSTEIN**) 888

MANGANERZ (*n*), **PRISMATOIDISCHES-**: manganite 889

MANGANERZ (*n*), **PYRAMIDALES-**: hausmannite 890

MANGANERZ, SCHWARZES- (*n*): hausmannite (see **HAUSMANNIT**) 891

MANGANERZ (*n*), **UNTEILBARES-**: psilomelane 892

MANGAN (*n*), **ESSIGSAURES-**: manganese acetate; $Mn(C_2H_3O_2)_2.4H_2O$; Sg. 1.6 893

MANGANFARBE (*f*): manganese colour (see **MANGANWEISS** and **MANGANVIOLETT**) 894

MANGAN (*n*), **FETTSAURES**: manganese sebate 895

MANGANGEHALT (*m*): manganese content 896

MANGANGLANZ (*m*): alabandite (see **ALABANDIT**) 897

MANGAN (*n*), **GOLDCHLORWASSERSTOFFSAURES-**: manganese chloraurate; $Mn(AuCl_4)_2.8H_2O$; or $12H_2O$ 898

MANGANHALTIG: manganiferous; containing manganese 899

MANGAN, HARZSAURES- (*n*): manganese resinate (see **MANGANOXYD, HARZSAURES-**) 900

MANGANHEPTOXYD (*n*): manganese heptoxide; Mn_2O_7 901

MANGAN, HOLZÖLSAURES- (*n*): manganese lignoleate 902

MANGANHYDROXYD (*n*): manganic (hydrate) hydroxide (*Manganum hydroxydatum*) $Mn(OH)_2$; Sg. 3.26 (see also **MANGANIT**) 903

MANGANHYDROXYDUL (*n*): manganous hydroxide 904

MANGANHYPEROXYD (*n*): (see **MANGANDIOXYD**) 905

MANGANICHLORID (*n*): manganic chloride 906

MANGANIG: manganous 907

MANGANIGSAUER: manganite of; combined with manganous acid 908

MANGANIGSÄURE (*f*): manganous acid 909

MANGANIHYDROXYD (*n*): manganic hydroxide (see **MANGANDIOXYD**) 910

MANGANIN (*m*): (83% Cu; 4% Ni; 13% Mn; or 58% Cu; 41% Ni; 10% Mn); manganin (a copper manganese alloy) 911

MANGANIPHOSPHAT (*n*): manganic phosphate 912

MANGANISALZ (*n*): manganic salt 913

MANGANISULFAT (*n*): manganic sulphate 914

MANGANIT (*m*): manganite, grey manganese ore, hydrated manganese oxide; $Mn_2O_3.H_2O$; Sg. 4.3 915

MANGANIVERBINDUNG (*f*): manganic compound 916

MANGANKARBID (*n*): manganese carbide (see **MANGANCARBID**) 917

MANGANKARBONAT (*n*): (see **MANGANCARBONAT**) 918

MANGANKIES (*m*): rhodonite, manganese spar (see **MANGANSPAT** and **RHODONIT**); Hauerite (see **HAUERIT**) 919

MANGANKIESEL (*m*): rhodonite (see **RHODONIT**) 920

MANGAN, KOHLENSAURES-: (*n*): manganese carbonate (see **MANGANCARBONAT**) 921

MANGANKUPFER (*n*): cupro-manganese, manganese copper (about 20-30% Mn., with sometimes up to 4% Fe) 922

MANGANKUPFEROXYD (*n*): Crednerite (mineral) (about 44% CuO; and 55% Mn_3O_4); Sg. 4.99 923

MANGANLAKTAT (*n*): manganese lactate; $Mn(C_3H_5O_3)_2.3H_2O$ 924

MANGANLEGIERUNG (*f*): manganese alloy 925

MANGANLEGIERUNG (*f*), **MAGNETISIERBARE-**: magnetizable manganese alloy (being cupro-manganese, see **MANGANKUPFER**, plus 3% or more of Al, As, B, Bi or Sb, with not less than 9% Mn) 926

MANGAN, LEINÖLSAURES- (*n*): manganese linoleate; $Mn(C_{18}H_{31}O_2)_2$ 927
MANGANMETALL (*n*): manganese metal 928
MANGANMETASILIKAT (*n*): manganese metasilicate; $MnSiO_3$; Sg. 3.35-3.58 929
MANGANNITRAT (*n*): manganese nitrate; $Mn(NO_3)_2+6H_2O$; Sg. 1.82 930
MANGANOCALCIT (*m*): manganocalcite (see MANGANSPAT) 931
MANGANOCOLUMBIT (*m*): manganocolumbite (see COLUMBIT) 932
MANGANOFERRUM (*n*): ferromanganese 933
MANGANOHYDROXYD (*n*): manganous hydroxide 934
MANGANOION (*n*): manganous ion 935
MANGANOKARBONAT (*n*): manganous carbonate; $MnCO_3$ (see MANGANCARBONAT) 936
MANGANOLEAT (*n*): manganese oleate (*Manganum oleinicum*); $Mn(C_{18}H_{33}O_2)_2$ 937
MANGAN (*n*), **ÖLSAURES-**: manganese oleate (see MANGANOLEAT) 938
MANGANOMANGANIT (*n*): mangano-manganic oxide; Mn_3O_4 (see MANGANOXYDOXYDUL) 939
MANGANOOXALAT (*n*): manganese oxalate (see MANGANOXYDUL, OXALSAURES-) 940
MANGANOSALZ (*n*): manganous salt 941
MANGANOSALZLÖSUNG (*f*): manganous salt solution 942
MANGANOSIT (*m*): manganosite (natural manganous oxide); MnO; Sg, 5.18 943
MANGANOSULFAT (*n*): manganous sulphate; $MnSO_4.4H_2O$; Sg. 2.107; Bp. 30°C. [944
MANGANOTANTALIT (*m*): manganotantalite (see TANTALIT) 945
MANGANOTANTALITH (*m*): manganotantalite (a form of tantalite in which most of the iron is replaced by manganese) (see TANTALIT) 946
MANGANOXALAT (*n*): manganese oxalate (see MANGANOXYDUL, OXALSAURES-) 947
MANGAN (*n*), **OXALSAURES-**: manganese oxalate (see MANGANOXYDUL, OXALSAURES-) 948
MANGANOXYD (*n*): (*Manganum oxydatum*): manganese (manganic) oxide; Mn_2O_3; Sg. 4.5 (see MANGANSESQUIOXYD) 949
MANGANOXYD (*n*), **HARZSAURES-**: manganese (manganic) resinate; $Mn(C_{20}H_{29}O_2)_2$ (an important quick-drier) 950
MANGANOXYDHYDRAT (*n*): (*Manganum hydroxydatum*): manganese (manganic) hydroxide, manganese hydrate; $[Mn(OH)_2]$; Sg. 3.26 (see also MANGANIT); Bistre, manganese brown 951
MANGANOXYDOXYDUL (*n*): mangano-manganic oxide, hausmannite; Mn_3O_4; Sg. 4.61 [952
MANGANOXYD, ROTES- (*n*): mangano-manganic oxide; Mn_3O_4; Sg. 4.61 953
MANGANOXYDSALZ (*n*): manganic salt 954
MANGANOXYDUL (*n*): manganous oxide (*Manganum oxydulatum*); MnO; Sg. 5.091 [955
MANGANOXYDUL (*n*), **BORSAURES-**: manganese borate (see MANGANBORAT) 956
MANGANOXYDUL (*n*), **ESSIGSAURES-**: manganese acetate (see MANGANAZETAT) 957
MANGANOXYDULHYDRAT (*n*): manganous hydroxide; $Mn_2O_3+H_2O$; Sg. 4.335 [958
MANGANOXYDUL (*n*), **KOHLENSAURES-**: manganese carbonate (see MANGANCARBONAT) 959
MANGANOXYDUL (*n*), **OXALSAURES-**: (*Manganum oxalicum*); manganese oxalate, manganous oxalate; $MnC_2O_4.5H_2O$; Sg. 2.4 960
MANGANOXYDULOXYD (*n*): mangano-manganic oxide; Mn_3O_4; Sg. 4.61 961
MANGANOXYDULSALZ (*n*): manganous salt 962
MANGANOXYDULSALZLÖSUNG (*f*): manganous salt solution 963
MANGANOXYDUL (*n*), **SCHWEFELSAURES-**: manganous sulphate (see MANGANSULFAT) 964
MANGANOXYDULVERBINDUNG (*f*): manganous compound 965
MANGANOXYDVERBINDUNG (*f*): manganic compound 966
MANGANPECHERZ (*n*): triplite (see TRIPLIT) 967
MANGANPEKTOLITH (*m*): (see SCHIZOLITH) 968
MANGANPEROXYD (*n*): manganese peroxide (see MANGANDIOXYD) 969
MANGANPHOSPHAT (*n*): manganese phosphate; manganous ortho-phosphate; $Mn_3(PO_4)_2.7H_2O$ 970
MANGANPRÄPARAT (*n*): manganese preparation 971
MANGANSACCHARAT (*n*): manganese saccharate 972
MANGAN (*n*), **SALPETERSAURES-**: manganese nitrate (see MANGANNITRAT) 973
MANGANSAUER: manganate of; combined with manganic acid 974
MANGANSÄURE (*f*): manganic acid; H_2MnO_4 975
MANGANSÄUREANHYDRID (*n*): manganic anhydride; manganese trioxide (see MANGAN-TRIOXYD) 976
MANGANSCHAUM (*m*): wad; bog manganese (10-20% water) (see WAD) 977
MANGANSCHLAMM (*m*): Weldon mud 978
MANGANSCHWARZ (*n*): manganese black (see MANGANSUPEROXYD) 979
MANGAN, SCHWEFELSAURES- (*n*): manganese sulphate (see MANGANSULFAT) 980
MANGANSELENID (*n*): manganese selenide; MnSe; Sg. 5.59 981
MANGANSESQUIOXYD (*n*): manganese sesquioxide (*Manganum oxydatum*) (see MANGAN-OXYD) 982
MANGANSILICID (*n*): manganese silicide; Mn_2Si; Sg. 6.2; MnSi; Sg. 5.24 983
MANGANSILICOFLUORID (*n*): manganese silicofluoride; $MnSiF_6+6H_2O$; Sg. 1.9038 [984
MANGANSPAT (*m*): manganese spar (manganese carbonate, $MnCO_3$, Sg. 3.61, rhodochrosite; manganese silicate, $MnSiO_3$, rhodonite, Sg. 3.57) 985
MANGANSTAHL (*m*): manganese steel (2-6% Mn; increases hardness of steel) 986
MANGANSULFAT (*n*): manganese sulphate (*Manganum sulfuricum*); $MnSO_4.4H_2O$; Sg. 2.107; Bp. 30°C. 987
MANGANSULFID (*n*): manganese sulphide 988
MANGANSULFÜR (*n*): manganous sulphide; MnS; Sg. 4.0 989
MANGANSUPEROXYD (*n*): manganese peroxide (see MANGANDIOXYD) 990
MANGANTONGRANAT (*m*): spessartite (see SPES-SARTIN) 991
MANGANTRIFLUORID (*n*): manganese trifluoride; MnF_3; Sg. 3.54 992
MANGANTRIOXYD (*n*): manganic anhydride, manganese trioxide; MnO_3 993
MANGAN (*n*), **UNTERPHOSPHORIGSAURES-**: manganese hypophosphite; $Mn(H_2PO_2)_2.H_2O$ 994
MANGANVERBINDUNG (*f*): manganese compound 995

MANGANVIOLETT (*n*): Nuremberg violet, manganese violet [a manganese (manganic) phosphate] $Mn_2(PO_4)_3$ 996

MANGANVITRIOL (*m*): manganese (sulphate) vitriol (see **MANGANSULFAT**) 997

MANGANWEISS (*n*): manganese white; manganous carbonate; $MnCO_3$ (see **MANGANCARBONAT**) 998

MANGANZINKSPAT (*m*): manganese-zinc spar; $(Mn,Zn)O.\ CO_2$ 999

MANGEL (*m*): lack; blemish; indigence; penury; want; deficiency; fault; defect, distress; (*f*): mangle; calender; rolling-press 000

MANGELENDE ZUSAMMENZIEHUNG (*f*): Asystole, heart failure due to dilation (Med.) 001

MANGELHAFT: deficient; lacking; defective; imperfect; inefficient; faulty; marred; incomplete; inchoate 002

MANGELHAFTIGKEIT (*f*): defectiveness, faultiness, imperfection, incompletion, inadequacy, insufficiency, scarcity, dearth, lack 003

MANGELHOLZ (*n*): (mangle)-roller 004

MANGELN: to be wanting; want; fail; lack; mangle; calender; roll 005

MANGELS: in default of 006

MANGELWURZEL (*f*): mangel-wurzel (*Beta maritima*) 007

MANGFUTTER (*n*): mixed fodder 008

MANGOBAUM (*m*): mango tree (*Mangifera indica*) 009

MANGOBLÄTTER (*n.pl*): mango leaves (see **MANGOBAUM**) 010

MANGOLD (*m*): (see **RÜBE**) 011

MANGOLDWURZEL (*f*): (see **MANGELWURZEL**) 012

MANGROVENEXTRAKT (*m*): mangrove extract 013

MANIE (*f*): mania; madness 014

MANIER (*f*): manner; fashion; way; style; habit; deportment 015

MANIERIERT: affected 016

MANIFEST: manifest-(ly); (*n*): manifesto; manifest (see **SCHIFFSMANIFEST**) 017

MANIHOTÖL (*n*): Manihot oil (from seeds of *Manihot glaziovii*); Sg. 0.926 018

MANILAHANF (*m*): Manilla hemp; Siam hemp; abaca (fibre from a plantain, *Musa textilis*) 019

MANILATAU (*n*): manilla rope 020

MANIPULIEREN: to manipulate; handle 021

MANJAK (*m*): (see **BARBADOSASPHALT**) 022

MANN (*m*): man; husband; male; male-servant 023

MANNA (*f* and *n*): manna (a sugary substance from *Fraxinus ornus*) (see also **HIMMELBROT**) 024

MANNAESCHE (*f*): flowering ash (*Fraxinus ornus*) 025

MANNAZUCKER (*m*): mannite; manna sugar; mannitol; $C_6H_8(OH)_6$; Sg. 1.52; Mp. 165-166°C.; Bp. 290-295°C. 026

MANNESBART (*n*): traveller's joy, wild clematis (*Clematis vitalba*) 027

MANNESREIFE (*f*): puberty (of age); maturity 028

MANNIGFACH: manifold; multifarious; various; different; varied; diverse; divers; sundry 029

MANNIGFALTIG: (see **MANNIGFACH**) 030

MANNIGFALTIGKEIT (*f*): variety; multiplicity; diversity; difference 031

MANNIT (*n*): mannite, mannitol, manna sugar (active principle of manna; see **MANNA**); $C_6H_8(OH)_6$; Sg. 1.52; Mp. 165/6°C.; Bp. 290/5°C. 032

MANNITESTER (*m*): mannite ester 033

MÄNNLICH: male; masculine; manly; bold; stout; valiant; brave 034

MÄNNLICHKEIT (*f*): manliness; manhood; virility 035

MANNLOCH (*n*): manhole 036

MANNLOCHBODEN (*m*): manhole-end (of a drum) 037

MANNLOCHBÜGEL (*m*): manhole (cross-bar) bridge 038

MANNLOCHDECKEL (*m*): manhole cover 039

MANNLOCHDICHTUNG (*f*): manhole-joint 040

MANNLOCHRING (*m*): manhole ring 041

MANNLOCHTÜR (*f*): manhole door 042

MANNLOCHVERPACKUNG (*f*): manhole packing, manhole (joints, jointing) joint rings 043

MANNLOCHVERSCHLUSS (*m*): manhole (door) cover 044

MANNOSE (*f*): mannose (one of the grape sugar group of sugars) (see **DEXTROSE**) 045

MANNOZUCKERSÄURE (*f*): mannosaccharic acid 046

MANNSBART (*m*): (see **BARTGRAS**) 047

MANNSCHAFT (*f*): crew; forces; staff; assistants; gang (body of)-men 048

MANNSCHAFT (*f*), **SCHIFFS-**: crew 049

MANNSCHAFTSRAUM (*m*): crew's quarters 050

MANNSCHAFTSROLLE (*f*): (muster)-roll 051

MANNSCHEU: misanthropic 052

MANNSCHEUE (*f*): misanthropy (hatred or fear of man) 053

MANNSKRAUT (*n*): Pasque flower (*Anemone pulsatilla*) 054

MANNSSTIMME (*f*): male voice 055

MANOMETER (*n*): manometer; pressure gauge 056

MANOMETER (*n*), **DIFFERENTIAL-**: differential pressure gauge 057

MANOMETERFEDER (*f*): pressure gauge spring 058

MANOMETER (*n*), **FEDER-**: spring pressure gauge 059

MANOMETERGEHÄUSE (*n*): pressure gauge case 060

MANOMETER (*n*), **METALL-**: (see **MANOMETER, FEDER-**) 061

MANOMETER (*n*), **PLATTENFEDER-**: (flat)-spring pressure gauge 062

MANOMETER (*n*), **QUECKSILBER-**: mercury pressure gauge 063

MANOMETERROHR (*n*): pressure gauge pipe, P.G. pipe 064

MANOMETERSTAND (*m*): pressure gauge (indication) position 065

MANOMETER (*n*), **WASSERDRUCK-**: hydraulic pressure gauge 066

MANÖVER (*n*): manœuvre 067

MANÖVRIEREN: to manœuvre 068

MANÖVRIERFÄHIGKEIT (*f*): manœuvring capacity (of ships) 069

MANSARDE (*f*): attic, mansarde (see also **TROCKENSTUHL**) 070

MANSARDENDACH (*n*): mansarde roof (a roof sloping at two angles, in order to give more room in the attic) 071

MANSARDENFENSTER (*n*): dormer (attic) window 072

MANSCHEN: to paddle, splash, meddle, dabble; mix-(up) 073

MANSCHETTE (*f*): cuff; flap; sleeve; collar (Tech.) 074
MANSCHETTENPRESSE (*f*): (leather)-collar press 075
MANTEL (*m*): mantlepiece; sheet-(ing); pall; envelope; cover; top; crown; mantle; casing; case; shell; jacket; body; housing; coping; cloak; covering 076
MANTELBLECH (*n*): shell-plate 077
MANTELFLÄCHE (*f*): superficies 078
MANTELHEIZUNG (*f*): jacket heating 079
MANTELKRÄHE (*f*): hooded-crow (a member of the genus *Corvus*) 080
MANTELKÜHLER (*m*): jacket cooler, shell type cooler 081
MANTELROHR (*n*): jacket pipe 082
MANTELSACK (*m*): valise 083
MANTELTAKEL (*n*): runner tackle (Naut.) 084
MANTEL (*m*), **WASSER-**: water jacket 085
MANTEL (*m*), **ZYLINDER-**: cylinder jacket 086
MANUAL (*n*): manual, keyboard; notebook, waste-book, journal 087
MANUFAKTUR (*f*): manufacture; manufactory 088
MANUSKRIPT (*n*): manuscript (actually any matter written by hand; now often used for any original draft, whether hand or type-written) 089
MANUTEKT (*n*): mixture of a solution of phenol and formaldehyde condensation products, with a cellulose ester solution, forming a liquid which is painted on to the hands and dries to an elastic coating; (substitute for operating gloves in surgery) 090
MAPPE (*f*): portfolio; case; map 091
MARANTASTÄRKE (*f*): arrowroot; maranta starch (see **STÄRKE**) 092
MÄRCHEN (*n*): fairy-tale; story; tale; legend; fable; myth; saga; fiction; tradition; rumour; superstition 093
MÄRCHENHAFT: fabulous; legendary; fictitious; mythical 094
MARDER (*m*): marten; pine marten (a member of the weasel family, *Mustela martes*) 095
MARDERFELL (*n*): marten(skin) fur 096
MÄRE (*f*): news; report; tradition; story; rumour; legend (see **MÄRCHEN**) 097
MARETIN (*n*): maretin; $C_8H_{11}N_3O$; Mp. 183.5°C. 098
MARGARETENBLUME (*f*): marguerite, daisy, Paris daisy (*Chrysanthemum frutescens*) 099
MARGARIN (*n*): margarine, butterine, oleomargarine (beef tallow freed from stearin; is distinct from **MARGARINE**, which see) 100
MARGARINE (*f*): margarine (obtained from Olein and milk; see **KIRNEN**) (according to German law margarine must contain 10% Sesame oil; see **SESAMÖL**) 101
MARGARINEABFALL (*m*): margarine (refuse) waste 102
MARGARINEFABRIKATION (*f*): margarine manufacture 103
MARGARINEKONSERVIERUNGSPRÄPARAT (*n*): margarine preservative 104
MARGARINEKONSISTENZ (*f*): consistency of margarine 105
MARGARINE (*f*), **SCHMELZ-**: melted margarine (from margarine by melting, for the purpose of preserving it for a greater length of time) 106
MARGARINIERUNG (*f*): margarinizing 107
MARGARINSAUER: margarate of; combined with margaric acid 108
MARGARINSÄURE (*f*): margaric acid 109
MARGARIT (*m*): margarite (a micaceous mineral); CaO,Al_2O_3,SiO_2,H_2O; Sg. 2.95-3.05 (see also **KALKGLIMMER**) 110
MARGASIT (*m*): margasite (see **MARGARIT**) 111
MARGOL (*n*): margol (a name given to two entirely different substances. First, a mixture of volatile fatty acid for margarine manufacture. Second, a remedy for syphilis containing antimony, arsenic and silver) 112
MARGOSAÖL (*n*): margosa oil (from seeds of *Melia azedarach*) 113
MARIALITH (*m*): marialite (a form of Wernerite); $Na_4Al_3Si_9O_{24}Cl$; Sg. 2.56 114
MARIENBAD (*n*): (hot)-water bath (Chem.); bain marie (water bath similar in principle to the ordinary glue-pot) 115
MARIENBLUME (*f*): ox-eye-(daisy) (*Chrysanthemum leucanthemum*) 116
MARIENDISTEL (*f*): milk thistle (*Silybum marianum*) 117
MARIENEIS (*n*): (see **MARIENGLAS**) 118
MARIENFÄDEN (*m.pl*): gossamer, filmy cobweb, floating spider's web 119
MARIENFLACHS (*m*): (soft)-feather grass (*Stipa pennata*) 120
MARIENGLAS (*n*): alabaster; selenite (see **ERDGLAS**); gypsum (see **GYPS**); mica; muscovite (see **MUSKOVIT**) 121
MARIENGLOCKE (*f*): (see **MARIENGLÖCKCHEN**) 122
MARIENGLÖCKCHEN (*n*) Canterbury bell (*Campanula medium*) 123
MARIENGRAS (*n*): feather-grass (*Stipa pennata*); corn spurrey (*Spergula arvensis*) 124
MARIENKÄFER (*m*): ladybird (*Coccinella*) 125
MARIENKRAUT (*n*): toad-flax, wild flax, yellow toad-flax (*Linaria vulgaris*); arnica (*Arnica montana*) 126
MARIENMANTEL (*m*): Common lady's mantle (*Alchemilla vulgaris*) 127
MARIENNESSEL (*f*): white horehound, common (hoarhound) horehound (*Marrubium vulgare*) 128
MARIENSCHELLE (*f*): Lily-of-the-Valley (*Convallaria majalis*) 129
MARIENSCHLÜSSEL (*m*): primula, primrose (*Primula vulgaris*) 130
MARIENSCHUH (*m*): lady's slipper, common slipper orchid, moccasin flower (*Cypripedium calceolus*) 131
MARIENVEILCHEN (*n*): (see **MARIENGLÖCKCHEN**) 132
MARIGNACIT (*m*): marignacite; $(Nb,Ta)_2O_5$, $(Ce,Fe)_2O_3$,$(Si,Th,Ti)O_2$,$(Na,K)_2O$, $(Ca,Mg,U)O$,F,FeS_2; Sg. 4.13 133
MARINBLAU: marine blue 134
MARINE (*f*): navy; marine 135
MARINEAKADEMIE (*f*): marine (academy) school 136
MARINEBLAU (*n*): marine blue 137
MARINEBRONZE (*f*): marine bronze (same as **MASCHINENBRONZE**, but for marine purposes) 138
MARINEETAT (*m*): navy estimates 139
MARINELEIM (*m*): marine glue 140
MARINEMINISTERIUM (*n*): Admiralty 141
MARINEOFFIZIER (*m*): Naval officer 142
MARINESOLDAT (*m*): bluejacket, marine 143
MARJORAM (*n*): marjoram (see **MAJORAN**) 144

MARJORAMÖL (*n*): marjoram oil (see **MAJORANÖL**) 145

MARJORAN (*m*), **WILDER-**: origanum (see **DOSTEN-KRAUT**) 146

MARK (*n*): marrow; core; pith; pulp; medulla (Med.); (*f*): Mark (German currency); border; district; boundary; marsh; (as a prefix = medullary) 147

MARKASIT (*m*): marcasite, white iron pyrites (natural iron sulphide); FeS_2; Sg. 4.85 (see also **EISENDISULFID**) 148

MARKASITGLANZ (*m*): tetradymite (see **TETRADY-MIT**) 149

MARKBEIN (*n*): marrow bone 150

MARKE (*f*): mark; ticket; stamp; label; brand; cheque; tally; token; type; sort; kind; counter; signature 151

MARKEN: to mark; fix a boundary 152

MARKENSCHUTZGESETZ (*n*): law of trade marks' registration; Trade Marks' Registration Act 153

MARKETENDER (*m*): sutler (type of camp-follower who sells provisions) 154

MARKETTE (*f*): virgin wax in cake form 155

MARKFETT (*n*): marrow-fat 156

MARKFLÜSSIGKEIT (*f*): spinal fluid 157

MARKGEWICHT (*n*): troy weight 158

MARKIEREN: to mark; stamp; label; brand; indicate; distinguish 159

MARKIERUNG (*f*): marking; branding; stamping 160

MARKISE (*f*): marquee; awning; tent 161

MARKKNOCHEN (*m*): (see **MARKBEIN**) 162

MARKSCHEIDE (*f*): limit, boundary 163

MARKSCHEIDEKUNST (*f*): art of surveying; mine surveying 164

MARKSCHEIDUNG (*f*): mine surveying; demarcation of a boundary 165

MARKSCHWAMM (*m*): medullary sarcoma (a fleshy tumour of the medulla, consisting of embryonic connective tissue) (Med.) 166

MARKSTRAHL (*m*): medullary ray (ray radiating from the core to the bark of tree trunks, and exposed by a radial section through the longitudinal axis of the trunk) 167

MARKSUBSTANZ (*f*): medullary substance 168

MARKT (*m*): market; mart; fair; market-place; business 169

MARKTBERICHT (*m*): market report, report on the state of the market (Commercial) 170

MARKTEN: to market; bargain; buy or sell 171

MARKTFÄHIG: marketable 172

MARKTGÄNGIG: current; market 173

MARKTGUT (*n*): market (wares) articles 174

MARKTKAUF (*m*): market-price; market purchase 175

MARKTLAGE (*f*): position (state) of the market 176

MARKTPLATZ (*m*): market-place 177

MARKTPREIS (*m*): market-price 178

MARKTREIF: marketable; ready for market 179

MARKTVERHÄLTNISSE (*n.pl*): position (state) of the market 180

MARKTWAGEN (*m*): lumber-waggon; market-cart 181

MARKTZETTEL (*m*): market report; market prices 182

MARKTZOLL (*m*): market-dues 183

MARKUNG (*f*): district, boundary; demarcation 184

MARK (*n*), **VERLÄNGERTES-**: Medulla oblongata (Med.) 185

MARKWÄHRUNG (*f*): rate of exchange of the Mark; Mark Standard 186

MARLEINE (*f*): marline (Naut.) 187

MARLEISEN (*n*): marline spike 188

MARLEN: to marl 189

MARLI (*m*): cat-gut; gauze of linen or cotton thread 190

MARLIEN (*f*): marline 191

MARLIN (*f*): marline 192

MARLING (*f*): marline 193

MARLPFRIEM (*m*): iron splicer 194

MARLREEP (*n*): slab-line 195

MARLSPIEKER (*m*): (see **MARLPFRIEM**) 196

MARLY (*m*): (see **MARLI**) 197

MARMELADE (*f*): marmalade (a preserve or jam made from oranges or lemons) 198

MARMELADENPULVER (*n*): marmalade powder (consists essentially of cane sugar, etc., with colouring matter) 199

MARMOR (*m*): marble; Sg. 2.65-2.84 (crystalline, granular limestone) (see **CALCIUMCARBONAT** and **CALCIT**) 200

MARMORADER (*f*): vein (streak) in marble 201

MARMORARBEIT (*f*): marble-cutting, work in marble 202

MARMORBAND (*m*): marbled binding (of books) 203

MARMORBILD (*n*): marble statue, bust or sculpture 204

MARMORBLOCK (*m*): block of marble 205

MARMORBRUCH (*m*): marble quarry 206

MARMORGRUBE (*f*): marble quarry 207

MARMORIEREN: to marble; vein; mottle; grain 208

MARMORKALK (*m*): marble lime 209

MARMORKIESEL (*m*): kind of hornstone 210

MARMORLILIE (*f*): fritillary (*Fritillaria meleagris*) 211

MARMORMEHL (*n*): marble meal 212

MARMORN: marble; of marble; marble-like 213

MARMORPAPIER (*n*): marbled paper 214

MARMORPLATTE (*f*): marble slab 215

MARMORSCHLEIFER (*m*): marble-cutter, marble-polisher; stone-mason 216

MARMORSTEIN (*m*): marble (see **MARMOR**) 217

MARMORTAFEL (*f*): marble slab 218

MARODE: tired; exhausted; weary 219

MARODIEREN: to pillage, maraud, plunder, sack, despoil 220

MAROKKO (*n*): Morocco 221

MAROKKOLEDER (*n*): morocco-(leather) 222

MARONE (*f*): sweet chestnut (*Castanea vulgaris*) 223

MAROQUIN (*m*): morocco-(leather) 224

MARQUARDMASSE (*f*): (see **MARQUARDT'SCHE MASSE**) 225

MARQUARDT'SCHE MASSE (*f*): Marquardt's mass (porcelain-like insulating mass for tubes for pyrometer thermo-elements; suitable up to 1600°C.) 226

MARRON (*n*): maroon 227

MARS (*m*): top (Naut.) 228

MARSBRASSE (*f*): topsail brace 229

MARSCH (*f*): bog; marsh; fen; swamp; (*m*): march-(ing); tramp 230

MARSCHBODEN (*m*): marsh-(y) land or soil 231

MARSCHIEREN: to march 232

MARSCHIG: marshy 233

MARSCHISCHE PROBE (*f*): Marsh's test 234

MARSCHLAND (*n*): moorland, marshy country, fen 235

MARSCHTURBINE (*f*): march turbine 236

MARS (*m*), **FOCK-** : fore top 237
MARS (*m*), **GROSSE-** : main top (Naut.) 238
MARSLEESEGEL (*n*) : top-gallant studding sail (Naut.) 239
MARSRAND (*m*) : top rim 240
MARSRELING (*f*) : top railing 241
MARSSCHOTEN (*f.pl*) : top sheets (Naut.) 242
MARSSEGEL (*n* and *pl*) : topsail-(s) 243
MARSSTENGE (*f*) : top mast 244
MARSTOPPNANT (*f*) : topsail lift 245
MARTENSIT (*n*) : martensite (state of steel, change temperature 200°C.); properties: magnetisability, high conductional resistance to heat and electricity, lower specific weight than either perlite or austenite, high hardness, flow limit and tensile, and low elongation under tensile test) 246
MARTERN : to martyr; torment; torture 247
MARTINOFEN (*m*) : Siemens-Martin furnace (lasts 600/900 heatings) (The process is mostly basic on the continent, though sometimes acid. Produces cast steel or mild steel) 248
MARTINPROZESS (*m*) : (Siemens)-Martin process (see **MARTINOFEN**) 249
MARTINSTAHL (*m*) : Martin steel; open-hearth steel (see **MARTINOFEN**) 250
MARTIRERGESCHICHTE (*f*) : Martyrology (the history of martyrs) 251
MARTIUSGELB (*n*) : Martius yellow 252
MÄRZBLÜMCHEN (*n*) : liverwort (*Hepatica*) 253
MÄRZBLÜMCHEN (*n*), **DREILAPPIGES-** : liverwort (see **HEPATICA**) 254
MÄRZBLUME (*f*) : Snowdrop (*Galanthus nivalis*) 255
MÄRZGLÖCKCHEN (*n*) : Spring Snowflake (*Leucoium vernum*) 256
MÄRZVEILCHEN (*n*) : Sweet Violet (*Viola odorata*) 257
MASCAGNIN (*m*) : mascagnine, mascagnite; $(NH_4O)_2SO_3,H_2O$; Sg. 1.75 258
MASCHE (*f*) : mesh; stitch; compartment; eye-(let hole) 259
MASCHEN : to net 260
MASCHENSIEB (*n*) : (mesh)-sieve, classifying (grading) sieve 261
MASCHENWEITE (*f*) : mesh-(width); width of mesh; size of mesh 262
MASCHENWERK (*n*) : network 263
MASCHIG : meshed; netted; reticulated; reticular; retiform; meshy 264
MASCHINE (*f*) : machine; engine; apparatus; (*pl*) : machinery 265
MASCHINE (*f*), **DREIFACHVERBUND-** : triple expansion engine 266
MASCHINE (*f*), **EINZYLINDER-** : single-cylinder engine 267
MASCHINELEHRE (*f*) : engineering 268
MASCHINELL : mechanical 269
MASCHINENARBEIT (*f*) : machine work 270
MASCHINENBAU (*m*) : mechanical engineering, engine (building) construction, machine construction 271
MASCHINENBAUER (*m*) : machine (engine) builder or constructor; machinist; millwright; mechanical engineer; mechanician 272
MASCHINENBAUINGENIEUR (*m*) : mechanical engineer 273
MASCHINENBAUMEISTER (*m*) : mechanical engineer 274
MASCHINENBETRIEB (*m*) : engine (management) driving; engine driven plant; engine drive 275
MASCHINENBRONZE (*f*) : machine bronze (80-90% Cu; 16-8% Sn; 4-2% Zn.) 276
MASCHINENDRUCK (*m*) : machine printing, machine pressing 277
MASCHINENELEMENTE (*n.pl*) : machine parts 278
MASCHINENFABRIK (*f*) : engine (factory) works 279
MASCHINENFERTIG : machine finished, completely machined, finished (ready for assembly without further machining or hand-work) (of machine parts) 280
MASCHINENFETT (*n*) : machine (grease) lubricant 281
MASCHINENFORMEREI (*f*) : machine moulding 282
MASCHINENFÜHRER (*m*) : engine (locomotive) driver 283
MASCHINENFUNDAMENT (*n*) : engine bed-(plate), engine foundation 284
MASCHINENGARN (*n*) : twist, machine cotton 285
MASCHINENGESTELL (*n*) : engine frame 286
MASCHINENHAUS (*n*) : engine (house) shed 287
MASCHINENKRAFT (*f*) : engine power, duty 288
MASCHINENLACK (*m*) : engine varnish 289
MASCHINEN (*f.pl*), **LANDWIRTSCHAFTLICHE-** : agricultural machinery 290
MASCHINENLEHRE (*f*) : engineering 291
MASCHINENMÄSSIG : mechanical; machine-like; automatic 292
MASCHINENMESSER (*n*) : machine cutter 293
MASCHINENÖL (*n*) : engine lubricating oil, machine oil 294
MASCHINENPAPIER (*n*) : machine made paper 295
MASCHINENPAPPE (*f*) : engine board (Paper) 296
MASCHINENRAHMEN (*m*) : engine frame 297
MASCHINENRAUM (*m*) : engine room (on board ship, etc.) 298
MASCHINENRICHTUNG (*f*) : longitudinal direction (of paper) 299
MASCHINENSATZ (*m*) : engine set 300
MASCHINENSÄULE (*f*) : column 301
MASCHINENSCHLOSSER(*m*) : engine fitter 302
MASCHINENSCHMIERER (*m*) : engine (lubricator) greaser 303
MASCHINENSCHREIBEN (*n*) : typewriting, typing 304
MASCHINENTEILE (*m.pl*) : machine parts 305
MASCHINENWEBSTUHL (*m*) : power loom 306
MASCHINENWERK (*n*) : machinery 307
MASCHINENZEICHNEN (*n*) : machine drawing 308
MASCHINENZEICHNUNG (*f*) : engine drawing, machine drawing 309
MASCHINERIE (*f*) : machinery 310
MASCHINE (*f*), **TANDEM-** : tandem engine 311
MASCHINE (*f*), **ZWILLINGS-** : twin engine 312
MASCHINE (*f*), **ZWILLINGSVERBUND-** : compound twin engine 313
MASCHINIEREN : to machine, do any sort of work by machine 314
MASCHINIST (*m*) : engineer; machinist; engine-driver 315
MASER (*f*) : speck-(le); spot; mark; vein; curl; streak; grain (of wood, etc.); (*pl*) : measles (Med.) 316
MASERBIRKE (*f*) : weeping birch (*Betula pendula*) 317
MASERHOLZ (*n*) : veined (grained) wood 318
MASERIG : speckled; streaked; grained; spotted; veined; curled 319
MASERLE (*f*) : maple (see **MASHOLDER**) 320
MASERN : to vein, streak, grain, spot, speckle, mark 321
MASERPAPIER (*n*) : speckled (grained) paper 322

MASERUNG (*f*): veining; graining 323
MASHOLDER (*m*): maple tree (*Acer campestre*) 324
MASKE (*f*): mask (Phot., etc.); disguise 325
MASKEN: to mask (Phot., etc.); back (photographic plate); disguise 326
MASS (*n*): measure; gauge; size; dimension; degree; limit; moderation; measurement; extent; (as a prefix=volumetric) 327
MASSANALYSE (*f*): volumetric analysis 328
MASSANALYTISCH: volumetric-(al)-(ly); titrimetric-(al)-(ly) 329
MASSE (*f*): mass; lump; bulk; block; assets; stock; dry sand (Founding); mallet (Sculpture); paste (Ceramics); pulp (Paper); substance; stuff; dough; manner; measure; proportion; way; mode; method 330
MASSEBESTIMMUNG (*f*): determination of mass 331
MASSEINHEIT (*f*): unit of measure-(ment) 332
MASSEL (*f*): pig; slab; bloom (Metal) 333
MASSELBRECHER (*m*): pig breaker, pig crusher (Metal) 334
MASSEN: considering (seeing) that, because, whereas; to measure 335
MASSENARTIKEL (*m.pl*): articles (goods) in bulk or made under mass production 336
MASSENBEWEGUNG (*f*): mass motion 337
MASSENBILDUNG (*f*): massing, formation of a mass 338
MASSENERZEUGUNG (*f*): mass production, production in bulk 339
MASSENFABRIKATION (*f*): mass production, production in bulk 340
MASSENGESTEIN (*n*): unstratified rock (see **ERUPTIVGESTEIN**) 341
MASSENGUSS (*m*): dry-sand casting 342
MASSENHERSTELLUNG (*f*): mass production, production in bulk 343
MASSENMOMENT (*n*): moment of inertia 344
MASSENWIRKUNG (*f*): mass (action) effect 345
MASSENWIRKUNGSGESETZ (*n*): law of mass (action) effect 346
MASSESCHLAMM (*m*): body slip (Ceramics) 347
MASSESCHLICKER (*m*): body (slip) paste (Ceramics) 348
MASSFLASCHE (*f*): measuring flask 349
MASSFLÜSSIGKEIT (*f*): standard solution 350
MASSFORMEL (*f*): standard formula 351
MASSGABE (*f*): measure; proportion 352
MASSGABE, NACH-: according to 353
MASSGEBEND: determinative; decisive; authoritative; standard; deciding 354
MASSGEBEND, FÜR-SEIN: to dictate; be the standard for 355
MASSGEBUNG (*f*): measure, limitation, proportion 356
MASSGEFÄSS (*n*): measuring vessel; graduator 357
MASSHOLDER (*m*): field maple (see **MASHOLDER**) 358
MASSICOT (*m*): massicot (Mineral, PbO; 93% lead) (see **BLEIOXYD, GELBES-**) 359
MASSIEREN: to massage 360
MÄSSIG: moderate; frugal; measured; temperate 361
MÄSSIGEN: to palliate; pacify; restrain; tranquillize; calm; solace; moderate; temper; be moderate; check; allay; appease; relieve; ease; assuage; alleviate; soothe; mitigate; mollify; abate 362
MÄSSIGKEIT (*f*): moderation; temperance; moderate character; frugality (see also **MÄSSIGEN**) 363
MÄSSIGUNG (*f*): moderation, etc. (see **MÄSSIGEN**) 364
MASSIV: massive; clumsy; unalloyed; solid-(ly); massively 365
MASSIVE (*n*): massiveness, solidity 366
MASSIVREIFEN (*m*): heavy (solid) tyre 367
MASSIV SCHMIEDEN: to forge in the solid 368
MASSKUNDE (*f*): metrology (the science of weights and measures) 369
MASSLIEBCHEN (*n*): daisy (*Bellis perennis*) 370
MASSLOS: beyond measure, boundless, immeasurable, exhorbitant 371
MASSNAHME (*f*): mode of action; precaution; measure; measuring 372
MASSNEHMUNG (*f*): (see **MASSNAHME**) 373
MASSOYÖL (*n*): massoy oil 374
MASSOYRINDE (*f*): massoy bark 375
MASSREGEL (*f*): measure; step; expedient 376
MASSRÖHRE (*f*): measuring (graduated) tube; burette 377
MASSSTAB (*m*): scale; rule; measure; standard; proportion 378
MASSSTOCK (*m*): measuring stick, carpenter's rule 379
MASSSYSTEM (*n*): measuring system; system of measurement 380
MASSSYSTEM (*n*), **ABSOLUTES-**: absolute system of measurement; G.C.S. system, Gramm-centimeter-second system 381
MASSTEIL (*m*): part by measure-(ment); volumetric part 382
MAST (*f*): mast; acorns; nuts; feeding; food; fattening; (*m*): pole; mast 383
MASTADER (*f*): hemorrhoidal vein 384
MASTBACKE (*f*): mast cheek 385
MASTBAND (*n*): mast hoop 386
MASTBAUM (*m*): mast, boom (Marine) 387
MASTBÜGEL (*m*): mast hoop 388
MASTDARM (*m*): rectum 389
MASTDARMBLASENFISTEL (*f*): fistula recto-vesicalis (an unnatural joining of the rectum and the bladder) (Med.) 390
MASTDARMBLUTFLUSS (*m*): rectal hæmorrhage (due to hæmorrhoids) (Med.) 391
MASTDARMBRUCH (*m*): hernia intestini recti, archocele, heterocele, rectal hernia (Med.) 392
MASTDARMDUSCHE (*f*): rectal (douche) irrigation (Med.) 393
MASTDARMENTZÜNDUNG (*f*): (see **PROKTITIS**) 394
MASTDARMFISTEL (*f*): rectal fistula, fistula ani (Med.) 395
MASTDARMFISSUR (*f*): fissura ani (Med.) 396
MASTDARMKATARRH (*m*): mild form of rectal inflammation (see **PROKTITIS**) (Med.) 397
MASTDARMKNOTEN (*m* and *pl*): hæmorrhoid-(s), pile-(s) (Med.) 398
MASTDARMKREBS (*m*): rectal cancer, cancer of the rectum (Med.) 399
MASTDARMPOLYPEN (*m.pl*): polypi of the rectum, rectal polypi (see **POLYP**) (Med.) 400
MASTDARMSCHEIDENFISTEL (*f*): fistula recto-vaginalis (an abnormal connection of the rectum with the vagina) (Med.) 401
MASTDARMSCHLEIMHAUT (*f*): mucous membrane of the rectum (Med.) 402
MASTDARMVORFALL (*m*): prolapsus ani (Med.) 403
MASTDUCHT (*f*): mast (main) thwart 404
MÄSTEN: to feed; fatten; grow fat 405

MASTEN: to fit (equip, furnish, provide) with a mast; become (grow) fat 406
MASTENKRAN (*m*): masting sheers 407
MASTENMACHER (*m*): mast maker 408
MASTESELSHAUPT (*n*): mast cap 409
MASTFISCHUNG (*f*): mast partner 410
MASTFUSS (*m*): heel of a mast, mast foot 411
MAST (*m*), **GENIETETER-**: riveted mast 412
MAST (*m*), **GESCHWEISSTER-**: welded mast 413
MASTHAUSUNG (*f*): mast housing 414
MAST (*m*), **HOLZ-**: wooden mast 415
MASTIKATOR (*m*): masticator; compressor (for compressing caoutchouc) 416
MASTISOL (*n*): mastisol (solution of **MASTIX** (which see), in benzene) 417
MASTIX (*m*): cement; (gum)-mastic; mastix; Lentisk; Pistachia galls (see **MASTIXHARZ**) 418
MASTIXBAUM (*m*): mastic tree, mastic shrub (*Pistacia lentiscus*) 419
MASTIXFIRNIS (*m*): mastic varnish 420
MASTIXHARZ (*n*): mastix; (gum)-mastic (resin from *Pistacia lentiscus*); Sg. 1.04-1.07; Mp. 95.5°C. 421
MASTIXHOLZ (*n*): mastic wood, balsam, lentisk (*Pistacia lentiscus*); Sg. 0.85 422
MASTIXKITT (*m*): mastic-(cement); Sg. 1.07 (equal parts of mastic and linseed oil) (mixture of sandstone, limestone, litharge and linseed oil varnish) 423
MASTIXKRAUT (*n*): teucrium, germander (*Teucrium*) 424
MASTIXÖL (*n*): mastic oil; Sg. 0.858 (from Mastic) (see also **MASTIX**) 425
MASTIZIEREN: to masticate; chew; (mastication is also a term for the preliminary treatment of caoutchouc by means of rollers, prior to vulcanizing) 426
MASTKEIL (*m*): mast wedge 427
MASTKLAMPE (*f*): mast clamp, mast cleat 428
MASTKOKER (*m*): mast trunk 429
MASTKORB (*m*): top, scuttle; bower (Naut.) 430
MASTKRAN (*m*): masting shears 431
MASTKRAUT (*n*): procumbent pearlwort (*Sagina procumbens*) 432
MASTLOCH (*n*): mast hole 433
MAST (*m*), **NAHTLOSER-**: seamless (weldless) mast 434
MASTSCHULTER (*f*): mast hound 435
MASTSPUR (*f*): mast step 436
MASTTOPP (*m*): masthead 437
MAST (*m*), **TREIDEL-**: mast for towing boats 438
MASTVIEH (*n*): cattle for fattening, fatted cattle 439
MASTWÄCHTER (*m*): look-out, top-man (Naut.) 440
MASTWURM (*m*): Ascaris (*Ascaris lumbricoides*) (an intestinal nematode, or parasitic roundworm) (Med.) 441
MAST (*m*), **ZUSAMMENGESETZTER-**: built up mast 442
MASUT (*m*): liquid residue after extraction (separation) of the petroleum in the fractional distillation process of oil 443
MATÉ (*f*): maté, Paraguay tea (*Folia maté*), Brazilian holly (*Ilex paraguaiensis*) 444
MATENSTRAUCH (*m*): (see **MATE**) 445
MATERIAL (*n*): material 446
MATERIALEINKAUFPREIS (*m*): price of material 447
MATERIALFEHLER (*m*): fault in material, flaw in material 448
MATERIALFESTIGKEIT (*f*): strength of material 449
MATERIALGESCHÄFT (*n*): (see **MATERIALWAREN-GESCHÄFT**) 450
MATERIALIEN (*n.pl*): material-(s) 451
MATERIALIENEINKAUFPREIS (*m*): price of material 452
MATERIALISMUS (*m*): materialism 453
MATERIALPROBE (*f*): test-piece, sample of material for testing purposes; material test-(ing) 454
MATERIALPROBIERMASCHINEN (*f.pl*): material testing machinery 455
MATERIALPRÜFINSTRUMENT (*n*): material testing instrument 456
MATERIALPRÜFUNG (*f*): material testing 457
MATERIALPRÜFUNGSAMT (*n*): bureau for material testing 458
MATERIALPRÜFUNGSMASCHINEN (*f.pl*): material testing machinery 459
MATERIALUNTERSUCHUNG (*f*): material (examination) testing 460
MATERIALWAREN (*f.pl*): drugs; groceries; Colonial produce 461
MATERIALWARENGESCHÄFT (*n*): grocer's shop; druggist's shop; Colonial produce trade 462
MATERIALWAREN-GROSSHANDLUNG (*f*): firm of wholesale druggists 463
MATERIE (*f*): matter; substance; stuff; subject; pus (Med.) 464
MATERIELL: material, actual, real, materially, actually, really 465
MATEZIT (*n*): (see **FICHTENZUCKER**) 466
MATHEMATIK (*f*): mathematics 467
MATHEMATIKER (*m*): mathematician 468
MATHEMATISCH: mathematical-(ly) 469
MATICOBLÄTTER (*n.pl*): (see **MATIKOBLÄTTER**) 470
MATIKOBLÄTTER (*n.pl*): matico leaves (*Folia matico* from *Piper angustifolium*) 471
MATIKOÖL (*n*): matico oil; Sg. 0.93 472
MATRATZE (*f*): mattress 473
MATRISIEREN: to damp (Paper) 474
MATRIZE (*f*): matrix; die; mould; bed; negative (Phot.) 475
MATRIZENSTÜCK (*n*): die-(plate) 476
MATROSE (*m*): sailor; mariner 477
MATSCH (*m*): mash; pulp; slush; mire; squash 478
MATSCHEN: to mash, squash, bruise, pulp, splash 479
MATT: mate (Chess); dull; dead; matt; (of colours, etc.); insipid; ground (of glass); faint; heavy; feeble; exhausted; weak; pale; (*n*): dullness (of colours); deadness (of surface); mate (Chess) 480
MATTBLAU: pale blue, dull blue; matt blue (as opposed to a glossy blue) 481
MATTBRENNE (*f*): matt pickle (for matting metal; consists of nitric acid of Sg. 1.38, sulphuric acid of Sg. 1.84, common salt and white vitriol in proportion 300 : 200 : 5 : 1 parts) (Electro-plating) 482
MATTE (*f*): mat; meadow 483
MATTEISEN (*n*): white pig (Iron) 484
MATTEN: to mate (Chess); deaden, dull (Surfaces); exhaust, make faint, weary 485
MATTENBLUME (*f*): Marsh Marigold (*Caltha palustris*) 486
MATTENZEUG (*n*): matting 487
MATTFARBE (*f*): matt (dull; dead) colour 488

MATTGESCHLIFFEN: ground; frosted (of glass) 489
MATTGLANZ (*m*): dull (matt) finish; dull lustre 490
MATTGLAS (*n*): ground (frosted) glass 491
MATTGLASUR (*f*): matt glaze 492
MATTGOLD (*n*): matt (dead; dull) gold 493
MATTHEIT (*f*): dullness; deadness; faintness; feebleness; weakness; dimness; lassitude 494
MATTIERBRENNE (*f*): (see **MATTBRENNE**) 495
MATTIEREN: to deaden; dull; tarnish; give matt (surface) finish to; dim; frost; grind (Glass) 496
MATTIERUNG (*f*): deadening, dulling, matting, giving a matt finish to 497
MATTIERUNGSLACK (*m*): matting varnish 498
MATTIGKEIT (*f*): exhaustion, fatigue 499
MATTLACK (*m*): matt (dull) varnish (varnish with a dead surface) 500
MATTROT: dull (pale; matt) red 501
MATTSALZ (*n*): matting salt 502
MATTSCHEIBE (*f*): frosted glass, ground-glass-(focussing)-screen (Photography) 503
MATTSCHLEIFEN: to grind; frost (Glass) 504
MATTWEISS: dull (matt) white 505
MATZ (*m*): (see **QUARK**) 506
MAUER (*f*): wall; (as a prefix=mural) 507
MAUERANKER (*m*): wall clamp-(iron) 508
MAUERARBEIT (*f*): brickwork, masonry 509
MAUERDACH (*n*): wall coping 510
MAUERFEST: firm as a rock 511
MAUERFRASS (*m*): decay, crumbling (of brickwork or masonry) 512
MAUERKALK (*m*): mortar 513
MAUERKITT (*m*): mortar 514
MAUERN: to wall in or up; build-(up) (Brickwork) 515
MAUERPFEFFER (*m*): stonecrop; wall pepper (*Sedum acre*) 516
MAUERRAUTE (*f*): wall rue (*Amesium ruta-muraria*) 517
MAUERSALPETER (*m*): wall saltpeter; calcium nitrate (see **CALCIUMNITRAT**) 518
MAUERSALZ (*n*): calcium acetate (see **CALCIUM-AZETAT**) 519
MAUERSAND (*m*): bricklayer's sand (for making mortar) 520
MAUERSTEIN (*m*): building (brick) stone 521
MAUERTRAUBE (*f*): white stonecrop (*Sedum album*) 522
MAUERWERK (*n*): brickwork; masonry; walls; walling; stonework 523
MAUERZIEGEL (*m*): building brick 524
MAUKEN (*n*): fermenting (Ceramics) 525
MAUL (*n*): mouth (of animals only); muzzle; jaws; tongue; mule 526
MAULBEERBAUM (*m*): mulberry tree (*Morus nigra* and *Morus alba*) 527
MAULBEERE (*f*): mulberry (fruit of **MAULBEER-BAUM**, which see) 528
MAULBEERFEIGENBAUM (*m*): (see **FEIGENBAUM, ÄGYPTISCHER-**) 529
MAULBEERHOLZ (*n*): mulberry wood; Sg. 0.55-0.9 530
MAULEN: to pout; sulk 531
MAULESEL (*m*): mule 532
MAULFAUL: taciturn 533
MAULHELD (*m*): braggart 534
MAULKLEMME (*f*): lockjaw; tetanus (Med.) 535
MAULKORB (*m*): muzzle 536
MAULSPERRE (*f*): lockjaw; tetanus (Med.) 537
MAULSTICH (*m*): half-hitch (Naut.) 538
MAULTIER (*n*): mule 539
MAUL- UND KLAUENSEUCHE (*f*): foot-and-mouth disease (of animals) 540
MAULWURF (*m*): mole (*Talpa europæa*) 541
MAULWURFSHAUFEN (*m*): mole-hill 542
MAURER (*m*): mason; bricklayer 543
MAURERARBEIT (*f*): masonry 544
MAURERISCH: masonic 545
MAUS (*f*): mouse (*Mus*) 546
MAUSE (*f*): moulting 547
MÄUSEDARM (*m*): (common)-chickweed (*Stellaria media*) 548
MÄUSEDORN (*m*): butcher's broom (*Ruscus aculeatus*) 549
MÄUSEDRECK (*m*): mouse dung 550
MAUSEFALLE (*f*): mouse-trap 551
MÄUSEFRASS (*m*): damage due to mice 552
MÄUSEGIFT (*n*): ratsbane; (white)-arsenic 553
MAUSEN: to moult; catch mice 554
MAUSEOHR (*n*): scorpion-grass (*Myosotis*) 555
MAUSERN: moult 556
MÄUSETYPHUSBAZILLE (*f*): mouse typhus bacillus 557
MÄUSEVERTILGUNGSMITTEL (*n*): mouse (destroyer) poison 558
MAUSFAHL: mouse coloured; mouse-grey 559
MAUSFARBEN: (see **MAUSFAHL**) 560
MAUS (*f*), **FELD-**: field mouse (*Mus sylvaticus*) 561
MAUSGRAU: (see **MAUSFAHL**) 562
MAUS (*f*), **HAUS-**: house mouse (*Mus musculus*) 563
MAUT (*f*): toll, custom, dues, excise, (import)-duty 564
MAXIMAL: maximum 565
MAXIMALWERTIGKEIT (*f*): maximum valence 566
MAXIME (*f*): maxim; adage; aphorism; axiom; saying 567
MAXIMUM (*n*): maximum 568
MAZERATION (*f*): maceration; infusion (Perfumery) 569
MAZERIEREN: to macerate; attenuate; steep 570
MAZISÖL (*n*): (see **MUSKATBLÜTENÖL**) 571
MECHANIK (*f*): mechanics; mechanism 572
MECHANIKER (*m*): mechanician; mechanic; mechanical engineer 573
MECHANISCH: mechanic-(al)-(ly); automatic-(ally) 574
MECHANISCHE FEUERUNG (*f*): mechanical (grate) stoker 575
MECHANISCHE ROSTBESCHICKUNG (*f*): mechanical (stoking) stoker 576
MECHANISMUS (*m*): mechanism 577
MECONSÄURE (*f*): meconic acid (obtained from opium); $OHC_5HO_2(COOH)_2.3H_2O$ [578
MEDAILLE (*f*): medal 579
MEDAILLENBRONZE (*f*): medal bronze (same as German small copper coins) (95% Cu: 4% Sn; 1% Zn.) 580
MEDAILLON (*n*): medallion 581
MEDIANTE (*f*): mediant (Mus.) 582
MEDIE (*f*): medium 583
MEDIEN (*n.pl*): (plural of **MEDIE**), media; mediums 584
MEDIKAMENT (*n*): medicament (a medicinal substance) 585
MEDIKAMENTÖS: medicinal-(ly) 586
MEDINAL (*n*): medinal (sodium salt of diethylbarbituric acid; administered to cause sleep) (see **NATRIUM, DIÄTHYLBARBITURSAURES-**) 587

MEDIOGARN (*n*): medio-twist (Textile) 588
MEDIUM (*n*): medium 589
MEDIZIN (*f*): medicine; physic 590
MEDIZINAL: medicinal, officinal 591
MEDIZINALBEHÖRDE (*f*): Board of Health 592
MEDIZINALGEWICHT (*n*): troy weight 593
MEDIZINALRAT (*m*): health (officer) inspector 594
MEDIZINALRINDE (*f*): medicinal (officinal) bark 595
MEDIZINALWAREN (*f.pl*): drugs, medicaments 596
MEDIZINER (*m*): medical (man) student; physician 597
MEDIZINFERTIG: medicated 598
MEDIZINGLAS (*n*): medicine glass; dispensing (measure) glass 599
MEDIZINIEREN: to take (medicine) physic; medicate 600
MEDIZINIERT: medicated 601
MEDIZINISCH: medicinal; officinal; medical 602
MEDIZINISCHE FAKULTÄT (*f*): medical faculty 603
MEDIZINISCHE SEIFE (*f*): medicated soap (*Sapo medicatus*) 604
MEDIZINKORK (*m*): medicine (cork) stopper 605
MEDULLAR: medullary 606
MEER (*n*): sea; ocean 607
MEERARM (*m*): arm of the sea; inlet; bight 608
MEERBUSEN (*m*): bay; gulf 609
MEERENGE (*f*): channel; straits 610
MEERESALGE (*f*): seaweed, fucus 611
MEERESBODEN (*m*): sea bottom 612
MEERESFLÄCHE (*f*): sea (surface) level 613
MEERESFLUT (*f*): high water 614
MEERESGRUND (*m*): sea bottom 615
MEERESOBERFLÄCHE (*f*): sea (surface) level 616
MEERESSTROM (*m*): under current, sea current 617
MEERFARBE (*f*): sea green 618
MEERFENCHEL (*m*): samphire (*Crithmum maritimum*) 619
MEERFERKEL (*n*): porpoise (*Phocœna communis*) 620
MEERFINGER (*m*): belemnite (skeleton of an extinct animal found in Jurassic and also Cretaceous rocks) 621
MEERFISCH (*m*): sea-fish 622
MEERGANZ (*f*): barnacle, bernicle goose (*Bernicla leucopsis*) 623
MEERGEWÄCHS (*n*): sea (marine) plant 624
MEERGRAS (*n*): thrift, sea-pink (*Armeria maritima*); sedge, sea weed (*Cladium mariscus*) 625
MEERGRÜN: sea-green 626
MEERHANDEL (*m*): maritime trade 627
MEERHOSE (*f*): waterspout 628
MEERIGEL (*m*): sea urchin (*Echinus*) 629
MEERKOHL (*m*): sea kale (*Crambe maritima*) 630
MEERKOHLWINDE (*f*): sea-side bindweed (*Convolvulus soldanella*) 631
MEERKOKOS (*m*): Seychelles fan-leaved palm; double cocoanut-palm (*Lodoicea sechellarum*) 632
MEERKRAPP (*m*): (see **KRAPP**) 633
MEERKRAUT (*n*): (jointed)-glasswort; marsh samphire; crab-grass; saltwort (*Salicornia herbacea*) 634
MEERLATTICH (*m*): sea lettuce (*Ulva*) 635
MEERLUFT (*f*): sea air; sea breeze 636
MEERMANNSTREU (*f*): (see **MEERSTRANDSMANN-STREU**) 637
MEERMELDE (*f*): orach (*Atriplex hortensis*) 638
MEERNELKE (*f*): thrift, sea pink (*Armeria maritima*) 639
MEERNESSEL (*f*): (see **HYDROZOEN**, **SIPHONO-PHOREN**, **HYDROMEDUSEN** and **QUALLE**) 640
MEERPFLANZE (*f*): (see **MEERGEWÄCHS**) 641
MEERPORTULAK (*m*): water purslane, sea purslane (*Halimus portulacoides*) (see **MEERMELDE**) 642
MEERRETTICH (*m*): horseradish (*Armoracia rusticana*) 643
MEERRETTIG (*m*): (see **MEERRETTICH**) 644
MEERSALINE (*f*): sea-salt; salt-garden 645
MEERSALZ (*n*): sea salt 646
MEERSAND (*m*): sea sand; silt 647
MEERSCHALUMINIT (*m*): meerschaluminite (see **KAOLINIT**) 648
MEERSCHAUM (*m*): sea foam; meerschaum (a magnesium silicate); $Mg_2Si_3O_8.2H_2O$; Sg. 0.9; pipe clay; sepiolite; talc 649
MEERSCHAUMERSATZ (*m*): meerschaum substitute (made from meerschaum waste with a binder; or without meerschaum, from calcined magnesia, infusorial earth and lime cream from lime slaking) 650
MEERSCHILF (*n*): sea reed (*Psamma arenaria*) 651
MEERSCHWAMM (*m*): sea-sponge 652
MEERSCHWEIN (*n*): porpoise (*Delphinus*) (see **MEERFERKEL**) 653
MEERSCHWEINSCHEN (*n*): guinea pig 654
MEERSCHWEINTRAN (*m*): porpoise oil (from *Delphinus phocœna*); Sg. 0.926 655
MEERSENF (*m*): purple sea-rocket (*Cakile maritima*) 656
MEERSTADT (*f*): sea-port-(town); sea-side town 657
MEERSTRANDSGRASNELKE (*f*): (see **MEERGRAS**) 658
MEERSTRANDSKIEFER (*f*): Scotch pine (*Pinus sylvestris*) 659
MEERSTRANDSMANNSTREU (*f*): sea holly, sea-eryngo (*Eryngium maritimum*) 660
MEERSTRANDSWINDE (*f*): (see **MEERKOHLWINDE**) 661
MEERWASSER (*n*): sea water 662
MEERWASSERECHT: fast to sea water 663
MEERWURZEL (*f*): (see **MEERSTRANDSMANNSTREU**) 664
MEERZWIEBEL (*f*): Scilla; squill; sea onion (bulb of *Urginea maritima*) 665
MEERZWIEBELESSIG (*m*): squill vinegar (from the bulb of *Urginea maritima*) 666
MEERZWIEBELEXTRAKT (*m*): squill extract 667
MEGABROMIT (*m*): megabromite (see **EMBOLIT**) 668
MEGALOGONER KUPHONSPAT (*m*): (see **BREW-STERIT**) 669
MEGERKRAUT (*n*): yellow bedstraw (*Gallium verum*) 670
MEGILP (*m*): (see **MACGILP**) 671
MEHL (*n*): meal; flour (*Farina*); dust; powder 672
MEHLARTIG: farinaceous; like flour; flour-like; floury; mealy 673
MEHLBAUM (*m*): (see **MEHLBEERBAUM**) 674
MEHLBEERBAUM (*m*): white beam tree (*Sorbus aria* and *Pirus aria*); common-(wild)-medlar (*Mespilus germanica*) 675
MEHLBEERE (*f*): haw; whortleberry (*Vaccinium myrtillus*) 676

MEHLBEUTEL (*m*): sifter; bolter; bolting sieve 677
MEHLBIRN (*f*): white beam tree (*Sorbus aria*) 678
MEHLDORN (*m*): (see MEHLFÄSSCHEN) 679
MEHLFASS (*n*): flour (keg, barrel) tub; meal tub 680
MEHLFÄSSCHEN (*n*): common-(wild)-medlar (*Mespilus germanica*) 681
MEHLFRÜCHTE (*f.pl*): cereals 682
MEHLGEBEND: farinaceous; yielding (giving) flour 683
MEHLGIPS (*m*): earthy gypsum (see GYPS) 684
MEHLIG: (see MEHLARTIG) 685
MEHLKASTEN (*m*): bolting hutch; flour (box) tub or bin 686
MEHLKLEISTER (*m*): (flour)-paste 687
MEHLKÖRPER (*m*): endosperm (of grain); grain (i.e., that portion of the ear only which is ground to flour) 689
MEHLKREIDE (*f*): earthy calcite (see CALCIT) 690
MEHLMUTTER (*f*): ergot 691
MEHLPULVER (*n*): meal powder; priming powder 692
MEHLSACK (*m*): flour sack 693
MEHLSCHMERGEL (*m*): white goosefoot (*Chenopodium album*) 694
MEHLSIEB (*n*): flour sieve 695
MEHLSPEISE (*f*): farinaceous food 696
MEHLSTAUB (*m*): flour dust 697
MEHLSUPPE (*f*): gruel 698
MEHLTAU (*m*): mildew (see MELTAU, HOPFEN-SCHIMMEL and ROSENSCHIMMEL) 699
MEHLTAUPILZ (*m*): mildew fungus (*Erysiphe*) 700
MEHLTEIG (*m*): paste, dough 701
MEHLZUCKER (*m*): brown (cask; powdered) sugar 702
MEHR: more 703
MEHRATOMIG: polyatomic 704
MEHRAUFWAND (*m*): extra, excess (of anything used) 705
MEHRBASISCH: polybasic 706
MEHRBETRAG (*m*): extra amount, surplus, excess 707
MEHRBLUMIG: many-flowered 708
MEHRDEUTIG: ambiguous 709
MEHRDEUTIGKEIT (*f*): ambiguity 710
MEHRDREHUNG (*f*): multirotation 711
MEHREN: to increase; multiply; make more; augment 712
MEHRENTEILS: mostly; for the most part 713
MEHRER (*m*): multiplier, factor 714
MEHRERE: several; numerous; various; divers; sundry; many; greater; farther 715
MEHRERE ABSÄTZE (*m.pl*): a multiplicity of stages 716
MEHRERLEI: various; divers; diverse 717
MEHRFACH: manifold; repeated; numerous; multiple 718
MEHRFÄLTIG: (see MEHRFACH) 719
MEHRFARBENFILTER (*m* and *n*): multi-colour filter (Phot.) 720
MEHRFARBIG: multi-coloured; polychromatic; pleochroic 721
MEHRFARBIGKEIT (*f*): polychromatism; polychromy; pleochroism 722
MEHRFLAMMIG: having more than one flame 723
MEHRGÄNGIG: multiple threaded (of screws, etc.) 724
MEHRGEBOT (*n*): outbidding 725
MEHRGLIEDRIG: having several (numerous, more) members; complex 726
MEHRHEIT (*f*): majority; plural-(ity) 727
MEHRJÄHRIG: a number of (several) years old 728
MEHRKERNIG: polynuclear; having more than one nucleus 729
MEHRLEISTUNG (*f*): extra (additional increased) duty or output; overload 730
MEHRMALIG: reiterated; repeated 731
MEHRMALS: repeatedly; several times; again and again 732
MEHRPHASENSTROM (*m*): polyphase (multiphase) current (Elect.) 733
MEHRPHASENSTROMERZEUGER (*m*): multi-phase (polyphase) current generator 734
MEHRPOLIG: multipolar 735
MEHRPREIS (*m*): extra price; additional (cost) price 736
MEHRQUANTIG: multiple 737
MEHRSEITIG: polygonal, having many sides 738
MEHRSILBIG: polysyllabic 739
MEHRSTROM-TURBINE (*f*): multi-flow turbine 740
MEHRSTUFIG: multi-stage 741
MEHRSTÜNDIG: of several hours' duration 742
MEHRUNG (*f*): increase; augmentation; multiplication 743
MEHRWERT (*m*): extra (additional) value; polyvalence 744
MEHRWERTIG: multi-(poly)-valent 745
MEHRZAHL (*f*): majority; plural-(ity) 746
MEHRZELLIG: multi-cellular 747
MEHRZYLINDRIG: multi-cylinder 748
MEIDEN: to avoid; shun 749
MEIDEROLÖL (*n*): tar-fat oil (used as substitute for mineral oil for machine lubrication) 750
MEIER (*m*): farmer; dairy-farmer; steward; bailiff 751
MEIEREI (*f*): farm; dairy farm 752
MEIEREIERZEUGNISSE (*n.pl*): dairy produce 753
MEIERGUT (*n*): farm; dairy farm; tenement 754
MEIERHOF (*m*): farm; dairy farm; tenement 755
MEILE (*f*): mile (German = about 5 English miles; French = a league = about 3 miles) 756
MEILE (*f*), **ABGESTECKTE-**: measured mile 757
MEILE (*f*), **DEUTSCHE-**: German mile (see MEILE) 758
MEILE (*f*), **FRANZÖSISCHE-**: French mile (see MEILE) 759
MEILENFAHRT (*f*): (trial)-run over a-(measured)-mile 760
MEILENGELD (*n*): mileage 761
MEILENSTEIN (*m*): milestone 762
MEILENZEIGER (*m*): signpost; milestone 763
MEILER (*m*): (circular)-pile; mound; stack; clamp (Brick); kiln; charcoal-kiln; heap 764
MEILERKOHLE (*f*): charcoal 765
MEILEROFEN (*m*): kiln furnace 766
MEILERVERFAHREN (*n*): pile charring; kiln process 767
MEIN: my; mine 768
MEINEID (*m*): perjury; false oath 769
MEINEIDIG: foresworn; perjured 770
MEINEN: to mean; think; suppose; fancy; reckon; signify 771
MEINEN, ERNSTLICH-: to be serious; act in good faith; have good intentions 772
MEINERSEITS: for (on) my part 773
MEINES WISSENS: as far as I know, to the best of my knowledge 774
MEINETHALBEN: for my (part) sake; for aught I care; as far as I am concerned 775

MEINETWEGEN : for my (part) sake ; for aught I care ; as far as I am concerned 776
MEINUNG (*f*) : opinion ; idea ; meaning ; intention ; mind ; notion ; thought ; view 777
MEINUNGSÄUSSERUNG (*f*) : expression of opinion 778
MEINUNGSFREI : unbiassed, unprejudiced, having an open mind 779
MEINUNGSKAUF (*m*) : speculative purchase 780
MEINUNGSSTREIT (*m*) : conflict of views 781
MEINUNGSWUT (*f*) : fanaticism 782
MEIONIT (*m*) : meionite (a form of Wernerite); $Ca_4Al_6Si_6O_{25}$; Sg. 2.7 783
MEIRAN (*m*) : marjoram (see MAJORAN) 784
MEISCH (*m*) : mash (see also MAISCH) 785
MEISCHBOTTICH (*m*) : vat ; mash tub ; fermenting (vat) tub 786
MEISCHE (*f*) : mash-(ing) 787
MEISCHEN : to mash ; mix 788
MEISCHKUFE (*f*) : vat, mash tub, fermenting (vat) tub 789
MEISE (*f*) : titmouse ; tit (*Parus*) 790
MEISSEL (*m*) : chisel ; pledget ; lint dressing (Med.) 791
MEISSELFÖRMIG : chisel-shaped 792
MEISSELN : to chisel 793
MEISSELSTAHL (*m*) : chisel steel 794
MEIST : mostly ; more often than not ; most ; almost 795
MEISTBIETENDE (*m*) : higher (highest) bidder 796
MEISTENS : generally ; for the most part ; most ; mostly 797
MEISTENTEILS : generally, for the most part ; most ; mostly 798
MEISTER (*m*) : chief ; master ; teacher 799
MEISTERHAFT : masterly ; skilful ; skilfully 800
MEISTERHAND (*f*) : master-hand 801
MEISTERLAUGE (*f*) : potash lye, caustic potash solution (see KALILAUGE and KALIUMOXYD-HYDRAT) 802
MEISTERSTÜCK (*n*) : masterpiece ; master stroke 803
MEISTERWERK (*n*) : masterpiece ; work of a master 804
MEISTERWURZ (*f*) : masterwort ; felon grass (*Radix imperatoriæ*) (*Imperatoria ostruthium* and *Peucedanum ostruthium*) 805
MEISTERWURZEL (*f*) : masterwort ; felon grass (*Radix imperatoriæ*) (*Imperatoria ostruthium* and *Peucedanum ostruthium*) 806
MEISTERWURZEL (*f*), SCHWARZE- : Astrantia (*Astrantia major*) 807
MEKKABALSAM (*m*): Mecca (balm) balsam ; Balm of Gilead (resin of *Balsamodendron gileadense*) ; often wrongly used to express Canada balsam (see KANADABALSAM) 808
MEKONIDIN (*n*) : meconidine (an opium alkaloid); $C_{21}H_{23}NO_4$ 809
MEKONINSÄURE (*f*) : Meconinic acid 810
MEKONSÄURE (*f*) : meconic acid (see MECON-SÄURE) (from *Papaver somniferum*) 811
MELAKONIT (*m*) : melaconite, black copper (natural copper oxide) (about 75% copper) ; CuO ; Sg. 6.0 (see also TENORIT) 812
MELANCHOLIE (*f*) : melancholy ; spleen ; melancholia (a form of insanity) (Med.) 813
MELANCHOLISCH : melancholy ; melancholic 814
MELANERZ (*n*), DIPRISMATISCHES- : Ilvaite ; Sg. 4.0 815
MELANERZ (*n*), HEMIPRISMATISCHES- : Gadolinite 816
MELANERZ (*n*), PRISMATOIDISCHES- : **Allanite**, orthite, cerite 817
MELANERZ (*n*), TETARTOPRISMATISCHES- : **Allanite**, orthite, cerite 818
MELANGE (*f*) : mixture 819
MELANGIEREN : to mix 820
MELANGLANZ (*m*) : melane glance, stephanite (Mineralogy) (see STEPHANIT) 821
MELANGLANZ (*m*), PRISMATISCHER- : stephanite (see SPRÖDGLASERZ) 822
MELANGLANZ (*m*), RHOMBOEDRISCHER- : polybasite 823
MELANGLIMMER (*m*), RHOMBOEDRISCHER- : cronstedtite, sideroschisolite Sg. 3.35 824
MELANIT (*m*) : melanite (a black form of andradite) ; $Ca_3(Al,Fe)_2Si_3O_{12}$; Sg. 3.5 to 3.9 825
MELANOCERIT (*m*) : melanocerite ; Sg. 4.13 [826
MELANTERIT (*m*) : melanterite (a hydrous ferrous sulphate) ; $FeSO_4+7H_2O$; Sg. 1.85 [827
MELAPHYR (*m*) : melaphyre, porphyry (black) (Mineralogy) 828
MELASSE (*f*) : molasses (from sugar) 829
MELASSE-ENTZUCKERUNG (*f*) : desaccharification of molasses, extraction of sugar from molasses 830
MELASSEFUTTER (*n*) : molasses feed 831
MELASSEMEHL (*n*) : molasses meal 832
MELDE (*f*) : Orach-(plant) (*Atriplex hortensis*) 833
MELDEBRIEF (*m*) : letter of advice 834
MELDEN : to announce ; inform ; mention ; present ; apply ; notify ; advise ; sue for ; apprise ; report 835
MELDUNG (*f*) : advertisement (see also MELDEN) 836
MELIEREN : to mix ; mottle ; mingle ; shuffle ; variegate 837
MELIERFASERN (*f.pl*) : mottling fibres (Paper) 838
MELIERPAPIER (*n*) : mottled paper 839
MELILITH (*m*) : melilite ; $(Al,Fe)_4(Ca,Mg)_{11}Na_2(SiO_4)_9$; Sg. 2.9 840
MELILOTE (*f*) : melilot (*Melilotus officinalis* and *M. vulgaris*) 841
MELILOTENKLEE (*m*) : melilot (*Melilotus officinalis* and *M. vulgaris*) 842
MELILOTSÄURE (*f*) : melilotic acid 843
MELINIT (*n*) : melinite (an explosive, similar to lyddite ; see LYDDIT) 844
MELINOPHAN (*m*) : melinophane ; $NaCa_2Be_2Si_3O_{10}F$; Sg. 3.0 845
MELIORIEREN : to ameliorate 846
MELIS (*m*) : lump sugar ; (coarse)-loaf sugar 847
MELISSE (*f*) : balm ; balm mint (*Melissa officinalis*) 848
MELISSENBLÄTTER (*n.pl*) : balm leaves (*Folia melissæ* from *Melissa officinalis*) 849
MELISSENGEIST (*m*) : balm spirit 850
MELISSENKRAUT (*n*) : balm ; balm mint (see MELISSE) 851
MELISSENÖL (*n*) : balm oil ; melissa oil ; lemon-balm oil (from *Melissa officinalis*) ; Sg. 0.91 852
MELISSINSÄURE (*f*) : melissic acid ; ($C_{29}H_{59}$·COOH) ; Mp. 91°C. 853
MELISZUCKER (*m*) : lump sugar ; (coarse)-loaf sugar 854
MELITRIOSE (*f*) : melitriose (see RAFFINOSE) 855
MELK : giving milk ; milch- 856
MELKEN : to milk ; drain ; draw off ; tap 857
MELLIT (*m*) : Mellite ; honey-stone (see HONIG-STEIN) ; $Al_2(CO)_{12}.18H_2O$; Sg. 1.6 [858

MELLITHSÄURE (*f*): mellitic acid (see HONIG-STEINSÄURE) 859
MELLITSÄURE (*f*): mellitic acid (see HONIG-STEINSÄURE) 860
MELODISCH: melodious, tuneful, musical 861
MELONE (*f*): melon (*Cucumis melo*) 862
MELONENBAUM (*m*): papaw (*Carica papaya*) 863
MELONENKERNÖL (*n*): melon-(seed) oil (from *Cucumis melo*); Sg. 0.928 864
MELONENKÜRBIS (*m*): pumpkin (*Cucurbita pepo*) 865
MELTAU (*m*): mildew (see MEHLTAU) 866
MELTAU (*m*), **FALSCHER-**: false mildew (*Peronospora*) 867
MELTAUPILZ (*m*): mildew fungus (*Erysiphe*) 868
MELUBRIN (*n*): melubrin; ($C_{11}H_{11}N_2O.NH.CH_2.SO_3Na$) 869
MEMBRAN (*f*): membrane, vibrating diaphragm, diaphragm 870
MEMBRANDIFFUSION (*f*): osmosis 871
MEMBRAN-PUMPE (*f*): membrane pump; diaphragm pump 872
MEMME (*f*): coward 873
MEMMENHAFT: cowardly 874
MEMORANDENBUCH (*n*): memorandum (note) book 875
MEMORIEREN: to memorize, commit to memory 876
MENACCANIT (*m*): menaccanite, ilmenite, titanic iron ore (a ferrous titanate); $FeTiO_3$; Sg. 4.89 877
MENDIPIT (*m*): mendipite; $2PbO,PbCl_2$; Sg. 7.0 878
MENDOZIT (*m*): mendozite; $Al_2(SO_4)_3.Na_2SO_4.24H_2O$; Sg. 1.88 879
MENGE (*f*): quantity; amount; crowd; multitude; plenty; abundance; mass; number 880
MENGEEINHEIT (*f*): unit of quantity 881
MENGEN: to mix; blend; interfere; mingle; shuffle (Cards); compound; join; associate; admix; intermix 882
MENGENBESTIMMUNG (*f*): quantitative (determination) analysis 883
MENGENUNTERSUCHUNG (*f*): quantitative analysis 884
MENGENVERHÄLTNIS (*n*): quantitative (relation; relationship; proportion; composition) ratio; in metal alloying = $\frac{\text{solid mixed crystals}}{\text{liquid melt}}$ 885
MENGER (*m*): mixer 886
MENGEREI (*f*): mixing; mingling; mixture 887
MENGFUTTER (*n*): mixed (grain) feed 888
MENGGESTEIN (*n*): conglomerate; (fragments of rock bound together by a matrix) 889
MENGKAPSEL (*f*): mixing capsule 890
MENGSEL (*n*): mixture; topsy-turvy; medley 891
MENGSPATEL (*m*): mixing spatula 892
MENGUNG (*f*): mixing; mixture; admixture (see MENGEN) 893
MENGUNGSVERHÄLTNIS (*n*): proportion of ingredients (see also MENGENVERHÄLTNIS) 894
MENHADENÖL (*n*): American fish oil, menhaden oil (from the Menhaden or Mossbanker fish, *Clupea menhaden* or *Alosa menhaden*); (Sg. 0.93) 895
MENILIT (*m*): menilite (brown opaque opal) (see OPAL) 896
MENINGOKOKKENSERUM (*n*): meningococcus serum 897
MENISKUS (*m*): meniscus (of liquids) 898
MENNIG (*m*): minium; red lead; (Pb_3O_4); (Sg. 8.94); (natural red lead oxide, about 91% Pb) 899
MENNIGE (*f*): minium; red lead; (Pb_3O_4) (Sg. 8.94) (natural red lead oxide, about 91% Pb) 900
MENNIGEKITT (*m*): red-lead-(cement) 901
MENNIG (*m*), **GELBER-**: chrome yellow (see CHROMGELB) 902
MENSCH (*m*): man; person; human being; (*n*): woman; hussy 903
MENSCHENÄHNLICH: human; man-like; anthropomorphous; anthropoid; resembling man 904
MENSCHENALTER (*n*): generation 905
MENSCHENART (*f*): race (of human beings); type 906
MENSCHENFEIND (*m*): misanthrope; misanthropist; man-hater; enemy of man 907
MENSCHENFEINDLICH: misanthropic-(al)-(ly) 908
MENSCHENFEINDLICHKEIT (*f*): misanthropy; hatred of mankind 909
MENSCHENFRESSEND: anthropophagous, cannibalistic 910
MENSCHENFREUND (*m*): philanthropist; friend of mankind (see also PHILANTHROP) 911
MENSCHENFREUNDLICH: philanthropic-(al)-(ly) 912
MENSCHENFREUNDLICHKEIT (*f*): philanthropy; love of mankind 913
MENSCHENGESCHLÄCHT (*n*): human (race, species) kind, mankind 914
MENSCHENHASS (*m*): (see MENSCHENFEINDLICHKEIT) 915
MENSCHENKÖRPER (*m*): human body 916
MENSCHENKUNDE (*f*): anthropology, the science of man 917
MENSCHENLEBEN (*n*): human life 918
MENSCHENLEHRE (*f*): anthropology (see MENSCHENKUNDE) 919
MENSCHENMÖGLICH: humanly possible 920
MENSCHENMORD (*m*): homicide, murder 921
MENSCHENMÖRDER (*m*): homicide, murderer 922
MENSCHENPOCKEN (*f.pl*): smallpox (Med.) 923
MENSCHENSATZUNG (*f*): human institution 924
MENSCHENSCHEU: (see MENSCHENFEINDLICH) 925
MENSCHENVERSTAND (*m*): common-sense; human understanding 926
MENSCHLICH: human; humane 927
MENSTRUATION (*f*): menstruation; menses (Med.) 928
MENSTRUATIONSBEFÖRDERENDES MITTEL (*n*): menstruative 929
MENSTRUATIONSENDE (*f*): change of life (of female) (Med.) 930
MENSTRUATIONSMITTEL (*n*): menstruative (Med.) 931
MENSTRUIEREN: to menstruate (Med.) 932
MENSTRUUM (*n*): solvent 933
MENSUR (*f*): measure; measuring (glass) vessel; duel; graduated glass and porcelain measuring instrument; rhythm (Mus.) 934
MENSURGLAS (*n*): measuring (graduated) glass 935
MENTHEN (*n*): menthene; ($C_{10}H_{18}$ or $C_{10}H_{16}$) (see also PINEN) 936
MENTHOL (*n*): menthol; peppermint camphor; hexahydrothymol (from *Mentha arvensis* and *M. piperita*); ($C_{10}H_{20}O$) (Sg. 0.89; Mp. 43°C.; Bp. 213°C.) 937

MENTHOL (*n*), **BALDRIANSAURES-** : menthol (valerianate) valerate (see **VALIDOL**) 938
MENTHOLINSCHNUPFPULVER (*n*) : menthol snuff, mentholin snuff 939
MENTHOLSEIFE (*f*) : menthol soap 940
MENTHOLSTIFT (*m*) : menthol pencil 941
MENTHON (*n*) : menthone (a constituent of peppermint oil) 942
MENTHYL (*n*), **ESSIGSAURES-** : menthyl acetate 943
MENTHYLVALERIANAT (*n*) : menthyl valerate 944
MEÖL (*n*) : (see **MOWRABUTTER**) 945
MEPHITISCH : mephitic ; pertaining to (poisonous, nauseous) noxious exhalations 946
MEPHITISCHES GAS (*n*) : carbonic acid (see **KOHLENSÄURE**) 947
MERAZETIN (*n*) : meracetin ; $(C_8H_6O_4Hg)$ (54% Hg) 948
MERCAPTAN (*n*) : mercaptan, ethyl hydrosulphide; $C_2H_5.SH$; Sg. 0.8391 ; Bp. 37°C. 949
MERCERISATION (*f*) : mercerization (Cotton) 950
MERCERISIEREN : to mercerize (Cotton) 951
MERCERISIERUNG (*f*) : mercerization ; mercerizing 952
MERCERISIERVERFAHREN (*n*) : mercerizing process 953
MERCUR (*m*) : mercury (see also **MERKUR** and **QUECKSILBER**) 954
MERCURIUS (*m*) : mercury, quicksilver (see **QUECKSILBER**) 955
MERGEL (*m*) : marl ; (Sg. about 1.75) (earthy calcium carbonate ; see **CALCIUMCARBONAT**) 956
MERGELABLAGERUNG (*f*) : marl bed 957
MERGELART (*f*) : kind (sort ; variety) of marl 958
MERGELARTIG : marly 959
MERGELBODEN (*m*) : marly soil 960
MERGELERDE (*f*) : marly soil ; loose marl 961
MERGELIG : marly 962
MERGELKALK (*m*) : marly limestone 963
MERGELN : to marl ; manure with marl 964
MERGELSCHIEFER (*m*) : marl (slate) shale 965
MERIDIAN (*m*) : meridian 966
MERIDIANSCHNITT (*m*) : vertical centre section 967
MERIDIANSPANNUNG (*f*) : vertical (pressure) stress-(taken)-on the centre line 968
MERK (*n*) : mark ; sign ; note ; observation 969
MERKANTIL : commercial, mercantile, merchant 970
MERKBAR : perceptible ; noticeable ; evident ; appreciable ; sensible 971
MERKBUCH (*n*) : note (minute) book ; waste book 972
MERKEN : to notice ; mark ; perceive ; observe ; remember ; note ; retain ; bear in mind 973
MERKENSWERT : noteworthy ; memorable ; remarkable 974
MERKEWOHL (*n*) : nota-bene (N.B.) 975
MERKLICH : appreciable ; perceptible ; sensible (see also **MERKBAR**) 976
MERKMAL (*n*) : characteristic ; mark ; sign ; indication ; token ; symptom ; memorandum 977
MERKUR (*m*) : mercury (see also **QUECKSILBER**) 978
MERKURBLENDE (*f*) : cinnabar (see **ZINNOBER**) 979
MERKURHORNERZ (*n*) : horn quicksilver ; calomel (Min.) (see **QUECKSILBERHORNERZ**) 980
MERKURIAL : mercurial 981
MERKURIALIEN (*n.pl*) : mercurials (Pharm.) 982
MERKURIAMMONIUMCHLORID (*n*) : mercuric-ammonium chloride, white precipitate, amino-mercuric chloride, ammoniated mercury-(chloride) ; NH_2HgCl (*Hydrargyrum præcipitatum album*) 983
MERKURIBENZOAT (*n*) : (see **QUECKSILBERBENZOAT**) 984
MERKURIBROMID (*n*) : (see **QUECKSILBERBROMID**) 985
MERKURICHLORAMID (*n*) : (see **AMINOQUECKSILBERCHLORID**) 986
MERKURICHLORID (*n*) : mercuric chloride (see **QUECKSILBERCHLORID**) 987
MERKURICYANID (*n*) : mercuric cyanide (see **QUECKSILBERCYANID**) 988
MERKURIEREN : to mercurize ; combine with mercury 989
MERKURIERUNG (*f*) : mercurization 990
MERKURIJODAT (*n*) : (see **QUECKSILBERJODAT**) 991
MERKURIJODID (*n*) : mercuric iodide (see **QUECKSILBERJODID**) 992
MERKURILAKTAT (*n*) : mercuric lactate ; $Hg(C_3H_5O_3)_2$ 993
MERKURINITRAT (*n*) : mercuric nitrate (see **QUECKSILBEROXYDNITRAT**) 994
MERKURIOXYD (*n*) : mercuric oxide (see **QUECKSILBEROXYD**) 995
MERKURIRHODANID (*n*) : (see **QUECKSILBERRHODANID**) 996
MERKURISALZ (*n*) : mercuric salt 997
MERKURISULFAT (*n*) : mercuric sulphate (see **QUECKSILBEROXYDSULFAT**) 998
MERKURISULFAT (*n*), **BASISCHES-** : basic mercuric sulphate (*Turpethum minerale*) ; $HgSO_4.2HgO$ 999
MERKURISULFID (*n*) : mercuric sulphide (see **QUECKSILBERSULFID**) 000
MERKURIVERBINDUNG (*f*) : mercuric compound 001
MERKUROBROMID (*n*) : (see **QUECKSILBERBROMÜR**) 002
MERKUROCHLORID (*n*) : mercurous chloride (see **QUECKSILBERCHLORÜR**) 003
MERKUROGRAPHIE (*f*) : mercurography (an old photographic process of development by means of mercury) 004
MERKURONITRAT (*n*) : mercurous nitrate (see **QUECKSILBEROXYDULNITRAT**) 005
MERKUROOXYD (*n*) : mercurous oxide (see **QUECKSILBEROXYDUL**) 006
MERKUROSALZ (*n*) : mercurous salt 007
MERKUROSULFAT (*n*) : mercurous sulphate (see **QUECKSILBEROXYDULSULFAT**) 008
MERKUROSULFID (*n*) : mercurous sulphide (see **QUECKSILBERSULFÜR**) 009
MERKUROVERBINDUNG (*f*) : mercurous compound 010
MERKURVERBINDUNG (*f*) : mercury compound (see **MERKURI-** and **MERKURO-**) 011
MERKWÜRDIG : noteworthy ; remarkable ; noticeable 012
MERKWÜRDIGERWEISE : strange to say 013
MERKWÜRDIGKEIT (*f*) : curiosity 014
MERKZEICHEN (*n*) : (see **MERKMAL**) 015
MEROCHINEN (*n*) : meroquinene 016
MEROTROPIE (*f*) : merotropy ; merotropism (a phenomenon of certain chemical compounds) 017
MEROXEN (*m*) : meroxene (see **BIOTIT**) 018
MESACONSÄURE (*f*) : mesaconic acid ; $HOCO.C_2H_3HC_2O_2H$; Mp. 202°C. 019

MESITIN (*m*): mesitite, breunnerite (see MESITINSPAT) 020
MESITINSPAT (*m*): mesitite; $2MgO,CO_2 + FeO,CO_2$; Sg. 3.35 021
MESITYLEN (*n*): mesitylene, tri-methyl benzene, (obtained from coal tar); C_9H_{12}; Sg. 0.8694; Mp. – 57.5°C.; Bp. 164.5°C. 022
MESITYLOXYD (*n*): mesityl oxide 023
MESOLITH (*m*): mesolite; $Al_2O_3,CaO,Na_2O,Si_6O_{12}+5H_2O$; Sg. 2.3 024
MESOTAN (*n*): mesotane; $C_9H_{10}O_4$; Sg. 1.2 025
MESOTHORIUM (*n*): mesothorium (a radio-active dissociation product of thorium) 026
MESOTYP (*m*): mesotype (see NATROLITH) 027
MESOWEINSÄURE (*f*): mesotartaric acid; $HOOCCH(OH).CH(OH)COOH.H_2O$; Mp. 142°C. 028
MESOXALSÄURE (*f*): mesoxalic acid 029
MESSAMT (*n*): mass; office (Ecclesias.) 030
MESSAPPARAT (*m*): measuring (apparatus) instrument; meter 031
MESSBAND (*n*): tape measure; measuring tape 032
MESSBAR: measurable 033
MESSBEREICH (*m*): measuring range 034
MESSBILDVERFAHREN (*n*): photogrammetry (see PHOTOGRAMMETRIE) 035
MESSBRÜCKE (*f*), **WHEATSTONESCHE-**: Wheatstone's bridge 036
MESSBUCH (*n*): (surveyor's) note book; mass book, missal 037
MESSDOSE (*f*): measuring cylinder 038
MESSDOSENKOLBEN (*m*): measuring cylinder piston 039
MESSE (*f*): fair; market; mass (Catholic Church); mess-(room) (Naut.) 040
MESSEN: to measure; gauge; contain; survey; span; compare 041
MESSER (*m*): measurer; surveyor; meter; gauge; (*n*): knife; cutter 042
MESSERBESTECK (*n*): knife-case, knife sheath 043
MESSERFABRIKANT (*m*): cutler 044
MESSERKLINGE (*f*): knife-blade 045
MESSERKOPF (*m*): cutter head 046
MESSERPUTZMASCHINE (*f*): knife cleaning machine 047
MESSERSCHÄRFER (*m*): knife sharpener 048
MESSERSCHEIDE (*f*): knife (case) sheath 049
MESSERSCHMIED (*m*): cutler 050
MESSERSCHMIEDWAREN (*f.pl*): cutlery 051
MESSERSCHMIEDWARENPAPIER (*n*): cutlery paper 052
MESSERSCHNEIDE (*f*): knife-edge 053
MESSERSPITZE (*f*): knife-point 054
MESSERSTIEL (*m*): knife-handle 055
MESSERWELLE (*f*): plane shaft 056
MESSFAHNE (*f*): surveyor's flag 057
MESSFLASCHE (*f*): measuring (bottle) flask 058
MESSFLÜSSIGKEIT (*f*): measuring (test) fluid 059
MESSGEFÄSS (*n*): measuring vessel; measure 060
MESSGERÄTE (*n.pl*): measuring (apparatus) instruments; utensils used in celebrating mass 061
MESSGESCHIRR (*n*): measuring (apparatus) instruments 062
MESSING (*n*): brass (*Aurichalcum*) (cast, Sg. 8.1) (see also GELBGUSS) 063
MESSINGANODE (*f*): brass anode 064
MESSINGBAD (*n*): brass bath (for brass coating) 065
MESSINGBESCHLAG (*m*): brass (binding, mounting) fitting 066
MESSINGBLECH (*n*): brass (sheeting) plate; sheet brass 067
MESSINGBLÜTE (*f*): aurichalcite (see AURICHALCIT) 068
MESSINGDECKEL (*m*): brass (lid, cover) cap 069
MESSINGDRAHT (*m*): brass wire 070
MESSINGEN: of brass; brass; brazen; to coat with brass 071
MESSINGGIESSER (*m*): brass founder 072
MESSINGGIESSEREI (*f*): brass foundry; brass (casting) founding 073
MESSINGHÜTTE (*f*): brass foundry 074
MESSINGLACK (*m*): brass varnish 075
MESSINGMUTTER (*f*): brass nut 076
MESSINGNETZ (*n*): brass (netting) gauze 077
MESSINGPROFIL (*n*): brass section 078
MESSINGROHR (*n*): brass tube 079
MESSINGSCHLAGLOT (*n*): brass solder 080
MESSINGSCHMIED (*m*): brazier 081
MESSING (*n*), **SCHMIEDBARES-**: malleable brass (60% Cu; 40% Zn) 082
MESSINGSCHRAUBE (*f*): brass screw 083
MESSINGSPÄNE (*m.pl*): brass (shavings; cuttings; turnings) filings 084
MESSINGSTANGE (*f*): brass rod 085
MESSINGWALZWERK (*n*): brass rolling mill 086
MESSINGWAREN (*f.pl*): brass ware 087
MESSINGWERK (*n*): brass foundry 088
MESSINSTRUMENT (*n*): measuring instrument, measuring apparatus, meter 089
MESSKEIL (*m*): inside caliper for pipes; measuring wedge 090
MESSKETTE (*f*): surveying chain; surveyor's chain 091
MESSKÖLBCHEN (*n*): small measuring flask 092
MESSKOLBEN (*m*): measuring flask 093
MESSKUNDE (*f*): mensuration 094
MESSKUNST (*f*): geometry; surveying 095
MESSLATTE (*f*): measuring (staff) rod 096
MESSMITTEL (*n*): measuring instrument; means of measurement 097
MESSPIPETTE (*f*): scale (measuring) pipette (with graduated scale) 098
MESSREIHE (*f*): series of measurements 099
MESSROHR (*n*): measuring tube; burette 100
MESSRUTE (*f*): surveyor's staff; measuring (surveying) rod 101
MESSSCHEIBE (*f*): quadrant, sextant 102
MESSSERIE (*f*): series of measurements 103
MESSTISCH (*m*): surveyor's-(plan)-table 104
MESSUNG (*f*): measuring; measurement; mensuration; surveying (see also MESSEN) 105
MESSVERFAHREN (*n*): method (process) of measurement 106
MESSWIDERSTAND (*m*): measuring resistance 107
MESSZYLINDER (*m*): measuring cylinder (see MENSUR) 108
MESTE (*f*): box (receptacle) for keeping condiments or spices 109
MET (*m*): mead; metheglin; hydromel (honey diluted with water and fermented) 110
METAAMIDOPHENOL (*n*): metaamidophenol 111
METAANTIMONSÄURE (*f*): metantimonic acid 112
METAARSENIG: metarsenious 113
METAARSENSÄURE (*f*): metarsenic acid 114
METABISULFIT (*n*): metabisulphite 115
METABLEISÄURE (*f*): metaplumbic acid 116

METABOLIE (*f*) : metabolism (a series of chemical changes, comprising the forming and breaking up of complex substances from simple ones) 117
METABOLISCH : metabolic 118
METABORSÄURE (*f*) : metaboric acid 119
METACHLORANILIN (*n*) : metachloroaniline 120
METACHROMBEIZE (*f*) : metachrome mordant (a mixture of potassium bichromate and ammonium sulphate) 121
METACHROMFARBE (*f*) : metachrome dye 122
METACINABRE (*m*) : (see **METACINNABARIT**) 123
METACINNABARIT (*m*) : metacinnabarite ; HgS ; Sg. 7.75 124
METAEISENOXYD (*n*) : metaferric oxide 125
METAFERRIHYDRAT (*n*) : metaferric hydroxide 126
METAFERRIN (*n*) : metaferrin (an iron preparation, compounded from 10% iron, 10% phosporic acid and albumin) 127
METAFERROSE (*f*) : metaferrose (solution of metaferrin) 128
METAKOHLENSÄURE (*f*) : metacarbonic acid (common carbonic acid) ; (H_2CO_3) 129
METAKRESOL (*n*) : metacresol (see **KRESOL**) 130
METAKRESOTINSÄURE (*f*) : metacresotic acid 131
METALL (*n*) : metal 132
METALLABFALL (*m*) : scrap metal ; metal scrap 133
METALLADER (*f*) : metallic (metalliferous) vein or seam 134
METALLÄHNLICH : metallic ; metalline 135
METALLAMID (*n*) : metallic amide 136
METALLANSTRICH (*m*) : metallic (metal) coating ; coating (painting) of metal 137
METALLARBEIT (*f*) : metal work 138
METALLARBEITER (*m*) : metal worker 139
METALLARTIG : metallic ; metalline 140
METALLASCHEN (*f.pl*) : metallic ashes 141
METALLAZID (*n*) : metallic azide 142
METALLBEARBEITER (*m*) : metal worker 143
METALLBEARBEITUNG (*f*) : metal working 144
METALLBESCHICKUNG (*f*) : alloyage (the act of mixing metals) 145
METALLBESCHLAG (*m*) : metal (sheathing) casing ; clamp, cleat, clasp ; metal binding 146
METALLBESCHREIBUNG (*f*) : metallography (branch of metallurgy describing structure of metals) 147
METALLBILDUNG (*f*) : alligation (see **LEGIERUNG**) 148
METALLBLATT (*n*) : sheet of metal 149
METALLBLÜTE (*f*) : aurichalcite (see **AURICHALCIT**) 150
METALLBÜCHSE (*f*) : metal box ; metal bush-(ing) 151
METALLBÜRSTE (*f*) : (steel)-wire brush 152
METALLCARBID (*n*) : metallic carbide ; carbide of a metal (see **CALCIUMCARBID** ; **BARYUM-CARBID** ; **BOROCARBID** and **MOLYBDÄN-CARBID**) 153
METALLCEMENT (*m* and *n*) : metal cement 154
METALLCHLORID (*n*) : metallic chloride 155
METALLCHLORWASSERSTOFFSÄURE (*f*) : metallic hydrochloride (compound of metal and hydrochloric acid) 156
METALLCHROMIE (*f*) : metallochrome ; metal colouring 157
METALLDAMPF (*m*) : metallic vapour 158
METALLDICHTUNG (*f*) : metal packing 159
METALLDICHTUNGSRING (*m*) : metal-(lic) packing ring 160
METALLDRAHT (*m*) : metal (filament) wire 161
METALLDRAHTLAMPE (*f*) : metal filament lamp 162
METALL (*n*), **EDLES-** : precious metal 163
METALLEN : metal ; made of metal 164
METALLFADEN (*m*) : metal filament 165
METALLFADENLAMPE (*f*) : metal filament lamp 166
METALLFARBE (*f*): metallic colour ; bronze colour (see **BROKATFARBE**) 167
METALLFÄRBUNG (*f*) : metal colouring ; metallochrome 168
METALLFLECK (*m*) : metallic spot (a spot which occasionally occurs on albumenized paper, due to iron impurities in the paper) 169
METALLFOLIE (*f*) : metal foil (see also **CALIN-LEGIERUNG**) 170
METALLGEGENSTAND (*m*) : metal object 171
METALLGEKRÄTZ (*n*) : waste (scrap) metal 172
METALLGEMISCH (*n*) : alloy ; metallic mixture 173
METALLGEWEBE (*n*) : wire cloth (Paper) 174
METALLGEWINNUNG (*f*) : metallurgy ; metal extraction 175
METALLGIESSEREI (*f*) : (metal)-foundry 176
METALLGLANZ (*m*) : metallic lustre 177
METALLGLÄNZEND : having metallic lustre 178
METALLGLAS (*n*) : enamel 179
METALLGOLD (*n*) : Dutch (foil) gold ; Dutch metal (imitation gold-lead from alloying copper and zinc) (see also **TOMBAK**) 180
METALLGRAU (*n*) : metal grey 181
METALLGUSS (*m*) : metal (founding) casting 182
METALLGUSSSTÜCK (*n*) : metal casting 183
METALLHALTIG : metalliferous ; containing metal 184
METALLHÜTTE (*f*) : metal smelting works ; smeltery (for any metal except iron) 185
METALLHYDROXYD (*n*) : metallic hydroxide 186
METALLIN (*n*) : metallin (a cobalt-copper-iron-aluminium alloy) (35% Co ; 30% Cu ; 25% Al ; 10% Fe) 187
METALLISCH : metallic 188
METALLISCHER DAMPF (*m*) : metallic (fume) vapour 189
METALLISIEREN : to metallize 190
METALLITÄT (*f*) : metallic nature 191
METALLKALK (*m*) : metallic (oxide) calc or calx ; calcined metal 192
METALLKAPSEL (*f*) : metal cap 193
METALLKARBID (*n*) : (see **METALLCARBID**) 194
METALLKARBONYL (*n*) : compound of metal and carbonyl 195
METALLKLAMPE (*f*) : metal (chock, cleat, clasp) clamp 196
METALLKÖNIG (*m*) : metallic (button) regulus 197
METALLKORN (*n*) : metal grain 198
METALLKÖRNER (*n.pl*) : granulated metal ; metal grains 199
METALLKUNDE (*f*) : metallurgy (the science of metal refining) 200
METALLLACK (*m*) : metal (lacquer) varnish 201
METALLLEGIERUNG (*f*) : metallic alloy 202
METALLLIDERUNG (*f*) : metal packing 203
METALLLÖSUNG (*f*) : metallic solution 204
METALLMIKROSCOP (*n*) : metal microscope (for microscopical examination of metals) 205
METALLMUTTER (*f*) : matrix (of ores) ; brass (metal) nut 206
METALLNEBEL (*m*) : atomized metal, metal mist (molten metal which is sprayed through a nozzle by means of high pressure and highly

heated gases or vapours) (see METALLSPRITZ-VERFAHREN) 207
METALLOCHROMIE (*f*): metallochrome; metal colouring 208
METALLOGRAPHIE (*f*): metallography (microscopical investigation of metals) 209
METALLOGRAPHISCH: metallographic 210
METALLOID (*n*): metalloid, non-metal 211
METALLORGANISCH: metalloorganic; organometallic 212
METALLOXYD (*n*): metallic oxide 213
METALLOXYDHYDRAT (*n*): metallic hydroxide 214
METALLPACKUNG (*f*): metallic packing 215
METALLPLATTE (*f*): metal sheet or plate 216
METALLPOLIERMITTEL (*n*): metal polish 217
METALLPROBE (*f*): test for metal; assay 218
METALLPUTZMITTEL (*n*): metal polish 219
METALLREINIGUNGSMITTEL (*n* and *pl*): metal cleaning agent-(s) 220
METALLRÜCKSTAND (*m*): metallic residue 221
METALLSÄGE (*f*): metal saw 222
METALLSALZ (*n*): metallic salt 223
METALLSCHLAUCH (*m*): (flexible)-metallic hose 224
METALLSCHLEIFMITTEL (*n*): metal grinding agent (Emery, etc.) 225
METALLSCHLIFF (*m*): metal filings 226
METALLSCHWAMM (*m*): metallic sponge 227
METALLSEIFE (*f*): metal scrap (sebates of heavy metals) 228
METALLSILBER (*n*): imitation silver foil; Britannia metal 229
METALLSILIZIUMVERBINDUNG (*f*): metal-silicon compound (see METILLUR) 230
METALLSPÄNE (*m.pl*): metal (shavings; filings; chips) turnings 231
METALLSPEKTRUM (*n*): spectrum of metal 232
METALLSPIEGEL (*m*): metallic mirror 233
METALLSPRITZVERFAHREN (*n*): metal spraying process (a process by Schoop, for obtaining solid metal precipitations or coatings on objects) (see METALLNEBEL) 234
METALLSTAUB (*m*): metallic (metal) dust 235
METALLSULFID (*n*): metallic sulphide 236
METALLTEIL (*m*): metallic (metal) part 237
METALLTREIBARBEIT (*f*): hammered (chased) metal-work, metal chasing 238
METALLTUCH (*n*): wire cloth (Paper) 239
METALLÜBERZUG (*m*): metallic (covering) coating (see METALLSPRITZVERFAHREN); metal (casing) plating; metallic wash (to act as an anti-rust agent) 240
METALL (*n*), UNEDLES-: base metal 241
METALLURG (*m*): metallurgist 242
METALLURGISCH: metallurgic-(al)-(ly) 243
METALLVERARBEITUNG (*f*): metal working 244
METALLVERSETZUNG (*f*): alloy-(ing) 245
METALLVORRAT (*m*): banking reserve, cash in hand 246
METALLWAREN (*f.pl*): metal wares; hardware 247
METALLWEISS (*n*): metal white 248
METALLWOLLE (*f*): metallic wool 249
METAMER: metameric 250
METAMERIE (*f*): metamerism; metamery 251
METAMINBLAU (*n*): metamine blue 252
METAMORPHOSE (*f*): metamorphosis; transformation, structural change; degeneration, development, evolution 253
METANILGELB (*n*): metanil (metaniline) yellow 254
METANILROT (*n*): metanil (metaniline) red 255
METANILSÄURE (*f*): metanilic acid (see AMIDO-BENZOLSULFOSÄURE) 256
METANITRANILIN (*n*): metanitraniline (see NITRANILIN) 257
METANITROHYDROXYBENZOESÄURE (*f*): meta-nitrohydroxybenzoic acid, asymmetric; meta-nitrosalicylic acid; $C_6H_3COOH:OH:NO_2$; Mp. 235°C. 258
METANITROORTHOANISIDIN (*n*): metanitro-orthoanisidine 259
METANITROPARATOLUIDIN (*n*): metanitropara-toluidine 260
METANITROTOLUOL (*n*): metanitrotoluene 261
METANTIMONIG: metantimonious 262
METANTIMONSÄURE (*f*): metantimonic acid 263
METAPHENYLENBLAU (*n*): metaphenylene blue 264
METAPHENYLENDIAMIN (*n*): metaphenylene-diamine (see PHENYLENDIAMIN) 265
METAPHOSPHORIG: metaphosphorous 266
METAPHOSPHORSAUER: metaphosphate of; combined with metaphosphoric acid 267
METAPHOSPHORSÄURE (*f*): metaphosphoric acid; glacial phosphoric acid; (HPO_3); Sg. 2.35 268
METAPHYSIK (*f*): metaphysics 269
METAPHYSISCH: metaphysical 270
METASÄURE (*f*): meta acid 271
METASTABIL metastable 272
METASTELLUNG (*f*): meta position 273
METATITANSÄURE (*f*): metatitanic acid; $TiO(OH)_2$ 274
METATOLUIDIN (*n*): metatoluidine 275
METATOLUYLENDIAMIN (*n*): metatoluylenediamine, diaminotoluene (see TOLUYLENDIAMIN) 276
METAVANADINSÄURE (*f*): metavanadic acid 277
METAVERBINDUNG (*f*): meta compound (substitution product from benzol) 278
METAWOLFRAMSÄURE (*f*): metatungstic (metawolframic) acid 279
METAXIT (*m*): metaxite, serpentine (see SERPENTIN) 280
METAXYLIDIN (*n*): metaxylidine (see XYLIDIN) 281
METAXYLOL (*n*): metaxylol (see XYLOL) 282
METAZENTRISCH: metacentric 283
METAZENTRISCHE HÖHE (*f*): metacentric height 284
METAZENTRUM (*m*): metacentre (the point in a floating body which determines its stability or otherwise) 285
METAZINNSÄURE (*f*): metastannic acid 286
METAZIRKONSÄURE (*f*): metazirconic acid 287
METAZUCKERSÄURE (*f*): metasaccharic acid 288
METEOR (*n*): meteor 289
METEOREISEN (*n*): meteoric iron 290
METEOROLOG (*m*): meteorologist 291
METEOROLOGIE (*f*): meteorology 292
METEOROLOGISCH: meteorological-(ly) 293
METEORSTAUB (*m*): meteoric dust 294
METEORSTEIN (*m*): meteoric stone; aërolite 295
METEORWASSER (*n*): meteoric water, rain-(water) 296
METER (*m*): metre (long measure); (*n*): meter (for gas, etc.) 297
METERKILOGRAMM (*n*): kilogram-metre 298
METERMASS (*n*): metre rule 299
METHACETIN (*n*): methacetin, acetanisidin; $CH_3OC_6H_4NH.C_2H_3O$; Mp. 127°C. 300
METHACRYLSÄURE (*f*): methacrylic acid 301
METHAN (*n*): methyl hydride; methane; hydrogen bicarbide; marsh gas; (CH_4); Sg. 0.559; Mp. −184°C.; Bp. −160°C. 302

METHANAL (*n*): (see FORMALDEHYD) 303
METHANOL (*m*): (new name for "methyl alcohol" English is "methanol") (see METHYLAL--KOHOL) 304
METHODE (*f*): method 305
METHODISCH: methodical-(ly) 306
METHYL (*n*): methyl (the alkyl group); (CH_3) 307
METHYLACETANILID (*n*): methyl acetanilide; $C_6H_5N(CH_3)(CO.CH_3)$; Mp. 102°C. 308
METHYLACETAT (*n*): methyl acetate; $CH_3COO.CH_3$; Sg. 0.925-0.9562; Mp. – 98°C. Bp. 54°C.-57.5°C. 309
METHYLACETON (*n*): methylacetone (mixture of methyl acetate and acetone, in proportion 1 : 2; is a liquid and contains about 75% of foregoing mixture) 310
METHYLAL (*n*): methylal; methylenedimethylester; methylenedimethylate; formal; $CH_2(O.CH_3)_2$; Sg. 0.855-0.872; Bp. 42°C. 311
METHYLALDEHYD (*n*): methyl aldehyde (see FORMALDEHYD) 312
METHYLALKOHOL (*m*): (*Alcohol methylicus*): wood naphtha, methyl alcohol, methylated spirit, wood alcohol, wood spirit; CH_4O; Sg. 0.8; Mp. – 97.8°C.; Bp. 65.75°C. (see also HOLZGEIST) 313
METHYL (*n*), AMEISENSAURES- : methyl formate; CH_3COOH; Sg. 0.973; Mp. – 99.75°C.; Bp. 32°C. 314
METHYLAMIDOCRESOLSULFAT (*n*): methylaminocresol sulphate 315
METHYLAMIN (*n*): methyl amine; CH_5N; Sg. 0.699; Bp. – 6.7°C. 316
METHYLANILIN (*n*): methyl aniline, monomethylaniline; C_7H_9N; Sg. 0.9863; Mp. about – 80°C.; Bp. 193.8°C. 317
METHYLANTHRACEN (*n*): methylanthracene 318
METHYLANTHRANILAT (*n*): methyl anthranilate (see NEROLIÖL, SYNTETISCHES-) 319
METHYLARSONSÄURE (*f*): methylarsonic acid 320
METHYLAT (*n*): methylate 321
METHYLÄTHER (*m*): methyl ether; C_2H_6O; Bp. – 23.7°C. 322
METHYLÄTHYLKETON (*n*): methyl ethyl ketone; C_4H_8O; Sg. 0.8255; Mp. – 85.9°C.; Bp. 79.6°C. 323
METHYL (*n*), BALDRIANSAURES- : methyl valerate 324
METHYLBENZOAT (*n*): methyl benzoate; niobe oil; essence niobe; $C_6H_5CO_2CH_3$; Sg. 1.09; Mp. – 12.3°C.; Bp. 198.6°C. 325
METHYL (*n*), BENZOESAURES- : methyl benzoate (see METHYLBENZOAT) 326
METHYLBENZYLANILIN (*n*): methylbenzylaniline 327
METHYLBLAU (*n*): methyl blue 328
METHYLBRENZKATECHIN (*n*): (see GUAJACOL) 329
METHYLBROMID (*n*): methyl bromide, bromomethane; (CH_3Br); Sg. 1.732; Mp. – 84°C.; Bp. 4.5°C. 330
METHYL (*n*), BUTTERSAURES- : methyl butyrate (see METHYLBUTYRAT) 331
METHYLBUTYRAT (*n*): methylbutyrate; $C_3H_7COO.CH_3$ 332
METHYL (*n*), CHLORESSIGSAURES- : methyl chloroacetate 333
METHYLCHLORID (*n*): methyl chloride; chloromethane (*Methylum chloratum*); (CH_3Cl); Sg. 0.9915; Mp. – 103.6°C.; Bp. – 24.1°C. 334
METHYL (*n*), CHLORKOHLENSAURES- : methyl chlorocarbonate 335
METHYL (*n*), CHLORSULFONSAURES- : methylchlorosulfonate (poison gas); CH_3ClSO_3 [336
METHYLCHLORÜR (*n*): methyl chloride (see METHYLCHLORID) 337
METHYLCONIIN (*n*): methyl(conine) coniine; $C_9H_{19}N$ 338
METHYLCROTONSÄURE (*f*): methylcrotonic acid, tiglic acid, crotonolic acid (see CRO--TONOLSÄURE) 339
METHYLCYCLOHEXANOL (*n*): methylcyclohexanol (see METHYLHEXALIN) 340
METHYLDIPHENYLAMIN (*n*): methyldiphenylamine (see DIPHENYLAMIN) 341
METHYLEN (*n*): methylene 342
METHYLENBICHLORID (*n*): methylene bichloride (see METHYLENCHLORID) 343
METHYLENBLAU (*n*): methylene blue; methylated thionine (a tar dyestuff for cotton and silk); $[C_6H_3N(CH_3)_2Cl.N.S.C_6H_3N(CH_3)_2]$ 344
METHYLENBROMID (*n*): methylene bromide; CH_2Br_2; Sg. 2.4985; Bp. 98.5°C 345
METHYLENCHLORID (*n*): methylene chloride; dichloromethane; methylene bichloride (*Methylenum chloratum*); Sg. 1.3778; Mp. – 96.7°C.; Bp. 41.6°C.; (CH_2Cl_2) 346
METHYLENCHLORÜR (*n*): (see METHYLENCHLORID) 347
METHYLENDIÄTHYLÄTHER (*m*): (see ÄTHYLAL) 348
METHYLENDIMETHYLÄTHER (*m*): (see METHYLAL) 349
METHYLENDIOXYZINNAMENYLAKRYLSÄURE (*f*): (see PIPERINSÄURE) 350
METHYLENGRUPPE (*f*): methylene group 351
METHYLENGUJAKOL (*n*): methylene guaiacol 352
METHYLENJODID (*n*): methylene iodide; diiodomethane; (CH_2I_2); Sg. 3.3326; Mp. 5.7°C.; Bp. 152°C. 353
METHYLENKREOSOT (*n*): methylene creosote 354
METHYLENPHENYLGLYKOLÄTHER (*m*): (see JASMAL) 355
METHYLEOSIN (*n*): methyl eosine 356
METHYL (*n*), ESSIGSAURES- : methyl acetate; $CH_3CO_2.CH_3$; Sg. 0.925; Mp. – 98.05°C.; Bp. 54.05°C. 357
METHYLFORMIAT (*n*): methyl formiate; HCOO.CH_3 (see METHYL, AMEISENSAURES-) 358
METHYLGALLUSÄTHERSÄURE (*f*): methyl ether of gallic acid; orthomethylgallic acid 359
METHYLGRÜN (*n*): methyl green 360
METHYLHARNSTOFF (*m*): methyl urea 361
METHYLHEXALIN (*n*): methylhexalin, methylcyclohexanol; $C_7H_{13}OH$; Bp. 160-180°C. 362
METHYLHEXYLCARBINOL (*n*): (see CAPRYLAL--KOHOL, SEKUNDÄR-) 363
METHYLHEXYLKETON (*n*): methylhexylketone 364
METHYLHYDRÜR (*n*): methyl hydride; methane (see METHAN) 365
METHYLIEREN: to methylate 366
METHYLISOBUTYRAT (*n*): methyl isobutyrate; $C_3H_7COOCH_3$ 367
METHYLISOPROPYLBENZOL (*n*): methylisopropylbenzene (see CYMOL) 368
METHYLJODID (*n*): methyl iodide; iodomethane; CH_3I; Sg. 2.2; Mp. – 65°C.; Bp. 44°C. 369

METHYL (*n*), KOHLENSAURES- : methyl carbonate 370
METHYL (*n*), MALONSAURES : methyl malate 371
METHYLNAPHTALIN (*n*): methyl naphthalene 372
METHYLORANGE (*n*): methyl orange 373
METHYLORANGELÖSUNG (*f*): methyl-orange solution (composed of 1 gram of methyl orange to 1 litre of water) 374
METHYLORTHOTOLUIDIN (*n*): methyl ortho--toluidine 375
METHYLPENTOSAN (*n*): methyl pentosane 376
METHYLPHENOL (*n*): methyl phenol (see KRESOL) 377
METHYLPHOSPHIN (*n*): methyl phosphine, primary methyl phosphine; $PH_2.CH_3$ 378
METHYLPHOSPHIN (*n*), PRIMÄRES- : primary methyl phosphine (see METHYLPHOSPHIN) 379
METHYLPHOSPHIN (*n*), SEKUNDÄRES- : secondary methyl phosphine (see DIMETHYLPHOSPHIN) 380
METHYLPHOSPHIN (*n*), TERTIÄRES : tertiary methyl phosphine (see TRIMETHYLPHOSPHIN) 381
METHYLPHOSPHONIUMJODID (*n*): methyl phosphonium iodide (see PHOSPHONIUMJODID) 382
METHYLPHOSPHOSÄURE (*f*): phosphomethylic acid; $CH_3.PO(OH)_2$ 383
METHYLPROPANOL (*n*): methyl propanol; $(CH_3)_2CH.CH_2.OH$ (see ISOBUTYLALKOHOL) 384
METHYLPROPIONAT (*n*): methyl propionate; $CH_3.CH_2COO.CH_3$ 385
METHYL (*n*), PROPIONSAURES- : methyl propionate (see METHYLPROPIONAT) 386
METHYLPROPYLKETON (*n*): methyl propyl ketone; $CH_3.CO.C_3H_7$ 387
METHYLPYRIDIN (*n*), ALPHA- : alpha methylpyridine; $C_5H_4.N(CH_3)$; Sg. 0.95 388
METHYLRHODANID (*n*): methyl rhodanide 389
METHYLRHODANÜR (*n*): methyl sulfocyanide 390
METHYLROT (*n*): methyl red (a tar dyestuff) 391
METHYL, SALICYLSAURES- : (*n*): methyl salicylate; artificial oil of wintergreen (see WINTER--GRÜNÖL, KÜNSTLICHES-) 392
METHYL, SCHWEFELSAURES- (*n*): methyl sulphate; dimethyl sulphate; $(CH_3)_2SO_4$; Sg. 1.352; Mp. – 10°C.; Bp. 188°C. 393
METHYLSENFÖL (*n*): methyl mustard oil 394
METHYLSULFID (*n*): methyl sulphide, dimethyl sulphide, methanethiomethane; $(CH_3)_2S$; Sg. 0.845; Mp. – 83.2°C.; Bp. 37.5°C. 395
METHYLSULFONAL (*n*): trional; sulfonethylmethane; $CH_3(C_2H_5)C(SO_2C_2H_5)_2$; Mp. 76°C. 396
METHYLTRIOXYBENZOPHENON (*n*): (see COTOIN) 397
METHYLVERBINDUNG (*f*): methyl compound (compounds derived direct from the radical; CH_3) 398
METHYLVIOLETT (*n*): methyl violet; $C_{24}H_{28}N_3Cl$ or $C_{25}H_{30}N_3Cl$ (see PYOKTANIN, BLAUES-) 399
METHYLWASSERSTOFF (*m*): methane (see METHAN) 400
METHYL (*n*), ZIMTSAURES- : methyl cinnamate; $C_6H_5CH.CHCO_2CH_3$; Sg. 1.042; Mp. 36°C.; Bp. 260°C. 401
METHYLZINNSÄURE (*f*): methylstannic acid 402
METILLUR (*f*): metal-silicon compound, produced in electric furnace; used for acid-proof chemical apparatus 403
METOCHINON (*n*): metoquinone (combination from METOL, which see, and hydroquinone; a photographic chemical) 404
METOL (*n*): metol, methyl-para-amino-meta-cresol sulphate (Photographic developer) 405
METOLENTWICKLER (*m*): metol developer (consists of metol, neutral sodium sulphite and water in the proportion of 1 : 10 : 100) (Photography) 406
METONISCHER ZYKLUS (*m*): metonic cycle (of the moon) 407
METRISCH : metric-(al)-(ly) 408
METROSKOP (*n*): (see MUTTERSPIEGEL) 409
METRUM (*n*): metre, measure (Poetry) 410
METZE (*f*): peck (Dry measure); prostitute 411
METZELBANK (*f*): shambles 412
METZELEI (*f*): butchery; slaughter; massacre 413
METZELN : to butcher; slaughter; massacre 414
METZEN : to slaughter 415
METZGE (*f*): shambles 416
METZGER (*m*): butcher 417
MEUCHELMORD (*m*): assassination 418
MEUCHELMÖRDER (*m*): assassin 419
MEUCHELN : to assassinate; plot; conspire 420
MEUCHLER (*m*): assassin; plotter; one who conspires 421
MEUCHLERISCH : insidiously; treacherously 422
MEUCHLINGS : insidiously; treacherously 423
MEUTE (*f*): mutiny 424
MEUTEREI (*f*): mutiny; sedition; riot 425
MEWE (*f*): sea-gull, mew, sea-mew, gull (member of the family, *Larus*) 426
MEZZOTINTOSTICH (*m*): mezzotint 427
MIARGYRIT (*m*): miargyrite; natural silver-antimony glance; AgSbS; Sg. 5.2 (about 38% Ag.) 428
MICA (*m*): mica (natural hydrous silicate as biotite, muscovite, lepidolite, lithionite, etc. (see also GLIMMER) 429
MICROBROMIT (*m*): microbromite (see EMBOLIT) 430
MIENE (*f*): mien; look; air; feature; bearing; appearance; expression; aspect; manner; carriage; demeanour; countenance; deportment 431
MIENENSPIEL (*n*): signs; mimicry; pantomimic gestures; pantomime; dumb-show 432
MIERE (*f*): chickweed (*Alsine*) (Bot.) 433
MIERSIT (*m*): miersite; AgI 434
MIETBEDINGUNGEN (*f.pl*): terms of (agreement) lease 435
MIETBESITZ (*m*): tenancy 436
MIETE (*f*): mite; rent; hire; stack 437
MIETEN : to hire; rent; lease; engage 438
MIETER (*m*): tenant; lodger; lessee 439
MIETKONTRAKT (*m*): tenure; lease; contract to rent; agreement 440
MIETLING (*m*): hireling; mercenary 441
MIETVERTRAG (*m*): (see MIETKONTRAKT) 442
MIGRÄNE (*f*): megrim, migraine, sick (nervous) headache (Med.) 443
MIGRÄNEPULVER (*n*): headache powder 444
MIGRÄNESTIFT (*m*): megrim (headache) pencil 445
MIGRÄNIN (*n*): trade name for a headache powder which is a mixture of Coffeine citrate and Antipyrin; (*Antipyrinum coffeino-citricum*) 446
MIKA (*m*): (see MICA) 447
MIKANIT (*m*): mikanit; name for mica plates stuck together with a shellac solution, to serve as insulating material 448

MIKROBRENNER (*m*): micro burner 449
MIKROCHEMIE (*f*): microchemistry 450
MIKROCHEMISCH: microchemical 451
MIKROFARAD (*n*): microfarad 452
MIKROKLIN (*m*): microcline (see **ORTHOKLAS**); $KAlS_3O_8$; Sg. 2.55 453
MIKROKOSMISCH: microcosmic 454
MIKROKOSMUS (*m*): macrocosm, microcosm (a little world) 455
MIKROLITH (*m*): microlite; $Ca_2Ta_2O_7$; Sg. 6.0 456
MIKROM (*n*): microhm 457
MIKROMETER (*n*): micrometer (astronomical instrument) 458
MIKROMETERSCHRAUBE (*f*): micrometer screw 459
MIKROMETRISCH: micrometric-(al) (see also **HELIOMETER**) 460
MIKROORGANISMUS (*m*); **MIKROORGANISMEN** (*pl*): microorganism-(s) 461
MIKROPHON (*n*): microphone, transmitter (an instrument for observing minute sounds) 462
MIKROPHONRELAIS (*n*): microphone relay 463
MIKROPHOTOGRAPHIE (*f*): microphotograph; microscopic-(al) photograph-(y); microphotography (the reduction of photographic positives to microscopical dimensions) 464
MIKROSKOP (*n*): microscope (optical instrument for magnifying small objects) 465
MIKROSKOPIE (*f*): microscopical examination of objects 466
MIKROSKOPIEREN: to examine microscopically; examine under the microscope 467
MIKROSKOPISCH: microscopic-(al)-(ly) 468
MIKROSKOPOBJEKTIV (*n*): microscope objective 469
MIKROVOLT (*n*): microvolt 470
MILARIT (*m*): milarite; $KHCa_2Al_2(Si_2O_5)_6$; Sg. 2.59 471
MILBE (*f*): mite; tick; wood-louse 472
MILCH (*f*): milk (Sg. 1.03); milt; soft-roe (of fish), juice; latex (see **LATEX**); emulsion 473
MILCHABKÜHLUNG (*f*): milk cooling 474
MILCHABSONDERUNG (*f*): secretion of milk, lactation (Physiol.) 475
MILCHABTREIBEND: antilactic; antigalactic 476
MILCHACHAT (*m*): milk-white agate (see **ACHAT**) 477
MILCHADER (*f*): lacteal vein 478
MILCHÄHNLICH: milk-like; milky; lacteal; like milk 479
MILCHARTIG: milky; lacteal; milk-like 480
MILCHBESTANDTEIL (*m*): constituent (part) of milk 481
MILCHDRÜSE (*f*): lacteal gland 482
MILCHDRÜSENSTÖRUNG (*f*): a disorder of lactation (general term covering various complaints such as Agalactia, Galactorrhœa, etc.) (Med.) 483
MILCHEN: to give milk 484
MILCHEXTRAKT (*m*): milk extract 485
MILCHFÄLSCHUNG (*f*): adulteration (watering) of milk 486
MILCHFARBIG: milk-coloured; milky 487
MILCHFETT (*n*): milk fat (from the milk of animals) 488
MILCHFIEBER (*n*): milk (ephemeral) fever (temperature rise after childbirth) (Med.) 489
MILCHGANG (*m*): lacteal duct 490
MILCHGEBEND: lacteal, yielding milk 491
MILCHGEFÄSS (*n*): lacteal (duct, vein) vessel; any vessel or container for milk 492
MILCHGESCHWULST (*f*): lacteous tumour 493
MILCHGLAS (*n*): milk (milky; breast) glass; opal glass (contains calcium phosphate, tin oxide or white arsenic) 494
MILCHGRAD (*m*): lactometric (galactometric) degree (of milk) 495
MILCHGÜTE (*f*): quality of milk 496
MILCHHARNEN (*n*): chyluria (the emission of chyle with the urine) (Med.) 497
MILCHHARNFLUSS (*m*): (see **MILCHHARNEN**) 498
MILCHICHT: milky; lacteal 499
MILCHIG: milky; lacteal; lacteous; emulsive; resembling milk 500
MILCHKANAL (*m*): lacteal duct (Med.) 501
MILCHKONDENSATIONSANLAGE (*f*): milk condensing plant 502
MILCHKONSERVIERUNG (*f*): milk preservation (as sterilization and pasteurization) 503
MILCHKOST (*f*): milk diet (Med.) 504
MILCHKÜGELCHEN (*n*): milk-(fat) globule 505
MILCH; KUNST- (*f*): artificial milk 506
MILCHKUR (*f*): milk cure 507
MILCHLAB (*n*): rennet (an enzyme) 508
MILCHMALTYL (*n*): milk maltyl 509
MILCHMESSER (*m*): galactometer; lactometer; milk gauge 510
MILCH (*f*), **PASTEURISIERTE-**: pasteurized milk 511
MILCHPASTEURISIERUNGSAPPARAT (*m*): (milk)-pasteurizing apparatus 512
MILCHPORZELLAN (*n*): milk (porcelain) glass 513
MILCHPRÄPARAT (*n*): milk preparation (as milk extract or milk powder, etc.) 514
MILCHPROBER (*m*): galactometer; lactometer; milk gauge; milk tester 515
MILCHPRÜFER (*m*): (see **MILCHPROBER**) 516
MILCHPRÜFUNG (*f*): milk (testing) analysis 517
MILCHPULVER (*n*): milk powder 518
MILCHPULVER (*n*), **KÜNSTLICHES-**: artificial milk powder (made from casein, butyrin, milk sugar and salts) 519
MILCHPULVER (*n*), **NATÜRLICHES-**: natural milk powder (evaporated cow's milk) 520
MILCHQUARZ (*m*): milk-(y) quartz (a milky-white variety of quartz) (see **QUARZ**) 521
MILCHRAHM (*m*): cream (of milk) 522
MILCHRUHR (*f*): infantile diarrhœa (Med.) 523
MILCHSAFT (*m*): milky juice; latex (see **LATEX**) (Bot.); chyle (a milky fluid produced during digestion of food) 524
MILCHSAFTIG: chylous; like chyle; containing chyle 525
MILCHSAFTIGER HARNABGANG (*m*): chyluria, chylous urine (Med.) 526
MILCHSAUER: lactate of; combined with lactic acid 527
MILCHSÄURE (*f*): lactic acid (*Acidum lacticum*); alpha-hydroxypropionic acid; ethylidene-lactic acid; $CH_3CHOHCOOH$; Sg. 1.25; Mp. 18°C.; Bp. 119°C. 528
MILCHSÄUREBAKTERIEN (*f.pl*): lactic acid bacteria (such as *Bacillus delbrucki*, *Bacillus bulgaricus*, *Bacillus acidi lactici*, etc.) 529
MILCHSÄUREESTER (*m*): lactic ester 530
MILCHSÄUREFERMENT (*n*): lactic ferment 531
MILCHSÄUREGÄRUNG (*f*): lactic fermentation 532
MILCHSÄURE (*f*), **GÄRUNGS-**: fermentation lactic acid; alpha-oxypropionic acid (see **MILCHSÄURE**) 533

MILCHSÄURENITRIL (*n*): lactic nitrile; $CH_3.CH(OH)CN$ 534
MILCHSAURES SALZ (*n*): lactate 535
MILCHSOMATOSE (*f*): milk somatose 536
MILCHSPEISE (*f*): milk diet (Med.) 537
MILCHSTERILISATOR (*m*): milk sterilizer 538
MILCH (*f*), **STERILISIERTE-**: sterilized milk 539
MILCHSTRASSE (*f*): milky way (Astron.) 540
MILCHTRANK (*m*): posset (hot milk curdled with a liquor, such as ale or wine) (for curing coughs or colds) 541
MILCHTRANSPORTKANNE (*f*): milk churn 542
MILCHTROCKENANLAGE (*f*): milk drying plant 543
MILCHUNTERSUCHUNG (*f*): milk (examination) analysis 544
MILCHUNTERSUCHUNGSAPPARAT (*m*): apparatus for milk (examination) analysis 545
MILCHVAKUUMVERDAMPFAPPARAT (*m*): vacuum evaporating apparatus (evaporator) for milk 546
MILCHVERDAMPFANLAGE (*f*): milk evaporating plant 547
MILCHVERMEHREND: galactagog (increasing the production of milk) 548
MILCHVERTREIBEND: anti- (lactic) galactic 549
MILCHWAGE (*f*): galactometer; lactometer; milk gauge 550
MILCHWARM: lukewarm; at milk temperature 551
MILCHWASSER (*n*): whey (the thin or watery part of milk) 552
MILCHWEIN (*m*): kumiss; koumiss (a beverage from fermentation of mare's milk) 553
MILCHWEISS: milk-white 554
MILCHWIRTSCHAFT (*f*): dairy 555
MILCHWURZ (*f*): milkwort (*Polygala vulgaris*) 556
MILCHZAHN (*m*): milk-tooth 557
MILCHZUCKER (*m*): lactose; sugar of milk; milk sugar (see **ROHRZUCKER** and **LAKTOSE**); $C_{12}H_{22}O_{11}$ 558
MILCHZUCKERFABRIKATION (*f*): manufacture of lactose 559
MILCHZUCKERLÖSUNG (*f*): milk-sugar solution 560
MILCHZUCKERSÄURE (*f*): saccharolactic (saccholactic; mucic) acid; $(OH)_4C_4H_4(COOH)_2$; Mp. 213°C. 561
MILD: mild; soft; gentle; kind; charitable; temperate; tender; light; benevolent; generous 562
MILDE: (see **MILD**); (*f*): mildness; clemency; generosity; charity; benevolence 563
MILDERN: to temper; mitigate; soften; correct; alleviate; moderate; abate; soothe; mollify; tone down; assuage; lessen; allay; quell; pacify; diminish; relieve 564
MILDERUNG (*f*): mitigation; alleviation (see **MILDERN**) 565
MILDERUNGSGRÜNDE (*m.pl*): extenuating circumstances 566
MILDERUNGSMITTEL (*n*): mitigant; demulcent; corrective; emollient; lenitive 567
MILDTÄTIG: charitable; liberal; kind; gentle 568
MILITÄR (*n*): soldiers; military; army 569
MILITÄRISCH: military; soldierly 570
MILLERIT (*m*): millerite, capillary pyrites (natural nickel sulphide) (about 64% Ni); NiS; Sg. 4.6 to 5.65 (see also **NICKELSULFID**) 571
MILLIAMPERE (*n*): milli-ampere 572
MILLIARDE (*f*): milliard; a thousand millions 573
MILLIGRAMM (*n*): milligram 574
MILLIVOLT (*n*): millivolt 575
MILZ (*f*): spleen (Human); melt (Animal) 576
MILZADER (*f*): splenetic vein 577
MILZBESCHWERDE (*f*): (see **MILZKRANKHEIT**) 578
MILZBRAND (*m*): malignant pustule (Human or Animal); anthrax (Med.) 579
MILZBRANDBAZILLUS (*m*): anthrax bacillus (*Bacillus anthracis*) (Med.) 580
MILZBRANDFALL (*m*): case of anthrax 581
MILZBRANDSPORE (*f*): anthrax spore 582
MILZDRÜSE (*f*): spleen 583
MILZENTZÜNDUNG (*f*): splenitis (Med.) 584
MILZKRANK: splenetic 585
MILZKRANKHEIT (*f*): hypochondriasis, spleen (disorder of the spleen) (Med.) 586
MILZKRAUT (*n*): spleenwort, common spleenwort (*Asplenium trichomanes*) 587
MILZSTECHEN (*n*): stitch (in the side) 588
MILZSUCHT (*f*): hypochondriasis, spleen (Med.) 589
MILZSÜCHTIG: splenetic, hypochondriacal 590
MIMETESIT (*m*): mimetesite; $Pb_5O_4As_3Cl_3$; Sg. 7.2; mimetite (see **MIMETIT** and **FLOCK-ENERZ**) 591
MIMETIT (*m*): mimetite; $Pb_4(PbCl)(AsO_4)_3$; Sg. 7.2 (see also **MIMETESIT** and **FLOCK-ENERZ**) 592
MIMIKRY (*f*): mimicry 593
MIMOSARINDE (*f*): golden wattle (a bark, rich in tannic acid, which is obtained from *Acacia pycnantha*) 594
MIMOSE (*f*): mimosa, sensitive plant (*Mimosa pudica* or *M. sensitiva*); wattle, Australian wattle (*Acacia binervata* and *A. mollissina*) 595
MIMOSEEXTRAKT (*m*): mimosa extract, wattle-bark extract (from *Acacia binervata* or *A. mollissina*) 596
MIMOSENGUMMI (*n*): gum arabic (see **GUMMI ARABICUM**) 597
MIMOSENRINDE (*f*): mimosa bark, black-wattle bark (from *Acacia binervata*, *A. pycnantha* or *A. mollissina*) 598
MINDER: less; smaller; inferior; minor; lesser 599
MINDERGUT: (of)-inferior-(quality) 600
MINDERHEIT (*f*): minority; inferiority 601
MINDERJÄHRIG: minor; under age 602
MINDERJÄHRIGKEIT (*f*): minority; infancy 603
MINDERN: to diminish; lessen; decrease; reduce; abate; narrow; fall; drop; lower; relax; slacken; deduct; rebate; remit; allow (see also **MILDERN**) 604
MINDERUNG (*f*): diminution; decrease; reduction; drop; fall (see **MINDERN**) 605
MINDERWERTIG: of lower (inferior) quality or valence; low-grade; low-value 606
MINDERWERTIGER BRENNSTOFF (*m*): low-value fuel; fuel having a low calorific value 607
MINDERZAHL (*f*): minority; lesser (smaller) number 608
MINDESTBETRAG (*m*): minimum-(amount) 609
MINDESTBIETENDER (*m*): lowest bidder 610
MINDESTENS: at-(the)-least 611
MINDESTMASS (*n*): minimum-(size) 612
MINE (*f*): mine (either the explosive or the working) 613

MINENGANG (*m*): passage; shaft; sap; adit; working; gallery (of a mine) 614
MINENGAS (*n*): mine gas (see GRUBENGAS) 615
MINENGRÄBER (*m*): miner; sapper 616
MINENLEGER (*m*): mine layer (Ship) 617
MINENPULVER (*n*): blasting powder 618
MINENSCHIFF (*n*): mine layer (Ship) 619
MINENSPERRE (*f*): mine barrier, minefield 620
MINENSUCHSCHIFF (*n*): mine sweeper (Ship) 621
MINENWERFER (*m*): trench mortar 622
MINENZÜNDMASCHINE (*f*): mine priming machine 623
MINERAL (*n*): mineral; fossil 624
MINERALBESTANDTEIL (*m*): mineral (inorganic) constituent 625
MINERALBODEN (*m*): mineral soil, mineral-bearing ground 626
MINERALCHEMIE (*f*): mineral (inorganic) chemistry 627
MINERALFARBE (*f*): mineral (colour) dye 628
MINERALFARBSTOFF (*m*): mineral dyestuff 629
MINERALFARBSTOFF, KÜNSTLICHER- (*m*): artificial mineral dyestuff; chemical colours 630
MINERALFARBSTOFF, NATÜRLICHER- (*m*): natural mineral dyestuff; earth colours 631
MINERALFETT (*n*): mineral fat or grease (such as vaseline, etc.) (see VASELIN) 632
MINERALFETTWACHS (*n*): mineral tallow; Hatchettite; Ozocerite (see OZOKERIT) 633
MINERALGELB (*n*): Cassel yellow (see CASSELER-GELB); mineral yellow (see WOLFRAM-SÄURE) 634
MINERALGERBEREI (*f*): mineral tanning, tawing (see WEISSGERBEREI) 635
MINERALGERBUNG (*f*): mineral (tannage) tanning (see WEISSGERBEREI) 636
MINERALGRAU (*n*): mineral (stone, silver) grey 637
MINERALGRÜN (*n*): mineral green (see KUPFER, ARSENIGSAURES-; and KUPFERKARBONAT) 638
MINERALHÄRTE (*f*): (see NICHTCARBONATHÄRTE) 639
MINERALHEFE (*f*): mineral yeast (a yeast culture obtained by fermenting sugar in the presence of ammonium salts) 640
MINERALIEN (*n.pl*): minerals 641
MINERALIENKUNDE (*f*): mineralogy; the science of minerals 642
MINERALIENSAMMLUNG (*f*): collection of minerals 643
MINERALISCH: mineral; mineralogical 644
MINERALKERMES (*m*): Kermes mineral; antimony sulphide (see KERMESIT) 645
MINERALKUNDE (*f*): mineralogy; the science of minerals 646
MINERALLEHRE (*f*): mineralogy (the science of minerals) 647
MINERALMOHR (*m*): ethiops mineral; black mercuric sulphide (amorphous HgS); Sg. 7.6 (see also EISENMOHR and MAGNETIT) 648
MINERALMOOR (*m.* and *n*): mineral-water marsh 649
MINERALMÜHLE (*f*): mineral mill 650
MINERAL (*n*), NIOBIUMHALTIGES-: mineral containing (columbium) niobium, niobium (columbium) mineral 651
MINERALOG (*m*): mineralogist 652
MINERALOGIE (*f*): mineralogy; the science of minerals 653
MINERALOGISCH: mineralogical 654
MINERALÖL (*n*): mineral (rock) oil (such as petroleum, see BERGÖL) (also various kinds of machine and cylinder oils) 655
MINERALÖLFIRNIS (*m*): mineral oil varnish 656
MINERALÖLRÜCKSTAND (*m*): mineral (rock) oil residue 657
MINERALPECH (*n*): mineral pitch; asphalt (see ASPHALT) 658
MINERALREICH (*n*): mineral kingdom; (as an adjective) rich in minerals 659
MINERALSALZ (*n*): mineral salt 660
MINERALSÄURE (*f*): mineral acid 661
MINERALSCHWARZ (*n*): mineral black (graphitic slate, or ground graphite) 662
MINERALTALG (*m*): mineral tallow; Hatchettite; Ozocerite (see OZOKERIT) 663
MINERALTRENNUNG (*f*): mineral separation 664
MINERALTURPENTINÖL (*n*): mineral turpentine oil 665
MINERALTURPETH (*m*): turpeth mineral (a basic mercuric sulphate) 666
MINERALVIOLETT (*n*): (see MANGANVIOLETT) 667
MINERALWACHS (*n*): mineral wax; Ozokerite; Ozocerite (see OZOKERIT); Ceresin (see ZERESIN) 668
MINERALWASSER (*n*): mineral (aerated) water; soda-water (as Apollinaris, Selters, Hunyadi János, etc.) 669
MINERALWASSERAPPARAT (*m*): mineral-water apparatus; soda fountain 670
MINERALWASSERSALZ (*n*): mineral-water salt 671
MINERBLAU (*n*): mineral blue (see KUPFER-KARBONAT; FERRIFERROCYANID and WOL-FRAMBLAU) 672
MINERKUNDE (*f*): mineralogy; the science of minerals 673
MINETTE (*f*): oolitic iron ore, minette (variety of BRAUNEISENSTEIN, which see) [high phosphorous content, and medium iron content, latter about 28-39%, obtained in Luxemburg and the neighbouring portion of France (Briey)] 674
MINIATUR (*f*) -BILD (*n*): miniature 675
MINIEREN: to mine; sap; undermine 676
MINIERWERKZEUG (*n*): miner's tools 677
MINIMAL- (as a prefix): minimum- 678
MINIMALBETRAG (*m*): minimum; lowest (rate) amount 679
MINIMALDRUCK (*m*): minimum pressure 680
MINIMALSPANNUNG (*f*): minimum (tension) pressure 681
MINIME (*f*): minim (Mus.) 682
MINIMUM (*n*): minimum 683
MINISTER (*m*): minister 684
MINISTERIELL: ministerial 685
MINISTERIUM (*n*): ministry; administration; government; clergy 686
MINISTRIEREN: to minister 687
MINIUM (*n*): minium (see MENNIGE) 688
MINUS: minus; (*n*): minus 689
MINUSBETRAG (*m*): minus quantity; discount (Exchange) 690
MINUSELEKTRIZITÄT (*f*): negative electricity 691
MINUSZEICHEN (*n*): minus sign (–) 692
MINUTE (*f*): minute 693
MINUTENZEIGER (*m*): minute hand (of watches) 694
MINZE (*f*): mint (*Mentha*) 695
MIRABILIT (*m*): mirabilite (see GLAUBERSALZ) 696
MIRBANESSENZ (*f*): essence of mirbane; nitrobenzene; $C_6H_5NO_2$ (see NITROBENZOL) 697

MIRBANÖL (*n*): mirbane oil (see MIRBANESSENZ) 698

MIRRHE (*f*): myrrh (see MYRRHE) 699

MIRTE (*f*): myrtle (see MYRTE) 700

MISANTHROPISCH: misanthropic (see also MANN-SCHEU) 701

MISBEFINDEN (*n*): indisposition, malaise (Med.) 702

MISCHBAR: miscible; mixable; capable of being mixed or compounded 703

MISCHBARKEIT (*f*): miscibility; mixability 704

MISCHEN: to mix; compound; mingle; interfere; blend; adulterate; fettle (Metal); shuffle (Cards); alloy (Metal); combine (Chem.); meddle; interpose 705

MISCHER (*m*): mixer 706

MISCHFARBE (*f*): mixed (combination) colour 707

MISCHFLASCHE (*f*): mixing (bottle) flask 708

MISCHFUTTER (*n*): mixed (fodder) feed 709

MISCHGAS (*n*): mixed gas; semi-watergas; Dowson gas; Mond gas (see MONDGAS) 710

MISCHGEFÄSS (*n*): mixing vessel 711

MISCHGESCHLÄCHT (*n*): half-breed 712

MISCHGLAS (*n*): mixing glass 713

MISCHHEFE (*f*): mixed (composite) yeast 714

MISCHINFEKTION (*f*): mixed infection 715

MISCHKRISTALL (*m*): mixed crystal; crystallized solid solution (Metal smelting) 716

MISCHLING (*m*): hybrid; cross-breed; mongrel 717

MISCHMASCH (*m*): medley; mixture; mess 718

MISCHMASCHINE (*f*): mixing (machine) mill; mixer 719

MISCHMETALL (*n*): alloy; mixed metal (Cerium-iron pyrophoric alloy) 720

MISCHMÜHLE (*f*): mixing mill; mixer 721

MISCHROHR (*n*): mixing tube 722

MISCHSÄURE (*f*): mixed acid (a mixture of H_2SO_4 and HNO_3 for nitration) 723

MISCHTROMMEL (*f*): mixing drum; drum mixer 724

MISCHUNG (*f*): mixture; mixing; composition; compounding; compound; blend; combination; alloy; combining 725

MISCHUNGSGEWICHT (*n*): combining weight 726

MISCHUNGSHAHN (*m*): mixing (tap) cock 727

MISCHUNGSKOHLE (*f*): mixing coal (Soda) 728

MISCHUNGSLÜCKE (*f*): miscibility (mixability) gap 729

MISCHUNGSRECHNUNG (*f*): alligation (Maths.) 730

MISCHUNGSREGEL (*f*): rule (law) of mixtures 731

MISCHUNGSVERHÄLTNIS (*n*): composition of a mixture, ratio of the ingredients in a mixture, mixing proportion 732

MISCHUNGSVERWANDSCHAFT (*f*): (chemical)-affinity 733

MISCHUNGSWÄRME (*f*): mixing (mixture) heat 734

MISCHVENTIL (*n*): mixing valve 735

MISCHVORRICHTUNG (*f*): mixing (apparatus, plant, arrangement) device 736

MISCHWALZWERK (*n*): mixing rollers 737

MISCHZYLINDER (*m*): mixing cylinder 738

MISPEL (*f*): medlar (*Mespilus germanica*) 739

MISPICKEL (*m*): arsenopyrite, mispickel (see ARSENKIES) 740

MISSACHTEN: to disregard; disdain; despise; undervalue; slight; neglect; disrespect **741**

MISSBEGRIFF (*m*): misconception, misunderstanding 742

MISSBEHAGEN: to (be) displease-(d); be inconvenient; dislike; be disagreeable **743**

MISSBILDUNG (*f*): deformity; malformation; misshapenness; monstrosity (Med.) 744

MISSBILDUNG (*f*) DES HERZENS: malformation of the heart, congenital heart disease (Med.) **745**

MISSBILLIGEN: to oppose; disapprove of; disallow; repudiate; condemn 746

MISSBILLIGUNG (*f*): disapprobation 747

MISSBRAUCH (*m*): misuse; abuse 748

MISSBRAUCHEN: to take in vain; trespass upon; abuse; misuse **749**

MISSDEUTEN: to misconstrue; misunderstand; misinterpret 750

MISSEN: to miss; dispense with; want; lack 751

MISSERFOLG (*m*): failure 752

MISSERNTE (*f*): crop failure; bad (crop) harvest 753

MISSETAT (*f*): crime; misdeed; offence 754

MISSETÄTER (*m*): criminal; malefactor; wrongdoer 755

MISSFALLEN: to be displeasing; dislike; (*n*): displeasure 756

MISSFÄLLIG: disagreeable; unpleasant; displeasing; offensive 757

MISSFARBIG: discoloured 758

MISSFÄRBUNG (*f*): discolouration 759

MISSGEBILDE (*n*): monster, monstrosity (Med.) 760

MISSGEBURT (*f*): miscarriage; abortion 761

MISSGESCHICK (*n*): mishap; misfortune; disaster; accident; fatality 762

MISSGESTALTET: deformed; misshapen 763

MISSGLÜCKEN: to miscarry, fail, be attended with ill success 764

MISSGRIFF (*m*): blunder; mistake 765

MISSHANDELN: to maltreat; illtreat; abuse 766

MISSHANDLUNG (*f*): maltreatment; cruelty; abuse; illtreatment; illusage 767

MISSKLANG (*m*): discord, dissonance 768

MISSLICH: embarrassing; doubtful; precarious; dubious; critical; uncertain; difficult; delicate; disagreeable 769

MISSLICHKEIT (*f*): dubiety; dubiousness; doubtfulness 770

MISSLINGEN: to miscarry; fail; be attended with ill success 771

MISSMUTIG: ill-humoured; discontented; sad; melancholy; peevish; despondent; dejected; cross; discouraged 772

MISSPICKEL (*m*): mispickel (see ARSENKIES) 773

MISSRATEN: to miscarry; fail; guess wrongly; give (bad) wrong advice 774

MISSSTAND (*m*): impasse; inconvenient (critical, embarrassing)-position or condition; inconvenience, impropriety, embarrassment 775

MISSSTIMMUNG (*f*): ill-humour; peevishness; discord; discontentment 776

MISSTON (*m*): discord; dissonance 777

MISSTRAUEN: to mistrust; distrust; suspect; (*n*): suspicion; distrust 778

MISSTRAUISCH: distrustful-(ly), suspicious-(ly) 779

MISSTRITT (*m*): fault; error 780

MISSVERHÄLTNIS (*n*): asymmetry; disproportion; disparity; inadequacy; disagreement **781**

MISSVERSTÄNDNIS (*n*): misunderstanding; difference; misconception; misapprehension; mistake **782**

MISSVERSTEHEN: to misunderstand; mistake; misinterpret 783
MISSWACHS (*m*): crop failure; deformed growth; deformation; malformation 784
MIST (*m*): manure; dung; midden; mist; fog 785
MISTBEET (*n*): hotbed (of manure) 786
MISTBEIZE (*f*): dung bate (Leather) 787
MISTEL (*f*): mistletoe (various varieties suit themselves to various trees and will not interchange) (*Viscum quercinum* or *Viscum album*) (see also RIEMENBLUME) 788
MISTELDROSSEL (*f*): missel thrush; mistletoe thrush (*Turdus merula*) 789
MISTELKRAUT (*n*): mistletoe (*Viscum quercinum*) (see MISTEL) 790
MISTEN: to manure; clean (Stables) 791
MISTGRUBE (*f*): midden; pit (hole) for dung 792
MISTIG: foggy, misty, cloudy, clouded 793
MISTJAUCHE (*f*): liquid manure; sewage; dung water 794
MISTKÄFER (*m*): scarab; dor-beetle, dung-beetle (*Geotrupes stercorarius*) 795
MISTRAL (*m*): (see MAGISTRAL) 796
MIT: with; by; at; in; upon; under; together; also; jointly; simultaneously; (as a prefix = con-; com-; secondary; auxiliary; co-; counter-; contributory; joint; additional; fellow; simultaneous) 797
MITARBEITEN: to collaborate; co-operate; assist 798
MITARBEITER (*m*): fellow worker; collaborator; contributor; assistant; colleague 799
MITBEGREIFEN: to include 800
MITBESITZ (*m*): joint (tenure) tenancy; joint possession; common property; joint ownership 801
MITBEWERBER (*m*): competitor; rival 802
MITBRINGEN: to bring (carry) along with 803
MITBÜRGER (*m*): fellow citizen 804
MITDASEIN (*n*): co-existence 805
MITESSER (*m.pl*): comedones, blackheads (Med.) 806
MITFÜHREN: to carry (take, lead) along with 807
MITGEFÜHL (*n*): sympathy 808
MITGEHILFE (*m*): colleague; collaborator; assistant; associate; accomplice 809
MITGENOSSENSCHAFT (*f*): co-partnership 810
MITGIFT (*f*): dowry 811
MITGLIED (*n*): member; fellow 812
MITGLIEDSCHAFT (*f*): membership; fellowship 813
MITHIN: therefore; of course; consequently 814
MITHÜLFE (*f*): support, aid, help, assistance, co-operation 815
MITINHABER (*m*): co-partner; co-proprietor 816
MITISGRÜN (*n*): (see SCHWEINFURTGRÜN) 817
MITLAUFEN: to run with, act in common with 818
MITLAUT (*m*): consonant 819
MITLEBEND: coeval, contemporaneous, contemporary 820
MITLEID (*n*): compassion; pity; sympathy 821
MITLEIDEN: to suffer with; sympathize; have (compassion) pity 822
MITLEIDSBEZEUGUNG (*f*): condolence 823
MITMACHEN: to take part (join) in; conform to 824
MITMEHRER (*m*): co-efficient 825
MITMENSCH (*m*): fellow-creature 826
MITNEHMEN: to take (with) along; take; share; weaken; wear; exhaust; maltreat; criticize; catch 827
MITNEHMER (*m*): carrier; driver; tappet; catch; gripper; engaging piece; dog; clutch; nose; etc. (something that carries something else along with it) (Mech.) 828
MITPÄCHTER (*m*): joint (lessee) tenant 829
MITRECHNEN: to reckon (calculate) upon, count in, add in, include in a calculation 830
MITREEDER (*m*): part-owner (of a ship) 831
MITREISSEN: to carry (draw) over; carry down; involve; precipitate; entrain; carry along; prime (Boilers); (*n*): priming (Boilers) 832
MITRHEDER (*m*): (see MITREEDER) 833
MITSCHULD (*f*): complicity 834
MITSCHULDIGE (*m*): accessory; accomplice 835
MITSCHWINGEN: to co-vibrate; vibrate (oscillate, swing) with or simultaneously 836
MITSEIN: to be co-existent with; be (out) gone; be with; be in the company of 837
MITSPRECHEN: to speak with; to have a say (in the matter) 838
MITSTIMMEN: to agree with; cast one's vote for 839
MITTAG (*m*): noon; midday; south; meridian 840
MITTÄGIG: southern; noon; midday; meridional 841
MITTÄGLICH: (see MITTÄGIG) 842
MITTAGSBLUME (*f*): Hottentot fig (*Mesembryanthemum edule*) 843
MITTAGSESSEN (*n*): midday meal; dinner 844
MITTAGSGEGEND (*f*): southern region 845
MITTAGSKREIS (*m*): meridian 846
MITTAGSLÄNGE (*f*): meridional longitude 847
MITTAGSLINIE(*f*): meridian-(line) 848
MITTAGSSTUNDE (*f*): noontide; dinner-hour 849
MITTE (*f*): middle; average; midst; centre; medium; mean; (used as prefix = middle or centre); (used as suffix = middle or centre of) 850
MITTEILBAR: contagious; communicable 851
MITTEILEN: to communicate-(with); inform; impart; give-(notice); advise 852
MITTEILEND: communicative 853
MITTEILUNG (*f*): communication; information; notice; intelligence; advice 854
MITTEL (*n*): midst; middle; mean; means; expedient; medium; average; mediation; remedy; medicine; wealth; ratio; vehicle (for drugs); mass (Mining); centre; (as a prefix = middle, moderate; central; mean; average; median; medium; intermediate; middling; meso-; mezzo-) 855
MITTELALTER (*n*): middle-age-(s) 856
MITTELALTERLICH: medieval 857
MITTEL (*n*), **ARITHMETISCHES-**: arithmetical mean 858
MITTELART (*f*): intermediate (kind, sort) variety 859
MITTELBAR: mediate; indirect; collateral-(ly) 860
MITTELBAUCHGEGEND (*f*): mesogastric region 861
MITTEL (*n*), **BERUHIGENDES-**: sedative, depressant, anodyne (Med.) 862
MITTEL (*n* and *pl*) **BILDEN**: to average, take (form) the average of 863
MITTELDECK (*n*): middle deck 864
MITTEL (*n*), **DICHTES-**: dense medium 865

MITTELDRUCK (*m*) : mean (average) pressure 866
MITTELDRUCKTURBINE (*f*) : intermediate pressure turbine 867
MITTELDRUCKZYLINDER (*m*) : intermediate pressure cylinder 868
MITTEL (*n*), **DÜNNES-** : rare medium 869
MITTELECK (*n*) : lateral summit (Cryst.) 870
MITTELECKE (*f*) : (see **MITTELECK**) 871
MITTELFARBE (*f*) : intermediate (secondary) colour 872
MITTELFEIN : medium fine 873
MITTELFEIN PULVERISIERT : medium finely pulverized 874
MITTELFELL (*n*) : mediastinum 875
MITTELFINGER (*m*) : middle finger 876
MITTELFLEISCH (*n*) : perineum (Anat.) 877
MITTELFUSS (*m*) : instep ; metatarsus (Anat.) 878
MITTELGEBIRGE (*n*) : hills, highlands 879
MITTEL (*n*), **GEOMETRISCHES-** : geometrical mean 880
MITTELGLIED (*n*) : intermediate (term) member (of a series) ; middle phalanx (Anat.) 881
MITTELGRAD (*m*) : comparative degree 882
MITTELGROSS : medium (middle) sized 883
MITTELGRÖSSE (*f*) : medium (average) size 884
MITTELGUT : second (rate) quality 885
MITTELGUT (*n*) : medium quality (wares, articles) material ; medium quality (value) coal (obtained from coal with stone content after breaking and washing, and has 20-30% ash ; 20-30% waste. That got from mines supplying rich coal has approx. 10% moisture ; 15-20% ash ; 16% volatile ; calorific value about 5650 cals.) 886
MITTELHAND (*f*) : palm ; metacarpus (Anat.) 887
MITTELHARTER STAHL (*m*) : medium hard steel 888
MITTELHAUT (*f*) : mesocarp (the pulp of fruits, lying between the skin and the stone) (Bot.) 889
MITTELKIEL (*m*) : centre keel, middle keel 890
MITTELKIEL (*m*), **DURCHLAUFENDER-** : centre-through-plate keel 891
MITTELKIELPLATTE (*f*) : centre-through plate 892
MITTELKIELSCHWEIN (*n*) : centre keelson 893
MITTELKLÜVER (*m*) : middle jib 894
MITTELKREIS (*m*) : equator 895
MITTELLAGE (*f*) : middle (intermediate) position 896
MITTELLANG : of medium length 897
MITTELLAUGE (*f*) : medium strong (weak) lye 898
MITTELLINIE (*f*) : median line ; axis ; equator ; centre line ; middle line 899
MITTELLOS : without means 900
MITTELMANN (*m*) : middleman (Commercial) 901
MITTELMÄSSIG : middling ; mediocre ; average ; moderate ; ordinary ; indifferent 902
MITTELMÄSSIGKEIT (*f*) : mediocrity 903
MITTELMAST (*m*) : main-mast (Naut.) 904
MITTELMEER (*n*) : Mediterranean sea 905
MITTELMEHL (*n*) : pollard 906
MITTELÖL (*n*) : middle oil (fractional distillation) 907
MITTELPARTEI (*f*) : Centre party, moderate party 908
MITTELPREIS (*m*) : average (mean) price 909
MITTELPUNKT (*m*) : centre-(point) ; focus ; cynosure 910
MITTELPUNKTSPROJEKTION (*f*) : gnomonic projection 911
MITTELRAUM (*m*) : interval ; intermediate space or chamber 912
MITTELS : by means of ; through 913
MITTELSALZ (*n*) : neutral salt 914
MITTELSCHLÄCHTIG : middle-shot (of water-wheels) 915
MITTELSCHLAG (*m*) : middle size ; intermediate (kind, sort) variety 916
MITTELSCHWER : medium heavy 917
MITTELSCHWERT (*n*) : centre board 918
MITTELSMANN (*m*) : intercessor ; mediator ; umpire ; referee ; third party 919
MITTELSPANT (*n*) : midship frame 920
MITTELST : by means of 921
MITTELSTAND (*m*) : middle class 922
MITTELSTÄNDIG : centre ; middle ; intermediate ; occupying a middle position 923
MITTELSTELLE (*f*) : intermediate (place) position 924
MITTELSTRASSE (*f*) : middle (course) way 925
MITTELSTUFE (*f*) : intermediate (degree) stage 926
MITTELTRÄGER (*m*) : centre girder, centre beam, or support 927
MITTELWAND (*f*) : partition-(wall) 928
MITTEL (*n*), **WASSERENTZIEHENDES-** : dehydrating agent 929
MITTELWELLE (*f*) : middle shaft 930
MITTELWERT (*m*) : mean (average) value 931
MITTELWORT (*n*) : participle 932
MITTELZAHL (*f*) : mean ; average 933
MITTELZEIT (*f*) : mean time 934
MITTELZEUG (*n*) : middle (second) stuff (Paper) 935
MITTELZUSTAND (*m*) : average (intermediate) state 936
MITTEN : in the (midst) middle ; amidst ; midway-(between) 937
MITTERNACHT (*f*) : midnight ; north 938
MITTERNÄCHTIG : midnight ; northern ; nocturnal 939
MITTLER : (used as a prefix—see **MITTEL** used as a prefix) ; (*m*) : intercessor ; mediator (see **MITTELSMANN**) 940
MITTLERWEILE : meanwhile ; in the meantime 941
MITTÖNEN : to resound (sound, ring) at the same time or simultaneously 942
MITTSCHIFFS : amidships 943
MITTSOMMER (*m*) : midsummer 944
MITUNTER : occasionally ; at times ; sometimes 945
MITURSACHE (*f*) : contributory cause 946
MITWELT (*f*) : contemporary ; present age 947
MITWERBEN : to rival ; compete 948
MITWIRKEN : to co-operate ; assist ; take part ; help ; contribute ; concur ; collaborate 949
MITWIRKEND : co-operative ; subsidiary ; contributory ; collateral 950
MITWIRKENDE FAKTOREN (*m.pl*) : contributory factors 951
MITWIRKUNG (*f*) : co-operation ; help ; assistance ; concurrence 952
MITWISSEN (*n*) : knowledge ; joint knowledge ; cognisance ; cognition 953
MITZÄHLEN : to include, reckon as a part, take into account, add in 954
MITZIEHEN : to draw (drag, carry, pull) along with 955
MIXTUR (*f*) : mixture 956
MNEMONIK (*f*) : mnemonics (memory cultivation) 957
MÖBEL (*n*) : furniture (either as a single piece or as a whole) 958

MÖBELHÄNDLER (*m*): furniture dealer 959
MÖBELLACK (*m*): cabinet (furniture) varnish 960
MÖBELPOLITUR (*f*): furniture polish 961
MÖBELTISCHLER (*m*): cabinet maker 962
MÖBELTISCHLEREI (*f*): cabinet (furniture) making; cabinet making works 963
MÖBELTRANSPORT (*m*): furniture removal 964
MÖBELWAGEN (*m*): furniture van 965
MOBIL: mobile; active; quick; nimble; moveable 966
MOBILIEN (*n.pl*): moveables; furniture 967
MOBILISIEREN: to mobilize, collect together 968
MÖBLIEREN: to furnish; fit up 969
MOCK (*m*): German steel (obtained by directly refining cast iron) 970
MOCKSTAHL (*m*): (see **MOCK**) 971
MODALITÄT (*f*): mode, manner 972
MODE (*f*): mode; vogue; style; fashion; custom; (as a prefix—modish; stylish; fancy; fashionable) 973
MODEARTIKEL (*m* and *pl*): fancy (goods) articles 974
MODEL (*m*): modulus; mould; form; pattern; matrix 975
MODELSCHNEIDER (*m*): mould cutter; moulder; pattern maker 976
MODELL (*n*): model; pattern; mould; form; sample; type; sort; kind; sitter (Photography) 977
MODELLACK (*m*): model varnish 978
MODELLIEREN: to model; mould; fashion; form 979
MODELLIERER (*m*): pattern maker; modeller 980
MODELLIERHOLZ (*n*): modelling tool (made of wood) (Art) 981
MODELLIERMASSE (*f*): modelling clay (plasticine, etc.); plastic mass (composed of clay, sand, magnesia and magnesium chloride with some borax) 982
MODELLIERTON (*m*): modelling clay 983
MODELLIERWACHS (*n*): modelling wax 984
MODELLMACHER (*m*): pattern (model) maker 985
MODELLTISCHLER (*m*): model maker, pattern maker 986
MODELLTISCHLEREI (*f*): model (pattern) making, pattern shop 987
MODELLZEICHNUNG (*f*): model drawing, drawing from a model (Art) 988
MODELN: to model; mould; figure-(cotton goods, etc.) 989
MODER (*m*): mould; putridity; decay; mustiness 990
MODERDUFT (*m*): mouldy (musty) smell 991
MODERERDE (*f*): mould; compost (see **KOMPOST**) 992
MODERHAFT: mouldy; musty; moulded; mouldering; putrid; decaying; rotten 993
MODERIG: (see **MODERHAFT**) 994
MODERN: to moulder; putrefy; rot; decompose; decay; (as an adjective—modern; fashionable) 995
MODERNISIEREN: to modernize, bring up to date 996
MODEWAREN (*f.pl*): fancy goods 997
MODEZEITUNG (*f*) fashion paper 998
MODIFIKATION (*f*): modification 999
MODIFIZIEREN: to modify; qualify 000
MODISCH: fashionable; modish; stylish 001
MODUL (*m*): modulus 002
MODULIEREN: to modulate 003
MODUS (*m*): mood (Grammar) 004
MOËLLON (*n*): moellon (see **DEGRAS**) 005
MÖGEN: to be able; may; like; choose; prefer; have a mind to; desire; wish 006
MÖGLICH: possible; potential; feasible; practicable 007
MÖGLICHENFALLS: possibly; if possible 008
MÖGLICHERWEISE: as far as (possible) practicable; perhaps, possibly 009
MÖGLICHKEIT (*f*): possibility; practicability; feasibility 010
MÖGLICHKEITSFÄLLE (*m.pl*): contingencies 011
MÖGLICHST: as much as possible; utmost 012
MOHN (*m*): common poppy (*Papaver rhoeas*); opium poppy (*Papaver somniferum*) 013
MOHNARTIG: papaverous; poppy-like 014
MOHNBLUME (*f*): opium poppy (*Papaver somniferum*) 015
MOHNKOPF (*m*): poppy (head, capsule) pod (*Fructus papaveris*) 016
MOHNÖL (*n*): poppy-(seed)-oil (*Oleum papaveris*); Sg. 0.92-0.93 (from *Papaver somniferum*) 017
MOHNSAAT (*f*): poppy seed 018
MOHNSAFT (*m*): poppy juice; opium 019
MOHNSAMEN (*m*): poppy seed 020
MOHNSÄURE (*f*): meconic acid; $OHC_5HO_2(COOH)_2.3H_2O$ 021
MOHNSIRUP (*m*): poppy syrup; poppy sirup 022
MOHNSTOFF (*m*): narcotine; $C_{19}H_{14}NO_4(OCH_3)_3$ (one of the papaverin group of isoquinoline derivates of opium; an opium alkaloid) 023
MOHR (*m*): black; ethiops (see **EISENMOHR** and **MAGNETIT**); watered fabric; moire; moreen; watered (stuff) effect; clouded effect; moor; negro; mohair 024
MOHRBAND (*n*): watered ribbon 025
MÖHRE (*f*): carrot (*Daucus carota*) 026
MOHRENFLECHTE (*f*): plica polonica, Polish ringworm (a scalp disease caused by lice and filth) (Med.) 027
MÖHRENSAFT (*m*): carrot juice 028
MOHRISCH: moorish 029
MOHRRÜBE (*f*): carrot (*Daucus carota*) 030
MOHRSCHES SALZ (*n*): Mohr's salt (ferrous ammonium sulphate) (see **AMMONIUMFERRO-SULFAT**) 031
MOHSSCHE SKALA (*f*): Mohs' scale (a scale of hardnesses for metal) 032
MOIRIEREN: to water; cloud (Fabrics) 033
MOISSANIT (*m*): moissanite; CSi 034
MOL (*n*): mole; mol; gramme-molecule (the number of grammes of a substance corresponding to its molecular weight); (as a prefix = molar) 035
MOLASSE (*f*): molasse, tertiary sandstone 036
MOLASSE-KOHLE (*f*): molasse coal (obtained from left bank of Rhine to the Alps from rocks of Molasse-formation) (70/77%C.; 5/5½% H.; 23.5/18% O.; 3-9% S.) 037
MOLCH (*m*): salamander; spiteful person 038
MOLE (*f*): mole; mol (see also **MOL**) 039
MOLEKEL (*n*): molecule 040
MOLEKUL (*n*): (see **MOLEKEL**) 041
MOLEKÜL (*n*): (see **MOLEKEL**) 042
MOLEKULAR: molecular 043
MOLEKULARANZIEHUNG (*f*): molecular attraction 044
MOLEKULARBEWEGUNG (*f*): molecular (movement) motion 045
MOLEKULARDRUCK (*m*): molecular pressure 046
MOLEKULARERNIEDRIGUNG (*f*): molecular reduction 047
MOLEKULARFORMEL (*f*): molecular formula 048
MOLEKULARGEWICHT (*n*): molecular weight 049

MOLEKULARGEWICHTSBESTIMMUNG (*f*): determination of molecular weight 050
MOLEKULARGRÖSSE (*f*): molecular (weight) magnitude 051
MOLEKULARREFRAKTION (*f*): molecular refraction 052
MOLEKULARROTATION (*f*): molecular rotation (see DREHUNGSVERMÖGEN, MOLEKULARES-) 053
MOLEKULARSTROM (*m*): molecular current 054
MOLEKULARWÄRME (*f*): molecular heat 055
MOLEKULARWIRKUNG (*f*): molecular (effect) action 056
MOLEKULARZUSAMMENSETZUNG (*f*): molecular composition 057
MOLEKULARZUSTAND (*m*): molecular (condition) state 058
MOLEKÜL, GRAMM- (*n*): gramme-molecule (see MOL) 059
MOLEKÜLVERBINDUNG (*f*): molecular compound 060
MOLENBRUCH (*m*): molar fraction 061
MOLETTE (*f*): small, unhardened steel roll (Calico printing) 062
MOLGEWICHT (*n*): molar weight 063
MOLGRÖSSE (*f*): molar magnitude 064
MOLKEN (*f* and *pl*): whey 065
MOLKENTRANK (*m*): (see MILCHTRANK) 066
MOLKENWESEN (*n*): dairy (work) farming 067
MOLKENWIRTSCHAFT (*f*): dairy 068
MOLKEREI (*f*): dairy 069
MOLKEREIGERÄTE (*n.pl*): dairy (apparatus, plant) appliances 070
MOLKEREIMASCHINEN (*f.pl*): dairy machinery 071
MOLKICHT: wheyey; like whey; containing (resembling) whey 072
MOLKIG: wheyey; wheyish; like whey; resembling whey 073
MOLL: minor (Music) 074
MÖLLER (*m*): ore plus limestone or other flux (fed (charged) to blast furnace) 075
MÖLLERN: to mix (ores and fluxes) (Metal) 076
MÖLLERUNG (*f*): mixture; charge (of ores and fluxes) (Metal) 077
MOLLETIEREN: to mill (coins, etc.) 078
MOLLIG: pleasant; cosy; snug; soft; plump (Leather) 079
MOLLIGEN (*m*): glycerine substitute for textile and paper industries 080
MOLLOSE (*f*): (see MOLLIGEN) 081
MOLLTONART (*f*): minor key (Music) 082
MOLLTONLEITER (*f*): minor scale (Music) 083
MOLLUSKE (*f*): mollusk; mollusc 084
MOLVERHÄLTNIS (*n*): molar ratio 085
MOLVOLUM (*n*): molar volume 086
MOLYBDÄN (*n*): molybdenum (Mo) 087
MOLYBDÄNBLAU (*n*): molybdenum blue 088
MOLYBDÄNBLEIERZ (*n*): Wulfenite (natural lead molybdate); $PbMoO_4$; Sg. 6.7-7 089
MOLYBDÄNBLEISPAT (*m*): Wulfenite (see MOLYB-DÄNBLEIERZ) 090
MOLYBDÄNBORID (*n*): molybdenum boride; Mo_3B_4; Sg. 7.11 091
MOLYBDÄNCARBID (*n*): molybdenum carbide; Mo_2C; Sg. 8.9; MoC; Sg. 8.4 092
MOLYBDÄNCARBONYL (*n*): molybdenum carbonyl; $Mo(CO)_6$; Sg. 1.96 093
MOLYBDÄNDIJODID (*n*): molybdenum diiodide; MoI_2; Sg. 4.3 094
MOLYBDÄNDIOXYDIFLUORID (*n*): molybdenum dioxydifluoride; MoO_2F_2; Sg. 3.494 [095
MOLYBDÄNDISULFID (*n*): molybdenum disulphide; MoS_2; Sg. 4.6 096
MOLYBDÄNDRAHT (*m*): molybdenum wire 097
MOLYBDÄNEISEN (*n*): (see FERROMOLYBDÄN) 098
MOLYBDÄNFADEN (*m*): molybdenum filament 099
MOLYBDÄNGLANZ (*m*): molybdenite (natural molybdenum sulphide); molybdenum glance; MoS_2; Sg. 4.6-4.9 (about 60% Mo) 100
MOLYBDÄNIT (*m*): (see MOLYBDÄNGLANZ) 101
MOLYBDÄNKARBID (*n*): (see MOLYBDÄNCARBID) 102
MOLYBDÄNKIES (*m*): (see MOLYBDÄNGLANZ) 103
MOLYBDÄNLAMPE (*f*): molybdenum filament lamp 104
MOLYBDÄNLEGIERUNG (*f*): molybdenum alloy 105
MOLYBDÄNMETALL (*n*): molybdenum metal 106
MOLYBDÄNOCHER (*m*): molybdite; molybdic ochre; MoO_3; Sg. 4.0-4.5 107
MOLYBDÄNOCKER (*m*): (see MOLYBDÄNOCHER) 108
MOLYBDÄNOXYD (*n*): molybdenum oxide; molybdite (see MOLYBDÄNOCHER) 109
MOLYBDÄNOXYTETRAFLUORID (*n*): molybdenum oxytetrafluoride; $MoOF_4$; Sg. 3.001 110
MOLYBDÄNPRÄPARAT (*n*): molybdenum preparation 111
MOLYBDÄNSALZ (*n*): molybdenum salt 112
MOLYBDÄNSAUER: molybdate of; combined with molybdic acid 113
MOLYBDÄNSÄURE (*f*): molybdic acid (*Acidum molybdaenicum*); $H_2MoO_4 + H_2O$; Sg. 3.1; molybdite (see MOLYBDÄNOCHER) 114
MOLYBDÄNSÄUREANHYDRID (*n*): molybdic anhydride (see MOLYBDÄNTRIOXYD) 115
MOLYBDANSÄUREHYDRAT (*n*): molybdic acid (see MOLYBDÄNSÄURE) 116
MOLYBDÄNSESQUISULFID (*n*): molybdenum sesquisulphide; Mo_2S_3; Sg. 5.9 117
MOLYBDÄNSILBER (*n*): tellurwismuth 118
MOLYBDÄNSILICID (*n*): molybdenum silicide; $MoSi_2$; Sg. 5.88-6.2 119
MOLYBDÄNSTAHL (*m*): molybdenum steel 120
MOLYBDÄNSULFID (*n*): molybdenum sulphide; MoS_2; Sg. 4.6 121
MOLYBDÄNSULFIDFREI: free from molybdenum sulphide 122
MOLYBDÄNTRIOXYD (*n*): molybdenum trioxide; MoO_3; Sg. 4.39 123
MOLYBDÄNVERBINDUNG (*f*): molybdenum compound 124
MOLYBDAT (*n*): molybdate 125
MOLYBDIT (*m*): (see MOLYBDÄNOCHER) 126
MOLZUSTAND (*m*): molecular (molar) state or condition 127
MOMENT (*n*): moment, impetus, momentum; instant 128
MOMENTAN: momentary; momentarily; instantaneously; immediately 129
MOMENTAUFNAHME (*f*): snap-shot, instantaneous (exposure) photograph 130
MOMENT (*n*), **AUFRICHTENDES-**: righting moment (Ships) 131
MOMENTBILD (*n*): instantaneous photograph; snap-shot (Photography) 132
MOMENTENKURVE (*f*): curve of moments 133
MOMENT (*n*), **KIPPENDES-**: tipping (capsizing) moment 134
MOMENTKLEMME (*f*): instantaneous clamp 135

MOMENTPHOTOGRAPHIE (*f*): instantaneous photograph, snap shot (Photography) 136
MOMENT (*n*), **STATISCHES-**: static moment 137
MOMENTVERSCHLUSS (*m*): instantaneous shutter (of cameras) 138
MONADE (*f*): monad; atom; molecule; (indivisible)-particle 139
MONADISCH: monadic 140
MONALKYLIERT: monalkylated 141
MONAT (*m*): month 142
MONATLICH: monthly; menstrual (Med.) 143
MONATSBERICHT (*m*): monthly report 144
MONATSFLUSS (*m*): menses; period; menstruation (Med.) 145
MONATSFRIST (*m*): period of a month 146
MONATSGEHALT (*m*): monthly (pay, wages) salary 147
MONATSGELD (*n*): (see **MONATSGEHALT**) 148
MONATSHEFT (*n*): monthly (part, number) edition (of periodical); monthly journal 149
MONATSSCHRIFT (*f*): (see **MONATSHEFT**) 150
MONATSWEISE: monthly 151
MONATSZEIT (*f*): (see **MONATSFLUSS**) 152
MONAZIT (*m*): monazite (natural cerium-lanthanum phosphate with traces of other constituents); $(Ce,La)PO_4$; Sg. 5.0-5.25 153
MONAZITSAND (*m*): Monazite sand (an enriched sand consisting largely of Monazite, Magnetite, Zirconium and Quartz) 154
MÖNCH (*m*): monk; friar; water outlet; sluice; lock 155
MÖNCHSKAPPE (*f*): monkshood; aconite; friar's cowl (*Aconitum napellus*) 156
MÖNCHSRHABARBER (*m*): monk's rhubarb (*Rumex alpinus*) 157
MÖNCHSRHABARBERWURZEL (*f*): Monk's rhubarb root 158
MOND (*m*): moon; month; lunette 159
MOND, ABNEHMENDER- (*m*): waning moon 160
MONDBAHN (*f*): moon's (orbit) path 161
MONDBESCHREIBUNG (*f*): selenography (description of the moon) 162
MONDFINSTERNIS (*f*): eclipse of the moon 163
MONDFLECKEN (*m*): macula (moon-spot) 164
MONDFÖRMIG: lunate; crescent-shaped 165
MONDGAS (*n*): Mond gas (after Mond the inventor of it) (by simultaneously passing air and water vapour over glowing coals); 50% N., 16% H., 25% CO. 166
MOND, HALBER- (*m*): half-moon; crescent 167
MONDHELL: moonlight; moonlit 168
MONDJAHR (*n*): lunar year 169
MONDLICHT (*n*): moonlight 170
MONDMILCH (*f*): agaric mineral 171
MONDPHASE (*f*): phase of the moon 172
MONDPROZESS (*m*): Mond process (of obtaining refined nickel direct from the ore) 173
MONDSCHEIBE (*f*): face (disc)of the moon 174
MONDSCHEIN (*m*): moonlight, moonshine 175
MONDSCHEINEFFEKT (*m*): moonlight effect 176
MONDSICHEL (*f*): crescent of the moon 177
MONDSTEIN (*m*): moonstone; selenite (a plagioclase feldspar) (see **ERDGLAS, SELENIT** and **ORTHOKLAS**) 178
MONDSUCHT (*f*): insanity, lunacy (Med.) 179
MONDSÜCHTIG: lunatic, insane (Med.) 180
MONDVIERTEL (*n*): quarter of the moon 181
MOND, VOLLER- (*m*): full moon 182
MONDZAHL (*f*): lunar epact (the age of the moon) 183
MONDZIRKEL (*m*): lunar cycle 184
MOND, ZUNEHMENDER- (*m*): waxing moon 185
MONEL-METALL (*n*): monel metal (named from its discoverer, Ambroise Monel, an American. Obtained in Canada from Monel ore which is a natural alloy of nickel and copper; 67% Ni; 28% Cu; 5% Mn and iron; also traces of silicon and carbon) 186
MONESIA (*f*): monesia (*Chrysophyllum glyciphloeum*) 187
MONESIAEXTRAKT (*m*): monesia extract 188
MONESIARINDE (*f*): monesia bark; Guaranham; Buranhem (bark of *Chryscphyllum glyciphloeum*) 189
MONHEIMIT (*m*): monheimite; $(Fe,Zn)CO_3$; Sg. 4.17 190
MONIEREN: to reprove; admonish; censure; warn; rebuke; forewarn; advise; counsel; caution; acquaint; notify; inform 191
MONKEYGRAS (*n*): (see **PIASSAVA, BRASILISCHE-**) 192
MONOAZOFARBSTOFF (*m*): (see **AZOFARBSTOFF**) 193
MONOBARYUMSILIKAT (*n*): monobarium silicate; $BaSiO_3$ 194
MONOBROMCAMPHOR (*m*): (see **BROMKAMPHER**) 195
MONOBROMISOVALERYLHARNSTOFF (*m*), **ALPHA-**: alpha-monobromoisovalerylurea (see **BRO-MURAL**) 196
MONOBROMKAMPFER (*m*): monobromated camphor (see **BROMKAMPHER**) 197
MONOBROMOANTIPYRIN (*n*): monobromoantipyrine (see **BROMOPYRIN**) 198
MONOBROMPROPIONSÄURE (*f*): monobromopropionic acid, alpha-monobromopropionic acid; $(CH_3CHBrCOOH)$; Sg. 1.69; Mp. 24.5°C.; Bp. 203°C. 199
MONOCALCIUMMETAPHOSPHAT (*n*): mono-calcium metaphosphate; $Ca(PO_3)_2$ 200
MONOCALCIUMPHOSPHAT (*n*): monobasic calcium phosphate; calcium biphosphate; primary calcium phosphate; acid-(calcium)-phosphate; superphosphate; $CaH_4(PO_4)_2.H_2O$; Sg. 2.22 201
MONOCALCIUMSILIKAT (*n*): monocalcium silicate; $CaO.SiO_2$ 202
MONOCARBONSÄURE (*f*): monocarboxylic acid 203
MONOCHLORAMIN (*n*): hydrazine (see **HYDRAZIN**) 204
MONOCHLORESSIGSÄURE (*f*): monochloroacetic acid; $ClCH_2CO_2H$; Sg. 1.4; Mp. 62.5°C.; Bp. 186°C. 205
MONOCHLORHYDRIN (*n*): monochlorohydrin, asymmetric propylene-chlorhydrin; chloropropyleneglycol; alpha-propylene-chlorohydrin; $CH_2Cl.CH(OH)CH_2(OH)$; Sg. 1.33 206
MONOCHROM: monochromatic; (*n*): monochrome 207
MONOCHROMASIE (*f*): monochromatism 208
MONOCHROMATISCH: monochromatic 209
MONOCYCLISCH: monocyclic 210
MONOGAMIE (*f*): monogamy (marriage to one wife) 211
MONOGRAM (*n*): monogram (artistically interwoven initials) 212
MONOGRAPHIE (*f*): monograph (description of one thing) 213
MONOHETEROATOMIG: monoheteratomic 214
MONOHYDRAT (*n*): monohydrate; (H_2O) [215
MONOHYDRATISCH: monohydric 216
MONOJODAZETON (*n*): (see **JODAZETON**) 217

MONOKALIUMPHOSPHAT (*n*): monobasic potassium phosphate, potassium diphosphate; KH_2PO_4; Sg. 2.34; Mp. 96°C. 218
MONOKALZIUMPHOSPHAT (*n*): monobasic calcium phosphate, monocalcium phosphate, calcium biphosphate, superphosphate, acid (calcium) phosphate; $CaH_4(PO_4)_2.2H_2O$; Sg. 2.22 219
MONOKETONFARBSTOFF (*m*): (see **OXYKETON-FARBSTOFF**) 220
MONOKLIN: monoclinic 221
MONOKLINISCH: (see **MONOKLIN**) 222
MONOLOG (*m*): monologue 223
MONOMETHYLPARAAMIDOPHENOL (*n*): monomethyl-para-aminophenol; $C_6H_4OH.NH.CH_3$ 224
MONONITRONAPHTALIN (*n*), **ALPHA-**: (see **NITRO-NAPHTALIN**) 225
MONONITROTOLUOL (*n*): mononitrotoluene (see **NITROTOLUOL**) 226
MONOPOL (*n*): monopoly 227
MONOSACCHARID (*n*): monosaccharide; $C_6H_{12}O_6$ (see **DEXTROSE** and **FRUCHTZUCKER**) 228
MONOSÄURE (*f*): monobasic acid 229
MONOSULFOSÄURE (*f*): monosulfonic acid; Cassella's F (Baeyer's F.; Mono F.) acid; beta-naphthyl-aminemonosulfonic (beta-naphthylamine-delta-sulfonic) acid; $C_{10}H_6(NH_2)SO_3H.2:7$ 230
MONOTAL (*n*): monotal; $C_6H_4O.CH_3.OOC.CH_2.OCH_3$ 231
MONOTHEISMUS (*m*): monotheism (belief in one God) 232
MONOTON: monotonous 233
MONOTONIE (*f*): monotony, tedium, sameness 234
MONOWOLFRAMCARBID (*n*): monowolfram (monotungsten) carbide; WC; Sg. 15.7 235
MONOXYBENZOL (*n*): (see **PHENOL**) 236
MONSTRÖS: monstrous, enormous, gigantic 237
MONSTROSITÄT (*f*): monstrosity 238
MONSTRUM (*n*): monster 239
MONTAG (*m*): Monday 240
MONTAGE (*f*): erection; fitting-up; assembling; mounting 241
MONTAGEWERKSTATT (*f*): erecting shop, assembling shop 242
MONTÄGLICH: on Mondays; every Monday 243
MONTANIN (*n*): montanin (name for a concentrated solution of fluosilicic acid) 244
MONTANINDUSTRIE (*f*): mining industry 245
MONTANPECH (*n*): montan pitch 246
MONTANWACHS (*n*): montan wax; mountain wax (extracted from Lignite) (see **ERDHARZ** and **BRAUNKOHLENBITUMEN**) 247
MONTANWACHSFABRIKATION (*f*): montan wax manufacture 248
MONTEJUS (*n* and *pl*): pressure (bulb, vessel) apparatus, montejus (see **DRUCKBIRNE**) 249
MONTEUR (*m*): erector; fitter 250
MONTICELLIT (*m*): monticellite; $CaMgSiO_4$; Sg. 3.1 251
MONTIEREN: to erect; fit up; assemble; mount; set up; equip 252
MONTIERUNG (*f*): erection (see **MONTIEREN**); uniform, equipment (Military) 253
MONTMORILLONIT (*m*): montmorillonite (mixture of Al_2O_3 and SiO_2 hydrogels) (see **MÜLLERIT**) 254
MONTPELLIERGELB (*n*): Turner's yellow, patent yellow, Montpellier yellow (a basic lead chloride); $PbCl_2.5PbO$ 255
MONTROYDIT (*m*): montroydite (natural mercury oxide); HgO 256
MONTUR (*f*): soldier's uniform, regimentals 257
MOOR (*n*): moor; fen; morass; marsh; bog; swamp 258
MOORBAD (*n*): mud bath 259
MOORBODEN (*m*): marshy (soil) land (see **MOOR-GRUND**) 260
MOORERDE (*f*): peaty soil 261
MOORGRUND (*m*): moorland, fens, marshy soil 262
MOORIG: marshy; boggy; swampy 263
MOORLAND (*n*): (see **MOORGRUND**) 264
MOOR (*n*), **TORF-**: peat bog 265
MOOS (*n*): moss (*Lichen*) 266
MOOSACHAT (*m*): moss agate 267
MOOSARTIG: moss-like; mossy 268
MOOSBANK (*f*): mossy bank 269
MOOSBEDECKT: moss-covered 270
MOOSBEERE (*f*): moorberry, cranberry, mossberry (*Oxycoccus palustris*) 271
MOOSBEWACHSEN: moss-covered 272
MOOSBLUME (*f*): marsh marigold (*Caltha palustris*) 273
MOOSGRÜN (*n*): moss green 274
MOOSICHT: (see **MOOSARTIG**) 275
MOOSIG: mossy 276
MOOS (*n*), **IRLÄNDISCHES-**: Irish moss (*Chondrus crispus*) 277
MOOS (*n*), **ISLÄNDISCHES-**: Iceland moss (*Cetraria islandica*) 278
MOOSKAPSEL (*f*): pyxidium, pyxis (Bot.) 279
MOOSKRAUT (*n*): club moss (*Lycopodium clavatum*) 280
MOOSKUNDE (*f*): muscology (the science of mosses) 281
MOOSLEHRE (*f*): (see **MOOSKUNDE**) 282
MOOSPFLANZEN (*f.pl*): mosses; *Bryophyta* (Bot.) (see **BRYOPHYTEN**) 283
MOOSPULVER (*n*): lycopodium-(powder) 284
MOOSROSE (*f*): moss rose (*Rosa muscosa*) 285
MOOSSTÄRKE (*f*): lichenin; moss starch (mostly obtained from Iceland moss by boiling water) 286
MOOSTORF (*m*): peat turf 287
MORAL (*f*): moral-(s); ethics 288
MORALISCH: moral 289
MORALISIEREN: to moralize 290
MORAST (*m*): bog; morass; fen; swamp; moor; marsh; quagmire 291
MORASTERZ (*n*): bog ore (see **WIESENERZ**) 292
MORASTIG: marshy; swampy; boggy 293
MORATORIUM (*n*): moratorium, extension of time, delay, respite (see also **KARENZ**) 294
MORCHEL (*f*): (edible)-mushroom, morel (*Morchella esculenta*) 295
MORCHELVERGIFTUNG (*f*): mushroom poisoning 296
MORD (*m*): murder; homicide; fratricide; manslaughter; suicide; matricide; sororicide; parricide 297
MORDANSCHLAG (*m*): attempted murder 298
MORDANZIEREN: to mordant 299
MORDEN: to-(commit)-murder; (*n*): murder; slaughter 300
MÖRDER (*m*): murderer 301
MÖRDERISCH: murderous 302
MÖRDERLICH: terrible; enormous; fearful 303
MORENOSIT (*m*): morenosite (a hydrous nickel sulphate in green acicular crystals) (see **NICKELSULFAT**) 304
MORES (*pl*): manners, dicipline, morals 305

MORGANATISCH : morganatic (marriage with one of inferior rank) 306
MORGEN (*m*): morn; morning; the East; Orient; acre; (as an adverb—to-morrow) 307
MORGENAUSGABE (*f*): morning edition 308
MORGENBROT (*n*): breakfast 309
MORGEND : of to-morrow 310
MORGENDÄMMERUNG (*f*): morning; daybreak; dawn 311
MORGENLAND (*n*): the East; Orient 312
MORGENLÄNDISCH : eastern; oriental 313
MORGENSTERN (*m*): morning star 314
MORGENTAU (*m*): morning dew 315
MORIN (*n*): morin (a dye from old fustic, *Morus tinctoria*); Sg. 1.041 316
MORINGA (*f*): moringa (*Moringa aptera*); horse-radish tree (*Moringa pterygosperma*) 317
MORINGAGERBSÄURE (*f*): moringatannic acid 318
MORINSÄURE (*f*): morinic acid 319
MORION (*m*): morion (see **QUARZ**) 320
MORPHIN (*n*): morphia; morphine; $C_{17}H_{19}NO_3.H_2O$; Mp. 254°C. 321
MORPHINACETAT (*n*): morphine acetate (see **MORPHIN, ESSIGSAURES-**) 322
MORPHINBROMMETHYLAT (*n*): (see **MORPHOSAN**) 323
MORPHINDIESSIGSÄUREESTER (*m*), **SALZSAURER-** : (see **HEROIN**) 324
MORPHIN (*n*), **ESSIGSAURES-** : morphine acetate; $C_{17}H_{19}NO_3.C_2H_4O_2.3H_2O$; Mp. 200°C. [325
MORPHINHYDROCHLORID (*n*): morphine hydrochloride; morphine muriate; $C_{17}H_{19}NO_3HCl.3H_2O$; Mp. 250°C. 326
MORPHIN, MEKONSAURES- (*n*): morphine (meconate) bimeconate; $(C_{17}H_{19}NO_3)_2.C_7H_4O_7.5H_2O$ 327
MORPHINNITRAT (*n*): morphine nitrate; $C_{17}H_{19}NO_3.HNO_3$ 328
MORPHIN (*n*), **SCHWEFELSAURES-** : morphine sulphate; $(C_{17}H_{19}NO_3)_2.H_2SO_4.5H_2O$; Mp. 250°C. 329
MORPHINSULFAT (*n*): morphine sulphate (see **MORPHIN, SCHWEFELSAURES-**) 330
MORPHIUM (*n*): morphine; morphia; morphin (see **MORPHIN**) 331
MORPHIUM, SALZSAURES- (*n*): morphia muriate (see **MORPHINHYDROCHLORID**) 332
MORPHIUMSUCHT (*f*): morphinomania, morphinism, morphine (opium) habit, (addiction to the taking of morphium) (Med.) 333
MORPHOLOGIE (*f*): morphology (the science of organic form) 334
MORPHOSAN (*n*): morphosan; $C_{17}H_{19}NO_3.CH_3.Br.H_2O$ 335
MORSCH : worm-eaten; rotten; decayed; decaying; tender (Fabrics) 336
MORSCHHEIT (*f*): rottenness 337
MORSEALPHABET(*n*): morse alphabet, morse code 338
MORSEAPPARAT (*m*): morse apparatus 339
MORSEKONUS (*m*): morse taper (of drills) 340
MORSELLE (*f*): lozenge 341
MÖRSER (*m*): mortar (gun, also apparatus for pounding) 342
MÖRSERBATTERIE (*f*): mortar battery (Military) 343
MÖRSERBLOCK (*m*): pestle 344
MÖRSERKEULE (*f*): pestle 345
MORSESCHLÜSSEL (*m*): morse (key) tapper 346
MORSESCHREIBER (*m*): morse printer 347
MORSESTREIFEN (*m*): morse slip 348
MORSETASTER (*m*): (morse)-key 349
MORSETELEGRAPH (*m*): morse telegraph 350
MORSETELEGRAPHIE (*f*): morse telegraphy 351
MORTALITÄT (*f*): mortality 352
MÖRTEL (*m*): plaster; mortar; cement; Sg. 1.35-1.75 353
MÖRTELARBEIT (*f*): plastering, cementing; plaster (cement) work; stucco (plastering) work 354
MÖRTELFABRIKATION (*f*): mortar manufacture 355
MÖRTEL (*m*), **FEUERFESTER-** : fireproof mortar, fireclay 356
MÖRTEL (*m*), **HYDRAULISCHER-** : hydraulic mortar, cement 357
MÖRTEL (*m*), **KALK-** : lime cement (see **LUFTMÖRTEL**) 358
MÖRTELKELLE (*f*): trowel 359
MÖRTELMASCHINE (*f*): mortar mixing machine 360
MÖRTELMÜHLE (*f*): mortar mill 361
MÖRTELSAND (*m*): mortar sand 362
MÖRTEL (*m*), **SCHAMOTTE-** : fireproof mortar, fireclay 363
MÖRTELTROG (*m*): hod 364
MÖRTELWÄSCHE (*f*): grout-(ing) 365
MORVENIT (*m*): morvenite (see **HARMOTOM**) 366
MOSAIK (*f*): mosaic-(work); (as a prefix—mosaic; tessellated) 367
MOSAISCH : mosaic 368
MOSANDRIT (*m*): mosandrite; $(Ce,La,Si,Ti,Zr)O_2$, $(Ce,Y)_2O_3$, CaO,Na_2O,F,H_2O; Sg. 3.0 369
MOSCHEE (*f*): mosque 370
MOSCHUS (*m*): musk (from *Moschus moschiferus*) 371
MOSCHUSBEUTEL (*m*): musk gland (of the male musk deer) 372
MOSCHUSKÖRNER (*n.pl*): Abelmoschus, musk mallow, ambrette, amber seed, musk seed (seeds of *Abelmoschus*) 373
MOSCHUSKÖRNERÖL (*n*): Abelmoschus oil (from seeds of Abelmoschus) 374
MOSCHUS (*m*), **KÜNSTLICHER-** : artificial musk (see **TRINITROBUTYLTOLUOL**) 375
MOSCHUSÖL (*n*): musk oil 376
MOSCHUSTIER (*n*): musk deer (*Moschus moschiferus*) 377
MOSCHUSWURZEL (*f*): musk root; sumbul root 378
MOSKITE (*f*): mosquito 379
MOSKITO (*m*): (see **MOSKITE**) 380
MOSKITOVERTILGER (*m*): mosquito killer 381
MOSKOVADE (*f*): muscovado 382
MOSSIT (*m*): mossite; $(Fe,Mn)(Nb,Ta)_2O_6$; Sg. 6.45 383
MOST (*m*): must; fruit juice 384
MOSTAPFEL (*m*): cider apple 385
MOSTBÜTTE (*f*): vat for must 386
MOSTEREIGERÄTE (*n.pl*): grape pressing (appliances) apparatus 387
MOSTKELTER (*f*): wine (cider) press 388
MOSTMESSER (*m*): must gauge 389
MOSTPRESSE (*f*): wine (cider) press 390
MOSTRICH (*m*): mustard 391
MOSTWAGE (*f*): must (hydrometer) aerometer 392
MOTIVIEREN : to cause; assign a reason for 393
MOTOR (*m*): motor; prime mover 394
MOTORBETRIEBSGEMISCH (*n*): motor driving-mixture, motor spirit 395
MOTORBOOT (*n*): motor (boat) launch 396

MOTORENÖL (*n*) : motor oil 397
MOTORFAHRRAD (*n*) : motor cycle 398
MOTORISCHER PROZESS (*m*) : motive process 399
MOTORLASTWAGEN (*m*) : motor (tractor) lorry 400
MOTOR MIT KURZSCHLUSSANKER (*m*) : squirrel cage (S.C.) motor 401
MOTOR MIT SCHLEIFRINGLÄUFER (*m*) : slip-ring (S.R.) motor 402
MOTORRAD (*n*) : motor cycle 403
MOTORSCHLEPPER (*m*) : motor tug 404
MOTORSCHLEPPWAGEN (*m*) : motor lorry 405
MOTORSPRIT (*m*) : motor spirit (mixture of alcohol and other combustible substances, for driving motors) 406
MOTORWAGEN (*m*) : motor (waggon) lorry, autocar, motorcar 407
MOTORZWEIRAD (*n*) : motor bicycle or cycle 408
MOTTE (*f*) : moth 409
MOTTENBOHRER (*m*), KLEINER- : small moth-borer (*Diatroea saccharalis* and *Diatroea canella*) (destroys the sugar cane) 410
MOTTENFRAS (*m*) : moth-holes (in clothes); damage caused by moths 411
MOTTENPULVER (*n*) : moth powder, moth killer 412
MOTTENSA(F)FRAN (*m*) : meadow saffron (*Colchicum autumnale*) 413
MOTTENSCHUTZMITTEL (*n*) : moth preventative 414
MOTTENSICHER : moth proof 415
MOTTIG : moth-eaten, full of moths 416
MOUSSIEREN : to effervesce ; fizz ; sparkle 417
MÖWE (*f*) : sea-gull ; mew (see MEWE) 418
MOWRABUTTER (*f*) : mowra (butter) oil (see BASSIAÖL) (from seeds of *Bassia longifolia*) ; Sg. 0.9175 ; Mp. 42°C. 419
MOWRAÖL (*n*) : Mowra oil (see BASSIAÖL) 420
MÜCKE (*f*) : midge ; gnat ; fly 421
MUCKEN : to sulk ; mutter ; grumble ; be capricious 422
MÜCKENFANG (*m*) : red, German catch-fly ; catchfly ; silene ; fly-bane (*Lychnis viscaria*) 423
MUCKER (*m*) : bigot ; hypocrite ; grumbler ; capricious (malicious, canting, sulky) person 424
MUCKEREI (*f*) : bigotry ; hypocrisy ; cant 425
MUCKISCH : sullen ; peevish ; capricious 426
MÜDE : tired ; weary ; fatigued 427
MÜDIGKEIT (*f*) : fatigue ; weariness ; lassitude ; languor ; languidness ; exhaustion 428
MUFF (*m*) : sleeve ; socket ; (coupling)-box ; clutch ; muff 429
MUFFE (*f*) : (see MUFF) 430
MUFFEL (*f*) : muffle 431
MUFFELBODEN (*m*) : muffle bottom 432
MUFFELDEKOR (*n*) : muffle decoration ; muffle colour (an overglaze colour for decorating porcelain, and employed in a muffle at a temperature of 700-850°C.) 433
MUFFELFARBE (*f*) : muffle colour (see MUFFEL-DEKOR) 434
MUFFELN : to muffle ; mumble ; mutter 435
MUFFELOFEN (*m*) : muffle furnace ; assay furnace 436
MUFFEN : to muffle ; have a musty smell 437
MUFFENROHR (*n*) : socketed pipe 438
MUFFE (*f*), VERBINDUNGS- : connecting sleeve ; socket (of pipes) 439
MUFFIG : rank ; musty ; mouldy ; sullen ; sulky ; pouting 440
MÜFFIG : (see MUFFIG) 441
MÜHE (*f*) : pains ; trouble ; toil ; care ; labour 442
MÜHELOS : without (trouble) care 443
MÜHEN : to trouble ; take pains ; toil 444
MÜHEVOLL : troublesome ; painful ; irksome ; difficult 445
MÜHEWALTUNG (*f*) : assiduity, trouble, pains, attention, endeavour, care, discharge of duties 446
MÜHLBACH (*m*) : mill (brook) stream 447
MÜHLBEUTEL (*m*) : bolter 448
MÜHLE (*f*) : (grinding)-mill 449
MÜHLENBAUER (*m*) : millwright 450
MÜHLENSTAUB (*m*) : mill dust 451
MÜHLE (*f*), VERTIKALE- : vertical mill, edge mill, grinding mill 452
MÜHLGERINNE (*n*) : mill race ; mill stream 453
MÜHLLAUF (*m*) : drum 454
MÜHLSTEIN (*m*) : millstone ; Sg. 2.48 455
MÜHLTEICH (*m*) : mill pond 456
MÜHLWELLE (*f*) : mill shaft 457
MÜHSAL (*n*) : difficulty ; trouble 458
MÜHSAM : troublesome ; difficult ; hard ; irksome ; toilsome ; painful ; painstaking ; assiduous 459
MÜHSELIG : toilsome ; painful ; weary ; hard ; difficult 460
MULDE (*f*) : tray ; trough ; basin ; bowl ; pig ; pig-mould (Metal) ; tub ; pannier ; drain ; ditch ; gutter ; channel ; cavity 461
MULDENBLEI (*n*) : pig-lead (Metal) 462
MULDENKIPPER (*m*) : tipping tray conveyor 463
MULDENROST (*m*) : trough grate, basket shaped grate, concave grate (of furnaces) 464
MULEGARN (*n*) : mule twist (Textile) 465
MULL (*n*) : dust ; dry mould ; rubbish ; waste ; refuse ; (*m*) : mull ; muslin 466
MÜLL (*n*) : dust ; dry mould ; rubbish ; waste ; refuse 467
MÜLLER (*m*) : miller 468
MÜLLEREI (*f*) : milling 469
MÜLLEREIMASCHINEN (*f.pl*) : flour-mill (grinding milling) machinery 470
MÜLLERGAZE (*f*) : bolting cloth 471
MÜLLERIT (*m*) : müllerite (see MONTMORILLONIT); Sg. 1.97 472
MULLICIT (*m*) : mullicite, vivianite (see VIVIANIT) 473
MÜLLVERBRENNUNG (*f*) : burning (incineration) of dust, rubbish and other household waste 474
MÜLLVERBRENNUNGSANLAGE (*f*) : dust destructor, dust and rubbish burning plant 475
MÜLLVERBRENNUNGSOFEN (*m*) : destructor furnace 476
MULM (*m*) : ore dust ; earthy ore ; dust ; mould ; rotten wood-(dust) 477
MULMEN : to powder ; pulverize ; crumble-(to dust) 478
MULMIG : dusty ; earthy ; decayed ; worm-eaten ; friable 479
MULSTERIG : (see MULSTIG) 480
MULSTIG : musty, fusty, mouldy 481
MULTIPLE : multiple 482
MULTIPLIKANDUS (*m*) : multiplicand 483
MULTIPLIKATION (*f*) : multiplication 484
MULTIPLIKATIONSFAKTOR (*m*) : factor of multiplication 485
MULTIPLIKATOR (*m*) : multiplier (Maths., Elect., etc.) ; amplifier (Elect.) 486
MULTIPLIKATORSPULE (*f*) : multiplying coil, multiplier coil, amplifier (Elect.) 487

MULTIPLIZIEREN : to multiply 488
MULTIPLUM (*n*): multiple 489
MUMIE (*f*): mummy (an embalmed human body); (bitumen and resin from embalmed bodies) 490
MUMIENBILDUNG (*f*): mummification 491
MUMIENHAFT : mummy-like 492
MUMIENKASTEN (*m*): mummy case 493
MUMIFIKATION (*f*): mummification 494
MUMIFIZIEREN : to mummify 495
MUMIFIZIERUNG (*f*): mummification 496
MUMIIN (*n*): mummy; bitumen and resin from embalmed bodies 497
MUMME (*f*): mum (strong beer); mask; masque 498
MUMMELN : to mumble, mutter 499
MUMMEN : to muffle, disguise 500
MUMMEREI (*f*): mummery 501
MUMMESPIEL (*n*): masquerade 502
MUMPS (*pl*): mumps; idiopathic, spontaneous, inflammation of the parotid gland (*Parotitis polymorpha*); cynanche parotidea, epidemic parotitis (Med.) 503
MUND (*m*): mouth; opening; vent; orifice 504
MUNDART (*f*): idiom; dialect 505
MÜNDEL (*m* and *f*): ward; pupil; minor 506
MUNDEN : to taste well; relish 507
MÜNDEN : to discharge; empty; open-(out) into; lead into; flow (fall) into 508
MUNDFÄULE (*f*): disease of the gums 509
MUNDHÖHLE (*f*): oral cavity 510
MÜNDIG : of age 511
MÜNDIGKEIT (*f*): majority (the age at which a child becomes an adult by law) 512
MUNDKLEMME (*f*): lockjaw; tetanus (Med.) 513
MUNDKRAUT (*n*): speedwell (*Veronica officinalis*) 514
MUNDLEIM (*m*): glue; gum (suitable for licking) 515
MÜNDLICH : oral-(ly); verbal-(ly); by word of mouth; *viva voce* 516
MUNDLOCH (*n*): discharge opening, outlet 517
MUNDPILLE (*f*): cachou 518
MUNDREINIGUNGSMITTEL (*n*): mouth-wash 519
MUNDROSE (*f*): hollyhock (*Althæa rosea*) 520
MUNDSCHLEIMHAUTENTZÜNDUNG (*f*): stomatitis, inflammation of the mouth (Med.) 521
MUNDSCHWAMM (*m*): mouth sponge; thrush (ulcerated mouth due to parasitic fungus, *Oidium albicans*) (Med.) 522
MUNDSPERRE (*f*): (see **MUNDKLEMME**) 523
MUNDSTÜCK (*n*): mouth-piece 524
MÜNDUNG (*f*): mouth; outlet; aperture; exit; orifice; estuary; discharge; opening; railway terminus; muzzle (Gun); nozzle 525
MÜNDUNGSQUERSCHNITT (*m*): cross sectional outlet opening, discharge (outlet) area 526
MUNDWASSER (*n*): gargle; mouth-wash 527
MUNITION (*f*): munition; ammunition 528
MUNITIONSAUFZUG (*m*): shell crane, shell davit (Naut.) 529
MUNITIONSFABRIK (*f*): munition factory 530
MUNITIONSFABRIKATION (*f*): ammunition or munition (making) manufacture 531
MUNITIONSFABRIKATIONSMASCHINEN (*f.pl*): munition making machinery 532
MUNITIONSRAUM (*m*): powder magazine 533
MUNITIONSWAGEN (*m*): ammunition waggon, caisson (Military) 534
MUNKELN : to whisper; spread by whispering 535
MÜNSTER (*n*): Minster; cathedral 536
MUNTER : lively; cheerful; fit; brisk; sound; healthy; vigorous; merry; awake; gay; allegro; vigilant; bright; blithe; sprightly 537
MUNTZMETALL (*n*): Muntz metal (an alloy of 60% copper and 40% zinc with traces of iron) 538
MÜNZAMT (*n*): mint 539
MÜNZANSTALT (*f*): mint 540
MÜNZBESCHICKUNG (*f*): alloyage 541
MÜNZDRUCKWERK (*n*): stamping press; coining mill; coin stamp 542
MÜNZE (*f*): coin-(age); medal; mint; money; mint (*Mentha*) (Bot.) 543
MÜNZEN : to mint; coin; stamp (Coinage); forge 544
MÜNZENBRONZE (*f*): bronze for coins (German small coins; 95% Cu; 4% Sn; 1% Zn) 545
MÜNZFÄLSCHER (*m*): coiner (of false money); one who utters false money 546
MÜNZFUSS (*m*): standard (of coinage) (amount of coinage per given quantity of gold) 547
MÜNZGEHALT (*m*): fineness (of coins); standard (of currency); alloy 548
MÜNZGEWICHT (*n*): standard weight 549
MÜNZKENNER (*m*): numismatist; numismatologist 550
MÜNZKUNDE (*f*): numismatics (the study of coins) 551
MÜNZPRÄGUNG (*f*): minting, coining 552
MÜNZPROBE (*f*): assay 553
MÜNZMASCHINE (*f*): minting machine 554
MÜNZSAMMLUNG (*f*): collection of coins 555
MÜNZSTEMPEL (*n*): die 556
MÜNZWÄHRUNG (*f*): standard (of coins) 557
MÜNZWERT (*m*): standard (of currency) 558
MÜNZWESEN (*n*): minting; coining; coinage; system of coinage 559
MÜNZWÜRDIGUNG (*f*): assay (of coins) 560
MÜNZZUSATZ (*m*): alloy (in coins) 561
MÜRBE : mellow; tender; crisp; soft; brittle; friable; short (Metal); pliable; decayed 562
MÜRBEN : to become (mellow, tender, brittle, friable) short; decay; rot 563
MÜRBIGKEIT (*f*): mellowness; tenderness; softness; brittleness; friability; shortness (Metal); dry-rot; suppleness; crispness; unsoundness; decay; pliability 564
MUREXVERFAHREN (*n*): Murex process (of oil treatment) 565
MURGAFETT (*n*): Murga fat (from *Garcinia morella*); Mp. 37°C. 566
MURIAZIT (*m*): muriazite (see **ANHYDRIT**) 567
MURMELN : to murmur; mutter; grumble; whisper 568
MURMELTIER (*n*): marmot (*Arctomys*) (a species of rodent) 569
MURREN : to murmur; grumble; mutter; growl; complain 570
MÜRRISCH : peevish; surly; morose; sullen 571
MUS (*n*): pulp; jam; marmalade; fruit (sauce) pulp (*Succus*); pap; stewed fruit 572
MUSARTIG : pulpy; thick; pappy; pulp-(pap)-like 573
MUSCARIN (*n*): muscarine (a ptomaine from putrid fish); $C_5H_{13}NO_2$ 574
MUSCHEL (*f*): mussel (Shellfish); external ear; shell; cockle; conch 575
MUSCHELBLUME (*f*): tropical duckweed; water lettuce (*Pistia stratiotes*) 576

MUSCHELFÖRMIG : shell-like ; conchoidal 577
MUSCHELGEHÄUSE (*n*) : shell-(covering) 578
MUSCHELGELD (*n*) : shell-money (Shells used as currency by the natives of the Indies) 579
MUSCHELGOLD (*n*) : shell gold ; ormolu 580
MUSCHELGRUS (*m*) : alluvial shell deposit 581
MUSCHELIG : shelly ; conchoidal 582
MUSCHELKALK (*m*) : shell (lime) limestone 583
MUSCHELKENNER (*m*) : conchologist 584
MUSCHELKUNDE (*f*) : conchology, the study of shells 585
MUSCHELLINIE (*f*) : conchoid (a plain curve) 586
MUSCHELMARMOR (*m*) : lumachel ; shell marble (see MARMOR) 587
MUSCHELMERGEL (*m*) : shell-marl (see MERGEL) 588
MUSCHELSCHALE (*f*) : cockle (mussel) shell 589
MUSCHELSCHIEBER (*m*) : three-port slide valve 590
MUSCHELSEIDE (*f*) : Byssus (fine flax or the material which is made therefrom), mussel silk (the fine beard of the mussel) 591
MUSCHELSILBER (*n*) : shell silver 592
MUSCHELTIER (*n*) : mollusk 593
MUSCOVIT (*m*) : (see MUSKOVIT) 594
MUSIG : pulpy ; pappy ; thick ; pulp-like ; pap-like 595
MUSIK (*f*) : music 596
MUSIKALIEN (*pl*) : music, compositions 597
MUSIKALISCH : musical 598
MUSIKANT (*m*) : musician 599
MUSIKER (*m*) : musician 600
MUSIKUS (*m*) : musician 601
MUSIV : mosaic 602
MUSIVARBEIT (*f*) : mosaic-(work) 603
MUSIVGOLD (*n*) : mosaic gold ; bisulphuret of tin ; stannic sulphide (see ZINNDISULFID) 604
MUSIVISCH : mosaic 605
MUSIVSILBER (*n*) : mosaic silver 606
MUSIZIEREN : to make music ; play on an instrument 607
MUSKAT (*m*) : nutmeg (see MUSKATNUSS) 608
MUSKATBALSAM (*m*) : nutmeg butter (see MUSKAT--NUSSÖL) 609
MUSKATBIRNE (*f*) : muscatel (musk) pear 610
MUSKATBLÜTE (*f*) : mace (*Myristica officinalis* and *Myristica fragans*) (see also MACISBLÜTE) 611
MUSKATBLÜTENÖL (*n*) : mace oil (*Oleum macidis*, from mace seeds) (see MUSKATBLÜTE) ; Sg. 0.92 ; Bp. 175-200°C. 612
MUSKATBLÜTÖL (*n*) : (see MUSKATBLÜTENÖL) 613
MUSKATBUTTER (*f*) : nutmeg butter (see MUSKAT--NUSSÖL) 614
MUSKATE (*f*) : nutmeg (see MUSKATNUSS) 615
MUSKATELLER (*m*) : muscatel ; muscat ; muscadine ; muscatelle ; muscadel (Muscadine grape) 616
MUSKATELLERAPFEL (*m*) : musk apple 617
MUSKATELLERKRAUT (*n*) : clary (*Salvia verbenaca*) 618
MUSKATELLERWEIN (*m*) : (see MUSKATENWEIN) 619
MUSKATENBAUM (*m*) nutmeg tree (*Myristica officinalis* or *M. fragans*) 620
MUSKATENBIRNE (*f*) : (see MUSKATBIRNE) 621
MUSKATENBLUME (*f*) : (see MUSKATBLÜTE) 622
MUSKATENBLÜTE (*f*) : (see MUSKATBLÜTE) 623
MUSKATENBUTTER (*f*) : nutmeg butter (see MUS--KATNUSSÖL) 624
MUSKATENNUSS (*f*) : nutmeg (see MUSKATNUSS) 625
MUSKATENWEIN (*m*) : muscatel-(wine) (from muscadine grape) 626
MUSKATNUSS (*f*) : nutmeg (*Nux moschata*) (*Nuces nucistæ*) (kernel of *Myristica officinalis* and *Myristica fragans*) 627
MUSKATNUSSBAUM (*m*) : nutmeg tree (*Myristica moschata*) 628
MUSKATNUSSLEBER (*f*) : nutmeg liver, passive (congestion) hyperæmia of the liver (uniform enlargment of the liver) (Med.) 629
MUSKATNUSSÖL (*n*) : essential oil of nutmeg ; nutmeg oil ; myristica oil (*Oleum myristica*) (from seeds of *Myristica officinalis*) ; Sg. 0.86-0.93 630
MUSKATÖL (*n*), **CALIFORNISCHES-** : **Californian** nutmeg oil (from *Tumion californicum*) 631
MUSKATROSE (*f*) : musk rose 632
MUSKEL (*m*) : muscle ; (as a prefix = muscular) 633
MUSKELANSTRENGUNG (*f*) : muscular exertion 634
MUSKELATROPHIE (*f*) : muscular atrophy, amyotrophic lateral sclerosis, chronic poliomyelitis, wasting palsy (wasting of the muscles) (Med.) 635
MUSKELBAU (*m*) : muscular structure ; structure (contexture, formation, composition, constitution) of the muscles 636
MUSKELEIWEISS (*n*) : myosin (an albuminoid obtained from muscles) 637
MUSKELFASER (*f*) : muscular fibre 638
MUSKELGEWEBE (*n*) : muscular tissue 639
MUSKELIG : muscular 640
MUSKELKONTRAKTION (*f*) : muscular contraction 641
MUSKELKRAFT (*f*) : muscular (force, power) strength 642
MUSKELKRANKHEIT (*f*) : muscular trouble ; disease of the muscles (such as ; myositis, inflammation of the muscle ; polymyositis or dermato-myositis, swelling of the muscle) (Med.) 643
MUSKELLÄHMUNG (*f*) : (see MUSKELATROPHIE) 644
MUSKELLEHRE (*f*) : myology (the science of the muscles) 645
MUSKEL (*m*), **RAUTENFÖRMIGER-** : rhomboid muscle 646
MUSKELRHEUMATISMUS (*m*) : muscular rheumatism (Med.) 647
MUSKELSINN (*m*) : kinaesthesis (Physiol.) 648
MUSKELSTÄRKE (*f*) : muscular strength 649
MUSKELSTOFF (*m*) : sarcosine ; methyl-glycocoll ; $CH_2(NHCH_3)COOH$; Mp. 210-215°C. [650
MUSKELZERGLIEDERUNG (*f*) : myotomy (the dissection of muscles) 651
MUSKELZERLEGUNG (*f*) : myotomy (the dissection of muscles) 652
MUSKELZUCKER (*m*) : inosite ; $C_6H_6(OH)_6$ [653
MUSKELZUSAMMENZIEHUNG (*f*) : muscular contraction 654
MUSKON (*n*) : Muskon (a keton extracted from musk) ; $C_{16}H_{30}O$ 655
MUSKOVIT (*m*) : Muscovite ; potassium mica ; $H_2KAl_3(SiO_4)_3$; Sg. 2.93 656
MUSKOWIT (*m*) : (see MUSKOVIT) 657
MUSKULÖS : muscular, sinewy, stalwart, strong, brawny, powerful 658
MUSS (*n*) : pulp ; jam ; marmalade ; fruit sauce ; necessity ; compulsion 659
MUSSE (*f*) : leisure ; ease 660

MUSSELIN (*m*): muslin 661
MUSSELINEN: muslin; of muslin 662
MÜSSEN: must; ought; to be obliged to; be forced to; have to; be under the necessity of; be bound to 663
MUSSESTUNDE (*f*): leisure hour 664
MUSSEZEIT (*f*): leisure (hour) time; spare time 665
MUSSIEREN: to effervesce; sparkle; froth; ferment; foam 666
MÜSSIG: idle; at leisure; disengaged 667
MÜSSIGGEHEN: to loiter; loaf; stroll; idle; be slothful 668
MUSTER (*n*): pattern; sample; example; specimen; standard; model; muster; exemplification; illustration; instance; style; design; paragon 669
MUSTERBEUTEL (*m*): sample bag 670
MUSTERBILD (*n*): ideal; paragon; example; model 671
MUSTERBUCH (*n*): pattern-book 672
MUSTERFLASCHE (*f*): sample bottle 673
MUSTERHAFT: standard; typical; exemplary 674
MUSTERKARTE (*f*): pattern card 675
MUSTERMÄSSIG: standard; typical; exemplary 676
MUSTERN: to figure (Fabrics); emboss (Paper); bring to shade (Dyeing); muster; review; make (models) patterns; inspect; sum up; find fault with; criticize; examine; censure; survey 677
MUSTERPLATZ (*m*): parade ground 678
MUSTERROLLE (*f*): muster roll 679
MUSTERSAMMLUNG (*f*): specimen collection; collection of samples 680
MUSTERSCHUTZ (*m*): patenting, protection (of a design) (see **GEBRAUCHSMUSTER**) 681
MUSTERWORT (*n*): paradigm (an example) 682
MUSTERZEICHNUNG (*f*): design 683
MUT (*m*): courage; spirit; mood; humour; disposition; fortitude; bravery; cheerfulness; valour; gallantry; fearlessness; intrepidity; heroism; boldness; daring; pluck; mettle; manhood; hardihood; audacity; resolution; spunk; dauntlessness 684
MUTFASSEN: to take courage 685
MUTHMANNSFLÜSSIGKEIT (*f*): Muthmann's liquid (see **ACETYLENTETRABROMID**) 686
MUTHMANNIT (*m*): muthmannite; (Ag, Au) Te 687
MUTIG: spirited, intrepid, etc. (see **MUT**) 688
MUTLOS: crest-fallen; discouraged; dejected; despondent; spiritless (see **MUT**) 689
MUTMASSEN: to guess; conjecture; suppose; presume; surmise 690
MUTMASSLICH: conjectural; supposed; probable; presumptive; presumable 691
MUTMASSLICHKEIT (*f*): probability 692
MUTTER (*f*): matron; mother; matrix; womb; uterus; nut-(box); female screw thread 693
MUTTERBAND (*n*): uterine ligament 694
MUTTERBIENE (*f*): Queen-(bee) 695
MUTTERBLUTFLUSS (*m*): menorrhagia (profuse menstruation); metrorrhagia (menstrual hæmorrhage other than at usual periods) (Med.) 696
MUTTEREISEN (*n*): nut iron 697
MUTTERERDE (*f*): mother-earth; native soil 698
MUTTERFASS (*n*): mother vat (Vinegar) 699
MUTTERFIEBER (*n*): puerperal fever (Med.) (see **KINDBETTFIEBER**) 700
MUTTERFORM (*f*): parent form (Biol.); master mould (Ceramics) 701
MUTTERFRÄSMASCHINE (*f*): nut shaping machine 702
MUTTERGESTEIN (*n*): gangue; matrix 703
MUTTERGUMMI (*n*): galbanum (see **GALBANHARZ**) 704
MUTTERHALTER (*m*): (see **MUTTERZÄPFCHEN**) 705
MUTTERHARZ (*n*): (see **GALBANHARZ**) 706
MUTTERKORN (*n*): ergot-(of rye) (*Claviceps purpurea*); spurred rye (*Secale cornutum*) 707
MUTTERKORNDIALYSAT (*n*): dialized ergot-(of rye) 708
MUTTERKORNÖL (*n*): ergot oil (from *Claviceps purpurea*); Sg. 0.93 709
MUTTERKRANKHEIT (*f*): hysteria (Med.) 710
MUTTERKRANZ (*m*): (see **MUTTERZÄPFCHEN**) 711
MUTTERKRAUT (*n*): feverfew (*Chrysanthemum parthenium*); southernwood (*Artemisia abrotanum*); wild camomile (*Matricaria chamomilla*); balm mint (*Melissa officinalis*); seà milkwort (*Glaux maritima*); motherwort (*Leonurus cardiaca*) 712
MUTTERKUCHEN (*m*): placenta; secundines; after-birth (Med.) 713
MUTTERKÜMMEL (*m*): cumin (see **KÜMMEL, RÖMISCHER-**) 714
MUTTERLAUGE (*f*): mother liquor; melt (of Metal) 715
MUTTERLAUGENSALZ (*n*): bath salt (evaporated mother liquor) 716
MUTTERLEIB (*m*): womb; uterus (Med.) 717
MUTTERMAAL (*n*): mole; mark; spot; birth-mark (see **MUTTERMAL**) 718
MUTTERMAL (*n*): mole, birth-mark (*Macula materna*) (*Nœvus maternus*) (Med.) 719
MUTTERMORD (*m*): matricide (the action) 720
MUTTERMÖRDER (*m*): matricide (the doer) 721
MUTTERNELKE (*f*): caryophyllus (*Caryophyllus aromaticus*) 722
MUTTERPECH (*n*): meconium (Med.) 723
MUTTERPFLASTER (*n*): lead plaster (see **BLEI-PFLASTER**) 724
MUTTERPLAGE (*f*): hysteria (Med.) 725
MUTTERRING (*m*): (see **MUTTERZÄPFCHEN**) 726
MUTTERSCHAF (*n*): ewe 727
MUTTERSCHAFT (*f*): motherhood; maternity 728
MUTTERSCHEIDE (*f*): vagina (Anat.) 729
MUTTERSCHLÜSSEL (*m*): wrench (for nuts); screwhammer; screw-key; screw-wrench; screw-spanner 730
MUTTERSCHNITT (*n*): Cæsarean section (see **KAISERSCHNITT**) 731
MUTTERSCHRAUBE (*f*): screw nut; nut-screw; female screw 732
MUTTERSPIEGEL (*m*): speculum uteri (a mirror for examining the vagina and the womb) (Med.) 733
MUTTERSPRACHE (*f*): mother-tongue; native language 734
MUTTERSTAND (*m*): maternity 735
MUTTERSTOFF (*m*): mother (parent) substance 736
MUTTERSUBSTANZ (*f*): (see **MUTTERSTOFF**) 737
MUTTERWEH (*n*): hysteria (Med.) 738
MUTTERWITZ (*m*): mother wit 739
MUTTERWURZ (*f*): mountain arnica (*Arnica montana*) (see also **FENCHEL, GEMEINER-** and **BÄRENFENCHEL**) 740

MUTTERWUT (*f*): nymphomania, uterine fury (Med.) 741
MUTTERZÄPFCHEN (*n*): pessary (*Pessarium*) (Med.) 742
MUTTERZELLE (*f*): mother (parent) cell 743
MUTTERZIMT (*m*): cinnamon (*Cinnamomum*) 744
MUTVOLL: courageous; impetuous; spirited (see MUT) 745
MUTWILLIG: wanton; mischievous; petulant; high-spirited; saucy; naughty 746
MÜTZE (*f*): cap; bonnet; cover 747
MÜTZENFÖRMIG: cap-shaped; mitriform (Bot.) 748
MÜTZENSCHIRM (*m*): peak (of a cap) 749
MYDRIATICUM (*n*): mydriatic (remedy for mydriasis or dilated pupil of the eye) 750
MYELITIS (*f*): myelitis (see RÜCKENMARKSENTZÜNDUNG) 751
MYKOLOGIE (*f*): mycology (the science of fungi) 752
MYOCARDITIS (*f*): myocarditis, inflammation of the walls of the heart (Med.) 753
MYRICYLALKOHOL (*m*): myricylic alcohol (from *Corypha cerifera*) 754
MYRIKAWACHS (*n*): myrtle wax (from *Myrica cerifera*) (see MYRTLEWACHS) 755
MYRISTICIN (*n*): Myristicin (constituent of MUSKATBLÜTENÖL, which see) 756
MYRISTICINSÄURE (*f*): myristicic acid; $C_{14}H_{28}O_2$; Mp. 53.8°C.; Bp. 121-5°C. 757
MYRISTINSÄURE (*f*): myristic acid (see MYRISTICINSÄURE) 758
MYROBALANE (*f*): myrobalane (see PFLAUME, TÜRKISCHE-) 759
MYROBALANENEXTRAKT (*m*): myrobalane extract 760
MYROBALANENÖL (*n*): myrobalane oil (from seeds of *Terminalia punctata*) 761
MYRONSAUER: myronate 762
MYRONSÄURE (*f*): myronic acid; $C_{10}H_{19}NS_2O_{10}$ (from the seeds of black mustard, *Brassica nigra*) 763
MYROSIN (*n*): myrosin (an enzyme from the seeds of black mustard) 764
MYRRHE (*f*): myrrh; Sg. 1.36 (see MYRRHENHARZ) 765
MYRRHENHARZ (*n*): myrrh (gum resin of *Commiphora myrrha*) (see MYRRHE) 766
MYRRHENÖL (*n*): myrrh oil (from the gum resin of *Commiphora myrrha*); Sg. 0.98-1.01; Bp. 220-235°C. 767
MYRRHENTINKTUR (*f*): myrrh tincture 768
MYRTE (*f*): myrtle; myrtus (*Myrtle communis*) 769
MYRTE (*f*), BRABANNTER-: sweet gale, bog-myrtle (*Myrica gale*) 770
MYRTENBEERE (*f*): myrtle berry 771
MYRTENLAUBE (*f*): myrtle grove 772
MYRTENÖL (*n*): myrtle oil; Sg. 0.9 773
MYRTENWACHS (*n*): myrtle wax (from *Myrica cerifera*); Sg. 0.99-1.014; Mp. 40-48°C. (see MYRTLEWACHS) 774
MYRTHENÖL (*n*): myrtle oil (see MYRTENÖL) 775
MYRTLEWACHS (*n*): myrtle wax (a vegetable wax from the berries of *Myrica cerifera* and others; Mp. 42-49°C.) 776
MYRTILLA-PASTILLEN (*f.pl*): myrtilla (pastilles) lozenges 777
MYSTERIÖS: mysterious 778
MYSTERIUM (*n*): mystery 779
MYSTIFIZIEREN: to mystify 780
MYSTISCH: mystic-(al) 781
MYSTIZISMUS (*m*): mysticism 782
MYTHE (*f*): myth, legend, fable, saga, tradition, superstition 783
MYTHISCH: mythical 784
MYTHOLOGIE (*f*): mythology (investigation of legends or tradition) 785
MYTHOLOGISCH: mythological 786
MYTILOTOSCIN (*n*): mytilotoscine (an animal alkaloid from poisonous mussels); $C_6H_{15}NO_2$ 787

N

NABE (*f*): nave; boss; hub (Mach.) 788
NABEL (*m*): navel; boss; hilium (Bot.) 789
NABELBINDE (*f*): umbilical bandage; navel bandage 790
NABELBRUCH (*m*): hernia; omphalocele, exomphalos (Med.) 791
NABELFÖRMIG: umbilicate, umbilical, navel-like 792
NABELIG: umbilicate 793
NABELKRAUT (*n*): navelwort; common navelwort; cotyledon; wall-pennywort (*Cotyledon umbilicus*) 794
NABELSCHNUR (*f*): navel (cord) string; umbilical cord 795
NABELSTRANG (*m*): umbilical cord (see NABELSCHNUR) 796
NABENBÜCHSE (*f*): wheel box 797
NABENKAPPE (*f*): axle cap 798
NACH: to; toward; after; according to; behind; for; of 799
NACHÄFFEN: to ape; imitate; mimic; copy 800
NACHAHMEN: to imitate; adulterate; counterfeit; copy; mimic; follow; forge 801
NACHARBEIT (*f*): copy; afterwork; check 802
NACHARTEN: to resemble 803
NACHBAR (*m*): neighbour 804
NACHBARLICH: neighbouring, neighbourly, vicinal 805
NACHBARSCHAFT (*f*): neighbourhood; district; vicinity 806
NACHBARSCHICHT (*f*): neighbouring (layer) stratum 807
NACHBAU (*m*): extension; addition; alteration (copy) of a building or construction 808
NACHBEHANDLUNG (*f*): after-treatment; further treatment; supplementary treatment; secondary (subsidiary) treatment 809
NACHBEIZEN: to sadden (Dye) 810
NACHBESSERN: to improve; touch up; repair; retouch 811
NACHBESTELLUNG (*f*): repeat order 812
NACHBEZAHLUNG (*f*): subsequent payment 813
NACHBILD (*n*): copy; imitation; counterfeit; fac-simile 814
NACHBILDEN: to copy; reproduce; counterfeit; imitate 815
NACHBLEIBEN: to remain behind; lag; survive; be detained 816
NACHBLEIBEND: residuary 817
NACHBOHREN: to re-bore; widen; re-drill 818
NACHBRENNEN: to re-distill, etc. (see BRENNEN) 819

NACHCHROMIEREN : to after-chrome (Dye) 820
NACHDATIEREN : to post-date 821
NACHDEM : afterward-(s); after; according to (whether); hereafter; thereafter; then 822
NACHDEM ——— **SO** : as ——— so (then) 823
NACHDENKEN : to meditate; follow; reflect; consider; (*n*): meditation; reflection; consideration 824
NACHDENKEND : meditative; pensive; reflective; thoughtful 825
NACHDENKLICH : thoughtfully, meditatively, pensively, reflectively; important 826
NACHDICHTEN : to re-tighten, re-pack; re-condense, re-compress; imitate (Poetry) 827
NACHDRUCK (*m*): energy; stress; emphasis; pirated (edition) copy; reprint; second press (of wine) 828
NACHDRUCKEN : to pirate; copy; reprint; counterfeit 829
NACHDRÜCKEN : to reprint; push; force 830
NACHDUNKELN : to fill up; sadden; darken (Dye) 831
NACHEIFER (*m*): emulation 832
NACHEIFERN : to emulate; rival 833
NACHEIFERUNG (*f*): emulation 834
NACHEILEN : to lag (Mech.); pursue, hasten after 835
NACHEILUNG (*f*): negative (advance) lead; lagging; exhaust lap (Mechanical) 836
NACHEINANDER : one after another; in turn; successively; in series; tandem (of two only) 837
NACHEINANDERFOLGEND : successive 838
NACHEN (*m*): boat; skiff 839
NACHERBE (*m*): residuary legatee 840
NACHERNTE (*f*): gleanings; second (harvest) crop 841
NACHERZÄHLEN : to retell; repeat; recite; recapitulate 842
NACHFÄLSCHEN : to counterfeit, forge 843
NACHFÄRBEN : to dye (stain) again; re-dye; re-stain 844
NACHFIXIEREN : to fix again; re-fix 845
NACHFOLGE (*f*): succession; imitation; reversion 846
NACHFOLGEN : to follow; succeed; imitate; revert 847
NACHFOLGEND : following; subsequent; consecutive; continuous; supplementary; secondary; subsidiary; auxiliary; successive 848
NACHFOLGER (*m*): successor-(s); imitator 849
NACHFORMEN : to imitate; copy 850
NACHFORSCHEN : to investigate; search (for); scrutinize; enquire after; seek; trace 851
NACHFORSCHER (*m*): investigator 852
NACHFORSCHUNG (*f*): investigation; research; scrutiny; search; quest 853
NACHFRAGE (*f*): enquiry; demand; request 854
NACHFRAGEN : to enquire (about) after; trace; request; care (worry, trouble) about 855
NACHFÜLLEN : to refill; replenish; fill up; add to 856
NACHGÄREN : to (be) subject-(ed) to after fermentation, cleanse; clear 857
NACHGÄRFASS (*n*): cleansing [settling; clearing; after (secondary) fermentation]; cask 858
NACHGÄRUNG (*f*): after-(secondary) fermentation; cleansing; clearing; settling 859
NACHGEBEN : to give (in) way; yield; grant; comply; relax 860
NACHGEBURT (*f*): after-birth 861
NACHGERBEN : to re-tan 862
NACHGESCHALTETE TURBINE(*f*): low pressure end of turbine; turbine coupled at the rear of other plant 863
NACHGESCHMACK (*m*): aftertaste; tang 864
NACHGIEBIG : yielding; indulgent; flexible 865
NACHGIESSEN : to refill; pour (again) after; take a cast of; cast from 866
NACHGILBEN : to re-gild 867
NACHGRÜBELN : to muse upon; ponder over; reflect upon; search one's memory for 868
NACHGUSS (*m*) : refilling; second pouring; after-mash; second wort; sparging-(water) (Brewing); recast; copy 869
NACHHALL (*m*): echo; reverberation 870
NACHHALLEN : to resound; reverberate; echo 871
NACHHER : after-(wards); hereafter; later; thereafter; subsequently 872
NACHHERIG : subsequent; posterior; later 873
NACHHOLEN : to recover; make up; retrieve; overtake 874
NACHHUT (*f*): rear guard 875
NACHJAGEN : to pursue; chase; run after 876
NACHKEHREN : to sweep up after 877
NACHKLANG (*m*): echo, reverberation, ring 878
NACHKLINGEN : to echo, reverberate, resound, ring 879
NACHKOCHEN : to boil again; continue boiling; reboil 880
NACHKOMME (*m*): successor; descendant; (*pl*): posterity; offspring; issue; future generation 881
NACHKOMMEN : to follow; act up to; comply with; overtake; perform; fulfil; execute; obey; carry out; meet (Wishes) 882
NACHKOMMENSCHAFT (*f*): posterity, offspring, issue, future generation 883
NACHKÜHLEN : to after-cool, re-cool 884
NACHKÜNSTELN : to imitate; make an imitation of (in an artistic manner); copy (works of art); counterfeit 885
NACHKUPFERN : to after-treat with copper (treatment of dyed cotton with copper salt) (see also **KUPFERBEIZE**) 886
NACHKUR (*f*): after-cure 887
NACHLASS (*m*): estate; inheritance; assets; legacy; reduction; relaxation; intermission; abatement; remission; rebate; discount; allowance; drawback; deduction; residuum; residue; remains 888
NACHLASSEN : to temper; anneal (Metal); slacken; abate; relax; deduct; leave behind; subside; give way; yield; allow; remit; mitigate; discontinue 889
NACHLASSEND : intermittent 890
NACHLÄSSIG : negligent; remiss; careless; lax; dilatory; neglectful; thoughtless; slothful; supine; indolent; unthinking; slovenly 891
NACHLÄSSIGKEIT (*f*): slovenliness; negligence; inaccuracy 892
NACHLAUF (*m*): after-run; second runnings; chase; chasing 893
NACHLAUFEN : to run after; pursue; chase 894
NACHLEBENDE (*m*): survivor, descendant 895
NACHLESE (*f*): pickings, gleanings 896
NACHLEUCHTEN : to shine after; phosporesce; (*n*): afterglow; phosphorescence 897

NACHLIEFERUNG (*f*): subsequent delivery; completion of an order; further (later) consignment 898
NACHMACHEN: to imitate; counterfeit; copy; mimic; forge 899
NACHMAHD (*f*): after-math 900
NACHMAISCHE (*f*): after-mash 901
NACHMALEN: to re-paint; copy-(an original painting); paint (from) after (Art) 902
NACHMALIG: subsequent; following 903
NACHMALS: subsequently; afterwards 904
NACHMEHL (*n*): pollard 905
NACHMESSEN: to re-measure; check-(measurements); re-survey 906
NACHMITTAG (*m*): afternoon 907
NACHMITTÄGIG: post-meridian 908
NACHMITTAGSFLUT (*f*): afternoon tide 909
NACHMÜHLENÖL (*n*): inferior olive oil, residual oil [obtained from the press residue (last pressing) of olives; a low grade olive oil] 910
NACHNAHME (*f*): a sort of "cash on delivery" in which the sender can obtain from the Post Office at the place of dispatch, payment for the goods, and the Post Office accept the responsibility of obtaining payment at the other end before parting with the goods; re-imbursement 911
NACHNAHME, UNTER-: C.O.D. (Cash on Delivery) (see **NACHNAHME**) 912
NACHNUANCIEREN: to shade-(colours) 913
NACHORDNEN: to (arrange) classify according to 914
NACHOXYDATION (*f*): after-oxydation 915
NACHPACKEN: to re-pack 916
NACHPORTO (*n*): extra (carriage, freight) postage 917
NACHPRÄGEN: to counterfeit (forge) coins; copy a stamp 918
NACHPRESSE (*f*): press rolls (Paper); re-pressing machine 919
NACHPRESSEN: to re-press 920
NACHPRÜFEN: to check; determine; find out 921
NACHPUTZEN: to repolish; finish 922
NACHRÄUMEN: to clear up after someone else 923
NACHRECHNEN: to recalculate, check; verify; audit, examine (Accounts, etc.) 924
NACHREDE (*f*): epilogue; report; rumour; rejoinder; calumny; slander 925
NACHREDEN: to repeat; slander; caluminate; malign; defame; vilify; abuse; libel; revile; traduce 926
NACHRICHT (*f*): news; information; advice; tidings; notice; account; report; intelligence 927
NACHRICHTEN: to re-adjust; (make) arrange-(ments) in conformity with; execute 928
NACHRICHTER (*m*): executioner 929
NACHRICHTLICH: for information 930
NACHRÜCKEN: to push (move) forward, advance, follow up 931
NACHRUF (*m*): notice; posthumous fame; report; call 932
NACHRUHM (*m*): posthumous fame 933
NACHSALZEN: to resalt 934
NACHSATZ (*m*): conclusion; minor proposition 935
NACHSCHALTEN: to connect (couple) at the outlet side 936
NACHSCHLAGEN: to look up; turn up; refer to; consult; resemble; counterfeit; strike again 937
NACHSCHLAGEWERK (*n*): reference work; work of reference 938
NACHSCHLEIFEN: to re-grind; slur (Mus.); drag along 939
NACHSCHLEPPEN: to drag after; trail; tow 940
NACHSCHLÜSSEL (*m*): master key; false key; pick-lock 941
NACHSCHMIEREN: to re-lubricate 942
NACHSCHREIBEN: to copy; write from dictation; transcribe 943
NACHSCHRIFT (*f*): P.S.; postscript; copy; transcript 944
NACHSCHUB (*m*): secondary (throw) thrust; subsequent (move, throw, thrust) push 945
NACHSCHÜTTEN: to pour (again) after; add; refill 946
NACHSEHEN: to look (after, to); see to; search; attend to; overlook; excuse; examine; indulge; overhaul 947
NACHSETZEN: to add; postpone; put (place) after; regard as inferior; pursue; follow; depreciate; slight 948
NACHSETZUNG (*f*): pursuit (see **NACHSETZEN**) 949
NACHSICHT (*f*): inspection; indulgence; forbearance; respite; period of grace 950
NACHSICHTSVOLL: forbearing; indulgent 951
NACHSILBE (*f*): suffix 952
NACHSPÜREN: to track, investigate, trace, enquire into, search after 953
NÄCHST: next-(to); nearest; latest; subsequent; proximate 954
NÄCHSTBESTE (*m*): next (second) best 955
NÄCHSTE: next; nearest; closest; proximate; neighbouring; (*m*): neighbour 956
NACHSTECHEN: to re-feed; re-pass; copy. (Engravings) 957
NACHSTEHEN: to be placed after; be inferior; be secondary; follow 958
NACHSTEHEND: following; below; subjoined; appended; undermentioned 959
NACHSTELLBAR: adjustable; regulable 960
NACHSTELLEN: to regulate; adjust; place (after) behind; append; subjoin; put back 961
NACHSTEMMEN: to recaulk; (*n*): recaulking 962
NÄCHSTENS: next time; shortly; next of all; soon 963
NACHSTERILISIEREN (*n*): re-sterilization; secondary sterilization; to re-sterilize 964
NACHSTEUER (*f*): re-assessment; additional (tax) duty 965
NÄCHSTFOLGEND: next, subsequent, following 966
NACHSTREBEN: to strive after, tend towards, have a tendency (towards) in the direction of, emulate, aspire to 967
NACHSUCHEN: to search (look) for or after, bespeak, enlist, solicit (of sympathy); enquire, seek, ask for, request; apply (make application) for 968
NACHT (*f*): night 969
NACHTANGRIFF (*m*): night attack (Mil.) 970
NACHTARBEIT (*f*): night (work) shift 971
NACHTBLAU (*n*): night blue 972
NACHTBLINDHEIT (*f*): night-blindness, nyctalopia (inability to see in a poor light) (Med.) 973
NACHTBLÜHEND: night-blooming (Bot.) 974
NACHTBLUME (*f*): nocturnal flower (see also **JASMIN, ARABISCHER-**) 975
NACHTBLÜTLER (*m.pl*): *Nyctaginaceæ* (Bot.) 976
NACHTDIENST (*m*): night (service) duty 977
NACHTEIL (*m*): disadvantage; detriment; prejudice; damage; drawback; inconvenience 978

NACHTEILIG: disadvantageous, damaging, detrimental, derogatory, harmful, hurtful, prejudicial, inconvenient 979
NACHTEULE (*f*): night owl 980
NACHTFROST (*m*): night frost 981
NACHTGLEICHE (*f*): equinox 982
NACHTHAUS (*n*): binnacle 983
NACHTHAUSLAMPE (*f*): binnacle lamp 984
NACHTHYAZINTHE (*f*): tuberose (*Polianthes tuberosa*) 985
NACHTIGALL (*f*): nightingale (*Daulias luscinia*) 986
NACHTISCH (*m*): dessert; fruits; sweets 987
NACHTKERZE (*f*): night light (see **NACHTRÖSCHEN**) 988
NACHTLAMPE (*f*): night light 989
NÄCHTLICH: nocturnal, nightly, gloomy 990
NACHTLICHT (*n*): night light 991
NACHTMAHL (*n*): supper 992
NACHTÖNEN: to echo; resound; reverberate 993
NACHTRAG (*m*): addendum; addition; supplement; appendix 994
NACHTRAGEN: to append; supply; supplement; add; harbour (a grudge); pay; post-(up) 995
NACHTRÄGLICH: supplementary; extra; later; additional; subsequent-(ly); afterwards; further 996
NACHTRIPPER (*m*): gleet (Med.) 997
NACHTRÖSCHEN (*n*): evening primrose (*Oenothera biennis*) 998
NACHTS: by night; at night-time; during the night 999
NACHTSCHADE (*m*): nighthawk, nightjar, goatsucker, fern-owl (*Caprimulgus europaeus*) (Orn.) 000
NACHTSCHATTEN (*m*): nightshade (*Solanum nigrum*) 001
NACHTSCHATTENGEWÄCHSE (*n.pl*): *Solanaceæ* (Bot.) 002
NACHTSCHICHT (*f*): night-shift 003
NACHTSCHWALBE (*f*): (see **NACHTSCHADE**) 004
NACHTSCHWEISS (*m*): night sweat (Med.) 005
NACHTSEHEN (*n*): day-blindness 006
NACHTVIOLE (*f*): dame's violet; rocket; damask violet (*Hesperis matronalis*) 007
NACHTVIOLENÖL (*n*): (see **ROTREPSÖL**) 008
NACHTWÄCHTER (*m*): night-watchman 009
NACHTWANDLER (*m*): somnambulist, sleep-walker, noctambulist 010
NACHÜBERHITZER (*m*): re-superheater, intermediate superheater, reheater (for superheating between stages) (Turbines) 011
NACHÜBERHITZUNG (*f*): re-superheating, intermediate superheating, re-heating (Turbines) 012
NACH-UND-NACH: gradually; by degrees 013
NACHWEICHEN (*n*): couching (Brewing) 014
NACHWEIS (*m*): proof; indication; information; direction; detection; reference; evidence; test-(for) 015
NACHWEISBAR: evident; demonstrable; detectable; assignable; capable of being proved 016
NACHWEISEN: to detect; show; demonstrate; prove; indicate; inform; check; establish; identify; refer to; point out; authenticate 017
NACHWEISER (*m*): pointer; index; director; indicator; demonstrator; checker 018
NACHWEISUNG (*f*): proof; direction; demonstration; detection; identification; reference; information; evidence; check; indication; enquiry 019
NACHWEISUNGSBÜRO (*n*): information (intelligence) bureau; enquiry office 020
NACHWELT (*f*): posterity 021
NACHWIRKUNG (*f*): after-(secondary)-effect 022
NACHWUCHS (*m*): after (new) growth 023
NACHWÜRZE (*f*): second wort 024
NACHZAHLEN: to pay off (outstanding amounts); pay the balance; settle 025
NACHZÄHLEN: to re-count; check; re-total; work out again; re-work (of figures) 026
NACHZEICHNEN: to re-draw, copy (by drawing) 027
NACHZEICHNUNG (*f*): correction (to drawings or etchings); copy (of a drawing) 028
NACHZIEHEN: to re-draw, pull (after, behind) out again; follow, go after; tighten (Nuts, etc.) 029
NACHZÜNDUNG (*f*): retard (late) ignition 030
NACKEN (*m*): neck; nape 031
NACKEND: naked; bare; plain; uncovered; nude 032
NACKENWIRBEL (*m*): cervical vertebra (Anat.) 033
NACKT: naked; bare; plain; uncovered; nude; openly 034
NACKTFRÜCHTIG: gymnocarpous (Bot.) 035
NACKTKEIMER (*m*): Acotyledon, acotyledonous plant, plant destitute of cotyledons 036
NACKTSAMENPFLANZE (*f*): gymnosperm (Bot.) 037
NACKTSAMIG: gymnospermous (Bot.) 038
NADEL (*f*): needle; pin 039
NADELARBEIT (*f*): needlework 040
NADELARTIG: acicular; needle-like 041
NADELBODEN (*m*): cast iron bottom of melting pan, fitted with projections (needles), for letting in the air (Bessemer and Thomas processes) 042
NADELDRAHT (*m*): pin-wire 043
NADELEISENERZ (*n*): needle iron ore; Lepidocrocite; göthite (see **LEPIDOKROKIT** and **GÖTHIT**) 044
NADELERZ (*n*): needle ore; aikinite; $Pb_2Cu_2Bi_2S_6$; Sg. 6.5 045
NADELFEILE (*f*): needle file 046
NADELFÖRMIG: acicular; needle-shaped 047
NADELFUTTERAL (*n*): needle case 048
NADELGEFÜGE (*n*): acicular structure (Steel) 049
NADELHÖLZER (*n.pl*): *Coniferæ*, conifers, pines, firs (Bot.) 050
NADELHOLZKOHLE (*f*): pine charcoal 051
NADELHOLZTEER (*m*): pine tar 052
NADELKOPF (*m*): pin's head 053
NADELÖHR (*n*): eye (of a needle) 054
NADELÖLER (*m*): needle lubricator 055
NADELPAAR (*n*), **ASTATISCHES-**: astatic needles 056
NADELSONDE (*f*): needle-(probe) (Surg.) 057
NADELSPITZE (*f*): needle point 058
NADELSTEIN (*m*): natrolite; needlestone (see **NATROLITH**) 059
NADELSTICH (*m*): Acupuncture, treatment by pricking with needles (Med.); pinhole; pin-prick 060
NADELTELEGRAPH (*m*): needle telegraph 061
NADELVENTIL (*n*): needle valve 062
NADELVERFAHREN (*n*): needle process- (of oil extraction by rubbing the fruit shells or skins against needle points) 063

NADELZEOLITH (*m*): needle zeolite; natrolite (see NATROLITH) 064
NADELZINNERZ (*n*): acicular cassiterite (see CASSITERIT) 065
NADIR (*m*): nadir 066
NADORIT (*m*): nadorite; $PbCl_2,SbO_2$; Sg. 7.0 067
NAEGIT (*m*): naegite; $(Ce,Si,Zr)O_2,(Nb_2,Ta_2)O_5,(U,Fe_2Y_2)O_3,(Ca,Mg)O,H_2O$; Sg. 4.1 [068
NAFALAN (*n*): nafalan (a soft, ointment-like mass obtained from the distillation of a Caucasian naphtha; Mp. 65-70°C.) 069
NAFALANSEIFE (*f*): nafalan soap 070
NAFTA (*f*): naphtha (see under NAPHTA) 071
NAGEL (*m*): nail; peg; pin; tack; stud; plug; spike 072
NAGELBANK (*f*): pin rack 073
NAGELBOHRER (*m*): gimlet; piercer 074
NÄGELCHEN (*n*): tack; clove, pink (see NELKE) 075
NÄGELEIN (*n*): tack; clove; pink (see NELKE) 076
NAGELEISEN (*n*): heading tool; nail iron 077
NAGELHANDEL (*m*): nail trade 078
NAGELKEIL (*m*): wedge 079
NÄGELKRANKHEIT (*f*): disease of the nails (Med.) 080
NAGELKRAUT (*n*): common burnet, great burnet (*Sanguisorba officinalis* or *Poterium officinale*) 081
NAGELN: to nail; pin; peg; spike; tack 082
NAGELSCHMIED (*m*): nail-smith; nail-maker 083
NAGELZANGE (*f*): pincers; pliers; nippers 084
NAGELZIEHER (*m*): nail-claw 085
NAGEN: to gnaw; nibble; bite 086
NAGER (*m*): rodent 087
NAGETIER (*n*): rodent 088
NAGYAGIT (*m*): nagyagite (a natural sulfotelluride of gold, antimony and lead; $Au_2Sb_2Pb_{10}Te_6S_{15}$; Sg. 6.8-7.2 089
NAHE: near; imminent; proximate; adjacent; close; neighbouring; almost; approximate-(ly) 090
NÄHE (*f*): neighbourhood; nearness; vicinity; presence; proximity; imminence; approach; closeness; environs; locality; propinquity; district 091
NAHE, ES LAG-: it was found necessary 092
NAHELEGEN: to place near to; give rise to, suggest (an idea) 093
NAHEN: to approach; draw near; approximate; be imminent 094
NÄHEN: to sew; do needlework; stitch; hem; mend; lace 095
NÄHER: nearer; closer 096
NÄHERE (*n*): details; information; data; particulars 097
NÄHERKOMMEN: to approach, send, forward 098
NÄHERN: to draw near; approach; approximate; be imminent 099
NÄHERUNG (*f*): approximation 100
NÄHERUNGSGLEICHUNG (*f*): approximate equation 101
NAHESTEHEND: standing near; proximate; nearly (connected) related; juxtaposed; neighbouring 102
NAHEZU: well-nigh; almost; nearly; approximately; practically 103
NÄHGARN (*n*): thread; twine; (sewing)-cotton 104
NÄHKUNST (*f*): (the art of) needlework 105
NÄHMASCHINE (*f*): sewing machine 106
NÄHMASCHINENLACK (*m*): varnish for sewing machines 107
NÄHMASCHINENÖL (*n*): machine oil for sewing machines 108
NÄHNADEL (*f*): (sewing)-needle 109
NÄHRBODEN (*m*): nutritive (nutrient) soil; fertilising agent 110
NÄHREN: to feed; nourish; keep; support; foster; nurse; obtain one's living; sustain oneself; suckle; cherish; maintain 111
NÄHRFLÜSSIGKEIT (*f*): nutrient (nutritive) liquid or fluid (see NÄHRSAFT) 112
NÄHRGELD (*n*): maintenance; keep; board 113
NÄHRHAFT: nutritive; nourishing; nutritious; rich; lucrative; productive; fertile; alimentary 114
NÄHRHEFE (*f*): yeast for domestic use 115
NÄHRKRAFT (*f*): nutritive power; fertility 116
NAHRLOS: poor; unfertile; without (sustenance) nourishment; unprofitable 117
NÄHRLÖSUNG (*f*): nutritive solution 118
NÄHRMITTEL (*n*): food; sustenance; nourishment; provisions; victuals 119
NÄHRPRÄPARAT (*n*): food preparation (any preparation for providing nourishment, such as patent foods, etc.) 120
NÄHRSAFT (*m*): chyle; nutrient juice (fluid from food in the intestines) (Med.); sap (Bot.) 121
NÄHRSALZ (*n*): nutritive salt 122
NÄHRSTAND (*m*): labouring (working) class 123
NÄHRSTOFF (*m*): nutritive (substance) material; fertilizer; manure; nutrient; food-(stuff); provisions; nourishment 124
NÄHRSTOFFGEHALT (*m*): nutrient (nutritive, food) content 125
NAHRUNG (*f*): nourishment; maintenance; nutrition; sustenance; nutriment; food; keep; livelihood 126
NAHRUNGSBREI (*m*): chyme (the food pulp in the stomach) (Med.) 127
NAHRUNGSFLÜSSIGKEIT (*f*): chyle; nutritive liquid (see NÄHRSAFT) 128
NAHRUNGSKANAL (*m*): alimentary canal (Anat.) 129
NAHRUNGSLOS: without means of sustenance (see NAHRLOS) 130
NAHRUNGSMITTEL (*n*): food; sustenance; provisions (see NÄHRMITTEL) 131
NAHRUNGSMITTELKUNDE (*f*): science of nutrition 132
NAHRUNGSROHR (*n*): alimentary canal (Anat.) 133
NAHRUNGSSAFT (*m*): chyle; nutrient juice (see NÄHRSAFT) (Med.); sap (Bot.) 134
NAHRUNGSSALZ (*n*): nutritive salt 135
NAHRUNGSSTOFF (*m*): nutritive substance; sustenance; food-(stuff); nourishment 136
NAHRUNGSTEILCHEN (*n*): nutritive element 137
NAHRUNGSWERT (*m*): nutritive value 138
NAHRUNGSWERTEINHEIT (*f*): unit of nutritive value 139
NÄHRWERTBERECHNUNG (*f*): calculation (computation, determination) of the nutritive value 140
NAHT (*f*): weld; seam; suture 141
NAHTLOS: weldless; seamless 142
NAHTSTREIFEN (*m*): edge strip 143
NÄHWACHS (*n*): sewing wax 144
NAIV: naïve; natural; ingenuous 145
NAIVETÄT (*f*): naiveté; ingenuousness 146
NAKRIT (*m*): nakrite, kaolinite; $Al_2O_3,2SiO_2,2H_2O$; Sg. 2.5 (see also KAOLINIT) 147

NAME (*m*): name; title; denomination; repute; reputation; appellation; epithet; designation; character; description; distinction; fame; exponent 148
NAMENBRETT (*n*): name plate 149
NAMENDEUTUNG (*f*): etymology (the science of words) 150
NAMENLOS: nameless; obscure; unknown; anonymous; inexpressible; dreadful; unspeakable; unnamed 151
NAMENPLATTE (*f*): name plate 152
NAMENSCHILD (*n*): name-plate 153
NAMENSUNTERSCHRIFT (*f*): signature 154
NAMENTLICH: namely; especially; particularly; nominal (ly) 155
NAMHAFT: famous; well-known; renowned; named; considerable; especial; specified 156
NÄMLICH: namely; that is-(to say); (i.e.); (e.g.); (self)-same; identical; to wit; particularly 157
NÄMLICHE (*m*): the same; identical 158
NANKING (*m*): nankeen 159
NANZIGERSÄURE (*f*): lactic acid (see MILCHSÄURE) 160
NAPF (*m*): bowl; basin; pan; cup; pot; porringer; dish 161
NÄPFCHENSTEIN (*m*): druid's stone; lapidarian sculpture; cup-shaped stone 162
NÄPFCHENTRAGEND: cupuliferous (Bot.) 163
NAPFFÖRMIG: bowl-shaped 164
NAPHTA (*f*): naphtha; petroleum (see BERGÖL) 165
NAPHTACEN (*n*): naphthacene 166
NAPHTACYLSCHWARZ (*n*): naphthacyl black 167
NAPHTALDEHYD (*n*): naphthaldehyde 168
NAPHTALIN (*n*): naphthalene; naphthalin; tar camphor; $C_{10}H_8$; Sg. 0.9673; Mp. 80.1°C.; Bp. 218°C. 169
NAPHTALINBLAU (*n*): naphthalene blue 170
NAPHTALINCAMPHOR (*m*): naphthalene camphor (see NAPHTALIN) 171
NAPHTALINDERIVAT (*n*): naphthalene derivate 172
NAPHTALINDICARBONSÄURE (*f*): (see NAPHTAL-SÄURE) 173
NAPHTALINGELB (*n*): naphthalene yellow 174
NAPHTALINKUGEL (*f*): naphthalene ball 175
NAPHTALINSÄURE (*f*): naphthalenic acid 176
NAPHTALINSEIFE (*f*): naphthalene soap 177
NAPHTALINSULFOSAUER: naphthalenesulfonate, combined with naphthalenesulfonic acid 178
NAPHTALINSULFOSÄURE (*f*): naphthalene sulfonic acid; $C_{10}H_8SO_3.H_2O$; Mp. 85-90°C. 179
NAPHTALINTETRACHLORID (*n*): naphthalene tetrachloride 180
NAPHTALINWASCHÖL (*n*): naphthalene washing oil 181
NAPHTALOL (*n*): naphthalol (see SALIZYLSÄURE-B-NAPHTHYLESTER) 182
NAPHTALSÄURE (*f*): naphthalic acid $[C_{10}H_6(CO_2H)_2]$ (see PHTALSÄURE, ORTHO-) 183
NAPHTAMINFARBE (*f*): naphthamine (colour) dye 184
NAPHTANTHRACHINON (*n*): naphthanthraquinone 185
NAPHTAPECH (*n*): naphtha pitch 186
NAPHTASALBE (*f*): naphtha ointment 187
NAPHTASEIFE (*f*): naphtha soap 188
NAPHTAZARIN (*n*): naphthazarin 189
NAPHTAZIN (*n*): naphthazine 190
NAPHTAZINBLAU (*n*): naphthazine blue 191
NAPHTAZURIN (*n*): naphthazurine 192
NAPHTEN (*n*): naphthene; C_nH_{2n} (a saturated hydrocarbon as HEXAHYDROBENZOL, CYCLO-HEXAN; CYCLOPENTAN, which see) 193
NAPHTENSÄURE (*f*): naphthenic acid 194
NAPHTHALSÄURE (*f*): (see NAPHTALSÄURE) 195
NAPHTHENOL (*n*): (see HEXALIN) 196
NAPHTHIONSÄURE (*f*): naphthionic acid; alpha-naphthylamine sulfonic acid: $C_{10}H_6(NH_2)SO_3H$: 1 : 4 197
NAPHTIDIN (*n*): naphthidine 198
NAPHTINDON (*n*): naphthindone 199
NAPHTIONAT (*n*): naphthionate 200
NAPHTIONSALZ (*n*): naphthionate 201
NAPHTIONSAUER: naphthionate, combined with naphthionic acid 202
NAPHTIONSÄURE (*f*): naphthionic acid (see NAPHTHIONSÄURE) 203
NAPHTOCHINON (*n*): naphthoquinone 204
NAPHTOCUMARIN (*n*): naphthocumarin (or coumarin) 205
NAPHTOESÄURE (*f*): naphthoic acid 206
NAPHTOGEN (*n*): naphthogene 207
NAPHTOINDIGOBLAU (*n*): naphthoindigo blue 208
NAPHTOL (*n*): naphthol; $C_{10}H_7OH$; (*a*) Sg. 1.0954 to 1.224; Mp. 96°C.; Bp. 279°C. (*b*) Sg. 1.217; Mp. 128°C.; Bp. 285-290°C. 209
NAPHTOLBLAU (*n*): naphthol blue 210
NAPHTOLDERIVAT (*n*): naphthol derivate 211
NAPHTOLDISULFOSAURES NATRON (*n*): (see R-SALZ) 212
NAPHTOLFARBSTOFF (*m*): naphthol (colour) dye 213
NAPHTOLGELB (*n*): naphthol yellow 214
NAPHTOLIEREN: to naphtholize or naphtholate 215
NAPHTOLMONOSULFOSÄURE (*f*): (see NEVILLE SÄURE and WINTHERS SÄURE) 216
NAPHTOLPECH (*n*): naphthol pitch 217
NAPHTOLSALICYLAT (*n*): naphthol salicylate; alphol 218
NAPHTOLSALOL (*n*): naphtholsalol (see BETOL) 219
NAPHTOLSULFOSÄURE (*f*): naphtholsulfonic acid 220
NAPHTOLSULFOSÄURE B, BETA-: (see BAYERSCHE SÄURE) 221
NAPHTOLSULFOSÄURE C, ALPHA-: (see CLEVES SÄURE) 222
NAPHTOLSULFOSÄURE NW, ALPHA-: (see NEVILLE SÄURE and WINTHERS SÄURE) 223
NAPHTOLSULFOSÄURE S, BETA-: (see SCHÄFFER-SCHE SÄURE) 224
NAPHTOL-WISMUT (*n*), **BETA-**: (*Bismutum naphtolicum*): β-bismuth naphtholate; $Bi_2O_2(OH).(C_{10}H_7O)$ 225
NAPHTOSÄURESCHWARZ (*n*): naphtho-acid black 226
NAPHTOYL (*n*): naphthoyl 227
NAPHTSULTAM (*n*): naphthosultam 228
NAPHTSULTON (*n*): naphthosultone 229
NAPHTYL (*n*): naphthyl 230
NAPHTYLAMIN (*n*): naphthyl amine; $(C_{10}H_9N)$; α, Sg. 1.1011; Mp. 50°C.; Bp. 300.8°C. β, Sg. 1.0614; Mp. 112°C.; Bp. 306.1°C. 231
NAPHTYLAMINCHLORHYDRAT (*n*): (alpha)-naphthylamine hydrochloride; $C_{10}H_7NH_2.HCl$ [232
NAPHTYLAMINSCHWARZ (*n*): naphthylamine black 233

NAPHTYLAMIN (*n*), **SCHWEFELSAURES-** : naphthylamine sulphate 234
NAPHTYLAMINSULFOSÄURE (*f*) : naphthylamine sulphonic acid 235
NAPHTYLAMINSULFOSÄURE BR., BETA- : beta-naphthylaminesulfonic acid, Broenner's acid (see **BRÖNNERSCHE SÄURE**) 236
NAPHTYLAMINSULFOSÄURE D, BETA- : beta-naphthylaminesulfonic acid (see **DAHLSCHE SÄURE**) 237
NAPHTYLAMINSULFOSÄURE F, BETA- : (see **F. ACID**) 238
NAPHTYLBLAU (*n*) : naphthyl blue 239
NAPHTYLBLAUSCHWARZ (*n*) : naphthyl blueblack 240
NAPHTYLENBLAU (*n*) : naphthylene blue; new blue; fast blue; cotton blue; (a tar dye); $Cl(CH_3)_2N$; $C_6H_3NOC_{10}H_6$ 241
NAPHTYLPHENYLAMIN (*n*) : naphthylphenylamine; $C_{10}H_7.NH.C_6H_5$; Mp. 62°C.; Bp. 226°C. 242
NAPHTYONAT (*n*) : naphthionate 243
NAPHTYRIDIN (*n*) : naphthyridine 244
NAPTHEN (*n*): naphthene; C_nH_{2n} (see **NAPHTEN**) 245
NARBE (*f*) : scar; mark; seam; grain (Paper, Leather, etc.); cicatrice; pit; stigma (Bot.) 246
NARBEN : to grain; mark; scar; seam; shagreen (Leather); (*m*) : grain; graining; shagreening (Leather); pitting 247
NARBENBILDUNG (*f*) : scarring; seaming; cicatrization 248
NARBENLEDER (*n*) : grain-(ed) leather 249
NARBENLOS : unscarred; unmarked 250
NARBENSEITE (*f*) : grain-(ed) side (of leather) 251
NARBIG : scarred; grained 252
NARBIGMACHEN : to nap; twill (Cloth); grain 253
NARCEIN (*n*) : narcein (an opium alkaloid of the papaverin group); $C_{20}H_{18}NO_5(CH_3O)_3$ [254
NARCISSE (*f*) : narcissus; daffodil (see **NARZISSE**) 255
NARCOFORM (*n*) : narcoform (a local anæsthetic; mixture of 60% ethyl chloride; 35% methyl chloride and 5% ethyl bromide) 256
NARCOPHIN (*n*) : narcophine (narcotine-morphine meconate) 257
NARCOTIN (*n*) : narcotine (see **MOHNSTOFF**) 258
NARDE (*f*) : nard; spikenard (a plant of the order *Valerianaceæ*) (see **NARDENKRAUT**) 259
NARDENKRAUT (*n*) : spikenard, nard (*Lavendula spica*) 260
NARDENSAME (*m*) : nutmeg flower (*Nigella sativa*) 261
NARKOSE (*f*) : narcotic, anæsthetic 262
NARKOTIKUM (*n*) : narcotic 263
NARKOTIL (*n*) : methylene chloride (a local anæsthetic) (see **METHYLENCHLORID**) 264
NARKOTIN (*n*) : narcotine (see **MOHNSTOFF**) 265
NARKOTISCH : narcotic 266
NARKOTISCHER EXTRAKT (*m*) : narcotic extract 267
NARKOTISCHES MITTEL (*n*) : narcotic; hypnotic; soporific (Med.) 268
NARKOTISIEREN : to narcotize 269
NARR (*m*) : fool; idiot; imbecile; madman; lunatic 270
NARRHEIT (*f*) : foolishness; madness; folly; absurdity; idiocy; inanity; feeblemindedness; imbecility; insanity 271
NÄRRISCH : foolish; mad; odd; funny; droll; crazy; ridiculous; absurd; idiotic; inane; feebleminded; insane; wild; strange; extravagant 272
NARZEIN (*n*) : (see **NARCEIN**) 273
NARZISSE (*f*) : narcissus; daffodil 274
NARZISSENÖL (*n*) : narcissus oil 275
NARZYL (*n*) : (see **ÄTHYLNARCEINCHLORHYDRAT**) 276
NASALLAUT (*m*) : nasal sound 277
NASCHWERK (*n*) : sweetmeats; dainties 278
NASCIEREND : nascent 279
NASE (*f*) : nose; snout; beak; spout; cam; catch; tappet; nozzle; hook 280
NÄSELN : to snivel; snuffle; speak through one's nose; smell; scent (of dogs) 281
NASENBEIN (*n*) : nose bone; bridge of (bone in) the nose 282
NASENBILDUNG (*f*) : rhinoplastic operation (an operation to replace loss of whole or part of the nose) 283
NASENBLUTEN (*n*) : Epistaxis, nose-bleeding (Med.) 284
NASENBUCHSTABE (*m*) : nasal (letter) sound 285
NASENFLÜGEL (*m*) : nostril; wing of the nose 286
NASENGEWÄCHS (*n*) : polypus in the nose (Med.) 287
NASENHÖHLE (*f*) : nasal cavity (in the nose bone) 288
NASENKEIL (*m*) : nose key, nose wedge 289
NASENKOPPE (*f*) : tip of the nose 290
NASENKRAUT (*n*) : snapdragon (*Antirrhinum majus*) 291
NASENKUPPE (*f*) : tip of the nose 292
NASENLAUT (*m*) : nasal sound 293
NASENLOCH (*n*) : nostril 294
NASENPOPEL (*m*) : mucus from the nose 295
NASENRACHENRAUM (*m*) : naso-pharynx (Med.) 296
NASENRÜCKEN (*m*) : bridge of the nose 297
NASENRÜMPFEN (*n*) : sneering; turning up one's nose 298
NASENSATTEL (*m*) : bridge of the nose 299
NASENSCHEIDEWAND (*f*) : bridge of the nose 300
NASENSCHLEIM (*m*) : mucus from the nose 301
NASENSPITZE (*f*) : tip of the nose 302
NASENTON (*m*) : nasal (sound) tone 303
NASEWEIS : pert; saucy; impertinent; inquisitive; self-sufficient; cheeky; impudent; insolent; forward 304
NASHORN (*n*) : rhinoceros 305
NASS : wet; humid; damp; moist; hydrous 306
NASSBLEICHE (*f*) : wet bleach 307
NÄSSE (*f*) : wet-(ness); moisture; humidity; damp-(ness) 308
NÄSSEN : to wet; moisten; water; humidify 309
NASSER WEG (*m*) : wet process 310
NASSFÄULE (*f*) : potato blight; damp-rot 311
NASSMÜHLE (*f*) : wet-(grinding)-mill 312
NASSPROBE (*f*) : wet test 313
NÄSSPROBE (*f*) : moisture-(determination)-test 314
NASSVERFAHREN (*n*) : wet process 315
NASTIN (*n*) : nastin (a glyceride from *Streptothrix leproides*) 316
NASTURAN (*m*) : pitchblende (see **URANPECHERZ**) 317
NASZIEREND : nascent 318
NATALKÖRNER (*n.pl*) : (see **GELBBEEREN, CHINESISCHE-**) 319

NATIONALISIEREN : to nationalize ; naturalize 320
NATIONALITÄT (*f*) : nationality 321
NATIONALÖKONOMIE (*f*) : political economy 322
NATIONALSCHULD (*f*) : national debt 323
NATIV : native, natural 324
NATRIUM (*n*) : sodium (*Natrium*) (Na) (see also NATRON) 325
NATRIUMACETAT (*n*) : (see NATRIUM, ESSIG-SAURES-) 326
NATRIUMALAUN (*m*) : sodium alum (*Alumen natricum*) ; $Na_2SO_4.Al_2(SO_4)_3+24H_2O$ [327
NATRIUMALUMINAT (*n*) : sodium aluminate ; $Na_2Al_2O_4$ or $Al_2(NaO)_6$; Mp. 1800°C. [328
NATRIUMALUMINIUMCHLORID (*n*) : (see ALUM-INIUMNATRIUMCHLORID) 329
NATRIUMAMALGAM (*n*) : sodium amalgam (sodium-mercury alloy) 330
NATRIUM, AMEISENSAURES- (*n*) : sodium formate ; $NaCHO_2$; Sg. 1.92 (see NATRIUMFORMIAT) 331
NATRIUMAMID (*n*) : sodium amide ; sodamide ; $NaNH_2$; Mp. 155°C. ; Bp. 400°C. (used in Indigo synthesis) 332
NATRIUM (*n*), **AMYLSCHWEFELSAURES-** : sodium amylsulphate 333
NATRIUM, ANHYDROMETHYLENCITRONENSAURES- (*n*) : sodium anhydromethylene citrate 334
NATRIUMANHYDROPERSULFAT (*n*) : anhydrous sodium persulphate ; $Na_2S_2O_9$ 335
NATRIUM, ANTIMONSAURES- (*n*) : sodium antimonate 336
NATRIUM, ARSANILSAURES- (*n*) : sodium arsanilate ; atoxyl ; sodium-aniline arsonate ; soamin ; sodium-aminophenyl-arsonate ; $C_6H_4NH_2(AsO.OH.ONa)$ 337
NATRIUMARSENIAT (*n*) : sodium arsenate (*Natrium arsenicicum*) ; $Na_3AsO_4.12H_2O$; Sg. 1.759 ; Mp. 85.5°C. 338
NATRIUMARSENIAT, NEUTRALES- (*n*) : neutral sodium arsenate (see NATRIUMARSENIAT) 339
NATRIUM, ARSENIGSAURES- (*n*) : sodium arsenite ; Na_2HAsO_3 ; Sg. 1.87 340
NATRIUM, ARSENIKSAURES- (*n*) : sodium arsenate (see NATRIUM, ARSENSAURES-) 341
NATRIUM, ARSENSAURES- (*n*) : sodium arsenate ; $Na_3AsO_4.12H_2O$ (see NATRIUMARSENIAT) 342
NATRIUMÄTHYLAT (*n*) : sodium ethylate 343
NATRIUM (*n*), **ÄTHYLSCHWEFELSAURES-** : sodium ethylsulphate 344
NATRIUMAUROTHIOSULFAT (*n*) : (see GOLDOXYD-ULNATRON, UNTERSCHWEFLIGSAURES-) 345
NATRIUMAZETAT (*n*) : (see NATRIUM, ESSIG-SAURES-) 346
NATRIUM (*n*), **AZETYL-P-AMIDOPHENYLARSIN-SAURES-** : (see AZETYLARSANILAT) 347
NATRIUMAZID (*n*) : sodium azide ; $NaN_3.H_2O$ 348
NATRIUM (*n*), **BALDRIANSAURES-** : sodium valerate 349
NATRIUMBENZOAT (*n*) : sodium benzoate ; $NaC_7H_5O_2$ 350
NATRIUM, BENZOESAURES- (*n*) : (see NATRIUM-BENZOAT) 351
NATRIUM (*n*), **BENZOSULFOSAURES-** : sodium benzosulphonate 352
NATRIUM (*n*), **BERNSTEINSAURES-** : sodium succinate 353
NATRIUMBICHROMAT (*n*) : (*Natrium bichromicum*) : sodium bichromate ; sodium acid chromate ; sodium dichromate ; $Na_2Cr_2O_7.2H_2O$; Sg. 2.52 354
NATRIUMBIFLUORID (*n*) : sodium bifluoride, sodium acid fluoride ; NaF.HF 355
NATRIUMBIKARBONAT (*n*) : (*Natrium bicarbonicum*) : sodium bicarbonate ; sodium acid carbonate ; baking soda ; hydrosodic carbonate ; $NaHCO_3$; Sg. 2.2 356
NATRIUMBIOXALAT (*n*) : (*Natrium bioxalicum*) : sodium binoxalate ; $NaHC_2O_4$ 357
NATRIUMBISULFAT (*n*) : sodium bisulphate, nitre cake, sodium acid sulphate ; $NaHSO_4$ (*Natrium bisulfuricum*) ; Sg. 2.435 ; Mp. 300°C. 358
NATRIUMBISULFIT (*n*) : (*Natrium bisulfurosum*) : sodium bisulphite, leucogen, sodium acid sulphite ; $NaHSO_3$; Sg. 1.48 359
NATRIUMBORAT (*n*) : sodium borate ; sodium biborate ; borax ; $Na_2B_4O_7.10H_2O$ 360
NATRIUM (*n*), **BORSAURES-** : sodium borate (see NATRIUMBORAT) 361
NATRIUMBRECHWEINSTEIN (*m*) : antimonyl sodium tartrate 362
NATRIUMBROMAT (*n*) : sodium bromate ; $NaBrO_3$; Sg. 3.339 ; Mp. 384°C. 363
NATRIUMBROMID (*n*) : sodium bromide ; NaBr ; Sg. 3.014 ; Bp. 1455°C. (*Natrium bromatum*) : $NaBr.2H_2O$; Sg. 2.18 ; Mp. 758°C. 364
NATRIUM (*n*), **BROMSAURES-** : sodium bromate see NATRIUMBROMAT) 365
NATRIUM (*n*), **BUTTERSAURES-** : sodium butyrate 366
NATRIUMCARBID (*n*) : sodium carbide ; Na_2C_2 ; Sg. 1.575 367
NATRIUMCARBONAT (*n*) : sodium carbonate, soda, washing soda, carbonic soda, soda crystals, sal soda, carbonate of soda ; Na_2CO_3, Sg. 2.476-2.5, Mp. 849°C. ; or $Na_2CO_3.10H_2O$, Sg. 1.45, Mp. 34°C., Bp. 106°C. ; (as a mineral see THERMONATRIT) 368
NATRIUMCHLORAT (*n*) : sodium chlorate (*Natrium chloricum*) ; $NaClO_3$; Sg. 2.289-2.49 ; Mp. 255°C. 369
NATRIUM (*n*), **CHLORCHROMSAURES-** : sodium chlorochromate 370
NATRIUMCHLORID (*n*) : sodium chloride ; common salt ; rock salt ; sea salt ; table salt (*Natrium chloratum*) ; NaCl ; Sg. 2.16-2.17 ; Mp. 804°C. to 850°C. ; Bp. 1490°C. 371
NATRIUM, CHLORSAURES- (*n*) : sodium chlorate (*Natrium chloricum*) (see NATRIUMCHLORAT) 372
NATRIUMCHROMAT (*n*) : (*Natrium chromicum*) : sodium chromate ; $Na_2CrO_4.10H_2O$; Sg. 2.723 ; Mp. 19.9°C. 373
NATRIUMCHROMAT, NEUTRALES- (*n*) : neutral sodium chromate (see NATRIUMCHROMAT) 374
NATRIUM (*n*), **CHROMSAURES-** : sodium chromate (see NATRIUMCHROMAT) 375
NATRIUMCITRAT (*n*) : (*Natrium citricum*) : sodium citrate ; $2Na_3C_6H_5O_7.11H_2O$ or $Na_3C_6H_5O_7.H_2O$ 376
NATRIUM (*n*), **CITRONENSAURES-** : sodium citrate (see NATRIUMCITRAT) 377
NATRIUMCYANAMID (*n*) : sodium cyanamide ; Na_2CN_2 378
NATRIUMCYANID (*n*) : (*Natrium cyanatum*) : sodium cyanide ; NaCN 379
NATRIUMCYANÜR (*n*) : sodium cyanide 380
NATRIUM (*n*), **DIÄTHYLBARBITURSAURES-** : sodium diethylbarbiturate, medinal, dormonal, (*Natrium diaethylbarbituricum*) (often used

in place of DIÄTHYLBARBITURSÄURE as sleeping medicine on account of its subcutaneous and rectal action) 381

NATRIUMDIHYDROPHOSPHAT (*n*): sodium phosphate, monobasic; sodium acid phosphate, monosodium phosphate; $NaH_2PO_4+H_2O$, Sg. 2.04; $NaH_2PO_4+2H_2O$, Sg. 1.9096 382

NATRIUM (*n*), DIMETHYLARSENSAURES- : sodium dimethylarsenate, sodium cacodylate (about 55% arsenic); $(CH_3)_2AsO_2Na.3H_2O$ [383

NATRIUMDIWOLFRAMAT (*n*): sodium ditungstate; $(Na_2W_3O_7)_2$ 384

NATRIUM, DOPPELKOHLENSAURES- (*n*): sodium bicarbonate (see NATRIUMBIKARBONAT) 385

NATRIUM (*n*), ESSIGSAURES- : (*Natrium aceticum*): sodium acetate; $NaC_2H_3O_2.3H_2O$; Sg. 1.4; Mp. 58°C. 386

NATRIUM (*n*), ESSIGWOLFRAMSAURES- : sodium acetotungstate 387

NATRIUMFERRICYANÜR (*n*): sodium ferricyanide (see FERRICYANNATRIUM) 388

NATRIUMFERRIOXALAT (*n*): ferric-sodium oxalate, iron sodium oxalate; $Na_3Fe(C_2O_4)_3.4\frac{1}{2}H_2O$ 389

NATRIUMFERROCYANÜR (*n*): sodium ferrocyanide (see FERROCYANNATRIUM) 390

NATRIUMFERROZYANID (*n*): (see FERROCYAN-NATRIUM) 391

NATRIUM (*n*), FETTSAURES- : sodium sebate 392

NATRIUMFLUORID (*n*): sodium fluoride (*Natrium fluoratum*); NaF; Sg. 2.766; Mp. 980°C. 393

NATRIUMFLUORID (*n*), SAURES- : (see NATRIUM-BIFLUORID) 394

NATRIUMFORMIAT (*n*): (*Natrium formicicum*): sodium formate; H.COONa; Sg. 1.92 395

NATRIUM (*n*), GERBSAURES- : sodium tannate 396

NATRIUMGOLDCHLORID (*n*): sodium chloraurate; sodium chloroaurate; sodium aurichloride; sodium-gold chloride (*Auronatrium chloratum*) (see GOLDSALZ) 397

NATRIUM (*n*), HIPPURSAURES- : sodium hippurate 398

NATRIUM (*n*), HOLZESSIGSAURES- : sodium pyrolignite 399

NATRIUMHYDRAT (*n*): caustic soda; sodium hydroxide; NaOH (see NATRIUMHY-DROXYD) 400

NATRIUMHYDRID (*n*): sodium hydride; NaH; Sg. 0.92 401

NATRIUMHYDROCARBONAT (*n*): sodium hydrocarbonate; $NaHCO_3$; Sg. 2.206 402

NATRIUMHYDROSULFAT (*n*): sodium hydrosulphate; $NaHSO_4$; Sg. 2.742 403

NATRIUMHYDROSULFID (*n*): sodium hydrosulphide, sodium sulfhydrate; $NaSH.2H_2O$ 404

NATRIUMHYDROSULFIT (*n*): sodium hydrosulphite; $Na_2S_2O_4$ 405

NATRIUMHYDROXYD (*n*): caustic soda; sodium hydroxide; sodium hydrate; NaOH; Sg. 2.13; Mp. 318°C. (*Natrium hydricum*) 406

NATRIUMHYPERJODAT (*n*): sodium periodate 407

NATRIUMHYPOCHLORIT (*n*): sodium hypochlorite, Eau de Labarraque; NaOCl 408

NATRIUMHYPOPHOSPHAT (*n*): sodium hypophosphate; $Na_4P_2O_6+10H_2O$; Sg. 1.832 409

NATRIUMHYPOPHOSPHIT (*n*): sodium hypophosphite; $NaH_2PO_2.H_2O$ 410

NATRIUMHYPOSULFAT (*n*): sodium (hyposulphate) dithionate; $Na_2S_2O_6.2H_2O$; Sg. 2.175 411

NATRIUMHYPOSULFIT (*n*): (*Natrium thiosulfuricum*: sodium hyposulphite, sodium thiosulphate, antichlor, hypo.; $Na_2S_2O_3$, Sg. 1.667; $Na_2S_2O_3+5H_2O$, Sg. 1.729-1.736, Mp. 48°C. 412

NATRIUM (*n*), INDIGOSCHWEFELSAURES- : sodium indigosulphate 413

NATRIUM (*n*), ISOBUTYLSCHWEFELSAURES- : sodium isobutylsulphate 414

NATRIUMJODAT (*n*): sodium iodate; $NaIO_3$; Sg. 4.277 415

NATRIUMJODID (*n*): sodium iodide (*Natrium jodatum*); NaI; Sg. 3.665; Mp. 653°C.; Bp 1350°C 416

NATRIUMKALIUM (*n*), WEINSAURES- : sodium-potassium tartrate (see SEIGNETTESALZ) 417

NATRIUM (*n*), KARBOLSAURES- : sodium (carbolate, phenolate) phenate; NaC_6H_4OH 418

NATRIUMKARBONAT (*n*): sodium carbonate (see NATRIUMCARBONAT) 419

NATRIUM (*n*), KOHLENSAURES- : sodium carbonate, soda (see NATRIUMCARBONAT) 420

NATRIUM (*n*), KRESOTINSAURES- : soda cresotate 421

NATRIUMLAKTAT (*n*): sodium lactate (*Natrium lacticum*); $C_3H_5O_3Na$ 422

NATRIUM, MANGANSAURES- (*n*): sodium manganate; $Na_2MnO_4.10H_2O$; Mp. 17°C. 423

NATRIUMMETALL (*n*): sodium metal 424

NATRIUMMETAPHOSPHAT (*n*): sodium metaphosphate; $NaPO_3$; Sg. 2.476 425

NATRIUMMETASULFIT (*n*): sodium metasulphite 426

NATRIUM (*n*), MILCHSAURES- : sodium lactate (see NATRIUMLAKTAT) 427

NATRIUM (*n*), MOLYBDÄNSAURES- : sodium molybdate (see NATRIUMMOLYBDAT) 428

NATRIUMMOLYBDAT (*n*): (*Natrium molybdænicum*); sodium molybdate; Na_2MoO_4 [429

NATRIUMMONOSULFID (*n*): (*Natrium sulfuratum*): sodium monosulphide, sodium sulphide, sodium sulfuret; Na_2S; Sg. 1.856-2.471 430

NATRIUM (*n*), NAPHTIONSAURES- : sodium naphthionate, alphasodium naphthylamine sulfonate; $NaC_{10}H_6(NH_2)SO_3.4H_2O$ [431

NATRIUMNITRAT (*n*): sodium nitrate; $NaNO_3$ (see NATRIUM, SALPETERSAURES-) 432

NATRIUMNITRIT (*n*): (*Natrium nitrosum*); sodium nitrite; $NaNO_2$; Sg. 2.157; Mp. 213°C. 433

NATRIUMNITROFERROCYANÜR (*n*): sodium (nitroferrocyanide, nitroprussiate) nitroprusside; $Na_2Fe(CN)_5NO.2H_2O$; Sg. 1.68 434

NATRIUMOLEAT (*n*): sodium oleate 435

NATRIUMOLEATLÖSUNG (*f*): sodium oleate solution 436

NATRIUM, OLEINSAURES- (*n*): sodium oleate 437

NATRIUMOXALAT (*n*): sodium oxalate (*Natrium oxalicum*); $Na_2C_2O_4$ 438

NATRIUM (*n*), OXALSAURES- : sodium oxalate (see NATRIUMOXALAT) 439

NATRIUMOXYD (*n*): sodium oxide (*Natrium oxydatum*); Na_2O; Sg. 2.27 440

NATRIUMOXYDHYDRAT (*n*): sodium hydroxide (see NATRIUMHYDROXYD) 441

NATRIUMPALMITAT (*n*): sodium palmitate 442

NATRIUMPALMITATLÖSUNG (*f*): sodium palmitate solution 443

NATRIUMPARAWOLFRAMAT (*n*): sodium (paratungstate) para-wolframate; $Na_{10}W_{12}O_{41}.28H_2O$ 444

NATRIUMPENTASULFID (*n*): sodium pentasulphide; Na_2S_5 445

NATRIUMPERBORAT (*n*): sodium perborate; Perborin (*Natrium perboricum*); $NaBO_3+4H_2O$ 446

NATRIUMPERCARBONAT (*n*): (*Natrium percarbonicum*): sodium percarbonate; $Na_2CO_4.1\frac{1}{2}H_2O$ or $Na_2CO_4.8H_2O$ 447

NATRIUMPERCHLORAT (*n*): (*Natrium perchloricum*): sodium perchlorate; $NaClO_4$; Mp. 482°C. 448

NATRIUMPERJODAT (*n*): sodium periodate; $NaIO_4$, Sg. 3.865; $NaIO_4.3H_2O$, Sg. 3.219 449

NATRIUMPERKARBONAT (*n*): (see **NATRIUM-PERCARBONAT**) 450

NATRIUMPERMANGANAT (*n*): (*Natrium permanganicum*): sodium permanganate, Condy's fluid; $NaMnO_4$ 451

NATRIUMPEROXYD (*n*): (*Natrium peroxydatum*): sodium peroxide; sodium superoxide; sodium binoxide; sodium dioxide; oxone; Na_2O_2; Sg. 2.8 452

NATRIUMPEROXYDHYDRAT (*n*): hydrated sodium peroxide; $Na_2O_2.8H_2O$ 453

NATRIUMPERSULFAT (*n*): (*Natrium persulfuricum*): sodium persulphate; $Na_2S_2O_8$ 454

NATRIUM (*n*), **PHENYLÄTHYLBARBITURSAURES-**: (see **LUMINALNATRIUM**) 455

NATRIUMPHOSPHAT (*n*), **BASISCHES-**: basic sodium phosphate, tribasic sodium phosphate, trisodium phosphate, tertiary sodium phosphate, sodium ortho-phosphate; $Na_3PO_4.12H_2O$; Sg. 1.62-1.645; Mp. 77°C. [456

NATRIUMPHOSPHAT (*n*), **NEUTRALES-**: neutral sodium phosphate (see **NATRIUM, PHOSPHORSAURES-**) 457

NATRIUMPHOSPHAT, PRIM.- (*n*): sodium phosphate (primary); monobasic sodiumphosphate; sodium acid phosphate; sodium biphosphate; monosodium phosphate; monosodium orthophosphate; $NaH_2PO_4.H_2O$; Sg. 2.04 458

NATRIUMPHOSPHAT, SEC.- (*n*): sodium phosphate (secondary); $Na_2HPO_4.12H_2O$ (see **NATRIUM, PHOSPHORSAURES-**) 459

NATRIUMPHOSPHAT, TERT.- (*n*): sodium phosphate (tertiary); $Na_3PO_4.12H_2O$ (see **NATRIUM-PHOSPHAT, BASISCHES-**) 460

NATRIUM (*n*), **PHOSPHORSAURES-**: (*Natrium phosphoricum*): disodium phosphate, dibasic sodium phosphate, disodium orthophosphate, hydrodisodic phosphate, secondary sodium phosphate; $Na_2HPO_4.12H_2O$; Sg. 1.524; Mp. 35°C. 461

NATRIUM (*n*), **PIKRAMINSAURES-**: sodium picramate 462

NATRIUMPIKRAT (*n*): sodium picrate; $C_6H_2.ONa(NO_2)_3$ 463

NATRIUM (*n*), **PIKRINSAURES-**: sodium picrate (see **NATRIUMPIKRAT**) 464

NATRIUMPLATINCHLORID (*n*): sodium chloroplatinate 465

NATRIUMPLUMBAT (*n*): sodium plumbate (*Natrium plumbicum*); Na_2PbO_4 466

NATRIUMPOLYSULFID (*n*): sodium polysulphide 467

NATRIUM (*n*), **PROPIONSAURES-**: sodium propionate 468

NATRIUM (*n*), **PROPYLSCHWEFELSAURES-**: sodium propylsulphate 469

NATRIUMPYROPHOSPHAT (*n*): sodium pyrophosphate; $Na_4P_2O_7.6H_2O$; Sg. 1.82; or $Na_4P_2O_7$; Sg. 2.45; Mp. 970°C. 470

NATRIUM, PYROPHOSPHORSAURES- (*n*): sodium pyrophosphate (see **NATRIUMPYROPHOSPHAT**) 471

NATRIUMRHODANÜR (*n*): (*Natrium rhodanatum*): sodium (sulfocyanide; thiocyanate) sulfocyanate, sodium (rhodanate) rhodanide; NaCNS; Mp. 287°C. 472

NATRIUMSACCHARAT (*n*): sodium saccharate 473

NATRIUM, SALICYLSAURES- (*n*): sodium salicylate; $NaC_7H_5O_3$ 474

NATRIUM (*n*), **SALPETERSAURES-**: sodium nitrate, Chili (saltpeter) nitre, nitratine, soda nitre, cubic (saltpeter) nitre, soda saltpeter; $NaNO_3$; Sg. 2.267; Mp. 316°C. 475

NATRIUM, SALPETRIGSAURES- (*n*): sodium nitrite (see **NATRIUMNITRIT**) 476

NATRIUMSALZ DER DIÄTHYLBARBITURSÄURE (*f*): (*Natrium diæthylbarbituricum*): (see **NATRIUM, DIÄTHYLBARBITURSAURES-**) 477

NATRIUMSALZ (*n*) **DER DIOXYWEINSÄURE** (*f*): sodium dioxytartrate; $C_4H_4O_8Na_2.3H_2O$ 478

NATRIUM (*n*), **SAURES CHROMSAURES-**: sodium bichromate, bichromate of soda (see **NATRIUMBICHROMAT**) 479

NATRIUM (*m*), **SAURES KOHLENSAURES-**: sodium bicarbonate, bicarbonate of soda (see **NATRIUMBICARBONAT**) 480

NATRIUM (*n*), **SAURES PHOSPHORSAURES-**: sodium biphosphate (see **NATRIUMPHOSPHAT, PRIM.-**) 481

NATRIUM (*n*), **SAURES SCHWEFELSAURES-**: sodium bisulphate (see **NATRIUMBISULFAT**) 482

NATRIUM (*n*), **SAURES SCHWEFLIGSAURES-**: sodium bisulphite (see **NATRIUMBISULFIT**) 483

NATRIUM (*n*), **SCHWEFELSAURES-**: sodium sulphate (see **NATRIUMSULFAT** and **GLAUBERSALZ**) 484

NATRIUM, SCHWEFLIGSAURES- (*n*): sodium sulphite (see **NATRIUMSULFIT**) 485

NATRIUM, SELENIGSAURES- (*n*): sodium selenite; Na_2SeO_3 486

NATRIUM, SELENSAURES- (*n*): sodium selenate 487

NATRIUMSILBERHYPOSULFIT (*n*): (double)-hyposulphite of sodium and silver; $AgNaS_2O_3$ or $Ag_2Na_4.3(S_2O_3)$ 488

NATRIUMSILICOFLUORID (*n*): sodium silicofluoride, sodium fluosilicate, salufer; Na_2SiF_6; Sg. 2.679 489

NATRIUMSILIKAT (*n*): sodium silicate, water glass, soluble glass (see **NATRONWASSERGLAS** and **WASSERGLAS**) 490

NATRIUMSTANNAT (*n*): (*Natrium stannicum*): sodium stannate, preparing salt (40-56% SnO_2); $Na_2SnO_3.3H_2O$ 491

NATRIUM (*n*), **STEARINSAURES-**: soda stearate 492

NATRIUM (*n*), **SULFANILSAURES-**: sodium sulfanilate 493

NATRIUMSULFANTIMONIAT (*n*): sodium sulfantimoniate 494

NATRIUMSULFAT (*n*): sodium sulphate (*Natrium sulfuricum*); $Na_2SO_4.7H_2O$, Sg. 1.46-1.49, Mp. 32.4°C, Na_2SO_4, Sg. 2.67, Mp. 888°C. (see **GLAUBERSALZ**) 495

NATRIUMSULFAT, CALC.- (*n*): sodium sulphate (calcined); Na_2SO_4 (see GLAUBERSALZ, CALCINIERTES-) 496

NATRIUMSULFAT (*n*), EINFACH GEWÄSSERTES-: monohydrated sodium sulphate; $Na_2SO_4.H_2O$ 497

NATRIUMSULFAT (*n*), EINHYDRATIGES-: monohydrated sodium sulphate; $Na_2SO_4.H_2O$ 498

NATRIUMSULFAT (*n*), NEUTRALES-: neutral sodium sulphate; Na_2SO_4; Sg. 2.67; Mp. 888°C. 499

NATRIUMSULFAT (*n*), WASSERFREIES-: anhydrous sodium sulphate (see NATRIUMSULFAT, NEUTRALES-) 500

NATRIUMSULFAT (*n*), ZEHNFACHGEWÄSSERTES-: Glauber's salt; $NaSO_4.10H_2O$ 501

NATRIUMSULFHYDRAT (*n*): sodium hydrosulphide (see NATRIUMHYDROSULFID) 502

NATRIUMSULFID (*n*): (*Natrium sulfuratum*): sodium sulphide, sodium monosulphide, sodium sulfuret; Na_2S; Sg. 1.856-2.471 503

NATRIUMSULFIT (*n*): (*Natrium sulfurosum*): sodium sulphite; Na_2SO_3, Sg. 2.63, Mp. 150°C.; $Na_2SO_3.7H_2O$, Sg. 1.594, Mp. 150°C. 504

NATRIUM (*n*), SULFOPHENOLSAURES-: sodium sulphophenolate 505

NATRIUMSUPEROXYD (*n*): sodium peroxide; Na_2O_2 (see NATRIUMPEROXYD) 506

NATRIUMSUPEROXYDHYDRAT (*n*): (see NATRIUMPEROXYDHYDRAT) 507

NATRIUMTETRABORAT (*n*): sodium tetraborate, borax, sodium biborate; $Na_2B_4O_7$; Sg. 2.367 508

NATRIUMTHIOSULFAT (*n*): sodium (thiosulphate) hyposulphite; $Na_2S_2O_3+5H_2O$ (*Natrium thiosulfuricum*): (see NATRIUMHYPOSULFIT) 509

NATRIUM (*n*), ÜBERBORSAURES-: sodium perborate (see NATRIUMPERBORAT) 511

NATRIUM, ÜBERCHLORSAURES- (*n*): sodium perchlorate (see NATRIUMPERCHLORAT) 512

NATRIUM (*n*), ÜBERKOHLENSAURES-: (see NATRIUMPERCARBONAT) 513

NATRIUM (*n*), ÜBERMANGANSAURES-: sodium permanganate (see NATRIUMPERMANGANAT) 514

NATRIUM (*n*), ÜBERSCHWEFELSAURES-: (see NATRIUMPERSULFAT) 515

NATRIUM, UNTERCHLORIGSAURES- (*n*): sodium hypochlorite (see NATRIUMHYPOCHLORIT) 516

NATRIUM, UNTERPHOSPHORIGSAURES-: (*n*): sodium hypophosphite (see NATRIUMHYPOPHOSPHIT) 517

NATRIUM, UNTERSCHWEFELSAURES- (*n*): sodium hyposulphate (see NATRIUMHYPOSULFAT) 518

NATRIUM, UNTERSCHWEFLIGSAURES- (*n*): sodium hyposulphite (see NATRIUMHYPOSULFIT) 519

NATRIUMURANAT (*n*): sodium uranate; $Na_2U_2O_7$ 520

NATRIUM, URANSAURES- (*n*): sodium uranate, yellow uranium oxide; uranium yellow; Na_2UO_4 521

NATRIUM, VANADINSAURES- (*n*): sodium vanadate; sodium ortho-vanadate; $Na_3VO_4.16H_2O$; or (Na_3VO_4; Mp. 866°C.) 522

NATRIUMVERBINDUNG (*f*): sodium compound 523

NATRIUMWASSERSTOFF (*m*): sodium hydride 524

NATRIUM (*n*), WEINSAURES-: sodium tartrate; $Na_2C_4H_4O_6.2H_2O$; Sg. 1.794 525

NATRIUMWEINSTEIN (*m*): sodium tartrate (see NATRIUM, WEINSAURES-) 526

NATRIUMWOLFRAMAT (*n*): sodium (wolframate) tungstate (*Natrium wolframicum*); Na_2WO_4; Sg. 4.179; Mp. 698°C.; $Na_2WO_4+2H_2O$; Sg. 3.245-3.259 527

NATRIUM, WOLFRAMSAURES- (*n*): sodium tungstate (see NATRIUMWOLFRAMAT) 528

NATRIUM (*n*), XANTHOGENSAURES-: sodium xanthogenate 529

NATRIUM (*n*), ZIMTSAURES-: sodium cinnamate, hetol; $C_6H_5.CH.CH.CO_2Na$; Mp. 133.5°C. 530

NATRIUM, ZINNSAURES- (*n*): sodium stannate (see NATRIUMSTANNAT) 531

NATRIUMZITRAT (*n*): sodium citrate (see NATRIUMCITRAT) 532

NATRIUM, ZITRONENSAURES-: sodium citrate (see NATRIUMCITRAT) 533

NATRIUMZYANAMID (*n*): sodium cyanamide (see NATRIUMCYANAMID) 534

NATRIUMZYANID (*n*): sodium cyanide (see NATRIUMCYANID) 535

NATROBOROCALCIT (*m*): natroborocalcite; $NaCaB_5O_9.6H_2O$; Sg. 1.7 536

NATROBOROKALZIT (*n*): (see BORONATROKALZIT) 537

NATROCALCIT (*m*): natrocalcite, Gaylussite (see GAYLÜSSIT) 538

NATROLITH (*m*): natrolite (a zeolite) (an aluminium-sodium meta-silicate); $Na_2Al_2Si_3O_{10}.2H_2O$; Sg. 2.23 539

NATRON (*n*): caustic soda; sodium hydroxide; soda; Na_2O; (Mineral $Na_2CO_3+10H_2O$); Sg. 1.45 (see also NATRIUMHYDROXYD and NATRIUM) 540

NATRONALAUN (*m*): soda alum (see NATRIUMALAUN); mendozite (see MENDOZIT) 541

NATRONAMMONSALPETER (*m*): sodium-ammonium nitrate 542

NATRONÄTZLAUGE (*f*): caustic soda (solution) lye (see ÄTZNATRONLAUGE) 543

NATRONCELLULOSE (*f*): soda (fibre) pulp (Paper) 544

NATRON, DOPPELTKOHLENSAURES- (*n*): bicarbonate of soda (see DOPPELTKOHLENSAURES NATRON) 545

NATRON (*n*), DOPPELTSCHWEFLIGSAURES-: (see NATRIUMBISULFIT) 546

NATRON, EINFACHKOHLENSAURES- (*n*): soda; carbonate of soda; neutral carbonate of soda (see NATRIUMCARBONAT) 547

NATRONFELDSPAT (*m*): soda feldspar, albite (see ALBIT) 548

NATRON (*n*), FIXIER-: fixing salt (see NATRIUMHYPOSULFIT) 549

NATRONGLAS (*n*): soda glass 550

NATRONGLIMMER (*m*): paragonite (see PARAGONIT) 551

NATRONHALTIG: containing soda; sodaic 552

NATRONHYDRAT (*n*): sodium hydroxide (see NATRIUMHYDROXYD) 553

NATRONHYPEROXYD (*n*): sodium peroxide (see NATRIUMPEROXYD) 554

NATRONITRE (*m*): (see NATRONSALPETER) 555

NATRONITRIT (*n*): (see NATRONSALPETER) 556

NATRONKALK (*m*): soda lime (by heating equal parts of sodium hydrate and caustic lime to red heat) 557

NATRON, KOHLENSAURES- (*n*): sodium carbonate (see **NATRIUMCARBONAT**); mineral—$Na_2CO_3 + 10H_2O$; Sg. 1.45 (see also **THERMONATRIT**) 558

NATRONLAUGE (*f*): soda lye; caustic soda solution (see **ÄTZNATRONLAUGE**); Na_2O; (*Liquor Natrii Caustici*); Sg. 1.17; 15% NaOH (used in aluminium obtaining process) 559

NATRONMESOTYP (*m*): natrolite (see **NATROLITH**) 560

NATRONMETALL (*n*): metallic sodium; sodium metal 561

NATRON (*n*), **NAPHTOLDISULFOSAURES-**: R-salt; sodium salt of beta-naphtholdisulfonic acid 562

NATRON (*n*), **NUKLEINSAURES-**: sodium nucleinate (*Natrium nucleinicum*) 563

NATRON (*n*), **OXALSAURES-**: sodium oxalate (see **NATRIUMOXALAT**) 564

NATRONPASTILLE (*f*): troche or ball of sodium bicarbonate (Pharm.) 565

NATRON (*n*), **PHOSPHORSAURES-**: sodium phosphate (s e **NATRIUMPHOSPHAT, SEC.**; and **NATRIUM, PHOPHORSAURES-**) 566

NATRONSALPETER (*m*): soda nitre; natronitrite (Min.); soda (Chile) saltpeter; $NaNO_3$; sodium nitrate (see **NATRIUMNITRAT**) 567

NATRON (*n*), **SALPETERSAURES-**: sodium nitrate (see **NATRONSALPETER** and **NATRIUM, SALPETER--SAURES-**) 568

NATRONSALZ (*n*): soda (sodium) salt 569

NATRON (*n*), **SALZSAURES-**: sodium chloride (see **NATRIUMCHLORID**) 570

NATRON (*n*), **SAURES SCHWEFLIGSAURES-**: (see **NATRIUMBISULFIT**) 571

NATRONSCHWEFELLEBER (*f*): sodium polysulphide 572

NATRON (*n*) **CHWEFLIGSAURES-** (s e **NATRIUM--SULFIT**) 573

NATRON (*n*), **SCHWEFLIGSAURES, WASSERFREIES-**: anhydrous sodium sulphite; Na_2SO_3 (see **NATRIUMSULFIT**) 574

NATRONSEIFE (*f*): soda soap 575

NATRONSTOFF (*m*): soda pulp (Paper) 576

NATRON (*n*), **UNTERSCHWEFLIGSAURES-**: (see **NATRIUMHYPOSULFIT**) 577

NATRONVERBINDUNG (*f*): sodium compound 578

NATRONWASSERGLAS (*n*): water-glass; soluble glass; sodium silicate; $Na_2SiO_3.9H_2O$; Mp. 48°C., Na_2SiO_3; Mp. 1018°C. 579

NATRONWASSERGLASLÖSUNG (*f*): sodium silicate solution (*Liquor Natrii silicici*) (Sg. 1.35, about 35% sodium silicate) 580

NATRONWEINSTEIN (*m*): sodium tartrate; $Na_2C_4H_4O_6.2H_2O$; Sg. 1.79 (see **KALIUM--NATRIUMTARTRAT**) 581

NATRON (*n*). **WOLFRAMSAURES-**: (see **NATRIUM-WOLFRAMAT**) 582

NATRONZELLSTOFF (*m*): soda (pulp) fibre (Paper) 583

NATRON (*n*), **ZWEIFACHKOHLENSAURES-**: sodium bicarbonate (see **NATRIUMBIKARBONAT**) 584

NATRUM (*n*): (see under **NATRIUM** and **NATRON**) 585

NATTERBLÜMCHEN (*n*): milkwort (*Polygala vulgaris*) 586

NATTERWURZ (*f*): bistort (*Polygonum bistorta*); cuckoopint (*Arum maculatum*) 587

NATTERZUNGE (*f*): adder's tongue (*Ophioglossum vulgatum*) 588

NATUR (*f*): nature; character; constitution; essence; quality; species; sort; kind; disposition; temperament; nakedness; matter; material world; universe; creature 589

NATURALIEN (*pl*): natural (products) specimens; natural history specimens 590

NATURALISIEREN: to naturalize, acclimatize, accustom 591

NATURALLEISTUNG (*f*): payment in kind 592

NATURALPHOTOGRAPHIE (*f*): naturalistic photography, impressionistic photography (in which the image is kept only as sharp as the eye sees it) 593

NATURBALSAM (*m*): natural balsam 594

NATURBEGEBENHEIT (*f*): natural phenomenon 595

NATURBESCHREIBER (*m*): naturalist, natural historian 596

NATURBESCHREIBUNG (*f*): natural history 597

NATURERSCHEINUNG (*f*): natural phenomenon 598

NATURERZEUGNIS (*n*): natural product-(ion) 599

NATURFARBIG: natural coloured; fair (of leather) 600

NATURFEHLER (*m*): inborn defect; natural defect 601

NATURFORSCHER (*m*): scientific investigator-(of nature); naturalist; natural philosopher 602

NATURFORSCHUNG (*f*): natural philosophy 603

NATURGAS (*n*): natural gas (main constituent is Methane) 604

NATURGEMÄSS: according to nature; natural; regular; legitimate; consistent with (conformable to) nature 605

NATURGESCHICHTE (*f*): natural history 606

NATURGESETZ (*n*): natural law 607

NATURGETREU (*n*): true to (nature) life 608

NATURHÄRTE (*f*): natural hardness 609

NATURHISTORIKER (*m*): natural historian 610

NATURKRAFT (*f*): natural (power) force 611

NATURKUNDE (*f*): natural (philosophy) science; physics 612

NATURLEHRE (*f*): natural (philosophy) science; physics 613

NATÜRLICH: natural-(ly); native; genuine; innate; true; certainly; of course; by all means; legitimate; unaffected; real; original; indigenous; characteristic; essential; normal; regular; bastard; illegitimate; ingenuous; artless; life-like 614

NATÜRLICHE GRÖSSE (*f*): life-size; natural size; full size 615

NATÜRLICHER FARBSTOFF (*m*): natural colouring matter (animal or vegetable dyes) 616

NATURPHILOSOPHIE (*f*): natural philosophy 617

NATURPRODUKT (*n*): natural product-(ion) 618

NATURRECHT (*n*): law of nature 619

NATURREICH (*n*): kingdom of nature; nature 620

NATURSPIEL (*n*): freak of nature 621

NATURSTAND (*m*): natural state 622

NATURTRIEB (*m*): natural instinct 623

NATURWIDRIG: unnatural, abnormal, preternatural, inhuman, unfeeling, brutal, artificial, affected, constrained 624

NATURWISSENSCHAFT (*f*): natural (physical) science; physics; natural (philosophy) history 625

NATURZUG (*m*): characteristic; peculiarity; distinguishing (trait) feature 626

NAUMANNIT (*m*): naumannite; $(Ag_2,Pb)Se$; Sg. 8.0 627
NAUTIK (*f*): navigation; nautical matters 628
NAUTISCH: nautical 629
NAVIGATION (*f*): navigation 630
NAVIGATIONSRAUM (*m*): chart room 631
NAVYBLAU: navyblue 632
NAXOSSCHMIRGEL (*m*): Naxos emery 633
NDILOÖL (*n*): (see TACAMAHACFETT) 634
NEAPELGELB (*n*): Naples yellow (a basic lead antimonate); $PbSb_2O_6$; 635
NEAPELROT (*n*): Naples red 636
NEBEL (*m*): mist, haze, fog, smoke, nebula 637
NEBELAPPARAT (*m*): atomizer 638
NEBELBANK (*f*): bank of (mist) fog 639
NEBELBILD (*n*): dissolving view (Phot.) 640
NEBELFLECK (*m*): nebula (a cluster of stars); film; clouded spot 641
NEBELFLÜGEL (*m*): side-wing 642
NEBELGESCHOSS (*n*): smoke shell (Military) 643
NEBELHAFT: misty; foggy; nebulous; smoky; hazy; mist-like; filmy; cloudy; clouded 644
NEBELHORN (*n*): fog-horn 645
NEBELIG: (see NEBELHAFT) 646
NEBELKASTEN (*m*): smoke-box 647
NEBELKRANZ (*m*): nebulous (ring) rim; nebulous corona 648
NEBELREGEN (*m*): drizzle, Scotch mist 649
NEBELSIGNAL (*n*): fog-signal 650
NEBELSIGNALANLAGE (*f*): fog signalling (equipment) installation 651
NEBELTOPF (*m*): smoke pot 652
NEBELTROMMEL (*f*): smoke drum 653
NEBEN: beside; with; near; by; together with; as well as 654
NEBENABSICHT (*f*): secondary intention; side-issue 655
NEBENACHSE (*f*): secondary axis; conjugate axis 656
NEBENAN: next-door, near by 657
NEBENARBEIT (*f*): secondary (auxiliary, additional) work; side-line 658
NEBENAUSGABE (*f*): incidental expense, incidental outlay; supplement-(ary edition) 659
NEBENAUSGANG (*m*): emergency exit, auxiliary (exit) outlet, side passage (of houses) 660
NEBENBAHN (*f*): branch line (Railways), secondary line (Railways); siding; small-gauge railway 661
NEBENBEDEUTUNG (*f*): additional (signification) meaning; secondary importance 662
NEBENBEGRIFF (*m*): subordinate idea 663
NEBENBESTANDTEIL (*m*): secondary (ingredient) constituent 664
NEBENBETRACHTUNG (*f*): secondary consideration 665
NEBENBEWEIS (*m*): additional proof 666
NEBENBINDUNG (*f*): secondary (second-order) union, bond or linkage 667
NEBENBLATT (*n*): supplement (of papers); stipula (floral leaf) (Bot.) 668
NEBENDURCHMESSER (*m*): conjugate diameter 669
NEBENEINANDER: side by side; together; in parallel 670
NEBENEINANDER LIEGEND: lying side by side, arranged side by side, arranged in parallel, set in (pairs) battery; contiguous 671
NEBENEINANDERSCHALTUNG (*f*): parallel connection 672
NEBENEINANDERSTELLEN: to juxtapose; compare 673
NEBENEINANDERSTELLUNG (*f*): juxtaposition; comparison 674
NEBENEINGANG (*m*): side entrance 675
NEBENEINKÜNFTE (*f.pl*): incidental (emoluments) profits; additional income; income from other sources 676
NEBENERBE (*m*): co-heir 677
NEBENERZEUGNIS (*n*): by-product 678
NEBENFARBE (*f*): secondary colour 679
NEBENFLÄCHE (*f*): secondary (face) surface 680
NEBENFLUSS (*m*): tributary-(river) 681
NEBENFRAGE (*f*): question of secondary importance; side-issue 682
NEBENGANG (*m*): by-way; lateral vein (Mining); side-aisle; side passage 683
NEBENGEBÄUDE (*n*): extension (wing) of a building; outhouse 684
NEBENGESCHMACK (*m*): tang; aftertaste 685
NEBENGESETZ (*n*): by-law 686
NEBENGEWÄCHS (*n*): excrescence 687
NEBENGEWINN (*m*): additional (secondary) yield, gain or profit 688
NEBENGEWINNUNGSANLAGE (*f*): by-product plant 689
NEBENGLEIS (*n*): side track (Railways) 690
NEBENGRUPPE (*f*): accessory group 691
NEBENINTERESSE (*f*): private interests 692
NEBENKETTE (*f*): side (subordinate) chain 693
NEBENKOSTEN (*pl*): incidental expenses 694
NEBENLINIE (*f*): side (branch, secondary) line; collateral line 695
NEBENLUFT (*f*): secondary air 696
NEBENMOND (*m*): mock moon, paraselene 697
NEBENNIERE (*f*): suprarenal gland 698
NEBENORDNUNG (*f*): co-ordination 699
NEBENPLANET (*m*): satellite (Astron.) 700
NEBENPRODUKT (*n*): by-product 701
NEBENPRODUKTE-RÜCKGEWINNUNGSANLAGE (*f*): by-product recovery plant 702
NEBENROHR (*n*): side (branch) tube or pipe; by-pass pipe 703
NEBENROLLE (*f*): secondary (subordinate) rôle 704
NEBENRÜCKSICHT (*f*): secondary consideration 705
NEBENSATZ (*m*): subordinate sentence; incidental proposition 706
NEBENSCHIFF (*n*): side aisle 707
NEBENSCHLIESSUNG (*f*): shunt (Elect.) 708
NEBENSCHLUSS (*m*): (see NEBENSCHLIESSUNG) 709
NEBENSCHLUSSERREGUNG (*f*): shunt excitation 710
NEBENSCHLUSSMOTOR (*m*): shunt motor 711
NEBENSCHLUSSREGLUNG (*f*): shunt regulation (Elect.) 712
NEBENSCHLUSSREGULATOR (*m*): shunt regulator 713
NEBENSCHLUSSSCHALTUNG (*f*): shunt connection (Elect.) 714
NEBENSCHLUSSSPULE (*f*): shunt coil 715
NEBENSCHLUSSSTROM (*m*): induction current 716
NEBENSCHLUSSSTROMKREIS (*m*): shunt circuit 717
NEBENSCHLUSSWICKELUNG (*f*): shunt winding 718
NEBENSCHOSS (*m*): shoot; sucker 719
NEBENSERIE (*f*): subordinate series 720
NEBENSONNE (*f*): mock sun, parhelion 721
NEBENSPESEN (*f.pl*): additional charges 722
NEBENSPIRALE (*f*): secondary coil 723

NEBENSTRASSE (*f*): side-turning; by-road; side-street 724
NEBENSTROM (*m*): induction current (Elect.); tributary-(river) 725
NEBENTEIL (*m*): accessory part 726
NEBENTITEL (*m*): sub-title 727
NEBENTON (*m*): secondary accent 728
NEBENTYPUS (*m*): secondary type 729
NEBENURSACHE (*f*): secondary cause 730
NEBENVERDIENST (*m*): incidental profit 731
NEBENWINKEL (*m*): adjacent angle 732
NEBENWIRKUNG (*f*): secondary effect 733
NEBENWORT (*n*): adverb 734
NEBENWÖRTLICH: adverbial 735
NEBENZWECK (*m*): subordinate (secondary) aim or object 736
NEBST: besides; including; (together)-with; in addition to; as well as 737
NEGATIV: negative; (*n*): negative, plate (Photography) (see also PLATTE) 738
NEGATIVBAD (*n*): negative bath (Phot.) 739
NEGATIVE (*f*): negative 740
NEGATIVIEREN: to deny; gainsay; adjure; disavow; abnegate; veto; negative 741
NEGATIVLACK (*m*): negative varnish 742
NEGATIVPAPIER (*n*): negative paper (Phot.) 743
NEGATIVRETOUCHE (*f*): negative retouching 744
NEGATIVSTÄNDER (*m*): drying rack (for photographic negatives) 745
NEGATIVWASCHKASTEN (*m*): negative (plate) washer (Phot.) 746
NEGERPULVER (*n*): negro powder (an explosive, 86 to 90% ammonium nitrate; 9 to 11% trinitrotoluene and 1 to 3% graphite 747
NEGOZIIEREN: to negotiate, discount (Bills, etc.) 748
NEHMER (*m*): receiver; recipient 749
NEID (*m*): envy; jealousy; grudge; ill-will; enviousness 750
NEIDEN: to envy, grudge, begrudge, be jealous of 751
NEIDENSWERT: enviable 752
NEIGE (*f*): slope; inclination; incline; decline; end; wane; dregs; sediment; rest; remains; remainder; bow; curtsey; acclivity; declivity; slant 753
NEIGEN: to slope; incline; lean; bend; dip; bow; curtsey; decline; wane; be biassed; tilt; cant; slant; lie at an angle; take an oblique direction 754
NEIGUNG (*f*): inclination; slope; dip; pitch; tendency; bias; taste; propensity; disposition; affection; decline; incline; acclivity; declivity; gradient; wane; bow; curtsey; incidence 755
NEIGUNGSEBENE (*f*): inclined plane 756
NEIGUNGSFLÄCHE (*f*): inclined plane 757
NEIGUNGSLOT (*n*): axis of incidence 758
NEIGUNGSWINKEL (*m*): angle of inclination 759
NEKROLOGIE (*f*): necrology, record of deaths 760
NEKROMANT (*m*): necromancer 761
NEKROMANTIE (*f*): necromancy, conjuring 762
NELKE (*f*): pink-(flower) (*Dianthus plumarius*); clove (*Caryophyllus aromaticus*); carnation (*Dianthus caryophyllus*) 763
NELKENBAUM (*m*): clove bush (*Eugenia caryophyllata, Eugenia aromatica* or *Caryophyllus aromaticus*) 764
NELKENBLÜTE (*f*): clove (blossom) flower; pink 765
NELKENFARBE (*f*): pink (see PINK) 766
NELKENÖL (*n*): clove oil (*Oleum Caryophyllorum*) (from dried buds of *Caryophyllus aromaticus*) (see GEWÜRZNELKENÖL) 767
NELKENPFEFFER (*m*): pimento; allspice; Jamaica pepper; pimenta (*Pimenta officinalis*) 768
NELKENPFEFFERÖL (*n*): pimenta oil; Sg. 1.05 769
NELKENPFEFFERWASSER (*n*): pimento water 770
NELKENRINDE (*f*): cinnamon; cassia (bark of *Cinnamomum cassia*) 771
NELKENSÄURE (*f*): eugenol; caryophyllic acid (see EUGENOL) 772
NELKENSTEIN (*m*): iolite (see IOLITH) 773
NELKENWURZEL (*f*): pink (clove) root (*Radix caryophyllatæ*) 774
NELKENZIMT (*m*): clove (Chinese, cassia-)cinnamon (*Cinnamomum cassia*) 775
NELKENZIMTÖL (*n*): cassia oil; chinese cinnamon oil; Sg. 1.054 776
NENNBETRAG (*m*): nominal (amount) value 777
NENNEN: to term; name; call; designate; denominate; style; mention; refer to; speak of; nominate 778
NENNER (*m*): denominator; nominator; namer 779
NENNER (*m*), KLEINSTER-: lowest common denominator, L.C.M. (Maths) 780
NENNFALL (*m*): nominative case 781
NENNLEISTUNG (*f*): nominal horse-power; nominal (load) output or duty 782
NENNUNG (*f*): denomination, nomination, mention, naming, designation 783
NENNWERT (*m*): nominal value 784
NENNWORT (*n*): noun; substantive 785
NEODYM (*n*): neodymium (Nd) 786
NEODYMCARBID (*n*): neodymium carbide; NdC_2; Sg. 5.15 787
NEODYMCHLORID (*n*): neodymium chloride; $NdCl_3$; Sg. 4.134 788
NEODYMOXYD (*n*): neodymium oxide; Nd_2O_3 789
NEODYMSULFAT (*n*): neodymium sulphate; $Nd_2(SO_4)_3+8H_2O$; Sg. 2.85 790
NEODYMSULFID (*n*): neodymium sulphide; Nd_2S_3; Sg. 5.179 791
NEOFORM (*n*): neoform, triiodophenol bismuth; $C_6H_2I_3.OBiO$ 792
NEOGLYZERIN (*n*): neoglycerine (a glycerine substitute; Sg. 1.29; freezing point about −25°C.) 793
NEOHEXAL (*n*): neohexal; secondary hexamethylenetetramine sulfosalicylate; $2[(CH_2)_6N_4].SO_3H.C_6H_3(OH)COOH$; Mp. 180.5°C. 794
NEOLITH (*m*): neolite; $Mg(Ca,Fe,Mn)O,Al_2O_3,SiO_2,H_2O$; Sg. 2.7 795
NEOLOG (*m*): neologist (one who introduces new words or doctrines) 796
NEON (*n*): Neon (Ne) 797
NEONBOGENLAMPE (*f*): neon arc lamp (filled with neon gas) 798
NEONGAS (*n*): neon gas (see NEON) 799
NEONGLIMMLAMPE (*f*): neon glow lamp 800
NEOSALVARSAN (*n*): neosalvarsan (an improved form of salvarsan) 801
NEOTANNYL (*n*): neotannyl (*Aluminium acetotannate*) 802
NEOTANTALIT (*m*): neotantalite; $Nb_2O_5,Ta_2O_5,FeO,MnO,K_2O,Na_2O,H_2O$ 803
NEOTYP (*m*): neotype (calcite plus BaO) (see CALCIT); Sg. 2.8 804

NEOVIOLON (*n*): neoviolon (a synthetic violet scent) 805
NEPHELIN (*m*): nephelite (a sodium-aluminium silicate); $4(Na_2O,Al_2O_3),9SiO_2$; Sg. 2.58 (nephelite of massive structure often termed Elaeolite) 806
NEPHRIT (*m*): jade, nephrite (natural metasilicate of iron, lime and magnesium); $Ca(Mg,Fe)_3(SiO_3)_4$; Sg. 3.1; actinolite (see AKTINO--LITH) 807
NEPOTISMUS (*m*): nepotism (favouritism to those of one's own kin) 808
NEPTUNBLAU (*n*): neptune blue 809
NEPTUNIT (*m*): neptunite; $(Na,K)_2(Fe,Mn)(Si,Ti)_5O_{12}$; Sg. 3.23 810
NERADOL (*n*): neradol (an artificial tanning extract) 811
NERALTEIN (*n*): neraltein; $C_2H_5O.C_6H_4NHCH_2.SO_3Na$ 812
NERNSTLAMPE (*f*): Nernst lamp (an electric glow lamp with magnesia filaments) 813
NEROL (*n*): nerol (an isomeric terpene alcohol); $C_{10}H_{17}.OH$; Sg. 0.883; Bp. 225.5°C. [814
NEROLIN (*n*): nerolin; naphthol methyl ether; $C_{10}H_7(OCH_3)$ (see YARA-YARA and BETA-NAPHTOL-ÄTHYLÄTHER) 815
NEROLIÖL (*n*): neroli; oil of orange flowers (*Oleum Aurantii florum*); Sg. 0.875 (from flowers of *Citrus vulgaris risso*) 816
NEROLIÖL (*n*), **SYNTETISCHES-**: synthetic neroli oil (Methyl anthranilate); $C_6H_4(NH_2)CO_2CH_3$; Sg. 1.17; Mp. 24.5°C.; Bp. 125°C. [817
NERV (*m*): nerve; fibre; sinew; vein (of leaf); filament; string 818
NERVENANFALL (*m*): attack of nerves 819
NERVENBAU (*m*): nervous structure; structure of the nerves; nervous system 820
NERVENBERUHIGUNGSMITTEL (*n*): nerve sedative, anodyne, nerve stilling agent (Med.) 821
NERVENENTZÜNDUNG (*f*): neuritis (inflammation of a nerve or nerves) (Med.) 822
NERVENFIEBER (*n*): nervous fever (Med.) 823
NERVENGEWEBE (*n*): nervous tissue; nervous plexus (a net work of nerves) 824
NERVENHAUT (*f*): retina (of the eye) (the internal tissue of the eye, being the expansion of the optic nerve) 825
NERVENKNOTEN (*m* and *pl*): Ganglion, knot; (*pl*) ganglia (Med.) 826
NERVENKRANK: nervous; neurotic 827
NERVENKRANKHEIT (*f*): nervous (breakdown) disorder; disease of (injury to) a nerve or nerves (Med.) 828
NERVENKUNDE (*f*): neurology (the science of the nerves) 829
NERVENLÄHMUNG (*f*): palsy (absence of feeling of power to regulate the muscles); paralysis (local or general dislocation of the nervous system) (Med.) 830
NERVENLEHRE (*f*): (see NERVENKUNDE) 831
NERVENLEIDEN (*n*): neuroses (a nervous disorder without known cause) (Med.) 832
NERVENLOS: without nerves; nerveless 833
NERVENMITTEL (*n*): neurotic 834
NERVENSAFT (*m*): nervous fluid 835
NERVENSCHLAG (*m*): paralysis, stroke (see NERVENLÄHMUNG) 836
NERVENSCHMERZ (*m*): neuralgia (pain of the nerves) (if of the facial nerves is termed tic-doloreux, and if of the sciatic nerve in the leg, sciatica) 837
NERVENSCHWACH: weak-nerved, nervous (see NERVÖS) 838
NERVENSTÄRKEND: neurotic; nerve-strengthening; tonic 839
NERVENSTILLEND: sedative, narcotic 840
NERVENSTILLENDES MITTEL (*n*): sedative, narcotic 841
NERVENSTOFF (*m*): nerve substance; neural substance 842
NERVENSYSTEM (*n*): nervous system 843
NERVENZENTRUM (*n*): nerve centre 844
NERVIG: nervous; sinewy; vigorous; strong; nervy; robust; forcible; well-strung (see NERVÖS) 845
NERVINUM (*n*): nerve remedy 846
NERVÖS: nervous, timid; timorous, frightened, fearful, shaky, neurotic (see NERVIG) 847
NERVUS SYMPATHICUS (*f*): sympathetic-(nervous) system (Med.) 848
NESSEL (*f*): nettle (*Urtica*) (see also RAMIE) 849
NESSELAUSSCHLAG (*m*): (see NESSELSUCHT) 850
NESSELFIEBER (*n*): nettle rash; urticaria (a skin disease) (Med.) 851
NESSELFRIESEL (*m*): (see NESSELFIEBER) 852
NESSELKRAUT (*n*): hemp nettle (*Galeopsis tetrahit*) 853
NESSELMAL (*n*): rash, nettlerash (the red swelling of nettlerash) 854
NESSELSUCHT (*f*): nettlerash, urticaria (a skin disease) 855
NESSELTUCH (*n*): muslin; material made of nettle fibre 856
NESSLERS REAGENS (*n*): Nessler's reagent (solution of potassium iodide and mercuric chloride; for detecting ammonia; NH_3) 857
NESTEL (*f*): filament, string, net, lace 858
NESTELLOCH (*n*): eyelet-(hole) 859
NESTELN: to string, lace 860
NESTELNADEL (*f*): bodkin 861
NETTHEIT (*f*): neatness; kindness; pleasantness; cleanness; fairness; prettiness; spruceness; niceness; whiteness (Sugar) 862
NETTO: net; nett; clear; neat 863
NETTOBETRAG (*m*): nett (clear) amount 864
NETTOEINNAHME (*f*): net receipts, net turn-over 865
NETTOGEWICHT (*n*): nett (net) weight 866
NETTOGEWINN (*m*): net (nett; clear) profit 867
NETTOWERT (*m*): net value 868
NETZ (*n*): net; netting; network; gauze; plexus; omentum; reticle (of telescope); system; lines (of pipes or wires); snare; caul, (membrane covering the head of some newly-born children) 869
NETZARTIG: reticulated; reticular; net-like 870
NETZBEIZE (*f*): oil mordant (Turkey red dyeing) 871
NETZBRUCH (*m*): epiplocele, hernia of the omentum (Med.) (see NETZHAUT) 872
NETZDARMBRUCH (*m*): (see NETZBRUCH) 873
NETZEBENE (*f*): space lattice (raumgitter) plane (Crystallography) 874
NETZEN: to wet; steep; moisten; soak; humidify; damp; sprinkle 875
NETZFARBE (*f*): paint for nets 876
NETZFASS (*n*): steeping (cask) tub 877
NETZFLOTTE (*f*): wetting-out liquid (Dye) 878
NETZFLÜGLER (*m.pl*): neuroptera (Entomology) 879
NETZFÖRMIG: reticulated; reticular; net-shaped 880

NETZHAUT (*f*): retina (of the eye); omentum; epiploon (membrane covering the intestines) 881
NETZHAUTENTZÜNDUNG (*f*): retinitis (inflammation of the retina of the eye) (Med.) 882
NETZ (*n*), **KLÜVERBAUM-**: jib-boom net 883
NETZNADEL (*f*): netting needle 884
NETZNEGATIV (*n*): line screen (Phot.) 885
NETZSCHWERTEL (*m*): Gladiole (*Gladiolus communis*) 886
NETZSTÄNDER (*m*): steeping tub (Paper) 887
NETZWERK (*n*): net-work 888
NEU: new-(ly); lately; fresh; modern; revised; recent; up-to-date; novel 889
NEUBLAU (*n*): Saxon blue, new blue (see **NAPHTYLENBLAU**) (also a name given to washing blue, see **WASCHBLAU**) 890
NEUBURGER KIESELKREIDE (*f*): Neuburg silicious chalk 891
NEUBURGER KIESELWEISS (*n*): Neuburg silicious whiting 892
NEUBURGER KREIDE (*f*): Neuburg chalk 893
NEUERER (*m*): innovator; neologist (one who introduces new words or doctrines) 894
NEUERFUNDEN: newly invented 895
NEUERN: to innovate; make (changes) innovations; introduce novelties; freshen up; revive; refresh; renew 896
NEUERUNG (*f*): innovation; new departure; novelty (in patents); revivifying; freshening-up 897
NEUGELB (*n*): new yellow; chrome yellow 898
NEUGESINNT: having up-to-date (modern) views, ideas or tendencies; progressive 899
NEUGESTALTUNG (*f*): re-organization 900
NEUGEWÜRZ (*n*): pimento; allspice; Jamaica pepper (*Pimenta officinalis*) 901
NEUGIER(-DE) (*f*): curiosity; inquisitiveness 902
NEUGIERIG: curious; inquisitive; anxious 903
NEUGOLD (*n*): Mannheim gold 904
NEUGRAU (*n*): (see **NIGRISIN**) 905
NEUHEIT (*f*): novelty; newness; freshness 906
NEUIGKEIT (*f*): news; novelty; innovation; change; new departure 907
NEUKUPFER (*n*): titanium (see **TITAN**) 908
NEULICH: lately; of late; late; recent-(ly) 909
NEULING (*m*): novice; freshman; stranger; neophyte; new hand; beginner; learner; new arrival; convert; proselyte; probationer 910
NEUNACHTELTAKT (*m*): 9/8 time (Mus.) 911
NEUNECK (*n*): nonagon (a nine sided plane) 912
NEUNECKIG: nonagonal (having nine sides) 913
NEUNFACH: ninefold 914
NEUNFACHE (*n*): nine times the amount 915
NEUNFÄLTIG: (see **NEUNFACH**) 916
NEUNFLÄCHIG: having nine (sides) faces 917
NEUNGLEICH (*n*): club-moss (*Lycopodium clavatum*) 918
NEUNHEIL (*n*): (see **NEUNGLEICH**) 919
NEUNMAL: nine times 920
NEUNWERTIG: nonavalent 921
NEURALGIE (*f*): neuralgia (Med.) 922
NEURASTHENIE (*f*): neurasthenia (Med.) 923
NEURIDIN (*n*): neuridine (a ptomaine from human corpses); $C_5H_{12}N_2$ 924
NEURIN (*n*): neurine (a ptomaine from cadaveric putrefaction); $C_5H_{13}NO$ 925
NEURODIN (*n*): neurodin, acetyl-para-oxyphenylurethan; $C_6H_4(CO_2CH_3)NH.CO_2.C_2H_5$; Mp. 87°C. 926
NEUROFEBRIN (*n*): neurofebrin (a mixture of neuronal and antifebrin) 927
NEURONAL (*n*): neuronal, bromodiethylacetamide; $Br(C_2H_5)_2C_2NHOH$; Mp. 66.5°C. 928
NEUROSF (*f*): neuralgia (Med.) 929
NEUROT (*n*): new red (Turkey red dyeing) 930
NEUROTROPIN (*n*): neurotropin, urotropin methylene-citrate 931
NEUROTVERFAHREN (*n*): new red process 932
NEUSILBER (*n*): white copper, Argentan, German silver, pack fong (a nickel alloy) (in plate form 60% Cu; 20% Ni; 20% Zn; if cast 52-63% Cu; 22-6% Ni; 26-31% Zn) 933
NEUTRAL: neutral; neuter 934
NEUTRALBLAU (*f*): neutral blue 935
NEUTRALFARBE (*f*): neutral (colour) tint (such as Pain's grey) 936
NEUTRALISATION (*f*): neutralization 937
NEUTRALISATIONSWÄRME (*f*): heat of neutralization 938
NEUTRALISIEREN: to neutralize 939
NEUTRALITÄT (*f*): neutrality 940
NEUTRALON (*n*): neutralon (an aluminium silicate, insoluble in H_2O); $Al_2Si_6O_{15}.2H_2O$ 941
NEUTRALSALZ (*n*): neutral salt 942
NEUTRALWEISS (*n*): neutral white 943
NEUTRUM (*n*): neuter 944
NEUWIEDERBLAU (*n*): blue verditer (see **KALKBLAU**) (blue basic copper carbonate); $2CuCO_3.Cu(OH)_2$ 945
NEUWIEDERGRÜN (*n*): Neuwied green; Paris green (see **PARISERGRÜN**) 946
NEUZEIT (*f*): modern times; present age; to-day 947
NEUZEITLICH: modern; up-to-date; recent; recently; latterly; nowadays 948
NEUZUSTELLUNG (*f*): dismantling and re-erection (of blast furnace brickwork, as well as repairs to air heaters, etc.); reconditioning 949
NEVILLE SÄURE (*f*): Neville acid, alpha-naphtholsulfonic acid; $C_{10}H_6(OH)(SO_3H)$ 1 : 4; Mp. 170°C. 950
NEWJANSKIT (*m*): newjanskite; (Os,Ir,Pt,Rh,Ru); Sg. 19.0. 951
NICCOLIT (*m*): niccolite (natural nickel arsenide, about 44% Ni); Ni(As,Sb); Sg. 7.5 see also **NICKELIN**) 952
NICHTACHTUNG (*f*): disregard; disrespect 953
NICHTANNAHME (*f*): non-acceptance 954
NICHTCARBONATHÄRTE (*f*): non-carbonate hardness (of water) (hardness caused by chlorides, nitrates or sulphates, which is not removable by boiling) 955
NICHTDASEIN (*n*): non-existence 956
NICHTDISSOCIIERT: undissociated; not dissociated 957
NICHTEISENHALTIG: non-ferrous 958
NICHTEISENLEGIERUNG (*f*): non-ferrous alloy 959
NICHTEISENMETALL (*n*): non-ferrous metal 960
NICHTEISERN: non-ferrous 961
NICHTEIWEISSARTIG: non-albuminous, non-albuminoid 962
NICHTELEKTROLYT (*n*): non-electrolyte 963
NICHTERSCHEINUNG (*f*): default; non-appearance 964
NICHTEXISTENZFÄHIG: incapable of existence 965
NICHTFLÜCHTIG: not volatile; non-volatile 966
NICHTGASFÖRMIG: non-gaseous 967
NICHTGERBSTOFF (*m*): non-tan-(nin) 968
NICHTGEWOLLT: undesired, unrequired, unwished for 969

NICHTIG: null; void; futile; idle; vain; invalid; transitory; empty 970

NICHTIGKEIT (*f*): nullity; invalidity; futility; nothingness 971

NICHTION (*n*): non-ion 972

NICHTLEITEND: non-conducting; dielectric 973

NICHTLEITENDES MATERIAL (*n*): non-conducting (dielectric) material; covering; lagging; non-conducting composition 974

NICHTLEITER (*m*): non-conductor; insulator 975

NICHTLIEFERUNG (*f*): non-delivery 976

NICHTMETALL (*n*): non-metal 977

NICHTMETALLISCH: non-metallic 978

NICHTMISCHBAR: immiscible; incapable of mixing 979

NICHTNORMAL: abnormal; not normal 980

NICHTROSTEND: stainless, non-corrosive, non-rusting, non-corroding 981

NICHTS: nothing; naught; zero; none; (*n*): tutti, impure zinc oxide 982

NICHTSCHMELZBAR: non-fusible; non-meltable 983

NICHTSEIN (*n*): non-existence 984

NICHTSNÜTZIG: valueless; worthless; useless 985

NICHTSSAGEND: meaningless; insignificant; non-committal 986

NICHTSTUN (*n*): idleness; inaction; idling; inactivity 987

NICHTS, WEISSES- (*n*): nihil album (sublimed zinc oxide) 988

NICHTSWÜRDIG: worthless; base; vile; futile; contemptible 989

NICHTSYNCHRON: not synchronous, asynchronous 990

NICHTUMWANDELBAR: inconvertible 991

NICHTVERGRÜNBAR: non-fading; will not turn green (usually in reference to black) 992

NICHTVOLLSTRECKUNG (*f*): non-performance 993

NICHTWÄSSERIG: non-aqueous 994

NICHTWESENTLICH: non-essential 995

NICHTWISSEN (*n*): ignorance; lack of knowledge 996

NICHTZAHLUNG (*f*): non-payment; non-settlement 997

NICHTZUCKER (*m*): non-sugar 998

NICKEL (*n*): nickel (*Nicolum*) (Ni) 999

NICKELACETAT (*n*): nickel acetate; $Ni(C_2H_3O_2)_2.4H_2O$; Sg. 1.7443 000

NICKELACETAT (*n*), **WASSERFREIES-**: anhydrous nickel acetate; $Ni(C_2H_3O_2)_2$; Sg. 1.798 [001

NICKELALUMINIUM (*n*): nickel-aluminium (used for thermo-elements) (Ni 98%; Al 2%) [002

NICKELALUMINIUMBRONZE (*f*): nickel-aluminium bronze (10% Cu; 20% Sn; 40% Ni; 30% Al; or 88% Cu; 10% Ni; 2% Al) [003

NICKELAMMONIUMCHLORID (*n*): nickel-ammonium chloride 004

NICKELAMMONIUMNITRAT (*n*): nickel-ammonium nitrate; $Ni(NO_3)_2.4NH_3.2H_2O$ [005

NICKELAMMONIUM (*n*), **SCHWEFELSAURES-**: (*Nicolo-Ammonium sulfuricum*): nickel-ammonium sulphate; $NiSO_4.(NH_4)_2SO_4.6H_2O$; Sg. 1.93 006

NICKELAMMONIUMSULFAT (*n*): nickel-ammonium sulphate (see NICKELAMMONIUM, SCHWEFELSAURES-) 007

NICKELANODE (*f*): nickel anode 008

NICKELANTIMONGLANZ (*m*): ullmannite (see ULLMANNIT) 009

NICKELANTIMONID (*n*): nickel antimonide; NiSb; Sg. 7.7 010

NICKELANTIMONKIES (*m*): ullmannite (see ULLMANNIT) 011

NICKELAPPARAT (*m*): nickel apparatus 012

NICKELARSENID (*n*): nickel arsenide; NiAs; Sg. 7.57; Ni_3As_2; Sg. 7.86 013

NICKELARSENIKKIES (*m*): gersdorffite (see NICKELGLANZ) 014

NICKELARSENKIES (*m*): gersdorffite (see NICKELGLANZ) 015

NICKELAUSKLEIDUNG (*f*): nickel lining 016

NICKELAZETAT (*n*): (*Nicolum aceticum*): nickel acetate; $Ni(C_2H_3O_2)_2$ 017

NICKELBAD (*n*): nickel bath (for nickel-coating) 018

NICKELBESCHLAG (*m*): nickel fitting, mounting or binding 019

NICKELBLÜTE (*f*): nickel bloom; nickel ochre; annabergite (see ERDKOBALT, GRÜNER-) 020

NICKELBORID (*n*): nickel boride; NiB; Sg. 7.39 021

NICKELBROMID (*n*): nickel bromide; $NiBr_2$ or $NiBr_2.3H_2O$; Sg. 4.64 022

NICKELBRONZE (*f*): nickel bronze (50% Cu; 25% Sn; 25% Ni) 023

NICKELCARBID (*n*): nickel carbide; Ni_3C 024

NICKELCARBONAT (*n*): nickel carbonate; $NiCO_3$ 025

NICKELCARBONAT (*n*), **BASISCHES-**: basic nickel carbonate; $2NiCO_3.3Ni(OH)_2.4H_2O$ 026

NICKELCARBONYL (*n*): nickel carbonyl, nickel tetracarbonyl; $Ni(CO)_4$; Sg. 1.3185; Mp. −25°C.; Bp. 43°C. 027

NICKELCHLORAT (*n*): nickel chlorate; $Ni(ClO_3)_2.6H_2O$ 028

NICKELCHLORID (*n*): nickel chloride; nickelic chloride; $NiCl_2$; Sg. 2.56 (*Nicolum chloratum*) 029

NICKELCHLORÜR (*n*): nickel-(ous) chloride; $NiCl_2.6H_2O$ 030

NICKELCHROM (*n*): nickel-chromium (used for thermo-elements); (90% Ni; 10% Cr) 031

NICKELCYANÜR (*n*): nickel cyanide; $Ni(CN)_2.4H_2O$ 032

NICKELEISEN (*n*): ferro nickel (an iron nickel alloy); (also found in nature; FeNi; Sg. 7.5) 033

NICKELERZ (*n*): nickel ore 034

NICKEL (*n*), **ESSIGSAURES-**: nickel acetate (see NICKELAZETAT) 035

NICKELFLUORID (*n*): nickel fluoride; NiF_2; Sg. 4.63 036

NICKELFLUSSEISEN (*n*): low carbon content nickel steel 037

NICKELGEFÄSS (*n*): nickel vessel 038

NICKELGEWINNUNG (*f*): nickel production 039

NICKELGLANZ (*m*): nickel glance; gersdorffite (natural nickel-arsenic sulphide); $NiAs_2.NiS_2$ 040

NICKEL (*n*), **GOLDCHLORWASSERSTOFFSAURES-**: nickel chloraurate; $Ni(AuCl_4)_2.8H_2O$ 041

NICKEL (*n*), **GUSS-**: cast nickel 042

NICKELGYMNIT (*m*): genthite, nickelgymnite (natural nickel-magnesium silicate); $(Ni,Mg)_4Si_3O_{10}.6H_2O$; Sg. 2.15 043

NICKELHALTIG: nickeliferous; containing nickel 044

NICKEL (*n*), **HARZSAURES-**: nickel resinate 045

NICKELHYDROXYD (*n*): nickel hydroxide (see NICKELOXYDHYDRAT) 046

NICKELHYDROXYDUL (*n*): nickelous hydroxide (see NICKELOXYDULHYDRAT) 047

NICKELIN (*m*): nickelin (the alloy) (56% Cu; 31% Ni; 13% Zn; or 68% Cu; 32% Ni); niccolite (the mineral) (see ROTNICKELKIES and NICCOLIT) 048
NICKELKARBONYL (*n*): nickel carbonyl (see NICKELCARBONYL) 049
NICKELKIES (*m*): millerite (see MILLERIT) 050
NICKELKOBALTKIES (*m*): linnaeite (see LINNEIT) 051
NICKELKOHLENOXYD (*n*): nickel carbonyl (see NICKELCARBONYL) 052
NICKEL (*n*), **KOHLENSAURES-**: nickel carbonate (see NICKELCARBONAT) 053
NICKELLEGIERUNG (*f*): nickel alloy (for electric resistances; 88% Ni; 4% Al; and 8% Cr) 054
NICKELMÜNZE (*f*): nickel coin (German = 75% Cu and 25% Ni) 055
NICKELNITRAT (*n*): nickel nitrate; $Ni(NO_3)_2 + 6H_2O$; Sg. 2.06; Mp. 56.7°C.; Bp. 136.7°C. 056
NICKELOCHER (*m*): nickel ochre; annabergite (see ERDKOBALT, GRÜNER-) 057
NICKELOCKER (*m*): nickel ochre, ochraniccoli, annabergite (see NICKELBLÜTE) 058
NICKELOXALAT (*n*): nickel oxalate 059
NICKEL (*n*), **OXALSAURES-**: nickel oxalate 060
NICKELOXYD (*n*): nickel-(ic) oxide; nickel sesquioxide; black nickel oxide; nickel peroxide (*Nicolum oxydatum*); Ni_2O_3; Sg. 4.83 061
NICKELOXYDHYDRAT (*n*): nickelic hydroxide; $Ni_2(OH)_6$ 062
NICKELOXYDUL (*n*): nickel protoxide; nickel (-ous) oxide; nickel monoxide; green nickel oxide; NiO; Sg. 6.7 (*Nicolum oxydulatum*) 063
NICKELOXYDULAMMONIAK, SCHWEFELSAURES- (*n*): (see NICKELAMMONIUM, SCHWEFELSAURES-) 064
NICKELOXYDULHYDRAT (*n*): nickelous hydroxide; $Ni(OH)_2$; Sg. 4.3 065
NICKELPHOSPHID (*n*): nickel phosphide; Ni_2P, Sg. 6.3; NiP_2, Sg. 4.62; NiP_3, Sg. 4.19 066
NICKEL (*n*), **SALPETERSAURES-**: nickel nitrate (see NICKELNITRAT) 067
NICKELSALZ (*n*): nickel salt (see NICKELAMMONIUM, SCHWEFELSAURES-) 068
NICKEL (*n*), **SCHWEFELSAURES-**: nickel sulphate (see NICKELSULFAT) 069
NICKELSELENAT (*n*): nickel selenate; $NiSeO_4 + 6H_2O$; Sg. 2.31 070
NICKELSELENID (*n*): nickel selenide; NiSe; Sg. 8.46 071
NICKELSESQUIOXYD (*n*): nickel sesquioxide (see NICKELOXYD) 072
NICKELSILICID (*n*): nickel silicide; Ni_2Si; Sg. 7.2 073
NICKELSILIKAT (*n*): nickel silicate 074
NICKELSPEISE (*f*): nickel speiss (Metal) 075
NICKELSPIESSGLANZ (*m*): ullmannite (see ULLMANNIT) 076
NICKELSTAHL (*m*): nickel steel 077
NICKELSTAHLGUSS (*m*): nickel-steel casting 078
NICKELSTEIN (*m*): nickel matt (Metal) 079
NICKELSTEIN (*m*), **SCHWERER-**: heavy nickel ore; (Ni_2S) 080
NICKELSULFAT (*n*): (*Nicolum sulfuricum*): nickel sulphate; $NiSO_4$; Sg. 3.42; $NiSO_4 + 7H_2O$; Sg. 1.98 Mp. 99°C. 081
NICKELSULFID (*n*): nickel sulphide; Ni_2S; Sg. 5.52; NiS; Sg. 4.6 082
NICKELSULFÜR (*n*): nickel-(ous) sulphide; NiS; Sg. 4.6 083
NICKELTHERMIT (*n*): nickel thermit 084
NICKELÜBERZUG (*m*): nickel-plating 085
NICKELVITRIOL (*m*): nickel sulphate (see NICKELSULFAT) 086
NICKELWISMUTGLANZ (*m*): grünanite 087
NICOTIN (*n*): nicotine (see NIKOTIN) 088
NIE: never; at no (time) period; not at any (time) period 089
NIEDER: low-(er); inferior; nether; secondary; down; mean; beneath; under; vulgar; subordinate 090
NIEDERBEUGEN: to depress, bend down 091
NIEDERBIEGEN: to deflect (bend) downwards; sag 092
NIEDERBLASEN: to blow (down) out (of a furnace or a boiler) 093
NIEDERBRECHEN: to break (tear; pull) down; crush; overcome; overwhelm; collapse; fail; give out 094
NIEDERBRENNEN: to burn down; reduce by combustion 095
NIEDERBRINGEN: to sink, lower; bring down 096
NIEDERDRUCK (*m*): low pressure; L.P. (usually below 15 Atms.) 097
NIEDERDRUCKANLAGE (*f*): low pressure (plant) installation 098
NIEDERDRÜCKEN: to depress; press (force; beat; keep; push; thrust) down; oppress; crush; overwhelm 099
NIEDERDRUCKENDE (*n*): low pressure end (Turbines) 100
NIEDERDRUCKTEIL (*m*): low pressure part 101
NIEDERFAHREN: to descend; drive down; ride down; decline; go down 102
NIEDERFALLEN: to fall down; settle; precipitate 103
NIEDERGANG (*m*): descent; decline setting -(of the sun); downstroke; subsidence; ebb 104
NIEDERGEHEN: to go down; descend; set; decline; subside; sink; settle; diminish; decrease; lull; wane; abate 105
NIEDERHOLER (*m*): downhaul 106
NIEDERHOLER (*m*), **LEESEGEL-**: studding sail downhaul 107
NIEDERHOLZ (*n*): under-growth, under-wood, coppice 108
NIEDERKOMMEN: to come down, descend; be confined, lie in (Med.) 109
NIEDERLAGE (*f*): warehouse; branch; agency; depot; defeat 110
NIEDERLAND (*n*): lowland-(s), low lying (land) country 111
NIEDERLASSEN: to settle; establish oneself; let down 112
NIEDERLEGEN: to resign; abdicate; deposit; lay (lie; put) down; warehouse; store 113
NIEDERREISSEN: to carry (drag; tear; pull) down; destroy; demolish 114
NIEDERSCHLAG (*m*): precipitate; deposit; precipitation; sediment; condensation; rainfall; downstroke 115
NIEDERSCHLAGBAR: precipitable; capable of being (deposited; condensed) precipitated 116
NIEDERSCHLAGEN: to precipitate; deposit; condense; alloy; quell; fell; beat (force;

thrust; cut; strike) down; refute; depress; deject; dishearten; (as an adjective=crest-fallen; dejected; disheartened; low-spirited; prostrated) 117

NIEDERSCHLAGEND: precipitant; sedative; depressing; depressant; discouraging; disheartening 118

NIEDERSCHLAG (*m*), GALVANISCHER-: galvanic deposit 119

NIEDERSCHLAGGEFÄSS (*n*): precipitating (settling; condensing) vessel 120

NIEDERSCHLAGSARBEIT (*f*): precipitation process; reduction process (as decomposition of the slag by means of iron, in tin or lead ore reduction process) (Metal) 121

NIEDERSCHLAGSWASSER (*n*): condensate, condensed water, condensation 122

NIEDERSCHLAGUNG (*f*): precipitation; condensation; dejection; quelling; beating down; refutation; depression (see NIEDER-SCHLAGEN) 123

NIEDERSCHLAGWASSER (*n*): condensation, condensate, condensed water 124

NIEDERSCHMELZEN: to melt down 125

NIEDERSCHREIBEN: to write down, note 126

NIEDERSPANNUNG (*f*): low tension 127

NIEDERSTAUCHEN: to rivet over; burr over; bead (Tubes) 128

NIEDERTRETEN: to tread (trample) down 129

NIEDERUNG (*f*): low (ground) country; flats 130

NIEDERVOLTIG: low-voltage 131

NIEDERWÄRTS: downward-(s) 132

NIEDRIG: low; base; mean; vulgar; common; lowly 133

NIERE (*f*): kidney (Physiology); nodule 134

NIERENABSONDERUNG (*f*): renal secretion (Med.) 135

NIERENBAUM (*m*): cashew tree (*Anacardium occidentale*) 136

NIERENBAUM (*m*), WESTINDISCHER-: cashew-nut tree (*Anacardium occidentale*) 137

NIERENBECKEN (*n*): pelvis of the kidney (Anat.) 138

NIERENBECKENENTZÜNDUNG (*f*): pyelitis (inflammation of the pelvis of the kidney) (Med.) 139

NIERENBESCHWERDE (*f*): kidney (complaint) trouble 140

NIERENENTZÜNDUNG (*f*): nephritis (inflammation of the kidneys) (Med.) 141

NIERENERZ (*n*): kidney ore 142

NIERENFÖRMIG: reniform; kidney-shaped; nodular 143

NIERENGRIES (*m*): kidney gravel (Med.) 144

NIERENHAUT (*f*): renal (capsule) membrane 145

NIERENKOLIK (*f*): renal colic (*Nephralgia Calculosa*) (Med.) 146

NIERENKRANK: nephritic 147

NIERENKRANKHEIT (*f*): kidney (trouble) disease 148

NIERENSTEIN (*m*): jade; spherulite; nephrite (see NEPHRIT); nephrolithiasis; renal calculus (a kidney concretion) (Med.); kidney stone (*Lapis nephriticus*) 149

NIERENWEH (*n*): nephralgia (a pain in the kidneys) (Med.) 150

NIESEKRAMPF (*m*): paroxysmal sneezing (Med.) 151

NIESEKRAUT (*n*): sneezewort (*Achillea ptarmica*) 152

NIESEMITTEL (*n*): sternutatory (see NIESMITTEL) 153

NIESEN: to sneeze; (*n*): sneezing; sternutation (Med.) 154

NIESEWURZ (*f*): (see NIESWURZ) 155

NIESKRAMPF (*m*): (see NIESEKRAMPF) 156

NIESMITTEL (*n*): errhine, sternutatory (an agent or drug for promoting sneezing) (Med.) 157

NIESSBRAUCH (*m*): usufruct 158

NIESSUNG (*f*), NUTZ-: usufruct 159

NIESWURZ (*f*): white hellebore; American hellebore (*Radix hellebori*) (*Veratrum viride*) 160

NIESWURZEL (*f*): (see NIESWURZ) 161

NIESWURZEL (*f*), GRÜNE-: green hellebore (*Radix hellebori* of *Helleborus viridis*) 162

NIESWURZ (*f*), SCHWARZE-: black hellebore, Christmas rose (*Helleborus niger*) 163

NIESWURZ (*f*), STINKENDE-: oxheal (*Helleborus Foetidus*) 164

NIESWURZ (*f*), WEISSE-: white hellebore (*Radix veratri* of *Veratrum album*) 165

NIET (*n*): rivet 166

NIETE (*f*): rivet; blank (in a lottery) 167

NIETEISEN (*n*): rivet iron 168

NIETEN: to rivet; (*n*): riveting 169

NIETER (*m*): riveter 170

NIETERWÄRMER (*m*): rivet heater 171

NIETE (*f*) ZIEHEN: to draw a blank (in a lottery) 172

NIETHAMMER (*m*): riveting hammer 173

NIETKOPF (*m*): rivet head 174

NIETLOCH (*n*): rivet hole 175

NIETLOCHKANTENRISS (*m*): crack springing (extending) from a rivet hole 176

NIETMASCHINE (*f*): riveting machine 177

NIETNAGEL (*m*): rivet 178

NIETNAHT (*f*): riveted seam 179

NIETOFEN (*m*): rivet-heating furnace 180

NIETREIHE (*f*): row of rivets 181

NIETSCHAFT (*m*): rivet shank or shaft 182

NIETSCHLIESSKOPF (*m*): rivet point 183

NIETSETZER (*m*): snap-cup 184

NIETSTEMPEL (*m*): riveting set 185

NIETUNG (*f*), DECK-: deck-plate riveting 186

NIETUNG (*f*), DOPPELTE-: double riveting 187

NIETUNG (*f*), EINFACHE-: single riveting 188

NIETUNG (*f*), EINREIHIGE-: single riveting 189

NIETUNG (*f*), EINSCHNITTIGE-: single-shear riveting 190

NIETUNG (*f*), LASCHEN-: butt (cover-plate) riveting 191

NIETUNG (*f*), MEHRREIHIGE-: multiple riveting 192

NIETUNG (*f*), MEHRSCHNITTIGE-: multiple-shear riveting 193

NIETUNG (*f*), SCHELLKOPF-: riveting snap, snap-riveting 194

NIETUNG (*f*), ZWEIREIHIGE-: double riveting 195

NIETUNG (*f*), ZWEISCHNITTIGE-: double-shear riveting 196

NIET (*n*), VERSENKTES-: countersunk rivet 197

NIETVORHALTER (*m*): holder-up (for riveting) 198

NIGERÖL (*n*): gingelli oil; gigily oil, Abyssinian camomile oil (from *Anthemis mysorensis*); Sg. 0.9255 199

NIGERPFLANZE (*f*): Abyssinian camomile (*Anthemis mysorensis* or *Guizotia oleifera*) 200

NIGRIN (*m*): nigrine (see RUTIL) 201

NIGRISIN (*n*): nigrisin (silver grey or black grey tar-dyestuff) 202

NIGROSIN (*n*): nigrosin (blue or black dyestuff) 203

NIKARAGUAHOLZ (*n*): (see FERNAMBUKHOLZ) 204

NIKOTIN (*n*): nicotine (*Nicotinum*) (an alkaloid from tobacco); $C_{10}H_{14}N_2$; Sg. 1.009-1.015 Bp. 247°C. 205

NIKOTINEXTRAKTIONSAPPARAT (*m*): nicotine extraction apparatus 206

NIKOTINHALTIG: containing nicotine 207

NIKOTINLÖSUNG (*f*): nicotine solution 208

NIKOTIN (*n*), **SALICYLSAURES-**: nicotine salicylate; $C_{10}H_{14}N_2.C_7H_6O_3$; Mp. 117.5°C. 209

NIKOTINSÄURE (*f*): nicotic (nicotinic) acid; meta-(beta)- pyridine carboxylic acid; $C_6H_5NO_2$; Mp. 228°C. 210

NILBLAU (*n*): Nile blue 211

NILPFERD (*n*): hippopotamus (*Hippopotamus amphibius*) 212

NIOB (*n*): niobium; columbium; (Nb) 213

NIOBEÖL (*n*): niobe oil; $C_6H_5CO_2CH_3$; methyl benzoate (see **METHYLBENZOAT**) 214

NIOBIT (*m*): niobite (see also **COLUMBIT**); $FeNb_2O_6$; Sg. 6.0 215

NIOBIUM (*n*): niobium, columbium 216

NIOBIUMHALTIG: containing niobium 217

NIOBPENTAFLUORID (*n*): niobium (columbium) pentafluoride; NbF_5; Sg. 3.293 218

NIOBPENTOXYD (*n*): niobium (columbium) pentoxide; Nb_2O_5; Sg. 4.47 219

NIOBSÄURE (*f*): niobic (columbic) acid 220

NIOBWASSERSTOFF (*m*): niobium (columbium) hydride 221

NIPPEL (*m*): nipple 222

NIPPFLUT (*f*): neap tide 223

NIRVANIN (*n*): nirvanin; $(C_2H_5)_2N:CH_2CONHOHCO_2CH_3.HCl$; Mp. 185°C. (employed as a local anæsthetic in 1 to 2% solutions) 224

NIRVANOL (*n*): nirvanol; $C_6H_5C_2H_5CCONHNHCO$ 225

NISCHE (*f*): niche 226

NISTEN: to nest; nestle 227

NITON (*n*): niton (Nt), radium emanation 228

NITRAGIN (*n*): nitrogenous (azotic) manure, or fertilizer (for bacteria) (see **BAKTERIEN-STICKSTOFFDÜNGER**) 229

NITRANILIN (*n*): nitraniline; $(C_6H_6O_2N_2)$; ortho.-; Sg. 1.44; Mp. 71.5°C.; meta.-; Sg. 1.43; Mp. 114°C.; Bp. 285°C.; para.-; Sg. 1.424; Mp. 147°C. 230

NITRANILINCHLORHYDRAT (*n*): nitraniline hydrochloride 231

NITRANILINROT (*n*), **PARA-**: para-nitraniline red 232

NITRANILINSULFOSÄURE (*f*): nitraniline sulfonic acid 233

NITRANISOL (*n*): nitranisol 234

NITRAT (*n*): nitrate 235

NITRATANLAGE (*f*): nitrate plant 236

NITRATIN (*m*): (see **NATRONSALPETER**) 237

NITRATION (*n*): nitrate ion 238

NITRATMASCHINEN (*f.pl*): nitrate machinery (see also **NITRIER**) 239

NITRID (*n*): nitride (metal compounds of nitrogen); aluminium nitride (see **ALUMINIUMNITRID**) 240

NITRIDRANDSCHICHT (*f*): nitride rim layer (formed on nitrification of iron) 241

NITRIERANLAGE (*f*): nitrating plant 242

NITRIERAPPARAT (*m*): nitrating apparatus 243

NITRIERBAUMWOLLE (*f*): nitrating cotton 244

NITRIEREN: to nitrify; nitrate; (*n*): nitration; nitrification (a method of case-hardening steel) 245

NITRIERGEFÄSS (*n*): nitrating vessel 246

NITRIERGEMISCH (*n*): nitrating mixture 247

NITRIERSÄURE (*f*): nitrating acid, mixed acid (a mixture of the concentrated solutions of H_2SO_4 (Sg. 1.85) and HNO_3 (Sg. 1.4) 248

NITRIERTOPF (*m*): nitrating pot 249

NITRIERUNG (*f*): nitration; nitrification (an oxidization process of nitrogenous organic matter to nitric or nitrous acid) 250

NITRIERUNGSHÄRTE (*f*): nitration hardness (of steel, reaches to a depth of about ½ m.m.) 251

NITRIERUNGSPRODUKT (*n*): nitration product 252

NITRIERVORSCHRIFT (*f*): nitrating (instructions) directions 253

NITRIERZENTRIFUGE (*f*): nitrating centrifuge 254

NITRIL (*n*): nitrile (cyanide of an alkyl radicle) 255

NITRIT (*n*): nitrite; sodium nitrite (see **NATRIUM-NITRIT**) 256

NITRITFABRIKATION (*f*): nitrite manufacture 257

NITRITGLÄTTE (*f*): nitrite litharge 258

NITRITMUTTERLAUGE (*f*): (sodium)-nitrite mother liquor or lye 259

NITROÄTHAN (*n*): nitroethane; $C_2H_5.NO_2$ [260

NITROBENZALDEHYD (*n*): nitrobenzaldehyde; $(C_7H_5O_3N)$; ortho.-; Mp. 46°C.; Bp. 153°C.; meta.-; Mp. 58°C.; Bp. 164°C.; para.-; Mp. 106°C. 261

NITROBENZOESÄURE (*f*): nitrobenzoic acid; $(C_7H_5O_4N)$: ortho.-; Mp. 147°C.; meta.-; Sg. 1.494; Mp. 141°C.; para.-; Mp. 238°C. 262

NITROBENZOESÄUREPROPYLESTER (*m*), **PARA-**: propyl-para-nitrobenzoate 263

NITROBENZOL (*n*): nitrobenzene; oil of mirbane; essence of mirbane; artificial oil of bitter almonds; $C_6H_5O_2N$; Sg. 1.199; Mp. 3.0-8.7°C.; Bp. 205-210°C. 264

NITROBENZOL (*n*), **SEHR SCHWERES-**: technical nitrotoluene; Sg. 1.17 265

NITROBENZYLCHLORID (*n*): nitrobenzyl chloride; $C_6H_4(NO_2).CH_2Cl$ 266

NITROCELLULOSE (*f*): nitrocellulose, gun-cotton; $C_6H_7O_5(NO_2)_3$ 267

NITROCELLULOSE (*f*), **DI-**: di-nitrocellulose (see **COLLODIUMWOLLE**) 268

NITROCELLULOSE (*f*). **HEXA-**: hexa-nitrocellulose (see **SCHIESSWOLLE**) 269

NITROCELLULOSELÖSUNG (*f*): nitrocellulose solution 270

NITROCELLULOSE (*f*), **TRI-**: tri-nitrocellulose (see **SCHIESSWOLLE**) 271

NITROCHINON (*n*): nitroquinone 272

NITROCHLORANILIN (*n*): nitrochloroaniline 273

NITROCHLORBENZOL (*n*): nitrochlorobenzol; nitrochlorobenzene; $NO_2.C_6H_4NH_2$ 274

NITROCOCUSSÄURE (*f*): nitrococcic acid 275

NITRODERIVAT (*n*): nitro derivate 276

NITROFARBSTOFF (*m*): nitro dyestuff 277

NITROFETTKÖRPER (*m*): aliphatic nitro (compound) substance 278

NITROFETTSÄURE (*f*): nitro fatty acid 279

NITROGLUKOSE (*f*): nitro-glucose 280

NITROGLUKOSEPAPIER (*n*): nitro-glucose paper (Phot.) 281

NITROGLYZERIN (*n*): nitroglycerine; glycerol (glyceryl) trinitrate; trinitrin; trinitroglycerine; glonoin oil; blasting oil; $C_3H_5(O.NO_2)_3$; Sg. 1.6; freezing point 8-13°C.; explosion point 180-260°C. 282

NITROGLYZERINFABRIKATION (*f*): nitroglycerine manufacture 283
NITROGLYZERINTRÖPFCHEN (*n*): nitroglycerine drop 284
NITROGRUPPE (*f*): nitro group 285
NITROHALOGENBENZOL (*n*): halonitrobenzene 286
NITROIN (*n*): nitroin (a mixture of H_2SO_4, HNO_3 and oxydating agents) 287
NITROINELEMENT (*n*): nitroin-(galvanic)-element 288
NITROKOHLENSTOFF (*m*): tetranitromethane 289
NITROKÖRPER (*m*): nitro (substance) compound 290
NITROKUPFER (*n*): nitro copper 291
NITROLSÄURE (*f*): nitrolic acid 292
NITROMETER (*n*): nitrometer (an instrument for determining the nitrogen in Chile saltpeter and superphosphates) 293
NITROMETHAN (*n*): nitromethane; CH_3O_2N; Sg. 1.1639; Bp. 101.5°C. 294
NITRON (*n*): nitron, diphenylendoanilohydrotriazole; $CN_4.(C_6H_5)_3.CH$ (a reagent used in the analysis of explosives) 295
NITRONAPHTALIN (*n*): nitronaphthalene, alpha-(mono)-nitronaphthalene; $C_{10}H_7.NO_2$; Sg. 1.33; Mp. 61°C.; Bp. 304°C. 296
NITRONAPHTALINSULFOSÄURE (*f*): nitronaphthalenesulfonic acid; Laurent's acid; $C_{10}H_6$ $(NO_2)(SO_3H)$ 1:5 297
NITROPHENOL (*n*): nitrophenol; $(C_6H_5O_3N)$; ortho.-; Sg. 1.66; Mp. 44.27°C.; Bp. 214°C.; meta.-; Sg. 1.49; Mp. 95.3°C.; Bp. 194°C.; para.-; Sg. 1.48; Mp. 113.8°C. [298
NITROPHENYLPROPIOLSAUER, ORTHO-: ortho-nitrophenylpropiolate, combined with ortho-nitrophenylpropiolic acid 299
NITROPHENYLPROPIOLSÄURE (*f*), **ORTHO-**: ortho-nitrophenylpropolic acid (see **PROPIOLSÄURE**) 300
NITROPRUSSIDNATRIUM (*n*): (*Natrium nitroferricyanatum*): sodium nitroferricyanide; (*Natrium nitroprussicum*), sodium nitroprussiate; $Na_4Fe_2(CN)_{10}(NO)_2.4H_2O$ 301
NITROPULVER (*n*): smokeless (gun)powder 302
NITRÖS: nitrous 303
NITROSALICYLSÄURE (*f*), **META-**: meta-nitrosalicylic acid, meta-nitrohydroxybenzoic acid; $C_6H_3COOH:OH:NO_2$; Mp. 235°C. 304
NITROSAMINROT (*n*): nitrosamine red 305
NITROSCHWEFELSÄURE (*f*): nitrosulphuric acid 306
NITROSE (*f*): nitrose (solution of nitrosylsulphuric acid in sulphuric acid) (Sulphuric acid process) 307
NITROSESÄURE (*f*): nitrose acid (solution of nitrosylsulphuric acid in sulphuric acid) (Sulphuric acid process) 308
NITROSIEREN: to introduce nitrose group into; treat with nitrous acid 309
NITROSISULFOSÄURE (*f*): nitrosisulfonic acid 310
NITROSOBASE (*f*): nitroso base 311
NITROSOBENZOL (*n*): nitrosobenzene 312
NITROSO-BETA-NAPHTOL (*n*): nitroso-beta-naphthol (see **NITROSONAPHTOL**) 313
NITROSOFARBSTOFF (*m*): nitroso-dyestuff (a class of synthetic tar dyestuffs) 314
NITROSOGRUPPE (*f*): nitroso group 315
NITROSOKOBALTWASSERSTOFFSÄURE (*f*): cobaltinitrous acid 316
NITROSOMETHYLURETHAN (*n*): nitrosomethylurethane; $CH_3-N(NO)-CO_2.C_2H_5$ [317
NITROSONAPHTOL (*n*): nitrosonaphthol; $C_{10}H_6$ $(NO)OH$; alpha-nitroso-alpha-naphthol; Mp. 152°C.; alpha-nitroso-beta-naphthol; Mp. 106°C. 318
NITROSOPRODUKT (*n*): nitroso product 319
NITROSORESORCIN (*n*): nitrosoresorcinol 320
NITROSOSULFOSÄURE (*f*): nitrososulphonic acid 321
NITROSTÄRKE (*f*): nitro starch; xyloidine (see **NITROCELLULOSE**) 322
NITROSULFONSÄURE (*f*): nitrosulphonic acid; nitrosylsulphuric acid 323
NITROSYLCHLORID (*n*): nitrosyl chloride; $NOCl$; Sg. 1.4165 324
NITROSYLSCHWEFELSÄURE (*f*): nitrosylsulphuric acid; nitrosulphonic acid 325
NITROTOLUIDIN (*n*): nitrotoluidine 326
NITROTOLUOL (*n*): (mono)-nitrotoluene; $(C_7H_7O_2N)$; ortho.-; Sg. 1.1629; Mp. −3.85°C.; Bp. 222.3°C.; meta.-; Sg. 1.168; Mp. 16.1°C.; Bp. 231°C.; para.-; Sg. 1.1232; Mp. 54°C.; Bp. 237.7°C. 327
NITROTOLUOLGEMISCH (*n*): nitrotoluene mixture 328
NITROTOLUOLSULFOSÄURE (*f*): nitrotoluenesulfonic acid 329
NITROVERBINDUNG (*f*): nitro compound 330
NITROXYLGRUPPE (*f*): nitroxyl (nitro) group 331
NITROXYLOL (*n*): mononitroxylene, dimethylnitrobenzene, nitroxylol, nitroxylene; $C_6H_3(CH_3)_2NO_2$; ortho.-; Sg. 1.14; Mp. 29°C.; Bp. 258°C.; meta.-; Sg. 1.135; Mp. 2°C.; Bp. 246°C.; para.-; Sg. 1.13; Bp. 240°C. 332
NITROZELLULOSE (*f*): nitro-cellulose (see **COLLODIUMWOLLE** and **SCHIESSWOLLE**) 333
NITROZIMTSÄUREÄTHYLESTER (*m*): ethyl nitrocinnamate 334
NITROZIMTSÄUREESTER (*m*): nitrocinnamic ester 335
NITROZIMTSÄURE, ORTHO- (*f*): ortho-nitrocinnamic acid 336
NIVEAU (*n*): level 337
NIVEAURÖHRE (*f*): levelling tube; level 338
NIVELLIEREN: to level 339
NIVELLIERGESTELL (*n*): level-(ling stand) (Phot.) 340
NIVELLIERINSTRUMENT (*n*): levelling (staff) instrument 341
NIVELLIERPLATTE (*f*): levelling slab (an adjustable slab of glass or similar material for use on a levelling stand) (Photography) 342
NIVELLIERUNG (*f*): levelling 343
NIVELLIERWAGE (*f*): water level; spirit level 344
NIVENIT (*m*): nivenite (see **URANPECHERZ**) 345
NIXBLUME (*f*): water lily (*Nuphar lutea* and *Nymphæa alba*) 346
NJAMPLUNGÖL (*n*): (see **TACAMAHACFETT**) 347
NOBEL: noble, dignified, magnanimous, generous, honorable, stately, lordly; bombastic 348
NOBELIT (*n*): nobelite (a safety explosive) 349
NOCHMALIG: reiterated, repeated, renewed 350
NOCK (*n*): yard arm, gaff pole (Naut.) 351
NOCKBÄNDSEL (*n*): head earing 352
NOCKE (*f*): boss; cam; lifter; tappet (Mech.) 353
NOCKEN (*m*): (see **NOCKE**) 354
NOCKENSTEUERUNG (*f*): cam gear 355
NOCKENWELLE (*f*): cam shaft 356
NOCKGORDING (*f*): leech line 357
NOCKKLAMPE (*f*): yard arm cleat 358
NOCK (*n*), **RAH-**: yard arm 359

NOCKTAKEL (*n*): yard arm tackle 360
NOMENKLATUR (*f*): nomenclature 361
NOMINALBETRAG (*m*): nominal (value) amount 362
NOMINALEFFEKT (*m*): nominal effect 363
NOMINALKERZE (*f*): nominal candle power 364
NOMINALWERT (*m*): nominal value 365
NOMINATIV (*m*): nominative (case) 366
NOMINELL: nominal-(ly) 367
NONIUSEINTEILUNG (*f*): Vernier scale (a scale invented by Pierre Vernier for reading finest of divisions of graduated scales) 368
NONNENNÄGELEIN (*n*): nutmeg flower (*Nigella sativa*) 369
NONOCARBOZYKLISCH: nonacarbocyclic 370
NONTRONIT (*m*): nontronite (a mixture of Fe_2O_3 and SiO_2 hydrogels); Sg. 2.1 371
NONYLALDEHYD (*n*): nonylic aldehyde 372
NONYLSÄURE (*f*): nonylic acid 373
NOPAL (*m*): nopal, cochineal cactus (*Nopalea coccinellifera* or *Opuntia coccinellifera*) 374
NOPALPFLANZE (*f*): (see NOPAL) 375
NOPALSCHILDLAUS (*f*): cochineal insect (*Coccus cacti* (see COCHENILLE) 376
NOPPE (*f*): nap; burl 377
NOPPEN: to nap; burl 378
NOPPENBEIZE (*f*): burl dye 379
NOPPENSCHWARZ (*n*): burl black 380
NOPPENTINKTUR (*f*): burling (tincture) liquid 381
NORDHÄUSER VITRIOLÖL (*n*): (see SCHWEFEL--SÄURE, RAUCHENDE-) 382
NÖRDLICH: northerly; northern; arctic 383
NORDLICHT (*n*): north (northern) light; aurora borealis 384
NORDOST (*m*): northeast 385
NORDPOL (*m*): north pole 386
NORDSTERN (*m*): north (pole)star 387
NORGESALPETER (*m*): calcium nitrate (produced from the nitrogen of the atmosphere) (see CALCIUMNITRAT) 388
NORGINE (*f*): norgine (a colloid from seaweed) 389
NORIT (*n*): Norit (a preparation of vegetable charcoal used for decolourizing and clearing sugar syrups) 390
NORLEUCIN (*n*): norleucine, d-α-amino-n-caproic acid; $CH_3(CH_2)_3CH(NH_2).CO_2H$ 391
NORM (*f*): standard; rule; model; (NORMEN (*pl*): standards; rules) 392
NORMAL: normal; standard; ordinary; common; regular; usual; natural; legitimate; vertical; perpendicular 393
NORMALBEDINGUNG (*f*): normal (stipulation) condition 394
NORMALDRUCK (*m*): normal pressure 395
NORMALE (*n*): standard 396
NORMAL-EICHUNGSKOMMISSION (*f*): Standards Bureau (similar to the English Standard Committee) 397
NORMALELEMENT (*n*): standard cell 398
NORMALESSIG (*m*): standard (proof) vinegar 399
NORMALEXPOSITION (*f*): normal exposure (Phot.) 400
NORMALFARBE (*f*): normal (standard) colour or dye 401
NORMALFLÜSSIGKEIT (*f*): normal (standard) solution 402
NORMALGESCHWINDIGKEIT (*f*) normal velocity, standard speed 403
NORMALGESETZ (*n*): general (rule) law 404
NORMAL GEWICHT UND MASS: standard weights and measures 405
NORMALISIEREN: to standardize 406
NORMALISIERUNG (*f*): standardization 407
NORMALITÄT (*f*): normality 408
NORMALKERZE (*f*): standard candle power 409
NORMALLÖSUNG (*f*): normal (standard) solution 410
NORMALMASS (*n*): standard measure 411
NORMALSÄURE (*f*): normal (standard) acid 412
NORMALSPUR (*f*): normal (standard) gauge (of Railways) 413
NORMALSPURIG: normal (standard) gauge (of Railways) 414
NORMALSTÄRKE (*f*): normal (standard) strength 415
NORMALUNG (*f*): standardization 416
NORMALWEINGEIST (*m*): proof spirit 417
NORMALZUSTAND (*m*): normal (standard) state or condition 418
NORMEN (*f.pl*): standards; rules (see NORM) 419
NORMEN: to standardize 420
NORMENSAND (*m*): standard sand (Cement) 421
NORMIEREN: to standardize; rule 422
NORMUNG (*f*): standardization 423
NORTHUPIT (*m*): northupite; $2(MgCO_3$. $Na_2CO_3.NaCl)$; Sg. 2.4 424
NOSEAN (*m*): nosean (see HAUYN) 425
NOSOPHEN (*n*): nosophene, tetraiodophenolphthalein; $C(C_6H_2I_2.OH)_2C_6H_4.CO.O$ 426
NOT (*f*): need; necessity; want; labour; danger; distress; urgency; exigency; compulsion; strait; difficulty; extremity; penury; emergency; poverty; privation; misery; destitution; trouble 427
NOTADRESSE (*f*): address in case of need 428
NOTANKER (*m*): sheet anchor; spare anchor 429
NOTAR (*m*): notary 430
NOTARBEIT (*f*): urgent work, necessary work 431
NOTAUSGANG (*m*): emergency exit 432
NOTAUSLASS (*m*): emergency (exit) outlet 433
NOTBEHELF (*m*): makeshift; stopgap 434
NOTBREMSE (*f*): emergency brake 435
NOTDRINGEND: urgent; pressing 436
NOTDURFT (*f*): exigency; necessity; need 437
NOTDÜRFTIG: scanty; needy; indigent 438
NOTE (*f*): note; bill 439
NOTEN (*f.pl*): music, compositions 440
NOTENBANK (*f*): bank of issue 441
NOTENDRUCK (*m*): music printing 442
NOTENGESTELL (*n*): music-stand, desk 443
NOTENPAPIER (*n*): music (manuscript) paper 444
NOTENSTECHER (*m*): music engraver 445
NOTENSYSTEM (*n*): staff, stave 446
NOTENUMLAUF (*m*): circulation of notes 447
NOTFALL (*m*): emergency; exigency; need; case of need; necessity 448
NOTFORM (*f*): standby pipe-(in the boshes, in case of necessity, for supplying hot air to blast furnaces) 449
NOTGAR: undertanned (Leather) 450
NOTGEDRUNGEN: necessarily; forced; compulsory; compulsorily; forcibly; by compulsion 451
NOTHAFEN (*n*): port of distress 452
NOTIEREN: to note; notify; quote 453
NOTIERUNG (*f*): quotation; memo; memorandum 454
NOTIFIZIEREN: to notify, acquaint, advise, instruct, announce, apprise 455
NÖTIGE (*n*): the necessary, the needful 456
NÖTIGEN: to be needful; necessitate; urge; force; compel; press; constrain; entreat 457

NÖTIG HABEN : to need ; be in need of ; necessitate ; want ; require ; have necessary 458
NOTIZ (*f*) : notice ; note ; cognisance ; memorandum ; advice ; notification ; news ; announcement ; information ; intelligence ; heed ; observation ; mention ; intimation ; warning ; review ; comments ; remarks 459
NOTIZBUCH (*n*) : note book ; memorandum book 460
NOTJAHR (*n*) : year of (famine) scarcity 461
NOTLEINE (*f*) : communication cord 462
NOTMAST (*m*) : jury-mast (Naut.) 463
NOTMITTEL (*n*) : expedient ; (make)-shift 464
NOTORISCH : notorious 465
NOTORITÄT (*f*) : notoriety 466
NOTRAH (*f*) : jury yard 467
NOTREGLER (*m*) : safety regulator 468
NOTRUDER (*n*) : jury (preventer) rudder (Naut.) 469
NOTSIGNAL (*n*) : distress signal 470
NOTSTAND (*m*) : case (state) of distress or need 471
NOTSTENGE (*f*) : jury topmast (Naut.) 472
NOTTÜR (*f*) : emergency exit 473
NOTWEHR (*f*) : self-defence 474
NOTZUCHT (*f*) : rape (Med.) 475
NOTZÜCHTIGEN : to rape ; violate ; ravish 476
NOVASPIRIN (*n*) : novaspirin (a substitute for aspirin) ; $C_{21}H_{16}O_{11}$ 477
NOVASUROL (*n*) : novasurol (contains 34% Hg) ; $C_{16}H_{16}N_2O_6NaHg$ 478
NOVICHTAN (*n*) : novichtan (a product from bitumenous mineral oils) 479
NOVOCAIN (*n*) : para-aminobenzoyldiethylaminoethanol, novocain (*Novocainum*) : $C_6H_4NH_2COO.C_2H_4.N(C_2H_5)_2.HCl$; Mp. 156°C. 480
NOVOKAIN (*n*) : (see **NOVOCAIN**) 481
NOVOKOL (*n*) : novokol (sodium guaiacolphosphate) 482
NOVOVIOL (*n*) : novoviol (a violet base for toilet soaps) 483
NOVOZON (*n*) : novozon (a registered trade name for Magnesium peroxide, used for medicinal purposes) 484
N-SUBSTANZ (*f*) : nitrogenous substance 485
NUANCE (*f*) : nuance ; shade ; tint ; modulation ; variation 486
NUANCIEREN : to shade-(off) ; modulate ; vary ; tint 487
NUANCIERFARBE (*f*) : changing colour 488
NÜCHTERN : sober ; temperate ; cool ; empty ; fasting ; flat ; insipid ; reasonable 489
NÜCHTERNHEIT (*f*) : temperance ; sobriety ; fasting ; calmness 490
NUCLEIN (*n*) : nuclein (a combination of nucleic acid and albumin, obtained from yeast) 491
NUCLEINSAUER : nucleinate, combined with nucleic acid 492
NUCLEINSÄURE (*f*) : nucleic acid (organic acid containing phosphorous and nitrogen) 493
NUDELN (*f.pl*) : vermicelli ; macaroni 494
NUKLEIN (*n*) : (see **NUCLEIN**) 495
NUKLEOTINPHOSPHORSÄURE (*f*) : (see **THYMIN-SÄURE**) 496
NULL : nought ; zero ; null ; nill ; (*f*) : cipher ; nought ; zero ; blank 497
NULLEITUNG (*f*) : (see **NULLLEITUNG**) 498
NULLENZIRKEL (*m*) : bow compasses, spring bow compasses, spring bows 499
NULLGRAD (*m*) : zero-(degree) 500
NULLITÄT (*f*) : nullity ; invalidity 501
NULLLAGE (*f*) : zero-(position) 502
NULLLEITUNG (*f*) : neutral feeder (Elect. three-wire system) 503
NULLPUNKT (*m*) : zero-(point) 504
NULLSETZEN : to nullify (Maths.) (to make an expression = 0) 505
NULLSETZUNG (*f*) : zero-expression (Maths.) (where an expression = 0) 506
NULLSTRICH (*m*) : zero-(mark) 507
NULL UND NICHTIG : null and void ; invalid 508
NULL UND NICHTIG MACHEN : to annul ; nullify ; make-(null and)-void ; cancel ; invalidate ; render (make) harmless ; make of no effect 509
NULLWERTIG : avalent ; non-valent 510
NUMEAIT (*m*) : numeaite ; $(Mg,Ni)SiO_3.H_2O$; Sg. 2.55 (see also **GARNIERIT**) 511
NUMERALE (*n*) : numeral adjective 512
NUMERIEREN : to number 513
NUMERISCH : numerical 514
NUMERO : number 515
NUMERUS (*m*) : number 516
NUMMER (*f*) : number ; cipher 517
NUMMERFOLGE (*f*) : numeral (order) sequence 518
NUMMERIEREN : to number 519
NUMMERIERUNG (*f*) : numbering 520
NUMMERN : (see **NUMERIEREN** and **NUMMER-IEREN**) 521
NUMMERSCHILD (*n*) : number plate (of motor cars) 522
NÜRNBERGERVIOLETT (*n*) : Nuremberg violet ; manganic phosphate (see **MANGANVIOLETT**) **523**
NUSS (*f*) : nut (*Semen*, *Fructus* or *Nuces*) ; walnut (see also under **WALNUSS**) ; notch 524
NUSSABRIEB (*m*) : smalls, nuts (the small pieces broken off nut coal during classification and unloading) **525**
NUSSBAUM (*m*) : nut tree ; walnut tree (*Juglans regia*) 526
NUSSBAUMHOLZ (*n*) : walnut wood (wood of European walnut tree, *Juglans regia*) (Sg. 0.63-0.71) 527
NUSSBAUMLACK (*m*) : walnut varnish 528
NUSSBEIZE (*f*) : nut mordant 529
NUSSBLÄTTER (*n.pl*) : walnut (nut) leaves (*Folia juglandis* of *Juglans regia*) 530
NUSSBOHNE (*f*) : ground-nut, ground pea-nut (*Arachis hypogæa*) 531
NUSSBOHNENKAFFEE (*m*) : ground nut coffee, roasted beans of *Arachis hypogæa* 532
NUSSBRAUN : nut-brown ; hazel ; bistre ; (as a noun, is (*n*)) 533
NUSSGELENK (*n*) : enarthrosis (Med.) 534
NUSSGRAS (*n*) Chufa-(tuber), earth almond (*Cyperus esculentus*) 535
NUSSKERN (*m*) : kernel-(of a nut) 536
NUSSKIEFER (*f*) : stone pine (*Pinus cembra*) 537
NUSSKOHLE (*f*) : nut coal ; nuts ; cobbles (size 10-80 m.m. mesh) 538
NUSSKOHLE (*f*), **GEWASCHENE-** : washed nuts, washed cobbles (Coal) 539
NUSSKÜMMEL (*m*) : pig-nut (*Bunium flexuosum*) 540
NUSSÖL (*n*) : walnut (nut) oil (*Oleum juglandis*) (from the seeds of *Juglans regia*) ; Sg. 0.919-0.93 541
NUSSÖL (*n*), **AMERIKANISCHES-** : (see **HICKORYÖL**) 542
NUSSPILZ (*m*) : truffle (see **TRÜFFEL**) 543
NUSSSCHALE (*f*) : nut-shell 544

NUSSSTAUDE (*f*): hazel-bush (see under **HASEL**) 545
NUSS, WÄLSCHE- (*f*): walnut (see **NUSSBAUM**) 546
NÜSTER (*f*): nostril (of horses) 547
NÜSTERGATT (*n*): limber hole 548
NUT (*f*): rabbet; furrow; groove; thick end of tapered (wedge) brick; channel; gutter; slot; keyway 549
NUTE (*f*): (see **NUT**) 550
NUTEN: to channel, groove, slot, furrow; key (Mech.) 551
NUTENEISEN (*n*): plough bit 552
NUTENFRÄSER (*m*): slot cutter, slotting machine 553
NUTEN HAUEN: to groove, slot 554
NUTENHAUER (*m*): plough bit 555
NUTENHOBEL (*m*): grooving (rabbeting) plane 556
NUTENKEIL (*m*): (sunk-)key 557
NUTENSTOSSMASCHINE (*f*): grooving machine, slotting machine 558
NUTENZIEHMASCHINE (*f*): slotting machine 559
NUTROSE (*f*): nutrose (sodium caseinate; contains about 65% albuminous matter together with mineral substances, water and non-nitrogenous substances) 560
NUTSCHAPPARAT (*m*): suction apparatus 561
NUTSCHE (*f*): suction (filter) strainer 562
NUTSCHENFILTER (*n*): (see **NUTSCHE**) 563
NUTSCHENTRICHTER (*m*): suction funnel 564
NUTSCHER (*m*): suckling 565
NUTSCHTRICHTER (*m*): suction funnel 566
NUTSTEIN (*m*): grooved (slotted) brick 567
NUT-UND-FEDERSTEIN (*m*): tapered brick; wedge brick 568
NUTZANWENDUNG (*f*): practical (applicability; application) utility 569
NUTZARBEIT (*f*): effective (useful) work 570
NUTZBAR: useful; available; applicable; practicable; fit for use; effective; advantageous; beneficial; lucrative; profitable; suitable 571
NUTZBARKEIT (*f*): usefulness, applicability, fitness, effectiveness, utility, serviceability, suitability 572
NUTZBRINGEND: beneficial, profitable 573
NUTZDRUCK (*m*): effective pressure (usually mean effective pressure) 574
NUTZE (*f*): useful; profitable; effective; lucrative 575
NÜTZE: (see **NUTZE**) 576
NUTZEFFEKT (*m*): efficiency; useful effect (η) 577
NUTZEFFEKT (*m*), **WIRTSCHAFTLICHER-**: economical efficiency (η_w) 578
NUTZEN: to be of use; use; make use of; apply; make suitable; fit; help; benefit; profit; serve; avail 579
NÜTZEN: (see **NUTZEN**) 580
NUTZGEFÄLLE (*n*): effective drop (usually "effective heat drop") 581
NUTZHOLZ (*n*): timber; building timber 582
NUTZLEISTUNG (*f*): useful (effective) work, duty or output; efficiency; useful effect (see also **NUTZEFFEKT**) 583
NÜTZLICH: useful; profitable; expedient; advantageous; conducive; helpful; beneficial; convenient; serviceable; favourable; opportune; salutary 584
NUTZLOS: unprofitable; useless 585
NUTZNIESSUNG (*f*): usufruct 586
NUTZPFERDESTÄRKE (*f*): effective horse power; E.H.P. 587
NUTZPFERDESTUNDE (*f*): effective horse-power-hour (eff. H.P./hr.) 588
NUTZREICH: profitable, beneficial, advantageous, lucrative 589
NUTZUNG (*f*): profit, employment, using, usufruct, produce, income, revenue, emolument, remuneration, pecuniary profit, application, utilization 590
NUTZWIRKUNG (*f*). (see **NUTZEFFEKT**) 591
NYCTALOPIE (*f*): nyctalopia, night-blindness (inability to see in a poor light) (Med.) 592
NYLANDERS REAGENS (*n*): Nylander's reagent (a solution of Bismuth subnitrate and Seignette salt in an 8% soda lye; for the determination of grape sugar in urine) 593
NYMPHENBLUME (*f*): water lily (*Nymphæa alba*) 594
NYMPHOMANIE (*f*): nymphomania (Med.) 595

O

OATH (*m*): declaration (Patents); oath 596
OBDUKTION (*f*): post-mortem examination 597
OBDUKTIONSHAUS (*n*) mortuary 598
OBEN: above; upstairs; overhead; on (high) top; before; on the surface; previously; aloft; uppermost; at the head 599
OBENAN: at the (head) top; in the (beginning) first place; primarily 600
OBENANGEGEBEN: as given above; aforesaid; above-mentioned 601
OBENDREIN: over and above; above all; into the bargain; furthermore 602
OBENERWÄHNT: above-mentioned; aforesaid 603
OBENGENANNT: (see **OBENERWÄHNT**) 604
OBENHIN: superficially; slightly 605
OBERAPPELLATIONSGERICHT (*n*): high court of appeal 606
OBERARM (*m*): upper arm (Anat.) 607
OBERAUFSEHER (*m*): superintendent, foreman, chief inspector 608
OBERBAU (*m*): super-structure 609
OBERBAUCH (*m*): Epigastric region (the upper part of the abdomen) (Anat.) 610
OBERBAU (*m*), **EISENBAHN-**: permanent way (Railways) 611
OBERBEIN (*n*): thigh (Anat.) 612
OBERDECK (*n*): upper deck 613
OBERFLÄCHE (*f*): surface; area; superficies 614
OBERFLÄCHENEINHEIT (*f*): unit of (surface) area 615
OBERFLÄCHENENTWICKELUNG (*f*): surface development 616
OBERFLÄCHENFÄRBUNG (*f*): surface colouring (colouring the made-up sheets of paper) 617
OBERFLÄCHENGESTEIN (*n*): eruptive rock, volcanic rock 618
OBERFLÄCHENHÄRTUNG (*f*): case-hardening (Metals) 619
OBERFLÄCHEN-KONDENSATION (*f*): surface condensing-(plant) 620
OBERFLÄCHENKONDENSATOR (*m*): surface condenser 621
OBERFLÄCHENLEIMUNG (*f*): surface sizing (Paper) 622
OBERFLÄCHENSPANNUNG (*f*): surface (tension) pressure 623

OBERFLÄCHENVERBRENNUNG (*f*): surface combustion [in which a gas-air mixture is practically completely burnt with a small amount of excess air (about ½%) when forced under high pressure through a granular refractory mass] 624
OBERFLÄCHENVEREDELUNG (*f*): surface enrichment (Metal) (see **ALITIERVERFAHREN**) 625
OBERFLÄCHENVERGASER (*m*): surface carburetter 626
OBERFLÄCHENWIRKUNG (*f*): surface (effect) action 627
OBERFLÄCHENZEMENTIERUNG (*f*): case hardening (Metal) 628
OBERFLÄCHLICH: superficial-(ly) 629
OBERFLÄCHLICHKEIT (*f*): superficiality 630
OBERGÄRIG: top-fermenting (Brewing) 631
OBERGÄRIGE HEFE (*f*): top yeast (see **OBERHEFE**) 632
OBERGÄRUNG (*f*): top fermentation (Brewing) 633
OBERGERICHT (*n*): high (supreme) court 634
OBERGESENK (*n*): top swage 635
OBERGEWICHT (*n*): top weight 636
OBERGRUND (*m*): top layer of soil 637
OBERHALB: above 638
OBERHAND (*f*): back of the hand; wrist, metacarpus (Anat.); upper-hand, supremacy 639
OBERHAUS (*n*): upperhouse, senate (equivalent to the House of Lords) 640
OBERHAUT (*f*): epidermis 641
OBERHÄUTCHEN (*n*): cuticle (*Cuticula*) 642
OBERHEFE (*f*): top yeast (very rapid and stormy ferment) 643
OBERINGENIEUR (*m*): chief engineer 644
OBERIRDISCH: aerial 645
OBERKANTE (*f*): upper edge 646
OBERKESSEL (*m*): steam drum-(s) (of water tube boilers); top or upper drum or drums, upper boiler 647
OBERKIEL (*m*): upper keel 648
OBERKIELSCHWEIN (*n*): rider keelson 649
OBERKÖRPER (*m*): upper part of the body 650
OBERLAND (*n*): oberland, highland, mountainous district 651
OBERLAST (*f*): top weight 652
OBERLASTIG: top-heavy 653
OBERLAUF (*m*): upper deck 654
OBERLEDER (*n*): vamp (Leather) 655
OBERLEHRER (*m*): head-master 656
OBERLEIB (*m*): upper part of the body 657
OBERLEITUNG (*f*): overhead (wire) wiring, aerial line or main 658
OBERLICHT (*n*): top-light; fan-light; sky-light 659
OBERLIPPE (*f*): upper lip 660
OBERMAAT (*m*): chief mate (Nautical) 661
OBERMATROSE (*m*): leading seaman 662
OBERRAH (*f*): upper yard (Naut.) 663
OBERRINDE (*f*): outer bark; upper crust 664
OBERSCHENKEL (*m*): thigh; upper part of the leg 665
OBERSCHERGANG (*m*): upper sheer strake 666
OBERSCHICHT (*f*): upper (layer) stratum 667
OBERSCHIFF (*n*): upper work, dead work (of ships) 668
OBERSCHLÄCHTIG: over-shot (of water wheels) 669
OBERSCHWELLE (*f*): architrave (Architecture); lintel 670
OBERSEGEL (*n* and *pl*): top sails 671
OBERST: chief; top; uppermost; supreme; highest; (*m*): Colonel 672
OBERSTÄNDIG: hypogynus (Bot.) 673
OBERSTEUERMANN (*m*): first mate (Naut.) 674
OBERSTIMME (*f*): treble, soprano (Mus.); upper part, upper voice (Mus.) 675
OBERTASSE (*f*): cup 676
OBERTEIG (*m*): upper dough (Brewing) 677
OBERTEIL (*m*): upper (portion) part 678
OBERTON (*m*): over tone; harmonic 679
OBERVERDECK (*n*): upper deck (Naut.) 680
OBERWASSER (*n*): surface water 681
OBERWERK (*n*): upper work, dead work (of ships) 682
OBERZUG (*m*): upper pass (of boilers) 683
OBGENANNT: above-named, above mentioned 684
OBIG: above; foregoing; former 685
OBJEKT (*n*): object 686
OBJEKTIV: objective; (*n*): objective; object-glass; lens (of a camera) 687
OBJEKTIVBRETT (*n*): lens board 688
OBJEKTIVDECKEL (*m*): lens cap (of cameras) 689
OBJEKTIV (*n*), **DOPPEL-**: double lens 690
OBJEKTIV (*n*), **EINFACHES-**: single lens, single landscape lens 691
OBJEKTIVFASSUNG (*f*): lens holder, lens tube 692
OBJEKTIVISCH: objective 693
OBJEKTIVLICHTKRAFT (*f*): rapidity of a lens 694
OBJEKTIVÖFFNUNG (*f*): lens aperture 695
OBJEKTIV (*n*), **ORTHOSKOPISCHES-**: orthoscopic lens 696
OBJEKTIVROHR (*n*): (see **OBJEKTIVFASSUNG**) 697
OBJEKTIVSATZ (*m*): casket lens 698
OBJEKTIVVERSCHLUSS (*m*): instantaneous shutter (of a camera) 699
OBJEKTTISCH (*m*): stand (Microscope) 700
OBJEKTTRÄGER (*m*): slide; stand; (object)-holder 701
OBLAST (*f*): burden; load 702
OBLATE (*f*): wafer; seal 703
OBLIEGEN: to be incumbent upon; attend to; be obliged to; apply to; be one's duty 704
OBLIGATION (*f*): bond; debenture 705
OBLIGATORISCH: obligatory 706
OBMANN (*m*): chief man; head (such as a Chairman; Inspector; President; Overseer; Umpire; Foreman-(of a jury, etc.)) 707
OBOE (*f*): oboe; hautboy 708
OBSCHWEBEN: to be (pending) imminent 709
OBSERVATORIUM (*n*): observatory 710
OBST (*n*): fruit 711
OBSTBAU (*m*): horticulture; fruit growing 712
OBSTBAUM (*m*): fruit tree 713
OBSTBRANNTWEIN (*m*): fruit (brandy) spirit 714
OBSTDARRE (*f*): fruit (drying) kiln 715
OBSTESSIG (*m*): fruit vinegar 716
OBSTGARTEN (*m*): orchard 717
OBSTKELTER (*f*): fruit press 718
OBSTLESE (*f*): fruit-gathering 719
OBSTMAHLAPPARAT (*m*): fruit grinding (pulping) apparatus 720
OBSTMOST (*m*): fruit juice; must 721
OBSTMÜHLE (*f*): fruit (pulping) mill 722
OBSTMUS (*n*): fruit sauce; jam; marmalade 723
OBSTPRESSE (*f*): fruit press 724
OBSTWEIN (*m*): fruit wine; cider 725
OBSTZEIT (*f*): fruit season 726
OBSTZUCKER (*m*): fruit sugar; levulose; fructose 727
OCHER (*m*): ochre; ocher (see **OCKER**) 728

OCHERIG: ocherous 729
OCHOCOBUTTER (*f*): ochocoa butter (from nuts of *Ochocoa gabonii*); Mp. 53°C. 730
OCHS (*m*): ox; bull-(ock) 731
OCHSE (*m*): (see **OCHS**) 732
OCHSENAUGE (*f*): bullseye 733
OCHSENBREMSE (*f*): gad-fly 734
OCHSENFLEISCH (*n*): beef 735
OCHSENGALLE (*f*): ox-gall (ox bile thickened) 736
OCHSENHAUT (*f*): neat's hide 737
OCHSENKLAUENFETT (*n*): neat's-foot oil (*Oleum bubulum*); Sg. 0.92 738
OCHSENLEDER (*n*): neat's leather 739
OCHSENSTALL (*m*): ox-stall; cowshed 740
OCHSENZUNGE (*f*): ox-tongue (also Bot., *Helminthia echioides*) 741
OCHSENZUNGENWURZEL (*f*): (see **ALKANNA**) 742
OCKER (*m*): ochre, earthy iron oxide (natural mixture of ferric hydroxide, clay and lime) (yellow, red and brown, but mostly refers to the yellow) (brown ferric oxide (Fe_2O_3); red ferric oxide (Fe_2O_3); yellow iron oxide, umber, sienna) (see also **OCHER**) 743
OCKERARTIG: ocherous; like ochre 744
OCKERFARBEN: ocherous 745
OCKER (*m*), **GELBER-**: yellow ochre (see **ERDE, GELBE-** and **OCKER**) 746
OCKERIG: ocherous 747
OCKER (*m*), **ROTER-**: red ochre (see **OCKER**) 748
OCTAN (*n*): octane; C_8H_{18} 749
OCTANOL (*n*): (see **CAPRYLALKOHOL, PRIMÄR-**) 750
OCTAVE (*f*): octave; diapason 751
OCTYLALDEHYD (*n*): octylic aldehyde 752
OCTYLALKOHOL (*m*): octylic alcohol (see **CAPRYL-ALKOHOL, PRIMÄR-**) 753
OD (*m*): odyle, odyl, odic force, odyllic force 754
ODE (*f*): ode, lyric poem 755
ÖDE: waste; desolate; deserted; (*f*): desert; waste; solitude 756
ÖDEM (*n*): oedema (Med.) 757
ODEM (*m*): breath 758
ODER ABER: or on the other hand, or on the contrary, or instead 759
ODERMENNIG (*m*): agrimony (*Agrimonia eupatoria*) 760
ODOL (*n*): odol 761
ODOMETER (*n*): odometer; pedometer 762
ODYENDYEBUTTER (*f*): Odyendyea butter (from seeds of *Odyendyea gabonensis*) 763
OERSTEDIT (*m*): oerstedite (see **MALAKON**) 764
OFEN (*m*): furnace; oven; stove; kiln 765
OFENBANK (*f*): seat by the stove or fireside; bench of ovens or furnaces 766
OFENBLOCK (*m*): battery (block, bench) of ovens 767
OFENBRUCH (*m*): tutty (Zinc); burst (rupture) of an oven or furnace 768
OFENBRUST (*f*): furnace front 769
OFENDARREN (*n*): kiln drying 770
OFEN, ELEKTRISCHER- (*m*): electric furnace 771
OFEN (*m*), **EXPERIMENTIER-**: experimenting furnace 772
OFEN (*m*), **FLAMM-**: reverberatory furnace 773
OFENGALMEI (*m*): cadmium (see **GALMEI**) 774
OFEN (*m*), **GASSCHMELZ-**: gas-(melting)-furnace 775
OFEN (*m*), **GEBLÄSE-**: blast (crucible) furnace; Hoskins' furnace; forced-draught furnace 776
OFEN (*m*), **GEFÄSS-**: furnace in which vessels are contained (such as muffle furnaces, crucible furnaces, etc.) 777
OFENGLANZ (*m*): stove (gloss) polish 778
OFEN (*m*), **GLÜH-**: annealing furnace 779
OFEN (*m*), **HERD-**: hearth furnace 780
OFEN (*m*), **HERDFLAMM-**: reverberatory furnace 781
OFEN (*m*), **HOCH-**: blast furnace 782
OFEN (*m*), **HOH-**: blast furnace 783
OFENKACHEL (*f*): earthen stove tile; Dutch tile 784
OFENKLAPPE (*f*): damper (of ovens) 785
OFENKRÜCKE (*f*): rake 786
OFEN (*m*), **KUPOL-**: cupola furnace 787
OFEN (*m*), **LABORATORIUMS-**: laboratory furnace 788
OFENLACK (*m*): stove varnish 789
OFENLOCH (*n*): furnace mouth 790
OFEN (*m*), **MUFFEL-**: muffle furnace 791
OFENPLATTE (*f*): oven (furnace) plate 792
OFENPOLITUR (*f*): stove polish 793
OFENPUTZMITTEL (*n*): stove polish 794
OFEN (*m*), **RETORTEN-**: retort furnace 795
OFEN (*m*), **REVERBERIER-**: reverberatory furnace 796
OFEN (*m*), **RÖHREN-**: tube furnace 797
OFEN (*m*), **RÖST-**: roasting furnace 798
OFENROST (*m*): grate 799
OFENSAU (*f*): furnace sow (Metal) 800
OFEN (*m*), **SCHACHT-**: shaft furnace; blast furnace 801
OFENSCHIRM (*m*): stove screen; firescreen 802
OFEN (*m*), **SCHMELZ-**: melting furnace 803
OFENSCHWARZ (*n*): black-lead, graphite (Plumbago) 804
OFENSCHWÄRZE (*f*): black-lead 805
OFENSOHLE (*f*): furnace bed 806
OFEN (*m*), **TIEGEL-**: crucible furnace 807
OFENTROCKEN: oven (kiln) dried 808
OFENTÜR (*f*): oven (furnace; fire) door 809
OFENTÜRE (*f*): (see **OFENTÜR**) 810
OFEN (*m*), **VERBRENNUNGS-**: combustion furnace 811
OFEN (*m*), **VERSUCHS-**: test frunace 812
OFENWAND (*f*): furnace wall 813
OFENZIEGEL (*m*): stove (fire) tile or brick 814
OFFEN: open; clear; clever; frank; sincere; candid; public; outspoken; vacant; uncovered; expanded; extended; unreserved; undisguised; exposed; unprotected; unobstructed; accessible; unenclosed; unrestricted; evident; apparent; debatable 815
OFFENBAR: manifest; obvious; plain; evident; apparent; clear; palpable; notorious 816
OFFENBAREN: to disclose; reveal; manifest; publish; make public 817
OFFENBARUNG (*f*): revelation; manifestation; disclosure 818
OFFENHERZIG: frank, candid, unreserved, sincere, artless, undissembling, open-hearted, hearty, cordial, honest, fair, generous, liberal, free 819
OFFENKUNDIG: notorious; well-known; evident; obvious; public 820
OFFENSICHTLICH: obviously, apparent-(ly), evident 821
OFFENSIVE (*f*): offensive; aggression; attack 822
OFFERIEREN: to offer, bid, quote, estimate, tender, contract 823
OFFERTE (*f*): offer; bid; quotation; estimate; tender 824
OFFERTENBRIEF (*m*): tender (covering) letter 825
OFFICIELL: official 826

OFFIZIAL (*m*): official, officer, functionary 827
OFFIZIELL: official 828
OFFIZIERDECK (*n*): quarter-deck 829
OFFIZIERSMESSE (*f*): ward room (Naut.) 830
OFFIZIN (*f*): dispensary; chemist's-shop; workshop; laboratory 831
OFFIZINELL: officinal; medicinal 832
OFFIZIÖS: officious 833
ÖFFNEN: to open; dissect; loosen; undo; untie; lay bare; show; exhibit; expand; explain; commence; enter upon; disclose; reveal; throw open; make easy of access 834
ÖFFNEND: opening; aperient (Med.) 835
ÖFFNER (*m*): opener; sley (the reed of a loom) (Textile) 836
ÖFFNERMASCHINE (*f*): opener (Textile) 837
ÖFFNUNG (*f*): opening; aperture; port; orifice; mouth; dissection; evacuation (Med.); incision 838
ÖFFNUNGSMITTEL (*n*): opening medicine; aperient (Med.) 839
ÖFFNUNGSSTROM (*m*): induced current; breaking current 840
OFFSETDRUCK (*m*): off-set print 841
OFTMALIG: frequent; repeated; reiterated 842
OHM (*n*): ohm (Elect.) 843
OHMSCHES GESETZ (*n*): Ohm's law 844
OHNE: without; besides; apart from; save; except; excluding; exclusive of; but for 845
OHNEDEM: apart from that; besides; without that; moreover; likewise 846
OHNEDIES: apart from this; besides; without this; moreover; likewise 847
OHNEHIN: besides; moreover; likewise 848
OHNMACHT (*f*): syncope, fainting (Med.) 849
OHNMÄCHTIG: unconscious; faint; feeble; weak; impotent; swooning; powerless 850
OHR (*n*): ear; handle; eye; lug; ring (Tech.) 851
ÖHR (*n*): ear; handle; eye; lug; catch; ring (Tech.); bog iron ore (see **BRAUNEISENERZ**) 852
OHRBLATT (*n*): lobe (of the ear) 853
OHRBLUTGESCHWULST (*f*) **VON GEISTESKRANKEN**: *Hæmatoma auris*, insane ear (Med.) 854
OHRCHEN (*n*): auricle (Bot.) 855
OHRDRÜSE (*f*): (see **OHRSPEICHELDRÜSE**) 856
OHRDRÜSENBRÄUNE (*f*): mumps (Med.) 857
OHRENARZT (*m*): aurist (ear specialist) (Med.) 858
OHRENBEULEN (*f.pl*): mumps (Med.) 859
OHRENBOLZEN (*m*): ring bolt 860
OHRENDRÜSE (*f*): (see **OHRSPEICHELDRÜSE**) 861
OHRENFLUSS (*m*): ear-discharge (Med.) 862
OHRENHÖHLE (*f*): aural cavity 863
OHRENKNORPEL (*m*): cartilage of the ear 864
OHRENKRANKHEIT (*f*): disease of the ear 865
OHRENNERV (*m*): auditory (acoustic) nerve 866
OHRENSCHMALZ (*n*): cerumen; ear wax 867
OHRENSCHMERZ (*m*): ear-ache; otalgia (Med.) 868
OHRENSPIEGEL (*m*): ear speculum, auricular speculum (a mirror for examining the ear passages) (Med.) 869
OHRENSPRITZE (*f*): ear syringe (Med.) 870
OHRENSTEIN (*m*): otolith 871
OHRENZWANG (*m*): (see **OHRENSCHMERZ**) 872
OHRFLÜGEL (*m*): lobe (of the ear) 873
OHRFÖRMIG: auriform; auricular; auriculate; ear-shaped 874
OHRREISSEN (*n*): (see **OHRENSCHMERZ**) 875
OHRSCHNECKE (*f*): cochlea (Anat.) 876
OHRSPEICHELDRÜSE (*f*): parotid gland (*Glandula parotis*) (Med.) 877
OHRSPEICHELDRÜSENENTZÜNDUNG (*f*): parotitis, inflammation of the parotid gland (Med.) 878
OHRSPEICHELDRÜSENENTZÜNDUNG (*f*) **EPIDEMI--SCHE-**: epidemic parotitis (*Parotitis epidemica*) (see also **MUMPS**) (Med.) 879
OHRSPEICHELDRÜSENENTZÜNDUNG (*f*), **IDIOPATHI--SCHE-**: idiopathic inflammation of the parotid gland (see **MUMPS**) 880
OHRSPEICHELDRÜSENENTZÜNDUNG (*f*), **SPONTANE-**: spontaneous inflammation of the parotid gland (see **MUMPS**) 881
OHRTRICHTER (*m*): ear trumpet 882
OHRTROMMEL (*f*): tympanum, ear-drum 883
OHRTROMPETE (*f*): Eustachian tube (Anat.); ear-trumpet 884
OHRWACHS (*n*): cerumen; ear wax 885
OHRWASSER (*n*): endolymph 886
OHRWURM (*m*): ear-wig (*Forficula auricularia*) 887
ÖHSE (*f*): ear; eye; lug; ring; handle; catch; hook; loop (Tech.) 888
OKER (*m*): ochre (see **OCKER**) 889
OKERIG: ocherous 890
OKKLUDIEREN: to occlude 891
OKKULTISMUS (*m*): occultism 892
ÖKONOM (*m*): economist; agriculturist 893
ÖKONOMIE (*f*): economy 894
ÖKONOMIK (*f*): economy 895
ÖKONOMISCH: economic-(al); agricultural 896
OKTAEDER (*n*): octahedron 897
OKTAEDRISCH: octahedral 898
OKTAV (*n*): octavo 899
OKTAVE (*f*): octave (Mus.) 900
OKTOCARBOCYCLISCH: octacarbocyclic 901
OKTOGON (*n*): octagon (an eight-sided figure) 902
OKTOGONAL: octagonal, eight-sided, eight-angled 903
OKTOGONISCH: octagonal 904
OKTOGYNISCH: octogynous (Bot.) 905
OKTONAPHTEN (*n*): octanaphthene 906
OKTYALDEHYD (*n*): octyaldehyde; (C_7H_{15}.COH) 907
OKULIEREN: to inoculate; graft 908
OKULIERMESSER (*n*): grafting knife 909
ÖL (*n*): oil (*Oleum*) 910
ÖLABSCHEIDER (*m*): oil separator; oil trap 911
ÖL (*n*), **ANTISEPTISCHES-**: antiseptic oil 912
ÖLARTIG: oily; oleaginous 913
ÖL, ÄTHERISCHES- (*n*): volatile oil; essential oil 914
ÖLAUSSCHEIDER (*m*): oil separator, oil extractor 915
ÖLAUSSTRICH (*m*): coat of oil; oil painting 916
ÖLBAD (*n*): oil bath; green liquor (Turkey red dyeing) 917
ÖLBADFÜLLUNG (*f*): oil bath filling 918
ÖLBAUM (*m*): olive tree (*Olea europæa*) 919
ÖLBAUM, ECHTER- (*m*): olive tree (*Olea europæa*) 920
ÖLBAUM, FALSCHER- (*m*): oleaster (*Elæagnus*) 921
ÖLBAUMGUMMI (*n*): elemi; gum elemi (a resin obtained from the elemi tree, *Amyris elemifera*) 922
ÖLBAUMHARZ (*n*): elemi; gum elemi (a resin obtained from the elemi tree, *Amyris elemifera* or *Canarium commune*) 923
ÖLBAUMHOLZ (*n*): olive tree wood (from *Olea europæa*); Sg. 0.87 924
ÖLBEERE (*f*): olive (fruit of *Olea europæa*) 925

ÖLBEHÄLTER (*m*): oil reservoir, oil-storage tank 926
ÖLBEIZE (*f*): oil mordant (for dyeing cotton) (Turkey red dyeing) 927
ÖLBILDEND: olefiant; oil-forming 928
ÖLBILDENDES GAS (*n*): olefiant gas; ethylene (see ÄTHYLEN) 929
ÖLBLASE (*f*): oil-(boiling)-copper (Varnish manufacture) 930
ÖLBLATT (*n*): oil leaf; olive leaf (see ÖLBAUM) 931
ÖLBLAU (*n*): oil blue; indigo copper; CuS (see also KUPFERINDIG) 932
ÖLBLEICHANLAGE (*f*): oil bleaching plant 933
ÖLBLEICHEN (*n*): oil bleaching 934
ÖL (*n*), BRAUSENDES-: effervescent oil 935
ÖLBRENNER (*m*): oil burner 936
ÖLCHROMVERBINDUNG (*f*): oil-chromium compound (by reaction of fat-soluble halogen compounds of chromium on linseed oil) 937
ÖLDAMPF (*m*): oil vapour 938
ÖLDESODORISIERUNG (*f*): oil (deodorizing) deodorization 939
ÖLDESTILLATIONSVERFAHREN (*n*): oil distilling process 940
ÖLDRASS (*m*): oil (foots) dregs 941
ÖLDRUCK (*m*): oleograph; oleography 942
ÖLDRUCKBILD (*n*): oleograph 943
OLEANDER (*m*): oleander, rose-bay (*Nerium oleander*) 944
OLEASTER (*m*): oleaster (see ÖLBAUM, FALSCHER-) 945
OLEAT (*n*): oleate 946
OLEFIN (*n*): olefin, alkylene, alkene (a product from the dry distillation of organic matter; having a formula; C_nH_{n}..) 947
OLEFINALKOHOL (*m*): olefin-(ic) alcohol 948
OLEFINHALOID (*n*): olefin halide 949
OLEFINISCH: olefinic 950
OLEFINKETON (*n*): olefin(-ic) ketone 951
OLEIN (*n*): olein (glyceride of oleic acid in fats); (often incorrectly used technically as a synonym for) oleic acid (see ELAINSÄURE) 952
OLEIN (*n*), DESTILLAT-: (see OLEIN, DESTILLIERTES-) 953
OLEIN (*n*), DESTILLIERTES-: distilled olein 954
OLEIN (*n*), SAPONIFIKAT-: (see OLEIN, SAPONIFIZIERTES-) 955
OLEIN (*n*), SAPONIFIZIERTES-: saponified olein 956
OLEINSAUER: oleate, combined with oleic acid 957
ÖLEINSÄURE (*f*): oleic acid; $C_{18}H_{34}O_2$ (see ELAINSÄURE) 958
OLEINSEIFE (*f*): olein soap 959
ÖLEN: to oil; lubricate; anoint 960
OLEOKRESOT (*n*): creosote oleate 961
OLEOMARGARIN (*n*): Oleomargarine (see MARGARIN); Sg. 0.929; Mp. 30-40°C. 962
OLEOSOLFARBE (*f*): oil-soluble (colour) dye 963
ÖLER (*m*): oiler, lubricator 964
ÖLERSATZ (*m*): oil substitute 965
OLEUM (*n*): fuming sulphuric acid (see SCHWEFELSÄURE, RAUCHENDE-); oil 966
ÖLEXTRAKTIONSANLAGE (*f*): oil extraction plant; deoiling plant 967
ÖLEXTRAKTOR (*m*): oil extractor 968
ÖLFABRIK (*f*): oil factory 969
ÖLFÄNGER (*m*): oil catcher, save-oil, drip plate, oil trap 970
ÖLFANGSCHALE (*f*): oil drip cup 971
ÖLFARBE (*f*): oil (paint) colour 972
ÖLFARBENSTIFT (*m*): oil colour pencil or crayon (a crayon of oil colour in a useable form, with which it is possible to paint or mix colours direct on the canvas) 973
ÖLFASS (*n*): oil (cask) drum 974
ÖLFEUERUNG (*f*): oil furnace; oil firing 975
ÖLFEUERUNGSANLAGE (*f*): oil firing plant 976
ÖLFILM (*f*): skin on oil; film (coating) of oil 977
ÖLFILTER (*m*): oil filter 978
ÖLFIRNIS (*m*): oil varnish 979
ÖLFIRNISBAUM (*m*): Candlenut tree (*Aleurites cordata*), Tung chou tree (*Elæococca vernicia*) 980
ÖLFIRNISBAUMÖL (*n*): (see ELÄOKOKKAÖL) 981
ÖLFLASCHE (*f*): (see FLORENTINER FLASCHE) 982
ÖLFLECK (*m*): oil (grease) spot 983
ÖLFLECKEN (*m*): (see ÖLFLECK) 984
ÖL (*n*), FLÜCHTIGES-: volatile oil; essential oil 985
ÖLGAR: chamois 986
ÖLGAS (*n*): oil gas (see BLAUGAS) 987
ÖLGASTEER (*m*): oil gas tar 988
ÖL (*n*), GEBLASENES-: oxydated oil 989
ÖL, GEBLEICHTES- (*n*): bleached oil 990
ÖL, GEHÄRTETES- (*n*): hardened oil; hydrogenated oil 991
ÖLGELÄGER (*n*): oil (foots; dregs) sediment 992
ÖLGEMÄLDE (*n*): oil painting; picture in oils (Art) 993
ÖLGLAS (*n*): lubricator glass 994
ÖLGRÜN (*n*): (see CHROMGRÜN) 995
ÖLGRUND (*m*): oil ground-(ing), first coat of oil 996
ÖLHAHN (*m*): lubricator cock or tap 997
ÖLHALTIG: oleiferous; containing oil 998
ÖLHANDEL (*m*): oil trade 999
ÖLHÄNDLER (*m*): oil (merchant) dealer 000
ÖLHÄRTUNG (*f*): hardening (hydrogenation) of oil 001
ÖLHÄRTUNGSANLAGE (*f*): oil hardening (hydrogenation) plant 002
ÖLHARZ (*n*): oleoresin (mixture of resin and essential oil from the same plant) 003
ÖL, HARZ- (*n*): resin oil (see HARZÖL) 004
ÖLHEFEN (*f.pl*): oil (foots; dregs) sediment 005
OLIBANUM (*n*): olibanum; gum frankincense; gum Thus (from various African species of *Boswellia* trees) 006
ÖLICHT: (see ÖLIG) 007
ÖLIG: oily; oleaginous; unctuous 008
OLIGISTE (*m*): oligiste (a natural iron oxide) (see EISENOXYD) 009
ÖLIGKEIT (*f*): oiliness 010
OLIGOKLAS (*m*): oligoclase (isomorphous mixture of albite and anorthite) (see ALBIT and ANORTHIT); Sg. 2.68; Mp. 1200-1240°C.; $2(Ca,Na_2)O,2Al_2O_3,9SiO_2$ 011
OLIVE (*f*): olive (sometimes refers to the tree and sometimes to the fruit) (see ÖLBAUM) 012
OLIVENBAUM (*m*): olive tree (*Olea europæa*); Sg. 0.87 013
OLIVENERZ (*n*): olivenite (see OLIVENIT) 014
OLIVENFARBE (*f*): olive colour 015
OLIVENHARZ (*n*): olive resin 016
OLIVENHOLZ (*n*): olive wood (see OLIVENBAUM) 017
OLIVENIT (*m*): olivenite (copper arsenate and hydroxide); $Cu(CuOH)AsO_4$; Sg. 4.25 018
OLIVENKERNÖL (*n*): olive-kernel oil; Sg. 0.916-0.92 019

OLIVENÖL (*n*): olive oil; sweet oil (*Oleum olivæ*) (*Oleum olivarum*); Sg. 0.92 (from fruit of olive tree, *Olea europæa*) 020
OLIVENÖLFETTSÄURE (*f*): olive oil (fatty) sebacic acid 021
OLIVENÖLSEIFE (*f*): olive oil soap 022
OLIVIN (*m*): olivine, hyalosiderite, chrysolite (a magnesium ortho-silicate); $(Mg,Fe)_2SiO_4$; Sg. 3.4 023
ÖLKANNE (*f*): oil can 024
ÖLKAUTSCHUK (*m*): vulcanized oil (a caoutchouc substitute) 025
ÖLKELTER (*f*): oil press 026
ÖLKITT (*m*): putty 027
ÖLKUCHEN (*m*): oil cake; linseed cake 028
ÖLKUCHENBRECHER (*m*): oil (linseed) cake crusher 029
ÖLKUGEL (*m*): oil (drop) globule 030
ÖLLACK (*m*): oil varnish; water-proof varnish 031
ÖLLAMPE (*f*): oil lamp 032
ÖLLEDER (*n*): chamois 033
ÖLLÖSLICH: oil-soluble, soluble in oil 034
ÖLMALEREI (*f*): oil-painting; painting in oils 035
ÖLMASCHINE (*f*): oil engine 036
ÖLMESSER (*m*): oleometer; oil hydrometer 037
ÖLMILCH (*f*): oil emulsion 038
ÖL (*n*), MINERAL-: mineral oil 039
ÖLMÜHLE (*f*): oil mill 040
ÖLNAPF (*m*): oil cup 041
ÖLNUSSBAUM (*m*): butternut tree 042
ÖLOCKER (*m*): yellow ochre 043
ÖL (*n*), OXYDIERTES-: oxydated oil 044
ÖLPALME (*f*): oil palm (*Elæis guineensis*) 045
ÖLPAPIER (*n*): oil-paper, oiled paper, grease paper 046
ÖLPRESSE (*f*): oil press 047
ÖLPRESSSCHMIERPUMPE (*f*): forced oil lubrication pump 048
ÖLPRESSVERFAHREN (*n*): oil-pressing process 049
ÖLPRÜFER (*m*): oil tester 050
ÖLPRÜFMASCHINEN (*f.pl*): oil testing machinery 051
ÖLPUMPE (*f*): oil pump 052
ÖL, PUTZ- (*n*): cleaning oil 053
ÖLRAFFINIERANLAGE (*f*): oil refining plant 054
ÖLREICH: oil-bearing 055
ÖLREINIGER (*m*): oil (purifier) filter 056
ÖLRETTICH (*m*): common radish (*Raphanus sativus*) 057
ÖL (*n*), ROH-: crude oil 058
ÖLRÜCKSTAND (*m*): oil residue 059
ÖLRUSS (*m*): oil spot, lamp black 060
ÖLSAAT (*f*): rape-seed, linseed (seeds of *Brassica napus*) 061
ÖLSAMEN (*m*): rape-seed; linseed (see ÖLSAAT) 062
ÖLSAMMLER (*m*): oil catcher, save-oil, oil trap; oil receiver (see FLORENTINER FLASCHE) 063
ÖLSATZ (*m*): oil (foots; sediment; lees) dregs 064
ÖLSAUER: oleate of; combined with oleic acid 065
ÖLSÄURE (*f*): oleic acid; $C_{18}H_{34}O_2$; Mp. 14°C. (see ELAINSÄURE) 066
ÖLSÄUREGLYZERID (*n*): olein (glyceride of oleic acid in fats) 067
ÖLSÄUREHALTIG: containing oleic acid 068
ÖLSAURES SALZ (*n*): oleate 069
ÖLSCHALENRETORTE (*f*): oil shale retort 070
ÖLSCHALTER (*m*): oil switch 071
ÖLSCHIEFER (*m*): oil shale 072
ÖLSCHLAGEN (*n*): oil pressing 073
ÖLSCHLÄGER (*m*): oil (miller) presser 074
ÖL (*n*), SCHMIER-: lubricating oil 075
ÖLSCHWARZ (*n*): oil black; lampblack 076
ÖLSEIFE (*f*): oil soap; soft soap 077
ÖLSODASEIFE (*f*): Castile soap 078
ÖLSPRITZE (*f*): lubricating (oil) syringe 079
ÖLSTEIN (*m*): oil stone 080
ÖLSTOFF (*m*): olein (see OLEIN) 081
ÖLSTRAHLGEBLÄSE (*n*): oil jet blower 082
ÖLSÜSS (*n*): glycerol; glycerine (see GLYZERIN) 083
ÖLTRESTER (*m.pl*): oil (marc) lees (a residue left over from pressing) 084
ÖLTUCH (*n*): oilcloth 085
ÖLUNG (*f*): oiling; lubrication; greasing; anointment; anointing; consecration 086
ÖLUNG, LETZTE- (*f*): extreme unction 087
ÖLUNTERSUCHUNG (*f*): oil testing 088
ÖLUNTERSUCHUNGSAPPARAT (*m*): oil testing apparatus 089
ÖLVENTIL (*n*): lubricating (oil) valve 090
ÖLVERFAHREN (*n*): oil-(flotation)-process 091
ÖLVERGOLDUNG (*f*): oil gilding 092
ÖLVORKOMMEN (*n*): oil resources (of a country) 093
ÖL (*n*), VULKANISIERTES-: vulcanized oil (see ÖLKAUTSCHUK) 094
ÖLWAGE (*f*): oleometer; oil hydrometer 095
ÖLWANNE (*f*): oil well 096
ÖL (*n*), WASSERLÖSLICHES-: water-soluble oil 097
ÖLWEIDE (*f*): oleaster (*Elæagnus*) 098
ÖLWEIDENGEWÄCHSE (*n.pl*): *Elæagnaceæ* (Bot.) 099
ÖLWEISS (*n*): oil white 100
ÖLZELLE (*f*): oil cell (of plants) 101
ÖLZEUG (*n*): oil-skin, (the material of which nautical oil-skins are made) 102
ÖLZUCKER (*m*): elæosaccharum (mixture of fine sugar powder and an essential oil, in proportion of 2 g. to 1 drop) 103
ÖLZWEIG (*m*): olive branch (the sign of peace) 104
OMINÖS: ominous 105
OMPHACIT (*m*): omphacite, augite; MgO, SiO_2,Al_2O_3; Sg. 3.27 106
ÖNANTHALDEHYD (*n*): œnanthic aldehyde (see ÖNANTHOL) 107
ÖNANTHÄTHER (*m*): œnanthic (enanthic) ether (see COGNAKÖL) 108
ÖNANTHOL (*n*): œnanthol, enanthol, heptanal, œnanthal, heptoic aldehyde, œnanthic aldehyde; $C_7H_{14}O$; Sg. 0.85; Bp. 154°C. 109
ÖNANTHSÄURE (*f*): œnanthic (enanthic) acid 110
ÖNANTHYLSÄURE (*f*): œnanthylic (enanthylic) acid 111
ÖNIG (*m*): (actually an abbreviation for Österreichischer Normenausschuss für Industrie und Gewerbe) (the title of an Austrian Bureau of Standards) 112
ÖNOLOGISCH: œnologic-(al) 113
ÖNOLOGISCHES PRODUKT (*n*): œnological product 114
ONONIN (*n*): ononine (a glycoside from *Ononis spinosa*) 115
ONYX (*m*): onyx, chalcedony (see ACHAT and CHALCEDON); Sg. 2.65 116
OOLITH (*m*): oolite (a form of limestone) (see ROGENSTEIN) 117
OOLITHISCH: oolitic; granular 118
OPAK: opaque 119
OPAL (*m*): opal [natural hydrated (silica) silicon dioxide]; $SiO_2.H_2O$; Sg. 2.25 120
OPALARTIG: opaline; opal-like 121

OPALBLAU (*n*): opal blue 122
OPAL (*m*), **EDEL-**: precious opal 123
OPALESZENZ (*f*): opalescence 124
OPALESZIEREN: to opalesce 125
OPALESZIEREND: opalescent 126
OPAL (*m*), **HALB-**: semi-opal 127
OPALIN-ALLOPHAN (*m*), **EUCHROMATISCHER-**: chrysocolla 128
OPALINALLOPHAN (*m*), **LAMPROCHROMATISCHER-**: allophane (see **ALLOPHAN**) 129
OPALISIEREN: to opalesce 130
OPALISIEREND: opalescent 131
OPALJASPIS (*m*): opal jasper (see **OPAL**) 132
OPALPLATTE (*f*): opal plate 133
OPER (*f*): opera-(house) 134
OPERATEUR (*m*): operator; surgeon-(conducting an operation) 135
OPERATION (*f*): operation 136
OPERATIONSFELD (*n*): field of operations; part to be operated upon (Med.) 137
OPERATIONSHANDSCHUH (*m*): operating glove (Surgery) 138
OPERATIV: operative 139
OPERIEREN: to operate; conduct (perform) an operation; function; work 140
OPERMENT (*n*): orpiment (see **ARSENTRISULFID** and **AURIPIGMENT**) 141
OPFER (*n*): offering; sacrifice; slaughter; victim; prey; quarry; martyr; immolation; oblation; surrender 142
OPFERN: to offer-(up); sacrifice; immolate; surrender; give up; kill; destroy; slaughter 143
OPFERTIER (*n*): sacrifice, victim, prey, quarry 144
OPFER (*n*), **ZUM-FALLEN**: to fall a prey to, become the (prey of) victim of, be the quarry of 145
OPHTHALMIA (*f*): ophthalmia, inflammation of the superficial structure of the eye (Med.) 146
OPIANSÄURE (*f*): opianic acid; dimethoxy-ortho-aldehydrobenzoic acid 147
OPIAT (*n*): opiate, anodyne, sedative, narcotic 148
OPIUM (*n*): opium (dried latex or milky juice obtained from the variety of poppy, *Papaver somniferum*); Sg. 1.34 149
OPIUMALKALOID (*n*): opium alkaloid 150
OPIUM (*n*), **DIALYSIERTES-**: dialysed opium (*Opium dialysatum*) 151
OPIUMTINKTUR (*f*): tincture of opium, laudanum (opium dissolved in alcohol) 152
OPODELDOC (*n*): opodeldoc 153
OPOPONAX (*n*): opoponax (a gum-resin) 154
OPOPONAXÖL (*n*): opoponax oil 155
OPPONIEREN: to oppose 156
OPTATIV: optative (expressive of desire) 157
OPTIEREN: to choose 158
OPTIK (*f*): optics (the science of light) 159
OPTIKER (*m*): optician 160
OPTIKUS (*m*): optician 161
OPTIMISMUS (*m*): optimism 162
OPTIMITÄT (*f*): excellence 163
OPTISCH: optic-(al)-(ly) 164
OPTISCH-AKTIV: optically active 165
OPTISCHE AKTIVITÄT (*f*): optical activity (capacity for rotating the plane of vibration of rectilinearly polarized light) (see **OPTISCHES DREHUNGSVERMÖGEN**) 166
OPTISCHER BRENNPUNKT (*m*): visual focus 167
OPTISCHER SENSIBILISATOR (*m*): optical sensitizer 168
OPTISCHES DREHUNGSVERMÖGEN (*n*): optical rotation (see **OPTISCHE AKTIVITÄT**) 169
OPTISCHES PYROMETER (*n*): optical pyrometer 170
ORAKEL (*n*): oracle 171
ORAKELHAFT: oracular, authoritative 172
ORANGE: orange (of dyes); (*f*): orange (fruit of *Citrus aurantium*) 173
ORANGEGELB (*n*): orange yellow 174
ORANGE-MENNIGE (*f*): orange minium 175
ORANGENBAUM (*m*): orange tree (*Citrus aurantium*) 176
ORANGENBLÄTTER (*n.pl*): orange leaves 177
ORANGENBLÜTE (*f*): orange blossom (see **ORANGE** and **POMERANZE**) 178
ORANGENBLÜTENÖL (*n*): (*Oleum aurantii florum*): orange blossom oil (see **NEROLIÖL**) 179
ORANGENBLÜTENWASSER (*n*): orange blossom water 180
ORANGENFARBE (*f*): orange-(colour) 181
ORANGENGELB: orange-yellow 182
ORANGENHOLZ (*n*): orange wood (Sg. 0.7-0.93) 183
ORANGENSAMENÖL (*n*): orange-(seed) oil (from *Citrus aurantium*) 184
ORANGENSCHALE (*f*): orange peel 185
ORANGENSCHALENÖL (*n*): orange peel oil (bitter) (see **POMERANZENSCHALENÖL, BITTERES-**) 186
ORANGENSCHALENÖL (*n*), **SÜSSES-**: orange peel oil (sweet) (see **POMERANZENSCHALENÖL, SÜSSES-**) 187
ORANGEOCKER (*m*): orange ochre 188
ORANGIT (*m*): orangite; ($ThSiO_4$); (Sg. 4.9) (see **THORIT**) 189
ORATORIUM (*n*): oratorio (sacred musical drama) 190
ORCANETTE (*f*): (see **ALKANNA**) 191
ORCHESTER (*n*): orchestra 192
ORCHIDEENÖL (*n*): Ylang-ylang oil (see **YLANG-YLANGÖL**) 193
ORCIN (*n*): orcin (see **ORZIN**) 194
ORDEN (*m*): order; insignia 195
ORDENSZEICHEN (*n*): insignia (sign; badge) of an order 196
ORDENTLICH: ordinary; regular; exact; orderly; steady; downright; proper-(ly); tidy; respectable 197
ORDER (*f*): order; command 198
ORDINALE (*n*): ordinal-(number) 199
ORDINALZAHL (*f*): ordinal-(number) 200
ORDINANZ (*f*): order, ordinance; orderly (Military) 201
ORDINÄR: ordinary; mean; common; vulgar; inferior 202
ORDINATE (*f*): ordinate (the vertical ruling of graph paper) 203
ORDINATION (*f*): ordination; investment 204
ORDINIEREN: to ordain; invest 205
ORDNEN: to order; regulate; settle; arrange; dispose; classify; organize; construe (of a sentence) 206
ORDNUNG (*f*): order-(ing); arrangement; disposition; classification; class; settling; settlement; magnitude (Astronomy) 207
ORDNUNGSGEMÄSS: orderly; regular; for the sake of regularity; methodical; according to order; duly; well 208
ORDNUNGSLOS: disorderly, untidy, irregular, confused, turbulent, unruly, riotous, disorganized, jumbled, topsy-turvy, upset 209
ORDNUNGSMÄSSIG: orderly; regular; methodical; according to order; duly; well 210
ORDNUNGSWIDRIG: irregular; disorderly (see **ORDNUNGSLOS**) 211

ORDNUNGSZAHL (*f*): ordinal number 212
OREXIN (*n*): orexine, phenyldihydroquinazoline; $C_{14}H_{12}N_2.2H_2O$ 213
OREXINCHLORHYDRAT (*n*): (*Orexinum hydrochloricum*): orexine hydrochloride; phenyldihydroquinazoline hydrochloride; $C_{14}H_{12}N_2.HCl.2H_2O$ 214
OREXIN (*n*): **GERBSAURES-**: (*Orexinum tannicum*): orexine tannate 215
ORGAN (*n*): organ; part; member; link; device; medium; mechanism; gear; valve; fitting; mounting; arrangement 216
ORGANISATOR (*m*): organiser 217
ORGANISCH: organic 218
ORGANISCHER FARBSTOFF (*m*): organic (colouring matter) dye 219
ORGANISIEREN: to organize 220
ORGANISIERUNG (*f*): organization, organizing 221
ORGANISMUS (*m*): organism 222
ORGANIST (*m*): organist 223
ORGANOGEL (*n*): organogel (the coagulated organic solution of a colloid) 224
ORGANOGRAPHIE (*f*): organography (description of the organs) (see **PFLANZENORGANO--GRAPHIE**) 225
ORGANOMETALL (*n*): organometallic compound 226
ORGANOSOL (*n*): organosol (a colloid dissolved in an organic solvent) 227
ORGANOTHERAPEUTISCH: organotherapeutic 228
ORGANOTHERAPEUTISCHES PRÄPARAT (*n*): organotherapeutical preparation 229
ORGEL (*f*): organ (the instrument) 230
ORGELBALG (*m*): organ bellows 231
ORGELBAUER (*m*): organ builder 232
ORGELCHOR (*n*): organ-loft 233
ORGELPFEIFE (*f*): organ-pipe 234
ORGELPUNKT (*m*): pedal note (Mus.) 235
ORGELREGISTER (*n*): organ-stop 236
ORGELSPIELER (*m*): organist 237
ORGELSTIMME (*f*): organ stop 238
ORGELSTIMMEN (*n*): re-voicing of an organ, tuning of an organ 239
ORGELZUG (*m*): set (series, row) of organ pipes; organ-stop 240
ORIENT (*m*): orient; East 241
ORIENTALISCH: oriental 242
ORIENTBLAU (*n*): orient blue 243
ORIENTIEREN: to orient; find one's way; set oneself right; find out; become acquainted with 244
ORIENTIEREN, SICH-: to make oneself acquainted with, obtain information regarding, prime oneself 245
ORIENTIERUNG (*f*): orientation; information 246
ORIGANUMKRAUT (*n*): Spanish marjoram (*Herba origani*) (*Origanum marjorana*) 247
ORIGANUMÖL (*n*): origanum oil (see also under **DOSTEN**); thyme oil; Sg. 0.92 248
ORIGINAL (*n*): original 249
ORIGINALAUFNAHME (*f*): original photograph 250
ORIGINALBOHRER (*m*): master tap 251
ORIGINALITÄT (*f*): originality 252
ORIGINALLITHOGRAPHIE (*f*): original lithograph (by the artist himself) 253
ORIGINALLITHOGRAPHIEDRUCK (*m*): artist's proof 254
ORIGINELL: original-(ly) 255
ORKAN (*m*): hurricane 256
ORLEAN (*n*): annatto (a vegetable dyestuff from the plant, *Bixa orellana*) 257
ORLOPBALKEN (*m*): orlop beam 258
ORLOPDECK (*m*): orlop deck 259
ORNAT (*m*): official (costume) dress; vestments 260
ORNITHIN (*n*): ornithine, diaminovalerianic acid; $NH_2.CH_2(CH_2)_2CH(NH_2).CO_2H$ 261
ORNITHOLOGIE (*f*): ornithology (the science of birds) 262
ORPHOL (*n*): (*Bismutum naphtolicum*): orphol, bismuth-beta-naphtholate; $Bi(C_{10}H_6OH)_3$ 263
ORPIMENT (*n*): orpiment (see **AURIPIGMENT** and **OPERMENT**) 264
ORSATAPPARAT (*m*): Orsat apparatus (for testing flue gases) 265
ORSEILLE (*f*): archil; orchil; orseille; persio; cudbear (colouring matter from lichen, *Roccella tinctoria*, etc.) (see **LACKMUSFLECHTE**) 266
ORSEILLEERSATZ (*m*): archil substitute 267
ORSEILLEEXTRAKT (*m*): archil extract (obtained by extraction with lime water and precipitation with acids) 268
ORSEILLE IM TEIG: real (actual) archil (obtained by fermenting the lichen, *Roccella tinctoria*, with ammonia) 269
ORSEILLEPRÄPARAT (*n*): archil preparation 270
ORSELLINSÄURE (*f*): orsellinic acid 271
ORSELLSÄURE (*f*): orsellic (diorsellinic; lecanoric) acid 272
ORT (*m*): place; region; location; locality; spot; point; locus (Maths.) 273
ORTBESCHREIBEND: topographical 274
ORTBESCHREIBUNG (*f*): topography (description of places) 275
ORTFEST: stationary; fixed 276
ORT (*m*), **GEOMETRISCHER-**: geometrical locus 277
ORTHIT (*m*): orthite, allanite; $CaF_2(AlOH)(AlCeFe)_2(SiO_4)_3$; Sg. 3.5 278
ORTHOAMEISENSÄURE (*f*): orthoformic acid; o-formic acid 279
ORTHOAMEISENSÄUREESTER (*m*): ortho-formic ester; $C_7H_{16}O_3$; Sg. 0.8971; Bp. 145.5°C. 280
ORTHOAMIDOBENZOESÄURE (*f*): orthoaminobenzoic acid, anthranilic acid; $C_6H_4(NH_2)(CO_2H)$; Mp. 144°C. 281
ORTHOANISIDIN (*n*): orthoanisidine 282
ORTHOANTIMONSÄURE (*f*): orthoantimonic acid 283
ORTHOARSENSÄURE (*f*): orthoarsenic acid; $H_3AsO_4.\frac{1}{2}H_2O$; Sg. 2.5; Mp. 35.5°C. 284
ORTHOBORSÄURE (*f*): orthoboric acid; H_3BO_3; Sg. 1.435; Mp. 184°C. 285
ORTHOCHINON (*n*): orthoquinone 286
ORTHOCHLORBENZALDEHYD (*n*): orthochlorobenzaldehyde 287
ORTHOCHLORTOLUOL (*n*): orthochlorotoluene 288
ORTHOCHROMATISCH: orthochromatic, sensitive to colours (usually of photographic plates) 289
ORTHOCHROMPLATTE (*f*): orthochromatic plate (Phot.) 290
ORTHOCYAMIN (*n*): orthocyamine 291
ORTHODOX: orthodox 292
ORTHOFORM (*n*): orthoform, methyl meta-amino-para-oxybenzoate; $C_6H_3.NH_2.OH.COOCH_3$; Mp. 142°C. 293
ORTHOGONAL: orthogonal 294
ORTHOGRAPHIE (*f*): orthography 295
ORTHOGRAPHISCH: orthographical-(ly) 296

ORTHOKIESELSÄURE (*f*): orthosilicic acid 297
ORTHOKLAS (*m*): orthoclase; potash feldspar (see **FELDSPAT**) (a potassium-aluminium silicate); $(K.Na)AlSi_3O_8$; Sg. 2.55 [298
ORTHOKLASPORPHYR (*m*): orthoclase porphry (a form of porphry in which there is little or no quartz; see **PORPHYR** and **QUARZPORPHYR**) 299
ORTHOKOHLENSÄURE (*f*): orthocarbonic acid 300
ORTHOKOHLENSÄUREESTER (*m*): ortho-carbonic ester; $C_9H_{20}O_4$; Sg. 0.9197; Bp. 159°C. [301
ORTHOKRESOL (*n*): orthocresol 302
ORTHOKRESOTINSÄURE (*f*): orthocresotic (orthocresotinic) acid 303
ORTHOMETHYLPHENOL (*n*): (see **KRESOL, ORTHO-**) 304
ORTHONITRANILIN (*n*): orthonitraniline 305
ORTHONITRANISOL (*n*): orthonitranisol 306
ORTHONITROBENZALDEHYD (*n*): orthonitrobenzaldehyde 307
ORTHONITROPHENOL (*n*): orthonitrophenol (see **NITROPHENOL**) 308
ORTHONITROTOLUOL (*n*): orthonitrotoluene (see **NITROTOLUOL**) 309
ORTHOOXALSÄUREDIPHENYLESTER (*m*): diphenyorthooxalate; $[(OH)_2H_5C_6.O]C.C[(OH)_2O.C_6H_5]$ 310
ORTHOPHENOLSULFOSÄURE (*f*): orthophenolsulphonic acid, Aseptol (see **ASEPTOL**) 311
ORTHOPHENYLENDIAMIN (*n*): orthophenylenediamine (see **PHENYLENDIAMIN**) 312
ORTHOPHOSPHORSÄURE (*f*): orthophosphoric acid; H_3PO_4; Sg. 1.88; Mp. 38.6°C. [313
ORTHOSALPETERSÄURE (*f*): orthonitric acid 314
ORTHOSALPETRIG: orthonitrous 315
ORTHOSÄURE (*f*): ortho acid 316
ORTHOSCHWARZ (*n*): ortho black 317
ORTHOSE (*f*): orthose (see **ORTHOKLAS**) 318
ORTHOSKOPISCHES OBJEKTIV (*n*): orthoscopic lens 319
ORTHOSTELLUNG (*f*): ortho position 320
ORTHOTITANSÄURE (*f*): ortho-titanic acid; $Ti(OH)_4$ 321
ORTHOTOLUIDIN (*n*): orthotoluidine (see **TOLUIDIN**) 322
ORTHOTOLUOLSULFAMID (*n*): orthotoluene sulfamide, toluenesulfoneamine; $CH_3C_6H_4SO_2NH_2$; Mp. 155°C. 323
ORTHOTOLUOLSULFOCHLORID (*n*): orthotoluene sulfochloride, toluene sulfonechoride; $CH_3C_6H_4SO_2Cl$ 324
ORTHOVERBINDUNG (*f*): ortho compound 325
ORTHOXYLIDIN (*n*): orthoxylidine (see **XYLIDIN**) 326
ORTHOXYLOL (*n*): orthoxylol (see **XYLOL**) 327
ORTISOMERIE (*f*): position isomerism 328
ÖRTLICH: local-(ly); topical-(ly) 329
ORTOL (*n*): ortol (a photographic chemical; methyl orthoaminophenolesulphate + hydroquinone, in proportion of 2 : 1) 330
ORTSBEHÖRDE (*f*): local authority 331
ORTSBESCHAFFENHEIT (*f*): locality; nature of a locality 332
ORTSCHAFT (*f*): place; locality; district; village; hamlet; township; cantonment; community 333
ORTSGEBRAUCH (*m*): local usage or custom 334
ORTSISOMERIE (*f*): place (position) isomerism 335
ORTSVERÄNDERUNG (*f*): change of (place) position 336
ORT UND STELLE, AN-: on the spot 337
ORZEIN (*n*): orcein (the actual colouring matter of archil; obtained from various species of lichen; see **ORSEILLE** and **LACKMUSFLECHTE**) 338
ORZIN (*n*): orcin, methyl-resorcinol, dioxytoluene, orcinol, (see also **ORSEILLE**); $C_7H_8O_2.H_2O$; Sg. 1.29; Mp. 56°C. Bp. 288.5°C. 339
OSCILLATORIAART (*f*): type (kind; sort) of oscillatoria or oscillaria (a genus of Confervoid algæ of the order *Oscillariaceæ* or *Oscillatoriaceæ* 340
ÖSE (*f*): ear; catch; lug; ring; eye; handle; hook; loop; shank, etc. (Tech.) 341
ÖSENBLATT (*n*): flange; tongue; lip, etc. (Tech.) 342
ÖSEN (*f.pl*), **RUDER-**: rudder braces, rudder gudgeons 343
ÖSFASS (*n*): boat bailer 344
OSMAT (*n*): osmate 345
OSMIRIDIUM (*n*): osmiridium (see **NEWJANSKIT**) 346
OSMIUM (*n*): osmium; (Os) 347
OSMIUMCHLORWASSERSTOFFSÄURE (*f*): osmium, chlorine and hydrogen acid; H_2OsCl_6 or H_3OsCl_6 348
OSMIUMFADEN (*m*): osmium filament 349
OSMIUM LAMPE (*f*): osmium lamp (an electric glow lamp, invented by Auer) 350
OSMIUMSALZ (*n*): osmium salt 351
OSMIUMSAUER: osmate, combined with osmic acid 352
OSMIUMSÄURE (*f*): osmic acid (erroneously employed for osmium tetroxide, see **ÜBEROSMIUMSÄURE**) 353
OSMIUMSÄUREANHYDRID (*n*): osmic anhydride (see **ÜBEROSMIUMSÄURE**) 354
OSMIUMSAURES SALZ (*n*): osmate 355
OSMIUMSUPEROXYD (*n*): osmium peroxide; OsO_4 (see **ÜBEROSMIUMSÄURE**) 356
OSMIUMTETROXYD (*n*): osmium tetroxide; OsO_4 (see **ÜBEROSMIUMSÄURE**) 357
OSMIUMTRIOXYD (*n*): osmium trioxide (not known in a free state); OsO_3 358
OSMIUMVERBINDUNG (*f*): osmium compound 359
OSMOSE (*f*): osmose; osmosis (diffusion through a membrane) 360
OSMOSEAPPARAT (*m*): osmosis apparatus (for molasses) 361
OSMOSEVERFAHREN (*n*): osmotic process 362
OSMOSIEREN: to osmose; osmosise; diffuse through a membrane 363
OSMOTISCH: osmotic-(ally) 364
OSMOTISCHER DRUCK (*m*): osmotic pressure (set up by osmosis) 365
OSRAM (*n*): osram 366
OSRAMLAMPE (*f*): Osram lampe (an electric glow-lamp invented by Dr. Auer) 367
OST: east; orient 368
OSTEN (*m*): east; orient 369
OSTEOLITH (*m*): osteolite, apatite (see **APATIT**) 370
OSTEOLOGIE (*f*): osteology (the science of bones) 371
OSTERBLUME (*f*): Passion flower (*Passiflora cœrulea*) 372
OSTERBLUME (*f*), **WEISSE-**: white wood anemone (*Anemone nemorosa*) 373
OSTERFEST (*n*): Easter; feast of the Passover 374
OSTERLILIE (*f*): Bermuda Easter lily (*Lilium Harrisii*, or *Lilium longiflorum eximium*) 375
OSTERLUZEI (*f*): birthwort (*Aristolochia clematitis*) 376

OSTERLUZEIGEWÄCHSE (*n.pl*): *Aristolochiaceæ* (Bot.) 377
OSTGRENZE (*f*): eastern (frontier) boundary 378
ÖSTLICH: easterly; eastern; oriental 379
OSTRANIT (*m*): ostranite (see **MALAKON**) 380
OSTSEE (*f*): Baltic sea 381
OSZILLIEREN: to oscillate, vibrate 382
OSZILLIEREND: oscillating, vibrating; oscillatory, vibratory 383
OSZILLIERUNG (*f*): oscillation, vibration 384
OTHÄMATOMA (*f*): *Hæmatoma auris*, insane ear (Med.) 385
OTTER (*f*): otter; adder; viper 386
OTTERBALG (*m*): viper's skin 387
OTTERNFELL (*n*): otter's skin 388
OTTERNGIFT (*n*): viper's (venom) poison 389
OTTRELITH (*m*): ottrelite; $H_2FeAl_2SiO_7$; Sg. 3.52 390
OUTREMER (*m*): (see **LASURIT**) 391
OVAL: oval, elliptical 392
OVALDREHBANK (*f*): oval lathe 393
OVALWERK (*n*): oval chuck (of lathes) 394
OVARIEN (*f.pl*): ovaries (egg cells) (Physiology); ovule sac (lower sack-like portion of carpel containing the ovule) (Bot.) 395
OWENIT (*m*): owenite; $H_{18}Fe_8(Al,Fe)_8Si_6O_{41}$; Sg. 3.2 396
OXALAT (*n*): oxalate 397
OXALÄTHER (*m*): oxalic ether; ethyl oxalate 398
OXALESSIGESTER (*m*): oxalacetic ester 399
OXALIT (*m*): oxalite; $(FeO,C_2O_3)_2,3H_2O$; Sg. 2.2 400
OXALSAUER: oxalate of; combined with oxalic acid 401
OXALSÄURE (*f*): oxalic acid (*Acidum oxalicum*): $CO_2H.CO_2H.2H_2O$ Sg. 1.65; Mp. 187°C. 402
OXALSÄUREANLAGE (*f*): oxalic acid plant 403
OXALSÄURE-DIÄTHYLESTER (*m*): oxalic diethylester; diethyl oxalate; $C_6H_{10}O_4$; Sg. 1.0793; Mp. – 40.6°C.; Bp. 184°C. 404
OXALSÄUREFABRIK (*f*): oxalic acid factory 405
OXALSÄURE (*f*), **FÜNFZIGSTEL-**: 1 : 50 oxalic acid, a 1 in 50 solution of oxalic acid (1.26 grams of oxalic acid crystals per litre) 406
OXALSÄURELÖSUNG (*f*): oxalic acid solution, (used for removing lines or figures from blue or white prints) 407
OXALSAURES SALZ (*n*): oxalate 408
OXALURSÄURE (*f*): oxaluric acid 409
OXAMETHAN (*n*): oxamethane 410
OXAMID (*n*): oxamide 411
OXAMINENTWICKLER (*m*): oxamine developer 412
OXAMINSCHWARZ (*n*): oxamine black 413
OXAPHOR (*n*): oxaphor (a 50% solution of oxycamphor (see **OXYKAMPFER**) 414
OXAZIN (*n*): oxazine, phenoxazine; $(C_6H_4)_2NOH$ 415
OXAZINFARBSTOFF (*m*): oxazine dye (a class of tar dyes) 416
OXHOFT (*n*): hogshead 417
OXON (*n*): oxone (sodium peroxide in tablet form) (see **NATRIUMPEROXYD**) 418
OXYAMMONIAK (*n*): oxyammonia; oxyammonium; hydroxylamine; oxammonium (see **HYDROXYLAMIN**) 419
OXYÄPFELSÄURE (*f*): hydroxymalic acid 420
OXYBENZALDEHYD (*n*): oxybenzaldehyde; $(C_7H_6O_2)$; ortho.-(salicylous acid); Sg. 1.152; Mp. about – 20°C.; Bp. 197°C.; meta.-; Mp. 108°C.; Bp. 240°C.; para.-; Mp. 116°C. 421
OXYBENZOESÄURE (*f*): (hydr-)oxybenzoic (salicylic) acid; $(C_7H_6O_3)$; ortho.-; Sg. 1.4835; Mp. 156-159°C.; meta.-; Sg. 1.473; Mp. 188°C.; para.- Sg. 1.468; Mp. 214°C. 422
OXYBENZOL (*n*): hydroxybenzene (see **PHENOL** and **CARBOLSÄURE**) 423
OXYBIAZOL (*n*): oxybiazole 424
OXYBIAZOLIN (*n*): oxybiazoline 425
OXYBUTTERSÄUREALDEHYD (*n*): oxybutyric aldehyde; aldol (see **ALDOL**) 426
OXYBUTYRALDEHYD (*n*): (see **ALDOL**) 427
OXYCARBONSÄURE (*f*): hydroxycarboxylic acid 428
OXYCELLULOSE (*f*): oxycellulose (see **OXYZELLU-LOSE**) 429
OXYCHINOLIN (*n*): hydroxyquinoline 430
OXYCHINOLIN, ALPHA- (*n*): alpha-hydroxyquinoline; Mp. 199.5°C. 431
OXYCHINON (*n*): hydroxyquinone 432
OXYCHLORKUPFER (*n*): copper oxychloride 433
OXYCHROMIN (*n*): oxychromine 434
OXYCYANQUECKSILBER (*n*): mercury oxycyanide (see **QUECKSILBEROXYCYANID**) 435
OXYD (*n*): (ic)-oxide 436
OXYDABEL: oxydizable; oxidizable 437
OXYDANTIE (*f*): oxydizer 438
OXYDARTIG: like an oxide 439
OXYDASE (*f*): oxydase (an enzyme which converts alcohol into acetic acid) 440
OXYDATION (*f*): oxydation (union with oxygen) 441
OXYDATIONSARTIKEL (*m*): oxydation style (Calico) 442
OXYDATIONSFÄHIG: capable of oxydation; oxydizable; oxidizable 443
OXYDATIONSFÄHIGKEIT (*f*): oxydizability 444
OXYDATIONSFLAMME (*f*): oxidizing flame 445
OXYDATIONSMITTEL (*n*): oxidizing (medium) agent 446
OXYDATIONSPOTENTIAL (*n*): oxydation potential 447
OXYDATIONSPRODUKT (*n*): oxidation product, product of oxidation 448
OXYDATIONSSTUFE (*f*): degree (stage) of oxidation 449
OXYDATIONSTERRASSE (*f*): oxydation terrace (Irrigation) 450
OXYDATIONSVORGANG (*m*): oxidation process 451
OXYDBESCHLAG (*m*): oxide coating 452
OXYDBRAUN (*n*): oxide brown 453
OXYDCHLORID (*n*): oxychloride 454
OXYDGELB (*n*): oxide yellow 455
OXYDGRAU (*n*): oxide grey 456
OXYDHALTIG: containing oxide-(s); oxidic; oxygenic 457
OXYDHAUT (*f*): film of oxide 458
OXYDHYDRAT (*n*): hydrated oxide; hydroxide; hydrate 459
OXYDIAMINOGEN (*n*): oxydiaminogen 460
OXYDIAMINSCHWARZ (*n*): oxydiamine black 461
OXYDIANILGELB (*n*): oxydianile yellow 462
OXYDIERBAR: oxidizable 463
OXYDIERBARKEIT (*f*): oxidizability 464
OXYDIEREN: to oxidize; oxydate 465
OXYDIERENDE FLAMME (*f*): (produced by a sufficiency of excess air); oxydizing (oxydating) flame 466
OXYDIERUNG (*f*): oxidation; oxydization 467
OXYDISCH: oxydic; oxygenic; containing oxide-(s) 468
OXYDOCKER (*m*): oxide ochre 469
OXYDOXYDUL (*n*): (-oso -ic) oxide 470

OXYDROT (*n*): Turkey red; purple oxide of iron; red oxide of iron (see **EISENOXYD**) 471
OXYDSCHICHT (*f*): layer of oxide 472
OXYDUL (*n*): (-ous) oxide; protoxide 473
OXYESSIGSÄURE (*f*): oxyacetic acid (see **GLYCOL-SÄURE**) 474
OXYGENIEREN: to oxygenize; oxygenate 475
OXYGENIERUNG (*f*): oxygenation 476
OXYKAMPFER (*m*): oxycamphor; $CHOHC_8H_{14}CO$; Mp. 204°C. 477
OXYKETON (*n*): hydroxy ketone 478
OXYKETONCARBONSÄURE (*f*): hydroxyketo-carboxylic acid 479
OXYKETONFARBSTOFF (*m*): hydroxy ketone dye (an important group of tar dyes) 480
OXYLIQUIT (*n*): oxyliquit (an explosive mixture of liquid air and charcoal powder) 481
OXYMEL (*n*): oxymel (dilute acetic acid and refined honey) 482
OXYNAPHTALIN (*n*): (mono)-hydroxynaphthalene (see **NAPHTOL**) 483
OXYNAPTOËSÄURE (*f*): oxynaphthoic acid 484
OXYNAROKTIN (*n*): oxynarcotine (an opium alkaloid); $C_{19}H_{14}NO_5(OCH_3)_3$ 485
OXYPHENYLCHINOLINDIKARBONSÄURE (*f*): oxyphenylquinolinedicarbonic acid; $C_{17}H_{11}NO_5.H_2O$ 486
OXYPHENYLPROPIONSÄURE (*f*), **L-α-AMINO-β-P-**: l-α-amino-β-p-oxyphenylpropionic acid (see **TYROSIN**) 487
OXYPROLIN (*n*): oxyproline, l-y-oxy-α-pyrrolidinecarbonic acid; $C_5H_9O_3N$ 488
OXYPROPIONSÄURE (*f*), **ALPHA-**: (see **MILCH-SÄURE, GÄRUNGS-**) 489
OXYPROPIONSÄURE (*f*), **L-α-AMINO-β-**: (see **SERIN**) 490
OXYSALZ (*n*): oxysalt 491
OXYSÄURE (*f*): oxyacid 492
OXYSCHWEFELSÄURE (*f*): oxysulphuric acid (see **CAROSCHE SÄURE**) 493
OXYSTEARINSÄURE (*f*): oxystearic acid 494
OXYTOLUOL (*n*): hydroxytoluene 495
OXYZELLULOSE (*f*): oxycellulose (from action of an oxydating agent on cellulose) 496
OZOBENZOL (*n*): ozobenzene 497
OZOKERIT (*m*): ozocerite; mineral wax; fossil wax; native paraffin; hydrocarbon; C_nH_{2n}; Sg. 0.85-0.95; Mp. 55-110°C. 498
OZON (*n*): ozone; (O_3); Mp. – 251.4°C.; Bp. – 112.3°C. 499
OZONAPPARAT (*m*): ozone apparatus 500
OZONDARSTELLUNGSMETHODE (*f*): method of producing (generating) ozone 501
OZONENTWICKLER (*m*): ozone generator 502
OZONENTWICKLUNG (*f*): ozone generation, generation of ozone 503
OZONERZEUGEND: ozoniferous; producing ozone; generating ozone 504
OZONGLIMMERRÖHRE (*f*): mica ozone tube 505
OZONHALTIG: ozoniferous; containing ozone 506
OZONISATOR (*m*): ozonizer 507
OZONISIEREN: to ozonize 508
OZONRÖHRE (*f*): ozone tube 509
OZONSAUERSTOFF (*m*): ozonized oxygen 510
OZONVENTILATOR (*m*): ozone (ventilator, blower) fan 511

P

PAAR (*n*): pair; couple; few; two; brace; match 512
PAAREN: to pair; couple; match; mate; copulate 513
PAARWEISE: in pairs; by twos; two and two; in couples 514
PACHNOLITH (*m*): pachnolite; $NaCaAlF_6.H_2O$; Sg. 2.95 515
PACHT (*f*): lease; tenure; rent 516
PACHTBRIEF (*m*): Lease; Conveyance 517
PACHTEN: to lease; rent 518
PACHTER (*m*): tenant; lessee 519
PÄCHTER (*m*): (see **PACHTER**) 520
PACHTGARTEN (*m*): allotment 521
PACHTGELD (*n*): rent 522
PACHTGUT (*n*): (see **PACHTHOF**) 523
PACHTHERR (*m*): lessor, landlord 524
PACHTHOF (*m*): leasehold (property, estate) farm 525
PACHTINHABER (*m*): lessee, tenant 526
PACHTKONTRAKT (*m*): lease; conveyance 527
PACHTUNG (*f*): leasing, renting; leasehold property 528
PACHTVERTRAG (*m*): Lease; conveyance 529
PACHTWEISE: on lease 530
PACHYMENINGITIS (*f*): pachymeningitis (subdural false membrane) (Med.) 531
PACK (*m*): pack; bundle; bale; packet; parcel 532
PACKEN: to bale; pack; stow; seize; grab; grasp 533
PACKER (*m*): packer 534
PACKESEL (*m*): pack-mule; baggage mule 535
PACKGARN (*n*): packing thread 536
PACKHAUS (*n*): warehouse; packing room 537
PACKHOF (*m*): warehouse; custom house; bonded store 538
PACKKORB (*m*): hamper 539
PACKLACK (*m*): sealing wax 540
PACKLEINEN (*n*): packing (canvas) linen, sacking 541
PACKLEINWAND (*f*): packing (canvas) linen; sacking 542
PACKMASCHINE (*f*): packing (bundling, baling) machine 543
PACKNADEL (*f*): packing needle 544
PACKPAPIER (*n*): wrapping (packing) paper; casing paper 545
PACKRAUM (*m*): packing room; hold; stowage; stuffing (packing) gland 546
PACKSTOFF (*m*): packing material 547
PACKTUCH (*n*): packing-(cloth) 548
PACKUNG (*f*): packing 549
PACKUNGSBOLZEN (*m*): gland bolt or stud, packing bolt or stud 550
PACKUNGSKOSTEN (*pl*): packing -(charges) 551
PACKUNGSLOS: without packing 552
PACKWAGEN (*m*): baggage waggon 553
PACKZEUG (*n*): packing-(material or tools) 554
PAJSBERGIT (*m*): Pajsbergite (see **RHODONIT**) 555
PAKET (*n*): packet; parcel; package 556
PAKETBEFÖRDERUNG (*f*): parcel delivery 557
PAKETBOOT (*n*): packet-(boat) 558
PAKETFÜLLMASCHINE (*f*): packet filling machine 559
PAKETPOST (*f*): parcel-post 560
PAKFONG (*n*): Pack Fong, German silver (a nickel alloy) (see **NEUSILBER**) 561

PAKT (*m*): pact, compact, agreement, covenant 562
PALÄOGRAPH (*m*): paleographer, paleographist 563
PALÄOGRAPHIE (*f*): palæography (the study of ancient writings) 564
PALÄOGRAPHISCH: paleographic-(al) 565
PALÄOLOGEN (*m*): paleologi 566
PALÄONTOLOG (*m*): paleontologist 567
PALÄONTOLOGIE (*f*): palæontology (the science of fossils) 568
PALÄONTOLOGISCH: paleontologic-(al) 569
PALATINROT (*n*): palatine red 570
PALAU (*n*): palau (a gold-palladium alloy, used as a substitute for platinum) 571
PÄLEISEN (*n*): scraping (hairing) knife (Tanning) 572
PÄLEN: to depilate, scrape off (remove hair) from skins; (*n*): depilation, hairing, peeling, scraping of hair from skins (Tanning) 573
PALETTE (*f*): pallet; palette 574
PALISANDERHOLZ (*n*): rosewood, South America (*Dalbergia nigra*); Sg. 0.7-1.0 575
PALL (*n*): pawl 576
PALLADICHLORWASSERSTOFFSÄURE (*f*): chloropalladic acid 577
PALLADIUM (*n*): palladium (Pd) 578
PALLADIUMBROMÜR (*n*): palladious bromide 579
PALLADIUMCHLORID (*n*): palladium (palladic) chloride; $PdCl_4$ (not known in a free state) 580
PALLADIUMCHLORIDLÖSUNG (*f*): palladic chloride solution 581
PALLADIUMCHLORÜR (*n*): palladium (bichloride) dichloride; palladious chloride; $PdCl_2$; Mp. 501°C. 582
PALLADIUMCHLORÜRLÖSUNG (*f*): palladious chloride solution 583
PALLADIUMCHLORWASSERSTOFF (*m*): chloropalladic acid 584
PALLADIUMDICHLORID (*n*): palladium (bichloride) dichloride; palladious chloride (see **PALLADIUMCHLORÜR**) 585
PALLADIUMJODÜR (*n*): palladious iodide; PdI_2 586
PALLADIUM, KOLLOIDALES- (*n*): colloidal palladium 587
PALLADIUMLEGIERUNG (*f*): palladium alloy 588
PALLADIUMMOHR (*m*): palladium black 589
PALLADIUMOXYD (*n*): palladic oxide 590
PALLADIUMOXYDUL (*n*): palladious oxide; PdO 591
PALLADIUMSALZ (*n*): palladium salt 592
PALLADIUMSCHWAMM (*m*): palladium sponge (by heating ammonium-palladic chloride) 593
PALLADIUMSCHWARZ (*n*): palladium black 594
PALLADIUMSILICID (*n*): palladium silicide; PdSi; Sg. 7.31 595
PALLADIUMWASSERSTOFF (*m*): palladium hydride 596
PALLADOCHLORWASSERSTOFFSÄURE (*f*): chloropalladious acid 597
PALLADOHYDROXYD (*n*): palladious hydroxide 598
PALLBETING (*f*): pawl bitts (Naut.) 599
PALLIATIV (*n*): palliative, lenitive 600
PALLISADE (*f*): palisade, fence 601
PALLISADENVERSCHANZUNG (*f*): enclosure, stockade 602
PALLISADENWURM (*m*): strongylus (genus of Nematoid worms, see also **SAUGWURM, TREMATODE, FADENWURM, BLASENWURM, BANDWURM**) (Med.) 603
PALLKLAMPE (*f*): pawl cleat (Naut.) 604
PALLRING (*m*): pawl rack 605
PALLSTÜTZE (*f*): pawl bitt 606
PALMACHRISTIÖL (*n*): (see **RICINUSÖL**) 607
PALMAPONEUROSE (*f*), **RETRAKTION DER-**: Dupuytren's contraction, contraction of the Palmar fascia (Med.) 608
PALMAROSAÖL (*n*): palmarosa oil (see **GERANIUMÖL**) 609
PALMBAUM (*m*): palm-(tree) 610
PALME (*f*): palm-(tree); palm-(of the hand) 611
PALMENGEWÄCHSE (*n.pl*): *Palmaceæ, Palmæ* (Bot.) 612
PALMENHOLZ (*n*): palmyra wood (see **PORKUPINEHOLZ**) 613
PALMENMEHL (*n*): sago (see **PALMENSTÄRKE**) 614
PALMENSTÄRKE (*f*): sago (from the sago palm) (see **PALMGRAUPEN**) 615
PALME (*f*), **SÜD-AMERIKANISCHE-**: South-American palm (*Ceroxylon andicolum*) 616
PALMETTOEXTRAKT (*m*): palmetto extract (an extract containing tannin, obtained from *Sabal palmetto*) 617
PALMETTOFASER (*f*): palmetto fibre (from *Sabal palmetto*) 618
PALMETTOPALME (*f*): palmetto palm, sabal (*Chamærops palmetto* and *Sabal palmetto*) 619
PALMFETT (*n*): palm (oil; grease; fat) butter; (*Oleum elaidis*) (from *Elæis guineensis*); Sg. 0.925; Mp. 27-43°C. 620
PALMGRAUPEN (*f.pl*): sago (from the sago palm, *Sagus rumphii* or *Sagus lævis*) 621
PALMITIN (*n*): palmitine (a glyceride of palmitic acid) 622
PALMITINSAUER: palmitate of, combined with palmitic acid 623
PALMITINSÄURE (*f*): (*Acidum palmitinicum*): palmitic (palmitinic) acid, cetylic acid; $C_{16}H_{32}O_2$; Sg. 0.847; Mp. 32 to 63°C.; Bp. 138°C. to 356°C. 624
PALMITINSÄURECETYLESTER (*m*): palmitic acid cetyl ester; Cetin (from Spermaceti); $C_{15}H_{31}COOC_{16}H_{13}$; Mp. 50°C.; Bp. 360°C. (see **WALRAT**) 625
PALMITINSÄURETHYMOLESTER (*m*): thymol palmitate 626
PALMKÄTZCHEN (*n*): catkin (of a palm tree) 627
PALMKERN (*m*): palm nut kernel 628
PALMKERNÖL (*n*): (*Oleum Elaidis*): palm nut (kernel) oil (from the fruit of *Elæis guineensis*); Sg. 0.92-0.97; Mp. 26-42.5°C. 629
PALMLILIE (*f*): yucca (Bot.) 630
PALMLILIE (*f*), **FADENTRAGENDE-**: Adam's needle (*Yucca filamentosa*) 631
PALMMEHL (*n*): sago (see **PALMGRAUPEN**) 632
PALMÖL (*n*): palm oil (see **PALMFETT**) (see also **PALMKERNÖL**) 633
PALMSONNTAG (*m*): Palm Sunday 634
PALMSTÄRKE (*f*): palm starch; sago (see **PALMGRAUPEN**) 635
PALMWACHS (*n*): palm wax (a vegetable wax from the South American palm, *Ceroxylon andicola*; Mp. 103.5°C.) 636
PALMWEIN (*m*): toddy, palm wine 637
PALMWOCHE (*f*): Passion Week 638
PALMYRAHOLZ (*n*): palmyra wood (see **PORKUPINEHOLZ**) 639
PALMZUCKER (*m*): palm sugar; jaggery (from various species of palm) (see **JAGARAZUCKER**) 640
PALMZWEIG (*m*): palm branch 641

PALPABEL: palpable, touchable, tangible, feelable; appreciable, noticeable, apparent 642
PALPATION (*f*): touching, feeling, touch, sensation of pressure 643
PALPITATION (*f*): palpitation, trembling, fear, emotion, agitation, trepidation, perturbation tremor, convulsion, quiver, throbbing, pulsation, flutter 644
PALPITIEREN: to palpitate, shake, tremble, pulsate, throb, quiver, shiver, beat, flutter, be agitated, go pit-a-pat 645
PALYXANDERHOLZ (*n*): Jacaranda wood (see PALISANDERHOLZ) 646
PANACEE (*f*): panacea, panpharmacon, universal remedy 647
PANAKOKOHOLZ (*n*): iron wood (see also EISEN-HOLZ) 648
PANAMARINDE (*f*): Quillai bark; Panama bark (see QUILLAYARINDE) 649
PANASCHIEREN: to streak with colour 650
PANCHROMATISCH: panchromatic 651
PANCHROMATISCHE PLATTE (*f*): panchromatic plate (Phot.) 652
PANCREAS (*m*): pancreas (Anat.) (see BAUCH-SPEICHELDRÜSE and PANKREAS) 653
PANCREASPRÄPARAT (*n*): pancreas preparation 654
PANCREATIN (*n*): (*Pancreatinum*): pancreatin (a mixture of enzymes from the pancreas of pigs and other warm blooded animals) 655
PANCREATISCH: pancreatic 656
PANDEMIE (*f*): wide-spread epidemic (such as the influenza scourge of 1915) 657
PANDEMISCH: wide-spread (of epidemics); spreading over whole continents 658
PANDERMIT (*m*): pandermite (Mineralogy); $Ca_2B_6O_{11}.3H_2O$; Sg. 2.35 659
PANEEL (*n*): panel, wainscot 660
PANEL (*n*): (see PANEEL) 661
PANIK (*f*): panic, scare 662
PANKREAS (*n*): pancreas (Anat.) (see BAUCH-SPEICHELDRÜSE and PANCREAS) 663
PANKREASDRÜSE (*f*): pancreatic gland; pancreas (see BAUCHSPEICHELDRÜSE) 664
PANKREATIN (*n*): pancreatine (see PANCREATIN) 665
PANKREATISCH: pancreatic 666
PANKREON (*n*): pancreon (a pancreas preparation) 667
PANKREONZUCKER (*m*): pancreon sugar 668
PANNETIERGRÜN (*n*): (see GUIGNETGRÜN) 669
PANORAMALINSE (*f*): panoramic lens 670
PANSE (*f*): paunch; first stomach of ruminants 671
PANSEN (*m*): (see PANSE) 672
PANTHEISMUS (*m*): pantheism (belief in the Universe as God) 673
PANTHEIST (*m*): pantheist 674
PANTHEISTISCH: pantheistic 675
PANTHER (*m*): (see PANTHERTIER and PANTHER-KATZE) 676
PANTHERKATZE (*f*): ocelot (*Felis pardalis*) 677
PANTHERTIER (*n*): panther (*Felis pardus*) 678
PANTOFFELBLUME (*f*): slipper-wort, calceolaria 679
PANTOPON (*n*): pantopon (used as a substitute for opium and morphine) 680
PANTRY (*f*): pantry 681
PANZER (*m*): armour; armour (plate) plating; Ironclad (Naut.) 682
PANZERBLECH (*n*): armour plate 683
PANZERBOLZEN (*m*): armour bolt 684
PANZERDECK (*n*): armour deck 685
PANZERFLECHTE (*f*): *Cetraria* (Bot.) 686
PANZERFLOTTE (*f*): ironclad fleet, fleet of ironclads 687
PANZERGÜRTEL (*m*): armour belt (Ships) 688
PANZERHINTERLAGE (*f*): armour wood-backing 689
PANZERMACHER (*m*): armourer, armour-maker 690
PANZERN: to armourplate; cover with armour-(plate) 691
PANZERPLATTE (*f*): armour plate 692
PANZERSCHIFF (*n*): ironclad-(ship) 693
PANZERSCHOTT (*n*): armour bulkhead 694
PANZERTIER (*n*): armadillo 695
PANZERTURM (*m*): armour-plated turret 696
PANZERUNG (*f*): armour plating 697
PÄONIE (*f*): peony (*Pæonia officinalis*) 698
PÄONIENBLÜTEN (*f.pl*): peony flowers (*Flores pæoniæ*) 699
PÄONIENSAMEN (*m*): peony seed 700
PÄONIENWURZEL (*f*): peony root 701
PAPAGEI (*m*): parrot 702
PAPAGEIGRÜN (*n*): parrot green (see KUPFER-ACETATARSENIAT) 703
PAPAIN (*n*): papain, papayotin, vegetable pepsin (an enzyme which is the active principle of *Carica papaya*) 704
PAPAVERAMIN (*n*): papaveramine (an opium alkaloid); $C_{21}H_{21}NO_5$ 705
PAPAVERIN (*n*): papaverine (an opium alkaloid) $(OCH_3)_2C_6H_3CH_2NC_5H_2C_4H_2(OCH_3)_2$; Mp. 147°C. 706
PAPAYOTIN (*n*): papayotin (see PAPAIN) 707
PAPEL (*f*): papule, pimple (a small, solid prominence of the skin) (Med.) 708
PAPIER (*n*): paper 709
PAPIERABGÄNGE (*m.pl*): waste paper 710
PAPIERAGENT (*m*): paper agent 711
PAPIERAUSSCHNITT (*m*): paper cutting, vignette 712
PAPIERBAHN (*f*): length-(of paper) 713
PAPIERBEARBEITUNGSMASCHINE (*f*): paper finishing machine 714
PAPIERBEREITUNG (*f*): paper-making 715
PAPIERBESCHNEIDEMASCHINE (*f*): paper triming machine 716
PAPIERBINDFADEN (*m*): paper string 717
PAPIERBLATT (*n*): sheet (leaf) of paper 718
PAPIERBOGEN (*m*): sheet (leaf) of paper 719
PAPIERBREI (*m*): paper pulp 720
PAPIERERN: paper; of paper 721
PAPIERFABRIK (*f*): paper (factory) mill 722
PAPIERFABRIKANT (*m*): paper (maker) manufacturer 723
PAPIERFABRIKATION (*f*): paper (making) manufacture 724
PAPIERFABRIKATIONSMASCHINEN (*f.pl*): paper-making machinery 725
PAPIERFARBEN (*f.pl*): paper colours 726
PAPIERFASER (*f*): paper fibre 727
PAPIERFILTER (*n*): paper filter 728
PAPIER (*n*), **FILTRIER-**: filter paper 729
PAPIERGELD (*n*): paper (currency) money; notes 730
PAPIERGESCHÄFT (*n*): stationery business, paper trade 731
PAPIERGRAS (*n*): papyrus, paper reed (*Papyrus antiquorum*) 732
PAPIERHANDEL (*m*): paper trade; stationery 733
PAPIERHÄNDLER (*m*): paper merchant; stationer 734

PAPIERHANDLUNG (*f*): stationer's shop 735
PAPIERKOHLE (*f*): paper coal (a type of lignite) 736
PAPIERKOPIE (*f*): paper copy; print (Phot.) 737
PAPIERKORB (*m*): waste-paper basket 738
PAPIERLADEN (*m*): stationer's shop 739
PAPIERLAGER (*n*): paper warehouse 740
PAPIERLEIM (*m*): paper size 741
PAPIER (*n*), **LÖSCH-**: blotting paper 742
PAPIERMACHÉ (*n*): paper mâché; papiermâché (a plastic mass of paper pulp with some binding material, or a number of layers of paper stuck together) 743
PAPIERMACHER (*m*): paper maker 744
PAPIERMÄKLER (*m*): bill broker 745
PAPIERMASCHINE (*f*): paper machine 746
PAPIERMASSE (*f*): (paper)-pulp (see **PAPIER-MACHÉ**) 747
PAPIERMÜHLE (*f*): paper mill 748
PAPIERMÜLLER (*m*): paper-maker 749
PAPIERPAPPE (*f*): paper board 750
PAPIERPERGAMENT (*n*): parchment paper (see under **PERGAMENT**) 751
PAPIER (*n*), **POSITIV-**: raw paper (Phot.) 752
PAPIERPRÜFUNG (*f*): paper testing 753
PAPIERROHSTOFF (*m*): raw material for paper-making 754
PAPIERROLLE (*f*): paper roll, roll of paper 755
PAPIERSACK (*m*): paper sack (sack made of paper) 756
PAPIERSCHERE (*f*): paper (shears) scissors 757
PAPIERSCHIRM (*m*): paper screen 758
PAPIERSCHNITZEL (*m*): paper cutting, paper scrap 759
PAPIERSORTE (*f*): type (sort; quality; kind) of paper 760
PAPIERSPEKULANT (*m*): stock-jobber 761
PAPIERSTAUDE (*f*): papyrus (see **PAPIERGRAS**) 762
PAPIERSTEIN (*m*): papyrolite 763
PAPIERSTEMPELPRESSE (*f*): paper embossing press 764
PAPIERSTOFF (*m*): paper pulp (see also **PAPIER-MACHÉ**) 765
PAPIERSTOFFFABRIK (*f*): paper pulp factory 766
PAPIERSTOFFGARN (*n*): paper twine 767
PAPIERSTRANG (*m*): paper strip 768
PAPIERSTREIFEN (*m*): paper strip 769
PAPIERTAPETE (*f*): wall-paper 770
PAPIERTEIG (*m*): (see **PAPIERMACHÉ**) 771
PAPIERUMLAUF (*m*): paper currency 772
PAPIERVALUTA (*f*): paper (exchange) value 773
PAPIERWÄHRUNG (*f*): (paper)-exchange value, value of a (bill, note) draft, paper (currency) money, bank-bill-(s); bank note-(s) 774
PAPIERWAREN (*f.pl*): paper (goods; wares) articles 775
PAPIERWÄSCHE (*f*): paper articles of attire (such as collars, cuffs and fronts) 776
PAPIERZEICHEN (*n*): watermark 777
PAPIERZEUG (*n*): stuff; (paper)-pulp 778
PAPIERZÜNDER (*m*): paper fuse 779
PAPILIONAZEEN (*f.pl*): *Papilionaceæ* (Bot.) 780
PAPILLE (*f*): papilla (of the skin) 781
PAPILLEN: papillary (resembling or having nipples) 782
PAPILLOTIEREN: to curl the hair; (*n*): hair-curling 783
PAPINIANISCHER TOPF (*m*): Papin's (autoclave) digester 784
PAPP (*m*): pap; paste 785
PAPPARBEIT (*f*): paste-(card)-board work 786
PAPPBAND (*m*): board-binding (for books) 787
PAPPBOGEN (*m*): pasteboard (cardboard) sheet 788
PAPPDECKEL (*m*): paste-(paper)-board-(cover) 789
PAPPE (*f*): paper board; paste board; card-board; millboard; paste; pap 790
PAPPEDICHTUNG (*f*): cardboard (pasteboard) packing; corrugated packing 791
PAPPE, GEKAUTSCHTE- (*f*): millboard; couched paper board 792
PAPPE, GELEIMTE- (*f*): paste board; sized paper board; Bristol board (the finest variety of paper board) 793
PAPPE (*f*), **GLANZ-**: glazed cardboard 794
PAPPEL (*f*): poplar-(tree) (*Populus alba*) 795
PAPPELART (*f*): variety (kind; sort) of poplar 796
PAPPELBAUM (*m*): (see **PAPPEL**) 797
PAPPELESPE (*f*): Aspen (*Populus tremula*) 798
PAPPELHOLZ (*n*): poplar wood (Sg. 0.4-0.6) (see **PAPPEL**) 799
PAPPELKNOSPEN (*f.pl*): poplar buds (*Gemmæ populi* of *Populus nigra*) 800
PAPPELKNOSPENÖL (*n*): poplar bud oil 801
PAPPELKRAUT (*n*): marsh mallow (*Althæa officinalis*) 802
PAPPELSTAUDE (*f*): (see **PAPPELKRAUT**) 803
PAPPELSTEIN (*m*): malachite (see **MALACHIT** and **ATLASERZ**) 804
PAPPELWEIDE (*f*): black poplar (*Populus nigra*) 805
PAPPEL (*f*), **WEISSE-**: white poplar (*Populus alba*); marsh mallow (*Althæa officinalis*) 806
PAPPEN: to paste; work with paste-(board); (as an adjective = of pasteboard) 807
PAPPENDECKEL (*m*): (see **PAPPDECKEL**) 808
PAPPENFABRIK (*f*): board mill (Paper) 809
PAPPENFORM (*f*): cardboard shape 810
PAPPENHERSTELLUNG (*f*): card-board manufacture 811
PAPPENLEIM (*m*): pasteboard glue 812
PAPPENMASCHINE (*f*): board machine (Paper) 813
PAPPENPRESSE (*f*): pasteboard (cardboard; millboard) press 814
PAPPENROLLE (*f*): cardboard roll (Paper) 815
PAPP, GEKAUTSCHTER- (*m*): (see **PAPPE, GEKAUT-SCHTE-**) 816
PAPPHÜLLE (*f*): pasteboard (cardboard) case or covering 817
PAPPHÜLSE (*f*): pasteboard (cardboard) case or covering 818
PAPPICHT: pappy 819
PAPPMASSE (*f*): (see **PAPIERMACHÉ**) 820
PAPPROHR (*n*): card-board tube 821
PAPPSCHACHTEL (*f*): pasteboard (cardboard) box 822
PAPPSORTE (*f*): type (sort, quality) of paper board 823
PAPRIKA (*f*): chilli; paprika; chili; chile; chilly; capsicum (*Fructus capsici*) (a sweet condiment made from pepper; from the pod of *Capsicum annuum*) 824
PAPRIKAÖL (*n*): capsicum (chili) oil (from *Capsicum annuum*); Sg. 0.92 825
PAPYRIN (*n*): parchment paper, vegetable parchment 826
PAPYROGRAPHIE (*f*): papyrogaphy (kind of photo-lithography) 827
PAPYROLITH (*m*): papyrolite 828
PAPYRUSSTAUDE (*f*): papyrus, biblus (*Cyperus papyrus*) 829

PARAAMIDOACETANILID (*n*): paraaminoacetanilide 830

PARA-AMIDOBENZOESÄURE-ÄTHYLESTER (*m*): ethyl-para-aminobenzoate (see **ANÄSTHESIN**) 831

PARAAMIDOPHENOL (*n*): para-aminophenol (a photographic chemical) (see **AMINOPHENOL**) 832

PARAAMIDOSALICYLSÄURE (*f*): paraaminosalicylic acid; $NH_2C_6H_3(OH)CO_2H$ 833

PARABANSÄURE (*f*): parabanic acid 834

PARABEL (*f*): fable; parable; simile; parabola; allegory; apologue; parabolic curve (on a graph) 835

PARABEL (*f*), **KUBISCHE-**: cubic parabola 836

PARABLAU (*n*): para blue 837

PARABOLISCH: parabolic; figurative; allegorical 838

PARABOLOID (*n*): paraboloid 839

PARABUTTER (*f*): (see **PARAPALMÖL**) 840

PARACENTESE (*f*): paracentesis, tapping, exploration, aspiration, drainage (Med.) 841

PARACIT (*m*): paracite (see **BORAZIT**) 842

PARACONSÄURE (*f*): paraconic acid 843

PARACYAN (*n*): paracyanogen 844

PARADE (*f*): parade; show; display; ceremony; ostentation; pomp; pageant 845

PARADESTÜCK (*n*): show-piece 846

PARADEVERSUCH (*m*): shop-window test (a test carried out with the object of showing off the plant to the best advantage, i.e., not an actual working condition test) 847

PARADIAMINSCHWARZ (*n*): paradiamine black 848

PARADICHLORBENZOL (*n*): paradichlorobenzene (see **DICHLORBENZOL, PARA-**) 849

PARADIEREN: to vaunt, make a show, flaunt, parade, display, show off 850

PARADIES (*n*): paradise; Eden 851

PARADIESAPFEL (*m*): tomato (*Lycopersicum esculentum*) 852

PARADIESBAUM (*m*): paradise tree; wild olive; bitter damson; mountain damson (*Simaruba officinalis*) 853

PARADIESFEIGE (*f*): banana; plantain (*Musa paradisiaca*) 854

PARADIESHOLZ (*n*): (see **ALOEHOLZ**) 855

PARADIESKÖRNER (*n.pl*): grains of paradise; guinea grains (*Amomum granum paradisi*) 856

PARADIESKÖRNERÖL (*n*): grains of paradise oil (from seeds of *Lecythis zabucajo*); Sg. 0.895 857

PARADIESPFLANZE (*f*): paradise plant (*Daphne mezereum*) 858

PARADIESVOGEL (*m*): bird of paradise (*Paradisea apoda*) 859

PARADIGMA (*n*): paradigm, example, model 860

PARADOX: paradoxical 861

PARADOXON (*n*): paradox 862

PARAFFIN (*n*): (*Paraffinum solidum*); paraffin scale, paraffin wax, ceresin; soft paraffin (*Paraffinum molle*); Sg. 0.88-0.89; Mp. 44-48°C.; hard paraffin (*Paraffinum durum*); Sg. 0.898-0.915°C. Mp. 52-58°C. (see also **ZERESIN** and **ERDWACHS**) 863

PARAFFINAL (*n*): "Paraffinal" (an emulsion of liquid paraffin and water in the proportion of 2 : 1) 864

PARAFFINAUGE (*f*): paraffin globule 865

PARAFFINBAD (*n*): paraffin bath 866

PARAFFINFABRIK (*f*): paraffin factory 867

PARAFFINFABRIKANLAGE (*f*): paraffin plant 868

PARAFFIN (*n*), **FESTES-**: paraffin wax, ceresin, solid paraffin; Mp. 77°C. (*Paraffinum solidum*) (Pharm.) (see **PARAFFIN, WEICH-** and **PARAFFIN, HART-**) 869

PARAFFIN (*n*), **FLÜSSIGES-**: (*Paraffinum liquidum*) liquid paraffin (see **PARAFFINÖL**) 870

PARAFFINGEWINNUNG (*f*): production of paraffin 871

PARAFFINHALTIG: containing paraffin 872

PARAFFIN (*n*) **HART-**: (*Paraffinum durum*): hard paraffin, paraffin wax; Sg. 0.9; Mp. 55°C. 873

PARAFFINIEREN: to wax with paraffin, to coat with paraffin, to (soak, steep) dip in paraffin 874

PARAFFINIERTES ZÜNDHOLZ (*n*): paraffin match 875

PARAFFININDUSTRIE (*f*): paraffin industry 876

PARAFFINKERZE (*f*): paraffin candle 877

PARAFFINMASSE (*f*): paraffin mass 878

PARAFFINÖL (*n*): (*Paraffinum liquidum*): paraffin oil, liquid paraffin; Sg. 0.88-0.895; Bp. over 360°C.; paraffin oil (obtained by fractional distillation has Sg. 0.87-0.925) 879

PARAFFINPAPIER (*n*): paraffin paper; tracing paper (paper made transparent with paraffin) 880

PARAFFINPRÄPARAT (*n*): paraffin preparation 881

PARAFFIN (*n*), **ROH-**: crude paraffin 882

PARAFFINSALBE (*f*): soft paraffin; petrolatum; paraffin ointment; vaseline; mineral fat; petroleum (ointment) jelly; Sg. 0.82-0.85; Mp. 46.5°C. 883

PARAFFINSCHMIERE (*f*): paraffin grease (for making tracing paper) 884

PARAFFINSCHUPPE (*f*): paraffin (flake) scale (see **PARAFFIN, WEICH-**) 885

PARAFFIN (*n*), **WEICH-**: soft paraffin, ceresin, paraffin scale; Sg. 0.885; Mp. 46°C. [886

PARAFORM (*n*): paraform (polymerized formaldehyde) (see **FORMALDEHYD**) 887

PARAFORMALDEHYD (*n*): (*Paraformaldehydum*): paraformaldehyde (see **FORMALDEHYD**) 888

PARAFUCHSIN (*n*): parafuchsine 889

PARAGENSIE (*f*): paragenesis, hybridism 890

PARAGONIT (*m*): paragonite (a form of mica); $(Na,K)_2O,2H_2O,3Al_2O_3,6SiO_2$; Sg. 2.78 [891

PARAGRAS (*n*): (see **PIASSAVA, BRASILISCHE-**) 892

PARAGUAY-TEE (*m*): (see **MATÉ**) 893

PARAGUMMI (*n*): para rubber 894

PARAGUMMIÖL (*n*): para-rubber oil (from seeds of *Hevea brasiliensis*) 895

PARAKAMPHERSÄURE (*f*): paracamphoric acid (see **KAMPHERSÄURE**) 896

PARAKAUTSCHUK (*m*): para rubber 897

PARAKAUTSCHUKÖL (*n*): (see **PARAGUMMIÖL**) 898

PARAKRESOL (*n*): paracresol (see **KRESOL, PARA-**) 899

PARAKRESSE (*f*): para cress (*Spilanthes oleracea*) 900

PARALAURIONIT (*m*): paralaurionite; $PbCl(OH)$; Sg. 6.05 901

PARALDEHYD (*n*): paraldehyde; $C_6H_{12}O_3$; Sg. 0.998; Mp. 10.5°C.; Bp. 124°C. 902

PARALLAKTISCH: parallactic-(al) (Astron.) 903

PARALLAXE (*f*): parallax (Astron.) 904

PARALLEL: parallel 905

PARALLELE (*f*): parallel-(line); comparison 906

PARALLELENZIEHER (*m*): parallel ruler 907

PARALLELEPIPEDON (*n*): parallepiped 908

PARALLELFÜHRUNG (*f*): guide(s); parallel motion 909
PARALLELITÄT (*n*): parallelity 910
PARALLELKREIS (*m*): parallel 911
PARALLELLINEAL (*n*): parallel ruler 912
PARALLELLINIE (*f*): parallel line 913
PARALLELOGRAMM (*n*): parallelogram 914
PARALLELOGRAMM (*n*), **KRÄFTE-**: force parallelogram, parallelogram of forces 915
PARALLELREISSER (*m*): marking gauge 916
PARALLELSCHALTER (*m*): parallel (multiple) switch 917
PARALLELSCHALTUNG (*f*): parallel coupling or connection 918
PARALLELSTELLE (*f*): parallel passage 919
PARALLELTRAPEZ (*n*): parallel trapezium, trapezoid (a trapezium having two of its sides parallel) 920
PARALLEPIPEDISCHER BAUSTEIN (*m*): (see **QUADER**) 921
PARALYSE (*f*): paralysis; paraplegia (Med.) 922
PARALYSE (*f*), **ALLGEMEINE-**: general (total) paralysis 923
PARALYSIEREN: to paralyse 924
PARALYSIS (*f*): paralysis, paraplegia (Med.) 925
PARALYSIS ASCENDANS ACUTA (*f*): acute ascending paralysis, Landry's paralysis (*Paralysis ascendans acuta*) (Med.) 926
PARALYSIS SPINALIS (*f*), **INTERMITTIERENDE-**: intermittent (paraplegia) spinal paralysis 927
PARALYSOL (*n*): lysol (see **LYSOL**) 928
PARAMAGNETISCH: paramagnetic 929
PARAMAGNETISMUS (*m*): paramagnetism 930
PARAMETER (*m*): parameter 931
PARAMIDOACETANILID (*n*): paraminoacetanilide (see **AMIDOACETANILID, PARA-**) 932
PARAMIDOPHENOL (*n*): paraminophenol (see **AMIDOPHENOL, PARA-**) 933
PARAMILCHSÄURE (*f*): paralactic acid (see **MILCHSÄURE**) 934
PARAMORFAN (*n*): paramorfan, dihydromorphine hydrochloride; $C_{17}H_{21}NO_3.HCl$ 935
PARANEPHRIN (*n*): paranephrine (blood-pressure raising constituent of suprarenal glands; used for stopping the flow of blood) 936
PARANITRACETANILID (*n*): para-nitroacetanilide; $NO_2.C_6H_4.NH.COCH_3$; Mp. 207°C. 937
PARANITRANILIN (*n*): paranitraniline (see **NITRA--NILIN, PARA-**) 938
PARANITROTOLUOL (*n*): paranitrotoluene (see **NITROTOLUOL, PARA-**) 939
PARANUSS (*f*): Brazil nut (*Bertholletia excelsa*) 940
PARANUSSÖL (*n*): brazil-nut oil (from *Bertholletia excelsa*) Sg. 0.918 941
PARAOXYBENZOËSÄURE (*f*): parahydroxybenzoic acid; $C_6H_4(OH)COOH.H_2O$; Mp. 210°C. 942
PARAPALME (*f*): cabbage palm (*Enterpe oleracea*) 943
PARAPALMÖL (*n*): cabbage palm oil (from the fruit of *Enterpe oleracea*) 944
PARAPHENOLSULFOSÄURE (*f*): paraphenolsulphonic acid; $C_6H_4.SO_3H.OH$ 945
PARAPHENYLENBLAU (*n*): paraphenylene blue 946
PARAPHENYLENDIAMIN (*n*): paraphenylenediamine (see **PHENYLENDIAMIN, PARA-**) 947
PARAPHIMOSIS (*f*): paraphimosis (disease of the penis) (Med.) 948
PARASÄURE (*f*): para acid, para-sulfaminobenzoic acid; Mp. about 288°C. 949
PARASITÄR: parasitic 950
PARASITEN (*m.pl*): parasites, bacteria, entozoa (Med.) 951
PARASTELLUNG (*f*): para position 952
PARAT: prepared; ready 953
PARATAKAMIT (*m*): paratakamite; $Cu_2Cl(OH)_3$; Sg. 3.74 954
PARATOLUIDIN (*n*): paratoluidine (see **TOLUIDIN, PARA-**) 955
PARATOLUOLSULFAMID (*n*): paratoluene sulfamide, toluenesulfoneamine; $CH_3C_6H_4SO_2NH_2$; Mp. 137°C. 956
PARATOLUOLSULFOCHLORID (*n*): paratoluene sulfochloride, toluene sulfonechloride; $CH_3C_6H_4SO_2Cl$; Mp. 69°C.; Bp. 145°C. [957
PARATOMER AUGITSPAT (*m*): amianthus, diopside, augite, pyroxene, acmite 958
PARAVERBINDUNG (*f*): para compound 959
PARAWEINSÄURE (*f*): paratartaric (racemic) acid; uvic acid; $C_2H_4O_2(COOH)_2.H_2O$; Sg. 1.7; Mp. 205°C. 960
PARAXYLIDIN (*n*): paraxylidine (see **XYLIDIN, PARA-**) 961
PARAXYLOL (*n*): paraxylol (see **XYLOL, PARA-**) 962
PARDON (*m*): pardon; quarter 963
PARDONNABEL: pardonable 964
PARDUNE (*f*): backstay (Naut.) 965
PAREIRAWURZEL (*f*): pareira root; velvet leaf; pareira brava (root of *Chondrodendron tomentosum*) 966
PARENCHYMZELLE (*f*): parenchyma cell (a six-sided cell containing chlorophyll, and forming the cellular tissue of plants) (Botany.) 967
PARENTHESE (*f*): parenthesis; parentheses; brackets 968
PARFÜM (*n*): perfume; scent 969
PARFÜMERIE (*f*): perfumery 970
PARFÜMERIEESSENZ (*f*): essence (solution of etheral oils in alcohol) 971
PARFÜMERIEETIKETT (*n*): perfume label 972
PARFÜMERIEETIKETTE (*f*): perfume label 973
PARFÜMERIEFABRIK (*f*): scent factory 974
PARFÜMERIEFABRIKANLAGE (*f*): scent manufacturing plant 975
PARFÜMERIEGRUNDSTOFF (*m*): perfumery materials; base for perfumes; materials from which scents are made 976
PARFÜMFLACON (*n*): scent (perfume) bottle or flask 977
PARFÜMIEREN: to scent; perfume 978
PARGASIT (*m*): pargasite (see **HORNBLENDE**) 979
PARI (*n*): par 980
PARI (*n*), **AL-**: at par 981
PARIEREN: to parry; stop; ward off; obey; bet; wager 982
PARISER: Paris (as an adjective) 983
PARISERBLAU (*n*): Paris (Chinese) blue; ferric-ferro cyanide; pure Berlin blue (see **BERLIN--ERBLAU**) 984
PARISERGELB (*n*): Paris yellow, lead chromate (see **BLEICHROMAT**) 985
PARISERGRÜN (*n*): Paris green; copper aceto-arsenite (see **KUPFERACETATARSENIAT**) 986
PARISERLACK (*m*): carmine lake (see **COCHENILLE**) 987
PARISERROT (*n*): Paris red (see **MENNIGE**) 988
PARISERSCHWARZ (*n*): Paris black 989
PARISIT (*m*): parisite; $Ce(OH,F),(Ce,La,Di)O.Co_2$; Sg. 4.25 990
PARITÄT (*f*): parity, equality, analogy, equivalence, sameness, similarity, correspondence 991
PARK (*m*): park 992

PARKESIEREN: to subject to Parkes' process (Metal); (*n*): the subjection to the Parkes' process 993
PARKES VERFAHREN (*n*): Parkes' process (extraction of silver from the product of preliminary purification of lead) 994
PARKETT (*n*): parquet (inlaid) floor-(ing); orchestra stalls (of a theatre) 995
PARKETTBODEN (*m*): parquet [inlaid-(wood)] flooring 996
PARKETTIEREN: to parquet (inlay) floors 997
PARKETTWACHS (*n*): parquet flooring wax 998
PARLAMENT (*n*): parliament 999
PARLAMENTARISCH: parliamentary 000
PARLAMENTSAKTE (*f*): act of parliament 001
PARODYN (*n*): parodyne (see **ANTIPYRIN**) 002
PAROTITIS (*f*): parotitis, inflammation of the parotid gland (Med.) 003
PARRY'SCHER TRICHTER (*m*): Parry's funnel (a cone, moveable downwards; for catching gas from blast furnaces; fitted at top) 004
PART (*m* and *n*): share; part; portion 005
PARTEI (*f*): party; part; faction; detachment; section; squad; side; following 006
PARTEIFÜHRER (*m*): party leader 007
PARTEIGÄNGER (*m*): partisan 008
PARTEIHAUPT (*m*): party leader 009
PARTEIISCH: prejudiced; partial; biased 010
PARTEILISCH: (see **PARTEIISCH**) 011
PARTEILOS: impartial; neutral; unprejudiced; unbiassed 012
PARTERRE (*n*): ground floor; parterre; pit (of a theatre) 013
PARTIALDRUCK (*m*): partial pressure 014
PARTIE (*f*): company; party; lot; quantity; batch; parcel; picnic game; match 015
PARTIELL: partial 016
PARTIEPREIS (*m*): wholesale price 017
PARTIEREN: to part; separate; divide; distribute; shuffle; cheat 018
PARTIESTOSS (*m*): finishing (winning) stroke 019
PARTIKEL (*f*): particle 020
PARTIKULAR: particular 021
PARTIKULARISIEREN: to particularize 022
PARTITIV: partitive 023
PARTITUR (*f*): score (Music) 024
PARTIZIP (*n*): participle 025
PARTIZIPIAL: participial 026
PARTIZIPIEREN: to participate, share, take part in 027
PARTIZIPIUM (*n*) **PRÄSENTIS**: present participle (Gram.) 028
PARTNERSCHAFT (*f*): partnership (Com.) 029
PARVOLIN (*n*): parvoline (a ptomaine from putrefying horse-flesh or mackerel); $C_9H_{13}N$ 030
PASS (*m*): pass, passport; narrows (Naut.); passage, defile; pace; fit 031
PASSAGIER (*m*): passenger; voyager; traveller 032
PASSAGIERSCHIFF (*n*): passenger (boat) ship 033
PASSAMT (*n*): passport (office) bureau 034
PASSATWIND (*m*): trade wind; monsoon 035
PASSAUERTIEGEL (*m*): a graphite crucible 036
PASSEINHEIT (*f*): unit of fit 037
PASSEN: to fit; suit; gauge; pass; be (becoming) appropriate; correspond; harmonize; be convenient 038
PASSEND: fit; suitable; just; appropriate; convenient; seasonable; opportune; fitting; proper 039
PASSIERBAR: passable 040
PASSIEREN: to pass (by or through); happen; take place; liquor (see **PASSIERMASCHINE**) 041
PASSIERGEWICHT (*n*): allowance, tolerance 042
PASSIERMASCHINE (*f*): liquoring machine (a machine for breaking up pulpy masses into as fine a distribution as possible, by forcing them through a sieve) 043
PASSIERSCHEIN (*m*): pass, permit 044
PASSIONIEREN: to be passionately (greatly, deeply) interested in, or fond of 045
PASSIONIERT: impassioned 046
PASSIONSBLUME (*f*): love-in-a-mist (*Nigella*); passion flower (*Passiflora coerulea*) (Bot.) 047
PASSIV: passive 048
PASSIVA (*n.pl*): liabilities, passive debts (see **PASSIVSCHULDEN**) 049
PASSIVISCH: passively 050
PASSIVITÄT (*f*): passivity 051
PASSIVSCHULDEN (*f.pl*): bills payable, debts, passive debts 052
PASSIVUM (*n*): passive voice (Gram.) 053
PASSLICH: fit; suitable; proper 054
PÄSSLICH: passable; tolerable 055
PASSROHR (*n*): make-up (piece) length, adaptor, template pipe 056
PASSSTÜCK (*n*): fitting (make-up) piece 057
PASSTOLERANZ (*f*): fitting tolerance 058
PASSUNG (*f*): fit, fitting, suitability, gauging 059
PASSUNGSSYSTEM (*n*): standard (for the fit of machine parts); standard gauge 060
PASSUS (*m*): passage (in a book, etc.) 061
PASTA (*f*): paste 062
PASTE (*f*): paste (*Pasta*) (also in reference to spurious jewellery) 063
PASTELFARBE (*f*): pastel (coloured-(ed) chalk; crayon 064
PASTEL-(L) (*m* and *n*): pastel 065
PASTELLMALEREI (*f*): pastel work, pastel drawing 066
PASTELLSTIFT (*m*): pastel, coloured chalk 067
PASTETCHENPFANNE (*f*): patty (pan) tin 068
PASTETE (*f*): pie; pastry 069
PASTETENBÄCKER (*m*): pastry cook 070
PASTEURISIERAPPARAT (*m*): pasteurizing apparatus 071
PASTEURISIEREN: to pasteurize 072
PASTILLE (*f*): pastille; medicated lozenge; troche (Pharm.) 073
PASTINAKE (*f*): parsnip 074
PASTORAL: pastoral; clerical 075
PASTÖS: pasty 076
PATCHOULIÖL (*n*): (see **PATSCHULIÖL**) 077
PATENT (*n*): patent: license; charter: commission (Military) 078
PATENTAMT (*n*): patent office (the official patent office only) 079
PATENTAMTLICHE GEBÜHR (*f*): patent office fee 080
PATENTANMELDUNG (*f*): patent application 081
PATENTANSPRUCH (*m*): patent claim 082
PATENTANWALT (*m*): patent (attorney) agent 083
PATENTBESCHREIBUNG (*f*): patent specification 084
PATENTBESTIMMUNG (*f*): patent (data, law) information 085
PATENTBLAU (*n*): patent blue 086
PATENTBRIEF (*m*): letters patent 087
PATENTBUREAU (*n*): patent office, patent agency 088
PATENTBÜRO (*n*): patent office, patent agency 089

PATENT (*n*), **DEFINITIVES-**: complete-(patent)-specification 090
PATENTFÄHIG: patentable 091
PATENTFÄHIGKEIT (*f*): patentability 092
PATENTGELB (*n*): patent yellow (see MONT--PELLIERGELB) 093
PATENTGESETZ (*n*): patent law 094
PATENTGESETZGEBUNG (*f*): patent legislation 095
PATENTGRÜN (*n*): patent green (see KUPFER--ACETATARSENIAT) 096
PATENTIEREN: to patent; take out a patent 097
PATENTIERTE (*m*): patentee 098
PATENTINHABER (*m*): patentee 099
PATENT (*n*), **KOMPLETTES-**: complete-(patent)-specification 100
PATENTLAND (*n*): patent country (country in which a patent is, or may be, taken out) 101
PATENTLISTE (*f*): patent (list) register 102
PATENTMEDIZIN (*f*): patent medicine 103
PATENT (*n*), **PROVISORISCHES-**: provisional-(patent)-specification 104
PATENTRECHT (*n*): patent right 105
PATENTREGISTER (*n*): register of patents 106
PATENTROT (*n*): patent red, vermillion (see ZINNOBER) 107
PATENTSALZ (*n*): patent salt (ammonium-antimony fluoride) 108
PATENTSCHRIFT (*f*): patent; patent specification 109
PATENTSCHUTZ (*m*): protection (by patent) 110
PATENTSCHWARZ (*n*): patent black 111
PATENTSUCHER (*m*): applicant (for a patent) 112
PATENTTRÄGER (*m*): patentee 113
PATENTVERFAHREN (*n*): patent (process) procedure 114
PATENTVERLÄNGERUNG (*f*): extension-(of duration)-of a patent 115
PATENTVERLÄNGERUNGSGESETZ (*n*): patent extension law 116
PATENTVERZINKUNG (*f*): zincing, zinc coating (in which the bath is given about 3% Al.; antirust agent for metal) 117
PATENTZEICHNUNG (*f*): patent drawing 118
PATERNOSTER (*m*): bucket elevator; paternoster 119
PATHETISCH: pathetic-(al)-(ly) 120
PATHOLOG (*m*): pathologist 121
PATHOLOGIE (*f*): pathology 122
PATHOLOGISCH: pathological 123
PATIENT (*m*): patient (Med.) 124
PATINA (*f*): green verditer, basic copper carbonate (a green coating formed on copper which is exposed to moist air) (see KUPFERCARBONAT, BASISCHES-) 125
PATINIERUNG (*f*): coating of copper with green verditer (see PATINA) 126
PATRIZE (*f*): punch 127
PATRONAT (*n*): patronage, advowson, living (Ecclesiastical) 128
PATRONE (*f*): cartridge; pattern; stencil; mandrel 129
PATRONENBÜCHSE (*f*): cartridge box 130
PATRONENFABRIK (*f*): cartridge factory 131
PATRONENHÜLSE (*f*): cartridge case 132
PATRONENPAPIER (*n*): cartridge paper (Paper) 133
PATRONENZYLINDER (*m*): cartridge case 134
PATRONIEREN: to make up into cartridges 135
PATRONIERMASCHINE (*f*): cartridge making machine 136
PATRONIT (*m*): patronite; Fe,Mo,S,Vd,Al_2O_3,SiO_2; Sg. 2.65 137
PATRONYMIKON (*n*): patronymic 138
PATSCHE (*f*): difficulty; dilemma; fix; mess; puddle; slush; mud; mire 139
PATSCHEN: to clap; dabble; slap; splash; smack; pour; patter (of rain) 140
PATSCHULIBLÄTTER (*n.pl*): patchouli leaves (*Folia patchouli* of *Pogostemon patchouli*) 141
PATSCHULIKRAUT (*n*): patchouli herb (*Pogostemon patchouli*) 142
PATSCHULIÖL (*n*): patchouli oil (*Oleum patchouli foliorum*) (an ethereal oil from leaves of *Pogostemon patchouli*); Sg. 0.925-0.995 143
PATT MACHEN: to put stalemate (Chess) 144
PATT SEIN: to be stalemate (Chess) 145
PATTINSONIEREN: to pattinsonize; (*n*): pattinsonizing (subjection to the Pattinson process) 146
PAUKE (*f*): kettle-drum; tympanum (Anat.); harangue; coffee-roaster 147
PAUKEN: to make a noise; cause a row 148
PAUKENFELL (*n*): drum-skin; membrana tympani (Anat.) 149
PAUKENGANG (*m*): ear duct 150
PAUPERISMUS (*m*): pauperism 151
PAUSCHALE (*f*): lot, lump sum, total 152
PAUSCHALSUMME (*f*): lump sum, sum total 153
PAUSCHEN: to smell; refine (Metal); melt; liquate; beat small; (*n*): refining; liquation 154
PAUSCHQUANTUM (*n*): bulk, total quantity, lot 155
PAUSCHSUMME (*f*): (see PAUSCHALSUMME) 156
PAUSCHT (*m* and *n*): post (Paper) 157
PAUSE (*f*): stop-(page); pause; tracing; break; cessation of work; suspension; interval; rest (Music); interruption 158
PAUSEN: to trace; caulk 159
PAUSIEREN: to stop; pause; cease; suspend; make a break; rest; interrupt 160
PAUSLEINEN (*n*): tracing cloth 161
PAUSLEINWAND (*f*): tracing cloth 162
PAUSPAPIER (*n*): tracing paper 163
PAVIAN (*m*): baboon 164
PAVILLON (*m*): pavilion 165
PEARCEIT (*m*): pearceite; $(Ag,Cu)_9AsS_6$; Sg. 6.15 166
PECH (*n*): pitch (*Pix*); cobbler's wax; bad luck 167
PECHART (*f*): sort (kind; variety) of pitch 168
PECHARTIG: like pitch; pitchy; bitumenous 169
PECHBAUM (*m*): (see PECHTANNE) 170
PECHBLENDE (*f*): pitchblende (natural uranium oxide); $U_3Pb_6O_{12}$; Sg. 7.03 (see also URANINIT and URANPECHERZ) 171
PECHDRAHT (*m*): shoemaker's (pitched) thread; cobbler's (waxed) thread 172
PECHEISENERZ (*n*): pitchblende (see PECH--BLENDE) 173
PECHERDE (*f*): bitumenous earth 174
PECHERZ (*n*): pitchblende (see PECHBLENDE) 175
PECHFACKEL (*f*): pitch torch; link 176
PECHFINSTER: pitch-dark 177
PECHGESCHMACK (*n*): pitchy taste 178
PECHGLANZ (*m*): pitchy lustre 179
PECHGLANZKOHLE (*f*): (see MOLASSE-KOHLE) 180
PECHGRANAT (*m*): colophonite 181
PECHHOLZ (*n*): resinous wood 182
PECHIG: pitchy; bitumenous 183
PECHKIEFER (*f*): pitch pine (see HARZKIEFER) 184
PECHKOHLE (*f*): pitch (bitumenous) coal 185

PECHNELKE (*f*): catch-fly (see MÜCKENFANG) 186
PECHÖL (*n*): oil of tar; tar oil; pitch oil; Sg. 0.86-0.87 187
PECHPFLASTER (*n*): asphalt; bitumen; tar-paving 188
PECH (*n*), SCHIFFS-: marine (caulking) pitch (from wood-tar) 189
PECH (*n*), SCHMIEDE-: forge pitch (a residue from the distillation of colophony) 190
PECH (*n*), SCHUSTER-: cobbler's wax (from wood tar) 191
PECHSCHWARZ: pitch-black 192
PECHSTEIN (*m*): pitch stone 193
PECHSTEINKOHLE (*f*): jet 194
PECH (*n*), STEINKOHLEN-: coal tar pitch (from distillation of coal tar) 195
PECHTANNE (*f*): pitch-pine (see HARZKIEFER); spruce (see FICHTE) 196
PECHTONNE (*f*): pitch (barrel, tun) cask 197
PECHURAN (*n*): pitchblende (see PECHBLENDE) 198
PEDAL (*n*): pedal 199
PEDALGUMMI (*n*): pedal rubber 200
PEDALKONUS (*m*): pedal cone 201
PEDANT (*m*): pedant 202
PEDANTERIE (*f*): pedantry 203
PEDANTISCH: pedantic-(al); pedagogic-(al) 204
PEDANTISMUS (*m*): pedantry 205
PEDELL (*m*): janitor; beadle 206
PEERD (*n*): foot rope (of yards) (Naut.) 207
PEERD (*n*), FOCKRAH-: fore yard foot-rope 208
PEGAMOID (*n*): pegamoid (in Germany the trade name of a leather substitute; in America the trade name of an aluminium paint) 209
PEGEL (*m*): water (depth) gauge 210
PEGELN: to gauge; sound; take soundings (Naut.); measure 211
PEGMATIT (*m*): pegmatite, giant granite (an ordinary granite of irregular texture, with large lumps of the constituent minerals) (see GRANIT) 212
PEGNIN (*n*): pegnin (a lactose-rennet ferment) 213
PEGU-KATECHU (*m*): (see KATECHU, BRAUNER-) 214
PEIL (*n*): (see PEGEL) 215
PEILEN: to sound (see PEGELN); take bearings, measure, take sun's altitude 216
PEILKOMPASS (*m*): variation compass, azimuth compass 217
PEILLOT (*n*): plummet (sounding-)lead 218
PEILROHR (*n*): sounding pipe 219
PEILSTOCK (*m*): sounding rod 220
PEIN (*f*): pain; torture; trouble; torment; agony 221
PEINIGEN: to torment; trouble; worry; torture; pain; upset; rack; plague 222
PEINLICH: painful; tormenting; penal; capital; painstaking; precise; criminal; careful; difficult; minute-(ly exact) 223
PEINLICHE KLAGE (*f*): summons for trespass 224
PEINLICHE UNTERSUCHUNG (*f*): trial on a capital charge 225
PEITSCHE (*f*): whip; cat-of-nine-tails 226
PEITSCHEN: to whip; lash; beat; scourge; flagellate; castigate 227
PEITSCHENFÖRMIG: flagelliform; whip-shaped 228
PEITSCHENHIEB (*m*): lash-(with a whip) 229
PEITSCHENRIEMEN (*m*): thong-(of a whip) 230
PEITSCHENSCHNUR (*f*): thong; whip-cord; **lash** 231
PEKORINUTS (*pl*): hickory nuts (from the tree, *Hicoria glabra*) 232
PEKTIN (*n*): pectin 233
PEKTINLÖSUNG (*f*): pectin solution 234
PEKTINSÄURE (*f*): pectic acid 235
PEKTINSTOFF (*m*): pectin-(substance) 236
PEKTINZUCKER (*m*): arabinose; pectinose; $C_5H_{10}O_5$; Mp. 159°C. 237
PEKTOLITH (*m*): pectolite; $NaHCa_2Si_3O_9$; Sg. 2.77 238
PEKUNIÄR: pecuniary 239
PELA (*n*): (see WACHS, CHINESISCHES-) 240
PELARGONSAUER: pelargonate of, combined with pelargonic acid 241
PELARGONSÄURE (*f*): pelargonic acid 242
PELLAGRA (*f*): pellagra (*Erythema pellagrosum*) (*an erythema of the skin*) (Med.) 243
PELLE (*f*): peel; skin; husk 244
PELLETIERIN (*n*): pelletierine 245
PELLIDOL (*n*): pellidol, diacetylaminoazotoluene (see DIAZETYLAMINOAZOTOLUOL) 246
PELZ (*m*): fur-(coat); pelt; skin (on liquids) 247
PELZEN: to graft; inoculate; of fur; furred; furry 248
PELZHANDEL (*m*): fur trade 249
PELZIG: furry; furred; having a skin; cottony (Bot.); fleecy 250
PELZREIS (*m*): graft 251
PENANG (*m*): (see KATECHUPALME) 252
PENDEL (*n* and *m*): pendulum 253
PENDELGEWICHT (*n*): balance (pendulum) weight 254
PENDELN: to swing; oscillate 255
PENDELSTAUER (*m*): balanced (swinging) ash dumping arrangement or back end arrangement (Mechanical stokers) 256
PENDELUHR (*f*): pendulum clock 257
PENDELUNG (*f*): oscillation 258
PENETRIEREN: to penetrate 259
PENIS (*m*): penis (Med.) 260
PENNIN (*m*): penninite, clinochlore; $Mg_2Al_2SiO_7.H_2O$: $Mg_3Si_2O_7.H_2O$; Sg. 2.69; ripidolite (see RIPIDOLITH) 261
PENSION (*f*): boarding house; board; boarding school; pension 262
PENSIONIEREN: to pension-(off); superannuate 263
PENSIONSLISTE (*f*). retired list 264
PENSUM (*n*): lesson, task 265
PENTACARBOCYCLISCH: pentacarbocyclic 266
PENTACHLORÄTHAN (*n*): pentachloroethane; C_2HCl_5; Sg. 1.685; Bp. 159°C. 267
PENTADECANSÄURE (*f*): pentadecoic acid 268
PENTAHYDRAT (*n*): pentahydrate; ($5H_2O$) 269
PENTAMETHYLBENZOL (*n*): pentamethylbenzene; $C_6H(CH_3)_5$ 270
PENTAMETHYLENIMID (*n*): (see PIPERIDIN) 271
PENTAN (*n*): pentane, amyl hydride, normal pentane, isopentane, petroleum ether (a limit hydrocarbon); C_5H_{12}; Sg. 0.634; Mp. – 130.8°C.; Bp. 35-40°C. (see also PETROLÄTHER) 272
PENTAN-NORMALTHERMOMETER (*n*): pentane normal thermometer 273
PENTATHIONSÄURE (*f*): pentathionic acid 274
PENTERPOL (*n*): penterpol (a cleaning preparation for washing, containing about 80% turpentine oil) 275
PENTINSÄURE (*f*): pentinic acid 276
PENTOSAN (*n*): pentosane 277
PENTOSANHALTIG: containing pentosane 278
PENTOSE (*f*): pentose 279

PENTOSURIE (*f*) : pentosuria 280
PEPSIN (*n*) : (*Pepsinum*) : pepsin (an enzyme from the mucous membrane of fresh pigs' or calves' stomachs) 281
PEPSINDRÜSE (*f*) : peptic gland 282
PEPSINESSENZ (*f*) : pepsin essence 283
PEPSINHALTIG : containing pepsin 284
PEPSINPRÄPARAT (*n*) : pepsin preparation 285
PEPSINWEIN (*m*) : pepsin wine 286
PEPTENZYM (*n*) : peptenzyme 287
PEPTISCH : peptic 288
PEPTON (*n*) : peptone (from digestion of albumen) 289
PEPTON (*n*), **ANIMALISCHES-** : animal peptone 290
PEPTONEISEN (*n*) : (see **EISENPEPTONAT**) 291
PEPTON (*n*), **FLEISCH-** : meat peptone (from digestion of red meat) 292
PEPTONISIEREN : to peptonize 293
PEPTON (*n*), **KASEIN-** : casein peptone 294
PEPTON (*n*), **PFLANZEN-** : vegetable peptone (from vegetable albumen) 295
PER : per ; for ; by (Book-keeping) ; on the principle of (see also **ÜBER-**) 296
PERACIDITÄT (*f*) : superacidity 297
PERAQUIN (*n*) : peraquin (a 30% peroxide of hydrogen) 298
PERAUTAN (*n*) : perautan (a mixture of polymerized formaldehyde with permanganate of potash) 299
PERBORAT (*n*) : perborate 300
PERBORAX (*m*) : sodium perborate, perborin ; $Na_2B_4O_8.10H_2O$ (see **NATRIUMPERBORAT**) 301
PERBORSÄURE (*f*) : perboric acid 302
PERBROMSÄURE (*f*) : perbromic acid 303
PERCARBONAT (*n*) : percarbonate 304
PERCHLORAT (*n*) : perchlorate 305
PERCHLORÄTHAN (*n*) : perchloroethane, carbon trichloride ; C_2Cl_6 ; Sg. 1.6298 ; Mp. 185°C. ; Bp. 185°C. 306
PERCHLORÄTHYLEN (*n*) : perchloroethylene, tetrachloroethylene ; carbon (dichloride) bichloride ; C_2Cl_4 ; Sg. 1.628 ; Bp. 119°C. 307
PERCHLORSÄURE (*f*) : perchloric acid ; $HClO_4$; Sg. 1.764-1.768 ; Bp. 16°C. ; Fraude's reagent 308
PERCHLORSÄUREANHYDRID (*n*) : (see **CHLORHEPT-OXYD**) 309
PERCHLORSÄUREHYDRAT (*n*) : perchloric acid, hydrated ; $HClO_4.H_2O$; Sg. 1.776-1.811 ; Mp. 50°C. 310
PERCHLORSÄUREMONOHYDRAT (*n*) : monohydrated perchloric acid ; $HClO_4.H_2O$; Sg. 1.776-1.811 ; Mp. 50°C. 311
PERCHROMSÄURE (*f*) : perchromic acid 312
PEREMTORISCH : peremptory ; peremptorily 313
PERENNIEREND : perennial (Bot.) 314
PERENNIERENDE PFLANZE (*f*) : perennial plant 315
PERFEKT : perfect 316
PERFEKTUM (*n*) : perfect tense 317
PERFORIEREN : to perforate 318
PERGAMENT (*n*) : parchment 319
PERGAMENTÄHNLICH : like parchment, parchment-like 320
PERGAMENTARTIG : like parchment, membranous 321
PERGAMENTBAND (*m*) : parchment binding 322
PERGAMENTEN : of parchment 323
PERGAMENTERSATZ (*m*) : imitation parchment ; parchment substitute (see **PERGAMENT-PAPIER, UNECHTES-**) 324
PERGAMENTHAUT (*f*) : parchment skin 325
PERGAMENTIEREN : to parchmentize 326
PERGAMENTIMITATION (*f*) : imitation parchment (see **PERGAMENTPAPIER, UNECHTES-**) 327
PERGAMENTLEIM (*m*) : parchment glue 328
PERGAMENTPAPIER (*n*) : parchment paper (see **PERGAMENTPAPIER, ECHTES-** and **UNECHTES-**) ; thick vellum 329
PERGAMENTPAPIER (*n*), **ECHTES-** : real parchment paper (from treatment of paper pulp with sulphuric acid or zinc chloride solution) 330
PERGAMENTPAPIER (*n*), **UNECHTES-** : imitation parchment paper (from sulphite cellulose sizes with aluminium sulphate or resin) 331
PERGAMENTPAPIER (*n*), **VEGETABILISCHES-** : vegetable parchment-(paper) 332
PERGAMENT (*n*), **TIERISCHES-** : animal parchment 333
PERGENOL (*n*) : pergenol (a hydrogen peroxide preparation in powder or tablet form, containing 12% hydrogen peroxide, 22% boric acid, sodium bitartrate and sodium perborate) 334
PERGLUTYL (*n*) : perglutyl (trade name for a hydrogen peroxide preparation for medical purposes) (consists of hydrogen peroxide, glycerine and gelatine) 335
PERGLYZERIN (*n*) : perglycerine (a glycerine substitute) 336
PERHYDRIEREN : to perhydrogenize 337
PERHYDRIT (*n*) : perhydrite (a hydrogen peroxide preparation from perhydrol with carbamide, contains over 30% of hydrogen peroxide) 338
PERHYDROL (*n*) : perhydrol (a very pure form of hydrogen peroxide, containing 30% by weight of H_2O_2) 339
PERIDOT (*m*) : peridot (see **OLIVIN**) 340
PERIKLAS (*m*) : periclase ; periclasite (natural magnesia oxide) ; MgO ; Sg. 3.8 341
PERIKLIN (*m*) : pericline (see **ALBIT**) 342
PERILLAÖL (*n*) : perilla oil (from *Perilla ocimoides* seeds) ; Sg. 0.93-0.945 343
PERILLASÄURE (*f*) : perillic acid 344
PERINEPHRITIS (*f*) : perinephritis (inflammation of the fibro-fatty envelope of the kidney) (Med.) 345
PERIODE (*f*) : period ; repetend (of a decimal) ; catamenia (Med.) ; phrase (Music) ; (see **RADIOAKTIVE PERIODE**) 346
PERIODENZAHL (*f*) : number of periods, periodicity (of electric current) 347
PERIODISCH : periodic-(al)-(ly) 348
PERIODISCHER DEZIMALBRUCH (*m*) : recurring decimal 349
PERIODISCHES SYSTEM (*n*) : periodic system 350
PERIODIZITÄT (*f*) : periodicity ; frequency 351
PERIODIZITÄTSGRAD (*m*) : degree of periodicity 352
PERIPATETIKER (*m*) : peripatetic 353
PERIPATETISCH : peripatetic 354
PERIPHERIE (*f*) : periphery 355
PERISKOP (*n*) : periscope 356
PERISKOPROHR (*n*) : periscope tube 357
PERISTALTIK (*f*) : peristalsis (the worm-like movement of the intestine) 358
PERISTALTIN (*n*) : peristaltin (a glucoside from the bark of *Cascara sagrada* ; a purgative) ; $C_{14}H_{18}O_8$ 359
PERISTALTISCH : peristaltic ; worm-like 360
PERITOMER ANTIMONGLANZ (*m*) : (see **SCHILF-GLASERZ**) 361
PERITOMES TITANERZ (*n*) : rutile (see **RUTIL**) 362

PERJODAT (*n*): periodate 363
PERJODSAUER: periodate; combined with periodic acid 364
PERJODSÄURE (*f*): periodic acid; $HIO_4.2H_2O$; Mp. 130-133°C. 365
PERKAGLYZERIN (*n*): perkaglycerine (a glycerine substitute) 366
PERKARBONAT (*n*): percarbonate 367
PERKOHLENSAUER: percarbonate 368
PERKOHLENSÄURE (*f*): percarbonic acid 369
PERKOLATION (*f*): percolation; displacement 370
PERKOLATOR (*m*): percolator 371
PERKRESAN (*n*): perkresan (a 50% cresol, water and soap mixture; a disinfectant) 372
PERKUSSION (*f*): percussion 373
PERKUSSIONSSATZ (*m*): percussion (detonating) composition; priming 374
PERKUSSIONSZÜNDHÜTCHEN (*n*): percussion cap 375
PERLARTIG: pearly; beadlike; pearl-like; nacreous 376
PERLASCHE (*f*): pearl-ash; refined potash; impure potassium carbonate (see **KALIUM-CARBONAT**) 377
PERLE (*f*): pearl (Sg. 2.65); bead; bubble; drop 378
PERLEN: to form bubbles or drops; sparkle; glisten-(like pearls); (as an adjective = pearl; of pearl) 379
PERLENÄHNLICH: (see **PERLARTIG**): 380
PERLENARTIG: (see **PERLARTIG**) 381
PERLENAUSTER (*f*): pearl oyster 382
PERLENDRUCK (*m*): pearl type (Printing) 383
PERLENFISCHEREI (*f*): pearl (fishery) fishing 384
PERLENGERSTE (*f*): pearl barley 385
PERLENGLANZ (*m*): pearly (nacreous) lustre 386
PERLENSAMEN (*m*): seed pearl 387
PERLFARBIG: pearl-coloured 388
PERLFÖRMIG: pearly, nacreous, bead-like, in pea crystals 389
PERLGLANZ (*m*): (see **PERLENGLANZ**) 390
PERLGLIMMER (*m*): margarite (Mineral) (see **MARGARIT**) 391
PERLGLIMMER (*m*), **AXOTOMER-**: pyrosmalite; Sg. 3.1 392
PERLGLIMMER (*m*), **HEMIPRISMATISCHER-**: margarite 393
PERLGRAUPEN (*f.pl*): pearl barley 394
PERLIG: pearly; nacreous 395
PERLIT (*n*): perlite (a fine mixture of **FERRIT** and **ZEMENTIT**, which see); (properties: magnetizability, lower specific weight than austenite. These are dependent upon the chemical composition and formation of its structural constituents) 396
PERLKOHLE (*f*): pearl (pea) coal (Size 4-10 mm.) 397
PERLMOOS (*n*): carrageen; pearl moss; Irish moss; Killeen; Pig wrack (*Chondrus crispus*) 398
PERLMUTTER (*f*): mother-of-pearl 399
PERLMUTTERARTIG: like mother-of-pearl 400
PERLMUTTERBLECH (*n*): crystallized tin-plate; moiré metallique 401
PERLMUTTERGLANZ (*m*): nacreous (mother-of-pearl) lustre 402
PERLMUTTERGLÄNZEND: pearly; of nacreous (mother-of-pearl) lustre 403
PERLMUTTERPAPIER (*n*): mother-of-pearl (nacreous) paper 404
PERLROHR (*n*): tube filled with beads 405
PERLSAGO (*m*): pearl sago 406
PERLSALZ (*n*): microcosmic salt; sodium ammonium (di)-phosphate (see **AMMONIUM-NATRIUMPHOSPHAT, SAURES-**); sodium phosphate (see **NATRIUM PHOSPHORSAURES-**) 407
PERLSAMEN (*m*): seed pearl 408
PERLSCHRIFT (*f*): pearl (Typ.) 409
PERLSINTER (*m*): opal (see **OPAL**) 410
PERLSPAT (*m*): pearl spar, dolomite (pearly or with a vitreous lustre) (see **DOLOMIT**) 411
PERLSTEIN (*m*): perlite; adularia (see **ADULAR**) 412
PERLSUCHT (*f*): murrain (of cattle) 413
PERLTEE (*m*): gunpowder tea 414
PERLWEISS (*n*): pearl white (see **WISMUTWEISS**) 415
PERMANENT: permanent 416
PERMANENTFARBE (*f*): permanent (colour, stain) dye 417
PERMANENTGRÜN (*n*): permanent green; chrome green (see **CHROMGRÜN**) 418
PERMANENTWEISS (*n*): permanent white, blanc fixé, barium sulphate; $BaSO_4$ 419
PERMANGANAT (*n*): permanganate 420
PERMEABEL: permeable 421
PERMEABILITÄT (*f*): permeability (μ) (Magnetism) 422
PERMEABILITÄTSCHWANKUNG (*f*): fluctuation in permeability 423
PERMEIEREN: to permeate 424
PERMONIT (*n*): permonite (a safety explosive consisting of Trinitrotoluene, ammonium nitrate and potassium perchlorate) 425
PERMUTATION (*f*): permutation 426
PERMUTIEREN: to permute (Maths.); exchange 427
PERMUTIERUNG (*f*): permutation 428
PERMUTIT (*m*): permutite (artificial zeolite, i.e., aluminium silicate; as natrolite, $Na_2Al_2Si_3O_{10}.2H_2O$., used for filtration) 429
PERMUTITVERFAHREN (*n*): permutite process 430
PERNAMBUKHOLZ (*n*): Brazil wood (see **FERNAM-BUKHOLZ**) 431
PERNICIÖS: pernicious 432
PERONIN (*n*): peronin, benzylmorphine hydrochloride (a narcotic) 433
PERONOSPORA (*f*): a grape-vine fungus (*Peronospora viticola*) 434
PEROWSKIT (*m*): perowskite (a rare isometric titanate); $(Ca,Fe)TiO_3$; Sg. 4.0 435
PEROXYD (*n*): peroxide (see **WASSERSTOFFSUPER-OXYD**) 436
PEROXYDOL (*n*): peroxydol, sodium perborate (see **NATRIUMPERBORAT**) (the term is most generally employed when it has reference to the use of sodium perborate as a bleaching agent) 437
PEROZID (*n*): perocid (mainly the crude sulphate of Cerite earths) 438
PERPENDIKEL (*n* and *m*): perpendicular-(line), vertical line, pendulum, upright 439
PERPENDIKULAR: perpendicular-(ly) 440
PERROTINE (*f*): perrotine (multicolour printing machine) 441
PERROTINENDRUCK (*m*): perrotine (multicolour) printing 442
PERSALZ (*n*): per-salt (as perborate, percarbonate, perchlorate, permanganate and persulphate) 443
PERSÄURE (*f*): peracid 444
PERSCHWEFELSÄURE (*f*): persulphuric acid 445

PERSENNING (*f*): tarpaulin 446
PERSIL (*n*): persil (a washing powder containing 10% sodium perborate) 447
PERSIMMONÖL (*n*): persimmon (date plum) oil (from *Diospyros virginiana*) 448
PERSIO (*m*): persio, red indigo (an archil preparation; see ORSEILLE) (a powder obtained by drying archil extract; see ORSEILLEEXTRAKT) 449
PERSISCH: Persian 450
PERSISCHROT (*n*): Persian red (see CHROMROT) 451
PERSON (*f*): person-(age); character; rôle; individual 452
PERSONAL (*n*): staff; assistants; complement; personnel; members; company; crew 453
PERSONALABGABE (*f*): poll-tax 454
PERSONALIEN (*pl*): personalities 455
PERSONENAUFZUG (*m*): passenger lift 456
PERSONENDAMPFER (*m*): passenger (boat) steamer 457
PERSONENNAME (*m*): noun personal 458
PERSONENSTEUER (*f*): poll tax 459
PERSONENVERWECHSELUNG (*f*): mistaken identity 460
PERSONENZUG (*m*): passenger train 461
PERSONENZUGLOKOMOTIVE (*f*): passenger engine 462
PERSONENZUGVERKEHR (*m*): passenger traffic (Railways) 463
PERSONIFIKATION (*f*): personification, impersonation 464
PERSONIFIZIEREN: to personify; impersonate 465
PERSONIFIZIERUNG (*f*): personification, impersonation 466
PERSÖNLICH: personal-(ly); in person 467
PERSÖNLICHKEIT (*f*): personality 468
PERSPEKTIV (*n*): perspective; telescope, opera-glass 469
PERSPEKTIVE (*f*): perspective 470
PERSPEKTIVE (*f*), PARALLEL-: parallel perspective 471
PERSPEKTIVISCH: perspective-(ly), prospective-(ly), in prospect, in perspective 472
PERSULFAT (*n*): persulphate 473
PERSULFOCYANSÄURE (*f*): persulphocyanic acid; perthiocyanic acid 474
PERSULFOMOLYBDÄNSÄURE (*f*): persulfomolybdic acid; thiopermolybdic acid 475
PERTINENZIEN (*pl*): appurtenances; adjuncts 476
PERTÜRKOL (*n*): pertürkol (trade name of a textile cleaning and grease removing agent) 477
PERUANISCH: Peruvian 478
PERUBALSAM (*m*): Peruvian balsam, black balsam (*Balsamum peruvianum*) (obtained from the tree *Myroxylon pereiræ*); Sg. 1.135-1.15 479
PERUBALSAM, KÜNSTLICHER- (*m*): synthetic Peru-(vian) balsam; perugene 480
PERUBALSAMÖL (*n*): Peru-(vian) balsam oil 481
PERUBALSAM (*m*), SYNTHETISCHER-: synthetic Peru balsam, perugene 482
PERÜCKENBAUM (*m*): young fustic (*Rhus cotinus*) 483
PERÜCKENSUMACH (*m*): young fustic (*Rhus cotinus*) 484
PERUGEN (*n*): perugene; synthetic Peru-(vian) balsam 485
PERUGENSEIFE (*f*): perugene soap 486
PERUOL (*n*): peruol (a 25% benzyl benzoate solution in castor oil) 487
PERURINDE (*f*): Peruvian bark; (yellow)-cinchona bark; Jesuit's bark; calisaya bark (bark of *Cinchona calisaya*) 488
PERUSKABIN (*n*): peruscabine, benzyl benzoate (see BENZOESÄUREBENZYLESTER) 489
PESSIMISTISCH: pessimistic 490
PEST (*f*): pest; blight; pestilence; plague (general term for any of the forms of plague; Bubonic, Oriental, Glandular, Inguinal, Indian, Levantine, Pali); septic pestilence, oriental typhus (Med.); nuisance 491
PESTARTIG: pestilential, contagious 492
PESTBEULE (*f*): plague spot 493
PESTFLECKEN (*m*): plague spot 494
PESTHAFT: pestilential 495
PESTILENZ (*f*): pestilence (see PEST) 496
PESTILENZIALISCH: pestilential 497
PESTILENZKRAUT (*n*): goat's rue (*Galega officinalis*) 498
PESTKRANK: plague-stricken 499
PETALINSPAT (*m*), PRISMATISCHER-: petalite 500
PETALIT (*m*): petalite (an aluminium-lithium metasilicate) (a lithium mineral of rare occurrence); $(Li,Na)Al(Si_2O_5)_2$; Sg. 2.45 501
PETARDE (*f*): petard 502
PETECHIEN (*pl*): petachiæ, purple spots on the skin due to minute hæmorrhages (Med.) 503
PETERSILIE (*f*): parsley (*Petroselinum sativum*) 504
PETERSILIENKRAUT (*n*): parsley-(herb); (*Herba petroselini*) 505
PETERSILIENÖL (*n*): parsley oil; Sg. 1.07 (distilled from the fruit of parsley; see PETERSILIE) 506
PETERSILIENSAMEN (*m*): parsley seed (see PETERSILIE) 507
PETERSILIENSCHIERLING (*m*): fool's parsley (*Aethusa cynapium*) 508
PETERSILIENWURZEL (*f*): parsley root (see PETERSILIE) 509
PETERSVOGEL (*m*): stormy petrel 510
PETIT (*f*): brevier (Typ.) 511
PETITGRAINÖL (*n*): petitgrain oil (*Oleum petitgrain*) (obtained from fruit, flowers and leaves of *Citrus bigaradia*); Sg. 0.89 512
PETITION (*f*): petition 513
PETITIONIEREN: to petition, request, address, pray, supplicate, entreat, sue, apply, solicit, appeal, crave, beg, ask, make application, prefer a request 514
PETREFAKT (*m*): fossil 515
PETREFAKTENKUNDE (*f*): palæontology (the science of fossils) (see PALÄONTOLOG) 516
PETRIFIZIEREN: to petrify, turn (change, convert) into stone, fossilize 517
PETRISCHALE (*f*): Petri dish 518
PETRISCHES SCHÄLCHEN (*n*): Petri dish 519
PETROGRAPHISCH: petrographic 520
PETROKLASTIT (*n*): petroclastite (a safety explosive, consisting of pitch, potassium bichromate, saltpeter and sulphur; only explodes in a confined space) 521
PETROLÄTHER (*m*): petroleum ether; Canadol; Sherwood oil; Sg. 0.65, Bp. 40-70°C. (see also PENTAN and KANADOL) 522
PETROLEUM (*n*): petroleum, petrol (see BERGÖL and ERDÖL); Sg. 0.88 523
PETROLEUMASPHALT (*m*): petroleum (pitch) asphalt (a black bituminous mass from the

residue of petroleum distillation); (see also ASPHALT) 524
PETROLEUMÄTHER (*m*): (see PETROLÄTHER and BENZIN) 525
PETROLEUMBEHÄLTER (*m* and *pl*): petroleum (petrol) reservoir-(s) or (storage)-tank-(s) 526
PETROLEUMBENZIN (*n*): petroleum benzine; safety (Danforth) oil; Sg. 0.667-0.707; Bp. 80-100°C. (see also BENZIN) 527
PETROLEUMDESTILLATIONSANLAGE (*f*): petroleum distilling plant 528
PETROLEUMDESTILLATIONSAPPARAT (*m*): petroleum distilling apparatus 529
PETROLEUMFARBE (*f*): petroleum colour 530
PETROLEUMFEUERUNG (*f*): petroleum furnace 531
PETROLEUMKOCHER (*m*): petroleum (cooker) cooking stove; oil stove 532
PETROLEUMMASCHINE (*f*): petrol-engine 533
PETROLEUMPRÜFER (*m*): petroleum tester 534
PETROLEUMRAFFINERIE (*f*): petroleum refinery 535
PETROLEUMRAFFINIERANLAGE (*f*): petroleum refining plant 536
PETROLEUMTANKANLAGE (*f*): petroleum tank-(plant) 537
PETROLEUMUNTERSUCHUNGSAPPARAT (*m*): petroleum testing apparatus 538
PETROLEUMWERK (*n*): petroleum works 539
PETROLKOKS (*m* and *pl*): petrol coke; oil coke; petroleum coke 540
PETROLPECH (*n*): petroleum pitch (see PETROLEUM-ASPHALT) 541
PETROLSÄURE (*f*): petrolic acid 542
PETSCHAFT (*n*): signet; seal 543
PETSCHE (*f*): drying (frame) room 544
PETSCHIER (*n*): signet; seal 545
PETSCHIEREN: to seal 546
PETTO (*m*), IN-HABEN: to retain, have (keep) in reserve, keep to oneself 547
PETZEN: to give information concerning, inform against 548
PETZIT (*m*): petzite (see TELLURGOLDSILBER); $(Ag,Au)_2Te$; Sg. 9.0 549
PFAD (*m*): path; way; footpath; track 550
PFADEISEN (*n*): socket, pivot 551
PFADGERECHTIGKEIT (*f*): right of way 552
PFADLOS: pathless; trackless 553
PFAFF (*m*): rivet stamp or driver; underlet (Brewing) (see also PFAFFE) 554
PFAFFE (*m*): priest; parson 555
PFAFFENBAUM (*m*): spindle tree; wahoo (*Euonymus atropurpureus*) 556
PFAFFENHÜTCHEN (*n*): wahoo (see PFAFFEN-BAUM) 557
PFAFFENSTIEL (*m*): trifle 558
PFAHL (*m*): pile; stake; stick; post; pole; pale; prop; picket 559
PFAHLBAU (*m*): fencing; paling; piling; lake dwelling (house or hut raised on piles above the water) 560
PFÄHLEN: to empale; prop; stake; picket; fence 561
PFAHLHECKE (*f*): paling; fencing; palisade; stockade 562
PFAHLMAST (*m*): pole mast (mast in one piece) 563
PFAHLRAMME (*f*): pile-driver 564
PFAHLSCHUH (*m*): foot, shoe (at the bottom of a post or pile) 565
PFAHLSTICH (*m*): bowline hitch (Naut.) 566
PFAHLWERK (*n*): paling; piling; fencing; palisade; stockade 567
PFAHLWURM (*m*): ship worm (*Teredo navalis*) 568
PFAHLWURZEL (*f*): taproot (see PFEILWURZEL) 569
PFAHLZAUN (*m*): (see PFAHLHECKE) 570
PFALZ (*f*): Palatinate 571
PFALZGRAFSCHAFT (*f*): Palatinate 572
PFAND (*m*): pledge; forfeit; pawn; gage; security; mortgage; deposit; guarantee; earnest; surety 573
PFANDBAR: capable of being pledged 574
PFÄNDBAR: distrainable 575
PFANDBELASTUNG (*f*): mortgage 576
PFANDBRIEF (*m*): mortgage; bond; bill of sale 577
PFÄNDEN: to seize; pledge; fine; distrain; take as a pledge 578
PFÄNDERSPIEL (*n*): game of forfeits 579
PFANDGEWÄHR (*f*): mortgage security 580
PFANDGLÄUBIGER (*m*): mortgagee 581
PFANDGUT (*n*): mortgage 582
PFANDHAUS (*n*): pawnshop; pawnbroker's 583
PFANDINHABER (*m*): mortgagee 584
PFANDKONTRAKT (*m*): mortgage; bill of sale; bond 585
PFANDLEIHER (*m*): pawnbroker 586
PFANDLÖSUNG (*f*): redemption of a pledge 587
PFANDRECHTLICH: hypothecary 588
PFANDSASS (*m*): mortgagee 589
PFANDSATZ (*m*): mortgage 590
PFANDSCHEIN (*m*): pawn ticket 591
PFÄNDUNG (*f*): distraint, seizure; packing (piece) block 592
PFANDVERSCHREIBUNG (*f*): mortgage deed 593
PFANDWEISE: by (pawn) mortgage 594
PFANNE (*f*): pan; copper; boiler; pantile; bearing; bush (Mech.); socket; acetabulum (the socket in the pelvis into which the head of the femur fits) (Anat.) 595
PFANNE (*f*), LAGER-: pillow, pillow bush, bearing (bush, step) brass 596
PFANNENFLICKER (*m*): tinker 597
PFANNENFÜGUNG (*f*): trochoid (Anat.) (a rotary or pivotal joint) 598
PFANNENGESTELL (*n*): trivet 599
PFANNENGRUBE (*f*): trochoidal cavity (cavity of the cotyle) 600
PFANNENHAUS (*n*): salt-works 601
PFANNENHEERD (*m*): salt-works 602
PFANNENKUCHEN (*m*): pancake; omelette; round flat lump of fæces-(floating on sewage water) 603
PFANNENMEISTER (*m*): salt-works inspector (see also PFÄNNER) 604
PFANNENSCHNAUZE (*f*): pan lip 605
PFANNENSTEIN (*m*): pan (fur) scale (Salt); $CaSO_4.Na_2SO_4$ and NaCl; boiler scale 606
PFANNENWERK (*n*): salt works 607
PFANNENZIEGEL (*m*): pantile 608
PFÄNNER (*m*): salt manufacturer; proprietor of a salt works 609
PFANNE (*f*), SPURLAGER-: step brass 610
PFARRACKER (*m*): glebe land 611
PFARRBEZIRK (*m*): parish 612
PFARRBUCH (*n*): parish register 613
PFARRE (*f*): living; parsonage; parish 614
PFARRER (*m*): parson; priest; cleric; clergyman; minister; incumbent 615
PFARRGUT (*n*): (see PFARRACKER and PFARR-WIESE) 616
PFARRHUFE (*f*): (see PFARRACKER and PFARR-WIESE) 617
PFARRKIRCHE (*f*): parish church 618

PFARRLEHEN (*n*): advowson; living 619
PFARRWIESE (*f*): glebe-meadow 620
PFAU (*m*): peacock 621
PFAUBLAU (*n*): peacock blue 622
PFAUENAUGE (*n*): peacock's eye; peacock butterfly (*Vanessa Io*); emperor moth (*Saturnia carpini*) 623
PFAUENFEDER (*f*): peacock's feather 624
PFEFFER (*m*): pepper 625
PFEFFER, GANZER- (*m*): whole pepper 626
PFEFFER, GESTOSSENER- (*m*): ground pepper 627
PFEFFERGEWÄCHSE (*n.pl*): *Piperaceæ* (Bot.) 628
PFEFFERIG: peppery 629
PFEFFERKORN (*n*): pepper-corn 630
PFEFFERKRAUT (*n*): peppergrass (*Lepidium*); summer savory (*Satureia hortensis*) (*Herba satureiæ*); stonecrop (*Sedum*) 631
PFEFFERKRAUTÖL (*n*): summer savory oil (from *Satureia hortensis*) 632
PFEFFERKUCHEN (*m*): gingerbread 633
PFEFFERKÜMMEL (*m*): cumin (see **KÜMMEL, RÖMISCHER-**) 634
PFEFFERLADE (*f*): pepper (spice) box 635
PFEFFERMINZE (*f*): peppermint (both plant and product) (*folia et herba Menthæ piperitæ*); brandy mint; lamb mint 636
PFEFFERMINZGERUCH (*m*): peppermint (odour) aroma 637
PFEFFERMINZÖL (*n*): peppermint oil (*Oleum menthæ piperitæ*) (an ethereal oil from *Mentha piperita*); Sg. 0.895-0.926 638
PFEFFERMINZÖL (*n*), **ENGLISCHES-**: English peppermint oil, Mitcham peppermint oil (the finest variety of peppermint oil) 639
PFEFFERMINZPFLANZE (*f*): peppermint plant (*Mentha piperita*) 640
PFEFFERMINZTROPFEN (*m.pl*): essence of peppermint 641
PFEFFERMÜHLE (*f*): pepper mill 642
PFEFFERMÜNZE (*f*): (see **PFEFFERMINZE**) 643
PFEFFERN: to pepper 644
PFEFFERÖL (*n*): (black)-pepper oil; Sg. 0.87-0.905 (extracted from *Piper nigrum*); savory oil (see **PFEFFERKRAUTÖL**) 645
PFEFFER (*m*), **SPANISCHER-**: annual capsicum (*Capsicum annuum*) 646
PFEFFERSTAUDE (*f*): pepper (tree) bush (see **PFEFFERSTRAUCH**) 647
PFEFFERSTEIN (*m*): peperine 648
PFEFFERSTRAUCH (*m*): pepper (tree) bush (*Piper*) 649
PFEIFE (*f*): pipe; whistle; fife; tube; (honey-comb)-cell 650
PFEIFE (*f*), **DAMPF-**: steam whistle, whistle alarm 651
PFEIFE (*f*), **GEDECKTE-**: stopped pipe (of organs) 652
PFEIFEN: to pipe; whistle 653
PFEIFENBAUM (*m*): lilac 654
PFEIFENBESCHLAG (*m*): mounting of a pipe 655
PFEIFENERDE (*f*): pipe clay (see **PFEIFENTON**) 656
PFEIFENFORM (*f*): pipe-shape; pipe-mould (of a tobacco-pipe) 657
PFEIFENFÖRMIG: tubular; pipe-shaped 658
PFEIFENKOPF (*m*): bowl (of a pipe) 659
PFEIFENROHR (*n*): pipe stem 660
PFEIFENSPITZE (*f*): mouth-piece (of a pipe) 661
PFEIFENSTRAUCH (*m*): Dutchman's pipe (*Aristolochia sipho*); mock orange; (white-)syringa (*Philadelphus coronarius*) 662
PFEIFENTON (*m*): pipeclay; kaolin (see **KAOLIN** and **BOLUS, WEISSER-**) 663
PFEIFE (*f*), **OFFENE-**: open pipe (of organs) 664
PFEIFER (*m*): piper; whistler 665
PFEIL (*m*): arrow; dart; shaft; bolt 666
PFEILEISEN (*n*): arrow-head 667
PFEILER (*m*): pillar; pier; upright; column; jamb; support; prop 668
PFEILERSPIEGEL (*m*): pier-glass 669
PFEILERSTEIN (*m*): basalt (see **BASALT**) 670
PFEILERWEITE (*f*): width between pillars; distance between columns 671
PFEILFÖRMIG: arrow-shaped; sagittate 672
PFEILGESCHWIND: (see **PFEILSCHNELL**) 673
PFEILGIFT (*n*): arrow poison (used by the Indians, etc., for poisoning the tips of their arrows) 674
PFEILHÖHE (*f*): rise (of an arch) 675
PFEILKÖCHER (*m*): quiver 676
PFEILKRAUT (*n*): arrowhead (Bot.) (*Sagittaria sagittæfolia*) 677
PFEILSCHNELL: swift as an arrow 678
PFEILSPITZE (*f*): arrow-head 679
PFEILSTEIN (*m*): belemnite (Geol.) 680
PFEILWURZ (*f*): taproot; arrowroot (*Maranta arundinacea*) 681
PFEILWURZEL (*f*): (see **PFEILWURZ**) 682
PFEILWURZELKLEBER (*m*): arrowroot moutant, permanent paste (Photography) (see **KLEBER**) 683
PFEILWURZELMEHL (*n*): arrow-root-(starch or flour) 684
PFEILZEICHEN (*n*): arrow-(sign) 685
PFENNIG (*m*): pfennig (smallest German coin, one hundred to the Mark) 686
PFENNIGKRAUT (*n*): (see **HELLERKRAUT**) 687
PFERCH (*m*): pen; fold; sheep-dung 688
PFERCHE (*f*): pen; fold 689
PFERCHEN: to pen; manure 690
PFERCHLAGER (*n*): sheep-pen, sheep-fold 691
PFERD (*n*): horse 692
PFERDEARZNEIKUNDE (*f*): veterinary surgery 693
PFERDEARZT (*m*): veterinary surgeon 694
PFERDEBAHN (*f*): horse tramway 695
PFERDEBÄNDIGER: horse-breaker, breaker-in of horses 696
PFERDEBESCHLAG (*m*): horse (shoeing) shoes 697
PFERDEBREMSE (*f*): horse-fly, gadfly (see **BREMSE**) 698
PFERDEDECKE (*f*): horse-rug 699
PFERDEDUNGMEHL (*n*): horse-dung meal 700
PFERDEFETT (*n*): horse-fat; Sg. 0.93-0.94; Mp. 20-38°C. 701
PFERDEHAAR (*n*): horse-hair 702
PFERDEHARNSÄURE (*f*): hippuric acid (see **HIPPURSÄURE**) 703
PFERDEHUF (*m*): horse's hoof 704
PFERDEKASTANIE (*f*): horse-chestnut (*Hippocastanum vulgare, Aesculus hippocastanum*) 705
PFERDEKRAFT (*f*): horse-power; H.P. (see also **PFERDESTÄRKE**) 706
PFERDEMARKT (*m*): horse-fair; mart-(for horses) 707
PFERDEMÄSSIG: equine 708
PFERDEMILCH (*f*): mare's milk 709
PFERDEMINZE (*f*): horsemint (*Monarda punctata*) 710
PFERDEMINZÖL (*n*): horsemint (monarda) oil (distilled from *Monarda punctata*); Sg. 0.92-0.94 711
PFERDEMIST (*m*): horse-dung 712
PFERDEPOLEI (*m*): (see **PFERDEMINZE**) 713
PFERDERENNEN (*n*): horse-racing 714

PFERDESCHWAMM (*m*): aphthæ (Vet.), wind-gall 715
PFERDESCHWEMME (*f*): horse-pond 716
PFERDESTÄRKE (*f*): (P.S.): horsepower (H.P.) 717
PFERDESTÄRKE (*f*), **EFFEKTIVE-**: (PS_e): effective horse power (EHP) 718
PFERDESTÄRKE (*f*), **INDIZIERTE-**: (PS_i): indicated horse power (IHP) 719
PFERDESTÄRKE (*f*), **NOMINELLE-**: nominal horse power 720
PFERDESTÄRKESTUNDE (*f*): horse-power-hour, H.P. hr. 721
PFERDESTRIEGEL (*f*): curry comb 722
PFERDETRÄNKE (*f*): horse-pond 723
PFERDEZOTE (*f*): fetlock 724
PFERDEZUCHT (*f*): horse breeding 725
PFERDIG: horse, horse-power, H.P., of — H.P. (the above word is usually prefixed by a number, thus: 2-PFERDIG=2 H.P.) 726
PFIFF (*m*) whistle; trick; knack; whiff 727
PFIFFIG: sly; crafty; cunning; shrewd; artful; wily; subtle; tricky 728
PFINGSTBLUME (*f*): peony (*Pæonia officinalis*) 729
PFINGSTEN (*n*): Whitsuntide; Pentecost 730
PFINGSTROSE (*f*): peony (see PÄONIE) 731
PFIRSCHE (*f*): (see PFIRSICH) 732
PFIRSICH (*m*): peach (*Prunus persica*) (both the tree and the fruit) 733
PFIRSICHÄTHER (*m*): peach essence 734
PFIRSICHBAUM (*m*): peach tree (*Prunus persica*) 735
PFIRSICHBLÜTE (*f*): peach blossom (of *Prunus persica* or *Amygdalus persica*) 736
PFIRSICHBRANNTWEIN (*m*): peach brandy 737
PFIRSICHE (*f*): (see PFIRSICH) 738
PFIRSICHHOLZ (*n*): peach wood 739
PFIRSICHKERN (*m*): peach (kernel) stone 740
PFIRSICHKERNÖL (*n*): peach kernel oil (from *Amygdalus persica*); Sg. 0.915-0.9285 [741
PFIRSICHÖL (*n*): peach oil 742
PFLANZBAR: plantable 743
PFLANZBÜRGER (*m*): planter; settler; colonist 744
PFLANZE (*f*): plant; (*pl*): plants; vegetation 745
PFLANZE (*f*), **ADER-**: moss 746
PFLANZE (*f*), **DROSSEL-**: fern 747
PFLANZEISEN (*n*): dibble 748
PFLANZE (*f*), **KNABEN-**: orchid 749
PFLANZEN: to plant; set-(up); erect; settle 750
PFLANZENALBUMIN (*n*): plant (albumin) albumen 751
PFLANZENALKALI (*n*): vegetable alkali 752
PFLANZENALKALOID (*n*): vegetable (plant) alkaloid 753
PFLANZEN (*f.pl*), **ALPEN-**: alpine plants 754
PFLANZENANATOMIE (*f*): plant anatomy 755
PFLANZENARTIG: vegetable; plant-like; phytoid 756
PFLANZENASCHE (*f*): vegetable (plant) ash-(es) 757
PFLANZENBASE (*f*): vegetable (plant) base or alkaloid 758
PFLANZEN (*f.pl*), **BEDECKTSAMIGE-**: *Angiospermæ* (Bot.) (see ANGIOSPERMEN) 759
PFLANZENBEET (*n*): bed-(of plants) 760
PFLANZENBESCHREIBUNG (*f*): description of (vegetation) plants; phytography 761
PFLANZENBLEICHEN (*n*): plant bleaching 762
PFLANZEN (*f.pl*), **BUSCH-**: bush plants 763
PFLANZENBUTTER (*f*): vegetable butter (see KOKOSBUTTER) 764
PFLANZENCHEMIE (*f*): phytochemistry; plant (vegetable) chemistry 765
PFLANZEN (*f.pl*), **EINKEIMBLÄTTRIGE-**: *Monocotyleæ* (Bot.) 766
PFLANZENEIWEISS (*n*): plant or vegetable (protein; albumin) albumen 767
PFLANZENERDE (*f*): vegetable mould; humus; garden mould (see HUMUS) 768
PFLANZENERNÄHRUNG (*f*): plant (fertilizing, manuring) nourishing or nourishment 769
PFLANZENERZIEHUNG (*f*): plant (culture) cultivation 770
PFLANZENEXEMPLAR (*n*): botanical specimen 771
PFLANZENFARBE (*f*): vegetable colour 772
PFLANZENFARBSTOFF (*m*): vegetable colouring matter, vegetable dye-(stuff) 773
PFLANZENFASER (*f*): vegetable (plant) fibre; cellulose (see ZELLULOSE) 774
PFLANZENFASERPAPIER (*n*): paper from vegetable fibre 775
PFLANZENFASERSTOFF (*m*): vegetable (gluten; cellulose) fibrin 776
PFLANZEN (*f.pl*), **FELD-**: field plants 777
PFLANZENFETT (*n*): vegetable fat 778
PFLANZENFIBRIN (*n*): (see PFLANZENFASERSTOFF) 779
PFLANZENFORSCHER (*m*): botanist 780
PFLANZENFRESSEND: herbivorous; graminivorous; plant-eating; phytophagous 781
PFLANZENGALLERTE (*f*): vegetable jelly (jelly from fruit, mosses or algæ) 782
PFLANZENGARTEN (*m*): botanic-(al) garden 783
PFLANZENGEOGRAPHIE (*f*): plant geography 784
PFLANZENGEWEBE (*n*): (see ZELLGEWEBE) 785
PFLANZENGRÜN (*n*): chlorophyl (see CHLOROPHYLL) 786
PFLANZENGUMMI (*n*): vegetable gum (see DEXTRIN) 787
PFLANZENHAARE (*n.pl*): plant hairs 788
PFLANZENHANDEL (*m*): kitchen (vegetable, nursery) gardening 789
PFLANZENHAUS (*n*): conservatory; greenhouse 790
PFLANZEN (*f.pl*), **HEIDE-**: heath plants 791
PFLANZENKASEIN (*n*): legumin; vegetable casein 792
PFLANZENKEIM (*m*): plant germ 793
PFLANZENKENNER (*m*): botanist 794
PFLANZENKENNTNIS (*f*): botany; knowledge of plants 795
PFLANZENKOHLE (*f*): charcoal 796
PFLANZENKÖRPER (*m*): vegetable substance 797
PFLANZENKRANKHEIT (*f*): plant disease 798
PFLANZENKUNDE (*f*): botany; phytology 799
PFLANZEN (*f.pl*), **LAUBABWERFENDE-**: deciduous plants 800
PFLANZENLAUGENSALZ (*n*): potash 801
PFLANZENLEBEN (*n*): vegetable life 802
PFLANZENLEHRE (*f*): botany; phytology 803
PFLANZENLEIM (*m*): gliadin; gluten; vegetable glue; aparatine (from hot treatment of starch with caustic soda) (see also COLLODIN) 804
PFLANZENMILCH (*f*): vegetable milk, latex (see LATEX) 805
PFLANZENMISCHLING (*m*): bastard plant 806
PFLANZENMORPHOLOGIE (*f*): plant morphologie (science of the structure of plants) 807
PFLANZEN (*f.pl*), **NACKTSAMIGE-**: *Gymnospermæ* (Bot.) 808

PFLANZEN (*f.pl*), NADELHOLZ- : coniferous plants 809

PFLANZENNÄHRSALZ (*n*): food (feeding; fertilizing) salt for plants 810

PFLANZENNÄHRSTOFF (*m*): plant (food, fertiliser) manure 811

PFLANZENÖL (*n*): vegetable oil 812

PFLANZENORGANOGRAPHIE (*f*): plant organography (science of the exterior structure of plants; description of the organs of plants) 813

PFLANZENPALÄONTOLOGIE (*f*): plant palæontology (the science of fossil plants) 814

PFLANZENPAPIER (*n*): botanical demy (Paper); Court plaster (see PFLASTER, ENGLISCHES-) 815

PFLANZENPATHOLOGIE (*f*): plant pathology (science of plant diseases) 816

PFLANZENPECH (*n*): vegetable pitch 817

PFLANZENPHYSIOLOGIE (*f*): plant physiology 818

PFLANZENPHYTOTOMIE (*f*): (plant)-phytotomy (plant anatomy, vegetable anatomy, dissection of plants) 819

PFLANZEN (*f.pl*), PRÄRIE- : prairie plants 820

PFLANZENREICH (*n*): vegetable kingdom 821

PFLANZENSAFT (*m*): vegetable (plant) juice; sap 822

PFLANZENSALZ (*n*): potash 823

PFLANZENSAUGER (*m*): plant sucker; parasitic plant 824

PFLANZENSÄURE (*f*): vegetable acid 825

PFLANZEN (*f.pl*), SAVANNEN- : savannha plants 826

PFLANZENSCHLEIM (*m*): mucilage 827

PFLANZENSCHMALZ (*n*): (see PFLANZENBUTTER and KOKOSÖL) 828

PFLANZENSCHUTZMITTEL (*n*): plant (protective; protector; preserver) preservative 829

PFLANZENSEIDE (*f*): vegetable silk (seed hairs of *Asclepias cornuti*; *Calotropis gigantea*, etc.) 830

PFLANZENSEKRET (*n*): plant secretion 831

PFLANZENSTEIN (*m*): phytolite 832

PFLANZEN (*f.pl*), STEPPEN- : steppe plants 833

PFLANZENSTOFF (*m*): vegetable matter 834

PFLANZENSYNONYMIK (*f*): plant synonyms 835

PFLANZENTALG (*m*): vegetable tallow (see PFLANZENWACHS; TALG, CHINESISCHER-; MALABARTALG; PINEYTALG; VATERIATALG; VIROLAFETT) 836

PFLANZENTERATOLOGIE (*f*): plant teratology (science of the malformation of plants or vegetable monstrosities) 837

PFLANZENTIER (*n*): zoophyte 838

PFLANZEN (*f.pl*), TROPISCHE- : tropical plants 839

PFLANZEN (*f.pl*), URWALD- : virgin forest plants 840

PFLANZENVERMEHRUNG (*f*): propagation of plants 841

PFLANZENWACHS (*n*): vegetable wax (see MYRICAWACHS; MYRTLEWACHS; MYRTENWACHS; KARNAUBAWACHS; CEREAWACHS; PALMWACHS; PELA; WACHS, CHINESISCHES-; WACHS, JAPANISCHES-) 842

PFLANZENWACHSTUM (*n*): vegetation; plant growth 843

PFLANZENWECHSEL (*m*): rotation of crops 844

PFLANZENWELT (*f*): vegetable kingdom; plant world 845

PFLANZEN (*f.pl*), WIESEN- : meadow plants 846

PFLANZENZELLE (*f*): vegetable (plant) cell 847

PFLANZENZELLENSTOFF (*m*): cellulose (see ZELLULOSE) 848

PFLANZENZUCHT (*f*): plant cultivation 849

PFLANZENZÜCHTUNG (*f*): plant (cultivation) culture 850

PFLANZEN (*f.pl*), ZWEIKEIMBLÄTTRIGE- : *Diocotyleæ* (Bot.) 851

PFLANZER (*m*): planter; settler; colonist; plantation owner 852

PFLANZLICH : vegetable 853

PFLANZORT (*m*): settlement 854

PFLANZUNG (*f*): planting; plantation; settlement; settling; colony; estate; colonization 855

PFLASTER (*n*): plaster; pavement; paving 856

PFLASTERBANDSTEIN (*m*): curbstone-(of a pavement) 857

PFLASTEREINFASSUNG (*f*): curbing (of pavements) 858

PFLASTER (*n*), EINGELEGTES- : flag-stones, flagged (tesselated) pavement 859

PFLASTER, ENGLISCHES- (*n*): Court plaster 860

PFLASTERER (*m*): pavior; plasterer 861

PFLASTERN : to plaster; pave 862

PFLASTERRAMME (*f*): ram-(mer) (for paving stones) 863

PFLASTERSTEIN (*m*): paving stone 864

PFLASTERSTÖSSEL (*m*): (see PFLASTERRAMME) 865

PFLASTERZIEGEL (*m*): paving (tile) brick 866

PFLATSCHEN : to pad (Calico) 867

PFLATSCHFARBE (*f*): padding liquor (Calico) 868

PFLAUME (*f*): plum; prune 869

PFLAUME (*f*), DAMASZENER- : damson (see KRIECHE and SYRISCHE PFLAUME) 870

PFLAUME, GETROCKNETE- (*f*): prune; dried plum (of *Prunus domestica*) 871

PFLAUMENBAUM (*m*): plum tree, wild plum (*Prunus domestica*) 872

PFLAUMENBAUMHOLZ (*n*): plum tree wood (Sg. 0.6-0.9) (see PFLAUMENBAUM) 873

PFLAUMENGEIST (*m*): (see SLIBOWOTZ) 874

PFLAUMENKERN (*m*): plum (kernel) stone 875

PFLAUMENKERNÖL (*n*): plum kernel oil (from *Prunus domestica* and *Prunus damascæna*); Sg. 0.92 876

PFLAUMENMUS (*n*): plum jam; stewed plums 877

PFLAUME (*f*), TÜRKISCHE- : Turkish plum (*Prunus cerasifera*) 878

PFLEGE (*f*): care; nursing; fostering; education; rearing; administration; superintendence; encouragement; cultivation 879

PFLEGEBEFOHLENE (*m* and *f*): ward 880

PFLEGEKIND (*n*): foster-child 881

PFLEGELOS : uncared for; neglected 882

PFLEGEN : to attend to; care for; tend; foster; nurse; cultivate; rear; be given to; be accustomed; (be)-use-(d) to; be wont; indulge; administer 883

PFLEGER (*m*): foster-parent; trustee; tutor; guardian 884

PFLICHT (*f*): duty; obligation 885

PFLICHTANKER (*m*): sheet anchor 886

PFLICHTBRUCH (*m*): disloyalty; perjury 887

PFLICHTENLEHRE (*f*): ethics; morals 888

PFLICHTFREI : free from obligation 889

PFLICHTIG : obliged, compelled by duty, bound; (often used as a suffix with same meanings) 890

PFLICHTLOS : disloyal, undutiful 891

PFLICHTMÄSSIG: according to the dictates of (duty) conscience 892
PFLICHTSCHULDIG: obligatory; in duty bound 893
PFLICHTVERLETZUNG (*f*): dereliction of duty, breach of duty 894
PFLICHTVERSÄUMNIS (*f*): neglect of duty 895
PFLOCK (*m*): plug; peg; pin 896
PFLOCKBOHRER (*m*): borer, piercer (Mineralogy) 897
PFLÖCKEN: to plug; peg; pin; fasten 898
PFLOCKSCHIESSEN (*n*): blasting (Mineralogy) 899
PFLÜCKEN: to pluck; pick; gather 900
PFLUG (*m*): plough; drag (Dredging) 901
PFLÜGBAR: arable 902
PFLUG (*m*), DAMPF-: steam plough 903
PFLUGEISEN (*n*): ploughshare 904
PFLÜGEN: to plough; drag 905
PFLUG (*m*), HAND-: hand plough 906
PFLUGLAND (*n*): arable land, ploughed land 907
PFLUGMOTOR (*m*): plough motor, motor for a plough 908
PFLUG (*m*), MOTOR-: motor plough 909
PLFUGSCHAR (*f*): ploughshare; (os)-vomer (Anat.) 910
PFLUGSCHARBEIN (*n*): (os-)vomer (Anat.) 911
PFORTADER (*f*): portal vein 912
PFORTE (*f*): gate; door; entrance; port-(al); port-hole 913
PFORTENDECKEL (*m*): port (cover) lid 914
PFORTENHANGER (*m*): port pendant 915
PFORTGAT (*n*): port-hole (Naut.) 916
PFÖRTNER (*m*): gatekeeper; doorkeeper; porter; pylorus (Anat.) 917
PFOSTE (*f*): (see PFOSTEN) 918
PFOSTEN (*m*): post; stake; pole; pier; jamb; scantling 919
PFRIEM (*m*): punch; awl; bodkin 920
PFRIEME (*f*): (see PFRIEM) 921
PFRIEMEN (*m*): (see PFRIEM); broom (*Cytisus scoparius*); (as a verb = to punch) 922
PFRIEMENGELD (*n*): primage (Com.) 923
PFRIEMENGRAS (*n*): feather-grass (*Stipa pennata*) 924
PFRIEMENHOLZ (*n*): broom (see PFRIEMEN) 925
PFRIEMENKRAUT (*n*): broom (*Cytisus scoparius*) 926
PFROPF (*m*): stopper; bung; plug; wad; cork; scion; graft; thrombus (Med.) 927
PFROPFEN: to stopper; stuff; cram; plug; bung; cork; graft; (*m*): stopper; bung; cork; graft; tap; plug; wad; thrombus (Med.) 928
PFROPFENZIEHER (*m*): corkscrew 929
PFROPFMESSER (*n*): grafting knife; grafter 930
PFROPFREIS (*n*): graft 931
PFUHL (*m*): slough; pool; puddle 932
PFÜHL (*m* and *n*): bolster; pillow; cushion 933
PFÜHLEN: to cushion; bed 934
PFUHLICHT: boggy, sloughy; marshy 935
PFUND (*n*): pound (Sterling or weight); lb.; £; (as a prefix—often means stout or large) 936
PFUNDGELD (*n*): poundage 937
PFUNDGEWICHT (*n*): poundage; weight in pounds (lbs.) 938
PFUNDHEFE (*f*): (see PRESSHEFE) 939
PFUNDLEDER (*n*): sole leather 940
PFUNDWEISE: by the pound; in (pounds) lbs., or £ 941
PFUSCHEN: to meddle; bungle; dabble; cheat; fudge; flash 942
PFUSCHER (*m*): bungler (see also PFUSCHEN) 943
PFÜTZE (*f*): puddle; slough; pool 944
PFÜTZEN: to pump water out of 945
PFÜTZIG: muddy; full of puddles 946
PHAGEDÄNISCHES WASSER (*n*): yellow mercurial lotion (Pharmacy) 947
PHAKOLITH (*m*): phakolite; $(CaNa_2K_2)Al_2Si_4O_{12}.6H_2O$; Sg. 2.1 948
PHANEROGAME (*f*): phanerogam (Bot). 949
PHÄNOMEN (*n*): phenomenon 950
PHANTASIE (*f*): fancy; fantasy; imagination; fantasia; improvization (Music) 951
PHANTASIEREICH: imaginative; inventive; (*n*): realm of fancy 952
PHANTASIEREN: to fancy; imagine; be delirious; rave; wander; indulge in fantasies; improvize (Music); invent; create out of one's fancies 953
PHANTASMAGORIE (*f*): phantasmagoria, illusive images, optical illusions 954
PHANTASMAGORISCH: phantasmagorial; illusive 955
PHANTAST (*m*): dreamer; visionary 956
PHANTASTISCH: fantastic; fanciful 957
PHARMAKOLITH (*m*): pharmacolite (a calcium arsenate); $HCaAsO_4.2H_2O$; Sg. 2.7 [958
PHARMAKOLOGIE (*f*): (*Materia medica*): pharmacology (the science of medicines) 959
PHARMAKOLOGISCH: pharmacological-(ly) 960
PHARMAKOPÖE (*f*): pharmacopeia 961
PHARMAKOSIDERIT (*m*): pharmacosiderite, cube ore (see also WÜRFELERZ); $3Fe_2O_3.2As_2O_5.12H_2O$ or $2FeAsO_4.Fe(OH)_3.5H_2O$; Sg. 2.95 962
PHARMAZEUT (*m*): pharmacist; pharmaceutist; druggist; apothecary 963
PHARMAZEUTIK (*f*): pharmaceutics 964
PHARMAZEUTISCH: pharmaceutical 965
PHARMAZEUTISCHE SPEZIALITÄT (*f*): pharmaceutical specialty 966
PHARMAZEUTISCHES PRODUKT (*n*): pharmaceutical product 967
PHARMAZIE (*f*): pharmacy 968
PHARYNGITIS (*f*): pharyngitis (inflammation of the pharynx) (Med.) 969
PHARYNX (*m*): pharynx (Anat.) 970
PHASE (*f*): phase; aspect, view; appearance, phasis 971
PHASENGESETZ (*n*): phase rule 972
PHASENLEHRE (*f*): doctrine of phases 973
PHASENREGEL (*f*): phase rule 974
PHASENZAHL (*f*): number of phases 975
PHASOTROPIE (*f*): phasotropy; phasotropism 976
PHELLANDREN (*n*): phellandren (a hydrocarbon of the methane group) 977
PHENACETIN (*n*): phenacetin (see PHENAZETIN) 978
PHENACETOLIN (*n*): phenacetolin (see PHENAZE-TOLIN) 979
PHENAKIT (*m*): phenakite, phenacite (Beryllium orthosilicate); Be_2SiO_4; Sg. 2.98 980
PHENAMINBLAU (*n*): phenamine blue 981
PHENANTHREN (*n*): phenanthrin, orthodiphenyleneethylene, phenanthrene (from coal tar, a by-product from the purification of anthracene); $C_{14}H_{10}$; Sg. 1.0631; Mp. 99.5°C.; Bp. 340°C. 982
PHENANTHRENCHINON (*n*): phenanthrenequinone; $C_6H_4CO_2.CO_2C_6H_4$; Sg. 1.405; Mp. 202°C. 983

PHENAZETIN (*n*): (*Phenacetinum*): phenacetin, aceto-para-phenetidine, acetphenitidine; $C_6H_4O.C_2H_5NH(CO.CH_3)$; Mp. 135°C. [984

PHENAZETOLIN (*n*): phenacetolin (from action of concentrated sulphuric acid and glacial acetic acid on Phenol) 985

PHENAZON (*n*): (*Phenazonum*): antipyrine, phenazone (see ANTIPYRIN) 986

PHENETIDIN (*n*): phenetidin, para-aminophenetol; $C_6H_4O.C_2H_5.NH_2$ 987

PHENETIDIN (*n*), ALPHA-BROMISOVALERYL-PARA-: alpha-bromo-isovaleryl-para-phenetidin, phenoval (see PHENOVAL) 988

PHENETIDIN (*n*), AMINOAZET-PARA-: aminoaceto-para-phenetidin (see PHENOKOLL) 989

PHENETIDIN (*n*), AZET-PARA-: (see PHENAZETIN) 990

PHENETIDIN, CITRONENSAURES- (*n*): phenitidin citrate 991

PHENETIDIN (*n*), GLYKOKOLL-PARA-: glycocoll-para-phenetidin (see PHENOKOLL) 992

PHENETIDIN, MILCHSAURES- (*n*): phenitidin lactate 993

PHENETOL (*n*): phenetol, phenyl ethyl ether; $C_6H_5.O.C_2H_5$; Sg. 0.971; Bp. 172°C. [994

PHENETOL (*n*), PARA-AMINO-: (see PHENETIDIN) 995

PHENETOL (*n*), PARA-NITRO-: para-nitrophenetol; $C_6H_4(NO_2)O.C_2H_5$ 996

PHENGIT (*m*): phengite (see MUSKOVIT) 997

PHENOCOLLUM (*n*), SALZSAURES-: phenocoll hydro-chloride (*Phenocollum hydrochloricum*) (see PHENOKOLL) 998

PHENOKOLL (*n*): (*Phenocollum*): phenocoll, aminoaceto-para-phenetidin, glycocoll- para-phenetidin; $C_6H_4(O.C_2H_5)NH.CO.CH_2.NH_2$ 999

PHENOKOLLCHLORHYDRAT (*n*): (*Phenocollum hydrochloricum*): phenocoll hydrochloride 000

PHENOL (*n*): (*Acidum carbolicum*): phenol; carbolic acid; phenic (phenylic) acid; phenyl hydrate; hydroxybenzene; C_6H_6O; Sg. 1.0489-1.0677; Mp. 43°C.; Bp. 181.5°C. (see also CARBOLSÄURE); [as a prefix = phenolate, phenoxide (of metals); phenol, phenolic] 001

PHENOLALUMINIUM (*n*): aluminium (carbolate; phenate; phenolate) phenoxide or phenylate 002

PHENOLAPPARAT (*m*): phenol apparatus 003

PHENOLÄTHER (*m*): (see PHENYLÄTHER) 004

PHENOLÄTHYLÄTHER (*m*): phenetol (see PHENETOL) 005

PHENOLBESTIMMUNG(*f*): determination of phenol 006

PHENOLCAMPHOR (*m*): phenol camphor 007

PHENOLCARBONSÄURE (*f*): phenolcarboxylic acid 008

PHENOLDARSTELLUNG (*f*): phenol manufacture 009

PHENOLDERIVAT (*n*): phenol derivate 010

PHENOLDISULFOSÄURE (*f*): phenol disulphonic acid; $C_6H_3OH(SO_3H_2)$ 011

PHENOLGEHALT (*m*): phenol content 012

PHENOLHOMOLOG (*m*): phenol homologue 013

PHENOLIN (*n*): phenolin (a disinfectant identical with Lysol) 014

PHENOLKALIUM (*n*): potassium (carbolate; phenate; phenolate; phenoxide) phenylate (see KALI, KARBOLSAURES-) 015

PHENOLKALZIUM (*n*): calcium (carbolate; phenate; phenolate; phenoxide) phenylate 016

PHENOLKARBONSÄURE (*f*): (see SALIZYLSÄURE) 017

PHENOLMETHYLÄTHER (*m*): phenol-methyl ether, anisol (see ANISOL); $C_6H_5.OCH_3$ 018

PHENOLMOLEKÜL (*n*): phenol molecule 019

PHENOLNATRIUM (*n*): sodium (carbolate; phenate; phenolate) phenoxide 020

PHENOLNATRIUMLÖSUNG (*f*): sodium phenolate solution 021

PHENOL (*n*), NICHT-: non-phenol 022

PHENOLPHTALEIN (*n*): (*Phenolphthaleinum*): phenolphthalein, dioxytriphenylcarbinol-carboxylic anhydride; $C(C_6H_4.OH)_2C_6H_4.CO.O.2H_2O$; Sg. 1.277; Mp. 251.5°C. 023

PHENOLQUECKSILBER (*n*): mercury (carbolate; phenate; phenolate) phenoxide 024

PHENOL (*n*), ROH-: crude phenol, crude carbolic acid; Sg. 1.055 (see PHENOL) 025

PHENOLSÄURE (*f*): phenolic acid 026

PHENOLSCHWEFELSÄURE (*f*): phenol-(phenyl-) sulphuric acid 027

PHENOLSULFOSÄURE (*f*): phenolsulfonic acid; sulfocarbolic acid; $C_6H_6SO_3$ 028

PHENOLUT (*n*): phenolut (a colloidal cresol solution with 40% crude cresol) 029

PHENOLVERGIFTUNG (*f*): phenol poisoning 030

PHENOLWISMUT (*m* and *n*): bismuth (carbolate; phenate; phenolate; phenoxide) phenylate; phenolbismuth (from interaction of phenol and bismuth hydroxide) 031

PHENOSTAL (*n*): phenostal (trade name for diphenyl ortho-oxalate; see ORTHOOXAL-SÄUREDIPHENYLESTER) 032

PHENOSULFOSÄURE (*f*): sulphophenic acid 033

PHENOTHIAZIN (*n*): phenothiazine (see THIO-DIPHENYLAMIN) 034

PHENOVAL (*n*): phenoval, alpha-bromo-isovaleryl-para-phenetidin; $(CH_3)_2.CH.CHBr.CO.NH.C_6H_4.OC_2H_5$; Mp. 149°C. 035

PHENOXAZIN (*n*): phenoxazine (see OXAZIN) 036

PHENYFORM (*n*): phenyform (a mixture of phenol and formaldehyde in powder form; an antiseptic) 037

PHENYL (*n*): phenyl; C_6H_5 038

PHENYLACETALDEHYD (*n*): phenylacetaldehyde, ethylalbenzene, toluyl aldehyde; $C_6H_5CH_2CHO$; Sg. 1.032; Mp. under – 10°C.; Bp. 193.5-206°C. 039

PHENYLACETAMID (*n*): phenylacetamide (see ACETANILID) 040

PHENYLACRYLSÄURE (*f*): (see ZIMTSÄURE) 041

PHENYLALANIN (*n*): phenylalanine, l-α-amino-β-phenylpropionic acid; $C_6H_5.CH_2.CH(NH_2).CO_2H$ 042

PHENYLALKOHOL (*m*): phenyl alcohol, phenol (see PHENOL) 043

PHENYLALPHANAPHTYLAMIN (*n*): phenyl-alpha-naphthylamine; $C_{10}H_7.NH.C_6H_5$; Mp. 62°C.; Bp. 226-335°C. 044

PHENYLAMEISENSÄURE (*f*): (see BENZOESÄURE) 045

PHENYLAMIDOAZOBENZOL (*n*): phenyl amino-azobenzene; $C_6H_5.N.N.C_6H_4.NH.C_6H_5$ [046

PHENYLAMIN (*n*): phenylamine (see ANILIN) 047

PHENYLAMINSCHWARZ (*n*): phenylamine black 048

PHENYLANILIN (*n*): (see DIPHENYLAMIN) 049

PHENYLARSENCHLORÜR (*n*): phenylarsenious chloride 050

PHENYLÄTHER (*m*): phenyl ether; $C_{12}H_{10}O$; Mp. 28°C.; Bp. 252-257°C. 051

PHENYLÄTHYLALKOHOL (*m*): phenylethylic (phenylethyl) alcohol; benzyl carbinol; $C_6H_5CH_2CH_2OH$; Sg. 1.024; Mp. -27°C.; Bp. 219°C. 052

PHENYLÄTHYLÄTHER (*m*): phenyl ethyl ether, phenetol (see **PHENETOL**) 053

PHENYLÄTHYLBARBITURSAURES NATRIUM (*n*): (see **LUMINALNATRIUM**) 054

PHENYLÄTHYLMALONYLHARNSTOFF (*m*): (see **LUMINAL**) 055

PHENYLBENZAMID (*n*): (see **BENZANILID**) 056

PHENYLBETANAPHTYLAMIN (*n*): phenyl-beta-naphthylamine; $C_{10}H_7.NH.C_6H_5$; Mp. 107.7°C.; Bp. 395-399.5°C. 057

PHENYLBORAT (*n*): phenyl borate; $C_6H_5BO_2$ [058

PHENYLBORCHLORID (*n*): phenylboron chloride 059

PHENYLBORSÄURE (*f*): phenylboric (borophenylic) acid; $(C_6H_5BO_2).(C_6H_5B_3O_5)$; Mp. 204°C. 060

PHENYLCHLORID (*n*): phenyl chloride, monochlorobenzene, monochlorobenzol; C_6H_5Cl; Sg. 1.106; Mp. -44.9°C.; Bp. 132°C. 061

PHENYLCHLOROFORM (*n*): phenylchloroform (see **BENZOTRICHLORID**) 062

PHENYLCINCHONINSÄURE (*f*): phenylcinchonic (phenylcinchonine; phenylquinoline carboxylic) acid; $C_6H_5C_9H_5NCO_2H$; Mp. 210°C. 063

PHENYLCYANAT (*n*): (see **KARBANIL**) 064

PHENYLCYANID (*n*): (see **BENZONITRIL**) 065

PHENYLDIHYDROCHINAZOLIN (*n*): phenyldihydroquinazoline (see **OREXIN**) 066

PHENYLDIHYDROCHINAZOLINCHLORHYDRAT (*n*): phenyldihydroquinazoline hydrochloride (see **OREXINCHLORHYDRAT**) 067

PHENYLDIMETHYLPYRAZOL (*n*): (see **ANTIPYRIN**) 068

PHENYLDIMETHYLPYRAZOLON (*n*): (*Pyrazolonum phenyldimethylicum*): antipyrine (see **ANTI-PYRIN**) 069

PHENYLDIMETHYLPYRAZOLON (*n*), **SALIZYL-SAURES-**: (*Pyrazolonum phenyldimethylicum salicylicum*): salipyrine (see **SALIPYRIN**) 070

PHENYLEN (*n*): phenylene; C_6H_4 (see also **ANTIPYRIN**) 071

PHENYLENBLAU (*n*): phenylene blue 072

PHENYLENBRAUN (*n*): phenylene brown 073

PHENYLENDIAMIN (*n*): phenylenediamine, diaminobenzene; $(C_6H_4)_2NH$ or $C_6H_4(NH_2)_2$; ortho.-; Mp. 102°C.; Bp. 258°C.; meta.-; Sg. 1.139; Mp. 63°C.; Bp. 287°C.; para.-; Mp. 140°C.; Bp. 267°C. 074

PHENYLENDIAMINCHLORHYDRAT (*n*): phenylenediamine hydrochloride; $C_6H_4(NH_2)_2.HCl$ 075

PHENYLENREST (*m*): phenylene residue 076

PHENYLESSIGSAUER: phenylacetate of, combined with phenylacetic acid 077

PHENYLESSIGSÄURE (*f*): phenylacetic acid; alphatoluic acid; $C_6H_5CH_2CO_2H$; Sg. 1.081; Mp. 76°C.; Bp. 262°C. 078

PHENYLESSIGSÄURENITRIL (*n*): phenylacetic acid nitrile (see **BENZYLCYANID**) 079

PHENYLFETTSÄURE (*f*): phenylated fatty acid 080

PHENYLGLYOXYL-ORTHO-KARBONSÄURE (*f*): phtalonic acid (see **PHTALONSÄURE**) 081

PHENYLHARNSTOFF (*m*): phenyl urea 082

PHENYLHYDRAZIN (*n*): phenylhydrazine; $C_6H_8N_2$; Sg. 1.0981; Bp. 19.6°C.; Mp. 243.5°C. [083

PHENYLHYDRAZINCHLORHYDRAT (*n*): phenylhydrazine hydrochloride; $C_6H_5HNNH_2.HCl$ 084

PHENYLHYDRAZINMONOSULFOSÄURE (*f*): phenylhydrazinemonosulphonic acid 085

PHENYLHYDRAZINSULFOSÄURE (*f*): phenylhydrazinesulfonic acid 086

PHENYLIEREN: to phenylate 087

PHENYLISOCYANAT (*n*): phenyl isocyanate, carbanile; C_7H_5ON; Sg. 1.0952; Bp. 166°C. 088

PHENYLISOSULFOCYANAT (*n*): phenyl isosulfocyanate, phenyl isothiocyanate, phenyl mustard oil, thiocarbanil, phenylthiocarbonimide; $C_6H_5.CNS$; Sg. 1.138; Mp. -21°C.; Bp. 221°C. 089

PHENYLJODIDCHLORID (*n*): phenyl iodochloride 090

PHENYLMETHYLKETON (*n*): phenyl methyl ketone; $C_6H_5.CO.CH_3$ 091

PHENYLMILCHSÄURE (*f*): phenyllactic acid 092

PHENYLNAPHTYLAMIN (*n*): phenylnaphthylamine (see **PHENYLALPHANAPHTYLAMIN** and **PHENYLBETANAPHTYLAMIN**) 093

PHENYLORTHOTOLUIDIN (*n*): phenyl orthotoluidine 094

PHENYLPARACONSÄURE (*f*): phenylparaconic acid 095

PHENYLPROPIOLSÄURE (*f*): phenylpropiolic acid; C_6H_5C: COOH; Mp. 136.5°C. 096

PHENYLPROPIONSÄURE (*f*): phenyl-propionic acid 097

PHENYLPROPIONSÄURE (*f*), **L-α-AMINO-β-**: (see **PHENYLALANIN**) 098

PHENYLSALICYLAT (*n*): phenyl salicylate, salol, salicylic acid phenyl ester; $C_6H_4OHCO_2C_6H_5$; Sg. 1.26; Mp. 41.9°C.; Bp. 172.5°C. 099

PHENYL, SALICYLSAURES- (*n*): phenyl salicylate; salol (see **PHENYLSALICYLAT**) 100

PHENYLSÄURE (*f*): phenol; phenylic acid (see **PHENOL** and **KARBOLSÄURE**) 101

PHENYLSCHWEFELSÄURE (*f*): phenylsulphuric acid; $C_6H_5O.SO_3H$ 102

PHENYLSENFÖL (*n*): phenyl mustard oil (see also **PHENYLISOSULFOCYANAT**): C_7H_5NS; Sg. 1.1289; Mp. -21°C.; Bp. 218.5°C. 103

PHENYLSILICIUMCHLORID (*n*): phenylsilicon chloride 104

PHENYLTRIBORAT (*n*): phenyl triborate; $C_6H_5B_3O_5$ 105

PHENYLURETHAN (*n*): phenyl urethane; $CO<\substack{NH_2\\OC_6H_5}$ 106

PHENYLWASSERSTOFF (*m*): phenyl hydride; benzene (see **BENZOL**) 107

PHILANTHROP (*m*): philanthropist (see also **MENSCHENFREUND**) 108

PHILANTHROPISCH: philanthropic 109

PHILLIPSIT (*m*): phillipsite; $(K_2,Ca)Al_2(SiO_3)_5.4H_2O$; Sg. 2.2 110

PHILOLOG (*m*): philologer; philologist 111

PHILOLOGIE (*f*): philology (the study of language) 112

PHILONOTIS (*f*): Philonotis moss 113

PHILOSOPH (*m*): philosopher 114

PHILOSOPHEM (*n*): theorem 115

PHILOSOPHENWOLLE (*f*): philosopher's wool; sublimed zinc oxide 116

PHILOSOPHIE (*f*): philosophy 117
PHILOSOPHIEREN: to philosophize 118
PHILOSOPHISCH: philosophical-(ly) 119
PHIMOSIS (*f*): phimosis (disease of the penis) (Med.) 120
PHIOLE (*f*): phial 121
PHLEGMA (*n*): phlegm; coolness; apathy 122
PHLEGMASIA DOLENS: *Phlegmasia (alba) dolens*, white leg (Med.) 123
PHLEGMATISCH: phlegmatic-(ally) 124
PHLOBAPHEN (*n*): bark tan, bark dye (Tanning material) 125
PHLOGISTISCH: phlogistic 126
PHLOGOPIT (*m*): phlogopite (a magnesium mica); $(H,K)_3MgAlSi_3O_{12}$; Sg. 2.82 127
PHLORCHINYL (*n*): phloroquinyl 128
PHLORETIN (*n*): phloretine 129
PHLOROGLUCID (*n*): phloroglucide 130
PHLOROGLUCIN (*n*): phloroglucinol (see **PHLOROGLUZIN**) 131
PHLOROGLUZIN (*n*): phloroglucine, phloroglucinol, trioxybenzene 1, 3, 5; $C_6H_3(OH)_3.2H_2O$; Mp. 200-219°C. 132
PHLOROL (*n*): phlorol; $C_8H_{10}O$ 133
PHOBROL (*n*): phobrol (a disinfectant) 134
PHOLERIT (*m*): pholerite (see **KAOLIN** and **KAOLINIT**) 135
PHONETIK (*f*): phonetics (the science of articulate sounds) 136
PHONETISCH: phonetic 137
PHÖNICIN (*n*): phenicin 138
PHONIK (*f*): acoustics (the science of sounds) 139
PHÖNIXSCHWARZ (*n*): phœnix black 140
PHONOGRAPHENWACHS (*n*): phonograph wax 141
PHONOGRAPHENWALZE (*f*): phonograph (record) cylinder 142
PHONOLITH (*m*): phonolite (an effusive rock, containing augite, hornblende, leucite, magnetic iron, nepheline and sanidine with about 14% potash) (see **KLINGSTEIN**) 143
PHONOLITHMEHL (*n*): phonolite meal, ground (crushed) phonolite (used as a fertilizer) 144
PHOSGEN (*n*): phosgene, carbonyl chloride (see **CHLORKOHLENOXYD**); $COCl_2$ 145
PHOSGENIT (*m*): phosgenite (natural chlorocarbonate of lead); $PbO,CO_2,PbCl_2$; Sg. 6.2 146
PHOSPHAT (*n*): phosphate 147
PHOSPHATACIDITÄT (*f*): phosphate acidity 148
PHOSPHATISCH: phosphatic 149
PHOSPHATMEHL (*n*): phosphate (meal) powder, ground phosphate 150
PHOSPHATMÜHLE (*f*): phosphate mill 151
PHOSPHENYLIG: phosphenylous 152
PHOSPHENYLSÄURE (*f*): phosphenylic acid 153
PHOSPHIN (*n*): chrysanilin, phosphine, phosphuretted hydrogen, hydrogen phosphide; PH_3; Sg. 1.185; Mp. – 133.5°C.; Bp. – 85°C. 154
PHOSPHINIGSÄURE (*f*): phosphinous acid 155
PHOSPHINSÄURE (*f*): phosphinic acid 156
PHOSPHOCERIT (*m*): phosphocerite (see **MONAZIT**) 157
PHOSPHONIUM (*n*): phosphonium; PH_4 158
PHOSPHONIUMCHLORID (*n*): phosphonium chloride; PH_4Cl; Mp. 26°C. 159
PHOSPHONIUMJODID (*n*): phosphonium iodide, methyl phosphonium iodide; $P(CH_3)_4I$ [160
PHOSPHOR (*m*): phosphorous; (P) 161
PHOSPHORARTIG: phosphorous; like phosphorous 162
PHOSPHORÄTHER (*m*): phosphoric ether; ethyl phosphate 163
PHOSPHORBASE (*f*): phosphorous base 164
PHOSPHORBASIS (*f*): phosphorous base 165
PHOSPHORBESTIMMUNG (*f*): determination of phosphorous 166
PHOSPHORBLEI (*n*): lead phosphide; pyromorphite (Mineral) (see **PYROMORPHIT**) 167
PHOSPHORBLEIERZ (*n*): pyromorphite (Mineral) (see **PYROMORPHIT**) 168
PHOSPHORBLEISPAT (*m*): pyromorphite (Mineral) (see **PYROMORPHIT**) 169
PHOSPHORBROMID (*n*): phosphorous bromide; phosphorous pentabromide; PBr_5 170
PHOSPHORBROMÜR (*n*): phosphorous (tri)-bromide; PBr_3; Sg. 2.495-2.923 171
PHOSPHORBRONZE (*f*): phosphor-bronze (84-90% Cu; 16-10% Sn; and trace of Zn; with up to 1% phosphorous) 172
PHOSPHORCALCIUM (*n*): calcium phosphide; photophor (see **CALCIUMPHOSPHID**) 173
PHOSPHORCHALCIT (*m*): phosphorchalcite (see also **LUNNIT**); $(CuOH)_3PO_4$; Sg. 3.9 174
PHOSPHORCHLORID (*n*): phosphorous (phosphoric) chloride; phosphoric perchloride; phosphorous pentachloride (*Phosphorous pentachloratus*); PCl_5; Sg. 3.6; Bp. 160-165°C. 175
PHOSPHORCHLORÜR (*n*): phosphorous (tri)-chloride (*Phosphorous trichloratus*); PCl_3; Sg. 1.47-1.613; Mp. – 111.8°C.; Bp. 76°C. 176
PHOSPHOREISEN (*n*): iron phosphide (see **EISENPHOSPHID**) 177
PHOSPHOREISENERZ (*n*): phosphorous iron ore 178
PHOSPHORESZENZ (*f*): phosphorescence 179
PHOSPHORESZIEREN: to phosphoresce 180
PHOSPHORESZIEREND: phosphorescent 181
PHOSPHORFLEISCHSÄURE (*f*): phosphocarnic acid 182
PHOSPHORGEHALT (*m*): phosphorous content 183
PHOSPHORGRUPPE (*f*): phosphorous group 184
PHOSPHORHALTIG: phosphorated; (containing)-phosphorous; phosphatic 185
PHOSPHORIG: phosphorous 186
PHOSPHORIGE SÄURE (*f*): phosphorous acid; H_3PO_3; Sg. 1.651 187
PHOSPHORIGSAUER: phosphite of; combined with phosphorous acid 188
PHOSPHORIGSÄURE (*f*): phosphorous acid (see **PHOSPHORIGE SÄURE**) 189
PHOSPHORIGSÄUREANHYDRID (*n*): phosphorous (trioxide) anhydride; P_2O_3 (see **PHOSPHORTRIOXYD**) 190
PHOSPHORISCH: phosphoric 191
PHOSPHORISIEREN: to phosphorate; phosphorize 192
PHOSPHORIT (*m*): phosphorite (an earthy variety of apatite) (natural calcium phosphate); $3Ca_3P_2O_8.CaF(Cl)_2$; Sg. 3.15 (see also **APATIT** and **TRICALCIUMPHOSPHAT**) 193
PHOSPHORIT-SUPERPHOSPHAT (*n*): mineral superphosphate, phosphorite superphosphate 194
PHOSPHORJODID (*n*): phosphorous (tri)-iodide 195
PHOSPHORJODÜR (*n*): phosphorous diiodide; P_2I_4 196
PHOSPHORKALK (*m*): calcium phosphide (see **CALCIUMPHOSPHID**) 197
PHOSPHORKALZIUM (*n*): (see **PHOSPHORCALCIUM**) 198

PHOSPHORKUPFER (*n*): copper alloy with 5-15% P; copper phosphide (see **KUPFERPHOSPHID**); Libethenite (see **LIBETHENIT**); Phosphorchalcite (see **PHOSPHORCHALCIT**) 199

PHOSPHORKUPFERERZ (*n*): Libethenite (see **LIBETHENIT**) 200

PHOSPHORLÖFFEL (*m*): deflagrating (phosphorous) spoon 201

PHOSPHORMETALL (*n*): metallic phosphide; phosphorous metal 202

PHOSPHORMILCHSAUER: lactophosphate (*Phospholacticum*) 203

PHOSPHORMOLYBDÄNSÄURE (*f*): phosphomolybdic acid (Sonnenschein's reagent); $H_3PO_4 : 12MoO_3$ 204

PHOSPHORNATRIUM (*n*): sodium phosphide 205

PHOSPHORNITRID (*n*): phosphorous nitride; P_3N_5; Sg. 2.51 206

PHOSPHOROGRAPHIE (*f*): phosphorography 207

PHOSPHORÖL (*n*): phosphorated oil (Pharm.) 208

PHOSPHORÖL (*n*), **BRAUSENDES-**: effervescent phosphorated oil 209

PHOSPHOROXYCHLORID (*n*): phosphorous oxychloride; phosphoryl chloride (*Phosphorous oxychloratus*); $POCl_3$; Sg. 1.68-1.712; Mp. 1.25°C.; Bp. 107.2°C. 210

PHOSPHOROXYD (*n*): phosphorous oxide (see **PHOSPHORPENTOXYD** and **PHOSPHORTRIOXYD**) 211

PHOSPHORPASTA (*f*): phosphorous paste 212

PHOSPHORPENTABROMID (*n*): phosphorous pentabromide (see **PHOSPHORBROMID**) 213

PHOSPHORPENTACHLORID (*n*): phosphorous pentachloride (see **PHOSPHORCHLORID**) 214

PHOSPHORPENTASULFID (*n*): (*Phosphorus pentasulfuratus*): phosphorous pentasulphide; P_2S_5; Sg. 2.03; Mp. 276°C.; Bp. 520°C. 215

PHOSPHORPENTOXYD (*n*): phosphorous pentoxide; phosphoric anhydride; P_2O_5; Sg. 2.387; Mp. 800°C. 216

PHOSPHORPROTEID (*n*): phosphoprotein 217

PHOSPHORSALZ (*n*): microcosmic salt; sodium-ammonium phosphate; $HNaNH_4PO_4.4H_2O$ 218

PHOSPHORSAUER: phosphate of; combined with phosphoric acid 219

PHOSPHORSÄURE (*f*): (ortho)-phosphoric acid; H_3PO_4; Sg. 1.884; Mp. 38.6°C. 220

PHOSPHORSÄUREANHYDRID (*n*): phosphoric anhydride; phosphorous pentoxide (see **PHOSPHORPENTOXYD**) 221

PHOSPHORSÄUREFABRIKANLAGE (*f*): phosphoric acid plant 222

PHOSPHORSESQUISULFID (*n*): phosphorous sesquisulphide; tetraphosphorous trisulphide; P_4S_3; Sg. 2.0; Mp. 172°C.; Bp. 407.8°C. 223

PHOSPHORSTAHL (*m*): phosphorous steel 224

PHOSPHORSUBSULFID (*n*): (see **PHOSPHORSESQUI-SULFID**) 225

PHOSPHORSULFID (*n*): phosphorous sulphide; P_4S_3; Sg. 2.03; or P_4S_7; Sg. 2.19; or P_4S_{10}; Sg. 2.09 226

PHOSPHORSULFOBROMID (*n*): phosphorous sulfobromide; $PSBr_3$; Sg. 2.85 227

PHOSPHORSULFOCHLORID (*n*): phosphorous sulfochloride; $PSCl_2$; Sg. 1.634-1.668 228

PHOSPHORTRIBROMID (*n*): phosphorous tribromide; PBr_3 (see **PHOSPHORBROMÜR**) 229

PHOSPHORTRICHLORID (*n*): phosphorous trichloride; PCl_3 (see **PHOSPHORCHLORÜR**) 230

PHOSPHORTRIOXYD (*n*): phosphorous trioxide, phosphorous anhydride; P_2O_3; Sg. 1.69-2.135 231

PHOSPHORTRISULFID (*n*): (*Phosphorus trisulfuratus*): phosphorous trisulphide; P_2S_3; Mp. 290°C.; Bp. 540°C. 232

PHOSPHORVERBINDUNG (*f*): phosphorous compound 233

PHOSPHORVERGIFTUNG (*f*): phosphorous poisoning; phossy-jaw (Med.) 234

PHOSPHORWASSERSTOFF (*m*): hydrogen phosphide; phosphoretted hydrogen; phosphorous hydride (see **PHOSPHIN** and also **PHOSPHORWASSERSTOFF, FESTER-**; **FLÜSSI-GER-**; and **-GAS**) 235

PHOSPHORWASSERSTOFF, FESTER- (*m*): solid hydrogen phosphide; P_4H_2 236

PHOSPHORWASSERSTOFF, FLÜSSIGER- (*m*): liquid hydrogen phosphide; P_2H_4; Sg. 1.012 [237

PHOSPHORWASSERSTOFFGAS (*n*): phosphoretted hydrogen gas; phosphine; PH_3; Sg. 0.744 238

PHOSPHORWASSERSTOFF, GASFÖRMIGER- (*m*): (see **PHOSPHORWASSERSTOFFGAS**) 239

PHOSPHORWEINSÄURE (*f*): phosphovinic acid 240

PHOSPHORWOLFRAMSÄURE (*f*): phosphotungstic (phosphowolframic) acid; $2(H_3PO_4)_2 WO_3 - 39H_2$ 241

PHOSPHORZINK (*n*): zinc phosphide; Zn_3P_2; Sg. 4.55 242

PHOSPHORZINN (*n*): tin phosphide (see **ZINNPHOS-PHID**) 243

PHOSPHORZÜNDHOLZ (*n*): phosphorous match 244

PHOSPHORZÜNDHÖLZCHEN (*n*): phosphorous match 245

PHOSPHURANYLIT (*m*): phosphuranylite; $(UO_2)_3(PO_4)_2.6H_2O$ 246

PHOTOCHEMIE (*f*): photochemistry 247

PHOTOCHEMIGRAPHIE (*f*): zincography, zinc etching (a photo-mechanical printing process on zinc) 248

PHOTOCHEMISCH: photochemical 249

PHOTOCHROMATIE (*f*): photochromatism (natural colour photography) 250

PHOTOCHROMIE (*f*): photography in natural colours; photochromatism; crystoleum painting (see **CHROMOPHOTOGRAPHIE**) 251

PHOTODYNAMISCH: photodynamic 252

PHOTOELEKTRISCH: photoelectric 253

PHOTOGALVANOGRAPHIE (*f*): photo-galvanography (process for the production of a printing block by electro-typing on a photograph) 254

PHOTOGEN (*n*): solar oil, kerosene (see **LEUCHT-PETROLEUM**) 255

PHOTOGLYPTIE (*f*): Woodbury-type printing (a photomechanical process of printing) 256

PHOTOGRAMM (*n*): photograph; photogram (a photogrammetrical record; see **PHOTO-GRAMMETRIE**) 257

PHOTOGRAMMETRIE (*f*): photogrammetry (the art of taking geodetical measurements or surveying, by the medium of photography) 258

PHOTOGRAPH (*m*): photographer 259

PHOTOGRAPHIE (*f*): photograph-(y); portrait-(ure) 260

PHOTOGRAPHIEFIRNIS (*m*): photographic varnish 261

PHOTOGRAPHIE (*f*) **OHNE OBJEKTIV** (*n*): pinhole photography 262

PHOTOGRAPHIEREN: to photograph 263

PHOTOGRAPHISCH: photographic-(ally) 264
PHOTOGRAPHISCHE CHEMIKALIEN (*f.pl*): photographic chemicals 265
PHOTOGRAPHISCHE LÖSUNG (*f*): photographic solution 266
PHOTOGRAPHISCHER APPARAT (*m*): camera; photographic apparatus 267
PHOTOGRAPHISCHES PAPIER (*n*): photographic paper 268
PHOTOGRAPHISCHE TROCKENPLATTE (*f*): photographic dry-plate 269
PHOTOGRAVÜRE (*f*): photogravure (from half-tone copper plate process) 270
PHOTOKOPIE (*f*): copy, print, positive (Photography) 271
PHOTOLITHOGRAPHIE (*f*): photolithography 272
PHOTOLITHOPHANE (*f*): photography on porcelain 273
PHOTOLYSE (*f*): photolysis 274
PHOTOMEKANISCHES VERFAHREN (*n*): photomechanical process 275
PHOTOMETER (*n*): photometer (a meter for measuring the strength of light, as an actinometer) 276
PHOTOMETRIE (*f*): (see LICHTMESSUNG) 277
PHOTOMETRISCH: photometric 278
PHOTOMINIATUR (*f*): crystoleum painting; minature photograph, photographic miniature 279
PHOTOPHYSIKALISCH: photophysical 280
PHOTOPLASTIGRAPHIE (*f*): photo-sculpture, photographic sculpture (in which photographic silhouettes serve as a guide during the clay-modelling process) 281
PHOTORELIEF (*n*): Woodbury-type (see PHOTO-GLYPTIE) 282
PHOTOSKULPTUR (*f*): photo-sculpture, photographic bas-relief (see PHOTOPLASTIGRAPHIE) 283
PHOTOSPHÄRE (*f*): photosphere (a sphere of light) 284
PHOTOSYNTHESE (*f*): photosynthesis 285
PHOTOTAXIS (*f*): phototaxis 286
PHOTOTELEGRAPHIE (*f*): photo-telegraphy (a process of transmitting photographs by means of telegraphy) 287
PHOTOTOPOGRAPHIE (*f*): photographic surveying 288
PHOTOTROPIE (*f*): phototropy; phototropism 289
PHOTOTROPISMUS (*m*): phototropism 290
PHOTOXYLOGRAPHIE (*f*): photography on wood blocks 291
PHOTOZINKOGRAPHIE (*f*): photozincography (see PHOTOCHEMIEGRAPHIE) 292
PHOTOZINKOTYPIE (*f*): photozincotype-(process) 293
PHRENIKUS (*m*): phrenic nerve (motor nerve for the diaphragm) (Med.) 294
PHRENOLOG (*m*): phrenologist 295
PHRENOLOGISCH: phrenological-(ly) 296
PHTALAMINSÄURE (*f*): phthalamic acid 297
PHTALAZIN (*n*): phthalazine 298
PHTALEIN (*n*): phthalein (see EOSIN and RHOD-AMIN) 299
PHTALID (*n*): phthalide 300
PHTALMONOPERSÄURE (*f*): monoperphthalic acid 301
PHTALONSÄURE (*f*): phthalonic acid 302
PHTALSAUER: phthalate of; combined with phthalic acid 303
PHTALSÄURE (*f*): (*Acidum phtalicum*): phthalic acid, naphthalic acid; ($C_8H_6O_4$); ortho.-; Sg. 1.585-1.593; Mp. 203-213°C.; meta.-; Mp. about 300°C.; (commercially the term is often erroneously used in reference to the anhydride; see PHTALSÄUREANHYDRID) 304
PHTALSÄUREANHYDRID (*n*): phthalic (acid) anhydride; $C_8H_4O_3$; Sg. 1.527; Mp. 128°C.; Bp. 277 to 284.5°C. 305
PHTALSÄURE (*f*), **HANDELS-**: commercial phthalic acid, phthalic anhydride (see PHTALSÄURE-ANHYDRID) 306
PHTHISIKER (*m*): (see PHTISIKER) 307
PHTISIKER (*m*): phthisical subject, one suffering from phthisis, consumptive (Med.) 308
PHULWARABUTTER (*f*): (see GHEABUTTER) 309
PHYKOCYAN (*n*): phycocyanogen (a green colouring matter from certain Algæ) 310
PHYKOLOG (*m*): phycologist (a student of Algæ) 311
PHYLLINSÄURE (*f*): phyllinic acid (from *Prunus lauro-cerasus*) 312
PHYSIK (*f*): physics; natural philosophy; tin composition (Dye) 313
PHYSIKALISCH: physical-(ly) 314
PHYSIKALISCH-CHEMISCH: physico-chemical 315
PHYSIKANT (*m*): natural philosopher 316
PHYSIKAT (*n*): physician's office 317
PHYSIKBAD (*n*): tin composition (Dye) (tin solution in *aqua regia*) 318
PHYSIKER (*m*): physicist; natural philosopher 319
PHYSIKSALZ (*n*): red spirit (Dye) 320
PHYSIKUS (*m*): physician 321
PHYSIOGNOM (*m*): physiognomist 322
PHYSIOGNOMIE (*f*): physiognomy 323
PHYSIOLOG (*m*): physiologist 324
PHYSIOLOGIE (*f*): physiology 325
PHYSIOLOGISCH: physiological-(ly) 326
PHYSISCH: physical 327
PHYSOSTIGMIN (*n*): physostigmine, eserine (alkaloid of calabar bean, *Physostigma venenosum*); $C_{15}H_{21}O_2N_3$; Mp. 102.5°C. [328
PHYSOSTIGMIN (*n*), **SALICYLSAURES-**: physostigmine salicylate (*Physostigminæ salicylas*) 329
PHYSOSTIGMIN (*n*), **SCHWEFELSAURES-**: physostigmine sulphate (*Physostigminæ sulfas*) 330
PHYSOSTIGMIN (*n*), **SCHWEFLIGSAURES-**: physostigmine sulphite 331
PHYSOSTOL (*n*): physostol (a 1% solution of physostigmine in olive oil) 332
PHYTIN (*n*): phytin (potassium-magnesium double salt of anhydroxymethylenediphosphoric acid) (contains about 23% phosphorous) 333
PHYTOCHEMIE (*f*): phytochemistry (see PFLANZ-ENCHEMIE) 334
PHYTOCHEMISCH: phytochemical 335
PHYTOLACCIN (*n*): phytolaccine (from the poke root, *Phytolacca decandra*) 336
PHYTOLIN (*n*): phytoline 337
PHYTONAL (*n*): phytonal 338
PHYTOPALÄONTOLOGIE (*f*): (see PFLANZEN-PALÄONTOLOGIE) 339
PHYTOPHTHORA (*f*): dry-rot (blight) of potatoes (*Phytophthora infestans*) 340
PHYTOPHYSIOLOGIE (*f*): (see PFLANZENPHY-SIOLOGIE) 341
PHYTOTOMIE (*f*): phytotomy (vegetable anatomy, dissection of plants) 342
PIANINO (*n*): upright piano-(forte); cottage piano-(forte) 343
PIANO (*n*): piano-(forte); (as an adjective and adverb—soft-(ly); piano) 344

PIASSABA (*f*): (see PIASSAVA) 345
PIASSAVA (*f*): para (monkey) grass 346
PIASSAVA (*f*), AFRIKANISCHE-: African piassava, bast fibre (from the palm, *Raphia vinifera*) 347
PIASSAVA (*f*), BRASILISCHE-: piassava, monkey grass, para-grass (the leaf fibre of the palm, *Attalea funifera*) 348
PIAUZIT (*m*): piauzite (a fossil resin) (Sg. 1.2) [349
PICHAPPARAT (*m*): pitching apparatus; pitching machine 350
PICHEN: to pitch; coat with pitch 351
PICHEREIMASCHINEN (*f.pl*): pitching machinery (Brewing) 352
PICHLEINWAND (*f*): tarpaulin 353
PICHURIMBOHNE (*f*): Brazil bean 354
PICHURIMTALGSÄURE (*f*): (see LAURINSÄURE) 355
PICK (*m*): thrust; cut; prick; puncture; stab; pick; tick (of a watch) 356
PICKE (*f*): pick-(axe) 357
PICKEL (*m*): pimple; pickle (Chrome tanning) 358
PICKELIG: pimply; blotched 359
PICKELN (*n*): pickling (Chrome tanning process); (as a verb = to pickle) 360
PICKEN: to pick; peck; tick (see also PICK) 361
PICKET (*n*): picquet (Card game); outpost; sentry; picket 362
PICOLINSÄURE (*f*): picolinic acid 363
PICOTIT (*m*): picotite, chrome spinel; (Fe,Mg) $O(Al,Cr,Fe)_2O_3$; Sg. 4.08 364
PICRAMINSAUER: picramate of, combined with picramic acid (see also PIKRAMINSAUER) 365
P.I-DIAGRAMM (*n*): p.i-diagram (pressure-total heat diagram) 366
PIEDESTAL (*n*): pedestal, foot, base 367
PIEDMONTIT (*m*): piedmontite, piemontite (see PIEMONTIT) 368
PIEK (*f*): peak, topping lift (Naut.) 369
PIEKBALKEN (*m*): panting beam 370
PIEKE (*f*): pike (the weapon) 371
PIEKEISEN (*n*): pike (tip) point 372
PIEKFALL (*n*): peak halliard 373
PIEKHÖLZER (*n.pl*): crotches 374
PIEKPFORTE (*f*): raft port 375
PIEKSCHAFT (*m*): pikestaff 376
PIEKSTÜCKE (*n.pl*): crotches 377
PIEKTANKS (*m.pl*): peak tanks 378
PIEMONTIT (*m*): piemontite; $HCa_2(Al,Mn,Si)_3O_{10}$; Sg. 3.4 379
PIETÄT (*f*): piety; devotion; reverence 380
PIETISMUS (*m*): devotion 381
PIETISTISCH: pious, sanctimonious 382
PIEZOELEKTRIZITÄT (*f*): piezoelectricity 383
PIGMENT (*n*): pigment 384
PIGMENTFARBSTOFF (*m*): pigment-(colouring matter) 385
PIGMENTFREI: free from pigment 386
PIGMENTHALTIG: pigmented; containing pigment 387
PIGMENTIEREN: to pigment; be pigmented 388
PIGMENTPAPIER (*n*): pigment paper (Phot.) 389
PIGMENTVERFAHREN (*n*): pigment process (Phot.) 390
PIGNOLFARBE (*f*): pignol dye 391
PIK (*n*): peak; pique; resentment; grudge; umbrage; offence; spite; displeasure; irritation; spades (Cards) 392
PIKANT: piquant; pungent; poignant; biting; spicy 393
PIKANTE (*n*): piquancy 394
PIKE (*f*): (see PIEKE and PIK) 395
PIKÉ (*m*): (see PIQUÉ) 396
PIKET-(T) (*n*): (see PICKET) 397
PIKIEREN: to pique, incite, goad, urge, instigate, spur, annoy, displease, offend, provoke, affront, incense, nettle, irritate, vex, sting, take (give) umbrage, bear grudge, give offence, resent 398
PIKNOMETER (*n*): pycnometer (a measuring instrument) 399
PIKOLIN (*n*): picolin, alpha-methyl-pyridine (a homologue of pyridine); C_6H_7N (see METHYLPYRIDIN, ALPHA-) 400
PIKPFAHL (*m*): picket; post 401
PIKRAMINSAUER: picramate of, combined with picramic acid 402
PIKRAMINSÄURE (*f*): picramic (picraminic) acid; dinitroaminophenol; $C_6H_2(NO_2)_2(NH_2)OH$; Mp. 168°C. 403
PIKRATPULVER (*n*): picrate powder (a smokeless gunpowder) (the base is usually the ammonium or potassium salt of picric acid) 404
PIKRINSAUER: picrate of, combined with picric acid 405
PIKRINSÄURE (*f*): picric (picronitric) acid; trinitrophenol (*Acidum picrinicum*); $C_6H_2.OH(NO_2)_3$; Sg. 1.77; Mp. 122.5°C. 406
PIKRINSÄUREAPPARAT (*m*): picric acid apparatus 407
PIKRINSÄURE (*f*), ROH-: crude picric acid 408
PIKROCROCIN (*n*): picrocrocin (the bitter principle of saffron); $C_{38}H_{66}O_{17}$ 409
PIKROLITH (*m*): picrolite (see SERPENTIN) 410
PIKROMERIT (*m*): picromerite; $K_2Mg(SO_4)_2.6H_2O$; Sg. 2.03 411
PIKROSMINSTEATIT (*m*), PRISMATISCHER-: picrosmine; $2MgO.SiO_2$; Sg. 2.62 412
PIKROTIN (*n*): picrotin (see PIKROTOXIN); $C_{15}H_{18}O_7$ 413
PIKROTOXIN (*n*): picrotoxin (the bitter principle of *Cocculus indicus*; a mixture of picrotoxinin and picrotin in the proportion of 55 : 45) (see PIKROTIN and PIKROTOXININ); Mp. 200°C. 414
PIKROTOXININ (*n*): picrotoxinin (see PIKRO-TOXIN); $C_{15}H_{16}O_6.H_2O$ 415
PIKRYL (*n*): picryl 416
PILEE (*f*): crushed sugar 417
PILEEZUCKER (*m*): crushed sugar 418
PILGER (*m*): pilgrim 419
PILGERN: to wander; make a pilgrimage 420
PILGERSCHRITTVERFAHREN (*n*): tube rolling process 421
PILGERSCHRITTWALZWERK (*n*): a rolling mill for making seamless tubes 422
PILLE (*f*): pill; pilule 423
PILLENBAUM (*m*): spider-flower (*Cleome viscosa*) 424
PILLENBAUMÖL (*n*): spider-flower oil (from seeds of *Cleome viscosa*) 425
PILLENDOSE (*f*): pill-box 426
PILLENDREHER (*m*): pill-maker; apothecary; *Lamellicornia* (Entomology) 427
PILLENGLAS (*n*): phial or glass tube (for pills) 428
PILLENKRAUT (*n*): *Pillularia* (Bot.) 429
PILLENMASCHINE (*f*): pill-(making)-machine 430
PILLENROLLER (*m*): pill roller (for giving the spherical form to pills) 431
PILLENSCHACHTEL (*f*): pill-box 432
PILOCARPIN (*n*): (*Pilocarpinum*): pilocarpine (an alkaloid from jaborandi leaves, of *Pilocarpus jaborandi*); $C_{11}H_{16}N_2O_2$; Mp. 34°C. 433

PILOCARPIN (*n*), **SALICYLSAURES-**: pilocarpin salicylate 434
PILOCARPIN (*n*), **SALPETERSAURES-**: (*Pilocarpinæ nitras*): pilocarpin nitrate 435
PILOCARPIN (*n*), **SALZSAURES-**: (*Pilocarpinæ hydrochloridum*): pilocarpin hydro-chloride 436
PILOKARPIN (*n*): pilocarpin (see **PILOCARPIN**) 437
PILOT (*m*): pilot; pilot fish; rudder fish (*Naucrates ductor*) 438
PILZ (*m*): fungus; fungosity; mushroom; agaric; upstart; species of (acotyledon) cellular cryptogam 439
PILZÄHNLICH: fungoid; like a (fungus) mushroom 440
PILZARTIG: (see **PILZHAFT**; **PILZIG** and **PILZÄHNLICH**) 441
PILZBESATZ (*m*): fungous content, amount of fungus present 442
PILZENZYM (*n*): fungus enzyme 443
PILZFADEN (*m*): (see **HYPHE**) 444
PILZFÖRMIG: fungiform; fungoid; mushroom-(shaped) 445
PILZHAFT: fungous; fungoid; spongy; mushroom-like; fungus-like; excrescent; ephemeral; upstart 446
PILZHYPHE (*f*): hypha (see **HYPHE**) 447
PILZIG: (see **PILZHAFT**) 448
PILZKUNDE (*f*): mycology (the science of fungi) 449
PILZMASSE (*f*): fungous mass 450
PILZSAMEN (*m*): mycelium (spawn of fungi) 451
PILZSÄURE (*f*): fungic acid 452
PILZSTOFF (*m*): fungine 453
PILZTÖTEND: fungicidal 454
PILZVERGIFTUNG (*f*): mushroom poisoning 455
PILZZOTTE (*f*): tangle (ribbon; streamer) of water fungi 456
PILZZUCKER (*m*): mushroom sugar 457
PIMARSÄURE (*f*): pimaric acid 458
PIMELINSÄURE (*f*): pimelic acid 459
PIMELITH (*m*): pimelite; $(Al,Ni)_2O_3,MgO,SiO_2$; Sg. 2.45 460
PIMENT (*n*): pimento; allspice; Jamaica pepper; pimenta; pimento bush (*Pimenta officinalis*) 461
PIMENTÖL (*n*): pimento (allspice) oil (from fruit of *Pimenta officinalis*); Sg. 1.045-1.055 [462
PIMPERNUSS (*f*): bladder-nut (*Staphylea*); pistachio (*Pistacia vera*); filbert 463
PIMPERNUSSBAUM (*m*): bladder-nut tree 464
PIMPINELLE (*f*): burnet saxifrage; common pimpernel (*Pimpinella saxifraga*) 465
PIMPINELLWURZEL (*f*): pimpernel root (*Radix pimpinellæ*) 466
PINACHROM (*n*): pinachrome (colour photograph) 467
PINACHROMIE (*f*): pinachrom process (see **PINATYPIE**) 468
PINAKEL (*m*): pinnacle 469
PINAKOL (*n*): pinacol (a photographic chemical) (see also **PINAKON**) 470
PINAKOLSALZ (*n*): pinacol salt (a 20% solution of sodium aminoacetate) (Phot.) 471
PINAKON (*n*): pinacol (see also **PINAKOL**); tetramethylethyleneglycol; $(CH_3)_2C(OH)$ 472
PINAKONBILDUNG (*f*): pinacol formation 473
PINANG (*m*): (see **KATECHUPALME**) 474
PINASSE (*f*): pinnace (Naut.) 475
PINATYPIE (*f*): pinatype (a process of colour photography) 476
PINCETTE (*f*): tweezers; nippers; forceps; pincers 477
PINEN (*n*): pinene, terebenthene (a constituent of **MUSKATBLÜTENÖL**, which see); $(C_{10}H_{16})$; Sg. 0.859; Bp. 156°C. 478
PINENCHLORHYDRAT (*n*): pinene hydrochloride, terpene hydrochloride, turpentine camphor, turpentine monohydrochloride, artificial camphor; $C_{10}H_{16}HCl$; Mp. 125°C.; Bp. 208°C. 479
PINEYTALG (*m*): piney tallow (see **MALABARTALG**) 480
PINGUIN (*m*): penguin 481
PINGUIT (*m*): pinguite (a mixture of Fe_2O_3 and SiO_2 hydrogels); Sg. 2.33 482
PINIE ((*f*): (stone)-pine (*Pinus pinea*); pine cone 483
PINIENKIEFER (*f*): (stone)-pine (*Pinus pinea*) 484
PINIT (*n*): pinitol; pinite (see **CORDIERIT**); pine sugar (see **FICHTENZUCKER**) 485
PINITGRANIT (*m*): granite, cordierite (see **GRANIT** and **CORDIERIT**) 486
PINITPORPHYR (*m*): granite, cordierite (see **GRANIT** and **CORDIERIT**) 487
PINK (*n*): pink (an underglaze colour used in ceramics, composed of a chromium-tin compound) 488
PINK-COLOUR (*n*): pink colour (see **PINK**) 489
PINKEN: to treat with pink salt; strike a light from flint and steel; hammer (on an anvil) 490
PINKGLASUR (*f*): pink glaze (Ceramics) (see **PINK**) 491
PINKSALZ (*n*): pink salt (ammonium stannic chloride) (see **AMMONIUMZINNCHLORID**) 492
PINNAYÖL (*n*): (see **TACAMAHACFETT**) 493
PINNE (*f*): pin; peg (Tech.); quill feather; thin side (of a hammer); (wire)-tack; pivot; centre-pin; (capstan)-spindle; tiller (Naut.) 494
PINNENBAUM (*m*): breast-beam (Weaving) 495
PINNENSÄGE (*f*): tenon saw 496
PINNE (*f*), **RUDER-**: rudder (helm) tiller, tiller 497
PINNOIT (*m*): pinnoite; $MgB_2O_4.3H_2O$; Sg. 3.35 498
PINOLIN (*n*): pinoline, resin spirit (see **HARZ-SPIRITUS**) 499
PINOTÖL (*n*): (see **PARAPALMÖL**) 500
PINSÄURE (*f*): pinic acid 501
PINSEL (*m*): pencil; brush (for painting) (Art) 502
PINSELEI (*f*): pencilling; daubing; stupidity 503
PINSELFÖRMIG: pencil-shaped 504
PINSELFÜHRUNG (*f*): (method of)-handling of a brush 505
PINSELN: to pencil; paint; daub; be (stupid) foolish 506
PINSELSCHIMMEL (*m*): any kind of *Pencillium* mould 507
PINSELSTRICH (*m*): stroke (touch) with a brush; brush-mark 508
PINUSHARZ (*n*): pine resin; colophony (see **KOLOPHONIUM**) 509
PINZETTE (*f*): (see **PINCETTE**) 510
PIONIER (*m*): pioneer 511
PIONIERARBEIT (*f*): pioneer work 512
PIONIERPATRONE (*f*): pioneer cartridge 513
PIONNIER (*m*): (see **PIONIER**) 514
PIPERAZIN (*n*): piperazine, piperazidine, diethylenediamine, ethyleneimine, pyrazine hexahydride; $C_4H_{10}N_2$; Mp. 104-107°C.; Bp. 145°C. 515

PIPERAZIN (*n*), **CHINASAURES-** : (see **SIDONAL**) 516
PIPERIDIN (*n*): piperidine; $C_5H_{11}N$; Sg. 0.8615; Mp. about – 17°C.; Bp. 105.8°C. 517
PIPERIN (*n*): piperine (an alkaloid); $C_{17}H_{19}NO_3$ (the active principle of pepper from *Piper nigrum*); Sg. 1.19; Mp. 128°C. 518
PIPERINSÄURE (*f*): piperinic acid, methylene-dioxycinnamenylacrylic acid, piperic acid; $C_{11}H_{10}O_4$ 519
PIPERONAL (*n*): piperonal (see **HELIOTROPIN**) 520
PIPERONALDARSTELLUNG (*f*): preparation of piperonal 521
PIPERONAL (*n*), **ROH-** : crude piperonal 522
PIPETTE (*f*): pipette (a measuring vessel) 523
PIPETTENGESTELL (*n*): pipette stand 524
PIPETTIEREN : to pipette; measure with a pipette; introduce into a pipette 525
PIPITZAHOINWURZEL (*f*): pipitzahuak root 526
PIPPS (*m*): (see **PIPS**) 527
PIPS (*m*): pip (a bird ailment of the nature of a cold) 528
PIQUÉ (*m*): piqué (a corded cotton fabric; usually white) 529
PIRAT (*m*): pirate 530
PIRENE (*f*): carbon tetrachloride, pirene (a fire extinguisher) (see **TETRACHLORKOHLENSTOFF**) 531
PISANG (*m*): plantain, pisang, banana (*Musa paradisiaca*) 532
PISSOIRÖL (*n*): lavatory oil 533
PISSOPHAN (*m*): pissophane; $(Fe_2O_3)_2SO_3.15H_2O$; Sg. 1.95 534
PISTAZIE (*f*): pistachio-(nut)-tree (*Pistacia vera*) 535
PISTAZIENÖL (*n*): pistachio-nut oil (from *Pistacia vera*); Sg. 0.9185 536
PISTIL (*n*): pestle; pistil (Bot.) 537
PISTON (*m* and *n*): piston; cornet-à-piston (Musical instrument) 538
PITAZIT (*m*): pitazite (a variety of epidot) (see **EPIDOT**) 539
PITCHPINEHOLZ (*n*): pitch pine wood (see **HARZ-KIEFER**) 540
PITRAL (*n*): pitral (a medicinal tar preparation) 541
PITTINERZ (*m*): pitch-blende (see **URANPECHERZ** and **URANINIT**) 542
PITTORESK : picturesque 543
PITTYLEN (*n*): pittylene (from the action of formaldehyde on pine tar) 544
PITUGLANDOL (*n*): (see **HYPOPHYSENEXTRAKT**) 545
PITUITRIN (*n*): pituitrin (see **HYPOPHYSEN-EXTRAKT**) 546
PIXAVON (*n*): pixavon (a liquid tar soap, being a mixture of Pittylene and Potash soap) 547
PJURI (*n*): (see **PURRÉE** and **INDISCHGELB**) 548
PLACK (*m*): plaque; patch; blot; plane-(surface) 549
PLACKEN : to ram; stamp; pester; flatten; toil; harass; drudge (see also **PLAGEN**); placard; (*m*): patch; placard (see **PLACK**) 550
PLACKER (*m*): pesterer; oppressor 551
PLACKEREI (*f*): drudgery; turmoil; toil; oppression (see **PLACKEN** and **PLAGEN**) 552
PLAGE (*f*): bother; trouble; vexation; drudgery; plague; torment; misery; nuisance; scourge; affliction 553
PLAGEN : to plague; bother; harass; trouble; worry; vex; toil; drudge; torment; oppress; pester; be a nuisance; annoy; scourge; afflict 554
PLAGEN, SICH- : to trouble, toil, work hard, be careful, take pains 555
PLAGER (*m*): tormenter 556
PLAGGE (*f*): sod, turf 557
PLAGGEN : to cut sods 558
PLAGIAR (*m*): plagiarist 559
PLAGIAT (*n*): plagiarism; plagiary 560
PLAGIATOR (*m*): (see **PLAGIAR**) 561
PLAGIOKLAS (*m*): plagioclase-(feldspar) (see **ALBIT, ANDESIN, ANORTHIT, BYTOWNIT, LABRADORIT, OLIGOKLAS**) 562
PLAGIONIT (*m*): plagionite; $5PbS.4Sb_2S_3$; Sg. 5.4 563
PLAIDIEREN : to plead 564
PLAKAT (*n*): placard; show-card; poster 565
PLAKATANZEIGER (*m*): hand-bill 566
PLAKATFARBE (*f*): lithographic (placard) ink or colour 567
PLAN (*m*): plane; plain; (ground)-plan; design; proposal; project; scheme; ground (Painting); (grass)-plot; lawn; (as an adjective and adverb = plain-(ly); plane; flat; smooth; clear; obvious; simple) 568
PLANDREHBANK (*f*): face plate lathe, surface lathe 569
PLANE (*f*): cover; tilt; awning 570
PLÄNE (*f*): plain 571
PLANEN : to plan; design; concert; plot; devise; scheme; contrive; concoct; propose; project; plane 572
PLANET (*n*): planet 573
PLANETARISCH : planetary 574
PLANETARIUM (*n*): orrery (an instrument for illustrating planetary movements) 575
PLANETENBAHN (*f*): orbit (of a planet) 576
PLANETENMASCHINE (*f*): (see **PLANETARIUM**) 577
PLANETENSTAND (*m*): aspect (of the planets) 578
PLANETENSYSTEM (*n*): planetary system 579
PLANETENZEICHEN (*n* and *pl*): sign-(s) of the planets 580
PLANFILM (*f*): flat film (Phot.) 581
PLANGEMÄSS : (see **PLANMÄSSIG**) 582
PLANHAMMER (*m*): planishing hammer 583
PLANIEREN : to plane; planish; level; smooth; glue; size (Paper) 584
PLANIERMASSE (*f*): size (Paper) 585
PLANIMETER (*n*): planimeter 586
PLANKE (*f*): board; plank 587
PLANKEISEN (*n*): caulking (iron) tool 588
PLÄNKELN : to harass; skirmish; snipe (Mil.) 589
PLANKEN : to board, plank 590
PLANKENGANG (*m*): strake (Naut.) 591
PLANKENLAGE (*f*): planking shift 592
PLANKENWERK (*n*): planking, boarding, wainscoting 593
PLANKONKAV : plano-concave 594
PLANKONKAVES GEBILDE (*n*): plano-concave form 595
PLANKONVEX : plano-convex 596
PLANKTON (*n*): the organisms which float about at random and which can live either in salt or fresh water 597
PLANLOS : without plans; purposeless; without design 598
PLANMACHER (*m*): planner; designer, projector, one who makes plans 599
PLANMÄSSIG : systematic-(al)-(ly); according to plan; concerted; arranged 600
PLANPARALLEL : plano-parallel (of surface) 601

PLANROST (*m*): horizontal grate, flat grate (of furnaces) 602
PLANROST (*m*), **HANDBESCHICKTER-**: hand-fired-(flat)-grate 603
PLANSCHEIBE' (*f*): chuck, face plate, surface place (of lathes) 604
PLANSCHEIBE (*f*), **UNIVERSAL-**: universal surface (face) plate 605
PLANSCHEN: to splash; dabble 606
PLANSPIEGEL (*m*): plane mirror 607
PLANSYMMETRISCH: planisymmetric-(al) 608
PLANTAGE (*f*): plantation 609
PLANTAGENBESITZER (*m*): plantation owner, planter 610
PLANTARIN (*n*): plantarin (trade name of a dressing and finishing preparation) 611
PLANUNG (*f*): planning, projecting (see also **PLANEN**) 612
PLAPPERHAFT: babbling; garrulous, talkative 613
PLASMA (*m*): chalcedony (see **CHALCEDON**) 614
PLASMOLYSE (*f*): plasmolysis (contraction of protoplasm in active cells, due to action of a re-agent) (Bot.) 615
PLASMOLYTISCH: plasmolytic 616
PLASMON (*n*): plasmon, sodium caseinate (mixture of moist casein with sodium bicarbonate) 617
PLASMONKAKAO (*m*): plasmon cocoa 618
PLASTIK (*f*): plastic art 619
PLASTISCH: plastic; soft; pliable; formative; easily moulded 620
PLASTISCHE MASSE (*f*): plastic mass (such as Bakelite; see **BAKELIT**. and others) 621
PLASTIZITÄT (*f*): plasticity 622
PLASTROTYL (*n*): plastrotyl (trade name for a plastic form of trinitrotoluene; see **TRINITRO-TOLUOL**) 623
PLATANE (*f*): plane tree (*Platanus occidentalis*) 624
PLATANENGEWÄCHSE (*n.pl*): *Platanaceæ* (Bot.) 625
PLATANENHOLZ (*n*): plane tree wood; Sg. 0.5 to 0.7 (see **PLATANE**) 626
PLATIN (*n*): platinum (Pt) 627
PLATINA (*f*): platina, old name for platinum (see **PLATIN**); platina (a zinc-copper alloy, with 54% zinc and 46% copper) 628
PLATINAMMONIUMCHLORID (*n*): chloride of platinum and ammonium; ammonium chloroplatinate (see **AMMONIUMPLATIN-CHLORID**) 629
PLATINAPPARAT (*m*): platinum apparatus 630
PLATINASBEST (*m*): platinized asbestos (asbestos soaked in $PtCl_4$ solution, dipped in ammonium chloride solution and heated; has a platinum content up to 25%) 631
PLATINBAD (*n*): platinum bath (for platinizing) 632
PLATINBARRE (*f*): platinum ingot 633
PLATINBLASE (*f*): platinum still 634
PLATINBLECH (*n*): platinum foil 635
PLATINBLENDE (*f*): platinum screen 636
PLATINCHLORID (*n*): platinum (platinic) chloride; platinum tetrachloride; $PtCl_4.5H_2O$; Sg. 2.43 637
PLATINCHLORÜR (*n*): platinous chloride; platinum (dichloride) bichloride; $PtCl_2$; Sg. 5.87 638
PLATINCHLORWASSERSTOFF (*m*): chloroplatinic acid; $H_2PtCl_6.H_2O$; Sg. 2.431 (see also **PLATINICHLORWASSERSTOFF**) 639
PLATINCHLORWASSERSTOFFSÄURE (*f*): (see **PLAT-INCHLORWASSERSTOFF**) 640
PLATINCYANBARIUM (*n*): barium platinocyanide (see **BARYUMPLATINCYANUR**) 641
PLATINCYANÜR (*n*): platinous cyanide; platinocyanide; cyanoplatinite; $Pt(CN)_2$ 642
PLATINCYANWASSERSTOFF (*m*): platinocyanic acid; cyanoplatinous acid; $H_2Pt(CN)_4$ 643
PLATINCYANWASSERSTOFFSÄURE (*f*): (see **PLAT-INCYANWASSERSTOFF**) 644
PLATINDRAHT (*m*): platinum wire 645
PLATINDRAHTHALTER (*m*): platinum-wire holder 646
PLATINDRUCK (*m*): platinotype process (Phot.) 647
PLATINE (*f*): plate; mill bar (Metal); flat bar; slab (billet) for metal sheets 648
PLATINELEKTRODE (*f*): platinum electrode 649
PLATINERSATZ (*m*): platinum substitute 650
PLATINERSATZELEKTRODE (*f*): platinum substitute electrode 651
PLATINERZ (*n*): platinum ore (metallic platinum plus iron, iridium, osmium, palladium, rhodium, ruthenium, etc.) 652
PLATINERZKORN (*n*): grain of platinum ore 653
PLATINFARBE (*f*): platinum colour 654
PLATINFOLIE (*f*): platinum foil 655
PLATINHALTIG: containing platinum; platiniferous 656
PLATINHYDROSOL (*n*): platinum hydrosol (colloidal platinum; see **HYDROSOL**) (by adding a few drops of a very diluted solution of hydrazine hydrate to a 1 in 1500 solution of $PtCl_4$) 657
PLATINICHLORID (*n*): platinic chloride (see **PLATINCHLORID**) 658
PLATINICHLORWASSERSTOFF (*m*): chloroplatinic acid; platinum (tetra)-chloride; platinic chloride; $PtCl_4$ (by dissolving metallic platinum in aqua regia; see **KÖNIGSWASSER** and **PLATINCHLORWASSERSTOFF**) 659
PLATINICHLORWASSERSTOFFSÄURE (*f*): (see **PLAT-INCHLORWASSERSTOFF**) 660
PLATINIEREN: to platinize; combine with platinum; cover with platinum; coat with platinum 661
PLATINIERTER ASBEST (*m*): platinized asbestos (see **PLATINASBEST**) 662
PLATINIRIDIUM (*n*): platinum-iridium; Pt+Ir (with copper, iron, rhodium and other constituents) Sg. 22.8 663
PLATINIRIDIUMLEGIERUNG (*f*): platinumiridium alloy (with an iridium content up to 10%) 664
PLATINISALZ (*n*): platinic salt 665
PLATINIT (*n*): platinite (nickel-iron alloy) 666
PLATINIVERBINDUNG (*f*): platinic compound 667
PLATINKALIUMCHLORÜR (*n*): potassium chloroplatinite (see **KALIUMPLATINOCHLORID**) 668
PLATINKAPSEL (*f*): platinum capsule (for coal analysis) 669
PLATIN (*n*), **KOLLOIDALES-**: colloidal platinum 670
PLATINLEGIERUNG (*f*): platinum alloy 671
PLATINLÖFFEL (*m*): platinum spoon 672
PLATIN (*n*), **LÖSLICHES-**: soluble platinum, colloidal platinum 673
PLATINLÖSUNG (*f*): platinum solution 674
PLATINMETALL (*n*): platinum metal 675
PLATINMETALL (*n*), **LEICHTES-**: light platinum metal (palladium, rhodium and ruthenium) 676

PLATINMETALL (*n*), **SCHWERES-**: heavy platinum metal (iridium, osmium and platinum) 677

PLATINMOHR (*m*): platinum (mohr) black (obtained by precipitation with organic reducing agents); Sg. 15.8-17.6 678

PLATINOCHLORID (*n*): platinous chloride (see **PLATINCHLORÜR**) 679

PLATINOCHLORWASSERSTOFF (*m*): chloroplatinous acid; $H_2PtCl_4.4H_2O$ 680

PLATINOCHLORWASSERSTOFFSÄURE (*f*): (see **PLATINOCHLORWASSERSTOFF**) 681

PLATINOCYANWASSERSTOFF (*m*): platinocyanic acid; cyanoplatinous acid (see **PLATIN-CYANWASSERSTOFF**) 682

PLATINOCYANWASSERSTOFFSÄURE (*f*): (see **PLATINCYANWASSERSTOFF**) 683

PLATINOID (*n*): platinoid (an alloy) (55.5% Cu; 22% Ni; 22% Zn; 0.5% W) 684

PLATINOVERBINDUNG (*f*): platinous compound 685

PLATINOXYD (*n*): platinum (platinic) oxide; PtO_2 686

PLATINOXYDUL (*n*): platinum (platinous) oxide; PtO 687

PLATINOXYDULVERBINDUNG (*f*): platinous compound 688

PLATINOXYDVERBINDUNG (*f*): platinic compound 689

PLATINPAPIER (*n*): platinotype paper (Phot.) 690

PLATINSALMIAK (*m*): ammonium chloroplatinate (see **AMMONIUMPLATINCHLORID**) 691

PLATINSALZ (*n*): platinum salt 692

PLATINSAUER: platinate of; combined with platinic acid 693

PLATINSÄURE (*f*): platinic acid 694

PLATINSCHALE (*f*): platinum dish 695

PLATINSCHIFFCHEN (*n*): platinum boat 696

PLATINSCHWAMM (*m*): platinum sponge; spongy platinum (Pt) (a porous mass obtained by heating ammonium chloroplatinate; see **AMMONIUMPLATINCHLORID**) 697

PLATINSCHWARZ (*n*): platinum black (see **PLATIN-MOHR**) 698

PLATINSELENID (*n*): platinum selenide; $PtSe_3$; Sg. 7.15; $PtSe_2$; Sg. 7.65 699

PLATINSILICID (*n*): platinum silicide; Pt_2Si; Sg. 13.8; $PtSi$; Sg. 11.63 700

PLATINSPATEL (*m*): platinum spatula 701

PLATINSPITZE (*f*): platinum point 702

PLATINSULFÜR (*n*): platinum sulphide; PtS; Sg. 8.897 703

PLATINTIEGEL (*m*): platinum crucible 704

PLATINTONBAD (*n*): platinum toning bath 705

PLATINVERBINDUNG (*f*): platinum compound 706

PLATINWASSERSTOFFCHLORID (*n*): (see **PLATIN-CHLORWASSERSTOFFSÄURE**) 707

PLATMENAGE (*f*): cruet-stand 708

PLATSCHEN: to splash 709

PLÄTSCHERN: to plash, splash, murmur, ripple 710

PLÄTTBAR: laminable, capable of being (scaled) laminated 711

PLATT: flat; smooth; level; even; plain; prostrate; low; horizontal; champaign (of an extent of country); dull; quite; absolutely; vulgar; (*n*): patois; dialect 712

PLATTBLANK: sleek (of leather) 713

PLATTBOGIG: elliptical 714

PLATTBOOT (*n*): (any)-flat bottomed boat 715

PLÄTTCHEN (*n*): lamella (Bot.); small plate; fillet (Arch.) 716

PLATTDEUTSCH: low-German; dialect-German; [also as a noun (*n*)] 717

PLATTE (*f*): plate (see also **NEGATIV**); sheet; slab; lamina; tray; salver; leaf; flag-stone; flaw (Textile); flat (portion) object; smoothing (flat) iron; swan-neck cover (Brick); plinth; plateau; sand-bank; plane 718

PLÄTTE (*f*): smoothing (flat) iron 719

PLATTE (*f*), **GANZE-**: wholeplate (size $8\frac{1}{2}'' \times 6\frac{1}{2}''$) (Photography) 720

PLATTE (*f*), **HOCHEMPFINDLICHE-**: highly sensitive plate (a photographic dry plate) 721

PLÄTTEISEN (*n*): (see **PLÄTTE**) 722

PLATTE (*f*), **MOMENT-**: snap-shot plate (photographic dry plate) 723

PLÄTTEN: to flatten; iron; laminate; scale; exfoliate 724

PLATTEN: to level; flatten; plate; laminate 725

PLATTENDRUCK (*m*): stereotype, stereotypography, stereotyping 726

PLATTENFORMAT (*n*): size of plate 727

PLATTENFÖRMIG: lamellar; lamelliform; laminated; plate-like; foliated; plate-shaped 728

PLATTENGANG (*m*): strake of plates or plating (Naut.) 729

PLATTENHALTER (*m*): plate holder (Phot.) 730

PLATTENHEBER (*m*): plate-lifter 731

PLATTEN, IN- BRECHEN: to laminate; flake off; scale; exfoliate 732

PLATTENKAUTSCHUK (*n*): sheet rubber 733

PLATTENKULTUR (*f*): plate culture 734

PLATTENKUPFER (*n*): sheet copper 735

PLATTENLAGE (*f*): layer of plates 736

PLATTENMETALL (*n*): sheet metal 737

PLATTENOBERFLÄCHE (*f*): surface of the plate 738

PLATTENPUTZPULVER (*n*), **GRÜNES-**: infusorial earth (Phot.) 739

PLATTENSCHRIFT (*f*): stereotype 740

PLATTENSTÄRKE (*f*): plate thickness 741

PLATTENSTEIN (*m*): flat (brick, stone) slab, (used as a filling material) 742

PLATTENSTOSS (*m*): butt (of plates) 743

PLATTENTROCKENSTÄNDER (*m*): drying rack (for photographic plates) 744

PLATTENTURM (*m*): plate (column) tower (a form of reaction tower in which perforated, ribbed plates are employed as the filling material; see **REAKTIONSTRUM**) 745

PLATTENWASCHKASTEN (*m*): (plate)-washing tank 746

PLATTENWECHSEL (*m*): changing of plates, plate changing (Phot.) 747

PLATTENWECHSELVORRICHTUNG (*f*): plate changing arrangement 748

PLATTENZÄHLER (*m*): plate counter, indicator (Phot.) 749

PLATTE (*f*), **PORTRÄT-**: portrait plate (a photographic dry-plate) 750

PLATTFORM (*f*): platform (used to describe anything flat like a platform, but not used for station platform) 751

PLATTGARN (*n*): embroidery cotton 752

PLÄTTGLOCKE (*f*): box-iron, smoothing iron 753

PLATTIEREN: to plate 754

PLATTIERT: plated 755

PLATTIERUNG (*f*): plate; plating 756

PLATTINE (*f*): (see **PLATINE**) 757

PLÄTTMITTEL (*n*): ironing agent or medium (any substance used for ironing washing) 758

PLATTNERIT (*m*): plattnerite; PbO_2; Sg. 8.55 759
PLÄTTÖL (*n*): ironing oil (for glossing washing) 760
PLÄTTPULVER (*n*): ironing powder (composed of borax, talc and tragacanth) (for ironing washing) 761
PLATTSCHIENEN (*f.pl*): plate-rails (Railways) 762
PLATTSTEIN (*m*): flag-(stone) 763
PLATTSTRECKEN: to stretch; dress (Felt); press; (*n*): pressing; stretching; dressing 764
PLÄTTWACHS (*n*): ironing wax (composed of Japan wax, paraffin and stearic acid in the proportion of 2 : 2 : 1; used for ironing washing) 765
PLÄTTWÄSCHE (*f*): ironed linen 766
PLÄTTWÄSCHE (*f*), **IMPRÄGNIERTE-**: impregnated ironed linen (ordinary ironed linen articles of attire which have been dipped in an amyl acetate and collodium solution to make them washable) 767
PLATTZANGE (*f*): pliers 768
PLATZ (*m*): place; town; room; spot; site; station; square; seat; locality; location; smash; (open)-space; stand; area; courtyard; (cab)-rank; grade; standing; position; situation; appointment; post; dwelling; residence; way; city; village; point; fort-(ress); bun; small cake; burst; explosion; break; rupture; crack 769
PLÄTZCHEN (*n*): small cake; bun; troche; pastille (*Rotulæ*) (Pharmacy); lozenge; tablet; small (town) place 770
PLATZEN: to burst; crack; crash; smash; break; rupture; explode 771
PLÄTZEN: to blaze-(trees); smack, pop, slap 772
PLATZKOKS (*m*): coke direct from the spot where it is unloaded 773
PLATZPATRONE (*f*): blank cartridge 774
PLATZREGEN (*m*): heavy shower of rain; sudden downpour 775
PLATZ-VERTRETUNG (*f*): local agency 776
PLATZWECHSEL (*m*): local bill 777
PLAUDERHAFT: talkative; loquacious; gossiping 778
PLAUSIBEL: plausible 779
PLEBEJER (*m*): plebeian 780
PLEBEJISCH: plebeian 781
PLEBISZIT (*n*): plebescite 782
PLEJADEN (*pl*): pleiades (cluster of stars in Taurus) (Astron.) 783
PLENARSITZUNG (*f*): full sitting (of any assembly of persons) 784
PLENUM (*n*): full (complete) assembly 785
PLEOCHROISMUS (*m*): pleochroism 786
PLEONASMUS (*m*): pleonasm, redundancy, diffuseness, tautology 787
PLEONAST (*m*): pleonaste, (a variety of Spinel); MgO,Al_2O_3; Sg. 3.6 788
PLEONASTISCH: pleonastic, redundant, diffuse, superfluous, tautological 789
PLESSIT (*m*): plessite; (Fe,Ni,Co); Sg. 7.55 790
PLEUELKOPF (*m*): connecting rod head 791
PLEUELSTANGE (*f*): connecting rod 792
PLEUELSTANGENSCHAFT (*m*): (body of the)-connecting rod 793
PLEURA (*f*): pleura (the serous membrane investing the lungs) (Med.) 794
PLEURAHÖHLE (*f*): pleural cavity (Med.) 795
PLEURITIS (*f*): pleurisy (inflammation of the pleura) (Med.) 796
PLINSE (*f*): omelette; pancake 797
PLINTH-(E) (*f*): plinth (Arch.) 798
PLINZE (*f*): (see **PLINSE**) 799
PLIOCEN: pliocene (Geology) 800
PLOMBE (*f*): plug; filling (Dental); leaden seal (for boxes, sacks, etc.) 801
PLOMBIEREN: to seal with lead; affix leaden seals; stop (Teeth) 802
PLÖTZLICH: sudden-(ly); instantaneous-(ly); abrupt-(ly) 803
PLÖTZLICHKEIT (*f*): suddenness; abruptness 804
PLUMBAT (*n*): plumbate 805
PLUMBISALZ (*n*): plumbic salt (see under **BLEI-**) 806
PLUMBOCALCIT (*m*): plumbocalcite; (Pb,Ca)O, CO_2; Sg. 2.75 807
PLUMBOSALZ (*n*): plumbous salt (see under **BLEI-**) 808
PLUMIERASÄURE (*f*): plumieric acid (from *Plumiera rubra*) 809
PLUMOSIT (*m*): plumosite; $Pb_2Sb_2S_5$; Sg. 5.6 810
PLUMP: bulky; clumsy; awkward; gross; coarse; blunt; heavy; ill-bred; (*m*): heavy fall; bump 811
PLUNDER (*m*): rubbish; lumber; trumpery; trash; rags; litter 812
PLÜNDERN: to plunder; pillage; sack 813
PLUNGER (*m*): plunger, piston, ram-(mer), ramrod 814
PLUNGERKOLBEN (*m*): piston trunk, plunger piston 815
PLUNGERPUMPE (*f*): plunger pump; ram (displacement) pump 816
PLUNSCHER (*m*): plunger 817
PLURAL (*m*): plural 818
PLURALITÄT (*f*): plurality, multiplicity 819
PLUS: plus (+) 820
PLÜSCH (*m*): plush 821
PLUSQUAMPERFEKTUM (*n*): pluperfect (Grammar) 822
PLUSZEICHEN (*n*): plus sign; (+) 823
PLUTONISCH: plutonian, infernal 824
PLUTONISCHES GESTEIN (*n*): (see **PLUTONIT**) 825
PLUTONISMUS (*m*): Plutonian theory 826
PLUTONIT (*n*): plutonic rock 827
PNEUMATIK (*m*): pneumatics, pneumatology (the science of air and other elastic fluids); pneumatic tyre 828
PNEUMATISCH: pneumatic 829
PNEUMONIE (*f*): pneumonia (inflammation of the lungs) (Med.) 830
PNEUMONIE (*f*), **BRONCHO-**: broncho (catarrhal, lobular) pneumonia (Med.) 831
PNEUMONIE (*f*), **CROUPÖSE-**: croupous (acute, lobar) pneumonia (Med.) 832
PÖBEL (*m*): rabble; mob 833
PÖBELHAFT: vulgar; plebeian; coarse; low 834
POCHE (*f*): beater, mallet, crusher, pounder, stamper 835
POCHEISEN (*n*): ram, foot (shoe) of a stamper 836
POCHEN to beat; knock; stamp; brag; boast; trust; pulverize; break, (Ore); throb; pound; crush 837
POCHER (*m*) stamp; ram; breaker; crusher; boaster; braggart 838
POCHERZ (*n*): ore (as mined) 839
POCHHAMMER (*m*) pounding hammer 840
POCHHERD (*m*): buddle 841
POCHMEHL (*n*): crushed (broken; pulverized) ore 842
POCHMÜHLE (*f*): stamp mill 843

POCHSCHLÄGEL (*m*): ore (hammer) breaker 844
POCHSCHUH (*m*): ram, foot (shoe) of a stamper 845
POCHSTEIGER (*m*): manager of a stamp mill 846
POCHSTEMPEL (*m*): stamper, stamping iron 847
POCHSTEMPELREIHE (*f*): stamp battery 848
POCHWERK (*n*): stamp mill 849
POCKE (*f*): pock-(mark); (*pl*) (POCKEN), smallpox; variola (Med.) (a contagious and eruptive fever) 850
POCKENGIFT (*n*): smallpox virus 851
POCKENGRUBE (*f*): pock-mark 852
POCKENHOLZ (*n*): pockwood, lignum vitæ, guaiacum wood (see GUAJAKHOLZ) 853
POCKENLYMPHE (*f*): vaccine lymph 854
POCKENNARBE (*f*): pock mark 855
POCKENNARBIG: pockmarked; pitted 856
POCKENSTEIN (*m*): variolite (Mineralogy) 857
POCKENWURZEL (*f*): (see CHINAWURZEL) 858
POCKHOLZ (*n*): pockwood; lignum vitæ; guaiacum wood (see GUAJAKHOLZ) 859
POCKHOLZLAGER (*n*): guaiacum-wood bearing, bearing made of lignum vitæ 860
PODAGRA (*n*): gout; podagra 861
PODEST (*m*): resting place, stage, platform; landing (of stairs) 862
PODIUM (*n*): platform, stage 863
PODOPHYLLIN (*n*) podophyllin-(resin) (an alkaloid from the root of *Podophyllum peltatum*) 864
PODOPHYLLINHARZ (*n*): podophyllin resin (see PODOPHYLLIN) 865
PODOPHYLLWURZEL (*f*): podophyllum root (*Radix podophylli* of *Podophyllum emodi*) 866
POESIE (*f*): poetry; poem 867
POETISCH: poetic-(al)-(ly) 868
POGAÖL (*n*): Poga oil (from *Poga oleosa*) 869
POINTIEREN: to give point to -(remarks); to point 870
POKAL (*m*): goblet; (loving)-cup; bumper 871
PÖKEL (*m*): pickle; brine (the liquid) 872
PÖKELFASS (*n*): pickling (salting) vat or tub 873
PÖKELFLEISCH (*n*): pickled (salted) meat 874
PÖKELHÄRING (*m*): pickled herring 875
PÖKELHERING (*m*): pickled herring 876
PÖKELN: to pickle; salt; corn 877
PÖKELROGEN (*m*): caviare (made from salted roes) 878
POKER (*m*): poker, pricker 879
POKULIEREN: to carouse; drink-(heavily) 880
POL (*m*) pole 881
POLABSTAND (*m*): polar distance 882
POLAR: polar 883
POLARISATION (*f*) polarization 884
POLARISATIONSAPPARAT (*m*) polarizing (polarization) apparatus 885
POLARISATIONSEBENE (*f*): plane of polarization 886
POLARISATIONSPRISMA (*n*): polarizer 887
POLARISATIONSSTROM (*m*): polarization (secondary) current (Elect.) 888
POLARISATOR (*m*): polarizer 889
POLARISIERBAR: polarizable 890
POLARISIEREN: to polarize 891
POLARISIERUNG (*f*): polarization; polarizing 892
POLARITÄT (*f*): polarity 893
POLARKREIS (*m*): polar circle; arctic zone 894
POLECK (*n*): summit (Cryst.) 895
POLECKE (*f*): (see POLECK) 896
POLEI (*f* and *m*): pennyroyal (*Mentha pulegium*) 897
POLEI, AMERIKANISCHE-(R)- (*m* and *f*): American pennyroyal; squaw mint (*Hedeoma pulegioides*) 898
POLEIÖL (*n*): pennyroyal oil; hedeoma oil (distilled from *Hedeoma pulegioides*); Sg. 0.935 899
POLEI, WILDE-(R)- (*f* and *m*): corn-mint (*Mentha arvensis*) 900
POLEMIK (*f*): polemic-(s); controversy 901
POLEMISCH: polemical; controversial 902
POLEN (*n*): poling (an oxydating purification of the metal in iron pans by the aid of green wooden poles) (Tin); Poland 903
POLEN: to pole (Metal) 904
POLEY (*m*): pennyroyal (see also POLEI) 905
POLHÖHE (*f*): latitude; elevation of the pole 906
POLIANIT (*m*): polianite (a manganese dioxide); MnO_2; Sg. 4.9 907
POLICE (*f*): insurance policy 908
POLIERAPPARAT (*m*): polishing apparatus 909
POLIERBAR: polishable 910
POLIERBÜRSTE (*f*): polishing brush 911
POLIEREISEN (*n*): burnisher 912
POLIEREN: to polish; burnish; smooth; brighten 913
POLIERER (*m*): polisher; burnisher 914
POLIERKALK (*m*): polishing chalk or lime, whiting 915
POLIERKREIDE (*f*): (see POLIERKALK) 916
POLIERLEDER (*n*): wash-leather 917
POLIERMITTEL (*n*): polishing (medium, agent) substance, polish 918
POLIERÖL (*n*): polishing oil 919
POLIERPAPIER (*n*): polishing paper; sandpaper 920
POLIERPASTA (*f*): polishing paste 921
POLIERPULVER (*n*): polishing powder 922
POLIERROT (*n*): colcothar; rouge; crocus; polishing red; jeweller's red or rouge 923
POLIERSCHEIBE (*f*): polishing disc 924
POLIERSCHIEFER (*m*): polishing slate (see TRIPEL) 925
POLIERSTAHL (*m*): burnisher 926
POLIERSTOCK (*m*): polishing stick; polishing anvil (of braziers) 927
POLIERWACHS (*n*): polishing wax 928
POLITIK (*f*), POLITIKA (*pl*): policy, politics 929
POLITIKER (*m*): politician 930
POLITISCH: politic-(al)-(ly) 931
POLITUR (*f*): polish-(ing); gloss; flatting-(varnish); French polish 932
POLITURFÄHIG: capable of being polished 933
POLITURLACK (*m*): flatting (polishing) varnish 934
POLIZE (*f*): insurance policy 935
POLIZEI (*f*): police; police station; constabulary 936
POLIZEIAMT (*n*): police station 937
POLIZEIBEAMTE (*m*): police officer, policeman 938
POLIZEIBEHÖRDE (*f*): police authorities 939
POLIZEIDIENER (*m*): policeman; constable 940
POLIZEIGERICHT (*n*): police-court 941
POLIZEILICH: police; of the police 942
POLIZEILICHE AUFSICHT (*f*): police supervision 943
POLIZEIORDNUNG (*f*): police regulation 944
POLIZEIWIDRIG: against (contrary to) police regulations 945
POLIZIST (*m*): policeman; constable 946
POLKLEMME (*f*): terminal 947
POLLANTIN (*n*): pollantin (an antitoxin for hay fever; obtained from the blood serum of

horses in Germany and from the pollen of various plants such as Golden rod, in America) 948
POLLEN (*m*): pollen (Bot.) 949
POLLENÜBERTRAGUNG (*f*): pollination (Bot.) 950
PÖLLER (*m*): bitt, bitton, bollard 951
PÖLLER (*m*), DOPPEL-: double bitton 952
POLLMEHL (*n*): coarse meal 953
POLLUCIT (*m*): pollucite (see POLLUX) 954
POLLUX (*m*): pollux; $H_2Cs_4Al_4(SiO_3)_9$; Sg. 3.0 955
POLONIUM (*n*): polonium (an element discovered by Curie; a radio-active substance obtained from pitch-blende) 956
POLONIUM-WISMUT (*m* and *n*): the name given to radio-active bismuth, obtained from pitch-blende 957
POLSTEIN (*m*): lode-stone, loadstone (a magnetic ore) 958
POLSTEINKRAFT (*f*): magnetic (attractive) power, attraction, magnetism 959
POLSTER (*n*): bolster; cushion; hassock; pad-(ding); upholstery 960
POLSTERFÖRMIG: pulvinate (Bot. and Arch.) 961
POLSTERN (*m*): pole-star, lode-star, loadstar 962
POLSTERN: to stuff, upholster, quilt, pad 963
POLSTERSTUHL (*m*): upholstered chair; easy chair 964
POLTERGEIST (*m*): hobgoblin; noisy fellow 965
POLTERKAMMER (*f*): lumber room 966
POLTERN: to make (a noise; an uproar; a clamour; a racket; a din; a hubbub) a tumult; scold; bounce; bluster; create a disturbance 967
POLWECHSEL (*m*): change (reversal) of poles 968
POLYBASIT (*m*): polybasite (silver sulphantimonite); $(As,Sb)(Ag,Cu)_9S_6$; Sg. 6.2 969
POLYCHROM: polychromatic, pleochroic; (*n*): polychrome; pyromorphite (see PYROMOR-PHIT) 970
POLYCHROMATISCHER FELDSPAT (*m*): labradorite 971
POLYCHROMSÄURE (*f*): polychromic acid 972
POLYCYCLISCH: polycyclic 973
POLYDISPERS: polydisperse 974
POLYEDER (*n*): polyhedron 975
POLYGALASÄURE (*f*): polygalic acid (from *Polygala senega*) 976
POLYGALYSAT (*n*): polygalysat (a preparation obtained by dialysis from the senega or snake root, *Polygala senega*) 977
POLYGAMISCH: polygamous 978
POLYGLOTTE (*f*): polyglot 979
POLYGONISCH: polygonal 980
POLYGYNISCH: polyginian (Bot.); polygynous 981
POLYHALIT (*m*): polyhalite (Mineral) (a natural potassium, calcium and magnesium sulphate from the Stassfurt district of Germany) (see ABRAUMSALZ); $(Ca,K_2,Mg)SO_4.H_2O$: Sg. 2.77 982
POLYHEDRISCH: polyhedral 983
POLYHETEROATOMIG: polyheteratomic 984
POLYHYPERJODAT (*n*): polyperiodate 985
POLYKIESELSÄURE (*f*): polysilicic acid 986
POLYKRAS (*m*): polycrase; $(Nb,Ta)_2O_5(Y,Er,Ce)_2O_3,FeO,(Ti,U)O_2,H_2O$; Sg. 4.9 987
POLYLAKTOL (*n*): Polylaktol (trade name of an iron-albumose preparation for increasing milk secretion) 988
POLYMER: polymeric 989
POLYMERIE (*f*): polymerism 990
POLYMERISATION (*f*): polymerization 991
POLYMERISATIONSNEIGUNG (*f*): tendency to polymerization 992
POLYMERISATIONSPRODUKT (*n*): polymerization product, product of polymerization 993
POLYMERISIERBAR: polymerizable 994
POLYMERISIEREN: to polymerize 995
POLYMERISIERTES ÖL (*n*): polymerized (blown, oxidized) oil, (oils, such as linseed or rape oil, which are made denser by having air blown through them when either in a cold or heated state) 996
POLYMERISIERUNG (*f*): polymerization, polymerizing 997
POLYMETHYLEN (*n*): polymethylene (see NAPTHEN) 998
POLYMIGNYT (*m*): polymignyte; $(Nb,Ta)_2O_5,(Y,Er,Ce,Fe,Di,La)_2O_3,(Ti,Sn,Th,Zr)O_2,(Ca,Fe)O,H_2O$; Sg. 4.8 999
POLYMOLYBDÄNSÄURE (*f*): polymolybdic acid 000
POLYMORPH: polymorphous; polymorphic 001
POLYMORPHIE (*f*): polymorphy; polymorphism 002
POLYP (*m*): polypus (a pedunculated growth springing from a mucous surface) (Med.); polype (Zoophyte) 003
POLYPEPTID (*n*): polypeptide 004
POLYPHONISCH: polyphonic 005
POLYSACCHARID (*n*): polysaccharide; $(C_6H_{10}O_5)_n$ (see STÄRKE, ZELLULOSE and DEXTRIN) 006
POLYTECHNIK (*f*): polytechnics 007
POLYTECHNISCH: polytechnic 008
POLYTHEISMUS (*m*): polytheism 009
POLYTHIONSÄURE (*f*): polythionic acid 010
POLYTROPE (*f*): polytropic (curve) line, entropy (pressure-volume, p-v) line or curve 011
POLYTROPISCH: polytropic 012
POLYURIA (*f*): polyuria, diabetes insipidus (Med.) 013
POMADE (*f*): pomade; pomatum 014
POMADENBÜCHSE (*f*): pomade (pomatum) pot 015
POMADIG: cool; phlegmatic; indifferent; free-and-easy 016
POMERANZE (*f*): orange (see also ORANGE) 017
POMERANZENÄHNLICH: orange-like 018
POMERANZENBAUM (*m*): orange tree (*Citrus vulgaris* and *Citrus aurantium*) 019
POMERANZENBAUM (*m*), BITTERFRÜCHTIGER-: bitter orange tree, Seville orange (*Citrus vulgaris*) 020
POMERANZENBAUM (*m*), SÜSSFRÜCHTIGER-: (sweet)-orange tree (*Citrus aurantium*) 021
POMERANZENBLÜTE (*f*): orange-blossom 022
POMERANZENBLÜTENÖL (*n*): neroli; orange-flower oil (see NEROLIÖL) 023
POMERANZENFARBIG: orange (yellow) coloured 024
POMERANZENGELB: orange-yellow; orange 025
POMERANZENKONFEKT (*n*): confection of oranges (*Confectio aurantii* or *Conditum aurantii*) 026
POMERANZENLIQUER (*m*): curaçao 027
POMERANZENÖL (*n*): orange oil (see POMERANZ-ENSCHALENÖL, BITTERES-) 028
POMERANZENSAFT (*m*): orange juice 029
POMERANZENSCHALE (*f*), EINGEMACHTE-: candied orange-peel 030
POMERANZENSCHALEN (*f.pl*): orange peels 031
POMERANZENSCHALENÖL (*n*): orange-(peel) oil (see POMERANZENSCHALENÖL, BITTERES- and SÜSSES-) 032

POMERANZENSCHALENÖL, BITTERES- (*n*): bitter orange-(peel) oil (*Oleum aurantii corticis amari*) (from skins of fruit of *Citrus vulgaris*); Sg. 0.844 033
POMERANZENSCHALENÖL, SÜSSES- (*n*): sweet orange-(peel) oil (*Oleum aurantii corticis*) (from skins of fruit of *Citrus aurantium*); Sg. 0.852 034
POMERANZEN-(SCHALEN) -SIRUP (*m*): syrup of orange 035
POMERANZENTINKTUR (*f*): tincture of orange 036
POMP (*m*): pomp; state 037
POMPEJANISCH: Pompeian 038
POMPEJANISCHROT (*n*): Pompeian red 039
POMPELMUSE (*f*): shaddock; Adam's apple; forbidden fruit; grape fruit (*Citrus decumana*) 040
POMPHAFT: pompous; stately; grand; ostentatious; majestic; showy; grandiloquent; grandiose; magniloquent 041
POMPÖS: pompous, magnificent, majestic, superb; splendid, august, dignified 042
PONCEAU (*n*): ponceau (of dyes) 043
PONTON (*m*): pontoon 044
POPANZ (*m*): bugbear 045
POPULÄR: popular 046
POPULARISIEREN: to popularize 047
POPULARITÄT (*f*): popularity 048
POPULÖS: populous, crowded, thickly populated 049
PORCELLAN (*n*): porcelain (see **PORZELLAN**) 050
PORCELLANERDE (*f*): kaolin, china clay (see **KAOLIN**) 051
PORE (*f*): pore; spiracle (breathing-hole; animal or vegetable) 052
PORENKAPAZITÄT (*f*): pore capacity 053
PORENVOLUMEN (*n*): pore volume 054
PORIG: porous; having (full of) pores 055
PORKUPINEHOLZ (*n*): porcupine wood, palmyra wood (wood of the cocoanut palm) 056
PORKUPINENHOLZ (*n*): (see **PORKUPINEHOLZ**) 057
PORÖS: porous; permeable; penetrable; percolable 058
POROSITÄT (*f*): porosity; porousness 059
PORPHYR (*m*): porphry, porphyry (an effusive rock, consisting of a compact mass containing crystals of feldspar, mica, quartz, etc.); Sg. red, 2.77 060
PORPHYRARTIG: porphyritic 061
PORPHYRFELSEN (*m*): porphyritic rock 062
PORPHYRGESTEIN (*n*): porphyritic rock 063
PORPHYRTUFF (*m*): tufaceous porphyry 064
PORREE (*m*): (see **WINTERPORREE** and **SOMMER-PORREE**) 065
PORSCH (*m*): marsh tea; wild rosemary (*Ledum palustre*) 066
PORSCHKRAUT (*n*): (see **PORSCH**) 067
PORST (*m*): (see **PORSCH**) 068
PORTIER (*m*): porter; doorkeeper 069
PORTION (*f*): share; portion; part; ration 070
PORTLANDKALK (*m*): Portland limestone 071
PORTLANDZEMENT (*m* and *n*): Portland cement (Sg. 1.3) 072
PORTLAND-ZEMENT-FABRIK (*f*): Portland-cement-works 073
PORTO (*n*): postage; carriage; (cost of)-transport-(charges) 074
PORTOFREI: postfree; postpaid 075
PORTOSATZ (*m*): (rate of)-postage, carriage or transport 076
PORTRAIT (*n*): portrait; likeness (see also **POR-TRÄT**) 077
PORTRAITIEREN: to portray 078
PORTRAITMALER (*m*): portrait painter 079
PORTRÄT (*n*): portrait (see also **PORTRAIT**) 080
PORTRÄTAUFNAHME (*f*): portraiture (Phot.) 081
PORTRÄTBÜSTE (*f*): bust portrait, head and shoulders (Phot.) 082
PORTRÄTOBJEKTIV (*n*): portrait lens 083
PORTUGALLOÖL (*n*): (see **PORTUGALÖL**) 084
PORTUGALÖL (*n*): Portugal (portugallo) oil; orange-peel oil (see **POMERANZENSCHALENÖL, SÜSSES-**) 085
PORTUGALWASSER (*n*): laurel water 086
PORTULAK (*m*): purslane (*Peplis portula*) 087
PORTWEIN (*m*): port wine, port (Sg. 1.0) 088
PORZELLAN (*n*): porcelain; china; Sg. 2.3 (a compact, white, translucent form of pottery) 089
PORZELLANAUFSATZ (*m*): porcelain (china) service 090
PORZELLANBECHER (*m*): porcelain (cup) bowl 091
PORZELLANBLAU: China-blue; (*n*): China blue 092
PORZELLANBREI (*m*): porcelain slip 093
PORZELLANBRENNEREI (*f*): porcelain factory 094
PORZELLANBRENNOFEN (*m*): porcelain kiln 095
PORZELLAN (*n*), **ECHTES-**: real porcelain (see **FELDSPATPORZELLAN**) 096
PORZELLANEINSATZ (*m*): porcelain liner (see also **EINSATZ**) 097
PORZELLANEMAILFARBE (*f*): porcelain enamel 098
PORZELLANEN: porcelain; of porcelain; china; of china 099
PORZELLANERDE (*f*): porcelain (china) clay; kaolin (see **KAOLIN**) 100
PORZELLANFABRIK (*f*): porcelain factory 101
PORZELLANFARBE (*f*): porcelain paint, porcelain colour (ground, coloured glass, for decorating porcelain) 102
PORZELLANFILTER (*n*): porcelain filter 103
PORZELLANGEFÄSS (*n*): porcelain (china) vessel or vase 104
PORZELLANGEGENSTAND (*m*): porcelain (article) object 105
PORZELLANGERÄT (*n*): porcelain (vessels) utensils 106
PORZELLANGERÄTSCHAFT (*f*): porcelain (vessels) utensils 107
PORZELLANGLASUR (*f*): porcelain glaze (consisting of kaolin, feldspar, chalk, calcite or marble, and sand) 108
PORZELLANGLÄTTE (*f*): (see **PORZELLANGLASUR**) 109
PORZELLANGLÜHAPPARAT (*m*): porcelain heating apparatus 110
PORZELLANGRIFF (*m*): porcelain handle 111
PORZELLANHAHN (*m*): porcelain cock 112
PORZELLAN (*n*), **HARTES-**: hard porcelain (see **FELDSPATPORZELLAN**) 113
PORZELLANINFUNDIERBÜCHSE (*f*): porcelain (infusion vessel, infusing vessel) digester 114
PORZELLAN (*n*) **ISOLATOR** (*m*): porcelain insulator 115
PORZELLAN (*n*), **JAPANISCHES-**: Japanese porcelain 116
PORZELLANJASPIS (*m*): porcelain jasper; porcelanite 117
PORZELLANKASSEROLLE (*f*): porcelain casserole 118
PORZELLANKITT (*m*): porcelain cement 119
PORZELLANKÜHLSCHLANGE (*f*): porcelain cooling-coil 120
PORZELLANLACK (*m*): porcelain varnish 121

PORZELLANMALEREI (*f*): porcelain painting 122
PORZELLANMANUFAKTUR (*f*): porcelain manufacture, porcelain factory 123
PORZELLANMASSE (*f*): porcelain (body) mass (see MARQUARDT'SCHE MASSE) 124
PORZELLANMÖRSER (*m*): porcelain mortar 125
PORZELLANMÖRTEL (*m*): pozzuolana mortar (see POZZOLANERDE) 126
PORZELLANMUFFEL (*f*): porcelain muffle 127
PORZELLANPLATTE (*f*): porcelain plate 128
PORZELLANRETORTE (*f*): porcelain retort 129
PORZELLANROHR (*n*): porcelain tube (for encasing thermo-elements) 130
PORZELLANRÖHRE (*f*): porcelain tube 131
PORZELLANSATTE (*f*): porcelain (bowl) dish 132
PORZELLANSCHALE (*f*): porcelain (bowl) dish 133
PORZELLANSCHIFFCHEN (*n*): porcelain boat 134
PORZELLANSCHMELZTIEGEL (*m*): porcelain crucible 135
PORZELLANSPAT (*m*): scapolite (see SKAPOLITH) 136
PORZELLANTEIL (*m*): piece of porcelain 137
PORZELLANTIEGEL (*m*): porcelain crucible 138
PORZELLANTON (*m*): porcelain (china) clay; kaolin (see KAOLIN) 139
PORZELLANTONUMSCHLAG (*m*): kaolin cataplasm; plaster splint 140
PORZELLANTRICHTER (*m*): porcelain funnel 141
PORZELLANTROMMEL (*f*): porcelain drum 142
PORZELLAN, UNECHTES- (*n*): earthenware 143
PORZELLANUNTERSATZ (*m*): porcelain (support) stand; porcelain (saucer) tray 144
PORZELLANWANNE (*f*): porcelain trough 145
PORZELLANWAREN (*f.pl*): porcelain wares 146
POSE (*f*): pose; quill 147
POSITIV: positive-(ly); (*n*): positive; print (Phot.) 148
POSITIVE (*f*): positive (Photography) 149
POSITIVLACK (*m*): positive varnish, crystal varnish (for varnishing transparencies, lantern-slides, etc., composed of gum dammar and benzene or alcohol, Canada balsam, sandarac and yellow shellac) 150
POSITIVRETOUCHE (*f*): print re-touching (Phot.) 151
POSITUR (*f*): posture, position 152
POSSE (*f*): antic; farce; burlesque; jest; fun; prank; drollery 153
POSSEN (*m*): trick; prank (see POSSE) 154
POSSENHAFT: comical; funny; droll; odd; ludicrous 155
POSSIERLICH: (see POSSENHAFT) 156
POST (*f*): post; news; mail; stage; post-office; entry; item; lot 157
POSTADRESSBUCH (*n*): postal directory 158
POSTADRESSE (*f*): postal address 159
POSTAMENT (*n*): pedestal; stand; base 160
POSTAMT (*n*): post-office 161
POSTANSTALT (*f*): post-office 162
POSTANWEISE (*f*): money order, postal order 163
POSTANWEISUNG (*f*): money order 164
POSTBEAMTE (*m*): postal (post-office) official 165
POSTBÜREAU (*n*): post office 166
POSTDAMPFER (*m*): mail boat, mail steamer 167
POSTDAMPFSCHIFF (*n*): mail boat, mail steamer 168
POSTDATIEREN: to post-date; (*n*): post-dating 169
POSTDIREKTION (*f*): General Post Office 170
POSTEN (*m*): post; place; station; sum; item; entry; amount; lot; parcel; batch (Metal); sentry; situation 171
POSTEXPEDITION (*f*): post office; parcel post 172
POSTFREI: postfree; postpaid 173
POSTGELD (*n*): postage 174
POSTIEREN: to post, place 175
POSTKARTE (*f*): post-card 176
POSTLAGERND: poste restante; to be called for at the post office 177
POSTMEISTER (*m*): post-master 178
POSTNACHNAHME (*f*): postal remittance (see NACHNAHME) 179
POSTPAPIER (*n*): letter paper, Bank (Paper) 180
POSTSCHEIN (*m*): Post-office receipt, post-office form 181
POSTSCHIFF (*n*): packet-(boat); mail steamer; steam-packet 182
POSTSKRIPTUM (*n*): postscript, PS 183
POSTSPARKASSE (*f*): post office savings bank 184
POSTSTRASSE (*f*): high road, main road 185
POSTULAT (*n*): postulate, conjecture, hypothesis, supposition, assumption, theory, assumed truth (*Postulatum*); self-evident (problem) position 186
POSTULIEREN: to postulate, presuppose, assume, conjecture; plead, entreat, solicit, supplicate, beseech 187
POSTVERBAND (*m*): (see POSTVEREIN) 188
POSTVERBINDUNG (*f*): postal communication 189
POSTVEREIN (*m*): postal union 190
POSTWENDEND: by return of post 191
POSTZEICHEN (*n*): postmark 192
POSTZUG (*m*): mail-train 193
POTASCHE (*f*): potash; potassium carbonate (see KALIUMCARBONAT and POTTASCHE) 194
POTENTIAL (*n*): potential 195
POTENTIALMESSUNG (*f*): potential measurement 196
POTENTIALNIVEAU (*n*): potential level 197
POTENTIALSPRUNG (*m*): potential difference (Elect.) 198
POTENTIELL: potential 199
POTENTIIEREN: to render potent 200
POTENZ (*f*): power 201
POTENZ (*f*), **DRITTE-**: third power, cube 202
POTENZIEREN: to raise to a (higher) power (Maths.) 203
POTENZIERUNG (*f*): involution 204
POTENZREIHE (*f*): exponential series (Maths.) 205
POTENZ (*f*), **ZWEITE-**: second power, square 206
POTTASCHE (*f*): potash; potassium carbonate (see POTASCHE and KALIUMCARBONAT) 207
POTTASCHEANLAGE (*f*): potash plant 208
POTTASCHEGEWINNUNG (*f*): manufacture (extraction) of potash 209
POTTASCHELÖSUNG (*f*): potash solution 210
POTTASCHEOFEN (*m*): potash furnace 211
POTTECHT: fast to potting (Dye) 212
POTTERDE (*f*): potter's clay 213
POTTFISCH (*m*): sperm whale; spermaceti whale; cachelot (*Physeter macrocephalus*) 214
POTTFISCHÖL (*n*): sperm oil; train oil; whale-(body) oil; Sg. 0.927 215
POTTHAKEN (*m*): pot-hook 216
POTTISCHE KRANKHEIT (*f*): Pott's disease, caries of the spine (a destructive disease affecting one or more vertebrae and their attendant cartilages) (Med.) 217
POTTLOT (*n*): blacklead; graphite 218
POUSSIEREN: to court; pay attention to; push; promote; encourage; patronize 219
POWELLIT (*m*): powellite; $Ca(Mo,W)O_4$; Sg. 4.5 220

POZZOLANERDE (*f*): pozzuolana (see also **PORZELLANMÖRTEL**) 221
PRÄ (*n*): preference 222
PRÄBABYLONISCH: pre-babylonian 223
PRACHT (*f*): splendour; state; magnificence; pomp; display; gorgeousness; ostentation 224
PRACHTAUSGABE (*f*): edition de luxe 225
PRACHTDRUCK (*m*): art printing 226
PRÄCHTIG: magnificent; gorgeous; splendid; sumptuous; ostentatious 227
PRACHTLOS: unostentatious 228
PRACHTTANNE (*f*): (American)-red fir (*Abies magnifica*) 229
PRACHTVOLL: (see **PRÄCHTIG**) 230
PRÄCIPITAT (*n* and *m*): precipitate (see **PRÄZIPITAT**) 231
PRÄCISION (*f*): precision (see also **PRÄZ-**) 232
PRÄCORDIALANGST (*f*): præcordial (anxiety) oppression (Med.) 233
PRÄDESTINIEREN: to predestine 234
PRÄDIKABEL: predicable 235
PRÄDIKABILITÄT (*f*): predicability 236
PRÄDIKAMENT (*n*): predicament (Log.) 237
PRÄDIKAT (*n*): predicate 238
PRÄDISPONIEREN: to predispose 239
PRÄDISPONIEREND: predisposing 240
PRÄEXISTENZ (*f*): pre-existence; former life 241
PRÄFEKT (*m*): prefect 242
PRÄFIX (*n*): prefix 243
PRÄGEANSTALT (*f*): mint 244
PRÄGEN: to stamp; mint; imprint; coin 245
PRÄGEPRESSE (*f*): stamping press 246
PRAGMATISCH: pragmatic-(al), meddling, interfering, intrusive, obtrusive, meddlesome 247
PRÄGNANT: with child, pregnant; important, fraught with meaning 248
PRÄGNANZ (*f*): pregnancy; weightiness, importance 249
PRÄGUNG (*f*): coining; minting; stamping; coinage; imprinting 250
PRAHLEN: to boast; vaunt; brag; be ostentatious; shine; be loud (of colours) 251
PRAHM (*m*): pontoon, lighter 252
PRAHMSPRITZE (*f*): fire-float 253
PRÄKLUSION (*f*): foreclosure (Law) 254
PRAKTIK (*f*): practice 255
PRAKTIKANT (*m*): practitioner; laboratory (student) worker 256
PRAKTIKER (*m*): expert; practician; practical man; experienced person 257
PRAKTISCH: practical-(ly); in practice 258
PRAKTISIEREN: to practise 259
PRAKTIZIEREN: to practise 260
PRÄLIMINAR: preliminary 261
PRÄLIMINARIEN (*pl*): preliminaries 262
PRALL: stout; tight; elastic; (*m*): impingement; recoil; rebound; ricochet; bounce; reflection; deflection (see also **PRELL-**) 263
PRALLEN: to bound; rebound; bounce; ricochet; impinge; be reflected; deflect; be deflected; recoil; baffle 264
PRALLHEIT (*f*): tightness; elasticity; tension 265
PRALLKÖRPER (*m*): (any body which acts as a deflector or dam), baffle, arrester, retarder, retarding device, deflector 266
PRALLKRAFT (*f*): resiliency; elasticity 267
PRALLLICHT (*n*): reflected light 268
PRALLPLATTE (*f*): baffle-(plate), deflecting (deflector) plate 269
PRALLSCHUSS (*m*): ricochet 270
PRALLWINKEL (*m*): angle of reflection 271
PRÄLUDIUM (*n*): prelude 272
PRÄMIE (*f*): premium; prize; bonus; bounty; royalty 273
PRÄMIENSCHEIN (*m*): premium bond 274
PRÄMIEREN: to give (present with; award) a prize; accord a prize or award; bargain for time 275
PRÄMIEVERTRAG (*m*): royalty agreement 276
PRÄMISSE (*f*): premise, proposition, ground, support, argument 277
PRANGEN: to shine; make a show; parade; vaunt 278
PRANKE (*f*): clutch 279
PRÄNUMERANDO: beforehand, in advance 280
PRÄNUMERANT (*m*): subscriber 281
PRÄNUMERIEREN: to subscribe 282
PRÄPARAT (*n*): preparation 283
PRÄPARATENGLAS (*n*): preparation (cylinder) glass; specimen tube 284
PRÄPARATENRÖHRCHEN (*n*): preparation (specimen) tube 285
PRÄPARIEREN: to prepare; adjust; trim; dissect (Med.) 286
PRÄPARIERSALZ (*n*): preparing salt; sodium stannate (see **NATRIUMSTANNAT**) 287
PRÄPARIERTES PAPIER (*n*): sensitized paper (Phot.) 288
PRÄPOSIT (*n*): (the name given to an explosive consisting mainly of potassium nitrate and sulphur) 289
PRÄPOSITION (*f*): preposition 290
PRÄROGATIVE (*f*): prerogative, exclusive privilege 291
PRASEM (*m*): prase; (green variety of) quartz (see **QUARZ**) 292
PRÄSENS (*n*): present tense 293
PRASENSTEIN (*m*): (see **PRASEM**) 294
PRÄSENT (*n*): gift; present 295
PRÄSENTANT (*m*): bearer (one who presents a cheque for payment); presenter (holder) (of a bill) 296
PRÄSENTIEREN: to present 297
PRÄSENTIERTELLER (*m*): salver; tray 298
PRÄSENZ (*f*): presence 299
PRASEODYM (*n*): praseodymium (Pr) 300
PRASEODYMCARBID (*n*): praseodymium carbide; PrC_2; Sg. 5.1 301
PRASEODYMCHLORID (*n*): praseodymium chloride; $PrCl_3$; Sg. 4.017-4.07; $PrCl_3+7H_2O$; Sg. 2.251 302
PRASEODYMDIOXYD (*n*): praseodymium dioxide; PrO_2; Sg. 5.978 303
PRASEODYM (*n*), **GOLDCHLORWASSERSTOFFSAURES-**: praseodymium chloraurate; $PrCl_3,AuCl_3+10H_2O$ 304
PRASEODYMOXYD (*n*): praseodymium oxide; Pr_2O_3; Sg. 6.88 305
PRASEODYMSELENAT (*n*): praseodymium selenate, $Pr_2(SeO_4)_3$; Sg. 4.3 306
PRASEODYMSULFAT (*n*): praseodymium sulphate; $Pr_2(SO_4)_3$; Sg. 3.72; $Pr_2(SO_4)_3+5H_2O$; Sg. 3.173; $Pr_2(SO_4)_3+8H_2O$; Sg. 2.817-2.819 307
PRASEODYMSULFID (*n*): praseodymium sulphide; Pr_2S_3; Sg. 5.042 308
PRASEODYMSUPEROXYD (*n*): praseodymium peroxide; PrO_2 309
PRASEOKOBALTSALZ (*n*): praseocobaltic salt 310
PRASEOLITH (*m*): praseolite (a decomposed form of Cordierite) (see **CORDIERIT**); Sg. 2.75 [311
PRÄSERVATIV (*n*): preservative 312
PRÄSERVIEREN: to preserve 313

PRÄSERVIERUNG (*f*): preserving; preservation 314
PRÄSERVIERUNGSMITTEL (*n*): preservative 315
PRÄSIDENT (*m*): president; chairman 316
PRÄSIDIEREN: to preside; take the chair 317
PRÄSIDIUM (*n*): presidency; chair 318
PRASSELN: to crackle; rustle; rattle 319
PRASSEN: to revel; carouse; debauch; gourmandize; spend; feast; riot 320
PRÄTENDENT (*m*): claimant; pretender 321
PRÄTERITUM (*n*): past-(definite)-tense, preterite (Gram.) 322
PRATZE (*f*): lug, bracket, claw 323
PRÄVALIEREND: prevailing; prevalent 324
PRAXIS (*f*): practice; exercise; conventional (usual) methods 325
PRÄZEDENS (*n*): precedent 326
PRÄZEDENZ (*f*): precedence 327
PRÄZIPITAT (*n*): precipitate 328
PRÄZIPITATION (*f*): precipitation 329
PRÄZIPITATIONSWÄRME (*f*): heat of precipitation 330
PRÄZIPITAT (*n*), ROTES-; PRÄZIPITAT (*m*), ROTER-: red precipitate (see QUECKSILBEROXYD) 331
PRÄZIPITAT (*m*), UNSCHMELZBARER WEISSER-: (see AMINOQUECKSILBERCHLORID) 332
PRÄZIPITAT (*n*), WEISSES-; PRÄZIPITAT (*m*), WEISSER-: white precipitate (see MERKURI--AMMONIUMCHLORID) 333
PRÄZIPITIEREN: to precipitate 334
PRÄZIPITIERGEFÄSS (*n*): precipitating vessel 335
PRÄZIPITIERMITTEL (*n*): precipitant 336
PRÄZIS: precise 337
PRÄZISIEREN: to define; make (be) precise 338
PRÄZISION (*f*): precision 339
PRÄZISIONSFERTIGGUSS (*m*): precision finished casting 340
PRÄZISIONSGUSS (*m*): precision casting 341
PRÄZISIONSMECHANIK (*f*): precision mechanics (such as clocks, measuring and recording instruments, etc.) 342
PRÄZISIONSWAGE (*f*): precision balance 343
PREDIGEN: to preach 344
PREDIGT (*f*): sermon; preaching; lecture 345
PREHNIT (*m*): Prehnite; $2CaO,Al_2O_3,3SiO_2,H_2O$; Sg. 2.88 346
PREIS (*m*): price; cost; prize; praise; value; rate; reward; glory 347
PREISABGABE (*f*): quotation (of prices) 348
PREISANGABE (*f*): quotation (of prices) 349
PREISANSATZ (*m*): quotation, price 350
PREISAUFGABE (*f*): subject (question) for competition 351
PREISAUSSTELLUNG (*f*): prize (exhibition) competition 352
PREISBEWERBER (*m*): competitor 353
PREISEN: to praise; commend; treasure 354
PREISERMÄSSIGUNG (*f*): reduction in prices, fall (drop) in prices, decline (decrease) in prices 355
PREISGEBEN: to give (away; over) up; expose; abandon; deliver up; prostitute; surrender; allow to leak out 356
PREISKOURANT (*m*): price list 357
PREISLISTE (*f*): price-list 358
PREISSATZ (*m*): estimate, valuation 359
PREISSCHRIFT (*f*): prize (essay) article 360
PREISSELBEERE (*f*): cranberry; bilberry (*Vaccinium*) 361
PREISSTAND (*m*): quotation, rate (on money market) 362
PREISSTEIGERUNG (*f*): advance, increase, rise (in price or prices) 363
PREISUNTERSCHIED (*m*): difference (change) in quotation 364
PREISVERZEICHNIS (*n*): price list 365
PREISWERT: valuable 366
PREISWÜRDIG: praiseworthy 367
PRELLE (*f*): tossing; rebounding 368
PRELLEN: to toss; cheat; make rebound; deflect; defraud; extort (see also PRALL-) 369
PRELLEREI (*f*): cheating; defrauding; extortion; swindling 370
PRELLHAMMER (*m*): sledge-hammer 371
PRELLKLOTZ (*m*): anvil 372
PRELLSTEIN (*m*): curbstone 373
PRELLWAND (*f*): deflector-(plate); baffle-(plate) 374
PREMIER JUS: (see FEINTALG) 375
PRESENNING (*f*): tarpaulin 376
PRESSANT: pressing, urgent 377
PRESSBALKEN (*m*): press table; platen; brake-beam; press-cheek 378
PRESSBAR: compressible 379
PRESSBAUM (*m*): (press)-lever 380
PRESSBAUSCH (*m*): printing frame pad (Phot.) 381
PRESSBEHÄLTER (*m*): press (accumulator) container or receiver; pressure (accumulator, container) receiver 382
PRESSBENGEL (*m*): (printing press)-bar 383
PRESSBEUTEL (*m*): filter bag (for presses); press-(ing) bag 384
PRESSDECKEL (*m*): tympan; printer's frame 385
PRESSE (*f*): press; gloss; stamper; pressure; dilemma; lustre; difficulty 386
PRESSE (*f*), HYDRAULISCHE-: hydraulic press 387
PRESSEN: to press; squeeze; urge; oppress; impress; strain; stamp; compress; gloss; dress; finish 388
PRESSE (*f*), ÖL-: oil press 389
PRESSE (*f*), VERTIKALE-: vertical press 390
PRESSFILTER (*n*): press (pressure) filter 391
PRESSFLÄCHE (*f*): press-(ing) surface, pressure (surface) area 392
PRESSFLÜSSIGKEIT (*f*): pressure (power) liquid or fluid; press-(ed) liquid or fluid 393
PRESSFORM (*f*): pressure (press) mould; pattern 394
PRESSFREIHEIT (*f*): freedom of the press 395
PRESSGAS (*n*): compressed gas; pressure gas, gas under pressure 396
PRESSGASLICHT (*n*): pressure gas-light (in which the gas is put under pressure by means of a pump) 397
PRESSGEFÄLLE (*n*): pressure head 398
PRESSGLANZ (*m*): gloss from pressing 399
PRESSGLAS (*n*): pressed glass 400
PRESSGUSSRIPPENROHR (*n*): cast iron pipe with pressed on ribs or fins 401
PRESSGUT (*n*): material to be pressed 402
PRESSHEFE (*f*): pressed (compressed) yeast, (pressed state of BRANNTWEINHEFE, which see) 403
PRESSIEREN: to be urgent, be pressing, be in a hurry; urge, press, importune, dun 404
PRESSKOHLE (*f*): pressed (charcoal) coal; briquette 405
PRESSKOLBEN (*m*): press (pressure) piston or ram 406
PRESSKOLBENQUERHAUPT (*m*): cross head 407
PRESSKORK (*m*): compressed cork 408

PRESSLING (*m*): pressed article; something pressed; (*pl*) pressed beet slices; expressed beet pulp (Sugar) 409
PRESSLUFT (*f*): compressed air; (as a prefix= air-blast; pneumatic) 410
PRESSLUFTABKLOPFER (*m*): pneumatic scaling (hammer; chipper) tool 411
PRESSLUFTANLAGE (*f*): compressed air plant 412
PRESSLUFTANTRIEB (*m*): pneumatic (compressed air) drive 413
PRESSLUFTARMATUR (*f*): compressed air mounting-(s) 414
PRESSLUFTBEWEHRUNG (*f*): compressed air mountings 415
PRESSLUFTBOHRMASCHINE (*f*): pneumatic (drilling machine; boring machine) drill 416
PRESSLUFTENTSTAUBUNGSANLAGE (*f*): compressed air dust (removing) extracting plant 417
PRESSLUFTFLASCHENZUG (*m*): pneumatic pulley block-(s) or tackle 418
PRESSLUFTGAS (*n*): pressure air gas, air gas under pressure 419
PRESSLUFTHAMMER (*m*): pneumatic hammer 420
PRESSLUFTMEISSEL (*m*): pneumatic chisel 421
PRESSLUFTMESSER (*m*): compressed air meter 422
PRESSLUFTMOTOR (*m*): compressed air motor 423
PRESSLUFTMOTORWINDE (*f*): compressed air motor (winch) capstan 424
PRESSLUFTNIETFEUER (*n*): riveting furnace or forge with air blast 425
PRESSLUFTNIETHAMMER (*m*): pneumatic riveting hammer 426
PRESSLUFTNIETUNG (*f*): pneumatic riveting 427
PRESSLUFTSANDSIEBMASCHINE (*f*): pneumatic sand screening machine 428
PRESSLUFTSCHLAUCH (*m*): compressed air hose 429
PRESSLUFTSTAB (*m*): compressed air pipe 430
PRESSLUFTSTAMPFER (*m*): compressed air (pneumatic) ram-(mer) 431
PRESSLUFTWERKZEUG (*n*): pneumatic tool-(s) 432
PRESSLUFTZULEITUNG (*f*): compressed air inlet pipe or piping 433
PRESSMASCHINE (*f*): calender 434
PRESSMÜHLE (*f*): pressing mill 435
PRESSÖL (*n*): fixed oil 436
PRESSPAHN (*m*): Presspahn (insulating cardboard) (Paper) 437
PRESSPAN (*m*): press board (Paper) 438
PRESSPAPIER (*n*): cardboard 439
PRESSPLATTE (*f*): pressplate; press table; platen; die 440
PRESSPLUNGER (*m*): press piston, ram or plunger 441
PRESSPUMPE (*f*): force (pressure) pump 442
PRESSPUMPETAUCHKOLBEN (*m*): force (pressure) pump plunger 443
PRESSPUMPWERK (*n*): force or press-(ure) pump, hydraulic pump; hydraulic pumping station 444
PRESSRING (*m*): press ring 445
PRESSRÜCKSTAND (*m*): expressed residue; residue from pressing 446
PRESSSACK (*m*): filter bag (for presses); press-(ing) bag 447
PRESSSAFT (*m*): expressed (press) juice 448
PRESSSÄULE (*f*): column; tie rod (of presses) 449
PRESSSCHILD (*n*): press shield 450
PRESSSCHMIERUNG (*f*): forced lubrication 451
PRESSSCHRAUBE (*f*): pressure screw 452
PRESSSCHUH (*m*): brake block 453
PRESSSCHWITZVERFAHREN (*n*): press sweating process 454
PRESSSITZ (*m*): medium force fit 455
PRESSSPAN (*m*): glazed cardboard; press-board (Paper) 456
PRESSSPINDEL (*f*): pressure (screw) spindle 457
PRESSSTANGE (*f*): press lever 458
PRESSSTEMPEL (*m*): (press)-die 459
PRESSSTIEL (*m*): brake post 460
PRESSSTRAHLTURBINE (*f*): pressure (reaction) turbine 461
PRESSTALG (*m*): pressed tallow; Sg. 0.937-0.952; Mp. 35-52°C. 462
PRESSTEIL (*m*): pressing, stamping; blank-(for pressing or stamping) 463
PRESSTORF (*m*): pressed peat 464
PRESSTUCH (*n*): filter (press) cloth 465
PRESSUNG (*f*): pressure; pressing; compression; impression; draught (of air through furnaces) 466
PRESSVERFAHREN (*n*): pressing process 467
PRESSWALZE (*f*): press roll 468
PRESSWAND (*f*): press cheek 469
PRESSWASSER (*n*): pressure water; (as a prefix= hydraulic) 470
PRESSWASSERANLAGE (*f*): hydraulic plant 471
PRESSWASSERLEITUNG (*f*): hydraulic (pressure water) pipe, piping, pipeline or main 472
PRESSWASSERPROBE (*f*): hydraulic (pressure)-test 473
PRESSWASSERPUMPE (*f*): hydraulic (pressure water) test pump 474
PRESSWASSERRÜCKFÜHRUNG (*f*): hydraulic pressure return 475
PRESSZIEGEL (*m*): pressed brick 476
PRESSZUCKER (*m*): compressed sugar 477
PRESSZWANG (*m*): censoring (restriction of the freedom) of the press 478
PRESSZYLINDER (*m*): press cylinder 479
PREUSSISCHBLAU (*n*): Prussian blue (see **BERLIN-ERBLAU**) 480
PRIAPISMUS (*m*): (see **RUTENKRAMPF**) 481
PRICEIT (*m*): priceite; $Ca_2B_6O_{11}.3H_2O$; Sg. 2.37 482
PRICKE (*f*): prick; prickle 483
PRICKEL (*m*): prickle; prickling 484
PRICKELN: to prick; prickle 485
PRICKELND: sharp; prickling; pungent; pricking 486
PRIESTER (*m*): priest; minister 487
PRIESTERLICH: priestly, sacerdotal 488
PRIESTERWEIHE (*f*): ordination; consecration 489
PRIMA (*f*): first of exchange (Bills of Exchange); highest class of a school; (as a prefix= prime; first-class; best quality; A1 quality) 490
PRIMAFORTE (*f*): first (prime) quality 491
PRIMAL (*n*): primal (a hair-dye, the active principle of which is toluylene diamine) 492
PRIMANER (*m*): senior form boy 493
PRIMÄR: primary; idiopathic (Med.) 494
PRIMÄRE KRISTALLE (*m.pl*): primary crystals (which dictate the arrangement of the constituent parts of steel structure, and which form on the change from the liquid state of the metal to the solid) 495
PRIMÄRKREIS (*m*): primary circuit (Elect.) 496
PRIMÄRLUFT (*f*): primary air 497
PRIMÄRSPULE (*f*): primary coil (Elect.) 498
PRIMÄRSTROM (*m*): primary current (Elect.) 499
PRIMAS (*m*): primate 500

PRIMASODA (*f*): refined (first quality) soda 501
PRIMAT (*n*): primacy 502
PRIMAVERAHOLZ (*n*): prima vera, white mahogany; Sg.0.6-0.8 503
PRIMAVISTA: at first sight 504
PRIMAWARE (*f*): prime (superior) article or goods 505
PRIMAWECHSEL (*m*): first-(bill)-of exchange 506
PRIME (*f*): prime; prima (Mus.) 507
PRIMEL (*f*): primrose (*Primula vulgaris*) 508
PRIMGELD (*n*): primage 509
PRIMITIV: primitive; original 510
PRIMULIN (*n*): primuline (an organic dye from the action of fuming sulphuric acid and sulphur on paratoluidine) 511
PRIMULINVERFAHREN (*n*): primuline process, diazotype printing (Phot.) 512
PRIMZAHL (*f*): prime number 513
PRINCIP (*n*): principle 514
PRINCIP, WIRKSAMES- (*n*): active principle (of ferments, etc.) 515
PRINZ (*m*): prince 516
PRINZIP (*n*): principle (see also PRINCIP) 517
PRINZIPAL (*m*): principal; head-master; chief; employer; head; manager; governor; master, etc. 518
PRINZIPIELL: principally, principal 519
PRINZIPIEN (*pl*): rudiments 520
PRINZIPIENREITER (*m*): one who is hard and fast to his principles 521
PRIOR (*m*): prior 522
PRIORIT (*m*): priorite (see BLOMSTRANDIN) 523
PRIORITÄT (*f*): priority; precedence 524
PRIORITÄTSAKTIE (*f*): preference share 525
PRIORITÄTSANLEIHE (*f*): mortgage-(loan) 526
PRIORITÄTSINTERVALL (*n*): interval of priority (time between announcement of a patent at home and the later announcement in another country) 527
PRIORITÄTSRECHT (*n*): right of priority 528
PRISE (*f*): prize; pinch; capture 529
PRISMA (*n*): prism 530
PRISMAÄHNLICH: prismoidal; prism-shaped; prismatic 531
PRISMAARTIG: (see PRISMAÄHNLICH) 532
PRISMAFÖRMIG: (see PRISMAÄHNLICH) 533
PRISMATIN (*m*): prismatite; $MgAl_2SiO_6$; Sg. 3.3 534
PRISMATISCH: prismatic 535
PRISMENFLÄCHE (*f*): prismatic face 536
PRISMENFÖRMIG: prismatic; prism-shaped 537
PRISMENGLAS (*n*): prismatic (prism) glass 538
PRISMENKANTE (*f*): prismatic edge 539
PRISMENSPEKTRUM (*n*): prismatic spectrum 540
PRITSCHE (*f*): bat; ferule; washing-floor (Alum); parapet; stillage, support (for drying purposes) (Dye); bench; plank bed 541
PRITSCHEN: to beat, bat, slap 542
PRIVAT: private 543
PRIVATANGELEGENHEIT (*f*): private (affair; matter) business 544
PRIVATDOZENT (*m*): University lecturer (receives fees but no salary) 545
PRIVATIM: in private, privately, in confidence, confidential-(ly), private and confidential 546
PRIVATISIEREN: to live (retired) privately 547
PRIVATISSIMUM (*n*): private university lecture 548
PRIVATLEBEN (*n*): private life 549
PRIVATMITTEILUNG (*f*): private (report) communication 550
PRIVATRECHT (*n*): civil law 551
PRIVATSTUNDE (*f*): private lesson 552
PRIVET (*n*): (water)-closet 553
PRIVILEGIEREN: to privilege 554
PRIVILEGIUM (*n*): privilege 555
PRO: per; pro; for 556
PROBAT: proved; tried; proof 557
PROBE (*f*): test; essay; test-piece; experiment; assay; sample; specimen; proof (Maths.); trial; probation; rehearsal 558
PROBEABDRUCK (*m*): proof-(print) sheet 559
PROBEABZUG (*m*): proof-(print) sheet 560
PROBEANLAGE (*f*): trial (test) plant 561
PROBEARBEIT (*f*): specimen, sample 562
PROBEBELASTUNG (*f*): test load 563
PROBEBLATT (*n*): (see PROBEBOGEN) 564
PROBEBOGEN (*m*): proof-sheet; pattern sheet; specimen 565
PROBE, BRUCH- (*f*): breaking test 566
PROBE, BRÜCKEN- (*f*): loading test on bridge 567
PROBEDRUCK (*m*): test-pressure; proof-(sheet) (Printing) 568
PROBE (*f*), DURCHSCHNITTS-: average sample 569
PROBEESSIG (*m*): proof vinegar 570
PROBEFAHRT (*f*): trial (trip) run 571
PROBEFAHRTGESCHWINDIGKEIT (*f*): trial speed 572
PROBEFAHRTSDAUER (*f*): duration of trial or trial (run) trip (of ships, etc.) 573
PROBEFÄRBUNG (*f*): test dyeing 574
PROBEFEST: (see PROBEHALTIG) 575
PROBEFLAMME (*f*): test flame 576
PROBEFLÄSCHCHEN (*n*): sampling bottle 577
PROBEFLASCHE (*f*): sample (test; sampling) bottle 578
PROBEFLÜSSIGKEIT (*f*): test liquid 579
PROBEGEWICHT (*n*): test (standard) weight 580
PROBEGLAS (*n*): sampling glass; test (sample) tube 581
PROBEGOLD (*n*): standard gold 582
PROBEGRUBE (*f*): trial (prospect) pit; examination pit 583
PROBEGUT (*n*): sample 584
PROBEHALTIG: proof; standard 585
PROBEJAHR (*n*): year of probation 586
PROBEKARTE (*f*): pattern card 587
PROBE, KETTEN- (*f*): chain test 588
PROBEKORN (*n*): assay button 589
PROBEKUNST (*f*): docimasy, docimacy, assaying (the art of assaying ore) 590
PROBELAST (*f*): test load 591
PROBELAUF (*m*): trial run, test 592
PROBELEHRLING (*m*): apprentice; probationer 593
PRÖBELN: to determine by trial and error (Maths.) 594
PROBELÖFFEL (*m*): assay spoon 595
PROBEMACHEN (*n*): testing; assaying 596
PROBEMASS (*n*): standard measure 597
PROBEMÄSSIG: according to sample 598
PROBEMÜNZE (*f*): proof coin 599
PROBEMUSTER (*n*): sample for testing 600
PROBEN: to test, prove, check, probe (see PRO-BIEREN) 601
PROBENADEL (*f*): touch needle 602
PROBENEHMEN (*n*): sampling 603
PROBENEHMER (*m*): sampler 604
PROBENGLAS (*n*): specimen (glass) tube 605
PROBENSTECHER (*m*): sampler; thief tube (Liquids); proof stick (Sugar) 606
PROBEPAPIER (*n*): test paper 607
PROBEPFAHL (*m*): test pipe 608

PROBER (*m*): tester; sampler; assayer; gauge 609

PROBERECHNUNG (*f*): trial calculation 610

PROBE, RECK- (*f*): elongation test 611

PROBEROHR (*n*): test tube; trial (tube) pipe (Mech.) 612

PROBERÖHRCHEN (*n*): (see PROBEROHR) 613

PROBERÖHRE (*f*): (see PROBEROHR) 614

PROBESÄURE (*f*): test (standard) acid 615

PROBESCHERBE (*f*): cupel; trial piece (Ceramics) 616

PROBESCHERBEN (*m*): (see PROBESCHERBE) 617

PROBESCHLAG (*m*): trial cruise (of sailing boats) 618

PROBESCHRIFT (*f*): draft; sample (specimen) of writing; trial (essay) article 619

PROBESILBER (*n*): standard silver 620

PROBESPIRITUS (*m*): proof spirit 621

PROBESTECHER (*m*): (see PROBENSTECHER) 622

PROBESTEIN (*m*): touchstone; sample stone 623

PROBESTOFF (*m*): sample 624

PROBESTUBE (*f*): laboratory; test room 625

PROBESTÜCK (*n*): test-piece; sample; specimen; pattern 626

PROBETIEGEL (*m*): assay crucible; cupel 627

PROBEVORRICHTUNG (*f*): testing arrangement; testing device 628

PROBEWAGE (*f*): assay balance 629

PROBE, WASSERDRUCK- (*f*): hydraulic-(pressure) test 630

PROBEWEINGEIST (*m*): proof spirit 631

PROBEZEIT (*f*): time (duration) of trial or test; probation, noviciate 632

PROBE, ZERREISS- (*f*): breaking (tensile) test 633

PROBEZIEHEN (*n*): sampling; taking of samples 634

PROBEZIEHER (*m*): sampler 635

PROBEZIEHUNG (*f*): sampling, taking of samples 636

PROBEZINN (*n*): standard tin 637

PROBE, ZUR-: as a trial; for a test; on (approbation) approval 638

PROBEZYLINDER (*m*): trial jar; test tube; test cylinder 639

PROBIERAPPARAT (*m*): testing (device) apparatus 640

PROBIERBLEI (*n*): assay (test) lead 641

PROBIERBLEIMASS (*n*): test (assay) lead measure 642

PROBIERBLEISIEB (*n*): test (assay) lead screen or sieve 643

PROBIERBLEITUTTE (*f*): lead assaying crucible 644

PROBIEREN: to test; try; prove; assay; sample; analyse; taste; rehearse (see PROBEN) 645

PROBIERER (*m*): analyst; tester; sampler; assayer 646

PROBIERGEFÄSS (*n*): assay-(ing) vessel 647

PROBIERGERÄT (*n*): assaying (testing) apparatus 648

PROBIERGERÄTSCHAFTEN (*f.pl*): (see PROBIERGERÄT) 649

PROBIERGEWICHT (*n*): assay weight 650

PROBIERGLAS (*n*): test (glass) tube 651

PROBIERGLÄSCHEN (*n*): (see PROBIERGLAS) 652

PROBIERGLÄTTE (*f*): test litharge 653

PROBIERGOLD (*n*): standard gold 654

PROBIERHAHN (*m*): test (try) cock; gauge cock 655

PROBIERHÄHNCHEN (*n*): (see PROBIERHAHN) 656

PROBIERKLUFT (*f*): assayer's tongs 657

PROBIERKORN (*n*): assay button 658

PROBIERKUNST (*f*): assaying; docimacy (see PROBEKUNST) 659

PROBIERLABORATORIUM (*n*): assay laboratory 660

PROBIERLÖFFEL (*m*): assay spoon 661

PROBIERMASCHINE (*f*): testing machine 662

PROBIERMETALL (*n*): test metal 663

PROBIERNADEL (*f*): touch needle 664

PROBIEROFEN (*m*): assay furnace 665

PROBIERPAPIER (*n*): test paper 666

PROBIERROHR (*n*): test tube 667

PROBIERRÖHRE (*f*): test tube 668

PROBIERRÖHRENGESTELL (*n*): test-tube (rack) stand 669

PROBIERSCHERBE (*f*): (see PROBIERSCHERBEN) 670

PROBIERSCHERBEN (*m*): cupel; assay test; trial piece (Ceramics) 671

PROBIERSTATION (*f*): testing shop 672

PROBIERSTEIN (*m*): touchstone 673

PROBIERTIEGEL (*m*): assay crucible; cupel 674

PROBIERTUTE (*f*): assay crucible 675

PROBIERTÜTE (*f*): (see PROBIERTUTE) 676

PROBIERTUTTE (*f*): (see PROBIERTUTE) 677

PROBIERVENTIL (*n*): test valve 678

PROBIERVORRICHTUNG (*f*): gauge; testing (device; apparatus) arrangement 679

PROBIERWAGE (*f*): assay balance 680

PROBIERZANGE (*f*): assayer's tongs 681

PROBLEM (*n*): problem 682

PROBLEMATISCH: problematic-(al); uncertain; questionable; dubious; doubtful 683

PROCENT: per cent. (see also PROZENT) 684

PROCHLORIT (*m*): prochlorite; $H_{12}Mg_7Si_4O_{27}Al_4$; Sg. 2.85 685

PROCURA (*f*): procuration 686

PRODUKT (*n*): produce, product 687

PRODUKTION (*f*): production 688

PRODUKTIV: productive-(ly) 689

PRODUKTIVGENOSSENSCHAFT (*f*): co-operative society 690

PRODUKTIVITÄT (*f*): productiveness 691

PRODUZENT (*m*): grower; producer 692

PRODUZIEREN: to produce; grow; yield; show; bring forward; supply; furnish 693

PROFAN: profane 694

PROFANIEREN: to profane 695

PROFESS (*m*): profession, vow-(s) 696

PROFESSION (*f*): profession, calling, trade 697

PROFESSIONIST (*m*): professional man 698

PROFESSOR (*m*): professor 699

PROFESSORAT (*n*): professorship 700

PROFESSORHAFT: professorial 701

PROFESSORISCH: professorial 702

PROFESSORSCHAFT (*f*): professorship 703

PROFESSUR (*f*): professorship 704

PROFIL (*n*): profile; side (view) section; (*pl*): sections, section irons 705

PROFILSKALA (*f*): edgewise scale 706

PROFILWALZE (*f*): section roll (for rolling metal sections) (Rolling mills) 707

PROFIT (*m*): profit, gain 708

PROFITIEREN: to profit, gain 709

PROFITLICH: profitable 710

PROGNOSE (*f*): prognosis (Med.) 711

PROGNOSTIKON (*n*): prognostic 712

PROGNOSTIZIEREN: to foretell, predict, prophesy, prognosticate, presage, augur, divine 713

PROGRAMM (*n*): programme 714

PROGRESSIV: progressive 715

PROHIBITIVZOLL (*m*): prohibition (prohibitive) duty 716

PROJEKT (*n*): project, scheme, proposal, plan 717
PROJEKTIEREN: to plan, project, scheme, propose, design, plot 718
PROJEKTIERUNG (*f*): planning, projecting (see PROJEKTIEREN) 719
PROJEKTIL (*n*): projectile, shell 720
PROJEKTION (*f*): projection 721
PROJEKTION (*f*), MITTELPUNKTS-: gnomonic projection 722
PROJEKTION (*f*), PARALLEL-: parallel projection 723
PROJEKTIONSAPPARAT (*m*): projecting apparatus, projector; optical lantern 724
PROJIZIEREN: to project; design; scheme; plot; plan 725
PROKLAMIEREN: to proclaim, announce, publish, promulgate, declare, advertise, make known, spread abroad, bruit 726
PROKTITIS (*f*): rectal inflammation (Med.) 727
PROKURA (*f*): power of attorney, procuration 728
PROKURA (*f*) ERTEILEN: to give a power of attorney 729
PROKURAFÜHRER (*m*): one who has a power of attorney 730
PROKURATRÄGER (*m*): (see PROKURAFÜHRER) 731
PROKURIST (*m*): (see PROKURAFÜHRER) 732
PROLETARIER (*m*): proletarian 733
PROLIN (*n*): proline, 1-pyrrolidine-carbonic acid; $C_5H_9O_2N$ 734
PRÖLLSCHE STEUERUNG (*f*): Proell's gear 735
PROLOG (*m*): prologue 736
PROLONGATION (*f*): prolongation (see PROLONGIEREN) 737
PROLONGATIONSGESCHÄFT (*n*): continuation business (Commerce) 738
PROLONGIEREN: to prolong, extend, continue, lengthen, protract, defer; renew (Bills) 739
PROMILLE: per thousand (usually shown by the sign‰) 740
PROMOVIEREN: to graduate; take a degree; promote 741
PROMPT: prompt-(ly), quick-(ly), ready, readily, punctual-(ly) 742
PROMPT BEZAHLEN: to pay prompt cash, pay ready money 743
PROMPTHEIT (*f*): promptness, promptitude, readiness, alacrity 744
PRONOMEN (*n*): pronoun 745
PRONYSCHER ZAUM (*m*): Prony brake 746
PROPAESIN (*n*): propaesin (see PROPÄSIN) 747
PROPAGANDA (*f*): propaganda 748
PROPAGIEREN: to propagate 749
PROPAGIERUNG (*f*): propagation; culture 750
PROPAN (*n*): propane; C_3H_8; Sg. 0.536; Bp. – 37°C. 751
PROPÄSIN (*n*): propaesin, propyl-para-aminobenzoate; $C_{10}H_{13}NO_2$ 752
PROPELLERFEDER (*f*): propeller feather 753
PROPELLERFLÜGEL (*m*): propeller blade 754
PROPELLERWELLE (*f*): propeller shaft 755
PROPELLERWELLENLAGER (*n*): propeller-shaft bearing 756
PROPELLERWIRKUNGSGRAD (*m*): propeller (propulsive) efficiency 757
PROPHET (*m*): prophet 758
PROPHETISCH: prophetic 759
PROPHEZEIEN: to prophesy, predict, foretell, divine, presage 760
PROPHYLAKTICUM (*n*): prophylactic (a preventive medicine) (Med.) 761
PROPHYLAKTIKUM MALLEBREIN (*n*): (see MALLEBREIN) 762
PROPIOLSÄURE (*f*): ortho-nitrophenylpropiolic acid, propiolic acid; $C_6H(NO_2)C:C.CO_2H$ 763
PROPIONALDEHYD (*n*): propionic (propylic, propyl) aldehyde; C_3H_6O; Sg. 0.8066; Bp. 48-49.5°C. 764
PROPIONAMID (*n*): propion amide 765
PROPIONITRIL (*n*): (see PROPIONSÄURE-NITRIL) 766
PROPIONSAUER: combined with propionic acid, propionate of 767
PROPIONSÄURE (*f*): propionic acid, metactonic acid, ethylcarbonic acid, methylacetic acid; $C_3H_6O_2$; Sg. 0.9871-0.992; Mp. – 19.3 to – 22°C.; Bp. 140°C. 768
PROPIONSÄUREANHYDRID (*n*): propionic anhydride; $C_6H_{10}O_3$; Sg. 1.0336; Bp. 167°C 769
PROPIONSÄURE-ÄTHYLESTER (*m*): propionic ethylester, ethyl propionate; $C_5H_{10}O_2$; Sg. 0.9125; Mp. – 72.6°C.; Bp. 99.1°C. 770
PROPIONSÄURE-NITRIL (*n*): propionic nitrile; C_3H_5N; Sg. 0.7882; Mp. – 103.5°C.; Bp. 97.1°C. 771
PROPIONYLCHLORID (*n*): propionyl chloride; C_3H_5OCl; Sg. 1.0646; Bp. 77.8-78.3°C. 772
PROPLATINUM (*n*): proplatinum (trade name of an alloy of 72% Ni; 24% Ag; 4% Bi) 773
PROPONAL (*n*): (*Acidum dipropylbarbituricum*): proponal, dipropylbarbituric acid, dipropylmalonylurea; $C_{10}H_{16}N_2O_3$; Mp. 145°C. 774
PROPONENT (*m*): mover; proposer 775
PROPONIEREN: to move; propose 776
PROPORTION (*f*): proportion 777
PROPORTIONAL: proportional 778
PROPORTIONALE (*f*), MITTLERE-: mean proportional 779
PROPORTIONALGRÖSSE (*f*): proportional 780
PROPORTIONALITÄT (*f*): proportionality 781
PROPORTIONALITÄTSGRENZE (*f*): proportional extension limit 782
PROPORTIONAL, MITTELBAR-: indirectly proportional 783
PROPORTIONAL, UMGEKEHRT-: inversely proportional 784
PROPORTIONAL, UNMITTELBAR-: directly proportional 785
PROPORTIONIEREN: to proportion; graduate; form (shape) symmetrically 786
PROPYLACETAT (*n*): propyl acetate; $CH_3COO.C_3H_7$ 787
PROPYLALDEHYD (*n*): propyl (propylic, propionic) aldehyde (see PROPIONALDEHYD) 788
PROPYLALKOHOL (*m*): propyl alcohol (*Alcohol propylicus*); $C_2H_5.CH_2.OH$; Sg. 0.808; Mp. – 127°C.; Bp. 97°C. 789
PROPYLALKOHOL (*m*), ISO-: pseudopropyl alcohol, isopropyl alcohol, dimethyl carbinol, secondary propyl alcohol; $(CH_3)_2:CH.OH$ or C_3H_8O; Sg. 0.7887; Bp. 82.1°C. 790
PROPYLALKOHOL (*m*), NORMALER-: normal propyl alcohol, ethyl carbinol (see ÄTHYLKARBINOL) 791
PROPYLALKOHOL (*m*), PRIMÄRER-: primary propyl alcohol, ethyl carbinol (see ÄTHYLKARBINOL) 792
PROPYLALKOHOL (*m*), SEKUNDÄRER-: secondary propyl alcohol, iso-propyl alcohol (see PROPYLALKOHOL, ISO-) 793
PROPYL (*n*), AMEISENSAURES-: propyl formate (see PROPYLFORMIAT) 794

PROPYLAMIN (*n*): propylamine 795
PROPYL (*n*), **BALDRIANSAURES-**: propyl valerate 796
PROPYL (*n*), **BENZOESAURES-**: propyl benzoate 797
PROPYLBROMID (*n*): propyl bromide 798
PROPYL (*n*), **BUTTERSAURES-**: propyl butyrate 799
PROPYLCHLORID (*n*): propyl chloride 800
PROPYL (*n*), **ESSIGSAURES-**: propyl acetate (see **PROPYLACETAT**) 801
PROPYLFORMIAT (*n*): propyl formate; $HCOO.C_3H_7$ 802
PROPYLJODID (*n*): propyl iodide; C_3H_7I; Sg. 1.7427; Mp. – 98.8°C.; Bp. 101.7°C. 803
PROPYLJODID (*n*), **ISO-**: isopropyl iodide; C_3H_7I; Sg. 1.7033; Mp. – 89/92°C. Bp. 88.6/88.9°C. 804
PROPYL (*n*), **PROPIONSAURES-**: propyl propionate 805
PROPYLSCHWEFELSAUER: propylsulphate of, combined with propylsulphuric acid 806
PROROGIEREN: to prorogue, adjourn 807
PROSILOXAN (*n*): prosiloxan (a silicon compound of the formula; $HSiOH$) 808
PROSOPIT (*m*): prosopite; $Al_2(F,OH)_6,CaF_2$; Sg. 2.89 809
PROSPEKT (*m*): prospect, prospectus, leaflet, brochure, syllabus 810
PROSTATE (*f*): prostate gland (Med.) 811
PROSTITUIEREN: to prostitute; expose 812
PROTALBINSÄURE (*f*): protalbinic acid (same derivation and use as **LYSALBINSÄURE**, which see) 813
PROTARGOL (*n*): protargol; silver proteinate (*Argentum proteinicum*): (silver content over 8%) 814
PROTEIN (*n*): protein, albuminous substance (see also **EIWEISS**) 815
PROTEINKÖRPER (*m*): protein-(substance) 816
PROTEINMEHL (*n*): protein flour 817
PROTEINSAUER: proteinate of 818
PROTEINSTOFF (*m*): protein-(substance) 819
PROTEINURIE (*f*): albuminuria (Med.) 820
PROTEOLYTISCH: proteolytic 821
PROTEST (*m*) **BEIBRINGEN**: to lodge a protest 822
PROTESTIEREN: to protest; object 823
PROTESTKOSTEN (*f.pl*): protest charges 824
PROTEST (*m*), **MIT- ZURÜCKKOMMEN**: to be returned dishonoured (Commerical) 825
PROTHORAX (*m*): prothorax 826
PROTOCATECHUALDEHYD (*n*): protocatechuic aldehyde; $C_7H_6O_3$ 827
PROTOCATECHUSÄURE (*f*): protocatechuic acid 828
PROTOGIN (*m*): protogine-(granite), alpine granite 829
PROTOGYNISCH: protogynous, proterogynous (having the stigma ready for fertilization before the pollen is ready to fertilize it) (Bot.) 830
PROTOKATECHUALDEHYD (*n*): (see **PROTOCATE-CHUALDEHYD**) 831
PROTOKATECHUSÄURE (*f*): protocatechuic acid 832
PROTOKOLL (*n*): protocol; record; minutes; report; certificate 833
PROTOKOLLANT (*m*): actuary; clerk; recorder 834
PROTOKOLLBUCH (*n*): minute book 835
PROTOKOLLIEREN: to record; register 836
PROTONUCLEIN (*n*): protonucleine 837
PROTOPIN (*n*): protopine (an opium alkaloid); $C_{20}H_{19}NO_5$ 838
PROTOPLASMA (*n*): protoplasm 839
PROTOYXD (*n*): protoxide 840
PROTOZOEN (*pl*): protozoa (unicellular organisms) 841
PROTYLIN (*n*): protylin (a phosphorous-albumen compound with about 80% albumen and 6% P_2O_5) 842
PROTZE (*f*): limber, carriage (of a gun) 843
PROTZWAGEN (*m*): gun-limber; gun-carriage 844
PROUSTIT (*m*): proustite, light ruby silver (sulpharsenite of silver) (see **ROTGILTIGERZ, LICHTES-**): Ag_3AsS_3; Sg. 5.55 845
PROVENCERÖL (*n*): a highly refined olive oil (see **OLIVENÖL** and **BAUMÖL**) 846
PROVIANT (*m*): victuals; provisions; store; supply; supplies; provender 847
PROVIANTHAUS (*n*): store-house (for provisions) 848
PROVIANTIEREN: to provide; provision; victual; supply; furnish; stock; store 849
PROVIANTRAUM (*m*): store-room 850
PROVIANTWESEN (*n*): commissariat 851
PROVISION (*f*): commission; provision; percentage 852
PROVISIONSABRECHNUNG (*f*): statement of commission 853
PROVISOR (*m*): dispenser (Pharm.) 854
PROVISORISCH: provisional; provisory 855
PROVISORIUM (*n*): provisional arrangement 856
PROVOZIEREN: to provoke 857
PROZEDUR (*f*): procedure; proceeding 858
PROZENT (*n*): per cent.; percentage 859
PROZENTGEHALT (*m*): percentage-(content) 860
PROZENTIG: percent-(age) 861
PROZENTISCH: (see **PROZENTIG**) 862
PROZENTSATZ (*m*): percentage; rate per cent. 863
PROZENTTEILUNG (*f*): percentage scale 864
PROZENTUALE GRÖSSE (*f*): the percentage; the amount per cent. 865
PROZESS (*m*): process; litigation; action; law-suit; case; (Legal)-proceeding-(s) 866
PROZESSAKTEN (*f.pl*): records of a case 867
PROZESSFÄHIG: actionable 868
PROZESSFÜHRER (*m*): litigant; plaintiff; counsel for the plaintiff 869
PROZESSIEREN: to litigate; bring an action 870
PROZESSKOSTEN (*f.pl*): costs of a case 871
PRUDEL (*m*): steam; ebullition; bubbling 872
PRUDELN: to boil, bubble, steam 873
PRÜFEISEN (*n*): probe 874
PRÜFEN: to test; try; prove; assay; taste; examine; check 875
PRÜFER (*m*): tester; taster; assayer; examiner 876
PRÜFPROTOKOLL (*n*): test (report) certificate 877
PRÜFSTEIN (*m*): touchstone; test 878
PRÜFUNG (*f*): test-(ing); trial; proof; assay; examination; temptation 879
PRÜFUNG (*f*), **BEI DER-**: on (under) test 880
PRÜFUNGSATTEST (*n*): test certificate 881
PRÜFUNGSATTEST, AMTLICHES- (*n*): official test certificate 882
PRÜFUNGSSCHEIN(*m*): test (report) certificate 883
PRÜFUNGSSCHRIFT (*f*): thesis (University) 884
PRÜFUNGSZEUGNIS (*n*): test certificate; diploma 885
PRÜGEL (*m*): club; cudgel; blow; thrashing; flogging; fighting 886
PRÜGELN: to beat; thrash; flog; fight 887

PRUNELLE (*f*): prunello, prunella (the name given to dried plums the skin of which has been removed) 888

PRUNELLENSALZ (*n*): sal (prunella) prunelle; fused (very fine) potassium nitrate (see KALIUMNITRAT) 889

PRUNELLSALZ (*n*): (see PRUNELLENSALZ) 890

PRUNK (*m*): pomp; show; state; parade; ostentation; splendour 891

PRUNKEN: to be ostentatious; make a show; parade 892

PSEUDO: pseudo; false 893

PSEUDOALKANNA (*f*): (see ALKANNA) 894

PSEUDOBUTYLALKOHOL (*m*) pseudobutyl alcohol, tertiary butyl alcohol; $(CH_3)_2COHCH_3$; Sg. 0.786; Mp. 25.5°C.; Bp. 83°C. 895

PSEUDOBUTYLCHLORID (*n*): pseudobutyl chloride 896

PSEUDOCONHYDRIN (*n*): pseudoconhydrine; $C_8H_{17}NO$; Mp. 101.5°C.; Bp. 230°C. 897

PSEUDOCUMIDIN (*n*): pseudocumidine (see CUMIDIN) 898

PSEUDOCUMOL (*n*): pseudocumene (see CUMOL) 899

PSEUDOCUMOLSULFOSÄURE (*f*): pseudocumene-sulfonic acid 900

PSEUDOEPHEDRIN (*n*): pseudoephedrine (see EPHEDRIN) 901

PSEUDOGLYZERIN (*n*): pseudo-glycerine (a mixture of starch sugar, dextrin, and $MgCl_2$) 902

PSEUDOHARNSTOFF (*m*): pseudourea 903

PSEUDOKATALYSATOR (*m*): pseudocatalyser 904

PSEUDOKATALYSE (*f*): pseudocatalysis 905

PSEUDOKATALYTISCH: pseudocatalytic 906

PSEUDOKUMOL (*n*): trimethylbenzene (see CUMOL) 907

PSEUDOLEUKÄMIE (*f*): lymphadenoma (*Anœmia lymphatica*) (enlargement of lymphatic glands) (Med.) 908

PSEUDOLÖSUNG (*f*): pseudo solution 909

PSEUDOMALACHIT (*m*): pseudomalachite; $(Cu.OH)_3PO_4$; Sg. 3.9 910

PSEUDOMERIE (*f*): pseudomerism 911

PSEUDOMORPH: pseudomorphous 912

PSEUDOMORPHOSE (*f*): pseudomorph-(osis) 913

PSEUDONYM: pseudonymous 914

PSEUDOPHIT (*m*): pseudophite (see PENNIN) 915

PSEUDOPROPYLALKOHOL (*m*): pseudopropyl alcohol (see PROPYLALKOHOL, ISO-) 916

PSEUDOSÄURE (*f*): pseudo acid 917

PSEUDOXANTHIN (*n*): pseudo-xanthine (an animal alkaloid from urine and flesh); $C_4H_5N_5O$ 918

PSILOMELAN (*m*): psilomelane (mainly manganese oxide), manganese hydrate (a manganese ore); $BaMnO_3(MnO_2,Mn_2O_3,K_2O,H_2O)$ or H_4MnO_5; Sg. 4.3 919

PSILOMELANGRAPHIT (*m*), UNTEILBARER-: asbolane 920

PSOASABSCESS (*m*): psoas abscess (an abscess in the sheath of the Psoas muscle) (Med.) 921

PSORIALAN (*n*): (name given to a salve obtained by action of margaric acid on yellow mercury oxide, and used for skin diseases; see SCHUPPENFLECHTE) 922

PSORIASIS (*f*): (see SCHUPPENFLECHTE) 923

PSYCHISCH: psychic-(al) 924

PSYCHOANALYTIKER (*m*): psychoanalyst 925

PSYCHOLOG (*m*): psychologist 926

PSYCHOLOGIE (*f*): psychology 927

PSYCHOLOGISCH: psychological 928

PTERIDOPHYTEN (*pl*): *Pteridophytes* (Bot.) 929

PUBERTÄT (*f*): puberty 930

PUBLIK: public 931

PUBLIKATION (*f*): publication (both the action and the article) 932

PUBLIKUM (*n*): public, audience 933

PUBLIZIEREN: to publish; promulgate; make (public) known; prove (Wills) 934

PUCHERIT (*m*): pucherite; $BiVO_4$; Sg. 6.25 935

PUDDELARBEITER (*m*): puddler (Metal) 936

PUDDELBETT (*n*): puddling furnace bed 937

PUDDELEISEN (*n*): puddled iron; puddling iron; puddle bar 938

PUDDELLUPPE (*f*): puddle-ball; bloom (Metal) 939

PUDDELMASCHINE (*f*): puddler 940

PUDDELN: to puddle; (*n*): puddling 941

PUDDELOFEN (*m*): puddling furnace (see FLAMM-OFEN) 942

PUDDELPROZESS (*m*): puddling process 943

PUDDELSCHLACKE (*f*): puddling slag (Steel works process by-product; 55-60% Fe) 944

PUDDELSOHLE (*f*): puddling furnace bed 945

PUDDELSPITZE (*f*): puddler's paddle 946

PUDDELSTAB (*m*): puddle bar 947

PUDDELSTAHL (*m*): puddled steel 948

PUDDELVERFAHREN (*n*): puddling process (see FLAMMOFENFRISCHEN) 949

PUDDELWALZE (*f*): puddle roll; blooming roll 950

PUDDELWALZWERK (*n*): puddle (puddling; blooming) rolls 951

PUDDELWERK (*n*): puddling works 952

PUDDINGPULVER (*n*): pudding powder 953

PUDEL (*m*): poodle; spaniel; miss; blunder; bungle 954

PUDELN: to miss; blunder; bungle 955

PUDER (*m*): powder; face powder 956

PUDERIG: powdery; powdered 957

PUDERN: to powder; pulverize 958

PUDERPAPIER (*n*): powder paper 959

PUDERQUAST (*m*): powder puff 960

PUDERSCHACHTEL (*f*): powder-box (for face powders) (see PULVERSCHACHTEL) 961

PUDERZUCKER (*m*): powdered sugar 962

PUERPERALFIEBER (*n*): puerperal fever, childbed fever, puerperal (surgical) septicæmia (a contagious fever in connection with child-birth) (Med.) 963

PUFF (*m*): blow; thump; puff 964

PUFFEN: to pop; puff; thump; beat; cuff 965

PUFFER (*m*): buffer 966

PUFFERBATTERIE (*f*): buffer battery 967

PUFFERBÜCHSE (*f*): buffer box 968

PUFFERFEDER (*f*): buffer spring 969

PUFFERHUB (*m*): buffer-stroke (Machinery) 970

PUFFERSCHEIBE (*f*): buffer disc 971

PUFFERSTÄNDER (*m*): buffer stop (Railway) 972

PUFFERSTANGE (*f*): buffer rod 973

PUFFERWEHR (*f*): buffer stop (Railway) 974

PUFFERWIRKUNG (*f*): buffer-(ing) action or effect 975

PUKALLSCHE MASSE (*f*): Pukall's mass (a hard, porous porcelain mass, used for making filters) 976

PULPA (*f*): pulp; fruit pulp (see MUS) 977

PÜLPENFÄNGER (*m*): pulp trap (for catching the coarse matter floating in the waste water from beet sugar factories) 978

PULS (*m*): pulse; pulsation; beat; throb 979

PULSADER (*f*): artery 980

PULSADERGESCHWULST (*f*): aneurism (dilation of an artery) (Med.) 981
PULSADER, GROSSE- (*f*): aorta 982
PULSADERKROPF (*m*): (see PULSADERGESCHWULST) 983
PULSATILLENBLÄTTER (*n.pl*): wind-flower leaves (see HAKELKRAUT) 984
PULSATILLENKAMPFER (*m*): (see ANEMONIN) 985
PULSGLAS (*n*): cryophorus 986
PULSHAMMER (*m*): cryophorus: water-hammer 987
PULSIEREN: to pulsate; beat; throb; palpitate 988
PULSOMETER (*n* and *m*): pulsometer 989
PULSSCHLAG (*m*): pulsation; beat (of pulse); throb; palpitation 990
PULT (*n*): desk 991
PULTDACH (*n*): lean-to roof 992
PULTFEUERUNG (*f*): step-grate furnace or firing; inclined grate (Boilers) 993
PULTOFEN (*m*): back-flame hearth (Metal) 994
PULVER (*n*): powder (*Pulvis*); gunpowder 995
PULVERARTIG: powdery; like powder 996
PULVERBÜCHSE (*f*): powder box, powder flask 997
PULVERDAMPF (*m*): powder smoke 998
PULVERFABRIK (*f*): powder factory 999
PULVERFASS (*n*): powder (barrel) keg 000
PULVERFLASCHE (*f*): powder (bottle) flask 001
PULVERFORM (*f*): powder form, form of powder 002
PULVERFÖRMIG: powdery; in the form of powder; small; fine; dusty; like dust; dust-like; floury 003
PULVERGLAS (*n*): powder bottle 004
PULVERHOLZ (*n*): black alder (*Alnus glutinosa*) 005
PULVERIG: powdery; dusty; pulverulent 006
PULVERISIERBAR: pulverable; capable of being pulverized; pulverizable 007
PULVERISIEREN: to pulverize; powder; pulverate; grind (to dust or powder); crush; triturate 008
PULVERISIERMASCHINE (*f*): pulverizing machine (*pl*): pulverizing machinery 009
PULVERISIERUNG (*f*): pulverization; pulveration; powdering; crushing; grinding; trituration 010
PULVERKAMMER (*f*): powder magazine 011
PULVERKORN (*n*): grain of powder 012
PULVERMAGAZIN (*n*): powder magazine 013
PULVERMASS (*n*): charge of powder 014
PULVERMEHL (*n*): priming-(powder) 015
PULVERMÖRSER (*m*): powder mortar 016
PULVERMÜHLE (*f*): powder mill 017
PULVERMÜLLER (*m*): powder maker 018
PULVERN: to pulverize; powder (see PULVER-ISIEREN) 019
PULVERPRESSE (*f*): powder press 020
PULVER (*n*), **RAUCHLOSES-**: smokeless (gun)-powder 021
PULVER (*n*), **RAUCHSCHWACHES-**: gunpowder which gives off very little smoke, nearly (practically) smokeless (gun)-powder 022
PULVERRINNE (*f*): train of (gun)-powder 023
PULVERSAND (*m*): drift sand 024
PULVERSATZ (*m*): powder composition 025
PULVERSCHACHTEL (*f*): powder-box (for any powders except face powders) (see PUDER-SCHACHTEL) 026
PULVERZUCKER (*m*): powdered sugar 027
PUMPBRUNNEN (*m*): well fitted with a pump 028
PUMPE (*f*): pump; inflator 029
PUMPE, ABDAMPFSTRAHL- (*f*): exhaust steam injector 030
PUMPE, ABDRÜCK- (*f*): hydraulic (pressure) test pump 031
PUMPE, ABESSINIER- (*f*): hollow ram pump 032
PUMPE, ABTEUF- (*f*): mine pump; bore-hole pump 033
PUMPE, ABWÄSSER- (*f*): sewage pump 034
PUMPE, AKKUMULATOR- (*f*): accumulator pump 035
PUMPE, AMMONIAK- (*f*): ammonia pump 036
PUMPE (*f*), **AMMONIAKWASSER-**: ammonia-water (gas water) pump 037
PUMPE, AUSZIEHBARE- (*f*): telescopic pump 038
PUMPE, AXIAL- (*f*): axial pump 039
PUMPE (*f*), **BALLAST-**: ballast pump 040
PUMPE (*f*), **BASTARD-**: bastard pump (a pump which sucks air and water simultaneously) 041
PUMPE, BAU- (*f*): contractor's pump 042
PUMPE, BEHÄLTER- (*f*): reservoir pump 043
PUMPE, BENZIN- (*f*): petrol pump 044
PUMPE, BERIESELUNGS- (*f*): water circulating or irrigating pump 045
PUMPE, BETRIEB-(S)- (*f*): working pump; work's pump 046
PUMPE, BILGE- (*f*): bilge pump 047
PUMPE, BOHRLOCH- (*f*): bore-hole pump (see PUMPE, ABTEUF-) 048
PUMPE, BRENNSTOFF- (*f*): fuel pump 049
PUMPE, BRUNNEN- (*f*): well pump 050
PUMPE (*f*), **CENTRIFUGAL-**: centrifugal pump 051
PUMPE, CORNISH-DAMPF- (*f*): Cornish steam pump 052
PUMPE, DAMPF- (*f*): steam pump 053
PUMPE, DÄMPFER- (*f*): dash-pot 054
PUMPE, DESTILLAT- (*f*): distillate (distilled water) pump 055
PUMPE (*f*), **DIAPHRAGMA-**: diaphragm (membrane) pump 056
PUMPE, DIFFERENTIAL- (*f*): differential pump 057
PUMPE, DIFFERENTIALEIL- (*f*): differential high-speed pump 058
PUMPE, DOCK- (*f*): dock pump 059
PUMPE, DOPPEL- (*f*): twin (tandem) pump; duplex pump 060
PUMPE, DOPPELSCHLEUDER- (*f*): twin (tandem; duplex; double) centrifugal pump 061
PUMPE, DOPPELSTRAHL- (*f*): double tube injector 062
PUMPE, DOPPELTURBINEN- (*f*): twin (tandem; duplex; double) turbine pump 063
PUMPE, DOPPELWIRKENDE- (*f*): double-acting pump 064
PUMPE, DOPPELWIRKENDE-DRUCK- (*f*): double-acting force pump 065
PUMPE, DOPPELWIRKENDE SAUG-UND-DRUCK- (*f*): double-acting suction and force pump 066
PUMPE, DOPPELWIRKENDE-VENTIL- (*f*): double-acting valve pump 067
PUMPE, DREHBEWEGUNG- (*f*): rotary (motion) pump 068
PUMPE, DREHKOLBEN- (*f*): semi-rotary (rotary; wing; vane) pump; pump with rotating piston 069
PUMPE, DREIFACHE TAUCHKOLBEN- (*f*): triple (three-throw) plunger pump 070
PUMPE, DRILLINGEIL- (*f*): three-throw high-speed pump 071

PUMPE, DRILLINGSPRESS- (*f*): three-throw (press; pressure) force pump 072
PUMPE, DRUCK- (*f*): force (compression; pressure; discharge) pump 073
PUMPE, DRUCKLUFT- (*f*): air-lift (compressed air) pump 074
PUMPE (*f*), DRUCKLUFTHEBE-: air-lift pump 075
PUMPE, DRUCKSTRAHL- (*f*): injector 076
PUMPE, DRUCKWASSERPRESS- (*f*): hydraulic pressure pump 077
PUMPE, DUPLEX- (*f*): duplex pump 078
PUMPE, EIL- (*f*): high-speed (express) pump 079
PUMPE, EINACHSIGE- (*f*): deep-well (prospecting) pump; single-barrel (lift) pump 080
PUMPE, EINFACHE DRUCKWASSER- (*f*): single-acting (hand) pressure pump 081
PUMPE, EINFACHE HANDPRESS- (*f*): single-acting hand pressure pump 082
PUMPE, EINFACHE SAUG- (*f*): ordinary (single-acting) suction pump 083
PUMPE, EINFACHWIRKENDE- (*f*): single-acting pump 084
PUMPE, EINSTUFIGE- (*f*): single-stage pump 085
PUMPE, EINZYLINDER- (*f*): single-cylinder pump 086
PUMPE, ELEKTRISCH-ANGETRIEBENE- (*f*): electrically driven pump 087
PUMPE, ELEKTRISCH-BETRIEBENE- (*f*): electrically driven pump 088
PUMPE, ELEKTRISCHE HAUSKAPSEL- (*f*): electrically driven rotary (domestic: house; house-service) pump 089
PUMPE, EVOLVENTEN- (*f*): involute pump 090
PUMPE, EXZENTER- (*f*): eccentric pump 091
PUMPE, FABRIK- (*f*): factory (works) pump 092
PUMPE, FAHRBARE BAU- (*f*): portable contractor's pump 093
PUMPE, FAHRBARE STOLLEN- (*f*): portable (mine-gallery) adit pump 094
PUMPE, FEDERPLATTEN- (*f*): diaphragm (membrane) pump 095
PUMPE, FEUERLÖSCH- (*f*): fire pump 096
PUMPE, FILTERPRESS- (*f*): filter press pump 097
PUMPE, FLÜGEL- (*f*): semi-rotary (wing) pump 098
PUMPE (*f*), FLÜSSIGKEITS-: liquid (liquor; lye) pump 099
PUMPE, FÖRDER- (*f*): delivery (supply) pump 100
PUMPE, FREISTEHENDE- (*f*): self-contained (independent) pump 101
PUMPE, FÜLL- (*f*): charging pump 102
PUMPE, FUSS- (*f*): foot pump 103
PUMPE, FUSSLUFT- (*f*): foot pump 104
PUMPE, GAS- (*f*): gas pump 105
PUMPE, GASDRUCK- (*f*): gas pressure pump 106
PUMPE, GEMENGE- (*f*): mixture (combined air and gas) pump 107
PUMPE, GEMISCH- (*f*): charging pump (see also PUMPE, GEMENGE-) 108
PUMPE, GENERATOR- (*f*): primary turbine; generator pump 109
PUMPE, GENTIL- (*f*): Gentil pump; centrifugal pump with self-starter 110
PUMPE, GLYZERIN- (*f*): glycerine pump 111
PUMPE, HALBLIEGENDE- (*f*): oblique (diagonal; inclined) pump 112
PUMPE (*f*), HAND-: hand pump 113
PUMPE, HANDDRUCK- (*f*): hand (pressure) force pump 114
PUMPE, HANDKOLBEN- (*f*): hand operated (piston; bucket) plunger pump 115
PUMPE, HANDLUFT- (*f*): hand air pump 116
PUMPE, HANDÖL- (*f*): hand oil pump 117
PUMPE, HANDPRESS- (*f*): hand operated press (force; pressure; test) pump 118
PUMPE, HANDVERDICHTUNGS- (*f*): hand operated compression pump 119
PUMPE (*f*), HARTBLEI-: hard-lead pump (for use with acids) 120
PUMPE, HARTGUMMI- (*f*): hard (India)-rubber pump 121
PUMPE, HAUS- (*f*): house (house-service; general-service; general-purposes) pump 122
PUMPE, HAUSKAPSEL (*f*): rotary (domestic; house; house-service) pump 123
PUMPE, HAUSWASSER- (*f*): domestic water pump 124
PUMPE (*f*), HEBE-: lift pump 125
PUMPE, HEBEL- (*f*): lever pump 126
PUMPE, HILFS- (*f*): standby (auxiliary) pump 127
PUMPE, HOCHDRUCK- (*f*): high-lift (high-pressure) pump 128
PUMPE, HOCHDRUCKSCHLEUDER- (*f*): high-lift (high-pressure) centrifugal pump 129
PUMPE, HOF- (*f*): yard (general-purposes) pump 130
PUMPE, HUB- (*f*): lift (lifting) pump 131
PUMPE (*f*), HYDRAULISCHE-: hydraulic pump 132
PUMPE, KALTWASSER- (*f*): cold water pump 133
PUMPE, KANAL- (*f*): sewage (drainage) pump 134
PUMPE, KANALISATIONS- (*f*): sewage (drainage) pump 135
PUMPE, KAPSEL- (*f*): rotary (wing; vane) pump (with rotating piston); valveless pump 136
PUMPE, KAPSELSOLE- (*f*): rotary brine pump 137
PUMPE (*f*), KESSELSPEISE-: boiler feed pump 138
PUMPE, KETTEN- (*f*): chain pump 139
PUMPE, KOLBEN- (*f*): piston (plunger; bucket) pump 140
PUMPE, KOLBENLOSE- (*f*): pistonless (plungerless) pump 141
PUMPE, KOLBENSOLE- (*f*): plunger (piston) brine pump 142
PUMPE, KOLBENSTUFEN- (*f*): differential pump; plunger stage pump 143
PUMPE; KOMPRESSIONS- (*f*): compression (force) pump 144
PUMPE, KONDENSAT- (*f*): condensate pump 145
PUMPE, KONDENSATORSCHLEUDER- (*f*): centrifugal condensed steam (condenser; wet-air) pump 146
PUMPE, KREISEL- (*f*): centrifugal pump 147
PUMPE, KUGELSPEISE- (*f*): spherical feed pump 148
PUMPE, KÜHLWASSER- (*f*): circulating (cooling) water pump 149
PUMPE, LADE- (*f*): charging pump 150
PUMPE (*f*), LAUGE-: lye pump (usually employed in the soap industry) 151
PUMPE, LAUGENSCHLEUDER- (*f*): centrifugal (brine) lye pump 152
PUMPE, LENS- (*f*): bilge pump 153
PUMPE, LENZ- (*f*): bilge pump 154
PUMPE, LIEGENDE- (*f*): horizontal pump 155
PUMPE, LUFT- (*f*): air pump; tyre (bicycle) pump 156
PUMPE, LUFT- MIT FUSS (*f*): foot pump (for tyres) 157
PUMPE (*f*), MAGEN-: stomach pump (Med.) 158
PUMPE, MAMMUT- (*f*): mammoth pump 159
PUMPE, MEHRFACHE HOCHDRUCKSCHLEUDER- (*f*): multi-stage high-lift (high-pressure) centrifugal pump 160

PUMPE, MEHRSTUFIGE- (*f*): multi-stage pump 161
PUMPE, MEMBRAN- (*f*): diaphragm (membrane) pump 162
PUMPE, MITTELDRUCKSCHLEUDER- (*f*): medium (intermediate) lift or pressure pump 163
PUMPEN: to pump; borrow; lend; accept (take) credit; give (request) credit 164
PUMPENANLAGE (*f*): pumping (plant) installation 165
PUMPENANORDNUNG (*f*): pump arrangement 166
PUMPENANTRIEB (*m*): pump drive 167
PUMPENANTRIEBSMOTOR (*m*): pump motor, motor for driving the pump 168
PUMPENARBEIT, INDIZIERTE- (*f*): indicated pump (performance; duty; output) horse-power 169
PUMPENART (*f*): type of pump 170
PUMPENBAGGER (*m*): pump dredger 171
PUMPENBAU (*m*): pump construction 172
PUMPENBAUART (*f*): type of pump; construction of pump 173
PUMPENBAUER (*m*): pump (constructor; maker) manufacturer 174
PUMPENDECKEL (*m*): pump cover-(plate) 175
PUMPENDIAGRAMM (*n*): pump diagram 176
PUMPENEIMER (*m*): pump bucket 177
PUMPENENTWÄSSERUNG (*f*): draining by pump-(ing); pumping 178
PUMPENFLÜGEL (*m*): pump impeller blade 179
PUMPENGEHÄUSE (*n*): pump (chamber) casing 180
PUMPENGESTÄNGE (*n*): pump rods 181
PUMPENHAHN (*m*): pump cock 182
PUMPENHAUS (*n*): pump house 183
PUMPENHUB (*m*): stroke-(of a pump) 184
PUMPE, NIEDERDRUCKSCHLEUDER- (*f*): low lift (pressure) centrifugal pump 185
PUMPE, NIEDERSCHLAGWASSER- (*f*): condensate (condensed water) pump 186
PUMPENINDIKATOR (*m*): pump indicator 187
PUMPENKAMMER (*f*): pump (clearance) chamber 188
PUMPENKASTEN (*m*): pump (case; casing; housing) chest 189
PUMPENKENNLINIE (*f*): pump diagram 190
PUMPENKLEID (*n*): (see PUMPENKASTEN) 191
PUMPENKOLBEN (*m*): pump (piston) plunger 192
PUMPENKONSTRUKTION (*f*): pump construction 193
PUMPENKÖRPER (*m*): pump body 194
PUMPENKÖRPERDECKEL (*m*): pump cover 195
PUMPENKURBEL (*f*): pump crank 196
PUMPENLEDER (*n*): washer; sucker (of a pump) 197
PUMPENLEITUNG (*f*): pump (main; water conduit) piping 198
PUMPENLIEFERUNG (*f*): pump (delivery) output 199
PUMPENMACHER (*m*): pump maker 200
PUMPENPFERDEKRAFT (*f*): pump horse-power 201
PUMPENRAD (*n*): pump impeller-(wheel) 202
PUMPENRAUM (*m*): pump (chamber) room 203
PUMPENRAUMFLUR (*f*): pump room floor 204
PUMPENREGULATOR (*m*): pump (regulator) governor 205
PUMPENSATZ (*m*): unit, set (of pumps) 206
PUMPENSAUGRAUM (*m*): pump suction chamber 207
PUMPENSCHACHT (*m*): pump shaft 208
PUMPENSCHACHTBAGGER (*m*): pump-(hopper)-dredger 209
PUMPENSCHAUBILD (*n*): pump diagram 210
PUMPENSCHLAG (*m*): stroke-(of a pump) 211
PUMPENSCHLAUCH (*m*): rubber connecting tube (for tyre pumps); flexible pump hose (for hand water pumps) 212
PUMPENSCHWENGEL (*m*): pump (swingle) handle 213
PUMPENSTANGE (*f*): pump rod 214
PUMPENSTEUERUNG (*f*): pump valve gear 215
PUMPENSTIEFEL (*m*): pump (cylinder) barrel 216
PUMPENSTUBE (*f*): pump room 217
PUMPENSUMPF (*m*): pump (well)sump 218
PUMPENTEIL (*m*): pump part 219
PUMPENVENTIL (*n*): pump valve 220
PUMPENWERK (*n*): pump parts 221
PUMPENWIDERSTAND (*m*): pump resistance 222
PUMPENZUBEHÖR (*n*): pump accessories 223
PUMPENZYLINDER (*m*): pump cylinder 224
PUMPE, ÖL- (*f*): oil pump 225
PUMPE, PETROLEUM- (*f*): petrol (petroleum) pump 226
PUMPE, PITCHER- (*f*): pitcher pump 227
PUMPE, PLUNGER- (*f*): plunger (ram; displacement) pump 228
PUMPE, PNEUMATIK- (*f*): pneumatic pump; foot pump or hand pump for tyres 229
PUMPE, PORZELLAN- (*f*): porcelain pump 230
PUMPE, PRESS- (*f*): force pump; press (pressure) pump; test pump 231
PUMPE (*f*): PRESSSCHMIER-: forced lubrication pump 232
PUMPE, PROBIER- (*f*): test pump 233
PUMPE, PRÜF- (*f*): hydraulic-(pressure) test pump 234
PUMPE, QUECKSILBERLUFT- (*f*): mercury air pump 235
PUMPE, RAMMROHR- (*f*): hollow ram pump 236
PUMPE, REINWASSER- (*f*): clear (clean; pure) water pump 237
PUMPE, RIEMEN- (*f*): belt driven pump 238
PUMPE, ROHRBRUNNEN- (*f*): artesian well pump 239
PUMPE, ROHRKOLBEN- (*f*): hollow piston pump 240
PUMPE, ROHWASSER- (*f*): dirty (untreated) water pump 241
PUMPE, ROTATIONS- (*f*): rotary pump (pump with rotary piston) 242
PUMPE (*f*), ROTIERENDE-: rotary pump 243
PUMPE, RUNDLAUFKOLBEN- (*f*): (see PUMPE, ROTATIONS-) 244
PUMPE (*f*), SALZWASSER-: brine pump, salt water pump, sea water pump 245
PUMPE, SAUG- (*f*): suction (lift) pump; bucket pump 246
PUMPE, SAUGENDE- (*f*): sucking (drawing; lifting; suction) pump 247
PUMPE, SAUGSTRAHL- (*f*): ejector 248
PUMPE (*f*), SÄURE-: acid pump 249
PUMPE, SÄUREMEMBRAN- (*f*): acid diaphragm (membrane) pump 250
PUMPE, SCHACHT- (*f*): shaft (well) pump 251
PUMPE, SCHAUFEL- (*f*): centrifugal pump 252
PUMPE, SCHAUFELRAD- (*f*): bucket wheel pump 253
PUMPE, SCHEIBENKOLBEN- (*f*): pump with solid piston 254
PUMPE, SCHIEBERGESTEUERTE- (*f*): pump with slide valve 255
PUMPE, SCHIFFS- (*f*): ship's pump 256
PUMPE (*f*), SCHIFFSLENZ-: bilge pump 257

PUMPE, SCHLAG- (*f*): hand oil pump (on motor cycles, etc.) 258
PUMPE, SCHLEUDER- (*f*): centrifugal pump 259
PUMPE, SCHMIER- (*f*): lubricating pump 260
PUMPE, SCHNELLLAUFENDE KOLBEN- (*f*): high-speed piston pump 261
PUMPE, SCHRAUBEN- (*f*): axial pump; screw pump 262
PUMPE, SCHRAUBENKOLBEN- (*f*): helical (helicoidal) piston pump 263
PUMPE, SCHWINGENDE TAUCHKOLBEN- (*f*): oscillating plunger pump 264
PUMPE, SCHWUNGRADDAMPF- (*f*): fly-wheel steam pump 265
PUMPE, SCHWUNGRADLOSE- (*f*): direct-acting pump-(without fly wheel) 266
PUMPE, SELBSTTÄTIGE- (*f*): self-acting (automatic) pump 267
PUMPE, SENK- (*f*): mine pump 268
PUMPE, SIMPLEX- (*f*): simplex pump 269
PUMPE, SOLE- (*f*): brine pump 270
PUMPE, SPEISE- (*f*): feed pump 271
PUMPE, SPIRITUS- (*f*): alcohol pump 272
PUMPE, SPÜL- (*f*): scavenging pump 273
PUMPE, STÄNDER- (*f*): stand pump 274
PUMPE, STEHENDE- (*f*): vertical pump 275
PUMPE, STOLLEN- (*f*): mine-gallery (adit) pump 276
PUMPE, STOSSWEISE WIRKENDE- (*f*): intermittent working pump 277
PUMPE, STRAHL- (*f*): injector; ejector 278
PUMPE, STRASSEN- (*f*): street (general-purposes) pump 279
PUMPE, STUFEN- (*f*): stage pump 280
PUMPE, STUFENKOLBEN- (*f*): differential pump 281
PUMPE, TAUCHKOLBEN- (*f*): plunger (ram; displacement) pump 282
PUMPE (*f*), TEER-: tar pump 283
PUMPE, TELESKOP- (*f*): telescopic pump 284
PUMPE (*f*), TIEFBOHR-: deep-well pump 285
PUMPE, TIEFBRUNNEN- (*f*): deep well pump 286
PUMPE, TOPF- (*f*): pitcher pump 287
PUMPE, TRAGBARE- (*f*): portable pump 288
PUMPE, TURBINEN- (*f*): turbine pump; turbo-pump 289
PUMPE (*f*), TURBO-: turbo-pump 290
PUMPE, UMLAUF- (*f*): circulating pump 291
PUMPE, UNTERDRUCKDAMPF- (*f*): steam vacuum pump 292
PUMPE, UNTERDRUCKLUFT- (*f*): vacuum air pump 293
PUMPE, VAKUUM- (*f*): vacuum pump 294
PUMPE, VENTILKOLBEN- (*f*): bucket valve piston pump 295
PUMPE, VERBUNDDAMPF- (*f*): compound steam pump 296
PUMPE, VERDICHTERSCHMIER- (*f*): compressor oil (lubricating) pump 297
PUMPE, VERDICHTUNGS- (*f*): compression pump 298
PUMPE, VERDRÄNGER- (*f*): plunger (ram; displacement) pump 299
PUMPE, VIERTAKT- (*f*): four-stroke pump 300
PUMPE, VOR- (*f*): auxiliary pump 301
PUMPE, WÄLZ- (*f*): circulating pump 302
PUMPE, WALZEN- (*f*): drum pump 303
PUMPE, WAND- (*f*): wall pump 304
PUMPE, WANDDAMPF- (*f*): wall steam pump 305
PUMPE, WARMWASSER-(*f*): hot (water) well pump 306
PUMPE, WASSERDRUCK- (*f*): water motor pump 307
PUMPE, WASSERDRUCKDREHBEWEGUNG- (*f*): water-motor, rotary-(motion) pump 308
PUMPE, WASSERPFOSTENENTLEER- (*f*): hydrant draining pump 309
PUMPE, WASSERSTRAHL- (*f*): water jet injector 310
PUMPE, WASSERUMLAUF- (*f*): water circulating pump 311
PUMPE, WASSERWERK- (*f*): waterwork's pump 312
PUMPE, WINDMOTOR- (*f*): windmill pump 313
PUMPE, WÜRGEL- (*f*): semi-rotary -(wing)-pump 314
PUMPE, ZAHNRAD- (*f*): gear-driven (geared) pump 315
PUMPE, ZENTRIFUGAL- (*f*): centrifugal pump 316
PUMPE, ZIRKULATIONS- (*f*): circulating pump 317
PUMPE, ZUBRINGER- (*f*): feeder (auxiliary fire-engine) pump 318
PUMPE, ZWEIACHSIGE- (*f*): double-barrel (lift) pump; deep-well (prospecting) pump 319
PUMPE, ZWEIFACHE HANDPRESS- (*f*): two-throw (double-acting) hand press, pressure or force pump 320
PUMPE, ZWILLING- (*f*): twin pump 321
PUMPE, ZYLINDROKONISCHESCHRAUBEN- (*f*): cylindroconic helical (helicoidal; screw) pump 322
PUMPMASCHINE (*f*): pumping engine 323
PUMPMASCHINE (*f*), VERBUND-: compound pumping engine 324
PUMPSTATION (*f*): pumping station 325
PUMPWAGE (*f*): aerometer-(for pump) 326
PUMPWERK (*n*): pumping (works) station; pump 327
PUNCH (*m*): punch (the drink) 328
PUNCHESSENZ (*f*): punch essence 329
PUNKT (*m*): point; period; dot; full-stop; heading; rule (in a set of rules); extent 330
PUNKTIEREN: to point; punctuate; dot (notes in music, etc.); draw a dotted line; stipple; tap (Med.); stipulate; enumerate 331
PUNKTIERFEDER (*f*): dotting pen 332
PUNKTIERNADEL (*f*): stipple, dotting (stippling) needle; tapping needle (Med.) 333
PUNKTIERRÄDCHEN (*n*): dotting wheel 334
PUNKTIERT: shown dotted, dotted (of drawings) 335
PUNKTIERTE LINIE (*f*): dotted line 336
PÜNKTLICH: punctual-(ly); exact; accurate; prompt-(ly); precise 337
PÜNKTLICHKEIT (*f*): punctuality 338
PUNKTSCHWEISSEN: to spot-weld; (*n*): spot-welding 339
PUNKTSTRICH (*m*): semicolon 340
PUNKTUM (*n*): period; full-stop 341
PUNKTUR (*f*): puncture 342
PUNSCH (*m*): punch 343
PUNSCHESSENZ (*f*): punch essence 344
PUNZE (*f*): hall-mark, stamp; punch 345
PUNZEN: to punch; stamp; (*m*): punch 346
PUNZIEREN: to hall-mark, stamp (Precious metals) 347
PUPILL (*m*): pupil; ward 348
PUPILLE (*f*): pupil (of the eye) 349
PUPILLENERWEITERNDES MITTEL (*n*): agent for enlarging the pupil of the eye 350
PUPILLENVERENGEND: contracting the pupil (of the eye) 351

PUPILLENVERENGUNG (*f*): contraction of the pupil-(of the eye) 352
PUPPE (*f*): doll; puppet; pupa; cocoon; chrysalis; larva; chess-man; float (Fishing); lay figure; core (of an electric cell) 353
PUPPEN: to change into a chrysalis 354
PUPPENHÜLLE (*f*): cocoon 355
PUPPENLARVE (*f*): larva 356
PUPPENWEISS (*n*): doll white 357
PUPPENZUSTAND (*m*): chrysalis (state) stage 358
PUR: mere, absolute, sheer; pure; real, genuine, undefiled, unpolluted, unmixed, unadulterated 359
PURGANS (*n*): (see **PURGANZ**) 360
PURGANZ (*f*): purgative; evacuant; cathartic; abstergent 361
PURGATIN (*n*): (trade name of a purge, the composition of which is trioxydiacetylanthraquinone) 362
PURGEN (*n*): purgen (trade name of a purge, the composition of which is Phenolphthalein) 363
PURGIER(*n*): purge, purgative, physic, cathartic; a scourer 364
PURGIERBAUM (*m*): black-alder (*Alnus glutinosa*); physic nut tree; South Sea laurel (*Croton tiglium*) 365
PURGIEREN: to purge; physic; scour; boil-off silk (with soap); cleanse; purify; clear; clarify; defecate; deterge; wash away 366
PURGIERGURKE (*f*): colocynth (fruit of *Citrullus colocynthis*) 367
PURGIERHARZ (*n*): scammony-(resin) (*Scammoniæ resina*) (from the root of *Convolvulus scammonia*) 368
PURGIERKASIE (*f*): (see **PURGIERKASSIA**) 369
PURGIERKASSIA (*f*): purging cassia; Indian laburnum; drumstick; pudding stick (dried fruit of *Cassia fistula*) 370
PURGIERKONFEKT (*n*): purgative confection 371
PURGIERKRAUT (*n*): hedge hyssop (*Gratiola*) 372
PURGIERLEIN (*n*): purging flax; cathartic (mountain) flax (*Linum catharticum*) 373
PURGIERMITTEL (*n*): purgative; purge; physic; cathartic; evacuant; aperient 374
PURGIERMOOS (*n*): Iceland moss (thallus of *Cetraria islandica*) 375
PURGIERNUSS (*f*): purging (physic) nut (*Curcas purgans*) 376
PURGIERPILLE (*f*): purging pill 377
PURGIERPULVER (*n*): purging powder, purgative 378
PURGIERSALZ (*n*): purgative salt 379
PURGIERTABLETT (*n*): aperient (purgative) tablet or lozenge 380
PURGIERWURZEL (*f*): purging root; rhubarb (*Rheum palmatum*); jalap (*Exogonium purga*) 381
PURIFIZIEREN: to purify; clear; purge; cleanse; refine; clarify; fine; defecate (Liquors) 382
PURINBASE (*f*): purine base 383
PURINDERIVAT (*n*): purine derivate 384
PURIST (*m*): purist 385
PUROSTROPHAN (*n*): (trade name for a crystallized, chemically pure form of strophantin) 386
PURPUR (*m*): purple 387
PURPURBLENDE (*f*), **PRISMATISCHE-**: Kermes-(mineral) (see **ANTIMONBLENDE**) 388
PURPUREOKOBALTVERBINDUNG (*f*): purpureo-cobaltic compound 389
PURPURERZ (*n*): (see **KIESABBRAND**) 390
PURPURFARBE (*f*): purple (colour) dye 391
PURPURFARBIG: purple-(coloured) 392
PURPURFIEBER (*n*): purples (Med.) 393
PURPURGLUT (*f*): purple glow, purple heat 394
PURPURHOLZ (*n*): violet wood; purpleheart wood (*Copaifera bracteata*); Sg. 0.81-1.0 or (*Peltogyne paniculata*); Sg. 0.85-1.0 [395
PURPURIG: purplish 396
PURPURKARMIN (*m*): murexide; purple carmine 397
PURPURLÜSTER (*m*): purple lustre (a gold-tin preparation, for porcelain decoration) (Ceramics) 398
PURPURMUSCHEL (*f*): purple shell 399
PURPURN: purple; of purple 400
PURPURROT: crimson; purple-red 401
PURPURSÄURE (*f*): purpuric acid 402
PURPURSCHWEFELSÄURE (*f*): sulphopurpuric acid 403
PURRÉE (*n*): Pjuri, Indian yellow (colouring matter obtained from the urine of cows fed with mango leaves) (the colouring principle is the magnesium salt of Euxanthinic acid) (see **INDISCHGELB**) 404
PURREN: to purr; poke (stir) a fire; rouse the watch (Naut.) 405
PURZELBAUM (*m*): somersault 406
PURZELN: to tumble; turn somersaults 407
PUSTEL (*f*): pimple; pustule (a small skin abscess, containing pus) (Med.) 408
PUSTEN: to pant; breathe; puff; snort; blow 409
PÜSTER (*m*): puff-ball 410
PUSTROHR (*n*): blow-pipe 411
PUTE (*f*): turkey-hen 412
PUTER (*m*): turkey-cock 413
PUTRESCIN (*n*): putrescine (a ptomaine from human corpses); $C_4H_{12}N_2$ 414
PUTSCH (*m*): riot (attempted revolt) 415
PÜTTINGBAND (*n*): futtock hoop (Naut.) 416
PÜTTINGSBOLZEN (*m*): chain bolt 417
PÜTTINGSWANTEN (*n.pl*): futtock rigging, futtock shrouds (Naut.) 418
PÜTTINGTAUE (*n.pl*): futtock rigging, futtock shrouds (Naut.) 419
PUTZ (*m*): plaster-(ing); attire; dress; ornament; finery; polish; toilet; rough-casting 420
PUTZBAUMWOLLE (*f*): cotton-waste 421
PUTZE (*f*): snuffers 422
PÜTZE (*f*): bucket 423
PUTZEISEN (*n*): (plasterer's)-trowel 424
PUTZEN: to clean; cleanse; burnish; polish; scour; trim; attire; dress; plaster; snuff (a wick); brush; wipe; be ornamental; be becoming; suit; adorn; rebuke; groom; prune; pluck (Birds); blow (Nose); (*m*): boss; projection 425
PUTZEREI (*f*): dressing, etc. (see **PUTZEN**) 426
PÜTZE (*f*), **TEER-**: tar bucket 427
PUTZHAHN (*m*): cleaning cock (Mech.) 428
PUTZHANDEL (*m*): milliner's (business) trade 429
PUTZHANDLUNG (*f*): milliner's shop 430
PUTZIG: droll; funny 431
PUTZKALK (*m*): polishing chalk 432
PUTZKASTEN (*m*): toilet (dressing) case 433
PUTZKRAM (*m*): millinery; finery 434
PUTZLAGE (*f*): coat of plaster 435
PUTZMACHEREI (*f*): millinery-(trade) 436
PUTZMACHERIN (*f*): milliner 437
PUTZMASCHINE (*f*): cleaning (scutching) machine 438

PUTZMATERIAL (*n*): polishing (cleaning) material 439
PUTZMAURER (*m*): plasterer 440
PUTZMITTEL (*n*): polishing agent, polish 441
PUTZÖL (*n*): cleaning oil 442
PUTZPASTE (*f*): polishing paste, polish 443
PUTZPRÄPARAT (*n*): polish; polishing preparation 444
PUTZPULVER (*n*): cleaning (polishing) powder 445
PUTZSCHICHT (*f*): coat (layer) of plaster 446
PUTZSEIFE (*f*): cleaning soap 447
PUTZSTEIN (*m*): cleaning stone, bath brick 448
PUTZSTUBE (*f*): dressing room 449
PUTZTISCH (*m*): dressing (toilet) table; cleaning table (Ceramics) 450
PUTZTUCH (*n*): cloth (for cleaning or polishing); cleaning (cloth; waste) rag 451
PUTZWAREN (*f.pl*): millinery 452
PUTZWASSER (*n*): scouring water (dilute acid) 453
PUTZWERG (*n*): tow, waste (for cleaning) 454
PUTZWOLLE (*f*): cotton waste-(for cleaning purposes) 455
PUTZWOLLENENTÖLUNG (*f*): deoiling of cotton waste, extraction of oil from cotton waste 456
PUTZWOLLENENTÖLUNGSANLAGE (*f*): cotton-waste (de-oiling) cleaning plant 457
PUTZWOLLENWÄSCHEREIANLAGE (*f*). cotton-waste washing (cleaning) plant 458
PUTZWOLLÖL (*n*): wool-waste (cotton-waste) oil 459
PUTZZEUG (*n*): cleaning (utensils) material 460
PUZZOLANE (*f*): pozzuolana (see POZZOLANERDE) 461
PUZZOLANERDE (*f*): (see POZZOLANERDE) 462
PUZZUOLANERDE (*f*): (see POZZOLANERDE) 463
PYÄMIE (*f*): pyæmia, purulent infection (Med.) 464
PYKNIT (*m*): pycnite (a columnar form of Topas) (see TOPAS); Sg. 3.5 465
PYKNOMETER (*n*): pycnometer (a measuring instrument) 466
PYLEPHLEBITIS (*f*): pylephlebitis, portal (thrombosis) phlebitis (Med.) 467
PYLORUSTEIL (*m*): pylorus (Anat.) 468
PYOHÄMIE (*f*): pyæmia, purulent infection (Med.) 469
PYOKTANIN (*n*): pyoktanin (a tar dye, either blue or yellow) 470
PYOKTANIN (*n*), **BLAUES-**: (*Pyoktaninum cæruleum*): blue pyoktanin (see METHYLVIOLETT) 471
PYOKTANIN (*n*), **GELBES-**: (*Pyoktaninum aureum*): yellow pyoktanin (see AURAMIN) 472
PYRAMIDE (*f*): pyramid 473
PYRAMIDENFÖRMIG: pyramidal 474
PYRAMIDENWÜRFEL (*m*): tetrahexahedron 475
PYRAMIDOL (*n*): pyramidol 476
PYRAMIDON (*n*): Pyramidon, dimethylamino-antipyrine, amidopyrine; $C_{13}H_{17}N_3O$; Mp. 106/107°C. 477
PYRAMIDON (*n*), **KAMPFERSAUERES-**: pyramidon camphorate 478
PYRAMIDON (*n*), **SALIZYLSAUERES-**: pyramidon salicylate 479
PYRAMINGELB (*n*): pyramine yellow 480
PYRARGYRIT (*m*): pyrargyrite, dark ruby silver (sulphantimonite of silver) (see ANTIMON-SILBERBLENDE and ROTGILTIGERZ, DUNKLES-) Ag_3SbS_3; Sg. 5.85 481
PYRAZIN (*n*): pyrazine; $C_4H_4N_2$ (see also ANTIPYRIN) 482
PYRAZOLIN (*n*): pyrazoline (see ANTIYPRIN) 483
PYRAZOLON (*n*): pyrazolone 484
PYRAZOLONFARBSTOFF (*m*): pyrazolone (colouring matter) dye (see TARTRAZIN) 485
PYRENOMYCETE (*f*): *Pyrenomycetes* (an order of *Ascomycetons*, see *Claviceps Purpurea*) (Fungi) 486
PYRHELIOMETER (*n*): pyrheliometer (see HELIO-THERMOMETER) 487
PYRIDIN (*n*): pyridine; C_5H_5N; Sg. 0.9893; Mp. -42°C.; Bp. 115.5°C. 488
PYRIDINBASE (*f*): pyridine base (general formula, $C_nH_{2n}-5N$) 489
PYRIDINDERIVAT (*n*): pyridine derivate 490
PYRIMIDINBASE (*f*): pyrimidine base 491
PYRIT (*m*): pyrites; pyrite; iron pyrites (see EISENKIES and EISENDISULFID) 492
PYRITABBRAND (*m*): pyrites cinder-(s) 493
PYRITISCH: pyritic 494
PYRITOFEN (*m*): pyrites (burner; oven) furnace 495
PYRO (*n*): pyro (see PYROGALLUSSÄURE) 496
PYROANTIMONSÄURE (*f*): pyroantimonic acid 497
PYROARSENSÄURE (*f*): pyroarsenic acid 498
PYROCATECHIN (*n*): pyrocatechol, pyrocatechin (see BRENZKATECHIN) 499
PYROCHLOR (*m*): pyrochlore (mainly a niobate of cerium metals) (see also MIKROLITH); $Nb_2O_5,TiO_2,ThO_2,CaO,Ce_2O_3,FeO,HgO,MgO,UO,F,Na_2O,H_2O$; Sg. 4.4 500
PYROELEKTRIZITÄT (*f*): pyro electricity 501
PYRO-ENTWICKLER (*m*): pyro developer (in which pyrogallol is the main developing principle: see also PYROGALLUSSÄURE) 502
PYROGALLOL (*n*): pyrogallol; pyrogallic acid (see BRENZGALLUSSÄURE) 503
PYROGALLOTRIACETAT (*n*): pyrogallotriacetate 504
PYROGALLUSSÄURE (*f*): pyrogallic acid; pyro; pyrogallol (see BRENZGALLUSSÄURE) 505
PYROGEN: pyrogenic; (*n*): pyrogene (a dye) 506
PYROGENFARBSTOFF (*m*): pyrogene (pyrogenic) dye-(stuff) 507
PYROKATECHIN (*n*): pyrocatechol (see BRENZ-CATECHIN) 508
PYROLSCHWARZ (*n*): pyrol black 509
PYROLUSIT (*m*): pyrolusite (natural hydrous manganese oxide); $MnO_2.H_2O$; Sg. 4.85 (see also BRAUNSTEIN) 510
PYROMETER (*n*): pyrometer (an instrument for measuring high temperatures over 575°F.) 511
PYROMETER (*n*), **OBTISCHES-**: optical pyrometer 512
PYROMETER (*n*), **QUECKSILBER-**: mercury pyrometer 513
PYROMETER (*n*), **THERMOELEKTRISCHES-**: thermoelectric pyrometer 514
PYROMETRIE (*f*): pyrometry 515
PYROMETRISCH: pyrometric 516
PYROMETRISCHER HEIZEFFEKT (*m*): pyrometric effect (see note to VERBRENNUNGSTEMPERA-TUR) 517
PYROMORPHIT (*m*): pyromorphite (lead chlorophosphate); $Pb_5Cl(PO_4)_3$; Sg. 6.9 (see also BRAUNBLEIERZ) 518
PYRONALBRAUN (*n*): pyronal brown 519
PYRONIN (*n*): pyronine 520
PYRONINFARBSTOFF (*m*): pyronine dye-stuff (see EOSIN and RHODAMIN) 521
PYROP (*m*): pyrope (a member of the garnet group); $Mg_3(Al,Cr,Fe)_2(SiO_4)_3$; Sg. 7.5 522

PYROPENTYLEN (*n*): pyropentylene (see CYCLO-PENTADIEN) 523
PYROPHANIT (*m*): pyrophanite; $MnTiO_3$; Sg. 4.54 524
PYROPHOR (*n*): pyrophorous 525
PYROPHORE LEGIERUNG (*f*): pyrophorous alloy (see AUERMETALL) 526
PYROPHORES METALL (*n*): pyrophorous metal 527
PYROPHORISCH: pyrophoric 528
PYROPHOSPHORSAUER: pyrophosphate of; combined with pyrophosphoric acid 529
PYROPHOSPHORSÄURE (*f*): pyrophosphoric acid 530
PYROPHOSPHORSAURES SALZ (*n*): pyrophosphate 531
PYROPHOTOGRAPHIE (*f*): enamel; enamelling (Phot.) 532
PYROPHYLLIT (*m*): pyrophyllite, pencil stone (natural hydrous aluminium silicate); $HAlSi_2O_6$; Sg. 2.85 533
PYROPISSIT (*m*): pyropissite (amorphous hydrocarbon) (similar in composition to Ozocerite; see OZOKERIT); Sg. 0.9 534
PYROSÄURE (*f*): pyro acid 535
PYROSCHLEIMSÄURE (*f*): pyromucic acid (see BRENZSCHLEIMSÄURE) 536
PYROSCHWEFELIG: pyrosulphurous 537
PYROSCHWEFELSÄURE (*f*): pyrosulphuric acid; disulphuric acid (see SCHWEFELSÄURE, RAUCHENDE-) 538
PYROSTILBIT (*m*): pyrostilbite, kermesite; Sb_2S_2O; Sg. 4.55 539
PYROSTILPNIT (*m*): pyrostilpnite; Ag_3SbS_3; Sg. 4.25 540
PYROSULFURYLCHLORID (*n*): pyrosulfuryl chloride; $S_2O_5Cl_2$; Sg. 1.844-1.872 541
PYROTECHNIK (*f*): pyrotechnics 542
PYROTECHNIKER (*m*): pyrotechnist 543
PYROTECHNISCH: pyrotechnic-(al) (appertaining to fireworks) 544
PYROTHEN (*n*): pyrothen (a disinfectant composed of cresol and sulphuric acid) 545
PYROWEINSÄURE (*f*): pyrotartaric acid (see BRENZWEINSÄURE) 546
PYROXEN (*m*): pyroxine, pyroxene (a metasilicate) (see DIOPSID, AUGIT, JADËIT, ENSTATIT, HYPERSTHEN, DIALLAG and HEDENBERGIT); Sg. about 3.2-3.9 547
PYROXYLIN (*n*): pyroxylin, guncotton (see SCHIESSBAUMWOLLE) 548
PYROXYLINLACK (*m*): pyroxylin varnish (nitrocellulose dissolved in a volatile solvent; for coating metal and paper) 549
PYRRHOSIDERIT (*m*): pyrrhosiderite (see GÖTHIT 55[illegible]
PYRRHOTIN (*m*): pyrrhotite, magnetic pyrites (an iron sulphide); $Fe_{11}S_{12}$; Sg. 4.59-4.65 (see also MAGNETKIES) 551
PYRRHOTIT (*m*): pyrrhotite (see MAGNETKIES and PYRRHOTIN) 552
PYRROL (*n*): pyrrole; C_4H_5N; Sg. 0.9481-0.975; Bp. 126.2°C-131°C. 553
PYRROLIDINCARBONSÄURE (*f*), L-: (see PROLIN) 554
PYRROLIDINCARBONSÄURE (*f*), L-Y-OXY-α-: (see OXYPROLIN) 555

Q

QUABBE (*f*): marsh, swamp, quagmire 556
QUABBELN: to shake; quiver; be flabby; feel sick; tremble 557
QUABB(E)LICHT: marshy; boggy, swampy; flabby; sick, unwell, trembling, shaky, shaking, queachy, moving, yielding; nauseated, queasy, qualmish 558
QUABBLIG: (see QUABBELICHT) 559
QUACKELBEERE (*f*): (see KRAMMETSBEERE) 560
QUACKSALBER (*m*): quack; charlatan; mountebank 561
QUACKSALBERN: to hang about; dance attendance on; practice quackery 562
QUADDEL (*f*): rash (of nettle-rash) (see NESSEL-MAL) 563
QUADER (*f*): block of stone; free (ashlar) stone; hewn (quartered) stone 564
QUADERFORMATION (*f*): chalk formation (Geol.) 565
QUADERGEBIRGE (*n*): hills (mountains) of chalk formation 566
QUADERPUTZ (*m*): imitation freestone facing of buildings 567
QUADERSANDSTEIN (*m*): sandstone of chalk formation 568
QUADERSANDSTEINFORMATION (*f*): sandstone of chalk formation 569
QUADERSTEIN (*m*): (see QUADER) 570
QUADRANGEL (*m*): quadrangle, square 571
QUADRANGULÄR: quadrangular, square 572
QUADRANGULIEREN: to square 573
QUADRANT (*m*): quadrant; quarter of a circle 574
QUADRANTEN-FEUCHTIGKEITSMESSER (*m*): quadrant hygrometer 575
QUADRANTOXYD (*n*): a compound of four atoms of metal to one atom of oxygen 576
QUADRAT (*n*): square (the figure, also Mathematics) (*Quadratus*); quadrat (Typ.); natural (*Quadratum*) (♮) (Music) 577
QUADRATEISEN (*n*): square (iron) bar; square bar iron 578
QUADRATFÖRMIG: square; quadrangular; quadrilateral 579
QUADRATISCH: quadratic; square; tetragonal (Cryst.) 580
QUADRATISCHE GLEICHUNG (*f*): quadratic equation 581
QUADRATMASS (*n*): (see FLÄCHENMASS) 582
QUADRATMETER (*m*): square meter; sq.m.; m^2 583
QUADRATPYRAMIDE (*f*): square pyramid 584
QUADRATSCHEIN (*m*): (see GEVIERTSCHEIN) 585
QUADRATSCHRIFT (*f*): Hebrew writing 586
QUADRATSPUNDUNG (*f*): square groove and tongue 587
QUADRATSUMME (*f*): sum of squares (Maths.) 588
QUADRATUR (*f*): quadrature 589
QUADRATWURZEL (*f*): square root ($\sqrt{\ }$) (Maths.) 590
QUADRATZAHL (*f*): square-(number); second power-(of a number); $(....^2)$ 591
QUADRATZENTIMETER (*n*): square centimeter; sq. cm.; cm^2 592
QUADRATZIFFER (*f*): the two end figures of a squared number 593
QUADRIEREN: to square; raise to the second power 594
QUADRILATERAL: quadrilateral, four-sided 595

QUADRINOM (*n*): expression containing four terms (Maths.) 596
QUADRIPARTITION (*f*): dividing (separating) into four parts 597
QUADRISYLLABUM (*n*): four-syllable word 598
QUADRONE (*f*): quadroon, quarteroon (offspring of a white and a terceroon) 599
QUADRUPEL: quadruple; four-fold 600
QUADRUPELALLIANZ (*f*): quadruple alliance, a four-powers alliance 601
QUAI (*n*): quay, landing-place, wharf (see KAI) 602
QUAIMAUER (*f*): (see KAIMAUER) 603
QUÄKER (*m*): quaker 604
QUÄKERHAFER (*m*): quaker-oats 605
QUAL (*f*): torment; pang; agony; pain; anguish; torture; distress; affliction; worry 606
QUÄLEN: to distress; afflict; worry; plague; torture; torment; cause agony; give pain; harass; vex; persecute; deaden (Colours) 607
QUALIFIKATION (*f*): qualification 608
QUALIFIKATIONSBERICHT (*m*): qualification report (a report in which a superior proposes an inferior for a post) (Mil.) 609
QUALIFIZIEREN: to qualify; fit; modify; classify (in order of quality); state the quality of 610
QUALIFIZIERUNG (*f*): classification-(according to quality) 611
QUALITÄT (*f*): quality 612
QUALITATIV: qualitative 613
QUALITÄTSARBEIT (*f*): quality work, work of first class quality 614
QUALITÄTSEISEN (*n*): first quality iron, special iron, pig for making best wrought iron [from siderite (see SPATEISENSTEIN) and charcoal] 615
QUALITÄTSKOEFFIZIENT (*m*): co-efficient of quality 616
QUALITÄTSREGULIERUNG (*f*): quality governing; regulating the mixture (Machinery, internal combustion) 617
QUALLE (*f*): jellyfish; sea-blubber; sea-nettle; medusa (see MEERNESSEL) 618
QUALLENPOLYP (*m*): medusa, hydromedusa, hydrozoa, siphonophore (*Coelentera*) 619
QUALM (*m*): vapour; (dense)-smoke; steam; exhalation 620
QUALMBAD (*n*): vapour bath 621
QUALMEN: to steam; (puff out)-smoke 622
QUALSTER (*m*): phlegm (Med.); currant (berry) eating lice (*Pentatoma baccarum*) 623
QUALSTERN: to bring up phlegm, spit, expectorate 624
QUANDEL (*m*): (kiln)-chimney or stack 625
QUANTENHAFT: pertaining to the quantum theory, or to quanta 626
QUANTENHYPOTHESE (*f*): quantum hypothesis 627
QUANTENTHEORIE (*f*): quantum theory 628
QUANTITÄT (*f*): quantity 629
QUANTITATIV: quantitative 630
QUANTITÄTSFAKTOR (*m*): factor of capacity, quantitative factor (of energy) 631
QUANTITIEREN: to measure quantity; (*n*): measurement of quantity 632
QUANTSWEISE: about; as it were 633
QUANTUM (*n*): quantum; quota; quantity; portion; amount; share 634
QUAPPE (*f*): lote, eel pout, burbot (*Lota vulgaris*) 635
QUARANTAINE (*f*): quarantine 636
QUARANTÄNE (*f*) **FLAGGE** (*f*): quarantine flag 637
QUARANTÄNEFRAGE (*f*): quarantine question 638
QUARANTÄNEHAFEN (*m*): quarantine port 639
QUARANTÄNEMASSREGEL (*f*): quarantine measures 640
QUARG (*m*): (see QUARK) 641
QUARK (*m*): curds; a sort of cream cheese; trash; excrement; rubbish; trifle 642
QUARKFARBE (*f*): (see KÄSEFARBE) 643
QUARKKÄSE (*f*): whey cheese 644
QUARKLEIM (*m*): casein (see KASEIN) 645
QUARKMOLKEN (*f.pl*): (see MOLKEN) 646
QUARRE (*n*): square (Mil.) 647
QUARRE (*f*): squalling child 648
QUARREN: to squall, cry, yell, scream, whine, grumble 649
QUARRI (*m*): *Euclea undulata* (Bot.) 650
QUART (*n*): quarto (Typ.); quart-(measure) (1.136 litres) 651
QUARTAL (*n*): quarter-(day); (as a prefix = quarterly) 652
QUARTALGERICHT (*n*): quarter sessions 653
QUARTALSÄUFER (*m*): dipsomaniac 654
QUARTANFIEBER (*n*): quartan ague (a fever which recurs every fourth day) 655
QUARTANT (*m*): quarto-(volume) 656
QUARTÄR: quarternary, fourfold, consisting of four; (*n*): quarternary 657
QUARTÄRFORMATION (*f*): quarternary formation, post-tertiary formation (Alluvium and Diluvium) (Geol.) 658
QUARTATION (*f*): parting, separating, separation (the separation of gold from silver by nitric acid) 659
QUARTBAND (*m*): quarto volume 660
QUARTE (*f*): quarter; fourth (Music) 661
QUARTERDECK (*n*): quarter deck 662
QUARTE (*f*), **REINE-**: perfect fourth (of the diatonic scale) (Mus.) 663
QUARTERMEISTER (*m*): quartermaster 664
QUARTERNÄR: quarternary (see QUARTÄR) 665
QUARTERONE (*f*): quarteroon (see QUADRONE) 666
QUARTETT (*n*): quartet; quartette (Music) 667
QUARTETT (*n*), **FLÖTEN-**: flute quartet (Flute, Violin, Viola and 'Cello) (but may also be applied to a quartet of four flutes) 668
QUARTETT (*n*), **FRAUEN-**: female-(voice)-quartet 669
QUARTETT (*n*), **GEMISCHTES-**: mixed-(voice)-quartet (a quartet of male and female voices) (usually Soprano, Contralto, Tenor and Bass) 670
QUARTETT (*n*), **HORN-**: horn quartet (for four horns) 671
QUARTETT (*n*), **KLAVIER-**: piano quartet (Piano, Violin, Viola and 'Cello) 672
QUARTETT (*n*), **MÄNNER-**: male-(voice)-quartet 673
QUARTETT (*n*), **STREICH-**: string quartet (Music) 674
QUARTE (*f*), **ÜBERMÄSSIGE-**: augmented fourth (Mus.) 675
QUARTE (*f*), **VERMINDERTE-**: diminished fourth (Mus.) 676
QUARTFORMAT (*n*): quarto-(size) 677
QUARTGEIGE (*f*): kit (old instrument) 678
QUARTIER (*n*): quarters; abode; billet; dwelling; lodging; quarter; fourth-(part); district; place; position; location; region; locality; territory; mercy; baulk (Billiards); watch (Naut.) 679

QUARTIEREN: to (furnish with)-quarter-(s), billet, lodge; divide (cut up) into four parts; (*n*): quartering; billeting 680
QUARTIERMEISTER (*m*): quarter-master 681
QUARTUHR (*f*): sand-glass (Naut.) 682
QUARZ (*m*): quartz; rock crystal; silex; silica; silicon dioxide; SiO_2; Sg. 2.65; Mp. 1700°C. 683
QUARZÄHNLICH: quartz-like; like quartz; quartzose; quartzitic 684
QUARZARTIG: (see QUARZÄHNLICH) 685
QUARZDRÜSE (*f*): crystalline quartz 686
QUARZFADEN (*m*): quartz thread 687
QUARZFELS (*m*): quartz rock; quartzite (see CHALCEDON) 688
QUARZFLUSS (*m*): coloured, crystalline quartz 689
QUARZGLAS (*n*): quartz-glass (see KIESELSÄURE and SILICIUMDIOXYD); Sg. 2.206 690
QUARZGLASGEFÄSS (*n*): quartz-glass vessel 691
QUARZGLAS QUECKSILBERTHERMOMETER (*n*): quartz-glass mercury thermometer 692
QUARZGLAS (*n*), UNDURCHSICHTIGES-: opaque quartz glass (from electrical melting of sand) 693
QUARZGLASWIDERSTANDSTHERMOMETER (*n*): quartz-glass resistance thermometer 694
QUARZGUT (*n*): quartz (appliances) utensils; appliances (utensils) made of quartz-glass; opaque quartz-glass (see QUARZGLAS, UN-DURCHSICHTIGES-) 695
QUARZIG: quartzy; quartzose 696
QUARZIN (*m*): quartzite, chalcedony (see CHALCEDON) 697
QUARZIT (*m*): quartzite (see CHALCEDON) 698
QUARZKORN (*n*): quartz grain, grain of quartz 699
QUARZMEHL (*n*): quartz powder 700
QUARZPORPHYR (*m*): quartz porphry (the common form of prophry; see PORPHYR and ORTHO-KLASPORPHYR) 701
QUARZ (*m*), PRISMATISCHER-: cordierite 702
QUARZRÖHRE (*f*): quartz pipe 703
QUARZSAND (*m*): quartz sand 704
QUARZSCHIEFER (*m*): quartz (siliceous) shale, slate or schist 705
QUARZSCHMELZMATERIAL (*n*): quartz material to be melted (i.e., usually rock-crystal; see BERGKRISTALL) 706
QUARZSINTER (*m*): siliceous sinter 707
QUARZTRACHYT (*m*): quartz trachyte 708
QUARZZERKLEINERUNGSANLAGE (*f*): quartz crushing plant 709
QUARZZIEGEL (*m*): quartz (Dinas) brick 710
QUASI: quasi; as (if) it were; as though; apparently; in a certain (manner) sense; to a certain extent 711
QUASIABITTER (*n*): (see QUASSIABITTER) 712
QUASIELASTISCH: quasi-elastic 713
QUASSIABITTER (*n*): quassin (see QUASSIN) 714
QUASSIAHOLZ (*n*): quassia-(wood); bitter-ash; bitterwood (*Quassia amara* or *Picrasma excelsa*) (used as a substitute for hops); Sg. 0.54 715
QUASSIENHOLZ (*n*): quassia-(wood) (see QUASSIA-HOLZ) 716
QUASSIN (*n*): quassin (from bark of *Picrasma excelsa* or *Quassia amara*) 717
QUAST (*m*): tassel; brush; tuft; mop; puff 718
QUASTE (*f*): (see QUAST) 719
QUASTIG: tufted; tasselled 720
QUAST (*m*), TEER-: tar brush 721

QUATEMBER (*m*): quarter-day; (as a prefix = quarterly) 722
QUATERNÄR: quarternary, composed of four parts 723
QUATSCH (*m*): squash; stuff and nonsense; twaddle; pap 724
QUATSCHEN: to (crush) squash; talk (rot) nonsense; splash 725
QUEBRACHIT (*n*): quebrachite; quebrachitol 726
QUEBRACHO (*n*): quebracho (bark of *Aspidosperma quebracho-blanco*) (used for tanning purposes) 727
QUEBRACHOEXTRAKT (*m*): quebracho extract (contains tannin and is used in Tanning industry) 728
QUEBRACHOHOLZ (*n*): quebracho wood (wood of *Aspidosperma quebracho*) (see QUEBRACHO) 729
QUEBRACHOHOLZBAUM (*m*): quebracho-wood tree (*Aspidosperma quebracho* and *Loxopterygium lorentzii*) 730
QUEBRACHORINDE (*f*): quebracho bark (bark of *Aspidosperma quebracho*) (see QUEBRACHO) 731
QUECKE (*f*): couch grass; quitch; graminis (*Agropyrum repens*); sand carex (*Carex arenaria*) 732
QUECKE (*f*), KLEINE-: creeping bent-grass, fine-top grass (*Agrostis alba*) 733
QUECKENGRAS (*n*): couch grass (see QUECKE) 734
QUECKENWURZEL (*f*): quitch (dog; quick) root (*Radix graminis*) (from *Agropyron repens*); Carex root (*Rhizoma caricis* of *Carex arenaria*) 735
QUECKSILBER (*n*): mercury; quicksilver; (Hg) (*Hydrargyrum*); Sg. 13.59; Mp. – 38.89°C.; Bp. 357.3°C. (see MERKURI- and MERKURO-) 736
QUECKSILBERANTIMONSULFID (*n*): mercury-antimony sulphide 737
QUECKSILBERATOXYL (*n*): (see QUECKSILBER, PARA-AMINOPHENYLARSINSAURES-) 738
QUECKSILBERAUFLAGE (*f*): coating of mercury 739
QUECKSILBERAZID (*n*): (see STICKSTOFFWASSER-STOFFSÄURE) 740
QUECKSILBERBAROMETER (*n*): mercury barometer 741
QUECKSILBERBENZOAT (*n*): mercuric benzoate; $Hg(C_7H_5O_2)_2.H_2O$; Mp. 165°C. 742
QUECKSILBERBOGEN (*m*): mercury arc 743
QUECKSILBERBRANDERZ (*n*): idrialite 744
QUECKSILBERBROMID (*n*): mercury (mercuric) bromide; $HgBr_2$; Sg. 5.738-6.604; Mp. 236.5°C.; Bp. 318°C. 745
QUECKSILBERBROMÜR (*n*): mercurous bromide; Hg_2Br_2 or $HgBr$; Sg. 7.307 746
QUECKSILBERCHLORID (*n*): perchloride of mercury; sublimate; muriate of quicksilver; sublimate of mercury; mercury (mercuric) chloride; corrosive sublimate; mercury bichloride (*Hydrargyrum bichloratum*); $HgCl_2$; Sg. 5.32-5.41; Mp. 265°C.-278°C.; Bp. 303-307°C. 747
QUECKSILBERCHLORIDLÖSUNG (*f*): mercuric chloride solution 748
QUECKSILBERCHLORÜR (*n*): mercurous (sub)-chloride; mercury monochloride; calomel; mercury chloride, mild; (*Hydrargyrum chloratum mite*); $HgCl$ or Hg_2Cl_2; Sg. 6.993 749

QUECKSILBERCHLORÜR (*n*), KOLLOIDALES- : colloidal mercurous chloride (see KALOMELOL) 750

QUECKSILBERCYANID (*n*): mercuric cyanide; $Hg(CN)_2$ (a disinfectant) (see CYANQUECK-SILBER) 751

QUECKSILBERCYANÜR (*n*): mercurous cyanide; mercury protocyanide; $Hg_2(CN)_2$ 752

QUECKSILBERDAMPF (*m*): mercury vapour 753

QUECKSILBERDAMPFLAMPE (*f*): mercury vapour lamp 754

QUECKSILBERDESTILLATIONSAPPARAT (*m*): mercury distilling apparatus 755

QUECKSILBERDIÄTHYL (*n*): mercury diethyl 756

QUECKSILBERDIMETHYL (*n*): mercury dimethyl 757

QUECKSILBERDIPHENYL (*n*): mercury diphenyl 758

QUECKSILBERDREHLUFTPUMPE (*f*): mercury rotating air pump 759

QUECKSILBERERZ (*n*): mercury ore 760

QUECKSILBERFADEN (*m*): mercury (thread) filament 761

QUECKSILBERFAHLERZ (*n*): tetrahedrite with mercury content 762

QUECKSILBERFARBE (*f*): mercury colour (mainly vermilion, see ZINNOBER) 763

QUECKSILBERFULMINAT (*n*): (see KNALLQUECK-SILBER) 764

QUECKSILBERGLIDIN (*n*): mercury glidine 765

QUECKSILBERGLIDINE (*f*): mercury glidine 766

QUECKSILBERGUHR (*f*): native sulphate of mercury 767

QUECKSILBERHALOGEN (*n*): mercury halide 768

QUECKSILBERHALTIG: mercurial; containing mercury 769

QUECKSILBERHORNERZ (*n*): horn quicksilver; native calomel (natural chloride of mercury) (see CHLORQUECKSILBER) 770

QUECKSILBERJODAT (*n*): mercuric iodate; $Hg(IO_3)_2$ 771

QUECKSILBERJODID (*n*): mercury (mercuric) iodide; mercury biniodide; HgI_2; Sg. 6.257 772

QUECKSILBERJODÜR (*n*): mercurous iodide; Hg_2I_2; Sg. 7.7 773

QUECKSILBERKAPSEL (*f*): (thermometer)-bulb 774

QUECKSILBERKATHODE (*f*): mercury cathode 775

QUECKSILBER, KOLLOIDALES- (*n*): colloidal mercury 776

QUECKSILBERLAKTAT (*n*): (see MERKURILAKTAT) 777

QUECKSILBERLAMPE (*f*): mercury lamp 778

QUECKSILBERLEBERERZ (*n*): hepatic cinnabar 779

QUECKSILBERLEGIERUNG (*f*): amalgam; mercury alloy 780

QUECKSILBERLICHT (*n*): mercury light 781

QUECKSILBERLICHTLAMPE (*f*): mercury (light) lamp 782

QUECKSILBERLÖSUNG (*f*): mercury solution 783

QUECKSILBERLUFTPUMPE (*f*): mercury air pump 784

QUECKSILBERMANOMETER (*n*): mercury pressure gauge 785

QUECKSILBERMASSE (*f*): mercury mass 786

QUECKSILBERMITTEL (*n*): mercurial 787

QUECKSILBERMOHR (*m*): ethiops mineral; black mercuric sulphide (see QUECKSILBERSULFID) 788

QUECKSILBERN: of mercury, of quicksilver, mercurial 789

QUECKSILBERNIEDERSCHLAG (*m*): red precipitate (see QUECKSILBEROXYD and PRÄZIPITAT, ROTER-) 790

QUECKSILBERNITRAT (*n*): mercury nitrate (see QUECKSILBEROXYDNITRAT and QUECKSILBER-OXYDULNITRAT) 791

QUECKSILBEROBERFLÄCHE (*f*): mercury surface 792

QUECKSILBEROXYCYANID (*n*): mercury (mercuric) oxycyanide; $Hg(CN)_2.HgO$; Sg. 4.43 [793

QUECKSILBEROXYD (*n*): mercuric oxide; red mercury oxide; red precipitate (*Hydrargyrum oxydatum*); HgO; Sg. 11.0-11.29 [794

QUECKSILBEROXYDACETAT (*n*): mercuric acetate; $Hg(C_2H_3O_2)_2$; Sg. 3.286 795

QUECKSILBEROXYD (*n*), CHLORSAURES- : mercuric chlorate; $(C_{24}H_{39}O_5)_2Hg$ 796

QUECKSILBEROXYD (*n*), GELBES- : (*Hydrargyrum oxydatum flavum*); yellow mercuric oxide; HgO; Sg. 11.03 797

QUECKSILBEROXYD (*n*), KOLLOIDALES- : colloidal mercuric oxide 798

QUECKSILBEROXYDNITRAT (*n*): mercuric nitrate (*Hydrargyrum nitricum oxydatum*); $Hg(NO_3)_2.8H_2O$; Mp. 6.7°C. 799

QUECKSILBEROXYD (*n*), ROTES- : (*Hydrargyrum oxydatum rubrum*): red mercuric oxide; HgO; Sg. 11.08-11.2 (see QUECKSILBER-OXYD) 800

QUECKSILBEROXYDSALZ (*n*): mercuric salt (see MERKURI-) 801

QUECKSILBEROXYDSULFAT (*n*): mercuric sulphate, mercury (bisulphate) persulphate; $HgSO_4$; Sg. 6.5 (see also QUECKSILBERSULFAT) 802

QUECKSILBEROXYDSULFAT (*n*), BASISCHES- : basic mercuric sulphate (*Turpethum minerale*); $HgSO_4.2HgO$; Sg. 6.44 803

QUECKSILBEROXYDUL (*n*): mercurous oxide; mercury binoxide; Hg_2O; Sg. 9.8 804

QUECKSILBEROXYDULACETAT (*n*): mercurous acetate; $Hg_2(C_2H_3O_2)_2$ 805

QUECKSILBEROXYDULNITRAT (*n*): mercurous nitrate (*Hydrargyrum nitricum oxydulatum*); $HgNO_3$ or $Hg_2(NO_3)_2$; Sg. 4.3; Mp. 70°C. 806

QUECKSILBEROXYDULNITRIT (*n*): mercurous nitrite; $HgNO_2$; Sg. 5.925 807

QUECKSILBEROXYDULSALZ (*n*): mercurous salt (see MERKURO-) 808

QUECKSILBEROXYDULSULFAT (*n*): mercurous sulphate; Hg_2SO_4; Sg. 7.121 809

QUECKSILBEROXYZYANID (*n*): (see QUECKSILBER-OXYCYANID) 810

QUECKSILBER (*n*), PARA-AMINO-PHENYLARSIN-SAURES- : mercury para-aminophenylarsinate; $[C_6H_4(NH_2)AsO_2(OH)]_2Hg$ 811

QUECKSILBERPFLASTER (*n*): mercurial plaster (Pharm.) 812

QUECKSILBER, PHOSPHORSAURES- (*n*): mercury phosphate 813

QUECKSILBERPILLE (*f*): blue pill 814

QUECKSILBERPRÄPARAT (*n*): mercury preparation 815

QUECKSILBERPRÄZIPITAT (*m* and *n*): (mercury)-precipitate 816

QUECKSILBERPRÄZIPITAT, GELBES- (*n*): yellow precipitate 817

QUECKSILBERPRÄZIPITAT, ROTES- (*n*): red precipitate (see PRÄZIPITAT, ROTES-) 818

QUECKSILBERPRÄZIPITAT (*m*), UNSCHMELZBARER WEISSER- : (see AMINOQUECKSILBERCHLORID) 819

QUECKSILBERPRÄZIPITAT (*m*), WEISSER- : (see MERKURIAMMONIUMCHLORID) 820
QUECKSILBERPUMPE (*f*) : mercury pump 821
QUECKSILBERPYROMETER (*n*) : mercury pyrometer 822
QUECKSILBERREINIGUNGSAPPARAT (*m*) : mercury purifying apparatus 823
QUECKSILBERREINIGUNGSFLASCHE (*f*) : mercury purifying flask 824
QUECKSILBERRESORBIN (*n*) : mercury resorbine (a mercurial ointment made with resorbine) 825
QUECKSILBERREST (*m*) : mercury residue 826
QUECKSILBERRHODANID (*n*) : mercury (mercuric) thiocyanate; mercuric (sulfocyanate; rhodanide) sulfocyanide; $Hg(SCN)_2$ [827
QUECKSILBERRHODANÜR (*n*) : mercurous (thiocyanate; sulfocyanate; sulfocyanide) rhodanide 828
QUECKSILBERRUSS (*m*) : stupp; mercurial soot 829
QUECKSILBERSALBE (*f*) : mercurial ointment 830
QUECKSILBERSALBE, GELBE- (*f*) : mercuric nitrate ointment; yellow mercurial ointment; $Hg(NO_3)_2.2H_2O$ 831
QUECKSILBER (*n*), SALICYLSAURES- : mercury salicylate 832
QUECKSILBERSALPETER (*m*) : mercury nitrate (see QUECKSILBEROXYDNITRAT and QUECK-SILBEROXYDULNITRAT) 833
QUECKSILBER, SALPETERSAURES- (*n*) : (see QUECK-SILBERSALPETER) 834
QUECKSILBERSALZ (*n*) : mercurial; mercury salt (see MERKURI- and MERKURO-) 835
QUECKSILBERSÄULE (*f*) : mercury column 836
QUECKSILBER, SCHWEFELSAURES- (*n*) : mercury sulphate (see QUECKSILBEROXYDSULFAT and QUECKSILBEROXYDULSULFAT) 837
QUECKSILBERSPAT (*m*) : horn quicksilver (see QUECKSILBERHORNERZ) 838
QUECKSILBERSPIEGEL (*m*) : mercury mirror 839
QUECKSILBERSUBLIMAT (*n*) : corrosive sublimate (see QUECKSILBERCHLORID) 840
QUECKSILBERSUCCINIMID (*n*) : mercury succinimide 841
QUECKSILBERSULFAT (*n*) : mercury sulphate (*Hydrargyrum sulfuricum*) (see QUECK-SILBEROXYDSULFAT and QUECKSILBER-OXYDULSULFAT) 842
QUECKSILBERSULFID (*n*) : mercury (mercuric) sulphide (*Hydrargyrum sulfuratum*); HgS; Sg. 7.63 (black=ethiops mineral; Sg. 7.67) (red=cinnabar; vermilion; Sg. 8.13) [843
QUECKSILBERSULFID (*n*) : ROTES- : red mercuric sulphide, cinnabar (see ZINNOBER and CHINESISCHROT) 844
QUECKSILBERSULFID, ROTES-, GEFÄLLT- : precipitated red mercuric sulphide; Sg. 8.129 [845
QUECKSILBERSULFID, ROTES-, SUBLIMIERT- : sublimed red mercuric sulphide; Sg. 8.146 [846
QUECKSILBERSULFID (*n*), SCHWARZES- : black mercuric sulphide (see METACINNABARIT and QUECKSILBERSULFID); Sg. 7.701-7.748 [847
QUECKSILBERTHERMOMETER (*n*) : mercury thermometer 848
QUECKSILBERVERBINDUNG (*f*) : mercury compound (see MERKURI- and MERKURO-) 849
QUECKSILBERVERDAMPFUNGSAPPARAT (*m*) : mercury evaporating apparatus, mercury vapourising apparatus 850
QUECKSILBERVERFAHREN (*n*) : mercury process 851
QUECKSILBERVERGIFTUNG (*f*) : mercurial poisoning 852
QUECKSILBERVERSTÄRKER (*m*) : mercury intensifier (Phot.) 853
QUECKSILBERVITRIOL (*m*) : mercuric sulphate (see QUECKSILBEROXYDSULFAT) 854
QUECKSILBERWANNE (*f*) : mercury trough 855
QUECKSILBERZITTERN (*n*) : trembling of the limbs due to mercurial poisoning 856
QUECKSILBERZYANID (*n*) : mercuric cyanide (see QUECKSILBERCYANID) 857
QUELL (*m*) : spring; well; source; fountain; authority 858
QUELLBAR : capable of swelling 859
QUELLBOTTICH (*m*) : steeping (tub) vat (see QUELLSTOCK) 860
QUELLDAUER (*f*) : duration of steeping 861
QUELLE (*f*) : spring; well; source; fountain; authority 862
QUELLE HABEN : to spring from; take its source from; rise; have its origin in; be caused by 863
QUELLEN : to spring; issue; flow; swell; well; soak; steep; tumify; (a)-rise from; absorb (suck up moisture from the air by the sap remaining in wood after drying); (*n*) : absorption; steeping, etc. 864
QUELLENBAUM (*m*) : traveller's tree (*Ravenala madagascariensis*) 865
QUELLENMÄSSIG : on good authority 866
QUELLENMOOS (*n*) : water-moss (*Fontinalis*) 867
QUELLER (*m*) : glass-wort (*Salicornia herbacea*) 868
QUELLGAS (*n*) : spring gas 869
QUELLGRUND (*m*) : swamp; quagmire; river-head 870
QUELLMESSER (*m*) : spring-gauge, brine-gauge 871
QUELLREIF : sufficiently (steeped) soaked 872
QUELLREIFE (*f*) : sufficiency of (steeping) soaking 873
QUELLSALZ (*n*) : spring (well) salt 874
QUELLSAND (*m*) : quicksand 875
QUELLSÄURE (*f*) : crenate 876
QUELLSOLE (*f*) : mother-liquor 877
QUELLSTOCK (*m*) : steep-(ing) (vat; tank) cistern (for malt) 878
QUELLUNG (*f*) : welling; swelling; tumefaction; soaking 879
QUELLUNGSWÄRME (*f*) : heat of (tumefaction) swelling 880
QUELLWASSER (*n*) : spring water 881
QUENDEL (*m*) : wild (carraway) thyme (*Herba serpylli* of *Thymus serpyllum*); lemon thyme (*Thymus serpyllum vulgaris*) 882
QUENDELKRAUT (*n*) : (*Herba serpylli*) : wild thyme, lemon thyme (see QUENDEL) 883
QUENDELÖL (*n*) : wild (lemon) thyme oil (from *Thymus serpyllum vulgaris*); Sg. 0.89-0.92 884
QUENSTEDIT (*m*) : quenstedite (a natural ferric sulphate); $Fe_2(SO_4)_3.10H_2O$ 885
QUER : cross; across; athwart; diagonal; oblique-(ly); transverse-(ly); perversely; contrary; queer; wrong; irrelevant 886
QUERACHSE (*f*) : transverse axis 887
QUERAXT (*f*) : twibill 888
QUERBALKEN (*m*) : traverse (transverse) beam; cross (beam) bar; transom 889
QUERBALKEN, GROSSER- (*m*) : architrave 890
QUERBAND (*n*) : cross-tie; traverse 891
QUERBAUM (*m*) : crossbar 892
QUERBEIL (*n*) : adze 893

QUERBINDE (*f*) : cross-tie ; traverse 894
QUERBORHUNG (*f*) : cross (hole) boring ; transverse borehole 895
QUERBUNKER (*m*) : cross bunker 896
QUERCIT (*n*) : quercitol ; quercite ; acorn sugar ; pentahydroxycyclohexane ; $C_6H_7(OH)_5$; Sg. 1.59 ; Mp. 234°C. 897
QUERCITRON (*n*) : quercitron (powdered bark of various species of oak, as, *Quercus nigra* ; *Q. coccinea* or *Q. tinctoria* ; contains up to 25% tannin and the dyestuff quercetine and quercitrine) 898
QUERCITRONEXTRAKT (*m*) : quercitron extract 899
QUERCITRONRINDE (*f*) : quercitron bark (bark of *Quercus coccinea tinctoria*) (see also QUERZITRONRINDE) 900
QUERDURCH : across ; transversely ; through (the middle) ; straight (across) through ; without turning to right or left 901
QUERDURCHMESSER (*m*) : transverse diameter 902
QUERDURCHSCHNITT (*m*) : (see QUERSCHNITT) 903
QUERE (*f*) : cross (oblique ; transverse ; diagonal) direction ; breadth 904
QUERFALL (*m*) : disappointment 905
QUERFASER (*f*) : transverse fibre 906
QUERFEDER (*f*) : transverse spring 907
QUERFELDEIN : (a)-cross country 908
QUERFELDEIN SCHLAGEN : to strike across country 909
QUERFLÜGEL (*m*) : transept (Arch.) 910
QUERFRAGE (*f*) : cross question-(ing) ; interrogation ; cross examination 911
QUERFRAGEN TUN : to cross-question, cross-examine 912
QUERGANG (*m*) : traverse, cross passage or corridor 913
QUER GEGENÜBER : opposite 914
QUERGRABEN (*m*) : traverse (Military) 915
QUERHAUPT (*n*) : cross-head-(beam) 916
QUERHOBELMASCHINE (*f*) : shaping machine 917
QUERHOLZ (*n*) : cross-bar ; transverse (beam) bar ; transom ; cross (end) grain (Wood) 918
QUERKALBEN (*n.pl*) : cross chocks 919
QUER, KREUZ UND- : upside down, topsy-turvy, in all directions 920
QUERL (*m*) : whorl (Bot.) ; egg-whisk (see also QUIRL) 921
QUERLATTE (*f*) : cross-(bar) lath 922
QUERLAUFEND : transverse 923
QUERLEN : to twirl, whirl round, whisk 924
QUERLINIE (*f*) : cross (diagonal) line 925
QUERNAHT (*f*) : circumferential seam ; cross (transverse ; butt) seam, weld or suture 926
QUERNAHTBRENNER (*m* and *pl*) : circumferential burner-(s) 927
QUERPROFIL (*n*) : cross (transverse) section 928
QUERRICHTUNG (*f*) : (in a) transverse direction 929
QUERRICHTUNG (*f*), IN DER- : transversely 930
QUERRIEGEL (*m*) : cross (rail ; bar) tie ; anchor ; cotter ; stretcher ; thwart 931
QUERRISS (*m*) : transverse crack ; cross-section ; fracture (Minerals) 932
QUERROHR (*n*) : (see QUERSIEDER) 933
QUERROHRKESSEL (*m*) : (see QUERSIEDERKESSEL) 934
QUERROHRVERSTEIFUNG (*f*) : Galloway (cross) tube stay-(ing) (of boilers) 935
QUERSÄGE (*f*) : cross cut saw 936
QUERSALING (*f*) : cross tree 937
QUERSATTEL (*m*) : side-saddle 938
QUERSCHIFF (*n*) : transept (Arch.) 939
QUERSCHIFFS : athwart (Ships) 940
QUERSCHLEIFEN : to grind-(wood) across the grain 941
QUERSCHNITT (*m*) : cross-cut ; cross-section ; transverse section ; cross sectional area ; area ; crosscut section ; diameter (of a cylinder or conic section) 942
QUERSCHOTT (*n*) : cross bulkhead 943
QUERSCHUBVORRICHTUNG (*f*) : cross-wise thrust gear ; transverse (lateral) thrust gear 944
QUERSCHWELLE (*f*) : transverse sleeper (see also QUERHOLZ) 945
QUERSCHWINGUNG (*f*) : transverse (lateral) vibration 946
QUERSIEDER (*m*) : Galloway tube, cross tube (of boilers) 947
QUERSIEDERKESSEL (*m*) : boiler with Galloway tube (either a two-flued Cornish boiler or a Lancashire boiler fitted with Galloway tubes) 948
QUERSINNIG : cross-grained; irritable; contrary; contentious 949
QUERSPANT (*n*) : square frame 950
QUERSTANGE (*f*) : tension (cross ; transverse) bar 951
QUERSTOLLEN (*m*) : cross (adit) gallery (Mining) 952
QUERSTRASSE (*f*) : cross-road 953
QUERSTRICH (*m*) : hyphen, dash ; break (Typ.) ; cross-line (of fractions) ; cancelling line (for fractions) ; ledger-line (Mus.) 954
QUERSTROM (*m*) : transverse (current) flow 955
QUERSTÜCK (*n*) : cross piece 956
QUERSTUTZEN (*m*) : cross-tube (of boilers) 957
QUERTRÄGER (*m*) : cross (transverse) bearer, beam, girder, tie, support, bearer-bar, grate-bar bearer 958
QUERÜBER : opposite, over against 959
QUERÜBER GEHEN : to cross over 960
QUERULIEREN : to be querulous, grumble, complain 961
QUERVERBAND (*m*) : transverse strengthening, stiffening or framing 962
QUERVERBINDUNG (*f*) : cross connection ; cross-tube (of boilers) 963
QUERVERSTEIFUNG (*f*) : diagonal (bar or stiffener) 964
QUERWAND (*f*) : transverse (partition) wall ; diaphragm 965
QUERWEG (*m*) : cross path 966
QUERWELLE (*f*) : transverse shaft 967
QUERWIND (*m*) : side wind 968
QUERZETIN (*n*) : quercetine (a yellow dye) (obtained from Quercitron bark, see QUERZITRONRINDE) 969
QUERZIT (*n*) : (see QUERCIT) 970
QUERZITRIN (*n*) : quercitrine (a yellow dye obtained from quercitron bark ; see QUERZITRONRINDE) ; $C_{21}H_{22}O_{12}.2H_2O$ 971
QUERZITRONRINDE (*f*) : quercitron bark (of *Quercus nigra* or *Q. tinctoria* ; used for dyeing) (see also QUERCITRON-) 972
QUERZUSAMMENZIEHUNG (*f*) : lateral contraction 973
QUESE (*f*) : water-blister 974
QUETSCHE (*f*) : wild plum (*Prunus domestica*) (Bot.) ; squeezer ; presser ; crusher ; wringer ; pincher ; crushing machine ; state of being crushed ; dilemma ; fix 975
QUETSCHEN : to pinch ; squeeze ; crush ; bruise ; mash ; contuse ; squash ; squeegee (Phot.) 976

QUETSCHER (*m*) : squeezer ; squeegee (Phot.) 977
QUETSCHGRENZE (*f*) : limit of (compressibility) compression 978
QUETSCHHAHN (*m*) : pinch-cock ; squeezing cock (rubber-tube clip) 979
QUETSCHHAHN, SCHRAUBEN- (*m*) : screw pinch-cock 980
QUETSCHMASCHINE (*f*) : crushing machine 981
QUETSCHMÜHLE (*f*) : crushing (bruising) mill 982
QUETSCHUNG (*f*) : squeezing, etc. ; contusion (see QUETSCHEN) 983
QUETSCHWALZE (*f*) : crushing roll ; roller squeegee 984
QUETSCHWERK (*n*) : crushing mill 985
QUETSCHWUNDE (*f*) : bruise, contusion 986
QUEUE (*f*) : cue (Billiards) 987
QUICK : quick, brisk, lively, animated, alert, sprightly, active 988
QUICKARBEIT (*f*) : amalgamation 989
QUICKBEIZE (*f*) : amalgamating fluid 990
QUICKBEUTEL (*m*) : amalgamating skin 991
QUICKBORN (*m*) : spring (of water) 992
QUICKBREI (*m*) : amalgam (see QUECKSILBER-LEGIERUNG) 993
QUICKEN : to amalgamate 994
QUICKERZ (*n*) : mercury (quicksilver) ore 995
QUICKFASS (*n*) : amalgamation cask 996
QUICKMETALL (*n*) : amalgamated metal 997
QUICKMÜHLE (*f*) : (ore)- amalgamating mill 998
QUICKWASSER (*n*) : quickening liquid ; mercury (mercurial) solution (solution of a mercuric salt ; for plating) 999
QUIDAM (*m*) : someone ; so-and-so (when referring to persons) 000
QUIEK (*m*) : squeak 001
QUIEKEN : to squeak ; squeal 002
QUILLAJABAUM (*m*) : quillai-(a) tree (*Quillaja saponaria*) 003
QUILLAJARINDE (*f*) : quillai-(a) bark (see QUILLAYARINDE) 004
QUILLAJASÄURE (*f*) : quillaic acid 005
QUILLAYAEXTRAKT (*m*) : quillai-(a) bark extract 006
QUILLAYARINDE (*f*) : quillai-(a) bark ; China (soap ; Panama ; Murillo) bark (of *Quillaja saponaria*) 007
QUINKAILLERIE (*f*) : ironmongery, hard-ware ; fancy (goods) wares 008
QUINTE (*f*) : fifth (Music) 009
QUINTENFORTSCHREITUNG (*f*) : consecutive fifths, succession of fifths (Mus.) 010
QUINTE (*f*), REINE- : perfect fifth (Mus.) 011
QUINTESSENZ (*f*) : quintessence 012
QUINTETT (*n*) : quintet (Mus.) 013
QUINTE (*f*), VERMINDERTE- : diminished fifth (Mus.) 014
QUIRL (*m*) : curl (measure of vortex) ; vortex ; twirling (whirling) device ; whorl (Bot.) ; fidgety person (see also QUERL) 015
QUIRLEN : to twirl ; turn ; whirl 016
QUITT : quits ; even ; rid 017
QUITTE (*f*) : quince (*Cydonia vulgaris* or *Pirus cydonia*) 018
QUITTENAPFEL (*m*) : quince 019
QUITTENBAUM (*m*) : quince tree (see QUITTE) 020
QUITTENBROT (*n*) : confection of quince 021
QUITTENGELB : quince-yellow 022
QUITTENKERN (*m*) : quince seed (of *Cydonia vulgaris*) 023
QUITTENMISPEL (*f*) : Cotoneaster, quince-leaved medlar (*Cotoneaster vulgaris*) 024
QUITTENÖLSÄURE (*f*) : quince oleic acid (isomer of ricinolic acd) (see RICINOLSÄURE) 025
QUITTENSAFT (*m*) : quince juice 026
QUITTENSAMEN (*m*) : (see QUITTENKERN) 027
QUITTENSAMENÖL (*n*) : quince seed oil (from *Cydonia vulgaris*) ; Sg. 0.922 028
QUITTENSCHLEIM (*m*) : quince emulsion 029
QUITTIEREN : to acknowledge (receipt) a bill ; quit ; abandon ; leave ; discharge ; settle 030
QUITTIERUNG (*f*) : receipt-(ing) of a bill 031
QUITTUNG (*f*) : receipt ; quittance ; acknowledgment 032
QUOTE (*f*) : quota ; share 033
QUOTIENT (*m*) : quotient 034
QUOTIEREN : to quote (Prices) ; bid ; tender ; offer ; estimate 035

R

RAA (*f*) : yard (Naut.) 036
RAA, GROSSE- (*f*) : main-yard (Naut.) 037
RAASEGEL (*n*) : yard-sail 038
RABATT (*m*) : rebate, discount, abatement, reduction, drawback 039
RABATTE (*f*) : border-(bed) ; facing (Tailoring) 040
RABATTIEREN : to discount, allow, reduce, abate 041
RABE (*m*) : raven 042
RABENAAS (*n*) : carrion 043
RABENGLIMMER (*m*) : zinnwaldite (see ZINNWALDIT) 044
RABENSCHWARZ : inky ; jet-black 045
RABENSTEIN (*m*) : gallows ; place of execution ; belemnite 046
RACEMISCH : racemic 047
RACEMISIEREN : to racemise 048
RACEMKÖRPER (*m*) : racemic (compound) substance 049
RACHE (*f*) : vengeance ; revenge ; vindication 050
RACHEN (*m*) : throat ; jaws ; mouth ; pharynx (Med.) ; fauces ; abyss 051
RÄCHEN : to avenge ; revenge ; retaliate ; vindicate ; requite 052
RACHENBEIN (*n*) : jaw bone 053
RACHENBLUME (*f*) : labiate flower (Bot.) 054
RACHENFÖRMIG : labiate (Bot.) 055
RACHENHÖHLE (*f*) : pharyngeal cavity (Anat.) 056
RACHENKRANKHEIT (*f*) : throat-disease, disease of the throat (Med.) 057
RACHSÜCHTIG : revengeful ; vindictive ; vengeful ; malicious ; malignant ; spiteful ; rancorous ; malevolent 058
RACK (*n*) : parrel, truss (of a yard) (Naut.) 059
RACKBAND (*n*) : truss hoop 060
RACKKETTE (*f*) : truss pendant 061
RACKNIEDERHOLER (*m*) : parrel downhaul 062
RACKSCHLITTEN (*m*) : parrel rib (Naut.) 063
RACKTAU (*n*) : parrel rope, truss (Naut.) 064
RAD (*n*) : wheel ; gear 065
RADACHSE (*f*) : axle (shaft) of a wheel ; axletree ; wheel-axle 066

RADARM (*m*): spoke 067
RADBANDAGE (*f*): tyre (of wheels) 068
RADBEWEGUNG (*f*): rotation; rotary motion 069
RADBODEN (*m*): wheel rim base; wheel cover plate 070
RADBÜCHSE (*f*): wheel (grease) box 071
RADDAMPFER (*m*): paddle steamer 072
RADDREHUNG (*f*): rotation; torsion; torque 073
RADE (*f*): corn-cockle; corn campion (*Agrostemma githago*) 074
RADEBRECHEN: to murder (a language); mutilate 075
RÄDELERZ (*n*): wheel ore; bournonite (see **BOURNONIT**) 076
RÄDELN: to riddle; wind (Silk) 077
RÄDELSFÜHRER (*m*): ring-leader 078
RADEMACHER (*m*): wheelwright 079
RADEN (*m*): corn-cockle (see **RADE**) 080
RÄDER (*m*): sieve, screen, riddler 081
RÄDERAUFZIEHPRESSE (*f*): wheel press 082
RÄDERFRÄSMASCHINE (*f*): gear cutting machine, wheel (cutting) milling machine 083
RÄDERGEHÄUSE (*n*): casing, housing; gear-case; paddle-box 084
RÄDERGETRIEBE (*n*): geared drive; gearing 085
RÄDERN: to sift; riddle; screen; fit wheels to 086
RÄDERPAAR (*n*): couple of wheels 087
RÄDERPRESSE (*f*): wheel press 088
RÄDERPRESSE (*f*), **DRUCKWASSER-**: hydraulic wheel press 089
RÄDERPRESSE (*f*), **PRESSWASSER-**: hydraulic wheel press 090
RÄDERSCHNEIDEMASCHINE (*f*): wheel cutting (gear cutting) machine 091
RÄDERTIER (*n*): rotifer, wheel animalcule 092
RÄDERTIERCHEN (*n*): rotifer; wheel animalcule 093
RÄDERÜBERSETZUNG (*f*): gear transmission; gear ratio 094
RÄDERWERK (*n*): wheel-(s); gearing; sifting apparatus; clock-work 095
RADFAHREN (*n*): cycling; (Verb): to cycle, ride a cycle 096
RADFAHRER (*m*): cyclist 097
RADFELGE (*f*): wheel rim; felly 098
RADFLANSCH (*m*): wheel flange 099
RADFÖRMIG: rotate (Bot.); wheel-shaped 100
RADGEHÄUSE (*n*): splasher, wheel cover (on locomotive wheels) 101
RADGESTELL (*n*): bogie frame (Railways) 102
RADIAL: radial-(ly) 103
RADIALLAGER (*n*): radial bearing 104
RADIALSCHUB (*m*): redial thrust (of bearings) 105
RADIALTURBINE (*f*): radial-flow turbine 106
RADIATOR (*m*): radiator (for heating purposes, and for motor cars) 107
RADIENWINKEL (*m*): angle formed by two radii 108
RADIEREN: to erase; etch 109
RADIERFIRNIS (*m*): etching varnish 110
RADIERGRUND (*m*): etching ground 111
RADIERGUMMI (*n*): (rubber)-eraser 112
RADIERKUNST (*f*): art of etching 113
RADIERMESSER (*n*): erasing (scraping) knife; scraper 114
RADIERNADEL (*f*): etching (needle) point 115
RADIERPULVER (*n*): pounce (a very fine powder) 116
RADIERUNG (*f*): erasure; erasing; etching 117
RADIERWASSER (*n*): erasing water (diluted aqua fortis) 118
RADIES (*n*): radish (*Raphanus sativus*) 119
RADIESCHEN (*n*): (see **RADIES**) 120
RADIIEREN: to radiate 121
RADIKAL: radical; (*m*): radical 122
RADIKALESSIG (*m*): radical vinegar; acetic or glacial acetic acid 123
RADIOAKTINIUM (*n*): radioactinium (a radio active substance) (RdAc) (atomic weight 226) 124
RADIOAKTIV: radioactive 125
RADIOAKTIVE PERIODE (*f*): radioactive period (the time in which the radioactivity of a substance diminishes by half) (see also **HALBWERTSZEIT**) 126
RADIOAKTIVER STOFF (*m*): radioactive substance 127
RADIOAKTIVE SUBSTANZ (*f*): radioactive substance 128
RADIOAKTIVITÄT (*f*): radioactivity 129
RADIOBLEI (*n*): radiolead (a radioactive substance) 130
RABIOBLEISALZ (*n*): radiolead salt 131
RADIOELEMENT (*n*): radioelement 132
RADIOGRAMM (*n*): radio-telegram 133
RADIOMETER (*n*): radiometer 134
RADIOTELEGRAPHIE (*f*): radio-telegraphy, wireless telegraphy 135
RADIOTELLUR (*n*): radiotellurium (a radioactive substance) 136
RADIOTHOR (*n*): radiothorium (a radioactive substance) (RTh or RdTh) (atomic weight 228) 137
RADIOTHORIUM (*n*): (see **RADIOTHOR**) 138
RADIOTHORIUMGEHALT (*m*): radiothorium content 139
RADIUM (*n*): radium (Ra) 140
RADIUMATOM (*n*): radium atom 141
RADIUMBROMID (*n*): radium bromide; $RaBr_2$ 142
RADIUMCHLORID (*n*): radium chloride; $RaCl_2$; Mp. 1650°C. 143
RADIUMELEMENT (*n*): radium element (RaEl) 144
RADIUMEMANATION (*f*): radium emanation, Niton (Nt) 145
RADIUMGEHALT (*m*): radium content 146
RADIUMHALTIG: containing radium 147
RADIUMJODID (*n*): radium iodide 148
RADIUMPRÄPARAT (*n*): radium preparation 149
RADIUMSALZ (*n*): radium salt 150
RADIUMSTRAHLEN (*m* and *pl*): radium rays; Becquerel rays (rays from radioactive substances) 151
RADIUMSTRAHLEN, α-: alpha rays, positively electrically charged radium rays 152
RADIUMSTRAHLEN, β-: beta rays, negative radium rays 153
RADIUMVERBINDUNG (*f*): radium compound 154
RADIUS (*m*): radius (both Anatomy and Mathematics) 155
RADIUSVEKTOR (*m*): radius vector 156
RADIZIEREN: to extract the root (Maths.); (*n*): extraction of the root 157
RADKAMMER (*f*): (wheel)-stage (Turbines) 158
RADKASTEN (*m*): paddle box 159
RADKRANZ (*m*): rim; shrouding; tyre (of wheel) 160
RADLAUF (*m*): travel; rotation; run 161
RADLINIE (*f*): cycloid (Geometry) 162
RADNABE (*f*): boss 163

RADREIBUNGSVERLUST (*m*): wheel friction loss (Turbines) 164
RADREIF (*m*): tyre; rim (of wheel) 165
RADREIFEN (*m*): (see **RADREIF**) 166
RADREIFENAUFZIEHPRESSE (*f*): tyre press (for forcing on tyres) 167
RADREIFPRESSE (*f*): tyre press 168
RADSATZ (*m*): pair of wheels-(complete with axles), complete wheel set 169
RADSCHAUFEL (*f*): vane; bucket; scoop; blade; sweep; paddleboard 170
RADSCHEIBE (*f*): circular disc, wheel disc 171
RADSCHLOSS (*n*): wheel-lock 172
RADSCHMIERE (*f*): wheel (axle) grease 173
RADSCHUH (*m*): brake; drag 174
RADSCHUTZHAUBE (*f*): protecting cover over wheel; wheel cover 175
RADSPEICHE (*f*): spoke (of a wheel) 176
RADSPERRE (*f*): drag; brake 177
RADSPUR (*f*): wheel track; rut 178
RADSTAND (*m*): wheel-base 179
RADTEILUNG (*f*): wheel pitch 180
RADTIEFE (*f*): depth of rim 181
RADWELLE (*f*): axle-(tree); wheel and axle 182
RADWIRKUNGSGRAD (*m*): wheel efficiency (Turbines) 183
RADZAHN (*m*): cog (of wheels) 184
RAFAELIT (*m*): rafaelite (see **PARALAURIONIT**) 185
RAFFEL (*f*): flax-comb 186
RAFFEN: to snatch (up) away; gather-(up); collect 187
RAFFGIER (*f*): rapacity, rapaciousness, ravenousness, avidity, voracity, voraciousness, avariciousness 188
RAFFGUT (*n*): stolen goods 189
RAFFHOLZ (*n*): fallen wood 190
RAFFINADE (*f*): refined sugar 191
RAFFINADKUPFER (*n*): refined copper, tough (tough-pitch) copper (99.5-99.8% copper) 192
RAFFINATION (*f*): refining 193
RAFFINATIONSERTRAG (*m*): rendement (Sugar) 194
RAFFINATIONSWERT (*m*): (see **RAFFINATIONS-ERTRAG**) 195
RAFFINATKUPFER (*n*): refined copper (see **RAFFINADKUPFER**) 196
RAFFINATSILBER (*n*): refined silver 197
RAFFINERIE (*f*): refinery; finesse 198
RAFFINEUR (*m*): refinery; refiner; refining machine (Paper) 199
RAFFINIERANLAGE (*f*): refinery (refining) plant 200
RAFFINIEREN: to refine; meditate (speculate) upon; (*n*): refining; meditation; speculation 201
RAFFINIERFEUER (*n*): refining (refinery) furnace or hearth 202
RAFFINIERHERD (*m*): (see **RAFFINIERFEUER**) 203
RAFFINIEROFEN (*m*): refining furnace 204
RAFFINIERT: refined; cunning; designing 205
RAFFINIERZINK (*n*): refined zinc 206
RAFFINOSE (*f*): raffinose, melitriose (a sugar); $C_{18}H_{32}O_{16}.2H_2O$ 207
RAGEN: to tower; project; stick out 208
RAH (*f*): yard (of sails) 209
RAHARM (*m*): yard-arm, quarter (of a yard) (Naut.) 210
RAHBÄNDSEL (*n*): roband 211
RAHEN (*f.pl*), **UNTER-**: lower yards 212
RAH (*f*), **GROSSE-**: main yard 213
RAHKETTE (*f*): yard chain 214
RAHLEIK (*n*): head rope 215
RAHLIEK (*n*): head rope 216
RAHM (*m*): cream; crust; frame; soot 217
RAHMÄHNLICH: creamy; creamlike 218
RAHMARTIG: creamy; creamlike 219
RAHMBILDUNG (*f*): cream formation 220
RAHMEISEN (*n*): chase-bar; iron frame 221
RAHMEN: to skim; remove cream from; frame; cream; form a cream; (*m*): frame; framing; compass; bounds; form: edge; scope; welt (of boots) 222
RAHMEN (*m*), **FAHRRAD-**: frame (of a cycle) 223
RAHMEN (*m*), **MOTORWAGEN-**: chassis (of motor car or lorry) 224
RAHMEN (*m*), **RUDER-**: rudder frame 225
RAHMENSÄGE (*f*): frame saw 226
RAHMENSPANT (*n*): web-frame 227
RAHMENSTRINGER (*m*): web stringer 228
RAHMERZ (*n*): foamy kind of wad 229
RAHMFARBIG: cream-coloured 230
RAHMGEWINNUNG (*f*): separation (of cream from milk) 231
RAHMIG: creamy; sooty 232
RAHMKAMMER (*f*): creamery, room for the storage of cream 233
RAHMKÄSE (*m*): cream cheese 234
RAHMKELLE (*f*): (cream)-skimmer 235
RAHMMESSER (*m*): creamometer 236
RAHMROLLE (*f*): sash-pulley 237
RAHRING (*m*): yard ring 238
RAHSCHIFF (*n*): square-rigged ship 239
RAHSEGEL (*n*): square sail 240
RAH (*f*), **VORBRAM-**: fore top gallant yard 241
RAIN (*m*): ridge; boundary; headland; border; edge 242
RAINFARN (*m*): tansy; buttons (*Tanacetum vulgare*) 243
RAINFARNKRAUT (*n*): tansy-(herb) (*Herba tanaceti*) 244
RAINFARNÖL (*n*): tansy oil; Sg. 0.925-0.955 [245
RAINWEIDE (*f*): privet 246
RAISONNIEREN: to reason 247
RAKEL (*f*): doctor (see **ABSTREICHMESSER**) 248
RAKELKITZE (*f*): faults in printing due to uneven doctor (Textile) 249
RAKETE (*f*): rocket (a firework) 250
RALLE (*f*): rail, land-rail, daker-hen, corn-crake (*Crex pratensis*) 251
RALSTONIT (*m*): ralstonite; $3(Na_2,Mg.Ca)F_2.4Al_2F_6,6H_2O$; Sg. 2.4 252
RAMBUTANTALG (*m*): rambutan tallow (from the seeds of *Nephelium lappaceum*); Sg. 0.893; Mp. 40°C. 253
RAMIE (*f*): Ramie; Chinese nettle (*Urtica nivea* or *Urtica boehmeria*) 254
RAMIEFASER (*f*): ramie fibre (bast fibre of **RAMIE**, which see) 255
RAMIEGARN (*n*): ramie yarn (see **RAMIE**) 256
RAMIFIZIEREN: to ramify 257
RAMIFIZIERUNG (*f*): ramification 258
RAMMBÄR (*m*): rammer, pile driver 259
RAMMBLOCK (*m*): ram-(mer); ram block; pile-driver 260
RAMME (*f*): ram-(mer); pile-driver 261
RAMMEN: to ram; beat down; drive 262
RAMMER (*m*): one who rams; rammer; pile-driver 263
RAMMKLOTZ (*m*): ram block; pile-driver 264
RAMMSTEVEN (*m*): ram-(stem) (of a ship) 265
RAMMVORICHTUNG (*f*): ramming apparatus 266
RAMPE (*f*): ramp; platform; footlights; landing; staircase 267

RAMPONIEREN : to damage ; spoil ; injure 268
RAMPONIERT : damaged (see RAMPONIEREN); disabled (of ships) 269
RAMSCH (m): lot ; bulk ; lump 270
RAMSCHE, IM- : in bulk ; in a lump ; in a lot 271
RAMTILLE (f) : (see NIGERPFLANZE) 272
RAND (m): edge ; brink ; rim ; margin ; brim ; border ; flange ; lip ; ledge 273
RANDAL (m): row ; disturbance ; brawl 274
RANDANMERKUNG (f): marginal (annotation) note 275
RANDBEMERKUNG (f): marginal (reference, annotation) note 276
RANDELN : to border ; edge ; rim ; mill ; brim ; engrail 277
RANDEN : (see RANDELN) 278
RÄNDEN : (see RANDELN) 279
RANDERKLÄRUNG (f) : (marginal)-annotation 280
RÄNDERN : (see RANDELN) 281
RÄNDERSCHEIBE (f) : edge (cutter) mill 282
RANDFARNE (m.pl) : *Cheilanthinae* (Bot.) 283
RANDGÄRUNG (f): rim fermentation (Brewing) 284
RANDIT (m) : randite ; $U(OH)_4.(CO_3)_6.Ca_5H_2O$ 285
RANDKOLBEN (m): ferret (Glass) 286
RANDPLANKE (f) : margin plank 287
RANDPLATTE (f) : margin plate 288
RANDSCHÄRFE (f) : marginal definition 289
RANDSCHICHT (f): rim layer (formed during nitrification of iron 290
RANDSOMHOLZ (n) : fashion piece 291
RANDSTÄNDIG : marginal 292
RANDSTEIN (m): curb-stone 293
RANDWEISUNG (f) : marginal reference 294
RANDZONE (f): outer (exterior) portion ; rim zone (Metal) 295
RANG (m): rank ; quality ; row ; tier ; dignity ; order ; grade ; class ; degree ; quality ; rate (of a ship) 296
RANGE (f): romp ; tom-boy ; (m): urchin : brat 297
RANGIERARBEITER (m): shunter (Railways) 298
RANGIERBAHNHOF (m): siding, goods station (Railways) 299
RANGIEREN : to shunt (Railways); arrange, place, classify, size, rank, grade 300
RANGIERGELEISE (n) : main line (Railways) 301
RANGIERLOKOMOTIVE (f) : shunting engine 302
RANGMÄSSIG : according to (degree, quality) rank 303
RANGORDNUNG (f) : order of precedence 304
RANGSCHIFF (n) : ship of the line (Naval) 305
RANK (m) : intrigue ; artifice ; trick ; plot 306
RANK : crank (of ships) ; winding, creeping 307
RANKE (f): runner; tendril; shoot (of plants) 308
RANKEN : to shoot ; creep ; run ; climb (of plants) ; (m) : tendril ; runner ; shoot 309
RANKENGEWÄCHS (n): runner ; creeper (of plants) 310
RÄNKESCHMIED (m) : plotter, intriguer 311
RÄNKE (m.pl) SCHMIEDEN : to plot, intrigue 312
RANKES SCHIFF (n) : top-heavy ship 313
RÄNKEVOLL : cunning ; intriguing 314
RANKHEIT (f) : crankness (of a ship) ; crookedness 315
RANULA (f): ranula (see FRÖSCHLEINGE--SCHWULST) (Med.) 316
RANUNKEL (f): ranunculus ; crowfoot ; gold-cup ; buttercup ; king-cup (*Ranunculus bulbosus*) (Bot.) 317

RÄNZEL (n) : satchel ; knapsack ; wallet 318
RANZEN (m) : satchel ; knapsack ; wallet ; (as a verb = to rove about) 319
RANZIDITÄT (f) : rancidity, rankness 320
RANZIG : rancid ; rank 321
RANZIGKEIT (f) : rancidness ; rancidity ; rankness 322
RANZIONIEREN : to-(hold to)-ransom, redeem 323
RAPID : rapid 324
RAPIDENTWICKLER (m) : rapid developer (Phot.) 325
RAPIDIN (n): rapidin (a mineral oil distillate); Sg. 0.75-0.8 ; Bp. 100-250°C. 326
RAPIDITÄT (f) : rapidity (Phot., etc.) 327
RAPINSÄURE (f) : rapinic acid ; $C_{17}H_{33}CO_2H$ [328
RAPP (m) : rape ; grape pomace 329
RAPPE (m): black horse ; (f): rape ; grape pomace 330
RAPPEL (m) : rage ; madness ; staggers 331
RAPPELN : to rattle, clatter ; be crazy ; make haste 332
RAPPORTIEREN : to report ; relate 333
RAPPORTRAD (n): driven wheel (cog wheel in gear with, and driven by, another cog wheel) 334
RAPPS (m) : (see RAPS) 335
RAPS (m): rape-(seed) ; winter rape ; cole seed (*Brassica napus*) ; summer rape, colza (*Brassica campestris*) 336
RAPSÖL (n): rape-(seed) oil ; colza oil ; Sg. 0.913-0.917 ; Mp. 17-22 °C. (from *Brassica campestris*) (see RÜBÖL) 337
RAPSSAAT (f) : colza ; rape seed ; napus (seeds of *Brassica napus*) 338
RAPUNZEL (f): lamb's lettuce (*Valerianella carinata*) 339
RAR : rare ; scarce 340
RARITÄT (f) : rarity ; curiosity 341
RASCH : quick ; swift ; prompt ; impetuous ; brisk ; speedy ; expeditious ; rash : hasty ; (m): rash ; serge (Cloth) 342
RASCHELN : to rustle 343
RASCHHEIT (f) : speediness ; promptness ; rashness (see RASCH) 344
RASCHIGRINGE (m.pl) : RASCHIGSRINGE (m.pl): Dr. Raschig's rings (a filling material for absorption towers, reaction vessels, distillers, dust extracting plants, etc.), filter rings (small hollow wrought iron cylinders) 345
RASCHLAUFEND : high-speed 346
RASEN : to rave ; bluster ; rage ; fume ; (m): turf ; lawn ; sod ; meadow ; (green)-sward 347
RASENBANK (f) : turf-(seat) ; grassy bank 348
RASENBLEICHE (f): grass (sun) bleaching ; bleaching (meadow) plot 349
RASEND : raging ; raving ; furious ; frantic ; mad-(ly) ; enraged 350
RASEND WERDEN : to become furious (enraged) 351
RASENEISENERZ (n): bog iron ore ; limonite ; brown (meadow) iron ore ; brown hematite ; Sg. 3.4 (see LIMONIT ; BRAUNEISENSTEIN and BRAUNEISENERZ) 352
RASENEISENSTEIN (m) : (see RASENEISENERZ) 353
RASENERZ (n): (see RASENEISENERZ) 354
RASENHOPFEN (m) : wild hop 355
RASENPLATZ (m) : grass plot ; greensward 356
RASENSTECHEN (n) : turf-cutting 357
RASENWALZE (f) : lawn roller 358
RASIERBECKEN (n) : shaving pot 359

RASIEREN: to raze; shave 360
RASIERFLECHTE (*f*): Herpes (*Herpes tonsurans*) 361
RASIERGRIND (*m*): scald (*Tinea favosa*) 362
RASIERMESSER (*n*): razor 363
RASIERPINSEL (*m*): shaving brush 364
RASIERSEIFE (*f*): shaving soap 365
RASIG: turfed; grassy 366
RASPE (*f*): rasp; grater 367
RASPEL (*f*): (see **RASPE**) 368
RASPELHAUER (*m*): rasp maker 369
RASPELHAUS (*n*): house of correction 370
RASPELN: to rasp; grate 371
RASPELSPÄNE (*m.pl*): raspings; filings 372
RASPIT (*m*): raspite; $PbWO_4$ 373
RASSE (*f*): race; breed; type (of yeast); culture; variety (Bot.) 374
RASSELN: rattle; clatter 375
RASSE (*f*), **REINE-**: thoroughbred type 376
RAST (*f*): stage; rest; repose; bosh-(es) (the bottom truncated cone-shaped portion or chamber of blast furnaces) 377
RASTE (*f*): stage (of a journey) 378
RASTEN: to rest; repose; halt; pause 379
RASTER (*m*): screen (Three-colour photography) 380
RASTER, COLOR- (*m*): colour screen (Three-colour photography) 381
RASTERPLATTE (*f*): screen plate; line (ruled) screen; colour screen (Three-colour photography) 382
RASTGÄRUNG (*f*): slow and incomplete fermentation (Brewing) 383
RASTLOS: restless-(ly); thoroughly; constant -(ly); continuous-(ly); incessant-(ly); ceaseless-(ly); without rest (see **RESTLOS**) 384
RASTLOS BEANTWORTEN: to answer satisfactorily 385
RASTLOS KLÄREN: to clear up thoroughly (a matter) 386
RASTORT (*m*): halt, resting (stopping) place 387
RASTRAL (*n*): pen for ruling music staves; music pen 388
RASTWINKEL (*m*): the angle of the boshes (see **RAST**) 389
RAT (*m*): counsel; advice; council; means; board; court; councillor; deliberation; expedient; prudence; senator; alderman; senate; consultation; remedy 390
RATA (*f*): rate; quota; instalment; proportion; part payment 391
RATANHIAGERBSÄURE (*f*): rhatania-tannic acid (obtained from rhatany root) 392
RATANHIARINDE (*f*): rhatany bark 393
RATANHIAWURZEL (*f*): payta; rhatany-(root) (*Radix ratanhiæ*) (of *Krameria triandra*, *Krameria ixina* or *Krameria argentia*) 394
RATE (*f*): (see **RATA**) 395
RATEN: to advise; counsel; guess; solve; help 396
RATENZAHLUNG (*f*): payment by instalments 397
RATFRAGEN: to consult 398
RATGEBER (*m*): counsellor; adviser; counsel; advocate 399
RATHAUS (*n*): town hall; senate 400
RATIFIZIEREN: to ratify 401
RATION (*f*): ration; portion; allowance 402
RATIONAL: rational 403
RATIONALISMUS (*m*): rationalism 404
RATIONALISTISCH: rationalistic 405
RATIONELL: rational 406
RATKAMMER (*f*): council-chamber 407
RÄTLICH: advisable; expedient; frugal (see **RATSAM**) 408
RATLOS: helpless; perplexed; without advice 409
RATMANN (*m*): councillor; alderman; senator 410
RATSAM: advisable; sparing; frugal; useful; thrifty; expedient; prudent 411
RATSAMKEIT (*f*): advisability; usefulness 412
RATSBEFEHL (*m*): order in council 413
RATSBESCHLUSS (*m*): decree; resolution 414
RATSCHE (*f*): ratchet 415
RATSCHLAG (*m*): advice; counsel; suggestion 416
RATSCHLAGEN: to consult; deliberate; decide 417
RATSCHLUSS (*m*): decree; resolution; decision 418
RÄTSEL (*n*): riddle; puzzle; enigma; mystery; problem 419
RÄTSELHAFT: enigmatical; mysterious; problematic-(al) 420
RATSGLIED (*n*): alderman; senator; member of a (senate) council; councillor 421
RATSSCHREIBER (*m*): Clerk to the Council, town clerk 422
RATSSITZUNG (*f*): council meeting 423
RATSSTUBE (*f*): council chamber 424
RATSVERLASS (*m*): order in council, decree 425
RATSVERSAMMLUNG (*f*): council meeting 426
RATSWAHL (*f*): (municipal)-election 427
RATTE (*f*): rat 428
RÄTTEN: to screen; sieve; riddle 429
RATTENFALLE (*f*): rat-trap 430
RATTENFÄNGER (*m*): rat-catcher; ratter 431
RATTENGIFT (*n*): rat poison (white arsenic) 432
RATTENPULVER (*n*): (see **RATTENGIFT**) 433
RATTENSCHWANZ (*m*): rat-tail file; rat's tail 434
RATTENTOD (*m*): rat (killer) poison 435
RATTENVERTILGUNGSMITTEL (*n*): rat (poison) destroyer; rat killer 436
RÄTTER (*m*): riddle; screen; sieve 437
RAUB (*m*): robbery; rapine; plunder; theft; prey; spoil; booty; rape; kidnapping 438
RAUBBEGIERDE (*f*): rapacity; rapaciousness 439
RAUBBEGIERIG: rapacious 440
RAUBEN: to rob, steal, plunder, pillage; ravish rape; thieve; deprive of 441
RÄUBER (*m*): robber; pirate 442
RAUB (*m*), **GEWALTSAMER-**: robbery with violence 443
RAUB (*m*), **JUNGFERN-**: rape 444
RAUB (*m*), **KINDER-**: kidnapping 445
RAUBMORD (*m*): murder for the sake of robbery 446
RAUBSUCHT (*f*): rapacity 447
RAUBTIER (*n*): wild beast, beast of prey 448
RAUBTIER WITTERUNG (*f*): wild beast's (beast of prey's) scent 449
RAUBVOGEL (*m*): bird of prey 450
RAUCH (*m*): smoke; fume; vapour; (as an adjective = coarse; shaggy; undressed; hairy; furred; rough) 451
RAUCHACHAT (*m*): smoky agate (see **ACHAT**) 452
RAUCHAPFEL (*m*): thorn apple 453
RAUCHARTIG: smoky 454
RAUCHBAD (*n*): vapour bath 455
RAUCHBEZUG (*m*): (see **RAUCHKLEID**) 456
RAUCHBLÄTTERIG: rough (coarse) leaved 457
RAUCHDARRE (*f*): smoke drying-kiln (for kiln-drying malt) 458

RAUCHE (*n*): hairiness; roughness; shagginess; rough side; fur-(side); severity 459
RAUCHEN: to smoke; fume; reek; (*n*): smoking; fuming 460
RAUCHEND: smoking; fuming; reeking 461
RAUCHENDE SALPETERSÄURE (*f*): fuming nitric acid (see under SALPETERSÄURE) 462
RAUCHENDE SCHWEFELSÄURE (*f*): fuming sulphuric acid (see under SCHWEFELSÄURE) 463
RAUCHENTWICKLUNG (*f*): smoke generation 464
RAUCHER (*m*): smoker (the man, or the compartment of a train) 465
RÄUCHER (*m*): fumigator 466
RÄUCHERESSENZ (*f*): aromatic essence 467
RÄUCHERESSIG (*m*): aromatic vinegar 468
RÄUCHERFASS (*n*): censer (used in the Catholic Church) 469
RÄUCHERIG: smoky; dingy 470
RÄUCHERKERZE (*f*): fumigating candle 471
RÄUCHERMITTEL (*n* and *pl*): disinfectant-(s) 472
RÄUCHERN: to fumigate; perfume; smoke; incense; fume; cure; dry 473
RÄUCHERPFANNE (*f*): fumigating pan (see RAUCHFASS) 474
RÄUCHERPULVER (*n*): fumigating powder 475
RÄUCHERUNG: fumigation; smoking (of fish and meat); incense burning 476
RÄUCHERWERK (*n*): perfume-(s); incense 477
RAUCHFANG (*m*): chimney; flue; uptake (of boilers); hood 478
RAUCHFANGKEHRER (*m*): chimney sweep 479
RAUCHFANGTRICHTER (*m*): smoke funnel, smoke uptake 480
RAUCHFASS (*n*): censer 481
RAUCHFLEISCH (*n*): smoked meat 482
RAUCHFREI: smokeless 483
RAUCHGAR: dressed with the hair on (Skins); well-smoked; thoroughly (cured) smoked 484
RAUCHGAS (*n*): flue gas (smoke or chimney gas); furnace gas; gas of combustion 485
RAUCHGASANALYSATOR (*m*): flue gas analyser (an instrument for the determination of the CO_2 content of flue gases) 486
RAUCHGASANALYSE (*f*): flue-gas analysis 487
RAUCHGASGESCHWINDIGKEITSMESSER (*m*): flue-gas (velocity meter) speed indicator 488
RAUCHGASPROBE (*f*): flue gas sample 489
RAUCHGASPRÜFER (*m*): flue gas (tester) testing instrument or apparatus 490
RAUCHGASSPEISEWASSERVORWÄRMER (*m*): flue-gas economiser 491
RAUCHGASUNTERSUCHUNG (*f*): flue gas (testing) analysis 492
RAUCHGASUNTERSUCHUNGSAPPARAT (*m*): flue gas testing apparatus (such as Orsat apparatus) 493
RAUCHGASVERWERTUNG (*f*): flue-gas utilization 494
RAUCHGASVORWÄRMER (*m*): flue gas (pre-heater, feed-water heater) economiser 495
RAUCHGASVORWÄRMUNG (*f*): flue gas (feed-water,-pre)-heating 496
RAUCHGASWÄRME (*f*), **UNBENUTZTE-**: (unexpended portion of the heat of the flue gases) waste heat of flue gases 497
RAUCHGELB: smoky-yellow 498
RAUCHHELM (*m*): smoke helmet 499
RAUCHIG: smoky 500
RAUCHKALK (*m*): magnesian limestone 501
RAUCHKAMMER (*f*): smoke box (of a locomotive) 502
RAUCHKANAL (*m*): flue 503
RAUCHKLAPPE (*f*): smoke (flue) damper 504
RAUCHKLEID (*n*): preventive covering (coating) on sails against soot and smoke 505
RAUCHLEDER (*n*): chamois leather; cordovan 506
RAUCHLOS: smokeless 507
RAUCHLOSES PULVER (*n*): smokeless (gun)-powder 508
RAUCHMANTEL (*m*): uptake (of boilers); mantel-piece (in a house) 509
RAUCHPFANNE (*f*): (see RAUCHFASS) 510
RAUCHPULVER (*n*): fumigating powder 511
RAUCHQUARZ (*m*): smoky quartz; cairngorm stone (see QUARZ) 512
RAUCHRING (*m*): smoke ring 513
RAUCHROHR (*n*): flue, smoke stack, funnel, chimney 514
RAUCHRÖHRE (*f*): flue (see RAUCHROHR) 515
RAUCHSCHIEBER (*m*): smoke (flue) damper 516
RAUCHSCHIEBERSTEUERUNG (*f*): damper gear 517
RAUCHSCHWACH: nearly (practically) smokeless, giving off very little smoke 518
RAUCHSCHWACHES PULVER (*n*): (gun)-powder which gives off little smoke, smokeless (gun)-powder 519
RAUCHSCHWARZ: black as soot 520
RAUCHSEGEL (*n*): smoke sail 521
RAUCHTABAK (*m*): tobacco 522
RAUCHTOPAS (*m*): smoky (quartz) topaz (brown or black variety of quartz); brown cairngorm; Sg. 2.65 (see QUARZ) 523
RAUCHVERBRENNUNG (*f*): smoke (consumption) burning 524
RAUCHVERGIFTUNG (*f*): smoke poisoning; flue-gas poisoning (from inhaling products of incomplete combustion) 525
RAUCHVERHÜTUNG (*f*): smoke prevention; smoke-abatement 526
RAUCHVERZEHRUNG (*f*): smoke consumption 527
RAUCHWACKE (*f*): cellular stone; red-land limestone (Geol.); dolomite (Min.) (see DOLOMIT) 528
RAUCHWAREN (*f.pl*): furs; peltry 529
RAUCHWARENHÄNDLER (*m*): trader in furs; fur merchant; furrier 530
RAUCHWERK (*n*): furs 531
RAUCHZIMMER (*m*): smoking room 532
RAUCHZUG (*m*): flue; pass (of a boiler) 533
RAUDE (*f*): scab; mange (of animals); rubbers (Vet.) 534
RÄUDE (*f*): mange (of animals); scab, itch; rubbers (Vet.) 535
RÄUDESEIFE (*f*): mangre soap 536
RÄUDIG: scabbed; scabby; mangy 537
RAUFBOLD (*m*): bully; brawler 538
RAUFE (*f*): hackle; rack; flax-comb 539
RAUFEN: to pull; pluck; scuffle; fight 540
RAUFSUCHT (*f*): brawling (bullying) tendencies; pugnacity 541
RAUFZANGE (*f*): tweezers 542
RAUH: rough; harsh; unglazed; raw; coarse; rude; hoarse; cold; bleak; piercing; inclement; chilly; inexperienced; unseasoned; uncooked; immature; green; unripe; crude; unfinished; sensitive; unmanufactured; sore; bare; rugged 543
RAUHARTIKEL (*m*): raised style (Calico) 544
RAUHBLÄTT-(E)-RIG: rough-leaved, asperifolious (Bot.) 545
RAUHE (*f*): roughness, etc. (see RAUH); moulting-time; (*n*): rough (see RAUHEN, AUS DEM—ARBEITEN) 546

RAUHEN: to nap (Cloth), tease, teasel, dress, card; roughen; moult (of birds) 547
RAUHEN, AUS DEM—ARBEITEN: to cut (work) out of the rough, rough-hew 548
RAUHFROST (*m*): rime; hoar-frost 549
RAUHHAARIG: rough-haired; hirsute (Bot.) 550
RAUHHEIT (*f*): roughness; harshness; hoarseness; rudeness; coarseness; acerbity (see **RAUH**) 551
RAUHMASCHINE (*f*): roughing machine (Textile) 552
RAUHREIF (*m*): (see **RAUHFROST**) 553
RAUHWACKE (*f*): (see **RAUCHWACKE**) 554
RAUHWEIZEN (*m*): bearded (awned) wheat 555
RAUKE (*f*): hedge-mustard (*Sisymbrium officinale*) 556
RAUKENKOHL (*m*): garden rocket (*Eruca sativa*) 557
RAUM (*m*): space; volume; room; hold (of ship); chamber; scope; area; opportunity; capacity 558
RÄUMAHLE (*f*): broach 559
RAUMANALYSE (*f*): volumetric analysis (Gases) 560
RAUMANKER (*m*): sheet anchor 561
RAUMBALKEN (*m*): hold beam 562
RAUMBALKENSTÜTZE (*f*): hold beam (support, pillar) stanchion 563
RAUMBESTIMMUNG (*f*): determination of volume 564
RAUMBILD (*n*): space diagram 565
RAUMCHEMIE (*f*): stereochemistry 566
RAUMCHEMISCH: stereochemical 567
RAUMDECK (*n*): orlop deck 568
RAUMEINHEIT (*f*): unit of (space) volume 569
RÄUMEN: to clear-(away); remove; clean; evacuate; quit; leave; decamp; retreat; give up; surrender; scavenge; scrape 570
RAUMGEBEN: to give way; make room; yield; grant 571
RAUMGEBILD (*n*): space diagram 572
RAUMGEBILDE (*n*): space diagram 573
RAUMGEHALT (*m*): volumetric content; content by volume; spatial content 574
RAUMGEOMETRIE (*f*): solid geometry 575
RAUMGEWICHT (*n*): volumetric weight 576
RAUMGITTER (*n*): "raumgitter"; space lattice (Cryst.) 577
RAUMGITTEREBENE (*f*): space lattice (raumgitter) plane (Cryst.) 578
RAUMGRÖSSE (*f*): volume; content; capacity; area 579
RÄUMIG: spacious, roomy, capacious 580
RAUMINHALT (*m*): capacity; volume; content; cubature; cubic (content) capacity 581
RAUMINHALTMESSER (*m*): volume meter 582
RAUMISOMERIE (*f*): stereoisomerism; spatial isomerism 583
RAUMKURVE (*f*): three dimensional curve 584
RAUMLEITER (*f*): hold ladder 585
RÄUMLICH: spatial; relating to (occupying) space; volumetric; roomy; spacious; capacious 586
RÄUMLICHE FLÄCHE (*f*): three dimensional (stereometrical) curve 587
RÄUMLICHKEIT (*f*): quality of occupying space; roominess; extension; space; specific volume; spaciousness; spatiality; extent 588
RAUM MACHEN: to make room; give way 589
RAUMMASS (*n*): measure of capacity; solid measure 590
RAUMMENGE (*f*): volume 591
RAUMMESSER (*m*): volume meter 592
RAUMMETER (*n*): cubic meter 593
RAUMNADEL (*f*): borer 594
RÄUMNADEL (*f*): scraping cutter; needle, nail (Min.) 595
RÄUMNADELMASCHINE (*f*): die slotting machine 596
RÄUMNADELZIEHMASCHINEN (*f.pl*): broaching machinery 597
RAUMSCHOTS SEGELN: to sail with flowing sheets, to sail off the wind (Naut.) 598
RAUMSTRINGER (*m*): hold stringer 599
RAUMSTÜTZE (*f*): hold (pillar) stanchion or support 600
RÄUMTE (*f*): offing (Marine) 601
RAUMTEIL (*m*): (part by)-volume; element of volume; volumetric part 602
RAUMTIEFE (*f*): depth of hold 603
RAUMUMFANG (*m*): cubic capacity; superficies, area 604
RÄUMUNG (*f*): clearing up; evacuation; scavenging 605
RAUMVERÄNDERUNG (*f*): change in volume 606
RAUMVERHÄLTNIS (*f*): proportion by volume; volume (volumetric) ratio or relation 607
RAUM (*m*) **VERMINDERTEN DRUCKES**: reduced pressure chamber 608
RAUMVERMINDERUNG (*f*): reduction of (capacity) volume 609
RAUMWEGERUNG (*f*): hold ceiling 610
RAUMZUSAMMENHANG (*m*): continuity 611
RAUNEN: to whisper; rumour 612
RAUPE (*f*): caterpillar; worm; whim; fancy 613
RAUPEN: to destroy (clear of) caterpillars 614
RAUPENFRASS (*m*): (caterpillar)-blight 615
RAUPENLEIM (*m*): caterpillar glue 616
RAUSCH (*m*): intoxication; carouse; carousal; drunkenness; rush-(ing); roar-(ing); rustling; murmur; rustle; pounded ore (Mining) 617
RAUSCHBEERE (*f*): cranberry (*Vaccinium oxycoccus*) 618
RAUSCHBEERGEWÄCHSE (*n.pl*): *Empetraceæ* (Bot.) 619
RAUSCHEN: to rush; murmur; rustle; whisper; gurgle; whistle; roar 620
RAUSCHGELB (*n*): orpiment (see **AURIPIGMENT** and **ARSENTRISULFID**) 621
RAUSCHGELB (*n*), **ROTES-**: realgar (see **ARSENDISULFID, ARSENROT** and **REALGAR**) 622
RAUSCHGOLD (*n*): Dutch gold; tinsel 623
RAUSCHLEDER (*n*): chamois leather 624
RAUSCHMITTEL (*n*): narcotic 625
RAUSCHROT (*n*): realgar (see **ARSENDISULFID, ARSENROT** and **REALGAR**) 626
RAUSCHSILBER (*n*): imitation silver foil; silver tinsel 627
RÄUSPERN: to clear one's throat 628
RAUTE (*f*): rhomb; rhombus; fillet; facet; diamond; rhomboid; lozenge; rue-(herb) (*Herba rutæ* of *Ruta graveolens*) 629
RAUTEN: to cut into facets 630
RAUTENFELD (*n*): lozenge (Heraldry) 631
RAUTENFLACH (*n*): rhombohedron 632
RAUTENFLÄCHE (*f*): rhombus; rhomb; facet 633
RAUTENFLÄCHNER (*m*): rhombohedron 634
RAUTENFÖRMIG: rhombic; diamond-shaped; rhomboidal; lozenge-shaped; quadrangular; like a facet 635

RAUTENGEWÄCHSE (*n.pl*): *Rutaceæ* (Bot.) 636
RAUTENGLAS (*n*): pane (square, facet) of glass 637
RAUTENÖL (*n*): oil of rue (from *Ruta graveolens*) 638
RAUTENSPAT (*m*): rhomb spar: dolomite (see DOLOMIT) 639
RAUTENSTEIN (*m*): jewel cut into facets 640
RAVISONÖL (*n*): ravison oil, black-sea rape oil (from seeds of *Brassica campestris*); Sg. 0.92; Mp. – 8°C. 641
RAYON (*m*): ray; radius 642
RAZEMOS: racemosus, clustered (-like grapes) 643
RAZOUMOFFSKIN (*m*): razoumoffskite (see MONT-MORILLONIT) 644
REAGENS (*n*): reagent (see REAGENZ) 645
REAGENSGLAS (*n*): test (tube) glass 646
REAGENSGLASHALTER (*m*): test tube holder 647
REAGENSPAPIER (*n*): test paper; reagent paper; litmus paper 648
REAGENSRÖHRE (*f*): test tube 649
REAGENTIEN (*n.pl*): reagents (plural of REAGENS) 650
REAGENTIENFLASCHE (*f*): reagent (bottle) flask 651
REAGENZ (*n*): reagent (see REAGENS) 652
REAGENZGLAS (*n*): test (tube) glass 653
REAGENZGLASHALTER (*m*): test (tube) glass holder 654
REAGENZIEN (*n.pl*): reagents (plural of REAGENZ) 655
REAGENZIENFLASCHE (*f*): reagent bottle 656
REAGENZPAPIER (*n*): test paper (see REAGENS-PAPIER) 657
REAGIEREN: to react 658
REAGIEREND: reactive; reacting 659
REAGIERGLAS (*n*): test (tube) glass 660
REAGIERGLASBÜRSTE (*f*): test tube brush 661
REAGIERGLASGESTELL (*n*): test tube stand 662
REAGIERGLASHALTER (*m*): test tube holder 663
REAGIERKELCH (*m*): test (glass) cup; reagent vessel 664
REAGIERZYLINDER (*m*): test tube 665
REAKTION (*f*): reaction 666
REAKTIONÄR: reactionary; (*m*): reactionary 667
REAKTIONSBAHN (*f*): path of a reaction; course of a reaction 668
REAKTIONSFÄHIG: reactive; capable of reacting 669
REAKTIONSFÄHIGKEIT (*f*): reactivity; reacting (capability) capacity 670
REAKTIONSFLÄCHE (*f*): reaction surface 671
REAKTIONSGEMISCH (*n*): reaction mixture 672
REAKTIONSGESCHWINDIGKEIT (*f*): reaction velocity 673
REAKTIONSGLAS (*n*): reaction vessel, test-tube 674
REAKTIONSGLEICHUNG (*f*): equation of a reaction 675
REAKTIONSLOS: reactionless 676
REAKTIONSLOSIGKEIT (*f*): non-reaction; absence of reaction 677
REAKTIONSMASSE (*f*): mass resulting from a reaction; reaction mass 678
REAKTIONSMITTEL (*n*): reagent 679
REAKTIONSOBERFLÄCHE (*f*): reaction surface 680
REAKTIONSRAD (*n*): reaction wheel 681
REAKTIONSRAUM (*m*): reaction space 682
REAKTIONSSTUFE (*f*): step (stage) of a reaction 683
REAKTIONSTURBINE (*f*): reaction turbine 684
REAKTIONSTURM (*m*): reaction tower or column (for cleaning gases, for absorption of gases in liquids and for regeneration of gases) 685
REAKTIONSVERLAUF (*m*): course of a reaction 686
REAKTIONSWIRKUNG (*f*): reaction effect 687
REAL: actual; real; really; actually; material: substantial; concrete; (*n*): stand (Typ.): (*m*): real (Spanish coin) 688
REALE (*n*): material; reality; historical (exact) science 689
REALGAR (*m*): realgar (natural arsenic monosulphide) (Mineral); AsS; Sg. 3.0; realgar (arsenic disulphide); As_2S_2; Sg. 3.4/3.6; Mp. 307-320°C.; Bp. 565°C. 690
REALISIEREN: to realize; sell out (Com.) 691
REALISMUS (*m*): realism 692
REALISTISCH: realistic 693
REALITÄT (*f*): reality; (*pl*): real estate, landed property 694
REALKREDIT (*m*): mortgage - (on landed property) 695
REALPHILOSOPHIE (*f*): metaphysics 696
REALSCHULE (*f*): modern (polytechnic; industrial; practical; scientific; high; technical) school 697
REALWERT (*m*): actual (real) value 698
REBE (*f*): vine 699
REBELLIEREN: to rebel, revolt, mutiny 700
REBENBLATT (*n*): vine leaf 701
REBENGEWÄCHSE (*n.pl*): *Vitaceæ* (Bot.) 702
REBENRUSS (*m*): vine (Frankfort) black (from carbonization of wine lees and vine wood) 703
REBENSAFT (*m*): grape juice; wine 704
REBENSCHWARZ (*n*): (see REBENRUSS) 705
REBENSTOCK (*m*): vine 706
REBHUHN (*n*): partridge 707
REBLAUS (*f*): *Phylloxera-(vastatrix)*; vine louse (a vine blight) 708
RECALESZENZ (*f*): cooling (Metal) 709
RECEIVER (*m*): receiver 710
RECENSENT (*m*): reviewer; critic (see also REZ-) 711
RECEPT (*n*): prescription; recipe 712
RECHEN (*m*): grating; grid; ratchet; rack; rake; screen; sieve; catcher; trap; strainer; sluice (for things floating in water); (as a verb = to rack; rake; screen; sieve; strain) 713
RECHENFEHLER (*m*): error in calculation; miscalculation 714
RECHENKNECHT (*m*): ready-reckoner 715
RECHENKUNST (*f*): arithmetic 716
RECHENMASCHINE (*f*): calculating machine, comptometer 717
RECHENMEISTER (*m*): arithmetician 718
RECHENPFENNIG (*m*): marker; counter 719
RECHENSCHAFT (*f*): account 720
RECHENSCHAFT GEBEN: to answer; give account of 721
RECHENSCHAFTSPFLICHTIG: responsible; accountable 722
RECHENSCHIEBER (*m*): calculating (slide) rule 723
RECHENSTAB (*m*): calculating (slide) rule 724
RECHENSTIFT (*m*): slate pencil 725
RECHENSTUNDE (*f*): arithmetic lesson 726
RECHENTAFEL (*f*): blackboard; multiplication (calculating; reckoning) table-(s), slate, etc. 727
RECHNEN: to reckon; rank; calculate; class; count; estimate; compute; charge;

esteem; deem; (*n*): arithmetic; mathematics 728
RECHNEN, AUF-: to count on; reckon upon; rely upon; base on 729
RECHNEN, MIT-: to count on; anticipate; reckon (upon) with 730
RECHNER (*m*): calculator; arithmetician 731
RECHNERISCH: mathematical-(ly) 732
RECHNUNG (*f*): calculation; reckoning; account; invoice; estimate; computation; arithmetic; calculus; mathematics; bill 733
RECHNUNG ABLEGEN: to render an account 734
RECHNUNG FÜHREN: to keep an account 735
RECHNUNG (*f*), **OFFENE-**: current (running) account 736
RECHNUNGSABNAHME (*f*): auditing (of accounts) 737
RECHNUNGSART (*f*): method (mode; manner) of calculating or reckoning 738
RECHNUNGSARTIKEL (*m*): item 739
RECHNUNGSAUFSEHER (*m*): auditor 740
RECHNUNGSAUSZUG (*m*): current account; extract of account 741
RECHNUNGSBELEG (*m*): voucher 742
RECHNUNGSBILANZ (*f*): balance of account 743
RECHNUNGSBUCH (*n*): account-book 744
RECHNUNGSFÜHRER (*m*): accountant; book-keeper 745
RECHNUNGSJAHR (*n*): financial year 746
RECHNUNGSLISTE (*f*): statement -(of account) 747
RECHNUNGSMÄSSIG: in accordance with (reckoning) calculation 748
RECHNUNGSPRÜFER (*m*): (see **RECHNUNGSAUF-SEHER**) 749
RECHNUNGSREST (*m*): balance of account 750
RECHNUNGSREVISOR (*m*): (see **RECHNUNGSAUF-SEHER**) 751
RECHNUNGSTAFEL (*f*): reckoning (calculating) table 752
RECHNUNGSWESEN (*n*): book-keeping, accountancy, accounts, auditing 753
RECHNUNG (*f*) **TRAGEN**: to take into account, bear in mind, take into one's calculations, make provision for 754
RECHNUNG UND GEFAHR: account and risk (Com.) 755
RECHT: correct; accurate; equitable; just; lawful; reasonable; actually; greatly; quite; truly; real; right; true; legitimate; genuine; own; fitting; agreeable; rightly; very; straight; (*n*): power; right; title; justice; privilege; law; claim; reason; (*pl*): duties; taxes 756
RECHT (*n*), **BÜRGERLICHES-**: civil law 757
RECHTE (*f*): right (side; party) hand 758
RECHTE (*n.pl*), **AUSGEHENDE-**: export duties 759
RECHTECK (*n*): rectangle 760
RECHTECKIG: rectangular 761
RECHTEN: to dispute; plead; litigate; remonstrate; contest 762
RECHTE SEITE (*f*): right (off) side; (in Germany and other countries the right side is the " near side "); correct side; top 763
RECHTFERTIG: just, fair, righteous, justified 764
RECHTFERTIGEN: to justify; exculpate; vindicate 765
RECHTFERTIGUNG (*f*): justification; vindication; exculpation 766
RECHTGEBEN: to give in; concede a point; grant; assent 767
RECHT (*n*), **GEMEINES-**: common law 768
RECHT (*n*), **GESCHRIEBENES-**: statute law 769
RECHTGLÄUBIG: orthodox 770
RECHTGLÄUBIGKEIT (*f*): orthodoxy 771
RECHTHABEN: to be right 772
RECHTHABER (*m*): disputant; one who dogmatizes; disputer 773
RECHTHABERISCH: dogmatic 774
RECHTLEHRIG: orthodox 775
RECHTLICH: judicial; upright; legal; honest; lawful; just; proper; legitimate; fair 776
RECHTLICHKEIT (*f*): rectitude; legality; honesty; probity; fairness; integrity 777
RECHTLINIG: rectilinear 778
RECHTLOS: lawless; unjust; illegal; illegitimate; unlawful; unfair 779
RECHTLOSIGKEIT (*f*): lawlessness; unjustness; illegality; illegitimacy; unlawfulness; unfairness 780
RECHT MACHEN, EINEM-: to suit (please) one 781
RECHTMÄSSIG: legal; legitimate; rightful; lawful; just; fair 782
RECHTMÄSSIGKEIT (*f*): legality; legitimacy; lawfulness; justness; fairness 783
RECHT, MIT-: rightly, justly, reasonably 784
RECHT (*n*), **PEINLICHES-**: penal code 785
RECHTS: to the right; at (on) the right 786
RECHTSABTRETUNG (*f*): making over (cession) of a right 787
RECHTSANSPRUCH (*m*): lawful (just; legitimate) claim 788
RECHTSANWALT (*m*): attorney; solicitor 789
RECHTSBEFLISSEN: studying law 790
RECHTSBEFLISSENER (*m*): law student 791
RECHTSBEFUGNIS (*f*): competence 792
RECHTSBEISTAND (*m*): counsel; legal (advisor) advice 793
RECHTSBESTÄNDIG: legal; valid; legitimate 794
RECHTSBEWEIS (*m*): legal proof 795
RECHTSCHAFFEN: upright; just; honest (of character); righteously; honestly 796
RECHTSCHAFFENHEIT (*f*): probity; honesty; integrity 797
RECHTSCHREIBUNG (*f*): spelling; orthography 798
RECHTSDREHEND: dextrorotatory 799
RECHTSDREHUNG (*f*): right-handed polarization; dextrorotation 800
RECHT SEIN, EINEM-: to be agreeable to, be content with, be satisfied with, be suited 801
RECHTSFALL (*m*): law-case; law-suit; case 802
RECHTSFÄLLIG WERDEN: to lose a (law-suit) action 803
RECHTSFRAGE (*f*): law question; point of law 804
RECHTSGANG (*m*): legal proceeding; the course of the law 805
RECHTSGÄNGIG: right-handed (of screws, etc.) 806
RECHTSGELEHRSAMKEIT (*f*): law, jurisprudence 807
RECHTSGELEHRTE (*m*): jurist; lawyer; one versed in the law 808
RECHTSGEMÄSS: according to law, in conformity with the law 809
RECHTSGEWINDE (*n*): right-handed thread 810
RECHTSGLEICHHEIT (*f*): legal equality 811
RECHTSGÜLTIG: (see **RECHTSKRÄFTIG**) 812
RECHTSHANDEL (*m*): law (case) suit; action; legal proceedings 813
RECHTSHÄNGIG: pending 814
RECHTSHILFE (*f*): legal assistance 815
RECHTSHÜLFE (*f*): legal assistance 816
RECHTSKRÄFTIG: legal; valid 817

RECHTSKUNDE (*f*): jurisprudence; the science of the law 818
RECHTSMILCHSÄURE (*f*): dextro-lactic acid 819
RECHTSMITTEL (*n*): legal remedy 820
RECHTSPFLEGE (*f*): administration of justice 821
RECHTSPOLARISATION (*f*): right-handed polarization; dextrorotation 822
RECHTSPRECHUNG (*f*): orthoepy; right pronunciation 823
RECHTSPROPELLER (*m*): right-handed propeller 824
RECHTSSÄURE (*f*): dextro acid 825
RECHTSSCHLUSS (*m*): decree; judgment 826
RECHTSSCHRAUBE (*f*): right-handed screw 827
RECHTSSPRUCH (*m*): verdict, sentence 828
RECHTSSTAND (*m*): jurisdiction; justification 829
RECHTSSTÄNDIG: justifiable 830
RECHTSSTREIT (*m*): case, law-suit, action, legal proceedings 831
RECHTSTITEL (*m*): title 832
RECHTSUCHER (*m*): claimant; plaintiff 833
RECHTSUM: to the right-about 834
RECHTSURKUNDE (*f*): legal document 835
RECHTSVERFAHREN (*n*): legal proceedings (see **RECHTSHANDEL**) 836
RECHTSVERFASSUNG (*f*): legal code, judicature 837
RECHTSVERSTÄNDIGE (*m*): jurist; one versed in the law; lawyer 838
RECHTSWEINSÄURE (*f*): dextrotartaric acid; $C_4H_6O_6$ 839
RECHTSWIDRIG: contrary to the law; unlawful; illegal 840
RECHTSWISSENSCHAFT (*f*): jurisprudence; science of law 841
RECHTWINKEL (*m*): rectangle, right-angle 842
RECHTWINKLIG: rectangular; right-angled 843
RECHTZEITIG: opportune; convenient; seasonable; timely; propitious; well-timed; appropriate 844
RECIPIENT (*m*): recipient; receiver (see also under **REZ-**) 845
RECK (*n*): rack; stretcher; horizontal (pole) bar 846
RECKBANK (*f*): rack; stretcher 847
RECKE (*f*): rack; stretcher; horizontal (pole) bar; (*m*): giant; hero 848
RECKEN: to stretch; extend; rack; draw (Metal); shingle; tilt 849
RECKHOLZ (*n*): stretcher; (*pl*): boot-trees 850
RECKWALZE (*f*): finishing roll (Metal) 851
RECKWALZWERK (*n*): finishing rolls (Metal) 852
RECKWERK (*n*): finishing rolls; wooden fence 853
REDAKTEUR (*m*): editor 854
REDAKTION (*f*): editorship; editorial (office) staff; editing; preparation; drawing up; (of documents) 855
REDE (*f*): speech; harangue; talk; discourse; oration; language; rumour; account; conversation; report; rhetoric 856
REDE, IN DIE- FALLEN: to interrupt 857
REDE (*f*), **IN- STEHEND**: in question, referred to, alluded to, under discussion 858
REDEKUNST (*f*): rhetoric; oratory 859
REDEN: to equip; fit out -(ships) 860
REDEN: to speak; talk; discourse; say; converse; (*n*): speaking; speech 861
REDENSART (*f*): expression; phrase-(ology); term; idiom; speech; figure of speech 862
REDESTEHEN: to give an account of; answer for 863
REDESTILLATION (*f*): re-fractionation, repeated fractionating 864
REDEWENDUNG (*f*): idiom, construction 865
REDE, ZUR- STELLEN: to question, call to account 866
REDIGIEREN: to edit (a newspaper) 867
REDLICH: honest; upright; fair; candid; just 868
REDLICHKEIT (*f*): probity; candour; honesty 869
REDMACHEN (*n*): modelling (free-hand engraving of ceramic wares when in a leather-hard state) 870
REDNER (*m*): speaker; orator 871
REDNERBÜHNE (*f*): (public)-platform; chair; pulpit 872
REDNERISCH: rhetorical 873
REDRUTHIT (*m*): redruthite; Cu_2S; Sg. 5.65 874
REDSELIG: talkative; loquacious 875
REDSELIGKEIT (*f*): loquacity 876
REDUCIERBAR: reducible (see also **REDUZ-**) 877
REDUCIEREN: to reduce 878
REDUCTION (*f*): reduction (Metals, etc.) 879
REDUKTION (*f*): reduction (of metals, etc.) 880
REDUKTION, DIREKTE- (*f*): direct reduction (splitting up of iron ores by glowing carbon in blast furnace work) 881
REDUKTION, INDIREKTE- (*f*): indirect reduction (splitting up of iron ores by carbon monoxide in blast furnace work) 882
REDUKTIONSFÄHIG: capable of reduction 883
REDUKTIONSFAKTOR (*m*): factor of reduction, reduction factor 884
REDUKTIONSFLAMME (*f*): reducing flame (see **REDUZIERENDE FLAMME**) 885
REDUKTIONSGETRIEBE (*n*): reduction gearing 886
REDUKTIONSKOEFFIZIENT (*m*): co-efficient of reduction 887
REDUKTIONSKRAFT (*f*): reducing power 888
REDUKTIONSMITTEL (*n*): reducing agent; reducer 889
REDUKTIONSMUFFE (*f*): reducing piece, reducing socket 890
REDUKTIONSOFEN (*m*): reduction furnace 891
REDUKTIONSPOTENTIAL (*n*): reducing (reduction) potential 892
REDUKTIONSROHR (*n*): reduction (pipe) piece, swaged pipe 893
REDUKTIONSRÖHRE (*f*): reduction (pipe) piece; swaged tube 894
REDUKTIONSROLLE (*f*): reduction roller 895
REDUKTIONSVENTIL (*n*): reduction (reducing) valve 896
REDUKTIONSZIRKEL (*m*): reduction compasses 897
REDUKTIONSZONE (*f*): reduction zone (in which the ores are split up in blast furnace work) 898
REDUZIERBAR: reducible (see also **REDUC-**) 899
REDUZIEREN: to reduce 900
REDUZIERENDE FLAMME (*f*): reducing flame (produced by an insufficiency of excess air) 901
REDUZIERSALZ (*n*): reducing salt 902
REDUZIERVENTIL (*n*): reducing valve 903
REE: ready all, ready about (Nautical command) 904
REEDE (*f*): roadstead (Naut.) (see **RHEDE**) 905
REEDEN: to equip (fit out) ships 906
REEDER (*m*): shipowner 907

REEDEREI (*f*): ownership (Ships); shipping (trade) interest; equipment (fitting out) of ships 908
REEDEREIFLAGGE(*f*): house flag 909
REEF (*n*): reef (see also **REFF**) 910
REEFTALJE (*f*), **MARS-**: topsail reef tackle 911
REELL: real; solid; honest; sound; fair; just; essential 912
REELLITÄT (*f*): solidity; honesty 913
REEP (*n*): rope 914
REEPSCHLÄGER (*m*): rope maker 915
REFERAT (*n*): abstract; review; report 916
REFERENT (*m*): reporter 917
REFERENZ (*f*): reference, testimonial 918
REFERIEREN: to report; relate 919
REFF (*n*): reef (see also **REEF**) 920
REFFBAND (*n*): reef band 921
REFFBANDLOCH (*n*): eyelet hole 922
REFFBÄNDSEL (*n*): reef band 923
REFFEN: to reef (a sail) 924
REFFKNOTEN (*m*): reef knot 925
REFFLEGEL (*m*): reef cringle 926
REFFLEINE (*f*): reef line 927
REFFTALJE (*f*): reef tackle 928
REFFZEISING (*f*): reef point 929
REFLEKTANT (*m*): enquirer, prospective client 930
REFLEKTIEREN: to reflect; consider; think of; have in (view) mind 931
REFLEKTIEREN AUF: to be a candidate for, to apply for, have in (prospect) view 932
REFLEKTIEREN ÜBER: to reflect upon, consider, con, study, turn over in one's mind, think of, muse upon 933
REFLEKTION (*f*): reflection 934
REFLEKTOR (*m*): reflector 935
REFLEKTORSCHIRM (*m*): reflector 936
REFLEX (*m*): reflection; reflex 937
REFLEXGALVANOMETER (*n*): reflex (reflection, reflecting) galvanometer 938
REFLEXION (*f*): reflection; reflex; contemplation 939
REFLEXIONSBILD (*n*): reflection-(figure or illustration) 940
REFLEXIONSEBENE (*f*): plane of reflection 941
REFLEXIONSGALVANOMETER (*n*): reflex (reflection; reflecting) galvanometer 942
REFLEXIONSGONIOMETER (*n*): reflecting goniometer 943
REFLEXIONSINTENSITÄT (*f*): intensity of reflection 944
REFLEXIONSSTRAHL (*m*): reflected ray 945
REFLEXIONSWASSERSTANDZEIGER (*m*): reflex water-gauge-(s) 946
REFLEXIONSWINKEL (*m*): angle of reflection 947
REFLEXLÄHMUNG (*f*): reflex paraplegia, partial urinary paraplegia (Med.) 948
REFORM (*f*): reform 949
REFORMATION (*f*): reformation, conversion 950
REFORMIEREN: to reform; convert 951
REFRAKTÄR: refractory 952
REFRAKTION(*f*): refraction. 953
REFRAKTIONSWINKEL (*m*): angle of refraction 954
REFRAKTOMETER (*n*): refractometer (an instrument for measuring refractive index) 955
REFRAKTOMETERHALTER (*m*): refractometer holder 956
REFRAKTOMETERPRISMA (*n*): refractometer prism 957
REFRAKTOMETRISCH: refractometric 958
REFRAKTOR (*m*): refractor; prismatic (telescope) glass 959
REFRIGERATOR (*m*): refrigerator 960
REFRIGERIEREN: to freeze; refrigerate 961
REG: lively, close, animated (see **REGE**) 962
REGAL (*n*): shelf; revenue; rack; bookcase; stand (Typ.); organ register; (as a prefix = regal, royal) 963
REGALIEREN: to regale 964
REGATTA (*f*): regatta 965
REGE: movable; stirring; moving; quick; active; lively; close; animated; extensive; brisk; zealous; industrious 966
REGEL (*f*): rule; precept; standard; principle; order; menses; menstruation (Med.) 967
REGELBAR: regulable 968
REGELEINRICHTUNG (*f*): regulating (apparatus, device) arrangement 969
REGEL, IN DER-: as a general rule; generally; usually 970
REGELLOS: anomalous; irregular; without order 971
REGELLOSIGKEIT (*f*): irregularity; anomaly 972
REGELMÄSSIG: regular-(ly); ordinary; uniform; constant; normal 973
REGELMÄSSIGKEIT (*f*): regularity; uniformity 974
REGELN:, to regulate; order; arrange 975
REGELRECHT: in accordance with (rule) precept; regularly; regular; orderly; correct; normal 976
REGELSPURIG: standard gauge (Railways) 977
REGELVENTIL (*n*), **GESTEUERTES-**: automatic regulating valve (actuated by valve gear) 978
REGELWIDRIG: contrary to rule; irregular 979
REGEN: to move; arouse; be aroused; be active; excite; stir; touch upon; mention; rise; (*m*): rain; shower; downpour 980
REGENBAD (*n*): shower bath 981
REGENBÖ (*f*): white squall (Naut.) 982
REGENBOGEN (*m*): rainbow 983
REGENBOGENFARBEN (*f.pl*): prismatic (rainbow; iridescent) colours 984
REGENBOGENFARBIG: iridescent; rainbow-coloured 985
REGENBOGENHAUT (*f*): iris (Anatomy) 986
REGENDACH (*n*): eaves 987
REGENDICHT: rainproof, waterproof, shower-proof 988
REGEN, DIE- BEZIEHUNGEN (*f.pl*): the close connection 989
REGENERATION (*f*): regeneration 990
REGENERATIONSVERFAHREN (*n*): regenerative process 991
REGENERATIV: regenerative 992
REGENERATIVFEUERUNG (*f*): regenerative furnace 993
REGENERATIVOFEN (*m*): regenerating furnace 994
REGENERATIVPRINZIP (*n*): regenerative principle 995
REGENERATIVVERFAHREN (*n*): regenerative process 996
REGENERATOR (*m*): regenerator 997
REGENERATOROFEN (*m*): regenerating furnace 998
REGENERIEREN: to regenerate 999
REGENERIERUNG (*f*): regeneration 000
REGENERIERUNGSMASSE (*f*): regenerating mass (for hardening steel castings) 001
REGENFANG (*m*): cistern; rain-water (tub) barrel 002
REGENFASS (*n*): rain-water (tub) barrel 003
REGENGALLE (*f*): water-gall 004
REGENGUSS (*m*): downpour (of rain); heavy shower 005

REGENHÖHE (*f*): (depth of)-rainfall 006
REGENLOS: rainless 007
REGENMANTEL (*m*): raincoat; waterproof-(coat); mackintosh; macintosh 008
REGENMASS (*n*): rain gauge 009
REGENMENGE (*f*): quantity of rain; (rain)-fall 010
REGENMESSER (*m*): rain gauge 011
REGENROCK (*m*): (see **REGENMANTEL**) 012
REGENSCHAUER (*m* and *n*): shower of rain 013
REGENSCHIRM (*m*): umbrella 014
REGENSCHIRMFÖRMIG: umbelliferous (Bot.) 015
REGENSCHIRMGESTELL (*m*): umbrella frame 016
REGENSCHNECKE (*f*): slug 017
REGENT (*m*): regent, administrator 018
REGENTROPFEN (*m*): rain drop 019
REGENVOGEL (*m*): plover 020
REGENWASSER (*n*): rain-water 021
REGENWETTER (*n*): rainy (weather) season; wet season 022
REGENWOLKE (*f*): rain cloud 023
REGENWURM (*m*): earthworm 024
REGENZEIT (*f*): rainy (wet) season 025
REGIE (*f*): administration; management 026
REGIE, IN EIGENER- AUSFÜHREN: to execute (carry out) in person or personally 027
REGIERBAR: governable, manageable 028
REGIEREN: to reign; guide; rule; govern; regulate; manage; administrate; sway; steer 029
REGIERER (*m*): ruler; governor 030
REGIERUNG (*f*): government; rule; administration; reign; management 031
REGIERUNGSADVOKAT (*m*): counsel for the Crown 032
REGIERUNGSASSESSOR (*m*): government assessor 033
REGIERUNGSBEZIRK (*m*): county (actually in Germany a Government District) 034
REGIERUNGSBLATT (*n*): government organ, official gazette 035
REGIERUNGSLOS: without law and order, anarchical 036
REGIERUNGSNOTE (*f*): government (bill) note 037
REGIEVERSCHLUSS (*m*): bond 038
REGIMENT (*n*): regiment; government; power 039
REGIMENT, DAS- FÜHREN: to rule, be master 040
REGINAVIOLETT (*n*): Regina violet 041
REGION (*f*): region; portion, part 042
REGISSEUR (*m*): manager 043
REGISTER (*n*): register; record; table of contents; index; list; damper-(plate); register (of an organ) 044
REGISTERTONNE (*f*): registered ton 045
REGISTERTONNENGEHALT (*m*): registered tonnage 046
REGISTERZUG (*m*): register (of an organ); coupler 047
REGISTRATOR (*m*): recorder; registrar 048
REGISTRATUR (*f*): registry; record (registrar's) office 049
REGISTRATURSCHRANK (*m*): filing cabinet 050
REGISTRATURSYSTEM (*n*): filing system 051
REGISTRIERAPPARAT (*m*): registering apparatus 052
REGISTRIEREN: to register; record; indicate 053
REGISTRIERINSTRUMENT (*n*): recording (indicating, registering) instrument, recorder 054
REGISTRIERKASSE (*f*): cash register 055
REGLER (*m*): regulator; governor; mixing valve 056
REGLISSE (*f*): liquorice-(liquid); marshmallow paste 057
REGNEN: to rain; pour 058
REGNERISCH: rainy 059
REGRESS (*m*): recourse; regress; remedy; appeal 060
REGRESSIV: regressive 061
REGSAM: active; agile; quick; brisk; nimble 062
REGSAMKEIT (*f*): activity; briskness; agility 063
REGULÄR: regular 064
REGULATIV (*n*): rule, regulation 065
REGULATOR (*m*): regulator; governor 066
REGULATORFEDER (*f*): governor spring 067
REGULATORKLAPPE (*f*): governor slide, governor valve 068
REGULATORKUGEL (*f*): governor ball 069
REGULATORSPINDEL (*f*): governor spindle 070
REGULATORSTANGE (*f*): governor rod 071
REGULATORWELLE (*f*): governor shaft 072
REGULIEREN: to regulate; adjust; govern 073
REGULIERHAHN (*m*): regulating cock 074
REGULIERSCHIEBER (*m*): regulating valve, regulating damper, regulating slide 075
REGULIERSCHRAUBE (*f*): regulating screw 076
REGULIERSTANGE (*f*): regulating rod 077
REGULIERUNG (*f*): regulation; regulating; governing 078
REGULIERVENTIL (*n*): regulating valve 079
REGULIERVORRICHTUNG (*f*): regulating apparatus, regulating device 080
REGULIERWIDERSTAND (*m*): compensating resistance, regulating resistance 081
REGULIN (*n*): reguline 082
REGULINISCH: reguline 083
REGULUS (*m*): regulus; matte (an artificial sulphide for reduction action) (Metallurgy) 084
REGUNG (*f*): emotion; stir; motion; moving; movement; agitation; excitement; impulse 085
REGUNGSKRAFT (*f*): power of movement; motive power 086
REGUNGSLOS: motionless; still 087
REH (*n*): roe; deer; (*f*): ribband (Naut.) 088
REHBOCK (*m*): roe-buck 089
REHBRAUN (*n*): fawn-(colour) 090
REHFAHL: fawn-coloured 091
REHFARBIG: fawn-coloured 092
REHFELL (*n*): doe-skin 093
REHHAUT (*f*): doe-skin 094
REHKALB (*n*): fawn; calf (of a deer) 095
REHLEDER (*n*): doe-skin; buck-skin 096
REHPOSTEN (*m.pl*): buck-shot 097
REIBAHLE (*f*): broach, reamer, reamering tool 098
REIBAPPARAT (*m*): grinding apparatus, triturating apparatus 099
REIBE (*f*): grater; rasp 100
REIBECHT: fast to rubbing 101
REIBEISEN (*n*): grater; rasp; reamering tool 102
REIBEN: to rub; graze; grind; triturate; reamer; rasp; scour; chafe; grate; abrase; pulverize; vex; annoy; provoke; torment; irritate 103
REIBEPULVER (*n*): abrasive powder 104
REIBER (*m*): grater; rubber; pestle; grinder 105
REIBEREI (*f*): exasperation; provocation; vexation; irritation; torment; annoyance; aggravation 106
REIBFLÄCHE (*f*): rubbing surface 107

REIBGUMMI (*n*): (rubber)-eraser 108
REIBKEULE (*f*): pestle 109
REIBMASCHINE (*f*): grinding machine 110
REIBRAD (*n*): friction wheel 111
REIBRADGETRIEBE (*n*): friction gear 112
REIBRADVORGELEGE (*n*): friction gear 113
REIBSCHALE (*f*): mortar 114
REIBSTEIN (*m*): grind-stone; grinding stone 115
REIBUNG (*f*): friction; rubbing; grinding; trituration; rasping; scouring; chafing; grating; abrasion; rub; difficulty; clash (see also REIBEREI) 116
REIBUNG (*f*), GLEITENDE-: sliding friction 117
REIBUNG (*f*), ROLLENDE-: rolling friction 118
REIBUNGSBEIWERT (*m*): coefficient of friction, (μ) 119
REIBUNGSBREMSE (*f*): friction brake 120
REIBUNGSDURCHMESSER (*m*): frictional diameter (of tubes) (where pressure drop is caused by tube friction alone exclusive of any other resistances) 121
REIBUNGSELEKTRIZITÄT (*f*): frictional electricity; galvanism 122
REIBUNGSFAKTOR (*m*): friction (rubbing) factor, coefficient of friction 123
REIBUNGSFLÄCHE (*f*): rubbing surface 124
REIBUNGSFREI: without friction, frictionless 125
REIBUNGSKOEFFIZIENT (*m*): coefficient of friction, (μ) 126
REIBUNGSKUPPELUNG (*f*): friction (coupling) clutch 127
REIBUNGSPROBE (*f*): friction test 128
REIBUNGSRAD (*n*): friction wheel (see REIBRAD) 129
REIBUNGSROLLE (*f*): running sheave, friction roller 130
REIBUNGSVERLUST (*m*): friction loss 131
REIBUNGSVORGELEGE (*n*): friction(-al) gear or gearing 132
REIBUNGSWÄRME (*f*): frictional heat 133
REIBUNGSWIDERSTAND (*m*): frictional resistance 134
REIBUNGSZAHL (*f*): coefficient of friction, (μ) 135
REIBUNGSZIFFER (*f*): (see REIBUNGSZAHL) 136
REIBUNG, WALZENDE- (*f*): rolling friction 137
REIBZÜNDHÖLZCHEN (*n*): friction match 138
REICH: large; rich; wealthy; abundant; abounding in; having a high content; (*n*): empire; realm; kingdom 139
REICHARDTIT (*m*): reichardtite (a natural magnesium sulphate) (see MAGNESIUM-SULFAT and BITTERSALZ) 140
REICHEN: to give; offer; pass; reach; extend; be sufficient; suffice; last; administer; bestow 141
REICHHALTIG: plentiful; copious; abundant; rich 142
REICHHALTIGKEIT (*f*): abundance; richness 143
REICHHEIT (*f*): richness, wealth, opulence 144
REICHLICH: rich; full; ample; copious; abundant; plentiful; large 145
REICHSANZEIGER (*m*): official (imperial) gazette 146
REICHSAUSGLEICHAMT (*n*): clearing (office) house 147
REICHSCHAUM (*m*): zinc crust, (rich in silver, etc., formed during Parkes' process) (Metal) 148
REICHSCHMELZEN (*n*): precious metal smelting 149
REICHSFOLGE (*f*): imperial succession 150
REICHSGESUNDHEITSAMT (*n*): Imperial Board of Health 151
REICHSHOFRAT (*m*): Aulic council (or a member of such council) 152
REICHSKAMMERGERICHT (*n*): High court, supreme court 153
REICHSKANZLER (*m*): Imperial Chancellor 154
REICHSLAND (*n*): Imperial territory 155
REICHSMARK (*f*): (Imperial)-Mark 156
REICHSMÜNZE (*f*): coin of the realm 157
REICHSRAT (*m*): senate; senator 158
REICHSSCHLUSS (*m*): Imperial decree 159
REICHSSTADT (*f*): (free)-Imperial city 160
REICHSSTREIT (*m*): legal proceedings 161
REICHSTAG (*m*): diet 162
REICHSVERFASSUNG (*f*): constitution of the empire 163
REICHSWÄHRUNG (*f*): Imperial currency 164
REICHTUM (*m*): wealth; riches; abundance; richness; affluence 165
REICHWEITE (*f*): range 166
REIF: ready; ripe; mature; (*m*): hoar-frost; rime; white-frost; collar; tyre; hoop; ring; bloom (of fruit) 167
REIFE (*f*): ripeness; maturity; age; puberty 168
REIFEISEN (*n*): tyre (hoop) iron 169
REIFFN: to ripen; mature; freeze; cover with hoar frost; hoop; put on a tyre; (*m*): collar; tyre; hoop; ring; circle 170
REIFEN (*m*), RAD-: rim (of a wheel) tyre 171
REIFLICH: mature 172
REIF (*m*), PNEUMATIK-: pneumatic tyre 173
REIFROST (*m*): mildew 174
REIF (*m*) VOLLGUMMI-: solid tyre 175
REIHE (*f*): series (Maths.); row; order; succession; suite; train; range; turn; rank; file; number; sequence 176
REIHE (*f*), ARITHMETISCHE-: arithmetical series 177
REIHE (*f*), DER- NACH: in series, in succession, successively, in turn, in sequence, in order, one after the other 178
REIHEN: to arrange (in series; in a row; in order); rank; file; range; baste; string (Beads); stitch; classify 179
REIHENANSATZ (*m*): progression (Maths.) (a series of members forming a progression), sequence, series 180
REIHENENTWICKLUNG (*f*): progression, progressive development 181
REIHENFOLGE (*f*): sequence; order; succession; series 182
REIHENMASCHINE (*f*): tandem-engine; series dynamo 183
REIHENSCHALTUNG (*f*): series connection; connection in series (Elect.) 184
REIHENWEISE: by turns; in succession; successively; in rows 185
REIHER (*m*): heron 186
REIHE (*f*), SINUS-: sine series (Trigonometry) 187
REILSEGEL (*n*): royal-(sail) 188
REIM (*m*): rhyme 189
REIMEN: to rhyme; make rhymes; versify; agree 190
REIMFALL (*m*): cadence 191
REIMSATZ (*m*): stanza, strophe 192
REIMSCHLUSS (*m*): (see REIMSATZ) 193
REIN: pure; clean; tidy; clear; sheer; purely; entirely; absolute-(ly); correct; genuine; blank; quite; dressed; innocent; plain; net; fair; sound; perfect; major (Music); purified; rectified 194

REINBLAU (*n*): pure blue 195
REINDARSTELLUNG (*f*): purification; rectification: presentation in a pure (rectified) condition 196
REINECLAUDE (*f*): Italian plum (*Prunus italica*) 197
REINE KASSE (*f*): net cash 198
REINERTRAG (*m*): net (clear) profit 199
REINGEIST (*m*): alcohol 200
REINGELB (*n*): pure yellow 201
REINGEWICHT (*n*): net weight 202
REINGEWINN (*m*): net (clear) gain, yield, or profit 203
REINGOLD (*n*): pure gold 204
REINHEIT (*f*): purity; clearness; clarity; etc.; fineness (see REIN) 205
REINHEITSGRAD (*m*): degree of (purity) fineness 206
REINIGEN: to purify; clean; clear; cleanse; scour (Silk); fine; refine (Metals, etc.); rectify (Spirits); purge; dress (Flax, etc.); clarify; disinfect 207
REINIGER (*m*): purifier; clarifier, etc. 208
REINIGUNG (*f*): purification; purgation; refining; rectification; menstruation; menses (Med.) (see also REINIGEN) 209
REINIGUNGSAPPARAT (*m*): purifier; purifying apparatus 210
REINIGUNGSBEHÄLTER (*m*): filter-(ing) tank 211
REINIGUNGSMASCHINE (*f*): cleaning machine 212
REINIGUNGSMASSE (*f*): purifying mass 213
REINIGUNGSMITTEL (*n*): purifier; purifying agent; cleanser; cleaning agent; clarifier; detergent; purge; purgative 214
REINIGUNGSTURM (*m*): purifying tower 215
REINIGUNGSWEG (*m*): excretory (duct) passage (Anat.); method of purification 216
REINIT (*m*): reinite; $FeWO_4$; Sg. 6.64 217
REINKOHLE (*f*): (combustible substance in coal or coke, i.e.) combustible; pure carbon 218
REINKULTIVIERUNG (*f*): pure (culture) cultivation (Bact.) 219
REINKULTUR (*f*): (see REINKULTIVIERUNG) 220
REINLICH: cleanly 221
REINLICHKEIT (*f*): cleanliness; neatness 222
REINNICKEL (*n*): pure nickel (about 99.5% nickel, rest cobalt) 223
REINSCHRIFT (*f*): fair copy 224
REINZUCHT (*f*): pure culture-(of yeast) 225
REINZUCHTAPPARAT (*m*): apparatus for pure yeast culture 226
REIS (*m*): rice (*Oryza sativa*); (*n*): twig; sprig; shoot; scion; sucker 227
REISBAU (*m*): rice-growing 228
REISBESEN (*m*): birch broom 229
REISBRANNTWEIN (*m*): rice spirit; arrack 230
REISBREI (*m*): rice pudding 231
REISBÜNDEL (*n*): fagot 232
REISBÜSCHEL (*n*): (see REISBÜNDEL) 233
REISE (*f*): journey; travel; voyage; trip; tour; run 234
REISEBESCHREIBUNG (*f*): travels, book of travels; itinerary of a tour 235
REISEBUCH (*n*): guide book, itinerary 236
REISEFERTIG: ready -(for a journey) 237
REISEKAMERA (*f*): holiday camera, tourist's camera 238
REISEKOFFER (*m*): travelling trunk 239
REISEKÜCHE (*f*): portable kitchen 240
REISEN: to journey; travel; go; leave; set out; tour; voyage; (*n*): peregrination; itinerancy; travelling, etc. 241
REISEND: touring, etc. (see REISEN); itinerant 242
REISENDE (*m*): traveller; passenger; commercial traveller; tourist; voyager; wayfarer; pilgrim; itinerant 243
REISEPASS (*m*): passport 244
REISEPOSTEN (*m*): situation as traveller 245
REISESPIEGEL (*m*): pocket mirror 246
REISESTATIV (*n*): tourist stand (for cameras) 247
REISEUNKOSTEN (*pl*): travelling expenses 248
REISEZUG (*m*): caravan 249
REISHOLZ (*n*): brushwood 250
REISICHT (*n*): brushwood; prunings; twigs 251
REISIG (*m* and *n*): brushwood; twigs; prunings; copse 252
REISIGE (*m*): trooper; horseman 253
REISKÖRPER (*m*): rice body (Med.) 254
REIS-MASCHINEN (*f.pl*): rice machinery 255
REISMEHL (*n*): rice flour 256
REISÖL (*n*): rice oil (from *Oryza sativa*) 257
REISPAPIER (*n*): rice paper (from *Aralia papyrifera*) 258
REISSAUS (*n*): flight 259
REISSAUSNEHMEN: to fly; take to flight; run away; decamp 260
REISSBLEI (*n*): graphite; blacklead (*Plumbago*); drawing (blacklead) pencil 261
REISSBLEITIEGEL (*m*): graphite crucible 262
REISSBRETT (*n*): drawing board 263
REISSCHLEIM (*m*): rice water 264
REISSEN: to tear; pull; rend; split; drag; sketch; trace; draw; burst; bruise; slit; crack; break; pain; scribe; etch; grave; mark; crush (Malt); rough-grind (Glass); rupture; (*n*): rending, etc.; colic (Med.); splitting (of wood along the fibre) 265
REISSEND: tearing; bursting; cracking, etc.; rapacious; rapid; wild; acute; rapidly 266
REISSER (*m*): gauge; sketcher 267
REISSER (*m*), **PARALLEL-**: surface gauge 268
REISSFEDER (*f*): drawing pen 269
REISSFESTIGKEIT (*f*): tensile strength; tenacity; resistance to tearing or breaking 270
REISSHAKEN (*m*): mortise-chisel 271
REISSKOHLE (*f*): charcoal; crayon (for drawing) 272
REISSKORN (*n*): artificial grain (Leather) 273
REISSMASCHINE (*f*): tearing machine 274
REISSNADEL (*f*): scribing (needle) tool, etching (needle) tool, graving tool, marking tool 275
REISSNAGEL (*m*): drawing pin 276
REISSPELZ (*m*): rice spelt 277
REISSPELZTEER (*m*): rice-spelt tar (extracted from rice spelt by dry distillation) 278
REISSSCHIENE (*f*): T-square (for drawing) 279
REISSSTIFT (*m*): drawing pencil (see also REISS-NADEL) 280
REISSTÄRKE (*f*): rice starch 281
REISSZEUG (*n*): drawing instruments; mathematical instruments; case of instruments 282
REISSZWECKE (*f*): drawing pin 283
REITBAHN (*f*): horse-ride; riding school 284
REITBAR: rideable, capable of being ridden; good (of roads) 285
REITEN: to ride 286
REITEND: mounted 287
REITER (*m*): rider; horseman; trooper; riddle; sieve; screen; cross-beam; (*f*): sieve; riddle; screen 288

REITEREI (*f*): riding; cavalry 289
REITERSTATUE (*f*): equestrian statue 290
REITKUNST (*f*): equestrian art; horsemanship 291
REITPFERD (*n*): saddle-horse 292
REITSCHULE (*f*): riding school 293
REITSTOCK (*m*): (sliding)-puppet (Mech.) 294
REITWEG (*m*): bridle-path 295
REIZ (*m*): charm; enticement; attraction; grace; tonic; stimulus; stimulant; irritant; irritation; incentive; provocation; exasperation; revivifier 296
REIZBAR: sensitive; irritable; susceptible; inflammable (Med.) 297
REIZEN: to irritate; provoke; exasperate; stimulate; charm; excite; allure; attract; tempt; revivify; entice; seduce; incite; induce; vivify 298
REIZEND: charming; lovely; beautiful; excellent 299
REIZGIFT (*n*): irritant poison 300
REIZLOS: non-irritant; non-irritating; non-stimulating; without charms; unattractive; insipid; unalluring 301
REIZMITTEL (*n*): irritant; stimulant; incentive; inducement 302
REIZSCHWELLENSTÄRKE (*f*): the strength of sound (the number of vibrations, which is just capable of affecting the sense of hearing, i.e., being heard) 303
REIZUNG (*f*): irritation; provocation, etc. (see REIZ and REIZEN) 304
REIZVOLL: charming 305
REIZWIRKUNG (*f*): stimulating (irritating) effect or action 306
REKAPITULIEREN: to recapitulate 307
REKLAMATION (*f*): claim 308
REKLAME (*f*): advertisement 309
REKLAMEANZEIGE (*f*): advertisement 310
REKLAME MACHEN: to advertise 311
REKLAMIEREN: to claim; reclaim; advertise 312
REKRISTALLISATION (*f*): recrystallization 313
REKRISTALLISATIONSVERSUCH (*m*): recrystallizing (recrystallization)-test 314
REKRISTALLISIEREN: to recrystallize 315
REKRUT (*m*): recruit 316
REKRUTIEREN: to recruit 317
REKRYSTALLISATION (*f*): recrystallization (see REKRISTALLISATION) 318
REKRYSTALLISIEREN: to recrystallize 319
REKTAL: rectal; of the rectum (Med.) 320
REKTIFIKATION (*f*): rectification (the process of eliminating one constituent of a fluid mixture, by distillation) 321
REKTIFIKATIONSAPPARAT (*m*): rectifying apparatus; rectifier 322
REKTIFIKATIONSKOLONNE (*f*): rectifying column 323
REKTIFIKATIONSPRINZIP (*n*): rectification principle 324
REKTIFIKATIONSSÄULE (*f*): rectifying column 325
REKTIFIKATIONSSYSTEM (*n*): rectifying (rectification) system 326
REKTIFIKATIONSVORRICHTUNG (*f*): rectifying or rectification (apparatus) arrangement 327
REKTIFIZIERAPPARAT (*m*): rectifying apparatus; rectifier 328
REKTIFIZIEREN: to rectify 329
REKTIFIZIERUNG (*f*): rectification 330
REKTILINEARLINSE (*f*): rectilinear lens 331
REKTOR (*m*): rector; headmaster 332
REKUPERATOR (*m*): recuperator 333
REKURS (*m*): (see REGRESS) 334
RELAIS (*n*): relay (Mech. and Elect.) 335
RELAISGLOCKE (*f*): relay bell 336
RELAISMAGNET (*m*): relay magnet 337
RELAISSCHALTUNG (*f*): relay connection 338
RELAISSTROM (*m*): relay circuit 339
RELATIV: relative-(ly); respective-(ly) 340
RELATIVUM (*n*): relative (Grammar) 341
RELAXIEREN: to relax 342
RELEGIEREN: to expel, relegate 343
RELIEF (*n*): relief, relievo 344
RELIEFDRUCK (*m*): Woodbury-type printing (see PHOTOGLYPTIE) 345
RELIEFDRUCKMASCHINE (*f*): relief printing machine 346
RELIGION (*f*): religion 347
RELIGIONSDULDUNG (*f*): religious toleration 348
RELIGIONSLEHRE (*f*): doctrine 349
RELIGIONSSATZ (*m*): dogma 350
RELIGIONSTRENNUNG (*f*): schism 351
RELIGIONSUNTERRICHT (*m*): religious instruction 352
RELIGIONSWISSENSCHAFT (*f*): theology 353
RELING (*f*): railing (on ships); bulwark 354
RELINGSTÜTZE (*f*): bulwark stanchion, railing (support) upright or column 355
RELIQUIE (*f*): relic 356
REMANENTER MAGNETISMUS (*m*): residual magnetism 357
REMANENZ (*f*): residual magnetism; residue, remainder 358
REMBRANDTBELEUCHTUNG (*f*): Rembrandt lighting (in photography to give a Rembrandt effect) 359
REMESSE (*f*): remittance (Com.) 360
REMIS: tied, drawn (of games) 361
REMISSE (*f*): (see REMESSE) 362
REMITTENT (*m*): remitter 363
REMITTIEREN: to remit; send; forward; return 364
REMONTEPFERD (*n*): remount 365
REMONTIEREN: to re-ascend; re-erect; bloom for long periods (of flowers); re-mount 366
RENDANT (*m*): cashier; paymaster 367
RENDEMENT (*n*): rendement; yield; output 368
RENDITA (*f*): extent of weighting-(of fabrics) 369
RENKEN: to bend; wrench; twist 370
RENNARBEIT (*f*): direct process (iron extraction from ore) 371
RENNAUTOMOBIL (*n*): racing car 372
RENNBAHN (*f*): (race)-course; track 373
RENNBOOT (*n*): racing boat 374
RENNEISEN (*n*): malleable iron (obtained by direct process) 375
RENNEN: to run; race; melt; smelt; extract metal (malleable iron) direct from the ore; bloom (Metal) 376
RENNFEUER (*n*): bloomery hearth (direct process extraction of malleable iron); smelting furnace 377
RENNFEUERVERFAHREN (*n*): bloomery process (direct process extraction of maleable iron) 378
RENNHERD (*m*): (see RENNFEUER) 379
RENNJACHT (*f*): racing yacht 380
RENNPFERD (*n*): race-horse 381
RENNSCHIFF (*n*): racing yacht; cutter 382
RENNSCHLACKE (*f*): direct-process slag (Iron) 383
RENNSTAHL (*m*): direct-process (natural) steel (made directly from the ore) 384
RENNSTEIN (*m*): gutter; sink 385
RENNTIER (*n*): reindeer 386

RENNTIERFLECHTE (*f*) : reindeer moss 387
RENNWAGEN (*m*) : racing car 388
RENOLFARBE (*f*) : renol dye 389
RENOMMEE (*n*) : renown 390
RENOMMIEREN : to boast ; brag ; bully ; spread about 391
RENOMMIERT : renowned ; well-known 392
RENOMMIST (*m*) : boaster ; braggart 393
RENOMMISTISCH : boastful ; bragging 394
RENOVIEREN : to renovate ; re-decorate ; renew 395
RENSSELAERIT (*m*) : Rensselaerit (a form of talcum) (see TALK) 396
RENTABEL : profitable ; lucrative 397
RENTABILITÄT (*f*) : profitableness 398
RENTAMT (*n*) : exchequer ; revenue (board) office 399
RENTE (*f*) : income ; rent ; revenue ; annuity ; pension ; interest ; (*pl*) : stocks 400
RENTIER (*m*) : one of independent means 401
RENTIEREN : to pay ; be profitable ; bring in rent or income 402
REPARATIONSSCHEIN (*m*) : reparation form 403
REPARATUR (*f*) : repair-(s) 404
REPARATURKASTEN (*m*) : repair (box) outfit 405
REPARATURKOSTEN (*f.pl*) : repair costs ; upkeep costs 406
REPARIEREN : to repair ; mend 407
REPASSIEREN : to boil off-(silk) a second time 408
REPETIEREN : to repeat 409
REPETIERUHR (*f*) : repeater-(watch) 410
REPETIONSKREIS (*m*) : repeating circle 411
REPONIEREN : to replace, put back, re-set ; set (Surgical) 412
REPOSITORIUM (*n*) : repository ; bookshelf 413
REPRÄSENTANT (*m*) : representative 414
REPRÄSENTIEREN : to represent 415
REPRESSALIEN (*f.pl*) : reprisals 416
REPRODUKTION (*f*) : reproduction 417
REPRODUKTIONSVERFAHREN (*n*) : reproduction process (process of reproducing drawings, paintings, manuscripts, etc.) 418
REPRODUZIERBAR : reproducible 419
REPRODUZIEREN : to reproduce 420
REPS (*m*) : rape (see also RAPS) 421
REPSKOHL (*m*) : rape (see also RAPS) 422
REPSÖL (*n*) : (see RÜBÖL and RAPSÖL) 423
REPUBLIK (*f*) : republic 424
REPUBLIKANISCH : republican 425
REQUIRIEREN : to requisition ; request ; require ; demand 426
REQUISIT (*n*) : requisite ; (*pl*) : properties ; props (Theatrical) 427
RESEDA (*f*) : mignonette (*Reseda odorata*) 428
RESEDA, GELBE- (*f*) : dyer's weed (*Reseda luteola*) 429
RESEDAÖL (*n*) : mignonette oil ; reseda oil (extracted from *Reseda odorata*) 430
RESEDASAMENÖL (*n*) : mignonette-(seed)-oil (from *Reseda odorata*) 431
RESERV : reserve, standby, spare, auxiliary (Mech.) 432
RESERVAGE (*f*) : resist ; reserve (Calico) 433
RESERVAGEARTIKEL (*m*) : resist (reserve) style (Calico) 434
RESERVAT (*n*) : reservation 435
RESERVE (*f*) : reserve (Military and General) ; resist ; reserve (Calico) ; stand-by ; spare ; auxiliary (Mech.) 436
RESERVEDEPLACEMENT (*n*) : reserve of buoyancy 437
RESERVEDRUCK (*m*) : resist (reserve) style ; resist printing (Calico) 438
RESERVEFONDS (*m*) : reserve fund 439
RESERVEGUT (*n*) : spares 440
RESERVELEITUNG (*f*) : reserve (spare, stand-by, duplicate) main or pipeline 441
RESERVEMITTEL (*n*) : resist ; reserve (Calico) 442
RESERVEMUSTER (*n*) : reference patterns (Textile) 443
RESERVESCHWIMMKRAFT (*f*) : reserve of buoyancy 444
RESERVETEIL (*m*) : spare part, replace part 445
RESERVIEREN : to reserve ; subject to action of a resist or reserve (Calico) 446
RESERVIERUNGSMITTEL (*n*) : resist ; reserve (Calico) 447
RESERVOIR (*n*) : tank ; reservoir 448
RESERVOIR (*n*), **WASSER-** : water (tank) reservoir 449
RESIDENZ (*f*) : residence 450
RESIDIEREN : to reside, live, dwell 451
RESINAT (*n*) : resinate 452
RESINATFARBE (*f*) : (see HARZFARBE) 453
RESINIT (*n*) : resinite (an artificial resin) 454
RESINOL (*n*) : resinol (an artificial resin) 455
RESISTENZ (*f*) : resistance 456
RESOLUTION (*f*) : resolution, resolve, decree, decision 457
RESONANZ (*f*) : resonance 458
RESONANZBODEN (*m*) : sound-(ing) board 459
RESONANZDECKE (*f*) : (see RESONANZBODEN) 460
RESONANZKASTEN (*m*) : sound-box 461
RESONANZLOCH (*n*) : sound hole 462
RESONANZSCHWINGUNG (*f*) : resonance (oscillation) vibration 463
RESONATOR (*m*) : resonator 464
RESORBIERBAR : resorptive ; re-absorptive 465
RESORBIEREN : to re-absorb ; resorb 466
RESORBIN (*n*) : resorbine (an emulsion composed of wax, water and almond oil, etc.) 467
RESORCIN (*n*) : resorcin ; resorcinol ; dioxybenzene, (meta-) ; $C_6H_4(OH)_2$; Sg. 1.27 ; Mp. 110-111.6°C. ; Bp. 271-280°C. (see also RESORZIN) 468
RESORCINFARBSTOFF (*m*) : resorcin-(ol) dye 469
RESORCINSEIFE (*f*) : resorcin-(ol) soap 470
RESORPTION (*f*) : resorption, re-absorption 471
RESORZIN (*n*) : (see RESORCIN) 472
RESORZINBENZOYLKARBONSÄURE-ÄTHYLESTER (*m*): ethyl resorcinbenzoylcarbonate ; Mp. 134°C. 473
RESORZINPHTALEIN (*n*) : resorcinphthalein (see FLUORESCEIN) 474
RESORZINREST (*m*) : resorcin-(ol) residue 475
RESPEKT (*m*) : respect 476
RESPEKTABEL : respectable 477
RESPEKTIEREN : to respect, reverence, esteem, regard, value, prize, revere ; honour (Bills, etc.) 478
RESPEKTIERLICH : respectable 479
RESPEKTIV : respective 480
RESPEKTIVE : respectively 481
RESPEKTTAGE (*m.pl*) : days of grace (Comm.) 482
RESPEKTWIDRIG : disrespectful 483
RESPIRATIONSAPPARAT (*m*) : breathing apparatus, apparatus for restoring respiration 484
RESPIRATIONSNAHRUNGSMITTEL (*n*) : respiratory food 485
RESPIRO (*n*) : grace ; respite (Com.) 486
REST (*m*) : residue ; arrears ; remainder ; rest ; outstanding (quantity) amount ; remains ; remnant ; balance ; difference 487

RESTANT (*m*): defaulter; one in arrears; (*pl*): arrears 488
RESTAURATION (*f*): restoration; restaurant; tavern 489
RESTEN: (see **RESTIEREN**) 490
RESTGLIED (*n*): remaining (term) member, remainder 491
RESTHÄRTE (*f*): (see **NICHTCARBONATHÄRTE**) 492
RESTIEREN: to remain; be left; be in arrear; rest; be (outstanding) overdue 493
RESTIEREND: outstanding, overdue 494
RESTITUTION (*f*): restitution 495
RESTITUTIONSFLUID (*n*): restitution fluid 496
RESTLOS: absolutely, completely; (actually means "without leaving a residue or a doubtful point") (see also **RASTLOS**) 497
RESTLOS BEANTWORTEN: to answer satisfactorily (a question) 498
RESTLOS KLÄREN: to clear up thoroughly (a question) 499
RESTSTRAHL (*m*): residual ray 500
RESTSTROM (*m*): residual current 501
RESTVALENZ (*f*): residual valence 502
RESULTANTE (*f*): resultant 503
RESULTAT (*n*): result; inference; consequence; deduction; outcome; effect; termination; end; issue; fruit 504
RESULTIEREN: to result; follow; issue; arise; rise; accrue; spring; originate; end; terminate; be derived; ensue 505
RESULTIEREND: resultant; resulting 506
RESUPINATION (*f*): resupination (the rotation of a flower on its stem to bring it into the correct position for polination) (Bot.) 507
RETARDIEREN: to retard, delay, check, impede, slow up, slacken speed; go slow, lose (of clocks) 508
RETENTIONSRECHT (*n*): lien (Law) 509
RETINIT (*m*): retinite (a fossil resin); Sg. 1.25 510
RETIRIEREN: to retire; retreat; fall back 511
RETORTE (*f*): retort (Mech. and Chem.) 512
RETORTENBAUCH (*m*): bulb (belly) of a retort 513
RETORTENGESTELL (*n*): retort stand 514
RETORTENGRAPHIT (*n*): retort graphite; carbon 515
RETORTENHALS (*m*): neck of a retort 516
RETORTENHALTER (*m*): retort (stand) holder 517
RETORTENHAUS (*n*): retort house 518
RETORTENKITT (*m*): retort cement 519
RETORTENKOHLE (*f*): retort (carbon) charcoal 520
RETORTENMÜNDUNG (*f*): mouth of a retort 521
RETORTENOFEN (*m*): retort furnace 522
RETORTENVERKOKUNG (*f*): retort coking 523
RETORTENVORSTOSS (*m*): adapter; condenser (Zinc) 524
RETORTENWAND (*f*): retort wall 525
RETOUCHE (*f*): retouching 526
RETOUCHIEREN: to retouch 527
RETOUCHIERLACK (*m*): retouching (medium) varnish 528
RETOUCHIERPULT (*n*): retouching desk (Phot.) 529
RETOURDAMPF (*m*): return steam 530
RETOUREN (*pl*): returns; empties 531
RETOUR-KONTO (*n*): return account 532
RETOURNIEREN: to return, send (go, give, turn) back, retrace, recur, revert; recompense, repay, requite; remit, transmit 533
RETOURÖL (*n*): recovered oil 534
RETOURWAREN (*f.pl*): returns (Goods) 535
RETRAKTION (*f*): retraction, contraction (Med.) 536
RETRATTE (*f*): re-draft (Com.) 537
RETTBAR: retrievable, remediable, deliverable, capable of being saved 538
RETTEN: to save; rescue; deliver; retrieve; help; remedy; salvage; preserve; escape; vindicate 539
RETTER (*m*): saviour; rescuer; deliverer; preserver 540
RETTICH (*m*): radish (*Raphanus sativus*) 541
RETTICHÖL (*n*): radish oil (seeds of *Raphanus sativus*) 542
RETTIG (*m*): (see **RETTICH**) 543
RETTUNG (*f*): delivery; rescue; salvage; escape; recovery; help 544
RETTUNGSAPPARAT (*m*): rescue apparatus 545
RETTUNGSBOJE (*f*): life buoy 546
RETTUNGSBOOT (*n*): lifeboat 547
RETTUNGSGÜRTEL (*m*): life belt 548
RETTUNGSLOS: irretrievable; irremediable 549
RETTUNGSMITTEL (*n*): resource; remedy; expedient 550
RETTUNGSRING (*m*): life buoy, life-belt 551
REUE (*f*): penitence; regret; repentance; remorse; contrition; sorrow 552
REUEN: to repent; regret; rue; be sorry-(for) 553
REUGELD (*n*): forfeit; penalty 554
REUIG: repentant; rueful; contrite; remorseful; penitent 555
REUMÜTIG: (see **REUIG**) 556
REUSE (*f*): weir-basket 557
REUTE (*f*): mattock; hoe; clearing (rooting) out 558
REUTEN: to dig (root) out; hoe 559
REUTER (*m*): sieve; riddler; riddle; screen 560
REUTERN: to screen; riddle; sieve 561
REVANCHE (*f*): revenge 562
REVANCHIEREN: to revenge, exact or obtain (revenge, indemnification, satisfaction); avenge, vindicate, retaliate, requite, take vengeance 563
REVERBERIEREN: to reverberate; reflect; radiate (of heat) 564
REVERBERIEROFEN (*m*): reverberatory furnace 565
REVERS (*m*): reverse; undertaking (in writing); obverse 566
REVERSIBEL: reversible (also of a colloid which can be transformed from a sol to a gel by the action of warmth, or from a gel to a sol by the action of water; see **SOL** and **GEL**) 567
REVERSIERWALZWERK (*n*): return (reversible) mill, roll or (rolling)-train 568
REVIDIEREN: to revise; examine; inspect (by the Customs officials); overhaul 569
REVIER (*n*): district, quarter, preserve, region, ward, circuit, territory, province 570
REVISION (*f*): overhauling (Machinery), examination (Customs); inspection (Customs, Boiler, etc.); review, revision 571
REVISIONSTÜR (*f*): inspection (examination) door 572
REVISOR (*m*): examiner, inspector, auditor 573
REVOLTIEREN: to revolt, rise, rebel 574
REVOLUTION (*f*): revolution, rebellion, mutiny, insurrection, revolt, rising 575
REVOLUTIONÄR: revolutionary, insurrectionary, rebellious, mutinous, insubordinate; (*m*): revolutionary 576

REVOLUTIONIEREN: to revolt; revolutionize 577
REVOLVEBLENDE (*f*): wheel (rotary) diaphragm (Phot.) 578
REVOLVERBANK (*f*): turret lathe 579
REVOLVERDREHBANK (*f*): revolver (turret) lathe 580
REVOZIEREN: to revoke, recant, retract, repeal, recall, annul, cancel, rescind, countermand, abolish, abrogate, make void 581
REVUE (*f*): review 582
REZENSENT (*m*): reviewer; critic 583
REZENSIEREN: to review; criticize 584
REZENSION (*f*): review; criticism; revision; critique 585
REZEPT (*n*): recipe; prescription; receipt; quittance 586
REZEPTIVITÄT (*f*): receptivity 587
REZEPTUR (*f*): receivership; dispensing, (of medicines) 588
REZESS (*m*): recess; treaty; arrangement; compact; arrears; (outstanding)-balance 589
REZIPE (*n*): recipe, prescription 590
REZIPIENT (*m*): receiver; recipient 591
REZIPIEREN: to receive 592
REZIPROK: reciprocal-(ly); converse-(ly) 593
REZIPROKER WERT (*m*): reciprocal value 594
RHABARBER (*m*): rhubarb (*Rheum officinale* or *Rheum palmatum*) 595
RHABARBER EXTRAKT (*m*): rhubarb extract 596
RHABARBERGELB (*n*): chrysophanic acid (see CHRYSOPHANSÄURE) 597
RHABARBERSÄURE (*f*): (see RHABARBERGELB) 598
RHABARBERWEIN (*m*): rhubarb wine 599
RHABARBERWURZEL (*f*): rhubarb root (*Radix rhei* of *Rheum officinale* or *Rheum palmatum*) 600
RHABDIT(*m*): rhabdite; $(Fe,Ni,Co)_3P$; Sg. 7.15 601
RHABDOPHAN (*m*): rhabdophane; $(Er,La,Y)PO_4.H_2O$; 602
RHACHITIS (*f*): rachitis, rickets (Med.) 603
RHAETIZIT (*m*): rhaetizite (a form of disthene); Al_2SiO_4; Sg. 3.63 604
RHAMNIN (*n*): rhamnin (a bitter principle obtained from the fruit of the purging buckthorn, *Rhamnus cathartica*) 605
RHAMNOKATHARTIN (*n*): rhamno-cathartin (a bitter principle obtained from the fruit of the purging buckthorn, *Rhamnus cathartica*) 606
RHAMNOSE (*f*): rhamnose 607
RHAPONTIKWURZEL (*f*): rhapontic root (*Radix rhapontici* of *Rheum rhaponticum*) 608
RHEDE (*f*): roadstead (Naut.) (see also REED-) 609
RHEDER (*m*): (see REEDER) 610
RHEIN (*m*): Rhine (the river); (*n*): dioxymethylanthraquinone (see DIOXYMETHYL-ANTHRACHINON) 611
RHEINGAUER ERDE (*f*): yellow ochre (see BOLUS, WEISSER-) 612
RHEINSÄURE (*f*): rheic (chrysophanic) acid (see CHRYSOPHANSÄURE) 613
RHEINWEIN (*m*): Rhine (Rhenish) wine; hock 614
RHEOSTAT (*m*): rheostat 615
RHETORIK (*f*): rhetoric-(s) 616
RHETORISCH: rhetorical 617
RHEUMATIN (*n*): rheumatine (see SALIZYLCHININ, SALIZYLSAURES-) 618
RHEUMATISCH: rheumatic 619
RHEUMATISMUS (*m*): rheumatism (Med.) 620
RHEUMATISMUSÖL (*n*): anti-rheumatic oil 621
RHISSANFARBE (*f*): rhissane colour 622
RHIZOMORPH: rhizomorphus, root-like 623
RHIZOPODEN (*f.pl*): rhizopoda (see also HELIO-ZOEN) (Med.) 624
RHODAFORM (*n*): rhodaform, methylhexamethylenetetramine rhodanide 625
RHODAMIN (*n*): rhodamine 626
RHODAN (*n*): thiocyanogen; sulphocyanogen; (as a prefix = thiocyanate of; thiocyano-; rhodanite; -sulfocyanide; -sulphocyanide; etc.) (see RHODANID and RHODANÜR) 627
RHODANALUMINIUM (*n*): aluminium thiocyanate (see ALUMINIUMRHODANÜR) 628
RHODANAMMONIUM (*n*): ammonium thiocyanate (see AMMONIUMRHODANÜR) 629
RHODANAT (*n*): thiocyanate; sulphocyanate (see RHODANÜR) 630
RHODANBARYUM (*n*): barium thiocyanate (see BARYUMRHODANÜR) 631
RHODANEISEN (*n*): iron (ferric) thiocyanate 632
RHODANEISENROT (*n*): ferric thiocyanate 633
RHODANEIWEISSVERBINDUNG (*f*): thiocyanogen-albumen compound 634
RHODANID (*n*): (an -ic-) thiocyanate, sulphocyanate, sulfocyanate, rhodanide, sulphocyanide or sulfocyanide 635
RHODANION (*n*): thiocyanogen ion (CNS′) 636
RHODANKALIUM (*n*): potassium (thiocyanate) rhodanide (see KALIUMRHODANÜR) 637
RHODANKALZIUM (*n*): calcium (thiocyanate) rhodanide (*Calcium rhodanatum*); $Ca(CNS)_2.3H_2O$ 638
RHODANKUPFER (*n*): cuprous (thiocyanate) rhodanide (see KUPFERRHODANÜR) 639
RHODANLÖSUNG (*f*): thiocyanate solution 640
RHODANMETALL (*n*): metallic thiocyanate 641
RHODANMETHYL (*n*): methyl thiocyanate 642
RHODANNATRIUM (*n*): sodium thiocyanate (see NATRIUMRHODANÜR) 643
RHODANNICKEL (*m*): nickel thiocyanate 644
RHODANSALZ (*n*): thiocyanate; sulphocyanate; sulfocyanide; sulphocyanide; rhodanide 645
RHODANSINAPIN (*n*): sinapin rhodanate; $C_{16}H_{23}NO_5HCNS$ 646
RHODANTONERDE (*f*): aluminium thiocyanate (see ALUMINIUMRHODANÜR) 647
RHODANÜR (*n*): (an -ous) thiocyanate, sulphocyanate, rhodanide, or sulphocyanide 648
RHODANVERBINDUNG (*f*): thiocyanogen compound (see RHODANÜR and RHODANID) 649
RHODANWASSERSTOFF (*m*): hydrogen thiocyanate; thiocyanic (sulphocyanic) acid; HSCN 650
RHODANWASSERSTOFFSÄURE (*f*): thiocyanic (sulphocyanic) acid; HSCN 651
RHODANZINN (*n*): stannic thiocyanate 652
RHODANZINNOXYD (*n*): stannic thiocyanate 653
RHODIN (*n*): rhodine 654
RHODINASÄURE (*f*): rhodinic acid 655
RHODINOL (*n*): rhodinol (similar to geraniol; a terpene alcohol obtained from rose oil) 656
RHODIUM (*n*): Rhodium (Rh) 657
RHODIUMCHLORWASSERSTOFFSÄURE (*f*): acid with rhodium, chlorine and hydrogen content; H_4RhCl_7 and H_3RhCl_6 658
RHODIUMSALZ (*n*): rhodium salt 659
RHODOCHROSIT (*m*): rhodochrosite (manganese protocarbonate); $MnCO_3$; Sg. 3.04—3.6 660
RHODONIT (*m*): rhodonite, manganese spar (Manganese metasilicate); $(Ca,Fe,Mn)(SiO_3)$; Sg. 3.6 (see also MANGANSPAT and KIESEL-MANGAN) 661

RHODULINBLAU (*n*): rhoduline blue 662
RHODULINROT (*n*): rhoduline red 663
RHOMBARSENIT (*m*): rhombarsenite, claudetite (see CLAUDETIT) 664
RHOMBENFÖRMIG: rhombic; rhomb-shaped 665
RHOMBISCH: rhombic 666
RHOMBOEDER (*n*): rhombohedron 667
RHOMBOEDERFÜLLSTEINE (*m.pl*): rhomboid filling stones 668
RHOMBOEDRISCH: rhombohedral 669
RHOMBOIDISCH: rhomboid; rhomboidal 670
RHOMBUS (*m*): rhombus 671
RHÖNIT (*m*): rhoenite; $Ca_3,Na_6K_6Al_4Mg_4Fe_2Si_6Ti_6O_{30}$; Sg. 3.55 672
RHUSESTER (*m*): rhus ester 673
RHUSLACK (*m*): Rhus varnish 674
RHUSOL (*n*): rhusol 675
RHUSOLAT (*n*): rhusolate 676
RHUSOLIN (*n*): rhusoline 677
RHUSOLLACK (*m*): rhusol varnish 678
RHUSOL, LEINÖLSAURES- (*n*): rhusol linoleate 679
RHUSOLLINOLEAT (*n*): rhusol linoleate 680
RHYTHMISCH: rhythmic-(al) 681
RHYTHMUS (*m*): rhythm 682
RICHTAPPARAT (*m*): trueing (straightening) apparatus 683
RICHTBAUM (*m*): pulley-beam 684
RICHTBLEI (*n*): plummet; plumb line; plumb-bob 685
RICHTE (*f*): direction; short cut 686
RICHTEN: to direct; address; set; adjust; adapt; turn; execute; arrange; dress; straighten; level-(up); true; make true; erect; raise; rise; settle; criticize; arbitrate; tend; (incline)-toward; bend; align; try; condemn; · judge; pass sentence; conform; agree; guide; umpire; carry out; accommodate oneself; put into practice; order; regulate; correct; aim at; point; trim (Sails); execute (a criminal); put (Questions) 687
RICHTEN, SICH- AUF: (see RICHTEN, SICH DANACH-) 688
RICHTEN, SICH DANACH-: to make one's arrangements (to suit) accordingly; prepare (make preparations) for; be guided by 689
RICHTEN, ZU GRUNDE-: to ruin, destroy 690
RICHTER (*m*): judge; justice; magistrate; arbiter; arbitrator; referee; umpire 691
RICHTERLICH: judicial-(ly) 692
RICHTIG: right; exact; just; true; fair; correct-(ly); regular; accurate; straight; certainly; sure enough 693
RICHTIG SITZEN: to sit true 694
RICHTIG WERDEN: to come to hand (of letters, etc.); come to (an understanding, an agreement) with 695
RICHTKORN (*n*): sight (of a rifle) 696
RICHTLATTE (*f*): rule; batten 697
RICHTLINIEN (*f.pl*): correct (directional) lines; most suitable methods; regulations; guide lines 698
RICHTLOT (*n*): (see RICHTBLEI) 699
RICHTMASCHINE (*f*): trueing (straightening) machine 700
RICHTMASS (*n*): standard; gauge; ruler 701
RICHTPLATTE (*f*): adjusting (straightening; trueing) table; trueing block; trueing plate 702
RICHTPLATZ (*m*): place of execution 703
RICHTSCHEIT (*n*): straight-edge; rule; level; ruler; batten 704
RICHTSCHNUR (*f*): plumb (chalk) line; level; rule (of conduct) 705
RICHTSCHRAUBE (*f*): adjusting screw 706
RICHTSPANT (*n*): principal (chief) frame 707
RICHTUNG (*f*): direction; bearing; aim; adjusting; course (see RICHTEN) 708
RICHTUNG (*f*) ANGEBEN: to take the lead, point the way, give the cue 709
RICHTUNG, GERADE- (*f*): alignment 710
RICHTUNGSÄNDERUNG (*f*): change of direction 711
RICHTUNGSLINIE (*f*): direction (guide) line 712
RICHTUNGSWINKEL (*m*): angle of elevation 713
RICHTWAGE (*f*): level 714
RICINELAIDINSÄURE (*f*): ricinelaidic acid (stereoisomer of ricinoleic acid; see RICINOLSÄURE); Mp. 52.5°C. 715
RICINOLSÄURE (*f*): ricinoleic acid; $C_{17}H_{32}(OH)COOH$; Sg. 0.945; Mp. 4.5°C.; Bp. 250°C. 716
RICINSÄURE (*f*): ricinic acid (isomer of ricinoleic acid, see RICINOLSÄURE); Mp. 81.5°C. 717
RICINUS (*m*): castor-oil plant, Palma Christi; oil plant (*Ricinus communis*) 718
RICINUSMEHL (*n*): castor flour 719
RICINUSÖL (*n*): castor-oil; ricinus oil (*Oleum ricini*); Sg. 0.915-0.965 (from seeds of castor-oil plant) 720
RICINUSÖLPRÄPARAT (*n*): castor-oil preparation 721
RICINUSÖLSÄURE (*f*): ricinoleic acid (see RICINOLSÄURE) (from saponification of castor oil) 722
RICINUSSAMEN (*m*): castor bean; mexico seed; ricinus (seeds of *Ricinus communis*) 723
RICKARDIT (*m*): rickardite; Cu_4Te_3; Sg. 7.54 724
RIEBECKIT (*m*): riebeckite; $NaFeSiO_3$; Sg. 3.4 725
RIECHBEIN (*n*): ethmoid-(bone) (Anat.) 726
RIECHEN: to scent; smell; perceive 727
RIECHEND: smelling; strong; redolent; odoriferous; odourous; perfumed; fragrant; aromatic; odorant; high (of meat) 728
RIECHEN NACH: to smell of 729
RIECHESSIG (*m*): aromatic vinegar 730
RIECHFLÄSCHCHEN (*n*): smelling bottle 731
RIECHMITTEL (*n*): scent 732
RIECHNERV (*m*): olfactory nerve (Anat.) 733
RIECHSALZ (*n*): smelling salts 734
RIECHSTOFF (*m*): perfume; scent 735
RIECHSTOFFANLAGE (*f*): perfume (scent) manufacturing plant 736
RIECHSTOFF (*m*), **BLÜTEN-**: flower perfume 737
RIECHSTOFFE, SYNTHETISCHE- (*m.pl*): synthetic perfumes and scents 738
RIECHSTOFF (*m*), **KÜNSTLICHER-**: artificial (synthetic) perfume 739
RIECHWAREN (*f.pl*): scents; perfumes 740
RIECHWASSER (*n*): scented water 741
RIED (*n*): reed; moor; marsh (see RIET) 742
RIEDGRAS (*n*): reed; sedge 743
RIEFE (*f*): groove; channel; furrow; fluting; chamfer; rifling; scoring 744
RIEFELN: (see RIEFEN) 745
RIEFEN: to score, scratch, cut, groove, rifle, chamfer, channel, flute 746
RIEFIG: grooved, channeled, pitted, fluted, furrowed, rifled, scored, scratched, chamfered 747
RIEGEL (*m*): cross bar; bar (of soap, etc.); rail; nogging piece; cross (beam) tie; latch; bolt; lock; sliding-bolt-(of doors); obstacle 748

RIEGELFEDER (*f*): lock spring 749
RIEGELHAKEN (*m*): latch; staple 750
RIEGELHOLZ (*n*): middle length timber; cross-bar (of a door) 751
RIEGELN: to lock; bolt; latch; bar 752
RIEGELWAND (*f*): framework; wall; partition 753
RIEGELWERK (*n*): framework 754
RIEMEN (*m*): fillet; beading (narrow floor board); strap; string; lace; thong; belt-(ing); band; brace; oar-(of boats) 755
RIEMEN: to row (a boat) 756
RIEMENANTRIEB (*m*): belt drive 757
RIEMENAUSRÜCKER (*m*): belt disengaging gear 758
RIEMENAUSRÜCKVORRICHTUNG (*f*): belt shifting arrangement, belt disengaging device, belt (shifter) disengager, belt adjuster 759
RIEMENBETRIEB (*m*): belt gearing, belt pulleys and shafting 760
RIEMENBLUME (*f*): yellow berried mistletoe (*Loranthus europæus*) (see also **MISTEL**) 761
RIEMENBLUMENGEWÄCHSE (*n.pl*): *Loranthaceæ* (Bot.) 762
RIEMENDOLLE (*f*): rowlock 763
RIEMENFETT (*n*): belt (fat; dressing) grease; lubricant for driving belts 764
RIEMENFÜHRER (*m*): belt guide 765
RIEMENFUSSBODEN (*m*): parquet (inlaid) flooring 766
RIEMENGABEL (*f*): belt guide, belt fork 767
RIEMEN (*m*), **GEKREUZTER-**: crossed belt 768
RIEMENGESCHWINDIGKEIT (*f*): belt velocity 769
RIEMENGRIFF (*m*): handle of an oar 770
RIEMENKALK (*m*): cyanite 771
RIEMENKLAMPEN (*f.pl*): rowlocks (Naut.) 772
RIEMENKRALLE (*f*): belt fastener 773
RIEMEN (*m*), **LAUF-**: (driving)-belt 774
RIEMENLEISTUNG (*f*): belt (capacity) duty 775
RIEMENLEITER (*m*): belt guide 776
RIEMENSCHAFT (*m*): handle of an oar 777
RIEMENSCHEIBE (*f*): belt pulley; sheave 778
RIEMENSCHEIBE (*f*), **FESTE-**: fixed pulley 779
RIEMENSCHEIBE (*f*), **LOSE-**: loose pulley 780
RIEMENSCHLAG (*m*): stroke-(of an oar) 781
RIEMENSCHRAUBE (*f*): belt fastener 782
RIEMENSPANNER (*m*): belt stretcher, belt tightener 783
RIEMENSPANNUNG (*f*): belt tension 784
RIEMENSPANNVORRICHTUNG (*f*): belt tightener, tightening arrangement or device 785
RIEMENSTELLVORRICHTUNG (*f*): belt shifting arrangement, belt disengaging device, belt (shifter) disengager, belt adjuster 786
RIEMENSTÜCK: queen closer (Brick) 787
RIEMEN (*m*), **TREIB-**: (driving)-belt 788
RIEMENTRIEB (*m*): belt (drive) driving, flexible drive 789
RIEMENVERBINDER (*m*): belt fastener 790
RIEMENWELLE (*f*): belt shaft 791
RIEMENWERK (*n*): straps; belting; harness 792
RIEMENZEUG (*n*): (see **RIEMENWERK**) 793
RIEMER (*m*): saddler, harness-maker 794
RIEMSCHEIBE (*f*): (see **RIEMENSCHEIBE**) 795
RIES (*n*): ream (of paper) 796
RIESE (*m*): giant; colossus 797
RIESEL (*m*): drizzle; trickle; ripple; rippling; sleet 798
RIESELANLAGE (*f*): irrigation plant; cooling plant, cooling tower (in which a liquid is cooled by trickling) 799
RIESELFELD (*n*): irrigation field (irrigated with sewage); (field of a) sewage farm 800
RIESELFLÄCHE (*f*): irrigation (irrigating) surface (of a sewage farm) 801
RIESELGUT (*n*): irrigation (sewage) farm 802
RIESELJAUCHE (*f*): sewage (for irrigation and fertilization purposes) 803
RIESELN: to ripple; purl; trickle; to cool by trickling (of water); drizzle; percolate; (*n*): trickling, etc. 804
RIESELTURM (*m*): (water)-cooling tower 805
RIESELWASSER (*n*): sewage-water, irrigation-water, percolated (drainage) water 806
RIESELWIESE (*f*): (see **RIESELFELD**) 807
RIESENHAFT: gigantic, colossal, enormous, immense, huge, monstrous, vast, cyclopean, herculean, prodigious 808
RIESENMÄSSIG: gigantic, colossal, enormous (see **RIESENHAFT**) 809
RIESENSCHILDKRÖTE (*f*): giant tortoise (*Chelonia mydas* and *Chelonia cahouana*) 810
RIESENSCHLANGE (*f*): python; anaconda; boa constrictor 811
RIESENWUCHS (*m*): macrosomatia (enlargement of the whole body) (Med.) 812
RIET (*n*): reed; moor; marsh (see **RIED**) 813
RIETGRAS (*n*): sedge 814
RIFF (*n*): reef; ridge 815
RIFFEL (*f*): polisher, rasp; corrugation, undulation, ripple, wave; flax-comb; chequering; rib; (*m*): chiding, scolding, reprimand 816
RIFFELBILDUNG (*f*): formation of (corrugations, undulations) waves 817
RIFFELBLECH (*n*): chequer plating, chequered plate; corrugated plate; corrugated sheet-(ing) 818
RIFFELGLAS (*n*): ribbed glass 819
RIFFELLÄNGE (*f*): length of (wave, corrugation) undulation 820
RIFFELN: to rib, corrugate, chequer, undulate, impart waves to -(tram rails, etc.); ripple, file, comb flax; rasp, polish; burnish; channel, flute, groove (see also **RIEFEN**); reprimand, scold, chide, upbraid 821
RIFFELWALZE (*f*): grooved roll 822
RIGORÖS: rigorous, severe, rigid 823
RIKOSCHETIEREN: to ricochet, glance off 824
RILLE (*f*): furrow; groove; chamfer; channel 825
RIMESSE (*f*): remittance, bill receivable (Com.) 826
RIND (*n*): neat; ox; (*pl*): cattle (with horns) 827
RINDE (*f*): rind; bark (*Cortex*); crust; scab 828
RINDENGEWEBE (*n*): cortical tissue 829
RINDENSCHICHT (*f*): cortical layer 830
RINDENTIER (*n*): crustacean 831
RINDERBRATEN (*m*): roast beef 832
RINDERFETT (*n*): beef (fat) suet (see **RINDFETT**) 833
RINDERMARKFETT (*n*): beaf-marrow-fat 834
RINDERTALG (*m*): beef (tallow) fat (see **RINDFETT**) 835
RINDFÄLLIG: peeling; barking; shedding bark 836
RINDFETT (*n*): beef (suet) fat (Sg. 0.92) 837
RINDFLEISCH (*n*): beef 838
RINDFLEISCHBRÜHE (*f*): beef-tea 839
RINDIG: crusty, scabby, covered with bark 840
RINDSBLUT (*n*): ox-blood 841
RINDSFETT (*n*): beef suet 842

RINDSGALLE (*f*): ox gall 843
RINDSHAUT (*f*): neat's (bullock's) hide 844
RINDSKLAUENFETT (*n*): neat's-foot oil 845
RINDSLEDER (*n*): neat's leather 846
RINDSTALG (*m*): beef (fat) tallow 847
RINDSVIEH (*n*): cattle 848
RINDSVIEHZUCHT (*f*): cattle breeding 849
RINDSZUNGE (*f*): ox-tongue 850
RING (*m*): ring; circle; band; hoop; flange; collar; coil; link; syndicate; trust; cycle (Mech.); halo 851
RINGALKOHOL (*m*): cyclic alcohol 852
RINGAMIN (*n*): cyclic amine 853
RINGBAHN (*f*): circular railway 854
RINGBILDUNG (*f*): ring formation 855
RINGBOLZEN (*m*): ring bolt 856
RINGDYNAMO (*f*): ring dynamo 857
RINGEL (*m*): ring-(let); circle; curl 858
RINGELBLUME (*f*): marigold (*Flores calendulæ*) 859
RINGELIG: annular; ring-like 860
RINGELN: to ring; curl; girdle; encircle; blaze (Trees) 861
RINGELTAUBE (*f*): ring-dove 862
RINGEN: to wrest-(le); wring; ring; curl; struggle; strive after; grapple-(with); (*n*): wrestling, etc. 863
RINGER (*m*): wrestler 864
RINGE, RASCHIG'S- (*m.pl*): (Doctor)-Rashig's rings (filling material for absorption towers, reaction vessels, distillers, dust extraction plants, etc.) 865
RINGFÖRMIG: ring-shaped; annular; cyclic 866
RINGHAKEN (*m*): ring hook 867
RINGKETON(*n*): cyclic ketone 868
RINGLAGER (*n*): ring bearing 869
RINGLEITUNG (*f*): ring main (of pipes) 870
RINGLOTTE (*f*): Italian plum (*Prunus italica*) 871
RINGMAUER (*f*): enclosing wall; city wall 872
RINGOFEN (*m*): circular (annular) kiln; rotary furnace 873
RINGROST (*m*): circular grate; annular grid 874
RINGS: around 875
RINGSCHICHT (*f*): annular layer 876
RINGSCHLITZ (*m*): annulus 877
RINGSCHMIERLAGER (*n*): ring lubricated bearing 878
RINGSCHMIERUNG (*f*): ring lubrication 879
RINGSPALT (*m*): annular clearance; annulus 880
RINGSPANNUNG (*f*): ring tension; circumferential (peripheral) pressure, tension or stress 881
RINGSPINNMASCHINE (*f*): ring spinning machine (Textile) 882
RINGSUM: around; round about 883
RINGSYSTEM (*n*): cyclic (ring) system 884
RINGUNGESÄTTIGT: containing an unsaturated ring; cyclically unsaturated (Chem.) 885
RINGVERBINDUNG (*f*): ring (cyclic) compound 886
RINGWICKELUNG (*f*): ring winding 887
RINGZAPFEN (*m*): ring pivot 888
RINKIT (*m*): rinkite; $(Si,Ti)O_2,(Ca,Ce,La)O, Na_2OF$; Sg. 3.46 889
RINMANNS GRÜN (*n*): Rinmann's green (see KOBALTGRÜN) 890
RINNE (*f*): channel; canal; furrow; groove; gutter (Building or Street); drain; sewer; trench; gully 891
RINNEIT (*m*): rinneit (Mineralogy); Sg. 2.34 892
RINNEN: to run; leak; flow; drip; drop; drain 893
RINNENEISEN (*n*): trough shaped section iron 894
RINNENFÖRMIG: channeled; fluted; grooved; furrowed 895
RINNSTEIN (*m*): sink; gutter-(stone) 896
RIPIDOLITH (*m*): ripidolite (see CHLORIT) 897
RIPPE (*f*): rib; timber (Building and Marine); web (Metal); corrugation 898
RIPPEN: to rib; flute; cord; frill; corrugate; (as a prefix = ribbed; fluted; grooved; costal; intercostal (Med.) 899
RIPPENBRUCH (*m*): fracture of a rib (Med.) 900
RIPPENFELL (*n*): (costal)-pleura (Anat.) 901
RIPPENFELLENTZÜNDUNG (*f*): inflammation of the pleura; pleurisy (Med.) 902
RIPPENFÜHRUNG (*f*): valve guide 903
RIPPENHAUT (*f*): (costal)-pleura (Anat.) 904
RIPPENKNORPEL (*m*): costal cartilage (Anat.) 905
RIPPENPULSADER (*f*): intercostal artery (Anat.) 906
RIPPENROHR (*n*): gilled (tube) pipe; ribbed (tube) pipe 907
RIPPENTRICHTER (*m*): ribbed (fluted) funnel 908
RIPPENWEH (*n*): stitch; pain in the side 909
RISIGALLUM (*n*): realgar 910
RISIKO (*n*): risk; peril; speculation; chance 911
RISKIEREN: to risk; speculate; chance; imperil; take a risk 912
RISÖRIT (*m*): risorit; $(Nb,Ta)_2O_5,(Ce,La,Er,Nd,Y)_2O_3,TiO_2,CaO$; Sg. 4.2 913
RISPE (*f*): panicle (Bot.) 914
RISS (*m*): fissure; cleft; crack; crevice; tear; gap; flaw; chink; rent; draft; design; drawing; sketch; outline; lash; projection (Geom.); shake (crack in tree-wood); fracture; furrow; lines (of a ship); laceration; plan; elevation; breath; stroke; blank 915
RISSE (*m.pl*): rupture, hernia (Med.) 916
RISSEBENE (*f*): plane of projection 917
RISSIG: fissured; rent; cracked; torn; full of flaws or cracks; sprung (of wood) 918
RISSSCHLAGZAHL (*f*): number of strokes to cause a crack to appear (Metal testing) 919
RISSTAFEL (*f*): plane of projection 920
RISSVERFAHREN (*n*): method of projection 921
RISSWUNDE (*f*): lacerated wound (Med.) 922
RIST (*m* and *n*): withers (Vet.); instep; wrist 923
RITT (*m*): ride; riding 924
RITTER (*m*): knight; cavalier 925
RITTERLICH: knightly; valiant; brave; chivalrous 926
RITTERMÄSSIG: (see RITTERLICH) 927
RITTERSPORN (*m*): larkspur (*Delphinium consolida, Delphinium pubescens, Delphinium staphisagria*) 928
RITTERSPORNBLÜTEN (*f.pl*): larkspur flowers (*Flores calcatrippæ*) 929
RITTINGERSCHAUFEL (*f*): Rittinger's blade (a blade ending radially) 930
RITTMEISTER (*m*): captain (of cavalry); riding master 931
RITUS (*m*): rite 932
RITZ (*m*): rift; crevice; crack; fissure; cleft; slit; tear; gap; flaw; scratch; chink 933
RITZE (*f*): (see RITZ) 934
RITZEL (*n*): pinion-(wheel); gear-(ed) wheel; toothed wheel; driver 935
RITZEL (*n*), ROHHAUT-: raw-hide pinion 936

RITZELWELLE (*f*): pinion shaft 937
RITZEN: to crack; scratch; slit; split; tear; scar; score; cut; grave in; engrave; etch; notch; fissure; lance; graze 938
RITZIG: scarred; scored; scratched 939
RITZMESSER (*n*): lancet 940
RIVALISIEREN: to rival, emulate, oppose, match, equal 941
RIZINUSÖL (*n*): castor oil (see **RICINUSÖL**) 942
ROBBA (*m* and *f*): seal (*Phoca*) 943
ROBBENFANG (*m*): sealing; seal-hunting 944
ROBBENFELL (*n*): seal-skin 945
ROBBENTRAN (*m*): seal oil; Sg. 0.925; Mp. 22-33°C. 946
ROBERTSKRAUT (*n*): herb Robert, stinking cranesbill (*Geranium robertianum*) 947
ROBINIAÖL (*n*): robinia oil 948
ROBURIT (*n*): roburite (a safety explosive; ammonium nitrate, dinitrobenzene, etc.) 949
ROCHE (*m*): ray; roach (Fish) (see **ROCHEN**); castle; rook (Chess piece) 950
ROCHELLESALZ (*n*): Rochelle salt; Seignette salt; potassium sodium tartrate (see **KALIUM-NATRIUMTARTRAT**) 951
RÖCHELN: to rattle-(in the throat); (*n*): rattling; rale; snoring; stertor (Med.); death-rattle 952
ROCHEN (*m*): ray; roach (Fish) (*Raja clavata, Raja batis* and *Trigon pastinaca* and *Leuciscus rutilus*); (as a verb = to castle (Chess)) 953
ROCHENLEBERÖL (*n*): roach (ray, thornback) liver oil (from *Trigon pastinaca, Raja batis* or *Raja clavata*) 954
ROCHENLEBERTRAN (*m*): (see **ROCHENLEBERÖL**) 955
ROCHIEREN: to castle (Chess) 956
ROCK (*m*): coat; robe; gown; skirt; petticoat; frock; dress 957
ROCKSCHOSS (*m*): coat-tail 958
RODEN: to root out; clear; grub-(up) 959
RODINAL (*n*): rodinal (a photographic developer with para-aminophenol as the developing principle) 960
ROGEN (*m*): (hard)-roe; spawn (of fish) 961
ROGENFISCH (*m*): female fish 962
ROGENSTEIN (*m*): oolitic lime; oolite; roe-stone (see **OOLITH**) 963
ROGENSTEINARTIG: oolitic 964
ROGGEN (*m*): rye (*Secale cereale*) 965
ROGGENBROT (*n*): rye bread 966
ROGGENDISTEL (*f*): eryngo: field eryngo (*Eryngium campestre*) 967
ROGGENKORNBRAND (*m*): rye smut (*Ustilago secalis*) (Bot.) 968
ROGGENMEHL (*n*): rye flour 969
ROGGENMUTTER (*f*): ergot (of rye); spurred rye (*Secale cornutum*) (of *Claviceps purpurea*) 970
ROGGENÖL (*n*): rye oil (from *Secale cereale*) 971
ROH: crude; rough; raw; coarse; gross; vulgar; unbleached (Linen); native (Min.); untanned (Leather); rude 972
ROHARBEIT (*f*): ore smelting (Metal) 973
ROHAUSCHNITT (*m*): crude tallow (obtained from the fatty tissue of animals, containing blood, muscles, etc.) 974
ROHBENZOL (*n*): crude benzol (benzole or benzene) (see **BENZOL**) 975
ROHBRAUNKOHLE (*f*): rough (brown coal) lignite (more earthy than "**BRAUNKOHLE**"; about 50-60% moisture and calorific value 2000/2300 cals.) 976
ROHEINNAHME (*f*): gross receipts 977
ROHEISEN (*n*): pig (crude) iron; Mp. 1000-1200°C. 978
ROHEISEN (*n*), **GRAUES-**: grey pig iron (grey fracture, slowly cooled, melting point 1200-1250°C. specific weight 7.0-7.3) 979
ROHEISEN, HALBIERTES- (*n*): mottled pig iron 980
ROHEISEN (*n*), **KALTERBLASENES-**: cold blast pig iron (300-400°C. air temperature) 981
ROHEISEN (*n*), **WARMERBLASENES-**: hot blast pig iron (600-800°C. air temperature) 982
ROHEISEN (*n*), **WEISSES-**: white pig iron [white fracture, quickly cooled (quenched), melting pt. 1100-1130°C., specific weight 7.5-7.8] 983
ROHERTRAG (*m*): gross (receipts, returns) profit 984
ROHERZ (*n*): raw (rough) ore 985
ROHERZEUGNIS (*n*): raw product 986
ROHFASERBESTIMMUNG (*f*): determination of crude fibre 987
ROHFRISCHEN (*n*): first (grey iron) refining or fining (Metal) 988
ROHFRISCHPERIODE (*f*): (re)-fining period or stage; boil-(ing) stage (Iron) 989
ROHFRUCHT (*f*): unmalted grain (Brewing) 990
ROHGANG (*m*): incomplete reduction and poor operation of blast furnace 991
ROHGLAS (*n*): raw (coarse) glass 992
ROHGUMMIMILCH (*f*): rubber latex 993
ROHGUSS (*m*): rough-(iron)-casting 994
ROHHAUT (*f*): raw (skin) hide 995
ROHHAUTGETRIEBE (*n*): raw-hide gearing, raw-hide pinion 996
ROHHAUTRITZEL (*n*): raw-hide pinion 997
ROHHAUTVORGELEGE (*n*): raw-hide gearing 998
ROHHEIT (*f*): roughness; crudeness, etc. (see **ROH**) 999
ROHIGKEIT (*f*): (see **ROHHEIT**) 000
ROHKERN (*m*): crude tallow (the fatty portions of animals); Sg. 0.943-0.952; Mp. 40-48.5°C. 001
ROHKNÜPPEL (*m*): rough billet (Metal) 002
ROHKOLLODIUM (*n*): plain collodion (Phot.) 003
ROHKUPFER (*n*): raw copper 004
ROHMATERIAL (*n*): raw material 005
ROHMETALL (*n*): crude (raw) metal 006
ROHMONTANWACHS (*n*): crude montan wax (see **MONTANWACHS**) 007
ROHNAPHTA (*f*): crude (petroleum) naphtha (see **NAPHTA**) 008
ROHNAPHTALIN (*n*): crude naphthalene 009
ROHNICKEL (*n*): (see **WÜRFELNICKEL**) 010
ROHOFEN (*m*): ore furnace 011
ROHÖL (*n*): crude oil 012
ROHPAPIER (*n*): plain paper 013
ROHPAPPE (*f*): raw board (Paper) 014
ROHPARAFFIN (*n*): crude paraffin 015
ROHPETROLEUM (*n*): crude (oil) petroleum 016
ROHPIKRINSÄURE (*f*): crude picric acid 017
ROHPLATIN (*n*): crude platinum (grains of platinum separated from the ore by washing) 018
ROHPRODUKT (*n*): raw product, raw material 019
ROHR (*n*): tube; pipe; flue; barrel; reed; cane, etc. (see **RÖHRE**) 020
ROHRABSCHNEIDER (*m*): tube cutter, pipe cutter 021
ROHRABSCHNITT (*m*): section of a tube, piece cut off a tube 022
ROHRAMMER (*f*): reed-bunting, water-sparrow, reed-sparrow, chuck, king bird, ring-bunting (*Emberiza schœniclus*) (Ornithology) 023
ROHRANORDNUNG (*f*): pipe (piping) arrangement or lay-out 024

ROHRANSCHLUSS (*m*): pipe connection 025
ROHRARMATUREN (*f.pl*): tube (accessories) fittings 026
ROHRAUSGLEICHER (*m*): pipe compensator, compensating pipe 027
ROHRBATTERIE (*f*): section (bank, nest) of tubes 028
ROHRBEFESTIGUNG (*f*): pipe fastening, pipe support, hanger, clip, etc. 029
ROHRBLATT (*n*): reed (Weaving) 030
ROHRBRUCH (*m*): pipe burst 031
ROHRBRUCHVENTIL (*n*): isolating (stop) valve; steam-pipe (main) isolating or stop valve 032
ROHRBUND (*n*): bundle of reeds (see also ROHR-BÜNDEL) 033
ROHRBÜNDEL (*n*): bank (nest, pass, section, element) of tubes (Boilers); bundle of tubes (Shipping specifications) (see also ROHR-BUND) 034
ROHRBÜRSTE (*f*): tube brush 035
ROHRBUSCH (*m*): reeds; sedge 036
RÖHRCHEN (*n*): tubule; reed (of organ); little tube or pipe 037
ROHRDACH (*n*): roof thatched with reeds, reed-thatch 038
ROHRDOMMEL (*f*): bittern (Ornithology) 039
RÖHRE (*f*): tube; pipe; nozzle; funnel; conduit; channel; spout; tunnel; shaft (see ROHR); valve (Wireless Tel.) 040
ROHREINWALZSTELLE (*f*): tube hole (in the tube plate of a boiler drum) 041
ROHRELEMENT (*n*): bank (nest; section; element; pass) of tubes 042
ROHREN: (of)-cane, (of)-reed; to cut (canes) reeds 043
RÖHREN: to roar, bellow 044
RÖHRENARTIG: tubular; fistular; tube-like 045
RÖHRENCASSIE (*f*): purging cassia, Indian laburnum, pudding (stick) pipe, drumstick (*Cassia fistula*) 046
RÖHRENDAMPFKESSEL (*m*): tubular-(steam)-boiler (see RÖHRENKESSEL) 047
ROHRENDANWÄRMEFEUER (*n*): tube end heating fire 048
RÖHRENFAHRT (*f*): (see RÖHRENGANG) 049
RÖHRENFÖRMIG: (see RÖHRENARTIG) 050
RÖHRENGANG (*m*): pipe-line, conduit, main 051
RÖHRENGESCHWÜR (*n*): fistula (Med.) 052
RÖHRENKASSIE (*f*): purging cassia (see RÖHREN-CASSIE) 053
RÖHRENKESSEL (*m*): (multi-)tubular boiler; (any type of)-tubular boiler or flue boiler 054
RÖHRENKLEMME (*f*): tube (clamp) clip 055
RÖHRENKONDENSATOR (*m*): surface condensor 056
RÖHRENKÜHLER (*m*): tubular (condenser) cooler 057
RÖHRENLIBELLE (*f*): tube level 058
RÖHRENLOT (*n*): pipe solder 059
RÖHRENMANNA (*f*): flake manna 060
RÖHRENNUDELN (*f.pl*): macaroni 061
RÖHRENOFEN (*m*): tube furnace 062
RÖHRENPAPIER (*n*): tube paper (Paper) 063
RÖHRENQUALLE (*f*): (see QUALLE, HYDROMEDUSEN and MEERNESSEL) 064
RÖHRENSENDER (*m*): valve transmitter (Wireless Tel.) 065
RÖHRENSICHERUNG (*f*): tubular fuse 066
RÖHRENSTRASSE (*f*): tube mill (Rolling mills) 067
RÖHRENTRÄGER (*m*): pipe (support) hanger or clip 068
RÖHRENWALZWERK (*n*): tube rolling mill 069
RÖHRENWERK (*n*): tubing; piping 070
ROHRFÖRMIG: tubular 071
ROHRGEIER (*m*): harpy, marsh-harrier, moor-buzzard, duck-hawk (*Circus aeruginosus*) (Ornithology) 072
ROHRGRUPPE (*f*): nest (bank, group, section) of tubes 073
ROHRHAKEN (*m*): wall hook, pipe support 074
ROHRHENNE (*f*): coot, water-fowl, moor-hen, gallinule, water-hen, spotted crake (*Gallinula chloropus* or *Crex porzana*) (Ornithology) 075
ROHRHUHN (*n*): (see ROHRHENNE) 076
RÖHRICH (*n*): reed-bed 077
RÖHRIG: tubular; fistular; (*n*): reed-bed 078
ROHR (*n*), INDISCHES-: bamboo, (*Bambusa*) (see BAMBUS) 079
ROHRKOLBEN (*m*): reed-mace (*Typha*) (Bot.) 080
ROHRKOLBENSCHILF (*n*): reed mace (*Typha*) 081
ROHRKOLBENSCHILF (*n*), BREITBLÄTTRIGES-: great reed mace (*Typha latifolia*) 082
ROHRKOLBENSCHILFFASER (*f*): (see TYPHAFASER) 083
ROHRKOLBENSCHILF (*n*), SCHMALBLÄTTRIGES-: lesser reed-mace (*Typha angustifolia*) 084
ROHRKRATZER (*m*): tube scrape 085
ROHRKRÜMMER (*m*): bend, bent pipe, elbow 086
ROHRLEGER (*m*): pipe fitter 087
ROHRLEITUNG (*f*): pipe line, main, conduit 088
ROHRLEITUNGS-ARMATUREN (*f.pl*): pipe-line (valves) fittings or mountings 089
ROHRMÜHLE (*f*): tube mill 090
ROHRPLAN (*m*): piping lay-out, arrangement of piping; tube (drawing, plan) section 091
ROHRPLATTE (*f*): tube plate (of boilers) 092
ROHRPOST (*f*): tube (pneumatic) post 093
ROHRPULVER (*n*): tubular powder 094
ROHRREINIGUNGSVORRICHTUNG (*f*): tube cleaning arrangement 095
ROHRRING (*m*): ring main (of pipes); ferrule (of condensers) 096
ROHRRING (*m*), KONDENSATOR-: condenser ferrule 097
ROHRSCHELLE (*f*): wall hook, pipe support, (pipe)-hanger 098
ROHRSCHLANGE (*f*): (tubular)-coil; worm; spiral 099
ROHRSCHMIED (*m*): barrel forger 100
ROHRSCHNEIDER (*m*): cutter (of sugar canes); tube-cutter 101
ROHRSCHRAUBSTOCK (*m*): tube vice 102
ROHRSCHUSS (*m*): section of tube header 103
ROHR (*n*), SPANISCHES-: Spanish (Italian) reed (*Arundo donax*) 104
ROHRSTAB (*m*): cane; bamboo 105
ROHRSTOCK (*m*): cane; bamboo 106
ROHRSTÖPSEL (*m*): tube plug, tube nipple, blind nipple 107
ROHRSTOSSE (*f*): reed-cutter; reed cutting apparatus 108
ROHRSTÖSSER (*m*): (see ROHRSTOSSE) 109
ROHRSTUHL (*m*): cane chair 110
ROHRTEILUNG (*f*): tube (spacing) pitching 111
ROHRVERBINDUNG (*f*): tube joint, pipe joint, pipe (tube) connection 112
ROHRVERSCHLUSS (*m*): tube cap; blind nipple; tube plug (of boiler tubes) 113
ROHRVERSCHRAUBUNG (*f*): tube (pipe) joint, screwed (joint) connection 114
ROHRWEITE (*f*): bore (pipe or tube); pipe (internal) diameter; i.d. 115
ROHRWERK (*n*): piping; tubing; reeds (of an organ) 116
ROHRWISCHER (*m*): tube brush 117

ROHRZANGE (*f*): pipe-wrench 118
ROHRZUCKER (*m*): cane sugar; saccharose; sucrose; $C_{12}H_{22}O_{11}$; Sg. 1.588; Mp. 189.2°C. 119
ROHRZUCKERSAFT (*m*): cane juice 120
ROHRZUCKERVERBINDUNG (*f*): cane sugar compound; saccharate 121
ROHSCHIENE (*f*): bloom, bar (flat iron 15-20 mm. thick and 50-100 mm. wide, rolled) (Metal) 122
ROHSCHLACKE (*f*): raw-(ore)-slag 123
ROHSCHMELZE (*f*): melt (Metal) 124
ROHSCHMELZEN (*n*): ore smelting 125
ROHSCHWEFEL (*m*): crude sulphur (see SCHWEFEL) 126
ROHSEIDE (*f*): raw silk 127
ROHSODA (*f*): crude soda; black ash (of Leblanc's process) (about 45% sodium carbonate, 30% calcium sulphide and 10% calcium hydroxide) 128
ROHSPAT (*m*): iron (spar) ore, siderite (see EISENSPAT) 129
ROHSPELZ (*m*): raw spelt (see SPELZ) 130
ROHSPIRITUS (*m*): raw spirit; crude spirit 131
ROHSTAHL (*m*): natural (crude; raw) steel 132
ROHSTEIN (*m*): coarse metal (Copper) (result of converter treatment) 133
ROHSTEINSCHMELZEN (*n*): coarse metal smelting; DEUTSCHER PROZESS (German process, with coke in shaft furnace); ENGLISCHER PROZESS (English process, without coal in reverberatory furnace) 134
ROHSTOFF (*m*): raw (crude) material or substance; raw product 135
ROHSTOFFE ZUR LEIMFABRIKATION: gluestock; raw material for glue manufacture 136
ROHTALG (*m*): crude tallow; Sg. 0.943-0.952; Mp. 40-48.5°C. 137
ROHWOLLE (*f*): raw wool 138
ROHZINK (*n*): crude (coarse) zinc, raw zinc (97-98% zinc; 2-3% lead and traces of iron, arsenic, antimony and cadmium) 139
ROHZINN (*n*): raw tin (97-98% tin) (see WERKZINN) 140
ROHZUCKER (*m*): raw (unrefined) sugar 141
ROIL (*n*): royal-(sail) 142
ROILSEGEL (*n*): royal-(sail) 143
ROLLBAR: voluble, capable of being rolled 144
ROLLBAUM (*m*): capstan; windlass 145
ROLLBEIN (*n*): trochlea (Anat.) 146
ROLLBEWEGUNG (*f*): rolling motion (of a ship) 147
ROLLE (*f*): roll; roller; caster; pulley; rôle; calender; scroll; mangle; part; barrel; cylinder; twist; coil; register; list; character; catalogue; inventory 148
ROLLEISEN (*n*): rolled iron 149
ROLLEN: to roll; calender; mangle; trundle; wheel; whirl; revolve; bind; enroll; level; smooth; press; impel; flatten; rotate; turn; gyrate; rock; tumble; flow; run 150
ROLLENBLECH (*n*): roll sheet iron 151
ROLLENBLEI (*n*): rolled (sheet) lead 152
ROLLENDE REIBUNG (*f*): roller bearing; rolling friction 153
ROLLENFÖRMIG: roll-shaped; cylindrical 154
ROLLENKUPFER (*n*): copper in rolls, sheet copper 155
ROLLENLAGER (*n*): roller bearing 156
ROLLENPRESSE (*f*): rolling press 157
ROLLENSCHNECKE (*f*): voluta 158
ROLLENTABAK (*m*): roll tobacco 159
ROLLENZUG (*m*): block and tackle; pulley-block 160
ROLLFASS (*n*): rolling cask (for polishing purposes) 161
ROLLFILM (*f*): roll film (Phot.) 162
ROLLFLASCHE (*f*): narrow-necked cylindrical bottle 163
ROLLGELD (*n*): cartage (transport) charges 164
ROLLGERSTE (*f*): winter barley 165
ROLLGUT (*n*): material to be rolled, rolled material 166
ROLLHOLZ (*n*): rolling pin 167
ROLLKASSETTE (*f*): roll holder (Phot.) 168
ROLLKLOBEN (*m*): pulley 169
ROLLMASCHINE (*f*): rolling machine 170
ROLLMATERIAL (*n*): rolling stock (Railways) 171
ROLLMESSING (*n*): rolled (sheet) brass 172
ROLLMÜHLE (*f*): rolling mill 173
ROLLSCHICHT (*f*): course (upright) on end (Bricks) 174
ROLLSCHLITTSCHUH (*m*): roller-skate 175
ROLLSITZ (*m*): rolling seat (of rowing boats) 176
ROLLSTEIN (*m*): boulder; garden-roller 177
ROLLSTUHL (*m*): chair on casters or rollers 178
ROLLVORHANG (*m*): curtain on rollers (Theatre); roller-blind 179
ROLLWAGEN (*m*): truck; trolley 180
ROMAN (*m*): novel; romance; tale; story; fiction 181
ROMANDICHTER (*m*): novelist 182
ROMANHAFT: fictitious; romantic 183
RÖMISCH: roman 184
RÖMISCHE BERTRAMWURZEL (*f*): pellitory; pyrethrum (*Radix anacyclus pyrethrum*) 185
RÖMISCHE KAMILLE (*f*): Roman camomile (*Anthemis nobilis*) 186
RÖMISCHE LICHTER (*n.pl*): roman lights (Fireworks) 187
RÖMISCHE MINZE (*f*): spearmint (*Mentha spicata*) 188
RÖMISCHER KÜMMEL (*m*): cumin (*Cuminum cyminum*) 189
RÖNTGENAPPARAT (*m*): Röntgen apparatus 190
RÖNTGENAUFNAHME (*f*): radiography, X-ray photograph 191
RÖNTGENPHOTOGRAPHIE (*f*): radio-photograph-(y), X-ray photograph-(y) 192
RÖNTGENROHR (*n*): Röntgen tube 193
RÖNTGENRÖHRE (*f*): Röntgen tube 194
RÖNTGENSCHIRM (*m*): fluorescent screen, Röntgen screen 195
RÖNTGENSTRAHLEN (*m.pl*): Röntgen (X)-rays 196
RÖNTGENTROCKENPLATTE (*f*): Röntgen dry plate 197
ROOBE (*pl*): fruit pulps (see MUS) 198
ROSA: rose-(coloured); (*f*): rose (Colour) 199
ROSALACK (*m*): rose lake 200
ROSAMIN (*n*): rosamine, benzorhodamine 201
ROSANILIN (*n*): rosaniline; triamido-tolyldiphenyl carbinol; $HOC(C_6H_3CH_3NH_2)(C_6H_4NH_2)_2$ 202
ROSANILINFARBSTOFF (*m*): rosaniline (colour) dye 203
RÖSCH: brittle; coarse; hard baked or roasted 204
RÖSCHEN: to age; cure; dig (trench or tunnel) 205
RÖSCHERZ (*n*): brittle silver ore; stephanite 206
RÖSCHGEWÄCHS (*n*): (see RÖSCHERZ) 207

ROSE (*f*): rose; rosette; erysipelas (Med.): starfish; sea-star 208
ROSELITH (*m*): roselite; $(CoCa)_3(AsO_4)_2$, $8H_2O$; Sg. 3.46 209
ROSELLAN (*m*): rosellane (a variety of anorthite) (see **ANORTHIT**); Sg. 2.7 210
ROSENADER (*f*): saphena 211
ROSENÄHNLICH: rose-like, rosaceous 212
ROSENARTIG: (see **ROSENÄHNLICH**) 213
ROSENBAUM (*m*): rosetree; rhododendron; rose-bay 214
ROSENBLÄTTER (*n.pl*): rose leaves 215
ROSENBRAND (*m*): blight on roses 216
ROSENBUSCH (*m*): rose bush 217
ROSENBUSCHIT (*m*): rosenbuschite; $2Na_2ZrO_2F_2.6CaSiO_3.Ti_2SiO_6$; Sg. 3.31 218
ROSENDISTEL (*f*): eryngo (*Eryngium*) 219
ROSENDORN (*m*): dog rose (*Rosa canina*); sweet briar-(rose) (*Rosa rubiginosa*) 220
ROSENESSENZ (*f*): attar of roses (see **ROSENÖL**) 221
ROSENFARBIG: rosy; rose-coloured; pink; roseate 222
ROSENGEWÄCHSE (*n.pl*): *Rosaceæ* (Bot.) 223
ROSENHOLDER (*m*): guelder rose; rose-elder; snowball tree (*Viburnum opulus*) 224
ROSENHOLZ (*n*): rosewood (*Rosa damascena* or *Rosa centifolia*, and wood thereof) 225
ROSENHOLZ (*n*), **AMERIKANISCHES-**: American rosewood, bois de citron (wood of *Amyris balsamifera*) 226
ROSENHOLZ (*n*), **JAMAIKA-**: Jamaica rosewood (see **ROSENHOLZ, AMERIKANISCHES-**) 227
ROSENHOLZÖL (*n*): rosewood oil 228
ROSENHONIG (*m*): honey of rose (Pharm.) 229
ROSENKNOPF (*m*): rose-bud 230
ROSENKOHL (*m*): Brussels sprouts 231
ROSENKONSERVE (*f*): confection of rose (Pharm.) 232
ROSENKRANZ (*n*): rosary; garland of roses 233
ROSENKUPFER (*n*): rose copper; rosette copper (see **ROTKUPFER**) 234
ROSENLACK (*m*): rose lake; rose madder 235
ROSENLORBEER (*m*): oleander; mountain rose (*Rosa*) 236
ROSENMALVE (*f*): rose-mallow, holly-hock (*Althœa rosea*) 237
ROSENÖL (*n*): oil (attar) of roses; rose oil (*Oleum rosæ*); Sg. 0.845-0.865; Bp. 229°C. (distilled from flowers of *Rosa damascena*) 238
ROSENÖL (*n*), **KÜNSTLICHES-**: artificial rose oil 239
ROSENÖLSURROGAT (*n*): rose-oil substitute 240
ROSENPAPPEL (*f*): (see **ROSENMALVE**) 241
ROSENQUARZ (*m*): rose quartz [an opaque, rose-coloured variety of quartz of the composition; $SiO_2(TiO_2)$]; Sg. 2.65 242
ROSENROT: rose-red; rose-coloured; rose (see **ROSENFARBIG**) 243
ROSENSCHIMMEL (*m*): mildew-(of roses) (*Sphærotheca pannosa*) 244
ROSENSPAT (*m*): rhodochrosite (see **RHODOCHROSIT**) 245
ROSENSTEIN (*m*): rose diamond 246
ROSENSTOCK (*m*): rose-bush; rose-tree 247
ROSENSTOCK, WILDER- (*m*): eglantine; sweet-briar; eglatere (*Rosa rubiginosa*) 248
ROSENSTRAUCH (*m*): (see **ROSENSTOCK**) 249
ROSENWASSER (*n*): rose water 250
ROSENZUCKER (*m*): (see **ROSENKONSERVE**) 251
ROSEOKOBALTSALZ (*n*): roseocobaltic salt 252
ROSEOLA (*f*): roseola, patchy erythema, rose rash (Med.) 253
ROSES METALL (*n*): Roses metal (a bismuth alloy, containing Bi,Pb and Sn in the proportion of 8 : 8 : 3) (Mp. 79°C.) 254
ROSETTE (*f*): rosette; register; draught regulator (Mech.) 255
ROSETTENARTIG: like a rosette 256
ROSETTENFENSTER (*m*): rose window 257
ROSETTENKUPFER (*n*): (see **ROSENKUPFER**) 258
ROSETTIEREN: to make rose-(tte) copper 259
ROSIEREN: to dye (pink) rose 260
ROSIERSALZ (*n*): rose salt (Dye); tin composition (solution of tin in aqua regia) 261
ROSIG: rosy; roseate; rose-coloured; red; flushed; ruddy; blushing 262
ROSINDULIN (*n*): rosinduline 263
ROSINE (*f*): raisin 264
ROSINFARBE (*f*): dark red, raisin colour, crimson 265
ROSINFARBIG: raisin-coloured 266
ROSINOL (*n*): rosinol, retinol, rosin oil (from fractional distillation of rosin above 360°C.); Sg. 0.98-1.11 267
ROSINROT (*n*): (see **ROSINFARBE**) 268
ROSMARIN (*m*): rosemary (*Rosmarinus officinalis*) 269
ROSMARINBLÄTTER (*n.pl*): rosemary leaves 270
ROSMARINBLÜTEN (*f.pl*): rosemary flowers 271
ROSMARINÖL (*n*): rosemary oil (*Oleum rosmarini*); Sg. 0.9-0.915 272
ROSOLROT (*n*): rosol red 273
ROSOLSÄURE (*f*): rosolic acid; para-rosolic acid; $C_{20}H_{16}O_3$; Mp. 270°C. 274
ROSS (*n*): horse; steed 275
ROSSARZNEIKUNST (*f*): veterinary surgery 276
ROSSBREMSE (*f*): horse (gad) fly 277
ROSSHAAR (*n*), **KÜNSTLICHES-**: artificial horse-hair 278
ROSSHUF (*m*): coltsfoot (*Tussilago farfara*) 279
ROSSKASTANIE (*f*): horse chestnut (*Aesculus hippocastanum*) (wood has Sg. 0.51-0.63) 280
ROSSKASTANIENGEWÄCHSE (*n.pl*): *Hippocastanaceæ* (Bot.) 281
ROSSKASTANIENÖL (*n*): horse-chestnut oil (from seeds of *Aesculus hippocastanum*) 282
ROSSSCHWEFEL (*m*): horse brimstone; impure greyish sulphur 283
ROST (*m*): rust; ferric hydroxide; $Fe(OH)_3$ (see **FERRIHYDROXID**); charge (Roasting); grate (of furnaces); grid-(iron); blight; smut; mildew (Bot.) 284
RÖSTABGANG (*m*): loss (of weight) due to roasting 285
RÖSTARBEIT (*f*): roasting-(process); smelting-(process) 286
ROST (*m*), **BAGASSE-**: bagasse (megass) grate 287
ROSTBALKEN (*m*): grate-(bar) bearer 288
ROSTBESCHICKUNGSANLAGE (*f*): grate feeding installation (such as conveyer plant, coal chutes, etc.) 289
ROST (*m*), **BEWEGLICHER-**: moveable (moving) grate, travelling grate, rotary (rotating) grate 290
ROSTBLECH (*n*): grid-plate 291
ROSTBRAND (*m*): rust; smut; mildew 292
ROST (*m*), **DREH-**: revolving grate 293
RÖSTE (*f*): ore to be roasted; roasting (place) charge; rettery; flaxhole; retting; steeping (Flax) [actually **ROTTE** from **VERROTTEN**—**VERFAULEN**, to rot. It is a method of obtaining bast fibre from the stems of plants by the addition of micro-organisms, bacteria

or fungus in a moist state, i.e., a rotting process] (Textile) 294

RÖSTEN: to burn; smelt; roast; torrefy; broil; grill; toast; rot, ret or steep (Flax and Hemp); (*n*): torrefaction; retting (rotting) process, etc.; roasting (an oxydating heating of ore) 295

ROSTEN: to rust; become rusty; corrode; (*n*): rusting; corrosion 296

ROSTENTFERNER (*m*): rust remover 297

RÖSTER (*m*): roaster 298

RÖSTERWERK (*n*): grid, grating 299

RÖSTE (*f*), **TAU-**: dew retting process (in which the flax is exposed to dew or rain, for obtaining bast fibre. Takes 3 weeks and more) 300

ROSTFARBE (*f*): rust colour 301

ROSTFELD (*n*): section (of sectional grates) 302

ROST (*m*), **FESTER-**: fixed grate 303

ROSTFEUERUNG (*f*): stoking, firing (furnaces), furnace fitted with a grate 304

ROSTFLÄCHE (*f*): grate (area) surface 305

ROSTFLÄCHENEINHEIT (*f*): unit of grate surface 306

ROSTFLECK (*m*): rust (iron) spot, mark, or mould 307

ROSTFLECKEN (*m*): (see ROSTFLECK) 308

ROSTFLECKENENTFERNER (*m*): rust (iron) stain remover 309

ROSTFLECKENWASSER (*n*): rust (iron) stain remover 310

RÖSTGAR (*n*): gas from roasting-(pyrites, etc.) 311

ROSTGELB (*n*): rust yellow (a buff-coloured pigment; hydrous iron oxide) (see under EISEN-) 312

RÖSTGUMMI (*n*): dextrin (see DEXTRIN) 313

RÖSTGUT (*n*): material to be roasted; roasted material; material to be (rotted) retted (Textile) 314

RÖSTHAUS (*n*): smeltery 315

RÖSTHÜTTE (*f*): smeltery 316

ROSTIG: rusty; corroded 317

RÖSTKASTEN (*m*): rotting box or case 318

RÖSTKASTEN (*m*), **SCHWIMMENDER-**: floating retting case or box (used in streams or ponds in Belgium for retting flax) 319

ROST (*m*), **KETTEN-**: chain grate 320

ROST (*m*), **KIPP-**: tipping grate 321

RÖSTKUFE (*f*): retting vat 322

RÖSTMALZ (*n*): roasted (black) malt 323

ROST (*m*), **MECHANISCHER-**: mechanical grate 324

ROST (*m*), **MULDEN-**: trough grate 325

RÖSTOFEN (*m*): roasting (oven; furnace) kiln; roaster; calciner 326

RÖSTPFANNE (*f*): heating pan, steam jacketed kettle (for heating starch to make dextrin); frying pan 327

RÖSTPOSTEN (*m*): roasting charge 328

RÖSTPROBE (*f*): calcination (assay) test 329

RÖSTPRODUKT (*n*): product of roasting; retting, etc. (see RÖSTEN) 330

RÖSTPROZESS (*m*): roasting (calcining) process 331

ROSTRAL (*n*): music pen 332

RÖSTREAKTIONSARBEIT (*f*): sulphatising (roasting) calcining process, roasting of lead ore in reverberatory furnace (formation of metallic lead and sulphurous acid) 333

RÖSTREDUKTIONSVERFAHREN (*n*): oxydizing (roasting) calcining process (converter process for lead) 334

ROST (*m*), **ROTIERENDER-**: rotary (rotating) grate 335

RÖSTRÜCKSTAND (*m*): residue from roasting 336

ROST (*m*), **RUND-**: circular grate 337

RÖSTSCHERBEN (*m*): roasting dish; scorifier 338

RÖSTSCHLACKE (*f*): slag from roasting 339

ROST (*m*), **SCHLACKEN-**: dumping grate, dumping bars 340

RÖSTSCHMELZEN (*n*): roasting and smelting (Metal) 341

ROST (*m*), **SCHRÄGER-**: inclined grate 342

ROST (*m*), **SCHÜTTEL-**: shaking grate; rocking-bar (grate) furnace 343

ROSTSCHUTZ (*m*): rust (prevention) preventative 344

ROSTSCHÜTZEND: protecting against rust, anti-rusting 345

ROSTSCHUTZFARBE (*f*): anti-rusting (anti-incrustant) paint 346

ROSTSCHUTZFETT (*n*): anti-rust fat (for coating polished metal surfaces) 347

ROSTSCHUTZMASSE (*f*): anti-incrustant, protector against rust 348

ROSTSCHUTZMITTEL (*n*): anti-rust agent; anti-corrosion agent; rust preventative; anti-incrustant 349

ROSTSCHUTZÖL (*n*): anti-rusting oil 350

ROSTSICHERHEIT (*f*): resistance to rust; non-rusting capacity 351

ROSTSPALTE (*f*): air-space (in a grate) 352

ROSTSTAB (*m*): grate bar; firebar; furnace bar 353

ROSTSTAB (*m*), **HOHL-**: hollow (grate)-bar 354

RÖSTSTAUB (*m*): dust of roasted ore 355

ROST (*m*), **STUFEN-**: step grate 356

ROSTTRÄGER (*m*): grate bar (firebar) bearer, bearer bar 357

ROST (*m*), **TREPPEN-**: step grate 358

RÖSTUNG (*f*): roasting; calcination; retting (Flax) (see RÖSTEN) 359

RÖSTVERFAHREN (*n*): roasting (retting) process (see RÖSTE) 360

ROST-VERHÜTEND: anti-rusting, anti-corrosive, rust preventing, anti-incrustant 361

RÖSTVORGANG (*m*): retting process (Flax) (see RÖSTE); roasting process (Ores) 362

RÖSTVORRICHTUNG (*f*): roasting (retting) apparatus, contrivance or device 363

ROT: russet; red; florid; ruddy; rubicund; (*n*): red; rouge; redness; flush 364

ROTAMESSER (*m*): (gas)-flow meter (used for measuring the strength of a current of gas or liquid, and thus indicating the quantity passing per hour) 365

ROTATION (*f*): rotation 366

ROTATIONSACHSE (*f*): axis of rotation 367

ROTATIONSAPPARAT (*m*): panoramic camera 368

ROTATIONSBEWEGUNG (*f*): rotational (rotary) motion 369

ROTATIONSDYNAMOMETER (*n*): rotation dynamometer 370

ROTATIONSFLÄCHE (*f*): surface of (rotation) revolution 371

ROTATIONSHOCHDRUCKGEBLÄSE (*n*): rotary, high-pressure blower 372

ROTATIONSKONSTANTE (*f*): rotation constant (Polarization) 373

ROTATIONSPUMPE (*f*): rotary pump 374

ROTATOR (*m*): tube mill (for dry grinding of clinker, etc.) 375

ROTAUGE (*n*): roach (Fish); red-eye; rud (*Leuciscus erythophthalmus*) (Ichthyology) 376

ROTÄUGIG: red-eyed 377

ROTBEIZE (*f*): red (mordant) liquor (Cotton dyeing) 378
ROTBLAU: reddish blue 379
ROTBLAUSAURES KALI (*n*): (see BLUTLAUGENSALZ, ROTES-) 380
ROTBLEIERZ (*n*): red lead ore; crocoite (natural lead chromate); $PbCrO_4$; Sg. 5.95 381
ROTBLEISPAT (*m*): (see ROTBLEIERZ) 382
ROTBRAUN: red-brown; bay; ruddy 383
ROTBRAUNSTEINERZ (*n*): rhodonite (see RHODONIT) 384
ROTBRUCH (*m*): red-short-(ness) (Metal) 385
ROTBRÜCHIG: red-short 386
ROTBUCHE (*f*): red beech (see BUCHE) 387
ROTBUCHENHOLZ (*n*): red beech wood (see BUCHE) 388
ROTDROSSEL (*f*): red-wing, swine-pipe, wind-thrush (*Turdus iliacus*) 389
RÖTE (*f*): red-(ness); madder; blush; red colour 390
ROTEICHE (*f*): Quebec oak; red oak (*Quercus rubra*) 391
ROTEICHENHOLZ (*n*): red oak wood 392
ROTEISEN (*n*): hematite pig 393
ROTEISENERZ (*n*): red iron ore; hematite (see ROTEISENSTEIN) 394
ROTEISENOCKER (*m*): red-(iron)-ocher; earthy hematite (see RÖTEL) 395
ROTEISENSTEIN (*m*): red iron ore; hematite (40-70% iron); Fe_2O_3; Sg. 5.25 (see also EISENOXYD and EISENGLANZ); red ochre; red iron ochre (see RÖTEL) 396
ROTEISENSTEIN (*m*), **DERBER-**: compact variety of hematite (see ROTEISENSTEIN and EISEN-GLANZ) 397
ROTEISENSTEIN (*m*), **MULMIGER-**: earthy variety of hematite (see ROTEISENSTEIN and EISEN-GLANZ) 398
RÖTEL (*m*): red (chalk) ocher (a soft ochrous variety of hematite) (see also ROTEISENSTEIN); reddle (a natural iron oxide); Fe_2O_3; red-wood (see FERNAMBUKHOLZ) 399
RÖTELN (*f.pl*): German measles; Rubella (Med.) 400
RÖTEN: to redden; become red; flush; blush; make red 401
RÖTEND: rubefacient 402
ROTERLE (*f*): alder (see ERLE) 403
ROTER PRÄZIPITAT (*m*): red precipitate; red mercuric oxide (see PRÄZIPITAT) 404
ROTE RUHR (*f*): dysentery (Med.) 405
ROTFAHL: tawny 406
ROTFARBIG: red-(coloured); russet-(coloured) 407
ROTFÄULE (*f*): red-rot (a fungus, *Trametes radiciperda*, which attacks fir trees) 408
ROTFEUER (*n*): red fire 409
ROTFICHTE (*f*): red fir-(tree) (*Pinus sylvestris*) 410
ROTFILTER (*m* and *n*): red filter (for three-colour printing or photography) 411
ROTFLECKIG: red-spotted 412
ROTFÖHRE (*f*): (see ROTFICHTE) 413
ROTGAR: tanned-(to a russet colour) 414
ROTGELB: orange; reddish yellow 415
ROTGERBER (*m*): tanner (see under GERB- and LOH-) 416
ROTGERBEREI (*f*): tanning; tannery 417
ROTGIESSER (*m*): brazier 418
ROTGIESSEREI (*f*): braziery 419
ROTGILTIGERZ (*n*), **DUNKLES-**: dark ruby silver ore, pyrargyrite, antimony-silver blende (about 60% silver); $3Ag_2S.Sb_2S_3$ (see also PYRARGYRIT) 420
ROTGILTIGERZ (*n*), **LICHTES-**: light ruby silver, proustite, arsenic silver blende; $3Ag_2S.As_2S_3$ (about 65% silver); Sg. 5.5 421
ROTGLÜHEND: red-hot; glowing red 422
ROTGLÜHHITZE (*f*): red heat 423
ROTGLUT (*f*): red heat 424
ROTGÜLTIG (*n*): red silver ore (see ROTGILTIG-) 425
ROTGÜLTIGERZ (*n*): (see ROTGÜLTIG-) 426
ROTGÜLTIGERZ, DUNKLES- (*n*): pyrargyrite (see ROTGILTIG-) 427
ROTGÜLTIGERZ, LICHTES- (*n*): proustite (see ROTGILTIG-) 428
ROTGUSS (*m*): red brass (84% Cu; 10% Sn; 6% Zn; for loco. ram rods); (88% Cu; 6% Zn; 6% Sn; for highly stressed machine parts); (see also TOMBAK); bronze, gun-metal (G.M.); braziery 429
ROTGUSSARMATUR (*f*): gun-metal fittings 430
ROTGUSSBEWEHRUNG (*f*): gun metal (fittings) working parts 431
ROTHEILWURZEL (*f*): (see TORMENTILLWURZEL) 432
ROTHIRSCH (*m*): stag; hart 433
ROTHOLZ (*n*): Brazil wood; camwood; red wood (see FERNAMBUKHOLZ) 434
ROTHOLZEXTRAKT (*m*): redwood extract 435
ROTHOLZLACK (*m*): redwood lake (the colouring principle of which is brasilein (see also BRASILEIN) 436
ROTIERAPPARAT (*m*): rotary apparatus 437
ROTIEREN: to rotate; revolve; gyrate 438
ROTIEREND: rotating; rotary; revolving; gyrating 439
ROTIERTROMMEL (*f*): rotary (rotating) drum 440
ROTKLEESAMENÖL (*n*): red clover seed oil 441
ROTKOHL (*m*): red cabbage 442
ROTKOHLE (*f*): red charcoal 443
ROTKUPFER (*n*): red copper-(ore); chalcotrichite; cuprite; ruby copper; Cu_2O; Sg. 5.95 (natural red copper oxide about 89% Cu) 444
ROTKUPFERERZ (*n*): (see ROTKUPFER) 445
ROTLAUF (*m*): erysipelas; dysentery (Med.) 446
ROTLAUFGÜRTEL (*m*): shingles (Med.) 447
ROTLAUFSERUM (*n*): erysipelas serum 448
RÖTLICH: reddish 449
ROTLIEGENDES (*n*): Lower Permian (shale) sand-stone (from Germany) (Geol.) 450
ROTMACHEN: to redden 451
ROTMACHEND: rubefacient 452
ROTNICKELKIES (*m*): niccolite (natural nickel arsenide); NiAs; Sg. 7.7 (see NICCOLIT) 453
ROTÖL (*n*): red oil; oleic acid (see OLEINSÄURE) 454
ROTOR (*m*): rotor (of turbines, etc.) 455
ROTOXYD (*n*): red mercury oxide (see QUECK-SILBEROXYD); red oxide, iron sesquioxide, red ferric oxide (see EISENOXYD) 456
ROT-RAUSCHGELB (*n*): realgar (see REALGAR) 457
ROTREPSÖL (*n*): rocket oil (from *Hesperis matronalis*); Sg. 0.928-0.934 458
ROTRÜSTER (*f*): (see ULME) 459
ROTSALZ (*n*): sodium acetate (see NATRIUM, ESSIGSAURES-); red potash, (potash coloured red by Fe_2O_3) 460
ROTSANDSTEIN (*m*): red sandstone (see also ROTLIEGENDES) 461
ROTSCHIMMEL (*m*): roan horse 462
ROTSCHLEIER (*m*): red fog (Phot.) (see also FARBSCHLEIER) 463

ROTSCHLINGE (*f*): mealy guelder rose, wayfaring tree (*Viburnum lantana*) 464
ROTSCHMIED (*m*): copper-smith 465
ROTSILBER (*n*): red silver ore (see ROTGILTIG-) 466
ROTSILBERERZ (*n*): red silver ore (see ROTGILTIG-) 467
ROTSILBERERZ, DUNKLES- (*n*): pyrargyrite (see ROTGILTIG-) 468
ROTSILBERERZ, LICHTES- (*n*): proustite (see ROTGILTIG-) 469
ROTSPAT (*m*): rhodonite (see RHODONIT) 470
ROTSPIESSGLANZ (*m*): kermesite (see ANTI--MONBLENDE) 471
ROTSPIESSGLANZERZ (*n*): (see ROTSPIESSGLANZ) 472
ROTSTEIN (*m*): red ochre; reddle (see RÖTEL); red (tile) brick 473
ROTSTIFT (*m*): red (pencil) crayon 474
ROTSUCHT (*f*): nettle-rash (Med.) 475
ROTTANNE (*f*): red fir (*Pinus sylvestris*); pitch pine (*Pinus abies* and *Pinus excelsa*); spruce (see FICHTE); red (Canadian) pine (*Pinus rubra*) 476
ROTTE (*f*): retting (Flax) (see RÖSTE); troop; gang; set; band; horde 477
ROTTEN: to ret or steep (Flax and hemp) (see RÖSTEN); plot; mutiny; combine 478
ROTULME (*f*): (see ULME) 479
ROTWÄLSCH (*n*): (see ROTWELSCH) 480
ROTWEIN (*m*): red wine 481
ROTWELSCH (*n*): gibberish; jargon; slang 482
ROT WERDEN: to become red; flush; blush; redden 483
ROTWILD (*n*): red deer; venison 484
ROTZ (*m*): mucous (from the nose); glanders 485
ROTZINKERZ (*n*): red zinc ore; zincite (natural red zinc oxide; about 80% zinc); ZnO; Sg. 5.55 (see ZINKIT) 486
ROUGE (*f*): rouge, face (grease) paint; red ferric oxide (see EISENOXYD) 487
ROULEAU (*n*): roller-blind; roll-(er) 488
ROULEAUDRUCK (*m*): roller printing (see WALZEN--DRUCK) 489
ROULEAUDRUCKMASCHINE (*f*): roller printing machine 490
ROULIEREN: to be in circulation; roll; circulate 491
ROUTE (*f*): way, route, road 492
ROUTINIERT: accustomed to routine, versed, drilled, experienced 493
ROWLANDIT (*m*): Rowlandite; $2Y_2O_3.3SiO_2$ or $(Fe,Mg)(Ce,La,Y)_2(YF)_2Si_4O_{14}$; Sg. 4.5 494
ROYAL (*n*): royal-(sail) 495
R-SALZ (*n*): R-salt (sodium salt of β- naphtholdisulfonic acid) 496
R.-SÄURE (*f*): R. acid $[C_{10}H_5(OH)(SO_3H)_2]$ β-naphtholdisulfonic acid 497
RÜBE (*f*): turnip; rape; beet. beetroot 498
RÜBE, GELBE- (*f*): carrot (*Daucus carota*) 499
RÜBELBRONZE (*f*): bronze (28.5-38.9% Cu; 18.2-40% Ni; 6.1-8.3% Al; 25.4-34.6% Fe) (in three different kinds); manganese bronze 500
RUBELLIT (*m*): rubellite, tourmaline (see TUR--MALIN) 501
RÜBENASCHE (*f*): beet ashes 502
RÜBENBREI (*m*): beet pulp 503
RÜBENBREIAPPARAT (*m*): root pulper (Beet sugar) 504
RÜBENERZEUGEND: beet-producing 505
RÜBENESSIG (*m*): beet-(root) vinegar 506
RÜBENFELD (*n*): turnip field 507
RÜBENFÖRMIG: turnip-shaped; napiform 508
RÜBENGERUCH (*m*): beet smell 509
RÜBENHAHNENFUSS (*m*): bulbous crowfoot (*Ranunculus bulbosus*) 510
RÜBENKOHL (*m*): rape (*Brassica rapa*) 511
RÜBENKRAUT (*n*): turnip tops 512
RÜBENMELASSE (*f*): beet-(root) molasses 513
RÜBENPOTTASCHE (*f*): potash from beetroot molasses 514
RÜBENROHZUCKER (*m*): raw beet sugar (see RÜBENZUCKER) 515
RÜBENSAFT (*m*): beet-(root) juice 516
RÜBENSAMEN (*m*): turnip (rape; beet; carrot) seed (see RÜBSEN) 517
RÜBENSCHLEMPE (*f*): spent wash, vinasse slops (a residual liquid from the distillation of beet molasses) (see also SCHLEMPE) 518
RÜBENSCHNITZEL (*m*): beet slice 519
RÜBENSIRUP (*m*): beet-(root) syrup 520
RÜBENSPIRITUS (*m*): beet spirit (from beet molasses) 521
RÜBENSTECHER (*m*): beet sampler 522
RÜBENZUCKER (*m*): beet sugar (see ROHRZUCKER) 523
RÜBENZUCKERFABRIK (*f*): beet sugar factory 524
RÜBE, ROTE- (*f*): beet; beetroot (*Beta vulgaris*) 525
RUBERYTHRINSÄURE (*f*): ruberythric acid (from *Rubia tinctorum*) 526
RÜBE, WEISSE- (*f*): turnip (*Brassica campestris*) 527
RUBIADINGLYKOSID (*n*): rubiadine glycoside (a glycoside from *Rubia tinctorum*) 528
RUBICHLORSÄURE (*f*): rubichloric acid (from *Rubia tinctorum*) 529
RUBIDINOL (*n*): (see HIMBEERÖL) 530
RUBIDIUM (*n*): rubidium (Rb) 531
RUBIDIUMALAUN (*m*): rubidium alum $[Rb_2SO_4.Al_2(SO_4)_3.24H_2O]$ 532
RUBIDIUMBROMID (*n*): rubidium bromide; $RbBr$; Sg. 2.78-3.21 533
RUBIDIUMCHLORID (*n*): rubidium chloride; $RbCl$; Sg. 2.209 534
RUBIDIUM (*n*), **GOLDCHLORWASSERSTOFFSAURES-**: rubidium chloraurate; $RbAuCl_4$ 535
RUBIDIUMHALTIG: containing rubidium 536
RUBIDIUMHYDRID (*n*): rubidium hydride; RbH 537
RUBIDIUMHYDROXYD (*n*): rubidium hydroxide; $RbOH$; Sg. 3.203 538
RUBIDIUMJODAT (*n*): rubidium iodate; $RbIO_3$; Sg. 4.559 539
RUBIDIUMJODID (*n*): rubidium iodide; RbI; Sg. 3.428-3.438; Mp. 641.5°C. 540
RUBIDIUMNITRAT (*n*): rubidium nitrate; $RbNO_3$; Sg. 3.0955 541
RUBIDIUMOXYD (*n*): rubidium oxide; Rb_2O; Sg. 3.72 542
RUBIDIUMPENTASULFID (*n*): rubidium pentasulphide; Rb_2S_5; Sg. 2.618 543
RUBIDIUMPERJODAT (*n*): rubidium periodate; $RbIO_3$; Sg. 3.918 544
RUBIDIUMPERMANGANAT (*n*): rubidium permanganate; $RbMnO$; Sg. 3.235 545
RUBIDIUMPLATINCHLORID (*n*): rubidium chloroplatinate 546
RUBIDIUMSELENAT (*n*): rubidium selenate; Rb_2SeO_4; Sg. 3.8995 547
RUBIDIUMSILICOFLUORID (*n*): rubidium (fluosilicate) silicofluoride; Rb_2SiF_6; Sg. 3.332 [548

RUBIDIUMSULFAT (*n*): rubidium sulphate; Rb_2SO_4; Sg. 3.6113 549

RUBIDIUMSUPEROXYD (*n*): rubidium peroxide; Rb_2O_2; Sg. 3.65; Rb_2O_3; Sg. 3.53 550

RUBIDIUMVERBINDUNG (*f*): rubidium compound 551

RUBIN (*m*): ruby (of dyes or colours); ruby (the gem) (Sg. 4.23); ruby; corundum-(of a deep red colour); Al_2O_3; Sg. 3.95 (see also ALUMINIUMOXYD) 552

RUBINBALAS (*m*): balas ruby; ruby spinel 553

RUBINBLENDE (*f*): pyrargyrite (see ROTGILTIG--ERZ, DUNKLES-); proustite; arsenical silver blende (see PROUSTIT) 554

RUBINBLENDE (*f*), HEMIPRISMATISCHE-: miargyrite 555

RUBINBLENDE (*f*), RHOMBOEDRISCHE-: pyrargyrite, antimonial silver blende 556

RUBINFARBIG: ruby-(coloured) 557

RUBINGLAS (*n*): ruby glass 558

RUBINGLIMMER (*m*): goethite (see GÖTHIT) 559

RUBINGRANAT (*m*): rock ruby (type of garnet) 560

RUBIN (*m*), KÜNSTLICHER-: artificial ruby (artificial corundum) 561

RUBINROT: ruby-(red) 562

RUBINSCHWEFEL (*m*): ruby sulphur; realgar (see ARSENDISULFID, ARSENROT and REALGAR) 563

RUBINSPINELL (*m*): ruby spinel 564

RÜBÖL (*n*): colza oil; rape-(seed) oil (*Oleum rapæ*); Sg. 0.915 (from *Brassica napus* and *Brassica rapa*) (see RAPSÖL) 565

RÜBÖLEXTRAKTIONSANLAGE (*f*): rape-(seed)-oil extraction plant 566

RÜBÖLKUCHEN (*m*): rape-seed cake 567

RUBRIK (*f*): column (of a newspaper, etc.); heading; side-heading; title; rubric 568

RÜBSAMEN (*m*): turnip (rape; carrot; beet) seed (see RÜBSEN) 569

RÜBSEN (*m*): rape seed (of *Brassica rapa*) 570

RÜBSENÖL (*n*): (see RAPSÖL) 571

RUCHBAR: notorious; well-known 572

RUCHBARKEIT (*f*): notoriety; publicity 573

RUCHLOS: wicked; vicious; profligate; nefarious; scentless; infamous 574

RUCHLOSIGKEIT (*f*): profligacy; wickedness 575

RUCHTBAR: (well)-known, notorious, public (see RUCHBAR) 576

RUCHTBARKEIT (*f*): notoriety 577

RUCK (*m*): jerk; pull; jolt; tug; shock 578

RÜCK: back 579

RÜCKANSICHT (*f*): rear (back) view, end view; reverse, obverse 580

RÜCKANSPRUCH (*m*): counter-claim 581

RÜCKANTWORT (*f*): reply, response, rejoinder 582

RÜCKÄUSSERUNG (*f*): reply (by return) 583

RÜCKBAR: movable 584

RÜCKBERUFUNGSSCHREIBEN (*n*): letter of recall 585

RÜCKBETAGEN: to postdate 586

RÜCKBEWEGUNG (*f*): retrogression, retrograde (return) movement or stroke 587

RÜCKBILDEN: to re-form; form again 588

RÜCKBLEIBSEL (*n*): remainder; residue; waste; sediment, etc. (see RÜCKSTAND) 589

RÜCKBLICK (*m*): retrospect 590

RÜCKBLICKEND: retrospective 591

RÜCKBUCHUNG (*f*): writing back (Com.) 592

RÜCKDRAHT (*m*): return wire 593

RÜCKEN: to move; crawl; march; pull; push; pass; stir; proceed; (*m*): back; rear; ridge; bridge (of nose); obverse; reverse 594

RÜCKENDECKUNG (*f*): parados (Military) 595

RÜCKENFELL (*n*): pleura (Med.) 596

RÜCKENFLOSSE (*f*): dorsal fin 597

RÜCKENHALT (*m*): reserve; support 598

RÜCKENHAUT (*f*): pleura (Med.) 599

RÜCKENLEHNE (*f*): chair-back 600

RÜCKENMARK (*n*): spinal cord (Med.) 601

RÜCKENMARKSAPOPLEXIE (*f*): spinal apoplexy, hæmatomyelia (*Hæmorrhagia Medullæ spinalis*) (hæmorrhage into the spinal cord) (Med.) 602

RÜCKENMARKSENTZÜNDUNG (*f*): myelitis (*Myelitis acuta*) (inflammation and partial softening of the spinal cord) 603

RÜCKENMARKSERSCHÜTTERUNG (*f*): concussion of spinal cord (Med.) 604

RÜCKENMARKSERWEICHUNG (*f*): myelomalacia (*Mollities medullæ spinalis*) (non-inflammatory white or simple softening of the spinal cord) (Med.) 605

RÜCKENMARKS, GRAUE DEGENERATION DER HINTERSTRÄNGE DES-: Locomotor ataxy; *Tabes dorsalis* (degeneration of posterior columns) (Med.) 606

RÜCKENMARKSKOMPRESSION (*f*): compression of spinal cord (Med.) 607

RÜCKENMARKS, MULTIPLE SKLEROSE DES-: disseminated sclerosis, insular sclerosis, multiple sclerosis of the spinal cord, multi-locular sclerosis (Med.) 608

RÜCKENMARKS, PRIMÄRE LATERALSKLEROSE DES-: primary lateral sclerosis (see SPINALPARALYSIS, SPASTISCHE-) (Med.) 609

RÜCKENMARKS, PRIMÄRE SKLEROSE DER SEITEN--STRÄNGE DES-: primary sclerosis of the lateral columns (see SPINALPARALYSIS, SPASTISCHE-) (Med.) 610

RÜCKENMARKSQUETSCHUNG (*f*): crushing of spinal cord (Med.) 611

RÜCKENMARKSZERREISSUNG (*f*): punctured spinal cord (Med.) 612

RÜCKEN, NÄHER-: to draw near; approach; advance 613

RÜCKENSPALTE (*f*): cleft spine, *Spina bifida* (as Spinal meningocele, Meningo-myelocele or Syringo-myelocele) (Med.) 614

RÜCKENWEH (*n*): lumbago; pain in the back (Med.) 615

RÜCKENWIRBEL (*m*): (dorsal)-vertebra (Anat.) 616

RÜCKENWIRBELBEIN (*n*): (see RÜCKENWIRBEL) 617

RÜCKEN, ZUSAMMEN-: to push together; draw together 618

RÜCKERINNERUNG (*f*): reminiscence; recollection; remembrance 619

RÜCKERSTATTUNG (*f*): restitution, restoration, return, reparation, recompense, repayment, recession 620

RÜCKFAHRKARTE (*f*): return ticket 621

RÜCKFAHRT (*f*): return-(journey or voyage) 622

RÜCKFAHRTKARTE (*f*): return ticket 623

RÜCKFALL (*m*): relapse; reversion 624

RÜCKFÄLLIG: revertible 625

RÜCKFLUSS (*m*): return flow, efflux 626

RÜCKFLUSSKÜHLER (*m*): return (reflux) condenser 627

RÜCKFORDERUNG (*f*): reclaiming; reclamation 628

RÜCKFRACHT (*f*): return freight 629

RÜCKFRAGE (*f*): demand; counter-question 630
RÜCKFÜHRSCHNECKE (*f*): return (spiral) worm 631
RÜCKGABE (*f*): return-(ing), restoration, giving back 632
RÜCKGANG (*m*): retirement; return; withdrawal; retrogression; retraction; decline; retreat; back (return) stroke or movement (Mach.); recoil 633
RÜCKGÄNGIG: retrograde; retrogressive; null and void 634
RÜCKGÄNGIGMACHEN: to cancel; annul; rescind; break off 635
RÜCKGEHEN: to go back, return, recede, retrograde, withdraw 636
RÜCKGEWINNBAR: recoverable 637
RÜCKGEWINNUNG (*f*): recovery; regaining 638
RÜCKGRAT (*m*): spine; backbone; spinal (vertebral) column 639
RÜCKGRATES, KRANKHEIT DES-: disease of the spine (Med.) 640
RÜCKGRATES, KRÜMMUNG DES-: curvature of the spine (Med.) 641
RÜCKGRATSGELENK (*n*): vertebra 642
RÜCKGRÄTSKRÜMMUNG (*f*): curvature of the spine (Med.) 643
RÜCKGRATSTIER (*n*): vertebrate 644
RÜCKGRATSVERKRÜMMUNG (*f*): curvature of the spine (Med.) 645
RÜCKGRATSVERKRÜMMUNG (*f*), **SEITLICHE-**: scoliosis, lateral curvature of the spine (Med.) 646
RÜCKHALT (*m*): support; reserve; stay; reservation; restraint 647
RÜCKHALTLOS: unreserved; frank; open; candid; outspoken; undissembling; honest; straight; artless; guileless; above-board; sincere 648
RÜCKKEHR (*f*): return 649
RÜCKKLANG (*m*): reverberation; resounding; echo 650
RÜCKKOHLUNG (*f*): re-carbonization 651
RÜCKKÜHLANLAGE (*f*): cooling plant (usually, water cooling plant) 652
RÜCKKÜHLEN: to re-cool 653
RÜCKKUNFT (*f*): return 654
RÜCKLADUNG (*f*): return-cargo 655
RÜCKLAGE (*f*): reserve funds (Com.) 656
RÜCKLAUF (*m*): recoil (Gun); recurrence; return (of a movement); back-(return) stroke (Mech.) 657
RÜCKLAUFEND: recurrent; retrogressive; retrograde 658
RÜCKLÄUFIG: (see **RÜCKLAUFEND**) 659
RÜCKLEITUNG (*f*): return pipe line, return main, return wire 660
RÜCKLIEFERUNG (*f*): return (of goods); restitution 661
RÜCKLINGS: backward; from behind 662
RÜCKMARSCH (*m*): retirement; retreat; return march 663
RÜCKPRALL (*m*): recoil; rebound; re-percussion; reaction 664
RÜCKPRÄMIE (*f*): return-(of)-premium 665
RÜCKPROVISION (*f*): commission on re-draft (Com.) 666
RÜCKRECHNUNG (*f*): return account 667
RÜCKREISE (*f*): return journey; homeward voyage (of ships) 668
RÜCKRUF (*m*): recall 669
RÜCKSCHEIN (*m*): reflection (Light); bond 670
RÜCKSCHLAG (*m*): recoil; rebound; retrogression; return; back-firing; back-stroke; reaction; striking-back 671
RÜCKSCHLAGEN (*n*): back-firing; lighting-back 672
RÜCKSCHLAGHEMMUNG (*f*): prevention of (striking back) back-firing 673
RÜCKSCHLAGVENTIL (*n*): check valve; non-return (back pressure) valve 674
RÜCKSCHLUSS (*m*): conclusion-(a posteriori); inference 675
RÜCKSCHREIBEN (*n*): answer; reply 676
RÜCKSCHRITT (*m*): retrogression; relapse; recession; retirement; backward movement 677
RÜCKSCHRITTSPARTEI (*f*): retrograde party 678
RÜCKSEITE (*f*): back; reverse; obverse 679
RÜCKSENDUNG (*f*): return-(consignment) 680
RÜCKSICHT (*f*): respect; regard; consideration; notice; discretion; deference; attention; motive 681
RÜCKSICHTLICH: considering 682
RÜCKSICHT, MIT- AUF: having due regard to; taking into account; bearing in mind; in conformity with; including 683
RÜCKSICHT (*f*), **OHNE- AUF**: excluding; exclusive of; regardless of; without (bearing in mind) taking into account; without (giving due consideration to) due regard to 684
RÜCKSICHTSLOS: regardless; indiscreet; without consideration 685
RÜCKSICHTSVOLL: discreet, considerate 686
RÜCKSITZ (*m*): back-seat 687
RÜCKSPRACHE (*f*): conference; reference; consultation 688
RÜCKSPRUNG (*m*): rebound; resilience 689
RÜCKSPRUNG TUN: to recant; withdraw; back out of 690
RÜCKSPRUNGVERFAHREN (*n*): rebound process (of hardness testing) (a small hammer falls from from a given height on the test piece and the hardness is calculated from the rebound) 691
RÜCKSTAND (*m*): residue; remainder; balance; arrears; refuse; remains; waste; scale; sediment 692
RÜCKSTÄNDIG: in arrears; overdue; outstanding; residual; waste; owing; backward 693
RÜCKSTÄNDIGE (*n*): outstanding amount, balance, remainder, residue, waste, scale 694
RÜCKSTÄNDLER (*m*): defaulter 695
RÜCKSTANDSRECHNUNG (*f*): balance account 696
RÜCKSTAU (*m*): back-pressure (or more correctly an accumulation of steam or pressure which, while it is not in itself back-pressure, produces back-pressure); damming-up 697
RÜCKSTELLIG: retrograde; retrogressive; null and void 698
RÜCKSTELLIGMACHEN: to cancel 699
RÜCKSTOSS (*m*): repulsion; recoil; back-stroke 700
RÜCKSTRAHLER (*m*): reflector 701
RÜCKSTRAHLUNG (*f*): reflection 702
RÜCKSTROM (*m*): return current 703
RÜCKTRATTE (*f*): re-draft; return bill 704
RÜCKTRITT (*m*): retreat; retirement; retrogression; resignation; return 705
RÜCKÜBERHITZER (*m*): reheat superheater 706
RÜCKUMWÄLZUNG (*f*): counter-revolution 707
RÜCKVERGÜTEN: to repay; compensate; reimburse; pay back; requite 708

RÜCKVERGÜTUNG (*f*): repayment; compensation; reparation; reimbursement 709
RÜCKVERSICHERUNG (*f*): re-insurance; guarantee 710
RÜCKWAND (*f*): back wall 711
RÜCKWÄRTS: backward-(s); back 712
RÜCKWÄRTSEXZENTER (*m*): backward eccentric 713
RÜCKWÄRTSFAHRT (*f*): return (of a movement); backward running, retrograde motion, reverse, backing 714
RÜCKWÄRTSGANG (*m*): retrograde motion, return (of a movement), backward running, reverse, backing 715
RÜCKWÄRTSHUB (*m*): back (return) stroke 716
RÜCKWÄRTSLAGE (*f*): reclining position; inclination 717
RÜCKWÄRTSLIEGEN: to recline; incline; slope backwards 718
RÜCKWÄRTSWIRKEN: to react; reciprocate; be retrospective 719
RÜCKWÄRTSWIRKEND: reciprocal; retrospective; reacting; reactive 720
RÜCKWECHSEL (*m*): return bill; re-draft; re-imbursement 721
RÜCKWEG (*m*): way back; return 722
RÜCKWEISE: intermittently; in jerks 723
RÜCKWIRKEN: to react; reciprocate; be retrospective 724
RÜCKWIRKEND: reciprocal; retrospective; reacting; reactive 725
RÜCKWIRKUNG (*f*): reaction 726
RÜCKZAHLUNG (*f*): repayment; re-imbursement 727
RÜCKZIELEND: reflective (Grammar) 728
RÜCKZOLL (*m*): drawback (Com.) 729
RÜCKZÖLLIG: entitled to drawback (Com.) 730
RÜCKZOLLSCHEIN (*m*): debenture 731
RÜCKZUG (*m*): retreat; return-(train) 732
RÜDE (*m*): (male)-dog; hound; male fox or wolf 733
RUDEL (*n*): flock; herd; stirring pole; whisk 734
RUDER (*n*): oar; rudder; helm 735
RUDERBANK (*f*): rowing (bench) seat, thwart 736
RUDERBESCHLAG (*m*): rudder irons 737
RUDERBLATT (*n*): blade-(of an oar) 738
RUDERBOOT (*n*): rowing boat 739
RUDERDOLLE (*f*): rowlock 740
RUDERDUCHT (*f*): thwart 741
RUDERER (*m*): rower, oarsman 742
RUDERFÜHRER (*m*): helmsman 743
RUDERFÜSSIG: web-footed 744
RUDERGÄNGER (*m*): helmsman, steersman, pilot, cox-(swain) 745
RUDERGAT (*n*): (see **RUDERDOLLE**) 746
RUDERGESCHIRR (*n*): steering gear 747
RUDERGESTÄNGE (*n*): steering gear 748
RUDERHAKEN (*m*): pintle 749
RUDERHAUS (*n*): wheel house (Ships) 750
RUDERHELM (*n*): helm, tiller (Naut.) 751
RUDERHERZ (*n*): main piece of a rudder 752
RUDERJOCH (*n*): rudder yoke 753
RUDERKLAMPE (*f*): thole-pin 754
RUDERKLUB (*m*): rowing club 755
RUDERKOKER (*m*): rudder case, rudder trunk 756
RUDERMASCHINE (*f*): steering engine 757
RUDERN: to row; paddle; steer (a boat); (*n*): rowing, etc. 758
RUDERÖSEN (*f.pl*): rudder braces, rudder gudgeons 759
RUDERPINNE (*f*): tiller 760
RUDERRAD (*n*): steering wheel; paddle-wheel 761
RUDERRADGEHÄUSE (*n*): paddle-box 762
RUDERSCHLAG (*m*): stroke (of an oar) 763
RUDERSPORT (*m*): rowing 764
RUDERSTANGE (*f*): gaff 765
RUDERSTEVEN (*m*): rudder post 766
RUDERTALJE (*f*): rudder-tackle 767
RUDERVEREIN (*m*): rowing club 768
RUF (*m*): calling; call; report; rumour; reputation; fame; vocation; renown; name; trade; repute; ejaculation; exclamation; shout; character; profession; summons; vogue; fashion 769
RUFEN: to call; cry (see **RUF**) 770
RÜFFELN: to reprimand, reprove, censure, reproach, blame, admonish, upbraid, chide 771
RUFIGALLUSSÄURE (*f*): rufigallic acid 772
RUFTASTER (*m*): bell-push 773
RÜGBAR: blamable; actionable 774
RÜGE (*f*): censure; blame; reproof; accusation; denouncement; fine 775
RÜGEN: to blame; reprove; censure; accuse; denounce; resent; take exception to; fine; reprimand; reproach; admonish; upbraid; chide; punish 776
RUHBÜTTE (*f*): storage (stock) vat or tub 777
RUHE (*f*): rest; repose; sleep; quiet; quiescence; peace; stagnation; calm; silence; fulcrum; hush; relaxation; self-possession; sang-froid 778
RUHELAGE (*f*): position of rest; stationary position 779
RUHELOS: restless; fluctuating; changeable 780
RUHEN: to rest; stop; pause; sleep; repose; be (static; quiescent; stagnant; latent) stationery; lie fallow 781
RUHEND: resting; quiescent; latent; stagnant; static (Elect.); fallow; uninvested (of money) 782
RUHEPLATZ (*m*): resting place; landing (of stairs) 783
RUHEPULVER (*n*): sedative powder 784
RUHEPUNKT (*m*): pause; point (of rest) stoppage or repose; fulcrum; centre of gravity 785
RUHESTAB (*m*): mahl-stick (Art.) 786
RUHESTAND (*m*): retirement; state of rest 787
RUHESTÖRUNG (*f*): breach of the peace, disturbance, nuisance 788
RUHESTROM (*m*): current of rest, closed circuit current (Elect.) 789
RUHESTROMBATTERIE (*f*): closed-circuit battery 790
RUHESTROMGLOCKE (*f*): closed-circuit bell 791
RUHEVOLL: restful, peaceful, serene, tranquil, quiescent 792
RUHEWINKEL (*m*): angle of repose 793
RUHIG: at rest; tranquil; calm; still; quiet; quiescent; serene; composed; peaceful 794
RUHM (*m*): fame; glory; praise; renown; reputation; honour; pride; boast 795
RUHMBEGIERIG: ambitious 796
RÜHMEN: to commend; praise; extol; glorify; boast; brag; pride oneself; celebrate; say (tell) of 797
RÜHMLICH: laudable, honourable, glorious 798
RUHMLOS: inglorious 799
RUHMREICH: glorious 800
RUHR (*f*): dysentery; diarrhœa (Med.) 801
RUHRANFALL (*m*): attack of (dysentery) diarrhœa (Med.) 802

RÜHRAPPARAT (*m*): stirrer; agitator; blunger (Ceramics) 803
RUHRARTIG: diarrhœtic (Med.) 804
RÜHRBAR: movable; susceptible; sensitive 805
RÜHREI (*n*): scrambled egg 806
RÜHREISEN (*n*): poker; iron (stirrer) agitator 807
RÜHREN: to touch; affect; impress; move (Feelings); stir; pole; agitate; beat; strike; turn (Soil); mix; whip (Cream) 808
RÜHREND: touching; affecting; moving; pathetic 809
RÜHRER (*m*): stirrer; agitator (Paper); beater; striker; blunger (Ceramics) 810
RÜHRFASS (*n*): churn; dolly tub (Metal) 811
RÜHRGEBLÄSE (*n*): agitator-(by means of a blower) 812
RÜHRHAKEN (*m*): rake; rabble; stirrer 813
RÜHRHARKE (*f*): (see **RÜHRKRÜCKE**) 814
RÜHRHOLZ (*n*): paddle; stirrer (of wood) 815
RÜHRIG: stirring; busy; agile; active; nimble 816
RÜHRIGKEIT (*f*): agility, activity, nimbleness 817
RÜHRKELLE (*f*): ladle; pot-ladle 818
RUHRKRAUT (*n*): Cudweed (*Gnaphalium*) 819
RÜHRKRÜCKE (*f*): rake; stirring (pole) crutch; mash rake (Brewing) 820
RÜHRKÜBEL (*m*): mixing trough 821
RÜHRLÖFFEL (*m*): ladle; pot-ladle 822
RÜHR-MICH-NICHT-AN (*n*): Touch-me-not, yellow balsam (*Impatiens noli-me-tangere*) (Bot.) 823
RÜHRMILCH (*f*): buttermilk 824
RUHRMITTEL (*n*): dysentery remedy 825
RUHRRINDE (*f*): Simar-(o)-uba bark; Paradise tree bark; Mountain (bitter) damson bark; Paraiba-(bark) (bark of *Simaruba officinalis*) 826
RUHR, ROTE- (*f*): dysentery (Med.) 827
RÜHRSCHAUFEL (*m*): trowel; pot-ladle 828
RÜHRSCHEIT (*n*): paddle; pole; rake; stirrer; spatula 829
RUHR (*f*), **SCHWARZE-**: melaena (*Dysenteria splenica*) (Med.) 830
RÜHRSPATEL (*m*): stirring spatula 831
RÜHRSTAB (*m*): stirrer; stirring rod (pole or paddle); mixer 832
RÜHRSTANGE (*f*): (see **RÜHRSTAB**) 833
RÜHRTROMMEL (*f*): revolving iron cylinder (for heating starch to make dextrin); mixing-drum (Concrete) 834
RÜHRUNG (*f*): stirring; agitation; emotion; feeling; touching; movement; bea·ing 835
RÜHRVORRICHTUNG (*f*): stirring (agitating) arrangement 836
RUHR, WEISSE- (*f*): diarrhœa (Med.) 837
RÜHRWERK (*n*): stirrer; agitator; blunger (Ceramics) 838
RUHRWURZEL (*f*): tormentil root (see **TORMEN-TILLWURZEL**); ipecacuhanna (of *Cephælis ipecacuanha*; calumba (of *Jateorhiza palmata*) 839
RUIN (*m*): ruin; decay; downfall 840
RUINE (*f*): ruins; ruin 841
RUINENHAFT: in ruins, crumbled, decayed, tumble-down 842
RUINENMARMOR (*m*): ruin marble 843
RUINIEREN: to ruin; spoil; decay; destroy 844
RUJAHOLZ (*n*): young fustic; Venetian sumac wood 845
RÜLLÖL (*n*): (see **LEINDOTTERÖL**) 846
RÜLLSAAT (*f*): (see **LEINDOTTERPFLANZE**) 847
RÜLPS (*m*): belch; eructation (Med.) 848
RÜLPSEN: to belch, eruct, eructate 849
RUM (*m*): rum 850
RUMÄTHER (*m*): rum ether; ethyl butyrate or formate; formic ether diluted with alcohol (see **AMEISENÄTHER**) 851
RUMBRENNEREI (*f*): rum distillery 852
RUMESSENZ (*f*): rum essence; ethyl butyrate or formate 853
RUMMEL (*m*): row; rumble; boom; uproar; rubbish; lumber; bulk; lump; lot; gross 854
RUMMELN: to rumble 855
RUMOR (*m*): noise; bustle; rumour 856
RUMOREN: to make a noise or bustle 857
RUMPELKAMMER (*f*): lumber room 858
RUMPELN: to rumble; rattle; turn (upside down) topsy-turvy 859
RUMPF (*m*): body; rump; trunk; torso; hull (Marine) 860
RÜMPFEN: to pucker; pout; curl up; wrinkle 861
RUMPF (*m*), **SCHIFFS-**: body of a ship 862
RUMPFSPRIT (*m*): double rum 863
RUND: round; circular; curved; rotund; spherical; plain; frank; even (of numbers) 864
RUNDBOGEN (*m*): circular arch, Roman arch 865
RUNDBRECHER (*m*): circular breaker 866
RUNDBRENNER (*m*): circular burner or nozzle 867
RUNDE (*f*): round 868
RÜNDE (*f*): rounding; roundness; rotundity; curve; curvature 869
RUNDEISEN (*n* and *pl*): rod, round bar(s) (Iron) 870
RUNDEN: to round-(off), even up; settle, finish; make round, curve 871
RUNDERHABEN: convex 872
RUNDFAHRT (*f*): circular (trip; voyage; drive; journey) tour 873
RUNDFEILE (*f*): round file (Tools) 874
RUNDFLASCHE (*f*): round (Florence) flask 875
RUNDFRAGE (*f*): circular letter, general enquiry (a letter or personal application to a number of people, asking for replies to certain questions or inviting them to give their experience on certain points) 876
RUNDGANG (*m*): revolution, rotation; rounds 877
RUNDGEMÄLDE (*n*): panorama 878
RUNDGESANG (*m*): round (Music) 879
RUNDHOBEL (*m*): hollow-nosed plane 880
RUNDHOHL: concave 881
RUNDHÖHLUNG (*f*): concavity 882
RUNDHOLZ (*n*): round timber 883
RUNDKELLE (*f*): round trowel 884
RUNDKIESEL (*m*): pudding stone 885
RUNDKOLBEN (*m*): round bottomed flask 886
RUNDLAUF (*m*): course, circulation, circular (movement, path, travel) course 887
RUNDLICH: circular; arched; rotund 888
RUNDREISE (*f*): circuit; circular (trip) tour 889
RUNDSÄGE (*f*): circular saw 890
RUNDSÄULE (*f*): cylindrical column; cylinder 891
RUNDSCHAU (*f*): review 892
RUNDSCHIEBER (*m*): piston valve, round slide valve (Mech.) 893
RUNDSCHIEBERSTEUERUNG (*f*): piston valve (gear) distribution 894
RUNDSCHREIBEN (*n*): circular-(letter); round robin 895
RUNDSCHRIFT (*f*): copper-plate writing, round-hand writing 896

RUNDSCHRIFTFEDER (*f*): copper-plate pen, round-hand pen 897
RUNDUM: round about, around 898
RUNDUNG (*f*): rounding; roundness; rotundity; arch; curvature; curve 899
RUNDWEG: plainly 900
RUNDZANGE (*f*): round nosed pliers 901
RUNDZIRKEL (*m*): calipers 902
RUNENSCHRIFT (*f*): runic (inscription) characters 903
RUNGE (*f*): pin; bolt; rung 904
RUNISCH: runic 905
RUNKEL (*f*): beet-(root) (*Beta vulgaris*, a variety of *Beta maritima*) 906
RUNKELRÜBE (*f*): (see **RUNKEL**) 907
RUNKELRÜBENZUCKER (*m*): beet sugar (see **ROHRZUCKER**) 908
RUNZEL (*f*): wrinkle; fold; pucker; crease; rumple 909
RUNZELIG: wrinkled; puckered; creased 910
RUNZELN: to wrinkle; rumple; pucker; crease; fold 911
RÜPEL (*m*): boor; lout 912
RUPFEN: to pluck; pick; pull; fleece 913
RUPIE (*f*): rupee (Indian coinage) 914
RUPPIG: shabby; tattered; ragged; niggardly; stingy; mean; paltry 915
RUPRECHTSKRAUT (*n*): herb Robert (*Geranium robertianum*) 916
RUSAÖL (*n*): ginger-grass oil (distilled from *Andropogon*) 917
RUSS (*m*): soot; lampblack; rust; carbon black 918
RUSSABBLÄSER (*m.pl*): soot blowers 919
RUSSABBLASVORRICHTUNG (*f*): sooting (valve) arrangement; soot blower 920
RUSSARTIG: soot-like 921
RUSSAUSBLASEROHR (*n*): sooting tube 922
RUSSBRAUN (*n*): bistre 923
RUSSBRENNAPPARAT (*m*): lampblack manufacturing apparatus 924
RUSSBRENNEREI (*f*): lampblack (manufacture) manufactory 925
RÜSSEL (*m*): trunk; nose; snout; nozzle; proboscis; muzzle 926
RUSSERZEUGUNG (*f*): lampblack manufacture 927
RUSSFARBE (*f*): lamp-black; bistre 928
RUSSIG: sooty 929
RUSSISCH: Russian 930
RUSSISCHES BILD (*n*): vignette on a black ground (Phot.) 931
RUSSISCHES GLAS (*n*): mica; Muscovy glass; Muscovite (see **MUSKOVIT**) 932
RUSSISCHGRÜN (*n*): Russian green 933
RUSSKOBALT (*m*): asbolite 934
RUSSSCHWARZ (*n*): lampblack 935
RUSSSORTE (*f*): kind (sort) of soot, lampblack or rust (see **RUSS**) 936
RÜSTBAUM (*m*): scaffold-pole 938
RÜSTBOCK (*m*): trestle; jack 939
RÜSTE (*f*): channel; setting (of the sun) 940
RÜSTE (*f*), **GROSSE-**: main channel 941
RÜSTEN: to prepare; equip; furnish; arm; fit-out; dress; prop; erect scaffolding; support; shore-(up) 942
RÜSTER (*f*): elm-(tree) (see **ULME**) 943
RÜSTERN: maple; elm; of elm 944
RÜSTERRINDE (*f*): elm bark 945
RÜSTHOLZ (*n*): shore; strut; support; prop 946
RÜSTIG: robust; vigorous; stout; active; strong 947
RÜSTUNG (*f*): armament; arming; equipment; preparation; scaffolding; armature (Elect.); fittings; apparatus; implements 948
RÜSTZEUG (*n*): scaffolding; set of tools; crane; implements 949
RUTE (*f*): rod; twig; switch; stick; wand; pole; yard; rod (Measure); penis (Med.) 950
RUTENKRAMPF (*m*): priapism (an affection of the penis) (Med.) 951
RUTENSEGEL (*n*): lateen (yard) sail 952
RUTHENIUM (*n*): Ruthenium; (Ru) 953
RUTHENIUMDIOXYD (*n*): ruthenium dioxide; RuO_2; Sg. 7.2 954
RUTHENIUMOXYD (*n*): ruthenium oxide; ruthenium sesquioxide; Ru_2O_3 955
RUTHENIUMOXYDUL (*n*): ruthenium monoxide; RuO 956
RUTHENIUMVERBINDUNG (*f*): ruthenium compound 957
RUTHERFORDIN (*n*): rutherfordine, uranyl carbonate; UCO_5; Sg. 4.8 958
RUTHERFORDIT (*m*): rutherfordite; $(Ti,U)O_2$, $(Ce,Y)_2O_3$; Sg. 5.62 959
RUTIL (*n*): rutile [natural Titanium (di)-oxide]; TiO_2; (90-98% TiO_2); Sg. 4.25; Mp. 1560°C. 960
RUTSCH (*m*): slip, slide, fall, landslip 961
RUTSCHBAHN (*f*): slide, chute 962
RUTSCHE (*f*): chute, gravity chute, slide 963
RUTSCHEN: to slide; slip; glide; shoot 964
RUTSCHPULVER (*n*): talc-(um) powder 965
RUTTE (*f*): (see **QUAPPE**) 966
RÜTTELAPPARAT (*m*): shaking apparatus (usually a sieve); winnowing machine (for grain) 967
RÜTTELN: to agitate; stir; toss; shake; jolt; rattle; jog; winnow (Grain) 968

S

SAAL (*m*): hall; assembly room; saloon; drawing-room 969
SAALGLÄTTE (*f*): floor polish 970
SAARBAUM (*m*): (see **SCHWARZPAPPEL**) 971
SAARBUCHE (*f*): (see **SCHWARZPAPPEL**) 972
SAAT (*f*): seed; young crops; green corn; sowing 973
SAATFELD (*n*): corn-field 974
SAATGUT (*n*): seeds; young crops 975
SAATWECHSEL (*m*): rotation of crops 976
SAATZEIT (*f*): sowing time 977
SABADILLESSIG (*m*): sabadilla vinegar (from sabadilla leaves) 978
SABADILLSAMEN (*m* and *pl*): sabadilla seed-(s); Indian barley caustic; Cevadilla (of *Asagraea officinalis*) 979
SABALOL (*n*): sabalol 980
SABBOTIEREN: to wreck 981
SÄBEL (*m*): sword; sabre; scimitar; cutlass 982
SÄBELFÖRMIG: ensiform; sword-shaped 983
SABINERBAUM (*m*): savin-(tree) (*Juniperus sabina*) 984
SABROMIN (*n*): sabromine [potassium salt of dibromobehenic acid; $Ca(C_{22}H_{41}Br_2O_2)_2$] 985
SACCHARAT (*n*): saccharate 986

SACCHARID (*n*): saccharide (one of the grape sugar group of sugars) 987
SACCHARIMETER (*n*): saccharometer; saccharimeter 988
SACCHARIN (*n*): saccharol; glycosine; saxin; sykose; glusimide; garantose; glusidum; saccharine; benzo-sulfinide; benzoylsulfonic imide; neo-saccharin; gluside; glycophenol; saccharinol; saccharinose; ortho-benzoic sulfimide; $C_6H_4COSO_2NH$; Mp. 224°C. 989
SACCHAROFERROL (*n*): saccharoferrol 990
SACCHAROIDISCH: saccharoid; saccharoidal 991
SACCHAROSE (*f*): saccharose (one of the cane sugar group of sugars) (see ROHRZUCKER) 992
SACHALINER KNÖTERICH (*m*): Sakhalin knotgrass; Saghalien knotgrass (from Saghalien, the Island off East Siberia) (*Polygonum sachalinense*) 993
SACHARIN (*n*): (see SACCHARIN) 994
SACHBEWEIS (*m*): concrete evidence, material facts, material proof 995
SACHDIENLICH: relevant 996
SACHE (*f*): matter; concern; affair; cause; business; thing; subject; case; lawsuit; (*pl*): goods; clothes; properties 997
SACHFÜHRER (*m*): advocate; attorney; agent; manager 998
SACHGEMÄSS: pertinent; appropriate; suitable; objective; serviceable; matter-of-fact 999
SACHKENNER (*m*): expert 000
SACHKENNTNIS (*f*): expert knowledge; practice; experience; practical knowledge 001
SACHKUNDE (*f*): (see SACHKENNTNIS) 002
SACHKUNDIG: expert; experienced; capable; versed; having practical (knowledge) experience 003
SACHLAGE (*f*): state of affairs; circumstances 004
SACHLICH: material; real 005
SÄCHLICH: neuter 006
SACHREGISTER (*n*): (subject)-index 007
SÄCHSISCH: Saxon 008
SÄCHSISCHBLAU (*n*): Saxon-(y) blue; cobalt blue (see BLAUFARBE) 009
SÄCHSISCHES BLAU (*n*): Saxon-(y) blue; cobalt blue (see BLAUFARBE) 010
SACHT: soft; light; easy; low; gentle; scoured (Silk) 011
SACHTE: (see SACHT) 012
SACHVERHALT (*m*): circumstances 013
SACHVERSTÄNDIG: expert; experienced; competent; having special knowledge; acquainted with 014
SACHVERSTÄNDIGE (*m*): expert; competent person; specialist 015
SACHVERSTÄNDIGER (*m*): specialist, expert, competent person 016
SACHWALTER (*m*): attorney; lawyer; counsel; manager 017
SACHWERT (*m*): actual (real) value 018
SACHWÖRTERBUCH (*n*): encyclopædia 019
SACK (*m*): sack; bag; pouch; pocket (also referring to fluids; Technical); cyst 020
SACKAUSKLOPFMASCHINE (*f*): sack (bag) dusting or cleaning machine 021
SACKAUSSTÄUBEMASCHINE (*f*): sack (bag) dusting or cleaning machine 022
SACKBAHN (*f*): loop line 023
SACKBAND (*n*): string (for tying bags) 024
SÄCKCHEN (*n*): little bag; purse 025
SÄCKEL (*m*): (see SÄCKCHEN) 026
SACKEN: to bag; put in sacks; sag; bulge; become baggy; subside; lower; settle 027
SACKFILTER (*n*): bag (sack) filter 028
SACKFÖRDERANLAGE (*f*): sack-handling plant 029
SACKFRUCHT (*f*): sporangium 030
SACKFÜLLAPPARAT (*n*): sack filling apparatus 031
SACKGASSE (*f*): blind alley; cul-de-sac 032
SACKLEINEN (*n*): sacking; sackcloth 033
SACKLEINWAND (*f*): (see SACKLEINEN) 034
SACKTRAGEND: sacciferous, having a sac 035
SACKTUCH (*n*): (see SACKLEINEN) 036
SACKVERSCHLIESSAPPARAT (*m*): sack (closing) shutting apparatus 037
SACK (*m*), WASSER-: water trap 038
SADEBAUM (*m*): savin-(tree) (*Juniperus sabina*) 039
SADEBAUMBLÄTTER (*n. pl*): savin leaves (*Folia sabinæ*) 040
SADEBAUMKRAUT (*n*): savin herb (*Herba savinæ*) 041
SADEBAUMÖL (*n*): savin oil; Sg. 0.92 042
SÄEMANN (*m*): sower 043
SÄEMASCHINE (*f*): sowing machine; drill plough; seeding machine (Agricultural) 044
SÄEN: to sow 045
SÄER (*m*): sower 046
SAFFIAN (*m*): morocco-(leather) 047
SAFFLOR (*m*): safflower (*Flores carthami*) (see FÄRBERDISTEL) (see also ZAFFER); also a cobalt colour 048
SAFFLORGELB (*n*): safflower (saffron) yellow (from *Carthamus tinctorius*) 049
SAFFLORIT (*m*): safflorite; $Co(AsS)_2$; Sg. 7.1 050
SAFFLORKARMIN (*n*): (see CARTHAMIN) 051
SAFFLORROT (*n*): safflower (saffron) red, carthamine (from *Carthamus tinctorius*) (see CARTHAMIN) 052
SAFFRAN (*m*): saffron (*Crocus sativus*) 053
SAFLORÖL (*n*): safflower oil (from *Carthamus tinctorius*); Sg. 0.927 054
SAFRAN (*m*): (see SAFFRAN) 055
SAFRANBRONZE (*f*): Tungsten bronze, Wolfram bronze 056
SAFRANGELB (*n*): saffron-yellow; saffranine 057
SAFRANHALTIG: containing saffron 058
SAFRANIN (*n*): safranine, aniline pink, aniline rose (a dyestuff) 059
SAFRANSURROGAT (*n*): saffron surrogate; saffron substitute; antinonnine 060
SAFROL (*n*): safrol, shikimol, allylpyrocatecholmethylene ether (the main constitutent of camphor oil); $C_3H_5C_6H_3O\ CH_2$; Sg. 1.096-1.107; Mp. 11°C.; Bp. 233°C. 061
SAFRONAL (*n*): safronal 062
SAFRONARTIG: saffron-like 063
SAFT (*m*): juice (*Succus*); sap; fluid; liquid; gravy; syrup; lymph; liquor 064
SAFTBEHÄLTER (*m*): nectary (the part of the flower which secretes or contains honey); sap-vessel 065
SAFTBRAUN (*n*): sap brown 066
SAFTGANG (*m*): sap passage or duct 067
SAFTGELB (*n*): Dutch pink, sap yellow (from dried Persian berries, *Rhamnus frangula*) 068
SAFTGRÜN (*n*): sap green (from unripe berries of *Rhamnus*) 069
SAFTHALTIG: sappy; sap bearing 070
SAFTIG: juicy; sappy; succulent 071
SAFTLEITUNG (*f*): sap conduction; sap passage or duct 072
SAFTLOS: sapless; juiceless 073

SAFTREICH : juicy ; sappy ; richly sap-bearing 074
SAFTZELLE (*f*) : lymph cell 075
SAGAPANGUMMI (*n*) : sagapenum 076
SAGAPENGUMMI (*n*) : sagapenum 077
SAGBAR : capable of being (expressed) uttered ; utterable 078
SAGE (*f*) : legend ; tradition ; saga ; saying ; rumour ; myth ; folk-lore 079
SÄGE (*f*) : saw 080
SÄGEARTIG : serrated ; saw-like ; serrate 081
SÄGEBLATT (*n*) : saw blade 082
SÄGEBLÄTTERIG : serretifoliate 083
SÄGEBOCK (*m*) : saw-bench, sawing trestle 084
SÄGEBOGEN (*m*) : saw bow or frame 085
SÄGEFEILE (*f*) : saw file 086
SÄGEFÖRMIG : serrate, serrated, saw-like 087
SÄGEGATTER (*n*) : saw-mill 088
SÄGEGESTELL (*n*) : saw frame 089
SÄGE (*f*), **HORIZONTAL-** : horizontal saw 090
SÄGE (*f*), **KREIS-** : circular saw 091
SÄGEMASCHINE (*f*) : sawing-machine 092
SÄGEMEHL (*n*) : sawdust 093
SÄGEMÜHLE (*f*) : sawmill 094
SAGEN : to say ; speak ; tell ; mention ; remark ; mean; declare ; render ; admit 095
SÄGEN : to saw ; cut 096
SAGENHAFT : traditional ; legendary ; fabulous ; mythical 097
SÄGENSCHÄRFER (*m*) : saw-sharpener 098
SÄGESCHNITT (*m*) : saw-cut 099
SÄGESPÄNE (*m.pl*) : sawdust, shavings (approx. 12% fixed carbon, 40% volatile matter, 0.7% ash ; calorific value, dry, 5000 Kcal. per kg., or as fired, with 50% moisture, 2250 Kcal/kg.) 100
SÄGESPÄNFEUERUNG (*f*) : saw-dust furnace ; saw-dust firing 101
SÄGE (*f*), **STICH-** : piercing saw 102
SÄGETISCH (*m*) : saw table 103
SÄGEWERK (*n*) : saw-mill 104
SÄGEZAHN (*m*) : saw-tooth ; (*pl*) : indentations 105
SAGO (*m*) : sago 106
SAGOPALME (*f*) : sago (palm) tree (*Sagus lævis* and *Cycas circinalis*) 107
SAGRADARINDE (*f*) : cascara sagrada 108
SAGROTAN (*n*) : sagrotan (a disinfectant) 109
SAHLBAND (*n*) : selvedge ; border ; wall 110
SAHLLEISTE (*f*) : (see **SAHLBAND**) 111
SAHNE (*f*) : cream 112
SAHNEN : to skim-(milk) 113
SAHNENKÄSE (*m*) : cream-cheese 114
SAHNIG : creamy 115
SAIDSCHÜTZER SALZ (*n*) : magnesium sulphate (see **MAGNESIUMSULFAT**) 116
SAIGER : perpendicular, upright 117
SAIGERN : to liquate (Metal) (see also **SEI-**) 118
SAISON (*f*) : season 119
SAITE (*f*) : string ; cord ; catgut 120
SAITENBEZUG (*m*) : set of strings 121
SAITENBRETT (*n*) : tail-piece (of musical instruments) 122
SAITENDRAHT (*m*) : wire-string (for pianos, mandolines, etc.) 123
SAITENINSTRUMENT (*n*) : stringed instrument 124
SAITIG : stringed 125
SAJODIN (*n*) : sajodine [potassium salt of mono-iodobehenic acid ; $(C_{22}H_{42}O_2I)_2Ca$; iodine content 25%] 126
SÄKULÄR : secular 127
SÄKULÄRISIEREN : to secularize 128

SALACETIN (*n*) : salacetine 129
SALACETOL (*n*) : salacetol, acetolsalicylic ester ; $C_6H_4(OH)CO_2CH_2COCH_3$; Mp. 71°C. 130
SALAMSTEIN (*m*) : corundum (see **KORUND**) 131
SALÄR (*n*) : salary, wages 132
SALAT (*m*) : lettuce ; salad 133
SALATBOHNE (*f*) : French bean (*Phaseolus vulgaris*) 134
SALATÖL (*n*) : salad oil ; olive oil (see **OLIVENÖL**) 135
SALATPFLANZE (*f*) : lettuce (*Lactuca*) 136
SALATRÜBE (*f*), **ROTE-** : common, red, garden beet root (*Beta vulgaris*) 137
SALAZETOL (*n*) : (see **SALACETOL**) 138
SALBADEREI (*f*) : twaddle ; gossip ; quackery 139
SALBADERN : to gossip 140
SALBAND (*n*) : (see **SAHLBAND**) 141
SALBE (*f*) : ointment ; salve ; unction ; unguent (*Ungentum*) 142
SALBEI (*f*) : sage (*Salvia*) 143
SALBEIBLÄTTER (*n.pl*) : sage leaves (*folia salviæ*) 144
SALBEIÖL (*n*) : sage oil ; salvia oil Sg. 0.92 (from leaves of *Salvia officinalis*) 145
SALBEMULL (*m*) : ointment (mull) muslin 146
SALBEN : to anoint ; rub with grease 147
SALBENBÜCHSE (*f*): ointment box 148
SALBENGRUNDLAGE (*f*) : ointment base 149
SALBENKRUKE (*f*) : ointment jar or pot 150
SALBENREIBMASCHINE (*f*) : ointment grinder 151
SALBENSPATEL (*m*) : spatula (for unctions) 152
SALBIG : unctuous 153
SALCREOLIN (*n*) : salcreolin 154
SALDIEREN : to balance (Accounts) 155
SALDO (*m*) : balance (of accounts) 156
SALDOVORTRAG (*m*) : balance carried forward 157
SALEN (*n*) : salene (methylglycolic ester and ethylglycolic ester of salicylic acid mixed in equal parts) 158
SALENAL (*n*) : salenal (an ointment containing 33⅓% salene ; see **SALEN**) 159
SALEP (*n*) : salep 160
SALEPKNOLLEN (*f. pl*) : salep root (*Tubera salep*, from various *Orchidaceæ* of the *Orphydineæ* group) 161
SALIBROMIN (*n*) : methyl dibromosalicylate ; $C_6H_2Br_2(OH)CO_2CH_3$ 162
SALICIN (*n*): salicine (*Salicinum*) ; $(C_{13}H_{18}O_7)$; Mp. 201°C. (see also undes **SALIZ-**) 163
SALICOL (*n*) : salicol (mixture of decomposed acetylsalicylic acid and 3-4% citric acid) 164
SALICOYLSÄURE (*f*) : salicylic acid (see **SALICYL-SÄURE**) (see also under **SALIZYL-**) 165
SALICYLALDEHYD (*n*) : salicylic aldehyde ; $C_6H_4(OH)CHO$, salicylous acid, ortho-oxybenzaldehyde ; Sg. 1.17 ; Mp. − 10°C. ; Bp. 196°C. 166
SALICYLAT (*n*) : salicylate 167
SALICYLÄTHER (*m*) : salicylic ether 168
SALICYLÄTHYLESTER (*m*) : salicylic ethylester 169
SALICYLESSIGSÄURE (*f*) : salicylacetic acid 170
SALICYLID (*n*): salicylic anhydride (see **SALICYL-SÄUREANHYDRID**) 171
SALICYLIEREN : to salicylate 172
SALICYLIERT : salicylated 173
SALICYLOSALICYLSÄURE (*f*) : (see **DIPLOSAL**) 174
SALICYLPRÄPARAT (*n*) : salicylate ; salicylic preparation 175
SALICYLSAUER : combined with salicylic acid ; salicylate of 176

SALICYLSÄURE (*f*): salicylic acid (see OXYBENZOE-SÄURE) (*Acidum salicylicum*) 177
SALICYLSÄUREAMYLESTER (*m*): salicylic amyl ester, amyl salicylate; $C_7H_5O_3:C_5H_{11}$; Sg. 1.045; Bp. 268-273°C. 178
SALICYLSÄUREANHYDRID (*n*): salicylic anhydride; C_6H_4COO; Mp. 156°C. 179
SALICYLSÄUREÄTHYLESTER (*m*): ethyl salicylate 180
SALICYLSÄUREBORNYLESTER (*m*): bornyl salicylate; $C_{10}H_{17}O.CO.C_6H_4.OH$ 181
SALICYLSÄUREMETHYLÄTHER (*m*): salicylic methyl ether 182
SALICYLSÄUREMETHYLESTER (*m*): methyl salicylate (see WINTERGRÜNÖL, KÜNSTLICHES-) 183
SALICYLSÄURESEIFE (*f*): salicylic acid soap 184
SALICYLURSÄURE (*f*): salicyluric acid 185
SALIFORMIN (*n*): hexamethylenetetramine salicylate; $(CH_2)_6N_4.C_6H_4(OH).COOH$ 186
SALIGEN (*n*): saligen 187
SALIMENTHOL (*n*): menthol salicylate 188
SALIN (*n*): salin (a dark-brown residue from the drying of the lye in the process of extracting potash from beech-wood ashes) (see SCHLEMPEKOHLE) 189
SALINE (*f*): saltern; salt (pit) works 190
SALINENEINRICHTUNG (*f*): saline plant 191
SALING (*f*): cross tree (Naut.) 192
SALINISCH: saline 193
SALIPYRIN (*n*): salipyrine phenyldimethylpyrazolon salicylate (*Antipyrinum salicylicum*); Mp. 92°C. 194
SALIT (*m*): salite (a variety of diopside); $CaMgSiO_3$; Sg. 3.4 195
SALIT (*n*): salite; bornyl salicylate (see SALICYLSÄUREBORNYLESTER) 196
SALIVIEREN: to salivate 197
SALIZYL-: (see under SALICYL-) 198
SALIZYLCHININ (*n*), **SALIZYLSAURES-**: salicylquinine salicylate; $C_6H_4(OH)CO_2.C_{20}H_{23}N_2O.C_6H_4(OH).CO_2H$; Mp. 179°C. 199
SALIZYLSÄUREÄTHYLESTER (*m*): ethyl salicylate 200
SALIZYLSÄURE-B-NAPHTHYLESTER (*m*): beta-naphthol salicylate; $C_7H_5O_3.C_{10}H_7$ 201
SALIZYLSÄURECHININESTER (*m*): (see SALOCHININ) 202
SALIZYLSÄUREMETHYLENAZETAT (*n*): salicylic acid methylene acetate (see INDOFORM) 203
SALIZYLSÄUREMETHYLESTER (*m*): methyl salicylate, artificial oil of wintergreen; $OHC_6H_4CO_2CH_3$; Sg. 1.1819; Mp. −8.3°C.; Bp. 223°C. 204
SALIZYLSÄUREPHENYLESTER (*m*): salicylic acid phenyl ester (see SALOL) 205
SALM (*m*): salmon 206
SALMIAK (*m*): sal ammoniac; muriate of ammonia; ammonium chloride (see AMMONIUMCHLORID); NH_4Cl; Sg. 1.52 207
SALMIAKFLÜSSIGKEIT (*f*): ammonia liquid (*Liquor ammonii caustici*) (solution of ammonia gas in water); Sg. 0.959 (about 10% NH_3) 208
SALMIAKGEIST (*m*): aqueous ammonia; liquid ammonia (*Liquor ammonii caustici*; about 30% NH_3); Sg. 0.895 (see also AMMONIAK) 209
SALMIAKLÖSUNG (*f*): solution of sal ammoniac 210
SALMIAKPASTILLE (*f*): sal ammoniac pastilles 211
SALMIAK (*m*), **PLATIN-**: ammonium chloroplatinate (see AMMONIUMPLATINCHLORID and PLATINSALMIAK) 212
SALMIAKSALZ (*n*): sal volatile; ammonium carbonate (see AMMONIUMKARBONAT); sal ammoniac; ammonium chloride (see AMMONIUMCHLORID) 213
SALMIAKSCHLACKE (*f*): sal ammoniac slag 214
SALMONIDEN GEWÄSSER (*n*): salmon water-(s), water containing salmon 215
SALMROT (*n*): salmon red 216
SALOCHININ (*n*): saloquinine, salicylic acid quinine ester, quinine salicylate; $C_6H_4(OH)CO_2.C_{20}H_{23}N_2O$ 217
SALOCOL (*n*): salocol (*Phenocollum salicylicum*); $C_6H_4(OC_2H_5)NH.CO.CH_2.NH_2.C_7H_6O_3$ 218
SALOCREOL (*n*): salocreol, creosote salicylate 219
SALOKOLL (*n*): (see SALOCOL) 220
SALOL (*n*): salol; phenyl salicylate; $C_6H_4OHCO_2C_6H_5$; Sg. 1.26; Mp. 41.9°C.; Bp. 172.5°C. 221
SALOLSEIFE (*f*): salol soap 222
SALOMONSSIEGEL (*n*): Solomon's seal (*Polygonatum multiflorum*) 223
SALON (*m*): saloon; drawing-room 224
SALONKAMERA (*f*): studio camera (Phot.) 225
SALOPHEN (*n*): acetyl-para-aminophenyl salicylate, salophen, acetaminosalol; $C_6H_4OH.COOC_6H_4NH.COCH_3$; Mp. 187°C. 226
SALOSANTAL (*n*): salosantal 227
SALPETER (*m*): saltpeter; nitre; potassium nitrate; KNO_3; Sg. 2.1 (see KALIUMNITRAT) 228
SALPETERARTIG: nitrous; saltpeter-like 229
SALPETERÄTHER (*m*): nitric ether; ethyl nitrate; $C_2H_5NO_3$; Sg. 1.116; Mp. 112°C.; Bp. 87.6°C. 230
SALPETERBILDUNG (*f*): nitrification 231
SALPETERDAMPF (*m*): nitrous fumes 232
SALPETERDRUSE (*f*): crystallized saltpeter 233
SALPETERDUNST (*m*): nitrous fumes 234
SALPETERERDE (*f*): nitrous earth 235
SALPETERESSIGSAUER: nitroacetate of; nitrate and acetate of 236
SALPETERFÜLLUNG (*f*): saltpeter (filling) content 237
SALPETERGAS (*n*): nitrous gas; saltpeter gas; gas from the manufacture of nitric acid; nitrous (nitric) oxide; (old name for)-nitrogen (see LACHGAS) 238
SALPETERGEIST (*m*): spirit of nitre; nitric acid (see SALPETERSÄURE) 239
SALPETERGEIST, VERSÜSSTER- (*m*): sweet spirit of nitre; ethyl nitrite; spirit of nitrous ether (see ÄTHYLNITRIT) 240
SALPETERGRUBE (*f*): saltpeter mine 241
SALPETERHALTIG: nitrous; nitric; containing saltpeter 242
SALPETERHALTIGER HÖLLENSTEIN (*m*): mitigated silver nitrate (Pharm.) 243
SALPETERHÜTTE (*f*): saltpeter works 244
SALPETERKESSEL (*m*): saltpeter boiler 245
SALPETER (*m*), **KONVERSIONS-**: conversion saltpeter (saltpeter obtained by conversion of sodium nitrate with KCl) 246
SALPETER, KUBISCHER- (*m*): Chili saltpeter; sodium nitrate (see NATRIUMNITRAT) 247
SALPETERLAGER (*n*): saltpeter deposit 248
SALPETERLUFT (*f*): nitrogen (Old name) 249
SALPETERMILCHSÄURE (*f*): ortho-nitrolactic acid 250
SALPETERPLANTAGE (*f*): saltpeter plantation 251
SALPETERSALZSAUER: nitromuriate of 252

SALPETERSALZSÄURE (*f*): nitrohydrochloric acid; aqua regia; nitromuriatic acid (see **KÖNIGS-WASSER**) 253

SALPETERSAUER: nitrate of; combined with nitric acid 254

SALPETERSÄURE (*f*): nitric acid; azotic acid; hydrogen nitrate (*Aqua fortis*); (*Acidum nitricum*); HNO_3; Sg. 1.522-1.559; Mp. $-41.3°C.$; Bp. 86°C. 255

SALPETERSÄUREANHYDRID (*n*): nitric anhydride; nitrogen pentoxide; N_2O_5 (see **STICK-STOFFPENTOXYD**) 256

SALPETERSÄUREANLAGE (*f*): nitric acid plant 257

SALPETERSÄUREÄTHER (*m*): nitric ether; ethyl nitrate (see **SALPETERÄTHER**) 258

SALPETERSÄUREDARSTELLUNG (*f*): production (output, manufacture) of nitric acid 259

SALPETERSÄUREESTER (*m*): nitric ester, erythrol tetranitrate; $C_4H_6(NO_3)_4$; Mp. 61°C. 260

SALPETERSÄUREFABRIKATION (*f*): nitric-acid manufacture 261

SALPETERSÄURE (*f*), **(ROTE)-RAUCHENDE-**: fuming nitric acid, nitrosonitric acid (saturated solution of N_2O_4 in HNO_3 having a Sg. over 1.45) 262

SALPETERSAURES KALI (*n*): (see **KALIUMNITRAT**) 263

SALPETERSAURES NATRON (*n*): sodium nitrate (see **NATRIUMNITRAT**) 264

SALPETERSAURES SALZ (*n*): nitrate 265

SALPETERSÄURETRIGLYZERID (*n*): (see **NITRO-GLYZERIN**) 266

SALPETERSÄUREVERFAHREN (*n*): nitric-acid process 267

SALPETERSCHWEFELSÄURE (*f*): nitrosylsulphuric acid; nitrosulphuric acid 268

SALPETERSIEDEREI (*f*): saltpeter (nitre) works 269

SALPETERSTOFF (*m*): nitrogen (see **STICKSTOFF**) 270

SALPETERSUPERPHOSPHAT (*n*): saltpeter superphosphate 271

SALPETERVERBRAUCH (*m*): saltpeter consumption, consumption (utilization, employment) of saltpeter 272

SALPETRICHT: saltpeter-like; nitrous 273

SALPETRIG: nitrous; containing saltpeter 274

SALPETRIGESÄURE (*f*): nitrous acid 275

SALPETRIGESCHWEFELSÄURE (*f*): nitrosylsulphuric acid; nitrosulphuric acid 276

SALPETRIGSAUER: nitrite of; combined with nitrous acid 277

SALPETRIGSÄURE (*f*): nitrous acid 278

SALPETRIGSÄUREANHYDRID (*n*): nitrous anhydride; N_2O_3 279

SALPETRIGSÄUREÄTHER (*m*): nitrous ether; ethyl nitrite (see **ATHYLNITRIT**) 280

SALPETRIGSÄUREÄTHYLESTER (*m*): (see **SALPET-RIGSÄUREÄTHER**) 281

SALPETRIGSAURES KALI (*n*): potassium nitrate; (KNO_2) 282

SALPETRIGSAURES SALZ (*n*): nitrite 283

SALSE (*pl*): fruit pulps 284

SALUSOL (*n*): salusol 285

SALUTIN (*n*): salutine 286

SALVARSAN (*n*): salvarsan, 606, six-o-six, arsenphenamine; $[HO(NH_2.HCl)C_6H_3As]_2.2H_2O$; diaminodihydroxyarsenobenzene hydrochloride 287

SALVE (*f*): salute; volley; round; salvo 288

SALVIEREN: to salvage, rescue 289

SALVIERUNG (*f*): salvage 290

SALVIERUNGSANLAGE (*f*): salvage plant 291

SALWEIDE (*f*): sallow willow (*Salix caprea*) 292

SALZ (*n*): salt; common salt; Sg. 2.13 (see **NATRIUMCHLORID**) 293

SALZABLAGERUNG (*f*): salt deposit 294

SALZADER (*f*): salt vein 295

SALZÄHNLICH: salt-like; like salt; haloid 296

SALZARTIG: salt-like; saline; salty 297

SALZÄTHER (*m*): muriatic ether; ethyl chloride; (Old chemical term; see **ÄTHYLCHLORID**) 298

SALZBAD (*n*): salt bath (for steel hardening, consists of barium chloride and potassium chloride) 299

SALZBASE (*f*): salifiable (salt) base 300

SALZ (*n*), **BASISCHES-**: basic salt 301

SALZBEREITUNG (*f*): salt making, salt preparation 302

SALZBERGWERK (*n*): salt mine 303

SALZBEZIRK (*m*): salt district 304

SALZBILDEND: salt forming; halogenous 305

SALZBILDER (*m*): halogen; salt-former; salt forming substance 306

SALZBILDNER (*m*): (see **SALZBILDER**) 307

SALZBILDUNG (*f*): salification; salt formation 308

SALZBILDUNGSFÄHIG: salifiable 309

SALZBLUMEN (*f.pl*): efflorescence of salt 310

SALZBRÜHE (*f*): brine; pickle 311

SALZBURGER VITRIOL (*m*): eagle vitriol 312

SALZDECKE (*f*): a cover impregnated with salts (to absorb poisonous gases) 313

A **SALZ** (*n*) **DER** *B* **SÄURE** (*f*): *A*-*B*-ate (see below *B* **SAURES** *A*) (Example: **NATRIUMSALZ DER DIOXYWEINSÄURE** = **DIOXYWEINSAURES NATRIUM** = sodium dioxytartrate or sodium salt of dioxytartaric acid) 314

SALZEN: to salt; season 315

SALZFASS (*n*): salt-box; salt-cellar 316

SALZFLEISCH (*n*): salt meat 317

SALZFLUSS (*m*): eczema (a skin disease) (Med.); saline flux 318

SALZGARTEN (*m*): salt garden (see **MEERSALINE**) 319

SALZGEHALT (*m*): salt content 320

SALZGEIST (*m*): spirit of salt; hydrochloric acid (see **SALZSÄURE**) 321

SALZGEIST, LEICHTER- (*m*): light spirit of salt; ethyl chloride (see **CHLORÄTHYL**) 322

SALZGEIST, LIBAVIUS' RAUCHENDER- (*m*): fuming liquor of Libavius; stannic chloride (see **ZINNCHLORID**) 323

SALZGEIST, SCHWERER- (*m*): heavy spirit of salt 324

SALZGEIST, VERSÜSSTER- (*m*): sweet spirit of salt 325

SALZGEMISCH (*n*): mixture of salts 326

SALZGEWINNUNG (*f*): production (output) of salt 327

SALZGLAZUR (*f*): salt glaze 328

SALZGRUBE (*f*): salt (mine) pit 329

SALZGURKE (*f*): pickled (gherkin) cucumber 330

SALZHALTIG: salt bearing; saliferous; containing salt; saline 331

SALZHANDEL (*m*): salt trade 332

SALZICHT: salty; saline; saltish 333

SALZIG: salty; briny; saline; saltish 334

SALZIGKEIT (*f*): saltiness; saltness; salinity 335

SALZKLUMPEN (*m*): lump of salt 336

SALZ (*n*), **KOCH-**: common (cooking) salt (see **NATRIUMCHLORID**) 337

SALZKORN (*n*): grain of salt 338

SALZKRAUT (*n*): salt wort (*Salsola kali* and *Salicornia herbacea*) 339
SALZKRISTALL (*m*): salt crystal 340
SALZKUCHEN (*m*): salt cake 341
SALZKUPFERERZ (*n*): atacamite (see ATAKAMIT) 342
SALZLAGER (*n*): salt (bed) deposit 343
SALZLAKE (*f*): brine 344
SALZLAUGE (*f*): brine; pickle 345
SALZLÖSUNG (*f*): brine; salt solution; solution of salt; saline solution 346
SALZMESSER (*m*): salinometer; salimeter; brine gauge 347
SALZMESTE (*f*): salt-box 348
SALZMÜHLE (*f*): salt mill 349
SALZNAPF (*m*): salt cellar 350
SALZ (*n*), NEUTRALES-: neutral salt 351
SALZNIEDERSCHLAG (*m*): saline deposit; salt deposit 352
SALZPAAR (*n*): pair of salts 353
SALZ, PALMITINSAURES- (*n*): palmitate 354
SALZPAPIER (*n*): salted paper (Phot.) 355
SALZPFANNE (*f*): salt (brine) pan 356
SALZPFANNENSTEIN (*m*): pan scale (Salt) 357
SALZPFLANZEN (*f.pl*): halophytes 358
SALZQUELLE (*f*): salt (saline) spring 359
SALZSAUER: hydrochloride of (Aniline and similar bases); chloride of (Metals, etc.); combined with hydrochloric acid 360
SALZSÄURE (*f*): hydrochloric (muriatic) acid (*Acidum hydrochloricum* or *muriaticum*) (see CHLORWASSERSTOFF); (HCl) 361
SALZSÄUREGAS (*n*): hydrochloric-acid gas; hydrogen chloride (see CHLORWASSERSTOFF) 362
SALZSÄUREHYDRAT (*n*): hydrochloric acid; $HCl.2H_2O$; Sg. 1.46; $HCl.H_2O$; Sg. 1.48 363
SALZ (*n*), SAURES-: acid salt, ———ate 364
SALZSAURES SALZ (*n*): chloride 365
SALZ, SCHWEFELSAURES- (*n*): sulphate 366
SALZSEE (*m*): salt lake 367
SALZSIEDEPFANNE (*f*): salt pan 368
SALZSIEDEREI (*f*): salt (making) works (see SALINE) 369
SALZSOLE (*f*): salt (spring) water; brine 370
SALZTROCKNER (*m*): salt dryer 371
SALZUNGER TROPFEN (*m.pl*): anti-rheumatic (tincture) drops 372
SALZWAGE (*f*): brine gauge; salinometer; salimeter 373
SALZWASSER (*n*): brine; salt water; saline 374
SALZWERK (*n*): salt work-(s); salt factory 375
SALZWIRKUNG (*f*): action (effect) of salt 376
SAMADERARINDE (*f*): Samadera bark 377
SAMARIUM (*n*): Samarium; (Sm) 378
SAMARIUMCARBID (*n*): samarium carbide; SmC_2; Sg. 5.86 379
SAMARIUMCHLORID (*n*): samarium chloride; $SmCl_3$; Sg. 4.27 380
SAMARIUM (*n*), GOLDCHLORWASSERSTOFFSAURES-: samarium chloraurate; $SmCl_3.AuCl_3+10H_2O$ 381
SAMARIUMOXYD (*n*): samarium oxide; Sm_2O_3; Sg. 8.347 382
SAMARIUMSULFAT (*n*): samarium sulphate; $Sm_2(SO_4)_3+8H_2O$; Sg. 2.93 383
SAMARSKIT (*m*): samarskite (a ferrous iron, cerium and uranium tantalate and niobate); $(U,Fe,Nb)_2O_3,(Sn,Th,W,Zr)O_2,(Ca,Ce,Cu,Fe,Mg,Y)O$; Sg. 5.7 384
SAMBAC (*m*): (see JASMIN, ARABISCHER-) 385
SAMBESISCHWARZ (*n*): zambesi black 386
SAME (*m*): seed (*Semen*); spawn; sperm; progeny 387
SAMEN (*m*): seed (*Semen*) 388
SAMENADER (*f*): spermatic vein 389
SAMENBEHÄLTER (*m*): seed-vessel, pericarp 390
SAMENBLÄSCHEN (*n*): spermatocyst 391
SAMENBRUCH (*m*): spermatocele 392
SAMENDRÜSE (*f*): testicle; prostate gland 393
SAMENERZEUGEND: spermatic 394
SAMENFACH (*n*): seed compartment 395
SAMENFADEN (*m*): spermatozoon 396
SAMENFLUSS (*m*): spermatorrhœa (a real or apparent discharge of seminal fluid) (Med.) 397
SAMENFLÜSSIGKEIT (*f*): seminal fluid 398
SAMENFÜHREND: seminiferous; spermatophorous 399
SAMENGANG (*m*): spermatic duct 400
SAMENGEHÄUSE (*n*): seed vessel, pericarp 401
SAMENHANDEL (*m*): seed trade 402
SAMENHÄNDLER (*m*): seed merchant 403
SAMENHAUT (*f*): episperm 404
SAMENHEFE (*f*): seed yeast 405
SAMENHÜLLE (*f*): perisperm 406
SAMENKEIM (*m*): germ; embryo 407
SAMENKELCH (*m*): seed cup 408
SAMENKERN (*m*): seed kernel; endosperm (Bot.); spermatic (sperm) nucleus (Physiol.) 409
SAMENKNOSPE (*f*): gemmule, ovule 410
SAMENKORN (*n*): seed; grain 411
SAMENKRONE (*f*): down 412
SAMENKUCHEN (*m*): placenta 413
SAMENLAPPEN (*m*): seed; lobe; cotyledon (Bot.) 414
SAMENLEHRE (*f*): spermatology 415
SAMENLEITER (*m*): spermatic duct 416
SAMENLOS: seedless 417
SAMENMANTEL (*m*): arillode-(of a seed), arillus 418
SAMENÖL (*n*): seed oil; cotton-seed oil (see BAUMWOLLSAMENÖL) 419
SAMENPFLANZEN (*f.pl*): *Phanerogamæ* (Bot.) 420
SAMENPRODUKTION (*f*): seed production 421
SAMENSAFT (*m*): seminal fluid 422
SAMENSTAUB (*m*): pollen (Plants) 423
SAMENSTRANG (*m*): seed-stalk; spermatic cord 424
SAMENTIERCHEN (*n*): spermatozoon 425
SAMENTRAGEND: seminiferous 426
SAMENTRÄGER (*m*): spermaphore 427
SAMENZELLE (*f*): spermatozoon; seminal cell 428
SAMENZUCKER (*m*): quercite; quercitol; acorn sugar; pentahydroxycyclohexane; $C_6H_7(OH)_5$; Sg. 1.58; Mp. 234°C. 429
SÄMEREI (*f*): seeds; (*pl*): kinds (classes; sorts) of seeds 430
SÄMIG: thick; viscous; viscid 431
SÄMISCH: chamois; soft; slimy 432
SÄMISCHGAR: chamois; oil tanned 433
SÄMISCHGERBEN (*n*): chamoising; oil tanning 434
SÄMISCHGERBER (*m*): chamois (tanner) dresser 435
SÄMISCHGERBEREI (*f*): chamois leather (tanning) tannery 436
SÄMISCHLEDER (*n*): chamois-(leather); wash-leather 437
SAMIT (*n*): carborundum 438
SAMMELBECKEN (*n*): reservoir; sump; basin, accumulator, receiver 439

SAMMELBOTTICH (*m*): starting (collecting) tub or vat (Brewing) 440
SAMMELBRUNNEN (*m*): pump well, hot well 441
SAMMELGEFÄSS (*n*): receiver; reservoir; collector; accumulator 442
SAMMELGLAS (*n*): preparation (specimen) tube; converging lens 443
SAMMELLINSE (*f*): convergent lens; positive lens; convex lens 444
SAMMELN: to collect; botanize; accumulate; assemble; gather; converge; concentrate; compose 445
SAMMELNAME (*m*): general (collective) term, generic term; collective noun 446
SAMMELPLATZ (*m*): collecting (meeting) place; dump 447
SAMMELRAUM (*m*): receiver; receptacle; collecting chamber; reception chamber 448
SAMMELROHR (*n*): collector; header 449
SAMMELSCHIENE (*f*): bus-bar (Elect.) 450
SAMMELWORT (*n*): collective (word) noun 451
SAMMET (*m*): velvet (see also SAMT) 452
SAMMETARTIG: velvet-like 453
SAMMETBLENDE (*f*): limonite (see LIMONIT) 454
SAMMETBRAUN (*n*): velvet brown (an iron pigment); bistre (see MANGANBRAUN) 455
SAMMETEN: velvet; velvety; of velvet 456
SAMMETERZ (*n*): azurite (see AZURIT) 457
SAMMETSCHWARZ (*n*): ivory black 458
SAMMLER (*m*): receiver; collector; compiler; accumulator; storage battery; gatherer 459
SAMMLUNG (*f*): accumulation; collection; assembly; gathering; compilation; rally 460
SAMMLUNGSGLAS (*n*): specimen (display) glass 461
SAMOL (*n*): menthol salicylate ointment (with about 25% menthol salicylate) 462
SAMT: together with; (*m*): velvet (see also SAMMET) 463
SAMTARTIG: velvety; velvet-like 464
SAMTBLUME (*f*): amaranth (*Amaranthus candatus*) 465
SAMTBRAUN (*n*): velvet brown (an iron pigment; bistre (see MANGANBRAUN) 466
SAMTGELB (*n*): zinc yellow (see ZINKGELB) 467
SÄMTLICH: all; entire; jointly; collectively; the whole of 468
SAMT UND SONDERS: one and all, altogether 469
SANAPHORIN (*n*): sanaphorine 470
SANATOGEN (*n*): sanatogen (sodium glycerino-phosphate 5% and casein 95%) 471
SANATOL (*n*): sanatol (a disinfectant) 472
SAND (*m*): sand; Sg. 1.39-1.8 473
SANDALE (*f*): sandal 474
SANDARACH (*m*): gum sandarach; realgar 475
SANDARAK (*m*): (see SANDARACH) 476
SANDARAKGUMMI (*n*): (gum)-sandarac-(resin) (from *Callitris quadrivalvis*) 477
SANDARAKHARZ (*n*): (see SANDARAKGUMMI) 478
SANDARAKLACK (*m*): sandarac varnish 479
SANDBAD (*n*): sand bath; arenation 480
SANDBADSCHALE (*f*): sand-bath dish 481
SANDBANK (*f*): sandbank 482
SANDBERG (*m*): sand dune 483
SANDBESTREUER (*m*): sand-(ing) strewing apparatus 484
SANDBINDEMITTEL (*n*): sand (binding agent) agglutinant (Casting) 485
SANDBODEN (*m*): sandy soil 486
SANDBÜCHSENBAUM (*m*): sand-box tree (*Hura crepitans*) 487
SANDEBENE (*f*): sandy plain 488
SANDEL (*m*): sandalwood (of *Pterocarpus santalinus*) 489
SANDELHOLZ (*n*): (see SANDEL) 490
SANDELHOLZÖL (*n*): sandalwood oil; santal oil; santalwood oil; sandal oil (from *Cæsalpinia echinata*); Sg. 0.95-0.98 491
SANDELHOLZ (*n*), WESTINDISCHES-: West Indian sandalwood, bois de citron (wood of *Amyris balsamifera*) 492
SANDELÖL (*n*): (see SANDELHOLZÖL) 493
SANDEN: to sand, sprinkle sand upon 494
SAND, FINER- (*m*): fine sand; grit 495
SANDFORM (*f*): sand mould (Founding) 496
SANDFORMER (*m*): sand moulder 497
SANDFORMEREI (*f*): sand moulding 498
SANDGEBLÄSE (*n*): sand-blast (machinery) apparatus 499
SANDGRIES (*m*): coarse sand; nne gravel 500
SAND, GROBER- (*m*): (see SANDGRIES) 501
SANDGRUBE (*f*): sand-pit; gravel-pit 502
SANDGRUND (*m*): sandbank (Naut.) 503
SANDGUSS (*m*): sand casting 504
SANDHAUFEN (*m*): sand-heap 505
SANDIG: sandy; gravelly; arenaceous; arenose; arenous 506
SANDKASTEN (*m*): sand box 507
SANDKORN (*n*): grain of sand 508
SANDKUCHEN (*m*): sponge cake 509
SANDMERGEL (*m*): sandy marl 510
SANDPAPIER (*n*): sand paper 511
SANDRIEDGRASWURZEL (*f*): carex root (of *Carex arenaria*); sand (star) sedge root 512
SANDSÄURE (*f*): sand acid; hydrofluosilicic acid; silicofluoric acid; H_2SiF_6 513
SANDSCHIEFER (*m*): schistous sandstone 514
SANDSEIFE (*f*): sand soap 515
SANDSIEB (*n*): sieve; screen; sand-riddling arrangement 516
SANDSTEIN (*m*): sandstone; Sg. 2.2 517
SANDSTOPFBÜCHSE (*f*): sand seal (Blast furnaces) 518
SANDTORTE (*f*): sponge cake 519
SANDUHR (*f*): sand-glass; hour-glass 520
SANDWÜSTE (*f*): sandy desert 521
SANDZUCKER (*m*): crude ground sugar; lactose (see LAKTOSE) 522
SANFT: gentle; soft; smooth; mild 523
SÄNFTE (*f*): sedan chair 524
SÄNFTIGEN: to soften; alleviate; mitigate; smooth 525
SANFTMÜTIG: mild; gentle; tender-hearted; meek 526
SANG (*m*): song 527
SÄNGER (*m*): songster; singer 528
SANGUARIN (*n*): sanguarine (an alkaloid from *Sanguinaria canadensis*) 529
SANGUARINSÄURE (*f*): sanguarinic acid (from *Sanguinaria canadensis*) 530
SANGUIFERRIN (*n*): sanguiferrine 531
SANGUINISPILLE (*f*): sang .nis pill 532
SANGUINOSE (*f*): sanguinose 533
SANIDIN (*m*): sanidine, glassy feldspar (Vitreous orthoclase); $KAlSi_3O_8$; Sg. 2.55 (see also ORTHOKLAS) 534
SANIKEL (*m*): sanicle (*Sanicula europœa*) 535
SANITÄR: sanitary 536
SANITÄT (*f*): sanitation; hygiene; health 537
SANITÄTER (*m*): stretcher bearer (equivalent to R.A.M.C. man in English army) 538
SANITÄTSKOLLEGIUM (*n*): board of health 539
SANITÄTSPFLEGE (*f*): sanitation 540
SANITÄTSWESEN (*n*): sanitary (affairs) matters 541

SANOLEUM (*n*) : sanoleum 542
SANTALOL (*n*) : santalol 543
SANTALÖL (*n*) : santal oil (see SANDELHOLZÖL) 545
SANTALÖLKAPSEL (*f*) : santal oil capsule 546
SANTEL (*m*) : sandal (see SANDEL) 547
SANTOL (*n*) : santol 548
SANTONIN (*n*) : santonin; $C_{15}H_{18}O_3$; Sg. 1.87; Mp. 169.5°C. (from *Artemisia pauciflora*) 549
SANTONINPRÄPARAT (*n*) : santonin preparation 550
SANTONINSÄURE (*f*) : santoninic acid 551
SANTONINZELTCHEN (*n*) : santonine (lozenge) tablet 552
SANTONSÄURE (*f*) : santonic acid 553
SANZA (*f*) : residue from olive oil extraction 554
SANZAÖL (*n*) : an oil obtained from the residue (Sanza) of oilve oil extraction 555
SAPANHOLZ (*n*) : Japan wood (of *Cæsalpinia sapan*) 556
SAPHIR (*m*) : sapphire; Sg. 3.99-4.0 (see also KORUND, EDLER- and SAPPHIR) 557
SAPHIRSPAT (*m*) : cyanite (see CYANIT) 558
SAPOCARBOL (*n*) : sapocarbol 559
SAPOCREOL (*n*) : sapocreol 560
SAPOFENA (*n*) : sapofena 561
SAPOKRESOL (*n*) : sapocresol 562
SAPONIFIZIEREN : to saponify; (*n*) : saponifying; saponification 563
SAPONIN (*n*) : saponine, saponin (from root of *Saponaria officinalis*); $C_{32}H_{52}O_{17}$ 564
SAPONIT TALK (*m*) : steatite (see SEIFENSTEIN) 565
SAPORVAL (*n*) : saporval 566
SAPOSALICYLSALBE (*f*) : saposalicylic ointment or salve 567
SAPOVASELINE (*f*) : sapovaseline 568
SAPPE (*f*) : sap; mine 569
SAPPEUR (*m*) : sapper 570
SAPPHIR (*m*) : sapphire (a blue corundum) (see KORUND, ALUMINIUMOXYD and SAPHIR); Sg. 3.99-4.0 571
SAPPHIRIN (*m*) : sapphirine; $Mg_5Al_{12}Si_2O_{27}$; Sg. 3.47 572
SAPROL (*n*) : saprol 573
SAPROSOL (*n*) : saprosol 574
SAPUCAJAÖL (*n*) : sapucaia-nut oil (from *Lecythis ollaria*) 575
SARDELLE (*f*) : sardine; anchovy 576
SARDELLENTRAN (*m*) : (see SARDINENÖL) 577
SARDINENÖL (*n*) : sardine oil, Japanese fish oil; Sg. 0.92-0.93 578
SARDINENÖL (*n*), JAPANISCHES- : Japanese sardine oil (from *Clupanodon melanosticta*) 579
SARDINIAN (*m*) : (see BLEISULFAT) 580
SARDONISCH : sardonic-(ally) 581
SARG (*m*) : coffin 582
SARGLACK (*m*) : coffin varnish 583
SARKASTISCH : sarcastic-(ally) 584
SARKIN (*n*) : sarkine (an animal alkaloid from urine and flesh); $C_5H_4N_4O$ 585
SARKOLITH (*m*) : sarkolite; $Al_2Na_2CaSiO_4$; Sg. 2.7 586
SARSAPARILLA (*f*) : sarsaparilla 587
SARSAPARILLAWURZEL (*f*) : sarsaparilla-(root) (*Radix sarsaparillæ* of a species of *Simalceæ*) 588
SASSAFRAS (*m*) : sassafras (*Sassafras variifolium*); ague tree; saxifrax 589
SASSAFRASHOLZ (*n*): sassafras (bark) wood; Saloop; Sg. 0.48; cinnamon wood 590
SASSAFRASÖL (*n*) : sassafras oil (from *Sassafras officinalis*); Sg. 1.08 591
SASSAFRASTEE (*m*) : sassafras tea 592
SASSAPARILLE (*f*) : sarsaparilla (see SARSAPARILLA) 593
SASSAPARILLENWURZEL (*f*) : (see SARSAPARILLA) 594
SASSOLIN (*m*) : sassoline, sassolite; $B(OH)_3$; Sg. 1.48 (see also BORAXSÄURE, PRISMATISCHE-) 595
SATANISCH : satanic-(al)-(ly) 596
SATINAGE (*f*) : glazing finish 597
SATINET (*m*) : satinet (half cotton and half wool) 598
SATINHOLZ (*n*) : satinwood (see ATLASHOLZ) 599
SATINIEREN : to glaze; satin; calender (Paper); burnish (Phot.); impart a glossy finish 600
SATINIERMASCHINE (*f*) : calender; rolling machine; burnishing machine; burnisher (Paper) 601
SATINIERTES ILLUSTRATIONS-PAPIER (*n*) : plate paper (Paper) 602
SATINOBER (*m*) : reddish-yellow ochre 603
SATINSPAT (*m*) : satin spar; gypsum (see GIPS) 604
SATINWEISS (*n*) : satin white 605
SATIRE (*f*) : satire 606
SATIRIKER (*m*) : satirist 607
SATIRISCH : satirical-(ly) 608
SATRAPOL (*n*) : satrapol 609
SATRAPOLADUROL (*n*) : satrapol adurol 610
SATROPOL HYDROCHINON (*n*) : satrapol hydroquinone 611
SATT : satiated; satisfied; replete; saturated; full; deep; weary 612
SATTDAMPF (*m*) : saturated steam 613
SATTE (*f*) : pan; bowl; dish 614
SATTEL (*m*) : bridge; saddle 615
SATTELFASS (*n*) : rider cask 616
SATTELFERTIG : ready to mount 617
SATTELFÖRMIG : saddle-shaped 618
SATTELN : to saddle 619
SATTELZEUG (*n*) : saddlery; harness 620
SATTHEIT (*f*) : satiety, repletion, fulness 621
SÄTTIGEN : to satiate; satisfy; saturate; sate; fill 622
SÄTTIGUNG (*f*) : satiety; repletion; satisfaction; saturation 623
SÄTTIGUNGSFÄHIG : capable of (saturation) being saturated 624
SÄTTIGUNGSGRAD (*m*) : degree of saturation 625
SÄTTIGUNGSKAPAZITÄT (*f*) : saturation capacity 626
SÄTTIGUNGSPUNKT (*m*) : saturation point 627
SÄTTIGUNGSTEMPERATUR (*f*) : saturation temperature; temperature of saturation (Steam) 628
SATTLER (*m*) : saddler, harness maker 629
SATTLERARBEIT (*f*) : saddler's work, saddlery, harness making 630
SATTLERWARE (*f*) : saddlery, harness 631
SATTSAM : sufficient-(ly); enough 632
SATTSAMKEIT (*f*) : sufficiency 633
SATURATIONSGEFÄSS (*n*) : saturator; saturation vessel 634
SATURATIONSÖL (*n*) : saturation oil 635
SATURATIONSSCHEIDUNG (*f*) : purification by carbonation (saturation of the juice with carbon dioxide) (Sugar) 636
SATURATIONSSCHLAMM (*m*) : sediment from carbonation (Sugar) 637
SATUREI (*f*) : savory (*Satureja*) 638
SATURIEREN : to saturate; carbonate (Sugar) 639

SATURNROT (*n*): minium; Saturn red (see BLEIMENNIGE) 640
SATURNZINNOBER (*m*): minium (see MENNIGE) 641
SATZ (*m*): sediment; deposit; precipitate; settlings; charge; batch; mixture; composition; dregs; grounds; pair (of wheels); plant; unit (of machinery); amount (fed; fired) supplied; nest; set (of tubes, machinery, etc.); theorem; proposition; principle; leap; jump; fry; young (of fish); price; rate; sentence; yeast; sediment (Brew.); pool; stake; statement (of mathematical question); position; point; thesis; period 642
SATZBRAUEN (*n*): brewing with cold malt extract 643
SATZKRÜCKE (*f*): yeast rake (Brew.) 644
SATZLEHRE (*f*): syntax (Grammar) 645
SATZSCHALE (*f*): settling dish 646
SATZWEISE: in (sentences) sets 647
SAU (*f*): sow (Metal); blot; drying kiln; hog; pig 648
SAUBER: clean; neat; pretty; elegant 649
SAUBERKEIT (*f*): fineness; cleanness 650
SÄUBERN: to cleanse; clean; purge 651
SAUBOHNE (*f*): broad bean (*Vicia faba*) (see also SOJABOHNE) 652
SAUBOHNENÖL (*n*): (see SOJABOHNENÖL) 653
SAUBORSTE (*f*): pig's bristle 654
SAUCIERE (*f*): sauce boat; gravy boat 655
SAUER: sour; acid; tart; hard; troublesome; peevish 656
SAUERAMPFER (*m*): sorrel (*Rumex*) 657
SAUERBAD (*n*): sour bath (Bleaching) 658
SAUERBRATEN (*m*): a German method of serving beef (beef soaked in vinegar and roasted with raisins) 659
SAUERBRÜHE (*f*): acid (sour) liquor; sulphite acid liquor (Paper) 660
SAUERBRUNNEN (*m*): mineral water spring; acid spring 661
SAUER CHROMSAUER: bichromate 662
SAUERDORNGEWÄCHSE (*n.pl*): *Berberidaceæ* (Bot.) 663
SAUERDORNWURZEL (*f*): (see BERBERITZEN-WURZEL) 664
SAUEREI (*f*): mess; piggishness 665
SAUERHONIG (*m*): oxymel (by boiling mixture of refined honey and acetic plant extracts) (see OXYMEL) 666
SAUERKLEE (*m*): wood sorrel (*Oxalis acetosella*) 667
SAUERKLEESALZ (*n*): salt of sorrel; acid potassium oxalate; potassium binoxalate; sal acetosella; essential salt of lemon; $KHC_2O_4.\frac{1}{2}H_2O$; Sg. 2.09 668
SAUERKLEESÄURE (*f*): oxalic acid (see OXAL-SÄURE) 669
SAUER KOHLENSAUER: bicarbonate 670
SAUERKRAUT (*n*): pickled cabbage 671
SÄUERLICH: acidulous; sourish 672
SÄUERLICHKEIT (*f*): acidity 673
SÄUERLING (*m*): sparkling mineral water 674
SAUERMACHEND: acidifying 675
SAUER MILCHSAUER: bilactate 676
SÄUERN: to sour; acidify; leaven (Bread) 677
SAUER OXALSAUER: ——acid oxalate, ——binoxalate 678
SAUER PHOSPHORSAUER: biphosphate 679
SAUER SCHWEFELSAUER: bisulphate 680
SAUER SCHWEFLIGSAUER: bisulphite 681
SAUERSTOFF (*m*): oxygen; (O) 682
SAUERSTOFFANLAGE (*f*): oxygen plant 683
SAUERSTOFFAPPARAT (*m*): oxygen apparatus (for poisoning and artificial respiration) 684
SAUERSTOFFARM: poor in oxygen; low oxygen content 685
SAUERSTOFFÄTHER (*m*): acetaldehyde (see ACETALDEHYD) 686
SAUERSTOFFATMUNGSAPPARAT (*m*): oxygen apparatus for artificial respiration; oxygen respirator 687
SAUERSTOFFBAD (*n*): oxygen bath 688
SAUERSTOFF BILDEND: forming oxygen, oxygen-forming 689
SAUERSTOFFENTWICKELUNG (*f*): generation of oxygen 690
SAUERSTOFFFÄNGER (*m*): oxygen absorbent 691
SAUERSTOFFFREI: free from oxygen 692
SAUERSTOFFGAS (*n*): oxygen gas (see SAUER-STOFF) 693
SAUERSTOFFGEHALT (*m*): oxygen content 694
SAUERSTOFFHALTIG: containing oxygen; oxydated; oxidized; burnt (of iron) 695
SAUERSTOFFMESSER (*m*): oxygen meter 696
SAUERSTOFFPOL (*m*): anode; oxygen pole 697
SAUERSTOFFREICH: rich in oxygen; high oxygen content 698
SAUERSTOFFSALZ (*n*): oxysalt; salt of an oxacid 699
SAUERSTOFFSÄURE (*f*): oxygen acid; oxacid 700
SAUERSTOFFSTRAHL (*m*): oxygen jet 701
SAUERSTOFFSTROM (*m*): oxygen current 702
SAUERSTOFFTRÄGER (*m*): oxygen carrier 703
SAUERSTOFFÜBERTRÄGER (*m*): oxygen carrier 704
SAUERSTOFFVERBINDUNG (*f*): oxygen compound 705
SAUERSTOFFWASSER (*n*): oxygen water 706
SAUERSTOFF ZIEHEND: extracting (using up) oxygen (such as welding flame in autogene welding) 707
SAUERTEIG (*m*): leaven 708
SAUERTÖPFISCH: cross; morose; peevish 709
SÄUERUNG (*f*): acidification; leavening 710
SÄUERUNGSFÄHIG: acidifiable 711
SAUERWASSER (*n*): sour water; sour (any dilute acid solution used in art); sparkling water 712
SAUER WEINSAUER: bitartrate, acid tartrate 713
SAUER WEINSTEINSAUER: bitartrate, acid tartrate 714
SAUERWERDEN (*n*): souring; acetification (of liquors) 715
SAUFEN: to drink (of animals); carouse 716
SÄUFER (*m*): hard drinker; drunkard; immoderate drinker 717
SAUGAPPARAT (*m*): aspirator; suction apparatus 718
SAUGEN: to suck (up); absorb; (*n*): suction; sucking; absorption 719
SÄUGEN: to suckle 720
SAUGEND: absorbent 721
SAUGER (*m*): aspirator; sucker; air (vacuum) pump 722
SÄUGETIER (*n*): mammal 723
SAUGFÄHIGKEIT (*f*): absorptive capacity 724
SAUGFESTIGKEIT (*f*): resistance to suction; suction strength 725
SAUGFILTER (*n*): suction (filter) strainer 726
SAUGFLASCHE (*f*): suction bottle 727
SAUGGAS (*n*): suction gas 728
SAUGGASANLAGE (*f*): suction gas plant 729
SAUGGASMOTOR (*m*): suction-gas motor 730

SAUGGLAS (*n*) : breast pump ; suction bottle 731
SAUGHAHN (*m*) : suction cock 732
SAUGHEBER (*m*) : siphon 733
SAUGHÖHE (*f*) : suction head ; suction height 734
SAUGKOLBEN (*m*) : valve piston 735
SAUGKORB (*m*) : strainer 736
SAUGLEITUNG (*f*) : suction (pipe ; piping) main ; pump suction (of pumps) 737
SÄUGLING (*m*) : infant ; suckling ; vacuum cleaner 738
SAUGMESSER (*m*) : vacuometer 739
SAUGPIPETTE (*f*) : suction pipette 740
SAUGPUMPE (*f*) : suction pump 741
SAUGROHR (*n*) : suction pipe or tube ; sucker 742
SAUGRÖHRE (*f*) : (see SAUGROHR) 743
SAUGSTUTZE (*f*) : suction branch 744
SAUGSTUTZEN (*m*) : suction branch (of pumps) 745
SAUGVENTIL (*n*) : suction valve 746
SAUGWARZE (*f*) : nipple ; teat 747
SAUGWIDERSTAND (*m*) : suction resistance 748
SAUGWÜRMER (*m.pl*) : trematoda, flukes (Tapeworms) (Med.) 749
SAUGZUG (*m*) : suction (induced) draught 750
SAUGZUGANLAGE (*f*) : suction draught plant 751
SÄULCHEN (*n*) : little column ; pillar ; post ; pile (Elect.) ; prism (Crystallography) 752
SÄULE (*f*) : stanchion ; support ; column ; pillar ; post ; pile (Elect.) ; prism (Crystallography) 753
SAULEDER (*n*) : pigskin 754
SÄULENACHSE (*f*) : prismatic axis (Cryst.) 755
SÄULENARTIG : columnar ; prismatic (Cryst.) 756
SÄULENFÖRMIG : (see SÄULENARTIG) 757
SÄULENFUSS (*m*) : pedestal ; base 758
SÄULENSCHAFT (*m*) : body (shaft) of a column 759
SÄULE (*f*), **WASSER-** : water column 760
SAUM (*m*) : seam ; selvedge ; border ; hem ; edge ; fringe ; list (Tin plate) 761
SÄUMEN : to hem ; border ; edge ; tarry ; delay ; stay ; square (Planks) 762
SÄUMIG : dilatory ; tardy ; negligent 763
SÄUMIGKEIT (*f*) : dilatoriness ; tardiness ; negligence 764
SÄUMNIS (*f*) : delay 765
SAUMPFANNE (*f*) : list pot (Tin plate) 766
SAUMTIER (*n*) : beast of burden 767
SAUMTOPF (*m*) : list pot (Tin plate) 768
SÄURE (*f*) : acid ; acidity ; sourness 769
SÄUREALIZARINSCHWARZ (*n*) : acid alizarine black 770
SÄUREAMID (*n*) : acid amide 771
SÄUREANHYDRID (*n*) : acid anhydride 772
SÄUREANZUG (*m*) : acid proof overall (Clothing) 773
SÄUREBAD (*n*) : acid bath 774
SAUREBALLON (*m*) : acid carboy 775
SÄUREBEIZE (*f*) : sour (Leather) 776
SÄUREBESTÄNDIG : stable (fast) to acids 777
SÄUREBILDEND : acid forming 778
SÄUREBILDNER (*m*) : acidifier ; acid former 779
SÄUREBRAUN (*n*) : acid brown 780
SÄUREBROMID (*n*) : acid bromide ; dibromide 781
SÄURECENTRIFUGE (*f*) : acid centrifuge 782
SÄURECHLORID (*n*) : acid chloride ; bichloride 783
SÄUREECHT : fast to acid 784
SÄUREENTWICKLER (*m*) : acid developer 785
SÄUREESTER (*m*) : ester ; acid ester (such as ESSIGESTER) 786
SÄUREFEST : acid proof 787
SÄUREFESTE AUSKLEIDUNG (*f*) : acid proof lining 788
SÄUREFESTER STEIN (*m*) : acid proof brick 789
SÄUREFLASCHE (*f*) : acid (bottle) carboy 790
SÄUREFLUORID (*n*) : acid fluoride ; bifluoride 791
SÄUREFREI : non-acid ; acid-free 792
SÄUREGEHALT (*m*) : acidity ; acid content 793
SÄUREGELB (*n*) : acid yellow 794
SÄUREGRAD (*m*) : degree of acidity 795
SÄUREGRÜN (*n*) : acid green 796
SÄUREHALOID (*n*) : acid halide 797
SÄUREHALTIG : acidiferous ; containing acid 798
SÄUREMESSER (*m*) : acidimeter 799
SÄUREMESSKUNST (*f*) : acidimetry 800
SÄUREPULSOMETER (*n*) : automatic montejus (see MONTEJUS and DRUCKAUTOMAT) 801
SÄUREPUMPE (*f*) : acid pump 802
SÄURERADIKAL (*n*) : acid radical 803
SAURE SCHLACKE (*f*) : acid slag (Siemens-Martin process) ; iron-manganese silicate slag (Bessemer process) (50-60% SiO_2 ; 20-30% FeO ; 15-20% MnO ; 1-3% CaO ; 1-3% Al_2O_3) 804
SÄURESCHLAUCH (*m*) : acid-proof (hose) tubing 805
SÄURESCHWARZ (*n*) : acid black 806
SAURES FLUORID (*n*) : bifluoride ; acid fluoride 807
SAURES KALIUMSALZ (*n*) : potassium acid salt, acid potassium-salt 808
SAURES KALIUMSALZ DER WEINSÄURE (*f*) : acid potassium salt of tartaric acid, potassium tartrate, Cremor tartari, cream of tartar ; potassium acid tartrate, potassium bitartrate (see CREMOR TARTARI) 809
SÄURESPALTUNG (*f*) : acid cleavage 810
SÄURESTÄNDER (*m*) : acid cistern 811
SÄURETRANSPORTKORB (*m*) : acid transport basket 812
SÄURETROG (*m*) : acid trough 813
SÄURETURM (*m*) : acid tower 814
SÄUREVENTIL (*n*) : acid valve 815
SÄUREWIDERSTEHEND : fast to acid ; acid-resisting 816
SÄUREWIDRIG : antacid (reducing acidity by increasing alkalinity) (Med.) 817
SÄUREZAHL (*f*) : acid number 818
SÄUREZENTRIFUGE (*f*) : acid centrifuge 819
SÄUREZUFUHR (*f*) : supply (addition) of acid 820
SAUROLO (*n*) : saurolo 821
SAUS (*m*) : storm ; riot ; whistle ; whiz ; rush 822
SÄUSELN : to rustle ; hum ; lisp ; buzz 823
SAUSEN : to rush ; hum ; bluster ; whistle ; whiz 824
SAUSEWIND (*m*) : blustering (gale) wind 825
SAUSTALL (*m*) : pigsty 826
SAVONA (*n*) : savona 827
SAVOYERKOHL (*m*) : savoy (Vegetable) 828
SCAMMONIUM (*n*) : scammony -(resin) (see SKAMMONIAHARZ) 829
SCANDIUM (*n*), **GOLDCHLORWASSERSTOFFSAURES-** : scandium chloraurate ; $3ScCl_3,2AuCl_3+21H_2O$ 830
SCANDIUMOXYD (*n*) : scandium oxide ; Sc_2O_3 ; Sg. 3.864 831
SCANDIUMSULFAT (*n*) : scandium sulphate ; $Sc_2(SO_4)_3$; Sg. 2.579 832
SCENERIE (*f*) : scenery 833
SCHAALSTEIN (*m*) : wollastonite 834
SCHÄBE (*f*) : awn ; chaff (of flax or hemp) ; scab ; itch 835
SCHABE (*f*) : woodlouse ; cockroach ; black-beetle 836
SCHABEEISEN (*n*) : scraper 837
SCHABEMESSER (*n*) : scraper ; scraping knife 838

SCHABEN: to shave; scrape; grate; rub 839
SCHABER (m): scraper; grater; skinner; scraping tool 840
SCHABEWERKZEUG (n): scraping tool, scraper 841
SCHÄBIG: shabby; mean; scabby 842
SCHÄBIGKEIT (f): shabbiness; meanness 843
SCHABINE (f) and (pl): waste from the manufacture of (foil) leaf metal 844
SCHABKUNST (f): mezzotint 845
SCHABLONE (f): pattern; model; template; stencil; mould; form 846
SCHABLONEARTIG: mechanical; according to pattern 847
SCHABLONEMÄSSIG: mechanical; according to pattern 848
SCHABSEL (n): scrapings; shavings; parings 849
SCHACH (n): chess; check; (m): chequer-work; trellis work; chequered portion (Blast furnaces, etc.) 850
SCHACHBRETT (n): chessboard; chequered board 851
SCHACHERN: to haggle 852
SCHACHMATT: check-mate 853
SCHACHT (m): shaft; pit; gorge; combustion chamber (of gas producer); well (of a lift); top truncated cone-shaped portion or chamber of a blast furnace 854
SCHACHTBAUARBEIT (f): sinking operations (of mine shafts) 855
SCHACHTEL (f): box; case 856
SCHACHTELDECKEL (m): box (case) cover or lid 857
SCHACHTELHALM (m): horsetail (*Herba equiseti* of *Equisetum arvense*) 858
SCHACHTELKRAUT (n): horsetail (*Herba equiseti* of *Equisetum arvense*) 859
SCHACHTELN: to box; pack; put in a box 860
SCHACHTELPAPPE (f): box-board (Paper) 861
SCHACHTOFEN (m): shaft (furnace) kiln; converter (see STÜCKOFEN) 862
SCHACHTRING (m): tubbing (for mines) 863
SCHACHTWINKEL (m): shaft angle (the angle of the upper truncated cone shaped portion of a blast furnace) 864
SCHADE: unfortunate; a pity; (m): damage; injury; prejudice; loss; hurt; defect; detriment 865
SCHÄDEL (m): cranium; skull 866
SCHÄDELLEHRE (f): phrenology 867
SCHADEN: to damage; harm; hurt; injure; prejudice; be hurtful; (m): damage; prejudice; loss; injury; defect; fault; hurt; detriment; (n): hurting; injuring; damaging 868
SCHADENBERECHNUNG (f): computation of damage; assessment of damage; statement of average 869
SCHADENERSATZ (m): damages; indemnification; compensation; reparation; indemnity 870
SCHADENERSATZPFLICHTIG: liable to pay compensation as damages 871
SCHADENFROH: malicious 872
SCHADHAFT: defective; damaged; spoiled; faulty 873
SCHADHAFTIGKEIT (f): defectiveness; faultiness 874
SCHÄDIGEN: to harm; injure 875
SCHÄDIGEND: harmful, injurious, detrimental 876
SCHÄDLICH: noxious; pernicious; injurious; dangerous; prejudicial; detrimental; unwholesome; offensive; harmful 877
SCHÄDLICHER RAUM (m): clearance (of a piston in a cylinder) 878
SCHÄDLICHKEIT (f): injuriousness 879
SCHADLOS: harmless; compensated; indemnified 880
SCHADLOS HALTEN: to compensate; indemnify 881
SCHADLOSHALTUNG (f): indemnification 882
SCHAF (n): sheep; ewe 883
SCHAFARTIG: sheep-like 884
SCHAFBEIN (n): sheep (bone) leg; bone ash 885
SCHAFBLATTERN (f): chicken-pox 886
SCHAFBOCK (m): ram 887
SCHAFDARMSEITE (f): catgut 888
SCHÄFER (m): shepherd 889
SCHÄFEREI (f): sheepfold; pen 890
SCHAFFELL (n): sheepskin; fleece 891
SCHAFFEN: to make; do; create; produce; procure; provide; get; furnish; bring; bring into being; take; work; be busy 892
SCHÄFFERSCHE SÄURE (f): Schæffer's acid, beta-naphthol-sulfonic acid; $C_{10}H_6(OH)SO_3H$ 2 : 6; Mp. 122°C. 893
SCHAFFLEISCH (n): mutton 894
SCHAFFNER (m): conductor; guard; manager; steward; agent 895
SCHAFFNERIN (f): stewardess; housekeeper; manageress 896
SCHAFFOT (n): scaffold 897
SCHAFFUNG (f): provision 898
SCHAFGARBE (f): yarrow; milfoil (*Achillæa millefolium*) 899
SCHAFGARBENBLÜTEN (f.pl): milfoil flowers (*Flores millefolii*) 900
SCHAFGARBENÖL (n): yarrow oil; milfoil oil 901
SCHAFHAUT (f): sheepskin; amnion (Anat.) 902
SCHAFHÄUTCHEN (n): amnion (Anat.) 903
SCHAFHÜRDE (f): pen; sheep-fold 904
SCHAFKÄSE (m): cheese from ewe's milk 905
SCHAFLEDER (n): sheepskin (Leather) 906
SCHAFMÄSSIG: sheepish 907
SCHAFMILCH (f): ewe's milk 908
SCHAFOTT (n): scaffold 909
SCHAFPELZ (n): fleece 910
SCHAFSCHERE (f): sheep shears 911
SCHAFSCHMIERE (f): sheep dip 912
SCHAFSCHUR (f): sheep-shearing 913
SCHAFSCHWEISS (m): yolk; suint 914
SCHAFT (m): shaft; handle; haft; shank; stock; trunk; stalk; body 915
SCHAFTALG (m): sheep's tallow (see HAMMELTALG) 916
SCHAFTBOHRER (m): screw tap 917
SCHAFTFRÄSER (m): end mill 918
SCHAFWASCHMITTEL (n): sheep dip 919
SCHAFWASSER (n): amniotic fluid (Anat.) 920
SCHAFWOLLE (f): sheep's wool 921
SCHAFZUCHT (f): sheep breeding 922
SCHAGRIN (n): shagreen 923
SCHAKAL (m): jackal 924
SCHAKE (f): link-(of a chain) 925
SCHÄKEL (m): shackle 926
SCHÄKEL, ANKER- (m): (anchor)-shackle 927
SCHÄKELBOLZEN (m): shackle bolt 928
SCHAKENKETTE (f): open link chain 929
SCHÄKER (m): jester; joker; wag; joke 930
SCHÄKEREI (f): jest; badinage; joke 931
SCHÄKERHAFT: jocose-(ly); joking-(ly); playful-(ly) 932
SCHÄKERN: to joke; jest 933
SCHAL: insipid; flat; stale 934
SCHÄLBE (f): heifer 935

SCHÄLCHEN (*n*): (small)-dish; capsule 936
SCHALE (*f*): dish; bowl; pan; cup; basin; saucer; scale (of a balance); husk; skin; peel; pod; shell; rind; bark; cover (of book); chill-mould (Founding); step, bush or brass (of bearings, etc.) 937
SCHÄLEN: to peel; skin; shell; pare; bark; decorticate; exfoliate; flay 938
SCHALENALGEN (*f.pl*): diatoms (small aquatic plants; single celled algæ) of the group *Diatomaceœ*) 939
SCHALENBLENDE (*f*): fibrous sphalerite (see SPHALERIT) 940
SCHALENEISENSTEIN (*m*): kidney (botryoidal) iron ore 941
SCHALENGUSS (*m*): chill casting 942
SCHALENGUSSFORM (*f*): chill 943
SCHALENHART: chilled (Founding) 944
SCHALENLEDERHAUT (*f*): chorion 945
SCHALENTRÄGER (*m*): tripod; dish support 946
SCHALFISCH (*m*): shell-fish 947
SCHALHEIT (*f*): staleness, insipidness 948
SCHALK (*m*): wag; rogue; knave 949
SCHALKHAFT: roguish; knavish; waggish 950
SCHALL (*m*): sound; resonance; acoustic-(s); peal; ring 951
SCHALLBODEN (*m*): sound-board 952
SCHALLDÄMPFER (*m*): sound damper; silencer 953
SCHALLEN: to sound; ring; resound 954
SCHALLGESCHWINDIGKEIT (*f*): sound velocity (Acoustics) 955
SCHALLGESCHWINDIGKEIT, ÜBER- (*f*): over-sound velocity 956
SCHALLGESCHWINDIGKEIT, UNTER- (*f*): under-sound velocity 957
SCHALLLEHRE (*f*): acoustics 958
SCHALLLOCH (*n*): sound-hole 959
SCHALLTRICHTER (*m*): trumpet, trumpet-piece, horn 960
SCHALLWELLE (*f*): sound wave 961
SCHALOTTE (*f*): shalot, shallot (*Allium ascalonicum*) 962
SCHALRAHMEN (*m*): frame-(work) (of ferro-concrete constructional work) 963
SCHALTANLAGE (*f*): switch-gear 964
SCHALTBRETT (*n*): switchboard 965
SCHALTEN: to switch; insert; connect (Elect.); rule; to have one's way; dispose; act 966
SCHALTEN, HINTEREINANDER-: to connect in series 967
SCHALTEN, IN REIHEN-: to connect in series 968
SCHALTEN, NEBENEINANDER-: to connect in parallel 969
SCHALTEN, PARALLEL-: to connect in parallel; to connect in multiple arc 970
SCHALTER (*m*): switch; circuit closer; commutator (Elect.); wicket; window; ruler; manager; ticket office window (Railway) 971
SCHALTERÖL (*n*): switch oil 972
SCHALTHEBEL (*m*): switch lever 973
SCHALTIER (*n*): crustacean 974
SCHALTJAHR (*n*): leap year 975
SCHALTSCHRANK (*m*): switch (cupboard, cabinet) case 976
SCHALTTAFEL (*f*): switch-board (Elect.) 977
SCHALTTAG (*m*): the extra day in a leap year 978
SCHALTUNG (*f*): connection (Elect.); disposal 979
SCHALTUNGSSCHEMA (*m*): diagram of connections 980
SCHALUNG (*f*): encasing, casing, sheeting, shell, framing, framework (see also EINSCHALUNG) 981
SCHÄLUNG (*f*): shelling; pealing; paring; barking; excoriation; desquamation (of skin, etc.) 982
SCHALUPPE (*f*): shallop 983
SCHALWAND (*f*): sheeting (of ferro-concrete constructional work) 984
SCHAM (*f*): shame; chastity; modesty; pudenda 985
SCHAMBEIN (*n*): pubis; os pubis (Anat.) 986
SCHÄMEN: to be ashamed; shame 987
SCHAMGANG (*m*): vagina (Anat.) 988
SCHAMGLIED (*n*): genital organ; sexual organ 989
SCHAMHAFT: modest; bashful; shamefaced; chaste 990
SCHAMLEISTE (*f*): groin 991
SCHAMLOS: shameless 992
SCHAMOTTE (*f*): grog; chamotte (Ceram.); fireclay; burnt clay 993
SCHAMOTTEFABRIK (*f*): fire-brick (fire-clay) factory, factory for (refractories) refractory material 994
SCHAMOTTEKAPSEL (*f*): fire-clay sagger (Ceram.) 995
SCHAMOTTEMASSE (*f*): fireclay 996
SCHAMOTTEMÖRTEL (*m*): fireproof cement; fireclay 997
SCHAMOTTESTEIN (*m*): firebrick; Sg. 2.2-2.7 998
SCHAMOTTETIEGEL (*m*): fireclay crucible 999
SCHAMOTTEWAREN (*f.pl*): fireproof goods; refractories; refractory (goods; wares) material 000
SCHAMOTTEZIEGEL (*m*): firebrick 001
SCHAMRÖTE (*f*): blush-(ing) for shame; blush of shame 002
SCHAMTEILE (*m.pl*): sexual organs 003
SCHANDE (*f*): dishonour; disgrace; shame; ignomy; infamy 004
SCHANDECK (*n*): gunwale 005
SCHANDECKEL (*m*): gunwale 006
SCHÄNDEN: to dishonour; disgrace; shame; violate 007
SCHANDFLECK (*m*): blot; stain; blemish 008
SCHÄNDLICH: disgraceful; shameful; infamous; scandalous 009
SCHÄNDUNG (*f*): violation, rape 010
SCHANDWÖRTER (*n.pl*): filthy (obscene) words or language 011
SCHANK (*m*): licensed house (for alcoholic liquors) 012
SCHÄNKBIER (*m*): draught beer 013
SCHANZE (*f*): entrenchment; chance; trench; sap; redoubt; pallisade 014
SCHANZEN: to strengthen a position by trenching or raising a parapet; (*n*): sapping; pioneer-work 015
SCHANZKLEID (*n*): bulwark 016
SCHANZKLEIDPFORTE (*f*): bulwark port 017
SCHANZKLEIDRELING (*f*): bulwark railing 018
SCHANZKLEIDSTÜTZE (*f*): bulwark stay 019
SCHAPINGMASCHINE (*f*): shaping machine 020
SCHAR (*f*): troop; host; multitude; crowd; herd; band; flock; plough-share 021
SCHARBEN: to chop 022
SCHARBOCK (*m*): scurvy (Med.); Pile-wort (*Ficaria ranunculoides*) (*Ranunculus ficaria*) (Bot.) 023
SCHARBOCKHEILEND: antiscorbutic 024

SCHARBOCKMITTEL (*n*): antiscorbutic 025
SCHAREN: to assemble; crowd or flock together 026
SCHARENWEISE: in (bands) troops 027
SCHARF: sharp; pungent; acrid; keen; acute; corrosive; hard; rigorous; astute; quick; biting; acid; stringent; severe; clear; in focus; lean (of body of a ship) 028
SCHÄRFE (*f*): sharpness; keenness; acridity; acuteness; edge; astuteness; severity; definition (Phot. and Art.) 029
SCHÄRFEN: to sharpen; whet; define 030
SCHARFFEUER (*n*): hard (sharp) fire 031
SCHARFFEUERDEKOR (*n*): sharp-fire decoration (see SCHARFFEUERFARBE) 032
SCHARFFEUEREMAIL (*n*): sharp fire enamel (Ceramics) 033
SCHARFFEUERFARBE (*f*): sharp fire (colour) paint (for decoration of hard porcelain) (an underglaze colour) 034
SCHARFGÄNGIG: sharp (of screws, etc.); sharp threaded; triangular threaded 035
SCHARFKANTIG: sharp-edged; having sharp edges 036
SCHARFKERBPROBE (*f*): sharp notch (test) test piece (Metal) 037
SCHARFRICHTER (*m*): executioner 038
SCHARFSAUER: very sour; strongly acid 039
SCHARFSCHMECKEND: acrid; pungent; tart; having a sharp taste; sharp 040
SCHARFSCHÜTZE (*m*): sharpshooter; sniper 041
SCHARFSICHTIG: keen (clear, quick) sighted 042
SCHARFSINN (*m*): acumen; astuteness; sagacity; penetration 043
SCHARFSINNIG: sagacious; astute 044
SCHARFWINKLIG: acute-angled 045
SCHARLACH (*m*): scarlet (fever or runner or colour); scarlatina (Med.) 046
SCHARLACHEN: scarlet 047
SCHARLACHFARBE (*f*): scarlet 048
SCHARLACHFARBEN: scarlet 049
SCHARLACHFIEBER (*n*): scarlet fever 050
SCHARLACHKÖRNER (*n.pl*): kermes-(grains) (dried female of the kermes insect, which yields a dye akin to carmine) (see KERMES-) 051
SCHARLACHROT: scarlet; bright-red (mercuric iodide) (see QUECKSILBERJODID) 052
SCHARLACHSALBE (*f*): scarlatina ointment 053
SCHARLACHWURM (*m*): cochineal insect (*Coccus cacti*) (see COCHENILLE) 054
SCHARNIER (*n*): joint; hinge 055
SCHÄRPE (*f*): sling (Med.); sash; scarf 056
SCHARPIE (*f*): lint 057
SCHARRE (*f*): scraper; rake 058
SCHARREN: to scrape; rake; scratch; (*m*): shambles 059
SCHARRWERK (*n*): scraping mechanism; scraper 060
SCHARTE (*f*): notch; fissure; sherd; nick; sawwort 061
SCHARTIG: notched; nicked; jagged 062
SCHATTEN (*m*): shade; shadow 063
SCHATTENBILD (*n*): shadow-picture; phantom; silhouette 064
SCHATTENDECK (*n*): shade deck 065
SCHATTENPALME (*f*): talipot palm (*Corypha umbraculifera*) 066
SCHATTENRISS (*m*): silhouette 067
SCHATTIEREN: to shade 068
SCHATTIERUNG (*f*): shading; shade; tint; lighting (Art) 069
SCHATZ (*m*): treasure; stock; wealth; store 070
SCHATZAMT (*m*): treasury; exchequer 071
SCHÄTZBAR: esteemed; estimable; valued; valuable; capable of (estimation) valuation; precious 072
SCHATZEN: to assess; tax 073
SCHÄTZEN: to value; estimate; consider; appraise; esteem; appreciate 074
SCHATZGRÄBER (*m*): treasure-hunter 075
SCHATZHAUS (*n*): storehouse; treasury 076
SCHATZKAMMER (*f*): storehouse; treasury; exchequer 077
SCHATZKISTCHEN (*n*): jewel casket 078
SCHAU (*f*): view; review; show 079
SCHAUARZT (*m*): coroner 080
SCHAUBILD (*n*): diagram; exhibit (Art); diagrammatic illustration 081
SCHAUBÜHNE (*f*): stage (of a theatre) 082
SCHAUDER (*m*): shudder; dread; terror; horror 083
SCHAUDERHAFT: dreadful; horrible 084
SCHAUDERIG: dreadful; horrible 085
SCHAUDERN: to shudder; shiver 086
SCHAUEN: to look-(at); examine; see; gaze; behold; view 087
SCHAUER (*m*): shudder; fit; thrill; shower; inspector; spectator; shed; shelter 088
SCHAUERVOLL: horrible 089
SCHAUFEL (*f*): shovel; paddle; scoop; bucket; blade (Turbine); ladle 090
SCHAUFELN: to shovel 091
SCHAUFELRAD (*n*): bucket (paddle) wheel (Ships, etc.); blade wheel (Turbines) 092
SCHAUFENSTER (*n*): show window 093
SCHAUGLAS (*n*): specimen (sample; display) glass 094
SCHAUKE (*f*), FLACHE-: punt 095
SCHAUKELKUVETTE (*f*): automatic rocker (Phot.); rocking (trough, dish) bath 096
SCHAUKELN: to rock; swing; shake 097
SCHAULINIE (*f*): curve (of a graph); diagrammatic illustration 098
SCHAULOCH (*n*): spy-hole 099
SCHAULUSTIG: curious 100
SCHAUM (*m*): foam; scum; froth; lather; head (Beer) 101
SCHAUMARTIG: frothy 102
SCHAUMBESTÄNDIG: capable of keeping a head 103
SCHAUMBESTÄNDIGKEIT (*f*): capacity for keeping a head 104
SCHAUMBILDUNG (*f*): formation of (foam, suds) froth; foaming, frothing, priming (of boilers), ebullition 105
SCHAUMBLASE (*f*): bubble (of froth or foam) 106
SCHÄUMEN: to froth; foam; sparkle; fizz (Wines); lather (Soap); skim; scum 107
SCHAUMERDE (*f*): aphrite (type of calcite) (see CALCIT) 108
SCHAUMERZ (*n*): soft or foamy wad (see WAD) 109
SCHAUMFÄHIGKEIT (*f*): capacity for forming (froth, foam) suds 110
SCHAUMGIPS (*m*): foliated gypsum (see GIPS) 111
SCHAUMGOLD (*n*): imitation (Dutch) gold; Dutch metal (see METALLGOLD) 112
SCHAUMHAUBE (*f*): head (Brewing) 113
SCHAUMIG: frothy; foamy 114
SCHAUMKALK (*m*): aragonite (see ARAGONIT) 115
SCHAUMKAUTSCHUK (*m*): rubber-sponge 116
SCHAUMKELLE (*f*): skimming (ladle) spoon; skimmer 117

SCHAUMLÖFFEL (*m*): skimming (ladle) spoon; skimmer 118
SCHAUMSCHWÄRZE (*f*): finely powdered animal charcoal 119
SCHAUMSEIFE (*f*): lathering soap 120
SCHAUMSPAT (*m*): analcite; analcime; $NaO.SiO_2+AlO_3Si_3O_2+H_2O$; Sg. 2.25 121
SCHAUMSTAND (*m*): head (Brewing) 122
SCHAUMTON (*m*): Fuller's earth (silica 53%, alumina 10%, iron oxide 8%, lime 0.5%, magnesia 1%, water 24%, etc.); Sg. 1.7-2.4 123
SCHAUMVERMÖGEN (*n*): capacity for forming (froth, foam) suds 124
SCHAUMWEIN (*m*): sparkling wine 125
SCHAUPLATZ (*m*): stage; theatre; scene 126
SCHAUSPIEL (*n*): play; drama; spectacle scene; sight 127
SCHAUSPIELARTIG: dramatic; spectacular; theatrical 128
SCHEBECKE (*f*): xebec (Mediterranean three-masted ship) 129
SCHECK (*m*): cheque 130
SCHECKBETRAG (*m*): value (amount) of cheque 131
SCHECKEN: to streak; mottle; spot; speckle 132
SCHECKIG: dappled; mottled; spotted; speckled; piebald 133
SCHECKSUMME (*f*): amount (value) of check 134
SCHEEL: askance; oblique; awry; with a squint; squint-eyed; jealous; envious; (*n*): wolfram (see WOLFRAM) 135
SCHEELBLEIERZ (*n*): scheeletite 136
SCHEELBLEISPAT (*m*): scheeletite 137
SCHEELERZ (*n*): scheelite (see SCHEELIT) 138
SCHEELE'SCHES GRÜN (*n*): Scheele's green (see KUPFER, ARSENIGSAURES-) 139
SCHEELE'SCHES SÜSS (*n*): glycerol; glycerine (see GLYZERIN) 140
SCHEELIT (*m*): scheelite (natural calcium tungstate); $CaWO_4$; Sg. 6.07 (see also CALCIUM-WOLFRAMAT) 141
SCHEELSÄURE (*f*): tungstic acid (see WOLFRAM-SÄURE) 142
SCHEELSUCHT (*f*): envy; jealousy 143
SCHEFFEL (*m*): bushel 144
SCHEIBE (*f*): slice; disc; cake (Wax, etc.); pane (Glass); comb (Honey); pulley; sheave; cut; wheel; dial; target 145
SCHEIBE, EXZENTRISCHE- (*f*): excentric-(wheel) disc 146
SCHEIBE (*f*), **GUSSEISEN-**: cast iron pulley 147
SCHEIBE (*f*), **KORK-**: cork pulley; cork sheet; cork disc 148
SCHEIBENANKER (*m*): disc armature 149
SCHEIBENFÖRMIG: disc-shaped 150
SCHEIBENGLAS (*n*): window glass 151
SCHIEBENHONIG (*m*): honey in the comb 152
SCHEIBENKUPFER (*n*): rose (rosette) copper (see ROTKUPFER) 153
SCHEIBENKUPPELUNG (*f*): plate coupling 154
SCHEIBENRAD (*n*): disc-(wheel) 155
SCHEIBENREISSEN (*n*): conversion into discs or rosettes (Metal) 156
SCHEIBENSCHIESSEN (*n*): target practice 157
SCHEIBENWACHS (*n*): cake-wax 158
SCHEIBENWASSERMESSER (*m*): disc water meter 159
SCHEIBGATT (*n*): sheave hole 160
SCHEIBIG: in discs or slices 161
SCHEIDBAR: separable; analysable 162
SCHEIDE (*f*): sheath; cover; vagina; boundary; border 163
SCHEIDEBÜRETTE (*f*): separating burette 164
SCHEIDEERZ (*n*): picked (screened) ore 165
SCHEIDEFLÜSSIGKEIT (*f*): separating (or parting) liquid 166
SCHEIDEGOLD (*n*): gold purified by (separation) parting 167
SCHEIDEGUT (*n*): material to be separated 168
SCHEIDEKAPELLE (*f*): cupel 169
SCHEIDEKOLBEN (*m*): separating flask; alembic 170
SCHEIDEKUNST (*f*): analytical chemistry 171
SCHEIDEKÜNSTLER (*m*): analytical chemist 172
SCHEIDELINIE (*f*): line of demarcation; boundary line 173
SCHEIDEMÜNZE (*f*): small coin 174
SCHEIDEN: to part-(with); separate; screen; sieve; cob (Ores); divide; analyse; decompose; pick; sort; sever; divorce; disjoin; refine; withdraw; depart; turn (of milk) 175
SCHEIDENSCHLEIM (*m*): vaginal mucus 176
SCHEIDEOFEN (*m*): almond (parting) furnace 177
SCHEIDEPFANNE (*f*): clarifier; defecating pan (Sugar) 178
SCHEIDESCHLAMM (*m*): defecation slime (Sugar) 179
SCHEIDESIEB (*n*): separating sieve 180
SCHEIDESILBER (*n*): parting silver 181
SCHEIDETRICHTER (*m*): separating funnel 182
SCHEIDEVORRICHTUNG (*f*): separating (apparatus) arrangement 183
SCHEIDEWAND (*f*): partition; barrier; mid-feather-wall; diaphragm; septum 184
SCHEIDEWASSER (*m*): nitric acid (used for separating) (see QUARTATION); aqua regis; aqua fortis (see SALPETERSÄURE) 185
SCHEIDEWEG (*m*): forked or cross roads 186
SCHEIDUNG (*f*): separation; parting; screening; dividing; analysing; decomposing; decomposition; picking; sorting; divorcing; turning (of milk); departing; departure; refining; withdrawing; withdrawal 187
SCHEIN (*m*): appearance; show; splendour; light; shine; lustre; bloom, (of oil, etc.); certificate; bond; pretext; bill; paper; document 188
SCHEINBAR: apparent; seeming; plausible; probable; apparently; seemingly; probably 189
SCHEINBARKEIT (*f*): plausibility; probability 190
SCHEINBEHELF (*m*): evasion 191
SCHEINEN: to appear; seem; shine 192
SCHEINFARBE (*f*): accidental colour 193
SCHEINFUSS (*m*): pseudopodium 194
SCHEINGOLD (*n*): spurious (false) gold 195
SCHEINGRUND (*m*): fallacy; apparent reason 196
SCHEINHEILIG: sanctimonious-(ly); hypocritical-(ly) 197
SCHEINSCHMAROTZER (*m*): epiphyta 198
SCHEINWERFER (*m*): searchlight; reflector; projector 199
SCHEISEGEL (*n*): skysail 200
SCHEIT (*n*): billet; log; block-(of wood) 201
SCHEITEL (*m*): summit; top; crown; origin of co-ordinates (Maths.); zenith; parting (of hair); point of an angle; angular point 202
SCHEITELBEIN (*n*): parietal bone (Anat.) 203
SCHEITELLINIE (*f*): vertical line 204

SCHEITELN : to part (Hair) 205
SCHEITELPUNKT (*m*) : zenith 206
SCHEITELRECHT : perpendicular ; vertical 207
SCHEITELWINKEL (*m*) : vertical angle ; azimuth ; (*pl*) : opposite (vertical) angles 208
SCHEITERHAUFEN (*m*) : funeral pile 209
SCHEITERN : to be wrecked ; be frustrated 210
SCHEITHOLZ (*n*) : billet wood, log split into billets 211
SCHEL : (see **SCHEEL**) 212
SCHELFEN : to peel-(off), exfoliate, skin, shell 213
SCHELFERN : to peel-(off), exfoliate, skin, shell 214
SCHELLACK (*m*) : shellac ; lac ; lacca (resin from sting of insect (*Coccus lacca*, on certain trees in East India) 215
SCHELLACKERSATZ (*m*) : shellac substitute 216
SCHELLACK, KÜNSTLICHER- (*m*) : artificial shellac 217
SCHELLACKPOLITUR (*f*) : French (shellac) polish 218
SCHELLACKWACHS (*n*) : shellac wax 219
SCHELLE (*f*) : bell ; handcuff ; manacle ; distance piece (in handcuff form ; for tubes) 220
SCHELLEN : to ring ; tingle 221
SCHELLFISCH (*m*) : cod ; haddock 222
SCHELLHAMMER (*m*) : large hammer 223
SCHELM (*m*) : rogue ; scoundrel 224
SCHELMISCH : roguish 225
SCHELTEN : to reprimand ; revile ; scold ; rebuke ; blame 226
SCHELTENSWERT : blamable 227
SCHEMA (*n*) : scheme ; sketch ; model ; pattern ; diagram ; form ; schedule ; proposal ; project ; lay-out (diagrammatic)-arrangement (Drawings) ; blank 228
SCHEMATISCH : schematic-(ally) ; diagrammatic-(ally) 229
SCHEMEL (*m*) : stool ; trestle ; bogie 230
SCHEMELWAGEN (*m*) : bogie truck 231
SCHEMEN (*m*) : phantom ; shadow 232
SCHENK (*m*) : retailer of liquor 233
SCHENKBIER (*n*) : draught beer 234
SCHENKE (*f*) : tavern ; public-house 235
SCHENKEL (*m*) : thigh ; femur (Anat.) ; side-(piece) ; shank ; leg ; flange (of section irons) 236
SCHENKELBEIN (*n*) : femur (Anat.) 237
SCHENKELBRUCH (*m*) : crural rupture 238
SCHENKELKNOCHEN (*m*) : femur (Anat.) 239
SCHENKELROHR (*n*) : elbow (Mech.) ; bent tube or pipe 240
SCHENKEN : to give ; fill ; present ; remit ; pour out ; retail (Liquor) ; give a donation 241
SCHENKGERECHTIGKEIT (*f*) : license to sell alcoholic liquors 242
SCHENKUNG (*f*) : present, gift, donation 243
SCHENKUNGSURKUNDE (*f*) : deed of gift 244
SCHERBEANSPRUCHUNG (*f*) : shearing (strain) stress ; shear 245
SCHERBEN (*m*) : (see **SHERBE**) 246
SCHERBENGEWÄCHS (*n*) : potted plant ; plant in a pot 247
SCHERBENKOBALT (*m*) : (*Cobaltum testaceum*) : native arsenic (see **ARSEN**) 248
SCHERBLOCK (*m*) : warping block 249
SCHERE (*f*) : scissors ; shears ; claws ; notch ; nick ; shear (Testing metal) 250
SCHEREN : (of stresses, etc.) to shear ; shave ; clip ; cut ; trouble ; plague ; tease ; vex ; bother ; clear off ; go ; quit ; sheer (of ships) 251
SCHERENKRAN (*m*) : shear legs 252
SCHEREN SCHLEIFEN : to grind scissors 253
SCHERENSCHNITT (*m*) : shear-(section or cut) 254
SCHERENSTAHL (*m*) : shear steel 255
SCHERER (*m*) : shearer 256
SCHEREREI (*f*) : bother ; trouble ; vexation 257
SCHERFESTIGKEIT (*f*) : shearing strength ; shear 258
SCHERFLEIN (*n*) : mite 259
SCHERGANG (*m*) : sheerstrake (Naut.) 260
SCHERGANG (*m*), **HAUPT-** : main sheerstrake 261
SCHERMASCHINE (*f*) : shearing machine, cutting machine, cutter 262
SCHERSPANNUNG (*f*) : shearing (strain) stress ; shear 263
SCHERWOLLE (*f*) : fleece ; sheared wool ; flock 264
SCHERZ (*m*) : joke ; fun ; sport ; jest ; rag ; raillery 265
SCHERZEN : to joke ; jest ; rag ; rail ; banter ; sport 266
SCHERZEND : jokingly 267
SCHERZHAFT : jocose ; funny ; jocular ; facetious 268
SCHEU : timorous ; shy ; bashful ; reserved ; timid ; coy ; (*f*) : timorousness ; timidity ; fear ; shyness. etc. 269
SCHEUCHE (*f*) : scarecrow 270
SCHEUCHEN : to scare ; frighten 271
SCHEUEN : to shy ; fear ; shun ; be timid ; avoid ; grudge ; dread ; be afraid 272
SCHEUER (*f*) : granary ; barn ; shed 273
SCHEUERFRAU (*f*) : charwoman 274
SCHEUERN : to wash ; scour ; rub ; char 275
SCHEUERPULVER (*n*) : scouring powder 276
SCHEUERSAND (*m*) : fine sand (as silver sand, for scouring purposes) 277
SCHEUNE (*f*) : granary ; barn ; threshing floor 278
SCHEUSAL (*n*) : monster 279
SCHEUSSLICH : horrid ; hideous ; horrible ; abominable ; filthy 280
SCHIBUTTER (*f*) : (see **SHEABUTTER**) 281
SCHICHT (*f*) : layer ; stratum ; bed ; part ; portion ; course ; charge (of furnace) ; batch ; shift (of work) ; film (of photographic plate) 282
SCHICHTBODEN (*m*) : mixing place (Metal) 283
SCHICHTEN : to arrange in (rows) layers ; bed ; stratify ; arrange ; stack ; charge ; pile-(up) 284
SCHICHTENGRUPPE (*f*) : formation ; strata group (Geology) 285
SCHICHTENKOHLE (*f*) : foliated (stratified) coal 286
SCHICHTENWEISE : stratified ; in (rows) layers 287
SCHICHTGESTEIN (*n*) : sedimentary rock ; stratified rock 288
SCHICHTSEITE (*f*) : film side (Phot.) 289
SCHICHTUNG (*f*) : stratification ; piling ; charging 290
SCHICHTWASSER (*n*) : ground water 291
SCHICKEN : to send ; dispatch ; happen or cause to happen ; be fit ; conform ; become ; suit ; be suitable ; accommodate-(oneself) ; prepare ; be proper ; convey ; be decorous ; behave 292
SCHICKLICH : proper ; fit ; becoming ; suitable ; convenient ; appropriate ; decorous 293
SCHICKSAL (*n*) : destiny ; fate ; fortune ; providence 294
SCHICKSALSSCHLAG (*m*) : stroke of misfortune, misfortune, catastrophe 295
SCHICKUNG (*f*) : dispensation ; providence 296

SCHIEBE (*f*) : shovel 297
SCHIEBEBÜHNE (*f*) : travelling platform, traverser 298
SCHIEBEFENSTER (*n*) : sash-window 299
SCHIEBEGITTER (*n*) : sliding grating; sliding lattice-work door (as of a lift) 300
SCHIEBEKARREN (*m*) : wheel-barrow 301
SCHIEBEN : to push; shove; slide; slip; cast; shift; transfer; put off; move 302
SCHIEBER (*m*) : slide (bar; valve; damper, etc.); carriage; shovel; pusher; damper; gate-valve; sliding door (of ashpits etc.); register 303
SCHIEBERDIAGRAMM (*n*) : slide valve diagram 304
SCHIEBERFEDER (*f*) : slide valve spring 305
SCHIEBERFLÄCHE (*f*) : valve face 306
SCHIEBERKASTEN (*m*) : slide-valve chest 307
SCHIEBERRAND (*m*) : valve face 308
SCHIEBERSPIEGEL (*m*) : slide valve face 309
SCHIEBERSTANGE (*f*) : valve spindle, or rod; damper rod 310
SCHIEBERSTANGENFÜHRUNG (*f*) : slide rod guide 311
SCHIEBERSTEUERUNG (*f*) : slide valve gear 312
SCHIEBERÜBERDECKUNG (*f*) : slide valve lap 313
SCHIEBERVENTIL (*n*) : slide valve; sluice valve; penstock (Mech.) 314
SCHIEBERWEG (*m*) : slide valve travel 315
SCHIEBETÜR (*f*) : sliding-door 316
SCHIEBEWEICHE (*f*) : sliding cross-over (of overhead rail transporter) 317
SCHIEBGEWICHT (*n*) : sliding weight 318
SCHIEBKARREN (*m*) : wheel-barrow 319
SCHIEBUNG (*f*) : pushing; sliding; slipping; shoving; manœuvre 320
SCHIEDSANALYSE (*f*) : umpire (deciding) analysis 321
SCHIEDSGERICHT (*n*) : court of reference; court of arbitration 322
SCHIEDSGERICHTLICHES VERFAHREN (*n*) : arbitration 323
SCHIEDSMANN (*m*) : referee; umpire; arbiter; arbitrator 324
SCHIEDSPRÜFUNG (*f*) : umpire test, arbitrary test 325
SCHIEDSRICHTER (*m*) : referee; umpire; arbiter; arbitrator; Justice of the Peace. 326
SCHIEDSRICHTERLICH : by means of arbitration 327
SCHIEDSSPRUCH (*m*) : award; decision 328
SCHIEDSURTEIL (*n*) : award; decision 329
SCHIEF : inclined; oblique; diagonal; crooked; slanting; askew; awry; amiss; ill 330
SCHIEFE (*f*) : slope; inclination; obliqueness; crookedness; inclined plane 331
SCHIEFER (*m*) : shale (Sg. 2.6); slate (Sg. 2.67-2.9); schist; flake; splinter; flaw 332
SCHIEFERBLAU : slate-blue 333
SCHIEFERBRUCH (*m*) : slate quarry 334
SCHIEFER BRUCHSTRICH (*m*) : slanting fractional line (as 1/2 instead of ½) 335
SCHIEFERDACH (*n*) : slate (tiled) roof 336
SCHIEFERFARBE (*f*) : slate colour 337
SCHIEFERGEBIRGE (*n*) : slate-(bearing) mountain-(s) or hill-(s) 338
SCHIEFERGRAU (*n*) : slate grey 339
SCHIEFERHAMMER (*m*) : slater's hammer 340
SCHIEFERIG : slaty; schistous; scaly; foliated; flaky 341
SCHIEFERKOHLE (*f*) : slate (slaty) coal; foliated (flaky) coal 342
SCHIEFERMEHL (*n*) : ground shale; crushed shale 343
SCHIEFERN : to exfoliate; shiver; scale (flake) off; laminate; lamellate 344
SCHIEFER, ÖL- (*m*) : oil (bearing)-shale 345
SCHIEFERÖL (*n*) : shale oil 346
SCHIEFERPAPIER (*n*) : slate paper 347
SCHIEFERPLATTE (*f*) : slab (plate) of slate 348
SCHIEFERSCHWARZ (*n*) : slate black 349
SCHIEFERSCHWELEREI (*f*) : shale distillery 350
SCHIEFERSPAT (*m*) : slate spar, (lamellar kind of calcite) 351
SCHIEFERSTEIN (*m*) : slate; shale; schist 352
SCHIEFERSTIFT (*m*) : slate pencil 353
SCHIEFERTALK (*m*) : talc slate; indurated talc 354
SCHIEFERTEER (*m*) : shale tar 355
SCHIEFERTEERÖL (*n*) : shale tar oil 356
SCHIEFERTON (*m*) : shale; slate clay 357
SCHIEFERUNG (*f*) : scaling (flaking) off; exfoliation 358
SCHIEFERWEISS (*n*) : fine, flaky type of white lead; slate white 359
SCHIEFES BILD (*n*) : a false impression 360
SCHIEFHEIT (*f*) : slope; inclination; slant; obliqueness; crookedness 361
SCHIEFLIEGEND : inclined; diagonal; sloping; oblique; slanting; askew 362
SCHIEFWINKLIG : oblique angled 363
SCHIELEN : to squint; leer; (*n*) : leering; squinting; squint; strabismus (Med.) 364
SCHIELEND : squinting; leering 365
SCHIEMANN (*m*) : boatswain's mate (Naut.) 366
SCHIEMANNSGARN (*n*): spun yarn 367
SCHIENBEIN (*n*) : tibia (Anat.); shin bone 368
SCHIENE (*f*) : rail; clout; bar; strip; slat; band (of iron); rim; hoop; tyre (of wheel); splint (Med.); bloom (Metal) 369
SCHIENEN : to splint 370
SCHIENENBAHN (*f*) : rail-(tram)-way 371
SCHIENENKOPF (*m*) : rail head 372
SCHIENENLASCHE (*f*) : fish plate (Railways) 373
SCHIENENNAGEL (*m*) : hook nail 374
SCHIENENRÄUMER (*m*) : rail guard 375
SCHIENENSTOSS (*m*) : rail joint 376
SCHIENENWEG (*m*) : (see SCHIENENBAHN) 377
SCHIENENWEITE (*f*) : gauge (Railway) 378
SCHIENE (*f*), ROHGEWALZTE- : bloom (Metal) 379
SCHIER : sheer; pure; almost; nearly; simply 380
SCHIERLING (*m*) : (spotted-)hemlock [*Herba conii-(maculati)* of *Conium maculatum*] 381
SCHIERLINGSAFT (*m*) : (spotted-)hemlock juice (from *Herba conii maculati*) 382
SCHIERLINGSFRUCHT (*f*) : (spotted-)hemlock seed (of *Herba conii maculati*) 383
SCHIERLINGSKRAUT (*n*) : (spotted-)hemlock (*Herba conii maculati*) (*Conium maculatum*) 384
SCHIERLINGSPFLANZE (*f*) : (spotted-)hemlock (*Conium maculatum*) 385
SCHIERLINGSTANNE (*f*) : (spotted-)hemlock (*Conium maculatum*) 386
SCHIERLINGSVERGIFTUNG (*f*) : Conium poisoning (from the Hemlock, *Conium maculatum*) (Med.) 387
SCHIESSBAUMWOLLE (*f*) : pyroxyline; trinitrocellulose; guncotton (see SCHIESSWOLLE and KOLLODIUMWOLLE) 388
SCHIESSBEDARF (*m*) : ammunition 389
SCHIESSEN : to shoot; fire; emit; dart; rush; flash; discharge; dash; (*n*) : shooting 390

SCHIESSGRABEN (*m*): trench 391
SCHIESSOFEN (*m*): bomb oven; tube furnace (an explosible oven for heating tubes) 392
SCHIESSPULVER (*n*): gunpowder 393
SCHIESSPULVER (*n*), **RAUCHLOSES-**: smokeless gunpowder 394
SCHIESSPULVER (*n*), **RAUCHSCHWACHES-**: gunpowder which gives off very little smoke, smokeless gunpowder 395
SCHIESSROHR (*n*): bomb tube (sealed tube wherein substances are heated under pressure) 396
SCHIESSRÖHRE (*f*): (see **SCHIESSROHR**) 397
SCHIESSCHEIBE (*f*): target 398
SCHIESSTOFF (*m*): gunpowder 399
SCHIESSWOLLE (*f*): guncotton (see **SCHIESS-BAUMWOLLE**) 400
SCHIESSWOLLEKOCHER (*m*): guncotton boiler 401
SCHIESSWOLLPULVER (*n*): guncotton powder 402
SCHIFF (*n*): ship; vessel; shuttle; nave 403
SCHIFFAHRT (*f*): navigation (see **SCHIFFFAHRT**) 404
SCHIFFBAR: navigable 405
SCHIFFBAU (*m*): shipbuilding 406
SCHIFFBAUER (*m*): shipbuilder, marine (naval) architect 407
SCHIFFBAUINGENIEUR (*m*): naval architect 408
SCHIFFBAUMEISTER (*m*): (master)-shipbuilder or shipwright 409
SCHIFFBAUPLATZ (*m*): shipyard, dockyard 410
SCHIFFBRUCH (*m*): shipwreck 411
SCHIFFBRÜCHIG: ship wrecked 412
SCHIFFBRÜCKE (*f*): pontoon (floating) bridge 413
SCHIFFCHEN (*n*): shuttle; keel (Bot.); little ship; boat 414
SCHIFFEN: to sail; navigate; ship; load; take aboard; send (by sea) 415
SCHIFFER (*m*): skipper; captain; master (Ships); sailor; mariner; navigator 416
SCHIFFERAUSDRUCK (*m*): nautical term 417
SCHIFFAHRER (*m*): sailor; mariner; navigator 418
SCHIFFAHRT (*f*): navigation; voyage; sea-journey; crossing 419
SCHIFFKUNDE (*f*): navigation; nautics 420
SCHIFFKUNST (*f*): (see **SCHIFFKUNDE**) 421
SCHIFFPUMPE (*f*): marine pump 422
SCHIFFSAGENTUR (*f*): shipping agency or agent 423
SCHIFFSARMIERUNG (*f*): ships equipment, armament 424
SCHIFFSBAUER (*m*): shipbuilder 425
SCHIFFSBEDÜRFNISSE (*n.pl*): naval stores; marine stores 426
SCHIFFSBEFEHLSHABER (*m*): captain; commander 427
SCHIFFSBEFRACHTUNGSVERTRAG (*m*): charter party 428
SCHIFFSBEUTE (*f*): prize 429
SCHIFFSBEWEGUNG (*f*): movement (motion) of a ship 430
SCHIFFSBODEN (*m*): ship's bottom 431
SCHIFFSDOCK (*n*): dock 432
SCHIFFSDOCKE (*f*): dock, basin 433
SCHIFFSEIL (*n*): cable, hawser, rope 434
SCHIFFSEINHEIT (*f*): marine unit 435
SCHIFFSGELEGENHEIT (*f*): shipping, freight, freighting opportunity 436
SCHIFFSGERIPPE (*n*): hull 437
SCHIFFSGLOCKE (*f*): ship's bell 438
SCHIFFSHAKEN (*m*): grapnel 439
SCHIFFSHILFSMASCHINE (*f*): (ship's) auxiliary engine; (*pl*): (ship's)-auxiliary machinery 440
SCHIFFSKÄFER (*m*): wood-eater, wood-beetle (*Lymexylon navale*) 441
SCHIFFSKAPITÄN (*m*): captain (of a ship) 442
SCHIFFSKESSEL (*m*): marine boiler 443
SCHIFFSKOMPASS (*m*): marine compass 444
SCHIFFSKÖRPER (*m*): hull, hulk 445
SCHIFFSKRAN (*m*): davit, derrick (Naut.) 446
SCHIFFSKÜCHE (*f*): cook-house, galley 447
SCHIFFSLADUNG (*f*): cargo 448
SCHIFFSLAFETTE (*f*): naval carriage 449
SCHIFFSLAST (*f*): cargo 450
SCHIFFSLEIM (*m*): marine glue 451
SCHIFFSLOG (*n*): ship's log 452
SCHIFFSMAKLER (*m*): ship-broker 453
SCHIFFSMANIFEST (*n*): ship's manifest (a document giving full information regarding cargo, passengers, crew, ports of call, etc.) 454
SCHIFFSMASCHINE (*f*): marine engine; ship's engine 455
SCHIFFSMASCHINENBAU (*m*): marine engineering 456
SCHIFFSMASCHINIST (*m*): ship's engineer, marine engineer 457
SCHIFFSMESSE (*f*): mess, mess room 458
SCHIFFSMITTELLINIE (*f*): centre line of the ship 459
SCHIFFSPECH (*n*): marine (ship's; caulking) pitch, (from wood-tar) 460
SCHIFFSPUMPE (*f*): marine (ship's) pump 461
SCHIFFSRAUM (*m*): hold 462
SCHIFFSREEDER (*m*): ship-owner, master 463
SCHIFFSREGISTER (*n*): code list 464
SCHIFFSRUMPF (*m*): hull of a ship 465
SCHIFFSSCHNABEL (*m*): head (nose) of a ship 466
SCHIFFSTAUFE (*f*): christening of a ship 467
SCHIFFSTURBINE (*f*): marine turbine 468
SCHIFFSVERDECK (*n*): deck-(of a ship) 469
SCHIFFSWELLE (*f*): propeller shaft 470
SCHIFFSWERFT (*f*): shipyard, shipbuilding yard 471
SCHIFFSWINDE (*f*): windlass, winch, capstan (Naut.) 472
SCHIFFSZEICHNUNG (*f*): lines (of a ship) 473
SCHIFFSZIMMERMANN (*m*): shipwright 474
SCHIFFWINDE (*f*): capstan 475
SCHIFTEN: to bind; join 476
SCHILD (*m*): shield; escutcheon; (*n*): badge; label; sign-(board); shell 477
SCHILDDRÜSE (*f*): thyroid gland (Anat.) 478
SCHILDDRÜSENESSENZ (*f*): thyroid solution (Pharm.) 479
SCHILDDRÜSENEXTRAKT (*m*): thyroid solution (Pharm.) 480
SCHILDERBLAU (*n*): (see **KASTENBLAU**) 481
SCHILDERN: to portray; paint; picture; depict; draw; to stand sentry 482
SCHILDFLECHTE (*f*): Parmelia (see **HIRNSCHÄDEL-MOOS**) 483
SCHILDFÖRMIG: scutiform; clypeiform; shield-shaped; thyroid (Anat.) 484
SCHILDKAMM (*m*): (tortoise)-shell comb 485
SCHILDKNORPEL (*m*): thyroid cartilage 486
SCHILDKRAUT (*n*): skullcap (*Herba scutellaria*) 487
SCHILDKRÖTE (*f*): tortoise; turtle; chelonian (*Testudinata*: *Chelonia*) 488
SCHILDKRÖTENÖL (*n*): tortoise oil (from *Chelonia mydas*); Sg. 0.9198 489
SCHILDKRÖTENSCHALE (*f*): tortoise shell 490

SCHILDKRÖTENSUPPE (*f*) : turtle soup 491
SCHILDLAUS (*f*) : cochineal insect (*Aspidiotus duplex*) 492
SCHILDPATT (*n*) : tortoise shell 493
SCHILDWACHE (*f*) : sentinel ; sentry ; guard ; sentry-go 494
SCHILF (*n*) : reed-(s) ; rush-(es) ; sedge 495
SCHILFDECKE (*f*) : rush mat 496
SCHILFERN : to exfoliate ; scale (flake) off 497
SCHILFGLASERZ (*n*) : Freieslebenite (see FREIESLE-BENIT) 498
SCHILFGRAS (*n*) : reeds 499
SCHILFIG : overgrown with rushes 500
SCHILFROHR (*n*) : reed (*Phragmites communis*) 501
SCHILLER (*m*) : irridescence ; glitter ; play of colours ; shimmer ; surface (metallic) colour ; splendour 502
SCHILLEREIDECHSE (*f*) : chameleon 503
SCHILLERFARBE (*f*) : shiller (changeable ; surface ; metallic) colour 504
SCHILLERFARBIG : irridescent ; showing colour play ; exhibiting shiller or metallic colour 505
SCHILLERGLANZ (*m*) : shiller ; coloured metallic lustre ; irridescent lustre 506
SCHILLERN : to irridesce ; show colour play ; change (vary) colour ; exhibit surface (metallic) colour or schiller ; glitter ; opalesce 507
SCHILLERND : irridescent ; glittering ; shimmering ; opalescent 508
SCHILLERQUARZ (*m*) : cat's-eye (Min.) 509
SCHILLERSEIDE (*f*) : shot silk 510
SCHILLERSPAT (*m*) : schiller spar (dceomposed bronzite) ; $3MgO,2SiO_2,2H_2O$; Sg. 2.7 511
SCHILLERSPAT (*m*), **DIATOMER-** : schiller spar, bastite (see SCHILLERSTEIN) 512
SCHILLERSPAT (*m*), **HEMIPRISMATISCHER-** : bronzite 513
SCHILLERSPAT (*m*), **PRISMATISCHER-** : diallage 514
SCHILLERSPAT (*m*), **PRISMATOIDISCHER-** : hypersthene 515
SCHILLERSTEIN (*m*) : schiller-spar, bastite ; $4(CaO.FeO.MgO)SiO_2+MgO.H_2O$; Sg. 2.7 (see also SCHILLERSPAT) 516
SCHILLERSTOFF (*m*) : iridescent (opalescent) substance ; esculine (see ÄSCULIN) 517
SCHIMMEL (*m*) : mildew ; mould ; grey horse 518
SCHIMMELIG : mouldy ; musty 519
SCHIMMELN : to mould ; become mouldy 520
SCHIMMELPILZ (*m*) : mould fungus ; mould (*Hyphomycetes*) (*Mycoderm oidium*) 521
SCHIMMER (*m*) : glint ; glimmer ; shimmer ; glitter ; shine ; sparkle 522
SCHIMMERLICHT (*n*) : glimmer ; glitter (see SCHIMMER) 523
SCHIMMERN : to glimmer ; shimmer ; glitter ; shine ; glisten ; sparkle ; glint 524
SCHIMPF (*m*) : abuse ; insult ; affront 525
SCHIMPFEN : to scold ; abuse ; call names ; insult ; use strong language ; grumble ; affront 526
SCHIMPFWORT (*n*) : invective ; strong word 527
SCHINDEL (*f*) : shingle ; splint (Med.) 528
SCHINDELDACH (*n*) : shingle roof 529
SCHINDEL (*f*), **HOLZ-** : wood-(en) shingle ; splint (Med.) 530
SCHINDELN : to cover with shingle ; splint (Med.) 531
SCHINDEN : to skin ; flay ; excoriate ; oppress ; exact ; be cruel 532
SCHINDEREI (*f*) : extortion ; excoriation ; flaying 533
SCHINKEN (*m*) : ham ; bacon 534
SCHINKENBEIN (*n*) : ham bone 535
SCHINKENKNOCHEN (*m*) : ham bone 536
SCHIPPE (*f*) : shovel ; skip ; scoop 537
SCHIRBEL (*m*) : bloom (Metal) (pieces broken off welded ingot with hammer) ; stamp 538
SCHIRM (*m*) : umbrella ; protection ; shade ; shelter ; screen ; guard ; umbel (Bot.) 539
SCHIRMBAUM (*m*) : (see CATAPPENBAUM) 540
SCHIRMEN : to shelter ; screen ; protect ; guard 541
SCHIRMER (*m*) : protector 542
SCHIRMMESSUNG (*f*) : screen (shutter) measurement (of water) 543
SCHIRMPFLANZE (*f*) : umbelliferous plant 544
SCHIRMSTATIV (*n*) : umbrella stand 545
SCHIRMWIRKUNG (*f*) : (three-colour)-screen effect (Photography) 546
SCHIRRMEISTER (*m*) : head man, foreman 547
SCHIZOLITH (*m*) : schizolite ; $3SiO_2.2(Ca,Fe,Mn)O.Na_2H_2O$; Sg. 3.0 548
SCHLABBERN : to slobber ; overflow ; gossip 549
SCHLABBERROHR (*n*) : overflow pipe 550
SCHLABBERVENTIL (*n*) : check valve ; overflow valve ; snifting valve 551
SCHLABBERWASSER (*n*) : overflow-(water) 552
SCHLACHT (*f*) : battle ; engagement ; fight ; action 553
SCHLACHTEN : to slaughter ; kill ; butcher 554
SCHLACHTHAUS (*n*) : slaughter-house 555
SCHLACHTHOF (*m*) : knacker's yard, slaughter-house 556
SCHLACHTHOFPRAXIS (*f*) : slaughter-house practice 557
SCHLACHTOPFER (*n*) : victim ; offering ; sacrifice 558
SCHLACHTSCHIFF (*n*) : warship, battleship, ship of the line 559
SCHLACKE (*f*) : scoria ; cinder ; dross ; slag ; clinker (Coal) ; refuse ; slack (Coal) ; sediment 560
SCHLACKEN : to remove (form) clinker or slag (Furnaces) 561
SCHLACKENARTIG : scoriaceous ; clinkery ; drossy 562
SCHLACKENAUGE (*f*) : cinder (notch) tap ; slag hole 563
SCHLACKENBILDUNG (*f*) : scorification ; formation of slag 564
SCHLACKEND : clinkering, caking 565
SCHLACKENEINSCHLUSS (*m*) : slag content ; presence of scoria (slag or cinders) 566
SCHLACKENEISEN (*n*) : cinder iron ; slag iron 567
SCHLACKENFALLPLATTE (*f*) : dumping-plate 568
SCHLACKENFORM (*f*) : nozzle-(tube) (for tapping or running off the liquid slag in blast furnace work) ; slag-block 569
SCHLACKENFRISCHEN (*n*) : pig boiling 570
SCHLACKENGANG (*m*) : cinder (fall) duct ; slag duct 571
SCHLACKENHAKEN (*m*) : slag hook, cinder or clinkering hook 572
SCHLACKENHALTIG : containing (slag) clinker 573
SCHLACKENHERD (*m*) : slag (hearth) furnace 574
SCHLACKENKASTEN (*m*) : cinder (ash) box 575
SCHLACKENKUCHEN (*m*) : cake (lump) of clinker or slag 576
SCHLACKENKÜHLVORRICHTUNG (*f*) : ash (clinker) cooling or quenching arrangement (of furnaces) 577

SCHLACKENLAVA (*f*) : scoriaceous lava 578
SCHLACKENLOCH (*n*) : cinder (notch) tap ; slag (hole) tap 579
SCHLACKENMEHL (*n*) : slag dust, ground slag 580
SCHLACKENOFEN (*m*) : slag furnace 581
SCHLACKENPUDDELN (*n*) : pig boiling 582
SCHLACKENROHEISEN (*n*) : cinder pig (Metal) 583
SCHLACKENRÖSTEN (*n*) : roasting of slag 584
SCHLACKENSAND (*m*) : slag sand (formed by granulation) 585
SCHLACKENSCHERBE (*f*) : scorifier 586
SCHLACKENSCHERBEN (*m*) : scorifier 587
SCHLACKENSPIESS (*m*) : cinder (slag) iron 588
SCHLACKENSPUR (*f*) : cinder (notch) tap ; slag hole 589
SCHLACKENSTAUB (*m*) : slack ; coal dust 590
SCHLACKENSTAUER (*m*) : fuel-retarder (of stokers) 591
SCHLACKENSTEIN (*m*) : slag brick 592
SCHLACKENSTEINFABRIK (*f*) : slag-brick works 593
SCHLACKENTRICHTER (*m*) : ash (funnel) hopper (of furnaces) 594
SCHLACKENWOLLE (*f*) : slag (mineral ; silicate) wool 595
SCHLACKENZACKEN (*m*) : cinder (front) plate 596
SCHLACKENZEMENT (*m*) : slag cement 597
SCHLACKENZINN (*n*) : prillion (tin obtained from slag) 598
SCHLACKIG : scoriaceous ; clinkery ; drossy 599
SCHLAF (*m*) : sleep ; repose ; rest 600
SCHLAFARZNEI (*f*) : narcotic ; soporific 601
SCHLAFBEFÖRDERND : soporific ; narcotic 602
SCHLAFBRINGEND : soporific ; narcotic 603
SCHLÄFE (*f*) : temple (Anat.) 604
SCHLAFEN : to sleep ; repose ; be quiescent ; rest ; lie dormant ; be non-active 605
SCHLÄFENADER (*f*) : temporal vein (Anat.) 606
SCHLÄFENBEIN (*n*) : temporal bone (Anat.) 607
SCHLÄFERIG : drowsy ; sleepy 608
SCHLAFF : flabby ; loose ; flaccid ; slack ; soft ; lax ; indolent ; remiss 609
SCHLAFFHEIT (*f*) : indolence, laxity, flaccidness, slackness 610
SCHLAFLOS : sleepless 611
SCHLAFLOSIGKEIT (*f*) : sleeplessness ; insomnia (Med.) 612
SCHLAFMITTEL (*n*) : soporific ; sleeping draught ; narcotic 613
SCHLAFMOHN (*m*) : opium poppy (*Papaver somniferum*) 614
SCHLAFSTÖRUNG (*f*) : sleep disorder (such as wakefulness, pervigilium, insomnia, disturbed or restless sleep) (Med.) 615
SCHLAFSUCHT (*f*) : somnolence ; lethargy ; trance (Med.) 616
SCHLAFTRANK (*m*) : sleeping draught ; soporific ; narcotic 617
SCHLAFTRUNK (*m*) : (see **SCHLAFTRANK**) 618
SCHLAFTRUNKEN : drowsy ; sleepy 619
SCHLAFZIMMER (*n*) : bedroom 620
SCHLAG (*m*) : blow ; stroke (of an oar, etc.) ; knock ; kick ; beat ; shock ; percussion ; apoplexy (Med.) ; kind ; coinage ; stamp ; sort ; felling (cutting) of wood ; layer ; bed ; lay ; turn (of a rope) ; impact ; rap 621
SCHLAGADER (*f*) : artery 622
SCHLAGARBEIT (*f*) : impact work 623
SCHLAGBAR : ready for (felling ; cutting) hewing down 624
SCHLAGBAUM (*m*) : turnstile 625
SCHLÄGEL (*m*) : mallet ; wooden hammer ; drumstick ; beater 626
SCHLAGELOT (*n*) : hard solder (see **SCHLAGLOT**) 627
SCHLAGEMPFINDLICHKEIT (*f*) : sensitiveness to shock (percussion or impact) 628
SCHLAGEN : to strike ; cuff ; kick ; beat (of oars, etc.) ; knock ; hit ; churn (Butter) ; press (Oil) ; put ; slay ; coin ; fight ; fell (hew ; cut) wood ; stamp ; hammer (of water) ; drive (Nails) ; warble ; concern ; belong ; impress ; dash ; lay ; coil (a rope) ; build ; construct ; throw (a bridge) ; flap (of sails) 629
SCHLAGEND : impressive ; striking ; explosive ; convincing 630
SCHLAGENDES WETTER (*n*) : fire damp 631
SCHLÄGER (*m*) : beater ; striker ; duellist 632
SCHLÄGEREI (*f*) : fighting ; row ; brawl 633
SCHLÄGERWALZE (*f*) : beater roll 634
SCHLAGFERTIG : absolutely complete ; ready to fight 635
SCHLAGFLUSS (*m*) : apoplexy (Med.) 636
SCHLAGGOLD (*n*) : beaten gold ; leaf gold 637
SCHLAGHOLZ (*n*) : undergrowth ; underwood 638
SCHLAG (*m*), **KALTER-** : a flash of lightning which does not cause fire when it strikes 639
SCHLAGLOT (*n*) : hard solder (pure copper, bronzes or brasses) 640
SCHLAGLOT (*n*), **GELBES-** : yellow hard solder (Brass filings and zinc) 641
SCHLAGLOT (*n*), **WEISSES-** : white hard solder (Brass filings + Tin + Zinc) 642
SCHLAGMANN (*m*) : stroke (a man in a rowing boat) 643
SCHLAGPROBE (*f*) : impact test or test piece 644
SCHLAGREGEN (*m*) : shower 645
SCHLAGSAAT (*f*) : hempseed 646
SCHLAGSAHNE (*f*) : whipped cream 647
SCHLAGSEITE (*f*) : lap-side (Naut.) ; the side of anything with which one strikes, striking face 648
SCHLAGSIEB (*n*) : precipitating sieve (water runs over a sieve and drives the heavy clay, containing particles of material to be regained, through in the form of mud) 649
SCHLAGSITZ (*m*) : tunking fit, wringing fit, driving fit (Mech.) 650
SCHLAGUHR (*f*) : striking clock 651
SCHLAGWASSER (*n*) : bilge-(water) 652
SCHLAGWASSERPLATTE (*f*) : wash plate 653
SCHLAGWEITE (*f*) : striking distance ; spark distance (of electric spark) 654
SCHLAGWELLE (*f*) : billow 655
SCHLAGWERK (*n*) : striking apparatus ; impact testing machine ; ram ; rammer 656
SCHLAGWETTER (*n*) : fire damp 657
SCHLAGWETTERSICHER : firedamp proof 658
SCHLAGWETTERZÜNDFÄHIG : capable of being ignited by fire-damp 659
SCHLAGWIDERSTAND (*m*) : impact (shock) resistance 660
SCHLAGZÜNDER (*m*) : percussion cap 661
SCHLAMM (*m*) : mud ; mire ; sludge ; slush ; silt ; slurry ; slime ; slip (Ceramic) ; smudge (Coal) (obtained from the Saar mines, 3 to 4 metres deep in the ground, exploited as a form of peat, mean average moisture 15 to 20% ; sent for several consecutive years from Saar and packed into huge basins with an area of several hectares and several metres deep ; allowed to dry out for several years

to obtain above result, about 18% ash, about 5,000 calories); waste washings (Coal); scum; sediment; tailings; ooze 662
SCHLAMMABBLASAPPARAT (*m*): blow-off (valve; cock) apparatus; mud-hole; scum cock 663
SCHLAMMABBLASVENTIL (*n*): blow-off valve; scum valve 664
SCHLAMMABLASS (*m*): mud-hole; blow-off (Boilers) 665
SCHLAMMABLASSAPPARAT (*m*): scum (cock) valve, blow-off (cock) valve, mud hole, any arrangement for letting mud or slime escape 666
SCHLAMMABLASSVENTIL (*n*): blow-off valve; scum valve 667
SCHLÄMMANALYSE (*f*): elutriation (washing) analysis 668
SCHLÄMMAPPARAT (*m*): elutriating (washing) apparatus 669
SCHLÄMMEN: to elutriate; levigate; wash; buddle (Mining); remove (cleanse from) mud 670
SCHLÄMMFASS (*n*): washing (tub) tank; dolly tub (Mining) 671
SCHLÄMMFLASCHE (*f*): elutriating flask 672
SCHLÄMMGEFÄSS (*n*): elutriating (reservoir) vessel 673
SCHLÄMMGLAS (*n*): elutriating glass 674
SCHLAMMHAHN (*m*): scum cock, blow-off cock 675
SCHLAMMIG: slimy; miry; sludgy; muddy; slushy 676
SCHLÄMMKREIDE (*f*): whiting; whitening; prepared (ground; powdered) chalk 677
SCHLAMMLOCH (*n*): mud hole 678
SCHLAMMLOCHTÜR (*f*): mud hole door 679
SCHLAMMPFÄNNCHEN (*n*): scum pan (Salt) 680
SCHLÄMMPROZESS (*m*): elutriating (washing) process 681
SCHLAMMROHR (*n*): scum pipe; blow-off pipe or bend (Boilers) 682
SCHLAMMSAMMLER (*m*): mud-drum (of boilers) 683
SCHLAMMSCHLICH (*m*): washed ore slime 684
SCHLÄMMSCHLICH (*m*): washed ore slime 685
SCHLAMMTEICH (*m*): settling trough, sediment trough 686
SCHLÄMMTRICHTER (*m*): elutriating funnel 687
SCHLAMMTROMMEL (*f*): mud drum (of Boilers) 688
SCHLÄMMUNG (*f*): elutriation; washing; levigation; buddling 689
SCHLAMMVENTIL (*n*): scum (mud) valve; blow-off valve (Boilers) 690
SCHLÄMMVORRICHTUNG (*f*): elutriating (washing) apparatus 691
SCHLÄMMZYLINDER (*m*): elutriating cylinder 692
SCHLAMPIG: slovenly; dirty 693
SCHLANGE (*f*): snake; serpent; hose; worm; coil 694
SCHLÄNGELN: to wind; meander 695
SCHLÄNGELND: serpentine; winding; meandering; sinuous 696
SCHLANGENFÖRMIG: serpentine; winding; meandering; sinuous; coiled 697
SCHLANGENHOLZ (*n*): snakewood 698
SCHLANGENKÜHLER (*m*): coil condenser 699
SCHLANGENLINIE (*f*): helical line 700
SCHLANGENROHR (*n*): coil; spiral tube 701
SCHLANGENRÖHRE (*f*): coil; spiral tube 702
SCHLANGENROHR-ÜBERHITZER (*m*): coil (bent tube) type superheater 703
SCHLANGENSTEIN (*m*): serpentine; ophite 704
SCHLANGENWINDUNG (*f*): coil 705
SCHLANGENWURZEL (*f*): snakeroot (*Radix actææ* of *Actæa racemosa*); birthwort; sangrel; snake weed; serpentaria (*Radix serpentariæ*) (root of *Aristolochia serpentaria*) 706
SCHLANGENWURZEL (*f*), VIRGINISCHE-: (*Radix serpentariæ*), Virginian snakeroot, birthwort, snakeweed, sangrel (*Aristolochia serpentaria*) 707
SCHLANK: slim; slender; lank-(y) 708
SCHLAPPE (*f*): slipper; slap; reverse; discomfiture; defeat; loss 709
SCHLAPPEN: to flap; hang-(down) 710
SCHLAPPERMILCH (*f*): curdled milk 711
SCHLAPPHUT (*m*): slouch hat 712
SCHLARAFFE (*m*): sluggard 713
SCHLARFE (*f*): slipper 714
SCHLAU: cunning; sly; sharp; knowing; crafty; cute; astute 715
SCHLAUCH (*m*): flexible (tube, pipe) hose; skin; drunkard; glutton; utricle 716
SCHLAUCHEN: to transfer (fill) by means of hose 717
SCHLAUCHPILZE (*m.pl*): *Ascomycetes* 718
SCHLAUCHSTÜCK (*n*): piece of flexible tube 719
SCHLAUCHVERBINDUNG (*f*): hose coupling 720
SCHLAUHEIT (*f*): astuteness; sharpness; craftiness 721
SCHLECHT: bad; poor; ill; base; mean; rotten; low; evil; vile 722
SCHLECHTDENKEND: evil-minded 723
SCHLECHTERDINGS: absolutely; utterly; by all means 724
SCHLECHTERDINGS NICHT: by no means 725
SCHLECHTHEIT (*f*): badness; rottenness; etc. (see SCHLECHT) 726
SCHLECHTHIN: merely; plainly; simply 727
SCHLECHTWEG: merely; plainly; simply 728
SCHLECHT WERDEN: to go bad, rot, become foul, degenerate, deteriorate; turn, putrefy 729
SCHLECKEN: to lick; lap; be dainty 730
SCHLECKERN: (see SCHLECKEN) 731
SCHLEGEL (*m*): mallet; maul; wooden hammer; beater; drumstick 732
SCHLEHDORN (*m*): (see SCHWARZDORN) 733
SCHLEHDORNBLÜTE (*f*): black-thorn flower 734
SCHLEHE (*f*): sloe (fruit and bush, *Prunus spinosa*) 735
SCHLEHE (*f*), GEFÜLLTE-: (see KRIECHE) 736
SCHLEHENBLÜTE (*f*): sloe blossom 737
SCHLEHENDORN (*m*): blackthorn (*Prunus spinosa*) 738
SCHLEHENFRUCHT (*f*): sloe berries (*Fructus acaciæ*) 739
SCHLEICHEN: to slink; creep; sneak 740
SCHLEICHENDES GIFT (*n*): slow poison 741
SCHLEICHER (*m*): sneak 742
SCHLEICHGUT (*n*): contraband 743
SCHLEICHHANDEL (*m*): smuggling; illicit trade 744
SCHLEICHWEG (*m*): by (secret) path 745
SCHLEIER (*m*): veil; lawn (Fabric); haze; cloudiness; mist; cloak; fog (Phot.); film 746
SCHLEIER (*m*), CHEMISCHER-: chemical fog (Photography) 747
SCHLEIERFREI: non-veiling, non-clouding, non-fogging; free from fog (Phot.) 748
SCHLEIERN: to cloak, veil, draw a veil over, cloud, fog 749
SCHLEIERTUCH (*n*): lawn 750

SCHLEIFBÜGEL (*m*): sliding bow (at the top of the contact standard of overhead electric trains) 751
SCHLEIFBÜRSTE (*f*): sliding brush (for rails); brush (Elect.) 752
SCHLEIFDAUMEN (*m*): (sliding)- contact cam or contract finger 753
SCHLEIFE (*f*): sledge; cravat; grinding mill; loop; noose; knot; slip-knot; train (of clothes); horseshoe curve; bend; slide; hairpin bend; guide; bow 754
SCHLEIFE (*f*), **AUSGLEICH-**: expansion (loop) bend (Piping) 755
SCHLEIFE (*f*), **DAMPF-**: steam circulation and automatic water return 756
SCHLEIFE (*f*), **DIAGRAMM-**: diagram loop (of engine indicator diagram) 757
SCHLEIFE (*f*), **KURBEL-**: crank guide 758
SCHLEIFEN: to grind; whet; hone; edge; polish; demolish; raze; sharpen; cut (Gems); trail; rub; drag; slip; slide; skid; rub down (Paint); skate; slur (Music) 759
SCHLEIFENBAHNHOF (*m*): loop (depôt) station (where the up train simply runs round a circle and is then the down train) 760
SCHLEIFENKURVE (*f*): loop 761
SCHLEIFEN LASSEN: to let slip; to have ground or sharpened 762
SCHLEIFENLEITUNG (*f*): loop (line) main (see **SCHLEIFLEITUNG**); pipe-line fitted with an expansion (loop) bend or bends 763
SCHLEIFENROHR (*n*): expansion (loop) bend 764
SCHLEIFENSCHALTUNG (*f*): loop connection 765
SCHLEIFENUMKEHRUNG (*f*): sweep reversion; loop flow-reversion (Turbines) 766
SCHLEIFENWICKLUNG (*f*): lap winding, multiple circuit winding 767
SCHLEIFER (*m*): grinder; cutter 768
SCHLEIFER (*m*), **KREUZKOPF-**: crosshead guide (slide) block 769
SCHLEIFE (*f*), **UMKEHR-**: reversing loop (of rope-ways) (the end station where the buckets run round a circle preparatory to starting on the return journey) 770
SCHLEIFFAHRT (*f*): dragging (of a balloon) 771
SCHLEIFFEDER (*f*): spring of brake shoe; sliding spring 772
SCHLEIFFINGER (*m*): contact finger (Elect.) 773
SCHLEIFFLÄCHE (*f*): sliding surface 774
SCHLEIFHEBEL (*m*): sliding lever 775
SCHLEIFKETTE (*f*): chain brake, chain for stopping railway waggons 776
SCHLEIFKLOTZ (*m*): grinding block 777
SCHLEIFKOHLE (*f*): carbon brush (Elect.) 778
SCHLEIFKONTAKT (*m*): sliding contact (Elect.) 779
SCHLEIFKUPPLUNG (*f*): friction (overload) coupling 780
SCHLEIFLACK (*m*): body varnish 781
SCHLEIFLACKÜBERZUG (*m*): coat-(ing) of body varnish 782
SCHLEIFLEITUNG (*f*): contact (loop) line (Elect.) 783
SCHLEIFMASCHINE (*f*): grinding machine, grinder 784
SCHLEIFMATERIAL (*n*): abrasive 785
SCHLEIFMITTEL (*n*): abrasive 786
SCHLEIFMÜHLE (*f*): grinding mill 787
SCHLEIFÖL (*n*): grinding oil, oil for grinding purposes 788
SCHLEIFPASTE (*f*): grinding paste, emery paste (for valves) 789
SCHLEIFPULVER (*n*): grinding (polishing) powder 790
SCHLEIFRAD (*n*): grinding (polishing) wheel 791
SCHLEIFRING (*m*): slip (collector) ring, packing ring (Mech.) 792
SCHLEIFRINGANKER (*m*): slip ring (rotor) armature (Elect.) 793
SCHLEIFRINGLÄUFER (*m*): slip-ring rotor (Elect.) 794
SCHLEIFSCHEIBE (*f*): grinding (disc) wheel 795
SCHLEIFSCHIENE (*f*): sliding contact (Elect.) 796
SCHLEIFSCHLAMM (*m*): swarf 797
SCHLEIFSCHUH (*m*): sliding shoe 798
SCHLEIFSEL (*n*): grindings 799
SCHLEIFSTAUB (*m*): grindings 800
SCHLEIFSTEIN (*m*): grind-(whet)-stone; hone; Sg. about 2.15 801
SCHLEIFSTOFF (*m*): ground pulp (Paper) 802
SCHLEIFTAG (*m*): skating day 803
SCHLEIFUNG (*f*): grinding; polishing; sharpening; cutting (Gems); dragging; trailing; slipping; sliding; demolishing; honing; whetting; razing 804
SCHLEIFVERSUCH (*m*): grinding test 805
SCHLEIM (*m*): slime; mucus; mucilage; phlegm; glutin 806
SCHLEIMARTIG: slimy; mucoid; glutinous; phlegmy; mucous; mucilaginous; viscous 807
SCHLEIMBILDEND: slime-forming 808
SCHLEIMDRÜSE (*f*): mucous gland; pituitary gland (*Glandula pituitaria*) (see **HYPOPHYSE**) 809
SCHLEIMEN: to generate mucous; purify 810
SCHLEIMFLUSS (*m*): blennorrhœa (Med.) 811
SCHLEIMHARZ (*n*): gum resin (see **GUMMIHARZ**) 812
SCHLEIMHAUT (*f*): mucous membrane (Med.) 813
SCHLEIMHAUT (*f*), **ENTZÜNDUNG DER- DES KOLONS**: colitis (inflammation of the mucous membrane of the colon) (Med.) 814
SCHLEIMHAUTERKRANKUNG (*f*): disease of the mucous membrane (Med.) 815
SCHLEIMIG: slimy; mucous; phlegmy; mucilaginous; glutinous; mucoid; viscous 816
SCHLEIMIGE GÄRUNG (*f*): mucous (viscous) fermentation 817
SCHLEIMIGKEIT (*f*): sliminess 818
SCHLEIMPILZ (*m*): slime (mould) fungus (*Myxomycetes*) 819
SCHLEIMSAUER: mucate of; saccharolactate of; combined with (saccharolactic) mucic acid 820
SCHLEIMSÄURE (*f*): mucic acid; saccharolactic acid; $C_6H_{10}O_8$; Mp. 213°C. 821
SCHLEIMSCHICHT (*f*): *Rete mucosum*, mucous layer-(of the skin), *Stratum mucosum* (lowermost layer of the epidermis) 822
SCHLEIMSTOFF (*m*): mucin 823
SCHLEIMTIER (*n*): mollusk 824
SCHLEIMZELLULOSE (*f*): mucocellulose 825
SCHLEIMZUCKER (*m*): levulose (see **LÄVULOSE**) 826
SCHLEISSE (*f*): splint-(er) 827
SCHLEISSEN: to split; slit; strip 828
SCHLEMMEN: to eat (greedily) like a glutton (see also **SCHLÄMM**) 829
SCHLEMMEREI (*f*): gluttony 830
SCHLEMPE (*f*): spent wash; vinasse slops (a residual liquid from distillation of molasses) 831

SCHLEMPEASCHE (*f*): salin; crude potash from beet vinasse (see SALIN) 832
SCHLEMPEKOHLE (*f*): (see SCHLEMPEASCHE) 833
SCHLEMPEKOHLE (*f*), **WEISSGEBRANNTE-**: calcined salin (see SCHLEMPEKOHLE and SALIN) (contains up to 60% potash) 834
SCHLEMPEMETHODE (*f*): method (process) of washing and compressing yeast scum into cakes 835
SCHLENDERN: to stroll; saunter; loiter; lounge 836
SCHLENKERN: to sling; fling; swing; dangle; toss 837
SCHLEPPBOOT (*n*): tug-boat 838
SCHLEPPDAMPFER (*m*): tug-(boat); steam tug 839
SCHLEPPE (*f*): train; trail; truck (Founding); felt board (Paper) 840
SCHLEPPEN: to tow; drag; trail; tug 841
SCHLEPPER (*m*): tug-boat, tug 842
SCHLEPPMÜHLE (*f*): drag-(stone) mill 843
SCHLEPPNETZ (*n*): drag net 844
SCHLEPPSCHIEBERSTEUERUNG (*f*): drag valve gear 845
SCHLEPPSCHIFF (*n*): tugboat, tug 846
SCHLEPPSEIL (*n*): tow-rope, hawser or cable; drag-rope 847
SCHLEPPTAU (*n*): (see SCHLEPPSEIL) 848
SCHLEPPWAGEN (*m*): lorry 849
SCHLEUDER (*f*): sling; centrifuge; separator 850
SCHLEUDERGESCHÄFT (*n*): undercutting, underselling 851
SCHLEUDERHONIG (*m*): strained honey 852
SCHLEUDERMASCHINE (*f*): centrifugal-(machine); centrifuge; hydro-extractor; catapult 853
SCHLEUDERN: to sling; fling; hurl; roll; cut prices; strain; centrifuge; hydro-extract; undersell; undercut-(prices) 854
SCHLEUDERPREIS (*m*): cut price 855
SCHLEUDERVERDICHTER (*m*): centrifugal compresser 856
SCHLEUDERWARE (*f*): rubbish, cheap and nasty goods, inferior articles or wares 857
SCHLEUNIG: quick; hasty; speedy; immediate 858
SCHLEUNIGKEIT (*f*): speediness 859
SCHLEUSE (*f*): sluice; sewer; flood-gate 860
SCHLEUSENGAS (*n*): sewer gas 861
SCHLEUSENSCHIEBER (*m*): sluice valve 862
SCHLEUSENTOR (*n*): lock (flood) gate; sluice-gate 863
SCHLEUSENTÜR (*f*): (see SCHLEUSENTOR) 864
SCHLEUSENVENTIL (*n*): sluice valve 865
SCHLEUSETOR (*n*): sluice gate; lock-gate 866
SCHLICH (*m*): byway; trick; concentrates; schlich; dressed ore (Metal); artifice 867
SCHLICHT: sleek; smooth; even; fine (of files); plain; straight (of wood); homely; simple 868
SCHLICHTAPPRETURMITTEL (*n*): dressing, size 869
SCHLICHTE (*f*): dressing; blackwash (Founding); skim (white) coat (Plastering); size (Fabrics) 870
SCHLICHTECHT: fast to sizing 871
SCHLICHTEN: to sleek (Leather); planish (Metal); size (Fabrics); mediate; smooth; plane; dress; settle; adjust; compose 872
SCHLICHTFEILE (*f*): smooth (smoothing) file 873
SCHLICHTHAMMER (*m*): planishing hammer 874
SCHLICHTHOBEL (*m*): smoothing plane; jointing plane 875
SCHLICHTLAUFSITZ (*m*): free fit 876
SCHLICHTLAUFSITZ (*m*), **WEITER-**: loose fit 877
SCHLICHTMASCHINE (*f*): finishing machine; smoothing machine (Textile) 878
SCHLICHTÖL (*n*): sizing oil 879
SCHLICHTPRÄPARAT (*n*): finishing (sizing; dressing) preparation 880
SCHLICHTSEIFE (*f*): sizing soap 881
SCHLICHTSTAHL (*m*): flat tool 882
SCHLICHTWALZE (*f*): finishing roll 883
SCHLICK (*m*): schlich; ore slime (Metal); mud; slime 884
SCHLICKER (*m*): paste; slip (Ceramics); dross (Metal) 885
SCHLIEFEN: to slip; creep 886
SCHLIERE (*f*): schliere; streak (in glass and igneous rocks) 887
SCHLIERENFREIES GLAS (*n*): glass free from streaks 888
SCHLIESSBAUM (*m*): beam; bar 889
SCHLIESSBLECH (*n*): bolt nab; closing plate (Stokers) 890
SCHLIESSE (*f*): peg; catch; pin; latch; anchor; locking-device 891
SCHLIESSEN: to shut; lock; close; bind; contract; embrace; conclude; wind up 892
SCHLIESSEN, SICH-: to vitrify (Ceramics); to decide, make up one's mind 893
SCHLIESSKOPF (*m*): rivet head 894
SCHLIESSLICH: final-(ly); ultimate-(ly); lastly 895
SCHLIESSMUSKEL (*m*): sphincter, obturator (Anat.) 896
SCHLIESSUNG (*f*): conclusion; closing; shutting 897
SCHLIESSUNGSDRAHT (*m*): loop wire 898
SCHLIESSUNGSSTROM (*m*): closing current (Elect.) 899
SCHLIFF (*m*): sharpening; grinding; polish; smoothness; grindings; cut (of gem); microscopical (metallographic) section or specimen 900
SCHLIFFSTOPFEN (*m*): ground stopper 901
SCHLIMM: bad; severe; sad; ill; sore; evil 902
SCHLINGE (*f*): sling; noose; loop; trap; snare; knot 903
SCHLINGEL (*m*): rogue; rascal; scoundrel; sluggard 904
SCHLINGEN: to wind; twist; twine; swallow; gulp; climb; creep; sling 905
SCHLINGERBEWEGUNG (*f*): rolling motion (of a ship) 906
SCHLINGERKIEL (*m*): bilge keel, drift keel 907
SCHLINGERN: to roll (Ship motion) 908
SCHLINGERND: rolling (motion of a ship) 909
SCHLINGERPARDUNE (*f*): preventer backstay 910
SCHLINGERPLATTE (*f*): wash plate 911
SCHLINGPFLANZE (*f*): creeper (of plants); climbing plant 912
SCHLINGSTRAUCH (*m*): climber, climbing plant (Bot.) 913
SCHLIPPE'SCHES SALZ (*n*): Schlippe's salt 914
SCHLITTEN (*m*): sledge; sleigh; cradle; truck; slide; runner; baseboard (Phot.); bed (of a planing machine, etc.); shoe (of a cross head) 915
SCHLITTENBALKEN (*m.pl*): sliding baulks (of a slipway) 916
SCHLITTENINDUKTOR (*m*): slide induction apparatus 917
SCHLITTENLÄUFER (*m*): skater 918

SCHLITTEN (*m*), STAPEL- : cradle (Shipbuilding) 919
SCHLITTSCHUH (*m*) : skate 920
SCHLITTSCHUHLAUFEN : to skate ; (*n*) : skating 921
SCHLITZ (*m*) : slot ; slit ; cleft ; fissure ; slash ; port (Mech.) 922
SCHLITZ, AUSPUFF- (*m*) : exhaust port 923
SCHLITZBRENNER (*m*) : batswing burner 924
SCHLITZEN : to slot ; slit ; split ; rip ; cleave 925
SCHLITZSTEUERUNG (*f*) : distribution by means of ports 926
SCHLOSS (*n*) : lock ; snap ; clasp ; castle ; palace 927
SCHLOSSE (*f*) : hailstone 928
SCHLOSSENSCHAUER (*m*) : hail-storm 929
SCHLOSSER (*m*) : locksmith ; fitter 930
SCHLOSSERARBEIT (*f*) : locksmith's (trade) work ; fitter's work 931
SCHLOSSEREI (*f*) : fitting shop ; locksmith's-(shop) 932
SCHLOSSERWERKSTATT (*f*) : locksmith's-(shop) ; fitter's shop 933
SCHLOSSFEDER (*f*) : lock spring 934
SCHLOSSRIEGEL (*m*) : lock bolt 935
SCHLOT (*m*) : soil pipe ; flue ; shaft ; smoke-stack ; chimney ; gutter ; channel 936
SCHLOTTE (*f*) : (see SCHLOT) 937
SCHLOTTER (*m*) : sediment from boiling (Salt) 938
SCHLOTTERIG : shaking ; loose ; flapping ; dangling 939
SCHLOTTERN : to flap ; dangle ; hang loose ; shake 940
SCHLUCHT (*f*) : gorge ; ravine ; defile ; cavity 941
SCHLUCHZEN : to sob ; hiccough 942
SCHLUCK (*m*) : gulp ; draught ; mouthful 943
SCHLUCKEN : to swallow ; gulp ; hiccough ; (*m*) : hiccough ; drink ; gulp ; draught ; (*n*) : swallowing, etc. 944
SCHLUCKENS, STÖRUNG DES- : deglutition, difficulty in swallowing (Med.) 945
SCHLUMMER (*m*) : slumber 946
SCHLUMMERN : to slumber 947
SCHLUMPIG : slovenly ; slatternly 948
SCHLUND (*m*) : pharynx ; throat ; gullet ; gulf ; precipice ; chasm ; abyss ; œsophagus; mouth (of a cave) 949
SCHLUNDKOPF (*m*) : upper pharynx 950
SCHLUPF (*m*) : slip (of pulleys) 951
SCHLÜPFEN : to slip ; slide ; glide 952
SCHLUPFEN : (see SCHLÜPFEN) 953
SCHLUPFLOCH (*n*) : loop-hole 954
SCHLUPFPULVER (*n*) : talcum powder (see MAGNES--IUMSILIKAT) 955
SCHLÜPFRIG : slippery ; lascivious ; obscene 956
SCHLÜPFRIG MACHEN : to grease ; lubricate ; make slip easily 957
SCHLÜPFUNG (*f*) : slip (Elect.) 958
SCHLÜPFUNGSKURVE (*f*) : slip curve (Elect.) 959
SCHLÜPFUNGSWIDERSTAND (*m*) : slip resistance (Elect.) 960
SCHLUPFWINKEL (*m*) : haunt ; lurking place 961
SCHLÜRFEN : to sip ; lap ; suck air (Pumps) 962
SCHLUSS (*m*) : conclusion ; end ; close-(ure) ; resolution 963
SCHLÜSSEL (*m*) : key ; screw-driver ; wrench ; switch (Elect.) 964
SCHLÜSSELBART (*m*) : ward ; bit (of a key) 965
SCHLÜSSELBEIN (*n*) : clavicle ; collar bone (Anat.) 966
SCHLÜSSELBLUME (*f*) : primrose ; cowslip (*Primula veris*) 967
SCHLÜSSELBUND (*m*) : bunch of keys 968
SCHLÜSSELFLECHTE (*f*) : Parmelia (see HIRN--SCHÄDELMOOS) 969
SCHLÜSSEL (*m*), GABEL- : fork spanner 970
SCHLÜSSELLOCH (*n*) : key-hole (in a door, or in the key itself) 971
SCHLÜSSELRING (*m*) : key-ring ; ring for keys 972
SCHLÜSSELWEITE (*f*) : clearance (opening) of a spanner or screw hammer 973
SCHLUSSFOLGE (*f*) : conclusion ; inference ; syllogism 974
SCHLUSSFOLGERUNG (*f*) : (see SCHLUSSFOLGE) 975
SCHLÜSSIG WERDEN : to come to a decision ; decide ; resolve ; make up one's mind 976
SCHLUSSLATERNE (*f*) : tail light (Railways and Motor cars, etc.) ; rear light 977
SCHLUSSRECHNUNG (*f*) : final account ; balance 978
SCHLUSSREDE (*f*) : epilogue 979
SCHLUSSSATZ (*m*) : conclusion ; conclusive remark 980
SCHLUSSSEITE (*f*) : rear (final) page 981
SCHLUSSSTEIN (*m*) : keystone 982
SCHLUSSWORT (*n*) : last word ; conclusion 983
SCHLUTTE (*f*) : alkekengi 984
SCHMACH (*f*) : disgrace ; offence ; ignominy 985
SCHMACHTEN : to pine ; yearn ; long for ; languish 986
SCHMÄCHTIG : slim ; slight ; slender ; lank 987
SCHMACHVOLL : ignominious ; disgraceful 988
SCHMACK (*m*) : sumac (see SUMACH) ; taste 989
SCHMACKEN : to sumac or treat with sumac 990
SCHMACKHAFT : savoury ; palatable ; tasty ; nice 991
SCHMACKIEREN : to sumac or treat with sumac 992
SCHMÄHEN : to slander ; inveigh ; revile ; abuse ; detract ; injure 993
SCHMÄHLICH : disgraceful ; vile ; scandalous ; ignominious 994
SCHMÄHREDE (*f*) : slander ; abuse 995
SCHMÄHSCHRIFT (*f*) : libel 996
SCHMAL : narrow ; slender ; scanty ; small 997
SCHMÄLEN : to scold ; chide 998
SCHMÄLERN : to narrow ; reduce ; abridge ; constrict ; swage ; taper ; lessen ; detract ; diminish 999
SCHMÄLERUNG (*f*) : swaging ; reduction ; constriction ; derogation ; diminution 000
SCHMALSEITE (*f*) : narrow side 001
SCHMALSPURBAHN (*f*) : narrow gauge railway 002
SCHMALSPURLOKOMOTIVE (*f*) : small-gauge locomotive, locomotive for small-gauge (narrow-gauge) railway 003
SCHMALTE (*f*) : smalt (see BLAUFARBE) 004
SCHMALZ (*n*) : lard (Sg. 0.95) ; grease ; fat ; suet (see SCHWEINEFETT) 005
SCHMALZARTIG : like lard ; greasy 006
SCHMALZ (*n*), BESTE SORTE : best (first quality) lard, neutral lard 007
SCHMALZBUTTER (*f*) : melted butter 008
SCHMALZEN : to grease ; lard (also written SCHMÄLZEN) 009
SCHMALZ (*n*), GERINGSTE SORTE : third quality lard, low quality lard, Western (prime) steam lard 010
SCHMALZIG : greasy ; fatty ; like lard 011
SCHMALZ (*n*), MITTELSORTE : middle (second) quality lard, cattle rendered lard, leaf lard 012

SCHMALZÖL (*n*): lard (oleo) oil; Sg. 0.915 [013
SCHMAND (*m*): slime; sludge; cream 014
SCHMAROTZER (*m*): parasite 015
SCHMAROTZERISCH: parasitic-(al) 016
SCHMAROTZERPFLANZE (*f*): parasitic plant 017
SCHMAUCH (*m*): smoke 018
SCHMAUCHEN: to smoke 019
SCHMAUS (*m*): feast; banquet; treat 020
SCHMECKEN: to taste; savour; relish 021
SCHMEER (*n*): fat; suet; grease 022
SCHMEERIG: fatty; greasy 023
SCHMEICHELEI (*f*): flattery; adulation 024
SCHMEICHELHAFT: complimentary; flattering 025
SCHMEICHELN: to flatter; coax; caress; fawn upon 026
SCHMEISSEN: to throw; fling; dash; deposit; strike 027
SCHMELZ (*m*): enamel; glaze; melt; melting; fusion; smalt 028
SCHMELZANLAGE (*f*): smelting plant 029
SCHMELZARBEIT (*f*): smelting-(process); enamelling; enamelled work 030
SCHMELZARBEITER (*m*): smelter; founder; enameller 031
SCHMELZBAR: fusible; meltable 032
SCHMELZBARKEIT (*f*): fusibility 033
SCHMELZBLAU (*n*): smalt (see **BLAUFARBE**) 034
SCHMELZBUTTER (*f*): melted butter 035
SCHMELZE (*f*): smelting; smeltery; fusion; melting; melt; fused mass; batch (Glass); ball (Soda); mill oil (Spinning) 036
SCHMELZE (*f*), **FLÜSSIGE-**: molten metal 037
SCHMELZEN: to melt; smelt; fuse; (*n*): smelting; melting; fusion 038
SCHMELZER (*m*): smelter; founder; melter 039
SCHMELZEREI (*f*): foundry; smelting works; smeltery 040
SCHMELZERZ (*n*): smelting ore 041
SCHMELZFARBE (*f*): vitrifiable pigment; enamel; majolica colour 042
SCHMELZFARBENBILD (*n*): enamel, pictures on majolica ware (Ceramics) 043
SCHMELZFEUER (*n*): refinery (Iron); smelting fire 044
SCHMELZFLUSS (*m*): melt; fused (molten) mass 045
SCHMELZGLAS (*n*): fusible glass; enamel 046
SCHMELZGLASUR (*f*): enamel 047
SCHMELZGUT (*n*): material (to be smelted) suitable for smelting purposes 048
SCHMELZHAFEN (*m*): smelting pot 049
SCHMELZHERD (*m*): smelting (hearth) furnace; front hearth of black-ash furnace (Soda) 050
SCHMELZHITZE (*f*): melting heat 051
SCHMELZHÜTTE (*f*): smelting works; foundry; smeltery 052
SCHMELZKEGEL (*m*): fusible cone (see **SEGER-KEGEL**) 053
SCHMELZKESSEL (*m*): melting (vessel; kettle) pan; caustic pan (Soda) 054
SCHMELZKUNST (*f*): smelting; enamelling 055
SCHMELZKURVE (*f*): fusion curve; curve of melting point 056
SCHMELZLINIE (*f*): (see **SCHMELZKURVE**) 057
SCHMELZLÖFFEL (*m*): ladle 058
SCHMELZMALEREI (*f*): enamel painting 059
SCHMELZMITTEL (*n*): flux 060
SCHMELZOFEN (*m*): smelting (melting) furnace; mill heating furnace 061
SCHMELZÖL (*n*): melting oil; mill oil (Spinning) 062
SCHMELZPERLE (*f*): blowpipe bead; enamel bead; bugle 063
SCHMELZPFANNE (*f*): melting pan (see **SCHMELZ-KESSEL**) 064
SCHMELZPOST (*f*): post; smelting charge 065
SCHMELZPROCESS (*m*): smelting (melting) process 066
SCHMELZPUNKT (*m*): fusing (fusion; melting) point 067
SCHMELZPUNKTBESTIMMUNG (*f*): determination of fusion point, or melting point 068
SCHMELZRAUM (*m*): hearth (Metal) 069
SCHMELZSTAHL (*m*): German steel; natural steel 070
SCHMELZTEMPERATUR (*f*): fusion (melting) temperature 071
SCHMELZTIEGEL (*m*): melting pot; crucible 072
SCHMELZTIEGELDECKEL (*m*): crucible cover; tile (Metal) 073
SCHMELZTIEGELHALTER (*m*): crucible (holder) support 074
SCHMELZTIEGELZANGE (*f*): crucible tongs 075
SCHMELZTOPF (*m*): melting pot 076
SCHMELZTRÖPFCHEN (*n*): blowpipe bead 077
SCHMELZÜBERZUG (*m*): enamelling, enamel (consists of quartz, borax, feldspar, magnesia and clay) (for cast iron and tin articles) 078
SCHMELZUNG (*f*): smelting; melting; fusion 079
SCHMELZWÄRME (*f*): heat of fusion; melting heat 080
SCHMELZWERK (*n*): foundry; smelting works; smeltery; enamelled work 081
SCHMELZWÜRDIG: suitable for smelting; workable 082
SCHMELZZEIT (*f*): time of (melting; smelting) fusion; journey (Glass) 083
SCHMELZZEUG (*n*): smelting tools 084
SCHMELZZONE (*f*): zone of (melting) fusion 085
SCHMER (*m*): fat; suet; grease 086
SCHMERGEL (*m*): emery (see **KORUND**) 087
SCHMERZ (*m*): pain; sorrow; grief; ache; misery 088
SCHMERZHAFT: painful; distressing; dolorous 089
SCHMERZLICH: painful; grievous 090
SCHMERZLOS: painless 091
SCHMERZSTILLEND: anodyne; having anodyne (pain-relieving) qualities 092
SCHMERZVOLL: painful 093
SCHMETTERLING (*m*): butterfly; lint 094
SCHMETTERLINGSBLÜTLER (*m.pl*): *Papilionaceæ* (Bot.) 095
SCHMETTERN: to blare; crash; clang; ring; smash; dash 096
SCHMIDT (*m*): (black)-smith 097
SCHMIED (*m*): (black)-smith 098
SCHMIEDBAR: forgeable; workable; malleable; ductile 099
SCHMIEDBARKEIT (*f*): malleability; ductility 100
SCHMIEDE (*f*): smithy; forge 101
SCHMIEDEAMBOSS (*m*): (forge)-anvil 102
SCHMIEDEARBEIT (*f*): smithing; forging; smith's work 103
SCHMIEDEBLASEBALG (*m*): forge bellows 104
SCHMIEDEEISEN (*n*): wrought iron (W.I.) (from puddling process); malleable iron 105
SCHMIEDEEISERN: (of)-wrought iron (W.I.); (of)-malleable iron 106
SCHMIEDEESSE (*f*): smith's forge 107
SCHMIEDEFEUER (*n*): forge-(fire); smith's forge or hearth 108

SCHMIEDEGEBLÄSE (*n*): forge bellows 109
SCHMIEDEHAMMER (*m*): sledge (smith's) hammer; forge (drop) hammer 110
SCHMIEDEHERD (*m*): forge-(hearth); smith's (hearth) forge 111
SCHMIEDEISEN (*n*): wrought iron (W.I.); malleable iron 112
SCHMIEDEKOHLE (*f*): forge coal, smith's coal (about 85% carbon) 113
SCHMIEDEN: to smith; forge; concoct; devise; hatch; fabricate 114
SCHMIEDEPECH (*n*): pitch (a residue in the distilling flask from the distillation of colophony) 115
SCHMIEDEPRESSE (*f*): forge (drop) press 116
SCHMIEDESTAHL (*m*): forged steel 117
SCHMIEDESTEMPEL (*m*): drop-stamp 118
SCHMIEDESTÜCK (*n*): forging 119
SCHMIEDEZANGE (*f*): smith's (forge) tongs 120
SCHMIEGE (*f*): bevel 121
SCHMIEGEN: to bend; bevel; incline; cling 122
SCHMIEGEN AN: to press close to; cling to 123
SCHMIEGEN UM: to twine around or about 124
SCHMIEGEN VOR: to cringe; submit; crouch before 125
SCHMIEGSAM: pliant; lithe; flexible; supple; submissive 126
SCHMIEGSAMKEIT (*f*): flexibility; submissiveness 127
SCHMIEGUNG (*f*): bevelling 128
SCHMIELE (*f*): bulrush 129
SCHMIERAPPARAT (*m*): grease box; oil cup; lubricator; lubricating apparatus; oiler 130
SCHMIERAPPARAT (*m*) MIT SICHTBAREM TROPF-ENFALL (*m*): sight-feed lubricator 131
SCHMIERBÜCHSE (*f*): grease box; oiler; oil cup; lubricator; axle box 132
SCHMIERE (*f*): (cart)-grease; lubricant; lubricating oil; lubricating grease; salve; ointment; smear; dip (for cattle) 133
SCHMIEREN: to smear; daub; grease; oil; lubricate; anoint; spread; butter; scribble; scrawl; thrash; adulterate (Liquors) 134
SCHMIERFETT (*n*): grease; lubricant; cart-grease 135
SCHMIERFLECK (*m*): grease spot; smear 136
SCHMIERGEFÄSS (*n*): lubricating box, grease box, oil cup, lubricator, etc. 137
SCHMIERHAHN (*m*): lubricating (oil, grease) cock, lubricator cock or tap 138
SCHMIERIG: smeary; greasy; oily; viscous; glutinous; dirty 139
SCHMIERKANNE (*f*): oil can 140
SCHMIERKÄSE (*m*): cream cheese 141
SCHMIERLEDER (*n*): oil dressed leather 142
SCHMIERMASSE (*f*): lubricant; ointment; liniment; unction; salve 143
SCHMIERMATERIAL (*n*): (see SCHMIERMASSE) 144
SCHMIERMITTEL (*n*): (see SCHMIERMASSE) 145
SCHMIERNUTE (*f*): grease channel, lubricating groove 146
SCHMIERÖL (*n*): lubricating oil; lubricant 147
SCHMIERPRESSE (*f*): lubricating press 148
SCHMIERPUMPE (*f*): lubricating pump; oil pump 149
SCHMIERRING (*m*): lubricating ring 150
SCHMIERSEIFE (*f*): soft soap 151
SCHMIERUNG (*f*): greasing; oiling; lubricating; smearing; lubrication 152
SCHMIERVASE (*f*): lubricating box, grease box, oil cup, lubricator 153
SCHMIERVORRICHTUNG (*f*): lubricating arrangement; lubricator 154
SCHMIERWOLLE (*f*): greasy wool 155
SCHMINKBOHNE (*f*): kidney (French) bean 156
SCHMINKE (*f*): grease paint; face paint; rouge 157
SCHMINKEN: to paint; rouge; make up 158
SCHMINKMITTEL (*n*): cosmetic 159
SCHMINKROT (*n*): rouge 160
SCHMINKWEISS (*n*): flake white; pearl white; bismuth white; basic nitrate or oxychloride of bismuth (see WISMUTSUBNITRAT) 161
SCHMINKWURZEL (*f*): (see ALKANNA) 162
SCHMIRGEL (*m*): emery (see KORUND) 163
SCHMIRGELFEILE (*f*): emery file 164
SCHMIRGELLEINEN (*n*): emery cloth 165
SCHMIRGELLEINWAND (*f*): emery cloth 166
SCHMIRGELN: to emery; rub with emery 167
SCHMIRGELPAPIER (*n*): emery paper 168
SCHMIRGELPULVER (*n*): emery (dust) powder 169
SCHMIRGELRAD (*n*): emery wheel 170
SCHMIRGELSCHEIBE (*f*): emery wheel 171
SCHMIRGELSCHLEIFRAD (*n*): emery wheel 172
SCHMIRGELTUCH (*n*): emery-cloth 173
SCHMISS (*m*): throw; strike; blow 174
SCHMITZEN: to lash; cut; whip 175
SCHMOLLEN: to pout; sulk 176
SCHMOREN: to stew 177
SCHMORTOPF (*m*): stew-(ing) pot 178
SCHMUCK: neat; trim; pretty; (*m*): ornament; decoration; finery; jewellery; frill; embellishment; adornment 179
SCHMÜCKEN: to dress; decorate; adorn; trim; embellish 180
SCHMUCKINDUSTRIE (*f*): ornament (jewellery) industry 181
SCHMUCKSTEIN (*m*): gem 182
SCHMUGGELN: to smuggle 183
SCHMUNZELN: to smirk; smile; simper 184
SCHMUTZ (*m*): dirt; smut; filth; soil; mess; muck; stain 185
SCHMUTZDECKE (*f*): top layer of a filter 186
SCHMUTZEN: to spot; stain; soil; dirty; foul; tarnish 187
SCHMUTZFÄNGER (*m*): mudguard 188
SCHMUTZFLECK (*m*): smut; spot; stain 189
SCHMUTZIG: dirty; filthy; smutty; soiled; foul; nasty; sordid 190
SCHMUTZKONKURRENZ (*f*): unfair competition 191
SCHMUTZPAPIER (*n*): waste paper 192
SCHMUTZWASSER (*n*): fouled water, dirty water 193
SCHMUTZWOLLE (*f*): wool in the yolk 194
SCHNABEL (*m*): beak; bill; nozzle; nose 195
SCHNABELFÖRMIG: beak-shaped; hooked 196
SCHNAKE (*f*): joke; gnat 197
SCHNALLE (*f*): clasp; buckle 198
SCHNALLEN: to clasp; buckle 199
SCHNALZEN: to clap; clash; smack; snap; cluck 200
SCHNAPPDECKE (*f*): extension (of gas mask) 201
SCHNAPPEN: to snap; snatch; grasp; grab; gasp 202
SCHNÄPPER (*m*): catch; latch; trigger 203
SCHNAPS (*m*): spirits; dram; liqueur; brandy 204
SCHNAPSFLASCHE (*f*): brandy (spirit) flask or bottle 205
SCHNARCHEN: to snore; snort; (*n*): snoring; stertor (Med.) 206

SCHNARCHVENTIL (*n*): snifting valve; over-pressure valve; air-inlet valve 207
SCHNATTERN: to chatter; cackle 208
SCHNAUBEN: to snort; blow; puff; pant 209
SCHNAUFEN: to wheeze; puff 210
SCHNAUMAST (*m*): snow-mast 211
SCHNAUZE (*f*): snout; muzzle; mouth; nose; nozzle; spout 212
SCHNECKE (*f*): snail; slug; cockle; fusee; worm (also Mech.); endless screw; spiral; volute; cochlea (Anat.); helice 213
SCHNECKENFÖRMIG: spiral; helical 214
SCHNECKENGANG (*m*): auger; snail's pace 215
SCHNECKENGETRIEBE (*n*): helical (worm) gear-(ing) 216
SCHNECKENGEWINDE (*n*): helix 217
SCHNECKENHAUS (*n*): snail's shell 218
SCHNECKENKLEE (*m*): snail clover 219
SCHNECKENLINIE (*f*): spiral (helical) line; conchoid 220
SCHNECKENRAD (*n*): worm (wheel) gear; helical gearing 221
SCHNECKENWELLE (*f*): worm shaft 222
SCHNEE (*m*): snow 223
SCHNEEBAHN (*f*): run (for sledges) 224
SCHNEEBALL (*m*): snow-ball 225
SCHNEEFLOCKE (*f*): snowflake 226
SCHNEEGESTÖBER (*n*): snow-storm 227
SCHNEEGIPS (*m*): snowy (foliated) gypsum 228
SCHNEEGLÖCKCHEN (*n*): snowdrop (*Galanthus nivalis*) 229
SCHNEEGRENZE (*f*): snow line 230
SCHNEEIG: snowy; snow-white; spotlessly white 231
SCHNEEKRISTALL (*m*): snow crystal 232
SCHNEEWEHE (*f*): snow-drift 233
SCHNEEWEISS (*n*): snow (zinc) white 234
SCHNEEWOLKE (*f*): snow cloud 235
SCHNEE, ZU-SCHLAGEN: to beat to a froth 236
SCHNEIDBACKEN (*f.pl*): screw dies 237
SCHNEIDBRENNER (*m*): cutting burner 238
SCHNEIDE (*f*): edge; knife-edge; cut; cutting-edge; keenness 239
SCHNEIDEBACKE (*f*): die 240
SCHNEIDEBRETT (*n*): cutting board 241
SCHNEIDEEISEN (*n*): tap (for screws); die-(plate) 242
SCHNEIDEEISENHALTER (*m*): die stock 243
SCHNEIDEKLUPPE (*f*): die stock 244
SCHNEIDELN: to prune, lop 245
SCHNEIDEMASCHINE (*f*): cutting machine; guillotine (for paper); slicing (shearing) machine 246
SCHNEIDEMÜHLE (*f*): cutting (sawing) mill; sawmill 247
SCHNEIDEN: to cut; carve; saw; lop; prune; adulterate; intersect; disect; bisect 248
SCHNEIDEN, GEWINDE-(*n*): to cut screws, to cut threads (Mechan.) 249
SCHNEIDER (*m*): cutter; tailor 250
SCHNEIDERARBEIT (*f*): tailoring; tailor's work 251
SCHNEIDERKREIDE (*f*): tailor's chalk; French chalk 252
SCHNEIDEZAHN (*m*): incisor 253
SCHNEIDIG: sharp; keen; plucky 254
SCHNEIDKANTE (*f*): cutting edge, sheared edge (Metal) 255
SCHNEIDKLUPPE (*f*): screw stock 256
SCHNEIDWERKZEUG (*n*): cutting tool(s) or machine(ry), cutter 257
SCHNEIEN: to snow 258
SCHNELL: fast; quick; sudden; rapid; swift; speedy 259
SCHNELLARBEITSTAHL (*m*): high-speed (tool) steel 260
SCHNELLAUFEND: high speed, fast running 261
SCHNELLBINDER (*m*): quick-setting cement 262
SCHNELLBLEICHE (*f*): quick (chemical) bleaching 263
SCHNELLBOHRMASCHINE (*f*): high-speed (boring) drilling machine 264
SCHNELLDREHBANK (*f*): high-speed lathe 265
SCHNELLDREHSTAHL (*m*): high speed (tool) steel 266
SCHNELLEN: to spring; let fly; fling; cheat; jerk; jump 267
SCHNELLENTWICKLER (*m*): rapid developer (Phot.) 268
SCHNELLESSIG (*m*): quick vinegar 269
SCHNELLESSIGBEREITUNG (*f*): quick vinegar process 270
SCHNELLFEUER (*n*): rapid (drum) fire 271
SCHNELLFINGERIG: dexterous; quick-fingered 272
SCHNELLFLÜSSIG: easily fusible 273
SCHNELLGÄRUNG (*f*): quick fermentation 274
SCHNELLIGKEIT (*f*): rapidity; velocity; speed; quickness; swiftness 275
SCHNELLIGKEITSMESSER (*m*): speed indicator; speedometer 276
SCHNELLKRAFT (*f*): elasticity 277
SCHNELLKRÄFTIG: elastic 278
SCHNELLLAUFEND: high-speed; fast running 279
SCHNELLLOT (*n*): fusible (alloy) metal; soft solder (Cadmium alloy; 2 parts Cd.; 4 parts Sn., 2 parts Pb.); Mp. 149°C. 280
SCHNELLPHOTOGRAPHIE (*f*): instantaneous photograph-(y); snapshot 281
SCHNELLPHOTOGRAPHIE (*f*), **AMERIKANISCHE-**: ferrotype (see **FERROTYPIE**) 282
SCHNELLPRESSE (*f*): mechanical press 283
SCHNELLPROBE (*f*): rapid test 284
SCHNELLRÖSTE (*f*): quick (chemical) retting (Flax) 285
SCHNELLSCHLUSSVENTIL (*n*): safety valve 286
SCHNELLSCHREIBEKUNST (*f*): stenography; shorthand 287
SCHNELLSCHRIFT (*f*): stenography; shorthand 288
SCHNELLTROCKNEND: siccative; quick-drying 289
SCHNELLZUG (*m*): fast (express) train 290
SCHNELLZUGLOKOMOTIVE (*f*): express engine 291
SCHNELLZÜNDER (*m*): quick match 292
SCHNEPFE (*f*): woodcock; snipe 293
SCHNEPPE (*f*): snout; spout; lip; nozzle 294
SCHNEUZEN: to snuff (Candle); blow (Nose) 295
SCHNIPPELN: to chip; cut (into shreds) to pieces 296
SCHNIPPISCH: pert; saucy; snappish 297
SCHNITT (*m*): cut-(ting); incision; slice; section (of drawing); edge (of book or knife); operation (Surgical); crop; (point of)-intersection (of lines) 298
SCHNITTBOHNE (*f*): French bean 299
SCHNITTBRENNER (*m*): batswing burner 300
SCHNITTE (*f*): cut; slice; chop; steak; sulphur match (for casks) 301
SCHNITTER (*m*): reaper; mower; corn cutter 302
SCHNITTFLÄCHE (*f*): area (surface) of a section; section-(al plane, surface or area); notched (cut) surface, (for notch tests of metal) 303
SCHNITTHANDEL (*m*): retail trade 304

SCHNITTHOLZ (*n*) : sawn timber 305
SCHNITTKANTE (*f*) : leading (foreward) edge (of propeller) 306
SCHNITTLAUCH (*m*) : chive ; garlic (*Allium schœnopraæum*) 307
SCHNITTLINIE (*f*) : line of intersection 308
SCHNITTPUNKT (*m*) : point of intersection 309
SCHNITTWAREN (*f.pl*) : draper's (dry) goods 310
SCHNITZ (*m*) : cut ; slice ; chip ; shred ; scrap ; clipping ; cutting ; section 311
SCHNITZARBEIT (*f*) : carving ; carved work 312
SCHNITZEL (*n*) : cut ; slice ; chip ; shred ; scrap ; clipping ; cutting ; section 313
SCHNITZELEI (*f*) : cutting ; carving (of wood) 314
SCHNITZELMASCHINE (*f*) : slicer ; slicing machine ; shredding machine 315
SCHNITZELN : to cut ; chip ; whittle ; carve (Wood) 316
SCHNITZEN : (see SCHNITZELN) 317
SCHNITZER (*m*) : cutter ; carver ; blunder 318
SCHNITZERN : to blunder ; make a mistake 319
SCHNITZLER (*m*) : cutter ; carver 320
SCHNÖDE : contemptuous ; contemptible ; base ; mean ; vile 321
SCHNOPPERN : to snuffle 322
SCHNÖRKEL (*m*) : scroll ; flourish 323
SCHNÜFFELN : to sniff ; pry ; snuffle ; smell 324
SCHNÜFFELVENTIL (*n*) : snifting (over-pressure) valve ; air-inlet valve 325
SCHNUPFARTIG : catarrhic ; catarrhal 326
SCHNUPFEN (*m*) : catarrh, cold (inflammation or congestion of mucous membrane) (Med.); (as a verb : to sniff, snuff, take snuff, inhale through the nose) 327
SCHNUPFENPULVER (*n*) : snuff powder 328
SCHNUPFENVERHÜTUNGSMITTEL (*n.* and *pl*) : catarrh (cold) preventative-(s) or prophylactic-(s) 329
SCHNUPFTABAK (*m*) : snuff 330
SCHNUPFTABAKSDOSE (*f*) : snuff-box 331
SCHNUPFTABAKSFABRIKANT (*m*) : snuff (maker), manufacturer 332
SCHNUPPERN : to snuffle ; sniff ; snuff (a candle) 333
SCHNUR (*f*) : cord ; string ; twine ; line ; tape ; band ; lace 334
SCHNÜRBAND (*n*) : lace ; string 335
SCHNÜRBODEN (*m*) : mould (stage) loft floor 336
SCHNÜREN : to string ; lace ; tie up ; strap ; constrict 337
SCHNURGERADE : upright ; straight ; (dead)-level 338
SCHNÜRKOLBEN (*m*) : flask with constricted neck 339
SCHNURLAUF (*m*) : groove (of a wheel) 340
SCHNÜRLOCH (*n*) : eyelet hole 341
SCHNÜRNADEL (*f*) : bodkin 342
SCHNURRBART (*m*) : moustache 343
SCHNURRE (*f*) : rattle ; story ; joke ; drollery 344
SCHNURREN : to hum ; rattle ; whirr ; buzz ; whiz ; purr ; fib 345
SCHNURRIG : droll ; funny ; waggish 346
SCHNURSCHEIBE (*f*) : grooved wheel or disc 347
SCHNURSTRACKS : straight ; direct-(ly) 348
SCHOBERN : to stack ; pile ; heap-(up) 349
SCHOCK (*n*) : sixty ; land-tax 350
SCHOFEL : wretched ; trashy ; mean ; paltry ; (*m*) : refuse ; trash 351
SCHOKOLADE (*f*) : chocolate 352
SCHOKOLADENBRAUN (*n*) : chocolate brown 353
SCHOKOLADENFABRIK (*f*) : chocolate (works) factory 354
SCHOKOLADENLACK (*m*) : chocolate varnish 355
SCHOKOLADENTAFEL (*f*) : cake of chocolate 356
SCHOLLE (*f*) : clod ; soil ; flounder ; plaice 357
SCHÖLLKRAUT (*n*) : celandine (*Chelidonium*) 358
SCHON : already ; yet ; ever ; now ; duly ; since ; indeed ; no doubt ; surely ; merely ; even ; alone 359
SCHÖN : beautiful ; lovely ; handsome ; fine ; fair 360
SCHÖNE (*f*) : fining ; isinglass ; fair one ; beauty 361
SCHONEN : to spare ; protest ; save ; forbear ; preserve from injury 362
SCHÖNEN : to fine ; refine ; clear ; clarify ; gloss ; brighten (Colours) ; beautify ; embellish 363
SCHONERBARK (*f*) : barquentine 364
SCHONERBAUM (*m*) : main boom of a schooner 365
SCHONERBRIGG (*f*) : brigantine 366
SCHONERGAFFEL (*f*) : main gaff of a schooner 367
SCHONERKUFF (*f*) : dutch-built schooner 368
SCHÖNFÄRBEN (*n*) : fine-colour dyeing ; garment dyeing 369
SCHÖNFÄRBER (*m*) : dyer in fine colours 370
SCHÖNFÄRBEREI (*f*) : garment dyeing 371
SCHÖNGRÜN (*n*) : Paris green ; copper aceto-arsenite 372
SCHÖNHEIT (*f*) : beauty ; loveliness ; fineness ; handsomeness ; fairness 373
SCHÖNHEITSESSENZ (*f*) : beauty essence 374
SCHÖNHEITSMITTEL (*n*) : aid to beauty ; cosmetic 375
SCHÖNHEITSWASSER (*n*) : beauty wash 376
SCHONUNG (*f*) : forbearance ; indulgence ; sparing 377
SCHONUNGSLOS : pitiless ; unsparing ; relentless 378
SCHÖNUNGSMITTEL (*n*) : fining agent ; brightening medium (for colours) 379
SCHOOSS (*m*) : lap ; tail (of a coat) ; womb ; shoot ; sprig 380
SCHOPF (*m*) : tuft ; shock or head of hair ; top (of trees) 381
SCHÖPFBRUNNEN (*m*) : (dipping)-well (of water) 382
SCHÖPFBÜTTE (*f*) : pulp (stuff) vat (Paper) 383
SCHÖPFEIMER (*m*) : bucket (of a well) 384
SCHÖPFEN : to draw ; scoop ; dip ; bale ; ladle ; create ; conceive ; obtain 385
SCHÖPFER (*m*) : creator ; maker ; author ; originator ; drawer (of water, etc.) ; scoop ; ladle ; bucket ; dipper 386
SCHÖPFERISCH : creative 387
SCHÖPFFORM (*f*) : paper mould (Paper) 388
SCHÖPFGEFÄSS (*n*) : scoop ; dipper ; ladle ; bucket 389
SCHÖPFHERD (*m*) : casting crucible 390
SCHÖPFKELLE (*f*) : scoop ; ladle 391
SCHÖPFLÖFFEL (*m*) : scoop ; ladle 392
SCHÖPFPROBE (*f*) : dipping test or sample (Thomas process) 393
SCHÖPFRAD (*n*) : bucket wheel 394
SCHÖPFRAHMEN (*m*) : deckle (Paper) 395
SCHÖPFTIEGEL (*m*) : ladle 396
SCHÖPFUNG (*f*) : creation 397
SCHOPPEN (*m*) : shed ; coach-house ; pint ; measure of beer or wine 398
SCHÖPS (*m*) : wether ; mutton ; simpleton 399
SCHÖPSENBRATEN (*m*) : roast mutton 400
SCHÖPSENFLEISCH (*n*) : mutton 401
SCHÖPSENKEULE (*f*) : leg of mutton 402

SCHÖPSENTALG (*m*): mutton tallow; prepared suet (see **HAMMELTALG**) 403
SCHORF (*m*): scab; scurf 404
SCHORFIG: scabby; scurfy 405
SCHÖRL (*m*): schorl; Sg. 3.17 (black tourmaline); $(Al,Fe,Mn,Mg)SiO_2.B_2O_3$ 406
SCHÖRLIT (*m*): pycnite (type of topaz) 407
SCHORNSTEIN (*m*): chimney; stack; flue; smokestack; funnel (of a ship) 408
SCHORNSTEINDECKEL (*m*): chimney (cover) cowl 409
SCHORNSTEIN (*m*), **KLAPP-**: hinged funnel, funnel which lowers (mostly in respect of tug-boats) 410
SCHORNSTEINMANTEL (*m*): chimney casing 411
SCHORNSTEINVERLUST (*m*): chimney (flue) loss 412
SCHORNSTEINZUG (*m*): chimney draught; natural draught 413
SCHOSS (*m*): lap; womb; tails (of coat); shoot; sprig 414
SCHOSSEN: to shoot-(forth) 415
SCHOSSFREI: free from taxes; exempt 416
SCHOSSGERINNE (*n*): channel; trough 417
SCHÖSSLING (*m*): shoot; sprout; sucker 418
SCHOSSPFLICHTIG: liable to taxes; taxable 419
SCHOSSREIS (*n*): sprig; shoot 420
SCHOTE (*f*): pod; husk; shell; sheet (Naut.); (*pl*): (green)-peas 421
SCHOTE (*f*), **GROSSE-**: main sheet 422
SCHOTENBLOCK (*m*): sheet block 423
SCHOTENDORN (*m*): acacia; black locust 424
SCHOTENERBSE (*f*): (green)-pea 425
SCHOTENGEWÄCHS (*n*): leguminous plant 426
SCHOTENKLAMPE (*f*): cleat-(of a sheet) 427
SCHOTENPFEFFER (*m*): capsicum; red pepper (*Capsicum fastigiatum*) 428
SCHOTENPFLANZE (*f*): leguminous plant 429
SCHOTENSTICH (*m*): sheet bend, sheet knot 430
SCHOTHORN (*n*): clew 431
SCHOTT (*n*): bulkhead (Naut.) 432
SCHOTTER (*m*): macadam; ballast; broken stone 433
SCHOTTERHERSTELLUNG (*f*): ballast (broken stone) manufacture 434
SCHOTT (*n*), **LÄNGS-**: longitudinal bulkhead 435
SCHOTTPLATTE (*f*): bulkhead plate 436
SCHOTT (*n*), **QUER-**: cross bulkhead 437
SCHOTTÜR (*f*): bulkhead door 438
SCHOTTVERSTEIFUNG (*f*): bulkhead (stiffener) stay 439
SCHOTT (*n*), **WASSERDICHTES-**: watertight bulkhead 440
SCHRAFFIEREN: to cross shade; hatch; line 441
SCHRAFFIERUNG (*f*): hatching, cross shading 442
SCHRAFFUR (*f*): shading, hatching, shaded or hatched portion of a drawing 443
SCHRAFFURPLATTE (*f*): line screen (Phot.) 444
SCHRÄG: inclined; slanting; oblique; sloping; bevel; diagonal 445
SCHRÄGE (*f*): inclination; slope; slant; bevel 446
SCHRAGEN (*m*): jack; trestle 447
SCHRÄGLÖTEN: to solder on the slant 448
SCHRÄGROST (*m*): inclined grate 449
SCHRÄGSTELLUNG (*f*): inclination 450
SCHRÄGWALZWERK (*n*): inclined rolling mill, Mannesmann-type rolling mill 451
SCHRÄMMASCHINE (*f*): (coal)-cutting machine 452
SCHRAMME (*f*): scratch; scar; cicatrice; slash 453
SCHRAMMEN: to scar; scratch; score; slash 454
SCHRAMMIG: scored; scarred; scratched 455
SCHRANK (*m*): cupboard; cabinet; safe; closet; press; case; locker 456
SCHRANKE (*f*): barrier; bar; limit; bound; restriction; (*pl*): lists (of tournaments) 457
SCHRÄNKEISEN (*n*): saw set 458
SCHRÄNKEN: to (put a)-cross; set (a saw) 459
SCHRANKEN, IN DIE-TRETEN: to enter the lists 460
SCHRANKENLOS: limitless; unlimited; boundless; unbounded 461
SCHRÄNKLEHRE (*f*): set gauge (for setting saws) 462
SCHRÄNKWEISE: crosswise 463
SCHRAPE (*f*): scraper; tracer 464
SCHRAPEN: to scrape 465
SCHRAPER (*m*): scraper; tracer 466
SCHRATSEGEL (*n*): fore and aft sail 467
SCHRAUBDECKEL (*m*): screw cap; screw-on lid 468
SCHRAUBE (*f*): (male)-screw; male-thread; worm; helice 469
SCHRAUBEN: to screw; worm; banter; mock; jeer 470
SCHRAUBENBOHRER (*m*): screw (auger) tap; twist drill 471
SCHRAUBENBOLZEN (*m*): screw bolt 472
SCHRAUBENBRUNNEN (*m*): screw aperture; screw trunk 473
SCHRAUBENDAMPFER (*m*): screw steamer 474
SCHRAUBENDREHBANK (*f*): screw cutting lathe 475
SCHRAUBENDREHER (*m*): screwdriver; wrench 476
SCHRAUBENFEDER (*f*): helical spring 477
SCHRAUBENFLÜGEL (*m*): blade of a screw or propeller; propeller blade 478
SCHRAUBENFÖRMIG: screw-shaped; helical; spiral 479
SCHRAUBENFUTTER (*n*): screw chuck 480
SCHRAUBENGANG (*m*): thread of a screw (either male or female); screwthread 481
SCHRAUBENGETRIEBE (*n*): worm gearing 482
SCHRAUBENGEWINDE (*n*): (see **SCHRAUBENGANG**) 483
SCHRAUBENGLAS (*n*): glass or jar with screw-(ed) neck 484
SCHRAUBENKLUPPE (*f*): screw stock 485
SCHRAUBENKUPPELUNG (*f*): screw coupling; screwed joint 486
SCHRAUBENLEHRE (*f*): screw gauge 487
SCHRAUBENLINIE (*f*): helical (line) curve 488
SCHRAUBENLOCH (*n*): screw hole 489
SCHRAUBENMUTTER (*f*): (screw)-nut; female (screw) thread 490
SCHRAUBENPRESSE (*f*): screw (fly) press 491
SCHRAUBENQUETSCHHAHN (*m*): screw (squeezing) pinch cock; screw clip 492
SCHRAUBENRAD (*n*): helical gear; worm wheel 493
SCHRAUBENRUTSCHE (*f*): screw-shaped (winding) gravity chute 494
SCHRAUBENSCHIFF (*n*): screw driven ship 495
SCHRAUBENSCHLEPPER (*m*): screw tug 496
SCHRAUBENSCHLÜSSEL (*m*): screw-(hammer; wrench; spanner) key 497
SCHRAUBENSCHLÜSSEL (*m*), **ENGLISCHER-**: monkey (spanner) wrench 498
SCHRAUBENSCHNEIDEBANK (*f*): screw-cutting lathe 499

SCHRAUBENSCHNEIDKLUPPE (*f*): screw stock 500
SCHRAUBENSCHNEIDMASCHINE (*f*): screw cutter, screw cutting machine 501
SCHRAUBENSENKER (*m*): countersink 502
SCHRAUBENSPINDEL (*f*): screw rod or spindle 503
SCHRAUBENSTEVEN (*m*): propeller (screw) post 504
SCHRAUBENSTOCK (*m*): vice; bench vice 505
SCHRAUBENSTOCKBACKE (*f*): vice cheek 506
SCHRAUBENUMSTEUERUNG (*f*): screw reversing gear 507
SCHRAUBENVERBINDUNG (*f*): screw coupling; screwed joint 508
SCHRAUBENWELLE (*f*): propeller shaft 509
SCHRAUBENWELLENTUNNEL (*m*): screw (propeller) shaft tunnel 510
SCHRAUBENWINDE (*f*): screw-jack 511
SCHRAUBENZIEHER (*m*): screw-driver 512
SCHRAUBENZWINGE (*f*): screw clamp 513
SCHRAUBE (*f*), PROPELLER-: propeller, screw propeller, screw 514
SCHRAUBSTOCK (*m*): vice 515
SCHRAUBSTOCKBACKEN (*f.pl*): vice jaws 516
SCHRAUBSTOCK (*m*), PARALLEL-: parallel vice 517
SCHRECK (*m*): fright; terror 518
SCHRECKEN: to be afraid; chill; frighten; terrify; (*m*): fright; horror; terror 519
SCHRECKENSHERRSCHAFT (*f*): reign of terror 520
SCHRECKLICH: awful; frightful; terrible; fearful; dreadful; terrific 521
SCHRECKSTEIN (*m*): malachite 522
SCHREI (*m*): cry; shout; scream; shriek; cake; bloom (Metal) 523
SCHREIBART (*f*): style-(of writing); spelling; method of writing (such as grammar, construction, etc.) 524
SCHREIBEN: to write; mark; communicate (by letter); record (Instruments); (*n*): epistle; letter; writing; writ; communication 525
SCHREIBER (*m*): writer; clerk; copyist; recording (gauge) meter; recorder; steam meter 526
SCHREIBERSIT (*m*): schreibersite; $(Fe,Ni,Co)_3P$; Sg. 7.15 527
SCHREIBFEDER (*f*): pen 528
SCHREIBFEHLER (*m*): error in writing; clerical error 529
SCHREIBKIES (*m*): marcasite (see MARKASIT) 530
SCHREIBMASCHINE (*f*): typewriter 531
SCHREIBMASCHINENPAPIER (*n*): typewriting paper (Paper) 532
SCHREIBMATERIALEN (*n.pl*): writing materials; stationery 533
SCHREIBPAPIER (*n*): writing paper 534
SCHREIBPULT (*n*): writing desk 535
SCHREIBTINTE (*f*): writing ink 536
SCHREIBWEISE (*f*): style; manner of writing 537
SCHREIBWERK (*n*): recording (registering) gear, device or apparatus, recorder 538
SCHREIBZEUG (*n*): writing materials 539
SCHREIEN: to cry; scream; shout; shriek; (*n*): crying; shouting, etc., cry (of tin) 540
SCHREIEND: loud (of colours); crying; screaming 541
SCHREIN (*m*): cabinet; chest; case; casket; press; shrine 542
SCHREINER (*m*): cabinet-maker; joiner 543
SCHREITEN: to stride; step; proceed; stalk 544
SCHRIFT (*f*): (hand)-writing; type; characters; paper; publication; calligraphy 545
SCHRIFTABSATZ (*m*): paragraph 546
SCHRIFTART (*f*): style (of letters or writing) 547
SCHRIFTERZ (*n*): sylvanite (see SYLVANIT) 548
SCHRIFTFÜHRER (*m*): secretary 549
SCHRIFTGIESSER (*m*): type founder 550
SCHRIFTGIESSEREI (*f*): type foundry 551
SCHRIFTGIESSERMETALL (*n*): type metal 552
SCHRIFTGOLD (*n*): sylvanite (see SYLVANIT) 553
SCHRIFTGRANIT (*m*): graphic granite 554
SCHRIFTGUSS (*m*): type founding 555
SCHRIFTJASPIS (*m*): jasper opal 556
SCHRIFTKASTEN (*m*): (letter)-case 557
SCHRIFTLEITER (*m*): editor 558
SCHRIFTLICH: in writing; written; in black and white; by letter 559
SCHRIFTMALEREI (*f*): lettering; sign-writing 560
SCHRIFTMÄSSIG: in writing; by letter 561
SCHRIFTMETALL (*n*): type metal (see LETTERN-METALL) 562
SCHRIFTMUTTER (*f*): matrix; type mould 563
SCHRIFTSCHNEIDER (*m*): cutter (of characters) 564
SCHRIFTSETZER (*m*): compositor; type-setter 565
SCHRIFTSTEIN (*m*): graphic granite 566
SCHRIFTSTELLER (*m*): author; writer 567
SCHRIFTTELLUR (*n*): graphic tellurium; sylvanite (see SYLVANIT) 568
SCHRIFTWECHSEL (*m*): interchange of letters; correspondence 569
SCHRIFTZEICHEN (*n*): character; letter 570
SCHRIFTZEUG (*n*): type metal (see LETTERN-METALL); writing materials 571
SCHRIFTZUG (*m*): (written)-character; flourish 572
SCHRITT (*m*): step; stride; pace 573
SCHRITT HALTEN: to keep (pace) step 574
SCHRITTWEISE: step by step; by steps or stages 575
SCHRÖCKINGERIT (*m*): schröckingerite; UC_2O_6. plus Aq. 576
SCHROFF: rough; steep; gruff; harsh; abrupt; rugged 577
SCHROFFHEIT (*f*): gruffness; abruptness; steepness 578
SCHRÖPFEN: to cup; scarify; (*n*): cupping (to relieve congestion) (Med.); scarification 579
SCHROT (*m*): cut; block; clipping; piece; (small)-shot; groats; selvedge; plumb bob; grist (Brewing) 580
SCHROTBEIL (*n*): chopping axe; chopping hammer 581
SCHROTEN: to chip; clip; cut in pieces; kibble; (rough)-grind; crush; bruise; roll; shoot (put) into the cellar (Casks); gobble-(food) 582
SCHRÖTER (*m*): handler of barrels; cellarman; stag beetle; woodcutter 583
SCHROTEREI (*f*): crushing room (for malt) (Brewing) 584
SCHROTFABRIK (*f*): shot factory 585
SCHROTHAMMER (*m*): chopping hammer 586
SCHRÖTLING (*m*) cutting; piece; blank; planchet (Minting) 587
SCHROTMEHL (*n*): coarse meal; groats 588
SCHROTMEISSEL (*m*): hot chisel 589
SCHROTMETALL (*n*): shot metal 590
SCHROTMÜHLE (*f*): grist (groat; bruising; kibbling; malt) mill 591
SCHROTSÄGE (*f*): cross cut saw 592
SCHROTT (*m* and *n*): old iron, scrap metal, de-tinned iron (waste) scrap (worked into the bloom, in puddling process, or used up in Siemens Martin process) 593

SCHROTTVERFAHREN (*n*) : scrap process (Siemens Martin process in which scrap metal is used) 594
SCHRUBBEN : to scrub ; scour 595
SCHRUBBER (*m*) : scrubber (for gases, etc.) 596
SCHRUBBERBESEN (*m*) : scrubbing brush 597
SCHRUBBFEILE (*f*) : coarse file ; rough file 598
SCHRUBBHOBEL (*m*) : jack plane 599
SCHRÜEN : to give the biscuit baking to (Porcelain) 600
SCHRUMPFEN : to shrink ; contract ; shrivel ; wrinkle ; crumple 601
SCHRUMPFEND : astringent (Med.) 602
SCHRUMPFEND WIRKEN : to cause shrinking ; to have an astringent action 603
SCHRUMPFIG : shrivelled ; crumpled 604
SCHRUMPFMASS (*n*) : contraction ; shrinkage (amount of-) 605
SCHRUMPFSITZ (*m*) : shrunk fit, heavy force fit, shrunk seating 606
SCHRUMPFUNG (*f*) : (see SCHRUMPFMASS) 607
SCHRUMPFUNGSDRUCK (*m*) : shrinkage pressure, pressure due to shrinking 608
SCHRUPPEN : to plane off 609
SCHUB (*m*) : push ; shove ; thrust ; throw ; shearing ; shear 610
SCHUB (*m*), AXIAL- : axial thrust 611
SCHUBFACH (*n*) : drawer 612
SCHUBFENSTER (*n*) : sash-(window) 613
SCHUBFESTIGKEIT (*f*) : shearing strength 614
SCHUBGESETZ (*n*) : law of displacement (due to thrust) 615
SCHUBKARREN (*m*) : wheel-barrow 616
SCHUBKRAFT (*f*) : shear ; thrust 617
SCHUBKURBEL (*f*) : crank 618
SCHUBLADE (*f*) : drawer 619
SCHUBLEERE (*f*) : sliding gauge ; slide gauge 620
SCHUBLEHRE (*f*) : (see SCHUBLEERE) 621
SCHUBMODUL (*m*) : modulus of thrust, thrust modulus 622
SCHUBSPANNUNG (*f*) : displacement (sliding) pressure, shearing stress, thrust 623
SCHUBVORGANG (*m*) : displacement process (due to thrust) 624
SCHUBWEISE : by (thrusts) shoves, or pushes ; gradually ; in batches 625
SCHÜCHTERN : timid ; shy ; coy ; bashful 626
SCHÜCHTERNHEIT (*f*) : timidity ; shyness ; coyness ; bashfulness 627
SCHUFTEN : to toil ; drudge 628
SCHUFTIG : rascally ; base ; mean ; shabby 629
SCHUH (*m*) : shoe 630
SCHUHBÜRSTE (*f*) : shoe-brush 631
SCHUHCREME (*f*) : shoe (polish) cream 632
SCHUHDRAHT (*m*) : twine ; shoemaker's (cobbler's) thread 633
SCHUHHORN (*n*) : shoehorn 634
SCHUHLEIM (*m*) : shoe glue 635
SCHUHMACHER (*m*) : shoemaker, cobbler 636
SCHUHMACHERWACHS (*n*) : cobbler's wax 637
SCHUHPUTZER (*m*) : shoe-black 638
SCHUHRIEMEN (*m*) : shoelace ; latchet 639
SCHUHSCHNALLE (*f*) : shoe buckle 640
SCHUHSCHWÄRZE (*f*) : blacking 641
SCHUHSOHLE (*f*) : sole 642
SCHUHWICHSE (*f*) : blacking 643
SCHULD (*f*) : fault ; guilt ; blame ; crime ; debt ; cause 644
SCHULDBRIEF (*m*) : any writing acknowledging indebtedness ; I.O.U. ; promissory note ; bond ; note of hand 645
SCHULDBUCH (*n*) : ledger ; account book 646
SCHULDEN : to owe ; be (culpable) to blame 647
SCHULDENFREI : free from debts 648
SCHULDENTILGUNGSFOND (*m*) : sinking fund 649
SCHULDFORDERUNG (*f*) : claim (of a debt) ; demand-(note) ; request for payment 650
SCHULDIG : due ; indebted ; owing ; at fault ; guilty ; culpable ; outstanding 651
SCHULDIGKEIT (*f*) : duty ; debt ; obligation 652
SCHULDIG SEIN : to be in debt, be indebted, owe, be guilty, be culpable, be the cause of 653
SCHULDLOS : guiltless ; blameless ; innocent 654
SCHULDNER (*m*) : debtor 655
SCHULDSCHEIN (*m*) : bond ; acknowledgment of indebtedness 656
SCHULDVERSCHREIBUNG (*f*) : (see SCHULDBRIEF) ; obligation 657
SCHULE (*f*) : school ; instruction-book (Music) 658
SCHULEN : to school ; teach 659
SCHÜLER (*m*) : scholar ; pupil 660
SCHULGEGENSTAND (*m*) : subject (of study in schools) 661
SCHULKREIDE (*f*) : school chalk 662
SCHULMEISTER (*m*) : school-master 663
SCHULTAFEL (*f*) : blackboard 664
SCHULTER (*f*) : shoulder 665
SCHULTERBEIN (*n*) : humerus (Anat.) 666
SCHULTERBLATT (*n*) : shoulder blade ; scapula (Anat.) 667
SCHULTER (*f*), MAST- : mast hound 668
SCHULZIT (*m*) : schulzite, geokronite ; Pb_5SbS_4 ; Sg. 6.0 669
SCHUND (*m*) : offal ; refuse ; rubbish ; filth ; muck (often applied to inferior goods) 670
SCHUNDGRUBE (*f*) : cess-pool 671
SCHÜPPCHEN (*n*) : flake ; small scale 672
SCHUPPE (*f*) : scale ; scurf ; flake 673
SCHÜPPE (*f*) : scoop ; shovel ; (*pl*) : spades (Cards) 674
SCHUPPEN : to scale off ; desquamate ; (*m*) : shed ; lean-to ; coachhouse 675
SCHÜPPENARTIG : squamous ; scaly ; flaky 676
SCHUPPENFISCH (*m*) : scaly fish 677
SCHUPPENFLECHTE (*f*) : psoriasis (an inflammatory disease) (Med.) 678
SCHUPPENGLÄTTE (*f*) : flake litharge 679
SCHUPPENSTEIN (*m*) : lepidolite (see LEPIDOLITH) 680
SCHUPPENTIER (*n*) : armadillo 681
SCHUPPIG : scaly ; squamous ; flaky ; scaled 682
SCHUPPIGKEIT (*f*) : scaliness ; flakiness 683
SCHUR (*f*) : shearing ; fleece ; clip ; annoyance 684
SCHÜRBEL (*m*) : bloom ; stamp (Metal) 685
SCHÜREISEN (*n*) : poker ; pricker ; fire-(hook) iron 686
SCHÜREN : to stir ; poke ; fire ; supply to the furnace ; feed (on to the grate) into the furnace ; stoke ; free the air spaces of a grate from ashes and clinker 687
SCHÜRER (*m*) : poker ; stoker ; grate 688
SCHÜRER, MECHANISCHER- (*m*) : mechanical (automatic) stoker or grate 689
SCHÜRFBOHRMASCHINE (*f*) : boring machine for prospecting work 690
SCHURFEN : to scratch ; scrape ; open (a mine) ; prospect 691
SCHÜRHAKEN (*m*) : poker ; pricker ; fire-(hook) iron 692
SCHURKISCH : rascally 693
SCHÜRLOCH (*n*) : stoke (fire) hole 694
SCHÜRPLATTE (*f*) : dead-plate (of grates) 695

SCHÜRSTANGE (*f*): poker, pricker, fire-iron 696
SCHURWOLLE (*f*): fleece (Wool) 697
SCHURZ (*m*): apron 698
SCHÜRZE (*f*): apron 699
SCHÜRZEN: to tie; tuck (gird) up; truss 700
SCHUSS (*m*): shot; blast; charge; shoot; weft (Fabrics); header; drum; box (Mech.); batch (Baking); section; portion 701
SCHÜSSEL (*f*): dish 702
SCHÜSSELZINN (*n*): pewter 703
SCHUSSFEST: bullet-proof 704
SCHUSSGARN *n*): weft, mule (Textile) 705
SCHUSSWEITE (*f*): range; distance 706
SCHUSSWUNDE) (*f*): gunshot wound 707
SCHUSTER (*m*): cobbler; shoemaker 708
SCHUSTERDRAHT (*m*): twine; thread 709
SCHUSTERN: to cobble 710
SCHUSTERPECH (*n*): cobbler's wax 711
SCHUSTERPFRIEM (*m*): punch 712
SCHUTT (*m*): refuse; rubbish 713
SCHÜTTBODEN (*m*): corn-loft; granary 714
SCHÜTTE (*f*): truss; bundle 715
SCHÜTTELFROST (*m*): rigor, shivering fit (Med.) 716
SCHÜTTELLÄHMUNG (*f*): shaking palsy, Parkinson's disease, paralysis agitans (the involuntary trembling of the body due to old age) (Med.) 717
SCHÜTTELN: to shake; agitate; churn; rock; jolt; riddle 718
SCHÜTTELROST (*m*): shaking (rocking) grate; rocking-bar furnace 719
SCHÜTTELRUTSCHE (*f*): shaking (rocking) chute 720
SCHÜTTELWERK (*n*): rocking (shaking) mechanism or apparatus 721
SCHÜTTELZYLINDER (*m*): shaking (stoppered) cylinder 722
SCHÜTTEN: to pour; shed; yield (Mining); shoot 723
SCHÜTTGELB (*n*): Dutch pink (see **SAFTGELB**) 724
SCHÜTTGUT (*n*): loose material (such as gravel, coal, etc.) 725
SCHÜTTHÖHE (*f*): amount (depth) of charge; thickness of fire 726
SCHÜTTLOCH (*n*): feed-(ing) hole; hopper-mouth 727
SCHÜTTUNG (*f*): pouring; shedding; extract-yielding materials (Brew.); bottoming (for roads) 728
SCHUTZ (*m*): protection; safeguard; shelter; defence; screen; fence; cover; guard; preserver 729
SCHUTZANSTRICH (*m*): protective (coating) covering with paint; coating with protective paint 730
SCHUTZBEIZE (*f*): reserve; resist (Calico) 731
SCHUTZBEIZENDRUCK (*m*): resist style 732
SCHUTZBLATTERN (*f.pl*): cow-pox 733
SCHUTZBLECH (*n*): guard plate 734
SCHUTZBRIEF (*m*): safe-conduct 735
SCHUTZBRILLE (*f*): protecting (spectacles) goggles; eye protectors 736
SCHUTZDECK (*n*): shelter deck 737
SCHUTZDECKE (*f*): protective (coating) cover 738
SCHÜTZE (*m*): archer; gunner; shot; (*m.pl*): flood-gate; dike; sluice 739
SCHÜTZEN: to protect; preserve; guard; safeguard; shelter; cover; defend; dam 740
SCHUTZENGEL (*m*): guardian angel 741
SCHÜTZENSCHAFT (*f*): rifle club 742
SCHUTZGATTER (*n*): portcullis; flood-gate; sluice 743
SCHUTZGLAS (*n*): (gauge)-glass protector 744
SCHUTZHÜLLE (*f*): protective (coat) covering; protecting (tube; envelope) case 745
SCHUTZIMPFUNG (*f*): vaccination; inoculation 746
SCHUTZKAPPE (*f*): protecting (cap) cover-(ing) 747
SCHUTZKOLLOID (*n*): protective colloid 748
SCHUTZKRAFT (*f*): protective power 749
SCHUTZLACK (*m*): protecting varnish 750
SCHÜTZLING (*m*): charge; protégé 751
SCHUTZLOS: unprotected; defenceless 752
SCHUTZMANN (*m*): policeman 753
SCHUTZMANTEL (*m*): protecting (case) jacket 754
SCHUTZMARKE (*f*): (registered) trademark 755
SCHUTZMASKE (*f*): face guard or mask 756
SCHUTZMASSE (*f*): resist (Calico) 757
SCHUTZMAUER (*f*): protecting wall; rampart; bulwark; dam 758
SCHUTZMITTEL (*n*): preventive; preservative; prophylactic (Med.); (Calico) resist 759
SCHUTZNETZ (*n*) **TORPEDO-**: torpedo netting 760
SCHUTZPAPP (*m*): resist; reserve (Calico) 761
SCHUTZPLATTE (*f*): protecting (baffle) plate 762
SCHUTZPOCKENGIFT (*n*): vaccine virus 763
SCHUTZRECHT (*n*): patent rights, copyright, protection 764
SCHUTZRING (*m*): guard ring 765
SCHUTZSALZLÖSUNG (*f*): protective salt solution (for gas masks) 766
SCHUTZSCHICHT (*f*): protective layer 767
SCHUTZSCHIRM (*m*): (protective)-screen 768
SCHUTZSTANGE (*f*): guardbar 769
SCHUTZSTOFF (*m*): protective material 770
SCHUTZTRICHTER (*m*): protecting funnel draining funnel (inverted to carry off condensed vapours) 771
SCHUTZVORRICHTUNG (*f*): protecting (device) contrivance or arrangement; protector 772
SCHUTZWACHE (*f*): (safe)-guard; escort 773
SCHUTZWAND (*f*): saving (retaining) wall 774
SCHUTZWIDERSTAND (*m*): protective resistance (Elect.) 775
SCHUTZWIRKUNG (*f*): protective effect 776
SCHUTZZOLL (*m*): protective duty 777
SCHWABBER (*m*): swab 778
SCHWACH: weak; faint; feeble; low; thin; light; poor; dim 779
SCHWÄCHE (*f*): weakness; faintness; feebleness; thinness; lightness; debility (Med.); infirmity 780
SCHWÄCHEN: to weaken; enfeeble; impair; seduce; make tender; depress; debilitate 781
SCHWACHHERZIG: faint-hearted 782
SCHWACHLASTPERIODE (*f*): period of light load 783
SCHWACHSÄUERLICH: weakly acid 784
SCHWACHSIEDEND: simmering; gently boiling 785
SCHWACHSINNIG: weak-minded 786
SCHWACHSTROM (*m*): weak (minimum) current (Elect.) 787
SCHWÄCHUNG (*f*): weakening; enfeebling; impairing; thinning; lightening; dimming; depressing 788
SCHWÄCHUNGSMITTEL (*n*): depressant 789
SCHWADEN (*m*): (fire; choke)-damp (Mining); suffocating vapour; swath; fumes; (any suffocating fumes) (see **WRASEN**) 790

SCHWADENFANG (*m*): ventilator; hood 791
SCHWADEN, FEURIGER- (*m*): fire-damp 792
SCHWAGER (*m*): brother-in-law 793
SCHWÄGERIN (*f*): sister-in-law 794
SCHWALBE (*f*): swallow 795
SCHWALBENSCHWANZ (*m*): dovetail 796
SCHWALBENSCHWANZBEFESTIGUNG (*f*): dovetail (joint) method of attachment 797
SCHWALBENWURZ-(EL) (*f*): swallowwort; swallowroot; vincetoxicum (Pharm.) (*Vincetoxicum officinale*) 798
SCHWALL (*m*): flood; billow; swell; throng; crowd 799
SCHWAMM (*m*): sponge; fungus; mushroom; tinder; tutty (Zinc); lot; mass 800
SCHWAMMARTIG: fungous; spongy; sponge-like 801
SCHWAMMBILDUNG (*f*): formation (growth) of fungus (on wood) 802
SCHWÄMMCHEN (*n*): thrush (a parasitic stomatitis, the mucous membrane, particularly of the mouth, being affected by the parasite, *Oidium albicans*) (Med.) 803
SCHWAMMFILTER (*n*): sponge filter 804
SCHWAMMGIFT (*n*): muscarine; mushroom poison 805
SCHWAMMGUMMI (*n*): rubber sponge 806
SCHWAMMHOLZ (*n*): decayed (spongy) wood 807
SCHWAMMIG: spongy; fungous; porous; bibulous (Paper) 808
SCHWAMMKOHLE (*f*): sponge charcoal 809
SCHWAMMKUPFER (*n*): copper sponge; spongy copper 810
SCHWAMMKÜRBISKERNÖL (*n*): loofah oil (from *Luffa aegyptica*); Sg. 0.925 811
SCHWAMMSTOFF (*m*): fungin 812
SCHWAMMTOD (*m*): fungicide; fungus destroyer 813
SCHWAN (*m*): swan 814
SCHWAND (*m*): shrinkage; contraction; waste; loss; disappearance; dwindling 815
SCHWANEN: to presage; prognosticate 816
SCHWANG (*m*): swing; vogue 817
SCHWANGER: pregnant with child 818
SCHWÄNGERN: to impregnate; become (be) pregnant-(with); saturate 819
SCHWANGERSCHAFT (*f*): pregnancy 820
SCHWÄNGERUNG (*f*): impregnation; saturation 821
SCHWANK: slender; flexible; wavering; (*m*): joke 822
SCHWÄNKBÜRSTE (*f*): bottle brush 823
SCHWÄNKEN: to rinse; wave; shake 824
SCHWANKEN: to vary; fluctuate; oscillate; shake; reel; rock; waver; be irresolute; hesitate; roll; stagger 825
SCHWANKEND: uncertain; irresolute; unsettled; fluctuating; unsteady (see SCHWANKEN) 826
SCHWANKHALLE (*f*): rinsing room (Brewing) 827
SCHWANKUNG (*f*): variation; fluctuation; oscillation; shaking; reeling; rocking; wavering; hesitation; uncertainty; staggering; irresolution 828
SCHWANZ (*m*): tail; train 829
SCHWÄNZER (*m*): truant; shirker 830
SCHWANZHAHN (*m*): stopcock with outlet through end of the key 831
SCHWANZHAMMER (*m*): tail hammer 832
SCHWANZSTERN (*m*): comet 833
SCHWANZWELLE (*f*): tail-end shaft 834
SCHWÄREN: to suppurate; fester; (*m*): ulcer 835
SCHWARM (*m*): swarm; throng; crowd; colony; cluster 836
SCHWÄRMEN: to swarm; revel; wander; rave; day-dream; imagine; go into rhapsodies over; fancy 837
SCHWÄRMER (*m*): visionary; fanatic; enthusiast; reveller; rioter; rover; dreamer; sphinx; firework; serpent 838
SCHWÄRMEREI (*f*): fanaticism; enthusiasm; day-dreaming 839
SCHWÄRMZEIT (*f*): swarming time 840
SCHWARTE (*f*): skin; rind; covering; crust; scalp; bark (of trees) 841
SCHWARZ: black; swarthy; dark; sombre; obscure; overcast; gloomy; (*n*): black-(colour) 842
SCHWARZÄUGIG: black-eyed 843
SCHWARZBASE (*f*): black base 844
SCHWARZBEIZE (*f*): black-(iron) liquor (Dyeing) 845
SCHWARZBLAU: very dark blue; blue-black 846
SCHWARZBLECH (*n*): black plate; untinned iron plate (ordinary iron plate, not tinned or coated with other anti-rust metal) 847
SCHWARZBLEI (*n*): black lead; graphite 848
SCHWARZBLEIERZ (*n*): black lead spar; carboniferous cerussite 849
SCHWARZBRAUN: dark-brown 850
SCHWARZBRAUNSTEIN (*m*): psilomelane (see PSILOMELAN) 851
SCHWARZBROD (*n*): German black bread 852
SCHWARZBRÜCHIG: black-short (Metal) 853
SCHWARZBRÜHE (*f*): (see HOLZESSIGSAURES EISEN) 854
SCHWARZDORN (*m*): black-thorn; sloe (*Prunus spinosa*) 855
SCHWÄRZE (*f*): black-(ing); printer's ink; swarthiness; blackness; obscurity; shadow (Phot.) (see SCHWARZ) 856
SCHWÄRZEN: to blacken; black; darken; obscure; become overcast 857
SCHWARZERLE (*f*): alder (see ERLE) 858
SCHWARZES WASSER (*n*): black mercurial lotion (Pharm.) 859
SCHWARZFARBE (*f*): black colour 860
SCHWARZFÄRBER (*m*): dyer in black 861
SCHWARZGALLIG: atrabilarious 862
SCHWARZGAR: black-tanned 863
SCHWARZGEBRANNT: kishy (Metal) 864
SCHWARZGELB: tawny; dark yellow 865
SCHWARZGRAU: olive; dark grey 866
SCHWARZGÜLTIGERZ (*n*): stephanite; polybasite (see STEPHANIT and POLYBASIT) 867
SCHWARZKALK (*m*): lump variety of hydraulic lime 868
SCHWARZKREIDE (*f*): black chalk 869
SCHWARZKÜMMEL (*m*): nutmeg plant and nutmeg flower (*Flores Nigella sativa* of *Nigella sativa*) 870
SCHWARZKÜMMELÖL (*n*): nutmeg oil (from seeds of *Nigella sativa*); Sg. 0.9248 871
SCHWARZKÜMMELSAMEN (*m* and *pl*): nutmeg (the fruit of *Nigella sativa*) 872
SCHWARZKUNST (*f*): mezzotint 873
SCHWARZKUPFER (*n*): black (coarse) copper; $4\,Cu + SO_2$ (about 94% Cu.) 874
SCHWARZKUPFERERZ (*m*): melaconite (see MELAKONIT) 875
SCHWÄRZLICH: blackish 876
SCHWARZLIEGEN: to be settled; be clear (of beer, etc.) 877

SCHWARZMANGANERZ (*n*): hausmannite (see HAUSMANNIT) 878
SCHWARZPAPPEL (*f*): black poplar (*Populus nigra*) 879
SCHWARZPECH (*n*): black (common) pitch 880
SCHWARZPULVER (*n*): black powder 881
SCHWARZSCHMELZ (*m*): black enamel 882
SCHWARZSEHER (*m*): pessimist 883
SCHWARZSENF (*m*): (see SENF, SCHWARZER-) 884
SCHWARZSENFÖL (*n*): black-mustard oil (from *Sinapis nigra*) 885
SCHWARZSILBERERZ (*n*): black silver; stephanite (see STEPHANIT) 886
SCHWARZSPIESSGLANZ (*m*): bournonite (see BOURNONIT) 887
SCHWARZSPIESSGLANZERZ (*n*): bournonite (see BOURNONIT) 888
SCHWARZSTREIFIG: black-streaked 889
SCHWARZTANNE (*f*): black-pine; Scotch fir 890
SCHWÄRZUNGSGRAD (*m*): degree of blackness 891
SCHWÄRZUNGSMITTEL (*n*): darkening agent (Phot.) 892
SCHWARZVITRIOL (*m*): black vitriol; impure ferrous sulphate 893
SCHWARZWASSERFIEBER (*n*): hæmoglobinuric fever, Blackwater fever (Med.) 894
SCHWARZ-WEISSKUNST (*f*): black and white (art) 895
SCHWARZ-WEISSKUNST (*f*), **LITHOGRAPHISCHE-**: lithographic black and white art, etching on stone 896
SCHWARZWURZ (*f*): symphytum (Pharm.); symphytum root (*Radix consolidæ*) 897
SCHWARZWURZEL (*f*): (see SCHWARZWURZ) 898
SCHWARZZINKERZ (*n*): franklinite (see FRANKLINIT) 899
SCHWATZEN: to chatter; prattle; babble; talk 900
SCHWATZHAFT: talkative; loquacious 901
SCHWATZIT (*m*): schwatzite; $Cu_{10}Hg_2Sb_4S_{13}$; Sg. 5.1 902
SCHWEBE (*f*): suspense; suspension; suspender; sling 903
SCHWEBEBAHN (*f*): suspended railway 904
SCHWEBEFÄHRE (*f*): overhead (suspended) ferry 905
SCHWEBEN: to be suspended; wave; hang; hover; float; be pending; to be in (suspense) suspension 906
SCHWEBESTOFF (*m*): matter (substance) in suspension or suspended 907
SCHWEDISCH: Swedish 908
SCHWEDISCHES GRÜN: Scheele's green; Swedish green (see KUPFER, ARSENIGSAURES-) 909
SCHWEDISCHES HÖLZCHEN (*n*): Swedish (safety) match 910
SCHWEDISCHES ZÜNDHOLZ (*n*): Swedish match 911
SCHWEFEL (*m*): sulphur (S); brimstone; Sg. 1.9-2.05; Mp. 120°C.; Bp. 444.6°C. 912
SCHWEFELALKALI (*n*): alkali sulphide 913
SCHWEFELALKOHOL (*m*): carbon disulphide (old chemical name for carbon disulphide) (see SCHWEFELKOHLENSTOFF) 914
SCHWEFELAMMON (*n*): sulphuret of ammonia; ammonium sulphide (see AMMONIUMSULFID) 915
SCHWEFELAMMONIUM (*n*): (see SCHWEFELAMMON) 916
SCHWEFELANTIMON (*n*): antimony sulphide (*Antimonium crudum*) (see ANTIMONSULFID) 917
SCHWEFELANTIMONSAUER: combined with thioantimonio (sulphoantimonic) acid; thioantimonate (sulphoantimonate) of 918
SCHWEFELANTIMONSÄURE (*f*): sulpho-(thio)-antimonic acid 919
SCHWEFELANTIMON-SCHWEFELNATRON (*n*): Schlippe's salt (see SCHLIPPE'SCHES SALZ) 920
SCHWEFELARSEN (*n*): sulphide of arsenic; realgar (see REALGAR) 921
SCHWEFELARSEN (*n*), **GELBES-**: (see AURIPIGMENT) 922
SCHWEFELARSENIG: sulpho-(thio)-arsenious 923
SCHWEFELARSENIK (*n*): sulphide of arsenic (see REALGAR) 924
SCHWEFELARSEN, ROTES- (*n*): sulphide of arsenic; realgar (see REALGAR) 925
SCHWEFELARSENSÄURE (*f*): sulpho-(thio)-arsenic acid 926
SCHWEFELART (*f*): kind (sort; type; variety) of sulphur 927
SCHWEFELARTIG: sulphurous 928
SCHWEFELÄTHER (*m*): ethyl (sulphuric) ether (see ÄTHYLÄTHER) 929
SCHWEFELBAD (*n*): sulphur bath 930
SCHWEFELBARYUM (*n*): barium sulphide (see BARYUMSULFID) 931
SCHWEFELBARYUMOFEN (*m*): barium sulphide (oven) furnace 932
SCHWEFELBESTIMMUNG (*f*): sulphur determination 933
SCHWEFELBLAUSÄURE (*f*): sulpho-(thio)-cyanic acid; CNHS 934
SCHWEFELBLEI (*n*): lead sulphide (see BLEISULFID and BLEIGLANZ) 935
SCHWEFELBLUMEN (*f.pl*): flowers of sulphur; sublimated sulphur; brimstone; sulphur flour; (S) 936
SCHWEFELBLÜTEN (*f.pl*): (see SCHWEFELBLUMEN) 937
SCHWEFELBRENNOFEN (*m*): sulphur kiln 938
SCHWEFELBROMÜR (*n*): sulphur monobromide; S_2Br_2; Sg. 2.629 939
SCHWEFELBROT (*n*): sulphur loaf 940
SCHWEFELCADMIUM (*n*): cadmium sulphide (see CADMIUMSULFID) 941
SCHWEFELCALCIUM (*n*): calcium sulphide (see CALCIUMSULFID) 942
SCHWEFELCHLORID (*n*): sulphur (chloride) dichloride; SCl_2 (see SCHWEFELDICHLORID) 943
SCHWEFELCHLORÜR (*n*): sulphur monochloride; S_2Cl_2; Sg. 1.68-1.701; Mp. –80° C.; Bp. 138°C. 944
SCHWEFELCHROM (*n*): chromium sulphide (see CHROMSULFID) 945
SCHWEFELCYAN (*n*): sulpho-(thio)-cyanogen; cyanogen sulphide 946
SCHWEFELCYANAMMONIUM (*n*): ammonium sulpho-(thio)-cyanate; NH_4SCN; Sg. 1.306; Mp. 159°C. 947
SCHWEFELCYANKALIUM (*n*): potassium sulpho-(thio)-cyanate (see KALIUMSULFOCYANAT) 948
SCHWEFELCYANMETALL (*n*): metallic sulpho-(thio)-cyanate 949
SCHWEFELCYANSÄURE (*f*): sulpho-(thio)-cyanic acid 950
SCHWEFELCYANWASSERSTOFFSÄURE (*f*): sulpho-(thio)-cyanic acid 951
SCHWEFELDAMPF (*m*): sulphur (sulphurous) vapour or fumes 952

SCHWEFELDICHLORID (*n*): sulphur dichloride; sulphur bichloride; (see also SCHWEFEL-CHLORID); SCl_2; Sg. 1.621; Mp. – 78°C.; Bp. 59°C. 953
SCHWEFELDIOXYD (*n*): sulphur dioxide; SO_2 (see SCHWEFLIGSÄUREANHYDRID) 954
SCHWEFELECHT: fast to (stoving) sulphurous acid (Dyeing) 955
SCHWEFELEISEN (*n*): iron (ferrous) sulphide (see EISENSULFÜR) 956
SCHWEFELEISEN (*n*), ANDERTHALB-: iron sesquisulphide (see FERRISULFID) 957
SCHWEFELEISEN, EINFACH- (*n*): (see EISENSULFÜR) 958
SCHWEFELEISEN, ZWEIFACH- (*n*): (see EISENDI-SULFID) 959
SCHWEFELERZ (*n*): sulphur ore 960
SCHWEFELESSIGSAUER: sulphacetate 961
SCHWEFELFARBE (*f*): sulphur colour 962
SCHWEFELFARBSTOFF (*m*): sulphur dye 963
SCHWEFELFORM (*f*): sulphur (brimstone) mould 964
SCHWEFELGALLIUM (*n*): gallium sulphide 965
SCHWEFELGEHALT (*m*): sulphur content 966
SCHWEFELGELB: sulphur-(yellow) coloured 967
SCHWEFELGERBUNG (*f*): sulphur tannage 968
SCHWEFELGERMANIUM (*n*): germanium sulphide 969
SCHWEFELGOLD (*n*): gold sulphide (see GOLD-SULFID) 970
SCHWEFELGRUBE (*f*): sulphur (pit) mine 971
SCHWEFELHALOGEN (*n*): sulphur halide 972
SCHWEFELHALTIG: sulphurous; containing sulphur 973
SCHWEFELHARNSTOFF (*m*): thiourea; sulphourea; sulfourea; thiocarbamide; CH_4N_2S; Sg. 1.406; Mp. 180.6°C. 974
SCHWEFELHOLZ (*n*): sulphur match; lucifer 975
SCHWEFELHÖLZCHEN (*n*): (see SCHWEFELHOLZ) 976
SCHWEFELHÜTTE (*f*): sulphur refinery 977
SCHWEFELIG: sulphurous 978
SCHWEFELINDIUM (*n*): indium sulphide 979
SCHWEFELJODÜR (*n*): sulphur moniodide 980
SCHWEFELKADMIUM (*n*): cadmium sulphide (see CADMIUMSULFID) 981
SCHWEFELKALIUM (*n*): potassium sulphide; liver of sulphur; hepar (see KALIUMSULFID) 982
SCHWEFELKALIUM (*n*), ZWEIFACHES-: potassium bisulphuret 983
SCHWEFELKALK (*m*): lime sulphide; calcium sulphide (see KALK, SCHWEFLIGSAURER-) 984
SCHWEFELKALZIUM (*n*): calcium sulphide (see CALCIUMSULFID) 985
SCHWEFELKAMMER (*f*): sulphur (stove) chamber 986
SCHWEFELKARBOLSÄURE (*f*): sulphocarbolic acid; phenolsulphonic acid; $C_6H_6SO_3$ [987
SCHWEFELKIES (*m*): pyrites (see PYRIT); ferrous sulphide (see EISENSULFID and EISENDISUL-FID) 988
SCHWEFELKIES, GEMEINER- (*m*): pyrite (see PYRIT) 989
SCHWEFELKIES, PRISMATISCHER- (*m*): marcasite (see MARKASIT) 990
SCHWEFELKOBALT (*m*): cobalt sulphide (see KOBALTSULFID) 991
SCHWEFELKOHLENSÄURE (*f*): sulpho-(trithio)-carbonic acid; H_2CS_3 992
SCHWEFELKOHLENSTOFF (*m*): carbon (bisulphide) disulphide; (CS_2) (*Carboneum sulfuratum*); Sg. 1.2634; Mp. – 110°C.; Bp. 46.3°C. 993
SCHWEFELKOLBEN (*m*): sulphur distilling retort 994
SCHWEFELKUCHEN (*m*): sulphur cake 995
SCHWEFELKUPFER (*n*): copper sulphide; Cu_2S (see KUPFERSULFID) 996
SCHWEFELLATWERGE (*f*): confection of sulphur 997
SCHWEFELLÄUTEROFEN (*m*): sulphur refining furnace 998
SCHWEFELLEBER (*f*): hepar; liver of sulphur (*Hepar sulfuris*); potassium sulphide; sulphuret of potash (see KALIUMSULFID) 999
SCHWEFELMAGNESIUM (*n*): magnesium sulphide (see MAGNESIUMSULFID) 000
SCHWEFELMANGAN (*n*): manganese sulphide 001
SCHWEFELMETALL (*n*): metallic sulphide 002
SCHWEFELMILCH (*f*): milk of sulphur; sulphur precipitate 003
SCHWEFELMILCHSEIFE (*f*): sulphur milk soap 004
SCHWEFELMOLYBDÄN (*n*): molybdenum sulphide (see MOLYBDÄNSULFID) 005
SCHWEFELMONOCHLORID (*n*): sulphur monochloride (see SCHWEFELCHLORÜR) 006
SCHWEFEL (*m*), MONOKLINER-: monoclinic sulphur; Mp. 120°C. 007
SCHWEFELN: to sulphur; sulphurate; sulphurize; treat (fumigate) with sulphur fumes 008
SCHWEFELNATRIUM (*n*): sodium sulphide (see NATRIUMSULFID) 009
SCHWEFELNATRIUMOFEN (*m*): sodium sulphide furnace 010
SCHWEFELNATRON (*n*): (see SCHWEFELNATRIUM) 011
SCHWEFELNICKEL (*m*): nickel sulphide (see NICKELSULFID) 012
SCHWEFELNIEDERSCHLAG (*m*): sulphur precipitate; milk of sulphur 013
SCHWEFELOFEN (*m*): sulphur burner 014
SCHWEFELOXYD (*n*): sulphur oxide (see SCHWEFEL-DIOXYD and SCHWEFELTRIOXYD) 015
SCHWEFELPHOSPHOR (*m*): phosphorous sulphide (see PHOSPHORSULFID) 016
SCHWEFELPILZE (*m.pl*): *Beggiatoa* (Fungus) 017
SCHWEFEL (*m*), PRÄZIPITIERTER-: sulphur precipitate, milk of sulphur, precipitated sulphur 018
SCHWEFEL (*m*), PULVERISIERTER-: powdered (pulverized) sulphur 019
SCHWEFELQUECKSILBER (*n*): mercury (mercuric) sulphide (see QUECKSILBERSULFID) 020
SCHWEFELQUELLE (*f*): sulphur (sulphurous) spring 021
SCHWEFEL (*m*), RAFFINIERTER-: refined sulphur 022
SCHWEFELRAUCH (*m*): sulphur fumes 023
SCHWEFELRÄUCHERUNG (*f*): sulphur fumigation 024
SCHWEFEL, RHOMBISCHER- (*m*): rhombic sulphur; Mp. 114.5°C. 025
SCHWEFELSALBE (*f*): sulphur ointment 026
SCHWEFELSALZ (*n*): sulphur (sulpho, thio) salt; sulphate 027
SCHWEFELSANDSEIFE (*f*): sulphur sand soap 028
SCHWEFELSAUER: sulphate of; combined with sulphuric acid 029
SCHWEFELSÄURE (*f*): sulphuric acid; H_2SO_4; oil of vitriol; Sg. 1.8204-1.854; Mp. 10.46°C.; Bp. 210-338°C. 030
SCHWEFELSÄUREANHYDRID (*n*): sulphuric anhydride; sulphur trioxide; SO_3 (see SCHWEFEL-TRIOXYD) 031

SCHWEFELSÄUREANLAGE (*f*): sulphuric acid plant 032
SCHWEFELSÄUREDIMETHYLESTER (*m*): dimethyl sulphate (see DIMETHYLSULFAT) 033
SCHWEFELSÄUREFABRIKATION (*f*): sulphuric acid manufacture 034
SCHWEFELSÄUREFREI: sulphuric-acid free, free from sulphuric acid 035
SCHWEFELSÄUREHYDRAT (*n*): sulphuric acid hydrate; $H_2SO_4+H_2O$; Sg. 1.783 036
SCHWEFELSÄUREKAMMER (*f*): sulphuric acid chamber 037
SCHWEFELSÄUREKONZENTRATIONSANLAGE (*f*): sulphuric acid concentration plant 038
SCHWEFELSÄUREMONOHYDRAT (*n*): sulphuric acid monohydrate (see SCHWEFELSÄUREHYDRAT) 039
SCHWEFELSÄURE, PYRO- (*f*): pyrosulphuric acid; $H_2S_2O_7$ 040
SCHWEFELSÄURE (*f*), RAUCHENDE-: oil of vitriol (*Acidum sulfuricum fumans*); fuming sulphuric acid, oleum (mixture of H_2SO_4 and SO_3); Sg. 1.86-1.9 (by cooling to 0°, pyrosulphuric acid crystals formed; $H_2S_2O_7$; decomposes at 35°C.) 041
SCHWEFELSAURES BLEIOXYD (*n*): lead vitriol 042
SCHWEFELSAURES KALI (*n*): potassium sulphate 043
SCHWEFELSAURES NATRON (*n*): sulphate of soda, sodium sulphate 044
SCHWEFELSAURE TONERDE (*f*): alumina sulphate (see ALUMINIUMSULFIT) 045
SCHWEFELSCHLACKE (*f*): sulphur dross 046
SCHWEFELSCHWARZ (*n*): sulphur black 047
SCHWEFELSEIFE (*f*): sulphur soap 048
SCHWEFELSELEN (*n*): selensulphur (a natural compound of sulphur and selenium) 049
SCHWEFELSESQUIOXID (*n*): sulphur sesquioxide; S_2O_3 050
SCHWEFELSILBER (*n*): silver sulphide; Ag_2S (see SILBERSULFID) 051
SCHWEFELSPIESSGLANZ (*m*): stibnite (see STIBNIT) 052
SCHWEFELSTANGE (*f*): roll of sulphur 053
SCHWEFELSTÜCK (*n*): piece of sulphur 054
SCHWEFELTHALLIUM (*n*): thallium sulphide (see THALLIUMSULFÜR) 055
SCHWEFELTONERDE (*f*): alumina sulphide (see ALUMINIUMSULFID) 056
SCHWEFELTONUNG (*f*): sulphur toning (Phot.) 057
SCHWEFELTRIOXYD (*n*): sulphur trioxide, sulphuric anhydride; SO_3; Sg. 1.81-1.98 [058
SCHWEFELUNG (*f*): sulphurization; sulphuration; fumigation with sulphur fumes 059
SCHWEFELVERBINDUNG (*f*): sulphur compound 060
SCHWEFELWASSER (*n*): sulphur water 061
SCHWEFELWASSERSTOFF (*m*): hydrogen sulphide; H_2S; Sg. 0.964-1.19; Mp. −83.8°C.; Bp. −60.2°C.; sulphuretted hydrogen 062
SCHWEFELWASSERSTOFF-AMOMNIAK (*n*): (see SCHWEFELAMMONIUM) 063
SCHWEFELWASSERSTOFFREST (*m*): hydrogen sulphide residue; sulphydryl; (SH′) 064
SCHWEFELWASSERSTOFFSÄURE (*f*): hydrosulphuric acid; H_2S (see SCHWEFELWASSERSTOFF) 065
SCHWEFELWEINSÄURE (*f*): sulphovinic acid (Old term); ethylsulphuric acid; sulfethylic acid; sulfovinic acid; monoethyl sulphate; $C_2H_5HSO_4$; Sg. 1.32; Bp. 280°C. 066
SCHWEFELWERK (*n*): sulphur refinery 067
SCHWEFELWISMUT (*n*): bismuth sulphide (see WISMUTSULFID) 068
SCHWEFELZINK (*m* and *n*): zinc sulphide (see ZINKSULFID) 069
SCHWEFELZINKWEISS (*n*): zincolith (pigment of chiefly zinc sulphide); lithopone (see LITHOPONE) 070
SCHWEFELZINN (*n*): tin sulphide (see ZINNSULFID and ZINNSULFÜR) 071
SCHWEFELZYAN (*n*): sulpho-(thio)-cyanogen; cyanogen sulphide 072
SCHWEFELZYANWASSERSTOFF (*m*): sulpho-(thio)-cyanic acid; CNHS 073
SCHWEFLIG: sulphurous 074
SCHWEFLIGE SÄURE (*f*): sulphurous acid (see SCHWEFLIGSÄURE) 075
SCHWEFLIGE SÄURE (*f*), WASSERFREIE, FLÜSSIGE-: anhydrous liquid sulphurous acid 076
SCHWEFLIGE SÄURE, WÄSSRIGE-: (*f*): hydrated sulphurous acid; aqueous sulphurous acid 077
SCHWEFLIGSAUER: sulphite of; combined with sulphurous acid 078
SCHWEFLIGSÄURE (*f*): sulphurous acid; H_2SO_3 079
SCHWEFLIGSÄUREANHYDRID (*n*): sulphurous anhydride; sulphur dioxide; SO_2; Sg. 1.434; Mp. −76.1°C.; Bp. −10°C. 080
SCHWEFLIGSÄUREGAS (*n*): sulphur dioxide (see SCHWEFLIGSÄUREANHYDRID) 081
SCHWEFLIGSÄUREWASSER (*n*): aqueous sulphurous acid 082
SCHWEFLIGWEINSÄURE (*f*): sulphovinous (sulfovinous; ethylsulphurous; ethylsulfurous) acid; $C_2H_5HSO_3$ 083
SCHWEIF (*m*): tail; train; warp (of fabrics) 084
SCHWEIFEN: to tail; curve; chamfer; rinse; ramble; stray 085
SCHWEIFSÄGE (*f*): fretsaw 086
SCHWEIFSTERN (*m*): comet 087
SCHWEIFUNG (*f*): curve; swell; rounding 088
SCHWEIGEN: to be silent; be quiet; (*n*): silence 089
SCHWEIGEN AUFERLEGEN: to impose silence 090
SCHWEIGSAM: silent; taciturn; reserved; quiet 091
SCHWEIGSAMKEIT (*f*): taciturnity; quietness 092
SCHWEIMEL (*m*): giddiness; dizziness 093
SCHWEIN (*n*): swine; pig; hog; sow 094
SCHWEINBROT (*n*): sow bread (*Cyclamen europæum*) 095
SCHWEINEFETT (*n*): lard; pig's (hog's) fat or grease; Sg. 0.94; Mp 28-50°C.; pork fat (*Adeps suillus*) 096
SCHWEINEFLEISCH (*n*): pork 097
SCHWEINEPEST (*f*): swine fever 098
SCHWEINEREI (*f*): piggishness; filthiness; dirtiness; filth; mess 099
SCHWEINESCHMALZ (*n*): lard; Sg. 0.94 (see SCHWEINEFETT) 100
SCHWEINESCHMER (*n*): (see SCHWEINEFETT) 101
SCHWEINFURTERGRÜN (*n*): Paris green; Schweinfurt-(h) green (see KUPFERACETATARSENIAT and PARISER GRÜN) 102
SCHWEINIGEL (*m*): hedgehog; filthy (dirty) fellow or urchin 103
SCHWEINISCH: piggish; swinish; filthy 104
SCHWEINSHAUT (*f*): pig (hog) skin 105
SCHWEINSLEDER (*n*): pig (hog) skin (Leather) 106
SCHWEISS (*m*): sweat; perspiration 107

SCHWEISSANLAGE (*f*): welding plant 108
SCHWEISSAPPARAT (*m*): welding apparatus 109
SCHWEISSARBEIT (*f*): welding 110
SCHWEISSBAR: capable of being welded 111
SCHWEISSBEFÖRDERND: diaphoretic; sudorific 112
SCHWEISSBRENNER (*m*): welding burner 113
SCHWEISSECHT: fast to perspiration 114
SCHWEISSEISEN (*n*): welding iron; weld iron; wrought iron (in which the W.I. is produced in pieces which weld together; Puddling process) 115
SCHWEISSEN: to weld; commence to melt; sweat; bleed; leak (of liquids); (*n*): welding 116
SCHWEISSER (*m*): welder 117
SCHWEISSERREGEND: diaphoretic; sudorific 118
SCHWEISSFEHLER (*m*): welding defect 119
SCHWEISSFLÜSSIGKEIT (*f*): sweat; perspiration 120
SCHWEISSGEWASCHEN: washed in the grease (of wool) 121
SCHWEISSHITZE (*f*): welding heat; white heat 122
SCHWEISSIG: perspiring; sweaty; bloody 123
SCHWEISS (*m*), KALTER-: cold sweat 124
SCHWEISSLOCH (*n*): (sweat; perspiration) pore 125
SCHWEISSMITTEL (*n*): diaphoretic; sudorific (an agent for causing sweating) 126
SCHWEISSNAHT (*f*): welded seam 127
SCHWEISSOFEN (*m*): welding (reheating) furnace 128
SCHWEISSPROZESS (*m*): welding process 129
SCHWEISSPUDER (*m*): sweat powder (anti-sweat powder) 130
SCHWEISSPULVER (*n*): welding powder; diaphoretic powder (powder producing sweating); sweat powder 131
SCHWEISSSCHLACKE (*f*): welding slag 132
SCHWEISSSCHMIEDEEISEN (*n*): weld iron (see SCHWEISSEISEN) 133
SCHWEISSSTAHL (*m*): weld (wrought) steel (in which the wrought iron is produced in pieces which weld together; Puddling process) 134
SCHWEISSSTELLE (*f*): position of weld; soldered (welded) seam; spot (place, seam) to be soldered or welded 135
SCHWEISSSTRASSE (*f*): welding machine 136
SCHWEISSTREIBEND: sudorific; diaphoretic 137
SCHWEISSTREIBENDES MITTEL (*n*): diaphoretic (Med.) 138
SCHWEISSUNG (*f*): weld-(ing) 139
SCHWEISSWALZEN (*f.pl*): roughing rolls 140
SCHWEISSWARM: welding hot 141
SCHWEISSWÄRME (*f*): welding heat 142
SCHWEISSWOLLE (*f*): wool (in the yolk) containing suint 143
SCHWEIZ (*f*): Switzerland 144
SCHWEIZER (*m*): Swiss 145
SCHWEIZERKÄSE (*m*): Swiss cheese; Gruyère cheese 146
SCHWELCHEN: to wither; air-dry (Brew.) 147
SCHWELCHMALZ (*n*): withered (air-dried) malt (Brew.) 148
SCHWELEN: to distil; break-up; smoulder; burn slowly 149
SCHWELGAS (*n*): gaseous product from distillation of brown coal; given off by decomposition of the bitumen (10-20% CO_2; 10-25% CH_4; 10-30% H.; 10-30% N.; 5-15% CO; plus heavy hydrocarbons, O and H_2S; calorific value about 1200-3600 calories) 150
SCHWELGEN: to revel; debauch; carouse 151
SCHWELGEREI (*f*): carousal; revelry; debauchery 152
SCHWELKEN: to wither; air-dry (Brew.) 153
SCHWELKOHLE (*f*): coal for distillation 154
SCHWELLBAR: capable of swelling 155
SCHWELLBEIZE (*f*): swelling liquor (Leather) 156
SCHWELLBRAUNKOHLE (*f*): lignite with high volatile content; lignite for distillation 157
SCHWELLE (*f*): step; sill; curb; beam; tie; joist; crossbar; ledge; threshold; sleeper (Railway); sole (Mine) 158
SCHWELLEN: to expand; distend; rise; swell; bulge; belly; plump (Leather); grow; tumify 159
SCHWELLGEWEBE (*n*): erectile tissue (Anat.) 160
SCHWELLKÖRPER (*m*): corpus cavernosum (Anat.) 161
SCHWELLKRAFT (*f*): plumping power (Leather) 162
SCHWELLMITTEL (*n*): plumping agent (Leather) 163
SCHWELLRETORTE (*f*): distilling retort 164
SCHWELLUNG (*f*): swelling; tumefaction 165
SCHWELOFEN (*m*): distilling oven, distiller, still 166
SCHWELPERIODE (*f*): duration of distillation process 167
SCHWELSCHACHT (*m*): retort, distilling hopper or chamber 168
SCHWELWASSER (*n*): water of distillation (from distillation of brown coal) 169
SCHWEMME (*f*): watering-(place); horse-pond 170
SCHWEMMEN: to wash; water; float; soak; deposit; flush (Sanitary) 171
SCHWEMMSTEIN (*m*): (vitrified)-stone float 172
SCHWEMMSYSTEM (*n*): flushing system (of removing human excrement from the neighbourhood of dwellings, as opposed to the old method of cartage) 173
SCHWENGEL (*m*): swingle (of a flail); (pump)-handle; swingle-tree (of waggons); crank (of a wheel); lever; bar (of a press, etc.); pendulum (of clocks) 174
SCHWENKEN: to flourish; wave; swing; rinse; brandish 175
SCHWER: heavy; hard; severe; difficult; thick; weighty; ponderous; great; strong 176
SCHWERATMIG: asthmatic; having difficulty in breathing 177
SCHWERBELADEN: heavily laden 178
SCHWERBENZOL (*n*): solvent naphtha (see SOLVENT-NAPHTHA) 179
SCHWERBLEIERZ (*n*): plattnerite; PbO_2; Sg. 9.4 180
SCHWERE (*f*): weight; hardness; heaviness; severity; difficulty; thickness; gravity 181
SCHWEREMESSER (*m*): barometer; aerometer; hydrometer 182
SCHWERERDE (*f*): heavy earth; baryta; (BaO) (see BARYUMOXYD) 183
SCHWERER LÖSLICH: (more)-difficultly soluble 184
SCHWERES WEINÖL (*n*): oil of wine (oily residue from ether preparation) 185
SCHWERFÄLLIG: heavy; clumsy; slack; dull 186
SCHWERFÄLLIGKEIT (*f*): clumsiness; heaviness 187
SCHWERFLÜCHTIG: not easily volatilized; difficultly volatile 188

SCHWERFLÜCHTIGKEIT (*f*): difficult volatility 189
SCHWERFLÜSSIG: difficultly (liquefiable) fusible; refractory; dissolved with difficulty; difficultly soluble; viscous; viscid; thickly liquid; flowing with difficulty; non-mobile 190
SCHWERFLÜSSIGKEIT (*f*): refractoriness 191
SCHWERFRUCHT (*f*): heavy (grain) cereals 192
SCHWERGEFRIERBAR: difficultly (freezable) congealable 193
SCHWERGUT (*n*): dead-weight 194
SCHWERHÖRIG: hard of hearing 195
SCHWERKRAFT (*f*): (force of) gravity; gravitation 196
SCHWERLEDER (*n*): sole leather 197
SCHWERLICH: scarcely; hardly; difficulty 198
SCHWERLÖSLICH: difficultly soluble 199
SCHWERMETALL (*n*): heavy metal 200
SCHWERMUT (*f*): melancholy; sadness; melancholia (a form of insanity) (Med,) 201
SCHWERMÜTIG: melancholy; sad 202
SCHWERÖL (*n*): heavy oil 203
SCHWERPUNKT (*m*): centre of gravity; (used figuratively = mainstay; basis; backbone) 204
SCHWERREDUZIERBAR: difficultly reducible 205
SCHWERSCHWARZ (*n*): weighted black (Dye) 206
SCHWERSPAT (*m*): heavy spar; barite; barium sulphate; $BaSO_4$ (see BARYUMSULFAT and BARYT); Sg. 4.55 207
SCHWERSPATMÜHLE (*f*): barite mill 208
SCHWERSTEIN (*m*): scheelite (see SCHEELIT) 209
SCHWERT (*n*): sword; centreboard (amidships; Naut.) 210
SCHWERTANTALERZ (*n*): tantalite (see TANTALIT) 211
SCHWERTBOHNE (*f*): kidney bean 212
SCHWERTBOOT (*n*): boat with centre board or leeboard 213
SCHWERTFISCH (*m*): sword-fish 214
SCHWERTFÖRMIG: sword-like 215
SCHWERTJACHT (*f*): centre board yacht 216
SCHWERTLILIE (*f*): iris; fleur-de-lis; fleur-de-luce; flag; orrice root (*Iris germanica*) 217
SCHWERWIEGEND: serious; weighty 218
SCHWESTER (*f*): sister 219
SCHWESTERMORD (*m*): sororicide 220
SCHWESTER CHIFF (*n*): sister ship 221
SCHWIEGERMUTTER (*f*): mother-in-law 222
SCHWIELE (*f*): callosity; hardness of skin; hard (structure) texture; weal; mark; welt (on the skin) 223
SCHWIELIG: callous; horny 224
SCHWIERIG: hard; difficult 225
SCHWIERIGKEIT (*f*): difficulty; hardness; trouble; obstacle 226
SCHWIMMBLASE (*f*): air bladder; swimming bladder; water wings 227
SCHWIMMDOCK (*n*): floating dock 228
SCHWIMMEBENE (*f*): plane of flotation 229
SCHWIMMEN: to swim; float 230
SCHWIMMER (*m*): swimmer; float 231
SCHWIMMERMESSER (*m*): float meter 232
SCHWIMMERSTANGE (*f*): float rod 233
SCHWIMMERVENTIL (*n*): float valve; ball valve 234
SCHWIMMFÄHIGKEIT (*f*): capacity for floating, buoyancy 235
SCHWIMMFEDER (*f*): fin 236
SCHWIMMFLOSSE (*f*): fin 237
SCHWIMMGERSTE (*f*): float barley; skimmings (Brew.) 238
SCHWIMMHAUT (*f*): web 239
SCHWIMMKIESEL (*m*): float stone; vitrified stone float 240
SCHWIMMKÖRPER (*m*): swimming (floating) body; float; buoyant body 241
SCHWIMMKRAFT (*f*): buoyancy 242
SCHWIMMKRAN (*m*): floating crane 243
SCHWIMMLINIE (*f*): flotation curve 244
SCHWIMMMETHODE (*f*): flotation method 245
SCHWIMMSAND (*m*): quicksand 246
SCHWIMMSEIFE (*f*): floating soap 247
SCHWIMMSTEIN (*m*): floatstone; vitrified stone float 248
SCHWIMMWAGE (*f*): hydrometer 249
SCHWINDEL (*m*): fraud; swindle; dizziness; vertigo; giddiness; lot; bag of tricks (Vulgarism) 250
SCHWINDELGESELLSCHAFT (*f*): bogus company 251
SCHWINDELN: to defraud; cheat; swindle; be (become) dizzy or giddy 252
SCHWINDEN (*n*): shrinkage (reduction of volume during cooling of the metal); contraction, etc. 253
SCHWINDEN: to shrink; dwindle; vanish; waste; wither; contract; disappear 254
SCHWINDLER (*m*): swindler; fraud; cheat 255
SCHWINDMASS (*n*): shrinkage; amount of (contraction; waste) shrinkage; amount of volumetric reduction 256
SCHWINDPUNKT (*m*): vanishing point 257
SCHWINDSUCHT (*f*): phthisis; consumption; tuberculosis (Med.) 258
SCHWINDSÜCHTIG: phthisical; consumptive; in consumption; suffering from tuberculosis 259
SCHWINDSUCHTSWURZEL (*f*): bugbane, black snakeroot (*Actæa racemosa*) 260
SCHWINDUNG (*f*): shrinkage (on solidification of molten metals, etc.); contraction; vanishing; waste; dwindling; disappearance; atrophy; wasteage; withering; non-stability 261
SCHWINDUNGSFÄHIGKEIT (*f*): shrinking (tendency) property 262
SCHWINDUNGSPUNKT (*m*): vanishing point 263
SCHWINGE (*f*): swingle; wing; pinion; fan; winnow 264
SCHWINGELGRAS (*n*): fescue-(grass) (*Festuca ovina*) 265
SCHWINGEN: to swing; rock; oscillate; vibrate; flourish; wave; brandish; winnow; swingle; centrifuge; rise; soar 266
SCHWINGEND: oscillatory, rocking, oscillating swinging 267
SCHWINGENDE WELLE (*f*): rocking shaft 268
SCHWINGFEDER (*f*): pinion 269
SCHWINGROST (*m*): rocking grate 270
SCHWINGSIEB (*n*): swinging (rocking) sieve 271
SCHWINGUNG (*f*): swinging; rocking; oscillation; vibration, etc. 272
SCHWINGUNGSBEWEGUNG (*f*): vibratory (oscillatory) movement or motion 273
SCHWINGUNGSDAUER (*f*): duration (period) of oscillation 274
SCHWINGUNGSEBENE (*f*): plane of vibration (Polarisation) 275
SCHWINGUNGSKNOTEN (*m*): point of rest, point of support 276
SCHWINGUNGSMITTELPUNKT (*m*): centre (axis) of oscillation 277
SCHWINGUNGSWEITE (*f*): extent of vibration 278

SCHWINGUNGSWELLE (*f*): undulation (wave) due to vibration or oscillation 279

SCHWINGUNGSZAHL (*f*): vibration number; rate of vibration 280

SCHWINGUNGSZEIT (*f*): time (duration) of vibration 281

SCHWIRREN: to whirr; buzz; whizz; centrifuge 282

SCHWITZBAD (*n*): steam (vapour) bath 283

SCHWITZE (*f*): sweat-(ing); perspiration 284

SCHWITZEN: to sweat; perspire 285

SCHWITZIG: sweaty; perspiring 286

SCHWITZMITTEL (*n*): sudorific; diaphoretic (agent for causing perspiration) 287

SCHWITZPULVER (*n*): diaphoretic powder 288

SCHWITZRÖSTE (*f*): steam (retting) rettery (Flax) (see also RÖSTE) 289

SCHWITZWASSER (*n*): perspiration, sweat, condensation (on metal, etc.) 290

SCHWÖDE (*f*): place where (or state of being) limed (Hides) 291

SCHWÖDEBREI (*m*): lime cream (Leather) 292

SCHWÖDEFASS (*n*): lime vat (Leather) 293

SCHWÖDEGRUBE (*f*): lime pit (Leather) 294

SCHWÖDEMASSE (*f*): lime cream (Leather) 295

SCHWÖDEN: to lime (Hides) 296

SCHWÖREN: to swear; take an oath 297

SCHWÖREN, FALSCH-: to commit perjury; perjure 298

SCHWÜL: sultry; close; oppressive; heavy (of atmosphere) 299

SCHWÜLE (*f*): sultriness; closeness; oppresiveness; heaviness (of atmosphere) 300

SCHWULST (*m*): swelling; tumour; bombast 301

SCHWUND (*m*): atrophy (Med.); wasteage; shrinkage; withering; disappearance 302

SCHWUNG (*m*): swing; vibration; activity; oscillation 303

SCHWUNGBEWEGUNG (*f*): vibratory (motion) movement 304

SCHWUNGBRETT (*n*): swing board 305

SCHWUNGGEWICHT (*n*): pendulum 306

SCHWUNGHAFT: swinging; lively 307

SCHWUNGKRAFT (*f*): vibrating power; centrifugal force; liveliness 308

SCHWUNGMASCHINE (*f*): centrifuge 309

SCHWUNGRAD (*n*): flywheel 310

SCHWUNGRING (*m*): revolving ring 311

SCHWUR (*m*): oath 312

SCHWÜRIG: suppurating 313

SCLEROSIS (*f*): sclerosis (Med.) 314

SCONTIEREN: to balance, square-(accounts) 315

SCONTO (*m*): allowance, discount 316

SCOPOLAMIN (*n*): scopolamine 317

SCOPOMORPHIN (*n*): scopomorphine 318

SCROFIN (*n*): scrofin 319

SEALINGPAPIERE (*n.pl*): sealing papers 320

SEBACINSAUER: sebate, sebacinate, combined with sebacic acid 321

SEBACINSÄURE (*f*): sebacic acid (see STEARIN and ELAINSÄURE) 322

SEBACYLSAUER: sebacylate 323

SECACORNIN (*n*): secacornin (a purified preparation of ergot) 324

SECERNIEREN: to secrete 325

SECHS: six 326

SECHSATOMIG: hexatomic 327

SECHSECK (*n*): hexagon 328

SECHSECKIG: hexagonal 329

SECHSERLEI: of six different kinds 330

SECHSERSALZ (*n*): magnesium chloride; $MgCl_2.6H_2O$ 331

SECHSFACH: sextuple; sixfold 332

SECHSFÄLTIG: (see SECHSFACH) 333

SECHSFLACH (*n*): hexahedron 334

SECHSFLÄCHIG: hexahedral 335

SECHSFLÄCHNER (*m*): hexahedron 336

SECHSGLIEDRIG: six-numbered 337

SECHSJÄHRIG: every six years; six years old, aged six 338

SECHSMALIG: six times; six-fold 339

SECHSMONATLICH: half-yearly 340

SECHSSÄURIG: hexacid 341

SECHSSEITIG: hexagonal 342

SECHSTÄGIG: six-day; of six days; every six days 343

SECHSTE: sixth 344

SECHSTEL (*n*): sixth 345

SECHSWERTIG: hexavalent 346

SECHSWINKLIG: hexangular; hexagonal 347

SECHSZÄHLIG: sixfold 348

SECHZEHN: sixteen 349

SECHZIG: sixty 350

SEDATIVSALZ (*n*): sedative salt; boracic acid; boric acid (see BORSÄURE); H_3BO_3 351

SEDATIVUM (*n*): sedative (Med.) 352

SEDIMENTGESTEIN (*n*): sedimentary rock, stratified rock 353

SEDIMENTIERBASSIN (*n*): sediment basin, settling basin 354

SEDIMENTIERKAMMER (*f*): settling (precipitating) chamber 355

SEDIMENTIERTEICH (*m*): settling trough, sediment trough 356

SEDIMENTIERUNG (*f*): settling, formation (precipitation) of sediment 357

SEDOBROL (*n*): sedobrol (contains about 50% NaBr.) 358

SEE (*f*): (inland)-sea; lake 359

SEEALGEN (*f.pl*): marine algæ 360

SEEBAD (*n*): sea bath; seaside resort; watering place 361

SEEBUCH (*n*): log-book (Naut.) 362

SEEERZ (*m*): lake (iron) ore (see WIESENERZ and MORASTERZ) 363

SEEFAHRER (*m*): mariner; navigator; voyager 364

SEEFAHRT (*f*): navigation; voyage 365

SEEGANG (*m*): sea 366

SEEGEWÄCHS (*n*): marine plant; seaweed (*Algæ*) 367

SEEGRAS (*n*): sea-weed; sea-wrack; sea tang (*Algæ*) 368

SEEHAFEN (*m*): harbour; sea-port 369

SEEHAHN (*m*): sluice cock 370

SEEHALTEND: sea-worthy 371

SEEHANDEL (*m*): sea (maritime) trade or commerce 372

SEEHUND (*m*): seal (*Phoca*) 373

SEEHUNDSTRAN (*m*): seal oil (see ROBBENTRAN) 374

SEE, IN—GEHEN: to go (put) to sea, to sail 375

SEEKABEL (*n*): submarine cable 376

SEEKARTE (*f*): chart (Naut.) 377

SEEKARTENPAPIERE (*n.pl*): chart papers (Paper) 378

SEEKOHL (*m*): sea kale (*Crambe maritima*) 379

SEEKOKOS (*m*): Seychelles fan-leaved palm; lodoicea; double cocoanut-palm (*Lodoicea sechellarum*) 380

SEEKRÄHE (*f*): cormorant (*Phalacrocorax*) 381

SEEKRANK: seasick 382

SEEKRANKHEIT (*f*): seasickness 383

SEEKRANKHEIT-GEGENMITTEL (*n*): seasickness (antidote; preventative; remedy) prophylactic 384

SEEKREBS (*m*): lobster 385

SEEKUH (*f*): sea-cow, dugong (*Halicore australis* and *Halicore indicus*); (*pl*): *Sirenia* (an order of marine animals) 386

SEEKUNDE (*f*): navigation; nautical knowledge 387

SEEKÜSTE (*f*): sea-coast; sea-board 388

SEELE (*f*): soul; mind; spirit; shaft (of blast furnace); core (of cables) 389

SEELENADEL (*m*): nobility of the mind 390

SEELENANGST (*f*): torture (anguish) of the mind 391

SEELENFORSCHER (*m*): psychologist 392

SEELENFRIEDEN (*m*): peace of mind; tranquility of mind; mental rest; mental respite 393

SEELENFROH: very glad 394

SEELENHIRT (*m*): shepherd of souls; pastor 395

SEELENKUNDE (*f*): psychology 396

SEELENLEHRE (*f*): (see SEELENKUNDE) 397

SEELENMESSE (*f*): requiem mass 398

SEELENRUHE (*f*): tranquility of mind; mental rest; peace of mind; mental respite 399

SEELEUTE (*pl*): mariners; seamen; seafaring people 400

SEELUFT (*f*): sea-air; sea-breeze 401

SEEMANN (*m*): mariner 402

SEEMASCHINIST (*m*): ship's engineer 403

SEEMÄSSIGE VERPACKUNG (*f*): seaworthy packing, packing for shipment 404

SEEMINE (*f*): sea mine; floating mine 405

SEEMOOS (*n*): carrageen; sea moss (*Chondrus crispus*) 406

SEENKUPFER (*n*), AMERIKANISCHES-: American lake copper 407

SEE, OFFENE- (*f*): open sea; main 408

SEEPFLANZE (*f*): marine plant 409

SEEREISE (*f*): sea voyage 410

SEESALZ (*n*): sea salt 411

SEESCHADEN (*m*): sea damage; average; damage by sea 412

SEESCHIFF (*n*): sea-going vessel 413

SEESCHLICK (*m*): sea ooze 414

SEESCHULE (*f*): naval (school) college 415

SEESEIDE (*f*): (see MUSCHELSEIDE) 416

SEESOLDAT (*m*): mariner; bluejacket 417

SEESTAAT (*m*): maritime (naval) power or country 418

SEETANG (*m*): seaweed; sea (wrack) tang (see TANG) 419

SEETIER (*n*): marine animal 420

SEETÜCHTIG: seaworthy 421

SEEUFER (*n*): sea-shore; fore-shore; strand 422

SEEUMGRENZT: sea-bound 423

SEEUNTÜCHTIG: unseaworthy 424

SEEWASSER (*n*): sea-water; brine 425

SEEWESEN (*n*): marine 426

SEEZUNGE (*f*): sole (Fish) 427

SEGEL (*n*): sail 428

SEGELAREAL (*n*): area (of sails) (Ships) 429

SEGELBOOT (*n*): sailing boat 430

SEGELFERTIG: ready to (sail) put off or put out 431

SEGELGARN (*n*): twine 432

SEGELGATT (*n*): sail room 433

SEGELHAKEN (*m*): sail hook 434

SEGELHANDSCHUH (*m*): sailmaker's palm 435

SEGELHEISS (*n*): hoist of a sail 436

SEGELKAMMER (*f*): sail room, sail loft 437

SEGELKLEID (*n*): cloth (of a sail) 438

SEGELKOJE (*f*): sail room, sail loft 439

SEGELLIEK (*n*): bolt rope 440

SEGELMACHER (*m*): sail-maker 441

SEGELMOMENT (*n*): moment of a sail 442

SEGELN: to sail, set sail, run 443

SEGELNADEL (*f*): sail needle 444

SEGELN, MIT HALBEM WINDE-: to tack, sail on a tack 445

SEGELPUNKT (*m*): centre of effort of the sails 446

SEGELSCHIFF (*n*): sailing (boat) vessel 447

SEGELSCHLITTEN (*m*): ice boat, ice yacht 448

SEGELSCHLITTENSPORT (*m*): ice boating 449

SEGELSTANGE (*f*): yard (of sails) (Naut.) 450

SEGELSYSTEMSCHWERPUNKT (*m*): centre of effort of the sails 451

SEGELTIEFE (*f*): depth of sail 452

SEGELTUCH (*n*): canvas, sail-cloth 453

SEGEN (*m*): benediction; blessing; bliss; grace; yield; abundance 454

SEGERKEGEL (*m*): Seger's fusible cone (cones of silicate mixture; about 6 cm. high by 2 cm. base) (a standard for fire resisting qualities of brick, etc.) consists of 59 numbers (No. 022=590°C. rising by thirties to 010=950°C.; 09=970°C. rising by twenties to 01=1130°C. to 1=1150°C. to 36=1850°C.) (These figures are only approximate) (see latest edition of "Hütte") 455

SEGERPORZELLAN (*n*): Segêr porcelain (a soft porcelain consisting of aluminium silicate, quartz and feldspar) 456

SEGER'SCHER KEGEL (*m*): (see SEGERKEGEL) 457

SEGMENT: segmental; (*n*): segment 458

SEGNEN: to bless; cross; give benediction 459

SEHE (*f*): pupil (of eye) 460

SEHEN: to see; look; behold; regard; observe; perceive; view; note; gather, etc.; (*n*): vision; sight; seeing 461

SEHENSWERT: worth seeing 462

SEHENSWÜRDIG: worth seeing 463

SEHENSWÜRDIGKEIT (*f*): anything worth seeing; sight; curiosity; spectacle; place of interest 464

SEHER (*m*): seer; prophet 465

SEHHÜGEL (*m*): optic thalami, Thalamus opticus (part of the brain) (Med.) 466

SEHLEHRE (*f*): optics 467

SEHNE (*f*): sinew; nerve; tendon; fibre (Metal); chord (Geom.); string 468

SEHNEISEN (*n*): fibrous iron 469

SEHNEN: to long; yearn-(for); (*n*): longing 470

SEHNENSCHNITT (*m*): (see TANGENTIALSCHNITT), section through tree trunk showing knots or irregular vein (grain) 471

SEHNERV (*m*): optic nerve 472

SEHNIG: sinewy; fibrous 473

SEHNLICH: longing-(ly); eager-(ly); passionate-(ly) 474

SEHNSUCHT (*f*): yearning; desire; longing; aspiration 475

SEHR: very (much; well) greatly 476

SEHROHR (*n*): telescope 477

SEHWEITE (*f*): horizon; visual distance; distance (length) of sight 478

SEICHT: low; flat; shallow; superficial 479

SEIDE (*f*): silk 480

SEIDEBLAU (*n*): silk blue 481

SEIDE, KUNST- (*f*): artificial (imitation) silk; art silk 482

SEIDE, KÜNSTLICHE- (*f*): artificial (imitation) silk; art. silk 483
SEIDEL (*n*): pint; half-litre; tankard (Beer) 484
SEIDELBAST (*m*): mezereon (*Daphne mezereum*); Paradise plant; Magell; Spurge flax; Olive spurge; Daphne; Dwarf bay; Wild pepper 485
SEIDELBASTGEWÄCHSE (*n.pl*): *Thymelæaceæ* (Bot.) 486
SEIDELBASTRINDE (*f*): mezereon bark; spurge flax bark 487
SEIDELBASTWURZEL (*f*): mezereon root (*Radix mezerei*) 488
SEIDEN: silk-(en); of silk 489
SEIDENABFALL (*m*): silk waste 490
SEIDENARBEIT (*f*): silk-work 491
SEIDENARTIG: silky 492
SEIDENASBEST (*m*): silky asbestos 493
SEIDENBAU (*m*): silk (seri-) culture; silkworm culture; breeding of silkworms 494
SEIDENFABRIK (*f*): silk (mill) manufactory 495
SEIDENFÄRBER (*m*): silk dyer 496
SEIDENFIBRIN (*n*): fibroin 497
SEIDENFLOR (*m*): silk gauze 498
SEIDENGARN (*n*): silk yarn; spun silk 499
SEIDENGAZE (*f*): silk gauze 500
SEIDENGLANZ (*m*): silky lustre 501
SEIDENGRÜN (*n*): silk green 502
SEIDENHANDEL (*m*): silk trade 503
SEIDENHOLZ (*n*): satinwood (see **ATLASHOLZ**) 504
SEIDENLEIM (*m*): sericin; silk glue 505
SEIDENPAPIER (*n*): tissue paper; small hands (Paper) 506
SEIDENPFLANZENGEWÄCHSE (*n.pl*): *Asclepiadaceæ* (Bot.) 507
SEIDENRAUPE (*f*): silkworm 508
SEIDENSCHMETTERLING (*m*): silkworm moth 509
SEIDENSCHREI (*m*): scroop of silk 510
SEIDENSPINNER (*m*): silk-spinner 511
SEIDENSPINNERPUPPE (*f*): chrysalis (of silk worm) 512
SEIDENSTICKEREI (*f*): silk-embroidery 513
SEIDENWAREN (*f.pl*): silks, silk (stuffs) goods 514
SEIDENWEBER (*m*): silk (thrower) weaver 515
SEIDENWURM (*m*): silk-worm 516
SEIDENZEUG (*n*): silk, silk material 517
SEIDENZUCHT (*f*): silk (seri)-culture, silkworm culture, breeding of silkworms 518
SEIDENZWIRNEN (*n*): silk-throwing 519
SEIDENZWIRNER (*m*): throwster 520
SEIDLING (*m*): cocoon (of silkworms) 521
SEIDLITZ PAPIERE (*n.pl*): Seidlitz papers 522
SEIDLITZPULVER (*n*): Seidlitz powder 523
SEIFE (*f*): soap; buddle (ore washing trough); silt; soap deposit (a sand deposit containing metals, in rivers, etc.); soapstock (Cotton seed oil refining) 524
SEIFE (*f*), **ANTISEPTISCHE-**: antiseptic soap 525
SEIFECHT: fast to soaping 526
SEIFEN: to soap; lather; scour; wash (Mining); to buddle (wash) ores 527
SEIFEN (*f.pl*): pups, soaps (Bricks); the boulders, rubble and general mass of water-worn material at the bottom of river valleys 528
SEIFENABFALL (*m*): soap scrap 529
SEIFENARBEIT (*f*): buddling; ore washing process 530
SEIFENARTIG: saponaceous; soapy 531
SEIFENASCHE (*f*): soap ashes 532
SEIFENBAD (*n*): soap bath 533
SEIFENBALSAM (*m*): soap liniment; opodeldoc 534
SEIFENBAUM (*m*): soap tree (*Sapindus saponaria*) 535
SEIFENBAUMFETT (*n*): soap-tree fat (from seeds of *Sapindus saponaria*) 536
SEIFENBAUMFRUCHT (*f*): soap berry (fruit of *Sapindus saponaria*) 537
SEIFENBEREITUNG (*f*): soap making 538
SEIFENBILDUNG (*f*): soap formation; saponification 539
SEIFENBLASE (*f*): soap bubble 540
SEIFENBRÜHE (*f*): soap suds 541
SEIFENERDE (*f*): Fuller's earth; saponaceous clay; marl 542
SEIFENERZ (*n*): alluvial ore; buddled ore 543
SEIFENFABRIK (*f*): soap factory 544
SEIFENFABRIKANT (*m*): soap (maker) manufacturer 545
SEIFENFABRIKATION (*f*): soap (making) manufacture 546
SEIFENFABRIKATIONSMASCHINEN (*f.pl*): soap-making machinery 547
SEIFENFARBEN (*f.pl*): soap colours 548
SEIFENFORM (*f*): soap frame 549
SEIFENFÜLLUNG (*f*): soap filling 550
SEIFENGOLD (*n*): placer gold 551
SEIFENKESSEL (*m*): soap boiler 552
SEIFENKOCHAPPARAT (*m*): soap boiling apparatus 553
SEIFENKRAUT (*n*): soap weed; soap wort (see **SEIFENWURZEL**) 554
SEIFENKUGEL (*f*): soap-ball 555
SEIFENKÜHLPRESSE (*f*): soap cooling press 556
SEIFENLAPPEN (*m*): washing glove 557
SEIFENLAUGE (*f*): soap (solution) suds or lye 558
SEIFENLEIM (*m*): soap (paste) glue; soap size (Paper) 559
SEIFENLÖSUNG (*f*): soap solution 560
SEIFENNAPF (*m*): soap dish 561
SEIFENNUSS (*f*): soap nut (nut of *Sapindus saponaria*); Sapindus nut 562
SEIFENPAPIER (*n*): soap paper 563
SEIFENPARFÜM (*n*): soap (perfume) scent 564
SEIFENPULVER (*n*): soap powder 565
SEIFENPULVERPACKMASCHINE (*f*): soap powder packing machine 566
SEIFENRIEGEL (*m*): bar of soap 567
SEIFENRINDE (*f*): soap bark (see **QUILLAYARINDE**) 568
SEIFENSCHABSEL (*n*): soap scrap-(s) 569
SEIFENSCHAUM (*m*): lather 570
SEIFENSCHMIERE (*f*): soap stuff (Leather) 571
SEIFENSIEDEN: to boil (make) soap; (*n*): soap (making) boiling 572
SEIFENSIEDER (*m*): soap boiler 573
SEIFENSIEDERASCHE (*f*): soap ashes 574
SEIFENSIEDEREI (*f*): soap works 575
SEIFENSIEDERLAUGE (*f*): caustic lye (see **ÄTZ-LAUGE**) 576
SEIFENSIEDUNG (*f*): soap boiling 577
SEIFENSPIRITUS (*m*): spirit of soap 578
SEIFENSTEIN (*m*): steatite; soapstone; $3MgO, 4SiO_2, H_2O$; Sg. 2.26 (see **STEATIT**) 579
SEIFENSTOFF (*m*): saponine (Chemical); $C_{32}H_{54}O_{18}$ (see **SAPONIN**) 580
SEIFENTAFEL (*f*): slab of soap 581
SEIFENTÄFELCHEN (*n*): cake of soap; tablet of soap 582
SEIFENTALG (*m*): soap tallow 583
SEIFENTELLERCHEN (*n*): soap dish 584
SEIFENTON (*m*): saponaceous clay 585
SEIFENWASSER (*n*): (soap)-suds; soap-(y) water 586

SEIFENWURZEL (*f*): soaproot; soapwort (*Saponaria officinalis*) 587
SEIFENWURZEL (*f*), **ROTE-**: soapwort root (*Radix saponariæ rubra* of *Saponaria officinalis*) 588
SEIFENZÄPFCHEN (*n*): suppository 589
SEIFENZINN (*n*): stream tin (tin obtained from rivers) 590
SEIFICHT: saponaceous; soapy 591
SEIFIG: (see **SEIFICHT**) 592
SEIGER: perpendicular; (*m*): plummet; perpendicular 593
SEIGERARBEIT (*f*): liquation-(process) (Metal) 594
SEIGERBLEI (*n*): liquation lead 595
SEIGERDÖRNER (*m.pl*): liquation dross (difficultly meltable tin-iron, tin-copper alloys, from liquation of raw tin) 596
SEIGERHERD (*m*): liquation hearth 597
SEIGERHÜTTE (*f*): liquation works 598
SEIGERN: to liquate; refine 599
SEIGEROFEN (*m*): liquation furnace 600
SEIGERPFANNE (*f*): liquation pan 601
SEIGERSCHLACKE (*f*): liquation slag 602
SEIGERSTÜCK (*n*): liquation cake 603
SEIGERUNG (*f*): liquation (Metal smelting) (caused by disintegration on solidification; separation, during cooling, of the alloys which are most difficult to melt, from the remaining solution of mother liquor. The position of this liquation remains liquid the longest; usually the core or top. The slower the cooling the greater the liquation. The place last to solidify, has the greatest amount of foreign matter. This results in tensile differences between the portions) 604
SEIGERWERK (*n*): liquation work-(s) 605
SEIGNETTESALZ (*n*): Seignette salt; Rochelle salt; sodium potassium tartrate (see **KALIUM-NATRIUMTARTRAT**) 606
SEIHE (*f*): filter; strainer; spent malt (Brew.) 607
SEIHEBODEN (*m*): strainer (perforated) bottom 608
SEIHEGEFÄSS (*n*): filtering (straining) vessel 609
SEIHEN: to filter; strain 610
SEIHEPAPIER (*n*): filter paper 611
SEIHER (*m*): filter; strainer 612
SEIHESACK (*m*): filtering bag; straining bag 613
SEIHESTEIN (*m*): filtering stone 614
SEIHETRICHTER (*m*): strainer; straining (filtering) funnel 615
SEIHETUCH (*n*): filter (straining) cloth 616
SEIHGEFÄSS (*n*): filtering (straining) vessel 617
SEIHKORB (*m*): strainer 618
SEIHSACK (*m*): filtering bag, straining bag 619
SEIHTUCH (*n*): filtering (straining) cloth 620
SEIL (*n*): cable; rope; line; cord 621
SEILBAHN (*f*): rope tramway; cableway, ropeway 622
SEILBRÜCKE (*f*): rope bridge 623
SEILEN: to equip (furnish) with ropes 624
SEILER (*m*): rope maker 625
SEILERWAREN (*f.pl*): cordage; ropes 626
SEILFETT (*n*): rope grease; cable lubricant 627
SEILFIRNIS (*m*): rope (cable) varnish 628
SEILFÖRDERANLAGE (*f*): ropeway 629
SEILKAUSCHE (*f*): grummet thimble 630
SEILKURVE (*f*): link polygon 631
SEILPOLYGON (*n*): funicular polygon 632
SEILROLLE (*f*): rope pulley 633
SEILSCHEIBE (*f*): rope pulley 634
SEILSCHMIERE (*f*): rope grease 635
SEILSPANNROLLE (*f*): (cable)-stretching (straining) pulley 636
SEILSPANNUNG (*f*): rope tension 637
SEILTRIEB (*m*): rope drive 638
SEILTROMMEL (*f*): winding drum (for ropes) 639
SEILWERK (*n*): ropes; cordage 640
SEIM (*m*): glutinous liquid; slime; honeycomb; mucilage 641
SEIMHONIG (*m*): strained honey 642
SEIMIG: glutinous; mucilaginous; slimy 643
SEIN: to be; exist; (as a pronoun): his; its; (*n*): being; existence 644
SEINERSEITS: for (on) his part 645
SEINETHALBEN: for his sake; on his behalf 646
SEINETWEGEN: (see **SEINETHALBEN**) 647
SEINIGE (*m.f* or *n*): his (own); (*n*): his property 648
SEINIGEN (*f.pl*): his (family) relations 649
SEISING (*f*): seizing 650
SEISMOGRAPH (*m*): seismograph (an instrument for recording vibrations) 651
SEIT: since 652
SEITAB: apart 653
SEITDEM: since; since then; ever since 654
SEITE (*f*): side; page (Book); face (Tech.); flank; party; folio 655
SEITENANMERKUNG (*f*): marginal note 656
SEITENANSICHT (*f*): side (elevation) view; profile 657
SEITENBAHN (*f*): branch (side) line 658
SEITENBUNKER (*m*): wing bunker 659
SEITENDAVIT (*m*): quarter davit 660
SEITENDRUCK (*m*): lateral pressure 661
SEITENECKE (*f*): lateral summit (Cryst.) 662
SEITENFLÄCHE (*f*): lateral face; facet (Cryst.) 663
SEITENGEBÄUDE (*n*): wing; extension (of buildings) 664
SEITENHÖHE (*f*): moulded depth (of ships) 665
SEITENKANTE (*f*): lateral edge 666
SEITENKETTE (*f*): side chain 667
SEITENLÄNGE (*f*): length of a side 668
SEITENLICHT (*n*): bullseye; side-light 669
SEITENLINIE (*f*): side line; collateral line 670
SEITENRISS (*m*): side (view) elevation 671
SEITENROHR (*n*): side (branch) pipe 672
SEITENS: on behalf of; on the part of 673
SEITENSCHMERZ (*m*): pleurodynia, intercostal myalgia (cramp or muscular rheumatism of the wall of the chest) (Med.) 674
SEITENSCHUB (*m*): axial (side, lateral) thrust 675
SEITENSCHWERT (*n*): lee board 676
SEITENSPRUNG (*m*): leap (jump) to one side 677
SEITENSTECHEN (*n*): stitch (in the side) 678
SEITENSTÜCK (*n*): sidepiece; counterpart 679
SEITENSTÜTZE (*f*): quarter stanchion 680
SEITENTAKEL (*n*): runner tackle 681
SEITENTISCH (*m*): side board 682
SEITENTÜR (*f*): side door 683
SEITENVERWANDT: cognate 684
SEITENWEG (*m*): by-path, by-way 685
SEITENWENDUNG (*f*): turning (aside), deflection 686
SEITENZAHL (*f*): number of pages; page number 687
SEITENZEIGER (*m*): index (of a book) 688
SEITHER: up till now; since then; hitherto 689
SEITLICH: (at the) side; lateral-(ly); collateral-(ly) 690
SEITWÄRTS: sideways; aside 691
SEKANTE (*f*): secant 692
SEKRET (*n*): secretion; privy 693

SEKRETÄR (*m*) : secretary 694
SEKRETION (*f*) : secretion 695
SEKRETZELLE (*f*) : secretion cell 696
SEKT (*m*): dry wine (particularly Champagne) 697
SEKTE (*f*) : sect; cult 698
SEKTION (*f*) : section (of boilers, etc.); dissection ; autopsy (Med.) 699
SEKTIONWUNDE (*f*) : post-mortem wound (Med.) 700
SEKTOR (*m*) : sector 701
SEKTORENVERSCHLUSS (*m*) : sector shutter (Phot.) 702
SEKTWEIN (*m*) : dry wine 703
SEKUNDÄR : secondary 704
SEKUNDÄRBAHN (*f*) : branch line (Railways); secondary line (Railways) 705
SEKUNDÄRKREIS (*m*) : secondary circuit (Elect.) 706
SEKUNDÄRLUFT (*f*) : secondary air 707
SEKUNDÄRSTRAHLUNG (*f*) : secondary radiation 708
SEKUNDE (*f*) : second (Time) 709
SELBE : same 710
SELBER : self 711
SELBIG : (the) same ; that 712
SELBST : self ; myself, etc. ; even 713
SELBSTACHTUNG (*f*) : self-esteem ; self-respect 714
SELBSTÄNDIG : independent ; self-dependent ; substantial ; self-contained 715
SELBSTÄNDIGE EINHEIT (*f*) : self-contained unit (Plant) 716
SELBSTANLAUF (*m*) : self-starter ; automatic starting device 717
SELBSTAUFOPFERUNG (*f*) : self-sacrifice ; self-devotion 718
SELBSTBEFLECKUNG (*f*) : masturbation (Med.) 719
SELBSTBEHERRSCHUNG (*f*) : self-control ; self-possession 720
SELBSTBESTÄUBUNG (*f*) : self-pollination (Bot.) 721
SELBSTBETRUG (*m*): self-delusion; self-deceit 722
SELBSTBEWUSST : self-conscious ; conceited 723
SELBSTBEWUSSTSEIN (*n*) : self-consciousness ; conceit 724
SELBSTBIOGRAPHIE (*f*) : autobiography 725
SELBSTDICHTEND : self-packing 726
SELBSTEIGEN : one's (own) personal property 727
SELBSTEINSTELLEND : self-aligning 728
SELBSTENTLADER (*m*) : automatic (discharger) discharging truck ; automatic (tip-cart) tipping tray 729
SELBSTENTWICKLER (*m*) : automatic generator (for carbonic acid) 730
SELBSTENTZÜNDLICH : spontaneously inflammable 731
SELBSTENTZÜNDUNG (*f*) : spontaneous (self) ignition ; spontaneous combustion 732
SELBSTERHALTUNG (*f*) : self-preservation 733
SELBSTERHITZUNG (*f*) : self-heating 734
SELBSTERKENNTNIS (*f*) : self-knowledge 735
SELBSTERNIEDRIGUNG (*f*) : self-humiliation ; self-debasement 736
SELBSTERREGEND : self-exciting 737
SELBSTERREGER (*m*) : automatic (self-acting) exciter 738
SELBSTERZEUGUNG (*f*) : autogenesis ; spontaneous (production) generation 739
SELBSTFARBE (*f*) : self (solid) colour 740
SELBSTGÄRUNG (*f*) : spontaneous fermentation 741
SELBSTGEFÄLLIG : self-complacent ; self-conceited ; self-sufficient 742
SELBSTGEFÄLLIGKEIT (*f*) : self-complacency ; self-conceit ; self-sufficiency 743
SELBSTGEFÜHL (*n*) : self-(consciousness) confidence 744
SELBSTGEMACHT : self (home) made 745
SELBSTGENÜGSAM : self-sufficient ; self-complacent 746
SELBSTGESPRÄCH (*n*) : soliloquy 747
SELBSTGIFT (*n*) : autotoxin 748
SELBSTGREIFER (*m*) : automatic grab 749
SELBSTHASS (*m*) : self-hatred 750
SELBSTHEIL (*n*) : (see BRUNELLE) 751
SELBSTHERRSCHAFT (*f*) : self-command ; autocracy 752
SELBSTHERRSCHER (*m*) : autocrat 753
SELBSTHILFE (*f*) : self-help 754
SELBSTHITZENDES SCHWEISSEN (*n*) : autogene (autogenous) welding 755
SELBSTINDUKTION (*f*) : automatic (self-acting) induction 756
SELBSTISCH : selfish 757
SELBSTKANTE (*f*) : selvedge ; list 758
SELBSTKLUG : conceited 759
SELBSTKOSTENPREIS (*m*): cost price (to oneself) 760
SELBSTLAUT (*m*) : vowel 761
SELBSTLAUTER (*m*) : vowel 762
SELBSTLOB (*n*) : self-praise 763
SELBSTLOS : unselfish 764
SELBSTLÖTUNG (*f*) : autogenic soldering 765
SELBSTMORD (*m*) : suicide ; felo-de-se 766
SELBSTMÖRDER (*m*) : felo-de-se, suicide 767
SELBSTMÖRDERISCH : suicidal 768
SELBSTÖLER (*m*) : automatic (self-acting) oiler or oil feeder 769
SELBSTPEINIGUNG (*f*) : mortification ; self-chastisement 770
SELBSTREDEND : self-evident 771
SELBSTREGISTRIEREND : self-registering, automatically registering 772
SELBSTREGULIEREND : regulating (adjusting) automatically ; self-acting 773
SELBSTREINIGEND : self-purifying (of water, by means of fungi, organisms, etc.) 774
SELBSTREINIGUNG (*f*) : self-purification 775
SELBSTRUHM (*m*) : vain-glory, self-praise 776
SELBSTSCHLUSSVENTIL (*n*) : automatic closing valve 777
SELBSTSCHRIFT (*f*) : autograph 778
SELBSTSPINNER (*m*) : self-acting spinning machine (Textile) 779
SELBSTSTÄNDIG : independent ; self-dependent ; substantial 780
SELBSTSUCHT (*f*) : selfishness ; egoism 781
SELBSTTÄTIG : automatic ; self-acting ; self-raising (Flour) 782
SELBSTTÄUSCHUNG (*f*) : self-delusion ; self-deceit 783
SELBSTÜBERWINDUNG (*f*) : self-victory 784
SELBSTUNTERBRECHER (*m*) : automatic switch ; cut-out ; interrupter (Elect.) 785
SELBSTVERACHTUNG (*f*) : self-contempt 786
SELBSTVERBRENNUNG (*f*) : spontaneous combustion 787
SELBSTVERDAUUNG (*f*) : autodigestion 788
SELBSTVERLEUGNUNG (*f*) : self-denial ; self-abnegation 789
SELBSTVERSTÄNDLICH : of course ; obviously ; self-evident 790
SELBSTVERTEIDIGUNG (*f*) : self-defence 791
SELBSTVERTRAUEN (*n*) : self-confidence, self-reliance 792

SELBST, VON-: automatically; spontaneously 793
SELBSTWIRKEND: automatic; self-acting 794
SELBSTZERSETZUNG (*f*): spontaneous decomposition 795
SELBSTZEUGUNG (*f*): spontaneous generation 796
SELBSTZUFRIEDENHEIT (*f*): self-contentment; self-complacency; self-conceit; self-sufficiency 797
SELBSTZÜNDER (*m*): self igniter; automatic lighter; pyrophorus 798
SELBSTZÜNDUNG (*f*): self-ignition, spontaneous (ignition) combustion 799
SELEKTION (*f*): selection 800
SELEKTIV: selective 801
SELEN (*n*): Selenium; (Se) 802
SELENBLEI (*n*): lead selenide; PbSe; Sg. 8.5; clausthalite (Min.) 803
SELENBLEIKUPFER (*n*): lead-copper selenide; $(Pb,Cu_2)Se$; Sg. 7.25 804
SELENBROMÜR (*n*): selenium monobromide; Se_2Br_2; Sg. 3.604 805
SELENBRÜCKE (*f*): selenium bridge 806
SELENCHLORID (*n*): selenium (selenic) chloride 807
SELENCHLORÜR (*n*): selenium monochloride; selenous chloride; (Se_2Cl_2); Sg. 2.906 [808
SELENCYANID (*n*): selenocyanate 809
SELENCYANKALIUM (*n*): potassium selenocyanate 810
SELENDIOXYD (*n*): selenium dioxide; SeO_2; Sg. 3.954 811
SELENEISEN (*n*): ferrous selenide; (FeSe) 812
SELENERZ (*n*): selenium ore 813
SELENHALOGEN (*n*): selenium halide 814
SELENHALTIG: containing selenium; seleniferous 815
SELENID (*n*): -ic selenide 816
SELENIG: selenious 817
SELENIGE SÄURE (*f*): selenious (selenous) acid; H_2SeO_3. Sg. 3.0066; H_2SeO_4, Sg. 2.608 to 2.95 818
SELENIGESÄUREANHYDRID (*n*): selenious anhydride; SeO_2 819
SELENIGSAUER: selenite of; combined with selenious acid 820
SELENIT (*m*): selenite (a transparent variety of gypsum) (see **ERDGLAS** and **GYPS-**) 821
SELENIUM (*n*): selenium (see **SELEN**) 822
SELENKUPFER (*n*): copper selenide; Cu_2Se; Sg. 6.9; Berzeline (Min.) 823
SELENKUPFERBLEI (*n*): copper-lead selenide, zorgite; $2PbSe.9Cu_2Se$; Sg. 5.6 (see also **ZORGIT**) 824
SELENMETALL (*n*): metallic selenide 825
SELENMONOSULFID (*n*): selenium monosulphide; SeS; Sg. 3.056 826
SELENOXYD (*n*): selenium oxide 827
SELENQUECKSILBER (*n*): mercury selenide, tiemannite; HgSe; Sg. 8.33 828
SELENQUECKSILBERBLEI (*n*): lerbachite 829
SELENSAUER: selenate of; combined with selenic acid 830
SELENSÄURE (*f*): selenic acid; H_2SeO_4; Sg. 2.608-2.95 831
SELENSÄUREHYDRAT (*n*): hydrated selenic acid, selenic acid hydrate; $H_2SeO_4+H_2O$; Sg. 2.356-2.627 832
SELENSCHLAMM (*m*): selenium (mud) slime; chamber sludge (a deposit in sulphuric acid works) 833
SELENSCHWEFELQUECKSILBER (*n*): onofrite 834
SELENSILBER (*n*): silver selenide; naumannite; Ag_2Se; Sg. 8.0 (see **NAUMANNIT**) 835
SELENSOL (*n*): selenium sol, colloidal (solution of) selenium 836
SELENÜR (*n*): -ous selenide 837
SELENWASSERSTOFF (*m*): hydrogen selenide; H_2Se; Sg. 2.12 838
SELENWASSERSTOFFSÄURE (*f*): hydroselenic acid 839
SELENWISMUTGLANZ (*m*): selenium-bismuth glance; Bi_2Se_3; Sg. 6.6 840
SELENZELLE (*f*): selenium cell 841
SELFAKTOR (*m*): self-acting spinning machine (for fine spinning; Textile) 842
SELIG: happy; blessed; deceased; late 843
SELIGKEIT (*f*): happiness, blessedness, bliss, beatitude, felicity, salvation 844
SELIGSPRECHUNG (*f*): beatification 845
SELLAIT (*n*): sellaite (natural magnesium fluoride); MgF_2; Sg. 2.97 846
SELLERIE (*m.* and *f*): celery (*Apium graveolens*) 847
SELLERIEÖL (*n*): celery oil; Sg. 0.87-0.895 [848
SELLERIESALZ (*n*): celery salt 849
SELLERIESAMEN (*m* and *pl*): celery seed-(s) 850
SELLERSBOCK (*m*): plummer block 851
SELTEN: rare; occasional; scarce; unusual; seldom 852
SELTENE ERDE (*f*): rare earth 853
SELTENHEIT (*f*): rarity; scarcity 854
SELTENSTRÄNGE (*f*): lateral columns (of the spinal cord) (Med.) 855
SELTERSWASSER (*n*): Seltzer water 856
SELTSAM: singular; curious; strange; unusual; odd 857
SEMICARBAZIDCHLORHYDRAT (*n*): semicarbazide hydrochloride 858
SEMICYCLISCH: semi-cyclic 859
SEMIDINUMLAGERUNG (*f*): semidine rearrangement 860
SEMIHYDRAT (*n*): semi-hydrate 861
SEMIMER: semimeric 862
SEMINAR (*n*): seminary 863
SEMIPERMEABILITÄT (*f*): semi-permeability 864
SEMIPERMEABEL: semi-permeable 865
SEMMEL (*f*): roll (of wheat bread) 866
SEMMELMEHL (*n*): wheat flour 867
SENARMONTIT (*m*): senarmontite; Sb_2S_3; Sg. 5.25 (see also **ANTIMONTRIOXYD**) 868
SENAT (*m*): senate 869
SENDELUFTLEITER (*m*): transmitting aerial (Wireless Tel.) 870
SENDEN: to send; forward; dispatch; consign; transmit 871
SENDER (*m*): transmitter (Tel.) 872
SENDERÖHRE (*f*): transmitting valve (for producing vibrations or waves) (Wireless Tel.) 873
SENDUNG (*f*): sending; dispatch; consignment; lot; shipment; parcel; mission; legation 874
SENEGALGUMMI (*n*): senegal gum, gum arabic (*Gummi arabicum*) 875
SENEGALINPUDER (*n*): senegalin powder 876
SENEGAWURZEL (*f*): senega root; snake root (*Radix senegæ* of *Polygala senega*) 877
SENESBAUM (*m*): senna-tree (see **SENNES-**) 878
SENF (*m*): mustard (*Sinapis alba* and *Sinapis nigra*) 879
SENFBRÜHE (*f*): mustard sauce 880
SENFBÜCHSE (*f*): mustard pot 881

SENFGEIST (*m*): mustard oil; oil of mustard (volatile); Sg. 1.018-1.029; Bp. 149°C. [882
SENFKORN (*n*): mustard seed-(grain) 883
SENFKRAUT (*n*): hedge mustard (*Sisymbrium officinale*) 884
SENFMEHL (*n*): ground (powdered) mustard 885
SENFÖL (*n*): mustard oil (see SCHWARZSENFÖL and WEISSSENFÖL) 886
SENFÖLESSIGSÄURE (*f*): 2, 4-thiazoledione 887
SENFÖL (*n*), **KÜNSTLICHES-**: artificial mustard oil, allyl iso-thiocyanate, allyl mustard oil; C_3H_5CNS; (a poison gas) 888
SENFPAPIER (*n*): mustard (paper) plaster; vesicatory 889
SENFPFLASTER (*n*): mustard plaster; vesicatory 890
SENFSAME (*m*): mustard seed 891
SENFSAMEN (*m*): mustard seed 892
SENFSÄURE (*f*): sinapic acid 893
SENF (*m*), **SCHWARZER-**: black mustard, red mustard (*Sinapis nigra*) 894
SENFUMSCHLAG (*m*): mustard plaster 895
SENF (*m*), **WEISSER-**: white mustard, yellow mustard, *Semen erucæ* (*Sinapis alba*) 896
SENGEN: to singe; scorch; parch; scald 897
SENKBLEI (*n*): plummet; plumb-line; sinker; plumb-bob; lead 898
SENKBODEN (*m*): false bottom; strainer (Brew.) 899
SENKE (*f*): swallow; probe (Surgical); lowering; low ground; sump; cesspool 900
SENKEL (*m*): plummet; lace; plumbline; plumbbob 901
SENKELNADEL (*f*): bodkin 902
SENKEN: to sink; lower; subside; countersink; depress; be depressed; settle; submerge; sag 903
SENKER (*m*): sinker 904
SENKER (*m*), **SCHRAUBEN-**: countersink 905
SENKGRUBE (*f*): cesspool; sump; sink; sewer 906
SENKKASTEN (*m*): caisson 907
SENKKÖRPER (*m*): sinker 908
SENKLOT (*n*): lead, solder 909
SENKNADEL (*f*): probe 910
SENKRECHT: vertical; upright; perpendicular 911
SENKRECHTE (*f*): vertical-(line), upright 912
SENKSCHNUR (*f*): plumbline 913
SENKSPINDEL (*f*): hydrometer; specific gravity spindle; areometer; density meter (for liquids) 914
SENKUJU (*f*): Japanese senkuju (*Angelica refracta*) 915
SENKUNG (*f*): sinking; lowering; inclination; falling; subsiding; hollow 916
SENKWAGE (*f*): hydrometer; specific gravity spindle; areometer; densimeter; density meter (for liquids); plumb rule 917
SENNALATWERGE (*f*): confection of senna (powdered senna leaves, sugar syrup, and tamarind pulp) 918
SENNESBAUM (*m*): senna-tree (see SENNESBLÄTTER) 919
SENNESBLÄTTER (*n.pl*): senna leaves [Tinnevelly senna (*Senna indica*) (Pharm.), from *Cassia angustifolia* or *Cassia elongata*; Alexandrian senna from *Cassia acutifolia*] 920
SENNESBLÄTTERPULVER (*n*): powdered senna leaves 921
SENNESFRÜCHTE (*f.pl*): senna pods 922
SENNESSTRAUCH (*m*): senna (see SENNESBLÄTTER) 923
SENSE (*f*): scythe 924
SENSENSCHÄRFER (*m*): hone, scythe sharpener 925
SENSIBEL: sensible; sensitive 926
SENSIBILISATOR (*m*): sensitizer (Phot.) 927
SENSIBILISATOR (*m*), **CHEMISCHER-**: chemical sensitizer 928
SENSIBILISATOR (*m*), **OPTISCHER-**: optical sensitizer 929
SENSIBILISIEREN: to sensitize-(Photographic plates) 930
SENSIBILISIERUNGSFARBSTOFF (*m*): sensitizer (Phot.) 931
SENSIBILISIERUNGSMAXIMUM (*n*): sensitizing (limit) maximum, maximum (limit of) sensitiveness 932
SENSIBILITÄT (*f*): sensibility; sensitiveness 933
SENSITOCOLORIMETER (*n*): colour sensitometer (for measuring the sensitiveness of colours) (Phot.) 934
SENTE (*f*): ribband, harpin (Naut.) 935
SEPARATOR (*m*): separator 936
SEPARIEREN: to separate 937
SEPIA (*f*): sepia; cuttle-fish (*Sepia officinalis*) 938
SEPIABRAUN (*n*): sepia-(brown) (from the fish, "*Sepia*") 939
SEPIENZEICHNUNG (*f*): sepia drawing 940
SEPIOLITH (*m*): sepiolite, meerschaum (a magnesium silicate); $2MgO.3SiO_2.2H_2O$ 941
SEPTICÄMIE (*f*): septicæmia, septic poisoning (Med.) 942
SEPTISCH: septic 943
SEQUESTRIEREN: to sequestrate; sequester 944
SERAIL (*n*): seraglio 945
SERICINSÄURE (*f*): (see MYRISTICINSÄURE) 946
SERICIT (*m*): sericite (compact potash mica) (see MUSKOVIT) 947
SERIE (*f*): series 948
SERIENSCHNITT (*m*): serial section 949
SERIKOSE (*f*): sericose 950
SERIN (*n*): serine, l-α-amino-β-oxypropionic acid; $CH_2(OH).CH(NH_2).CO_2H$ 951
SERÖS: serous; watery 952
SEROSITÄT (*f*): serosity 953
SERPENTIN (*m*): serpentine-(stone) (a magnesium silicate); $H_4(MgFe)_3Si_2O_9$; Sg. 2.6 954
SERPENTINASBEST (*m*): fibrous serpentine; amianthus (see AMIANT) 955
SERPENTINSTEATIT (*m*), **PRISMATISCHER-**: (see SERPENTIN) 956
SERPENTINSTEIN (*m*): serpentine-(stone) (see SERPENTIN) 957
SERUM (*n*): serum 958
SERUMGLAS (*n*): serum glass 959
SERUMTHERAPIE (*f*): serum therapeutics 960
SERVIETTE (*f*): serviette; napkin 961
SERVIETTENPAPIERE (*n.pl*): serviette papers (Paper) 962
SERVIS (*n*): service; plate 963
SERVOMOTOR (*m*): servomotor 964
SESAM (*m*): sesame (*Sesamum orientale*) 965
SESAM (*m*), **INDISCHER-**: Indian sesame (*Sesamum indicum*) 966
SESAM (*m*), **LEVANTINER-**: Levant sesame (*Sesamum orientale*) 967
SESAMÖL (*n*): sesame oil; gingelly oil; benne oil; gigily oil; teel oil (from *Sesamum orientale*); Sg. 0.923; Mp. 26-32°C. 968
SESAMÖL (*n*), **DEUTSCHES-**: (see LEINDOTTERÖL) 969
SESAMSAAT (*f*): sesame seed 970

SESQUITERPEN (*n*): sesquiterpene 971
SESSEL (*m*): easy-chair; stool; seat 972
SESSHAFT: settled; stationery; residing; resident; sedentary; established 973
SETTIESEGEL (*n*): settie sail 974
SETZBETT (*n*): settling bed 975
SETZBORD (*m*): wash board (Boats) 976
SETZBOTTICH (*m*): settling tub or vat 977
SETZEN: to sit-(down); set; place; put; plant; sieve; jig (Tech.); wager; settle; be deposited; precipitate; subside; sink; cross; traverse; leap; substitute (Maths.); lay; compose; fix; assume; write; be given that; incorporate; perch 978
SETZEN, (FÜR- SETZEN): to express—as—(Maths.), to write, substitute (Maths.) 979
SETZEN, IN GANG- : to put into (action) commission 980
SETZER (*m*): setter; tamper; compositor 981
SETZERFEHLER (*m*): printer's error or compositor's error 982
SETZHAKEN (*m*): cant hook, crow 983
SETZHAMMER (*m*): set hammer 984
SETZKASTEN (*m*): settling tank (water passes through continuously and is kept washing backward and forward. The difference in specific weight causes the coal to rise above the waste and flow off with the water leaving the waste behind) 985
SETZLATTE (*f*): level rule 986
SETZLING (*m*): slip; young (tree) plant 987
SETZMACHINE (*f*): settling machine (for coal washing) 988
SETZMEISSEL (*m*): setter 989
SETZUNG (*f*): expression (Maths.) 990
SETZWEGER (*m*): spirketting 991
SETZZAPFEN (*m*): suppository 992
SEUCHE (*f*): pestilence; plague; infectious (contagious) disease; epidemic 993
SEUFZEN: to sigh 994
SEVENBAUM (*m*): savin (*Juniperus sabina*) 995
SEVENKRAUT (*n*): (see SEVENBAUM) 996
SEXTANT (*m*): sextant (Naut.) 997
SEXUELL: sexual 998
SEZERNIEREN: to secrete 999
SEZIEREN: to dissect 000
SHAPINGMASCHINE (*f*): shaping machine 001
SHEABUTTER (*f*): shea butter (from seeds of *Bassia parkii*); Sg. 0.9176; Mp. 23-31°C. 003
SHEDDACH (*n*): shed-roof 004
SHERARDISIEREN (*n*): zincing process by Sherard, Sherardizing (process for coating metal with iron-zinc alloy above which is a fine coat of pure zinc) (excellent anti-rust treatment) 005
SHERBE (*f*): potsherd; piece; fragment; (flower)-pot; crock; scorifier; cupel 006
SHIKIMISÄURE (*f*): shikimic acid 007
SHOLE (*f*), RUDER- : rudder sole 008
SIAKTALG (*m*): (see BALAMTALG) 009
SICCATIV: siccative; quick-drying; (*n*): siccative; quick-drier (see SIKKATIV) 010
SICCATIVEXTRAKT (*m*): siccative extract 011
SICCATIVPULVER (*n*): siccative powder 012
SICCOLIN (*n*): siccolin 013
SICCOLINEUMLACK (*m*): siccolineum varnish 014
SICH: oneself; him (her; it; them)-self or selves; one another; each other 015
SICHEL (*f*): sickle; crescent 016
SICHELFÖRMIG: crescent-shaped 017
SICHER: safe; certain; reliable; secure; true; sure; firm 018
SICHER GEHEN: to go carefully, be on the safe side, look before one leaps 019
SICHERHEIT (*f*): safety; reliability; security; surety; guarantee; certainty; firmness 020
SICHERHEITSAPPARAT (*m*): safety apparatus 021
SICHERHEITSFAKTOR (*m*): factor of safety 022
SICHERHEITSGRAD (*m*): factor (degree) of safety 023
SICHERHEITSKOEFFIZIENT (*m*): co-efficient of safety; factor of safety 024
SICHERHEITSLAMPE (*f*): safety lamp 025
SICHERHEITSLEITER (*f*): safety (emergency ladder 026
SICHERHEITSMASSREGEL (*f*): safety (expedient) measure 027
SICHERHEITSMUTTER (*f*): safety nut, lock nut 028
SICHERHEITSÖL (*n*): safety oil 029
SICHERHEITSRAND (*m*): safe (margin, border) edge 030
SICHERHEITSREGLER (*m*): safety (governor) regulator 031
SICHERHEITSRÖHRE (*f*): safety tube 032
SICHERHEITSSCHIENE (*f*): safety rail, guide rail 033
SICHERHEITSSPRENGSTOFF (*m*): safety explosive 034
SICHERHEITSTREPPE (*f*): safety staircase, emergency staircase 035
SICHERHEITSTRICHTERROHR (*n*): safety tube 036
SICHERHEITSVENTIL (*n*): safety valve 037
SICHERHEITSVENTIL (*n*) -AUSBLASELEITUNG (*f*): safety valve escape pipe 038
SICHERHEITSVENTIL (*n*), DOPPELTWIRKENDES- : double-acting safety valve (i.e., a pair of valves) 039
SICHERHEITSWINDE (*f*): safety winch 040
SICHERHEITSZÜNDER (*m*): safety detonator 041
SICHERHEITSZÜNDHÖLZCHEN (*n*): safety match 042
SICHERHEITSZÜNDSCHNUR (*f*): safety fuse 043
SICHERLICH: undoubtedly; surely; certainly 044
SICHERN: to secure; guarantee; ensure; safeguard 045
SICHERSTELLUNG (*f*): security, guarantee 046
SICHERUNG (*f*): securing; security; safety device; catch; fastening; fuse; cut-out (Elect.) 047
SICHERUNG (*f*), BLEI- : lead (cut-out) fuse 048
SICHT (*f*): sight (Com., etc.) 049
SICHTBAR: visible; evident; conspicuous 050
SICHTBARKEIT (*f*): visibility; conspicuousness 051
SICHTBARLICH: visibly 052
SICHTEN: to sift; winnow; screen 053
SICHTER (*m*): sifter; sieve; screen 054
SICHTLICH: visible; evident; perceptible 055
SICHTWECHSEL (*m*): bill payable at sight 056
SICKERN; to ooze; trickle; drop; percolate; drain 057
SICKERWASSER (*n*): ground water 058
SIDERINGELB (*n*): siderine yellow; iron chromate; $Fe_2(CrO_4)_3$ (see EISENCHROMAT) 059
SIDERIT (*m*): siderite, spathic iron, sidérose, chalybite (a ferrous carbonate); $FeCO_3$; Sg. 3.8 (see also EISENSPAT) 060
SIDEROSTHENLUBROSE (*f*): siderosthene-lubrose; (trade name for a protective paint for iron) 061

SIDIO-QUARZGLASGERÄT (n): sidio-quartz glass utensils 062
SIDIO-QUARZGLASGERÄTE (n.pl.): sidio-quartz glass utensils 063
SIDONAL (n): sidonal, piperazine quinate (*Piperazinum chinicum*); Mp. 170°C. 064
SIDONAL (n), NEU-: New Sidonal (25% quinic acid and 75% quinic acid anhydride) 065
SIE: she; her; it; they; them; you 066
SIEB (n): sieve; sifter; screen; strainer; bolter (for flour); riddle 067
SIEBANLAGE (f): screening plant 068
SIEBBEIN (n): ethmoid bone (Anat.) 069
SIEBBODEN (m): sieve (perforated) bottom 070
SIEBEN: to sift; sieve; strain; bolt (Flour); riddle; screen; (as an adjective = seven) 071
SIEBENBLÄTTERIG: seven-leaved, heptaphyllous 072
SIEBENECK (n): heptagon 073
SIEBENECKIG: heptagonal 074
SIEBENFACH: seven-fold 075
SIEBENFÄLTIG: seven-fold 076
SIEBENGESTIRN (n): pleiades (Astronomy) 077
SIEBENGLIEDRIG: seven-membered 078
SIEBENMAL: seven times 079
SIEBENTE: seventh 080
SIEBENWERTIG: septi-(hepta-)valent 081
SIEBEREI (f): screening, sieving; riddling; place of screening or sieving 082
SIEBFÖRMIG: sieve-(like) shaped 083
SIEBMASCHINE (f): sifting (screening) machine 084
SIEBMEHL (n): coarse flour; siftings 085
SIEBPLATTE (f): sieve plate 086
SIEBSATZ (m): set of sieves 087
SIEBSCHALE (f): dish with perforated bottom 088
SIEBSEL (n): siftings 089
SIEBSTAUB (m): siftings 090
SIEBTEHALB: six and a half 091
SIEBTROMMEL (f): drum sieve; riddling drum; revolving screen; screening drum 092
SIEBTUCH (n): bolting cloth 093
SIEBWASSER (n): back water (Paper) 094
SIEBZEHN: seventeen 095
SIEBZIG: seventy 096
SIECH: sick-(ly); languishing; infirm 097
SIECHEN: to sicken; languish; become infirm 098
SIEDE (f): boiling; seething 099
SIEDEABFALL (m): boiling (scum) sediment 100
SIEDEAPPARAT (m): boiling apparatus 101
SIEDEBLECH (n): boiling plate 102
SIEDEERSCHEINUNG (f): boiling; ebullition 103
SIEDEGEFÄSS (n): boiling vessel; pan; boiler; copper, etc. 104
SIEDEGRAD (m): boiling point 105
SIEDEHAUS (n): boiling house 106
SIEDEHITZE (f): boiling heat 107
SIEDEKESSEL (m): boiling vessel; pan; boiler; copper, etc. 108
SIEDEKOLBEN (m): boiling flask 109
SIEDEKURVE (f): boiling (point) curve 110
SIEDELAUGE (f): boiling (liquor) lye 111
SIEDELEI (f): settlement 112
SIEDELN: to settle 113
SIEDEN: to boil; seethe; brew (Beer); refine; (n): boiling; ebullition; seething 114
SIEDEN, IM ALAUN-: to alum 115
SIEDEN LASSEN: to allow to boil; to let boil 116
SIEDEPFANNE (f): boiling (evaporating) pan 117
SIEDEPUNKT (m): boiling point 118
SIEDEPUNKTBESTIMMUNG (f): determination of boiling point 119
SIEDEPUNKTSERHÖHUNG (f): increase (raising) of boiling point 120
SIEDER (m): boiler 121
SIEDEREI (f): boiling (room) house; refinery (Sugar) 122
SIEDEROHR (n): boiling (distilling) tube; water tube (of boilers as opposed to steam tube) 123
SIEDERÖHRE (f): (see SIEDEROHR) 124
SIEDESALZ (n): common salt; refined (table) salt (see NATRIUMCHLORID) 125
SIEDESOLE (f): brine 126
SIEDEVERZUG (m): delay in boiling 127
SIEDEZEIT (f): boiling season; boiling (time) period 128
SIEDKESSEL (m): boiling vessel; pan; boiler; copper, etc. 129
SIEDLER (m): settler 130
SIEG (m): victory; triumph; win; conquest; gain 131
SIEGEL (n): seal (of wax; lead, etc.) 132
SIEGELERDE (f): Lemnian earth; terra sigillata (see BOLUS, WEISSER-) 133
SIEGELLACK (m and n): sealing wax 134
SIEGELMARKE (f): seal-(stamp) 135
SIEGELN: to seal 136
SIEGELRING (m): signet ring 137
SIEGELWACHS (n): (soft)-sealing wax 138
SIEGEN: to conquer; triumph; be victorious; win; vanquish 139
SIEGENIT (m): siegenite (an explosive) 140
SIEGREICH: victorious; triumphant 141
SIEGWURZ (f): gladiole (*Gladiolus communis*) 142
SIEKEN: to seam; crease 143
SIEL (n): dyke, lock, sluice; sewer 144
SIELWASSER (n): dyke (lock) water 145
SIEMENSGAS (n): producer gas 146
SIENAERDE (f): sienna; terra sienna (a clay from California, coloured by iron and manganese oxide) 147
SIENNA (f): sienna (the colour); also the name of a clay (see SIENAERDE) 148
SIGNAL (n): signal 149
SIGNALAPPARAT (m): signalling apparatus 150
SIGNALBUCH (n): signal (code) book (Marine) 151
SIGNALEMENT (n): description (of a person) 152
SIGNALFARBE (f): signal colour 153
SIGNALFLAGGE (f): signal flag 154
SIGNALGLOCKE (f): signal bell 155
SIGNALISIEREN: to signal; signalize 156
SIGNALLAMPE (f): signal lamp 157
SIGNALLATERNE (f): signal (lamp) lantern 158
SIGNALMAST (m): signal mast 159
SIGNALRAKETE (f): signal rocket 160
SIGNALROT (n): (see BLEIMENNIGE) 161
SIGNALSCHEIBE (f): disc signal 162
SIGNATUR (f): signature; stamp; mark; brand 163
SIGNIEREN: to sign; stamp; mark; brand 164
SIGNIERTINTE (f): marking ink 165
SIKKATIV (n): siccative; (quick)-drier (for varnishes and paints); manganese borate (see MANGANBORAT) (see also SICCATIV) 166
SIKKATIVEXTRAKT (m): siccative extract 167
SIKKATIVFIRNIS (m): siccative varnish 168
SIKKATIV (n), FLÜSSIGES-: liquid siccative 169
SIKKATIVMASSE (f): siccative 170
SIKKATIVPULVER (n): siccative powder 171
SILBE (f): syllable 172
SILBENMASS (n): quantity; metre 173
SILBENRÄTSEL (n): charade 174

SILBER (*n*): silver; (Ag); plate (for table use, etc.) 175
SILBERACETAT (*n*): silver acetate; $AgC_2H_3O_2$ 176
SILBERADER (*f*): vein of silver 177
SILBERALBUMINAT (*n*): silver albuminate 178
SILBERAMALGAM (*n*): silver amalgam; Ag,Hg; Sg. 13.9 179
SILBERANTIMONGLANZ (*m*): silver-antimony glance, miargyrite (see **MIARGYRIT**) 180
SILBERANTIMONID (*n*): silver antimonide; Ag_2Sb 181
SILBERANTIMONSULFID (*n*): silver-antimony sulphide; Ag_3SbS_3 182
SILBER (*n*), **APFELSAURES-**: (see **SILBERMALAT**) 183
SILBERAPPARAT (*m*): silver apparatus 184
SILBERARBEIT (*f*): silver-work 185
SILBERARBEITER (*m*): silver-smith 186
SILBERARTIG: silvery; argentine 187
SILBERÄTZSTEIN (*m*): lunar caustic; silver nitrate (see **SILBERNITRAT**) 188
SILBERAZID (*n*): silver azide; silver hydroazoate; AgN_3 189
SILBERBAD (*n*): silver bath (Phot.) 190
SILBER (*n*), **BALDRIANSAURES-**: (see **SILBER-VALERIANAT**) 191
SILBERBARRE (*f*): silver (bar) ingot 192
SILBERBARREN (*m*): (see **SILBERBARRE**) 193
SILBERBAUM (*m*): silver tree (*arbor Dianæ*) (Chem.) 194
SILBERBELEGUNG (*f*): silver coating 195
SILBERBERGWERK (*n*): silver mine 196
SILBERBICHROMAT (*n*): silver bichromate; $Ag_2Cr_2O_7$; Sg. 4.77 197
SILBERBLATT (*n*): silver (foil) leaf 198
SILBERBLECH (*n*): silver (foil) plate 199
SILBERBLENDE (*f*): proustite; pyrargyrite (see **PROUSTIT** and **PYRARGYRIT**) 200
SILBERBLICK (*m*): silver "blick"; silver brightening or fulguration 201
SILBERBRENNEN (*n*): silver refining 202
SILBERBROMAT (*n*): silver bromate; $AgBrO_3$; Sg. 5.104-5.206 203
SILBERBROMID (*n*): silver bromide; AgBr; Sg. 6.473; Mp. 427°C. (*Argentum bromatum*) 204
SILBERBROMÜR (*n*): silver sub-bromide 205
SILBERBUTYRAT (*n*): silver butyrate (see **SILBER-BUTYRAT, NORMAL-**; and **SILBERISOBUTYRAT**) 206
SILBERBUTYRAT (*n*), **NORMAL-**: normal silver butyrate; $AgCH_3(CH_2)_2CO_2$ 207
SILBERCAPROAT (*n*): silber caproate; $AgC_6H_{11}O_2$ 208
SILBERCAPRYLAT (*n*): silver caprylate; $AgC_8H_{15}O_2$ 209
SILBERCARBONAT (*n*): silver carbonate; Ag_2CO_3 210
SILBERCHLORAD (*n*): silver chloride; AgCl (see **SILBERCHLORID**) 211
SILBERCHLORAT (*n*): silver chlorate; $AgClO_3$; Sg. 4.4-4.43 212
SILBERCHLORID (*n*): silver chloride; AgCl; Sg 5.553-5.57; Mp. 451°C. 213
SILBERCHLORÜR (*n*): silver subchloride 214
SILBERCHROMAT (*n*): silver chromate; Ag_2CrO_4; Sg. 5.625 215
SILBERCITRAT (*n*): silver citrate; $AgC_6H_5O_7$ (see **ITROL**) 216
SILBER (*n*), **CITRONSAURES-**: silver citrate (see **SILBERCITRAT**) 217
SILBERCYANID (*n*): silver cyanide: AgCN; Sg. 3.95 218
SILBERCYANÜR (*n*): silver cyanide 219
SILBERDITHIONAT (*n*): silver dithionate: $Ag_2S_2O_6+2H_2O$ 220
SILBERDRAHT (*m*): silver wire 221
SILBERDRUCK (*m*): silver print (Phot.) 222
SILBERERZ (*n*): silver ore (see **STEPHANIT**) 223
SILBERERZ, RUSSIGES- (*n*): black silver ore 224
SILBERESSIGSALZ (*n*): silver acetate (see **SILBER-ACETAT**) 225
SILBER (*n*), **ESSIGSAURES-**: silver acetate (see **SILBERACETAT**) 226
SILBERFADEN (*m*): silver thread 227
SILBERFAHLERZ (*n*): argentiferous tetrahedrite (see **WEISSGÜLTIGERZ**) 228
SILBERFÄLLUNGSMITTEL (*n*): silver precipitant 229
SILBERFARBE (*f*): silver-(colour) 230
SILBERFARBEN: silvery; silver-coloured 231
SILBERFARBIG: silvery; silver-coloured 232
SILBER (*n*), **FETTSAURES-**: silver sebate 233
SILBERFLUORID (*n*): silver fluoride; AgF; Sg. 5.852 234
SILBERFLUORÜR (*n*): silver subfluoride 235
SILBERFOLIE (*f*): silver foil 236
SILBERFORMIAT (*n*): silver formate; $AgCHO_2$ 237
SILBERGEHALT (*m*): silver content 238
SILBERGERÄT (*n*): silver plate 239
SILBERGESCHIRR (*n*): silver plate 240
SILBERGEWINNUNG (*f*): silver extraction; mining of silver 241
SILBERGLANZ (*m*): silver glance, argentite, acanthite (natural silver sulphide); Ag_2S (about 85% silver); Sg. 7.25 (see also **SILBERSULFID**); silver (lustre, glimmer, glitter, sheen) gleam 242
SILBERGLÄNZEND: silvery 243
SILBERGLANZERZ (*n*): argentite, silver glance (see **SILBERGLANZ**) 244
SILBERGLAS (*n*): argentite (see **SILBERGLANZ**) 245
SILBERGLASERZ (*n*): (see **SILBERGLANZ**) 246
SILBERGLÄTTE (*f*): (silver)-litharge (litharge after quick cooling, yellow coloured) (see **BLEIGLÄTTE**) 247
SILBERGLIMMER (*m*): muscovite; common mica (see **MUSKOVIT**) 248
SILBERGLYKOLAT (*n*): silver glycolate; $AgCH_2OH_2CO_2$ 249
SILBERGOLD (*n*): argentiferous gold; electrum (see **ELEKTRUM**) 250
SILBER (*n*), **GOLDCHLORWASSERSTOFFSÄURES-**: silver chloraurate; $AgAuCl_4$ 251
SILBERGRAPHIT (*m*): (see **GRAPHIT**) 252
SILBERGRAU: silver grey; (*n*): silver-grey (the colour) 253
SILBERGRUBE (*f*): silver mine 254
SILBERHALTIG: argentiferous; having a silver content; containing silver 255
SILBERHORNERZ (*n*): horn silver; cerargyrite (see **CERARGYRIT**) 256
SILBERHÜTTE (*f*): silver (works) foundry 257
SILBERIG: silvery 258
SILBERISOBUTYRAT (*n*): silver isobutyrate; $(CH_3)_2CHCO_2Ag$ 259
SILBERJODAT (*n*): silver iodate; $AgIO_3$; Sg. 5.525 260
SILBERJODID (*n*): silver iodide; AgI; Sg. 5.674; Mp. 556°C. 261
SILBERJODÜR (*n*): silver subiodide 262

SILBER (*n*), **KOHLENSAURES-**: silver carbonate (see **SILBERCARBONAT**) 263

SILBER, KOLLOIDALES- (*n*): colloidal silver; collargol; argentum credé 264

SILBERKOPIE (*f*), **POSITIVE-**: positive silver copy (Phot.) 265

SILBERKUPFERGLANZ (*m*): silver-copper glance; stromeyerite (see **STROMEYERIT**) 266

SILBERLAKTAT (*n*): silver lactate; $AgCH_3CHOHCOO$ or $AgC_3H_5O_3.H_2O$ (see **ACTOL**) 267

SILBER, LEBENDIGES- (*n*): quicksilver (*Argentum vivum*) (see **QUECKSILBER**) 268

SILBERLEGIERUNG (*f*): silver alloy 269

SILBERLÖSEND: silver-solvent 270

SILBERLÖSUNG (*f*): silver solution 271

SILBERLOT (*n*): silver solder (about 25% Cu, and 75% Ag) 272

SILBERMALAT (*n*): silver malate; $Ag_2C_2H_3OH(COO)_2$ 273

SILBERMALONAT (*n*): silver malonate; $Ag_2CH_2(COO)_2$ 274

SILBERMESSER (*m*): argentometer (an instrument for showing the number of grains per ounce of silver nitrate in a pure silver bath) (Phot.) 275

SILBER (*n*), **MILCHSAURES-**: silver lactate; $AgCH_3CHOHOCO$ (see **ACTOL**) 276

SILBERMÜNZE (*f*): silver coin 277

SILBERN: silver; of silver; (as a verb = to silver; silver-plate; paint with silver paint) 278

SILBERNIEDERSCHLAG (*m*): silver precipitate 279

SILBERNITRAT (*n*): silver nitrate, Lunar caustic (*Lapis infernalis*) (*Argentum nitricum*); $AgNO_3$; Sg. 4.35; Mp. 218°C. 280

SILBERNITRIT (*n*): silver nitrite; $AgNO_2$; Sg. 4.453 281

SILBER, NUCLEINSAURES- (*n*): silver nucleinate 282

SILBEROLEAT (*n*): silver oleate; $AgC_{18}H_{33}CO_2$ 283

SILBERÖNANTHAT (*n*): silver oenanthate; $AgC_9H_{17}O_2$ 284

SILBERÖNANTHYAT (*n*): silver oenanthyeate; $AgC_7H_{13}O_2$ 285

SILBERORTHOARSENAT (*n*): silver orthoarsenate; Ag_3AsO_4; Sg. 6.657 286

SILBERORTHOPHOSPHAT (*n*): silver orthophosphate; Ag_3PO_4; Sg. 6.37; Mp. 849°C. (see **SILBERPHOSPHAT**) 287

SILBEROXALAT (*n*): silver oxalate; $Ag_2C_2O_4$ 288

SILBER (*n*), **OXALSAURES-**: silver oxalate (see **SILBEROXALAT**) 289

SILBEROXYCHLORID (*n*): silver oxychloride; $AgClO$ 290

SILBEROXYD (*n*): silver oxide; Ag_2O; Sg. 7.521 291

SILBEROXYDAMMONIAK (*n*), **SALPETERSAURES-**: silver-ammonium nitrate; $2NH_3AgNO_3$ (contains 77.5% silver nitrate, and 22.5% ammonia) 292

SILBEROYXD (*n*), **CHROMSAURES-**: (see **SILBERCHROMAT**) 293

SILBEROXYD (*n*), **CITRONENSAURES-**: (see **SILBERCITRAT**) 294

SILBEROXYD, ESSIGSAURES- (*n*): silver acetate (see **SILBERACETAT**) 295

SILBEROXYD (*n*), **SALTPETERSAURES-**: (see **SILBERNITRAT**) 296

SILBEROXYDUL (*n*): silver suboxide 297

SILBERPALMITAT (*n*): silver palmitate; $AgC_{16}H_{31}O_2$ 298

SILBER (*n*), **PALMITINSAURES-**: silver palmitate (see **SILBERPALMITAT**) 299

SILBERPAPIER (*n*): silver paper or silvered paper 300

SILBERPAPPEL (*f*): white poplar (see **PAPPEL**) 301

SILBERPARALAKTAT (*n*): silver paralactate: $AgCH_3CHOHCOO$ 302

SILBERPHOSPHAT (*n*): silver phosphate; Ag_3PO_4; Sg. 6.37-7.324; Mp. 849°C. 303

SILBER (*n*), **PHOSPHORSAURES-**: silver phosphate (see **SILBERPHOSPHAT**) 304

SILBERPLATTIERT: silver plated 305

SILBERPLATTIERUNG (*f*): silver plating 306

SILBERPROBE (*f*): silver (test) assay 307

SILBERPROPIONAT (*n*): silver propionate; $AgC_3H_5O_2$ 308

SILBER, PROTEINSAURES- (*n*): silver proteinate (*Argentum proteinicum*) 309

SILBERPUTZ (*m*): silver polish 310

SILBERPUTZSEIFE (*f*): silver (polishing, cleaning) soap 311

SILBERREICH: having high silver content; rich in silver 312

SILBERSALPETER (*m*): silver nitrate (see **SILBERNITRAT**) 313

SILBER, SALPETERSAURES- (*n*): (see **SILBERSALPETER**) 314

SILBERSALZ (*n*): silver salt (Dye), sodium salt of anthraquinone monosulphonic acid 315

SILBERSCHAUM (*m*): thin leaves of silver 316

SILBERSCHEIDUNG (*f*): silver (separation) refining 317

SILBERSCHLAGLOT (*n*): silver solder (see **SILBERLOT**) 318

SILBERSCHLAMM (*m*): silver tailings 319

SILBERSCHMELZHITZE (*f*): silver melting-point, degree of heat (temperature) at which silver melts (i.e., about 960.5°C.) 320

SILBERSCHMIED (*m*): silversmith 321

SILBERSCHMIERKUR (*f*): Collargol ointment cure (the external application of collargol ointment; see also **COLLARGOLSALBE**) 322

SILBERSCHWÄRZE (*f*): (earthy)-argentite; silver glance (see **SILBERGLANZ**) 323

SILBER (*n*), **SCHWEFELSAURES-**: silver sulphate (see **SILBERSULFAT**) 324

SILBERSCHWEFLIG: argentosulphurous 325

SILBERSELENID (*n*): silver selenide; Ag_2Se [326

SILBERSELENIT (*n*): silver selenite; Ag_2SeO_3; Sg. 5.9297 327

SILBERSPAT (*m*): cerargyrite (see **CERARGYRIT**) 328

SILBERSPIEGEL (*m*): silver mirror 329

SILBERSPIESSGLANZ (*m*): dyscrasite; antimonial silver (natural silver antimonide) (see **ANTIMONSILBER**) 330

SILBERSTAHL (*m*): silver steel 331

SILBER (*n*), **STEARINSAURES-**: silver stearate; $AgC_{18}H_{35}O_2$ 332

SILBERSTOFF (*m*): silver (brocade) cloth 333

SILBERSTUFE (*f*): silver-ore 334

SILBERSULFAT (*n*): silver sulphate; Ag_2SO_4; Sg. 5.4-5.45 335

SILBERSULFID (*n*): silver sulphide; Ag_2S; Sg. 6.85-7.32; Mp. 842°C. 336

SILBERTANNENÖL (*n*): pine oil 337

SILBERTARTRAT (*n*): silver tartrate; $Ag_2C_4H_4O_6$ 338

SILBERTIEGEL (*m*): silver crucible 339

SILBERVALERIANAT (*n*): silver valerianate; $Ag(CH_3)_2CHCH_2CO_2$ 340

SILBERVITRIOL (*m*): silver sulphate (see SILBERSULFAT) 341
SILBERWAREN (*f.pl*): silver ware 342
SILBERWEISS (*n*) silver-white 343
SILBERZEUG (*n*): silver-(plate) ware 344
SILBERZITRAT (*n*): (see ITROL and SILBERCITRAT) 345
SILICIUM (*n*): silicon; (Si) (see SILIZIUM) 346
SILICOFLUORID (*n*): silicofluoride; fluosilicate (see SILIZIUMFLUORID) 347
SILICOFLUORWASSERSTOFFSÄURE (*f*): (see KIESELFLUORWASSERSTOFFSÄURE) 348
SILICOMANGAN (*n*): silicomanganese 349
SILICON (*n*): silicone 350
SILIKA (*f*): silica 351
SILIKASTEIN (*m*): silica brick; Sg. 2.3 2.6 352
SILIKAT (*n*): silicate 353
SILIKATGESTEIN (*m*): silicate rock 354
SILIKOAMEISENSÄURE (*f*): silicoformic acid 355
SILIKON (*n*): silicone 356
SILIKOSPIEGEL (*m*): ferromanganese silicon (blast furnace product) (65-68% Fe; 20% Mn; 10-12 % Si; 2-2.5% C.; 0.18% P.) (produced in electric furnace 50-75% Mn; 20-35% Si; 0.65-1%C; 0.05% P: 0.02 % S.) 357
SILIT (*n*): silicon carbide (SiC); Sg. 3.12-3.2 [358
SILIZID (*n*): silicide 359
SILIZIUM (*n*): silicon; (Si) (see SILICIUM) 360
SILIZIUMAMEISENSÄURE (*f*): silicoformic acid 361
SILIZIUMÄTHAN (*n*): silicoethane 362
SILIZIUMBORID (*n*): silicon boride; SiB_3, Sg. 2.52; SiB_6, Sg. 2.47 363
SILIZIUMBROMID (*n*): silicon bromide 364
SILIZIUMBROMOFORM (*n*): silicon bromoform, silicobromoform; $SiHBr_3$; Sg. 2.5-2.7 365
SILIZIUMBRONZE (*f*): silicon-bronze (91-99.9% Cu; 0.03-9. % Sn; 0.03-0.05% Si) 366
SILIZIUMBRONZEDRAHT (*m*): silicon-bronze wire 367
SILIZIUMCARBID (*n*): silicon carbide; SiC; Sg. 3.12-3.2 (see SILIZIUMKARBID) 368
SILIZIUMCHLORID (*n*): silicon chloride; $SiCl_4$ (see SILIZIUMTETRACHLORID) 369
SILIZIUMCHLOROBROMID (*n*): silicon chlorobromide; $SiClBr_3$; Sg. 2.432 370
SILIZIUMCHLOROFORM (*n*): silicochloroform; silicon chloroform; $SiHCl_3$; Sg. 1.34-1.65 371
SILIZIUMDIOXYD (*n*): silicon dioxide; SiO_2; siliceous earth; silica; silicic anhydride 372
SILIZIUMEISEN (*n*): iron silicide 373
SILIZIUMFLUORID (*n*): fluosilicate; silicofluoride; SiF_4 374
SILIZIUMFLUORVERBINDUNG (*f*): fluosilicate 375
SILIZIUMFLUORWASSERSTOFF (*m*): fluosilicic acid; H_2SiF_6 376
SILIZIUMFLUORWASSERSTOFFSÄURE (*f*): (see SILIZIUMFLUORWASSERSTOFF) 377
SILIZIUMHALTIG: siliceous; containing silicon 378
SILIZIUMHEXACHLORID (*n*): silicon hexachloride; Si_2Cl_6; Sg. 1.58 379
SILIZIUMJODID (*n*): silicon iodide 380
SILIZIUMJODOFORM (*n*): silicon iodoform, silicoiodoform; $SiHI_3$; Sg. 3.286-3.314 381
SILIZIUMKARBID (*n*): silicon carbide; SiC; carborundum; Sg. about 3.15 382
SILIZIUMKOHLENSTOFF (*m*): (see SILIZIUMKARBID) 383
SILIZIUMKREIDE (*f*): silicious chalk 384
SILIZIUMKUPFER (*n*): silicon copper (see KUPROSILICIUM); copper silicide 385
SILIZIUMMAGNESIUM (*n*): magnesium silicide 386
SILIZIUM-MANGANSTAHL (*m*): silicon-manganese steel 387
SILIZIUMMETALL (*n*): metallic silicide 388
SILIZIUMMETHAN (*n*): silicon tetrahydride; silicomethane 389
SILIZIUMMONOXYD (*n*): silicon monoxide; SiO 390
SILIZIUMNITRID (*n*) silicon nitride; SiN, Sg. 3.17; Si_2N_4, Sg. 3.44; Si_2N_3, Sg. 3.64 [391
SILIZIUMOXALSÄURE (*f*): silicooxalic acid 392
SILIZIUMOXYD (*n*): silicon dioxide; SiO_2 393
SILIZIUMOXYDHYDRAT: hydrated iodide of silicon; silicic acid 394
SILIZIUMSTAHL (*m*): silicon steel 395
SILIZIUMSTEIN (*m*): silica brick 396
SILIZIUMSULFID (*n*): silicon sulphide; SiS; Sg. 1.853 397
SILIZIUMTETRABROMID (*n*): silicon tetrabromide; $SiBr_4$; Sg. 2.8124 398
SILIZIUMTETRACHLORID (*n*): silicon tetrachloride; silicon chloride; $SiCl_4$; Sg. 1.485-1.524; Mp. – 89°C.; Bp. 56.9°C. 399
SILIZIUMTETRAHYDRÜR (*n*): silicon tetrahydride (see SILIZIUMMETHAN) 400
SILIZIUMVERBINDUNG (*f*): silicon compound 401
SILIZIUMWASSERSTOFF (*m*): silicon hydride; hydrogen silicide; Si_2H_6; Sg. >1.0 402
SILL (*n*): coaming, sill 403
SILLIMANIT (*m*): sillimanite, fibrolite (an aluminium silicate; also a new refractory material); Al_2SiO_5 or $Al_2O_3.SiO_2$; Sg. 3.24 (see ALUMINIUMSILIKAT) 404
SILO (*m*): silo; bin (Brew.); bunker 405
SIMONYIT (*m*): simonyite (see BLÖDIT) 406
SIMPEL: simple; plain 407
SIMS (*m*): shelf; cornice; moulding 408
SIMULTAN: simulatneous-(ly) 409
SINALBIN (*n*): sinalbin (a glycoside from white mustard); $C_{30}H_{44}N_2S_2O_{16}$ or $C_{30}H_{42}N_2S_2O_{15}$ 410
SINAPINSÄURE (*f*): sinapic acid (see SENFSÄURE) 411
SINAPOL (*n*): sinapol (a washing and fulling preparation) 412
SINGEN: to sing; chant 413
SINGULO-SILIKATSCHLACKE (*f*): singulo silicate slag (such as $2CaO.SiO_2$) 414
SINIGRIN (*n*): (see KALIUM, MYRONSAURES-) 415
SINKEN: to sink; subside; decrease; fall; drop; descend; go down; submerge (of submarines) 416
SINN (*m*): mind; sense; feeling; inclination; meaning 417
SINNBILD (*n*): symbol; emblem; allegory 418
SINN (*m*), **DREH-**: direction of rotation 419
SINNEN: to think; reflect; meditate; muse; speculate 420
SINNENWELT (*f*): external world 421
SINNENWERKZEUG (*n*): organ of sense 422
SINNESÄNDERUNG (*f*): change of mind 423
SINNESART (*f*): character; disposition 424
SINNFÄLLIG: thoughtful, deliberate 425
SINNGEDICHT (*n*): epigram 426
SINNGEMÄSS: of course; obviously; naturally; patent 427
SINNGETREU: faithful (of reproduction of anything, or translation) 428

SINNIG : sensible ; ingenious ; thoughtful ; judicious 429

SINNLICH : sensual ; sensuous ; sensitive ; sentient 430

SINNLICHKEIT (*f*) : sensuality 431

SINNLOS : senseless ; devoid of meaning 432

SINNPFLANZE (*f*) : sensitive plant 433

SINNREICH : witty ; ingenious ; judicious 434

SINNSPRUCH (*m*) : sentiment ; motto ; maxim 435

SINNVERWANDT : synonymous 436

SINNWIDRIG : absurd 437

SINOIDE (*f*) : sinusoid 438

SINTER (*m*) : cinder ; iron dross ; sinter ; stalactites 439

SINTERKOHLE (*f*) : non-caking (sintering) coal ; non-clinkering coal 440

SINTERN : to form sinter ; sinter ; drop ; trickle ; incrust ; vitrify ; cake ; clinker 441

SINTERQUARZ (*m*) : siliceous sinter 442

SINTERRÖSTVERFAHREN (*n*) : sintering process (Roasting process for lead) 443

SINTERSCHLACKE (*f*) : clinker 444

SINTERUNG (*f*) : sintering ; formation of sinter ; dropping ; trickling ; incrustation ; vitrifaction ; caking ; clinkering 445

SINTERWASSER (*n*) : sinter-forming water (impregnated with mineral matter) 446

SINUS (*m*) : sine (Maths.) 447

SINUSBUSSOLE (*f*) : sine compass, sine galvanometer 448

SINUSLINIE (*f*) : sine curve 449

SINUSSATZ (*m*) : theorem of the sine (Maths.) 450

SINUS VERSUS EINES BOGENS : versed sine of an arc 451

SIPHONIEREN : to siphon 452

SIPHONOPHOREN (*pl*) : siphonophores (such as Portuguese man-of-war ; type of jellyfish) 453

SIPPE (*f*) : kin ; kindred ; tribe 454

SIPPSCHAFT (*f*) : (see SIPPE) 455

SIRENE (*f*) : siren 456

SIRENENSCHALLTRICHTER (*m*) : siren (trumpet) horn 457

SIRENENTON (*m*) : siren sound 458

SIRENENVENTIL (*n*) : siren valve 459

SIRIUS (*m*) : artifical horse-hair (Textile) 460

SIRIUSGELB (*n*) : sirius yellow 461

SIRIUSMETALLAMPE (*f*) : sirius metal lamp (a type of Tungsten lamp) 462

SIROLIN (*n*) : siroline (a solution of thiocol in orange-syrup in proportion of 1 : 14) 463

SIRUP (*m*) : treacle ; syrup (*Sirupus*) (a thickly viscous solution of sugar in aqueous or alcoholic liquids) 464

SIRUPARTIG : syrupy ; treacly ; thick ; viscous 465

SIRUPDICK : (see SIRUPARTIG) 466

SIRUPHALTIG : syrupy ; containing syrup 467

SIRUPKONSISTENZ (*f*) : consistency of syrup 468

SIRUPÖS : syrupy ; treacly 469

SIRUPPFANNE (*f*) : syrup pan 470

SISALHANF (*m*) : Sisal hemp (Textile) (from the *Agave rigida*) 471

SISMONDIN (*m*) : sismondine, sismondite (a black variety of chloritoid) 472

SITTE (*f*) : custom ; manner ; practice ; use ; usage ; habit ; (*pl*) : morals ; manners ; customs 473

SITTENLEHRE (*f*) : ethics ; moral philosophy 474

SITTENLEHRER (*m*) : moralist 475

SITTENLOS : immoral 476

SITTENVERDERBNIS (*n*) : corruption-(of morals) ; demoralization 477

SITTER (*m*) : first futtock ; lower futtock (Naut.) 478

SITTIG : modest ; well-mannered 479

SITTLICH : moral 480

SITTLICHKEIT (*f*) : morality 481

SITTLICHKEITSGEFÜHL (*n*) : morality ; moral sense 482

SITTSAM : reserved ; chaste ; modest ; decent 483

SITTSAMKEIT (*f*) : chastity ; decency ; modesty 484

SITZ (*m*) : seat ; chair ; fit ; see (of a bishop) ; residence ; seating ; setting (of brickwork, etc.) ; set 485

SITZARBEIT (*f*) : sedentary occupation 486

SITZEN : to sit ; fit ; be (imprisoned) incarcerated ; rest ; be (fitted) attached ; be seated ; set ; reside 487

SITZEND : sitting ; sedentary ; residing ; having one's seat-(at) 488

SITZ (*m*), **ENGSTER-** : snug fit 489

SITZFLÄCHE (*f*) : seat, seating, face or surface 490

SITZUNG (*f*) : sitting ; session 491

SITZUNGSBERICHT (*m*) : report of the (proceedings) session 492

SKALA (*f*) : scale (graduated on instruments ; thermometers, etc.) 493

SKALENBELEUCHTUNGSSPIEGEL (*m*) : scale illuminating mirror 494

SKALENINTERVALL (*n*) : scale (division) interval 495

SKALENOEDER (*n*) : scalenohedron 496

SKALENRICHTIGKEIT (*f*) : scale correctness 497

SKALENROHR (*n*) : scale tube 498

SKALPIEREN : to scalp 499

SKAMMONIAHARZ (*n*) : scammony-(resin) 500

SKAMMONIAWURZEL (*f*) : scammony root (*Radix scammoniæ* of *Convolvulus scammonia*) 501

SKAMMONIENHARZ (*n*) : scammony-(resin) 502

SKAMMONIUM (*n*) : scammony resin 503

SKANDAL (*m*) : scandal ; row ; noise 504

SKANDIUM (*n*) : Scandium ; (Sc) 505

SKAPOLITH (*m*) : scapolite, wernerite ; $Ca_4Al_6Si_6O_{25}$: $Na_4Al_3Si_9O_{24}Cl$; Sg. 2.68 ; Mp. 1120-1140°C. 506

SKARABAEUS (*m*) : scarabeus, scarab 507

SKARIOL (*m*) : prickly lettuce (*Lactuca scariola*) 508

SKATOL (*n*) : skatol, beta-methyl-indol ; C_9H_9N ; Mp. 94°C. 509

SKEISEGEL (*n*) : skysail 510

SKELETT (*n*) : skeleton (Anat.) 511

SKELETTIEREN : to become a skeleton, become as thin as a rake, become all skin and bone 512

SKEPTISCH : sceptic-(al) 513

SKIFF (*n*) : skiff 514

SKIZZE (*f*) : sketch ; rough draft 515

SKIZZENBUCH (*n*) : sketch book 516

SKIZZIERBLOCK (*m*) : sketch block, sketching pad 517

SKIZZIEREN : to sketch ; draft ; (*n*) : sketching ; planning ; plotting ; roughing-out ; drafting ; designing 518

SKLAVE (*m*) : slave 519

SKLAVISCH : servile ; slavish 520

SKLEROKLAS (*m*) : scleroclase ; $PbAs_2S_4$; Sg. 5.39 521

SKLEROSE (*f*) : sclerosis (Med.) 522

SKLEROSE (*f*), **LATERAL-** : lateral sclerosis (see also SPINALPARALYSIS, SPASTISCHE-) 523

SKLEROSE (*f*), **MULTIPLE**: multiple sclerosis (Med.) 524

SKLEROSE (*f*), **PRIMÄRE-**: primary (idiopathic) sclerosis (see also SPINALPARALYSIS, SPASTISCHE-) 525

SKLEROTIUM (*n*): sclerotic (cornea of eye) 526

SKOLECIT (*m*): scolecite (a zeolite); $CaAl_2Si_3O_{10}.3H_2O$; Sg. 2.28 527

SKOLEZIT (*m*): scolezite, needlestone, lime mesotype; $CaO.SiO_2+AlO_3.SiO$; Sg. 2.25 528

SKOPOLAMIN (*n*): scopolamine (*Scopolaminum*); hyoscine (*Hyoscinum*) (alkaloid from seeds of *Hyoscyamus niger*); $C_{17}H_{21}NO_4$; Mp. 50-59°C. 529

SKORBUT (*m*): scurvy 530

SKORODIT (*m*): scorodite (a hydrous ferric arsenate); $FeAsO_4.2H_2O$; Sg. 3.2 531

SKROFEL (*f*): scrofula 532

SKROFULÖS: scrofulous 533

SKROPHULOSE (*f*): scrofula, struma (a tubercular affection of the lymphatic glands) (Med.) 534

SKRUBBER (*m*): scrubber (for gas) 535

SKRUBBERREINIGUNG (*f*): scrubbing (of gases) 536

SKRUPEL (*m*): scruple 537

SKRUPULÖS: scrupulous 538

SKULL (*n*): scull (Naut.); oar 539

SKULLBOOT (*n*): sculling boat 540

SKULLEN: to scull (Naut.), to row 541

SKUTTERUDIT (*m*): skutterudite (a natural arsenic-cobalt ore); $CoAs_3$ (about 21% Cobalt); Sg. 6.65 (see also TESSERALKIES) 542

SLIBOWITZ (*m*): (see SLIBOWOTZ) 543

SLIBOWOTZ (*m*): slibowotz (a brandy made in Serbia from plums) 544

SLIP (*m*): slip (of a propeller) 545

SLIPTROSSE (*f*): slip rope 546

SLIWOWITZ (*m*): (see SLIBOWOTZ) 547

SLUP (*f*): sloop (Naut.) 548

SMALTE (*f*): smalt; (Sg. 2.44); cobalt blue (see BLAUFARBE); cobalt silicate (see KOBALTSILIKAT) 549

SMALTIN (*m*): smaltine, smaltite (a cobalt arsenide); $CoAs_2$; Sg. 6.55 (see also SPEISKOBALT) 550

SMALTIT (*m*): smaltite (see SPEISKOBALT) 551

SMARAGD (*m*): emerald (a variety of Beryl, see BERYLL); Sg. 2.7 552

SMARAGDEN: emerald; of emerald 553

SMARAGDFARBEN: emerald; of emerald 554

SMARAGDGRÜN (*n*): Guignet's green; viridian; emerald (chrome) green; $Cr_2(OH)_6$. (a hydrated chromic oxide) (see GUIGNET-(S)-GRÜN) 555

SMARAGDIT (*m*): smaragdite (a variety of hornblende) (see HORNBLENDE) 556

SMARAGDMALACHIT (*m*): euchroite; $Cu_4O.AsO_5+7H_2O$ (about 47% copper oxide, 34% arsenic acid and 19% water) 557

SMARAGDMALACHIT (*m*), **RHOMBOEDRISCHER-**: dioptase 558

SMARAGD (*m*), **PRISMATISCHER-**: euclase; Sg. 3.05 559

SMARAGD (*m*), **RHOMBOEDRISCHER-**: phenakite (glucine 45%, silica 55%); $G_2O.SiO_2$; Sg. 2.98 560

SMARAGDSPAT (*m*): amazonite; amazon stone; green feldspar (see MIKROKLIN) 561

SMARAGD (*m*), **SYNTHETISCHER-**: synthetic emerald (a glass + beryllium oxide) 562

SMILAXART (*f*): *Smilaceæ* (Bot.) 563

SMIRGEL (*m*): emery (a mineral of finely granular Corundum and other minerals such as Magnetite; see KORUND and MAGNETIT) 564

SMITHSONIT (*m*): smithsonite (a zinc carbonate) (when in honeycombed masses is known as "Dry-bone ore"); $ZnCO_3$; Sg. 4.33 (see also ZINKSPAT) 565

S.M.-STAHL (*m*): Siemens-Martin steel 566

SO: so; thus; then; such; as 567

SOBALD: as soon as 568

SOCIETÄT (*f*): society; company 569

SOCKE (*f*): sock 570

SOCKEL (*m*): pedestal; base; support; stand; trestle; tray; saucer; foot; socle; plinth (Arch.) 571

SOCKEN: to crystallize out; contract (of metals) 572

SOD (*m*): sod; turf; lump of peat 573

SODA (*f*): soda; sodium carbonate (see NATRIUMCARBONAT); $Na_2CO_3.10H_2O$; Sg. 1.458-1.493; Bp. 106°C. 574

SODAASCHE (*f*): soda ash; Na_2CO_3; crude sodium carbonate 575

SODABRECHER (*m*): soda crusher 576

SODA (*f*), **CALCINIERTE-**: anhydrous sodium carbonate, soda ash, calcined soda (see SODA, KALZINIERTE-) 577

SODAFABRIK (*f*): soda (alkali) factory or works 578

SODAFABRIKATION (*f*): soda manufacture 579

SODAFABRIKEINRICHTUNG (*f*): soda plant 580

SODAHALTIG: containing soda 581

SODA (*f*), **KALZINIERTE-**: soda ash, calcined soda, anhydrous sodium carbonate 582

SODA, KAUSTISCHE- (*f*): caustic soda; sodium hydroxide (see NATRIUMHYDRAT) 583

SODAKESSEL (*m* and *pl*): soda pan-(s) 584

SODA (*f*), **KOHLENSAURE-**: carbonate of soda 585

SODAKRISTALLE (*m.pl*): soda crystals 586

SODAKÜPE (*f*): soda vat 587

SODALAUGE (*f*): soda lye 588

SODALITH (*m*): sodalite; $Na_4(AlCl)Al_2(SiO_4)_3$; Sg. 2.3 589

SODALÖSUNG (*f*): soda solution 590

SODAMEHL (*n*): sodium carbonate monohydrate (powdered); $Na_2CO_3.H_2O$ 591

SODANN: then; in (such; that) case 592

SODAOFEN (*m*): soda (black-ash) furnace (Alkali manufacture) 593

SODARÜCKSTAND (*m*): soda residue 594

SODASALZ (*n*): soda (sodium) salt; sodium carbonate; soda ash (see SODAASCHE and NATRIUMCARBONAT) 595

SODASALZ, CALCINIERTES- (*n*): sodium carbonate; soda ash (see SODAASCHE) 596

SODASCHMELZE (*f*): black ash (45% sodium carbonate, 30% calcium sulphide and 10% calcium hydroxide) 597

SODASEE (*m*): soda lake 598

SODASEIFE (*f*): soda soap 599

SODASTEIN (*m*): caustic soda (see NATRIUMHYDRAT) 600

SODATROMMEL (*f*): soda drum 601

SODAWASSER (*n*): soda water 602

SODBRENNEN (*n*): heartburn; waterbrash; pyrosis (Med.) 603

SODIEREN: to wash (treat) with soda 604

SOEBEN: just; just now 605

SOFERN: as far as; inasmuch as; in case 606

SOFORT: immediately; instantly; at once; forthwith 607

SOFORTIG: instantaneous; immediate 608

SOFTENINGÖL (*n*): softening oil 609
SOGAR: even 610
SOGENANNT: so-called 611
SOGGEN: to precipitate in crystal form; crystallize out 612
SOGGEPFANNE (*f*): crystallizing pan 613
SOGGESALZ (*n*): common salt (see NATRIUM-CHLORID) 614
SOGLEICH: immediately; at once; instantly; forthwith 615
SOHLBAND (*n*): gangue; matrix (Mining) 616
SOHLE (*f*): sole; floor; bottom; sill; splint; bed (of streams) 617
SOHLE (*f*), **DOCK-**: floor (bottom) of a dock 618
SOHLEN: to sole 619
SOHLENFILZ (*m*): sole felt, felt for soles 620
SOHLENLEDER (*n*): sole leather 621
SOHLLEDER (*n*): sole leather 622
SOHLPLATTE (*f*): sole-plate 623
SOJA (*f*): soy bean (*Soja hispida*; see also *Dolichos sesquipedalis*); soy sauce 624
SOJABOHNE (*f*): soy bean (see under SOJA) 625
SOJABOHNENÖL (*n*): soy oil, soja bean oil, soya-bean oil, Chinese bean oil, bean oil (from *Soja hispida* or *Soja japonica*); Sg. 0.926; Mp. 22-31°C. 626
SOJAÖL (*n*): (see SOJABOHNENÖL) 627
SOL (*n*): sol (a colloidal solution or extremely fine suspension of a colloid in water or in an organic solvent), the disperse phase of a colloid 628
SOLANEEN (*f.pl*): *Solanaceæ* (Bot.) 629
SOLANIDIN (*n*): solanidine; ($C_{26}H_{41}NO_2$) 630
SOLANIN (*n*): solanine; ($C_{42}H_{75}NO_{15}$) (an alkaloid from *Solanum dulcamara*) 631
SOLARISATION (*f*): solarization (see SOLARISIER-UNG) (of photographic plates) 632
SOLARISIERUNG (*f*): solarization, reversal (in photography, due to the extreme action of light which causes the parts which should be dark in the negative to come up light and vice versa) 633
SOLARÖL (*n*): solar oil, photogene, Kerosene (see LEUCHTPETROLEUM) 634
SOLARSTEARIN (*n*): stiffened lard 635
SOLBEHÄLTER (*m*): brine (container; reservoir) cistern or tank 636
SOLBOHRLOCH (*n*): salt well 637
SOLCH: such; such a 638
SOLCHENFALLS: in such-(a)-case 639
SOLCHER: such; such a 640
SOLCHERGESTALT: in such a (way) manner 641
SOLCHERLEI: such-(kind) 642
SOLD (*m*): pay 643
SOLDAT (*m*): soldier 644
SOLDATESKA (*f*): soldiery 645
SOLE (*f*): salt (water) spring; brine 646
SOLEBAD (*n*): brine (tank) reservoir 647
SOLEKÜHLER (*m*): salt-water (brine) cooler 648
SOLENOID (*n*): solenoid 649
SOLESCHICHT (*f*): stratum of brine 650
SOLID: steady (of character) 651
SOLIDBLAU (*n*): solid blue 652
SOLIDGRÜN (*n*): solid green (nitroso dyestuff; dinitroresorcinol) 653
SOLIDITÄT (*f*): solidity, steadiness 654
SOLIFERSEIFE (*f*): solifer soap 655
SOLL (*n*): debit 656
SOLLDURCHMESSER (*m*): theoretical (nominal) diameter 657
SOLLEN: to be; be (supposed; destined) intended for; be (obliged) permitted to 658
SÖLLER (*m*): balcony; terrace; platform; loft 659
SOLLWERT (*m*): theoretical (nominal) value; anticipated value; debit value 660
SOLORINSÄURE (*f*): solorinic acid 661
SOLPFANNE (*f*): brine pan 662
SOLQUELLE (*f*): brine (salt) spring or well 663
SOLSALZ (*n*): well (spring) salt 664
SOLUROL (*n*): solurol (Thymic acid) (see THYMIN-SÄURE) 665
SOLUTIN (*n*): solutin (a water-proof tar and asphalt covering mass) 666
SOLUTOL (*n*): solutol (a disinfectant) 667
SOLVATATION (*f*): conversion into a sol (Colloidal chemistry) 668
SOLVATISIEREN: to convert into a sol (Colloidal chemistry) (see SOL) 669
SOLVENT: solvent 670
SOLVENTNAPHTA (*f*): solvent naphtha; Sg. 0.875; Bp. 160°C. 671
SOLVENZ (*f*): solvency 672
SOLVENZIE (*f* and *n*): solvent 673
SOLVEOL (*n*): solveol (water-soluble cresol): Sg. 1.155 674
SOLVIEREN: to dissolve 675
SOLVIEREND: solvent; dissolving 676
SOLWAGE (*f*): salinometer; brine gauge 677
SOMATOSE (*f*): somatose 678
SOMIT: so; therefore; consequently; thus 679
SOMMER (*m*): summer 680
SOMMERBIRNE (*f*): early pear 681
SOMMEREICHE (*f*): oak tree (*Quercus pedunculata*); Sg. 0.7-1.05 682
SOMMERFÄDEN (*m.pl*): gossamer 683
SOMMERFLECK (*m*): freckle 684
SOMMERGERSTE (*f*): spring barley 685
SOMMERPORREE (*m*): leek (*Allium ampeloprasum*) 686
SOMMERPUNKT (*m*): summer solstice 687
SOMMERSAAT (*f*): spring corn 688
SOMMERSPROSSE (*f*): freckle; (*pl*): freckles, lentigo, ephelides (Med.) (see also SONNEN-FLECKEN) 689
SOMMERWEIZEN (*m*): spring wheat (*Triticum æstivum*); summer wheat 690
SOMMERZEIT (*f*): summer time 691
SOMMERZEUG (*n*): summer (light) dress-material 692
SOMMERZWIEBEL (*f*): onion (*Allium cepa*) 693
SOMNOFORM (*n*): somnoform (a mixture of ethyl chloride, methyl chloride and ethyl bromide in proportion 60:35:5) (an anæsthetic) 694
SONACH: so; accordingly; therefore; thus; consequently 695
SONATE (*f*): sonata (Music) 696
SONDE (*f*): sound; probe; plummet 697
SONDER: without 698
SONDERABDRUCK (*m*): reprint; special (separate) print or impression 699
SONDERAUSGABE (*f*): special (separate) edition 700
SONDERBAR: singular; strange; peculiar; odd; curious; marvellous; remarkable 701
SONDERFALL (*m*): exception-(al case) 702
SONDERGLEICHEN: unequalled 703
SONDERLICH: special; particular; material-(ly) 704
SONDERLING (*m*): odd (whimsical; singular; queer) character or person 705
SONDERN: to separate; part; segregate; sort; sever; sunder; disjoin; (as a conjunction but; on the contrary) 706

SONDERS, SAMT UND- : one and all; each and every one 707
SONDERSTELLUNG (*f*): special (separate; unique; exceptional; exclusive) position 708
SONDERUNG (*f*): separation; sundering; segregation; disjoining; sorting; severing; parting 709
SONDIEREN : to sound; probe 710
SONNE (*f*): sun 711
SONNEN : to sun; expose to (bask in) the sun 712
SONNENAUGE (*n*): adularia 713
SONNENBAHN (*f*): ecliptic; cycle of the sun 714
SONNENBLUME (*f*): sunflower (*Helianthus annuus*) 715
SONNENBLUMENÖL (*n*): sunflower oil; Sg. 0.93 (from *Helianthus annuus*) 716
SONNENBRONZE (*f*): a cobalt-copper-aluminium alloy. (40-60% Co; 30-40% Cu; 10% Al.) 717
SONNENFINSTERNIS (*f*): eclipse of the sun; solar eclipse 718
SONNENFLECK (*m*): sun-spot; freckle 719
SONNENFLECKEN (*m.pl*): Ephelis, sunburn (a pigmentary discolouration of the skin) (Med.) 720
SONNENGEFLECHT (*n*): solar plexus 721
SONNENHAFT : sunny; radiant 722
SONNENHELL : bright as (with) the sun 723
SONNENHITZE (*f*): solar heat 724
SONNENHÖHE (*f*): altitude (of the sun) 725
SONNENJAHR (*n*): solar year 726
SONNENKRAUT (*n*): sundew (*Drosera otundifolia*) 727
SONNENLICHT (*n*): sunlight 728
SONNENNÄHE (*f*): perihelion 729
SONNENRÖSCHEN (*n*): rock-rose (*Helianthemum*) 730
SONNENROSENÖL (*n*): sunflower oil (see SONNEN-BLUMENÖL) 731
SONNENSCHEIBE (*f*): the sun('s disc) 732
SONNENSCHIRM (*m*): sunshade; parasol 733
SONNENSEGEL (*n*): awning (Ships) 734
SONNENSEGELSTÜTZE (*f*): awning stanchion (Ships) 735
SONNENSPEKTRUM (*n*): solar spectrum 736
SONNENSTÄUBCHEN (*n*): mote; atom 737
SONNENSTEIN (*m*): sunstone; aventurine feldspar 738
SONNENSTICH (*m*): sunstroke, heatstroke, insolation (Med.) 739
SONNENSTILLSTAND (*m*): solstice 740
SONNENSTRAHL (*m*): sunbeam; ray of sunlight; solar ray 741
SONNENSTRAHLUNG (*f*): solar radiation 742
SONNENSYSTEM (*n*): solar system 743
SONNENTAU (*m*): sundew (*Drosera rotundifolia*) (Botany) 744
SONNENUHR (*f*): sundial 745
SONNENWÄRME (*f*): solar heat, heat of (from) the sun 746
SONNENWENDE (*f*): solstice; heliotrope 747
SONNIG : sunny 748
SONNTAG (*m*): Sunday 749
SONNTÄGIG : dominical 750
SONST : else; besides; otherwise; usually; in other respects; formerly; moreover 751
SONSTIG : former; other; remaining; sundry 752
SONSTWO : elsewhere 753
SOOR (*f*): thrush (see SCHWÄMMCHEN) (Med.) 754
SOPHOL (*n*): sophol (silver formonucleinate) 755
SORBENS (*n*): (see ADSORBENS) 756
SORBET (*n*): sherbet 757
SORBETT (*n*): sherbet 758
SORBIN (*n*): sorbin (a sugar obtained from the berries of the mountain ash, "*Sorbus aucuparia*") 759
SORBINROT (*n*): sorbine red 760
SORBINSÄURE (*f*): sorbic acid 761
SORGE (*f*): care; sorrow; concern; apprehension; fear; worry; anxiety 762
SORGEN : to care; sorrow; grieve; concern oneself; be anxious; fear; worry; provide for; apprehend 763
SORGEN (DA) FÜR : to ensure (that) 764
SORGFALT (*f*): care; carefulness; solicitude; diligence 765
SORGFÄLTIG : careful -(ly); solicitous; diligent 766
SORGLICH : solicitous; anxious; careful; apprehensive 767
SORGLOS : careless; indolent; reckless; unconcerned; negligent 768
SORGLOSIGKEIT (*f*): indolence; unconcern; negligence; carelessness 769
SORPTION (*f*): (see ADSORPTION) 770
SORREN : to seize, lash 771
SORTE (*f*): sort; kind; brand; quality; type; species 772
SORTIERAPPARAT (*m*): sorting apparatus (Paper) 773
SORTIEREN : to sort; size; assort; screen; sift; pick; arrange 774
SORTIERMASCHINE (*f*): sorting machine 775
SORTIERUNG (*f*): assorting (Paper) 776
SORTIMENT (*n*): assortment, set 777
SO-UND-SOVIEL : so-and-so-much; such-and-such an amount; a given (amount) quantity 778
SOUPLESEIDE (*f*): souple silk 779
SOUPLIEREN : to souple (Silk) 780
SOUVERÄNES MITTEL : sovereign remedy 781
SOVIEL : as much as; so much; as far as 782
SOWIE : as also; as well as; besides; as; together with 783
SOWJET : soviet (Russia) 784
SOWOHL : as well as 785
SOYA (*f*): soy (see SOJA) 786
SOYABOHNENÖL (*n*): soya-bean oil (see SOJA) 787
SOZIOLOGIE (*f*): sociology 788
SOZIUS (*m*): partner 789
SOZOJODOL (*n*): (see SOZOJODOLSÄURE) 790
SOZOJODOLSALZ (*n*): sozoiodol salt 791
SOZOJODOLSÄURE (*f*): sozoiodic acid (*Sozojodol acidium*), diiodo-para-phenolsulphonic acid; $C_6H_2I_2(OH)SO_3H$ 792
SOZOLSÄURE (*f*): sozolic acid; O-phenol-sulphonic acid (see ASEPTOL) 793
SPACHTEL (*f*): spatula; smoother 794
SPÄHEN : to spy; pry; explore; search; glance intently 795
SPAKE (*f*): hand spike, bar 796
SPAKE (*f*), BRATSPILL- : windlass bar 797
SPALIER (*n*): trellis; fence 798
SPALIEROBST (*n*): wall fruit 799
SPALT (*m*): cleavage; clearance; rent; split; crack; slit; cleft; fissure; rift; fracture; space; opening; chink; chip (of wood) 800
SPALTALGEN (*f.pl*): *Diatomaceæ*; diatoms 801

SPALTBAR: cleavable; separable; divisible; capable of being (cloven) split; fissible 802
SPALTBARKEIT (*f*): cleavage; cleavability; divisibility; capacity for being split 803
SPALTBREITE (*f*): width of (slit) crack; clearance 804
SPALTE (*f*): clearance; split; crack, etc. (see SPALT); division; column (of print) 805
SPALTEBENE (*f*): cleavage plane 806
SPALTEN: to split; slit; crack; rend; cleave; break up; decompose; dissociate; separate; divide; fissure; fracture; rupture; disjoin 807
SPALTENWEISE: in columns (of print) 808
SPALTFLÄCHE (*f*): cleavage (area; surface) face or plane 809
SPALTFRUCHT (*f*): cremocarp, schizocarp 810
SPALTFÜSSER (*m*): fissipeds, schizopods, Copepoda 811
SPALTHOLZ (*n*): chopped wood; firewood 812
SPALTHUFER (*m*): ruminant, animal with cloven hoofs 813
SPALTIG: split; cleavable; cracked, etc.; columned (Printing) 814
SPALTIG, GUT-: easily cleavable, easily split 815
SPALTIG, LEICHT-: easily cleavable, easily split 816
SPALTIG, SCHWER-: difficultly cleavable, difficult to split 817
SPALTKÖRPER (*m*): cleavage (substance) product 818
SPALTMÜNDUNG (*f*): stoma (Botany) 819
SPALTÖFFNUNG (*f*): stoma (Botany); crack (see also SPALTE) 820
SPALTPILZE (*m.pl*): fission fungi; *Schizomycetes* 821
SPALTPRODUKT (*n*): cleavage (fission) product 822
SPALTQUERSCHNITT (*m*): cross sectional opening 823
SPALTRISS (*m*): cleavage (clearance) crack 824
SPALTROHR (*n*): split tube; collimator (of spectroscope) 825
SPALTSÄGE (*f*): cleavage saw 826
SPALTSPIEL (*n*): (amount of) clearance; play (Turbines, etc.) 827
SPALTSTÜCK (*n*): fragment; detached portion 828
SPALTUNG (*f*): (see SPALTEN); scission; disunion; division; disension; fission; dissociation; split; rupture; schism 829
SPALTUNGSFLÄCHE (*f*): cleavage (face) plane 830
SPALTUNGSKRYSTALL (*m*): cleavage crystal 831
SPALTUNGSPRODUKT (*n*): cleavage (fission) product; dissociation product 832
SPALTUNGSPROZESS (*m*): cleavage process 833
SPALTUNGSRICHTUNG (*f*): cleavage direction 834
SPALTUNGSVORGANG (*m*): cleavage process 835
SPALTVERLUST (*m*): clearance loss -(es) (Turbines, etc.) 836
SPALTWEITE (*f*): clearance, gauge, pitch; air-space (of grates) 837
SPALZE (*f*): chaff; husk; beard (of ears of cereals); awn 838
SPAN (*m*): shaving; chip; boring; splinter; shred; veneer (Wood) 839
SPÄNEFILTER (*m*): filter (with shavings as the filter bed) 840
SPÄNEN: to suckle; wean 841
SPANFASS (*n*): chip cask (Brewing) 842
SPÄNFASS (*n*): (see SPANFASS) 843
SPANGE (*f*): buckle; clasp; stay-bolt; bracelet; bangle 844
SPANGELEISEN (*n*): crystalline pig iron 845
SPANGELIG: spangled; glistening 846
SPANGLIG: spangled; glistening 847
SPANGRÜN (*n*): verdigris (see KUPFERACETAT, BASISCHES-) 848
SPANISCH: spanish; double-dutch 849
SPANISCHE FLIEGEN (*f.pl*): Spanish flies; cantharides 850
SPANISCHE KREIDE (*f*): (see SIEFENSTEIN) 851
SPANISCHER KLEE (*m*): red clover (*Trifolium*) 852
SPANISCHER PFEFFER (*m*): capsicum; red pepper; cayenne pepper (*Capsicum annuum*) 853
SPANISCHE SODA (*f*): barilla (impure sodium carbonate) 854
SPANISCHES ROHR (*n*): Spanish (Bengal; Italian) cane (*Arundo donax*) 855
SPANISCHWEISS (*n*): Spanish white, bismuth white (see WISMUTSUBNITRAT) 856
SPANN (*m*): instep 857
SPANNADER (*f*): sinew 858
SPANNBAR: ductile; tensile 859
SPANNBARKEIT (*f*): extensibility; ductility 860
SPANNE (*f*): span; stretch 861
SPANNEN: to span; stretch; strain; tighten; pull-up; increase tension or pressure of; put under pressure; superheat; cock (a gun); pinch; be exciting; stress; tie; fetter; shackle; be intent; suspend; brace; be interesting 862
SPANNER (*m*): stretcher, tightener 863
SPANNHAMMER (*m*): stretching hammer 864
SPANNHÜLSE (*f*): collar (Mechanical) 865
SPANNKNORPEL (*m*): thyroid cartilage 866
SPANNKRAFT (*f*): tension; strain; stress; elasticity; expansibility; extensibility; potential energy 867
SPANNKRÄFTIG: elastic 868
SPANNMASCHINE (*f*): (see RICHTMASCHINE) 869
SPANNPLATTE (*f*): face plate, fixing plate, setting-up plate (Lathes) 870
SPANNRAHMEN (*m*): frame 871
SPANNRING (*m*): expansion ring, stretching ring, tightening (spring) ring 872
SPANNROLLE (*f*): tension pulley, idler 873
SPANNSÄGE (*f*): frame saw 874
SPANNSCHRAUBE (*f*): tightening (stretching; tension; straining; fixing) screw 875
SPANNSEIL (*n*): shackle; guy-rope 876
SPANNSTRICK (*m*): tether 877
SPANNUNG (*f*): tension, etc. (see SPANNEN); potential; suspense; attention; difference; strained relations; discord; stress; voltage (Elect.) 878
SPANNUNG (*f*), **ELEKTRISCHE-**: voltage, tension 879
SPANNUNGSABFALL (*m*): potential drop 880
SPANNUNGSDIFFERENZ (*f*): potential difference 881
SPANNUNGSEINHEIT (*f*): unit of tension 882
SPANNUNGSFREI: unstressed 883
SPANNUNGSGRAD: degree of (stress) tension 884
SPANNUNGSMESSER (*m*): voltmeter (Elect.) 885
SPANNUNGSREIHE (*f*): electromotive series; contact series 886
SPANNUNGSTHEORIE (*f*): strain (tension) theory (Organic Chemistry) 887
SPANNUNGSUNTERSCHIED (*m*): difference in (stress) tension; potential difference (Elect.) 888

SPANNUNGSVERLUST (*m*): tension loss; voltage loss (Elect.) 889
SPANNUNGSZEIGER (*m*): tension (pressure) indicator; voltmeter (Elect.) 890
SPANT (*n*): frame 891
SPANTABSTAND (*m*): pitching (spacing) of the frames, distance between the frames 892
SPANT (*n*), **BALANCIER-**: balance frame 893
SPANTBOLZEN (*m*): frame bolt 894
SPANTEISEN (*n*): frame angle iron 895
SPANTENRISS (*m*): body-plan 896
SPANT (*n*), **HÖLZERNES-**: timbers 897
SPANTWINKELEISEN (*n*): frame angle iron 898
SPARASSOL (*n*): a crystallized product of metabolism of the fungus *Sparassis ramosa*; $C_{10}H_{12}O_4$; Mp. 67.5°C. 899
SPARBRENNER (*m*): economical burner 900
SPARDECK (*n*): spar deck 901
SPARDECKSCHIFF (*n*): spar decked ship 902
SPAREN: to save; reserve; spare; to be economical with 903
SPARGEL (*m*): (edible)-asparagus (*Asparagus officinalis*) 904
SPARGELBEET (*n*): asparagus bed 905
SPARGELD (*n*): savings (of money) 906
SPARGELSAMENÖL (*n*): asparagus seed oil (from seeds of *Asparagus officinalis*); Sg. 0.928 907
SPARGELSTEIN (*m*): asparagus stone; sort of apatite (see **APATIT**) 908
SPARGELSTOFF (*m*): asparagine (see **ASPARAGIN**) 909
SPARKALK (*m*): flooring plaster (slower setting than Plaster of Paris), estrich gypsum, Keene's cement 910
SPARKAPSEL (*f*): economy sagger (Ceram.) 911
SPARKASSE (*f*): savings bank 912
SPÄRLICH: sparse; scanty; scarce; sparing; economical; parsimonious; frugal 913
SPARÖLER (*m*): economical lubricator 914
SPARREN (*m*); spar; rafter 915
SPARRENHOLZ (*n*): rafter timber 916
SPARRHOLZ (*n*): ends (of building timber) 917
SPARRWERK (*n*): rafters 918
SPARSAM: economical; sparing; thrifty; close; saving; parsimonious; frugal 919
SPARSAMKEIT (*f*): parsimony; thrift; economy 920
SPARTEIN (*n*): sparteine ($C_{15}H_{26}N_2$); (Bp. 181°C); (an alkaloid from *Spartium scoparium*) 921
SPARTGRAS (*n*): esparto grass (*Stipa tenacissima*) 922
SPASS (*m*): jest; joke; sport; fun 923
SPASSEN: to jest; joke; sport; make (poke) fun; play 924
SPASSHAFT: droll; jocular; jocose; sportive; joking; jesting 925
SPASTISCH: spastic (Med.) 926
SPAT (*m*): spar; spavin 927
SPÄT: late 928
SPATARTIG: spathic; sparry 929
SPATEISENSTEIN (*m*): siderite; spathic iron ore; black band ore (see **SIDERIT**); ($FeCO_3$; (about 25-40% iron)); (see **EISENKARBONAT** and **EISENOXYDULCARBONAT**) 930
SPATEL (*m and pl.*): spatula (Med.) 931
SPATELFÖRMIG: spatulate; spatula-shaped 932
SPATEN (*m*): spade 933
SPÄTERNTE (*f*): late harvest 934
SPATIG: spavined; sparry; spathic 935
SPÄTSOMMER (*m*): latter part of summer; beginning of autumn 936
SPATSTEIN (*m*): selenite (see **SELENIT**) 937
SPATZ (*m*): sparrow 938
SPÄTZÜNDUNG (*f*): retard-(ed) (late) ignition 939
SPAZIEREN: to stroll; walk; promenade 940
SPAZIERSTOCK (*m*): walking-stick 941
SPECHT (*m*): woodpecker 942
SPECIELL: special; specific; particular-(ly) 943
SPECIES (*f*): species; specie; herbs; drugs; simples (Pharm.) 944
SPECIFISCH: specific 945
SPECK (*m*): lard; Sg. 0.95; bacon; fat; blubber (of the whale); (as a prefix = amyloid; fatty) 946
SPECKÄHNLICH: like lard; fatty 947
SPECKARTIG: like lard; fatty 948
SPECKARTIGE DEGENERATION (*f*): waxy degeneration, amyloid (albuminoid) disease (Med.) 949
SPECKHAUT (*f*): buffy coat (Physiology); fat membrane 950
SPECKIG: lardy; fat; fatty; flinty (Barley); heavy (Bread) 951
SPECKKUCHEN (*m*): oil cake 952
SPECKÖL (*n*): lard oil; Sg. 0.915 953
SPECKSTEIN (*m*): steatite; talc; potstone; soapstone (*creta hispanica*) (a hydrated magnesian silicate; $3MgO.4SiO_2.H_2O$; Sg. 2.7); (see also **TALK**) 954
SPECKSTEINARTIG: steatitic 955
SPECKSTEINTALK (*m*): steatite: $3MgO,4SiO_2,H_2O$; Sg. 2.65 956
SPECKSTOFF (*m*): lardy (amyloid) substance 957
SPECKSUBSTANZ (*f*): (see **SPECKSTOFF**) 958
SPECULUM (*n*): speculum (an instrument for exploring deep-seated parts of the body), [such as aural (optical, nasal, throat, rectal, vaginal) speculum] (Med.) 959
SPEDIEREN: to dispatch; consign; send; forward; transmit 960
SPEDITEUR (*m*): shipping agent-(s); forwarding agent; carrier 961
SPEER (*m*): spear; lance 962
SPEERKIES (*m*): spear pyrites (type of marcasite); (see **MARKASIT**) 963
SPEICHE (*f*): spoke (of wheels); spike; radius (Anat.) 964
SPEICHEL (*m*): saliva; spittle (Med.) 965
SPEICHELBEFÖRDERND: sialagog; promoting flow of saliva 966
SPEICHELDRÜSE (*f*): salivary gland (Med.) 967
SPEICHELFISTEL (*f*): salivary fistula (of the duct of the parotid gland) (Med.) 968
SPEICHELFLUSS (*m*): salivation; profuse flow of saliva; Ptyalism (Med.) 969
SPEICHELFLÜSSIGKEIT (*f*): saliva 970
SPEICHELKASTEN (*m*): saliva chamber 971
SPEICHELMITTEL (*n*): sialagogue; (see **SPEICHEL-TREIBENDES MITTEL**) 972
SPEICHELN: to spit 973
SPEICHELSTEIN (*m*): salivary calculus (usually phosphate of lime concretions of Stenson's duct or duct of the parotid gland, the ducts of the submaxillary or sublingual glands) (Med.) 974
SPEICHELSTOFF (*m*): ptyalin (an enzyme) 975
SPEICHELTREIBENDES MITTEL (*n*): sialagogue (agent for increasing the secretion of saliva) (Med.) 976

SPEICHEN: to fit (equip; furnish) with spokes 977
SPEICHER (*m*): loft; warehouse; storehouse; granary; elevator; container; accumulator (Mech.); receiver; reservoir; storage tank 978
SPEICHERANLAGE (*f*): accumulator (plant) installation; (grain)-elevators or lofts 979
SPEICHER (*m*), **GETREIDE-**: grain elevator 980
SPEICHERN: to store; accumulate 981
SPEIEN: to vomit; spit; spew 982
SPEIGATT (*n*): scupper-(hole) 983
SPEIGATTKLAPPE (*f*): scupper hole leather, scupper leather 984
SPEIGATTROHR (*n*): scupper pipe 985
SPEILER (*m*): skewer 986
SPEILERN: to skewer 987
SPEISE (*f*): food; board; nourishment; dish; meal; bell (gun) metal; mortar; speiss (Metal) 988
SPEISEAPPARAT (*m*): feed -(ing) apparatus 989
SPEISEBESTANDTEIL (*m*): food constituent 990
SPEISEBREI (*m*): chyme 991
SPEISEESSIG (*m*): household vinegar 992
SPEISEFETT (*n*): nutrient fat; refined fat; fat suitable for human consumption 993
SPEISEGANG (*m*): alimentary canal 994
SPEISEGATTER (*n*): feeding door, feed opening 995
SPEISEGELB (*n*): pale yellow speiss 996
SPEISEHAHN (*m*): feed cock 997
SPEISEHAUS (*n*): restaurant; eating-house 998
SPEISEKAMMER (*f*): larder; pantry 999
SPEISEKANAL (*m*): alimentary canal 000
SPEISEKARTE (*f*): menu; bill of fare 001
SPEISEKOBALT (*m*): smaltite; (see **SMALTIT**) 002
SPEISELEITUNG (*f*): feed (pipe, piping) main; feeder (Elect.) 003
SPEISEMASSE (*f*): ration 004
SPEISEN: to eat; take food; dine; sup; feed; board; nourish 005
SPEISEÖL (*n*): salad (olive; sweet) oil; edible oil 006
SPEISEORDNUNG (*f*): diet 007
SPEISEPUMPE (*f*): feed pump 008
SPEISEROHR (*n*): feed (supply) pipe 009
SPEISERÖHRE (*f*): feed (supply) pipe; esophagus; Oesophagus (Anat.) 010
SPEISESAAL (*m*): dining (room) hall 011
SPEISESAFT (*m*): chyle 012
SPEISESALZ (*n*): table (common; cooking) salt; (see **SALZ**) 013
SPEISESCHRANK (*m*): larder; pantry; safe (for food) 014
SPEISETALG (*m*): refined tallow (suitable for human consumption) (see **FEINTALG**) 015
SPEISEVENTIL (*n*): feed valve 016
SPEISEVORRICHTUNG (*f*): feed -(ing) (arrangement) apparatus, feeding device (for boilers, etc.) 017
SPEISEWALZE (*f*): feed -(ing) roll 018
SPEISEWASSER (*n*): feed-(water) 019
SPEISEWASSERAUSTRITT (*m*): feed (water) outlet 020
SPEISEWASSERBEHÄLTER (*m*): feed tank (sometimes also, hot-well) 021
SPEISEWASSEREINTRITT (*m*): feed inlet 022
SPEISEWASSERENTGASUNG (*f*): feed water degassing, extraction of gases from feed water, feed water distillation 023
SPEISEWASSERENTLÜFTUNG (*f*): feed water deaeration 024
SPEISEWASSERFILTER (*m*): feed water filter 025
SPEISEWASSERMESSER (*m*): feed water meter 026
SPEISEWASSERREGLER (*m*): feed water regulator 027
SPEISEWASSERREGULATOR (*m*): feed water regulator 028
SPEISEWASSERREINIGER (*m*): feed water purifier 029
SPEISEWASSERREINIGUNG (*f*): feed water purification 030
SPEISEWASSERTANK (*m*): feed (water) tank 031
SPEISEWASSERTEMPERATUR (*f*): feed (water) temperature 032
SPEISEWASSERVENTIL (*n*): feed (water) check valve 033
SPEISEWASSERVORWÄRMER (*m*): pre-heater, feed water heater, economiser 034
SPEISEWASSER (*n*), **ZUSATZ-**: make-up feed-(water) 035
SPEISEZÄHNE (*m.pl*): feeding (cogs) teeth (Mech.) 036
SPEISEZETTEL (*m*): menu, bill of fare 037
SPEISEZIMMER (*n*): dining room; coffee room (of a hotel) 038
SPEISEZUCKER (*m*): table (refined) sugar 039
SPEISKOBALT (*m*): smaltite (natural cobalt arsenide) $(Co,Fe,Ni)As_2$; Sg. 6.75 (see also **SMALTIN**) 040
SPEISSUNG (*f*): splice (of ropes) 041
SPEISUNG (*f*): feed -(ing); eating; dining; boarding; supply; feed 042
SPEKTAKEL (*m*): noise; uproar; spectacle 043
SPEKTAKELN: to make a noise; cause an uproar; create a scene 044
SPEKTRAL: spectral 045
SPEKTRALANALYSE (*f*): spectrum analysis 046
SPEKTRALANALYTISCH: spectrometric; spectroscopic 047
SPEKTRALAPPARAT (*m*): spectroscopic apparatus (an apparatus for measuring refractive index) 048
SPEKTRALBEOBACHTUNG (*f*): spectroscopic observation 049
SPEKTRALFARBE (*f*): spectral (spectrum) colour 050
SPEKTRALLINIE (*f*): spectral (spectrum) line 051
SPEKTRALPROBE (*f*): spectrum test 052
SPEKTRALROHR (*n*): spectral (spectrum) tube 053
SPEKTRALRÖHRE (*f*): (see **SPEKTRALROHR**) 054
SPEKTRALTAFEL (*f*): spectral (spectrum) chart 055
SPEKTREN (*n.pl*): spectra (plural of **SPEKTRUM**) 056
SPEKTROMETER (*n*): spectrometer (an instrument for measuring the refractive index) 057
SPEKTROMETRISCH: spectrometric 058
SPEKTROSKOP (*n*): spectroscope (an instrument for examination of spectra) 059
SPEKTRUM (*n*): spectrum 060
SPEKULANT (*m*): speculator 061
SPEKULATION (*f*): speculation, venture, risk, bargain 062
SPEKULIEREN: to speculate, venture, risk, chance 063
SPELZ (*m*): spelt (*Triticum spelta*) 064
SPELZE (*f*): husk; chaff; awn; beard (of ears of cereals) 065
SPELZWEIZEN (*m*): spelt; awned wheat (*Triticum spelta*) 066
SPENDE (*f*): gift 067
SPENDEN: to spend; distribute 068

SPENDIEREN: to spend; donate; give 069
SPERBEERE (*f*): service berry (*Amelanchier canadensis*) 070
SPERBER (*m*): sparrow-hawk 071
SPERLING (*m*): sparrow 072
SPERMA (*n*): semen; sperm 073
SPERMAZET (*n*): spermaceti (*Cetaceum*); Sg. 0.945 074
SPERMAZETI (*n*): (see **SPERMAZET**) 075
SPERMAZETÖL (*n*): spermaceti oil 076
SPERMIN (*n*): spermine ($C_5H_{14}N_2$) 077
SPERMÖL (*n*): sperm oil (mainly dodecatyl oleate $C_{17}H_{33}COOC_{12}H_{25}$); Sg. 0.88 (from Cachalot or Sperm whale, *Physeter macrocephalus*) 078
SPERRBAUM (*m*): bar; barrier; turnpike 079
SPERRE (*f*): dam, barrier, blockade, blockage, obstruction, bar, impediment, barricade, drag, prohibition, embargo 080
SPERREN: to confine; bar; blockade; stop; close; shut-(up); prohibit; space out; resist; spread out; open wide; gape; place an embargo on; refuse; struggle; impede; seal 081
SPERRFLÜSSIGKEIT (*f*): sealing liquid 082
SPERRHAHN (*m*): stop-cock 083
SPERRHAKEN (*m*): catch 084
SPERRHOLZ (*n*): ply wood (see also **SPARRHOLZ**) 085
SPERRHOLZHERSTELLUNG (*f*): manufacture of ply wood 086
SPERRKIES (*m*): marcasite (see **MARKASIT**) 087
SPERRKLINKE (*f*): catch, pawl 088
SPERRKREUZ (*n*): turnstile 089
SPERRRAD (*n*): cog (ratchet) wheel; rack wheel 090
SPERRVENTIL (*n*): stop valve 091
SPERRWEIT: gaping; wide open 092
SPERRYLITH (*m*): Sperrylite (platinum arsenide); $PtAs_2$; Sg. 10.6 093
SPESEN (*f.pl*): charges; costs; expenses 094
SPESSARTIN (*m*): spessartite, spessartine $(Fe,Mn)_3(Al,Fe)_2(SiO_4)_3$; Sg. 4.0 095
SPEZEREI (*f*): grocery; spices 096
SPEZIALFALL (*m*): special case 097
SPEZIALHARTSTEINGUT (*n*): special hard stone ware; Sg. 2.6 098
SPEZIALISIEREN: to specialize 099
SPEZIALISIERUNG (*f*): specialization, specializing 100
SPEZIALIST (*m*): specialist 101
SPEZIALITÄT (*m*): speciality 102
SPEZIALWAGEN (*m*): special type of (wagon, carriage, truck) car, etc. (see under **WAGEN**) 103
SPEZIES (*f*): species; specie; herbs; drugs; simples (Pharm.) 104
SPEZIFIKUM (*n*): specific 105
SPEZIFISCH: specific 106
SPEZIFISCHES GEWICHT (*n*): specific (weight) gravity 107
SPHALERIT (*m*): sphalerite, zinc blende, black-jack (natural zinc sulphide); ZnS; Sg. 4.05 108
SPHÄRE (*f*): sphere 109
SPHÄRISCH: spherical -(ly) 110
SPHÄROKRISTALL (*m*): sphero- (spherical) crystal 111
SPHÄROSIDERIT (*m*): spherosiderite, clay (band) ironstone (a variety of spathic iron ore) (see **SIDERIT**, **TONEISENSTEIN** and **SPATEISENSTEIN**) 112
SPHEN (*m*): titanite, sphene (a silicate and titanate of Calcium); $CaSiTiO_5$; Sg. 3.5 (see also **TITANIT**) 113
SPICKEN: to smoke; lard 114
SPICKÖL (*n*): spike oil (see **SPIKÖL**) 115
SPIEGEL (*m*): (see also **SPEIGELFASER**); looking-glass; mirror; speculum (see **SPECULUM**); polished (silver; silky) surface (of liquids, fabrics, etc.); any flat surface; stern (of ship) 116
SPIEGELBEPLATTUNG (*f*): stern plating 117
SPIEGELBILD (*n*): reflected (likeness) image 118
SPIEGELBLANK: polished; bright 119
SPIEGELBRONZE (*f*): mirror bronze (68.21% Cu; and 31.79% Sn for mirrors of astronomical instruments) 120
SPIEGELEISEN (*n*): ferromanganese, specular (cast-) iron, iron manganate, hematite of metallic appearance, spiegel-eisen, mirror iron (pig iron with 10-40% manganese and 5% carbon content) 121
SPIEGELFABRIKATION (*f*): looking-glass (mirror; reflector) manufacture 122
SPIEGELFASER (*f*): medullary ray; silver (silky) grain 123
SPIEGELFENSTER (*n*): plate-glass window 124
SPIEGELFLÄCHE (*f*): surface (of liquids, fabrics, etc.); (any) flat surface 125
SPIEGELFLOSS (*n*): spiegeleisen (see **SPIEGEL-EISEN**); ferromanganese 126
SPIEGELFOLIE (*f*): tinfoil; silvering (on a mirror) 127
SPIEGELGALVANOMETER (*n*): mirror galvanometer 128
SPIEGELGIESSEREI (*f*): plate-glass factory 129
SPIEGELGLAS (*n*): plate glass; mirror glass 130
SPIEGELGLASPLATTE (*f*): bevel plate (photography; on which photographs are squeegeed) 131
SPIEGELHECK (*n*): square stern 132
SPIEGELIG: mirror-like; specular 133
SPIEGELMETALL (*n*): speculum metal; mirror metal (tin-copper alloy; an average of 66% Cu and 34% Sn) 134
SPIEGELN: to (be) reflect -(ed); shine: glitter 135
SPIEGELPLATTE (*f*): stern plate 136
SPIEGELSCHNITT: a radial section through the longitudinal axis of a tree trunk 137
SPIEGELSTEIN (*m*): specular stone 138
SPIEGELUNG (*f*): reflection; mirage 139
SPIEKE (*f*): (spike-) lavender; spikenard (*Lavandula spica*; *Inula squarrosa* and *Aralia racemosa*) 140
SPIEKER (*m*): nail, spike 141
SPIEKERHAUT (*f*): wood sheathing 142
SPIEKERN: to spike, nail 143
SPIEKÖL (*n*): oil of spike (see **LAVENDELÖL** and **SPIKÖL**) 144
SPIEL (*n*): game; sport; play; performance; pack of cards; working; action (Mech.); stroke (of engines); play; clearance; any mechanical movement 145
SPIELART (*f*): variety (Biology); manner of playing; type of clearance (Mech.) 146
SPIELEN: to play; trifle; sparkle (Gems); to have (clearance) play; to work (Mech.); act; perform; sport; disport 147
SPIELGLÜCK (*n*): chance; luck at play 148
SPIELKARTE (*f*): playing-card 149
SPIELKASTEN (*m*): musical-box 150
SPIELMANN (*m*): spielmann; musician 151

SPIELMARKE (*f*): counter 152
SPIELPAPIERE (*n.pl*): speculative stock 153
SPIELRAUM (*m*): margin; latitude; elbow-room clearance; play (Mech.); play-room; free (motion) movement; tolerance; (amount of)-error 154
SPIELSUCHT (*f*): gambling fever 155
SPIELWAREN (*f.pl*): toys; playthings 156
SPIELZEUG (*n*): toy -(s); plaything -(s) 157
SPIERE (*f*): boom, spar (Naut.) 158
SPIERE (*f*), **FOCK-**: fore yard boom 159
SPIERE (*f*), **LEESEGEL-**: studding sail boom 160
SPIERE (*f*), **VORBRAMLEESEGEL-**: fore topgallant studding sail boom 161
SPIESS (*m*): lance; spear; spit; pike 162
SPIESSEN: to pierce; spear; spit 163
SPIESSGLANZ (*m*): stibnite (see **STIBNIT**); antimony (see also **ANTIMON**) 164
SPIESSGLANZARTIG: antimonial 165
SPIESSGLANZASCHE (*f*): antimony ash 166
SPIESSGLANZBLEIERZ (*n*) bournonite (see **BOURNONIT**) 167
SPIESSGLANZBLENDE (*f*): kermesite; antimony blend (see **KERMESIT**) 168
SPIESSGLANZBLUMEN (*f.pl*): flowers of antimony 169
SPIESSGLANZBLÜTE (*f*): (see **SPIESSGLANZERZ**) 170
SPIESSGLANZBUTTER (*f*): antimony- (butter) trichloride; $SbCl_3$ (a caustic liquid); Sg. 1.35 171
SPIESSGLANZERZ (*n*): antimony ore (*Stibium oxydatum album*) (see **ANTIMONOXYD**) 172
SPIESSGLANZERZ (*n*), **GRAUES-**: stibnite (see **STIBNIT**) 173
SPIESSGLANZERZ (*n*), **SCHWARZES-**: bournite (see **BOURNONIT**) 174
SPIESSGLANZERZ, (*n*), **WEISSES-**: valentinite (see **VALENTINIT**) 175
SPIESSGLANZHALTIG: antimonial; containing antimony 176
SPIESSGLANZKERMES (*m*): kermesite (see **KER-MESIT**) 177
SPIESSGLANZKÖNIG (*m*): antimony regulus; (Sb); (see **ANTIMON**) 178
SPIESSGLANZLEBER (*f*): liver of antimony (*hepar antimonii*) 179
SPIESSGLANZMITTEL (*n*): antimonial remedy 180
SPIESSGLANZMOHR (*m*): aethiops antimonialis (preparation of mercury and antimony sulphides) (Pharm.) 181
SPIESSGLANZOCKER (*m*): antimony (antimonial) ochre (see **ANTIMONOCKER**) 182
SPIESSGLANZOXYD (*n*): antimony trioxide; Sb_2O_3; Sg. 5.2-5.67 183
SPIESSGLANZSAFFRAN (*m*): crocus of antimony 184
SPIESSGLANZSÄURE (*f*): antimonic acid 185
SPIESSGLANZSCHWEFEL (*m*): antimony sulphide; Sb_2S_3; Sg. 4.56; Mp. 546°C 186
SPIESSGLANZSILBER (*n*): dyscrasite (see **DYS-KRASIT**) 187
SPIESSGLANZWEIN (*m*): antimonial wine 188
SPIESSGLANZWEINSTEIN (*m*): tartrated antimony (Pharm.) 189
SPIESSGLANZWEISS (*n*): antimony (trioxide) white (see **SPIESSGLANZOXYD**) 190
SPIESSGLANZZINNOBER (*m*): kermesite (see **KER-MESIT**) 191
SPIESSGLAS (*n*): antimony (see **ANTIMON**) 192
SPIESSIG: spear-like; speary; stibnite-like crystallization 193
SPIESSKOBALT (*m*): smaltite (see **SMALTIT**) 194
SPIKÖL (*n*): spike oil; lavender spike oil (an inferior **LAVENDELÖL**, which see) (from *Lavandula spica*); Sg. 0.9-0.92 195
SPILL (*n*): capstan windlass 196
SPILLE (*f*): pin; peg 197
SPILLMOTOR (*m*): capstan or windlass (driving) motor 198
SPINALAPOPLEXIE (*f*): spinal apoplexy (see **RÜCKENMARKSAPOPLEXIE**) 199
SPINALLÄHMUNG (*f*): spinal paralysis (Med.) 200
SPINALLÄHMUNG (*f*), **AKUTE-ERWACHSENER-**: acute spinal paralysis of adults, acute atrophic spinal paralysis, acute inflammation of grey anterior horns, (*Poliomyelitis anterior acuta*) (Med.) 201
SPINALLÄHMUNG (*f*), **SUBAKUTE-ERWACHSENER-**: chronic atrophic spinal paralysis, subacute and chronic inflammation of grey anterior horns (*Poliomyelitis anterior subacuta et chronica*) (Med.) 202
SPINALPARALYSIS (*f*), **SPATISCHE-**: spastic spinal paralysis (*Paralysis Spinalis Spastica*), idiopathic (primary) lateral sclerosis or sclerosis of the lateral columns (Med.) 203
SPINAT (*m*): spinach (*Spinacia oleracea*) 204
SPIND (*n*): locker 205
SPINDEL (*f*): spindle; axle; pivot; peg; pin; spire; mandrel; mandril 206
SPINDELBANK (*f*): bank of spindles (Textile) 207
SPINDLEBAUM (*m*): spindle-tree; maple; wahoo (*Euonymus europæus* and *Euonymus atropurpureus*) (see **HUNDSBAUM**) 208
SPINDELBAUMÖL (*n*): spindle-tree oil (from *Euonymus europæus*) 209
SPINDELFÖRMIG: fusiform; spindle-shaped 210
SPINDELHOLZ (*n*): spindle-tree wood 211
SPINDELÖL (*n*): spindle oil 212
SPINDELPRESSE (*f*): screw-press 213
SPIND (*n*), **WERKZEUG-**: locker 214
SPINELL (*m*): spinel; $MgAl_2O_4$; Sg. 3.5-3.62 215
SPINELL (*m*), **EDLER-**: spinel (natural magnesium aluminate); $MgO.\ Al_2O_3$; Sg. 3.62 216
SPINNAKER (*m*): spinnaker 217
SPINNE (*f*): spider 218
SPINNEN: to spin; twist; plot 219
SPINNENARTIG: spidery; spider-like 220
SPINNENGEWEBE (*n*): cobweb; spider's web 221
SPINNENWEBE (*f*): cobweb; spider's web 222
SPINNER (*m*): millowner, spinner 223
SPINNEREI (*f*): spinning; spinning (trade) works or mill 224
SPINNFASER (*f*): spinning (textile) fibre; combing fibre 225
SPINNMASCHINE (*f*): spinning machine (Textile) 226
SPINNÖL (*n*): spinning oil (Textile) 227
SPINNWEBE (*f*): cobweb; spider's web 228
SPINNWEBENARTIG: cobweb-like; arachnoid 229
SPINNWOLLE (*f*): spinning wool; raw wool for spinning 230
SPION (*m*): spy; scout 231
SPIONIEREN: to spy; scout; (*n*): scouting; spying 232
SPIRALBOHRER (*m*): twist drill 233
SPIRALE (*f*): spiral; coil; helice; helical line 234
SPIRALFEDER (*f*): spiral-spring 235

SPIRALFÖRMIG : helical ; spiral ; coiled 236
SPIRALIG : helical -(ly) ; spiral -(ly) 237
SPIRALLINIE (*f*) : spiral line 238
SPIRALPUMPE (*f*) : spiral pump 239
SPIRALROHR (*n*) : spiral (pipe) tube 240
SPIRILLE (*f*) : *Spirillum rugula* (a fungus) 241
SPIRITLÖSLICHE FARBE (*f*) : colour soluble in alcohol 242
SPIRITUS (*m*) : spirit -(s), (rectified, Sg. 0.82) ; (pl. **SPIRITUOSA** or **SPIRITUOSEN**) ; Ethyl alcohol (see **ÄTHYLALKOHOL** ; see also **SPRIT**) 243
SPIRITUSARTIG : alcoholic ; spirituous 244
SPIRITUSBLAU (*n*) : spirit blue 245
SPIRITUSBRENNEREI (*f*) : distillery 246
SPIRITUSFABRIK (*f*) : distillery ; spirit works or factory 247
SPIRITUSGERUCH (*m*) : spirit odour 248
SPIRITUSLACK (*m*) : spirit varnish 249
SPIRITUSLAMPE (*f*) : spirit lamp 250
SPIRITUSMOTOR (*m*) : spirit motor 251
SPIRITUSWAGE (*f*) : spirit level ; alcoholometer 252
SPIROMETER (*n*) : spirometer (an instrument for measuring the capacity of the chest, i.e., after the deepest inspiration, the total amount of air expelled by the deepest expiration) (Med.) 253
SPIROYLIG : spiroylous 254
SPIROYLIGE SÄURE (*f*) : spiroylous acid ; salicylaldehyde (see **SALICYLALDEHYD**) 255
SPIROYLSÄURE (*f*) : salicylic acid (see **SALICYL-SÄURE**) 256
SPIRSÄURE (*f*) : (see **SPIROYLSÄURE**) 257
SPITAL (*n*) : hospital ; infirmary 258
SPITZ : pointed ; taper ; circular ; acute ; (*m*) : paddle (Metal) ; spitz, pomeranian dog 259
SPITZBERG (*m*) : peak (of mountains) 260
SPITZBLATTERN (*f.pl*) : chicken-pox (Med.) 261
SPITZBOGEN (*m*) : splayed circle (Metal section) 262
SPITZBOHRER (*m*) : awl, bit, drill 263
SPITZBUBE (*m*) : rogue ; thief ; sharper ; pick-pocket 264
SPITZE (*f*) : lace ; point ; mouth-piece ; top ; spire ; summit ; apex ; pinnacle ; vertex ; tip ; head ; forefront ; cusp ; peak ; pen-nib ; peak-load (of machinery) ; taper 265
SPITZEN : to sharpen ; point ; sprout (of grain) ; prick up (the ears) 266
SPITZENBELASTUNG (*f*) : peak load 267
SPITZENDREHBANK (*f*) : centre lathe 268
SPITZENENTLADUNG (*f*) : point discharge (Elect.) 269
SPITZENGLAS (*n*) : reticulated glass 270
SPITZENHANDEL (*m*) : lace-trade 271
SPITZENKLÖPPEL (*m*) : bobbin (for lace-making) 272
SPITZENPAPIER (*n*) : lace paper 273
SPITZENWIRKUNG (*f*) : needle effect 274
SPITZFEILE (*f*) : pointed file, taper file 275
SPITZFINDIG : crafty ; subtle ; cunning 276
SPITZGLAS : sediment glass ; any tall, conical glass 277
SPITZHAUFEN (*m*) : couch (Brew.) 278
SPITZHUND (*m*) : spitz ; pomeranian dog 279
SPITZIG : pointed ; tapering ; biting ; sharp ; cunning ; acute ; not uniform (of colours) ; keen ; poignant 280
SPITZKOLBEN (*m*) : tapered flask 281
SPITZKUGEL (*f*) : conical bullet 282
SPITZMALZ (*n*) : chit malt 283
SPITZMORCHEL (*f*) : mushroom (*Morchella*) 284
SPITZNAME (*m*) : nickname 285
SPITZSÄULE (*f*) : obelisk 286
SPITZTRICHTER (*m*) : tail funnel 287
SPITZWINKLIG : acute-angled 288
SPITZZAHN (*m*) : eye-tooth 289
SPITZ ZULAUFEND : tapering 290
SPLEISSE (*f*) : splint ; splinter ; shard ; shiver ; splice 291
SPLEISSEN : to splice ; split ; cleave ; shiver 292
SPLINT (*m*) : peg ; pin ; sap ; split-pin ; cotter -(pin) ; splint ; sapwood ; laburnum 293
SPLINTBOLZEN (*m*) : key bolt 294
SPLINTER (*m*) : splinter ; scale ; splint ; shard ; shiver 295
SPLINTHOLZ (*n*) : sapwood (the outer light rings of tree trunks, which portions of the wood are still in the process of growing) 296
SPLISSEN : to splice 297
SPLISSHORN (*n*) : splicing fid (see **MARLPFRIEM**) 298
SPLISSUNG (*f*) : splice, splicing 299
SPLITS (*pl*) : scones, splits (Bricks) 300
SPLITTER (*m*) : splinter ; scale ; splint ; shard ; shiver 301
SPLITTERBSE (*f*) : split-pea 302
SPLITTERIG : splintery 303
SPLITTERKOHLE (*f*) : splint coal 304
SPLITTERN : to splinter ; shatter ; shiver ; scale ; split 305
SPLITTERRICHTEN : to be hypercritical ; censure ; carp ; split hairs 306
SPLITTERWAND (*f*) : splinter screen (on war vessels) 307
SPODIUM, SCHWARZES- (*n*) : bone-black (see **KNOCHENKOHLE**) 308
SPODIUM, WEISSES- (*n*) : bone-ash (see **KNOCHEN-ASCHE**) 309
SPODUMEN (*m*) : spodumene, hiddenite, kunzite (a lithium-aluminium metasilicate) $LiAl(SiO_3)_2$; Sg. 3.14 310
SPONTAN : spontaneous-(ly) 311
SPORADISCH : sporadic 312
SPORE (*f*) : spore 313
SPORENBILDEND : spore-forming 314
SPORENBILDUNG (*f*) : spore formation 315
SPORENHALTIG : containing spores 316
SPORN (*m*) : spur ; incentive ; stimulus 317
SPORNRÄDCHEN (*n*) : rowel (of a spur) 318
SPORNSTREICHS : instantaneously ; at top speed ; immediately 319
SPORT (*m*) : sport (takes the place of the suffix "-ing" in English in words referring to sport such as "boating," etc.) 320
SPORTEL (*f*) : perquisite ; fee 321
SPORTELN : to yield fees 322
SPOTT (*m*) : ridicule ; derision ; raillery ; irony ; satire 323
SPOTTBILD (*n*) : caricature 324
SPOTTBILLIG : dirt-cheap ; very (extremely) cheap 325
SPOTTDRESSEL (*f*) : mocking-bird 326
SPÖTTELN : to jeer ; gibe 327
SPOTTEN : to mock ; ridicule ; scoff ; deride 328
SPÖTTER (*m*) : mocker ; scoffer 329
SPOTTGEBOT (*n*) : trifling sum ; ridiculously low offer ; cut offer 330
SPOTTGEDICHT (*n*) : satire 331
SPOTTGELD (*n*) : trifling sum ; ridiculously low price ; cut price 332
SPÖTTISCH : scornful ; ironical ; mocking ; scoffing ; derisive ; ironical ; satirical-(ly) 333

SPOTTNAME (*m*): nickname 334
SPOTTPREIS (*m*): (very) low price; (under) cut price 335
SPOTTREDE (*f*): irony 336
SPOTTSCHRIFT (*f*): satire 337
SPOTTWEISE: mockingly; ironically 338
SPOTTWOHLFEIL: extremely (dirt) cheap 339
SPRACHE (*f*): speech; language; tongue; discussion; conference 340
SPRACHEIGENHEIT (*f*): idiom 341
SPRACHFEHLER (*m*): grammatical error (in speech) 342
SPRACHFERTIG: voluble; having a good command of language-(s) 343
SPRACHFERTIGKEIT (*f*): volubility; command of language-(s) 344
SPRACHFORSCHER (*m*): philologer, philologist, linguist 345
SPRACHFORSCHUNG (*f*): philology 346
SPRACHKENNER (*m*): linguist 347
SPRACHKUNDE (*f*): knowledge of language-(s); philology 348
SPRACHLEHRE (*f*): grammar 349
SPRACHLEHRER (*m*): language (master) teacher 350
SPRACHLICH: grammatical; lingual 351
SPRACHLOS: speechless 352
SPRACHLOSIGKEIT (*f*): aphasia, defect of speech (Med.) 353
SPRACHREINIGER (*m*): purist 354
SPRACHROHR (*n*): speaking tube 355
SPRACHSCHATZ (*m*): vocabulary; thesaurus 356
SPRACHSCHNITZER (*m*): blunder (error) in grammar 357
SPRACHSTÖRUNG (*f*): speech (disorder) defect (Med.) 358
SPRACHWIDRIG: ungrammatical 359
SPRACHWISSENSCHAFT (*f*): science of languages 360
SPRATZEN: to spurt; spit 361
SPRATZKUPFER (*n*): copper rain (from surface of the molten metal) 362
SPRECHART (*f*): idiom; manner of speaking; dialect 363
SPRECHEN: to speak; say; talk; discuss; converse; confer; pass-(sentence); (*n*): speaking, etc. 364
SPRECHEN, FREI-: to discharge; absolve; pronounce not guilty 365
SPRECHEN, SCHULDIG-: to declare (pronounce) guilty 366
SPRECHSAAL: forum; hall for speaking; name of a German technical paper in connection with the glass industry 367
SPREELATTE (*f*): sheer batten (Naut.) 368
SPREITEN: to spread-(out); extend 369
SPREIZEN: to prop up; spread (stretch) open; sprawl; boast; expand; force (apart) open; jack up; raise up by driving a support under; open against pressure 370
SPREIZFEDER (*f*): expanding spring 371
SPREIZGEWINDE (*f*): expanding screw-thread 372
SPREIZKEIL (*m*): expanding (key, cotter) wedge 373
SPREIZRING (*m*): expanding ring 374
SPREIZWIRKUNG (*f*): expanding action 375
SPRENGAPPARAT (*m*): sparger; sprinkling apparatus 376
SPRENGARBEIT (*f*): blasting 377
SPRENGBOCK (*m*): truss 378
SPRENGEISEN (*n*): cracking ring (Glass) 379
SPRENGEL (*m*): district; diocese 380
SPRENGEN: to blast; spring; explode; burst-(open); blow-up; scatter; sprinkle; water; drizzle; gallop; dash 381
SPRENGFLÜSSIGKEIT (*f*): explosive liquid 382
SPRENGGELATINE (*f*): explosive (blasting) gelatine 383
SPRENGGRABEN (*m*): mine 384
SPRENGGRANATE (*f*): high explosive (H.E.) shell or grenade 385
SPRENGKANNE (*f*): watering-can 386
SPRENGKAPSEL (*f*): detonating cap; detonator; blasting cap 387
SPRENGKOHLE (*f*): cracking coal (Glass) 388
SPRENGKRAFT (*f*): explosive (power) force 389
SPRENGKRÄFTIG: powerfully explosive 390
SPRENGLADUNG (*f*): bursting (explosive) charge 391
SPRENGLOCH (*n*): blast-hole 392
SPRENGLUFT (*f*): (see **OXYLIQUIT**) 393
SPRENGMITTEL (*n*): explosive 394
SPRENGÖL (*n*): nitroglycerine; blasting oil; nitroleum; trinitroglycerine; trinitrin; glonoin oil; glyceryl trinitrate; $CH_2NO_3.CHNO_3.CH_2NO_3$; Sg. 1.6; freezes at 13°C; explodes at 260°C 395
SPRENGÖL (*n*), **NOBELSCHES-**: Nobel's blasting oil (see **NITROGLYZERIN**) 396
SPRENGPATRONE (*f*): explosive (blasting) cartridge 397
SPRENGPULVER (*n*): blasting powder 398
SPRENGSALPETER (*m*): saltpeter blasting powder 399
SPRENGSCHNUR (*f*): fuse 400
SPRENGSCHUSS (*m*): shot 401
SPRENGSTOFF (*m*): explosive 402
SPRENGTECHNIK (*f*): explosive manufacture; technology of explosives 403
SPRENGUNG (*f*): exploding; blasting; explosion; bursting; sprinkling 404
SPRENGWEDEL (*m*): sprinkler 405
SPRENGWERK (*n*): truss; scaffolding, framing 406
SPRENGWIRKUNG (*f*): explosive (action) effect 407
SPRENGZÜNDER (*m*): fuse 408
SPRENKEL (*m*): spot; speckle 409
SPRENKELIG: spotted; speckled; mottled; variegated 410
SPRENKELMANIER (*f*): (see **SPRITZMANIER**) 411
SPRENKELN: to speckle; spot; mottle; sprinkle 412
SPRENKLER (*m*): sparger; sprinkler 413
SPREU (*f*): chaff 414
SPRICHWORT (*n*): saying; proverb; adage 415
SPRICHWÖRTLICH: proverbial-(ly) 416
SPRIESSE (*f*): stay, prop, support 417
SPRIESSEN: to sprout; germinate; shoot 418
SPRIETSEGEL (*n*): sprit sail 419
SPRINGBOCK (*m*): wild (mountain) goat 420
SPRINGBRUNNEN (*m*): fountain; waterspout; geyser; jet of water 421
SPRINGEN: to crack; break; burst; leap; spring; bound; jump; spout 422
SPRINGER (*m*): jumper, etc. (see **SPRINGEN**); knight (Chess) 423
SPRINGFEDER (*f*): spring 424
SPRINGFEDERWAGE (*f*): spring balance 425
SPRINGFLUT (*f*): spring-tide 426
SPRINGGURKE (*f*): squirting cucumber (*Memordica elaterium*) (see **ELATERIN**) 427

SPRINGGURKENEXTRAKT (*m*): elaterium (a sediment from juice of *Ecballium elaterium*) 428
SPRINGKOLBEN (*m*): Bologna flask 429
SPRINGKRAFT (*f*): (power of) recoil; springiness; elasticity 430
SPRINGKRÄFTIG: springy; elastic 431
SPRINKLER (*m*): sprinkler 432
SPRINKLERVERFAHREN (*n*): sprinkler process (of spraying waste water on filter beds) 433
SPRIT (*m*): spirit-(s) (see also SPIRITUS-) 434
SPRITDRUCKEN (*n*): spirit printing (Calico) 435
SPRITFARBE (*f*): spirit colour (Calico) 436
SPRITFARBSTOFF (*m*): spirit soluble (colour) dye 437
SPRITHALTIG: spirituous; containing spirit; fortified (of wine) 438
SPRITLACK (*m*): spirit varnish 439
SPRITLÖSLICH: spirit-soluble, soluble in spirit 440
SPRITZAPPARAT (*m*): aerograph, spraying apparatus 441
SPRITZBEWURF (*m*): rough plastering; rough cast 442
SPRITZBÜCHSE (*f*): squirt, syringe, sprayer 443
SPRITZE (*f*): syringe; sprayer; squirt; fire-engine 444
SPRITZEN: to squirt; spray; spatter; syringe; inject; spout; spirt; throw (with hose); aerograph 445
SPRITZENHAUS (*n*): fire-station; engine-house 446
SPRITZFARBE (*f*): colour for spraying 447
SPRITZFLASCHE (*f*): wash-(ing) bottle (for precipitates) 448
SPRITZGURKE (*f*): squirting cucumber (*Ecballium agreste*) 449
SPRITZGUSS (*m*): die casting 450
SPRITZKOPF (*m*): spraying nozzle 451
SPRITZKORK (*m*): sprinkler (cork) stopper 452
SPRITZKRANZ (*m*): sparger (Brew.) 453
SPRITZMANIER (*f*): spraying process (a stiff brush, filled with lithographic ink, is rubbed over a wire gauze at a little distance from the stone and sprays a fine coating of colour on to it) 454
SPRITZMITTEL (*n*): injection 455
SPRITZNUDELN (*f.pl*): vermicelli 456
SPRITZRING (*m*): throwing ring, thrower (Oil lubrication) 457
SPRITZRÖHRE (*f*): syringe 458
SPRITZTON (*m*): tone obtained by spraying process (see SPRITZMANIER) 459
SPROCK: decayed, brittle, perished, rotten (of wood) 460
SPRÖDE: brittle; short; hard; shy; prim; coy; reserved; prudish; fragile 461
SPRÖDGLANZERZ (*n*): brittle silver ore; stephanite (see STEPHANIT); polybasite 462
SPRÖDGLASERZ (*n*): (see SPRÖDGLANZERZ) 463
SPRÖDIGKEIT (*f*): brittleness; shortness; reserve; shyness; primness; prudery; coyness; prudishness 464
SPROSS (*m*): sprout; sprig; shoot; offshoot; offspring; scion; germ; step; degree 465
SPROSSE (*f*): (see SPROSS) 466
SPROSSE (*f*), LEITER-: ladder step or rung 467
SPROSSEN: to sprout; shoot; spring; germinate; descend (from) 468
SPROSSENBIER (*n*): spruce beer 469
SPROSSENEXKRAKT (*m*): spruce essence 470
SPROSSPILZ (*m*): budding fungus; yeast fungus (*Saccharomyces cerevisiæ*) 471
SPROTTE (*f*): sprat 472
SPROTTENÖL (*n*): sprat oil (from *Clupea sprattus*); Sg. 0.927 473
SPROTTENTRAN (*m*): (see SPROTTENÖL) 474
SPRUCEBIER (*n*): spruce beer (brewed in Canada from the shoots of *Picea nigra*) 475
SPRUCH (*m*): motto; saying; text; verdict; sentence; decree 476
SPRUDEL (*m*): fountain; hot spring; geyser; well 477
SPRUDELN: to bubble; spout; spring 478
SPRUDELSALZ (*n*): Karlsbad salt 479
SPRUDELSTEIN (*m*): deposit from hot springs 480
SPRÜHEN: to emit; vomit; spray; spit; scatter; sprinkle; drizzle; scintillate 481
SPRÜHKUPFER (*n*): copper rain (from surface of molten metal) 482
SPRUNG (*m*): jump; leap; spring; bound; crack; flaw; fault; fissure; rift; sheer (of a deck) 483
SPUCKE (*f*): spittle; saliva 484
SPUCKEN: to spit 485
SPUCKNAPF (*m*): spittoon 486
SPUK (*m*): spectre; ghost; spook; apparition 487
SPUKEN: to (be) haunt-(ed); make a noise 488
SPÜLBAD (*n*): rinsing (washing) bath 489
SPULBAUM (*m*): spindle-tree; maple (*Euonymous europæus*) 490
SPULE (*f*): spool; quill; coil (Elect.); bobbin 491
SPÜLEIMER (*m*): rinsing (washing) pail 492
SPÜLEN: to rinse; flush; wash; swill 493
SPULEN: to wind (on a spool); coil; reel; spool 494
SPULENANKER (*m*): slipring (rotor) armature 495
SPULE (*f*), SEKUNDÄR-: secondary coil 496
SPÜLFLÜSSIGKEIT (*f*): rinsing liquid 497
SPÜLGEFÄSS (*n*): rinsing vessel 498
SPÜLHALLE (*f*): rinsing room (Brewing) 499
SPÜLICHT (*n*): slops; dishwater; hog-wash; spent wash (Distilling) 500
SPÜLJAUCHE (*f*): wash-up water; washing water polluted water which has been used for washing purposes in chemical and industrial works) 501
SPÜLMASCHINE (*f*): rinsing machine 502
SPÜLNAPF (*m*): rinsing bowl; slop basin 503
SPÜLPRÄPARAT (*n*): rinsing preparation 504
SPÜLTOPF (*m*): rinsing (pot) jar 505
SPÜLWASSER (*n*): rinsing (flushing: dish-) water; swill; wash 506
SPUND (*m*): plug; bung-(hole); stopper; tongue (Woodwork); hole; socket 507
SPÜNDEN: to plug; bung; stopper; cask (Liquors); tongue and groove (Woodwork) 508
SPUNDHOBEL (*m*): match (plough) plane; rebating plane 509
SPUNDLOCH (*n*): bung-hole 510
SPÜNDUNG (*f*): rabbet, rebate 511
SPUNDVOLL: brimful 512
SPUR (*f*): track; trace; footstep; trail; vestige; scent; channel; gutter; spore (of plants); rut; spur; permanent way (Railways) 513
SPURARBEIT (*f*): concentration (Metal) 514
SPUREN: to concentrate (Metal) 515
SPÜREN: to track; trace; trail; feel; notice; perceive; follow the scent 516
SPURENWEISE: in traces 517

SPÜRHUND (*m*): blood-hound 518
SPURKRANZ (*m*): wheel flange 519
SPURLAGER (*n*): step bearing 520
SPURLOS: trackless; without a trace 521
SPURPFANNE (*f*): step 522
SPURSCHLACKE (*f*): concentration slag 523
SPURSTEIN (*m*): concentration (metal) matt (result of re-roasting; 70%-75% copper) 524
SPURWEITE (*f*): gauge (of a track, railways, etc.); gauge of permanent way 525
SPURZAPFEN (*m*): vertical journal, pivot 526
SPUTA (*n.pl*): (plural of SPUTUM, which see) 527
SPUTEN: to be quick; speed; (make) haste 528
SPUTUM (*n*): sputum (Med.) 529
S-ROHR (*n*): S-tube 530
STAAR (*m*): cataract (Med.); starling 531
STAARNADEL (*f*): couching needle (Med.) 532
STAAROPERATION (*f*): couching for a cataract 533
STAARSTECHEN (*n*): couching for the cataract (Med.) 534
STAAT (*m*): state; pomp; ceremony 535
STAATENBUND (*m*): confederacy; treaty 536
STAATLICH: national 537
STAATSAMT (*n*): public office 538
STAATSANGEHÖRIGKEIT (*f*): nationality 539
STAATSANGELEGENHEIT (*f*): affairs of state 540
STAATSANLEIHE (*f*): government (state) loan 541
STAATSBÜRGER (*m*): citizen 542
STAATSEINKÜNFTE (*f.pl*): revenue; finances 543
STAATSGESETZ (*n*): constitutional law 544
STAATSLEHRE (*f*): political science 545
STAATSPAPIER (*n*): state-paper; (*pl*): bonds; stocks; public (funds) securities; Government (bonds; stock) securities 546
STAATSRAT (*m*): state (council) councillor; privy council-(lor) 547
STAATSSCHATZ (*m*): exchequer 548
STAATSSCHULD (*f*): national debt 549
STAATSVERBRECHEN (*n*): political crime 550
STAATSVERBRECHER (*m*): political criminal 551
STAATSVERFASSUNG (*f*): government, constitution 552
STAATSVERWALTUNG (*f*): government, administration 553
STAATSWISSENSCHAFT (*f*): politics, political (science) economy 554
STAATSWOHL (*n*): common (good) weal, public benefit 555
STAATSZEITUNG (*f*): gazette 556
STAB (*m*): rod: bar; stick; staff-(officers) 557
STABBAKTERIE (*f*): bacillus 558
STÄBCHEN (*n*): small rod (bar, stick or staff); bacillus; bougie 559
STÄBCHENFÖRMIG: rod (bar) shaped 560
STABEISEN (*n*): bar iron 561
STABEISENHÄRTEOFEN (*m*): bar heating furnace 562
STABFÖRMIG: rod (bar) shaped 563
STABIL: stable; non-active; inactive; stationary (sometimes used as a prefix to express land plant as opposed to marine) 564
STABILITÄT (*f*): stability 565
STABILITÄTSGRENZE (*f*): range (limit) of stability 566
STABILITÄTSHEBELARM (*m*): stability lever 567
STABILITÄTSKURVE (*f*): stability curve 568
STABILITÄTSRECHNUNG (*f*): stability computation 569
STABKRANZ (*m*): corona radiata (Biology) 570
STABMAGNET (*m*): bar magnet (magnetized bar) 571
STABSOFFIZIER (*m*): staff-officer (Military) 572
STABTHERMOMETER (*n*): thermometer with scale on stem (not separate) as in a clinical thermometer, usually uncased thermometer 573
STACHEL (*m*): thorn; spike; spine; prong; prickle; sting (of insects) (the cause, not the effect) 574
STACHELBEERE (*f*): gooseberry (*Ribes grossularia*) 575
STACHELBEERSTRAUCH (*m*): gooseberry-bush 576
STACHELDRAHT (*m*): barbed wire 577
STACHELIG: thorny; prickly; pungent; biting 578
STACHELMOHN (*m*): prickly poppy (*Argemone grandiflora*) 579
STACHELN: to prick; sting; spur; prod; goad; stimulate 580
STACHELROCHE (*m*): thornback, ray (*Raia clavata*), (Icht) 581
STACHELSCHWEIN (*n*): porcupine; hedge-hog 582
STACHELSCHWEINHOLZ (*n*): porcupine wood (from Palmyra palm, *Borassus flabelliformis*) 583
STACHELWALZWERK (*n*): toothed rolls (Mech.) 584
STACKET (*n*): stockade; palisade; railing; fence; paling 585
STADEL (*m*): shed; open kiln; stall (for roasting ores) 586
STADELRÖSTUNG (*f*): open-kiln (stall) roasting -(process) 587
STADIUM (*n*): stage; phase; state 588
STADT (*f*): town; city 589
STADTABGEORDNETE (*m*): town-councillor; member on a borough council 590
STADTABGEORDNETER (*m*): town-councillor; a member on a borough council 591
STADTADEL (*m*): resident nobility 592
STADTAMT (*n*): municipal office; mayoralty (Germany) 593
STADTAMTMANN (*m*): mayor 594
STADTANWALT (*m*): recorder 595
STADTBAHN (*f*): municipal (light) railway or tramway 596
STADTBANN (*m*): district (belonging to a town); town bounds or boundaries 597
STADTBEKANNT: generally known: talk of the town 598
STADTBEWOHNER (*m*): citizen; inhabitant (of a town) 599
STADTBEZIRK (*m*): ward; district; quarter 600
STADTBRIEF (*m*): letter sent by local post 601
STADTBUCH (*n*): municipal register 602
STADTBÜRGER (*m*): citizen 603
STADTBÜRGERRECHT (*n*): civic rights 604
STADTFESTE (*f*): citadel 605
STADTFESTUNG (*f*): citadel 606
STADTGEBIET (*n*): township 607
STADTGEGEND (*f*): quarter 608
STADTGEMEINDE (*f*): municipality; corporation 609
STADTGERICHT (*n*): magistrate's (recorder's) court 610
STADTGESPRÄCH (*n*): talk of the town 611
STADTGUT (*n*): municipal property; commons 612
STADTHAUS (*n*): town hall 613

STÄDTISCH: town, municipal, city, borough, belonging to a town 614
STADTKUNDIG: notorious 615
STADTLAUGE (*f*): town sewage, waste water from towns 616
STADTLEBEN (*n*): town life 617
STADTMILIZ (*f*): civic (militia) guard 618
STADTPHYSIKUS (*m*): town physician 619
STADTPOST (*f*): local (town) post 620
STADTRAT (*m*): town (municipal) council-(lor) 621
STADTRECHT (*n*): civic rights; municipal bye-laws 622
STADTRICHTER (*m*): judge; recorder; sheriff 623
STADTSCHREIBER (*m*): town-clerk 624
STADTSCHREIBEREI (*f*): town clerk's office; rolls 625
STADTSCHULE (*f*): board school 626
STADTSCHULKOMMISSION (*f*): school-board 627
STADTSCHULRAT: inspector (of board schools); member of school-board 628
STADTSTEUER (*f*): rates 629
STADTTEIL (*m*): ward; quarter 630
STADTVERORDNETE (*m*): town-councillor 631
STADTVERORDNETENVERSAMMLUNG (*f*): council meeting; meeting of the town council 632
STADTVERORDNETENVORSTEHER (*m*): chairman of town council; mayor 633
STADTVERORDNETER (*m*): town councillor 634
STADTVIERTEL (*n*): quarter; ward 635
STADTVOGT (*m*): provost; bailiff 636
STADTVOGTEI (*f*): prison (of a town) 637
STADTWAGE (*f*): public scales 638
STADTWAGEN (*m*): hackney cab 639
STADTWAPPEN (*n*): city arms 640
STADTWESEN (*n*): civic (municipal) matters, concerns or affairs 641
STAFFEL (*f*): step; degree 642
STAFFELEI (*f*): easel 643
STAFFELROST (*m*): step (stage) grate 644
STAFFIEREN: to equip; trim; garnish; prepare; furnish; supply; dress 645
STAG (*m*): stay 646
STAGAUGE (*n*): stay collar 647
STAGNIEREN: to stagnate 648
STAGSEGEL (*n*): stay-sail 649
STAG, ÜBER . . . . GEHEN: to stay 650
STAHL (*m*): steel; Sg. 7.73-7.9; Mp. 1375°C 651
STAHLABFALL (*m*): steel (waste) scrap 652
STAHLARBEIT (*f*): steel (work) process 653
STAHLARBEITER (*m*): steel worker 654
STAHLARTIG: steel-like; steely; chalybeate 655
STAHLARZNEI (*f*): chalybeate; medicine containing iron 656
STAHLBARRE (*f*): steel bar 657
STAHLBEREITUNG (*f*): steel making 658
STAHLBLAU: steel-blue; steel-coloured 659
STAHLBLECH (*n*): steel (sheet) plate; sheet steel 660
STAHLBLECHMANTEL (*m*): steel (plate) casing 661
STAHLBRONZE (*f*): (extra hardened)-gunmetal (about 90% Cu and 10% Sn; sometimes also a brass with a manganese and iron content) 662
STAHLBRUNNEN (*m*): chalybeate (iron-water) spring 663
STAHLDRAHT (*m*): steel wire 664
STAHLDREHSPÄNE (*m.pl*): steel turnings or shavings 665
STAHL (*m*), **EDEL-**: the best sorts of steel such as crucible steel, electro steel and special steels, including nickel steel, chromium steel, manganese steel, tungsten steel, vanadium steel, molybdenum steel 666
STAHLEISEN (*n*): steely iron; Siemens Martin pig iron 667
STÄHLEN: to steel; harden; make into steel; temper 668
STÄHLERN: steel; of steel 669
STAHLERZ (*n*): steel ore; siderite (see **SIDERIT**) 670
STAHLFASS (*n*): steel barrel 671
STAHLFEDER (*f*): steel (pen) spring 672
STAHLFLASCHE (*f*): steel (cylinder) bottle 673
STAHLFORMGUSS (*m*): steel mould casting 674
STAHLFRISCHFEUER (*n*): steel (refining furnace or hearth) finery 675
STAHLGEFÄSS (*n*): steel vessel 676
STAHL (*m*), **GESCHMIEDETER-**: forged steel 677
STAHL (*m*), **GEWALZTER-**: rolled steel 678
STAHLGRAU: steel-grey 679
STAHLGUSS (*m*): steel casting; tough cast iron 680
STAHLGUSSARMATUREN (*f.pl*): cast-steel (fittings) mountings, C.S. fittings or mountings 681
STAHLGUSSKOKILLE (*f*): (lasts up to 400 castings) (see **KOKILLE**) 682
STAHLGUSSVENTIL (*n*): cast steel valve 683
STAHLHAHN (*m*): steel cock 684
STAHLHALTIG: chalybeate; containing iron 685
STAHLHÄRTUNG (*f*): steel hardening 686
STAHLHÜTTE (*f*): steel works 687
STAHLKOBALT (*m*): smaltite (see **SMALTIT**) 688
STAHLKOHLEN (*n*): steel formation from wrought iron by carbonisation 689
STAHLKONSTRUKTION (*f*): constructional steel-work; steel (building) construction 690
STAHLMITTEL (*n*): medicine containing iron; chalybeate 691
STAHLMOBEL (*n*): steel furniture 692
STAHLMOLETTE (*f*): (see **MOLETTE**) 693
STAHLMÖRSER (*m*): steel mortar 694
STAHL (*m*), **NICHTROSTENDER-**: stainless (non-rusting, non-corroding) steel 695
STAHL (*m*), **V2A NICHTROSTENDER-**: Krupp's V2A stainless chrome-nickel steel, of austenite variety, 20% chromium and about 7% nickel 696
STAHLPERLE (*f*): steel bead 697
STAHLPLATTE (*f*): steel plate or sheet 698
STAHLPRÄPARAT (*n*): iron preparation (Pharm.) 699
STAHLPUDDELN (*n*): steel puddling 700
STAHLQUELLE (*f*): chalybeate (iron-water) spring 701
STAHLROHR (*n*): steel tube 702
STAHL (*m*), **RUHIGER-**: quiet steel (in which the solidification takes place without movement and gas bubbles are not formed) 703
STAHL (*m*), **SCHNELLDREH-**: high speed (tool) steel 704
STAHLSEIL (*n*): steel wire rope 705
STAHLSPAN (*m*): steel (shavings; filings; chips) turnings 706
STAHLSTANGE (*f*): steel bar 707
STAHLSTECHEN (*n*): steel engraving; engraving on steel 708
STAHLSTECHEREI (*f*): (see **STAHLSTECHEN**) 709
STAHLSTEIN (*m*): siderite (see **SIDERIT**) 710
STAHLSTICH (*m*): steel engraving 711
STAHLTROMMEL (*f*): steel drum 712

STAHLWAREN (*f.pl*): hard-ware, iron or steel goods 713
STAHLWASSER (*n*): chalybeate (iron) water 714
STAHLWEIN (*m*): iron wine 715
STAHLWERK (*n*): steel works 716
STAHLWERKSANLAGE (*f*): steel-works plant 717
STAKET (*n*): stockade; railing; palisade; paling; fence 718
STALL (*m*): stable; stall; sty 719
STALLBUTTER (*f*): dairy butter 720
STALLEN: to stable; put up; instal 721
STALLUNG (*f*): stabling 722
STAMM (*m*): stem; trunk; stalk; stock; staff; tribe; origin; race 723
STAMMAKTIEN (*f.pl*): common stock shares; ordinary shares; original share 724
STAMMÄLTERN (*pl*): progenitors; ancestors 725
STAMMBAUM (*m*): pedigree; genealogical tree 726
STAMMELN: to stammer; stutter; falter 727
STAMMEN: to originate from; spring from; descend 728
STAMMERBE (*m*): lineal descendant; heir-at-law 729
STAMM-FABRIKATION (*f*): head office and works 730
STAMMFARBE (*f*): primary colour 731
STAMMFOLGE (*f*): line of descent 732
STAMMGUT (*n*): ancestral (family) estate 733
STÄMMIG: robust; strong; sturdy; massive; with a stem 734
STAMMKAPITAL (*n*): capital; stock-fund 735
STAMMKÖRPER (*m*): parent (substance; compound) body 736
STAMMLEHEN (*n*): fee-simple 737
STAMMLINIE (*f*): trunk (main) line; lineage 738
STAMMNAME (*f*): generic name 739
STAMMPRIORITÄTSAKTIE (*f*): preference share 740
STAMMREIS (*n*): sucker 741
STAMMROLLE (*f*): register 742
STAMMSITZ (*m*): ancestral (seat) home 743
STAMMSUBSTANZ (*f*): parent substance; substance of origin 744
STAMMSYLBE (*f*): root-syllable 745
STAMMTAFEL (*f*): genealogical table 746
STAMMVERMÄCHTNIS (*n*): entail 747
STAMMVERMÖGEN (*n*): capital 748
STAMMVERWANDT: kindred; akin; cognate 749
STAMMVERWANDTSCHAFT (*f*): kinship; affinity 750
STAMMWORT (*n*): root 751
STAMMWÜRZE (*f*): original wort 752
STAMMWURZEL (*f*): principal root 753
STAMPFBEWEGUNG (*f*): pitching motion (of ships) 754
STAMPFE (*f*): stamp-(er); pestle; ram-(mer); punch; beater; crushing; pounding; stamping 755
STAMPFEISEN (*n*): iron pestle 756
STAMPFEN: to stamp; punch; beat; ram; pound; crush; pulp; mash; bruise; express (Oil); pitch (Ship's motion) 757
STAMPFEND: pitching (motion of a ship) 758
STAMPFER (*m*): stamper; pestle; ram-(mer); punch; beater; crusher; pounder, etc. 759
STAMPFGANG (*m*): crushing mill 760
STAMPFHAUFEN (*m*): batch (Paper) 761
STAMPFKLOTZ (*m*): pile driver; ram-(mer) 762
STAMPFMASCHINE (*f*): pulping machine 763
STAMPFMÜHLE (*f*): crushing mill; stamping mill 764
STAMPFREITEN: to pitch while at anchor (Naut.) 765
STAMPFSTAG (*n*): martingale stay 766
STAMPFSTEVEN (*m*): straight stem (Naut.) 767
STAMPFSTOCK (*m*): martingale boom 768
STAMPFWERK (*n*): stamping (crushing; breaking) mill 769
STAMPFZUCKER (*m*): crushed sugar 770
STAND (*m*): stand; position; level; place; state; condition; height; station; rank; rate of exchange; class; standing; platform; housing (Rolling mills) 771
STANDARD-LÖSUNG (*f*): standard solution 772
STANDARTE (*f*): ensign, standard (Marine); brush (of fox) 773
STANDBAROMETER (*n*): stand (stationary) barometer 774
STANDBILD (*n*): statue 775
STANDBILDERBRONZE: art bronze for statues (see **KUNSTBRONZE**) 776
STÄNDCHEN (*n*): serenade 777
STANDER (*m*): pendant 778
STÄNDER (*m*): pedestal; bearer; support; stanchion; pillar; post; stand-(ard); jar; cistern; water-tub; tank; container; upright; housing (of mill); pole; stator (Elect.) 779
STANDESERÖFFNUNG (*f*): opening of parliament 780
STANDFÄHIG: stable 781
STANDFESTIGKEIT (*f*): stability 782
STANDFLASCHE (*f*): bottle (flask) for standing in a fixed place 783
STAND (*m*), **FÜHRER-**: driver's platform; foot-plate (Railways) 784
STANDGEFÄSS (*n*): stock tub (Brewing); vessel (jar) for standing in a fixed place; show (storage) jar 785
STANDHAFT: steady; firm; constant; steadfast; persevering; stable 786
STANDHAFTIGKEIT (*f*): steadiness, constancy, resoluteness, firmness, steadfastness, perseverance, stoicism 787
STANDHALTEN: to hold one's ground; be firm; stand up to, be suitable for, withstand (certain conditions) 788
STAND (*m*), **HEIZER-**: stoker's (stoking) platform or floor; firing floor 789
STÄNDIG: constant; fixed; firm; stationary; permanent; regular; continual 790
STAND, IN-SETZEN: to enable; put in a position to 791
STANDKUGEL (*f*): stationary bulb 792
STANDLEHRE (*f*): statics (Maths.) 793
STANDLINIE (*f*): base-(line) 794
STANDMÖRSER (*m*): mortar with a firm base on which to stand 795
STANDORT (*m*): stand; station; habitat (Botany) 796
STANDPLATZ (*m*): (see **STANDORT**) 797
STANDPUNKT (*m*): standpoint; point of view; standard; position 798
STANDRECHT (*n*): martial law; court martial 799
STANDREDE (*f*): harangue 800
STANDRISS (*m*): elevation (Art) 801
STANDROHR (*n*): standpipe; pipe or hose of fire engine 802
STANDWIND (*m*): steady (breeze) wind (Naut.) 803
STANGE (*f*): stick; rod; pole; bar; perch; roll (Sulphur); ingot (Metal); stake 804
STANGENBLEI (*n*): bar lead 805
STANGENDRAHT (*m*): bar wire 806

STANGENEISEN (*n*): bar iron 807
STANGENGOLD (*n*): bar (ingot) gold 808
STANGENKUPFER (*n*): bar copper 809
STANGENLACK (*m*): stick-lac 810
STANGENSCHWEFEL (*m*): stick (roll) sulphur 811
STANGENSEIFE (*f*): bar soap 812
STANGENSILBER (*n*): bar (ingot) silver 813
STANGENSTEIN (*m*): pycnite; columnar topaz (see PYKNIT) 814
STANGENTABAK (*m*): roll tobacco; twist 815
STANGENZINN (*n*): bar tin 816
STANGENZIRKEL (*m*): beam compasses 817
STANGE (*f*), **SCHIEBER-**: valve spindle 818
STÄNKERN: to stink; quarrel; search 819
STANNIBROMID (*n*): stannic bromide (see ZINN-BROMID) 820
STANNICHLORID (*n*): stannic chloride; tin tetrachloride (see ZINNCHLORID) 821
STANNICHLORWASSERSTOFFSÄURE (*f*): chlorostannic acid (see ZINNCHLORWASSERSTOFF-SÄURE) 822
STANNICHROMAT (*n*): stannic chromate; tin chromate; $Sn(CrO_4)_2$ 823
STANNIHYDROXYD (*n*): stannic hydroxide 824
STANNIJODID (*n*): stannic iodide (see ZINNJODID) 825
STANNIN (*m*): stannite, tin pyrites (a copper-tin-iron sulphide; Cu_2S,FeS,SnS_2 or Cu_2FeSnS_4; Sg. 4.4 826
STANNIOL (*n*): tinfoil (thin tin plates or leaf tin) 827
STANNIOXYD (*n*): stannic oxide; stannic anhydride; tinstone; tin ash; tin dioxide; tin peroxide; flowers of tin; SnO_2; Sg. 6.75; Mp. 1127°C 828
STANNIPHOSPHID (*n*): stannic phosphide; tin phosphide; Sn_2P_2; Sg. 6.6 829
STANNIREIHE (*f*): stannic series 830
STANNISULFID (*n*): stannic sulphide; mosaic gold; artificial gold; SnS_2; Sg. 4.5 831
STANNIVERBINDUNG (*f*): stannic compound 832
STANNOAZETAT (*n*): tin acetate; stannous acetate (*Stannum aceticum*) 833
STANNOBROMID (*n*): stannous bromide (see ZINNBROMÜR) 834
STANNOCHLORID (*n*): stannous chloride; tin salt; tin chloride; tin protochloride; tin dichloride; tin bichloride; $SnCl_2$, Mp. 249°C, Bp. 620°C; or $SnCl_2.2H_2O$, Sg. 2.71, Mp. 37.7°C 835
STANNOCHLORWASSERSTOFFSÄURE (*f*): chlorostannous acid 836
STANNOCHROMAT (*n*): stannous chromate; tin chromate; $SnCrO_4$ 837
STANNOHYDROXYD (*n*): stannous hydroxide 838
STANNOJODID (*n*): stannous iodide (see ZINN-JODÜR) 839
STANNOJODWASSERSTOFFSÄURE (*f*): iodostannous acid 840
STANNOOXALAT (*n*): stannous oxalate; SnC_2O_4 841
STANNOOXYD (*n*): stannous oxide; tin protoxide; tin monoxide; SnO; Sg. 6.3 842
STANNOSALZ (*n*): stannous salt 843
STANNOSULFAT (*n*): stannous sulphate; $SnSO_4$ 844
STANNOSULFID (*n*): stannous sulphide (see ZINNSULFÜR) 845
STANNOTARTRAT (*n*): stannous tartrate; $SnC_4H_4O_6$ 846
STANNOVERBINDUNG (*f*): stannous compound 847
STANZE (*f*): stamping; stamp: punch; embossing machine 848
STANZEN: to stamp; emboss; punch 849
STANZLOCH (*n*): punched hole 850
STANZMASCHINE (*f*): perforating machine, punching machine, stamping machine, slotting machine 851
STANZPRESSE (*f*): stamping press 852
STANZTEILE (*m.pl*): stampings; blanks (for stamping) 853
STAPEL (*m*): staple; warehouse; store; heap; pile; stack; dump; slips; stock; slipway (Shipbuilding) 854
STAPELBAR: capable of being (heaped) piled up; subject to staple laws 855
STAPELKEIL (*m*): launching wedge 856
STAPELKLOTZ (*m*): wedge block (of slipways) 857
STAPELLAUF (*m*): slipway (Shipbuilding), launch-(ing) 858
STAPELN: to store; warehouse; heap (pile) up; stack 859
STAPELPLATZ (*m*): ship-yard 860
STAPELSCHLITTEN (*m*): (launching)- cradle (Shipbuilding) 861
STAPELSTÜTZE (*f*): shore, prop 862
STAPELUNG (*f*): storing; storage 863
STAPHYLOM (*n*): staphyloma (of the cornea or of the sclerotic) (Med.) 864
STAR (*m*): cataract (Med.); starling 865
STARANIS (*m*): star anise (see STERNANIS) 866
STARK: strong; mighty; powerful; heavy; fat; thick; loud; large; hard; stout; keen; corpulent; robust; big; stentorious; wide-(ly); great-(ly); sharp-(ly); considerably; considerably 867
STÄRKE (*f*): starch; $(C_6H_{10}O_5)_x$; Sg. 0.95-1.5 (*Amylum*); strength; thickness; force; vigour; corpulency; stoutness 868
STÄRKEART (*f*): variety of starch 869
STÄRKEARTIG: amyloid; amylaceous; starchy 870
STÄRKEBLAU (*n*): starch blue 871
STÄRKEFABRIK (*f*): starch factory 872
STÄRKEGLANZ (*m*): starch gloss (for glossing washing) 873
STÄRKEGRAD (*m*): intensity; degree of strength 874
STÄRKEGUMMI (*n*): dextrin; starch gum; British gum; $C_{12}H_{20}O_{10}$ (see DEXTRIN) 875
STÄRKEHALTIG: starchy; containing starch 876
STÄRKEKLEISTER (*m*): starch paste; starch moutant (see KLEBER) 877
STÄRKEKORN (*n*): starch (grain) granule; grain of starch 878
STÄRKEKÖRNCHEN (*n*): granule of starch 879
STÄRKEKÖRNERRASTER (*m*): screen of starch grains (Phot.) 880
STÄRKEMEHL (*n*): starch (powder; flour) 881
STÄRKEMEHLÄHNLICH: amyloid; amylaceous; starchy 882
STÄRKEMITTEL (*n*): tonic; restorative; strengthener; strengthening remedy 883
STÄRKEN: to starch; strengthen; thicken; stiffen; stimulate; refresh; restore; confirm; corroborate 884
STÄRKEND: corroborative; strengthening; restorative; tonic; stimulating 885
STÄRKESIRUP (*m*): glucose; starch syrup 886
STÄRKEWASSER (*n*): starching liquid 887
STÄRKEWEIZEN (*m*): emmer; starch wheat 888

STÄRKEZUCKER (*m*): (dextro-) glucose; starch sugar (*Saccharum amylaceum*); $C_6H_{12}O_6$; (see DEXTROSE) 889
STARKFARBIG: strongly (highly; loudly) coloured 890
STARKLEIBIGKEIT (*f*): corpulence; stoutness 891
STARKMUSKELIG: muscular 892
STARKSTROM (*m*): strong (intense, excess) current (Elect.) 893
STÄRKUNGSMITTEL (*n*): tonic; restorative; stimulant; strengthener; strengthening remedy 894
STARKWANDIG: thick-walled 895
STARKWIRKEND: efficacious; drastic; powerful 896
STARR: rigid: stiff; stubborn; numb; inflexible 897
STARRE (*f*): rigor, stiffness, rigidity, numbness, stubbornness, obstinacy 898
STARREN: to be (rigid) stiff; be numb; stare; be (benumbed) chilled; stiffen 899
STARRHEIT (*f*): rigidity; stiffness; numbness; obstinacy; inflexibility 900
STARRKÖPFIG: headstrong; obstinate; stubborn; hot-headed; inflexible 901
STARRKRAMPF (*m*): tetanus; lock-jaw (Med.) 902
STARRLEINEN (*n*): buckram 903
STARRLEINWAND (*f*): buckram 904
STARRSINN (*m*): stubbornness; obstinacy; hot-headedness 905
STARRSINNIG: stubborn; obstinate; hot-headed 906
STARRSUCHT (*f*): catalepsy; torpor (Med.) 907
STASSFURTIT (*m*): stassfurtite, boracite (see BORAZIT) 908
STÄT: constant; fixed; stable; firm; continuous 909
STÄTIG: constant; fixed; stable; continual; continuous; invariable 910
STÄTIGKEIT (*f*): constancy; stability; continuity 911
STATIK (*f*): statics 912
STATION (*f*): station; stage 913
STATIONÄR: stationary 914
STÄTISCH: restive 915
STATISCH: static 916
STATIST (*m*): mute; dumb-(person) 917
STATISTIK (*f*): statistics 918
STATIV (*n*): stand; support 919
STATOR (*m*): stator (Elect.) 920
STATT: in place of; instead of; in lieu of; (*f*): stead; place; lieu 921
STÄTTE (*f*): place; room; stead 922
STATTEN, VON-GEHEN: to take place; pass off; go off; proceed; succeed; prosper 923
STATTEN, ZU-KOMMEN: to be of (advantage) use; stand in good stead 924
STATTFINDEN: to take place; occur; happen; result 925
STATTGEHABT: previous 926
STATTHABEN: to take place; occur; happen; result 927
STATTHAFT: allowable; permissible; legal; lawful; admissible 928
STATTHALTER (*m*): governor; viceroy 929
STATTLICH: stately; splendid; fine; regal; portly 930
STATTLICHKEIT (*f*): stateliness; splendour 931
STATUE (*f*): statue 932
STATUIEREN: to lay down, set (an example), maintain 933
STATUR (*f*): stature, size 934
STATUT (*n*): statute, regulation 935
STAU (*m*): retention, accumulation, storage, damming, pocket, banking up, restriction, stowage 936
STAUB (*m*): dust; powder; pollen (Bot.) 937
STAUBARTIG: dustlike; powdery 938
STAUBBEUTEL (*m*): anther (Bot.) 939
STAUBBINDUNG (*f*): laying the dust 940
STÄUBCHEN (*n*): particle; mote; atom 941
STAUBDICHT: dustproof 942
STAUBECKEN (*n*): storage basin; reservoir 943
STAUBEN: to be dusty 944
STÄUBEN: to dust; powder; spray; scatter; raise dust 945
STÄUBERN: to dust; drizzle; drift; drive away 946
STAUBFÄNGER (*m*): dust (grit) catcher 947
STAUBFEIN: very fine; as fine as dust 948
STAUBFÖRMIG: powdery, small, fine, dusty 949
STAUBGEFÄSS (*n*): stamen (Bot.) 950
STAUBGEHALT (*m*): dust content (of gases, etc.) 951
STAUBIG: dusty; powdery 952
STAUBKALK (*m*): air-slaked lime 953
STAUBKAMM (*m*): tooth-comb 954
STAUBKOHLE (*f*): coal dust; powdered (fuel) coal; pulverized coal (size 0-$\frac{3}{4}$ mm.) 955
STAUBKOKS (*m*): coke dust (9% CO_2 content at furnace, calorific value less than 5,000 cals.) pulverized coke 956
STAUBÖL (*n*): dust oil 957
STAUBREGEN (*m*): drizzle (of rain); Scotch mist 958
STAUBSAMMLER (*m*): dust (catcher) collector; grit catcher 959
STAUBSAUGER (*m*): vacuum cleaner 960
STAUBSCHREIBER (*m*): recording dust meter 961
STAUBSIEB (*n*): dust sieve 962
STAUBTEE (*m*): tea dust 963
STAUBWOLKE (*f*): cloud of dust 964
STAUCHBAR: malleable 965
STAUCHEN: to knock; beat; compress (by a blow); stem; jolt; toss; pile up (Flax); expand-(tubes); shorten; to rivet over; burr over; bead; clench; hammer (press) down; up-set (Rivets) 966
STAUCHKALIBER (*n*): up-set pass (Rolling mills) 967
STAUCHSTICH (*m*): up-set pass (Rolling mills) 968
STAUCHUNG (*f*): knocking; beating, etc. (see STAUCHEN) 969
STAUDE (*f*): shrub; bush 970
STAUDENARTIG: bushy, shrubby 971
STAUDENGEWÄCHS (*n*): herbaceous plant-(s) 972
STAUDENSALAT (*m*): cabbage lettuce (*Lactuca scariola sativa*) 973
STAUDRUCK (*m*): back pressure; static pressure (the pressure of a fluid which has been dammed); dynamic pressure (the pressure which causes a fluid to dam up) 974
STAUE (*f*): restriction; dam; baffle; accumulation; pocket; obstruction; blockage; stagnant water 975
STAUEN (*n*): stowing, stowage (of cargo); accumulating, damming, pocketing 976
STAUEN: to restrict; baffle; dam; bank up; choke; pocket; accumulate; store; block up; form a pocket; stow; congest; obstruct 977

STAUFFERBÜCHSE (*f*): Stauffer lubricator 978
STAUFLANSCH (*m*): reducing flange; throttle; diaphragm (of a steam meter) (see STAURAND) 979
STAUFREI: unrestricted 980
STAUFREI, NICHT-: restricted 981
STAUGERÄT (*n*): static measuring apparatus 982
STAUHÖHE (*f*): head (of water) 983
STAUINHALT (*m*): capacity, water capacity, reservoir capacity, content (of reservoirs) 984
STAUKÖRPER (*m*): baffle plate; diaphragm; dam 985
STAUNEN: to be astonished; stare; be surprised; wonder; marvel; (*n*): astonishment; surprise; wonder 986
STAUNENSWERT: wonderful; surprising; astonishing; marvellous 987
STAU (*m*), **NORMALER-**: normal water level (of reservoirs) 988
STAUPE (*f*): rod; whip; scourge 989
STAUPENDELBRÜCKE (*f*): balanced ash dumping arrangement; swinging ash dumping arrangement (of mech. stokers) 990
STAURAND (*m*): throttle diaphragm, reducing flange (Steam meters) (a diaphragm in a pipe, having a hole in it less than the internal diameter of the pipe) 991
STAUROLITH (*m*): staurolite (natural ferrous iron aluminium silicate); $HFeAl_5Si_2O_{13}$; Sg. 3.65 992
STAUSCHEIBE (*f*): baffle plate; diaphragm; dam 993
STAUUNG (*f*): pocket; obstruction; congestion; accumulation; pocketing; rise (of water); stowage (of cargo); damming; restriction; stowing 994
STAUVORRICHTUNG (*f*): dumping arrangement, back-end arrangement (of mechanical stokers) 995
STAUVORRICHTUNG (*f*) **FÜR WANDERROSTE** (*m.pl*): back-end arrangement for mechanical stokers 996
STAUWASSER (*n*): stagnant water 997
STAUWEIHER (*m*): mill-pond; reservoir 998
STEAMKASTEN (*m*): steam chest 999
STEARAT (*n*): stearate 000
STEARIN (*n*): stearine [glycerylstearic ester (*Stearinum*)]; $C_{57}H_{110}O_6$; Sg. 0.8621; Mp. 71.5°C 001
STEARINFABRIKATION (*f*): stearine manufacture 002
STEARINFABRIKATIONSANLAGE (*f*): stearine manufacturing plant 003
STEARINGOUDRON (*m*): stearine pitch (a pitch-like residue left in the retort from the distillation of fatty acids) 004
STEARINKERZE (*f*): stearine candle 005
STEARINKUCHEN (*m*): stearine cake 006
STEARINÖL (*n*): stearine oil; oleic acid (see ELAINSÄURE) 007
STEARINPECH (*n*): stearine pitch (see STEARIN-GOUDRON) 008
STEARINSAUER: stearate of, combined with stearic acid 009
STEARINSÄURE (*f*): (*Acidum stearinicum*); stearic (stearinic) acid, cetylacetic acid, stearophanic acid; $CH_3(CH_2)_{16}CO_2H$ or $C_{18}H_{36}O_2$; Sg. 0.843 to 0.941; Mp. 69.3 to 71.3°C; Bp. 232-291°C 010
STEARINSAURESSALZ (*n*): stearate 011
STEARINSEIFE (*f*): stearin soap 012
STEARINTEER (*m*): candle tar; stearin pitch; candle pitch; palm pitch; Mp. 140-170°C (residue from distillation of fatty acids) 013
STEAROPTEN (*n*): stearoptene (a paraffin); $C_{16}H_{34}$; Mp. 36.5°C 014
STEATIT (*m*): steatite, talc, soapstone (magnesium silicate); $3MgO,4SiO_2.H_2O$; Sg. 2.65 (see also SPECKSTEIN) 015
STEATOM (*n*): steatoma (a sebaceous cyst) (Med.) 016
STECHAPFEL (*m*): thorn apple (*Datura Stramonium*); stramonium (Pharm.); Christ's thorn (see JUDENBAUM) 017
STECHAPFELBLÄTTER (*n.pl*): stramonium leaves 018
STECHAPFELKRAUT (*n*): stramonium (*herba Stramonii*); stramonium leaves 019
STECHBEITEL (*m*): mortice chisel 020
STECHEN (*n*): breaking up and sorting roughly, without much care (Coal direct from the mine) (see the verb STECHEN) 021
STECHEN: to stick; prick; sting; pierce; cut; engrave; tap; spire; spout; incline; kill; feed material into a roll; work (use) a roll (Rolling mills); break up and sort roughly (Coal) 022
STECHEND: penetrating; piercing; pungent; stinging 023
STECHER (*m*): pricker; proof-stick (Sugar); sampler; engraver 024
STECHFLIEGE (*f*): (any) stinging fly; gad-fly (see BREMSE) 025
STECHGINSTER (*m*): furze (*Ulex europæus*) 026
STECHHEBER (*m*): thief tube 027
STECHKOLBEN (*n*): pipette 028
STECHKUNST (*f*): (the art of) engraving 029
STECHPALME (*f*): Christ's thorn (*Paliurus*); holly-tree (*Ilex aquifolium*) 030
STECHPALMENHOLZ (*n*): holly-(tree)-wood (of *Ilex aquifolium*); Sg. 0.75-0.95 031
STECHPROBE (*f*): touchstone test 032
STECHROCHE (*m*): roach (*Trigon pastinaca*) (Icht.) 033
STECKDOSE (*f*): wall-plug (Elect.) 034
STECKEN: to stick; hide; stay; remain; set; fix; put; place; breakdown; (*m*): staff; rod; stick 035
STECKENDOSE (*f*): wall plug (Elect.) 036
STECKENPFERD (*m*): hobby horse 037
STECKER (*m*): plug (of electric wall plug) 038
STECKKAPSEL (*f*): push-on cap 039
STECKKAPSELGLAS (*n*): glass tube with push-on cap 040
STECKLING (*m*): layer; shoot 041
STECKNADEL (*f*): pin 042
STEG (*m*): bridge; path; stay; cross-(bar) piece; stud (of a chain); strap; bar; web (of metal sections) 043
STEGKETTE (*f*): stud-link chain 044
STEGKREUZ (*n*): turn-stile 045
STEGREIF (*m*): impromptu; extemporization 046
STEHBOLZEN (*m*): stay-(bolt) 047
STEHBÜTTE (*f*): stock tub (Brewing) 048
STEHEN: to stand; become; be; be (vertical) upright; fit; stop; come to a standstill; answer for 049
STEHEND: standing; upright; vertical; stationary; stable; stagnant 050
STEHENDES GUT (*n*): standing rigging (Nautical) 051

STEHKOLBEN (*m*): flat-bottomed flask 052
STEHLAGER (*n*): pedestal bearing; plummer block 053
STEHLEN: to steal; rob; pilfer; thieve; (*n*): stealing; robbing; pilfering; thieving 054
STEIERISCH: Styrian 055
STEIERMARK (*f*): Styria 056
STEIF: rigid; stiff; firm; precise; formal; stubborn; inflexible; pedantic 057
STEIFE (*f*): stiffening; stiffener; prop; shore; stay; stiffness; starch; buttress 058
STEIFEN: to stiffen; prop; stay; starch; buttress 059
STEIFHEIT (*f*): stiffness; rigidity; formality; pedantry; stubbornness; inflexibility 060
STEIFKÖPFIG: stubborn; stiff-necked 061
STEIFLEINEN (*n*): buckram 062
STEIFLEINWAND (*f*): buckram 063
STEIFUNG (*f*): stiffening; starching 064
STEIG (*m*): path 065
STEIGBÜGEL (*m*): stirrup; stapes (Anat.) 066
STEIGE (*f*): steps; ladder; stair-(case); score; hen-roost 067
STEIGEN: to mount; climb; rise; ascend; descend; increase; advance; progress; improve 068
STEIGERN: to increase; raise; heighten; enhance 069
STEIGEROHR (*n*): riser; rising (ascending) pipe 070
STEIGERUNG (*f*): increase; gradation; rise; raising; heightening; enhancing; enhancement; comparison (Grammar) 071
STEIGERUNG (*f*), **TEMPERATUR-**: increase of temperature 072
STEIGRAUM (*m*): space (in which to rise) to permit of rising 073
STEIGROHR (*n*): riser; rising (ascending) pipe or tube 074
STEIGSTEIN (*m*): step (stone) brick (for climbing up the insides of chimneys, shafts, etc.) 075
STEIGUNG (*f*): rise; rising; increase; incline; pitch (of a screw); gradient; ascent 076
STEIGUNGSMESSER (*m*): pitchometer (an instrument for determining the pitch of a screw-propeller) 077
STEIL: steep; precipitous 078
STEILROHRKESSEL (*m*): (inclined) vertical tube boilers 079
STEILSCHRIFT (*f*): vertical writing 080
STEIN (*m*): stone; rock; matt (Metal); gravel; calculus; concretion (Medical); piece; chessman; pawn; brick; voussoir (of an arch) 081
STEINABFÄLLE (*m.pl*): stone chips 082
STEIN (*m*), **ABGERUNDETER-**: bull-nosed brick 083
STEINÄHNLICH: stony; stone-like 084
STEINALAUN (*m*): rock alum (see **ALAUN**) 085
STEINALT: ancient; as old as the hills 086
STEINARBEIT (*f*): stone work; metal smelting 087
STEINARTIG: stony; stone-like 088
STEINAUFLÖSEND: lithontriptic, stone-dissolving 089
STEINAUFLÖSENDES MITTEL (*n*): lithontriptic; calculus solvent (for dissolving urinary calculi) (Med.) 090
STEINAUFLÖSUNGSMITTEL (*n*): (see **STEINAUF-LÖSENDES MITTEL**) 091
STEINBILD (*n*): statue 092
STEINBOCK (*m*): wild (mountain) goat 093
STEINBOHRER (*m*): rock drill 094
STEINBRECH (*m*): saxifrage (*Saxifraga*) 095
STEINBRECHER (*m*): stone-(breaker) crusher 096
STEINBRECHGEWÄCHSE (*n.pl*): *Saxifragaceæ* (Bot.) 097
STEINBRECHMASCHINE (*f*): (see **STEINBRECHER**) 098
STEINBRUCH (*m*): quarry 099
STEINBRUCHMASCHINE (*f*): quarrying machine 100
STEINBÜHLERGELB (*n*): barium (chromate) yellow (see **BARIUMCHROMAT**) 101
STEINBUTTER (*f*): rock butter 102
STEIN DER WEISEN (*m*): philosopher's stone 103
STEINDRUCK (*m*): lithograph-(y); lithographic printing 104
STEINDRUCKER (*m*): lithographer 105
STEINDRUCKEREI (*f*): lithography; lithographic printing -(works) 106
STEINDRUCKFARBE (*f*): lithographic ink 107
STEINEICHE (*f*): oak (see **WINTEREICHE**) 108
STEINERN: stone; of stone; stony; rocky; rock 109
STEIN (*m*), **FEUERFESTER-**: firebrick; Sg. 2.2-2.7 110
STEINFLACHS (*m*): amianthus; mountain flax; asbestos (see **ASBEST**) 111
STEINFRUCHT (*f*): stone fruit 112
STEIN, GEBRANNTER- (*m*): brick 113
STEINGRAU: stone-grey; (*n*): stone-grey (the colour) 114
STEINGRAVÜRE (*f*): (see **STEINRADIERUNG**) 115
STEINGRUBE (*f*): stone (pit) mine 116
STEINGRÜN (*n*): terre verte (Colour) (see **GRÜNE ERDE**) 117
STEINGUT (*n*): stoneware; white ware; Sg. 1.3-1.6; earthenware (porous clay wares which are glazed) 118
STEINGUT (*n*), **ENGLISCHES-**: English white ware (in which the glazes are transparent; made from plastic clay; consist of 35-69% clay and 35-60% quartz) 119
STEINHART: hard as stone 120
STEINHOLZ (*n*): xylolite 121
STEINHOLZARBEIT (*f*): xylolite work 122
STEINICHT: stony; rocky; of stone; stone 123
STEINIG: stony; rock-(y); (of) stone 124
STEINIGEN: to lapidate, stone 125
STEINIGUNG (*f*): lapidation, stoning 126
STEINKENNER (*m*): lithologist 127
STEINKITT (*m*): stone cement 128
STEINKLEE (*m*): melilot (*Melilotus officinalis*); white clover (*Trifolium repens*) 129
STEINKOHLE (*f*): mineral (steam) coal; anthracite; coal; pit-coal; fossil-coal; stone-coal 130
STEINKOHLE, MINDERE- (*f*): sub-bituminous coal 131
STEINKOHLENASCHE (*f*): coal ashes 132
STEINKOHLENBENZIN (*n*): benzene; benzol; benzole; coal naphtha (see **BENZOL**) 133
STEINKOHLENBERGWERK (*n*): coal mine 134
STEINKOHLENGAS (*n*): coal gas 135
STEINKOHLENKOKS (*m*): coke 136
STEINKOHLENKREOSOT (*n*): (see **PHENOL**) 137
STEINKOHLENSCHICHT (*f*): coal seam 138
STEINKOHLENSCHLACKE (*f*): cinders (from coal) 139
STEINKOHLENSTAUB (*m*): coal dust 140
STEINKOHLENTEER (*m*): coal tar (by-product of destructive distillation of coal) 141
STEINKOHLENTEERBLASE (*f*): coal-tar still 142

STEINKOHLENTEERESSENZ (*f*) : first light oil 143
STEINKOHLENTEERKAMPHER (*m*) : white tar; naphthalene; $C_{10}H_8$ (see **NAPHTALIN**) 144
STEINKOHLENTEERKREOSOT (*n*) : (coal tar)- creosote; carbolic acid; Phenol (see **PHENOL** and **CARBOLSÄURE**) 145
STEINKOHLENTEERÖL (*n*) : coal-tar oil; naphtha; heavy naphtha; Sg. 0.925; Bp. 160-220°C 146
STEINKOHLENTEERÖL, LEICHTES- : light oil 147
STEINKOHLENTEERÖL, SCHWERES- : heavy oil; creosote oil 148
STEINKOHLENTEERPECH (*n*) : coal tar pitch (from refining of coal tar; forms residue) 149
STEINKOHLENVERKOHLUNG (*f*) : coking (of coal) 150
STEINKOHLENVERKOKUNG (*f*) : coking (of coal) 151
STEINKRAUT (*n*) : stonecrop (*Sedum*) 152
STEINKRUG (*m*) : stone jar 153
STEINKUNDE (*f*) : lithology, the science of stones 154
STEINLAGE (*f*) : layer or course of brickwork 155
STEINMALZ (*n*) : glassy (vitreous) malt 156
STEINMARK (*m*) : clay (see **KAOLINIT**); Sg. 2.5 157
STEINMEHL (*n*) : crushed stone 158
STEINMEISSEL (*m*) : stone chisel 159
STEINMETZ (*m*) : stone mason 160
STEINMÖRTEL (*m*) : concrete; cement; Portland cement 161
STEINNUSS (*f*) : ivory nut; corozo nut (*Phytelephas macrocarpa*) 162
STEINOBST (*n*) : stone fruit 163
STEINÖL (*n*) : petroleum; rock-oil (Old term for **ERDÖL**, which see) 164
STEINPAPPE (*f*) : roofing (fabric; paper) board; carton pierre (see **DACHPAPPE**) 165
STEINPECH (*n*) : stone (hard) pitch 166
STEINPFLASTER (*n*) : stone pavement 167
STEINPILZ (*m*) : (*Boletus edulis*); champignon; edible Boletus 168
STEINPLATTE (*f*) : slab of stone 169
STEINPORZELLAN (*n*) : hard porcelain 170
STEINQUITTE (*f*) : cotoneaster (Bot.) (*Cotoneaster*) 171
STEINRADIERUNG (*f*) : etching on stone 172
STEINREICH : stony; full of stones; (*n*) : mineral kingdom 173
STEINRIFF (*n*) : ridge; reef 174
STEINRÖSTEN (*n*) : regulus (matt) roasting 175
STEINSALZ (*n*) : rock salt; Sg. 2.18 (see **NATRIUM-CHLORID**); halite (see **HALIT**) 176
STEINSALZLAGER (*n*) : rock salt (bed) deposit 177
STEINSCHLAG (*m*) : broken stone 178
STEINSCHLEIFER (*m*) : lapidary 179
STEINSCHNEIDEN (*n*) : stone-cutting; lapidation 180
STEINSCHNEIDER (*m*) : lapidary, stone-cutter, jeweller 181
STEINSCHNITT (*m*) : lithotomy (removal of stone from the bladder by incision) (Med.) 182
STEINSCHRAUBE (*f*) : stone bolt 183
STEINSCHRIFT (*f*) : inscription on stone 184
STEINSTOPFEN (*m*) : stone stopper 185
STEINTAFEL (*f*) : stone slab 186
STEINWAREN (*f.pl*) : stoneware 187
STEINWURF (*m*) : stone's throw 188
STEINZEICHNUNG (*f*) : lithograph, drawing on stone 189
STEINZEIT (*f*) : stone age 190
STEINZEUG (*n*) : stoneware (Ceram.); hardware; Sg. 2.47-2.65 191
STEINZEUGFLIESSE (*f*) : hardware paving brick or tile 192
STEINZEUGPUMPE (*f*) : hardware pump (for gases) 193
STEISS (*m*) : rump; buttocks 194
STEISSBEIN (*n*) : coccyx (Anat.) 195
STEISSBEINSCHMERZ (*m*) : Coccygodyna (a neuralgic pain of or near the coccyx) (Med.) 196
STEK (*m*) : bend; hitch, knot 197
STELLBAR : adjustable; moveable 198
STELLBOTTICH (*m*) : fermenting vat (Brew.); settling (vat) tub (Dyeing) 199
STELLDICHEIN (*n*) : appointment; rendezvous 200
STELLE (*f*) : place, (also of decimals); position; spot; situation; point; employment; office 201
STELLE-GESUCH (*n*) : situation wanted 202
STELLEN : to put; place; impose; superimpose; set; adjust; regulate; stop; check; furnish; station; arrange; come forward; dissemble; feign; hold up; trim (Sails); supply; shade (Colours); place oneself; appear; prove to be; stand; pretend (to be) 203
STELLENGESUCH (*n*) : application for a situation 204
STELLENGESUCHE (*n.pl*) : applications for situations; situations wanted 205
STELLENWEISE : in places; at certain points; here and there 206
STELLENZAHL (*f*) : index (Maths.) 207
STELLE (*f*), **OFFENE-** : vacancy (vacant situation) 208
STELLESUCHENDE (*m*) : candidate, applicant for employment 209
STELLHAHN (*m*) : regulating cock 210
STELLIN (*n*) : stellin (trade term for a purified motor spirit) 211
STELLING (*f*) : stage, staging 212
STELLKEIL (*m*) : wedge bolt; adjustable wedge 213
STELLMACHER (*m*) : cartwright; wheelwright 214
STELLMUTTER (*f*) : lock-(ing) nut; adjusting nut 215
STELLRING (*m*) : loose collar; adjustable collar; adjusting ring 216
STELLSCHRAUBE (*f*) : set screw; adjusting screw 217
STELLSTIFT (*m*) : adjusting pin 218
STELLSTÜCK (*n*) : adjusting (distance) piece 219
STELLUNG (*f*) : placing, etc. (see **STELLEN**); situation; position; arrangement; attitude; disposition; constellation; posture 220
STELLUNGSISOMERIE (*f*) : position (place) isomerism 221
STELLVERTRETER (*m*) : deputy; substitute; representative; proxy; (when used in combination means "vice-") 222
STELLVERTRETUNG (*f*) : substitution; deputizing; representation; proxy 223
STELZE (*f*) : stilt; wooden leg 224
STEMMEISEN (*n*) : chisel; mortise chisel; double bevelled chisel; caulking (tool) iron 225
STEMMEN : to notch (Trees); caulk; chisel; stem; prop; oppose; resist 226
STEMMHAMMER (*m*) : caulking hammer 227

STEMMKANTE (*f*): caulking (caulked) edge 228
STEMMMEISSEL (*m*): caulking tool 229
STEMMNAHT (*f*): caulked seam 230
STEMMSETZE (*f*): caulking tool 231
STEMMTOR (*n*): lock gate 232
STEMMUNG (*f*): caulking 233
STEMPEL (*m*): mark; stamp; stamper; die; punch; pestle; pounder; brand; piston; plunger; ram-(mer) 234
STEMPELFARBE (*f*): marking ink 235
STEMPELGEBÜHR (*f*): stamp-duty 236
STEMPELMARKE (*f*): mark; stamp 237
STEMPELN: to stamp; mark; brand 238
STEMPELZEICHEN (*n*): stamp; mark; brand; hall-mark (Precious metals) 239
STENGE (*f*): (top)-mast 240
STENGE (*f*), GROSSE-: main top mast 241
STENGEGUT (*n*): topmast rigging 242
STENGEL (*m*): stem; stalk; pole 243
STENGELIG: columnar (Mineralogy) 244
STENGELKOHLE (*f*): columnar coal 245
STENGESTAG (*n*): stay (Naut.) 246
STENOGRAPHIE (*n*): stenography, shorthand 247
STEPHANIT (*m*): stephanite (silver sulphantimonite); Ag_5SbS_4; Sg. 6.25 248
STEPHANSKÖRNER (*n.pl*): stavesacre seeds (seeds of *Delphinium staphisagria*) 249
STEPPDECKE (*f*): quilt-(ed) cover 250
STEPPE (*f*): steppes; desert; heath 251
STEPPEN: to quilt; stitch 252
STEPPENWIESE (*f*): prairie 253
STERBEN: to die; fade-(away); disappear; be effaced; (*n*): dying; death 254
STERBLICH: mortal 255
STERBLICHKEIT (*f*): mortality; death-rate 256
STEREOAUFNAHME (*f*): stereoscopic photograph 257
STEREOCHEMIE (*f*): stereochemistry (science of the spacial relation of atoms to a molecule) 258
STEREOCHEMISCH: stereochemical 259
STEREOISOMERIE (*f*): stereoisomerism 260
STEREOMETRIE (*f*): stereometry; solid geometry 261
STEREOMETRISCH: stereometric-(al) 262
STEREOSKOPAUFNAHME (*f*): stereoscopic photograph-(y) 263
STEREOTYPE (*f*): stereotype 264
STEREOTYPIE (*f*): stereotype 265
STEREOTYPIE-PAPIERE (*n.pl*): stereotyping papers (Paper) 266
STEREOTYPIEREN: to stereotype 267
STERIL: sterile; sterilized 268
STERILISATION (*f*): sterilization, disinfection 269
STERILISATOR (*m*): sterilizer 270
STERILISIERAPPARAT (*m*): sterilizing apparatus 271
STERILISIEREN: to sterilize 272
STERILISIERMITTEL (*n*): sterilizing agent, sterilizer 273
STERILISIERUNG (*f*): sterilization 274
STERILITÄT (*f*): sterility; barrenness 275
STERISCH: steric; spatial 276
STERN (*m*): star; asterisk; pupil (of eye) 277
STERNANEMONE (*f*): Star anemone (*Anemone hortensis*) 278
STERNANIS (*m*): star anise (*Illicium anisatum*) 279
STERNANISÖL (*n*): star anise oil; Sg. 0.985 280
STERNBERGIT (*n*): sternbergite, iron-silver glance; $AgFe_2S_3$; Sg. 4.23 281
STERNBESCHREIBUNG (*f*): astrography 282
STERNBILD (*n*): constellation 283
STERNBÜCHSE (*f*): stern bush 284
STERNCHEN (*n*): small star; asterisk 285
STERNDEUTER (*m*): astrologer 286
STERNFÖRMIG: stellate; star-shaped; stelliform; stellar 287
STERNFORSCHER (*m*): astrologer; astronomer 288
STERNGRUPPE (*f*): group (cluster) of stars 289
STERNHELL: starlight 290
STERNHIMMEL (*m*): starry firmament 291
STERNIG: starry, star-like 292
STERNJAHR (*n*): sidereal year 293
STERNKARTE (*f*): astronomical chart 294
STERNKENNER (*m*): astronomer 295
STERNKUNDE (*f*): astronomy 296
STERNRAD (*n*): stern-wheel 297
STERNROHR (*n*): telescope 298
STERNSAPPHIR (*m*): star sapphire 299
STERNSCHNUPPE (*f*): meteor; shooting star; falling star 300
STERNWARTE (*f*): observatory 301
STERNZEIT (*f*): sidereal time 302
STERROMETALL (*n*): sterro metal (similar to delta metal) (see DELTAMETALL) 303
STET: constant; stable; continuous; fixed; firm 304
STETHOGRAPH (*m*): stethograph (an instrument for recording the movements of the chest) (Med.) 305
STETHOMETER (*n*): stethometer (an instrument for measuring the movement of the chest) (Med.) 306
STETHOSKOP (*n*): stethoscope (an instrument for auscultation of the chest) (Med.) 307
STETIG: constant; stable; continuous; fixed; firm; invariable; continual; regular; even-(ly); consistent-(ly) 308
STETIGKEIT (*f*): constancy; stability; continuity 309
STETIGKEITSGRENZE (*f*): proportional extension limit 310
STETS: ever; invariable; invariably; continually, regularly, always, constantly, consistently 311
STEUER (*n*): helm; rudder; (*f*): duty; tax; revenue; impost 312
STEUERACHSE (*f*): steering axle 313
STEUERAMT (*n*): customs (office) house; revenue office 314
STEUERANLAGE (*f*): imposition of a tax; steering gear 315
STEUERAUSCHLAG (*m*): assessment 316
STEUERBAR: assessable; rateable; taxable; dutiable 317
STEUERBARKEIT (*f*): liability to taxation 318
STEUERBEAMTE (*m*): tax collector; revenue (customs) officer 319
STEUERBEAMTER (*m*): (see STEUERBEAMTE) 320
STEUERBEDIENTE (*m*): exciseman; customs officer 321
STEUERBEFEHL (*m*): edict or parliamentary bill concerning taxes 322
STEUERBEHÖRDE (*f*): treasury (board or office) 323
STEUERBEWEGUNG (*f*): distributing valve motion 324
STEUERBEWILLIGUNG (*f*): right to grant or refuse taxes 325
STEUERBORD (*n*): starboard (Naut.) 326
STEUERBORDSEITE (*f*): starboard side 327

STEUERBORDSWACHE (*f*): starboard-watch 328
STEUERBORDWACHE (*f*): starboard watch 329
STEUERBRÜCKE (*f*): bridge (helmsman or steersman's post) 330
STEUERBUCH (*n*): rate-book 331
STEUEREINNEHMER (*m*): tax (collector) gatherer; receiver of taxes 332
STEUERENDE (*n*): stern; poop (Naut.) 333
STEUERERHEBUNG (*f*): collection (levying) of taxes 334
STEUERERLASS (*m*): exemption from (taxation) duty; remission of (taxes) duty 335
STEUERFREI: duty (tax)-free; exempt 336
STEUERFREIHEIT (*f*): exemption (from duty or taxes) 337
STEUERGELD (*n*): tax money 338
STEUERGESTÄNGE (*n*): valve gear rods 339
STEUERHAUS (*n*): wheel house (Ships) 340
STEUERKASSE (*f*): treasury; receiver's office 341
STEUERKASSENSCHEIN (*m*): bank-note 342
STEUERKOLLEGIUM (*n*): revenue board 343
STEUERKOMPASS (*m*): common compass 344
STEUERKUFE (*f*): rudder runner (of ice boats) 345
STEUERKUFE (*f*), **EISJACHT-**: ice yacht rudder runner 346
STEUERLASTIG: down by the stern (of ships) 347
STEUERMANN (*m*): steersman; pilot; helmsman; coxswain; cox 348
STEUERMANNSMAAT (*m*): second mate (Naut.) 349
STEUERMARKE (*f*): revenue stamp 350
STEUERN: to control; steer; regulate; pilot; navigate; stop; prevent; check; help; aid; actuate; direct a course; contribute; pay-(taxes); operate 351
STEUERNAGEL (*m*): linch-pin 352
STEUERORGAN (*n*): distributing valve (Turbines) 353
STEUERPFLICHTIG: assessable; rateable; taxable 354
STEUERPFLICHTIGE (*m*): ratepayer; anyone who is taxable 355
STEUERRAD (*n*): (steering)-wheel; helm 356
STEUERRADTROMMEL (*f*): steering wheel barrel 357
STEUERREEP (*n*): tiller-rope (Naut.) 358
STEUERRUDER (*n*): rudder; helm (Naut.) 359
STEUERSATZ (*m*): rate 360
STEUERSCHALTER (*m*): controller, controlling switch, automatic (starter) starting gear 361
STEUERSCHEIN (*m*): tax receipt (for payment); rate receipt 362
STEUERSCHIEBER (*m*): by-pass (easing) slide or gate valve 363
STEUERUNG (*f*): controller, control, controlling (gear) device; control gear; steering gear; valve gear (Pumps); distributor; distribution; distributing gear; regulator; actuation (Mech.); steering (see **STEUERN**): driving (mechanism) gear; regulation (of temperatures, etc.) 365
STEUERUNG, AUSKLINK- (*f*): trip gear 366
STEUERUNG, AUSLASS- (*f*): exhaust gear 367
STEUERUNG, AUSPUFF- (*f*): distribution by means of exhaust ports 368
STEUERUNG, DOPPELSCHIEBER- (*f*): double ended piston valve gearing 369
STEUERUNG (*f*), **DRUCKKNOPF-**: push-button (control, controller) controlling device 370
STEUERUNG, EINLASS- (*f*): admission gear 371
STEUERUNG, EXZENTER- (*f*): eccentric valve gear 372
STEUERUNGSANTRIEB (*m*): valve gear 373
STEUERUNG, SCHIEBER- (*f*): slide valve gear 374
STEUERUNG, SCHLITZ- (*f*): distribution by means of ports 375
STEUERUNGSDAUMEN (*m*): cam 376
STEUERUNGSDIAGRAMM (*n*): distribution diagram 377
STEUERUNGSEINRICHTUNG (*f*): distributing arrangement (Turbines); steering arrangement or gear 378
STEUERUNGSGESTÄNGE (*n*): valve gear 379
STEUERUNGSGETRIEBE (*n*): cam shaft driving mechanism; steering gear 380
STEUERUNGSHAHN (*m*): distributing cock 381
STEUERUNGSHEBEL (*m*): linkwork (Mech.); starting (distributing) lever; valve-gear lever 382
STEUERUNGSHÖCKER (*m*): cam 383
STEUERUNGSKNAGGE (*m*): cam 384
STEUERUNGSNOCKEN (*m*): cam 385
STEUERUNGSWELLE (*f*): cam shaft; gear (gearing; geared) shaft 386
STEUERUNG, VENTIL- (*f*): valve gear; valve distribution 387
STEUERUNG, ZÜND- (*f*): ignition distribution 388
STEUERUNG, ZWANGLÄUFIGE- (*f*): governor controlled valve gear; positive controlled valve gear 389
STEUERUNG, ZWEISEITIGE- (*f*): distribution both sides (of the piston) 390
STEUERVERWALTUNG (*f*): revenue administration 391
STEUERVORGANG (*m*): process of distribution (Turbines) 392
STEUERVORRICHTUNG (*f*): controlling device 393
STEUERWELLE (*f*): cam shaft; eccentric shaft 394
STEUERWESEN (*n*): taxation; tax (affairs) matters 395
STEUERZAHLER (*m*): tax payer 396
STEUERZETTEL (*m*): bill of taxes; tax paper; assessment; receipt for rates or taxes 397
STEUERZUSCHLAG (*m*): addition to rates or taxes; increase in taxation 398
STEVEN (*m*): stem (Naut.) 399
STEVENROHR (*n*): stern tube 400
STEVENROHRBÜCHSE (*f*): stern bush 401
STEVENSTÜTZE (*f*): stem prop 402
STIBIOTANTALIT (*m*): stibiotantalite (Ta,Nb) SbO_4; Sg. 5.98-7.37 403
STIBLITH (*m*): stiblite; $H_2Sb_2O_5$; Sg. 5.2 404
STIBNIT (*m*): stibnite, antimony glance (antimony trisulphide); Sb_2S_3; Sg. 4.55 (see also **ANTIMONGLANZ**) 405
STICH (*m*): bend; knot; hitch; puncture; stitch; stab; prick; thrust; sting; stitch (shooting pain); engraving; tinge; cast (Colours); tapping; tap hole; tapped metal (Metal); pass (Rolling mills) 406
STICHAUGE (*f*): tap hole 407
STICHE, IM-LASSEN: to leave in the lurch 408
STICHEISEN (*n*): tapping bar (Metal) 409
STICHEL (*m*): burin; graver; style; stamp; graving tool 410

STICHELN: to prick; puncture; stitch; jeer; sneer; sew; be sarcastic; satirize 411
STICHFLAMME (*f*): thin (fine; pointed) flame; blow-pipe flame; impinging flame 412
STICH HALTEN: to stand the test 413
STICHHALTIG: proof; valid 414
STICHHÖHE (*f*): rise (of an arch) 415
STICHIG: tinged (Colours) 416
STICHKULTUR (*f*): stab culture 417
STICHLOCH (*n*): tap hole 418
STICHLOCHSTOPFMASCHINE (*f*): machine for closing the molten iron tapping hole of a blast furnace 419
STICHMASS (*n*): gauge, calipers, template, pattern 420
STICHPFROPF (*n*): tap hole plug 421
STICHPROBE (*f*): sample obtained by tapping or (pricking; piercing); tapped metal assay (Metal); pricking test (Brew.) 422
STICHSÄGE (*f*): lock saw; fret-saw; keyhole saw 423
STICHWEIN (*m*): sample wine 424
STICHWORT (*n*): catchword; cue; sarcasm; distinctive (name) term 425
STICKDAMPF (*m*): suffocating vapour; choke damp 426
STICKDUNST (*m*): (see STICKDAMPF) 427
STICKEN: to embroider; stitch; choke; suffocate 428
STICKENHUSTEN (*m*): choking cough 429
STICKEREI (*f*): fancywork; embroidery; needlework 430
STICKGAS (*n*): nitrogen (gas); carbon-dioxide (see STICKSTOFF) 431
STICKKOHLENSTOFF (*m*): nitrogen carbide; carbon nitride 432
STICKLUFT (*f*): nitrogen; azote (see STICK-STOFF); close air 433
STICKMUSTER (*n*): embroidery pattern 434
STICKOXYD (*n*): nitric oxide; (NO); Sg. 1.269 435
STICKOXYDUL (*n*): nitrous oxide; laughing gas; N_2O; Sg. 1.2257-1.53 liquid; Sg. 0.937 gas; liquid Mp. – 102°C; Bp. – 89.8°C 436
STICKRAHMEN (*m*): embroidering frame 437
STICKSTOFF (*m*): nitrogen (N) (*Nitrogenium*); azote (Az) (always used in French chem. formulæ instead of "N"); Sg. (gas) 0.9674; (liquid), 0.804; (solid) 1.0265; Mp. – 210.5°C; Bp. – 195.5°C; (as a prefix: nitrogen, nitrogenous, nitric, nitrous, azote, azotic, nitride of, azide of) 438
STICKSTOFFAMMONIUM (*n*): ammomium nitride 439
STICKSTOFFANLAGE (*f*): nitrogen plant 440
STICKSTOFFBENZOYL (*n*): benzoyl (azide) nitride 441
STICKSTOFFBESTIMMUNG (*f*): determination of nitrogen 442
STICKSTOFFBESTIMMUNGSAPPARAT (*m*): apparatus for determination of nitrogen; nitrogen determining apparatus 443
STICKSTOFFBOR (*n*): boron nitride 444
STICKSTOFFCHLORID (*n*): sal ammoniac; ammonium chloride; NH_4Cl; Sg. 1.52-1.532 445
STICKSTOFFCYANTITAN (*n*): titanium cyanonitride; Ti_5CN_4; Sg. 5.29 446
STICKSTOFFDIOXYD (*n*): nitrogen dioxide; (NO_2) 447
STICKSTOFFDÜNGER (*m*): nitrogenous (fertilizer) manure 448
STICKSTOFFDÜNGERANLAGE (*f*): nitrogenous (fertilizer) manure plant 449
STICKSTOFFFREI: non-nitrogenous; free of nitrogen 450
STICKSTOFFGAS (*n*): nitrogen gas (see STICKSTOFF) 451
STICKSTOFFGEHALT (*m*): nitrogen content 452
STICKSTOFFGLYZERIN (*n*): nitroglycerine 453
STICKSTOFFGLYZERINANLAGE (*f*): nitroglycerine plant 454
STICKSTOFFHALOGEN (*n*): nitrogen halide 455
STICKSTOFFHALTIG: nitrogenous; containing nitrogen; azotic 456
STICKSTOFFHALTIGE ABFÄLLE (*m.pl*): azotic (nitrogenous) waste 457
STICKSTOFFKALK (*m*): (see KALKSTICKSTOFF) 458
STICKSTOFFKALOMEL (*n*): mercurous (azide) nitride 459
STICKSTOFFKALZIUM (*n*): calcium nitride (see CALCIUMNITRID) 460
STICKSTOFFKOHLENOXYD (*n*): carbonyl (azide) nitride; carbodiazide 461
STICKSTOFFLITHIUM (*n*): lithium nitride 462
STICKSTOFFMAGNESIUM (*n*): magnesium nitride 463
STICKSTOFFMETALL (*n*): metallic nitride 464
STICKSTOFFNATRIUM (*n*): sodium nitride 465
STICKSTOFFOXYD (*n*): nitrogen oxide (see STICK-OXYD) 466
STICKSTOFFOXYDUL (*n*): nitrous oxide (see STICKOXYDUL) 467
STICKSTOFFOXYDULGAS (*n*): nitrous oxide gas; laughing gas (see STICKOXYDUL) 468
STICKSTOFFPENTASULFID (*n*): nitrogen pentasulphide; N_2S_5; Sg. 1.901 469
STICKSTOFFPENTOXYD (*n*): nitrogen pentoxide; nitric anhydride; N_2O_5; Mp. 30°C; Bp. 45-50°C 470
STICKSTOFFQUECKSILBER (*n*): mercury nitride 471
STICKSTOFFQUECKSILBEROXYDUL (*n*): mercurous (azide) nitride 472
STICKSTOFFREICH: rich in nitrogen; having high nitrogen content; highly nitrogenous 473
STICKSTOFFSAMMLER (*m*): nitrogen (collector) storing plant; leguminous plant 474
STICKSTOFFSELENID (*n*): nitrogen selenide (NSe) 475
STICKSTOFFSILBER (*n*): silver nitride 476
STICKSTOFFSILIZID (*n*): silicon nitride; nitrogen silicide 477
STICKSTOFFSULFID (*n*): nitrogen sulphide (NS) 478
STICKSTOFFTETRASULFID (*n*): nitrogen tetrasulphide; N_2S_4; Sg. 2.1166 479
STICKSTOFFTETROXYD (*n*): nitrogen tetroxide; N_2O_4; Sg. 1.4903 480
STICKSTOFFTITAN (*n*): titanium nitride 481
STICKSTOFFTRIOXYD (*n*): nitrogen trioxide, nitrogen anhydride; N_2O_3; Sg. 1.453 482
STICKSTOFFVANADIN (*n*): vanadium nitride 483
STICKSTOFFVERBINDUNG (*f*): nitrogen compound 484
STICKSTOFFWASSERSTOFF (*m*): hydrogen (nitride) trinitride; HN_3; nitrogen hydride 485
STICKSTOFFWASSERSTOFFSAUER: hydroazoate of; combined with hydroazoic acid 486
STICKSTOFFWASSERSTOFFSÄURE (*f*): hydronitric (hydroazoic) acid 487

STICKSTOFFWASSERSTOFFSÄUREPHENYLESTER (*m*): phenyl (azide) hydroazoate 488
STICKSTOFFWASSERSTOFFSAURES SILBER (*n*): (see SILBERAZID) 489
STICKWERK (*n*): fancywork; embroidery; needlework 490
STICKWETTER (*n.pl*): after-damp; choke-damp; foul air (Mining) 491
STIEBEN: to be scattered; drizzle; drive; fly about (like dust); be dusty; drift (Snow) 492
STIEFEL (*m*): boot; shoe; case; tube; barrel 493
STIEFELABSATZ (*m*): heel (of a boot) 494
STIEFELBÜRSTE (*f*): boot-brush, blacking-brush 495
STIEFELLACK (*m*): shoe polish 496
STIEFEL, PUMPEN- (*m*): (pump) barrel; pump body 497
STIEFELSCHAFT (*m*): leg (of a boot) 498
STIEFELSCHWÄRZE (*f*): blacking 499
STIEFELWICHSE (*f*): blacking 500
STIEFMÜTTERCHEN (*n*): pansy; heart's ease (*Viola tricolor*) 501
STIEFMÜTTERCHENKRAUT (*n*): pansy herb (*herba violæ tricoloris* of *Viola tricolor*) 502
STIEFMÜTTERCHENTEE (*m*): pansy herb (*herba violæ tricoloris* of *Viola tricolor*) 503
STIEGE (*f*): ladder; stile; stairs; staircase; steps; score 504
STIEL (*m*): stem; stalk (of flowers); shaft; haft; handle; stick 505
STIELEICHE (*f*): (see SOMMEREICHE) 506
STIELEN: to furnish with a haft or handle; fit a handle 507
STIER (*m*): steer; ox; bull 508
STIEREN: to stare; look surprised 509
STIER, JUNGER- (*m*): bullock 510
STIFT (*m*): pencil; crayon; stump; pin; snag; tack; peg; bolt; stud; tag; apprentice; (*n*): monastery; seminary; foundation; chapter; cathedral 511
STIFTEN: to found; originate; make; establish; tack; cause; excite; bring about; institute 512
STIFTER (*m*): founder; originator 513
STIFTFARBE (*f*): pastel; coloured crayon; pencil colour 514
STIFTSCHRAUBE (*f*): set (bolt) screw 515
STIFTUNG (*f*): foundation; institution; origination; establishment 516
STIFTUNGSTAG (*m*): anniversary (of a foundation) 517
STIGMA (*n*): stigma 518
STIL (*m*): style (Bot.) 519
STILART (*f*): (see STIL) 520
STILBEN (*n*): stilben; $C_{14}H_{12}$; Sg. 0.9708; Mp. 124°C; Bp. 307°C 521
STILBENGELB (*n*): stilben yellow 522
STILBIT (*m*): stilbite, desmine $(Na_2,Ca)Al_2Si_6O_{16}.6H_2O$; Sg. 2.15 523
STILETT (*n*): stiletto 524
STILISIEREN: to write (with due regard to style) 525
STILL: still; quiet; dull; stagnant; inanimate; silent; calm; quiescent; tranquil; peaceful; secret 526
STILLE (*f*): stillness; quietness; stagnation; inanimation; silence; dullness; calmness; quiescence; tranquillity; peacefulness; secrecy 527
STILLEN: to still; suckle; quiet; silence; allay; alleviate; stop; gratify; calm; appease; quench; become quiescent; pause 528
STILLEND: quieting; calming; sedative; allaying 529
STILLENDE (*f*): nursing mother 530
STILLHALTEN: to be quiet; be still; keep still; pause; stop 531
STILLINGIAFETT (*n*): stillingia fat (covering the mesocarp of seeds of *Stillingia sebifera*) 532
STILLINGIAÖL (*n*): stillingia oil; tallow-seed oil from seeds of *Stillingia sebifera*): Sg. 0.9432 533
STILLLEBEN (*n*): still-life (Art) 534
STILLMITTEL (*n*): sedative; narcotic 535
STILLSCHWEIGEN: to be silent; be quiet; (*n*): silence 536
STILLSCHWEIGEND: tacit; silent 537
STILLSCHWEIGENDS: tacitly; silently 538
STILLSETZEN: to stop, shut down, bring to a standstill, put out of commission, take off the load (of plant) 539
STILLSETZUNG (*f*): stoppage, shutting down 540
STILLSTAND (*m*): standstill; pause; stop; cessation; stoppage; shutting down; (sometimes used to imply " breakdown ") 541
STILLSTEHEN: to become stationary; stand still; pause; stop; stagnate; cease working 542
STILLUNG (*f*): stilling; pause; appeasing; quieting; alleviation (see STILLEN); lactation (Med.) 543
STILLUNGSMITTEL (*n*): sedative; narcotic 544
STILPNOMELAN (*m*): stilpnomelane; Sg. 3.0 545
STILPNOSIDERIT (*n*): stilpnosiderite (an Fe_2O_3 hydrogel) (see BRAUNEISENERZ); Sg. 3.35 546
STIMMBAND (*n*): vocal cord 547
STIMME (*f*): voice; part (Music); vote (Politics); sound 548
STIMMEN: to be in tune; tune; accord; agree; balance; dispose; vote; induce; sound; (*n*): voting; agreeing; tuning 549
STIMMENMEHRHEIT (*f*): voting majority 550
STIMMGABEL (*f*): tuning fork 551
STIMMRECHT (*n*): suffrage rights 552
STIMMUNG (*f*): tuning; key; mood; frame of mind; tone; humour; disposition 553
STIMULANS (*n*): stimulant 554
STIMULANTIA (*n.pl*): stimulants (plural of STIMULANS) 555
STIMULATORENGÄRUNG (*f*): stimulative fermentation 556
STIMULIEREN: to stimulate 557
STIMULIERUNG (*f*): stimulation 558
STINKASAND (*m*): asafoetida (*Ferula foetida*) 559
STINKASANT (*m*): (see STINKASAND) 560
STINKBAUMÖL (*n*): (see JAVAOLIVENÖL) 561
STINKEN: to stink; smell; be foetid; be rancid 562
STINKEND: stinking; foetid; smelling; rancid 563
STINKFLUSS (*m*): foetid fluorspar; bituminous fluoride 564
STINKFLUSSSPAT (*m*): (see STINKFLUSS) 565
STINKHARZ (*n*): asafoetida (gum resin from roots of *Ferula foetida*) 566
STINKIG: stinking; foetid; smelling 567
STINKKALK (*m*): anthraconite; bituminous limestone (see STINKSTEIN) 568

STINKKOHLE (*f*): foetid coal; bituminous coal 569
STINKMERGEL (*f*): foetid marl; bituminous marl 570
STINKNASE (*f*): Ozaena, chronic (atrophic, foetid) Rhinitis (Med.) 571
STINKQUARZ (*m*): foetid quartz; bituminous quartz 572
STINKRAUM (*m*): gas chamber 573
STINKRAUMPROBE (*f*): gas-chamber test 574
STINKSPAT (*m*): foetid fluorspar; bituminous fluoride 575
STINKSTEIN (*m*): foetid stone; bituminous limestone (a bituminous shale found at Seefeld in the Tyrol) 576
STIPENDIUM (*m*): stipend; pension; foundation 577
STIRN (*f*): brow; forehead; front; face; crown; vertex (of arches) 578
STIRNARTERIE (*f*): frontal artery (Anat.) 579
STIRNBOGEN (*m*): frontal arch 580
STIRNE (*f*): brow; forehead; crown; vertex (of arches) 581
STIRNFLÄCHE (*f*): face (front) of an arch 582
STIRNFRÄSER (*m*): face and side cutter 583
STIRNFRÄSMASCHINE (*f*): face cutting machine, face and side cutting machine 584
STIRNGETRIEBE (*n*): spur pinion 585
STIRNHAMMER (*m*): front hammer 586
STIRNLOCKE (*f*): fore-lock 587
STIRNRAD (*n*): spur wheel; spur gear 588
STIRNRADGETRIEBE (*n*): spur gear-(ing) 589
STIRNSEITE (*f*): front; face; facade (Arch.) 590
STIRNWAND (*f*): front wall; face; facade 591
STIRNZAPFEN (*m*): gudgeon 592
STIRRHOLZ (*n*): wooden stirrer 593
STÖBERIG: flying about; drizzling; drifting; blowing about 594
STÖBERN: to rummage (fly; hunt) about; drift (of snow); drizzle (of rain) 595
STÖBERWETTER (*n*): sleet; snow; drizzle 596
STOCHERN: to stir (poke) a fire; prick; pick 597
STÖCHIOMETRIE (*f*): stoichiometry 598
STÖCHIOMETRISCH: stoichiometric-(al) 599
STOCHMEISSEL (*m*): pricker (for coal, a firing tool) 600
STOCK (*m*): stick; haft; stock; butt; handle; stump; block; trunk; post; staff; story (of buildings); clamp (Brick); floor; stem; (as a prefix = stone; completely; entirely; utterly) 601
STOCKAMBOSS (*m*): stock-anvil 602
STOCK, ANKER- (*m*): (anchor) stock 603
STOCKBLIND: stone blind 604
STOCKDEGEN (*m*): sword-stick 605
STOCKEN: to stop; falter; hesitate; slacken; decay; stagnate; mould; congeal; curdle; coagulate; stick; become (musty) mildewed; cease; solidify 606
STOCKERZ (*n*): ore in large lumps 607
STOCKFINSTER: pitch-dark 608
STOCKFISCH (*m*): stockfish; dried cod; cod, ling, hake (*Gadus morrhua*), (Icht.); blockhead; fool 609
STOCKFISCHLEBERÖL (*n*): (see LEBERTRAN) 611
STOCKFISCHLEBERTRAN (*m*): cod liver oil (see STOCKFISCHLEBERÖL and LEBERTRAN) 612
STOCKFLECK (*m*): mildew; mouldy stain 613
STOCKFLECKIG: mouldy; spotted with mould; musty; mildewed 614
STOCKHAUS (*n*): block-house 615
STOCKIG: musty; mouldy; stubborn; obstinate 616
STÖCKIG: (see STOCKIG) 617
STOCKLACK (*m*): stick-lac; varnish for (walking)-sticks 618
STOCKPUNKT (*m*): solidifying (solidification) point 619
STOCKS (*pl*): stocks (English Government stocks only) 620
STOCKSCHERE (*f*): stock shears 621
STOCKSCHLACKE (*f*): shingling slag 622
STOCKSTILL: stockstill; motionless 623
STOCKTAUB: stone deaf; deaf as a post 624
STOCKUNG (*f*): hesitation (see STOCKEN) 625
STOCKWERK (*n*): floor; story (of a building) 626
STOCKZAHN (*m*): grinder; molar (of teeth) 627
STOFF (*m*): matter; stuff; material; fabric; substance; pulp (Paper); subject-(matter) 628
STOFFANSATZ (*m*): anabolism (Biology) 629
STOFFAUSTAUSCH (*m*): exchange of material 630
STOFFBILDUNG (*f*): formation of a substance 631
STOFFBÜTTE (*f*): stuff chest (Paper) 632
STOFFDRUCK (*m*): (see ZEUGDRUCK) 633
STOFFFABRIK (*f*): pulp mill (Paper) 634
STOFFFÄNGER (*m*): save-all (Paper) 635
STOFFFÄRBUNG (*f*): pulp colouring (Paper) 636
STOFFGATTUNG (*f*): kind (class; sort) of material or substance 637
STOFF (*m*), GOLDGELBER-: canary medium (Phot.) 638
STOFFKUFE (*f*): stuff vat (Paper) 639
STOFFLEHRE (*f*): chemistry 640
STOFFLICH: material 641
STOFFLOS: unsubstantial; immaterial 642
STOFFMENGE (*f*): quantity (amount) of material or substance 643
STOFFMILCH (*f*): pulp milk (thinned paper pulp) 644
STOFFMÜHLE (*f*): pulping engine; hollander; stuff engine (Paper) 645
STOFFPATENT (*n*): chemical patent; patent on chemical substances; material patent 646
STOFFRAHMEN (*m*): filter disc 647
STOFFTEILCHEN (*n*): particle of matter 648
STOFFVERBRAUCH (*m*): consumption of (material) substance or matter 649
STOFFWECHSEL (*m*): (inter)-change of substance; metabolism 650
STOFFWECHSELPRODUKT (*n*): product of metabolism 651
STOFFWECHSELPROZESS (*m*): metabolic process 652
STOFFWECHSELVORGANG (*m*): metabolic process 653
STOFFZAHL (*f*): number of (materials) substances 654
STÖHNEN: to groan; (*n*): groan-(ing) 655
STOLLEN: to prop; furnish with posts; soften (Skins); (*m*): gallery; heading; tunnel; working; drift; slice of bread and butter; prop; post 656
STOLLENBAU (*m*): heading; working (of mines) 657
STOLLENHOLZ (*n*): sixth section, log of wood cut into six parts 658
STOLLENSCHACHT (*m*): shaft of mines 659
STOLPERN: to stumble; blunder; trip 660

STOLZ: proud; arrogant; presumptuous; haughty; lofty; stately; superb; (*m*): pride; arrogance, etc. 661
STOLZIEREN: to strut; be arrogant 662
STOLZIT (*m*): stolzite; $PbWO_4$; Sg. 8.0 663
STOMACHALE (*n*): stomachic (an agent for increasing the secretion of the stomach) 664
STOMACHALMITTEL (*n*): (see STOMACHALE) 665
STOMACHICUM (*n*): (see STOMACHALE) 666
STOPFBÜCHSE (*f*): stuffing (box) gland; packing gland 667
STOPFBÜCHSENBRILLE (*f*): stuffing box gland 668
STOPFBÜCHSENPACKUNG (*f*): gland packing; stuffing box packing 669
STOPFBÜCHSE (*f*), **STEVENROHR-**: stern-tube gland 670
STÖPFEL (*m*): plug; stopper; cork 671
STOPFEN to stop; stuff; darn; fill; constipate; be costive; caulk (Ships); stem; obstruct; check; cram; (*m*): stopper (Brick); plug; cork 672
STOPFEND: (see STOPFEN); styptic; astringent 673
STOPFGARN (*n*): darning (yarn) wool 674
STOPFMITTEL (*n*): styptic; astringent (Med.) 675
STOPFNADEL (*f*): darning needle 676
STOPFROHR (*n*): stopper pipe, sleeve (Brick) 677
STOPFVENTIL (*n*): stop valve 678
STOPFWERG (*n*): oakum (for caulking purposes) 679
STOPPEL (*f*): stubble 680
STOPPELFELD (*n*): stubble-field 681
STOPPEN: to stop 682
STOPPER (*m*): stopper 683
STOPPUHR (*f*): stop-watch 684
STÖPSEL (*m*): plug; stopper; cork 685
STÖPSELFLASCHE (*f*): stoppered bottle 686
STÖPSELHAHN (*m*): stopper cock 687
STÖPSELN: to plug; stopper; cork 688
STÖPSELSCHALTER (*m*): plug commutator; plug switch 689
STÖPSELSCHNUR (*f*): plug cord 690
STÖPSELSICHERUNG (*f*): plug fuse 691
STÖPSELUMSCHALTER (*m*): plug commutator 692
STÖR (*m*): sturgeon 693
STORAX (*m*): storax (balsam), styrax (from bark of *Liquidambar orientale*); Sg. 0.89-1.1; Bp. 150-300°C 694
STORAXBAUM (*m*): Styrax tree (*Liquidambar orientale*) 695
STORAXÖL (*n*): storax oil; styrax oil; Sg. 0.89-1.1; Bp. 150-300°C 696
STORCH (*m*): stork 697
STORCHSCHNABEL (*m*): pantograph; crane's (stork's) bill; (also plant, "*Geranium*") 698
STÖREN: to upset; disturb; derange; annoy; trouble; interfere with; stir; poke; interrupt; perturb 699
STÖRFREIHEIT (*f*): freedom from interruption (Wireless Tel.) 700
STORNIEREN: to cancel; delete 701
STÖRRIG: troublesome; refractory; obstinate 702
STÖRRIGKEIT (*f*): obstinacy 703
STÖRUNG (*f*): derangement; trouble; disorder; disturbance; interruption; alteration; conversion; perturbation 704
STOSS (*m*): impact; impulse; push; blow; thrust; stroke; shock; percussion; jolt; recoil; collision; bump; pile; file (Papers); joint; blast (of trumpet); stab; kick; butt-(joint); chock; hammer (in pipelines) 705
STOSSBEWEGUNG (*f*): forward (projectile) motion; ramming (action) movement 706
STOSSBLECH (*n*): butt strap 707
STOSSBOLZEN (*m*): butt bolt 708
STÖSSEL (*m*): ram; stamper; pestle; pounder; knocker 709
STOSSEN: to push; bump; hit; ram; kick; knock; pound; pulverise; join; dash; thrust; recoil; stamp; butt; meet 710
STÖSSER (*m*): pounder; knocker; ram; pestle; stamper; tappet; cam; stop (of a door, valve, etc.) 711
STOSSERDE (*f*): stiff clay 712
STOSSFUGE (*f*): (upright)-joint 713
STOSSKANTE (*f*): butt edge 714
STOSSMASCHINE (*f*): shaping (grooving) machine 715
STOSS (*m*), **PLANKEN-**: plank butt 716
STOSSPLATTE (*f*): butt plate; butt strap 717
STOSSPUNKT (*m*): point of impact 718
STOSSRING (*m*): locking ring 719
STOSSVOGEL (*m*): bird of prey 720
STOSSWEISE: jerkily; in a jolting manner; by starts or jerks 721
STOSSWELLE (*f*): percussion wave 722
STOSSWIND (*m*) gust (of wind) 723
STOSSWINKEL (*m*): angle of incidence; bosom piece 724
STOSSZÜNDER (*m*): percussion fuse 725
STOTTERN: to stutter; falter; stammer; hesitate 726
STRAAK (*m*): sheer 727
STRAAKGEWICHT (*n*): sheer iron 728
STRAAKLATTE (*f*): batten for drawing curves; sheer lath 729
STRABISMUS (*m*): strabismus, squint (Med.) 730
STRACKS: straightway; immediately 731
STRAFBAR: punishable; culpable; guilty; criminal 732
STRAFBARKEIT (*f*): culpability 733
STRAFE (*f*): punishment; fine; penalty; chastisement; correction 734
STRAFEN: to punish; fine; penalize; reprove; rebuke; chastise; correct; reprehend 735
STRAFF: tense; stretched; under tension; tight; rigid; extended 736
STRAFFÄLLIG: culpable; punishable 737
STRAFFHEIT (*f*): tension; tightness; rigidity 738
STRAFGELD (*n*): fine; penalty; forfeit 739
STRAFGESETZ (*n*): penal law 740
STRÄFLICH: severe-(ly); criminal; wrong; punishable; culpable; reprehensible 741
STRAFLOS: innocent; guiltless; unpunished; exempt from punishment 742
STRAFLOSIGKEIT (*f*): impunity 743
STRAHL (*m*): ray; beam; jet; shaft; lightning flash; radius; (straight)-line (Graphs); spoke (Wheel) 744
STRAHLAPPARAT (*m*): jet apparatus; injector; pump; blower, etc. 745
STRAHLASBEST (*m*): plumose asbestos (see ASBEST) 746
STRAHLBARYT (*m*): Bologna stone; radiated barite 747

STRAHLBLENDE (*f*): Sphalerite (see SPHALERIT) 748
STRAHLDÜSE (*f*): spray-(ing) nozzle, jet: ejector-(nozzle) 749
STRAHLEN: to radiate; emit rays; beam 750
STRAHLENART (*f*): kind of rays 751
STRAHLENBRECHEND: refractive; refracting 752
STRAHLENBRECHUNG (*f*): refraction 753
STRAHLENBRECHUNGSMESSER (*m*): refractometer (an instrument for determining the refractive index) 754
STRAHLENBÜNDEL (*n*): pencil of rays 755
STRAHLENBÜSCHEL (*m*): pencil of rays 756
STRAHLEND: radiating; radiant 757
STRAHLENDE WÄRME (*f*): radiant (radiating) heat 758
STRAHLENFIGUR (*f*): radiating figure 759
STRAHLENFÖRMIG: radiate-(d) 760
STRAHLENGLIMMER (*m*): striated mica (see GLIMMER) 761
STRAHLENKEGEL (*m*): cone of rays 762
STRAHLENKUPFER (*n*): clinoclasite (see STRAHL-ERZ) 763
STRAHLENMESSER (*m*): radiometer; actinometer (instrument for measuring actinic power of sun's rays) 764
STRAHLENPILZ (*m*): *Actinomyces* 765
STRAHLENSTEIN (*m*): actinolite; amianthus (see STRAHLSTEIN) 766
STRAHLENWERFEN (*n*): radiation 767
STRAHLER (*m*): radiator 768
STRAHLERZ (*n*): clinoclasite; $6CuO, As_2O_5, 3H_2O$; Sg. 4.3 769
STRAHLGIPS (*m*): fibrous gypsum (see GIPS) 770
STRAHLIG: radiated (Mineralogy); radiant 771
STRAHLKEIL (*m*): belemnite (skeleton of extinct animal allied to the cuttle-fish) (Palæontology) 772
STRAHLKIES (*m*): marcasite (see MARKASIT) 773
STRAHLPUMPE (*f*): jet pump; injector 774
STRAHLPUNKT (*m*): radiating (radiant) point 775
STRAHLQUARZ (*m*): fibrous quartz (see QUARZ) 776
STRAHLSCHÖRL (*m*): radiated tourmaline (see TURMALIN) 777
STRAHLSTEIN (*m*): nephrite, actinolite; amphibole, amianthus (see NEPHRIT and AMPHIBOL) Sg. 3.15 778
STRAHLUNG (*f*): radiance; radiation 779
STRAHLUNGSVERMÖGEN (*n*): radiating power 780
STRAHLUNGSWÄRME (*f*): (heat of) radiation; radiant heat 781
STRAHLUNTERBRECHER (*m*): jet-interrupter 782
STRAHLWÄRME (*f*): radiant heat 783
STRAHLZEOLITH (*m*): stilbite (see STILBIT) 784
STRÄHNE (*f*): hank; skein; strand; wisp 785
STRAITSZINN (*n*): straits tin (silverwhite in colour) (from Straights Settlements) 786
STRALSTEIN (*m*): nephrite (see NEPHRIT and STRAHLSTEIN) 787
STRAMM: tight; tense; rigid; robust; strict; stretched; straight; at attention (Military) 788
STRAMPELN: to kick; struggle 789
STRAND (*m*): strand; bank; shore 790
STRANDEN: to be stranded; be (run) aground; be shipwrecked 791
STRANDNELKE (*f*): sea lavender (*Limonium*) 792
STRANDPFLANZE (*f*): seaside (littoral) plant 793
STRANFA (*f*): straw yarn (from chemical treatment of straw; used for sack-making, etc.) 794
STRANG (*m*): skein; rope; cord; halter; string; (pipe)- line; strand; shaft 795
STRANGERKRANKUNG (*f*): degeneration (Med.) 796
STRANGERKRANKUNG DES RÜCKENMARKS: degeneration of the spinal cord (Med.) 797
STRANGFARBIG: dyed in the yarn 798
STRANGPRESSE (*f*): wire cutting press; trace press 799
STRANGULIEREN: to strangle; (*n*): strangulation 800
STRAPAZE (*f*): toil; hardship; fatigue 801
STRASS (*n*): (see MAINZER FLUSS) 802
STRASSE (*f*): mill, machine; street; (high)-way; (high)-road; strait (Naut.); (as a suffix = mill (Mach.)) 803
STRASSENAUFSEHER (*m*): road surveyor 804
STRASSENBAHN (*f*): tramway 805
STRASSENBAHNSCHIENE (*f*): tramline, tramway rail 806
STRASSENBAHNWAGEN (*m*): tram-car,. street-railway car 807
STRASSENBAU (*m*): road-making 808
STRASSENBELEUCHTUNG (*f*): street lighting 809
STRASSENBETONIERUNG (*f*): road-reinforcement 810
STRASSENFAHRZEUG (*n*): road vehicle 811
STRASSENPFLASTER: paving 812
STRASSENRAUB (*m*): highway-robbery 813
STRASSENSPRENGUNG (*f*): street (road) watering 814
STRASSENTRANSPORT (*m*): road transport 815
STRASSENWALZE (*f*): road-roller, steam-roller 816
STRATIFIZIEREN: to stratify 817
STRÄUBEN: to bristle; resist; ruffle-up; strive against 818
STRÄUBIG: rough; coarse (of wool); rebellious; unwilling; resisting 819
STRAUBIG: (see STRÄUBIG) 820
STRAUCH (*m*): bush; shrub 821
STRAUCHARTIG: shrublike 822
STRAUCHELN: to stumble; fail; blunder 823
STRAUCHIG: bushy: shrubby 824
STRAUCHWERK (*n*): underwood; undergrowth; shrubs; bushes; scrub 825
STRAUSS (*m*): bouquet; nosegay; bunch; crest; bush; tuft; combat; strife; ostrich 826
STRAUSSFEDER (*f*): ostrich feather 827
STREBE (*f*): stay; prop; buttress; post; strut; brace; diagonal; leg 828
STREBEN: to strive; struggle; press; tend (towards or away from); endeavour; aspire; (*n*): endeavour; effort; tendency 829
STREBEPFEILER (*m*): buttress 830
STREBSAM: industrious; aspiring; active; ambitious; assiduous 831
STRECKBALKEN (*m*): cross (joist) beam; binding piece; transverse beam 832
STRECKBAR: extensible; malleable; ductile 833
STRECKE (*f*): section; distance; finite line; line of given length (Geom.); space; stretch; extent; distance; drawing (Metal); drift (Mining); tract 834

STRECKE (*f*), **EBENE-**: level, level run, level stretch 835
STRECKEN: to distend; extend; increase; stretch; flatten; spread; draw-(out); roll (Metals, etc.); lengthen; elongate; put down; lay (a keel) 836
STRECKENBOGEN (*m*): frame (for mines) 837
STRECKENFÖRDERUNG (*f*): aerial (cableway) ropeway; transporter, conveyor 838
STRECKENLAST (*f*): distributed (partial) load (on girders, etc.) 839
STRECKENTEILUNG (*f*): dividing up (division) of a given length, section, line or distance (Geom.) 840
STRECKENZUSATZMASCHINE (*f*): booster (Elect.) 841
STRECKER (*m*): stretcher; extensor (Med.); flattener (Glass) 842
STRECKGRENZE (*f*): yield point; limit of elasticity; elastic limit; upper yield limit 843
STRECKMETALL (*n*): expanded metal 844
STRECKOFEN (*m*): flattening furnace (Glass) 845
STRECKSTAHL (*m*): rolled steel 846
STRECKWERK (*n*): rolls; rolling mill (Metal) 847
STREICH (*m*): stroke; prank; trick; blow; stripe; lash 848
STREICHBAR: strokeable; plastic 849
STREICHBÜRSTE (*f*): paint (varnish) brush 850
STREICHE (*f*): spatula 851
STREICHEISEN (*n*): skimming iron, skimmer (for molten iron) 852
STREICHELN: to stroke; caress 853
STREICHEN: to stroke; rub; whet; strike (flags or matches); paint; varnish; coat; erase; cancel; mould (Brick); card (Wool) spread (Butter); scrape (Skins); stain or coat (Paper); move; rush; sweep; pass -(over); come in contact with; impinge; roam; rove; wander; run (in a given direction); touch; smooth; graze; lower (Sails); scrub (of gases on surfaces) 854
STREICHFÄHIGKEIT (*f*): smearing capacity (i.e., about the consistency of butter) 855
STREICHFARBE (*f*): stain-(ing colour) (Paper) 856
STREICHFLÄCHE (*f*): striking surface (Matches); surface of contact (Gases, etc.) 857
STREICHHOLZ (*n*): match 858
STREICHHÖLZCHEN (*n*): match 859
STREICHHOLZPAPIER (*n*): match paper 860
STREICHINSTRUMENT (*n*): stringed instrument 861
STREICHKAPPE (*f*): friction (striking) cap 862
STREICHKASTEN (*m*): colour tub (Dyeing) 863
STREICHKONTAKT (*m*): rubbing contact 864
STREICHMASS (*n*): measure 865
STREICHMASSE (*f*): friction composition (for matches) 866
STREICHMUSTER (*m*): stained paper pattern (Paper) 867
STREICHOFEN (*m*): reverberatory furnace 868
STREICHRIEMEN (*m*): razor strop 869
STREICHSTEIN (*m*): hone; touchstone 870
STREICHTORF (*m*): pressed peat 871
STREICHZÜNDHOLZ (*n*): match 872
STREIF (*m*): band; streak; strip; tape; stripe; strap; track; vein 873
STREIFEN: to stripe; striate; streak; flute; channel; strip; graze; wander; ramble; strip off; roam; sweep over; scrub (Gases over heating surface); touch upon 874
STREIFIG: streaked; striated; streaky; veined; striped 875
STREIFKASTEN (*m*): settling tank, clearing tank 876
STREIK (*m*): strike (of employees against employers) 877
STREIT (*m*): strife; combat; contest; dispute; debate; fight; quarrel; fray; disagreement; difference; discord 878
STREITBAR: valiant; warlike; contestable; questionable; disputable 879
STREITEN: to struggle; fight; contest; combat; question; oppose; dispute; contend; debate; quarrel; differ; disagree 880
STREITFERTIG: ready for the fray 881
STREITFRAGE (*f*): question at issue; matter of dispute; debateable (point) question 882
STREITHANDEL (*m*): litigation, law-suit, action 883
STREITIG: disputed; questionable; disputable 884
STREITKRÄFTE (*f.pl*): opposing forces 885
STREITPUNKT (*m*): the question at issue; the debateable (question) point; point of dissension 886
STREITSACHE (*f*): disputable (question) matter; action; law-suit 887
STRENG: severe; rigorous; strict; harsh; rigid 888
STRENGE (*f*): severity; rigour; rigidity 889
STRENGFLÜSSIG: refractory; difficult to fuse 890
STRENGGENOMMEN: strictly speaking 891
STRENGIT (*m*): strengite; H_2FePO_6; Sg. 2.87 892
STRENGLOT (*n*): hard solder (see **SCHLAGLOT**) 893
STREU (*f*): litter 894
STREUBLAU (*n*): powder blue (the coarsest form of Cobalt blue; see **BLAUFARBE**) 895
STREUDÜSE (*f*): spraying nozzle, atomizer, distributing nozzle, diffuser 896
STREUEN: to spray; strew; scatter 897
STREUFARBE (*f*): powdered colour 898
STREUGLANZ (*m*): brass powder 999
STREUGOLD (*n*): gold dust 900
STREUHÄRTEPULVER (*n*): case-hardened powder 901
STREUKUPFER (*n*): copper rain (from surface of molten metal) 902
STREUPULVER (*n*): dusting (strewing) powder; a mixture of boracic powder and French chalk (for bed sores); lycopodium; bleaching powder 903
STREUSAND (*m*): sand for sanding floors 904
STREUUNGSBEREICH (*m*): region (range) of distribution 905
STREUZUCKER (*m*): powdered sugar 906
STRICH (*m*): stroke; streak; line; dash; train (of powder); grain (of wood); nap (of cloth); batch (of bricks); direction; course; flock; brood; tract; region; hyphen; comma; bar (Music); notch; flaw (of glass); flight (migration) of birds 907
STRICHELN: to streak; hatch; shade; break; (chain)- dot 908
STRICHKULTUR (*f*): streak culture 909
STRICHPUNKT (*m*): semicolon 910
STRICHPUNKTIERT: chain dotted (of lines on drawings) 911
STRICHVOGEL (*m*): bird of passage 912
STRICHZEIT (*f*): time for the migration of birds 913

STRICK (*m*): cord; string; rope; line; snare; halter 914
STRICKARBEIT (*f*): knitting 915
STRICKEN: to knit; net; reticulate 916
STRICKFÖRMIG: ropelike; cordlike; restiform (Anat.) 917
STRICKGARN (*n*): knitting (yarn) wool 918
STRICKLEITER (*f*): rope-ladder 919
STRICKNADEL (*f*): knitting-needle 920
STRICKWERK (*n*): ropes, cordage; reticulated work, network 921
STRICKWOLLE (*f*): knitting wool 922
STRIEGEL (*f*): curry-comb 923
STRIEGENER ERDE (*f*): kaolin (see BOLUS, WEIS-SER-) 924
STRIEME (*f*): stripe, weal, streak 925
STRIKT: strict 926
STRINGER (*m*): stringer 927
STRINGER (*m*), BALKEN-: beam stringer 928
STRINGERWINKEL (*m*): stringer angle 929
STRIPPER (*m*): stripper (for stripping the KOKILLE from the casting) 930
STRITTIG: questionable; contested; in dispute 931
STROH (*n*): straw 932
STROHARBEIT (*f*): straw-work; thatching 933
STROHBLUME (*f*): immortelle; everlasting (*Helichrysum bracteatum*) (Bot.); artificial straw flower 934
STROHBUND (*n*): truss (of straw) 935
STROHBUTTER (*f*): straw butter (from animals fed on straw) 936
STROHDACH (*n*): thatched roof 937
STROHDECKE (*f*): straw (covering) casing; straw-mat 938
STROHERN: straw; of straw 939
STROHFARBIG: straw-coloured 940
STROHFASERGARN (*n*): straw yarn (see STRANFA) 941
STROHFEILE (*f*): straw file 942
STROHFLACHS (*m*): raw flax (see FLACHS) 943
STROHGEFLECHT (*n*): straw-plaiting 944
STROHGELB: straw (yellow) coloured 945
STROHHALM (*m*): a straw 946
STROHHÜLSE (*f*): straw (covering; case; cover) envelope, (for bottles) 947
STROHHUT (*m*): straw hat 948
STROHHUTFARBE (*f*): straw-hat dye 949
STROHHUTLACK (*m*): straw-hat varnish 950
STROHHUTREINIGUNGSMITTEL (*n*): straw-hat (cleaner; cleaning material) cleaning agent (such as salts of lemon) 951
STROHHÜTTE (*f*): thatched hut 952
STROHIG: straw-like; like straw 953
STROHKESSEL (*m*): straw boiler (Paper) 954
STROHLAGER (*n*): straw (litter) bed 955
STROHMANN (*m*): dummy (Whist); scarecrow 956
STROHMATTE (*f*): straw-mat 957
STROHPAPIER (*n*): straw paper (Paper) 958
STROHPAPPE (*f*): straw board (Paper) 959
STROHSEIL (*n*): rope of twisted straw 960
STROHSTOFF (*m*): straw (pulp) stuff (Paper) 961
STROHWEIN (*m*): straw wine 962
STROHWISCH (*m*): straw wisp 963
STROHZELLSTOFF (*m*): (see ESPARTOGRAS) 964
STROHZEUG (*n*): straw (stuff) pulp (Paper) 965
STROLCH (*m*): loafer; tramp 966
STROM (*m*): stream; flow; current (Elect.); torrent; flood 967
STROMAB: down-stream 968
STROMABLEITUNG (*f*): shunt 969
STROMABNEHMER (*m*): collector, brush (Elect.) 970
STROMABWÄRTS: down stream 971
STROMABWEICHUNG (*f*): current (variation) fluctuation 972
STROMANKER (*m*): stream anchor 973
STROMANZIEGER (*m*): current indicator (see STROMZEIGER) 974
STROMART (*f*): type of current (Elect.) 975
STROMAUF: up-stream 976
STROMAUFWÄRTS: up-stream 977
STROMBETT (*n*): river-bed 978
STROMDICHTE (*f*): current density 979
STROMDICHTIGKEIT (*f*): current density 980
STROMDURCHGANG (*m*): passage of current 981
STROMEINHEIT (*f*): unit of current 982
STRÖMEN: to flow; pour; stream; pass; flood; sweep 983
STROM (*m*), ERD-: terrestrial current; earth current 984
STROMERZEUGER (*m*): current generator 985
STROMEYERIT (*m*): stromeyerite (natural sulphide of copper and silver, about 30% Cu. and 50% Ag); $(Ag, Cu)_2S$; Sg. 6.25 986
STROMFELD (*n*): field (of current) 987
STROMGESCHWINDIGKEIT (*f*): velocity of current 988
STROMGESCHWINDIGKEITSMESSER (*m*): current (flow) meter; hydrometer 989
STROM (*m*), GLEICHGERICHTETER-: rectified current 990
STROMGLEICHRICHTER (*m*): current rectifier 991
STROMINDIKATOR (*m*): current indicator (see STROMZEIGER) 992
STROMINDUKTION (*f*): induction of current; electromagnetic induction 993
STROM (*m*), INDUZIERENDER-: inducing current 994
STROM (*m*), INTERMITTIERENDER-: intermittent current 995
STROMKREIS (*m*): circuit (Elect.) 996
STROMKREIS (*m*), ÄUSSERER-: external circuit 997
STROMKREIS (*m*), GESCHLOSSENER-: closed circuit 998
STROMKREIS (*m*), INDUZIERENDER-: inducing (induction) circuit 999
STROMKREIS (*m*), INDUZIERTER-: induced circuit 000
STROMKREIS (*m*), INNERER-: inner circuit 001
STROMKREIS (*m*), MAGNETISCHER-: magnetic circuit 002
STROMKREIS, OFFENER- (*m*): open circuit; broken circuit 003
STROMKREIS SCHLIESSEN: to close the circuit 004
STROMKURVE (*f*): current curve 005
STROMLAUF (*m*): flow of current; circulation 006
STROMLEITER (*m*): conductor (Elect.) 007
STROMLEITUNG (*f*): conduction (Elect.) 008
STROMLOS: without current 009
STROMLOSIGKEIT (*f*): absence of current 010
STROMMENGE (*f*): amount (strength) of current (Elect.) 011
STROMMESSER (*m*): ammeter; amperemeter (Elect.); current (flow) meter (Water meters) 012
STROM (*m*), NEGATIVER-: negative current 013
STROM (*m*), OFFENER-: open current 014
STROM (*m*), POSITIVER-: positive current 015

STROM (*m*), **PRIMÄR-**: primary current 016
STROM (*m*), **PRIMÄRER-**: primary current 017
STROMPRÜFER (*m*): current detector 018
STROMQUELLE (*f*): source of current (Elect.) 019
STROMREGLER (*m*): current regulator 020
STROMREGULATOR (*m*): current regulator 021
STROMRICHTUNG (*f*): direction of current 022
STROMSAMMLER (*m*): collector 023
STROMSCHLUSS (*m*): closing (of a current) 024
STROMSCHLÜSSEL (*m*): switch (Elect.) 025
STROMSCHWANKUNG (*f*): current fluctuation 026
STROM (*m*), **SEKUNDÄRER-**: secondary current 027
STROMSPANNUNG (*f*): current (potential) voltage 028
STROMSTÄRKE (*f*): current (intensity) strength 029
STROMSTOSS (*m*): current (pulsation) impulse; water-hammer 030
STROM (*m*), **THERMO-**: thermo-electric current 031
STROMUMKEHRER (*m*): commutator (Elect.); current reverser 032
STROMUMKEHRUNG (*f*): reversal of current 033
STRÖMUNG (*f*): flow; pouring; streaming; flowing; flood; stream; current; sweeping; flux; circulation 034
STRÖMUNGSENERGIE (*f*): flow energy, kinetic energy 035
STRÖMUNGSMESSER (*m*): flow-meter 036
STROMUNTERBRECHER (*m*): interrupter; circuit breaker (Elect.) 037
STROMVERBRAUCH (*m*): current consumption 038
STROMVERLUST (*m*): current loss 039
STROMWECHSLER (*m*): commutator (Elect.) 040
STROMWEISE: in torrents; in a stream 041
STROMWENDER (*m*): commutator (Elect.); current reverser 042
STROMZEIGER (*m*): ammeter; amperemeter (Elect.); current indicator 043
STROMZEIGER (*m*), **WEICHEISEN-**: moving iron ammeter 044
STROMZINN (*n*): stream tin; grain cassiterite (see **CASSITERIT**) 045
STROMZUFÜHRUNG (*f*): current conducting; current conduction; current supply; current lead 046
STRONTIAN (*m*): strontia; strontium (see **STRONTIUM**) 047
STRONTIANERDE (*f*): strontia; (SrO); strontium oxide (see **STRONTIUMOXYD**) 048
STRONTIANHALTIG: containing strontia; having a strontia content 049
STRONTIANHYDRAT (*n*): strontium hydrate (see **STRONTIUMHYDROXYD**) 050
STRONTIANIT (*m*): strontianite (natural strontium carbonate); $SrCO_3$; Sg. 3.7 (see also **STRONTIUMCARBONAT**) 051
STRONTIANSALPETER (*m*): strontium nitrate (see **STRONTIUMNITRAT**) 052
STRONTIANSALZ (*n*): strontia (strontium) salt 053
STRONTIANWASSER (*n*): strontia water 054
STRONTIANZUCKER (*m*): strontium sucrate 055
STRONTIUM (*n*): strontium; (Sr); (see also **STRONTIAN**) 056
STRONTIUMACETAT (*n*): strontium acetate; $Sr(C_2H_3O_2)_2.\frac{1}{2}H_2O$ 057
STRONTIUMACETAT (*n*), **WASSERFREIES-**: anhydrous strontium acetate; $Sr(C_2H_3O_2)_2$; Sg. 2.099 058
STRONTIUMAZETAT (*n*): strontium acetate (see **STRONTIUMACETAT**) 059
STRONTIUMBROMAT (*n*): strontium bromate; $Sr(BrO_3)_2+H_2O$; Sg. 3.773 060
STRONTIUMBROMID (*n*): strontium bromide; $SrBr_2$, Sg. 4.216; $SrBr_2+6H_2O$, Sg. 2.358 061
STRONTIUMCARBID (*n*): strontium carbide; SrC_2; Sg. 3.19 062
STRONTIUMCARBONAT (*n*): strontium carbonate; $SrCO_3$; Sg. 3.62 063
STRONTIUMCHLORAT (*n*): strontium chlorate; $Sr(ClO_3)_2$; Sg. 3.152 064
STRONTIUMCHLORID (*n*): strontium chloride; $SrCl_2+6H_2O$, Sg. 1.954-1.964; $SrCl_2$, Sg. 3.054 065
STRONTIUM (*n*), **CHLORSAURES-**: strontium chlorate 066
STRONTIUMCHROMAT (*n*): strontium chromate; $SrCrO_4$; Sg. 3.353 067
STRONTIUMDIOXYD (*n*): strontium dioxide (see **STRONTIUMPEROXYD**) 068
STRONTIUM (*n*), **ESSIGSAURES-**: strontium acetate 069
STRONTIUMFLUORID (*n*): strontium fluoride; SrF_2; Sg. 2.44 070
STRONTIUMHYDRAT (*n*): strontium hydroxide (see **STRONTIUMHYDROXYD**) 071
STRONTIUMHYDROXYD (*n*): strontium hydroxide; $Sr(OH)_2+8H_2O$, Sg. 1.396; $Sr(OH)_2$ Sg. 3.625 072
STRONTIUMJODID (*n*): strontium iodide; SrI_2; Sg. 4.549 073
STRONTIUM (*n*), **KOHLENSAURES-**: strontium carbonate (see **STRONTIUMCARBONAT**) 074
STRONTIUMLAKTAT (*n*): strontium lactate; $Sr(C_3H_5O_3)_2.H_2O$ 075
STRONTIUMMETASILIKAT (*n*): strontium metasilicate; $SrSiO_3$; Sg. 3.91 076
STRONTIUM (*n*), **MILCHSAURES-**: strontium lactate (see **STRONTIUMLAKTAT**) 077
STRONTIUMMOLYBDAT (*n*): strontium molybdate; $SrMoO_4$; Sg. 4.145 078
STRONTIUMNITRAT (*n*): strontium nitrate; $Sr(NO_3)_2$; Sg. 2.93; Mp. 645°C 079
STRONTIUMNITRIT (*n*): strontium nitrite; $Sr(NO_2)_2$; Sg. 2.867 080
STRONTIUMORTHOSILIKAT (*n*): strontium orthosilicate; Sr_2SiO_4; Sg. 3.84 081
STRONTIUMOXALAT (*n*): strontium oxalate 082
STRONTIUM (*n*), **OXALSAURES-**: strontium oxalate 083
STRONTIUMOXYD (*n*): strontium oxide; SrO; Sg. 4.34 084
STRONTIUMOXYDHYDRAT (*n*): strontium hydroxide; $[Sr(OH)_2+8H_2O.]$ (see **STRONTIUM-HYDROXYD**) 085
STRONTIUMPERBORAT (*n*): strontium perborate 086
STRONTIUMPEROXYD (*n*): strontium peroxide, strontium dioxide; SrO_2; Sg. 0.456 087
STRONTIUMSALICYLAT (*n*): strontium salicylate; $Sr(C_7H_5O_3)_2.2H_2O$ 088
STRONTIUM (*n*), **SALIZYLSAURES-**: (see **STRONTIUMSALICYLAT**) 089
STRONTIUM (*n*), **SALPETERSAURES-**: strontium nitrate (see **STRONTIUMNITRAT**) 090
STRONTIUMSALZ (*n*): strontium salt 091
STRONTIUM (*n*), **SCHWEFELSAURES-**: strontium sulphate (see **STRONTIUMSULFAT**) 092
STRONTIUMSULFAT (*n*): strontium sulphate; $SrSO_4$; Sg. 3.71-3.91; Mp. 1605°C 093

STRONTIUMSULFID (*n*): strontium sulphide; SrS; Sg. 3.72 094
STRONTIUMTARTRAT (*n*): strontium tartrate 095
STRONTIUMWASSERSTOFF (*m*): strontium hydride 096
STRONTIUM (*n*), WEINSAURES- : strontium tartrate 097
STRONTIUMWOLFRAMAT (*n*): strontium wolframate; strontium tungstate; $SrWO_4$; Sg. 6.184 098
STRONTIUM, WOLFRAMSAURES- (*n*): (see STRONTIUMWOLFRAMAT) 099
STROPHANTHUSÖL (*n*): strophanthus oil (from seeds of *Strophantus hispidus*); Sg. 0.925 100
STROPHANTIN (*n*): strophantine (a glycoside from *Strophantus kombe*) 101
STROPHANTUSSAMEN (*m.pl*): strophantus seeds (*semen strophanti*) (from *Strophantus kombe*) 102
STROPHE (*f*): strophe 103
STROPP (*m*) (also *n*): strap (*m*); strop (*n*) 104
STROTZEN: to swell (up); be (swollen) exuberant; abound in 105
STROTZEND: strutting; exuberant 106
STRUDEL (*m*): vortex; eddy; whirlpool 107
STRUDELN: to whirl; eddy; gush 108
STRUKTUR (*f*): structure; composition; texture 109
STRUKTURFARBE (*f*): structure colour 110
STRUKTURFORMEL (*f*): structural formula 111
STRUKTURIDENTISCH: structurally identical 112
STRUKTURLOS: amorphous; structureless 113
STRUKTURVERÄNDERUNG (*f*): structural (change; fluctuation; variation) alteration 114
STRUMA (*m*): goitre (Med.) 115
STRUMPF (*m*): stocking; hose; (incandescent) mantle 116
STRUMPFBAND (*n*): garter 117
STRUMPFFABRIK (*f*): stocking factory 118
STRUMPFHÄNDLER (*m*): hosier 119
STRUMPFWAREN (*f.pl*): hosiery 120
STRUMPFWARENPAPIER (*n*): hosiery paper (Paper) 121
STRUMPFWEBER (*m*): stocking-maker 122
STRUNK (*m*): trunk; stump; stock; stalk; stem 123
STRUPPIG: shaggy; rough; bristly 124
STRÜVERIT (*m*): strüverite; $FeO.(Ta,Na)_2O_5.4TiO_2$ 125
STRUVIT (*m*): struvite; $(NH_4)MgPO_4.6H_2O$; Sg. 1.7 126
STRYCHNIN (*n*): strychnine (*Strychninum*); $C_{21}H_{22}O_2N_2$; Mp. 268°C (an alkaloid from the seeds of *Strychnos nux vomica*) 127
STRYCHNINGETREIDE (*n*): strychnine (cereals, grain) corn 128
STRYCHNIN (*n*), SALPETERSAURES- : strychnine nitrate (*Strychninae nitras*) 129
STRYCHNIN (*n*), SCHWEFELSAURES- : strychnine sulphate (*Strychninae sulphas*) 130
STRYCHNINVERGIFTUNG (*f*): strychnine poisoning 131
STRYCHNOSSAMENÖL (*n*): strychnos seed oil (from seeds of *Strychnos nux vomica*) 132
STUBBFETT (*n*): stubb fat (see STUPPFETT) 133
STUBE (*f*): apartment; chamber; room 134
STUCCATUR (*f*): stucco 135
STUCCATURARBEIT (*f*): stucco-work 136
STUCK (*m*): stucco-work 137
STÜCK (*n*): piece; parcel; lump; bit; lot; fragment; part; bloom (Metal); piece (of artillery); number; head (of cattle) 138
STÜCKARBEIT (*f*): piece-work 139
STUCKATEUR (*m*): stucco worker 140
STUCKATUR (*f*): stucco 141
STÜCKCHEN (*n*): particle; small (part) portion; little piece; bit; morsel 142
STÜCKELN: to break (cut) into pieces 143
STÜCKEN: to patch; piece together; cut into pieces 144
STÜCKENZUCKER (*m*): lump (crushed) sugar 145
STÜCKERZ (*n*): lump ore 146
STÜCKFÄRBER (*m*): piece dyer 147
STÜCKFARBIG: dyed in the piece 148
STÜCKFORM (*f*): gun-mould 149
STÜCKGIESSEREI (*f*): gun foundry 150
STÜCKGIPS (*m*): plaster of Paris 151
STÜCKGUT (*n*): piece goods; gun metal 152
STÜCKKOHLE (*f*): lump coal; cherry coal; coarse lump coal (size over 80 mm.) 153
STÜCKLISTE (*f*): detail list 154
STÜCKMETALL (*n*): gun metal 155
STÜCKOFEN (*m*): bloomery furnace (see SCHACHTOFEN) 156
STÜCKPFORTE (*f*): port-(hole) (for a gun) 157
STUCKPREIS (*m*): price (each) per piece 158
STÜCKPULVER (*n*): gun-powder 159
STÜCKREICH: screened, sieved, riddled 160
STÜCKWAREN (*f.pl*): piece goods 161
STÜCKWEISE: piecemeal; bit by bit; by retail 162
STUDENT (*m*): student; pupil; scholar 163
STUDIEN (*pl.* of STUDIUM): studies 164
STUDIEREN: to study 165
STUDIERLAMPE (*f*): reading-lamp 166
STUDIERSTUBE (*f*): study 167
STUDIUM (*n*): study 168
STUFE (*f*): step; degree; grade; rank; stage 169
STUFEN: to step, grade, rank 170
STUFENARTIG: by steps or stages; gradual-(ly); stagewise 171
STUFENERZ (*n*): graded (piece) ore 172
STUFENFOLGE (*f*): gradation; succession of (progression by) stages or steps 173
STUFENGANG (*m*): gradation; succession of (progression by) stages, steps or degrees 174
STUFENGESETZ (*n*): law of (stages) progression 175
STUFENLEITER (*f*): scale (of stages or progression) 176
STUFENRAD (*n*): stepped wheel 177
STUFENREGEL (*f*): law of (stages) progression 178
STUFENROST (*m*): step-grate 179
STUFENSCHEIBE (*f*): step-(ped) pulley, grooved pulley 180
STUFENVENTIL (*n*): stage valve, step valve 181
STUFENWEISE: stagewise; by (steps; degrees) stages; gradually 182
STUFENWEISES ERHITZEN (*n*): stage (re)-heating (Steam plant) 183
STUFENZAHL (*f*): number of stages 184
STUHL (*m*): stool; seat; chair; stool; excreta; faeces; feces; excrement; motion (Med.) 185
STUHLBEFÖRDERND: laxative; aperient (Med.) 186
STUHLFLECHTER (*m*): maker of cane chair seats 187

STUHLGANG (*m*): stool; motion; faeces; excreta; excrement (Med.) 188
STUHLKAPPE (*f*): chair-cover 189
STUHLLEHNE (*f*): chair-back 190
STUHLMACHER (*m*): chair-maker 191
STUHL (*m*), SCHIENEN-: rail chair 192
STUHL (*m*), TROMMEL-: drum-stool 193
STUHLVERHALTUNG (*f*): constipation (Med.) 194
STUHLZÄPFCHEN (*n*): suppository (Med.) 195
STUKATUR (*f*): stucco 196
STUKATURARBEIT (*f*): stucco-work 197
STULPE (*f*), FLAMMEN-: upturned (inverted, upside down) flame duct 198
STÜLPEN: to invert; turn (upside down) inside out; place (over) upon; upturn 199
STÜLPWAND (*f*): retaining wall 200
STUMM: dumb; mute; silent; speechless 201
STUMME (*m* and *f*): mute; dumb person 202
STUMMEL (*m*): stump; end 203
STUMMHEIT (*f*): speechlessness; silence; dumbness 204
STUMPEN (*m*): stump 205
STÜMPER (*m*): duffer; bungler 206
STÜMPEREI (*f*): bungling 207
STÜMPERHAFT: bungling, unskilful, ignorant 208
STUMPF (*m*): end; stump; trunk; (as an adj. = blunt; dull; obtuse) 209
STUMPFECKIG: blunt-(cornered) 210
STUMPFEN: to blunt, dull 211
STUMPFHEIT (*f*): bluntness; dullness 212
STUMPFKANTIG: blunt-(edged) 213
STUMPFKEGEL (*m*): truncated cone 214
STUMPFLÖTEN: to butt-solder, solder end to end 215
STUMPFNASE (*f*): snub nose 216
STUMPFSINN (*m*): stupidity; imbecility; dullness (of intellect); mental (stupor) torpor (Med.) 217
STUMPFSTOSSSCHWEISSUNG (*f*): butt welding 218
STUMPFWINKLIG: obtuse-angled 219
STUMPIG: stumpy 220
STUNDE (*f*): hour; time; lesson 221
STUNDENGLAS (*n*): hour-glass 222
STUNDENWEISE: hourly; by the hour 223
STÜNDLICH: hourly; per hour; every hour 224
STUNDUNG (*f*): respite 225
STUPP (*f*): mercurial root; stupp 226
STUPPFETT (*n*): a fatty hydrocarbon mixture obtained from the refining of STUPP; stupp fat 227
STURM (*m*): storm; alarm; fury; tumult; tempest; assault; attack 228
STURMDECK (*n*): awning deck (Ships) 229
STÜRMEN: to storm; roar; assault; attack 230
STURMGLOCKE (*f*): alarm-bell 231
STURMHUT (*m*): aconite; monkshood (*Aconitum napellus*) 232
STURMHUTKRAUT (*n*): aconite herb (*herba aconiti napelli*) (from *Aconitum napellus*) 233
STÜRMISCH: stormy; turbulent; boiling; impetuous; tempestuous 234
STURMKLÜVER (*m*): fore topmast staysail 235
STURMLEITER (*f*): scaling-ladder, storming-ladder 236
STURMSCHRITT (*m*): the double (Military) 237
STURMSIGNAL (*n*): storm signal 238
STURMVOGEL (*m*): stormy-petrel 239
STURMWETTER (*n*): stormy weather 240
STURMWIND (*m*): hurricane, gale, squall 241
STURMZÜNDER (*m*): fusee 242
STURZ (*m*): drop; fall; decrease; plunge; overthrow; business failure; waterfall; ruin; slab; plate (Iron) 243
STURZACKER (*m*): freshly ploughed land 244
STURZBAD (*n*): plunge (shower)-bath; douche 245
STURZBLECH (*n*): plate iron 246
STÜRZE (*f*): cover; lid 247
STÜRZEN: to plunge; overturn; throw; hurl; overthrow; dump; rush; dash; precipitate; fall; pour; ruin; to plough for the first time; break (Land) 248
STURZFLAMME (*f*): reverberatory flame 249
STUTE (*f*): mare 250
STUTENMILCH (*f*): mare's milk 251
STÜTZBALKEN (*m*): underpinning; support-(ing beam) 252
STÜTZE (*f*): support; prop; stay; buckstay; stanchion; bearing; bearer; leg; shore 253
STUTZEN: to top; lop; clip; stop; trim; curtail; crop; be (nonplussed) perplexed; (*m*): connecting (branch) pipe; (flanged)-socket; pocket (for thermometers); a low (glass) cylinder 254
STÜTZEN: to support; prop; stay; rest; lean (upon); be based (rely) upon; underpin; shore; buttress 255
STÜTZFLÄCHE (*f*): supporting surface 256
STÜTZGEWEBE (*n*): supporting tissue 257
STUTZIG: startled; nonplussed; perplexed; staggered 258
STÜTZMAUER (*f*): supporting wall, abutment, buttress 259
STÜTZMITTEL (*n*): supporting (medium) agent 260
STÜTZPLATTE (*f*): bracket plate 261
STÜTZPUNKT (*m*): bearing (supporting) surface; point of support; fulcrum 262
STÜTZROLLENLAGER (*n*): roller supports (for pipe lines) 263
STUTZSCHWANZ (*m*): bobtail 264
STÜTZSUBSTANZ (*f*): supporting substance 265
STUTZUHR (*f*): mantel clock 266
STÜTZWEITE (*f*): span 267
STYPHNINSÄURE (*f*) styphnic acid 268
STYPTICIN (*n*): stypticin, cotarnine hydrochloride (as the name implies, a styptic or blood stanching agent) (see COTARNIN, SALZSAURES-) 269
STYPTICUM (*n*): styptic 270
STYPTISCH: styptic 271
STYPTIZIN (*n*): stypticin (*Cotarninum hydrochloricum*) (see COTARNIN, SALZSAURES-) 272
STYPTOL (*n*): styptol (*Cotarninum phtalicum*) (see COTARNIN, PHTALSAURES-) 273
STYRACIN (*n*): styracin 274
STYRAKOL (*n*): styracol (*Styracolum*), guaiacol cinnamate; $C_{16}H_{14}O_3$; Mp. 130°C 275
STYRAX (*n*): styrax (see STORAX) 276
STYRAXSEIFE (*f*): storax soap (see STORAX) 277
STYROL (*n*): styrol, styrene, cinnamol, cinnamene, styrolene, vinylbenzene, phenylethylene; C_8H_8; Sg. 0.9073; Bp. 146°C 278
SUBCHLORÜR (*n*): subchloride 279
SUBERINSÄURE (*f*): suberic acid 280
SUBHALOID (*n*): subhalide; subhaloid 281
SUBHASTATION (*f*): auction-(sale) 282

SUBJEKT (*n*): subject 283
SUBKUTAN: subcutaneous-(ly), hypodermically 284
SUBLAMIN (*n*): Sublamin (trade name for a mercuric sulphate and ethylene diamine compound); $HgSO_4.2C_2H_8N_2.2H_2O$ 285
SUBLIMAT (*n*): sublimate; mercuric chloride; corrosive sublimate (see QUECKSILBERCHLOR-ID and ÄTZSUBLIMAT) 286
SUBLIMATIONSWÄRME (*f*): heat of sublimation 287
SUBLIMATPASTILLE (*f*): sublimate pastille 288
SUBLIMATSEIFE (*f*): sublimate soap 289
SUBLIMIERAPPARAT (*m*): subliming (sublimating) apparatus 290
SUBLIMIERBAR: sublimable 291
SUBLIMIEREN: to sublimate; sublime 292
SUBLIMIERGEFÄSS (*n*): sublimation vessel 293
SUBLIMIEROFEN (*m*): subliming furnace 294
SUBLIMIERUNG (*f*): sublimation 295
SUBMIKROSKOPISCH: submicroscopical (an ultra-microscopical particle which can be made visible); under the microscope 296
SUBMITIEREN: to submit, send in, forward, tender 297
SUBNORMALE (*f*): subnormal 298
SUBOXYD (*n*): suboxide 299
SUBSKRIBENT (*m*): subscriber 300
SUBSKRIBIEREN: to subscribe 301
SUBSKRIPTION (*f*): subscription 302
SUBSPECIES (*f*): subspecies (a subdivision of a species) 303
SUBSTANTIV: substantive-(ly); (*n*): noun, substantive 304
SUBSTANZ (*f*): substance; matter; body; principle; stuff; material; subject-(matter) 305
SUBSTANZ (*f*), TIERISCHE-: animal (substance) matter 306
SUBSTITUENT (*m*): substituent; substitute 307
SUBSTITUIERBAR: replaceable; capable of being substituted 308
SUBSTITUIEREN: to substitute; replace 309
SUBSTITUIERUNG (*f*): substitution; replacing 310
SUBSTITUT (*m*): substitute 311
SUBSTRAT (*n*): substratum; foundation 312
SUBTANGENTE (*f*): subtangent 313
SUBTRAHIEREN: to subtract; deduct 314
SUBTRAHIERMASCHINE (*f*): subtracting machine 315
SUBTRAKTION (*f*): subtraction 316
SUBTRAKTIVE METHODE (*f*): subtractive method (of three colour printing) 317
SUBVENIEREN: to assist; stay; support; help; relieve; supply; provide 318
SUBVENTIONIEREN: to subsidize; grant; aid; help; supply 319
SUCCINIMID (*n*): succinimide (see BERNSTEIN-SÄURE) 320
SUCCINOABIETINOLSÄURE (*f*): succinoabietinic acid; $C_{40}H_{60}O_5$ 321
SUCCINYLCHLORID (*n*): succinyl chloride 322
SUCCINYLSALICYLSÄURE (*f*): (see DIASPIRIN) 323
SUCCUSPRÄPARAT (*n*): (see LAKRITZENPRÄPARAT) 324
SUCHE (*f*): search 325
SUCHEN: to search; look for; seek; try; want; demand; attempt; explore 326
SUCHER (*m*): searcher; seeker; probe 327
SUCHSPINDEL (*f*): exploring spindle 328
SUCHT (*f*): sickness; disease; epidemic; mania (Med.) 329
SUCHT, FALLENDE- (*f*): epilepsy (Med.) 330
SUCHT, GELBE- (*f*): yellow jaundice (Med.) 331
SÜCHTIG: suppurative; [as a suffix, means: suffering from (a disease); greedy (eager, anxious) for] 332
SUD (*m*): boil-(ing); brew-(ing); gyle; amount brewed; decoction; mordant (Dyeing); seething 333
SÜD (*m*): south 334
SUDELEI (*f*): filth-(iness); soiling 335
SUDELHAFT: slovenly; soiled; dirty 336
SUDELN: to soil; dirty; sully; make filthy; daub 337
SÜDEN (*m*): south 338
SUDFETT (*n*): bone (fat) grease (fat or grease obtained from bones by boiling) 339
SÜDFRÜCHTE (*f.pl*): tropical fruits 340
SUDHAUS (*n*): boiling (brewing) house 341
SUDLER (*m*): filthy (dirty) person; sloven 342
SÜDLICH: south; southern 343
SÜDOST: south-east 344
SUDPFANNE (*f*): boiling pan (for brine) 345
SÜDPOL (*m*): South Pole 346
SUDSALZ (*n*): boiled salt (crystallizes from boiled brine) (almost pure NaCl) 347
SUDSEIFENBAD (*n*): soap bath; suds 348
SÜDWAL (*m*): southern (Australian) whale (*Balaena Australis*) 349
SÜDWÄRTS: southward 350
SUDWERK (*n*): boiling (brewing) apparatus 351
SUDWESEN (*n*): brewing 352
SÜDWEST: south-west 353
SUDZEIT (*f*): duration of brewing or boiling 354
SÜFFIG: bibulous; palatable 355
SUGGESTION (*f*): suggestion 356
SÜHNE (*f*): atonement; expiation 357
SÜHNEN: to atone for; expiate 358
SÜHNUNG (*f*): atonement; expiation 359
SUKZESSIV: successive 360
SUKZINIMID (*n*): succinimide (see BERNSTEIN-SÄURE) 361
SULCHLONITHÄRTE (*f*): (see NICHTCARBON-ATHÄRTE) 362
SULFAMID (*n*): sulfamide 363
SULFAMINBRAUN (*n*): sulfamine brown 364
SULFAMINSÄURE (*f*): sulfamic acid 365
SULFANILSAUER: sulfanilate, combined with sulfanilic acid 366
SULFANILSÄURE (*f*): (*Acidum sulfanilicum*); sulfanilic acid, para-animobenzol-sulphonic acid; $C_6H_4(NH_2)SO_3H$ 367
SULFANILSÄURELÖSUNG (*f*), SALZSAURE-: solution of sodium sulfanilate crystals (in proportion of 2 grams to 1 litre water) 368
SULFANTIMONIAT (*n*): sulfantimoniate 369
SULFANTIMONIG: sulph- (thio)- antimonious 370
SULFANTIMONSÄURE (*f*): sulph- (thio)- antimonic acid 371
SULFARSENIG: sulph- (thio)- arsenious 372
SULFARSENSÄURE (*f*): sulph- (thio)- arsenic acid 373
SULFAT (*n*): sulphate; sodium sulphate (see NATRIUMSULFAT and GLAUBERSALZ) 374
SULFATANLAGE (*f*): sulphate plant 375
SULFATISIEREN: to sulphatize 376
SULFATKESSEL (*m*): sulphate boiler 377
SULFATOFEN (*m*): sulphate furnace 378
SULFATPFANNE (*f*): sulphate pan 379

SULFATSTOFFFABRIK (*f*): sulphate pulp mill (Paper) 380
SULFATSTOFF-SEALINGS: sulphate sealings (Paper) 381
SULFATWASSER (*n*): sulphate water, water containing sulphates 382
SULFATZELLSTOFF (*m*): sulphate fibre (Paper) 383
SULFCARBAMINSÄURE (*f*): thiocarbamic acid; β-thiocarbamic acid; $HOCSNH_2$ 384
SULFCARBANIL (*n*): thiocarbanil; phenyl iso-thiocyanate 385
SULFHYDRAT (*n*): hydrosulphide 386
SULFHYDRID (*n*): hydrosulphide 387
SULFID (*n*): sulphide (preceded by -ic) 388
SULFIDAL (*n*): sulfidal (colloidal sulphur; 80%S) 389
SULFIEREN: to sulphonate; (*n*): sulphonation 390
SULFIN (*n*): sulphonium; sulphine; SH_3' 391
SULFINSÄURE (*f*): sulphinic acid 392
SULFIT (*n*): sulphite 393
SULFITCELLULOSE (*f*): sulphite pulp (Paper) 394
SULFITLAUGE (*f*): sulphite (liquor) lye 395
SULFITLAUGENANLAGE (*f*): sulphite lye plant 396
SULFITLAUGE (*f*), SAURE-: acid sulphite solution (sodium bisulphite solution) 397
SULFITPAPPE (*f*): sulphite board (Paper) 398
SULFITSTOFFFABRIK (*f*): sulphite pulp mill (Paper) 399
SULFITVERFAHREN (*m*): sulphite process 400
SULFITZELLSTOFF (*m*): sulphite fibre (Paper) 401
SULFKOHLENSÄURE (*f*): sulpho- (trithio)- carbonic acid; thionocarbonic acid; HOCSOH 402
SULFOBASE (*f*): sulphur base 403
SULFOBASIS (*f*): sulphur base 404
SULFOBENZOESAUER: sulfobenzoate 405
SULFOBORIT (*m*): sulphoborite; $2\ MgSO_4.2Mg_2B_2O_5.9H_2O$; Sg. 2.45 406
SULFOCARBAMID (*n*): (see SULPHOCARBAMID) 407
SULFOCARBANILID (*n*): sulfocarbanilide 408
SULFOCARBOLSAUER: sulfophenate 409
SULFOCYAN (*n*): thio- (sulpho)- cyanogen 410
SULFOCYANÄTHYL (*n*): ethyl (sulphocyanate) thio-cyanate (see ÄTHYL, SULFOCYANSAURES-) 411
SULFOCYANEISEN (*n*): iron (ferric) thiocyanate 412
SULFOCYANKALIUM (*n*): potassium sulpho- (thio)- cyanate 413
SULFOCYANSAUER: sulpho- (thio-) cyanate of; combined with sulpho- (thio-) cyanic acid 414
SULFOCYANSÄURE (*f*): sulpho- (thio-) cyanic acid 415
SULFOCYANVERBINDUNG (*f*): sulpho- (thio-) cyanate 416
SULFOGRUPPE (*f*): sulphonic (sulpho) group 417
SULFOHYDRAT (*n*): hydrosulphide 418
SULFOKARBOLSÄURE (*f*): sulphocarbolic acid 419
SULFOKARBONSÄURE (*f*): sulpho- (trithio-) carbonic acid; H_2CS_3 420
SULFOKOHLENSÄURE (*f*): sulpho- (trithio-) carbonic acid; H_2CS_3 421
SULFOMONOPERSÄURE (*f*): permonosulphuric acid; H_2SO_5; Mp. 45°C (see CARO'SCHE SÄURE) 422
SULFONAL (*n*): sulfonal, diethylsulphone dimethylmethane, sulphone methane, sulfone methane; $(CH_3)_2C(SO_2C_2H_5)_2$; Mp. 125°C 423
SULFONAPHTOL (*n*): sulfonaphthol 424
SULFONAZURIN (*n*): sulfonazurine 425
SULFONCARBONSÄURE (*f*): sulphonocarboxylic (sulphonecarboxylic) acid 426
SULFONIEREN: to sulphonate 427
SULFONSÄURE (*f*): sulphonic acid 428
SULFONSÄURECARBONSÄURE (*f*): sulphocarboxylic acid; $R''(SO_3H)CO_2H$ 429
SULFONSÄURESCHWARZ (*n*): sulfonic acid black 430
SULFOPHENOLSAUER: sulphophenolate 431
SULFORHODAMIN (*n*): sulphorhodamine, sulphothiamine 432
SULFORÖL (*n*): sulphurous carbon oil 433
SULFOSALIZYLSAUER: sulfosalicylate, thiosalicylate 434
SULFOSALZ (*n*): sulpho (thio) salt 435
SULFOSÄURE (*f*): sulpho acid; sulphonic acid; sulphacid 436
SULFOSOT (*n*): sulfosot (sulphurated, water-soluble creosote) 437
SULFOSOTSIRUP (*m*): sulfosot syrup 438
SULFOXYLCHLORID (*n*): sulfoxyl chloride 439
SULFOZYANAT (*n*): thio- (sulpho)- cyanate 440
SULFTHIOKOHLENSÄURE (*f*): thiolthionocarbonic acid; HOCSSH 441
SULFÜR: protosulphide; sulphide (preceded by -ous) 442
SULFURIEREN: to sulphurize: sulphonate 443
SULFUROLSCHWARZ (*m*): sulfurol black 444
SULFURYLCHLORID (*n*): sulfuryl chloride; SO_2Cl_2; Sg. 1.67-1.708 445
SULFURYLHYDROXYLCHLORID (*n*): sulfuryl hydroxylchloride; $SO_2Cl.OH$; Sg. 1.776-1.784 446
SÜLL (*n*): sill, coaming 447
SULPHINSALZ (*n*): sulphonium (sulphine) salt 448
SULPHOHARNSTOFF (*m*): thiocarbamide (see THIOCARBAMID) 449
SULPHURÖL (*n*): sulphurous oil (extracted from the press residue of olives) 450
SULPHURYLCHLORID (*n*): (see SULFURYLCHLORID) 451
SÜLZE (*f*): jelly; gelatine; brine; pickled meat 452
SULZE (*f*): (see SÜLZE) 453
SÜLZEN: to salt; pickle 454
SULZFLEISCH (*f*): pickled meat 455
SULZIG: gelatinous 456
SÜLZIG: (see SULZIG) 457
SUMACH (*m*): sumac (*Rhus glabra*) 458
SUMACHBAUM (*m*): sumach tree (*Rhus acuminata, Rhus succedanea, Rhus sylvestris* or *Rhus vernicifera*) 459
SUMACHEXTRAKT (*m*): sumac extract 460
SUMACHGEWÄCHSE (*n.pl*): *Terebinthaceæ* (Bot.) 461
SUMACHWACHS (*n*): (see JAPANWACHS) 462
SUMBULWURZEL (*f*): sumbul (musk) root 463
SUMMAND (*m*): term of a sum (Maths.) 464
SUMMARISCH: summary 465
SUMMATION (*f*): summation, total, addition 466
SUMME (*f*): total; amount; sum; deduction 467
SUMMEN: to total; add up; sum; cast; buzz; hum 468

SUMMEN (*n*): humming, buzzing; head-noises, noises in the ears (Med.) 469
SUMMENGLEICHUNG (*f*): summation equation 470
SUMMENWIRKUNG (*f*): combined (total) effect, or action 471
SUMMER (*m*): buzzer (Engineering, etc.) 472
SUMMIEREN: to total; sum; add up; cast 473
SUMPF (*f*): bog; marsh; swamp; pit; sump; pool; fen; moor 474
SUMPFBODEN (*m*): swamp; bog; marshy-ground 475
SUMPFEISENSTEIN (*m*): bog iron ore (see RASEN-EISENERZ) 476
SUMPFERZ (*n*): bog iron ore, limonite (see BRAUNEISENSTEIN) 477
SUMPFESCHEL (*m*): zaffre (see ESCHEL) 478
SUMPFGAS (*n*): marsh gas; CH_4 (see METHAN) 479
SUMPFGEGEND (*f*): fen; marshy country 480
SUMPFICHT: marshy, swampy, boggy 481
SUMPFIG: marshy, boggy, swampy 482
SUMPFLAND (*n*): bog-land, marshy land, moor 483
SUMPFNELKE (*f*): water (purple) avens (*Geum rivale*) 484
SUMPFPORSCH (*m*): marsh tea (*Ledum palustre*) 485
SUMPFPORST (*m*): marsh tea (*Ledum palustre*) 486
SUMPFSILGE (*f*): marsh parsley (*Peucedanum palustre*) 487
SUMPFWASSER (*n*): sump-water; bog-water 488
SUMPFZYPRESSE (*f*): cypress (*Taxodium distichum*) 489
SUND (*m*): strait; sound 490
SÜNDE (*f*): sin 491
SÜNDENBEKENNTNIS (*n*): confession (of sins) 492
SÜNDENBOCK (*m*): scapegoat 493
SÜNDENERLAS (*m*): remission of sins; absolution 494
SÜNDER (*m*): sinner; culprit; delinquent 495
SÜNDFLUT (*f*): the flood (deluge) of the Bible 496
SÜNDHAFT: sinful; peccable 497
SÜNDIG: (see SÜNDHAFT) 498
SÜNDIGEN: to sin; offend 499
SÜNDLICH: sinful 500
SUPERFIZIELL: superficial 501
SUPEROXYD (*n*): peroxide; superoxide 502
SUPERPHOSPHAT (*n*): superphosphate; hypophosphate (see MONOCALCIUMPHOSPHAT and CALCIUM, PHOSPHORSAURES-) 503
SUPERPHOSPHATANLAGE (*f*): superphosphate plant 504
SUPERPONIEREN: super-(impose) pose 505
SUPERPOSITION (*f*): superimposement, superimposing, superposing 506
SUPERSATURIEREN: to supersaturate 507
SUPPE (*f*): soup 508
SUPPENLAUCH (*m*): (see SCHNITTLAUCH) 509
SUPPENNAPF (*m*): soup (dish) tureen; porringer 510
SUPPLEMENTWINKEL (*m*): supplement-angle 511
SUPPLIK (*f*): supplication, suit, application, request, petition 512
SUPPLIKANT (*m*): petitioner, suitor, applicant 513
SUPPLIZIEREN: to supplicate, apply, petition-(for), request, sue 514
SUPPORT (*m*): rest, support 515
SUPPOSITORIE (*f*): suppository (Med.) 516
SUPPOSITORIENPRESSE (*f*): suppository mould 517
SUPRAMINSCHWARZ (*n*): supramine black 518
SUPRARENIN (*n*): suprarenin (see ADRENALIN) 519
SUPREMAT (*n*): supremacy 520
SURREN: to hum; buzz 521
SURROGAT (*n*): substitute 522
SURROGATSTOFF (*m*): substitute; pulp substitute (Paper) 523
SUSPENDIEREN: to suspend 524
SUSPENSION (*f*): suspension (coarsely dispersed solid particles floating in fluids); hanging, depending 525
SUSPENSOID (*n*): suspensoid; granular colloid; (colloidal distribution of solid substances in fluids) 526
SÜSS: sweet 527
SÜSSBIER (*n*): sweet beer 528
SÜSSE (*f*): sweetness 529
SÜSSEN: to sweeten; dulcify; soften 530
SÜSSERDE (*f*): beryllia; glucina; (BeO) 531
SÜSSHOLZ (*n*): liquorice-(root) 532
SÜSSHOLZEXTRAKT (*m*): liquorice extract (see LAKRITZENSAFT) 533
SÜSSHOLZEXTRAKTPRÄPARAT (*n*): (see LAKRITZEN-PRÄPARAT) 534
SÜSSHOLZSAFT (*m*): liquorice (extract) juice 535
SÜSSHOLZWURZEL (*f*): liquorice root (*Radix liquiritiæ* of *Glycyrrhiza glabra*) 536
SÜSSHOLZZUCKER (*m*): glycyrrhizic acid 537
SÜSSIGKEIT (*f*): sweetness; suavity; sweets 538
SÜSSKLEE (*m*): sainfoin (*Hedysarum onobrychis*) 539
SÜSSLICH: sweetish 540
SÜSSMANDELÖL (*n*): sweet almond oil (see MANDELÖL) (*Oleum amygdalarum expressum*) (from *Prunus amygdalus dulcis*); Sg. 0.918 541
SÜSSSTOFF (*m*): sweet-(ening) substance 542
SÜSSWASSER (*n*): fresh water 543
SUSZEPTIBILITÄT (*f*): susceptibility (κ) (Magnetism) 544
SWINTER (*m*): suint (see WOLLSCHWEISS) 545
SYENIT (*m*): greenstone; syenite (Geol.) 546
SYENITGRANIT (*m*): granite (see GRANIT) 547
SYLVANERZ (*n*): sylvanite (see SYLVANIT) 548
SYLVANIT (*m*): sylvanite, sylvane (gold-silver telluride); $(Au,Ag)Te_2$; Sg. 8.1 549
SYLVESTER (*m*): New Year's Eve 550
SYLVESTERABEND (*m*): New Year's Eve 551
SYLVIN (*m*): sylvite, sylvine (natural potassium chloride, from Stassfurt district of Germany); KCl; Sg. 1.987 (see also LEOPOLDIT) 552
SYLVINIT (*m*): sylvinite (sylvite and rock salt; see SYLVIN and STEINSALZ) 553
SYLVINSÄURE (*f*): sylvic (abietic) acid (see ABIETINSÄURE) 554
SYMBIOSE (*f*): symbiosis (an intimate relationship between plants and animals or between plants, in which each absorbs nourishment from the other) 555
SYMBOL (*n*): symbol; sign 556
SYMBOLISCH: symbolic-(al)-(ly) 557
SYMMETRIE (*f*): symmetry 558
SYMMETRIEACHSE (*f*): axis of symmetry 559
SYMMETRIEEBENE (*f*): plane of symmetry 560
SYMMETRISCH: symmetric; symmetrical; symmetrically 561
SYMPATHETISCH: sympathetic-(ally) 562

SYMPATHIE (*f*): sympathy 563
SYMPATHISIEREN: to sympathize 564
SYMPHONIE (*f*): symphony 565
SYMPTOM (*n*): symptom 566
SYNAERESIS (*f*): synaeresis 567
SYNALLAGMATISCH: synallagmatic; reciprocal (of a contract between two or more parties) 568
SYNCHRON: synchronous 569
SYNCHRONMOTOR (*m*): synchronous motor 570
SYNCHRON, NICHT-: not synchronous, asynchronous 571
SYNDIKAT (*n*): syndicate 572
SYNERGISMUS (*m*): synergism; mutual action 573
SYNGENIT (*m*): syngenite; $K_2Ca(SO_4)_2.H_2O$ (see KALUSZIT); Sg. 2.6 574
SYNONYM: synonymous; (*n*): synonym 575
SYNONYMA (*n.pl*): synonyms (plural of SYNONYM) 576
SYNONYMISCH: synonymous-(ly) 577
SYNOVITIS (*f*): synovitis (Med.) 578
SYNOVITIS (*f*), ACUTE-: acute synovitis, inflammation of the synovial membrane (Med.) 579
SYNTETISCH: synthetic; artificial 580
SYNTHESE (*f*): synthesis 581
SYNTHESIEREN: to synthesize 582
SYNTHETISCH: synthetic 583
SYNTONIN (*n*): syntonine, muscle fibrin, parapeptone 584
SYRGOL (*n*): syrgol (a colloidal silver compound) 585
SYRINGASÄURE (*f*): syringic acid 586
SYRINGOMYELIE (*f*): syringomyelia, syringomyelitis, hydromyelia, central gliomatosis, *Hydrorrhachis interna* (Med.) 587
SYRISCHE PFLAUME (*f*): Syrian plum (*Prunus syriaca*) (see also KRIECHE) 588
SYRUP (*f*): (see SIRUP) 589
SYSTEM (*n*): system 590
SYSTEMATISCH: systematic; systematical; systematically 591
SYSTEMATISIEREN: to systematize 592
SYSTEMSCHWERPUNKT (*m*): centre of gravity 593
SZENE (*f*): scene 594
SZIENTIFISCH: scientific 595
SZINTILLIEREN: to scintillate 596

T

TABAK (*m*): tobacco (*Nicotiana tabacum*); snuff 597
TABAKBLATT (*n*): tobacco-leaf 598
TABAKDÜNGER (*m*): tobacco manure 599
TABAKEXTRAKT (*m*): tobacco extract 600
TABAKKAUEN (*n*): tobacco-chewing 601
TABAKÖL (*n*): tobacco oil 602
TABAKPAPIER (*n*): tobacco paper (Paper) 603
TABAKRAUCHEN (*n*): tobacco smoking 604
TABAKSAMENÖL (*n*): tobacco-seed oil 605
TABAKSASCHE (*f*): tobacco ashes 606
TABAKSBAU (*m*): tobacco-growing 607
TABAKSBEUTEL (*m*): tobacco-pouch 608
TABAKSBLATT (*n*): tobacco leaf 609
TABAKSBRÜHE (*f*): tobacco juice 610
TABAKSDAMPF (*m*): tobacco smoke 611
TABAKSDOSE (*f*): snuff box; tobacco box 612
TABAKSFABRIK (*f*): tobacco factory 613
TABAKSHÄNDLER (*m*): tobacconist 614
TABAKSKAMPHER (*m*): nicotianin(-e) 615
TABAKSPFLANZUNG (*f*): tobacco plantation 616
TABAKSRAUCH (*m*): tobacco smoke 617
TABAKSSCHNUPFEN (*n*): snuff-taking 618
TABAKSSTANGE (*f*): cake (bar; roll) of tobacco 619
TABAKVERGIFTUNG (*f*): tobacco (nicotine) poisoning 620
TABASCHIR (*m*): tabashir; tabasheer 621
TABAXIR (*m*): (see TABASCHIR) 622
TABELLARISCH: tabular 623
TABELLARISCHE ZUSAMMENSTELLUNG (*f*): tabulated summary, schedule 624
TABELLARISIEREN: to tabulate; index; summarize 625
TABELLE (*f*): table; summary; synopsis; index; list; register; schedule 626
TABLEAU (*n*): board 627
TABLETT (*n*): tablet; lozenge 628
TABLETTENGLAS (*n*): tablet tube; glass tube for containing tablets 629
TABLETTENMASCHINE (*f*): tablet (making; compressing) machine 630
TABORINDIKATOR (*m*): Tabor indicator 631
TACAMAHACFETT (*n*): tacamahac fat (from seeds of *Calophyllum inophyllum*) 632
TACHHYDRIT (*m*): tachhydrite; $CaMg_2Cl_6$, $12H_2O$; Sg. 1.95 633
TACHOGRAPH (*m*): tachograph; recording (speed meter) tachometer; velocity recorder 634
TACHOMETER (*n*): tachometer, speed meter, speedometer, velocity meter 635
TADEL (*m*): fault; blame; reproof; blemish; censure 636
TADELHAFT: faulty, censurable, blamable 637
TADELLOS: faultless; blameless 638
TADELN: to criticize; blame; reprove; censure; find fault with 639
TADELNSWERT: censurable 640
TADELSÜCHTIG: censorious; critical 641
TADLER (*m*): fault-finder 642
TAFEL (*f*): table; tablet; plate; slab; sheet; cake; pane; slate; blackboard; chart; index; diagram 643
TAFELARTIG: tabular 644
TAFELAUFSATZ (*m*): table (appointments) appurtenances 645
TAFELBIER (*n*): table beer 646
TAFELBLEI (*n*): sheet lead 647
TÄFELCHEN (*n*): tablet; small table (see TAFEL) 648
TAFELFARBE (*f*): local (topical) colour (Calico) 649
TAFELFÖRMIG: tabular 650
TAFELGESCHIRR (*n*): table (service) ware or appurtenances 651
TAFELGLAS (*n*): sheet glass 652
TAFELLACK (*m*): shellac 653
TAFELMESSING (*n*): sheet brass 654
TAFELN: to dine; feast; sup; eat; sit at table 655
TÄFELN: to wainscot; floor; board; panel 656
TAFELÖL (*n*): table oil; salad oil; olive oil (refined) 657
TAFELQUARZ (*m*): tabular quartz (see QUARZ) 658
TAFELSALZ (*n*): table salt (see SALZ) 659
TAFELSCHIEFER (*m*): roofing slate 660
TAFELSERVIS (*n*): table service; plate 661
TAFELSILBER (*n*): table-silver 662
TAFELSPAT (*m*): wollastonite (see WOLLASTONIT) 663

TAFELSTEIN (*m*): table diamond 664
TÄFELUNG (*f*): panelling, wainscotting 665
TAFELWAGE (*f*): counter (platform) scales 666
TÄFELWERK (*n*): panelling, boarding, wainscot 667
TAFELZEUG (*n*): table linen 668
TAFFET (*m*): taffeta; sarcenet 669
TAFFONAL (*n*): taffonal (a solution of resin in benzol) 670
TAFFT (*m*): (see **TAFFET**) 671
TAFT (*m*): (see **TAFFET**) 672
TAFTEN: of taffeta 673
TAFTPAPIER (*n*): satin paper 674
TAFTWEBER (*m*): satin weaver 675
TAG (*m*): day 676
TAGARBEIT (*f*): day work; day shift 677
TAGEARBEIT (*f*): day-work; day-shift 678
TAGEBAU (*m*): open working 679
TAGEBLATT (*n*): daily (news)-paper 680
TAGEBUCH (*n*): diary; journal; day-book 681
TAGEDIEB (*m*): idler; loiterer 682
TAGELANG: for days; all day long 683
TAGELOHN (*m*): daily wages; day's pay 684
TAGELÖHNER (*m*): day-labourer; worker on daily wages or pay 685
TAGEN: to dawn; become light; sit (of assemblies) 686
TAGESANBRUCH (*m*): dawn; daybreak 687
TAGESLICHT (*n*): daylight 688
TAGESORDNUNG (*f*): order of the day 689
TAGESPREIS (*m*): current price 690
TAGESZEIT (*f*): daytime; hour (of the day) 691
TÄGLICH: daily; quotidian; diurnal; diurnally 692
TAGSATZUNG (*f*): sitting; meeting 693
TAGSCHICHT (*f*): day work, day shift 694
TAGUNG (*f*): convention, meeting 695
TAGWASSER (*n*): surface water 696
TAGWEISE: by days 697
TAILLE (*f*): waist; figure; bodice 698
TAKEL (*n*): tackle 699
TAKELAGE (*f*): rig (of ships), rigging 700
TAKELMEISTER (*m*): rigger 701
TAKELN: to tackle; rig 702
TAKELUNG (*f*): rig, rigging, tackle (Naut.) 703
TAKELWERK (*n*): tackle; rigging 704
TAKELWERKSTATT (*f*): rigging loft 705
TAKLER (*m*): rigger 706
TAKT (*m*): time (Music); measure; beat; tact; stroke (Mech.) 707
TAKTIK (*f*): tactics 708
TAKTMÄSSIG: in time; on the beat 709
TAKTSTRICH (*m*): bar (Music) 710
TAL (*n*): valley; vale; dale (see also **THAL**) 711
TALAR (*m*): gown; robe 712
TALBUCHE (*f*): red beech 713
TALCUM (*n*): talc; talcum; French chalk (see **TALK**) 714
TALENGE (*f*): narrow pass into or out of a valley 715
TALENT (*n*): talent; gift 716
TALERKÜRBISÖL (*n*): (see **TELFAIRIAÖL**) 717
TALG (*m*): tallow; Sg. 0.94; Mp. 43-48° C.; (*Sebum*); fat; grease; suet; Sebum (Anat.) (see also **FETT** and **RINDFETT**) 718
TALGÄHNLICH: like tallow; sebaceous 719
TALGART (*f*): kind of tallow 720
TALGARTIG: tallowy; sebaceous 721
TALGBAUM (*m*): tallow tree; Chinese tallow tree 722
TALGBAUM, WESTAFRIKANISCHER- (*m*): West African tallow tree (*Pentadesma butyracea*) 723
TALGBROT (*n*): tallow cake 724
TALG (*m*), **CHINESISCHER-**: Chinese tallow (a vegetable tallow from the seeds of *Sapium sebiferum*; Mp. 44.5 °C.) 725
TALGDRÜSE (*f*): sebaceous gland 726
TALGDRÜSENAUSSCHWITZUNG (*f*): seborrhœa, stearrhœa, steatorrhœa (a skin affection) (Med.) 727
TALGEN: to tallow; grease; produce tallow 728
TALGFETT (*n*): stearine (see **STEARIN**) 729
TALGFETTSÄURE (*f*): stearic acid (see **STEARINSÄURE**) 730
TALGGRIEBEN (*f.pl*): tallow cracklings; greaves 731
TALGGRIEBENBROT (*n*): tallow (dripping) cake 732
TALGIG: tallowy; suety 733
TALGLICHT (*n*): tallow candle 734
TALGLORBEERBAUM (*m*): tallow laurel (*Tetranthera laurifolia*) 735
TALGÖL (*n*): tallow oil 736
TALGOL (*n*): talgol (a hardened oil); Mp. 35-40°C. 737
TALGPRÄPARAT (*n*): tallow preparation 738
TALGRUND (*m*): bottom of a valley 739
TALGSAMENÖL (*n*): (see **STILLINGIAÖL**) 740
TALGSÄURE (*f*): stearic acid (see **STEARINSÄURE**), sebacic acid (see **SEBACINSÄURE**) 741
TALGSCHMELZANLAGE (*f*): tallow rendering plant; tallow melting plant 742
TALGSCHMELZEN (*n*): tallow rendering 743
TALGSEIFE (*f*): tallow soap; lard soap 744
TALGSTEIN (*m*): soapstone; steatite (see **STEATIT**) 745
TALGSTOFF (*m*): stearin; stearine (see **STEARIN**) 746
TALGTROG (*m*): (tallow)-mould 747
TALGZELLE (*f*): sebaceous cell 748
TALIPOT (*f*): (see **SCHATTENPALME**) 749
TALJE (*f*): tackle 750
TALJEBLOCK (*m*): tackle block 751
TALJE (*f*), **BOOTS-**: boat tackle 752
TALJELÄUFER (*m*): tackle fall 753
TALJEREEP (*n*): lanyard 754
TALK (*m*): talc; talcum; French chalk (see **SPECKSTEIN**); soapstone; steatite (see **SEIFENSTEIN** and **STEATIT**); (see also **MEERSCHAUM**); potstone (natural hydrous magnesium silicate); $3MgO.4SiO_2.H_2O$; Sg. 2.75 755
TALKARTIG: talcous 756
TALKERDE (*f*): magnesite; magnesia; magnesium oxide (see **MANGANESIUMOXYD** and **MAGNESIT**) 757
TALKGLIMMER (*m*): lamellar talc 758
TALKGLIMMER (*m*), **HEMIPRISMATISCHER-**: mica (see **GLIMMER**) 759
TALKGLIMMER (*m*), **RHOMBOEDRISCHER-**: biotite 760
TALK (*m*), **SAPONIT-**: steatite (see **SEIFENSTEIN**) 761
TALKSCHIEFER (*m*): slaty talc; talcous slate 762
TALKSPAT (*m*): magnesite (see **MAGNESIT** and **MAGNESITSPAT**) 763
TALK (*m*), **SPECKSTEIN-**: (see **STEATIT**) 764
TALKSTEIN (*m*): soapstone; steatite (see **SEIFENSTEIN** and **STEATIT**) 765
TALKSTEINPACKUNG (*f*): soapstone (steatite) packing 766

TALKUM (*n*): talc; talcum; French chalk (see TALK) 767
TALLIANIN (*n*): tallianine 768
TALLÖL (*n*): liquid resin (a waste product from the manufacture of cellulose from pine wood) 769
TALMIGOLD (*n*): Tombac (see TOMBAK) 770
TALOSCHLEIMSÄURE (*f*): talomucic acid 771
TALSPERRE (*f*): barrage; dam-(across a valley) 772
TALSPERRENBAU (*m*): barrage (dam) construction 773
TALSPERRENSCHIEBER (*m*): sluice-valve; sluice; penstock 774
TAMARACKHOLZ (*n*): tamarack wood, hackmatack, American larch; Sg. 0.63 775
TAMARINDE (*f*): tamarind; Indian date (*Tamarindus indica*) 776
TAMARINDENKONSERVE (*f*): tamarind preserve 777
TAMARINDENMUS (*n*): tamarind (jam) pulp 778
TAMARINDENWEIN (*m*): tamarind wine 779
TAMARISKENGEWÄCHSE (*n.pl*): *Tamaricaceæ* (Bot.) 780
TAMBOUR (*m*): drummer; drum 781
TAMBOURLAGER (*n*): drum bearing 782
TAMBOURMAJOR (*m*): drum-major 783
TANARGAN (*n*): tanargan, tanargentan (a tannin-silver-albumen compound, with about 10% Ag.) 784
TANARGENTAN (*n*): (see TANARGAN) 785
TAND (*m*): bauble; trinket; trifle; toy; idle talk 786
TÄNDELN: to dawdle; trifle; toy; play 787
TANG (*m*): seaweed; seawrack (*Algæ*) 788
TANGASCHE (*f*): kelp (the ash of seaweed) 789
TANGENTE (*f*): tangent 790
TANGENTIAL: tangential 791
TANGENTIALEBENE (*f*): tangent plane 792
TANGENTIALKEIL (*m*): tangent wedge 793
TANGENTIALKRAFT (*f*): tangential force 794
TANGENTIALSCHNITT (*m*): tangential section (a longitudinal section through a tree trunk) 795
TANGENTIALSPANNUNG (*f*): tangential (stress) pressure 796
TANGENTIALTURBINE (*f*): tangential flow turbine 797
TANGENTIALWIDERSTAND (*m*): tangential resistance 798
TANGIEREN: to touch-(upon); be tangent to 799
TÄNIOL (*n*): taniol (a mixture of Sebirol, turpentine oil, and dithymol salicylate) 800
TANK (*m*): tank 801
TANKANLAGE (*f*): tank plant; tanks 802
TANKDAMPFER: tank ship (Naut.) (for oil and other fluids) 803
TANK (*m*), WASSER-: water tank 804
TANNAL (*n*): aluminium tannate 805
TANNALBIN (*n*): tannalbine 806
TANNAT (*n*): tannate 807
TANNE (*f*): fir-(tree) (*Pinus sylvestris*); pine (*Pinus picea*) (see also FICHTE and KIEFER) 808
TANNEN: fir; of fir; pine; of pine; deal 809
TANNENAPFEL (*m*): fir cone 810
TANNENBAUM (*m*): fir (pine) tree (see TANNE) 811
TANNENBAUMKRISTALL (*m*): arborescent crystal 812
TANNENDUFT (*m*): pine odour 813
TANNENGEWÄCHSE (*n.pl*): *Abietaceae* (Bot.) 814
TANNENHAIN (*m*): pine (grove) wood 815
TANNENHARZ (*n*): fir (pine) resin 816
TANNENHARZSÄURE (*f*): abietic acid, abietinic acid; $C_{44}H_{64}O_5$; Mp. 182°C. 817
TANNENHOLZ (*n*): fir (pine) wood; deal; Sg. 0.48-0.66 818
TANNENNADEL (*f*): pine needle 819
TANNENNADELÖL (*n*): pine needle oil (see KIEFERN-NADELÖL) 820
TANNENSAMENÖL (*n*): pine seed oil (from seeds of *Pinus picea*) 821
TANNENWALD (*m*): fir (pine) wood or forest 822
TANNENZAPFEN (*m*): fir cone 823
TANNENZAPFENÖL (*n*): fir cone oil 824
TANNIEREN: to tan; mordant with tannic acid (Dye.) 825
TANNIGEN (*n*): tannigen, acetyltannin, diacetyltannin (more accurately a mixture of acetyltannin and diacetyltannin) (see DIACETYL-TANNIN) 826
TANNIN (*n*): tannic acid; tannin; digallic acid; gallotannic acid; $C_{10}H_{14}O_9$ 827
TANNINALBUMINAT (*n*): tannin albuminate 828
TANNINBLEISALBE (*f*): lead tannate ointment 829
TANNINEXTRAKT (*m*): tannin extract 830
TANNINEXTRAKT—FABRIKATIONSANLAGE (*f*): tannin extracting plant 831
TANNINLEDERSCHWARZ (*n*): tannin leather black 832
TANNINLÖSUNG (*f*): tannin solution 833
TANNINSALBE (*f*): tannin ointment 834
TANNINSEIFE (*f*): tannin soap 835
TANNINSTOFF (*m*): tannin; tannic acid (see TANNIN) 836
TANNINVERFAHREN (*n*): tannin process (Phot.) 837
TANNISMUT (*n*): tannismut 838
TANNOBROMIN (*n*): tannobromine 839
TANNOFORM (*n*): tannoform 840
TANNOPIN (*n*): tannopin 841
TANNOTHYMOL (*n*): tannothymol 842
TANNYL (*n*): tannyl 843
TANOCOL (*n*): tanocol 844
TANTAL (*n*): tantalum; (Ta) 845
TANTALCHLORID (*n*): tantalum chloride; $TaCl_5$; Sg. 3.68 846
TANTALDIOXYD (*n*): tantalum dioxide; TaO_2 847
TANTALERZ (*n*): tantalum ore; tantalite (see TANTALIT) 848
TANTALFLUORID (*n*): tantalum fluoride; TaF_5; Sg. 4.744 849
TANTALIT (*m*): tantalite (natural iron-manganese tantalate); Ta_2FeO_6; Sg. 6.7 to 7.5 850
TANTALLAMPE (*f*): tantalum lamp 851
TANTALOXYD (*n*): (Ta_2O_5); tantalum oxide; tantalum pentoxide; Sg. 7.53 (see TANTAL-DIOXYD; TANTALPENTOXYYD and TANTAL-TETROXYD) 852
TANTALPENTOXYD (*n*): tantalum pentoxide; Ta_2O_5; Sg. 7.53 853
TANTALPRÄPARAT (*n*): tantalum preparation 854
TANTALSÄURE (*f*): tantalic acid (see TANTAL-OXYD); H_3TaO_4 855
TANTALSILICID (*n*): tantalum silicide; $TaSi_2$; Sg. 8.83 856
TANTALTETROXYD (*n*): tantalum tetroxide Ta_2O_4 857
TANTALVERBINDUNG (*f*): tantalum compound 858
TANZ (*m*): dance; row; brawl 859

TANZBEWEGUNG (*f*): Brownian movement; dance (dancing) movement 860
TANZBODEN (*m*): dancing-floor; ball-room 861
TANZEN: to dance 862
TAPET (*n*): carpet 863
TAPETE (*f*): tapestry; wallpaper; hanging; tapetum (Anat.) 864
TAPETENLEIM (*m*): paperhanging (glue) paste (paste for ordinary wall-paper, and glue for raised paper or lincrusta) 865
TAPETENPAPIER (*n*): wall paper; long elephant (Paper) 866
TAPEZIEREN: to paper; hang with tapestry; upholster 867
TAPEZIERER (*m*): paper-hanger; upholsterer 868
TAPEZIERUNG (*f*): papering (House decoration); upholstering 869
TAPFER: brave; valiant; courageous; heroic; valorous; gallant 870
TAPFERKEIT (*f*): bravery; valour; courage; heroism; gallantry 871
TAPIOLIT (*m*): tapiolite; $(Ta,Nb)_2FeO_6$; Sg. 7.35 872
TAPISSERIEGARN (*n*): carpet yarn 873
TAPPEN: to grope; fumble 874
TÄPPISCH: awkward; clumsy 875
TARA (*f*): tare 876
TARANTEL (*f*): tarantula 877
TARIERBECHER (*m*): tare cup 878
TARIEREN: to tare 879
TARIERSCHROT (*n* and *m*): tare shot 880
TARIERWAGE (*f*): tare balance 881
TARIF (*m*): tariff; schedule; rate 882
TARIRIFETT (*n*): tariri fat (from seeds of *Picramnia camboita*) 883
TARIRINSÄURE (*f*): taririnic acid; $C_{17}H_{31}CO_2H$; Mp. 50.5°C. 884
TARTARALIN (*n*): tartaralin 885
TARTARISIEREN: to tartarize 886
TARTRABARIN (*n*): tartrabine 887
TARTRAZIN (*n*): tartrazin (a yellow tar dye-stuff, from action of two parts of phenylhydrazine-monosulphonic acid on one part of dioxytartaric acid); $C_{16}H_{12}N_4O_8$ 888
TARTRONSÄURE (*f*): tartronic acid 889
TASCHE (*f*): pocket; pouch; bursa (Med.) 890
TÄSCHELKRAUT (*n*): (see **HELLERKRAUT**) 891
TÄSCHELKRAUTSAMENÖL (*n*): pennycress oil (from *Thlaspi arvense*) 892
TASCHENAUSGABE (*f*): pocket edition 893
TASCHENBUCH (*n*): pocket-book 894
TASCHENFORMAT (*n*): pocket size 895
TASCHENKREBS (*m*): (ordinary)-crab 896
TASCHENMESSER (*n*): pocket-knife; pen-knife 897
TASCHENSPIEL (*n*): jugglery 898
TASCHENTUCH (*n*): pocket handkerchief 899
TASCHENUHR (*f*): (pocket)-watch 900
TASCHENWÖRTERBUCH (*n*): pocket dictionary 901
TÄSCHNER (*m*): wallet (trunk) maker 902
TASSE (*f*): cup; dish (Phot.) 903
TASSENROT (*n*): (see **CARTHAMIN**) 904
TASTATUR (*f*): keys; set of keys 905
TASTE (*f*): key (of piano or other musical instrument) 906
TASTEN: to touch; grope; feel; palpate (Med.) 907
TASTENINSTRUMENT (*n*): keyed instrument; instrument fitted with (having) keys 908
TASTER (*m*): feeler; antenna; calipers; key (Telegraph); compasses 909
TASTER (*m*), **AUSSEN-**: outside calipers 910
TASTER, GEGENSPRECH- (*m*): duplex key (Telegraph) 911
TASTER (*m*), **INNEN-**: inside calipers 912
TASTER (*m*), **MORSE-**: morse key 913
TASTSINN (*m*): sense of touch 914
TAT (*f*): deed; act; action; doing; fact; exploit; achievement 915
TATBESTAND (*m*): facts of the case; fact; reality; actuality 916
TATENDRANG (*m*): desire for action 917
TATENLOS: inactive; indolent 918
TÄTER (*m*): doer; perpetrator; author; culprit 919
TÄTIG: active; busy 920
TÄTIGKEIT (*f*): activity; occupation; function; action 921
TAT, IN DER-: indeed; in fact; as a matter of fact; sure (true) enough 922
TATKRAFT (*f*): energy 923
TATKRÄFTIG: energetic 924
TÄTLICH: violent; actual 925
TÄTLICHKEIT (*f*): (act of) violence 926
TÄTOWIEREN: to tattoo 927
TATSACHE (*f*): fact; matter of fact; reality; actuality; truth (of the matter) 928
TATSÄCHLICH: in reality, as a matter of fact, in point of fact, actual-(ly), real-(ly) 929
TATZE (*f*): paw; claw; cam (Tech.) 930
TAU (*m*): dew; thaw; (*n*): cable; rope; cord; tow 931
TAUANKER (*m*): tow anchor 932
TAUB: deaf; empty; barren; numb 933
TAUBE (*f*): dove; pigeon 934
TAUBENSCHLAG (*m*): dove-cot; pigeon-house 935
TAUBHEIT (*f*): deafness 936
TAUBKOHLE (*f*): anthracite (see **ANTHRAZIT**) 937
TAUBNESSEL (*f*): blind nettle, dead-nettle 938
TAUBNESSELBLÜTEN (*f.pl*): nettle flowers (*flores lamii albi*) 939
TAUCHAPPARAT (*m*): diving apparatus 940
TAUCHBAR: submersible 941
TAUCHBATTERIE (*f*): plunge battery 942
TAUCHBOOT (*n*): submersible boat (for example driven by steam on the surface and by electricity under water) 943
TAUCHEN: to dip; steep; plunge; dive; immerse; submerge; sink; duck 944
TAUCHER (*m*): diver; plunger (Mech.); sinker (Mech.); dipper (Phot.) 945
TAUCHERANZUG (*m*): diving (dress) suit 946
TAUCHERAPPARAT (*m*): diving apparatus 947
TAUCHERGLOCKE (*f*): diving bell 948
TAUCHERKOLBEN (*m*): plunger (Eng.) 949
TAUCHFLUID (*n*): dip; dipping fluid (for incandescent gas mantles) 950
TAUCHKOLBEN (*m*): trunk-(piston), plunger 951
TAUCHKORN (*n*): sinker (Brew.) 952
TAUCHKUVETTE (*f*): dipping (trough) bath (Phot.) 953
TAUCHLACK (*m*): dipping varnish 954
TAUCHROHR (*n*): submerged (submersible) tube 955
TAUCHSPINDEL (*f*): tail (of a thermometer) 956
TAUCHZYLINDER (*m*): plunger (colorimeter); plunge cylinder 957
TAUEN: to thaw; dew; taw (Hides); tow 958
TAUENDE (*n*): junk (of a rope); rope's-end 959
TAUE (*n.pl*), **PÜTTING-**: futtock rigging; futtock shrouds (Naut.) 960
TAUFEN: to baptize; christen; duck 961
TAUFSTEIN (*m*): font 962

TAUGEN : to be (good) fit ; of value ; be of use ; be suitable ; be apt 963
TAUGEN, NICHTS- : to be (worthless) valueless ; be good for nothing ; be inept 964
TAUGLICH : fit ; good ; able ; apt 965
TAUGLICHKEIT (*f*) : aptness, aptitude, fitness, ability, suitability, value, propriety 966
TAUIG : dewy 967
TAUKARDEEL (*n*) : rope strand 968
TAULAST (*f*) : cable room (Marine), storeage room for cables ; towing load 969
TAU (*n*), LINKSGESCHLAGENES- : left hand rope 970
TAUMEL (*m*) : transport ; frenzy ; giddiness ; intoxication ; reeling ; staggering ; vertigo 971
TAUMELIG : reeling ; staggering ; giddy 972
TAUMELKORN (*n*) : darnel (*Lolium temulentum*) 973
TAUMELN : to be (giddy) intoxicated ; be in a (transport) frenzy ; to reel ; stagger 974
TAUPUNKT (*m*) : dew point 975
TAURACK (*n*) : truss parrel 976
TAU (*n*), RECHTSGESCHLAGENES- : right hand rope 977
TAURISCIT (*m*) : tauriscite (a hydrous ferrous sulphate) ; $FeSO_4+7H_2O$ 978
TAUROCHOLSÄURE (*f*) : taurocholic acid ; choleic (choleinic ; choliaic) acid ; sulphocholeic acid ; $C_{26}H_{45}NSO_7$ (from the bile of the ox) 979
TAURÖSTE (*f*) : dew retting process (Flax) see (RÖSTE, TAU-) 980
TAUSCH (*m*) : exchange ; barter 981
TAUSCHEN : to exchange ; barter 982
TÄUSCHEN : to delude ; deceive ; cheat ; disappoint ; be deceived 983
TÄUSCHEND : delusive ; illusory 984
TÄUSCHENDE ÄHNLICHKEIT (*f*) : striking (similarity) resemblance 985
TAUSCHEREI (*f*) : exchange ; barter 986
TAUSCHERSETZUNG (*f*) : double decomposition 987
TAUSCHHANDEL (*m*) : exchange ; barter 988
TÄUSCHUNG (*f*) : delusion ; deception ; illusion ; disappointment 989
TAUSEELE (*f*) : core (heart) of a rope 990
TAUSEND : thousand 991
TAUSENDFACH : thousand-fold 992
TAUSENDFÄLTIG : thousand-fold 993
TAUSENDGÜLDENKRAUT (*n*) : lesser centaury (*Erythræa centaurium*) (*herba centauri minoris*) 994
TAUSENDSCHÖN (*n*) : love-lies-bleeding (*Amaranthus caudatus*) 995
TAUSENDSCHÖNCHEN (*n*) : pansy 996
TAUSENDSTEIN (*m*) : (new method of expressing the quality of gold in parts of a thousand ; 14 carat equals $\frac{14}{24}\times 1000$) 997
TAUSTOPPER (*m*) : rope stopper 998
TAUTOMER : tautomeric 999
TAUTOMERIE (*f*) : tautomerism 000
TAUTROPFEN (*m*) : dew-drop 001
TAUWERK (*n*) : rigging (Naut.) ; cordage 002
TAUWETTER (*n*) : thaw 003
TAXAMETERUHR (*f*) : taximeter-(dial) 004
TAXATION (*f*) : taxation 005
TAXBAUM (*m*) : yew-(tree) (*Taxus baccata*) 006
TAXE (*f*) : tax ; rate ; duty ; price 007
TAXIEREN : to tax ; rate ; value ; assess ; appraise ; price 008
TAXUS (*m*) : yew-(tree) 009
TEAKBAUM (*m*) : teak, Java (*Tectona grandis*) 010
TEAKHOLZ (*n*) : teak wood (see TEAKBAUM) ; Sg. 0.6 011
TECHNIK (*f*) : technic ; technique ; arts ; technics ; technology 012
TECHNIKER (*m*) : technist ; technologist 013
TECHNIKUM (*n*) : technical institution 014
TECHNISCH : technical-(ly) 015
TECHNISCHER LEITER (*m*) : technical manager ; chief engineer 016
TECHNOLOG (*m*) : technologist 017
TECHNOLOGIE (*f*) : technology (the science of producing finished articles or products from raw materials, suitable for human use), (either chemical or mechanical) 018
TECHNOLOGISCH : technological 019
TECKEL (*m*) : German terrier 020
TEE (*m*) : tea (*Thea chinensis*) 021
TEE (*m*), ABFÜHRENDER- : (*Species laxantes*) : laxative tea 022
TEEBAUM (*m*) : tea plant (*Thea chinensis*) 023
TEE (*m*), HARNTREIBENDER- : (*Species diureticæ*) : diuretic tea 024
TEEMISCHUNG (*f*) : tea ; specific (*Species*) (Pharm.) (mixture of parts of plants, either together or with other substances) 025
TEER (*m*) : tar ; Sg. 1.02 ; cart-grease 026
TEERACETONLÖSUNG (*f*) : tar-acetone solution 027
TEERARM : having a poor (low) tar content 028
TEERARTIG : tarry 029
TEERASPHALT (*m*) : coal-tar pitch ; tar asphalt (residue from coal-tar distillation) 030
TEERAUSSCHEIDER (*m*) : tar separator 031
TEERBAUM (*m*) : Scotch pine (*Pinus sylvestris*) 032
TEERBLASE (*f*) : tar boiler (wrought iron vessel for fractional distillation of tar) 033
TEERBRENNEREI (*f*) : tar factory 034
TEERBÜCHSE (*f*) : grease-box ; tar-box 035
TEERDAMPF (*m*) : tar (fumes) vapour 036
TEERDESTILLATION (*f*) : tar (distilling) distillation 037
TEERDESTILLATIONSANLAGE (*f*) : tar distilling plant 038
TEEREXTRAKTION (*f*) : tar (extracting) extraction 039
TEEREN : to tar ; form (cover with) tar ; (*n*) : tarring 040
TEERENTWÄSSERUNGSANLAGE (*f*) : water extraction plant for tar 041
TEEREXTRAKTIONSAPPARAT (*m*) : tar extractor 042
TEERFARBE (*f*) : coal-tar (colour) dye 043
TEERFARBSTOFF (*m*) : (see TEERFARBE) 044
TEERICHT : tar-like ; tarry 045
TEERIG : tarry ; covered with tar 046
TEERKESSEL (*m*) : tar boiler 047
TEERLACK (*m*) : tar varnish 048
TEEROFEN (*m*) : tar (kiln) furnace 049
TEERÖL (*n*) : oil of tar ; coal-tar oil ; Sg. 0.867 050
TEERÖLANLAGE (*f*) : oil of tar (coal-tar oil) plant 051
TEERÖLFEUERUNG (*f*) : tar oil furnace, tar oil firing, tar oil fuel 052
TEERPAPIER (*n*) : tar-(red) paper 053
TEERPAPPE (*f*) : tar board (see DACHPAPPE) 054
TEERPRODUKT (*n*) : tar product 055

TEERPRÜFER (*m*): tar tester 056
TEERPUMPE (*f*): tar pump 057
TEERPÜTZE (*f*): tar bucket 058
TEERQUAST (*m*): tar brush 059
TEERRÜCKSTAND (*m*): tar residue (usually from distillation) 060
TEERSÄURE (*f*): tar acid; any phenol from coal tar 061
TEERSCHWEFELSEIFE (*f*): tar-sulphur soap 062
TEERSCHWELAPPARAT (*m*): tar distilling apparatus 063
TEERSEIFE (*f*): tar (coal-tar) soap 064
TEERWASSER (*n*): tar water; ammoniacal liquor (Gas) 065
TEER, WASSERGAS- (*m*): water-gas tar; Sg. 1.1 (by-product from carburetted water gas) 066
TEERWERG (*n*): tarred oakum 067
TEESAMENÖL (*n*): tea seed oil (from *Camellia theifera*); Sg. 0.922 068
TEESAMENÖL (*n*), CHINESISCHES-: Chinese tea seed oil 069
TEESTRAUCH (*m*), ECHTER-: tea plant (*Camellia theifera*) 070
TEICH (*m*): ditch; pond; pool; tank 071
TEICHBINSE (*f*): bulrush (*Scirpus lacustris*) 072
TEICHGITTER (*n*): grating 073
TEICHWASSER (*n*): pond-water 074
TEICHZAPFEN (*m*): sluice, lock 075
TEIG (*m*): paste; dough 076
TEIGARTIG: pasty; doughy 077
TEIGFARBE (*f*): dough colour 078
TEIGFORM (*f*): form of dough or paste 079
TEIGIG: doughy; dough-like; pasty; mellow 080
TEIGMULDE (*f*): (dough)-trough 081
TEIGWARE (*f*): paste article; dye in paste form 082
TEIL (*m*): part; element; portion; share; division; component; party (Legal); tome; volume; piece; constituent 083
TEILAPPARAT (*m*): dividing (head) apparatus 084
TEILBAR: divisible; separable 085
TEILBARKEIT (*f*): divisibility 086
TEILCHEN (*n*): particle; element 087
TEILCHENVERSCHIEBUNG (*f*): displacement of particles 088
TEILDRUCK (*m*): partial pressure 089
TEILEN: to part; divide; separate; participate; share; graduate; to pitch; space out 090
TEILER (*m*): divider; divisor (Maths.) 091
TEILER (*m*), GEMEINSCHAFTLICHER-: common divisor 092
TEILFUGE (*f*): joint 093
TEILHABEN: to partake; share; take part; participate 094
TEILHABER (*m*): participator; partner 095
TEILHABER (*m*), DIRIGIERENDER-: managing partner 096
TEILHABER (*m*), STILLER-: sleeping partner 097
TEILHAFT: participating; sharing; participant; partaking 098
TEILHAFTIG: (see TEILHAFT) 099
TEILHAFT MACHEN: to communicate, share 100
TEILKRAFT (*f*): component (of pressure) 101
TEILKREIS (*m*): pitch circle 102
TEILMASCHINE (*f*): dividing machine 103
TEILNAHME (*f*): sympathy; interest; participation 104
TEILNEHMEN: to participate; take part; take an interest in 105
TEILNEHMER (*m*): participant; part-owner 106
TEILREAKTION (*f*): partial reaction 107
TEILS: in part; partly 108
TEILS, GROSSEN-: in a large measure 109
TEILS, GRÖSSTEN-: for the (most part) major portion 110
TEILSTRICH (*m*): graduation mark; pitch 111
TEILUNG (*f*): separation; division; sharing; graduation; scale; partition; pitch; pitching; distribution; spacing out 112
TEILUNGSEBENE (*f*): plane of (division) joint 113
TEILUNGSGESETZ (*n*): law of partition 114
TEILUNGSKOEFFIZIENT (*m*): distribution coefficient 115
TEILUNGSZAHL (*f*): dividend 116
TEILUNGSZEICHEN (*n*): division sign (: or sometimes ÷) 117
TEILUNGSZUSTAND (*m*): state of division 118
TEILVORGANG (*m*): partial process 119
TEILWEISE: partial-(ly); in part 120
TEILZAHL (*f*): quotient 121
TEILZIRKEL (*m*): dividers (Drawing instrument) 122
TEIL, ZUM-: in part; partly 123
TEIN (*n*): theine (see KAFFEIN) 124
TEINGEHALT (*m*): Theine (caffeine) content 125
TEINT (*m*): complexion 126
TEKHOLZ (*n*): teak wood 127
TEKTUR (*f*): slip-(for sticking on) (such as a price slip or an errata slip) 128
TELEFONMUNDSTÜCK (*n*): telephone mouthpiece (see TELEPHON-) 129
TELEGRAPH (*m*): telegraph 130
TELEGRAPHENDRAHT (*m*): telegraph wire 131
TELEGRAPHENISOLATOR (*m*): telegraph insulator 132
TELEGRAPHENKABEL (*n*): telegraph cable 133
TELEGRAPHENLEITUNG (*f*): telegraph line or wire 134
TELEGRAPHENLINIE (*f*): telegraph-line or wire 135
TELEGRAPHENSCHLÜSSEL (*m*): telegraphic key 136
TELEGRAPHENSTANGE (*f*): telegraph pole 137
TELEGRAPHENSTATION (*f*): telegraph station or office 138
TELEGRAPHIE (*f*): telegraphy 139
TELEGRAPHIEREN: to telegraph, wire 140
TELEGRAPHISCH: telegraphic-(ally) 141
TELEMOTOR (*m*): telemotor 142
TELEOBJEKTIV (*n*): tele-photographic lens 143
TELEPATHIE (*f*): telepathy 144
TELEPHON (*n*): telephone, 'phone 145
TELEPHONAMT (*n*): telephone exchange 146
TELEPHON-ANSCHLUSS (*m*): telephone number, telephonic connection 147
TELEPHONAPPARAT (*m*): telephone apparatus 148
TELEPHONHAKEN (*m*): telephone hook 149
TELEPHONHÖRER (*m*): telephone receiver or ear-piece 150
TELEPHONIE (*f*): telephony 151
TELEPHONIEREN: to telephone, to 'phone 152
TELEPHONISCH: telephonic 153
TELEPHONLEITUNG (*f*): telephone line or cable 154
TELEPHONNETZ (*n*): telephone system, network of telephone lines or cables 155
TELEPHONSTATION (*f*): telephone exchange 156
TELEPHONZENTRALE (*f*): telephone exchange 157
TELESKOP (*n*): telescope 158
TELFAIRIAÖL (*n*): telfairia oil (from seeds of *Telfairia pedata*); Sg. 0.918 159
TELFAIRIASÄURE (*f*): telfairic acid (from TELFAIRIAÖL, which see); $C_{17}H_{31}CO_2H$; Mp. 48°C. 160

TELLER (*m*): plate; dish; palm (of hand) 161
TELLERFÖRMIG: plate-shaped 162
TELLERFUSS (*m*): plate-base (of columns, etc.); plate-shaped base 163
TELLERROT (*n*): (see **CARTHAMIN**) 164
TELLERVENTIL (*n*): disc valve 165
TELLUR (*n*): tellurium; (Te) 166
TELLURALKYL (*n*): alkyl telluride 167
TELLURBLEI (*n*): lead telluride; altaite (see **ALTAIT**) 168
TELLURDIÄTHYL (*n*): diethyl telluride 169
TELLURDIOXYD (*n*): tellurium dioxide; TeO_2; Sg. 5.66-5.85 170
TELLURERZ (*n*): tellurium ore 171
TELLURGLANZ (*m*): tellurium glance; nagyagite (see **NAGYAGIT**) 172
TELLURGOLDSILBER (*n*): gold silver telluride; Petzite (see **PETZIT**); $(Au.Ag)_2Te$; Sg. 8.83 173
TELLURHALOGEN (*n*): tellurium halide 174
TELLURIG: tellurous 175
TELLURIGE SÄURE (*f*): tellurous acid: H_2TeO_3 176
TELLURIGSAUER: tellurite, combined with tellurous acid 177
TELLURIT (*m*): tellurite (see **TELLURDIOXYD**) 178
TELLURIUM (*n*): tellurium 179
TELLURMETALL (*n*): metallic telluride 180
TELLUROCHER (*m*): tellurite (Min.) (see **TELLURDI-OXYD**) 181
TELLUROXYD (*n*): tellurium oxide 182
TELLURSAUER: tellurate of; combined with telluric acid 183
TELLURSÄURE (*f*): telluric acid; $H_2TeO_4+2H_2O$ or H_2TeO_4; Sg. 3.441 184
TELLURSÄUREHYDRAT (*n*): telluric acid, hydrated; $Te(OH)_6$; Sg. 3.053-3.071 185
TELLURSILBER (*n*): silver telluride; Ag_2Te; Sg. 8.25; petzite (Min.) (see **PETZIT**) 186
TELLURTRIOXYD (*n*): tellurium trioxide; TeO_3; Sg. 5.087 187
TELLURVERBINDUNG (*f*): tellurium compound 188
TELLURWASSERSTOFF (*m*): hydrogen telluride; tellurium hydride; H_2Te; Sg. 2.57 189
TELLURWISMUT (*n*): bismuth telluride (see also **MOLYBDÄNSILBER**) 190
TEMPEL (*m*): temple; church 191
TEMPERAMENT (*n*): temperament 192
TEMPERATUR (*f*): temperature 193
TEMPERATURÄNDERUNG (*f*): temperature (change; alteration; fluctuation) variation 194
TEMPERATURBEOBACHTUNG (*f*): temperature observation 195
TEMPERATURBEREICH (*m*): temperature range 196
TEMPERATURDIFFERENZ (*f*): temperature difference 197
TEMPERATUREINFLUSS (*m*): temperature influence 198
TEMPERATURERHÖHUNG (*f*): temperature increase 199
TEMPERATURGEFÄLLE (*n*): temperature drop 200
TEMPERATURGRAD (*m*): degree (of temperature) 201
TEMPERATUR (*f*), **KRITISCHE-**: critical temperature 202
TEMPERATURMESSER (*m*): thermometer; pyrometer; temperature measurer 203
TEMPERATURMESSUNG (*f*): temperature measurement; (temperature)-reading 204
TEMPERATURMITTEL (*n*): mean temperature; average temperature 205
TEMPERATURREGLER (*m*): temperature regulator, thermo-regulator 206
TEMPERATURREGLUNG (*f*): temperature regulation; thermo-regulation; temperature regulating (thermo-regulating) device 207
TEMPERATURSCHWANKUNG (*f*): temperature (variation) fluctuation 208
TEMPERATURVERÄNDERUNG (*f*): temperature (change; variation) fluctuation 209
TEMPERGUSS (*m*): malleable-iron casting 210
TEMPERIEREN: to temper; anneal (Metal); chill; cool; reduce (bring) to the correct temperature (Brewing) 211
TEMPERKOHLE (*f*): temper carbon; graphite 212
TEMPERN: to temper; anneal 213
TEMPEROFEN (*m*): annealing (oven) furnace 214
TEMPERPROZESS (*m*): (see **GLÜHFRISCHEN**) 215
TEMPERSTAHL (*m*): temper steel; annealed steel 216
TEMPORÄR: temporary 217
TENAKEL (*n*): filtering frame; copyholder (Printing) 218
TENDENZ (*f*): tendency 219
TENDER (*m*): tender (Ship) 220
TENDERBREMSE (*f*): tender brake 221
TENDERLOKOMOTIVE (*f*): tender (engine) locomotive 222
TENDERMASCHINE (*f*): tender (engine) locomotive 223
TENNANTIT (*m*): tetrahedrite, fahlore, tennantite, grey copper ore (copper-arsenic sulphide); $Cu_8As_2S_7$; Sg. 4.4-4.85 (see **ARSENFAHLERZ** and **ARSENIKFAHLERZ**) 224
TENNE (*f*): (threshing)-floor (of a barn) 225
TENORIT (*m*): tenorite, melaconite (natural black copper oxide); CuO; Sg. 6.0 (see **MELA-KONIT**) 226
TEPHROIT (*m*): tephroite (a member of the chrysolite group); Mn_2SiO_4; Sg. 4.0 227
TEPPICH (*m*): carpet; rug; tapestry; blanket; cover (for table) 228
TEPPICHFILZPAPIER (*n*): carpet felt paper 229
TEPPICHWIRKER (*m*): carpet-maker; rug manufacturer 230
TERATOLITH (*m*): teratolite, bole (see **BOL**) 231
TERATOLOGIE (*f*): teratology (the science of animal or vegetable monstrosities) 232
TERBINERDE (*f*): terbia; terbium oxide 233
TERBIUMCHLORID (*n*): terbium chloride; $TbCl_3$; Sg. 4.35 234
TEREBINSÄURE (*f*): terebic acid 235
TEREPHTALSÄURE (*f*): terephthalic acid 236
TERMIN (*m*): term; time; period; fixed (appointed) day, date or time 237
TERMINOLOGIE (*f*): terminology 238
TERMINWEISE: by instalments; at fixed periods 239
TERMITE (*f*): white ant (popular but incorrect name); termite (insects of the family, *Termitidæ*) 240
TERNÄR: ternary; composed of three parts 241
TERNEBLECH (*n*): terne-plate 242
TERPANGRUPPE (*f*): methane group 243
TERPENCHEMIE (*f*): terpene chemistry 244
TERPENFREI: terpeneless; free from terpene; terpene-free 245
TERPENGRUPPE (*f*): terpene group 246
TERPENTIN (*m*): turpentine; turps; $C_{10}H_{16}$; Sg. 0.87; Bp. 155-180°C. 247
TERPENTINART (*f*): kind of turpentine 248

TERPENTINARTIG : like turpentine ; terebinthine 249
TERPENTINFIRNIS (*m*) : turpentine ; crude turpentine (obtained from various coniferous trees, *Pinus*) 250
TERPENTINGEIST (*m*) : spirit (oil) of turpentine (see TERPENTIN) 251
TERPENTINHARZ (*n*) : crude turpentine ; turpentine resin (see TERPENTINFIRNIS) 252
TERPENTIN (*m*), **KANADISCHER-** : (*Balsamum Canadense*) : Canadian turpentine (see KANADABALSAM) 253
TERPENTINKIEFER (*f*) : pitch pine (see HARZ-KIEFER) 254
TERPENTINÖL (*n*) : oil (spirit) of turpentine ; Sg. 0.87 (see TERPENTIN) 255
TERPENTINÖLERSATZ (*m*) : oil (spirit) of turpentine substitute 256
TERPENTINPECH (*n*) : turpentine pitch 257
TERPENTINSALBE (*f*) : turpentine ointment 258
TERPENTINSEIFE (*f*) : turpentine soap 259
TERPENTINSPIRITUS (*m*) : spirit (oil) of turpentine (see TERPENTIN) 260
TERPENTINSURROGAT (*n*) : turpentine substitute 261
TERPIN (*n*) : terpene 262
TERPINEOL (*n*) : terpineol 263
TERPINHYDRAT (*n*) : hydrated terpene ; terpene hydrate 264
TERPINOL (*n*) : terpinol 265
TERPINYLACETAT (*n*) : terpinyl acetate (see TERPINYL, ESSIGSAURES-) 266
TERPINYL (*n*), **ESSIGSAURES-** : terpinyl acetate ; $C_{10}H_{17}CO_2CH_3$; Sg. 0.97 ; Mp. – 50°C. ; Bp. 220°C. 267
TERPOCHIN (*n*) : terpochine 268
TERRAIN (*m*) : ground 269
TERRASSE (*f*) : terrace (see also OXYDATIONS-TERRASSE) 270
TERRESTRISCH : terrestrial 271
TERRINE (*f*) : tureen ; bowl 272
TERRITORIUM (*n*) : territory 273
TERTIÄR : tertiary 274
TERZERONE (*m*) : terceroon (see also QUADRONE) 275
TESSERALKIES (*m*) : skutterudite (see SKUTTERU-DIT) ; smaltite (see SMALTIT) 276
TESSERALSYSTEM (*n*) : isometric system 277
TEST (*m*) : test ; test furnace ; cupel-(lation furnace) ; graphite ; indicator ; molybdenite (see MOLYBDÄNOCHER) 278
TESTAMENT (*n*) : testament ; last will 279
TESTAMENTARISCH : by will, testamentary 280
TESTAMENTSVOLLSTRECKER (*m*) : executor (of a a will) 281
TESTASCHE (*f*) : bone ash (see KNOCHENASCHE) 282
TESTATOR (*m*) : testator 283
TESTIEREN : to (make) leave a will ; testify 284
TETANIE (*f*) : tetanilla, tetany, idiopathic muscular spasm (Med.) 285
TETANISIEREN : to tetanize 286
TETANUSANTITOXIN (*n*) : tetanus antitoxin, anti-tetanus serum 287
TETANUSHEILSERUM (*n*) : anti-tetanus serum 288
TETARTOPRISMATISCHES MELANERZ (*n*) : allanite, orthite, cerite 289
TETRABORSÄURE (*f*) : tetraboric acid 290
TETRABROMÄTHAN (*n*) : tetrabromoethane (see BROMACETYLEN) 291
TETRABROMFLUORESCEIN (*n*) : tetrabromofluorescine, eosin (colouring matter obtained from fluorescine and used in orthochromatic photographic work) 292
TETRABROMFLUORESCEINKALIUM (*n*) : erythrosine, potassium salt of tetraiodofluorescine (see JODEOSIN) 293
TETRABROMKOHLENSTOFF (*m*) : carbon tetrabromide ; CBr_4 294
TETRACHLORÄTHAN (*n*) : tetrachloroethane, acetylene tetrachloride ; $C_2H_2Cl_4$; Sg. 1.58 ; Mp. – 36°C. ; Bp. 147.2°C. 295
TETRACHLORÄTHYLEN (*n*) : tetrachloroethylene, carbon (dichloride) bichloride, tetrachloroethene ; $(CCl_2)_2$; Sg. 1.6 ; Mp. – 19°C. ; Bp. 118.5°C. 296
TETRACHLORCHINON (*n*) : tetrachloroquinone (see CHLORANIL) 297
TETRACHLORKOHLENSTOFF (*m*) : carbon tetrachloride ; CCl_4 ; Sg. 1.6326 ; Mp. – 22.95°C. ; Bp. 76.8°C. 298
TETRACHLORKOHLENSTOFFSEIFE (*f*) : tetrachloride soap 299
TETRACHLORPHTALSÄURE (*f*) : tetrachlorophthalic acid ; $C_6Cl_4(CO_2H)_2$ 300
TETRACHLORSILICIUM (*n*) : silicon tetrachloride ; $SiCl_4$ 301
TETRACHLORZINN (*n*) : tin tetrachloride ; $SnCl_4$ 302
TETRACYANOL (*n*) : tetracyanol 303
TETRADYMIT (*m*) : tetradymite ; Bi_2Te_2S ; Sg. 7.45-7.8 304
TETRAEDER (*n*) : tetrahedron 305
TETRAEDRISCH : tetrahedral 306
TETRAËDRIT (*m*) : tetrahedrite ; fahlore ; $Cu_8Sb_2S_7$; Sg. 4.4-4.85 307
TETRAFLUORKOHLENSTOFF (*m*) : carbon tetrafluoride ; CF_4 308
TETRAHEXAEDER (*m*) : tetrahexahedron 309
TETRAHYDRAT (*n*) : tetrahydrate ; $(4H_2O)$ 310
TETRAJODÄTHYLEN (*n*) : (see DIJODOFORM) 311
TETRAJODKOHLENSTOFF (*m*) : carbon tetriodide ; CI_4 312
TETRALIN (*n*) : tetralin, tetrahydronaphthalin (used as a substitute for turpentine oil) 313
TETRALUTION (*f*) : tetralution (a carbon tetrachloride soluble in H_2O) 314
TETRAMETHYLÄTHYLENGLYKOL (*n*) : (see PINAKON) 315
TETRAMETHYLDIAMINOBENZHYDROL (*n*) : tetramethyldiaminobenzhydrol, Michler's hydrol, tetramethyldiaminodiphenylcarbinol ; $(CH_3)_2N.C_6H_4CH(OH).C_6H_4N(CH_3)_2$; Mp. 96°C. 316
TETRAMETHYLPHOSPHONIUMHYDROXYD (*n*) : tetra-methyl phosphonium hydroxide ; $P(CH_3)_4.OH$ 317
TETRAMULSION (*f*) : tetramulsion 318
TETRAOXYPHTALOPHENANHYDRID (*n*) : (see FLUORESCEIN) 319
TETRAPOL (*n*) : tetrapol 320
TETRATHIONSÄURE (*f*) : tetrathionic acid ; $H_2S_4O_6$ 321
TETRAVANADINSÄURE (*f*) : tetravanadic acid 322
TETRINSÄURE (*f*) : tetrinic acid 323
TEUER : dear ; costly ; beloved ; expensive 324
TEUERUNG (*f*) : dearth ; famine ; rise (increase) in price ; dearness 325
TEUERUNGSZUSCHLAG (*m*) : increase in price, increase due to rise in prices 326
TEUFEL (*m*) : devil 327
TEUFELEI (*f*) : devilry ; devilishness ; diabolical machinations 328

TEUFELSABBIS (*m*): devil's bit; primrose scabious (*Scabiosa succisa*) 329
TEUFELSABBISWURZEL (*f*): succise root (see TEUFELSABBIS) 330
TEUFELSDRECK (*m*): asafœtida (gum resin from *Ferula narthex*) 331
TEUFLISCH: devilish; diabolical 332
TEXASIT (*m*): texasite; $Ni_3H_2CO_4.4H_2O$; Sg. 2.65 333
TEXELN: to adze 334
TEXEN (*n*): texene 335
TEXT (*m*): text, substance, words, wording, lecture, (subject)-matter 336
TEXTIL: textile 337
TEXTILBETRIEB (*m*): textile works 338
TEXTILINDUSTRIE (*f*): textile industry 339
TEXTILMASCHINEN (*f.pl*): textile machinery 340
TEXTILÖL (*n*): textile oil 341
TEXTILPULVER (*n*): textile powder 342
TEXTILSEIFE (*f*): textile soap 343
TEXTILSEIFEN-KOCHAPPARAT (*m*): textile soap boiler 344
TEXTILTECHNIK (*f*): textile technics 345
TEXTILWAREN (*f.pl*): textile goods, textiles 346
TEXTILWERK (*n*): textile works 347
TEXTOSE (*f*): textose 348
THAL (*n*): valley (see also TAL) 349
THALBUCHE (*f*): red beech (*Fagus sylvatica*) 350
THALENGE (*f*): narrow pass into or out of a valley 351
THALLICHLORAT (*n*): thallic chlorate 352
THALLIION (*n*): thallic ion 353
THALLINSULFAT (*n*): thalline sulphate 354
THALLINUM (*n*), GERESAURES-: thallinum tannate 355
THALLINUM (*n*), SCHWEFELSAURES-: thallinum sulphate 356
THALLINUM (*n*), WEINSAURES-: thallinum tartrate 357
THALLIOXYD (*n*): thallic oxide (see THALLIUM-OXYD) 358
THALLISALZ (*n*): thallic salt 359
THALLISULFAT (*n*): thallic sulphate 360
THALLIUM (*n*): thallium; (Tl) 361
THALLIUMACETAT (*n*): thallium acetate 362
THALLIUMALAUN (*m*): thallium alum (see ALAUN) 363
THALLIUMBROMÜR (*n*): thallous (thallium) bromide; TlBr; Sg. 7.54 364
THALLIUMCHLORID (*n*): thallium (thallic) chloride; $TlCl_3 - H_2O$ 365
THALLIUMCHLORÜR (*n*): thallous chloride; TlCl; Sg. 7.02 366
THALLIUMCHLORÜRCHLORID (*n*): thallium chloro-chloride; $3TlCl.TlCl_3$; Sg. 5.9 367
THALLIUMGLAS (*n*): thallium glass 368
THALLIUMHYDROXYD (*n*): thallic hydroxide 369
THALLIUMHYDROXYDUL (*n*): thallous hydroxide 370
THALLIUMJODID (*n*): thallium (thallic) iodide 371
THALLIUMJODÜR (*n*): thallous iodide; TlI; Sg. 7.056-7.072 372
THALLIUM (*n*), KOHLENSAURES-: thallium carbonate 373
THALLIUMOXYD (*n*): thallium (thallic) oxide; Tl_2O_3; Sg. 5.56 374
THALLIUMOXYDUL (*n*): thallous oxide; Tl_2O 375
THALLIUMOXYDULCARBONAT (*n*): thallous carbonate; Tl_2CO_3; Sg. 7.11 376
THALLIUMOXYDULCHLORAT (*n*): thallous chlorate; $TlClO_3$; Sg. 5.047 377
THALLIUMOXYDULNITRAT (*n*): thallous nitrate; $TlNO_3$; Sg. 5.3-5.8 378
THALLIUMOXYDULPERCHLORAT (*n*): thallous perchlorate; $TlClO_4$; Sg. 4.89 379
THALLIUMOXYDULPHOSPHAT (*n*): thallous phosphate; Tl_3PO_4; Sg. 6.89 380
THALLIUMOXYDULSULFAT (*n*): thallous sulphate; Tl_2SO_4; Sg. 6.77 381
THALLIUM (*n*), SALICYLSAURES-: thallium salicylate 382
THALLIUM (*n*), SCHWEFELSAURES-: thallium sulphate 383
THALLIUMSELENAT (*n*): thallium selenate; Tl_2SeO_4; Sg. 6.875 384
THALLIUMSULFÜR (*n*): thallous sulphide; Tl_2S; Sg. 8.0 385
TALLIUMVERBINDUNG (*f*): thallium compound 386
THALLIUM (*n*), WEINSAURES-: thallium tartrate 387
THALLIVERBINDUNG (*f*): thallic compound 388
THALLOCHLORAT (*n*): thallous chlorate (see THALLIUMOXYDULCHLORAT) 389
THALLOCHLORID (*n*): thallous chloride (see THALLIUMCHLORÜR) 390
THALLOION (*n*): thallous ion 391
THALLOJODAT (*n*): thallous iodate (see THALLIUM-JODÜR) 392
THALLOPHYTEN (*pl*): *Thallophytes* (including Algæ, Fungi and Lichens) (Bot.) 393
THALLOSALZ (*n*): thallous salt 394
THALLOVERBINDUNG (*f*): thallous compound 395
THAPSIASÄURE (*f*): thapsic acid 396
THEBAIN (*n*): thebaine, para-morphine; $C_{17}H_{15}NO(OCH_3)_2$; Mp. 193°C. 397
THEBAINCHLORHYDRAT (*n*): thebaine hydrochloride; $C_{19}H_{21}NO_3.HCl.H_2O$ 398
THEIN (*n*): theine (see KAFFEIN) 399
THEMA (*n*): theme; topic; subject; essay 400
THENARDIT (*m*): thenardite (natural sodium sulphate); Na_2SO_4; Sg. 2.7 (see also NATRIUMSULFAT) 401
THENARDSBLAU (*n*): Thenards blue (see KOBALT-BLAU) 402
THEOBROMIN (*n*): theobromine, dimethylxanthine (the active principle of cocoa); $C_7H_8O_2N_4$; Mp. 329°C. 403
THEOBROMINNATRIOACETAT (*n*): theobromine-sodium acetate (see AGURIN) 404
THEOBROMINNATRIOSALICYLAT (*n*): theobromine-sodium salicylate 405
THEOBROMINNATRIUMACETAT (*n*): theobromine-sodium acetate (see AGURIN) 406
THEOBROMINSÄURE (*f*): theobromic acid; $C_{64}H_{128}O_2$ 407
THEOCIN (*n*): theocine 408
THEODOLIT (*m*): theodolite 409
THEOLACTIN (*n*): theolactine 410
THEOLOG (*m*): theologian; divine 411
THEOLOGIE (*f*): theology; divinity 412
THEOLOGISCH: theological-(ly) 413
THEOPHONIN (*n*): theophonine 414
THEOPHYLLIN (*n*): theophylline 415
THEORETIKER (*m*): theorist 416
THEORETISCH: theoretical-(ly) 417
THEORIE (*f*): theory 418
THERAPEUTISCH: therapeutic 419
THERAPIE (*f*): therapeutics 420
THERAPINSÄURE (*f*): therapinic acid; $C_{13}H_{19}CO_2H$ 421
THERME (*f*): hot-spring 422
THERMIOL (*n*): thermiol 423

THERMISCH : thermal 424
THERMIT (*n*): thermite (mixture of metallic oxide and aluminium); ($Fe_2O_3 + Al_2$) etc. 425
THERMOCHEMIE (*f*) : thermochemistry 426
THERMOCHEMISCH : thermochemical 427
THERMODYNAMISCH : thermodynamic-(al) 428
THERMOELEKTRISCH : thermoelectric 429
THERMOELEKTRIZITÄT (*f*) : thermo-electricity 430
THERMOELEMENT (*n*) : thermoelement, thermocouple (of pyrometers) 431
THERMOKETTE (*f*) : thermoelement 432
THERMOLYSE (*f*) : thermolysis 433
THERMOMETER (*n*) : thermometer 434
THERMOMETERHALTER (*m*): thermometer (pocket) holder 435
THERMOMETERKUGEL (*f*) : thermometer bulb 436
THERMOMETER (*n*), QUECKSILBER- : mercury thermometer 437
THERMOMETERROHR (*n*) : thermometer tube; capillary tube 438
THERMOMETERRÖHRE (*f*) : thermometer tube, capillary tube 439
THERMOMETERSKALA (*f*) : thermometer (thermometric) scale 440
THERMOMETRIE (*f*) : thermometry (measurement of temperature with a thermometer) 441
THERMOMETRISCH : thermometric 442
THERMONATRIT (*m*) : thermonatrite (a natural sodium carbonate); $Na_2CO_3.H_2O$; Sg. 1.55 443
THERMOREGLER (*m*) : thermo-regulator, temperature regulator 444
THERMOSÄULE (*f*) : thermopile 445
THERMOSFLASCHE (*f*) : thermos flask 446
THERMOSTAT (*m*) : thermostat 447
THERMOSTROM (*m*) : thermo-electric current 448
THESE (*f*) : thesis 449
THESPESIAÖL (*n*) : mallow (musk-mallow, rose-mallow) oil (from *Hibiscus populneus*) 450
THIAMIN (*n*) : thiamine 451
THIAZIN (*n*) : thiazine, phenothiazine (see THIODIPHENYLAMIN) 452
THIAZINBRAUN (*n*) : thiazine brown 453
THIAZINFARBSTOFF (*m*) : (see THIONINFARBSTOFF) 454
THIAZOLGELB (*n*) : thiazol yellow 455
THIOANTIMONSÄURE (*f*) : thioantimonic acid 456
THIOARSENIGSÄURE (*f*) : thioarsenious acid 457
THIOARSENSÄURE (*f*) : thioarsenic acid 458
THIOCARBAMID (*n*) : thiocarbamide, thiourea; CH_4N_2S 459
THIOCARBAMINSÄURE (*f*) : thiocarbamic acid; thiolcarbamic acid; $HSCONH_2$ 460
THIOCARBANILID (*n*) : thiocarbanilide; $CS(NH.C_6H_5)_2$ 461
THIOCHROMOGEN (*n*) : thiochromogene 462
THIOCOL (*n*) : thiocol 463
THIOCRESOL (*n*) : thiocresol 464
THIOCYAN (*n*) : thiocyanogen 465
THIOCYANKALIUM (*n*) : potassium thiocyanate (see KALIUMSULFOCYANAT) 466
THIOCYANSAUER : thiocyanate of; combined with thiocyanic acid 467
THIOCYANSÄURE (*f*) : thiocyanic acid 468
THIOCYANVERBINDUNG (*f*) : thiocyanate 469
THIODIN (*n*) : thiodine 470
THIODIPHENYLAMIN (*n*) : thiodiphenylamine, thiazine, phenothiazine; $(C_6H_4)_2NHS$ 471
THIOFLAVIN (*n*) : thioflavine 472
THIOGENOL (*n*) : thiogenol 473
THIOGENSCHWARZ (*n*) : thiogene black 474
THIOGERMANIUMSÄURE (*f*) : thiogermanic acid 475
THIOHARNSTOFF (*m*) : thio urea; sulfourea; sulphourea; thiocarbamide; CH_4N_2S ; Sg. 1.4 ; Mp. 180°C. 476
THIOKOHLENSÄURE (*f*) : thiocarbonic acid; trithio-carbonic acid; H_2CS_3 477
THIOLAN (*n*) : thiolan 478
THIOLSÄURE (*f*) : thiolic acid 479
THIOLSEIFE (*f*) : thiol soap 480
THIONCARBAMINSÄURE (*f*) : thionocarbamic acid 481
THIONINFARBSTOFF (*m*) : thionine dye, thiazine dye (a class of tar dyes) 482
THIONKOHLENSÄURE (*f*) : thionocarbonic acid 483
THIONKOHLENTHIOLSÄURE (*f*) : thiolthionocarbonic acid; HOCSSH 484
THIONSÄURE (*f*) : thionic acid 485
THIONSCHWARZ (*n*) : thione black 486
THIONYLBROMID (*n*) : thionyl bromide, sulphur oxybromide; $SOBr_2$; Sg. 2.61-2.68 487
THIONYLCHLORID (*n*) : thionyl chloride, sulphur oxychloride; $SOCl_2$; Sg. 1.676 488
THIOPHEN (*n*) : thiophen; C_4H_4S ; Sg. 1.0705 ; Bp. 84°C. 489
THIOPHENOL (*n*) : thiophenol; C_6H_6S ; Sg. 1.078 ; Bp. 169.5°C. 490
THIOPHORFARBE (*f*) : thiophor dye 491
THIOPHOSPHORSÄURE (*f*) : thiophosphoric acid 492
THIOPHTALID (*n*) : thiophthalide 493
THIOPHTEN (*n*) : thiophthene 494
THIO-P-TOLUIDIN (*n*) : thio-para-toluidine; $NH_2(CH_3)C_6H_3.S.C_6H_3(CH_3)NH_2$ 495
THIOSALZ (*n*) : thio salt 496
THIOSÄURE (*f*) : thio acid 497
THIOSCHWEFELSÄURE (*f*) : thiosulphuric acid; $H_2S_2O_3$ 498
THIOSINAMIN (*n*) : thiosinnamine, allylsulphurea; $CSNH_2NHC_3H_5$ (from action of ammonia on mustard oil) (see also ALLYLSULFOCARBAMID) 499
THIOSULFAT (*n*) : thiosulphate; hyposulphite 500
THIOSULFOSÄURE (*f*) : thiosulphonic acid 501
THIOVERBINDUNG (*f*) : thio compound 502
THIOXINSCHWARZ (*n*) : thioxine black 503
THIOZINNSÄURE (*f*) : thio-stannic acid 504
THOMASEISEN (*n*) : Thomas iron; basic iron 505
THOMASMEHL (*n*) : Thomas meal; ground (pulverized) basic slag from Thomas process (used as agricultural manure) 506
THOMASPHOSPHATMEHL (*n*) : Thomas phosphate meal (a fertilizer) (see THOMASMEHL) 507
THOMASPROZESS (*m*) : Thomas process. [This process most in use on the continent. For pig iron with high phosphorous content.] 508
THOMASSCHLACKE (*f*) : basic slag (Thomas process) [(47-49% CaO ; 18-23% P_2O_5 ; 6-10% FeO ; 6% Fe_2O_3 ; 6-8% MnO ; 7% SiO_2 ; 1-3% MgO ; 0.6-1.5% CaS). It is pulverized into THOMASMEHL, which see, and used as an agricultural manure.] 509
THOMASSCHLACKENMÜHLE (*f*) : Thomas (basic) slag mill 510
THOMASSTAHL (*m*) : Thomas (basic) steel 511
THOMSONIT (*m*) : Thomsonite (a zeolite); $(Na_2.Ca)O.Al_2O_3.2SiO_2.2\frac{1}{2}H_2O$ or $2(Ca,Na_2)Al_2Si_2O_8.2H_2O$; Sg. 2.36 512
THOMSONOLITH (*m*) : thomsonolite; $Na_2CaAlF_6.H_2O$; Sg. 2.95 513
THOR (*n*) : gate; thorium; (Th) 514

THORCARBID (*n*): thorium carbide; ThC_2: Sg. 8.96 515
THORERDE (*f*): thoria; thorium oxide 516
THORERZ (*n*): thorium ore; monazite (see **MONAZIT**) 517
THORIANIT (*m*): thorianite (natural thorium and uranium oxide, with various other constituents); $(Th.U)O_2$; Sg. 8.0-9.7 [518
THORIT (*m*): thorite, orangite (thorium silicate); $ThSiO_4.2H_2O$; Sg. 4.9-5.2 (see also **ORANGIT**) 519
THORIUM (*n*): thorium 520
THORIUM-AKTINIUM (*n*): thorium-actinium (the name given to radio-active thorium) 521
THORIUMANHYDRID (*n*): thorium anhydride (see **THORIUMDIOXYD**) 522
THORIUMBORID (*n*): thorium boride; ThB_4; Sg. 7.5; ThB_6; Sg. 6.4 523
THORIUMBROMID (*n*): thorium bromide; $ThBr_4$; Sg. 5.67 524
THORIUMCARBID (*n*): thorium carbide; ThC_2; Sg. 8.96-10.15 525
THORIUMCHLORID (*n*): thorium chloride; $ThCl_4$; Sg. 4.59 526
THORIUMDIOXYD (*n*): thorium dioxide; ThO_2; Sg. 9.88 527
THORIUMEMANATION (*f*): thorium emanation (a radio-active substance) 528
THORIUMMETASULFAT (*n*): thorium metasulphate; $Th(PO_3)_4$; Sg. 4.08 529
THORIUMNIEDERSCHLAG (*m*): thorium precipitate 530
THORIUMNITRAT (*n*): thorium nitrate; $Th(NO_3)_4-6H_2O$ 531
THORIUMOXYD (*n*): thorium oxide; ThO_2; Sg. 9.87 532
THORIUMPRÄPARAT (*n*): thorium preparation 533
THORIUM (*n*), **SALPETERSAURES-**: thorium nitrate (see **THORIUMNITRAT**) 534
THORIUMSALZ (*n*): thorium salt 535
THORIUMSILICID (*n*): thorium silicide; $ThSi_2$; Sg. 7.96 536
THORIUMSULFAT (*n*): thorium sulphate; $Th(SO_4)_2$, Sg. 4.225; $Th(SO_4)_2-9H_2O$, Sg. 2.767 537
THORIUMSULFID (*n*): thorium sulphide; ThS_2; Sg. 6.7-6.8 538
THORIUMVERBINDUNG (*f*): thorium compound 539
THROMBOSE (*f*): thrombosis 540
THROSTLEGARN (*n*): water twist (Textile) 541
THUJAÖL (*n*): thuja oil; Sg. 0.92 (from leaves of *Thuja occidentalis*) 542
THUJON (*n*): thujon 543
THÜR (*f*): door 544
THÜRE (*f*): door 545
THYMEN (*n*): thymene 546
THYMIAN (*m*): thyme (*Herba thymi*) (*Thymus vulgaris*) (wild = *Thymus serpyllum*) (see also **FELDTHYMIAN**) 547
THYMIANEXTRAKT (*m*): thyme extract 548
THYMIANKRAUT (*n*): thyme-(herb) (see **THYMIAN**) 549
THYMIANKRAUTFLUIDEXTRAKT (*m*): (*Extractum herbæ Thymi vulgaris fluidum*): thyme fluid extract 550
THYMIANÖL (*n*): oil of thyme; thyme oil (*Oleum thymi*); Sg. 0.9-0.95 (from *Thymus vulgaris*) 551
THYMIN (*n*): thymine; HN.OC.HN.CO.C.CH. CH_3 (a pyrimidine base) 552
THYMINSÄURE (*f*): thymic acid, solurol; $C_{30}H_{46}N_4O_{15}.2P_2O_5$ 553
THYMOCHINON (*n*): thymoquinone 554
THYMOL (*n*): thymol, thyme camphor, thymic acid, isopropyl-meta-cresol; $(CH_3)_2CH.C_6H_3(CH_3)OH$; Sg. 0.98; Mp. 49°C.; Bp. 232°C. (see also **THYMIANÖL**) 555
THYMOLJODAT (*n*): thymol iodate 556
THYMOLPALMITAT (*n*): thymol palmitate 557
THYMOLSEIFE (*f*): thymol soap 558
THYMOTAL (*n*): thymotal (thymol carbon) 559
THYMUSDRÜSE (*f*): thymus gland (Med.) 560
THYREOIDINTABLETTEN (*f.pl*): thyreoidin tablets 561
THYRESOL (*n*): thyresol 562
TIDE (*f*): tide (of the sea) 563
TIEF: deep; low; profound; heavy (Breathing); far; extreme; dark (of colours) 564
TIEFÄUGIG: hollow-eyed 565
TIEFBAUTEN (*m.pl*): underground (works) constructions, works (constructions) below ground level 566
TIEFBLAU (*n*): deep blue; dark blue 567
TIEFBLICK (*m*): insight; penetration; penetrating glance 568
TIEFBOHRER (*m*): auger; deep-(well)-boring machine 569
TIEFBOHRMASCHINE (*f*): deep-(well) boring machine 570
TIEFBOHRUNG (*f*): deep boring 571
TIEFBRUNNEN (*m*): deep well 572
TIEFBRUNNENWASSER (*n*): deep-well water 573
TIEFDENKEND: profound, penetrating 574
TIEFDENKER (*m*): profound thinker 575
TIEFDRUCK (*m*): low pressure 576
TIEFE (*f*): deep; depth; deepness; depth; profundity 577
TIEFEBENE (*f*): lowland; plain 578
TIEFE (*f*), **GEMALLTE-**: moulded depth 579
TIEFEINDRINGEND: penetrating 580
TIEFEN: to deepen; lower; take sounding (Naut.) 581
TIEFENGESTEIN (*n*): plutonic rock (see **PLUTONIT**) 582
TIEFENLEHRE (*f*): depth gauge 583
TIEFENLINIE (*f*): current of a river 584
TIEFENREGLER (*m*): depth (regulator) governor (for submarines) 585
TIEFENSCHÄRFE (*f*): sharpness of the shadows (Photo., etc.) 586
TIEFGANG (*m*): draught (of ships or vessels) 587
TIEFGANGMARKEN (*f.pl*): draught marks (Ships) 588
TIEFGELB: deep yellow 589
TIEFHAMMER (*m*): hollowing hammer 590
TIEFKÜHLUNG (*f*): low (intense) cooling; refrigeration; cooling to low temperature 591
TIEFLIEGEND: deep-seated; sunken 592
TIEFLOT (*n*): deep-sea lead (Naut.) 593
TIEFROT: deep red 594
TIEFRUND: concave 595
TIEFRÜNDE (*f*): concavity 596
TIEFSCHÄFTIG: of the low-warp (Weaving) 597
TIEFSCHWARZ: deep black 598
TIEFSEEKABEL (*n*): deep-sea cable 599
TIEFSINN (*m*): profound thought, thoughtfulness; melancholy 600
TIEFSINNIG: pensive; melancholy; profound; thoughtful; serious 601
TIEFTEMPERATURGEWINNUNG (*f*): low temperature carbonization, low temperature production 602

TIEFTON (*m*): secondary accent; deep (sound) tone 603
TIEFTÖNEND: deep-sounding 604
TIEGEL (*m*): crucible; pot; stewpan; saucepan 605
TIEGELBRENNER (*m*): crucible maker 606
TIEGELBRENNOFEN (*m*): crucible oven 607
TIEGELDECKEL (*m*): crucible (lid) cover 608
TIEGELFLUSSSTAHL (*m*): crucible cast steel 609
TIEGELFORM (*f*): crucible mould 610
TIEGELFORMEREI (*f*): crucible moulding 611
TIEGELFUTTER (*n*): crucible lining 612
TIEGELGIESSEREI (*f*): casting in crucibles 613
TIEGELGUSS (*m*): crucible casting 614
TIEGELGUSSSTAHL (*m*): crucible-(cast) steel 615
TIEGELGUSSSTAHLGIESSEN (*n*): casting of crucible steel 616
TIEGELHOHLFORM (*f*): crucible mould 617
TIEGELOFEN (*m*): crucible furnace 618
TIEGELPROBE (*f*): crucible test 619
TIEGELSCHMELZVERFAHREN (*n*): crucible process 620
TIEGELSTAHL (*m*): crucible steel 621
TIEGELUNTERSATZ (*m*): crucible (support) stand 622
TIEGELZANGE (*f*): crucible tongs 623
TIEKBAUM (*m*): (see TEAKBAUM) 624
TIEMANNIT (*m*): tiemannite (mercury selenide); HgSe; Sg. 8.33 625
TIER (*n*): animal; beast; cattle; brute 626
TIERART (*f*): kind (species) of animal 627
TIERARZNEI (*f*): veterinary surgery; veterinary remedy 628
TIERARZNEIMITTEL (*n*): veterinary remedy 629
TIERARZT (*m*): veterinary surgeon 630
TIERBESCHREIBUNG (*f*): zoology; zoography 631
TIERBLASE (*f*): bladder (of animals) 632
TIERBUDE (*f*): menagerie 633
TIERCHEMIE (*f*): animal chemistry 634
TIERCHEN (*n*): animalcule; tiny (creature) animal 635
TIERFASER (*f*): animal fibre 636
TIERFETT (*n*): animal fat 637
TIERFIBRIN (*n*): animal fibrin 638
TIERGARTEN (*m*): Zoo; Zoological garden 639
TIERGATTUNG (*f*): genus of animals 640
TIERGESCHICHTE (*f*): natural history 641
TIERGIFT (*n*): animal poison; venom 642
TIERHEILKUNDE (*f*): veterinary science 643
TIERISCH: bestial; animal; brutish 644
TIERKEIM (*m*): animal (germ) embryo 645
TIERKENNER (*m*): zoologist 646
TIERKOHLE (*f*): animal charcoal 647
TIERKÖRPER (*m*): animal body 648
TIERKREIS (*m*): zodiac 649
TIERKREISZEICHEN (*n*): sign of the Zodiac 650
TIERKUNDE (*f*): zoology 651
TIERLEBEN (*n*): animal life 652
TIERLEHRE (*f*): zoology 653
TIERLEIM (*m*): animal (glue) size; gluten (Paper) 654
TIERMALER (*m*): animal painter 655
TIERMILCH (*f*): animal milk 656
TIERÖL (*n*): animal (bone) oil (*Oleum animale*); Dippel's oil; Sg. 0.94 657
TIERPFLANZE (*f*): zoophyte 658
TIERQUÄLEREI (*f*): cruelty to animals 659
TIERREICH (*n*): animal kingdom 660
TIERSCHAU (*f*): cattle show 661
TIERSCHUTZVEREIN (*m*): Society for the prevention of cruelty to animals 662
TIERSTEIN (*m*): zoolite 663
TIERSTOFF (*m*): animal substance 664
TIERVERSUCH (*m*): experiment on an animal 665
TIERWELT (*f*): animal world 666
TIERWESEN (*n*): animal; brute nature 667
TIERZELLE (*f*): animal cell 668
TIGER (*m*): tiger 669
TIGERAUGE (*f*): tiger-eye 670
TIGERFELL (*n*): tiger skin 671
TIGERHAUT (*f*): tiger-skin 672
TIGERN: to spot, speckle, stripe, variegate, mottle 673
TIGLINSÄURE (*f*): tiglinic (tiglic) acid; $C_5H_8O_2$; Mp. 64-5°C.; Bp. 198-5°C. 674
TILGBAR: capable of being (eradicated; cancelled; paid; deleted) destroyed (see TILGEN) 675
TILGEN: to destroy; annul; cancel; eradicate; blot out; erase; settle; delete; pay (discharge), (Debts); extinguish 676
TILGUNG (*f*): cancellation; annulment; extinction (see TILGEN) 677
TINGIBEL: stainable 678
TINGIEREN: to stain; dye 679
TINKAL (*m*): tincal (a natural sodium borate) (see also BORAX); $Na_2B_4O_7.10H_2O$; Sg. 1.75-1.81 680
TINKTORIELL: tinctorial 681
TINKTUR (*f*): tincture (aqueous or alcoholic, thinly-liquid extract of animal or vegetable matter) (Pharm.) 682
TINKTURENPRESSE (*f*): tincture press 683
TINKTURPRESSE (*f*): tincture press 684
TINTE (*f*): ink; tint; mess; pickle 685
TINTENARTIG: inky; like ink 686
TINTENEXTRAKT (*m*): ink extract 687
TINTENFABRIKANT (*m*): ink maker 688
TINTENFASS (*n*): inkstand; inkwell 689
TINTENFASSFEDER (*f*): fountain pen 690
TINTENFISCH (*m*): cuttlefish (*Sepia*) 691
TINTENFISCHSCHWARZ (*n*): sepia (Pigment) 692
TINTENFLASCHE (*f*): ink-bottle 693
TINTENFLECK (*m*): ink (mark; spot) stain; blot 694
TINTENGLAS (*n*): ink glass 695
TINTENGUMMI (*n* and *m*): ink eraser 696
TINTENKOPIERVERFAHREN (*n*): iron gallate process (Phot.); ink process (Art) 697
TINTENKRUG (*m*): ink jar or bottle 698
TINTENLÖSCHER (*m*): blotter 699
TINTENPUDER (*m*): ink powder 700
TINTENPULVER (*n*): ink powder 701
TINTENSTEIN (*m*): inkstone 702
TINTENTABELLE (*f*): ink (tablet) pellet 703
TINTIG: inky 704
TINTOMETER (*n*): tintometer 705
TIPP (*m*): tap 706
TIPPEN: to tap; strike gently; to tip 707
TIRANN (*m*): tyrant (see TYRANN) 708
TISANE (*f*): ptisan, diet drink, (Med.) 709
TISCH (*m*): table; board 710
TISCHBIER (*n*): table beer 711
TISCHBLATT (*n*): leaf (of a table) 712
TISCHBOHRMASCHINE (*f*): bench drilling-machine 713
TISCHCHEN (*n*): tablet; little table 714
TISCHDECKE (*f*): table (cover) cloth 715
TISCHEN: to dine; sit at table 716
TISCHGÄNGER (*m*): boarder 717
TISCHGEBET (*n*): grace 718
TISCHGEDECK (*n*): table linen 719
TISCHGELD (*n*): board 720
TISCHGERÄT (*n*): table (utensils; ware) appointments 721

TISCHGESCHIRR (*n*): (see TISCHGERÄT) 722
TISCHGESTELL (*n*): trestle 723
TISCHGLOCKE (*f*): (table)-bell 724
TISCHKASTEN (*m*): drawer (in a table) 725
TISCHKONTAKT (*m*): desk (table) push or contact switch 726
TISCHLER (*m*): joiner; carpenter; cabinet-maker 727
TISCHLERARBEIT (*f*): joinery; joiner's work; carpentry; cabinet-making 728
TISCHLEREI (*f*): carpentry, joinery, joiner's work, cabinet-making 729
TISCHLERGESELL (*m*): journeyman (carpenter) joiner 730
TISCHLERHANDWERK (*n*): joinery, joiner's trade, carpentry, cabinet-making 731
TISCHLERHOBEL (*m*): joiner's plane 732
TISCHLERLEIM (*m*): joiner's glue 733
TISCHLERWERKZEUG (*n*): carpenter's (joiner's) tools 734
TISCHMESSER (*n*): table-knife 735
TISCHSTATIV (*n*): table stand; studio stand (Phot.) 736
TISCHTELEPHON (*n*): table (desk) telephone 737
TISCHTUCH (*n*): table-cloth 738
TISCHVENTILATOR (*m*): desk fan; table fan 739
TISCHWEIN (*m*): table wine 740
TISCHZEIT (*f*): meal-time 741
TISCHZEUG (*n*): table linen 742
TITAN (*n*): titanium; (Ti) 743
TITANAMMONIUMFORMIAT (*n*): titanium-ammonium formate; $Ti(HCO_2)_3.3Ti(HCO_2)_2OH.2NH_4HCO_2.H_2O$ 744
TITANAMMONIUMOXALAT (*n*): titanium-ammonium oxalate; $TiO(NH_4.C_2O_4)_2.H_2O$ 745
TITANAMMONOXALAT (*n*): titanium-ammonium oxalate (see TITANAMMONIUMOXALAT) 746
TITANCARBID (*n*): titanium carbide; TiC; Sg. 4.25 747
TITANCHLORID (*n*): titanium (tetrachloride) chloride; $TiCl_4$; Sg. 1.76; Mp. −25°C.; Bp. 136.4°C. 748
TITANDERIVAT (*n*): titanium derivate 749
TITANDIJODID (*n*): titanium diiodide; TiI_2; Sg. 4.3 750
TITANDIOXYD (*n*): titanium dioxide; TiO_2; Sg. 4.13-4.25; Mp. 1560°C. 751
TITANEISEN (*n*): titaniferous iron; titanic iron ore; $FeTiO_3(+Fe_2O_3)$; Sg. 4.89; ferrous titanate; Washingtonite (see WASHINGTONIT); Ilmenite (see ILMENIT); menaccanite (about 37% iron) 752
TITANEISENERZ (*n*): ilmenite; titanic iron ore (see TITANEISEN and ILMENIT) 753
TITANEISENSAND (*m*): ilmenite; titaniferous iron sand (see ILMENIT) 754
TITANEISENSTEIN (*m*): ilmenite; titanic iron ore; Iserine (see TITANEISEN) 755
TITANERZ (*n*): titanium (titanic) ore; ilmenite (see ILMENIT and TITANEISEN) 756
TITANERZ (*n*), PERITOMES-: rutile (see RUTIL) 757
TITANERZ (*n*), PRISMATISCHES-: sphene 758
TITANERZ (*n*), PYRAMIDALES-: anatase (see ANATAS) 759
TITANFLUORID (*n*): titanium fluoride; TiF_4; Bp. 284°C. 760
TITANFLUORWASSERSTOFFSÄURE (*f*): fluotitanic acid; H_2TiF_6 761
TITANFORMIAT (*n*): titanium formate 762
TITANGLAS (*n*): titanium glass (by melting SiO_2 with titanic acid in a reducing atmosphere) 763
TITANHALOGEN (*n*): titanium halide 764
TITANHALTIG: titaniferous; containing titanium 765
TITANISALZ (*n*): titanic salt 766
TITANIT (*m*): titanite (see also SPHEN); $CaSiTiO_5$; Sg. 3.5 767
TITANIUMSALZ (*n*): titanium salt 768
TITANKALIUMOXALAT (*n*): titanium-potassium oxalate; $TiO(KC_2O_4)_2.2H_2O$ 769
TITANLÖSUNG (*f*): titanium solution 770
TITANMETALL (*n*): titanium metal 771
TITANNITRID (*n*): titanium nitride; Ti_2N_2; Sg. 5.1-5.18 772
TITANOFLUORWASSERSTOFFSÄURE (*f*): fluotitanous acid 773
TITANOSALZ (*n*): titanous salt 774
TITANOXYD (*n*): titanium oxide (see TITANDIOXYD and TITANSESQUIOXYD) 775
TITANOXYD (*n*), BLAUES-: blue titanium oxide; $2Ti_2O_3.3TiO_2$ 776
TITANOXYDHYDRAT (*n*): titanic hydroxide 777
TITANOXYDHYDRAT (*n*), SCHWARZES-: (black)-titanium hydroxide; $Ti(OH)_3$ 778
TITANOXYDULHYDRAT (*n*): titanous hydroxide 779
TITANPHOSPHID (*n*): titanium phosphide; TiP; Sg. 3.95 780
TITANPRÄPARAT (*n*): titanium preparation 781
TITANSALZ (*n*): titanium salt (see TITANISALZ and TITANOSALZ) 782
TITANSAUER: titanate of; combined with titanic acid 783
TITANSÄURE (*f*): titanic acid; TiO_2 784
TITANSÄUREANHYDRID (*n*): titanic anhydride; TiO_2 785
TITANSÄURELAKTAT (*n*): titanic acid lactate; $TiO_2(C_3H_5O_3)_4$ 786
TITANSESQUIOXYD (*n*): titanium sesquioxide; Ti_2O_3 787
TITANSILICID (*n*): titanium silicide; $TiSi_2$; Sg. 4.02 788
TITANSILOXYD (*n*): (see TITANGLAS) 789
TITANSULFAT (*n*): titanium sulphate; $TiO.SO_4$ 790
TITANTETRACHLORID (*n*): titanium tetrachloride; $TiCl_4$; Sg. 1.76; Bp. 136°C. 791
TITANTETRAFLUORID (*n*): titanium tetrafluoride; TiF_4; Sg. 2.798-2.833 792
TITANTHERMIT (*m*): titanium thermit 793
TITANTRICHLORID (*n*): titanium trichloride; $TiCl_3$ 794
TITANVERBINDUNG (*f*): titanium compound 795
TITANYLSULFAT (*n*): (see TITANSULFAT) 796
TITEL (*m*): title; claim 797
TITELBLATT (*n*): title page 798
TITER (*n*): titre; titer 799
TITERFLÜSSIGKEIT (*f*): standard solution; test solution 800
TITERSTELLUNG (*f*): standardization; establish- of titre 801
TITRAGE (*f*): titration, analysis 802
TITRIERANAYLASE (*f*): titration (volumetric) analysis 803
TITRIERAPPARAT (*m*): titrating (volumetric) apparatus 804
TITRIEREN: to titrate; test 805
TITRIERFLÜSSIGKEIT (*f*): titrating (standard) solution 806
TITRIERGERÄTE (*n.pl*): titrating apparatus 807
TITRIERMETHODE (*f*): titration method 808
TITRIERSÄURE (*f*): titrating (standard) acid 809
TITRIERUNG (*f*): titration 810

TITRIERVORRICHTUNG (*f*): titrating apparatus 811
TITRIMETRISCH: titrimetric-(al)-(ly), volumetric-(al)-(ly) 812
TITULAR: titular 813
TITULIEREN: to entitle; term; call; name 814
TOBEN: to rage; rave; storm; roar; rant 815
TOBEND: boisterous; raging 816
TOBSUCHT (*f*): delirium; madness; insanity; raving; fury; mania (Med.) 817
TOCHTER (*f*): daughter; (as a prefix = secondary; branch; filial) 818
TOCHTERANSTALT (*f*): branch establishment 819
TOCHTERLAND (*n*): colony 820
TOCHTERLOGE (*f*): branch lodge (Freemasonry) 821
TOCHTERSTAAT (*m*): colony 822
TOD (*m*): death; decease; demise (see also TOT) 823
TODBRINGEND: fatal; deadly; lethal; mortal 824
TODDY (*m*): toddy (see PALMWEIN) 825
TODESANGST (*f*): death-agony; mortal fright 826
TODESBLÄSSE (*f*): deathy-pallor 827
TODESFALL (*m*): death, demise, decease, case of death 828
TODESGABE (*f*): fatal (lethal) dose 829
TODESGEFAHR (*f*): death peril 830
TODESGRAUEN (*n*): death horror 831
TODESKAMPF (*m*): agony; death-struggle 832
TODESNOT (*f*): death-peril 833
TODESSCHRECKEN (*m*): fear (terror) of death 834
TODESSCHWEISS (*m*): death-sweat 835
TODESSTOSS (*m*): death blow 836
TODESSTRAFE (*f*): capital punishment; death penalty 837
TODESURTEIL (*n*): death-sentence 838
TODGEBOREN: still-born 839
TÖDLICH: fatal; deadly; lethal; mortal 840
TÖDLICHKEIT (*f*): mortality, deadliness 841
TODMÜDE: tired to death, dog-tired 842
TODSTILL: lifeless, motionless, still as death, dead-still, immovable 843
TODSÜNDE (*f*): deadly (mortal) sin 844
TOILETTE (*f*): toilet; dressing (toilet) table 845
TOILETTENSPIEGEL (*m*): looking-glass; pier glass 846
TOILETTSEIFE (*f*): toilet soap 847
TOKAIER (*m*): Tokay-(wine) 848
TOLAN (*n*): tolan; $C_{14}H_{10}$; Mp. 60°C. 849
TOLANROT (*n*): tolan red 850
TOLEDOBLAU (*n*): toledo blue 851
TOLERANT: tolerant 852
TOLERANZ (*f*): tolerance; limits of error 853
TOLIDIN (*n*): tolidine 854
TOLIDIN (*n*), **ORTHO-**: orthotolidine; $(CH_3.C_6H_3.NH_2)_2$; Mp. 128°C. 855
TOLL: mad; furious; crazy; raging; frantic; foolish; nonsensical; reckless 856
TOLLBEERE (*f*): belladonna; deadly-nightshade berry (see also BELLADONNA) 857
TOLLEN: to rave; be crazy 858
TOLLHAUS (*n*): madhouse; asylum 859
TOLLHÄUSLER (*m*): lunatic; maniac 860
TOLLHEIT (*f*): madness; frenzy; rage; fury; mad trick; foolhardiness; temerity; rashness; recklessness 861
TOLLKERBEL (*m*): hemlock; wild chervil 862
TOLLKIRSCHE (*f*): belladonna; deadly nightshade (see also BELLADONNA) 863
TOLLKIRSCHENÖL (*n*): (see BELLADONNAÖL) 864
TOLLKOPF (*m*): mad-cap; hot-headed person; dare-devil 865
TOLLKÖPFIG: mad-cap; hot-headed; dare-devil 866
TOLLKORN (*n*): darnel; cockleweed (see TAUMELKORN) 867
TOLLKÖRNER (*n.pl*): thorn-apple seeds (*Semen strammonii* of *Datura strammonium*) 868
TOLLKRAUT (*n*): belladonna; thorn apple; deadly nightshade (*Datura strammonium*) (see also BELLADONNA) 869
TOLLKÜHN: foolhardy; rash; madcap; dare-devil; reckless 870
TOLLKÜHNHEIT (*f*): foolhardiness; rashness; temerity; recklessness 871
TOLLPATSCH (*m*): lout; booby; blockhead; clown; awkward fellow; dolt; idiot 872
TÖLPEL (*m*): (see TOLLPATSCH) 873
TÖLPELEI (*f*): awkwardness; clownishness 874
TÖLPELKRANKHEIT (*f*): mumps (see MUMPS) 875
TÖLPISCH: clumsy; awkward; clownish 876
TOLRÜBE (*f*): white bryony (*Bryonia alba*) 877
TOLSUCHT (*f*): madness 878
TOLUBALSAM (*m*): opobalsam, Tolu balsam, Tolu resin (*Balsamum tolutanum*) (from *Toluifera Balsamum*); Sg. 1.2; Mp. 63°C. 879
TOLUCHINON (*n*): toluquinone 880
TOLUIDIN (*n*): toluidine; (C_7H_9N); ortho.—Sg. 0.9986; Bp. 199.7°C.; meta.—Sg. 0.9986; Bp. 203.3°C.; para.—Sg. 0.9538; Mp. 45°C.; Bp. 200.4°C. 881
TOLUIDINBLAU (*n*): toluidin blue 882
TOLUIDINSULFOSÄURE (*f*): toluidin sulfonic acid 883
TOLUNITRIL (*n*): tolunitrile 884
TOLUNITRIL (*n*), **PARA-**: para-tolunitrile; $C_6H_4(CH_3)NC$ 885
TOLUOL (*m*): toluol; toluene; phenylmethane; methylbenzene; C_7H_8; Sg. 0.8845; Mp. −92.4°C.; Bp. 110.7°C. 886
TOLUOLSULFAMID (*n*): toluene sulfamide (see ORTHOTOLUOLSULFAMID and PARATOLUOLSULFAMID) 887
TOLUOLSULFOCHLORID (*n*): toluene sulfochloride (see PARATOLUOLSULFOCHLORID and ORTHOTOLUOLSULFOCHLORID) 888
TOLUOLSÜSS (*n*): saccharin-(e) (see SACCHARIN) 889
TOLUSIRUP (*m*): tolu sirup 890
TOLUTINKTUR (*f*): tolu tincture 891
TOLUYLALDEHYD (*n*): (see PHENYLACETALDEHYD) 892
TOLUYLENDIAMIN (*n*): toluylene diamine; $C_6H_3(CH_3)(NH_2)_2$; meta.—Mp. 99°C.; Bp. 280°C.; para.—Bp. 274°C. 893
TOLUYLENDIAMIN (*n*), **SCHWEFELSAURES-**: toluylenediamine sulphate 894
TOLUYLENDIAMINSULFOSÄURE (*f*): toluylenediamine sulfacid 895
TOLUYLENROT (*n*): toluylene red 896
TOLUYLENSCHWARZ (*n*): toluylene black 897
TOLUYLSÄURE (*f*): toluylic (toluic) acid; $(C_8H_8O_2)$; ortho.—Sg. 1.0621; Mp. 104°C.; Bp. 259°C.; meta.—Sg. 1.0543; Mp. 109°C.; Bp. 263°C.; para.—Mp. 180°C.; Bp. 275°C. 898
TOLYLSCHWARZ (*n*): tolyl black 899
TOMATE (*f*): tomato (*Lycopersicum esculentum*) 900
TOMBAK (*n*): (red brass) tombac (a copper-zinc alloy) (82-89% Cu; 18-11% Zn.) 901

TON (*m*): clay; argil (see KAOLIN); sound; key; tone; tint; tune; accent; fashion; pitch; stress; disposition 902
TONADER (*f*): clay seam 903
TONART (*f*): kind of clay; key; pitch; tune; mode; tone; sound, etc. 904
TONARTIG: clayey; argillaceous 905
TONBAD (*n*): toning bath (Phot.) 906
TONBEARBEITUNG (*f*): clay-working 907
TONBEARBEITUNGSMASCHINEN (*f.pl*): clay-working machinery 908
TONBEIZE (*f*): red liquor (Dyeing); aluminium acetate (see ALUMINIUMACETAT) 909
TONBESCHLAG (*m*): coating of clay 910
TONBILDNEREI (*f*): ceramics 911
TONBODEN (*m*): clay soil 912
TONBREI (*m*): clay slip (Ceramics) 913
TONDECKE (*f*): clay cover 914
TONEISENSTEIN (*m*): clay ironstone; clay band; Sphærosiderite (see SPHÄROSIDERIT) 915
TONEN: to tone (Phot.); (*n*): toning (Phot.) 916
TÖNEN: to sound; chime; ring; tone; tint; intone; (*n*): ringing; sounding; resounding; clang; chime 917
TONERDE (*f*): alumina; clay; clayey (argillaceous) soil (see ALUMINIUMOXYD); alumina silicate; $Fe_2-Al_2O_3$ 918
TONERDE, AMEISENSAURE- (*f*): alumina formiate; aluminium formate; $Al(CO_2H)_3$ 919
TONERDEBEIZE (*f*): red liquor (Dyeing) (see ALUMINIUMACETAT) 920
TONERDE, CHLORSAURE- (*f*): alumina chlorate; aluminium chlorate (*Aluminium chloricum*); $Al(ClO_3)_3$ 921
TONERDE, ESSIGSAURE- (*f*): alumina acetate; aluminium acetate (*Aluminium aceticum*); $Al(C_2H_3O_2)_3$ 922
TONERDEHALTIG: aluminiferous; containing alumina 923
TONERDE, HARZSAURE- (*f*): alumina resinate; aluminium resinate; $Al(C_{44}H_{62}O_5)_3$ 924
TONERDEHYDRAT (*n*): alumina hydrate, aluminium hydrate, aluminium hydroxide, hydrated alumina, precipitated aluminium oxide; $Al_2O_3.xH_2O$; Sg. 2.3 925
TONERDE (*f*), KIESELSAURE-: (*Aluminium silicicum*): an aluminium silicate, Kaolin, Argilla, Porcelain clay, terra alba, China clay, bolus alba or white bole, Kaolinite; $Al_2O_3.2SiO_2.2H_2O$ 926
TONERDE (*f*), MILCHSAURE-: (*Aluminium lacticum*): aluminium lactate; $Al(C_3H_5O_3)_3$ 927
TONERDE, NAPHTOLSULFOSAURE- (*f*): alumina sulphonaphtholate 928
TONERDENATRON (*n*): sodium aluminate (see NATRIUMALUMINAT) 929
TONERDE, ÖLSAURE- (*f*): alumina oleate; aluminium oleate; $Al(C_{18}H_{33}O_2)_3$ 930
TONERDE (*f*), OXALSAURE-: aluminium oxalate (see ALUMINIUMOXALAT) 931
TONERDE (*f*), SALPETERSAURE-: aluminium nitrate (see ALUMINIUMNITRAT) 932
TONERDE (*f*), SCHWEFELSAURE-: aluminium sulphate; $Al_2(SO_4)_3$; Sg. 2.7; $Al_2(SO_4)_3.18H_2O$; Sg. 1.6 (see also ALUMINIUM-SULFAT) 933
TÖNERN: clay; of clay; argillaceous; clayey; earthen 934
TONFARBE (*f*): clay colour 935
TONFARBIG: clay-coloured 936
TON (*m*), FEUERFESTER-: fireclay 937
TONFIXIERBAD (*n*): (combined)-toning and fixing bath (Phot.) 938
TONFIXIERBAD (*n*,) NEUTRALES-: neutral toning and fixing bath (Phot.) 939
TONFIXIERBADPATRONE (*f*): toning and fixing bath cartridge (Phot.) 940
TONFIXIERBAD (*n*), SAURES-: acid toning and fixing bath (Phot.) 941
TONFIXIERSALZ (*n*): toning and fixing salt (Phot.) 942
TONGESCHIRR (*n*): earthenware; pottery 943
TONGIPS (*m*): argillaceous (clayey) gypsum (see GIPS) 944
TONGRUBE (*f*): clay-pit 945
TONHALTIG: argillaceous; containing clay 946
TONHÖHE (*f*): pitch (of sound) 947
TONICUM (*n*): tonic; stimulant (Med.) 948
TONIG: argillaceous; clayey 949
TONIGENOL (*n*): tonigenol 950
TONINDUSTRIE (*f*): clay industry; pottery (ceramic) industry 951
TONINDUSTRIELLER (*m*): clay worker 952
TONISCH: tonic 953
TONISCHES MITTEL (*n*): tonic (Med.) 954
TONISIEREN: to stimulate, tone up, act as a tonic 955
TONISIEREND: having a tonic effect, stimulating 956
TONKABOHNE (*f*): tonka bean; tonquin bean; coumarouna bean; snuff bean (bean of *Dipteryx oppostifolia*) 957
TONKALK (*m*): argillocalcite; clayey limestone 958
TONKEGEL (*m*): clay cone 959
TONKUNST (*f*): phonetics; music; science of sound 960
TONLAGER (*n*): bed (stratum) of clay 961
TONLEITER (*f*): gamut; scale (of sound) 962
TONMASSE (*f*): paste (Ceramics); clay mass 963
TONMERGEL (*m*): clay marl 964
TONMÜHLE (*f*): clay (pug) mill 965
TÖNNCHEN (*n*): keg; small cask 966
TONNE (*f*): cask; tun (Capacity); barrel; keg; ton (of 1,000 kg.) (Weight) 967
TONNENBOJE (*f*): tun buoy 968
TONNENGEHALT (*m*): tonnage, tons burden 969
TONNENGELD (*n*): tonnage-(money) 970
TONNENGEWICHT (*n*): tonnage-(weight) 971
TONNENMASS (*n*): tonnage-(measurement); shipping measurement (in shipping tons of 40 cu. ft. per ton) 972
TONNENWEISE: by (casks, tuns, barrels, kegs) tons 973
TONOL (*n*): tonol 974
TONPAPIER (*n*): tinted paper; (*pl*): tinted writings (Paper) 975
TON (*m*), PLASTISCHER-: plastic clay, kaolin (see BOLUS, WEISSER- and KAOLIN) 976
TONREINIGER (*m*): stone separator (Ceramics) 977
TONRETORTE (*f*): clay retort 978
TONRÖHRE (*f*): clay (tube) pipe 979
TONSANDSTEIN (*m*): clayey (argillaceous) sandstone 980
TONSALZ (*n*): toning salt 981
TONSATZ (*m*): composition (Music) 982
TONSCHÄLCHEN (*n*): clay dish 983
TONSCHÄLCHENFORM (*f*): clay dish mould 984
TONSCHICHT (*f*): layer (stratum; bed) of clay 985
TONSCHIEFER (*m*): argillite; clay slate 986
TONSCHLAMM (*m*): clay slip 987

TONSCHNEIDER (*m*): pug mill; clay mill 988
TONSEIFE (*f*): aluminous soap 989
TONSILBE (*f*): accented syllable 990
TONSTEIN (*m*): clay stone 991
TONSTEINGUT (*n*): clay stone white ware 992
TONSUBSTANZ (*f*): aluminium silicate, clay-substance or mass) (see ALUMINIUMSILIKAT) 993
TONTIEGEL (*m*): clay crucible 994
TONTIEGELFORM (*f*): clay crucible mould 995
TONUNG (*f*): toning (Phot.) 996
TONUS (*m*): tone, state (of the body) 997
TONWAREN (*f.pl*): earthenware; pottery (see TÖPFERWARE) 998
TONWERT (*m*): tonal value, tonal quality 999
TONZEICHEN (*n*): accent 000
TONZELLE (*f*): clay cell 001
TONZIEGEL (*m*): clay tile 002
TONZUSCHLAG (*m*): aluminous flux (Metal) 003
TONZYLINDER (*m*): clay cylinder 004
TOP (*m*): top; mast-head (Naut.) 005
TOPAS (*m*): topaz; topas; $Al_2(F,OH)_2SiO_4$ or $5Al_2SiO_5.Al_2SiF_{10}$; Sg. 3.55 006
TOPASFLUSS (*m*): artificial topaz 007
TOPASSCHÖRLIT (*m*): columnar topaz; pycnite (see PYKNIT) 008
TOPF (*m*): pot; jar; crock 009
TOPFBUTTER (*m*): crock-butter 010
TÖPFCHEN (*n*): little pot or jar 011
TOPFDECKEL (*m*): pot (cover) lid 012
TOPFEN (*m*): (see QUARK) 013
TÖPFER (*m*): potter 014
TÖPFERARBEIT (*f*): potter's work; pottery 015
TÖPFEREI (*f*): pottery; ceramics 016
TÖPFEREIMASCHINEN (*f.pl*): ceramic (pottery) machinery 017
TÖPFERERDE (*f*): potter's (earth) clay; argil 018
TÖPFERERZ (*n*): alquifou; potter's ore 019
TÖPFERFARBE (*f*): pottery (ceramic) colour 020
TÖPFERGESCHIRR (*n*): pottery; crockery 021
TÖPFERGUT (*n*): (see TÖPFERGESCHIRR) 022
TÖPFERHANDWERK (*n*): potter's trade; pottery 023
TÖPFERN: earthen 024
TÖPFEROFEN (*m*): potter's kiln 025
TÖPFERSCHEIBE (*f*): potter's wheel 026
TÖPFERTON (*m*): potter's clay 027
TÖPFERWARE (*f*): pottery; crockery; potter's ware (see TONWAREN) 028
TÖPFERZEUG (*n*): pottery; earthenware 029
TOPFFRUCHT (*f*): sapucaia nut (fruit of *Lecythis ollaria*) 030
TOPFGIESSEREI (*f*): pot casting; making crucible steel 031
TOPFGLASUR (*f*): earthenware glaze; alquifou 032
TOPFMÜHLE (*f*): barrel mill 033
TOPFSCHERBE (*f*): potsherd 034
TOPFSTEIN (*m*): soapstone; potstone (see SEIFENSTEIN) 035
TOPISCH: topical 036
TOPMAST (*m*): top-mast (Naut.) 037
TOPOGRAPHIE (*f*): topography 038
TOPOGRAPHISCH: topographical-(ly) 039
TOPOPHOTOGRAPHIE (*f*): photographic surveying 040
TOPP (*m*): masthead, top (Naut.) 041
TOPPEN: to trim (the yards) (Naut.) 042
TOPPLATERNE (*f*): top lantern 043
TOPPNANT (*f*): lift 044
TOPPNANT (*f*), **MARS-**: topsail lift 045
TOPPSEGEL (*n*): top sail 046
TOPPSEGEL (*n*), **GAFFEL-**: gaff top sail 047
TOR (*m*): fool; (*n*): gate; portal; door; gateway 048
TORF (*m*): peat; turf; Sg. 0.6-1.33 049
TORFARTIG: peaty; like peat 050
TORFASCHE (*f*): peat ashes 051
TORF, BAGGER- (*m*): dredged peat 052
TORFBASTFASER (*f*): peat bast fibre 053
TORFBODEN (*m*): peat-soil; peat-storage floor 054
TORFBRIKETT (*n*): peat briquette 055
TORFEISENERZ (*n*): bog iron ore (see RASENEISENERZ) 056
TORF, FASER- (*m*): fibrous peat 057
TORFFASER (*f*): peat fibre; Sg. 1.339 058
TORFFEUER (*n*): peat fire 059
TORFGAGELSTRAUCH (*m*): sweet gale, bog-myrtle (*Myrica gale*) 060
TORFGESCHMACK (*m*): peaty (flavour) taste 061
TORFGRUBE (*f*): place where peat is cut 062
TORF, HAND- (*m*): hand-cut peat 063
TORFIG: peaty 064
TORFKOHLE (*f*): peat (charcoal; coal; carbon) fuel 065
TORFLAGER (*n*): peat (yard) bog 066
TORFLÜGEL (*m*): wing (of a double gate) 067
TORF, MASCHINEN- (*m*): machine-cut peat 068
TORFMASSE (*f*): peat 069
TORFMEHL (*n*): peat meal 070
TORF, MODER- (*m*): mouldy peat 071
TORFMOOS (*n*): peat moss; sphagnum moss 072
TORFMÜLL (*n*): peat dust, peat mould (see TORFMULL) 073
TORFMULL (*n*): peat dust; peat (mould) refuse (the upper grass layer of peat beds is separated into the grass fibre, "Peat straw" and the mould attached to it, "Peat mould") 074
TORFÖL (*n*): peat (turf) oil 075
TORF, PECH- (*m*): pitch peat 076
TORF, PRESS- (*m*): pressed peat 077
TORFRAUCHGESCHMACK (*m*): flavour of peat smoke 078
TORF, SPECK- (*m*): bitumenous peat 079
TORFSTECHEN: to cut peat 080
TORFSTEIN (*m*): peat brick; peat briquette 081
TORF, STICH- (*m*): dug peat 082
TORF, STREICH- (*m*): moulded peat 083
TORFSTREU (*f*): peat straw, peat litter, peat fibre (see TORFMULL) 084
TORFTEER (*m*): peat (turf) tar 085
TORF, WURZEL- (*m*): fibrous peat 086
TORHEIT (*f*): folly; foolishness; silliness; foolery 087
TÖRICHT: silly; foolish; absurd 088
TORKEL (*f*): wine-press 089
TORMENTILLGERBSÄURE (*f*): tormentilla tannic acid (from *Potentilla tormentilla*) 090
TORMENTILLWURZEL (*f*): (*Rhizoma tormentillæ*): tormentilla root, blood root, potentilla root (root of *Potentilla tormentilla*) (see BLUTWURZEL) 091
TORNISTER (*m*): knapsack; pack 092
TORPEDOBOOT (*n*): torpedo boat 093
TORPEDOBOOTZERSTÖRER (*m*): torpedo boat destroyer, destroyer 094
TORPEDOJÄGER (*m*): torpedo boat destroyer. destroyer 095
TORPEDOLUFTPUMPE (*f*): torpedo air pump 096
TORPEDOMUNITION (*f*): torpedo ammunition 097
TORPIDITÄT (*f*): torpor (an inactive bodily and mental condition) 098

TORQUIEREN: to twist; subject to (torsion) torque 099
TORSIOGRAPH (*m*): torsiograph (an instrument for reading the critical number of revolutions, degree of inequality, etc.) 100
TORSION (*f*): torsion; torsional stress; twist; torque 101
TORSIONSBEANSPRUCHUNG (*f*): torsional (strain) stress, torque 102
TORSIONSDYNAMOMETER (*n*): torsion dynamometer 103
TORSIONSFESTIGKEIT (*f*): torsional strength; torque 104
TORSIONSGALVANOMETER (*n*): torsion galvanometer 105
TORSIONSINDIKATOR (*m*): torsion indicator, torsion meter 106
TORSIONSKRAFT (*f*): torsional (power)-force 107
TORSIONSMOMENT (*n*): torsional moment, moment of torsion, torque 108
TORSIONSSCHWINGUNG (*f*): torsional variation or vibration (the critical number of revolutions) 109
TORSIONSSPANNUNG (*f*): torsional stress 110
TORSIONSSTEIFIGKEIT (*f*): torsional strength, resistance to torsional stresses 111
TORSIONSWAGE (*f*): torsion balance 112
TORSIONSWINKEL (*m*): angle of torsion 113
TORT (*m*): injury; wrong; tort (Legal) 114
TORTE (*f*): tart; cake 115
TORTENBÄCKER (*m*): pastry-cook 116
TORULAN (*n*): torulan 117
TORWEG (*m*): gateway 118
TOSEN: to rage; roar 119
TOT: dead; lifeless; immovable; motionless (see also **TOD**) 120
TOTAL: total-(ly); complete-(ly) 121
TOTALREFLEKTOMETER (*n*): a type of refractometer (an instrument for measuring refractive index) 122
TOTBRENNEN: to overburn; kill; burn to death 123
TÖTEN: to kill; deaden; soften; pass (Time); destroy; annihilate; murder 124
TOTENACKER (*m*): graveyard; churchyard; burial-ground 125
TOTENAMT (*n*): requiem mass 126
TOTENBAHRE (*f*): bier 127
TOTENBLASS: pallid, deathly pale 128
TOTENBLUME (*f*): marigold (*Calendula*) 129
TOTENEULE (*f*): screech-owl 130
TOTENFARBE (*f*): livid colour; colour of death 131
TOTENGERIPPE (*n*): skeleton 132
TOTENGERUCH (*m*): cadaveric odour; smell of a corpse 133
TOTENGERÜST (*n*): scaffold 134
TOTENGLOCKE (*f*): passing bell 135
TOTENGRÄBER (*m*): grave-digger 136
TOTENKOPF (*m*): caput mortuum; death's head; colcothar (see **COLCOTHAR**) 137
TOTENSTARRE (*f*): rigor mortis (stiffening of the muscles after death) (Med.) 138
TOTENSTILLE (*f*): deathly (absolute) silence 139
TOTENUHR (*f*): death-watch-(beetle) 140
TOTENWAGEN (*m*): hearse 141
TÖTER (*m*): killer; extinguisher; murderer 142
TOTER GANG (*m*): end play (Machinery) 143
TOTES GEWICHT (*n*): dead weight 144
TOTES WASSER (*m*): still water 145
TOTGEBOREN; still-born 146
TOTGEBRANNT: overburned; dead 147
TOTGEGERBT: overtanned 148
TOTGEKOCHT: overboiled 149
TOTHOLZ (*n*): dead wood 150
TOTLAGE (*f*): dead centre (of crank) 151
TOTMAHLEN (*n*): overgrinding; overbeating (Paper) 152
TOTSCHLAG (*m*): murder; homicide; manslaughter 153
TÖTUNG (*f*): killing, murder, homicide, manslaughter; mortification 154
TOUR (*f*): turn; tour; revolution; round 155
TOURENZAHL (*f*): number of revolutions; r.p.m. (revolutions per minute) 156
TOURENZÄHLER (*m*): speed (indicator) counter, speedometer 157
TOURILL (*n*): Woulf-(f) bottle 158
TOURNANTÖL (*n*): rank olive oil; lowest grade of olive oil (skimmings from the top of juice from pressed olives) 159
TOXIKOLOG (*m*): toxicologist 160
TOXIKOLOGIE (*f*): toxicology (the science of poisons and their antidotes) 161
TOXIKOLOGISCH: toxicological-(ly) 162
TOXIKOLOGISCHE CHEMIE (*f*): the chemistry of poisons, toxicological chemistry 163
TOXISCH: toxic; poisonous 164
TOXIZITÄT (*f*): toxicity 165
TRAB (*m*): trot 166
TRABANT (*m*): satellite 167
TRABEN: to trot; (*n*): trotting 168
TRACHT (*f*): costume; mode; fashion; dress; course; pregnancy; litter (of animals); load; charge 169
TRACHTEN: to try; strive; attempt; tend to; aspire to; endeavour; (*n*): endeavour; attempt, etc. 170
TRÄCHTIG: pregnant (of animals) 171
TRÄCHTIGKEIT (*f*): pregnancy; gestation (of animals) 172
TRACHYT (*m*): trachyte 173
TRACHYTTUFF (*m*): trachytic tuff 174
TRAGACANTH (*m*): tragacanth (see **TRAGANT**) 175
TRAGACHSE (*f*): carrying axle, supporting axle 176
TRAGANT (*m*): tragacanth (*Gummi tragacanthæ*) (gum from *Astragalus gummifer* or *Astragalus verus*) 177
TRAGANTGUMMI (*n*): gum tragacanth (*Gummi traganth*) 178
TRAGANTIN (*n*): tragantine 179
TRAGANTSCHLEIM (*m*): tragacanth 180
TRAGBAHRE (*f*): stretcher; litter; hand-barrow 181
TRAGBALKEN (*m*): (supporting)-beam 182
TRAGBAND (*n*): strap; sling; suspender; belt (of belt conveyor) 183
TRAGBAR: portable; productive; fruitful; bearing; fertile; wearable; fast to wear (Dyeing) 184
TRAGE (*f*): litter; stretcher; frame; hand-barrow 185
TRÄGE: inert; inactive; idle; lazy; sluggish; dull; slow 186
TRAGECHT: fast to wear 187
TRAGEN: to carry; bear; produce; wear; dress; support; yield; suffer; endure; be pregnant 188
TRAGENDES GEWÖLBE (*n*): saving arch (Brickwork) 189
TRÄGER (*m*): bearer; porter; support; carrier; wearer; bracket; column; stanchion; girder; joist; stay; upright; beam; buckstay 190

TRÄGERPLATTE (*f*): rider plate, bearer plate 191
TRÄGERSÄGE (*f*): joist (I-beam, girder) saw 192
TRÄGERSTRASSE (*f*): joist mill (Rolling mills) 193
TRAGFÄHIGKEIT (*f*): buoyancy; tonnage; burden; bearing strength; productiveness; capacity; carrying capacity 194
TRAGFEDER (*f*): bearing spring 195
TRÄGHEIT (*f*): inertia laziness; idleness; dullness; slowness; sluggishness; indolence 196
TRÄGHEITSMOMENT (*n*): moment of inertia 197
TRAGISCH: tragic-(al); tragically 198
TRAGKETTE (*f*): supporting (suspension) chain 199
TRAGKRAFT (*f*): carrying capacity; supporting power; bearing strength 200
TRAGÖDIE (*f*): tragedy 201
TRAGPLATTE (*f*): supporting (bearer) plate 202
TRAGRAD (*n*): trailing wheel 203
TRAGRIEMEN (*m*): main-braces 204
TRAGRING (*m*): supporting ring (metal ring supporting blast furnace shaft or upper truncated cone shaped portion of blast furnace) 205
TRAGWEITE (*f*): range; significance; bearing; importance; distance of carry 206
TRAGZAPFEN (*m*): journal, bearing (neck) journal 207
TRAKTAMENT (*n*): treatment; treat; allowance; pay 208
TRAKTAT (*m*): treatise; treaty; tract 209
TRAKTIEREN: to treat; entertain; to stand treat; regale 210
TRAMBAHN (*f*): tramway; tram; tramway line; tramway system 211
TRAMPELN: to trample; stamp 212
TRAN (*m*): blubber; train oil; Sg. 0.92; fish oil 213
TRANBRENNEREI (*f*): blubber (train oil) boiling works 214
TRANCHIEREN: to carve 215
TRÄNE (*f*): tear; drop; globule 216
TRÄNEND: in tears; weeping; lachrymose 217
TRÄNENDRÜSE (*f*): lachrymal gland 218
TRÄNENFISTEL (*f*): lachrymal fistula 219
TRÄNENFLÜSSIGKEIT (*f*): lachrymal fluid 220
TRÄNENLOS: tearless 221
TRÄNENWEIDE (*f*): weeping willow (*Salix babylonica*) 222
TRANGERUCH (*m*): train oil (blubber) smell or odour 223
TRANIG: greasy; like train oil 224
TRANK (*m*): drink; beverage; potion; draft (of medicine); decoction 225
TRÄNKCHEN (*n*): draft; physic; poison; decoction 226
TRÄNKE (*f*): watering-place water-(ing) (of animals) 227
TRÄNKEN: to water (Cattle) to saturate; pickle; steep; moisten; soak; impregnate; suckle; give to drink; (*n*): impregnation; soaking, etc. 228
TRÄNKFLÜSSIGKEIT (*f*): impregnating, pickling (wood preserving) fluid or liquid 229
TRÄNKTROG (*m*): impregnating (pickling) trough (for wood) 230
TRÄNKUNG (*f*): saturation; impregnation; steeping (see **TRÄNKEN**) 231
TRANLEDER (*n*): leather; dressed with train oil 232
TRANRAFFINERIEANLAGE (*f*): train-oil (blubber) refining plant, fish oil refining plant 233
TRANRAFFINIERANLAGE (*f*): fish oil refining plant; blubber (train oil) refining plant 234
TRANRAFFINIEREN (*n*): train oil (blubber) refining 235
TRANSCHMIERE (*f*): daubing (for leather) 236
TRANSEIFE (*f*): train-oil soap; black soap 237
TRANS-FORM (*f*): trans-form 238
TRANSFORMATOR (*m*): transformer 239
TRANSFORMATORENÖL (*n*): transformer oil 240
TRANSFORMATORENSÄULE (*f*): transformer column 241
TRANSFORMATORÖL (*n*): transformer oil 242
TRANSFORMIEREN: to convert, transform, change; (*n*): transforming 243
TRANSIEDER (*m*): blubber (train-oil) boiler 244
TRANSIEDEREI (*f*): blubber (train oil) boiling works 245
TRANSITO (*m*): transit; passage 246
TRANSLATIONSEBENE (*f*): translation plane (crystallographic sliding plane) 247
TRANSMISSION (*f*): transmission; belt drive; (as a prefix=mechanically driven; belt driven) 248
TRANSMISSIONSANLAGE (*f*): (transmission)-shafting (plant) installation 249
TRANSMISSIONSDYNAMOMETER (*n*): belt dynamometer 250
TRANSMISSIONSÖL (*n*): shafting and gearing oil 251
TRANSMISSIONSWELLE (*f*): shafting; line shafting; connecting (driving; transmission) shaft 252
TRANSOMPLATTE (*f*): transom plate 253
TRANSPARENT: transparent 254
TRANSPARENTBILD (*n*): diapositive, transparency (Phot.) 255
TRANSPARENTFARBE (*f*): transparent colour 256
TRANSPARENTSEIFE (*f*): transparent soap 257
TRANSPARENZ (*f*): transparency 258
TRANSPECK (*m*): blubber 259
TRANSPONIEREN: to transpose (Music, etc.) 260
TRANSPORT (*m*): transport; conveyance 261
TRANSPORTABEL: portable 262
TRANSPORTANLAGE (*f*): conveying (conveyor) plant; transporter plant 263
TRANSPORTBAND (*n*): conveyor belt 264
TRANSPORTDAMPFER (*m*): transport steamer, transport 265
TRANSPORTEUR (*m*): conveyor; transporter; carrier; shipper; protractor (Geometry) 266
TRANSPORTGEFÄSS (*n*): transporting vessel 267
TRANSPORTGESELLSCHAFT (*f*): shipping company; transport company; carrier 268
TRANSPORTIEREN: to transport; convey; carry; transfer 269
TRANSPORTKORB (*m*): transport basket 270
TRANSPORTVERFAHREN (*n*): transfer carbon process (Phot.) 271
TRANSSTELLUNG (*f*): trans position 272
TRANSTEARIN (*n*): blubber (fish oil) stearine (see **STEARIN**) 273
TRANSVERSAL: transverse 274
TRANSVERSALE (*f*): transverse 275
TRANSVERSALMASSSTAB (*m*): transverse scale 276
TRANSZENDENT: transcendent-(al)-(ly) 277
TRAPEZ (*n*): trapeze; trapezium 278
TRAPEZGEWINDE (*n*): buttress thread 279
TRAPEZOEDER (*n*): trapezohedron 280
TRAPEZREGEL (*f*): trapezoidal rule 281
TRAPP (*m*): trap (Mineral); heavy (loud) step 282
TRAPPARTIG: trappean; trap-like; of the nature of trap 283

TRAPPE (*f*): footstep; trap; bustard 284
TRAPPELN: to tramp(le); stump; patter 285
TRAPPEN: to tramp; stamp; walk heavily 286
TRAPPPORPHYR (*m*): trachyte; trachytic porphyry 287
TRAPPSAND (*m*): coarse gravel 288
TRASS (*m*): trass, calcareous tufa 289
TRASSANT (*m*): drawer (of a bill) 290
TRASSAT (*m*): drawee (Com.) 291
TRASSIEREN: to draw (a bill) 292
TRASSIERTER WECHSEL (*m*): bill-(of exchange); draft 293
TRATTE (*f*): bill-(of exchange); draft 294
TRAUBE (*f*): grape; bunch of grapes; cluster: raceme (Bot.) 295
TRAUBENABFALL (*m*): husks (marc) of grapes 296
TRAUBENÄHNLICH: grape-like; like a bunch of grapes; botryoidal 297
TRAUBENARTIG: (see TRAUBENÄHNLICH) 298
TRAUBENBEERE (*f*): grape 299
TRAUBENBLEI (*n*): mimetite (see MIMETIT); pyromorphite (see PYROMORPHIT) 300
TRAUBENFÖRMIG: botryoidal (Bot.); racemose; aciniform; grapelike; of the form of grapes 301
TRAUBENKAMM (*m*): stalk 302
TRAUBENKERN (*m*): grape (seed) stone 303
TRAUBENKERNÖL (*n*): grape seed oil; Sg. 0.92-0.956 304
TRAUBENKIRSCHE (*f*): bird cherry (*Prunus padus*) 305
TRAUBENKIRSCHENRINDE (*f*): bird-cherry tree bark 306
TRAUBENKRAUT (*n*): American wormseed, Chenopodium, Jerusalem oak (*Chenopodium ambrosioides*) 307
TRAUBENKRAUT, MEXIKANISCHES- (*n*): American wormseed (*Herba botyros mexicanæ*), (see TRAUBENKRAUT and JESUITENTEE) 308
TRAUBENKRAUTÖL (*n*): American wormseed oil, Chenopodium oil -(from fruit of *Chenopodium ambrosioides*); Sg. 0.975 309
TRAUBENLESE (*f*): vintage 310
TRAUBENMOST (*m*): grape must 311
TRAUBENMUS (*n*): grape must 312
TRAUBENSAFT (*m*): grape juice; wine 313
TRAUBENSAUER: racemate of; combined with racemic acid 314
TRAUBENSÄURE (*f*): para-tartaric acid; racemic acid; uvic acid; inactive tartaric acid; $C_4H_6O_6.H_2O$; Sg. 1.697; Mp. 205.5°C. 315
TRAUBENVITRIOL (*m*): crystallized ferrous sulphate (see FERROSULFAT) 316
TRAUBENZUCKER (*m*): dextrose; grape sugar; $C_6H_{12}O_6$ (see DEXTROSE) 317
TRAUEN: to trust; confide in; rely on; marry; dare; venture; risk 318
TRAUER (*f*): mourning; sorrow; affliction; grief 319
TRAUERBIRKE (*f*): weeping birch (*Betula pendula*) 320
TRAUERN: to mourn; sorrow; grieve 321
TRAUERSPIEL (*n*): tragedy; drama 322
TRAUERWEIDE (*f*): weeping willow (*Salix babylonica*) 323
TRAUERZUG (*m*): funeral procession 324
TRAUFE (*f*): eaves; drip; gutter; trough 325
TRAUFEFASS (*n*): water-tub 326
TRÄUFELN: to trickle; drip; drop 327
TRAUFEN: to drop; drip 328
TRÄUFEN: to drip; drop 329
TRAUFERINNE (*f*): gutter 330
TRAUFERÖHRE (*f*): gutter-pipe 331
TRAUFESTEIN (*m*): gutter 332
TRAUFWASSER (*n*): rainwater 333
TRAULICH: intimate; snug; cosy; cordial; trustingly; familiar; comfortable 334
TRAUM (*m*): dream; fancy; vision; imagination 335
TRAUMATICIN (*n*): traumaticin (solution of india-rubber in chloroform) 336
TRÄUMEN: to dream; imagine; fancy 337
TRÄUMER (*m*): dreamer; visionary 338
TRÄUMEREI (*f*): reverie; fancy; dreaming; imagination 339
TRÄUMERISCH: visionary; fanciful 340
TRAURIG: sad; wretched; melancholy; sorrowful, doleful; mournful 341
TRAURIGKEIT (*f*): grief, etc. (see TRAURIG) 342
TRAURING (*m*): wedding-ring 343
TRAUSCHEIN (*m*): marriage (license) certificate 344
TRAUT: dear; cosy 345
TRAUUNG (*f*): marriage-(ceremony) 346
TRAUUNGSSCHEIN (*m*): marriage certificate 347
TRAVERSE (*f*): traverse 348
TRAVESTIE (*f*): travesty 349
TRAVESTIEREN: to travesty 350
TREBER (*f*): spent (remains) residue; husks; grains (Brew.); ground malt; malt dust; draff (spent malt or the marc of grapes) 351
TREFFEN: to hit; strike; meet; find; provide; take-(measures): arrive at; come to-(decision); happen; befall; (*n*): combat; meeting; battle; fight 352
TREFFEND: appropriate; suitable; striking 353
TREFFER (*m*): hit; prize; one who hits 354
TREFFLICH: admirable; excellent; choice 355
TREFFLICHKEIT (*f*): excellence 356
TREFFPUNKT (*m*): point of (impingement; meeting) impact 357
TREFOL (*n*): trefol 358
TREFOLIN (*n*): trefolin 359
TREIBANKER (*m*): driving anchor 360
TREIBARBEIT (*f*): cupellation; hammered (chased) work; chasing 361
TREIBASCHE (*f*): cupel (bone) ashes (see KNOCHENASCHE) 362
TREIBBRÜHE (*f*): old liquor 363
TREIBEBEET (*n*): hot-bed 364
TREIBEIS (*n*): floating (drifting; drift) ice 365
TREIBEISEN (*n*): white pig (Iron) 366
TREIBEN: to drive; actuate; impel; press; propel; urge; drift; float; shoot; sprout; carry on; do; work at; cupel; sublime; work (hammer; emboss; chase) metals; raise (Dough or Hides); practice (Occupations); force (Plants); put forth (Leaves); circulate; ferment; act as a diuretic; blow (Cement); beat (Game); move; push 367
TREIBENDES MEDIUM (*n*): driving medium 368
TREIBEN, MIT DEM HAMMER- : to chase 369
TREIBEN, VOR ANKER- : to drag the anchor 370
TREIBEN, WILLENSLOS- : to float about at random 371
TREIBER (*m*): driver; drover; propeller; impeller; refiner (Metals); beater; task-master 372
TREIBER (*m*), PACKUNGS- : packing driver 373
TREIBESAND (*m*): quicksand 374
TREIBHAUS (*n*): hot-house 375
TREIBHERD (*m*): cupellation furnace; refining (furnace) hearth 376

TREIBHOLZ (*n*): driftwood 377
TREIBKETTE (*f*): driving chain 378
TREIBKRAFT (*f*): motive (driving) power 379
TREIBLADUNG (*f*): propelling charge 380
TREIBMASCHINE (*f*): driving engine; prime mover 381
TREIBMITTEL (*n*): purgative (Med.); propellant; motive (driving) medium, agent or fluid 382
TREIBOFEN (*m*): refining furnace 383
TREIBÖL (*n*): motor oil; driving oil (for oil engines); fuel oil (for boilers) 384
TREIBPROZESS (*m*): cupellation 385
TREIBRIEMEN (*m*): driving belt 386
TREIBRIEMENFETT (*n*): belt grease; belt dressing; belt lubricant 387
TREIBSCHERBEN (*m*): cupel 388
TREIBSCHWEFEL (*m*): native sulphur (see SCHWEFEL) 389
TREIBSEIL (*n*): driving rope 390
TREIBSITZ (*m*): driving fit 391
TREIBWELLE (*f*): driving shaft 392
TREIDELLEINE (*f*): tracking rope, tow (towing) rope 393
TREIDELMAST (*m*): mast for towing boats 394
TREIDELN: to track, tow 395
TREIDELWEG (*m*): tow (towing) path 396
TREISEGEL (*n*): try-sail 397
TREISEGEL (*n*), VOR-: fore trysail 398
TREMATODEN (*f.pl*): trematoda, flukes (Tapeworms) (Med.) 399
TREMOLIT (*m*): tremolite, asbestos (a fibrous variety of amphibole); $CaMg_3(SiO_3)$; Sg. 2.95 (see also AMPHIBOL) 400
TRENNBAR: separable; divisible (see TRENNEN) 401
TRENNBARKEIT (*f*): divisibility (see TRENNEN) 402
TRENNEN: to separate; divide; sever; sunder; resolve; cleave; split-(up); decompose; part; disjoin 403
TRENNSÄGE (*f*): cleavage saw 404
TRENNUNG (*f*): separation; division; severing; sundering; parting; cleaving; cleavage; severance; decomposition 405
TRENNUNGSFLÄCHE (*f*): cleavage (parting) plane, surface or area 406
TRENNUNGSPLATTE (*f*): division (diaphragm, baffle) plate 407
TRENNUNGSPRODUKT (*n*): separation product 408
TRENNUNGSSTELLE (*f*): point of separation (into separate units of the high and low pressure parts of turbines); break (the point at which portions of a ruptured piece of metal have parted company) 409
TRENNUNGSVERFAHREN (*n*): separation process 410
TRENNUNGSWAND (*f*): partition; diaphragm; baffle plate; division plate 411
TREPPE (*f*): stairs; staircase; flight of stairs; steps; step 412
TREPPENABSATZ (*m*): landing (in a house) 413
TREPPENFÖRMIG: stepped 414
TREPPENGELÄNDER (*n*): baluster; balustrade 415
TREPPENLÄUFER (*m*): stair carpet 416
TREPPENROST (*m*): step-grate 417
TREPPENROSTFEUERUNG (*f*): step-grate furnace 418
TREPPENSTUFE (*f*): step; stair 419
TRESOR (*m*): treasury; safe 420
TRESORSCHEIN (*m*): treasury (bill) note 421
TRESORSTAHL (*m*): steel for safes 422
TRESPE (*f*): brome grass (*Bromus squarrosus*) 423
TRESTER (*m.pl*): residue; grounds; husks (grape marc; olive husks, etc.) (see also TREBER) 424
TRESTERBRANNTWEIN (*m*): grape-marc brandy 425
TRESTERWEIN (*m*): grape-marc wine; afterwine 426
TRETEN: to step; tread; go; enter; trample; actuate a (treadle) foot-pump 427
TRETGEBLÄSE (*n*): treadle (foot) bellows or blower 428
TRETKONTAKT (*m*): floor contact 429
TREU: true; faithful; loyal; honest; tenacious 430
TREUBRUCH (*m*): faithlessness; breach of faith; perfidy 431
TREUBRÜCHIG: faithless; untrue; perfidious 432
TREUE (*f*): faithfulness; sincerity; faith; uniformity; loyalty; fidelity; honesty 433
TREUHERZIG: true-hearted; sincere 434
TREULICH: honest, faithful-(ly) 435
TREULOS: faithless (see TREUBRÜCHIG) 436
TRI (*n*): (see TRICHLORÄTHYLEN) 437
TRIACETIN (*n*): triacetine 438
TRIAKISOKTAEDER (*n*): triagonal; trisoctahedron; triakisoctahedron 439
TRIÄTHYLAMIN (*n*): triethylamine; $C_6H_{15}N$; Sg. 0.7277; Bp. 88.9°C. 440
TRIÄTHYLPHOSPHIN (*n*): triethylphosphine 441
TRIAZOJODID (*n*): triazoiodide; iodine azide; (N_3I) 442
TRIAZOLSCHWARZ (*n*): triazol black 443
TRIBOLUMINESZENZ (*f*): tribolumin essence 444
TRIBROMANILIN (*n*): aniline tribromide 445
TRIBROMBUTYLALKOHOL (*m*), TERTIÄRER-: tertiary tribromobutylalcohol, brometon; $C_4H_7OBr_3$ (77% Bromine) 446
TRIBROMESSIGSÄURE (*f*): tribromacetic acid 447
TRIBROMPHENOL (*n*): tribromophenol (see BROMOL) 448
TRIBROMPHENOLWISMUT (*m* and *n*): (*Bismutum tribromphenolicum*): xeroform, bismuth (tribromophenate) tribromophenolate; $Bi_2O_3(C_6H_2Br_3OH)$ 449
TRIBUNE (or TRIBÜNE) (*f*): rostrum; platform; gallery; pulpit 450
TRIBUT (*m*): tribute 451
TRIBUTÄR: tributary 452
TRICALCIUMARSENIAT (*n*): (see CALCIUM, ARSENSAURES-) 453
TRICALCIUMPHOSPHAT (*n*): tribasic calcium phosphate; tricalcium orthophosphate; tricalcic phosphate; tertiary calcium phosphate; precipitated calcium phosphate; rock phosphate; $Ca_3(PO_4)_2$; Sg. 3.18; Mp. 1550°C. 454
TRICALCOL (*n*): tricalcol 455
TRICARBALLYLSÄURE (*f*): tricarballylic acid 456
TRICHLORALDEHYD (*n*): trichloroacetic aldehyde (see CHLORAL) 457
TRICHLORÄTHYLEN (*n*): trichlorethylene; ethylene trichloride; C_2HCl_3; Sg. 1.471; Mp. –70°C.; Bp. 86°C. 458
TRICHLORBUTTERSÄURE (*f*): trichlorobutyric acid 459
TRICHLORCHINON (*n*): trichloroquinone 460
TRICHLORESSIGSAUER: trichloracetate, combined with trichloracetic acid 461
TRICHLORESSIGSÄURE (*f*): trichloroacetic acid; $CCl.CO_2H$; Sg. 1.6302; Mp. 57°C.; Bp. 196.5°C. 462

TRICHLORISOPROPYLALKOHOL (*m*): trichloroisopropyl alcohol 463
TRICHLORMETHAN (*n*): trichloromethane (see CHLOROFORM) 464
TRICHLORMILCHSÄURE (*f*): trichlorolactic acid 465
TRICHROMSÄURE (*f*): trichromic acid 466
TRICHTER (*m*): funnel; trumpet-(piece); infundibulum (Bot.); hopper (of mechanical grates, etc.) 467
TRICHTEREINLAGE (*f*): filter cone 468
TRICHTERFÖRMIG: funnel-shaped; infundibular; infundibuliform 469
TRICHTERHALS (*m*): neck of a funnel 470
TRICHTERHALTER (*m*): funnel holder 471
TRICHTERKOLBEN (*m*): funnel flask 472
TRICHTERN: to pour through a funnel 473
TRICHTERROHR (*n*): funnel (tube) pipe 474
TRICHTERSTATIV (*n*): funnel stand 475
TRICHTERWANDUNG (*f*): funnel wall 476
TRICYANCHLORID (*n*): tricyanyl chloride; cyanuryl chloride 477
TRICYANSÄURE (*f*): tricyanic (cyanuric) acid (see CYANURSÄURE) 478
TRIDYMIT (*m*): tridymite (a natural silicon dioxide); SiO_2; Sg. 2.311 (see also SILICIUM-DIOXYD) 479
TRIEB (*m*): impetus; driving force; impulse; drive; inclination; instinct; shoot; sprout; driving; drifting; drift 480
TRIEBKRAFT (*f*): driving (motive; propelling) power or force; capacity for forming a head (of ferments) 481
TRIEBRAD (*n*): driving wheel 482
TRIEBROHR (*n*): sliding tube; telescopic tube; lens tube 483
TRIEBSTANGE (*f*): driving rod 484
TRIEBWERK (*n*): driving (machine; machinery) mechanism; motor; (driving)-gear; gearing; transmission; shafting 485
TRIEBWERKSTEILE (*m.pl*): driving (working, wearing) parts 486
TRIEFEN: to drop; trickle; drip; run; be bleary (of the eye) 487
TRIFERRINMALTYL (*n*): triferrine maltyl 488
TRIFERROCARBID (*n*): (see EISENCARBID) 489
TRIFERROL (*n*): triferrol 490
TRIFT (*f*): drift; pasture; pasturage; herd; drove 491
TRIFTIG: weighty; substantial; sound; valid; cogent; adrift 492
TRIFTIGKEIT (*f*): validity; cogency 493
TRIGASVERFAHREN (*n*): Delensky's method of gas production 494
TRIGEMIN (*n*): trigemine 495
TRIGONOMETRIE (*f*), EBENE-: plane trigonometry 496
TRIGONOMETRIE (*f*), SPHÄRISCHE-: spherical trigonometry 497
TRIGONOMETRISCH: trigonometric-(al)-(ly) 498
TRIHYDRAT (*n*): trihydrate; ($3H_2O$) 499
TRIISOBUTYLAMIN (*n*): triisobutylamine 500
TRIJODMETHAN (*n*): triiodomethane (see JODO-FORM) 501
TRIKALIUMPHOSPHAT (*n*): tripotassium phosphate; tribasic potassium phosphate 502
TRIKLIN: triclinic 503
TRIKLINISCH: triclinic 504
TRIKRESOLSEIFE (*f*): tricresol soap 505
TRILLER (*m*): trill; quaver; shake 506
TRILLERN: to trill; shake (Music) 507
TRIM (*m*): trim (of a ship) 508
TRIMÄNDERUNG (*f*): change of trim (Naut.) 509
TRIMELLITHSÄURE (*f*): trimellitic acid 510
TRIMESINSÄURE (*f*): trimesic acid 511
TRIMETHYLAMIN (*n*): trimethylamine; C_3H_9N; Sg. 0.662 Bp. 3.2/3.8°C. 512
TRIMETHYLAMINCHLORHYDRAT (*n*): trimethylamine hydrochloride; $N(CH_3)_3HCl$ 513
TRIMETHYLBENZOL (*n*): tri-methyl benzene (see MESITYLEN) 514
TRIMETHYLCARBINOL (*n*): trimethylcarbinol 515
TRIMETHYLENBROMID (*n*): trimethylene bromide 516
TRIMETHYLPHOSPHIN (*n*): trimethyl phosphine, tertiary methyl phosphine; $P(CH_3)_3$; Bp. 40°C. 517
TRIMETHYLPHOSPHINOXYD (*n*): **Trimethyl phosphine oxide**; $(CH_3)_3PO$ 518
TRIMETHYLPYROGALLOL (*n*): trimethylpyrogallol 519
TRIMETHYLSULFINJODID (*n*): trimethylsulfine iodide 520
TRIMMEN: to trim 521
TRIMMER (*m*): trimmer (of coal, etc.) 522
TRIMMOMENT (*n*): moment of trim 523
TRIMOLYBDÄNSÄURE (*f*): trimolybdic acid 524
TRIMORPHIE (*f*): trimorphism 525
TRINATRIUMPHOSPHAT (*n*): tribasic sodium phosphate; trisodium phosphate; sodium orthophosphate; $Na_3PO_4+12H_2O$; Sg. 1.63; Mp. 77°C. 526
TRINITRANILIN (*n*): trinitraniline 527
TRINITRIN (*n*): (see NITROGLYZERIN) 528
TRINITROBENZOESÄURE (*f*): trinitrobenzoic acid 529
TRINITROBENZOL (*n*): trinitrobenzene 530
TRINITROBUTYLTOLUOL (*n*): trinitrobutyltoluene; $CH_3.C_6H.(NO_2)_3.C(CH_3)_3$; Mp. 96-97°C. 531
TRINITROCHLORBENZOL (*n*): trinitrochlorobenzene 532
TRINITROGLYZERIN (*n*): (see NITROGLYZERIN) 533
TRINITROKRESOL (*n*): trinitrocresol; $C_7H_4(OH)(NO_2)_3$ 534
TRINITROMETHAN (*n*): trinitromethane (see CHLORPIKRIN) 535
TRINITROPHENOL (*n*): trinitrophenol (see PIKRIN-SÄURE) 536
TRINITROPHENOLNATRIUM (*n*): sodium picrate (see NATRIUMPIKRAT) 537
TRINITROTOLUOL (*n*): trinitrotoluene 538
TRINKBAR: drinkable; potable 539
TRINKBECHER (*m*): drinking cup, vessel or glass 540
TRINKEN: to drink; absorb 541
TRINKGEFÄSS (*n*): drinking vessel 542
TRINKGELD (*n*): tip; gratuity 543
TRINKGESCHIRR (*n*): drinking (vessels) ware; glasses 544
TRINKGLAS (*n*): (drinking)-glass 545
TRINKLIED (*n*): drinking song 546
TRINKSTUBE (*f*): tap-room; bar-parlour 547
TRINKWASSER (*n*): drinking water 548
TRIO-GROBSTRASSE (*f*): three-high breaking down mill (Rolling mills) 549
TRIONAL (*n*): trional (see METHYLSULFONAL) 550
TRIOWALZE (*f*): three-high rolls (Rolling mills) 551
TRIOWALZWERK (*n*): three-high rolling mill 552
TRIOXYBENZOESÄURE (*f*): (see GALLAPFEL-SÄURE) 553

TRIOXYBENZOL (*n*): pyrogallic acid; trioxybenzol (see **BRENZGALLUSSÄURE** and **PYROGALLOL**); trioxybenzene (see **PHLOROGLUZIN**) 554

TRIOXYDIAZETYLANTHRACHINON (*n*): trioxydiacetylanthraquinone 555

TRIOXYMETHYLEN (*n*): trioxymethylene 556

TRIPEL (*m*): tripoli; rottenstone; Sg. 1.98 (a type of Kieselguhr); tripolite; opal (see **OPAL**) 557

TRIPELN: to trip 558

TRIPELPHOSPHAT (*n*): triple phosphate 559

TRIPELSALZ (*n*): triple salt 560

TRIPELSTEIN (*m*): tripoli stone (see **TRIPEL**) 561

TRIPHAN (*m*): triphane, spodumene (see **SPODUMEN**); Sg. 3.15 562

TRIPHANSPAT (*m*), **AXOTOMER-**: prehnite; $2CaO.SiO_2+AlO_3SiO_2$; Sg. 3 563

TRIPHANSPAT (*m*), **PRIMATISCHER-**: spodumene 564

TRIPHENYLAMIN (*n*): triphenylamine; $C_{18}H_{15}N$; Mp. 127°C. 565

TRIPHENYLCARBINOL (*n*): triphenyl carbinol; $C_{19}H_{16}O$; Mp. 162.5°C. 566

TRIPHENYLMETHAN (*n*): triphenylmethane; $C_{19}H_{16}$; Sg. 1.0166; Mp. 92°C.; Bp. 359°C. 567

TRIPHENYLMETHANFARBSTOFF (*m*): triphenylmethane dye-stuff 568

TRIPHENYLPHOSPHAT (*n*): triphenyl phosphate; $(C_6H_5)_3PO_4$; Mp. 53°C.; Bp. 245°C. 569

TRIPHENYLSILIZIUMCHLORID (*n*): triphenylsilicon chloride 570

TRIPHYLIN (*m*): triphyline, triphylite (a lithium phosphate); $(Fe,Mn)LiPO_4$; Sg. 3.5 571

TRIPLIEREN: to treble 572

TRIPLIT (*m*): triplite (a natural fluophosphate of iron and manganese); $(Mn,Fe)PO_4$, $(Mn,Fe)F_2$; Sg. 3.6 573

TRIPPEL (*m*): tripoli (see **TRIPEL**) 574

TRIPPER (*m*): clap; gonorrhœa (Med.) 575

TRIPPERGIFT (*n*): gonorrhœal virus (Med.) 576

TRIPPERRHEUMATISMUS (*m*): rheumatoid arthritis, gonorrhœal (arthritis) rheumatism (Med). 577

TRIPPKEIT (*m*): trippkeite; $CuAs_2O_4$ 578

TRISALYT (*n*): trisalite 579

TRISÄURE (*f*): tribasic acid 580

TRISULFAMINSÄURE (*f*): trisulfamic acid 581

TRITHIONSÄURE (*f*): trithionic acid; $H_2S_3O_6$ [582

TRITONSCHWARZ (*n*): triton black 583

TRITT (*m*): step; footstep; tread; track; footprint; kick; trace; pace 584

TRITTBRETT (*n*): step (of a bus, tram, coach, etc.) 585

TRITTGEBLASE (*n*): treadle (foot) bellows or blower 586

TRIUMPHIEREN: to triumph; conquer, gain a victory, win 587

TRIVALIN (*n*): trivalin 588

TRIVIAL: trite; trivial 589

TRIVIALNAME (*m*): popular name 590

TRIWOLFRAMCARBID (*n*): triwolfram (tritungsten) carbide; W_3C; Mp. above 2700°C. 591

TROCHOIDE (*f*): trochoid 592

TROCKEN: dry; barren; arid; anhydrous (Chem.) 593

TROCKENANLAGE (*f*): drying plant 594

TROCKENAPPARAT (*m*): dryer; desiccator; drying apparatus 595

TROCKENAPPARAT, ROTIERENDER-: rotary dryer 596

TROCKENBATTERIE (*f*): dry battery (Elect.) 597

TROCKENBLECH (*n*): drying plate 598

TROCKENBODEN (*m*): drying (loft) floor 599

TROCKENBRETT (*n*): drying board 600

TROCKENCHLOR (*n*): dry chemicking (Bleaching) 601

TROCKENDOCK (*n*): dry dock 602

TROCKENE DESTILLATION (*f*): dry distillation (heating with the exclusion of air, being the separation of the gaseous elements from the carbon) 603

TROCKENELEMENT (*n*): dry (cell) battery (Elect.) 604

TROCKENE NADEL (*f*): dry-point (Etching) 605

TROCKENEXTRAKT (*m*): dry extract 606

TROCKENFÄULE (*f*): dry rot 607

TROCKENFILZ (*m*): drying felt 608

TROCKENFIRNIS (*m*): japan; siccative varnish 609

TROCKENFUTTER (*n*): dry (feed) forage or fodder 610

TROCKENGEHALT (*m*): dry (solid) content 611

TROCKENGERÜST (*n*): drying (rack) frame 612

TROCKENGESTELL (*n*): drying (stand) frame 613

TROCKENGLAS (*n*): drying glass 614

TROCKENGUT (*n*): material to be dried 615

TROCKENHAUS (*n*): drying house 616

TROCKENHEFE (*f*): dried yeast 617

TROCKENHEIT (*f*): dryness; drought; barrenness; aridity 618

TROCKENKAMMER (*f*): drying (room) chamber 619

TROCKENKASTEN (*m*): drying (oven) chest; drier 620

TROCKENLEGEN (*n*): drainage 621

TROCKENLEGUNG (*f*): drainage 622

TROCKENMALEREI (*f*): pastel (drawing) painting 623

TROCKENMASCHINE (*f*): drying machine; drier (Paper, Textile, etc.) 624

TROCKENMASS (*n*): dry measure 625

TROCKENMITTEL (*n*): siccative; drier; drying agent 626

TROCKENOFEN (*m*): drying (kiln) oven, or stove (Brickworks, etc.) 627

TROCKENÖL (*n*): drier; siccative (for oils) 628

TROCKENPLATTE (*f*): drying plate; dry-plate (Phot.) (glass coated with a sensitive silver salt emulsified in gelatine) 629

TROCKENPLATTENLACK (*m*): dry-plate varnish 630

TROCKENPLATZ (*m*): drying place 631

TROCKENPRÄPARAT (*n*): dry preparation; drying (preparation) medium; drier 632

TROCKENPROBE (*f*): dry (assay) test 633

TROCKENPUDDELN (*n*): dry puddling 634

TROCKENPULVER (*n*): drying powder 635

TROCKENRAUM (*m*): drying room; drier; drying (space) chamber 636

TROCKENREINIGUNG (*f*): dry cleaning 637

TROCKENROHR (*n*): drying (tube) pipe 638

TROCKENRÜCKSTAND (*m*): dry residue 639

TROCKENSCHÄLCHEN (*n*): drying capsule 640

TROCKENSCHALE (*f*): drying dish 641

TROCKENSCHRANK (*m*): drying (oven) chest; drier; drying closet, chamber, cupboard or room 642

TROCKENSCHWINDUNG (*f*): shrinkage due to drying 643

TROCKENSTÄNDER (*m*): drying-rack 644

TROCKENSTUBE (*f*): drying (stove) room 645
TROCKENSTUHL (*m*): drying arrangement of printing machine 646
TROCKENSUBSTANZ (*f*): dry substance; solid matter 647
TROCKENTREBER (*f*): dried grains (Brewing) 648
TROCKENTROMMEL (*f*): rotary dryer; drying drum; drum drier 649
TROCKENTUNNEL (*m*): tunnel drier 650
TROCKENTURM (*m*): turret drier (Leather); drying tower 651
TROCKENVORGANG (*m*): drying process 652
TROCKENVORRICHTUNG (*f*): drying (arrangement; apparatus) contrivance or device 653
TROCKENWALZE (*f*): drying (roller) roll 654
TROCKENZYLINDER (*m*): drying cylinder 655
TROCKNE (*f*): dryness 656
TROCKNE DESTILLATION (*f*): carbonization, dry distillation 657
TROCKNEN: to dry; desiccate; air; season (Wood) 658
TROCKNEND: drying; siccative 659
TROCKNENDES ÖL (*n*): drying oil 660
TROCKNER (*m*): drier 661
TROCKNIS (*f*): dryness 662
TROCKNUNG (*f*): drying; desiccation 663
TRODDEL (*f*): tassel; tuft 664
TRÖDEL (*m*): rubbish; old clothes; lumber; rags 665
TRÖDELBUDE (*f*): rag-shop; second-hand clothes shop 666
TRÖDELHAFT: trumpery; dawdling 667
TRÖDELMANN (*m*): tally-man 668
TRÖDELN: to dawdle; sell (second-hand articles; old clothes) rags 669
TRÖDLER (*m*): second-hand (old clothes) dealer 670
TROEGERIT (*m*): (see **TRÖGERIT**) 671
TROG (*m*): trough; hod; tray 672
TROGAPPARAT (*m*): trough apparatus; trough battery (Elect.) 673
TROGBATTERIE (*f*): trough battery (Elect.) 674
TROG, DIELEKTRISCHER- (*m*): dielectric trough 675
TRÖGELCHEN (*n*): little trough 676
TRÖGERIT (*m*): troegerite; $(UO_2)_3,As_2O_5,12H_2O$; Sg. 3.25 677
TROGSTECHER (*m*): stirrer (Sugar) 678
T-ROHR (*n*): T-tube; T-pipe 679
TROLIT (*m*): trolite (see **EISENSULFÜR**); Mp. 1197°C. 680
TROLLEN: to toddle; roll; decamp; pack; go away 681
TROMMEL (*f*): drum; tympanum (Anat.); barrel 682
TROMMELANKER (*n*): drum armature 683
TROMMELDARRE (*f*): drum kiln 684
TROMMEL (*f*), **EISERNE-**: iron drum 685
TROMMELFELL (*n*): drum-(head) skin; tympanic membrane; tympanum (Anat.) 686
TROMMELFEUER (*n*): drum fire 687
TROMMELHAUT (*f*): (see **TROMMELFELL**) 688
TROMMEL (*f*), **INDIKATOR-**: paper cylinder of an indicator 689
TROMMELKLÖPFEL (*m*): drumstick 690
TROMMELKLÖPPEL (*m*): drumstick 691
TROMMELKONVERTER (*m*): drum converter 692
TROMMELMÜHLE (*f*): drum mill; grinding mill 693
TROMMELN: to drum; beat a drum 694
TROMMELSCHLAG (*m*): drumbeat; roll; beating (rolling) of a drum 695
TROMMELSCHLÄGEL (*m*): drumstick 696
TROMMELSCHLÄGER (*m*): drummer 697
TROMMELSIEB (*n*): mixing sieve; drum sieve 698
TROMMELSTOCK (*m*): drumstick 699
TROMMELWICKELUNG (*f*): drum winding 700
TROMPETE (*f*): trumpet; (eustachian)-tube (Anat.) 701
TROMPETE, EUSTACHISCHE- (*f*): eustachian tube (Anat.) 702
TROMPETEN: to (sound a) trumpet 703
TROMPETENSTICH (*m*): sheep-shank 704
TRONA (*f*): natural soda (obtained from the African soda lakes in lower Egypt. Believed to be, by inversion of the letters, the origin of the word **NATRON**); $Na_2CO_3.NaHCO_3.2H_2O$; Sg. 2.15 705
TRONAVIOLETT (*n*): tronaviolet 706
TROOSTIT (*m*): troostite (a zinc or manganese orthosilicate); $(Zn,Mn)_2SiO_4$; Sg. 4.1; (*n*): troostite (microscopically structureless form of perlite) 707
TROPACOCAIN (*n*): tropacocaine (an alkaloid from *Erythroxylon coca*); $C_{15}H_{19}NO_2$ 708
TROPACOCAINCHLORHYDRAT (*n*): tropacocaine hydrochloride; $C_{15}H_{19}NO_2.HCl$; Mp. 271°C. 709
TROPAKOKAIN (*n*): (see **TROPACOCAIN**) (*Tropacocainum*); Mp. 49°C. 710
TROPÄOLIN (*n*): tropeoline 711
TROPASÄURE (*f*): tropic acid 712
TROPEN (*f.pl*): tropics 713
TROPENFRUCHT (*f*): tropical fruit 714
TROPENGEWÄCHS (*n*): tropical vegetation 715
TROPENPFLANZE (*f*): tropical plant 716
TROPF (*m*): dunce; simpleton; booby 717
TROPFBAD (*n*): douche; shower-bath 718
TROPFBAR: liquid; fluid; capable of forming drops 719
TROPFBARFLÜSSIG: liquid 720
TROPFBARKEIT (*f*): liquidity; capability of forming drops 721
TROPFBERNSTEIN (*m*): liquid amber (see **STORAX**) 722
TROPFBRETT (*n*): (plate)-drainer 723
TRÖPFCHEN (*n*): little drop; dunce; simpleton; booby; ninny 724
TRÖPFCHENKULTUR (*f*): drop culture 725
TRÖPFELFETT (*n*): drippings; dripping 726
TRÖPFELN: to drip; trickle; drop 727
TRÖPFELPFANNE (*f*): dripping pan; dripper 728
TRÖPFELWERK (*n*): drying (graduation) house (Salt) 729
TROPFEN: to drop; drip; trickle; (*m.pl*): drops (Med.); (*m*): drop; globule; tear 730
TROPFENFALL (*m*): gutter; eaves; drip 731
TROPFENFÖRMIG: drop-shaped 732
TROPFENMIXTUR (*f*): drops (Med.) 733
TROPFENWEISE: drop by drop; dropwise; by drops 734
TROPFENZÄHLER (*m*): drop bottle 735
TROPFFASS (*n*): hogshead (Sugar) 736
TROPFFEUER (*n*): rain of fire, such as golden rain (Fireworks) 737
TROPFFLASCHE (*f*): drop bottle 738
TROPFGLAS (*n*): drop glass (pipette or drop bottle) 739
TROPFHAHN (*m*): drop cock 740
TROPFHAUS (*n*): curing house (Sugar) 741
TROPFKALK (*m*): drop chalk; calcium carbonate (see **CALCIUMCARBONAT**) 742
TROPFKANTE (*f*): list (Tin plate) 743
TROPFKASTEN (*m*): save-all (Paper) 744

TROPFKÖRPER (*m*): filter-bed (circular pile of coke in sewage works, through which the waste water is allowed to percolate) (see also FÜLLKÖRPER) 745
TROPFNASS: dripping wet 746
TROPFPFANNE (*f*): dripping pan 747
TROPFRINNE (*f*): gutter 748
TROPFROHR (*n*): drop-(ping) tube 749
TROPFSCHWEFEL (*m*): drop (torrefied) sulphur 750
TROPFSTEIN (*m*): dripstone; stalagmite; stalactite; travertine (natural calcium carbonate) (see CALCIUMCARBONAT) 751
TROPFSTEINARTIG: stalactitic 752
TROPFTRICHTER (*m*): dropping funnel 753
TROPFWÄSSER (*n.pl*): drops (Med.) 754
TROPFWEIN (*m*): droppings; leakings (of wine) 755
TROPFZINK (*n*): drop zinc 756
TROPHÄE (*f*): trophy 757
TROPHONIN (*n*): trophonine 758
TROPIN (*n*): tropin 759
TROPISCH: tropical 760
TROPON (*n*): tropon 761
TROSS (*m*): baggage; baggage train (Military); crowd 762
TROSSE (*f*): tow-rope, hawser 763
TROSSE (*f*), **ANKER-**: anchor cable 764
TROSSELGARN (*n*): water twist (Textile) 765
TROSSE (*f*), **STAHL-**: steel hawser 766
TROST (*m*): comfort; consolation 767
TRÖSTEN: to console; comfort 768
TRÖSTLICH: consolatory, consolable 769
TROSTLOS: inconsolable; disconsolate; comfortless 770
TROSTREICH: comforting, consolatory 771
TRÖSTUNG (*f*): consolation, comfort 772
TROTTE (*f*): wine press; bruising mill 773
TROTTEL (*m*): idiot; cretin 774
TROTTEN: to trot; trudge 775
TROTTOIR (*n*): pavement; sidewalk 776
TROTYL (*n*): trotyl 777
TROTZ: in spite (defiance) of; despite; (*m*): defiance; spite; scorn; insolence 778
TROTZ BIETEN: to defy; brave 779
TROTZDEM: nevertheless; notwithstanding 780
TROTZEN: to be defiant; bid defiance; brave; be obstinate (refractory) sulky; boast of; presume; defy; dare 781
TROTZIG: defiant; defying; insolent; daring; presumptuous 782
TROTZKÖPFIG: stubborn; defiant 783
TRUB (*m*): dregs; sediment (Brewing); deposit; lees; thickness (Wine); cloudiness 784
TRUBBIER (*n*): beer from sediment bag; cloudy beer 785
TRÜBE: turbid; dim; cloudy; muddy; dull; thick; sad; gloomy; troubled; (*f*): turbidity; slime (Metal); cloudiness (Liquid) 786
TRUBEL (*m*): tumult; trouble 787
TRÜBEN: to trouble; dim; dull; cloud; make turbid; make gloomy 788
TRÜBEND: clouding 789
TRÜBEND, NICHT-: non-clouding 790
TRUBSACK (*m*): filter (sediment) bag (Brewing) 791
TRÜBSAL (*f* and *n*): trouble; affliction; tribulation; gloom 792
TRÜBSELIG: sad; troubled; woeful; afflicted; gloomy; miserable 793
TRÜBSELIGKEIT (*f*): adversity; misery; calamity 794
TRÜBSINN (*m*): depression; gloominess; gloom; melancholy; sadness; dejection; misery 795
TRÜBUNG (*f*): turbidity; making turbid 796
TRÜFFEL (*f*): truffle (*Tuber æstivum*) 797
TRUG (*m*): fraud; deception; deceit; fallacy 798
TRUGBILD (*n*): illusion; phantom; mirage 799
TRÜGEN: to deceive; cheat; be deceitful; be deceptive; delude; be mistaken 800
TRÜGEND: deceptive, illusory, deceitful, delusive, fallacious 801
TRÜGERISCH: (see TRÜGEND) 802
TRUGGEBILDE (*n*): illusion; phantom 803
TRUGGEWEBE (*n*): tissue of (deceit) lies 804
TRUGLOS: artless 805
TRUGLOSIGKEIT (*f*): artlessness 806
TRUGWERK (*n*): deception, illusion 807
TRUHE (*f*): chest; trunk 808
TRUMEAU (*m*): pier 809
TRUMEAUSPIEGEL (*m*): pier-glass 810
TRÜMMER (*n.pl*): ruins; remains; wreck; fragments; rubbish 811
TRÜMMERACHAT (*m*): brecciated agate (see ACHAT) 812
TRÜMMERGESTEIN (*n*): breccia; conglomerate 813
TRUMPF (*m*): trump 814
TRUMPFAS (*n*): ace of trumps (Cards) 815
TRUMPFEN: to trump 816
TRUNK (*m*): drink; draft; potion; drinking; liquor; trunk 817
TRUNKEN: drunk; intoxicated 818
TRUNKENHEIT (*f*): drunkenness; intoxication 819
TRUNKMASCHINE (*f*): trunk engine 820
TRUPP (*m*): troop; set; flock; band; gang 821
TRUPPE (*f*): troop; company; body; troupe; (*pl*): troops; forces 822
TRUPPENTRANSPORTSCHIFF (*n*): troop-transport 823
TRÜSCHE (*f*): (see QUAPPE) 824
TRUTHAHN (*m*): turkey-(cock) 825
TRUTHENNE (*f*): turkey-(hen) 826
TRUTZ (*m*): offence; offensive 827
TRUTZBÜNDNIS (*n*): offensive alliance 828
TRUXILLIN (*n*), **ALPHA-**: alpha truxillin (see ISATROPYLCOCAIN) 829
TRYBOL (*n*): trybol 830
TRYPSALIN (*n*): trypsaline 831
TRYPSOGEN (*n*): trypsogene 832
TRYPTISCH: tryptic 833
TRYPTOPHAN (*n*): tryptophane, l-α-amino-β-indolepropionic acid; $NH.CH_2.CH(NH_2).CO_2H$ 834
T-STÜCK (*n*): T; tee; T-piece 835
T-TEER (*m*): (TIEFTEMPERATURTEER); low temperature tar (coal tar distilled at low temperature) 836
TUBE (*f*): tube 837
TÜBE (*f*): tube 838
TUBEN: (plural of TUBUS (*m*)=tube); (as a prefix=tube; tubular) 839
TUBENFÜLLMASCHINE (*f*): tube filling machine 840
TUBERCULIN (*n*): tuberculin (see TUBERKULIN) 841
TUBERKEL (*f*): tubercle; (as a prefix=tubercular; tubercle) 842
TUBERKELBAZILLUS (*m*): bacillus of tuberculosis 843

TUBERKULIN (*n*): tuberculin (an antituberculosis agent) (a glycerine broth culture of the tubercle bacillus, evaporated at 100°C. to one tenth of its original volume; for subcutaneous injection in cases of tuberculosis) (Med.) 844
TUBERKULOL (*n*): tuberculol 845
TUBERKULÖS: tubercular 846
TUBERKULOSE (*f*): tuberculosis 847
TUBERKULOSEMITTEL (*n*): tuberculosis remedy 848
TUBOCURARIN (*n*): tubocurarine (an alkaloid from CURARE, which see); $C_{19}H_{21}O_4N$ 849
TUBULIERT: tubulated 850
TUBULÖS: tubular; tubulous 851
TUBULUS (*m*): tubule; tubulation; tubulare 852
TUBUS (*m*): tube 853
TUCH (*n*): fabric; material; cloth; shawl; handkerchief 854
TUCHARTIG: like cloth 855
TUCHBAUM (*m*): beam; roller 856
TUCHBLAU (*n*): cloth blue 857
TUCHBEREITER (*m*): cloth dresser 858
TUCHEN: cloth; of cloth 859
TUCHFABRIK (*f*): cloth (factory) mill 860
TUCHFABRIKANT (*m*): cloth manufacturer 861
TUCHFARBE (*f*): cloth dye 862
TUCHFÄRBER (*m*): cloth-dyer 863
TUCHFÄRBEREI (*f*): cloth dyeing 864
TUCHGEWÖLBE (*n*): draper's shop 865
TUCHHALLE (*f*): draper's hall; cloth hall 866
TUCHHANDEL (*m*): cloth trade; drapery trade 867
TUCHHÄNDLER (*m*): cloth merchant; draper 868
TUCHHANDLUNG (*f*): cloth trade; draper's shop 869
TUCHLADEN (*m*): draper's shop 870
TUCHMACHER (*m*): cloth (worker) maker; clothier 871
TUCHNADEL (*f*): breast-pin; clothes peg 872
TUCHNAPF (*m*): colour cup 873
TUCHPRESSE (*f*): cloth press 874
TUCHRAHMEN (*m*): tenter; cloth-frame 875
TUCHRASCH (*m*): serge 876
TUCHRAUHER (*m*): friezer; napper; carder 877
TUCHREISENDE (*m*): cloth salesman; commercial traveller in cloth 878
TUCHREISENDER (*m*): (see TUCHREISENDE) 879
TUCHREST (*n*): remnant (Cloth) 880
TUCHSARSCHE (*f*): serge 881
TUCHSCHERE (*f*): cloth shears 882
TUCHSCHERMASCHINE (*f*): friezing (cloth-shearing) machine 883
TUCHSCHROTE (*f*): list of cloth 884
TUCHSTOPFER (*m*): fine-drawer (Cloth) 885
TÜCHTIG: capable; able; effective; strong; skilful; sound; apt; fit; qualified; solid; hard; large 886
TÜCHTIGKEIT (*f*): excellence, efficiency, thoroughness 887
TUCHWALKE (*f*): fulling (Cloth) 888
TUCHWALKER (*m*): fuller (Cloth) 889
TUCHWARE (*f*): drapery 890
TUCHWEBER (*m*): cloth weaver 891
TUCHWEBEREI (*f*): cloth manufacture; cloth factory; weaving 892
TÜCKE (*f*): malice; spite; trick; insidiousness 893
TÜCKISCH: insidious; malignant; malicious; spiteful 894
TUFFARTIG: tufaceous 895
TUFFKALK (*m*): tufaceous limestone; travertine (see CALCIUMCARBONAT) 896
TUFFSTEIN (*m*): tuff; tufa; travertine 897
TUFFSTEINARTIG: tufaceous 898
TUGEND (*f*): virtue; morality 899
TUGENDGESETZ (*n*): moral law 900
TUGENDHAFT: virtuous 901
TUGENDLEHRE (*f*): morals; ethics 902
TUGENDLEHRER (*m*): moralist 903
TUGENDREICH: virtuous 904
TÜLL (*m*): net-work 905
TÜLLE (*f*): spout; nozzle; socket 906
TÜLLSPITZE (*f*): point-lace 907
TULPE (*f*): tulip (*Tulipa*) 908
TULPENBAUM (*m*): tulip tree; whitewood; Canary (*Liriodendron tulipifera*); Sg. 0.45-0.6 909
TULPENZWIEBEL (*f*): tulip bulb 910
TUMENOL (*n*): tumenol 911
TUMESZIEREN: to swell; tumefy 912
TUMMELN: to exercise; bestir oneself; keep moving; bustle; hurry; tumble 913
TUMMLER (*m*): tumbler; porpoise; dolphin 914
TÜMPEL (*m*): pool; tymp (Blast furnaces) 915
TUMULT (*m*): tumult; bustle; noise; uproar 916
TUMULTUARISCH: tumultuous-(ly) 917
TUN: to do; make; take; perform; put; act; (*n*): doing; action; conduct 918
TÜNCHE (*f*): whitewash; plaster; parget; rough-cast; limewash 919
TÜNCHEN: to plaster; rough-cast; whitewash; parget; limewash; (*n*): whitewashing 920
TÜNCHER (*m*): whitewasher 921
TÜNCHFARBE (*f*): plastering colour 922
TÜNCHMASCHINE (*f*): white-washing machine 923
TÜNCHSCHICHT (*f*): finishing coat (Plastering) 924
TÜNCHWERK (*n*): whitewashing; pargeting 925
TUNGÖL (*n*): tung oil (see ELÄOKOKKAÖL) 926
TUNGSTEIN (*m*): scheelite (see SCHEELIT) 927
TUNKE (*f*): dip; pit; steep 928
TUNKEN: to dip; soak; steep 929
TUNLICH: feasible; practicable; convenient 930
TUNLICHKEIT (*f*): practicability; convenience; feasibility 931
TUNNEL (*m*): tunnel 932
TUNNEL (*m*), PROPELLERWELLEN-: propeller shaft tunnel 933
TUNNELTÜR (*f*): tunnel door 934
TUNNEL (*m*), WELLEN-: propeller shaft tunnel 935
TUPF (*m*): spot 936
TÜPFEL (*m*): spot; dot; point; jot 937
TÜPFELIG: dotted; spotted; speckled; mottled 938
TÜPFELMETHODE (*f*): spot (drop) method 939
TÜPFELN: to spot; speckle; stipple; dot; mottle 940
TÜPFELPROBE (*f*): spot (drop) test 941
TUPFEN: to dab; touch; tip 942
TÜPFEN: (see TUPFEN) 943
TUPFER (*m*): tampon; pledget; swab 944
TÜR (*f*): door 945
TÜRANGEL (*f*): door-hinge 946
TÜRBESCHLAG (*m*): door (mountings) fittings 947
TURBINE (*f*): turbine 948
TURBINE (*f*), AXIAL-: axial turbine 949
TURBINE (*f*), DAMPF-: steam turbine 950
TURBINE (*f*), EINSTUFIGE-: single stage turbine 951

TURBINE (*f*), **MEHRSTUFIGE-** : multi-stage turbine 952
TURBINENGEHÄUSE (*n*) : turbine housing 953
TURBINENKÖRPER (*m*) : turbine body or housing 954
TURBINENLAGER (*n*) : turbine bearing 955
TURBINENPUMPE (*f*) : turbine pump, turbo-pump 956
TURBINENRAD (*n*) : turbine wheel 957
TURBINENSCHAUFEL (*f*) : turbine blade, vane or bucket 958
TURBINE (*f*), **PARTIAL-** : partial turbine 959
TURBINE (*f*), **RADIAL-** : radial flow turbine 960
TURBINE (*f*), **REAKTIONS-** : reaction turbine 961
TURBINE (*f*), **TANGENTIAL-** : tangential flow turbine 962
TURBINE (*f*), **UMSTEUERBARE-** : reversible turbine 963
TURBINE (*f*), **VOLL-** : full (supply) admission turbine 964
TURBINE (*f*), **WASSER-** : water turbine 965
TURBODYNAMO (*f*) : turbodynamo 966
TURBOGEBLÄSE (*n*) : turbo blower 967
TURBOGENERATOR (*m*) : turbo-generator 968
TURBOKOMPRESSER (*m*) : turbo-compressor 969
TÜRE (*f*) : door 970
TÜRFELD (*n*) : door panel 971
TÜRFLÜGEL (*m*) : wing (of a folding door) 972
TURGESZENT : turgescent, swelling, swollen 973
TURGIT (*m*) : turgite, hydrohematite; $Fe_4O_5(OH)_2$; Sg. 4.14 974
TÜRGRIFF (*m*) : door (knob) handle 975
TURICOL (*n*) : turicol 976
TÜRKEI (*f*) : Turkey (Geography) 977
TÜRKIS (*m*) : turquoise (hydrous aluminium phosphate); $(Al_2O_3)_2P_2O_5.5H_2O$; Sg. 2.75 978
TÜRKISBLAU (*n*) : turquoise blue 979
TÜRKISCH : Turkish 980
TÜRKISCHE BOHNE (*f*) : scarlet runner (*Phaseolus multiflorus*) 981
TÜRKISCHER WEIZEN (*m*) : maize; Indian corn (*Zea mays*) 982
TÜRKISCHES KORN (*n*) : (see **TÜRKISCHER WEIZEN**) 983
TÜRKISCHROT (*n*) : Turkey red (Dyeing) 984
TÜRKISCHROTFÄRBEREI (*f*) : Turkey-red dyeing-(works) 985
TÜRKISCHROTÖL (*n*) : Turkey-red oil (a watery solution of oil mordant from castor oil), sulfonated castor oil 986
TÜRKISCHROTÖL-RÜHRWERK (*n*) : turkey red oil stirring apparatus 987
TÜRKLINKE (*f*) : door-latch 988
TÜRKLOPPER (*m*) : (door)-knocker 989
TÜRKONÖL (*n*) : turkon oil 990
TÜRKONTAKT (*m*) : door contact 991
TURM (*m*) : tower; turret; spire; steeple; castle; rook (Chess) 992
TURMALIN (*m*) : tourmaline; turmaline; $H_2O.Na_2O.MgO.Al_2O_3.B_2O_3.SiO_2$ (an alumina silicate); Sg. 2.94-3.24; Mp. 1000-1100°C. (see also **SCHÖRL** and **INDIGOLITH**) 993
TURMALINKRISTALL (*m*) : tourmaline crystal 994
TURMALINPLATTE (*f*) : tourmaline plate (Polarization) 995
TURMALINZANGE (*f*) : tourmaline tongs 996
TÜRMCHEN (*n*) : turret 997
TURMDECKSCHIFF (*n*) : turret ship 998
TURMDREHMASCHINE (*f*) : turret turning engine 999
TÜRMEN : to tower; pile up; rise-(to a height) 000
TURMFÖRMIG : turreted; tower-like 001
TURMGESCHÜTZ (*n*) : turret gun 002
TURMHOCH : towering, very high 003
TURMKNOPF (*m*) : ball (of a steeple) 004
TURMPANZER (*m*) : turret armour 005
TURMSCHIFF (*n*) : turret ship 006
TURMSPITZE (*f*) : spire 007
TURMUHR (*f*) : turret clock; clock in a (tower) steeple 008
TURNEN : to practice gymnastics; give a gymnastic display 009
TURNER (*m*) : gymnast 010
TURNERSGELB (*n*) : Turner's yellow (see **MONTPELLIERGELB**) 011
TURNKUNST (*f*) : gymnastics 012
TURNLEHRER (*m*) : master (teacher) of gymnastics 013
TURNMEISTER (*m*) : (see **TURNLEHRER**) 014
TURNVEREIN (*m*) : gymnastic club 015
TURPENTINÖL (*n*) : turpentine-(oil) 016
TÜRPFOSTEN (*m*) : door-post 017
TURPITHWURZEL (*f*) : turpeth root (*Radix turpethi*) 018
TÜRRAHMEN (*m*) : door frame 019
TÜRRIEGEL (*m*) : (door)-bolt 020
TÜRSCHLOSS (*n*) : door lock 021
TÜRSCHLÜSSEL (*m*) : door-key 022
TÜRSCHWELLE (*f*) : door (sill, ledge) step, threshold 023
TÜRSTOCK (*m*) : (see **TÜRZARGE**) 024
TURTELTAUBE (*f*) : turtle-dove 025
TÜRZARGE (*f*) : door frame 026
TÜRZIEHER (*m*) : door spring, any type of door closer 027
TUSCALINROT (*n*) : tuscaline red 028
TUSCH (*m*) : Indian ink; flourish (of trumpets) 029
TUSCHE (*f*) : Indian ink 030
TUSCHELN : to whisper 031
TUSCHEN : to paint with Indian ink or water colours; tinge (usually applied to wash drawing) 032
TUSCHFARBE (*f*) : water colours 033
TUSCHPINSEL (*m*) : ink brush 034
TUSSOL (*n*) : tussol 035
TUTE (*f*) : assay crucible; glass cylinder; paper bag (usually with special reference to such of conical shape) 036
TÜTE (*f*) : (see **TUTE**) 037
TUTEN : to toot; sound a horn 038
TÜTENDRUCK (*m*) : paper-bag printing 039
TÜTENMASCHINE (*f*) : paper bag machine 040
TÜTENPAPIER (*n*) : bag paper, paper for bags 041
TUTTE (*f*) : assay-(ing) crucible; glass cylinder; paper bag (see **TUTE**) 042
TYALID (*n*) : tyalide 043
TYLCALSIN (*n*) : tylcalsin 044
TYNDALLISATION (*f*) : a fractionized sterilization of ampules containing living bacteria, in which the ampules are heated to 60°C. for 2 hours per day for a number of days 045
TYP (*m*) : type, kind, sort 046
TYPE (*f*) : type, kind, sort, style 047
TYPENMETALL (*n*) : type metal 048
TYPENMOLEKÜL (*n*) : type molecule 049
TYPENMUSTER (*n*) : standard sample 050
TYPENTHEORIE (*f*) : type theory 051
TYPENWASCHLAUGE (*f*) : type wash-(ing fluid) 052
TYPHAFASER (*f*) : reed-mace fibre (fibre of *Typha latifolia* and *Typha angustifolia*) 053

TYPHOSE ERSCHEINUNGEN (*f.pl*): typhoid state (Med.) 054
TYPHUS ABDOMINALIS (*m*): typhoid (enteric, gastric, pythogenic) fever 055
TYPHUSBAZILLUS (*m*): typhus bacillus 056
TYPHUS (*m*), **FLECK-**: (see **KRIEGSPEST**) 057
TYPISCH: typical-(ly) 058
TYPOGRAPHIE (*f*): typography (a method of raised printing) 059
TYPOGRAPHISCH: typographical 060
TYPOPHORSCHWARZ (*n*): typephore black 061
TYPUS (*m*): type 062
TYRANN (*m*): tyrant; despot 063
TYRANNEI (*f*): tyranny; despotism 064
TYRANNISCH: tyrannical; despotic 065
TYRANNISIEREN: to tyrannize-(over) 066
TYROLERGRÜN (*n*): Tyrol green (see **GRÜNERDE**) 067
TYROSIN (*n*): para-hydroxyphenylalanine, tyrosine; $HO.C_6H_4.CH_2.CH(NH_2)CO_2H$; l-α-amino-β-p-oxyphenylpropionic acid; $HO.CH_2.CH(NH_2).CO_2H$ 068

U

ÜBEL: bad; badly; evil; sick-(ly); ill; wrong; unwell; (*n*): evil; ill; disease; misfortune; injury; sickness; nausea 069
ÜBELANGEBRACHT: misplaced 070
ÜBELBEFINDEN (*n*): indisposition 071
ÜBEL DARAN SEIN: to be badly off; be in a bad (case) way 072
ÜBELKEIT (*f*): nausea; sickness 073
ÜBELKLANG (*m*): dissonance 074
ÜBELLAUNIG: ill-humoured 075
ÜBELLAUNIGKEIT (*f*): ill-humour 076
ÜBELNEHMEN: to take (ill) amiss 077
ÜBELNEHMERISCH: touchy 078
ÜBELRIECHEND: foul-(bad; ill) smelling; foul; fœtid 079
ÜBELSTAND (*m*): nuisance; inconvenience; disadvantage; evil; bad state of things; impropriety 080
ÜBELTAT (*f*): offence; misdeed; crime 081
ÜBELTÄTER (*m*): malefactor; evil-doer 082
ÜBELWOLLEN (*n*): malevolence 083
ÜBEN: to practice; exert; do; exercise; execute 084
ÜBER: over; on; above; upon; besides; about; concerning; by way of; across; via; beyond; more than; less-(than); during; past; on account of 085
ÜBERALL: everywhere; all over 086
ÜBERANTWORTEN: to consign; deliver 087
ÜBERÄSCHERN: to overlime (Leather) 088
ÜBERAUS: exceedingly; extremely; excessively 089
ÜBERBAUEN: to build over; erect on top of 090
ÜBERBEIN (*n*): bony excrescence; spavin 091
ÜBERBELICHTUNG (*f*): over-exposure (Phot.) 092
ÜBERBIETEN: to outbid; surpass; outdo; bid (offer) too much 093
ÜBERBILDEN: to over-refine 094
ÜBERBINDEN: to bind (upon) over 095
ÜBERBLEIBSEL (*n*): remainder; residue; remnant; rest; waste 096
ÜBERBLICK (*m*): synopsis; survey 097
ÜBERBLICKEN: to survey; over-look 098
ÜBERBORSAUER: perborate; combined with perboric acid 099
ÜBERBORSÄURE (*f*): perboric acid; HBO_4 100
ÜBERBORSAURES SALZ (*n*): perborate 101
ÜBERBREITEN: to spread (cover) over 102
ÜBERBRENNEN: to overburn, etc. (see **BRENNEN**) 103
ÜBERBRINGEN: deliver; transmit; present; bear (bring)-(over) 104
ÜBERBRINGER (*m*): bearer (Com.) 105
ÜBERBROMSÄURE (*f*): perbromic acid; $HBrO_4$ 106
ÜBERBRÜCKEN: to bridge over 107
ÜBERBÜRDEN: to over-burden; overload 108
ÜBERCHLORSAUER: perchlorate of; combined with perchloric acid 109
ÜBERCHLORSÄURE (*f*): perchloric acid (see **PERCHLORSÄURE**) 110
ÜBERCHLORSÄUREANHYDRID (*n*): perchloric anhydride (see **CHLORHEPTOXYD**) 111
ÜBERCHLORSAURES SALZ (*n*): perchlorate 112
ÜBERCHROMSÄURE (*f*): perchromic acid; $HCrO_4$ 113
ÜBERDAUERN: to outlast 114
ÜBERDECKE (*f*): coverlet (of a bed) 115
ÜBERDECKEN: to cover (spread) over; overlap 116
ÜBERDECKUNG (*f*): lap, overlapping; covering 117
ÜBERDEM: besides; moreover 118
ÜBERDENKEN: to reflect (meditate) upon (*n*); meditation; reflection 119
ÜBERDESTILLIEREN: to over-distil; distil over 120
ÜBERDIES: besides; moreover 121
ÜBERDRUCK (*m*): absolute pressure; pressure above vacuum (i.e., 1 Atms, above gauge pressure); excess (excessive) pressure; overprint; cover-printing (Calico) 122
ÜBERDRUCKARTIKEL (*m*): cover-print style (Calico) 123
ÜBERDRUCKWIRKUNG (*f*): reaction effect 124
ÜBERDRÜSSIG: disgusted; tired; weary; satiated 125
ÜBERECK: across; diagonal-(ly) 126
ÜBERECKS: (see **ÜBERECK**) 127
ÜBEREILEN: to overtake; precipitate; hurry 128
ÜBEREILT: premature; precipitate; rash; inconsiderate; (over)-hasty 129
ÜBEREILUNG (*f*): hastiness; precipitation 130
ÜBEREINANDER: one upon another; superposed; superimposed 131
ÜBEREINANDER ANORDNEN: to superimpose, to superpose 132
ÜBEREINANDER LAGERN: to stack; place one upon another; superpose; super-impose 133
ÜBEREINANDERLAGERUNG (*f*): superposition; superimposement; superposing 134
ÜBEREINKOMMEN: to agree; conform; (*n*): agreement; conformity 135
ÜBEREINKUNFT, NACH- (*f*): according to (agreement) contract 136
ÜBEREINSTIMMEN: to agree; correspond; harmonize; check; dye to pattern (Dyeing); to be unanimous; concur; accord; be of one mind; mean the same thing 137
ÜBEREINSTIMMEND: agreeing; concordant; consistent; conformable; in accordance 138

ÜBEREINSTIMMUNG (*f*): agreement; correspondence; harmony; concordance; consistence; conformity; unanimity 139

ÜBEREINTREFFEN: to correspond, agree (see also ÜBEREINSTIMMEN) 140

ÜBERELASTISCHE BEANSPRUCHUNG (*f*): over-elastic stress 141

ÜBEREUTEKTOID: hypereutectic (of steel) 142

ÜBEREXPOSITION (*f*): over-exposure (Phot.) 143

ÜBERFAHREN: to drive (pass) over; ride (sail) over; convey (take) over or across; cover 144

ÜBERFAHRT (*f*): crossing, passage; ferry 145

ÜBERFALL (*m*): surprise; fall (Water); invasion; attack; ambush; hold-up; overflow; weir (of water) 146

ÜBERFALLEN: to surprise; fall upon; overtake; fall over; ambush; attack; invade; hold up 147

ÜBERFANGEN: to flash; cover; case (Glass) 148

ÜBERFANGGLAS (*n*): flashed glass (Phot.) 149

ÜBERFÄRBEARTIKEL (*m*): cross-dyed style (Calico) 150

ÜBERFÄRBEN: to cross-dye; over-dye; over-(colour) stain; top; fill-up; pad (Cotton warp); dye on a mordant 151

ÜBERFEDER (*f*): ink holder 152

ÜBERFEIN: superfine; overfine; over-refined 153

ÜBERFETTEN: to superfat (Soap); overstuff (Leather) 154

ÜBERFETTET: superfatted 155

ÜBERFEUERN: to overfire; overheat 156

ÜBERFIRNISSEN: to varnish; coat (cover) with varnish 157

ÜBERFLIESSEN: to overflow; flow over 158

ÜBERFLÜGELN: to outflank; surpass 159

ÜBERFLUSS (*m*): overflow; abundance; plenty; surfeit; profusion; superfluity; redundance 160

ÜBERFLÜSSIG: waste (Water, etc.); superfluous; profuse; overflowing; excess; abundant; plentiful; redundant 161

ÜBERFLUTEN: to overflow; flood; swamp 162

ÜBERFORMEN: to mould on an inside mould (Ceramics) 163

ÜBERFÜHRBAR: convertible; transferable; transportable 164

ÜBERFUHRE (*f*): passage; transport 165

ÜBERFÜHREN: to convert; convey; transport; transfer; lead (carry) over or across; (by)-pass; convince; convict; transform 166

ÜBERFÜHREND: convincing 167

ÜBERFÜHRUNG (*f*): conviction; crossing; conversion; transformation 168

ÜBERFÜHRUNGSZAHL (*f*): transference number 169

ÜBERFÜLLE (*f*): excess; repletion; plethora; superabundance 170

ÜBERFÜLLEN: to overstock; overload; overcharge; overfill; surfeit 171

ÜBERFÜTTERUNG (*f*): overfeeding, etc. (see ÜBERFÜLLEN) 172

ÜBERGABE (*f*): delivery; handing-over; surrender; yielding-(up) 173

ÜBERGANG (*m*): conversion; turn; transition; change; passage; crossing; transfer; blending, or shading-off (Colours) 174

ÜBERGANGSFORM (*f*): transitional (shape) form 175

ÜBERGANGSPUNKT (*m*): transition point; place of transition 176

ÜBERGANGSSTADIUM (*n*): transition-(al) state 177

ÜBERGANGSSTÄHLE (*m.pl*): transition steels (steels which change from Austenite to Martensite state on slight heating or cold working) 178

ÜBERGANGSSTELLE (*f*): transition point; place of transition 179

ÜBERGANGSTEMPERATUR (*f*): transition temperature 180

ÜBERGANGSZEIT (*f*): transitional period, duration of (time occupied by) transition 181

ÜBERGANGSZUSTAND (*m*): transition (transitional) state 182

ÜBERGAR: overdone; too hot (of a furnace); dry (of copper); over-refined (of metals) 183

ÜBERGEBEN: to deliver; commit; surrender; vomit; yield-(up); hand over 184

ÜBERGEHEN: to pass (run; go) over or by; change; overflow; omit; over-look; turn; shade; blend (of colours); refer to; obtain; arrive at; fix; decide; determine; transfer; convert; cross; peruse; be converted (Chem.); (*n*): conversion, etc. 185

ÜBERGEHEND: transitive 186

ÜBERGEWICHT (*n*): over-weight; preponderance, superiority 187

ÜBERGIESSEN: to douche; irrigate (Med.); transfer; spill; pour (on) over 188

ÜBERGIPSEN: to parget; plaster 189

ÜBERGLASEN: to glaze-(over); overglaze; vitrify 190

ÜBERGLASUNG (*f*): overglaze; overglazing; glaze; vitrification 191

ÜBERGLASUR (*f*): overglaze 192

ÜBERGLASURFARBE (*f*): over-glaze colour (Ceramics) 193

ÜBERGOLDEN: to gild 194

ÜBERGOLDUNG (*f*): gilding; gilt 195

ÜBERGREIFEN: encroach; overlap; (*n*): overlapping; encroaching 196

ÜBERGUSS (*m*): anything poured over; crust; icing (Sugar) 197

ÜBERGUT: too good; above (the normal) standard 198

ÜBERHAND (*f*): upper hand; lead; command 199

ÜBERHANDNAHME (*f*): prevalence; taking of the upper hand; (too great)-increase 200

ÜBERHAND NEHMEN: to take the upper hand; prevail; become too powerful; increase to too great an extreme 201

UBERHÄNGEND: raking (Naut.); overhanging 202

ÜBERHÄUFEN: overstock; overload; overwhelm; accumulate 203

ÜBERHAUPT: in general; on the whole; at all; generally; altogether 204

ÜBERHEBEN: to excuse; exempt; boast; be (proud) conceited; strain; dispense with; lift over 205

ÜBERHELFEN: to help (through) over; tide over 206

ÜBERHITZE (*f*): superheat; waste heat 207

ÜBERHITZEN: to overheat; superheat 208

ÜBERHITZER (*m*): superheater 209

ÜBERHITZER-ENTWÄSSERUNG (*f*): superheater drain 210

ÜBERHITZER (*m*), GETRENNTER-: independent-(ly fired) superheater 211

ÜBERHITZUNG (*f*): superheat-(ing); (often loosely employed when "total temperature" is meant) 212

ÜBERHOHLEN: to outstrip; overtake; list; lie over; heel over (Naut.); lurch; overhaul 213
ÜBERIMPFEN: to vaccinate; innoculate (from one to another) 214
ÜBERIRDISCH: heavenly; superterrestrial; overland; supernatural; celestial 215
ÜBERJODSÄURE (*f*): periodic acid; $HIO_4.2H_2O$; Mp. 131.5°C. 216
ÜBERKALKEN: to overlime 217
ÜBERKÄLKEN: to overlime 218
ÜBERKALTEN: to undercool; supercool 219
ÜBERKIPPEN: to overturn; tip over 220
ÜBERKISTE (*f*): outside (case) box 221
ÜBERKLEBEN: to paste (on) over; glue over; stick over 222
ÜBERKLEISTERN: (see ÜBERKLEBEN) 223
ÜBERKLOTZEN: to slop-pad (Calico) 224
ÜBERKOCHEN: to boil over; (*n*): priming (of boilers); boiling-over 225
ÜBERKOHLENSAUER: percarbonate of; combined with percarbonic acid 226
ÜBERKOHLENSÄURE (*f*): percarbonic acid 227
ÜBERKOHLENSAURES SALZ (*n*): percarbonate 228
ÜBERKOMMEN: to get; receive; seize; attack; obtain 229
ÜBERKOMPENSIEREN: to over-compensate 230
ÜBERLADEN: to overcharge; overload; surcharge; overburden; surfeit 231
ÜBERLAPPT GESCHWEISST: lap-welded 232
ÜBERLAPPTLÖTEN: to lap-solder 233
ÜBERLAPPTSCHWEISSEN: to lap-weld 234
ÜBERLAPPUNG (*f*): lap, lapping, overlap-(ping) 235
ÜBERLAPPUNGSNIETUNG (*f*): lap-riveting 236
ÜBERLASSEN: to leave to; abandon; give up; resign to; commit; yield 237
ÜBERLAST (*f*): overload; overweight; surcharge; surfeit 238
ÜBERLASTEN: to overload, overburden, overcharge 239
ÜBERLASTFÄHIGKEIT (*f*): overload capacity 240
ÜBERLÄSTIG: importunate; troublesome 241
ÜBERLASTVENTIL (*n*): overload valve 242
ÜBERLAUF (*m*): by-pass; overflow; net profit 243
ÜBERLAUFEN: to by-pass; run over; overflow; annoy; desert; overrun; importune 244
ÜBERLAUFROHR (*n*): by-pass pipe; overflow pipe 245
ÜBERLEBEN: to survive; outlive 246
ÜBERLEGEN: to consider; deliberate; reflect; lie over; heel over; lurch; list (Naut.); (*adj.*): prevalent; superior; pre-eminent 247
ÜBERLEGENHEIT (*f*): superiority; prevalence; pre-eminence 248
ÜBERLEGUNG (*f*): consideration; deliberation; reflection 249
ÜBERLEITEN: to lead (conduct; pass) over; transfuse (Blood) 250
ÜBERLESEN: to peruse; read (over) through 251
ÜBERLIEFERUNG (*f*): surrender; transmission; delivery; tradition 252
ÜBERLISTEN: to outwit; overreach; (take by) surprise 253
ÜBERLÖST: overgrown (Malt) 254
ÜBERMACHEN: to make over; leave; will; transmit; send; forward 255
ÜBERMACHT (*f*): predominance; superiority; superior (power) force; advantage (in point of strength) 256
ÜBERMALEN: to re-paint; paint over 257
ÜBERMANGANSAUER: permanganate of; combined with permanganic acid 258
ÜBERMANGANSÄURE (*f*): permanganic acid; $HMnO_4$ 259
ÜBERMANGANSAURES KALI (*n*): permanganate of potash; $KMnO_4$; Sg. 2.7 260
ÜBERMANGANSAURES SALZ (*n*): permanganate 261
ÜBERMANNEN: to overpower; overcome 262
ÜBERMASS (*n*): excess; superfluity; superabundance 263
ÜBERMÄSSIG: excessive; immoderate; extended; augmented (Theory of harmony) (see also QUARTE) 264
ÜBERMENSCHLICH: superhuman 265
ÜBERMITTELN: to send, forward, transmit 266
ÜBERMUT (*m*): exuberance; arrogance; haughtiness; presumption 267
ÜBERMÜTIG: exuberant; arrogant; haughty; presumptuous; wanton 268
ÜBERNAHME (*f*): taking over, acceptance 269
ÜBERNÄHREN: to overfeed 270
ÜBERNATÜRLICH: supernatural, not natural 271
ÜBERNEHMEN: to take over; receive; take charge (possession) of; accept; assume; enter upon-(responsibilities); overdo; overeat; overcharge; undertake 272
ÜBEROSMIUMSÄURE (*f*): perosmic acid; perosmic anhydride; perosmic oxide; osmic acid; osmic anhydride; osmium tetroxide; (*Acidum osmicum*); OsO_4; Sg. 8.89; Mp. 20°C.; Bp. 100°C. (see OSMIUMSUPEROXYD) 273
ÜBEROXYD (*n*): peroxide 274
ÜBEROXYDIEREN: to peroxidize 275
ÜBEROXYDIERUNG (*f*): peroxidation 276
ÜBERPFLATSCHEN: to pad (Calico) 277
ÜBERPOLEN: to overpole 278
ÜBERPROBE (*f*): over-proof spirit 279
ÜBERPROBEWEINGEIST (*m*): over-proof spirit 280
ÜBERPRODUKT (*n*): by-product; residual product 281
ÜBERPRODUKTION (*f*): surplus (over) production 282
ÜBERRAGEN: to tower above 283
ÜBERRASCHEN: to surprise; take unawares 284
ÜBERRASCHEND: astonishing; surprising 285
ÜBERRASCHUNG (*f*): astonishment; surprise 286
ÜBERRECHNEN: to count over; work over; check; calculate 287
ÜBERREDEN: to persuade; talk over 288
ÜBERREICHEN: to present, hand over, deliver 289
ÜBERREIF: over-ripe 290
ÜBERREISSEN: to prime; draw over; (*n*): priming (Boilers) 291
ÜBERREIZEN: to over-excite; over-irritate 292
ÜBERREST (*m*): remainder; residue; remains; scraps; waste; remnant; (*pl*): bones (Human) 293
ÜBERROCK (*m*): overcoat; frock coat 294
ÜBERRUTHENIUMSÄURE (*f*): perruthenic acid 295
ÜBERSÄEN: to strew 296
ÜBERSALZEN: to salt too much 297
ÜBERSATT: surfeited; glutted; super-saturated; satiated 298
ÜBERSÄTTIGEN: to surfeit; supersaturate; satiate 299
ÜBERSÄTTIGUNG (*f*): satiety; supersaturation; surfeit; glut 300
ÜBERSAUER: too (sour) acid; having more than two parts of acid to one part of base (Salts) 301

ÜBERSÄUERN : to peroxidize ; over-acidify 302
ÜBERSAURES OXALSAURES KALI (*n*) : potassium tetroxalate 303
ÜBERSCHATTEN : to overshadow ; cast a shadow upon 304
ÜBERSCHÄTZEN : to over-estimate ; overrate ; over-value 305
ÜBERSCHAUEN : to look over ; overlook ; survey 306
ÜBERSCHÄUMEN : to froth over ; foam over 307
ÜBERSCHICKEN : to send, transmit, remit 308
ÜBERSCHIESSEN : to shift (of ship's cargoes) ; (*n*) : shifting 309
ÜBERSCHIFFEN : to sail over ; ship ; transport 310
ÜBERSCHLAG (*m*) : approximate (estimate ; calculation) computation ; somersault ; facing ; cuff 311
ÜBERSCHLAGEN : to estimate (calculate) approximately ; omit ; pass over ; neglect ; compute ; become lukewarm ; become moderately warm ; turn over ; turn a somersault 312
ÜBERSCHLAGRECHNUNG (*f*) : estimate ; estimation ; computation ; approximate calculation 313
ÜBERSCHMELZEN : to enamel ; superfuse 314
ÜBERSCHMIEREN : to daub ; smear ; besmear 315
ÜBERSCHREIBEN : to superscribe ; entitle ; inscribe ; head ; address ; direct (an envelope, etc.) 316
ÜBERSCHREITEN : to overstep ; go beyond ; exceed (a given limit or point) ; step over ; transgress ; violate 317
ÜBERSCHREITUNG (*f*) : transgression ; violation ; overstepping 318
ÜBERSCHRIFT (*f*) : superscription ; inscription ; heading ; address ; direction (of letters, etc.) 319
ÜBERSCHUSS (*m*) : excess ; surplus ; residue 320
ÜBERSCHÜSSIG : surplus ; excessive ; excess ; remaining 321
ÜBERSCHÜTTEN : to pour (in) over 322
ÜBERSCHWÄNGERN : to over-impregnate 323
ÜBERSCHWÄNGLICH : (super)-abundant, exceeding 324
ÜBERSCHWÄNZEN : to sparge (Brew.) ; (*n*) : sparging 325
ÜBERSCHWEFELSAUER : persulphate of ; combined with persulphuric acid 326
ÜBERSCHWEFELSÄURE (*f*) : persulphuric acid ; $H_2S_2O_8$ (only known as a solution in a free state) 327
ÜBERSCHWEFELSAURES SALZ (*n*) : persulphate 328
ÜBERSCHWELLEN : to overplump (Leather) 329
ÜBERSCHWEMMEN : to overflow ; flood ; deluge ; inundate 330
ÜBERSCHWENGLICH : excessive ; exuberant ; exceeding 331
ÜBERSEEISCH : transmarine ; oversea ; transmaritime 332
ÜBERSEGELN : to run down (one ship of another) 333
ÜBERSEHEN : to overlook ; omit ; oversee ; peruse ; supervise ; look over ; survey ; review ; view ; précis ; summarize ; abstract 334
ÜBERSENDEN : to send ; transmit ; consign ; pass on ; remit 335
ÜBERSENDUNG (*f*) : transmission ; shipment ; conveyance ; consignment 336
ÜBERSETZEN : to translate ; transport ; ferry across (Rivers) ; overcharge ; top (Dyeing) ; ship (pass ; leap) over ; cross ; transmit 337
ÜBERSETZER (*m*) : translator 338
ÜBERSETZUNG (*f*) : translation ; version ; (gear)-ratio ; speed reduction (of reduction gearing) ; transmission gear (Mech.) ; transposition 339
ÜBERSICHT (*f*) : survey ; review ; abstract ; synopsis ; view ; insight ; précis ; summary 340
ÜBERSICHTIGKEIT (*f*) : squinting 341
ÜBERSICHTLICH : clear ; easily visible ; capable of being seen at a glance 342
ÜBERSICHTSPLAN (*m*) : plan (of site) 343
ÜBERSICHTSTABELLE (*f*) : table ; tabular summary 344
ÜBERSIEDELN : to migrate ; emigrate 345
ÜBERSIEDEN : to boil over ; (*n*) : extra brew (Brewing) 346
ÜBERSILBERN : to silver ; silverplate ; plate 347
ÜBERSINNLICH : abstract ; metaphysical ; transcendent 348
ÜBERSPANNEN : to overstrain ; force ; span ; exaggerate 349
ÜBERSPANNT : overstrained ; eccentric ; exaggerated 350
ÜBERSPANNUNG (*f*) : overvoltage ; overtension (Elect.) ; exaggeration, etc. 351
ÜBERSPINNEN : to cover ; spin over (Wire, etc.) 352
ÜBERSPRINGEN : to jump (leap) over ; spring across ; skip ; omit ; pass abruptly ; alternate 353
ÜBERSPRITZEN : to spray (spurt ; squirt) over 354
ÜBERSPÜLEN : to overflow ; wash ; drench 355
ÜBERSTÄNDIG : over-mature , over-ripe 356
ÜBERSTEHEN : to overcome ; endure ; stand out ; project ; surmount 357
ÜBERSTEIGEN : to exceed ; overcome ; surmount ; overstep ; overflow ; step over 358
ÜBERSTEIGERN : to outbid ; raise the price 359
ÜBERSTEIGGEFÄSS (*n*) : overflow vessel 360
ÜBERSTIMMEN : to outvote ; overrule 361
ÜBERSTRAHLEN : irradiate ; outshine ; illuminate ; shine upon 362
ÜBERSTREICHEN : to coat ; paint (spread) over ; apply (Paint) ; pass over ; scrub 363
ÜBERSTREUEN : to strew (over) ; sprinkle (over) 364
ÜBERSTRÖMEN : to by-pass ; overflow ; inundate ; run (flow) over 365
ÜBERSTRÖMVENTIL (*n*) : overflow (by-pass) valve 366
ÜBERSTÜLPEN : to invert (over something) ; upturn ; turn upside down 367
ÜBERSTUNDE (*f*) : over-time 368
ÜBERSTÜRZEN : to overturn ; act rashly ; hurry ; upset ; tip (knock) over 369
ÜBERSTÜRZT : precipitate ; hasty 370
ÜBERSULFID (*n*) : persulphide 371
ÜBERSYNCHRON : hypersynchronous 372
ÜBERTAKELN : to over-rig (of ships) 373
ÜBERTÄUBEN : to deafen ; stun ; drown (a noise) ; dumbfound 374
ÜBERTEUERN : to overcharge 375
ÜBERTRAG (*m*) : transport ; transfer ; carry forward ; brought forward 376
ÜBERTRAGBAR : transferable ; infectious (Disease) 377
ÜBERTRAGEN : to transfer ; transmit ; transport ; transcribe ; translate ; carry (over) ;

charge; confer; commission; carry forward; bring forward 378
ÜBERTRAGER (*m*): carrier; transferrer; transmitter; transporter; transcriber; translator; relay (Mech.) 379
ÜBERTRAGUNG (*f*): transfer; transference; transmission; transporting; transcription; translation; transport; charging; conferring; commissioning 380
ÜBERTREFFEN: to excel; surpass; exceed; outdo 381
ÜBERTREIBEN: to exaggerate; exceed; overdo; force (drive) over; be immoderate 382
ÜBERTREIBKÜHLER (*m*): efflux condenser (in which the condensate is led away) (as opposed to return flow condenser) 383
ÜBERTRETEN: to overstep; step (pass; go) over; transgress; exceed; trespass 384
ÜBERTRETUNG (*f*): trespass; transgression 385
ÜBERTRIEBEN: exaggerated; excessive; immoderate 386
ÜBERTRITT (*m*): change; transfer 387
ÜBERTROCKNEN: to over-dry 388
ÜBERTÜNCHEN: to whitewash; gloss over; plaster; parget 389
ÜBERVOLL: superabundant 390
ÜBERVORTEILEN: to overreach; defraud; impose upon 391
ÜBER-VULKANISATION (*f*): over-vulcanization (the region in which the mechanical properties of vulcanized caoutchouc quickly decrease again) 392
ÜBERWACHEN: to supervise; watch over 393
ÜBERWACHSEN: to overgrow; grow over; outgrow 394
ÜBERWALLEN: to boil over; overflow 395
ÜBERWÄLTIGEN: to overcome; overpower; vanquish 396
ÜBERWEISEN: to convince, assign, cede, make over, send, transmit (Payment), endorse (Bills) 397
ÜBERWEISSEN: to whiten, whitewash, gloss over 398
ÜBERWEISUNGSSPESEN (*f.pl*): inspection (costs) charges 399
ÜBERWERFEN: to overthrow; throw (cast) over 400
ÜBERWICHTIG: preponderating, outweighing 401
ÜBERWIEGEN: to outweigh; overbalance; surpass 402
ÜBERWIEGEND: preponderating (see ÜBERWIEGEN) 403
ÜBERWINDEN: to surmount; overcome; conquer; vanquish; subdue 404
ÜBERWINTERN: to winter; spend the winter; keep through the winter 405
ÜBERWÖLBEN: to vault; arch over 406
ÜBERWURF (*m*): roughcast; dressing gown; bush (Mech.); hasp; (anything placed or thrown over) 407
ÜBERWURFFLANSCH (*m*): screwed on flange 408
ÜBERWURFMUTTER (*f*): screwed (nut)-cap; screw-cap 409
ÜBERZAHL (*f*): odds; surplus; greater number 410
ÜBERZÄHLIG: supernumerary 411
ÜBERZEUGEN: to convince; persuade; make (certain) sure 412
ÜBERZEUGT: certain; convinced; sure; persuaded 413
ÜBERZEUGUNG (*f*): conviction; persuasion; certainty 414
ÜBERZIEHEN: to coat; case; cover; line; encase; overlay; plate; incrust; put (on) over; become (coated) covered 415
ÜBERZUCKERN: to sugar; ice; candy (Peel) 416
ÜBERZUG (*m*): coat; coating; covering; skin; cover; incrustation; crust; lining; plating; case; casing 417
ÜBERZUGLACK (*m*): coating varnish 418
ÜBERZUGSGALVANOPLASTIK (*f*): (see GALVANO-PLATTIERUNG) 419
ÜBERZUGSLACK (*m*): coating varnish 420
ÜBLICH: customary; usual; general 421
ÜBRIG: over; remaining; left; to spare; superfluous; residual 422
ÜBRIGBLEIBEN: to remain over; be left; survive 423
ÜBRIGENS: besides; however; furthermore; moreover 424
ÜBRIGGEBLIEBEN: remaining; residual 425
ÜBUNG (*f*): practice; use; dexterity; exercise; discipline; routine; drill-(ing) 426
U-EISEN (*n*): channel iron 427
UFER (*n*): bank; coast; beach; shore 428
UFERBAHN (*f*): coastal railway or railway along a river bank 429
UFERBAUKUNST (*f*): dike and bank building and (fortifying) strengthening 430
UFERBEZIRK (*m*): sea coast; coastal district 431
UFERLÄUFER (*m*): sand-piper (*Actitis*) 432
UFERLOS: boundless; shoreless 433
UFERPFLANZEN (*f.pl*): river bank or seashore plants 434
UFERSCHUTZBAUTEN (*pl*): embankment 435
UFERSCHWALBE (*f*): sand martin 436
UFERSEITE (*f*): bank; water's edge 437
U-FÖRMIG: U-shaped 438
UHR (*f*): watch; clock; meter; hour; dial (of measuring instruments); o'clock 439
UHRENÖL (*n*): watch (clock) oil 440
UHRFEDER (*f*): watch-spring; clock-spring 441
UHRGEHÄUSE (*n*): watch-case; clock-case 442
UHRGLAS (*n*): watch glass; clock glass 443
UHRMACHER (*m*): watch (clock) maker 444
UHRMACHERWERKZEUG (*n*): watch (clock) maker's tools 445
UHRSCHLÜSSEL (*m*): watch key 446
UHRWERK (*n*): clock-work; works (of a clock or watch) 447
UHRZEIGER (*m*): hand (of a watch or clock) 448
UHRZIFFERBLATT (*n*): clock dial 449
ULEXIT (*m*): ulexite; $BaCaB_5O_9.6H_2O$; Sg. 1.73 450
ULLMANNIT (*m*): ullmannite (a nickel-antimony sulphide); NiSbS; Sg. 6.35 451
ULME (*f*): elm (*Ulmus campestris*); Sg. 0.6-0.85 452
ULMENBAUM (*m*): elm-(tree) 453
ULMENGEWÄCHSE (*n.pl*): *Ulmaceæ* (Bot.) 454
ULMENHOLZ (*n*): elm wood (see ULME) 455
ULMENRINDE (*f*): elm bark (*Cortex ulmi*) 456
ULMINSAUER: ulmate, combined with ulmic acid 457
ULMINSÄURE (*f*): ulmic acid 458
ULMINSAURES SALZ (*n*): ulmate 459
ULOTHRIXFÄDCHEN (*n* and *pl*): ulothrix streamer (of a genus of Confervoid algæ, *Ulotrichaceæ*) 460
ULTRAFILTER (*m*): (see GALLERTFILTER) 461
ULTRAFILTRATION (*f*): ultra (fine) filtration 462
ULTRAMARIN (*n*): ultramarine (blue); $Na_4(NaS_3.Al)Al_2(SiO_4)_3$ 463

ULTRAMARINERSATZ (*m*): ultramarine substitute 464
ULTRAMARINGELB (*n*): ultramarine yellow, barium yellow; $BaCrO_4$ (see BARIUM-CHROMAT) 465
ULTRAMARIN (*m*), GELBER-: barium yellow (see BARIUMCHROMAT) 466
ULTRAMARINSCHWARZ (*n*): ultramarine black 467
ULTRAMIKROSKOPISCH: ultramicroscopical (see SUBMIKROSKOPISCH and AMIKROSKOPISCH) 468
ULTRAROT: ultra (infra) red 469
ULTRAVIOLETTE STRAHLEN (*f.pl*): ultraviolet rays 470
UM: round; at; by; about; over; up; for; to; in order to; near; past; after 471
UMÄNDERN: to change; alter; convert; remodel; exchange; transform; alternate 472
ÜMÄNDERUNG (*f*): alternation (etc., see UMÄNDERN) 473
UMARBEITEN: to recast; work over; convert; work up; remodel; change 474
UMBAU (*m*): rebuilding, re-erection, conversion 475
UMBAUEN: to convert, re-build, re-erect 476
UMBETTEN: to re-bed; transfer to another bed 477
UMBIEGEN: to bend (round) over; to double back; flange 478
UMBILDEN: to transform; remodel; reform 479
UMBINDEN: to tie round; bind; rebind 480
UMBLASEN: to blow (round; down) over 481
UMBÖRDELN: to border, flange, edge 482
UMBRA (*f*): umber (see MANGANBRAUN); Cologne earth (see BRAUNKOHLENPULVER) 483
UMBRAERDE (*f*): umber; Cologne earth (see BRAUNKOHLENPULVER) 484
UMBRA (*f*), GEBRANNTE-: burnt umber (by heating raw umber; see MANGANBRAUN) 485
UMBRA (*f*), KÖLNISCHE-: Cologne earth (see BRAUNKOHLENPULVER) 486
UMBRECHEN: to break (up) down 487
UMBRINGEN: to destroy; kill; murder 488
UMDECKEN: to re-cover 489
UMDESTILLIEREN: to rectify 490
UMDREHEN: to rotate; turn (round) about; twirl; impart a rotary motion to; revolve 491
UMDREHEND: rotatory; revolving 492
UMDREHUNG (*f*): revolution; turn; rotation; twirling 493
UMDREHUNGSACHSE (*f*): axis of (rotation) revolution 494
UMDREHUNGSBEWEGUNG (*f*): rotary motion 495
UMDREHUNGSGESCHWINDIGKEIT (*f*): velocity of rotation 496
UMDREHUNGSPUNKT (*m*: centre of rotation 497
UMDREHUNGSREGLER (*m*): revolution governor 498
UMDREHUNGSZAHL (*f*): number of revolutions; speed; velocity 499
UMDREHUNGSZAHLEN, MIT—VON ? BIS ? (*f.pl*): with speed regulation from ? to ? 500
UMDRUCK (*m*): reprint (as from artist's original stone etching on to another stone, for mass production) 501
UMDRUCKEN: to reprint 502
UMFALLEN: to fall down; topple over 503
UMFANG (*m*): circumference; extent; compass; circuit; range; periphery; scope 504
UMFANGEN: to surround; embrace; encircle 505
UMFANGSGESCHWINDIGKEIT (*f*): peripheral (circumferential) speed or velocity 506
UMFANGSKRÄFTE (*f.pl*): peripheral forces 507
UMFÄRBEN: to re-dye 508
UMFASSEN: to embrace; comprise; span 509
UMFASSEND: extensive; comprehensive; copious 510
UMFLECHTEN: to twist (round) about; to weave (plait) again 511
UMFORMEN: to remodel; transform (Elect. etc.); convert 512
UMFORMER (*m*): transformer; converter (Elect.) 513
UMFORMER (*m*), ROTIERENDER-: rotary converter 514
UMFORMUNGSARBEIT (*f*): deformation (Rolling Mills) 515
UMFÜHREN: to lead round; by-pass 516
UMFÜHRUNG (*f*): by-pass-(ing) 517
UMFÜLLEN: to transfer (Fluids); decant; re-rack (Brewing); transfuse 518
UMGANG (*m*): circuit; rotation; turn; round; company; society; intercourse; procession; association 519
UMGÄNGLICH: sociable 520
UMGÄNGLICHKEIT (*f*): sociability 521
UMGEBEN: to surround; enclose 522
UMGEBUNG (*f*): environment; surroundings; environs; neighbourhood; company 523
UMGEGEND (*f*): environment; surroundings; environs; neighbourhood; country 524
UMGEHEN: to by-pass; work; be in action; go (round) about; associate; deal; manage; avoid; evade; elude; circumvent; circumambulate; circumnavigate 525
UMGEHEND: by return-(of post) 526
UMGEHUNG (*f*): by-pass-(ing), etc. (see UMGEHEN) 527
UMGEKEHRT: inverted; inverse-(ly); opposite; reverse; converse; vice versa; on the contrary 528
UMGEKEHRTE FÄLLUNG (*f*): reverse precipitation 529
UMGEKEHRTE REAKTION (*f*): reverse reaction 530
UMGESTALTEN: to transform; change; remodel; recast; reform; transmute; transmogrify 531
UMGESTÜLPT: inverted; upturned; overturned 532
UMGEWANDELT: transformed; converted; changed; inflected (of Grammar) 533
UMGIESSEN: to decant; recast; transfer (anything fluid by pouring) 534
UMGRENZEN: to circumscribe; define; limit; bound 535
UMGÜRTEN: to gird, girdle, encircle, surround 536
UMGUSS (*m*): decantation; recast-(ing); transfer 537
UMHANG (*m*): wrap; shawl; veil; curtain 538
UMHAUEN: to fell; hew (cut) down 539
UMHER: about; around; here and there 540
UMHERLAUFEN: to wander (ramble; rove; stray) about 541
UMHERWANDELN: (see UMHERLAUFEN) 542
UMHERZIEHEN: (see UMHERLAUFEN) 543
UMHIN: but; otherwise; about 544
UMHÜLLEN: to cover; jacket; case; wrap; envelope; shroud 545
UMHÜLLUNG (*f*): covering; jacket-(ing); casing; wrapping; wrapper; envelope; shroud-(ing); case; housing 546

UMKANTEN: to tilt; cant 547
UMKEHR (*f*): return; reversion; turning back 548
UMKEHRBAR: invertible; reversible 549
UMKEHREN: to invert; change; reverse; convert; turn (round; back; about; over) inside out 550
UMKEHRWALZWERK (*n*): return (reversible) roll, mill or train (Rolling mills) 551
UMKIPPEN: to invert; tip over 552
UMKLAMMERN: to embrace; clasp; cling to; bracket 553
UMKLEIDUNG (*f*): travesty; change of dress 554
UMKOMMEN: to spoil; perish; die 555
UMKREIS (*m*): circuit; extent; circumference; compass; range; circle; periphery 556
UMKREISEN: to rotate (revolve) around or about; turn; circle 557
UMKRISTALLISIEREN: to re-crystallize; (*n*): re-crystallization 558
UMKRÜCKEN: to rake; rabble; mash (Brew.) 559
UMKRYSTALLISIEREN: to re-crystallize; (*n*): re-crystallization 560
UMLADEN: to tranship 561
UMLAGERN: to re-arrange; surround; besiege; beleaguer 562
UMLAGERUNG (*f*): rearrangement; surrounding; besieging 563
UMLAUF (*m*): rotation; circulation; revolution; cycle; by-pass 564
UMLAUFEN: to rotate; circulate; revolve; by-pass; run (over) down 565
UMLAUFRICHTUNG (*f*): direction of rotation; direction of circulation 566
UMLAUFSZAHL (*f*): number of revolutions, speed, velocity 567
UMLAUFUNG (*f*): revolution 568
UMLAUFVENTIL (*n*): by-pass valve 569
UMLAUF (*m*), **WASSER-**: water circulation; circulating system (for water) 570
UMLAUFZÄHLER (*m*): revolution counter, tachometer, speedometer 571
UMLEGEN: to shift; change; put on; relay; repair; turn (Malt); surround; lay over 572
UMLENKEN: to turn (round) about, steer round 573
UMLIEGEND: surrounding; circumjacent 574
UMMANTELUNG (*f*): casing 575
UMMAUERN: to brick (round, up) in, encase (surround) with masonry 576
UMNEBELN: to encase (surround) with mist, befog, dim, cloud 577
UMNIETEN: to rivet 578
UMPACKEN: to repack; pack afresh; renew the packing 579
UMPFLANZEN: to transplant; plant (round) about 580
UMRANGIEREN: to turn (Railway waggons on a turntable) 581
UMRECHNEN: to convert (Maths.; as from English to Metric) 582
UMRECHNUNGSFAKTOR (*m*): conversion factor 583
UMREISEN: to travel round, journey (voyage, sail) round, circumnavigate 584
UMREISSEN: to pull (down) over; demolish 585
UMRINGEN: to encircle; surround; encompass; ring 586
UMRISS (*m*): sketch; outline; contour 587
UMROLLMASCHINE (*f*): rewinding machine (Paper) 588
UMRÜHREN: to stir-(up); pole; work; puddle; (*n*): stirring-(up); puddling (Iron); poling (Copper) 589
UMSATZ (*m*): turnover; exchange; sale; transaction; business; return; conversion; reversion; double decomposition (Chem.) 590
UMSCHAFFEN: to transform; remodel (see **UMGESTALTEN**) 591
UMSCHALTEN (*n*): change-(ing) over, etc. (see the verb) 592
UMSCHALTEN: to switch over; reverse; commutate (Elect.) 593
UMSCHALTER (*m*): switchboard; (current)-reverser; commutator (Elect.) 594
UMSCHALTSTÖPSEL (*m*): switch plug 595
UMSCHAU (*f*): review; survey 596
UMSCHAUEN: to look (about) back or round; prospect 597
UMSCHAUFELN: to turn over with a shovel 598
UMSCHICHTIG: alternately; in layers 599
UMSCHIFFEN: to trans-ship; circumnavigate 600
UMSCHLAG (*m*): cover-(ing); wrapper; wrapping; envelope; poultice; cataplasm; sale; hem; facing; collar; turn-(ing); sudden change; transition; fomentation 601
UMSCHLAGBOHRER (*m*): centre-bit 602
UMSCHLAGEN: to throw on (a wrap or shawl); wrap (up) about; apply; turn (about) over; knock down; change; overturn; upset; decompose; poultice; foment; envelop 603
UMSCHLAGETUCH (*n*): wrapper; shawl 604
UMSCHLAGPAPIER (*n*): wrapping paper 605
UMSCHLAGSPUNKT (*m*): transition point 606
UMSCHLEIERN: to veil; envelop; wrap; cover 607
UMSCHLIESSEN: to envelop, enclose, embrace, surround 608
UMSCHMELZEN: to refound; remelt; recast 609
UMSCHNÜREN: to relace; bind (tie) with cord 610
UMSCHREIBEN: to rewrite; circumscribe; transcribe; describe (a circle round); paraphrase 611
UMSCHREIBUNG (*f*): transcription; paraphrase 612
UMSCHÜTTELN: to shake; agitate; shake up; stir up; mix 613
UMSCHÜTTEN: to pour into another vessel; decant 614
UMSCHWEIF (*m*): digression; round about way 615
UMSCHWENKEN: to turn round; rotate 616
UMSCHWUNG (*m*): revolution; rotation 617
UMSEGELN: to circumnavigate 618
UMSEHEN: to look (back) about or around; be circumspect; prospect 619
UMSETZEN (*n*): puddling (the pushing of various lots of iron towards the rear seal wall or back bridge wall) 620
UMSETZEN: to transpose; change; convert; transplant; sell; exchange; place in a fresh position 621
UMSETZUNG (*f*): transposition; conversion; change, etc.; double decomposition 622
UMSICHT (*f*): prospect; circumspection; panorama 623
UMSICHTIG: circumspect 624
UMSONST: gratis; free; in vain; for nothing; aimlessly; causelessly 625

UMSPANNEN: to span; fathom; change horses 626

UMSPÜLEN: to wash- (over; round); to play (about) around; to sweep-(over) (by gases, etc.); to scrub (by gases, etc.) 627

UMSTAND (*m*): fact; condition: circumstance; situation (*pl.*); conditions; ceremonies; formalities; particulars 628

UMSTÄNDEN (*m.pl.*), **UNTER SONST GLEICHEN-**: other conditions being equal 629

UMSTAND, KRITISCHER- (*m*): critical condition 630

UMSTÄNDLICH: circumstantial; bothersome; ceremonious; complex; detailed 631

UMSTANDSWORT (*n*): adverb 632

UMSTECHEN: to stir up; turn; re-engrave 633

UMSTEHEND: surrounding; following; next (*pl*); bystanders 634

UMSTEIGEN: to change (Trains, etc.) 635

UMSTELLBAR: reversible; transposable; invertible; variable 636

UMSTELLEN: to reverse; transpose; invert; surround; beset 637

UMSTEUERBAR: reversible 638

UMSTEUERN: to reverse (Machinery) 639

UMSTEUERUNG (*f*): reversing gear 640

UMSTEUERVENTIL (*n*): change-over valve (valve for changing direction of flow of the gas and air) (Siemens Martin process) 641

UMSTIMMEN: to change the tone; alter a person's views; change (alter) the action or effect of (usually in reference to a photographic developer) 642

UMSTIMMUNGSMITTEL (*n*): alterative 643

UMSTOSSEN: to abolish; cancel; overthrow; knock (over) down; annul; overturn 644

UMSTREUEN: to strew (about) 645

UMSTRÖMEN: to flow round; by-pass 646

UMSTÜLPEN: to invert; upturn; overturn; turn upside down 647

UMSTURZ (*m*): overthrow; downfall 648

UMSTÜRZEN: to upset, overthrow; fall (over) down; subvert 649

UMTAUSCH (*m*): exchange 650

UMTAUSCHEN: to exchange 651

UMTRIEB (*m*): circulation; intrigue 652

UMTUN: to put (around) on; enquire; seek; look for 653

UMWALZEN: to revolve; rotate; revolutionize; turn (roll) round 654

UMWÄLZEND: revolutionary; epoch-making 655

UMWÄLZPUMPE (*f*): rotary pump 656

UMWANDELBAR: transformable; convertible 657

UMWANDELBARKEIT (*f*): transformability; convertibility 658

UMWANDELN: to transform; be transformed; convert; change; inflect; conjugate (Grammar) 659

UMWANDLER (*m*): converter; transformer (Elect.) 660

UMWANDLUNG (*f*): transformation; conversion; change; inflection (Grammar); metamorphosis 661

UMWANDLUNGSFÄHIG: convertible; transformable 662

UMWANDLUNGSPRODUKT (*n*): transformation product 663

UMWANDLUNGSPUNKT (*m*): transformation point; transition point 664

UMWANDLUNGSTHEORIE (*f*): transformation theory 665

UMWANDLUNGSWÄRME (*f*): transformation heat 666

UMWECHSELN: to change; alter; alternate; transform; exchange 667

UMWEG (*m*): detour; roundabout way; evasion 668

UMWENDEN: to turn (over; upside down) about; invert 669

UMWERFEN: to overthrow; upset; knock (over) down 670

UMWICKELN: to wrap round; cover; enfold 671

UMWICKELUNG (*f*): wrapping; casing; covering 672

UMZÄUNUNG (*f*): hedge; stockade; fence; enclosure 673

UMZIEHEN: to cover; draw round; pull down; move; change (clothes, etc.); cloud 674

UMZUG (*m*): procession; wandering; removal; change 675

UMZÜNGELN: to envelop; play about (of flames); surround; encircle 676

UNABÄNDERLICH: invariable; unalterable; unchangeable; immutable 677

UNABHÄNGIG: independent-(ly) 678

UNABHÄNGIGKEIT (*f*): independence 679

UNABHELFLICH: irremediable 680

UNABLÄSSIG: uninterrupted; incessant-(ly); unceasing-(ly); continual-(ly); constant-(ly) 681

UNABSEHBAR: incalculable; immeasurable 682

UNABSICHTLICH: unintentional-(ly) 683

UNABSORBIERT: unabsorbed 684

UNABWENDBAR: inevitable 685

UNACHTSAM: negligent; inadvertent; careless; inattentive 686

UNÄHNLICH: dissimilar; unlike 687

UNAKTINISCH: non-actinic 688

UNAL (*n*): unal (a solid form of Rodinal: see **RODINAL**) (Phot.) 689

UNANGEFOCHTEN: unchallenged; undisputed; unquestioned 690

UNANGEMESSEN: inadequate; unsuitable 691

UNANGENEHM: disagreeable; unpleasant 692

UNANNEHMBAR: unacceptable; disagreeable; unpleasant 693

UNANSEHNLICH: mean; insignificant 694

UNANWENDBAR: inapplicable, unsuitable, impracticable 695

UNART (*f*): bad (habits) manners; naughtiness; ill (bad; improper) behaviour 696

UNATEMBAR: irrespirable 697

UNAUFGEFORDERT: unprovoked 698

UNAUFHALTSAM: irresistible; incessant; continual 699

UNAUFLÖSBAR: insoluble 700

UNAUFLÖSBARKEIT (*f*): insolubility 701

UNAUFLÖSLICH: insoluble 702

UNAUFLÖSLICHKEIT (*f*): insolubility 703

UNAUSBLEIBLICH: infallible; certain 704

UNAUSDEHNBAR: inexpansible; non-ductile 705

UNAUSFÜHRBAR: impracticable; impossible 706

UNAUSFÜHRLICH: incomplete; lacking in detail 707

UNAUSGEFÜHRT: undischarged, not (executed)-carried out 708

UNAUSGEMACHT: uncertain; undecided 709

UNAUSGESETZT: uninterrupted; continual 710

UNAUSLÖSCHBAR: inextinguishable; indelible (of ink) 711

UNAUSLÖSCHLICH: (see **UNAUSLÖSCHBAR**) 712

UNAUSSPRECHLICH: unspeakable; inexpressible; unutterable; ineffable 713

UNAUSSTEHLICH : intolerable ; insufferable 714
UNAUSWEICHLICH : inevitable 715
UNBALANZ (*f*), (DYNAMISCHER)- : (dynamic) out-of-balance 716
UNBÄNDIG : intractable ; unmanageable ; indomitable 717
UNBEACHTET : exclusive of, not taking into account, neglecting, unnoticed 718
UNBEARBEITET : unwrought (Metals) ; undressed ; raw (Hides); untilled (Land) ; unworked ; rough 719
UNBEDACHTSAM : imprudent ; inconsiderate ; indiscreet 720
UNBEDECKT : uncovered 721
UNBEDEUTEND : unimportant ; insignificant ; trifling 722
UNBEDINGT : unconditional ; unqualified ; unlimited ; unrestricted ; implicit ; absolute(ly) ; without fail 723
UNBEFANGEN : unconcerned ; unprejudiced ; impartial ; natural ; ingenuous 724
UNBEFLECKT : unblemished ; unstained ; immaculate ; pure 725
UNBEFRIEDIGEND : unsatisfactory ; insufficient 726
UNBEFUGT : incompetent 727
UNBEGREIFLICH : incomprehensible, inconceivable 728
UNBEGRENZT : unlimited, unconditioned, unbounded, boundless 729
UNBEHAGEN (*n*) : discomfort ; uneasiness ; dislike 730
UNBEHÜLFLICH : clumsy ; awkward ; helpless ; unwieldy 731
UNBEHUTSAM : incautious ; imprudent ; unwary 732
UNBEKANNT : unacquainted ; unknown ; ignorant 733
UNBEKÜMMERT : unconcerned ; careless 734
UNBELASTET : unencumbered ; free ; not loaded 735
UNBELEBT : lifeless ; inanimate 736
UNBEMERKBAR : unnoticeable ; imperceptible 737
UNBENANNT : anonymous ; unnamed ; indefinite (Maths.) ; innominate (Anat.) 738
UNBENOMMEN : still permissible ; permitted ; not prohibited 739
UNBEQUEM : uncomfortable ; inconvenient ; incommodious 740
UNBERECHTIGT : unjustified 741
UNBERICHTIGT : uncorrected ; not settled 742
UNBESCHÄDIGT : undamaged ; safe ; unhurt 743
UNBESCHÄFTIGT : unemployed, not working, out of work ; (*m.pl*) : unemployed 744
UNBESCHRÄNKT : unlimited ; undefined ; unbounded ; unconditioned ; absolute 745
UNBESCHREIBLICH : indescribable ; inexpressible 746
UNBESCHÜTZT : unprotected 747
UNBESTÄNDIG : fickle ; unstable ; changeable ; inconstant 748
UNBESTÄNDIGKEIT (*f*) : fickleness ; instability ; inconstancy 749
UNBESTIMMT : indefinite ; indeterminate ; undetermined ; vague ; uncertain ; undecided 750
UNBESTRITTEN : unquestioned ; undoubted ; uncontested ; undisputed 751
UNBETEILIGT : not concerned ; non-participating 752
UNBETRÄCHTLICH : inconsiderable 753
UNBETRETEN : untrodden ; unbeaten 754
UNBEUGSAM : inflexible 755
UNBEUGSAMKEIT (*f*) : inflexibility 756
UNBEWAFFNET : unarmed ; unaided ; naked (of eye) 757
UNBEWANDERT : not (versed) experienced in 758
UNBEWEGLICH : fixed ; immovable ; motionless ; inflexible ; still ; quiescent 759
UNBEWOHNBAR : uninhabitable 760
UNBEWUSST : unaware of, unknown, unconscious of 761
UNBEZAHLT : unpaid 762
UNBEZWEIFELT : undoubted 763
UNBIEGSAM : inflexible, rigid 764
UNBILDE (*f*) : wrong ; injury ; injustice ; inclemency ; unfairness 765
UNBILL (*f*) : (see UNBILDE) 766
UNBILLIG : unjust ; unfair ; unreasonable ; inequitable 767
UNBRAUCHBAR : useless ; waste 768
UNBRENNBARER STOFF (*m*) : incombustible matter 769
UNDANK (*m*) : ingratitude 770
UNDANKBAR : ungrateful 771
UNDECYLENSÄURE (*f*) : undecylenic acid 772
UNDEHNBAR : inextensible ; non-ductile ; unmalleable ; non-malleable 773
UNDENKBAR : unimaginable ; unbelievable ; incredible 774
UNDEUTLICH : indistinct ; unsharp ; unintelligible ; inarticulate 775
UNDICHT : leaky ; pervious ; not tight ; permeable 776
UNDICHTHEIT (*f*) : leak-(iness) ; perviousness ; permeability ; leakage 777
UNDIENLICH : unserviceable ; unsuitable ; unfit 778
UNDISSOZIIERT : undissociated 779
UNDULDSAM : intolerant 780
UNDURCHDRINGLICH : impermeable ; impenetrable ; impervious ; inscrutable 781
UNDURCHFÜHRBAR : impracticable ; not feasible 782
UNDURCHLÄSSIG : impervious ; impermeable 783
UNDURCHSICHTIG : opaque ; not transparent 784
UNDURCHSICHTIGKEIT (*f*) : opacity 785
UNEBEN : uneven ; unequal ; rough ; rugged 786
UNEBENHEIT (*f*) : inequality ; unevenness ; roughness ; ruggedness 787
UNECHT : false ; spurious ; counterfeit ; unreal ; artificial ; improper ; not genuine ; illegitimate ; fugitive ; not fast ; loose(Colours) 788
UNEDEL : ignoble ; mean ; base 789
UNEGAL : unequal ; uneven ; rough ; rugged 790
UNEGALITÄT (*f*) : unevenness ; inequality ; roughness ; ruggedness 791
UNEHRE (*f*) : disgrace ; dishonour 792
UNEHRLICH : dishonest ; dishonourable 793
UNEIGENTLICH : figurative ; not literal ; not (true ; proper ; in a strict sense ; strictly) real 794
UNEINGESCHRÄNKT : unconditioned ; unlimited ; unrestricted ; unrestrained 795
UNEINIG : disagreeing ; not in harmony ; at variance ; disunited ; discordant 796
UNELASTISCH : inelastic 797
UNEMPFÄNGLICH : unreceptive ; unsusceptible 798
UNEMPFINDLICH : insensible ; insensitive ; unsusceptible ; cold ; indifferent 799

UNEMPFINDLICH GEGEN : not (influenced) affected by, unaffected (uninfluenced) by 800
UNEMPFINDLICHKEIT (*f*): insensibility, anæsthesia (Med.) 801
UNENDLICH : endless ; infinite ; eternal 802
UNENDLICH VIEL : infinite number ; infinite (amount) quantity 803
UNENTBEHRLICH : indispensable ; necessary 804
UNENTGELTLICH : free ; gratis ; gratuitous-(ly) 805
UNENTHALTSAM : incontinent 806
UNENTSCHIEDEN : undecided, undetermined 807
UNENTSCHLOSSEN : undecided, irresolute 808
UNENTWEGT : firm-(ly) 809
UNENTWICKELT : undeveloped 810
UNENTWIRRBAR : inextricable 811
UNENTZÜNDBAR : uninflammable 812
UNERBITTLICH : inexorable 813
UNERFAHREN : inexperienced 814
UNERFORSCHLICH : inscrutable 815
UNERGIEBIG : unyielding ; unprofitable 816
UNERGRÜNDLICH : unfathomable ; inscrutable ; impenetrable 817
UNERHEBLICH : insignificant ; inconsiderable 818
UNERHÖRT : unheard of 819
UNERKENNLICH : indiscernible ; nurecognizable ; ungrateful 820
UNERKLÄRBAR : inexplicable ; unaccountable 821
UNERLÄSSLICH : indispensable ; irremissible 822
UNERLAUBT : illegal ; illicit ; not (allowed) permitted 823
UNERMESSLICH : immense ; immeasurable ; vast 824
UNERMÜDLICH : indefatigable 825
UNERÖRTERT : unexplained ; undecided 826
UNERREICHBAR : unattainable ; inaccessible 827
UNERSÄTTLICH : insatiable ; greedy 828
UNERSCHLOSSEN : untapped, not opened up 829
UNERSCHÖPFLICH : inexhaustible 830
UNERSETZBAR : irreparable ; not replaceable ; irretrievable 831
UNERSETZLICH : (see UNERSETZBAR) 832
UNERSTEIGBAR : insurmountable ; inaccessible 833
UNERTRÄGLICH : insufferable ; intolerable 834
UNERWÄHNT : unmentioned 835
UNERWARTET : unexpected ; unanticipated ; unlooked for ; unforeseen ; unawares 836
UNFÄHIG : incapable ; unfit ; unable 837
UNFALL (*m*): disaster ; misfortune ; mischance ; accident 838
UNFEHLBAR : unfailing ; infallible ; certain ; sure 839
UNFERN : near ; not far from 840
UNFÖRMIG : deformed ; disproportionate 841
UNFREI : unpaid ; excluding, exclusive of 842
UNFREI, FRACHT- : exclusive of freight 843
UNFREIWILLIG : involuntary, involuntarily 844
UNFRIEDE (*f*): discord, dissention, enmity 845
UNFRUCHTBAR : sterile, unfruitful, barren, not fertile 846
UNFRUCHTBARKEIT (*f*): barrenness, sterility (Med.) ; fruitlessness 847
UNFUG (*m*): disorder ; mischief ; misdemeanour 848
UNFÜHLBAR : imperceptible ; impalpable 849
UNGANGBAR : impracticable ; not feasible ; unusual 850
UNGAR : underdone ; not done ; undertanned (Leather) 851
UNGASTLICH : inhospitable 852
UNGEACHTET : although ; notwithstanding ; nevertheless ; in spite of ; excluding 853
UNGEAHNT : unexpected ; unanticipated ; undreamt of 854
UNGEBAHNT : unbeaten, trackless, untrodden 855
UNGEBAUT : uncultivated 856
UNGEBETEN : unbidden, unasked, uninvited 857
UNGEBILDET : uneducated, coarse ; uncivilized 858
UNGEBLEICHT : unbleached. 859
UNGEBRANNT : unburnt 860
UNGEBRANNTES : unburnt matter ; residue 861
UNGEBRÄUCHLICH : not customary, unusual 862
UNGEBRAUCHT : unused ; residual 863
UNGEBREMSTE WELLE (*f*): undamped wave 864
UNGEBÜHRLICH : indecent ; improper ; unbecoming 865
UNGEBUNDEN : unbound ; loose ; dissolute ; prose ; uncombined (Elements) ; unrestrained ; free 866
UNGEBUNDENE WÄRME (*f*): free heat 867
UNGEDÄMPFTE WELLE (*f*): undamped wave 868
UNGEFÄHR : about ; approximate-(ly) (*n*): chance ; hazard 869
UNGEFÄHRDET : safe ; unendangered 870
UNGEFÄLLIG : disagreeable ; disobliging 871
UNGEFÄRBT : undyed ; uncoloured 872
UNGEFORDERT: unasked (uncalled) for 873
UNGEFÜGIG : not pliant ; not flexible ; not pliable 874
UNGEGERBT : untanned 875
UNGEGOREN : unfermented 876
UNGEGRÜNDET : unfounded ; false 877
UNGEHALTEN : indignant ; angry 878
UNGEHEISSEN : unasked ; unbidden ; spontaneous ; voluntarily 879
UNGEHEIZT : unheated 880
UNGEHEMMT : unchecked ; unarrested 881
UNGEHEUER : huge ; enormous ; immense (*n*): monster 882
UNGEHINDERT : unhindered ; unchecked ; unprevented ; unrestricted 883
UNGEHOBELT : unplaned, rough ; unpolished, uncouth 884
UNGEHOFFT : unhoped for, unexpected 885
UNGEHÖRIG : improper ; unsuitable 886
UNGEKOCHT : unboiled 887
UNGEKRÄNKT : uninjured ; not (grieved) vexed 888
UNGEKÜNSTELT : artless ; unaffected ; natural 889
UNGELADEN : uninvited ; unloaded ; not charged 890
UNGELÄUTERT : unrefined ; unpurified 891
UNGELEGEN : inopportune ; unseasonable ; inconvenient ; incommodious 892
UNGELEHRT : unlearned ; illiterate 893
UNGELEIMT : unglued ; unsized (Paper) 894
UNGELENK : awkward ; stiff ; inflexible ; intractable 895
UNGELÖSCHT : unslaked ; unquenched 896
UNGEMACH (*n*): trouble ; adversity ; hardship ; discomfort 897
UNGEMÄSSIGT : immoderate ; not moderated 898
UNGEMEIN : uncommon ; extraordinary ; exceedingly 899
UNGEMESSEN : boundless ; unmeasured ; unlimited 900
UNGEMISCHT : unmixed 901
UNGEMÜNZT : uncoined 902
UNGEMÜTLICH : unpleasant ; uncomfortable 903

UNGENANNT : unnamed ; anonymous ; innominate (Anat.) 904
UNGENAU : inaccurate ; inexact 905
UNGENAUIGKEIT (*f*) : inaccuracy ; inexactitude 906
UNGENEIGT : disinclined ; unfriendly ; unwilling ; unfavourably inclined 907
UNGENIESSBAR : unpalatable ; uneatable 908
UNGENÜGEND : insufficient ; inadequate 909
UNGENÜGSAM : insatiable ; greedy 910
UNGENUTZT : unused ; unemployed ; waste 911
UNGEORDNET : inordinate 912
UNGEPRÜFT : untried ; unexamined ; not tested 913
UNGERADE : uneven ; not straight ; odd (of numbers) 914
UNGERATEN : ill-bred ; depraved 915
UNGERECHNET : not reckoned ; not counted ; not included 916
UNGERECHT : unjust 917
UNGEREGELT : irregular 918
UNGEREIMT : unrhymed ; absurd ; incongruous 919
UNGEREINIGT : unpurified 920
UNGERN : unwillingly 921
UNGERÖSTET : unroasted ; not calcined ; not subjected to the retting process (Flax) 922
UNGERUFEN : uncalled 923
UNGESALZEN : unsalted ; insipid ; fresh ; unseasoned 924
UNGESÄTTIGT : unsaturated ; not satiated 925
UNGESÄUERT : not acidified ; unleavened (Bread) 926
UNGESÄUMT : seamless ; immediate ; prompt 927
UNGESCHÄLT : unpeeled ; unhusked ; not (pared) shelled 928
UNGESCHEHEN : undone ; unaccomplished 929
UNGESCHICHTET : unstratified 930
UNGESCHICKT : awkward ; inept 931
UNGESCHLACHT : rude ; uncouth 932
UNGESCHLIFFEN : rude ; uncut ; unpolished ; rough ; coarse 933
UNGESCHMÜCKT : unadorned 934
UNGESCHWÄCHT : not weakened, unimpaired 935
UNGESEHEN : unseen, unperceived, unnoticed 936
UNGESELLIG : unfriendly, unsociable 937
UNGESETZLICH : unlawful, illegal 938
UNGESTALT : misshapen ; deformed 939
UNGESTILLT : unappeased 940
UNGESTÖRT : undisturbed 941
UNGESTÜM : violent ; impetuous ; vehement ; (*m* and *n*) : violence ; vehemence ; impetuosity 942
UNGESUCHT : unsought ; unaffected 943
UNGESUND : unwholesome, unhealthy 944
UNGETADELT : uncensured ; unblamed 945
UNGETEILT : undivided, ungraduated 946
UNGETROCKNET : undried 947
UNGEÜBT : inexperienced 948
UNGEÜBTE (*m*) : inexperienced person, novice ; (*pl*) : inexperienced 949
UNGEWASCHEN : unwashed 950
UNGEWISS : uncertain ; indefinite ; doubtful 951
UNGEWISSENHAFT : unconscientious 952
UNGEWITTER (*n*) : thunderstorm ; tempest 953
UNGEWOGEN : unfavourable 954
UNGEWÖHNLICH : unusual ; extraordinary ; uncommon 955
UNGEWOHNT : unaccustomed ; unusual ; unwonted 956
UNGEZÄUMT : unbridled ; unrestrained ; unchecked 957
UNGEZIEFER (*n*) : vermin 958
UNGEZIEFERVERTILGUNGSMITTEL (*n*) : vermin (killer) destroyer 959
UNGEZÜGELT : unbridled ; unrestrained ; unchecked 960
UNGEZWUNGEN : unconstrained ; natural ; voluntarily 961
UNGHWARIT (*m*) : unghwarite (an Al_2O_3 and SiO_2 hydrogel) 962
UNGIFTIG : non-poisonous 963
UNGILTIG : invalid ; null ; void 964
UNGLASIERT : not glazed ; matt ; unglazed 965
UNGLAUBE (*m*) : incredulity ; unbelief ; disbelief 966
UNGLÄUBLICH : incredible 967
UNGLEICH : unlike ; unequal ; uneven ; odd ; not uniform ; dissimilar 968
UNGLEICHARTIG : dissimilar ; heterogeneous 969
UNGLEICHE WINKELEISEN (*n.pl*) : unequal angles 970
UNGLEICHFÖRMIG : unsymmetrical ; unlike ; not uniform ; different ; variable 971
UNGLEICHHEIT (*f*) : dissimilarity ; unevenness ; inequality ; disparity 972
UNGLEICHMÄSSIG : not uniform ; unsymmetrical ; disproportionate ; dissimilar 973
UNGLEICHSEITIG : having unequal sides ; scalene (triangle) (Geometry) 974
UNGLEICHWINKELIG : having unequal angles 975
UNGLÜCK : bad (ill) luck ; misfortune ; accident ; disaster ; calamity 976
UNGLÜCKLICH : unfortunate ; ill-fated ; unlucky 977
UNGLÜCKLICHERWEISE : unfortunately 978
UNGLÜCKSELIG : calamitous ; unhappy ; unfortunate 979
UNGLÜCKSFALL (*m*) : accident ; disaster 980
UNGLÜCK, ZUM- : unfortunately ; unhappily ; as ill-luck would have it 981
UNGNADE (*f*) : disgrace ; disfavour 982
UNGNADIAÖL (*n*) : ungnadia oil (from seeds of *Ungnadia speciosa*) ; Sg. 0.912 983
UNGNÄDIG : unkind ; ungracious 984
UNGÜLTIG : void ; invalid ; null 985
UNGUNST (*f*) : unfavourableness ; disfavour 986
UNGÜNSTIG : unfavourable ; unpropitious 987
UNGUT : amiss ; ill 988
UNGÜTIG : unfriendly ; unkind 989
UNHALTBAR : untenable ; not durable ; frivolous 990
UNHALTIG : not containing any ; having no content of 991
UNHÄMMERBAR : not malleable 992
UNHEIL (*n*) : harm ; calamity ; mischief 993
UNHEILBAR : incurable 994
UNHEILIG : profane ; unholy 995
UNHEILSAM : noxious ; unwholesome 996
UNHEIMLICH : dismal ; creepy ; haunted ; uncomfortable ; uneasy 997
UNHEIZBAR : not heatable 998
UNHÖFLICH : impolite ; uncivil 999
UNHÖRBAR : inaudible 000
UNHYGIENISCH : unhygienic, not hygienic 001
UNIERT : united 002
UNIFARBE (*f*) : plain (uniform ; self) colour (Dyeing) 003
UNIFÄRBEN : to dye (solid ; uniformly) a self (uniform) shade 004
UNIKUM (*n*) : single (unique) example or copy ; sole existing (example) copy 005
UNINTERESSANT : uninteresting 006

UNIPOLAR: unipolar 007
UNIPOLARDYNAMO (*f*): unipolar dynamo 008
UNIPOLARITÄT (*f*): unipolarity 009
UNIPOLARMASCHINE (*f*): unipolar dynamo 010
UNISTÜCKWARE (*f*): plain (uniform; self) shade piece goods 011
UNITARISCH: unitary; unitarian 012
UNIVERSAL: universal 013
UNIVERSALARZNEI (*f*): universal remedy; cure-all; sovereign remedy 014
UNIVERSALEISEN (*n.pl*): flitchplates 015
UNIVERSALGELENK (*n*): universal joint 016
UNIVERSALLACK (*m*): universal varnish 017
UNIVERSALMITTEL (*n*): universal remedy; cure-all; sovereign remedy 018
UNIVERSALREZEPT (*n*): universal (prescription) remedy; cure-all; sovereign (prescription) remedy 019
UNIVERSALSCHRAUBENSCHLÜSSEL (*m*): universal screw spanner, universal screw wrench, universal screw hammer, monkey (spanner) wrench 020
UNIVERSALSCHWARZ (*n*): universal black 021
UNIVERSITÄT (*f*): university 022
UNIVERSUM (*n*): universe 023
UNKE (*f*): toad; frog 024
UNKEN: to croak 025
UNKENNBAR: unrecognizable; undiscernible 026
UNKENNTLICH: (see **UNKENNBAR**) 027
UNKENNTNIS (*f*): ignorance; lack of knowledge 028
UNKEUSCH: lascivious 029
UNKLAR: confused; turbid; not clear 030
UNKLAREN, IM- LASSEN: to leave (in the dark) without instructions 031
UNKLUG: foolish; stupid; indiscreet; imprudent 032
UNKLUGHEIT (*f*): indiscretion; imprudence 033
UNKONTROLLIERBAR: uncontrollable 034
UNKORRIGIERT: uncorrected 035
UNKOSTEN (*f.pl*): charges; expenses; costs; disbursements; outlay; outgoings 036
UNKRAUT (*n*): weed-(s); tare-(s) 037
UNKRAUTSAMEN (*m* and *pl*): weed seeds 038
UNKRAUTVERTILGUNGSMITTEL (*n*): weed-killer 039
UNKRISTALLISIERBAR: uncrystallizable 040
UNKUNDIG: unacquainted (with); ignorant of 041
UNLÄNGST: recently; of late; not long ago 042
UNLAUTER: sordid; impure; unclean 043
UNLEIDLICH: insufferable; intolerable; impatient 044
UNLENKSAM: unmanageable 045
UNLESBAR: illegible 046
UNLESERLICH: illegible 047
UNLEUGBAR: undeniable; unquestionable; incontrovertible; indisputable; irrefutable 048
UNLIEB: unpleasant; objectionable; disagreeable 049
UNLIEBSAM: unpleasant; objectionable; disagreeable 050
UNLÖSBAR: insoluble; indissoluble; not soluble; insolute; not solvable (Maths.) 051
UNLÖSCHBAR: unquenchable 052
UNLÖSLICH: insoluble; not solvable; insolute 053
UNLÖSLICHE (*n*): insolvent matter, insolubility 054
UNLÖSLICHKEIT (*f*): insolubility 055
UNLUST (*f*): displeasure; dislike; disgust 056
UNLUSTIG: disinclined; sad; reluctant; dull; heavy 057
UNMASS (*n*): immoderateness 058
UNMASSE (*f*): large (great; enormous) quantities 059
UNMASSGEBLICH: inauthoritative; humble 060
UNMÄSSIG: intemperate; immoderate; excessive 061
UNMASSSTÄBLICH: not to scale 062
UNMENSCH (*m*): monster; brute; barbarian 063
UNMENSCHLICH: inhuman; brutal; barbarian; superhuman 064
UNMERKLICH: imperceptible 065
UNMESSBAR: immeasurable 066
UNMISCHBAR: immiscible 067
UNMISCHBARKEIT (*f*): immiscibility 068
UNMITTELBAR: immediate; direct-(ly) 069
UNMITTELBARKEIT (*f*): directness 070
UNMÖBLIERT: unfurnished 071
UNMODERN: old-fashioned; out-of-date; not modern 072
UNMÖGLICH: impossible 073
UNMÖGLICHKEIT (*f*): impossibility 074
UNMÜNDIG: minor; under age 075
UNMÜNDIGKEIT (*f*): minority 076
UNMUT (*m*): discontent; dejection; sadness 077
UNNACHAHMBAR: inimitable 078
UNNACHAHMLICH: inimitable 079
UNNACHGIEBIG: inflexible; unyielding 080
UNNACHLÄSSLICH: unyielding, irremissible, unabating, indispensable 081
UNNACHSICHTIG: unrelenting 082
UNNAHBAR: inaccessible; unapproachable 083
UNNATÜRLICH: unnatural; uncommon; affected 084
UNNENNBAR: inexpressible; unutterable; ineffable 085
UNNÖTIG: unnecessary; needless 086
UNNÜTZ: useless; vain; naughty; unprofitable; fruitless 087
UNORDENTLICH: disordered; disorderly; untidy; inordinate 088
UNORDNUNG (*f*): litter; disorder; confusion 089
UNORGANISCH: inorganic 090
UNPAARWERTIG: of odd valence 091
UNPARTEIISCH: impartial; unbiassed 092
UNPARTEIISCHE (*m*): unbiassed person 093
UNPARTEILICHKEIT (*f*): impartiality 094
UNPASS: unwell; indisposed 095
UNPASSEND: unsuitable; unfit; improper 096
UNPÄSSLICH: unwell; indisposed; ill 097
UNPLASTISCH: non-plastic 098
UNPOLIERT: unpolished; rough 099
UNPOLITISCH: impolitic 100
UNPRESSBAR: incompressible 101
UNPRESSBARKEIT (*f*): incompressibility 102
UNPRODUKTIV: unproductive 103
UNRAT (*m*): trash; refuse; excretion; dirt; garbage; excrement; offal; dross (Metal) 104
UNRECHT: wrong; unjust; unfair; undue; false; (*n*): injustice; wrong; error; fault; injury 105
UNRECHTLICH: unjust; dishonest 106
UNRECHTMÄSSIG: illegal; unlawful; illegitimate 107
UNREDLICH: dishonourable; dishonest; unfair; unjust 108
UNREGELMÄSSIG: abnormal; irregular; anomalous 109
UNREIF: immature; unripe 110

UNREIFE (*f*): immaturity; unripeness 111
UNREIN: unclean; impure 112
UNREINHEIT (*f*): impurity; uncleanness 113
UNREINIGKEIT (*f*): impurity; uncleanness 114
UNREINLICH: uncleanly 115
UNREIZBAR: not sensitive; non-irritable 116
UNRETTBAR: irrecoverable; irretrievable 117
UNRICHTIG: false; erroneous; wrong; unjust; incorrect 118
UNRUHE (*f*): uneasiness; restlessness; turbulence; unrest; trouble; commotion; balance (of watch) 119
UNRUHIG: restless; troubled; turbulent; uneasy 120
UNRUHIGES FLUSSEISEN (*n*): restless mild steel (in which, during solidification, gas is constantly generated, forming bubbles in the ingot, which bubbles weld together when the ingot is worked) 121
UNRUND: untrue, out of truth, not (round) circular 122
UNSAGBAR: unspeakable 123
UNSÄGLICH: unspeakable 124
UNSANFT: harsh; rough; not soft 125
UNSATINIERT: unglazed (Paper) 126
UNSATINIERTE FARBPAPIERE (*n.pl*): unglazed colours (Paper) 127
UNSAUBER: dirty; sordid; unclean-(ly); squalid; impure; slovenly 128
UNSCHÄDLICH: innocuous; harmless; safe 129
UNSCHÄDLICHES LICHT (*n*): safe light (Phot., etc.) 130
UNSCHÄTZBAR: invaluable; inestimable; priceless 131
UNSCHEINBAR: unsightly; insignificant (plain) looking; dull; tarnished; simple 132
UNSCHICKLICH: unbecoming; unfit; inconvenient; improper; indecent 133
UNSCHLITT (*m*): suet; tallow (see TALG and RINDFETT) 134
UNSCHLÜSSIG: irresolute; undecided; wavering 135
UNSCHMACKHAFT: insipid; unpalatable; tasteless; unsavoury 136
UNSCHMELZBAR: infusible; not meltable; not fusible 137
UNSCHMELZBARKEIT (*f*): infusibility 138
UNSCHULD (*f*): purity; innocence 139
UNSCHULDIG: innocent; not guilty; guiltless; pure; undefiled; guileless 140
UNSCHWEISSBAR: not weldable, unweldable 141
UNSCHWEISSBARER STAHL (*m*): harsh steel, unweldable steel 142
UNSCHWER: easy; not difficult 143
UNSEIFBAR: unsaponifiable 144
UNSEIFBARE (*n*): unsaponifiable matter 145
UNSELIG: unhappy; unfortunate; vicious; fatal 146
UNSERTHALBEN: on our behalf; in our name; for our sake 147
UNSICHER: unsafe; insecure; unsteady; uncertain; unreliable; unstable; dubious 148
UNSICHERHEIT (*f*): uncertainty; unreliability (see UNSICHER) 149
UNSICHTBAR: invisible 150
UNSINN (*m*): nonsense; madness; absurdity 151
UNSINNIG: nonsensical; mad; absurd; ridiculous; insane 152
UNSITTE (*f*): bad (habit) custom 153
UNSITTLICH: indecent; immoral; unmannerly; indecorous 154
UNSPALTBAR: uncleavable 155
UNSPALTIG: uncleavable, impossible to split 156
UNSTÄT: variable; fluctuating; unsteady; unstable; inconstant; restless; unsettled 157
UNSTÄTIG: (see UNSTÄT) 158
UNSTÄTIGKEIT (*f*): fluctuation; variation; instability (see UNSTÄT) 159
UNSTATTHAFT: illicit; inadmissible; unlawful 160
UNSTATTHAFTIGKEIT (*f*): impropriety (see UN-STATTHAFT) 161
UNSTERN (*m*): misfortune; unlucky star; disaster; calamity 162
UNSTETIG: uneven, irregular 163
UNSTÖRBAR: imperturbable 164
UNSTRÄFLICH: blameless; irreproachable 165
UNSTRECKBAR: not (extensible; malleable) ductile 166
UNSTREITIG: indisputable; unquestionable; indubitable; incontestable 167
UNSTUDIERT: unlettered; unstudied; illiterate 168
UNSULFIERT: unsulphonated 169
UNSUMME (*f*): immense (enormous) sum 170
UNSYMMETRISCH: unsymmetrical 171
UNTADELHAFT: blameless, irreproachable, unblamable 172
UNTADELIG: unexceptional 173
UNTAT (*f*): misdeed; crime 174
UNTÄTIG: inert; dormant; inactive; indolent; idle 175
UNTÄTIGKEIT (*f*): inertness; inactivity; indolence; inaction; idleness 176
UNTAUGLICH: unsuitable; unfit; useless 177
UNTEILBAR: indivisible; inseparable 178
UNTEILBARKEIT (*f*): indivisibility 179
UNTEILHAFTIG: not sharing in; having no part in 180
UNTEILNEHMEND: indifferent (see UNTEIL-HAFTIG) 181
UNTEN: beneath; under-(neath); below; at the bottom 182
UNTENAN: last; at the (foot; bottom) end 183
UNTENBENANNT: hereafter mentioned; mentioned below; undermentioned 184
UNTENHIN: downwards; down 185
UNTENLIEGEND: inferior; lying below 186
UNTENSTEHEND: given below; under-mentioned 187
UNTER: below; under-(neath):; beneath; among; between; during; in the midst of; less (than); betwixt; amidst 188
UNTERABTEILEN: to sub-divide 189
UNTERABTEILUNG (*f*): sub-division 190
UNTER ANDERN: amongst other things; amongst others 191
UNTERARM (*m*): forearm 192
UNTERAUGENHÖHLEN: infra-orbital 193
UNTERBAU (*m*): foundation; sub-structure 194
UNTERBAUCH (*m*): hypogastrium 195
UNTERBAUCHBRUCH (*m*): hypogastrocele (Med.) 196
UNTERBAUEN: to build under; underpin; build a sub-structure 197
UNTERBEAMTER (*m*): inferior (subordinate) official or officer 198
UNTERBEHÖRDE (*f*): inferior court 199
UNTERBEINKLEIDER (*n.pl*): drawers; pants 200
UNTERBELICHTUNG (*f*): under-exposure (Phot.) 201

UNTERBIETEN: to outbid; bid below; be outdone by 202
UNTERBILANZ (*f*): deficit 203
UNTERBINDEN: to ligature; tie up; bandage 204
UNTERBINDUNG (*f*): ligature; bandaging 205
UNTERBLEIBEN: to cease; be left undone; not occur; be omitted; be obviated; be done away with 206
UNTERBRECHEN: to interrupt; stop; break; suspend; disconnect; cut out 207
UNTERBRECHER (*m*): interrupter; breaker (Elect.); cut-out 208
UNTERBRECHERFEDER (*f*): interrupting spring 209
UNTERBRECHERGEHÄUSE (*n*): contact breaker case (Motor cars) 210
UNTERBRECHERSPITZE (*f*): interruptor point, interruptor tip 211
UNTERBRECHER (*m*), **STROM-**: contact breaker 212
UNTERBRECHUNG (*f*): interruption; stop; break; intermission 213
UNTERBRECHUNGSSTROM (*m*): interrupted current 214
UNTERBREITEN: to submit 215
UNTERBRINGEN: to give lodging to; provide for; dispose of; invest; sell; place 216
UNTERBROCHEN: interrupted; broken; intermittent 217
UNTERBROMIG: hypobromous 218
UNTERBROMIGE SÄURE (*f*): hypobromous acid 219
UNTERBROMIGSAUER: hypobromite of; combined with hypobromous acid 220
UNTERCHLORIG: hypochlorous 221
UNTERCHLORIGE SÄURE (*f*): hypochlorous acid; (HOCl) 222
UNTERCHLORIGSAUER: hypochlorite of; combined with hypochlorous acid 223
UNTERCHLORIGSAURES SALZ (*n*): hypochlorite 224
UNTERCHLORSÄURE (*f*): hypochloric acid 225
UNTERDECK (*n*): lower deck 226
UNTERDES: meanwhile; in the meantime 227
UNTERDESSEN: (see **UNTERDES**) 228
UNTERDRUCK (*m*): negative pressure; vacuum 229
UNTERDRÜCKEN: to suppress; repress; oppress; depress 230
UNTERE: under; inferior; lower; nether 231
UNTEREINANDER: together; confusedly; among one another 232
UNTEREINTEILUNG (*f*): subdivision; section 233
UNTERESSIGSAUER: subacetate of 234
UNTEREUTEKTOID: hypoeutectic (of steel) 235
UNTEREXPOSITION (*f*): under-exposure (Phot.) 236
UNTERFANGEN: to venture; dare; undertake; underpin (Building) 237
UNTERFLÄCHE (*f*): base; under-surface 238
UNTERFORM (*f*): subspecies; sub-variety; inferior (species) variety 239
UNTERFUTTER (*n*): lining 240
UNTERGANG: setting (of sun); fall; ruin; decline; going down; sinking 241
UNTERGÄRIG: fermented from below; bottom-fermenting; bottom-fermented 242
UNTERGÄRIGE HEFE (*f*): bottom yeast (Brew.) 243
UNTERGÄRUNG (*f*): bottom fermentation (Brew.) 244
UNTERGATTUNG (*f*): subgenus; lower type 245
UNTERGEBEN: inferior; subordinate; to subject; submit; place under; commit 246
UNTERGEHEN: to set (of the sun); go (down) under; become extinct; perish; decline; submerge; founder; go under; sink 247
UNTERGEORDNET: subservient; secondary; subordinate; inferior; minor 248
UNTERGESCHOBEN: substituted; forged; spurious; suppositious 249
UNTERGESCHOSS (*n*): ground floor 250
UNTERGESENK (*n*): bottom (lower) swage 251
UNTERGESTELL (*n*): under-carriage; supporting structure; chassis; carriage; foundation; substructure 252
UNTERGLASUR (*f*): under-glaze 253
UNTERGLASURFARBE (*f*): under-glaze-(colour) (Ceramics) 254
UNTERGRABEN: to undermine; to dig in; corrupt 255
UNTERGRÄDIG: underproof 256
UNTERGRÄTEN-: infraspinous 257
UNTERGRUND (*m*): underground; subsoil; bottom (Dyeing); bottom (first) print (Calico) 258
UNTERGRUNDBAHN (*f*): underground railway 259
UNTERGRUNDFARBE (*f*): bottom colour (Dyeing) 260
UNTERHALB: below; under; beyond 261
UNTERHALTEN: to support; entertain; converse; discourse; sustain; maintain; keep (up) under; keep; amuse-(oneself) 262
UNTERHALTEND: entertaining 263
UNTERHALTUNG (*f*): maintenance; support; upkeep; entertainment; conversation; sustenance; amusement 264
UNTERHALTUNGSKOSTEN (*f.pl*): working expenses, maintenance, upkeep-(charges), cost of (maintenance) upkeep, running costs 265
UNTERHANDELN: to negotiate; treat; mediate; arbitrate; transact 266
UNTERHAUS (*n*): House of Commons 267
UNTERHAUT (*f*): underskin; (as a prefix = subcutaneous-) 268
UNTERHEFE (*f*): bottom yeast (slow and quiet ferment) 269
UNTERHÖHLEN: to undermine 270
UNTERHOLZ (*n*): underwood, undergrowth 271
UNTERIRDISCH: underground; subterranean; subterraneous 272
UNTERJOCHEN: to subjugate; subdue 273
UNTERJODIG: hypoiodous 274
UNTERJODIGSAUER: hypoiodite of; combined with hypoiodous acid 275
UNTERJODIGSÄURE (*f*): hypoiodous acid; 2IOH 276
UNTERKANTE (*f*): lower edge 277
UNTERKEHLE (*f*): double chin; gullet 278
UNTERKESSEL (*m*): mud drum-(s) (of water tube boilers-), lower drums, bottom drums, lower boiler 279
UNTERKIEFER (*m*): inferior maxilla; lower jaw (Anat.) 280
UNTERKIEL (*m*): lower keel 281
UNTERKINN (*n*): double chin 282
UNTERKOMMEN: to find (shelter; lodging) employment; (*n*): situation; employment, lodging 283
UNTERKRIECHEN: to crawl under 284
UNTERKRIEGEN: to overcome; get the better of 285

UNTERKÜHLEN: to undercool; cool intensely; cool below condensation temperature (of a liquid) 286
UNTERKUNFT (*f*): shelter; lodging 287
UNTERLAGE (*f*): bearing; support; case; bed; foundation; trestle; stand; stay; groundwork; substratum; subsoil; lining 288
UNTERLAGEN (*f.pl*): data; particulars; enclosures 289
UNTERLAGSPLATTE (*f*): sole plate; tie (bearing; base; foundation; bed) plate 290
UNTERLAGSRING (*m*): supporting ring 291
UNTERLAGSSCHEIBE (*f*): collar, washer, rivet plate 292
UNTERLAND (*n*): lowland-(s) 293
UNTERLASS (*m*): cessation; intermission 294
UNTERLASSEN: to cease; discontinue; leave off; fail; omit; neglect 295
UNTERLAUFEN: to occur; creep in; fill; extravasate; (*n*): extravasated; bloodshot; filled 296
UNTERLAUGE (*f*): underlye; spent lye (Soap) 297
UNTERLEFZE (*f*): under-lip 298
UNTERLEGEN: to lay (put) under; inferior; subordinate 299
UNTERLEGSCHEIBE (*f*): washer (see **UNTERLAGSSCHEIBE**) 300
UNTERLEIB (*m*): abdomen; belly 301
UNTERLEIBSENTZÜNDUNG (*f*): peritonitis (Med.) 302
UNTERLEIBSKRANKHEIT (*f*): stomachic (abdominal) trouble; disorder of the bowels (Med.) 303
UNTERLIEGEN: to be liable; be subject; succumb; lie under; be (inferior) subordinate 304
UNTERLIEK (*n*): foot rope 305
UNTERLIPPE (*f*): under-lip 306
UNTERLÖST: insufficiently grown (Malt) 307
UNTERMASTEN (*m.pl*): lower masts 308
UNTERMAUERN: to brick under; underpin; support on brickwork 309
UNTERMENGEN: to intermix 310
UNTERMINIEREN: to undermine, sap 311
UNTERMISCHEN: to intermingle, intermix 312
UNTERNEHMEN: to undertake; be enterprising; attempt; (*n*): enterprise; undertaking 313
UNTERNEHMEND: enterprising; venturesome; bold 314
UNTERNEHMUNG (*f*): undertaking; venture; enterprise 315
UNTERNEHMUNGSGEIST (*m*): energy; enterprise; go 316
UNTERNEHMUNGSLUST (*f*): energy; enterprise; go 317
UNTERNEHMVERBAND (*m*): syndicate; trust; etc. (any kind of business combination) 318
UNTERORDNEN: to subordinate 319
UNTERPFAND (*n*): pledge; mortgage; security; pawn 320
UNTERPHOSPHORIG: hypophosphorous 321
UNTERPHOSPHORIGE SÄURE (*f*): hypophosphorous acid; H_3PO_2; Sg. 1.493; Mp. 26.5°C. 322
UNTERPHOSPHORIGSAUER: hypophosphite of; combined with hypophosphorous acid 323
UNTERPHOSPHORIGSÄURE (*f*): hypophosphorous acid (see **UNTERPHOSPHORIGE SÄURE**) 324
UNTERPHOSPHORIGSAURES SALZ (*n*): hypophosphite 325
UNTERPHOSPHORSÄURE (*f*): hypophosphoric acid; $H_4P_2O_6$ 326
UNTERPROBE (*f*): under-proof spirit 327
UNTERPROBEWEINGEIST (*m*): under-proof spirit 328
UNTERREDUNG (*f*): conversation; conference; interview; discourse 329
UNTERRICHT (*m*): instruction 330
UNTERRICHTEN: to instruct; teach; inform; prime 331
UNTERRICHTEND: instructive 332
UNTERRICHTSANSTALT (*f*): academy; school 333
UNTERRINDE (*f*): under (bark) crust 334
UNTERSAGEN: to prohibit; forbid; interdict 335
UNTERSALPETERSÄURE (*f*): hyponitric acid; nitrogen peroxide (see **STICKSTOFFTETROXYD**) 336
UNTERSALPETRIG: hyponitrous 337
UNTERSALPETRIGSAUER: hyponitrite of; combined with hyponitrous acid 338
UNTERSATZ (*m*): stand; support; saucer; stay; assumption; foundation; trestle; tray; substructure; base; socle; plinth (Arch.) 339
UNTERSCHALE (*f*): saucer 340
UNTERSCHÄTZEN: to under-rate; under-value 341
UNTERSCHEIDBAR: discernible; distinguishable 342
UNTERSCHEIDEN: to discern; distinguish; differentiate; discriminate 343
UNTERSCHEIDUNGSMARKIERUNG (*f*): distinctive marking 344
UNTERSCHEIDUNGSVERMÖGEN (*n*): discriminative (faculty) power 345
UNTERSCHENKEL (*m*): lower leg; lower thigh 346
UNTERSCHIEBEN: to push under; forge; substitute; suppose 347
UNTERSCHIEBUNG (*f*): (see **UNTERSCHIEBEN** and **DUBLETTEN**) 348
UNTERSCHIED (*m*): difference; partition; distinction; discrimination 349
UNTERSCHIEDLICH: diverse; different 350
UNTERSCHIFF (*n*): lower work, quick work (of ships) 351
UNTERSCHLÄCHTIG: under-shot (of water wheels) 352
UNTERSCHLAG (*m*): shelf (of a deck) 353
UNTERSCHLAGEN: to embezzle; purloin; cross (the arms); intercept-(letters) 354
UNTERSCHLEIF (*m*): embezzlement; fraud; defalcation 355
UNTERSCHREIBEN: to sign; subscribe; underwrite 356
UNTERSCHRIFT (*f*): subscription; signature 357
UNTERSCHUBFEUERUNG (*f*): underfeed (grate; stoker) furnace; underfeed firing- (method) 358
UNTERSCHUBROST (*m*): underfeed (grate) stoker 359
UNTERSCHÜREN: to (under)-stoke, (under)-feed (a fire) 360
UNTERSCHWEFELSAUER: hyposulphate (dithionate) of; combined with hyposulphuric (dithionic) acid 361
UNTERSCHWEFELSÄURE (*f*): hyposulphuric (dithionic) acid 362
UNTERSCHWEFLIG: hyposulphurous 363
UNTERSCHWEFLIGSAUER: thiosulphate; hyposulphite of; combined with hyposulphurous acid 364
UNTERSCHWEFLIGSÄURE (*f*): hyposulphurous acid; thiosulphuric acid 365
UNTERSEEBOOT (*n*): U-boat; submarine 366
UNTERSEEISCH: submarine 367

UNTERSEEKABEL (n): submarine cable 368
UNTERSEGEL (n.pl): lower sails 369
UNTERSETZEN: to put (set) under 370
UNTERSETZSCHERBE (f): crucible stand; plate or saucer for placing under something 371
UNDERSETZT: thick-set; squat 372
UNTERSIEGELN: to seal; attach a seat to 373
UNTERSINKEN: to sink; descend; dive; go (to the bottom) down; fall 374
UNTERST: undermost; lowest 375
UNTERSTECKEN: to place (put) under or amongst; incorporate 376
UNTERSTEHEN: to dare; venture; pertain 377
UNTERSTELLEN: to place (put; stand) under; take (shelter) cover 378
UNTERSTEMMEN: to support, prop-(up), shore-(up) 379
UNTERSTOCK (m): underback (Brew.) 380
UNTERSTREICHEN: to underline; mark 381
UNTERSTÜTZEN: to assist; support; aid; uphold; prop-(up); underpin 382
UNTERSTÜTZUNGSKLOBEN (m): supporting block 383
UNTERSUCHEN: to enquire-(into); inspect; probe; investigate; examine; search; scrutinize; test; survey 384
UNTERSUCHUNG (f): investigation; examination; enquiry; research; test; inspection 385
UNTERSUCHUNGSCHEMIKER (m): research chemist 386
UNTERSUCHUNGSGEGENSTAND (m): object (matter) investigated or to be (investigated; tested) examined 387
UNTERSUCHUNGSLABORATORIUM (n): research laboratory 388
UNTERSUCHUNGSMITTEL (n): means of (research; examination) investigation 389
UNTERTAN: subject to; dependent on; (m): subject; vassal 390
UNTERTÄNIG: subject; submissive; obedient; subjective; humble 391
UNTERTASSE (f): saucer 392
UNTERTAUCHEN: to submerge; immerse; dive; (n): submersion 393
UNTERTAUCHUNG (f): submersion; immersion; diving 394
UNTERTEIG (m): underdough (Brew.) 395
UNTERTEIL (m): under (lower; inferior) part or portion 396
UNTERTEILEN: to subdivide; split up 397
UNTERTROMMEL (f): lower (mud) drum (of boilers) 398
UNTERWACHSEN: interlarded 399
UNTERWALD (m): undergrowth; underbush; scrub; brushwood; underwood 400
UNTERWANTEN (n.pl.): lower (rigging) shrouds 401
UNTERWÄRTS: downwards; underneath 402
UNTERWEGS: on (by) the way or road 403
UNTERWEGSVERLUST (m): transmission loss 404
UNTERWEIBIG: hypogynous 405
UNTERWEISEN: to teach; instruct 406
UNTERWERFEN: to subject-(to); submit-(to); subdue 407
UNTERWIND (m): forced-draught 408
UNTERWINDEN: to venture, hazard, dare, presume 409
UNTERWINDFEUERUNG (f): forced-draught furnace 410
UNTERWÜHLEN: to undermine 411
UNTERWÜRFIG: obsequious; submissive; subject-(ive) 412
UNTERZEICHNEN: to underwrite; sign; subscribe; execute (Deeds, etc.); ratify 413
UNTERZEICHNETE (m): subscriber 414
UNTERZEUG (n): underclothing 415
UNTERZIEHEN: to draw (lay) under; put (on) under-(neath); undergo; undertake 416
UNTERZUNGEN-: sublingual 417
UNTERZUG (m): carling (of a deck) 418
UNTIEFE (f): shallowness; shallows; sands; sandbank 419
UNTIER (n): monster 420
UNTILGBAR: indelible; irredeemable; inextinguishable 421
UNTRENNBAR: inseparable 422
UNTRENNBARKEIT (f): inseparability 423
UNTREU: faithless; perfidious; unfaithful; disloyal 424
UNTRINKBAR: undrinkable 425
UNTRÖSTBAR: inconsolable 426
UNTRÖSTLICH: inconsolable 427
UNTRÜGLICH: infallible; unmistakable; sure; certain 428
UNTÜCHTIG: inept; incapable; unsuitable; unfit 429
UNTUGEND (f): vice; fault; bad habit 430
UNTULICH: not feasible; impracticable 431
UNÜBERLEGT: rash; inconsiderate 432
UNÜBERSETZBAR: untranslatable 433
UNÜBERSTEIGLICH: insurmountable 434
UNÜBERTREFFLICH: unrivaled; unsurpassable; incomparable 435
UNÜBERWINDLICH: invincible; unconquerable 436
UNUMGÄNGLICH: unavoidable; indispensable; inevitable; unsociable 437
UNUMSCHRÄNKT: absolute; unlimited; unbounded 438
UNUMSTÖSSLICH: irrefutable; irrefragable; inviolable 439
UNUMWUNDEN: plain; frank; candid; straightforward 440
UNUNTERBROCHEN: unbroken; uninterrupted; continuous-(ly) 441
UNUNTERRICHTET: uninstructed 442
UNUNTERSCHEIDBAR: indistinguishable 443
UNUNTERSUCHT: uninvestigated 444
UNUNTERZEICHNET: unsigned 445
UNVERÄNDERLICH: invariable; unchangeable; constant; immutable; unalterable 446
UNVERÄNDERT: unaltered; unchanged; constant; unvaried 447
UNVERANTWORTLICH: irresponsible; inexcusable 448
UNVERARBEITET: unwrought; not (made up) worked 449
UNVERÄUSSERLICH: inalienable 450
UNVERBESSERLICH: incorrigible; perfect; incapable of being improved 451
UNVERBINDLICH: not binding; disobliging 452
UNVERBLÜMT: plain 453
UNVERBRANNT: unburned 454
UNVERBRANNTES (n): unburnt matter 455
UNVERBRENNBAR: incombustible 456
UNVERBRENNBARKEIT (f): incombustibility 457
UNVERBRENNLICH: incombustible 458
UNVERBRENNLICHKEIT (f): incombustibility 459
UNVERBRÜCHLICH: inviolable 460
UNVERBUNDEN: unconnected; uncombined; unobliged 461

UNVERBÜRGT: unconfirmed; unwarranted; not authentic 462
UNVERDAULICH: indigestible 463
UNVERDAULICHKEIT (*f*): indigestibility 464
UNVERDAUUNG (*f*): indigestion 465
UNVERDERBLICH: incorruptible 466
UNVERDERBLICHKEIT (*f*): incorruptibility 467
UNVERDICHTBAR: incondensable; incompressible 468
UNVERDICHTBARKEIT (*f*): incondensability; incompressibility 469
UNVERDIENT: undeserved; unmerited 470
UNVERDORBEN: unspoiled; undamaged; sound 471
UNVERDROSSEN: unwearied; patient; unremitting; indefatigable 472
UNVERDROSSENHEIT (*f*): assiduity 473
UNVEREINBAR: incompatible; irreconcilable; incongruous 474
UNVERFÄLSCHT: unadulterated; genuine 475
UNVERFAULBAR: imputrescible; unputrefiable 476
UNVERFROREN: cool; unabashed 477
UNVERGÄNGLICH: not perishable 478
UNVERGÄRBAR: unfermentable 479
UNVERGESSLICH: unforgetable; memorable 480
UNVERGLAST: unvitrified; unglazed 481
UNVERGLEICHBAR: incomparable; matchless 482
UNVERGLEICHLICH: incomparable; matchless 483
UNVERGOLTEN: unrewarded 484
UNVERGRÜNBAR: ungreenable (Dyeing) 485
UNVERGRÜNLICH: ungreenable (Dyeing) 486
UNVERHALTEN: unconcealed; open; free 487
UNVERHÄLTNISMÄSSIG: disproportionate-(ly) 488
UNVERHINDERT: unimpeded; unhindered; unhampered; unchecked 489
UNVERHOHLEN: unconcealed; open 490
UNVERKÄUFLICH: unsaleable; unmarketable 491
UNVERKENNBAR: unmistakable; evident; obvious; plain; indubitable; indubitably 492
UNVERKÜRZT: unabridged 493
UNVERLETZBAR: invulnerable; inviolable 494
UNVERLETZT: unimpaired; unhurt 495
UNVERLIERBAR: inalienable 496
UNVERLÖSCHBAR: inextinguishable 497
UNVERMEIDLICH: unavoidable; inevitable 498
UNVERMENGT: unmixed; pure; unalloyed 499
UNVERMERKT: unperceived; unnoticed 500
UNVERMINDERT: undiminished; unabated 501
UNVERMISCHBAR: immiscible 502
UNVERMISCHT: unmixed; pure; unalloyed 503
UNVERMÖGEN (*n*): impotence; inability; incapacity 504
UNVERMÖGEND: impotent; unable; incapable 505
UNVERMUTET: unexpected; unawares 506
UNVERNEHMLICH: inaudible 507
UNVERNUNFT (*f*): absurdity; unreasonableness; irrationality 508
UNVERNÜNFTIG: irrational; absurd; silly 509
UNVERÖFFENTLICHT: unpublished 510
UNVERRICHTET: unperformed; undone 511
UNVERRÜCKT: unmoved; fixed 512
UNVERSCHULDET: unmerited; innocent; undeserved; not (in debt) indebted 513
UNVERSEHENS: unexpectedly; unawares 514
UNVERSEHRBAR: inviolable 515
UNVERSEHRT: undamaged; uninjured; unhurt; safe 516
UNVERSEIFBAR: unsaponifiable 517
UNVERSIEGBAR: inexhaustible 518
UNVERSIEGELT: unsealed 519
UNVERSILBERT: unsilvered; not silverplated 520
UNVERSÖHNLICH: implacable; irreconcilable 521
UNVERSÖHNT: unreconciled 522
UNVERSORGT: unprovided for 523
UNVERSTAND (*m*): indiscretion; imprudence; want of (sense) judgment 524
UNVERSTÄNDIG: imprudent; unwise; foolish; injudicious 525
UNVERSTÄNDLICH: incomprehensible; unintelligible 526
UNVERSTEUERBAR: not dutiable; not assessible; not taxable 527
UNVERSTEUERT: duty unpaid; untaxed; not assessed for taxation 528
UNVERSUCHT: untried 529
UNVERTILGBAR: undestroyable; indelible 530
UNVERTRÄGLICH: unsociable; incompatible; quarrelsome 531
UNVERTRÄGLICHKEIT (*f*): unsociability; incompatibility 532
UNVERWAHRT: unpreserved; unguarded 533
UNVERWANDT: unmoved; unrelated 534
UNVERWEILT: without delay 535
UNVERWELKLICH: imperishable 536
UNVERWERFLICH: unexceptionable 537
UNVERWESLICH: undecaying; imputrescible; incorruptible 538
UNVERWÜSTLICH: indestructible 539
UNVERZAGT: intrepid; undismayed; undaunted 540
UNVERZEIHLICH: unpardonable 541
UNVERZOLLT: duty unpaid 542
UNVERZÜGLICH: instant; immediate; without delay 543
UNVERZWEIGT: unbranched 544
UNVOLLENDET: unfinished; incomplete 545
UNVOLLKOMMEN: imperfect; incomplete 546
UNVOLLSTÄNDIG: incomplete; defective 547
UNVORBEREITET: unprepared 548
UNVORHERGESEHEN: unforeseen; unlooked for; unanticipated; unexpected 549
UNVORSÄTZLICH: unpremeditated; unintentional 550
UNVORSICHTIG: careless; incautious; inconsiderate; improvident 551
UNVORTEILHAFT: unprofitable; disadvantageous 552
UNWÄGBAR: imponderable; unweighable 553
UNWAHR: false; untrue 554
UNWAHRHEIT (*f*): untruth; falsehood; lie 555
UNWAHRSCHEINLICH: improbable; unlikely 556
UNWANDELBAR: invariable; unchangeable; unalterable; immutable; constant 557
UNWEGSAM: trackless; pathless; impassable; impracticable 558
UNWEIT: not far from; not far away 559
UNWERT: unworthy 560
UNWESEN (*n*): nuisance; disorder; confusion; disturbance 561
UNWESENTLICH: inconsiderable; unessential; immaterial; unsubstantial 562
UNWETTER (*n*): storm; bad weather 563
UNWICHTIG: unimportant; insignificant; negligible; inconsiderable 564
UNWIDERLEGBAR: irrefutable (see **UNLEUGBAR**) 565
UNWIDERLEGLICH: (see **UNLEUGBAR**) 566

UNWIDERRUFLICH : irrevocable 567
UNWIDERSETZLICH : irresistible 568
UNWIDERSPRECHLICH : incontestable, unquestionable, irrefragable, undeniable 569
UNWIDERSTEHLICH : irresistible 570
UNWIEDERBRINGLICH : not restorable ; irrecoverable 571
UNWILLE (*m*) : resentment ; displeasure ; anger ; indignation 572
UNWILLIG : resentful ; indignant 573
UNWILLKÜRLICH : involuntary 574
UNWIRKSAM : inactive ; void ; inefficient ; ineffective ; ineffectual ; inefficacious 575
UNWIRSCH : brusque ; cross 576
UNWIRTBAR : barren ; inhospitable ; dreary ; desert 577
UNWISSEND : ignorant ; unknowing ; unskilled 578
UNWISSENTLICH : unknowingly ; unconsciously 579
UNZAHL (*f*) : enormous (tremendous ; immense) number 580
UNZÄHLBAR : innumerable 581
UNZÄHLIG : innumerable 582
UNZE (*f*) : ounce 583
UNZEIT (*f*) : wrong (unsuitable) time 584
UNZEITIG : untimely ; immature ; inopportune ; unseasonable 585
UNZERBRECHLICH : unbreakable 586
UNZERLEGBAR : indecomposable 587
UNZERREISSBAR : untearable ; indestructible 588
UNZERSETZBAR : indecomposable 589
UNZERSETZT : undecomposed 590
UNZERSTÖRBAR : indestructible 591
UNZERSTÖRBARKEIT (*f*) : indestructibility 592
UNZERSTÖRT : undestroyed 593
UNZERTEILT : undivided 594
UNZERTRENNBAR : inseparable 595
UNZERTRENNLICH : inseparable 596
UNZIEHBAR : not ductile 597
UNZIEMLICH : unbecoming ; unseemly 598
UNZÜCHTIG : obscene ; unchaste ; lascivious ; prostitute ; lewd 599
UNZUFRIEDEN : dissatisfied ; discontented ; malcontented 600
UNZUGÄNGLICH : inaccessible 601
UNZULÄNGLICH : inadequate ; insufficient 602
UNZULÄSSIG : inadmissible 603
UNZUREICHEND : insufficient 604
UNZUSAMMENDRÜCKBAR : incompressible 605
UNZUSAMMENHÄNGEND : unconnected 606
UNZUTRÄGLICH : unhealthy ; disadvantageous 607
UNZUTREFFEND : incorrect-(ly) ; wrong-(ly) ; inapplicable 608
UNZUVERLÄSSIG : unreliable ; uncertain ; untrustworthy ; precarious 609
UNZWECKMÄSSIG : inexpedient ; unsuitable 610
UNZWEIDEUTIG : unambiguous ; unequivocal 611
UNZWEIFELHAFT : undoubted ; indubitable ; indubitably 612
UPASBAUM (*m*) : upas tree (*Antiaris toxicaria*) 613
UPASGIFT (*n*) : upas poison 614
ÜPPIG : luxuriant ; rank ; rich ; luxurious ; wanton 615
UR (*m*) : aurochs (*Bos primigenus*) 616
URAHN (*m*) : ancestor ; great-(grandfather), grandmother 617
URALIT (*m*) : uralite (a pseudomorphous variety of amphibole) (see **AMPHIBOL**) ; Sg. 3.15 [618
URALT : ancient ; very old 619
URÄLTERN (*pl*) : great-grandparents ; ancestors 620
URALTERS : from time immemorial 621
URAMIE (*f*) : uremia 622
URAN (*n*) : uranium ; (U) 623
URANACETAT (*n*) : uranium acetate (see **URAN, ESSIGSAURES-**) 624
URANAT (*n*) : uranate 625
URANCARBID (*n*) : uranium carbide 626
URANCHLORID (*n*) : uranium chloride 627
URANDIOXYD (*n*) : uranium dioxide ; UO_2 (see **URANOXYDUL**) 628
URANERZ (*n*) : uranium ore 629
URAN (*n*), **ESSIGSAURES-** : uranium acetate, uranyl acetate ; $UO_2(C_2H_3O_2)_2.2H_2O$ 630
URANGELB (*n*) : uranium yellow (see **URANOXYD, GELBES-**) 631
URANGLAS (*n*) : uranium glass 632
URANGLIMMER (*m*) : uranium mica ; uranite (see **URANIT**) 633
URANHALTIG : uraniferous ; containing uranium 634
URANHEXAFLUORID (*n*) : uranium hexafluoride ; UF_6 ; Sg. 4.68 ; Mp. 69.5°C. ; Bp. 56°C. 635
URANIABLAU (*n*) : urania blue 636
URANIN (*n*) : uranine, uranine yellow ; $Na_2C_{20}H_{10}O_5$ 637
URANINIT (*m*) : uraninite, pitch-blende (see **PECHBLENDE** and **URANPECHERZ**) ; Sg. 6.4-9.7 638
URANINITRAT (*n*) : uranic (uranyl) nitrate ; $UO_2(NO_3)_2+6H_2O$; Sg. 2.8 ; Mp. 60°C. ; Bp. 118°C. 639
URANIOXYD (*n*) : uranic oxide 640
URANIT (*m*) : uranite ; $(U_2O_2)Cu.2PO_4.8H_2O$; Sg. 3.5 641
URANIVERBINDUNG (*f*) : uranic (uranyl) compound 642
URANKALKCARBONAT (*n*) : uranothallite (see **URANOTHALLIT**) 643
URANKALKKARBONAT (*n*) : uranothallite (see **URANOTHALLIT**) 644
URANKOPIERVERFAHREN (*n*) : uranium copying process, uranium printing process (Photo.) 645
URANNATRIUMACETAT (*n*) : uranium-sodium acetate ; $UO_2(C_2H_3O_2)_2.NaC_2H_3O_2$ 646
URAN-NATRIUMCHLORID (*n*) : uranium-sodium chloride ; $UCl_4.2NaCl$; Mp. 390°C. 647
URANNITRAT (*n*) : uranium nitrate (see **URANI-NITRAT**) 648
URANOCHALCIT (*m*): uranochalcite ; U_3O_4, $(Ca,Cu,Fe)O,SO_3,H_2O$; Sg. 3.2 649
URANOCIRIT (*m*) : uranocirite ; $Ba(UO_2)_2(PO_4)_2.8H_2O$; Sg. 3.5 650
URANOCKER (*m*) : uranium ochre ; $(U_2O_3)3SO_3,14H_2O$ 651
URANOHYDROXYD (*n*) : uranous hydroxide 652
URANOLITH (*m*) : meteorite 653
URANOOXYD (*n*) : uranium (uranous) oxide ; UO_2 654
URANOPHAN (*m*) : uranophane ; $CaU_2Si_2O_{11}.6H_2O$; Sg. 3.88 655
URANOPILIT (*m*) : uranopilite ; $8UO_3.CaO.2SO_3.25H_2O$; Sg. 3.85 656
URANOREIHE (*f*) : uranous series 657
URANOSALZ (*n*) : uranous salt 658
URANOSPHÄRIT (*m*) : uranosphaerite ; $Bi_2O_3.2UO_3.3H_2O$; Sg. 6.4 659
URANOSPINIT (*m*) : uranospinite ; $Ca(UO_2.AsO_4)_2$; Sg. 3.45 660

URANOSULFAT (*n*): uranium (uranous) sulphate; $U(SO_4)_2.4H_2O$ 661
URANOTHALLIT (*m*): uranothallite; $U(CO_3)_2.2CaCO_3.10H_2O$ (see also LIEBIGIT) 662
URANOTHORIT (*m*): uranothorite; $USiO_4.ThSiO_4$; Sg. 4.13 663
URANOTIL (*m*): uranotile; $CaU_2Si_2O_{11}.6H_2O$; Sg. 3.88 664
URANOURANAT (*n*): uranous uranate (see URAN-OXYDOXYDUL) 665
URANOVERBINDUNG (*f*): uranous compound 666
URANOXYD (*n*): uranium (uranic) oxide; UO_3, Sg. 5.2; U_3O_8, Sg. 7.31 667
URANOXYD (*n*), GELBES-: yellow uranium oxide, sodium uranate, uranium yellow; Na_2UO_4 668
URANOXYDNITRAT (*n*): uranic nitrate, $UO_2.(NO_3)_2+6H_2O$; Sg. 2.807 669
URANOXYDOXYDUL (*n*): uranoso-uranic oxide; uranous uranate; U_3O_8; Sg. 7.31 670
URANOXYD (*n*), SALPETERSAURES-: uranic nitrate (see URANOXYDNITRAT) 671
URANOXYDSULFAT (*n*): uranic sulphate; $UO_2.SO_4+3H_2O$; Sg. 3.28 672
URANOXYDUL (*n*): uranium dioxide; uranium (uranous) oxide; (UO_2); Sg. 10.15-10.95; Mp. 2176°C. 673
URANOXYDULOXYD (*n*): uranoso-uranic oxide; uranous uranate; U_3O_8; Sg. 7.31 674
URANOXYDUL, SALPETERSAURES- (*n*): uranous nitrate 675
URANOXYDULSALZ (*n*): uranous salt 676
URANPECHBLENDE (*f*): uraninite; pitchblende; U_3O_4(Pb.Fe.Ag.Mg.Bi.Ca.SiO_2 and other constituents); Sg. 8.9 677
URANPECHERZ (*n*): uraninite; pitchblende (see URANPECHBLENDE, also PECHBLENDE and URANINIT) 678
URANPENTACHLORID (*n*): uranium pentachloride; UCl_5 679
URAN (*n*), PHOSPHORSAURES-: uranium phosphate (see URANYLPHOSPHAT) 680
URANPRÄPARAT (*n*): uranium preparation 681
URAN (*n*), SALPETERSAURES-: uranium nitrate 682
URANSAUER: uranate of, combined with uranic acid 683
URANSÄURE (*f*): uranic acid 684
URANSÄUREANHYDRID (*n*): uranic anhydride; (UO_3) 685
URANSTRAHLEN (*m.pl*): uranium rays 686
URANSUPEROXYD (*n*): uranium peroxide; $UO_4.2H_2O$ 687
URANTETRABROMID (*n*): uranium tetrabromide; UBr_4; Sg. 4.838 688
URANTETRACHLORID (*n*): uranium tetrachloride, UCl_4 689
URANTETRAJODID (*n*): uranium tetraiodide; UI_4; Sg. 5.6; Mp. 500°C. 690
URANTRICHLORID (*n*): uranium trichloride; UCl_3 691
URANTRIOXYD (*n*): uranium trioxide; UO_3; Sg. 5.2 692
URANVERBINDUNG (*f*): uranium compound 693
URANVERSTÄRKER (*m*): uranium intensifier (Phot.) 694
URANVITRIOL (*m*): uranochalcite, uranium vitriol, uranium vitriolic ore (a hydrous uranium sulphate); U_3O_4(Fe,Ca,Cu)O, SO_3,H_2O; Sg. 3.19 (see URANOCHALCIT) 695
URANYLACETAT (*n*): uranium (uranyl) acetate (see URAN, ESSIGSAURES-) 696
URANYLAZETAT (*n*): uranyl acetate; $UO_2(C_2H_3O_2)_2.2H_2O$ 697
URANYLNITRAT (*n*): uranium (uranyl) nitrate; $UO_2(NO_3)_2+6H_2O$; Mp. 59.5°C. 698
URANYLPHOSPHAT (*n*): uranium (uranyl) phosphate; $UO_2(HPO_4)_2+4H_2O$ 699
URANYLSULFAT (*n*): uranyl sulphate; $UO_2SO_4.3H_2O$ 700
URAO (*m*): urao (natural hydrous sodium carbonate) (see TRONA) 701
URATSTEIN (*m*): gravel, bladder stone 702
URBAR: tillable; arable 703
URBARMACHUNG (*f*): cultivation 704
URBESTANDTEIL (*m*): prime (original; fundamental) constituent 705
URBEWOHNER (*m*): original inhabitant; primeval inhabitant; aborigine 706
URBILD (*n*): original; prototype 707
URETHAN (*n*): urethane 708
URFARBE (*f*): primary colour 709
URFELS (*m*): primitive rock 710
URFORM (*f*): original form; prototype 711
URGEBIRGE (*n*): primitive (rock) mountains 712
URGESTEIN (*n*): primitive rock 713
URGEWICHT (*n*): standard (original) weight 714
URGRANIT (*m*): primitive granite 715
URGRUND (*m*): original cause; foundation; basis 716
URGRÜNSTEIN (*m*): gabbro (see GABBRO) 717
URHEBER (*m*): founder; originator; author 718
URICEDIN (*n*): uricedine 719
URIFORM (*n*): uriform 720
URIN (*m*): urine 721
URINABSATZ (*m*): urinary sediment 722
URINGLAS (*n*): urine test-tube 723
URINOL (*n*): urinol 724
URINSTEIN (*m*): urinary calculus 725
URKALK (*m*): primary (primitive) limestone 726
URKOMISCH: extremely comical 727
URKRAFT (*f*): primitive power; original force 728
URKRÄFTIG: hearty; very powerful 729
URKUNDE (*f*): document; diploma; record; voucher; deed 730
URKUNDEN: to attest; testify; prove 731
URKUNDLICH: documentary; authentic 732
URLAND (*n*): primordial land 733
URLAUB (*m*): leave of absence; pass; furlough (Mil.) 734
URLÄUTER (*m*): sod oil; degras (Leather) (see DEGRAS) 735
URMASS (*n*): standard measure 736
URMETALLE (*pl*): component metals of an alloy 737
URNE (*f*): urn 738
UROCOL (*n*): urocol 739
UROGOSAN (*n*): urogosan 740
U-RÖHRE (*f*): U-tube 741
UROL (*n*): urol 742
UROLIN (*n*): uroline 743
UROPHERIN (*n*): uropherine 744
UROSIN (*n*): urosin 745
UROTROPIN (*n*): urotropin (see HEXAMETHYL-ENTETRAMIN) 746
URPLÖTZLICH: very sudden-(ly) 747
URPREIS (*m*): original price; cost price; manufacturer's price 748
URQUELL (*m*): origin; fountain-head; source 749
URSACHE (*f*): cause; reason; reasoning 750
URSÄCHLICH: causative; causal 751
URSÄURE (*f*): —— uric acid 752

URSCHRIFT (*f*): original document 753
URSOL (*n*): ursol 754
URSPRACHE (*f*): primitive language 755
URSPRUNG (*m*): origin; source 756
URSPRÜNGLICH: original-(ly); first; primitive; primary; primordial; primarily 757
URSTOFF (*m*): primary (primitive) matter; element; basic substance 758
URSTOFFLEHRE (*f*): atomic theory (theory of the primary matter forming the elements) 759
URSTOFFLICH: elementary 760
URSUBSTANZ (*f*): primary (primitive) matter; element; basic substance 761
URTEER (*m*): crude tar (tar gained at 300-500°C. without being exposed to higher temperatures) 762
URTEIL (*n*): judgment; verdict; opinion; decision; proposition (Logic); sentence 763
URTEILEN: to judge; decide; form an opinion; pass sentence 764
URTEILSKRAFT (*f*): (power of) judgment 765
URTEILSSPRUCH (*m*): verdict; judgment; sentence; decision 766
URTEILSVERMÖGEN (*n*): (power of) judgment 767
URTIER (*n*): protozoon 768
URTITER (*m*): original titer; titrimetric standard 769
URTITERSUBSTANZ (*f*): standard titrimetric substance 770
URTYPUS (*m*): original type 771
URURÄLTERN (*pl*): great-great-grandparents; progenitors, ancestors 772
URWELLEN: to double (Metal) 773
URWELT (*f*): primeval (primitive) world 774
URWESEN (*n*): primeval (primitive; original) being 775
URWÜCHSIG: native; original; blunt; rough 776
URZELLE (*f*): ovum; primitive cell 777
URZEUGUNG (*f*): abiogenesis 778
URZUSTAND (*m*): primitive (original) condition 779
USNINSÄURE (*f*): usnic acid 780
USURPIEREN: to usurp 781
UTENSILIEN (*n.pl*): utensils; implements 782
UTROVAL (*n*): utroval 783
UVINSÄURE (*f*): uvic acid; para-tartaric acid (see TRAUBENSÄURE) 784
UVIOLÖL (*n*): uviol oil 785
UVITINSÄURE (*f*): uvitic acid; $C_9H_8O_4$ 786
UWAROWIT (*m*): uvarovite; $Ca_3Cr_2(SiO_4)_3$; Sg. 3.55 787

V

VACCINIEREN: to vaccinate 788
VACUA (*n.pl*): vacua (plural of vacuum) 789
VACUUM (*n*): vacuum (see VAKUUM) 790
VAGABUNDIERENDER STROM (*m*): galvanic current (that current set up by the contact of two metals) 791
VAGINAKAPSELN (*f.pl*): vaginal capsules (*Globuli*) (Pharm.) 792
VAGINALKUGEL (*f*): (see VAGINALKAPSELN) 793
VAGUS (*m*): vagus, pneumogastric nerve (Med.) 794
VAKANZ (*f*): holiday; vacation; vacancy 795
VAKUUM (*n*): vacuum; negative pressure 796
VAKUUMAPPARAT (*m*): vacuum apparatus 797
VAKUUMBLITZABLEITER (*m*): vacuum (lightning) protector 798
VAKUUMBREMSE (*f*): vacuum brake 799
VAKUUMEINDAMPFAPPARAT (*m*): vacuum (evaporator) evaporating apparatus 800
VAKUUMGÄRUNG (*f*): vacuum fermentation (Brew.) 801
VAKUUMGLÜHLAMPE (*f*): vacuum incandescent lamp 802
VAKUUMKNETMASCHINE (*f*): vacuum kneading machine 803
VAKUUMMANOMETER (*n*): vacuum gauge, vacuum manometer 804
VAKUUMMESSER (*m*): vacuum gauge 805
VAKUUMMETER (*n*): vacuum (meter) gauge; condenser gauge 806
VAKUUMMISCHMASCHINE (*f*): vacuum mixing machine 807
VAKUUMPFANNE (*f*): vacuum pan 808
VAKUUMPUMPE (*f*): vacuum pump 809
VAKUUMREINIGUNGSAPPARAT (*m*): vacuum (cleaner) cleaning apparatus 810
VAKUUMROHR (*n*): vacuum tube 811
VAKUUMRÖHRE (*f*): vacuum tube 812
VAKUUMTROCKENANLAGE (*f*): vacuum drying plant 813
VAKUUMTROCKENAPPARAT (*m*): vacuum (drying apparatus) drier 814
VAKUUMTROCKENSCHRANK (*m*): vacuum (drying closet; drying chamber) drier 815
VAKUUMTROCKENTROMMEL (*f*): vacuum (drying drum) drier 816
VAKUUMVERDAMPFAPPARAT (*m*): vacuum evaporator 817
VAKZINE (*f*): vaccine 818
VAKZINIEREN: to vaccinate 819
VALENTINIT (*m*): valentinite; white antimony; Sb_2O_3; Sg. 5.7 (see ANTIMONTRIOXYD) 820
VALENZ (*f*): valency; valence 821
VALENZLEHRE (*f*): doctrine of valence 822
VALERALDEHYD (*n*): valeraldehyde; $C_5H_{10}O$; Sg. 0.8185; Bp. 103.4°C. 823
VALERALDEHYD (*n*), ISO-: isovaleraldehyde; $C_5H_{10}O$; Sg. 0.8041; Bp. 92°C. 824
VALERALDOXIM (*n*): valeraldoxime

$$C_4H_9C\begin{cases}H\\NOH\end{cases}$$

825
VALERIANAT (*n*): valerianate, valerate 826
VALERIANÖL (*n*): valerian oil (see BALDRIAN) 827
VALERIANSÄURE (*f*): valeric (valerianic) acid (*Acidum valerianicum*); $C_5H_{10}O_2$; Sg. 0.9415; Mp. – 58.5°C.; Bp. 186.4°C. [828
VALERIANSÄUREAMYLÄTHER (*m*): amyl valerate; apple essence; apple oil; amyl valerianate; $C_4H_9CO_2C_5H_{11}$; Sg. 0.88; Bp. 203.7°C. [829
VALERIANSÄURE-ÄTHYLESTER (*m*): ethyl (valerat) valerianate; $C_7H_{14}O_2$; Sg. 0.8765; Bp. 144.6°C. 830
VALERIANSÄURE (*f*), ISO-: iso-valerianic acid, iso-valeric acid; $(CH_3)_2CHCH_2CO_2H$; Mp. – 36°C.; Bp. 173.7°C. 831
VALERIANSÄUREMENTHOLESTER (*m*): menthol valerate (see VALIDOL) 832
VALERIANSÄURESALZ (*n*): valerate; valeric acid salt 833

VALERYDIN (*n*): valerydin 834
VALERYLCHLORID (*n*): valeryl chloride 835
VALET (*n*): farewell 836
VALIDOL (*n*): (*Mentholum valerianicum*): validol, menthol valerate 837
VALIN (*n*): (see AMINOISOVALERIANSÄURE) 838
VALISAN (*n*): Brovalol (an American term) 839
VALKASA (*n*): valkasa 840
VALOFIN (*n*): valofin 841
VALONEA (*f*): valonea 842
VALONEAEXTRAKT (*m*): valonea extract (from *Quercus ægilops*) 843
VALUTA (*f*): value; exchange 844
VALUTA-STEUER (*f*): value (*ad valorem*) tax 845
VALYL (*n*): valyl 846
VANAD (*n*): vanadium; (V): (see also VANADIUM) 847
VANADAT (*n*): vanadate 848
VANADIN (*n*): vanadium; (V)(see VANADIUM) 849
VANADINBLEIERZ (*n*): vanadinite (see VANADINIT) (natural lead chlorovanadate); $Pb_5Cl(VO_4)_3$; Sg. 7.0 850
VANADINBLEISPAT (*m*): vanadinite (see VANADINIT) 851
VANADINCHLORID (*n*): vanadium chloride, divanadyl tetrachloride, hypovanadic hydrochloride; $(VO)_2Cl_4.5H_2O$; Sg. 3.23 852
VANADINIT (*m*): vanadinite; $Pb_5ClV_3O_{12}$ or $3Pb_3V_2O_8.PbCl_2$; Sg. 7.0 853
VANADINOXYTRICHLORID (*n*): vanadium oxy-(tri)-chloride; $VOCl_3$ 854
VANADINPENTAFLUORID (*n*): vanadium pentafluoride; VF_5 855
VANADINPENTOXYD (*n*): vanadium pentoxide; V_2O_5; Sg. 3.49; Vanadic acid (see VANADINSÄURE) 856
VANADINSALZ (*n*): vanadium salt (either -ic or -ous) 857
VANADINSAUER: vanadate of; combined with vanadic acid 858
VANADINSÄURE (*f*): vanadic acid; HVO_3 (see also VANADINSÄUREANHYDRID) 859
VANADINSÄUREANHYDRID (*n*): vanadic anhydride; vanadium pentoxide; V_2O_5; Sg. 3.357; Mp. 658°C. 860
VANADINSTAHL (*m*): vanadium steel 861
VANADINSTICKSTOFF (*m*): vanadium nitride 862
VANADINVERBINDUNG (*f*): vanadium compound 863
VANADISALZ (*n*): vanadic salt 864
VANADIUMCARBID (*n*): vanadium carbide; VC; Sg. 5.36 865
VANADIUMDICHLORID (*n*): vanadium (dichloride) bichloride; VCl_2; Sg. 3.28 866
VANADIUMDIOXYD (*n*): vanadium dioxide; V_2O_2; Sg. 3.64 867
VANADIUMOXYBROMID: vanadium oxybromide; VOBr; Sg. 4.0 868
VANADIUMOXYD (*n*): vanadium oxide 869
VANADIUMOXYDICHLORID (*n*): vanadium oxydichloride; $VOCl_2$; Sg. 2.88 870
VANADIUMOXYDIFLUORID (*n*): vanadium oxydifluoride; VOF_2; Sg. 3.396 871
VANADIUMOXYTRICHLORID (*n*): vanadium oxytrichloride; $VOCl_3$; Sg. 1.836-1.854 [872
VANADIUMOXYTRIFLUORID (*n*): vanadium oxytrifluoride; VOF_3; Sg. 2.459 873
VANADIUMPENTAFLUORID (*n*): vanadium pentafluoride; VF_5; Sg. 2.177 874
VANADIUMPENTASULFID (*n*): vanadium pentasulphide; V_2S_5; Sg. 3.0 875
VANADIUMPENTOXYD (*n*): vanadium pentoxide; V_2O_5; Sg. 3.49-3.357; Mp. 658°C. 876
VANADIUMSÄURE (*f*): vanadic acid, vanadium pentoxide; V_2O_5 (see VANADIUMPENTOXYD) 877
VANADIUMSESQUIOXYD (*n*): vanadium sesquioxide; V_2O_3; Sg. 4.72-4.87 878
VANADIUMSILICID (*n*): vanadium silicide; VSi_2, Sg. 4.42; V_2Si; Sg. 5.48 879
VANADIUMSULFAT (*n*): vanadium sulphate, vanadyl sulphate; $VO_2SO_3.2H_2O$ or $V_2O_2(SO_4)_2.4H_2O$ 880
VANADIUMSULFID (*n*): vanadium sulphide (see VANADIUMPENTASULFID) 881
VANADIUMTETRACHLORID (*n*): vanadium tetrachloride; VCl_4; Sg. 1.836-1.865 882
VANADIUMTETRAFLUORID (*n*): vanadium tetrafluoride; VF_4; Sg. 2.975 883
VANADIUMTETROXYD (*n*): vanadium tetroxide; V_2O_4 884
VANADIUMTRICHLORID (*n*): vanadium trichloride; VCl_3; Sg. 3.0 885
VANADIUMTRIFLUORID (*n*): vanadium trifluoride; VF_3; Sg. 3.363 886
VANADIUMTRIOXYD (*n*): vanadium trioxide; V_2O_3; Sg. 4.87 887
VANADOFORM (*n*): vanadoform 888
VANADOSALZ (*n*): vanadous salt 889
VANADYLSALZ (*n*): vanadyl salt 890
VANDYCKBRAUN (VAN-DYCK-BRAUN) (*n*): vandyke brown; Van Dyke brown 891
VANILLASÄURE (*f*): vanillic acid (from *Vanilla planifolia*) 892
VANILLE (*f*): vanilla 893
VANILLENBOHNE (*f*): vanilla bean (fruit of *Vanilla planifolia*) 894
VANILLENESSENZ (*f*): vanilla essence 895
VANILLENKAMPFER (*m*): vanillin (see VANILLIN) 896
VANILLENPFLANZE (*f*): vanilla plant (*Liatris odoratissima* and *Vanilla planifolia*) 897
VANILLENZUCKER (*m*): vanilla sugar 898
VANILLIN (*n*): vanillin; $C_6H_3(CHO)(OCH_3)OH$; Mp. 80.5°C.; Bp. 285°C.; vanillic aldehyde; methylprotocatechuic aldehyde 899
VANILLINSÄURE (*f*): vanillic acid; $C_8H_8O_4$ [900
VANILLINZUCKER (*m*): vanillin sugar 901
VANILLON (*n*): vanillon 902
VANTHOFFIT (*m*): vanthoffite; $3Na_2SO_4.MgSO_4$; Sg. 2.7 903
VAPOCAIN (*n*): vapocain 904
VAREC (*m*): kelp (the ashes of seaweed) 905
VARIABEL: variable 906
VARIABILITÄT (*f*): variability 907
VARIABLE (*f*): variable 908
VARIANTE (*f*): alternative; variant 909
VARIATION (*f*): variation 910
VARIATIONSKOMPASS (*m*): variation compass 911
VARIETÄT (*f*): variety 912
VARIIEREN: to vary 913
VASE (*f*): vase 914
VASELIN (*n*): vaseline; mineral fat; petrolatum; petroleum jelly; Sg. 0.83; Mp. 46-5°C. (see also VASELINE) 915
VASELINANLAGE (*f*): vaseline plant 916
VASELINE (*f*): vaseline, petrolatum; (Sg. 0.82-0.88; Mp. 33-48°C.); *Adeps mineralis, Adeps petrolei* (see also VASELIN) 917
VASELIN (*n*), KUNST-: Vaseline, petroleum jelly (*Unguentum paraffini*) [solution of paraffin and ceresin in VASELINÖL (*Paraffinum liquidum*)] 918

VASELINÖL (*n*): vaseline oil 919
VASELINSEIFE (*f*): vaseline soap 920
VASENOL (*n*): vasenol 921
VASOGEN (*n*): vasogene 922
VASOGENPRÄPARAT (*n*): vasogene preparation 923
VATER (*m*): father 924
VATERIATALG (*m*): vateria tallow (see MALABARTALG) 925
VATERLAND (*n*): native (country) land; fatherland; Germany 926
VATERLÄNDISCH: national; native 927
VATERLANDSFREUND (*m*): patriot 928
VATERLANDSLIEBE (*f*): patriotism 929
VÄTERLICH: paternal; fatherly 930
VATERLOS: fatherless; orphaned 931
VATERMORD (*m*): parricide 932
VATERSTADT (*f*): native town, birthplace 933
VATERTEIL (*m*): patrimony 934
VATERUNSER (*n*): The "Our Father"; The Lord's Prayer 935
VAUCANSONSCHE KETTE (*f*): pitch chain 936
VAUQUELINIT (*m*): vauquelinite; 3(PbCu)O, Cr_2O_3; Sg. 5.65 937
VEGETABILIEN (*f.pl*): herbal (vegetable) drugs 938
VEGETABILISCH: vegetable; herbal 939
VEGETATION (*f*): vegetation 940
VEILCHEN (*n*): violet (*Viola*) 941
VEILCHENBLAU: violet 942
VEILCHENBLÜTEN (*f.pl*): violet flowers (*Flores violæ odoratæ*) 943
VEILCHENBLÜTENESSENZ (*f*): extract of violet 944
VEILCHENBLÜTENÖL (*n*): violet flower oil 945
VEILCHEN, DREIFARBIGES- (*n*): pansy (*Viola tricolour*) 946
VEILCHENHOLZ (*n*): violet wood 947
VEILCHENKETON (*n*): violet ketone 948
VEILCHENÖL, SYNTHETISCHES-: (*n*) synthetic (artificial) violet oil 949
VEILCHENSTEIN (*m*): iolite; water sapphire; cordierite (see CORDIERIT) 950
VEILCHENSTRAUSS (*m*): bouquet (nosegay; bunch) of violets 951
VEILCHENWURZ (*f*): iris (orris) root (*Rhizoma iridis* of *Iris germanica* or *Iris pallida*) (*Rhizoma iridis florentinæ* of *Iris florentina*) 952
VEILCHENWURZEL (*f*): (see VEILCHENWURZ) 953
VEILCHENWURZELÖL (*n*): iris (orris) oil (see IRISÖL) 954
VEITSBOHNE (*f*): (see VITSBOHNE) 955
VEITSTANZ (*m*): minor chorea; St. Vitus' Dance (Med.) 956
VEKTOR (*m*): vector 957
VEKTORANALYSIS (*f*): vector analysis 958
VEKTORIELL: vectorial 959
VEKTOR (*m*), RADIUS-: radius vector 960
VELIN (*n*): vellum 961
VELINFORM (*f*): wove mould (Paper) 962
VELINPAPIER (*n*): wove (vellum) paper 963
VELOCIPED (*m*): velocipede, cycle 964
VELOCIPEDBAU (*m*): cycle construction 965
VELOUTIEREN: to flock (Paper) 966
VENE (*f*): vein 967
VENENBLUT (*n*): venous blood 968
VENENHÄUTCHEN (*n*): choroid membrane 969
VENENSTEIN (*m*): phlebolite; vein stone 970
VENERIE (*f*): veneral disease 971
VENERISCH: venereal 972
VENETIANER KREIDE (*f*): (see SEIFENSTEIN) 973
VENETIANISCHROT (*n*): Venetian rod (see VENEZIANISCHROT) 974
VENEZIANERKREIDE (*f*): (see VENETIANER KREIDE) 975
VENEZIANERWEISS (*n*): Venetian white 976
VENEZIANISCHROT (*n*): Venetian red (pure iron oxide) 977
VENÖS: venous 978
VENTIL (*n*): valve 979
VENTIL, ABSPERR- (*n*): stop valve 980
VENTILATIONSANLAGE (*f*): ventilating plant 981
VENTILATIONSMASCHINE (*f*): blower, fan, blast engine, blowing engine, ventilator; fan engine 982
VENTILATOR (*m*): ventilator; blower; fan 983
VENTILATORSCHAUFEL (*f*): fan blade 984
VENTILDECKEL (*m*): valve cover 985
VENTIL (*n*), DOPPELSITZ-: double-seat valve 986
VENTIL, DROSSEL- (*n*): throttle valve 987
VENITL (*n*), ENTLASTETES-: balanced valve, equilibrium valve 988
VENTIL (*n*), ETAGEN-: multiple seat valve, stage valve, step valve 989
VENTILFÄNGER (*m*): catcher 990
VENTIL (*n*), FEDERBELASTETES-: spring loaded valve 991
VENTILGEGENGEWICHT (*n*): valve balance weight 992
VENTILGEHÄUSE (*n*): valve (chamber) body, or chest 993
VENTIL (*n*), GESTEUERTES-: geared valve 994
VENTILGEWICHT (*n*): valve weight 995
VENTIL (*n*), GEWICHTSBELASTETES-: weight loaded valve, dead-weight valve 996
VENTIL (*n*), GLOCKEN - bell-shaped valve 997
VENTILHEBEL (*m*): valve lever 998
VENTILHEBER (*m*): valve lifter 999
VENTIL (*n*) HILFS-: auxiliary valve 000
VENTILHUB (*m*): lift of a valve 001
VENTILIEREN: to ventilate 002
VENTILKASTEN (*m*): valve chest 003
VENTILKEGEL (*m*): valve cone 004
VENTIL (*n*), KEGEL-: conical valve 005
VENTILKLAPPE (*f*): valve flap, or clack 006
VENTIL (*n*), KLAPPEN-: clack valve 007
VENTIL (*n*), KLAPPERNDES-: chattering valve 008
VENTILKOLBEN (*m*): valve piston 009
VENTILKUGEL (*f*): valve ball 010
VENTIL (*n*), KUGEL-: spherical valve, ball-valve 011
VENTILLOS: valveless 012
VENTILMASCHINE (*f*): poppet-valve engine 013
VENTILNOCKEN (*m*): valve lifter 014
VENTILORGAN (*n*): valve part 015
VENTILRÖHRE (*f*): valve (Wireless Tel.) 016
VENTILSCHEIBE (*f*): valve disc 017
VENTIL (*n*), SICHERHEITS-: safety valve 018
VENTILSITZ (*m*): valve seat 019
VENTILSITZFLÄCHE (*f*): valve seating 020
VENTILSITZSCHLEIFEN (*n*): valve reseating 021
VENTIL (*n*), SPEISE-: feed valve, feed check valve 022
VENTILSPINDEL (*f*): valve spindle 023
VENTILSTANGE (*f*): valve spindle, valve rod 024
VENTILSTELLSCHRAUBE (*f*): valve adjusting screw 025
VENTILSTEUERUNG (*f*): valve gear; poppet-valve distributing gear (Steam Engines) 026
VENTIL (*n*), STUFEN-: multiple-seat valve, stage valve, step valve 027
VENTILSTUTZEN (*m*): valve branch 028
VENTILTELLER (*m*): valve disc 029
VENTIL (*n*), UNGESTEUERTES-: self-acting valve, automatic (ungeared) valve 030
VENTIL (*n*), ZYLINDRISCHES-: cylindrical valve 031

VERABFOLGEN: to deliver; surrender; hand over; remit; send 032
VERABREDEN: to appoint; agree upon; concert; bespeak 033
VERABREICHEN: to dispense; deliver 034
VERABSÄUMEN: to neglect 035
VERABSCHEUEN: to detest; abhor; abominate 036
VERABSCHIEDEN: to take leave; dismiss; discharge 037
VERACCORDIEREN: to come to an agreement on 038
VERACHTEN: to scorn; despise; disdain; contemn; disparage 039
VERÄCHTLICH: contemptuous; contemptible; despicable 040
VERACHTUNG (*f*): scorn; contempt 041
VERALLGEMEINERN: to generalize 042
VERALLGEMEINERUNG (*f*): generalization 043
VERALTEN: to grow (old) obsolete; become (antiquated) out of date 044
VERALTET: antiquated; obsolete; ancient; out-of-date 045
VERÄNDERLICH: fluctuating; variable; temporary; changeable; unsteady; fickle; alterable 046
VERÄNDERLICHE (*f*): variable 047
VERÄNDERLICHKEIT (*f*): variability; unsteadiness; changeability; fluctuation; fickleness 048
VERÄNDERN: to vary; change; alter; concentrate (Ores); modify; transmogrify 049
VERÄNDERUNG (*f*): variation; change; alteration; modification 050
VERÄNDERUNGSPRODUKT (*n*): product of change 051
VERANKERN: to tie, grapple, anchor, fasten, stay 052
VERANKERUNG (*f*): fastening, anchoring, staying, grappling, tie 053
VERANLASSEN: to cause; give rise to; occasion; induce; engage 054
VERANLASSUNG (*f*): cause; occasion; inducement; motive 055
VERANSCHAULICHEN: to illustrate, show 056
VERANSCHLAGEN: to estimate; value; rate; tax 057
VERANSTALTEN: to manage; arrange; prepare 058
VERANTWORTEN: to justify; answer for; account for; be responsible for 059
VERANTWORTLICH: responsible; answerable; accountable 060
VERARBEITEN: to work-(up); treat; manufacture; throttle; reduce-(pressure), (of steam); to deal with 061
VERARBEITUNGSFÄHIGKEIT (*f*): workability 062
VERARGEN: to find fault with 063
VERARMEN: to become poor; decrease in value; impoverish 064
VERASCHEN: to ash; incinerate 065
VERASCHUNG (*f*): ashing; incineration 066
VERÄSTELUNG (*f*): ramification 067
VERÄSTUNG (*f*): ramification 068
VERATRIN (*n*): veratrine; $C_{37}H_{53}NO_{11}$; Mp. 180°C. 069
VERATRINSULFAT (*n*): veratrine sulphate; $(C_{32}H_{49}NO_9)_2.H_2SO_4$ 070
VERATROL (*n*): veratrol (see **BRENZCATECHINDI-METHYLÄTHER**) 071
VERATRON (*n*): veratron 072
VERATRUMSÄURE (*f*): veratric acid; $C_9H_{10}O_4$ 073
VERAUSGABEN: to lay out; expend; pay; disburse 074
VERÄUSSERLICH: alienable 075
VERÄUSSERN: to alienate 076
VERBAND (*m*): bandage; dressing (Med.); bandaging; fastening; connection; union; binding; association; join; bond (Masonry) 077
VERBANDMATERIAL (*n*): bandages, dressings, any binding material 078
VERBAND (*m*), **SCHIFFS-**: framing (of a ship) 079
VERBANDSFORMEL (*f*): (Formula made up by Internationalen Verband der Dampfkessel-Überwachungs-Vereine, for the determination of the heat value of fuel from its chemical constituents: $h=81C+290(H-\frac{1}{8}O)+25S-6W$, equals absolute heat value in WE per kg. 080
VERBANDSTOFF (*m*): binding material; surgical bandage; antiseptic dressing 081
VERBANDSTOFF (*m*), **ANTISEPTISCHER-**: antiseptic dressing (Med.) 082
VERBANDWATTE (*f*): compressed wadding 083
VERBANNEN: to banish; outlaw; exile; proscribe; interdict 084
VERBAUEN: to build up; brick up; enclose 085
VERBAUERN: to rusticate 086
VERBEISSEN: to stifle; suppress; forbear; brook 087
VERBEIZEN: to overbate (Leather) 088
VERBENAÖL (*n*): verbena oil; East Indian verbena oil; lemongrass oil (see **GRASÖL**) 089
VERBENE (*f*): vervain (*Verbena officinalis*) 090
VERBERGEN: to hide; secrete; conceal 091
VERBESSERN: to improve; better; correct; reform; amend; ameliorate 092
VERBESSERUNG (*f*): improvement; increase; amelioration; amendment; reformation; betterment; reform; correction 093
VERBESSERUNGSMITTEL (*n*): corrective 094
VERBEUGEN: to bow; incline 095
VERBIEGEN: to warp; bend wrongly; strain 096
VERBIETEN: to forbid; prohibit; interdict; inhibit 097
VERBINDEN: to combine; bind-(up); bandage; dress (Med.); link; oblige; connect; join; unite; tie-(up); interconnect 098
VERBINDLICH: binding; obliging; obliged; obligatory; firm (of prices) 099
VERBINDLICHEN DANK (*m*): best (many) thanks; thank you very much 100
VERBINDLICHKEIT (*f*): kindness; obligation; civility; liability 101
VERBINDUNG (*f*): compound; combination; conjunction; union; combining; connection; joining; binding; joint; fastening; bond; alliance; communication; society; blending (of colours, etc.) 102
VERBINDUNG (*f*), **CHEMISCHE-**: chemical combination; chemical compound 103
VERBINDUNG, IN DIESER-: in this connection 104
VERBINDUNGSBOLZEN (*m*): holding (down) bolt 105
VERBINDUNGSFORM (*f*): form of combination 106
VERBINDUNGSGANG (*m*): connection passage or duct; binding (strake) strip (Naut.) 107
VERBINDUNGSGEWICHT (*n*): combining weight 108
VERBINDUNGSGLEICHUNG (*f*): combining equation 109

VERBINDUNGSGLIED (*n*): connecting (link) member 110
VERBINDUNGSKEIL (*m*): key (for keying to a shaft, etc.), connecting key 111
VERBINDUNGSKLAMMER (*f*): brace; clip; clamp; bracket 112
VERBINDUNGSKRAFT (*f*): combining power 113
VERBINDUNGSLINIE (*f*): line of communication; connecting line 114
VERBINDUNGSMITTEL (*n*): binding substance or medium; binder 115
VERBINDUNGSMOLEKÜL (*n*): molecule of compound 116
VERBINDUNGSMUFFE (*f*): junction box 117
VERBINDUNGSPLATTE (*f*): connecting plate 118
VERBINDUNGSPUNKT (*m*): junction; juncture; join 119
VERBINDUNGSROHR (*n*): connecting (pipe) tube 120
VERBINDUNGSRÖHRE (*f*): (see VERBINDUNGSROHR) 121
VERBINDUNGSSCHLAUCH (*m*): connecting (tube) tubing; connecting hose; flexible connection 122
VERBINDUNGSSCHRAUBE (*f*): connecting screw 123
VERBINDUNGSSTANGE (*f*): connecting rod 124
VERBINDUNGSSTELLE (*f*): junction; point of juncture; join 125
VERBINDUNGSSTREBEN (*n*): affinity (Chem.) 126
VERBINDUNGSSTRICH (*m*): hyphen 127
VERBINDUNGSSTÜCK (*n*): connecting piece 128
VERBINDUNGSSTUFE (*f*): stage of combination 129
VERBINDUNGSVERHÄLTNIS (*n*): combining proportion 130
VERBINDUNGSVOLUMEN (*n*): combining volume 131
VERBINDUNGSWÄRME (*f*): heat of combination 132
VERBINDUNGSWERT (*m*): combining value 133
VERBINDUNGSWORT (*n*): conjunction 134
VERBINDUNGSZEICHEN (*n*): hyphen 135
VERBISSEN: crabbed; sullen 136
VERBITTEN: to decline; object to; deprecate 137
VERBITTERN: to embitter 138
VERBLASEN: to blow (Glass and Metals); dilute (Colours) 139
VERBLEIBEN: to remain; continue; abide 140
VERBLEICHEN: to fade; expire; grow pale; die; bleach (Phot.) 141
VERBLEIEN: to lead; line (coat) with lead 142
VERBLEIUNG (*f*): leading; lead lining 143
VERBLENDEN: to dazzle; blind; delude; to face (with material); beguile 144
VERBLENDER (*m*): face brick (Ceramics) 145
VERBLOCKUNG (*f*): locking, fastening, latching 146
VERBLÜFFEN: to confuse; dumbfound; stupefy; startle; surprise 147
VERBLÜHEN: to wither; fade; decay 148
VERBLÜMEN: to disguise 149
VERBLÜMT: figurative 150
VERBLUTEN: to bleed to death; to stop (cease) bleeding 151
VERBOLZEN: to bolt 152
VERBORGEN: to lend; sell on credit; (as an adj.=latent; secret; hidden; concealed) 153
VERBOT (*n*): prohibition; interdiction; inhibition 154
VERBRÄMEN: to trim; border; adorn 155
VERBRANNT: over-pickled (Metal) (when the object loses its metallic lustre due to too long a pickling treatment); burnt-out; over-exposed (Phot.) (see VERBRENNEN) 156
VERBRAUCH (*m*): use; consumption 157
VERBRAUCHEN: to use; consume; employ; utilize; absorb 158
VERBRAUCHER (*m*): consumer; user 159
VERBRAUCHSGEGENSTAND (*m*): commodity; article of consumption 160
VERBRAUCHSTELLE (*f*): point of (consumption) employment; place where used 161
VERBRAUCHT: spent; used up; stale; worn out; consumed 162
VERBRAUSEN: to subside; stop (cease) fermenting 163
VERBRECHEN: to break; commit an offence; perpetrate a crime; (*n*): offence; guilt; crime 164
VERBRECHER (*m*): delinquent; criminal; offender; convict 165
VERBREITEN: to disseminate; diffuse; distribute; spread; divulge; propagate 166
VERBREITERN: to broaden; widen 167
VERBREITERUNG (*f*): broadening; widening 168
VERBREITUNG (*f*): dissemination; diffusion; distribution; spreading; range; extension; circulation (of newspapers) 169
VERBREITUNGSZENTRUM (*n*): centre of distribution 170
VERBRENNBAR: combustible 171
VERBRENNBARKEIT (*f*): combustibility 172
VERBRENNEN: to burn; bake; scald; scorch; tan; to burn (away) out; over-pickle; (see VERBRANNT); fuse up 173
VERBRENNLICH: inflammable; combustible 174
VERBRENNLICHE (*n*): combustible-(matter) 175
VERBRENNLICHE LUFT (*f*): inflammable air (obsolete name for hydrogen) 176
VERBRENNLICHKEIT (*f*): inflammability; combustibility 177
VERBRENNUNG (*f*): combustion; burning; cremation; fusing; baking; scorching 178
VERBRENNUNG (*f*), RAUCHFREIE-: smokeless combustion 179
VERBRENNUNGSANALYSE (*f*): combustion analysis 180
VERBRENNUNGSERZEUGNIS (*n*): product of combustion 181
VERBRENNUNGSGASE (*n.pl*): combustion gases; gases of combustion 182
VERBRENNUNGSGLAS (*n*): combustion glass 183
VERBRENNUNGSINTENSITÄT (*f*): intensity of combustion 184
VERBRENNUNGSKAMMER (*f*): combustion chamber 185
VERBRENNUNGSKONTROLLE (*f*): combustion control 186
VERBRENNUNGSKRAFTMASCHINE (*f*): internal combustion engine 187
VERBRENNUNGSLÖFFEL (*m*): combustion spoon 188
VERBRENNUNGSLUFT (*f*): air for combustion 189
VERBRENNUNGSMASCHINE (*f*): internal combustion engine 190
VERBRENNUNGSMATERIAL (*n*): material for combustion, combustible; fuel 191
VERBRENNUNGSOFEN (*m*): combustion furnace 192
VERBRENNUNGSPRODUKT (*n*): product(s) of combustion 193

VERBRENNUNGSRAUM (*m*) : combustion chamber ; furnace 194
VERBRENNUNGSROHR (*n*) : combustion tube 195
VERBRENNUNGSRÜCKSTAND (*m*) : residue on (ignition) combustion 196
VERBRENNUNGSSCHIFFCHEN (*n*) : combustion boat 197
VERBRENNUNGSTEMPERATUR (*f*) : combustion temperature (temperature obtained on combustion of a body with theoretical amount of air) 198
VERBRENNUNGSTHEORIE (*f*) : combustion theory ; theory of combustion 199
VERBRENNUNGSTURBINE (*f*) : combustion turbine 200
VERBRENNUNGSVERSUCH (*m*) : combustion (experiment) test 201
VERBRENNUNGSVORGANG (*m*) : process of combustion ; combustion process 202
VERBRENNUNGSWÄRME (*f*) : heat of combustion 203
VERBRENNUNGSZONE (*f*) : zone of combustion 204
VERBRIEFEN : to confirm (secure) by writing ; to give in black and white ; charter 205
VERBRINGEN : to waste ; spend ; pass ; consume 206
VERBRÜHEN : to scald 207
VERBUNDDYNAMO (*f*) : compound dynamo 208
VERBUNDEN : bound ; joined ; coupled ; connected ; compound, etc. (see BINDEN) 209
VERBÜNDEN : to associate ; ally 210
VERBUNDMASCHINE (*f*) : compound engine 211
VERBUNDSTÜCK (*n*) : fitting ; coupling ; make-up piece 212
VERBUNDVAKUUMPUMPE (*f*) : compound vacuum pump 213
VERBÜRGEN : to warrant ; guarantee ; act as (security) surety ; go bail 214
VERBUTTERN : to convert into butter ; churn 215
VERDACHT (*m*) : distrust ; suspicion 216
VERDÄCHTIG : suspected ; suspicious 217
VERDAMMEN : to damn ; condemn 218
VERDÄMMEN : to dam 219
VERDAMPFAPPARAT (*m*) : evaporator ; vapourizer ; evaporating apparatus 220
VERDAMPFEN : to evaporate ; vapourize (see ABDAMPFEN and EINDAMPFEN) 221
VERDAMPFER (*m*) : evaporator 222
VERDAMPFLEISTUNG (*f*) : output ; duty ; quantity of steam generated ; evaporation (when given as total) ; rating (when given per unit of heating surface) 223
VERDAMPFOBERFLÄCHE (*f*) : evaporating surface 224
VERDAMPFPFANNE (*f*) : evaporating pan 225
VERDAMPFSCHALE (*f*) : evaporating dish 226
VERDAMPFUNG (*f*) : evaporation ; vapourization 227
VERDAMPFUNGSFÄHIG : capable of evaporation 228
VERDAMPFUNGSPFANNE (*f*) : evaporating pan 229
VERDAMPFUNGSPUNKT (*m*) : vapourization point 230
VERDAMPFUNGSRÜCKSTAND (*m*) : residue on evaporation 231
VERDAMPFUNGSVERMÖGEN (*n*) : evaporative capacity 232
VERDAMPSUNGSVERSUCH (*m*) : evaporation (steam) test 233
VERDAMPFUNGSWÄRME (*f*) : heat of vapourization ; latent heat ; heat of evaporation (see WÄRME, VERDAMPFUNGS-) 234
VERDAMPFUNGSZAHL (*f*) : (see VERDAMPFZIFFER) 235
VERDAMPFUNGSZIFFER (*f*) : factor of evaporation ; evaporative rating (lbs. of water evaporated per lb. of fuel) 236
VERDAMPFZIFFER (*f*) : coefficient of evaporation ; rating (see VERDAMPFUNGSZIFFER) 237
VERDANKEN : to have to thank ; to owe (thanks) ; be indebted to 238
VERDAUEN : to digest ; be digested ; be digestible 239
VERDAULICH : digestible 240
VERDAUUNG (*f*) : digestion 241
VERDAUUNGSDAUER (*f*) : duration of digestion 242
VERDAUUNGSEINGEWEIDE (*n*) : digestive tract 243
VERDAUUNGSFÄHIG : digestible 244
VERDAUUNGSFÄHIGKEIT (*f*) : digestibility 245
VERDAUUNGSFERMENT (*n*) : proteolytic ferment, enzyme 246
VERDAUUNGSFLÜSSIGKEIT (*f*) : digestive fluid ; gastric juice 247
VERDAUUNGSGESCHÄFT (*n*) : digestive process 248
VERDAUUNGSMITTEL (*n*) : digestive-(remedy) 249
VERDAUUNGSOFEN (*m*) : digestive oven 250
VERDAUUNGSROHR (*n*) : alimentary canal 251
VERDAUUNGSSAFT (*m*) : digestive fluid ; gastric juice 252
VERDAUUNGSSCHWÄCHE (*f*) : weak digestion ; indigestion 253
VERDAUUNGSSTÖRUNG (*f*) : disorder of digestion 254
VERDAUUNGSVORGANG (*m*) : digestive process 255
VERDECK (*n*) : deck ; cover 256
VERDECKEN : to cover ; mask ; conceal ; hide 257
VERDENKEN : to find fault with ; blame for ; misconstrue 258
VERDERBEN : to spoil ; ruin ; damage ; decay ; corrupt, (*n*) : destruction ; perdition ; corruption ; decay ; ruin ; damage ; spoilation 259
VERDERBLICH : perishable ; injurious ; destructive ; pernicious ; corruptive 260
VERDERBUNG (*f*) : spoilation, ruination, corruption, damage, decay, rotting 261
VERDEUTLICHEN : to elucidate ; explain 262
VERDEUTSCHEN : to explain ; make clear ; translate into German 263
VERDICHTBAR : condensable 264
VERDICHTEN : to condense ; liquefy (Vapours, etc.) ; compress ; pack ; concentrate ; consolidate ; densify 265
VERDICHTER (*m*) : compressor ; condenser 266
VERDICHTUNG (*f*) : condensation ; liquefaction ; compression ; concentration ; consolidation ; packing ; density 267
VERDICHTUNGSAPPARAT (*m*) : condenser ; compressor ; condensing (compressing) apparatus 268
VERDICHTUNGSGRAD (*m*) : degree of condensation ; compression) concentration 269
VERDICHTUNGSHUB (*m*) : compression stroke 270
VERDICHTUNGSRING (*m*) : packing ring 271
VERDICHTUNGSRÖHRE (*f*) : condensing (liquefying) tube ; liquefier ; condenser 272
VERDICHTUNGSSCHEIBE (*f*) : packing disc 273
VERDICHTUNGSWÄRME (*f*) : heat of (condensation) compression 274
VERDICKEN : to concentrate ; inspissate ; thicken ; condense ; coagulate ; curdle ; become viscous ; curd 275

VERDICKUNG (*f*): thickening; condensation; concentration; inspissation; coagulation; curdling; paste (Calico) 276
VERDICKUNGSMITTEL (*n*): thickener; thickening medium 277
VERDIENEN: to deserve; earn; make; gain; merit 278
VERDIENST (*n*): deserts; merit; profit; earnings; reward; gain 279
VERDINGEN: to hire out; contract for; charter 280
VERDINGUNG (*f*): invitation to tender; auction; contract; agreement; hiring; chartering 281
VERDOLMETSCHEN: to interpret, translate 282
VERDOPPELN: to double; duplicate 283
VERDORBEN: spoiled; ruined; wasted; depraved; corrupted 284
VERDORBENHEIT (*f*): rottenness, decay, corruption, putrefaction 285
VERDORREN: to wither; dry up 286
VERDRAHTEN: to wire 287
VERDRÄNGBAR: displaceable 288
VERDRÄNGEN: to displace; remove; drive (out) away; supplant 289
VERDRÄNGUNG (*f*): displacement; removal 290
VERDREHEN: to distort; twist; wrench; warp; contort; pervert 291
VERDREIFACHEN: to treble; triplicate 292
VERDREIFACHUNG (*f*): trebling; triplication 293
VERDRIESSEN: to vex; annoy; grieve; trouble; fret 294
VERDRIESSLICH: vexed; morose; tiresome; irksome; peevish; fretful 295
VERDROSSEN: loath; slow; indolent; lazy; vexed; annoyed; grieved; averse; reluctant 296
VERDRUCKEN: to misprint; use up in printing 297
VERDRÜCKEN: to crush; crumple; mash; overpress 298
VERDRUSS (*m*): ill-will; annoyance; anger; trouble; vexation 299
VERDUFTEN: to evaporate 300
VERDUNKELN: to obscure; darken; eclipse 301
VERDÜNNBAR: rarefiable; capable (of dilution) of being thinned 302
VERDÜNNEN: to dilute (Liquids); attenuate (Wort, etc.); thin; rarefy (Gases); break down (Spirits) 303
VERDÜNNT: dilute-(d); thin; rare; attenuated; thinned; rarefied; adulterated 304
VERDÜNNUNG (*f*): dilution; thinning; rarefaction; attenuation; evacuation; exhaustion; exhausting (of air); adulteration 305
VERDÜNNUNGSGESETZ (*n*): dilution law 306
VERDÜNNUNGSMITTEL (*n*): diluent; attenuant (Med.) 307
VERDÜNNUNGSWÄRME (*f*): heat of dilution 308
VERDUNSTBAR: vapourizable; capable of (being evaporated) evaporation; vaporizable 309
VERDUNSTEN: to vapourize; evaporate; exhale; volatilize 310
VERDÜNSTEN: to vapourize; evaporate; exhale 311
VERDUNSTUNG (*f*): evaporation; vapourization; volatilization 312
VERDÜNSTUNG (*f*): evaporation; vapourization 313
VERDUNSTUNGSFÄHIGKEIT (*f*): evaporative capacity 314
VERDUNSTUNGSKÄLTE (*f*): cold due to evaporation 315
VERDUNSTUNGSWÄRME (*f*): heat of evaporation 316
VERDURSTEN: to die of thirst 317
VERDÜSTERN: to darken 318
VERDUTZEN: to nonplus; puzzle; abash 319
VEREDELN: to dress; improve; finish; refine (Metals. etc.); elevate; ennoble; enrich; purify; cultivate (Plants) 320
VEREDELUNGSPROZESS (*m*): improving process (for rags and waste) 321
VEREHREN: to reserve; honour; admire; adore; venerate 322
VEREIDEN: to administer (bind by) an oath 323
VEREIN (*m*): union; club; society; company; association; confederation; federation 324
VEREINBAR: reconcilable 325
VEREINBAREN: to reconcile; agree (decide) upon; join; accord 326
VEREINBARKEIT (*f*): compatibility 327
VEREINBARUNG (*f*): reconciliation; decision; agreement 328
VEREINFACHEN: to simplify; reduce (Maths.) 329
VEREINFACHUNG (*f*): simplification; reduction 330
VEREINIGEN: to combine; unite; join; collect; accord; (make)-agree; ally; associate; concert 331
VEREINIGUNG (*f*): union; association; combination; club; agreement; alliance; concert 332
VEREINZELN: to separate; detach; isolate; dismember; individualize 333
VEREINZELT: separate; detached; isolated; sporadically 334
VEREISEN: to freeze up; turn to ice; coagulate 335
VEREITELN: to thwart; baulk; defeat; frustrate; disappoint; baffle 336
VEREITERN: to suppurate (Med.) 337
VERENGEN: to swage; narrow; contract; taper; restrict 338
VERENGERN: (see VERENGEN) 339
VERENGUNG (*f*): contraction, swaging, narrowing, restriction 340
VERERBEN: to hand down; transmit; receive as a legacy; inherit; bequeath; devolve upon 341
VERERBT: hereditary 342
VERERBUNG (*f*): bequeathing; inheritance 343
VERESTERN: to esterify 344
VERESTERUNG (*f*): esterification 345
VEREWIGEN: to perpetuate; immortalize 346
VEREWIGT: deceased; late 347
VERFAHREN: to proceed; act; transport; export; deal; treat; work; manage; muddle; mistake the road; blunder; convey 348
VERFAHREN (*n*): process; means; procedure; proceeding; mode; method; manner; management; conduct 349
VERFAHREN, BASISCHES- (*n*): basic process (Metal) 350
VERFAHRENPATENT (*n*): process patent 351
VERFAHREN, SAURES- (*n*): acid process (Metal) 352
VERFAHRUNGSART (*f*): mode (method) of procedure; process; treatment 353
VERFAHRUNGSWEISE (*f*): (see VERFAHRUNGSART) 354

VERFALL (*m*) : decline ; decay ; deterioration ; lapse ; expiration ; ruin ; disuse ; fall 355
VERFALLEN : to decline ; decay ; deteriorate ; lapse ; be forfeited ; expire ; fall ; hit ; chance ; go to rack and ruin ; become due 356
VERFÄLSCHEN : to adulterate ; debase ; falsify ; counterfeit ; forge ; interpolate 357
VERFÄLSCHER (*m*) : coiner ; forger ; falsifier ; adulterator ; counterfeiter 358
VERFÄLSCHUNGSMITTEL (*n*) : adulterant 359
VERFANGEN : to operate ; be of avail ; have effect ; be (caught) ensnared 360
VERFÄNGLICH : deceitful ; insidious ; captious ; dangerous 361
VERFÄRBEN : to fade ; discolour ; change colour ; dye poorly ; use up in dyeing 362
VERFÄRBUNG (*f*) : fading ; discolouration 363
VERFASSEN : to compose ; write ; prepare ; compile 364
VERFASSER (*m*) : author ; writer 365
VERFASSUNG (*f*) : composition ; writing ; condition ; constitution ; state ; situation ; compilation ; preparation 366
VERFASSUNGSMÄSSIG : constitutional 367
VERFAULBAR : putrescible 368
VERFAULEN : to decompose ; rot ; putrefy 369
VERFAULT : rotten ; putrid 370
VERFECHTEN : to defend ; advocate ; dispute 371
VERFEHLEN : to fail ; miss ; be unsuccessful ; mistake the road 372
VERFEHLT : misplaced ; unsuccessful 373
VERFEINERN : to improve ; fine ; refine 374
VERFERTIGEN : to prepare ; manufacture ; make ; compose ; fabricate 375
VERFESTIGUNG (*f*) : strengthening, increased strength (due to external influences) 376
VERFETTUNG (*f*) : fatty degeneration 377
VERFILZEN : to felt ; mat ; clog up (with reference to lubrication) 378
VERFILZUNGSFÄHIGKEIT (*f*) : felting (matting) property 379
VERFINSTERN : to obscure ; darken ; eclipse 380
VERFLACHEN : to (become) level ; flatten 381
VERFLECHTEN : to interlace ; involve ; interweave 382
VERFLECHTUNG (*f*) : complexity (see VERFLECHTEN) 383
VERFLIEGEN : to fly (off) away ; vanish ; volatilize ; evaporate 384
VERFLIEGEND : evanescent ; volatile 385
VERFLIESSEN : to pass ; go ; elapse ; flow (off) away ; blend ; subside (Water) ; expire 386
VERFLUCHEN : to curse ; execrate 387
VERFLÜCHTIGBAR : volatilizable 388
VERFLÜCHTIGEN : to volatilize 389
VERFLÜCHTIGUNG (*f*) : volatilization 390
VERFLUCHUNG (*f*) : curse ; execration ; malediction 391
VERFLÜSSIGEN : to liquefy ; condense (Gases or Vapours) 392
VERFLÜSSIGUNG (*f*) : liquefaction ; condensation (of gases or vapours) 393
VERFLÜSSIGUNGSANLAGE (*f*) : liquifying plant 394
VERFLÜSSIGUNGSAPPARAT (*m*) : liquefying (condensing) apparatus ; liquefier ; condenser 395
VERFLÜSSIGUNGSMITTEL (*n*) : liquefacient 396
VERFOLGEN : to pursue ; prosecute ; follow (up) ; carry on ; continue ; persecute 397
VERFRACHTEN : to freight ; charter (a vessel) 398
VERFRACHTUNG (*f*) : conveyance, chartering, charter (of ships), freighting 399
VERFRISCHEN : to fine ; refine (Metal) 400
VERFRÜHT : premature 401
VERFÜGBAR : available ; at disposal 402
VERFÜGEN ÜBER : to dispose (of) ; have at one's (command) disposal ; order ; arrange ; have available 403
VERFÜGUNG (*f*) : disposal ; service ; command ; order ; arrangement ; disposition ; availability 404
VERFÜHREN : to mislead ; corrupt ; seduce ; lead astray ; convey ; transport ; tempt ; allure 405
VERFÜHRERISCH : tempting ; seductive ; alluring 406
VERFÜLLEN : to fill 407
VERFUTTERN : to feed ; overfeed 408
VERFÜTTERN : (see VERFUTTERN) 409
VERFÜTTERUNGSVERSUCH (*m*) : feeding experiment 410
VERGÄLLEN : to denature ; denaturize (process of making substances unsuitable for human consumption) ; gall ; embitter ; (*n*) : denaturing 411
VERGANGEN : past ; gone ; last 412
VERGÄNGLICH : transient ; perishable ; finite ; transitory 413
VERGÄRBAR : attenuable ; fermentable 414
VERGÄREN : to ferment ; attenuate 415
VERGÄRUNGSGRAD (*m*) : degree of (attenuation) fermentation 416
VERGÄRUNGSMESSER (*m*) : zymometer ; zymoismeter (instrument for determining degree and heat of fermentation of liquids) 417
VERGASBAR : vapourizable ; gasifiable 418
VERGASEN : to gasify ; vapourize ; distil (Coal) 419
VERGASER (*m*) : vapourizer ; gasifier ; carburetter (Motors) 420
VERGASERDÜSE (*f*) : carburetter nozzle 421
VERGASERHEBEL (*m*) : carburetter lever 422
VERGASERKOLBEN (*m*) : carburetter piston 423
VERGASERSCHWIMMER (*m*) : carburetter float 424
VERGASERVENTIL (*n*) : carburetter valve 425
VERGASUNGSKAMMER (*f*) : gasifying chamber 426
VERGATTERN : to screen off (by grating or latticework) 427
VERGEBEN : to forgive ; give ; bestow ; pardon ; misdeal (Cards) ; remit 428
VERGEBENS : in vain ; vainly 429
VERGEBLICH : in vain ; vain ; vainly ; futile ; fruitless ; useless 430
VERGEGENWÄRTIGEN : to represent ; realize ; imagine ; figure 431
VERGEHEN : to vanish ; cease ; pass ; fade ; elapse ; perish ; exercise ; transgress ; commit a fault ; make a mistake ; decay ; lose 432
VERGEHEN (*n*) : fault ; transgression ; offence ; crime ; error 433
VERGEISTIGEN : to spiritualize 434
VERGELBEN : to turn yellow 435
VERGELTEN : to reward ; return ; repay ; requite ; recompense ; retaliate ; remunerate 436
VERGELTUNG (*f*) : reward ; remuneration ; retaliation (see VERGELTEN) 437
VERGESSEN : to forget ; omit ; neglect ; (*n*) : forgetfulness ; forgetting 438
VERGESSENHEIT (*f*) : oblivion ; forgetfulness 439
VERGESSLICH : oblivious ; forgetful 440

VERGEUDEN: to waste; squander; spend; lavish 441
VERGEWISSERN: to confirm; assure; convince; satisfy; make sure; certify; ascertain 442
VERGIESSEN: to cast (run) in; cast badly; shed; spill 443
VERGIFTEN: to poison; infect 444
VERGILBEN: to yellow, turn yellow, fade (Phot., etc.) 445
VERGILBT: yellowed 446
VERGIPSEN: to plaster 447
VERGISSMEINNICHT (*n*): forget-me-not; germander 448
VERGITTERN: to grate, lattice; rail off; screen 449
VERGLASBAR: vitrifiable 450
VERGLASEN: to glaze; vitrify 451
VERGLASUNG (*f*): glaze; glazing; vitrifaction 452
VERGLEICH (*m*): comparison; agreement; contract; arrangement; accord 453
VERGLEICHBAR: comparable 454
VERGLEICHEN: to compare; come to terms; agree; settle; accord 455
VERGLEICHEND: comparative 456
VERGLEICHENDE UNTERSUCHUNG (*f*): comparative test 457
VERGLEICHLICH: comparable 458
VERGLEICHPRISMA (*n*): comparison (comparative) prism 459
VERGLEICHSVERSUCH (*m*): comparative test 460
VERGLEICHSWEISE: as (by way of) a comparison, by agreement 461
VERGLEICHUNG (*f*): comparison, parallel 462
VERGLEICHUNGSWEISE: comparatively, in comparison; (*f*): method (mode) of comparison 463
VERGLIMMEN: to die out; cease glowing 464
VERGLÜHBRAND (*m*): biscuit baking (Ceramics) 465
VERGLÜHEN: to bake; cease glowing; cool down; biscuit-bake (Ceramics); (*n*): biscuit-baking (Ceramics) 466
VERGLÜHOFEN (*m*): biscuit kiln (Ceramics) 467
VERGNÜGEN: to please; gratify; delight; be (delighted; gay) cheerful; enjoy oneself 468
VERGNÜGT: pleased; happy; delighted 469
VERGNÜGUNG (*f*): pleasure; sport; amusement; delight; diversion 470
VERGNÜGUNGSREISENDER (*m*): tourist; a person travelling for pleasure 471
VERGOLDEN: to gild; gold-plate 472
VERGOLDET: gold-plated; gilt; gilded 473
VERGOLDUNG (*f*): gold-plating; gilding; gilt 474
VERGOLDUNG (*f*), **GALVANISCHE-**: electro-gilding 475
VERGOLDUNGSWACHS (*n*): gilder's wax 476
VERGOLDUNGSWASSER (*n*): quickening liquid 477
VERGÖNNEN: to grant; permit; allow; vouchsafe 478
VERGOREN: fermented; finished (Brew.) 479
VERGÖTTERN: to deify; idolize; adore; make a god of 480
VERGRABEN: entrench; bury; inter 481
VERGREIFEN: to mistake; violate; steal; embezzle; attack; buy (up) out; miss the mark 482
VERGRIFFEN: out of print; sold out, etc. 483
VERGRÖSSERN: to enlarge; increase; exaggerate; magnify; grow larger 484
VERGRÖSSERUNG (*f*): enlargement (of photos., etc.); increase; aggrandisement; exaggeration; magnification 485
VERGRÖSSERUNGSAPPARAT (*m*): enlarging (camera) apparatus (Phot.) 486
VERGRÖSSERUNGSGLAS (*n*): magnifying glass; microscope 487
VERGRÖSSERUNGSKRAFT (*f*): magnifying power 488
VERGRÖSSERUNGSLINSE (*f*): magnifying lens 489
VERGRÖSSERUNGSSKALA (*f*): magnifying scale 490
VERGRÜNEN: to turn green; fade; lose green colour 491
VERGRÜNLICH: greenable (Dyeing) 492
VERGÜNSTIGEN: to concede; permit; allow 493
VERGUSSMASSE (*f*): filling mass, cement, grouting 494
VERGÜTEN: to compensate; make (amends) good; allow; repair; (re)-pay; reimburse; improve (Metal) 495
VERGÜTIGEN: (see **VERGÜTEN**) 496
VERGÜTIGUNG (*f*): (see **VERGÜTUNG**) 497
VERGÜTUNG (*f*): reimbursement; crediting; reparation; compensation; improvement; strengthening (Steel) (quick cooling followed by re-heating) 498
VERHAFT (*m*): arrest, custody, seizure, apprehension, imprisonment 499
VERHAFTEN: to arrest; apprehend; take (into custody) in charge; imprison; seize 500
VERHALLEN: to die (fade; pass) away (of sound) 501
VERHALT (*m*): condition; state; fact 502
VERHALTEN: to behave; act; be; retain; suppress; (with)-hold; keep back; (*n*): behaviour; conduct; retention; suppression 503
VERHÄLTNIS (*n*): ratio; rate; relation; proportion; situation; circumstance; condition; connection 504
VERHÄLTNISANZEIGER (*m*): exponent (Maths.) 505
VERHÄLTNIS, DAMPF- (*n*): steam condition; condition (state) of the steam 506
VERHÄLTNISGLEICHHEIT (*f*): proportion; comparative equality 507
VERHÄLTNISMÄSSIG: comparatively; in comparison; proportional-(ly); proportionate; commensurate; commensurable; comparative 508
VERHÄLTNISWIDRIG: disproportionate 509
VERHÄLTNISWORT (*n*): preposition 510
VERHÄLTNISZAHL (*f*): proportional number; coefficient of proportion 511
VERHALTUNG (*f*): behaviour; conduct; retention; suppression; concealment 512
VERHANDELN: to transact; sell; try; negotiate; treat; deal; discuss 513
VERHANDLUNG (*f*): proceeding; debate; transaction; selling; negotiation 514
VERHÄNGEN: to hang; proclaim; decree; destine 515
VERHÄNGNIS (*n*): fate; destiny; fatality 516
VERHÄNGNISGLAUBE (*m*): fatalism 517
VERHÄNGNISVOLL: fatal 518
VERHARREN: to persist, remain, persevere 519
VERHÄRTEN: to indurate; harden 520
VERHÄRTUNG (*f*): induration; hardening; obdurateness 521
VERHARZEN: to resinify; become resinous; resin; convert into resin (by polymerization) 522

VERHASST: odious; hated; hateful 523
VERHAUEN: to hack; prune; cut (up) down; lop 524
VERHEEREN: to devastate; desolate; lay waste; destroy 525
VERHEHLEN: to conceal; hide; camouflage; dissemble 526
VERHEIMLICHEN: to disguise; secrete; keep secret; conceal 527
VERHEIRATEN: to marry; wed; enter into matrimony; give in marriage 528
VERHEISSEN: to promise 529
VERHEIZEN: to burn; consume; fire (feed; supply) coal on to the grate 530
VERHELFEN: to help; raise to; assist in; procure 531
VERHETZEN: to incite; instigate 532
VERHINDERN: to prevent; obstruct; hinder; debar from; preclude; impede; inhibit 533
VERHINDERUNG (*f*): obstruction; obstacle; hindrance; impediment; prevention; preventative; inhibition 534
VERHOLEN: to warp (a ship) 535
VERHOLZEN: to become wood; lignify 536
VERHÖR (*n*): hearing; trial; examination 537
VERHÖREN: to hear; try; examine; question; not to hear; to make a mistake in hearing 538
VERHORNEN: to become horn-(y) 539
VERHÜLLEN: to veil; cover-(over) 540
VERHUNGERN: to starve; hunger; famish 541
VERHUNZEN: to spoil; bungle; make a mess of 542
VERHÜTEN: to prevent; avert; obviate; hinder 543
VERHÜTEND: preventive 544
VERHÜTTEN: to smelt; work (Ores) 545
VERHÜTTUNG (*f*): smelting; working (of metals) 546
VERHÜTUNG (*f*): prophylaxis (Med.); prevention 547
VERHÜTUNGSMITTEL (*n* and *pl*): preventative, prophylactic (Med.) 548
VERIMPFEN: to innoculate; transmit (by contagion of inoculation) 549
VERIRREN: to stray; go astray; err; lose one's way; mistake the road 550
VERIRRUNG (*f*): aberration 551
VERJAGEN: to expel; drive (off; out) away; chase-(away) 552
VERJÄHREN: to grow old 553
VERJAUCHEN: to putrefy 554
VERJAUCHUNG (*f*): putrefaction 555
VERJUBELN: to squander 556
VERJÜNGEN: to reduce; constrict; narrow; swage; taper; diminish; renew; renovate; rejuvenate 557
VERJÜNGUNG (*f*): reduction; taper; constriction; swage; swaging; diminution; rejuvenation 558
VERKALKBAR: calcinable 559
VERKALKEN: to calcine; calcify 560
VERKALKUNG (*f*): calcination 561
VERKANNT: mistaken 562
VERKAPPEN: to mask; disguise 563
VERKAPSELN: to put on (seal with) capsules 564
VERKAPSELUNGSMASCHINE (*f*): capsuling machine; (*pl*): capsuling (machines) machinery 565
VERKÄSEN: to become caseous or cheesy 566
VERKÄSUNG (*f*): caseation (Med.) 567
VERKAUF (*m*): sale 568
VERKAUFEN: to sell; dispose of 569
VERKÄUFER (*m*): vendor 570
VERKÄUFLICH: marketable; saleable; venal; capable of being bribed 571
VERKAUF MIT ABRECHNUNG, ZUM-: on sale or return 572
VERKAUFSPREIS (*m*): sales (selling) price 573
VERKAUFSRECHNUNG (*f*): account-sales 574
VERKEHR (*m*): traffic; trade; commerce; communication; intercourse 575
VERKEHREN: to traffic; trade; communicate; associate; have intercourse with; pervert; invert; reverse; associate; visit; turn 576
VERKEHRSMITTEL (*n*): conveyance 577
VERKEHRSSICHER: safe for (commerce) traffic 578
VERKEHRSSICHERER SPRENGSTOFF (*m*): safety explosive 579
VERKEHRT: wrong; up-side-down; inverted; reverse-(d); perverse; perverted; absurd 580
VERKEILEN: to wedge; key (Mech.) 581
VERKEILUNG (*f*): wedging, keying 582
VERKENNEN: to misunderstand; mistake; misapprehend 583
VERKETTEN: to form into a chain; to link-(together) 584
VERKETTUNG (*f*): linkage; concatenation; connection; bond 585
VERKETTUNGSFÄHIGKEIT (*f*): ability for forming chains; linkage capacity 586
VERKIESELN: to silicify 587
VERKIESELUNG (*f*): silicification 588
VERKIESEN: to ballast; gravel 589
VERKITTEN: to lute; cement; seal 590
VERKITTUNG (*f*): luting; cementing 591
VERKLAGEN: to accuse; sue; inform 592
VERKLAGER (*m*): plaintiff; informer; accuser 593
VERKLAGTE (*m* or *f*): accused, defendant 594
VERKLAMMERN: to clamp; to bracket or brace; dowel 595
VERKLAMMERUNG (*f*): dowelling 596
VERKLÄREN: to transfigure, glorify, transform 597
VERKLATSCHEN: to slander; calumniate 598
VERKLEBEN: to agglutinate; gum; glue; cement; lute; stick-(together); plaster-(over) 599
VERKLEIBEN: (see **VERKLEBEN**) 600
VERKLEIDEN: to case; line; face; mask; disguise; wainscot; incrust 601
VERKLEIDUNG (*f*): casing; lining; facing; disguise; incrustation 602
VERKLEINERN: to decrease; diminish; reduce; taper; swage; disparage; belittle; underrate; detract; lessen 603
VERKLEINERUNG (*f*): decrease; diminution; reduction; disparagement; defamation; underrating; detraction 604
VERKLEINERUNGSMASSSTAB (*m*): scale of reduction 605
VERKLEINERUNGSZIRKEL (*m*): reduction compasses 606
VERKLEISTERN: to paste (up); glue (up); make into paste 607
VERKLEISTERT: pasty 608
VERKNETEN (*n*): kneading; agglomeration 609
VERKNISTERN: to decrepitate 610
VERKNISTERUNG (*f*): decrepitation 611
VERKNÖCHERN: to ossify; form bone 612
VERKNÖCHERUNG (*f*): ossification 613

VERKNORPELN : to become cartilaginous ; form gristle 614
VERKNORPELT : cartilaginous 615
VERKNORPELUNG (*f*) : chondrification 616
VERKNÜPFEN : to bind ; tie ; connect ; join ; unite ; combine 617
VERKNÜPFUNG (*f*) : union ; connection ; combination 618
VERKOBALTEN : to cobalt, coat with cobalt (Metal) 619
VERKOBALTUNG (*f*). ELEKTROLYTISCHE- : electrolytic cobalting (as a rust preventative) 620
VERKOCHEN : to concentrate ; boil (down) away ; overboil ; boil too much ; overheat ; spoil (injure) by boiling 621
VERKOHLEN : to carbonize ; char ; change (convert) into coal 622
VERKOHLUNG (*f*) : carbonization ; charring 623
VERKOHLUNGSOFEN (*m*) : carbonizing furnace 624
VERKOKEN : to coke ; carbonize 625
VERKOKUNG (*f*) : coking-(process) ; carbonization 626
VERKOKUNGSOFEN (*m*) : coke oven 627
VERKOMMEN : to degenerate ; perish ; decay ; die ; starve ; go to rack and ruin 628
VERKORKEN : to cork-(up) 629
VERKORKMASCHINE (*f*) : corking machine 630
VERKÖRPERN : to embody 631
VERKREIDEN : to calcify 632
VERKREIDUNG (*f*) : calcification 633
VERKRÜMELN : to crumble 634
VERKRÜMMEN : to crook ; bend ; curve ; hook 635
VERKRÜPPELN : to (become) cripple-(d) ; stunt 636
VERKÜHLEN : to cool (off) down 637
VERKÜNDEN : to announce ; publish ; foretell ; predict ; promulgate 638
VERKÜNDIGEN : (see VERKÜNDEN) 639
VERKÜPBAR : capable of being used as a vat dye 640
VERKÜPEN : to reduce a vat dye (before dyeing) 641
VERKUPFERN : to copper ; copper-plate ; coat with copper 642
VERKUPPELN : to couple ; connect 643
VERKÜRZEN : to abridge ; curtail ; shorten ; diminish ; abbreviate 644
VERKÜRZUNG (*f*) : shortening, abridgment, abbreviation 645
VERLACHEN : to mock ; laugh at 646
VERLADEN : to (un)-load ; transship ; freight ; transport ; (*n*) : (un)-loading ; transshipment ; transport 647
VERLADUNG (*f*) : (un)-loading ; transshipment ; freight ; lading 648
VERLAG (*m*) : publishing house ; publication ; publishing ; expenses ; disbursement ; outlay ; advance 649
VERLAGERN : to shift ; move ; displace ; rearrange 650
VERLAGERUNG (*f*) : shifting ; movement ; displacement ; rearrangement ; moving 651
VERLAGSBUCHHÄNDLER (*m*) : publisher 652
VERLAGSBUCHHANDLUNG (*f*) : publishing (house) firm ; publishers 653
VERLAGSHÄNDLER (*m*) : publisher 654
VERLAGSHANDLUNG (*f*) : publisher's ; publishing (house) firm 655
VERLAGSRECHT (*n*) : copyright 656
VERLANGEN : to ask ; wish for ; demand ; desire ; want ; long for ; require ; request ; (*n*) : desire ; request ; longing 657

VERLÄNGERN : to lengthen ; protract ; extend ; prolong ; stretch ; produce (Maths.) ; replenish (Liquor, etc.) 658
VERLÄNGERTES MARK (*n*) : Medulla oblongata (Med.) 659
VERLÄNGERUNG (*f*) : lengthening ; extension ; prolongation ; elongation 660
VERLANGSAMEN : to slow (up) down ; retard 661
VERLANGSAMUNG (*f*) : retardation 662
VERLASCHEN : to fish-(together), lash 663
VERLASCHUNG (*f*) : fishing, lashing 664
VERLASS (*m*) : reliance ; trust ; inheritance ; succession 665
VERLASSEN : abandoned ; forsaken ; destitute ; to leave ; abandon ; give up ; depart from ; forsake ; depend (upon) ; rely (upon) ; trust ; throw aside ; dump 666
VERLASSENHEIT (*f*) : abandonment, destitution 667
VERLASSENSCHAFT (*f*) : bequest, inheritance 668
VERLÄSSIG : reliable ; trustworthy 669
VERLÄSTERN : to slander, calumniate 670
VERLAUB (*m*) : permission 671
VERLAUF (*m*) : course (of a curve, etc.) ; lapse (of time) ; expiration 672
VERLAUF (*m*) DER LINIEN : tapering of the lines (of ships) 673
VERLAUFEN : to proceed ; follow a course ; pass (of time) ; elapse ; happen ; be scattered ; go astray ; take place ; subside (of water) ; blend (of colours) ; extend ; expire 674
VERLAUT (*m*) : rumour ; report 675
VERLAUTEN : to be reported ; be (heard) said ; be (rumoured) hinted 676
VERLEBEN : to live ; pass ; spend (of time) 677
VERLEBT : decrepit ; used up 678
VERLEGEN : puzzled ; spoiled ; stale ; confused ; embarrassed ; to lay ; retard ; obstruct ; transfer ; delay ; transport ; (mis-)place ; publish (Books, etc.) ; become (choked) stopped up 679
VERLEGENHEIT (*f*) : perplexity ; embarrassment ; difficulty 680
VERLEGENHEIT, IN- BRINGEN : to place in a difficult (awkward) position ; embarrass ; puzzle ; perplex 681
VERLEGER (*m*) : publisher ; bookseller 682
VERLEIDEN : to spoil ; make (render) distasteful or disagreeable ; disgust 683
VERLEIHEN : to lend ; confer ; bestow ; invest with ; grant ; give to 684
VERLEITEN : to seduce ; mislead 685
VERLERNEN : to forget ; unlearn 686
VERLESEN : to read (wrong) aloud ; pick 687
VERLETZEN : to damage ; injure ; infringe ; offend ; hurt ; violate ; wound 688
VERLETZLICH : vulnerable 689
VERLETZLICHKEIT (*f*) : vulnerability, capacity for being (injured) damaged 690
VERLEUGNEN : to disavow ; deny ; disown 691
VERLEUGNUNG (*f*) : denial ; abnegation 692
VERLEUMDEN : to slander ; asperse ; defame ; calumniate ; detract 693
VERLIEBEN : to become enamoured of ; fall in love with ; take a fancy to 694
VERLIEBT : amorous ; in love with ; fond of ; enamoured 695
VERLIEREN : to lose ; disappear ; be lost ; go astray ; get lost ; suffer loss 696
VERLIESS (*n*) : dungeon 697

VERLOBEN : to become (betrothed) engaged; affiance 698
VERLOCKEN : to seduce; entice; mislead; lead astray 699
VERLOGEN : untruthful 700
VERLOHNEN : to be worth; pay; requite 701
VERLOOSEN : to draw lots; raffle 702
VERLOSCHEN : dead; extinct; gone out 703
VERLÖSCHEN : to extinguish; to be extinguished; go out; expire; die 704
VERLÖTEN : to solder-(up); close up by adhesion (Med.) 705
VERLUST (*m*): loss; waste; escape; detriment; damage 706
VERLUST (*m*), DRUCK- : pressure loss 707
VERLUST DURCH ABZIEHENDE GASE (*m*): chimney loss (of combustion); flue loss 708
VERLUST (*m*), KRAFT- : power loss 709
VERLUST (*m*), LADUNGS- : loss of charge 710
VERLUSTLOS : free from losses; ideal (when applied to engineering plant) 711
VERLUSTQUELLE (*f*): source of loss 712
VERLUST (*m*), REIBUNGS- : friction loss 713
VERLUST (*m*), SPANNUNGS- : tension loss; loss of voltage 714
VERLUST (*m*), STROM- : loss of current 715
VERLUST (*m*), WÄRME- : heat loss 716
VERMACHEN : to make over to; bequeath; devise; shut; stop 717
VERMÄCHTNIS (*n*): legacy; bequest 718
VERMÄCHTNISERBE (*m*): legatee 719
VERMAHLEN : to grind-(up); crush; pulverize 720
VERMÄHLEN : to marry; give in marriage; espouse 721
VERMAHNEN : to admonish; exhort; remind 722
VERMALEDEIEN : to curse; execrate 723
VERMALEDEIUNG (*f*): malediction; imprecation 724
VERMANNIGFALTIGEN : to duplicate; multiply; diversify 725
VERMAUERN : to wall (brick, build) up 726
VERMEHREN : to increase; multiply; propagate; enlarge; grow; rise; augment 727
VERMEHRUNG (*f*): increase; propagation; addition; augmentation; multiplication; enlargement 728
VERMEIDEN : to obviate; evade; avoid; shirk; forbear; shun ; elude 729
VERMEIDLICH : avoidable; avertable 730
VERMEINEN : to mean; suppose; think 731
VERMEINTLICH : supposed; supposititious; pretended; presumptive 732
VERMELDEN : to announce; tell; advise; inform 733
VERMENGEN : to mix; blend; mingle; confuse; confound; meddle with 734
VERMERKEN : to perceive; note; observe 735
VERMESSEN : to measure-(wrongly); venture; dare; protest; size-(up); survey (Land); bold; audacious; presumptuous 736
VERMESSEN (*n*): surveying 737
VERMESSENHEIT (*f*): temerity; presumption; boldness; audacity 738
VERMESSINGEN : to brass 739
VERMESSUNG (*f*): surveying; survey; measurement; measuring; mismeasurement 740
VERMESSUNGSDECK (*n*): tonnage deck 741
VERMESSUNGSGESETZ (*n*): tonnage law 742
VERMESSUNGSKUNDE (*f*): geodesy; mensuration 743
VERMESSUNGSLÄNGE (*f*): length for tonnage (Ships), tonnage length 744
VERMICULIT (*m*): [MAGNESIAGLIMMER (which see), plus Na_2O]; vermiculite; Sg. 2.3 745
VERMIETEN : to let; hire out; lease 746
VERMIETER (*m*): lessee; hirer; letter 747
VERMILLON (*m*): vermilion (see ZINNOBER) 748
VERMINDERN : to lessen; reduce; diminish; minimize; decrease; decline; grow less 749
VERMINDERT : diminished (Harmony) (see QUARTE) 750
VERMINDERUNG (*f*): reduction; decrease; diminution (see VERMINDERN) 751
VERMISCHEN : to mix; adulterate; blend; mingle; alloy; amalgamate 752
VERMISCHT : mixed; miscellaneous; mingled; adulterated; blended; alloyed; amalgamated 753
VERMISCHUNG (*f*): mixture; mixing; alloy; blend-(ing); adulteration; miscellany; medley; amalgamation 754
VERMISSEN : to miss; regret; want; lack 755
VERMITTELN : to arrange; reconcile; adjust; mediate; interpose; negotiate; arbitrate; intercede 756
VERMITTELST : by means (the aid; the help) of 757
VERMITTELUNG (*f*): means; agency; interposition; mediation; conveyance; adjustment; intercession 758
VERMODERN : to moulder; decay 759
VERMÖGE : according to; by virtue of 760
VERMÖGEN : to induce; be able; be capable of; have (power) capacity for; to find it possible to; have influence; (*n*): capability; capacity; faculty; power; ability; fortune; wealth; property 761
VERMÖGEND : wealthy; rich; propertied; powerful 762
VERMÖGENSRECHNUNG (*f*): Capital account 763
VERMÖGENSSTEUER (*f*): property tax 764
VERMUTEN : to think; suppose; suspect; conjecture; assume; presume; expect; anticipate 765
VERMUTLICH : probable; probably; likely; presumably 766
VERMUTUNG (*f*): conjecture; surmise; supposition; guess; opinion; conclusion; thought; assumption 767
VERNACHLÄSSIGBAR : negligible; negligibly 768
VERNACHLÄSSIGEN : to neglect; disregard; slight 769
VERNACHLÄSSIGUNG (*f*): negligence; neglect 770
VERNAGELN : to nail up 771
VERNARBEN : to (be) cicatrize-(d); to scar 772
VERNARRTHEIT (*f*): infatuation 773
VERNEHMEN : to observe; perceive; distinguish; understand; hear; see; notice; learn; try; examine; interrogate; (*n*): intelligence; report; hearing; understanding 774
VERNEHMLICH : distinct; audible; perceptible; intelligible 775
VERNEHMUNG (*f*): trial; examination 776
VERNEINEN : to negative; deny 777
VERNEINEND : negative 778
VERNICKELUNG (*f*): nickel-plating; nickeling 779
VERNICHTEN : to annihilate; destroy; annul; cancel 780
VERNICHTUNG (*f*): destruction, annihilation, cancellation 781
VERNICHTUNGSANLAGE (*f*): destructor-(plant) 782

VERNICHTUNGSPULVER (*n*): destructive powder (for weeds, insects and vermin) 783
VERNICKELN: to nickel; nickel-plate 784
VERNICKELUNGSSALZ (*n*): nickel-plating salt 785
VERNIETEN: to rivet; clinch 786
VERNUNFT (*f*): reason; intelligence; intellect; judgment; sense; rationality; logic 787
VERNUNFTGEMÄSS: logically, according to reason, rationally 788
VERNÜNFTIG: sensible; rational; reasonable; logical; intelligent; judicious 789
VERNUNFTLEHRE (*f*): logic 790
VERNUNFTLOS: illogical, irrational, unreasonable 791
VERNUNFTMÄSSIG: rational, reasonable, logical 792
VERNUNFTSCHLUSS (*m*): logical (rational) conclusion, rationalism, syllogism 793
VERNUNFTWIDRIG: irrational, illogical, opposed to (against, contrary to) reason or common sense 794
VERNUNFTWISSENSCHAFT (*f*): logic, philosophy 795
VERÖDEN: to lay (become) waste; desolate; devastate 796
VERÖFFENTLICHEN: to advertise; publish; announce 797
VERÖFFENTLICHUNG (*f*): publication; advertisement; announcement; (*pl*): literature 798
VERONAL (*n*): veronal 799
VERONALNATRIUM (*n*): (see **NATRIUM, DIÄTHYL-BARBITURSAURES-**) 800
VERONESERGELB (*n*): Verona yellow (see **CASSELERGELB**) 801
VERONESERGRÜN (*n*): Verona (Veronese) green (green earth from Verona) (see **GRÜNERDE**) 802
VERORDNEN: to order; decree; institute; appoint; prescribe (Med.); dispose; enact; ordain 803
VERORDNUNG (*f*): statute; order; ordinance; enactment; decree; appointment; prescription (Med.) 804
VERORDNUNGEN (*f.pl*): statute laws, statutes 805
VERPACHTEN: to let; lease 806
VERPACKEN: to pack-(up); bale; bag 807
VERPACKFLASCHE (*f*): packing bottle 808
VERPACKUNG (*f*): packing; lining; casing; bagging, etc. 809
VERPACKUNGSMACHINE (*f*): packing machine 810
VERPASSEN: to lose; miss; pass 811
VERPESTEN: to taint; poison; infest 812
VERPESTET: infected with pest, tainted 813
VERPFÄHLEN: to empale; palisade; fence 814
VERPFÄNDEN: to pledge; pawn; mortgage; hypothecate 815
VERPFLANZEN: to transplant; transmit 816
VERPFLEGEN: to tend; take care of; nurse; foster; maintain; provide for 817
VERPFLICHTEN: to bind; oblige; engage 818
VERPFLICHTEND: obligatory 819
VERPFLICHTUNG (*f*): duty; obligation; liability; engagement 820
VERPFUSCHEN: to bungle 821
VERPICHEN: to pitch; tar; coat (cover) with pitch or tar 822
VERPLATINIEREN: to platinize 823
VERPLATINIERUNG (*f*): platinization 824
VERPÖNEN: to forbid; dissuade by threats 825
VERPROVIANTIEREN: to provision 826
VERPROVIANTIERUNG (*f*): supply, provisioning 827
VERPUFFEN: to deflagrate; puff (off) out; detonate; fire; explode; crackle; crepitate 828
VERPUFFUNG: deflagration; detonation; explosion; decrepitation; crepitation; crackling 829
VERPUFFUNGSAPPARAT (*m*): explosion apparatus 830
VERPUFFUNGSDRUCK (*m*): detonation (explosion, firing, combustion) pressure 831
VERPUFFUNGSERSCHEINUNG (*f*): explosion phenomenon 832
VERPUFFUNGSPROBE (*f*): deflagration test 833
VERPUFFUNGSRÖHRE (*f*): explosion (tube) pipe 834
VERPUFFUNGSSPANNUNG (*f*): combustion (firing) pressure (of internal combustion engines) 835
VERPUPPEN: to change into the chrysalis state 836
VERPUTZEN: to plaster 837
VERQUICKEN: to quicken; amalgamate; coat copper and its alloys with gold or silver 838
VERQUICKUNG (*f*): quickening, amalgamating 839
VERQUICKUNGSLÖSUNG (*f*): quickening (amalgamating) solution 840
VERRAMMELN: to barricade; block (up) 841
VERRAT (*m*): treason; treachery 842
VERRATEN: to betray 843
VERRAUCHEN: to evaporate; go up in smoke; smoke 844
VERRÄUCHERN: to smoke; cure 845
VERRECHNEN: to calculate; reckon; miscalculate; be mistaken 846
VERRECHNUNGSSCHECK (*m*): cheque in settlement of account; crossed cheque 847
VERREIBEN: to triturate; rub to a fine powder; to grind fine; to rub (grind) away; spread (spoil) by rubbing 848
VERREIBUNG (*f*): trituration (a fine powder produced by constant rubbing of a drug with milk sugar) 849
VERREISSEN: to tear; wear out 850
VERRENKEN: to sprain, dislocate 851
VERRICHTEN: to execute; do; perform; carry out 852
VERRICHTUNG (*f*): execution; performance; affair; business; function 853
VERRIEGELN: to bolt; bar 854
VERRINGERN: to lessen; reduce; decrease; diminish; deteriorate; abate 855
VERRINNEN: to pass (of time); to run off 856
VERROSTEN: to rust 857
VERROTTEN: to rot 858
VERRUCHT: nefarious; infamous; godless; profligate 859
VERRÜCKEN: to disturb; derange; displace; remove; dislocate 860
VERRÜCKT: insane; crazy; mad; deranged 861
VERRUF (*m*): disgrace; discredit; infamy 862
VERRUFEN: infamous; to decry; cry down 863
VERS (*m*): verse 864
VERSAGEN: to refuse; miss-(fire); deny; promise; fill (a situation); jib; fail (to work); refuse to (act) function; to be (non-effective) of no avail 865
VERSAGER (*m*): failure; misfire; one who (refuses) objects; a person on whom a medicine fails to act 866
VERSALZEN: to oversalt; mar 867
VERSAMMELN: to assemble; meet; congregate 868

VERSAMMLUNG (*f*) : meeting; congress; assembly; congregation; convention 869
VERSAND (*m*): export-(ation); shipping; shipment; consignment; dispatch 870
VERSANDANWEISUNG (*f*): forwarding instructions 871
VERSANDART (*f*): method of dispatch 872
VERSANDBIER (*n*): export beer 873
VERSANDEN: to (cover with) sand 874
VERSANDFLASCHE (*f*): bottle (cylinder) for dispatch; carboy 875
VERSANDGLÄSER (*n.pl*): glasses for dispatch 876
VERSANDSKOSTEN (*pl*): transport-(ation) (charges) costs; carriage; freight 877
VERSATZ (*m*): mixing; treatment; compound; pawning; pledge; addition; alloy; batch; (Ceramic); layer; layaway (Leather); gobbing; (Mining) 878
VERSATZNIETUNG (*f*): staggered riveting 879
VERSATZUNG (*f*): mixing (see **VERSATZ**); staggered pitching; staggering (of rivets; tubes, etc.); spigot; dovetail; skew notch 880
VERSAUERN: to turn sour 881
VERSÄUERN: to acidify; make (too) sour 882
VERSÄUMEN: to miss; neglect; fail 883
VERSBAU (*m*): versification 884
VERSCHAFFEN: to procure; secure; supply; acquire; provide; furnish 885
VERSCHALEN: to cover, board up, case; fish-(together) 886
VERSCHALLEN: to die away (of sound) 887
VERSCHALUNG (*f*): casing 888
VERSCHALUNG, HOLZ- (*f*): boarding (surrounding or enclosing something) 889
VERSCHÄMT: modest; bashful; ashamed 890
VERSCHANZEN: to barricade; entrench; fortify 891
VERSCHÄRFEN: to sharpen; intensify 892
VERSCHÄUMEN: to cease (stop) foaming 893
VERSCHEIDEN: to die; expire 894
VERSCHENKEN: to give away; retail; make a present 895
VERSCHERZEN: to (throw; trifle) fling away; forfeit 896
VERSCHEUCHEN: to scare (frighten) away 897
VERSCHICKEN: to send away; consign; transport 898
VERSCHIEBBAR: movable; sliding; displaceable; capable of being postponed 899
VERSCHIEBEN: to move; shift; remove; displace; postpone; defer; delay; procrastinate 900
VERSCHIEBUNG (*f*): movement; shifting; displacement; postponement; delay-(ing); slip; dislocation (Geol.) 901
VERSCHIEDEN: different; differing; various; deceased; divers; diverse; several 902
VERSCHIEDENARTIG: of various (kinds; species) sorts; varied 903
VERSCHIEDENBLÄTTERIG: various-leaved (Bot.) 904
VERSCHIEDENFARBIG: variegated; of various (different) colours 905
VERSCHIEDENHEIT (*f*): variety; diversity; difference 906
VERSCHIEDENTLICH: variously; severally 907
VERSCHIESSEN: to fade; discolour; to shoot (away) off; discharge; shade off (Colours) 908
VERSCHIFFEN: to ship; consign; export 909
VERSCHIMMELN: to grow (become) mouldy; moulder 910
VERSCHIMMELT: mouldy 911
VERSCHLACKEN: to scorify; slag; reduce (be reduced) to slag or scoria; clinker 912
VERSCHLACKUNG (*f*): scorification; clinkering; formation of clinker 913
VERSCHLACKUNGSPROBE (*f*): scorification assay 914
VERSCHLACKUNGSZONE (*f*): clinkering zone, zone in which the formation of clinker takes place (of furnaces) 915
VERSCHLAFEN: sleepy; drowsy; to oversleep; sleep (off) away 916
VERSCHLAG (*m*): compartment; partition-(wall); crate 917
VERSCHLAGEN: lukewarm; tepid; cunning; wily; sly; crafty; shrewd; to warm; take the chill off; cool; become lukewarm; make a difference; matter; board (nail; close) up; partition; recoin; drive (throw) away; lose (damage; spoil) by striking 918
VERSCHLÄMMEN: to silt up; become filled (choked) with mud 919
VERSCHLECHTERN: to make worse; debase; impair; deteriorate; spoil; degrade 920
VERSCHLECHTERUNG (*f*): debasement; deterioration; degeneration; spoilation; degradation 921
VERSCHLEIERN: to veil; draw the veil over; cloud (of beer); fog (of photographs) 922
VERSCHLEIERUNG (*f*): veiling; clouding, fogging (of photographic plates, etc.) 923
VERSCHLEIMEN: to silt up; choke up with (slime; phlegm; mud; mucous) pus; suppurate; foul 924
VERSCHLEIMUNG (*f*): silting (choking) up (with slime; phlegm; mucous; mud or pus); suppuration 925
VERSCHLEISS (*m*): sale; retail; wear and tear; polish; consumption 926
VERSCHLEISSEN: to sell; polish; wear away; use up; consume 927
VERSCHLEISSFEST: impervious to wear and tear 928
VERSCHLEISSFESTIGKEIT (*f*): (capacity of)-resistance to wear and tear 929
VERSCHLEPPEN: to carry off; misplace; protract; drag out; delay; retard 930
VERSCHLEPPUNG (*f*): protraction; dragging, etc.; delay-(ing); retardation 931
VERSCHLEUDERN: to squander; throw (fling) away; sell dirt-cheap; under-sell; dissipate; waste 932
VERSCHLIESSBAR: capable of being (shut; locked) closed; sealable; lockable 933
VERSCHLIESSEN: to shut; lock; close; seal; stop; reserve 934
VERSCHLIMMERN: to make (grow; become) worse; deteriorate; degenerate 935
VERSCHLINGEN: to entangle; twist; intertwine; entwine; swallow; devour; gulp down 936
VERSCHLUCKEN: to gulp down; swallow; absorb 937
VERSCHLUSS (*m*): closing; stopping; stoppage; closure; seal; fastening; shutting (locking) device; fastener; stopper; cover; snap; lock; clasp; trap (for ashes, water, etc.); bond (Customs; of goods); shutter (Photo.); hopper (for ashes; water, etc.); confinement; custody; imprisonment; connection; sealing 938

VERSCHLUSSDECKEL (*m*): seal, sealing (closing) cover 939

VERSCHLUSSHAKEN (*m*): locking (coupling) hook 940

VERSCHLUSS (*m*), HERMETISCHER-: hermetical seal 941

VERSCHLUSSKAPSEL (*f*): closing capsule 942

VERSCHLUSSKÖLBCHEN (*n*): sealed flask; pressure flask 943

VERSCHLUSSSCHRAUBE (*f*): closing (locking) screw 944

VERSCHLUSSSTÜCK (*n*): stopper; plug; closing piece 945

VERSCHMACHTEN: to faint; languish 946

VERSCHMÄHEN: to scorn; disdain; despise 947

VERSCHMÄLERN: to diminish; swage; narrow; constrict; reduce 948

VERSCHMELZEN: to fuse; melt; smelt; melt (together) away; blend; merge; coalesce 949

VERSCHMELZUNG (*f*): melting; fusion; smelting; coalescence; blending 950

VERSCHMERZEN: to get over; bear patiently; cease (stop) grieving over or about; brook 951

VERSCHMIEREN: to lute; smear; daub; plaster (up) 952

VERSCHMITZT: subtle; artful; sly; crafty; cunning; shrewd 953

VERSCHMUTZEN: to dirty; soil; foul 954

VERSCHNEIDEN: to cut; prime; clip; castrate; adulterate; blend 955

VERSCHNEIDUNG (*f*): castration; adulteration 956

VERSCHNITT (*m*): cuttings; chips; adulteration; blending 957

VERSCHNITTSORTE (*f*): type (kind, sort) of adulteration; adulterated (type, kind) sort 958

VERSCHNITTWEIN (*m*): adulterated wine 959

VERSCHOLLEN: lost; missing; unknown; deceased 960

VERSCHONEN: to spare; excuse; forbear 961

VERSCHÖNERN: to beautify; embellish 962

VERSCHRÄNKEN: to cross (the arms); interlace 963

VERSCHRAUBEN: to screw (up or on); to screw wrongly 964

VERSCHRAUBUNG (*f*): screwing, screw-(ed) coupling, screwed joint 965

VERSCHREIBEN: to prescribe; order; miswrite; transfer (make a mistake) in writing; use up in writing; assign; give (promise; bind oneself) in writing 966

VERSCHREIBUNG (*f*): prescription; order; note; bond; assignment; obligation; promise (in black and white) 967

VERSCHREIEN: to decry; cry down 968

VERSCHROBEN: tangled; distorted; queer; eccentric; preposterous; perverse 969

VERSCHRUMPFEN: to contract; shrink 970

VERSCHUB (*m*): procrastination; delay 971

VERSCHULDEN: to owe, merit, be guilty 972

VERSCHULDET: indebted; in debt; merited; embarrassed; guilty; encumbered 973

VERSCHÜTTEN: to shed; spill; choke; bury; fill up 974

VERSCHWEIGEN: to suppress; to keep secret; conceal 975

VERSCHWELEN: to smoulder; burn slowly; distil (Tar, etc.) 976

VERSCHWELUNG (*f*): distillation (Tar, etc.) 977

VERSCHWENDEN: to squander; waste; dissipate; be (prodigal) spendthrift 978

VERSCHWENDER (*m*): squanderer; prodigal; spendthrift 979

VERSCHWIEGEN: discreet; reserved; secretive; taciturn; silent 980

VERSCHWIEGENHEIT (*f*): secrecy; discretion; reserve; taciturnity; silence 981

VERSCHWIMMEN: to blend; run into one another (Colours); vanish; dissolve 982

VERSCHWINDEN: to vanish; disappear 983

VERSCHWINDEND GERING: infinitely small; extremely small; practically nil; infinitesimal 984

VERSCHWINDEND KLEIN: (see VERSCHWINDEND GERING) 985

VERSCHWOMMEN: indistinct; indefinite; blurred; fuzzy (of Photos.) (see VERSCHWIMMEN) 986

VERSCHWÖREN: to conspire; forswear; swear; adjure; take an oath; plot 987

VERSCHWÖRUNG (*f*): conspiracy; plot 988

VERSEGELUNG (*f*): sailing out of (deviating from) a given course 989

VERSEHEN: to furnish; supply; provide; attend to; perform; conduct; expect; be mistaken in; be aware; invest with; be in error; make a mistake; provide oneself; (*n*): error; mistake; oversight; blunder 990

VERSEIFBAR: saponifiable 991

VERSEIFEN: to saponify 992

VERSEIFUNG (*f*): saponification 993

VERSEIFUNGSFASS (*n*): saponification tun 994

VERSEIFUNGSPRODUKT (*n*): saponification product 995

VERSEIFUNGSZAHL (*f*): saponification (value) number 996

VERSENDEN: to forward; send; export; transmit; consign; dispatch; convey; expedite 997

VERSENDER (*m*): shipper; sender; exporter; consigner 998

VERSENGEN: to scorch; singe 999

VERSENKEN: to sink; submerge 000

VERSENKER (*m*): countersink; sinker 001

VERSETZBAR: mixable; miscible; capable of being mixed, compounded, etc. 002

VERSETZEN: to stop; allay; tamp; displace; obstruct; translocate; transplant; remove; transpose; pledge; pawn; mortgage; misplace; be stopped up; change course; mix; treat; alloy; compound; handle (Hides); reply; rejoin 003

VERSETZEN, MIT-: to add to; replace by; substitute 004

VERSETZUNG (*f*): compounding, etc.; alligation (Maths.); permutation; retention (Med.) 005

VERSICHERER (*m*): Insurer; Insurance Company 006

VERSICHERN: to insure; assure; assert; affirm; make sure of; ascertain; ensure; underwrite; convince 007

VERSICHERUNG (*f*): insurance; assurance; affirmation; conviction 008

VERSICHERUNGSAGENT (*m*): insurance agent, underwriter 009

VERSICHERUNGSANSTALT (*f*): insurance office 010

VERSICHERUNGSBEDINGUNGEN (*f.pl*): conditions (terms) of insurance 011

VERSICHERUNGSGEBÜHR (*f*): insurance premium 012

VERSICHERUNGSGELD (*n*): insurance (money) premium 013
VERSICHERUNGSGESELLSCHAFT (*f*): underwriters; insurance company 014
VERSICHERUNGSNEHMER (*m*): insured person; one who takes out a policy 015
VERSICHERUNGSSCHEIN (*m*): insurance policy; bond 016
VERSICHERUNGSURKUNDE (*f*): policy 017
VERSICKERN LASSEN: to allow to percolate or drain away 018
VERSICKERUNG (*f*): percolation 019
VERSIEDEN: to boil away 020
VERSIEGELTES SCHREIBEN (*n*): depositum; pli-cachete; sealed document (a sealed manuscript of an invention which one does not at the moment wish to patent. Placed in the hands of a notary, or technical paper in Germany with the object of being able to prove date of invention should a patent be taken out later by someone else, and to be able to claim an interest in the patent) 021
VERSIEGEN: to be (drained) exhausted; dry up; decay 022
VERSIERT: versed 023
VERSILBERN: to silver-plate; electro-plate; silver; realize; turn into (money) cash; discount 024
VERSILBERT, GALVANISCH-: electro-plated 025
VERSILBERUNG (*f*): (silver)-plating; silvering 026
VERSILBERUNG (*f*), **GALVANISCHE-**: electro-plating 027
VERSILBERUNGSBAD (*n*): silver bath 028
VERSILBERUNGSFLÜSSIGKEIT (*f*): silvering fluid 029
VERSINKEN: to sink 030
VERSINNBILDEN: to represent; symbolize 031
VERSINNBILDLICHEN: to represent; symbolize 032
VERSINNLICHEN: to represent, symbolize 033
VERSKLAVUNG (*f*): enslavement 034
VERSMASS (*n*): measure (of poetry) 035
VERSÖHNEN: to reconcile; appease; propitiate; mediate; placate; pacify 036
VERSÖHNLICH: forgiving; placable; reconcilable 037
VERSORGEN: to provide-(for); supply; maintain; sustain 038
VERSORGEN, (MIT-): to provide-(with) 039
VERSORGUNG (*f*): provision; supply; maintenance; situation 040
VERSPÄTEN: to be late; delay; retard; be overdue (of trains, boats, etc.) 041
VERSPERREN: to block; obstruct; bar; barricade 042
VERSPIEKERN: to nail, spike 043
VERSPIELEN: to gamble-(away) 044
VERSPLEISSEN: to splice 045
VERSPOTTEN: to deride; ridicule; scoff at; mock 046
VERSPRECHEN: to promise; engage; be engaged; make a slip of the tongue; (*n*): promise 047
VERSPRENGEN: to scatter; disperse 048
VERSPRITZEN: to squirt; spurt; shed; spill 049
VERSPUNDEN: to plough; bung, cork up, stopper 050
VERSPÜNDEN: to bung; cork up; stopper 051
VERSPÜREN: to feel; perceive; notice 052
VERSTÄBEN: to chamfer 053
VERSTÄHLEN: to steel; to (coat; cover) case with steel; edge with steel; convert into steel 054
VERSTAND (*m*): intelligence; sense; understanding; reason; judgment; appreciation 055
VERSTANDESKRAFT (*f*): intellectual (power) faculty 056
VERSTANDESLOS: senseless; nonsensical 057
VERSTANDESSCHÄRFE (*f*): sagacity, penetration, perspicacity 058
VERSTÄNDIG: intelligent; sensible; reasonable; clever; judicious 059
VERSTÄNDIGEN: to inform; arrange; agree; come to terms 060
VERSTÄNDLICH: comprehensible; intelligible 061
VERSTÄNDNIS (*n*): comprehension; intelligence; agreement; arrangement; understanding 062
VERSTÄNDIGUNG (*f*): arrangement; explanation; understanding 063
VERSTÄRKEN: to strengthen; succour; increase; amplify; thicken; stay; stiffen; reinforce; fortify; concentrate; intensify (Photo.) 064
VERSTÄRKER (*m*): intensifier (Photo.); amplifier (Wireless Tel.); strengthener; stiffener; stay 065
VERSTÄRKERRÖHRE (*f*): (valve)-amplifier (Wireless Tel.) 066
VERSTÄRKUNG (*f*): strengthening; strengthener; stiffening; stiffener; re-inforcement; stay; succour 067
VERSTÄRKUNG (*f*), **LOKALE-**: local intensification (Photo., etc.) 068
VERSTÄRKUNGSBAND (*n*): strengthening band or hoop, hoop iron 069
VERSTÄRKUNGSGRAD (*m*): degree of intensification; degree of amplification (Wireless Tel.) 070
VERSTÄRKUNGSRIPPE (*f*): angle iron (hoop, ring, rib, support, strengthener) 071
VERSTÄRKUNGSSCHIRM (*m*): intensifying screen (Photo.) 072
VERSTATTEN: to allow; permit 073
VERSTÄUBEN: to convert into dust; dust away; scatter like dust; atomize; pulverize 074
VERSTÄUBER (*m*): spray; atomizer; diffuser; sprayer; vapourizer 075
VERSTAUCHEN: to strain; sprain; dislocate 076
VERSTAUEN: to stow 077
VERSTAUUNG (*f*): stowage 078
VERSTECHEN: to adulterate; blend 079
VERSTECK (*n*): place of concealment; ambush; hiding place 080
VERSTECKEN: to hide; conceal; lurk 081
VERSTECKT: hidden; concealed; insincere 082
VERSTEHEN: to conceive; understand; appreciate; comprehend; gather; consent; agree; mean; have an understanding; be skilled (good) at 083
VERSTEHEN, SICH- MIT: to include; consent; agree 084
VERSTEIFEN: to reinforce, stiffen, strengthen 085
VERSTEIFUNG (*f*): stiffener, stiffening, reinforcement 086
VERSTEIGERER (*m*): auctioneer 087
VERSTEIGERN: to put up for auction; sell by auction 088
VERSTEIGERUNG (*f*): auction-(sale) 089
VERSTEINERN: to petrify 090
VERSTEINERUNG (*f*): petrifaction 091
VERSTEINUNG (*f*): devitrification (Metal) 092
VERSTELLBAR: adjustable; movable 093

VERSTELLEN : to adjust; move; actuate; shift; transpose; misplace; disguise; disfigure; obstruct; dissemble; replace; substitute (be substituted) for; take the place of; oust 094
VERSTELLKRAFT (*f*): actuating (moving) power or force 095
VERSTELLT : fictitious; feigned; simulated 096
VERSTEMMEN : to caulk 097
VERSTEMMER (*m*): caulker 098
VERSTEUERN : to pay (taxes) duty on; charge (taxes) duty on; assess for taxation 099
VERSTEUERT : duty-paid; taxed 100
VERSTIMMT : out of (humour) tune 101
VERSTOCKEN : to harden; be obdurate; grow stubborn; rot; mould 102
VERSTOHLEN : furtive; stealthy; clandestine; secret-(ly) 103
VERSTOPFEN : to stop(up); block; choke; obstruct; clog; constipate (Med.) 104
VERSTOPFUNG (*f*): blockage; stoppage; choking; clogging, impediment; obstruction; constipation; costiveness (Med.) 105
VERSTORBEN : late; deceased; defunct 106
VERSTÖREN : to trouble; confuse; disturb; derange; disperse; dissipate 107
VERSTOSSEN : to reject; repel; cast (out) off; offend; make a mistake; blunder; drive away; disown; expel; repudiate 108
VERSTREBEN : to strut 109
VERSTREBUNG (*f*): strut, strutting 110
VERSTREICHEN : to stop (fill) up; spread over; expire; elapse; pass (away) 111
VERSTREUEN : to scatter; disperse 112
VERSTRICKEN : to entangle; ensnare 113
VERSTÜMMELN : to maim; mutilate; mangle 114
VERSTUMMEN : to become (dumb; mute) silent 115
VERSUCH (*m*): assay; test; experiment; trial; effort; attempt; research; essay 116
VERSUCHEN : to try; test; sample; taste; endeavour; experience; experiment; attempt; essay; tempt; seduce; entice 117
VERSUCHER (*m*): tempter; seducer; experimenter; tester 118
VERSUCHPLAN (*m*): test program 119
VERSUCHSANLAGE (*f*): test (experimental) plant 120
VERSUCHSANORDNUNG (*f*): test arrangement; test plant lay-out; method of test 121
VERSUCHSDAUER (*f*): duration of test; length of test 122
VERSUCHSFEHLER (*m*): experimental error 123
VERSUCHSFELD (*n*): field for experiment 124
VERSUCHSLABORATORIUM (*n*): experimental laboratory 125
VERSUCHSMATERIAL (*n*): experimental material 126
VERSUCHSMETHODE (*f*): experimental method 127
VERSUCHSREIHE (*f*): series of (experiments) tests 128
VERSUCHSSERIE (*f*): (see VERSUCHSREIHE) 129
VERSUCHSSTAB (*m*): test (rod) bar 130
VERSUCHSSTADIUM (*n*): research (test) stage 131
VERSUCHSSTATION (*f*): test station 132
VERSUCHSSTELLE (*f*): test station; place of test 133
VERSUCHSVERHÄLTNISSE (*n.pl*): test conditions 134
VERSUCHUNG (*f*): temptation; seduction; experiment 135
VERSUDELN : to soil; sully; dirt; daub; mess 136
VERSUMPFEN : to become (marshy) boggy 137
VERSÜSSEN : to sweeten; edulcorate; oversweeten; purify by washing 138
VERSÜSSTER SALPETERGEIST (*m*): sweet spirit of nitre; spirit of nitrous ether (Pharm). 139
VERTÄFELUNG (*f*): panelling, wainscotting 140
VERTAGEN : to adjourn 141
VERTÄUEN : to moor, tie up (a boat or ship); (*n*): mooring 142
VERTAUSCHEN : to exchange; interchange; change; barter; permute; transmute; transmogrify 143
VERTAUSCHUNG (*f*): exchange; interchange; change; bartering; permutation 144
VERTEIDIGEN : to defend; maintain 145
VERTEIDIGER (*m*): defendant; defender 146
VERTEIDIGUNG (*f*): defence; protection; advocacy; apology 147
VERTEILEN : to distribute; disperse; divide; diffuse; disseminate; dispense; assign; apportion; spread 148
VERTEILER (*m*): distributor 149
VERTEILUNG (*f*): distribution; repartition (see VERTEILEN) 150
VERTEILUNGSGESETZ (*n*): law of distribution 151
VERTEILUNGSHAHN (*m*): branch cock, distributing cock 152
VERTEILUNGSKRAFT (*f*): distributing (power) force, condensing force 153
VERTEILUNGSMITTEL (*n*): distribution agent 154
VERTEILUNGSPLATTE (*f*): division plate, baffle, partition 155
VERTEILUNGSROHR (*n*): distribution tube or pipe; blast main (of blast furnaces) 156
VERTEILUNGSSATZ (*m*): distribution (principle) law 157
VERTEILUNGSSCHIENE (*f*): distributing (bar) busbar (Elect.) 158
VERTEILUNGSVERMÖGEN (*n*): inductive capacity (Elect.) 159
VERTEUERN : to raise (or rise) in price; enhance 160
VERTIEFEN : to deepen; bury oneself (in a book); be (absorbed; lost) deeply engaged; become more intense 161
VERTIEFUNG (*f*): deepening; cavity; depression; groove; dip; indentation; hollow 162
VERTIKAL : vertical-(ly) 163
VERTIKALACHSE (*f*): vertical axis 164
VERTIKALE (*f*): vertical line, upright 165
VERTIKALGALVANOMETER (*n*): balance galvanometer 166
VERTIKALSCHNITT (*m*): vertical section 167
VERTILGEN : to eradicate; exterminate; extirpate; destroy 168
VERTILGUNG (*f*): extermination, destruction, extirpation, eradication 169
VERTRAG (*m*): agreement; contract; treaty; bargain; convention 170
VERTRAGEN : to tolerate; endure; suffer; bear; agree; harmonize; be (consistent) compatible 171
VERTRÄGLICH : compatible; amicable; consistent; sociable 172
VERTRÄGLICHKEIT (*f*): compatibility; amicability; sociability 173
VERTRÄGLICHKEITSBEDINGUNG (*f*): compatibility condition 174

VERTRAGSDAUER (*f*): duration of contract, contract period 175
VERTRAGSMÄSSIG: in accordance with (contract)-agreement, as agreed, (as) stipulated, (as) arranged 176
VERTRAUEN: to (en)-trust; rely upon; confide-(in); (*n*): trust; confidence 177
VERTRAUEN SCHENKEN: to repose confidence in 178
VERTRAUENSSACHE (*f*): private and confidential matter 179
VERTRAULICH: confidential; intimate; familiar 180
VERTRAUT: conversant; intimate; proficient; familiar 181
VERTREIBEN: to dispel; drive (chase; pass) away; sell; soften (Colours); expel; disperse; dissipate; exile; banish 182
VERTRETBAR: replaceable; deputizable 183
VERTRETBARKEIT (*f*): replaceability; deputizability 184
VERTRETEN: to replace; represent; deputize; intercede for; take the place of; obstruct; strain; stretch; act for 185
VERTRETER (*m*): deputy; representative; intercessor; substitute; agent 186
VERTRETERBÜRO (*n*): agency 187
VERTRETUNG (*f*): agency; representation; intercession 188
VERTRIEB (*m*): market; sale; disposal 189
VERTROCKNEN: to dry up; drain; wither; fade; dessicate 190
VERTROCKNUNG (*f*): drying up; dessication 191
VERTRÖSTEN: to soothe; console; appease 192
VERTUN: to waste; lavish; dissipate; squander; spend 193
VERÜBELN: to take (amiss) ill 194
VERÜBEN: to commit; perpetrate 195
VERUNEHREN: to dishonour; defame; disgrace 196
VERUNEINIGEN: to quarrel; disunite; disjoin; disassociate; disagree; set at variance; cause discord; fall out 197
VERUNGLIMPFEN: to discredit; disparage; slander; calumniate; asperse; defame 198
VERUNGLÜCKEN: to come to grief; meet with an accident; be wrecked; perish 199
VERUNREINIGEN: to soil; vitiate; pollute; contaminate; defile; foul 200
VERUNREINIGUNG (*f*): pollution; impurity; contamination; soiling 201
VERUNSTALTEN: to deform; disfigure 202
VERUNTREUEN: to defraud; embezzle 203
VERUNZIEREN: to deface; disfigure; mar 204
VERURSACHEN: to produce; incur; cause; induce; bring about 205
VERURTEILEN: to sentence; condemn; doom 206
VERVEIN (*n*): verveine 207
VERVIELFACHEN: to multiply; copy; duplicate 208
VERVIELFÄLTIGEN: to multiply; copy; duplicate 209
VERVIELFÄLTIGUNG (*f*): multiplying, copying, duplicating 210
VERVOLLKOMMNEN: to improve; perfect; ground oneself 211
VERVOLLSTÄNDIGEN: to complete 212
VERVOLLSTÄNDIGUNG (*f*): completion; complement 213
VERWACHSEN: deformed; overgrown; choked with weeds; to grow (over; together) crooked; coalesce; be overgrown 214
VERWAHREN: to preserve; keep; guard 215
VERWAHRLOSEN: to neglect; spoil; slight 216
VERWAHRUNG (*f*): preservation; custody; keeping; protest 217
VERWAIST: destitute; orphan; deserted; abandoned; forsaken 218
VERWALTEN: to administer; manage 219
VERWALTER (*m*): stewart, agent (for estates, etc.), manager, administrator 220
VERWALTUNG (*f*): administration; management; Board of Directors 221
VERWALTUNGSRAT (*m*): board of directors 222
VERWANDELBAR: convertible; transformable 223
VERWANDELBARKEIT (*f*): convertibility; transformability 224
VERWANDELN: to convert; transmute; transform; transmogrify; change; turn; metamorphose; alter 225
VERWANDLUNG (*f*): metamorphosis, etc. (see VERWANDELN) 226
VERWANDT: related; akin; kin; cognate; allied; analogous; consanguineous; affiliated; connected 227
VERWANDTE (*m.* and *f*): relation; relative 228
VERWANDTSCHAFT (*f*): relationship; affinity; kindred; connection; tie 229
VERWANDTSCHAFTLICH: allied; congenital 230
VERWANDTSCHAFTSLEHRE (*f*): doctrine of affinity 231
VERWARNEN: to warn 232
VERWASCHEN: faded; washed-out; to wash(away) out 233
VERWÄSSERN: to dilute; water; weaken 234
VERWECHSELN: to exchange; confound; mix up 235
VERWEGEN: forward; rash; daring; audacious; temerarious; precipitate 236
VERWEGENHEIT (*f*): temerity; audacity; presumption; foolhardiness 237
VERWEHEN: to blow (away) over 238
VERWEHREN: to forbid; refuse; prevent; hinder; prohibit 239
VERWEICHLICHT: effeminate 240
VERWEIGERN: to decline; refuse; deny 241
VERWEILEN: to tarry; stay; sojourn; delay 242
VERWEISEN: to refer; banish; rebuke; expel; reprimand; reproach; censure; reprove 243
VERWEISUNG (*f*): reference; banishment; reproof 244
VERWELKEN: to fade; wither; decay 245
VERWENDBAR: applicable; available; utilizable 246
VERWENDEN: to apply; bestow upon; employ; use; utilize; invest; spend; turn away; intercede 247
VERWENDUNG (*f*): employment; use; appropriation; intercession; application; utilization 248
VERWERFEN: to reject; refuse; repudiate misplace; abandon; warp; dislocate (Mining) 249
VERWERFUNG (*f*): rejection; refusal; repudiation; slip; throw; fault; dislocation (Mining) 250
VERWERTBAR: conservable; utilizable; convertible (into money) 251
VERWERTEN: to conserve; turn to account; utilize; effectively (use) employ 252
VERWERTEN FÜR: to use for 253

VERWERTETE WÄRME (*f*): conserved heat; effective (utilizable) heat 254
VERWERTUNG (*f*): conservation; utilization (see VERWERTEN): 255
VERWESEN: to decay; moulder; rot; decompose; putrefy; manage; administer 256
VERWESUNG (*f*): decay; decomposition; administration 257
VERWESUNGPROZESS (*m*): process of (decomposition; decay) putrefaction 258
VERWICHEN: former; past; last 259
VERWICKELN: to involve; entangle; complicate; intrigue 260
VERWICKELT: complicated; involved 261
VERWICKELUNG (*f*): complexity; plot; complication 262
VERWIEGEN: to weigh-(out or wrongly) 263
VERWILLIGEN: to grant; permit; allow; consent 264
VERWINDEN: to recover from; get over 265
VERWIRKLICHEN: to realize; bring to (perfection) maturity; make (concrete) real 266
VERWIRREN: to (en)-tangle; embarrass; perplex; complicate; puzzle; distract; confound; confuse 267
VERWIRRT: complex; complicated; confused; distracted; puzzled; disturbed; perplexed 268
VERWISCHEN: to efface; cancel; nullify; blot (wipe) out or away; smudge; smear 269
VERWITTERBARKEIT (*f*): weathering capacity (of rocks or stones) 270
VERWITTERN: to effloresce; weather; be weather-beaten 271
VERWITTERT: weathered 272
VERWITTERTER KALK (*m*): air-slaked lime (see CALCIUMHYDROXYD) 273
VERWITTERUNG (*f*): efflorescence; weathering 274
VERWÖHNEN: to pamper; spoil 275
VERWORREN: complex; complicated; confused; distracted; puzzled; disturbed; perplexed; intricate 276
VERWUNDBAR: vulnerable 277
VERWUNDEN: to wound 278
VERWUNDERN: to wonder; astonish; surprise; admire 279
VERWÜNSCHEN: to damn; curse; imprecate; execrate; bewitch 280
VERWÜNSCHT: damned; cursed; bewitched 281
VERWÜSTEN: to lay waste; devastate; waste; squander; destroy; desolate 282
VERZAGEN: to despair; despond; be (faint-hearted) timid 283
VERZÄHLEN: to miscount 284
VERZAHNEN: to cog; indent; tooth; gear 285
VERZAHNUNG (*f*): gear, gearing, toothing 286
VERZAHNUNG (*f*), KONISCHE-: conical (bevel) gear or gearing; angle (angular) gear or gearing 287
VERZAHNUNG (*f*), LIEGENDE-: horizontal toothing 288
VERZAHNUNG (*f*), STEHENDE-: upright (vertical) toothing 289
VERZAHNUNG (*f*), ZYLINDRISCHE-: cylindrical (gear) gearing 290
VERZAPFEN: to mortise; join; retail (Liquors); tap 291
VERZAPFUNG (*f*): mortising; mortice joint 292
VERZEHREN: to consume; spend; waste 293
VERZEICHNEN: to record; catalogue; note; specify; draw wrongly; quote; set (write) down 294
VERZEICHNIS (*n*): register; list; index; inventory; invoice; catalogue; schedule 295
VERZEICHNUNG (*f*): distortion (Phot.) 296
VERZEIHEN: to excuse; pardon; forgive; remit 297
VERZEIHLICH: excusable; pardonable 298
VERZERREN: to distort 299
VERZICHT (*m*): renunciation; resignation 300
VERZICHTEN: to renounce 301
VERZIEHEN: to distort; warp; buckle; withdraw; disappear; move; delay; tarry 302
VERZIEHEN (*n*): distortion, warping (of wood) 303
VERZIEREN: to embellish; ornament; decorate; illustrate; adorn 304
VERZIERUNG (*f*): embellishment; ornament; decoration; adornment 305
VERZINKEN: to galvanize; zinc; coat (line) with zinc; dovetail; joggle 306
VERZINKEREI (*f*): galvanizing-(works) 307
VERZINKTES EISEN (*n*): galvanized iron 308
VERZINKUNG (*f*): galvanizing; zincing; zinc coating; dovetailing; joggling 309
VERZINNEN: to tin; line (coat) with tin 310
VERZINNTES EISENBLECH (*n*): tin-plate 311
VERZINNUNG (*f*): tinning 312
VERZINSEN: to pay (yield) interest; put out at interest 313
VERZINSUNG (*f*): interest; placing out at interest; investment; payment (calculation) of interest on 314
VERZÖGERER (*m*): retarder; retarding agent; restrainer (Phot.); dawdler; procrastinator 315
VERZÖGERN: to retard; defer; postpone; adjourn; delay; procrastinate; prolong; protract; linger; dawdle; loiter; suspend 316
VERZÖGERUNG (*f*): retardation; postponement; delay; procrastination 317
VERZÖGERUNGSMITTEL (*n*): retarding agent (Phot., etc.) 318
VERZOLLBAR: dutiable; taxable 319
VERZOLLEN: to charge (pay) duty or toll on 320
VERZOLLT: duty paid 321
VERZUCKERN: to sugar; candy; saccharify 322
VERZUCKERUNG (*f*): candying; sugaring; saccharification 323
VERZÜCKT: enraptured; entranced 324
VERZUCKUNG (*f*): convulsion 325
VERZUG (*m*): delay; retardation 326
VERZUNDERN: to cover with scale (Metal); flake; exfoliate 327
VERZWEIFELN: to despair 328
VERZWEIFELT: desperate 329
VERZWEIGEN: to ramify; branch; fork; bifurcate 330
VERZWEIGUNG (*f*): ramification; branching; forking; bifurcation 331
VERZWEIGUNGSSTELLE (*f*): branch; fork; bifurcation 332
VERZWILLINGT: twinned 333
VESALVIN (*n*): vesalvine 334
VESEN (*m*): raw spelt (see ROHSPELZ) 335
VESUVIAN (*m*): vesuvianite, californite; $HCa_2Al(SiO_4)_2$; Sg. 3.4 (see IDOKRAS) 336
VESUVIN (*n*): vesuvin 337
VETERINÄRPRAXIS (*f*): veterinary practice 338
VETIVERÖL (*f*): (see IVARANCHUSAÖL) 339
VETIVERWURZEL (*f*): vetiver root (see IVARAN-CHUSAWURZEL) (*Radix vetiveris*) 340
VEXIEREN: to puzzle; vex; tease 341

VEXIEREREI (*f*): vexation 342
V-FÖRMIG: V-shaped 343
VIBRATION (*f*): vibration 344
VIBRIEREN: to vibrate 345
VIBROGRAM (*m*): vibrogram, vibrograph (record) diagram 346
VIBROGRAPH (*m*): vibrograph (an instrument for recording vibrations) 347
VIBRONA (*n*): vibrona 348
VIBURNUMRINDE (*f*): black-haw bark (of *Viburnum prunifolium*) 349
VICEADMIRAL (*m*): vice-admiral 350
VICEKÖNIG (*m*): viceroy 351
VICEKONSUL (*m*): vice-consul 352
VICEPRESIDENT (*m*): vice-president 353
VIDALSCHWARZ (*n*): vidal black 354
VIEH (*n*): beast; cattle; brute; stock 355
VIEHARZNEI (*f*): veterinary (medicine) remedy 356
VIEHARZNEIKUNDE (*f*): veterinary (surgery) science 357
VIEHARZT (*m*): veterinary surgeon 358
VIEHDÜNGER (*m*): stable manure 359
VIEHFUTTER (*n*): fodder; forage 360
VIEHHÄNDLER (*m*): cattle-dealer 361
VIEHISCH: brutal; bestial; beastly 362
VIEHMARKT (*m*): cattle-market 363
VIEHSALZ (*n*): rock salt; cattle salt 364
VIEHSEUCHE (*f*): murrain 365
VIEHSTALL (*m*): stable, stall, cow-house 366
VIEHSTAND (*m*): stock-(of cattle) 367
VIEHWEIDE (*f*): pasturage 368
VIEHZUCHT (*f*): cattle-breeding, cattle-rearing 369
VIEL: much; many; (as a prefix=multi-; poly-) 370
VIELDEUTIG: ambiguous 371
VIELECK (*n*): polygon 372
VIELECKIG: polygonal; many angled 373
VIELERLEI: multifarious; various; many; diverse; divers 374
VIELFACH: manifold; various; multifarious 375
VIELFACHE (*n*): multiple 376
VIELFACH-TELEGRAPHIE (*f*): multiplex telegraphy 377
VIELFÄLTIG: manifold; various; multifarious 378
VIELFÄLTIGKEIT (*f*): variety; multiplicity 379
VIELFARBIG: variegated; polychromatic; many-coloured 380
VIELFLACH (*n*): polyhedron 381
VIELFLÄCHIG: polyhedral; many-sided 382
VIELFÖRMIG: polymorphous; multiform 383
VIELFRÄSSIG: gluttonous; voracious 384
VIELGLIEDERIG: polynominal; of many parts 385
VIELGÖTTEREI (*f*): polytheism 386
VIELHEIT (*f*): plurality 387
VIELKRYSTALLPROBE (*f*): multicrystal test or test piece 388
VIELLEICHT: possibly; perhaps 389
VIELMALIG: reiterated; often repeated 390
VIELMEHR: rather; more; much more 391
VIELPOLIG: multipolar 392
VIELSAGEND: significant; expressive 393
VIELSÄURIG: polybasic; polyacid 394
VIELSEIT (*n*): polyhedron 395
VIELSEITIG: polyhedral; many-sided 396
VIELSILBIG: polysyllabic 397
VIELTEILIG: polynominal; of many parts 398
VIELVERSPRECHEND: (very)-promising; having (great) possibilities 399
VIELWERTIG: polyvalent; multivalent 400
VIELZELLIG: multicellular 401
VIER: four 402
VIERATOMIG: tetratomic 403
VIERBASISCH: tetrabasic 404
VIERBEINIG: four-legged 405
VIERDRITTELKOHLENSAURES NATRIUM (*n*): ($Na_2CO_3.2NaHCO_3.3H_2O$); trona (see also TRONA) 406
VIERECK (*n*): quadrangle; square 407
VIERECKIG: quadrangular; square; four-cornered 408
VIERERLEI: of four different kinds 409
VIERERSALZ (*n*): magnesium chloride; $MgCl_2.4H_2O$ 410
VIERFACH: quadruple; fourfold; quadruplex; quaternary 411
VIERFACH-CHLORKOHLENSTOFF (*m*): carbon tetrachloride (see TETRACHLORKOHLENSTOFF) 412
VIERFACHEXPANSIONSMASCHINE (*f*): quadruple expansion engine 413
VIERFACH OXALSAUER: tetroxalate 414
VIERFÄLTIG: fourfold; quadruple; quaternary; quadruplex 415
VIERFLACH (*n*): tetrahedron 416
VIERFLÄCHIG: tetrahedral; four-faced 417
VIERFLÄCHNER (*m*): tetrahedron 418
VIERFLÜGLIG: four-bladed 419
VIERFUSS (*m*): four-footed stand 420
VIERFÜSSIG: four-footed; quadruped 421
VIERGLIEDRIG: four-membered; tetragonal (Cryst.); quadrinominal (Maths.) 422
VIERJÄHRIG: quadrennial; four years old 423
VIERKANT (*n*): square 424
VIERKANTEISEN (*n*): square (iron) bar-(s) (Metal) 425
VIERKANTFEILE (*f*): square file 426
VIERKANTIG: quadrangular; square; four-cornered 427
VIERKURBELIGE DAMPFMASCHINE (*f*): quarter-crank engine 428
VIERMALIG: four times repeated 429
VIERMASTBARK (*f*): four-masted bark 430
VIERMASTVOLLSCHIFF (*n*): four-masted ship 431
VIERRÄDERIG: four-wheeled 432
VIERSÄURIG: tetracid 433
VIERSEIT (*n*): quadrilateral 434
VIERSEITIG: quadrilateral; square; four-sided 435
VIERSILBIG: quadrisyllabic; four-syllable 436
VIERTÄGIG: quartan; four days old 437
VIERTAKT (*m*): four-(stroke; cycle) throw 438
VIERTAKTMOTOR (*m*): four (cycle) stroke motor 439
VIERTE: fourth 440
VIERTEILEN: to quarter 441
VIERTEL (*n*): quarter; fourth 442
VIERTELJAHR (*n*): quarter (three months) 443
VIERTELJÄHRLICH: quarterly; every three months 444
VIERTELPLATTE (*f*): quarter-plate (Phot.) 445
VIERTELSTUNDE (*f*): quarter of an hour 446
VIERTENS: fourthly 447
VIERUNDZWANZIGFLÄCHNER (*m*): icositetrahedron 448
VIERWANDIG: four-walled 449
VIERWEGEHAHN (*m*): four-way cock 450
VIERWEGHAHN (*m*): four-way cock 451
VIERWERTIG: tetravalent; quadrivalent 452

VIERWERTIGKEIT (*f*): tetravalence; quadrivalence 453
VIERZÄHLIG: fourfold 454
VIERZEHN: fourteen 455
VIERZEHN TAGE (*m.pl*): fourteen days; a fortnight 456
VIERZIG: forty 457
VIGNETTE (*f*): vignette (Phot.) 458
VIGNETTIERDECKEL (*m*): vignetting board (Phot.) 459
VIGNETTIEREN: to vignette 460
VILLIAUMIT (*m*): villiaumite; NaF; Sg. 2.8 (see FLUORNATRIUM) 461
VINYLBENZOL (*n*): (see STYROL) 462
VINYLCYANÜR (*n*): vinyl cyanide 463
VINYLTRICHLORID (*n*): vinyl trichloride (see ÄTHYLENCHLORID) 464
VIOFLOR (*n*): vioflor 465
VIOFORM (*n*): vioform 466
VIOLAN (*m*): violane; $(Ca,Fe,Mg,Mn,Na_2)O,Al_2O_3,SiO_2$; Sg. 3.2 467
VIOLAQUERCITRIN (*n*): viola quercitrine (from *Viola tricolor*) 468
VIOLETT (*n*): violet (Colour) 469
VIOLETT: violet-(coloured) 470
VIOLETTHOLZ (*n*): violet wood, Myall (*Acacia pendula*) 471
VIOLETTLACK (*m*): violet lake 472
VIOLETT-SCHWARZ: violet black 473
VIOLETTSTICHIG: violet-tinged 474
VIOLINE (*f*): violin 475
VIOLINESCHLÜSSEL (*m*): treble-clef 476
VIOLON (*m*): bass-viol 477
VIOLONCELL (*n*): violoncello 478
VIPER (*f*): viper, adder 479
VIPERNGIFT (*n*): viper's (poison) venom 480
VIRGINISCH: Virginian 481
VIRIDINLACK (*m*): viridin lake 482
VIROL (*n*): virol (Pharm.) 483
VIROLAFETT (*n*): virola fat (a vegetable fat from the seeds of *Myristica sebifera*; Mp. 45-50°C.) 484
VIRTUELL: virtual-(ly) 485
VIRTUOS (*m*): virtuoso 486
VISCOS: viscous (see VISKOS) 487
VISETHOLZ (*n*): fustic; Sg. 0.63 (from *Rhus cotinus*) 488
VISETHOLZEXTRAKT (*m*): fustic extract 489
VISIER (*n*): visor; sight (of a gun) 490
VISIEREN: to visa; aim; gauge 491
VISIERGRAUPEN (*f.pl*): tinstone, cassiterite (see CASSITERIT) 492
VISIERSCHEIBE (*f*): focussing screen (Phot.) 493
VISION (*f*): vision 494
VISITATION (*f*): visitation; search; (Customs)-; examination 495
VISITENKARTE (*f*): visiting card 496
VISITIEREN: to visit; search; probe; examine 497
VISITKARTE (*f*): Carte-de-visite (Phot.) (size about $4'' \times 2\frac{1}{2}''$) 498
VISKOS: viscous 499
VISKOSIMETER (*n*): viscosimeter 500
VISKOSITÄT (*f*): viscosity 501
VISKOSITÄTSMITTEL (*n*): viscosity (agent) medium 502
VITAMIN (*n*): vitamin 503
VITOLINSCHWARZ (*n*): vitoline black 504
VITREOSIL (*n*): quartz-glass (made from quartz sand; Sg. 2.2; Mp. 2000°C.) 505
VITRIOL (*m*): vitriol 506
VITRIOL (*m*), ADLER-: (see VITRIOL, DOPPEL-) 507
VITRIOLARTIG: vitriolic; like vitriol 508
VITRIOLÄTHER (*m*): ethyl ether (see ÄTHYLÄTHER) 509
VITRIOL (*m*), BLAUER-: blue vitriol (see KUPFERVITRIOL) 510
VITRIOLBLEIERZ (*n*): anglesite (see ANGLESIT) 511
VITRIOLBLEISPAT (*m*): anglesite (see ANGLESIT) 512
VITRIOL (*m*), DOPPEL-: double vitriol (mixed crystallization of EISEN and KUPFERVITRIOL, which see) 513
VITRIOLERZ (*n*): vitriol ore (see VITRIOLKIES) 514
VITRIOLFLASCHE (*f*): carboy 515
VITRIOL, GRÜNER- (*m*): green vitriol (see EISENOXYDULSULFAT) 516
VITRIOLHALTIG: vitriolic; containing vitriol 517
VITRIOLIN (*n*): vitrioline 518
VITRIOLKIES (*m*): marcasite (see MARKASIT) 519
VITRIOLÖL (*n*): oil of vitriol; fuming sulphuric acid (see SCHWEFELSÄURE) 520
VITRIOLÖL, FESTES- (*n*): sulphur trioxide; sulphuric acid (see SCHWEFELSÄURE) 521
VITRIOLSÄURE (*f*): vitriolic acid; sulphuric acid (see SCHWEFELSÄURE) 522
VITRIOLSCHIEFER (*m*): alum slate 523
VITSBOHNE (*f*): kidney bean (*Phaseolus vulgaris*) 524
VIVIANIT (*m*): vivianite (hydrous ferrous phosphate); $3FeO,2P_2O_5,8H_2O$; Sg. 3.0-6.0) see BLAUE ERDE) 525
VIVISEKTION (*f*): vivisection 526
VIZEADMIRAL (*m*): vice-admiral (see also VICE-) 527
VLIES (*n*): fleece; fleecy matter 528
VLIESS (*n*): (see VLIES) 529
VOGEL (*m*): bird; fowl; target (for shooting) 530
VOGELAMBER (*m*): spermaceti (see SPERMAZET) 531
VOGELBAUER (*m*): bird cage 532
VOGELBEERBAUM (*m*): service tree; mountain ash (*Sorbus aucuparia* or *Pyrus aucuparia*) 533
VOGELBEERE (*f*): service berry; mountain ash berry (*Sorbus aucuparia*) 534
VOGELBEERENÖL (*n*): service berry oil (from seeds of *Sorbus aucuparia*) 535
VOGELDUNST (*m*): small (buck) shot 536
VOGELEI (*n*): bird's egg 537
VOGELFÄNGER (*m*): fowler; bird-catcher 538
VOGELFLINTE (*f*): fowling piece 539
VOGELFUTTER (*n*): bird-seed 540
VOGELHÄNDLER (*m*): bird-seller, dealer in birds, bird-fancier 541
VOGELHAUS (*n*): aviary 542
VOGELHECKE (*f*): aviary, breeding cage 543
VOGELKIRSCHE (*f*): bird-cherry (*Prunus padus*) 544
VOGELKUNDE (*f*): ornithology 545
VOGELLEIM (*m*): bird-lime 546
VOGELMIST (*m*): bird dung or droppings 547
VOGELNEST (*n*): bird's nest 548
VOGELPERSPEKTIVE (*f*): bird's eye view 549
VOGELSCHEUCHE (*f*): scarecrow 550
VOGLIT (*m*): voglite; $(Ca,Cu)CO_3,U(CO_3)_2.H_2O$ 551
VOGT (*m*): magistrate; steward; constable; governor; bailiff; judge; prefect 552
VOGTEI (*f*): prefecture; bailiwick 553
VOKAL (*m*): vowel 554
VOKALMUSIK (*f*): vocal music 555
VOLANTLAGER (*n*): flywheel bearing 556

VOLBORTHIT (*m*): volborthite; (Cu,Ca,Ba) V_2O_5,H_2O; Sg. 3.5 557
VOLK (*n*): folk; nation; people; men; forces; herd; flock; crowd; mob; multitude 558
VOLKARM: unpopulated 559
VÖLKERRECHT (*n*): International law, law of nations 560
VÖLKERSCHAFT (*f*): tribe, nation, people 561
VOLKLOS: unpopulated, deserted, vacant 562
VOLKREICH: populous 563
VOLKSBLATT (*n*): newspaper 564
VOLKSFEST (*n*): national (public) festival, or holiday 565
VOLKSHERRSCHAFT (*f*): democracy 566
VOLKSJUSTIZ (*f*): lynch-law 567
VOLKSLIED (*n*): folk song 568
VOLKSMÄSSIG: popular 569
VOLKSTÜMLICH: national; popular 570
VOLKSWIRTSCHAFT (*f*): political economy 571
VOLL: full; flush; solid; whole; entire; gross (of prices); complete; compact; plump (of leather); perfect 572
VOLLAKTIE (*f*): fully paid up share 573
VOLLAUF: (in) plenty; plentifully; (in) abundance; abundantly; in full; fully; ample; amply 574
VOLLAUSGEZOGENE LINIE (*f*): full (unbroken) line (usually in drawings) 575
VOLLBAHN (*f*): full gauge railway. main (trunk) line 576
VOLLBELASTUNG (*f*): full load 577
VOLLBLUT (*n*): thorough-bred 578
VOLLBLÜTIG: plethoric; full-blooded 579
VOLLBLÜTIGKEIT (*f*): plethora; full-bloodedness 580
VOLLBRINGEN: to execute; accomplish; achieve; perform 581
VOLLBÜRTIG: of (full) age 582
VOLLDAMPF (*m*): full steam 583
VOLLDECKSCHIFF (*n*): three-deck vessel or ship; three-decker 584
VOLLDRUCK (*m*): full pressure 585
VOLLENDEN: to complete; terminate; finish; perfect; end; die; accomplish 586
VOLLENDET: perfect; accomplished 587
VOLLENDS: wholly; altogether; quite; entirely; completely; absolutely 588
VOLLFÜHREN: to execute; carry out; perform; accomplish 589
VOLLGEHALT (*m*): full value or capacity 590
VOLLGILTIG: of full value; sufficient 591
VOLLHALTIG: standard; of full value 592
VOLLHEIT (*f*): fulness; plenitude 593
VOLLHUB (*m*)-(VENTIL) (*n*): full-lift-(valve) 594
VÖLLIG: entire-(ly); full-(y); complete-(ly); total-(ly); whole; wholly; quite 595
VÖLLIGKEIT (*f*): fulness; plenitude; corpulence; fineness (beauty and utility of a ship's lines) 596
VÖLLIGKEITSGRAD (*m*): block-coefficient (of a ship); coefficient of fineness 597
VOLLJÄHRIG: of age 598
VOLLJÄHRIGKEIT (*f*): majority; man's estate 599
VOLLKOMMEN: perfect; ideal; entire; complete; accomplished; consumate 600
VOLLKOMMENE FLÜSSIGKEIT (*f*): perfect liquid (a liquid free of friction) 601
VOLLKRAFT (*f*): energy; vigour 602
VOLLLASTBETRIEB, (NORMALER)- (*m*): (normal)-full load 603
VOLLMACHT (*f*): power of attorney; authority; authorization; right 604
VOLLMILCH (*f*): whole milk; full-cream milk 605
VOLLMOND (*m*): full moon 606
VOLLMUNDIG: having good body; full (of liquors) 607
VOLLPIPETTE (*f*): pipette to deliver one definite quantity; pipette without graduated scale 608
VOLLSCHIFF (*n*): full-rigged ship 609
VOLLSCHMIEDEN: to forge in the solid 610
VOLLSCHWARZ (*n*): full black 611
VOLLSPANT (*n*): watertight frame 612
VOLLSTÄNDIG: complete-(ly); entire-(ly); total-(ly); full-(y); integral 613
VOLLSTRECKEN: to carry out; execute; perform 614
VOLLSTRECKEND: executive 615
VOLLWICHTIG: weighty; of full weight; competent 616
VOLLZÄHLIG: complete 617
VOLLZAHLUNG (*f*): payment in full 618
VOLLZIEHEN: to consummate; execute; put into effect; carry out; perform; accomplish 619
VOLT (*n*): volt (Elect.) 620
VOLTAGOMETER (*n*): voltagometer 621
VOLTAISCH: voltaic 622
VOLTAMETER (*n*): voltameter 623
VOLTAMPERE (*n*): voltampere 624
VOLTASCHE SÄULE (*f*): voltaic battery 625
VOLTIGIEREN: to vault 626
VOLTMETER (*n*): voltmeter 627
VOLUM (*n*): volume 628
VOLUMÄNDERUNG (*f*): volumetric (alteration; fluctuation) change 629
VOLUMEINHEIT (*f*): unit of volume 630
VOLUMEN (*n*): volume 631
VOLUMENÄNDERUNG (*f*): volumetric (alteration, fluctuation) change 632
VOLUMENGEWICHT (*n*): volumetric weight 633
VOLUMENOMETER (*n*): volume meter 634
VOLUMETRISCH: volumetric 635
VOLUMGESETZ (*n*): law of volumes 636
VOLUMGEWICHT (*n*): volumetric weight; weight per unit of volume 637
VOLUMINA (*n.pl*): volumes 638
VOLUMINÖS: voluminous 639
VOLUMMESSER (*m*): volumeter 640
VOLUMPROZENT (*n*): percent by volume 641
VOLUMTEIL (*m*): part by volume 642
VOLUMVERGRÖSSERUNG (*f*): volumetric increase 643
VOLUMVERHÄLTNIS (*n*): volumetric (relation) proportion 644
VOLUMVERMEHRUNG (*f*): volumetric increase 645
VOLUMZUNAHME (*f*): volumetric increase 646
VOLUMZUWACHS (*m*): volumetric increase 647
VOMIEREN: to vomit 648
VOMITARTARIN (*n*): tartar emetic substitute 649
VOMITIV (*n*): emetic 650
VON: of; from; about; upon; on; by; von (Title) 651
VONNÖTEN: necessary; in need; needful 652
VOR: before; formerly; of; from; for; on; since; ago; prior to; in preference to; inlet (Mech.) 653
VORABEND (*m*): eve 654
VORABENTSCHÄDIGUNG (*f*): part payment, prepayment 655
VORAHNUNG (*f*): premonition; presentiment 656
VORALLEM: primarily; first of all 657

VORÄLTERN (*pl*): forefathers, ancestors, progenitors 658
VORAN: before; on; in front; at the head; in advance 659
VORANGEHEN: to go before, lead the way, take the lead, go in advance, precede, proceed 660
VORANORDNEN: to arrange (fit; couple; connect) on the inlet side of (Plant) 661
VORANSCHICKEN: to send before; premise; observe beforehand 662
VORANSCHLAG (*m*): estimate 663
VORANZEIGE (*f*): preliminary notice 664
VORARBEIT (*f*): preparation; preparatory (preliminary) treatment or work 665
VORARBEITEN: to prepare-(work); subject to preliminary treatment; surpass 666
VORARBEITER (*m*): foreman 667
VORAUF: on; before 668
VORAUS: in advance; before-(hand); in anticipation 669
VORAUSBERECHNUNG (*f*): estimation; preliminary calculation 670
VORAUSBESTIMMEN: to stipulate 671
VORAUSBEZAHLEN: to pay in advance; prepay 672
VORAUSGEHEN: to go in advance; precede 673
VORAUSGESETZT: provided; presupposed; assumed; assuming 674
VORAUSHABEN: to have (the better of) an advantage over 675
VORAUS, IM-: in advance 676
VORAUSNEHMEN: to anticipate 677
VORAUSSAGE (*f*): prediction 678
VORAUSSAGEN: to foretell; predict; prophesy 679
VORAUSSCHICKEN: to send before; premise 680
VORAUSSEHEN: to foresee; anticipate 681
VORAUSSETZEN: to assume; presuppose; suppose; stipulate 682
VORAUSSETZUNG (*f*): hypothesis; condition; stipulation; assumption; supposition 683
VORAUSSICHT (*f*): foresight; forethought; expectation; anticipation 684
VORAUSSICHTLICH: prospective; probable; most probably; presumably 685
VORBAD (*n*): preliminary bath 686
VORBAU (*m*): projection 687
VORBAUEN: to prevent; obviate; guard against; project; build (before) in front of 688
VORBAUUNG (*f*): prevention; prophylaxis (Med.) 689
VORBAUUNGSMITTEL (*n*): prophylactic (Med.) 690
VORBEARBEITUNG (*f*): preliminary (working) treatment 691
VORBEDACHT (*m*): forethought; premeditation 692
VORBEDACHT, MIT- (*m*): deliberately; on purpose; premeditated; with malice aforethought 693
VORBEDENKEN: to think (consider; deliberate) beforehand 694
VORBEDEUTEN: to forebode; portend; omen 695
VORBEDEUTUNG (*f*): foreboding; premonition; omen 696
VORBEDINGUNG (*f*): preliminary (condition) stipulation; (*pl*): essentials 697
VORBEHALT (*m*): proviso; reservation; restriction 698
VORBEHALTEN: to reserve the right; keep back; condition 699
VORBEHANDLUNG (*f*): preliminary treatment 700
VORBEI: past; done; by; gone; over; finished 701
VORBEI PASSIEREN: to pass (over) by 702
VORBEIZEN: to mordant previously (Dyeing) 703
VORBELICHTUNG (*f*): preliminary (illumination) exposure 704
VORBEMERKUNG (*f*): preliminary remark 705
VORBEREITEN: to prepare 706
VORBEREITUNG (*f*): preparation 707
VORBEREITUNGSMASCHINE (*f*): preparing machine (Textile) 708
VORBERICHT (*m*): introduction; preamble; preface 709
VORBESCHEID (*m*): summons; citation 710
VORBESCHEIDEN: to summons; cite 711
VORBESPRECHUNG (*f*): preliminary discussion 712
VORBESTIMMEN: to predetermine; predestinate 713
VORBEUGEN: to prevent; obviate; bend forward; stoop 714
VORBEUGEND: precautionary; preventive; prophylactic 715
VORBEUGUNG (*f*): prevention; prophylaxis (Med.) 716
VORBEUGUNGSMITTEL (*n*): preventive; prophylactic (Med.) 717
VORBILD (*n*): (proto)-type; pattern; model; emblem; representation 718
VORBILDEN: to prepare; school; typify; represent 719
VORBINDEN: to tie (put) on 720
VORBLOCKEN (*n*): preliminary rolling; blooming (Rolling mills) 721
VORBOHREN: to bore, drill, perforate 722
VORBOTE (*m*): harbinger, forerunner, messenger 723
VORBRAMPARDUNE (*f*): fore top gallant backstay 724
VORBRAMRAH (*f*): fore topgallant yard 725
VORBRAMSEGEL (*n*): fore topgallant sail 726
VORBRECHER (*m*): preliminary crusher (Sugar cane) 727
VORBRENNE (*f*): preliminary pickle (a mixture of nitric acid of Sg. 1.33, lampblack and cooking salt in proportion of 200 : 1.5 : 1 parts by weight) (Electroplating) 728
VORBRINGEN: to produce; adduce; bring forward; utter 729
VORDACH (*n*): penthouse 730
VORDATIEREN: to antedate 731
VORDEM: formerly 732
VORDER: anterior; front; fore 733
VORDERACHSE (*f*): front axle, leading axle, guiding axle 734
VORDERANSICHT (*f*): front (view) elevation 735
VORDERARM (*m*): fore-arm 736
VORDERBLENDE (*f*): front diaphragm, diaphragm in front of a lens (Phot.) 737
VORDERDECK (*n*): forecastle deck (Naut.) 738
VORDERFINGER (*m*): forefinger 739
VORDERFUSS (*m*): instep 740
VORDERGEBÄUDE (*n*): extension (projection; front part) of a building; porch 741
VORDERGLIED (*n*): antecedent; front rank 742
VORDERGRUND (*m*): foreground; front 743
VORDERHAND: provisionally; for the present 744
VORDERHERD (*m*): forehearth 745
VORDERKANTE (*f*): fore edge 746
VORDERLIEK (*n*): fore leech (Naut.) 747
VORDERLUKE (*f*): fore hatch 748
VORDERRAD (*n*): leading (front) wheel 749

VORDERRAUM (*m*): fore hold (Naut.) 750
VORDERSATZ (*m*): antecedent; (*pl*): antecedents; premises 751
VORDERSCHIFF (*n*): fore (body) part (of a ship) 752
VORDERSEITE (*f*): face; front-(side); obverse; front face 753
VORDERSPANT (*n*): fore frame 754
VORDERST: foremost 755
VORDERSTEVEN (*m*): stem 756
VORDERTEIL (*m*): forehead; front; fore-part; head; nose; bow (of a ship); van-(guard) 757
VORDERTREFFEN (*n*): van-(guard), first line (Military) 758
VORDERWAND (*f*): front-(wall); breast (of a blast furnace); face (of a building) 759
VORDERWÜRZE (*f*): first wort (Brew.) 760
VORDERZAHN (*m*): front tooth 761
VORDRÄNGEN: to push (press) forward 762
VORDRINGEN: to push (press) forward; advance; (*n*): advance; advancing 763
VORDRUCK (*m*): proof; first impression; first (bottom) printing (Calico) 764
VORDRUCKEN: to prefix 765
VORDRUCKWALZE (*f*): dandy roll (Paper) 766
VOREILEN: to outrun; hasten-(forward); anticipate; (*n*): lead (Mechanical) 767
VOREILENDER GENERATOR (*m*): leading generator 768
VOREILIG: rash; forward 769
VOREILUNG (*f*): lead (Mechanical) 770
VOREILUNGSWINKEL (*m*): angle of lead 771
VOREILWINKEL (*m*): angle of lead 772
VOREINGENOMMEN: prejudiced; prepossessed 773
VORELTERN (*pl*): ancestors, forefathers 774
VOREMPFINDUNG (*f*): premonition; foreboding; misgiving; anticipation 775
VORENTHALTEN: to withold; keep back; detain 776
VORERINERRUNG (*f*): preface; preamble; premonition 777
VORERNTE (*f*): early harvest 778
VORERST: firstly; before (first of) all; for the present; for the nonce 779
VORERWÄHNT: aforementioned; above-mentioned; aforesaid 780
VORFAHR (*m*): ancestor; predecessor; forefather; progenitor 781
VORFALL (*m*): case; event; happening; accident; incident; occurrence; prolapse (Med.) (*Prolapsus*) 782
VORFALLEN: to occur; happen; prolapse (Med.) 783
VORFÄRBUNG (*f*): preliminary (staining; colouring) dyeing; grounding; bottoming (Dyeing) 784
VORFECHTER (*m*): champion 785
VORFEILE (*f*): bastard file 786
VORFEUERUNG (*f*): projecting furnace; extended furnace; Dutch oven (furnace) 787
VORFINDEN: to meet with; find; come upon; be (forthcoming) found 788
VORFORDERN: to summon (before); cite 789
VORFRISCHEN (*n*): preliminary refining (partial refining and then the removal of the slag) (Metal) 790
VORFÜHREN: to produce; bring (before) out; introduce; demonstrate; lead before 791
VORFÜHRUNG (*f*): demonstration 792
VORGANG (*m*): process; proceeding; procedure; transaction; event; occurrence; precedent; precedence; example; reaction (Chem.); first runnings (Distilling) 793
VORGÄNGER (*m*): predecessor; antecessor; leader 794
VORGÄNGIG: preceding; previous 795
VORGEBÄUDE (*n*): extension (projection) of a building; vestibule; porch; entrance; lobby 796
VORGEBEN: to pretend; feign; suggest; give odds; advance; (*n*): pretext; pretence 797
VORGEBIRGE (*n*): promontory; headland; cape 798
VORGEBLICH: pretended; so-called; would-be 799
VORGEFASST: preconceived 800
VORGEFÜHL (*n*): misgiving; presentiment; premonition 801
VORGEHEN: to precede; proceed; go (before) on; happen; have the preference; be fast (Clocks) 802
VORGELEGE (*n*): gearing; reduction gear-(ing) 803
VORGELEGE (*n*), **RÄDER-**: wheel gearing 804
VORGELEGE (*n*), **REIBRAD-**: friction gear 805
VORGELEGE (*n*), **TRIEBSTOCK-**: pin-wheel gear(-ing) 806
VORGEMACH (*n*): anteroom; antechamber; vestibule 807
VORGEMELDET: above-mentioned, aforementioned 808
VORGENANNT: above-mentioned, aforementioned 809
VORGESCHICHTLICH: prehistoric 810
VORGESCHMACK (*m*): foretaste 811
VORGESCHOBENER POSTEN (*m*): advanced post 812
VORGESETZTE (*m*): chief; head; superior; overseer 813
VORGESTERN: day before yesterday 814
VORGLÜHEN: annealing; preliminary heating 815
VORGLÜHOFEN: annealing (furnace) oven; biscuit kiln (Ceramics) 816
VORGREIFEN: to anticipate; forestall 817
VORHABEN: to purpose; intend; wear; have on; be engaged in; design; (*n*): design; purpose; plan; intention; idea; project 818
VORHALLE (*f*): lobby; antechamber; vestibule; porch; entrance-hall 819
VORHALTEN: to reproach; hold before; last; hold out; charge with 820
VORHAND (*f*): wrist 821
VORHANDEN: existing; present; at (on) hand; available; ready; extant 822
VORHANDENSEIN: to exist; be (extant) available; (*n*): existence; presence; availability 823
VORHANG (*m*): (drop)-curtain 824
VORHÄNGESCHLOSS (*n*): padlock 825
VORHAUPT (*n*): forehead 826
VORHAUT (*f*): foreskin 827
VORHELLING (*f*): launch-(ing) slip 828
VORHER: before; previously 829
VORHERBEDENKEN: to premeditate 830
VORHERBESTIMMEN: to predetermine; predestine 831
VORHERBESTIMMUNG (*f*): predestination 832
VORHERD (*m*): forehearth 833
VORHERGEHEN: to precede; go before 834
VORHERGEHEND: previous; preceding; foregoing; former; prior 835

VORHERIG : previous ; preceding ; foregoing ; former ; prior 836
VORHERSAGEN : to prophesy ; predict ; foretell 837
VORHERSCHEN : to prevail ; predominate 838
VORHERSEHEN : to foresee ; anticipate 839
VORHIN : previously ; before ; lately ; recently ; heretofore ; a short while ago ; just now 840
VORHOF (*m*) : porch ; vestibule ; outercourt ; forecourt ; entry ; entrance 841
VORHOLBEWEGUNG (*f*) : forward motion 842
VORHUT (*f*) : (van)-guard 843
VORIG : preceding ; last ; former ; previous 844
VORJAHR (*n*) : previous year 845
VORKALKULATOR (*m*) : estimator 846
VORKAUF (*m*) : forestalling, pre-emption 847
VORKAUFEN : to forestall, pre-empt 848
VORKEHR (*f*) : precaution ; provision 849
VORKEHREN : to prepare ; predispose ; turn (to the front) forwards ; take precautions 850
VORKEHRUNG (*f*) : precaution ; provision ; precautionary (safety) measure ; predisposition 851
VORKENNTNIS (*n*) : preliminary knowledge ; (*pl*) : elements ; rudiments 852
VORKOMMEN : to occur ; happen ; seem ; appear be (present ; presented) admitted ; crop up ; be found ; come before ; (*n*) : occurrence ; presence 853
VORKOST (*f*) : provisions 854
VORKÜFEN : to bottom with a vat dye 855
VORKÜHLUNG (*f*) : pre-cooling ; preliminary cooling 856
VORLADEN : to summon ; cite 857
VORLADUNG (*f*) : summons ; citation 858
VORLAGE (*f*) : gas header ; receiver ; condenser (Zinc) ; proposal ; matter ; bill ; extension piece ; original ; pattern ; copy 859
VORLAGERN : to extend (before) in front of 860
VORLÄNGST : long (ago) since 861
VORLASS (*m*) : first runnings ; first light-oil (Coal tar) ; foreshot (Whisky) 862
VORLASSEN : to admit ; give precedence to ; give access to 863
VORLASTIG : down by the head (of ships) 864
VORLAUF (*m*) : first runnings ; first light-oil (Coal tar) ; foreshot (Whisky) 865
VORLÄUFER (*m*) : forerunner ; precursor ; sign ; indication ; spur of a mountain 866
VORLÄUFIG : preliminary ; provisional ; previous ; meantime ; for the present ; for the time being ; just now 867
VORLAUT : forward ; noisy ; hasty ; inconsider ate 868
VORLEBEN (*n*) : past ; past career ; record ; antecedents 869
VORLEESEGEL (*n*) : fore studding sail 870
VORLEGEN : to apply ; lay before ; put on ; submit ; place before ; propose ; exhibit ; display 871
VORLEGESCHLOSS (*n*) : padlock 872
VORLEIK (*n*) : fore leech (Naut.) 873
VORLESEN : to deliver ; recite ; read (before ; out) aloud 874
VORLESUNG (*f*) : lecture ; recitation ; reading 875
VORLETZT : last but one ; penultimate 876
VORLIEBE (*f*) : fondness ; preference ; predilection 877
VORLIEGEN : to exist ; be (perceptible) found ; to be ; to have (before one) received ; be prior to (of patents) ; lie (before) in front 878
VORLIEGEND : present, in question, under consideration, existing 879
VORLIEK (*n*) : fore leech (Naut.) 880
VORLUKE (*f*) : fore hatch 881
VORMACHEN : to put (do) before ; do as an example ; deceive 882
VORMAISCHAPPARAT (*m*) : foremashing apparatus (Brew.) 883
VORMAISCHEN : to foremash (Brew.) 884
VORMAISCHER (*m*) : foremasher (Brew.) 885
VORMALS : formerly ; heretofore ; previously 886
VORMANN (*m*) : foreman 887
VORMARS (*m*) : fore top 888
VORMARSBRASSE (*f*) : fore topsail brace 889
VORMARSPARDUNE (*f*) : fore topmast backstay 890
VORMARSSEGEL (*n*) : fore topsail 891
VORMARSSPIERE (*f*) : fore topyard boom 892
VORMARSSTENGE (*f*) : fore topmast 893
VORMERKUNG (*f*) : note ; notice 894
VORMILCH (*f*) : foremilk ; colostrum 895
VORMITTAG (*m*) : forenoon ; morning 896
VORMUND (*m*) : trustee ; guardian ; tutor 897
VORMUNDSCHAFT (*f*) : trusteeship ; guardianship 898
VORN : before ; in front 899
VORNAME (*m*) : christian name 900
VORNE : before ; in front 901
VORNEHM : aristocratic ; distinguished ; noble 902
VORNEHMEN : to undertake ; take (up) ; purpose ; intend ; take in hand ; (*n*) : purpose ; intention 903
VORNEHMLICH : especially ; chiefly ; preferably ; mostly ; particularly ; principally 904
VORNEHMST : foremost ; capital ; chief ; principal ; excellent 905
VORNE, VON- : from the (commencement ; start) beginning 906
VORNIERE (*f*) : head kidney 907
VORNIERENGANG (*m*) : Wolffian duct 908
VOROBERLEESEGEL (*n*) : fore topmast studding sail 909
VOROFFERTE (*f*) : preliminary quotation 910
VORPIEK (*f*) : fore peak (Naut.) 911
VORPIEKFALL (*n*) : fore peak halliard 912
VORPLATZ (*m*) : vestibule ; esplanade 913
VORPOSTEN (*m*) : outpost 914
VORPREDIGEN : to preach to ; lecture 915
VORPROBE (*f*) : preliminary test 916
VORPRÜFUNG (*f*) : preliminary (examination) testing 917
VORPUMPE (*f*) : wort pump ; circulator (Brew.) 918
VORRAGEN : to be prominent ; stand out ; excel 919
VORRAH (*f*) : fore yard (Naut.) 920
VORRANG (*m*) : superiority ; precedence ; preeminence 921
VORRAT (*m*) : store ; stock ; supply ; provision 922
VORRATFLASCHE (*f*) : stock bottle 923
VORRATGEFÄSS (*n*) : stock pot ; stock vessel ; reservoir 924
VORRATGRUBE (*f*) : storage pit ; bunker 925
VORRÄTIG : on hand ; in stock ; stored up 926
VORRATLÖSUNG (*f*) : stock solution 927
VORRATSEIWEISS (*n*) : supply protein 928
VORRATSFLASCHE (*f*) : stock bottle 929

VORRATSGRUBE (*f*) : storage pit ; bunker 930
VORRATSHAUS (*n*) : magazine, store-house 931
VORRATSKAMMER (*f*) : store-room 932
VORRATSLÖSUNG (*f*) : stock solution 933
VORRATSRAUM (*m*) : store-room 934
VORRATSTURM (*m*) : storage tower ; bunker 935
VORRAUM (*m*) : coking chamber (of furnaces) ; antichamber 936
VORRECHT (*n*) : privilege ; prerogative 937
VORRECHTSAKTIE (*f*) : preference share 938
VORREDE (*f*) : preamble ; preface ; prologue, introduction : introductory remarks 939
VORREDNER (*m*) : last speaker 940
VORREIBEN : to grind beforehand 941
VORREINIGUNG (*f*) : preliminary refining (removal of foreign metal except silver from raw lead ; see **WERKBLEI**) 942
VORRICHTEN : to prepare ; arrange ; fit up ; equip 943
VORRICHTUNG (*f*) : arrangement ; equipment ; apparatus ; mechanism ; preparation ; device ; contrivance ; gear ; appliance ; dressing 944
VORROIL (*n*) : fore royal-(sail) (Naut.) 945
VORROILBAUCHGORDING (*f*) : fore royal buntline 946
VORROILBRASSE (*f*) : fore royal brace 947
VORROILFALL (*n*) : fore royal halliard 948
VORROILGEITAU (*n*) : fore royal clew line 949
VORROILPARDUNE (*f*) : fore royal backstay 950
VORROILSEGEL (*n*) : fore royal-(sail) 951
VORROILSTENGE (*f*) : fore royal mast 952
VORRÜCKEN : to progress ; push forward ; advance ; reproach 953
VORRÜSTE (*f*) : fore channel 954
VORSAAL (*m*) : antechamber ; anteroom ; (entrance)-hall 955
VORSAGEN : to dictate, rehearse, repeat 956
VORSATZ (*m*) : design ; purpose ; intention ; screen ; anything placed before 957
VORSÄTZLICH : intentional ; deliberate 958
VORSATZ, MIT- : intentionally ; deliberately 959
VORSCHALTEN : to connect ; couple (at the inlet side) ; cut in ; introduce into an electric circuit ; put (connect) in series with 960
VORSCHALTRAD (*n*) : additional (extra ; auxiliary) wheel 961
VORSCHALTSTUFE (*f*) : auxiliary stage ; high pressure stage (Turbines) 962
VORSCHALTTURBINE (*f*) : auxiliary turbine ; high pressure (end) turbine ; any turbine connected to the inlet side of another 963
VORSCHALTUNG (*f*) : connection ; introduction ; addition ; coupling (at the inlet side) 964
VORSCHALTWIDERSTAND (*m*) : rheostat resistance ; external resistance box (of Wattmeter) ; series resistance 965
VORSCHEIN (*m*) : appearance 966
VORSCHEIN, ZUM- KOMMEN : to appear ; be (found out) discovered ; come to light 967
VORSCHIEBEN : to push forward ; advance 968
VORSCHIESSEN : to advance-(money) ; to circulate (Brew.) ; shoot forward 969
VORSCHIFF (*n*) : fore body (of a ship), fore part (of a ship) 970
VORSCHLAG (*m*) : proposal ; proposition ; flux (Metal) ; fusion ; suggestion ; motion 971
VORSCHLAGEN : to suggest ; propose ; prefix ; move ; put before 972
VORSCHLAGHAMMER (*m*) : sledge hammer 973
VORSCHMACK (*m*) : foretaste ; most prominent (predominant) flavour 974
VORSCHMECKEN : to predominate (of taste) 975
VORSCHNEIDEN : to rough hew (Wood) 976
VORSCHNELL : hasty ; rash ; premature ; forward 977
VOR-SCHONERSEGEL (*n*) : fore schoonersail 978
VORSCHREIBEN : to dictate ; direct ; prescribe ; write out ; order ; specify 979
VORSCHRIFT (*f*) : recipe ; prescription ; instruction-(s) ; direction-(s) ; copy ; order ; command ; precept ; injunction ; specification 980
VORSCHRIFTSMÄSSIG : as (directed) prescribed ; according to (directions) prescription 981
VORSCHUB (*m*) : furtherance ; aid ; assistance ; pushing forward ; feeding (ram ; device ; worm ; plunger) arrangement ; feed ; advance (Mech.) 982
VORSCHUBGESCHWINDE (*n*) **AM KETTERNOST** (*m*) : speed of chain grate stoker ; rate of feed 983
VORSCHUBGESCHWINDIGKEIT (*f*) : speed of (advance) travel (of mechanical grates) 984
VORSCHUB (*m*), **SCHNECKEN-** : worm feeder 985
VORSCHUH (*m*) : vamp ; upper leather ; toe-cap 986
VORSCHULE (*f*) : preparatory school 987
VORSCHUSS (*m*) : advance (Money) 988
VORSCHÜTZEN : to allege ; pretend ; plead ; dam-(up) 989
VORSCHWEBEN : to float-(before) ; have in mind ; hover-(before) 990
VORSEGEL (*n.pl*) : head sails 991
VORSEHEN : to foresee ; provide for ; take (care) heed 992
VORSEHUNG (*f*) : providence 993
VORSETZBLATT (*n*) : fly-leaf 994
VORSETZEN : to place (set ; put) before ; prefix ; intend ; propose ; resolve (decide) upon 995
VORSETZER (*m*) : anything placed in front of something else, such as a guard, shield, protector, shroud, screen 996
VOR SICH GEHEN : to happen ; take place ; occur (of a process) 997
VORSICHT (*f*) : providence ; foresight ; (pre)-caution ; care ; prudence ; circumspection 998
VORSICHTIG : careful ; cautious ; prudent ; circumspect 999
VORSICHTSMASSREGEL (*f*) : precautionary (safety) measure 000
VORSIEDEN : to subject to preliminary boiling 001
VORSILBE (*f*) : prefix 002
VORSITZ (*m*) : presidency ; chairmanship 003
VORSITZEN : to preside 004
VORSITZENDER (*m*) : president ; chairman 005
VORSITZENDER-STELLVERTRETER (*m*) : vice-president 006
VORSKEISEGEL (*n*) : fore sky sail 007
VORSORGE (*f*) : forethought ; precaution 008
VORSPANNMASCHINE (*f*) : pilot engine 009
VORSPANNTURBINE (*f*), **HOCHDRUCK-** : high pressure auxiliary turbine ; high pressure end 010
VORSPIEGELN : to pretend ; show in a wrong light ; play to ; deceive 011
VORSPIEGELUNG (*f*) : illusion 012
VORSPIEL (*n*) prologue ; prelude 013
VORSPRECHEN : to pay a visit ; make a call ; speak (in front of) before 014
VORSPRINGEN : to project ; spring out ; be prominent ; stand out ; jut out 015
VORSPRUNG (*m*) : projection ; advantage ; prominence ; advance ; lead ; start 016
VORSTADT (*f*) : suburb 017

VORSTAND (*m*): security; bail; executive committee; directorate; superior; director; chief; head 018
VORSTECKEN: to place (insert; put; stick) before; propose; appoint; prefix 019
VORSTECKER (*m*): peg; pin; split-pin; pawl 020
VORSTECKSTIFT (*m*): cotter pin, split-pin 021
VORSTEHEN: to stand (out) before; project; precede; preside over; administer; direct; be prominent; superintend 022
VORSTEHEND: preceding; above 023
VORSTEHENDE (*n*): foregoing 024
VORSTEHER (*m*): manager; superintendent; head; chief; director; inspector; president 025
VORSTEHERDRÜSE (*f*): prostate gland 026
VORSTELLEN: to place before; demonstrate; visualize; (re)-present; personate; imagine; introduce 027
VORSTELLIGMACHEN: to remonstrate 028
VORSTELLPLATTE (*f*): dead-plate (of grates) 029
VORSTELLUNG (*f*): introduction; conception; (re)-presentation; remonstrance; idea; notion 030
VORSTELLUNGSBILD (*n*): imaginary picture 031
VORSTENGE (*f*): fore top mast 032
VORSTENGEWANTEN (*n.pl*): fore mast rigging 033
VORSTEVEN (*m*): stem (Naut.) 034
VORSTOSS (*m*): edging; adapter; lap; projection; lug 035
VORSTRECKEN: to lend; credit; advance; loan; stretch forward 036
VORSTRICHMASSE (*f*): priming-(mass) 037
VORSTUFE (*f*): primer; first step 038
VORSUD (*m*): first boiling 039
VORTAKEL (*n*): fore tackle 040
VORTEIL (*m*): advantage; benefit; profit; boon; gain; interest 041
VORTEILHAFT: advantageous; preferable; favourable; profitable; beneficial 042
VORTIEGEL (*m*): forehearth (Tin); lead pot 043
VORTRAG (*m*): lecture; address; recital; exposition; discourse; statement; enunciation; report; delivery; balance (Commerce); carry forward 044
VORTRAGEN: to carry (forward) before; lecture; state; propose 045
VORTREFFLICH: superior; excellent; prime; splendid 046
VORTREFFLICHKEIT (*f*): superiority; excellence 047
VORTREISEGEL (*n*): fore trysail 048
VORTRETEN: to stand out; step forward 049
VORTRITT (*m*): precedence 050
VORTROCKNEN: to dry beforehand 051
VORTUN: to place (put) before; surpass; push oneself forward 052
VORTURBINE (*f*): auxiliary turbine (usually high pressure end of turbine) (see VORSCHALT-TURBINE) 053
VORÜBER: past; done; gone; over; finished 054
VORÜBERGEHEN: to pass (by) away 055
VORÜBERGEHEND: transitory; transient; temporary 056
VORUNTERLEESEGEL (*n*): lower fore studding sail 057
VORUNTERSUCHUNG (*f*): preliminary (investigation) test 058
VORURTEIL (*m*): bias; prejudice; condemnation 059
VORURTEILEN: to judge without hearing the evidence; be (biassed) prejudiced 060
VORVERÖFFENTLICHUNG (*f*): prior publication; prior art (of patents) 061
VORVERSUCH (*m*): preliminary (experiment; trial) test 062
VORWACHE (*f*): outpost; advanced sentry 063
VORWACHS (*m*): bee glue; propolis (resinous matter collected by bees from plants) 064
VORWALKEN: to scour (Cloth) 065
VORWALTEN: to predominate; prevail 066
VORWALZE (*f*): blooming roll; roughing roll; bloom roll (Rolling mills); (*pl*): blooming train 067
VORWALZEN: to rough down; bloom 068
VORWALZWERK (*n*): blooming (roughing) rolls, mill or train 069
VORWAND (*m*): pretence; pretext; (*f*): front wall; façade 070
VORWÄRMAPPARAT (*m*): preheating apparatus 071
VORWÄRMEN: to preheat; warm; anneal 072
VORWÄRMER (*m*): preheater; economizer; feed-(water)-heater 073
VORWÄRMESTUFE (*f*): preheating stage 074
VORWÄRMETEMPERATUR (*f*): preheating temperature 075
VORWÄRMOFEN (*m*) preheating (annealing) oven or furnace 076
VORWÄRMUNG (*f*): preheating 077
VORWÄRMZONE (*f*): preheating zone 078
VORWÄRTS: onward; on; forward 079
VORWÄRTSEXZENTER (*m*): foreward eccentric 080
VORWÄRTSGANG (*m*): forward motion 081
VORWÄRTS GEHEN: to drive or to go (ahead) forward (Naut., etc.), to advance, proceed 082
VORWÄRTSTURBINE (*f*): foreward turbine 083
VORWÄSCHE (*f*): preliminary washing 084
VORWEG: before-(hand) 085
VORWEGNEHMEN: to anticipate; take beforehand 086
VORWEIN (*m*): first runnings 087
VORWEISEN: to produce; exhibit 088
VORWELT (*f*): antiquity; ancients; former age; ancestors; progenitors 089
VORWENDEN: to allege; pretend 090
VORWERFEN: to cast in one's face; cast up at one; throw before; reproach with; upbraid with; project (push) forward; (*n*): reproach-(ing); forward motion 091
VORWIEGEN: to preponderate; outweigh 092
VORWIEGEND: predominantly; predominating; especially; chiefly; by preference 093
VORWISSEN (*n*): foreknowledge 094
VORWITZIG: inquisitive; pert; meddlesome; forward; prying 095
VORWORT (*n*): preface; foreword; introduction; preposition; proem; preamble 096
VORWURF (*m*): blame; reproach; object; subject 097
VORZÄHLEN: to enumerate; count before 098
VORZEICHEN (*n*): omen; token; indication sympton; sign (Maths.); prognostication 099
VORZEICHNEN: to define, make clear, illustrate, draw as an example, line out, rough out, sketch, plan, draft 100
VORZEICHNUNG (*f*): pattern; copy (of drawings); example; sketch 101
VORZEIGEN: to present; exhibit; show; produce 102

VORZEIT (*f*): olden times; by-gone days 103
VORZEITIG: untimely; premature 104
VORZIEHBAR: preferable 105
VORZIEHEN: to prefer; show preference for; draw forward 106
VORZUG (*m*): preference; advantage; merit; virtue; privilege; precedence; superiority; priority; prerogative 107
VORZÜGLICH: choice; superior; preferable; excellent; exquisite; distinguished; chiefly; especially 108
VORZÜGLICHKEIT (*f*): superiority; excellence 109
VORZUGSAKTIE (*f*): preference share 110
VORZUGSDIVIDENDE (*f*): preferential dividend 111
VORZUGSWEISE: preferably; especially; pre-eminently; principally; chiefly 112
VORZÜNDUNG (*f*): pre-ignition; premature (advanced) ignition 113
VORZUSEHEND: provided, arranged for 114
VOUTENDECKE (*f*): arched (floor) ceiling; vaulted roof 115
VULKAN (*m*): vulcan; volcano 116
VULKANFASER (*f*): vulcanized fibre 117
VULKANFIBER (*f*): vulcanized fibre 118
VULKANGAS (*n*): volcanic gas 119
VULKANGLAS (*n*): volcanic glass; obsidianite 120
VULKANISCH: volcanic 121
VULKANISCHES GESTEIN (*n*): volcanic rock 122
VULKANISIERAPPARAT (*m*): vulcanizer; vulcanizing apparatus 123
VULKANISIEREN: to vulcanize 124
VULKANISIERKESSEL (*m*): vulcanizing boiler 125
VULKANISIERUNG (*f*): vulcanizing; vulcanization 126
VULKANIT (*n*): vulcanite 127
VULKANÖL (*n*): vulcan oil; mineral lubricating (sperm) oil 128

W

WAAGE (*f*): (see WAGE) 129
WABE (*f*): honeycomb 130
WABENARTIG: honeycombed 131
WACH: awake; wakeful; waking 132
WACHE (*f*): watch (Naut., etc.); guard house; guard; sentry; picket 133
WACHEFEUER (*n*): camp-fire; watch-fire 134
WACHEHAUS (*n*): guard-house 135
WACHEN: to watch; wake; be awake; guard; stand sentry; be on (sentry-go) guard or picket duty 136
WACHMITTEL (*n*): antisoporific 137
WACHOLDER (*m*): juniper (*Juniperus communis*) 138
WACHOLDERBAUM (*m*): juniper tree (*Juniperus communis*) 139
WACHOLDERBEERE (*f*): juniper berry (*Fructus juniperi*) 140
WACHOLDERBEERÖL (*n*): oil of juniper; juniper-berry oil; Sg. 0.87 141
WACHOLDERBRANNTWEIN (*m*): Holland gin 142
WACHOLDERHOLZ (*n*): juniper wood (see WACH-OLDER); Sg. 0.5-0.7 143
WACHOLDERHOLZÖL (*n*): juniper wood oil 144
WACHOLDERMUS (*n*): (*Succus Juniperi inspissatus*): elderberry pulp 145
WACHOLDERÖL (*n*): juniper oil (see WACHOLDER-BEERÖL and WACHOLDERHOLZÖL) 146
WACHOLDERÖL, BRENZLICHES- (*n*): oil of cade (see CADEÖL) 147
WACHOLDERSPIRITUS (*m*): spirit of juniper 148
WACHOLDERSPITZEN (*f.pl*): juniper tops 149
WACHOLDERSTRAUCH (*m*): juniper tree (*Juniperus communis*) 150
WACHS (*n*): wax; Sg. 0.965; (*m*): growth 151
WACHSABDRUCK (*m*): wax impression 152
WACHSÄHNLICH: wax like; waxy 153
WACHSALAUN (*m*): crystallized alum (see ALAUN) 154
WACHSALKOHOL (*m*): wax alcohol 155
WACHSAM: vigilant; watchful 156
WACHSART (*f*): variety of wax 157
WACHSARTIG: wax-like; waxy 158
WACHSBAUM (*m*): bayberry tree; wax myrtle; candleberry myrtle (see WACHSMYRTE) 159
WACHSBEIZE (*f*): wax mordant 160
WACHSBILD (*n*): wax (model) figure 161
WACHSBILDNEREI (*f*): wax modelling 162
WACHSBLEICHANLAGE (*f*): wax bleaching plant 163
WACHSBLEICHE (*f*): wax (bleaching) bleaching works 164
WACHSBLEICHEN: wax bleaching 165
WACHSBODEN (*m*): wax cake 166
WACHSBOTTICH (*m*): roching cask (Alum) 167
WACHS (*n*), **CHINESISCHES-**: Chinese wax; pela (from an insect on the tree, *Fraxinus chinensis*) 168
WACHSDRÜSE (*f*): ceruminous gland 169
WACHSEN: to grow; increase; swell (Lime); sprout (Malting) 170
WÄCHSEN: to wax 171
WÄCHSERN: waxen; (of)-wax 172
WACHSFARBE (*f*): wax colour (oil-soluble colour or dye) 173
WACHSFARBIG: wax-coloured 174
WACHSFIGUR (*f*): wax-figure 175
WACHSFIRNIS (*m*): wax varnish 176
WACHSGAGEL (*m* and *f*): wax myrtle (*Myrica cerifera*) (see WACHSBAUM) 177
WACHSGAGELSTRAUCH (*m*): (see WACHSMYRTE) 178
WACHSGELB: wax yellow; yellow as wax 179
WACHSHANDEL (*m*): wax trade 180
WACHS (*n*), **JAPANISCHES-**: Japan wax (a vegetable wax from the seeds of *Rhus vernicifera* and others; Mp. 52.5°C.) 181
WACHSKERZCHEN (*n*): wax taper; wax match 182
WACHSKERZE (*f*): wax candle 183
WACHSKITT (*m*): wax cement 184
WACHSKOHLE (*f*): pyropissite (see PYROPISSIT) 185
WACHSKUCHEN (*m*): wax cake 186
WACHSLEINEN (*n*): oilcloth 187
WACHSLEINWAND (*f*): oil-cloth 188
WACHSLICHT (*n*): wax (candle) light 189
WACHSMALEREI (*f*): encaustic painting 190
WACHSMYRTE (*f*): wax myrtle (*Myrica cerifera*) (see WACHSBAUM) 191
WACHSÖL (*n*): wax oil 192
WACHSOPAL (*m*): wax opal 193
WACHSPALME (*f*): wax palm (*Corypha cerifera*) (see also CARNAUBAPALME) 194
WACHSPAPIER (*n*): wax paper 195
WACHSPFLASTER (*n*): cerate (*Ceratum*) 196

WACHSPRÄPARAT (*n*): wax preparation 197
WACHSSALBE (*f*): cerate; ointment 198
WACHSSCHEIBE (*f*): cake of wax; wax tablet 199
WACHSSCHMELZANLAGE (*f*): wax (rendering) melting plant 200
WACHSSCHMELZEN (*n*): wax (melting) rendering 201
WACHSSEIFE (*f*): wax soap 202
WACHSSTOCK (*m*): wax candle; wax taper 203
WACHSTAFEL (*f*): wax tablet 204
WACHSTAFFET (*m*): waxed (oiled) silk; oilsilk 205
WACHSTUCH (*n*): oilcloth; oil-skin; waxcloth 206
WACHSTUM (*n*): growth; increase 207
WACHSWAREN (*f.pl*): wax (goods) wares 208
WACHSZÜNDHOLZ (*n*): wax match 209
WACHT (*f*): watch; guard 210
WACHTEL (*f*): quail 211
WÄCHTER (*m*): guard; warder; watchman; male nurse 212
WACKELIG: rickety; shaky; shaking; tottering 213
WACKELN: to waggle; rock; shake; totter 214
WACKER: brave; stout; true; valiant; gallant 215
WAD (*m*): wad, bog manganese; MnO,MnO_2,H_2O; Sg. 3.0-3.5 216
WADE (*f*): calf of the leg 217
WADENBEIN (*n*): fibula (Anat.) 218
WADERZ (*n*): wad, bog manganese, asbolite (see WAD) 219
WADGRAPHIT (*m*), SCHAUMARTIGER-: wad 220
WAFFE (*f*): arms; weapon 221
WAFFENDIENST (*m*): military service 222
WAFFENFABRIK (*f*): arms factory 223
WAFFENFETT (*n*): rifle oil or fat; any oil, grease or fat for small arms 224
WAFFENHAUS (*n*): armoury; arsenal 225
WAFFENÖL (*n*): rifle oil 226
WAFFENRÜSTUNG (*f*): armament; arming 227
WAFFENSCHMIED (*m*): armourer 228
WAFFENSTILLSTAND (*m*): armistice, truce, cessation of hostilities 229
WAFFNEN: to arm 230
WÄGBAR: ponderable; weighable 231
WÄGBARKEIT (*f*): ponderability; weighability 232
WAGE (*f*): balance; weigh-bridge; scales; weighing machine; level; hydrometer; ærometer; densimeter 233
WAGEARM (*m*): balance arm 234
WAGE, AUTOMATISCHE- (*f*): automatic (balance) weighing apparatus 235
WAGEBALKEN (*m*): balance (arm) beam (Scales) 236
WAGE, BRÜCKEN- (*f*): weigh-bridge 237
WAGE (*f*), DEZIMAL-: decimal balance 238
WAGE (*f*), FEDER-: spring balance 239
WAGE (*f*), FLÜSSIGKEITS-: hydrometric balance (see HYDROMETRISCHE WAGE) 240
WAGE (*f*), GAS-: gas balance 241
WAGEGEHÄUSE (*n*): balance case 242
WÄGEGLAS (*n*): weighing glass 243
WÄGEGLÄSCHEN (*n*): weighing glass 244
WAGEHALSIG: bold; rash; daredevil; foolhardy 245
WAGE (*f*), HYDROSTATISCHE-: hydrometer, aerometer (for liquids) 246
WAGE (*f*), LAUFGEWICHTS-: balance with poise or sliding weight, steelyard 247
WAGEN: to venture; hazard; risk; chance; dare; (*m*): waggon; carriage; conveyance; coach; car; cart; truck; vehicle, etc. 248
WÄGEN: to weigh; ponder; balance; poise 249
WAGENACHSE (*f*): axletree 250
WAGENBAUER (*m*): cartwright, carriage builder 251
WAGENDECKE (*f*): tarpaulin; waggon; cover; hood 252
WAGENDEICHSEL (*f*): pole (of a waggon) 253
WAGENFABRIK (*f*): carriage (waggon) works; scale (balance) factory 254
WAGENFEDER (*f*): carriage spring 255
WAGENFETT (*n*): axle-grease; cart-grease 256
WAGENFETTANLAGE (*f*): cart-grease plant 257
WAGENGESTELL (*n*): car frame, chassis (of automobiles) 258
WAGENGLEIS (*n*): cart (rut) track 259
WAGENHAUS (*n*): coach-house 260
WAGENKASTEN (*m*): carriage body 261
WAGENLACK (*m*): carriage varnish 262
WAGENLADUNG (*f*): cartload, truckload, load 263
WAGENMACHER (*m*): cartwright, coach (carriage) builder 264
WAGENÖL (*n*): waggon oil 265
WAGENPFERD (*n*): cart-horse 266
WAGENRAD (*n*): waggon (carriage) wheel 267
WAGENREMISE (*f*): coach-house 268
WAGEN (*m*), SCHLACKEN-: slag wagon 269
WAGENSCHMIERE (*f*): cart-grease; axle-grease 270
WAGENSCHOPPEN (*m*): carriage-shed 271
WAGENSPUR (*f*): cart (rut) track 272
WÄGEPIPETTE (*f*): weighing pipette 273
WAGE (*f*), POST-: postal scales 274
WÄGER (*m*): weigher 275
WAGERECHT: horizontal; level 276
WAGERECHTE (*f*): horizontal (line) 277
WÄGERÖHRCHEN (*n*): weighing tube 278
WAGESCHALE (*f*): balance (scale) pan 279
WÄGESCHALE (*f*): weighing dish 280
WÄGESCHIFFCHEN (*n*): weighing boat 281
WAGE (*f*), ZÄHL-: recording balance 282
WAGGON (*m*): (railway)-carriage; car; coach; waggon; wagon 283
WAGGONBAU (*m*): wagon (carriage) construction 284
WAGGONFABRIK (*f*): carriage (wagon) works 285
WAGGONLACK (*m*): carriage varnish; railway carriage or waggon varnish 286
WAGGONWAGE (*f*): weigh-bridge 287
WAGLICH: hazardously 288
WAGNER (*m*): wheelwright; cartwright 289
WAGNERIT (*m*): wagnerite; $MgF_2,MgPO_4$; Sg. 3.0 290
WAGNIS (*n*): hazard; risk; chance; venture 291
WAGRECHT: horizontal 292
WAGSCHALE (*f*): scale (balance) pan 293
WÄGUNG (*f*): weighing 294
WAGZIMMER (*n*): weighing (balance) room 295
WAHL (*f*): choice; selection; option; election 296
WÄHLBAR: eligible 297
WÄHLBARKEIT (*f*): eligibility 298
WÄHLEN: to choose; elect; prefer; select 299
WÄHLER (*m*): chooser; voter; elector 300
WÄHLERISCH: fastidious; particular; dainty 301
WAHLFÄHIG: having a vote; eligible for a vote; qualified to vote 302

WAHLLOS: without choice; haphazard 303
WAHLRECHT (*n*): franchise; suffrage 304
WAHLSPRUCH (*m*): device; motto . 305
WAHLSTIMME (*f*): suffrage; vote 306
WAHLVERWANDSCHAFT (*f*): elective affinity; congeniality 307
WAHN (*m*): error; illusion; delusion; folly; fancy; presumption 308
WAHNBILD (*n*): illusion; phantom 309
WÄHNEN: to imagine; suppose; fancy; presume; think 310
WAHNSINN (*m*): delirium; frenzy; insanity; madness; monomania (Med.) 311
WAHNSINNIG: mad; frantic; insane 312
WAHR: real; true; genuine; veritable 313
WAHREN: to look after; keep; preserve; guard 314
WÄHREN: to continue; last; endure 315
WÄHREND: while; whilst; whereas; during; for 316
WAHRHAFT: true; truthful; veracious; genuine; sure; sincere; real; regular 317
WAHRHAFTIG: really; truly; verily; true; truthful; positive 318
WAHRHAFTIGKEIT (*f*): truth; veracity 319
WAHRHEIT (*f*): truth-(fulness); fact; verity; reality 320
WAHRLICH: really; truly; surely; verily 321
WAHRNEHMBAR: perceptible 322
WAHRNEHMEN: to perceive; observe; notice; profit by; take care of; attend to 323
WAHRNEHMUNG (*f*): observation; perception; attention 324
WAHRSAGEN: to predict; prophesy; divine 325
WAHRSAGER (*m*): prophet; diviner; fortune-teller; soothsayer 326
WAHRSAGEREI (*f*): fortune-telling 327
WAHRSAGUNG (*f*): divination, fortune-telling 328
WAHRSCHEINLICH: probable; probably; likely; plausible 329
WAHRSCHEINLICHKEIT (*f*): probability; likelihood 330
WAHRSCHEINLICHKEITSRECHNUNG (*f*): calculation of probabilities; sizing up of chances 331
WAHRSPRUCH (*m*): verdict 332
WAHRUNG (*f*): vindication; support 333
WÄHRUNG (*f*): currency; value; coinage; standard; duration 334
WÄHRWOLF (*m*): were-wolf 335
WAHRZEICHEN (*n*): sign; token 336
WAID (*m*): woad (see **FÄRBERWAID**) 337
WAISE (*f*): orphan 338
WAISENHAUS (*n*): orphanage 339
WAL (*m*): whale 340
WALD (*m*): wood; forest 341
WALDAHORN (*m*): sycamore (*Acer pseudo-platanus*) 342
WALDANEMONE (*f*): wood anemone (*Anemone silvestris*) 343
WALDBRAND (*m*): forest fire 344
WALDBRUDER (*m*): anchorite; hermit 345
WÄLDCHEN (*n*): thicket, grove, copse 346
WALDEN'SCHE UMKEHRUNG (*f*): Walden inversion 347
WALDGEWÄCHS (*n*): forest plant 348
WALDGOTT (*m*): sylvan deity; satyr 349
WALDHORN (*n*): French horn; buglehorn 350
WALDIG: woody, wooded 351
WALDKIRSCHE (*f*): wild cherry (*Prunus padus*) 352
WALDKULTUR (*f*): afforestation 353
WALDLERCHE (*f*): wood-lark 354
WALDMALVE (*f*): mallow (*Malva sylvestris*) 355
WALDMEISTER (*m*): woodruff (*Asperula odorata*) 356
WALDMEISTERÄTHER (*m*): woodruff ether 357
WALDREBE (*f*): clematis (*Clematis vitalba*) 358
WALDRECHT (*n*): forest laws 359
WALDRISS (*m*): (see **KERNRISS**) 360
WALDRÖSCHEN (*n*): wood anemone, white wood anemone (*Anemone nemorosa*) 361
WALDSCHWAMM (*m*): tree fungus (of any kind) 362
WALDSTROM (*m*): wood-torrent 363
WALDUNG (*f*): woods; forest; woodland 364
WALDWOLLE (*f*): pine wool 365
WALDWOLLÖL (*n*): pine needle oil (see **KIEFERN-NADELÖL**) 366
WALDWOLLPRODUKT (*n*): pine needle product 367
WALFISCH (*m*): whale 368
WALFISCH (*m*), **GRÖNLÄNDISCHER-**: Greenland whale (*Balaena mysticetus*) 369
WALFISCHSPECK (*m*): whale blubber 370
WALFISCHTRAN (*m*): train oil; whale oil; sperm oil; Sg. 0.93 371
WALKBLAU (*n*): fulling blue 372
WALKE (*f*): fulling; fulling (machine) mill (Textile) 373
WALKEN: to full (Cloth); felt 374
WALKER (*m*): fuller 375
WALKERDE: Fuller's earth (see **FULLERERDE**) 376
WALKERDISTEL (*f*): fuller's teasel (*Dipsacus fullonum*) 377
WALKERERDE (*f*): Fuller's earth (see **FULLERERDE**) 378
WALKERTON (*m*): Fuller's earth (see **FULLERERDE**) 379
WALKEXTRAKT (*m*): fulling extract 380
WALKFETT (*n*): fuller's oil; fulling oil, fat or grease 381
WALKFETTGEWINNUNGSANLAGE (*f*): fulling oil extraction plant 382
WALKMÜHLE (*f*): fulling mill 383
WALKMÜLLER (*m*): fuller 384
WALKÖL (*n*): fulling oil 385
WALKSCHWARZ (*n*): fulling black 386
WALKSEIFE (*f*): fuller's soap; fulling soap 387
WALKTON (*m*): Fuller's earth (see **FULLERERDE**) 388
WALKTROG (*m*): fulling trough 389
WALL (*m*): wall; dam; rampart; coast; bank 390
WALLACH (*m*): Wallachian; gelding; castrated horse 391
WALLEN: to bubble; boil up; simmer; heave; wave; undulate 392
WÄLLEN: to simmer; boil 393
WALLFAHRER (*m*): pilgrim 394
WALLFAHRT (*f*): pilgrimage 395
WALLFAHRTEN: to go on a pilgrimage 396
WALLFISCH (*m*): (see **WALFISCH**) 397
WALLNUSS (*f*): walnut (fruit of *Juglans regia*) (see also **WALNUSS**) 398
WALLNUSSÖL (*n*): (see **NUSSÖL**) 399
WALLNUSSSCHALEN (*f.pl*): walnut shells (*Cortex nucis juglandis*) 400
WALLONENSCHMIEDE (*f*): Walloon process 401
WALLRATH (*m*): (see **WALRAT**) 402
WALLROSS (*n*): walrus 403
WALLSTEIN (*m*): dam of a blast furnace 404
WALLSTEINPLATTE (*f*): dam plate 405
WALLUNG (*f*): ebullition; boiling; undulation; agitation; excitement; motion 406

WALNUSS (*f*): walnut (fruit of *Juglans regia*) (see under **NUSS**) 407
WALNUSSBAUM (*m*): walnut tree (*Juglans regia*) (see **NUSSBAUM**) 408
WALNUSSBAUM, GRAUER- (*m*): butternut tree (*Juglans cinerea*) 409
WALNUSSBAUM, WEISSER- (*m*): hickory (*Hicoria*) 410
WALNUSSHOLZ (*n*): walnut wood (see **NUSSBAUM** and **WALLNUSS, SCHWARZER**) 411
WALNUSSÖL (*n*): walnut oil (see **NUSSÖL**) 412
WALNUSSSCHALEN (*f.pl*): walnut shells (*Cortex nucis juglandis*) 413
WALNUSS (*m*), **SCHWARZER-**: black walnut, American walnut (*Juglans nigra*); Sg. 0.5-0.9 414
WALÖL (*n*): whale oil (northern from the Greenland whale, *Balaena mysticetus* and southern from the whale *Balaena australis*); Sg. 0.916-0.93 415
WALPURGIN (*m*): walpurgin; $As_4Bi_{10}O_{28}(U_3O_6)$. $10H_2O$; Sg. 5.75 416
WALPURGIS NACHT (*f*): Walpurgis night; night of 1st May 417
WALRAT (*m*): spermaceti; sperm (*Cetaceum*); Sg. 0.94 (Cetyl palmitate) 418
WALRATÖL (*n*): sperm oil; Sg. 0.93 419
WALROSS (*n*): walrus; (*Phoca*) seal 420
WALROSSTRAN (*m*): (see **ROBBENTRAN**) 421
WALTEN: to manage; govern; rule; dispose; domineer 422
WALTRAN (*m*): whale oil; train oil (see **WALÖL**) 423
WALZBLECH (*n*): rolled plate 424
WALZBLEI (*n*): sheet (rolled) lead 425
WALZE (*f*): roll; roller; cylinder; drum; (*pl*): train of rolls 426
WALZE (*f*), **GEZAHNTE-**: cogged (toothed) roll or roller 427
WALZEISEN (*n*): rolled (drawn) iron; pin; axle 428
WALZE (*f*), **KANNELIERTE-**: grooved roll 429
WALZEN: to waltz; roll; mill [see also **WALZE** (*pl.*)] 430
WÄLZEN: to roll; revolve; turn over; wallow; welter 431
WALZENAPPARAT (*m*): rolling frame (Dyeing) 432
WALZENBRECHER (*m*): roller crusher (for coal, etc.) 433
WALZENBUND (*n*): body of roll (Rolling mills) 434
WALZENDRUCK (*m*): cylinder printing; roller printing (Textile); roll (roller) pressure 435
WALZENDRUCKMASCHINE (*f*): roller printing machine 436
WALZENFÖRMIG: cylindrical 437
WALZENGATTER (*n*): roller saw-mill, saw mill with cylinders 438
WALZENGLAS (*n*): cylinder glass 439
WALZENKESSEL (*m*): cylindrical boiler 440
WALZENMASSE (*f*): roller composition 441
WALZENMÜHLE (*f*): roller mill 442
WALZENQUETSCHE (*f*): roller squeezer (for soap) 443
WALZENREIBMASCHINE (*f*): roller (grinder) grinding machine 444
WALZENSINTER (*m*): rolling mill (sinter) scale (60-75% Fe) (steelworks process byproduct) 445
WALZENSTAHL (*m*): steel for rolls 446
WALZENSTÄNDER (*m*): housing (for rolls) 447
WALZENSTRASSE (*f*): rolling (mill) machine, roll-train 448
WALZENSTRECKE (*f*): roll train; rolling mill 449
WALZENTROCKNER (*m*): drum (rotary) drier 450
WALZENUMDREHUNG (*f*): revolution of the roll-(er) 451
WALZENZUG (*m*): set (train) of rolls 452
WALZENZUGSMASCHINE (*f*): rolling-mill engine 453
WALZER (*m*): waltz; valse 454
WALZGERÜST (*n*): (roll)-housing 455
WALZHART: hard rolled 456
WALZMESSING (*n*): rolled brass (same as **GELB-GUSS**, but less zinc constituent, see **GELBGUSS**) 457
WALZSTAHL (*m*): rolled steel 458
WALZSTRASSE (*f*): rolling (train) mill; blooming mill 459
WALZWERK (*n*): roll train; rolling (blooming) mill (Metal); crushing (mill) rolls (Sugar) 460
WALZWERKSANLAGE (*f*): rolling-mill (crushing-mill) plant 461
WALZZINN (*n*): rolled (sheet) tin 462
WAMME (*f*): paunch; belly; dewlap 463
WAMMS (*m*): jerkin; doublet; waistcoat 464
WAND (*f*): wall; septum; partition; side; cheek; shell (of any sort of vessel) 465
WANDBEWURF (*m*): plastering 466
WANDDICKE (*f*): wall thickness 467
WANDEL (*m*): variation; change; traffic; trade; life; conduct; behaviour 468
WANDELBAR: variable; convertible; changeable; perishable; inconstant 469
WANDELBARKEIT (*f*): variability; changeableness; inconstancy 470
WANDELLOS: unalterable; unchangeable; invariable 471
WANDELN: to vary; change; transform; convert; wander; walk; trade; travel; live 472
WANDELSTERN (*m*): planet 473
WANDELUNG (*f*): transformation; change 474
WANDERBLOCK (*m*): erratic block (Geol.) 475
WANDERER (*m*): wanderer; nomad; vagrant; traveller 476
WANDERJAHR (*n*): year of travelling 477
WANDERLEBEN (*n*): wanderer's life, vagrant (nomad's) life 478
WANDERN: to wander; walk; go; travel; migrate 479
WANDERROST (*m*): travelling grate-(stoker) 480
WANDERROSTFEUERUNG (*f*): travelling grate-(stoker); mechanical stoker; chain grate-(stoker) 481
WANDERSCHAFT (*f*): travels; tour; peregrination 482
WANDERUNG (*f*): travels; walking; migration 483
WANDERUNGSGESCHWINDIGKEIT (*f*): migration velocity 484
WANDERUNGSSINN (*m*): migration direction 485
WANDKONSOLE (*f*): wall bracket 486
WANDKRAN (*m*): wall crane 487
WANDKRAUT (*n*): wall pellitory (*Herba parietariæ*) (*Parietaria officinalis*) 488
WANDLAGER (*n*): wall bearing 489
WANDLEUCHTER (*m*): sconce 490
WANDMALEREI (*f*): mural painting 491
WANDNIEDERSCHLAG (*m*): wall condensation (of pipes, etc.) 492
WANDPLATTE (*f*): wall plate 493
WANDPUTZ (*m*): plastering 494
WANDSPIEGEL (*m*): pier-glass 495

WANDSTÄRKE (*f*): (wall)-thickness 496
WANDTAFEL (*f*): blackboard; wall-chart 497
WANDTELEPHON (*n*): wall telephone 498
WANDUHR (*f*): wall-clock 499
WANDUNG (*f*): partition; wall-(surface) 500
WANDVENTILATOR (*m*): wall fan 501
WANGE (*f*): cheek 502
WANKELHAFT: inconstant; unsteady; unsettled; vacillatory; changeable; fickle 503
WANKELMUT (*m*): vacillation; fickleness; inconstancy 504
WANKELMÜTIG: (see WANKELHAFT) 505
WANKEN: to waver; totter; stagger; reel; shake 506
WANN: when; then 507
WANNE (*f*): vat; trough; well; tub; bath-tub; coop; fan 508
WANNEN: to winnow; fan 509
WANNE (*f*), ÖL-: oil well 510
WANTAUGE (*f*): eye of a shroud 511
WANTEN (*n.pl*): shrouds (rigging) (Naut.) 512
WANTEN (*n.pl*), PÜTTINGS-: futtock rigging, futtock shrouds (Naut.) 513
WANTENSCHRAUBE (*f*): rigging screw 514
WANTKLAMPE (*f*): shroud cleat 515
WANTTAU (*n*): shroud (Naut.) 516
WANZE (*f*): bug 517
WANZIG: buggy; full of (overrun with) bugs 518
WAPPEN (*m*): escutcheon; (coat of) arms 519
WAPPENKUNDE (*f*): heraldry; blazonry 520
WARDEIN (*m*): mint warden 521
WARE (*f*): ware; article; manufacture; (*pl*): goods; merchandize 522
WARENGESCHÄFT (*n*): merchant's business 523
WARENLAGER (*n*): store; shop; warehouse 524
WARENSTEMPEL (*m*): trademark; manufacturer's mark on goods 525
WARENZEICHEN (*n*): trade mark 526
WARM: hot; warm 527
WARMBLÜTER (*m.pl*): warm-blooded animals 528
WARMBLÜTIG: warm (hot)-blooded 529
WARMBRÜCHIG: hot-short; red-short (Metal) 530
WARMBRUNNEN (*m*): hot spring 531
WÄRME (*f*): warmth; heat 532
WÄRMEABGABE (*f*): heat (transmitted) given off; heat loss 533
WÄRMEABLEITUNG (*f*): heat (transmission) conduction 534
WÄRMEABSORPTION (*f*): heat absorption 535
WÄRMEÄNDERUNG (*f*): change of temperature; heat change; thermal variation 536
WÄRME, ANSTELL- (*f*): initial (commencing; starting) temperature (Brew.) 537
WÄRMEÄQUIVALENT (*n*): heat equivalent, thermal equivalent 538
WÄRMEAUFNAHME (*f*): heat absorption 539
WÄRMEAUFNAHMEFÄHIGKEIT (*f*): heat absorptivity 540
WÄRMEAUFWAND (*m*): heat consumption 541
WÄRMEAUSDEHNUNG (*f*): expansion due to heat 542
WÄRMEAUSDEHNUNGSZAHL (*f*): coefficient of (heat) expansion 543
WÄRMEAUSGLEICH (*m*): heat balance 544
WÄRMEAUSGLEICHPRINZIP (*n*): regenerative principle, principle of heat (exchange) balance 545
WÄRMEAUSNUTZUNG (*f*): heat efficiency; thermal efficiency 546
WÄRMEAUSSTRAHLUNG (*f*): heat radiation 547
WÄRMEAUSTAUSCH (*m*): heat exchange; heat interchange 548
WÄRMEAUSTAUSCHKÖRPER (*m*): heat exchanging body; heat exchanger 549
WÄRMEBEHANDLUNG (*f*): heat treatment 550
WÄRMEBEIWERT (*m*): heat coefficient; thermal coefficient 551
WÄRMEBEWEGUNG (*f*): heat motion 552
WÄRMEBILANZ (*f*): heat balance 553
WÄRMEBILDUNG (*f*): formation (production) of heat 554
WÄRMEBINDUNG (*f*): heat (absorption) consumption 555
WÄRMEDICHTE (*f*): heat density; quantity of heat per unit of volume 556
WÄRMEDURCHGANG (*m*): heat transmission 557
WÄRMEDURCHGANGSZAHL (*f*): amount (coefficient) of heat transmission 558
WÄRMEDURCHLÄSSIG: diathermic; diathermanous; transcalent; permeable to heat; conductive to heat; capable of transmitting heat 559
WÄRMEEFFEKT (*m*): heat effect; thermal effect 560
WÄRMEEINHEIT (*f*): heat unit (kilogram calorie); thermal unit; (1 W.E. = 3.968 B.T.U.); (1 W.E. per Kg. = 9/5ths B.T.U. per lb.) 561
WÄRMEEINSTRAHLUNG (*f*): (heat) radiation; (heat)-absorption 562
WÄRMEEINWIRKUNG (*f*): action of heat (see WÄRMEEINSTRAHLUNG) 563
WÄRMEELEKTRIZITÄT (*f*): thermo-electricity 564
WÄRMEENTBINDUNG (*f*): heat (disengagement) release 565
WÄRMEENTWICKELUNG (*f*): evolution of heat 566
WÄRMEERSPARNIS (*f*): heat (saving) economy; saving in heat; heat conservation 567
WÄRMEERZEUGEND: heat-producing; heat-generating 568
WÄRMEERZEUGER (*m*): heat generator; heat (producer) producing substance 569
WÄRMEGEBEND: giving heat, exothermic, exothermal 570
WÄRMEGEFÄLLE (*n*): heat drop 571
WÄRMEGRAD (*m*): temperature; degree of heat 572
WÄRMEINHALT (*m*): heat content 573
WÄRMEKAPAZITÄT (*f*): heat (calorific) capacity 574
WÄRMEKRAFTLEHRE (*f*): thermo-dynamics 575
WÄRMEKRAFTMASCHINE (*f*): heat engine 576
WÄRMELEHRE (*f*): theory of heat 577
WÄRMELEITEND: heat-conducting 578
WÄRMELEITER (*m*): thermal (heat) conductor 579
WÄRMELEITFÄHIGKEIT (*f*): heat (thermal) conductivity 580
WÄRMELEITUNG (*f*): heat conduction; thermal conduction 581
WÄRMELEITUNGSVERMÖGEN (*n*): heat (thermal) conductivity 582
WÄRMELEITVERMÖGEN (*n*): thermal (heat) conductivity 583
WÄRMEMECHANISCH: heat-mechanical 584
WÄRMEMENGE (*f*): heat quantity 585
WÄRMEMESSER (*m*): thermometer; calorimeter; any heat (meter) measuring apparatus 586
WÄRMEMESSINSTRUMENT (*n*): calorimeter; thermometer (instrument for measuring heat) 587
WÄRMEMESSUNG (*f*): calorimetry; measurement of heat 588
WÄRMEMOTOR (*m*): heat motor; heat engine 589
WÄRMEN: to warm; heat; become warm 590
WÄRME (*f*), NUTZBARE-: effective heat 591

WÄRMEOFEN (*m*): heating (tempering, hardening, annealing) furnace 592
WÄRMEPLATTE (*f*): warming plate; hot-plate 593
WÄRMEQUANTUM (*n*): heat quantity 594
WÄRMEQUELLE (*f*): heat source 595
WÄRMESAMMLER (*m*): heat accumulator 596
WÄRMESATZ (*m*): heat law; law of thermodynamics 597
WÄRMESCHRANK (*m*): warming cabinet; drying cupboard 598
WÄRMESCHUTZ (*m*): heat conservation; heat (insulation) insulator 599
WÄRMESCHUTZMASSE (*f*): non-conducting composition; lagging; heat conserver; insulation; insulator 600
WÄRMESCHUTZMITTEL (*n*): heat conserver; (heat)-insulator; non-conducting composition; insulating material; non-conductor 601
WÄRMESCHWANKUNG (*f*): heat (variation) fluctuation 602
WÄRMESCHWINGUNG (*f*): heat vibration 603
WÄRMESPANNUNG (*f*): heat stress; stress (due to) set up by heat 604
WÄRMESPEICHER (*m*): regenerator; heat accumulator 605
WÄRMESPEKTRUM (*n*): heat (thermal) spectrum 606
WÄRMESTAUUNG (*f*): heat pocket 607
WÄRMESTOFF (*m*): thermogene; caloric 608
WÄRMESTRAHL (*m*): heat ray 609
WÄRMESTRAHLUNG (*f*): heat radiation 610
WÄRMESUMME (*f*): heat sum 611
WÄRMETECHNISCH: heat technical 612
WÄRMETECHNOLOGIE (*f*): heat technology 613
WÄRMETHEORIE (*f*): heat theory 614
WÄRMETÖNUNG (*f*): heat tone 615
WÄRMETRAGEND: conveying (carrying) heat; caloriferous 616
WÄRMEÜBERGANG (*m*): heat (transfer) transmission 617
WÄRMEÜBERGANGSWERT (*m*): coefficient of heat transfer 618
WÄRMEÜBERTRAGUNG (*f*): heat (transfer) transmission 619
WÄRME (*f*), VERBORGENE-: latent heat (see WÄRME, VERDAMPFUNGS-) 620
WÄRMEVERBRAUCH (*m*): heat consumption; temperature drop 621
WÄRME (*f*), VERDAMPFUNGS-: heat of (vapourization) evaporation, (a general term which, unless otherwise qualified, is usually held to mean "Total heat of evaporation." It is, however, often loosely employed for latent heat, see WÄRME, VERBORGENE-) 622
WÄRMEVERDICHTUNG (*f*): heat density; quantity of heat per unit of volume 623
WÄRMEVERLUST (*m*): heat loss 624
WÄRMEVERMÖGEN (*n*): heat capacity 625
WÄRMEVERTEILUNG (*f*): heat distribution 626
WÄRMEVERZEHREND: consuming heat, endothermic, endothermal 627
WÄRMEVORGANG (*m*): thermal (heat) phenomenon or process 628
WÄRMEVORRAT (*m*): store (supply, stock) of heat 629
WÄRMEWERT (*m*): calorific (heat) value 630
WÄRMEWIRKUNG (*f*): thermal (action) effect 631
WÄRMEWIRTSCHAFT (*f*): heat (thermo) economy 632
WÄRMEZUFUHR (*f*): heat (addition) supply 633
WÄRMEZUSTAND (*m*): thermal condition 634
WÄRMFLASCHE (*f*): warming (hot-water) bottle 635
WARMGEPRESST: hot-pressed 636
WARMGEZOGEN: hot-drawn 637
WARMLAUF (*m*): overheating 638
WARM LAUFEN: to run hot, become hot, heat, overheat 639
WARMLAUFEN (*n*): running hot, getting hot, heating, overheating 640
WARMLUFTANLAGE (*f*): hot-air plant 641
WÄRMOFEN (*m*): (see GLÜHOFEN) 642
WÄRMPLATTE (*f*): warming (hot) plate 643
WARMPRESSEN: to hot-press; (*n*): hot-pressing 644
WARMRECKEN: to hot-draw; hot-work (Metal) 645
WARMWASSERBEHÄLTER (*m*): hot-water tank 646
WARMWASSERBEREITER (*m*): geyser; calorifier 647
WARMWASSERBEREITUNGSANLAGE (*f*): hot-water plant 648
WARMWASSERHEIZUNG (*f*): (hot-) water heating (-apparatus) 649
WARMWASSERLEITUNG (*f*): hot-water piping 650
WARMWASSERSYSTEM (*n*): hot-water system 651
WARNEN: to warn; caution; admonish 652
WARNUNG (*f*): warning; caution; admonition 653
WARNUNGSSCHILD (*n*): warning board, danger notice board 654
WARNUNGSTAFEL (*f*): warning board, danger notice board 655
WARPANKER (*m*): grapnel 656
WARTE (*f*): lookout; observatory; watchtower; observation post 657
WARTEN: to wait; nurse; attend to; care for; stay; tarry 658
WÄRTER (*m*): warder; keeper; nurse; attendant; waiter 659
WÄRTERIN (*f*): wardress; attendant; keeper; nurse 660
WARTHIT (*m*): warthite (see BLÖDIT) 661
WARTUNG (*f*): attendance; nursing; tending; waiting; waiters; attendants; attention 662
WARUM: why; wherefore 663
WÄRZCHEN (*n*): (see PAPILLE) 664
WARZE (*f*): wart; nipple; teat; pin; mastoid; boss; knob; lug; projection 665
WARZENÄHNLICH: wartlike; mamillary; papillary 666
WARZENARTIG: (see WARZENÄHNLICH) 667
WARZENFÖRMIG: wartlike; wartshaped; mamillary; papillary 668
WARZENKRAUT (*n*): warty plant; marigold (*Calendula officinalis*) 669
WARZIG: warty 670
WAS: which; what; that 671
WASCHANLAGE (*f*): washing plant 672
WASCHANSTALT (*f*): laundry; washing establishment 673
WASCHAPPARAT (*m*): washer; washing apparatus 674
WASCHBAR: washable 675
WASCHBECKEN (*m*): washing (washhand) basin 676
WASCHBERGE (*m.pl*): waste washings (Coal mining) 677
WASCHBLAU (*n*): dolly blue, washing blue; blue; blueing (a blue-tinged starch for washing purposes), any blue for blueing washing 678

WÄSCHE (*f*) washing; clothes; linen; wash 679
WASCHECHT: fast to washing; washable 680
WÄSCHEGLANZ (*m*): washing gloss 681
WASCHEN: to wash; buddle; scour; scrub (of gases, etc.) 682
WASCHER (*m*): washer; scourer; scrubber (of gases, etc.) 683
WÄSCHEREI (*f*): laundry; washhouse; gossip 684
WÄSCHEREIMASCHINEN (*f.pl*): laundry machinery 685
WÄSCHESCHRANK (*m*): linen-cupboard 686
WÄSCHEWACHS (*n*): wax for washing 687
WÄSCHEZEICHENTINTE (*f*): marking ink 688
WASCHFASS (*n*): washing tub 689
WASCHFLASCHE (*f*): wash (washing) bottle 690
WASCHFRAU (*f*): washerwoman; laundress 691
WASCHGOLD (*n*): placer gold 692
WASCHHAFT: gossipping, talkative 693
WASCHHAUS (*n*): washhouse 694
WASCHKASTEN (*m*): washing tank; print washer, plate washer (Phot.) 695
WASCHKESSEL (*m*): boiler; copper (for washing purposes) 696
WASCHKLAMMER (*f*): clothes-peg 697
WASCHKRYSTALL (*n*): soda (washing) crystals; $Na_2CO_3.10H_2O$ (see **SODA**) 698
WASCHLAPPEN (*m*): dish-cloth 699
WASCHLEDER (*n*): washleather; chamois leather 700
WASCHMASCHINE (*f*): washing machine 701
WASCHMITTEL (*n*): detergent; washing agent; lotion (Med.) 702
WASCHÖL (*n*): washing oil 703
WASCHPRÄPARAT (*n*): washing preparation, detergent 704
WASCHPROBE (*f*): assay of washed (buddled) ore 705
WASCHPRODUKT (*n*): product of washing; washings (Coal) 706
WASCHSEIFE (*f*): washing soap 707
WASCHTROMMEL (*f*): washing (cylinder) drum 708
WASCHTURM (*m*): washing tower 709
WASCHUNG (*f*): washing; wash; lotion 710
WASCHWANNE (*f*): washing-tub 711
WASCHWASSER (*n*): washing-water; water for washing purposes 712
WASCHWIRKUNG (*f*): cleansing (action) effect 713
WASCHWURZEL (*f*): soapwort (*Saponaria officinalis*) 714
WASCHZINN (*n*): stream tin (see **STROMZINN**) 715
WASEN (*m*): greensward; turf 716
WASHINGTONIT (*m*): washingtonite; $FeTiO_3$; Sg. 4.9 717
WASSER (*n*): water; H_2O; (*Aqua*); (Sg. 1.0; generally the standard on which specific gravities are based) 718
WASSER (*n*), **ABGEZOGENES-**: distilled liquor 719
WASSERABLASS (*m*): drainage; draining 720
WASSERABLEITUNG (*f*): drainage; draining 721
WASSERABSCHEIDER (*m*): water (separator) extractor 722
WASSERABSCHEIDUNG (*f*): water separation 723
WASSERABSTOSSEND: repelling water 724
WASSERABTRITT (*m*): water-closet (Sanitation); water (exit, outlet) discharge 725
WASSERÄHNLICH: watery; like water; water-like; aqueous 726
WASSERAMPFER (*m*): water dock (*Rumex aquaticus*) 727
WASSERANALYSE (*f*): water analysis 728
WASSERANZIEHEND: hydroscopic; hygroscopic; attracting water or moisture 729
WASSERARM: having low (water) moisture content 730
WASSERARMATUREN (*f.pl*): water fittings 731
WASSER, AROMATISCHES- (*n*): aromatic (distilled) water (solutions of mixtures of volatile vegetable substances in water (Pharm.)) 732
WASSERARTIG: aqueous; watery; like water; water-like 733
WASSERAUFNAHME (*f*): water absorption 734
WASSERAUFSAUGUNGSVERMÖGEN (*n*): water absorptivity, capillary attraction 735
WASSERAUSSCHEIDUNG (*f*): water separation 736
WASSERAUSSPÜLING (*f*): rinsing with water 737
WASSERAUSTRITT (*m*): water (outflow) exit; elimination of water; efflux of water 738
WASSERBAD (*n*): water bath 739
WASSERBALG (*m*): serous cyst (Med.) 740
WASSERBAU (*m*): hydraulics 741
WASSERBAUKUNST (*f*): hydraulic engineering 742
WASSERBAUTEN (*m.pl*): hydraulic plant; hydraulic (engineering) constructional work 743
WASSERBEHÄLTER (*m*): water tank; reservoir; cistern; water back (Brew.) 744
WASSERBEIZE (*f*): water mordant 745
WASSERBERÜHRT: submerged (of boiler heating surface; i.e., that portion of the heating surface (touched) in direct contact with water) 746
WASSERBESTÄNDIG: stable in (resistant to) water; waterproof 747
WASSER (*n*), **BETRIEBS-**: operating fluid; water used for driving purposes 748
WASSERBILDUNG (*f*): formation of water 749
WASSERBLASE (*f*): bubble; vessel for heating water; blister; vesicle (Med.) 750
WASSERBLAU: sea-blue; blue like water; (*n*): sea-blue; water-blue 751
WASSERBLEI (*n*): molybdenum glance; molybdenite; MoS_2 (see **MOLYBDÄNGLANZ**) 752
WASSERBLEIOCHER (*m*): molybdite; molybdic ocher (see **MOLYBDÄNOCHER**) 753
WASSERBLEISÄURE (*f*): molydic acid (see **MOLYBDÄNSÄURE**) 754
WASSERBRUCH (*m*): hydrocele; swelling of the scrotum (Med.) 755
WASSERDAMPF (*m*): steam; water vapour 756
WASSERDAMPFBAD (*n*): steam (vapour) bath 757
WASSERDESTILLATION (*f*): distillation of water 758
WASSERDESTILLATIONSAPPARAT (*m*): water distilling (apparatus) plant 759
WASSER (*n*), **DESTILLIERTES-**: distilled water (see also **WASSER, AROMATISCHES-**) 760
WASSERDICHT: waterproof; water-tight; impervious to water 761
WASSERDICHTE (*f*): watertightness; imperviousness to water; water-density 762
WASSERDICHTES PAPIER (*n*): drying Royals (Paper) 763
WASSERDICHTIGKEIT (*f*): imperviousness to water; waterproof capacity 764
WASSERDICHTMACHEN (*n*): waterproofing 765
WASSERDRUCK (*m*): water-pressure; hydraulic pressure 766
WASSERDRUCKPROBE (*f*): water-pressure (hydraulic) test 767
WASSERDUNST (*m*): water vapour 768
WASSERDURCHBRUCH (*m*): water (leak) burst (of pipes) 769

WASSERDURCHLÄSSIG: permeable (to water); porous 770
WASSERDURCHLÄSSIGKEIT (*f*): permeability-(to water); filtrative capacity; porosity 771
WASSERECHT: fast to water 772
WASSERECHTMACHUNGSMITTEL (*n*): water-resisting agent; waterproofing material 773
WASSEREIMER (*m*): (water)-pail 774
WASSERENTEISENUNG (*f*): extraction of iron from water 775
WASSERENTHÄRTUNG (*f*): water softening 776
WASSERENTZIEHEND: dehydrating, removing (extracting) water 777
WASSERENTZIEHUNG (*f*): dehydration; extraction (removal) of water 778
WASSERERGUSS (*m*): watery effusion; œdema 779
WASSERFALL (*m*): water-fall; cascade; cataract 780
WASSERFARBE (*f*): water-colour 781
WASSERFASS (*n*): water-tub 782
WASSERFEDER (*f*): water featherfoil; water violet (*Hottonia palustris*) 783
WASSERFENCHEL (*m*): water fennel (*Foeniculum vulgare*) 784
WASSERFENCHELÖL (*n*): water-fennel oil 785
WASSERFEST: water-tight; waterproof 786
WASSERFIRNIS (*m*): water varnish 787
WASSERFLÄCHE (*f*): water level; surface (sheet) of water 788
WASSERFLASCHE (*f*): water-bottle 789
WASSERFLECK (*m*): water stain 790
WASSER, FLIESSENDES- (*n*): running water 791
WASSERFLUT (*f*): flood, inundation, deluge 792
WASSERFÖRMIG: watery; like water 793
WASSERFREI: anhydrous; free from water 794
WASSERFÜHREND: water-bearing 795
WASSERFÜHRUNG (*f*): water supply (either the amount or the method of supply) 796
WASSERGANG (*m*): waterway; aqueduct; drain 797
WASSERGAS (*n*): water gas 798
WASSERGASANLAGE (*f*): watergas plant 799
WASSERGEFÄSS (*n*): water vessel; lymphatic vessel (Med.) 800
WASSERGEFLÜGEL (*n*): water-fowl 801
WASSERGEHALT (*m*): water (moisture) content 802
WASSERGEKÜHLT: water-cooled 803
WASSERGEKÜHLTES LAGER (*n*): water-cooled bearing 804
WASSER (*n*), GESÄTTIGTES-: saturated water (usually; saturated solution) 805
WASSERGESCHWULST (*f*): hygroma; oedema 806
WASSERGEWÄCHS (*n*): aquatic plant 807
WASSERGLAS (*n*): water-glass (see NATRON-WASSER-GLAS and KALIWASSERGLAS); isinglass; silicate of potash and soda (see DOPPELWASSERGLAS) 808
WASSERGLASANLAGE (*f*): silicate (silica) plant 809
WASSERGLASFARBE (*f*): silica colour 810
WASSERGLASPRÄPARAT (*n*): silica preparation 811
WASSERGLASSEIFE (*f*): water-glass soap 812
WASSERGRABEN (*m*): ditch; trench; channel; moat 813
WASSER (*n*), GURGEL-: gargle 814
WASSERHAHN (*n*): water (cock) tap 815
WASSERHALTEND: water (retaining) containing 816
WASSERHALTENDE KRAFT (*f*): water capacity 817
WASSERHALTIG: aqueous; hydrous; hydrated; containing water; water-bearing 818
WASSERHART: impervious to water; waterproof; water-tight 819
WASSERHARZ (*n*): Burgundy pitch; Burgundy resin (from *Abies excelsa*) 820
WASSERHAUT (*f*): hyaloid membrane; amnion 821
WASSERHEILKUNDE (*f*): hydropathy; hydrotherapeutics 822
WASSERHELL: transparent; clear as water 823
WASSERHOSE (*f*): water-spout 824
WASSERHUHN (*n*): water-hen 825
WASSER (*n*), HYGROSKOPISCHES-: hygroscopic water 826
WÄSSERIG: watery; hydrous; aqueous; serous (Med.); insipid; flat 827
WÄSSERIGE FEUCHTIGKEIT (*f*): aqueous humour (Med.) 828
WASSERJUNGFER (*f*): dragon-fly (*Libellula*) 829
WASSERKALK (*m*): hydraulic (water) lime 830
WASSERKAMMER (*f*): drum, header, water-space (of boilers) 831
WASSERKANNE (*f*): water (watering) can, ewer 832
WASSERKAPAZITÄT (*f*): water capacity 833
WASSERKASTEN (*m*): cistern 834
WASSERKIES (*m*): marcasite; FeS_2 (see MARKASIT) 835
WASSERKITT (*m*): hydraulic cement 836
WASSERKLAR: clear as water 837
WASSER (*n*), KÖLNISCHES-: Eau de Cologne 838
WASSERKOPF (*m*): hydrocephalus; Hydrops capitis; water on the brain (Med.) 839
WASSERKRAFT (*f*): water power 840
WASSERKRAFTANLAGE (*f*): water power plant; hydraulic power plant 841
WASSERKRAFTLEHRE (*f*): hydrodynamics 842
WASSERKRAN (*m*): water faucet 843
WASSERKREBS (*m*): noma, Cancrum oris (a gangrenous ulceration of the mouth) (Med.) 844
WASSERKRESSE (*f*): water-cress (*Nasturtium officinale*) 845
WASSERKRISTALL (*m*): rock crystal 846
WASSERKRUG (*m*): water (pot) jug; pitcher 847
WASSERKÜHLKASTEN (*m*): water block 848
WASSERKÜHLUNG (*f*): water (jacket) cooling 849
WASSERKUNST (*f*): hydraulics 850
WASSERLAUF (*m*): conduit; stream; channel; water-course; gutter; drain 851
WASSER, LAUFENDES- (*n*): running water 852
WASSERLEITUNG (*f*): water (pipe; piping) main; aqueduct; conduit; channel 853
WASSER (*n*), LEITUNGS-: water from the mains (usually from the town mains, but sometimes referring to water taken from the pipelines of power plant) 854
WASSERLEITUNGSROHR (*n*): water pipe; water main 855
WASSERLILIE (*f*): water-lily (*Nymphœa alba*) 856
WASSERLINIE (*f*): water line 857
WASSERLINIENPLAN (*m*): half breadth plan 858
WASSERLINIENRISS (*m*): half breadth plan 859
WASSERLINSE (*f*): duckweed (*Lemna minor*) 860
WASSERLOS: without water; waterless 861
WASSERLÖSLICH: water soluble 862
WASSERLÖSLICHE FETTE (*n.pl*): water soluble fats 863
WASSERLÖSLICHE ÖLE (*n.pl*): water soluble oils 864
WASSERLUFTPUMPE (*f*): water vacuum pump 865
WASSERMAISCHE (*f*): mash; aqueous infusion 866

WASSERMALEREI (*f*): water colour painting 867
WASSERMANGEL (*m*): shortage (want; deficiency) of water; water (shortage) famine 868
WASSERMANTEL (*m*): water jacket 869
WASSERMASS (*n*): water gauge 870
WASSERMELONENÖL (*n*): water-melon oil (from *Cucumis citrullus*); Sg. 0.925 871
WASSERMESSER (*m*): water (gauge) meter; hydrometer; hygrometer; aerometer; densimeter (for liquids) 872
WASSERMESSER (*m*), **AMMONIAK-**: ammonia-water (gas liquor) meter 873
WASSERMESSER (*m*), **KONDENS-**: condensate water meter 874
WASSERMESSER (*m*), **SPEISE-**: feed water meter 875
WASSERMESSFLÜGEL (*m*): water measuring (hydrometric) vane or propeller, current meter 876
WASSERMESSUNGSLEHRE (*f*): hydrometry 877
WASSERMÖRTEL (*m*): hydraulic mortar; cement 878
WASSERMOTOR (*m*): hydraulic motor 879
WASSERMÜHLE (*f*): water-mill 880
WASSER (*n*), **MUND-**: mouth wash 881
WÄSSERN: to water; hydrate; soak; irrigate 882
WASSERNABEL (*m*): marsh pennywort (*Hydrocotyle*) 883
WASSERNUSS (*f*): water (caltrop) nut (*Trapa natans*) 884
WASSEROPAL (*m*): water opal; hyalite; hydrophane; adularia; avanturine feldspar (see **HYALIT**) 885
WASSERPAPIER (*n*): waterleaf (Paper) 886
WASSERPFLANZE (*f*): water-plant 887
WASSERPLAN (*m*): hydroplane 888
WASSERPOCKEN (*f.pl*): chickenpox (Med.) 889
WASSERPROBE (*f*): water (hydraulic) test; water sample 890
WASSERPRÜFER (*m*): water tester 891
WASSERPUMPE (*f*): water (hydraulic) pump 892
WASSERPYROMETER (*n*): water pyrometer 893
WASSERQUELLE (*f*): spring, source 894
WASSERRAD (*n*): water-wheel 895
WASSERRADSCHAUFEL (*f*): paddle of a water wheel, bucket 896
WASSERRADZELLE (*f*): bucket 897
WASSERRATTE (*f*): water-rat 898
WASSERREICH: abounding in water 899
WASSERREINIGUNG (*f*): water purification 900
WASSERREINIGUNGSANLAGE (*f*): water-purifying plant 901
WASSERREST (*m*): water residue; hydroxyl 902
WASSERRINNE (*f*): (water)-channel; gutter 903
WASSERROHR (*n*): water (pipe) tube 904
WASSERRÖHRE (*f*): (see **WASSERROHR**) 905
WASSERRÖHRENKESSEL (*m*): water-tube boiler 906
WASSERROHRKESSEL (*m*): water-tube boiler 907
WASSERRÖSTE (*f*): retting process of flax in ditches, trenches or streams; for obtaining bast fibre (Textile) (see **RÖSTE**) 908
WASSERSACK (*m*): water pocket 909
WASSERSÄULE (*f*): water column 910
WASSERSCHADEN (*m*): damage caused by (water) flood 911
WASSERSCHEIDE (*f*): watershed 912
WASSERSCHEU (*f*): fear of water; hydrophobia; rabies (Med.) 913
WASSERSCHIEBER (*m*): penstock; sluice-valve 914
WASSERSCHIERLING (*m*): cowbane; water hemlock (*Cicuta virosa*) 915
WASSERSCHLAG (*m*): water hammer-(ing) (in pipes, etc.) 916
WASSERSCHLANGE (*f*): coil (for water); water-hose; bladderwort (*Utricularia*); water-snake 917
WASSERSCHLUSS (*m*): water seal; trap 918
WASSERSEGEL (*n*): water sail 919
WASSERSILBER (*n*): quicksilver; mercury (see **QUECKSILBER**) 920
WASSERSPIEGEL (*m*): water level; surface (sheet) of water 921
WASSERSPRITZE (*f*): syringe; squirt; sprinkler 922
WASSERSTAG (*n*): bobstay 923
WASSERSTAND (*m*): water-level; water-gauge; constant water-level device 924
WASSERSTÄNDER (*m*): water-tub 925
WASSERSTANDSANZEIGER (*m*): water-gauge 926
WASSERSTANDS-ENTWÄSSERUNG (*f*): water-gauge drain 927
WASSERSTANDSGLAS (*n*): (water)-gauge glass 928
WASSERSTANDSHAHN (*m*): (water)-gauge cock 929
WASSERSTANDSLINIE (*f*): water-line, water-level 930
WASSERSTANDSREGLER (*m*): water-level regulator; water-gauge 931
WASSERSTANDSROHR (*n*): gauge glass, water-gauge pipe 932
WASSERSTANDSSCHUTZGLAS (*n*): gauge glass protector 933
WASSERSTANDSSCHUTZVORRICHTUNG (*f*): water-gauge protectors 934
WASSERSTANDSZEIGER (*m*): water-level pointer (indicator); water gauge 935
WASSERSTÄRKE (*f*), (**ZUGSTÄRKE IN** m.m.)-: (draught in m.m.)- water-gauge (W.G.) 936
WASSERSTEIN (*m*): scale, incrustation, deposit, sediment 937
WASSERSTERILISIERUNGSMITTEL (*n*): water (sterilizer) sterilizing agent 938
WASSERSTOFF (*m*): hydrogen; (H) 939
WASSERSTOFFANLAGE (*f*): hydrogen plant 940
WASSERSTOFFARM: low hydrogen content 941
WASSERSTOFFENTWICKELUNG (*f*): generation (production; evolution) of hydrogen 942
WASSERSTOFFFREI: free from hydrogen 943
WASSERSTOFFGAS (*n*): hydrogen gas; H_2; Sg. 0.0695; Mp. -259°C.; Bp. -252°C. 944
WASSERSTOFFGASANLAGE (*f*): hydrogen gas plant 945
WASSERSTOFFGEHALT (*m*): hydrogen content 946
WASSERSTOFFHALTIG: hydrogenous; containing hydrogen 947
WASSERSTOFFHYPEROXYD (*n*): hydrogen peroxide; peroxide of hydrogen (see **WASSERSTOFFSUPEROXYD**) 948
WASSERSTOFFION (*n*): hydrogen ion 949
WASSERSTOFFIONENKONZENTRATION (*f*): concentration of hydrogen ions 950
WASSERSTOFFJON (*n*): hydrogen ion 951
WASSERSTOFFKALIUM (*n*): potassium hydride 952
WASSERSTOFFKNALLGAS (*n*): detonating gas (see **KNALLGAS**) 953
WASSERSTOFFMESSER (*m*): hydrogen meter 954
WASSERSTOFFMONOXYD (*n*): hydrogen monoxide, water; H_2O; (Aqua) 955
WASSERSTOFFPALLADIUM (*n*): palladium hydride 956

WASSERSTOFFPEROXYD (*n*): peroxide of hydrogen; hydrogen peroxide (see WASSERSTOFF-SUPEROXYD); H_2O_2; Sg. 1.452-1.4995) 957
WASSERSTOFFPERSULFID (*n*): hydrogen persulphide; (H_2S_2; Sg. 1.376); (H_2S_3; Sg. 1.496) 958
WASSERSTOFFREICH: having high hydrogen content 959
WASSERSTOFFSALZ (*n*): hydrogen salt 960
WASSERSTOFFSAUER: combined with a hydracid 961
WASSERSTOFFSÄURE (*f*): hydracid 962
WASSERSTOFFSTROM (*m*): hydrogen current 963
WASSERSTOFFSULFID (*n*): hydrogen sulphide; sulphuretted hydrogen; H_2S; Sg. 1.19; Mp. – 83°C.; Bp. – 60°C. 964
WASSERSTOFFSUPEROXYD (*n*): hydrogen peroxide (*Hydrogenium peroxydatum*); H_2O_2; Mp. – 2°C.; Bp. 70.7°C. (see WASSERSTOFF-PEROXYD) 965
WASSERSTOFFSUPEROXYDANLAGE (*f*): peroxide of hydrogen plant 966
WASSERSTOFFVERBINDUNG (*f*): hydrogen compound 967
WASSERSTOFFWERTIGKEIT (*f*): hydrogen valence 968
WASSERSTOFFZÜNDMASCHINE (*f*): hydrogen (Döbereiner's) lamp 969
WASSERSTRAHL (*m*): jet of water; water-jet 970
WASSERSTRAHLGEBLÄSE (*n*): water-jet blast 971
WASSERSTRAHLLUFTPUMPE (*f*): water-jet vacuum pump 972
WASSERSTROM (*m*): current of water 973
WASSERSTURZ (*m*): water-fall 974
WASSERSUCHT (*f*): dropsy (Med.) 975
WASSERSÜCHTIG: dropsical; suffering from dropsy 976
WASSERSUPPE (*f*): water-gruel 977
WASSERTALK (*m*): brucite (see BRUCIT) 978
WASSERTECHNIK (*f*): hydraulics 979
WASSERTIEFE (*f*): depth of water; draught (of ships) 980
WASSERTIER (*n*): aquatic (water) animal 981
WASSER, TOTES- (*n*): still water 982
WASSERTREIBEND: hydragog (Med.) 983
WASSERTREIBENDES MITTEL (*n*): hydragogue (a purge causing a watery discharge) (Med.) 984
WASSERTROG (*m*): water-trough 985
WASSERTROPFEN (*m*): drop of water 986
WASSERTURBINE (*f*): water turbine, turbine tube cleaner 987
WASSER (*n*), ÜBERHITZTES-: superheated water 988
WASSERUMLAUF (*m*): water circulation 989
WASSERUNDURCHLÄSSIG: watertight, impermeable 990
WÄSSERUNG (*f*): hydration; soaking; watering; irrigation 991
WASSERUNLÖSLICH: insoluble in water 992
WASSERUNTERSUCHUNG (*f*): water analysis 993
WASSERVENTIL (*n*): water-valve 994
WASSERVERBRAUCH (*m*): water consumption 995
WASSERVERDRÄNGUNG (*f*): (water)-displacement (of ships) 996
WASSERVERDUNSTUNGSFÄHIGKEIT (*f*): evaporative capacity 997
WASSERVERGOLDUNG (*f*): water gilding 998
WASSERVERLUST (*m*): water loss 999
WASSERVERMÖGEN (*n*): water capacity 000
WASSERVERSCHLUSS (*m*): water seal 001
WASSERVERSORGUNG (*f*): water works; water supply 002
WASSERVERUNREINIGUNG (*f*): water (contamination) impurities 003
WASSERWAGE (*f*): (water)-level; hydrometer; hydrostatic balance; spirit level (employed erroneously) 004
WASSERWANNE (*f*): water (pneumatic) trough 005
WASSERWEG (*m*): waterway (usually referring to navigable streams); water (passage; opening) area 006
WASSERWEIDE (*f*): osier (*Salix*) 007
WASSERWERK (*n*): waterworks 008
WASSERWERKSANLAGE (*f*): waterworks plant 009
WASSERWERT (*m*): water equivalent 010
WASSERWERTSWÄRME (*f*): water equivalent heat (the heat measurable by thermometer) 011
WASSER (*n*), WOHLRIECHENDES-: perfume, scent 012
WASSERZEICHEN (*n*): watermark (Paper) 013
WASSERZEMENT (*m*): hydraulic cement 014
WASSERZERSETZUNG (*f*): decomposition of water 015
WASSERZUFLUSS (*m*): water (feed) supply 016
WÄSSRIG: watery; hydrous; aqueous; serous (Med.); flat; insipid 017
WATEN: to wade 018
WATERGARN (*n*): water twist (Textile) 019
WATERMASCHINE (*f*): water machine (Textile) 020
WATERPROOFFIRNIS (*m*): water-proof varnish 021
WATSCHELN: to waddle; toddle 022
WATT (*n*): watt (Elect.) 023
WATTE (*f*): cotton-wool (from *Gossypium barbadense*); wadding; padding; wad 024
WATTEARTIG: wadding-like, like cotton-wool 025
WATTEBAUSCH (*m*): wadding plug, cotton-wool plug 026
WATTEPFROPF (*m*): plug (wad) of cotton-wool or wadding 027
WATTESCHICHT (*f*): layer of (cotton-wool) wadding 028
WATTIEREN: to wad; pad 029
WATTMETER (*n*): watt meter 030
WATTSTUNDE (*f*): watt hour 031
WAU (*m*): weld; yellow-weed; dyer's weed (*Reseda luteola*) 032
WAUSAMENÖL (*n*): weld-seed-oil (from *Reseda luteola*) 033
WAVELLIT (*m*): wavellite (hydrous aluminium phosphate); $Al_3(F,OH)P_2O_4.7H_2O$; Sg. 2.35 (see also LASIONIT) 034
WEBE (*f*): web; weft 035
WEBELEINE (*f*): ratline 036
WEBELEINSTECK (*m*): clove hitch 037
WEBEN: to weave 038
WEBER (*m*): weaver 039
WEBERBAUM (*m*): warping loom 040
WEBERDISTEL (*f*): fuller's teasel; fuller's thistle (*Dipsacus fullonum*) 041
WEBEREI (*f*): weaving-(mill); tissue; texture 042
WEBEREIMASCHINE (*f*): weaving machine (Textile) 043
WEBERSCHIFFCHEN (*n*): shuttle 044
WEBERSPULE (*f*): shuttle; spool 045
WEBERZETTEL (*m*): warp 046
WEBSTERIT (*m*): websterite; $Al_2O_3,SO_3,9H_2O$; Sg. 1.65 047
WEBSTUHL (*m*): loom (Textile) 048

WECHSEL (*m*): alternation (Elect.); change; vicissitude; fluctuation; exchange; bill; draft; rotation of crops; junction; joint; cock; tap 049
WECHSELBANK (*f*): exchange-bank 050
WECHSELBAR: changeable 051
WECHSELBEWEGUNG (*f*): intermittent motion, alternating movement 052
WECHSELBEZIEHUNG (*f*): drawing (Commerce); correlation 053
WECHSELBRIEF (*m*): bill (of exchange) 054
WECHSELFALL (*m*): alternative; vicissitude; dilemma 055
WECHSELFARBIG: iridescent; changing colour 056
WECHSELFELD (*n*): alternating (Elect.) field 057
WECHSELFIEBER (*n*): intermittent fever 058
WECHSELGEBÜHR (*f*): exchange 059
WECHSELGESPRÄCH (*n*): dialogue 060
WECHSELGETRIEBE (*n*): variable speed gear, change-gear 061
WECHSELHAHN (*m*): change cock 062
WECHSELHANDEL (*m*): banking 063
WECHSELKASTEN (*m*): changing box (Phot.) 064
WECHSELKURS (*m*): rate of exchange 065
WECHSELN: to alternate; change; exchange; interchange; shift 066
WECHSEL (*m*), OFFENER-: blank cheque 067
WECHSELPREIS (*m*): rate of exchange 068
WECHSEL, PRIMA- (*m*): first (bill) of exchange 069
WECHSELRÄDER (*n.pl*): change (wheels) gear-(ing) 070
WECHSELRECHNUNG (*f*): bank (banking, banker's) account 071
WECHSELSACK (*m*): changing bag (Phot.) 072
WECHSELSEITIG: alternate; reciprocal; interchangeable; mutual 073
WECHSELSTÄNDIG: alternate 074
WECHSELSTEIN (*m*): glazed tile of brick 075
WECHSELSTROM (*m*): alternating current; (A.C.) 076
WECHSELSTROMERZEUGER (*m*): alternator (Elect.) 077
WECHSELSTROMGLOCKE (*f*): magneto bell 078
WECHSELSWEISE: alternately; reciprocally; interchangeably; mutually; by turns 079
WECHSELVENTIL (*n*): change valve 080
WECHSELVERHÄLTNIS (*n*): reciprocal (relation) proportion 081
WECHSELWEIBCHEN (*n*): contact knob, contact stud 082
WECHSELWEISE: alternately; mutually; interchangeably; reciprocally 083
WECHSELWINKEL (*m.pl*): alternate angles 084
WECHSELWIRKUNG (*f*): reciprocal (action) effect 085
WECHSELZERSETZUNG (*f*): double decomposition 086
WECHSLER (*m*): money-changer; banker 087
WECK (*m*): small loaf; roll 088
WECKE (*f*): (see WECK) 089
WECKEN: to wake; awake; waken; arouse; rouse; (*m*): small loaf; roll 090
WECKER (*m*): alarm-(clock) 091
WECKUHR (*f*): alarm clock 092
WEDEL (*m*): brush; fan 093
WEDELN: to wag; fan 094
WEDER (——NOCH): neither (——nor) 095
WEG (*m*): way; walk; route; street; road; passage; path; process; method; manner; course; distance; travel; stroke (Mech.) 096
WEG: away; gone 097
WEGARBEITEN: to work (away) off; remove; get rid of (by working) 098
WEGÄTZEN: to eat away; etch away; corrode; remove by caustics 099
WEGBEGEBEN: to withdraw; go away; retire; depart 100
WEGBEIZEN: to eat away; etch away; corrode; remove by caustics 101
WEGBEKOMMEN: to remove; carry (take) away; get away 102
WEGBLEIBEN: to remain (stay) away or out; be (elided) omitted 103
WEGBRINGEN: to remove; carry (take; bring) away 104
WEGDRÄNGEN: to push (force; hustle) away 105
WEGE (*m*), AUF NASSEM-: by (means of) the wet process 106
WEGE (*m*), AUF TROCKNEM-: by (means of) the dry process 107
WEGEBAU (*m*): road-making 108
WEGEBAUMATERIAL (*n*): road making material 109
WEGEDORN (*m*): purging blackthorn: waythorn (*Rhamnus catharticus*) 110
WEGELAGERN: to waylay 111
WEGEMESSER (*m*): odometer; pedometer 112
WEGEN: regarding; on account of; due to; because (by reason; by virtue) of 113
WEGER (*m*) and (*pl*): ceiling 114
WEGER (*m*), BALK-: (beam)-clamp 115
WEGERICH (*m*): plantain; (*Plantago*) 116
WEGERICHKRAUT (*n*): plantain (*Herba plantaginis*) 117
WEGERUNG (*f*): ceiling 118
WEGERUNG (*f*), LOSE-: loose ceiling 119
WEGERUNGSPLANKE (*f*): ceiling (board) plank 120
WEGERUNG (*f*), ZWISCHENDECK-: between deck ceiling 121
WEGESCHRANKE (*f*): barrier 122
WEGFALL (*m*): suppression; omission; elimination; obviation 123
WEGFALLEN: to be omitted; cease; drop 124
WEGFALLENDES FUSSENDE (*n*): waste; crop (of cast ingots, Metal) 125
WEGFAULEN: to rot (away) off 126
WEGFEILEN: to file (off) away 127
WEGFLIEGEN: to fly (off) away 128
WEGFLIESSEN: to flow away 129
WEGFRESSEN: to eat (away) up; corrode; remove by caustics; devour 130
WEGGIESSEN: to pour away 131
WEGHALTEN: to keep off; withdraw 132
WEGHAUEN: to cut off, lop, trim 133
WEGKARREN: to cart away 134
WEGKEHREN: to sweep away, turn (away) off 135
WEGKOMMEN: to come off; get away; get lost; escape 136
WEGLÄNGE (*f*): length of path or travel 137
WEGLASSEN: to omit; leave out; neglect 138
WEGLEITEN: to lead (off) away; remove; drain; evacuate 139
WEGLEITEND: efferent 140
WEGLOCKEN: to entice (decoy) away 141
WEGMACHEN: to make (away) off; remove; withdraw; efface 142
WEGNAHME (*f*): seizure; removal; deduction 143
WEGNEHMEN: to take away; seize; remove; capture; deduct 144
WEGRADIEREN: to rub out; erase 145
WEGRÄUMEN: to clear away; remove 146

WEGREIBEN : to rub (off; away) out; delete; eliminate 147
WEGREISEN : to set off on a journey or voyage; depart 148
WEGREISSEN : to tear (pull) away; drag away 149
WEGSÄGEN : to saw (off) away 150
WEGSAM : penetrable; pervious 151
WEGSAMKEIT (*f*) : penetrability; perviousness 152
WEGSCHABEN : to scrape (away) off 153
WEGSCHAFFEN : to eliminate; remove 154
WEGSCHAFFUNG (*f*) : elimination; removal 155
WEGSCHAUFELN : to shovel away 156
WEGSCHEREN : to shear (off) away, cut (off) away; go away; sheer (off) away 157
WEGSCHLEIFEN : to grind off 158
WEGSCHMELZEN : to melt away 159
WEGSCHNEIDEN : to cut (away) off, lop, trim 160
WEGSCHÜTTEN : to pour away, throw away 161
WEGSCHWEMMEN : to float (wash) away 162
WEGSEIN : to be away; be (gone; lost) absent 163
WEGSENGEN : to singe off 164
WEGSETZEN : to leap; put (away) aside 165
WEGSPÜLEN : to (rinse) wash away 166
WEGSTREICHEN : to cut out; cross out; strike out; delete; cancel; erase; annul 167
WEGTRAGEN : to carry away; remove; transport 168
WEGTUN : to put (aside) away; dismiss; remove 169
WEGWÄLZEN : to roll (away) off 170
WEGWASCHEN : to (rinse) wash away 171
WEGWEISEN : to turn (send) away 172
WEGWEISER (*m*) : signpost; guide; finger-post; directory; rack rope (between the shrouds) (Naut.) 173
WEGWERFEN : to throw (aside) away; reject; abase oneself 174
WEGWERFEND : disparaging 175
WEGZERREN : to tear (drag) away 176
WEGZIEHEN : to depart; pull (off) away; draw (march) away; move 177
WEH : woe; pain; ache; sore; painful; aching; (*n*) : woe; grief; pang; ache; pain; soreness; (*pl*) : labour-pains 178
WEHE (*f*) : grief; pain; ache 179
WEHEN : to wave; blow 180
WEHENMITTEL (*n*) : ecbolic 181
WEHKLAGE (*f*) : lament; wail; lamentation; wailing 182
WEHKLAGEN : to lament; wail 183
WEHMUT (*f*) : melancholy; sadness; sorrowfulness 184
WEHMÜTIG : melancholy; sad; sorrowful; mournful 185
WEHR (*f*) : defence; bulwark; weapon; (*n*) : weir; dam; dike 186
WEHRE (*f*) : defence 187
WEHREN : to check; restrain; prevent; keep; resist; defend oneself 188
WEHR (*n*), FEUERUNGS- : bridge wall (of a furnace) 189
WEHR (*n*), KOHLEN- : auxiliary bridge wall (in a furnace) 190
WEHRKRONE (*f*) : crown or top of a weir, or bridge wall of a furnace 191
WEHRLOS : unprotected; unarmed; defenceless 192
WEHRLOSIGKEIT (*f*) : unprotectedness; defencelessness 193
WEHRSTAND (*m*) : the military 194
WEHTUN : to give pain; ache 195
WEIB (*n*) : wife; woman 196
WEIBERHAFT : womanish, effeminate 197
WEIBERSTIMME (*f*) : female voice; treble 198
WEIBIG : (a botanical suffix meaning petals or styles) 199
WEIBISCH : womanish, effeminate 200
WEIBLICH : feminine; womanly; female 201
WEICH : soft; weak; mild; gentle; tender; mellow 202
WEICHBLEI (*n*) : soft (refined) lead 203
WEICHBLEIBEND : non-hardening; remaining soft 204
WEICHBOTTICH (*m*) : steeping tub 205
WEICHBRAND (*m*) : soft (place) brick 206
WEICHBRAUNSTEIN (*m*) : pyrolusite; MnO_2; Sg. 4.85 (see PYROLUSIT) 207
WEICHBÜTTE (*f*) : cistern; steep tank 208
WEICHDAUER (*f*) : duration of (soaking) steeping 209
WEICHE (*f*) : softness; switch; siding; soaking pit (leather); (*pl*) : groin; side; flank; points (Railways) 210
WEICHEN : to soak; soften; steep; retreat; yield; recede; give ground; make way 211
WEICHEND : easier 212
WEICHENHEBEL (*m*) : switch lever 213
WEICHENPLATTE (*f*) : base plate for railway line points 214
WEICHENSTELLER (*m*) : switchman, pointsman 215
WEICHENZUNGE (*f*) : switch tongue 216
WEICHFASS (*n* : steeping tub 217
WEICHFEUERN : to melt down 218
WEICHFLOSS (*n*) : porous white pig (Metal) 219
WEICHGUMMI (*n*) : soft rubber 220
WEICHHARZ (*n*) : oleoresin; soft resin (mixture of resin and essential oil from the same plant) 221
WEICHHEIT (*f*) : softness; weakness; mildness; gentleness; mellowness 222
WEICHHERZIG : tender-heated, soft 223
WEICHHOLZ (*n*) : soft wood 224
WEICHKUFE (*f*) : steeping vat 225
WEICHLEDER (*n*) : soft leather 226
WEICHLICH : soft; tender; effeminate; indolent; delicate; weak 227
WEICHLICHKEIT (*f*) : softness; tenderness; indolence; effeminacy; delicacy; weakness 228
WEICHLING (*m*) : weakling; voluptuary 229
WEICHLOT (*n*) : soft solder (see LÖTZINN) 230
WEICHLÖTEN : to soft-solder 231
WEICHMACHEN : to soften; (*n*) : softening 232
WEICHMANGANERZ (*n*) : pyrolusite; polianite (see PYROLUSIT and POLIANIT) 233
WEICHMETALL (*n*) : soft metal 234
WEICHPARAFFIN (*n*) : soft paraffin (see PARAFFIN) 235
WEICHPORZELLAN (*n*) : soft porcelain 236
WEICHPORZELLAN (*n*), ENGLISCHES- : English soft porcelain, bone porcelain, bone china (see KNOCHENPORZELLAN) 237
WEICHPORZELLAN (*n*), FRANZÖSISCHES- : French soft porcelain (translucent frit porcelain) 238
WEICHPORZELLAN (*n*), NATÜRLICHES- : bone porcelain (see KNOCHENPORZELLAN) 239
WEICHSELBAUM (*m*) : cherry tree (see WEICHSELKIRSCHE) 240
WEICHSELKIRSCHE (*f*) : mahaleb cherry, agrigot (*Prunus mahaleb* or *Cerasus mahaleb*) 241

WEICHSELROHR (*n*): cherry-wood (tobacco)-pipe-stem (made from the wood of *Prunus mahaleb*) 242
WEICHSELROHRKULTUR (*f*): cultivation of cherry wood 243
WEICHSELROHRPFLANZUNG (*f*): cherry-wood plantation 244
WEICHSTOCK (*m*): steeping (tub; tank) cistern 245
WEICHTEIL (*m*): soft part 246
WEICHTIER (*n*): mollusk 247
WEICHWASSER (*n*): steeping water; steep; soaking liquor 248
WEICHWERDEN: to become soft; (*n*): softening 249
WEID (*m*): woad (plant, *Isatis tinctoria* and also product from fermentation of leaves of this plant; contains a small percentage of Indigo): (*f*): hunting 250
WEIDE (*f*): pasture; pasturage; willow (see WEIDENBAUM); osier (*Salix viminalis*) 251
WEIDE (*f*), BAND-: osier (*Salix viminalis*) 252
WEIDE (*f*), KORB-: osier (see WEIDE, BAND-) 253
WEIDELAND (*n*): pasture (grazing) land 254
WEIDEN: to feed; graze; pasture; delight in 255
WEIDENBAST (*m*): willow bast 256
WEIDENBAUM (*m*): willow tree (*Salix alba*); Sg. 0.43-0.63 257
WEIDENBITTER (*n*): salicin (see SALICIN) 258
WEIDENBOHRER (*m*): goat-moth caterpillar (see HOLZBOHRER) 259
WEIDENBUSCH (*m*): willow bush 260
WEIDENGEWÄCHSE (*n.pl*): *Salicaceæ* (Bot.) 261
WEIDENHOLZ (*n*): willow (wood) (see WEIDEN-BAUM) 262
WEIDENKOHLE (*f*): willow charcoal 263
WEIDENKORB (*m*): willow basket 264
WEIDENRINDE (*f*): willow bark 265
WEIDENRUTE (*f*): willow (twig; stick) switch 266
WEIDICHT (*n*): willow grove; osier bed 267
WEIDLICH: thoroughly; sound-(ly); stout; valiant 268
WEIDMANN (*m*): sportsman; hunter; huntsman 269
WEIDWERK (*n*): game; hunt; hunting; chase 270
WEIFE (*f*): reel; spool 271
WEIFEN: to wind; reel 272
WEIGEMESSER (*n*): grounding tool (Etching) 273
WEIGERN: to decline; refuse; deny 274
WEIGERUNG (*f*): refusal; denial 275
WEIHE (*f*): sanction; ordination; inauguration; consecration; kite (Bird) 276
WEIHEN: to consecrate; devote; ordain 277
WEIHER (*m*): pool; (fish)-pond; reservoir 278
WEIHEVOLL: solemn 279
WEIHNACHTEN (*f.pl*): Christmas 280
WEIHNACHTSABEND (*m*): Christmas Eve 281
WEIHNACHTSWURZEL (*f*): black hellebore; Christmas rose (*Helleborus niger*) 282
WEIHRAUCH (*m*): incense; olibanum; frankincense (see OLIBANUM) 283
WEIHRAUCHHARZ (*n*): frankincense; incense resin; gum thus 284
WEIHRAUCHÖL (*n*): olibanum oil (from gum thus); Sg. 0.88 285
WEIL: since; because; while; as (long as); during 286
WEILCHEN: little (short) while 287
WEILE (*f*): (space of)-time; while 288
WEILEN: to stay; spend; tarry 289
WEILER (*m*): hamlet 290
WEIN (*m*): wine; vine (*Vitis*) 291
WEINÄHNLICH: like wine; wine-like 292
WEINAPFEL (*m*): wine apple 293
WEINART (*f*): kind (sort; type) of wine 294
WEINARTIG: vinous; like wine 295
WEINBAU (*m*): viniculture; vine or wine (growing; cultivation) culture 296
WEINBAUER (*m*): grape grower; vine (wine) grower 297
WEINBEERE (*f*): grape 298
WEINBEERÖL (*n*): grape-seed oil, wine-(grape)-stone oil; (from grapes; *Vitis vinifera*) (Sg. 0.93) (see also COGNAKÖL) 299
WEINBERG (*m*): vineyard 300
WEINBLATT (*n*): vine leaf 301
WEINBLATTLAUS (*f*): vine fretter (*Phylloxera*) 302
WEINBLUME (*f*): bouquet (of wine); vine blossom 303
WEINBLÜTE (*f*): vine blossom 304
WEINBOHRER (*m*): vine borer (*Prionus laticollis*) 305
WEINBUKETT (*n*): boquet (of wines) 306
WEINBÜTTE (*f*): wine coop, wine tub 307
WEINDROSSEL (*f*): red-wing (*Turdus iliacus*) 308
WEINEN: to whine; cry; lament; weep; (*n*): crying; whining 309
WEINEREI (*f*): crying; lamenting; whining 310
WEINERLICH: whining; lachrymose; tearful 311
WEINERNTE (*f*): vintage 312
WEINERZEUGEND: wine-producing 313
WEINESSIG (*m*): (wine)-vinegar 314
WEINFABRIK (*f*): winery; wine manufactory 315
WEINFARBE (*f*): wine colour 316
WEINFARBSTOFF (*m*): colouring matter (for wine) 317
WEINFASS (*n*): wine-cask 318
WEINFLASCHE (*f*): wine-bottle 319
WEINGAR: fermented 320
WEINGARE MAISCHE (*f*): mash (Distilling) 321
WEINGARTEN (*m*): vineyard 322
WEINGÄRTNER (*m*): grape (vine) grower 323
WEINGÄRUNG (*f*): fermentation of wine 324
WEINGEGEND (*f*): wine-district; wine growing country 325
WEINGEHALT (*m*): wine content 326
WEINGEIST (*m*): spirit of wine; alcohol (see ALKOHOL) 327
WEINGEISTFIRNIS (*m*): spirit varnish 328
WEINGEISTHALTIG: alcoholic; containing (alcohol) spirit of wine 329
WEINGEISTIG: alcoholic; spirituous 330
WEINGEISTMESSER (*m*): alcoholometer (a hydrometer for determining the alcoholic strength of liquids) 331
WEINGEIST, VERSÜSSTER- (*m*): spirit of nitrous ether (see ÄTHYLNITRIT) 332
WEINGLAS (*n*): wine-glass 333
WEINHALTIG: containing wine 334
WEINHANDEL (*m*): wine trade 335
WEINHÄNDLER (*m*): wine-merchant 336
WEINHAUS (*n*): wine-tavern; wine-vaults 337
WEINHEFEN (*f.pl*): wine (yeast) lees 338
WEINIG: vinous; like wine 339
WEINKAMM (*m*): rape; grape pomace 340
WEINKELLER (*m*): wine (vaults) cellar; wine tavern 341
WEINKELTER (*f*): wine press 342
WEINKERNÖL (*n*): grape-seed oil (see WEIN-BEERÖL) 343

WEINKÜFER (*m*): wine cooper 344

WEINLAGER (*n*): wine (store) storage, wine cellar 345

WEINLAND (*n*): vine-growing country; wine district 346

WEINLAUB (*n*): vine-leaves, vine foliage 347

WEINLESE (*f*): vintage; grape harvest 348

WEINMESSER (*m*): oenometer; vinometer (a hydrometer to determine the alcoholic strength of wine) 349

WEINMET (*m*): vinous hydromel 350

WEINMOST (*m*): grape (juice) must 351

WEINMOTTE (*f*): vine moth (*Tortricina ambiguella*) 352

WEINMUTTER (*f*): yeast 353

WEINÖL (*n*): (o)-enanthic ether; oil of wine (see COGNAKÖL) 354

WEINÖL, SCHWERES- (*n*): (heavy)-oil of wine 355

WEINPALME (*f*): toddy palm; wine palm (*Caryota urens*) 356

WEINPRESSE (*f*): wine-press 357

WEINPROBE (*f*): sample (sampling) of wine; wine-sample; wine-sampling 358

WEINPRÜFUNG (*f*): wine (sampling) testing 359

WEINRANKE (*f*): vine-branch; vine tendril 360

WEINREBE (*f*): vine (*Vitis vinifera*) 361

WEINREBENARTIG: sarmentaceous 362

WEINREBENGEWÄCHSE (*n.pl*): *Ampelidæ* (Bot.) 363

WEINREBENSCHWARZ (*n*): Frankfort black (see REBENSCHWARZ) 364

WEINREICH: vinous; rich (abounding) in vines 365

WEINROSE (*f*): eglantine (*Rosa rubiginosa*) 366

WEINROT: wine-red; ruby; claret red 367

WEINSAUER: tartrate of; combined with tartaric acid 368

WEINSÄUERLICH: tart; sourish; acid 369

WEINSÄURE (*f*): tartaric acid; $C_4H_6O_6$; Sg. 1.7598; Mp. 169°C. (see also WEINSTEINSÄURE) 370

WEINSÄURE-DIÄTHYLESTER (*m*): tartaric diethylester; $C_8H_{14}O_6$; Sg. 1.2059; Bp. 280°C. 371

WEINSAURES SALZ (*n*): tartrate 372

WEINSCHENKE (*f*): wine tavern 373

WEINSCHÖNE (*f*): wine fining 374

WEINSCHRÖTER (*m*): cellar man 375

WEINSPRIT (*m*): brandy 376

WEINSTEIN (*m*): potassium acid tartrate (see KALIUM, SAURES WEINSTEINSAURES-); tartar argol; cream of tartar; tartar acid potassium tartrate 377

WEINSTEINARTIG: tartaric; tartarous; like tartar 378

WEINSTEINBILDUNG (*f*): formation of tartar, tartarization 379

WEINSTEINERSATZ (*m*): tartar substitute; acid sodium sulphate 380

WEINSTEINFABRIKATIONSANLAGE (*f*): tartaric acid manufacturing plant 381

WEINSTEIN, GEREINIGTER- (*m*): (purified)-cream of tartar (*Cremor tartari*) 382

WEINSTEINKOHLE (*f*): black flux 383

WEINSTEINPRÄPARAT (*n*): acid sodium sulphate (see NATRIUMBISULFAT) 384

WEINSTEINRAHM (*m*): cream of tartar (*Cremor tartari*) 385

WEINSTEIN, ROHER- (*m*): crude tartar; argol; wine stone 386

WEINSTEINSALZ (*n*): salt of tartar; potassium carbonate 387

WEINSTEINSAUER: tartrate of; combined with tartaric acid 388

WEINSTEINSÄURE (*f*): tartaric acid, dextrotartaric acid, dioxysuccinic acid; $CO_2H(CH(OH))_2CO_2H.H_2O$; Sg. 1.76; Mp. 170°C. (see also WEINSÄURE) 389

WEINSTOCK (*m*): (grape)-vine (*Vitis vinifera*) 390

WEINSTUBE (*f*): wine shop 391

WEINSUPPE (*f*): wine soup 392

WEINTRAUBE (*f*): grape or bunch of grapes 393

WEINTREBER (*f.pl*): marc; lees (of grapes) 394

WEINTRESTER (*m.pl*): marc; lees (of grapes) 395

WEINUNTERSUCHUNG (*f*): wine (examination; investigation) testing 396

WEINUNTERSUCHUNGSAPPARAT (*m*): wine testing apparatus 397

WEINVERFÄLSCHUNG (*f*): wine adulteration 398

WEINWAGE (*f*): vinometer; oenometer (a hydrometer to determine the alcoholic strength of wine) 399

WEISE: wise; prudent; sage; (*m*): sage; philosopher; (*f*): manner; mode; method; way; fashion; melody 400

WEISE (*f*), IN WIRKSAMER-: in an effective manner, effectively, efficaciously, efficiently 401

WEISEN: to point out; direct; indicate; show; teach 402

WEISER (*m*): guide; sign-(post); indicator; hand; pointer; Queen-bee 403

WEISHEIT (*f*): wisdom; knowledge; prudence 404

WEISHEITSZAHN (*m*): wisdom-tooth 405

WEISLICH: wisely; prudently 406

WEISS: white; clean; blank; (*n*): white 407

WEISSAGEN: to prophesy; predict; divine 408

WEISSAGER (*m*): prophet; diviner 409

WEISSAGUNG (*f*): prophecy, divination, prediction 410

WEISSAKAZIENÖL (*n*): locust oil (from *Robinia pseudacacia*) 411

WEISSÄTZUNG (*f*): white discharge (Calico) 412

WEISSBAD (*n*): white liquor bath (Turkey red dyeing) 413

WEISSBIER (*n*): white (pale) beer, Weissbier (a top-fermentation, pale-coloured, acid-tasting beer from mixed malts and water containing salt and gypsum; 1 to 3.5% alcohol by weight) 414

WEISSBLAU: whitish-blue 415

WEISSBLECH (*n*): tin-plate; tinned iron plate or tinned plate of any other metal (to act as an anti-rust agent) 416

WEISSBLECHABFÄLLE (*m.pl*): tin (plate) scrap 417

WEISSBLECHWAREN (*f.pl*): tin ware 418

WEISSBLEICHE (*f*): full bleach 419

WEISSBLEIERZ (*n*): white lead ore, cerussite (a natural lead carbonate) (77.5% lead); $PbCO_3$; Sg. 6.5 (see CERUSSIT and BLEICARBONAT) 420

WEISSBLÜTIGKEIT (*f*): leucocythemia 421

WEISSBROT (*n*): white (wheaten) bread 422

WEISSBRÜCHIG: having a white (pale; light) fracture 423

WEISSBRÜHE (*f*): dègras (see DEGRAS, GERBERFETT and LEDERFETT) 424

WEISSBUCHE (*f*): hornbeam (see HORNBAUM) 425

WEISSBUCHENHOLZ (*n*): hornbeam (see HORNBAUM) 426

WEISSDORN (*m*): hawthorn; whitethorn (*Crataegus oxyacantha*) 427

WEISSDORNARTEN (*f.pl*): species of whitethorn (*Crataegus*) 428
WEISSDORNGEBÜSCH (*n*): whitethorn bush 429
WEISSDORNHECKE (*f*): whitethorn hedge 430
WEISSE (*f*): whiteness; whitewash; white; (*m* and *f*): white person 431
WEISSE GLUT (*f*): white heat 432
WEISSEISEN (*n*): white iron 433
WEISSE MAGNESIA (*f*): magnesia (magnesium) alba; basic magnesium carbonate; $4MgCO_3.Mg(OH)_2.5H_2O$; Sg. 2.18 434
WEISSEN: to whitewash; whiten; bleach; refine (Iron); (*n*): whitening 435
WEISSERDE (*f*): terra alba; superfine hardening 436
WEISSER FLUSS (*m*): leucorrhea 437
WEISSER KUPFERSTEIN (*m*): white metal (Copper) 438
WEISSER LEIM (*m*): gelatine 439
WEISSER VITRIOL (*m*): white vitriol; zinc sulphate (see **ZINKSULFAT**) 440
WEISSERZ (*n*): white ore, siderite (arsenopyrite plus Ag.) (see **ARSENKIES+AG.**); Sg. 6.0 441
WEISSES EISENBLECH (*n*): tin plate 442
WEISSES NICHTS (*n*): nihil album; zinc oxide (see **ZINKOXYD**) 443
WEISSFÄRBEN: to bleach; paint white or colour white 444
WEISSFÄRBER (*m*): bleacher 445
WEISSFEUER (*n*): white fire 446
WEISSGAR: tawed; tanned 447
WEISSGERBEN: to taw; tan 448
WEISSGERBER (*m*): tawer; tanner 449
WEISSGERBERDEGRAS (*n*): dègras (see **DEGRAS**) 450
WEISSGERBEREI (*f*): tawing; tanning; tawery; (alum)-tannery 451
WEISSGERBERFETT (*n*): dègras (see **DEGRAS**) 452
WEISSGILTIG (*m*): argentiferous tetrahedrite (see **SILBERFAHLERZ**, **FAHLERZ** and **WEISSGÜLTIGERZ**) 453
WEISSGLÜHEN: to heat white hot; to heat to (incandescence) white heat; (*n*): white heat; incandescence; heating to white heat 454
WEISSGLÜHEND: white-hot; incandescent 455
WEISSGLÜHHITZE (*f*): white heat; incandescence 456
WEISSGLUT (*f*): white heat; incandescence 457
WEISSGOLD (*n*): platinum (see **PLATIN**) 458
WEISSGRAU: light-grey 459
WEISSGÜLTIGERZ (*n*): polytelite; argentiferous tetrahedrite (see **FAHLERZ**, **SILBERFAHLERZ** and **FREIBERGIT**) 460
WEISSGUSS (*m*): white metal (see **WEISSMETALL**) 461
WEISSHOLZBAUM (*m*): whitewood, Canary (*Liriodendron tulipifera*); Sg. 0.45-0.6 462
WEISSKALK (*m*): crude calcium acetate; pyrolignite of lime; fat lime; $Ca(C_2H_3O_2)_2.H_2O$ 463
WEISSKIEFER (*f*): white (American) pine (*Pinus strobus*) 464
WEISSKIES (*m*): arsenopyrite (see **ARSENOPYRIT**) 465
WEISSKLEESAMENÖL (*n*): white clover-seed-oil 466
WEISSKOCHEN: to degum (Silk) 467
WEISSKRAM (*m*): linen-drapery 468
WEISSKUPFER (*n*): domeykite; native copper arsenide (see **DOMEYKIT**); white copper; Pakfong (see **NEUSILBER**) 469
WEISSLACK (*m*): white (lac) varnish 470
WEISSLEDER (*n*): white (tawed) leather 471
WEISSLICH: whitish 472
WEISSLOT (*n*): soft solder (see **LÖTZINN**) 473
WEISSMACHEN: to whiten; bleach 474
WEISSMEHL (*n*): white (wheat) flour 475
WEISSMESSING (*n*): white brass (20-50% Cu; 50-80% Zn; whitish yellow colour; used for buttons, candelsticks, etc.) 476
WEISSMETALL (*n*): Babbitt metal; white metal; German silver (high duty; 90% Sn; 80% Sb; 2% Cu; middle duty 76.7% Sn; 15.5% Sb; 7.8% Cu; low duty; 85% Sn; 10% Sb; 5% Cu); zinc alloys also go by this name as well as tin alloys 477
WEISSNICKELERZ (*n*): Rammelsbergite; chloanthite; white nickel ore; $(Ni,Co,Fe)As_2$; Sg. 7.1 478
WEISSNICKELKIES (*m*): (see **WEISSNICKELERZ** and **ARSENNICKELKIES**) 479
WEISSOFEN (*m*): refining furnace 480
WEISSPAPP (*m*): white resist (Calico) 481
WEISSPAPPEL (*f*): white poplar; abele (see **PAPPEL**) 482
WEISSROT: whitish red 483
WEISSSENF (*m*): (see **SENF, WEISSER-**) 484
WEISSSENFÖL (*n*): white-mustard oil (from *Sinapis alba*) 485
WEISSSIEDEKESSEL (*m*): blanching copper 486
WEISSSIEDEN: to blanch 487
WEISSSPIESSGLANZ (*m*): valentinite; white antimony (a natural antimony oxide) (see **VALENTINIT**) 488
WEISSSPIESSGLANZERZ (*n*): valentinite; antimony bloom; white antimony (see **ANTIMONTRIOXYD**) 489
WEISSSTEIN (*m*): white metal (Copper) 490
WEISSSTRAHLIG: white; radiated 491
WEISSSTRAHLIGES EISEN (*n*): white iron 492
WEISSSTUCK (*m*): white stucco 93
WEISSSUD (*m*): blanching (solution) 494
WEISSSYLVANERZ (*n*): sylvanite (see **SYLVANIT**) 495
WEISSTANNE (*f*): silver fir 496
WEISSTELLUR (*n*): sylvanite (see **SYLVANIT**) 497
WEISSWAREN (*f.pl*): white (linen) goods 498
WEISSWARM: white-hot 499
WEISSWEIN (*m*): white wine 500
WEISSWURZEL (*f*): Solomon's seal (*Polygonatum officinale*) 501
WEISUNG (*f*): direction; order; instruction; injunction 502
WEIT: far-(off); wide; remote; distant; afar; great; large; much; considerably 503
WEITAUS: by far 504
WEITE (*f*): distance; length; range; width; extent; remoteness 505
WEITEN: to widen; distend; expand 506
WEITER: further; farther; additional; on; collateral; else; go on! proceed! 507
WEITERBEHANDLUNG (*f*): subsequent treatment 508
WEITERES, OHNE-: off hand; without consideration 509
WEITERFÄRBEN: to continue dyeing 510
WEITERGEBEN: to pass (hand) on 511
WEITERGEHEND: progressive 512
WEITERGERBEN: to re-tan 513
WEITERUNG (*f*): prolixity; length; verbosity; amplitude 514

WEITERUNGEN (*f.pl*): complications; formalities 515
WEITGEHEND: far-reaching 516
WEITGETRIEBEN: extended; intensive 517
WEITHALSFLASCHE (*f*): wide-necked bottle 518
WEITHALSIG: wide-necked 519
WEITHIN: far off; wide extent 520
WEITLÄUFIG: widespread; roomy; scattered; distant; straggling; diffuse; detailed; circumstantial 521
WEITLÄUFIGKEIT (*f*): prolixity; verbosity 522
WEITLOCHIG: large (holed) pored 523
WEITMASCHIG: wide (coarse) meshed 524
WEITMÜNDIG: wide-mouthed 525
WEITREICHEND: far-reaching; copious; extensive 526
WEITRINGIG: wide-ringed, coarse-ringed 527
WEITSCHWEIFIG: diffuse; circuitous; verbose; circumstantial 528
WEITSCHWEIFIGKEIT (*f*): prolixity; verbosity 529
WEITSICHTIG: far-sighted 530
WEITTRAGEND: important; carrying to a distance 531
WEITUNG (*f*): widening; width 532
WEITWINKELOBJEKTIV (*n*): wide-angle lens 533
WEIZEN (*m*): wheat (*Triticum sativum*) 534
WEIZENACKER (*m*): wheat field 535
WEIZENKLEBER (*m*): wheat gluten 536
WEIZENKLEIE (*f*): wheat bran 537
WEIZENKORN (*n*): wheat-(grain) 538
WEIZENMALZ (*n*): wheat malt 539
WEIZENMEHL (*n*): wheat flour 540
WEIZENMELTAU (*m*): wheat mildew (*Erysiphe graminis*) 541
WEIZENÖL (*n*): wheat oil (from *Triticum vulgare*); Sg. 0.9245 542
WEIZEN (*m*), SPELZ-: spelt; awned wheat (*Triticum spelta*) 543
WEIZENSTÄRKE (*f*): wheat starch 544
WEIZENSTROH (*n*): wheat straw 545
WEIZEN, TÜRKISCHER- (*m*): maize; Indian corn (see MAIS) 546
WELCHER: who; which; what; some; that; any 547
WELCHER AUCH: whosoever 548
WELCHERLEI: of what kind 549
WELCHES AUCH: whatsoever 550
WELDONSCHLAMM (*m*): Weldon mud (mixture of manganese and calcium manganites from Manganese dioxide recovery process; contains 60-70% manganese dioxide) 551
WELK: faded; withered; flaccid; languid 552
WELKBODEN (*m*): withering (air-drying) floor (Brew.) 553
WELKEN: to fade; decay; wither; (air)-dry; wilt 554
WELKMALZ (*n*): withered (air-dried) malt 555
WELLBALKEN (*m*): axle beam (see WELLBAUM) 556
WELLBAUM (*m*): axletree; shaft; arbour 557
WELLBLECH (*n*): corrugated sheet-(ing); corrugated (iron) plate (Metal) 558
WELLBLECHHAUS (*n*): corrugated iron shed; hut of corrugated sheeting 559
WELLE (*f*): wave; (propeller)-shaft; axle; roller; roll; arbour; faggot; bundle; stria (in glass); billow 560
WELLE (*f*), BIEGSAME-: flexible shaft 561
WELLE (*f*), GEBREMSTE-: damped wave 562
WELLE (*f*), GEDÄMPFTE-: damped wave 563
WELLEN: to simmer; boil; corrugate; weld; roll 564
WELLENARTIG: undulatory; wave-like 565
WELLENBERG (*m*): wave-(crest); mountainous wave 566
WELLENBEWEGUNG (*f*): undulatory motion; wave motion 567
WELLENBOCKBEFESTIGUNG (*f*): strut palms (of a propeller shaft) 568
WELLENENDE (*n*): shaft end 569
WELLENFLÄCHE (*f*): wave surface 570
WELLENFÖRMIG: wavy; undulatory; undulating 571
WELLENHÖHE (*f*): height of waves 572
WELLENLAGER (*n*): shaft bearing 573
WELLENLÄNGE (*f*): wave length; length of waves 574
WELLENLEITUNG (*f*): shafting 575
WELLENLINIE (*f*): wavy (wave) line; undulation; corrugation 576
WELLENMESSER (*m*): ondameter (Wireless tel.) 577
WELLENSCHWINGUNG (*f*): undulation 578
WELLENSTRANG (*m*): (line)-shaft 579
WELLENTAL (*n*): trough (of a wave); wave hollow 580
WELLENTHEORIE (*f*): wave theory 581
WELLENZAPFEN (*m*): shaft end, axle end, axle arm 582
WELLENZUG (*m*): set (series) of waves (Wireless Tel.) 583
WELLER (*m*): mud; loam 584
WELLE (*f*), UNGEBREMSTE-: undamped wave 585
WELLE (*f*), UNGEDÄMPFTE-: undamped wave 586
WELLIG: wavy; undulatory; undulating 587
WELLPACKPAPIER (*n*): corrugated paper 588
WELLPAPIER (*n*): corrugated paper 589
WELLPAPPE (*f*): corrugated (paper) cardboard 590
WELLPAPPENKARTON (*m*): corrugated cardboard box 591
WELLRAD (*n*): arbour wheel; wheel and axle 592
WELLRIPPE (*f*): curved (spoke) rib 593
WELLROHR: corrugated (tube; flue) pipe 594
WELLROHRSCHUSS (*m*): corrugated flue (ring) section 595
WELSCH: foreign; Welsh 596
WELSCHZWIEBEL (*f*): (see WINTERPORREE) 597
WELT (*f*): world; universe 598
WELTALL (*n*): universe 599
WELTALTER (*n*): age of the world; age 600
WELTAUGE (*f*): hydrophane; white opal (see HYDROPHAN) 601
WELTAUSSTELLUNG (*f*): international exhibition; world's fair 602
WELTBERÜHMT: world (far) famed; world-famous 603
WELTBESCHREIBUNG (*f*): cosmography (science of the constitution of the earth) 604
WELTBÜRGERLICH: cosmopolitan 605
WELTGEGEND (*f*): region; district; quarter (of the world); cardinal point 606
WELTGESCHICHTE (*f*): history (of the world or universe) 607
WELTGÜRTEL (*m*): zone (of the world) 608
WELTHANDEL (*m*): world commerce 609
WELTKLUG: worldly-wise; shrewd; politic; prudent 610
WELTKÖRPER (*m*): heavenly body 611
WELTKUGEL (*f*): (terrestrial)-globe or sphere 612
WELTKUNDIG: public; notorious 613
WELTLICH: mundane; worldly; secular; temporal; civil; profane 614
WELTLICHKEIT (*f*): worldliness; secularity 615

WELTMEER (*n*): ocean 616
WELTPOSTVEREIN (*m*): Postal Union 617
WELTRUF (*m*): world-wide reputation 618
WELTSPRACHE (*f*): universal language 619
WELTSTADT (*f*): metropolis 620
WELTTEIL (*m*): part (quarter) of the world; continent; region; district 621
WELTWEISE (*m*): philosopher 622
WELTWEISHEIT (*f*): philosophy 623
WENDE (*f*): era; epoch; turning-(point) 624
WENDEKREIS (*m*): tropic 625
WENDELTREPPE (*f*): winding (stairway) stairs 626
WENDEN: to turn; rotate; wind 627
WENDEN, SICH -AN: to apply (refer) to; address; turn to; communicate with 628
WENDEPUNKT (*m*): turning point; point of inflexion 629
WENDER (*m*): clearer, winder (Textile) 630
WENDUNG (*f*): turn-(ing) 631
WENIG: little; few; some 632
WENIGER: less; fewer 633
WENIGKEIT (*f*): littleness; smallness; small (amount) number 634
WENIGSTE: least; fewest 635
WENIGSTENS: at least 636
WENN: when; if; whenever 637
WENNAUCH: although; even (if) though 638
WENNGLEICH: (see WENNAUCH) 639
WENNSCHON: (see WENNAUCH) 640
WER: who; which; he who; whoever; whosoever 641
WER -AUCH: whoever; whosoever 642
WERBEN: to sue; court; levy; recruit; apply 643
WERBESCHRIFTEN (*pl*): pamphlets, catalogues, etc. (printed matter for use in prospecting for business) 644
WERBUNG (*f*): courting; recruiting; competition (Com.); application 645
WERDEGANG (*m*): development 646
WERDEN: to be; become; turn; grow 647
WERFEN: to throw; fling; cast; project; apply oneself; warp; distort; become cloudy (Beer) 648
WERFEN (*n*): distortion, warping (of wood) 649
WERFT (*f*): woof; weft; wharf; dock-(yard); shipyard 650
WERG (*n*): oakum; tow (hemp fibre impregnated with tar) 651
WERK (*n*): work-(s); (power)-station; mechanism; apparatus; stuff (Paper); pig of raw lead; factory; mill; deed 652
WERKBANK (*f*): (work)-bench 653
WERKBLEI (*n*): raw lead (product of roasting) 654
WERKBOTTICH (*m*): stuff Vat (Paper) 655
WERKBÜTTE (*f*): stuff vat (Paper) 656
WERKFÄHIGKEIT (*f*): workability 657
WERKFÜHRER (*m*): foreman; superintendent 658
WERKHOLZ (*n*): timber 659
WERK, INS- RICHTEN: to carry out; execute; put into practice 660
WERK (*n*), LEBENDIGES-: quick work, lower work (of ships) 661
WERKLEUTE (*pl*): hands; workmen; labourers 662
WERKMEISTER (*m*): foreman; work's manager 663
WERKPROBE (*f*): sample of metal 664
WERKSILBER (*n*): silver extracted from lead ore 665
WERKSTATT (*f*): work-(shop) room; studio; laboratory 666
WERKSTÄTTE (*f*): work-(shop) room; studio; laboratory 667
WERKSTATTZEICHNUNG (*f*): working drawing 668
WERKSTEIN (*m*): quarry stone; freestone 669
WERKSTOFF (*m*): material 670
WERKSTÜCK (*n*): piece to be worked or treated; ore to be smelted; ingot; free-stone, etc. 671
WERKTAG (*m*): working day 672
WERKTÄGIG: commonplace; workaday 673
WERKTÄTIG: operative; acting; practical; working 674
WERKTÄTIGKEIT (*f*): industry 675
WERKTISCH (*m*): bench; working-table; work-table 676
WERK (*n*), TOTES-: dead work, upper work (of ships) 677
WERKZEUG (*n*): implement; tool-(s); instrument 678
WERKZEUGHALTER (*m*): tool holder; tool-chest 679
WERKZEUGKASTEN (*m*): tool chest 680
WERKZEUGMASCHINE (*f*): machine tool 681
WERKZEUGSCHRANK (*m*): tool (chest) cabinet 682
WERKZEUGSPIND (*n*): locker 683
WERKZEUGSTAHL (*m*): tool steel 684
WERKZINK (*n*): raw zinc 685
WERKZINN (*n*): raw tin (see ROHZINN); (97-98% tin) 686
WERMUT (*m*): wormwood (*Herba absynthii* of *Artemisia absinthium*); vermuth; bitters; absinthe 687
WERMUTÖL (*n*): wormwood oil, oil of wormwood, absinthe oil (from leaves of *Artemisia absinthium*); Sg. 0.925/0.955 688
WERMUTSCHNAPS (*m*): vermuth; absinthe 689
WERNERIT (*m*): wernerite, scapolite; $Ca_4Al_6Si_6O_{25}.Na_4Al_3Si_9O_{24}Cl$; Sg. 2.68 690
WER -NUR: whoever 691
WERST (*f*): Verst; Russian mile 692
WERT: worth; valuable; valued; dear; worthy; respected; (*m*): value; price; merit; rate; worth; (*pl*): figures, etc. 693
WERTANGABE (*f*): statement of value 694
WERTBESTIMMUNG (*f*): valuation; evaluation 695
WERTGESCHÄTZT: esteemed; valued 696
WERTHALTEN: to esteem; value; have regard for 697
-WERTIG (*suffix*): valent; of —— valency (see EINWERTIG, ZWEIWERTIG and DREIWERTIG) 698
WERTIGKEIT (*f*): valence; validity 699
WERTPAPIER (*n*): security (valuable paper); (*pl*): stocks; securities; bonds 700
WERTPAPIERBÖRSE (*f*): stock exchange 701
WERTPAPIERMARKT (*m*): stock market 702
WERTSACHEN (*f.pl*): valuables 703
WERTSCHÄTZEN: to esteem highly; value; rate (regard) highly 704
WERTVERHÄLTNIS (*n*): relative value 705
WERTVOLL: valuable 706
WERTZEICHENPAPIER (*n*): bond-(paper) 707
WESEN (*n*): affairs; matters; being; essence; concerns; property; nature; character; condition; manner; means; substance; gist (of a matter) 708
WESENTLICH: intrinsic; substantial-(ly); considerable; appreciable; appreciably; real; essential; material-(ly); decided 709

WESENTLICHEN, IM- : in (general) the main; in (particular) essential; mainly; particularly; generally; essentially; intrinsically; for the (most part) major portion 710
WESHALB : wherefore; why; for what reason 711
WESPE (*f*) : wasp 712
WESPE (*f*), HOLZ- : wood wasp (*Vespa arborea*) 713
WEST (*m*) : west-(wind); occident 714
WESTE (*f*) : waistcoat; vest 715
WESTFALIT (*m*) : westfalite 716
WESTLICH : west; occidental; western; westward; westerly 717
WESTWÄRTS : westward 718
WESWEGEN : why 719
WETT : quits; even 720
WETTBEWERB (*m*) : rivalry; competition 721
WETTBEWERBSKLAUSEL (*f*) : competition clause (usual in Germany before the war, in contracts for situations, and prevented one from accepting a similar position with any other firm, or of starting up on one's own, in certain places defined in the agreement) 722
WETTE (*f*) : wager; bet; competition; rivalry 723
WETTEIFER (*m*) : competition; emulation; rivalry; contention 724
WETTEIFERN : to compete; contend; vie; emulate; rival 725
WETTEN : to wager; bet 726
WETTER (*n*) : weather; storm; damp (Mining); (*m*) : better; wagerer; betting-man 727
WETTERABLEITER (*m*) : lightning conductor 728
WETTERBESTÄNDIG : weather-proof, weather-resisting 729
WETTERDACH (*n*) : penthouse; eaves 730
WETTERDYNAMIT (*n*) : wetterdynamite; blasting dynamite (52% nitroglycerin; 34% soda crystals; 14% Kieselgur) 731
WETTERECHT : weather-proof; fast to exposure; resistant to weather 732
WETTERFAHNE (*f*) : weather-vane; weather-cock 733
WETTERFEST : weatherproof; fast to exposure; resistant to weather 734
WETTERFULMINIT (*n*) : wetterfulminite; blasting fulminite 735
WETTERGLAS (*n*) : weatherglass; barometer 736
WETTERHAHN (*m*) : weather-cock 737
WETTERKUNDE (*f*) : meteorology 738
WETTERLAMPE (*f*) : miner's lamp; hurricane lamp 739
WETTERLAUNISCH : changeable, moody, irritable, peevish 740
WETTERN : to storm; thunder and lighten 741
WETTERPROPHET (*m*) : weather prophet; clerk of the weather 742
WETTERPROPHEZEIUNG (*f*) : weather (forecasts) forecasting 743
WETTERSCHADEN (*m*) : damage-(due to the weather) 744
WETTERSICHER : weather-proof, weather-resisting 745
WETTERSTEIN (*m*) : belemnite (skeleton of extinct animal similar to the cuttle-fish) 746
WETTERWENDISCH : changeable; moody; fickle; irritable 747
WETTERWOLKE (*f*) : thunder-cloud 748
WETTKAMPF (*m*) : contest; combat 749
WETTLAUF (*m*) : race 750
WETTRENNEN (*n*) : race-(s) 751
WETTSEGELN (*n*) : yacht racing 752
WETTSTREIT (*m*) : contest; emulation; contention 753
WETZEN : to hone; whet; sharpen 754
WETZSTEIN (*m*) : hone; whetstone; (Sg. white, 2.88); oilstone 755
WHEWELLIT (*m*) : whewellite; $CaC_2O_4.H_2O$ [756
WICHSE (*f*) : polish; blacking; polishing wax 757
WICHSEN : to polish; wax; glaze; clean; black (Boots) 758
WICHT (*m*) : fellow; wight; creature 759
WICHTIG : weighty; important; serious 760
WICHTIGKEIT (*f*) : weightiness; importance; consequence; important (weighty; serious) matter 761
WICKE (*f*) : vetch; tare (*Vicia sativa*) 762
WICKEL (*m*) : roll; roller; curl paper; cigar filler 763
WICKELN : to roll (wrap; wind) up; twist; enwrap; swathe; enfold; encase 764
WICKELUNG (*f*) : wrapping; casing; winding (and Elect.) 765
WICKENSTROH (*n*) : vetch straw 766
WICKE (*f*), SPANISCHE- : scarlet lupin-(e) 767
WIDDER (*m*) : ram; wether; Aries, the ram (first sign of the Zodiac) 768
WIDDERN : to turn (Malt) 769
WIDDERSCHIFF (*n*) : ram 770
WIDDERSCHWÄRMER (*m*) : burnet moth (*Zygæna*) 771
WIDER : against; contrary (opposed; opposite; counter) to; (as a prefix = counter-; contra-; re-; anti-; gain-; with-; dis-; etc.) 772
WIDERDRUCK (*m*) : reaction; counterpressure; back pressure 773
WIDERFAHREN : to happen to; befall 774
WIDER, FÜR UND- ; pros and cons 775
WIDERHAARIG : perverse; cross-grained 776
WIDERHAKEN (*m*) : barb; beard 777
WIDERHALL (*m*) : (re)-echo 778
WIDERHALLEN : to (re)-echo 779
WIDERHALT (*m*) : hold, grip, support, resistance 780
WIDERHALTEN : to hold out, withstand, resist, support, hold against 781
WIDERLAGER (*n*) : supporting wall, abutment, buttress; thrust-block 782
WIDERLAGERSTEIN (*m*) : skewback (Bricks) 783
WIDERLEGEN : to disprove; refute; contradict; deny; confute 784
WIDERLEGLICH : refutable 785
WIDERLICH : repulsive; disagreeable; repugnant; nauseous; offensive 786
WIDERN : to be (repulsive; disagreeable; repugnant; nauseous) offensive 787
WIDERNATÜRLICH : preternatural; unnatural 788
WIDERPART (*m*) : counterpart; opposition; adversary 789
WIDERRATEN : to dissuade from 790
WIDERRECHTLICH : unlawful, illegal 791
WIDERREDE (*f*) : objection; contradiction; opposition 792
WIDERRUFEN : to revoke; recall; retract; recant; disavow 793
WIDERRUFLICH : revocable 794
WIDERSACHER (*m*) : opponent; adversary; antagonist 795
WIDERSCHEIN (*m*) : reflection; reverberation 796
WIDERSETZEN : to oppose; resist 797
WIDERSETZLICH : resistable; refractory 798
WIDERSINNIG : contradictory; nonsensical; absurd; inconsistent 799

WIDERSPENSTIG: stubborn: refractory; obstinate 800
WIDERSPIEGELN: to reflect 801
WIDERSPIEL (n): opposition; contrary 802
WIDERSPRECHEN: to contradict; gainsay 803
WIDERSPRUCH (m): contradiction; variance; opposition; gainsaying 804
WIDERSTAND (m): resistance; opposition 805
WIDERSTAND (m), FESTER-: fixed resistance (Elect.) 806
WIDERSTANDSBRÜCKE (f): resistance bridge 807
WIDERSTANDSFÄHIG: capable of resistance; resistant 808
WIDERSTANDSFÄHIGKEIT (f): resistance (property) capacity; resistivity; capability of resisting 809
WIDERSTANDSHÖHE (f): height of resistance 810
WIDERSTANDSKASTEN (m): resistance box 811
WIDERSTANDSKRAFT (f): resistance 812
WIDERSTANDSKURVE (f): resistance curve 813
WIDERSTANDSLINIE (f): line of resistance 814
WIDERSTANDSMOMENT (n): moment of resistance 815
WIDERSTANDSROLLE (f): resistance coil, choking coil 816
WIDERSTANDSSPULE (f): resistance coil, choking coil 817
WIDERSTANDSTHERMOMETER (n): resistance thermometer 818
WIDERSTANDSVERMÖGEN (n): (capacity for, or capability of) resistance 819
WIDERSTANDSZAHL (f): coefficient of resistance 820
WIDERSTAND (m), VERSTELLBARER-: variable resistance (Elect.) 821
WIDERSTEHEN: to withstand; resist; be repugnant; oppose 822
WIDERSTOSS (m): countershock 823
WIDERSTREBEN: to resist; oppose; strive (strain; struggle) against; (n): resistance 824
WIDERSTREITEN: to resist, oppose, contradict, militate against, contest, combat 825
WIDERWÄRTIG: adverse; disagreeable; repugnant; contrary 826
WIDERWILLE (m): antipathy; repugnance; aversion; ill-will; reluctance 827
WIDERWILLIG: vexatious; reluctant 828
WIDMANNSTÄTTENSCHE STRUKTUR (f): Widmannstätten's structure (where the various perlite fields in the steel show straight line borders) 829
WIDMEN: to dedicate; devote; apply 830
WIDMUNG (f): dedication; devotion; application 831
WIDRIG: adverse; contrary; repugnant; nauseous; loathsome; offensive 832
WIDRIGENFALLS: in the contrary case 833
WIE: how; like; as 834
WIE -AUCH: however 835
WIEBEL (m): weevil (a plant eating beetle of the family, *Curculionidæ*) 836
WIED (m): woad (see WEID) 837
WIEDER: back; again; in return; anew; afresh 838
WIEDERABDRUCK (m): reprint 839
WIEDERABDRUCKEN: to republish; reprint 840
WIEDERANFANGEN: to begin again; commence afresh; recommence 841
WIEDERANKÜPFEN: to renew (acquaintance). rebutton on 842
WIEDERANNAHME (f): re-adoption 843
WIEDERANSTELLUNG (f): reinstatement 844
WIEDERANWACHSEN (n): further increase 845
WIEDERAUFBAUEN: to rebuild; reconstruct 846
WIEDERAUFKOMMEN: to recover 847
WIEDERBELEBEN: to revive; char; re-burn (Sugar); put new life into; resuscitate; revivify; liven-up 848
WIEDERBELEBUNG (f): revival; resuscitation; charring; boom; general improvement 849
WIEDERBELEBUNGSAPPARAT (m): life restoring (resuscitating) apparatus 850
WIEDERBELEBUNGSMITTEL (n): restorative 851
WIEDERBEZAHLEN: to pay back; repay; reimburse 852
WIEDERBILDEN: to reform; form again 853
WIEDERBRINGEN: to bring back; restore; return; retrieve 854
WIEDERERLANGEN: to regain; recover 855
WIEDERERSTATTUNG (f): return, restitution, repayment, restoration 856
WIEDERERZEUGEN: to regenerate; reproduce 857
WIEDERERZEUGUNG (f): regeneration; reproduction 858
WIEDERFÄLLEN: to re-precipitate 859
WIEDERFÄRBEN: to re-dye; re-colour 860
WIEDERFORDERN: to demand (back) the return; reclaim 861
WIEDERGABE (f), PHOTOGRAPHISCHE-: photographic reproduction 862
WIEDERGEBEN: to return; translate; reproduce; illustrate; show; depict; render; give back; transcribe 863
WIEDERGEBURT (f) regeneration; rejuvenation; re-birth 864
WIEDERGENESUNG (f): recovery (of health) 865
WIEDERGEWINNEN: to regain; recover 866
WIEDERGEWINNUNG (f): recovery; regaining; recuperation 867
WIEDERGEWINNUNGSANLAGE (f): recovery (recuperator) plant 868
WIEDERGEWINNUNGSAPPARAT (m): recovery apparatus, recuperator 869
WIEDERHALL (m): reverberation; echo; resonance 870
WIEDERHERSTELLEN: to reproduce; replace; re-erect; re-install; re-adjust; restore; revive; revivify; re-animate; repair; patch up; refit; redintegrate; re-establish; re-instate; regenerate; reconstruct 871
WIEDERHERSTELLUNG (f): re-erection; restoring, etc. (see WIEDERHERSTELLEN) 872
WIEDERHERVORBRINGEN: to regenerate; reproduce 873
WIEDERHOLEN: to repeat; recur; fetch (bring) back; reiterate; echo; match; reproduce 874
WIEDERHOLUNG (f): repetition; recurrence; reiteration; reproduction 875
WIEDERKÄUEN: to ruminate; chew the cud 876
WIEDERKÄUER (m): ruminant; cud-chewing animal 877
WIEDERKEHR (f): recurrence; return 878
WIEDERKEHREN: to recur; return 879
WIEDERSCHEIN (m): reflection; reflex 880
WIEDERSEHEN (n): renewed meeting; to meet again 881
WIEDERSTOSS (m): countershock; impact 882
WIEDERSTRAHL (m): reflected ray 883
WIEDERSTRAHLEN: to reflect 884
WIEDERTÄUFER (m): anabaptist 885
WIEDERUM: again; once again; in return; afresh; anew 886
WIEDERVEREINIGEN: to reconcile; reunite 887

WIEDERVEREINIGUNG (*f*): reunion; reconcilement; reconciliation 888
WIEDERVERGELTEN: to retaliate; requite; return; recompense; repay 889
WIEDERVERGELTUNG (*f*): retaliation; requital; return; recompense 890
WIEDERVERKAUF (*m*): resale 891
WIEDERVERKÄUFER (*m*): middleman 892
WIEDERVERSTÄRKEN: to replenish; strengthen again 893
WIEDERVERWANDELN: to reconvert 894
WIEDERWÄSSERUNG (*f*): rehydration; re-(watering) soaking 895
WIEGE (*f*): cradle 896
WIEGEMESSER (*n*): chopping (mincing) knife 897
WIEGEN: to rock; weigh 898
WIEGENLIED (*n*): lullaby 899
WIEHERN: to neigh 900
WIEKE (*f*): pledget; probe; tent (Surgical) 901
WIEN (*n*): Vienna 902
WIENER: Viennese; Vienna 903
WIENERÄTZPULVER (*n*): Vienna (caustic) paste; potassium hydroxide and lime 904
WIENERBLAU (*n*): Vienna blue (see KOBALTBLAU) 905
WIENERISCH: Viennese; Vienna 906
WIENER KALK (*m*): (see WIENERPUTZKALK) 907
WIENERLACK (*m*): carmine lake 908
WIENERPUTZKALK (*m*): Vienna-(polishing) chalk 909
WIENERROT (*n*): Vienna red (see CHROMROT) 910
WIESE (*f*): meadow 911
WIESEL (*n*): weasel 912
WIESENERZ (*n*): meadow ore; meadow (bog) iron ore; limonite (see BRAUNEISENSTEIN) 913
WIESENFLACHS (*m*): purging flax (*Linum catharticum*) 914
WIESENGRÜN (*n*): Paris green (see KUPFERACE--TATARSENIAT) 915
WIESENGRUND (*m*): meadow land 916
WIESENHEU (*n*): meadow hay 917
WIESENKALK (*m*): freshwater (spring) limestone 918
WIESENKLEE (*m*): red clover (*Trifolium pratense*) 919
WIESENKNÖTERICH (*m*): snakeweed; bistort (*Polygonum bistorta*) 920
WIESENKULTUR (*f*): meadow cultivation 921
WIESENLEIN (*m*): purging flax (*Linum catharticum*) 922
WIEVIEL: how much; how many 923
WIEWOHL: (al-)though 924
WIIKIT (*m*): wiikite; $(Nb,Ta)_2O_5,(Ce,Sc,Y)_2O_3$, $UO_3,(Si,Th,Ti,Zr)O_2,FeO$; Sg. 4.85 925
WILD: fierce; wild; savage; unruly; barbarous; (*n*): game, deer; venison 926
WILDBAD (*n*): a bath (watering place) situated in a wild-(mountainous)- region: Wildbad (the name of a town in the Black Forest, Germany) 927
WILDE (*m.* and *f*): savage; wilderness; wildness 928
WILDENTENFETT (*n*): wild-duck fat; Sg. 0.93; Mp. 25-34°C. 929
WILDER SENF (*m*): hedge mustard (*Sisymbrium officinale*) 930
WILDES FLEICH (*n*): proud flesh 931
WILDES GESTEIN (*n*): non-metalliferous rock 932
WILDGESCHMACK (*m*): game-like taste; taste of (wild birds) game 933
WILDHEIT (*f*): savageness; wildness 934
WILDKIRSCHE (*f*): wild cherry (*Prunus avium*) 935
WILDKIRSCHENRINDE (*f*): wild cherry bark 936
WILDLEDER (*n*): deerskin; buckskin 937
WILDNIS (*f*): wilderness; desert 938
WILDPRET (*n*): venison; deer; game 939
WILDSCHWEINFETT (*n*): boar's fat; Sg. 0.96; Mp. 45°C. 940
WILLE (*m*): will; intent-(ion); design; permission; mind; purpose; object 941
WILLEMIT (*m*): willemite (zinc ortho-silicate); Zn_2SiO_4; Sg. 4.05 942
WILLEN (*m*): (see WILLE) 943
WILLENSFREIHEIT (*f*): free will 944
WILLENSKRAFT (*f*): will-power; strength of will; volition 945
WILLENSLOS: random 946
WILLENSMEINUNG (*f*): desire; intention; wish; will; pleasure; object; idea 947
WILLFAHREN: to comply with; yield to; agree to; carry out; gratify; comply; grant 948
WILLFÄHRIG: ready; willing; compliant; complaisant 949
WILLIG: ready; willing; voluntary 950
WILLIGEN: to consent; signify one's (readiness; preparedness) willingness; agree to; assent; grant 951
WILLKOMMEN: welcome; (*n.* and *m.*): welcome 952
WILLKÜR (*f*): arbitrariness; free-will; option; discretion; volition 953
WILLKÜRLICH: arbitrary; optional; voluntary 954
WIMMELN: to teem; crowd; swarm 955
WIMMERN: to whine; moan; whimper 956
WIMPEL (*m*): pennant; streamer 957
WIMPELLEINE (*f*): pennant halliard 958
WIMPER (*f*): eyelash 959
WIMPERND: winking; ciliated; ciliary 960
WIND (*m*): wind; air; breeze; draught; blast (of a blower); scent (see LUFT) 961
WINDBESTÄUBUNG (*f*): wind-pollination (Bot.) 962
WINDBLÄSER (*m*): blower; blast apparatus 963
WINDBÜCHSE (*f*): air-gun 964
WINDDARM (*m*): colon (Anat.) 965
WINDDRUCK (*m*): wind (air) pressure; blast pressure; draught 966
WINDDÜRRE: air-dried 967
WINDE (*f*): windlass; capstan; worm; screw; winch; pulley; binding; winding; coil; stria (Glass) 968
WINDEBOCK (*m*): lifting jack 969
WINDEEISENEINSATZ (*m*): tap frame 970
WINDEI (*n*): addled egg 971
WINDEISEN (*n*): stock and dies 972
WINDELN: to swathe 973
WINDEN: to wind; wriggle; coil; wring; draw (haul) up 974
WINDERHITZER (*m*): air heater, regenerator, hot-blast stove 975
WINDERHITZUNG (*f*): air-heating; blast-heating 976
WINDERHITZUNGSAPPARAT (*m*): air heater; regenerator; hot blast stove 977
WINDFAHNE (*f*): (weather)-vane; weather-cock 978
WINDFORM (*f*): tuyère; twyer; nozzle (of Blast furnaces) 979
WINDFRISCHEN (*n*): air refining (Metal) (Bessemer and Thomas process producing "FLUSSEISEN" and "FLUSSSTAHL," which see) 980

WIND (*m*), **FRISCHER-** : fresh wind, breeze, gale 981
WINDFRISCHVERFAHREN (*n*), **BASISCHES-** : Thomas process (see WINDFRISCHEN) 982
WINDEGESCHWINDIGKEIT (*f*) : velocity of the wind 983
WINDGESCHWINDIGKEITSMESSER (*m*) : anemometer; wind-gauge (instrument for measuring the velocity of the wind) 984
WINDGESCHWULST (*f*) : Subcutaneous Emphysema (distention of the areolar tissue with gas or air (Med.) 985
WIND (*m*), **GUTER-** : fair wind 986
WINDHAUBE (*f*) : air cover, air seal; cover (seal) of an air duct 987
WINDHUND (*m*) : greyhound 988
WINDIG : windy; breezy; shaky; visionary 989
WINDKASTEN (*m*) : wind (air) chest; air box; air chamber; tuyère box (Metal) 990
WINDKESSEL (*m*) : air chamber, air vessel, air receiver 991
WINDLEITUNG (*f*) : air duct; blast (pipe) main (Metal); tuyère; twyer 992
WINDLEITUNGSROHR (*n*) : blast (main) pipe; tuyère; air-duct; twyer 993
WIND (*m*), **MÄSSIGER-** : moderate breeze or wind 994
WINDMESSER (*m*) : anemometer, wind gauge (see ANEMOMETER) 995
WINDMOTOR (*m*) : wind motor 996
WINDMÜHLE (*f*) : wind-mill 997
WINDOFEN (*m*) : wind (air) furnace; blast furnace 998
WINDPOCKEN (*f.pl.*) : chicken-pox (Med.) 999
WINDPRESSUNG (*f*) : wind (air) pressure; blast pressure; draught 000
WINDREEP (*n*) : top rope (Naut.) 001
WINDROHR (*n*) : tuyère (blast) pipe; air duct; twyer 002
WINDROSE (*f*) : wind rose, anemone (*Anemone pulsatilla*) 003
WINDSAMMLER (*m*) : air reservoir 004
WINDSCHADEN (*m*) : damage caused by wind 005
WINDSCHICHTMASCHINE (*f*) : wind (screening) sifting machine 006
WINDSCHIEF : warped; bent; twisted (by the wind) 007
WINDSCHIRM (*m*) : wind-screen 008
WIND (*m*), **SCHLECHTER-** : foul wind 009
WINDSEITE (*f*) : leeward; lee-side; windward-(side) 010
WINDSICHTER (*m*) : wind sifter 011
WINDSORSEIFE (*f*) : Windsor soap 012
WINDSTÄRKE (*f*) : force (strength) of the wind 013
WINDSTILL : calm 014
WINDSTILLE (*f*) : calm 015
WINDSTOSS (*m*) : gust (of wind) 016
WIND (*m*), **STÜRMISCHER-** : stormy wind, gale 017
WINDSTUTZEN (*m*) : branch air-duct, blast-pipe, nozzle 018
WINDSUCHT (*f*) : Tympanites, flatulency (distension of abdomen due to gas) (Med.) 019
WINDTREIBEND : carminative; antispasmotic 020
WINDTROCKEN : air-dried; air-dry 021
WINDTROCKNEN : to wind (air) dry 022
WINDTROCKNUNG (*f*) : blast drying (Blast furnace operation); air drying 023
WINDUNG (*f*) : coil; helice; winding; spiral; twist; haulage; turn 024
WIND (*m*), **UNGÜNSTIGER-** : foul wind 025
WINDWEHE (*f*) : (snow)-drift 026
WINDZUG (*m*) : air current; ventilator 027
WINK : wink; nod; hint; sign 028
WINKEL (*m*) : angle; corner; nook 029
WINKELABSTAND (*m*) : angular distance; distance between angles 030
WINKELBEWEGUNG (*f*) : angular motion 031
WINKELBLATT (*n*) : local paper 032
WINKEL (*m*), **DREH-(UNGS)-** : angle of torsion 033
WINKELEISEN (*n*) : angle-(iron) 034
WINKELEISEN, GLEICHSCHENKLICHES- (*n*) : equal angles 035
WINKELEISEN, UNGLEICHSCHENKLICHES- (*n*) : unequal angles 036
WINKEL (*m*), **ENTGEGENGESETZTER-** : contrary (opposite) angle 037
WINKELFÖRMIG : angular 038
WINKELGESCHWINDIGKEIT (*f*) : angular velocity 039
WINKELHAHN (*m*) : angle cock 040
WINKELHAKEN (*m*) : composing stick 041
WINKELHEBEL (*m*) : cranked lever; bell-crank lever 042
WINKELIG : angular; (as suffix=-angled; -cornered) 043
WINKELLINIE (*f*) : diagonal 044
WINKELMASS (*n*) : carpenter's square 045
WINKELMESSER (*m*) : goniometer (Surveying) (an instrument for measuring angles) 046
WINKEL (*m*), **NEBEN-** : adjacent angle, contiguous angle 047
WINKELRAD (*n*) : bevel wheel 048
WINKELRECHT : rectangular; right-angled 049
WINKEL (*m*), **RECHTER-** : right angle 050
WINKELRING (*m*) : angle ring, flange ring 051
WINKEL (*m*), **SCHEITEL-** : opposite angle, vertical angle 052
WINKEL (*m*), **SCHIEFER-** : oblique angle 053
WINKELSCHIENE (*f*) : angle iron 054
WINKELSPIEGEL (*m*) : goniometer (Surveying) 055
WINKEL (*m*), **SPITZER-** : acute angle 056
WINKEL (*m*), **STUMPFER-** : obtuse angle 057
WINKEL (*m*), **UNIVERSAL-** : universal square 058
WINKELVERZAHNUNG (*f*) : conical gear, bevel (angle) gear 059
WINKEL (*m*), **WECHSEL-** : alternate angle 060
WINKELWULSTEISEN (*n*) : bulb angle-(iron) 061
WINKELZUG (*m*) : pretext; shift; evasion 062
WINKEN : to wink; nod; hint; beckon; (give a)-sign 063
WINKLIG : angular; (as suffix=-angled; -cornered) 064
WINNETON (*n*) : winneton 065
WINSELN : to whine; whimper; moan 066
WINTER (*m*) : winter 067
WINTERAPFEL (*m*) : winter apple 068
WINTERARBEIT (*f*) : winter work 069
WINTERBAUM (*m*) : winter berry (North American)-black alder-(shrub) (*Ilex verticillata*) 070
WINTERBIRNE (*f*) : winter pear 071
WINTEREICHE (*f*) : oak tree (*Quercus sessiliflora*); Sg. 0.7-1.05 072
WINTERGERSTE (*f*) : winter-barley 073
WINTERGETREIDE (*n*) : winter corn 074
WINTERGRÜN (*n*) : wintergreen (*Pyrola minor*); periwinkle (*Vinca*); myrtle; ivy 075
WINTERGRÜN, DOLDENFÖRMIGES- (*n*) : chimaphila leaves (*Herba pirolæ umbellatæ*) 076
WINTERGRÜNKRAUT (*n*) : periwinkle (evergreen) leaves (*Herba vincæ pervincæ*) 077

WINTERGRÜNÖL (*n*): oil of wintergreen (see GAULTHERIAÖL) 078
WINTERGRÜNÖL, KÜNSTLICHES- (*n*): methyl salicylate; artificial oil of wintergreen; $C_6H_4(OH)COOCH_3$; Sg. 1.185; Mp. – 8.3°C.; Bp. 222°C. 079
WINTERHAFT: wintry 080
WINTERKLEIDUNG (*f*): winter (clothes) wear 081
WINTERLICH: (see WINTERHAFT) 082
WINTERMÄSSIG: (see WINTERHAFT) 083
WINTERPORREE (*m*): leak (*Allium porrum*) 084
WINTERQUARTIER (*n*): winter quarters 085
WINTERSAAT (*f*): winter corn 086
WINTERWEIZEN (*m*): winter wheat (*Triticum*) 087
WINTERZEIT (*f*): winter-(time) 088
WINTHERS SÄURE (*f*): Winther's acid (see NEVILLE SÄURE) 089
WINZER (*m*): vine dresser 090
WINZIG: tiny; minute; small; petty; diminutive 091
WIPFEL (*m*): summit; top (of trees) 092
WIPPBRÜCKE (*f*): bascule bridge 093
WIPPE (*f*): whip; counterpoise; gibbet; balancing; seesaw; clipping (from coins); lever (Rolling mills) 094
WIPPEN: to whip; balance; poise 095
WIPPER (*m*): tipper; waggon tipper; clipper 096
WIPPERENSTAUBUNG (*f*): extraction of dust by tipping 097
WIR: we 098
WIRBEL: vortex; eddy; whirlpool; whirl-wind; whirl; vertebra; vertigo; intoxication; spigot; button; collar; sheave; swivel; crown (of the head) 099
WIRBELBEIN (*n*): vertebra (Anat.) 100
WIRBELBEWEGUNG (*f*): vortex (whirling) motion 101
WIRBELIG: whirling; eddying 102
WIRBELKNOCHEN (*m*): vetrebra (Anat.) 103
WIRBELLOS: invertebrate; without eddy 104
WIRBELN: to whirl; eddy; spin; warble; roll (beat) of a drum 105
WIRBELSÄULE (*f*): vertebral (spinal) column; spine (Anat.) 106
WIRBELSTROM (*m*): eddy current, whirlpool, vortex 107
WIRBELSTURM (*m*): cyclone; tornado 108
WIRBELTIER (*n*): vertebrate; (*pl*): vertebrata 109
WIRBELWIND (*m*): whirlwind; eddy 110
WIRKEN: to work; act; (have) effect; perform; produce; operate; knead; weave; be practicable 111
WIRKEND, EINSEITIG-: acting on one side only 112
WIRKLICH: actual; real; truly; true; authentic 113
WIRKLICHKEIT (*f*): reality; actuality; truth; authenticity 114
WIRKSAM: effective; efficient; active; efficacious-(ly); operative; practicable 115
WIRKSAMES PRINZIP (*n*): active principle (of ferments, etc.) 116
WIRKSAMKEIT (*f*): effectiveness; efficiency; activity; strength; agency; efficacy 117
WIRKSAMKEIT, IN -SETZEN: to throw into gear 118
WIRKSTUHL (*m*): loom 119
WIRKTISCH (*m*): kneading bench 120
WIRKUNG (*f*): effect; action; agency; operation; efficacy 121
WIRKUNG (*f*), **HYPNOTISCHE-**: hypnotic (action) effect 122
WIRKUNGSART (*f*): method of operation; kind of action 123
WIRKUNGSGRAD (*m*): (η); efficiency; effect 124
WIRKUNGSGRAD, ANLAGE- (*m*): total (overall) efficiency; efficiency of the (plant) installation 125
WIRKUNGSGRAD, ELEKTRISCHER- (*m*): electrical efficiency 126
WIRKUNGSGRAD, GESAMT- (*m*): (η); total (overall) efficiency 127
WIRKUNGSGRAD, HYDRAULICHER- (*m*): hydraulic efficiency; (η_h) 128
WIRKUNGSGRADKURVE (*f*): efficiency curve 129
WIRKUNGSGRAD, MECHANISCHER- (*m*): (η_m); mechanical efficiency 130
WIRKUNGSGRAD, THERMISCHER- (*m*): (η_t); thermal (heat) efficiency 131
WIRKUNGSGRAD, THERMODYNAMISCHER- (*m*): thermo-dynamic efficiency 132
WIRKUNGSGRADVERBESSERUNG (*f*): improvement in efficiency 133
WIRKUNGSGRAD, VOLUMETRISCHER- (*m*): volumetric efficiency (Compressors) 134
WIRKUNGSGRAD, WIRTSCHAFTLICHER- (*m*): (η_w); economical efficiency 135
WIRKUNGSKRAFT (*f*): efficiency; efficacy; effective force; working power 136
WIRKUNGSKREIS (*m*): sphere of (action) activity 137
WIRKUNGSLOS: inefficient; inactive; ineffective; ineffectual 138
WIRKUNGSSPHÄRE (*f*): sphere of (action) activity 139
WIRKUNGSWEISE (*f*): method of operation; mode of acting; action 140
WIRKUNGSWERT (*m*): effective value; strength (of acids, etc.) 141
WIRR: tangled; confused 142
WIRRBUND (*n*): truss of straw 143
WIRRE (*f*): disorder; confusion 144
WIRRSEIDE (*f*): silk (waste) refuse 145
WIRRSTROH (*n*): short straw; straw ends 146
WIRRWARR (*m*): muddle; hubbub; jumble; confusion 147
WIRT (*m*): landlord; host; innkeeper 148
WIRTEL (*m*): whorl 149
WIRTELFÖRMIG: verticillate 150
WIRTSCHAFT (*m*): hotel; tavern; public-house; inn; household; housekeeping; economy; inn-keeping; husbandry 151
WIRTSCHAFTEN: to manage; keep (house) an inn 152
WIRTSCHAFTLICH: household; economical-(ly); thrifty 153
WIRTSCHAFTLICHER NUTZEFFEKT (*m*): economical efficiency; (η_w) 154
WIRTSCHAFTLICHKEIT (*f*): economy; thrift 155
WIRTSHAUS (*n*): hotel; tavern; inn; public-house 156
WISCH (*m*): mop; rag; whisk; wisp; wipe; piece of waste paper; trash; rub 157
WISCHEN: to mop; wipe; slip (whisk) away; rub 158
WISCHER (*m*): duster; wiper; stump (Art) 159
WISCHLAPPEN (*m*): rag; floor-cloth 160
WISCHTUCH (*n*): duster 161
WISERIN (*m*): (see XENOTIM) 162
WISMUT (*m* and *n*): bismuth; (Bi); [Mineral = Bi(As)] 163

WISMUTACETAT (*n*): bismuth acetate; BiO. $C_2H_3O_2$ 164
WISMUTARTIG: bismuthic, bismuthal, like bismuth 165
WISMUT (*n*), BALDRIANSAURES-: bismuth valerate 166
WISMUT (*n*), BASISCH GALLUSSAURES-: bismuth subgallate (see DERMATOL) 167
WISMUTBENZOAT (*n*): bismuth benzoate; $Bi(C_7H_5O_2)_3$ 168
WISMUTBLAU (*n*): bismuth blue (from precipitation of bismuth chloride solution with potassium ferrocyanide solution) 169
WISMUTBLENDE (*f*): bismuth blende; eulytite; $2Bi_2O_3.3SiO_2$; Sg. 6.1 170
WISMUTBLÜTE (*f*): bismuth ochre; bismite; Bi_2O_3; bismuth trioxide (see WISMUTOCKER) 171
WISMUTBROMID (*n*): bismuth bromide 172
WISMUTCARBONAT (*n*): bismuth (sub)-carbonate; bismuth oxycarbonate; $Bi_2O_3.CO_2.H_2O$. Sg. 6.86 173
WISMUTCHLORID (*n*): bismuth (tri)-chloride; $BiCl_3$; Sg. 4.56; Mp. 227°C. (*Bismutum chloratum*) 174
WISMUTCHROMAT (*n*): bismuth chromate; $Bi_2O_3.2CrO_3$ 175
WISMUTEN: to bismuth-solder 176
WISMUTERZ (*n*): bismuth ore 177
WISMUTGALLAT, BASISCHES- (*n*): (see DERMATOL) 178
WISMUT (*n*), GERBSAURES-: bismuth tannate (from action of tannic acid on bismuth hydroxide) 179
WISMUTGLANZ (*m*): bismuth glance, bismuthinite, bismuth ochre; Bi_2S_3; Sg. 6.5 (see BISMUTIN) (see also WISMUTTRISULFID) 180
WISMUTGLANZ, PRISMATISCHER- (*m*): aikinite 181
WISMUTGOLD (*n*): maldonite; Au_2Bi; Sg. 8.95 (see also MALDONIT) 182
WISMUTHALTIG: bismuthic, bismuthal, containing bismuth 183
WISMUT (*n*), HARZSAURES-: bismuth resinate 184
WISMUTHYDRAT (*n*): bismuth (tri)-hydrate, hydrated bismuth oxide (see WISMUTHYDR-OXYD) 185
WISMUTHYDROXYD (*n*): bismuth (hydroxide) hydrate; $Bi(OH)_3$ (see also WISMUTHYDRAT); (*Bismutum hydroxydatum*) 186
WISMUTJODID (*n*): bismuth iodide; bismuth triiodide; BiI_3; Sg. 5.65; Mp. 408°C. [187
WISMUT (*n*), KOHLENSAURES-: bismuth carbonate (see WISMUTCARBONAT) 188
WISMUTLEGIERUNG (*f*): bismuth alloy 189
WISMUTLOT (*n*): bismuth solder 190
WISMUTMAGISTER (*n*): bismuth magistery (see WISMUTSUBNITRAT) 191
WISMUTNIEDERSCHLAG (*m*): bismuth precipitate; bismuth oxynitrate 192
WISMUTNITRAT (*n*): bismuth (tri)-nitrate; $Bi(NO_3)_3+5H_2O$; Sg. 2.78; Mp. 74°C. [193
WISMUTNITRAT (*n*), BASISCHES-: basic bismuth nitrate, bismuth subnitrate (see WISMUT-SUBNITRAT) 194
WISMUTOCKER (*m*): bismuth ochre; bismite (a natural bismuth oxide); Bi_2O_3; Sg. 4.5 195
WISMUTOLEAT (*n*): bismuth oleate (oleic acid and bismuth trioxide) 196
WISMUTOXYBROMID (*n*): bismuth oxybromide; BiOBr; Sg. 8.082 197
WISMUTOXYCARBONAT (*n*): (see WISMUTCAR-BONAT) 198
WISMUTOXYCHLORID (*n*): bismuth oxychloride; BiOCl; Sg. 7.717; pearl white; bismuthyl chloride (*Bismutum oxychloratum*) 199
WISMUTOXYD (*n*): bismuth oxide; bismuth trioxide; Bi_2O_3; Sg. 8.15; Mp. 820°C. (*Bismutum oxydatum*) 200
WISMUTOXYDHYDRAT (*n*): bismuth hydroxide (see WISMUTHYDROXYD and WISMUTHYDRAT) 201
WISMUTOXYFLUORID (*n*): bismuth oxyfluoride; BiOF; Sg. 7.5 202
WISMUTOXYJODID (*n*): bismuth oxyiodide; BiOJ; Sg. 7.922 203
WISMUTPENTOXYD (*n*): (see WISMUTSÄUREAN-HYDRID) 204
WISMUT (*n*), PHOSPHORSAURES-: bismuth phosphate 205
WISMUT (*n*), SALICYLSAURES-: bismuth salicylate 206
WISMUT (*n*), SALPETERSAURES-: bismuth nitrate (see WISMUTNITRAT) 207
WISMUTSALZ (*n*): bismuth salt 208
WISMUTSAUER: bismuthate; combined with bismuthic acid 209
WISMUTSÄURE (*f*): bismuthic acid 210
WISMUTSÄUREANHYDRID (*n*): bismuth acid anhydride, bismuth pentoxide; Bi_2O_5 211
WISMUTSAURES SALZ (*n*): bismuthate 212
WISMUTSILBERERZ (*n*): bismuth-silver ore; $Ag_{10}Bi$ 213
WISMUTSPAT (*m*): bismuth spar; $Bi_2O_3.CO_2$; Sg. 7.2 214
WISMUTSUBGALLAT (*n*): (*Bismuthum subgallicum*): bismuth subgallate (see DERMATOL) 215
WISMUTSUBNITRAT (*n*): (*Bismutum subnitricum*): magistery of bismuth; bismuth subnitrate, basic bismuth nitrate, bismuth white $BiO(NO_3)$ or $[Bi(NO_3)_3.2Bi(OH)_3]$; Sg. 4.93 216
WISMUTSULFID (*n*): bismuth sulphide; Bi_2S_3; Sg. 7.39; bismuth trisulphide 217
WISMUTTELLUR (*n*): telluric bismuth; tetradymite (see TETRADYMIT) 218
WISMUTTRIBROMID (*n*): bismuth tribromide; $BiBr_3$; Sg. 5.604 219
WISMUTTRIBROMPHENOLAT (*n*): (see XEROFORM) 220
WISMUTTRICHLORID (*n*): bismuth trichloride; $BiCl_3$; Sg. 4.56 221
WISMUTTRIFLUORID (*n*): bismuth trifluoride; BiF_3; Sg. 5.32 222
WISMUTTRIJODID (*n*): bismuth triiodide, bismuth iodide; BiI_3; Sg. 5.65-5.82; Mp. 408°C. 223
WISMUTTRIOXYD (*n*): bismuth trioxide, bismuth oxide; Bi_2O_3; Sg. 8.15-8.8; Mp. 820°C. 224
WISMUTTRISULFID (*n*): bismuth trisulphide; Bi_2S_3; Sg. 7.39 225
WISMUTVANADINAT (*n*): bismuth vanadinate, pucherite (see PUCHERIT) 226
WISMUTVERBINDUNG (*f*): bismuth compound 227
WISMUTWEISS (*n*): bismuth white, pearl white, Spanish white (Bismuth subnitrate); $BiO(NO_3)$ 228
WISPEL (*f*): medlar (*Pyrus germanica*); (*n*): old corn measure 229
WISPERN: to whisper 230

WISSBEGIER (*f*): inquisitiveness; curiosity 231
WISSBEGIERDE (*f*): inquisitiveness; curiosity 232
WISSBEGIERIG: inquisitive; curious 233
WISSEN: to know; (*n*): learning; knowledge 234
WISSEN, MIT-: on purpose; consciously; deliberately; wilfully; knowingly 235
WISSENSCHAFT (*f*): knowledge; learning; intelligence; science 236
WISSENSCHÄFTLER (*m*): scientist 237
WISSENSCHAFTLICH: scientific-(ally) 238
WISSENSCHAFTLICHKEIT (*f*): scientific nature 239
WISSENSCHAFTSLEHRE (*f*): theory of science 240
WISSENS, MEINES-: as far as (I know) my knowledge carries me; to the best of my knowledge 241
WISSENSZWEIG (*m*): branch of knowledge 242
WISSENTLICH: deliberate, wilful, knowing, conscious, on purpose 243
WITHERIT (*m*): witherite (a natural barium carbonate); $BaO.CO_2$; Sg. 4.25 244
WITTERN: to smell; scent; thunder 245
WITTERUNG (*f*): scent (of animals); trail; weather 246
WITTERUNGSEINFLUSS (*m*): atmospheric influence 247
WITTERUNGSKUNDE (*f*): meteorology 248
WITTERUNGSLEHRE (*f*): meteorology 249
WITTICHENIT (*m*): wittichenite; Cu_3BiS_3; Sg. 4.3 250
WITT'SCHE SCHEIBE (*f*): Witt'-(s) plate 251
WITWE (*f*): widow 252
WITWER (*m*): widower 253
WITZ (*m*): joke; wit-(tiness); sense 254
WITZBLATT (*n*): humorous (comic) paper 255
WITZBOLD (*m*): wit; wag 256
WITZELEI (*f*): witticism; quibble 257
WITZIG: witty; humorous; funny; ingenious 258
WITZIGEN: to (make wiser) teach sense 259
WITZWORT (*n*): pun; witticism; sally 260
WO: where 261
WOBEI: whereby; whereat 262
WOCHE (*f*): week; (*pl.*): confinement 263
WOCHENAUSWEIS (*m*): weekly (statement)-return 264
WOCHENBETT (*n*): childbed (Med.) 265
WOCHENFLUSS (*m*): lochia (the discharge following childbirth) (Med.) 266
WOCHENSCHRIFT (*f*): weekly 267
WÖCHENTLICH: weekly 268
WÖCHNERIN (*f*): woman in childbed 269
WODURCH: by (what) which-(means); whereby; through (due to) which 270
WOFERN: provided; so far as; in case; if 271
WOFÜR: for (what) which; wherefore 272
WOGE (*f*): billow; wave 273
WOGEGEN: against which 274
WOGEN: to fluctuate; wave; surge 275
WOHER: whence 276
WOHIN: whither; where 277
WOHL: well; indeed; probably; (*n*): weal; welfare; good; benefit; health; well-being 278
WOHLAN: well; come on 279
WOHLANGEBRACHT: well-timed 280
WOHLANSTÄNDIGKEIT (*f*): decency; decorum 281
WOHLAUF: well; now then; come on; cheer up 282
WOHLBEDACHT: well-considered; deliberate 283
WOHLBEDÄCHTIG: well-considered; deliberate 284
WOHLBEFINDEN (*n*): good health; well-being 285
WOHLBEGRÜNDET: well-founded 286
WOHLBEHAGEN (*n*): comfort 287
WOHLBEHALTEN: in good condition; safe (and sound) 288
WOHLBEKANNT: well-known 289
WOHLBELEIBT: corpulent 290
WOHLBELEIBTHEIT (*f*): corpulency 291
WOHLBELESEN: well-read 292
WOHLBESETZT: well-filled 293
WOHLBEWUSST: conscious; well-known 294
WOHLDEFINIERT: well-defined 295
WOHLERFAHREN: experienced; expert; skilled 296
WOHLERGEHEN (*n*): prosperity; welfare; well-being 297
WOHLERHALTEN: in good condition; well-preserved 298
WÖHLERIT (*m*): wöhlerite (natural siliconiobate) $(Ca,Fe,Mn,Na_2)O,Nb_2O_5,SiO_2,ZrO_2$; Sg. 3.4 299
WOHLERWOGEN: well-considered 300
WOHLFAHRT (*f*): welfare; happiness; prosperity 301
WOHLFEIL: cheap 302
WOHLFEILHEIT (*f*): cheapness 303
WOHLGEBAUT: well constructed; well-built 304
WOHLGEBILDET: well-formed; well-educated; well-made 305
WOHLGEFALLEN (*n*): pleasure; liking; delight; satisfaction 306
WOHLGEFÄLLIG: pleasant; pleased; pleasing; agreeable 307
WOHLGELITTEN: popular 308
WOHLGEMEINT: well-meant; well-intended 309
WOHLGEMUT: merry; cheerful; cheery; (*m*): origanum (see DOSTENKRAUT) 310
WOHLGENÄHRT: well-nourished; well-fed 311
WOHLGEORDNET: well-arranged 312
WOHLGERATEN: perfect; well-done; well-bred 313
WOHLGERUCH (*m*): perfume; fragrance 314
WOHLGESCHMACK (*m*): agreeable (pleasant) taste or flavour 315
WOHLGESINNT: well-disposed; well-intentioned 316
WOHLGESTALT (*f*): well-built figure (of the human form) 317
WOHLGEÜBT: skilled; practised 318
WOHLGEWOGEN: well-inclined; benevolent 319
WOHLHABEND: well-off; well-to-do; wealthy; opulent 320
WOHLHABENHEIT (*f*): opulence 321
WOHLKLANG (*m*): harmony; euphony 322
WOHLLAUT (*m*): harmony; euphony 323
WOHLLEBEN (*n*): luxury; high-living; good cheer 324
WOHLMEINEND: well-meaning; kind 325
WOHLREDEND: well-spoken; eloquent 326
WOHLRIECHEND: sweet-smelling; fragrant 327
WOHLSCHMECKEND: savoury; palatable 328
WOHLSEIN (*n*): well-being; good health; prosperity 329
WOHLSTAND (*m*): prosperity; welfare 330
WOHLTAT (*f*): benefit; kindness; good (action) deed; blessing; charity 331
WOHLTÄTER (*m*): benefactor 332

WOHLTÄTIG : beneficial ; benevolent ; wholesome ; beneficient ; charitable 333
WOHLTÄTIGKEIT (*f*) : beneficence ; benevolence ; charity 334
WOHLTUN : to benefit ; do good ; give pleasure ; be charitable 335
WOHLVERLEI (*m*) : Arnica (*Arnica montana*) 336
WOHLVERSEHEN : well-provided 337
WOHLWEISLICH : wisely ; very wisely 338
WOHLWOLLEN : to be well-(kindly)-disposed towards ; wish well ; (*n*) : benevolence ; favour ; good-will 339
WOHLWOLLEND : well-intentioned ; well-disposed ; benevolent 340
WOHNEN : to dwell ; live ; reside ; abide ; lodge ; stay 341
WOHNHAFT : resident ; residing ; dwelling ; living 342
WOHNHAUS (*n*) : dwelling-(house) 343
WOHNLICH : habitable ; comfortable 344
WOHNORT (*m*) : residence ; dwelling-place ; habitation ; abiding-place 345
WOHNPLATZ (*m*) : residence ; dwelling place ; habitation ; abiding-place 346
WOHNSITZ (*m*) : residence ; domicile 347
WOHNSTUBE (*f*) : living (sitting) room 348
WOHNUNG (*f*) : residence ; habitation ; dwelling ; mansion ; apartments ; lodging ; house 349
WOHNUNGSANZEIGER (*m*) : directory 350
WÖLBEN : to arch ; vault ; curve 351
WÖLBSTEIN (*m*) : arch brick ; voussoir 352
WÖLBSTEIN, END- (*m*) : end arch-brick 353
WÖLBSTEIN, SEITEN- (*m*) : side arch-brick 354
WÖLBUNG (*f*) : arch ; vault-(ing) ; curvature 355
WOLF (*m*) : wolf ; lupus (Med.) ; chafing ; bloom ; lump (Metal) (see LUPPE) 356
WOLFACHIT (*m*) : wolfachite ; $Ni(As,S,Sb)_2$; Sg. 6.4 357
WÖLFISCH : wolfish 358
WOLFRAM (*m* and *n*) : wolframite (see WOL-FRAMIT) ; tungsten ; wolfram (W) (a grey powder) ; Sg. 18.77 ; Mp. 3267°C. ; [Mineral = $(Fe.Mn)O.WO_3$] 359
WOLFRAMAT (*n*) : wolframate ; tungstate 360
WOLFRAMBLAU (*n*) : wolfram (tungsten) blue, mineral blue ; W_2O_5 361
WOLFRAMBLEIERZ (*n*) : stolzite (see STOLZIT) ; lead tungstate (see BLEIWOLFRAMAT) 362
WOLFRAMCARBID (*n*) : tungsten (wolfram) carbide ; (WC), Sg. 15.7 ; (W_2C), Sg. 16.06 363
WOLFRAMCHLORID (*n*) : tungsten chloride 364
WOLFRAMDIJODID (*n*) : wolfram (tungsten) diiodide ; WI_2 ; Sg. 6.9 365
WOLFRAMDIOXYD (*n*) : tungsten (wolfram) dioxide ; WO_2 366
WOLFRAMDISULFID (*n*) : wolfram (tungsten) disulphide ; (WS_2) ; Sg. 7.5 367
WOLFRAMERZ (*n*) : tungsten ore 368
WOLFRAMFADEN (*m*) : tungsten filament 369
WOLFRAMIT (*m*) : wolframite (natural iron-manganese tungstate) ; $(Fe.Mn)WO_4$ (about 76% WO_3) ; Sg.7.25 370
WOLFRAMKARBID (*n*) : (see WOLFRAMCARBID) 371
WOLFRAMLAMPE (*f*) : tungsten lamp 372
WOLFRAMMETALL (*n*) : tungsten metal ; metallic tungsten 373
WOLFRAMOCHER (*m*) : tungstite ; tungstic (wolfram) ochre ; (WO_3) ; Sg. 5.5 374
WOLFRAMOXYD (*n*) : tungsten oxide (see WOL-FRAMDIOXYD) 375
WOLFRAMPHOSPHID (*n*) : wolfram (tungsten) phosphide ; (WP), Sg. 8.5 ; (WP_2), Sg. 5.8 376
WOLFRAMPRÄPARAT (*n*) : tungsten preparation 377
WOLFRAMSALZ (*n*) : tungsten salt 378
WOLFRAMSAUER : tungstate of ; combined with tungstic acid 379
WOLFRAMSÄURE (*f*) : tungstic (wolframic) acid ; $W_2O_5(OH)_2$ or $(H_2WO_4+H_2O)$ 380
WOLFRAMSÄUREANHYDRID (*n*) : tungstic anhydride 381
WOLFRAMSÄURESALZ (*n*) : tungstate ; tungstic acid salt 382
WOLFRAMSILICID (*n*) : tungsten (wolfram) silicide ; WSi_2 ; Sg. 9.4 383
WOLFRAMSTAHL (*m*) : tungsten steel ; wolfram steel 384
WOLFRAMTETRAJODID (*n*) : wolfram (tungsten) tetriodide ; WI_4 ; Sg. 5.2 385
WOLFRAMTRIOXYD (*n*) : tungstic acid ; (WO_3) ; Sg. 6.84 ; tungsten (wolfram) trioxide 386
WOLFSBERGIT (*m*) : wolfsbergite, chalcostibite (antimonial copper glance) ; $SbCuS_2$; Sg. 4.9 387
WOLFSBOHNE (*f*) : lupine (*Lupinus polyphyllus*) 388
WOLFSGRAU : wolf-gray 389
WOLFSKIRSCHE (*f*) : belladonna (*Atropa belladonna*) 390
WOLFSMILCH (*f*) : wolf's milk ; spurge (*Euphorbia helioscopia*) 391
WOLFSPELZ (*m*) : wolf's (fur) skin 392
WOLFSSTAHL (*m*) : natural steel 393
WOLFSWURZ (*f*) : wolf's bane ; aconite (see EISENHUT) 394
WOLKE (*f*) : cloud 395
WOLKEN : to cloud ; become cloudy 396
WOLKENANGRIFF (*m*) : cloud-gas attack 397
WOLKENARTIG : cloud-like 398
WOLKENBLENDE (*f*) : cloud stop (Phot.) 399
WOLKENBRUCH (*m*) : cloud-burst 400
WOLKIG : clouded ; cloudy 401
WOLLABGANG (*m*) : waste wool 402
WOLLARBEIT (*f*) : wool-work ; wool-dressing 403
WOLLARTIG : woolly ; wool-like 404
WOLLASTONIT (*m*) : wollastonite (a calcium silicate) ; $CaSiO_3$; Sg. 2.85 (see CALCIUM-METASILIKAT) 405
WOLLBLUMEN (*f.pl*) : mullen (*Flores verbasci*) 406
WOLLE (*f*) : wool 407
WOLLEN : of wool ; woollen ; to want ; be willing ; will ; wish ; have a mind to ; (*n*) : will ; volition 408
WOLLENATLAS (*m*) : satin cloth 409
WOLLENFABRIK (*f*) : wool-factory 410
WOLLENFABRIKATION (*f*) : (woollen)-cloth manufacture 411
WOLLENGARN (*n*) : wollen yarn 412
WOLLENGEWERBE (*n*) : wool trade 413
WOLLENSORTEN (*f.pl*) : wools ; sorts (kinds) of wools 414
WOLLENWEBER (*m*) : wool weaver 415
WOLLE, PHILOSOPHISCHE- (*f*) : philosopher's wool ; zinc oxide (see ZINKOXYD) 416
WOLLFÄRBER (*m*) : wool dyer 417
WOLLFARBIG : dyed in the wool 418
WOLLFETT (*n*) : wool (grease) fat (see ADEPS LANAE) 419
WOLLFETT (*n*), GEREINIGTES- : Lanolin, purified wool (fat) grease, Lanain, Laniol, Lanesin, Lanichol, Lanum 420

WOLLFETTGEWINNUNGSANLAGE (*f*): wool (fat) grease extraction plant 421
WOLLFETTPECH (*n*): pitch from wool grease 422
WOLLGRASFASER (*f*): cotton-grass fibre (fibre of *Eriophorum vaginatum*) 423
WOLLHANDEL (*m*): wool-trade 424
WOLLIG: wool-like; woolly 425
WOLLKÄMMER (*m*): wool-comber 426
WOLLKRAUT (*n*): mullen; mullein (*Verbascum thapsus*) 427
WOLLMARKT (*m*): wool market 428
WOLLÖL (*n*): wool oil 429
WOLLPULVER (*n*): flock 430
WOLLSCHMIERE (*f*): suint; wool yolk (see also WOLLFETT) 431
WOLLSCHUR (*f*): (sheep)-shearing-(time) 432
WOLLSCHWARZ (*n*): wool black 433
WOLLSCHWEISS (*m*): suint; (wool)-yolk 434
WOLLSCHWEISSASCHE (*f*): suint (yolk) (pot)-ash 435
WOLLSCHWEISSEFTT (*n*): wool grease (see WOLL-FETT); Sg. 0.94; Mp. 41°C. 436
WOLLSTAUB (*m*): flock 437
WOLLSTICKEREI (*f*): wool work; crewel work; tapestry 438
WOLLUST (*f*): lust; luxury; pleasure; carnality; sensuality; bliss; voluptuousness 439
WOLLÜSTIG: lustful; carnal-(minded); lascivious; libidinous; voluptuous; sensuous 440
WOLLWÄSCHEREI (*f*): wool (washery) washing 441
WOMIT: wherewith; by which 442
WOMÖGLICH: if (where) possible 443
WONNE (*f*): pleasure; bliss; delight; rapture 444
WONNEREICH: delightful; delicious; blissful 445
WONNESCHAUER (*m*): thrill of delight 446
WONNIG: blissful; delightful 447
WOODS METALL (*n*): Wood's metal (1 part Cadmium; 4 parts bismuth; 4 parts tin; 2 parts lead) (Mp. 60/70°C.) 448
WOOTZSTAHL (*m*): Wootz steel 449
WORAN: whereat; whereon; by (at; in) what; whereby 450
WORAUF: whereupon; whereon; on (upon) what or which 451
WORAUS: from (by) what or which; wherefrom 452
WOREIN: into (what) which; whereinto 453
WORFELN: to fan; winnow 454
WORFSCHAUFEL (*f*): fan; winnow 455
WORGEN: to vomit, retch, belch-(out), choke (see also WÜRGEN) 456
WORIN: in (what) which, wherein 457
WORP (*f*): transom 458
WORT (*n*): word; expression; term 459
WORTABLEITUNG (*f*): etymology; derivation 460
WORTARM: mute; silent; taciturn; reserved; reticent; curt; sparing (short) of words; laconic 461
WORTAUSDRUCK (*m*): phraseology 462
WORTBRÜCHIG: faithless 463
WÖRTERBUCH (*n*): dictionary; vocabulary; thesaurus; glossary; lexicon 464
WORTFOLGE (*f*): sequence (order) of words 465
WORTFÜGUNG (*f*): syntax; construction of sentences 466
WORTFÜHRER (*m*): speaker; spokesman 467
WORTGEPRÄNGE (*n*): bombast; gasconade; magniloquence; brag 468
WORTGETREU: literal 469
WORTKARG: laconic; curt; brief (see also WORTARM) 470
WORTKUNDE (*f*): etymology 471
WORTLAUT (*m*): text; wording 472
WÖRTLICH: literal; verbatim; verbal 473
WORTREGISTER (*n*): vocabulary; list of words 474
WORTREICH: loquacious; garrulous; talkative; chattering; fluent; voluble; copious; eloquent; verbose; rich in words 475
WORTSCHATZ (*m*): vocabulary 476
WORTSCHWALL (*m*): bombast; gasconade; magniloquence; brag 477
WORTSPIEL (*n*): pun; play on words 478
WORTSTREIT (*m*): wordy contention; contention regarding words; battle of words 479
WORTVERBINDUNG (*f*): context; connection (of) between words 480
WORTWECHSEL (*m*): interchange of words; dispute; contention; argument 481
WORTZEICHEN (*n*): catchword; trade name 482
WORÜBER: about (of; upon; on) which or what; whereat; whereof; whereon; whereupon 483
WORUNTER: among which; under which or what 484
WOSELBST: where 485
WOULFFSCHE FLASCHE (*f*): Woulff bottle 486
WOULFSCHE FLASCHE (*f*): Woulff bottle 487
WOVON: of (what; which) whom; whereof 488
WOVOR: of (for) which; before (what) which; whereat 489
WOZU: why; whereat; whereto; to-(for) what or which; wherefore 490
WRACK (*n*): wreck; refuse 491
WRASEN (*m* and *pl*): a mixture of air, brown-coal (lignite) dust and water; foul-air; fumes; smoke-fumes (special kind of fumes of the above composition) (see also SCHWADEN) (sometimes loosely applied to any vapour, such as the clouds of steam from a laundry, etc.) 492
WRICKEN: to scull (with one oar); wriggle 493
WRICKRIEMEN (*m*): scull (Naut.). oar 494
WRINGEN: to wring 495
WUCHER (*m*): usury 496
WUCHERN: to practice usury; luxuriate; pullulate; proliferate (Biology); multiply; grow rapidly 497
WUCHERUNG (*f*): proliferation (Biology) (see WUCHERN) 498
WUCHS (*m*): growth; stature; form; grain of wood 499
WUCHT (*f*): burden; force; weight; fulcrum (Mech.); live (load) force; (*Vis viva*) 500
WUCHTEN: to weigh heavy 501
WÜHLEN: to agitate; rummage; dig; rake; root; stir 502
WÜHLER (*m*): agitator 503
WULFENIT (*m*): wulfenite (natural lead molybdate); $PbMoO_4$; Sg. 6.85 504
WULST (*m*): roll; pad-(ding); elevation; swelling; enlargment; bulb 505
WULSTEISEN (*n*): bulb iron 506
WULSTKIEL (*m*): bulb keel 507
WULSTIG: swelled; padded; stuffed; swollen; tumid; puffy; bulbous; enlarged 508
WULSTWINKELEISEN (*n*): bulb angle iron 509
WUND: chafed; sore; wounded 510
WUNDARZNEI (*f*): remedy for wounds; styptic; surgery 511
WUNDARZNEIKUNDE (*f*): surgery 512
WUNDARZNEIKUNST (*f*): surgery 513
WUNDARZT (*m*): surgeon 514

WUNDÄRZTLICH: surgical 515
WUNDBALSAM (*m*): vulnerary balsam 516
WUNDE (*f*): sore; wound; injury; hurt 517
WUNDER (*n*): wonder; miracle; marvel 518
WUNDERBAR: wonderful; strange; miraculous; marvellous 519
WUNDERBAUM (*m*): locust-tree (*Robinia pseudacacia*); castor-oil plant (*Ricinus communis*) 520
WUNDERERDE (*f*): lithomarge (a soft unctuous variety of Kaolin) 521
WUNDERERSCHEINUNG (*f*): miraculous phenomenon 522
WUNDERGLAUBE (*f*): belief in miracles 523
WUNDERGROSS: prodigious 524
WUNDERHÜBSCH: charming; very pretty 525
WUNDERKERZE (*f*): sparkler 526
WUNDERKIND (*n*): child (infant) prodigy 527
WUNDERKRAFT (*f*): miraculous (power) force 528
WUNDERLAMPE (*f*): magic lamp 529
WUNDERLICH: singular; odd; strange; whimsical 530
WUNDERN: to surprise; wonder; astonish; marvel; be astounded 531
WUNDERPFEFFER (*m*): allspice (*Pimenta officinalis*) 532
WUNDERSALZ (*n*): Glauber's salt; sal mirabile (see GLAUBERSALZ) 533
WUNDERSAM: wonderful 534
WUNDERSCHÖN: very (wonderfully) beautiful or fine 535
WUNDERTAT (*f*): miracle; wonderful deed 536
WUNDERTÄTIG: miraculous 537
WUNDERTIER (*n*): monster 538
WUNDERVOLL: wonderful; marvellous; miraculous 539
WUNDKRAUT (*n*): golden rod-(leaves) (*Solidago virgaurea*); Woundwort (*Stachys*) 540
WUNDPULVER (*n*): vulnerary powder 541
WUNDSALBE (*f*): vulnerary (salve) ointment 542
WUNDSCHERE (*f*): surgical scissors 543
WUNDSCHRECK (*m*): shock (due to a wound) (Med.) 544
WUNDSCHWAMM (*m*): surgeon's agaric; German tinder; amadou (*Polyporus fomentarius*) 545
WUNDSTEIN (*m*): copper aluminate 546
WUNDWASSER (*n*): wound lotion 547
WUNSCH (*m*): desire; wish 548
WÜNSCHELRUTE (*f*): divining rod 549
WÜNSCHEN: to wish (for); desire 550
WÜNSCHENSWERT: desirable 551
WUNSCHGEMÄSS: in accordance with your wishes, as desired 552
WÜRDE (*f*): dignity; degree; honour; office 553
WÜRDELOS: undignified 554
WÜRDERN: to estimate; tax; value 555
WÜRDEVOLL: dignified; grave 556
WÜRDIG: worthy; deserving 557
WÜRDIGEN: to value; estimate; appraise; appreciate; prize; favour; deign; vouchsafe 558
WÜRDIGKEIT (*f*): value; worth; merit 559
WURF (*m*): cast; throw; litter; brood; projection (Mech.) 560
WURFANKER (*m*): kedge 561
WURFBESCHICKER (*m*): sprinkler stoker, automatic stoker (which imitates mechanically the firing of fuel on to the grate by means of a shovel) 562
WÜRFEL (*m*): cube; dice; block; die 563
WÜRFELALAUN (*m*): cubic alum (see ALAUN) 564
WÜRFELBECHER (*m*): dice-box 565
WÜRFELECK (*n*): cubic summit; corner of a cube 566
WÜRFELECKE (*f*): (see WÜRFELECK) 567
WÜRFELERZ (*n*): pharmacosiderite; cube ore; $3Fe_2O_3.2As_2O_5.12H_2O$ (see PHARMAKOSIDERIT) 568
WÜRFELFÖRMIG: cubic-(al) 569
WÜRFELGIPS (*m*): anhydrite (see ANHYDRIT) 570
WÜRFELIG: cubic-(al); checkered 571
WÜRFELINHALT (*m*): cubic contents 572
WÜRFELN: to checker; play at dice 573
WÜRFELNICKEL (*n* and *m*): cube nickel; raw nickel (98-99% nickel; 1-1.5% cobalt and 0.3-0.5% iron) 574
WÜRFELSALPETER (*m*): cubic (nitre) saltpeter; sodium nitrate (see NATRIUMNITRAT) 575
WÜRFELSCHIEFER (*m*): argillite; clay slate (see KAOLIN) 576
WÜRFELSPAT (*m*): anhydrite (see ANHYDRIT) 577
WÜRFELSPIEL (*m*): dice-(game) 578
WÜRFELSTEIN (*m*): boracite (see BORACIT) 579
WÜRFELZAHL (*f*): cube (Maths.) 580
WÜRFELZEOLITH (*m*): analcite; analcime (see ANALCIM) 581
WÜRFELZUCKER (*m*): cube sugar 582
WURFFEUERUNG (*f*): sprinkler stoker (see WURFBESCHICKER) 583
WURFGESCHOSS (*n*): projectile; missile 584
WURFKRAFT (*f*): projectile force 585
WURFLEHRE (*f*): ballistics 586
WURFLINIE (*f*): trajectory; curve or line of (flight) projection 587
WURFMINE (*f*): trench-mortar bomb or shell; flying pig 588
WURFRAUCHKÖRPER (*n*): smoke bomb 589
WURFSCHAUFEL (*f*): winnowing-blade 590
WURFSCHEIBE (*f*): quoit; discus 591
WURFSPIESS (*m*): javelin 592
WURFWEITE (*f*): range 593
WÜRGELPUMPE (*f*): rotary pump 594
WÜRGEN: to destroy; strangle; choke; gulp; retch; throttle; swallow 595
WÜRGERISCH: murderous 596
WURM (*m*): worm; reptile; snake; maggot; vermin; vermiform process (Anat.) 597
WURMABTREIBEND: anthelmintic; vermifuge (Med.) 598
WURMABTREIBUNGSMITTEL (*n*): vermifuge; anthelmintic (Med.) 599
WURMÄHNLICH: wormlike; helical 600
WURMANTRIEB (*m*): worm drive (Mech.) 601
WURMARTIG: wormlike 602
WURMARZNEI (*f*): anthelmintic; vermifuge (Med.) 603
WURMEN: to annoy; vex 604
WURMFARN (*m*): male fern (*Aspidium athamanticum*) 605
WURMFARNWURZEL (*f*): male fern root (*Rhizoma pannæ* of *Aspidium athamanticum*) 606
WURMFÖRMIG: vermiform; vermicular; worm-shaped 607
WURMFORTSATZ (*m*): vermiform process (Anat.) 608
WURMFRASS (*m*): worm-holes (in wood); worm-eaten condition (of wood) 609
WURMFRÄSSIG: worm-eaten 610
WURMIG: worm-eaten, wormy 611
WURMLOCH (*n*): worm-hole 612
WURMMEHL (*n*): worm-dust (from worm-eaten wood) 613

WURMMITTEL (*n*): anthelmintic; vermifuge (Med.) 614
WURMRAD (*n*): worm wheel 615
WURMRINDE (*f*): worm bark 616
WURMSAMEN (*m*): wormseed (of *Chenopodium ambrosioides*, or *Artemisia pauciflora*) 617
WURMSAMENBITTER (*n*): santonin (see SANTONIN) 618
WURMSAMENÖL (*n*): wormseed oil (Levant, Sg. 0.93; Baltimore, Sg. 0.975) 619
WURMSAMENÖL, AMERIKANISCHES- (*n*): oil of (American) Baltimore wormseed; chenopodium oil; goose-foot oil (from *Chenopodium ambrosioides*); Sg. 0.975 620
WURMSTICHIG: worm-eaten; doubtful; fishy; questionable 621
WURMTABLETT (*n*): worm tablet 622
WURMTANG (*m*): worm moss 623
WURMTREIBEND: anthelmintic; vermifuge 624
WURMWIDRIG: anthelmintic; vermifuge 625
WURST (*f*): sausage; roll; pad-(ding) 626
WURSTDARM (*m*): sausage skin 627
WURSTFLEISCH (*n*): sausage meat 628
WURSTGIFT (*n*): sausage poison 629
WURSTKRAUT (*n*): marjoram (*Origanum majorana*) 630
WURSTSTEIN (*m*): pudding stone 631
WURSTSUPPE (*f*): pudding broth 632
WURTZIT (*m*): wurtzite (a natural zinc sulphide); (Zn,Fe)S; Sg. 4.0 (see also ZINKSULFID and BLENDE) 633
WURZ (*f*): herb; wort; root 634
WÜRZE (*f*): condiment; spice; seasoning; wort (Brew.) 635
WÜRZEBRECHEN (*n*): breaking of the wort 636
WÜRZEBÜCHSE (*f*): spice-box 637
WÜRZEKÜHLER (*m*): wort cooler (Brew.) 638
WURZEL (*f*): root [of plant (*Radix* or *Rhizoma*) and mathematics $\sqrt{\ }$]; carpus (of the hand); tarsus (of the foot) 639
WURZELARTIG: root-like; rhizomorphus 640
WURZEL (*f*) AUSZIEHEN: to extract the root 641
WURZELBAUM (*m*): mangrove (*Rhizophora mangle*) 642
WURZELBILDEND: rhizophorous 643
WURZELBLATT (*n*): radical (leaf arising from the top of the root) (Bot.) 644
WURZELBLÜTLER (*m.pl*): *Rhizantheæ* 645
WURZELBOHRER (*m*): root borer (*Castnia licus*) (destroys the sugar cane) 646
WURZELBUCHSTABE (*m*): radical letter 647
WÜRZELCHEN (*n*): radicle; little root 648
WURZELDRUCK (*m*): endosmotic pressure 649
WURZELEXPONENT (*m*): radical (exponent) index 650
WURZELFASER (*f*): root fibre 651
WURZELFRESSEND: rhizophagous 652
WURZELFRÜCHTIG: rhizocarpous 653
WURZELFÜSSER (*m.pl*): rhizopoda 654
WURZELGEWÄCHS (*n*): edible root; tuber 655
WURZELGRÖSSE (*f*): radical (Maths.) 656
WURZELKEIM (*m*): radicle (Bot.) 657
WURZELKNÖLLCHEN (*n*): root nodule; tubercle 658
WURZELKNOLLEN (*m*): tuber 659
WURZELKNOTEN (*m*): node 660
WURZELLAUS (*f*): *Phylloxera* 661
WURZELLOS: arrhizous 662
WURZELMÜNDIG: rhizostomatous 663
WURZELN: to take root; be rooted; root 664
WURZELPILZE (*m.pl*): *Rhizomorphæ* 665
WURZELRANKE (*f*): tendril; sucker; runner 666
WURZELSCHNEIDEMASCHINE (*f*): root (cutter) cutting machine 667
WURZELSPROSSEND: stoloniferous 668
WURZELSTOCK (*m*): rhizome; rootstock 669
WURZELTRAGEND: rhizophorous 670
WURZELWERK (*n*): roots 671
WURZELWORT (*n*): radical word 672
WURZELZAHL (*f*): root number 673
WURZELZEICHEN (*n*): radical sign (Maths.) 674
WÜRZEN: to season; spice; (*n*): seasoning 675
WÜRZENELKE (*f*): clove (see GEWÜRZNELKE) 676
WÜRZESPIEGEL (*m*): surface of the wort (Brew.) 677
WÜRZHAFT: spicy, aromatic 678
WÜRZIG: aromatic; spicy 679
WÜRZLOS: unspiced; unseasoned 680
WÜRZNÄGELEIN (*n*): clove (see GEWÜRZNELKE) 681
WÜRZNELKE (*f*): clove (see GEWÜRZNELKE) 682
WÜRZPFANNE (*f*): wort (kettle) boiler 683
WÜRZSIEDEPFANNE (*f*): wort (brewing) boiler or copper 684
WÜRZUNG (*f*): seasoning 685
WÜRZWEIN (*m*): spiced wine 686
WUST (*m*): rubbish; chaos; mess; filth; confused mass 687
WÜST: desert; wild; waste; desolate; dissolute 688
WÜSTE (*f*): desert; waste; wilderness; desolation 689
WÜSTEN: to waste; desolate; destroy 690
WÜSTENSAND (*m*): desert sand; Sg. 2.4 691
WUT (*f*): rage; madness; fury; mania; delirium (Med.) 692
WÜTEN: to rage; rave; fume; foam; chafe 693
WÜTEND: furious; raging; raving; fuming; chafing; foaming 694
WÜTIG: mad; raging; furious 695
WUTSEUCHE (*f*): rabies 696

X

XANTHALIN (*n*): xanthaline (an opium alkaloid); $C_{37}H_{36}O_9$ 697
XANTHIN (*n*): xanthine, ureous acid, dioxypurine (an animal alkaloid from urine and flesh); $C_5H_4N_4O_2$ 698
XANTHINDERIVAT (*n*): xanthin derivate 699
XANTHISCH: xanthic 700
XANTHIUMBLÄTTER (*n.pl*): xanthium leaves (*Herba Xanthi spinosæ*) 701
XANTHOCREATININ (*n*): xanthocreatinine (an animal alkaloid from fresh meat); $C_5H_{10}N_4O$ 702
XANTHOGENSAUER: xanthogenate, combined with xanthogenic (xanthic) acid 703
XANTHOGENSÄURE (*f*): xanthic (xanthogenic) acid 704
XANTHOGENSAURES KALI (*n*): potassium xanthogenate (see KALIUMXANTHOGENAT) 705
XANTHOGENSAURES KALIUM (*n*): potassium xanthogenate 706
XANTHOKOBALTCHLORID (*n*): xanthocobaltic chloride 707
XANTHOKON (*m*): xanthocone (from Freiberg, in Saxony), xanthaconite (Mexico);

Ag. about 64%; As. 15%; S. 21%; with perhaps a trace of Fe; Sg. 5.17 708
XANTHON (n): xanthone; $C_6H_4.O.CO.C_6H_4$ 709
XANTHONFARBSTOFF (m): xanthone colouring matter 710
XANTHOPHYLLIT (m): xanthophyllite; chrysophane; $2(Mg.Ca)_3SiO_8+3Al_2O_3.SiO_2$; Sg. 3.0 (see CHRYSOPHAN) 711
XANTHOPROTEINSÄURE (f): xanthoproteic acid 712
XANTINE (f): xantine 713
X-BEINIG: knock-kneed 714
X-BELIEBIG: any 715
XENOLITH (m): xenolite; $2Al_2O_3,3SiO_2$; Sg. 3.58 716
XENON (n): xenon; (X) 717
XENOTIM (m): xenotime (yttrium phosphate); $(Y,Ce)PO_4$; Sg. 4.5 718
XERASE (f): xerase (a yeast preparation) 719
XERES (m): sherry 720
XERESWEIN (m): sherry 721
XEROFORM (n): xeroform; bismuth (tribromophenate) tribromophenolate; $Bi_2O_3(C_6H_2Br_3OH)$ (see TRIBROMPHENOLWISMUT) 722
XIMENIAÖL (n): ximenia oil (from seeds of *Ximenia americana*) 723
X-MAL: x-times, a number of times, over and over again 724
X-STRAHLEN (m.pl): Röntgen (X) rays 725
XYLAM (n): xylam (a disinfectant) 726
XYLAN (n): xylane 727
XYLENOL (n): xylenol, dimethylphenol (from coal tar); $C_8H_{10}O$; Sg. 1.036; Mp. 26/75°C.; Bp. 211/225°C. 728
XYLIDIN (n): xylidine, amino-xylene, amino-dimethylbenzene; $C_8H_{11}N$; ortho.—Sg. 0.982; Bp. 212/215°C.; meta.—Sg. 0.9184; Bp. 215°C.; para.—Sg. 0.98; Mp. 15°C.; Bp. 215°C. 729
XYLIDINROT (n): xylidine red 730
XYLOCHINON (n): xyloquinone 731
XYLOGRAPH (m): xylographer, wood-engraver 732
XYLOGRAPHIE (f): xylography; wood engraving 733
XYLOIDIN (n): xyloidine (see NITROCELLULOSE) 734
XYLOL (n): xylol, xylene, o.m.or p. dimethylbenzene; C_8H_{10}; ortho.—Sg. 0.8633; Mp. −27.1°C.; Bp. 141°C.; meta.—Sg. 0.8642; Mp. −54.8°C.; Bp. 139.2°C.; para.—Sg. 0.8612; Mp. 15°C.; Bp. 137.5°C. (see DIMETHYLBENZOL) 735
XYLOLITH (m): xylolite 736
XYLOPHOTOGRAPHIE (f): photography on wood; photograph on wood 737
XYLORCIN (n): xylorcin-(ol) 738
XYLOSE (f): xylose (see HOLZZUCKER and KIRSCH-GUMMI) 739
XYLYLSÄURE (f): xylic acid 740

Y

YACHT (f): yacht 741
YACHTAGENTUR (f): yacht agency 742
YACHTBAU (m): yacht building 743
YACHTBAUER (m): yacht builder 744
YACHTBEIBOOT (n): yacht lifeboat 745
YACHTREGATTA (f): yacht racing or regatta 746
YACHTWESEN (n): yachting 747
YACHTWETTSEGELN (n): yacht racing or regatta 748
YAK (m): yak (*Bos grunniens*) 749
YAMSWURZEL (f): yam (*Dioscorea batatas*) 750
YAMWURZEL (f): (see YAMSWURZEL) 751
YARA-YARA (n): Yara-yara, β-naphtholmethyl ether, nerolin; $C_{10}H_7OCH_3$; Mp. 72°C. Bp. 274°C. 752
YAWL (f): yawl (Naut.) 753
YERBASTRAUCH (m): Paraguay tea; maté (*Ilex paraguaiensis*) (*Herba santa* of *Eriodictyon glutinosum*) 754
YERBAZIN (n): Yerbazine 755
YERPRÄPARAT (n): yer preparation 756
YLANGIN (n): ylangine 757
YLANG-YLANG-ÖL (n): ylang-ylang oil (*Oleum anonæ*) or (*Oleum Unonæ*); Sg. 0.95 758
YOGHURT (m): yoghurt 759
YOGHURTFERMENT (n): yoghurt ferment 760
YOGHURTPRÄPARAT (n): yoghurt preparation 761
YOHIMBIN (n): yohimbine (an alkaloid from the bark of *Corynanthe yohimbe*); $C_{23}H_{32}N_2O_4$; Mp. 231°C. 762
YOHIMBINDERIVAT (n): yohimbine derivate 763
YOHIMBINHYDROCHLORID (n): yohimbine hydrochloride; $C_{23}H_{32}N_2O_4.HCl$ 764
YOHIMBIN (n), SALZSAURES-: (see YOHIMBIN-HYDROCHLORID) 765
YOHYDROL (n): trade name for YOHIMBIN-HYDROCHLORID, which see 766
YPERBAUM (m): Dutch elm-tree 767
YPSILONEULE (f): Y-moth (*Noctua gamma*) 768
YSOP (m): hyssop (*Hyssopus officinalis*) 769
YSOPKRAUT (n): hyssop herb or leaves (*Herba hyssopi*) 770
YSOPÖL (n): oil of hyssop; hyssop oil (from *Hyssopus officinalis*); Sg. 0.93 771
YTTERBIUM (n): ytterbium; (Yb) 772
YTTERBIUMCARBONAT (n): ytterbium carbonate; $Yb_2(CO_3)_3+4H_2O$; Sg. 3.67 773
YTTERBIUMCHLORID (n): ytterbium chloride; $YbCl_3+6H_2O$; Sg. 2.575 774
YTTERBIUM (n), GOLDCHLORWASSERSTOFFSAURES-: ytterbium chloraurate; $YbCl_3.AuCl_3+9H_2O$ 775
YTTERBIUMNITRAT (n): ytterbium nitrate; $Yb(NO_3)_3+4H_2O$; Sg. 2.682 776
YTTERBIUMOXYD (n): ytterbium oxide; Yb_2O_3; (Sg. 9.175 ?) 777
YTTERBIUMSELENAT (n): ytterbium selenate; $Yb_2(SeO_4)_3$; Sg. 4.14; $Yb_2(SeO_4)_3+8H_2O$; Sg. 3.49 778
YTTERBIUMSULFAT (n): ytterbium sulphate; $Yb_2(SO_4)_3$; Sg. 3.62-3.793; $Yb_2(SO_4)_3+8H_2O$; Sg. 3.28-3.286 779
YTTERERDE (f): yttria; Y_2O_3 (see YTTRIUM-OXYD) 780
YTTERHALTIG: yttric; yttrious; containing yttrium 781
YTTERSPAT (m): xenotime; yttrium spar (see XENOTIM) 782
YTTRIALITH (m): yttrialite (natural thorium and yttrium silicate, with various other constituents) $(Y,Ca,La,Al)_2O_3,UO_3,(Si,Th)O_2,(Ca,Fe,Pb)O$; Sg. 4.6 783
YTTRIUM (n): yttrium; (Y) or (Yt) 784
YTTRIUMACETAT (n): yttrium acetate; $Yt(C_2H_3O_2)_3.8H_2O$ 785

YTTRIUMCARBID (*n*): yttrium carbide; YC_2; Sg. 4.13 786
YTTRIUMCARBONAT (*n*): yttrium carbonate; $Yt_2(CO_3)_3.3H_2O$ 787
YTTRIUMCHLORID (*n*): yttrium chloride; $YCl_3.6H_2O$; Sg. 2.58 788
YTTRIUM (*n*), GOLDCHLORWASSERSTOFFSAURES-: yttrium chloraurate; $YtCl_3,2AuCl_3+16H_2O$ 789
YTTRIUMNITRAT (*n*): yttrium nitrate; $Y(NO_3)_3.4H_2O$ 790
YTTRIUMOXYD (*n*): yttrium oxide; Y_2O_3; Sg. 5.046 (see also YTTERERDE) 791
YTTRIUMPRÄPARAT (*n*): yttrium preparation 792
YTTRIUMPYROPHOSPHAT (*n*): yttrium pyrophosphate; $Y_4(P_2O_7)_3$; Sg. 3.059 793
YTTRIUMSULFAT (*n*): yttrium sulphate; $[Y_2(SO_4)_3$; Sg. 2.612]; $[Y_2(SO_4)_3+8H_2O$; Sg. 2.54-2.558] 794
YTTRIUMVERBINDUNG (*f*): yttrium compound 795
YTTROCERIT (*m*): yttrocerite; $(Y,Ca)O,Ce_2O_3,F$; Sg. 3.4 796
YTTROFLUORIT (*n*): yttrofluorite; $(Y_2.Ca_3)F_6$; Sg. 3.55 797
YTTROILMENIT (*m*): Samarskite (see SAMARSKIT) 798
YTTROKRASIT (*m*): yttrocrasite; $Y_2O_3,(Th,Ti)O_2,H_2O$; 799
YTTROTANTALIT (*m*): yttrotantalite; $(Y.Ca.Fe.UO)TaO$; Sg. 5.65 800
YTTROTITANIT (*m*): yttrotitanite; $SiO_2.TiO_2.Al_2O_3.Fe_2O_3.CaO.YO.CeO$; Sg. 3.6 801
YTTROZERIT (*m*): yttrocerite; $(Ca.Y)O.Ce_2O_3.Fe$; Sg. 3.4 802
YUKKA (*f*): Yucca (*Yucca*) (Bot.) (either *Yucca Aloifolia* or *Yucca filamentosa*) (see PALMLILIE) 803
YUKKAARTEN (*f*): varieties of yucca (see YUKKA) 804
YUKKABLÜTEN (*f.pl*): yucca flowers (see YUKKA) 805
YUKKAMOTTE (*f*): Yucca moth (*Pronuba yuccasella*) 806
YUKKASELLAWEIBCHEN (*n*): female Yucca moth 807
YUMBEHOABAUM (*m*): Yohimbe tree (an African tree, *Corynanthe yohimbe*) 808

Z

Refer also to C for words not found under Z

ZACKE (*f*): prong; projection; tooth; process; twig; dent; bough; edging; scollop; plate (Metal) 810
ZÄCKE (*f*): tick (see ZECKE) 811
ZACKEN: to notch; indent; tooth; scollop; (*m*): (see ZACKE) 812
ZACKIG: notched; toothed; indented; jagged; serrated; scolloped; pointed; pronged 813
ZAFFER (*m*): zaffre; cobalt-(ic) oxide; (CO_2O_3), containing slight impurities 814
ZAGELEISEN (*n*): slab iron 815
ZAGEL (*m*), ROH-: puddle bar (iron) 816
ZAGEN: to be (timorous) afraid; tremble; fear; be faint-hearted 817
ZAGHAFT: timid; afraid; faint-hearted 818
ZAGHAFTIGKEIT (*f*): fairt-heartedness; timorousness; timidity 819
ZÄH: viscous; viscid; tenacious; tough; stingy 820
ZÄHE: (see ZÄH) 821
ZÄHEISEN (*n*): toughened iron 822
ZÄHFESTIG: tenacious 823
ZÄHFESTIGKEIT (*f*): tenacity 824
ZÄHFLÜSSIG: viscous; refractory 825
ZÄHHEIT (*f*): viscidity, viscosity, tenacity, toughness, semi-liquidity, clamminess, stickiness, glutinosity, gummosity, mucosity, adhesiveness, sliminess 826
ZÄHIG: viscous; viscid; tenacious; tough; gelatinous; glutinous; mucilaginous; clammy 827
ZÄHIGKEIT (*f*): viscosity; viscidity; toughness; tenacity; clamminess (see also ZÄHHEIT) 828
ZÄHIGKEITSFAKTOR (*m*): factor of viscosity, viscosity factor 829
ZÄHIGKEITSKOEFFIZIENT (*m*): coefficient of viscosity 830
ZÄHIGKEITSMODUL (*m*): modulus of viscosity, factor of viscosity, coefficient of viscosity 831
ZÄHIGKEITSZAHL (*f*): coefficient of viscosity; viscosity factor 832
ZÄHKUPFER (*n*): tough-pitch copper 833
ZAHL (*f*): numeral; number; cipher; (as a suffix=coefficient; number) 834
ZÄHLAPPARAT (*m*): counter; counting apparatus 835
ZÄHLBAR: computable; numerable; calculable 836
ZAHLBAR: due; payable 837
ZAHLEN: to pay 838
ZÄHLEN: to count; number; reckon; enumerate; tell; compute; calculate 839
ZAHLENFOLGE (*f*): numerical order 840
ZAHLENGRÖSSE (*f*): numerical quality 841
ZAHLENLEHRE (*f*): arithmetic 842
ZAHLENMÄSSIG: numerically 843
ZAHLENWERT (*m*): numerical value 844
ZAHLER (*m*): payer 845
ZÄHLER (*m*): numerator (Maths.); counter; teller; marker; recorder; meter 846
ZAHL (*f*), GANZE-: integer number 847
ZAHL (*f*), GEBROCHENE-: fractional number 848
ZAHL (*f*), GERADE-: even number 849
ZAHLLOS: numberless; innumerable 850
ZAHLMEISTER (*m*): paymaster; treasurer; cashier 851
ZAHLREICH: numerous 852
ZAHLUNG (*f*): payment; paying 853
ZÄHLUNG (*f*): counting; numbering; computation; telling; calculation; reckoning; enumeration 854
ZAHL (*f*), UNGERADE-: uneven number, odd number 855
ZAHLUNGSBEDINGUNGEN (*f.pl*): terms of payment 856
ZAHLUNGSEINSTELLUNG (*f*): bankruptcy, suspension of payment 857
ZAHLUNGSFÄHIG: solvent (Com.); able to pay 858
ZAHLUNGSFRIST (*f*): time for payment 859
ZAHLUNGSSCHEIN (*m*): receipt (Com.) 860
ZAHLUNGSUNFÄHIG: insolvent (Com.); unable to pay 861
ZAHLUNGSUNFÄHIGKEIT (*f*): insolvency 862
ZAHLUNGSVERMÖGEN (*n*): solvency 863

ZÄHLVORRICHTUNG (*f*): counter; register; indicator 864
ZÄHLWERK (*n*): counter, teller 865
ZAHLWERT (*m*): numerical value 866
ZAHLWORT (*n*): numeral; number 867
ZAHLZEICHEN (*n*): numeral; figure; cipher 868
ZAHM: tame; gentle; domestic-(ated) 869
ZÄHMBAR: tamable 870
ZÄHMEN: to tame; domesticate; control; restrain; break in; subjugate; train; bridle; restrain 871
ZAHN (*m*): tooth; cog; prong; projection; process; dent; point 872
ZAHNARZNEI (*f*): dentistry 873
ZAHNARZT (*m*): dentist; dental surgeon 874
ZAHNÄRZTLICH: dental 875
ZAHNÄRZTLICHE PRÄPARÄTE (*n.pl*): dental preparations 876
ZAHNBEIN (*n*): dentine 877
ZAHNBOHRER (*m*): tooth-drill 878
ZAHNBÜRSTE (*f*): tooth-brush 879
ZÄHNELN: to indent; tooth; notch; serrate; scollop 880
ZAHNEMAIL (*n*): dental enamel 881
ZAHNEN: to indent; tooth; serrate; notch; cut one's teeth; (*n*): dentition 882
ZAHNFÄULE (*f*): caries (rottenness) of the teeth 883
ZAHNFLEISCH (*n*): gum-(s) 884
ZAHNFLEISCHGESCHWÜR (*n*): Parulis, gumboil, alveolar abscess (Med.) 885
ZAHNFORM (*f*): tooth form or section 886
ZAHNFORMEL (*f*): dental formula 887
ZAHNFÖRMIG: tooth-shaped; tooth-like; odontoid; dentiform 888
ZAHNFORTSATZ (*m*): odontoid; alveolar process 889
ZAHNHEILKUNDE (*f*): dental (science) surgery; dentistry 890
ZAHNHÖHE (*f*): length of tooth 891
ZAHNHÖHLE (*f*): socket (of a tooth) 892
ZAHNIG (ZÄHNIG): toothed, dentate, serrate, notched, pronged, cogged 893
ZAHNKEIM (*m*): dental pulp 894
ZAHNKETTE (*f*): sprocket chain, gear chain 895
ZAHNKETTENTRIEB (*m*): sprocket (chain) drive 896
ZAHNKITT (*m*): dental (cement) filling 897
ZÄHNKLAPPERN (*n*): chattering of the teeth 898
ZÄHNKNIRSCHEN (*n*): grinding (gnashing) of the teeth 899
ZAHNKRANZ (*m*): spur gear 900
ZAHNKUPPELUNG (*f*): geared coupling 901
ZAHNLÄNGE (*f*): length (depth) of a tooth 902
ZAHNLATWERGE (*f*): tooth paste; dentifrice 903
ZAHNLOS: toothless; without cogs; edentate 904
ZAHNLÜCKE (*f*): space (gap) between teeth 905
ZAHNPASTA (*f*): tooth-paste; dentifrice 906
ZAHNPASTE (*f*): tooth-paste; dentifrice 907
ZAHNPROFIL (*n*): tooth form or section 908
ZAHNPULPA (*f*): dental pulp 909
ZAHNPULVER (*n*): tooth-powder; dentifrice 910
ZAHNRAD (*n*): cog wheel; gear wheel; sprocket wheel; cogged (toothed) wheel 911
ZAHNRADBAHN (*f*): rack railway, funicular railway 912
ZAHNRÄDERGETRIEBE (*n*): toothed gear 913
ZAHNRÄDERWERK (*n*): gearing, cogs 914
ZAHNRADGETRIEBE (*n*): gear, gearing, toothed gearing 915
ZAHNRAD (*n*), KONISCHES-: bevel wheel 916
ZAHNRADSCHNEIDEMASCHINE (*f*): gear cutting machine 917
ZAHNRADVORGELEGE (*n*): reduction gearing 918
ZAHNREISSEN (*n*): toothache, odontalgia (Med.) 919
ZAHNSCHIENE (*f*): rack rail 920
ZAHNSCHMELZ (*m*): dental enamel; encaustum 921
ZAHNSCHMERZEN (*m.pl*): tooth-ache; odontalgia (Med.) 922
ZAHNSCHMERZMITTEL (*n*): tooth-ache (cure) remedy 923
ZAHNSCHMERZSTILLER (*m*): toothache remedy 924
ZAHNSTANGE (*f*): indented bar, rack 925
ZAHNSTANGENGETRIEBE (*n*): rack gear 926
ZAHNSTANGENTRIEB (*m*): rack and pinion-(gear or drive) 927
ZAHNSTEIN (*m*): tartar (on human teeth) 928
ZAHNSTOCHER (*m*): toothpick 929
ZAHNUNG (*f*): dentition (Med.); serration; toothing; gearing; cogs 930
ZAHNWASSER (*n*): tooth wash; dentifrice 931
ZAHNWEH (*n*): tooth-ache (Med.); odontalgia 932
ZAHNWEHHOLZ (*m*): prickly ash (*Zanthoxylum*) 933
ZAHNWEINSTEIN (*m*): tartar (on human teeth) 934
ZAHNWERK (*n*): gearing 935
ZAHNWURZ (*f*): toothwort (*Dentaria bulbifera*; *Lathræa squamaria*) 936
ZAHNZEMENT (*m*): dental cement 937
ZÄHPOLEN: to toughen metal by poling 938
ZAIN (*m*): ingot; pig; bar; rod 939
ZAINEISEN (*n*): bar iron; ingot iron 940
ZAINEN: to draw out; stretch; make (cast) into ingots (Metal) 941
ZAINFORM (*f*): ingot mould 942
ZAINMETALL (*n*): the raw product for making bronze colour; ingot metal 943
ZAINSCHMELZE (*f*): melt of brass and tombac (see MESSING and TOMBAK) (used as a substitute for leaf gold) 944
ZAINSILBER (*n*): ingot silver 945
ZANGE (*f*): (pair of) pliers; pincers; forceps; nippers; tweezers; tongs 946
ZÄNGEARBEIT (*f*): shingling (Metal) 947
ZANGE (*f*), BRENNER-: (gas) burner pliers 948
ZANGE (*f*), FLACH-: flat pliers 949
ZÄNGELCHEN (*n*): nippers; tweezers; pincers 950
ZÄNGEN: to shingle (Metal) 951
ZÄNGEN (*n*): shingling (Metal) (the hammering of the blooms into square blocks and welding them) 952
ZANGENFÖRMIG: forcipate 953
ZANGENGRIFF (*m*): handle of (pliers; pincers; forceps) tongs, etc. 954
ZÄNGER (*m*): shingler (Metal) 955
ZÄNGESCHLACKE (*f*): shingling slag (Metal) 956
ZÄNGEWALZE (*f*): shingling roll (Metal) 957
ZANK (*m*): quarrel; contention; altercation; row; dispute; discord 958
ZANKAPFEL (*m*): bone of contention 959
ZANKEN: to quarrel; contend; dispute; wrangle 960
ZÄNKEREI (*f*): quarrel; contention; altercation 961
ZANKHAFT: quarrelsome; captious 962
ZÄNKISCH: quarrelsome; captious 963
ZANKLUST (*f*): quarrelsomeness 964

ZÄPFCHEN (*n*): small (pin) peg; uvula (Anat.); suppository 965
ZÄPFCHENDRÜSE (*f*): uvular gland (Med.) 966
ZÄPFCHENENTZÜNDUNG (*f*): inflammation of the uvula (Med.) 967
ZÄPFCHENSCHNITT (*m*): staphylotomy 968
ZAPFEN (*m*): plug; peg; pin; spigot; pivot; bung; tap; tenon (Wood-working); cone (of pine tree, etc.); axle; trunnion; journal; gudgeon; pintle; to draw (tap) liquids 969
ZAPFENBAUM (*m*): conifer 970
ZAPFENBOHRER (*m*): tap-borer 971
ZAPFENFÖRMIG: conical; cone-shaped; peg-shaped 972
ZAPFENFRÜCHTIG: coniferous 973
ZAPFENKORN (*n*): ergot 974
ZAPFENLAGER (*n*): bush; collar; bearing; socket; pillow; block; chocks; plummer-block; step (journal) bearing 975
ZAPFENLOCH (*n*): bung (tap; peg) hole; mortice (Wood working) 976
ZAPFENREIBUNG (*f*): journal (gudgeon) friction 977
ZAPFEN (*m*), RUDER-: rudder pintle 978
ZAPFEN (*m*), STEHENDER-: pivot 979
ZAPFENSTREICH (*m*): retreat; tattoo (Military) 980
ZAPFENTRAGEND: coniferous 981
ZAPFENWEIN (*m*): leaked wine; tapped wine 982
ZAPFER (*m*): tapster; drawer 983
ZAPON (*m*): varnish (see ZAPONLACK) 984
ZAPONLACK (*m*): varnish composed of pyroxylin and amyl acetate 985
ZAPPELIG: fidgety; writhing 986
ZAPPELN: to fidget; writhe; struggle; flounder; kick 987
ZAR (*m*): Czar 988
ZARATIT (*m*): zaratite; $Ni_3H_2CO_4,4H_2O$; Sg. 2.65 989
ZARGE (*f*): edge; rim; border; brim; case; frame 990
ZART: delicate; tender; soft; fine 991
ZARTHEIT (*f*): delicacy; tenderness; softness; fineness 992
ZÄRTLICH: fond; tender-(ly); soft; delicate 993
ZÄRTLICHKEIT (*f*): fondness; tenderness; softness; delicacy 994
ZASER (*f*): filament; fibre; funicle; hair; tendril; twine; thread 995
ZÄSERCHEN (*n*): fibril; capillament 996
ZASERIG: filamentous; fibrous; filaceous; filiform; thread-like; fibrillous; wiry; stringy; capillary; funicular; capilliform; hairy 997
ZASERN: to teaze; unravel; unwind; untwist 998
ZÄSIUM (*n*): cesium; (Cs) (see CÄSIUM) 999
ZASPEL (*f*): hank; skein 000
ZAUBER (*m*): magic; witchcraft; spell; charm; enchantment; incantation; conjuring 001
ZAUBEREI (*f*): magic; witchcraft; spell; scorcery; wizardry; charms 002
ZAUBERER (*m*): wizard; magician; sorcerer; conjurer 003
ZAUBERHAFT: magic; enchanting; charming; bewitching 004
ZAUBERHASEL (*f*): witch-hazel; winter-bloom; striped alder; tobacco-wood (*Hamamelis virginiana*) (see HAMAMELIS) 005
ZAUBERHASELBLÄTTER (*n.pl*): witch hazel leaves 006
ZAUBERIN (*f*): witch; sorceress; enchantress 007
ZAUBERISCH: (see ZAUBERHAFT) 008
ZAUBERKRAFT (*f*): magic power 009
ZAUBERKUNST (*f*): witchcraft, magic-(art) 010
ZAUBERLATERNE (*f*): magic lantern 011
ZAUBERN: to indulge in witchcraft, practise magic, enchant, charm, conjure, bewitch 012
ZAUBERPHOTOGRAPHIE (*f*): magic (photograph) photography; spirit (photograph) photography 013
ZAUBERSTAB (*m*): magic wand 014
ZAUBERWURZEL (*f*): mandrake (*Mandragora*) 015
ZAUDERHAFT: tardy; dilatory; tarrying; hesitating; lingering 016
ZAUDERN: to hesitate; delay; tarry; linger; loiter 017
ZAUM (*m*): rein; bridle 018
ZÄUMEN: to bridle; restrain; check; curb 019
ZAUMLOS: unbridled; unrestricted; unchecked; unrestrained 020
ZAUM (*m*), PRONYSCHER-: Prony brake 021
ZAUN (*m*): hedge; fence; fencing 022
ZAUNDRAHT (*m*): fence wire 023
ZÄUNEN: to hedge; fence 024
ZAUNGITTER (*n*): wire fencing; fence netting 025
ZAUNLATTICH (*m*): (see SKARIOL) 026
ZAUNLATTICH (*m*), WILDER-: (see SKARIOL) 027
ZAUNPFAHL (*m*): fence (picket) post 028
ZAUNREBE (*f*): bryony (*Bryonia alba*) 029
ZAUNRÜBE (*f*): bryony (*Bryonia alba*) 030
ZAUNRÜBENWURZEL (*f*): bryony root; bryonia; brionia (*Radix bryoniæ* of *Bryonia alba* or *Bryonia dioica*) 031
ZAUSELN: to drag; pull; pick; tug 032
ZAUSEN: to drag; pull; pick; tug 033
ZEBRAHOLZ (*n*): palmyra wood (see PORKUPINE-HOLZ) 034
ZECHE (*f*): bill; reckoning; score; mine; mining company; colliery; share; drinking-bout 035
ZECHEN: to booze; tipple; drink; carouse 036
ZECHENBESITZER (*m*): colliery owner 037
ZECHENKOKS (*m*): mine coke (7-15% ash content and about 15% moisture) 038
ZECHFREI: scot-free 039
ZECHSTEIN (*m*): zechstein, upper permian (from Stassfurt; contains carnallite, gypsum, rock salt, etc.) 040
ZECKE (*f*): tick (*Melophagus ovinus*) 041
ZEDER (*f*): cedar (*Pinus cedrus*, *Cedrus libani*, *Juniperus virginiana* and *Larix europœa*) 042
ZEDERBLATTÖL (*n*): cedar leaf oil (from *Juniperus virginiana*); Sg. 0.886 043
ZEDERHOLZÖL (*n*): cedar wood oil (from *Juniperus virginiana*); Sg. 0.95 044
ZEDERNHOLZ (*n*): cedar wood (see CEDERNHOLZ) 045
ZEDERÖL (*n*): cedar oil (see ZEDERBLATTÖL and ZEDERHOLZÖL) 046
ZEDIEREN: to transfer; cede; assign 047
ZEDRATÖL (*n*): citron oil (see CEDRATÖL) 048
ZEDROÖL (*n*): citron oil (see CEDRATÖL) 049
ZEH (*m*): toe; stick; knot (of ginger); clove (of garlic) 050
ZEHE (*f*): (see ZEH) 051
ZEHENSPITZE (*f*): tip-toe; tip of the toe, toe tip 052
ZEHN: ten 053
ZEHNBASISCH: decabasic 054
ZEHNECK (*n*): decagon 055
ZEHNFACH: tenfold 056

ZEHNFÄLTIG : tenfold 057
ZEHNJÄHRIG : decennial 058
ZEHNMALIG : ten times repeated 059
ZEHNTE : tenth ; (*m*) : tithe 060
ZEHNTEILIG : of ten parts 061
ZEHNTEL (*n*) : tenth 062
ZEHNTELNORMALALKALI (*n*) : 1 : 10 normal alkali, one in ten normal alkali 063
ZEHNTELNORMALPERMANGANAT (*n*) : 1 : 10 normal permanganate, a 1 in 10 solution of normal permanganate of potash (3.165 grams $KMnO_4$ per litre) 064
ZEHNTELNORMALSÄURE (*f*) : 1 : 10 normal acid, one in ten normal (standard) acid 065
ZEHREN : to consume ; waste ; feed ; live ; eat and drink 066
ZEHREND : consumptive, wasting 067
ZEHRFIEBER (*n*) : hectic fever 068
ZEHRUNG (*f*) : consumption ; waste ; priming (Explosives) ; expenses ; provisions 069
ZEICHEN (*n*) : sign ; symbol ; mark ; brand ; stamp ; token ; symptom ; signal ; indication ; reference (on a letter) ; ensign (Naut.) 070
ZEICHENBOGEN (*m*) : drawing sheet, sheet of drawing paper 071
ZEICHENBRETT (*n*) : drawing board 072
ZEICHENBUCH (*n*) : drawing book 073
ZEICHENBUREAU (*n*) : drawing office 074
ZEICHENDEUTER (*m*) : auger ; astrologer 075
ZEICHENDREIECK (*n*) : set square 076
ZEICHENERKLÄRUNG (*f*) : explanation of (symbols ; marks) signs, etc. 077
ZEICHENFEDER (*f*) : lettering (drawing) pen, mapping pen 078
ZEICHENKREIDE (*f*) : crayon ; chalk 079
ZEICHENKUNST (*f*) : art of drawing 080
ZEICHENPAPIER (*n*) : drawing paper 081
ZEICHENREGEL (*f*) : drawing (constructional) rule 082
ZEICHENSETZUNG (*f*) : punctuation 083
ZEICHENSPRACHE (*f*) : language of signs ; symbolic notation ; dumb language 084
ZEICHENSTIFT (*m*) : drawing pencil 085
ZEICHENSTUNDE (*f*) : drawing lesson 086
ZEICHENTINTE (*f*) : drawing ink ; marking ink 087
ZEICHENTISCH (*m*) : drawing (bench) table, drawing board 088
ZEICHNEN : to draw ; plot ; mark ; delineate ; design ; stamp ; brand ; sign ; subscribe (to loans) ; record (of meters) 089
ZEICHNER (*m*) : designer ; draughtsman ; drawer ; signatory ; subscriber (to loans) ; signer 090
ZEICHNERISCH : diagrammatic ; graphic 091
ZEICHNUNG (*f*) : drawing ; delineation ; diagram ; plan ; sketch ; pattern ; marking ; signature ; subscription (to loans) 092
ZEICHNUNGSSCHRANK (*m*) : drawing cabinet 093
ZEIGEFINGER (*m*) : fore (index ; first) finger 094
ZEIGEN : to point-(out) ; indicate ; show ; exhibit ; demonstrate ; appear ; become evident ; prove ; turn out 095
ZEIGER (*m*) : pointer ; index; hand ; indicator ; needle ; index finger ; bearer ; presenter 096
ZEIGER-GALVANOMETER (*n*) : dial galvanometer 097
ZEIGERTELEGRAPH (*m*) : dial telegraph 098
ZEIGERTELEGRAPHIE (*f*) : dial telegraphy 099
ZEIHEN : to accuse ; charge ; tax with ; convict ; impeach 100
ZEILE (*f*) : line ; row 101
ZEIN (*m*) : ingot ; pig ; bar ; rod (see ZAIN) 102
ZEISIG (*m*) : siskin ; greenfinch (*Fringilla spinus*) 103
ZEISIGGRÜN : light-yellowish green ; siskin-(green) coloured 104
ZEISING (*f*) : seizing 105
ZEISING (*f*), REFF- : reef point (Naut.) 106
ZEIT (*f*) : time ; period ; duration ; season ; era ; tide ; tense 107
ZEITABSCHNITT (*m*) : interval ; period (of time) ; epoch 108
ZEITALTER (*n*) : age ; generation ; era 109
ZEITANGABE (*f*) : date 110
ZEITAUFNAHME (*f*) : time exposure (Phot.) 111
ZEITAUFWAND (*m*) : time occupied ; loss of time 112
ZEITBEGEBENHEIT (*f*) : event 113
ZEITBEGRIFF (*m*) : conception of time 114
ZEITBIEGUNG (*f*) : conjugation 115
ZEITDAUER (*f*) : period (space) of time ; duration 116
ZEITE (*f*) : spout 117
ZEITEINHEIT (*f*) : unit of time 118
ZEITEINTEILUNG (*f*) : division of time 119
ZEITERSPAREND : time-saving 120
ZEITFOLGE (*f*) : chronological (order) sequence 121
ZEITFORM (*f*) : tense (Gram.) 122
ZEITFRAGE (*f*) : topic of the day 123
ZEITGEIST (*m*) : spirit of the age 124
ZEITGEMÄSS : up-to-date ; seasonable ; in accordance with current (present-day) practice 125
ZEITGENOSS (*m*) : contemporary 126
ZEITGENOSSE (*m*) : (see ZEITGENOSS) 127
ZEITHAFEN (*m*) : tidal harbour 128
ZEITHER : hitherto 129
ZEITHERIG : modern ; current ; present-day ; recent 130
ZEITIG : in good time ; timely ; early ; mature ; ripe 131
ZEITIGEN : to ripen ; mature ; bring to a head 132
ZEITIGUNG (*f*) : maturity 133
ZEITLANG (*f*) : time ; for a time 134
ZEITLAUF (*m*) : period ; juncture ; lapse of time 135
ZEITLEBENS : for (ever) life 136
ZEITLICH : temporal ; temporary ; secular ; earthly 137
ZEITLICHT (*n*). time-light (a flash-light mixture which burns away very slowly) (Phot.) 138
ZEITLOHN (*m*) : hourly rate of pay 139
ZEITLÖHNER (*m*) : a person engaged at an hourly rate of pay 140
ZEITLOSE (*f*) : meadow saffron (*Colchicum autumnale*) ; daisy 141
ZEITMASS (*n*) : (measure of) time ; measure ; time ; period 142
ZEITMESSER (*m*) : chronometer 143
ZEITMESSUNG (*f*) : chronometry ; measurement of time 144
ZEITORDNUNG (*f*) : (arrangement in) chronological order 145
ZEITPUNKT (*m*) : moment ; epoch ; instant 146
ZEITRAUBEND : occupying (consuming ; wasting) time 147
ZEITRAUM (*m*) : period ; space (of time) 148
ZEITRECHNUNG (*f*) : chronology ; era 149
ZEITSCHRIFT (*f*) : periodical ; magazine ; journal 150
ZEITSPANNE (*f*) : space (period) of time ; span 151
ZEITTEILCHEN (*n*) : (short) interval of time 152

ZEITUNG (*f*) : newspaper ; periodical ; journal ; gazette 153
ZEITUNGSABONNAMENT (*m*) : subscription 154
ZEITUNGSANNONCE (*f*) : advertisement, announcement 155
ZEITUNGSARTIKEL (*m*) : article 156
ZEITUNGSAUSGABE (*f*) : issue ; edition ; newspaper office 157
ZEITUNGSAUSSCHNITT (*m*) : cutting 158
ZEITUNGSBEILAGE (*f*) : newspaper supplement ; advertising (supplement) leaflet (enclosed in a newspaper or periodical) ; circular 159
ZEITUNGSDRUCK (*m*) : newspaper printing 160
ZEITUNGSEXPEDITION (*f*) : publishing office 161
ZEITUNGSPAPIER (*n*) : newspaper 162
ZEITUNGSSCHREIBER (*m*) : reporter ; journalist ; gazetteer 163
ZEITUNGSWESEN (*n*) : journalism ; newspaper trade ; the press 164
ZEITUNTERSCHIED (*m*) : difference (variation) in time 165
ZEITVERHÄLTNISSE (*n.pl*) : state of affairs 166
ZEITVERLAUF (*m*) : lapse of time 167
ZEITVERLUST (*m*): loss (waste) of time 168
ZEITVERSCHWENDUNG (*f*) : waste of time 169
ZEITVERTREIB (*m*) : pastime ; amusement ; diversion 170
ZEITVERTREIBEND : amusing ; diverting 171
ZEIT, VOR KURZER- : a short (time) while ago 172
ZEITWEILIG : present ; actual ; temporary ; at times ; for the time being 173
ZEITWEISE : occasionally ; temporarily ; periodically 174
ZEITWORT (*n*) : verb (Gram.) 175
ZEITZÜNDER (*m*) : time fuse 176
ZEIT, ZUR- : at the time ; at present 177
ZELASTERGEWÄCHSE (*n.pl*) : *Celastraceæ* (Bot.) 178
ZELEBRIEREN : to celebrate 179
ZELLÄHNLICH : cellular ; cell-like 180
ZELLCHEN (*n*) : small cell ; cellule 181
ZELLE (*f*) : cell ; cubicle ; bucket (Eng.) 182
ZELLENART (*f*) : sort (type ; kind) of cell 183
ZELLENARTIG : cellular ; cell-like 184
ZELLENBILDEND : cell-forming 185
ZELLENBILDUNG (*f*) : cell formation 186
ZELLENFASER (*f*) : cell fibre 187
ZELLENFLÜSSIGKEIT (*f*) : cell fluid ; cellular (juice) fluid 188
ZELLENFÖRMIG : cellular ; in cell form 189
ZELLENGANG (*m*) : cellular duct 190
ZELLENGEHALT (*m*) : cell content-(s) 191
ZELLENGEWEBE (*n*) : cellular tissue ; parenchyma 192
ZELLENHALTIG : cellular ; containing cells 193
ZELLENINHALT (*m*) : cell contents-(s) 194
ZELLENKERN (*m*) : cell nucleus 195
ZELLENKIES (*m*) : cellular pyrites 196
ZELLENKORALLEN (*f.pl*) : *Celleporæ* 197
ZELLENPFLANZEN (*f.pl*) : *Cellulares* 198
ZELLENRAD (*n*) : bucket wheel 199
ZELLENSAFT (*m*) : cell (fluid) juice 200
ZELLENSCHALTER (*m*) : cell switch 201
ZELLENSCHMELZ (*m*) : a goldsmith's enamel, Émail cloisonné 202
ZELLENSTOFF (*m*) : cellulose (see ZELLULOSE) 203
ZELLENWAND (*f*) : cell wall 204
ZELLE (*f*), PORÖSE- : porous cell 205
ZELLE (*f*), WASSERDICHTE- : watertight cell 206
ZELLGEWEBE (*n*) : cellular tissue ; parenchyma 207
ZELLGEWEBSENTZÜNDUNG (*f*) : inflammation of cellular tissue ; phlegmon 208
ZELLIG : celled ; cellular ; cellulous 209
ZELLINNERE (*n*) : interior of a cell 210
ZELLKERN (*n*) : cytoblast 211
ZELLKNOTEN (*m*) : cellular node 212
ZELLMEMBRAN (*f*) : celluar membrane 213
ZELLSAFT (*m*) : cell (fluid) juice, cellular fluid 214
ZELLSTOFF (*m*) : protoplasm ; cellulose ; $C_6H_{10}O_5$; chemical fibre ; Sg. 1.45 215
ZELLSTOFFFABRIK (*f*) : cellulose factory 216
ZELLSTOFFWATTE (*f*) : cellulose wadding 217
ZELLULOID (*n*) : celluloid ; xylonite (solid solution of nitrocellulose in camphor) 218
ZELLULOIDFABRIK (*f*) : celluloid factory 219
ZELLULOIDLACK (*m*) : celluloid varnish (a solution of celluloid in amyl acetate, acetone or ether) 220
ZELLULOID- ROHSTOFF (*m*) : celluloid raw material 221
ZELLULOIDWAREN (*f.pl*) : celluloid goods 222
ZELLULOSE (*f*) : cellulose ; $C_6H_{10}O_5$; chemical fibre ; Sg. 1.45 223
ZELLULOSEAZETAT (*n*) : cellulose acetate, sericose (see AZETYLZELLULOSE) 224
ZELLULOSEESTER (*m*) : cellulose acetate (see AZETYLZELLULOSE) 225
ZELLULOSEEXTRAKT (*m*) : cellulose extract 226
ZELLULOSEFILZ (*m*) : cellulose felt 227
ZELLULOSELÖSUNG (*f*) : cellulose solution 228
ZELLWAND (*f*) : cell wall 229
ZELOT (*m*) : zealot ; fanatic 230
ZELT (*n*) : tent ; pavilion ; awning 231
ZELTCHEN (*n*) : tablet ; lozenge 232
ZELTSCHNEIDER (*m*) : tent-maker 233
ZEMENT (*m*) : cement 234
ZEMENTATION (*f*) : cementation (case hardening by heating in a bath of barium chloride, potassium chloride and calcium cyanamide) 235
ZEMENTBREI (*m*) : grout (semi-liquid cement) 236
ZEMENTFABRIK (*f*) : cement works 237
ZEMENTFARBE (*f*) : cement colour ; colour for cement 238
ZEMENTIEREN : to (convert) cement (see ZEMENTA-TION) 239
ZEMENTIERFASS (*n*) : precipitation (cementing) vat 240
ZEMENTIERKASTEN (*m*) : converting chest 241
ZEMENTIERMITTEL (*n*) : cementing agent 242
ZEMENTIEROFEN (*m*) : converting furnace (Metal) 243
ZEMENTIERUNG (*f*) : cementation ; cementing 244
ZEMENTIERVERFAHREN (*n*) : cementing (cementation) process (annealing of metal, in which carbon enters to a certain depth of the metal to be annealed) 245
ZEMENTIT (*n*) : cementite (a hard compound of iron and iron carbide) 246
ZEMENTKALKSTEIN (*m*) : hydraulic limestone 247
ZEMENTKUPFER (*n*) : precipitated copper ; copper precipitate 248
ZEMENTMÜHLE (*f*) : cement mill 249
ZEMENTPRÜFUNG (*f*) : cement testing 250
ZEMENTSILBER (*n*): precipitated silver ; silver precipitate 251
ZEMENTSTAHL (*m*) : cementation (cement ; converted) steel 252
ZEMENTTIEGEL (*m*) : cement crucible 253
ZEMENTWAREN (*f.pl*) : cement ware-(s) 254
ZENIT (*m*) : zenith, climax 255
ZENSIEREN : to criticize ; censor 256
ZENSUR (*f*) : censorship ; report 257

ZENTESIMAL: centesimal 258
ZENTESIMALTHERMOMETER (*n*): centigrade thermometer 259
ZENTIFOLIE (*f*): cabbage rose (*Rosa centifolia*) 260
ZENTIGRAMM (*n*): centigram 261
ZENTIMETER (*n*): centimetre 262
ZENTNER (*m*): hundredweight; cwt.; quintal (50 kilogrammes) 263
ZENTNERLAST (*f*): heavy weight; hundredweight (cwt.) load 264
ZENTRAL: central 265
ZENTRALBEWEGUNG (*f*): central movement 266
ZENTRALBLATT (*n*): central (paper; organ) journal 267
ZENTRALBLENDE (*f*): central diaphragm (diaphragm between two lenses) (Phot.) 268
ZENTRALE (*f*): median-(line); generating station; power station 269
ZENTRALHEIZUNG (*f*): central heating-(plant), centralized heating 270
ZENTRALISIEREN: to centralize 271
ZENTRALIN (*n*): zentralin, dimethyldiphenylurea; $(CH_3)_2(C_6H_5)_2CON_2$; Mp. 120°C. 272
ZENTRALPUNKT (*m*): central point; centre 273
ZENTRALSCHMIERAPPARAT (*m*): central lubricating apparatus 274
ZENTRALSCHMIERUNG (*f*): central lubrication 275
ZENTRIEREN: to centre 276
ZENTRIEREND: centring 277
ZENTRIERMASCHINE (*f*): centring (centering) machine 278
ZENTRIERUNG (*f*): centring 279
ZENTRIFUGALABSCHEIDER (*m*): centrifugal separator 280
ZENTRIFUGALKRAFT (*f*): centrifugal force 281
ZENTRIFUGALMOMENT (*n*): centrifugal moment 282
ZENTRIFUGALPUMPE (*f*): centrifugal pump 283
ZENTRIFUGALREGULATOR (*m*): centrifugal (regulator) governor, Watt governor 284
ZENTRIFUGALSCHMIERAPPARAT (*m*): centrifugal lubricator 285
ZENTRIFUGALZERSTÄUBER (*m*): centrifugal atomizer 286
ZENTRIFUGE (*f*): centrifuge; sling; hydro-extractor; centrifugal 287
ZENTRIFUGENKUPPLUNG (*f*): centrifuge coupling 288
ZENTRIFUGENTROMMEL (*f*): centrifugal drum 289
ZENTRIFUGIEREN: to centrifuge; centrifugalize 290
ZENTRIFUGIERMASCHINE (*f*): centrifugal machine 291
ZENTRIFUGIERUNG (*f*): centrifugal (action) effect 292
ZENTRIPETALKRAFT (*f*): centripetal force 293
ZENTRISCH: central; concentric 294
ZENTRIWINKEL (*m*): angle at centre; angle formed by two radii of a circle 295
ZENTRUM (*n*): centre; bull's eye 296
ZENTRUMBOHRER (*m*): centre bit 297
ZEOLITH (*m*): zeolite (hydrous silicate of aluminium plus sodium and/or calcium) 298
ZEOLITH (*m*), KÜNSTLICHER-: artificial zeolite (see PERMUTIT and NATROLITH) 299
ZEPTER (*n*): sceptre 300
ZER (*n*): cerium; (Ce) (see also CER) 301
ZERARBEITEN: to prepare (destroy) by working; work; toil; overwork; crumble 302
ZERASIN (*n*): cerasin 303
ZERASINHALTIGES GUMMI (*n*): cerasin bearing gum (such as Cherry gum) 304
ZERAT (*n*): cerate (*Ceratum*) 305
ZERÄTZEN: to destroy by the aid of caustics; corrode 306
ZERBERSTEN: to burst (asunder) into fragments 307
ZERBRECHEN: to break; shatter; crack; fracture; rupture 308
ZERBRECHER (*m*): breaker 309
ZERBRECHLICH: fragile; breakable; brittle; frail 310
ZERBRECHLICHKEIT (*f*): fragility; brittleness, etc. 311
ZERBRECHUNG (*f*): breaking 312
ZERBRECHUNGSFESTIGKEIT (*f*): breaking (resistance) strain 313
ZERBRÖCKELN: to crumble 314
ZERBRÖCKELND: crumbling; friable 315
ZERDRÜCKEN: to crush; squash 316
ZERDRÜCKUNG (*f*): crushing; squashing; crushing (stress) load 317
ZERDRÜCKUNGSFESTIGKEIT (*f*): crushing resistance 318
ZEREALIEN (*f.pl*): cereals 319
ZEREBRAL: cerebral 320
ZEREBRALSYSTEM (*n*): cerebral system 321
ZEREBROSPINAL: cerebrospinal 322
ZEREISEN (*n*): cerium iron, Auer's metal (see AUERMETALL) 323
ZEREMONIE (*f*): ceremony 324
ZEREMONIELL: ceremonial 325
ZEREMONIÖS: ceremonious 326
ZERESIN (*n*): ceresin, mineral wax, cerin; ozocerite, solid paraffin wax; Sg. 0.918-0.922; Mp. 62-80°C. (see also PARAFFIN, FESTES-) 327
ZERFAHREN: to run (knock) down; crush by driving over; fly (burst) asunder; spoil (injure; ruin) roads by (careless) driving; thoughtless; unsteady; inattentive; absent-minded; disconnected 328
ZERFALL (*m*): dissociation; disintegration; decomposition; decay; ruin; delapidation 329
ZERFALLEN: to dissociate; disintegrate; decompose: fall to pieces; decay 330
ZERFALLGESCHWINDIGKEIT (*f*): decomposition velocity 331
ZERFALLPRODUKT (*n*): dissociation (decomposition) product 332
ZERFALLSPRODUKT (*n*): (see ZERFALLPRODUKT) 333
ZERFASERN: to separate into fibres; teaze out 334
ZERFETZEN: to tear (cut) to pieces; slash; shred; tatter; mangle 335
ZERFLEISCHEN: to lacerate 336
ZERFLIEGEN: to disperse; fly asunder 337
ZERFLIESSBAR: deliquescent; liquefiable 338
ZERFLIESSEN: to melt; deliquesce; (be) dissolve-(d); run (Colours); liquefy 339
ZERFLIESSEND: deliquescent 340
ZERFLIESSLICH: deliquescent 341
ZERFLIESSUNG (*f*): deliquescence; liquefaction 342
ZERFRESSEN: to corrode; cauterize; eat away; gnaw 343
ZERFRESSEND: corrosive; septic 344
ZERFRESSUNG (*f*): corrosion; cauterization; sepsis 345
ZERGEHEN: to dwindle; melt; disperse; dissolve; deliquesce 346

ZERGLIEDERN : to dismember ; decompose ; dissect ; analyse 347
ZERGLIEDERUNG (*f*) : dismemberment ; anatomy ; dissection ; decomposition ; analysis 348
ZERHACKEN : to hack (cut) to pieces ; mince 349
ZERIN (*n*) : cerin (see ZERESIN) 350
ZERISULFATABSCHWÄCHER (*m*) : ceric sulphate reducer (Phot.) 351
ZERIT (*m*) : cerite (see CERIT) 352
ZERITERDE (*f*) : cerite earth (contains cerium, lanthanum and neodymium) (see CERIT) 353
ZERIUM (*n*) : cerium ; (Ce) (see CER) 354
ZERKLEINERN : to disintegrate ; reduce to small pieces ; comminute ; masticate ; grind ; crush 355
ZERKLEINERUNG (*f*) : disintegration ; grinding, etc. 356
ZERKLEINERUNGSAPPARAT (*m*) : disintegrator 357
ZERKLEINERUNGSMASCHINE (*f*) : grinding (pulverizing) machine ; crusher ; crushing (machine) mill ; disintegrator 358
ZERKLÜFTET : fissured ; cleft ; riven ; segmented 359
ZERKLÜFTUNG (*f*) : cleavage, fission, fissure, segmentation 360
ZERKNACKEN : to crack 361
ZERKNALLEN : to explode 362
ZERKNICKEN : to crack ; snap 363
ZERKNIRSCHTHEIT (*f*) : penitence, remorse, crushed state 364
ZERKNISTERN : to decrepitate 365
ZERKNITTERN : to crumple ; rumple ; crush ; ruffle 366
ZERKOCHEN : to boil to pieces ; spoil by overboiling 367
ZERKRATZEN : to scratch 368
ZERKRÜMELN : to crumble 369
ZERLASSEN : to liquefy ; melt ; dissolve 370
ZERLEGBAR : divisible (Maths.) ; decomposable ; portable ; demountable ; capable of being dismantled 371
ZERLEGEN : to decompose ; cut (carve ; split) up ; take (to pieces) apart ; resolve (Forces) ; dissect ; analyse ; demount ; dismantle (Mech.) ; separate (into constituent parts) ; divide ; sunder 372
ZERLEGEN (*n*) : decomposition (used in the abstract as " decomposition of force ") 373
ZERLEGUNG (*f*) : decomposition ; resolution (of forces), etc. 374
ZERLÖCHERN : to punch ; perforate 375
ZERMAHLEN : to pulverize ; grind ; triturate ; powder ; granulate ; crush 376
ZERMALMEN : to grind ; bruise ; crush ; pulverize ; triturate 377
ZERMÜRBEN : to decay ; rot ; become friable 378
ZERNAGEN : to gnaw ; corrode 379
ZERNICHTEN : to destroy, annihilate 380
ZERNIEREN to blockade, invest, besiege 381
ZERNITRAT (*n*) : cerium nitrate ; $Ce(NO_3)_3.6H_2O$ 382
ZEROSIN (*n*) : (see ZERESIN) 383
ZEROVERBINDUNG (*f*) : cerous compound 384
ZEROXYD (*n*) : cerium oxide ; CeO_2 385
ZERPLATZEN : to burst (asunder) ; crack ; rupture ; explode 386
ZERQUETSCHEN : to squash ; crush ; bruise ; mash 387
ZERRBILD (*n*) : caricature 388
ZERREIBBAR : friable ; triturable 389
ZERREIBEN : to triturate ; pulverize ; powder ; granulate ; grind 390
ZERREIBLICH : friable ; triturable 391
ZERREIBLICHKEIT (*f*) : friability 392
ZERREIBUNG (*f*) : pulverization ; trituration ; granulation 393
ZERREISSBAR : tearable ; breakable 394
ZERREISSEN : to tear ; break ; rupture ; rend ; lacerate ; wear out 395
ZERREISSFESTIGKEIT (*f*) : tensile strength (Metal) 396
ZERREISSPROBE (*f*) : tensile test 397
ZERREISSUNG (*f*) : rupture ; laceration ; rending 398
ZERREN : to pull ; tear ; drag ; tug ; lug ; haul ; worry (a bone) 399
ZERRENNBODEN (*m*) : slag bottom (Metal) 400
ZERRENNEN : to fine ; refine (Metal) 401
ZERRENNER (*m*) : refiner (Metal) 402
ZERRENNFEUER (*n*) : refining fire (Metal) 403
ZERRENNHERD (*m*) : refining furnace 404
ZERRINNEN : to dissolve ; melt ; fade away ; vanish ; disappear ; dwindle 405
ZERRÜHREN : to mix (stir ; beat) up ; mash ; pulp 406
ZERRUSIT (*m*) : cerrusite ; $(PbO.CO_2)$; cerusite, cerussite (natural lead carbonate, about 77.5% lead ; Sg. 6.5) 407
ZERRÜTTEN : to undermine-(health) ; derange, destroy ; shatter, disorder ; disturb ; ruin ; unsettle ; distrust 408
ZERRÜTTUNG (*f*) : derangement ; destruction ; disorder ; disturbance ; distraction ; confusion 409
ZERSÄGEN : to saw to pieces 410
ZERSCHELLEN : to shiver ; shatter ; smash ; be (smashed) broken up ; be wrecked ; be dashed to pieces 411
ZERSCHLAGEN : beaten ; disappointed ; to shiver ; shatter ; smash ; crush ; break in pieces ; bruise ; beat ; be disappointed ; be beaten ; batter ; be dispersed 412
ZERSCHMELZEN : to melt ; dissolve 413
ZERSCHMETTERN : to shatter ; smash ; dash to pieces 414
ZERSCHNEIDEN : to shred ; cut (up) to pieces ; mince ; carve 415
ZERSETZBAR : decomposable 416
ZERSETZBARKEIT (*f*) : decomposability 417
ZERSETZEN : to decompose ; disintegrate ; dissociate ; break up into constituent parts ; decay 418
ZERSETZER (*m*) : decomposing agent ; decomposer 419
ZERSETZLICH : unstable ; liable to decompose 420
ZERSETZLICHKEIT (*f*) : instability ; liability to decomposition 421
ZERSETZUNG (*f*) : decomposition ; dissolution 422
ZERSETZUNGSDESTILLATION (*f*) : destructive distillation (see CRACKPROZESS) 423
ZERSETZUNGSERSCHEINUNG (*f*) : phenomenon of decomposition 424
ZERSETZUNGSERZEUGNIS (*f*) : product of decomposition 425
ZERSETZUNGSFÄHIG : capable of decomposing 426
ZERSETZUNGSKUNST (*f*) : analysis 427
ZERSETZUNGSMITTEL (*n*) : decomposing agent 428
ZERSETZUNGSPFANNE (*f*) : decomposing pan 429
ZERSETZUNGSPRODUKT (*n*) : decomposition product 430
ZERSETZUNGSPROZESS (*m*) : process of decomposition 431
ZERSETZUNGSPUNKT (*m*) : decomposing point 432

ZERSETZUNGSSPANNUNG (*f*): decomposition voltage 433
ZERSETZUNGSVORGANG (*m*): decomposition process 434
ZERSETZUNGSWÄRME (*f*): decomposition heat 435
ZERSETZUNGSWIDERSTAND (*m*): electrolytic resistance 436
ZERSETZUNGSZELLE (*f*): decomposition cell 437
ZERSPALTEN: to cleave; split up; crack 438
ZERSPLITTERN: to splinter; split; shiver; break up; dissipate; scatter; fritter away 439
ZERSPRENGEN: to blow-up; burst; shatter; spring; explode; split 440
ZERSPRINGEN: to explode; burst; crack; rupture; break; fly into pieces 441
ZERSTAMPFEN: to pound; pulverize 442
ZERSTAUBEN: to convert into dust 443
ZERSTÄUBEN: to comminute; reduce to dust; pulverize; diffuse; atomize; spray; flower (Mercury) 444
ZERSTÄUBUNG (*f*): spraying, atomizing, atomization 445
ZERSTÄUBUNGSAPPARAT (*m*): atomizer; spray; diffusor; pulverizer; vapourizer 446
ZERSTIEBEN: to spray; scatter as dust; be scattered as dust; vanish 447
ZERSTÖRBAR: destructible; perishable 448
ZERSTÖRBARKEIT (*f*): destructibility 449
ZERSTÖREN: to destroy; ruin; interrupt; break down; demolish; wreck; frustrate; overthrow 450
ZERSTÖREND: disruptive 451
ZERSTÖRUNG (*f*): destruction; ruin; disruption; break-down; demolition; havoc 452
ZERSTÖRUNGSTRIEB (*m*): vandalism 453
ZERSTÖRUNGSWUT (*f*): vandalism 454
ZERSTOSSEN: to pulverize; pound; powder; bruise; shatter; triturate 455
ZERSTREUEN: to disseminate; scatter; disperse; diffuse; divert; sow; dissipate; distract; shed; spread; disband; distribute; dispel; strew; broadcast 456
ZERSTREUT: abstracted; dispersed (see ZERSTREUEN); absent-minded 457
ZERSTREUUNG (*f*): abstractedness; absence of mind; dispersion; diffusion; diversion; dispersal; dissipation 458
ZERSTREUUNGSLINSE (*f*): divergent (concave) lens, negative lens 459
ZERSTREUUNGSPUNKT (*m*): point of divergence 460
ZERSTREUUNGSVERMÖGEN (*n*): dispersive power 461
ZERSTÜCKELN: to dismember; cut (divide) into pieces 462
ZERSTÜCKEN: (see ZERSTÜCKELN) 463
ZERSTÜMMELN: to mutilate 464
ZERTEILBAR: separable, divisible 465
ZERTEILBARKEIT (*f*): divisibility 466
ZERTEILEN: to separate; divide; disperse; dismember; disunite; split up; dissolve; resolve; analyse 467
ZERTEILUNG (*f*): division; resolution; dissolution; separation; analysis 468
ZERTEILUNGSMITTEL (*n*): resolvent 469
ZERTIFIKAT (*n*): certificate 470
ZERTRENNEN: to unravel; separate; unstitch; rip-(up) 471
ZERTRENNLICH: separable 472
ZERTRETEN: to crush under foot, tread down, trample 473
ZERTRÜMMERN: to shatter; break (up) down; destroy; go to rack and ruin; demolish 474
ZERTRÜMMERUNG (*f*): destruction; demolition; decay 475
ZERUSSIT (*m*): cerussite (see CERUSSIT) 476
ZERVELATWURST (*f*): saveloy 477
ZERWÜRFNIS (*n*): strife; dissension; discord; difference; dispute 478
ZERZUPFEN: to pull to pieces; touse 479
ZESSIEREN: to cease 480
ZESSION (*f*): transfer; cession 481
ZESSIONSURKUNDE (*f*): deed of assignment, transfer 482
ZETTEL (*m*): bill; placard; label; ticket; card; slip of paper; check; note; handbill; warp; poster 483
ZETTELANSCHLAGEN (*n*): bill-posting 484
ZETTELBAUM (*m*): warp-beam 485
ZETTELGARN (*n*): warp, water (Textile) 486
ZETTELKATALOG (*m*): card (catalogue) index 487
ZEUG (*n*): material; stuff; implements; pulp (Paper); metal; yeast (Brew.); cloth; fabric; matter; utensils; lumber; trash; tackle; goods; things 488
ZEUGBAUM (*m*): cloth-beam 489
ZEUGBÜTTE (*f*): stuff chest; pulp vat (Paper) 490
ZEUGDRUCK (*m*): calico (cloth) printing; textile printing 491
ZEUGDRUCKEREI (*f*): print works; calico printing 492
ZEUGE (*m*): witness; deponent 493
ZEUGELINIE (*f*): generating line 494
ZEUGEN: to generate; engender; procreate; beget; breed; produce; testify; depose; witness 495
ZEUGENAUSSSAGE (*f*): deposition; statement (by a witness) 496
ZEUGENVERHÖR (*n*): examination (of witnesses) 497
ZEUGFABRIK (*f*): woollen mill; cloth factory 498
ZEUGFÄNGER (*m*): stuff catcher (Paper) 499
ZEUGFÄRBEREI (*f*): cloth (calico) dyeing 500
ZEUGGEBEN (*n*): the addition of yeast (Brew.); to add yeast 501
ZEUGHANDEL (*m*): textile trade 502
ZEUGHAUS (*n*): armoury; arsenal 503
ZEUGKASTEN (*m*): stuff chest (Paper); tool chest 504
ZEUGLUMPEN (*m.pl*): rags (Paper) 505
ZEUGMANUFAKTUR (*f*): cloth manufacture 506
ZEUGNIS (*n*): testimonial; testimony; evidence; deposition; witness; certificate 507
ZEUGNIS (*n*) ABLEGEN: to give evidence, depose 508
ZEUGNISABSCHRIFT (*f*): copy of testimonial 509
ZEUGSCHMIED (*m*): armourer; tool-smith 510
ZEUGSICHTER (*m*): pulp strainer (Paper) 511
ZEUGUNG (*f*): generation; production; procreation; breeding 512
ZEUGUNGSFÄHIG: capable of (generation) procreation 513
ZEUGUNGSFLÜSSIGKEIT (*f*): semen; seminal fluid 514
ZEUGUNGSGLIED (*n*): genital (organ) member 515
ZEUGUNGSMITTEL (*n*): aphrodisiac 516
ZEUGUNGSSTOFF (*m*): semen 517
ZEUGUNGSTEILE (*m.pl*): genitive (sexual) organs 518
ZEUGWANNE (*f*): yeast tub (Brew.) 519
ZEUGWAREN (*f.pl*): textiles; textile (fabrics) goods 520

ZEUNERIT (*m*): zeunerite; $Cu(UO_2)_2(AsO_4)_2$; Sg. 3.53 521
ZIBEBE (*f*): cubeb (*Piper cubeba*); raisin 522
ZIBET (*m*): civet 523
ZIBETBAUM (*m*): East Indian durian tree (*Durio zibethinus*) 524
ZIBETH (*m*): civet 525
ZICHORIE (*f*): chicory; succory; endive (*Cichorium intybus*) 526
ZICHORIENWURZEL (*f*): chicory root 527
ZICKZACK (*m*): zig-zag 528
ZICKZACKBLITZ (*m*): forked lightening 529
ZICKZACKLINIE (*f*): zig-zag (jagged) line 530
ZICKZACKNIETUNG (*f*): zig-zag (staggered) riveting 531
ZIEGE (*f*): she-goat; goat 532
ZIEGEL (*m*): tile, (Sg. 1.82); brick (common or red brick), Sg. 1.4-1.6 533
ZIEGELARTIG: tegular; like a tile 534
ZIEGELBAU (*m*): brick construction 535
ZIEGELBRAND (*m*): tile (brick) burning; batch (Bricks or Tiles) 536
ZIEGELBRENNEN (*n*): tile (brick) burning 537
ZIEGELBRENNER (*m*): tile (brick) burner or maker 538
ZIEGELBRENNEREI (*f*): tile (brick) works or kiln; tile (brick) burning 539
ZIEGELBRENNOFEN (*m*): brick kiln 540
ZIEGELDACH (*n*): tiled roof 541
ZIEGELDACHFÖRMIG: imbricated 542
ZIEGELDECKER (*m*): bricklayer, tiler 543
ZEIGELEI (*f*): brickworks; brickyard; brick-kiln 544
ZIEGELEIEINRICHTUNG (*f*): plant for brickworks 545
ZIEGELEIMASCHINEN (*f.pl*): brick-making machinery, machinery for brickworks 546
ZIEGELERDE (*f*): clay for bricks 547
ZIEGELERZ (*n*): tile ore (cuprite+limonite) (see ROTKUPFERERZ and LIMONIT) 548
ZIEGELFARBIG: brick-coloured; brick-red 549
ZIEGEL (*m*), FEUERFESTER-: firebrick (Sg. 2.2-2.7) 550
ZIEGELHÜTTE (*f*): brick kiln; brick works 551
ZIEGELMAUER (*f*): brick wall 552
ZIEGELMEHL (*n*): brick dust 553
ZIEGELOFEN (*m*): brick kiln 554
ZIEGELPRESSE (*f*): tile press 555
ZIEGELROT: brick-red; brick-coloured 556
ZIEGELSTEIN (*m*): brick; tile (see ZIEGEL) 557
ZIEGELSTREICHER (*m*): brick-maker 558
ZIEGELTON (*m*): clay for bricks 559
ZIEGENARTIG: capriform; goat-like 560
ZIEGENAUGE (*f*): goat grass (*Aegilops ovata*) 561
ZIEGENBOCK (*m*): he-goat 562
ZIEGENBUTTER (*f*): goat's butter 563
ZIEGENFELL (*n*): goatskin 564
ZIEGENFÜSSIG: capriped, goat-footed 565
ZIEGENHAAR (*n*): goat's hair 566
ZIEGENKÄSE (*m*): goat's milk cheese 567
ZIEGENLAB (*n*): rennet 568
ZIEGENLEDER (*n*): goatskin; kid (Leather) 569
ZIEGENMILCH (*f*): goat's milk 570
ZIEGENPETER (*m*): mumps (see MUMPS) 571
ZIEGENSTEIN (*m*): bezoar (a morbid concretion in the stomach of ruminants) 572
ZIEGENWEIDE (*f*): goat willow (*Salix caprea*) 573
ZIEGLER (*m*): tile (brick) maker 574
ZIEHARM (*m*): handle; arm; crank 575
ZIEHBANK (*f*): draw bench 576
ZIEHBAR: ductile, extensible 577
ZIEHBARKEIT (*f*): ductility 578
ZIEHBRÜCKE (*f*): draw-bridge 579
ZIEHBRUNNEN (*m*): well 580
ZIEHEISEN (*n*): draw plate 581
ZIEHEN: to pull; draw; drag; tug; educate; reap; rear; bring up; breed; raise; cultivate; extract (Drugs); mould (Candles); rifle (Guns); floor; couch (Grain); penetrate; soak; stretch; move; warp; incline; extrude (Tubes); be ropy; attract; (*n*): ropiness (Liquids); twinge; move; traction; draft; drawing; extrusion (Tubes); hauling; haulage 582
ZIEHEN, IN DIE LÄNGE-: to drag (draw) out; protract 583
ZIEHEN, IN ZWEIFEL-: to question; doubt 584
ZIEHENLASSEN: to let (allow to) draw; infuse 585
ZIEHEN, NUTZEN-: to reap (advantage) profit 586
ZIEHEN, SICH-: to distort, warp 587
ZIEHEN, ZU RATE-: to consult 588
ZIEHFEDER (*f*): drawing pen 589
ZIEHFENSTER (*n*): sash window 590
ZIEHHARMONIKA (*f*): accordion; concertina 591
ZIEHKIND (*n*): foster child 592
ZIEHKLINGE (*f*): draw plate, scraper 593
ZIEHKLOBEN (*m*): pulley 594
ZIEHKRAFT (*f*): tractive (drawing) force or power; attraction 595
ZIEHMESSER (*n*): draw knife 596
ZIEHÖL (*n*): oil for drawing 597
ZIEHPFLANZE (*f*): nursery plant 598
ZIEHPFLASTER (*n*): vesicatory 599
ZIEHPROBE (*f*): sample drawn or for drawing out; extrusion sample or test 600
ZIEHSCHLEIFE (*f*): slip knot 601
ZIEHSEIL (*n*): tow-(ing) rope 602
ZIEHSTANGE (*f*): draw-bar; rifling-rod 603
ZIEHSTRANG (*m*): trace 604
ZIEHTAG (*m*): moving day 605
ZIEHTEILE (*m.pl*): blanks (for drawing) (Metal) 606
ZIEHUNG (*f*): draft (see ZIEHEN): drawing; attraction 607
ZIEHWEG (*m*): towing path 608
ZIEL (*n*): object; goal; end; aim; butt; target; mark; limit; time; term; scope 609
ZIELEN: to aim; tend to 610
ZIELENDES ZEITWORT (*n*): transitive verb 611
ZIELLOS: aimless-(ly) 612
ZIELPUNKT (*m*): bull's eye; mark; aim; goal 613
ZIELSCHEIBE (*f*): target; mark; butt 614
ZIEMEN: to suit; become; be becoming; be fitting; be suitable 615
ZIEMLICH: moderate; fair; rather; tolerable; considerable; pretty 616
ZIER (*f*): grace; ornament; decoration 617
ZIERAT (*m*): decoration; ornament; finery 618
ZIERBLUME (*f*): ornamental plant 619
ZIERDE (*f*): ornament; honour; grace; decoration 620
ZIEREN: to ornament; grace; decorate; adorn; be (coy; prim) affected 621
ZIEREREI (*f*): affectation; airs-(and graces) 622
ZIERLICH: graceful; dainty; pretty; elegant; neat; fine; smart; nice 623
ZIERLICHER KONTRAPUNKT (*m*): florid counterpoint (Music) 624
ZIERPFLANZE (*f*): ornamental plant 625
ZIEST (*m*): wound-wort (*Stachys sylvatica*) 626

ZIFFER (*f*): cipher; digit; number; figure; cryptograph; (as prefix=coefficient) 627
ZIFFERBLATT (*n*): dial-(plate) (of a watch or gauge); face (of a watch) 628
ZIFFERSCHRIFT (*f*): cypher; code 629
ZIGARETTE (*f*): cigarette 630
ZIGARETTENPAPIER (*n*): cigarette paper (Paper) 631
ZIGARETTENSPITZE (*f*): cigarette holder 632
ZIGARRE (*f*): cigar 633
ZIGARRENSPITZE (*f*): cigar holder 634
ZIGEUNER (*m*): gipsy 635
ZIGEUNERWEISE (*f*): gipsy tune 636
ZILLE (*f*): skiff 637
ZIMMER (*n*): chamber; room; apartment 638
ZIMMERARBEIT (*f*): carpenter's work 639
ZIMMERAUFNAHME (*f*): indoor (portraiture) photograph 640
ZIMMERDECKE (*f*): ceiling 641
ZIMMERGERÄT (*n*): furniture 642
ZIMMERHANDWERK (*n*): carpentry; carpenter's trade 643
ZIMMERHOF (*m*): timber-yard 644
ZIMMERHOLZ (*n*): timber 645
ZIMMERMÄDCHEN (*n*): chambermaid 646
ZIMMERMANN (*m*): carpenter 647
ZIMMERMANNSSTICH (*n*): timber hitch 648
ZIMMERMEISTER (*m*): master carpenter 649
ZIMMERN: to build of wood; timber; square 650
ZIMMEROFEN (*m*): stove 651
ZIMMERTEMPERATUR (*f*): room temperature 652
ZIMMERWERK (*n*): carpentry; carpenter's work 653
ZIMMET (*m*): cinnamon (see ZIMT) 654
ZIMMT (*m*): cinnamon (see ZIMT) 655
ZIMPERLICH: coy; squeamish; prim 656
ZIMPERN: to be (coy; squeamish) prim; simper 657
ZIMT (*m*): cinnamon (see ZIMT AUS CEYLON; ZIMT AUS CHINA and ZIMTBAUM) 658
ZIMTALDEHYD (*n*): cinnamic aldehyde; cinnamaldehyde; C_9H_8O; Sg. 1.0497; Mp. about – 7.5°C.; Bp. 209.5°C. 659
ZIMTALKOHOL (*m*): cinnamic alcohol; $C_9H_{10}O$; Sg. 1.0338; Mp. 33°C.; Bp. 254.5°C. 660
ZIMT AUS CEYLON: Ceylon cinnamon (from *Cinnamomum zeylanicum*) 661
ZIMT AUS CHINA: Cassia bark (see CASSIA) (from *Cinnamomum cassia*) 662
ZIMTBAUM (*m*): cinnamon tree (*Cinnamomum zeylanicum*) 663
ZIMTBLUME (*f*): cinnamon (cassia) flower 664
ZIMTBLÜTE (*f*): cinnamon (cassia) flower or bud 665
ZIMTBLÜTENÖL (*n*): cassia bud oil (see CASSIAÖL) 666
ZIMTBRAUN: cinnamon-brown 667
ZIMTCARBONSÄURE (*f*): carboxycinnamic acid 668
ZIMTCASSIAÖL (*n*): (see CASSIAÖL) 669
ZIMT, CHINESISCHER- (*m*): cassia bark (see CASSIA) 670
ZIMTDERIVAT (*n*): cinnamon derivate 671
ZIMTFARBE (*f*): cinnamon (brown) colour 672
ZIMTFARBIG: cinnamon coloured 673
ZIMTFISTULA (*f*): purging cassia (*Cassia fistula*) 674
ZIMTKANEEL (*m*): canella bark (see KANEEL) 675
ZIMTKASSIA (*f*): cassia bark (see CASSIA) 676
ZIMTKASSIENÖL (*n*): oil of cassia (see CASSIAÖL) 677
ZIMTÖL (*n*): oil of cinnamon (from *Cinnamomum aromaticum*); Sg. 1.04 678
ZIMTRINDE (*f*): cinnamon-(bark) 679
ZIMTRÖSCHEN (*n*): mock orange (*Philadelphus coronarius*) 680
ZIMTSAUER: cinnamate of; combined with cinnamic acid 681
ZIMTSÄURE (*f*): cinnamic acid, cinnamylic acid, beta-phenylacrylic acid; $C_6H_5CHCHCOOH$; Sg. 1.248; Mp. 133°C.; Bp. 300°C. (see also CINNAMYLSÄURE) 682
ZIMTSÄUREÄTHYLESTER (*m*): ethyl cinnamate, cinnamic ethylester, cinnamylic ether; $C_{11}H_{12}O_2$; Sg. 1.0537; Mp. 12°C.; Bp. 271°C. 683
ZIMTSÄUREMETHYLESTER (*m*): cinnamic methylester, methyl cinnamate 684
ZIMTSÄURE-M-KRESOLESTER (*m*): meta-cresol cinnamate 685
ZIMTSÄURE-NITRIL (*n*): cinnamic nitrile; C_9H_7N; Sg. 1.037; Mp. 11°C.; Bp. 255°C. 686
ZIMTSTEIN (*m*): essonite; cinnamon stone; Hessonite (see GROSSULAR) 687
ZIMTTINKTUR (*f*): tincture of cinnamon 688
ZIMTWASSER (*n*): cinnamon water 689
ZIMT, WEISSER- (*m*): canella bark (see KANEEL) 690
ZINCKENIT (*m*): zinckenite; $PbSb_2S_4$; Sg. 5.32 691
ZINDER (*m*): cinder 692
ZINEOL (*m*): cineole (see EUKALYPTOL) 693
ZINK (*n*): zinc; (Zn); Sg. 7.142; Mp. 419°C.; Bp. 918°C.; spelter 694
ZINKACETAT (*n*): zinc acetate; $Zn(C_2H_3O_2)_2.3H_2O$; Sg. 1.72; Mp. 235-257°C. 695
ZINKACETAT (*n*), WASSERFREIES-: anhydrous zinc acetate; $Zn(C_2H_3O_2)_2$; Sg. 1.84; Mp. 241.5°C. 696
ZINKALAUN (*m*): aluminium-zinc sulphate, zinc alum 697
ZINKAMMONIUMCHLORID (*n*): zinc-ammonium chloride; $ZnCl_2.5NH_3.H_2O$ 698
ZINK (*n*), APFELSAURES-: zinc mallate; $ZnC_4H_4O_5.3H_2O$ 699
ZINKARA (*f*): an (anti-corrosion agent) anti-corrosive; rust (preventative) preventer (Metal) (an oxide of zinc and sodium) 700
ZINKARSENAT (*n*): zinc arsenate; $Zn_3(AsO_4)_2$; Sg. 4.913 701
ZINKARTIG: zinc-like; zincy 702
ZINKASCHE (*f*): zinc (ash) dross; oxide of zinc (see ZINKOXYD) 703
ZINKÄTHYL (*n*): zinc ethyl; $C_4H_{10}Zn$; Sg. 1.182; Mp. – 28°C.; Bp. 118°C. 704
ZINKÄTHYLSULFAT (*n*): zinc ethyl sulphate; $Zn(C_2H_5.SO_4)_2.2H_2O$ 705
ZINKÄTZUNG (*f*): zinc etching, zincography 706
ZINK (*n*), BALDRIANSAURES-: zinc valerate 707
ZINKBARYT (*m*), BRACHYTYPER-: willemite 708
ZINKBARYT (*m*), PRISMATISCHER-: Smithsonite (see ZINKGLAS) 709
ZINKBESCHLAG (*m*): zinc (covering) coating 710
ZINKBICHROMAT (*n*): zinc (dichromate) bichromate; $ZnCO_2O_7$ 711
ZINKBLECH (*n*): zinc (galvanised) sheet-(ing) plate 712
ZINKBLENDE (*f*): black jack, zinc blende, sphalerite (zinc sulphide); ZnS (about 67% Zn); Sg. 4.0 (see SPHALERIT) 713
ZINKBLUMEN (*f.pl*): flowers of zinc; zinc oxide; ZnO (see ZINKOXYD) 714

ZINKBLÜTE (*f*) : hydrozincite, zinc bloom (natural basic zinc carbonate) ; $ZnCO_3.2Zn(OH)_2$; Sg. 3.65 715

ZINK, BORSAURES- (*n*): zinc borate ; $ZnBO_3$ [716

ZINKBROMAT (*n*): zinc bromate ; $Zn(BrO_3)_2$; Sg. 2.566 ; Mp. 100°C. 717

ZINKBROMID (*n*): zinc bromide ; $ZnBr_2$; Sg. 3.643-4.219 ; Mp. 394°C. ; Bp. 650°C. 718

ZINKBUTTER (*f*) : butter of zinc ; zinc chloride ; $ZnCl_2$ (see ZINKCHLORID) 719

ZINKCARBONAT (*n*): zinc carbonate ; $ZnCO_3$; Sg. 4.44 ; an anti-corrosion agent (Metal) (see ZINKKARBONAT) 720

ZINKCHLORAT (*n*) : zinc chlorate ; $Zn(ClO_3)_2.6H_2O$; Mp. 60°C. 721

ZINKCHLORID (*n*) : zinc chloride ; $ZnCl_2$; Sg. 2.91 ; Mp. 262°C. ; Bp. 730°C. 722

ZINKCHLORIDLAUGE (*f*) : zinc chloride (solution) lye 723

ZINK (*n*), CHROMSAURES- : zinc chromate ; $(ZnCrO_4.7H_2O)$ 724

ZINK, CHRYSOPHANSAURES- (*n*) : zinc chrysophanate 725

ZINKCYANÜR (*n*) : zinc cyanide ; $Zn(CN)_2$ 726

ZINKDACH (*n*) : zinc roof 727

ZINKDAMPF (*m*) : zinc vapour 728

ZINKDESTILLIEROFEN (*m*) : zinc-(distillation) furnace 729

ZINKDICHROMAT (*n*) : (see ZINKBICHROMAT) 730

ZINKE (*f*) : spike ; prong ; lug ; tooth 731

ZINKEISENERZ (*n*) : franklinite (see FRANKLINIT) 732

ZINKEISENSPAT (*m*) : ferruginous calamine ; $(Zn,Fe)CO_3$; Sg. 4.3 733

ZINKEISENSTEIN (*m*) : franklinite (see FRANKLINIT) 734

ZINKEN (*m*) : (see ZINKE) ; (as an *adj.* : zinc, of zinc) ; (as a *verb* : galvanize, zinc ; tooth, fit with prongs) 735

ZINKENIT (*m*): zincenite ; $PbSb_2S_4$; Sg. 5.33 736

ZINKENWEITE (*f*) : width between prongs (of coal fork) 737

ZINKERZ (*n*) : zinc ore (see ZINKSPAT and ZINKFAHLERZ) 738

ZINKERZ, ROTES- (*n*) : zincite (see ZINKIT) 739

ZINK (*n*), ESSIGSAURES- : zinc acetate ; $Zn(C_2H_3O_2)_2.3H_2O$; Sg. 1.72 ; Mp. 235/257°C. (see ZINKACETAT) 740

ZINKFAHLERZ (*n*) : antimonious tetrahedrite (plus about 9% zinc) (see ANTIMONFAHLERZ) 741

ZINKFARBE (*f*) : zinc colour 742

ZINKFEILSPÄNE (*m.pl*) : zinc filings 743

ZINKFLUAT (*n*) : zinc fluosilicate 744

ZINKFLUORID (*n*) : zinc fluoride ; ZnF_2 ; Sg. 4.84 ; Mp. 734°C. 745

ZINKFOLIE (*f*) : zinc foil 746

ZINKFORMIAT (*n*) : zinc formate ; $Zn(CHO_2)_2.2H_2O$ 747

ZINKGEKRÄTZ (*n*) : zinc dross ; oxide of zinc (see ZINKOXYD) 748

ZINKGELB (*n*) : zinc yellow (a mixture of zinc chromate, potassium bichromate and zinc oxide), $3(ZnCrO_4).K_2Cr_2O_7.ZnO$; sometimes also zinc chromate, $(ZnCrO_4.7H_2O)$ 749

ZINKGEWINNUNG (*f*) : zinc (production) extraction 750

ZINKGLAS (*n*) : Smithsonite electric calamine ; siliceous calamine (a siliceous zinc oxide) ; $Zn_2O.SiO_2$; Sg. 3.4 (see KIESELZINKERZ) 751

ZINKGLASERZ (*n*) : (see ZINKGLAS) 752

ZINK (*n*), GOLDCHLORWASSERSTOFFSAURES- : zinc chloraurate ; $Zn(AuCl_4)_2+8H_2O$. or $12H_2O$ 753

ZINKGRAU (*n*) : zinc grey (a mixture of ZINKOXYD and ZINKSTAUB, which see) 754

ZINKGRÜN (*n*) : zinc green (a mixture of zinc yellow and Prussian blue) 755

ZINKHALTIG : containing zinc ; zinciferous 756

ZINK, HARZSAURES- (*n*) : zinc resinate 757

ZINKHOCHÄTZUNG (*f*) : (see PHOTOZINKOTYPIE) 758

ZINKHÜTTE (*f*) : zinc works 759

ZINKHYDROSULFIT (*n*) : zinc hydrosulphite 760

ZINKIG : spiked ; pronged 761

ZINKISCH : zincky 762

ZINKIT (*m*) : zincite, red zinc ore (a natural zinc oxide) ; $(Zn,Mn)O$; Sg. 5.55 (see also ROTZINKERZ) 763

ZINKJODAT (*n*) : zinc iodate ; $Zn(IO_3)_2$ 764

ZINKJODID (*n*) : zinc iodide ; ZnI_2 ; Sg. 4.696 ; Mp. 446°C. ; Bp. 624°C. 765

ZINKKALK (*m*) : zinc dross (see ZINKOXYD) 766

ZINK (*n*), KARBOLSAURES- : zinc (carbolate) phenolate 767

ZINKKARBONAT (*n*) : zinc carbonate ; $ZnCO_3$ (see ZINKCARBONAT and ZINK, KOHLENSAURES-) 768

ZINKKARBONAT (*n*), BASISCHES- : basic zinc carbonate (formed by exposure of zinc to moist air) 769

ZINKKIESEL (*m*) : siliceous calamine (see KIESELZINKERZ) 770

ZINKKIESELERZ (*n*) : (see ZINKKIESEL) 771

ZINKKOHLENBATTERIE (*f*) : zinc-carbon battery 772

ZINK, KOHLENSAURES- (*n*) : zinc carbonate ; $(ZnCO_3)$; Sg. 4.42-4.45 ; Mp. loses CO_2 at 300°C. 773

ZINKLAKTAT (*n*) : zinc lactate ; $Zn(C_3H_5O_3)_2.3H_2O$ 774

ZINK, LEINÖLSAURES- (*n*) : zinc linoleate 775

ZINKMETASILIKAT (*n*) : zinc metasilicate ; $ZnSiO_3$; Sg. 3.42-3.86 776

ZINKMETHYL (*n*) : zinc methyl ; C_2H_6Zn ; Sg. 1.386 ; Mp. – 40°C. ; Bp. 46°C. 777

ZINK (*n*), MILCHSAURES- : zinc lactate (see ZINKLAKTAT) 778

ZINKNAFALAN (*n*) : zinc nafalane 779

ZINK (*n*), NAPHTALINSULFOSAURES- : zinc sulfonaphthalate 780

ZINKNITRAT (*n*) : zinc nitrate ; $Zn(NO_3)_2.6H_2O$; Sg.2 .065 ; Mp. 36.4°C. ; Bp. 131°C. 781

ZINKOFEN (*m*) : zinc furnace 782

ZINKOFENBRUCH (*m*) : cadmia ; tutty 783

ZINKOGRAPHIE (*f*) : (see ZINKÄTZUNG) 784

ZINKOLITH (*m*) : (see LITHOPON) 785

ZINK (*n*), ÖLSAURES- : zinc oleate ; $Zn(C_{18}H_{33}O_2)_2$ 786

ZINKORTHOSILIKAT (*n*) : zinc orthosilicate ; Zn_2SiO_4 ; Sg. 3.7 787

ZINK (*n*), OXALSAURES- : zinc oxalate ; $ZnC_2O_4.2H_2O$; Sg. 2.58 788

ZINKOXYD (*n*) : zinc oxide ; ZnO ; Sg. 5.78 ; zinc white ; Chinese white ; flowers of zinc ; nix alba (see also WEISSES NICHTS) 789

ZINKOXYDANLAGE (*f*) : zinc oxide plant 790

ZINKOXYDNATRON (*n*) : oxide of zinc and sodium ; zinc-sodium oxide ; anti-corrosion agent ; rust preventative 791

ZINKOXYD (n), ROTES- : zincite (see ZINKIT) 792
ZINKPECHERZ (n) : sphalerite (see SPHALERIT) 793
ZINKPERBORAT (n) : zinc perborate ; $ZnBo_3.H_2O$ 794
ZINKPERHYDROL (n) : zinc perhydrol 795
ZINKPERMANGANAT (n): zinc permanganate ; $Zn(MnO_4)_2.2H_2O$ 796
ZINKPEROXYD (n) : zinc peroxide (see ZINK-SUPEROXYD) 797
ZINKPHENAT (n) : zinc phenate ; $Zn(C_6H_5O)_2$ 798
ZINKPHOSPHAT (n) : zinc phosphate ; $Zn_3(PO_4)_2$, Sg. 3.998 ; $Zn_3(PO_4)_2.8H_2O$, Sg. 3.109 [799
ZINKPHOSPHID (n) : zinc phosphide ; Zn_3P_2, Sg. 4.55 ; ZnP_2, Sg. 2.97 800
ZINKPHOSPHIT (n) : zinc phosphite : $ZnHPO_3.2\frac{1}{2}H_2O$ 801
ZINK (n), PHOSPHORSAURES- : zinc phosphate (see ZINKPHOSPHAT) 802
ZINKPLATTE (f) : zinc plate 803
ZINKPOL (m) : zinc pole ; cathode 804
ZINKPRÄPARAT (n) : zinc preparation 805
ZINKRAUCH (m) : zinc fume 806
ZINKSALBE (f) : zinc ointment 807
ZINK, SALIZYLSAURES- : zinc salicylate ; $Zn(C_6H_4CO_2)_2.3H_2O$ 808
ZINK (n), SALPETERSAURES- : zinc nitrate (see ZINKNITRAT) 809
ZINKSALZ (n) : zinc salt 810
ZINKSCHAUM (m) : zinc scum 811
ZINKSCHNITZEL (n.pl) : zinc filings 812
ZINKSCHWAMM (m) : cadmia ; tutty 813
ZINK (n), SCHWEFELSAURES- : zinc sulphate (see ZINKSULFAT) 814
ZINK (n), SCHWEFLIGSAURES- : zinc sulphite 815
ZINKSELENID (n) : zinc selenide ; ZnSe ; Sg. 5.42 816
ZINKSILICOFLUORID (n) : zinc fluosilicate ; $ZnSiF_6.6H_2O$ 817
ZINKSILIKAT (n) : zinc silicate 818
ZINKSPAT (m) : zinc ore, Sg. 4.45, ($ZnCO_3$) (zinc carbonate), Smithsonite (52% zinc) ; $2ZnO.SiO_2.H_2O$ (hydrous zinc silicate), calamine (54% zinc) 819
ZINKSPINELL (m) : gahnite, zinc spinel ; ZnO, Al_2O_3 ; Sg. 4.34 820
ZINKSTAB (m) : zinc rod 821
ZINKSTAUB (m): zinc dust; zinc powder; (70-90% Zn) (from reduction process of zinc in muffle furnace) 822
ZINK (n), STEARINSAURES- : zinc stearate ; $Zn(C_{18}H_{35}O_2)_2$ 823
ZINKSUBGALLAT (n) : zinc (sub)-gallate (mixture of zinc oxide and gallic acid in proportion of 44 : 56) 824
ZINKSUBKARBONAT (n) : zinc subcarbonate, precipitated zinc carbonate ; $2ZnCO_3.3Zn(OH)_2$ 825
ZINKSULFAT (n) : zinc sulphate ; $ZnSO_4.7H_2O$; Sg. 2.015 ; Mp. 50°C. ; $ZnSO_4$; Sg. 3.49 [826
ZINKSULFAT (n), WASSERFREIES- : anhydrous zinc sulphate ; Sg. 3.4 827
ZINKSULFID (n) : zinc sulphide ; ZnS ; Sg. 3.98-4.06 ; Mp. 1049°C. 828
ZINKSULFIT (n) : zinc sulphite ; $ZnSO_3.2H_2O$ [829
ZINKSUPEROXYD (n) : zinc peroxide ; ZnO_2 830
ZINKTITANAT (n) : zinc titanate ; $ZnTiO_3$; Sg. 3.17 831
ZINKTRIPELSALZ (n) : zinc triple salt 832
ZINK (n), ÜBERMANGANSAURES- : zinc permanganate (see ZINKPERMANGANAT) 833
ZINKVERBINDUNG (f) : zinc compound 834
ZINKVERGIFTUNG (f) : zinc poisoning (Med.) 835
ZINKVITRIOL (m) : zinc sulphate, white vitriol, (also found in a natural state) ; $ZnSO_4$, $7H_2O$; Sg. 2.05 (see also ZINKSULFAT) 836
ZINKVITRIOLANLAGE (f) : white vitriol plant 837
ZINK (n), WEINSAURES- : zinc tartrate 838
ZINKWEISS (n) : zinc white ; zinc oxide ; ZnO (see ZINKOXYD) 839
ZINKWERK (n): zinc works 840
ZINKWOLLE (f) : flowers of zinc (see ZINKOXYD) 841
ZINKZINNAMALGAM (n) : zinc-tin amalgam (zinc, tin and mercury in proportion of 25 : 25 : 50) 842
ZINKZITRAT (n) : zinc citrate ; $Zn_3(C_6H_5O_7)_2.2H_2O$ 843
ZINN (n) : tin ; (Sn) ; (*Stannum*) 844
ZINNACETAT (n) : tin acetate (see STANNO-AZETAT) 845
ZINNADER (f) : tin (vein) lode 846
ZINNAFTER (m) : tin ore refuse 847
ZINNAMALGAM (n) : tin amalgam (a mercury-tin alloy) 848
ZINNAMMONIUMCHLORID (n) : pink salt ; chloride of tin and ammonia (see PINKSALZ) 849
ZINNARTIG : stannous, tin-like 850
ZINNASCHE (f) : tin ash, stannic anhydride ; SnO_2 ; Sg. 6.75 ; Mp. 1127°C. 851
ZINNBAD (n) : tin bath 852
ZINNBAUM (m) : Arbor Jovis, tin tree (Chem.) 853
ZINNBEIZE (f) : tin (spirit) mordant 854
ZINNBERGWERK (n) : tin mine 855
ZINNBLATT (n) : tin foil 856
ZINNBLECH (n) : sheet tin ; tin plate 857
ZINNBLENDE (f) : tin blende 858
ZINNBLUMEN (f.pl) : flowers of tin (see ZINN-ASCHE) 859
ZINNBROMID (n) : tin (stannic) bromide ; $SnBr_4$; Sg. 3.322 860
ZINNBROMÜR (n) : stannous bromide ; tin dibromide ; $SnBr_2$; Sg. 5.117 861
ZINNBROMWASSERSTOFFSAUER : bromostannate (stannibromide) of ; combined with bromostannic acid 862
ZINNBROMWASSERSTOFFSÄURE (f) : bromostannic acid 863
ZINNBRONZE (f) : (see ZINNDISULFID) 864
ZINNBUTTER (f) : butter of tin ; stannic chloride (see ZINNCHLORID and STANNICHLORID) 865
ZINNCHLORID (n) : tin (stannic) chloride ; tin tetrachloride ; $SnCl_4$; Sg. 2.279 ; Mp. -33°C. ; Bp. 114°C. (used for silk dyeing) 866
ZINNCHLORÜR (n) : stannous chloride ; tin (dichloride ; protochloride) salt ; $SnCl_2.2H_2O$; Sg. 2.7 (see STANNOCHLORID) 867
ZINNCHLORWASSERSTOFFSAUER : chlorostannate (stannichloride) of ; combined with chlorostannic acid 868
ZINNCHLORWASSERSTOFFSÄURE (f) : chlorostannic acid ; $H_2SnCl_6.6H_2O$. ; Sg. 1.925 [869
ZINNCHROMAT (n) : (see STANNICHROMAT and STANNOCHROMAT) 870
ZINNDIBROMID (n) : tin dibromide, stannous bromide ; $SnBr_2$; Sg. 5.117 871
ZINNDICHLORID (n) : tin dichloride ; stannous chloride ; $SnCl_2+2H_2O$; Sg. 2.7 872
ZINNDIOXYD (n) : tin dioxide (see ZINNASCHE) 873
ZINNDIPHENYLCHLORID (n) : diphenylstannic chloride 874
ZINNDISULFID (n) : mosaic gold ; artificial gold ; stannic sulphide ; SnS_2 ; Sg. 4.5 875

ZINNDOSE (*f*): tin box 876

ZINNDRAHT (*m*): tin wire 877

ZINNE (*f*): pinnacle; battlement; point 878

ZINNEN: to tin; tin; of tin; pewter 879

ZINNENFÖRMIG: crenated, scalloped, crenellated, turretted, battlemented, pinnacled, pointed 880

ZINNERN: tin; pewter; of (pewter) tin 881

ZINNERZ (*n*): cassiterite; tin ore (see **ZINNOXYD**); tin stone (see **ZINNSTEIN** and **AUSSTRICH**) 882

ZINN (*n*), **ESSIGSAURES-**: tin acetate (see **STANNO-AZETAT**) 883

ZINNFEILICHT (*n*): tin filings 884

ZINNFEILSPÄNE (*m.pl*): tin filings 885

ZINNFOLIE (*f*): tin foil 886

ZINNGEKRÄTZ (*n*): tin (sweepings) refuse; tin dross 887

ZINNGERÄT (*n*): tin (pots and pans) vessels; pewter 888

ZINNGESCHIRR (*n*): (see **ZINNGERÄT**) 889

ZINNGESCHREI (*n*): cry (crackling) of tin; tin-cry (due to crystalline structure) 890

ZINNGIESSER (*m*): tin founder; pewterer 891

ZINNGLASIERT: tin-glazed 892

ZINNGLASUR (*f*): tin (glazing) glaze 893

ZINNGRUBE (*f*): tin mine 894

ZINNGRÜN (*n*): tin green (see **GENTELES GRÜN**) 895

ZINNHALTIG: stanniferous; containing tin 896

ZINNHYDROXYD (*n*): tin (stannic) hydroxide 897

ZINNHYDROXYDUL (*n*): stannous hydroxide 898

ZINNJODID (*n*): tin (stannic) iodide; SnI_4; Sg. 4.696 899

ZINNJODÜR (*n*): stannous iodide; SnI_2 900

ZINNKALK (*m*): calcined tin 901

ZINNKIES (*m*): stannite; tin pyrites (see **STANNIN**); $Cu_2S.FeS.SnS_2$.; Sg. 4.4 902

ZINNKNIRSCHEN (*n*): cry (crackling) of tin; tin-cry (see **ZINNGESCHREI**) 903

ZINNKRÄTZE (*f*): tin dross 904

ZINNKREISCHEN (*n*): (see **ZINNKNIRSCHEN**) 905

ZINN (*n*), **KRYSTALL-**: (see **STANNOCHLORID**) 906

ZINNLEGIERUNG (*f*): tin alloy 907

ZINNLÖSUNG (*f*): tin solution; tin spirit (Dyeing) 908

ZINNLOT (*n*): tin solder 909

ZINN (*n*), **MILCHSAURES-**: tin lactate 910

ZINNMONOXYD (*n*): (see **STANNOOXYD**) 911

ZINNOBER (*m*): cinnabar; natural mercury sulphide; HgS; Sg. 8.15 (see **QUECK-SILBERSULFID**); vermilion (see **CHINES-ISCHROT**) 912

ZINNOBERERZ (*n*): cinnabar (natural mercury sulphide) (see **ZINNOBER**) 913

ZINNOBER (*m*), **GEMAHLENER-**: vermilion (see **ZINNOBER**) 914

ZINNOBER (*m*), **GRÜNER-**: (see **CHROMGRÜN**) 915

ZINNOBERROT (*n*): vermilion (see **ZINNOBER**) 916

ZINNOBERSPAT (*m*): crystallized cinnabar (see **ZINNOBERERZ**) 917

ZINNOFEN (*m*): tin furnace 918

ZINN (*n*), **OXALSAURES-**: tin oxalate (see **STANNO-OXALAT**) 919

ZINNOXYD (*n*): tin (stannic) oxide; SnO_2; Sg. 6.95 (see **STANNIOXYD**) 920

ZINNOXYDHYDRAT (*n*): tin hydroxide (stannic or stannous hydroxide) 921

ZINNOXYDNATRON (*n*): sodium stannate (see **NATRIUMSTANNAT**) 922

ZINNOXYDUL (*n*): stannous oxide; SnO; Sg. 6.3 (see **STANNOOXYD**) 923

ZINNOXYDULSALZ (*n*): stannous salt 924

ZINNOXYDULVERBINDUNG (*f*): stannous compound 925

ZINNOXYDVERBINDUNG (*f*): stannic compound 926

ZINNPAUSCHE (*f*): tin slag 927

ZINNPEROXYD (*n*): (see **ZINNASCHE**) 928

ZINNPEST (*f*): tin plague (decomposition of tin to a powder on organ pipes and tinware in unheated rooms) 929

ZINNPFANNE (*f*): tin pot (Tinning) 930

ZINNPHOSPHID (*n*): tin phosphide; SnP_3; Sg. 4.1; Sn_4P_3; Sg. 5.18 931

ZINNPLATTE (*f*): tin (sheet) plate 932

ZINNPRÄPARAT (*n*): tin preparation 933

ZINNPROBE (*f*): tin sample; tin (assay) test 934

ZINNPROTOXYD (*n*): tin protoxide (see **STANNO-OXYD**) 935

ZINNQUECKSILBER (*n*): (see **ZINNAMALGAM**) 936

ZINN (*n*), **RAFFINIERTES-**: refined tin (99.6-99.9% tin) 937

ZINNREICH: rich in tin 938

ZINNRHODANÜR (*n*): tin sulfocyanide 939

ZINNROHR (*n*): tin pipe 940

ZINNRÖHRE (*f*): tin pipe 941

ZINN (*n*), **SALPETERSAURES-**: tin nitrate 942

ZINNSALZ (*n*): tin salt; tin dichloride; stannous chloride; $SnCl_2.2H_2O$ (see **ZINNCHLORÜR**) 943

ZINNSAND (*m*): granular tin 944

ZINNSAUER: stannate of; combined with stannic acid 945

ZINNSAUM (*m*): selvedge; list of tin 946

ZINNSÄURE (*f*): stannic acid; H_2SnO_3 947

ZINNSÄUREANHYDRID (*n*): stannic anhydride; tin dioxide (see **STANNIOXYD**) 948

ZINNSÄURE, META- (*f*): metastannic acid; $SnO(OH)_2$ 949

ZINNSÄURE, ORTHO- (*f*): orthostannic acid; $Sn(OH)_4$ 950

ZINNSEIFE (*f*): stream tin (see **STROMZINN**) 951

ZINNSELENÜR (*n*): stannous selenide; SnSe; Sg. 6.179 952

ZINNSODA (*f*): sodium stannate (see **NATRIUM-STANNAT**) 953

ZINNSTEIN (*m*): cassiterite (78.6% tin) (SnO_2); tin stone, tin ash, stannic oxide, stannic anhydride, flowers of tin, tin peroxide, tin dioxide; Sg. 6.6-6.9 (see **ZINNOXYD**) 954

ZINNSULFAT (*n*): (see **STANNOSULFAT**) 955

ZINNSULFID (*n*): tin (stannic) sulphide; SnS_2; Sg. 4.51 (see **STANNISULFID**) 956

ZINNSULFÜR (*n*): stannous sulphide; SnS; Sg. 5.03-5.08 957

ZINNSUPEROXYD (*n*): (see **ZINNASCHE**) 958

ZINNTARTRAT (*n*): (see **STANNOTARTRAT**) 959

ZINNTELLURÜR (*n*): stannous telluride; SnTe; Sg. 6.478 960

ZINNTETRABROMID (*n*): tin tetrabromide; stannic bromide; $SnBr_4$; Sg. 3.322-3.349 961

ZINNTETRACHLORID (*n*): tin tetrachloride; stannic chloride; $SnCl_4$; Sg. 2.229-2.2788 962

ZINNTETRAFLUORID (*n*): tin tetrafluoride; stannic fluoride; SnF_4; Sg. 4.78 963

ZINNTETRAJODID (*n*): tin tetraiodide, stannic iodide; SnJ_4; Sg. 4.696 964

ZINNVERBINDUNG (*f*): tin compound 965

ZINNVERGIFTUNG (*f*): tin poisoning 966

ZINNWALDIT (*m*): zinnwaldite (see also **LEPIDO-LITH**); $(Li,K)(F,OH)_2FeAl_3Si_5O_{16}$; Sg. 3.0 967

ZINNWAREN (*f.pl*): tin-ware 968
ZINNWÄSCHE (*f*): buddled tin 969
ZINS (*m*): rent; due; tribute; (*pl*): interest 970
ZINSBAR: tributary, liable to pay rent or interest 971
ZINSBUCH (*n*): rent-roll; rental 972
ZINSEN: to pay (yield) interest or rent 973
ZINSESZINSEN (*pl*): compound interest 974
ZINSFREI: rent-free, free of interest 975
ZINSFUSS (*m*): rate of interest, discount 976
ZINSPFLICHTIG: tributary, liable to pay rent or interest 977
ZINSTAG (*m*): rent-day 978
ZINSUNTERSCHIED (*m*): difference of interest 979
ZIPFEL (*m*): tip; end; point; top; lappet; corner 980
ZIPFELIG: having (tips; ends; points) corners 981
ZIRBELDRÜSE (*f*): pineal body (see EPIPHYSE) 982
ZIRBELFICHTE (*f*): Swiss stone pine (*Pinus cembra*) 983
ZIRBELKIEFER (*f*): (see ZIRBELFICHTE) 984
ZIRBELNUSS (*f*): cedar nut (edible nut from Swiss stone pine, *Pinus cembra*) 985
ZIRKA: about; almost; approximately 986
ZIRKEL (*m*): pair of compasses; circle 987
ZIRKELBEIN (*n*): leg of compasses 988
ZIRKELFÖRMIG: circular 989
ZIRKELIT (*m*): zirkelite; $(Ca,Fe)(Th,Ti,Zr)_2O_5$; Sg. 4.74 990
ZIRKELLINIE (*f*): circular line; curve 991
ZIRKELN: to circle; measure with a pair of compasses 992
ZIRKELSPITZE (*f*): point of a pair of compasses 993
ZIRKEL (*m*), STANGEN-: beam compasses 994
ZIRKON (*n*): zirconium, (Zr); zircon (Mineral) (natural zirconium silicate), $ZrO_2.SiO_2$, Sg. 4.4 995
ZIRKONACETAT (*n*), BASISCHES-: basic zirconium acetate; $Zr(C_2H_3O_2)_3.OH$ 996
ZIRKONCARBID (*n*): zirconium carbide; ZrC and ZrC_2 997
ZIRKONCARBONAT (*n*), BASISCHES-: basic zirconium carbonate; $3ZrO_2CO_2.6H_2O$ 998
ZIRKONCHLORID (*n*): zirconium (tetra)-chloride; $ZrCl_4$; Bp. 400°C. 999
ZIRKONCHLORID (*n*), BASISCHES-: (see ZIRKON-OXYCHLORID) 000
ZIRKONDIOXYD (*n*): zirconium dioxide; ZrO_2 001
ZIRKONERDE (*f*): zirconia; zirconium oxide; ZrO_2 (see ZIRKON) 002
ZIRKONFLUORID (*n*): zirconium fluoride; ZrF_4; Sg. 4.433 003
ZIRKONGLAS (*n*): zirconium glass 004
ZIRKONHYDROXYD (*n*): zirconium hydroxide; $Zr(OH)_4$; Sg. 3.25 005
ZIRKONIUM (*n*): (see ZIRKON) 006
ZIRKONIUMVERBINDUNG (*f*): zirconium compound 007
ZIRKONLICHT (*n*): zircon light 008
ZIRKONNITRAT (*n*): zirconium nitrate; $Zr(NO_3)_4.5H_2O$ 009
ZIRKONORTHOPHOSPHAT (*n*): (see ZIRKON-PHOSPHAT) 010
ZIRKONOXYCHLORID (*n*): zirconium oxychloride, basic zirconium chloride; $ZrOCl_2.8H_2O$ 011
ZIRKONOXYD (*n*): zirconium oxide; zirconic anhydride; ZrO_2; Sg. 5.489-5.75; Mp. 2500°C. 012
ZIRKONPHOSPHAT (*n*): zirconium (ortho)-phosphate, basic zirconium phosphate; $5ZrO_2(P_2O_5)_4.8H_2O$ 013
ZIRKONPHOSPHAT (*n*), BASISCHES-: (see ZIRKON-PHOSPHAT) 014
ZIRKONPHOSPHID (*n*): zirconium phosphide; ZrP_2; Sg. 4.77 015
ZIRKONPRÄPARAT (*n*): zirconium preparation 016
ZIRKON (*n*), SALPETERSAURES-: zirconium nitrate (see ZIRKONNITRAT) 017
ZIRKONSÄURE (*f*): zirconic acid; Baddeleyite (see BADDELEYIT) 018
ZIRKONSILICID (*n*): zirconium silicide; $ZrSi_2$; Sg. 4.88 019
ZIRKONSTAHL (*m*): zirconium steel 020
ZIRKONSULFAT (*n*): zirconium sulphate; $Zr(SO_4)_2.4H_2O$ 021
ZIRKONTETRACHLORID (*n*): (see ZIRKONCHLORID) 022
ZIRKONTETRAFLUORID (*n*): zirconium tetrafluoride; ZrF_4; Sg. 4.433 023
ZIRKULAR: circular; (*n*): circular 024
ZIRKULIEREN: to circulate 025
ZIRPEN: to chirp 026
ZISCHELN: to whisper 027
ZISCHEN: to sizz-(le); hiss; fizz; whiz; (*n*): hissing; fizzing 028
ZISSOIDE (*f*): cissoid 029
ZISTERNE (*f*): cistern 030
ZITAT (*n*): citation 031
ZITIEREN: to cite; summon 032
ZITRAL (*n*): (see TERPENALDEHYD) 033
ZITRAT (*n*): citrate 034
ZITRATLÖSLICH: citrate soluble 035
ZITRIN (*n*): citrine (transparent; light yellow) (see QUARZ) 036
ZITRONAT (*n*): candied lemon peel 037
ZITRONBARTGRAS (*n*): lemon grass (*Andropogon citratus*) 038
ZITRONE (*f*): citron; lemon 039
ZITRONELLÖL (*n*): citronella oil (see CITRONEL-LAÖL) 040
ZITRONELLSÄURE (*f*): citronellic acid 041
ZITRONENÄTHER (*m*): citric ether 042
ZITRONENBAUM (*m*): lemon tree (*Citrus medica var. Acida*) 043
ZITRONENESSENZ (*f*): lemon essence 044
ZITRONENFARBE (*f*): lemon colour (see ZITRONEN-GELB) 045
ZITRONENGELB (*n*): lemon yellow, citrine, chrome yellow (see CHROMGELB and BLEI-CHROMAT); zinc yellow (see ZINKGELB) 046
ZITRONENGRAS (*n*): lemon grass; citron grass (*Andropogon schoenanthus*) 047
ZITRONENHOLZ (*n*): lemon (citron) wood; (Sg. 0.7-0.8) 048
ZITRONENKRAUT (*n*): Southernwood (*Artemisia abrotanum*) (see ZITRONENMELISSE) 049
ZITRONENMELISSE (*f*): balm-(mint), (*Melissa officinalis*) 050
ZITRONENÖL (*n*): lemon oil (*Oleum citri*) (see CEDRATÖL) 051
ZITRONENSAFT (*m*): lemon juice 052
ZITRONENSAUER: citrate of; combined with citric acid 053
ZITRONENSÄURE (*f*): citric acid; oxytricarballylic acid; $(CO_2HCH_2)_2C(OH)CO_2H$; Sg. 1.5_; Mp. 153°C. 054
ZITRONENSÄURELÖSLICH: soluble in citric acid 055
ZITRONENSCHALE (*f*): lemon peel 056
ZITRONENSCHALENÖL (*n*): lemon peel oil 057

ZITRONENWASSER (*n*): lemon water; lemonade 058
ZITRONSAUER: citrate of; combined with citric acid 059
ZITRONSÄURE (*f*): citric acid (see ZITRONENSÄURE) 060
ZITROVANILLE (*f*): citrovanilla 061
ZITRULLE (*f*): watermelon (*Citrullus vulgaris*) 062
ZITTERAAL (*m*): electric eel 063
ZITTERESPE (*f*): aspen; poplar (*Populus tremula*) 064
ZITTERN: to tremble; shake; quiver; vibrate 065
ZITTERPAPPEL (*f*): aspen (see ZITTERESPE) 066
ZITTERWURZEL: zedoary (see ZITWER) 067
ZITTWER (see ZITWER) 068
ZITWER (*m*): zedoary plant (*Curcuma zedoaria*); ginger (*Zingiber officinalis*); aconite; sweet flag (*Acorus calamus*) 069
ZITWERKRAUT (*n*): tarragon (*Artemisia dracunculus*) 070
ZITWERÖL (*n*): oil of zedoary; Sg. 0.922-1.01 071
ZITWERSAMEN (*m*): zedoary seed (see ZITWER); santonica; wormseed; Levant wormseed (*Semen cinæ*) (of *Artemisia pauciflora*) 072
ZITWERWURZEL (*f*): zedoary plant (see ZITWER) (*Rhizoma zedoariæ* of *Curcuma zedoaria*) 073
ZITZ (*m*): chintz; calico 074
ZITZE (*f*): teat; nipple 075
ZITZENFÖRMIG: mammillary 076
ZITZENTIER (*n*): mammal 077
ZIVIL: civil 078
ZIVILINGENIEUR (*m*): civil engineer 079
ZIZYPHUS (*m*): zizyphus (see JUDENDORN) 080
ZOBEL (*m*): sable 081
ZOBELPELZ (*m*): sable-(fur) 082
ZOBER (*m*): tub 083
ZOBTENFELS (*m*): gabbro (see GABBRO) 084
ZÖGERN: to delay; hesitate; tarry; wait; loiter; linger 085
ZÖGERUNG (*f*): delay; hesitation 086
ZÖGLING (*m*): pupil; scholar 087
ZOISIT (*m*): zoisite, a dimorphous modification of Epidot; $HCa_4Al_3Si_3O_{13}$; Sg. 3.3 088
ZÖLESTIN (*m*): celestite (see CÖLESTIN) 089
ZOLL (*m*): custom; duty; toll; customs (import; export) duty; inch (Measure) 090
ZOLLAMT (*n*): customhouse; Customs 091
ZOLLBAR: dutiable 092
ZOLLBEAMTE (*m*): Custom's (Revenue) officer 093
ZOLLBEAMTER (*m*): (see ZOLLBEAMTE) 094
ZOLLBEDIENTE (*m*): (see ZOLLBEAMTE) 095
ZOLLEINNEHMER (*m*): tax-gatherer; receiver (collector) of taxes; revenue officer 096
ZOLLEN: to pay duty or toll 097
ZOLLFREI: duty-free 098
ZOLLFREIHEIT (*f*): exemption from duty 099
ZOLLGEBÜHR (*n*): Custom's (duty) dues 100
ZOLLHAUS (*n*): Custom-house 101
ZOLLPFLICHTIG: dutiable 102
ZOLLRECHNUNG (*f*): Custom's invoice 103
ZOLLSATZ (*m*): tariff rate 104
ZOLLSTAB (*m*): inch rule 105
ZOLLSTOCK (*m*): carpenter's rule, foot rule 106
ZOLLTARIF (*m*): Custom's tariff 107
ZOLLVEREIN (*m*): tariff union; Custom's union; Custom's federation 108
ZOLLVERSCHLUSS (*m*): bond 109
ZONE (*f*): zone 110
ZONE (*f*), GEMÄSSIGTE-: temperate zone 111
ZONE (*f*), KALTE-: frigid zone 112
ZONE (*f*), TROPEN-: tropical zone 113
ZOOCHEMIE (*f*): zoochemistry 114
ZOOCHEMISCH: zoochemical 115
ZOOLOG (*m*): zoologist 116
ZOOLOGIE (*f*): zoology 117
ZOPF (*m*): top (of tree); tuft; tress; pigtail; cue; pedantry; formalities; ribbon; streamer (Sewage disposal) 118
ZOPFEN: to lop (off all the unsuitable wood of trees) 119
ZORGIT (*m*): zorgite; Se(Cu₂.Pb); Sg. 7.25 120
ZORN (*m*): anger; rage; wrath; passion 121
ZORNIG: angry; wrathful; passionate 122
ZÖRULEUM (*n*): ceruleum, sky-blue (a cobalt colour; cobalt stannate) 123
ZOSTER (*f*): zoster, zona (Med.) 124
ZOTTE (*f*): tuft; tangle; rag; ribbon; streamer; tail (Sewage disposal) 125
ZOTTEL (*f*): (see ZOTTE) 126
ZOTTELWOLLE (*f*): shaggy wool 127
ZOTTENHAUT (*f*): chorion (Embryology) 128
ZOTTIG: matted; shaggy; ragged 129
ZU: to; for; in; at; on; too; toward; closed 130
ZUBEHÖR (*m* & *n*): accessories; belongings; appurtenances; fittings 131
ZUBEHÖRTEILE (*m.pl*): accessories, parts, spare parts, spares 132
ZUBEKOMMEN: to get closed; to get (extra) in addition 133
ZUBENANNT: surnamed 134
ZUBER (*m*): tub 135
ZUBEREITEN: to prepare; dress; finish; adjust 136
ZUBINDEN: to bind up, tie up 137
ZUBRENNEN: to close by heating; roast; calcine; cauterize 138
ZUBRINGEN: to bring; spend; stay; add to 139
ZUBRÜHEN: to add boiling water in mashing (Brew.) 140
ZUBUSSE (*f*): contribution; supply 141
ZUBÜSSEN: to contribute; supply; spend 142
ZUCHT (*f*): breed; breeding; race; rearing; cultivation; education; culture; propriety; discipline; modesty 143
ZÜCHTEN: to grow; rear; breed 144
ZÜCHTER (*m*): grower; raiser; breeder 145
ZUCHTHAUS (*n*): house of correction 146
ZÜCHTIG: chaste; discreet; modest; virtuous 147
ZÜCHTIGEN: to punish; chastise; correct; discipline; castigate 148
ZÜCHTIGKEIT (*f*): chastity, modesty, pudicity, virtue, continence 149
ZÜCHTIGUNG (*f*): chastisement, correction, punishment, chastening, castigation 150
ZUCHTLOS: undisciplined; disorderly 151
ZUCHTLOSIGKEIT (*f*): insubordination 152
ZUCHTMEISTER (*m*): taskmaster 153
ZUCHTVIEH (*n*): cattle for breeding purposes 154
ZUCKEN: to jerk; twitch; palpitate; be spasmodic; stir; shrug (the shoulders) 155
ZÜCKEN: to draw (a sword) 156
ZUCKER (*m*): sugar; Sg. 0.61 (see ROHRZUCKER, MILCHZUCKER, FRUCHTZUCKER, etc.); (as a chemical prefix = sucrate, saccharate) 157
ZUCKERÄHNLICH: sugar-like; saccharoid 158
ZUCKERAHORN (*m*): sugar maple (*Acer saccharinum*) 159
ZUCKERART (*f*): kind (sort; type; variety) of sugar 160

ZUCKERARTIG : sugar-like ; saccharoid ; saccharine ; sugary 161
ZUCKERÄTHER (*m*) : sugar ether 162
ZUCKERAUSBEUTE (*f*) : rendement ; sugar-yield 163
ZUCKERBÄCKER (*m*) : confectioner 164
ZUCKERBÄCKEREI (*f*) : confectionery ; confectioner's shop 165
ZUCKERBACKWERK (*n*) : confectionery ; sweetmeat ; pastry 166
ZUCKERBARYT (*n*) : barium (sucrate) saccharate 167
ZUCKERBAU (*m*) : sugar-cane cultivation 168
ZUCKERBESTIMMUNG (*f*) : sugar determination 169
ZUCKERBILDUNG (*f*) : saccharification ; formation of sugar ; glycogenesis 170
ZUCKERBROT (*n*) : sugar-loaf ; sweet bread 171
ZUCKERBÜCHSE (*f*) : sugar-basin ; (sugar)-caster 172
ZUCKERBUSCH (*m*) : sugar bush ; honey flower (*Protea mellifera*) 173
ZUCKERCOULEUR (*f*) : sugar colour, caramel, burnt sugar, sugar colouring 174
ZUCKERDICKSAFT (*m*) : treacle ; molasses (see MELASSE) 175
ZUCKERDOSE (*f*) : sugar-basin ; (sugar)-caster 176
ZUCKERERBSE (*f*) : sugar-plum ; sugar-pea (Bot.) (*Pisum sativum saccharatum*) 177
ZUCKERERDE (*f*) : sugar boiler's clay ; animal charcoal for sugar refining 178
ZUCKERERZEUGUNG (*f*) : sugar production or output 179
ZUCKERFABRIK (*f*) : sugar (factory) refinery 180
ZUCKERFABRIKANT (*m*) : sugar refiner 181
ZUCKERFORM (*f*) : sugar mould 182
ZUCKERGÄRUNG (*f*) : sugar fermentation 183
ZUCKERGEBACKENE (*n*) : sweetmeats 184
ZUCKERGEHALT (*m*) : sugar content 185
ZUCKERGEHALTMESSER (*m*) : saccharimeter (a polariscope to test solutions of sugar by means of polarized light) 186
ZUCKERGEHALTSWAGE (*f*) : saccharometer (a hydrometer for determining the specific gravity of sugar solutions) 187
ZUCKERGEIST (*m*) : rum 188
ZUCKERGEWINNUNG (*f*) : sugar (manufacture) extraction 189
ZUCKERGUSS (*m*) : sugar icing 190
ZUCKERHALTIG : saccharine ; saccharated ; containing sugar 191
ZUCKERHANDEL (*m*) : sugar trade 192
ZUCKERHARNEN (*n*) : glycosuria ; sugary condition of urine (Pathology) 193
ZUCKERHARNRUHR (*f*) : diabetes-(mellitus) ; sweet diabetes ; grape sugar in the urine (Med.) 194
ZUCKERHEFEN (*f.pl*) : sugar refuse 195
ZUCKERHIRSE (*f*) : sugar grass ; sweet shaloo (*Sorghum saccharatum*) 196
ZUCKERHONIG (*m*) : treacle ; molasses (see MELASSE) 197
ZUCKERHUT (*m*) : sugar-loaf 198
ZUCKERIG : saccharine ; sugary 199
ZUCKERIN (*n*) : saccharin (see SACCHARIN) 200
ZUCKERINDUSTRIE (*f*) : sugar industry 201
ZUCKERKALK (*m*) : sugar-lime ; calcium (sucrate) saccharate 202
ZUCKERKALKENTWICKLER (*m*) : calcium sucrate (calcium saccharate) developer 203
ZUCKERKAND (*m*) : sugar-candy 204
ZUCKERKANDIS (*m*) : sugar candy 205
ZUCKERKESSEL (*m*) : sugarboiler 206
ZUCKERKISTENHOLZ (*n*) : Havana cedar wood ; sugar box-wood (*Cedrela guianensis*) 207
ZUCKERKOHLE (*f*) : sugar charcoal 208
ZUCKERKONVENTION (*f*) : sugar convention 209
ZUCKERKORN (*n*) : grain of sugar 210
ZUCKERKOULEUR (*f*) : (see ZUCKERCOULEUR) 211
ZUCKERKRANKE (*m*) : diabetic (Med.) 212
ZUCKERKRANKHEIT (*f*) : diabetes ; diabetes mellitus (see ZUCKERHARNRUHR) 213
ZUCKER, KRYSTALL- (*m*) : crystallized sugar 214
ZUCKERKÜPE (*f*) : sugar vat 215
ZUCKERLOSE HARNRUHR (*f*) : polyuria, diabetes insipidus (Med.) 216
ZUCKERLÖSUNG (*f*) : sugar solution ; syrup 217
ZUCKERMASCHINEN (*f.pl*) : sugar machinery 218
ZUCKERMEHL (*n*) : powdered sugar 219
ZUCKERMELDE (*f*) : garden orache ; mountain spinach (*Atriplex hortensis*) 220
ZUCKERMELONE (*f*) : sweet melon (*Cucumis melo*) 221
ZUCKERMESSER (*m*) : saccharimeter 222
ZUCKERMESSKUNST (*f*) : saccharimetry 223
ZUCKERMESSUNG (*f*) : saccharimetry 224
ZUCKERMILBE (*f*) : sugar louse (*Glyciphagus sacchari*) 225
ZUCKERMÜHLE (*f*) : sugar mill 226
ZUCKERN : to sugar ; sweeten ; candy ; edulcorate 227
ZUCKER (*m*), NITRIERTER- : nitro-glucose (see NITROGLUKOSE) 228
ZUCKERPALME (*f*) : sugar palm (*Arenga saccharifera*) 229
ZUCKERPFLANZUNG (*f*) : sugar plantation ; sugar estate 230
ZUCKERPILZ (*m*) : amadou (*Polyporus fomentarius*) 231
ZUCKERPLANTAGE (*f*) : (see ZUCKERPFLANZUNG) 232
ZUCKERPLÄTZCHEN (*n*) : small sugared bun or cake 233
ZUCKERPRÄMIE (*f*) : sugar bounty 234
ZUCKERPROBE (*f*) : sugar (sample) test 235
ZUCKERRAFFINERIE (*f*) : sugar refinery 236
ZUCKERROHR (*n*) : sugar-cane (*Saccharum officinarum*) 237
ZUCKERRÖHRCHEN (*n*) : sugar tube (small funnel used for sugar determination) 238
ZUCKERROHR, CHINESISCHES- (*n*) : sugar grass (*Sorghum saccharatum*) (see ZUCKERHIRSE) 239
ZUCKERROHR-RÜCKSTÄNDE (*m.pl*) : megass ; bagasse (the waste sugar-cane, which is used as fuel) 240
ZUCKERROHRSAFT (*m*) : cane juice 241
ZUCKERROSE (*f*) : red rose 242
ZUCKERRÜBE (*f*) : sweet turnip ; beetroot ; sugar beet (*Beta vulgaris*) 243
ZUCKERRÜBENESSIG (*m*) : beetroot ((sugar)-beet) vinegar 244
ZUCKERRÜBENMELASSE (*f*) : beetroot (beet) molasses 245
ZUCKERRÜBENSAFT (*m*) : beetroot ((sugar)-beet) juice 246
ZUCKERRÜBENZUCKER (*m*) : beet sugar (see ROHR-ZUCKER) 247
ZUCKERRUHR (*f*) : diabetes ; diabetes mellitus (see ZUCKERHARNRUHR) 248
ZUCKERSAFT (*m*) : sugar cane juice or syrup 249
ZUCKERSATZ (*m*) : molasses (see MELASSE) 250
ZUCKERSAUER : saccharate (oxalate) of ; combined with saccharic (oxalic) acid 251

ZUCKERSÄURE (*f*): saccharic (oxalic) acid; $C_6H_{10}O_8$ 252
ZUCKERSAURES SALZ (*n*): sucrate; saccharate 253
ZUCKERSCHALE (*f*): sugar-basin 254
ZUCKERSCHAUM (*m*): powdered animal charcoal (see ZUCKERERDE) 255
ZUCKERSCHLAMM (*m*): lime scum (Beet sugar) 256
ZUCKERSCHLEUDER (*f*): sugar (centrifugal) centrifuge 257
ZUCKERSCHOTENBAUM (*m*): honey locust (*Gleditschia triacanthos*) 258
ZUCKERSIEDEN (*n*): sugar (refining) boiling 259
ZUCKERSIEDER (*m*): sugar (refiner) boiler 260
ZUCKERSIEDEREI (*f*): sugar refinery 261
ZUCKERSIRUP (*m*): molasses (see MELASSE) 262
ZUCKERSIRUP (*m*), WEISSER-: (*Sirupus simplex*) white sugar sirup 263
ZUCKERSTEIN (*m*): granular albite; NaO. $Si_3O_2+AlO_3.Si_3O_2$; Sg. 2.59 264
ZUCKERSTOFF (*m*): saccharine-(matter) (see also SACCHARIN) 265
ZUCKERSTRONTIAN (*m*): strontium (sucrate) saccharate 266
ZUCKERTINKTUR (*f*): (see ZUCKERCOULEUR) 267
ZUCKERUNTERSUCHUNG (*f*): sugar (examination; testing) analysis 268
ZUCKERUNTERSUCHUNGSAPPARAT (*m*): sugar testing apparatus 269
ZUCKERVERBINDUNG (*f*): saccharate; sugar compound; sucrate 270
ZUCKERWERK (*n*): sweetmeats; confectionery 271
ZUCKERWURZEL (*f*): skirret (*Sium sisarum*) 272
ZUCKERWURZEL (*f*), ARABISCHE-: chufa-(tuber) (*Cyperus esculentus*) 273
ZUCKERZANGE (*f*): sugar-tongs 274
ZUCKRIG: saccharine; sugary 275
ZUCKUNG (*f*): twitch; spasm; convulsion; fit 276
ZUDÄMMEN: to dam-(up) 277
ZUDECKEN: to close-(up); cover 278
ZUDEM: besides; furthermore; moreover 279
ZUDRÄNGEN: to crowd; throng; intrude; squeeze (force) one's way towards 280
ZUDREHEN: to screw up; shut (by means of a handwheel, or by any other means involving a rotary motion) 281
ZUDRINGLICH: intrusive; obtrusive; importunate 282
ZUDRINGLICHKEIT (*f*): importunity 283
ZUDRÜCKEN: to close (by the application of pressure); shut; push (force; press) together 284
ZUEIGNEN: to dedicate; arrogate; appropriate; attribute; assume 285
ZUEIGNUNG (*f*): appropriation; dedication 286
ZUEILEN: to hurry (hasten, fly, rush, dash, scurry) towards 287
ZUERKENNEN: to award; adjudge; allow; admit; adjudicate; confer upon 288
ZUERKENNUNG (*f*): award; adjudication; judgment 289
ZUERST: first of all; at first; above all; first-(ly) 290
ZUFALL (*m*): accident; chance; incident; hazard 291
ZUFALLEN: to fall to the lot of; be affected by; shut; fall upon 292
ZUFÄLLIG: accidental; casual; incidental; chance; unessential 293
ZUFÄLLIGERWEISE: as it happens; by accident 294
ZUFÄLLIGKEIT (*f*): accidental occurrence, contingency, incident 295
ZUFLIESSEN: to flow (in) towards 296
ZUFLUCHT (*f*): recourse; refuge; shelter 297
ZUFLUCHTSORT (*m*): place of refuge, shelter, asylum 298
ZUFLUCHTSSTÄTTE (*f*): place of refuge; shelter; asylum 299
ZUFLUSS (*m*): flow; influx; affluence; confluence; flux; resources; supply; tributary (of rivers) 300
ZUFLUSSROHR (*n*): supply (feed) pipe 301
ZUFLÜSTERN: to whisper to 302
ZUFOLGE: owing to; in consequence of; according to; due to; pursuant to 303
ZUFÖRDERN: to supply (forward) to 304
ZUFRIEDEN: satisfied; contented; happy; pleased 305
ZUFRIEDENHEIT (*f*): contentment; satisfaction; pleasure; happiness 306
ZUFRIEDENSTELLEN: to satisfy; content 307
ZUFRIEDENSTELLEND: satisfactory 308
ZUFRIEREN: to freeze-(up); congeal 309
ZUFÜGEN: to inflict; cause (a wound); add; do 310
ZUFÜGUNG (*f*): infliction; addition 311
ZUFUHR (*f*): supply; delivery; supplies; importation; conveyance; transport; addition 312
ZUFÜHREN: to supply; lead to; convey; transport; feed; import; conduct; add; bring 313
ZUFÜHRUNG (*f*): supply; inlet; feed; conduction; conveyance; lead (Elect.) (see ZU-FÜHREN) 314
ZUFÜHRUNGSDRAHT (*m*): supply wire, lead (Elect.) 315
ZUG (*m*): expedition; lineament; flock; troop; crowd; herd; (tractive)-stress; tensile-(strain); move; impulse; draught; train; disposition; procession; traction; pull; tug; drawing; passage; progress; motion; range (of mountains); stretch (of leather); trait; feature; line; stroke; pulley; piston; flue; duct; pass (of boilers) 316
ZUGABE (*f*): extra; surplus; addition; supplement 317
ZUGANG (*m*): access; admittance; entrance; avenue; admission; approach 318
ZUGÄNGLICH: approachable; accessible 319
ZUGÄNGLICH, SCHWER-: difficult of access; inaccessible 320
ZUGBEANSPRUCHUNG (*f*): tractive (strain) stress, tensile stress, tensile 321
ZUGBELASTUNG (*f*): traction load 322
ZUGBOLZEN (*m*): set pin 323
ZUGBRÜCKE (*f*): draw-bridge 324
ZUGEBEN: to give in; add; admit; grant; permit; allow; consent; concede 325
ZUGEGEN: present 326
ZUGEHEN: to approach; go (on) up to; happen; come to pass; meet; arrive; come; close; be received; come to hand 327
ZUGEHÖREN: to belong to; appertain to 328
ZUGEHÖRIG: belonging (pertaining; appertaining) to; appropriate; proper; accessory; accompanying 329
ZÜGEL (*m*): bridle; rein-(s) 330
ZUGELASTIZITÄT (*f*): elasticity under traction 331
ZÜGELLOS: unbridled, unrestricted, unchecked, licentious, dissolute 332

ZÜGELN : to bridle, check, curb, restrict 333
ZUGEMÜSE (*n*) : greens, vegetables ; (*pl*) : herbs 334
ZUGENANNT : surnamed 335
ZUGESELLEN : to associate with, mix with, accompany 336
ZUGESPITZT : tapered ; acuminate ; pointed 337
ZUGESTÄNDNIS (*n*) : concession, admission 338
ZUGESTEHEN : to grant ; admit ; concede ; permit ; agree ; allow ; consent 339
ZUGETAN : devoted ; attached 340
ZUGFEDER (*f*) : draw spring 341
ZUGFESTIGKEIT (*f*) : tensile strength ; tenacity 342
ZUGGARN (*n*) : drag-net 343
ZUGHAKEN (*m*) : coupling hook (on Railway coaches and trucks) 344
ZUGIESSEN : to pour (in ; on) to ; add ; fill up 345
ZÜGIG : pliant ; flexible (Leather) 346
ZUG, KAMIN- (*m*) : chimney (natural) draught 347
ZUGKANAL (*m*) : flue ; air-duct 348
ZUGKLAPPE (*f*) : damper (in flues) 349
ZUGKNOPF (*m*) : pull contact, draw knob 350
ZUGKONTAKT (*m*) : draw knob, pull contact 351
ZUGKRAFT (*f*) : traction ; tension ; tractive (force) power ; attraction 352
ZUG, KÜNSTLICHER- : mechanical draught 353
ZUGLEICH : together ; at the same time ; simultaneously ; at once 354
ZUGLOCH (*n*) : draught (vent ; air) hole or opening 355
ZUGMESSER (*m*) : draught gauge ; blast (meter ; indicator) gauge 356
ZUGMITTEL (*n*) : attraction ; vesicant ; vessicatory (Med.) 357
ZUGMUFFEL (*f*) : continuous muffle (Ceramics) 358
ZUG, NATÜRLICHER- : natural draught 359
ZUGOFEN (*m*) : wind furnace 360
ZUGÖFFNUNG (*f*) : draught (vent ; air) hole or opening 361
ZUGPFERD (*n*) : draught-horse 362
ZUGPFLASTER (*n*) : vessicatory ; blistering plaster 363
ZUGREGLER (*m*) : draught regulator ; damper 364
ZUGREIFEN : to fall to (Eating) ; help oneself ; seize ; grasp ; lay (take) hold of 365
ZUGROHR (*n*) : draught (vent ; air) pipe 366
ZUGRUNDELEGEN : to base (upon) on 367
ZUGRUNDELEGUNG (*f*) : basis ; the basing upon 368
ZUGRUNDELEGUNG, UNTER- (*f*) : basing upon ; on the basis of 369
ZUGSALBE (*f*) : resin cerate 370
ZUG (*m*), SAUG- : suction (induced) draught 371
ZUGSEIL (*n*) : (hauling)-rope or cable ; hawser 372
ZUGSPANNUNG (*f*) : tractive force ; tensile stress 373
ZUGSTANGE (*f*) : draw-bar, drawing bar, tie rod ; piston rod ; valve rod 374
ZUGSTÄRKE (*f*) : draught 375
ZUGUSS (*m*) : addition ; infusion 376
ZUGUTEKOMMEN : to be benefited by 377
ZUGUTEMACHEN : to work-(up) (of ores, etc.) 378
ZUGUTERLETZT : finally ; lastly 379
ZUGVOGEL (*m*) : bird of passage ; migratory bird 380
ZUGWAGEN (*m*) : tractor 381
ZUGWEISE : in (flocks ; crowds ; troops) herds 382
ZUHAUEN : to rough-hew ; fight tooth and nail ; hit hard 383
ZUHEILEN : to heal (close) up ; consolidate (Med.) 384
ZUHEILUNG (*f*) : consolidation (Med.) 385
ZUHILFENAHME (*f*) : aid ; employment ; utilization 386
ZUHORCHEN : to listen ; attend ; pay attention ; hearken 387
ZUHÖREN : to listen ; attend ; pay attention ; hearken 388
ZUHÖRER (*m*) : auditor ; hearer ; (*pl*) : audience 389
ZUHÖRERSCHAFT (*f*) : audience 390
ZUKITTEN : to cement up 391
ZUKLEBEN : to stick (glue ; gum) up 392
ZUKLEISTERN : to paste (stick) up 393
ZUKLINKEN : to latch 394
ZUKNÖPFEN : to button up 395
ZUKNÜPFEN : to tie up 396
ZUKOMMEN : to become ; come (up) to ; be due ; belong to ; be (suitable) fit ; fall to one's (share) lot ; behove 397
ZUKUNFT (*f*) : future ; futurity 398
ZUKÜNFTIG : next ; to come ; future 399
ZULAGE (*f*) : increase ; addition ; allowance 400
ZULANGEN : to suffice ; help oneself ; reach 401
ZULÄNGLICH : sufficient ; enough ; competent 402
ZULÄNGLICHKEIT (*f*) : sufficiency ; competence 403
ZULASSEN : to admit ; allow ; grant ; permit ; leave (shut ; unopened) closed ; turn on supply of ; leave room for 404
ZULÄSSIG : admissible ; allowable ; permissible 405
ZULÄSSIGE BELASTUNG (*f*) : admissible (safe) load 406
ZULÄSSIGKEIT (*f*) : admissibility 407
ZULASSUNG (*f*) : admission 408
ZULAUF (*m*) : inlet, supply ; concourse, crowd 409
ZULAUFEN : to run ; flock ; crowd ; flow towards ; run towards ; gravitate 410
ZULAUFLEITUNG (*f*) : inlet (duct) piping 411
ZULEGEN : to put to ; add ; gain (in weight) ; take ; procure ; get ; cover ; attribute ; ascribe ; impute ; assign 412
ZULEIMEN : to glue (gum) up ; stick up 413
ZULEITEN : to lead (conduct ; supply) to 414
ZULEITUNG (*f*) : supply-(ing) ; conduction ; lead ; supply-pipe 415
ZULEITUNGSROHR (*n*) : inlet (supply) pipe ; feed pipe 416
ZULETZT : finally ; lastly ; at last 417
ZULÖTEN : to solder up 418
ZUMACHEN : to shut ; close 419
ZUMAL : especially ; particularly ; chiefly ; seeing that 420
ZUMAUERN : to brick (up) in ; wall up ; set-(boilers) 421
ZUMEIST : mostly ; for the most part 422
ZUMESSEN : to attribute ; ascribe ; impute 423
ZUMISCHEN : to admix ; mix with ; (*n*) : admixture 424
ZUMISCHSTOFF (*m*) : admixture 425
ZUMUTEN : to expect (from) ; require (from) ; demand (from) 426
ZUMUTUNG (*f*) : imputation ; requirement ; desire ; expectation ; demand 427
ZUNÄCHST : first (of all) ; nearest ; next ; next of all 428
ZUNAGELN : to nail up 429

ZUNÄHEN: to stitch (sew) up 430
ZUNAHME (*f*): increase; improvement; progress-(ion); increment 431
ZUNAHMEVORSCHRIFTEN (*f.pl*): instructions for the permissible increase (Steel tests) 432
ZUNAME (*m*): surname 433
ZÜNDAKKUMULATOR (*m*): ignition accumulator 434
ZÜNDAPPARAT (*m*): primer; priming apparatus; ignition apparatus 435
ZÜNDBAND (*n*): quick match; fuse 436
ZÜNDBAR: inflammable 437
ZÜNDBARKEIT (*f*): inflammability 438
ZÜNDEN: to set fire to; take (catch) fire; ignite; kindle; prime; inflame 439
ZUNDER (*m*): tinder; occasion; cause; forge scale; touchwood 440
ZÜNDER (*m*): igniter; lighter; fuse; cinder; match; detonator; primer 441
ZÜNDER (*m*), ELEKTRISCHER-: electric fuse 442
ZUNDERN: to burn (away) out (of firebars, etc.); fuse up 443
ZÜNDERPULVER (*n*): fuse powder 444
ZÜNDERSATZ (*m*): fuse composition 445
ZUNDERSCHWAMM (*m*): spunk; touchwood; tinder fungus; red-rot fungus (*Trametes radiciperda*) 446
ZÜNDFLÄMMCHEN (*n*): pilot (flame) light 447
ZÜNDGEWÖLBE (*n*): combustion (ignition) arch (of furnaces) 448
ZÜNDHEBEL (*m*): sparking lever 449
ZÜNDHOLZ (*n*): match 450
ZÜNDHÖLZCHEN (*n*): match 451
ZÜNDHÖLZCHEN, SCHWEDISCHES- (*n*): Swedish (safety) match 452
ZÜNDHOLZFABRIK (*f*): match factory 453
ZÜNDHOLZMASCHINEN (*f.pl*): match making machinery 454
ZÜNDHOLZMASSE (*f*): match composition 455
ZÜNDHOLZPAPIER (*n*): match paper 456
ZÜNDHOLZSATZ (*m*): match composition 457
ZÜNDHOLZSCHACHTEL (*f*): match box 458
ZÜNDHOLZSCHACHTELMASCHINEN (*f.pl*): matchbox making machinery 459
ZÜNDHÜTCHEN (*n*): percussion (detonating) cap 460
ZÜNDHÜTCHENSATZ (*m*): priming (composition) 461
ZÜNDKAPSEL (*f*): percussion (detonating) cap 462
ZÜNDKERZE (*f*): wax (vesta) match; sparking plug 463
ZÜNDKIRSCHE (*f*): ignition pellet; pellet primer 464
ZÜNDLOCH (*n*): touch-hole 465
ZÜNDMASSE (*f*): ignition composition 466
ZÜNDMITTEL (*n*): primer; igniting agent 467
ZÜNDÖFFNUNG (*f*): ignition (port) opening 468
ZÜNDPAPIER (*n*): touch paper 469
ZÜNDPILLE (*f*): ignition pellet; pellet primer 470
ZÜNDPULVER (*n*): priming-(powder) 471
ZÜNDSATZ (*m*): priming-(composition) 472
ZÜNDSCHALTBRETT (*n*): igniter switchboard 473
ZÜNDSCHNUR (*f*): match; fuse; quick-match 474
ZÜNDSCHWAMM (*m*): German tinder; amadou (*Polyporus fomentarius*); spunk; partially decayed wood 475
ZÜNDSTELLE (*f*): place of ignition 476
ZÜNDSTIFT (*m*): ignition pin 477
ZÜNDSTOFF (*m*): primer; igniting agent; inflammable material 478
ZÜNDUNG (*f*): priming; ignition; igniting 479
ZÜNDUNGSGEWÖLBE (*n*): combustion arch 480
ZÜNDUNGSVERSAGER (*m*): miss-fire 481
ZÜNDVORRICHTUNG (*f*): igniter; ignition device; sparking device 482
ZÜNDWAREN (*f.pl*): inflammable goods 483
ZUNEHMEN: to increase; improve; advance; augment; prosper; grow; take more; rise 484
ZUNEIGEN: to tend; incline towards; have a (tendency) propensity; conduce; trend; gravitate towards 485
ZUNEIGUNG (*f*): attachment; inclination; propensity; proclivity; bent; turn; leaning; predisposition; predilection; proneness; propension; propendency; sympathy 486
ZUNFT (*f*): corporation; guild; fraternity; profession; company 487
ZUNFTGELEHRTER (*m*): professional man 488
ZÜNFTIG: incorporated; belonging (pertaining) to a guild or corporation 489
ZUNGE (*f*): tongue; language; sole-(fish); projection; lug, distance-piece 490
ZÜNGELCHEN: lip; tab; tongue; languet (of an organ pipe) 491
ZÜNGELN: to lick (of flames) 492
ZÜNGELND: lambent 493
ZUNGENBAND (*n*): fillet, ligament of the tongue 494
ZUNGENFERTIG: loquacious, talkative, garrulous, voluble, glib, fluent, wordy, verbose, diffuse, rambling, prolix 495
ZUNGENFORMIG: tongued, tongue-shaped 496
ZUNGENSCHIENE (*f*): points (Railway lines) 497
ZUNGENVORFALL (*m*): macroglossia (enlargement of the tongue) (Med.) 498
ZUNGENZÄPFCHEN (*n*): epiglottis (Physiol.) 499
ZUNICHTE: ruined; undone (see also ZUNICHTE MACHEN) 500
ZUNICHTE MACHEN: to ruin; undo; destroy; annul; nullify; cancel; reverse; neutralize; repeal; revoke; rescind; dissolve; set aside; quash 501
ZUOBERST: uppermost 502
ZUORDNEN: to associate 503
ZUPACKEN: to pack up 504
ZUPFEN: to pluck; pull; tug; pick 505
ZUPFLASTERN: to plaster up; fill in a hole (gap) in paving 506
ZUPFLEINWAND (*f*): lint 507
ZUPFROPFEN: to cork (stop) up 508
ZUPIPETTIEREN: to add to the pipette 509
ZURAMMEN: to close (seal) by ramming or by means of a ram 510
ZURATEN: to advise in favour 511
ZURAUNEN: to whisper to; suggest to; (*n*): whispering; suggestion 512
ZURECHNEN: to add (to an account); impute, blame, hold (responsible) accountable for 513
ZURECHNUNG (*f*): imputation 514
ZURECHNUNGSFÄHIG: responsible; accountable 515
ZURECHT: right; in (time) order; in (good) condition 516
ZURECHTBRINGEN: to put (set) right; arrange 517
ZURECHTHELFEN: (see ZURECHTBRINGEN) 518
ZURECHTKOMMEN: to make do with, get on well 519
ZURECHTLEGEN: to arrange; put (set) right 520
ZURECHTMACHEN: to prepare; get ready; adjust 521

ZURECHTWEISEN: to show (demonstrate) the right way; correct; set to rights 522
ZUREDEN: to speak to; persuade; encourage; exhort 523
ZUREICHEN: to hand to; suffice; be (enough) sufficient 524
ZURICHTEN: to prepare; finish; dress; make ready; leaven; fit 525
ZURIEGELN: to bolt (up) 526
ZÜRNEN: to be angry 527
ZURREN: to seize, lash 528
ZURRING (*f*): seizing, lashing 529
ZURÜCK: back; behind; backward-(s) 530
ZURÜCKBEBEN: to recoil; start back 531
ZURÜCKBEGEBEN: to return; go back 532
ZURÜCKBEGEHREN: to wish (desire) to return; to demand back 533
ZURÜCKBEGLEITEN: to escort back; show out 534
ZURÜCKBEHALTEN: to retain; detail; reserve; keep back 535
ZURÜCKBEKOMMEN: to recover; get back 536
ZURÜCKBERUFEN: to recall; call back 537
ZURÜCKBEUGEN: to bend back 538
ZURÜCKBEZAHLEN: to repay; pay back; reimburse; discharge (settle; satisfy; liquidate) debts; quit (a score); balance (square) an account; pay off; refund 539
ZURÜCKBEZIEHEN: to be reflected; to refer to 540
ZURÜCKBEZÜGLICH: reflective (Grammar) 541
ZURÜCKBIEGEN: to bend back 542
ZURÜCKBLEIBEN: to remain behind; lag; be (slow) late; be inferior; fall short 543
ZURÜCKBLEIBEND: residual, etc. 544
ZURÜCKBLICKEN: to review; recall; look back 545
ZURÜCKBRINGEN: to bring back; reclaim; undeceive; reduce (Arithmetic) 546
ZURÜCKDATIEREN: to antedate; back-date 547
ZURÜCKDENKEN: to reflect; recall; think of the past; imagine back 548
ZURÜCKDISKONTIEREN: to re-discount; discount again 549
ZURÜCKDRÄNGEN: to repress; push (press; force; drive) back 550
ZURÜCKDREHEN: to turn back 551
ZURÜCKDÜRFEN: to be allowed to return 552
ZURÜCKEILEN: to hurry (hasten) back 553
ZURÜCKERBITTEN: to beg (ask) back 554
ZURÜCKERHALTEN: to recover; get back 555
ZURÜCKERINNERN: to remember; recollect; bring back to mind 556
ZURÜCKEROBERN: to re-conquer 557
ZURÜCKERSTATTEN: to refund; restore; return 558
ZURÜCKFAHREN: to spring back; recoil; return; rebound 559
ZURÜCKFALLEN: to revert; relapse; fall back 560
ZURÜCKFLIESSEN: to ebb; recede; flow back 561
ZURÜCKFORDERN: to reclaim; demand back 562
ZURÜCKFÜHRBAR: attributable; traceable; reducible 563
ZURÜCKFÜHREN: to attribute to; trace (lead) back; reduce; re convey 564
ZURÜCKGABE (*f*): restitution; restoration; return; reddition; rendition; re-investment; rehabilitation 565
ZURÜCKGEBEN: to restore; return; refund; surrender; give back (see ZURÜCKGABE) 566
ZURÜCKGEHEN: to return; deteriorate; retract; decline; fall; go back; recede; retrogress 567
ZURÜCKGELEITEN: to lead back 568
ZURÜCKGEWINNEN: to regain; recover; win back 569
ZURÜCKGEWINNUNG (*f*): regaining; recovery; recuperation 570
ZURÜCKGEZOGEN: retired; secluded 571
ZURÜCKGEZOGENHEIT (*f*): retirement 572
ZURÜCKGREIFEN: to return; revert; reconsider; refer to 573
ZURÜCKHALTEN: to detain; retain; repress; keep back; reserve; be reserved; keep aloof; drag 574
ZURÜCKHALTEND: reserved; shy; backward; behind the times 575
ZURÜCKKAUFEN: to buy back 576
ZURÜCKKEHREN: to return; recur; go (come) back 577
ZURÜCKKOMMEN: to return; recover; retrieve; revert to; come back 578
ZURÜCKKUNFT (*f*): return 579
ZURÜCKLASSEN: to leave-(behind); allow to return; abandon 580
ZURÜCKLAUFEN: to run (flow) back; ebb; recoil; retrograde 581
ZURÜCKLEGEN: to shelve; lay aside; put by; cover (Distance); reserve 582
ZURÜCKLEITEN: to lead (up) back to; reduce to 583
ZURÜCKLENKEN: to turn back 584
ZURÜCKLEUCHTEN: to light back; be reflected 585
ZURÜCKLIEFERN: to deliver up; send back; return 586
ZURÜCKNAHME (*f*): recall; revocation; resumption; withdrawal; taking back 587
ZURÜCKNEHMEN: to retract; recall; take back; cancel; revoke; withdraw 588
ZURÜCKPRALLEN: to recoil; rebound; be reflected 589
ZURÜCKRECHNEN: to reckon back; to draw upon (Com.) 590
ZURÜCKREISEN: to return; travel back; journey back 591
ZURÜCKROLLEN: to roll back 592
ZURÜCKRÜCKEN: to move (set; push) back 593
ZURÜCKRUDERN: to row back; backwater 594
ZURÜCKRUFEN: to recall; call back 595
ZURÜCKSCHAFFEN: to render (take; give) back; cause to be returned 596
ZURÜCKSCHALLEN: to resound; reverberate 597
ZURÜCKSCHAUDERN: to start back (recoil) with horror 598
ZURÜCKSCHAUEN: to reflect; review; look back 599
ZURÜCKSCHIEBEN: to push back; repulse; postpone; defer 600
ZURÜCKSCHIFFEN: to ship (row) back; send back by boat 601
ZURÜCKSCHLAGEN: to repel; repulse; drive (beat; strike) back; turn back (Bedclothes); light back; back-fire; (*n*): back-firing 602
ZURÜCKSCHLAGEND: back-firing 603
ZURÜCKSCHLEPPEN: to drag back 604
ZURÜCKSCHLEUDERN: to hurl back 605
ZURÜCKSCHLIESSEN: to reason (use an argument) a posteriori 606
ZURÜCKSCHNEIDEN: to cut (down) back (Plants, etc.) 607
ZURÜCKSCHNELLEN: to dart (cast) back 608
ZURÜCKSCHREIBEN: to write back; reply 609
ZURÜCKSEHEN: to look back; reflect; meditate; consider; bethink; remember; recall; review 610

ZURÜCKSEHNEN: to desire (long, wish) to return or be back 611
ZURÜCKSETZEN: to put back; set back; slight; neglect; reduce; disregard 612
ZURÜCKSETZUNG (*f*): putting back; neglect; slight; disregard-(ing) 613
ZURÜCKSINKEN: to sink (fall) back 614
ZURÜCKSPIEGELN: to reflect 615
ZURÜCKSPRINGEN: to jump (spring) back; recoil; rebound; be reflected 616
ZURÜCKSPÜLEN: to wash (rinse) back 617
ZURÜCKSTEIGEN: to rise again 618
ZURÜCKSTOSSEN: to push back; repel; drive (force) back; repulse 619
ZURÜCKSTOSSEND: repulsive; repellent 620
ZURÜCKSTOSSUNG (*f*): repulsion (see ZURÜCK-STOSSEN) 621
ZURÜCKSTOSSUNGSKRAFT (*f*): repulsive power; power of repulsion 622
ZURÜCKSTRAHLEN: to reflect; radiate (of heat) 623
ZURÜCKSTRÖMEN: to flow (stream) back 624
ZURÜCKTITRIEREN: to titrate back; re-titrate 625
ZURÜCKTREIBEN: to drive back; repel; force back; check 626
ZURÜCKTRETEN: to recede; retire; go back; light back; back-fire 627
ZURÜCKWEICHEN: to recede; recoil; retire; give way; step back; go back 628
ZURÜCKWEISEN: to decline; send back; refuse; reject; repudiate 629
ZURÜCKWERFEN: to drive back; throw back; reflect 630
ZURÜCKZAHLEN: to pay back, repay, reimburse, discharge (settle) a debt, quit (a score), balance (square) an account, pay off, satisfy, liquidate, refund 631
ZURÜCKZIEHEN: to recede; retire; draw back; retract; renounce; withdraw; retreat 632
ZURUF (*m*): call; shout; acclamation; cheer 633
ZURUNDEN: to round (off) 634
ZURÜSTEN: to equip; prepare; fit out; arm 635
ZURZEIT: at the time 636
ZUSAGE (*f*): promise; consent 637
ZUSAGEN: to agree; consent; accept; promise; please; answer; assert; suit 638
ZUSAMMEN: together; (con)-jointly; altogether 639
ZUSAMMENARBEITEN: to collaborate; work in conjunction with; (*n*): collaboration 640
ZUSAMMENBACKEN: to stick together; cake; clinker 641
ZUSAMMENBALLEN: to ball; conglomerate; agglomerate 642
ZUSAMMENBAU (*m*): assembling; assembly (of machine parts); putting together 643
ZUSAMMENBAUEN: to set in battery; build together; put (fit) together; assemble; erect 644
ZUSAMMENBERUFEN: to convoke, convene, call together, summon together, convocate 645
ZUSAMMENBRECHEN: to break up; cave (fall) in; collapse 646
ZUSAMMENBRINGEN: to bring together; collect; amass; accumulate; assemble; agglomerate 647
ZUSAMMENDRÜCKBAR: compressible 648
ZUSAMMENDRÜCKBARKEIT (*f*): compressibility 649
ZUSAMMENDRÜCKBARKEITSKOEFFIZIENT (*m*): coefficient of compressibility 650
ZUSAMMENDRÜCKEN: to compress; press together; force together 651
ZUSAMMENDRÜCKUNG (*f*): compression 652
ZUSAMMENFAHREN: to shrink; travel together 653
ZUSAMMENFALLEN: to cave (fall) in; collapse; coincide; converge 654
ZUSAMMENFALTEN: to fold (together) up 655
ZUSAMMENFASSEN: to collect; comprise; sum up; summarize; comprehend; grasp 656
ZUSAMMENFLUSS (*m*): conflux; confluence; concourse 657
ZUSAMMENFÜGEN: to unite; join; combine; couple; pair; put together 658
ZUSAMMENFÜGUNG (*f*): conjunction; combining; combination 659
ZUSAMMENGEHÖRIG: correlated; corresponding; belonging together; homologous 660
ZUSAMMENGESCHLIFFEN: ground together; machined to make a face to face joint 661
ZUSAMMENGESETZT: compound (Sections); composed (Girders); put (fitted) together; erected; assembled; composite; built up; compounded 662
ZUSAMMENGESETZTE PROFILE (*n.pl*): compound (composed) sections (Rolling mill work) 663
ZUSAMMENGESETZTE WELLE (*f*): composed (built-up) shaft (made up of various parts) 664
ZUSAMMENGIESSEN: to pour together; cast in one piece (Metal) 665
ZUSAMMENHALT (*m*): cohesion; unity 666
ZUSAMMENHALTBAR: cohesive; comparable; capable of holding (hanging) together 667
ZUSAMMENHALTEN: to hold (hang) together; cohere; compare 668
ZUSAMMENHANG (*m*): connection; relation; consistency; coherence; cohesion; conjunction 669
ZUSAMMENHANGEN: (see ZUSAMMENHÄNGEN) 670
ZUSAMMENHÄNGEN: to be (related) connected; cohere; hang together 671
ZUSAMMENHÄNGEND: composite, hanging (sticking, clinging) together 672
ZUSAMMENHÄUFEN: to heap (together) up; accumulate; aggregate 673
ZUSAMMENKITTEN: to cement together 674
ZUSAMMENKLEBEN: to glue (stick) together; agglutinate 675
ZUSAMMENKOMMEN: to meet; assemble 676
ZUSAMMENKUNFT (*f*): meeting; convention; assembly; conference 677
ZUSAMMENLAGERUNG (*f*): assembly 678
ZUSAMMENLAUFEN: to converge; curdle; coagulate; run (flock) together; blend 679
ZUSAMMENLEIMEN: to glue together 680
ZUSAMMENLEIMUNG (*f*): gluing (sticking) together; agglutination 681
ZUSAMMENLÖTEN: to solder (together) up 682
ZUSAMMENPASSEN: to suit, fit-(together) 683
ZUSAMMENPRESSEN: to press (force) together; compress 684
ZUSAMMENREIBEN: to grind (rub) together 685
ZUSAMMENRÜCKEN: to push together; crowd 686
ZUSAMMENRUFEN: to convoke; convocate; convene; call (summon) together 687
ZUSAMMENSCHLEIFEN: to grind together; machine for a face to face joint 688
ZUSAMMENSCHMELZEN: to melt (together; down); fuse; clinker; alloy 689

ZUSAMMENSCHRAUBEN: to screw together 690
ZUSAMMENSCHRUMPFEN: to shrink; contract; wrinkle; shrivel 691
ZUSAMMENSCHÜTTELN: to shake (down) together 692
ZUSAMMENSCHÜTTEN: to mix; pour together 693
ZUSAMMENSCHWEISSEN: to braze, weld together 694
ZUSAMMENSETZEN: to compose; put (fit) together; combine; compound; construct; consist; build up (of various parts); assemble; erect; range side by side; pile up; compile 695
ZUSAMMENSETZUNG (f): composition; combination; compounding; building up; synthesis; compound; complication; assembly structure 696
ZUSAMMENSTELLEN: to tabulate; summarize; put together; assemble; schedule; group; combine; juxtapose; compile; plot (on a graph); confront; compare 697
ZUSAMMENSTELLUNG (f): tabulation; summary; assembly; grouping; juxtaposition; table; schedule; combination; collation; compilation; classification 698
ZUSAMMENSTIMMEN: to agree; tune together; harmonize; concur; accord; be consistent (compatible; commensurate); correspond; tally; consort with; dovetail; match; conform 699
ZUSAMMENSTIMMEND: concordant; consonant; agreeing; congruous; harmonious; in accord; accordant 700
ZUSAMMENSTOSS (m): collison; encounter; join-(t); butt 701
ZUSAMMENSTOSSEN: to collide; encounter; adjoin 702
ZUSAMMENSTOSSEND: contiguous; adjacent 703
ZUSAMMENTREFFEN: to coincide; meet; concur; accord 704
ZUSAMMENTROCKNEN: to dry (shrivel) up 705
ZUSAMMENWACHSEN: to coalesce; grow together 706
ZUSAMMENZIEHBAR: to contract; abridge; draw together; assemble; shrink; tighten; gather 707
ZUSAMMENZIEHBARKEIT (f): contractibility; astringent action 708
ZUSAMMENZIEHEND: contractive; astringent; costive (Med.) 709
ZUSAMMENZIEHENDES MITTEL (n): astringent 710
ZUSAMMENZIEHUNG (f): contraction; shrinkage; concentration 711
ZUSATZ (m): admixture; addition; appendix; codicil; supplement; extra; margin; corollary; alloy; flux (Metal) 712
ZUSATZGENERATOR (m): booster (Elect.) 713
ZUSÄTZLICH: additional-(ly); auxiliary; supplementary 714
ZUSATZLUFT (f): secondary air (for combustion) 715
ZUSATZMASCHINE (f): booster (Elect.) 716
ZUSATZMETALL (n): alloy 717
ZUSATZMITTEL (n): supplementary (auxiliary; additional) agent 718
ZUSATZPATENT (n): addition to a patent 719
ZUSATZTRANSFORMATOR (m): booster transformer (Elect.) 720
ZUSATZVENTIL (n): auxiliary (by-pass) valve; additional (stand-by; spare) valve 721
ZUSCHALTEN: to switch on, make connection, to close the circuit 722
ZUSCHÄRFEN: to sharpen; point; bevel 723
ZUSCHAUEN: to look (at) on; behold; regard 724
ZUSCHAUER (m): spectator 725
ZUSCHICKEN: to send (forward; dispatch; remit; consign) to; ship (post) to 726
ZUSCHIEBEN: to shut; push to; close 727
ZUSCHLAG (m): increase; addition; flux; extra; margin; adjudication 728
ZUSCHLAGEN: to add; increase; slam (Doors); close-(up); shut (with force); adjudicate; adjudge; strike 729
ZUSCHLÄGER (m): striker 730
ZUSCHLIESSEN: to lock up; close; seal 731
ZUSCHMIEREN: to smear (daub) over 732
ZUSCHMELZEN: to seal; close by melting 733
ZUSCHNALLEN: to buckle (strap) up 734
ZUSCHNEIDEN: to cut to suit, to cut down to, to trim 735
ZUSCHNÜREN: to lace (up) 736
ZUSCHRAUBEN: to screw (up), screw on 737
ZUSCHREIBEN: to attribute; credit; impute; dedicate; assign; add; inform; communicate; ascribe; write to 738
ZUSCHRIFT (f): communication; letter; dedication; favour 739
ZUSCHÜREN: to stoke (feed) a fire 740
ZUSCHUSS (m): addition; supply; contribution 741
ZUSCHÜTTEN: to add; pour (on; in) to; fill up with 742
ZUSCHWÖREN: to swear to 743
ZUSEHEN: to look (on) at, suffer 744
ZUSEHENS: noticeably; visibly 745
ZUSENDEN: to send (dispatch; forward; consign; remit) to; ship (post) to 746
ZUSETZEN: to add; alloy; mix; obstruct; contribute; stop up 747
ZUSICHERN: to assure; promise; asseverate 748
ZUSIEGELN: to seal (up) 749
ZUSPERREN: to bar, barricade 750
ZUSPITZEN: to point; taper; sharpen 751
ZUSPITZUNG (f): point-(ing); taper-(ing); acumination 752
ZUSPRECHEN: to agree with; speak to; adjudicate; adjudge; encourage; comfort; console 753
ZUSPRUCH (m): encouragement; call; address; consolation; comfort 754
ZUSPÜNDEN: to bung (up) 755
ZUSTAND (m): condition; state; circumstances; situation 756
ZUSTANDEBRINGEN: to bring about; accomplish; effect 757
ZUSTANDEKOMMEN: to come about; happen; recur 758
ZUSTANDEKOMMEN (n): occurrence 759
ZUSTÄNDIG: appropriate; suitable; belonging to; competent; pertinent; appertaining 760
ZUSTÄNDIGKEIT (f): competence; property; appurtenance 761
ZUSTANDSÄNDERUNG (f): change of state 762
ZUSTANDSDIAGRAMM (n): phase diagram; diagram of state; solidification diagram (of metal smelting) 763
ZUSTANDSFORMEL (f): formula of state 764
ZUSTANDSGLEICHUNG (f): equation of state 765
ZUSTANDSVERSCHIEBUNG (f): change in (displacement of) state 766
ZUSTECKEN: to pin together 767

ZUSTEHEN : to belong; pertain; become; be incumbent upon 768
ZUSTELLEN : to present; deliver; shut; close; pay; hand to 769
ZUSTIMMEN : to agree; consent; acquiesce; assent; accede; concur; approve; endorse 770
ZUSTIMMUNG (*f*) : consent; agreement; acquiescence; assent; concordance; compliance; approval; ratification 771
ZUSTOPFEN : to stop (stuff) up; pack; close; cork; mend; stopper; bung 772
ZUSTÖPSELN : to cork; stopper 773
ZUSTOSSEN : to happen; occur; befall 774
ZUSTRÖMEN : to flow (in)-to; stream (in)-to 775
ZUTAT (*f*) : addition; complement; ingredient; necessaries 776
ZUTEILEN : to allot; assign; distribute; impart; apportion 777
ZUTRAGEN : to carry (bring; convey) to; happen; chance; report; conduce 778
ZUTRÄGLICH : beneficial; good; profitable; conducive; useful; wholesome 779
ZUTRÄGLICHKEIT (*f*) : conduciveness; usefulness 780
ZUTRAUEN : to confide in; credit with; expect of; trust (to); rely upon 781
ZUTRAUEN (*n*) : confidence; trust 782
ZUTRAULICH : confiding; confidential 783
ZUTRAULICHKEIT (*f*) : trust 784
ZUTREFFEN : to correspond; agree; prove correct; happen; take place 785
ZUTRETEN : to enter; mingle 786
ZUTRITT (*m*) : admittance; access; accession; joining; entry; admission 787
ZUTULICH : affectionate; attached; attentive; obliging; officious; insinuating 788
ZUTULICHKEIT (*f*) : attachment 789
ZUTUN : to add; put to; furnish; shut; close; (*n*) : co-operation; agency; aid; instrumentality; doing 790
ZUVERLÄSSIG : authentic; reliable; certain; sure; trustworthy; authoritative 791
ZUVERLÄSSIGKEIT (*f*) : reliability; authenticity; infallibility 792
ZUVERSICHT (*f*) : reliance; confidence; certainty; trust 793
ZUVERSICHTLICH : certain; positive; confident; sure 794
ZUVOR : previously; before; beforehand; first; heretofore; formerly 795
ZUVORKOMMEN : to prevent; anticipate 796
ZUVORKOMMEND : obliging; anticipating; complaisant; polite; courteous 797
ZUVORTUN : to outdo; surpass; excel 798
ZUWACHS (*m*) : growth; increment; increase; accretion; accession; augmentation 799
ZUWACHSEN : to grow over; increase; accrue; augment; become overgrown 800
ZUWACHSZONE (*f*) : annual ring (Trees) 801
ZUWACHSZONENBILDUNG (*f*) : the formation of annual rings (Trees) 802
ZUWEGE BRINGEN : to bring about, effect, accomplish 803
ZUWEILEN : sometimes 804
ZUWEISEN : to show (direct) to; recommend 805
ZUWIDER : against; contrary to; repugnant; offensive; averse 806
ZUWIDERHANDELN : to act in contravention; contravene 807
ZUZEITEN : at times 808
ZUZIEHEN : to call in; consult; invite; incur; pull to; draw (to) together; catch (a complaint) 809
ZUZIEHUNG (*f*) : incurring, help, drawing, assistance 810
ZWACKEISEN (*n*) : pincers 811
ZWACKEN : to pinch; tease; cheat 812
ZWANG (*m*) : coercion; force; compulsion; violence; constraint 813
ZWÄNGEN : to coerce; force; press 814
ZWANGLÄUFIG : forced (Water circulation, etc.); positive (Mech.) 815
ZWANGLÄUFIG BEWEGEN : to actuate positively, to regulate (of valves) 816
ZWANGLÄUFIGER WASSERUMLAUF (*m*) : forced water circulation 817
ZWANGLÄUFIGE VENTILSTEUERUNG (*f*) : regulated valve gear 818
ZWANGLÄUFIG HINDURCHTREIBEN : to force through (of water, etc.) 819
ZWANGLOS : unconstrained 820
ZWANGSCHIENE (*f*) : guard rail (Railways) 821
ZWANGSMITTEL (*n*) : means of (constraint) restraint; coercive measure 822
ZWANGSSYNDIKAT (*n*) : compulsory syndicate 823
ZWANGSWEISE : compulsory; by force; forcible; compulsorily; forcibly 824
ZWANGVORSTELLUNG (*f*) : mental obsession, imperative ideas (Med.) 825
ZWANZIG : twenty; a score 826
ZWANZIGFLACH (*n*) : icosohedron 827
ZWANZIGSTE : twentieth 828
ZWANZIGSTEL (*n*) : twentieth-(part); 1/20th 829
ZWAR : indeed; it is true; truly; to be sure; namely; certainly; no doubt 830
ZWARK (*m*) : (see QUARK) 831
ZWECK (*m*) : end; purpose; design; object; aim; peg; pin; tack; intention 832
ZWECKDIENLICH : efficient; serviceable 833
ZWECKE (*f*) : peg; pin; tack; nail 834
ZWECKEN : to (peg; pin; tack) nail; aim at; have for its (aim; object; goal) purpose; be of use 835
ZWECKENTSPRECHEND : necessary (suitable) for the purpose; appropriate 836
ZWECKLOS : useless; aimless 837
ZWECKMÄSSIG : appropriate; judicious; suitable; expedient; preferable; advantageous 838
ZWECKMÄSSIGERWEISE : advantageous-(ly); preferably 839
ZWECKMÄSSIGSTEN, AM- : preferably; most advantageously 840
ZWECKS : for the purpose of; with the object (intention) of 841
ZWECKWIDRIG : unsuitable; inexpedient; inappropriate; injudicious 842
ZWEI : two; (as a prefix=bi-; di-) 843
ZWEIACHSIG : biaxial 844
ZWEIARMIG : two-armed 845
ZWEIATOMIG : diatomic 846
ZWEIBASISCH : dibasic 847
ZWEIBASISCHE SÄURE (*f*) : dibasic acid 848
ZWEIDECKER (*m*) : two-decker 849
ZWEIDECKSCHIFF (*n*) : two decker, two decked ship 850
ZWEIDEUTIG : equivocal; ambiguous 851
ZWEIDEUTIGKEIT (*f*) : ambiguity; equivocation 852
ZWEI DRITTEL GESÄTTIGTES CALCIUMPHOSPHAT (*n*) : (see DICALCIUMPHOSPHAT) 853

ZWEIERLEI: different; of two different kinds or sorts 854
ZWEIFACH: double; twofold; twice; two times; (in chemistry, prefix=bi-; di-) 855
ZWEIFACHBASISCH: dibasic 856
ZWEIFACH CHROMSAUER: bichromate of 857
ZWEIFACHSAURES KALZIUMPHOSPHAT (*n*): monobasic calcium phosphate (see **MONOCALCIUM-PHOSPHAT**) 858
ZWEIFACHSCHWEFELSAUER: bisulphate 859
ZWEIFACHSCHWEFLIGSAUER: bisulphite 860
ZWEIFÄLTIG: double, twofold 861
ZWEIFARBIG: two-coloured; two-colour 862
ZWEIFEL (*m*): doubt; uncertainty; question 863
ZWEIFELHAFT: questionable; doubtful; uncertain; dubious 864
ZWEIFEL HEGEN: to doubt; entertain (have) doubts 865
ZWEIFELLOS: indubitable; doubtless; unquestionably 866
ZWEIFELN: to doubt; entertain (have) doubts; question 867
ZWEIFEL, OHNE-: doubtless; indubitably; unquestionably 868
ZWEIFELSGRUND (*m*): cause for doubt or question 869
ZWEIFELSOHNE: doubtless; without (doubt) question 870
ZWEIFELSUCHT (*f*): scepticism, dubiety 871
ZWEIFELSÜCHTIG: sceptic-(al) 872
ZWEIFLAMMROHRKESSEL (*m*): two-flued Cornish boiler 873
ZWEIFLER (*m*): sceptic, doubter 874
ZWEIFÜSSIG: having two feet; bipedal 875
ZWEIG (*m*): branch; twig; bough; sprig; spur (of hills); department 876
ZWEIGDRAHT (*m*): shunt wire 877
ZWEIGIG: branched; bifurcated; forked 878
ZWEIGLEITUNG (*f*): branch (main) pipe-line 879
ZWEIGLIEDRIG: two-membered; binomial (Maths.) 880
ZWEIGROHR (*n*): branch pipe 881
ZWEIGRUTSCHE (*f*): bifurcated chute 882
ZWEIGWIDERSTAND (*m*): shunt resistance 883
ZWEIHALSIG: two-necked 884
ZWEIHÄNDIG: two-handed 885
ZWEIHÄUFIG: dioecious 886
ZWEIHEIT (*f*): dualism, duality 887
ZWEIHÖRNIG: bicornute, bicornous, having two horns 888
ZWEIHUFIG: bisulcate, cloven-hoofed 889
ZWEIJÄHRIG: of (for) two years; biennial; two years old 890
ZWEIKAMMERSYSTEM (*n*): two chamber system 891
ZWEIKAMPF (*m*): duel; single combat 892
ZWEIKAPSELIG: bicapsular; bivalvular 893
ZWEIKEIMIG: dicotyledonous (Bot.) 894
ZWEIKERNCHINON (*n*): binuclear quinone 895
ZWEIKNOSPIG: bigeminate 896
ZWEIKÖPFIG: two-headed 897
ZWEILAPPIG: dicoccous (Bot.), bilobate 898
ZWEILEIBIG: amphibious 899
ZWEILEITERSYSTEM (*n*): two wire system 900
ZWEILIPPIG: bilabiate 901
ZWEIMAL: twice 902
ZWEIMALIG: reiterated; twice done 903
ZWEIMASTER (*m*): brig, two-master (Naut.) 904
ZWEIMONATLICH: bi-monthly; every two months 905
ZWEIPFÜNDIG: of (weighing) two pounds 906
ZWEIPHASENSTROM (*m*): diphase (two phase) current 907
ZWEIPHASENSTROMANKER (*m*): diphase (two phase) armature 908
ZWEIPHASENSTROMKREIS (*m*): diphase (two phase) circuit 909
ZWEIPHASIG: diphase; two-phase 910
ZWEIPLATTIG: bilamellar 911
ZWEIPOLIG: bipolar 912
ZWEIRAD (*n*): bicycle 913
ZWEIREIHIG: distichous 914
ZWEIRIPPIG: binervate, double ribbed, bicarinate 915
ZWEISAMENLAPPIG: dicotyledonous (Bot.) 916
ZWEISÄURIG: diacid 917
ZWEISCHALIG: bivalvular 918
ZWEISCHENKELIG: double-branch; having two (legs) branches 919
ZWEISCHNEIDIG: two-edged 920
ZWEISCHRAUBENDAMPFER (*m*): twin-screw steamer 921
ZWEISEITIG: bilateral; two-sided 922
ZWEISILBIG: disyllabic, of two syllables 923
ZWEISITZIG: double-seated, having two seats 924
ZWEISPALTIG: bifid, double columned 925
ZWEISPITZIG: bicuspid 926
ZWEISPRACHIG: bilingual 927
ZWEISPURIG: double-tracked 928
ZWEISTIMMIG: two-voiced, in two voices (Music) 929
ZWEISTROM-TURBINE (*f*): double-flow turbine 930
ZWEISTÜNDIG: two-hourly, every two hours, lasting two hours, of two hours duration 931
ZWEITÄGIG: two-day, of two days, every two days, two days old 932
ZWEITAKT (*m*): two-stroke 933
ZWEITAKTMOTOR (*m*): two-stroke motor 934
ZWEITAKTVERFAHREN (*n*): two-stroke cycle 935
ZWEITE: second 936
ZWEITEILIG: two-part; bifurcated; bipartite; bisected 937
ZWEITENS: secondly 938
ZWEIVIERTELPAUSE (*f*): minim rest (Music) 939
ZWEIVIERTELTAKT (*m*): two-four ($\frac{2}{4}$) time; half common time (Music) 940
ZWEIWANDIG: two-walled; double walled 941
ZWEIWEGEHAHN (*m*): two-way cock 942
ZWEIWEGHAHN (*m*): two-way cock 943
ZWEIWEIBIG: digynous; bigamous 944
ZWEIWERTIG: bivalent 945
ZWEIWERTIGE SÄURE (*f*): dibasic acid 946
ZWEIWERTIGES ELEMENT (*n*): dyad 947
ZWEIWERTIGKEIT (*f*): bivalence 948
ZWEIWUCHS (*m*): rickets 949
ZWEIZACKIG: double-pointed, two-pronged, bifurcated 950
ZWEIZÄHLIG: double; twofold; binary; dual 951
ZWEIZÄHNIG: two-toothed, bidentate 952
ZWEIZEHIG: didactyle 953
ZWEIZEILIG: distichous 954
ZWEIZINKIG: two-pronged; bifurcated 955
ZWEIZÜNGIG: double-tongued 956
ZWERCH: across, obliquely, diagonally, awry 957
ZWERCHFELL (*n*): diaphragm (Anatomy) 958
ZWERCHPFEIFE (*f*): fife (Instrument) 959
ZWERG (*m*): dwarf; pigmy 960
ZWERGARTIG: dwarfish 961
ZWERGBAUM (*m*): dwarf-tree 962
ZWERGMISPEL (*f*): Cotoneaster (Bot.) (*Cotoneaster*) 963

ZWETSCHE (*f*): plum (see ZEWETSCHENBAUM) 964
ZWETSCHE, GEDÖRRTE- (*f*): prune 965
ZWETSCHENBAUM (*m*): plum tree, wild plum (*Prunus œconomica* or *Prunus domestica*) 966
ZWETSCHGENBAUM (*m*): (see ZWETSCHENBAUM) 967
ZWICK (*m*): nip; pinch 968
ZWICKAUERGELB (*n*): chrome yellow (see CHROM-GELB and BLEICHROMAT) 969
ZWICKAUERGRÜN (*n*): (see SCHWEINFURTGRÜN) 970
ZWICKBOHRER (*m*): gimlet 971
ZWICKEL (*m*): try-cock; test-cock; wedge; gusset 972
ZWICKEN: to pinch; worry; nip 973
ZWICKER (*m*): pince-nez 974
ZWICKMÜHLE (*f*): mill 975
ZWIEBACK (*m*): biscuit; rusk 976
ZWIEBEL (*f*): onion (*Allium cepa*); bulb 977
ZWIEBELARTIG: bulbous 978
ZWIEBEL (*f*), ASKALONISCHE-: shalotte (see SCHALOTTE) 979
ZWIEBELFÖRMIG: bulbous; bulbiform 980
ZWIEBEL (*f*), GEMEINE-: (common) onion (see ZWIEBEL) 981
ZWIEBELGEWÄCHS (*n*): bulbous plant; bulb 982
ZWIEBELLAUCH (*m*): spring onion (*Allium schoenoprasum*) 983
ZWIEBELMARMOR (*m*): cipolin 984
ZWIEBELROT: onion-red 985
ZWIEBELSAFT (*m*): onion juice 986
ZWIEBELSCHALE (*f*): onion (peel) skin 987
ZWIEBELTRAGEND: bulbiferous 988
ZWIELICHT: twilight 989
ZWIESELIT (*m*): triplite (see TRIPLIT) 990
ZWIESPALT (*m*): discord; schism; difference; dissention; quarrel; disunion 991
ZWIETRACHT (*f*): dissension; discord 992
ZWIETRÄCHTIG: discordant 993
ZWILLICH (*m*): ticking; drill 994
ZWILLING (*m*): twin; (*pl*): Castor and Pollux (the constellation) 995
ZWILLINGSACHSE (*f*): twinning axis 996
ZWILLINGSARTIG: of twin (nature) form 997
ZWILLINGSBILDUNG (*f*): twin formation 998
ZWILLINGSBLÜTIG: geminiflorous 999
ZWILLINGSDAMPFMASCHINE (*f*): twin engine, twin-cylinder engine, tandem engine 000
ZWILLINGSEBENE (*f*): twin plane 001
ZWILLINGSKERNE (*m.pl*): twin nuclei 002
ZWILLINGSKRYSTALL (*m*): twin crystal 003
ZWILLINGSMASCHINE (*f*): twin engine 004
ZWILLINGSPUMPE (*f*): twin pump 005
ZWILLINGSSCHRAUBE (*f*): twin screw 006
ZWILLINGSSCHRAUBENDAMPFER (*m*): twin-screw steamer 007
ZWINGE (*f*): clamp; vice; hoop; ferrule; press 008
ZWINGE (*f*), LEIM-: glue press 009
ZWINGEN: to force; compel; conquer; subdue; constrain; coerce 010
ZWINGEND: cogent, etc. 011
ZWINGER (*m*): wedge; prison 012
ZWINKERN: to twinkle; wink 013
ZWIRN (*m*): thread; yarn 014
ZWIRNBAND (*n*): tape 015
ZWIRNEN: to twine; twist; throw (silk); (as an adjective = of thread) 016
ZWIRNFADEN: thread 017
ZWIRNMASCHINE (*f*): thread machine 018
ZWISCHEN: between; among; amongst; betwixt; (as a prefix = incidental; intermediate; mid-) 019
ZWISCHENAKT (*m*): interlude; interval 020
ZWISCHENBAD (*n*): intermediate bath 021
ZWISCHENDAMPFENTNAHME (*f*): bleeding; inter-stage (intermediate) leak-off (Turbines) 022
ZWISCHENDECK (*n*): between deck, steerage (Naut.) 023
ZWISCHENDECKBALKEN (*m*): between deck beam 024
ZWISCHENDURCH: through; at intervals 025
ZWISCHENFALL (*m*): incident; episode 026
ZWISCHENFÖRDERER (*m*): intermediate (auxiliary) conveyor or transporter 027
ZWISCHENGANG (*m*): intermediate strake 028
ZWISCHENGEFÄSS (*n*): intermediate vessel 029
ZWISCHENGLASURFARBE (*f*): between-glaze colour (Ceram.) 030
ZWISCHENGLASURMALEREI (*f*): painting between glazes 031
ZWISCHENGLIED (*n*): intermediate member; connecting-link; go-between 032
ZWISCHENHÄNDLER (*m*): middleman (Com.); mediator 033
ZWISCHENKÖRPER (*m*): intermediate (substance) body 034
ZWISCHENKÜHLER (*m*): intermediate cooler 035
ZWISCHENKUNFT (*f*): intervention 036
ZWISCHENLAGE (*f*): intermediate position; intermediate layer; interposition 037
ZWISCHENLIEGEND: interposing, intervening, intermediate 038
ZWISCHENMAHL (*n*): collation 039
ZWISCHENMAUER (*f*): partition, party-wall 040
ZWISCHENPLATTE (*f*): intermediate (intercostal) plate 041
ZWISCHENPRODUKT (*n*): intermediate product 042
ZWISCHENRAUM (*m*): intermediate (chamber) space; interval; interstice 043
ZWISCHENREAKTION (*f*): intermediate reaction 044
ZWISCHENREDE (*f*): digression; interlocution; interruption 045
ZWISCHENSATZ (*m*): parenthesis, incident 046
ZWISCHENSCHALTEN: to intercalate; inter-connect; fit between 047
ZWISCHENSCHIEBEN (*n*): interpolation, insertion, intercalation 048
ZWISCHENSENTE (*f*): intermediate ribband 049
ZWISCHENSPANT (*n*): intermediate frame 050
ZWISCHENSPIEL (*n*): interlude; intermezzo 051
ZWISCHENSTADIUM (*n*): intermediate stage 052
ZWISCHENSTEIN (*m*): blue metal; blue stone (Copper) 053
ZWISCHENSTRINGER (*m*): intercostal stringer 054
ZWISCHENSTÜCK (*n*): intermediate piece; distance piece 055
ZWISCHENSTUFE (*f*): intermediate stage 056
ZWISCHENTON (*m*): intermediate (medium) shade (Colours); intermediate tone (Sound) 057
ZWISCHENTRÄGER (*m*): intermediary; cross tie; transverse (intermediate) girder 058
ZWISCHENÜBERHITZER (*f*): intermediate super-heater 059
ZWISCHENÜBERHITZUNG (*f*): intermediate super-heating 060
ZWISCHENUMSTAND (*m*): incident 061
ZWISCHENWAND (*f*): partition; diaphragm; baffle-(wall); division wall 062
ZWISCHENWEITE (*f*): interval; distance between 063
ZWISCHENWELLE (*f*): intermediate shaft 064

ZWISCHENZEIT (*f*): interim; interval 065
ZWISCHENZELLRAUM (*m*): intercellular space 066
ZWISCHENZUSTAND (*m*): intermediate state 067
ZWIST (*m*): dispute; discord; disagreement; difference; disunion; dissension 068
ZWISTIG: quarrelling; discordant 069
ZWITSCHERN: to chirp; twitter 070
ZWITTER (*m*): hybrid; mongrel; hermaphrodite; bastard 071
ZWITTERARTIG: hybrid, androgynous 072
ZWITTERBLÜTIG: hermaphrodite, androgynous (Bot.) 073
ZWITTERPFLANZE (*f*): bastard plant 074
ZWÖLF: twelve; a dozen 075
ZWÖLFFACH: twelve-fold 076
ZWÖLFFÄLTIG: twelve-fold 077
ZWÖLFFINGERDARM (*m*): duodenum (Anat.) 078
ZWÖLFFLACH (*n*): dodecahedron 079
ZWÖLFFLÄCHIG: dodecahedral 080
ZWÖLFMAL: twelve times 081
ZWÖLFMÄNNIG: dodecandrian (Bot.) 082
ZWÖLFSEITIG: dodecahedral 083
ZWÖLFTE: twelfth 084
ZWÖLFTEILIG: duodecimal 085
ZWÖLFTEL (*n*): twelfth part; ($\frac{1}{12}$) 086
ZWÖLFTELFORMAT (*n*): duodecimo 087
ZYAN (*n*): cyanogen (see CYAN) 088
ZYANE (*f*): (corn)-blue bottle; common cornflower (*Centaurea cyanus*) 089
ZYANGAS (*n*): cyanogen gas (see CYAN) 090
ZYGOMA (*n*): zygoma, cheek-bone 091
ZYGOPHYLLEEN (*f.pl*): *Zygophylleæ* (Bot.) 092
ZYGOT (*n*): zygote (of cells) 093
ZYKLADEN (*f.pl*): Cyclades (Islands in the Greek Archipelago) 094
ZYKLAMEN (*n*): cyclamen, sow-bread (*Cyclamen europæum*) 095
ZYKLIDE (*f*): cyclide 096
ZYKLISCH: cyclic 097
ZYKLOIDE (*f*): cycloid 098
ZYKLOIDENVERZAHNUNG (*f*): cycloidal toothing 099
ZYKLON (*m*): cyclone; (also a blower used for dry process dust extraction from coal); cyclone (blower) fan 100
ZYKLOP (*m*): Cyclops 101
ZYKLOPENHAFT: cyclopean 102
ZYKLUS (*m*): circle; cycle; series 103
ZYLINDER (*m*): cylinder (Mech.); top hat; lamp chimney 104
ZYLINDERÄHNLICH: cylindric-(al) 105
ZYLINDERARTIG: cylindric-(al) 106
ZYLINDERBOHRMASCHINE (*f*): cylinder boring-machine 107
ZYLINDERDAMPFKESSEL (*m*): any cylindrical boiler, (usually Lancashire or Cornish boiler) 108
ZYLINDERDREHBANK (*f*): slide lathe 109
ZYLINDERFETT (*n*): cylinder grease 110
ZYLINDERFLÄCHE (*f*): cylindrical surface 111
ZYLINDERFÖRMIG: cylindrical 112
ZYLINDERHEMMUNG (*f*): cylinder escapement 113
ZYLINDERKRATZE (*f*): cylinder carder 114
ZYLINDERMANGEL (*f*): calender; lack of cylinders 115
ZYLINDERMANTEL (*m*): jacket, cylinder surface or case 116
ZYLINDERN: to calender 117
ZYLINDERÖL (*n*): cylinder oil 118
ZYLINDER (*m*), PUMPEN-: pump-cylinder 119
ZYLINDERSTOPFBÜCHSE (*f*): cylinder stuffing (gland) box 120
ZYLINDERWALZE (*f*): cylinder (cylindrical) roll 121
ZYLINDRIEREN: to calender 122
ZYLINDRISCH: cylindrical 123
ZYLINDROID (*n*): cylindroid 124
ZYMARIN (*n*): zymarine (a glycoside from *Apocynum cannabinum*); Mp. 144°C. 125
ZYMIN (*n*): zymine 126
ZYMOL (*n*): zymol, methyl isopropyl-benzol; $CH_3.C_6H_4.C_3H_7$; Sg. 0.87; Bp. 175°C. 127
ZYMOLOGIE (*f*): zymology 128
ZYMOSAN (*n*): zymosane (a disinfectant; fluosilicic acid containing formaldehyde) 129
ZYNISCH: cynical, misanthropic-(al) 130
ZYPERGRAS (*n*): cypress grass (*Cyperus fuscus*) 131
ZYPERGRASARTIG: cyperaceous 132
ZYPERN (*n*): Cyprus (Geog.) 133
ZYPERWURZ (*f*): galingale (*Cyperus longus*) 134
ZYPRESSE (*f*): cypress (*Cupressus sempervirens*) 135
ZYPRESSE (*f*), DEUTSCHE-: tamarisk (*Tamarix*) 136
ZYPRESSENGEWÄCHSE (*n.pl*): *Cupressaceæ* (Bot.) 137
ZYPRESSENNUSS (*f*): cypress cone 138
ZYPRESSENWOLFSMILCH (*f*): Cypress spurge (*Euphorbia cyparissias*) 139
ZYPRESSE (*f*), SUMPF-: (see ZYPRESSE, VIRGINISCHE-) 140
ZYPRESSE (*f*), VIRGINISCHE-: deciduous cypress (*Taxodium distichum*) 141
ZYPRISCH: Cyprian, of Cyprus 142
ZYSTE (*f*): cyst (Med.) 143

LIST OF ATOMIC WEIGHTS, SPECIFIC GRAVITIES, MELTING POINTS AND BOILING POINTS OF ELEMENTS

Chemical Sign.	No.	German Term.	English Term.	Atomic Weight.	General Data.
A.	144	ARGON	Argon		(see Ar.)
Ad.	145	ALDEBARANIUM	Aldebaranium	172	(see Yb)
Ag.	146	SILBER	Silver	107.88	(*Argentum*) Sg. 10.53; Mp. 961.5°C.; Bp. 1955°C.
Al.	147	ALUMINIUM	Aluminium	27.1	Sg. 2.7; Mp. 658°C.; Bp. 2300°C.
Ar.	148	ARGON	Argon	39.88	(Gas: liquifies at – 186.1°C.)
As.	149	ARSEN	Arsenic	74.96	Sg. 4.7-5.7; Mp. 850°C.
Au.	150	GOLD	Gold	197.2	(*Aurum*) Sg. 19.2; Mp. 1062°C.; Bp. 2530°C.
B.	151	BOR	Boron	10.82	Sg. 2.45; Mp. 2350°C.
Ba.	152	BARIUM	Barium	137.4	Sg. 3.78; Mp. 850°C.; Bp. 950°C.
Be.	153	BERYLLIUM	Beryllium	9.02	Sg. 1.85; Mp. 1280°C.
Bi.	154	WISMUT	Bismuth	209.0	Sg. 9.747; Mp. 268°C.; Bp. 1420°C.
Br.	155	BROM	Bromine	79.92	Sg. 3.1883; Mp. – 7.3°C.; Bp. 58.7°C.
Bv.	156	BREVIUM	Brevium	234	
C.	157	KOHLENSTOFF	Carbon	12.0	(*Carboneum*): boils at 50-60 cm. pressure; liquid at 40 cm. pressure.
Ca.	158	CALCIUM	Calcium	40.07	Sg. 1.5446; Mp. 805°C.
Cb.	159	COLUMBIUM	Columbium	93.5	(see Nb.)
Cd.	160	CADMIUM	Cadmium	112.4	Sg. 8.642; Mp. 321°C.; Bp. 766°C.
Ce.	161	CERIUM	Cerium	140.2	Sg. 6.92; Mp. 645°C.
Cl.	162	CHLOR	Chlorine	35.46	Sg. 2.491; Mp. – 102°C.; Bp. – 33.6°C.
Co.	163	KOBALT	Cobalt	58.97	Sg. 8.5; Mp. 1478°C.
Cp.	164	KASSIOPEIUM	Cassiopeium	174	(see Lu.)
Cr.	165	CHROM	Chromium	52.0	Sg. 6.92; Mp. 1520°C.; Bp. 2200°C.
Cs.	166	CÄSIUM	Cesium(Caesium)	132.8	Sg. 1.87; Mp. 28.5°C.; Bp. 670°C.
Cu.	167	KUPFER	Copper	63.57	(*Cuprum*) Sg. 8.96; Mp. 1083°C.; Bp. 2310°C.
Dy.	168	DYSPROSIUM	Dysprosium	162.5	
Em.	169	EMANATION	Emanation	222	(see Nt.)
Er.	170	ERBIUM	Erbium	167.7	
Eu.	171	EUROPIUM	Europium	152.0	
F.	172	FLUOR	Fluorine	19.0	Sg. 1.14 compared with air; Mp. – 223°C.; Bp. – 187°C.
Fe.	173	EISEN	Iron	55.84	(*Ferrum*) Sg. 7.85-7.88; Mp. 1530°C.; Bp. 2450°C.
Ga.	174	GALLIUM	Gallium	69.9	Sg. 5.96; Mp. 30.15°C.
Gd.	175	GADOLINIUM	Gadolinium	157.3	
Ge.	176	GERMANIUM	Germanium	72.5	Sg. 5.469; Mp. 958°C.
Gl.	177	GLUCINUM	Glucinum	9.02	(see BERYLLIUM) (Be.)
H.	178	WASSERSTOFF	Hydrogen	1.008	Sg. 0.06949; Mp. – 259°C.; Bp. – 252°C.
He.	179	HELIUM	Helium	4.0	Sg. 1.98 compared with hydrogen; Mp. – 269°C.; Bp. – 268.75°C.
Hg.	180	QUECKSILBER	Mercury	200.6	(*Hydrargyrum*) Sg. 13.59; Mp. – 38.85°C.; Bp. 357.33°C.
Ho.	181	HOLMIUM	Holmium	163.5	
I.	182	JOD	Iodine	126.92	Sg. 4.98; Mp. 114.2°C.; Bp. 184°C.
In.	183	INDIUM	Indium	114.8	Sg. 7.362; Mp. 155°C.; Bp. 700°C.
Io.	184	IONIUM	Ionium	230	
Ir.	185	IRIDIUM	Iridium	193.1	Sg. 15.86-22.42; Mp. 1950-2250°C.
J.	186	JOD.	Iodine	126.92	(see I.)
K.	187	KALIUM	Potassium	39.10	Sg. 0.8621; Mp. 63.5°C.; Bp. 757.5°C.
Kr.	188	KRYPTON	Krypton	82.9	Mp. – 169°C.; Bp. – 151.7°C.
La.	189	LANTHAN	Lanthanum	139.0	Sg. 6.154; Mp. 810°C.
Li.	190	LITHIUM	Lithium	6.94	Sg. 0.534; Mp. 186°C.; Bp. over 1400°C.

Chemical Sign.	No.	German Term.	English Term.	Atomic Weight.	General Data.
Lu.	191	LUTETIUM	Lutetium (Luteoium)	175.0	(see Cp.)
Mg.	192	MAGNESIUM	Magnesium	24.32	Sg. 1.69-1.75; Mp. 650°C.; Bp. 1120°C.
Mn.	193	MANGAN	Manganese	54.93	Sg. 7.42; Mp. 1260°C.; Bp. 1900°C.
Mo.	194	MOLYBDÄN	Molybdenum	96.0	Sg. 8.56; Mp. 2410°C.
N.	195	STICKSTOFF	Nitrogen	14.008	Sg. (gas) 0.96737; (liquid) 0.804, (solid) 1.0265; Mp. – 210.5°C.; Bp. – 195.5°C.
Na.	196	NATRIUM	Sodium	23.0	Sg. 0.9712; Mp. 97.6°C.; Bp. 877.5°C.
Nb.	197	NIOBIUM	Niobium	93.5	(see Cb.) Sg. 7.06; Mp. 1760°C.
Nd.	198	NEODYM	Neodymium	144.3	Sg. 6.9563; Mp. 840°C.
Ne.	199	NEON	Neon	20.2	(Gas): Mp. 840°C.
Ni.	200	NICKEL	Nickel	58.68	(*Nicolum*) Sg. 8.63; Mp. 1450°C.
Nt.	201	NITON	Niton	222	(Radium emanation) (Gas)
O.	202	SAUERSTOFF	Oyxgen	16.0	(*Oxygenium*) Sg. 1.10535; Mp. – 227°C.; Bp. – 182.5°C.
Os.	203	OSMIUM	Osmium	190.9	Sg. 22.48; Mp. 2500°C.
P.	204	PHOSPHOR	Phosphorous	31.04	Sg. 1.83; Mp. 44.4°C.; Bp. 287°C.
Pb.	205	BLEI	Lead	207.2	(*Plumbum*) Sg. 11.34; Mp. 327°C.; Bp. 1525°C.
Pd.	206	PALLADIUM	Palladium	106.7	Sg. 11.4-11.9; Mp. 1550°C.
Pr.	207	PRASEODYM	Praseodymium	140.9	Sg. 6.4754; Mp. 940°C.
Pt.	208	PLATIN	Platinum	195.2	Sg. 21.16; Mp. 1753°C.
R.	209	RUBIDIUM	Rubidium	85.5	Sg. 1.532; Mp. 39°C.; Bp. 696°C.
Ra.	210	RADIUM	Radium	226.0	Mp. 700°C.
Rb.	211	RUBIDIUM	Rubidium	85.5	(see R.)
Rh.	212	RHODIUM	Rhodium	102.9	Sg. 12.1; Mp. 1970°C.
Ru.	213	RUTHENIUM	Ruthenium	101.7	Sg. 8.6; Mp. over 1950°C.
S.	214	SCHWEFEL	Sulphur	32.07	Sg. 2.046; Mp. 116.5°C.; Bp. 444.6°C.
Sa.	215	SAMARIUM	Samarium	150.4	(see Sm.)
Sb.	216	ANTIMON	Antimony	121.8	(*Stibium*) Sg. 6.69; Mp. 630°C.; Bp. 1300°C.
Sc.	217	SCANDIUM (SKANDIUM)	Scandium	45.1	Mp. 1200°C.
Se.	218	SELEN	Selenium	79.2	Sg. 4.27; Mp. 217°C.; Bp. 690°C.
Si.	219	SILICIUM	Silicon	28.3	Sg. 2.49; Mp. 1420°C.; Bp. 3500°C.
Sm.	220	SAMARIUM	Samarium	150.4	Mp. 1300°C.
Sn.	221	ZINN	Tin	118.7	(*Stannum*) Sg. 7.298; Mp. 232°C.; Bp. 1450-1600°C.
Sr.	222	STRONTIUM	Strontium	87.6	Sg. 2.54; Mp. 900°C.
Ta.	223	TANTAL	Tantalum	181.5	Sg. 14.49; Mp. 2900°C.
Tb.	224	TERBIUM	Terbium	159.2	
Te.	225	TELLUR	Tellurium	127.5	Sg. 6.15; Mp. 449°C.; Bp. 1390°C.
Th.	226	THORIUM	Thorium	232.1	Sg. 11.12; Mp. over 1700°C.
Ti.	227	TITAN	Titanium	48.1	Sg. 4.5; Mp. 1795°C.
Tl.	228	THALLIUM	Thallium	204.4	Sg. 11.85; Mp. 302°C.; Bp. 1280°C.
Tm.	229	THULIUM	Thulium	169.4	
Tu.	230	THULIUM	Thulium	169.4	
U.	231	URAN	Uranium	238.2	Sg. 18.685; Mp. 800°C.
V.	232	VANADIUM	Vanadium	51.0	Sg. 6.025; Mp. 1730°C.
W.	233	WOLFRAM	Tungsten	184.0	Sg. 18.77; Mp. 3267°C.
X.	234	XENON	Xenon	130.2	Sg. 4.42; Mp. – 140°C.; Bp. – 109.1°C.
Xe.	235	XENON	Xenon	130.2	(see X.)
Y.	236	YTTRIUM	Yttrium	88.7	Sg. 3.8; Mp. 1250°C.
Yb.	237	YTTERBIUM	Ytterbium (Neoytterbium)	173.5	(see Ad.); Mp. 1800°C.
Yt.	238	YTTRIUM	Yttrium	88.7	(see Y.)
Zn.	239	ZINK	Zinc	65.37	Sg. 7.142; Mp. 419°C.; Bp. 918°C.
Zr.	240	ZIRKONIUM	Zirconium	90.6	Sg. 4.15-6.4; Mp. 1500-2350°C.

See also under Abbreviations.

ABBREVIATIONS

See also List of Atomic Weights, as single letters and combinations of letters sometimes represent chemical terms.

Abbreviation.	No.	German.	English.	Remarks.
a/	241	AN	on	As in CHECK A/LONDON, check on London
a/	242	AM	on, on the	
a- (as a prefix)	243	A-	anti-, non-	
a.	244	AUS	of	
a.	245	AN, AM	on	
a.	246	AR	are	
a.	247	AMIDO-	amido	
a.	248	ASYMMETRISCH	asymmetric-(al)-(ly)	
A.	249	ALKOHOL	alcohol	
A.	250	AMPERE	ampere	Unit of electric current
Ä.	251	ÄTHER	ether	
a.a.O.	252	AN ANDERN ORTEN; AN ANGEFÜHRTEN ORTEN, AN ANGEGEBEN ORTEN	elsewhere; in the place cited	
Abb.	253	ABBILDUNG	illustration, diagram, portrait, cut, figure	
abs.	254	ABSOLUT	absolute	
Abschn.	255	ABSCHNITT	section	
absol.	256	ABSOLUT	absolute	
Absolv.	257	ABSOLVENT	one who has completed his studios at college	
Abt.	258	ABTEILUNG	department, part, portion	
Ac.	259	AKTINIUM	actinium	a radioactive substance; atomic weight 226
AcEm.	260	AKTINIUMEMANATION	actinium emanation	a radioactive substance; atomic weight 218
a.d.	261	AN DER (etc. according to context)	at the, on, on the, to the	
Adr.	262	ADRESSE	address	
Ae.	263	ÄTHER	ether	
A.G.	264	ATOMGEWICHT	atomic weight	
A.G.	265	AKTIENGESELLSCHAFT	joint stock Co.	
Ah.	266	AMPERESTUNDE	ampere hour	
akad.	267	AKADEMISCH	academical-(ly)	
AKO	268	ARSENSAURES KOBALT-OXYDUL	Cobaltous arsenate	See KOBALTOXYDUL, ARSENSAURES-
Akt.-Ges	269	AKTIENGESELLSCHAFT	joint stock company	
Alk.	270	ALKOHOL	alcohol	
alkal.	271	ALKALISCH	alkaline	
alkoh.	272	ALKOHOLISCH	alcoholic	
Am.	273	AMERIKANISMUS	Americanism	
Amp.	274	AMPERE	ampère, amp., ampères, amps.	
An.	275	ANMERKUNG	remark, note	
anerk.	276	ANERKANNT	recognized	
Anfr.	277	ANFRAGE-(N)	enquiry, enquiries	
Ang.	278	ANGEBOT-(E)	application(s), offer(s)	
Angeb.	279	ANGEBOT-(E)	application(s), offer(s)	
angew.	280	ANGEWANDT	used, employed; applied	
Anh.	281	ANHYDRID	anhydride	
Anl.	282	ANLAGE-(N)	installation-(s); enclosure-(s)	
Anm.	283	ANMERKUNG	remark, note	
Ann.	284	ANNALEN	annals	(usually refers to ANNALEN DER CHEMIE or ANNALEN DER PHYSIK)
anorg.	285	ANORGANISCH	inorganic	
Anspr.	286	ANSPRUCH	requirement	

Abbreviation.	No.	German.	English.	Remarks.
App.	287	APPARAT	apparatus	
aq.	288	WASSER (*Aqua*)	water ; H_2O	
asymm.	289	ASYMMETRISCH	asymmetric-(al)	
at.	290	ATMOSPHÄRE	atmospheres	
At.-Gew.	291	ATOM GEWICHT	atomic weight	
äth.	292	ÄTHERISCH	etheral	
Atm.	293	ATMOSPHÄRE-(N)	atmosphere(s), atm., atms.	
Atm. abs.	294	ATMOSPHÄRE ABSOLUT	absolute pressure in atmospheres	
Atü.	295	ATMOSPHÄRE ÜBERDRUCK	atmospheres absolute pressure	
a.u.a.	296	AUCH UNDER ANDERN	also among others	
Aufl.	297	AUFLAGE	edition	
Aufst.	298	AUFSTELLUNG	statement	
Ausl.	299	AUSLAND, AUSLÄNDISCH	foreign ; export	
ausschl.	300	AUSSCHLIESSLICH	excluding, exclusive	
ausw.	301	AUSWAGE-(N)	quantity (quantities) weighed out	
ausw.	302	AUSWÄRTIG	foreign	
autom.	303	AUTOMATISCH	automatic	
A.W.	304	AMPEREWINDUNG	ampere winding	
ä.W.	305	ÄUSSERE WEITE	outside diameter, o.d.	
AZ.	306	ACETYLZAHL	acetyl number	
B.	307	BILDUNG	formation ; training, education, experience	
Bd.	308	BAND	volume	
Bé.	309	(GRADE) BAUMÉ	(degrees) Baumé	
ber.	310	BERECHNET	calculated	
Ber.	311	BERICHT	report	(sing. or plur.)
Berat. Ing.	312	BERATENDER INGENIEUR	Consulting Engineer	
bes.	313	BESORGT	attends to, takes charge of	
bes.	314	BESTIMMT	definite, intended for	
Best.	315	BESTIMMUNG	determination	
Best.	316	BESTELLUNG	order	
Bestz.	317	BESTELLZETTEL	order form	
betr.	318	BETREFFEND (BETREFFENDE, BETREFFENDER, BETREFFENDES, etc.)	concerned, said ; in question	at the heading of letters = reference, referring to, concerning
bez.	319	BEZIEHUNGSWEISE	or, respectively	
bez. auf	320	BEZOGEN AUF	based on, corresponding to, in conformity with	
bezw.	321	BEZIEHUNGSWEISE	or, respectively	
Bild.	322	BILDUNG	education, training, experience, character	
Bohrg.	323	BOHRUNG	bore	
Br.	324	BREITE	latitude	
b.w.	325	BITTE WENDEN	p.t.o., please turn over	
Bz.	326	BESTELLZETTEL	order form	
bzgl.	327	BEZÜGLICH	respecting, relative, with reference to, regarding,	
Bzl.	328	BENZOL	Benzole, Benzol	Commercial Benzene
bzw.	329	BEZIEHUNGSWEISE	or, respectively	
C.	330	CELSIUS	Centigrade	
C.	331	COULOMB	Coulomb	Unit of electric transfer
C.	332	ELEKTRISCHE KAPAZITÄT	electrical capacity	
c.	333	SPEZIFISCHE WÄRME	specific heat	
ca.	334	CIRCA, ZIRKA	about, approximately, approx., nearly	
Cal.	335	(KILOGRAMM)-CALORIE	kilogram calorie(s), large calorie	
cal.	336	(GRAMM)-CALORIE	gram calorie(s), small calorie	
calc.	337	CALCINIERT	calcined	
cbcm.	338	KUBIKZENTIMETER	cubic centimeter	
ccm.	339	KUBIKZENTIMETER	cubic centimetre	
cg.	340	ZENTIGRAMM	centigram(s)	

Abbreviation.	No.	German.	English	Remarks.
chem.	341	CHEMISCH	chemical-(ly)	
Chlf.	342	CHLOROFORM	Chloroform	
Cie.	343	COMPAGNIE, GESELL--SCHAFT	Company, Co., Coy.	
C.I.F.	344	C.I.F.	c.i.f.	of shipping
cl.	345	ZENTILITER	centilitre	
cm.	346	ZENTIMETER	centimetre	
corr.	347	CORRIGIERT	corrected	
Cos.	348	KOSINUS	cosine	Trigonometry
c_p.	349	SPEZIFISCHE WÄRME BEI KONSTANTEM DRUCK	specific heat at constant pressure	
cpl.	350	COMPLETT	complete	
cr.	351	COURANT	of this month, current, instant, curt., inst.	
Crist.	352	KRYSTALL(E)	Crystal(s)	
Crist.	353	KRYSTALLISATION	Crystallization	
crist.	354	KRYSTALLISIERT	crystallized	
crist.	355	KRISTALLINISCH	crystalline	
c_v.	356	SPEZIFISCHE WÄRME BEI KONSTANTEM VOLUMEN	specific heat at constant volume	
Cy.	357			Often used in German chemical formulæ in place of CN in cyano-gen compounds
d.	358	DER, DEN, DES, DEM, DIE, DAS	the, to the, of the etc.	according to the context
d.	359	DIESES	of this-(month), current, instant, curt., inst.	
d.	360	RECHTSDREHEND	dextrorotatory	
D.	361	DICHTE	density, specific gravity	
D.	362	DAMPFSCHIFF	steamship, ss .	
D.A.5	363	DEUTSCHES ARZNEIBUCH, 5TE. AUSGABE	German Pharmacopœia, 5th Edition	
D.A.B.	364	DEUTSCHES APOTHEKER--BUCH, *Pharmaco--poea Germanica*	German Pharmacopœia	
D.A.B.5	365	DEUTSCHES ARZNEI BUCH V. AUFLAGE	German Pharmacopœia 5th edition	
D.Ap.V	366	DEUTSCHES APOTHEKER--BUCH V. AUSGABE	German Pharmacopœia 5th edition	
Darst.	367	DARSTELLUNG	preparation	
D.A.V.	368	DEUTSCHES ARZNEIBUCH V. AUFLAGE	German Pharmacopœia 5th edition	
d. Bl.	369	DIESES BLATTES	of this paper	
DD.	370	DAMPFDICHTE	vapour density	
D.D.	371	DICHTEN	densities	
DE.	372	DIELEKTRIZITÄTSKON--STANTE	dielectric constant	
desgl.	373	DESGLEICHEN	the same, idem, ditto, similar, such-like	
dest.	374	DESTILLATION	distillation	
dest.	375	DESTILLIERT	distilled	
dg.	376	DEZIGRAMM	decigram(s)	
dgl.	377	DERGLEICHEN	the like, such-like	
dgl.	378	DESGLEICHEN, FERNER, DITO	ditto, the same (idem) (id.)	
d.h.	379	DAS HEISST	namely, that is, that is to say, (*id est*), i.e., which means	
d.i.	380	DAS IST	(*id est*), that is, namely	
Diff.	381	DIFFERENZ	difference	
Dipl.-Ing.	382	DIPLOM INGENIEUR	diploma engineer	
Diss.	383	DISSOZIATION, DISSOCIA--TION	dissociation	
Diss.	384	DISSERTATION	dissertation	
d.J.	385	DIESES JAHRES	of this year, of the present year	
dl.	386	DEZILITER	decilitre	

Abbreviation.	No.	German.	English.	Remarks.
dm.	387	DEZIMETER	decimetre	
d.M.	388	DIESES MONATS	of this month, inst., curt., instant, current	
D.P.a.	389	DEUTSCHE PATENTANMEL-DUNG	German patent application	
drgl.	390	DERGLEICHEN	the like, such-like	
D.R.G.M.	391	DEUTSCHES REICHS GE-SCHÜTZTES MUSTER	German State protected material	
D.R.P.	392	DEUTSCHES REICHSPATENT	German State Patent	
d.s.	393	DASS SIND	i.e., that is, namely	
ds.	394	DIESES-(MONATS)	this, of this (month), current, instant, curt., inst.	
dsgl.	395	DESGLEICHEN	ditto, likewise, such-like	
Dutz.	396	DUTZEND	dozen	
dz.	397	DOPPELZENTNER (100 kg.)	hundredweight, cwt. (of 100 kilograms)	
E.	398	ERSTARRUNGSPUNKT	freezing point, solidification point	
E.	399	ELEKTROMOTORISCHE KRAFT	electromotive force	
E.E.	400	ENTROPIEEINHEIT	entropy unit	
Eg.	401	EISESSIG	glacial acetic acid	
e.h.	402	EHRENHALBER	(in titles) honorary ; as an honour ; honoured with (on account of)	
eig.	403	EIGENE (etc. according to context)	own	
einschl.	404	EINSCHLIESSLICH	including, inclusive of	
Einw.	405	EINWIRKUNG	influence, action, effect	
EMK.	406	ELEKTROMOTORISCHE KRAFT	electromotive force, EMF	
entspr.	407	ENTSPRECHEND	corresponding	
Entw.	408	ENTWICKELUNG	development, evolution	
Er.	409	ERSTARRUNGSPUNKT or ERSTARRT	solidification point or solidifies	
erb.	410	ERBITTEN	beg, request	
erb.	411	ERBETEN	requested	
erfahr.	412	ERFAHRENER	experienced	
Erg.-Bd.	413	ERGÄNZUNGSBAND	supplementary (volume) number	
erh.	414	ERHITZT	heated	
Erh.	415	ERHITZUNG	heating	
Erst. P.	416	ERSTARRUNGSPUNKT	solidification point	
erw.	417	ERWÄRMT	warmed, heated	
ev.	418	EVENTUELL	eventually, perhaps, in the event of	
evang.	419	EVANGELISCH	protestant (religion)	
event.	420	EVENTUELL	eventually, eventual, perhaps, in the event of	
evtl.	421	EVENTUELL	eventually, perhaps, in the event of	
Ew.	422	EUER	Your (in forms of address)	
Exped.	423	EXPEDITION	office (of a newspaper)	
Extr.	424	EXTRAKT	extract	
EZ.	425	ESTERZAHL	ester number	
F.	426	FEINSTE SORTE	finest grade	
F.	427	FUSIONSPUNKT	fusion point, melting point, m.p.	
F.	428	FAHRENHEIT	Fahrenheit, Fahr., F.	
F.	429	FARAD	Farad	Unit of electric capacity
f.	430	FEST	solid	
f.	431	FEIN	fine	
f.	432	FOLGENDE	the following, onwards	
f.	433	FÜR	for	
f.a.B.	434	FREI AN BORD	f.o.b., free on board	
Fernspr.	435	FERNSPRECHER	telephone, Tel.	
F.f.	436	FORTSETZUNG FOLGT	to be continued	
ff.	437	SEHR FEIN	very fine	

Abbreviation.	No.	German.	English.	Remarks.
ff.	438	UND FOLGENDE	*et seq.*, and following	
f.f.	439	SEHR FEIN	very fine	
f.f.	440	UND FOLGENDE	*et seq.*, and following	
fff.	441	SEHR FEIN	very fine	
fff.	442	UND FOLGENDE	*et seq.*, and following	
FFS.	443	FFEINER SAFFLOR	very fine grade smalt	
Fig.	444	FIGUR	figure	
fl.	445	FLÜSSIG	liquid, fluid	
fl.	446	GULDEN	florins	coinage
Fl.	447	FLÜSSIGKEIT	liquid, fluid	
Fll.	448	FLÜSSIGKEITEN	liquids, fluids	
fm.	449	FESTMETER	theoretical cubic metre (of wood)	
folg.	450	FOLGEND	following	
fr.	451	FRANKEN	francs	coinage
freundl.	452	FREUNDLICHST	kindly	
Fr.-Off.	453	FRANKO OFFERTEN	post paid applications	
Fr.-Offerten	454	FRANKO-OFFERTEN	post-paid (postage paid) applications	
FS.	455	FEINER SAFFLOR	fine grade smalt	
fum.	456	FUMAROID	fumaroid	
Fussb.	457	FUSSBODEN	floor	
g.	458	GRAMM	gram(s)	
G.	459	GESELLSCHAFT	Company, Co., Coy.	
gasf.	460	GASFÖRMIG	gasiform, gaseous	
gcal.	461	GRAMMKALORIE	gram calories	
GCS.	462	GRAMM-ZENTIMETER-SEKUNDE	Gram centimeter-second, CGS	absolute system of measurement
G.E.	463	GEWICHTSEINHEIT	Imperial weight, unit of weight, Standard of Weight	
geb.	464	GEBOREN	born	
geb.	465	GEBUNDEN	bound	
geb.	466	GEBILDET	educated	
Gebr.	467	GEBRÜDER	Brothers, Bros.	
gef.	468	GEFÄLLIG, GEFÄLLIGST	obliging, if you please	
gef.	469	GEFUNDEN	found	
gefl.	470	GEFÄLLIG, GEFÄLLIGST	obliging, if you please, esteemed	
Gef.P.	471	GEFRIERPUNKT	freezing point	
geg.	472	GEGEN	against	
gegr.	473	GEGRÜNDET	founded	
gel.	474	GELÖST	dissolved	
gem.	475	GEMAHLEN	ground, powdered, pulverized	
ges.	476	GESETZLICH	by law	
ges.	477	GESÄTTIGT	saturated	
Ges.	478	GESELLSCHAFT	Company, Coy., Co., Society, Soc.	
ges. gesch.	479	GESETZLICH GESCHÜTZT	patented, protected by law	
Gew.	480	GEWICHT	weight, gravity	
gew.	481	GEWÖHNLICH	common, ordinary, general	
Gew.-%	482	GEWICHTS PROZENT	percentage by weight	
Gew.-T.	483	GEWICHTSTEIL	part by weight	
GG.	484	GASGEWINDE	gas-thread	
Ggw.	485	GEGENWART	presence	
Gl.	486	GLEICHUNG	equation	
G.m.b.H.	487	GESELLSCHAFT MIT BE-SCHRÄNKTER HAF-TUNG (HAFTPFLICHT)	Company with limited liability, Limited (liability) company, Company Limited, Co. Ld.	
gr.	488	GRANULIERT	granulated	
gr.	489	GRAMM	gram-(s), g.	
grm.	490	GRAMM	grams, g.	
gründl.	491	GRÜNDLICH	entirely, fundamentally basically	
H.	492	HOHEIT	Highness	
H.	493	HÄRTE	hardness	

Abbreviation.	No.	German.	English.	Remarks.
H.	494	HÖHE	height, altitude	
H.	495	HABEN	credit	
H.	496	HENRY	Henry	Unit of electric induction
H.	497	HEIZWERT	calorific value	
h.	498	HOCHSCHMELZEND	high-melting	
h.	499	HOCH	high	
h.	500	HEILIG	holy, sacred	
h.	501	HEISS	hot	
h.	502	STUNDE	hour	
ha.	503	HEKTAR	hectare	
H.D.	504	HOCHDRUCK	H.P., high pressure	
Hekt.	505	HEKTOLITER	hectolitre	
Herst.	506	HERSTELLUNG	production, construction, manufacture	
HK.	507	HEFNER-KERZE-(N)	Hefner candle-(s)	
hl.	508	HEKTOLITER	hectolitre	
Hl.	509	HALBLEDER	half-leather	
H.M.	510	HOHLMASS	imperial measure, standard of measure	
holl.	511	HOLLÄNDISCH	Dutch	
/Hr.	512	STUNDE	hour	
hydr.	513	HYDRIERT	hydrated	
i.	514	IST	is	
i.	515	IMIDO	imido	
i.	516	IN	in	
i.	517	IM	in the	
I.	518	ELEKTRISCHE STROM--STÄRKE	strength of electrical current	
Ia.	519	PRIMA	first class, first quality	
id.	520	IDEM	the same, ditto, idem	
i.D.	521	IM DAMPF	in vapour, in the form of vapour	
i.J.	522	IM JAHRE	in the year	
inact.	523	INACTIV	inactive	
Ing.	524	INGENIEUR	engineer	
Inh.	525	INHALT	capacity, content	
inkl.	526	INKLUSIV	inclusive, including, incl.	
i.V.	527	IM VAKUUM	in a vacuum	
I.W.	528	INNERE WEITE	inside (internal) diameter, i.d., width in the clear	
J.	529			Used in all German chemical formulæ in place of I in iodine compounds
J.	530	JOULE	Joule	Unit of electric work
J.	531	MECHANISCHES WÄRME--ÄQUIVALENT	mechanical heat equivalent	
J.	532	JAHR	year	
Jahrg.	533	JAHRGANG	year	
Jg.	534	JAHRGANG	year	
J.Z.	535	JODZAHL	Iodine number	
——k.	536	——KEIT		
k.	537	KALT	cold	
k.	538	KONSTANT	constant	
k.	539	KÖNIGLICH	royal	
k.	540	KÄISERLICH	Imperial	
K.	541	KÖNIGLICH	royal	
K.	542	KÄISERLICH	Imperial	
K.	543	ELEKTRISCHE DISSOCIA--TIONSKONSTANTE	electric dissociation constant	
K.	544	KALORIE	calorie	
Kal. (see Cal; cal.)	545	KILOGRAM KALORIE	Kilogram calorie	
kalz.	546	KALZINIERT	calcined	
Kap.	547	KAPITEL	chapter	
kathl.	548	KATHOLISCH	catholic (religion)	
Kcal. or kcal.	549	KILOGRAMMKALORIE	kilogram calories	
Kg. or kg.	550	KILOGRAMM	kg., kilograms	
kgl.	551	KÖNIGLICH	royal	

Abbreviation.	No.	German.	English.	Remarks.
Kl.	552	KLASSE	class	
km.	553	KILOMETER	kilometre	
k.M.	554	KOMMENDEN MONATS	of next month, prox., proximo	
Kn.	555	KNOTEN	knot	marine; 1-60th of a degree of latitude
KOH.	556	KOHLENSAURES KOBALT-OXYDUL	cobaltous carbonate	(See KOBALTOXYDUL, KOHLENSAURES-)
konst.	557	KONSTANT	constant	
konz.	558	KONZENTRIERT	concentrated	
korr.	559	KORRIGIERT	corrected	
Kp.	560	KOCHPUNKT, SIEDEPUNKT	boiling point, b.p.	
kpl.	561	KOMPLETT	complete	
kr.	562	KRAN	hoist, crane	
Krist.	563	KRISTALLISATION	crystallization	
Krist.	564	KRISTALL-(E)	crystal-(s)	
krist.	565	KRISTALLINISCH	crystalline	
krist.	566	KRISTALLISIERT	crystallized	
Krit. Temp.	567	KRITISCHE TEMPERATUR	critical temperature	
kryst.(see krist.)	568			
kVA.	569	KILOVOLTAMPERE	Kilovoltampere	
KW. or kW.	570	KILOWATT	kilowatt, K.W.	one thousand watts
K.Wh. or kWh.	571	KILOWATTSTUNDE	Kilowatt hour	equals 864 W.E.
KW.-stoff.	572	KOHLENWASSERSTOFF	hydrocarbon	
L.	573	SELBSTINDUKTIONSKO-EFFIZIENT	inductivity (Elect.)	
——l.	574	——LICH		
l.	575	LIES	read	
l.	576	LÖSLICH	soluble	
l.	577	LITER	litre	liquid measure
l.	578	LINKSDREHEND	levorotatory	
10 l.	579	ZEHNFACHE LÄNGE	ten-fold length, i.e., a bar having a length equal to ten times its diameter, for elongation test of metal	
L=10d.	580	L=10D.	length equals ten times the diameter	in elongation test of metals
lab.	581	LABIL	labile, unstable	
landw.	582	LANDWIRTSCHAFTLICH	agricultural	
langj.	583	LANGJÄHRIG	for many years	
lebensl.	584	LEBENSLAUF	career	
Lfd.	585	LAUFEND	current, running, consecutive, regular	
lfd. Nr.	586	LAUFENDE NUMMER	serial number; current number; running number (in a series or arithmetical progression), column or line where these are numbered consecutively	
lg.	587	LANG	long, in length	
Lg.	588	LIGROIN	ligroin	See BENZIN
linksdr.	589	LINKSDREHEND	levorotatory	
ll.	590	LEICHT LÖSLICH	easily soluble, readily soluble	
lösl.	591	LÖSLICH	soluble	
Lösl.	592	LÖSLICHKEIT	solubility	
Lsg.	593	LÖSUNG	solution	
Lsgg.	594	LÖSUNGEN	solutions	
lt.	595	LAUT	in accordance with, as per	
l.w.	596	LICHTE WEITE	inside diameter, width in the clear	
m.	597	MIT	with	
m.	598	MEIN (etc. according to context)	my, mine, of my	
m.	599	MINUTE	minute	
m.	600	METER	metre	
m.	601	META	meta	Chemical

Abbreviation.	No.	German.	English.	Remarks.
m./	602	MEIN, ETC.	mine, etc. (according to context)	
M.	603	MARK	Mark	coinage
M.	604	METER	Metre	long measure
M.	605	MASSE	Mass	
M.	606	MITTEL-(SORTE)	medium-(grade)	
M.	607	MONAT	Month	
mA.	608	MILLIAMPERE	milliampere	one thousandth part of an ampere
mal.	609	MALEINOID	maleinoid	
Masch.	610	MASCHINE	machine, engine	
max.	611	MAXIMAL, MAXIMUM	maximum	
m.b.H.	612	MIT BESCHRÄNKTER HAF-TUNG (HAFTPFLICHT)	with limited liability, Limited, Lim., Ld., Ltd.	
M.D.	613	MITTELDRUCK	intermediate pressure	
m.E.	614	MEINES ERACHTENS	in my opinion	
M.E.	615	MÜNZEEINHEIT	coinage	
M.E.	616	MACHE-EINHEIT	Mache unit	of radio active substances
mechan.	617	MECHANISCH	mechanical-(ly)	
med.	618	MEDIZINISCH	medical	
mg.	619	MILLIGRAMM	milligram(s)	
m.G.	620	MIT GOLDSCHNITT	with gilt edges	
M.G.	621	MOLEKULARGEWICHT	molecular weight	
Milchz.	622	MILCHZUCKER	milk-sugar, lactose	
Mill.	623	MILLION	million	
min.	624	MINIMAL, MINIMUM	minimum	
min.	625	MINUTE	minute	
Min.	626	MINUTE(N)	minute(s)	
Mitt.	627	MITTEILUNG	communication, report	
Mk	628	MARK	Mark	
Mk.	629	MARKEN	marks	German currency
mkr.	630	MIKROSKOPISCH	microscopic(al)	
ml.	631	MILLILITER	millilitre	
Mm. or mm.	632	MILLIMETER	millimetre	
Mod.	633	MODEL	model	
mögl.	634	MÖGLICH	possible	
Mol.	635	MOLEKÜL(E), MOLEKUL	molecule(s)	
Mol-.Gew.	636	MOLEKULARGEWICHT	molecular weight	
Mol.-Refr.	637	MOLEKULARREFRAKTION	molecular refraction	
Mol. Vol.	638	MOLVOLUM	molar volume	
Mon.	639	MONOGRAPHIE	monograph	
MS.	640	MITTLERER SAFFLOR	medium grade smalt	
m.W.	641	MEINES WISSENS	to the best of my knowledge, as far as I know	
MW	642	MEGAWATT	Megawatt	
n.	643	NUTZEFFEKT, WIRKUNGS-GRAD	efficiency	
n.	644	GESAMTWIRKUNGSGRAD	total efficiency, overall efficiency	
n.	645	NETTO	ne(t)t	
n.	646	NÖRDLICH	northern	
n.	647	NORMAL	normal	
n.	648	NEUTRAL	neutral	
N.	649	NORD, NORDEN	north	
N.	650	NACHTS	at night, p.m.	
N.	651	NACHMITTAGS	afternoon, p.m.	
N.	652	LEISTUNG	output; power-(factor), duty, load, usually horse power, H.P. or HP	
N.	653	WIRKUNGSGRAD; GES-AMT WIRKUNGSGRAD	efficiency, total (overall) efficiency	
Nachf.	654	NACHFOLGER	successors	
natürl. Grösse	655	NATÜRLICHE GRÖSSE	full size; life size	
N.D.	656	NIEDERDRUCK	L.P., low pressure	
Nd.	657	NIEDERSCHLAG	precipitate	
Ndd.	658	NIEDERSCHLÄGE	precipitates	

Abbreviation.	No.	German.	English.	Remarks.
Ne.	659	EFFEKTIVE LEISTUNG	effective output (power or duty), usually effective horse power, E.H.P.	
neutr.	660	NEUTRAL	neutral	
n.F.	661	NEUE FOLGE	new series	
N_h.	662	HYDRAULISCHER WIRK-UNGSGRAD	hydraulic efficiency	
Ni.	663	INDIZIERTE LEISTUNG	indicated output (power or duty), usually indicated horse power, I.H.P.	
niedr.	664	NIEDRIG	low	
NK.	665	NORMALKERZE	standard candle power	
Nm.	666	NACHMITTAGS	p.m., afternoon	
N_m.	667	MECHANISCHER WIRK-UNGSGRAD	mechanical efficiency	
n.M.	668	NÄCHSTEN MONATS	of next month, prox., proximo	
norm.	669	NORMAL	normal	
norm.	670	NORMEN	standards	
Nr.	671	NUMMER	number, No.	
Nro	672	NUMERO	number	
N_t.	673	THERMISCHER WIRKUNGS-GRAD	thermal (heat) efficiency	
N_w.	675	WIRTSCHAFTLICHER NUTZ-EFFEKT	economical efficiency	
N_w.	676	WIRTSCHAFTLICHER WIRK-UNGSGRAD	economical efficiency	
o.	678	ODER	or	
o.	679	OBEN	above	
o.	680	ORDINÄR	ordinary grade	
o.	681	ORTHO	ortho	(Chemical)
O.	682	OST	east	
O.	683	ORDINÄRE SORTE	normal grade	
od.	684	ODER	or	
O.D.	685	OPTISCHES DREHUNGSVER-MÖGEN	optical rotation	
o.dgl.	686	ODER DERGLEICHEN	or such-like, or the like, or similar	
o.drgl.	687	ODER DERGLEICHEN	or the like, or such-like, or similar	
Off. Nr.	688	OFFERTE NUMMER	quotation number	
opt. akt.	689	OPTISCH AKTIV	optically active	
ordentl.	690	ORDENTLICH	ordinary	
org.	691	ORGANISCH	organic	
OS.	692	ORDINÄR SAFFLOR	ordinary grade smalt	
Österr.	693	ÖSTERREICHISCH	Austrian	
p.	694	PARA	para	(Chemical)
p.	695	PRO	per	
Pa.	696	PROTAKTINIUM	protactinium	(a radioactive substance, atomic weight 230)
p.A.	697	PER ADDRESSE	care of, c/o	
PAe.	698	PETROLEUMÄTHER	petroleum ether	
Pat.	699	PATENT	patent, pat.	
-pctig.	700	-PROZENTIG	per cent.	
PE.	701	PASSEINHEIT	unit of fit	
pers.	702	PERSÖNLICH	private, personal	
Pf.	703	PFUND	pound	weight or sterling
Pf.	704	PFENNIG	pfennigs	
Pf.	705	PFERDE	horse, horse-power, H.P., HP.	
Pfg.	706	PFENNIG	pfennigs	
Ph. Belg.	707	*Pharmakopoea Belgica*	Belgian Pharmacopœia	
Ph.G.V.	708	*Pharmacopoea Germanica* V. AUFLAGE	German Pharmacopœia 5th. edition	

Abbreviation.	No.	German.	English.	Remarks.
Ph. Helvet.	709	*Pharmakopoea Helve-tica*	Swiss Pharmacopœia	
P.K.	710	PFERDEKRAFT	horse power, H.P., IP.	
PKO.	711	PHOSPHORSAURES KOBALT-OXYDUL	cobaltous phosphate	See KOBALTOXYDUL, PHOSPHORSAURES
p.m.	712	PRO MINUTE	per minute, per min.	
PO.	713	KOBALTOXYDUL; PROT-OXYD	cobaltous oxide	See KOBALTOXYDUL
P.P. (*praetermissis praemittendis*)	714	MIT AUSLASSUNG DES VORZUSCHICKENDEN	omitting address and other prefatory remarks	
prakt.	715	PRAKTISCH	practical-(ly)	
prim.	716	PRIMÄR	primary	
Proc.	717	PROCENT, PROCENTIG	percent., percentage, %	
Prod.	718	PRODUKT	product	
promov.	719	PROMOVIERT	graduated	at the University
Proz.	720	PROZENT, PROZENTIG	percent., percentage, %	
PS.	721	POSTSKRIPTUM	Postscript, PS.	
P.S.	722	PFERDESTÄRKE	horse power, H.P., HP.	
P.Se	723	EFFEKTIVE PFERDESTÄRKE	effective horse power, effective H.P.	
P.Sh.	724	PFERDESTÄRKESTUNDE	horse-power hour, HP/hr.	
P.S./h.	725	PFERDESTÄRKE/STUNDE	H.P./hr., horse power hour	
P.Si.	726	INDIZIERTE PFERDESTÄRKE	indicated horse power, I.H.P.	
P.T. (*praetermisso titulo*)	727	MIT AUSLASSUNG DES TITELS	omitting the title	
pulv.	728	PULVERISIERT	powdered, ground, pulverized	
Q.	729	WÄRMEMENGE	heat quantity	
Q.	730	ELEKTRIZITÄTSMENGE	electrical quantity	
q.	731	QUADRAT	square	
qcm.	732	QUADRATCENTIMETER	square centimetre, sq. cm.	
qdm.	733	QUADRATDECIMETER	square decimeter, sq. dm.	
qm.; q.M.	734	QUADRAT METER	square metres, sq.m.	
qmm.	735	QUADRATMILLIMETER	square millimetre, sq.mm.	
Q.S.	736	QUECKSILBERSTAND	mercury column	
q.v.	737	*Quod vide*	which see	
r.	738	RECHTSDREHEND	dextrorotatory	
R.	739	RÉAUMUR	Réaumur	
R.	740	RING	cycle, cyclic	
R.	741	REFERAT	review, abstract	
R.	742	ELEKTRISCHER WIDER-STAND	electrical resistance	
rac.	743	RACEMISCH	racemic	
RaEl.	744	RADIUMELEMENT	radium element	
RaEm.	745	RADIUMEMANATION	radium emanation	See EMANATION in list of Atomic Weights
raff.	746	RAFFINIERT	refined	
rauch.	747	RAUCHEND	fuming	
rd.	748	RUND	about, approx., approximately, nearly	
rechtsdr.	749	RECHTSDREHEND	dextrorotatory	
resp.	750	RESPEKTIVE	respectively, or, or rather	
rez.	751	REZIPROK	reciprocal-(ly); conversely	
Rhld.	752	RHEINLAND	Rhineland	
Rk.	753	REAKTION	reaction	
Rkk.	754	REAKTIONEN	reactions	
RKO.	755	REINES KOBALTOXYD	pure cobalt oxide	See KOBALTOXYD
rm.	756	RAUMMETER	cubic meter	
Rm.	757	REICHSMARK	Mark	coinage
R.P.	758	REICHS PATENT	State (Imperial) patent	
Russ.	759	RUSSISCH	Russian	
S.	760	SÄURE	acid	

Abbreviation.	No.	German.	English.	Remarks.
S.	761	SEITE	page	
S.	762	SANKT	St., Saint	
S.	763	SIEMENS	Siemens' ohm	
S.	764	ENTROPIE	entropy	
S.	765	SEKUNDE	second	
s.	766	SIEHE	see	
s.	767	SYMMETRISCH	symmetric-(al)-(ly)	
s.	768	SEKUNDE	second	
/S.	769	STUNDE	hour	
s.a.	770	SIEHE AUCH	see also	
Sa.	771	SUMMA.	together, total	
schm.	772	SCHMELZEND	melting	
schm.	773	SCHMILZT	melts	
Schmp.	774	SCHMELZPUNKT	melting point	
Sch.P.	775	SCHMELZPUNKT	melting point	
schr.	776	SCHRIFT, SCHRIFTLICH (from) SCHREIBEN	writing, in writing, to write	
Schweizer.	777	SCHWEIZERISCH	Swiss	
Sd.	778	SIEDEPUNKT	boiling point, b.p.	
sd.	779	SIEDEND	boiling	
sd.	780	SIEDET	boils	
s.d.	781	SIEHE DIES	see this, q.v., *quod vide*, which see	
sec.	782	SEKUNDE	second	
sec.	783	SEKUNDÄR	secondary	
s.G.	784	SPEZIFISCHES GEWICHT	specific (weight) gravity, sp.gr.	
s.g.	785	SOGENANNT	so-called	
S.K.	786	SEGER-KEGEL	Segar cone	See SEGERKEGEL
sll.	787	SEHR LEICHT LÖSLICH	very easily (readily) soluble	
Sm.	788	SCHMELZPUNKT, or SCHMILZT	melting point, or melts	
S.M.	789	SEINE MAJESTÄT	His (or Her) Majesty, (H.M.)	
s.o.	790	SIEHE OBEN	see above (*vide supra*)	
sof.	791	SOFORT, SOFORTIGEN	immediate	
sog.	792	SOGENANNT	so-called	
sogen.	793	SOGENANNT	so-called	
Söh.	794	SÖHNE	Sons	
sp.	795	SPINDEL	spindle	
spec.	796	SPECIFISCH	specific	See under spez.
spez.	797	SPEZIFISCH	specific	
spez.	798	SPEZIELL	especially	
spez. Gew.	799	SPEZIFISCHES GEWICHT	specific (weight) gravity	
Sp. G.	800	SPEZIFISCHES GEWICHT	specific (weight) gravity	
spr.	801	(from) SPRECHEN	to speak	
Sr.	802	SEINER	(title) His	
s.S.	803	SIEHE SEITE	see page	
SS.	804	SÄUREN	acids	
St.	805	STUNDE	hour	
St.	806	SANKT	St.,Saint	
St.	807	STÜCK	item, each, in number, pieces	
St.A.	808	STAMMAKTIEN	common stock, ordinary shares	
städl.	809	STÄDLICH	municipal	
Std.	810	STUNDE	hour	
Std.	811	STUNDEN	hours	
Stde.	812	STUNDE	hour	
Stdn.	813	STUNDEN	hours	
Stell.	814	STELLUNG	situation, post	
Str.	815	STRASSE	street	
s.u.	816	SIEHE UNTEN	see below	
subcut.	817	SUBKUTAN	subcutaneous-(ly); hypodermic-(ally)	
subl.	818	SUBLIMIERT	sublimes	
sublim.	819	SUBLIMIERT	sublimes	
s.W.	820	SPEZIFISCHE WÄRME	specific heat	

Abbreviation.	No.	German.	English.	Remarks.
swl.	821	SEHR WENIG LÖSLICH	very difficultly soluble	
s.w.u.	822	SIEHE WEITER UNTEN	see below	
symm.	823	SYMMETRISCH	symmetric-(al)	
SZ.	824	SÄUREZAHL	acid number	
s. Zt.	825	SEINER ZEIT	at that time, then	
t.	826	TONNE(N)	ton(s)	of 1,000 kilograms
T.	827	TONNEN	tons	
T.	828	TAUSEND	thousand	
T.	829	TEILE	parts	
T.Adr.	830	TELEGRAMM ADRESSE, TELEGRAPHISCHE ADRESSE	telegraphic address, tel. adr.	
techn.	831	TECHNISCH	technical-(ly)	
Telegr.	832	TELEGRAMM	telegram	
Telegr.-Adr.	833	TELEGRAMM ADRESSE	telegraphic address	
Temp.	834	TEMPERATUR	temperature	
tert.	835	TERTIÄR	tertiary	
ThEm.	836	THORIUMEMANATION	thorium emanation	a radio-active substance· atomic weight 220
Tl(e).	837	TEIL(E)	part(s)	
Tr.	838	(GRAD) TRALLES	(degrees) Tralles	
Tragk.	839	TRAGKRAFT	carrying capacity, load	
Tri.	840	TRICHLORÄTHYLEN	trichloroethylene	
Tw.	841	TWADDLE (GRADE)	Twaddle (Degrees)	
u.	842	UND	and	
u.	843	UNTEN	below	
u.	844	UNTER	among, under	
u.a.	845	UNTER ANDERN	(*inter alia*), among others, amongst others, moreover, furthermore	
u.a.m.	846	UND ANDERE MEHR	and others	
u.a.m.	847	UND ANDERES MEHR	and so forth, and so on	
u.ä.m.	848	UND ÄHNLICHES MEHR	and the like, and suchlike, and so on, and so forth and so on	
u.a.O.	849	UND ANDERE ORTE	and elsewhere	
u.a.O.	850	UNTER ANDEREN ORTEN	among other places	
Üb.	851	ÜBERTRAG	carried forward, c/f.; brought forward, b/f	
u.d.f.	852	UND DIE FOLGENDE	and those following	
u.dgl.	853	UND DERGLEICHEN	and such-like	
u.dgl. m.	854	UND DERGELICHEN MEHR	and such-like	
u.E.	855	UNSERES ERACHTENS	in our opinion	
Ueb.	856	ÜBERTRAG	brought forward, b/f.; carried forward, c/f.	
u.f.	857	UND FOLGENDE	and the following, onwards	
u.Mk.	858	UNTER DEM MIKROSKOP	under the microscope	
Uml/Min.	859	UMLAUFUNGEN PRO MIN-UTE	revolutions per minute, r.p.m.	
ung.	860	UNGEFÄHR	about, approximately	
unk.	861	UNKORRIGIERT	uncorrected	
unl.	862	UNLÖSLICH	insoluble	
uns.	863	UNSYMMETRISCH	unsymmetric-(al)	
unt.	864	UNTER	among, under, below	
Unters.	865	UNTERSUCHUNG	examination, investigation	
unt. Zers.	866	UNTER ZERSETZUNG	with decomposition	
unveränd.	867	UNVERÄNDERT	unchanged	
unveränd.	868	UNVERÄNDERLICH	unchangeable, invariable, constant	
U.p.M.	869	UMLAUFUNGEN PRO MIN-UTE	revolutions per minute, r.p.m.	
u.s.f.	870	UND SEINE FRÜHERE, etc. (according to context)	and its previous (number, volume, edition, article, etc.)	
u.s.f.	871	UND SO FORT	and so on, and so forth, etc.	

Abbreviation.	No.	German.	English.	Remarks.
u.s.F.	872	UND SEINE FORTSETZUNG	and its continuation, and its following number or volume	
u.s.w.	873	UND SO WEITER	and so forth, and so on, etcetera, etc.	
u.U.	874	UNTER UMSTÄNDEN	on occasion, occasionally, under certain conditions, in certain (circumstances) cases, when required	
u.ü.V.	875	UNTER ÜBLICHEN VORBE-HALT	E. & O.E.; errors and omissions excepted; with the usual reservations	
u.Z.	876	UNTER ZERSETZUNG	with decomposition	
u.Zers.	877	UNTER ZERSETZUNG	with decomposition	
u. zw.	878	UND ZWAR	i.e., *id est*, that is, namely	
v.	879	BENACHBART, NAHE	neighbouring, near, vicinal	
v.	880	VON	from, of etc.	
v.	881	VON	von (titles)	
V.	882	VORMITTAGS	a.m., in the forenoon or morning	
V.	883	VORKOMMEN	presence, occurrence	
V.	884	VOLT	volts	electric pressure unit
V.A.	885	VOLLAKTIE	fully paid up share	
v.Chr.	886	VOR CHRISTI	before Christ, B.C.	
V.d.I.	887	VEREIN DEUTSCHER INGEN-IEURE	(the title of the leading German engineering society)	
v.d.L.	888	VOR DEM LÖTROHR	before the blow-pipe	
Ver.	889	VEREINIGT	United	
Verb.	890	VERBINDUNG	compound	
Verb.	891	VERBAND	Union	
Verbb.	892	VERBINDUNGEN	compounds	
verd.	893	VERDÜNNT	dilute, diluted	
Verf.	894	VERFASSER	author	
Verf.	895	VERFAHREN	process	
vergl.	896	VERGLEICHE	refer, compare, cf., see	
verh.	897	VERHEIRATET	married	
Verh.	898	VERHÄLTNISSE	conditions	
Verl.	899	VERLAG	publishing house, publisher	
Vers.	900	VERSUCH	assay, test, experiment, trial, attempt	
Vf.	901	VERFASSER	author	
Vff.	902	VERFASSER	authors	
vgl.	903	VERGLEICHE	compare, see, refer, cf.	
v.H.	904	VOM HUNDERT, PROZENT	per hundred, per cent., %	
Visc.	905	VISCOSITÄT	viscosity	
v.J.	906	VORIGEN JAHRES	of last year	
v.J.	907	VOM JAHRE	of the year	
v.Jr.	908	VORIGEN JAHRES	of last year	
v.Jr.	909	VOM JAHRE	of the year	
v.M.	910	VORIGEN MONATS	of last month, ultimo, ult.	
Vm.	911	VORMITTAGS	morning, a.m.	
Vol.	912	VOLUM, VOLUMEN	Volume, Vol.	
Vol.-Gew.	913	VOLUMGEWICHT, VOLU-METRISCHES GEWICHT	volumetric weight (specific gravity)	
Vol.-%	914	VOLUMPROZENT	percentage by volume, volumetric percentage	
Vol.-T.	915	VOLUMTEIL	part by volume	
Vorm.	916	VORMITTAGS	morning, a.m.	
vorm.	917	VORMALS	previously, late, formerly	
Vortr.	918	VORTRAG	c.f., carried forward; b.f., brought forward	
v.T.	919	VON TAUSEND, ‰, PRO MILLE	per thousand	

Abbreviation.	No.	German.	English.	Remarks.
v.u.	920	VON UNTEN	from beneath, from below, from the bottom	
VZ.	921	VERSEIFUNGSZAHL	saponification number, saponification value	
w.	922	WERTEN	valued, esteemed	In the phrase IHRES WERTEN SCHREIBENS
w.	923	WARM	warm, hot	
W.	924	WASSER	water	
W.	925	WATT	Watt	unit of electric power
W.E.	926	WÄRMEEINHEIT	heat unit, calorie	1 W.E. equals 3.968 B.T.U. 1 W.E. per k.g. equals 9/5ths. B.T.U. per lb.
wf	927	WASSERFREI	anhydrous	
wl.	928	WENIG LÖSLICH	difficultly soluble, only slightly soluble	
wlösl.	929	WASSERLÖSLICH	water soluble	
w.o.	930	WEITER OBEN	above	
Wrkg.	931	WIRKUNG	effect, action	
wss.	932	WÄSSERIG	watery, aqueous, hydrous	
Wwe.	933	WITWE	widow	
XP	934	EILBOTE BEZAHLT	express messenger paid	Telegrams
xte.	935	X-TE	x-th	
z.	936	ZU, ZUR, ZUM	to, at, by, for	
Z.	937	ZEITSCHRIFT	journal, periodical	
Z.	938	ZEIT	time	
Z.	939	ZOLL	inch	
Z.	940	ZEILE	line	
za.	941	ZIRKA	about, approx., approximately, nearly	
zahlr.	942	ZAHLREICH	numerous	
z.B.	943	ZUM BEISPIEL	*exempli gratia*, e.g., for example, for instance, say, namely, as a case in point	
z.E.	944	ZUM EXEMPEL	for example, *exempli gratia*, e.g., for instance, as a case in point	
Zeitschr.	945	ZEITSCHRIFT	publication, periodical, journal	
zers.	946	ZERSETZEND	decomposing	
zers.	947	ZERSETZT	decomposes	
Zers.	948	ZERSETZUNG	decomposition	
Zeugn.	949	ZEUGNIS-(SE)	testimonial-(s); reference-(s)	
Zf.	950	ZINSFUSS	percentage (rate) of interest	
z.H.	951	ZU HÄNDEN	for the attention of, care of	
Ziv. Ing.	952	ZIVIL INGENIEUR	Civil Engineer	
zl.	953	ZIEMLICH LÖSLICH	fairly soluble	
Zle.	954	ZEILE	line	
Zp.	955	ZERSETZUNGSPUNKT	decomposition point	
z.T.	956	ZUM TEIL	partly, in part	
Ztg.	957	ZEITUNG	journal, newspaper	
z.Th.	958	ZUM TEIL	in part, partly	
Zus.	959	ZUSATZ	addition	In connection with patents
Zus.	960	ZUSAMMENSETZUNG	composition	
zus.	961	ZUSAMMEN	together, totalling	
Zus.-P.	962	ZUSATZPATENT	addition to a patent	
zw.	963	ZWAR	true, no doubt	
zw.	964	ZWISCHEN	between	
zwl.	965	ZIEMLICH WENIG LÖSLICH	rather difficultly soluble, only slightly soluble	
Zyl.	966	ZYLINDER	cylinder	
z.Z.	967	ZUR ZEIT.	at the time, at present, acting	
z.Zt.	968	ZUR ZEIT	at the time, at that time, then, at present, acting	

SIGNS AND SYMBOLS

Abbreviation.	No.	German.	English.	Remarks.
1/50	969	FÜNFZIGSTEL	1 : 50	one in fifty; chemical method of expressing the strength of a dilution
,,	970	DESGLEICHEN, DITO	ditto, id., idem, the same	
—— -(Strom)	971	GLEICHSTROM	Direct Current. D.C.	
~ -(Strom)	972	WECHSELSTROM	alternating current, A.C.	
∽	973	ÄHNLICHKEITSZEICHEN	sign of similarity	
∽	974	UNGEFÄHR	approximately	
≌ or ≈	975	UNGEFÄHR GLEICH	approximately equal	
...°	976	GRAD	degree(s)	
° ig.	977	—GRADIG	of —— degrees	
——° igem	978	—GRADIGEM	of (a temperature of) —— degrees	
°Tr.	979	GRADE TRALLES	degrees Tralles	
°Tw.	980	GRADE TWADDLE	degrees Twaddle	
%-ig.	981	PROZENTIG	percent	
%-ige Lsg.	982	PROZENTIGE-(LÖSUNG)	% : per cent-(solution)	
‰	983	PROMILLE	per thousand	
α	984	WÄRMEAUSDEHNUNGS-KOEFFIZIENT	coefficient of heat expansion	
r	985	VERDAMPFUNGSWÄRME	heat of vapourization, latent heat	
μ	986	MIKRON	micron	a millionth part of a metre
μ	987	REIBUNGSZAHL	coefficient of friction	
μ	988	MAGNETISCHE DURCHLÄSSIGKEIT	permeability	Magnetism
μ F	989	MIKROFARAD	mikrofarad	one millionth part of a Farad
μ μ	990	MIKROMILLIMETER	micromillimetre	0.000001 m.m.
v	991	GESCHWINDIGKEIT	velocity	
ω	992	WINKELGESCHWINDIGKEIT	angular velocity	
Ω	993	OHM	Ohm	unit of electric resistance
MΩ	994	MEGOHM	Megohm	
ζ	995	WIDERSTANDSZAHL FÜR FLÜSSIGKEITSSTRÖMUNG	coefficient of resistance to the flow of fluids	
ℋ	996	MAGNETISCHE AUFNAHMEFÄHIGKEIT	susceptibility	Magnetism
:	997	VERHÄLTNIS	ratio	
:	998	GLEICH	equals	
÷	999	VERHÄLTNIS	ratio	
÷	000	DIVISIONSZEICHEN or TEILUNGSZEICHEN	division sign	
. or ×	001	MULTIPLIZIERUNGSZEICHEN	multiplication signs	
x	002	UNBEKANNTE ZAHL	unknown quantity	Algebra
ϕ or Φ	003	DURCHSCHNITT	diameter	
ϕ or Φ	004	MITTLERER DURCHMESSER	mean diameter	
>	005	GRÖSSER ALS	greater than	
<	006	KLEINER ALS	smaller than	
+	007	PLUS; DAS POSITIVE ZEICHEN; DAS POSITIVE VORZEICHEN	plus; the positive sign	

Abbreviation.	No.	German.	English.	Remarks.
—	008	MINUS; DAS NEGATIVE ZEICHEN; DAS NEGA-TIVE VORZEICHEN	minus; the negative sign	
—	009	BIS, 1 BIS 4	to; 1 to 4	
± or ∓	010	PLUS ODER MINUS	plus or minus	
=	011	GLEICH	equals	
/ (as 1/2)	012	TEILUNGSZEICHEN	division sign	
/ (as 1/2)	013	SCHIEFER BRUCHSTRICH	slanting fractional line	
/ (as 1/2)	014	VERHÄLTNIS	ratio	
□	015	QUADRAT-	square-, sq., $—^2$.	
Φ	016	ENTROPIE	entropy	
y	017	ZWEITE UNBEKANNTE ZAHL	second unknown quantity	Algebra

BOTANICAL SECTION.

In botanical terms the letters " i " and " y " are very often interchangeable, as *Pirus* or *Pyrus*; *Silvestris* or *Sylvestris*. Often too, " x " and " z " and " i " and " j " are interchangeable. One should therefore remember to try these various alternatives when looking up these terms.

Where the German or English is not given, the Botanical term itself is usually employed. If the German or English term is lacking it may be taken that the Botanical term is the one commonly in vogue.

Botanical Term.	No.	English Term.	German Term.
Abelmoschos esculentus	018	Okra ; gumbo	ROSENPAPPEL (*f*)
Abelmoschos moschatus	019	Musk	BISAMSTRAUCH (*m*); BISAM (*m*)
Abies	020	Fir, fir-tree (See also under *Picea* and *Pinus*)	TANNE (*f*); WEISSTANNE (*f*)
Abies alba	021	White spruce (See *Abies pectinata* and *Pinus picea*)	EDELTANNE (*f*); WEISSTANNE (*f*)
Abies amabilis	022	(North American) purple fir, red fir	PURPUR-TANNE (*f*)
Abies balsamea	023	Balsam fir; Canadian balm of Gilead fir (See also *Abies subalpina*)	BALSAMTANNE (*f*); GUMMI--FICHTE (*f*)
Abies bifida	024	(See *Abies momi*)	
Abies brachyphylla	025	Japanese spruce (See *Abies homolepsis*)	
Abies bracteata	026	Californian fir; bearded fir	GRANNEN-TANNE (*f*)
Abies canadensis	027	Hemlock spruce (See *Picea alba* and *Tsuga canadensis*)	HEMLOCKRINDEBAUM (*m*); KANA--DISCHE TANNE (*f*): HEM--LOCKTANNE (*f*)
Abies cephalonica	028	Greek fir	GRIECHISCHE TANNE (*f*)
Abies cilicica	029	Asian fir	ZILIZISCHE TANNE (*f*)
Abies concolor	030	Grey fir	GLEICHFARBIGE TANNE (*f*)
Abies douglasi	031	Douglas fir; Oregon pine (See *Pseudotsuga Douglasii* and *Pseudotsuga taxifolia*)	
Abies excelsa	032	Spruce, pine, white fir; Norway spruce fir	FICHTE (*f*); PECHTANNE (*f*); ROTTANNE (*f*)
Abies firma	033	(See *Abies momi*)	
Abies fraseri	034	Fraser's fir	FRASERS-TANNE (*f*)
Abies grandis	035	Giant fir	RIESEN-TANNE (*f*)
Abies homolepsis	036	Nikka fir; Japanese fir (See *Abies brachyphylla*)	NIKKA-TANNE (*f*)
Abies larix	037	Larch (See under *Pinus larix*)	
Abies magnifica	038	(Californian) silver fir	PRACHT-TANNE (*f*)
Abies mariena	039	(See *Picea nigra*)	
Abies mariesi	040	Maries fir	MARIES-TANNE (*f*)
Abies momi	041	Momi fir (See *Abies firma*)	MOMI-TANNE (*f*)
Abies nobilis	042	Silver fir (See *Abies pectinata*)	SILBER-TANNE (*f*)
Abies nordmanniana	043	Nordmann's fir; Caucasus fir	NORDMANNS-TANNE (*f*)
Abies numidica	044	(North)-African fir	NUMIDISCHE TANNE (*f*)
Abies pectinata	045	Fir; silver fir; common silver fir, deal tree (See *Abies alba*; *Abies picea*; *Pinus picea*; *Pinus pectinata* · *Abies nobilis*)	TANNE (*f*); EDELTANNE; WEISS--TANNE (*f*)
Abies picea	046	Silver fir (See *Picea abies* and *Abies pectinata*)	EDELTANNE (*f*)
Abies pinsapo	047	Spanish fir	SPANISCHE TANNE (*f*)

Botanical Term.	No.	English Term.	German Term.
Abies sachaliensis	048	(Yesso)-silver fir (See *Picea ajanensis*)	SACHALIN-TANNE (*f*)
Abies sibirica	049	Siberian fir	SIBIRISCHE TANNE (*f*)
Abies spectabilis	050	Himalayan fir	HIMALAJA-TANNE (*f*)
Abies subalpina	051	North American balsam fir (See *Abies balsamea*)	WESTLÄNDISCHE BALSAM-TANNE (*f*)
Abies torano	052	(See *Picea polita*)	
Abies veitchi	053	Veitch's fir (See *Picea Veitchi*)	VEITCHS-TANNE (*f*)
Abietaceae	054	Abietaceæ (See *Araucaria, Abies, Picea, Pinus, Larix, Tsuga, Pseudotsuga* and *Cedrus*)	
Abroma angustum	055	Cocoa mallow	KAKAOMALVE (*f*)
Abronia fragans	056	Sand verbena	SANDEISENKRAUT (*n*)
Abrus precatorius	057	Wild liquorice; paternoster, crab's eyes, weather plant, rosary pea	PATERNOSTERERBSE (*f*); PATERNOSTERKRAUT (*n*)
Abutilon esculentum	058	Indian mallow	INDIANISCHE MALVE (*f*); SAMT-PAPPEL (*f*); SCHMUCKMALVE (*f*); BASTARDEIBISCH (*m*)
Acacia	059	Acacia	AKAZIE (*f*)
Acacia arabica	060	(Indian)-babul tree (See *Acacia nilotica* and *Acacia vera*)	BABULBAUM (*m*); SSANT (*m*); SONT (*m*); BABUL (*m*); KIKAR (*m*)
Acacia armata	061	Australian acacia, Kangaroo-thorn	KÄNGURUHDORN (*m*)
Acacia binervata	062	(Australian)-wattle (See *Acacia mollisima* and also under *Mimosa*)	MIMOSE (*f*)
Acacia catechu	063	Catechu acacia	KATECHUAKAZIE (*f*)
Acacia falcata	064	Bastard myall	BASTARDMYALLHOLZ (*n*)
Acacia fistula	065	(See *Acacia seyal*)	
Acacia giraffæ	066	Camel-thorn	KAMELDORN (*m*)
Acacia homalophylla	067	Myallwood; violet wood	MYALLHOLZ (*n*); VEILCHENHOLZ (*n*); BLAUES EBENHOLZ (*n*)
Acacia horrida	068	Whitethorn-acacia	WEISSDORNAKAZIE (*f*); KAPSCHO-TENDORN (*m*)
Acacia melanoxylon	069	Blackwood tree	SCHWARZERNUTZHOLZBAUM (*m*)
Acacia mollisima	070	(See *Acacia binervata*)	
Acacia nilotica	071	(See *Acacia arabica*)	
Acacia pendula	072	(Australian)-myall-(wood) tree	MYALLHOLZBAUM (*m*)
Acacia pycnantha	073	Golden wattle-(tree)	MIMOSA (*f*); MIMOSORINDEBAUM (*m*); MIMOSE (*f*); MIMOSEN-RINDEBAUM (*m*)
Acacia senegal	074	Senegal acacia (See *Acacia verek*)	HASCHAB (*m*); VEREK (*m*)
Acacia seyal	075	Shittah tree (See *Acacia fistula*)	SSOSSAR (*m*)
Acacia vera	076	(See *Acacia arabica*)	
Acacia verek	077	(See *Acacia senegal*)	
Acanthus ilicifolius	078	Tropical bear's-breech	TROPISCHE BÄRENKLAU (*f*)
Acanthus mollis	079	Bear's breech, bear's foot	BÄRENKLAU (*f*); LÖWENKLAU (*f*); WEICHE BÄRENKLAU (*f*)
Acanthus spinosus	080	Bear's-breech	BÄRENKLAU (*f*)
Acer	081	Maple	AHORN (*m*)
Aceraceæ	082	(See *Acerineæ*)	
Aceras anthropophora	083	Man orchis, green-man orchis	ORCHIDEE (*f*); ORCHE (*f*); MEN-SCHENORCHE (*f*); MENSCHEN-ÄHNLICHE ORCHE (*f*); OHN-HORN (*n*); MENSCHENÄHN-LICHES OHNHORN (*n*)
Acer californicum	084	Californian maple (See *Negundo californicum*)	KALIFORNISCHER AHORN (*m*)
Acer campestre	085	Maple, field maple, common maple	MASSHOLDER (*m*); FELDAHORN (*m*); MASHOLDER (*m*); MAS-ERLE (*f*); KREUZBAUM (*m*); MASSELLER (*m*); NORDISCHER MASSHOLDER (*m*)

Botanical Term.	No.	English Term.	German Term.
Acer circinatum;	086	Round-leaved maple	RUNDBLÄTTRIGER AHORN (*m*)
Acer cordifolium	087	(See *Acer tataricum*)	
Acer crataegifolium	088	Whitethorn-leaved maple	WEISSDORNBLÄTTRIGER AHORN (*m*)
Acer dasycarpum	089	Silver (white) maple, swamp maple	WEISSER AHORN (*m*); SILBER-AHORN (*m*); GEISSBAUM (*m*)
Acer ginnala	090	Manchurian maple	MANDSCHURISCHER AHORN (*m*)
Acer grandidentatum	091	(See *Acer saccharinum*)	
Acer hybridum	092	Hybrid maple; bastard maple	BASTARDAHORN (*m*)
Acerineæ	093	Acerineæ (See *Acer*)	AHORNGEWÄCHSE (*n.pl*); ACER-AZEEN (*f.pl*); ACERINEEN (*f.pl*)
Acer macrophyllum	094	Large-leaved maple	GROSSBLÄTTRIGER AHORN (*m*)
Acer monspessulanum	095	French maple (See *Acer trilobatum*)	FRANZÖSISCHER AHORN (*m*)
Acer negundo	096	American maple, box elder (See *Negundo aceroides*)	ESCHENBLÄTTRIGER AHORN (*m*)
Acerosæ	097	Conifers	NADELHÖLZER (*n.pl*)
Acer palmatum	098	Fan maple	FÄCHERAHORN (*m*)
Acer pennsylvanicum	099	Pennsylvanian maple	PENNSYLVANISCHER AHORN (*m*)
Acer platanoides	100	Norway maple	SPITZAHORN (*m*); SPITZBLÄTT-RIGER AHORN (*m*)
Acer pseudo-platanus	101	Sycamore; great maple; Scotch plane	BERGAHORN (*m*); STUMPFBLÄTT-RIGER AHORN (*m*); GEMEINER BERGAHORN (*m*); WEISSER AHORN (*m*); SYKOMORE (*f*); WALDAHORN (*m*)
Acer rubrum	102	Red-flowered maple; scarlet maple (See also *Acer dasycarpum*)	ROTBLÜTIGER AHORN (*m*); ROTER AHORN (*m*)
Acer saccharinum	103	Sugar maple; American (bird's-eye)-maple	ZUCKERAHORN (*m*); BERGAHORN (*m*); AMERIKANISCHER AHORN (*m*); VOGELAUGENAHORN (*m*)
Acer tataricum	104	Tartar maple	TATARISCHER AHORN (*m*)
Acer trilobatum	105	Trilobate maple	DREILAPPIGER AHORN (*m*)
Acetabularia	106	Acetabularia, green seaweed	GRÜNALGEN (*f.pl*); SCHIRM-GRÜNALGEN (*f.pl*)
Achillea ageratum	107	Sweet Maudlin	
Achillea atrata	108	Black milfoil	SCHWÄRZLICHE SCHAFGARBE (*f*)
Achillea millefolium	109	Yarrow; millefoil, milfoil	GARBE (*f*); SCHAFGARBE (*f*); LÄMMERSCHWÄNZCHEN (*n*); GEMEINE SCHAFGARBE (*f*)
Achillea moschata	110	Musk milfoil	MOSCHUSSCHAFGARBE (*f*); IVA (*n*); IVAKRAUT (*n*)
Achillea ptarmica	111	Sneezewort, the pearl, goose-tongue	NIESEKRAUT (*n*); WEISSER DOR-ANT (*m*); DEUTSCHER BERT-RAM (*m*); SUMPFGARBE (*f*); WEISSER RAINFARN (*m*); BE-RUFSKRAUT (*n*)
Achorion schoenleinii	112		FAVUSPILZ (*m*)
Achras sapota	113	Bullt tree; Sapodilla plum (See *Sapota achras*)	SAPOTILLBAUM (*m*); NISPERO (*m*); MISPELBOOM (*m*)
Achyrophorus maculatus	114	(See *Hypochœris maculata*)	ALPENFERKELKRAUT (*n*)
Aciphylla	115	Bayonet plant; Spear grass	
Acokanthera spectabilis	116	(See *Toxicophlœa spectabilis*)	NATAL-ACOCANTHERA (*f*)
Aconitum lycoctonum	117	Yellow wolf's-bane	GELBER EISENHUT (*m*); WOLFSEI-SENHUT (*m*)
Aconitum napellus	118	Monk's-hood; aconite; wolfs-bane; blue rocket; friar's cowl; wolfbane; blue wolfs-bane	ACONITKNOLLE (*f*); EISENHUT (*m*); EISENHÜTCHEN (*n*), MÖNCHSKAPPE (*f*); AKONIT (*m*); STURMHUT (*m*); VENUS-WAGEN (*m*); WOLFSWURZ (*f*)
Acorus calamus	119	Sweet-flag; sweet-grass; sweet-cane; calmus; calamus; sweet-sedge (See *Calamus aromaticus*)	KALMUS (*m*); MAGENWURZ (*f*)
Acorus gramineus	120	Japanese calamus, small calamus (See *Acorus spuriosus*)	KLEINER KALMUS (*m*)
Acorus maculatum	121	Cuckoo-pint	MAGENWURZ (*f*)

Botanical Term.	No.	English Term.	German Term.
Acorus spuriosus	122	Japanese calamus, myrtle-grass (See *Acorus gramineus*)	JAPANISCHER KALMUS (*m*)
Acrocomia sclerocarpa	123	Gru-gru palm; Macaw tree; great macaw palm	MACAWBAUM (*m*); MACAHUBA (*m*); MACOYA (*m*)
Acrostichum alicorne	124	(See *Platycerium aleicorne*)	
Acrostichum barbatum	125	Elephant's-ear fern (See *Acrostichum crinitum*)	ZEILFARN (*m*)
Acrostichum crinitum	126	(See *Acrostichum barbatum*)	
Actæa racemosa	127	Snake-root, black snake-root, bugbane	SCHLANGENWURZEL (*f*); NORD-AMERIKANISCHES CHRISTOPHSKRAUT (*n*); KLAPPERSCHLANGENWURZEL (*f*); SCHWINDSUCHTSWURZEL (*f*)
Actæa spicata	128	Baneberry; herb Christopher; toad-root	CHRISTOPHSKRAUT (*n*); SCHWARZKRAUT (*n*); CHRISTOPHSWURZ (*f*); ÄHRENTRAGENDES CHRISTOPHSKRAUT (*n*)
Actinella grandiflora	129	American dwarf sunflower	AMERIKANISCHE ZWERGSONNENBLUME (*f*)
Actinocarpus damasonium	130	Damasonium (See *Damasonium stellatum*)	
Actinomeris squarrosa	131	North American sunflower	NORDAMERIKANISCHE SONNENBLUME (*f*)
Actinomyces	132		STRAHLENPILZ (*m*)
Actinotus helianthi	133	Flannel flower	
Adansonia digitata	134	Baobab tree; monkey bread; sour gourd	AFFENBROTBAUM (*m*); BAOBAB (*m*); VINKA (*f*); TABALDIE (*f*); DINNA (*f*); MOWANA (*f*); MBUJU (*m*)
Adansonia gregorii	135	Sour gourd tree	SAUREGURKENBAUM (*m*)
Adenanthera bicolor	136	Barbados pride	
Adenophora	137	Adenophora, gland bell-flower	DRÜSENTRÄGER (*m* and *pl*)
Adenorachis	138	(See *Amelanchier vulgaris*)	
Adiantum Capillus-veneris	139	Maiden-hair-(fern)	KRULLFARN (*m*); FRAUENHAAR (*n*); HAARFARN (*m*); GEMEINES FRAUENHAAR (*n*)
Adiantum cuneatum	140	South American Maiden-hair fern	SÜDAMERIKANISCHER KRULLFARN (*m*); SÜDAMERIKANISCHER HAARFARN (
Adlumia cirrhosa	141	Alleghany vine; Climbing fumitory	
Adonis	142	Pheasant's-eye	ADONISRÖSCHEN (*n*); TEUFELSAUGE (*n*)
Adonis æstivalis	143	Pheasant's eye	ADONISRÖSCHEN (*n*)
Adonis autumnalis	144	(Corn)-pheasant's eye; (autumn)-pheasant's eye; Adonis flower; red chamomile	BLUTROTBLÄTTRIGES ADONISRÖSCHEN (*n*)
Adonis vernalis	145	(Spring)-pheasant's eye; Ox-eye	ZITRONENGELBES ADONISRÖSCHEN (*n*)
Adoxa moschatellina	146	Moscatel	BISAMKRAUT (*n*); MOSCHUSKRAUT (*n*)
Aecidiacei	147	Aecidiacei, fungus of living plants	PILZE DER LEBENDEN PFLANZEN; ÄZIDIEN (*pl*)
Aecidium	148	Cluster-cup	ROSTPILZ (*m*); BECHER-ROST (*m*)
Aecidium allii	149	Garlic cluster cup	LAUCH-BECHERROST (*m*)
Aecidium berberidis	150	Barberry-(parasitic cup)-fungus; barberry cluster-cup	BERBERITZEN-BECHERROST (*m*)
Aecidium dracontii	151	Arum cluster cup	ARON-BECHERROST (*m*)
Aecidium galii	152	Bedstraw-cluster-cup	LIEBFRAUENBETTSTROH-BECHERROST (*m*)
Aecidium grossulariæ	153	Gooseberry cluster cup	STACHELBEEREN-BECHERROST (*m*)
Aecidium leucospermum	154	Anemone cluster-cup	ANEMONEN-BECHERROST (*m*)
Aecidium menthæ	155	Mint cluster cup	MINZEN-BECHERROST (*m*)
Aecidium orchidearum	156	Orchis cluster-cup	ORCHIDEEN-BECHERROST (*m*)
Aecidium ranunculacearum	157	Buttercup cluster cup; crowfoot cluster cup	RANUNKEL-BECHERROST (*m*)
Aecidium tragopogonis	158	Goatsbeard cluster cup	BOCKSBART-BECHERROST (*m*)
Aecidium urticæ	159	Nettle cluster-cup	NESSEL-BECHERROST (*m*)
Aecidium violæ	160	Violet-cluster cup	VIOLETTEN-BECHERROST (*m*)

Botanical Term.	No.	English Term.	German Term.
Aegilops ovata	161	Hard grass ; goat grass (See *Triticum durum*)	ZIEGENAUGE (*f*)
Aegle marmelos	162	Bael fruit tree	BELANUSSBAUM (*m*)
Aegopodium podagraria	163	Gout-weed ; bishopweed ; gont-weed ; herb Gerard	GÄNSESTREUZEL (*n*) ; GIERSCH (*m*); GEISSFUSS (*m*)
Aeranthus	164	(See *Angræcum*)	
Aerides odoratum	165	Air-plant	LUFTBLUME (*f*)
Aeschynanthus atrosanguinea	166	Blush-wort	
Aeschynomene aspera	167	(Indian)-shola pith (solah pith) plant	SHOLASTRAUCH (*m*); SOLA--STRAUCH (*m*)
Aesculus alba	168	(See *Pavia alba*)	
Aesculus carnea	169	Red horse chestnut (See *Aesculus spectabilis* and *Aesculus pavia*)	ROTE ROSSKASTANIE (*f*) ; ROTER ROSSKASTANIENBAUM (*m*)
Aesculus chinensis	170	Chinese horse chestnut	CHINESISCHE ROSSKASTANIE (*f*)
Aesculus flava	171	Sweet buckeye (See *Aesculus lutea*)	
Aesculus floribunda	172	(See *Aesculus carnea*)	
Aesculus hippocastanum	173	Horse chestnut (See *Hippocastanum vulgare*)	ROSSKASTANIE (*f*) ; GEMEINE ROSSKASTANIE (*f*) ; PFERDE--KASTANIE (*f*) ; PFERDENUSS (*f*)
Aesculus humilis	174	Dwarf horse chestnut	NIEDRIGE ROSSKASTANIE (*f*)
Aesculus lutea	175	Yellow-flowered horse-chestnut (See *Pavia lutea*)	GELBBLÜHENDE ROSSKASTANIE (*f*)
Aesculus parviflora	176	Small-flowered horse-chestnut (See *Pavia alba*)	KLEINBLÜTIGE ROSSKASTANIE (*f*)
Aesculus pavia	177	Red-flowered horse-chestnut, red buckeye (See *Pavia rubra* and also *Aesculus carnea*)	ROTBLÜHENDE ROSSKASTANIE (*f*)
Aesculus rubicunda	178	(See *Aesculus carnea*)	
Aesculus rubra	179	(See *Pavia rubra*)	
Aesculus spectabilis	180	(See *Aesculus carnea*)	
Aethalium septicum	181	Flower-of-tan-(fungus)	
Aethlonema coridifolium	182	Lebanon candytuft; Candy mustard	LIBANON-BAUMSENF (*m*)
Aethusa cynapium	183	Fool's parsley	PETERSILIENSCHIERLING (*m*); HUNDSGLEISSE (*f*) ; GARTEN--SCHIERLING (*m*); HUNDSPE--TERSILIE (*f*) ; KATZENPETER--LEIN (*m*) ; HUNDSEPPICH (*m*)
Aframomum	184	(See *Amomum melegueta*)	
Agalmyla staminea	185	Scarlet root-blossom	
Agapanthus alba	186	White African lily	WEISSE LIEBESBLUME (*f*)
Agapanthus elobosus	187	Dwarf African lily	KLEINE LIEBESBLUME (*f*)
Agapanthus giganteus	188	Giant African lily	GROSSE LIEBESBLUME (*f*)
Agapanthus umbellatus	189	(blue)-African lily ; agapanthus ; Cape lily	LIEBESBLUME (*f*); SCHMUCK--LILIE (*f*); BLAUE LIEBES--BLUME (*f*)
Agaricaceæ	190	Agaricaceæ	AGARIKAZEEN (*f.pl*); BLÄTTER--SCHWÄMME (*m.pl*)
Agaricus	191	Agaric, leaf fungus, gill fungus	HERBSTMUSSERON (*m*) ; BLÄTTER--SCHWAMM (*m*); BLÄTTER--PILZ (*m*)
Agaricus adiposus	192		SCHUPPENPILZ (*m*)
Agaricus albus	193	(See *Polyporus officinalis* and *Boletus laricis*)	
Agaricus arvensis	194	Field agaric, meadow agaric, meadow mushroom, horse mushroom	SCHAFCHAMPIGNON (*n* and *m*)
Agaricus cæsareus	195	Imperial mushroom (See *Amanita cæsarea*)	KAISERSCHWAMM (*m*)
Agaricus campestris	196	Mushroom, common mushroom, edible mushroom, field agaric (See *Psalliota*)	CHAMPIGNON (*n* and *m*)
Agaricus chirurgorum	197	(See *Polyporus officinalis* and *Boletus laricis*)	WUNDSCHWAMM (*m*)

Botanical Term.	No.	English Term.	German Term.
Agaricus esculentus	198	Edible agaric, nail fungus (See *Collybia*)	ESSBARER NAGELSCHWAMM (*m*)
Agaricus fascicularis	199	(See *Hypholoma*)	BÜSCHELSCHWAMM (*m*)
Agaricus furcatus	200	Green agaric	GRÜNLING (*m*)
Agaricus gambosus	201	St. George's mushroom (See *Tricholoma*)	MAISCHWAMM (*m*)
Agaricus graveolens	202		GLÜHSCHWAMM (*m*)
Agaricus igneus	203		(AUSTRALISCHER)-LAMPEN--SCHWAMM (*m*)
Agaricus lampas	204		
Agaricus melleus	205	Hallimasche, fir-tree rot; fir-tree fungus; fir-tree gangrene; gill fungus; leaf fungus (See *Armillaria*)	BLÄTTERSCHWAMM (*m*); HONIG--PILZ (*m*); BLÄTTERPILZ (*m*); HALLIMASCH (*m*)
Agaricus muscarius	206	Toadstool, fly agaric (See *Amanita*)	FLIEGENPILZ (*m*); GIFTIGER PILZ (*m*)
Agaricus mutabilis	207	(See *Pholiota*)	STOCKSCHWAMM (*m*)
Agaricus noctilucens	208		(BORNEO)-NACHTLEUCHTE (*f*)
Agaricus olearius	209		OLIVENSCHWAMM (*m*)
Agaricus oreades	210		NELKENBLÄTTERSCHWAMM (*m*)
Agaricus ostreatus	211	Oyster mushroom (See *Pleurotus*)	BUCHENPILZ (*m*); DREHLING (*m*)
Agaricus pantherinus	212		PANTHERSCHWAMM (*m*)
Agaricus phalloides	213		KNOLLENBLÄTTERSCHWAMM (*m*)
Agaricus pomonæ	214		POMONASCHWAMM (*m*)
Agaricus pratensis	215		WIESENSCHWAMM (*m*)
Agaricus procerus	216	(See *Lepiota*)	PARASOLSCHWAMM (*m*)
Agaricus prunulus	217	(See *Clitopilus*)	MAISCHWAMM (*m*); MUSSERON (*m*); HERBSTMUSSERON (*m*)
Agaricus rubescens	218	(See *Amanita*)	GRAUER FLIEGENSCHWAMM (*m*); PERLENSCHWAMM (*m*)
Agaricus scorodonius	219		LAUCHSCHWAMM (*m*)
Agaricus silvaticus	220		WALDCHAMPIGNON (*n* and *m*)
Agathæa cœlestis	221	Cape aster, blue marguerite	KAPASTER (*f*)
Agathis australis	222	Kauri pine (See *Dammara australis*)	KAURIFICHTE (*f*); DAMMAR--FICHTE (*f*)
Agave americana	223	American aloe, Mexican soap-plant, century plant	HUNDERTJÄHRIGE ALOE (*f*); MAGUEY, METL
Agave heteracantha	224	Tampico-fibre tree	TAMPIKOFASERBAUM (*m*); ISTLE (*f*); IXTLE (*f*)
Agave rigida	225	Mexican fibre	INDIAFASER (*f*); CHELEM, SACCI, HENEGUEN
Agave vivipara	226	Bombay aloe-hemp	BOMBAY-ALOEHANFBAUM (*m*)
Ageratum mexicanum	227	Floss-flower, bastard agrimony	BASTARD-ODERMENNIG (*m*)
Aglaonema	228	Poison-dart	
Agraphis non-scriptus	229	Blue-bell	
Agraphis nutans	230	English blue-bell (See *Scilla nutans*)	
Agrimonia eupatoria	231	Agrimony; common agrimony; liverwort (See *Agrimonia officinalis*)	LEBERKLETTE (*f*); ODERMENNIG (*m*); GEMEINER ODERMENNIG (*m*); STEINWURZ (*f*); ACKER--MENNIG (*m*); HEIL ALLER WELT (*n*); HEILANDSTEE (*m*)
Agrimonia odorata	232	Scented agrimony (See *Agrimonia eupatoria*)	WOHLRIECHENDER ODERMENNIG (*m*)
Agrimonia officinalis	233	(See *Agrimonia eupatoria*)	
Agropyron repens	234	(See *Agropyrum repens*)	
Agropyrum acutum	235	(See *Agropyrum repens*)	
Agropyrum barbata	236	(See *Agropyrum repens*)	
Agropyrum caninum	237	Dog-grass; fibrous agropyrum	HUNDSWEIZEN (*m*)
Agropyrum junceum	238	(See *Agropyrum repens*)	QUECKE (*f*); MEER-ZWECKEN (*m*); SEEZWECKEN (*m*)
Agropyrum littorale	239	(See *Agropyrum repens*)	
Agropyrum repens	240	Dog's-grass; quick-grass; quickens; couch grass, triticum, dog-grass, graminis; knot-grass; wheat-grass; dog-wheat; witch-grass; quitch-grass (See *Triticum repens*)	GRASWURZEL (*f*); KAMMGRAS (*n*); QUECKE (*f*); QUECKEN--GRAS (*n*); HUNDSWEIZEN (*m*); PÄDENGRAS (*n*); ZWECKEN (*m*)

Botanical Term.	No.	English Term.	German Term.
Agrostemma coronaria	241	(See *Lychnis coronaria*)	
Agrostemma flos-Jovis	242	(See *Lychnis flos-Jovis*)	
Agrostemma githago	243	Corn cockle; corn campion (See *Lychnis githago*)	RADE (*f*); RADEN (*m*); KORN-RADE (*f*); ACKERKRONE (*f*); KORNROSE (*f*)
Agrostis	244	Bent grass, cloud grass, spear grass	WINDHALM (*m*); STRAUSSGRAS (*n*)
Agrostis alba	245	Creeping bent grass, fine-top grass (See *Agrostis vulgaris*)	KLEINE QUECKE (*f*); FIORINGRAS (*n*); RIEDGRAS (*n*)
Agrostis alba stolonifera	246	Fiorin-(grass) (See *Agrostis stolonifera latifolia*)	FIORINGRAS (*n*)
Agrostis anemagrostis	247	(See *Agrostis spica-venti*)	
Agrostis canina	248	Rhode Island bent grass	HUNDSSTRAUSSGRAS (*n*)
Agrostis interrupta	249	(See *Agrostis spica-venti* and *Apera interrupta*)	
Agrostis nebulosa	250	cloud-grass, Spanish bent grass	NEBELGRAS (*n*)
Agrostis nigra	251	(See *Agrostis alba*)	
Agrostis pumila	252	(See *Agrostis alba*)	
Agrostis rubra	253	Red bent-grass	HERDENGRAS (*n*)
Agrostis scabra	254	Tickle-grass; fly-away grass; windward spiked grass; silky bent grass	HAARGRAS (*n*)
Agrostis setacea	255	Bristle agrostis; bristle bent grass	
Agrostis spica venti	256	Tickle-grass; fly-away grass; windward spiked grass; silky bent grass; silky agrostis (See *Apera spica-venti*)	WINDHALM (*m*); SCHLINGGRAS (*n*); TAUGRAS (*n*); GROSSE MEDDEL (*f*)
Agrostis stolonifera latifolia	257	(See *Agrostis alba stolonifera*)	
Agrostis vulgaris	258	Bent grass; red top; creeping bent grass; wiry grass (See *Agrostis alba*)	GEMEINES STRAUSSGRAS (*n*); KLEINE MEDDEL (*f*)
Ailanthus glandulosa	259	Ailanthus; tree of heaven; tree of the gods; varnish tree	GÖTTERBAUM (*m*); GEMEINER GÖTTERBAUM (*m*); BAUM DES HIMMELS (*m*)
Ailanthus procera	260	(See *Ailanthus glandulosa*)	
Aira alpina	261	(See *Aira cæspitosa*)	ALPENSCHMIELE (*f*)
Aira cæspitosa	262	Tufted hair grass; tufted aira	GEMEINE SCHMIELE (*f*); RASEN-SCHMIELE (*f*)
Aira canescens	263	Grey aira; grey bent-grass (See *Weingærtneria*)	HAFERSCHMIELE (*f*)
Aira caryophyllea	264	Silvery aira; silvery bent-grass; silvery hair-grass	SCHMIELE (*f*)
Aira flexuosa	265	Wavy aira; wavy bent-grass	HAFERSCHMIELE (*f*)
Aira lævigata	266	(See *Aira cæspitosa*)	
Aira montana	267	(See *Aira flexuosa*)	
Aira præcox	268	Early aira; early bent-grass; early hairgrass	
Aira setacea	269	(See *Aira flexuosa*)	
Aira uliginosa	270	(See *Aira flexuosa*)	
Aizoon hispanicum	271	Spanish evergreen	SPANISCHES IMMERGRÜN (*n*)
Ajuga chamæpitys	272	Ground pine; yellow bugle	
Ajuga genevensis	273	Erect bugle, gout ivy	
Ajuga pyramidalis	274	Pyramidal bugle (See *Ajuga genevensis*)	
Ajuga reptans	275	Common bugle; creeping bugle	GÜNSEL (*m*); KRIECHENDER GÜN-SEL (*m*)
Albugo candida	276	White rust (See *Cystopus candidus*)	KETTENSPOR-EIPILZ (*m*)
Alcea	277	(See *Althæa*)	
Alchemilla alpina	278	Alpine Lady's mantle; alpine alchemil	ALPENFRAUENMANTEL (*m*)
Alchemilla argentea	279	(See *Alchemilla alpina*)	
Alchemilla arvensis	280	Parsley piert; field lady's mantle; field alchemil; piert parsley	FELDFRAUENMANTEL (*m*)
Alchemilla conjuncta	281	(See *Alchemilla alpina*)	
Alchemilla hybrida	282	(See *Alchemilla vulgaris*)	
Alchemilla montana	283	(See *Alchemilla vulgaris*)	

Botanical Term.	No.	English Term.	German Term.
Alchemilla phanes	284	(See *Alchemilla arvensis*)	
Alchemilla vulgaris	285	Common Lady's mantle; common alchemil	MARIENMANTEL (*m*); FRAUEN-MANTEL (*m*); SINAU (*m*)
Aleurites cordata	286	Candle-nut tree, Tung chou (See *Elæococca vernicia*)	TUNGBAUM (*m*); ÖLFIRNISBAUM (*m*)
Aleurites moluccana	287	Candlenut (See *Aleurites triloba*)	BANKULNUSS (*f*); LICHTNUSS (*f*); CANDLENUSSBAUM (*m*); KER-ZENNUSS (*f*); LACKBAUM (*m*)
Aleurites triloba	288	Candlenut, candle-berry (See *Aleurites moluccana*)	BANKULNUSS (*f*); KERZENBEERE (*f*)
Algæ	289	Algæ; seaweed (See *Characeæ, Chlorophyceæ, Confervaceæ, Conjugatæ, Cyanophyceæ, Diatomaceæ, Florideæ, Phæophyceæ, Protococcoideæ, Rhodophyceæ, Siphoneæ, Volvocineæ*)	ALGE (*f*)
Alhagi camelorum	290	(Asian)-camel's thorn (See *Alhagi maurorum*)	ALHAGISTRAUCH (*m*); MANNA-KLEE (*m*)
Alhagi maurorum	291	(See *Alhagi camelorum*)	
Alisma	292	Water plantain, deil's spoons	FROSCHLÖFFEL (*m*); KRÄHEN-ZEHE (*f*)
Alismaceæ	293	Water plantains	WASSERLIESCHE (*n.pl*); ALISMA-ZEEN (*f.pl*); FROSCHLÖFFEL-PFLANZEN (*f.pl*); WASSER-LIESCHGEWÄCHSE (*n.pl*)
Alisma lanceolatum	294	Spear-leaved water plantain (See *Alisma plantago*)	
Alisma natans	295	Water plantain	FROSCHLÖFFEL (*m*)
Alisma plantago	296	Great water-plantain, thrum-wort (See *Alisma lanceolatum*)	WASSERWEGERICH (*m*); FROSCH-LÖFFEL (*m*)
Alisma ranunculoides	297	Lesser water-plantain	FROSCHLÖFFEL (*m*)
Alisma repens	298	(See *Alisma ranunculoides*)	
Alkanna tinctoria	299	Dyer's alkanet (See *Anchusa tinctoria*)	FÄRBEROCHSENZUNDE (*f*); AL-KANNA (*f*); ROTE SCHLANGEN-WURZEL (*f*); ALKANNAWUR-ZELPFLANZE (*f*)
Allanblackia stuhlmannii	300	Mkani, East-African fat-tree	OSTAFRIKANISCHER FETTBAUM (*m*)
Alliaria officinalis	301	Jack-by-the-hedge; sauce-alone; mustard garlic; garlic mustard (See *Sisymbrium alliaria*)	KNOBLAUCHHEDERICH (*m*); LAUCHHEDERICH (*m*)
Allium ampeloprasum	302	leek; wild leek (See *Allium porrum*)	PORREE (*m*); SOMMERPORREE (*m*)
Allium arenarium	303	(See *Allium scorodoprasum*)	
Allium ascalonicum	304	Shalot, shallot	SCHALOTTE (*f*); ESCHLAUCH (*m*); ASKALONISCHE ZWIEBEL (*f*)
Allium bulbiferum	305	(See *Allium scorodoprasum*)	
Allium carinatum	306	(See *Allium oleraceum*)	
Allium cepa	307	Onion	GEMEINE ZWIEBEL (*f*); BOLLE (*f*); SOMMERZWIEBEL (*f*); KÜCH-ENZWIEBEL (*f*)
Allium cepa aggregatum	308	Potato onion	
Allium cepa proliferum	309	Tree onion, Egyptian onion	BAUMZWIEBEL (*f*); ÄGYPTISCHE ZWIEBEL (*f*)
Allium compactum	310	(See *Allium vineale*)	
Allium complanatum	311	(See *Allium oleraceum*)	
Allium fistulosum	312	shalot; shallot; Welsh onion, Ciboul onion	SCHLOTTENZWIEBEL (*f*); SCHNITT-ZWIEBEL (*f*); WINTERZWIE-BEL (*f*); EWIGE ZWIEBEL (*f*); KLÖWEN (*m*); JAKOBSLAUCH (*m*); JOHANNISLAUCH (*m*); RÖHRENLAUCH (*m*)
Allium holmense	313	(See *Allium scorodoprasum*)	
Allium neapolitanum	314	Daffodil garlic	
Allium oleraceum	315	Field leek; garlic leek; field garlic; cabbage garlic	KOHLLAUCH (*m*)
Allium ophioscorodon	316	Pearl onion	PERLZWIEBEL (*f*)

Botanical Term.	No.	English Term.	German Term.
Allium porrum	317	Leek (See *Allium ampeloprasum*)	LAUCH (*m*); PORREE (*m*); BORREE (*m*); WINTERPORREE (*m*); WELSCHZWIEBEL (*f*); GE-MEINER LAUCH; SPANISCHER LAUCH; ASCHLAUCH; FLEISCH-LAUCH (*m*)
Allium sativum	318	Garlic	KNOBLAUCH (*m*); LAUCHKNOB-LAUCH (*m*)
Allium sativum var. Ophioscorodon	319	Garden rocambole, pearl onion	SCHLANGENLAUCH (*m*); ROGGEN-BOLLE (*f*)
Allium schoenoprasum	320	Chives; Spring onion	SCHNITTLAUCH (*m*); BREISLAUCH (*m*); GRASLAUCH (*m*); HOHL-LAUCH (*m*); JAKOBSLAUCH (*m*); JOHANNISLAUCH (*m*); SUPPENLAUCH (*m*); ZWIEBEL-LAUCH (*m*)
Allium scorodoprasum	321	Rocambole; sand leek; Spanish garlic	ZWIEBEL (*f*); ROCKENBOLLE (*f*); SCHLANGENKNOBLAUCH (*m*)
Allium sibiricum	322	(See *Allium schoenoprasum*)	
Allium sphærocephalum	323	Round-headed leek	
Allium triquetrum	324	Triquetrous leek	
Allium ursinum	325	Ramsons	BÄRENLAUCH (*m*)
Allium victorialis	326	Wild mandrake, serpent's garlic	WILDER ALRAUN (*m*); BERG-AL-RAUN (*m*); SIEGWURZ (*f*); ALLERMANNSHARNISCH (*m*); NETZWURZELIGER LAUCH (*m*); SCHLANGENLAUCH (*m*)
Allium vineale	327	Crow garlick	
Allosorus crispus	328	Rock brakes; rock brake; rock bracken; parsley fern	
Alnus alnobetula	329	alpine alder	DROSSEL (*f*)
Alnus americana	330	American alder (See *Alnus serratula*)	AMERIKANISCHE ERLE (*f*)
Alnus cordata	331	Cordate-leaved alder	HERZBLÄTTRIGE ERLE (*f*)
Alnus glutinosa	332	Alder; black alder; common alder; Aar	(KLEBRIGE)-ERLE (*f*); ELLER (*f*); ROTERLE (*f*); SCHWARZERLE (*f*); ELSE (*f*); ERLENBAUM (*m*); LÄUSEBAUM (*m*); PUL-VERHOLZ (*n*); PURGIERBAUM (*m*); HOHLKIRSCHE (*f*)
Alnus incana	333	White alder, Northern alder	WEISSERLE (*f*); GRAUERLE (*f*); NORDISCHE GRAUERLE (*f*)
Alnus japonica	334	Japanese alder	JAPANISCHE ERLE (*f*)
Alnus oregona	335	Oregon alder (See *Alnus rubra*)	OREGON ERLE (*f*)
Alnus rubra	336	Red alder	ROTERLE (*f*)
Alnus serrulata	337	American alder	HASELERLE (*f*)
Aloe hepatica	338	hepatic aloe, curacao	LEBERALOE (*f*); MATTE ALOE (*f*); CURASSAO-ALOE (*f*); NATAL-ALOE (*f*)
Aloe lucida	339	Cape aloe	KAPALOE (*f*); GLÄNZENDE ALOE (*f*); CURASSAO-ALOE (*f*)
Aloe marginata	340	Pearl aloe	PERLENALOE (*f*)
Aloe socotrina	341	(See *Aloe vulgaris*)	SANSIBARALOE (*f*); OSTAFRIKAN-ISCHE ALOE (*f*); BARBADOS-ALOE (*f*)
Aloe vulgaris	342	Aloe-(s) (See *Aloe socotrina*)	ALOË (*f* and *pl*)
Aloexylon agallochum	343	Eagle wood (See *Aquilaria agallocha*)	ALOEHOLZBAUM (*m*); CALAMBAC-BAUM (*m*)
Alopecurus agrestis	344	Slender foxtail	ACKERFUCHSSCHWANZ (*m*)
Alopecurus alpinus	345	Alpine foxtail	ALPENFUCHSSCHWANZ (*m*)
Alopecurus bulbosus	346	Fox-tail grass	KOLBENGRAS (*n*)
Alopecurus geniculatus	347	Marsh foxtail (See *Alopecurus bulbosus*)	GEKNIETER FUCHSSCHWANZ (*m*)
Alopecurus pratensis	348	Meadow foxtail	KOLBENGRAS (*n*); WIESENFUCHS-SCHWANZ (*m*); FUCHS-SCHWANZ (*m*)

Botanical Term.	No.	English Term	German Term.
Aloysia citriodora	349	Lemon-scented verbena; sweet-scented verbena, herb Louisa, lemon plant (See *Lippia citriodora* and *Verbena officinalis*)	ZITRONENKRAUT (*n*); PUNSCH-PFLANZE (*f*)
Alpinia galanga	350	(See *Alpinia officinarum*)	GALGANTWURZEL (*f*)
Alpinia officinarum	351	Galangal; colic root; Chinese ginger (See *Alpinia galanga*)	GALANGAWURZEL (*f*); GALGANT (*m*)
Alsine	352	Chickweed, alsine	MIERE (*f*); MEIRICH (*m*)
Alsine cherleri	353	Clovewort (See *Arenaria cherleri*)	ALPENMEIRICH (*m*)
Alsine stricta	354	(See *Arenaria uliginosa*)	STEIFER MEIRICH (*m*)
Alsine tenuifolia	355	(See *Arenaria tenuifolia*)	FEINBLÄTTRIGER MEIRICH (*m*)
Alsine verna	356	(See *Arenaria verna*)	FRÜHLINGSMEIRICH (*m*)
Alstonia scholaris	357	Devil's tree	TEUFELSBAUM (*m*)
Alstrœmeria peregrina	358	Inca lily	INKALILIE (*f*)
Alternanthera	359	(See *Telanthera*)	
Althæa frutex	360	(See *Hibiscus syriacus*)	
Althæa hirsuta	361	Hispid althæa	
Althæa officinalis	362	Marsh mallow; marsh althæa; guimauve	ALTHEE (*f*); PAPPELKRAUT (*n*); PAPPELSTAUDE (*f*); WEISSE PAPPEL (*f*); IBISCH (*m*); GEMEINER EIBISCH (*m*)
Althæa rosea	363	Hollyhock (See *Malva alcea*)	HERBSTROSE (*f*); MUNDROSE (*f*); ROSENPAPPEL (*f*); ROSEN-MALVE (*f*); PAPPELROSE (*f*); STOCKROSE (*f*); MALVE (*f*)
Alyssum argenteum	364	(See *Alyssum calycinum*)	
Alyssum calycinum	365	Small alyssum, sweet alyssum, sweet alysson, gold-dust, mad-wort, gold-basket, golden tuft (See *Alyssum argenteum*)	STEINKRAUT (*n*)
Alyssum maritimum	366	Sweet alyssum	
Alyssum saxitile	367	Gold-basket	GOLDKÖRNCHEN (*n*); FELSEN-STEINKRAUT (*n*)
Amanita	368	(See *Agaricus rubescens* and *Agaricus muscarius*)	
Amanita cæsarea	369	(See *Agaricus cæsareus*)	
Amanita muscaria	370	Fly amanita, fly agaric	FLIEGENSCHWAMM (*m*)
Amaranthus caudatus	371	Amaranth; cockscomb; love-lies-bleeding; prince's feathers	TAUSENDSCHÖN (*n*); SAMTBLUME (*f*); SAMMETBLUME (*f*); AMARANT (*m*); FUCHS-SCHWANZ (*m*)
Amaranthus hypochondriacus	372	Prince's-feather	
Amaranthus oleraceus	373	Vegetable amaranth (See *Euxolus oleraceus*)	GEMÜSEAMARANT (*m*)
Amaranthus tricolor	374	Indian amaranth; velvet flower	TAUSENDSCHÖN (*n*); PAPAGEIEN-FEDER (*f*)
Amaryllideæ	375	Amaryllideæ	AMARYLLIDEEN (*f.pl*)
Amaryllis	376	Agave, American aloe (See *Agave americana*)	NARZISSENLILIE (*f*)
Amaryllis aurea	377	(See *Lycoris aurea*)	
Amaryllis belladonna	378	Belladonna lily; Cape belladonna lily	MEXIKANISCHE LILIE (*f*); BELLA-DONNA-AMARYLLIS (*f*)
Amaryllis formosissima	379	(See *Sprekelia formosissima*)	
Amaryllis lutea	380	(See *Sternbergia lutea*)	
Amaryllis rigida	381	Sisal hemp agave	
Amaryllis sarniensis	382	Guernsey lily	
Amelanchier canadensis	383	June-berry; Service berry	KANADISCHE FELSENMISPEL (*f*); SPERBEERE (*f*)
Amelanchier vulgaris	384	Snow-mespilus, grape-pear (See *Aronia rotundifolia* and also *Pirus*)	GEMEINE TRAUBENBIRNE (*f*); GEMEINE FELSENBIRNE (*f*); ENGLISCHE MISPEL (*f*)
Amentaceæ	385	Catkins; amentaceous plants	AMENTACEEN (*f.pl*); JULIFLOREN (*f.pl*); KÄTZCHENTRÄGER (*m.pl*)

Botanical Term.	No.	English Term.	German Term.
Amesium Ruta-muraria	386	Wall rue	MAUERRAUTE (*f*)
Amicia zygomeris	387	(See *Zygomeris flava*)	
Ammophila arenaria	388	Common sea-reed, sea-bent, mat-grass (See *Arundo arenaria*)	BORSTENGRAS (*n*)
Ammophila arundinacea	389	Sea-reed, sand-reed (See *Psamma arenaria*)	STRANDHAFER (*m*); SANDROHR (*n*); SANDSCHILF (*n*); SAND--HALM (*m*)
Ammophila baltica	390	Baltic mat-grass, Baltic sea-reed	OSTSEEROHR (*n*)
Amomum cardamomum	391	(See *Cardamomum racemosum*)	SIAM KARDAMOMENPFLANZE (*f*)
Amomum grana paradisi	392	Grains of paradise plant; guinea grains plant (See *Amomum melegueta*)	PARADIESKÖRNER (*pl*)
Amomum maximum	393	(See *Cardamomum majus*)	
Amomum melegueta	394	Melegueta pepper plant; Guinea grains plant (See *Amomum grana paradisi* and *Elettaria cardamomum*)	MELEGUETA PFEFFERSTAUDE (*f*); PARADIESKÖRNERPFLANZE (*f*); GUINEAKÖRNERPFLANZE (*f*); MALAGETTAPFEFFER (*m*)
Amorpha fruticosa	395	Bastard indigo	FALSCHER INDIGO (*m*); STRAUCH--IGE UNFORM (*f*)
Ampelidæ	396	Ampelidæ, vines	WEINREBENGEWÄCHSE (*n.pl.*)
Ampelopsis hederacea	397	(See *Ampelopsis quinquefolia*)	
Ampelopsis heterophylla	398	(See *Vitis heterophylla humulifolia*)	
Ampelopsis quinquefolia	399	Virginia creeper (See *Quinaria* and *Vitis quinquefolia* and *Parthenocissus quinquefolia*)	WILDER WEIN (*m*); JUNGFERN-REBE (*f*); DOLDENREBE (*f*); KANADISCHE REBE (*f*); GE--MEINE DOLDENREBE (*f*); GE--MEINE JUNGFERNREBE (*f*)
Ampelopsis tricuspidata	400	(See *Ampelopsis Veitchii* and *Vitis inconstans*)	
Ampelopsis veitchii	401	(Veitch's)-virginia creeper (See *Vitis inconstans*)	VEITCHII (WEIN)-REBE (*f*)
Amygdalopsis Lindleyi	402	(See *Prunus triloba*)	
Amygdalus communis	403	Almond tree; bitter almond (See *Prunus amygdalus*)	MANDELBAUM (*m*); BITTER--MANDEL (*f*)
Amygdalus communis persica	404	Almond-peach (See *Prunus amygdalus var. persica*)	MANDELPFIRSICH (*m*)
Amygdalus nana	405	dwarf almond (See *Prunus nana*)	ZWERGMANDEL (*f*)
Amygdalus orientalis	406	(See *Prunus orientalis*)	
Amygdalus pedunculata	407	(See *Prunus triloba*)	
Amygdalus persica	408	Peach (See *Prunus persica* and *Persica vulgaris*)	GEMEINER PFIRSICH (*m*); PFIR--SCHE (*f*); PFIRSICH (*m*); PFIRSICHBAUM (*m*); PFIR--SICHE (*f*)
Amyris balsamifera	409	Jamaican rosewood tree	JAMAIKA ROSENHOLZBAUM (*m*); AMERIKANISCHES ROSENHOLZ--BAUM (*m*); WESTINDISCHES SANDELHOLZBAUM (*m*); BAL--SAMPFLANZE (*f*); BALSAM--BAUM (*m*); SALBENBAUM (*m*)
Amyris elemifera	410	Elemi tree	
Anacamptis pyramidalis	411	(See *Orchis pyramidalis*)	PYRAMIDENORCHE (*f*); HUNDS--WURZ (*f*)
Anacardium occidentale	412	Cashew-(nut)-tree	ELEFANTENLAUS (*f*); NIEREN--BAUM (*m*); KASCHU-ACAJOU--BAUM (*m*)
Anacharis alsinastrum	413	Canadian pondweed (See *Elodea canadensis*)	WASSERPEST (*f*); WASSERMYRTE (*f*); WASSERTHYMIAN (*m*)
Anacharis canadensis	414	(See *Anacharis alsinastrum*)	
Anacyclus officinarum	415	(See *Anacyclus pyrethrum*)	RINGBLUME (*f*); DEUTSCHER BER--TRAM (*m*); DEUTSCHE BERTRAMWURZEL (*f*); BER--TRAM-SPEICHELWURZEL (*f*); BERTRAMZAHNWURZEL (*f*)

Botanical Term.	No.	English Term.	German Term.
Anacyclus pyrethrum	416	Pyrethrum; pellitory; lungwort, Spanish chamomile	BERTRAM (*m*); BERTRAMRINGBLUME (*f*); ECHTE BERTRAMWURZEL (*f*); ECHTE JOHANNISWURZEL (*f*); RÖMISCHE BERTRAMWURZEL (*f*); RÖMISCHE JOHANNISWURZEL (*f*); BERTRAMKAMILLE (*f*)
Anagallis arvensis	417	Scarlet pimpernel; poor man's weather-glass; common pimpernel; weather-glass; shepherd's weather-glass; corn pimpernel	KOLDERKRAUT (*n*); GAUCHHEIL (*n*)
Anagallis tenella	418	Bog pimpernel	GAUCHHEIL (*n*)
Anagyris fœtida	419		STINKSTRAUCH (*m*)
Ananas sativa	420	Pineapple (See *Bromelia ananas*)	ANANAS (*f*); ANASSA (*f*); NANAS (*f*)
Anaphalis margaritacea	421	Pearly everlasting, pearly immortelle (See *Antennaria margaritacea*)	VIRGINISCHE IMMORTELLE (*f*); PAPIERBLUME (*f*)
Anastatica hierochontica	422	(See *Anastatica hierochuntina* and *Odontospermum*)	
Anastatica hierochuntina	423	Rose of Jerico	WEIHNACHTSROSE (*f*); HUFKRAUT (*n*); ROSE VON JERICHO (*f*)
Anchusa arvensis	424	Bugloss (See *Lycopsis arvensis*)	
Anchusa officinalis	425	Common alkanet; oxtongue; bugloss	OCHSENZUNGE (*f*)
Anchusa sempervirens	426	Green alkanet; evergreen alkanet	
Anchusa tinctoria	427	Alkanna, alkanet (See *Alkanna tinctoria*)	ALKANNA (*f*)
Andira araroba	428		ANGELIM (*m*)
Andira inermis	429	Partridge-wood tree	ANGELIM (*m*); REBHUHNHOLZBAUM (*m*)
Andromeda floribunda	430	North-American lavender heath	NORDAMERIKANISCHE LAVENDELHEIDE (*f*)
Andromeda japonica	431	Japanese lavender-heath	JAPANISCHE LAVENDELHEIDE (*f*); ASEBU (*f*); BASUIBOKU (*f*)
Andromeda polifolia	432	Marsh andromeda; lavender heath, lavender heather	LAVENDELHEIDE (*f*); ROSMARINHEIDE (*f*); FALSCHER PORST (*m*)
Andropogon	433	Gingergrass; beard-grass	GINGERGRAS (*n*); INGWERGRAS (*n*); BARTGRAS (*n*); MANNSBART (*m*)
Andropogon ciliatum	434	(See *Andropogon citratus*)	
Andropogon citratus	435	Lemongrass (See *Andropogon Nardus*)	LIMONGRAS (*n*); LEMONGRAS (*n*); NARDENBARTGRAS (*n*); ZITRONENBARTGRAS (*n*)
Andropogon ischæmum	436	Cock's-foot grass	HÜHNERFUSSGRAS (*n*); BLUTHIRSE (*f*)
Andropogon ivaranchusa	437	Vetiver-(root)	VETIVERWURZEL (*f*); IVARANKUSAWURZEL (*f*)
Andropogon laniger	438	Camel grass; ginger-grass	KAMELGRAS (*n*); KAMELHEU (*n*); GINGERGRAS (*n*)
Andropogon muricatus	439	(See *Andropogon ivaranchusa*)	
Andropogon nardus	440	(See *Andropogon citratus*)	NARDENBARTGRAS (*n*)
Andropogon schœnanthus	441	Citron grass	ZITRONENGRAS (*n*); ZITRONENBARTGRAS (*n*)
Andropogon sorghum	442	(See *Sorghum vulgare* and *Sorghum caffrorum*)	
Andropogon squarrosus	443	(See *Andropogon ivaranchusa*)	
Androsace	444	Sea-navelwort	MANNSSCHILD (*m*)
Androsæmum officinale	445	Tutsan (See *Hypericum androsæmum*)	HYPERICUM (*n*); MANNSBLUT (*n*); BLUTHEIL (*n*); GRUNDHEIL (*n*)
Anemone	446	Anemone; wind flower	WINDBLUME (*f*); WINDRÖSCHEN (*n*)
Anemone blanda	447	(Greek)-blue wind flower	
Anemone coronaria	448	Garden anemone	GARTENANEMONE (*f*)

Botanical Term.	No.	English Term.	German Term.
Anemone hepatica	449	Yellow (golden) trefoil; hepatica; liverwort (See also *Marchantia polymorpha* and *Hepatica*)	GOLDKLEE (*m*); LEBERANEMONE (*f*); DREILAPPIGES MÄRZ-BLÜMCHEN (*n*)
Anemone hortensis	450	Star anemone	STERNANEMONE (*f*)
Anemone japonica	451	Japanese anemone; Japanese wind-flower	JAPANISCHE ANEMONE (*f*)
Anemone nemorosa	452	Wood anemone; white wood anemone	WEISSE OSTERBLUME (*f*); APRIL-BLUME (*f*); WALDRÖSCHEN (*n*)
Anemone pavonina	453	Peacock anemone	PFAUENANEMONE (*f*)
Anemone pratensis	454	Meadow anemone	
Anemone pulsatilla	455	Pasque flower, wind flower	KÜCHENSCHELLENKRAUT (*n*); HAKELKRAUT (*n*); MANNS-KRAUT (*n*); BEISSWURZ (*f*); KÜCHENSCHELLE (*f*); KUH-SCHELLE (*f*)
Anemone ranunculoides	456	Yellow (golden) trefoil	GELBE OSTERBLUME (*f*)
Anemone silvestris	457	Wood anemone; snowdrop anemone	WALDANEMONE (*f*); OSTER-BLUME (*f*); OSTERSCHALE (*f*)
Anethum graveolens	458	Anethum; garden dill (See *Peucedanum graveolens*)	DILL (*m*); GURKENKRAUT (*n*); GARTENDILL (*m*); KÜMMER-LINGSKRAUT (*n*)
Anethum sowa	459	Sowa dill	SOWADILL (*m*)
Angelica anomala	460	Japanese biyakushi	BIYAKUSHI (*f*)
Angelica archangelica	461	Angelica; garden angelica; Holy Ghost (See *Archangelica officinalis*)	ENGELWURZ (*f*); ENGELWURZEL (*f*); BRUSTWURZEL (*f*)
Angelica officinalis	462	(See *Angelica archangelica*)	
Angelica refracta	463	Japanese Senkuju	SENKUJU (*f*)
Angelica sylvestris	464	Wild angelica	WALDANGELIKA (*f*)
Angiospermæ	465	Angiospermæ (Plants having a seed-vessel)	ANGIOSPERMEN (*f.pl*); BEDECKT-SAMIGE PFLANZEN (*f.pl*)
Angræcum fragans	466	Faham leaves; faham tea, Bourbon tea (See *Aeranthus*)	FAHAMBLÄTTER (*n.pl*); FAHAM-TEE (*m*); BOURBONTEE (*m*)
Anona	467	Cherimoya	FLASCHENBAUM (*m*)
Anonaceæ	468	Anonaceæ	FLASCHENBÄUME (*m.pl*); ANONA-ZEEN (*f.pl*)
Anona cherimolia	469	Cherimoya tree; Jamaica apple	FLASCHENBAUM (*m*); CHERI-MOYA (*f*)
Anona muricata	470	Custard apple; sour-sop	BRASILIEN CHERIMOYA (*f*); STACHEL-ANONE (*f*); SAUER-SACK (*m*)
Anona palustris	471	Marsh anona	SUMPF-ANONE (*f*)
Anona reticulata	472	Custard apple; bullock's heart	OCHSENHERZ (*n*)
Anona rhizantha	473	Flowering-rooted anona (See *Geanthemum rhizanthum*)	WURZELBLÜTIGE ANONE (*f*)
Anona squamosa	474	Sugar apple; sweet-sop; custard apple	WESTINDIEN STRAUCHARTIGE CHERIMOYA (*f*); ZUCKER-APFEL (*m*); ZIMTAPFELBAUM (*m*); ATHEAPFELBAUM (*m*); SÜSSSACK (*m*); ZUCKERAPFEL-BAUM (*m*); CURROSSOL (*m*); ZIMTAPFEL (*m*); RAHMAPFEL (*m*); GEWÜRZAPFEL (*m*)
Antennaria dioica	475	Mountain cudweed; mountain everlasting; mountain antennaria; cat's-foot; Cat's-ear; mouse-ear hawkweed (See *Gnaphalium dioicum*)	KATZENPFÖTCHEN (*n*); ENGEL-BLUME (*f*); PAPIERBLUME (*f*)
Antennaria margaritacea	476	Pearl antennaria (See *Anaphalis margaritacea*)	
Antennaria tomentosa	477	Cat's-ear	FERKELKRAUT (*n*)
Anthemis	478	Camomile	KAMILLE (*f*); ACKERKAMILLE (*f*)
Anthemis arvensis	479	Corn (chamomile) camomile	KUHDILL (*m*); ACKERKAMILLE (*f*); UNECHTE KAMILLE (*f*)
Anthemis cotula	480	Dog's camomile; mayweed; stinking camomile; stink mayweed; fetid (fœtid) camomile	HUNDSKAMILLE (*f*); FELDHUNDS-KAMILLE (*f*)

Botanical Term.	No.	English Term.	German Term.
Anthemis maritima	481	(See *Anthemis arvensis*)	
Anthemis nobilis	482	Chamomile, camomile, Roman camomile; common camomile (See also *Matricaria chamomilla*)	(RÖMISCHE)-KAMILLE (*f*)
Anthemis tinctoria	483	Yellow camomile	FÄRBERKAMILLE (*f*)
Anthericum liliago	484	St. Bernard's lily	GRASLILIE (*f*); ZAUNLILIE (*f*)
Anthistiria ciliata	485	Kangaroo grass (See *Themeda*)	KÄNGURUHGRAS (*n*)
Anthistiria vulgaris	486	(See *Anthistiria ciliata*)	
Anthoxanthum	487	Vernal grass	RUCHGRAS (*n*)
Anthoxanthum odoratum	488	Sweet vernal-grass	GOLDGRAS (*n*); RUCHGRAS (*n*)
Anthriscus	489	Chervil; bastard parsley	KÄLBERKROPF (*m*); KLETTEN-KERBEL (*m*)
Anthriscus cerefolium	490	Garden chervil, salad chervil (See *Scandix cerefolium* and *Chærophyllum temulentum*)	GARTENKERBEL (*m*); KERBEL (*m*)
Anthriscus sylvestris	491	Wild chervil; beaked parsley (See *Chærophyllum sylvestre*)	WIESENKERBEL (*m*); HAFERROHR (*n*); PFERDEKÜMMEL (*m*) KÄLBERROHR (*n*)
Anthriscus vulgaris	492	Chervil, common chervil (See *Chærophyllum anthriscus*)	GEMEINER KERBEL (*m*) HUNDSKERBEL (*m*)
Anthyllis barba-jovis	493	Jupiter's beard	JUPITERBART (*m*)
Anthyllis vulneraria	494	Lady's finger; common kidney vetch	WUNDKLEE (*m*); TANNENKLEE (*m*); WUNDBLUME (*f*); WOLL-BLUME (*f*)
Antiaris saccidora	495		SACKBAUM (*m*)
Antiaris toxicaria	496	Upas tree	GIFTBAUM (*m*); ANTIAR (*m*); UPASBAUM (*m*); ANTSCHEE (*f*)
Antirrhinum majus	497	Great snapdragon	LÄMMERKRAUT (*n*); LÖWENMAUL (*n*); NASENKRAUT (*n*); GROSSES LÖWENMAUL (*n*); GROSSER DORANT (*m*)
Antirrhinum orontium	498	Lesser snapdragon	KLEINER DORANT (*m*); FELD-LÖWENMAUL (*n*)
Apargia autumnalis	499	(See *Leontodon autumnalis*)	
Apargia hispida	500	(See *Leontodon hispidus*)	
Apargia taraxaci	501	(See *Leontodon taraxacum*)	
Apera interrupta	502	(See *Agrostis spica-venti*)	UNTERBROCHENER WINDHALM (*m*)
Apera spica-venti	503	(See *Agrostis spica-venti*)	GEMEINER WINDHALM (*m*)
Apetalæ	504	Apetalæ	BLUMENKRONENLOSE PFLANZEN (*f.pl*)
Apios fortunei	505	Japanese ground-nut; Japanese wild bean	JAPANISCHE KOLBENWICKE (*f*)
Apios tuberosa	506	Ground-nut; wild bean	KOLBENWICKE (*f*); ERDBIRNE (*f*)
Apium graveolens	507	Celery	EPPICH (*m*); GEMEINE SELLERIE (*f*)
Apium inundatum	508	Lesser apium	
Apium nodiflorum	509	Procumbent apium	
Apocynum androsæmifolium	510	Dog's bane	HUNDSKOHL (*m*); HUNDSWOLLE (*f*)
Apocynum cannabinum	511	American hemp	AMERIKANISCHER HANF (*m*)
Apollonias canariensis	512	Canary wood	KANARIENHOLZ (*n*)
Aponogeton fenestralis	513	Cape pondweed (See *Ouvirandra fenestralis*)	GITTERPFLANZE (*f*)
Aporanthus trifoliastrum	514	(See *Trigonella purpurascens*)	
Aporo-a dioica	515	Burmese kokra-wood tree	
Aquifoliaceæ	516	Aquifoliaceæ (See *Ilicineæ*)	AQUIFOLIAZEEN (*f.pl*); STECH-PALMENARTIGE GEWÄCHSE (*n.pl*)
Aquilaria agallocha	517	Agallocha-wood tree, Eagle-wood tree (See *Aloeoxylon agallochum*)	AGALLOCHEHOLZBAUM (*m*); AD-LERHOLZBAUM (*m*); ALOE-HOLZBAUM (*m*); PARADIES-HOLZBAUM (*m*)
Aquilaria malaccensis	518	Eagle-wood tree	ADLERBAUM (*m*); ADLERHOLZ-BAUM (*m*); ASPALATHAHOLZ-BAUM (*m*); RHODISER DORN-HOLZ (*n*)
Aquilegia canadensis	519	Common American columbine	KANADISCHE AKELEI (*f*)
Aquilegia chrysantha	520	Golden columbine	GOLDENE AKELEI (*f*)

Botanical Term.	No.	English Term	German Term.
Aquilegia cærulea	521	Rocky mountain columbine	
Aquilegia pyrenaica	522	Dwarf, blue columbine	
Aquilegia vulgaris	523	Columbine ; common columbine	AKELEI (*f*) ; AGLEI (*f*) ; AQUIL-EJA (*f*); GEMEINE AKELEI (*f*)
Arabis albida	524	White rock-cress	WEISSE GÄNSEKRESSE (*f*) ; WEISSER GÄNSEKOHL (*m*)
Arabis alpina	525	Alpine rockcress	ALPENGÄNSEKRESSE (*f*)
Arabis ciliata	526	Fringed rockcress	
Arabis hirsuta	527	Hairy rockcress	
Arabis perfoliata	528	Glabrous rockcress ; tower mustard	
Arabis petræa	529	Northern rockcress	
Arabis stricta	530	Bristol rockcress	
Arabis thaliana	531	Thale cress ; Thale rockcress ; wall-cress (See *Sisymbrium thaliana*)	RAUTENSENF (*m*)
Arabis turrita	532	tower cress ; tower rockcress	
Arachis hypogæa	533	Pea-nut, ground-nut, earth-nut, pig-nut ; ground pea-nut ; earth pea (See *Bunium flexuosum*)	ERDEICHEL (*f*) ; ERDNUSS (*f*) ; KNOLLWICKE (*f*) ; NUSSBOHNE (*f*) ; ERDBOHNE (*f*) ; ERD-MANDEL (*f*) ; ERDPISTAZIE (*f*) ; MANDUBIBOHNE (*f*)
Aralia japonica	534	(See *Fatsia japonica*)	
Aralia nudicaulis	535	North-American green aralia, North-American sarsaparilla root	NORDAMERIKANISCHE SASSAPAR-ILLENWURZEL (*f*)
Aralia papyrifera	536	Rice-paper plant (See *Tetrapanax papyrifera* and *Fatsia papyrifera*)	REISPAPIERPFLANZE (*f*)
Aralia quinquefolia	537	(See *Panax quinquefolium*)	
Aralia racemosa	538	Spikenard	SPIKE (*f*)
Aralia sieboldii	539	(See *Fatsia japonica*)	
Aralia spinosa	540	Angelica tree	ANGELIKABAUM (*m*)
Aralia trifolia	541	Dwarf ginseng (See *Panax trifolium*)	
Araucaria araucana	542	(See *Araucaria imbricata*)	ANDENTANNE (*f*)
Araucaria bidwillii	543	Bunya-Bunya tree	BUNYA-BUNYA-BAUM (*m*)
Araucaria brasiliensis	544	Brasilian araucaria	BRAZILISCHE ARAUKARIE (*f*) ; PINHEIRD
Araucaria columnaris	545	Columnar araucaria	SÄULENZYPRESSE (*f*)
Araucaria cunninghamii	546	Moreton-Bay pine	MORETONFICHTE (*f*)
Araucaria excelsa	547	Norfolk Island pine	NORFOLKTANNE (*f*)
Araucaria imbricata	548	Chile pine, monkey-puzzle ; araucaria (See *Araucaria araucana*)	ARAUKARIE (*f*) ; SCHMUCK-TANNE (*f*) ; CHILE-SCHMUCK-TANNE (*f*) ; CHILIFICHTE (*f*)
Arbor vitæ	549	(See *Thuja occidentalis*)	
Arbutus menziesii	550	(Californian)-madroña	SANDBEERE (*f*)
Arbutus unedo	551	Arbutus ; strawberry tree	ERDBEERBAUM (*m*)
Arbutus uva ursi	552	(See *Arctostaphylos uva ursi*)	
Archangelica officinalis	553	Angelica (See *Angelica archangelica*)	ENGELWURZ (*f*) ; ANGELIKA (*f*) ; BRUSTWURZEL (*f*)
Archontophœnix	554	(See *Seaforthia* and *Ptychosperma*)	AUSTRALISCHE PALME (*f*)
Arctium lappa	555	Burdock-(root) ; bardana ; clotbur ; common burdock	KLETTENWURZEL (*f*) ; KLETTE (*f*)
Arctium lappa edulis	556	Japanese gobo	JAPANISCHE SKOZONERE (*f*)
Arctium lappa glabra	557	(See *Arctium lappa*)	
Arctium majus	558	Great burdock	GROSSE KLETTENWURZEL (*f*)
Arctium minus	559	Lesser burdock (See *Arctium lappa*)	KLEINE KLETTENWURZEL (*f*)
Arctostaphylos alpina	560	Highland bearberry ; black bearberry	
Arctostaphylos glauca	561	Manzanita	
Arctostaphylos uva-ursi	562	Bear-berry ; common bearberry (See *Arbutus uva ursi*)	BÄRENTRAUBE (*f*)
Arctotis acaulis	563	Arctotis	BÄRENOHR (*n*)
Ardisia crenata	564	Spear-flower	SPITZBLUME (*f*)

Botanical Term.	No.	English Term.	German Term.
Areca catechu	565	Pinang, betel-nut palm	KATECHUPALME (*f*); PINANG (*m*); PENANG (*m*); AREKAPALME (*f*); BETELNUSSPALME (*f*)
Areca oleracea	566	(West Indian) cabbage palm (See *Oreodoxa oleracea*)	WEST-INDISCHE KOHLPALME (*f*)
Aregma bulbosum	567	(See *Phragmidium bulbosum*)	
Arenaria alpina	568	Alpine sandwort	ALPENSANDKRAUT (*n*)
Arenaria balearica	569	Sandwort	BALEARISCHES SANDKRAUT (*n*)
Arenaria biflora	570	two-flowered sandwort	ZWEIBLÜTIGES SANDKRAUT (*n*)
Arenaria cherleri	571	Cyphel (See *Alsine cherleri*)	
Arenaria ciliata	572	Fringed sandwort	GEWIMPERTES SANDKRAUT (*n*)
Arenaria graminifolia	573	Grass-leaved sandwort	GRASBLÄTTRIGES SANDKRAUT (*n*)
Arenaria grandiflora	574	Large-flowered sandwort	GROSSBLÜTIGES SANDKRAUT (*n*)
Arenaria montana	575	Sandwort (See *Arenaria alpina*)	
Arenaria multicaulis	576	Many-stemmed sandwort	VIELSTENGELIGES SANDKRAUT (*n*)
Arenaria norwegica	577	Norwegian sandwort	NORWEGISCHES SANDKRAUT (*n*)
Arenaria peploides	578	Sea purslane (See *Atriplex portulacoides*)	
Arenaria serpyllifolia	579	Thyme-leaved sandwort	QUENDELBLÄTTRIGES SANDKRAUT (*n*)
Arenaria tenuifolia	580	Fine-leaved sandwort (See *Alsine tenuifolia*)	
Arenaria trinervis	581	Three-nerved sandwort (See *Mœhringia trinervis*)	DREINERVIGE SPELLE (*f*)
Arenaria uliginosa	582	Bog sandwort (See *Alsine stricta*)	
Arenaria verna	583	Vernal sandwort (See *Alsine verna*)	
Arenga saccharifera	584	Sugar palm	ZUCKERPALME (*f*); MOLUKKISCHE ZUCKERPALME (*f*); GOMUTI--PALME (*f*); SAGWIREPALME (*f*)
Argania sideroxylon	585	Argan tree; Morocco argan tree	ARGANBAUM (*m*); EISENHOLZ--BAUM (*m*)
Argemone grandiflora	586	Prickly poppy, devil's fig, Mexican poppy	STACHELMOHN (*m*)
Aria	587	(See *Pirus aria*)	
Arisœma dracontium	588	(See *Arum dracontium*)	
Aristolochia clematitis	589	Birthwort	OSTERLUZEI (*f*); GEMEINE OSTER--LUZEI (*f*); WALDREBENHOHL--WURZ (*f*)
Aristolochia macrophylla	590	(See *Aristolochia sipho*)	
Aristolochia serpentaria	591	Virginian (snake-wort) snake-root, Sangrel, snakeweed, birthwort	VIRGINISCHE SCHLANGENWURZEL (*f*)
Aristolochia sipho	592	Dutchman's pipe	PFEIFEN-OSTERLUZEI (*f*); PFEI--FENSTRAUCH (*m*)
Aristotelia maqui	593	(Chilean)-maqui shrub	MAQUISTAUDE (*f*); CHILENISCHER JASMIN (*m*)
Armenia vulgaris	594	(See *Prunus armeniaca*)	
Armeniaca	595	(See *Prunus armeniaca*)	
Armeria maritima	596	Thrift; sea-pink; common thrift (See *Armeria vulgaris* and *Statice armeria*)	MEERNELKE (*f*); MEERGRAS (*n*); MEERSTRANDSGRASNELKE (*f*); HORNKRAUT (*n*); SEEGRAS (*n*); SEENELKE (*f*)
Armeria plantaginea	597	Plantain thrift	GRASNELKE (*f*)
Armeria vulgaris	598	(See *Armeria maritima*)	GRASNELKE (*f*); SANDNELKE (*f*); GRASBLUME (*f*); MEERGRAS (*n*); SEEGRAS (*n*); MEER--NELKE (*f*)
Armillaria	599	(See *Agaricus melleus*)	
Armoracia amphibium	600	(See *Nasturtium amphibium*)	MEERRETTICH (*m*)
Armoracia rusticana	601	Horse radish (See *Cochlearia armoracia*)	MEERRETTICH (*m*); MEERRETTIG (*m*); GREEN (*n*)

Botanical Term.	No.	English Term.	German Term.
Arnica montana	602	Arnica ; mountain arnica	WOHLVERLEIH (*m*) ; WOHLVERLEI (*m*) ; LAUGENKRAUT (*n*) ; FALL-KRAUT (*n*) ; MARIENKRAUT (*n*) ; MUTTERWURZ (*f*) ; ENGELKRAUT (*n*) ; MÖNCHSWURZ (*f*) ; JOHANNISBLUME (*f*) ; BERGWOHLVERLEI (*m*) ; BERGWOHLVERLEIH (*m*) ; FALLBLUMENKRAUT (*n*) ; JOHANNISBLUMENKRAUT (*n*) ; ENGELBLUMENKRAUT (*n*)
Arnoseris pusilla	603	Swine's succory ; dwarf arnoseris ; lamb's succory	LAMMKRAUT (*n*)
Aronia rotundifolia	604	(See *Amelanchier vulgaris* and also *Pirus*)	
Arrhenatherum avenaceum	605	Oat-grass ; false oat	KNOLLHAFER (*m*) ; RISPE (*f*) ; HOHER GLATTHAFER (*m*) ; WIESENHAFER (*m*) ; FRANZÖSISCHES RAIGRAS (*n*)
Arrhenatherum bulbosum	606	Oat-grass	KNOLLGRAS (*n*) ; KNOLLHAFER (*m*)
Arrhenatherum elatius	607	French rye-grass	FRANZÖSISCHES RAIGRAS (*n*)
Arrhenatherum nodosum	608	Oat-grass	KNOLLGRAS (*n*)
Artanthe elongata	609	Matico shrub (See *Piper angustifolium*)	MATIKOSTAUDE (*f*)
Artemisia abrotanum	610	Southernwood ; maiden's ruin ; old man ; lad's love ; boy's love ; old woman	EBERRAUTE (*f*) ; MUTTERKRAUT (*n*) ; STABWURZ (*f*) ; EBERREIS (*m*) ; ABERRAUTE (*f*) ; HOFRAUTE (*f*) ; HOFMANNS BAUM (*m*) ; ABRANDKRAUT (*n*) ; ZITRONENKRAUT (*n*) ; ZITRONELLE (*f*)
Artemisia absinthum	611	Wormwood ; absinth	ABSINTH (*m*) ; ABSYNTH (*m*) ; BEIFUSS (*m*) ; WERMUT (*m*)
Artemisia campestris	612	Field artemisia	HARTKRAUT (*n*)
Artemisia dracunculus	613	Tarragon	ZITWERKRAUT (*n*) ; DRAGUNBEIFUSS (*m*) ; ESTRAGON (*m*)
Artemisia maritima	614	Sea artemisia ; sea wormwood	SEEWERMUT (*m*) ; SEEBEIFUSS (*m*) ; MEERWERMUT (*m*) ; MEERBEIFUSS (*m*)
Artemisia mutellina	615		EDELRAUTE (*f*)
Artemisia pauciflora	616	Levant wormseed, Santonica	WURMSAMENKRAUT (*n*)
Artemisia pontica	617	Grey-leaved artemisia	RÖMISCHER BEIFUSS (*m*)
Artemisia vulgaris	618	Mugwort ; common artemisia	BEIFUSS (*m*) ; GEMEINER BEIFUSS (*m*) ; MUTTERKRAUT (*n*) ; HARTKRAUT (*n*)
Arthrolobium ebracteatum	619	(See *Ornithopus ebracteatus*)	
Artocarpus incisa	620	Bread-fruit tree	AFFENBROTBAUM (*m*) ; GEMEINER BROTFRUCHTBAUM (*m*) ; BROTBAUM (*m*)
Artocarpus integrifolia	621	Jack-fruit tree	JAKHOLZ (*n*) ; JAKHOLZBAUM (*m*) ; JAQUEIRAHOLZ (*n*) ; INDISCHER BROTBAUM (*m*) ; JAQUEIRA (*f*)
Arum dracontium	622	Green dragon (See *Arisæma dracontium*)	WUNDERKNOLLE (*f*)
Arum dracunculus	623	Common dragon (See *Dracunculus vulgaris*)	DRACHENWURZ (*f*) ; SCHLANGENKRAUT (*n*)
Arum maculatum	624	Cuckow-pint, arum, lords-and-ladies, cuckoo-pint ; wake-robin	ARON (*m*) ; NATTERWURZ (*f*) ; ARONSWURZEL (*f*) ; ARONSWURZ (*f*) ; AARONSWURZ (*f*) ; ARONSSTAB (*m*) ; AARONSSTAB (*m*) ; ZEHRWURZ (*f*) ; GEMEINER ARONSSTAB (*m*) ; ESELSOHR (*n*) ; AASBLUME (*f*) ; GEFLECKTER DEUTSCHER INGWER (*m*)
Arum tenuifolium	625	Narrow-leaved arum	
Arundinaria alpina	626	Alpine bamboo (See also BAMBUSA)	ALPEN-BAMBUS (*m*)

Botanical Term.	No.	English Term.	German Term.
Arundinaria japonica	627	Japanese bamboo (See *Bambusa metake*)	JAPANISCHER BAMBUS (*m*)
Arundo	628	Reed, sedge (See also *Ammophila, Calamagrostis, Calamus* and *Phragmites*)	RIET (*n*); ROHR (*n*); SCHILF (*m* and *n*)
Arundo arenaria	629	(See *Ammophila arenaria*)	STRANDHAFER (*m*); SANDROHR (*n*); SANDHALM (*m*); SAND-SCHILF (*m* and *n*); SCHALMEIENROHR (*n*)
Arundo calamagrostis	630	(See *Calamagrostis lanceolata*)	
Arundo donax	631	Spanish reed; Italian reed	SPANISCHES ROHR (*n*); ITALIENISCHES ROHR (*n*); PFAHLROHR (*n*); SCHALMEIENROHR (*n*); KLARINETENROHR (*n*)
Arundo phragmites	632	(See *Phragmites communis*)	
Asagræa officinalis	633	Sabadilla (See *Schoenocaulon officinale*)	
Asarum arifolium	634	Wild ginger	WILDER INGWER (*m*)
Asarum canadense	635	Canadian asarum, sweet-scented asarum	WOHLRIECHENDER ASARUM (*m*); KANADISCHE SCHLANGENWURZ (*f*)
Asarum europæum	636	Asarabacca, hazel (cabaric) root; hazel-wort	HASELWURZ (*f*); LEBERKRAUT (*n*); WILDE NARDE (*f*)
Aschion	637	Non-edible truffle (See *Tuber excavatum*)	UNECHTE TRÜFFEL (*f*)
Asclepiadaceæ	638	Asclepiadaceæ	SEIDENPFLANZENGEWÄCHSE (*n.pl*)
Asclepias acida	639	False Soma plant (See *Sarcostemma acida* and also *Ephedra vulgaris*)	UNECHTE HEILIGE SOMAPFLANZE (*f*)
Asclepias cornuti	640	Virginian swallow-wort (See *Asclepias syriaca*)	SCHWALBENWURZ (*f*); SEIDEN-PFLANZE (*f*)
Asclepias curassavica	641	Wild ipecacuanha	WILDE IPECACUANHA (*f*)
Asclepias gigantea	642	(See *Calotropis gigantea*)	
Asclepias incarnata	643	Swallow-wort (See *Asclepias cornuti*)	
Asclepias syriaca	644	(See *Asclepias cornuti*)	
Asclepias tuberosa	645	Pleurisy root; butterfly weed	
Ascotricha chartarum	646	Mildew, paper mildew	PAPIER-MELTAU (*m*)
Aspalatus ebenus	647	Green ebony	GRÜNEBENHOLZ (*n*); UNECHTES POCKHOLZ (*n*); GRÜNHERZ-HOLZBAUM (*m*)
Asparagus medeoloides	648	(See *Asparagus officinalis*; *Medeola asparagoides* and *Myrsiphyllum asparagoides*)	
Asparagus officinalis	649	(Common) asparagus	SPARGEL (*m*)
Aspergillus	650	Club mould (See also *Pencillium*)	SCHIMMEL-SCHLAUCHPILZ (*m*)
Aspergillus fumigatus	651	Barley club-mould	GERSTENKOLBENSCHIMMEL (*m*)
Aspergillus glaucus	652	Club mould (See *Aspergillus herbariorum* and *Eurotium glaucum*)	KOLBENSCHIMMEL (*m*); GEMEINER KOLBENSCHIMMEL (*m*)
Aspergillus herbariorum	653	Greyish-green club-mould (See *Aspergillus glaucus* and *Eurotium glaucum*)	GRAUGRÜNER KOLBENSCHIMMEL (*m*)
Aspergillus oryzæ	654	Saké club-mould, Japanese rice-wine club-mould	SAKÉKOLBENSCHIMMEL (*m*); REIS-WEIN-KOLBENSCHIMMEL (*m*); REISBIER - KOLBENSCHIMMEL (*m*)
Asperifoliaceæ	655	Asperifoliaceæ	RAUHBLÄTTRIGE PFLANZEN (*f.pl*)
Asperugo procumbens	656	Madwort; procumbent madwort	NIEDERLIEGENDER SCHLANGEN-ÄUGLEIN (*m*)
Asperula cynanchica	657	Squinancywort	
Asperula odorata	658	Woodruff	WALDMEISTER (*m*); MAIKRAUT (*n*); MÖSCH (*m*); GEMEINER WALDMEISTER (*m*)
Asphodeline lutea	659	(See *Asphodelus luteus*)	
Asphodeline taurica	660	White oriental asphodel (See *Asphodelus taurica*)	WEISSER ORIENTALISCHER ASPHO-DILL (*m*)

Botanical Term.	No.	English Term.	German Term.
Asphodelus albus	661	Asphodel, Silver rod, King's spear	DRECKLILIE (*f*); GOLDLILIE (*f*); AFFODILL (*m*); ASPHODILL (*m*); WEISSER ASPHODILL (*m*)
Asphodelus luteus	662	Yellow asphodel (See *Asphodeline lutea*)	GELBER ASPHODILL (*m*)
Asphodelus neglectus	663	(See *Asphodelus albus*)	
Asphodelus ramosus	664	(See *Asphodelus albus*)	
Asphodelus taurica	665	(See *Asphodeline taurica*)	
Aspidistra elatior	666	Parlour palm, aspidistra	ASPIDISTRA- (BLUME) (*f*)
Aspidistra lurida	667	Variegated aspidistra, aspidistra; parlour palm (See *Aspidistra elatior* and *Plectogyne variegata*)	GEFLECKT-BLÄTTRIGE ASPIDISTRA -(BLUME) (*f*)
Aspidium	668	Fern, male fern, shield fern, aspidium, filixmas, wood-fern (See *Lastrea Filix-mas* and *Dryopteris filix-mas marginalis*; also *Aspidium filix mas*)	FARN (*m*); FARREN (*m*); FARNKRAUT (*n*); FARRENKRAUT (*n*); FARNWURZEL (*f*); JOHANNISWURZ (*f*); JOHANNISWURZEL (*f*); SCHILDFARN (*m*)
Aspidium aculeatum	669	Prickly shield-fern (See *Polystichum aculeatum*)	STACHELIGER SCHILDFARN (*m*)
Aspidium athamanticum	670	Male fern	WURMFARN (*m*)
Aspidium cristatum	671	Crested shield-fern (See *Nephrodium cristatum* and *Lastrea cristata*)	
Aspidium filix mas	672	Male fern (See *Aspidium*)	MÄNNLICHES FARNKRAUT (*n*); FARNKRAUTMÄNNCHEN (*n*); JOHANNISWURZEL (*f*); WALDFARN (*m*); WURMFARN (*m*); TEUFELSKLAUE (*f*)
Aspidium lonchitis	673	Holly fern	
Aspidium oreopteris	674	Mountain shield fern	BERGSCHILDFARN (*m*)
Aspidium rigidum	675	Rigid shield-fern	STEIFER SCHILDFARN (*m*)
Aspidium spinulosum	676	Broad shield-fern	BREITER SCHILDFARN (*m*)
Aspidium thelypteris	677	Marsh shield-fern	SUMPFSCHILDFARN (*m*)
Aspidosperma quebracho-(blanco)	678	(Chilean) white quebracho (See also *Loxopterygium lorentzii*)	(WEISSER)- QUEBRACHOHOLZBAUM (*m*); WEISSER QUEBRACHO-RINDEBAUM (*m*)
Aspidosperma vargasii	679	West-Indian box tree	WESTINDISCHER BUCHSBAUM (*m*)
Aspidosperma vergacci	680	Yellow quebracho-wood tree	GELBER QUEBRACHOHOLZBAUM (*m*)
Asplenium	681	Spleenwort fern	MILZFARN (*m*); STREIFENFARN (*m*); STRICHFARN (*m*)
Asplenium Adiantum-nigrum	682	Black spleenwort	SCHWARZER STREIFENFARN (*m*)
Asplenium ceterach	683	Spleenwort (See *Ceterach officinarum* and *Scolopendrium ceterach*)	MILZFARN (*m*)
Asplenium filix-fœmina	684	Lady fern	FALSCHER WURMFARN (*m*); WEIBLICHER STREIFENFARN (*m*)
Asplenium fontanum	685	Rock spleenwort	QUELLENMILZFARN (*m*)
Asplenium germanicum	686	German spleenwort	DEUTSCHER STREIFENFARN (*m*)
Asplenium lanceolatum	687	Lanceolate spleenwort	LANZETTLICHER MILZFARN (*m*)
Asplenium marinum	688	Sea spleenwort	MEERSTREIFENFARN (*m*)
Asplenium ruta-muraria	689	Wall rue, tentwort	MAUERRAUTE (*f*)
Asplenium septentrionale	690	Forked spleenwort	NÖRDLICHER MILZFARN (*m*)
Asplenium trichomanes	691	Common spleenwort	MILZKRAUT (*n*); ROTES FRAUENHAAR (*n*); ATHON (*m*); MILZFARN (*m*); ROTER WIDENTON (*m*); BRAUNSTIELIGER MILZFARN (*m*)
Asplenium viride	692	Green spleenwort	GRÜNER STREIFENFARN (*m*)
Aster alpinus	693	Alpine aster, star-wort	ALPENASTER (*f*); STERNBLUME (*f*)
Aster amellus	694	Virgil's aster, Italian star-wort	VIRGILSASTER (*f*); ITALIENISCHE STERNBLUME (*f*)
Aster chinensis	695	China aster (See *Callistephus chinensis*)	ASTER (*f*); CHINESISCHE ASTER (*f*)
Aster linosyris	696	(See *Chrysocoma linosyris*)	
Aster tradescanti	697	Michaelmas daisy	GROSSBLÄTTERIGE STERNBLUME (*f*); HERBSTASTER (*f*)

Botanical Term.	No.	English Term.	German Term.
Aster tripolium	698	Sea aster	**SEEASTER** (*f*); **STRANDASTER** (*f*); **SUMPFASTER** (*f*)
Astilbe Japonica	699	Japanese saxifrage	**JAPANISCHE ASTILBE** (*f*)
Astracus stellatus	700	Star-fungus (See *Geaster hygrometricus*)	**STERNPILZ** (*m*)
Astragalus alpinus	701	Alpine astragal; alpine milk-vetch	**ALPEN-KAFFEEWICKE** (*f*)
Astragalus bœticus	702	Milk vetch	**KAFFEEWICKE** (*f*)
Astragalus danicus	703	Purple astragal (See *Astragalus hypoglottis*)	
Astragalus glycyphyllos	704	Sweet milk vetch; wild liquorice; liquorice vetch	**LAKRITZENWICKE** (*f*)
Astragalus gummifer	705	(See *Astragalus tragacantha*)	
Astragalus hypoglottis	706	Purple milk vetch (See *Astragalus danicus*)	
Astragalus tragacantha	707	Tragacanth plant; gum dragon plant	**TRAGANTPFLANZE** (*f*); **TRAGANT** (*m*)
Astragalus verus	708	(See *Astragalus tragacantha*)	
Astrantia major	709	Astrantia; master wort	**SCHWARZE MEISTERWURZEL** (*f*); **SCHWARZE MEISTERWURZ** (*f*); **ASTRANTIE** (*f*); **TALSTERN** (*m*); **GROSSER TALSTERN** (*m*); **STERNDOLDE** (*f*)
Astrantia minor	710	Small astrantia	**ASTRANZ** (*f*); **KLEINER TALSTERN** (*m*); **ALPENSTERN** (*m*); **STRÄNZE** (*f*); **OSTRANZ** (*f*); **KLEINER ALPENSTERN** (*m*)
Astrocaryum vulgare	711	Tucuma palm	**TUCUMAPALME** (*f*)
Atragene	712	(See *Clematis*)	
Atriplex calotheca	713	Arrow-leaved orache	**PFEILBLÄTTRIGE MELDE** (*f*)
Atriplex hastatum	714	Broad-leaved orache	**SPIESSBLÄTTRIGE MELDE** (*f*)
Atriplex hortensis	715	Orach; garden orache; mountain spinach (See *Halimus portulacoides*)	**MELDE** (*f*); **MEERMELDE** (*f*); **ARROCHE** (*f*); **MEERPORTULAK** (*m*); **ZUCKERMELDE** (*f*); **GARTENMELDE** (*f*); **WILDER SPINAT** (*m*)
Atriplex litorale	716	Littoral orache	**UFERMELDE** (*f*)
Atriplex nitens	717	Shining orache	**GLÄNZENDE MELDE** (*f*)
Atriplex patula	718	Common orache	**MELDE** (*f*); **AUSGEBREITETE MELDE** (*f*)
Atriplex pedunculata	719	Stalked orache (See *Obione pedunculata*)	
Atriplex portulacoides	720	Sea urslane	**PORTULAKMELDE** (*f*)
Atriplex rosea	721	Frosted orache	**STERNMELDE** (*f*)
Atriplex tartaricum	722	Tartar orache	**TARTARISCHE MELDE** (*f*)
Atropa belladonna	723	Dwale; deadly nightshade; belladonna; banewort	**BELLADONNA** (*f*); **TEUFELSKIRSCHE** (*f*); **WOLFSKIRSCHE** (*f*); **TOLLKRAUT** (*n*); **GEMEINE TOLLKIRSCHE** (*f*); **WOLFSMUT** (*f*)
Attalea cohune	724	Cohune palm	**COHUNE** (*f*); **COHUNEPALME** (*f*)
Attalea funifera	725	Brazilian coquilla-nut palm, bast palm	**PIASSAVA** (*f*); **PIKABAHANFBAUM** (*m*); **CHIQUICHIQUI-PALME** (*f*); **PIASSABEPALME** (*f*)
Aubrietia deltoidea	726	Purple rock cress; aubrietia	**GRIECHISCHE AUBRIETIA** (*f*)
Aucuba japonica	727	Japanese aucuba, Aoki	**JAPANISCHE AUKUBE** (*f*); **AOKI** (*f*)
Auklandia costus	728	Putchock	**KOSTWURZ** (*f*)
Auricularia auricula judæ	729	Jew's ear fungus (See *Hirneola auricula-Judas*)	**JUDASOHR** (*n*); **JUDASOHRPILZ** (*m*)
Avena brevis	730	Short oats	**KURZER HAFER** (*m*)
Avena elatior	731	French rye-grass (See *Arrhenatherum avenaceum*)	**FRANZÖSISCHES REIGRAS** (*n*)
Avena fatua	732	Wild oat(s)	**WILDER HAFER** (*m*)
Avena flavescens	733	Yellow oat-grass (See *Trisetum flavescens*)	**GOLDHAFER** (*m*)
Avena nuda	734	Naked oats	**NAKTHAFER** (*m*)
Avena orientalis	735	Tartarian oats	

Botanical Term.	No.	English Term.	German Term.
Avena pratensis	736	Perennial oat	TRIFTHAFER (*m*); BERGHAFER (*m*)
Avena pubescens	737	Meadow oats	RAINHAFER (*m*): WEICHHAARIGER WIESENHAFER (*m*)
Avena sativa	738	Common oats	HAFER (*m*)
Avena strigosa	739	Bristle-pointed oats	
Averrhoa bilimbi	740	Cucumber tree	GURKENBAUM (*m*)
Averrhoa carambola	741	Coromandel gooseberry; carambola	BAUMSTACHELBEERE (*f*)
Avicennia tomentosa	742	Mangrove tree	MANGROVE (*f*)
Avoira elais	743	(See *Elæis guineensis*)	
Azalea indica	744	Chinese dwarf azalea	CHINESISCHE NIEDRIGE AZALIE (*f*)
Azalea mollis	745	Japanese azalea	JAPANISCHE AZALIE (*f*)
Azalea pontica	746	Tall azalea; yellow-flowered azalea (See *Rhododendron flavum*)	HOHE AZALIE (*f*); GELBBLÜTIGE AZALIE (*f*)
Azalea procumbens	747	Azalea (See *Loiseleuria procumbens*)	FELSENSTRAUCH (*m*); AZALIE (*f*)
Azalea sinensis	748	Chinese azalea	CHINESISCHE AZALIE (*f*)
Azalea viscosa	749	North American azalea (See *Rhododendron viscosum*)	NORD-AMERIKANISCHE AZALIE (*f*); KLEBRIGE AZALEE (*f*)
Ballota fœtida	750	Horehound	STINKENDE TAUBNESSEL (*f*); GOTTVERGESS (*m*)
Ballota nigra	751	Black horehound (See also *Marrubium vulgare*)	BALLOTE (*f*); SCHWARZE BALLOTE (*f*); SCHWARZER ANDORN (*m*)
Balsamita vulgaris	752	Ale-cost, costmary (See *Chrysanthemum balsamita*)	
Balsamocarpum brevifolium	753	Algarobilla tree (See *Cæsalpinia brevifolia*)	ALGAROBILLABAUM (*m*)
Balsamodendron	754	Balsam bush (See *Commiphora*)	BALSAMSTAUDE (*f*); BALSAMSTRAUCH (*m*)
Bambusa	755	(See *Phyllostachys* and also *Dendrocalamus*)	
Bambusa arundinacea	756	(See *Bambusa metake*)	
Bambusa metake	757	Japanese bamboo (See *Arundinaria japonica*)	BAMBUS (*m*); GEMEINES BAMBUSROHR (*n*)
Bambusa vulgaris	758	(See *Bambusa metake*)	
Banksia australis	759	(Australian) honeysuckle tree	AUSTRALISCHE BANKSIA (*f*)
Baphia nitida	760	(African) camwood tree; redwood tree	KAMBIERBAUM (*m*); KAMHOLZBAUM (*m*); AFRIKANISCHER ROTHOLZBAUM (*m*); AFRIKANISCHES SANDELHOLZ (*n*); CAMBALHOLZBAUM (*m*); ANGOLAHOLZ (*n*)
Barbarea vulgaris	761	yellow rocket; upland cress; common wintercress; American wintercress	
Barosma	762	Bucco-leaf tree; Buchu tree	BUCCOBLÄTTERBAUM (*m*)
Bartsia alpina	763	Alpine bartsia	
Bartsia odontites	764	Red Bartsia	
Bartsia viscosa	765	Viscid Bartsia	
Bassia	766	Butter-tree; Indian butter-tree	BUTTERBAUM (*m*)
Bassia butyracea	767	Ghee butter-tree (See *Illipe butyracea*)	BUTTERBAUM (*m*); INDISCHER BUTTERBAUM (*m*); PHULWARABAUM (*m*)
Bassia latifolia	768	Mahwa butter-tree; moa tree (See also *Illipe latifolia*)	MAHWABAUM (*m*); MADHUKABAUM (*m*)
Bassia longifolia	769	Bassia tree	BASSIA (*f*)
Bassia parkii	770	Shea butter-tree (See *Illipe Parkii* and *Butyrospermum Parkii*)	
Batatas edulis	771	Sweet potato (See *Ipomœa batatas*)	BATATE (*f*)
Begonia nitida	772	Winter-flowering begonia; fibrous rooted begonia	SCHIEFBLATT (*n*)
Begonia rex	773	Rex begonia, ornamental-leaved begonia	

Botanical Term.	No.	English Term.	German Term.
Begonia semperflorens	774	Winter-flowering begonia; fibrous-rooted begonia	
Bellis perennis	775	Daisy; common daisy	MASSLIEBCHEN (*n*); MASSLIEB (*n*); SAMTRÖSCHEN (*n*); TAU-SENDSCHÖN (*n*); GÄNSE-BLUME (*f*)
Benthamia fragifera	776	(Nepaul) benthamia	BENTHAMIA (*f*)
Benzoin benzoin	777	Spice bush (See *Lindera benzoin* and *Styrax*)	BENZOELORBEER (*m*)
Benzoin odoriferum	778	Spicebush	BENZOELORBEER (*m*)
Berberideæ	779	Berberideæ	SAUERDORNGEWÄCHSE (*n.pl*); SAUERDÖRNER (*m.pl*); BERBER-IDAZEEN (*f.pl*)
Berberis aquifolium	780	Evergreen barberry (See *Mahonia aquifolium*)	GEMEINE MAHONIE (*f*)
Berberis aristata	781	Barberry; berberis; yellow-flowered barberry	BERBERITZE (*f*)
Berberis asiatica	782	Asiatic barberry	ASIATISCHER SAUERDORN (*m*)
Berberis buxifolia	783	Box-leaved barberry (See *Berberis dulcis*)	BUCHSBAUM-SAUERDORN (*m*)
Berberis cretica	784	Japanese barberry, Jaundice berry (See *Berberis thunbergii*)	
Berberis darwinii	785	Orange-yellow-flowered barberry	DARWINS-SAUERDORN (*m*)
Berberis dulcis	786	I ox-leaved barberry (See *Berberis buxifolia*)	
Berberis japonica	787	Long-leaved barberry (See also *Ilex japonica* and *Mahonia japonica*)	JAPANISCHE MAHONIE (*f*)
Berberis sinensis	788	Chinese barberry	CHINESISCHER SAUERDORN (*m*)
Berberis thunbergii	789	Low-growing barberry	THUNBERGS-SAUERDORN (*m*)
Berberis vulgaris	790	Barberry; pepperidge-bush; pipperidge-bush	GEMEINER SAUERDORN (*m*); BERBERITZE (*f*); ESSIGDORN (*m*); BERBESBEERE (*f*); GE-MEINER BERBERITZENSTRAUCH (*m*); SAUERACH (*m*)
Bertholletia excelsa	791	Brazil-nut (tree)	PARANUSS (*f*); PARANUSSBAUM (*m*); BRASILIANISCHER NUSS-BAUM (*m*); BERTHOLLETIA (*f*)
Beta maritima	792	Sea beet; mangel-wurzel	MANGELWURZEL (*f*); MANGEL-RÜBE (*f*); MANGOLDWURZEL (*f*)
Beta vulgaris	793	Beet(root)	BETE (*f*); ROTE RÜBE (*f*); SALAT-RÜBE (*f*); RUNKELRÜBE (*f*); ZUCKERRÜBE (*f*); MANGOLD (*n*)
Beta vulgaris cicla	794	Mangold	MANGOLD (*n*); BETE (*f*); BEISS-KOHL (*m*); RÖMISCHER SPINAT (*m*); RÖMISCHER KOHL (*m*)
Beta vulgaris macrorrhiza	795	Mangel-wurzel	MANGOLD (*n*)
Beta vulgaris rapa	796	Beetroot	RUNKELRÜBE (*f*)
Betonica officinalis	797	Betony (See *Stachys betonica*)	BETONIENKRAUT (*n*); ZIEST (*m*)
Betula	798	Birch, Queen of the woods	BIRKE (*f*)
Betula alba	799	Birch; white birch; common birch	BIRKE (*f*); GEMEINE BIRKE (*f*); MAIBAUM (*m*); WEISSBIRKE (*f*)
Betula alba pendula	800	Weeping birch	HÄNGEBIRKE (*f*)
Betula alba purpurea	801	Pendulous birch; purple-leaved birch	PURPURBLÄTTRIGE BIRKE (*f*)
Betula alnus	802	(See *Alnus glutinosa*)	
Betula carpinifolia	803	Japanese birch	JAPANISCHE BIRKE (*f*)
Betula costata	804	Beech-leaved birch	WEISSBUCHENBLÄTTRIGE BIRKE (*f*)
Betula daurica	805	Siberian birch	SIBERISCHE BIRKE (*f*)
Betula fruticosa	806	Bush birch	STRAUCHBIRKE (*f*)
Betula glandulosa	807	Dwarf birch	ZWERGBIRKE (*f*)

Botanical Term.	No.	English Term.	German Term.
Betula glutinosa	808	Common birch (See *Betula alba*)	BIRKE (*f*)
Betula lanulosa	809	(See *Betula nigra*)	
Betula lenta	810	Birch tree; North American sugar birch, cherry birch	BIRKE (*f*); ZUCKERBIRKE (*f*)
Betula lutea	811	(See *Betula lenta*)	
Betula nana	812	Dwarf birch	ZWERGBIRKE (*f*)
Betula nigra	813	Red birch	SCHWARZBIRKE (*f*); ROTBIRKE (*f*)
Betula odorata	814	(See *Betula alba*)	
Betula papyracea	815	Papery birch (See *Betula papyrifera*)	
Betula papyrifera	816	Paper birch (See *Betula papyracea*)	PAPIERBIRKE (*f*)
Betula pendula	817	Weeping birch (See *Betula verrucosa*)	MASERBIRKE (*f*)
Betula pubescens	818	Common birch	BIRKE (*f*); RUCHBIRKE (*f*); MAIENBAUM (*m*); GERUCH--BIRKE (*f*); RUCHBIRKE (*f*); NORDISCHE BIRKE (*f*); HARZ--BIRKE (*f*); MASERBIRKE (*f*); MOORBIRKE (*f*); RAUHBIRKE (*f*); STEINBIRKE (*f*); WEISS--BIRKE (*f*); WINTERBIRKE (*f*)
Betula pumila	819	Dwarf birch	ZWERGBIRKE (*f*)
Betula utilis	820	Himalayan papery birch	ASIATISCHE PAPIERBIRKE (*f*); BHOJPATRABIRKE (*f*); CHURJ--BIRKE (*f*)
Betula verrucosa	821	Birch; weeping birch (See *Betula pendula*)	WEISSBIRKE (*f*); BIRKE (*f*); HÄNGEBIRKE (*f*); TRAUER--BIRKE (*f*)
Bidens cernua	822	Bur-marigold	PRIESTERLAUS (*f*); ZWEIZAHN (*m*); PRIESTERLÄUSEKRAUT (*n*)
Bidens tripartita	823	Three-cleft bidens	
Bignonia capreolata	824	Trumpet flower, bignonia	BIGNONIE (*f*); TROMPETEN--BLUME (*f*)
Bignoniaceæ	825	Bignoniaceæ	BIGNONIAZEEN (*f.pl*); TROM--PETENBLÜTLER (*m.pl*)
Bignonia radicans	826	Indian jasmine (See *Tecoma radicans*)	WURZELNDE KLETTERTROMPETE (*f*)
Biota orientalis	827	(See *Thuja orientalis*)	ORIENTALISCHER LEBENSBAUM (*m*)
Bixa orellana	828	Orlean, anetta, arnatto, rocoe	ORLEAN (*m*)
Blechnum boreale	829	Brazilian tree-fern (See *Lomaria spicant*)	
Blechnum spicant	830	Hard fern; blechnum	GEMEINER RIPPENFARN (*m*)
Blitum capitatum	831	Strawberry spinach	ERDBEERSPINAT (*m*)
Blumenbachia lateritia	832	(See *Loasa lateritia*)	BLUMENBACHIA (*f*)
Blysmus compressus	833	Broad blysmus	
Blysmus rufus	834	Narrow blysmus	
Bœhmeria nivea	835	Grass cloth plant, China grass (See *Urtica nivea*)	RAMIE (*f*); CHINAGRAS (*n*); RHEAHANF (*m*); RHEA (*f*)
Boletus	836	Boletus	RÖHRENSCHWAMM (*m*); RÖHREN--PILZ (*m*); SCHWAMM (*m*)
Boletus badius	837	Sweet-chestnut fungus	MARONENPILZ (*m*)
Boletus bovinus	838	Cow-spunk	KUHPILZ (*m*)
Boletus castaneus	839	chestnut fungus	KASTANIENPILZ (*m*)
Boletus cervinus	840	Boletus (See *Elaphomyces*)	HIRSCHBRUNST (*f*)
Boletus chirurgorum	841	Surgeon's agaric (See *Agaricus chirurgorum*)	FEUERSCHWAMM (*m*)
Boletus edulis	842	Champignon, edible boletus	STEINPILZ (*m*); HERRENPILZ (*m*)
Boletus fomentarius	843	(See *Fomes fomentarius* and *Polyporus fomentarius*)	
Boletus granulatus	844		SCHMERLING (*m*)
Boletus igniarius	845	(See *Polyporus igniarius*)	FEUERSCHWAMM (*m*)
Boletus laricis	846	Female agaric (See *Agaricus albus*, *Agaricus chirurgorum* and *Polyporus officinalis*)	LÄRCHENSCHWAMM (*m*)

Botanical Term.	No.	English Term.	German Term.
Boletus luridus	847		HEXENPILZ (*m*)
Boletus luteus	848	Butter-fungus	BUTTERPILZ (*m*)
Boletus pachypus	849		DICKFUSS (*m*)
Boletus polyporus	850	Boletus (See *Polyporus*)	LÖCHERPILZ (*m*); LÖCHER--SCHWAMM (*m*)
Boletus satanas	851	Satan's fungus	BLUTPILZ (*m*); SATANSPILZ (*m*)
Boletus scaber	852	Birch fungus	BIRKENPILZ (*m*); KAPUZINER--PILZ (*m*)
Boletus tomentosus	853		ZIEGENLIPPE (*f*)
Boletus variegatus	854	Sand fungus	SANDPILZ (*m*)
Bombax ceiba	855	(See *Eriodendron anfractuosum*)	WOLLBAUM (*m*)
Bombax pentandrum	856	(See *Eriodendron anfractuosum*)	
Bongardia rauwolfii	857	Bongardia	
Borago laxiflora	858	Corsican borage	
Borago longifolia	859	Long-leaved borage	
Borago officinalis	860	Borage; bee bread; common borage	BORETSCH (*m*); GURKENKRAUT (*n*); BORRETSCH (*m*); GEMEI--NER BORETSCH (*m*)
Borago orientalis	861	Cretan borage	
Borassus æthiopum	862	Ethiopian fan palm; deleb palm	DELEBPALME (*f*)
Borassus flabelliformis	863	Palmyra palm	FÄCHERPALME (*f*); STACHEL--SCHWEINHOLZPALME (*f*); SCHIRMPALME (*f*); PALMYRA--PALME (*f*); TODDYPALME (*f*); LONTARPALME (*f*); WEIN--PALME (*f*)
Boswellia serrata	864	Boswellia	SALAIBAUM (*m*)
Botrychium lunaria	865	Moon-wort-(fern); botrychium; moon-fern	MONDRAUTE (*f*); WALPURGIS--KRAUT (*n*); RIPPENFARN (*m*)
Bovista nigrescens	866	Black egg-shaped puff-ball (See *Lycoperdon bovista*)	SCHWÄRZLICHER EIERSTÄUBLING (*m*); BOVIST (*m*)
Brachypodium pinnatum	867	Heath false-brome	FEDERSCHWINGEL (*m*); GEFIE--DERTE ZWENKE (*f*); FEDER--ZWENKE (*f*)
Brachypodium sylvaticum	868	Slender false-brome	ZWENKE (*f*); FEDERSCHWINGEL (*m*)
Brassica adpressa	869	Hoary brassica	
Brassica alba	870	White mustard; cultivated mustard (See *Sinapis alba*)	
Brassica campestris	871	Turnip; swede; wild navew; common turnip; smooth-leaved summer-rape; rape; colza; Swedish turnip; swede turnip; field brassica; Black-sea rape (See *Brassica rapa*)	RAPS (*m*); WEISSE RÜBE (*f*); RÜBENKOHL (*m*)
Brassica caulorapa	872	Turnip-rooted cabbage; Kohl-rabi; cole-rabi; underground Kohlrabi (See *Brassica napobrassica*)	KOHLRABI (*m*); KOHLRÜBE (*f*)
Brassica gemmifera	873	(See *Brassica oleracea gemmifera*)	
Brassica glauca	874	Yellow Indian rape	GELBER INDISCHER RAPS (*m*)
Brassica gongylodes	875	Surface Kohlrabi	OBERKOHLRABI (*m*)
Brassica monensis	876	Isle-of-Man brassica	
Brassica muralis	877	Wall brassica	
Brassica napobrassica	878	(See *Brassica caulorapa*)	
Brassica napus	879	Rape; Coleseed; (rough-leaved)-winter-rape; rapeseed	RAPS (*m*); REPS (*m*); RAPSKOHL (*m*)
Brassica nigra	880	Black mustard	SCHWARZER SENF (*m*)
Brassica oleracea	881	(Common)-cabbage; wild cab-bage; cole; cale; kale; borecole; broccoli; Brussels sprouts; cauliflower	KOHL (*m*); KAPPES (*m*)
Brassica oleracea acephala	882	Kale, borecole	WINTERKOHL (*m*); GARTENKOHL (*m*)
Brassica oleracea botrytis	883	Cauliflower	KARFIOL (*m*); BLUMENKOHL (*m*)
Brassica oleracea botrytis asparagoides	884	Broccoli	BLUMENKOHL (*m*); KARVIOL (*m*); SPARGELKOHL (*m*); KÄSEKOHL (*m*)

Botanical Term.	No.	English Term.	German Term.
Brassica oleracea bullata	885	Savoy (See *Brassica sabauda*)	WIRSINGKOHL (m); SAVOYERKOHL (m)
Brassica oleracea capitata	886	Cabbage, common cabbage	KOPFKOHL (m); WEISSKRAUT (n); KABBES (m); KAPPES (m); KABIS (m); KRAUT (n); ROT-KRAUT (n); WEISSKOHL (m); BUTTERKRAUT (n); FILDER-KRAUT (n); ZENTNERKRAUT (n); YORKERKRAUT (n); OCHSENHERZ (n)
Brassica oleracea Caulo-rapa	887	Kohl Rabi (See *Brassica caulorapa* and *Brassica napobrassica*)	
Brassica oleracea costata	888	Portugal cabbage, Couve Tronchuda	
Brassica oleracea gemmifera	889	Brussels sprouts	ROSENKOHL (m)
Brassica oleracea sylvestris	890	Wild cabbage	WILDER KOHL (m)
Brassica rapa	891	(Common)-turnip (See *Brassica campestris*)	FUTTERRÜBE (f); RÜBE (f); WASSERRÜBE (f); RÜBSEN (m)
Brassica rutabaga	892	Swede; rutabaga	KOHLRÜBE (f)
Brassica sabauda	893	Savoy (See also *Brassica oleracea bullata*)	SAVOYERKOHL (m); WIRSING (m); WELCHER KOHL (m); HERZ-KOHL (m); BÖRSCH (m)
Brassica sinapis	894	Charlock; wild mustard (See *Sinapis arvensis* and *Brassica sinapistrum*)	
Brassica sinapistrum	895	Charlock; wild mustard (See *Brassica sinapis*)	ACKERSENF (m); HEDERICH (m)
Brassica tenuifolia	896	Sand brassica; rocket	
Brassica vulgaris	897	Wild cabbage	KUHKOHL (m); EWIGER KOHL (m); BAUMKOHL (m); BLATTKOHL (m)
Brayera anthelmintica	898	Kousso-(plant), cusso-(plant), Brayera (See *Hagenia abyssinica*)	KOSOSTRAUCH (m)
Briza	899	Briza; quaking grass	LIEBESGRAS (n); ZITTERGRAS (n); AMOURETTENGRAS (n)
Briza erecta	900	Mexican quaking grass (See *Briza rotundata*)	MEXIKANISCHES ZITTERGRAS (n)
Briza maxima	901	South-European quaking grass	SÜDEUROPÄISCHES ZITTERGRAS (n)
Briza media	902	Quaking grass; briza	ZITTERGRAS (n)
Briza minor	903	Lesser quaking grass; briza	KLEINES ZITTERGRAS (n)
Briza rotundata	904	(See *Briza erecta*)	
Bromelia ananas	905	Pineapple (See *Ananas sativa*)	ANANAS (f)
Bromelia karatas	906	Fibrous pineapple (See *Karatas plumieri*)	FASERANANAS (f)
Bromelia silvestris	907	Bromelia	BROMELIA (f); ISTLESTAUDE (f)
Bromus arvensis	908	Field brome; taper field brome-grass	KLEINE ACKERTRESPE (f)
Bromus asper	909	Hairy brome; hairy wood brome-grass	WALDTRESPE (f)
Bromus brizæformis	910	Transcaucasian quaking grass	TRANSKAUKASISCHE TRESPE (f)
Bromus commutatus	911	Tumid brome-grass (See *Bromus multiflorus* and *Serrafalcus commutatus*)	
Bromus diandrus	912	Upright annual brome-grass	AUFRECHTE JÄHRLICHE TRESPE (f)
Bromus erectus	913	Upright brome; upright perennial brome-grass	AUFRECHTE TRESPE (f); AUSDAU-ERNDE TRESPE (f)
Bromus giganteus	914	Tall brome; giant brome (See *Festuca gigantea*)	GROSSE TRESPE (f)
Bromus inermis	915		WEHRLOSE TRESPE (f)
Bromus madritensis	916	Compact brome	
Bromus mango	917	Chilian breadplant	CHILI BROTPFLANZE (f)
Bromus maximus	918	Great brome	GROSSE TRESPE (f)
Bromus mollis	919	Lop-(grass); bobs; soft brome-grass	WEICHE TRESPE (f)
Bromus multiflorus	920	(See *Bromus commutatus* and *Serrafalcus commutatus*)	VIELBLÜTIGE TRESPE (f)

Botanical Term.	No.	English Term.	German Term.
Bromus pratensis	921	Meadow brome-grass	WIESENTRESPE (*f*)
Bromus racemosus	922	Smooth brome-grass	
Bromus schraderi	923	Prairie grass (See *Ceratochloa pendula*)	PRÄRIEGRAS (*n*); HORNSCHWINGEL (*m*)
Bromus secalinus	924	Smooth Rye brome-grass	ROGGENTRESPE (*f*); KORNTRESPE (*f*); GROSSE ACKERTRESPE (*f*); DORT (*m*); TÖBERICH (*m*)
Bromus squarrosus	925	Corn brome-grass	TRESPE (*f*); KORNTRESPE (*f*)
Bromus sterilis	926	Barren brome-(grass)	
Bromus velutinus	927	Downy Rye brome-grass	
Brosimum alicastrum	928	Bread-nut tree	BROTNUSSBAUM (*m*)
Brosimum Aubletti	929	Leopard wood; snakewood; speckled (letter) wood; Tigerwood (See also *Jacaranda cærulea*)	SCHLANGENHOLZ (*n*); MUSKATHOLZ (*n*); LETTERNHOLZ (*n*); BUCHSTABENHOLZ (*n*); TIGERHOLZ (*n*); SCHLANGENWURZEL (*f*); SCHLANGENHOLZBAUM (*m*); MUSKATHOLZBAUM (*m*); LETTERNHOLZBAUM (*m*); FASANENHOLZBAUM (*m*); TIGERHOLZBAUM (*m*)
Brosimum galactodendron	930	Cow-tree (See *Galactodendron utile*)	KUHBAUM (*m*); MILCHBAUM (*m*)
Broussonetia papyrifera	931	(See *Broussonetia tinctoria*)	JAPANISCHER PAPIERMAULBEERBAUM (*m*)
Broussonetia tinctoria	932	Paper mulberry (See *Morus tinctoria*)	PAPIERMAULBEERBAUM (*m*)
Brugiera gymnorrhiza	933	Mangrove tree	MANGROVE (*f*)
Brugmansia bicolor	934	(See *Datura sanguinea*)	
Brugmansia candida	935	(See *Datura arborea*)	
Brya ebenus	936	Green ebony; red ebony (See *Ebenus cretica* and *Jacaranda ovalifolia*)	GRÜNEBENHOLZ (*n*); UNECHTES POCKHOLZ (*n*); ROTEBENHOLZBAUM (*m*); GRENADILLHOLZBAUM (*m*); KOKOSHOLZBAUM (*m*); WESTINDISCHER EBENHOLZBAUM (*m*); AMERIKANISCHER EBENHOLZBAUM (*m*); WESTINDISCHER GRENADILLHOLZBAUM (*m*); GRÜNER EBENHOLZBAUM (*m*)
Bryanthus taxifolius	937	(See *Dabœcia taxifolia*)	
Bryonia alba	938	White bryony; black-berried bryony	HUNDSKIRSCHE (*f*); TOLLRÜBE (*f*); ZAUNREBE (*f*); ZAUNRÜBE (*f*); GEMEINE ZAUNRÜBE (*f*); HUNDSRÜBE (*f*); GICHTRÜBE (*f*)
Bryonia dioica	939	Red bryony; common bryony; red-berried bryonia	ROTFRÜCHTIGE ZAUNRÜBE (*f*)
Bryopogon	940		MOOSBART (*m*)
Bunium bulbocastanum	941	Ground nut (See *Carum bulbocastanum*)	ERDNUSS (*f*); ERDKASTANIE (*f*); KASTANIENKÜMMEL (*m*)
Bunium esculentum	942	Common pig-nut (See *Bunium flexuosum*)	
Bunium flexuosum	943	Ground-nut; pig-nut; hawk-nut; earth-chestnut; earthnut (See *Arachis hypogœa*)	ERDEICHEL (*f*); ERDNUSS (*f*); NUSSKÜMMEL (*m*)
Bupleurum aristatum	944	Narrow buplever	
Bupleurum falcatum	945	Falcate buplever	
Bupleurum rotundifolium	946	Throw-wax; hair's ear; buplever	
Bupleurum tenuissimum	947	Slender buplever	
Bursa bursa-pastoris	948	Shepherd's purse (See *Capsella Bursa-pastoris*)	HIRTENTASCHE (*f*); HIRTENTÄSCHLEIN (*n*)
Butea frondosa	949	(Indian or Bengal) Dhak tree or Palas tree	MALABARISCHER LACKBAUM (*m*); KINOBAUM (*m*); DHAKBAUM (*m*); OSTINDISCHER KINOBAUM (*m*); PULASKINOBAUM (*m*); PALASABAUM (*m*)

Botanical Term.	No.	English Term.	German Term.
Butomus umbellatus	950	Flowering rush ; butome	WASSERLIESCH (*m*) ; WASSER--VIOLE (*f*) ; BLUMENBINSE (*f*) ; SCHWANENBLUME (*f*)
Butyrospermum Parkii	951	Butter tree (See *Illipe Parkii*)	SCHIBAUM (*m*) ; BUTTERBAUM (*m*)
Buxus balearica	952	Balearic box-tree	BALEARISCHER BUCHSBAUM (*m*)
Buxus japonica	953	(Japanese)-box tree	JAPANISCHER BUCHSBAUM (*m*)
Buxus microphylla	954	Dwarf, small-leaved box tree	KLEINBLÄTTIGER (BUXBAUM) BUCHSBAUM (*m*)
Buxus sempervirens	955	Box, boxwood, box tree	BUXBAUM (*m*) ; BUCHSBAUM (*m*) ; GEMEINER BUCHSBAUM (*m*) ; ECHTER BUCHSBAUM (*m*)
Buxus suffruticosa	956	(English)-box tree	ENGLISCHER BUCHSBAUM (*m*)
Cactaceæ	957	Cactaceæ, cacti	KAKTEEN (*f.pl*)
Cactus pereskia	958	West Indian cactus	WESTINDISCHER KAKTUS (*m*)
Cactus serpentinus	959	Serpentine cactus	SERPENTIN-KAKTUS (*m*)
Cæoma pinitorquum	960	Pine rust (See *Melampsora tremulæ*)	KIEFERNDREHER (*m*)
Cæsalpinia bijuga	961	Brazil wood ; redwood	FERNAMBUKHOLZ (*n*) ; WEST--INDISCHER ROTHOLZBAUM (*m*)
Cæsalpinia brasiliensis	962	Brazil wood ; redwood	BRASILIENHOLZ (*n*) ; ROTHOLZ (*n*) ; FERNAMBUKHOLZ (*n*)
Cæsalpinia brevifolia	963	Algarobilla bush (See *Balsamocarpum brevifolium*)	ALGAROBILLASTRAUCH (*m*)
Cæsalpinia coriaria	964	Dividivi tree	DIVIDIVIBAUM (*m*) ; LIBIDIBI--BAUM (*m*)
Cæsalpinia crista	965	Brazil wood ; redwood	
Cæsalpinia echinata	966	Brazil wood, redwood, Mexican redwood	BRASILIENHOLZ (*n*) ; ROTHOLZ (*n*); FERNAMBUKHOLZ (*n*) ; MAR--THENHOLZ (*n*)
Cæsalpinia pipai	967	Pipi-pod tree	
Cæsalpinia pulcherrima	968	Barbados pride	
Cæsalpinia sappan	969	Sappan wood, Brazil redwood	INDISCHES ROTHOLZBAUM (*m*) ; SAPPANHOLZ (*n*) ; ROTHOLZ (*n*) ; BRASILIENHOLZ (*n*) ; BAHIAROTHOLZ (*n*) ; PERNAM--BUKHOLZ (*n*) ; FERNAMBUK--HOLZ (*n*) ; BRASILETTHOLZ (*n*) ; SAPPANHOLZBAUM (*m*)
Cakile maritima	970	Purple sea-rocket ; sea-rocket ; sea-cress	MEERSENF (*m*)
Calamagrostis epigeios	971	Wood smallreed	LANDROHRGRAS (*n*)
Calamagrostis lanceolata	972	Purple smallreed (See *Arundo calamagrostis*)	RIEDGRAS (*n*) ; TEICHROHRGRAS (*n*)
Calamagrostis stricta	973	Narrow smallreed	FEDERGRAS (*n*) ; ROHR (*n*) ; RIET (*n*)
Calamintha acinos	974	Basil thyme ; field calaminth ; field balm ; calamint (See *Melissa calamintha* and *Thymus calamintha*)	KALAMINTHE (*f*)
Calamintha clinopodium	975	Wild basil ; hedge calaminth (See *Clinopodium vulgare* and also *Satureia clinopodium*)	BERGMINZE (*f*)
Calamintha coccinea	976	Scarlet calamint (See *Cunila coccinea*)	SCHARLACHROTE KALAMINTHE (*f*)
Calamintha nepeta	977	Lesser calamint	
Calamintha officinalis	978	Common calamint ; common basil (See *Satureia calamintha*)	BERGMELISSE (*f*) ; BERGMINZE (*f*)
Calamus aromaticus	979	Officinal sweet-flag (See *Acorus calamus*)	
Calamus draco	980	(See *Dæmonorops draco*)	
Calamus rotang	981	Calamus ; rattan ; reed-palm	ROHR (*n*) ; ROHRPALME (*f*) ; ROTANGPALME (*f*) ; SCHILF--PALME (*f*) ; ROTTANGPALME (*f*)
Calamus scipionum	982	Calamus ; malacca cane	

Botanical Term.	No.	English Term.	German Term.
Calamus tenuis	983	Calamus ; rattan cane	
Calamus viminalis	984	Calamus ; rattan cane	
Calandrinia nitida	985	Pink-flowered calandrinia	CALANDRINIA (*f*)
Calanthe macrolaba	986	White-flowered, evergreen calanthe orchid	
Calanthe veratrifolia	987	White-flowered, evergreen calanthe orchid	
Calanthe versicolor	988	Lilac-flowered, evergreen calanthe orchid	
Calanthe vestita	989	white-flowered, deciduous calanthe orchid	
Calathea	990	Calathea	
Calceolaria	991	Slipper-wort, calceolaria	PANTOFFELBLUME (*f*) ; KALZEO--LARIE (*f*)
Calcitrapa hippophæstum	992	(See *Centaurea calcitrapa*)	
Calendula officinalis	993	Common pot marigold ; marygold ; old marigold ; marigold	RINGELBLUME (*f*) ; KALENDEL (*f*) ; TOTENBLUME (*f*) ; WARZEN--KRAUT (*n*) ; GOLDBLUME (*f*)
Calendula pluvialis	994	Small Cape marigold	KAP-RINGELBLUME (*f*)
Calla palustris	995	Calla ; bog arum	SCHLANGENKRAUT (*n*) ; KALLA (*f*) ; DRACHENWURZ (*f*) ; ROTER WASSERPFEFFER (*m*) ; SUMPFSCHLANGENKRAUT (*n*) ; SCHWEINSOHR (*n*) ; SCHLANG--ENWURZ (*f*) ; SUMPF--SCHWEINSWURZ (*f*)
Callistemon speciosus	996	Bottle brush tree (See *Metrosideros speciosa*)	AUSTRALISCHER CALLISTEMON (*m*)
Callistephus chinensis	997	China aster (See *Aster chinensis*)	CHINESISCHE ASTER (*f*)
Callitriche aquatica	998	Water starwort, star-grass	WASSERSTERN (*m*)
Callitriche verna	999	Water starwort (See *Callitriche aquatica*)	
Callitris arborea	000	Clanwilliam cedar	CLANWILLIAM-ZEDER (*f*)
Callitris juniperoides	001	South-African cedar	SÜDAFRIKANISCHE ZEDER (*f*) ; WACHOLDERARTIGE SANDARAK-ZYPRESSE (*f*) ; CEDERBOOM (*m*)
Callitris quadrivalvis	002	(North African)-sandarac tree ; sandarach tree (See *Thuja articulata*)	SANDARAKBAUM (*m*) ; SANDARAK--ZYPRESSE (*f*)
Calluna vulgaris	003	Common heath ; ling ; heather (See *Erica vulgaris*)	SANDHEIDE (*f*) ; HEIDE (*f*) ; HEIDEKRAUT (*n*) ; BESENKRAUT (*n*) ; BESENHEIDE (*f*) ; IMMER--SCHÖNKRAUT (*n*)
Calluna vulgaris alba	004	White heather	WEISSE SANDHEIDE (*f*) ; WEISSES GEMEINES HEIDEKRAUT (*n*)
Calocera viscosa	005		HIRSCHSCHWÄMMCHEN (*n*)
Calochortus albus	006	White mariposa lily	
Calochortus apiculatus	007	Yellow-flowered mariposa lily	
Calochortus cæruleus	008	Blue mariposa lily	
Calochortus lilacinus	009	Lilac mariposa lily	
Calochortus nitidus	010	White mariposa lily	
Calochortus pulchellus	011	Yellow-flowered mariposa lily	
Calophyllum calaba	012	Green-fruited calophyllum	
Calophyllum inophyllum	013	Red-fruited calophyllum ; poonwood tree	GUMMIAPFEL (*m*) ; SCHÖNBLATT (*n*)
Calophyllum tacamahaca	014	Tacamahac tree	TACAMAHACABAUM (*m*) ; CALA--BABALSAMBAUM (*m*) ; MAR--IENBALSAMBAUM (*m*)
Calopogon pulchellus	015	(North American)-tuberous-rooted orchid	
Calotropis gigantea	016	Akund-yercum ; Mudar plant ; Mudar tree, Ak bush, Yerkum bush (See *Asclepias gigantea*)	MUDARBAUM (*m*) ; AKSTAUDE (*f*); YERKUMSTRAUCH (*m*) ; MAD--ARSTRAUCH (*m*) ; SODOMSAP--FEL (*m*) ; OSCHERSTRAUCH (*m*)
Caltha leptosepala	017	White-flowered marsh marigold	

Botanical Term.	No.	English Term.	German Term.
Caltha palustris	018	Marsh marigold	KUHBLUME (*f*); MATTENBLUME (*f*); MOOSBLUME (*f*); DOTTERBLUME (*f*); BUTTERBLUME (*f*); SCHMIRGEL (*m*); FETTBLUME (*f*)
Caltha palustris floreplено	019	Double marsh marigold	
Caltha purpurascens	020	Purple-stemmed marsh marigold	
Calycanthus	021	Spice bush	GEWÜRZSTRAUCH (*m*)
Calycanthus fertilis	022	Many-flowered spice bush, Carolina allspice	REICHBLÜHENDER GEWÜRZSTRAUCH (*m*)
Calycanthus floridus	023	Carolina allspice	WOHLRIECHENDER GEWÜRZSTRAUCH (*m*); AMERIKANISCHER GEWÜRZSTRAUCH (*m*)
Calycanthus lævigatus	024	(See *Calycanthus fertilis*)	
Calycanthus macrophyllus	025	(See *Calycanthus occidentalis*)	
Calycanthus occidentalis	026	Californian sweet-scented shrub	GROSSBLÄTTRIGER GEWÜRZSTRAUCH (*m*); OCCIDENTALISCHER GEWÜRZSTRAUCH (*m*)
Calycanthus præcox	027	Winter sweet (See *Chimonanthus fragrans*)	JAPANISCHER GEWÜRZSTRAUCH (*m*)
Calycifloræ	028	Calycifloræ	KELCHBLÜTLER (*m.pl*)
Calystegia	029	Bind-weed (See *Convolvulus*)	
Calystegia dahurica	030	Rose-flowered calystegia	
Calystegia occidentalis	031	Double, pink-flowered calystegia	
Calystegia pubescens	032	Pink-flowered calystegia	
Calystegia sepium	033	Common bindweed (See *Convolvulus sepium*)	ZAUNWINDE (*f*); DEUTSCHE PURGIERWINDE (*f*); DEUTSCHE SKAMMONIE (*f*)
Calystegia soldanella	034	Sea bells	
Calystegia villosa	035	Trailing, primrose-flowered calystegia	
Camassia cusicki	036	Pale-blue-flowered camassia	
Camassia esculenta	037	(North American)-edible camassia	ESSBARE CAMASSIA (*f*)
Camassia howellii	038	Lilac-flowered camassia	
Camassia leichtlini	039	Creamy-white-flowered camassia	
Camelina sativa	040	Cameline; gold of pleasure	LEINDOTTER (*m*); ÖLDOTTER (*m*); RAPSDOTTER (*m*); RILLSAAT (*f*); DOTTERKRAUT (*n*); RÜLLSAAT (*f*); FLACHSDOTTERPFLANZE (*f*); LEINDOTTERPFLANZE (*f*); DOTTER (*m*); SCHMALZBLUME (*f*)
Camellia japonica	041	Common camellia; Japanese camellia	JAPANISCHE ROSE (*f*); JAPANISCHE (KAMELLIE) KAMELIE (*f*)
Camellia oleifera	042	White-flowered camellia	WEISSE (KAMELLIE) KAMELIE (*f*)
Camellia reticulata	043	Chinese, pink-flowered camellia	CHINESISCHE (KAMELLIE) KAMELIE (*f*)
Campanula	044	Bell-flower; Venus's looking-glass (See *Specularia hybrida*)	FRAUENSPIEGEL (*m*); GLOCKENBLUME (*f*)
Campanula glauca	045	Japanese Kokko shrub	JAPANISCHER KOKKOSTRAUCH (*m*)
Campanula glomerata	046	Clustered bell-flower; clustered campanula	
Campanula hederacea	047	Ivy campanula (See *Wahlenbergia hederacea*)	
Campanula hybrida	048	Corn campanula (See *Specularia hybrida*)	
Campanula latifolia	049	Giant bell-flower; giant campanula	
Campanula medium	050	Canterbury bell	MARIENGLÖCKCHEN (*n*); MARIENVEILCHEN (*n*); MARIENGLOCKE (*f*); MARIETTENVEILCHEN (*n*)
Campanula patula	051	Spreading campanula	
Campanula persicæfolia	052	Peach-leaved campanula	
Campanula pyramidalis	053	Chimney campanula, pyramidal bell-flower	
Campanula rapunculoides	054	Creeping bell-flower; creeping campanula	

Botanical Term.	No.	English Term.	German Term.
Campanula rapunculus	055	Garden rampion, rampion ; rampion campanula ; Coventry rape (See also *Phyteuma orbiculare*)	RAPUNZELGLOCKENBLUME (*f*)
Campanula rotundifolia	056	Hairbell ; harebell ; bluebell of -(Scotland)	
Campanula speculum	057	Venus's looking-glass (See *Specularia speculum*)	FRAUENSPIEGEL (*m*)
Campanula trachelium	058	Nettle-leaved bell-flower ; nettle-leaved campanula ; throat-wort (See *Trachelium cœruleum*)	HALSKRAUT (*n*)
Camphora officinarum	059	Camphor tree (See *Cinnamomum camphora*)	KAMPFERBAUM (*m*)
Campsis chinensis	060	Chinese jasmine	CHINESISCHE KLETTERTROMPETE (*f*)
Campsis radicans	061	(See *Tecoma radicans* and *Bignonia radicans*)	
Cananga odorata	062	Cananga ; Ilang-ilang tree (See *Fitzgeraldia*)	CANANGABAUM (*m*) ; ILANG--ILANG-BAUM (*m*)
Canarium commune	063	Canary-nut tree	KANARIENNUSSBAUM (*m*)
Canarium paniculatum	064	Colophane-wood tree	COLOPHANHOLZBAUM (*m*)
Canavalia	065		KRIMPBOHNE (*f*) ; KANAVALIE (*f*)
Canella alba	066	Canella (See *Winterana canella*)	KANELLA (*f*) ; CANELLARINDEN--BAUM (*m*) ; WEISSER ZIMT--BAUM (*m*) ; KANELBAUM (*m*) ; FALSCHE WINTERSRINDEN--BAUM (*m*)
Cannabis gigantea	067	Giant hemp	RIESENHANF (*m*)
Cannabis indica (Actually the dried tops of the female plants ; is a pharmaceutical term)	068	Indian hemp ; Beng ; Bhang ; Bang ; Hashish, Hasheesh (See *Cannabis sativa*)	INDISCHER HANF (*m*)
Cannabis sativa	069	Fimble (male) hemp ; common hemp (See *Galeopsis tetrahit*)	FIMMEL (*m*) ; HANF (*m*) ; FENNEL (*m*) ; HANFPFLANZE (*f*) ; GE--MEINER HANF (*m*)
Canna edulis	070	Indian shot ; canna	WESTINDISCHER ARROWROOT--BAUM (*m*) ; BLUMENROHR (*n*) ; ADEIRASTAUDE (*f*)
Cantharellus aurantiacus	071	Orange-coloured chantarelle	ORANGEFARBENER FALTEN--SCHWAMM (*m*)
Cantharellus cibarius	072	Edible mushroom ; chantarella ; yellow merulius ; chantarelle	EIERPILZ (*m*) ; PFIFFERLING (*m*) ; REHGEISS (*m*) ; KANTHARELLE (*f*) ; ESSBARER FALTEN--SCHWAMM (*m*) ; RÖTLING (*m*) ; GEHLCHEN (*n*) ; EIERSCHWAMM (*m*)
Capparidaceæ	073	Caparidaceæ (See *Capparis*)	KAPERNGEWÄCHSE (*n.pl*)
Capparis amygdalina	074	(West Indian)-white-flowered caper-(tree)	KAPERNSTRAUCH (*m*)
Capparis decidua	075	Arabian caper	ARABISCHER CAPPARIS (*m*)
Capparis coriacea	076		SIMULOBAUM (*m*)
Capparis odoratissima	077	Violet-flowered caper-(tree)	WOHLRIECHENDER KAPERN--STRAUCH (*m*)
Capparis spinosa	078	Caper-(bush or tree)	KAPER (*m*) ; GEMEINER KAPERN--STRAUCH (*m*)
Capsella Bursa-pastoris	079	Shepherd's purse ; Capsell (See *Bursa bursa-pastoris*)	HIRTENTASCHE (*f*) ; HIRTENTÄSCH--LEIN (*n*) ; HIRTENTÄSCHEL--KRAUT (*n*) ; TÄSCHELKRAUT (*n*)
Capsicum acuminatum	080	Long cayenne pepper	LANGER CAYENNEPFEFFER (*m*)
Capsicum annuum	081	Annual capsicum, cayenne pepper (See *Capsicum grossum*)	SPANISCHER PFEFFER (*m*) ; BEISS--BEERE (*f*) ; TASCHENPFEFFER (*m*) ; SCHOTENPFEFFER (*m*) ; PAPRIKA (*f*)
Capsicum baccatum	082	Cayenne pepper	CAYENNEPFEFFER (*m*) ; GOLD--PFEFFER (*m*) ; GUINEAPFEFFER (*m*) ; CHILLIKRAUT (*n*)

Botanical Term.	No.	English Term.	German Term.
Capsicum fasciculatum	083	Red cluster pepper	ROTER PFEFFER (*m*); CHILE-PFEFFER (*m*); CAYENNE-PFEFFER (*m*)
Capsicum fastigiatum	084	African pepper; bird pepper; Capsicum; Cayenne pepper; Chillies; Red pepper	BEISSBEERE (*f*)
Capsicum grossum	085	Guinea pepper; bell-pepper; red pepper (See *Capsicum annuum*)	GUINEAPFEFFER (*m*)
Carapa guianensis	086	Guiana carapa (See *Carapa procera*)	CARAPABAUM (*m*); ANDIROBA-BAUM (*m*)
Carapa procera	087	(See *Carapa guianensis*)	
Cardamine amara	088	Large bittercress	BITTERKRESSE (*f*)
Cardamine bulbifera	089	Coral-root; bulbiferous bitter-cress	SCHAUMKRAUT (*n*)
Cardamine diphylla	090	(See *Dentaria diphylla*)	
Cardamine hirsuta	091	Hairy bitter-cress	
Cardamine impatiens	092	Narrow-leaved bittercress	
Cardamine pratensis	093	Common bittercress; cuckoo flower; lady's smock; meadow bittercress	KUCKUCKSSCHAUMKRAUT (*n*); (GEMEINE)-WIESENKRESSE (*f*); GAUCHBLUME (*f*); SCHAUM-KRAUT (*n*)
Cardamomum longum	094	(See *Cardamomum majus*)	
Cardamomum majus	095	(See *Elettaria* and *Amomum maximum*)	JAVAKARDAMOMENPFLANZE (*f*); CEYLONKARDAMOMENPFLANZE (*f*); JAPANISCHE KAR-DAMOMENPFLANZE (*f*)
Cardamomum minus	096	(See *Elettaria cardamomum*)	
Cardamomum racemosum	097	(See *Amomum cardamomum*)	
Carduus	098	Thistle	DISTEL (*f*)
Carduus acanthoides	099	Welted thistle	
Carduus acaulis	100	Dwarf thistle	
Carduus arvensis	101	Creeping thistle	
Carduus crispus	102	Horse-thistle (See *Cirsium arvensis*)	KRATZDISTEL (*f*)
Carduus eriophorus	103	Woolly-(headed) thistle; friar's-thistle; friar's-crown	
Carduus heterophyllus	104	Melancholy thistle	
Carduus lanceolatus	105	Spear thistle	
Carduus marianus	106	Milk thistle, lady's thistle	MARIENDISTEL (*f*)
Carduus nutans	107	Musk thistle	BISAMDISTEL (*f*); ESELSDISTEL (*f*)
Carduus palustris	108	Marsh thistle	
Carduus personata	109	bur-thistle	KLETTENDISTEL (*f*)
Carduus pratensis	110	Meadow thistle	
Carduus pycnocephalus	111	Slender thistle	
Carduus tuberosus	112	tuberous thistle	
Carex acuta	113	Acute carex	SCHARFES RIEDGRAS (*n*)
Carex alpina	114	Alpine carex (See *Carex brizoides*)	ALPENRIEDGRAS (*n*)
Carex ampullacea	115	Bottle carex, ampule carex	AMPULLENRIEDGRAS (*n*)
Carex arenaria	116	Creeping carex; sand carex	QUECKE (*f*); SANDSEGGE (*f*); ROTE QUECKE (*f*); SANDRIED-GRAS (*n*); DEUTSCHE SARSA-PARILLE (*f*)
Carex atrata	117	Black carex	SCHWARZES RIEDGRAS (*n*)
Carex axillaris	118	Axillary carex	AXILLÄRES RIEDGRAS (*n*)
Carex baccans	119	Red-berried, Indian carex	INDISCHES RIEDGRAS (*n*)
Carex baldensis	120	Mount Baldo carex	TIROLER-SEGGE (*f*)
Carex brizoides	121	Alpine carex	ALPENGRAS (*n*); RASCH (*m*); WALDHAAR (*n*); SEEGRAS (*n*); ZITTERGRASARTIGE SEGGE (*f*)
Carex buxbaumii	122	Buxbaum's carex	BUXBAUMS RIEDGRAS (*n*)
Carex cæspitosa	123	Tufted carex (See *Carex vulgaris*)	
Carex canescens	124	Whitish carex	WEISSES RIEDGRAS (*n*)
Carex capillaris	125	Capillary carex	KAPILLÄRES RIEDGRAS (*n*)
Carex digitata	126	Fingered carex	GEFINGERTES RIEDGRAS (*n*)

Botanical Term.	No.	English Term.	German Term.
Carex dioica	127	Diœcious carex	DIÖZISCHES RIEDGRAS (*n*); ZWEI-HÄUSIGES RIEDGRAS (*n*)
Carex distans	128	Distant carex	
Carex divisa	129	Divided carex	
Carex echinata	130	Star-headed carex	
Carex elongata	131	Elongated carex	AUSGEDEHNTES RIEDGRAS (*n*)
Carex extensa	132	Long-bracted carex	
Carex filiformis	133	Slender carex	
Carex flava	134	Yellow carex	GELBES RIEDGRAS (*n*); GELBE SEGGE (*f*)
Carex glauca	135	Glaucous carex	MEERGRÜNE SEGGE (*f*)
Carex hirta	136	Hairy carex	HAARIGES RIEDGRAS (*n*)
Carex humilis	137	Dwarf carex	ZWERGRIEDGRAS (*n*)
Carex incurva	138	Curved carex	
Carex lagopina	139	Hare's-foot carex	HASENPFOTEN-SEGGE (*f*)
Carex leporina	140	Oval carex	
Carex limosa	141	Mud carex	SCHILFGRAS (*n*)
Carex montana	142	Mountain carex	RUCHGRAS (*n*)
Carex muricata	143	Prickly carex	
Carex pallescens	144	Pale carex	BLEICHES RIEDGRAS (*n*)
Carex paludosa	145	Marsh carex	SUMPFRIEDGRAS (*n*)
Carex panicea	146	Carnation carex; carnation grass	
Carex paniculata	147	Panicled carex	
Carex pauciflora	148	Few-flowered carex	
Carex pendula	149	British, drooping carex; pendulous carex	HÄNGENDES RIEDGRAS (*n*)
Carex pilulifera	150	Pill-headed carex	PILLENKÖPFIGES RIEDGRAS (*n*)
Carex præcox	151	Vernal carex	
Carex pseudocyperus	152	Cyperus-like carex	
Carex pulicaris	153	Flea carex	
Carex punctata	154	Dotted carex	
Carex remota	155	Remote carex (See also *Kobresia caricina*)	ENTFERNTÄHRIGE SEGGE (*f*)
Carex rupestris	156	Rock carex	
Carex saxatilis	157	Russet carex	
Carex strigosa	158	Thin-spiked carex	
Carex sylvatica	159	Wood carex	WALDRIEDGRAS (*n*)
Carex tomentosa	160	Downy carex	
Carex vesicaria	161	Bladder carex	BLASENRIEDGRAS (*n*)
Carex vulgaris	162	Common carex, sedge, bur-flag	RIEDGRAS (*n*); SANDSEGGE (*f*); SCHILFGRAS (*n*)
Carex vulpina	163	Fox carex	FUCHSBRAUNE SEGGE (*f*)
Carica papaya	164	Papaw tree; melon tree	MELONENBAUM (*m*); MAMMONA (*f*); PAPAYBAUM (*m*)
Carissa arduina	165	African carissa	
Carissa carandas	166	Evergreen Christ's thorn	
Carissa grandiflora	167	Natal plum	
Carlina acaulis	168	Carline thistle	KARLSDISTEL (*f*); WETTERDISTEL (*f*); SONNENDISTEL (*f*); ENGLISCHE DISTEL (*f*)
Carlina vulgaris	169	Carline thistle, common carline thistle	EBERWURZ (*f*)
Carpinus betulus	170	Hornbeam; yoke elm	HORNBAUM (*m*); HAINBUCHE (*f*); HAGEBUCHE (*f*); WEISSBUCHE (*f*); JOCHBAUM (*m*); HEISTER (*m*); HAARBUCHE (*f*); BIRKEN-HEISTER (*m*)
Carpinus orientalis	171	Oriental hornbeam	SÜDLÄNDISCHER HORNBAUM (*m*); SÜDLÄNDISCHE HAINBUCHE (*f*)
Carthamus tinctor'	172	safflower; dyer's saffron	FÄRBERDISTEL (*f*); FARBENDISTEL (*f*); WILDER SAFRAN (*m*); DEUTSCHER SAFRAN (*m*); FALSCHER SAFRAN (*m*); BÜRSTEN-KRAUT (*n*); BASTARD SAFRAN (*m*); SAFLOR (*m*)
Carum ajowan	173	Ajowan (See *Ptychotis ajowan*)	AJOWANKRAUT (*n*)
Carum bulbocastanum	174	Tuberous carum; bulbous caraway	KNOLLENKÜMMEL (*m*)

Botanical Term.	No.	English Term.	German Term.
Carum carui	175	Caraway-(plant); common caraway	KÜMMEL (*m*)
Carum carvi	176	(See *Carum carui*)	
Carum petroselinum	177	Parsley; common parsley; carum (See *Petroselinum sativum*)	PETERSILIE (*f*)
Carum segetum	178	Corn parsley; corn carum (See *Petroselinum segetum*)	
Carum verticillatum	179	Whorled (carum) parsley; whorled caraway	
Carya alba	180	White hickory (See also *Hicoria*)	HICKORY (*f*); WEISSE HICKORY (*f*); HICKORYNUSS (*f*)
Carya amara	181	Hickory; swamp hickory	HICKORY (*f*); BITTERE HICKORY (*f*); BITTERNUSS (*f*); HICK-ORYBAUM (*m*)
Carya minima	182	(See *Carya amara*)	
Carya myristicæfolia	183	Nutmeg hickory	
Carya olivæformis	184	Pecan-nut hickory; Illinois nut hickory (See also *Carya tomentosa*)	PEKANNUSSBAUM (*m*); ILLINOIS-NUSSBAUM (*m*)
Carya tomentosa	185	White-heart hickory; mocker nut tree (See also *Carya olivæformis*)	VEXIERNUSSBAUM (*m*); VEXIER-NUSS (*f*)
Caryocar butyrosum	186	Souari nut-(tree) (See *Pekea butyrosa*)	BOCKNUSS (*f*); MANDELAHORN (*m*); BUTTERAHORN (*m*); PEKEANUSS (*f*)
Caryocar nuciferum	187	Souari nut-(tree)	BOCKNUSS (*f*); KOPFNUSS (*f*); BUTTERNUSSBAUM (*m*); SOU-ARINUSSBAUM (*m*)
Caryocedrus	188	(See *Juniperus drupacea*)	
Caryophyllæ	189	Caryophyllæ (See *Caryophyllus, Dianthus, Silene, Lychnis, Agrostemma*)	NELKENGEWÄCHSE (*n.pl*)
Caryophyllus aromaticus	190	Caryophyllus; clove; clove tree (See *Eugenia aromatica* and *Jambosa caryophyllus*)	MUTTERNELKE (*f*); NELKE (*f*); NELKENBAUM (*m*); KREIDE-NELKE (*f*); GEWÜRZNELKEN-BAUM (*m*)
Caryota mitis	191	Fish-tail palm, caryota	
Caryota sobolifera	192	Fish-tail palm; caryota	
Caryota urens	193	Wine palm; caryota palm; toddy palm	BRENNPALME (*f*); WEINPALME (*f*)
Casimiroa edulis	194	Edible Mexican apple	WEISSE SAPOTE (*f*)
Cassia absus	195	Absus cassia	CHICHONPFLANZE (*f*); CHICHIM-KASSIE (*f*)
Cassia acutifolia	196	Cassia; Alexandria senna shrub (See SENNESBLÄTTER)	SENNESSTRAUCH (*m*); SENNES-BLÄTTERSTRAUCH (*m*)
Cassia angustifolia	197	Cassia (See SENNESBLÄTTER)	KASSIABAUM (*m*); SENNESBAUM (*m*)
Cassia auriculata	198	Cassia-bark tree	CASSIARINDEBAUM (*m*); TURWAR-RINDEBAUM (*m*)
Cassia caryophyllata	199	Cinnamon	NELKENZIMT (*m*)
Cassia chamæchrista	200	Partridge pea	
Cassia cinnamomea	201	Cassia	HOLZKASSIE (*f*)
Cassia elongata	202	Cassia; long leaved senna shrub (See SENNESBLÄTTER)	SENNESBAUM (*m*)
Cassia fistula	203	Purging Cassia, drumstick, Indian laburnum, pudding stick	FISTELKASSIE (*f*); PURGIER-KASIE (*f*); PURGIERKASSIA (*f*); ZIMTFISTULA (*f*); RÖHR-ENKASSIE (*f*)
Cassine crocea	204	Saffron-wood tree (See *Elæodendron croceum* and *Crocoxylon excelsum*)	HOHER SAFRONHOLZBAUM (*m*)
Castanea pumila	205	Dwarf-chestnut; chincapin; chinkapin; chinquapin	ZWERG-KASTANIE (*f*); NORD-AMERIKANISCHER CHINCAPIN-BAUM (*m*)
Castanea sativa	206	Chestnut; edible chestnut	KASTANIE (*f*); EDLE KASTANIE (*f*); EDELKASTANIE (*f*)

Botanical Term.	No.	English Term.	German Term.
Castanea vesca	207	Chestnut; sweet chestnut; Spanish chestnut (See *Castanea vulgaris*)	KASTANIE (*f*); EDELKASTANIE (*f*); ESSBARE KASTANIE (*f*)
Castanea vulgaris	208	Sweet chestnut (See *Castanea vesca*)	KASTANIE (*f*); MARONE (*f*); ECHTER KASTANIENBAUM (*m*); MARONENBAUM (*m*); KÄSTEN-BAUM (*m*)
Castanopsis chrysophylla	209	Golden-leaved castanopsis	GOLDBLÄTTRIGE EICHENKASTANIE (*f*)
Castanospermum australe	210	(Australian)-Moreton bay chestnut-(tree)	AUSTRALISCHER KASTANIENBAUM (*m*); BOHNENBAUM (*m*)
Casuarinæ	211	Casuarinæ, Botany-bay-wood plants, blackwood plants	KASUARINEEN (*f.pl*)
Casuarina equisetifolia	212	(New South Wales)-beef wood tree; blackwood plant; Botany-bay-wood plant	KEULENBAUM (*m*); KASUARINE (*f*); OCHSENFLEISCHHOLZ-BAUM (*m*); RINDFLEISCH-HOLZBAUM (*m*); EISENHOLZ-BAUM (*m*); SUMPFEICHE (*f*)
Casuarina fraseriana	213	Australian she-oak	WEIBLICHE EICHE (*f*)
Casuarina quadrivalvis	214	New-South-Wales swamp-oak	SUMPFEICHE (*f*)
Casuarina stricta	215	Beef-wood tree, Australian mahogany tree	BEEFWOODBAUM (*m*); AUSTRALISCHER MAHAGONIBAUM (*m*); KASUARINENHOLZBAUM (*m*); PFERDEFLEISCHHOLZBAUM (*m*)
Casuarina torulosa	216	Australian river-oak; forest-oak; Botany-bay oak	FLUSSEICHE (*f*); WALDEICHE (*f*)
Catalpa bignonioides	217	Catalpa; bean tree, Indian bean	(GEMEINER)-TROMPETENBAUM (*m*); CATALPABAUM (*m*); ZIGARRENBAUM (*m*)
Catalpa kæmpferi	218	Japanese bean tree; Japanese catalpa	JAPANISCHER TROMPETENBAUM (*m*)
Catalpa longissima	219	Catalpa	ANTILLENEICHE (*f*)
Catalpa ovata	220	Japanese catalpa	JAPANISCHER TROMPETENBAUM (*m*)
Catalpa syringæfolia	221	(See *Catalpa bignonioides*)	BUNGES TROMPETENBAUM (*m*)
Catha edulis	222	Arabian tea	KATHA (*f*); CATHASTRAUCH (*m*)
Caucalis anthriscus	223	Hedge parsley	
Caucalis arvensis	224	Spreading caucalis; goldilocks	
Caucalis daucoides	225	Small caucalis; bur-parsley	
Caucalis latifolia	226	Broad caucalis	
Caucalis nodosa	227	Knotted caucalis	
Ceanothus americanus	228	New Jersey tea, mountain sweet, Californian lilac, (American)-redroot (See also *Lachnanthes tinctoria*)	AMERIKANISCHE SÄCKELBLUME (*f*); ROTWURZEL (*f*); SECKEL-BLUME (*f*)
Ceanothus herbacus	229	(See *Ceanothus americanus*)	
Ceanothus ovatus	230	(See *Ceanothus americanus*)	
Cedrela guianensis	231	Cedar, Havanna (cigar-box) wood; West Indian cedar	ZIGARRENKISTENHOLZ (*n*); ZUCK-KERKISTENHOLZ (*n*); ZEDRO-BAUM (*m*)
Cedrela odorata	232	Cedar, Havanna (cigar-box) wood	ZIGARRENKISTENHOLZ (*n*); MEXIKANISCHE ZEDER (*f*); SPANISCHE ZEDER (*f*); ZEDRELA-HOLZBAUM (*m*); SPANISCHER ZEDERNHOLZBAUM (*m*); ZUCK-ERKISTENHOLZBAUM (*m*); ZIGARRENKISTENHOLZBAUM (*m*)
Cedrus atlantica	233	African cedar, Mount Atlas cedar, Fountain tree (See also *Cedrus libani*)	ATLAS-ZEDER (*f*); LIBANONZEDER (*f*)
Cedrus deodora	234	Himalayan cedar; deodar	HIMALAJA-ZEDER (*f*)
Cedrus libani	235	Cedar-(of Lebanon) (See also *Cedrus atlantica*)	(ECHTE)-ZEDER (*f*); LIBANON-ZEDER (*f*)
Celosia cristata	236	Common cockscomb	HAHNENKAMM (*m*)
Celtis australis	237	Lote wood; European nettletree	TRIESTER HOLZ (*n*); ZÜRGELBAUM (*m*); SÜDLICHER ZÜRGEL-BAUM (*m*)
Celtis kraussiana	238	South African stinkwood tree	STINKWOODBAUM (*m*); CAMDE-BOOBAUM (*m*)

Botanical Term.	No.	English Term.	German Term.
Celtis occidentalis	239	Lote wood, hackberry	TRIESTER HOLZ (*n*); ZÜRGEL-BAUM (*m*); NORDAMERIKAN-ISCHER ZÜRGELBAUM (*m*); WESTLICHER ZÜRGELBAUM (*m*)
Centaurea americana	240	Basket flower	
Centaurea aspera	241	Guernsey centaurea	
Centaurea benedicta	242	Blessed thistle, holy thistle (See *Cnicus benedictus*)	
Centaurea calcitrapa	243	Common star-thistle (See *Calcitrapa hippophæstum*)	BITTERBLÄTTRIGE FLOCKENBLUME (*f*)
Centaurea cyanus	244	(Corn)-bluebottle; common corn-flower	KORNBLUME (*f*); ZYANE (*f*); KORNFLOCKENBLUME (*f*); CYANE (*f*); TREMSE (*f*)
Centaureæ	245	Centaureæ (See *Cnicus* and *Calcitrapa*)	ZYANEEN (*f.pl*)
Centaurea montana	246	Mountain centaurea	BERGFLOCKENBLUME (*f*)
Centaurea moschata	247	Sweet-sultan	BISAMFLOCKENBLUME (*f*); MOS-CHUSBLUME (*f*)
Centaurea nigra	248	Black knapweed; hardhead	
Centaurea rhaponticum	249	(See *Rhaponticum cynaroides*)	
Centaurea scabiosa	250	Great knapweed, greater knap-weed	
Centaurea solstitialis	251	Yellow star-thistle; yellow cen-taurea (See also *Chlora perfoliata*)	GELBE FLOCKENBLUME (*f*)
Centranthus ruber	252	Red valerian; spur valerian; centranth; German lilac (See *Valeriana montana*)	SPORNBLUME (*f*)
Centrolobium robustum	253	Arariba, Zebra-wood tree	ZEBRAHOLZBAUM (*m*)
Centunculus minimus	254	Bastard pimpernel: chaffweed; centuncle	
Cephælis ipecacuanha	255	Ipecacuanha	KOPFBEERE (*f*); GOLDERDWURZEL (*f*)
Cephalanthera ensifolia	256	Narrow helleborine	
Cephalanthera pallens	257	White helleborine	
Cephalanthera rubra	258	Red helleborine	
Cephalanthus occidentalis	259	(Western)-button wood	KOPFBLUME (*f*)
Cephalotaxus pedunculata	260	Japanese cluster-flowered yew	KOPFEIBE (*f*); (KURZBLÄTTRIGE)-SCHEINEIBE (*f*)
Cerastium alpinum	261	Alpine mouse-ear chickweed; alpine cerast; snow-plant, snow-in-summer	BERGHORNKRAUT (*n*); ALPEN-HORNKRAUT (*n*)
Cerastium arvense	262	Field mouse-ear chickweed; field cerast	HORNKRAUT (*n*); ACKERHORN-KRAUT (*n*)
Cerastium sylvaticum	263	wood cerast	WALDHORNKRAUT (*n*)
Cerastium tomentosum	264	Snow plant; Snow-in-summer	SCHNEEKRAUT (*n*); KRÄUTLEIN PATENTIA (*n*); FILZIGES HORNKRAUT (*n*)
Cerastium trigynum	265	Starwort cerast	
Cerastium viscosum	266	Viscid mouse-ear chickweed	KLEBRIGES HORNKRAUT (*n*)
Cerastium vulgatum	267	Mouse-ear chickweed; common cerast	GEMEINES HORNKRAUT (*n*)
Cerasus acida	268	Sour cherry; bush cherry; Morello cherry (See *Prunus acida*)	SAUERKIRSCHE (*f*); STRAUCH-WEICHSEL (*f*); SAUERWEICH-SEL (*f*); WEICHSELKIRSCHE (*f*)
Cerasus avium	269	Wild cherry	HOLZKIRCHE (*f*); BERGKIRSCHE (*f*); VOGELKIRSCHE (*f*)
Cerasus camproniana	270	Wild cherry	HOLZKIRCHE (*f*)
Cerasus communis	271	(See *Prunus cerasus*)	
Cerasus japonica	272	(See *Prunus japonica*)	
Cerasus mahaleb	273	(See *Prunus mahaleb*)	
Cerasus mollis	274	Soft-haired cherry (See *Prunus mollis*)	WEICHHAARIGE KIRSCHE (*f*)
Cerasus padus	275	(See *Prunus padus*)	
Cerasus virginiana	276	American grape cherry	AMERIKANISCHE TRAUBENKIR-SCHE (*f*)
Cerasus vulgaris	277	(See *Prunus cerasus*)	
Ceratochloa pendula	278	(See *Bromus Schraderi*)	

Botanical Term.	No.	English Term.	German Term.
Ceratonia siliqua	279	Carob bean, carob tree; locust-(bean)-tree; bean tree; St. John's bread; (See also *Siliqua dulcis*)	HÜLSENBAUM (*m*); KAROBE (*f*); JOHANNISBROT (*n*); KARUBE (*f*); BOCKSHORNBAUM (*m*)
Ceratophylleæ	280	Ceratophylleæ (See *Ceratophyllum*)	HÖRNERBLÄTTER (*n.pl*)
Ceratophyllum demersum	281	Hornwort, common hornwort, morass-weed	HORNBLATT (*n*)
Cerbera tanghin	282	(See *Tanghinia venenifera*)	
Cercis siliquastrum	283	Judas tree	JUDASBAUM (*m*); LIEBESBAUM (*m*); GEMEINER JUDASBAUM (*m*)
Cercocarpus ledifolius	284	Bay mahogany	
Cereus fimbriatus	285	Pink-flowered (cactus) cereus, torch thistle, night-flowering cereus	SÄULENKAKTUS (*m*); SCHLANGEN--FACKELDISTEL (*f*); FACKEL--DISTEL (*f*)
Cereus flagelliformis	286	Pink-flowered, creeping (cactus) cereus	PEITSCHENKAKTUS (*m*); SCHLAN--GENKAKTUS (*m*)
Cereus geminatus	287	bud-bearing (cactus) cereus	CARDON (*m*); HECHO (*m*)
Cereus grandiflorus	288	Climbing (cactus) cereus; night-flowering cactus; night-blooming cereus (See *Cereus nycticalus*)	KÖNIGIN DER NACHT (*f*)
Cereus nycticalus	289	Queen-of-the-night-(cactus) (See *Cereus grandiflorus*)	KÖNIGIN (*f*) DER NACHT.
Cereus senilis	290	Grey-headed torch-thistle	GREISENHAUPT (*n*)
Cereus speciosissimus	291	torch thistle	SÄULENKAKTUS (*m*); FACKEL--DISTEL (*f*)
Ceriops Candolleana	292	Mangrove tree	MANGROVE (*f*)
Ceropegia candelabrum	293	Ceropegia	LEUCHTERBAUM (*m*)
Ceroxylon andicolum	294	South American palm; wax palm; Peruvian wax palm (See also *Klopstockia cerifera*)	SÜD-AMERIKANISCHE PALME (*f*); WACHSPALME (*f*)
Ceroxylon exorrhiza	295	Aerial-rooted wax palm	WACHSPALME (*f*); ANDENPALME (*f*)
Ceterach officinarum	296	Hart's tongue, ceterach (See *Asplenium ceterach* and *Scolopendrium ceterach*)	
Cetraria islandica	297	Iceland moss	ISLÄNDISCHES MOOS (*n*); HIRSCHHORNFLECHTE (*f*); PURGIERMOOS (*n*)
Chænomeles chinensis	298	(See *Cydonia chinensis*)	
Chænorrhinum minus	299	(See *Linaria minor*)	
Chærophyllum anthriscus	300	Burr chervil (See *Anthriscus vulgaris*)	
Chærophyllum bulbosum	301	Bulbous-rooted chervil	KNOLLIGER KÄLBERKROPF (*m*)
Chærophyllum sylvestre	302	Wild chervil; beaked parsley; cow weed (See *Anthriscus sylvestris*)	KÄLBERKROPF (*m*)
Chærophyllum temulentum	303	Rough chervil (See *Anthriscus cerefolium* and *Scandix cerefolium*)	
Chætomium	304	Bristle-mould	FREIFRÜCHTIGER KERN-SCHLAUCH--PILZ (*m*)
Chætomium chartarum	305	Paper bristle mould	PAPIER-KERNSCHLAUCHPILZ (*m*)
Chætomium griseum	306	Grey bristle mould	GRAUER KERNSCHLAUCHPILZ (*m*)
Chætomium murorum	307	Wall bristle mould	WAND-KERNSCHLAUCHPILZ (*m*)
Chamæcyparis lawsoniana	308	(See *Cupressus lawsoniana*)	LAWSONS-SCHEINZYPRESSE (*f*)
Chamæcyparis nutkatensis	309	(See *Cupressus nutkatensis*)	SITKAZYPRESSE (*f*)
Chamæcyparis obtusa	310	Japanese fire cypress	FEUERZYPRESSE (*f*); HINOKI--ZYPRESSE (*f*)
Chamæcyparis obtusa aurea	311	Golden Hinoki cypress	GOLDFARBIGE FEUERZYPRESSE (*f*)
Chamæcyparis obtusa nana	312	Dwarf Hinoki cypress	ZWERG-FEUERZYPRESSE (*f*)
Chamæcyparis obtusa pendula	313	Drooping Hinoki cypress	HÄNGENDE FEUERZYPRESSE (*f*)
Chamæcyparis pisifera	314	Japanese Sawara cypress	SAWARAZYPRESSE (*f*)
Chamæcyparis sphæroidea	315	(See *Cupressus thyiodes*)	ZEDERZYPRESSE (*f*)
Chamænerion angustifolium	316	Willow herb	ANTONSKRAUT (*n*)
Chamærops humilis	317	Dwarf palm, fan palm, African hair palm, European palm	ZWERGPALME (*f*)

Botanical Term.	No.	English Term.	German Term.
Chamærops palmetto	318	Palmetto palm ; sabal (See *Sabal palmetto*)	PALMETTOPALME (*f*)
Characeæ	319	(See *Nitella syncarpa*)	ARMLEUCHTERALGEN (*f.pl*)
Cheiranthus cheiri	320	Common wallflower, gilliflower	GOLDLACK (*m*); LACK (*m*); LACKVIOLE (*f*); MAUER-BLUME (*f*); GELBVEIGELEIN (*n*); GELBE VIOLE (*f*)
Cheiranthus fruticulosus	321	Wall-flower	LACKBLUME (*f*)
Chelidonium majus	322	Common celandine ; greater celandine ; swallow-wort	SCHELLKRAUT (*n*)
Chelone barbata	323	(See *Pentstemon barbatus*)	ROTBLÜTIGE SCHILDBLUME (*f*)
Chelone obliqua	324	Shell-flower	SCHILDBLUME (*f*)
Chenopodium	325	Goose-foot	GÄNSEFUSS (*m*); MELDE (*f*)
Chenopodium album	326	White goosefoot	MEHLSCHMERGEL (*m*); WEISSER GÄNSEFUSS (*m*)
Chenopodium altissimum	327	Grass-leaved goose-foot	HOHER GÄNSEFUSS (*m*)
Chenopodium ambrosioides	328	Goosefoot ; sweet-scented goose-foot	JESUITENTEE (*m*); TRAUBEN-KRAUT (*n*); PIMENTKRAUT (*n*); AMERIKANISCHES WURM-SAMENKRAUT (*n*); WOHL-RIECHENDER GÄNSEFUSS (*m*)
Chenopodium bonus-Henricus	329	Good-king-Henry ; mercury goosefoot ; allgood ; wild spinach	GUTER HEINRICH (*m*); WILDER SPINAT (*m*)
Chenopodium botrys	330	Oak-leaved goosefoot	EICHENBLÄTTRIGER GÄNSEFUSS (*m*)
Chenopodium ficifolium	331	Fig-leaved goosefoot	FEIGENBLÄTTRIGER GÄNSEFUSS (*m*)
Chenopodium glaucum	332	Glaucus goosefoot	MEERGRÜNER GÄNSEFUSS (*m*)
Chenopodium hybridum	333	Maple-leaved goosefoot	UNECHTER GÄNSEFUSS (*m*)
Chenopodium murale	334	Nettle-leaved goosefoot	MAUERGÄNSEFUSS (*m*)
Chenopodium olidum	335	Stinking goosefoot (See *Chenopodium vulvaria*)	
Chenopodium opulifolium	336	Snow-ball leaved goosefoot	SCHNEEBALLBLÄTTRIGER GÄNSE-FUSS (*m*)
Chenopodium polyspermum	337	Many-seeded goosefoot	VIELSAMIGER GÄNSEFUSS (*m*)
Chenopodium quinoa	338	Quinoa-(plant)	PERUANISCHER REIS (*m*); REIS-MELDE (*f*); QUINOAMELDE (*f*); CHILENISCHE GÄNSE-FUSS (*m*)
Chenopodium rubrum	339	Red goosefoot	ROTER GÄNSEFUSS(*m*)
Chenopodium urbicum	340	Upright goosefoot	STEIFER GÄNSEFUSS (*m*)
Chenopodium vulvaria	341	Stinking goosefoot (See *Chenopodium olidum*)	STINKENDER GÄNSEFUSS (*m*); BOCKSMELDE (*f*)
Chimonanthus fragrans	342	Winter sweet ; Japan allspice (See *Calycanthus præcox*)	
Chiococca racemosa	343	(See *Symphoricarpus racemosus*)	
Chionanthus virginica	344	(North American)-fringe tree	VIRGINISCHE SCHNEEBLUME (*f*)
Chloranthus officinalis	345		PFLAUMENPFEFFER (*m*)
Chlora perfoliata	346	Yellow-wort ; yellow centaury (See also *Centaurea solstitialis*)	
Chlorophora tinctoria	347	fustic wood, cuba wood (See *Manchura tinctoria* and *Maclura tinctoria*)	GELBHOLZ (*n*); FUSTIKHOLZ (*n*); GELBES BRASILHOLZ (*n*); ECHTES GELBHOLZ (*n*); FUSTIK (*m*); ALTER FUSTIK (*m*); ECHTER FUSTIK (*m*)
Chlorophyceæ	348	Green algæ, green seaweed	GRÜNALGEN (*f.pl*)
Chloroxylon swietenia ;	349	Satinwood	SATINHOLZ (*n*); ATLASHOLZ (*n*); FEROLEHOLZ (*n*); SEIDENHOLZ (*n*)
Chondrus crispus	350	Iceland moss, chondrus, carrageen, carrigeen, carragheen-(moss), Irish moss ; pearl moss ; Killeen ; pig-wrack ; rock-salt moss (See *Sphærococcus crispus*)	GALLERTMOOS (*n*); KNORPEL-TANG (*m*); IRLÄNDISCHES MOOS (*n*); PERLMOOS (*n*); KNORPEL-ROTALGE (*f*); IRI-SCHES MOOS (*n*); KARRAGHEEN (*n*)
Choripetalæ	351	Choripetalæ	GETRENNTBLUMENKRONENBLÄTT-RIGE PFLANZEN (*f.pl*)

Botanical Term.	No.	English Term.	German Term.
Chrysanthemum balsamita var. tanacetoides	352	(See *Balsamita vulgaris* and *Tanacetum balsamita*)	GRIECHISCHE MINZE (*f*); BALSAM-KRAUT (*n*); MARIENBLATT (*n*); RÖMISCHER SALBEI (*m*); FRAUENMINZE (*f*); PFEFFER-BLATT (*n*)
Chrysanthemum carinatum	353	Tricolor daisy (See *Chrysanthemum tricolor*)	EINJÄHRIGE GOLDBLUME (*f*)
Chrysanthemum coronarium	354	Crown daisy	
Chrysanthemum frutescens	355	Paris daisy; common marguerite	MARGARETENBLUME (*f*)
Chrysanthemum inodorum	356	Scentless chrysanthemum (See *Matricaria inodora*)	FALSCHE KAMILLE (*f*)
Chrysanthemum leucanthemum	357	Ox-eye-(daisy)	MARIENBLUME (*f*); GROSSES MASSLIEB (*n*)
Chrysanthemum parthenium	358	Feverfew (See *Pyrethrum parthenium*)	MUTTERKRAUT (*n*); MUTTER-KRAUTSKAMILLE (*f*)
Chrysanthemum segetum	359	Corn marigold; yellow oxeye	WUCHERBLUME (*f*)
Chrysanthemum tricolor	360	(See *Chrysanthemum carinatum*)	
Chrysanthemum vulgare	361	(See *Tanacetum vulgare*)	
Chrysocoma linosyris	362	Goldy locks (See *Aster linosyris*)	GOLDHAAR (*n*)
Chrysophyllum cainita	363	West Indian star apple	STERNAPFELBAUM (*m*); GOLD-BLATT (*n*)
Chrysophyllum glyciphlœum	364	Monesia	MONESIA (*f*)
Chrysosplenium alternifolium	365	Alternate-leaved chrysosplene; golden saxifrage	GOLDSTEINBRECH (*m*); GOLD-MILZ (*f*); GOLDMILZKRAUT (*n*); STEINKRESSE (*f*)
Chrysosplenium oppositifolium	366	Golden saxifrage; opposite-leaved chrysosplene	MILZKRAUT (*n*)
Chytridiei	367	(See *Chytridium*)	OHNFADENEIPILZE (*m.pl*)
Chytridium	368	Chytridium fungus	OHNFADENEIPILZ (*m*)
Cicendia pusilla	369	Dwarf cicendia	
Cicendia filiformis	370	Slender cicendia	
Cicer arietinum	371	Bengal (Indian) chick pea	KICHERERBSE (*f*); KICHERLING (*m*)
Cichorium intybus	372	Chicory; Succory; Endive	KRAUSSALAT (*m*); CICHORIE (*f*); FELDWEGWART (*m*); ZICHORIE (*f*)
Cicuta virosa	373	Cowbane; water hemlock	WASSERSCHIERLING (*m*); PAR-ZENKRAUT (*n*)
Cinchona calisaya	374	Cinchona tree, yellow-bark cinchona (See *Cinchona ledgeriana*)	GELBES CHINAHOLZ (*m*); CHINA-RINDENBAUM (*m*); CINCHONA-BAUM (*m*)
Cinchona ledgeriana	375	(See *Cinchona calisaya*)	
Cinchona officinalis	376	Cinchona wood; cinchona tree; pale-bark cinchona	CHINAHOLZ (*n*); CHINARINDEN-BAUM (*m*); BLASSES CHINA-HOLZ (*n*)
Cinchona succiruba	377	Cinchona tree, red-bark cinchona tree	ROTES CHINAHOLZ (*n*)
Cincinalis dealbata	378	(See *Nothochlæna dealbata*)	
Cineraria cruenta	379	Cineraria (See *Senecio cruentus*)	ASCHENKRAUT (*n*); ASCHEN-PFLANZE (*f*)
Cineraria maritima	380	(See *Senecio cineraria*)	MEERSTRANDS-ASCHENKRAUT (*n*)
Cinnamomum camphora	381	(See *Camphora officinarum*)	
Cinnamomum cassia	382	Cassia; (Chinese)-cinnamon	CASSIA (*f*); KASSIA (*f*); KASSIE (*f*)
Cinnamomum ceylonicum	383	(See *Cinnamomum Zeylanicum*)	
Cinnamomum Zeylanicum	384	(Ceylon)-cinnamon	ZIMTBAUM (*m*)
Circæa alpina	385	Alpine circæa	ALPENHEXENKRAUT (*n*)
Circæa lutetiana	386	Common circæa; enchanter's nightshade (See *Mandragora vernalis*)	HEXENKRAUT (*n*)
Cirsium arvensis	387	Creeping thistle (See *Carduus crispus*)	KRATZDISTEL (*f*)
Cirsium lanceolatum	388	Spear thistle (See *Cnicus lanceolatus*)	WEGDISTEL (*f*)
Cissampelos pareira	389	Velvet leaf	FALSCHES PAREIRAWURZELKRAUT (*n*); FALSCHES GRIESWURZEL-KRAUT (*n*)
Cissus antarctica	390	(See *Vitis antarctica*)	WILDE WEINREBE (*f*); KLIMME (*f*)

Botanical Term.	No.	English Term.	German Term.
Cistus crispus	391	Cistus ; rock rose	CISTROSE (*f*) ; ZISTROSE (*f*)
Cistus heterophyllus	392	Algerian cistus	ALGERISCHE ZISTROSE (*f*)
Cistus villosus	393	Cistus ; rock rose	ZISTENRÖSCHEN ; ZISTROSE (*f*) ; CISTROSE (*f*)
Citrullus colocynthis	394	Colocynth ; bitter gourd ; bitter cucumber ; bitter apple, coloquintida	KOLOQUINTE (*f*) ; PURGIERGURKE (*f*) ; ALHANDAL (*m*) ; POMA--QUINTE(*f*)
Citrullus vulgaris	395	Water-melon (See *Cucumis melo*)	WASSERMELONE (*f*) ; ZITRULLE (*f*)
Citrus aurantium	396	Sweet orange	SÜSSFRÜCHTIGER POMERANZEN--BAUM (*m*) ; ORANGENBAUM (*m*) ; POMERANZENBAUM (*m*) ; POMERANZE (*f*)
Citrus aurantium bergamia	397	Bergamot orange (See *Citrus aurantium var. bergamia*)	BERGAMOTTE (*f*)
Citrus aurantium bigaradia	398	Seville orange (See *Citrus vulgaris*)	
Citrus aurantium japonica	399	(See *Citrus japonica*)	
Citrus aurantium lusitanica	400	Portuguese orange	
Citrus aurantium melitensis	401	Blood-orange	
Citrus aurantium myrtefolia	402	Myrtle-leaved orange	
Citrus aurantium var. bergamia	403	Bergamot (See *Citrus aurantium bergamia*)	
Citrus aurantium variegata	404	Variegated orange	
Citrus decumana	405	Shaddock ; Adam's apple ; forbidden fruit ; grape fruit ; pomelo, pompelmous, pampelmoes	POMPELMUSE (*f*) ; POMELMUS (*m*) ; POMPELMUS (*m*)
Citrus japonica	406	Kumquat (See *Citrus aurantium japonica*)	ZWERGPOMERANZE(*f*)
Citrus limetta	407	Lime ; sweet lime (See *Citrus medica limetto*)	SÜSSE LIMONE (*f*) ; LIMETTE (*f*)
Citrus limonum	408	Lemon ; citron (See *Citrus medica limonum*)	SAURE LIMONE (*f*) ; FALSCHE ZITRONE (*f*) ; LIMONE (*f*)
Citrus lumia	409	Sweet lemon	LUMIE (*f*)
Citrus margarita	410	Pearl lemon	
Citrus medica	411	Lemon wood ; citron tree ; median lemon	ZITRONENHOLZ (*n*) ; LIMONE (*f*) ; ECHTE ZITRONE (*f*) ; ZITRONEN--BAUM (*m*)
Citrus medica limetto	412	(See *Citrus limetta*)	
Citrus medica limonum	413	(See *Citrus limonum*)	
Citrus medica var. acida	414	West Indian lime ; Tahiti lime ; Mandarin lime ; Sour Rangpur lime (See *Citrus limetta*)	ZITRONENBAUM (*m*)
Citrus nobilis	415	mandarin, mandarin orange ; kid-gloved orange ; tangerine	MANDARINE (*f*)
Citrus nobilis major	416	Mandarin orange	MANDARINE (*f*)
Citrus nobilis tangerana	417	Tangerine orange	MANDARINE (*f*)
Citrus peretta	418	Orange tree	PERETTENBAUM (*m*)
Citrus trifoliata	419	Three-leaved citron	DREIBLATTZITRONE (*f*)
Citrus triptera	420	(See *Citrus trifoliata*)	
Citrus vulgaris	421	Seville orange (See *Citrus aurantium bigaradia*)	BITTERFRÜCHTIGER POMERANZEN--BAUM (*m*)
Cladium mariscus	422	Common sedge ; prickly cladium	MEERGRAS (*n*)
Cladonia bellidiflora	423	Coral moss	KORALLENMOOS (*n*) ; SÄULEN--FLECHTE(*f*)
Cladonia rangiferina	424	Lapland moss ; reindeer moss	KORALLENMOOS (*n*) ; RENNTIER--FLECHTE (*f*)
Cladrastis lutea	425	(See *Cladrastis tinctoria*)	
Cladrastis tinctoria	426	(American)-yellow wood tree	NORDAMERIKANISCHES GELB--HOLZ (*n*)
Clavata	427	(See *Lycopodium clavatum*)	
Claviceps purpurea	428	Fungus of ergot ; ergot (See *Secale cornutum*)	MUTTERKORN (*n*) ; HUNGERKORN (*n*) ; STIELKOPF-SCHLAUCH--PILZ (*m*) ; HAHNENSPORN (*m*)

Botanical Term.	No.	English Term.	German Term.
Claytonia perfoliata	429	Winter purslane; perfoliate claytonia	DURCHWACHSENE CLAYTONIE (*f*)
Claytonia virginiana	430	Virginian Claytonia	VIRGINISCHE CLAYTONIE (*f*)
Clematis	431	Clematis (See *Atragene*)	WALDREBE (*f*); ALPENREBE (*f*)
Clematis flammula	432	Sweet-scented wild clematis	BRENNWALDREBENKRAUT (*n*); FEUERKRAUT (*n*)
Clematis orientalis	433	Yellow Indian virgin's bower	ORIENTALISCHE WALDREBE (*f*)
Clematis vitalba	434	Traveller's joy; bindwith; wild clematis; old man's beard	MANNESBART (*n*); DEUTSCHE WALDREBE (*f*)
Clematis viticella	435	Blue clematis	BLAUE WALDREBE (*f*)
Cleome pentaphylla	436	Spider flower	PILLENBAUM (*m*)
Clerodendron fragrans	437	(See *Clerodendron siphonanthus* and *Volkameria fragrans*)	
Clerodendron siphonanthus	438	(East Indian)-tube flower (See *Clerodendron fragrans* and *Volkameria fragans*)	LOSBAUM (*m*); SCHICKSALSBAUM (*m*)
Clethra acuminata	439	White alder	LAUBHEIDE (*f*); SCHEINELLER (*f*)
Clethra alnifolia	440	Sweet pepper-bush	GESÄGTBLÄTTRIGE LAUBHEIDE (*f*); SCHEINELLER (*f*)
Clianthus dampier	441	Glory pea	PRACHTBLUME (*f*)
Clianthus puniceus	442	Glory vine; parrot's beak	PRACHTBLUME (*f*)
Clinopodium vulgare	443	Field basil (See *Calamintha clinopodium* and also *Satureia clinopodium*)	KÖNIGSKRAUT (*n*)
Clitopilus	444	(See *Agaricus prunulus*)	
Clusia flava	445	(Jamaica)-balsam tree; Lecluse; Clusia	KLUSIE (*f*); AFFENAPFEL (*m*)
Cnicus benedictus	446	Holy thistle, blessed thistle (See *Centaurea benedicta*)	KARDOBENEDIKTENKRAUT (*n*)
Cnicus lanceolatus	447	Spear plum thistle (See *Cirsium lanceolatum*)	HEILDISTEL (*f*); BENEDIKTEN--KRAUT (*n*)
Coccoloba uvifera	448	(West Indian)-seaside grape tree	SEETRAUBE (*f*); MEERTRAUBE (*f*)
Cocculus indicus	449	Indian berries; staves acre seed	FISCHKÖRNER (*n.pl*); KOKKELS--KÖRNER (*n.pl*); LAUSEKÖRNER (*n.pl*)
Cocculus palmatus	450	(See *Jateorrhiza palmata*)	
Cochlearia anglica	451	English scurvy grass	ENGLISCHES LÖFFELKRAUT (*n*)
Cochlearia armoracia	452	Horse-radish (See *Armoracia rusticana*)	MEERRETTIG (*m*)
Cochlearia danica	453	Danish scurvy grass	DÄNISCHES LÖFFELKRAUT (*n*)
Cochlearia officinalis	454	Common scurvy grass; spoon-wort	LÖFFELKRAUT (*n*)
Cocos butyracea	455	Cocoanut (coconut) palm	KÖNIGSPALME (*f*)
Cocos nucifera	456	Cocoa-nut palm; cocoa-nut tree; coconut palm	KOKOSPALME (*f*); KOKOS (*m*); KOKOSBAUM (*m*); KOKOSNUSS--BAUM (*m*)
Codium bursa	457	Sea-purse	FILZ-GRÜNALGE (*f*); HOHLKU--GELIGE-FILZGRÜNALGE (*f*)
Coffea arabica	458	Coffee shrub, coffee tree	KAFFEEBAUM (*m*)
Coffea liberica	459	Coffee shrub, coffee tree	KAFFEEBAUM (*m*)
Coix lachryma	460	Job's tears-(grass)	HIOBSTRÄNEN (*f.pl*)
Cola acuminata	461	Cola, guru, Soudan coffee; African cola-nut (guru-nut) tree	GURUNUSSBAUM (*m*)
Colchicum autumnale	462	Meadow saffron; wild saffron; colchicum, meadow crocus; autumn crocus	KUHBUTTER (*f*); HERBSTZEITLOSE (*f*); MOTTENSAFRAN (*m*); MOTTENSAFFRAN (*m*); ZEIT--LOSE (*f*)
Coleosporium senecionis	463	Groundsel rust	GRINDKRAUT-SCHEIDENSPOR-ROSTPILZ (*m*)
Collybia	464	(See *Agaricus esculentus*)	
Colutea arborescens	465	Bladder senna	(DEUTSCHER)-BLASENSTRAUCH (*m*)
Colutea cruenta	466	(See *Colutea orientalis*)	
Colutea orientalis	467	Oriental bladder senna	ORIENTALISCHER BLASENSTRAUCH (*m*)
Comarum palustre	468	Purple-wort; marsh cinquefoil (See *Potentilla palustris*)	
Commiphora	469	(See *Balsamodendron*)	

Botanical Term.	No.	English Term.	German Term.
Compositæ	470	Compositæ	KORBBLÜTLER (*m.pl*)
Comptonia	471	(See *Myrica comptonia*)	
Conferva bombycina	472	Hair weed	WASSERFADEN (*m*)
Confervaceæ	473	Green thread algæ	FADENALGEN (*f.pl*)
Coniferæ	474	Conifers	KONIFERE (*f.pl*); ZAPFENBÄUME (*m.pl*); NADELHÖLZER (*n.pl*); ZAPFENTRÄGER (*m.pl*)
Coniomycetes	475	Dust fungi	STAUBBRANDPILZE (*m.pl*)
Conium maculatum	476	Hemlock; poison hemlock; cowbane; spotted hemlock; common hemlock	SCHIERLING (*m*); SCHIERLINGS-KRAUT (*n*); FLECKSCHIERLING (*m*); SCHIERLINGSPFLANZE (*f*); SCHIERLINGSTANNE (*f*)
Conjugatæ	477	Conjugatæ	JOCHALGEN (*f.pl*)
Conopodium denudatum	478	Earthnut; pig-nut; conopodium	ERDNUSS (*f*)
Convallaria majalis	479	Lily of the valley; May lily; May blossom; Park lily	MAIGLÖCKCHEN (*n*); MAIBLUME (*f*); MAIBLUMENKRAUT (*n*); MARIENSCHELLE (*f*)
Convallaria multiflora	480	Solomon's seal (See *Polygonatum multiflorum*)	SALOMONSSIEGEL (*n*); GROSSE MAIBLUME (*f*)
Convallaria polygonatum	481	Solomon's seal	SALOMONSSIEGEL (*n*)
Convolvulus arvensis	482	Small bindweed; field bindweed; lesser bindweed	FELDWINDE (*f*)
Convolvulus scammonia	483	Scammony (bindweed) convolvulus	SKAMMONIUMWURZEL (*f*); PURGIERWURZEL (*f*)
Convolvulus scoparius	484	Shrubby bindweed; Lignum Rhodii; rosewood shrub	BESENWINDE (*f*)
Convolvulus sepium	485	Great bindweed; convolvulus; calystegia; larger bindweed; greater bindweed; bear-bind	HECKENWINDE (*f*)
Convolvulus soldanella	486	Sea-(side) bindweed, sea withwind	MEERKOHLWINDE (*f*); MEER-STRANDSWINDE (*f*); MEER-STRANDWINDE (*f*)
Convolvulus tricolor	487	Dwarf convolvulus	DREIFARBIGE WINDE (*f*)
Conyza squarrosa	488	(See *Inula conyza*)	
Copaifera bracteata	489	Purple wood; purple heart; amaranth	PURPURHOLZ (*n*); LUFTHOLZ (*n*); AMARANTHOLZ (*n*); VIOLETT-HOLZ (*n*); VEILCHENHOLZ (*n*); BLAUES EBENHOLZ (*n*); AMARANTHOLZBAUM (*m*)
Copaifera lansdorfii	490	Copaiva balsam (copaiba balsam) tree	KOPAIVABAUM (*m*)
Copernica cerifera	491	Carnauba palm; (Brazilian)-wax palm (See *Corypha cerifera*)	CARNAUBAPALME (*f*)
Coprinus comatus	492	Maned mushroom	TINTLING (*m*); TINTENBLÄTTER-SCHWAMM (*m*)
Coptis trifolia	493	Gold-thread herb; three-leaved gold-thread	
Corallina officinalis	494	Coralline; hornwrack	KORALLENMOOS (*n*); KORALLEN-RINDE (*f*); KORALLEN-KALK-ALGE (*f*)
Corallorhiza innata	495	Spurless coralroot	KORALLENWURZEL (*f*)
Corchorus capsularis	496	Jute; corchorus	INDISCHER FLACHS (*m*); JUTE-HANF (*m*)
Corchorus japonica	497	(See *Kerria japonica*)	
Corchorus olitorius	498	Jute; corchorus; Jew's mallow	INDISCHER FLACHS (*m*)
Cordia gerascanthus	499	Dominica rosewood; St. Lucia wood; prince wood, Spanish elm	CYPERNHOLZ (*n*); ROSENHOLZ (*n*); SANKT LUCIENHOLZ (*n*)
Coreopsis	500	(American)-tickseed	WANZENBLUME (*f*); MÄDCHEN-AUGE (*n*)
Coriandrum sativum	501	Coriander; common coriander	KÜMMEL (*m*)
Coriaria myrtifolia	502	Ink-plant (See *Rhus coriaria*)	GERBERSTRAUCH (*m*)
Cormus domestica	503	(See *Sorbus domestica*)	
Cornus alba	504	Tartar privet (See *Cornus tartarica*); white privet	TARTARISCHER HARTRIEGEL (*m*); WEISSER HARTRIEGEL (*m*)
Cornus amomum	505	(See *Cornus sericea*)	

Botanical Term.	No.	English Term.	German Term.
Cornus mas	506	Cornel, cornelian cherry, dog-wood	KORNELKIRSCHE (*f*); DÜRLITZE (*f*); (GELBER)-HARTRIEGEL (*m*); CORNELIUS-KIRSCHE (*f*)
Cornus mascula	507	(See *Cornus mas*)	
Cornus pubescens	508	soft-haired cornel	WEICHHAARIGER HARTRIEGEL (*m*)
Cornus sanguinea	509	Dog-berry; prickwood; cornel; dog-wood; wild cherry; cornelian cherry (See *Ligustrum vulgare*)	HARTRIEGEL (*m*); KORNELLE (*f*); HARTBAUM (*m*); KORNELBAUM (*m*); GEMEINER HARTRIEGEL (*m*); HORNKIRSCHE (*f*); KORNELKIRSCHE (*f*); HORN-STRAUCH (*m*)
Cornus sericea	510	Silky-haired cornel (See *Cornus amomum*)	SEIDENHAARIGER HARTRIEGEL (*m*)
Cornus suecica	511	Dwarf cornel	SCHWEDISCHE HERLITZE (*f*)
Cornus tatarica	512	Tartar privet (See *Cornus alba*)	TATARISCHER HARTRIEGEL (*m*)
Coronilla coronata	513	Crown vetch, sickle wort (See *Coronilla varia*)	KRONENWICKE (*f*); SCHAFLINSE (*f*); KRONWICKE (*f*)
Coronilla emerus	514	Scorpion senna	SKORPIONSKRONWICKE (*f*); GAR-TEN-KRONENWICKE (*f*); GROSSES PELTSCHEN (*n*)
Coronilla varia	515	(See *Coronilla coronata*)	
Corrigiola littoralis	516	Strapwort	STRANDLING (*m*)
Cortaderia argentea	517	(See *Gynerium argenteum*)	
Corydalis claviculata	518	White climbing fumitory	WEISSE KLETTERNDE HOHLWURZ (*f*)
Corydalis lutea	519	Yellow fumitory	GELBE HOHLWURZ (*f*)
Corylus americana	520	American hazel	AMERIKANISCHER HASELSTRAUCH (*m*)
Corylus avellana	521	Hazel; hazel-nut	HASELNUSS (*f*); HASEL (*f*); HASELSTRAUCH (*m*); HAZEL (*f*); HASELNUSSSTRAUCH (*m*); LAMBERTNUSSBAUM (*m*); WALDHASELNUSS (*f*); NUSS-STAUDE (*f*)
Corylus colurna	522	Tree hazel	BAUM-HASELSTRAUCH (*m*); BAUM-HASELNUSS (*f*); TÜRKISCHE HASELNUSS (*f*); DICKNUSS (*f*)
Corylus ferox	523	Himalayan hazel	HIMALAJA-HASELNUSS (*f*); BEWEHRTE HASELNUSS (*f*)
Corylus heterophylla	524	Asian hazel	OSTASISCHE HASELNUSS (*f*)
Corylus lamberti	525	Lambert's filbert (See *Corylus tubulosa*)	
Corylus mandschurica	526	Manchurian hazel	MANDSCHURISCHE HASELSTRAUCH (*m*)
Corylus maxima	527	(See *Corylus tubulosa*)	LAMBERTS-HASELNUSS (*f*)
Corylus rostrata	528	North-American hazel	SCHNABEL-HASELSTRAUCH (*m*); GESCHNÄBELTE HASELNUSS (*f*)
Corylus tubulosa	529	Cob-nut, filbert, Kentish cob, great cob, Lambert's filbert (See *Corylus maxima*)	LAMBERTNUSSBAUM (*m*); BLUT-NUSS (*f*); LAMPERTS-HASEL-STRAUCH (*m*); LAMBERTNUSS (*f*); LAMBERTSNUSS (*f*); LANGBARTSNUSS (*f*)
Corynanthe yohimbe	530	Yohimbe tree	YUMBEHOABAUM (*m*)
Coryneum beijerinckii	531	Gumming fungus	
Corypha cerifera	532	Wax palm (See *Copernica cerifera*)	WACHSPALME (*f*)
Corypha gebauga	533	Java fan palm	JAVA FÄCHERPALME (*f*)
Corypha taliera	534	Indian fan palm	INDISCHE FÄCHERPALME (*f*)
Corypha umbraculifera	535	Talipot palm; Ceylon fan palm	SCHATTENPALME (*f*); TALIPOT (*f*)
Costus arabica	536	Costus	KOSTWURZ (*f*)
Costus speciosus	537	Hindu tree of heaven; Kushta	KUSHTABAUM (*m*); OSTINDISCHE KOSTWURZ (*f*)
Cotinus cotinus	538	Venetian sumac, young fustic (See *Rhus cotinus*)	FÄRBERBAUM (*m*): FISETHOLZ (*n*); FISTELHOLZ (*n*); VISETHOLZ (*n*)
Cotoneaster	539	Cotoneaster, quince-leaved medlar, rose-box (See *Cotoneaster vulgaris*)	STEINQUITTE (*f*); QUITTENMIS-PEL (*f*); ZWERGMISPEL (*f*)

Botanical Term.	No.	English Term.	German Term.
Cotoneaster integerrima	540	Common cotoneaster	GEMEINE STEINQUITTE (*f*)
Cotoneaster microphylla	541	Cotoneaster; small-leaved cotoneaster	KLEINBLÄTTRIGE STEINQUITTE (*f*)
Cotoneaster thymifolia	542	(See *Cotoneaster microphylla*)	
Cotoneaster vulgaris	543	(See *Cotoneaster*)	
Cotula coronopifolia	544	Buck's-horn	LAUGENBLUME (*f*)
Cotyledon agavoides	545	(See *Cotyledon californica*)	
Cotyledon californica	546	Californian cotyledon	KALIFORNISCHES NABELKRAUT (*n*)
Cotyledon grandiflora	547	Mexican cotyledon	MEXIKANISCHES NABELKRAUT (*n*)
Cotyledon spinosum	548	Chinese cotyledon	CHINESISCHES NABELKRAUT (*n*)
Cotyledon umbilicus	549	Common navelwort; wall-pennywort; pennyleaf; pennywort	NABELKRAUT (*n*)
Coumarouna odorata	550	(See *Dipterix odorata*)	TONKABOHNE (*f*)
Couroupita guianensis	551	(American)-cannon-ball tree	KANONENKUGELBAUM (*m*)
Crambe maritima	552	Sea-kale	MEERKOHL (*m*); SEEKOHL (*m*)
Crassula	553	Houseleek (See *Sedum dasyphyllum*)	DICKBLATT (*n*)
Crassulacea	554	(See *Sempervivum tectorum* and *Sedum*)	KRASSULAZEEN (*f.pl*); DICK-BLATTGEWÄCHSE (*n.pl*)
Crassula muscosa	555	(See *Tillæa muscosa*)	
Cratægus chamæmespilus	556	Dwarf medlar (See also under *Sorbus* and *Pirus*)	ZWERGMISPEL (*f*)
Cratægus coccinea	557	Scarlet-fruited whitethorn, scarlet thorn	SCHARLACHDORN(*m*); SCHARLACH-FRÜCHTIGER WEISSDORN (*m*); ROTDORN (*m*)
Cratægus crus-galli	558	Cockspur thorn (See also *Cratægus punctata*)	HAHNENDORN (*m*); HAHNEN-SPORNWEISSDORN (*m*)
Cratægus glandulosa	559	(See *Cratægus virginica* and *Cratægus rotundifolia*)	
Cratægus grandiflora	560	Large-flowered whitethorn	GROSSBLÜTIGER WEISSDORN (*m*)
Cratægus lobata	561	(See *Cratægus grandiflora*)	
Cratægus lucida	562	(See *Cratægus crus galli*)	
Cratægus mollis	563	Soft-haired whitethorn	WEICHHAARIGER WEISSDORN (*m*)
Cratægus odoratissima	564	Crimean cratægus	WOHLRIECHENDER WEISSDORN (*m*)
Cratægus oxyacantha	565	Whitethorn; hawthorn; haythorn; May (See *Mespilus oxyacantha*)	WEISSDORN (*m*); HAGEDORN (*m*); ZWEIGRIFFLICHER WEISSDORN (*m*)
Cratægus punctata	566	Spotted-fruited (speckled-fruited) whitethorn	PUNKTIERTFRÜCHTIGER WEISS-DORN (*m*)
Cratægus rotundifolia	567	Round-leaved whitethorn	RUNDBLÄTTRIGER WEISSDORN (*m*)
Cratægus tomentosa	568	Single-flowered whitethorn	EINBLÜTIGER WEISSDORN (*m*)
Cratægus torminalis	569	(See *Sorbus torminalis* and *Pirus torminalis*)	
Cratægus uniflora	570	(See *Cratægus tomentosa*)	
Cratægus virginica	571	Virginian whitethorn	DRÜSENTRAGENDER WEISSDORN (*m*)
Crepis	572	Hawksbeard	PIPPAU (*m*)
Crepis barbata	573	(See *Tolpis barbata*)	BART-PIPPAU (*m*)
Crepis biennis	574	Rough (crepis) hawk's-beard	ZWEIJÄHRIGER PIPPAU (*m*)
Crepis fœtida	575	Fœtid (fetid) crepis or hawk's-beard	STINKENDER PIPPAU (*m*)
Crepis hieracioides	576	Hawkweed crepis; hawk's-beard	HABICHTSKRAUTÄHNLICHER PIP-PAU (*m*)
Crepis paludosa	577	Marsh (crepis) hawk's-beard	SUMPF-PIPPAU (*m*)
Crepis taraxifolia	578	Beaked (crepis) hawk's-beard	KUHBLUMENBLÄTTRIGER PIPPAU (*m*)
Crepis virens	579	Smooth (crepis) hawk's-beard	GRÜNER PIPPAU (*m*)
Crescentia cujete	580	Calabash tree, gourd tree	KÜRBISBAUM (*m*)
Crinum	581	Cape coast lily, Cape lily	
Crinum asiaticum	582	White-and-green-flowered crinum	ASIATISCHE HAKENLILIE (*f*)
Crithmum maritimum	583	Samphire; sea samphire, sea fennel	MEERFENCHEL (*m*)
Crocoxylon excelsum	584	(See *Elæodendron croceum* and *Cassine crocea*)	SAFRANHOLZBAUM (*m*)
Crocus aureus	585	Crocus, yellow crocus	KROKOS (*m*): GELBER KROKOS (*m*)
Crocus imperati	586	Italian wild crocus	ITALIENISCHER KROKOS (*m*)
Crocus nudiflor	587	Autumnal crocus	HERBSTKROKUS (*m*)

Botanical Term.	No.	English Term.	German Term.
Crocus sativus	588	Saffron, French (Spanish) saffron; saffron crocus	KROKOS (*m*); SAFRAN (*m*); SAFFRAN (*m*); ECHTER SAFRAN (*m*); HERBSTSAFRAN (*m*)
Crocus vernus	589	Crocus; Spring crocus; purple crocus	KROKOS (*m*); FRÜHLINGSKROKUS (*m*)
Crotalaria juncea	590	Taag; Sunn; Shunum; sunhemp; Bengal hemp; Madras hemp; Sunn hemp	MADRASHANF (*m*); BENGALISCHER HANF (*m*); KLAPPERSCHOTE (*f*)
Croton eleuteria	591	Eluthera, sweet-wood	KASKARILLBAUM (*m*); SCHAKERILLBAUM (*m*)
Croton pavana	592	Tilly seed tree	KREBSBLUME (*f*)
Croton pseudochina	593	Colpachi	KASKARILLRINDENBAUM (*m*)
Croton tiglium	594	Croton, South sea laurel (See *Tiglium officinale*)	PURGIERKROTON (*m*); TIGLIBAUM (*m*)
Crozophora tinctoria	595	Litmus plant, turnsole	LACKMUSKRAUT (*n*); LACKMUS-PFLANZE (*f*)
Cruciferæ	596	Cruciferæ	KREUZBLÜTLER (*m.pl*); KRUZIFEREN (*f.pl*)
Cryptogamæ	597	Cryptogamia, flowerless plants, cellular plants	NICHTBLÜHENDE PFLANZEN (*f.pl*); BLÜTENLOSE PFLANZEN (*f.pl*); KRYPTOGAMEN (*f.pl*)
Cryptogamma crispa	598	Rock-brake, curled brake	KRYPTOGAME (*f*)
Cryptomeria japonica	599	Japanese cryptomeria, Japanese cedar	JAPANISCHE ZEDER (*f*); KRYPTOMERIE (*f*)
Cucubalus baccifer	600	(See *Silene cucubalus*)	
Cucumis anguinus	601	(See *Trichosanthes anguina*)	
Cucumis anguria	602	West Indian gherkin	WESTINDISCHE GURKE (*f*); BRASILISCHE GURKE (*f*)
Cucumis dipsaceus	603	Gherkin; hedgehog gourd	GURKE (*f*); PFEFFERGURKE (*f*)
Cucumis melo	604	Melon, sweet melon (See *Citrullus vulgaris*)	MELONE (*f*); ZUCKERMELONE (*f*)
Cucumis sativus	605	Cucumber	GURKE (*f*)
Cucurbita lagenaria	606	Gourd-bottle, bottle-gourd, trumpet gourd (See *Lagenaria vulgaris*)	FLASCHENKÜRBIS (*m*); KÜRBISFLASCHE (*f*)
Cucurbita pepo	607	Pumpkin; gourd; melon pumpkin	KÜRBIS (*m*); MELONENKÜRBIS (*m*)
Cucurbita pepo ovifera	608	Vegetable marrow	KÜRBIS (*m*); PFEBE (*f*); FLASCHENAPFEL (*m*)
Cudrania	609	Indian yellow-wood	INDISCHES GELBHOLZ (*n*)
Cuminum cyminum	610	Cumin, cummin	KREUZKÜMMEL (*m*); RÖMISCHER KÜMMEL (*m*); MUTTERKÜMMEL (*m*); PFEFFERKÜMMEL (*m*)
Cunila coccinea	611	(See *Calamintha coccinea*)	
Cunila mariana	612	Maryland dittany	
Cupressaceæ	613	Cupressaceæ (See *Thuja, Thujopsis, Cupressus, Libocedrus, Biota, Chamæcyparis* and *Juniperus*)	ZYPRESSENARTIGE ZAPFENTRÄGER (*m.pl*); ZYPRESSENGEWÄCHSE (*n.pl*)
Cupressus distichum	614	Virginian cypress	VIRGINISCHE ZYPRESSE (*f*); SUMPFZYPRESSE (*f*)
Cupressus funebris	615	(See *Cupressus pendula*)	
Cupressus glauca	616	Glaucus cypress	MEERGRÜNE ZYPRESSE (*f*)
Cupressus lawsoniana	617	Lawson's cypress (See *Chamæcyparis lawsoniana*)	LAWSONS-ZYPRESSE (*f*); OREGON-ZYPRESSE (*f*); LEBENSBAUM-ZYPRESSE (*f*)
Cupressus lusitanica	618	Drooping cypress (See *Cupressus pendula*)	
Cupressus macnabiana	619	Shrubby cypress	STRAUCHZYPRESSE (*f*)
Cupressus macrocarpa	620	Monterey cypress	MONTEREYZYPRESSE (*f*)
Cupressus nutkatensis	621	Alaska cypress (See *Chamæcyparis nutkatensis*)	NUTKA-ZYPRESSE (*f*)
Cupressus obtusa	622	Hinoki cypress	HINOKI-ZYPRESSE (*f*)
Cupressus pendula	623	Drooping cypress (See *Cupressus funebris*)	TRAUERZYPRESSE (*f*)
Cupressus pisifera	624	Japanese cypress	SAWERA-ZYPRESSE (*f*)
Cupressus pyramidalis	625	Pyramidal cypress	SÄULENZYPRESSE (*f*); PYRAMIDEN-ZYPRESSE (*f*)

Botanical Term.	No.	English Term.	German Term.
Cupressus sempervirens	626	Cypress, evergreen cypress	ZYPRESSE (*f*); ECHTE ZYPRESSE (*f*); IMMERGRÜNE ZYPRESSE (*f*); GEMEINE ZYPRESSE (*f*)
Cupressus thyiodes	627	White cedar (See *Chamæcyparis sphæroidea* and *Retinospora ericoides*)	ZEDER-ZYPRESSE (*f*)
Cupressus torulosa	628	Himalaya cypress	HIMALAJAZYPRESSE (*f*)
Cupuliferæ	629	Cupuliferæ	BECHERFRÜCHTLER (*m.pl*), NÄPFCHENFRÜCHTLER (*m.pl*)
Curcas purgans	630	Physic nut; purging nut; red ebony	PURGIERNUSS (*f*); GRANADIL--HOLZ (*n*); GRANATILLHOLZ (*n*); GRENADILLHOLZ (*n*)
Curculigo recurvata	631	Weevil-plant	RÜSSELLILIE (*f*)
Curcuma longa	632	Curcuma; turmeric; curry	GILBWURZ (*f*); GILBWURZEL (*f*); KURKUMA (*f*)
Curcuma zedoaria	633	Zedoary plant	ZITWER (*m*); ZITWERWURZEL (*f*); ZITTERWURZEL (*f*)
Cuscuta	634	Dodder	KLEBE (*f*); SEIDE (*f*)
Cuscuta epilinum	635	Flax dodder	FLACHSSEIDE (*f*)
Cuscuta epithymum	636	Lesser dodder	KLEESEIDE (*f*)
Cuscuta europæa	637	Greater dodder	NESSELSEIDE (*f*); VOGELSEIDE (*f*); TEUFELSZWIRN (*m*); KLEBE (*f*); RANGE (*f*)
Cuscuta trifolii	638	Clover dodder	DREIBLÄTTRIGE KLEESEIDE (*f*)
Cusparia trifoliata	639	(See *Cuspa trifoliata* and *Galipea cusparia*)	
Cuspa trifoliata	640	(See *Galipea cusparia*)	CUSPABAUM (*m*)
Cyanophyceæ	641	Blue algæ, blue seaweed	CYANOPHYZEEN (*f.pl*); BLAU--GRÜNE ALGEN (*f.pl*)
Cyathus vernicosus	642	Cornbells	KORNBECHERPILZ (*m*)
Cycadaceæ	643	Cycadaceæ, palm ferns	PALMFARNE (*m.pl*); PALMEN--FARNE (*m.pl*); ZAPFENPAL--MEN (*f.pl*)
Cycas circinalis	644	(See *Sagus lævis*)	FARNPALME (*f*); PALMFARN (*m*); UNECHTE SAGOPALME (*f*); INDISCHER PALMFARN (*m*)
Cycas revoluta	645	(Chinese)-sago palm	CHINESISCHE SAGOPALME (*f*); JAPANISCHER PALMFARN (*m*)
Cycas rumphii	646	Malayan palm fern	MALAIISCHER PALMFARN (*m*)
Cycas seemanni	647	Fiji palm fern	FIDSCHI-PALMFARN (*m*)
Cycas thouarsii	648	Madagascar palm fern	MADAGASSISCHER PALMFARN (*m*)
Cyclamen europæum	649	Sow-bread; cyclamen	SCHWEINBROT (*n*); ERDBROT (*n*); ALPENVEILCHEN (*n*); ZYKLA--MEN (*n*)
Cydonia alpina	650	Alpine quince	ALPENQUITTE (*f*)
Cydonia chinensis	651	China quince	CHINESISCHE SCHEINQUITTE (*f*)
Cydonia japonica	652	Japan quince (See *Pirus japonica*)	JAPANISCHE SCHEINQUITTE (*f*)
Cydonia vulgaris	653	Quince-(tree) (See *Pirus cydonia*)	GEMEINE QUITTE (*f*); QUITTE (*f*); QUITTENBAUM (*m*)
Cymbalaria cymbalaria	654	Cymbalaria (See *Cymbalaria muralis* and *Linaria cymbalaria*)	MAUERLEINKRAUT (*n*); VENUS--NABEL (*m*); MAUER-ZYMBER--KRAUT (*n*)
Cymbalaria elatinoides	655	(See *Linaria elatine*)	
Cymbalaria muralis	656	(See *Cymbalaria cymbalaria*)	
Cynanchum vincetoxicum	657	Swallowwort (See *Vincetoxicum officinale*)	GIFTWENDE (*f*)
Cynara cardunculus	658	Cardoon-(thistle)	KARDONE (*f*)
Cynara scolymus	659	Globe artichoke	GARTENDISTEL (*f*)
Cynodon dactylon	660	Bermuda grass; Cynodon; dog's-tooth; creeping cynodon (See also *Cynosurus echinatus*)	BERMUDAGRAS (*n*); HUNDS--HIRSE (*f*); HUNDSZAHN (*m*)
Cynoglossum montanum	661	Green hound's-tongue	BERGHUNDSZUNGE (*f*)
Cynoglossum officinale	662	Hound's tongue; common hound's tongue	HUNDSZUNGE (*f*)
Cynomorium coccineum	663	Orobanche; broom rape; fungus melitensis	MALTESERPILZ (*m*)
Cynosurus cristatus	664	Crested dog's-tail grass, gold-seed, cock's-comb grass	KAMMGRAS (*n*)

Botanical Term.	No.	English Term.	German Term.
Cynosurus echinatus	665	Dog's-tooth grass; rough dog's-tail grass	RAUHES KAMMGRAS (*n*)
Cyperaceæ	666	Cyperaceæ (See *Blysmus, Carex, Cyperus, Cladium, Eleocharis, Eriophorum, Kobresia, Rhynchospora, Scirpus* and *Schœnus*)	HALBGRÄSER (*n.pl*); SCHEIN-GRÄSER (*n.pl*); SAUERGRÄSER (*n.pl*); ZYPERGRÄSER (*n.pl*)
Cyperus alternifolius	667	Umbrella palm	MADAGASKAR-CYPERGRAS (*n*)
Cyperus esculentus	668	Chufa-(tuber); earth almond	ERDMANDEL (*f*); NUSSGRAS (*n*); ARABISCHE ZUCKERWURZEL (*f*)
Cyperus fuscus	669	Brown cyperus	ZYPERGRAS (*n*)
Cyperus longus	670	Galingale	ZYPERWURZ (*f*)
Cyperus papyrus	671	Papyrus, biblus, Egyptian paper plant, papyrus plant (See *Papyrus antiquorum*)	PAPYRUSSTAUDE (*f*): PAPIER-STAUDE (*f*)
Cypripedium calceolus	672	Lady's slipper; common slipper orchid; moccasin flower	MARIENSCHUH (*m*); FRAUEN-SCHUH (*m*)
Cystopteris bulbifera	673	Bulbiferous bladder fern	KNOLLENTRAGENDER BLASENFARN (*m*)
Cystopteris fragilis	674	(Brittle)-bladder fern	ZERBRECHLICHER BLASENFARN (*m*)
Cystopteris montana	675	Mountain bladder-fern	GEBIRGS-BLASENFARN (*m*)
Cystopteris regia	676	Imperial bladder fern	KÖNIGLICHER BLASENFARN (*m*)
Cystopus candidus	677	Cabbage white rust, white rust, crucifer white rust (See *Albugo candida*)	WEISSER ROST (*m*)
Cystopus cubicus	678	Goat's-beard white rust	BOCKSBART-KETTENSPOREIPILZ (*m*)
Cystopus portulacæ	679	Purslane white rust	PORTULAK-KETTENSPOREIPILZ (*m*)
Cystopus spinulosus	680	Thistle white rust	DISTEL-KETTENSPOREIPILZ (*m*)
Cytisus albus	681	White-flowered broom (See *Cytisus leucanthus*)	WEISSBLÜTIGER GEISSKLEE (*m*)
Cytisus ardoini	682	Ardoino's broom	ARDOINOS-GEISSKLEE (*m*)
Cytisus austriacus	683	Yellow-flowered broom; Austrian broom	ÖSTERREICHISCHER BOHNEN-STRAUCH (*m*)
Cytisus biflorus	684	Deciduous broom	ZWEIBLÜTIGER BOHNENSTRAUCH (*m*)
Cytisus canariensis	685	Genista (See *Genista*)	KANARIEN-BOHNENSTRAUCH (*m*)
Cytisus hirsutus	686	Hairy broom	HAARIGER BOHNENSTRAUCH (*m*)
Cytisus laburnum	687	Common laburnum (See *Laburnum vulgare*)	GOLDREGEN (*m*)
Cytisus leucanthus	688	White-flowered broom (See *Cytisus albus*)	WEISSBLUMIGER BOHNENSTRAUCH (*m*); BOHNENBAUM (*m*); GEISSKLEE (*m*)
Cytisus nigricans	689	Yellow-flowered broom	SCHWARZER BOHNENSTRAUCH (*m*)
Cytisus purpureus	690	Purple broom (See also *Viborgia purpurea*)	EDELREIS (*n*); ROTBLÜTIGER GOLDREGEN (*m*)
Cytisus scoparius	691	Broom; greenbroom, hogweed, bannal; common broom (See *Genista* and *Sarothamnus scoparius* and *Spartium scoparium*)	BESENGINSTER (*m*); BESENKRAUT (*n*); GINSTER (*m*); PFRIEMEN (*m*); PFRIEMENKRAUT (*n*)
Cytisus triflorus	692	(See *Cytisus hirsutus*)	
Daboecia polifolia	693	St. Dabeoc's heath; Cunnamara heath (See *Menziesia polifolia*)	
Daboecia taxifolia	694	Yew-leaved heath (See *Bryanthus taxifolius*)	
Dacrydium cupressinum	695	New Zealand dacrydium	TRÄNENEIBE (*f*); SCHUPPENEIBE (*f*)
Dacrydium franklinii	696	Huonpine; Tasmanian Huonpine	HUONFICHTE (*f*)
Dacrydium taxoides	697	New Caledonian dacrydium	
Dactylis cæspitosa	698	(See *Festuca flabellata* and *Poa flabellata*)	TUSSOCKGRAS (*n*); KNAULGRAS (*n*)
Dactylis glomerata	699	Cock's-foot grass; orchard-grass	KNÄULGRAS (*n*)
Dæmonorops draco	700	Dragon's blood palm (See *Calamus draco*)	DRACHENBLUTPALME (*f*)

Botanical Term.	No.	English Term.	German Term.
Dahlia superflua	701	Dahlia (See *Georgina juarezi*)	DAHLIA (*f*)
Dalbergia latifolia	702	West-Indian rosewood, Blackwood	WESTINDISCHES ROSENHOLZ (*n*); WESTINDISCHER ROSENHOLZ-BAUM (*m*)
Dalbergia nigra	703	Jacaranda (palisander) wood; (Brazilian)-rosewood (See also *Jacaranda cærulea*)	JAKARANDAHOLZ (*n*); PALISANDERHOLZ (*n*); ROSENHOLZ-BAUM (*m*)
Damasonium stellatum	704	Star damasonium; star fruit	
Dammara australis	705	Dammar pine, kauri pine (See *Agathis australis*)	KAURIFICHTE (*f*)
Dammara orientalis	706	Molucca dammar pine	DAMMARFICHTE (*f*)
Danaë racemosa	707	(See *Ruscus racemosus*)	GEMEINER TRAUBENDORN (*m*); ALEXANDRINISCHER LORBEER (*m*)
Daphne cneorum	708	Garland-flower	WOHLRIECHENDER SEIDELBAST (*m*)
Daphne laureola	709	Spurge laurel	KELLERHALS (*m*); LORBEERKRAUT (*n*); LORBEERSEIDELBAST (*m*)
Daphne mezereum	710	Mezereon, paradise plant, Oliver spurge, magell, wild pepper. spurge flax, dwarf bay	SEIDELBAST (*m*); KELLERHALS (*m*); GEMEINER SEIDELBAST (*m*); PARADIESPFLANZE (*f*); PFEFFERBEERE (*f*)
Darlingtonia californica	711	Californian pitcher-plant	SUMPFDARLINGTONIE (*f*)
Datura arborea	712	Angel's trumpet (See *Brugmansia candida*)	CHILE-STECHAPFEL (*m*)
Datura metel	713	Egyptian henbane	HERZBLÄTTRIGER STECHAPFEL (*m*)
Datura sanguinea	714	Orange-red-flowered datura (See *Brugmansia bicolor*)	BLUTROTGESTREIFTBLÜTIGER STECHAPFEL (*m*)
Datura stramonium	715	Thorn-apple; stramony, stramonium; apple Peru, Jamestown weed	STECHAPFEL (*m*); KRÖTENMELDE (*f*); IGELKOLBEN (*m*); RAUH-APFEL (*m*)
Daucus carota	716	Carrot; wild carrot	KAROTTE (*f*); MÖHRE (*f*); MOHR-RÜBE (*f*); GELBE RÜBE (*f*)
Delphinium ajacis	717	Larkspur, rocket larkspur	RITTERSPORN (*m*)
Delphinium consolida	718	Field larkspur	LERCHENKLAUE (*f*)
Delphinium exaltum	719	American delphinium; American larkspur	AMERIKANISCHER RITTERSPORN (*m*)
Delphinium staphysagria	720	Stavesacre	SCHARFER RITTERSPORN (*m*); LÄUSEKRAUT (*n*); RATTEN-PFEFFER (*m*); WOLFSKRAUT (*n*); STEPHANSKRAUT (*n*)
Dendrocalamus giganteus	721	Giant bamboo (See also *Bambusa*)	RIESENBAMBUS (*m*)
Dentaria bulbifera	722	Coralwort, toothwort	ZAHNWURZ (*f*)
Dentaria diphylla	723	(North American)-pepper root; tooth cress; two-leaved tooth cress (See *Cardamine diphylla*)	NORDAMERIKANISCHE PFEFFER-WURZ (*f*)
Desmodium canadense	724	Tick trefoil	BÜSCHELKRAUT (*n*); FESSELHÜLSE (*f*)
Desmodium gyrans	725	Telegraph plant (See *Pleurolobus*)	WANDELKLEE (*m*)
Dianthus alpinus	726	Alpine pink	ALPENNELKE (*f*)
Dianthus arenarius	727	Sand pink	SANDFEDERNELKE (*f*)
Dianthus armeria	728	Deptford pink	BÜSCHELNELKE (*f*); RAUHE NELKE (*f*)
Dianthus barbatus	729	Sweet William	FEDERNELKE (*f*); RAUHE NELKE (*f*); BARTNELKE (*f*)
Dianthus cæsius	730	Cheddar pink; mountain pink; carnation	BERGNELKE (*f*); PFINGSTNELKE (*f*)
Dianthus carthusianorum	731	Carthusian pink	KARTÄUSERNELKE (*f*)
Dianthus caryophyllus	732	Clove pink; carnation	NELKE (*f*); GARTENNELKE (*f*)
Dianthus chinensis	733	Chinese pink, Indian pink	CHINESISCHE NELKE (*f*)
Dianthus deltoides	734	Maiden pink	HEIDENELKE (*f*); BLUTS-TRÖPFCHEN (*n*)
Dianthus glacialis	735	Glacier pink	GLETSCHERNELKE (*f*)
Dianthus plumarius	736	Common pink	NELKE (*f*); GARTENFEDERNELKE (*f*)

Botanical Term.	No.	English Term.	German Term.
Dianthus plumosus	737	Common pink	NELKE (*f*)
Dianthus prolifer	738	Proliferous pink (See *Tunica prolifera*)	
Dianthus silvester	739	Wood pink	WALDNELKE (*f*)
Dianthus superbus	740	Fringed pink	PRACHTFEDERNELKE (*f*)
Diatomaceæ	741	Diatoms	KIESELALGEN (*f.pl*); BAZILLAR-IEN (*f.pl*); SPALTALGEN (*f.pl*)
Dicentra canadensis	742	bleeding heart	KANADISCHES JUNGFERNHERZ (*n*)
Dicentra cucullaria	743	Dutchman's breeches	
Dicentra spectabilis	744	Bleeding heart; Chinaman's breeches, seal flower, squirrel corn (See *Dielytra*)	HÄNGENDES HERZ (*n*); FLAM-MENDES HERZ (*n*)
Dichopsis gutta	745	Gutta-percha tree	GUTTAPERCHABAUM (*m*)
Dicotyleæ	746	Dicotyleæ	ZWEIKEIMBLÄTTRIGE PFLANZEN (*f.pl*)
Dictamnus fraxinella	747	Dittany; fraxinella (See *Origanum dictamnus*)	DIPTAM (*m*)
Dieffenbachia seguina	748	(West Indian)-dumb cane	WESTINDISCHE DIEFFENBACHIE (*f*)
Dielytra	749	(See *Dicentra*)	
Digitalis purpurea	750	Foxglove, digitalis, purple fox-glove, fairy glove; finger flower; elves-gloves; our-Lady's-gloves; bloody-fingers; deadmen's-bells	FINGERHUT (*m*); ROTER FINGER-HUT (*m*)
Digraphis arundinacea	751	Reed grass (See *Phalaris arundinacea*)	AUSDAUERNDES GLANZGRAS (*n*)
Dionæa muscipula	752	Venus's fly-trap	VENUSFLIEGENFALLE (*f*); FLIE-GENKLAPPE (*f*)
Dioscorea batatas	753	Yam (See *Dioscorea sativa*)	JAMPFLANZE (*f*); MEHLWURZEL (*f*); YAMSWURZEL (*f*); YAM-WURZEL (*f*)
Dioscorea sativa	754	(See *Dioscorea batatas*)	
Diospyros ebenum	755	Ebony-(tree)	EBENHOLZ (*n*); EBENHOLZ-DATTEL-PFLAUME (*f*)
Diospyros hirsuta	756	(Indian)-calamander-(wood)- tree	KOROMANDELHOLZBAUM (*m*); KALAMANDERHOLZBAUM (*m*); BUNTES EBENHOLZ (*n*)
Diospyros kaki	757	Date plum; Japanese date plum	DATTELPFLAUME (*f*)
Diospyros lotus	758	Green ebony; date plum; lotus; common date palm	GRÜNEBENHOLZ (*n*); UNECHTES POCKHOLZ (*n*); DATTEL-PFLAUME (*f*); LOTUSPFLAUME (*f*)
Diospyros texana	759	Persimmon; Mexican date plum	DATTELPFLAUME (*f*)
Diospyros virginiana	760	Persimmon; American date plum	DATTELPFLAUME (*f*); PERSIM-MONBAUM (*m*)
Diotis maritima	761	Cotton weed; weed cotton grass; cotton sedge; weed cotton sedge	
Dipsaceæ	762	Dipsaceæ (See *Dipsacus*, *Knautia* and *Scabiosa*)	KARDENGEWÄCHSE (*n.pl*)
Dipsacus fullonum	763	Teasel; fuller's thistle, fuller's teasel; fuller's-weed	WEBERDISTEL (*f*); KRATZDISTEL (*f*); KARDE (*f*); RAUHKARDE (*f*); WALKERDISTEL (*f*)
Dipsacus pilosus	764	Small teasel; shaggy teasel	RAUHE KARDENDISTEL (*f*)
Dipsacus sylvestris	765	Wild teasel	WILDE KARDENDISTEL (*f*)
Dipterix odorata	766	Tonka bean; Tonquin bean; (See *Coumarouna odorata*)	TONKABAUM (*m*)
Dipterocarpus trinervis	767	Gurjun balsam tree	ZWEIFLÜGELNUSSBAUM (*m*)
Dipteryx oppositifolia	768	Tonquin bean, tonka bean, coumarouna bean, snuff bean (See *Dipterix*) (See also *Coumarouna odorata* and *Faba de Tonca*)	TONKABOHNE (*f*)
Dirca palustris	769	Leather-wood	SUMPF-LEDERHOLZ (*n*); SUMPF-SEIDELBAST (*m*); MÄUSEHOLZ (*n*)

Botanical Term.	No.	English Term.	German Term.
Discorea elephantipes	770	Elephant's foot (See *Testudinaria elephantipes*)	ELEFANTENFUSS (*m*)
Dodecatheon meadia	771	Shooting star; American cowslip (See *Jeffreyi*)	GÖTTERBLUME (*f*)
Dodecatheon media frigidum	772	Dwarf American cowslip, dwarf dodecatheon	ZWERG-GÖTTERBLUME (*f*)
Dodecatheon meadia lancifolium	773	Large-leaved dodecatheon	ANZETTBLÄTTRIGE GÖTTER-BLUME (*f*)
Dolichos sesquipedalis	774	Asparagus bean, soy bean, hyacinth bean (See also *Soja hispida*)	SOJA (*f*); SOJABOHNE (*f*)
Dorema ammoniacum	775	Ammoniac plant	AMMONIAKPFLANZE (*f*)
Doronicum pardalianches	776	Great leopard's bane; panther strangler	KRAFTWURZEL (*f*); DORANT-WURZEL (*f*); DORANT (*m*); SCHWINDELWURZEL (*f*)
Doronicum plantagineum	777	Plantain-leaved leopard's bane (See *Senecio doronicum*)	GEMSKRAUTWURZEL (*f*); GEMS-WURZ (*f*)
Doryanthes excelsa	778	(Australian)-gigantic lily, (Australian)-giant lily, spear lily	AUSTRALISCHE RIESENLILIE (*f*)
Dothidea ribis	779	Currant fungus; gooseberry fungus	STACHELBEEREN-PUSTELLAGER-PILZ (*m*)
Draba aizoides	780	Yellow draba, yellow whitlow grass	IMMERGRÜNE HUNGERBLUME (*f*)
Draba hirta	781	Rock draba, rock whitlow grass	FELSEN-HUNGERBLUME (*f*)
Draba incana	782	Hoary draba, hoary whitlow grass	
Draba muralis	783	Wall draba; wall whitlow grass	WAND-HUNGERBLUME (*f*)
Draba verna	784	Common whitlow-grass (See *Erophila verna*)	HUNGERBLÜMCHEN (*n*); FRÜH-LINGSHUNGERBLUME (*f*)
Dracœna draco	785	Dragon tree, dragon-blood tree (See also *Dœmonorops draco*)	DRACHENBAUM (*m*); DRACHEN-BLUTBAUM (*m*); DRACHEN-LILIE (*f*); DRACHENPALME (*f*)
Dracocephalum canescens	786	Dragon's head	DRACHENKOPF (*m*)
Dracocephalum moldavica	787	Moldavian balm	MOLDAUDRACHENKOPFKRAUT (*n*); TÜRKISCHES MELISSENKRAUT (*n*)
Dracunculus vulgaris	788	Common dragon-plant, snake plant (See *Arum dracunculus*)	
Drimys winteri	789	Winter's-bark shrub	WINTERSRINDE (*f*); WINTERS-RINDENBAUM (*m*); WINTERS-GEWÜRZRINDENBAUM (*m*)
Drosera	790	Sundew; youth wort	SONNENTAU (*m*)
Drosera anglica	791	English sundew	LANGBLÄTTRIGER SONNENTAU (*m*); ENGLISCHER SONNENTAU (*m*)
Drosera longifolia	792	Oblong sundew (See *Drosera anglica*)	
Drosera rotundifolia	793	Round-leaved sundew; common sundew	SONNENKRAUT (*n*)
Dryas octopetala	794	White dryas; mountain avens	GEMEINE SILBERWURZ (*f*); ALPEN-GAMANDER (*m*)
Dryobalanops camphora	795	(Borneo)-camphor tree (See *Laurus camphora*)	BORNEOKAMPFERBAUM (*m*)
Dryopteris filix-mas marginalis	796	Fern, male fern (See *Aspidium* and *Lastrea Felix-mas*)	FARN (*m*); FARNKRAUT (*n*); JOHANNISWURZ (*f*); JOHANNIS-WURZEL (*f*)
Durio zibethinus	797	(East Indian)-durian tree	ZIBETBAUM (*m*)
Ebenus cretica	798	Grenada wood; red ebony (See *Brya ebenus*)	ROTEBENHOLZ (*n*)
Ecballium agreste	799	Squirting cucumber (See *Memordica elaterium*)	SPRITZGURKE (*f*)
Ecballium elaterium	800	(See *Memordica elaterium*)	
Eccremocarpus scaber	801	Eccremocarpus, Chilian glory-flower, eccremocarpus vine	SCHÖNREBE (*f*); HÄNGEFRUCHT (*f*)
Echinocactus williamsii	802	(Mexican)-echinocactus; hedgehog plant, hedgehog cactus	IGELKAKTUS (*m*)

Botanical Term.	No.	English Term.	German Term.
Echinops	803	Globe thistle (See *Echinops sphærocephalus*)	
Echinopsis cinnabarina	804	Echinopsis-(cactus); sea hedge-hog cactus	SEEIGELKAKTUS (*m*)
Echinops sphærocephalus	805	Globe thistle	KUGELDISTEL (*f*)
Echium plantagineum	806	Purple echium, purple viper's bugloss	PURPURNER NATTERKOPF (*m*)
Echium vulgare	807	Viper's bugloss; blueweed	GEMEINER NATTERNKOPF (*m*); BLAUER HEINRICH (*m*)
Edwardsia grandiflora	808	(See *Sophora tetraptera*)	
Elæagnus	809	Oleaster; wild olive	ÖLWEIDE (*f*); OLEASTER (*m*): FALSCHER ÖLBAUM (*m*)
Elæagnus angustifolia	810	Narrow-leaved oleaster	SCHMALBLÄTTRIGE ÖLWEIDE (*f*)
Elæagnus argentea	811	Silver oleaster, beef-suet tree, buffalo-berry, rabbit-berry (See *Shepherdia argentea*)	SILBERÖLWEIDE (*f*): SILBERWEISSE BÜFFELBEERE (*f*)
Elæagnus longipes	812	Japanese wild olive, Japanese oleaster	LANGSTIELIGE ÖLWEIDE (*f*)
Elæagnus parvifolia	813	Small-leaved oleaster	KLEINBLÄTTRIGE ÖLWEIDE (*f*)
Elæis guineensis	814	Oil palm (See *Avoira elais*)	ÖLPALME (*f*); AFRIKANISCHE ÖLPALME (*f*)
Elæococca vernicia	815	Candlenut tree, Tung chou (See *Aleurites cordata*)	ÖLFIRNISBAUM (*m*): TUNGBAUM (*m*)
Elæodendron croceum	816	(See *Cassine crocea* and *Crocoxylon excelsum*)	
Elæodendron orientale	817	Olive shrub	ÖLSTRAUCH (*m*)
Elaphomyces granulatus	818	Stag-truffle (See *Boletus cervinus*)	HIRSCHSTREULING (*m*): HIRSCH-TRÜFFEL (*m*): HIRSCH-TRÜFFELPILZ (*m*): HIRSCH-BRUNST (*m*); GEMEINER HIRSCHTRÜFFEL (*m*)
Elatinaceæ	819	Elatinaceæ	ELATINAZEEN (*f.pl*); TÄNNELGE-WÄCHSE (*n.pl*)
Elatine hexandra	820	Water-pepper; water-wort; six-stamened elatine	SECHSMÄNNIGE ELATINE (*f*)
Elatine hydropiper	821	Water-pepper; water-wort; eight-stamened elatine	WASSERPFEFFER (*m*); ACHTMÄN-NIGE ELATINE (*f*)
Eleocharis acicularis	822	Lesser spike-rush (See *Heleocharis acicularis*)	NADELFÖRMIGE TEICHSIMSE (*f*)
Eleocharis multicaulis	823	Many-stalked spike-rush	VIELSTENGELIGE TEICHSIMSE (*f*); TEICHRIED (*n*); TEICHBINSE (*f*)
Eleocharis palustris	824	Common spike-rush	SUMPFRIED (*n*); SUMPF-TEICH-SIMSE (*f*)
Eleocharis tuberosa	825	Tuberous spike-rush	WASSERKASTANIE (*f*)
Elettaria cardamomum	826	(Indian)-cardamom (See *Amomum melegueta*)	KARDAMOM (*m*); MALABARISCHE KARDAMOMENPFLANZE (*f*); KLEINE KARDAMOMENPFLANZE (*f*)
Elettaria repens	827	Cardamon	KARDAMOM (*m*)
Eleusine	828	Crab grass; jara grass	JARAGRAS (*n*)
Eleusine coracana	829	African millet	AFRIKANISCHE HIRSE (*f*); KORA-KANGRAS (*n*)
Eleusine indica	830	Wire grass	INDISCHES KORAKANGRAS (*n*)
Elodea canadensis	831	Canadian pondweed; water thyme; water weed (See *Anacharis alsinastrum*)	WASSERPEST (*f*)
Elymus arenarius	832	Lyme-grass; upright sea lyme-grass	SANDHAARGRAS (*n*)
Elymus europæus	833	Lyme-grass, rye-grass	HAARGRAS (*n*)
Elymus geniculatus	834	Pendulous sea lyme-grass	GEKNIETES HAARGRAS (*n*)
Embeliacea	835	Embeliacea; edible Embelia; edible Aembelia	MYRSINAZEEN (*f.pl*)
Empetrum nigrum	836	Crowberry; common crowberry; black-fruited crowberry, crake-berry, black-berried heath (See also *Vaccinium*)	SCHWARZE RAUSCHBEERE (*f*); KRÄHENBEERE (*f*)
Empetrum nigrum rubrum	837	Red-fruited crowberry	ROTE RAUSCHBEERE (*f*)
Encephalartos caffer	838	Kaffir bread plant; caffre bread	BROTPALMFARN (*m*); KAFFERN-BROTPALMFARN (*m*)

Botanical Term.	No.	English Term.	German Term.
Endophyllum	839	Cluster-cup	REIHENSPORENROST(*m*)
Entada scandens	840	Sea-bean	MEERBOHNE (*f*)
Enteromorpha	841	Pipe-weed	DARMMEERLATTICH (*m*); DARM-ALGE (*f*)
Ephedra americana	842	American sea-grape	AMERIKANISCHER MEERTRÄUBEL (*m*)
Ephedra andina	843	Andes sea-grape	ANDENMEERTRÄUBEL (*m*)
Ephedra distachya	844	(Russian)-joint fir; sea-grape	MEERTRÄUBCHEN (*n*); (ZWEI-ÄHRIGER)-MEERTRÄUBEL (*m*)
Ephedra fragilis	845	Fragile sea-grape	BRÜCHIGER MEERTRÄUBEL (*m*)
Ephedra major	846	Large sea-grape	GRÖSSERER MEERTRÄUBEL (*m*)
Ephedra nebrodensis	847	(See *Ephedra major*)	
Ephedra vulgaris	848	Holy Soma plant (See also *Asclepias acida*)	HEILIGE SOMAPFLANZE (*f*)
Epilobium alpinum	849	Alpine (epilobe) willow-herb	ALPEN-WEIDENRÖSCHEN (*n*)
Epilobium alsinefolium	850	Chickweed (epilobe) willow-herb	MEIRICHBLÄTTRIGE WEIDEN-RÖSCHEN (*n*)
Epilobium angustifolium	851	Willow herb; rose-bay; French willow; willow epilobe; bay willow; narrow-leaved willow herb	ANTONSKRAUT (*n*); FEUERKRAUT (*n*); ST. ANTONIUSKRAUT (*n*)
Epilobium hirsutum	852	Great willow herb; great epilobe; codlins and cream	HAARIGES WEIDENRÖSCHEN (*n*)
Epilobium montanum	853	broad-leaved (epilobe) willow-herb	BERG-WEIDENRÖSCHEN (*n*)
Epilobium palustre	854	Marsh (epilobe) willow-herb	SUMPF-WEIDENRÖSCHEN (*n*)
Epilobium parviflorum	855	Hoary (epilobe) willow-herb; small-flowered willow-herb	KLEINBLÜTIGES WEIDENRÖSCHEN (*n*)
Epilobium roseum	856	Pale (epilobe) willow-herb	ROSA-WEIDENRÖSCHEN (*n*)
Epilobium tetragonum	857	Square (epilobe) willow-herb	
Epimedium alpinum	858	Barrenwort; Bishop's hat	BISCHOFSMÜTZE (*f*); ALPEN-SOCKENBLUME(*f*)
Epipactis latifolia	859	Broad epipactis	BREITBLÄTTRIGE EPIPACTIS (*f*)
Epipactis palustris	860	Marsh epipactis	SUMPF-EPIPACTIS (*f*)
Epiphyllum truncatum	861	Snipped Indian fig	BLATTKAKTUS (*m*)
Epipogum aphyllum	862	Leafless epipogum	BLATTLOSER EPIPOGUM (*m*)
Episcea cupreata	863	Copper-leaved episcea	KUPFERIGE EPISCEA (*f*)
Equisetum arvense	864	Field horsetail; field equisetum; paddock-pipes; shave grass	KANNENKRAUT (*n*); SCHEUER-KRAUT (*n*); ACKER-SCHACH-TELHALM (*m*); DUWOK (*m*)
Equisetum hyemale	865	rough horse-tail; Dutch rush, shave grass	POLIERHEU (*n*); WINTERSCHACH-TELHALM (*m*); KANNENKRAUT (*n*); ZINNKRAUT (*n*)
Equisetum limosum	866	Smooth horse-tail	SCHLAMMSCHACHTELHALM (*m*)
Equisetum littorale	867	Bog horse-tail	SUMPFSCHACHTELHALM (*m*)
Equisetum maximum	868	Horse-tail; great horse-tail, fox-tailed asparagus	SCHACHTELHALM (*m*); SCHACH-TELKRAUT (*n*); GROSSSCHEI-DIGER SCHACHTELHALM (*m*)
Equisetum palustre	869	Marsh-horse-tail, horse pipe, horse willow, toad broom, toad pipe, shave grass	SCHACHTELHALM (*m*); SUMPF-SCHACHTELHALM (*m*)
Equisetum pratense	870	Shady horse-tail	WIESENSCHACHTELHALM (*m*)
Equisetum sylvaticum	871	Wood horse-tail	WALDSCHACHTELHALM (*m*)
Equisetum telmateia	872	Great horse-tail (See *Equisetum maximum*)	
Equisetum trachyodon	873	Long horse-tail	RAUHZÄHNIGER SCHACHTELHALM (*m*)
Equisetum variegatum	874	Variegated horse-tail	BUNTER SCHACHTELHALM (*m*)
Eragrostis ægyptica	875	Love grass, feather grass, Egypt-ian love grass	
Eranthis hyemalis	876	Winter aconite	WINTERWOLFSKRAUT (*n*); WIN-TERCHRISTWURZ (*f*); WIN-TERLING (*m*)
Erica arborea	877	Tree-heath; briar-wood bush	BRUYERHOLZSTRAUCH (*m*)
Erica carnea	878	Flesh-coloured heath; Mediter-ranean heath	FLEISCHFARBIGE BRUCHHEIDE (*f*); FLEISCHFARBIGES HEIDEKRAUT (*n*); ALPENHEIDEKRAUT (*n*)
Erica ciliaris	879	Ciliated heath; Dorset heath	

Botanical Term.	No.	English Term.	German Term.
Erica cinerea	880	Fine-leaved heath ; bell-heather, Scotch heather	GRAUE HEIDE (*f*)
Erica herbacea	881	(See *Erica carnea*)	
Erica Mackaiana	882	Mackay's heath	MACKAY'S-BRUCHHEIDE (*f*)
Erica mediterranea	883	Irish heath	
Erica tetralix	884	Cross-leaved heath	SUMPF-BRUCHHEIDE (*f*)
Erica vagans	885	Cornish heath	
Erica vulgaris	886	(See *Calluna vulgaris*)	HEIDEKRAUT (*n*) ; HEIDE (*f*)
Erigeron acris	887	Fleabane ; blue fleabane	BERUFKRAUT (*n*); FLOHKRAUT (*n*); BLAUE DÜRRWURZ (*f*)
Erigeron alpinus	888	Dwarf, mountain fleabane ; alpine erigeron	ALPENFLOHKRAUT (*n*)
Erigeron annuus	889	Annual fleabane (See *Stenactis annua*)	EINJÄHRIGE DÜRRWURZ (*f*)
Erigeron aurantiacus	890	Orange daisy	ORANGEROTBLÜTIGES FLOHKRAUT (*n*)
Erigeron canadensis	891	Scabious ; horseweed ; fleabane ; Canadian erigeron	BERUFKRAUT (*n*) ; FLOHKRAUT (*n*)
Erigeron speciosus	892	American fleabane ; handsome fleabane (See *Stenactis*)	BLAUBLÜTIGES FLOHKRAUT (*n*)
Erinus alpinus	893	Erinus	ALPEN-ERINUS(*m*)
Eriocaulon septangulare	894	Jointed pipe-wort	KUGELBINSE (*f*) ; SIEBENKANTIGE KUGELBINSE (*f*)
Eriodendron anfractuosum	895	Kapok tree (See also *Bombax ceiba* and *Bombax pentandrum*)	KAPOKBAUM (*m*)
Eriophorum alpinum	896	Alpine cotton-grass	ALPENWOLLGRAS (*n*)
Eriophorum polystachion	897	Common cotton-grass	GEMEINES WOLLGRAS (*n*); BIN--SENSEIDE (*f*)
Eriophorum vaginatum	898	Hare's-tail cotton-grass ; sheathing cotton-grass	WOLLGRAS (*n*) ; SCHEIDIGES WOLL-GRAS (*n*)
Eritrichium nanum	899	Fairy forget-me-not, Fairy Borage	ZWERGVERGISSMEINNICHT (*n*)
Erodium cicutarium	900	Hemlock stork's-bill ; erodium ; stork's-bill ; common erodium; heron's-bill	REIHERSCHNABEL (*m*)
Erodium maritimum	901	Sea erodium	MEER-REIHERSCHNABEL (*m*)
Erodium moschatum	902	Musk erodium	BISAMSTORCHSCHNABELKRAUT (*n*); MOSCHUSSTORCHSCHNABEL--KRAUT (*n*)
Erophila verna	903	(See *Draba verna*)	HUNGERBLÜMCHEN (*n*)
Eruca sativa	904	Rocket-(salad herb), garden rocket ; hedge mustard	RAUKENKOHL (*m*); RAUKE (*f*); SAATRANKE(*f*)
Ervum lens	905	Lentil plant (See *Lens esculenta*)	LINSE (*f*)
Eryngium campestre	906	Field eryngo	ROGGENDISTEL (*f*) ; BRACHDISTEL (*f*) ; MANNSTREU (*f*)
Eryngium maritimum	907	Sea holly ; sea eryngo	MEERMANNSTREU (*f*); MEER--STRANDSMANNSTREU (*f*); MEERWURZEL (*f*)
Erysibe	908	(See *Erysiphe*)	
Erysimum cheiranthoides	909	(Worm - seed) - treacle - mustard, alpine wallflower, hedge mustard	HEDERICH (*m*)
Erysimum officinale	910	(See *Sisymbrium officinale*)	
Erysimum orientale	911	Hare's-ear	MORGENLÄNDISCHER HEDERICH (*m*)
Erysiphe	912	Mildew fungus, blight (See *Erysibe*)	MELTAUPILZ (*m*); MELTAU (*m*); MEHLTAUPILZ (*m*); MEHLTAU (*m*); ECHTER MELTAU (*m*)
Erysiphe graminis	913	Grass mildew, wheat mildew (See *Oidium monilioides*)	WEIZENMELTAU (*m*)
Erysiphe martii	914	Pea mildew (See *Erysiphe pisi*)	
Erysiphe pisi	915	Pea mildew (See *Erysiphe martii*)	ERBSENMELTAU (*m*)
Erythraea centaurium	916	Common centaury ; lesser centaury, blush wort	TAUSENDGÜLDENKRAUT (*n*)

Botanical Term.	No.	English Term.	German Term.
Erythrina corallodendron	917	Coral wood; coral tree; coral flower; Jamaican bean-tree	KORALLENHOLZ (*n*); KORALLEN-BLUME (*f*); KORALLEN-STRAUCH (*m*)
Erythrina crista-galli	918	(Brazilian)-cockscomb; coral tree	HAHNENKAMM-KORALLENSTRAUCH (*m*); KORALLENBAUM (*m*)
Erythronium americanum	919	Yellow American dog's-tooth violet; yellow adder's tongue	AMERIKANISCHE ZAHNLILIE (*f*); AMERIKANISCHER HUNDSZAHN (*m*)
Erythronium dens-canis	920	(European)-dog's tooth violet; rose-coloured dog's-tooth violet	HUNDSZAHN (*m*); GEMEINE ZAHN-LILIE (*f*); ROTLING (*m*)
Erythronium grandiflorum	921	Yellow dog's-tooth violet	GROSSBLÜTIGE ZAHNLILIE (*f*)
Erythrophlœum guineense	922	(West African)-red water tree	AFRIKANISCHER ROTWASSERBAUM (*m*); SAFFYBAUM (*m*); SYFTY-BAUM (*m*)
Erythrophlœum laboucherii	923	(Australian)-red water tree	AUSTRALISCHER ROTWASSERBAUM (*m*)
Erythroxylon coca	924	Coca bush; coca-(shrub); coca tree; spadic bush	KOKA (*f*); KOKASTRAUCH (*m*)
Eucalyptus amygdalina	925	Peppermint tree	PFEFFERMINZBAUM (*m*)
Eucalyptus botryoides	926	Bastard mahogany	BASTARD-MAHAGONIE (*f*)
Eucalyptus citriodora	927	Citron-scented gum	WEISSRINDIGE SCHÖNMÜTZE (*f*)
Eucalyptus corynocalyx	928	Corynocalix, sweet eucalyptus	ZUCKERIGER EUCALYPTUS (*m*)
Eucalyptus diversicolor	929	White gum, karri	KARRI (*m*); WEISSER GUMMI-BAUM (*m*)
Eucalyptus emarginata	930	Australian mahogany tree (See *Eucalyptus marginata*)	JARRAHHOLZBAUM (*m*); AUSTRA-LISCHES MAHAGONI (*n*); AUS-TRALISCHER MAHAGONIBAUM (*m*)
Eucalyptus globulus	931	Blue gum, blue eucalyptus; Australian gum	BLAUER GUMMIBAUM (*m*); BLAUER EUCALYPTUS (*m*); EISENVEILCHENBAUM (*m*); FIEBERHEILBAUM (*m*)
Eucalyptus leucoxylon	932	White ironbark tree	EISENRINDENBAUM (*m*)
Eucalyptus marginata	933	Mahogany gum; yarrow tree; jarrah (See *Eucalyptus emarginata*)	JARRAH (*m*); FALSCHER MAHA-GONIBAUM (*m*)
Eucalyptus melanophloia	934	Silver-leaved ironbark tree	SILBERBLÄTTRIGER EISENRINDEN-BAUM (*m*)
Eucalyptus microcorys	935	Australian oak, tallow-wood tree	AUSTRALISCHE EICHE (*f*)
Eucalyptus occidentalis	936	Eucalyptus; malleto-bark tree	MALLETRINDEBAUM (*m*); MALLETO-RINDEBAUM (*m*)
Eucalyptus paniculata	937	Red ironbark tree	ROTER EISENRINDENBAUM (*m*)
Eucalyptus resinifera	938	Red mahogany; forest mahogany	KINOSAFTBAUM (*m*); LERPBAUM (*m*)
Eucalyptus rostrata	939	Red gum; red eucalyptus	ROTER GUMMIBAUM (*m*); ROTER EUCALYPTUS (*m*)
Eucharis grandiflora	940	Amazon lily	AMAZON-LILIE (*f*)
Eugenia aromatica	941	Clove, caryophyllus; fruiting myrtle (See *Caryophyllus aromaticus*)	GEWÜRZNELKE (*f*); KIRSCHMYRTE (*f*)
Eugenia caryophyllata	942	(See *Eugenia aromatica*)	
Eugenia Jambos	943	Rose apple	KIRSCHMYRTE (*f*)
Eugenia pimenta	944	Jamaica pepper; Allspice tree (See *Pimenta officinalis*)	
Eulalia japonica	945	(See *Miscanthus japonica*)	
Euonymus alata	946	Japanese spindle tree	GEFLÜGELTER SPINDELBAUM (*m*)
Euonymus americana	947	American spindle tree	AMERIKANISCHER SPINDELBAUM (*m*)
Euonymus atro-purpureus	948	Wahoo; spindle tree; burning-bush	PFAFFENHÜTCHEN (*n*); SPINDEL-BAUM (*m*); PFAFFENBAUM (*m*); PURPURBLÜTIGER SPIN-DELBAUM (*m*)
Euonymus bungeana	949	Bunge's spindle tree	BUNGES SPINDELBAUM (*m*)
Euonymus europœus	950	Spindletree	PFAFFENHÜTCHEN (*n*); DEUTSCHER BUCHSBAUM (*m*); SPINDEL-BAUM (*m*); SPULBAUM (*m*)
Euonymus hamiltoniana	951	Hamilton's spindle tree	HAMILTONS-SPINDELBAUM (*m*)
Euonymus japonica	952	Japanese spindle tree	JAPANISCHER SPINDELBAUM (*m*)

Botanical Term.	No.	English Term.	German Term.
Euonymus latifolia	953	Broad-leaved spindle tree	BREITBLÄTTRIGER SPINDEL--BAUM (*m*)
Euonymus micrantheus	954	(See *Euonymus bungeana*)	
Euonymus nana	955	Dwarf spindle tree	ZWERGIGER SPINDELBAUM (*m*)
Euonymus vulgaris	956	(See *Euonymus europæus*)	
Eupatorium cannabinum	957	Hemp-agrimony ; Eupatory ; dyer's-weed	WASSERDOST (*m*); WASSERHANF (*m*); WASSERSENF (*m*); HIRSCH--KLEE (*m*)
Eupatorium perfoliatum	958	Indian sage ; ague-weed ; fever-root; thoroughwort ; boneset ; feverwort	KUNIGUNDENKRAUT (*n*) ; WASSER--DOSTEN (*m*)
Eupatorium purpureum	959	Trumpet weed	PURPURROTBLÜTIGES ALPKRAUT (*n*)
Euphorbia	960	spurge ; wolfsmilk, poinsettia, caper spurge	WOLFSMILCH (*f*)
Euphorbia amygdaloides	961	Wood spurge	MANDELBLÄTTRIGE WOLFSMILCH (*f*)
Euphorbiaceæ	962	Euphorbiaceæ (See *Euphorbia*, *Buxus* and *Mercurialis*)	WOLFSMILCHGEWÄCHSE (*n.pl*)
Euphorbia cyparissias	963	Cypress spurge	ZYPRESSENWOLFSMILCH (*f*)
Euphorbia esula	964	Leafy spurge	BLATTREICHE WOLFSMILCH (*f*)
Euphorbia exigua	965	Dwarf spurge	ZWERGWOLFSMILCH (*f*)
Euphorbia helioscopia	966	Sun spurge ; wart-wort ; churn-staff ; cat's-milk ; wolf's milk	WOLFSMILCH (*f*) ; SONNENWOLFS--MILCH (*f*)
Euphorbia hiberna	967	Irish spurge	IRLÄNDISCHE WOLFSMILCH (*f*)
Euphorbia lathyris	968	caper spurge	KLEINES SPRINGKRAUT (*n*) ; MAUL--WURFSKRAUT (*n*)
Euphorbia longana	969	(See *Nephelium longana*)	
Euphorbia paralias	970	Sea spurge	MEERWOLFSMILCH (*f*)
Euphorbia peplis	971	Purple spurge	PURPURNE WOLFSMILCH (*f*)
Euphorbia peplus	972	Petty spurge	KLEINE WOLFSMILCH (*f*)
Euphorbia pilosa	973	Hairy spurge	HAARIGE WOLFSMILCH (*f*)
Euphorbia pilulifera	974	Australian snake-weed ; Australian cat's-hair	AUSTRALISCHE WOLFSMILCH (*f*)
Euphorbia platyphyllos	975	Broad spurge	BREITBLÄTTRIGE WOLFSMILCH (*f*)
Euphorbia pulcherrima	976	(See *Poinsettia pulcherrima*)	
Euphorbia segetalis	977	Portland spurge	
Euphrasia officinalis	978	Eye-bright ; euphrasy ; euphrasia	AUGENTROST (*m*)
Euplectella aspergillum	979	Glass sponge ; Venus' flower-basket	GIESKANNENSCHWAMM (*m*)
Eurotium glaucum	980	Herbarium mould (See *Aspergillus glaucus* and *Aspergillus herbariorum*)	
Eurotium herbariorum	981	(See *Eurotium glaucum*)	
Euryale amazonica	982	(See *Victoria regia*)	
Eusyce	983	(See *Ficus carica*)	
Euterpe oleracea	984	Assai palm ; cabbage tree ; cabbage palm	KOHLPALME (*f*) ; PARAPALME (*f*)
Eutuber	985	Edible truffel (See *Tuber æstivum*)	ECHTE TRÜFFEL (*f*)
Euxolus oleraceus	986	(See *Amaranthus oleraceus*)	
Evonymus	987	(See *Euonymus*)	
Excœcaria sebifera	988	Chinese tallow tree	CHINESISCHER TALGBAUM (*m*)
Exoascus	989	Witch knot fungus	FREISCHLAUCH-SCHEIBENPILZ (*m*)
Exoascus carpini	990	Hornbeam witch knot fungus (See *Taphria carpini*)	HAINBUCHEN-FREISCHLAUCH--SCHEIBENPILZ (*m*)
Exoascus insititiæ	991	Bullace witch knot fungus	SCHLEHEN-FREISCHLAUCH-SCHEI--BENPILZ (*m*)
Exoascus pruni	992	Plum fungus ; plum witch knot fungus	PFLAUMEN-FREISCHLAUCH-SCHEI--BENPILZ (*m*)
Exoascus turgidus	993	Birch witch knot fungus	BIRKEN-FREISCHLAUCH-SCHEIBEN--PILZ (*m*) ; HEXENBESENPILZ (*m*)
Exogonium purga	994	Jalap	JALAPA (*f*) ; JALAPE (*f*) ; JALAPPE (*f*) ; PURGIERWURZEL (*f*) ; JALAPENWINDE (*f*)

Botanical Term.	No.	English Term.	German Term.
Faba alba	995	White bean	WEISSE BOHNE (*f*)
Faba calabarica	996	(See *Physostigma venenosum*)	
Faba de Tonca	997	(See *Dipteryx oppositifolia*)	
Faba Ignatii	998	(See *Strychnos Ignatii*)	
Faba vulgaris	999	Horse bean ; broad bean (See *Vicia faba*)	FUTTERBOHNE (*f*)
Fagara clavaherculis	000	Prickly ash, yellow-wood, angelica tree (See *Xanthoxylum americanum*)	GELBHOLZ (*n*)
Fagara flava	001	Yellow-wood tree ; satin-wood tree	SEIDENHOLZBAUM (*m*) ; SATIN--HOLZBAUM (*m*)
Fagara pterota	002	Jamaica ironwood (See *Zanthoxylum pterota*)	JAMAIKA-EISENHOLZ (*n*)
Fagopyrum esculentum	003	Buck wheat (See *Polygonum fagopyrum*)	BUCHWEIZEN (*m*)
Fagus americana latifolia	004	(See *Fagus ferruginea*)	
Fagus castanea	005	(See *Castanea sativa*)	
Fagus cuprea	006	Copper beech	KUPFERBUCHE (*f*)
Fagus ferruginea	007	American red beech, copper beech	AMERIKANISCHE ROTBUCHE (*f*) : ROTBUCHE (*f*)
Fagus pendula	008	(See *Fagus sylvatica pendula*)	
Fagus pumila	009	(See *Castanea pumila*)	
Fagus purpurea	010	Purple beech	BLUTBUCHE (*f*)
Fagus silvatica	011	Beech, common beech, red beech	BUCHE (*f*) ; ROTBUCHE (*f*) ; GEMEINE BUCHE (*f*)
Fagus sylvatica asplenifolia	012	Fern-leaved beech	FARNBLÄTTRIGE BUCHE (*f*)
Fagus sylvatica pendula	013	Weeping beech	TRAUERBUCHE (*f*) ; HÄNGEBUCHE (*f*)
Fagus sylvatica purpurea	014	Purple-leaved beech	BLUTBUCHE (*f*)
Fagus sylvatica quercoides	015	Oak-leaved beech	EICHENBLÄTTRIGE BUCHE (*f*)
Fagus sylvatica variegata	016	Variegated-leaved beech	BUNTBLÄTTRIGE BUCHE (*f*)
Fatsia aralia papyrifera	017	(See *Fatsia papyrifera*)	
Fatsia japonica	018	Japanese fatsia ; Japanese aralia ; fig-leaf palm (See *Aralia sieboldii*)	JAPANISCHE FATSIE (*f*)
Fatsia papyrifera	019	Formosan fatsia ; Chinese rice paper plant (See also *Fatsia aralia papyrifera* ; *Aralia papyrifera* and *Tetrapanax papyrifer*)	REISPAPIERPFLANZE (*f*)
Ferolia guianensis	020	Satin-wood tree	SATINHOLZBAUM (*m*) ; ATLASHOLZ--BAUM (*m*)
Feronia elephantum	021	Elephant's-apple tree ; feronia	ELEFANTENAPFELBAUM (*m*)
Ferula communis	022	Giant fennel	RUTENKRAUT (*n*) ; GEMEINES STECKENKRAUT (*n*)
Ferula fœtida	023	Stinking fennel (See *Ferula narthes*)	STINKASAND (*m*) ; STINKASANT (*m*)
Ferula galbaniflua	024	Galbanum-resin plant, galbanum	GALBAN (*m*)
Ferula narthes	025	(See *Ferula fœtida*)	
Ferula sumbul	026	Musk, musk-root	SUMBULWURZEL (*f*) ; MOSCHUS--WURZEL (*f*)
Festuca arundinacea	027	Fescue-(grass) (See also *Poa*)	ROHRSCHWINGEL (*m*)
Festuca bromoides	028	Barren fescue-grass (See *Festuca myurus*)	
Festuca calamaria	029	(See *Festuca sylvatica*)	
Festuca duriuscula	030	Hard fescue-grass	HARTER SCHWINGEL (*m*)
Festuca elatior	031	Tall fescue-(grass) ; meadow fescue-(grass) (See *Festuca pratensis*)	WIESEN-SCHWINGEL (*m*)
Festuca flabellata	032	Tussock (Tussac) grass (See *Dactylis cæspitosa* and *Poa flabellata*)	TUSSOCKGRAS (*n*)
Festuca gigantea	033	Giant fescue-grass, great fescue-grass (See *Bromus giganteus*)	RIESEN-SCHWINGEL (*m*)
Festuca heterophylla	034	Various-leaved fescue	VERSCHIEDENBLÄTTRIGER SCHWINGEL (*m*)
Festuca loliacea	035	Spiked fescue	STACHELIGER SCHWINGEL (*m*)

Botanical Term.	No.	English Term.	German Term.
Festuca myurus	036	Rat's-tail fescue-(grass) (See *Festuca bromoides*)	MÄUSESCHWANZ-SCHWINGEL (*m*)
Festuca ovina	037	Sheep's fescue-grass (See *Festuca vivipara*)	RUCHGRAS (*n*); SCHAF-SCHWINGEL (*m*)
Festuca ovina tenuifolia	038	Fine-leaved sheep's fescue-grass	FEINBLÄTTRIGES RUCHGRAS (*n*)
Festuca pratensis	039	Meadow fescue (See *Festuca elatior*)	SCHWINGELGRAS (*n*); WIESEN-SCHWINGEL (*m*)
Festuca rubra	040	Red fescue	ROTER SCHWINGEL (*m*)
Festuca sciuroides	041	Squirrel-tailed fescue grass	EICHHORNSCHWANZIGER SCHWINGEL (*m*)
Festuca sylvatica	042	Reed fescue-(grass); wood fescue-grass (See *Festuca calamaria*)	WALD-SCHWINGEL (*m*)
Festuca uniglumis	043	One-glumed fescue-(grass); single-glumed fescue-grass	EINSPELZIGER SCHWINGEL (*m*)
Festuca vivipara	044	(See *Festuca ovina*)	
Ficaria verna	045	(See *Ranunculus ficaria*)	FEIGWURZ (*f*)
Ficus aoa	046	Samoa banyan tree	AOA-BANYANBAUM (*m*)
Ficus bengalensis	047	Banyan-(tree); Indian fig; prickly pear; banian-(tree) (See *Opuntia Ficus-Indica*)	BANIANE (*m*); INDISCHE FEIGE (*f*); BENGALISCHER BANYANBAUM (*m*)
Ficus benjamina	048	Malayan banyan tree	MALAIISCHER BANYANBAUM (*m*)
Ficus carica	049	Fig-tree	FEIGENBAUM (*m*); ECHTE FEIGE (*f*)
Ficus elastica	050	Gum tree; rubber tree; India-rubber plant; East India rubber tree	GUMMIBAUM (*m*); KAUTSCHUK-FEIGENBAUM (*m*); ASSAM-KAUTSCHUKBAUM (*m*)
Ficus Indica	051	Banyan (See *Opuntia Ficus-Indica*)	
Ficus laccifera	052	Indian lac fig-tree	INDISCHER LACKFEIGENBAUM (*m*)
Ficus macrophylla	053	Australian banyan	AUSTRALISCHER BANYANBAUM (*m*)
Ficus palmata	054	Hand-leaved fig	HANDBLÄTTRIGE FEIGE (*f*)
Ficus religiosa	055	Indian sacred fig-tree; pipul; Pipal; Peepul; Bo tree	HEILIGER FEIGENBAUM (*m*); BO-BAUM (*m*); PIPUL-BAUM (*m*); BODHI-BAUM (*m*)
Ficus retusa	056	Chinese banyan tree	CHINESISCHER BANYANBAUM (*m*)
Ficus saussureana	057	Cow-tree	KUHBAUM (*m*)
Ficus sycomorus	058	Egyptian fig tree	ÄGYPTISCHER FEIGENBAUM (*m*); MAULBEERFEIGENBAUM (*m*); SYKOMORENFEIGENBAUM (*m*); ECHTE SYKOMORE (*f*); MAULBEERFEIGE (*f*)
Ficus urostigma	059		MÖRDERFEIGE (*f*); WÜRZFEIGE (*f*)
Filago gallica	060	Narrow filago	SCHIMMELKRAUT (*n*)
Filago germanica	061	Common cudweed; everlasting; goldylocks; immortelle; filago	IMMERSCHÖN (*n*)
Filago minima	062	Field filago	KLEINES SCHIMMELKRAUT (*n*); FELD-SCHIMMELKRAUT (*n*)
Fistulina hepatica	063	Fistulina -(edible)-fungus; liver agaric	LEBERPILZ (*m*); ZUNGENPILZ (*m*); BLUTSCHWAMM (*m*); ZUNGEN-SCHWAMM (*m*); LEBER-SCHWAMM (*m*)
Fitzgeraldia	064	(See *Cananga odorata*)	
Fitzroya patagonica	065	Winter's bark	PATAGONISCHE ALERCEZYPRESSE (*f*)
Flaveria contrayerba	066	Peruvian flaveria	FLAVERIE (*f*)
Florideæ	067	Flowering algæ (See *Rhodophyceæ*)	BLÜTENTANGE (*m.pl*)
Fœniculum capillaceum	068	Fennel	FENCHEL (*m*)
Fœniculum commune	069	Mediterranean fennel	GEMEINER FENCHEL (*m*)
Fœniculum dulce	070	Sweet fennel, Florence fennel	SÜSSER FENCHEL (*m*)
Fœniculum officinale	071	Fennel	FENCHEL (*m*); GEMEINER FENCHEL (*m*); MUTTERWURZ (*f*)
Fœniculum tingitanum	072	African fennel	AFRIKANISCHER FENCHEL (*m*)
Fœniculum vulgare	073	Fennel; stinking fennel; evil-smelling fennel; common fennel	FENCHEL (*m*); STINKENDER FENCHEL (*m*)

Botanical Term.	No.	English Term.	German Term.
Fomes fomentarius	074	(See *Boletus fomentarius* and *Polyporus fomentarius*)	
Forsythia	075	Forsythia, Japanese golden bell tree	FORSYTHIE (*f*)
Fragaria	076	Strawberry	ERDBEERE (*f*)
Fragaria chilensis	077	Chilean strawberry, Chili strawberry	CHILIERDBEERE (*f*)
Fragaria chilœnsis grandiflora	078	Pine strawberry	GROSSBLÜTIGE CHILENISCHE ERDBEERE (*f*)
Fragaria collina	079	Green pine strawberry	HÜGELERDBEERE (*f*); KNACKERDBEERE (*f*)
Fragaria elatior	080	Hautboy strawberry, hautbois strawberry	HOCHSTENGELIGE ERDBEERE (*f*)
Fragaria vesca	081	Wood strawberry; wild strawberry; hautboy	WILDE ERDBEERE (*f*)
Fragaria virginiana	082	Virginian strawberry, scarlet strawberry	VIRGINISCHE ERDBEERE (*f*)
Frangula caroliniana	083	(See *Rhamnus caroliniana*)	KAROLINISCHER FAULBAUM (*m*)
Frankenia lœvis	084	Smooth sea-heath	FRANKENIE (*f*); FRANKENIENGEWÄCHS (*n*)
Fraxinus americana	085	(American)-white ash	WEISSESCHE (*f*)
Fraxinus anomala	086	Utah ash	UTAH-ESCHE (*f*)
Fraxinus aurea	087	Golden ash	GOLDENE ESCHE (*f*)
Fraxinus chinensis	088	Chinese ash	CHINESISCHE ESCHE (*f*)
Fraxinus crispa	089	Curl-leaved ash	GEKRÄUSELTBLÄTTRIGE ESCHE (*f*)
Fraxinus elonza	090	Elonza ash	ELONZA-ESCHE (*f*)
Fraxinus excelsior	091	(Common)-ash	(GEMEINE)-ESCHE (*f*); EDEL-ESCHE (*f*)
Fraxinus mandschurica	092	Manchurian ash	MANSCHURISCHE ESCHE (*f*)
Fraxinus nigra	093	Black ash	SCHWARZE ESCHE (*f*)
Fraxinus oregona	094	Oregon ash	OREGON-ESCHE (*f*)
Fraxinus ornus	095	Flowering ash, Manna ash	MANNAESCHE (*f*); GEMEINE BLUMENESCHE (*f*); MANNA-FLECHTE (*f*)
Fraxinus parvifolia	096	Small-leaved ash	KLEINBLÄTTRIGE ESCHE (*f*)
Fraxinus pendula	097	Weeping ash	TRAUERESCHE (*f*)
Fraxinus quadrangulata	098	Blue ash	BLAUE ESCHE (*f*)
Fraxinus rotundifolia	099	Round-leaved flowering ash	RUNDBLÄTTRIGE ESCHE (*f*)
Fraxinus syriaca	100	Syrian ash	SYRISCHE ESCHE (*f*)
Fraxinus tamarixifolia	101	Mastic-leaved ash	MASTIXBLÄTTRIGE ESCHE (*f*)
Fraxinus viridis	102	Green ash	GRÜNESCHE (*f*)
Fraxinus xanthoxyloides	103	Yellow-wood-leaved ash	GELBHOLZBLÄTTRIGE ESCHE (*f*)
Freycinetia Banksii	104	Kiekie-(shrub)	KLETTER-SCHRAUBENPALME (*f*)
Fritillaria imperialis	105	Crown imperial	KAISERMANTEL (*m*); KÖNIGS-KRONE (*f*)
Fritillaria meleagris	106	Fritillary; common fritillary; snake's head-(fritillary)	MARMORLILIE (*f*); SCHACH-BLUME (*f*)
Fuchsia coccinea	107	Common fuchsia, ear drops, lady's ear drops	FUCHSIA (*f*): FUCHSIE (*f*)
Fucus	108	Seaweed, sea-tang; bladder-wrack	SEETANG (*m*); MEERESALGE (*f*); MEERTANG (*m*); LEDERTANG (*m*)
Fucus crispus	109	(See *Chrondrus crispus*)	
Fucus spinosus	110	White seaweed	LEDERTANG (*m*)
Fucus vesiculosus	111	Bladder-wrack, sea-wrack, sea-weed	BLASENTANG (*m*)
Fumaria capreolata	112	Pink-flowered fumitory	RANKENDER ERDRAUCH (*m*)
Fumariaceœ	113	Fumariaceæ (See *Corydalis* and *Fumaria*)	ERDRAUCHGEWÄCHSE (*n.pl*)
Fumaria officinalis	114	(Common)-fumitory; fumaria	ERDRAUCH (*m*); FELDRAUCH (*m*); FELDRAUTE (*f*); GRÜNWURZEL (*f*)
Fungus chirurgorum	115	(See *Globaria bovista*)	
Gagea lutea	116	Yellow gagea; yellow star of Bethlehem	GELBSTERN (*m*)
Gagea sylvatica	117	Wood yellow star of Bethlehem	WALDGOLDSTERN (*m*)
Galactodendron utile	118	Cow-tree (See *Brosimum galactodendron*)	KUHBAUM (*m*)

Botanical Term.	No.	English Term.	German Term.
Galanthus latifolius	119	Broad-leaved snowdrop	SCHNEEGLÖCKCHEN (*n*); FRÜH--LINGSGLOCKE (*f*)
Galanthus nivalis	120	Snowdrop, fair maids of February	SCHNEEGLÖCKCHEN (*n*); MÄRZ--BLUME (*f*); MÄRZBECHER (*m*)
Gale	121	(See *Myrica gale*)	
Galega officinalis	122	Goat's rue; blue-flowered galega	GEISSRAUTE (*f*); PESTILENZ--KRAUT (*n*)
Galega officinalis alba	123	Goat's rue; white-flowered galega	WEISSE GEISSRAUTE (*f*)
Galeobdolon luteum	124	Archangel (See *Lamium galeobdolon*)	GELBE TAUBNESSEL (*f*); GOLD--NESSEL (*f*)
Galeopsis ladanum	125	Red hemp-nettle	ACKER-DAUN (*m*)
Galeopsis ochroleuca	126	Downy hemp-nettle	HANFNESSEL (*f*); DAUN (*m*); HOHLZAHN (*m*)
Galeopsis tetrahit	127	Common hemp-nettle, iron-wort	HANFNESSEL (*f*); LIEBERSCHES KRAUT (*n*); NESSELKRAUT (*n*); HOHLZAHN (*m*)
Galipea cusparia	128	Angostura (See *Cuspa trifóliata*)	ANGOSTURABAUM (*m*)
Galium	129	galium	LABKRAUT (*n*); LIEBKRAUT (*n*)
Galium anglicum	130	Wall galium	ENGLISCHES LABKRAUT (*n*)
Galium aparine	131	Cleavers; Goose-grass; bed-straw; catch-weed; clivers; hariff; scratch-weed	KLEBKRAUT (*n*)
Galium boreale	132	Northern galium	NORDISCHES LABKRAUT (*n*)
Galium cruciatum	133	Crosswort; maywort; may-weed; mugweed	KREUZLABKRAUT (*n*)
Galium mollugo	134	Hedge galium, great hedge bed-straw	GEMEINES LABKRAUT (*n*)
Galium palustre	135	Marsh galium	SUMPF-LABKRAUT (*n*)
Galium saxatile	136	Heath galium	HEIDE-LABKRAUT (*n*)
Galium tricorne	137	Corn galium	KORN-LABKRAUT (*n*)
Galium uliginosum	138	Swamp galium	MOOR-LABKRAUT (*n*)
Galium verum	139	Yellow bedstraw; lady's bed-straw; scratch weed	MEGERKRAUT (*n*); LIEBFRAUEN--BETTSTROH (*n*); LABKRAUT (*n*)
Galtonia candicans	140	White galtonia; giant hyacinth	WEISSE GALTONIE (*f*); RIESEN-HYAZINTHE (*f*)
Garcinia gambogia	141	Gamboge-(tree)	KAMBODSCHA-GUMMIGUTTBAUM (*m*)
Garcinia mangostana	142	Mangosteen tree	MANGOSTANBAUM (*m*); MANGO--STINBAUM (*m*)
Garcinia morella	143	Ceylon gamboge tree (See *Garcinia gambogia*)	CEYLON-GUMMIGUTTBAUM (*m*)
Gardenia citriodora	144	Citron-scented Gardenia (See *Mitriostigma axillaris*)	ZITRONENDUFTENDER GELBSCHO--TENBAUM (*m*)
Gardenia florida	145	Jasmine; yellow jasmine	GELBER JASMIN (*m*)
Gastridium lendigerum	146	Nitgrass; awned nitgrass	
Gaultheria procumbens	147	Wintergreen, mountain tea; tea-berry; box-berry; partridge-berry; deer-berry; creeping wintergreen, Canada tea	BERGTEE (*m*)
Geanthemum rhizanthum	148	(See *Anona rhizantha*)	
Geaster hygrometricus	149	Earth-star fungus (See *Astracus stellatus*)	WETTER-ERDSTERN (*m*)
Gelidium corneum	150	White seaweed	HART-ROTALGE (*f*)
Gelsemium sempervirens	151	Carolina jasmine; gelsemium; yellow jasmine; wild wood-bine (See *Jasminum nudiflorum*)	GIFTJASMIN (*m*); GELSEMIEN (*n*)
Genipa americana	152	(West Indian)-genipap tree	MARMELDOSE (*f*)
Genista	153	Broom, furze, whin, gorse (See *Cytisus scoparius* and *Cytisus canariensis*)	GINSTER (*m*); ERDPHRIEM (*m*)
Genista anglica	154	Needle whin; needle furze; petty whin; needle genista	ENGLISCHER GINSTER (*m*)
Genista canariensis	155	Rosewood	ROSENHOLZ (*n*)
Genista hispanica	156	Spanish gorse	SPANISCHER GINSTER (*m*)
Genista pilosa	157	Hairy genista	BEHAARTER GINSTER (*m*)

Botanical Term.	No.	English Term.	German Term.
Genista tinctoria	158	Woad; dyer's weed; dyer's broom; green-weed; green-broom; green-wood; dyer's genista; dyer's greenweed	FÄRBERGINSTER (*m*)
Gentiana acaulis	159	Old gentianella	STENGELLOSER ENZIAN (*m*); BLAUER FINGERHUT (*m*)
Gentiana amarella	160	Autumn gentian; felwort	GEERKRAUT (*n*); HERBST-GEER-KRAUT (*n*)
Gentiana asclepiadea	161	Swallow-wort	SPEERENSTICH (*m*)
Gentiana bavarica	162	Bavarian gentian	BAYRISCHER ENZIAN (*m*)
Gentiana campestris	163	Field gentian	FELDENZIAN (*m*)
Gentiana cruciata	164	Cross-wort	KREUZKRAUT (*n*); KREUZWURZ (*f*); KREUZ-ENZIAN (*m*)
Gentiana lutea	165	Gentian, yellow gentian; bitter-root; bitter-wort	ENZIAN (*m*); BITTERWURZEL (*f*); ENZIANWURZEL (*f*); BITTER-WURZ (*f*)
Gentiana nivalis	166	Small gentian	SCHNEE-GEERKRAUT (*n*)
Gentiana pneumonanthe	167	Marsh gentian, windflower	LUNGENBLUME (*f*); LUNGEN-ENZIAN (*m*)
Gentiana verna	168	Little gentian; spring gentian	FRÜHLINGS-GEERKRAUT (*n*)
Georgina juarezi	169	Cactus dahlia	KAKTUSDAHLIA (*f*); GEORGINE (*f*); ZWERGGEORGINE (*f*); LILIPUTGEORGINE (*f*); RÖHR-ENBLÜTIGE DAHLIA (*f*)
Geranium	170	Geranium; crane's-bill; stork's-bill	KRANICHSCHNABEL (*m*); STORCH-SCHNABEL (*m*)
Geranium columbinum	171	Long-stalked geranium	TAUBEN-STORCHSCHNABEL (*m*)
Geranium dissectum	172	Dove's-foot; cut-leaved geranium	SCHLITZBLÄTTRIGER STORCH-SCHNABEL (*m*)
Geranium lucidum	173	Shining geranium	GLÄNZENDER STORCHSCHNABEL (*m*)
Geranium maculatum	174	Geranium; crane's-bill, stork's-bill; (American)-alum root	FLECKSTORCHSCHNABEL (*m*)
Geranium molle	175	Dove's-foot crane's-bill; dove's-foot geranium	WEICHER STORCHSCHNABEL (*m*)
Geranium phæum	176	Dusky geranium	ROTBRAUNER STORCHSCHNABEL (*m*)
Geranium pratense	177	Geranium; meadow crane's-bill; meadow geranium	KRANICHHALS (*m*)
Geranium pusillum	178	Small-flowered geranium	ZWERG-STORCHSCHNABEL (*m*)
Geranium pyrenaicum	179	Mountain geranium	PYRENÄISCHER STORCHSCHNABEL (*m*)
Geranium robertianum	180	Stinking crane's-bill; herb-robert; bird's-eye	ROBERTSKRAUT (*n*); RUPRECHTS-KRAUT (*n*)
Geranium rotundifolium	181	Round-leaved geranium	RUNDBLÄTTRIGER STORCH-SCHNABEL (*m*)
Geranium sanguineum	182	Blood-red geranium; crane's-bill	BLUTROTER STORCHSCHNABEL (*m*)
Geranium sylvaticum	183	Wood geranium, crane's-bill	STORCHSCHNABEL (*m*)
Geum canadense	184	Blood-root	KANADISCHE NELKENWURZ (*f*)
Geum rivale	185	water avens; purple avens	SUMPFNELKE (*f*)
Geum urbanum	186	Wood avens; common avens; herb bennet	NELKENWURZ (*f*); IGELKRAUT (*n*)
Gingko biloba	187	Ginkga-tree, gingko tree, maiden-hair tree (See *Salisburya adiantifolia*)	JAPANISCHER NUSSBAUM (*m*); (ZWEILAPPIGER)-FÄCHERBLATT-BAUM (*m*)
Ginkgo	188	(See *Gingko*)	
Githago segetum	189	(See *Lychnis githago*)	
Gladiolus communis	190	(Common)-gladiolus; gladiole, corn flag, sword lily	NETZSCHWERTEL (*m*); SIEGWURZ (*f*); SCHWERTLILIE (*f*)
Gladiolus palustris	191	Swamp corn flag	SUMPF-SIEGWURZ (*f*); SCHWERTEL (*m*); ALLERMANNSHARNISCH (*m*)
Gladiolus primulinus	192	Maid of the Mist, primrose-coloured corn-flag	PRIMELFARBIGE SCHWERTLILIE (*f*)
Gladiolus pyramidatus	193	(See *Watsonia rosea*)	PYRAMIDEN-SCHWERTLILIE (*f*)
Glaucium flavum	194	Yellow horned poppy (See also *Glaucium luteum*)	GELBER HORNMOHN (*m*)
Glaucium glaucium	195	Horn poppy	HORNMOHN (*m*)

Botanical Term.	No.	English Term.	German Term.
Glaucium luteum	196	Horn-(ed) poppy; sea poppy; yellow horned poppy (See also *Glaucium flavum*)	HORNMOHN (*m*)
Glaux maritima	197	Sea milkwort; black saltwort; milk tare	MUTTERKRAUT (*n*); MILCHKRAUT (*n*)
Glechoma hederacea	198	Ground-ivy; gill; ale-hoof (See *Nepeta glechoma*)	GUNDERMANN (*m*); GUNDEL--REBE (*f*)
Gleditschia aquatica	199	Water locust	WASSER-GLEDITSCHIE (*f*)
Gleditschia inermis	200	(See *Gleditschia aquatica*)	WEHRLOSE GLEDITSCHIE (*f*)
Gleditschia macracantha	201	(See *Gleditschia triacanthos*)	GROSSDORNIGE GLEDITSCHIE (*f*)
Gleditschia monosperma	202	(See *Gleditschia aquatica*)	EINSAMIGE GLEDITSCHIE (*f*)
Gleditschia triacanthos	203	Holly, honey locust	CHRISTUSDORN (*m*); GLEDIT--SCHIE (*f*); DREIDORNIGE GLEDITSCHIA (*f*); ZUCKER--SCHOTENBAUM (*m*)
Globaria bovista	204	Giant puff-ball (See *Lycoperdon giganteum*)	KUGELIGER RIESENSTÄUBLING (*m*); WOLFSRAUCH (*m*); WUNDSCHWAMM (*m*)
Globularia	205	Globe daisy	KUGELBLUME (*f*)
Glyceria	206	Sweet-grass	ENTENGRAS (*n*); SÜSSGRAS (*n*); SCHWADEN (*m*)
Glyceria aquatica	207	Reedy sweet-grass (See *Poa aquatica*)	WASSER-SCHWADEN (*m*)
Glyceria Borreri	208	Borrer's sweet-grass (See *Schlerochloa Borreri*)	BORRERS-SCHWADEN (*m*)
Glyceria distans	209	Reflexed sweet-grass (See *Poa distans* and *Schlerochloa distans*)	ZERSTREUTLIEGENDER SCHWADEN (*m*)
Glyceria fluitans	210	Floating sweet-grass (See *Poa fluitans*)	HIMMELSTAU (*m*); MANNAGRAS (*n*); FLUTENDER SCHWADEN (*m*)
Glyceria loliacea	211	Dwarf rigid sweet-grass (See *Triticum loliaceum*)	RAIGRASÄHNLICHER SCHWADEN (*m*)
Glyceria maritima	212	Creeping sea sweet-grass (See *Schlerochloa maritima*)	MEERSTRANDS-SCHWADEN (*m*)
Glyceria procumbens	213	Procumbent sweet-grass (See *Schlerochloa procumbens*)	LIEGENDER SCHWADEN (*m*)
Glyceria remota	214	Remote sweet-grass	ENTFERNTBLÜTIGER SCHWADEN (*m*)
Glyceria rigida	215	Hard sweet-grass (See *Schlerochloa rigida*)	STEIFER SCHWADEN (*m*)
Glycine chinensis	216	(See *Wistaria chinensis*)	
Glycine fructescens	217	(See *Wistaria fructescens*)	
Glycine sinensis	218	(Chinese)-glycine (See *Wistaria chinensis*)	
Glycyrrhiza glabra	219	Liquorice plant, Spanish liquorice	SÜSSHOLZ (*n*); LAKRITZE (*f*); SÜSSHOLZWURZEL (*f*); LAK--RITZENWURZEL (*f*); SÜSS--HOLZSTAUDE (*f*); LAKRITZ--ENSTAUDE (*f*)
Glyptostrobus heterophyllus	220	(See *Taxodium heterophyllum*)	
Gnaphalium	221	Cudweed	RUHRKRAUT (*n*)
Gnaphalium dioicum	222	(See *Antennaria dioica*)	
Gnaphalium leontopodium	223	Edelweiss (See *Leontopodium alpinum*)	EDELWEISS (*n*)
Gnaphalium luteo-album	224	Cat's-foot flower; Jersey cud-weed	(ROTES)-KATZENPFÖTCHEN (*n*); FALBES RUHRKRAUT (*n*)
Gnaphalium supinum	225	Dwarf cudweed	ALPEN-RUHRKRAUT (*n*)
Gnaphalium sylvaticum	226	Wood cudweed	WALD-RUHRKRAUT (*n*)
Gnaphalium uliginosum	227	Marsh cudweed	TEICH-RUHRKRAUT (*n*); SUMPF--RUHRKRAUT (*n*)
Gnetacea	228	Joint firs; gnetacea (See *Ephedra distachya*)	GNETUMGEWÄCHS (*n*)
Gnetum	229	(See *Ephedra distachya*)	
Goldfussia	230	(See *Strobilanthes*)	GOLDFUSSIE (*f*)
Gomphia decora	231	Button flower; South American button flower	KNOPFBLUME (*f*); GOMPIE (*f*)
Gonolobus condurango	232	Condurango bark (See *Marsdenia condurango*)	

Botanical Term.	No.	English Term.	German Term.
Goodyera repens	233	Creeping goodyera, rattlesnake plantain, adder's violet	KRIECHENDES SPALT-KNABEN--KRAUT (*n*)
Gossypium acuminatum	234	Cotton plant	BAUMWOLLSTAUDE (*f*)
Gossypium bahma	235	Egyptian cotton	ÄGYPTISCHE BAUMWOLLSTAUDE (*f*)
Gossypium barbadense	236	Cotton plant; Barbados cotton plant; common cotton-(wool) plant	BAUMWOLLSTAUDE (*f*); BARBA--DOS-BAUMWOLLSTAUDE (*f*)
Gossypium herbaceum	237	Cotton plant	BAUMWOLLSTAUDE (*f*)
Gouania domingensis	238	Jamaican chaw-stick; Gouania	GOUANIE (*f*)
Grammatophyllum speciosum	239	Letter leaf	SCHRIFTBLATT (*n*)
Gratiola	240	Hedge hyssop (See also *Hyssopus officinalis*)	GNADENKRAUT (*n*); PURGIER--KRAUT (*n*)
Gravia paradoxa	241	(See *Monstera acuminata*)	
Guaiacum officinale	242	Pockwood tree, lignum-vitæ; guaiacum (See *Lignum sanctum*)	POCKHOLZ (*n*); FRANZOSENHOLZ (*n*); GUAJAKBAUM (*m*); POCK--HOLZBAUM (*m*)
Guajacum	243	(See *Guaiacum*)	
Guarea grandifolia	244	(West Indian)-alligator-wood tree (See also *Melia*)	GROSSBLÄTTRIGE GUAREA (*f*)
Guatteria laurifolia	245	White-lancewood tree	LORBEERBLÄTTRIGE GUATTERIE (*f*)
Guatteria virgata	246	Lancewood tree	GUATTERIE (*f*)
Guazuma ulmifolia	247	(American)-guazuma	ULMENBLÄTTRIGER GUAZUMA--BAUM (*m*)
Gymnocladus canadensis	248	Kentucky coffee-tree	KANADISCHER SCHUSSERBAUM (*m*)
Gymnocladus chinensis	249	Soap tree	CHINESISCHER SCHUSSERBAUM (*m*)
Gymnogramma chrysophylla	250	Naked fern	NACKTFARN (*m*)
Gymnospermæ	251	Gymnospermæ	NACKTSAMIGE PFLANZEN (*f.pl*)
Gymnosporangium juniperinum	252	Juniper cluster-cup	WACHOLDER-GITTERROST (*m*)
Gymnothrix latifolia	253	(See *Pennisetum latifolium*)	
Gynerium argenteum	254	Pampas grass (See *Cortaderia argentea*)	PAMPASGRAS (*n*)
Gypsophila muralis	255	Wall gypsophila; chalk-wort	MAUER-GIPSKRAUT (*n*)
Gyromitra esculenta	256	Mushroom	LORCHEL (*f*); FALTENMORCHEL (*f*)
Habenaria	257	Orchis	ORCHIDEE (*f*); HABENARIE-ORCHE (*f*)
Habenaria albida	258	Small orchis	WEISSLICHE HABENARIE-ORCHE (*f*)
Habenaria bifolia	259	Lesser butterfly orchis; Rein orchis	KLEINE SCHMETTERLINGORCHE (*f*)
Habenaria chlorantha	260	Great butterfly orchis	GROSSE SCHMETTERLINGORCHE (*f*)
Habenaria conopsea	261	Fragrant orchis	WOHLRIECHENDE ORCHE (*f*)
Habenaria intacta	262	Dense-spiked orchis	UNVERLETZTE ORCHE (*f*)
Habenaria viridis	263	Frog orchis; green orchis	GRÜNE HABENARIE-ORCHE (*f*)
Hæmanthus	264	Blood-flower; arnica; blood lily; Red Cape tulip	BLUTBLUME (*f*)
Hæmanthus toxicaria	265	African bloow-flower	GIFT-BLUTBLUME (*f*); HÄMAN--THUS (*m*)
Hæmatoxylon campechianum	266	Campeachy wood, logwood-(tree)	BLAUHOLZ (*n*); CAMPECHEHOLZ (*n*); BLUTHOLZ (*n*); BLUT--BAUM (*m*); BLAUHOLZBAUM (*m*)
Hagenia abyssinica	267	Cusso; Koussa; Brayera (See *Brayera anthelmintica*)	KOSSOBAUM (*m*); KUSSOBAUM (*m*); KUSSALA (*f*)
Halesia tetraptera	268	Snowdrop tree; Silver-bell	SCHNEEGLÖCKCHENBAUM (*m*)
Halidrys siliquosa	269	Sea-oak	MEEREICHEN-TANG (*m*)
Halimus portulacoides	270	Sea purslane (See *Atriplex hortensis*)	MEERPORTULAK (*m*)
Halymenia	271	(See *Iridæa edulis* and *Schizy--menia edulis*)	
Hamamelis virginiana	272	Witch-hazel, winter-bloom, striped alder, tobacco-wood	ZAUBERHASEL (*f*)
Harpagophytum procumbens	273	(Cape)-grapple plant	LIEGENDES WOLLSPINNENFRUCHT--KRAUT (*n*)
Hedeoma pulegioides	274	Squaw mint, American penny-royal	AMERIKANISCHE POLEI (*f*); AMERIKANISCHER POLEI (*m*)

Botanical Term.	No.	English Term.	German Term.
Hedera helix	275	Ivy ; bindwood ; common ivy	EPHEU (*m*) ; EFEU (*m*) ; GEMEINER EFEU (*m*) ; IMMERGRÜN (*n*)
Hedysarum alhagi	276	Agul	MANNAKLEE (*m*)
Hedysarum gyrans	277	French honeysuckle, shaking saintfoin	OSTINDISCHER SCHWINGKLEE (*m*)
Hedysarum onobrychis	278	Cock's head, common saintfoin (See *Onobrychis sativa*)	SÜSSKLEE (*m*)
Helenium autumnale	279	(North American)-helenium ; autumnal helenium ; Helen flower, sneeze-wort, sneezeweed (See *Inula helenium*)	ALANT (*m*)
Heleocharis	280	(See *Eleocharis*)	
Helianthemum canadense	281	Rock-rose, frost-wort, frost-weed	KANADISCHES SONNENRÖSCHEN (*n*)
Helianthemum canum	282	Hoary rock-rose	SONNENGÜNSEL (*m*)
Helianthemum fumana	283	Wild rose	HEIDENROSE (*f*)
Helianthemum guttatum	284	Spotted rock-rose (See *Tuberaria guttata*	GETÜPFELTES SONNENRÖSCHEN (*n*)
Helianthemum polifolium	285	White rock-rose	PULVRIGES SONNENRÖSCHEN (*n*)
Helianthemum pulverulentum	286	(See *Helianthemum polifolium*)	
Helianthemum vulgare	287	Common rock rose ; sun rose	SONNENRÖSCHEN (*n*)
Helianthus annuus	288	Common-(annual)-sunflower	SONNENBLUME (*f*) ; SONNENROSE (*f*)
Helianthus multiflorus	289	Sunflower	VIELBLÜTIGE SONNENBLUME (*f*)
Helianthus tuberosus	290	Jerusalem artichoke ; (Italian)-girasole	SAUBROT (*n*)
Helichrysum arenarium	291	Everlasting, immortelle, yellow everlasting	IMMERSCHÖN (*n*) ; IMMORTELLE (*f*)
Helichrysum bracteatum	292	Everlasting, immortelle	STROHBLUME (*f*)
Heliopsis lævis	293	Smooth heliopsis	SONNENAUGE (*n*)
Heliotropium	294	Vanilla heliotrope (See also *Vanilla*)	VANILLEN-HELIOTROP (*m*) ; VA--NILLENSTRAUCH (*m*)
Heliotropium peruvianum	295	Peruvian heliotrope ; heliotrope ; turnsole ; girasole ; cherry-pie (See also *Tournefortia*)	HELIOTROP (*m*)
Helipterum	296	Immortelle-flower, Australian everlasting (See *Waitzia aurea*)	IMMORTELLE (*f*)
Helleborus	297	Winter aconite ; hellebore ; Christmas rose ; Lenten rose	NIESWURZ (*f*)
Helleborus fœtidus	298	Stinking hellebore ; bear's foot ; fœtid (fetid) hellebore ; setter-wort ; oxheal	STINKENDE NIESWURZ (*f*)
Helleborus niger	299	Christmas rose ; black hellebore	SCHWARZE NIESWURZ (*f*) ; SCHNEEROSE (*f*) ; WEIHNACHTS--WURZEL (*f*)
Helleborus odorus	300	Fragrant hellebore	WOHLRIECHENDE NIESWURZ (*f*)
Helleborus orientalis	301	Lenten rose	ORIENTALISCHE NIESWURZ (*f*)
Helleborus viridis	302	Green hellebore ; bear's-foot ; (See also *Veratrum viride*)	GROSSE NIESWURZ (*f*)
Helminthia echioides	303	Ox-tongue	OCHSENZUNGE (*f*)
Helodea canadensis	304	(See *Elodea candensis* and *Anacharis alsinastrum*)	
Helvella crispa	305	Common helvel-(fungus) ; turban-top	LORCHEL (*f*)
Helvella leucophæna	306	Yellow turban-top	HERBSTMORCHEL (*f*)
Hemibasidii	307	Brand fungus (See *Tilletia* and *Ustilago*)	BRANDPILZE (*m. pl*) ; HEMIBASI--DIEN (*f.pl*)
Hepatica	308	Liverwort (See *Anemone hepatica* and *Marchantia polymorpha*)	LEBERBLUME (*f*) ; DREILAPPIGES MARZBLÜMCHEN (*n*) ; LEBER--MOOS (*n*)
Hepatica angulosa	309	(See *Hepatica*)	
Hepatica nobilis	310	Liver-leaf, liver-wort, crystal-wort	LEBERBLÜMCHEN (*n*)
Hepatica triloba	311	Liverwort	LEBERKRAUT (*n*)
Heracleum sphondylium	312	(Giant)-cow parsnip ; hog-weed ; Heracleum ; cartwheel flower	BÄRENKLAU (*f*)

Botanical Term.	No.	English Term.	German Term.
Herminium monorchis	313	Musk orchis	EINKNOLLIGE RAGWURZ (*f*); RAG--ORCHE (*f*)
Herniaria glabra	314	Rupture-wort	HORNKRAUT (*n*); BRUCHKRAUT (*n*); TAUSENDKORN (*n*)
Hesperis matronalis	315	Dame's violet; damask violet; double-(white)-rocket; rocket; sweet rocket; Dame's rocket; night-smelling rocket	NACHTVIOLE (*f*); MATRONEN--BLUME (*f*)
Hesperis tristis	316	Night-scented stock	KILTE (*f*)
Heuchera americana	317	Alum root; American sanicle	HEUCHERA (*f*)
Hevea brasiliensis	318	Para India-rubber plant	PARAGUMMIPFLANZE (*f*); PARA--KAUTSCHUKBAUM (*m*)
Hibiscus	319	Mallow; musk mallow, rose mallow, hemp mallow, Malabar rose; blacking plant	EIBISCH (*m*)
Hibiscus abelmoschus	320	Hibiscus (See *Malva moschata*)	MOSCHUS-EIBISCH (*m*)
Hibiscus rosa-sinensis	321	Blacking plant	ROSEN-EIBISCH (*m*); CHINESI--SCHER EIBISCH (*m*)
Hibiscus syriacus	322	Syrian mallow; Rose of Sharon (see *Althæa frutex*)	SYRISCHER ROSENEIBISCH (*m*)
Hicoria	323	Hickory (See also *Carya*)	WEISSER WALNUSSBAUM (*m*); HICKORYNUSS (*f*)
Hicoria glabra	324	Hickory-nut (See *Juglans glabra*)	FERKELNUSS (*f*); GLATTE HICK--ORYNUSS (*f*)
Hieracium	325	Hawkweed: golden mouse-ear	HABICHTSKRAUT (*n*)
Hieracium alpinum	326	Alpine hawkweed	ALPENHABICHTSKRAUT (*n*)
Hieracium cerinthoides	327	Honeywort hawkweed	CERINTHEÄHNLICHES HABICHTS--KRAUT (*n*)
Hieracium murorum	328	Wall hawkweed	MAUER-HABICHTSKRAUT (*n*)
Hieracium pilosella	329	Mouse-ear hawkweed	MÄUSEOHR-HABICHTSKRAUT (*n*)
Hieracium prenanthoides	330	Prenanth hawkweed	PRENANTHESÄHNLICHES HABICHTS--KRAUT (*n*)
Hieracium sabaudum	331	Savoy hawkweed	SAVOYER-HABICHTSKRAUT (*n*)
Hieracium umbellatum	332	Umbellate hawkweed	WIESEN-HABICHTSKRAUT (*n*)
Hierochlœ borealis	333	(Northern)-holy grass; northern sacred grass	NORDISCHES MARIENGRAS (*n*)
Hierochlœ odorata	334	Sweet-scented holy-grass; sweet-scented sacred grass	WOHLRIECHENDES MARIENGRAS (*n*)
Himanthalia lorea	335	Sea-thong	RIEMENTANG (*m*)
Himantoglossum hircinum	336	(See *Orchis hircina*)	
Hippocastanum vulgare	337	(See *Aesculus hippocastanum*)	
Hippocrepis comosa	338	Horse-shoe vetch; hippocrepis	HUFEISENKLEE (*m*)
Hippomane mancinella	339	Machineel-(tree)	MANZANILLABAUM (*m*)
Hippophaë rhamnoides	340	Sea buckthorn; sallow-thorn	SANDDORN (*m*); SEEDORN (*m*); GEMEINER (SANDDORN) SEE--DORN (*m*)
Hippuris vulgaris	341	Mare's-tail	GEMEINER TANNENWEDEL (*m*)
Hirneola auricula-Judas	342	Judas' ear-(fungus); Jew's ear (See *Auricularia auricula judæ*)	
Holcus lanatus	343	Yorkshire fog: woolly holcus; variegated soft grass	WOLLIGES HONIGGRAS (*n*)
Holcus mollis	344	Soft holcus; soft grass, velvet-grass	HONIGGRAS (*n*)
Holosteum umbellatum	345	Umbellate holosteum	DOLDENBLÜTIGE SPURRE (*f*)
Hordeum jubatum	346	Squirrel-tail grass	EICHHORNSCHWANZÄHNLICHE GERSTE (*f*)
Hordeum maritimum	347	Sea barley	STRAND-GERSTE (*f*)
Hordeum murinum	348	Wall barley; waybent; squirrel-tail grass	MÄUSEGERSTE (*f*); MAUERGERSTE (*f*)
Hordeum pratense	349	Meadow barley	WIESEN-GERSTE (*f*)
Hordeum sylvaticum	350	Wood barley	WALD-GERSTE (*f*)
Hordeum vulgare	351	Barley	GERSTE (*f*)
Hordeum zeocrithon	352	Battledore-barley	PFAUENGERSTE (*f*)
Horminium pyrenaicum	353	Horminium clary (See *Salvia horminium*)	DRACHENMAUL (*n*)
Hottonia palustris	354	Water-violet; feather foil	WASSERFEDER (*f*)
Hoya bella	355	(See *Hoya carnosa*)	

Botanical Term.	No.	English Term.	German Term.
Hoya carnosa	356	Wax-flower, honey-plant (See *Hoya bella*)	PORZELLANBLUME (*f*)
Humulus lupulus	357	Hop; hops	HOPFEN (*m*); HECKENHOPFEN (*m*)
Hura crepitans	358	Sand-box tree	SANDBÜCHSENBAUM (*m*)
Hutchinsia petræa	359	Rock Hutchinsia	STEIN-GEMSKRESSE (*f*)
Hyacinthus amethystinus	360	Spanish hyacinth	SPANISCHE HYAZINTHE (*f*)
Hyacinthus orientalis	361	Common hyacinth	HYAZINTHE (*f*), ORIENTALISCHE HYAZINTHE (*f*)
Hyacinthus orientalis albulus	362	Roman hyacinth	WEISSE ORIENTALISCHE HYAZINTHE (*f*); RÖMISCHE HYAZINTHE (*f*)
Hydnum erinaceum	363	Satyr's beard-(fungus)	IGELSCHWAMM (*m*)
Hydnum repandum	364	Urchin of the woods-(fungus)	STOPPELSCHWAMM (*m*)
Hydrangea hortensis	365	Common hydrangea	HORTENSIE (*f*): GARTEN WASSER--STRAUCH (*m*): HORTENSIA (*f*)
Hydrangea opuloides	366	(See *Hydrangea hortensis*)	
Hydrastis canadensis	367	Goldenseal; turmeric	GELBES BLUTKRAUT (*n*); GELB--WURZ (*f*)
Hydrocharideæ	368	Hydrocharideæ (See *Hydrocharis* and *Stratiotes*)	NIXENKRÄUTER (*n.pl*)
Hydrocharis asiatica	369	Oriental frog-bit	ÖSTLICHER FROSCHBISS (*m*)
Hydrocharis Morsus-ranæ	370	Frog-bit	GEMEINER FROSCHBISS (*m*)
Hydrocotyle asiatica	371	Flukewort, flukeweed, sheep's bane	WASSERNABEL (*m*); ÖSTLICHER WASSERNABEL (*m*)
Hydrocotyle vulgaris	372	Penny-wort; marsh penny-wort; white-rot, flukeweed	WASSERNABEL (*m*); GEMEINER WASSERNABEL (*m*)
Hygodium palmatum	373	Climbing fern	KLETTERFARN (*m*)
Hymenæa courbaril	374	(See *Robinia pseudacacia*)	
Hymenophyllum tunbridgense	375	Filmy fern; Tunbridge fern	TUNBRIDGER HAUTFARN (*m*)
Hyoscyamus niger	376	Henbane; stinking nightshade; hyoscyamus	BILSENKRAUT (*n*)
Hypericum androsæmum	377	Tutsan, sweet amber (See *Androsæmum officinale*)	KONRADSKRAUT (*n*)
Hypericum calycinum	378	Rose of Sharon, Aaron's beard; large flowered St. John's wort; large-flowered hypericum	GROSSKELCHIGES HARTHEU (*n*)
Hypericum dubium	379	Imperforate hypericum	UNBESTIMMTES HARTHEU (*n*)
Hypericum elodes	380	Marsh hypericum; marsh St. John's wort	SUMPF-HARTHEU (*n*)
Hypericum hirsutum	381	Hairy St. John's wort; hairy hypericum	RAUHHAARIGES HARTHEU (*n*)
Hypericum humifusum	382	Trailing hypericum	NIEDERLIEGENDES HARTHEU (*n*)
Hypericum linarifolium	383	Flax-leaved hypericum	FLACHSBLÄTTRIGES HARTHEU (*n*); LEINKRAUTBLÄTTRIGES HART--HEU (*n*)
Hypericum montanum	384	Mountain St. John's wort; mountain hypericum	BERG-HARTHEU (*n*)
Hypericum perforatum	385	St. John's wort; perforated St. John's wort; common hypericum; All-Saints-wort	HARTHEU (*n*); JOHANNISKRAUT (*n*)
Hypericum pulchrum	386	Small upright St. John's wort; slender hypericum	SCHÖNES HARTHEU (*n*)
Hypericum quadrangulum	387	Square-stalked St. John's wort; square-stalked hypericum	VIERKANTIGES HARTHEU (*n*)
Hyphæne thebaica	388	Egyptian doom-palm	DOUMPALME (*f*)
Hypholoma	389	(See *Agaricus fascicularis*)	
Hypochæris glabra	390	Glabrous hypochære	KAHLES FERKELKRAUT (*n*)
Hypochæris maculata	391	Spotted hypochære (See *Achyrophorus maculatus*)	GEFLECKTES FERKELKRAUT (*n*)
Hypochæris radicata	392	Long-rooted cat's ear; common cat's ear	KURZWURZELIGES FERKELKRAUT (*n*)
Hyppophæ	393	(See *Hippophaë*)	
Hyssopus officinalis	394	Hyssop (See also *Gratiola*)	GNADENKRAUT (*n*); ISOP (*m*); YSOP (*m*)
Hysterium macrosporum	395	Pine fungus (See *Hysterium nervisequium*)	RITZENSCHORF (*m*)
Hysterium nervisequium	396	(See *Hysterium macrosporum*)	
Iberis amara	397	Bitter Candytuft	BITTERE SCHLEIFERBLUME (*f*)
Iberis coronaria	398	Candytuft, rocket candytuft	BAUMSENF (*m*); BAUERNSENF (*m*)

Botanical Term.	No.	English Term.	German Term.
Iberis sempervirens	399	Candytuft; common, evergreen candytuft	BAUMSENF (*m*); IMMERGRÜNE SCHLEIFERBLUME (*f*)
Iberis sempervirens corifolia	400	Dwarf candytuft	ZWERG-SCHLEIFERBLUME (*f*)
Iberis umbellata	401	(Common)-candytuft	DOLDEN-SCHLEIFERBLUME (*f*)
Ignatia multiflora	402	(See *Strychnos Ignatii*, and also *Faba Ignatii*)	
Ilex aquifolium	403	(Common)-holly	STECHPALME (*f*); HÜLSENPALME (*f*); CHRISTDORN (*m*); GE--MEINER HÜLSSTRAUCH (*m*); HULST (*m*)
Ilex decidua	404	Deciduous holly	SOMMERGRÜNER HÜLSSTRAUCH (*m*)
Ilex ferox	405	Hedgehog holly	HÜLSSTRAUCH (*m*); DORNIGER HÜLSSTRAUCH (*m*)
Ilex glabra	406	Ink berry	KAHLER HÜLSSTRAUCH (*m*)
Ilex japonica	407	(See *Berberis japonica* and *Mahonia japonica*)	
Ilex paraguayensis	408	Brazilian holly; Paraguay tea; mate, maté, Brazil tea	MATE (*f*); MATENSTRAUCH (*m*)
Ilex verticillata	409	(North American)-black alder-(shrub); winter berry	WINTERBAUM (*m*); QUIRLIGER HÜLSSTRAUCH (*m*)
Ilicineæ	410	(See *Aquifoliaceæ* and *Ilex*)	ILICINEEN (*f.pl*)
Illecebrum verticillatum	411	Illecebrum	QUIRLBLÜTIGE KNORPELBLUME (*f*)
Illicium anisatum	412	Star anise, aniseed tree	STERNANIS (*m*)
Illicium religiosum	413	Japanese star anise	JAPANISCHER STERNANIS (*m*)
Illipe butyracea	414	(See *Bassia butyracea*)	
Illipe latifolia	415	(See *Bassia latifolia*)	
Illipe malabrorum	416	Illipe tree; Illupie tree	ILLIPÉBAUM (*m*); GALLERTBAUM (*m*)
Illipe Parkii	417	(See *Butyrospermum Parkii*)	
Impatiens biflora	418	(See *Impatiens fulva*)	ZWEIBLÜTIGES SPRINGKRAUT (*n*)
Impatiens fulva	419	Orange balsam (See *Impatiens biflora*)	GELBBRAUNES SPRINGKRAUT (*n*)
Impatiens Noli-me-tangere	420	Yellow balsam; touch-me-not	RÜHR-MICH-NICHT-AN (*n*)
Impatiens sultani	421	Zanzibar balsam	OSTAFRIKANISCHES SPRINGKRAUT (*n*); SULTANS-SPRINGKRAUT (*n*)
Imperatoria ostruthium	422	Masterwort, felon (grass) wort	KAISERWURZ (*f*); MEISTERWURZ (*f*); MEISTERWURZEL (*f*)
Imphee holcus saccharatus	423	Sweet soft-grass	AFRIKANISCHES ZUCKERROHR (*n*)
Indigofera tinctoria	424	Indigo plant	INDIGOSTRAUCH (*m*)
Inga vera	425	Indian kokra-wood tree	GRENADILLHOLZBAUM (*m*); KOKOS--HOLZBAUM (*m*); ROTEBEN--HOLZBAUM (*m*)
Inula conyza	426	Plowman's spikenard; ploughman's spikenard (See *Conyza squarrosa* and *Inula squarrosa*)	
Inula crithmoides	427	Golden samphire	FENCHELÄHNLICHER ALANT (*m*)
Inula dysenterica	428	Fleabane (inule) inula (See *Pulicaria dysenterica*)	RUHR-FLOHKRAUT (*n*)
Inula helenium	429	Elecampane; inula; scab-wort; horse-heal; elf-wort (See *Helenium autumnale*)	ALANT (*m*)
Inula Oculus Christi	430	Christ's eye	CHRISTUS-AUGE (*n*)
Inula pulicaria	431	Small fleabane (See *Pulicaria vulgaris*)	
Inula salicina	432	Willow leaved (inule) inula	WEIDENBLÄTTRIGES FLOHKRAUT (*n*)
Inula squarrosa	433	Ploughman's spikenard (See *Inula conyza* and *Conyza squarrosa*)	SPIKE (*f*)
Ipomœa batatas	434	Sweet potato (See *Batatas*)	BATATE (*f*)
Ipomœa coccinea	435	Scarlet ipomœa, morning glory, American bell-bind, moon-creeper	ROTE BATATE (*f*)
Ipomœa purga	436	Jalap	PURGIERWURZEL (*f*)

Botanical Term.	No.	English Term.	German Term.
Iridæa edulis	437	Brown dulse; edible dulse (See *Halymenia* and *Schizymenia edulis*)	HAUTALGE (*f*); HAUTTANG (*m*); SEEBAND (*n*)
Iris florentina	438	Florentine iris; orris plant (See also *Iris germanica*)	
Iris fœtidissima	439	Stinking iris; gladwyn; blue iris; gladwin; gladdon; fetid (fœtid) iris; roast-beef plant	STINKENDE SCHWERTLILIE (*f*)
Iris germanica	440	German iris, flag iris (See also *Iris florentina*)	FLAMME (*f*); HIMMELSLILIE (*f*)
Iris pallida	441	(See *Iris florentina*)	BLASSE SCHWERTLILIE (*f*)
Iris Pseud-acorus	442	Yellow (water)-flag; fleur-de-lis; flower-de-luce, iris, flower-de-lis; corn-flag; yellow iris; fleur-de-luce	ILGE (*f*); WASSERSCHWERTLILIE (*f*)
Iris robinsoniana	443	(See *Moræa robinsoniana*)	ROBINSONS-SCHWERTLILIE (*f*)
Isatis tinctoria	444	Dyer's woad; Indian indigo plant; woad plant	FÄRBERWAID (*m*); WAID (*m*), WIED (*m*)
Isoetes lacustris	445	Quillwort, european quillwort	SUMPF BRACHSENKRAUT (*n*)
Isolepsis	446	Mud-rush (See also *Scirpus*)	SIMSE (*f*)
Isolepsis fluitans	447	Floating mud-rush	FLUTENDE SIMSE (*f*)
Isolepsis gracilis	448	(See *Scirpus nodosus* and *Scirpus tenella*)	ZARTE SIMSE (*f*)
Isolepsis holoschœnus	449	Round cluster-headed mud rush (See *Scirpus holoschœnus*)	KNOPFGRASARTIGE SIMSE (*f*)
Isolepsis Savii	450	Savi's mud-rush (See *Scirpus Savii*)	SAVIS-SIMSE (*f*)
Isolepsis setacea	451	Bristle-stemmed mud-rush (See *Scirpus setaceus*)	
Itea virginica	452	Virginian itea; virginian rosemary-willow	ROSMARINWEIDE (*f*); VIRGINISCHER MOORSTRAUCH (*m*)
Jacaranda brasiliana	453	Jacaranda (Palisander) wood; (Brazilian)-rosewood; Amaranth wood	JAKARANDAHOLZ (*n*); PALISANDERHOLZ (*n*); PURPURHOLZ (*n*)
Jacaranda cœrulea	454	Mimosa-leaved ebony tree; green ebony tree (See also *Brosimum aubletti* and *Dalbergia nigra*)	JAKARANDABAUM (*m*)
Jacaranda ovalifolia	455	Green ebony-(tree) (See *Brya ebenus*)	GRÜNEBENHOLZ (*n*); UNECHTES POCKHOLZ (*n*)
Jacobæa elegans	456	(See *Senecio Jacobæa* and *Senecio elegans*)	
Jacquinia armillaris	457	West-Indian bracelet-wood	JACQUINIA (*f*)
Jambosa caryophyllus	458	(See *Caryophyllus aromaticus* and *Eugenia aromatica*)	
Jasione montana	459	Sheep's scabious; sheep's-bit	BERG-JASIONE (*f*)
Jasminum fruticans	460	Shrubby trefoil; yellow jasmine	FRUCHTTRAGENDER JASMIN (*m*)
Jasminum grandiflorum	461	Jessamine, jasmine	JASMIN (*m*); GROSSBLÜTIGER JASMIN (*m*)
Jasminum nudiflorum	462	Winter (jessamine) jasmine; yellow (jessamine) jasmine (See *Gelsemium sempervirens*)	JASMIN (*m*); GELBER JASMIN (*m*)
Jasminum officinale	463	White (jessamine) jasmine	JASMIN (*m*); WEISSER JASMIN (*m*)
Jasminum revolutum	464	Yellow-flowering, evergreen (jessamine) jasmine	GELBBLÜHENDER IMMERGRÜNER JASMIN (*m*)
Jasminum sambac	465	White-flowering, evergreen (Jessamine) jasmine	SAMBAC (*m*); ARABISCHER JASMIN (*m*); NACHTBLUME (*f*)
Jateorrhiza palmata	466	Calumba (See *Cocculus palmatus*)	KOLOMBOWURZEL (*f*)
Jatropha curcas	467	Physic nut	PURGIERNUSS (*f*)
Jatropha manihot	468	Manioc tree, cassava tree, tapioca tree (See also *Manihot utilissima*)	
Jeffreyi	469	(See *Dodecatheon media*)	
Jonesia asoca	470	Asoca tree; Indian asoca tree	ASOKABAUM (*m*)
Joxylon pomiferum	471	(See *Maclura aurantiaca*)	
Jubœa spectabilis	472	Honey palm	HONIGPALME (*f*)

Botanical Term.	No.	English Term.	German Term.
Juglans cinerea	473	Grey walnut, butternut tree; white walnut tree	GRAUNUSS (*f*); GRAUER WALNUSSBAUM (*m*); BUTTERNUSS (*f*); ASCHGRAUER NUSSBAUM (*m*)
Juglans glabra	474	(See *Hicoria glabra*)	
Juglans nigra	475	American walnut; black walnut	SCHWARZE WALNUSS (*f*); AMERIKANISCHER NUSSBAUM (*m*); SCHWARZER NUSSBAUM (*m*)
Juglans oblonga	476	(See *Juglans cinerea*)	
Juglans regia	477	Walnut tree	WALNUSSBAUM (*m*); NUSSBAUM (*m*); WELSCHER NUSSBAUM (*m*); GEMEINER WALNUSSBAUM (*m*); ECHTER WALNUSSBAUM (*m*); KÖNIGLICHER WALNUSSBAUM (*m*)
Junceæ	478	Junceæ (See *Juncus, Luzula* and *Narthecium*)	HALBGRÄSER (*n.pl*)
Juncus	479	Rush	BINSE (*f*)
Juncus acutus	480	Sharp rush	SCHARFE BINSE (*f*)
Juncus alpinus	481	Alpine rush	ALPEN-BINSE (*f*)
Juncus articulatus	482	Jointed rush	GEGLIEDERTE BINSE (*f*)
Juncus balticus	483	Baltic rush	BALTISCHE BINSE (*f*)
Juncus biglumis	484	Two-flowered rush	ZWEIBLÜTIGE BINSE (*f*)
Juncus bufonius	485	Toad rush	KRÖTENBINSE (*f*)
Juncus capitatus	486	Capitate rush	KOPFBINSE (*f*); KOPFBLÜTIGE BINSE (*f*)
Juncus castaneus	487	Chestnut rush	KASTANIENBINSE (*f*)
Juncus communis	488	Common rush	GEMEINE BINSE (*f*)
Juncus compressus	489	Round-fruited rush	RUNDFRÜCHTIGE BINSE (*f*); ZUSAMMENGEDRÜCKTE BINSE (*f*)
Juncus conglomeratus	490	Common rush; button-weed	KNÄUELBINSE (*f*); KNOPFBINSE (*f*)
Juncus effusus	491	Soft rush	WEICHE BINSE (*f*); FLATTERBINSE (*f*)
Juncus effusus spiralis	492	(See *Juncus effusus*)	FLATTERBINSE (*f*)
Juncus filiformis	493	Thread rush	FADEN-BINSE (*f*)
Juncus gerardi	494	Mud rush	SCHLAMMBINSE (*f*)
Juncus glaucus	495	Hard rush	MEERGRÜNE BINSE (*f*)
Juncus maritimus	496	Sea rush	MEERBINSE (*f*); MEERSTRANDSBINSE (*f*)
Juncus obtusiflorus	497	Obtuse rush	STUMPFBLÜTIGE BINSE (*f*)
Juncus pygmæus	498	Dwarf rush	ZWERGBINSE (*f*)
Juncus squamosus	499	Heath rush	MOOSBINSE (*f*)
Juncus supinus	500	Bog rush	SUMPF-BINSE (*f*)
Juncus sylvaticus	501	Great woodrush (See also *Luzula sylvatica*)	WALDBINSE (*f*)
Juncus tenuis	502	Slender rush	ZARTE BINSE (*f*)
Juncus trifidus	503	Highland rush	DREISPITZIGE BINSE (*f*)
Juniperus	504	Juniper	KRONAWETTBAUM (*m*); WACHOLDER (*m*)
Juniperus bermudiana	505	Bermuda cedar, pencil cedar	FLORIDA-ZEDERNHOLZBAUM (*m*); BERMUDISCHER SADEBAUM (*m*)
Juniperus capensis	506	(See *Thuja juniperoides*)	
Juniperus communis	507	Juniper; common juniper (See *Oxycedrus*)	WACHOLDER (*m*); GEMEINER WACHOLDER (*m*); MACHANDELBAUM (*m*); KRANEWIT (*m*)
Juniperus drupacea	508	Syrian juniper (See *Caryocedrus*)	PFLAUMEN-WACHOLDER (*m*)
Juniperus macrocarpa	509	Large fruited Juniper	GROSSFRÜCHTIGER WACHOLDER (*m*)
Juniperus nana	510	Dwarf juniper	ZWERG-WACHOLDER (*m*)
Juniperus oxycedrus	511	Cade (See *Oxycedrus*)	KADDIG (*m*); KADDICH (*m*); ZEDERNWACHOLDER (*m*)
Juniperus phœnicea	512	Phœnician juniper	ZYPRESSENWACHOLDER (*m*); ROTFRÜCHTIGER SADEBAUM (*m*)
Juniperus procera	513	East-African cedar	OSTAFRIKANISCHE ZEDER (*f*); BERGWACHOLDER (*m*); AFRIKANISCHER SADEBAUM (*m*)

Botanical Term.	No.	English Term.	German Term.
Juniperus sabina	514	Rock cedar; shrubby red cedar; savin	SEVENBAUM (*m*); SADEBAUM (*m*); SEPENBAUM (*m*); GEMEINER SADEBAUM (*m*); SABINER-BAUM (*m*); SABINE (*f*); SÄBENBAUM (*m*)
Juniperus sabina cupressifolia	515	Cypress-leaved savin	ZYPRESSENBLÄTTRIGER SADEBAUM (*m*)
Juniperus sabina tamariscifolia	516	Tamarisk-leaved savin	TAMARISKENBLÄTTRIGER SADE-BAUM (*m*)
Juniperus virginiana	517	Virginian juniper; pencil (red)-cedar; cigar-box cedar	BLEISTIFTHOLZ (*n*); FALSCHES ZEDERNHOLZ (*n*); VIRGIN-ISCHER SADEBAUM (*m*); ZIGARRENKISTCHENHOLZ (*n*); VIRGINISCHER WACHOLDER (*m*); VIRGINISCHE ZEDER (*f*); ROTE ZEDER (*f*)
Kadsura	518	Kadsura	KUGELFADEN (*m*)
Kalmia angustifolia	519	American laurel, mountain laurel, sheep laurel, swamp laurel	KLEINBLÄTTRIGE KALMIE (*f*); LORBEER-ROSE (*f*)
Kalmia latifolia	520	Mountain laurel; American laurel; calico bush	KALMIE (*f*); LÖFFELBAUM (*m*)
Karatas plumieri	521	(See *Bromelia karatas*)	
Kerria japonica	522	Kerria, Japanese rose, Jew's mallow (See *Corchorus japonica*)	JAPANISCHE HONIGROSE (*f*); RANUNKELSTRAUCH (*m*); JAPANISCHE KERRIE (*f*)
Khaya senegalensis	523	Senegal mahogany, African mahogany	AFRIKANISCHES MAHAGONI (*n*); AFRIKANISCHER MAHAGONI-BAUM (*m*)
Klopstockia cerifera	524	Wax palm (See also *Ceroxylon andicolum*)	
Knautia arvensis	525	Field scabious	KNOPFKRAUT (*n*); KRÄTZKRAUT (*n*); KOPFGRINDKRAUT (*n*)
Kniphofia aloides	526	Red-hot poker plant (See *Tritoma uvaria*)	KNIPHOFIA (*f*); HYAZINTHENALOE (*f*)
Kobresia caricina	527	Kobresia (See also *Carex remota*)	KOBRESIE (*f*)
Kochia scoparia	528	Summer cypress, belvedere, mock cypress	KOCHIE (*f*)
Kœleria cristata	529	Crested Kœleria	KÖLERIE (*f*); KAMMSCHMIELE (*f*)
Krameria ixina	530	Savanilla rhatany	RATANHIAWURZEL (*f*)
Krameria triandra	531	Peruvian rhatany	RATANHIAWURZEL (*f*); PERU-RATANHIAWURZEL (*f*)
Labiatæ	532	Labiatæ, labiate plants (See *Mentha, Nepeta, Salvia, Melittis, Thymus, Origanum, Galeopsis, Stachys, Ajuga, Scutellaria*)	RACHENBLÜTLER (*m.pl*)
Laburnum alpinum	533	Alpine laburnum, Scotch laburnum	ALPENGOLDREGEN (*m*)
Laburnum ramentaceum	534	Dwarf laburnum	NIEDRIGER GOLDREGEN (*m*)
Laburnum vulgare	535	Common laburnum, golden chain (See *Cytisus laburnum*)	GEMEINER GOLDREGEN (*m*); ECHTER GOLDREGEN (*m*); GOLDREGEN (*m*); WELSCHE LINSE (*f*)
Lachnanthes tinctoria	536	(American)-redroot (See also *Ceanothus americanus*)	
Lactarius	537	Orange-agaric	REIZKER (*m*)
Lactuca alpina	538	Purple-flowered lettuce (See *Mulgedium alpinum*)	ALPENLATTICH (*m*)
Lactuca muralis	539	Wall lettuce	MAUERLATTICH (*m*)
Lactuca perennis	540	Perennial lettuce	BLAUER LATTICH (*m*)
Lactuca saligna	541	Least lettuce; willow lettuce	WEIDENBLÄTTRIGER LATTICH (*m*)
Lactuca sativa	542	Common lettuce (See *Lactuca scariola sativa*)	SALATPFLANZE (*f*); LATTICH (*m*)
Lactuca scariola	543	Prickly lettuce	ZAUNLATTICH (*m*); WILDER ZAUN-LATTICH (*m*); SKARIOL (*m*); LEBERDISTEL (*f*)

Botanical Term.	No.	English Term.	German Term.
Lactuca scariola sativa	544	Cabbage lettuce (See *Lactuca sativa*)	KOPFLATTICH (*m*); KOPFSALAT (*m*); KRAUTSALAT (*m*); STAUDENSALAT (*m*); GARTEN-SALAT (*m*); LATTICH (*m*)
Lactuca tuberosa	545	Blue-flowered lettuce	KNOLLEN-LATTICH (*m*)
Lactuca virosa	546	Acrid lettuce; lettuce opium	GIFTLATTICH (*m*)
Lagenaria vulgaris	547	Bottle-gourd, trumpet gourd (See *Cucurbita lagenaria*)	FLASCHENKÜRBIS (*m*)
Lagerstrœmia Flos-Reginæ	548	Queen's flower	KÖNIGINBLUME (*f*)
Lagerstrœmia Indica	549	(Chinese)-crape myrtle, Indian lilac	INDISCHE LAGERSTROEMIE (*f*)
Lagetta lintearia	550	West Indian lace-bark; Lagetta (See *Linodendron lagetta*)	WESTINDISCHER SPITZENBAUM (*m*)
Lagurus ovatus	551	Hare's-tail grass	HASENSCHWANZ (*m*)
Laminaria digitata	552	Tangle; seaweed; oarweed; edible tangle	FLÜGELTANG (*m*)
Laminaria saccharina	553	Tangle-seaweed	RIEMENTANG (*m*)
Lamium album	554	White dead-nettle	TAUBNESSEL (*f*)
Lamium amplexicaule	555	Henbit	STENGELUMFASSENDE TAUB-NESSEL (*f*)
Lamium galeobdolon	556	Archangel; dead-nettle; yellow archangel (See *Galeobdolon luteum*)	GELBE TAUBNESSEL (*f*)
Lamium maculatum	557	Dwarf dead nettle; spotted dead-nettle	GEFLECKTE TAUBNESSEL (*f*)
Lamium purpureum	558	Red dead-nettle; purple-flowered dead-nettle	ROTE TAUBNESSEL (*f*)
Languncularia racemosa	559	Mangrove tree	MANGROVE (*f*)
Lantana camara	560	(West Indian)-lantana; Jamaica mountain sage; Surinam tea plant; Camara	WANDELBLÜTE (*f*)
Lapageria albiflora	561	(See *Lapageria rosea alba*)	WEISSBLÜTIGE LAPAGERIE (*f*)
Lapageria rosea	562	Chilean bell-flower	CHILENISCHE LAPAGERIE (*f*)
Lapageria rosea alba	563	White-flowered Chilean bell-flower (See *Lapageria albiflora*)	WEISSBLÜTIGE CHILENISCHE LAPA-GERIE (*f*)
Lapsana communis	564	Nipplewort; lapsane; cuckoo-sorrel, dock-cress	RAINKOHL (*m*)
Larix americana	565	American larch, Hackmatack, black larch, tamarack (See *Larix pendula* and *Larix occidentalis*)	AMERIKANISCHE LÄRCHE (*f*)
Larix decidua	566	(See *Larix europæa*)	GEMEINE LÄRCHE (*f*)
Larix europæa	567	Larch; deciduous larch, larch-fir (See *Pinus Larix*; *Larix decidua*; *Larix excelsa*)	LÄRCHE (*f*); LÄRCHENTANNE (*f*); GEMEINE LÄRCHE (*f*); STEIN-LÄRCHE (*f*); JOCHLÄRCHE (*f*); EUROPÄISCHE ZEDER (*f*)
Larix excelsa	568	(See *Larix europæa*)	
Larix japonica	569	(See *Larix leptolepsis*)	
Larix ledebourii	570	Russian larch	RUSSISCHE LÄRCHE (*f*)
Larix leptolepsis	571	Japanese larch (See *Larix japonica*)	JAPANISCHE LÄRCHE (*f*)
Larix occidentalis	572	(See *Larix americana*)	
Larix pendula	573	Black larch, Hackmatack (See *Larix americana*)	HÄNGELÄRCHE (*f*)
Larix sibirica	574	Siberian larch	SIBIRISCHE LÄRCHE (*f*)
Larrea mexicana	575	American creosote plant	KREOSOTSTRAUCH (*m*)
Laserpitium	576	Laser-wort (See *Silphium*)	LASERPFLANZE (*f*); LASERKRAUT (*n*)
Lasiandra	577	(See *Tibouchina elegans*)	
Lastrea cristata	578	(See *Nephrodium cristatum* and *Aspidium cristatum*)	
Lastrea dilatata	579	(See *Nephrodium spinulosum*)	
Lastrea Felix-mas	580	Male fern (See *Aspidium* and *Dryopteris filix-mas marginalis*, and *Nephrodium filix-mas*)	JOHANNISWURZEL (*f*)
Lastrea oreopteris	581	Sweet mountain fern (See *Nephrodium montanum*)	

Botanical Term.	No.	English Term.	German Term.
Lastrea recurva	582	(See *Nephrodium æmulum*)	
Lastrea thelypteris	583	Marsh fern (See *Nephrodium thelypteris*)	
Latania	584	Bourbon palm	SAMTPALME (*f*)
Lathrœa squamaria	585	Toothwort	ZAHNWURZ (*f*)
Lathyrus aphaca	586	Yellow vetchling	RANKEN-PLATTERBSE (*f*)
Lathyrus hirsutus	587	Rough pea	RAUHHAARIGE PLATTERBSE (*f*)
Lathyrus latifolius	588	(See *Lathyrus sylvestris platyphyllus*)	BREITBLÄTTRIGE PLATTERBSE (*f*)
Lathyrus macrorrhizus	589	Tuberous pea	GROSSWURZELIGE PLATTERBSE (*f*)
Lathyrus magellanicus	590	Lord Anson's pea	LORD ANSON'S PLATTERBSE (*f*)
Lathyrus maritimus	591	Sea pea	MEERSTRANDS-PLATTERBSE (*f*)
Lathyrus niger	592	Black pea	SCHWARZE PLATTERBSE (*f*)
Lathyrus nissolia	593	Grass vetchling	BLATTLOSE PLATTERBSE (*f*)
Lathyrus odoratus	594	Sweet pea	WOHLRIECHENDE PLATTERBSE (*f*)
Lathyrus palustris	595	Marsh pea	SUMPF-PLATTERBSE (*f*)
Lathyrus pratensis	596	Common meadow vetchling; meadow pea	PLATTERBSE (*f*); WIESEN-PLATT-ERBSE (*f*)
Lathyrus sylvestris-(platyphyllus)	597	Everlasting pea (See *Lathyrus latifolius*)	(BREITBLÄTTRIGE)-WALD-PLATT-ERBSE (*f*)
Lathyrus tingitanus	598	Tangier pea	TANGERISCHE PLATTERBSE (*f*)
Lathyrus tuberosus	599	Earth-nut; ground-nut; heath-pea; heath nut; earth pea	ERDEICHEL (*f*)
Laurelia Bovæ-Zelandiæ	600	(New-Zealand)-laurelia	LAURELIE (*f*)
Laurentia erinoides	601	(African)-laurentia	LAURENTIE (*f*)
Laurus camphora	602	Camphor laurel; (Chinese or Japanese)-camphor laurel (See *Dryobalanops camphora*)	KAMPFERBAUM (*m*)
Laurus nobilis	603	Bay-tree; laurel-tree; sweet laurel; sweet bay tree; noble laurel; poet's laurel, victor's laurel	LORBEER (*m*); EDLER LORBEER-BAUM (*m*)
Lavandula officinalis	604	Lavender (See *Lavandula vera*)	LAVENDEL (*m*)
Lavandula spica	605	Spikenard, nard, spike lavender	NARDENKRAUT (*n*); SPIKE (*f*)
Lavandula vera	606	(See *Lavandula officinalis*)	LAVENDEL (*m*)
Lavatera arborea	607	Tree-mallow; shrub lavatera	MALVENBAUM (*m*)
Lavatera olbia	608	Tree lavatera, tree mallow	LAVATERE (*f*)
Lavatera trimestris	609	Rose tree-mallow, rose lavatera	PAPPELROSE (*f*)
Lawsonia alba	610	Henna plant	HENNA (*f*); HENNASTRAUCH (*m*)
Lecanora tartarea	611	Cudbear; tartarean moss; European tartarean moss (See also *Roccella tinctoria*)	LACKFLECHTE (*f*)
Lecythea	612	Rust	ROST (*m*)
Lecythea caprearum	613	Birch rust; goat-willow rust (See *Uredo betulina* and *Melampsora betulina*)	BIRKENROST (*m*)
Lecythea populina	614	Poplar rust	PAPPELROST (*m*)
Lecythea rosæ	615	Rose rust	ROSENROST (*m*)
Lecythea ruborum	616	Bramble rust	BROMBEERROST (*m*)
Lecythea saliceti	617	Willow rust	WEIDENROST (*m*)
Lecythes ollaria	618	(Brazilian)-sapucaia-nut tree	TOPFBAUM (*m*); TOPFFRUCHT-BAUM (*m*); BRASILISCHER SAPUCAJANUSSBAUM (*m*)
Lecythes zabucajo	619	(Brazilian)-sapucaia-nut tree	ZABUCAJONUSSBAUM (*m*); ANNA-HOLZBAUM (*m*); SAPUCAJA-NUSSBAUM (*m*)
Ledum latifolium	620	Labrador tea	BREITBLÄTTRIGER PORST (*m*); MOTTENKRAUT (*n*); WANZEN-KRAUT (*n*)
Ledum palustre	621	Marsh tea; wild rosemary, marsh rosemary	PORSCH (*m*); PORST (*m*); MOTT-ENKRAUT (*n*); WILDER ROS-MARIN (*m*); PORSCHKRAUT (*n*); KIENPORST (*m*); SUMPF-PORSCH (*m*); SUMPFPORST (*m*)
Lemna arrhiza	622	Rootless duckweed (See *Wolfia arrhiza*)	WURZELLOSE WASSERLINSE (*f*)
Lemna gibba	623	Gibbous duckweed	WASSERLINSE (*f*)

Botanical Term.	No.	English Term.	German Term.
Lemna minor	624	Duckweed ; lesser duckweed	MEERLINSE (*f*) ; WASSERLINSE (*f*) ; KLEINE WASSERLINSE (*f*)
Lemna polyrrhiza	625	Many-rooted duckweed ; greater duckweed (See *Spirodela polyrrhiza*)	VIELWURZELIGE WASSERLINSE (*f*)
Lemna trisulca	626	Single-rooted duckweed ; ivy-leaved duckweed	ENTENGRÜTZE (*f*)
Lens esculenta	627	Lentil (See *Ervum lens*)	LINSE (*f*)
Lentibulariæ	628	Lentibulariæ (See *Pinguicula* and *Utricularia*)	FETTKRÄUTER (*n.pl*)
Leonotis leonurus	629	Lion's tail, lion's ear (See *Leonurus cardiaca*)	SCHARLACHROTE LEONOTIS (*f*) : LÖWENSCHWANZ (*m*)
Leontice Alberti	630	Lion's turnip	TRAPP (*m*)
Leontice leontopetalum	631	Lion's leaf	LÖWENBLUMENBLATT (*n*) ; LÖWEN--TRAPP (*m*)
Leontodon autumnalis	632	Autumnal hawkbit (See *Leontodon taraxacum* and *Apargia autumnalis*)	KUHBLUME (*f*)
Leontodon hirtus	633	Lesser hawkbit	KLEINHAARIGE KUHBLUME (*f*)
Leontodon hispidus	634	Common hawkbit	STEIFHAARIGE KUHBLUME (*f*)
Leontodon taraxacum	635	Dandelion (See also *Leontodon autumnalis*)	LÖWENZAHN (*m*) ; KUHBLUME (*f*)
Leontopodium alpinum	636	Edelweiss ; lion's foot (See *Gnaphalium leontopodium*)	EDELWEISS (*n*)
Leonurus cardiaca	637	Motherwort (See *Leonotis leonurus*)	HERZGESPANN (*n*) ; LÖWEN--SCHWANZ (*m*) ; MUTTERKRAUT (*n*)
Lepanthes sanguinea	638	West Indian, red-flowered lepanthes orchid	LEPANTHES-ORCHE (*f*)
Lepargyrea argentea	639	(See *Elægnus argentea*)	
Lepidium	640	Peppergrass ; cress ; pepperwort	PFEFFERGRAS (*n*) ; KRESSE (*f*) ; PFEFFERKRAUT (*n*)
Lepidium campestre	641	Mithridate pepperwort	FELDKRESSE (*f*)
Lepidium draba	642	Hoary cress	STENGELUMFASSENDE KRESSE (*f*)
Lepidium latifolium	643	Dittander	BREITBLÄTTRIGE KRESSE (*f*)
Lepidium ruderale	644	Narrow-leaved cress	SCHUTT-KRESSE (*f*)
Lepidium sativum	645	Common cress ; pepperwort ; garden cress	PFEFFERGRAS (*n*) ; KRESSE (*f*) ; GARTENKRESSE (*f*)
Lepidium Smithii	646	Smith's cress	SMITHS-KRESSE (*f*)
Lepidozamia peroffskyana	647	(See *Macrozamia denisonii*)	
Lepiota	648	(See *Agaricus procerus*)	
Leptamnium virginianum	649	Beechdrops	KREBSWURZ (*f*)
Lespedeza capitata	650	Bush clover ; Japanese clover	KOPFLESPEDEZIE (*f*)
Leucadendron argenteum	651	Wittebroom ; (Cape)-silver-tree	KAP-SILBERBAUM (*m*)
Leucocrinum montanum	652	(Colorado)-sand lily ; Californian soapwort	COLORADO-SANDLILIE (*f*)
Leucoium æstivum	653	Summer snowflake, summer snowdrop	SOMMER-KNOTENBLUME (*f*)
Leucoium vernum	654	Spring snowflake ; snowdrop	MÄRZGLÖCKCHEN (*n*) ; MÄRZ--BECHER (*m*)
Leucojum	655	(See under *Leucoium*)	
Leucopogon richei	656	(See *Leucopogon verticillatus*)	
Leucopogon verticillatus	657	(Australian)-native currant (See *Leucopogon richei*)	AUSTRALISCHE JOHANNISBEERE (*f*)
Levisticum officinale	658	Lovage ; sea parsley (See *Ligusticum scoticum* and *Ligusticum levisticum*)	LIEBSTÖCKEL (*m*) ; LIEBSTOCK (*m*) ; GEBRÄUCHLICHER LIEB--STÖCKEL (*m*)
Lewisia rediviva	659	(North American)-bitter root ; spatlum ; bitterwort	KALIFORNISCHE AUFERSTEHUNGS--PFLANZE (*f*)
Leycestera formosa	660	Himalayan honeysuckle ; flowering nutmeg	SCHÖNE LEYCESTERE (*f*)
Liatris elegans	661	Button snake root ; blazing star	PRACHTSCHARTE (*f*)
Liatris odoratissima	662	Vanilla plant, deer's tongue	VANILLENPFLANZE (*f*)
Liatris scariosa	663	Rattlesnake's master ; snake-root	PRACHTSCHARTE (*f*)
Libocedrus decurrens	664	Californian cedar ; Incense cedar ; white cedar	FLUSSZEDER (*f*) ; KALIFORNISCHE ZEDER (*f*) ; SPITZSCHUPPE (*f*) ; KALIFORNISCHE FLUSSZEDER (*f*) ; SCHUPPENZEDER (*f*)

Botanical Term.	No.	English Term.	German Term.
Libocedrus doniana	665	(New Zealand)-kawa tree	NEUSEELÄNDISCHE SCHUPPEN--ZEDER (*f*)
Lichen	666	Moss, lichen	FLECHTE (*f*)
Lichen prunastri	667	Blackthorn lichen	SCHLEHDORNFLECHTE (*f*)
Lichen quercinum	668	Oak moss	EICHENMOOS (*n*)
Licmophora flabellata	669	fan diatom	FÄCHERKIESELALGE (*f*)
Lignum sanctum	670	Holy wood (See *Guaiacum officinale*)	HEILIGENHOLZ (*n*)
Lignum vitæ	671	(See *Guaiacum officinale*)	
Ligusticum levisticum	672	(See *Levisticum officinale*)	
Ligusticum scoticum	673	Scottish lovage ; Scotch lovage (See *Levisticum officinale*)	LIEBSTOCK (*m*) ; LIEBSTÖCKEL (*m*)
Ligustrum chinense	674	Chinese privet (See *Ligustrum sinense* and *Ligustrum villosum*)	CHINESISCHE RAINWEIDE (*f*)
Ligustrum japonicum	675	Japanese privet	JAPANISCHE RAINWEIDE (*f*)
Ligustrum lucidum	676	Cordate-leaved privet	HERZBLÄTTRIGE RAINWEIDE (*f*)
Ligustrum ovalifolium	677	Oval-leaved privet	EIBLÄTTRIGE RAINWEIDE (*f*)
Ligustrum sinense	678	(See *Ligustrum chinense*)	
Ligustrum villosum	679	(See *Ligustrum chinense*)	
Ligustrum vulgare	680	Privet (See *Cornus sanguinea*)	HARTRIEGEL (*m*) ; KORNGÄRTE (*f*) ; LIGUSTER (*m*) ; RAIN--WEIDE (*f*) ; HECKEN-RAIN--WEIDE (*f*) ; GEISSHOLZ (*n*)
Lilium	681	Lily	LILIE (*f*)
Lilium auratum	682	Golden-rayed Japanese lily	GOLDLILIE (*f*) ; GOLDBANDLILIE (*f*)
Lilium bulbiferum	683	Orange lily (See *Lilium croceum*)	FEUERLILIE (*f*)
Lilium candidum	684	Madonna lily ; St. Joseph's lily ; Bourbon lily ; common white lily	WEISSE LILIE (*f*)
Lilium chalcedonicum	685	Greek lily ; Turk's cap lily	BRENNENDE LILIE (*f*) ; SCHAR--LACH-LILIE (*f*)
Lilium croceum	686	Orange lily (See *Lilium bulbiferum*)	SAFRAN-LILIE (*f*)
Lilium elegans	687	Elegant lily	ZIERLICHE LILIE (*f*)
Lilium giganteum	688	Giant lily	RIESENLILIE (*f*)
Lilium Harrisii	689	Bermuda Easter lily (See *Lilium longiflorum exim--ium*)	
Lilium lancifolium	690	(See *Lilium speciosum*)	LANZETTBLÄTTRIGE LILIE (*f*)
Lilium longiflorum	691	Long-flowered Japanese lily	LANGBLÜTIGE LILIE (*f*)
Lilium longiflorum eximium	692	Bermuda Easter lily (See also *Lilium Harrisii*)	OSTERLILIE (*f*)
Lilium martagon	693	Turk's-cap lily ; martagon lily	KÖNIGSLILIE (*f*) ; GELBWURZ (*f*) ; GOLDLILIE (*f*) ; GOLDWURZ (*f*) ; TÜRKENBUNDLILIE (*f*)
Lilium pardalinum	694	Panther lily	PANTHERLILIE (*f*)
Lilium pomponium	695	Scarlet pompone lily	POMPONISCHE LILIE (*f*)
Lilium speciosum	696	Japanese lily (See *Lilium lancifolium*)	JAPANISCHE LILIE (*f*) ; PRACHT--LILIE (*f*)
Lilium superbum	697	Swamp lily	SUMPFLILIE (*f*) ; STOLZE LILIE (*f*)
Lilium tenuifolium	698	Siberian lily	SIBERISCHE LILIE (*f*)
Lilium testaceum	699	Nankeen lily, reddish-yellow lily	ROT-GELBE LILIE (*f*)
Lilium tigrinum	700	Tiger lily	FEUERLILIE (*f*)
Lilium washingtonia	701	Californian lily	KALIFORNISCHE LILIE (*f*)
Limnanthemum nymphæo--ides	702	Nymphæa-like water lily ; limnanth ; shield-shaped limnanth (See *Limnanthemum peltatum*)	SEEROSENARTIGE SEETANNE (*f*) ; TAUCHE (*f*) ; SUMPFROSE (*f*)
Limnanthemum peltatum	703	Fringed water lily ; fringed buckbean ; marsh-flower (See *Limnanthemum nymphæo--ides* and *Villarsia nymphæo--ides*)	
Limonium	704	Sea lavender (See *Statice limonium*)	STRANDNELKE (*f*)
Limosella aquatica	705	mudwort ; limosel	GEMEINER SCHLÄMMLING (*m*)

Botanical Term.	No.	English Term.	German Term.
Linaria alpina	706	Alpine toadflax	ALPENLEINKRAUT (*n*)
Linaria cymbalaria	707	Ivy-leaved toadflax; Mother o millions; cymbalaria (See *Cymbalaria cymbalaria*)	ZYMBELKRAUT (*n*)
Linaria elatine	708	Sharp-pointed toad-flax (See *Cymbalaria elatinoides*)	SCHARFES LEINKRAUT (*n*)
Linaria minor	709	Lesser (linaria) toadflax (See *Chœnorrhinum minus*)	KLEINES LEINKRAUT (*n*)
Linaria pelisseriana	710	Pelliser's linaria; Jersey toadflax	PELLISERS-LEINKRAUT (*n*)
Linaria repens	711	Pale linaria; creeping toadflax	KRIECHENDES LEINKRAUT (*n*)
Linaria spuria	712	round-leaved (linaria) toadflax; male fluellen	RUNDBLÄTTRIGES LEINKRAUT (*n*)
Linaria supina	713	Supine linaria	LIEGENDES LEINKRAUT (*n*)
Linaria vulgaris	714	Yellow toadflax; toadflax; wild flax; common linaria	LEINKRAUT (*n*); MARIENKRAUT (*n*); GEMEINES LEINKRAUT (*n*); MARIENFLACHS (*m*)
Lindelofia spectabilis	715	Himalayan lungwort	HIMALAJA-LUNGENWURZ (*f*)
Lindera benzoin	716	Benjamin bush, spice bush (See *Benzoin benzoin*)	FIEBERBUSCH (*m*); GEWÜRZ-BUSCH (*m*)
Linnæa borealis	717	Linnæa, twin flower	MOOSRAUDE (*f*)
Linociera ligustrina	718	Jamaica rosewood	JAMAIKA-ROSENHOLZ (*n*)
Linodendron lagetta	719	(See *Lagetta lintearia*)	
Linum angustifolium	720	Flax plant; pale flax	FLACHS (*m*); LEINPFLANZE (*f*)
Linum catharticum	721	Purging flax; cathartic flax; mountain flax	PURGIERLEIN (*m*); WIESEN-FLACHS (*m*); WIESENLEIN (*m*)
Linum perenne	722	Perennial flax	AUSDAUERNDE LEINPFLANZE (*f*)
Linum trigynum	723	(See *Reinwardtia trigyna*)	
Linum usitatissimum	724	Common flax; linseed	FLACHS (*m*); LEINPFLANZE (*f*); LEIN (*m*); SCHLIESSLEIN (*m*)
Linum vulgare	725	(See *Linum usitatissimum*)	
Liparis lœselii	726	Fen orchis; two-leaved liparis	GLANZORCHE (*f*)
Lippia citriodora	727	Scented verbena; lemon plant (See *Aloysia citriodora* and *Verbena officinalis*)	ZITRONENKRAUT (*n*); ZITRONEN-STRAUCH (*m*)
Liquidambar asplenifolia	728	(See *Myrica asplenifolia*)	MILZFARNBLÄTTRIGER AMBER-BAUM (*m*)
Liquidambar styraciflora	729	Sweet-gum, liquidambar; gum-tree, red-gum (See *Nyssa multiflora*)	STORAXBAUM (*m*); AMBERBAUM (*m*); SATINNUSSBAUM (*m*)
Liriodendron tulipifera	730	Tulip tree, white wood tree, yellow poplar	TULPENBAUM (*m*); WEISSHOLZ-BAUM (*m*); GELBE PAPPEL (*f*); LILIENBAUM (*m*); GEMEINER TULPENBAUM (*m*)
Listera cordata	731	Lesser twayblade	HERZBLÄTTRIGES ZWEIBLATT (*n*)
Listera ovata	732	Twayblade; common twayblade	EIBLÄTTRIGES ZWEIBLATT (*n*)
Lithospermum arvense	733	Corn Gromwell; bastard alkanet	ACKER-STEINSAME (*m*)
Lithospermum officinale	734	Gromwell; common gromwell (See *Lithospermum prostatum*)	STEINSAME (*m*)
Lithospermum prostatum	735	(See *Lithospermum officinale*)	AUSGESTRECKTER STEINSAME (*m*)
Lithospermum purpureo-cœruleum	736	Creeping gromwell	BERG-STEINSAME (*m*)
Littorella juncea	737	Marsh littorel; marshweed	SUMPF-STRANDLING (*m*)
Littorella lacustris	738	Shore-weed; littorel	MEER-STRANDLING (*m*)
Livistona australis	739	Australian fan palm	AUSTRALISCHE FÄCHERPALME (*f*); AUSTRALISCHE LIVISTONIE (*f*)
Livistona humilis	740	Australian, dwarf fan palm	NIEDRIGE AUSTRALISCHE FÄCHER-PALME (*f*); ZWERG-LIVIS-TONIE (*f*)
Livistona rotundifolia	741	Round-leaved Saribu palm	RUNDBLÄTTRIGE SARIBUPALME (*f*)
Lloydia alpina	742	Mountain spider-wort (See *Lloydia serotina*)	ALPEN-LLOYDIE (*f*)
Lloydia serotina	743	Mountain Lloydia (See *Lloydia alpina*)	SPÄTBLÜHENDE LLOYDIE (*f*)
Loasa lateritia	744	Chili nettle (See *Blumenbachia lateritia*)	LOASE (*f*)
Lobaria pulmonaria	745	(See *Sticta pulmonaria*)	LUNGENMOOS (*n*); GRUBEN-FLECHTE (*f*); LUNGENFLECHTE (*f*)

Botanical Term.	No.	English Term.	German Term.
Lobelia cardinalis	746	Scarlet lobelia, cardinal flower	SCHARLACHBLÜTIGE LOBELIE (*f*)
Lobelia dortmanna	747	Water lobelia	WASSER-LOBELIE (*f*)
Lobelia inflata	748	Indian tobacco	VIOLETTBLÜTIGE LOBELIE (*f*)
Lobelia urens	749	Acrid lobelia	WESTEUROPÄISCHE LOBELIE (*f*)
Lochnera rosea	750	(See *Vinca rosea*)	
Lodoicea sechellarum	751	Seychelles fan-leaved palm; lodoicea; double cocoanut-palm	MEERKOKOS (*m*); SEFKOKOS (*m*); DOPPELKOKOS (*m*)
Loiseleuria procumbens	752	Loiseleuria, alpine azalea, trailing azalea (See *Azalea procumbens*)	NIEDERLIEGENDES FELSEN-RÖSCHEN (*n*)
Lolium arvense	753	Darnel; rye-grass (See *Lolium temulentum*)	TAUMELKORN (*n*); LOLCH (*m*)
Lolium italicum	754	Italian rye-grass	ITALIENISCHES RAIGRAS (*n*)
Lolium perenne	755	Rye-grass, red darnel	LOLCH (*m*); RAIGRAS (*n*); RAY-GRAS (*n*)
Lolium temulentum	756	Darnel-(grass) (See *Lolium arvense*)	TAUMELKORN (*n*)
Lomaria spicant	757	Hard fern, deer fern (See *Blechnum boreale*)	GEMEINER RIPPENFARN (*m*)
Lomatophyllum aloiflorum	758	Bourbon aloe	AFRIKANISCHE ALOE (*f*)
Lonicera	759	Honeysuckle	GEISSBLATT (*n*); ZÄUNLING (*m*)
Lonicera caprifolium	760	Perfoliate honeysuckle, goat-leaf honeysuckle	JE LÄNGER, JE LIEBER; ITALIEN-ISCHES GEISSBLATT (*n*)
Lonicera fragantissima	761	Yellow jasmine honeysuckle	STARKDUFTENDES GEISSBLATT (*n*)
Lonicera periclymenum	762	Woodbine; honeysuckle; com-mon honeysuckle	GEISSBLATT (*n*); WALDGEISS-BLATT (*n*); JELÄNGERJELIE-BER (*n*)
Lonicera sempervirens	763	Evergreen honeysuckle	IMMERGRÜNES GEISSBLATT (*n*)
Lonicera xylosteum	764	Fly honeysuckle, flag honey-suckle, bird cherry	ROTES GEISSBLATT (*n*); HUNDS-KIRSCHE (*f*); HECKENKIR-SCHE (*f*)
Lophodermium pinastri	765	Pine fungus	KIEFERNPILZ (*m*)
Loranthaceæ	766	Loranthaceæ (See *Loranthus* and *Viscum*)	RIEMENPFLANZEN (*f.pl*)
Loranthus europæus	767	Yellow-berried mistletoe (See also *Viscum*)	RIEMENBLUME (*f*)
Lotus angustissimus	768	Slender lotus	SCHMALER LOTUSKLEE (*m*)
Lotus corniculatus	769	(Common)-bird's-foot trefoil; common lotus; fingers-and-toes	LOTUSKLEE (*m*); HORNKLEE (*m*); SCHOTENKLEE (*m*); FLÜGEL-ERBSE (*f*)
Lotus major	770	Greater bird's-foot trefoil	LOTUSKLEE (*m*)
Loxopterygium lorentzii	771	(Mexican)-red quebracho (See also *Aspidosperma quebracho*)	(ROTER)-QUEBRACHOHOLZBAUM (*m*)
Luffa acutangula	772	(West Indian)- loofah plant	LÄNGSRIPPIGFRÜCHTIGE LOOFAH (*f*); WESTINDISCHE LOOFAH (*f*)
Luffa ægyptiaca	773	Towel-gourd; loofah plant	ÄGYPTISCHE LOOFAH (*f*)
Luffa cylindrica	774	Dishcloth plant; snake gourd; luffa sponge; sponge cucum-ber	SCHWAMMKÜRBIS (*m*)
Lunaria annua	775	Lunaria; honesty (See *Lunaria biennis*)	
Lunaria biennis	776	Money flower, satin flower (See *Lunaria annua*)	MONDKRAUT (*n*); MONDVIOLE (*f*)
Lupinus	777	Lupine; lupin	LUPINE (*f*)
Lupinus albus	778	White (lupin) lupine	WEISSE LUPINE (*f*)
Lupinus arboreus	779	Tree lupine	BAUMLUPINE (*f*)
Lupinus nanus	780	Dwarf lupine	ZWERGLUPINE (*f*)
Lupinus polyphyllus	781	Lupine, lupin	LUPINE (*f*); WOLFSBOHNE (*f*)
Luzula arcuata	782	Curved woodrush	MARBEL (*f*)
Luzula campestris	783	Field wood-rush, glow-worm grass	HAINBINSE (*f*); HAINSIMSE (*f*); HASENBROT (*n*); HEINSIMSE (*f*)
Luzula pilosa	784	Hairy wood-rush	BEHAARTE HAINSIMSE (*f*)
Luzula spicata	785	Spiked woodrush	SPIKEN-HAINSIMSE (*f*)
Luzula sylvatica	786	Great woodrush (See also *Juncus sylvaticus*)	WALD-MARBEL (*f*)

Botanical Term.	No.	English Term.	German Term.
Lychnis	787	Crow-flower, campion	LICHTNELKE (*f*)
Lychnis alba	788	(See *Lychnis vespertina*)	WEISSE LICHTNELKE (*f*)
Lychnis alpina	789	Alpine lychnis	ALPENLICHTNELKE (*f*)
Lychnis chalcedonica	790	Cross of Jerusalem (See *Lychnis coronaria*)	BRENNENDE LIEBE (*f*); JERUSALEMSBLUME (*f*); CHALZEDONISCHE LICHTNELKE (*f*); FEUERNELKE (*f*)
Lychnis coronaria	791	Scarlet campion; flower of Constantinople; cross of Malta; cross of Jerusalem; scarlet lychnis (See also *Silene*)	JERUSALEMSBLUME (*f*); FEUERNELKE (*f*); SCHARLACHBLUME (*f*); KRANZRADE (*f*); VEXIERNELKE (*f*)
Lychnis dioica	792	(See *Lychnis diurna*)	
Lychnis diurna	793	Red campion; pink campion; red robin; red lychnis (See *Lychnis dioica*)	ROTE LICHTNELKE (*f*); TAGBLÜHENDE LICHTNELKE (*f*)
Lychnis Flos-cuculi	794	Ragged robin, meadow lychnis	KUCKUCKSBLUME (*f*); KUCKUCKSSPEICHEL (*m*)
Lychnis flos-Jovis	795	Flower of Jove (See *Lychnis jovis*)	JUPITERLICHTNELKE (*f*)
Lychnis githago	796	Corn cockle (See *Papaver rhœas* and *Agrostemma githago*)	KORNRADE (*f*); KORNROSE (*f*)
Lychnis jovis	797	(See *Lychnis flos jovis*)	
Lychnis tomentosa	798	(See *Lychnis coronaria*)	
Lychnis vespertina	799	White campion; evening campion; white lychnis (See *Lychnis alba* and *Melandryum album*)	ABENDBLÜHENDE LICHTNELKE (*f*); ABENDLICHTNELKE (*f*)
Lychnis viscaria	800	Red, German catch-fly; fly-bane; silene; catch-fly; viscid lychnis (See *Viscaria alba*)	MÜCKENFANG (*m*); PECHNELKE (*f*)
Lycium barbarum	801	Box-thorn, Duke of Argyll's tea tree, African tea tree (See *Lycium sinense*)	TEUFELSZWIRN (*m*); BOCKSDORN (*m*)
Lycium europæum	802	Spiny box-thorn	EUROPÄISCHER BOCKSDORN (*m*)
Lycium halmifolium	803	(See *Lycium barbarum*)	
Lycium sinense	804	(See *Lycium barbarum*)	CHINESISCHER BOCKSDORN (*m*)
Lycoperdon bovista	805	Puff ball; egg-shaped puff-ball (See *Bovista nigrescens*)	
Lycoperdon gemmatum	806	Puff-ball	GEKÖRNTER STÄUBLING (*m*); GEMEINER STÄUBLING (*m*)
Lycoperdon giganteum	807	(See *Globaris bovista*)	
Lycopersicum esculentum	808	Tomato, love-apple	PARADIESAPFEL (*m*); LIEBESAPFEL (*m*)
Lycopodium alpinum	809	Savin-leaved club-moss; alpine club-moss	ALPENBÄRLAPP (*m*)
Lycopodium annotinum	810	Interrupted club-moss	SPROSSENDER BÄRLAPP (*m*)
Lycopodium clavatum	811	Lycopodium, common club-moss; witch meal, vegetable sulphur; stag-horn moss	ERDSCHWEFEL (*m*); BÄRLAPP (*m*); BÄRLAPPKRAUT (*n*); HEXENMEHL (*n*); MOOSKRAUT (*n*); NEUNGLEICH (*n*); NEUNHEIL (*n*); KOLBENMOOS (*n*); SCHLANGENKRAUT (*n*); SCHLANGENMOOS (*n*); KEULENFÖRMIGERBÄRLAPP (*m*)
Lycopodium dendroideum	812	Fir club moss (See *Lycopodium selago*)	
Lycopodium denticulatum	813	(See *Selaginella kraussiana*)	
Lycopodium inundatum	814	Marsh club-moss	SUMPFBÄRLAPP (*m*)
Lycopodium selago	815	Fir club-moss (See *Lycopodium dendroideum*)	TANNENBÄRLAPP (*m*); SEMUST (*m*)
Lycopsis arvensis	816	Bugloss; small bugloss	ACKER-WOLFSAUGE (*f*)
Lycopus europæus	817	Gypsy-wort	GEMEINER WOLFSTRAPP (*m*)
Lycoris aurea	818	Golden lily (See *Amaryllis aurea*)	GOLDENE LILIE (*f*); JAPANISCHER GOLDENER RITTERSTERN (*m*)
Lygodium palmatum	819	Climbing fern	KLETTERFARN (*m*)

Botanical Term.	No.	English Term.	German Term.
Lysimachia nemorum	820	Yellow pimpernel; wood loosestrife; wood lysimachia	WALDPFENNIGKRAUT (*n*); HAIN-GILBWEIDERICH (*m*)
Lysimachia nummularia	821	Money-wort; creeping-jenny; herb-two-pence	PFENNIGKRAUT (*n*)
Lysimachia thyrsiflora	822	Tufted lysimachia; tufted loosestrife (See *Naumburgia thyrsiflora*)	
Lysimachia vulgaris	823	Common lysimachia; loosestrife; yellow loosestrife	GEMEINES PFENNIGKRAUT (*n*); FRIEDLOS (*m*); FELBERICH (*m*); GILBWEIDERICH (*m*)
Lythrum hyssopifolia	824	Hyssop-leaved purple loosestrife; hyssop lythrum	YSOPBLÄTTRIGER WEIDERICH (*m*)
Lythrum salicaria	825	Purple loosestrife, willow weed, common loosestrife	GEMEINER WEIDERICH (*m*); BLUT-KRAUT (*n*)
Maba ebenus	826	Ceylon ebony tree	MAKASSAR-EBENHOLZBAUM (*m*)
Machærium schomburghii	827	Tiger-wood	SCHLANGENHOLZ (*n*)
Machærium violaceum	828	Mexican king-wood tree	MEXIKANISCHER KÖNIGSHOLZ-BAUM (*m*)
Maclura aurantiaca	829	Osage orange; bow-wood tree (See *Joxylon pomiferum*)	OSAGE-ORANGE (*f*)
Maclura tinctoria	830	Old fustic (See *Chlorophora tinctoria*)	ALTER FUSTIK (*m*)
Macrochloa tenacissima	831	(See *Stipa tenacissima*)	
Macropiper excelsum	832	(See *Piper excelsum*)	
Macropiper latifolium	833	Ava; Kava	KAVA (*f*); AVA (*f*)
Macrozamia denisonii	834	Giant fern palm, Swan river fern palm (See *Lepidozamia peroffskyana*)	SPITZSCHUPPENPALMFARN (*m*)
Madia sativa	835	Madia oil plant	MADIAÖLPFLANZE (*f*); ÖLMADIE (*f*)
Magnolia acuminata	836	(American)-cucumber tree	GURKENMAGNOLIE (*f*)
Magnolia conspicua	837	Yulan (See *Magnolia yulan*)	
Magnolia glauca	838	Swamp sassafras; beaver tree; sweet bay	GRAUGRÜNE MAGNOLIE (*f*)
Magnolia grandiflora	839	Large-flowered magnolia	GROSSBLUMIGE MAGNOLIE (*f*)
Magnolia obovata	840	Obovate magnolia (See *Magnolia purpurea*)	
Magnolia purpurea	841	Purple magnolia (See *Magnolia obovata*)	ROTE MAGNOLIE (*f*)
Magn ia tripetala	842	Umbrella tree (See *Magnolia Umbrella*)	SCHIRM-MAHAGONIE (*f*)
Magnolia umbellata	843	(See *Magnolia tripetala*)	
Magnolia umbrella	844	(See *Magnolia tripetala*)	
Magnolia yulan	845	Yulan (See *Magnolia conspicua*)	LILIEN-MAHAGONIE (*f*)
Mahonia aquifolium	846	(See *Berberis aquifolium*)	GLANZBLÄTTRIGE MAHONIE (*f*)
Mahonia japonica	847	(See *Berberis japonica* and *Ilex japonica*)	JAPANISCHE MAHONIE (*f*)
Maianthemum bifolium	848	(See *Smilacina bifolia*)	ZWEIBLÄTTRIGES SCHATTENBLÜM-CHEN (*n*)
Maianthemum convallaria	849	Twin-leaved (bi-foliate) lily of-the-valley (See *Simlacina bifolia*)	SCHATTENBLÜMCHEN (*n*)
Majorana hortensis	850	(See *Origanum majorana*)	WURSTKRAUT (*n*); MAIRAN (*m*)
Malachium aquaticum	851	Water chickweed (See *Stellaria aquatica*)	GEMEINER WASSERDARM (*m*); WASSERMIERE (*f*)
Malaxis paludosa	852	Bog orchis	SUMPF-WEICHORCHE (*f*)
Malcomia maritima	853	Virginian stock	MALCOMIE (*f*)
Mallotus philippinensis	854	Indian kamala tree (See *Rottlera tinctoria*)	KAMALABAUM (*m*)
Malope trifida	855	Large-flowered mallow-wort	DREISPALTIGE MALOPE (*f*)
Malpighia glabra	856	Barbados cherry	BARBADOSKIRSCHENBAUM (*m*)
Malpighia mexicana	857	Mexican malpighia (See *Stigmaphyllon ciliatum*)	MALPIGHIE (*f*); AZEROLENBAUM (*m*)
Malus angustifolia	858	Narrow-leaved apple	SCHMALBLÄTTRIGER APFEL (*m*)
Malus astracanica	859	Astrachan apple	WACHSAPFEL (*m*); EISAPFEL (*m*)

Botanical Term.	No.	English Term.	German Term.
Malus baccata	860	Siberian crab apple; berried apple	BEERENAPFEL (*m*)
Malus communis	861	Apple-tree (See also *Pirus malus*)	GEMEINER APFELBAUM (*m*)
Malus coronaria	862	American crab apple; garland apple	WOHLRIECHENDER APFEL (*m*)
Malus floribunda	863	Free-flowering apple	REICHBLÜTIGER APFEL (*m*)
Malus halliana	864	Hallian apple	HALLS-APFEL (*m*)
Malus prunifolia	865	Plum-leaved apple	PFLAUMENBLÄTTRIGER APFEL (*m*)
Malus ringo	866	Ringo apple	RINGO-APFEL (*m*)
Malus rivularis	867	Oregon crab apple	UFERAPFEL (*m*)
Malus spectabilis	868	Showy apple; Chinese apple (See *Pirus sinensis*)	CHINESISCHER APFEL (*m*); PRACHT-APFEL (*m*)
Malus toringo	869	Toringo apple (See *Sorbus toringo*)	TORINGO-APFEL (*m*)
Malva alcea	870	Vervain mallow (See *Althæa rosea*)	ROSENPAPPEL (*f*)
Malva moschata	871	Musk mallow (See *Hibiscus abelmoschus*)	MOSCHUS-MALVE (*f*)
Malva neglecta	872	Neglected mallow (See *Malva rotundifolia*)	UBERSEHENE MALVE (*f*)
Malva rotundifolia	873	Dwarf mallow, round-leaved mallow (See *Malva neglecta*)	KÄSEPAPPEL (*f*)
Malvastrum capense	874	False mallow; Cape false mallow	KAP-SCHEINMALVE (*f*); FLEISSIGES LIESCHEN (*n*)
Malva sylvestris	875	Common mallow, wild mallow, wild malva	WALDMALVE (*f*); KÄSEPAPPEL (*f*); MALVE (*f*); MALWE (*f*)
Mammea americana	876	Mammee apple; West Indian wild apricot; mammee tree	MAMAIBAUM (*m*)
Mammillaria	877	Thimble cactus, elephant's-tooth cactus, nipple cactus	ZITZENKAKTUS (*m*); WARZENKAKTUS (*m*)
Manchura tinctoria	878	Fustic wood, Cuba wood (See *Chlorophora tinctoria*)	GELBHOLZ (*n*); FUSTIKHOLZ (*n*)
Mandevilla	879	Chili jasmine	CHILE-JASMIN (*m*)
Mandragora	880	Mandrake (See also *Podophyllum peltatum*)	ZAUBERWURZEL (*f*)
Mandragora autumnalis	881	Autumn mandrake	MANDRAGORA (*f*)
Mandragora vernalis	882	Spring mandrake; mandragora, mandrake, devil's apple; enchanter's nightshade; Judæan wormwood (See *Circæa lutetiana*)	HEXENKRAUT (*n*); MANDRAGORA (*f*)
Mangifera indica	883	Mango tree	MANGOBAUM (*m*)
Manihot Aipi	884	(See *Manihot utilissima*)	
Manihot utilissima	885	Manioc tree, cassava tree, tapioca tree, cassava bush, cassado bush (See *Jatropha manihot*)	MANIOKSTRAUCH (*m*)
Maranta arundinacea	886	Arrow-root	PFEILWURZEL (*f*); PFEILWURZ (*f*); MAGENWURZEL (*f*); ARONSWURZEL (*f*)
Marasmius oreades	887	Oread; fairy-ring champignon; fairy-ring mushroom; champignon	HERBSTMUSSERON (*m*); KRÖSLING (*m*); NELKENSCHWAMM (*m*); LAUCHSCHWAMM (*m*)
Marattia fraxinea	888	Marattia-(fern); ash-leaf fern	KAPSELFARN (*m*); MARATTIA-FARN (*m*)
Marchantia polymorpha	889	Liverwort (See also *Anemone hepatica* and *Hepatica*)	LEBERBLUME (*f*); LEBERBLÜMCHEN (*n*); KOPFFRUCHTLEBERMOOS (*n*); SCHLIESSFRUCHTLEBERMOOS (*n*); LEBERKRAUT (*n*); STRAHLKOPFLEBERMOOS (*n*); VIELGESTALTIGES STRAHLKOPFLEBERMOOS (*n*)
Margyricarpus setosus	890	Pearl fruit-(shrub), pearl-berry	PERLBEERE (*f*)
Marica northiana	891	Toad-cup lily	NORTHS-MARICALILIE (*f*)
Marrubium vulgare	892	White horehound, common hoarhound (See also *Ballota nigra*)	ANDORN (*m*); ANDORNKRAUT (*n*); MARIENNESSEL (*f*)

Botanical Term.	No.	English Term.	German Term.
Marsdenia condurango	893	Condurango-bark tree (See *Gonolobus condurango*)	CONDURANGORINDENBAUM (*m*); GEIERRINDENBAUM (*m*)
Marsilea macropus	894	Nardoo; long-stalked marsilia; long-stalked marsilea	KLEEFARN (*m*); PILLENFRUCHT-FARN (*m*)
Martynia fragrans	895	Unicorn plant, elephant's trunk	WOHLRIECHENDE MARTYNIE (*f*)
Mathiola annua	896	Ten-week stock (See *Matthiola*)	
Matricaria chamomilla	897	Wild (chamomile) camomile; German camomile; common matricary (See also *Anthemis nobilis*)	ACKERKAMILLE (*f*); MUTTER-KRAUT (*n*)
Matricaria inodora	898	Corn feverfew; scentless May-weed; scentless matricary; spear-thistle; double may-weed (See *Chrysanthemum inodorum*)	GERUCHLOSES MUTTERKRAUT (*n*); GERUCHLOSE KAMILLE (*f*)
Matricaria parthenium	899	Feverfew; wild feverfew; vivey-vew; flirtwort; featherfall	FIEBERKRAUT (*n*)
Matthiola annua	900	Ten week stock, intermediate stock	EINJÄHRIGE LEVKOJE (*f*)
Matthiola incana	901	Hoary stock; common stock; Queen stock; Gilliflower, wall-flower-leaved stock, Brompton stock; stock- gilliflower	LEVKOJESTOCK (*m*); LEVKOJE (*f*)
Matthiola sinuata	902	Sea stock	LEVKOJE (*f*); MEER-LEVKOJE-STOCK (*m*)
Matthiola tristis	903	Night-scented stock; melancholy gentleman	NACHTDUFTENDE LEVKOJE (*f*)
Mauritia flexuosa	904	South American fan palm; Buriti palm	GEBOGENE BURITIPALME (*f*); MORICHE (*f*); MIRITI (*f*)
Mauritia vinifera	905	South American fan palm; Buriti palm	WEIN-BURITIPALME (*f*)
Meconopsis cambrica	906	Yellow Welsh poppy, Himalayan poppy, prickly poppy	WESTEUROPÄISCHER MOHNLING (*m*)
Meconopsis nepalensis	907	Nepaul poppy	NEPAUL-MOHNLING (*m*)
Meconopsis wallichii	908	Satin poppy	HELLBLAUER MOHNLING (*m*)
Medeola asparagoides	909	(See *Asparagus officinalis*)	
Medicago	910	Snail clover	SCHNECKENKLEE (*m*)
Medicago arborea	911	Moon trefoil	MONDKLEEBLATT (*n*)
Medicago denticulata	912	Toothed medick	GEZÄHNTER SCHNECKENKLEE (*m*)
Medicago echinus	913	Crown of thorns, Calvary clover	DORNIGER SCHNECKENKLEE (*m*)
Medicago falcata	914	Sickle medick, sickle podded medick	SICHELKLEE (*m*); SCHWEDISCHE LUZERNE (*f*)
Medicago lupulina	915	Black medick; yellow clover; hop; nonsuch; hop clover, hop trefoil	HOPFENKLEE (*m*)
Medicago maculata	916	Spotted medick	GEFLECKTER SCHNECKENKLEE (*m*)
Medicago minima	917	Bur medick	KLEINSTER SCHNECKENKLEE (*m*); GELBBLÜHENDER SCHNECKEN-KLEE (*m*)
Medicago sativa	918	Lucern; purple medick; Lucerne; Alfalfa; Spanish trefoil; French clover; medic; medick; purple medic; Brazilian clover; Chilean clover	EWIGER KLEE (*m*); LUZERNE (*f*); MONATSKLEE (*m*); SPARGEL-KLEE (*m*)
Medicago scutellata	919	Snail clover	SCHNECKENKLEE (*m*)
Melaleuca leucadendron	920	Cajeput tree	KAJEPUTBAUM (*m*); WEISSBAUM (*m*)
Melaleuca minor	921	Cajeput tree; small cajeput tree	KLEINER KAJEPUTBAUM (*m*)
Melampsora betulina	922	(See *Lecythea caprearum*)	SCHWARZSPORROST (*m*)
Melampsora lini	923	Flax rust	LEINROST (*m*)
Melampsora tremulæ	924	Aspen rust (See *Cæoma pinitorquum*)	ZITTERPAPPELROST (*m*)
Melampyrum arvense	925	Purple cow-wheat	FELDKUHWEIZEN (*m*); FELD-WACHTELWEIZEN (*m*)
Melampyrum cristatum	926	Crested cow-wheat	KAMM-WACHTELWEIZEN (*m*)

Botanical Term.	No.	English Term.	German Term.
Melampyrum pratense	927	Yellow cow-wheat; common cow-wheat	KUHWEIZEN (*m*); WIESEN-WACH--TELWEIZEN (*m*)
Melampyrum sylvaticum	928	Small-flowered cow-wheat	WALDKUHWEIZEN (*m*); WALD--WACHTELWEIZEN (*m*)
Melandryum album	929	(See *Lychnis vespertina*)	
Melia	930	Bead tree, Indian lilac	
Melia azedarach	931	Bread tree (See also *Guarea*)	PATERNOSTERBAUM (*m*); ZEDA--RACHBAUM (*m*)
Melianthus major	932	Great Cape honey-flower	GROSSER HONIGSTRAUCH (*m*)
Melilotus alba	933	White melilot; white trefoil	HONIGKLEE (*m*); RIESENKLEE (*m*)
Melilotus arvensis	934	Field melilot	FELDMELILOTE (*f*)
Melilotus cœrulea	935	Blue melilot	SCHABKLEE (*m*)
Melilotus officinalis	936	Melilot; yellow melilot; common melilot	STEINKLEE (*m*); MELILOTE (*f*); MELILOTENKLEE (*m*); HONIG--KLEE (*m*); MOTTENKRAUT (*n*); SCHOTENKLEE (*m*)
Melilotus vulgaris	937	(See *Melilotus officinalis*)	STEINKLEE (*m*)
Melissa calamintha	938	(See *Calamintha acinos*)	KALAMINTHE (*f*)
Melissa officinalis	939	Balm; balm mint	MELISSE (*f*); MELISSENKRAUT (*n*); MUTTERKRAUT (*n*); ZI--TRONENKRAUT (*n*); ZITRONEN--MELISSE (*f*)
Melittis melissophyllum	940	Bastard balm	MELISSENBLÄTTRIGER BIENEN--SAUG (*m*); IMMENBLATT (*n*)
Melocactus communis	941	Melon cactus, Turk's head, Turk's cap cactus, melon thistle, Pope's head	MELONENDISTEL (*f*); TÜRKEN--KOPF (*m*)
Memordica elaterium	942	Squirting cucumber (See *Ecballium elaterium* and also *Momordica*)	SPRINGGURKE (*f*)
Menispermaceæ	943	Menispermaceæ	MONDSAMENGEWÄCHSE (*n.pl*)
Menispermum canadense	944	Moon creeper, moon seed	KANADISCHER KOCKELSKÖRNER--BAUM (*m*)
Menispermum cocculus	945	Moon seed-(tree)	KOCKELSKÖRNERBAUM (*m*)
Menispermum palmatum	946	Moon-seed	MONDSAME (*f*)
Mentha	947	Mint; pennyroyal	MINZE (*f*)
Mentha aquatica	948	Capitate mint; hairy mint; water mint; Bergamot mint	WASSERMINZE (*f*)
Mentha arvensis	949	Corn mint	ACKERMINZE (*f*); WILDE POLEI (*f*); WILDER POLEI (*m*)
Mentha crispa	950	Curled mint	KRAUSEMINZE (*f*)
Mentha piperita	951	Pepper-mint	PFEFFERMINZ (*m*); PFEFFER--MÜNZ (*m*)
Mentha pulegium	952	Penny-royal	POLEI (*f* and *m*)
Mentha rotundifolia	953	Round-leaved mint	RUNDBLÄTTRIGE MINZE (*f*)
Mentha sativa	954	Whorled hairy mint	KULTIVIERTE MINZE (*f*)
Mentha spicata	955	Spear mint (See *Mentha viridis*)	KRAUSEMINZE (*f*)
Mentha sylvestris	956	Horse mint	WALDMINZE (*f*)
Mentha viridis	957	Spear mint, Lamb mint (See *Mentha spicata*)	KRAUSEMINZE (*f*); RÖMISCHE MINZE (*f*)
Menyanthes trifoliata	958	Buckbean, bog bean, marsh trefoil, bog trefoil, water trefoil (See *Trifolium*)	BITTERKLEE (*m*); FIEBERKLEE (*m*); (*m*); WASSERKLEE (*m*)
Menziesia cœrulea	959	Blue menziesia	BLAUE MENZIESIE (*f*)
Menziesia polifolia	960	St. Dabeoc's heath (See *Dabœcia polifolia*)	GAMANDERBLÄTTRIGE MENZIESIE (*f*)
Mercurialis annua	961	Annual mercury	EINJÄHRIGES BINGELKRAUT (*n*)
Mercurialis perennis	962	Mercury; dog's mercury	AUSDAUERNDES BINGELKRAUT (*n*)
Merendera	963	Pyrenean meadow saffron	ÖSTLICHE ZEITLOSE (*f*)
Mertensia maritima	964	Mertensia; sea mertensia	MEERSTRANDS-MERTENSIE (*f*)
Mertensia pulmonarioides	965	(See *Mertensia virginica*)	LUNGENKRAUTÄHNLICHE MER--TENSIE (*f*)
Mertensia virginica	966	Virginian cowslip (See *Mertensia pulmonarioides*)	VIRGINISCHE MERTENSIE (*f*)
Merulius destruens	967	(See *Merulius lacrymans*)	
Merulius lachrymans	968	(See *Merulius lacrymans*)	

Botanical Term.	No.	English Term.	German Term.
Merulius lacrymans	969	Wood fungus; common dry-rot; dry-rot fungus (See *Polyporus umbellatus*)	(ECHTER)-HAUSSCHWAMM (*m*); (ECHTER)-HOLZSCHWAMM (*m*)
Mesembryanthemum crystallinum	970	Ice plant	EISPFLANZE (*f*); EISKRAUT (*n*); EISBLUME (*f*)
Mesembryanthemum edule	971	Hottentot fig, fig marigold, mid-day flower	MITTAGSBLUME (*f*); ZASERBLUME (*f*)
Mespilus domestica	972	(See *Mespilus germanica*)	
Mespilus germanica	973	Medlar; common-(wild)-medlar (See *Mespilus domestica* and *Pirus germanica*)	MISPEL (*f*); MEHLBEERBAUM (*m*); MEHLDORN (*m*); MEHL-FÄSSCHEN (*n*); DEUTSCHE MISPEL (*f*); WISPEL (*f*)
Mespilus oxyacantha	974	(See *Cratægus oxyacantha*)	
Mesua ferrea	975	Ironwood tree	EISENHOLZBAUM (*m*)
Metrosideros buxifolia	976	Aki; New-Zealand Lignum vitæ	BUCHBLÄTTRIGER EISENMASS-BAUM (*m*)
Metrosideros robusta	977	(New Zealand)-rata tree	NEUSEELÄNDISCHE EICHE (*f*)
Metrosideros speciosa	978	(See *Callistemon speciosus*)	
Metrosideros vera	979	Rata tree	RATABAUM (*m*); EISENHOLZBAUM (*m*)
Metroxylon sagu	980	(East Indian)-sago palm	OST-INDISCHE SAGO-PALME (*f*)
Meum athamanticum	981	Bald-money; spicknel; spignel-(meu)	BÄRENFENCHEL (*m*); MUTTER-WURZ (*f*)
Microcachrys tetragona	982	(Tasmanian)-strawberry-fruited cypress	SCHUPPENBLÄTTRIGE STIEL-FRUCHTEIBE (*f*)
Microglossa albescens	983	Shrubby star-wort	
Microsphæra	984	(See *Microsphæria*)	
Microsphæra berberidis	985	Berberry blight	BERBERITZENKORALLENMELTAU (*m*)
Microsphæra grossulariæ	986	Gooseberry blight	STACHELBEEREN-KORALLENMEL-TAU (*m*)
Microsphæria	987	Blight (See *Microsphæra*)	KORALLENMELTAU (*m*)
Mikania scandens	988	Parlour ivy, German ivy	SCHNELLEFEU (*m*); SOMMER-EPHEU (*m*); SOMMEREFEU (*m*)
Milium effusum	989	Millet-grass; spreading millium (See also *Panicum miliaceum*)	GEMEINES FLATTERGRAS (*n*)
Mimosa pudica	990	Mimosa; sensitive plant; humble plant (See also under *Acacia*)	MIMOSE (*f*)
Mimosa sensitiva	991	Mimosa; larger mimosa; sensi-tive plant	MIMOSE (*f*); SINNPFLANZE (*f*)
Mimulus cardinalis	992	Cardinal flower	MASKENBLUME (*f*); LARVEN-BLUME (*f*); LOCHBLUME (*f*)
Mimulus cupreus	993	(See *Mimulus luteus*)	KUPFERIGE GAUKLERBLUME (*f*)
Mimulus luteus	994	Common monkey flower; yellow mimulus (See *Mimulus cupreus*)	GELBE GAUKLERBLUME (*f*); AFFEN-BLUME (*f*)
Mimulus maculosus	995	Spotted mimulus	MASKENBLÜTLER (*m*)
Mimulus moschatus	996	Common musk	MOSCHUSBLUME (*f*); GELBBLÜT-IGES MOSCHUSKRAUT (*n*)
Mirabilis jalapa	997	Marvel of Peru	FALSCHE JALAPE (*f*); WUNDER-BLUME (*f*); VIERUHRBLUME (*f*)
Miscanthus japonica	998	Zebra-striped rush (See *Eulalia japonica*)	JAPANISCHES SEIDENGRAS (*n*); JAPANISCHES STIELBLÜTEN-GRAS (*n*); JAPANISCHES STIELÄHRCHENGRAS (*n*)
Mitchella repens	999	Winter-clover; checkerberry, squaw-vine; partridge-berry, deer berry, chequer berry	KRIECHENDE REBHUHNBEERE (*f*)
Mitella diphylla	000	Mitre-wort, bishop's cap	MITTELLE (*f*); BISCHOFSMÜTZE (*f*)
Mitraria coccinea	001	(Scarlet)-mitre-pod, mitre flower	SCHARLACHROTE MÜTZENBLUME (*f*)
Mitriostigma axillaris	002	Citron-scented Gardenia (See *Gardenia citriodora*)	
Mœhringia trinervis	003	(See *Arenaria trinervis*)	
Mœnchia erecta	004	Upright mœnchia	AUFRECHTE MÖNCHIE (*f*)

Botanical Term.	No.	English Term.	German Term.
Mohria caffrorum	005	Frankincense	WEIHRAUCHFARN (*m*)
Molinia cœrulea	006	Bent-grass, Indian grass, lavender grass	BENTHALM (*m*)
Momordica balsamina	007	Balsam (balm) apple (See also *Memordica*)	BALSAMAPFEL (*m*)
Monarda didyma	008	Oswego tea, sweet bergamot	ZWILLING-MONARDE (*f*)
Monarda fistulosa	009	Wild bergamot, Oswego mint, bee balm, horsemint	HOHLSTENGELIGE MONARDE (*f*); PFERDEMINZE (*f*)
Monarda punctata	010	Horsemint	PFERDEMINZE (*f*); PFERDEPOLEI (*m*); GEFLECKTE PFERDE--MINZE (*f*)
Monocotyleœ	011	Monocotyleæ	EINKEIMBLÄTTRIGE PFLANZEN (*f.pl*)
Monotropa hypopitys	012	Yellow bird's-nest; monotrope	FICHTENSPARGEL (*m*)
Monotropa uniflora	013	Tobacco-pipe	OHNBLATT (*n*)
Monstera acuminata	014	Shingle plant (See *Monstera tenuis* and *Gravia paradoxa*)	LANGSPITZIGES FENSTERBLATT (*n*)
Monstera deliciosa	015	Delicious monstera; edible-fruited monstera (See *Philodendron pertusum*)	ESSBARFRÜCHTIGES FENSTERBLATT (*n*)
Monstera tenuis	016	(See *Monstera acuminata*)	SCHLANKES FENSTERBLATT (*n*)
Montia fontana	017	Water chickweed; blinks (See *Montia minor*)	UFERMONTIE (*f*)
Montia minor	018	Lesser montia (See *Montia fontana*)	UFERMONTIE (*f*)
Montia rivularis	019	River montia	BACHMONTIE (*f*)
Morœa bicolor	020	Butterfly iris	ZWEIFARBIGE MORAEA (*f*)
Morœa robinsoniana	021	Wedding flower (See *Iris robinsoniana*)	ROBINSONS-MORAEA (*f*)
Morchella esculenta	022	Common morel-(fungus); edible mushroom	MORCHEL (*f*)
Morina	023	Whorl flower	QUIRLBLUME (*f*)
Morinda citrifolia	024	Citron-leaved Indian mulberry	ZITRONBLÄTTRIGE INDISCHE MAULBEERE (*f*); SURINGI--WURZEL (*f*)
Moringa aptera	025	Moringa (See *Moringa pterygosperma*)	MORINGA (*f*)
Moringa pterygosperma	026	Ben-nut tree; (Indian)-horse-radish tree	MORINGA (*f*); ECHTER BENNUSS--BAUM (*m*)
Morus alba	027	Mulberry, white mulberry	MAULBEERBAUM (*m*); MAUL--BEERE (*f*); WEISSER MAUL--BEERBAUM (*m*)
Morus nigra	028	Mulberry, black mulberry, common mulberry	MAULBEERBAUM (*m*); MAUL--BEERE (*f*); SCHWARZER MAUL--BEERBAUM (*m*)
Morus pendula	029	Weeping mulberry	HÄNGENDER MAULBEERBAUM (*m*)
Morus rubra	030	Red mulberry	ROTER MAULBEERBAUM (*m*)
Morus tinctoria	031	Dyer's mulberry; old fustic; yellow Brazil wood (See *Broussonetia tinctoria*)	FÄRBERMAULBEERBAUM (*m*); GELBHOLZ (*n*)
Mucor mucedo	032	Mucor mould	KOPFSCHIMMEL (*m*); MIST-KOPF--SCHIMMEL (*m*)
Muehlenbeckia complexa	033	Fiddle-leaved muehlenbeckia, native Australian ivy	MÜHLENBECKIE (*f*)
Mulgedium alpinum	034	(See *Lactuca alpina*)	ALPEN-MILCHLATTICH (*m*)
Murraya exotica	035	Chinese box tree	CHINA BUCHSBAUM (*m*)
Musa paradisiaca	036	Banana; plantain; pisang	PARADIESFEIGE (*f*); PISANG (*m*)
Musa sapientum	037	Banana plant	BANANE (*f*)
Musa textilis	038	Abaca plantain; Manilla plantain	ABACA (*f*); MANILAHANFPFLANZE (*f*)
Muscari comosum	039	Tassel hyacinth	SCHOPFBLÜTIGE PERL-HYAZINTHE (*f*)
Muscari comosum monstrosum	040	Feather hyacinth	FEDERHYAZINTHE (*f*); FEDER--PERLHYAZINTHE (*f*)
Muscari moschatum	041	Musk hyacinth, musk grape hyacinth	MOSCHUSHYAZINTHE (*f*); MOS--CHUS-PERLHYAZINTHE (*f*); BISAM-PERLHYAZINTHE (*f*)

Botanical Term.	No.	English Term.	German Term.
Muscari racemosum	042	Grape hyacinth, starch hyacinth	TRAUBENHYAZINTHE (*f*); TRAU--BIGE PERLHYAZINTHE (*f*); TRÄUBEL (*m*)
Muscus cranii humani	043	(See *Parmelia perlata*)	
Mycetoma	044	Fungus-foot	MADURAFUSS (*m*)
Mycoderma aceti	045	Vinegar plant, vinegar fungus (See *Ulvina aceti*)	KAHMPILZ (*m*); ESSIGMUTTER (*f*)
Myosotis	046	Scorpion-grass; mouse-ear; forget-me-not	MAUSEOHR (*n*); VERGISSMEIN--NICHT (*n*)
Myosotis arvensis	047	Field scorpion-grass; field forget-me-not	ACKERMAUSEOHR (*n*); ACKER-VERGISSMEINNICHT (*n*)
Myosotis collina	048	Early scorpion-grass; early forget-me-not	HÜGEL-MAUSEOHR (*n*); HÜGEL--VERGISSMEINNICHT (*n*)
Myosotis palustris	049	Forget-me-not; (Great water)-scorpion-grass; water myosote; mouse-ear scorpion grass	MAUSEOHR (*n*); SUMPF-MAUSE--OHR (*n*); SUMPF-VERGISS--MEINNICHT (*n*)
Myosotis sylvatica	050	Wood (myosote) scorpion-grass; wood forget-me-not	WALDMAUSEOHR (*n*); WALD--VERGISSMEINNICHT (*n*)
Myosotis versicolor	051	Changing scorpion-grass; changing forget-me-not	WECHSELFARBIGES MAUSEOHR (*n*); WECHSELFARBIGES VERGISS--MEINNICHT (*n*)
Myosurus	052	Mousetail	MÄUSESCHWANZ (*m*)
Myosurus minimus	053	Common mousetail	KLEINSTER MÄUSESCHWANZ (*m*)
Myrica acris	054	Myrica; bay; gale; candle-berry myrtle	GAGEL (*m*)
Myrica asplenifolia	055	Fern-leaved myrtle (See *Liquidambar asplenifolia*)	FARNBLÄTTRIGER GAGELSTRAUCH (*m*); MILZFARNBLÄTTRIGER GAGEL (*m*)
Myrica caroliniensis	056	(See *Myrica cerifera*)	
Myrica cerifera	057	Wax-myrtle; bayberry; candle-berry myrtle	WACHSMYRTE (*f*); WACHSBAUM (*m*); WACHSGAGELSTRAUCH (*m*); GAGEL (*f*); KERZEN--BEERSTRAUCH (*m*); WACHS--GAGEL (*f*); LICHTMYRTE (*f*)
Myrica comptonia	058	(See *Myrica asplenifolia*)	
Myrica gale	059	Bog myrtle; sweet gale	TORFGAGELSTRAUCH (*m*); BRA--BANNTER MYRTE (*f*); ECHTER GAGEL (*m*)
Myrica palustris	060	(See *Myrica gale*)	
Myricaria germanica	061	(See *Tamarix germanica*)	BIRZSTRAUCH (*m*)
Myriophyllum	062	Water milfoil	TAUSENDBLATT (*n*)
Myriophyllum spicatum	063	Spiked water-milfoil	ÄHRENBLÜTIGES TAUSENDBLATT (*n*)
Myriophyllum verticillatum	064	Whorled water-milfoil	QUIRLBLÜTIGES TAUSENDBLATT (*n*)
Myristica fragans	065	Nutmeg (tree); mace	MUSKATBLÜTE (*f*); MACIS (*f*); MAZIS (*f*); MUSKATENBAUM (*m*); MUSKATNUSSBAUM (*m*)
Myristica moschata	066	Nutmeg tree	MUSKATNUSSBAUM (*m*)
Myristica officinalis	067	(See *Myristica fragans*)	
Myristica otoba	068	American nutmeg tree, American mace tree	OTOBABAUM (*m*); OKTOBABAUM (*m*); OKUBABAUM (*m*)
Myroxylon peruvianum	069	Peru balsam tree, Peruvian balsam tree	PERUBALSAMBAUM (*m*)
Myrrhis odorata	070	Sweet Cicely; myrrh	MYRRHE (*f*)
Myrsiphyllum asparagoides	071	(See *Asparagus officinalis*)	
Myrtus communis	072	Common myrtle	MYRTE (*f*)
Myxomycetes	073	Slime fungus, slime mould	SCHLEIMPILZ (*m*)
Naiadaceæ	074	Naiadaceæ, Naiades	NAJADACEEN (*f.pl*)
Naias flexilis	075	Slender naias; slender naiad	BIEGSAMES NIXENKRAUT (*n*)
Naias graminea	076	Grassy (naias) naiad	GRASSBLÄTTRIGES NIXENKRAUT (*n*)
Naias marina	077	Holly-leaved (naias) naiad	GROSSES NIXENKRAUT (*n*)
Naias minor	078	Small naiad, small naias	KLEINES NIXENKRAUT (*n*)
Najas	079	(See *Naias*)	
Nandina domestica	080	Heavenly bamboo	GARTEN-NANDINE (*f*)
Napoleona imperialis	081	(African)-belvisia; Imperial napoleona	KAISER-NAPOLEONSBLUME (*f*)

Botanical Term.	No.	English Term.	German Term.
Narcissus bicolor	082	Two-coloured narcissus (See *Narcissus pseudo-narcissus bicolor*)	ZWEIFARBIGE NARZISSE (*f*)
Narcissus biflorus	083	Peerless primrose; two-flowered narcissus	ZWEIBLÜTIGE NARZISSE (*f*)
Narcissus bulbocodium	084	Hoop-(ed)-petticoat daffodil (See *Narcissus corbularia*)	FEENLILIE (*f*); KNOLLIGE NAR-ZISSE (*f*)
Narcissus corbularia	085	(See *Narcissus bulbocodium*)	
Narcissus cyclamineus	086	Cyclamen-flowered daffodil	ZYKLAMENBLÜTIGE NARZISSE (*f*)
Narcissus incomparabilis	087	Chalice-cup daffodil	UNVERGLEICHLICHE NARZISSE (*f*)
Narcissus jonquilla	088	Jonquil	BINSENNARZISSE (*f*); JONQUILLE (*f*)
Narcissus juncifolius	089	Rush-leaved daffodil	BINSENBLÄTTRIGE NARZISSE (*f*)
Narcissus moschatus	090	Musk daffodil	MOSCHUSNARZISSE (*f*)
Narcissus odorus	091	Campernel	WOHLRIECHENDE NARZISSE (*f*); CAMPERNELLE (*f*)
Narcissus odorus plenus	092	Queen Anne's Jonquil	DOPPELBLÜTIGE CAMPERNELLE (*f*)
Narcissus papyraceus	093	Paper narcissus	PAPIERNARZISSE (*f*)
Narcissus poeticus	094	Poet's narcissus, poet's daffodil, pheasant's-eye narcissus	DICHTER-NARZISSE (*f*); ECHTE NARZISSE (*f*)
Narcissus pseudo-narcissus	095	Daffodil; Lent-lily; daffy-down-dilly; narcissus	GELBE MÄRZBLUME (*f*); HOR-NUNGSBLUME (*f*); PAPPEN-BLUME (*f*); NARZISSE (*f*); NARCISSE (*f*); GELBE NAR-ZISSE (*f*); FEENLILIE (*f*)
Narcissus pseudo-narcissus bicolor	096	(See *Narcissus bicolor*)	
Narcissus pseudo-narcissus capax plenus	097	Lemon double daffodil	ZITRONENGELBE DOPPELBLÜTIGE NARZISSE (*f*)
Narcissus pseudo-narcissus cernuus plenus	098	White double daffodil	WEISSE DOPPELBLÜTIGE NARZISSE (*f*)
Narcissus pseudo-narcissus lobularis plenus	099	Yellow double daffodil	GELBE DOPPELBLÜTIGE NARZISSE (*f*)
Narcissus pseudo-narcissus telemonius plenus	100	(Yellow) double English daffodil	GELBE DOPPELBLÜTIGE ENGLISCHE NARZISSE (*f*)
Narcissus tazetta	101	Tazetta narcissus; Polyanthes daffodil	TAZETTE (*f*)
Nardus stricta	102	Mat-grass	BORSTENGRAS (*n*)
Narthecium ossifragum	103	Bog asphodel, Lancashire asphodel	KUPFERWURZEL (*f*); GRASLILIE (*f*)
Nasturtium amphibium	104	Great watercress; amphibious yellow-cress (See *Armoracia amphibium* and *Sisymbrium amphibium*)	WASSERKRESSE (*f*)
Nasturtium capucine	105	Trophy-wort, Indian cress	KAPUZINERKRESSE (*f*)
Nasturtium officinale	106	Water cress, common water-cress (See also *Tropæolum minus* and *Sisymbrium nasturtium*)	MEERRETTICH (*m*); BRUNNEN-KRESSE (*f*)
Nasturtium palustre	107	Marsh watercress	SUMPF-BRUNNENKRESSE (*f*)
Nasturtium sylvestre	108	Creeping watercress, creeping yellow-cress (See *Sisymbrium sylvestre*)	WALDBRUNNENKRESSE (*f*)
Nasturtium terrestre	109	Annual yellow-cress (See *Sisymbrium terrestre*)	EINJÄHRIGE BRUNNENKRESSE (*f*)
Naumburgia thyrsiflora	110	Tufted loosestrife (See *Lysimachia thyrsiflora*)	STRAUSSBLÜTIGE NAUMBURGIE (*f*)
Nectandra puchurz major	111	Brazilian muscatel	PICHURIMBOHNE (*f*)
Nectandra rodiœi	112	Bebeeru, greenheart tree; bibiri	BAYBEERENBAUM (*m*)
Nectria cinnabarina	113	Canker fungus (See *Tubercularis vulgaris*)	ZINNOBERROTE NEKTRIE (*f*)
Negundium americanum	114	Ash-leaved maple (See *Negundo aceroides*)	
Negundo aceroides	115	Ash-leaved maple; box-elder (See *Negundium americanum* and *Negundo fraxinifolium*)	ESCHENBLÄTTRIGER AHORN (*m*)
Negundo californicum	116	(See *Acer californicum*)	
Negundo fraxinifolium	117	(See *Negundo aceroides*)	
Neillia amurensis	118	Nine bark (See *Spiræa amurensis*)	TRAUBENSPIERE (*f*)

Botanical Term.	No.	English Term.	German Term.
Nelumbium luteum	119	Chinese water lily; Egyptian bean; sacred bean; water bean	GELBBLÜHENDE NELUMBO (*f*)
Nelumbo lutea	120	(See *Nelumbium luteum*)	
Nemophila	121	Californian blue-bell	HAINBLUME (*f*); HAINSCHÖN-CHEN (*n*)
Neottia Nidus-avis	122	Bird's-nest-(orchid); bird's-nest-(orchis)	ORCHIDEE (*f*); ORCHE (*f*); VOGEL-NEST (*n*); GEMEINE NESTWURZ (*f*)
Nepenthes	123	Pitcher plant	KANNENSTRAUCH (*m*)
Nepeta cataria	124	Cat-mint; catnip; field balm; balm-mint	BERGMINZE (*f*); KATZENMINZE (*f*); KATZENMELISSE (*f*)
Nepeta glechoma	125	Ground-ivy; haymaids; ale-hoof; (See *Glechoma hederacea*)	GUNDERMANN (*m*); GUNDELREBE (*f*)
Nepeta hederacea	126	Ground-ivy (See *Nepeta glechoma* and *Glechoma hederacea*)	
Nephelium lappaceum	127	Burdock-like nephelium	RAMBUTAN (*m*)
Nephelium longana	128	(Indian)-longan tree (See *Euphorbia longana*)	LONGANBAUM (*m*)
Nephelium mutabile	129	Nephelium	PULASSAN (*m*)
Nephrodium aemulum	130	Hay-scented buckler fern (See *Lastrea recurva*)	NIERENFARN (*m*); HEUDUFTEN-DER NIERENFARN (*m*)
Nephrodium cristatum	131	Crested buckler fern (See *Aspidium cristatum* and *Lastrea cristata*)	KAMMFÖRMIGER NIERENFARN (*m*)
Nephrodium dryopteris	132	Oak fern	EICHENFARN (*m*)
Nephrodium filix-mas	133	Male fern, buckler fern (See *Lastrea filix-mas*)	WURMFARN (*m*)
Nephrodium fragrans	134	Fragrant wood fern	WOHLRIECHENDER NIERENFARN (*m*)
Nephrodium montanum	135	Mountain buckler fern (See *Lastrea oreopteris*)	BERG-NIERENFARN (*m*)
Nephrodium patens	136	Spreading wood fern	AUSGEDEHNTER NIERENFARN (*m*)
Nephrodium phegopteris	137	Beech fern (See *Phegopteris polypodioides*)	BUCHENFARN (*m*)
Nephrodium rigidum	138	Rigid buckler-fern	STEIFER NIERENFARN (*m*)
Nephrodium robertianum	139	Stork's-bill fern	STORCHSCHNABELFARN (*m*)
Nephrodium spinulosum	140	Prickly shield fern (See *Lastrea dilatata*)	DORNIGER NIERENFARN (*m*)
Nephrodium thelypteris	141	Female buckler fern, lady fern; marsh shield fern (See *Lastrea thelypteris*)	SUMPF-NIERENFARN (*m*)
Nephrolepsis cordifolia tuberosa	142	ladder fern; tuberous-rooted heart-shaped ladder fern	KNOLLENTRAGENDER NIEREN-SCHUPPENFARN (*m*)
Nerine sarniensis	143	Guernsey lily	GUERNSEYLILIE (*f*)
Nerium oleander	144	Common oleander; rose-bay	OLEANDER (*m*)
Nertera depressa	145	Bead plant, fruiting duckweed	KORALLENBEERE (*f*)
Neviusa alabamiensis	146	Alabama snow wreath	TRAUBENKERRIE (*f*)
Nicandra physaloides	147	Apple-of-Peru	PERUANISCHE SCHLUTTENARTIGE GIFTBEERE (*f*)
Nicotiana tabacum	148	Common tabacco plant	TABAK (*m*)
Nierembergia rivularis	149	Cup flower	BACH-NIEREMBERGIA (*f*)
Nigella damascena	150	Love-in-a-mist, love-in-a-puzzle, fennel-flower, devil-in-a-bush, love-in-the-mist, ragged lady	PASSIONSBLUME (*f*); BRAUT IN GRÜNEN; BRAUT IN HAAREN; GRETCHEN IM BUSCH; JUNG-FER IN GRÜNEN; JUNGFER IN HAAREN; KAPUZINERKRAUT (*n*); JUNGFER IM GRÜN (*f*)
Nigella sativa	151	Nutmeg flower	SCHWARZKÜMMEL (*m*); NARDEN-SAME (*m*); RÖMISCHER KORI-ANDER (*m*); SCHWARZER KORIANDER (*m*); NONNEN-NÄGELEIN (*n*)
Nigritella angustifolia	152	Small-leaved cabbage-rose	SCHMALLBLÄTTRIGES KOHLRÖS-CHEN (*n*); BRÄNDLI (*n*); KUHBRÄNDLI (*n*); KAMM-BLÜMLE (*n*); SCHWARZ-STÄNDEL (*m*)

Botanical Term.	No.	English Term.	German Term.
Nipa fruticans	153	Attap-(palm); East Indian attap	ATTAP-PALME (*f*); ATAPPALME (*f*); NIPAPALME (*f*)
Nitella syncarpa	154	Smooth chara (See *Characeæ*)	ARMLEUCHTERALGE (*f*)
Nolana prostrata	155	Chilian bell flower	GLOCKENWINDE (*f*)
Nopalea coccinellifera	156	Nopal; cochineal cactus (See *Opuntia coccinellifera*)	NOPAL (*m*); NOPALPFLANZE (*f*)
Nothochlæna dealbata	157	Cloak fern (See *Cincinalis dealbata*)	FILZFARN (*m*)
Nothochlæna nivea	158	Silver maiden-hair fern	SCHNEEWEISSER FILZFARN (*m*)
Nothoclæna	159	(See *Nothochlæna*)	
Notholæna	160	(See *Nothochlæna*)	
Notochlæna	161	(See *Nothochlæna*)	
Nuphar lutea	162	Yellow water lily; brandy bottle (See also *Nymphæa alba*)	NIXBLUME (*f*)
Nuttallia cerasiformis	163	Oso-berry tree	KIRSCHENARTIGE NUTTALLIA (*f*)
Nyctagineæ	164	Nyctagineæ, night-blooming plants	NACHTBLÜTLER (*m.pl*)
Nyctanthes arbor-tristes	165	Tree of sadness	BAUM (*m*) DER TRAURIGKEIT (*f*)
Nymphæa alba	166	White water lily (See also *Nuphar lutea*)	KOLBWURZ (*f*); NYMPHENBLUME (*f*); NIXBLUME (*f*); MUMMEL (*m*); NIZENBLUME (*f*); PLUMPE (*f*); WASSERLILIE (*f*); WASSERROSE (*f*)
Nymphæaceæ	167	Water-lilies	WASSERROSEN (*f.pl*); WASSER-LILIEN (*f.pl*)
Nymphæa lotus	168	Egyptian lotus	LOTUSBLUME (*f*)
Nyssa aquatica	169	Tupelo tree	TUPELOBAUM (*m*)
Nyssa multiflora	170	Black-gum; sour-gum; gum-tree (See *Nyssa villosa* and *Liquidambar styraciflora*)	
Nyssa silvatica	171	(North American)-mountain tupelo tree	BERG-TUPELOBAUM (*m*)
Nyssa villosa	172	Black-gum; sour-gum; yellow-gum (See *Nyssa multiflora*)	
Obione pedunculata	173	Salt-bush, salt orache, stalked orache (See *Atriplex pedunculata*)	SALZBUSCH (*m*); SALZMELDE (*f*); KEILMELDE (*f*); STIELFRÜCHT-IGE SALZMELDE (*f*)
Ochropsora sorbi	174	Mountain-ash yellow rust	EBERESCHEN-GELBROST (*m*)
Ocimum	175	(See *Ocymum*)	
Ocotea bullata	176	South African stinkwood tree	STINKHOLZBAUM (*m*); KAP-LOR-BEERBAUM (*m*); KAP-WAL-NUSSBAUM (*m*)
Ocymum basilicum	177	Sweet basil; basil	BASILIENKRAUT (*n*); HIRNKRAUT (*n*)
Ocymum minimum	178	(Chilean)-bush basil	CHILE-BASILIENKRAUT (*n*); KLEINSTES BASILIENKRAUT (*n*)
Ocymum sanctum	179	Holy basil	HEILIGES BASILIENKRAUT (*n*)
Ocymum viride	180	(West African)-fever plant	GRÜNES BASILIENKRAUT (*n*)
Odontoglossum	181	Violet scented orchid; almond-scented orchid	ODONTOGLOSSUM-ORCHE (*f*)
Odontospermum	182	(See *Anastatica hierochuntina*)	
Oenanthe crocata	183	Water-hemlock; hemlock œnanthe; hemlock water-dropwort	SAFRAN-REBENDOLDE (*f*)
Oenanthe fistulosa	184	(Common)-water-dropwort; tubular water-dropwort; common œnanthe	PFERDESAAT (*f*); REBENDOLDE (*f*)
Oenanthe phellandrium	185	Fine-leaved œnanthe	ROSSFENCHEL (*m*); FEINBLÄTT-RIGER ROSSFENCHEL (*m*)
Oenanthe pimpinelloides	186	Parsley œnanthe	PIMPERNELLÄHNLICHE REBEN-DOLDE (*f*)
Oenocarpus batava	187	Wine palm	MOSTPALME (*f*)
Oenothera acaulis	188	Stemless evening primrose (See *Oenothera taraxicifolia*)	STENGELLOSES NACHTRÖSCHEN (*n*)

Botanical Term.	No.	English Term.	German Term.
Oenothera biennis	189	Evening primrose	NACHTKERZE (*f*); NACHTRÖSCHEN (*n*); RAPUNZEL (*f*)
Oenothera fruticosa	190	Sun-drops; shrubby evening primrose	NACHTKERZE (*f*); STRAUCHIGE NACHTKERZE (*f*)
Oenothera taraxicifolia	191	White-flowering evening primrose; dandelion-leaved evening primrose (See *Oenothera acaulis*)	WEISS-BLÜHENDES NACHTRÖSCHEN (*n*)
Oidium balsamii	192	Turnip mildew	RÜBENMELTAU (*m*)
Oidium monilioides	193	Cereal mildew (See *Erysiphe graminis*)	GETREIDEMELTAU (*m*)
Oidium tuckeri	194	Grape-vine mildew	REBENMELTAU (*m*)
Oldfieldia africana	195	African oak; African teak	AFRIKANISCHE EICHE (*f*)
Olea aquifolia	196	Holly-leaved olive	STECHPALMENBLÄTTRIGER ÖL-BAUM (*m*)
Oleaceæ	197	Oleaceæ	ÖLBAUMARTIGE GEWÄCHSE (*n.pl*)
Olea europæa	198	Olive, olive tree, wild olive, oleaster	ÖLBAUM (*m*); OLIVENBAUM (*m*); ECHTER ÖLBAUM (*m*); WILDER ÖLBAUM (*m*); EUROPÄISCHER ÖLBAUM (*m*)
Olea fragrans	199	(See *Osmanthus fragrans*)	
Oleandra articulata	200	(African)-oleander fern; jointed oleander fern	AFRIKANISCHER OLEANDERFARN (*m*)
Oleandra nodosa	201	Knotty oleander fern	KRIECHENDER OLEANDERFARN (*m*)
Olea oleaster	202	(See *Olea europæa* and *Elæagnus*)	
Olearia haastii	203	New Zealand daisy bush	HAASTS-OLEARIE (*f*)
Olearia stellulata	204	Victorian snow-bush	STERN-OLEARIE (*f*)
Omphalea triandra	205	(Jamaica)-omphalea	OMPHALEA (*f*)
Omphalobium lambertii	206	Zebra-wood tree	ZEBRAHOLZBAUM (*m*)
Omphalodes linifolia	207	Venus's navelwort	LEINBLÄTTRIGES GEDENKEMEIN (*n*)
Omphalodes luciliæ	208	Rock navel-wort	FELSEN-GEDENKEMEIN (*n*)
Omphalodes verna	209	Creeping forget-me-not	MÄNNERTREU (*f*); MANNSTREU (*f*); FRÜHLINGS-GEDENKEMEIN (*n*)
Oncidium papilio	210	(West Indian)-butterfly plant; butterfly orchid	SCHMETTERLING-ORCHE (*f*)
Onobrychis sativa	211	Saintfoin; sainfoin, cock's-head-(clover) (See *Hedysarum onobrychis*)	ESPARSETTE (*f*); SCHILDKLEE (*m*); KATZENKOPFKLEE (*m*)
Onoclea germanica	212	Ostrich fern (See *Struthiopteris germanica*)	PERLFARN (*m*)
Onoclea sensibilis	213	Sensitive fern	REIZBARER PERLFARN (*m*)
Ononis antiquorum	214	Spinous rest-harrow (See *Ononis spinosa*)	
Ononis arvensis	215	Trailing rest-harrow; common rest-harrow, cammock	GEMEINE HAUHECHEL (*f*)
Ononis natrix	216	Goat-root	GELBE HAUHECHEL (*f*)
Ononis reclinata	217	Irish rest-harrow; spreading rest-harrow; small rest-harrow	LIEGENDE HAUHECHEL (*f*)
Ononis spinosa	218	Spinous rest-harrow	HAUHECHEL (*f*)
Onopordon acanthium	219	Cotton thistle; Scotch thistle	FRAUENDISTEL (*f*); MARIEN-DORN (*m*); KREBSDISTEL (*f*)
Onosma echioides	220	Golden drop	LOTWURZ (*f*)
Ophioglossum vulgatum	221	Adder's-tongue-(fern), adder's spear, serpent's tongue	NATTERZUNGE (*f*); SCHLANGEN-ZUNGE (*f*)
Ophiopogon	222	Snake's-beard	SCHLANGENBART (*m*)
Ophiorrhiza mangora	223	Snake-root	SCHLANGENWURZEL (*f*)
Ophioxylon serpentinum	224	Serpent-wood	SCHLANGENHOLZ (*n*)
Ophrys	225	Orchis	ORCHIDEE (*f*); ORCHE (*f*); FRAU-ENTRÄNE (*f*)
Ophrys apifera	226	Bee orchis	BIENEN-ORCHIS (*f*); BIENEN-ORCHE (*f*)
Ophrys arachnites	227	Late spider orchis	INSEKTENORCHE (*f*); SPÄTE SPINNENORCHE (*f*)
Ophrys aranifera	228	Spider orchis	SPINNENORCHE (*f*)
Ophrys muscifera	229	fly orchis	FLIEGENORCHE (*f*)
Ophrys tenthredinifera	230	Sawfly orchis	FLIEGENKRAUT (*n*)

Botanical Term.	No.	English Term.	German Term.
Oplismenus burmannii variegatus	231	Variegated panicum (See *Panicum variegatum*)	
Opuntia coccinellifera	232	Indian fig, cochineal cactus, nopal (See *Nopalea coccinellifera*)	NOPAL (*m*); NOPALPFLANZE (*f*); KOSCHENILLENBAUM (*m*)
Opuntia Ficus-Indica	233	Indian fig; prickly pear, barbary fig (See *Ficus Bengalensis* and *Ficus Indica*)	INDISCHE FEIGE (*f*)
Orchideæ	234	Orchideæ (See *Orchis, Ophrys, Habenaria, Aceras, Neottia*)	KUCKUCKSBLÜTLER (*m.pl*); ORCHI-DEEN (*f.pl*)
Orchis	235	Orchis	HODE (*f*); KNABENKRAUT (*n*); KUCKUCKSBLUME (*f*); ORCHI-DEE (*f*); RAGWURZ (*f*); ORCHE (*f*)
Orchis conopsea	236	(See *Habenaria conopsea*)	
Orchis coriophora	237	Bug orchis	WANZENORCHE (*f*)
Orchis foliosa	238	Madeira orchis	MADEIRAORCHE (*f*)
Orchis globosa	239	Globe orchis; round spiked orchis	KUGELORCHE (*f*)
Orchis hircina	240	Lizard orchis (See *Himantoglossum hircinum*)	EIDECHSENORCHE (*f*); BOCKS-ORCHE (*f*); BOCKS-RIEMEN-ZUNGE (*f*); BOCKS-ROLL-ZUNGE (*f*)
Orchis latifolia	241	Marsh orchis (See also *Orchis maculata*)	GLÜCKSHÄNDCHEN (*n*); JOHAN-NISHÄNDCHEN (*n*)
Orchis laxiflora	242	Loose orchis, Guernsey orchis	LOSBLÜTIGE ORCHE (*f*)
Orchis maculata	243	Spotted-(palmate)-orchis (See also *Orchis latifolia*)	GEFLECKTES KNABENKRAUT (*n*); FLECKENORCHE (*f*); GLÜCKS-HÄNDCHEN (*n*); CHRISTUS-HÄNDCHEN (*n*); MUTTER-GOTTESHÄNDCHEN (*n*)
Orchis mascula	244	Early-(purple)-orchis; dead-man's-fingers	MÄNNLICHES KNABENKRAUT (*n*)
Orchis militaris	245	Military orchis	KNABENKRAUT (*n*); SOLDATEN-ORCHE (*f*); HELMKNABEN-KRAUT (*n*)
Orchis morio	246	Green-winged-(meadow)-orchis	HARLEKIN (*m*); NARRENORCHE (*f*)
Orchis pallens	247	Pale orchis	BLASSORCHE (*f*); BLAUES KNAB-ENKRAUT (*n*)
Orchis purpurea	248	Lady orchis	PURPURNE ORCHE (*f*); DAMEN-ORCHE (*f*)
Orchis pyramidalis	249	Pyramidal orchis (See *Anacamptis pyramidalis*)	PYRAMIDENORCHE (*f*); HUNDS-WURZ (*f*)
Orchis sambucina	250	Elder-scented orchis	HOLUNDERDUFTENDES KNABEN-KRAUT (*n*); HOLUNDER-ORCHE (*f*)
Orchis simia	251	Simian orchis	AFFENORCHE (*f*)
Orchis spectabilis	252	North American orchis	PRACHTORCHE (*f*)
Orchis tridentata	253	Three-toothed orchis	DREIZAHNORCHE (*f*)
Orchis ustulata	254	Dark-winged-(dwarf)-orchis	BRANDORCHE (*f*)
Oreodoxa oleracea	255	(West Indian)-cabbage palm (See *Areca oleracea*)	WESTINDISCHE KOHLPALME (*f*)
Oreodoxa regia	256	King palm	KÖNIGSPALME (*f*)
Origanum dictamnus	257	Dittany of Crete (See *Dictamnus fraxinella*)	
Origanum marjorana	258	Sweet marjoram; garden marjoram; knotted marjoram (See *Majorana hortensis*)	MAJORAN (*m*); MEIRAN (*m*); ORIGANUMKRAUT (*n*); WURST-KRAUT (*n*)
Origanum onites	259	Pot marjoram (See *Majorana*)	WURSTKRAUT (*n*)
Origanum vulgare	260	Marjoram; common marjoram; wild marjoram	DOST (*m*); DOSTEN (*m*); MAJOR-AN (*m*); MEIRAN (*m*); WOHL-GEMUT (*m*)
Ornithogalum nutans	261	Drooping star-of-Bethlehem	NICKENDER MILCHSTERN (*m*)
Ornithogalum pyrenaicum	262	Spiked star of Bethlehem	PYRENAIISCHER MILCHSTERN (*m*); VOGELMILCH (*f*)

Botanical Term.	No.	English Term.	German Term.
Ornithogalum umbellatum	263	Common star of Bethlehem, onion plant; six-o'clock-flower	MILCHSTERN (*m*); DOLDIGER MILCHSTERN (*m*); DOLDEN--TRAUBIGER MILCHSTERN (*m*)
Ornithopus ebracteatus	264	Sand bird's-foot (See *Arthrolobium ebracteatum*)	SAND-KLAUENSCHOTE (*f*)
Ornithopus perpusillus	265	Birdsfoot; common birdsfoot	KLEINE KLAUENSCHOTE (*f*)
Ornithopus sativus	266	Serradilla, trefoil	SAATVOGELFUSS (*m*)
Orobanche cœrulea	267	Blue broomrape	BLAUE SOMMERWURZ (*f*)
Orobanche caryophyllacea	268	Clove-scented broom-rape	GEWÜRZNELKENDUFTENDE SOM--MERWURZ (*f*)
Orobanche cruenta	269	(Sainfoin)-broom-rape	BLUTROTE SOMMERWURZ (*f*)
Orobanche elatior	270	Tall broom-rape	HOHE SOMMERWURZ (*f*)
Orobanche hederœ	271	Ivy broom-rape	EFEU-SOMMERWURZ (*f*)
Orobanche major	272	Broom rape; great broom-rape	GROSSE SOMMERWURZ (*f*)
Orobanche minor	273	(Clover)-broom-rape; lesser broom-rape	KLEINE SOMMERWURZ (*f*); KLEE--TEUFEL (*m*)
Orobanche pruinosa	274	(Bean)-broom-rape	BOHNENTOD (*m*)
Orobanche ramosa	275	Branched broom-rape	HANF-SOMMERWURZ (*f*)
Orobanche rapsum	276	Greater broom-rape	RAPS-SOMMERWURZ (*f*)
Orobanche rubens	277	(Lucern)-broom-rape	LUZERN-SOMMERWURZ (*f*); RÖT--LICHE SOMMERWURZ (*f*)
Orobanche rubra	278	Red broom-rape	ROTE SOMMERWURZ (*f*)
Orobus lathyroides	279	(See *Vicia orobus* and *Vicia orobioides*)	STACHELSPITZIGE PLATTERBSE (*f*)
Orontium aquaticum	280	Golden club	GOLDKEULE (*f*); GOLDKOLBEN (*m*)
Oryza sativa	281	Rice-(grass)	REIS (*m*)
Osmanthus americanus	282	American devil's-wood tree	AMERIKANISCHER TEUFELSHOLZ--BAUM (*m*)
Osmanthus aquifolium	283	Holly-leaved olive	STECHPALMENBLÄTTRIGER SCHEINÖLBAUM (*m*)
Osmanthus fragrans	284	Fragrant olive (See *Olea fragrans*)	WOHLRIECHENDER ÖLBAUM (*m*); WOHLRIECHENDER SCHEINÖL--BAUM (*m*)
Osmunda regalis	285	Flowering fern; royal fern	RISPENFARN (*m*)
Ostrya carpinifolia	286	Hop-hornbeam; Caucasian iron-wood	GEMEINE HOPFENBUCHE (*f*)
Ostrya virginica	287	Ironwood; hop-hornbeam; American ironwood	EISENHOLZ (*n*); HOPFENBUCHE (*f*)
Othonna crassifolia	288	African rag-wort	AFRIKANISCHE RAGWURZ (*f*)
Ouvirandra fenestralis	289	Lattice leaf (lace leaf) plant (See *Aponogeton fenestralis*)	FENSTER-GITTERKRAUT (*n*); WAS--SERJAMS (*m*)
Oxalideœ	290	Oxalideæ	SAUERKLEEGEWÄCHSE (*n.pl*)
Oxalis acetosella	291	Wood sorrel; clover sorrel; shamrock	SAUERKLEE (*m*); KUCKUCKS--KLEE (*m*); GAUCHBROT (*n*); HASENKOHL (*m*); KLEESALZ--KRAUT (*n*); HASENAMFER (*m*); HAINAMFER (*m*)
Oxalis cernua	292	Bermuda butter-cup; drooping wood-sorrel	HÄNGENDER SAUERKLEE (*m*); NICKENDER SAUERKLEE (*m*)
Oxalis corniculata	293	Procumbent oxalis	GEHÖRNTER SAUERKLEE (*m*)
Oxycedrus	294	(See *Juniperus communis* and *Juniperus oxycedrus*)	
Oxycoccus macrocarpus	295	American cranberry (See *Vaccinium macrocarpum*)	GROSSFRÜCHTIGE MOOSBEERE (*f*)
Oxycoccus palustris	296	cranberry, bogberry, mossberry, moor-berry (See *Vaccinium oxycoccos*)	MOOSBEERE (*f*); KRANBEERE (*f*); KRANICHBEERE (*f*)
Oxydendron arboreum	297	Sorrel tree	SAUERBAUM (*m*)
Oxyria reniformis	298	Mountain sorrel; kidney sorrel; oxyrea	NIERENAMPFER (*m*)
Oxytropis campestris	299	Yellow oxytrope	FELD-SPITZKIEL (*m*)
Oxytropis montana	300	Mountain oxytrope	BERG-SPITZKIEL (*m*)
Oxytropis uralensis	301	Purple oxytrope; uralian oxy-trope	URAL-SPITZKIEL (*m*)
Pachyma cocos	302	Indian bread-(fungus); Tucka-hoe-(fungus)	LÖCHERSCHWAMM (*m*)
Padina paronia	303	Peacock weed	PFAUENTANG (*m*)

Botanical Term.	No.	English Term.	German Term.
Padus avium	304	(See *Prunus padus* and *Prunus avium*)	
Padus lauro-cerasus	305	(See *Prunus lauro-cerasus*)	KIRSCHLORBEER (*m*)
Padus oblonga	306	(See *Prunus virginiana*)	
Padus virginiana	307	(See *Prunus serotina*)	
Pœonia moutan	308	Tree peony	STRAUCH-PÄONIE (*f*)
Pœonia officinalis	309	Peony, common herbaceous peony, pæony, piony	PÄONIE (*f*); PFINGSTROSE (*f*); PFINGSTBLUME (*f*)
Paliurus aculeatus	310	Christ's thorn; garland thorn; Jew's thorn	STECHAPFEL (*m*); JUDENBAUM (*m*)
Panax quinquefolium	311	Ginseng; five-leaved ginseng (See also *Aralia quinquefolia*)	FÜNFBLÄTTRIGE KRAFTWURZ (*f*)
Panax trifolium	312	Dwarf ginseng; three-leaved ginseng (See *Aralia trifolia*)	DREIBLÄTTRIGE KRAFTWURZ (*f*)
Pancratium maritimum	313	Sea daffodil, Mediterranean lily	MEERSTRANDS-KRAFTZWIEBEL (*f*)
Pandanus odoratissimus	314	Screw pine	SCHRAUBENBAUM (*m*)
Panicum	315	Panic grass	BORSTENHIRSE (*f*); KNOLLEN-HIRSE (*f*); TURNIPGRAS (*n*)
Panicum crus-galli	316	Cockspur panicum, cockspur panick grass, panic grass	HADELGRAS (*n*); HÜHNERHIRSE (*f*); STACHELHIRSE (*f*)
Panicum glabrum	317	Glabrous panicum	KAHLE BORSTENHIRSE (*f*)
Panicum glaucum	318	Glaucous panicum (See *Setaria glauca*)	GELBHAARIGE BORSTENHIRSE (*f*); GELBHAARIGER FENNICH (*m*)
Panicum maximum	319	Guinea grass	GUINEAGRAS (*n*)
Panicum miliaceum	320	Millet grass (See also *Milium effusum*)	HIRSE (*f*); RISPENHIRSE (*f*)
Panicum molle	321	Para grass; soft panicum	PARAGRAS (*n*); WEICHE BORS-TENHIRSE (*f*)
Panicum sanguinale	322	Fingered panicum	BLUTHIRSE (*f*); BLUTFENNICH (*m*); FINGERHIRSE (*f*)
Panicum spectabile	323	Angola grass; showy panicum	ANGOLAGRAS (*n*); PRACHT-BORS-TENHIRSE (*f*)
Panicum variegatum	324	(See *Oplismenus burmannii variegatus*)	VERSCHIEDENBLÄTTRIGE BORS-TENHIRSE (*f*)
Panicum verticillatum	325	Rough panicum (See *Setaria verticillata*)	QUIRLBLÄTTRIGE BORSTENHIRSE (*f*)
Panicum viride	326	Green panicum; bottle-grass (See *Setaria viridis*)	GRÜNE BORSTENHIRSE (*f*)
Papaveraceœ	327	Papaveraceæ (See *Glaucium*, *Chelidonium*, *Meconopsis* and *Papaver*)	MOHNGEWÄCHSE (*n.pl*)
Papaver alpinum	328	Alpine poppy	ALPENMOHN (*m*)
Papaver argemone	329	Pale poppy	SANDMOHN (*m*)
Papaver dubium	330	Long-headed poppy	ZWEIFELHAFTE KLATSCHROSE (*f*)
Papaver glaucum	331	Tulip poppy; sea-green poppy	MEERGRÜNER MOHN (*m*)
Papaver hybridum	332	Rough poppy	KRUMMBORSTIGER MOHN (*m*); BASTARD-MOHN (*m*)
Papaver nudicaule	333	Iceland poppy	ARKTISCHER MOHN (*m*)
Papaver orientale	334	Oriental poppy	ORIENTALISCHER MOHN (*m*)
Papaver pavoninum	335	Peacock poppy	PFAUENMOHN (*m*)
Papaver rhœas	336	Field poppy; Shirley poppy, common red poppy; corn poppy (See *Lychnis githago*)	KORNROSE (*f*); KLAPPERROSE (*f*); KLATSCHROSE (*f*); MOHN (*m*)
Papaver rupifragum	337	Spanish poppy	SPANISCHER MOHN (*m*); FELSEN-MOHN (*m*)
Papaver somniferum	338	White poppy, opium poppy; mawseed	MOHNBLUME (*f*); MOHN (*m*)
Papilionaceœ	339	Papilionaceæ	PAPILIONACEEN (*f.pl*)
Papyrus antiquorum	340	(Egyptian)-paper reed, paper grass, paper plant, papyrus plant (See *Cyperus papyrus*)	PAPIERGRAS (*n*); PAPIERSTAUDE (*f*)
Paradisea liliastrum	341	St. Bruno's lily	PARADIESBAUM (*m*)
Parietaria	342	Wall pellitory; wall pelletory	GLASKRAUT (*n*)
Parietaria officinalis	343	Wall pellitory	GLASKRAUT (*n*); WANDKRAUT (*n*); GEBRÄUCHLICHES GLAS-KRAUT (*n*)

Botanical Term.	No.	English Term.	German Term.
Parinarium excelsum	344	Rough-skinned plum	WESTAFRIKANISCHE GRAUE PFLAUME (*f*)
Parinarium macrophyllum	345	(Guinea)-gingerbread plum	HONIGKUCHENPFLAUME (*f*)
Paris quadrifolia	346	Herb Paris; truelove; oneberry	EINBEERE (*f*)
Parmelia perlata	347	Parmelia (See *Muscus cranii humani*)	HIRNSCHÄDELMOOS (*n*); SCHILD-FLECHTE (*f*)
Parmentiera cerifera	348	Candle-tree	KERZENBAUM (*m*)
Parnassia palustris	349	Common Parnassus grass; grass of Parnassus	SUMPF-HERZBLATT (*n*)
Parochetus communis	350	Shamrock pea, blue-flowered shamrock	BLAUBLÜTIGER SAUERKLEE (*m*)
Paronychia argentea	351	Whitlow-wort, nail-wort	NAGELKRAUT (*n*)
Parottia jacquemontiana	352	Persian iron-wood	ROSENROTER TRANSKAUKASISCHER EISENHOLZBAUM (*m*)
Parthenocissus quinquefolia	353	(See *Vitis quinquefolia*, *Ampelopsis quinquefolia* and *Quinaria*)	
Pasania cuspidata	354	(See *Quercus cuspidata*)	
Passiflora cærulea	355	Passion flower	PASSIONSBLUME (*f*); OSTER-BLUME (*f*)
Passiflora maliformis	356	Water lemon; apple-formed passion flower	APFELFÖRMIGE PASSIONSBLUME (*f*)
Passiflora princeps	357	Scarlet passion flower (See *Tacsonia manicata*)	BLUTROTE PASSIONSBLUME (*f*)
Pastinaca sativa	358	Parsnip; common parsnip (See *Peucedanum sativum*)	PASTINAKE (*f*)
Paullinia sorbilis	359	Cascalotte	CASCALOTTE (*f*)
Pavetta borbonica	360	(Bourbon)-wild jasmine	WILDER BOURBON-JASMIN (*m*)
Pavia alba	361	(See *Aesculus parviflora*)	WEISSE ROSSKASTANIE (*f*)
Pavia lutea	362	(See *Aesculus lutea*)	GELBE ROSSKASTANIE (*f*)
Pavia rubra	363	(See *Aesculus pavia*)	ROTE ROSSKASTANIE (*f*)
Pedicularis palustris	364	Red rattle; marsh louse-wort	LÄUSEKRAUT (*n*)
Pedicularis sylvatica	365	Lousewort; red rattle	LÄUSEKRAUT (*n*)
Pedilanthus tithymaloides	366	Jew-bush, slipper spurge	CANDELILLASTRAUCH (*m*); SANTO DOMINGO IPECACUANHA (*f*)
Peireskia aculeata	367	(See *Pereskia aculeata*)	
Pekea butyrosa	368	(See *Caryocar butyrosum*)	
Pelargonium	369	Pelargonium, geranium	PELARGONIE (*f*); GERANIE (*f*)
Pelargonium capitatum	370	Rose-scented geranium	KOPF-PELARGONIE (*f*)
Pelargonium citriodorum	371	Citron-scented geranium	ZITRONEN-PELARGONIE (*f*)
Pelargonium crispum	372	Lemon-scented geranium	ZITRONENDUFTENDE PELARGONIE (*f*)
Pelargonium denticulatum filicifolium	373	Fern-leaved geranium	FARNBLÄTTRIGE GERANIE (*f*)
Pelargonium fragrans	374	Nutmeg-scented geranium	WOHLRIECHENDE PELARGONIE (*f*)
Pelargonium grandiflorum	375	Regal geranium, regal pelargonium	GROSSBLUMIGE PELARGONIE (*f*); BLUMISTEN-PELARGONIE (*f*)
Pelargonium hederæfolium	376	Ivy-leaved geranium (See *Pelargonium peltatum*)	EFEU-PELARGONIE (*f*)
Pelargonium inquinans	377	Scarlet geranium	BESCHMUTZTE PELARGONIE (*f*)
Pelargonium peltatum	378	(See *Pelargonium hederæfolium*)	SCHILD-PELARGONIE (*f*)
Pelargonium quercifolium	379	Oak-leaved geranium	EICHENBLÄTTRIGE PELARGONIE (*f*)
Pelargonium radula	380	Balsam-scented geranium	BÜRSTEN-PELARGONIE (*f*)
Pelargonium roseum	381	Rosy stork's-bill; rosy pelargonium	ROSEN-PELARGONIE (*f*); ROSEN-GERANIE (*f*); ROSENKRAUT (*n*)
Pelargonium tomentosum	382	Peppermint-scented geranium	MINZENDUFTENDE PELARGONIE (*f*)
Pelargonium tricolor	383	Three-coloured pelargonium	DREIFARBIGE PELARGONIE (*f*)
Pelargonium triste	384	Night-smelling geranium	NACHTDUFTENDE PELARGONIE (*f*)
Pelargonium zonale	385	Zonal geranium, horseshoe geranium	ZONAL-PELARGONIE (*f*); GÜRTEL-PELARGONIE (*f*)
Pelecyphora asseliformis	386	Hatchet cactus	MEXIKANISCHER KAKTUS (*m*)
Pellæa	387	Cliff brake fern (See *Platyloma* and also *Pteris*)	GLIEDERSTIELFARN (*m*)
Peltandra virginica	388	Arrow-arum (See also *Richardia africana*)	VIRGINISCHER ARON (*m*)
Peltigera	389	Dog lichen, ground liverwort	HUNDSFLECHTE (*f*)
Peltiphyllum peltatum	390	Peltate-leaved Californian shield-leaf	KALIFORNISCHES SCHILDBLATT (*n*)

Botanical Term.	No.	English Term.	German Term.
Pencillium	391	Mildew (See also *Aspergillus*)	PINSELSCHIMMEL (*m*)
Penicillium crustaceum	392	(See *Penicillium glaucum*)	BLAUGRÜNER PINSELSCHIMMEL (*m*)
Penicillium glaucum	393	Bread mould, mildew; green mould (See *Penicillium crustaceum*)	PINSELSCHIMMEL (*m*); SCHIMMEL--PILZ (*m*)
Pennisetum latifolium	394	Pennisetum (See *Gymnothrix latifolia*)	PINSELGRAS (*n*)
Pentadesma butyracea	395	African tallow tree	WESTAFRIKANISCHER TALGBAUM (*m*)
Pentastemon	396	(See *Pentstemon*)	
Pentstemon barbatus	397	Beard tongue (See *Chelone barbata*)	FÜNFFADEN (*m*)
Peperomia	398	Pepper-elder	ZWERGPFEFFER (*m*)
Peplis portula	399	Water purslane (See also *Portulaca oleracea*)	PORTULAK (*m*); GEMEINER AFTER--QUENDEL (*m*); BACHBURZEL (*f*); WASSER-PORTULAK (*m*); ZIPFELKRAUT (*n*)
Pereskia aculeata	400	Barbados gooseberry (See *Peireskia aculeata*)	BARBADOS-STACHELBEERE (*f*)
Peridermium pini	401	Fir-tree cluster-cup, bladder rust (See *Peridermium strobi*)	BLASENROST (*m*); KIEFERN--BLASENROST (*m*)
Peridermium strobi	402	(See *Peridermium pini*)	
Perilla ocimoides	403	Perilla	PERILLA (*f*)
Periploca græca	404	Silk-vine	GRIECHISCHE SCHLINGE (*f*)
Peristeria elata	405	Dove-orchid, dove-flower	HEILIGEN-GEIST BLUME (*f*); BLUME DES HEILIGEN GEISTES (*f*)
Pernettya mucronata	406	Prickly heath	STACHELSPITZIGE PERNETTYE (*f*); MYRTENKRÜGLEIN (*n*)
Peronospora	407	Mildew, false mildew, mould	FALSCHER MELTAU (*m*); TRAU--BENSPOR-EIPILZ (*m*)
Peronospora dipsaci	408	Teasel mould	WEBERKARDEN-TRAUBENSPOREI--PILZ (*m*)
Peronospora effusa	409	Spinach mould	SPINAT-TRAUBENSPOREIPILZ (*m*)
Peronospora ficariæ	410	Figwort mould (See *Peronospora sordida*)	FEIGENWURZ-TRAUBENSPOREI--PILZ (*m*)
Peronospora gangliformis	411	Lettuce mould	LATTICH-TRAUBENSPOREIPILZ (*m*)
Peronospora infestans	412	Potato (mould) mildew (See *Phytophthora infestans*)	KARTOFFEL-TRAUBENSPOREIPILZ (*m*); KARTOFFELKRANKHEIT (*f*)
Peronospora lamii	413	Dead-nettle mould	TAUBNESSEL-TRAUBENSPOREIPILZ (*m*)
Peronospora myosotidis	414	Forget-me-not mould	VERGISSMEINNICHT-TRAUBENSPOR--EIPILZ (*m*)
Peronospora nivea	415	Parsnip mould	PASTINAKEN-TRAUBENSPOREIPILZ (*m*)
Peronospora obliqua	416	(See *Peronospora rumicis*)	
Peronospora parasitica	417	Crucifer mould, cabbage mould	KRUZIFEREN - TRAUBENSPOREI--PILZ (*m*); KOHL-TRAUBEN--SPOREIPILZ (*m*)
Peronospora pygmæa	418	Anemone mould	ANEMONEN-TRAUBENSPOREIPILZ (*m*)
Peronospora rumicis	419	Dock mould (See *Peronospora obliqua*)	AMPFER-TRAUBENSPOREIPILZ (*m*)
Peronospora schachtii	420	Beet-root mould	ZUCKERRÜBEN-TRAUBENSPOREI--PILZ (*m*)
Peronospora schleideniana	421	Onion mould	ZWIEBEL-TRAUBENSPOREIPILZ (*m*)
Peronospora sordida	422	(See *Peronospora ficariæ*)	
Peronospora trifoliorum	423	Clover mould	KLEE-TRAUBENSPOREIPILZ (*m*)
Peronospora urticæ	424	Nettle mould	NESSEL-TRAUBENSPOREIPILZ (*m*)
Peronospora viciæ	425	Pea mould	ERBSEN-TRAUBENSPOREIPILZ (*m*)
Peronospora viticola	426	Grape-vine fungus; vine mildew (See *Plasmospora viticola* and *Plasmopara viticola*)	PERONOSPORA (*f*)
Persea gratissima	427	Avocado-pear, avocado, avigato, avocato, alligator-pear	ADVOKATENBIRNE (*f*); AVOCATO--BIRNE (*f*); ADVOKATEN--FRUCHT (*f*); AGUACATE (*f*); ALLIGATORBIRNE (*f*)

Botanical Term.	No.	English Term.	German Term.
Persica vulgaris	428	(See *Amydalus persica*)	
Petasites fragrans	429	Winter heliotrope	WOHLRIECHENDE PESTWURZ (*f*); WINTER-HELIOTROP (*m*)
Petasites vulgaris	430	Butter-bur; bog rhubarb; pestilence-wort, goat's-rue, wood mallow (See *Tussilago petasites*)	PESTWURZ (*f*); ROSSPAPPEL (*f*)
Petroselinum sativum	431	Parsley (See *Carum petroselinum*)	PETERSILIE (*f*)
Petroselinum segetum	432	(See *Carum segetum*)	GETREIDEN-PETERSILIE (*f*)
Petunia violocea	433	Petunia	PETUNIE (*f*)
Peucedanum cervaria	434	Sulphur-wort	SCHWEFELWURZ (*f*); HIRSCH-HAARSTRANG (*m*); HIRSCH-WURZ (*f*)
Peucedanum graveolens	435	Dill (See *Anethum graveolens*)	HOCHKRAUT (*n*)
Peucedanum officinale	436	Sulphur weed; hog's fennel; hogfennel, sow fennel	ECHTER HAARSTRANG (*m*) SAU-FENCHEL (*m*)
Peucedanum oreoselinum	437	Mountain celery; mountain parsley	BERGSELLERIE (*f*); BERGPETER-SILIE (*f*); BERG-HAARSTRANG (*m*); GRUNDHEIL (*n*)
Peucedanum ostruthium	438	Masterwort	MEISTERWURZEL (*f*)
Peucedanum palustre	439	Marsh parsley; milk parsley (See *Thysselinum palustre*)	SUMPFSILGE (*f*); SUMPF-HAAR-STRANG (*m*)
Peucedanum sativum	440	Parsnip (See *Pastinaca sativa*)	HAARSTRANG (*m*); ROSSKÜMMEL (*m*); SAU-FENCHEL (*m*)
Peziza	441	Cup fungus	BECHERLING (*m*)
Peziza acruginosa	442	Green rot	GRÜNFÄULE (*f*)
Peziza ciborioides	443	(See *Sclerotinia trifoliorum*)	
Peziza reticulata	444	Reticulated cup fungus	NETZBECHERLING (*m*)
Peziza venosa	445	Veined cup fungus	ADERBECHERLING (*m*)
Phædranassa	446	Queen lily	KÖNIGIN-LILIE (*f*)
Phæophyceæ	447	Brown seaweed, brown algæ	BRAUNALGEN (*f.pl*); TANGE (*m.pl*)
Phalænopsis amabilis	448	Indian butterfly-plant, moth orchid	FLIEGENDE-TAUBE ORCHE (*f*)
Phalaris arundinacea	449	Phlaris-(grass) (See *Digraphis arundinacea*)	ROHRARTIGES GLANZGRAS (*n*)
Phalaris arundinacea variegata	450	Lady's garters; ribbon grass; gardener's garters	VERSCHIEDENBLÄTTRIGES ROHR-ARTIGES- GLANZGRAS (*n*)
Phalaris canariensis	451	Canary-grass	KANARIENGRAS (*n*); KANARIEN-GLANZGRAS (*n*)
Phallus impudicus	452	Wood-witch; stink-horn-(fungus)	RUTENMORCHEL (*f*)
Phanerogamæ	453	Phanerogamous plants, flowering plants, phænogamous plants, vascular plants	SAMENPFLANZEN (*f.pl*); BLÜTEN-PFLANZEN (*f.pl*)
Phaseolus caracella	454	Snail flower, Indian snail flower	SCHNECKENBOHNE (*f*)
Phaseolus multiflorus	455	Climbing scarlet runner, scarlet runner bean	LAUFBOHNE (*f*); TÜRKISCHE BOHNE (*f*)
Phaseolus nanus	456	Dwarf bean	KRIECHBOHNE (*f*)
Phaseolus perennis	457	Kidney-bean	SCHMINKBOHNE (*f*)
Phaseolus vulgaris	458	Kidney bean; haricot bean; dwarf French bean	BRECHBOHNE (*f*); SALATBOHNE (*f*); VITSBOHNE (*f*); VEITS-BOHNE (*f*)
Phegopteris polypodioides	459	(See *Nephrodium phegopteris*)	
Phellandrium aquaticum	460	Water chervil	WASSERKERBEL (*m*)
Phellomyces	461	Potato rot	KARTOFFELFÄULE (*f*)
Philadelpheæ	462	Philadelpheæ	PFEIFENSTRÄUCHER (*m.pl*)
Philadelphus coronarius	463	Mock orange, (white)-syringa, wild jasmine	PFEIFENSTRAUCH (*m*); ZIMT-RÖSCHEN (*n*)
Phillyrea latifolia	464	Mock privet, jasmine box (See *Phyllirea*)	STEINLINDE (*f*)
Philodendron pertusum	465	(See *Monstera deliciosa*)	
Phleum alpinum	466	Alpine cat's-tail; alpine timothy grass	ALPENLIESCHGRAS (*n*)
Phleum arenarium	467	Sand Timothy-grass	SANDLIESCHGRAS (*n*)
Phleum asperum	468	Rough Timothy-grass	ACKER-LIESCHGRAS (*n*)
Phleum pratense	469	Timothy-grass; herds'-grass; cat's-tail-(grass)	LIESCHGRAS (*n*); LIESCHE (*f*); WIESEN-LIESCHGRAS (*n*)
Phlomis herbaventi	470	Jerusalem sage	BRANDKRAUT (*n*)

Botanical Term.	No.	English Term.	German Term.
Phlox maculata	471	Wild sweet-william; spotted phlox	GEFLECKTE FLAMMENBLUME (*f*)
Phlox paniculata	472	Panicled phlox	RISPIGE FLAMMENBLUME (*f*); HERBSTFLIEDER (*m*)
Phlox subulata	473	Moss pink; ground pink	FLAMMENBLUME (*f*)
Phœnix dactylifera	474	Date palm	DATTELBAUM (*m*); DATTELPALME (*f*)
Phœnix sylvestris	475	Indian-(date)-palm	INDISCHE DATTELPALME (*f*); WALD-DATTELPALME (*f*)
Pholiota	476	(See *Agaricus mutabilis*)	
Phormium tenax	477	New Zealand flax, common flax lily	FLACHSLILIE (*f*)
Photinia arbutifolia	478	Californian May-bush	ERDBEERBAUMBLÄTTRIGE GLANZ-MISPEL (*f*)
Photinia japonica	479	Japanese medlar tree; quince tree; loquat tree	JAPANISCHE GLANZMISPEL (*f*)
Photinia serrulata	480	Chinese hawthorn	CHINESISCHE GLANZMISPEL (*f*)
Phragmidium acuminatum	481	Burnet brand	BIBERNELL-QUERWANDSPOREN-ROST (*m*)
Phragmidium bulbosum	482	Brand, bramble brand (See *Aregma bulbosum*)	BROMBEER-QUERWANDSPOREN-ROST (*m*)
Phragmidium gracile	483	Raspberry brand	HIMBEER-QUERWANDSPORENROST (*m*)
Phragmidium mucronatum	484	Rose brand	ROSEN-QUERWANDSPORENROST (*m*)
Phragmidium obtusatum	485	Strawberry brand	ERDBEER-QUERWANDSPORENROST (*m*)
Phragmites communis	486	Reed, common reed (See *Arundo phragmites*)	SCHILFROHR (*n*); PFAHLROHR (*n*); ROHR (*n*); SCHILF (*n*)
Phycomyces	487	Water fungus	WASSERPILZ (*m*)
Phyllactinia guttata	488	Hazel blight	STACHELMELTAU (*m*)
Phyllirea	489	(See *Phillyrea*)	
Phyllostachys	490	Whangee cane (See *Bambusa*)	WHANGEEROHR (*n*)
Phyllostachys castillonis	491	Variegated bamboo	PFEFFERROHR (*n*)
Phyllostachys nigra	492	Black bamboo; Whangee cane	WANGHEEROHR (*n*); SCHWARZER BAMBUS (*m*)
Physalis alkekengi	493	Alkekengi; winter cherry; ground cherry; bladder herb	SCHLUTTE (*f*); JUDENKIRSCHE (*f*); KORALLENKIRSCHE (*f*)
Physalis peruviana-(edulis)	494	Cape gooseberry, Peruvian Cape gooseberry	ANANASKIRSCHE (*f*); PERUAN-ISCHE BLASENKIRSCHE (*f*); SCHLOTTE (*f*); ERDBEERTO-MATE (*f*); SCHLUTTE (*f*); KAP-STACHELBEERE (*f*)
Physocalymna floribundum	495	Tulip wood, Brazilian rosewood	ECHTES ROSENHOLZ (*n*); BRASIL-IANISCHES ROSENHOLZ (*n*)
Physospermum cornubiense	496	Bladder-seed; Cornish physosperm	BLASENSAME (*m*)
Physostegia virginiana	497	False dragon-head	FALSCHER DRACHENKOPF (*m*)
Physostigma venenosum	498	Calabar bean, ordeal bean (See *Faba calabarica*)	KALABARBOHNE (*f*)
Phytelephas macrocarpa	499	Ivory nut, corozo nut-(tree), vegetable ivory tree or palm	ELEFANTENNUSS (*f*); STEINNUSS (*f*); ELFENBEINNUSS (*f*)
Phyteuma nigra	500	Black rampion	SCHWARZE TEUFELSKRALLE (*f*)
Phyteuma orbiculare	501	Round-headed rampion, horned rampion (See also *Campanula rapunculus*)	RUNDKÖPFIGE TEUFELSKRALLE (*f*)
Phyteuma spicatum	502	Spiked rampion	SPICKEL (*m*); RAPUNZEL (*f*); WIESENKOHL (*m*)
Phytolacca acinosa	503	Indian poke	NACHTSCHATTEN (*m*)
Phytolacca decandra	504	Poke; pigeon-berry; poke-weed; garget; red-ink plant; Virginian poke	KERMESBEERE (*f*); SCHARLACH-BEERE (*f*); SCHMINKBEERE (*f*)
Phytophthora fagi	505	Beech fungus (See *Phytophthora omnivora*)	BUCHEN-TRAUBENSPOREIPILZ (*m*)
Phytophthora infestans	506	(See *Peronospora infestans*)	
Phytophthora omnivora	507	(See *Phytophthora fagi*)	BUCHENKEIMLINGSKRANKHEIT (*f*)
Picea	508	Fir (See also under *Abies* and *Pinus*)	FICHTE (*f*)

Botanical Term.	No.	English Term.	German Term.
Picea abies	509	Norway spruce, spruce (See *Abies picea*)	BALSAMTANNE (*f*)
Picea ajanensis	510	Yesso fir (See *Abies sachaliensis*)	JESO-FICHTE (*f*); AJANFICHTE (*f*)
Picea alba	511	Canadian-(white)-fir; white spruce (See *Abies canadensis*)	SCHIMMELFICHTE (*f*) KANADISCHE WEISSFICHTE (*f*)
Picea engelmanni	512	Engelmann's fir	ENGELMANN-FICHTE (*f*)
Picea excelsa	513	Spruce, pine, white fir; Norway spruce fir (See *Picea vulgaris*)	FICHTE (*f*); PECHTANNE (*f*); ROTTANNE (*f*); GEMEINE FICHTE (*f*); ROTFICHTE (*f*)
Picea glehni	514	Glehn's fir	GLEHNS FICHTE (*f*)
Picea hondoensis	515	Japanese fir; Hondo fir	HONDO-FICHTE (*f*)
Picea morinda	516	Himalayan fir (See *Pinus smithiana*)	HIMALAJA-FICHTE (*f*); TRÄNEN-FICHTE (*f*)
Picea nigra	517	Canadian-(black)-fir; black spruce (See *Abies mariena*)	SCHWARZFICHTE (*f*)
Picea obovata	518	Siberian fir	SIBIRISCHE FICHTE (*f*)
Picea omorica	519	Omorica fir	OMORIKA-FICHTE (*f*)
Picea orientalis	520	Oriental fir	ORIENT-FICHTE (*f*); SAPINDUS-FICHTE (*f*)
Picea polita	521	Japanese fir (See *Abies torano*)	STACHELFICHTE (*f*); TIGER-SCHWANZFICHTE (*f*)
Picea pungens	522	Prickly fir	STECHFICHTE (*f*)
Picea rubra	523	Red fir	ROTFICHTE (*f*)
Picea schrenkiana	524	Schrenk's fir	SCHRENKS FICHTE (*f*)
Picea sitchensis	525	Sitka fir, Alaska fir	SITKAFICHTE (*f*)
Picea succinifera	526	Amber fir (An extinct variety of the Upper Cretaceous period)	BERNSTEINFICHTE (*f*)
Picea veitchi	527	(See *Abies veitchi*)	
Picea vulgaris	528	(See *Picea excelsa*)	
Picræna excelsa	529	(See *Picrasma excelsa*)	
Picrasma excelsa	530	(See *Simaruba amara, Simaruba officinalis* and *Quassia amara*) (See also *Picræna excelsa*)	JAMAIKA UNECHTES QUASSIAHOLZ (*n*)
Picris hieracioides	531	Hawkweed picris; hawkweed ox-tongue	HABICHTSKRAUTÄHNLICHER BITTERICH (*m*)
Pieroma	532	(See *Tibouchina elegans*)	
Pilea muscosa	533	Stingless nettle, pistol plant, artillery plant	PISTOLENPFLANZE (*f*)
Pilocarpus pinnatifolius	534	Pilocarpus; pinnate-leaved cap-fruit	JABORANDI (*m*); JABORANDI-STRAUCH (*m*)
Pilularia globulifera	535	Pillwort; creeping pillwort (See *Pilularia pilulifera*)	
Pilularia pilulifera	536	Pillwort (See *Pilularia globulifera*)	PILLENFARN (*m*)
Pimelea	537	Rice-flower	REISBLUME (*f*)
Pimenta acris	538	Black cinnamon tree; wild clove; bitter allspice-tree	WESTINDISCHER BAYÖLBAUM (*m*); BAYRUMBAUM (*m*)
Pimenta officinalis	539	Pimento bush; Pimento; pimenta; Jamaica pepper; allspice (See *Eugenia pimenta*)	NELKENPFEFFER (*m*); NEUGEWÜRZ (*n*); PIMENT (*n*); WUNDERPFEFFER (*m*)
Pimpinella anisum	540	Anise, aniseed	ANIS (*m*); ANGIS (*m*)
Pimpinella magna	541	Greater pimpinel, great burnet saxifrage	PFEFFERWURZ (*f*); PIMPERNELL-WURZEL (*f*)
Pimpinella saxifraga	542	Burnet saxifrage; common pimpinel	BIBERNELL (*f*); BIBERNELLE (*f*); PIMPINELLE (*f*)
Pinguicula alpina	543	Alpine butterwort	ALPEN-FETTKRAUT (*n*)
Pinguicula lusitanica	544	Pale butterwort	LUSITANISCHES FETTKRAUT (*n*)
Pinguicula vulgaris	545	Butterwort; common butterwort: bog violet	GEMEINES FETTKRAUT (*n*)
Pinus	546	Fir (See also under *Abies* and *Picea*)	FÖHRE (*f*); KIEFER (*f*); KIEN-FÖHRE (*f*)
Pinus abies	547	Spruce, pine, white fir	FICHTE (*f*); PECHTANNE (*f*); ROTTANNE (*f*)

Botanical Term.	No.	English Term.	German Term.
Pinus australis	548	Pitch pine, Georgia pine, swamp pine ; southern pine ; Oregon pine	AMERIKANISCHE TERPENTINKIE-FER (*f*); AMERIKANISCHE HARZKIEFER (*f*); PITCHPINE (*f*); PECHBAUM (*m*); PECHTANNE (*f*)
Pinus austriaca	549	Black pine ; Austrian pine (See *Pinus laricio* and *Pinus nigra austriaca*)	SCHWARZFÖHRE (*f*); SCHWARZ-KIEFER (*f*)
Pinus balfourniana	550	Hickory pine	HICKORYFICHTE (*f*)
Pinus banksiana	551	Scrub pine	STRAUCHKIEFER (*f*)
Pinus bungeana	552	Lace bark pine	BUNGES KIEFER (*f*)
Pinus cedrus	553	Cedar	(ECHTE)-ZEDER (*f*)
Pinus cembra	554	Stone pine, cembra pine ; Swiss pine ; Swiss stone pine	ZIRBELKIEFER (*f*); ARVE (*f*); ZIRBELFICHTE (*f*); NUSS-KIEFER (*f*); ZIRBE (*f*)
Pinus echinata	555	Carolina pine ; short-leaved pine	CAROLINAKIEFER (*f*); KURZNADE-LIGE KIEFER (*f*)
Pinus excelsa	556	Spruce ; pine ; Bhotan pine	FICHTE (*f*); PECHTANNE (*f*); ROTTANNE (*f*); HIMALAJA-KIEFER (*f*); TRÄNENKIEFER (*f*)
Pinus halepensis	557	Jerusalem pine	ALEPPOKIEFER (*f*)
Pinus insignis	558	Monterey pine	MONTERAY-KIEFER (*f*)
Pinus jeffreyi	559	Jeffrey's pine	JEFFREYSKIEFER (*f*)
Pinus koraiensis	560	Korean pine	KOREA-KIEFER (*f*)
Pinus lambertiana	561	Lambert's pine, giant pine; sugar pine	RIESENKIEFER (*f*); ZUCKERKIE-FER (*f*)
Pinus laricio	562	Black pine ; Austrian pine ; Corsican pine (See *Pinus austriaca*)	SCHWARZFÖHRE (*f*); SCHWARZ-KIEFER (*f*)
Pinus larix	563	Larch (See *Larix decidua*;. *Larix excelsa* ; *Larix europæa*)	LÄRCHE (*f*); LÄRCHENTANNE (*f*)
Pinus leucodermis	564	White pine	WEISSRINDIGE KIEFER (*f*); SCHLANGENHAUTKIEFER (*f*)
Pinus montana	565	Mountain pine	BERGKIEFER (*f*); LEGFÖHRE (*f*); KRUMMHOLZKIEFER (*f*); LACK-HOLZ (*n*); LATSCHE (*f*); KNIEHOLZ (*n*); LEGKIEFER (*f*); KNIEHOLZKIEFER (*f*)
Pinus montana pumilio	566	Knee pine ; dwarf mountain pine (See *Pinus mughus*)	KRUMMHOLZBAUM (*m*); KRUMM-HOLZFICHTE (*f*); KRUMM-HOLZKIEFER (*f*); LATSCHE (*f*); LATSCHENKIEFER (*f*); KNIEHOLZ (*n*); KNIEHOLZ-KIEFER (*f*); LEGFÖHRE (*f*)
Pinus montana uncinata	567	Uncinate knee pine	LATSCHENKIEFER (*f*) MIT SCHILDER HACKENFÖRMIG GEKRÜMMT
Pinus monticicola	568	(North-west American)-Weymouth pine	WESTLÄNDISCHE WEYMOUTHS-KIEFER (*f*)
Pinus mughus	569	(See *Pinus montana pumilio*)	
Pinus nigra austriaca	570	(See *Pinus austriaca*)	
Pinus palustris	571	Pitch (fir) pine, southern pine, Georgia pine, swamp pine, Oregon pine ; long-leaved pine	PITCHPINE (*f*); AMERIKANISCHE TERPENTINKIEFER (*f*); PECH-KIEFER (*f*); AMERIKANISCHE HARZKIEFER (*f*); HARZ-FICHTE (*f*); HARZKIEFER (*f*); HARZTANNE (*f*); LANGNADE-LIGE KIEFER (*f*)
Pinus pectinata	572	Fir, silver fir (See *Abies pectinata*)	TANNE (*f*); EDELTANNE (*f*); WEISSTANNE (*f*); EDELFICHTE (*f*)
Pinus peuce	573	Balkan spruce	KUMELISCHE KIEFER (*f*)
Pinus picea	574	(See *Abies pectinata*)	
Pinus pinaster	575	Pinaster, cluster pine	STERNKIEFER (*f*); SEEFÖHRE (*f*); SEEKIEFER (*f*); IGELFÖHRE (*f*); STRANDKIEFER (*f*)
Pinus pinea	576	Pine ; stone pine, umbrella pine	PINIE (*f*); PINIENKIEFER (*f*)
Pinus ponderosa	577	Yellow pine	YELLOWPINE (*f*); GELBKIEFER (*f*)

Botanical Term.	No.	English Term.	German Term.
Pinus pumila	578	Dwarf pine	ZWERGKIEFER (*f*)
Pinus pumilio	579	(See *Pinus montana pumilio*)	
Pinus rigida	580	Pitch pine ; Georgia pine ; swamp pine ; southern pine ; Oregon pine	AMERIKANISCHE TERPENTINKIE--FER (*f*) ; AMERIKANISCHE HARZKIEFER (*f*) ; PITCHPINE (*f*) ; PECHKIEFER (*f*)
Pinus rubra	581	Red fir (See *Pinus sylvestris*)	
Pinus sabineana	582	North-American pine ; Sabine pine	SABINES KIEFER (*f*)
Pinus smithiana	583	(See *Picea morinda*)	
Pinus strobus	584	American yellow pine, white pine, Weymouth pine	WEYMOUTHKIEFER (*f*) ; WEIMUT--KIEFER (*f*) ; WEIMUTSKIEFER (*f*) ; STROBE (*f*)
Pinus sylvestris	585	Scotch pine, Scotch fir, red fir ; white pine ; northern (common) pine ; yellow fir	KIEFER (*f*) ; FACKELFÖHRE (*f*) ; FÖHRE (*f*) ; TEERBAUM (*m*) ; WEISSKIEFER (*f*) ; GEMEINE KIEFER (*f*) ; MEERSTRANDS--KIEFER (*f*) ; KERNHOLZ (*n*) ; WEISSFÖHRE (*f*)
Piper angustifolium	586	Narrow-leaved pepper plant (See *Artanthe elongata*)	SCHMALBLÄTTRIGER PFEFFER (*m*) ; SCHMALBLÄTTRIGE PFEFFER--STAUDE (*f*)
Piper betle	587	Betel pepper ; betel leaf	KAUPFEFFER (*m*)
Piper cubeba	588	Cubeb pepper ; Java pepper	KUBEBE (*f*) ; KUBEBENPFEFFER (*m*) ; ZIBEBE (*f*)
Piper excelsum	589	Lofty pepper plant (See *Macropiper excelsum*)	HOHE PFEFFERSTAUDE (*f*)
Piper nigrum	590	Black pepper	PFEFFERSTAUDE (*f*) ; PFEFFER--STRAUCH (*m*)
Piptanthus nepalensis	591	Nepaul laburnum	NEPAUL-BOHNENBAUM (*m*)
Piratinera guianensis	592	Leopard wood ; snakewood ; speckled (letter) wood	SCHLANGENHOLZ (*n*) ; MUSKAT--HOLZ (*n*) ; LETTERNHOLZ (*n*) ; BUCHSTABENHOLZ (*n*) ; TIGER--HOLZ (*n*)
Pirus	593	(See also *Malus* ; *Pyrophorum* ; *Sorbus* ; *Aria* ; *Amelanchier* ; *Aronia* ; *Adenorachis* ; *Cydonia* and *Mespilus*)	
Pirus amygdaliformis	594	Almond-leaved pear-(tree) (See *Pirus salicifolia*)	MANDELBLÄTTRIGER BIRNBAUM (*m*)
Pirus arbutifolia	595	Chokeberry	ROTFRÜCHTIGE APFELBEERE (*f*)
Pirus aria	596	White beam tree (See *Sorbus aria*)	MEHLBEERBAUM (*m*)
Pirus aucuparia	597	Mountain ash ; rowan tree ; quicken tree (See *Sorbus aucuparia*)	EBERESCHE (*f*)
Pirus baccata	598	Siberian crab	BEERENAPFEL (*m*)
Pirus chamæmespilus	599	Bastard medlar (See also under *Cratægus* and *Sorbus*)	BERG-MEHLBEERE (*f*) ; ZWERG-MEHLBEERE (*f*)
Pirus communis	600	Pear tree ; wild pear tree (See also *Pirus sinensis* and *Pyrophorum communis*)	BIRNBAUM (*m*) ; HOLZBIRNBAUM (*m*) ; WILDER BIRNBAUM (*m*)
Pirus coronaria	601	Sweet-scented crab	WOHLRIECHENDER APFEL (*m*)
Pirus cydonia	602	(See *Cydonia vulgaris*)	
Pirus dasyphylla	603	Felt-leaved apple	FILZBLÄTTRIGER APFEL (*m*)
Pirus domestica	604	Service tree	SPERBERBAUM (*m*) ; SPEIERLING (*m*) ; SPIERAPFEL (*m*)
Pirus elæagrifolia	605	Olive-leaved pear-(tree)	ÖLBAUMBLÄTTRIGER BIRNBAUM (*m*)
Pirus germanica	606	(See *Mespilus germanica*)	
Pirus heterophylla	607	Various-leaved pear-(tree)	VERSCHIEDENBLÄTTRIGER BIRN--BAUM (*m*)
Pirus hybrida	608	(See *Malus astracanica*)	
Pirus japonica	609	Japanese quince (See *Cydonia japonica*)	
Pirus malus	610	Apple tree ; crab-(apple)-tree ; wild apple tree (See also *Malus communis*)	APFELBAUM (*m*) ; HOLZAPFEL--BAUM (*m*) ; HAGAPFEL (*m*) ; WILDER APFELBAUM (*m*)

Botanical Term.	No.	English Term.	German Term.
Pirus malus sempervirens	611	Evergreen crab	IMMERGRÜNER HOLZAPFELBAUM (*m*)
Pirus nigra	612	Black-fruited aronia; black aronia	SCHWARZFRÜCHTIGE APFELBEERE (*f*)
Pirus nivalis	613	Snow pear	SCHNEEBIRNBAUM (*m*)
Pirus prunifolia	614	Siberian crab; plum-leaved apple	PFLAUMENBLÄTTRIGER APFEL (*m*)
Pirus pumila	615	Dwarf apple, bush apple	STRAUCHAPFEL (*m*)
Pirus salicifolia	616	Willow-leaved pear-(tree) (See *Pirus amygdaliformis*)	WEIDENBLÄTTRIGER BIRNBAUM (*m*)
Pirus sinensis	617	Snow pear, sand pear, Chinese pear-(tree) (See also *Pirus communis* and *Malus spectabilis*)	CHINESISCHER BIRNBAUM (*m*)
Pirus sorbus	618	(See *Sorbus domestica*)	
Pirus sylvestris	619	Smooth-leaved apple	GLATTBLÄTTRIGER APFEL (*m*)
Pirus toringo	620	Toringo crab, Japanese crab	TORINGO-APFEL (*m*)
Pirus torminalis	621	Wild service tree; beam tree (See *Sorbus torminalis* and *Cratægus torminalis*)	ELSE (*f*); ELZBEERE (*f*)
Pirus vulgaris	622	Quince (See *Pirus cydonia* and *Cydonia vulgaris*)	
Pistacia lentiscus	623	Mastic shrub; mastic tree	MASTIXBAUM (*m*)
Pistacia terebinthus	624	(Cyprus)-turpentine tree	TEREBINTHE (*f*); TERPENTIN--PISTAZIE (*f*)
Pistacia vera	625	Pistachio-nut tree, pistacio tree, pistacia tree	PISTAZIE (*f*); PIMPERNUSS (*f*); ECHTE PISTAZIE (*f*)
Pistia stratiotes	626	Tropical duckweed; water lettuce	MUSCHELBLUME (*f*)
Pisum arvense	627	Field pea	FUTTERERBSE (*f*)
Pisum elatius	628	Crown pea, mummy pea (See also *Pisum sativum umbellatum*)	
Pisum sativum	629	Garden pea	ERBSE (*f*)
Pisum sativum saccharatum	630	Sugar pea	ZUCKERERBSE (*f*)
Pisum sativum umbellatum	631	Crown pea (See also *Pisum elatius*)	KRONERBSE (*f*)
Pithecolobium dulce	632	Sweet pithecolobium; monkey's ear-ring	KAMATSCHILRINDEBAUM (*m*)
Pittosporum	633	Australian box tree	AUSTRALISCHER BUCHSBAUM (*m*)
Pittosporum crassifolium	634	Parchment bark	KLEBSAME (*f*)
Plantago	635	Plantain	KÖNIGSPISANG (*m*); WEGERICH (*m*); WEGEBREIT (*m*)
Plantago coronopus	636	Buck's-horn plantain	SCHLITZBLÄTTRIGER WEGERICH (*m*); GROSSES HIRSCHHORN (*n*)
Plantago lanceolata	637	Ribwort plantain; cocks-and-hens; rib-grass	LANZETTLICHER WEGERICH (*m*)
Plantago major	638	Great plantain; greater plantain; greater ribgrass	GROSSER WEGERICH (*m*)
Plantago maritima	639	Seaside plantain; sea-plantain; sea rib-grass	MEERSTRANDS-WEGERICH (*m*)
Plantago media	640	Hoary plantain; hoary ribgrass	MITTLERER WEGERICH (*m*)
Plantago psyllium	641	Fleawort-seed; fleabane-seed; flea seed	FLOHSAMEN (*m*)
Plasmodiophora alni	642	Alder fungus	ERLENPILZ (*m*)
Plasmodiophora brassicæ	643	Cabbage-(club-root)-fungus	FINGERSEUCHE (*f*); KOHLHERNIE (*f*); KROPFSEUCHE (*f*)
Plasmopara viticola	644	(See *Peronospora viticola*)	
Plasmospora viticola	645	Vine mildew (See *Peronospora viticola*)	FALSCHER MELTAU (*m*)
Platanus acerifolia	646	London plane (See *Platanus orientalis acerifolia*)	
Platanus lobada	647	(See *Platanus occidentalis*)	
Platanus occidentalis	648	Plane tree; sycamore; button-ball tree; water-birch; button-wood; Western plane (See *Platanus lobaba*)	PLATANE (*f*); ABENDLÄNDISCHE PLATANE (*f*); WASSERBUCHE (*f*); AMERIKANISCHE PLATANE (*f*)

Botanical Term.	No.	English Term.	German Term.
Platanus orientalis	649	Plane tree ; common plane tree ; oriental plane tree (See *Platanus palmata*)	PLATANE (*f*); MORGENLÄN-DISCHE PLATANE (*f*); ORIEN-TALISCHE PLATANE (*f*)
Platanus orientalis acerifolia	650	London plane (See *Platanus acerifolia*)	MORGENLÄNDISCHE PLATANE (*f*); AHORNBLÄTTRIGE PLATANE (*f*)
Platanus palmata	651	(See *Platanus orientalis*)	
Platanus vulgaris	652	Plane tree	PLATANE (*f*); GEMEINE PLATANE (*f*)
Platycerium alcicorne	653	Common elk's-horn fern, stag's-horn fern (See *Acrostichum alicorne*)	ELCHHORNFARN (*m*); FLACH-HORNFARN (*m*)
Platycodon grandiflorum	654	Chinese bell-flower	CHINESISCHE GLOCKENBLUME (*f*)
Platylobium	655	Flat pea	PLATTSCHOTE (*f*)
Platyloma	656	(See *Pellœa*)	
Platystemon californicus	657	Californian poppy, cream cups	KALIFORNISCHER MOHN (*m*)
Plectogyne variegata	658	(See *Aspidistra lurida*)	
Pleurolobus	659	(See *Desmodium gyrans*)	
Pleurotus	660	(See *Agaricus ostreatus*)	
Plocaria lichenoides	661	Jaffna moss; Agar-agar; Agal-agal ; Ceylon moss	AGAR-AGAR MOOS (*n*); JAFFNA-MOOS (*n*)
Plumbagineæ	662	Plumbagineæ (See *Statice* and *Armeria*)	BLEIWURZELPFLANZEN (*f.pl*)
Plumbago capensis	663	(Cape)-leadwort	KAP-BLEIWURZ (*f*)
Plumiera rubra	664	Frangipanni plant	FRANGIPANI (*f*)
Poa	665	Meadow-grass (See also *Festuca* and *Dactylis cæspitosa*)	RISPENGRAS (*n*); VIEHGRAS (*n*)
Poa alpina	666	Alpine meadow grass	ALPEN-RISPENGRAS (*n*)
Poa annua	667	Annual meadow-grass	STRASSENGRAS (*n*); BESENGIN-STER (*m*); BLAUGRAS (*n*); EINJÄHRIGES RISPENGRAS (*n*)
Poa aquatica	668	Reed meadow-grass (See *Glyceria aquatica*)	WASSER-RISPENGRAS (*n*)
Poa arachnifera	669	Texas blue grass	TEXAS-RISPENGRAS (*n*); TEXAS-BLAUGRAS (*n*)
Poa bulbosa	670	Bulbous meadow-grass	ZWIEBELIGES RISPENGRAS (*n*)
Poa compressa	671	Flattened meadow-grass	PLATTHALM-RISPENGRAS (*n*)
Poa distans	672	Reflexed meadow-grass (See *Glyceria distans* and *Schlerochloa distans*)	ENTFERNTES RISPENGRAS (*n*)
Poa flabellata	673	Tufted meadow-grass ; Tussock grass (See *Dactylis cæspitosa* and *Festuca flabellata*)	TUSSOCKGRAS (*n*)
Poa fluitans	674	Floating meadow-grass, manna grass (See *Glyceria fluitans*)	MANNAGRAS (*n*); FLUTENDES RISPENGRAS (*n*)
Poa laxa	675	Wavy meadow-grass	LOSES RISPENGRAS (*n*)
Poa loliacea	676	Darnel meadow-grass	TAUMELKORN-RISPENGRAS (*n*)
Poa maritima	677	Sea meadow-grass	MEERSTRANDS-RISPENGRAS (*n*)
Poa memoralis	678	Wood meadow-grass	HAIN-RISPENGRAS (*n*)
Poa pratensis	679	Spear grass ; blue grass ; June grass ; smooth meadow grass ; Kentucky blue grass ; smooth-stalked meadow-grass	RISPENGRAS (*n*); WIESEN-RIS-PENGRAS (*n*); ANGERGRAS (*n*)
Poa procumbens	680	Procumbent meadow-grass	LIEGENDES RISPENGRAS (*n*)
Poa rigida	681	Hard meadow-grass	STEIFES RISPENGRAS (*n*)
Poa serotina	682	Late meadow grass	SPÄTES RISPENGRAS (*n*)
Poa trivialis	683	Rough-stalked meadow-grass, variegated meadow-grass	GEMEINES RISPENGRAS (*n*); RAUHES RISPENGRAS (*n*)
Podocarpus totara	684	Totara pine	TOTARABAUM (*m*)
Podophyllum peltatum	685	May-apple ; mandrake ; mandragora ; wild lemon ; duck-foot (See also under *Mandragora*)	FUSSBLATT (*n*); ALRUNE (*f*); ALRAUN-(E) (*f*); MANDRA-GORA (*f*); AMERIKANISCHE WILDE LIMONE (*f*); WILDE ZITRONE (*f*); ENTENFUSS (*m*); MAIAPFEL (*m*); MANDRAKE (*f*)
Podosphæra kunzei	686	Blight, plum blight	MELTAU (*m*); PFLAUMENMELTAU (*m*)

Botanical Term.	No.	English Term.	German Term.
Pogostemon patchouli	687	Patchouli shrub; patchouli herb	PATSCHULIKRAUT (*n*)
Poinsettia pulcherrima	688	Mexican Christmas flower (See also *Euphorbia pulcherrima*)	MEXIKANISCHE WEIHNACHTS-BLUME (*f*)
Polemonium cæruleum	689	Blue Jacob's ladder; Greek valerian	HIMMELSLEITER (*f*)
Polianthes tuberosa	690	Tuberose	NACHTHYAZINTHE (*f*); TUBEROSE (*f*)
Polycarpon tetraphyllum	691	Four-leaved polycarp	VIERBLÄTTRIGES NAGELKRAUT (*n*)
Polygala	692	Milkwort	KREUZBLUME (*f*); MILCHBLUME (*f*)
Polygala senega	693	(North American)-senega-(root); snake-root	SENEGAWURZEL (*f*); RAMSEL (*m*); KLAPPERSCHLANGENWURZEL (*f*)
Polygala vulgaris	694	Milkwort; common sea milkwort; procession flower	KRANZBLUME (*f*); KREUZBLUME (*f*); KREUZKRAUT (*n*); MILCH-WURZ (*f*); NATTERBLÜM-CHEN (*n*); BITTERES KREUZ-KRAUT (*n*)
Polygonatum multiflorum	695	Solomon's seal, common Solomon's-seal; David's harp (See *Convallaria multiflora*)	SALOMONSSIEGEL (*m*); GROSSE MAIBLUME (*f*)
Polygonatum officinale	696	Angular Solomon's-seal	WEISSWURZEL (*f*)
Polygonatum verticillatum	697	Whorled Solomon's-seal	QUIRLBLÄTTRIGE WEISSWURZ (*f*)
Polygonum alpinum	698	Alpine polygonum	ALPENKNÖTERICH (*m*)
Polygonum amphibium	699	Amphibion polygonum	ORTSWECHSELNDER KNÖTERICH (*m*)
Polygonum aviculare	700	Knot-grass; knotweed	KNÖTERICH (*m*); VOGELKNÖTERICH (*m*)
Polygonum bistorta	701	Bistort; snakeweed; snakeroot; adder's-wort	WIESENKNÖTERICH (*m*); NATTER-WURZ (*f*); GANSAMFER (*m*); NATTERWURZEL (*f*)
Polygonum convolvulus	702	Climbing persicaria; climbing polygonum; black bindweed	WINDENKNÖTERICH (*m*)
Polygonum cuspidatum	703	Japanese polygonum	JAPANISCHER KNÖTERICH (*m*)
Polygonum dumetorum	704	Copse polygonum, false buck-wheat	HECKENKNÖTERICH (*m*)
Polygonum fagopyrum	705	Buck-wheat (See *Fagopyrum*)	GRICKEN (*m*); HEIDEKORN (*n*); BUCHWEIZEN (*m*)
Polygonum hydropiper	706	Common water-pepper; smart-weed; lakeweed	WASSERPFEFFER (*m*)
Polygonum lapathifolium	707	Pale polygonum	ACKERKNÖTERICH (*m*); AMPFER-BLÄTTERIGER KNÖTERICH (*m*)
Polygonum maritimum	708	Sea knot-grass	MEERKNÖTERICH (*m*)
Polygonum minus	709	Slender polygonum	KLEINER KNÖTERICH (*m*)
Polygonum orientale	710	Oriental polygonum	ORIENTALISCHER KNÖTERICH (*m*)
Polygonum persicaria	711	Persicaria polygonum; spotted knot-grass, spotted persicaria	GEMEINER KNÖTERICH (*m*); SCHARFER KNÖTERICH (*m*); PFIRSICHBLÄTTERIGER KNÖT-ERICH (*m*); FLOHKRAUT (*n*)
Polygonum sachalinense	712	Sakhalin knotgrass; Saghalien knotgrass	SACHALINER KNÖTERICH (*m*) RIESENKNÖTERICH (*m*)
Polygonum tinctorium	713	Indigo plant	FÄRBERKNÖTERICH (*m*)
Polygonum viviparum	714	Viviparous polygonum	SPITZKEIMENDER KNÖTERICH (*m*)
Polypodium alpestre	715	Alpine polypodium, alpine poly-pody	ALPENTÜPFELFARN (*m*)
Polypodium dryopteris	716	Oak fern	TÜPFELFARN (*m*)
Polypodium phegopteris	717	Beech polypody (See *Polypodium vulgare*)	
Polypodium vulgare	718	Common polypody; beech poly-pody (See *Polypodium phegopteris*)	KORALLENWURZEL (*f*); ENGEL-SÜSS (*m*); GEMEINER TÜPFEL-FARN (*m*)
Polypogon	719	Beard grass	BARTGRAS (*n*); MANNSBART (*m*); BÜRSTENGRAS (*n*); VIELBART (*m*); TAUSENDBART (*m*)
Polypogon littoralis	720	Beard grass; perennial beard grass	AUSDAUERNDES BÜRSTENGRAS (*n*)
Polypogon monspeliensis	721	Beard grass; annual beard grass	EINJÄHRIGES BÜRSTENGRAS (*n*)
Polyporus	722	Polyporus (See *Boletus polyporus*)	PORLING (*m*); LÖCHERPILZ (*m*)

Botanical Term.	No.	English Term.	German Term.
Polyporus betulinus	723	Razor-strop fungus; beech fungus	BUCHENPILZ (*m*); LÖCHERPILZ (*m*); BIRKENLÖCHERSCHWAMM (*m*)
Polyporus borealis	724	Pine-wood fungus	FICHTENHOLZSCHWAMM (*m*)
Polyporus dryadeus	725	Oak-wood fungus	EICHENHOLZSCHWAMM (*m*)
Polyporus fomentarius	726	Amadou, German tinder; surgeon's agaric (See *Fomes fomentarius* and *Boletus fomentarius*)	FEUERSCHWAMM (*m*); WUNDSCHWAMM (*m*); BUCHENPILZ (*m*); BUCHENSCHWAMM (*m*); ZUNDERSCHWAMM (*m*); ECHTER FEUERSCHWAMM (*m*); ZUCKER-PILZ (*m*)
Polyporus frondosus	727		KLAPPERSCHWAMM (*m*)
Polyporus hartigii	728		WEISSTANNENSCHWAMM (*m*)
Polyporus igniarius	729	Touchwood fungus (See *Boletus igniarius*)	WEIDENSCHWAMM (*m*)
Polyporus officinalis	730	Purging agaric, female agaric (See *Agaricus albus, Agaricus chirurgorum* and *Boletus laricis*)	LÄRCHENSCHWAMM (*m*); LÄRCHENPILZ (*m*)
Polyporus ovinus	731		SCHAFEUTER (*n*)
Polyporus pinicola	732	Pine fungus	FICHTENSCHWAMM (*m*)
Polyporus sulfureus	733	Red rot	ROTFÄULE (*f*)
Polyporus tuberaster	734		TUBERASTER (*m*)
Polyporus umbellatus	735	Dry-rot (See *Merulius lacrymans*)	HAUSSCHWAMM (*m*); EICHHASE (*f*); EICHPILZ (*m*); HASEL-SCHWAMM (*m*)
Polyporus vaporarius	736	White wood fungus	WEISSER (HOLZSCHWAMM) HAUS-SCHWAMM (*m*); UNECHTER (HAUSSCHWAMM) HOLZSCHWAMM (*m*); LOHBEETLÖCHERSCHWAMM (*m*); PORENHAUSSCHWAMM (*m*)
Polystichum aculeatum	737	(See *Aspidium aculeatum*)	STACHELIGER PUNKTFARN (*m*)
Polytrichum commune	738	Common polytrichum-(moss)	GEMEINES WIDERTONMOOS (*n*); HAARMOOS (*n*); FILZMÜTZE (*f*)
Pontederia cordata	739	Pickerel weed	HERZBLÄTTRIGE PONTEDERIE (*f*)
Populus	740	Poplar	PAPPEL (*f*)
Populus alba	741	Silver (white) poplar; Abele (tree)	SILBERPAPPEL (*f*): WEISSPAPPEL (*f*): WEISSE PAPPEL (*f*)
Populus angulata	742	North-American poplar	FLÜGELPAPPEL (*f*)
Populus angustifolia	743	Small-leaved poplar	SCHMALBLÄTTRIGE PAPPEL (*f*)
Populus balsamifera	744	Tacamahac, balsam poplar, cottonwood tree	BALSAMPAPPEL (*f*)
Populus canadensis	745	Cotton wood, Canadian poplar, poplar (See *Populus deltoidea* and *Populus monilifera*)	KANADISCHE PAPPEL (*f*); PAPPEL (*f*); KANADISCHE WALD-PAPPEL (*f*)
Populus candicans	746	Ontario poplar (See *Populus ontariensis*)	
Populus canescens	747	Grey poplar	GRAUE PAPPEL (*f*)
Populus deltoidea	748	(See *Populus canadensis*)	
Populus deltoidea erecta	749	Necklace poplar	
Populus euphratica	750	Euphrates poplar	EUPHRATPAPPEL (*f*)
Populus fremonti	751	Californian poplar	KALIFORNISCHE PAPPEL (*f*)
Populus italica	752	Italian poplar (See *Populus pyramidalis*)	
Populus laurifolia	753	Siberian poplar	LORBEERPAPPEL (*f*)
Populus macrophylla	754	(See *Populus ontariensis*)	
Populus monilifera	755	Canadian poplar; poplar; cotton wood (See *Populus canadensis*)	KANADISCHE PAPPEL (*f*); PAPPEL (*f*); KANADISCHE WALD-PAPPEL (*f*); ROSENKRANZ-PAPPEL(*f*)
Populus nigra	756	Black poplar (See *Populus pyramidalis*)	SCHWARZPAPPEL (*f*); SCHWARZE PAPPEL (*f*); ITALIENISCHE PAPPEL (*f*); PAPPELWEIDE (*f*); SAARBAUM (*m*); SAARBUCHE (*f*)
Populus nigra pyramidalis	757	Lombardy poplar (See *Populus pyramidalis*)	

Botanical Term.	No.	English Term.	German Term.
Populus ontariensis	758	Ontario poplar (See *Populus candicans* and *Populus macrophylla*)	ONTARIO-PAPPEL (*f*)
Populus pyramidalis	759	Italian poplar ; Lombardy poplar (See *Populus nigra* and *Populus italica*)	ITALIENISCHE PAPPEL (*f*) PYRA-MIDENPAPPEL (*f*); CHAUS-SEEPAPPEL (*f*)
Populus simoni	760	Chinese poplar	CHINESISCHE PAPPEL (*f*)
Populus suaveolens	761	Sweet-scented poplar	WOHLRIECHENDE PAPPEL (*f*)
Populus tremula	762	Aspen	ZITTERPAPPEL (*f*); ASPE (*f*); ESPE (*f*); PAPPELESPE (*f*)
Populus tremuloides	763	American aspen	AMERIKANISCHE ZITTERPAPPEL (*f*)
Populus trichocarpa	764	Hairy-fruited poplar; Californian poplar	HAARFRÜCHTIGE PAPPEL (*f*)
Populus tristis	765	Dark-leaved poplar	DUNKELBLÄTTRIGE PAPPEL (*f*)
Portulaca grandiflora	766	(Brazilian)-sun plant	BRASILISCHER PORTULAK (*m*)
Portulaca oleracea	767	Common purslane, purslain (See also *Peplis portula*)	PORTULAK (*m*); PURZELKRAUT (*n*); WURZELKRAUT (*n*); GEMEINER PORTULAK (*m*)
Potamogeton	768	Pond-weed,frog lettuce	LAICHKRAUT (*n*); SAMKRAUT (*n*)
Potamogeton acutifolius	769	Acute pondweed	SCHARFBLÄTTRIGES LAICHKRAUT (*n*)
Potamogeton crispus	770	Curly pondweed	GEKRÄUSELTES LAICHKRAUT (*n*); KRAUSEN-LAICHKRAUT (*n*)
Potamogeton densus	771	Opposite pondweed	DICHTBLÄTTRIGES LAICHKRAUT (*n*)
Potamogeton heterophyllus	772	Various-leaved pondweed	VERSCHIEDENBLÄTTRIGES LAICH-KRAUT (*n*)
Potamogeton lucens	773	Shining pondweed	SPIEGELNDES LAICHKRAUT (*n*)
Potamogeton natans	774	Pondweed; floating pondweed; broad pondweed	SCHWIMMENDES LAICHKRAUT (*n*)
Potamogeton obtusifolius	775	Obtuse pondweed	STUMPFBLÄTTRIGES LAICHKRAUT (*n*)
Potamogeton pectinatus	776	Fennel pondweed	KAMMFÖRMIGES LAICHKRAUT (*n*)
Potamogeton perfoliatus	777	Perfoliate pondweed	DURCHWACHSENBLÄTTRIGES LAICHKRAUT (*n*)
Potamogeton prœlongus	778	Long pondweed	LANGES LAICHKRAUT (*n*)
Potamogeton pusillus	779	Small pondweed; slender pondweed	ZWERG-LAICHKRAUT (*n*)
Potentilla	780	Cinquefoil	FINGERKRAUT (*n*)
Potentilla anserina	781	Silver weed; goose grass	GÄNSERICH (*m*); GÄNSEKRAUT (*n*); GÄNSEFINGERKRAUT (*n*)
Potentilla argentea	782	Silver weed; hoary potentil; hoary cinquefoil	SILBER-FINGERSTRAUCH (*m*)
Potentilla fragariastrum	783	Barren strawberry; strawberry-leaved cinquefoil	ERDBEERBLÄTTRIGER FINGER-STRAUCH (*m*)
Potentilla fruticosa	784	Shrubby (potentil) cinquefoil	GEMEINER FINGERSTRAUCH (*m*); GEMEINES FINGERKRAUT (*n*)
Potentilla palustris	785	Marsh (potentil) cinquefoil (See *Comarum palustre*)	SUMPF-FINGERSTRAUCH (*m*)
Potentilla reptans	786	Creeping cinquefoil (See *Tormentilla reptans*)	HAINFINGERKRAUT (*n*)
Potentilla rupestris	787	Rock (potentil) cinquefoil	FELSEN-FINGERSTRAUCH (*m*)
Potentilla tormentilla	788	Blood-root; tormentil (See *Tormentilla officinalis*)	BLUTWURZ (*f*); BLUTWURZEL (*f*); ROTWURZEL (*f*); RUHRWURZ (*f*)
Potentilla verna	789	Spring cinquefoil	FRÜHLINGS-FINGERSTRAUCH (*m*)
Poterium muricatum	790	Prickly salad burnet, muricated salad burnet	STACHELIGER BECHERSTRAUCH (*m*)
Poterium officinale	791	(See *Sanguisorba officinalis*)	PIMPERNELL (*m*); GROSSER BECH-ERSTRAUCH (*m*)
Poterium sanguisorba	792	Lesser burnet; common salad burnet; garden burnet	KLEINER BECHERSTRAUCH (*m*)
Poterium spinosum	793	Spiny burnet	DORNIGER BECHERSTRAUCH (*m*)
Primula amœna	794	Caucasian primrose	KAUKASISCHE PRIMEL (*f*)
Primula auricula	795	Auricula; mountain cowslip; bear's ear	GAMSWURZ (*f*)
Primula elatior	796	Oxlip	HOHE PRIMEL (*f*)
Primula farinosa	797	Bird's-eye primrose	MEHLIGE PRIMEL (*f*)

Botanical Term.	No.	English Term.	German Term.
Primula glutinosa	798	Primrose, blue spike	BLAUER SPEIK (*m*); KLEBRIGE PRIMEL (*f*)
Primula obconica	799	Japanese primrose	BECHER-PRIMEL (*f*); GIFT-PRIMEL (*f*)
Primula officinalis	800	Paigle; cowslip (See *Primula veris*)	SCHLÜSSELBLUME (*f*)
Primula variabilis	801	polyanthus	
Primula veris	802	Paigle; cowslip; fairy cup; petty mullein; oxlip (See *Primula officinalis*)	LERCHENBLUME (*f*)
Primula vulgaris	803	Primrose, primula; common primrose; wild primrose	HIMMELSCHLÜSSEL (*m*); MARIEN-SCHLÜSSEL (*m*); PRIMEL (*f*)
Pringlea antiscorbutica	804	Kerguelen cabbage	KERGUELENKOHL (*m*)
Prionium palmita	805	(South African)-palmite rush	PALMITSCHILF (*m*)
Prosopis glandulosa	806	Common mezquite	PROSOPIS (*f*); GEMEINER SÜSS-HÜLSENBAUM (*m*)
Prosopis pubescens	807	Screw bean; curly mezquite	HAARIGER SÜSSHÜLSENBAUM (*m*)
Protea mellifera	808	Honey flower, sugar bush	ZUCKERBUSCH (*m*)
Protococcoideæ	809	Globe algæ	KUGELALGEN (*f.pl*); PROTOKOK-KOIDEEN (*f.pl*)
Protococcus nivalis	810	Red alga	SCHNEEALGE (*f*)
Protococcus viridis	811	Yellowish-green globe alga	KUGELALGE (*f*)
Prunella vulgaris	812	Self-heal	BRUNELLE (*f*); GAUCHHEIL (*n*); BRAUNELLE (*f*); SELBSTHEIL (*n*); GOTTHEIL (*n*)
Prunus	813	Bird cherry; apricot; cherry; almond; peach; plum; damson	AHLKIRSCHE (*f*); APRIKOSE (*f*); KIRSCHE (*f*); MANDEL (*f*); PFIRSICH (*m*); PFLAUME (*f*); QUETSCHE (*f*); ZWETSCHE (*f*)
Prunus acida	814	(See *Cerasus acida*)	FRÄNKISCHE WUCHERKIRSCHE (*f*)
Prunus amygdalus	815	Almond tree (See *Amygdalus communis*)	MANDELBAUM (*m*); BITTERMAN-DEL (*f*); GEMEINE MANDEL (*f*)
Prunus amygdalus var. amera	816	Bitter almond	BITTERMANDEL (*f*)
Prunus amygdalus var. dulcis	817	Sweet almond	SÜSSE MANDEL (*f*)
Prunus amygdalus var. fragilis	818	Sweet almond	KRACHMANDEL (*f*)
Prunus amygdalus var. persica	819	(See *Amygdalus communis persica*)	GEMEINER PFIRSICH (*m*)
Prunus angustifolia	820	(See *Prunus insititia*)	CHIKASA-KIRSCHE (*f*)
Prunus armeniaca	821	Apricot; (Persian)-misch-misch; musch-musch	GEMEINE APRIKOSE (*f*); MISCH-MASCH (*m*); MARILLE (*f*); ALBERGE (*f*); APRIKOSEN-BAUM (*m*)
Prunus avium	822	Wild cherry; bird cherry; cherry tree; gean cherry; heart cherry (See *Prunus padus*)	VOGELKIRSCHE (*f*); WALDKIRSCHE (*f*); WILDKIRSCHE (*f*); SÜSS-KIRSCHE (*f*); KIRSCHBAUM (*m*); HOLZKIRSCHENBAUM (*m*); BAUERNKIRSCHE (*f*); MAIKIRSCHE (*f*); HERZKIR-SCHE (*f*); ZWIESEL (*m*)
Prunus cerasifera	823	Turkish plum; Myrobalan plum; Bedda nut; cherry plum	KIRSCHPFLAUME (*f*); TÜRKISCHE PFLAUME (*f*); MYROBALANE (*f*)
Prunus cerasus	824	Wild cherry; dwarf cherry; gean; Morella cherry (See also *Prunus pseudo-cerasus*)	BAUMWEICHSEL (*f*); SAUER-KIRSCHE (*f*); MORELLE (*f*); BAUERNWEICHSELKIRSCHE (*f*); WEICHSELKIRSCHE (*f*); BERN-STEINKIRSCHE (*f*); WILDER KIRSCHBAUM (*m*); WALD-KIRSCHBAUM (*m*); GLAS-KIRSCHE (*f*)
Prunus chamæceras	825	(See *Prunus fructicosa*)	
Prunus coccumilio	826	Neapolitan plum	NEAPOLITANISCHE PFLAUME (*f*)
Prunus communis	827	Damson, Damascene plum; sloe; blackthorn (See also *Prunus domestica*)	

Botanical Term.	No.	English Term.	German Term.
Prunus dasycarpa	828	Plum apricot	PFLAUMENAPRIKOSE (*f*)
Prunus demissa	829	Dwarf bird cherry	NIEDRIGE AHLKIRSCHE (*f*)
Prunus domestica	830	Plum tree ; wild plum (See also *Prunus communis* and *œconomica*)	PFLAUME (*f*); PFLAUMENBAUM (*m*); ZWETSCHE (*f*); ZWETSCHGE (*f*); ZWETSCHKE (*f*); ZWETSCHGENBAUM (*m*); ZWETSCHENBAUM (*m*); HAUSPFLAUME (*f*); QUETSCHE (*f*)
Prunus fructicosa	831	Dwarf cherry	ZWERGKIRSCHE (*f*); OSTHEIMER KIRSCHE (*f*); ZWERGKIRSCHBAUM (*m*)
Prunus incana	832	Grey-leaved dwarf cherry	GRAUBLÄTTRIGE ZWERGKIRSCHE (*f*)
Prunus insititia	833	Plum tree ; wild bullace (See *Prunus angustifolia*)	KIRSCHENPFLAUME (*f*); KRIECHPFLAUME (*f*); KRIECHE (*f*); KRIECHENPFLAUME (*f*); HAFERSCHLEHE (*f*); SPILLING (*f*); DAMASZENER PFLAUME (*f*); JOHANNISPFLAUME (*f*); GEFÜLLTE SCHLEHE (*f*); CHIKASAKIRSCHE (*f*); HAFERPFLAUME (*f*); PFLAUMENSCHLEHE (*f*)
Prunus italica	834	Italian plum, greengage	REINECLAUDE (*f*); RINGLOTTE (*f*); REINECLAUDENPFLAUME (*f*)
Prunus japonica	835	Japanese cherry	JAPANISCHE KIRSCHE (*f*)
Prunus lanceolata	836	(See *Prunus pennsylvanica*)	
Prunus lauro-cerasus	837	Cherry laurel (See *Padus lauro-cerasus*)	KIRSCHLORBEER (*m*); ECHTER KIRSCHLORBEER (*m*)
Prunus lusitanica	838	Portugal laurel	PORTUGIESISCHER KIRSCHLORBEER (*m*); LUSITANISCHER KIRSCHLORBEER (*m*)
Prunus mahaleb	839	Mahaleb cherry, St. Lucia cherry, rock cherry (*See Prunus odorata*)	WEICHSEL (*f*); FELSENKIRSCHE (*f*); WEICHSELKIRSCHE (*f*); LUCIENHOLZBAUM (*m*); TÜRKISCHE WEICHSELKIRSCHE (*f*); MAHALEBKIRSCHE (*f*); ST. LUCIENKIRSCHE (*f*); STEINWEICHSEL (*f*)
Prunus maritima	840	Seaside cherry (See *Prunus pubescens*)	STRANDKIRSCHE (*f*)
Prunus mollis	841	(See *Prunus nigra* and *Cerasus mollis*)	KANADISCHE KIRSCHE (*f*)
Prunus monticola	842	Mountain plum	GEBIRGSPFLAUME (*f*)
Prunus nana	843	(See *Amygdalus nana*)	ZWERGMANDEL (*f*)
Prunus nigra	844	Canadian cherry (See *Prunus mollis*)	KANADISCHE KIRSCHE (*f*)
Prunus odorata	845	(See *Prunus mahaleb*)	
Prunus œconomica	846	(See *Prunus domestica*)	
Prunus orientalis	847	Oriental almond	MORGENLÄNDISCHE MANDEL (*f*)
Prunus orthosepala	848	Texas cherry	TEXASKIRSCHE (*f*)
Prunus padus	849	Bird cherry ; hagberry ; wild cherry, grape cherry (See *Prunus avium*)	FAULBAUM (*m*); TRAUBENKIRSCHE (*f*); AHLKIRSCHE (*f*); GEMEINE AHLKIRSCHE (*f*); HECKENKIRSCHE (*f*); PADDELKIRSCHE (*f*); VOGELKIRSCHE (*f*); WALDKIRSCHE (*f*)
Prunus pennsylvanica	850	Pennsylvanian cherry	PENNSYLVANISCHE KIRSCHE (*f*)
Prunus persica	851	(See *Amygdalus persica*)	
Prunus persica lævis	852	Nectarine	NEKTARIE (*f*)
Prunus prostrata	853	Prostrate cherry	NIEDERGESTRECKTE ZWERGKIRSCHE (*f*)
Prunus pseudocerasus	854	Japanese cherry	JAPANISCHE KIRSCHE (*f*)
Prunus pubescens	855	(See *Prunus maritima*)	STRANDKIRSCHE (*f*)
Prunus pumila	856	Small cherry	KLEINE KIRSCHE (*f*)
Prunus racemosa	857	(See *Prunus avium* and *Prunus padus*)	TRAUBENKIRSCHE (*f*)
Prunus rubra	858	(See *Prunus virginiana*)	

Botanical Term.	No.	English Term.	German Term.
Prunus semperflorens	859	Ever-blooming cherry	IMMERBLÜHENDE KIRSCHE (*f*); ALLERHEILIGENKIRSCHE (*f*)
Prunus serotina	860	Late-blooming bird cherry (See *Padus virginiana*)	SPÄTBLÜHENDE AHLKIRSCHE (*f*)
Prunus serrulata	861	(See *Prunus pseudocerasus*)	SÄGEZAHNBLÄTTRIGE KIRSCHE (*f*)
Prunus sibirica	862	Siberian cherry	SIBIRISCHE KIRSCHE (*f*)
Prunus spinosa	863	Sloe; black-thorn; damson, plum, bullace	SCHLEHDORN (*m*); SCHWARZDORN (*m*); SCHLEHENPFLAUME (*f*); SCHLEHE (*f*)
Prunus subcordata	864	Broad-leaved plum	BREITBLÄTTRIGE PFLAUME (*f*)
Prunus susquehanæ	865	(See *Prunus pumila*)	
Prunus syriaca	866	Syrian plum	SYRISCHE PFLAUME (*f*); DAMAS-ZENER PFLAUME (*f*)
Prunus tomentosa	867	Tomentous cherry; rough-leaved cherry	FILZBLÄTTRIGE KIRSCHE (*f*); FILZIGE KIRSCHE (*f*)
Prunus triflora	868	Chinese cherry; Chinese plum	DREIBLÜTIGE KIRSCHE (*f*); DREI-BLÜTIGE PFLAUME (*f*)
Prunus triloba	869	Chinese peach	DREILAPPIGER PFIRSICH (*m*); MANDELAPRIKOSE (*f*); -RÖSCHENMANDEL (*f*)
Prunus utahensis	870	Utah plum	UTAH-PLFAUME (*f*)
Prunus virginiana	871	Virginian bird cherry (See *Padus oblonga*)	VIRGINISCHE AHLKIRSCHE (*f*)
Psalliota	872	(See *Agaricus campestris*)	
Psamma arenaria	873	Sea-reed; marram-(grass); sea matweed (See *Ammophila arundinacea*)	MEERSCHILF (*n*)
Pseudotsuga douglasii	874	Douglas spruce, red fir (See *Pseudotsuga taxifolia* and *Abies douglasi*)	DOUGLASTANNE (*f*)
Pseudotsuga japonica shirasawa	875	Japanese douglas spruce	JAPANISCHE DOUGLASTANNE (*f*)
Pseudotsuga mucronata	876	Douglas spruce, red fir	DOUGLASTANNE (*f*)
Pseudotsuga taxifolia	877	Yew-leaved Douglas spruce (See *Pseudotsuga Douglasii* and *Abies douglasi*)	EIBENBLÄTTERIGE DOUGLAS-TANNE (*f*)
Psidium cattleianum	878	(Chinese)-strawberry guava	ERDBEER-GUAYAVE (*f*)
Psidium guava	879	West Indian guava	GUAVA (*f*); GUAYAVE (*f*); GUA-YABE (*f*)
Psidium guayava	880	Guava; large fruited Guava (See *Psidium pyriferum*)	GROSSFRÜCHTIGE GUAYAVE (*f*)
Psidium pyriferum	881	Pear-shaped guava (See *Psidium guayava*)	BIRNFÖRMIGE GUAYAVE (*f*)
Psoralea esculenta	882	Pomme de prairie plant	ESSBARER DRÜSENKLEE (*m*)
Psychotria ipecacuanha	883	Ipecacuanha plant (See *Uragoga ipecacuanha*)	ECHTE BRECHWURZEL (*f*); BRECH-VEILCHEN (*n*)
Ptæroxylon utile	884	Sneeze-wood	NIESHOLZ (*n*)
Ptelea trifoliata	885	Hop-tree	HOPFENBAUM (*m*); KLEEBAUM (*m*); (GEMEINER)-LEDER-STRAUCH (*m*); KLEESTRAUCH (*m*)
Pteris	886	(See also *Pellæa*)	SAUMFARN (*m*)
Pteris aquilina	887	Brake, bracken, brachen; fern	SAUMFARN (*m*); ADLERFARN (*m*)
Pteris aquilina esculenta	888	Tasmanian edible fern	ESSBARER SAUMFARN (*m*)
Pteris serrulata	889	Spider fern	GESÄGTER SAUMFARN (*m*)
Pteris tremula	890	Trembling fern	ZITTER-SAUMFARN (*m*)
Pterocarpus erinaceus	891	African rosewood	AFRIKANISCHER ROSENHOLZBAUM (*m*)
Pterocarpus indicus	892	Burmese rosewood	PADOUKHOLZBAUM (*m*); PADUK-HOLZBAUM (*m*)
Pterocarpus santalinus	893	Red sandalwood	ROTES SANDELHOLZ (*n*); KALIA-TURHOLZ (*n*)
Pterocarya fraxinifolia	894	Ash-leaved caucasian walnut	KAUKASISCHE FLÜGELNUSS (*f*); FLÜGELNUSSBAUM (*m*)
Pterocarya stenoptera	895	Narrow-winged winged-nut tree	CHINESISCHE FLÜGELNUSS (*m*)
Pterospermum indicum	896	Amboina-wood tree	AMBOINAHOLZBAUM (*m*)
Ptychosperma elegans	897	Australian feather palm (See *Seaforthia elegans* and *Archontophœnix*)	AUSTRALISCHE FEDER-PALME (*f*)

Botanical Term.	No.	English Term.	German Term.
Ptychotis ajowan	898	(See *Carum ajowan*)	
Ptychotis coptica	899	Ajowan (See *Trachyspermum copticum*)	ADIOWANSAMEN (*m.pl*); AJOWAN-KÜMMEL (*m*)
Puccinia apii	900	Celery brand	SELLERIE-ROST (*m*)
Puccinia arundinacea	901	Reed brand	ROHR-ROST (*m*); RIED-ROST (*m*)
Puccinia asparagi	902	Asparagus rust	SPARGEL-ROST (*m*)
Puccinia buxi	903	Box brand	BUXBAUM-ROST (*m*)
Puccinia clandestina	904	Scabius brand	ABBISWURZEL-ROST (*m*)
Puccinia clinopodii	905	Clinopodium brand	CLINOPODIUM-ROST (*m*)
Puccinia coronata	906	Crown rust, coronated mildew	KRONENROST (*m*)
Puccinia coronifera	907	Oat brand	HAFER-KRONENROST (*m*)
Puccinia dispersa	908	Brown rust	BRAUN-ROST (*m*)
Puccinia fallens	909	Vetch brand	WICKEN-ROST (*m*)
Puccinia fusca	910	Anemone brand	ANEMONEN-ROST (*m*)
Puccinia glechomatis	911	Ground-ivy brand	GUNDERMANN-ROST (*m*)
Puccinia glumarum	912	Yellow rust	GELB-ROST (*m*)
Puccinia graminis	913	Blight, mildew, rubigo, rust, corn mildew, wheat rust, wheat fungus, black rust of cereals (caused by spores of *Aecidium berberidis* (which see) blown on to the wheat fields)	ECHTER MEHLTAU (*m*), MEHLTAU (*m*), MELTAU (*m*); ZWANZIG-ZELLSPOREN-ROST (*m*); GE-TREIDEN-SCHWARZROST (*m*)
Puccinia helianthi	914	Sunflower rust	SONNENBLUMEN-ROST (*m*)
Puccinia malvacearum	915	Hollyhock disease	ROSENMALVEN-ROST (*m*)
Puccinia menthæ	916	Mint brand	MINZEN-ROST (*m*)
Puccinia phlei pratensis	917	Grass rust	GRASROST (*m*)
Puccinia porri	918	Leek brand	LAUCHROST (*m*)
Puccinia prunorum	919	Plum-tree brand	PFLAUMENROST (*m*)
Puccinia rubigo vera	920	Wheat rust, red rust	WEIZENROST (*m*)
Puccinia saxifragarum	921	Saxifrage brand	STEINBRECH-ROST (*m*)
Puccinia simplex	922	Cereal dwarf rust	ZWERGROST (*m*)
Puccinia sparsa	923	Goat's-beard brand (See *Ustilago tragopogonis pratensis*)	
Puccinia suaveolens	924	Thistle brand (See *Puccinia syngenesiarum*)	DISTELROST (*m*)
Puccinia syngenesiarum	925	(See *Puccinia suaveolens*)	
Puccinia vincæ	926	Periwinkle brand	WINTERGRÜN-ROST (*m*)
Puccinia violarum	927	Violet brand	VIOLETTEN-ROST (*m*)
Pulicaria dysenterica	928	Fleabane; common fleabane (See *Inula dysenterica*)	BERUFKRAUT (*n*); RUHRKRAUT (*n*)
Pulicaria vulgaris	929	Small fleabane	GEMEINES FLOHKRAUT (*n*); FLOH-KRAUT (*n*); CHRISTINCHEN-KRAUT (*n*)
Pulmonaria angustifolia	930	Blue cowslip	KLEINBLÄTTRIGES LUNGENKRAUT (*n*)
Pulmonaria officinalis	931	Lungwort, Sage of Bethlehem	LUNGENKRAUT (*n*); GRUBEN-FLECHTE (*f*)
Pulmonaria saccharata	932	Pink-flowered lungwort	ZUCKERIGES LUNGENKRAUT (*n*)
Punica granatum	933	Pomegranate tree	GRANAT (*m*); GRANATBAUM (*m*); GRANATE (*f*)
Puschkinia libanotica	934	(See *Puschkinia scilloides*)	
Puschkinia scilloides	935	Striped squill (See *Puschkinia libanotica*)	PUSCHKINIE (*f*)
Pyrethrum parthenium	936	Fever few (See *Chrysanthemum parthenium*)	FIEBERKRAUT (*n*); MUTTER-KRAUT (*n*)
Pyrola media	937	Intermediate wintergreen	MITTLERES WINTERGRÜN (*n*)
Pyrola minor	938	Common wintergreen	WINTERGRÜN (*n*); KLEINES WIN-TERGRÜN (*n*)
Pyrola rotundifolia	939	Larger wintergreen	RUNDBLÄTTRIGES WINTERGRÜN (*n*)
Pyrola secunda	940	Serrated wintergreen	SEITENBLÜTIGES WINTERGRÜN (*n*)
Pyrola uniflora	941	One-flowered wintergreen	EINBLÜTIGES WINTERGRÜN (*n*)
Pyrophorum communis	942	(See *Pirus communis*)	
Pyrus	943	(See under *Pirus*)	
Pyxidanthera barbulata	944	Pine-barren beauty	
Quassia amara	945	Bitter ash; bitterwood (See *Simaruba amara* and *Picrasma excelsa*)	AMERIKANISCHER QUASSIAHOLZ-BAUM (*m*)

Botanical Term.	No.	English Term.	German Term.
Quercus	946	Oak	EICHE (*f*); EICHENBAUM (*m*)
Quercus ægilops	947	White-haired oak; Greek evergreen oak (See *Quercus pubescens* and *Quercus macrolepsis*)	WEISSHAARIGE EICHE (*f*); KNOPPEREICHE (*f*)
Quercus alba	948	(American)-white oak	WEISSEICHE (*f*)
Quercus ballota	949	Barbary oak; sweet-acorn oak; Ballota oak	BALLOTA-EICHE (*f*)
Quercus castanea	950	Chestnut oak (See *Quercus prinus*)	KASTANIENEICHE (*f*)
Quercus cerris	951	Austrian (Burgundy) oak; Turkey oak	ÖSTERREICHISCHE EICHE (*f*); BURGUNDEREICHE (*f*); ZERREICHE (*f*)
Quercus cinerea	952	(North American)-upland willow oak	WEIDEN-EICHE (*f*)
Quercus coccifera	953	(See *Quercus coccinea*)	KERMESEICHE (*f*)
Quercus coccinea	954	Dyer's oak, scarlet oak (See *Quercus coccifera*)	SCHARLACHEICHE (*f*)
Quercus coccinea tinctoria	955	Quercitron, dyer's oak	FÄRBEREICHE (*f*)
Quercus conferta	956	(See *Quercus esculus*)	DICKFRÜCHTIGE EICHE (*f*)
Quercus cuspiaata	957	Evergreen Japanese oak (See *Pasania cuspidata*)	IMMERGRÜNE JAPANISCHE EICHE (*f*); ZUGESPITZTE SÜDEICHE (*f*)
Quercus dentata	958	Asian oak	KAISEREICHE (*f*)
Quercus esculus	959	Thick-fruited oak (See *Quercus conferta*)	DICKFRÜCHTIGE EICHE (*f*)
Quercus fastigiata	960	Cypress oak	PYRAMIDENFÖRMIGE EICHE (*f*)
Quercus filicifolia	961	Fern-leaved oak	FARNBLÄTTERIGE EICHE (*f*)
Quercus glandulifera	962	Japanese oak	DRÜSENZÄHNIGE EICHE (*f*)
Quercus ilex	963	Ilex, holm (holly) oak, evergreen holm	STECHEICHE (*f*); IMMERGRÜNE EICHE (*f*); HAGEEICHE (*f*)
Quercus intermedia	964	Intermediate oak; Norwood oak	NORWOOD-EICHE (*f*); MITTLERE EICHE (*f*)
Quercus libani	965	Lebanon oak	LIBANON-EICHE (*f*)
Quercus lusitanica	966	Portuguese oak	PORTUGIESISCHE EICHE (*f*)
Quercus macranthera	967	Persian oak	PERSISCHE EICHE (*f*)
Quercus macrocarpa	968	North-American oak; large-fruited oak, burr oak	GROSSFRÜCHTIGE EICHE (*f*)
Quercus macrolepsis	969	Arcadian oak (See *Quercus ægilops*)	ARKADISCHE EICHE (*f*)
Quercus palustris	970	Bog oak; marsh oak	SUMPFEICHE (*f*)
Quercus pedunculata	971	Oak, common oak, British oak, forest oak (See also *Quercus robur*)	SOMMEREICHE (*f*); FRÜHEICHE (*f*); STIELEICHE (*f*)
Quercus pendula	972	Weeping oak	TRAUEREICHE (*f*)
Quercus phellos	973	Willow-oak	WEIDENEICHE (*f*)
Quercus prinus	974	(See *Quercus castanea*)	
Quercus pubescens	975	(See *Quercus ægilops*)	SCHWARZEICHE (*f*); WEICHFLAUMIGE EICHE (*f*)
Quercus robur	976	(Common)-oak, British oak, forest oak; black oak (See also *Quercus sessiliflora* and *Quercus pedunculata*)	SOMMEREICHE (*f*); FRÜHEICHE (*f*); STIELEICHE (*f*)
Quercus rubra	977	(American)-red oak; Quebec oak	ROTEICHE (*f*); ROTE EICHE (*f*)
Quercus salicina	978	(Japanese)-evergreen willow oak	WEIDENBLÄTTRIGE EICHE (*f*)
Quercus serrata	979	Serrated oak	SÄGEBLÄTTERIGE EICHE (*f*)
Quercus sessiliflora	980	Oak; mountain oak; chestnut oak; Durmast oak; sessile-fruited oak	STEINEICHE (*f*); WINTEREICHE (*f*); TRAUBENEICHE (*f*); FRÜHEICHE (*f*)
Quercus suber	981	Cork oak; cork tree	KORKEICHE (*f*); KORKBAUM (*m*)
Quercus tinctoria	982	Dyer's oak; quercitron	FARBENEICHE (*f*)
Quercus toza	983	Pyrenean oak	PYRENÄEN-EICHE (*f*)
Quercus vallonea	984	Gall-nut oak	VALONENEICHE (*f*)
Quillaja saponaria	985	Quillai-(a) tree	QUILLAJABAUM (*m*)
Quinaria	986	(See *Ampelopsis quinquefolia*; *Parthenocissus quinquefolia* and *Vitis quinquefolia*)	
Quisqualis indica	987	Rangoon creeper	INDISCHE FADENRÖHRE (*f*)

Botanical Term.	No.	English Term.	German Term.
Radiola millegrana	988	Flaxseed ; allseed	TAUSENDKÖRNIGER ZWERG-LEIN (*m*)
Ræstelia	989	Ræstelia, cluster-cup (See *Ræstelia*)	
Ramondia pyrenaica	990	Rosette mullein	RAMONDIE (*f*)
Ranunculus	991	Crowfoot ; butter-cup, gold-cup, king-cup ; ranunculus, fair maids of Kent, fair maids of France	HAHNENFUSS (*m*) ; BUTTERBLUME (*f*) ; RANUNKEL (*f*) ; GARTEN-BUTTERBLUME (*f*)
Ranunculus aconitifolius flore pleno	992	Fair maids of France	EISENHUTBLÄTTRIGER HAHNEN-FUSS (*m*)
Ranunculus acris	993	Upright crowfoot ; meadow crowfoot ; buttercup ; field buttercup	SCHARFER HAHNENFUSS (*m*)
Ranunculus acris flore pleno	994	Bachelor's buttons, double buttercup	DOPPELBLÜTIGER SCHARFER HAHNENFUSS (*m*)
Ranunculus alpestris	995	Alpine crow's-foot, alpine ranunculus	VORALPENHAHNENFUSS (*m*)
Ranunculus aquatilis	996	Water crowfoot ; common water buttercup ; water-fennel ; water-ranunculus	WASSERHAHNENFUSS (*m*)
Ranunculus arvensis	997	Corn crowfoot	ACKER-HAHNENFUSS (*m*)
Ranunculus asiaticus	998	Asiatic ranunculus	ASIATISCHER HAHNENFUSS (*m*) ; KUGELRANUNKEL (*f*)
Ranunculus auricomus	999	Wood crowfoot ; goldilocks	GOLDKOPF (*m*) ; GOLDGELBER HAHNENFUSS (*m*)
Ranunculus bulbosus	000	Bulbous crowfoot ; cuckoo-bud ; king's-cup ; butter-cup, gold-cup ; butter-flower ; king-cup ; bulbous buttercup	RANUNKEL (*f*) ; BUTTERBLUME (*f*) ; DOTTERBLUME (*f*) ; KUGELRANUNKEL(*f*)
Ranunculus chærophyllos	001	(See *Ranunculus flabellatus*)	
Ranunculus ficaria	002	Pilewort ; lesser celandine ; fig-wort ranunculus (See *Ficaria verna*)	LÄMMERBLUME (*f*) ; FEIGEN-WURZ (*f*) ; SCHARBOCK (*m*) ; SCHARBOCKSKRAUT (*n*)
Ranunculus flabellatus	003	Fine-leaved ranunculus (See *Ranunculus chærophyllos*)	FEINBLÄTTRIGER HAHNENFUSS (*m*) ; FÄCHERHAHNENFUSS (*m*)
Ranunculus flammula	004	Lesser spearwort	KLEINER SUMPF-HAHNENFUSS (*m*) ; BRENNENDER HAHNENFUSS (*m*)
Ranunculus hederaceus	005	Ivy crowfoot ; ivy-leaved crowfoot	HAARKRAUT (*n*) ; FROSCHKRAUT (*n*) ; EFEUBLÄTTRIGER WASSER-HAHNENFUSS (*m*)
Ranunculus hirsutus	006	Hairy ranunculus	HAARIGER HAHNENFUSS (*m*)
Ranunculus lingua	007	Great spearwort	GROSSER SUMPF-HAHNENFUSS (*m*) ; ZUNGENBLÄTTRIGER HAHNEN-FUSS (*m*)
Ranunculus lyallii	008	Rockwood lily, New Zealand water-lily	LYALLS-HAHNENFUSS (*m*)
Ranunculus montanus	009	mountain crow-foot, mountain ranunculus	BERGHAHNENFUSS (*m*)
Ranunculus ophioglossifolius	010	Snake-tongue ranunculus	NATTERZUNGENBLÄTTRIGER HAHNENFUSS (*m*)
Ranunculus parviflorus	011	Small-flowered ranunculus	KLEINBLÜTIGER HAHNENFUSS (*m*)
Ranunculus repens	012	Creeping crowfoot ; creeping buttercup	KRIECHENDER HAHNENFUSS (*m*)
Ranunculus sceleratus	013	Celery-leaved crowfoot ; celery-leaved ranunculus	BLASENZIEHENDER HAHNENFUSS (*m*)
Raphanus caudatus	014	Rat-tail radish	SCHWANZRETTICH (*m*)
Raphanus maritimus	015	Sea radish	MEER-RETTICH (*m*)
Raphanus raphanistrum	016	Jointed charlock ; white charlock ; wild radish	WILDER RAPS (*m*) ; ACKERRETTICH (*m*) ; HEDERICH (*m*)
Raphanus sativus	017	Radish, common radish	RADIES (*n*) ; RADIESCHEN (*n*) ; RETTICH (*m*) ; GARTENRET-TICH (*m*) ; ÖLRETTICH (*m*)
Raphia vinifera	018	(West African)-bamboo palm ; wine palm	WESTAFRIKANISCHE BAMBUS-PALME (*f*) ; WEINPALME (*f*) ; RAPHIA (*f*) ; RAPHIE (*f*)
Raphiolepsis indica	019	Indian hawthorne (See *Rhaphiolepsis indica*)	CHINESISCHE RHAPHIOLEPSIS (*f*)
Ravenala madagascariensis	020	Traveller's tree	QUELLENBAUM (*m*)

Botanical Term.	No.	English Term.	German Term.
Reinwardtia trigyna	021	Winter flax, East Indian flax (See *Linum trigynum*)	REINWARDTIE (*f*); WINTER-FLACHS (*m*)
Reseda alba	022	White mignonette	WEISSER WAU (*m*)
Reseda lutea	023	Wild mignonette; cut-leaved mignonette	GELBER WAU (*m*)
Reseda luteola	024	Weld; yellowweed; dyer's weed; dyer's rocket; woold; mignonette	WAU (*m*); GILBE (*f*); HARN-KRAUT (*n*); GELBKRAUT (*n*); FÄRBERWAU (*m*); GELBE RESEDA (*f*)
Reseda odorata	025	Mignonette; fragrant mignonette	RESEDA (*f*); GARTEN-RESEDA (*f*)
Retinospora ericoides	026	(See *Cupressus thyiodes*)	
Rhamnus	027	Buckthorn	KREUZDORN (*m*); FAULBAUM (*m*)
Rhamnus alnifolia	028	Alder-leaved buckthorn (See *Rhamnus purshiana*)	ERLENBLÄTTRIGER KREUZDORN (*m*)
Rhamnus alpina	029	Alpine buckthorn	ALPENKREUZDORN (*m*)
Rhamnus caroliniana	030	Carolina buckthorn (See *Frangula caroliniana*)	KAROLINISCHER FAULBAUM (*m*)
Rhamnus catharticus	031	Buckthorn; purging buckthorn; hartshorn; waythorn; common buckthorn (See *Rhamnus spinosa*)	HIRSCHDORN (*m*); WEGEDORN (*m*); FÄRBERBEERE (*f*); FÄR-BERKREUZDORN (*m*); KREUZ-DORN (*m*); GEMEINER KREUZ-DORN (*m*); HIRSCHBEERDORN (*m*); RAINBEERE (*f*)
Rhamnus frangula	032	Water elder, black alder, alder buckthorn, Avignon (Persian) berry; berry-bearing alder	FAULBAUM (*m*); KREUZHOLZ (*n*); KREUZDORN (*m*); KREUZ-BEERSTRAUCH (*m*); GELB-BEERSTRAUCH (*m*); ERLEN-FAULBAUM (*m*); KREUZ-BEERE (*f*); PULVERHOLZ-BAUM (*m*)
Rhamnus infectoria	033	Dyer's buckthorn	FÄRBERKREUZDORN (*m*)
Rhamnus pumila	034	Alpine (mountain) buckthorn; rock (dwarf) buckthorn (See *Rhamnus rupestris*)	FELSENFAULBAUM (*m*)
Rhamnus purshiana	035	Californian buckthorn; Pursh's buckthorn (See *Rhamnus alnifolia*)	PURSH'S FAULBAUM (*m*)
Rhamnus rupestris	036	(See *Rhamnus pumila*)	FELSENFAULBAUM (*m*)
Rhamnus spinosa	037	(See *Rhamnus catharticus*)	GEMEINER KREUZDORN (*m*)
Rhaphiolepsis indica	038	(See *Raphiolepsis indica*)	
Rhapis humilis	039	Ground rattan cane	ZWERG-STECKENPALME (*f*)
Rhaponticum cynaroides	040	Swiss knapweed (See *Centaurea rhaponticum*)	RÜBENDISTEL (*f*)
Rheum officinale	041	(See *Rheum palmatum*)	GEBRÄUCHLICHER RHABARBER (*m*)
Rheum palmatum	042	Medicinal rhubarb (See *Rheum officinale*)	PURGIERWURZEL (*f*); RHABARBER (*m*); ECHTER RHABARBER (*m*)
Rheum rhababerum	043	Rhubarb (See *Rheum undulatum*)	WELLIGER RHABARBER (*m*)
Rheum rhaponticum	044	Dwarf rhubarb, edible rhubarb, garden rhubarb	STUMPFER RHABARBER (*m*); PONTISCHER RHABARBER (*m*)
Rheum undulatum	045	(See *Rheum rhababerum*)	WELLIGER RHABARBER (*m*)
Rhexia virginica	046	Meadow beauty; deer grass	RHEXIA (*f*); REHGRAS (*n*)
Rhinanthus crista-galli	047	Yellow rattle; common cockscomb; common rattle	KLAPPERTOPF (*m*)
Rhipsalis cassytha	048	Mistletoe cactus; Cassytha	FADEN-GEISSELKAKTUS (*m*)
Rhizocarpæ	049	Water-ferns	WASSERFARNE (*m.pl*)
Rhizophora	050	Mangrove tree; mangle-tree	LEUCHTERBAUM (*m*); MANGLE-BAUM (*m*)
Rhizophora mangle	051	Horseflesh mahogany (See *Robinia panacoco*)	PFERDEFLEISCHHOLZ (*n*); BOL-LETRIEHOLZ (*n*); MANGROVE (*f*); WURZELBAUM (*m*); MANGLEBAUM (*m*)
Rhizophora mucronata	052	Mangrove tree	MANGROVE (*f*); MANGROVEN-RINDEBAUM (*m*)
Rhodiola rosea	053	(See *Sedum rhodiola*)	
Rhododendron	054	Rhododendron, mountain rose, alpine rose	ALPENROSE (*f*); ALMRAUSCH (*m*); GEISSSCHADEN (*m*); ALPEN-RAUSCH (*m*); ROSENLORBEER (*m*)

Botanical Term.	No.	English Term.	German Term.
Rhododendron chamæcistus	055	(See *Rhodothamnus chamæcistus*)	
Rhododendron flavum	056	(See *Azalea pontica*)	
Rhododendron hirsutum	057	Alpine rose	RAUHHAARIGE ALPENROSE (*f*); ALPENRAUSCH (*m*); SCHNEE-ROSE (*f*); DONNERROSE (*f*); ALPBALSAM (*m*)
Rhododendron viscosum	058	Swamp honeysuckle (See *Azalea viscosa*)	
Rhodophyceæ	059	Red algæ, red seaweed (See *Florideæ*)	ROTALGEN (*f.pl*)
Rhodorastrum dahuricum	060	Dahuri azalea; Siberian azalea (See also *Azalea*)	DAHURISCHE AZALIE (*f*); DAHURISCHE AZALEE (*f*)
Rhodothamnus chamæcistus	061	Ground cistus (See *Rhododendron chamæcistus*)	ZWERG-RADROSE (*f*)
Rhodotypus kerrioides	062	Japanese false kerria; white kerria	SCHEINKERRIE (*f*); KAIMAS-STRAUCH (*m*); JAMBUKI-STRAUCH (*m*)
Rhodymenia palmata	063	Purple dulse	ROSEN-ROTALGE (*f*)
Rhus aromatica	064	sweet-scented sumach (See *Rhus crenata*)	WOHLRIECHENDER SUMACH (*m*)
Rhus canadensis	065	(See *Rhus thyphina*)	AMERIKANISCHES FISETHOLZ (*n*)
Rhus coriaria	066	Tanner's (sumac) sumach (See *Coriaria myrtifolia*)	GERBERSTRAUCH (*m*); GERBER-SUMACH (*m*); LEDERSTRAUCH (*m*); LEDERBAUM (*m*); SUMACH (*m*); SCHMACK (*m*)
Rhus cotinoides	067	American smoke tree	AMERIKANISCHER PERÜCKEN-SUMACH (*m*)
Rhus cotinus	068	Wig tree; smoke tree; young fustic; smoke plant; yellow fustic; Venetian sumach (See *Cotinus cotinus*)	FISETHOLZ (*n*); FISTELHOLZ (*n*); PERÜCKENSUMACH (*m*); PE-RÜCKENSTRAUCH (*m*); PE-RÜCKENBAUM (*m*); GELBHOLZ (*n*); FUSTIKHOLZ (*n*); JUNGER FUSTIK (*m*)
Rhus crenata	069	(See *Rhus aromatica*)	KERBBLÄTTRIGER SUMACH (*m*)
Rhus glabra	070	shumac, sumac; tanner's sumac, sumach, smooth sumach (See *Rhus viridifolia*)	FÄRBERBAUM (*m*); GERBER-STRAUCH (*m*); GLATTER SUMACH (*m*)
Rhus succedanea	071	Japan wax tree	TALG-SUMACH (*m*)
Rhus thyphina	072	Canadian sumach, stag's-horn sumach (See *Rhus canadensis*)	ESSIGSUMACH (*m*); ESSIGBAUM (*m*)
Rhus toxicodendron	073	(North American)-poison oak	GIFTSUMACH (*m*); GIFTEFEU (*m*)
Rhus trilobata	074	Trilobate sumach	DREILAPPIGER SUMACH (*m*)
Rhus venenata	075	Poison elder	FIRNIS-SUMACH (*m*)
Rhus vernicifera	076	Lacquer tree; varnish tree	LACKBAUM (*m*)
Rhus viridifolia	077	(See *Rhus glabra*)	GRÜNBLÄTTRIGER SUMACH (*m*)
Rhynchospermum jasminoides	078	Chinese ivy, Chinese jasmine (See *Trachelospermum jasminoides*)	JASMINARTIGER SCHNABELSAME (*m*)
Rhynchospora alba	079	White beak-sedge	WEISSE MEERSIMSE (*f*)
Rhynchospora fusca	080	Brown-beak-sedge	BRAUNE MEERSIMSE (*f*)
Ribes album	081	White currant (See *Ribes rubrum album*)	WEISSE JOHANNISBEERE (*f*)
Ribes alpinum	082	Alpine currant; mountain currant	ALPEN-JOHANNISBEERE (*f*)
Ribes americanum	083	American currant	AHLBEERE (*f*); REICHBLÜTIGE JOHANNISBEERE (*f*)
Ribes atropurpureum	084	Dark-purple currant (See *Ribes petræum*)	
Ribes aureum	085	Yellow buffalo currant; Missouri currant, buffalo currant (See also *Ribes leiobotrys*)	GELBE JOHANNISBEERE (*f*)
Ribes cereum	086	Waxy-leaved currant	WACHSDRÜSIGE JOHANNISBEERE (*f*)
Ribes cynosbati	087	Dog bramble; dog currant	BORSTENFRÜCHTIGE STACHEL-BEERE (*f*)
Ribes diacantha	088	Twin-prickled currant	STACHELIGE JOHANNISBEERE (*f*)
Ribes divaricatum	089	Straggling white gooseberry	SPARRIGE STACHELBEERE (*f*)
Ribes fasciculatum	090	Bundled currant	BÜSCHELIGE JOHANNISBEERE (*f*)

Botanical Term.	No.	English Term.	German Term.
Ribes gordonianum	091	Gordon's currant; Gordonian currant	GORDONS JOHANNISBEERE (*f*)
Ribes gracile	092	(See *Ribes niveum*)	
Ribes grossularia	093	Gooseberry; wild gooseberry	STACHELBEERE (*f*); KRATZBEERE (*f*); ECHTE STACHELBEERE (*f*); RIBITZEL (*n*)
Ribes grossularia uva crispa	094	Smooth-berried gooseberry	KAHLBEEREN-STACHELBEERE (*f*); KAHLBEERIGE STACHELBEERE (*f*)
Ribes hirtellum	095	Fine-haired gooseberry; American gooseberry	KURZHAARIGE STACHELBEERE (*f*)
Ribes irriguum	096	Water-loving currant	WASSERLIEBENDE STACHELBEERE (*f*)
Ribes lacustre	097	Lake currant	SUMPF-JOHANNISBEERE (*f*)
Ribes leiobotrys	098	Buffalo currant (See also *Ribes aureum*)	KAHLTRAUTE JOHANNISBEERE (*f*)
Ribes menziesi	099	Menzies red currant	MENZIES-STACHELBEERE (*f*)
Ribes multiflorum	100	Many-flowered currant	VIELBLÜTIGE JOHANNISBEERE (*f*)
Ribes nigrum	101	Black currant (See *Ribes 'idum*)	AHLBEERE (*f*); SCHWARZE JO--HANNISBEERE (*f*); GICHTBEERE (*f*); PFEFFERBEERE (*f*); WAN--ZENBEERE (*f*)
Ribes niveum	102	Snowy currant (See *Ribes gracile*)	WEISSBLÜHENDE STACHELBEERE (*f*)
Ribes olidum	103	(See *Ribes nigrum*)	
Ribes orientale	104	Oriental currant	BEHAARTFRÜCHTIGE JOHANNIS--BEERE (*f*)
Ribes petræum	105	Rock currant (See *Ribes atropurpureum*)	FEOSEN-JOHANNISBEERE (*f*)
Ribes rotundifolium	106	Round-leaved currant	RUNDBLÄTTRIGE STACHELBEERE (*f*)
Ribes rubrum	107	Red currant	AHLBEERE (*f*); ROTE JOHANNIS--BEERE (*f*); GEMEINE JOHANNISBEERE (*f*)
Ribes rubrum album	108	(See *Ribes album*)	
Ribes sanguineum	109	Common flowering currant	BLUTROTE JOHANNISBEERE (*f*); ROTE JOHANNISBEERE (*f*)
Richardia africana	110	Lily of the Nile; white arum; Calla lily, arum lily (See *Zantedeschia æthiopica* and also *Peltandra virginica*)	ARON (*n* and *m*); ARUM (*m*); ARONLILIE (*f*)
Ricinus communis	111	Castor-oil plant, oil plant, palma christi, croton-seed plant, tilly-seed plant, castor-bean plant	RICINUS (*m*); GEMEINER WUNDER--BAUM (*m*); RIZINUS (*m*); PURGIERKÖRNERBAUM (*m*); PALMA CHRISTI (*f*); RÖMISCHE BOHNE (*f*); INDISCHE BOHNE (*f*); SONNENKORN (*n*); HÖL--LENFEIGE (*f*); SCHAFLAUS (*f*); ÖLKAFFEE (*m*); POMADEN--BOHNE (*f*)
Rivinia humilis	112	Rouge plant, blood berry, rouge berry (See also *Phytolacca*)	ZWERG-RIVINIE (*f*); BLUTBEERE (*f*)
Robinia hispida	113	Rose acacia	BRAUNROTE ROBINIE (*f*)
Robinia panacoco	114	Horseflesh mahogany (See *Rhizophora mangle*)	PFERDEFLEISCHHOLZ (*n*); BOL--LETRIEHOLZ (*n*); ROBINIE (*f*)
Robinia pseudoacacia	115	Locust, false acacia; locust tree (See *Hymenæa courbaril*)	ROBINIE (*f*); FALSCHE AKAZIE (*f*); HEUSCHRECKENBAUM (*m*); HÜLSENBAUM (*m*); SCHEIN--ROBINIE (*f*); KUGELROBINIE (*f*); KUGELAKAZIE (*f*); WEISSAKAZIE (*f*); VIRGIN--ISCHER SCHOTENDORN (*m*); WUNDERBAUM (*m*)
Robinia umbraculifera	116	(See *Robinia pseudoacacia*)	KUGELAKAZIE (*f*)
Roccella tinctoria	117	Tartarean moss, orchil (See also *Lecanora tartarea*)	LACKFLECHTE (*f*); LACKMUS--FLECHTE (*f*); ORSEILLE--FLECHTE (*f*); FÄRBERFLECHTE (*f*)

Botanical Term.	No.	English Term.	German Term.
Rodgersia podophylla	118	Rodgers' bronze-leaf	HANDBLÄTTRIGE RODGERSIE (*f*)
Roella	119	South African harebell	ROËLLE (*f*)
Roemeria hybrida	120	Roemerea ; Roemeria	MITTELMEER-MOHN (*m*)
Rœstelia	121	Rœstellia ; cluster-cup (See *Rœstelia*)	GITTER-ROST (*m*)
Rœstelia cancellata	122	Pear-tree cluster-cup	BIRN-GITTERROST (*m*)
Rœstelia cornuta	123	Mountain ash cluster cup ; horn-like cluster-cup	EBERESCHEN-GITTERROST (*m*)
Rœstelia lacerata	124	Hawthorn cluster-cup, lacerated cluster-cup	WEISSDORN-GITTERROST (*m*)
Rœstelia penicillata	125	Apple cluster-cup	APFEL-GITTERROST (*m*)
Romanzoffia sitchensis	126	Sitcha water-leaf	SUBARKTISCHES WASSERBLATT (*n*)
Romneya coulteri	127	Tree poppy, white bush	ROMNEYA (*f*)
Romulea columnæ	128	Romulea	ROMULEA (*f*)
Rosa	129	Rose	ROSE (*f*)
Rosa agrestis	130	Field rose	ZAUN-ROSE (*f*)
Rosa alpina	131	Alpine rose (See *Rosa inermis* and *Rosa pendulina*)	ALPENROSE (*f*)
Rosa arvensis	132	(White) field rose, Ayrshire rose (See *Rosa repens*)	WEISSE ROSE (*f*)
Rosa austriaca	133	(See *Rosa gallica*)	
Rosa banksiæ	134	Banksian rose	BANKS-ROSE (*f*)
Rosa bourboniana	135	Bourbon rose	BOURBON-ROSE (*f*)
Rosa bracteata	136	Macartney rose (See *Rosa microphylla*)	KLEINBLÄTTRIGE ROSE (*f*)
Rosa bractescens	137	Bracteated downy rose ; Macart--ney's rose	MACARTNEY-ROSE (*f*)
Rosa cæsia	138	Glaucous-leaved rose	BLAUGRÜNE ROSE (*f*)
Rosa canina	139	(Common)-dog rose ; hip-tree ; hep-tree ; hep-bramble ; hep-brier ; hep-briar ; wild-brier, hedge rose, wild rose	HUNDSROSE (*f*) ; HAGEROSE (*f*) ; HECKENROSE (*f*) ; HORNROSE (*f*) ; WILDE ROSE (*f*)
Rosa canina cortifolia	140	(See *Rosa coriaria*)	
Rosa canina glaucescens	141	Milky-green rose (See *Rosa glauca*)	MEERGRÜNE ROSE (*f*)
Rosa carolina	142	Carolina rose (See *Rosa pennsylvanica*)	KAROLINEN-ROSE (*f*)
Rosaceæ	143	Rosaceæ (See *Prunus*, *Spiræa*, *Rosa*, *Cratægus*, *Cotoneaster*, *Mespilus*, *Pirus*, *Sanguisorba*, *Agrimonia*, *Alchemilla*, *Rubus*, *Fragaria*, *Potentilla*)	ROSAZEEN (*f.pl*) ; ROSENBLÜTIGE GEWÄCHSE (*n.pl*)
Rosa centifolia	144	Cabbage rose (See *Rosa gallica centifolia*)	HUNDERTBLÄTTRIGE ROSE (*f*) ; ZENTIFOLIE (*f*) ; CENTIFOLI--ENROSE (*f*)
Rosa centifolia bifera	145	(See *Rosa damascena*)	
Rosa centifolia muscosa	146	(See *Rosa muscosa*)	
Rosa cinnomomea	147	Cinnamon rose (See *Rosa majalis*)	ZIMMTROSE (*f*)
Rosa coriaria	148	Leather-leaved rose (See *Rosa canina cortifolia*)	
Rosa damascena	149	Damascene rose, damask rose ; monthly rose (See *Rosa centifolia bifera*)	DAMASCENER ROSE (*f*) ; MONATS--ROSE (*f*)
Rosa elliptica	150	Elliptical rose	STARKRIECHENDE ROSE (*f*)
Rosa ferruginea	151	Red-leaved rose (See *Rosa rubrifolia*)	ROTBLÄTTRIGE ROSE (*f*)
Rosa fœtida	152	(See *Rosa lutea*)	
Rosa gallica	153	Damask rose ; red rose ; Provence rose ; French rose (See *Rosa austriaca*)	ESSIGROSE (*f*)
Rosa gallica centifolia	154	Cabbage rose (See *Rosa centifolia*)	
Rosa gallica muscosa	155	Moss rose (See *Rosa muscosa*)	
Rosa gallica provincialis	156	Provence rose	PROVENCERROSE (*f*)
Rosa glauca	157	(See *Rosa canina glaucescens*)	

Botanical Term.	No.	English Term.	German Term.
Rosa grevillei	158	Greville's Rose (See *Rosa multiflora grevillei*)	GREVILLES-ROSE (*f*)
Rosa hemisphærica	159	Hemispherical rose	SCHWEFEL-ROSE (*f*)
Rosa hibernica	160	Irish rose	IRLÄNDISCHE ROSE (*f*)
Rosa indica	161	Indian rose, China rose, monthly rose	INDISCHE ROSE (*f*); MONATSROSE (*f*)
Rosa indica fragrans	162	Tea rose (See *Rosa odorata*)	WOHLRIECHENDE INDISCHE ROSE (*f*)
Rosa indica minima	163	Fairy rose, Miss Lawrence's rose (See *Rosa lawrenciana*)	KLEINE INDISCHE ROSE (*f*); LAWRENCES-ROSE
Rosa indica sanguinea	164	Crimson China rose	BLUTROTE INDISCHE ROSE (*f*)
Rosa indica var. semperflorens	165	Monthly rose; ever-blowing rose	MONATSROSE (*f*)
Rosa inermis	166	(See *Rosa alpina*)	WEHRLOSE ROSE (*f*)
Rosa inodora	167	Faintly-scented sweet-briar	GERUCHLOSE WEINROSE (*f*)
Rosa involuta	168	Prickly unexpanded rose	
Rosa lævigata	169	Cherokee rose	CHEROKEE-ROSE (*f*)
Rosa lawrenciana	170	(See *Rosa indica minima*)	
Rosa lucida	171	Bright-leaved rose	LICHTE-ROSE (*f*)
Rosa lutea	172	Austrian briar; Austrian yellow brier (See *Rosa fœtida*)	GELBE ROSE (*f*)
Rosa majalis	173	(See *Rosa cinnamomea*)	
Rosa micrantha	174	Small-flowered sweet-briar	KLEINBLÜTIGE WEINROSE (*f*)
Rosa microphylla	175	Small-leaved rose (See *Rosa sempervirens* and *Rosa bracteata*)	KLEINBLÄTTRIGE ROSE (*f*)
Rosa mollis	176	(See *Rosa villosa*)	
Rosa moschata	177	Musk rose	MOSCHUSROSE (*f*)
Rosa multiflora grevillei	178	Seven sisters rose (See *Rosa grevillei*)	
Rosa multiflora-(polyantha)	179	Many-flowered rose; polyantha rose	BÜSCHELROSE (*f*)
Rosa muscosa	180	Moss rose	MOOSROSE (*f*)
Rosa nitida	181	Glossy-leaved rose (See *Rosa rubrispina*)	GLANZBLÄTTRIGE ROSE (*f*)
Rosa noisettiana	182	Noisette rose	NOISETTE-ROSE (*f*)
Rosa odorata	183	(See *Rosa indica fragrans*)	WOHLRIECHENDE ROSE (*f*)
Rosa pendulina	184	Drooping rose (See *Rosa alpina*)	
Rosa pennsylvanica	185	(See *Rosa carolina*)	
Rosa persica	186	Persian rose	PERSISCHE ROSE (*f*)
Rosa pimpinellifolia	187	Scotch rose; burnet rose; pimpinella-leaved rose (See *Rosa spinosissima*)	PIMPERNELL ROSE (*f*)
Rosa pomifera	188	Dog-rose, apple rose	APFELROSE (*f*)
Rosa repens	189	Creeping rose (See *Rosa arvensis*)	KRIECHROSE (*f*)
Rosa rubella	190	Red-fruited dwarf rose	ROT-FRÜCHTIGE ZWERG-ROSE (*f*)
Rosa rubifolia	191	(See *Rosa setigera*)	
Rosa rubiginosa	192	Sweet-briar; eglantine; eglatere; sweet-brier; sweet-briar rose	WEINROSE (*f*)
Rosa rubrifolia	193	Red-leaved rose (See *Rosa ferruginea*)	
Rosa rubrispina	194	(See *Rosa nitida*)	
Rosa rugosa	195	Japanese rose; wrinkled rose	RUNZELIGE ROSE (*f*)
Rosa semperflorens	196	(See *Rosa indica var. semperflorens*)	
Rosa sempervirens	197	Evergreen rose (See *Rosa microphylla*)	IMMERGRÜNE ROSE (*f*)
Rosa sepium	198	Small-leaved sweet-briar	KLEINBLÄTTRIGE WEINROSE (*f*)
Rose setigera	199	Prairie rose; bristle-bearing rose (See *Rosa rubifolia*)	PRÄRIEROSE (*f*); BORSTENROSE (*f*)
Rosa spinosissima	200	Scotch rose, burnet rose (See *Rosa pimpinellifolia*)	
Rosa systyla	201	Close-styled dog-rose	HUNDS-SCHLINGROSE (*f*)
Rosa tomentosa	202	downy-leaved dog-rose	FILZIGE ROSE (*f*)
Rosa trachyphylla	203	Coarse-haired rose	RAUHHAARIGE ROSE (*f*)
Rosa turbinata	204	Top-shaped-calyxed rose	KREISELROSE (*f*)

Botanical Term.	No.	English Term.	German Term.
Rosa villosa	205	Downy rose; soft-leaved round-fruited rose (See *Rosa mollis*)	WEICHHAARIGE ROSE (*f*)
Rosa virginiana	206	Virginian rose	VIRGINISCHE ROSE (*f*)
Rosmarinus officinalis	207	Wild rosemary; marsh rosemary; moorwort	LAVENDELHEIDE (*f*); HEIDE--BIENENKRAUT (*n*); ROSMAR--INHEIDE (*f*)
Rottlera tinctoria	208	(Indian)-kamala plant (See also *Mallotus philippinensis*)	KAMALAPFLANZE (*f*)
Rubia munjista	209	East Indian madder	OSTINDISCHER KRAPP (*m*)
Rubia peregrina	210	Wild madder	WILDER KRAPP (*m*); MITTEL--MEERISCHER KRAPP (*m*)
Rubia tinctorum	211	Madder	FÄRBERRÖTE (*f*); KRAPP (*m*); LANDRÖTE (*f*); KRAPPPFLANZE (*f*)
Rubus biflorus	212	Two-flowered raspberry (See *Rubus leucodermis*)	ZWEIBLÜTIGE HIMBEERE (*f*)
Rubus cæsius	213	Dewberry; bramble	ACKERBEERE (*f*)
Rubus chamæmorus	214	Cloudberry; mountain-bramble; knotberry; ground mulberry	MULTEBEERE (*f*)
Rubus cratægifolius	215	Japanese raspberry; cratægus-leaved raspberry	WEISSDORNBLÄTTRIGE HIMBEERE (*f*)
Rubus deliciosus	216	(Rocky mountain)-bramble; delicious bramble (See also *Rubus occidentalis*)	LIEBLICHE HIMBEERE (*f*)
Rubus fruticosus	217	Blackberry; bramble-(berry); shrubby bramble (See *Rubus plicatus*, *Rubus sulcatus* and *Rubus ulmifolius*)	KRATZBEERE (*f*); BROMBEERE (*f*); BRAMBEERE (*f*); BROMBEERGE--BÜSCH (*n*); FALTIGE BROM--BEERE (*f*)
Rubus idæus	218	Raspberry	ROTFRÜCHTIGE HIMBEERE (*f*); HIMBEERE (*f*); GELBFRÜCH--TIGE HIMBEERE (*f*)
Rubus idæus loganii	219	Loganberry	LOGANBEERE (*f*)
Rubus incisus	220	Incised bramble	EINGESCHNITTENE HIMBEERE (*f*)
Rubus laciniatus	221	American blackberry, cut-leaved blackberry	SCHLITZBLÄTTRIGE BROMBEERE (*f*)
Rubus leucodermis	222	White-skinned raspberry (See *Rubus biflorus*)	WEISSRINDIGE HIMBEERE (*f*)
Rubus nutkanus	223	Nootka-sound bramble, salmon berry	NUTKA-HIMBEERE (*f*)
Rubus occidentalis	224	Virginian raspberry (See also *Rubus deliciosus* and *Rubus odoratus*)	NORDAMERIKANISCHE HIMBEERE (*f*)
Rubus odoratus	225	(North American)-sweet-scented raspberry; Virginian raspberry; purple-flowering raspberry	WOHLRIECHENDE HIMBEERE (*f*)
Rubus phœnicolasius	226	Wineberry	ROTBORSTIGE HIMBEERE (*f*)
Rubus plicatus	227	Plaited bramble (See *Rubus fruticosus*)	FALTIGE BROMBEERE (*f*)
Rubus purpureus	228	(See *Rubus triphyllus*)	
Rubus rosæfolius flore simplici	229	Strawberry-raspberry; rose-leaved raspberry	ROSENBLÄTTRIGE HIMBEERE (*f*)
Rubus rosifolius	230	(See *Rubus rosæfolius flore simplici*)	
Rubus saxatilis	231	Stone rubus	STEINBEERE (*f*)
Rubus sorbifolius	232	(See *Rubus rosæfolius flore simplici*)	
Rubus spectabilis	233	Salmon berry, showy raspberry	PRÄCHTIGE HIMBEERE (*f*)
Rubus strigosus	234	Wild red raspberry	BORSTIGE HIMBEERE (*f*)
Rubus sulcatus	235	(See *Rubus fruticosus*)	
Rubus triphyllus	236	Three-leaved raspberry (See *Rubus purpureus*)	DREIBLÄTTRIGE HIMBEERE (*f*)
Rubus ulmifolia flore pleno	237	Daisy-flowered bramble	DOPPELBLÜTIGE ULMENBLÄTTRIGE BROMBEERE (*f*)
Rubus ulmifolius	238	Common bramble; elm-leaved bramble (See *Rubus fruticosus*)	ULMENBLÄTTRIGE BROMBEERE (*f*)
Rudbeckia	239	Cone-flower	SONNENHUT (*m*)

Botanical Term.	No.	English Term.	German Term.
Ruellia macracantha	240	Christmas pride	
Rumex	241	Sorrel ; dock	AMPFER (*m*)
Rumex abyssinicus	242	Abyssinian dock	ABESSINISCHER AMPFER (*m*)
Rumex acetosa	243	Sorrel, dock ; common sorrel, garden sorrel	AMPFER (*m*) ; SAUERAMPFER (*m*) ; GROSSER AMPFER (*m*)
Rumex acetosella	244	Sheep's sorrel	KUHZUNGE (*f*) ; FELDAMPFER (*m*); KLEINER AMPFER (*m*)
Rumex alpinus	245	Monk's rhubarb ; *Radix rhabarbari monachorum*	MÖNCHSRHABARBER (*m*) ; ALPEN--AMPFER (*m*) ; GRINDWURZ (*f*)
Rumex aquaticus	246	Smooth-fruited dock ; water dock ; red dock	WASSERAMPFER (*m*)
Rumex arifolius	247	Arum-leaved sorrel (See *Rumex montanus*)	ARONBLÄTTERIGER AMPFER (*m*)
Rumex conglomeratus	248	Clustered dock	GEKNÄUELTER AMPFER (*m*)
Rumex crispus	249	Curled dock ; curly dock ; yellow dock	KRAUSBLÄTTERIGER AMPFER (*m*)
Rumex hydrolapathum	250	Great water dock	RIESENWASSERAMPFER (*m*)
Rumex maritimus	251	Golden dock	MEERAMPFER (*m*)
Rumex maximus	252	Great dock	RIESENAMPFER (*m*)
Rumex montanus	253	Mountain sorrel ; arum-leaved sorrel (See *Rumex arifolius*)	BERGAMPFER (*m*) ; ARONBLÄT--TERIGER AMPFER (*m*)
Rumex obtusifolius	254	Dock ; broad-(leaved) dock	STUMPFBLÄTTERIGER AMPFER (*m*)
Rumex patientia	255	Patience dock, herb patience	GARTENAMPFER (*m*) ; ENGLISCHER SPINAT (*m*) ; GEMÜSEAMPFER (*m*)
Rumex pratensis	256	Meadow dock	WIESENAMPFER (*m*)
Rumex pulcher	257	Fiddle dock	SCHÖNER AMPFER (*m*)
Rumex sanguineus	258	Red dock, bloodwort ; bloody-veined dock ; red-veined dock	HAINAMPFER (*m*)
Ruppia maritima	259	Sea ruppia	RUPPIE (*f*) ; MEERFADEN (*m*) ; SEESTRANDS-MEERFADEN (*m*) ; WASSERRIEMEN (*m*)
Ruscus aculeatus	260	Butcher's broom ; knee holly	STACHELIGER MÄUSEDORN (*m*)
Ruscus hypoglossum	261	Tongue-leaved broom	ZUNGENBLÄTTRIGER MÄUSEDORN (*m*)
Ruscus hypophyllum	262	Double tongue	MÄUSEDORN (*m*) ; BREITBLÄT--TRIGER MÄUSEDORN (*m*)
Ruscus racemosus	263	Alexandrian laurel (See *Danaë racemosa*)	ALEXANDRINISCHER LORBEER (*m*)
Rutaceæ	264	Rutaceæ	RAUTEN (*f.pl*) ; RAUTENGEWÄCHSE (*n.pl*)
Ruta graveolens	265	Rue ; herb of (repentance) grace	RAUTE (*f*)
Sabal blackburniana	266	Thatch palm, fan palm	BLACKBURNS-SABALPALME (*f*)
Sabal palmetto	267	Cabbage palm ; cabbage palmetto ; sabal, palmetto palm (See *Chamærops palmetto*)	PALMETTOPALME (*f*) ; KOHL--PALME (*f*) ; SOMBREROBLÄT--TERBAUM (*m*)
Sabbatia campestris	268	Rose pink, American centaury	FELD-SABBATIA (*f*)
Sabina	269	(See *Juniperus sabina*)	
Saccharomyces cerevisiæ	270	Yeast fungus ; alcohol ferment ; yeast plant ; beer (ferment) yeast (See HEFENPILZ)	SPROSSPILZ (*m*) ; BRANNTWEIN--HEFE (*f*) ; HEFENPILZ (*m*) ; HEFENPFLANZE (*f*) ; HEFE--PILZ (*m*) ; BIERHEFE (*f*)
Saccharomyces ellipsoideus	271	Wine (ferment) yeast	WEINHEFE (*f*)
Saccharomyces mycoderma	272	Mycoderm	KAHMPILZ (*m*)
Saccharum officinarum	273	Sugar-cane	ZUCKERROHR (*n*)
Sagina glabra	274	Pearl-wort, pearl weed	MASTKRAUT (*n*)
Sagina linnæi	275	Alpine pearlwort	FELSENMASTKRAUT (*n*)
Sagina maritima	276	Sea pearlwort	MEERMASTKRAUT (*n*) ; STRAND--MASTKRAUT (*n*)
Sagina nodosa	277	Knotted spurrey ; knotted pearlwort	KNOTIGES MASTKRAUT (*n*)
Sagina procumbens	278	Procumbent pearlwort	MASTKRAUT (*n*) ; LIEGENDES MASTKRAUT (*n*)
Sagittaria sagittifolia	279	Arrowhead	PFEILKRAUT (*n*)
Sagus lævis	280	Sago palm (See *Sagus rumphii* and *Cycas circinalis*)	FACKELPALME (*f*) ; SAGOPALME (*f*)

Botanical Term.	No.	English Term.	German Term.
Sagus rumphii	281	Sago palm (See *Sagus lævis*)	FACKELPALME (*f*)
Saintpaulia ionanthe	282	African violet	USAMBARA-VEILCHEN (*n*)
Salicaceæ	283	Salicaceæ (See *Salix* and *Populus*)	WEIDENGEWÄCHSE (*n.pl*)
Salicornia	284	Glasswort, marsh samphire	KALIPFLANZE (*f*); GLASSCHMALZ (*m*)
Salicornia herbacea	285	Jointed glasswort; glasswort; marsh samphire; saltwort; crab-grass (See also *Salsola kali*)	GLASKRAUT (*n*); MEERKRAUT (*n*); SALZKRAUT (*n*); QUELLER (*m*); KRAUTIGER GLASSCHMALZ (*m*)
Salicornia radicans	286	Creeping glasswort	KRIECHENDER GLASSCHMALZ (*m*)
Salisburya adiantifolia	287	(See *Gingko biloba*)	ZWEILAPPIGER FÄCHERBLATT-BAUM (*m*)
Salix	288	Willow	WEIDE (*f*)
Salix acuminata	289	Long-leaved Sallow willow	LANGBLÄTTRIGE SALWEIDE (*f*)
Salix acutifolia	290	Sharp-leaved willow (See *Salix pruinosa*)	SCHARFBLÄTTERIGE WEIDE (*f*): SPITZBLÄTTERIGE WEIDE (*f*)
Salix adenophylla	291	(North-American)-hairy-leaved willow	EUKALYPTUSWEIDE (*f*); DRÜSEN-BLÄTTRIGE WEIDE (*f*)
Salix alba	292	White willow; golden willow; common willow; Huntingdon willow	WEISSWEIDE (*f*); SILBERWEIDE (*f*); KOPFWEIDE (*f*); FELBE (*f*); BITTERWEIDE (*f*); DOTTERWEIDE (*f*)
Salix alba cœrulea	293	Bat willow, cricket-bat willow (See *Salix alba*)	
Salix alba vitellina	294	(See *Salix pendula*)	
Salix ambigua	295	Ambiguous willow	ZWEIFELHAFTE WEIDE (*f*)
Salix amygdalina	296	Almond willow; French willow; almond-leaved willow (See *Salix triandra*)	MANDELWEIDE (*f*)
Salix andersoniana	297	Green mountain sallow willow	ANDERSONS-WEIDE (*f*)
Salix angustifolia	298	Little tree willow (See *Salix arbuscula*)	SPITZBLÄTTRIGE WEIDE (*f*)
Salix aquatica	299	Water-willow	WASSERWEIDE (*f*)
Salix arbuscula	300	(See *Salix angustifolia*)	SCHWARZWEIDE (*f*)
Salix arenaria	301	Downy mountain willow	SAND-WEIDE (*f*)
Salix argentea	302	Silky sand willow	SILBERWEIDE (*f*)
Salix aurita	303	Round-eared willow	OHRWEIDE (*f*)
Salix babylonica	304	Weeping willow (See *Salix pendula*)	TRÄNENWEIDE (*f*); TRAUERWEIDE (*f*)
Salix bicolor	305	Shining dark-green willow (See *Salix laurina* and *Salix violacea*)	ZWEIFARBIGE WEIDE (*f*)
Salix borreriana	306	Borrer's willow, dark upright willow	BORRERS-WEIDE (*f*)
Salix cœsia	307	Blue-green willow	BLAUGRÜNE WEIDE (*f*)
Salix caprea	308	Sallow willow; sallow; great sallow; goat willow; great round-leaved sallow willow	SALE (*f*); SALWEIDE (*f*); GE-MEINE SALWEIDE (*f*); PALM-WEIDE (*f*); GEMEINE SAAL-WEIDE (*f*); ZIEGENWEIDE (*f*)
Salix caprea pendula	309	Kilmarnock willow	KILMARNOCK-WEIDE (*f*); HÄNG-ENDE SALWEIDE (*f*)
Salix carinata	310	Folded-leaved willow	
Salix cinerea	311	Sallow willow; sallow; grey sallow	SALE (*f*); SALWEIDE (*f*); WERFT-WEIDE (*f*)
Salix cotinifolia	312	Quince-leaved sallow willow	
Salix cuspidata	313	Cuspidate willow (See *Salix pentandra*)	
Salix daphnoides	314	(See *Salix prœcox*)	
Salix daphnoides angustifolia	315	(See *Salix pruinosa*)	
Salix davalliana	316	Davallian willow	
Salix decipiens	317	White Welsh willow, varnished willow (See *Salix fragilis*)	
Salix dicksoniana	318	Broad-leaved mountain willow	DICKSONS-WEIDE (*f*); BREIT-BLÄTTRIGE BERGWEIDE (*f*)
Salix discolor	319	Blue-grey willow	BLAUGRAUE WEIDE (*f*)

Botanical Term.	No.	English Term.	German Term.
Salix doniana	320	Don's willow	DONS-WEIDE (*f*)
Salix fissa	321	(See *Salix purpurea*)	
Salix fœtida	322	Fishy willow	STINKENDE WEIDE (*f*)
Salix forbyana	323	Fine basket willow, fine basket osier	
Salix forsteriana	324	Glaucous mountain sallow willow	FORSTERS-WEIDE (*f*)
Salix fragilis	325	Crack willow ; withy wintergreen (See *Salix decipiens* and *Salix fragilissima*)	KNACKWEIDE (*f*) ; BRUCHWEIDE (*f*)
Salix fragilissima	326	(See *Salix fragilis*)	
Salix fusca	327	Brownish dwarf willow	BRAUNE ZWERG-WEIDE (*f*)
Salix glauca	328	Glaucous mountain willow	BLAUGRÜNE BERG-WEIDE (*f*)
Salix grisea	329	Silky-haired willow (See *Salix sericea*)	SEIDENHAARIGE WEIDE (*f*)
Salix hastata	330	Apple-leaved willow	APFELBLÄTTRIGE WEIDE (*f*)
Salix helix	331	Rose willow	ECHTE BACHWEIDE (*f*)
Salix herbacea	332	Herbaceous willow ; dwarf willow; least willow	KRAUTIGE WEIDE (*f*)
Salix hermaphrodita	333	(See *Salix pentandra*)	
Salix hirta	334	Hairy-branched willow	HAARIGGEZWEIGTE WEIDE (*f*)
Salix hoffmanniana	335	Hoffmann's willow, short-leaved triandrous willow	HOFFMANNS-WEIDE (*f*)
Salix incubacea	336	Trailing silk willow	SEIDEN-WEIDE (*f*)
Salix lambertiana	337	Boyton willow	LAMBERTS-WEIDE (*f*)
Salix lanata	338	Woolly willow ; woolly broad-leaved willow	WOLL-WEIDE (*f*)
Salix lanceolata	339	Sharp-leaved triandrous willow (See *Salix undulata*)	KORBWEIDE (*f*)
Salix lapponum	340	Downy willow	LAPPLÄNDISCHE WEIDE (*f*)
Salix laurina	341	(See *Salix bicolor*)	ZWEIFARBIGE WEIDE (*f*) ; LOR-BEERWEIDE (*f*)
Salix livida	342	Livid dwarf willow	SCHWARZBLAUE ZWERGWEIDE (*f*)
Salix longifolia	343	Long-leaved willow	LANGBLÄTTRIGE WEIDE (*f*)
Salix lucida	344	Shiny willow	GLANZWEIDE (*f*)
Salix medemi	345	Medem's willow	MEDEMS-WEIDE (*f*)
Salix mollissima	346	(See *Salix smithiana*)	
Salix monandra	347	(See *Salix purpurea*)	
Salix myricoides	348	(North American)-myrtle-leaved willow (See *Salix rigida*)	GAGELWEIDE (*f*)
Salix myrsinites	349	Whortle willow, green whortle-leaved willow	HEIDELBEERBLÄTTRIGE WEIDE (*f*)
Salix nigra	350	North-American black willow	SCHWARZE WEIDE (*f*)
Salix nigricans	351	Dark broad-leaved willow	DUNKELE BREITBLÄTTRIGE WEIDE (*f*); SCHWARZWEIDE (*f*)
Salix nitens	352	Shining-leaved wiuow	GLANZWEIDE (*f*)
Salix oleifolia	353	Olive-leaved willow	ÖLBAUMBLÄTTRIGE WEIDE (*f*)
Salix pendula	354	Weeping willow (See *Salix babylonica* and *Salix alba vitellina*)	TRAUERWEIDE (*f*)
Salix pentandra	355	Sweet bay-leaved willow, bay-leaved willow (See *Salix hermaphrodita*)	LORBEERWEIDE (*f*)
Salix petiolaris	356	Dark long-leaved willow	DUNKELE LANGBLÄTTRIGE WEIDE (*f*)
Salix phylicifolia	357	Tea-leaved willow (See *Salix radicans*)	SCHWARZWEIDE (*f*)
Salix polaris	358	Polar willow	POLARWEIDE (*f*) ; GLETSCHER-WEIDE (*f*)
Salix præcox	359	Mezereon willow (See *Salix daphnoides*)	REITWEIDE (*f*) ; KELLERHALS-WEIDE (*f*)
Salix prostrata	360	Prostrate willow	LIEGENDE WEIDE (*f*)
Salix pruinosa	361	Caspian willow (See *Salix acutifolia*)	KASPISCHE WEIDE (*f*) ; SPITZ-BLÄTTRIGE WEIDE (*f*)
Salix prunifolia	362	Plum-leaved willow	PFLAUMENBLÄTTRIGE WEIDE (*f*)
Salix purpurea	363	Red osier ; purple osier ; purple willow ; rose willow ; bitter willow, bitter purple willow (See *Salix fissa* and *Salix monandra*)	PURPURWEIDE (*f*)

Botanical Term.	No.	English Term.	German Term.
Salix radicans	364	Tea-leaved willow (See *Salix phylicifolia*)	TEEBLÄTTERIGE WEIDE (*f*)
Salix repens	365	Creeping willow ; common dwarf willow	ZWERGWEIDE (*f*) ; KRIECHENDE WEIDE (*f*)
Salix reticulata	366	Alpine willow ; reticulate willow ; wrinkled-leaved willow	GLETSCHERWEIDE (*f*) ; NETZ-ADERIGE WEIDE (*f*)
Salix retusa	367	Stunted willow	GESTUTZTE WEIDE (*f*)
Salix rigida	368	(See *Salix myricoides*)	
Salix rosmarinifolia	369	Rosemary-leaved willow	ROSMARINBLÄTTRIGE WEIDE (*f*)
Salix rubra	370	Green-leaved osier	ROTWEIDE (*f*)
Salix rupestris	371	Silky rock sallow willow	FELSEN-SALWEIDE (*f*)
Salix russelliana	372	Bedford willow (A cross between *Salix alba* and *Salix fragilis*)	RUSSELS-WEIDE (*f*)
Salix sericea	373	(See *Salix grisea*)	SEIDENHAARIGE WEIDE (*f*)
Salix serpyllifolia	374	Thyme-leaved willow	THYMIANBLÄTTERIGE WEIDE (*f*)
Salix smithiana	375	Silky-leaved osier (See *Salix mollissima*)	SMITHS-WEIDE (*f*)
Salix sphacelata	376	Withered-pointed sallow willow	
Salix stipularis	377	Auricled osier	OHRWEIDE (*f*)
Salix tenuifolia	378	Thin-leaved willow	DÜNNBLÄTTRIGE WEIDE (*f*)
Salix tetrapla	379	Four-ranked willow	VIERBLÄTTRIGE WEIDE (*f*)
Salix triandra	380	Osier ; long-leaved triandrous willow (See *Salix amygdalina*)	MANDELWEIDE (*f*)
Salix undulata	381	(See *Salix lanceolata*)	
Salix vacciniifolia	382	Bilberry-leaved willow	VACCINIUM-BLÄTTRIGE WEIDE (*f*)
Salix venulosa	383	Veiny-leaved willow	GEADERTBLÄTTERIGE WEIDE (*f*)
Salix viminalis	384	Osier, water willow, withy, osier willow	KORBWEIDE (*f*) ; FLECHTWEIDE (*f*) ; HANFWEIDE (*f*) ; KREBS-WEIDE (*f*) ; BANDWEIDE (*f*) ; WASSERWEIDE (*f*)
Salix violacea	385	(See *Salix bicolor*)	
Salix vitellina	386	Golden osier, yellow willow	GOLDENE FLECHTWEIDE (*f*) ; GELBE WEIDE (*f*) ; DOTTER-WEIDE (*f*)
Salix wulfeniana	387	Wulfenian willow	WULFENS-WEIDE (*f*)
Salpiglossis sinuata	388	Scalloped tube-tongue	TROMPETENZUNGE (*f*)
Salsola kali	389	Saltwort ; prickly saltwort (See also *Salicornia herbacea*)	KALI (*n*) ; SALZKRAUT (*n*) ; GE-MEINES SALZKRAUT (*n*)
Salvia argentea	390	Silver clary	SILBERSALBEI (*f* and *m*)
Salvia glutinosa	391	Glutinous clary ; Jupiter's distaff	KLEBERIGE SALBEI (*f*) ; KLEBE-RIGER SALBEI (*m*)
Salvia hians	392	Cashmere sage	KASHMIR-SALBEI (*f* and *m*)
Salvia horminum	393	Horminum clary (See *Horminium pyrenaicum*)	HORMINIUM-SALBEI (*f* and *m*)
Salvia officinalis	394	Sage ; red sage ; white sage	SALBEI (*m* and *f*); SALBEISTRAUCH (*m*)
Salvia pratensis	395	Meadow clary ; meadow sage	WIESENSALBEI (*f* and *m*)
Salvia sclarea	396	Clary (See *Sclarea vulgaris*)	SCHARLACHKRAUT (*n*)
Salvia silvestris	397	Wood clary	WALD-SALBEI (*f* and *m*)
Salvia verbenaca	398	English clary ; wild sage	MUSKATELLERKRAUT (*n*) ; SALBEI (*f* and *m*) ; WILDE SALBEI (*f*) ; WILDER SALBEI (*m*)
Salvia verticillata	399	Whorl-flowered clary	QUIRLBLÜTIGE SALBEI (*f*); QUIRL-BLÜTIGER SALBEI (*m*)
Sambucus	400	Marsh elder, elder tree	FACKELBAUM (*m*) ; HOLUNDER (*m*) ; FLIEDER (*m*) ; HOLDER (*m*)
Sambucus canadensis	401	Elder, Canadian elder	HOLDER (*m*) ; HOLLUNDER (*m*) ; HOLUNDER (*m*) ; KANADISCHER HOLUNDER (*m*)
Sambucus ebulus	402	Danewort ; dwarf elder	KRAUTHOLUNDER (*m*) ; FLIEDER (*m*)
Sambucus nigra	403	Elder ; common elder	HOLLUNDER (*m*) ; SCHWARZER FLIEDER (*m*) ; SCHWARZER HOLUNDER (*m*)

Botanical Term.	No.	English Term.	German Term.
Sambucus nigra foliis aureis	404	Golden elder	GOLDENER HOLUNDER (*m*); GELB-LICHBLÄTTRIGER SCHWARZER HOLUNDER (*m*)
Sambucus nigra laciniata	405	Cut-leaved elder	GEFIEDERTBLÄTTRIGER SCHWARZER HOLUNDER (*m*)
Sambucus racemosa	406	Scarlet-berried elder	ROTFRÜCHTIGER HOLUNDER (*m*); TRAUBENHOLUNDER (*m*); HIRSCH-HOLUNDER (*m*)
Samolus repens	407	Tasmanian water pimpernel	KRIECHENDE BUNGE (*f*)
Samolus valerandi	408	Brook-weed	SALZ-BUNGE (*f*)
Sanguinaria canadensis	409	Blood-root; puccoon; turmeric; red-root	BLUTWURZ (*f*); BLUTWURZEL (*f*)
Sanguisorba officinalis	410	Greater burnet; great burnet; common burnet (See *Poterium officinale*)	NAGELKRAUT (*n*); BLUTKRAUT (*n*)
Sanicula europœa	411	Wood sanicle	SANIKEL (*m*)
Sanseviera cylindrica	412	Angola hemp, bow-string hemp	ANGOLAHANF (*m*); BAJONETT-PFLANZE (*f*)
Sanseviera zeylanica	413	(East Indian)-moorva	MOORVAPFLANZE (*f*)
Santalum album	414	Sandalwood; East Indian sandalwood	(WEISSES)-SANDELHOLZ (*n*); SANDELHOLZBAUM (*m*)
Santalum citrini	415	Yellow sandalwood	GELBSANDELHOLZ (*n*); GELBES SANDELHOLZ (*n*)
Santolina chamœcyparissus	416	Lavender cotton	UNECHTE ZYPRESSE (*f*); GARTEN-ZYPRESSE (*f*); ZYPRESSEN-KRAUT (*n*); MEERWERMUT (*m*); HEILIGENPFLANZE (*f*)
Santolina maritima	417	Sea lavender cotton	BAUMWOLLKRAUT (*n*); HEILIGEN-KRAUT (*n*); MEERSTRANDS-KRAUT (*n*); MEERWERMUT (*m*)
Sapindus saponaria	418	(South American)-soap berry tree; soap tree	SEIFENBEERBAUM (*m*)
Saponaria ocymoides	419	Rock soap-wort	BASILIENARTIGES SEIFENKRAUT (*n*)
Saponaria officinalis	420	Soapwort; fuller's herb; hedge pink	MADENKRAUT (*n*); SEIFENWURZEL (*f*); WASCHWURZEL (*f*); (ECHTES)-SEIFENKRAUT (*n*)
Saponaria officinalis flore pleno	421	Double soapwort	DOPPELBLÜTIGES SEIFENKRAUT (*n*)
Sapota achras	422	Bully tree; sapodilla plum (See *Achras sapota*)	
Sarcostemma acida	423	(See *Asclepias acida* and also *Ephedra*)	UNECHTE HEILIGE SOMAPFLANZE (*f*)
Sargassum bacciferum	424	Floating gulf-weed	BEEREN-TANG (*m*)
Sarmienta repens	425	Chilian pitcher-flower	CHILENISCHE BECHERBLUME (*f*)
Sarothamnus scoparius	426	Broom (See *Cytisus scoparius* and *Spartium scoparium*)	BESENGINSTER (*m*); BESENKRAUT (*n*)
Sarracenia flava	427	Trumpet-leaf	SARRAZENIE (*f*)
Sarracenia purpurea	428	North American pitcher-plant; side-saddle flower; ' huntsman's horn, Indian cup	JÄGERMÜTZE (*f*)
Sassafras officinale	429	Sassafras wood	FENCHELHOLZ (*n*)
Satureia calamintha	430	(See *Calamintha officinalis*)	
Satureia clinopodium	431	(See *Calamintha clinopodium* and *Clinopodium vulgare*)	WIRBELDOST (*m*)
Satureia hortensis	432	Summer savory	BOHNENKRAUT (*n*); PFEFFER-KRAUT (*n*); KALBYSOPKRAUT (*n*); GARTENKÖLLE (*f*)
Satureia montana	433	Winter savory	BERG-BOHNENKRAUT (*n*)
Sauromatum guttatum	434	Monarch of the East	EIDECHSENWURZ (*f*)
Saussurea alpina	435	Alpine saussurea; alpine saw-wort	ALPEN-SCHARTE (*f*)
Saussurea japonica	436	Saw-wort (See *Seratula japonica*)	SCHARTE (*f*); JAPANISCHE SCHARTE (*f*)
Saxifraga	437	Saxifrage	BÄRENWURZEL (*f*); STEINBRECH (*m*)
Saxifraga aizoides	438	Yellow saxifrage	GEWIMPERTER STEINBRECH (*m*)

Botanical Term.	No.	English Term.	German Term.
Saxifraga aizoon	439	Silver saxifrage	SILBERMINZE (*f*); TRAUBEN-BLÜTIGER STEINBRECH (*m*)
Saxifraga cæspitosa	440	Tufted saxifrage	
Saxifragaceæ	441	Saxifragaceæ (See *Saxifrageæ*)	
Saxifraga ceratophylla	442	(See *Saxifraga trifurcata*)	
Saxifraga cernua	443	Drooping saxifrage	HÄNGENDER STEINBRECH (*m*)
Saxifraga geum	444	Kidney saxifrage	PORZELLANBLÜMCHEN (*n*)
Saxifraga granulata	445	Meadow saxifrage	WIESENSTEINBRECH (*m*); KÖRNIGER STEINBRECH (*m*)
Saxifraga granulata flore pleno	446	Double meadow saxifrage	DOPPELBLÜTIGER WIESENSTEIN-BRECH (*m*)
Saxifraga hirculus	447	Marsh saxifrage	SUMPFSTEINBRECH (*m*); MOOR-STEINBRECH (*m*)
Saxifraga hypnoides	448	Cut-leaved saxifrage, Dovedale moss	GESCHLITZTBLÄTTRIGER STEIN-BRECH (*m*)
Saxifraga nivalis	449	Alpine saxifrage	ALPENSTEINBRECH (*m*)
Saxifraga oppositifolia	450	Purple saxifrage; opposite-leaved saxifrage	GEGENBLÄTTRIGER STEINBRECH (*m*); BLAUES STEINMOOS (*n*)
Saxifraga peltata	451	Umbrella plant	
Saxifraga rivularis	452	Brook saxifrage	FLUSS-STEINBRECH (*m*)
Saxifraga sarmentosa	453	Mother of thousands; Aaron's beard; sailor plant	JUDENBART (*m*)
Saxifraga stellaris	454	Star saxifrage	STERN-STEINBRECH (*m*)
Saxifraga tridactylites	455	Rue-leaved saxifrage	DREIFINGERIGER STEINBRECH (*m*)
Saxifraga trifurcata	456	Stag's-horn rockfoil (See *Saxifraga ceratophylla*)	DREIFURCHIGER STEINBRECH (*m*)
Saxifraga umbrosa	457	St. Patrick's cabbage; London pride; None-so-pretty	PORZELLANBLÜMCHEN (*n*)
Saxifrageæ	458	Saxifrageæ (See *Chrysosplenium* and *Saxifraga*)	STEINBRECHGEWÄCHSE (*n.pl*)
Scabiosa arvensis	459	Field scabious	FELD-SKABIOSE (*f*)
Scabiosa atropurpureum	460	Pincushion flower, mournful widow, sweet scabious	KRÄTZKRAUT (*n*); PUPPUR-SKA-BIOSE (*f*); SCHWARZROTE SKABIOSE (*f*); STERNKOPF (*m*)
Scabiosa canescens	461	Hoary scabious	GRAUE SKABIOSE (*f*)
Scabiosa caucasica	462	Caucasian scabious	KAUKASISCHE SKABIOSE (*f*)
Scabiosa columbaria	463	Small scabious	TAUBEN-SKABIOSE (*f*)
Scabiosa succisa	464	Devil's bit; primrose scabious	TEUFELSABBIS (*m*); TEUFELS-ABBISWURZEL (*f*)
Scandix cerefolium	465	Garden chervil, salad chervil (See *Anthriscus cerefolium* and *Chærophyllum temulentum*)	GARTEN-NADELKERBEL (*f*)
Scandix pecten-(veneris)	466	Shepherd's needle; Venus's comb; needle chervil	NADELKERBEL (*f*); VENUSKAMM (*m*); KAMMFÖRMIGE NADEL-KERBEL (*f*)
Scheuchzeria palustris	467	Marsh scheuchzeria	SUMPF-BLASENBINSE (*f*)
Schinopsis lorentzii	468	Quebracho-wood tree	QUEBRACHOHOLZBAUM (*m*)
Schinus molle	469	Peruvian mastic tree	PERUANISCHER PFEFFERBAUM (*m*); AMERIKANISCHER MASTIX-BAUM (*m*)
Schizandra chinensis	470	Chinese schizandra	SPALTKÖPFCHEN (*n*); CHINE-SISCHES SPALTKÖPFCHEN (*n*)
Schizanthus	471	Fringe flower, butterfly flower	SPALTBLUME (*f*)
Schizomycetes	472	Fission fungus; fission plant	SPALTPILZ (*m*)
Schizophragma hydrangeoides	473	Climbing hydrangea	KLETTERNDE HYDRANGEA (*f*)
Schizophyceæ	474	Fission algæ	SPALTALGEN (*f.pl*)
Schizostylis coccinea	475	Crimson flag, caffre lily	KAFFER-LILIE (*f*)
Schizymenia edulis	476	(See *Halymenia* and *Iridæa edulis*)	
Schlerochloa borreri	477	(See *Glyceria borreri*)	
Schlerochloa distans	478	(See *Poa distans* and *Glyceria distans*)	
Schlerochloa maritima	479	(See *Glyceria maritima*)	
Schlerochloa procumbens	480	(See *Glyceria procumbens*)	
Schlerochloa rigida	481	(See *Glyceria rigida*)	
Schœnocaulon officinale	482	Sabadilla (See *Asagræa officinalis*)	SABADILLA (*f*); MEXIKANISCHES LÄUSEKRAUT (*n*)

Botanical Term.	No.	English Term.	German Term.
Schœnus nigricans	483	Bog rush ; black schœnus	SCHWÄRZLICHES KOPFRIED (*n*)
Sciadopitys verticillata	484	Umbrella pine, parasol fir tree	SCHIRMTANNE (*f*)
Scilla	485	(See also *Urginea maritima*)	MEERZWIEBEL (*f*); BLAUSTERN (*m*); ZILLE (*f*)
Scilla amœna	486	Pleasing scilla (See *Scilla messeniaca*)	SCHÖNER BLAUSTERN (*m*)
Scilla autumnale	487	Autumnal squill	HERBST-ZILLE (*f*)
Scilla bifolia	488	Two-leaved squill	ZWEIBLÄTTRIGER BLAUSTERN (*m*)
Scilla campanulata	489	Light-blue Spanish scilla	HASENGLÖCKCHEN (*n*)
Scilla cernua	490	Drooping scilla (See *Scilla festalis cernua*)	
Scilla festalis cernua	491	Drooping scilla (See *Scilla cernua*)	NICKENDER BLAUSTERN (*m*)
Scilla hispanica	492	Spanish squill	ENDYMION-BLAUSTERN (*m*)
Scilla maritima	493	Squill, sea onion (See *Urginea maritima*)	MEERZWIEBEL (*f*)
Scilla messeniaca	494	(See *Scilla amœna*)	
Scilla nonscripta	495	Blue-bell	UNBESCHRIEBENER BLAUSTERN (*m*)
Scilla nutans	496	Wild hyacinth ; bluebell ; wild bluebell ; bluebell squill (See *Agraphis nutans*)	GLOCKENBLUME (*f*)
Scilla sibirica	497	Siberian squill	SIBIRISCHE ZILLE (*f*)
Scilla verna	498	Vernal squill ; Spring squill	FRÜHLINGS-ZILLE (*f*)
Scirpus	499	(See also *Isolepsis*)	SIMSE (*f*)
Scirpus acicularis	500	Needle scirpus, needle club grass	NADEL-SIMSE (*f*)
Scirpus cæspitosus	501	Tufted scirpus ; scaly-stalked club-rush	RASEN-SIMSE (*f*)
Scirpus carinatus	502	Blunt-edged scirpus, blunt-edged club-rush	KIELFÖRMIGE SIMSE (*f*)
Scirpus compressus	503	compressed club-rush	ZUSAMMENGEDRÜCKTE SIMSE (*f*)
Scirpus fluitans	504	Scirpus ; floating scirpus	FLUTENDE SIMSE (*f*)
Scirpus glaucus	505	(See *Scirpus lacustris*)	MEERGRÜNE SIMSE (*f*)
Scirpus holoschœnus	506	Clustered scirpus (See *Isolepsis holoschœnus*)	KNOPFGRASARTIGE SIMSE (*f*)
Scirpus lacustris	507	Bull rush ; bast ; lake scirpus ; great club-rush (See *Scirpus glaucus*)	BINSE (*f*) ; SEE-SIMSE (*f*)
Scirpus lacustris tabernœmontani zebrina	508	Variegated porcupine rush	VERSCHIEDENBLÄTTRIGE SEE-SIMSE (*f*)
Scirpus maritimus	509	Sea scirpus ; salt-marsh club-rush	MEERSTRANDS-SIMSE (*f*)
Scirpus multicaulis	510	Many-stalked scirpus	VIELSTENGELIGE SIMSE (*f*)
Scirpus natalensis	511	Natal club-rush	NATAL-SIMSE (*f*)
Scirpus nodosus	512	Club rush (See *Isolepsis gracilis*)	ZARTE SIMSE (*f*)
Scirpus palustris	513	Creeping scirpus	SUMPF-SIMSE (*f*)
Scirpus parvulus	514	Small scirpus	KLEINE SIMSE (*f*)
Scirpus pauciflorus	515	Few-flowered scirpus, chocolate-headed club-rush	GERINGBLÜTIGE SIMSE (*f*)
Scirpus pungens	516	Sharp scirpus ; sharp club-rush	SCHARFE SIMSE (*f*)
Scirpus radicans	517	Rooting club-rush	WURZELNDE SIMSE (*f*)
Scirpus rufus	518	Brown club-rush	BRAUNE SIMSE (*f*)
Scirpus savii	519	Savi's scirpus (See *Isolepsis savii*)	SAVIS-SIMSE (*f*)
Scirpus setaceus	520	Bristle scirpus (See *Isolepsis setacea*)	BORSTENFÖRMIGE SIMSE (*f*)
Scirpus supinus	521	Prostrate club-rush	LIEGENDE SIMSE (*f*)
Scirpus sylvaticus	522	Wood scirpus ; wood club-rush	WALD-SIMSE (*f*)
Scirpus tenella	523	(See *Isolepsis gracilis*)	ZARTE SIMSE (*f*)
Scirpus triqueter	524	Triangular scirpus ; triangular club-rush	DREIECKIGE SIMSE (*f*)
Sclarea vulgaris	525	(See *Salvia sclarea*)	
Scleranthus	526	Knawel, German knot-grass	KNÄUEL (*m*) ; KNÄUL (*m*); KNAUEL (*m*)
Scleranthus annuus	527	Annual knawel	EINJÄHRIGER KNAUEL (*m*)
Scleranthus perennis	528	Perennial knawel	AUSDAUERNDER KNAUEL (*m*)
Scleria latifolia	529	Knife-grass	MESSERGRAS (*n*)
Scleroderma aurantiacum	530	(See *Scleroderma vulgare* and also *Bovista*)	

Botanical Term.	No.	English Term.	German Term.
Scleroderma vulgare	531	False truffel (See *Scleroderma auranticum* and also *Bovista*)	POMERANZENHÄRTLING (*m*); GEMEINER HARTBOVIST (*m*); HARTKARTOFFELBOVIST (*m*); FALSCHE TRÜFFEL (*f*); FELL-STREULING (*m*)
Sclerotinia trifoliorum	532	Clover canker (See *Peziza ciborioides*)	KLEEKRANKHEITSPILZ (*m*); KLEE-KREBS (*m*)
Scolopendrium ceterach	533	(See *Scolopendrium vulgare*; *Asplenium ceterach* and *Ceterach officinarum*)	
Scolopendrium officinarum	534	(See *Scolopendrium vulgare*)	
Scolopendrium vulgare	535	Hart's-tongue-(fern) (See *Scolopendrium officinarum*)	HIRSCHZUNGE (*f*); GEMEINE HIRSCHZUNGE (*f*)
Scolymus hispanicus	536	Golden thistle, Spanish oyster plant	GOLDENE DISTEL (*f*)
Scorzonera hispanica	537	Viper's grass	NATTERGRAS (*n*)
Scrofulariaceæ	538	Scrofulariaceæ (See *Scrophularineæ*)	
Scrophularia	539	Figwort	KNOTENWURZ (*f*); KROPFWURZ (*f*); BRAUNWURZ (*f*)
Scrophularia aquatica	540	Water figwort	WASSER-KNOTENWURZ (*f*)
Scrophularia nodosa	541	Knotted figwort	KNOTENWURZ (*f*)
Scrophularia scorodonia	542	Balm-leaved figwort	KNOBLAUCHARTIGE KNOTENWURZ (*f*)
Scrophularia vernalis	543	Yellow figwort	FRÜHLINGS-KNOTENWURZ (*f*)
Scrophularineæ	544	Scrophularineæ (See *Scrophularia, Digitalis, Verbascum, Melampyrum, Euphrasia, Pedicularis*)	PERSONATEN (*f.pl*)
Scutellaria	545	Skull-cap; helmet flower	SCHILDKRAUT (*n*); HELMKRAUT (*n*); SCHILDTRÄGER (*m*)
Scutellaria galericulata	546	Common skull-cap; greater skull-cap	SCHILDKRAUT (*n*); GROSSER SCHILDTRÄGER (*m*)
Scutellaria lateriflora	547	Skull-cap; side-flowering skull-cap	SEITENBLÜTIGES HELMKRAUT (*n*)
Scutellaria minor	548	Lesser skull-cap	KLEINER SCHILDTRÄGER (*m*)
Seaforthia elegans	549	Illawarra palm; Australian feather palm (See *Archontophœnix* and also *Ptychosperma elegans*)	ILLAWARRA-PALME (*f*)
Secale cereale	550	Rye	ROGGEN (*m*); ECHTER ROGGEN (*m*)
Secale cornutum	551	Spurred rye (See *Claviceps purpurea*)	MUTTERKORN (*n*); HUNGERKORN (*n*); HAHNENSPORN (*m*)
Sedum	552	Stonecrop	STEINKRAUT (*n*); HAUSWURZ (*f*); HAUSWURZEL (*f*); HENNE (*f*); FETTE HENNE (*f*)
Sedum acre	553	Wall pepper; stonecrop; biting stonecrop; wallwort	MAUERPFEFFER (*m*); PFEFFER-KRAUT (*n*)
Sedum album	554	White stonecrop	KNORPELKRAUT (*n*); MAUER-TRAUBE (*f*)
Sedum anglicum	555	English sedum	TRIPMADAM (*n*)
Sedum dasyphyllum	556	Thick-leaved sedum (See *Crassula*)	DICKBLÄTTRIGES STEINKRAUT (*n*)
Sedum reflexum	557	Reflex-leaved (sedum) stonecrop	ZURÜCKGEKRÜMMTE FETTHENNE (*f*)
Sedum rhodiola	558	Rose-root; midsummer-men (See *Sedum roseum* and *Rhodiola rosea*)	ROSENWURZ (*f*)
Sedum roseum	559	(See *Sedum rhodiola*)	ROSENROTE FETTHENNE (*f*)
Sedum rupestre	560	Rock (sedum) stonecrop	FELSEN-STEINKRAUT (*n*)
Sedum sexangulare	561	Tasteless (sedum) stonecrop	SECHSKANTIGES STEINKRAUT (*n*)
Sedum telephium	562	Live-long; orpine; (tuberous)-stonecrop	HENNE (*f*)
Sedum villosum	563	Hairy (sedum) stonecrop	LANGHAARIGES STEINKRAUT (*n*)
Selaginella kraussiana	564	Tree club moss, creeping moss (See *Lycopodium denticulatum*)	SCHUPPENBLATT-BÄRLAPP (*m*)
Selaginella selaginoides	565	Lesser club-moss; common selaginella	WIMPERZÄHNIGE SELAGINELLE (*f*)
Selago	566	(See *Lycopodium selago*)	

Botanical Term.	No.	English Term.	German Term.
Semecarpus anacardium	567	Malacca kidney bean	OSTINDISCHER MERKFRUCHT-BAUM (*m*)
Sempervivum arachnoideum	568	Cobweb house-leek	ÜBERSPONNENE HAUSWURZ (*f*); SPINNWEBENHAUSLAUCH (*m*); SPINNENWURZ (*f*)
Sempervivum globiferum	569	(See *Sempervivum soboliferum*)	KUGELFÖRMIGE HAUSWURZ (*f*)
Sempervivum soboliferum	570	Hen-and-chickens house-leek (See *Sempervivum globiferum*)	SPROSSENDE HAUSWURZ (*f*)
Sempervivum tectorum	571	Houseleek	DACHWURZ (*f*); HAUSLAUF (*m*); HAUSLAUCH (*m*); HAUSWURZ (*f*); HAUSWURZEL (*f*)
Senebiera coronopus	572	Wart-cress; common senebiera	KRÄHENFUSS (*m*)
Senebiera didyma	573	Lesser senebiera	KLEINER KRÄHENFUSS (*m*)
Senecio aquaticus	574	Marsh ragwort; water senecio	WASSER-KREUZKRAUT (*n*)
Senecio campestris	575	Field senecio	ACKERGRINDKRAUT (*n*); FELD-KREUZKRAUT (*n*)
Senecio cineraria	576	Dusty miller (See *Cineraria maritima*)	MEERSTRANDS-ASCHENKRAUT (*n*)
Senecio cruentus	577	(See *Cineraria cruenta*)	BUNTES KREUZKRAUT (*n*)
Senecio doronicum	578	Leopard's bane (See *Doronicum plantagineum*)	
Senecio elegans	579	(See *Senecio jacobœa* and *Jacobœa elegans*)	
Senecio erucifolius	580	Narrow-leaved senecio	SCHMALBLÄTTRIGES KREUZKRAUT (*n*)
Senecio jacobœa	581	Ragwort; common ragwort (See *Senecio elegans* and *Jacobœa elegans*)	JAKOBS-KREUZKRAUT (*n*)
Senecio macroglossus	582	Cape ivy	KAP-EPHEU (*m*)
Senecio mikanioides	583	German ivy	DEUTSCHER EPHEU (*m*)
Senecio paludosus	584	Fen senecio	MOOR-KREUZKRAUT (*n*)
Senecio palustris	585	Marsh senecio	SUMPFGRINDKRAUT (*n*); SUMPF-KREUZKRAUT (*n*)
Senecio saracenicus	586	Broad-leaved senecio	BREITBLÄTTRIGES KREUZKRAUT (*n*); UFER-KREUZKRAUT (*n*)
Senecio squalidus	587	Squalid senecio	BALDGREIS (*m*)
Senecio sylvaticus	588	Mountain groundsel; wood senecio	WALDGRINDKRAUT (*n*); WALD-KREUZKRAUT (*n*)
Senecio viscosus	589	Viscous senecio	KLEBERIGES GRINDKRAUT (*n*); KLEBERIGES KREUZKRAUT (*n*)
Senecio vulgaris	590	Groundsel; senecio	KREUZKRAUT (*n*); JAKOBSKRAUT (*n*); KRÖTENKRAUT (*n*); GRINDKRAUT (*n*); GEMEINES KREUZKRAUT (*n*); BALDGREIS (*m*); GREISKRAUT (*n*)
Sequoia gigantea	591	Wellingtonia, giant redwood; Mammoth tree (See *Wellingtonia gigantea* and *Washingtonia*)	WELLINGTONIE (*f*); MAMMUT-BAUM (*m*); RIESENMAMMUT-BAUM (*m*)
Sequoia sempervirens	592	Californian evergreen redwood tree	KÜSTENMAMMUTBAUM (*m*)
Serapias cordigera	593	Tongue-flowered orchid	HERZORCHE (*f*); SCHWERTORCHE (*f*)
Seratula japonica	594	(See *Saussurea japonica* and also *Serratula*)	
Seriana triternata	595	Supple jack	
Serrafalcus commutatus	596	(See *Bromus commutatus*)	
Serratula tinctoria	597	Saw-wort; dyer's saw-wort (See also *Seratula*)	FÄRBERDISTEL (*f*); FÄRBER-SCHARTE (*f*)
Sesamum indicum	598	Benne, sesame, bene, oil-plant, vangloe, tilseed, teel-seed	INDISCHER SESAM (*m*)
Sesamum orientale	599	Sesame	SESAM (*m*); ORIENTALISCHER SESAM (*m*)
Seseli hippomarathrum	600	Horse-fennel	PFERDE-SESEL (*m*)
Seseli libanotis-(montana)	601	Mountain seseli	SESEL (*m*); BERG-SESEL (*m*); BERG-HEILWURZ (*f*)
Setaria ambigua	602	Ambiguous fox-tail grass	TÄUSCHENDE BORSTENHIRSE (*f*)
Setaria glauca	603	Glaucous bristle-grass (See *Panicum glaucum*)	

Botanical Term.	No.	English Term.	German Term.
Setaria italica	604	Italian bristle-grass; Italian millet	KOLBENHIRSE (*f*); ITALIENISCHE BORSTENHIRSE (*f*)
Setaria verticillata	605	Rough bristle-grass (See *Panicum verticillatum*)	
Setaria viridis	606	Green bristle-grass (See *Panicum viride*)	
Shepherdia argentea	607	(See *Elægnus argentea* and *Lepargyrea argentea*)	
Sherardia arvensis	608	Field-madder	ACKER-SHERARDIE (*f*)
Shorea robusta	609	(North Indian)-Sal tree	SAL-BAUM (*m*); SALHARZBAUM (*m*); SAULBAUM (*m*)
Sibthorpia candida	610	(See *Sibthorpia europæa*)	
Sibthorpia europæa	611	Cornish moneywort (See *Sibthorpia candida*)	EUROPÄISCHE SIBTHORPIA (*f*)
Sibthorpia peregrina	612	Madeira moneywort	MADEIRA SIBTHORPIA (*f*)
Sideroxylon inerme	613	Iron wood-(tree)	EISENHOLZ (*n*); EISENHOLZBAUM (*m*)
Silaus pratensis	614	Pepper-saxifrage	WIESEN-SILAU (*m*)
Silene acaulis	615	Moss campion; dwarf silene; moss pink, cushion pink (See also *Lychnis*)	STENGELLOSES LEIMKRAUT (*n*)
Silene alpestris	616	Alpine catchfly	ALPENLEIMKRAUT (*n*)
Silene anglica	617	English catchfly	ENGLISCHES LEIMKRAUT (*n*)
Silene armeria	618	Sweet-william catchfly	GARTEN-LEIMKRAUT (*n*)
Selene conica	619	Striated silene	KEGEL-LEIMKRAUT (*n*)
Silene cucubalus	620	Bladder campion; white bottle (See *Silene inflata*; *Silene venosa* and *Silene vulgaris*; also *Cucubalus baccifer*)	BLASIGES LEIMKRAUT (*n*); BEER--ENNELKE (*f*); HÜHNERLIESCH (*n*); HÜHNERBISS (*m*)
Silene gallica	621	Small-flowered silene	FRANZÖSISCHES LEIMKRAUT (*n*)
Silene inflata	622	Bladder campion (See *Silene cucubalus*)	
Silene maritima	623	Sea-(bladder)-campion; white-flowered sea-campion	KÜSTEN-LEIMKRAUT (*n*)
Silene maritima flore pleno	624	Witch's thimble	DOPPELBLÜTIGES MEERSTRANDS-LEIMKRAUT (*n*)
Silene noctiflora	625	Night-blooming catchfly; night-flowering silene	NACHTBLÜHENDES LEIMKRAUT (*n*)
Silene nutans	626	Nottingham catchfly	NICKENDES LEIMKRAUT (*n*)
Silene otites	627	Spanish silene	OHRLÖFFEL-LEIMKRAUT (*n*)
Silene pennsylvanica	628	American wild pink	PENNSYLVANISCHES LEIMKRAUT (*n*)
Silene quinquevulnera	629	Catchfly	LEIMKRAUT (*n*)
Silene schafta	630	Persian catchfly	PERSISCHES LEIMKRAUT (*n*)
Silene venosa	631	(See *Silene cucubalus*)	TAUBENKROPF (*m*)
Silene vespertina	632	Evening catchfly	NACHTBLÜHENDES LEIMKRAUT (*n*)
Silene virginica	633	Fire pink	VIRGINISCHES LEIMKRAUT (*n*); FEUER-NELKE (*f*)
Silene viscosa	634	Catchfly	KLEBERIGES LEIMKRAUT (*n*)
Silene vulgaris	635	(See *Silene cucubalus*)	GEMEINES LEIMKRAUT (*n*)
Siliqua dulcis	636	St. John's Bread (See *Ceratonia siliqua*)	JOHANNISBROT (*n*)
Silphium	637	(See *Laserpitium*)	SILPHIE (*f*)
Silphium laciniatum	638	Compass plant	GESCHLITZTE SILPHIE (*f*)
Silphium perfoliatum	639	Cup plant	DURCHWACHSENE SILPHIE (*f*)
Silybum marianum	640	Milk thistle, holy thistle, blessed thistle	MARIENDISTEL (*f*)
Simaruba amara	641	Bitter damson, mountain damson (See *Simaruba officinalis* and *Quassia amara*)	BITTERESCHE (*f*)
Simaruba excelsa	642	(See *Picrasma excelsa*)	
Simaruba officinalis	643	Paradise tree, wild olive, bitter damson, mountain damson, Paraiba (See *Simaruba amara*)	PARADIESBAUM (*m*); RUHR--RINDENBAUM (*m*)
Simethis bicolor	644	Variegated simethis	ZWEIFARBIGE SIMETHIS (*f*)
Sinapis alba	645	White mustard; yellow mustard (See *Brassica alba*)	MOSTRICH (*m*); WEISSSENF (*m*)

Botanical Term.	No.	English Term.	German Term.
Sinapis arvensis	646	Charlock ; wild mustard (See *Brassica sinapistrum*)	ACKERSENF (*m*)
Sinapis nigra	647	Black mustard, red mustard	SCHWARZSENF (*m*)
Sinningia speciosa	648	Gloxinia	GLOXINIE (*f*)
Siphoneæ	649	Tubular algæ	SCHLAUCHALGEN (*f.pl*)
Sison amomum	650	Hedge sison ; bastard stone parsley	
Sisymbrium alliaria	651	Jack-by-the-hedge ; Sauce- alone (See *Alliaria officinalis*)	
Sisymbrium amphibium	652	(See *Nasturtium amphibium*)	SCHWIMMANDER (*m*)
Sisymbrium irio	653	London rocket	
Sisymbrium nasturtium	654	(See *Nasturtium officinale*)	
Sisymbrium officinale	655	Hedge mustard ; charlock (See *Erysimum officinale*)	HEDERICH (*m*); RAUKE (*f*); SENFKRAUT (*n*); WILDER SENF (*m*)
Sisymbrium sophia	656	Flixweed	FEINBLÄTTERIGER RAUTENSENF (*m*)
Sisymbrium sylvestre	657	(See *Nasturtium sylvestre*)	WALD-RAUTENSENF (*m*)
Sisymbrium terrestre	658	(See *Nasturtium terrestre*)	RAUTENSENF (*m*)
Sisymbrium thalianum	659	Common thale cress (See *Arabis thaliana*)	
Sisyrinchium angustifolium	660	Blue-eyed grass ; narrow-leaved sisyrinchium ; satin flower, rush lily	KLEINBLÜTIGER RÜSSELSCHWERTEL (*m*)
Sisyrinchium grandiflorum	661	Spring bell, purple snowdrop	GROSSBLÜTIGER RÜSSELSCHWER-TEL (*m*)
Sitopyros	662	(See *Triticum vulgare*)	
Sium angustifolium	663	Lesser sium	MERK (*m*); SCHMALBLÄTTRIGER MERK (*m*)
Sium latifolium	664	Broad sium ; water parsnip	BREITER MERK (*m*)
Sium sisarum	665	Skirret	MERK (*m*); WASSERMERK (*m*); ZUCKERWURZEL (*f*)
Smilacina bifolia	666	(See *Maianthemum convallaria*)	
Smilacina racemosa	667	False spikenard	FALSCHE SPIKNARDE (*f*)
Smilacina stellata	668	Star-flowered lily-of-the-valley	
Smilax aspera	669	Prickly ivy	STACHELIGE STECHWINDE (*f*)
Smilax glauca	670	Sarsaparilla ; Carrion flower ; sea-green smilax	MEERGRÜNE STECHWINDE (*f*); SARSAPARILLE (*f*)
Smilax medica	671	Sarsaparilla shrub (See *Smilax officinalis* and *Smilax ornata*)	SARSAPARILLEPFLANZE (*f*); SAR-SAPARILLE (*f*)
Smilax officinalis	672	(See *Smilax medica*)	
Smilax ornata	673	(See *Smilax medica*)	ZIER-STECHWINDE (*f*); SASSA-PARILLENWURZEL (*f*)
Smilax sassaparilla	674	(See *Smilax medica*)	
Smyrnium olusatrum	675	Alexanders	
Soja hispida	676	Soy bean (See also *Dolichos sesquipedalis*)	SOJA (*f*); SOJABOHNE (*f*); CHINE-SISCHE BOHNE (*f*); SAUBOHNE (*f*)
Solanaceæ	677	Solanaceæ (See *Solaneæ*)	
Solaneæ	678	Solaneæ, solanaceæ (See *Solanum*, *Hyoscyamus*, *Verbascum* and *Atropa*)	NACHTSCHATTENGEWÄCHSE (*n.pl*)
Solanum capsicastrum	679	Star capsicum-(nightshade); winter cherry ; Jerusalem cherry	NACHTSCHATTEN VON JERICHO (*m*)
Solanum crispum	680	Potato tree ; curled nightshade	GEKRÄUSELTER NACHTSCHATTEN (*m*)
Solanum dulcamara	681	bitter-sweet ; woody nightshade ; dulcamara	ALPRANKEN (*m*); BITTERSÜSS (*n*); NACHTSCHATTEN (*m*); BITTER-SÜSSNACHTSCHATTENGEWÄCHS (*n*)
Solanum jasminoides	682	Jasmine nightshade	JASMINARTIGER NACHTSCHATTEN (*m*)
Solanum melongena	683	Aubergine ; mad apple ; egg plant	EIERFRUCHT (*f*)
Solanum nigrum	684	Black nightshade ; black-berried nightshade	NACHTSCHATTEN (*m*)

Botanical Term.	No.	English Term.	German Term.
Solanum sodomeum	685	Dead-sea apple tree; Sodom apple tree	SODOMS-APFEL (*m*)
Solanum tuberosum	686	Potato	KARTOFFELPFLANZE (*f*); KARTOFFEL (*f*)
Soldanella alpina	687	Blue moon-wort	TRODDELBLUME (*f*); ALPENGLÖCKCHEN (*n*)
Solenostemma argel	688	Argel; arghel	MEKKA-SENNA (*f*)
Solidago virga-aurea	689	(See *Solidago virgaurea*)	
Solidago virgaurea	690	Golden rod	GOLDRUTE (*f*); WUNDKRAUT (*n*)
Sollya heterophylla	691	Australian bluebell creeper	
Sonchus arvensis	692	Corn sowthistle	ACKER-SAUDISTEL (*f*); ACKER-GÄNSEDISTEL (*f*)
Sonchus asper	693	Sowthistle	SAUDISTEL (*f*); GÄNSEDISTEL (*f*)
Sonchus oleraceus	694	Sowthistle; common milkthistle; common sowthistle	KOHLARTIGE SAUDISTEL (*f*); GLATTE GÄNSEDISTEL (*f*)
Sonchus palustris	695	Marsh sowthistle	SUMPF-SAUDISTEL (*f*)
Sophora japonica	696	Sophore, Chinese pagoda tree, Waifa	SOPHORE (*f*); SCHNURBAUM (*m*); JAPANISCHER SCHNURBAUM (*m*); SCHUSSERBAUM (*m*); NATALKÖRNERBAUM (*m*)
Sophora tetraptera	697	New Zealand laburnum (See *Edwardsia grandiflora*)	ROSENKRANZHÜLSE (*f*)
Sophronia	698	(See *Sophronitis*)	
Sophronitis cernua	699	Scarlet-flowered orchid, drooping orchid	SCHARLACH-BLÜTIGE ORCHE (*f*) HÄNGENDE ORCHE (*f*)
Sorbus aria	700	White beam tree (See *Pirus aria*)	MEHLBEERBAUM (*m*); MEHLBIRN (*f*); GEMEINE MEHLBIRNE (*f*)
Sorbus aucuparia	701	Mountain ash (See *Pirus aucuparia*)	VOGELBEERE (*f*); EBERESCHE (*f*); GEMEINE EBERESCHE (*f*); VOGELBEERBAUM (*m*)
Sorbus chamæmespilus	702	(See *Cratægus chamæmespilus* and under *Pirus*)	ZWERGMISPEL (*f*)
Sorbus domestica	703	Service tree (See *Pirus sorbus* and *Cormus domestica*)	SPEIERLING (*m*); SPERBERBAUM (*m*); ZAHME EBERESCHE (*f*); ELSBEERBAUM (*m*); BEERESCHE (*f*); ECHTER SPIERLING (*m*); SPERBAUM (*m*); SCHIERLINGSBAUM (*m*)
Sorbus toringo	704	(See *Malus toringo*)	
Sorbus torminalis	705	Wild service tree; beam tree (See *Pirus torminalis* and *Cratægus torminalis*)	ELSE (*f*); ELZBEERE (*f*); ELSBEERBAUM (*m*); ELSBEERE (*f*)
Sorghum caffrorum	706	Kaffir corn; caffre corn (See *Andropogon sorghum*)	DURRAGRAS (*n*); SORGHUMGRAS (*n*)
Sorghum saccharatum	707	Sugar grass; sweet shaloo	CHINESISCHES ZUCKERROHR (*n*); ZUCKERHIRSE (*f*)
Sorghum vulgare	708	Indian millet; Guinea corn; doura; durra, dhurra, dora, negro-corn; broom corn; caffre corn (See *Andropogon sorghum*)	MOHRENHIRSE (*f*); MOORHIRSE (*f*)
Sorosporium	709	Smut	BRAND (*m*)
Sorosporium scabies	710	Potato smut	KARTOFFELBRAND (*m*)
Soymida febrifuga	711	Coromandel redwood; East-Indian mahogany	COROMANDEL-ROTHOLZBAUM (*m*)
Sparassis crispa	712	Trunk fungus	STRUNKSCHWAMM (*m*); ECHTER ZIEGENBART (*m*)
Sparaxis	713	African harlequin flower	SCHLITZSCHWERTEL (*m* and *n*)
Sparganium minimum	714	Small sparganium; small bur-reed	KLEINER IGELKOLBEN (*m*)
Sparganium ramosum	715	Branched bur-reed; branched sparganium	GEZWEIGTER IGELKOLBEN (*m*)
Sparganium simplex	716	Simple bur-reed; simple sparganium	IGELKOLBEN (*m*)
Sparmannia africana	717	African hemp (See *Vossianthus*)	AFRIKANISCHER HANF (*m*); SÜDAFRIKANISCHE SPARMANNIE (*f*)
Spartina stricta	718	Cord-grass	STEIFES BESENGRAS (*n*)

Botanical Term.	No.	English Term.	German Term.
Spartium junceum	719	Spanish broom; yellow Spanish broom	SPANISCHER PFRIEMEN (*m*); GEMEINER PFRIEMEN (*m*)
Spartium scoparium	720	Broom (See *Sarothamnus scoparius* and also *Cytisus scoparius*)	BESENSTRAUCH (*m*); PFRIEMEN (*m*); PFRIEMENHOLZ (*n*); PFRIEMENSTRAUCH (*m*); GEMEINER BESENSTRAUCH (*m*); PFRIEMENKRAUT (*n*)
Specularia hybrida	721	Venus's looking-glass (See *Campanula hybrida*)	FRAUENSPIEGEL (*m*)
Specularia speculum	722	(See *Campanula speculum*)	FRAUENSPIEGEL (*m*)
Spergula arvensis	723	Corn spurrey	MARIENGRAS (*n*); KNÖTERICH (*m*); SPARK (*m*); SPERK (*m*); FELDSPARK (*m*)
Spergularia marina	724	Sandwort spurrey (See *Spergularia rubra*)	SCHUPPENMIERE (*f*)
Spergularia rubra	725	Sandwort; sandspurrey (See *Spergularia marina*)	ROTE SCHUPPENMIERE (*f*)
Sphærococcus crispus	726	(See *Chondrus crispus*)	
Sphærotheca	727	Blight	EINSCHLAUCH-MELTAU (*m*)
Sphærotheca castagnei	728	Hop mildew (See *Sphærotheca humuli*)	HOPFENSCHIMMEL (*m*); HOPFENMEHLTAU (*m*); HOPFENMELTAU (*m*)
Sphærotheca humuli	729	(See *Sphærotheca castagnei*)	
Sphærotheca pannosa	730	Mildew-(of roses)	ROSENSCHIMMEL (*m*); MEHLTAU (*m*); ROSENMEHLTAU (*m*); ROSENMELTAU (*m*)
Sphagnum	731	Sphagnum moss; bog moss	BLEICHMOOS (*n*); TORFMOOS (*n*)
Spigelia marilandica	732	Indian pink; pink root; worm-grass; Maryland pink-root; Carolina pink	WURMKRAUT (*n*)
Spilanthes oleracea	733	Para cress	PARAKRESSE (*f*); FLECKBLUME (*f*)
Spinacia oleracea glabra	734	Summer spinach	(SOMMER)-SPINAT (*m*)
Spinacia oleracea spinosa	735	Prickly spinach; winter spinach	(WINTER)-SPINAT (*m*)
Spinifex	736	(See *Triodia irritans*)	
Spiræa amurensis	737	(See *Neillia amurensis*)	
Spiræa aruncus	738	Goat's beard	GEISSBART (*m*)
Spiræa filipendula	739	Dropwort	SPIERSTAUDE (*f*)
Spiræa filipendula flore pleno	740	Double dropwort	DOPPELBLÜTIGE SPIERSTAUDE (*f*)
Spiræa lobata	741	Queen of the prairies (See *Spiræa ulmaria*)	SPIERSTRAUCH (*m*)
Spiræa salicifolia	742	Willow spiræa	WEIDENBLÄTTRIGER SPIERSTRAUCH (*m*)
Spiræa tomentosa	743	Steeple-bush, hardhack	GELBFILZIGER SPIERSTRAUCH (*m*)
Spiræa ulmaria	744	Meadow-sweet; queen of the meadow	KRAMPFKRAUT (*n*); GEISSBART (*m*); GEISSFUSS (*m*)
Spiranthes æstivalis	745	Lady's-tresses	SOMMER-WENDELORCHE (*f*)
Spiranthes autumnalis	746	Fragrant lady's tresses	HERBST-WENDELORCHE (*f*); DREHLING (*m*)
Spirillum rugula	747		SPIRILLE (*f*)
Spirodela polyrrhiza	748	(See *Lemna polyrrhiza*)	
Spondias	749	Hog-plum	SCHWEINSPFLAUME (*f*)
Spondias dulcis	750	Sweet Otaheite apple	TAHITI-APFEL (*m*); GOLDPFLAUME (*f*); SÜSSE BALSAMPFLAUME (*f*); CYTHERA-APFEL (*m*)
Spondias lutea	751	Golden apple; Jamaica plum; hog plum	GELBE BALSAMPFLAUME (*f*); GELBE CIRUELA (*f*); GELBPFLAUME (*f*); SCHWEINSPFLAUME (*f*)
Spondias mangifera	752	Mango-bearing hog-plum	MANGOPFLAUME (*f*)
Sprekelia formosissima	753	Jacobean lily (See *Amaryllis formosissima*)	JAKOBSLILIE (*f*)
Stachys arvensis	754	Corn woundwort; field woundwort	FELDZIEST (*m*)
Stachys betonica	755	Betony; wood betony (See *Betonica officinalis*)	BETONIENKRAUT (*n*)
Stachys germanica	756	Downy woundwort (See *Stachys lanata*)	
Stachys lanata	757	Lamb's ear	WOLLZIEST (*m*)
Stachys palustris	758	Marsh woundwort	SUMPFZIEST (*m*)

Botanical Term.	No.	English Term.	German Term.
Stachys recta	759	Erect woundwort	BERUFKRAUT (*n*); AUFRECHTER ZIEST (*m*)
Stachys sieboldii	760	(See *Stachys tuberifera*)	
Stachys sylvatica	761	Hedge woundwort; hedge-nettle	ZIEST (*m*)
Stachys tuberifera	762	Chinese artichoke (See *Stachys sieboldii*)	KNOLLENZIEST (*m*)
Stapelia asterias	763	Stapelia, African toad-flower, Carrion flower, star-fish flower	AASBLUME (*f*)
Stapelia bufonia	764	Toad-flower	KRÖTEN-AASBLUME (*f*)
Staphylea	765	Bladder-nut	PIMPERNUSS (*f*); KLAPPERNUSS (*f*); BLASENNUSS (*f*)
Staphylea bolanderi	766	(North American)-bladder-nut tree	BLASENNUSS (*f*); NORDAMERI-KANISCHE PIMPERNUSS (*f*)
Staphylea bumalda	767	(Japanese)-bladder-nut tree	JAPANISCHE PIMPERNUSS (*f*)
Staphylea colchica	768	Bladder-nut tree	KOLCHISCHE PIMPERNUSS (*f*)
Staphylea pinnata	769	Job's tears; bladder-nut tree; St. Anthony's nut	GEFIEDERTE PIMPERNUSS (*f*); ÖLNUSS (*f*); TOTENKÖPFCHEN-STRAUCH (*m*)
Staphylea trifolia	770	Bladder-nut tree	DREIZÄHLIGE PIMPERNUSS (*f*)
Statice armeria	771	(See *Armeria maritima*)	GRASNELKE (*f*)
Statice auriculæfolia	772	Rock statice	GAMSWURZBLÄTTRIGER WIDER-STOSS (*m*)
Statice limonium	773	Sea lavender; common sea lavender; sea pink (See *Limonium*)	ECHTER WIDERSTOSS (*m*)
Statice maritima	774	Sea thrift (See *Armeria maritima*)	MEERSTRANDS-WIDERSTOSS (*m*)
Statice reticulata	775	Matted statice	NETZ-WIDERSTOSS (*m*)
Stellaria aquatica	776	Water starwort; water stich-wort (See *Malachium aquaticum*)	WASSERMIERE (*f*); GEMEINER WASSERDARM (*m*)
Stellaria graminea	777	Lesser stichwort; lesser star-wort	GRASMIERE (*f*)
Stellaria holostea	778	Stellaria, starwort, stichgrass, stichwort, greater stichwort; great starwort	KRAGENBLUME (*f*); GROSSBLU-MIGE MIERE (*f*)
Stellaria media	779	Common chickweed	HÜHNERBISS (*m*); MÄUSEDARM (*m*); HÜHNERDARM (*m*); VOGELMIERE (*f*); MIERE (*f*); STERNMIERE (*f*); HÜHNER-MYRTE (*f*); HÜHNER-SCHWARM (*m*)
Stellaria nemorum	780	Wood stichwort; wood starwort	HAINMIERE (*f*)
Stellaria palustris	781	Marsh starwort; marsh stich-wort	SUMPFMIERE (*f*)
Stellaria paminea aurea	782	Yellow-leaved lesser stichwort	GELBBLÄTTRIGE STERNMIERE (*f*)
Stellaria uliginosa	783	Bog starwort; bog stichwort	SUMPFMIERE (*f*)
Stenactis annua	784	(See *Erigeron annuus*)	
Stenotaphrum glabrum variegatum	785	Variegated grass	GRUBENGRAS (*n*)
Stephanotis floribunda	786	Clustered wax flower, Madagascar jasmine; Madagascar chaplet flower	MADAGASKAR-STEPHANOTIS (*f*)
Sternbergia lutea	787	Winter daffodil; yellow star-flower (See *Amaryllis lutea*)	
Sticta pulmonaria	788	Lungwort (See *Lobaria pulmonaria*)	LUNGENFLECHTE (*f*)
Stigmaphyllon ciliatum	789	Golden vine (See *Malpighia mexicana*)	
Stigmatophyllvm	790	(See *Stigmaphyllon*)	
Stipa arenaria	791	Esparto grass, Alfa, Halfa (See *Stipa tenacissima* and also *Stupa*)	REIHERGRAS (*n*); ESPARTOGRAS (*n*)
Stipa pennata	792	Feather-grass; soft feather grass	MARIENGRAS (*n*); MARIENFLACHS (*m*); PFRIEMENGRAS (*n*)
Stipa tenacissima	793	Esparto grass, Alfa, Halfa (See *Macrochloa tenacissima* and *Stipa arenaria*)	SPANISCHES PFRIEMENGRAS (*n*); ESPARTOGRAS (*n*); HALFAGRAS (*n*)

Botanical Term.	No.	English Term.	German Term.
Stokesia cyanea	794	Stokes' aster	STOKES-ASTER (*f*)
Stratiotes aloides	795	Water soldier; crab's claw; water aloe	WASSERSCHERE (*f*); WASSERSCHER (*m*)
Strelitzia	796	Bird of Paradise flower; bird's-tongue flower	VOGELZUNGENBLUME (*f*); STRELITZIE (*f*)
Streptocarpus polyanthus	797	Cape primrose	DREHFRUCHT (*f*)
Strobilanthes	798	Cone-head (See *Goldfussia*)	
Strobilanthes flaccidifolius	799	Flabby-leaved cone-head; Assam indigo plant	ASSAMINDIGOPFLANZE (*f*)
Strophantus hispidus	800	Strophantus (See *Strophantus kombé*)	KOMBÉSTRAUCH (*m*); ONAGE--STRAUCH (*m*); INEESTRAUCH (*m*)
Strophantus kombé	801	Strophantus (See *Strophantus hispidus*)	STROPHANTUS (*m*)
Struthiopteris germanica	802	(See *Onoclea germanica*)	DEUTSCHER STRAUSSFARN (*m*)
Strychnos colubrina	803	Snakewood-(tree)	SCHLANGENHOLZ (*n*); SCHLANGEN--HOLZBAUM (*m*)
Strychnos ignatii	804	St. Ignatius' bean (See *Ignatia multiflora* and also *Faba ignatii*)	IGNATIUSBOHNE (*f*); IGNATIUS--BAUM (*m*)
Strychnos nux vomica	805	Nux vomica-(tree)	KRÄHENAUGE (*n*); PURGIERNUSS (*f*); BRECHNUSSBAUM (*m*); KRÄHENAUGENBAUM (*m*)
Strychnos potatorum	806	Clearing-nut plant	PURGIERNUSS (*f*); KLÄRNUSS--BAUM (*m*); ATSCHIERBAUM (*m*)
Strychnos tieuté	807	Upa plant	UPASTRAUCH (*m*)
Stryphnodendron barbatimao	808	Barbatimao-bark tree	BARBATIMAORINDEBAUM (*m*)
Stupa	809	(See *Stipa*)	
Stylophorum diphyllum	810	Celandine poppy	ZWEIBLÄTTRIGER GRIFFELTRÄGER (*m*)
Styrax benzoin	811	Benzoin tree; benjamin tree	BENZOEBAUM (*m*)
Styrax obassia	812	Storax	STORAXBAUM (*m*)
Suæda fruticosa	813	Seablite; shrubby suæda	GÄNSEFÜSSCHEN (*n*)
Suæda maritima	814	Seablite; herbaceous suæda	MEERSTRANDSGÄNSEFÜSSCHEN (*n*)
Subularia aquatica	815	Awl-wort	WASSER-PFRIEMENKRESSE (*f*)
Sutherlandia frutescens	816	Cape bladder senna	KAP-BLASENSENNA (*f*)
Swainsona coronillifolia	817	Darling River pea; Swainson's pea	AUSTRALISCHE SWAINSONA (*f*)
Swartzia	818	(Guiana)-bully tree	SWARTZIA (*f*)
Swartzia tormentosa	819	Iron wood, South American iron wood	(SÜDAMERIKANISCHES)-EISENHOLZ (*n*)
Sweertia	820	(See *Swertia*)	
Swertia chirata	821	Chirata, chiretta	CHIRETTA (*f*)
Swertia chirayita	822	Chirata; chiretta	CHIRETTA (*f*)
Swertia perennis	823	Marsh fel-wort	AUSDAUERENDE SWEERTIE (*f*); TARANT (*f*)
Swietenia mahagoni	824	mahogany tree	MAHAGONI (*n*); MAHAGONIHOLZ (*n*); MAHAGONIBAUM (*m*)
Sycomorus	825	(See *Ficus sycomorus*)	
Sympetalæ	826	Sympetalæ	VERWACHSENBLUMENKRONBLÄT--TRIGE PFLANZEN (*f.pl*)
Symphoricarpus occidentalis	827	Wolf berry	WESTLICHE SCHNEEBEERE (*f*)
Symphoricarpus racemosus	828	(North American)-snowberry shrub; snowberry tree (See also *Chiococca racemosa*)	GEMEINE SCHNEEBEERE (*f*)
Symphoricarpus vulgaris	829	Indian currant; coral berry	SCHNEEBEERE (*f*)
Symphyandra pendula	830	Pendulous bell-flower	HÄNGENDE GLOCKENBLUME (*f*)
Symphytum officinale	831	Comfrey; common comfrey	BEINWURZ (*n*); SCHWARZWURZ (*f*); BEINWELL (*m*)
Symphytum tuberosum	832	Tuberous comfrey	KNOLLIGER BEINWELL (*m*); WALL--WURZ (*f*); SCHWARZWURZEL (*f*)
Syringa	833	Lilac; pipe tree	LILA (*m*); LILAK (*m*); FLIEDER (*m*)
Syringa chinensis	834	Rouen lilac	CHINESISCHER FLIEDER (*m*)
Syringa emodi	835	Indian lilac	EMODI-FLIEDER (*m*)
Syringa japonica	836	Japanese syringa, Japanese lilac	JAPANISCHER FLIEDER (*m*)
Syringa josikœa	837	Hungarian lilac	UNGARISCHER FLIEDER (*m*)

Botanical Term.	No.	English Term.	German Term.
Syringa oblata	838	Round-leaved syringa	RUNDBLÄTTRIGER FLIEDER (*m*)
Syringa persica	839	Persian syringa ; Persian lilac	PERSISCHER FLIEDER (*m*)
Syringa synensis	840	(See *Syringa chinensis*)	
Syringa vulgaris	841	Syringa, lilac, common lilac	SYRINGENHOLZ (*n*) ; (GEMEINER)-FLIEDER (*m*); GEMEINER (WILDER) JASMIN (*m*); SPA--NISCHER (TÜRKISCHER) HO--LUNDER (*m*) ; LILA (*m*); LILAK (*m*) ; SPANISCHER FLIEDER (*m*)
Tabernæmontana coronaria	842	East Indian rose bay, Adam's apple	ADAMSAPFELBAUM (*m*)
Tabernæmontana dichotoma	843	Eve's apple ; tree of knowledge	BAUM (*m*) DER ERKENNTNIS (*f*); EVAAPFELBAUM (*m*)
Tabernæmontana utilis	844	Demerara milk-tree ; Hya-hya tree	DEMERARA-MILCHBAUM (*m*); HYA-HYA-BAUM (*m*)
Tacca pinnatifida	845	Otaheite salap plant	TAHITI ARROWROOT-PFLANZE (*f*)
Tacsonia manicata	846	Blood-red passion flower (See *Passiflora princeps*)	
Tagetes erecta	847	African marigold	STUDENTENBLUME (*f*) ; AFRIKA--NISCHE SAMTBLUME (*f*)
Tagetes lucida	848	Mexican marigold	SAMTBLUME (*f*) ; MEXIKANISCHE SAMTBLUME (*f*)
Tagetes patula	849	French marigold	FRANZÖSISCHE SAMTBLUME (*f*)
Taiwana cryptomerioides	850	Formosa cedar	FORMOSAZEDER (*f*)
Tamarindus indica	851	Tamarind tree, Indian date	TAMARINDE (*f*); SAUERDATTEL (*f*)
Tamarix	852	Tamarisk	DEUTSCHE ZYPRESSE (*f*) ; TAMA--RISKE (*f*)
Tamarix anglica	853	(See *Tamarix gallica*)	
Tamarix gallica	854	Common tamarisk, manna plant (See *Tamarix anglica* and *Tamarix pentandra*)	FÜNFMÄNNIGE TAMARISKE (*f*)
Tamarix germanica	855	German tamarisk (See *Myricaria germanica*)	DEUTSCHER BIRZSTRAUCH (*m*)
Tamarix mannifera	856	Tamarisk ; manna tree	MANNABAUM (*m*)
Tamarix pentandra	857	(See *Tamarix gallica*)	FÜNFMÄNNIGE TAMARISKE (*f*)
Tamarix sinensis	858	Chinese tamarisk	CHINESISCHE TAMARISKE (*f*)
Tamarix tetranda	859	Four-stamened tamarisk	VIERMÄNNIGE TAMARISKE (*f*)
Tamus communis	860	Black bryony ; our Lady's seal	GEMEINE SCHMERWURZ (*f*)
Tanacetum balsamita	861	Alecost ; costmary (See *Chrysanthemum balsamita var. tanacetoides* and *Balsamita vulgaris*)	MARIENBLATT (*n*)
Tanacetum vulgare	862	Wild tansy ; buttons ; alecost, wild agrimony, goose-grass (See *Chrysanthemum vulgare*)	GÄNSERICH (*m*) ; RAINFARN (*m*)
Tanghinia venenifera	863	Tanghinin almond ; Tangwi tree (See *Cerbera tanghin*)	TANGHIBAUM (*m*) ; TANGWIBAUM (*m*) ; GERICHTSBAUM (*m*)
Taphria carpini	864	(See *Exoascus carpini*)	
Taraxacum dens-leonis	865	Dandelion (See *Taraxacum officinale*)	LÖWENZAHN (*m*)
Taraxacum officinale	866	Dandelion (See *Taraxacum dens-leonis*)	LÖWENZAHN (*m*)
Taxaceæ	867	Taxaceæ (See *Gingko*, *Taxus* and *Cephalotaxus*)	EIBENGEWÄCHSE (*n.pl*)
Taxodium distichum	868	Deciduous cypress; black cypress; American black cypress	SUMPFZYPRESSE (*f*); VIRGI--NISCHE ZYPRESSE (*f*) ; AMERI--KANISCHE SUMPFZYPRESSE (*f*) ; SUMPFZEDER (*f*)
Taxodium heterophyllum	869	Small Chinese embossed cypress (See *Glyptostrobus heterophyllus*)	WASSERZYPRESSE (*f*); CHINE--SISCHE SUMPFZYPRESSE (*f*)
Taxodium mucronatum	870	Montezuma cypress	MONTEZUMAZYPRESSE (*f*) ; MEXI--KANISCHE SUMPFZYPRESSE (*f*)
Taxus aurea	871	(See *Taxus baccata aurea*)	
Taxus baccata	872	Yew-(tree); Irish yew-(tree); common yew tree (See *Taxus fastigiata*)	TAXUS (*m*); EIBE (*f*); IBENBAUM (*m*); EIBENBAUM (*m*); GEMEI--NE EIBE (*f*); EVENBAUM (*m*); GEMEINER TAXUSBAUM (*m*)

Botanical Term.	No.	English Term.	German Term.
Taxus baccata aurea.	873	Golden yew-(tree)	GOLDENE EIBE (*f*); GOLDENER EIBENBAUM (*m*)
Taxus canadensis	874	Ground hemlock; American yew	AMERIKANISCHE EIBE (*f*)
Taxus cephalotaxus	875	(See *Taxus baccata*)	
Taxus cuspidata	876	Japanese yew	JAPANISCHE EIBE (*f*)
Taxus erecta	877	Fulham yew	AUFRECHTE EIBE (*f*)
Taxus fastigiata	878	Irish yew (See *Taxus baccata*)	
Taxus globosa	879	Mexican yew	MEXIKANISCHE EIBE (*f*)
Tecoma australis	880	Australian wonga-wonga vine	AUSTRALISCHE WONGA-WONGA REBE (*f*)
Tecoma ipe	881	South American bowwood; pao d'arco	BOGENHOLZ (*n*)
Tecoma leucoxylon	882	Green-heart tree	GRÜNHERZHOLZBAUM (*m*); GRÜNES EBENHOLZ (*n*); GELBES EBEN--HOLZ (*n*); BASTARD-GUAJAK (*m*); BRAUNES EBENHOLZ (*n*)
Tecoma radicans	883	Common trumpet flower; trumpet creeper, Moreton Bay trumpet jasmine (See *Bignonia radicans* and *Campsis radicans*)	VIRGINISCHER JASMIN (*m*)
Tecophilæ	884	Chilian crocus	CHILENISCHER KROKOS (*m*)
Tectona grandis	885	Teak, teak tree	TEAKHOLZ (*n*); TIKHOLZ (*n*); TEAKBAUM (*m*); TIEKBAUM (*m*); INDISCHE EICHE (*f*)
Teesdalia nudicaulis	886	Common teesdalia	NACKTSTENGELIGE TEESDALIE (*f*)
Telanthera	887	Joy-weed (See *Alternanthera*)	WECHSELKÖLBCHEN (*n*)
Terebinthaceæ	888	Terebinthaceæ	SUMACHGEWÄCHSE (*n.pl*)
Terminalia chebula	889	Chebula tree; Myrobalan tree	MYROBALANENBAUM (*m*)
Testudinaria elephantipes	890	Elephant's foot; tortoise plant; Hottentot bread (See *Discorea elephantipes*)	SCHILDKRÖTENPFLANZE (*f*)
Tetragonia expansa	891	New Zealand spinach	NEUSEELÄNDISCHER SPINAT (*m*)
Tetrapanax papyrifera	892	(See *Aralia papyrifera* and *Fatsia papyrifera*)	PAPIERPANAX (*m*)
Teucrium botrys	893	Cut-leaved germander	GESCHLITZTBLÄTTRIGER GAMAN--DER (*m*)
Teucrium chamædrys	894	Wall germander	WANDGAMANDER (*m*); GEMEINER GAMANDER (*m*)
Teucrium marum	895	Teucrium; germander; cat-thyme	GAMANDER (*m*); MASTIXKRAUT (*n*); KATZENKRAUT (*n*); MARUMKRAUT (*n*); MOSCHUS--KRAUT (*n*)
Teucrium polium	896	Penny-royal germander	POLEI-GAMANDER (*m*)
Teucrium scordium	897	Water germander	LACHENKNOBLAUCH (*m*)
Teucrium scorodonia	898	Wood sage; germander; wood germander	GAMANDER (*m*)
Thalictrum alpinum	899	Alpine meadow rue	ALPEN-WIESENRAUTE (*f*); WIESEN--KRAUT (*n*)
Thalictrum anemonoides	900	Rune anemone	ANEMONEN-WIESENRAUTE (*f*)
Thalictrum flavum	901	(Common)-meadow rue; yellow meadow rue	GARTENRAUTE (*f*)
Thalictrum majus	902	Greater meadow rue	GROSSE WIESENRAUTE (*f*)
Thalictrum minus	903	Lesser meadow rue	KRÖTENDISTEL (*f*); KLEINE WIE--SENRAUTE (*f*)
Thamnocalamus falconeri	904	Himalayan bamboo	HIMALAJA-BAMBUS (*m*)
Thea bohea	905	Small tea shrub; bohea tea plant	KLEINER TEE (*m*); BOHEA-TEE (*m*)
Thea chinensis	906	(See *Thea sinensis*)	
Thea sinensis	907	Tea plant (See *Thea chinensis*)	TEE (*m*); CHINESISCHER TEE (*m*)
Thea viridis	908	Large tea shrub	GROSSER TEE (*m*); GRÜNER TEE (*m*)
Theloschistes	909	(See *Xanthoria parietina*)	
Themeda	910	(See *Anthistiria ciliata*)	
Theobroma cacao	911	Cacao tree, cocoa tree	KAKAOBAUM (*m*); SCHOKOLADEN--BAUM (*m*)

Botanical Term.	No.	English Term.	German Term.
Thesium linophyllum	912	Bastard toadflax	VERNEINKRAUT (*n*); LEINBLATT (*n*); BERGFLACHS (*m*)
Thevetia ahouai	913	Ahouai tree	AHOVAIBAUM (*m*); SCHELLEN-BAUM (*m*)
Thlaspi alpestre	914	Alpine pennycress	ALPENPFENNIGKRAUT (*n*)
Thlaspi arvense	915	Penny-cress; mithridate mustard	PFENNIGKRAUT (*n*); TÄSCHEL-KRAUT (*n*)
Thlaspi perfoliatum	916	Perfoliate penny-cress	DURCHWACHSENES PFENNIGKRAUT (*n*); DURCHWACHSENES TÄSCHELKRAUT (*n*)
Thrinax argentea	917	Silver thatch palm, broom palm	SILBERPALME (*f*)
Thrinax parviflora	918	Royal palmetto palm	KÖNIGLICHE-PALMETTOPALME (*f*)
Thuia	919	Arbor vitæ, lignum vitæ, tree of life (See *Thuja occidentalis*)	THUJA (*f*); LEBENSBAUM (*m*)
Thuja articulata	920	(See *Callitris quadrivalvis*)	
Thuja dolobrata	921	(See *Thujopsis dolobrata*)	
Thuja gigantea	922	Giant arbor vitæ (See *Thuja plicata*)	RIESEN-LEBENSBAUM (*m*)
Thuja japonica	923	Japanese arbor vitæ	JAPANISCHER LEBENSBAUM (*m*)
Thuja juniperoides	924	Cape juniper; cederboom (See *Juniperus capensis*)	CEDERBOOM (*m*)
Thuja lobbii	925	(See *Thuja plicata* and *Thuja gigantea*)	
Thuja occidentalis	926	(Canadian)-white cedar, arbor-vitæ, American arbor-vitæ (See *Thuia* and *Thuya*)	THUJA (*f*); GEWÖHNLICHER LE-BENSBAUM (*m*); ABENDLÄN-DISCHER LEBENSBAUM (*m*); AMERIKANISCHER LEBENSBAUM (*m*); GEMEINER LEBENSBAUM (*m*)
Thuja orientalis	927	Oriental arbor-vitæ, Chinese arbor-vitæ (See *Biota orientalis*)	MORGENLÄNDISCHER LEBENS-BAUM (*m*)
Thuja plicata	928	Red cedar, Canoe cedar (See *Thuja lobbii* and *Thuja gigantea*)	GEFALTETER LEBENSBAUM (*m*)
Thuja standishi	929	(See *Thuja japonica*)	
Thujopsis dolobrata	930	(See *Thuja dolobrata*)	HIBER (*f*); BEILBLÄTTRIGE HIBER (*f*); BEILSCHUPPE (*f*); HIBA-LEBENSBAUM (*m*)
Thuya	931	(See *Thuja*)	
Thuyopsis	932	(See *Thujopsis*)	
Thymelœaceœ	933	Thymelæaceæ	SEIDELBASTGEWÄCHSE (*n.pl*)
Thymus calamintha	934	Common calamint (See *Calamintha acinos*)	KALAMINTE (*f*)
Thymus citriodorus	935	Lemon thyme (See *Thymus serphyllum vulgaris*)	ZITRONEN-THYMIAN (*m*)
Thymus serphyllum	936	Wild thyme, wild caraway	FELDTHYMIAN (*m*); FELDKÜMMEL (*m*); QUENDEL (*m*); QUENDEL-KRAUT (*n*); GRUNDEL (*m*); FELDPOLEI (*f* and *m*); HÜHNER-POLEI (*f* and *m*)
Thymus serphyllum citri-dorus	937	(See *Thymus citriodorus*)	
Thymus serphyllum citri-odorus argenteus	938	Fraser's silver-leaved thyme	SILBERBLÄTTERIGER THYMIAN (*m*)
Thymus serphyllum citri-odorus aureum	939	Golden-leaved thyme	GOLDBLÄTTERIGER THYMIAN (*m*)
Thymus serphyllum coccineus	940	Crimson-flowered thyme	KARMESINBLÜTIGER THYMIAN (*m*)
Thymus serphyllum lanugino-sus	941	Woolly-leaved thyme	WOLLIGBLÄTTERIGER THYMIAN (*m*)
Thymus serphyllum vulgaris	942	Lemon thyme (See *Thymus citriodorus*)	
Thymus vulgaris	943	Thyme; garden thyme	DOST (*m*); DOSTEN (*m*); THYMIAN (*m*)
Thyrsacanthus rutilans	944	Thyrse flower	ORANGENROTE THYRSUSBLUME (*f*)
Thysselinum palustre	945	(See *Peucedanum palustre*)	
Tiarella cordifolia	946	Foam-flower, false mitre-wort	TIARELLE (*f*)

Botanical Term.	No.	English Term.	German Term.
Tibouchina elegans	947	Brazilian spider-flower (See *Lasiandra* and *Pieroma*)	BRASILIENISCHE SPINNENBLUME (*f*)
Tiglium officinale	948	(See *Croton tiglium*)	
Tigridia pavonia	949	Orange-flowered tiger-flower, Tiger iris	PFAUENLILIE (*f*); TIGERBLUME (*f*); TIGERLILIE (*f*)
Tilia	950	Lime tree	LINDE (*f*)
Tilia alba	951	White (silver) lime	SILBERLINDE (*f*)
Tilia americana	952	(North American)-black lime, bass wood tree	SCHWARZE LINDE (*f*)
Tilia argentea	953	Hungarian silver lime	UNGARISCHE SILBERLINDE (*f*)
Tilia cordata	954	Lime tree, linden tree, winter lime (See *Tilia europæa* and *Tilia cordata*)	LINDE (*f*); WINTERLINDE (*f*)
Tilia europæa	955	Lime tree, Linden tree (See *Tilia cordata* and *Tilia vulgaris*)	LINDE (*f*)
Tilia grandifolia	956	Summer lime; large leaved lime (See *Tilia platyphyllos*)	SOMMERLINDE (*f*); GROSSBLÄT--TRIGE LINDE (*f*)
Tilia heterophylla	957	Various-leaved silver lime, bee tree	VERSCHIEDENBLÄTTRIGE SILBER--LINDE (*f*)
Tilia mandschurica	958	Manchurian silver lime	MANDSCHURISCHE SILBERLINDE (*f*)
Tilia parvifolia	959	Winter lime; small leaved lime	WINTERLINDE (*f*); KLEINBLÄT--TRIGE LINDE (*f*); STEINLINDE (*f*)
Tilia persiflora	960	(See *Tilia cordata*)	
Tilia platyphyllos	961	Summer lime; large leaved lime (See *Tilia grandifolia*)	SOMMERLINDE (*f*); GROSSBLÄT--TRIGE LINDE (*f*)
Tilia pubescens	962	White-haired lime	WEISSHAARIGE LINDE (*f*)
Tilia tomentosa	963	(See *Tilia argentea*)	
Tilia ulmifolia	964	Winter lime; small leaved lime; Linn; Linden; Basswood tree	WINTERLINDE (*f*); KLEINBLÄT--TRIGE LINDE (*f*); STEINLINDE (*f*)
Tilia vulgaris	965	(See *Tilia europæa* and *Tilia cordata*)	
Tillæa muscosa	966	Mossy tillæa (See *Crassula muscosa*)	MOOSARTIGE TILLAE (*f*)
Tillandsia usneoides	967	(Tropical American)-tree moss, Spanish moss, old man's beard (See *Vriesia*)	GREISENBART (*m*)
Tilletia caries	968	Stinking wheat smut; bunt, brand, rust, bladder brand (See also *Ustilago tritici*)	STEINBRAND (*m*); SCHMIERBRAND (*m*); STINKBRAND (*m*)
Tilletia tritici	969	(See *Tilletia caries*)	
Todea barbara	970	Crape fern	ELEFANTENFARN (*m*)
Todea superba	971	Filmy fern, Prince of Wales' feather fern	TODEAFARN (*m*)
Tofieldia palustris	972	Scottish asphodel	NORDISCHE TOFIELDIE (*f*)
Tolpis barbata	973	Yellow garden hawkweed (See *Crepis barbata*)	
Toluifera pereiræ	974	Balsam tree	BALSAMBAUM (*m*)
Tordylium maximum	975	Great hartwort	ZIRMET (*m*)
Tormentilla erecta	976	(See *Potentilla tormentilla* and *Tormentilla officinalis*)	
Tormentilla officinalis	977	Common tormentil (See *Potentilla tormentilla*)	
Tormentilla reptans	978	Trailing tormentil (See *Potentilla reptans*)	
Torreya californica	979	Californian nutmeg	KALIFORNISCHE NUSSEIBE (*f*); KALIFORNISCHE MUSKATNUSS (*f*); WILDE MUSKATNUSS (*f*)
Torreya taxifolia	980	Stinking cedar, stinking yew	STINKEIBE (*f*)
Tournefortia	981	Summer heliotrope (See also *Heliotropium*)	SOMMER-HELIOTROP (*m*)
Toxicophlœa spectabilis	982	Winter sweet (See *Acokanthera spectabilis*)	
Trachelium cœruleum	983	Blue throat-wort (See *Campanula trachelium*)	NESSELBLÄTTRIGE GLOCKEN--BLUME (*f*)

Botanical Term.	No.	English Term.	German Term.
Trachelospermum jasminoides	984	(See *Rhynchospermum jasmin-oides*)	JASMINARTIGER SCHNABELSAME (*m*)
Trachyspermum copticum	985	Ajowan (*See Ptychotis coptica*)	
Tradescantia virginiana	986	(See *Tradescantia virginica*)	
Tradescantia virginica	987	Spiderwort, flower of a day (See *Tradescantia virginiana*)	VIRGINISCHE TRADESKANTIE (*f*)
Tragopogon glaber	988	Purple-flowered goat's-beard	PURPURBLÜTIGER BOCKSBART (*m*)
Tragopogon minor	989	Smaller goat's-beard	KLEINER BOCKSBART (*m*)
Tragopogon porrifolius	990	Salsify ; salsafy ; purple salsify ; vegetable oyster	SALSIFI (*f*) ; LAUCHBLÄTTRIGER BOCKSBART (*m*)
Tragopogon pratensis	991	Field goat's-beard ; yellow goat's beard ; Johnnie-go-to-bed-at-noon ; meadow salsify	BOCKSBART (*m*) ; WIESEN-BOCKS-BART (*m*) ; HAFERWURZ (*f*)
Trametes radiciperda	992	Tinder fungus ; red-rot fungus	ZUNDERSCHWAMM (*m*) ; ROT-FÄULE (*f*) ; FICHTENROT-FÄULE (*f*) ; SENKRÖHREN-SCHWAMM (*m*)
Trametes suaveolens	993	Sweet-smelling tinder fungus	WOHLRIECHENDER SENKRÖHREN-SCHWAMM (*m*)
Trapa bicornis	994	(See *Trapa natans*)	ZWEIHÖRNIGE WASSERNUSS (*f*)
Trapa bispinosa	995	Singhara nut plant	ZWEIDORNIGE WASSERNUSS (*f*)
Trapa natans	996	Water chestnut ; Jesuits' nut ; water caltrops, Singara nut (See *Trapa bicornis*)	JESUITENNUSS (*f*) ; WASSERNUSS (*f*)
Trichilia emetica	997	Mafura	MAFURA (*f*) ; MAFURRAH (*f*) ; AFRIKANISCHER MAFUREIRA-BAUM (*m*)
Trichilia glandulosa	998	New South Wales rosewood	DRÜSENTRAGENDE TRICHILIA (*f*)
Trichobasis	999	Rust	ROST (*m*)
Trichobasis betæ	000	Beet rust	ZUCKERRÜBENROST (*m*)
Trichobasis fabæ	001	Bean rust	BOHNENROST (*m*)
Trichobasis fallens	002	(See *Uromyces apiculata*)	
Trichobasis galii	003	Bedstraw rust	LIEBFRAUENBETTSTROHROST (*m*)
Trichobasis labiatarum	004	Mint rust	MINZENROST (*m*)
Trichobasis petroselini	005	Parsley rust	PETERSILIENROST (*m*)
Trichobasis polygonorum	006	(See *Uromyces polygoni*)	
Trichobasis rubigo vera	007	Corn rust	GETREIDEROST (*m*)
Trichobasis suaveolens	008	Sweet-smelling rust ; thistle rust	WOHLRIECHENDER ROST (*m*) ; DIS-TELROST (*m*)
Tricholoma	009	(See *Agaricus graveolens*)	
Trichomanes radicans	010	Cup goldilocks ; Killarney fern ; bristle fern	HAARFARN (*m*)
Trichosanthes anguina	011	Viper gourd, snake gourd, serpent cucumber (See *Cucumis anguinus*)	HAARBLUME (*f*)
Tricyrtis hirta	012	Japanese toad lily	JAPANISCHE KRÖTENLILIE (*f*)
Trientalis americana	013	Star-flower	AMERIKANISCHES SCHIRMKRAUT (*n*) ; SIEBENSTERN (*m*)
Trientalis europœa	014	Common trientale ; chickweed wintergreen	SCHIRMKRAUT (*n*)
Trifolium	015	Trefoil ; clover (See *Menyanthes trifoliata*)	KLEE (*m*) ; KOPFKLEE (*m*) ; DREI-BLÄTT (*n*) ; DREIBLÄTTRIGER KLEE (*m*) ; LEBERKRAUT (*n*) ; LEBERKLEE (*m*)
Trifolium agrarium	016	Golden clover	GOLDKLEE (*m*)
Trifolium alexandrinum	017	Egyptian clover, berseem	ALEXANDRINER-KLEE (*m*)
Trifolium alpestre	018	Wood clover	WALDKLEE (*m*)
Trifolium arvense	019	Hare's-foot trefoil ; hare's-foot clover	HASENKLEE (*m*) ; HASENPFÖTCHEN (*n*) ; KATZENKLEE (*m*) ; ACKER-KLEE (*m*) ; MÄUSEKLEE (*m*)
Trifolium badium	020	Leather-brown clover	LEDERBRAUNER KLEE (*m*)
Trifolium bocconi	021	Boccone's clover	BOCCONES-KLEE (*m*)
Trifolium filiforme	022	Slender clover	FADENFÖRMIGER KLEE (*m*)
Trifolium fragiferum	023	Strawberry clover	ERDBEERKLEE (*m*) ; BLASENKLEE (*m*)
Trifolium glomeratum	024	Clustered clover	KOPFKLEE (*m*)
Trifolium hybridum	025	Alsike clover	BASTARD-KLEE (*m*) ; SCHWE-DISCHER KLEE (*m*)

Botanical Term.	No.	English Term.	German Term.
Trifolium incarnatum	026	Crimson clover	INKARNATKLEE (*m*)
Trifolium lupinaster	027	Lupinaster	LUPINENKLEE (*m*)
Trifolium maritimum	028	Sea clover	MEER-KLEE (*m*)
Trifolium medium	029	Marl-grass; zigzag clover; meadow clover	MITTLERER KLEE (*m*)
Trifolium minus	030	Lesser (trefoil) clover	KLEINER KLEE (*m*)
Trifolium ochroleucum	031	Sulphur-coloured trefoil; sulphur clover	BLASSGELBER KLEE (*m*); ROSEN-KLEE (*m*)
Trifolium pratense	032	Purple clover; red clover	FUTTERKLEE (*m*); WIESENKLEE (*m*)
Trifolium procumbens	033	True hop clover; hop clover; hop trefoil	LIEGENDER KLEE (*m*); LIEGENDER HOPFENKLEE (*m*)
Trifolium repens	034	White clover; Dutch clover; St. Patrick's clover; shamrock	STEINKLEE (*m*); WEISSKLEE (*m*); WIESENKLEE (*m*)
Trifolium repens purpureum	035	Scotch shamrock	ROTER WIESENKLEE (*m*)
Trifolium resupinatum	036	Reversed clover	PERSISCHER KLEE (*m*); SCHABDER (*m*)
Trifolium rubens	037	Red clover	ROTER KLEE (*m*)
Trifolium scabrum	038	Rough clover	RAUHER KLEE (*m*)
Trifolium spadiceum	039	Hop clover; hop trefoil	HOPFENKLEE (*m*); BRAUNER KLEE (*m*)
Trifolium stellatum	040	Starry clover	STERN-KLEE (*m*)
Trifolium striatum	041	Soft knotted trefoil; knotted clover	GESTREIFTER KLEE (*m*)
Trifolium strictum	042	Upright clover	AUFRECHTER KLEE (*m*)
Trifolium subterraneum	043	Subterranean clover	UNTERIRDISCHER KLEE (*m*)
Trifolium suffocatum	044	Suffocated clover	ERSTICKTER KLEE (*m*)
Trifragmium	045	Complex brand (See *Triphragmium*)	DREIWANDSPOREN-ROST (*m*)
Trifragmium ulmariæ	046	Meadow-sweet brand	SPIERSTAUDEN-DREIWANDSPOREN-ROST (*m*)
Triglochin maritimum	047	Sea arrowgrass	MEERSTRANDS-DREIZACK (*m*)
Triglochin palustre	048	Marsh arrowgrass	SUMPF-DREIZACK (*m*)
Trigonella fœnum-græcum	049	Fenugreek; fœnugreek	KUHHORNKLEE (*m*)
Trigonella purpurascens	050	Bird's-foot trigonel; fenugreek (See *Aporanthus trifoliastrum*)	BOCKSHORNKLEE (*m*)
Trillium erectum	051	Birth-root	AUFRECHTER DRILLING (*m*)
Trillium erythrocarpum	052	Painted Wood lily	ROTFRÜCHTIGER DRILLING (*m*)
Trillium grandiflorum	053	Wake robin	DRILLING (*m*)
Trinia glauca	054	Sea-green honewort	BLAUGRÜNE TRINIE (*f*)
Trinia vulgaris	055	Common trinia; honewort	GEMEINE TRINIE (*f*)
Triodia irritans	056	(Australian)-porcupine grass; spinifex (See *Spinifex*)	STACHELKOPFGRAS (*n*); DREIZAHN-GRAS (*n*)
Triosteum	057	Fever-root; feverwort; horse gentian	DREISTEINWURZEL (*f*)
Triosteum perfoliatum	058	American feverwort	AMERIKANISCHE DREISTEINWURZEL (*f*); DURCHWACHSEN-BLÄTTRIGE DREISTEINWURZEL (*f*)
Triphragmium	059	Brand (See *Trifragmium*)	BRAND (*m*); ROST (*m*)
Trisetum flavescens	060	(See *Avena flavescens*)	GOLDHAFERGRAS (*n*); KLEINER WIESENHAFER (*m*)
Tristania neriifolia	061	Oleander-leaved tristania; scrub box-tree; brush box-tree	TRISTANIE (*f*); BUSCH-BUCHS-BAUM (*m*); LORBEERMYRTE (*f*)
Triticum æstivum	062	Spring wheat, summer wheat	SOMMERWEIZEN (*m*)
Triticum amyleum	063	Starch wheat	STÄRKE-WEIZEN (*m*)
Triticum caninum	064	Dog's tail grass	HUNDSQUECKE (*f*)
Triticum durum	065	Hard wheat (See *Aegilops ovata*)	HARTER WEIZEN (*m*); GERSTEN-WALCH (*m*)
Triticum loliaceum	066	(See *Glyceria loliacea*)	
Triticum monococcum	067	Single-grained wheat; spelt	EINKORN (*n*); PETERSKORN (*n*)
Triticum polonicum	068	Polish wheat	POLNISCHER WEIZEN (*m*)
Triticum repens	069	Couch-grass (See *Agropyrum repens*)	GRASWURZEL (*f*); LIEGENDE QUECKE (*f*)

Botanical Term.	No.	English Term.	German Term.
Triticum sativum	070	(Common)-wheat (See *Triticum vulgare*)	WEIZEN (*m*)
Triticum sativum turgidum	071	(See *Triticum turgidum*)	
Triticum spelta	072	Spelt; awned wheat	SPELZ (*m*); SPELZWEIZEN (*m*); IGELWEIZEN (*m*); REISDINKEL (*m*)
Triticum turgidum	073	Turgid wheat; cone wheat; mummy wheat (See *Triticum sativum turgidum*)	ENGLISCHER WEIZEN (*m*)
Triticum vulgare	074	Wheat (See *Triticum sativum* and also *Sitopyros*)	WEIZEN (*m*)
Tritoma uvaria	075	African flame-flower, torch lily, club lily (See *Kniphofia aloides*)	TRAUBIGE HYAZINTHENALOE (*f*)
Trollius europæus	076	Globe flower	KUGELRANUNKEL (*f*)
Tropæolum aduncum	077	Canary creeper (See *Tropæolum canariense* and *Tropæolum peregrinum*)	CANARIEN-KAPUZINERKRESSE (*f*); PERU-KAPUZINERKRESSE (*f*)
Tropæolum canariense	078	(See *Tropæolum aduncum*)	CANARIEN-KAPUZINERKRESSE (*f*)
Tropæolum majus	079	Great nasturtium, tall nasturtium	GROSSE KAPUZINERKRESSE (*f*)
Tropæolum minus	080	(Small)-nasturtium; Indian cress, dwarf nasturtium (See also *Nasturtium officinale* and *Sisymbrium nasturtium*)	KAPUZINERKRESSE (*f*); SPA--NISCHE KRESSE (*f*)
Tropæolum peregrinum	081	Canary creeper (See *Tropæolum aduncum*)	
Tropæolum polyphyllum	082	Yellow rock Indian cress	VIELSPALTIGE KAPUZINERKRESSE (*f*); VIELBLÄTTRIGE KAPUZIN--ERKRESSE (*f*)
Tropæolum speciosum	083	Flame-flowered nasturtium, flame flower	SECHSLAPPIGE KAPUZINERKRESSE (*f*)
Tsuga	084	Hemlock; hemlock spruce	HEMLOCKTANNE (*f*); SCHIER--LINGSTANNE (*f*)
Tsuga americana	085	(See *Tsuga canadensis*)	
Tsuga araragi	086	(See *Tsuga sieboldi*)	
Tsuga canadensis	087	Canadian hemlock; hemlock spruce (See *Tsuga americana* and *Abies canadensis*)	KANADISCHE SCHIERLINGSTANNE (*f*)
Tsuga caroliniana	088	Caroline hemlock	KAROLINEN - SCHIERLINGSTANNE (*f*)
Tsuga diversifolia	089	Variegated hemlock	VERSCHIEDENBLÄTTRIGE SCHIER--LINGSTANNE (*f*)
Tsuga mertensiana	090	Merten's hemlock	MERTENS-SCHIERLINGSTANNE (*f*)
Tsuga pattoniana	091	Patton's hemlock, Californian hemlock spruce	PATTONS-SCHIERLINGSTANNE (*f*)
Tsuga sieboldi	092	Araragi hemlock, Japanese hem-lock spruce (See *Tsuga araragi*)	ARARAGI-SCHIERLINGSTANNE (*f*)
Tuber æstivum	093	(Edible)-truffle (See *Eutuber*)	TRÜFFEL (*f*); NUSSPILZ (*m*); SOMMERTRÜFFEL (*f*); DEUTSCHE TRÜFFEL (*f*)
Tuberaria guttata	094	(See *Helianthemum guttatum*)	
Tubercularia vulgaris	095	(See *Nectria cinnabarina*)	
Tuber excavatum	096	Non-edible truffle; hollow wood truffle (See *Aschion*)	HOHLTRÜFFEL (*f*); HOLZTRÜFFEL (*f*)
Tulipa	097	Tulip	TULPE (*f*)
Tulipa acuminata	098	Turkish tulip	TÜRKISCHE TULPE (*f*)
Tulipa sylvestris	099	Wild tulip; yellow-flowered tulip	WALDTULPE (*f*)
Tunica prolifera	100	(See *Dianthus prolifer*)	
Tussilago farfara	101	Coltsfoot	ROSSHUF (*m*); HUFLATTIG (*m*); HUFLATTICH (*m*); BRUSTLAT--TICH (*m*); HIRSCHLATTIG (*m*)
Tussilago petasites	102	Butterbur (See *Petasites vulgaris*)	PESTWURZ (*f*); GIFTWURZ (*f*); SCHWEISSWURZ (*f*); NEUN--KRAFTWURZ (*f*)

Botanical Term.	No.	English Term.	German Term.
Typha	103	Reed-mace ; club rush	ROHRKOLBENSCHILF (*m* and *n*); KOLBENSCHILF (*m* and *n*); ROHRKOLBEN (*m*)
Typha angustifolia	104	Lesser reed-mace ; lesser cat's-tail	SCHMALBLÄTTRIGER ROHRKOLBEN--SCHILF (*m*); SCHMALBLÄT--TRIGES ROHRKOLBENSCHILF (*n*)
Typha latifolia	105	Cat's tail ; club-rush ; cat-o'-nine-tails ; great cat's-tail ; great reed-mace (erroneously termed " bullrush ")	BREITBLÄTTRIGER ROHRKOLBEN--SCHILF (*m*); BREITBLÄT--TRIGES ROHRKOLBENSCHILF (*n*)
Ulex compositus	106	(See *Ulex europæus*)	
Ulex europæus	107	Gorse ; whin ; (spring)-furze ; goss ; common furze ; Irish furze	HEIDEBUSCH (*m*); GEMEINER HECKSAME (*m*); GASPELDORN (*m*); STECHGINSTER (*m*)
Ulex europæus flore pleno	108	Double furze	DOPPELBLÜTIGER HECKSAME (*m*)
Ulex nanus	109	Autumnal furze ; dwarf furze	KLEINER HECKSAME (*m*)
Ulex strictus	110	Irish furze	AUFRECHTER HECKSAME (*m*)
Ulmus	111	Elm	ULME (*f*); KORKULME (*f*); RÜSTER (*f*)
Ulmus alata	112	Wahoo elm, winged elm	FLÜGELULME (*f*)
Ulmus americana	113	American elm (See *Ulmus fulva*)	FUCHSULME (*f*); AMERIKANISCHE ULME (*f*); WEISSULME (*f*)
Ulmus campestris	114	Elm, common elm, English elm, small-leaved elm	FELDRÜSTER (*f*); ULME (*f*); ROTRÜSTER (*f*); ROTULME (*f*); KORKULME (*f*); FELDULME (*f*); GEMEINE ULME (*f*); RÜSTER (*f*); LEINBAUM (*m*); IPER (*m*)
Ulmus campestris pendula	115	Weeping elm	TRAUERULME (*f*)
Ulmus campestris purpurea	116	Purple-leaved-(common)-elm	ROTBLÄTTRIGE FELDULME (*f*)
Ulmus campestris suberosa	117	(See *Ulmus suberosa*)	
Ulmus campestris variegata	118	Variegated elm	VERSCHIEDENBLÄTTRIGE FELD--ULME (*f*)
Ulmus crassifolia	119	Thick-leaved elm	DICKBLÄTTRIGE ULME (*f*)
Ulmus effusa	120	Soft leaved elm (See *Ulmus pedunculata*)	BASTRÜSTER (*f*); FLATTERULME (*f*); FLATTERRÜSTER (*f*)
Ulmus fulva	121	Slipper-elm (See *Ulmus americana*)	FUCHSULME (*f*)
Ulmus glabra	122	Wych elm (See *Ulmus montana* and *Ulmus nitens*)	GLATTBLÄTTRIGE ULME (*f*)
Ulmus glabra cornubiensis	123	Cornish elm (See *Ulmus campestris*)	
Ulmus glabra pendula	124	Weeping wych elm	HÄNGENDE HASELRÜSTER (*f*)
Ulmus major	125	Dutch elm (See *Ulmus campestris*)	
Ulmus microphylla	126	(See *Ulmus pumila*)	ZWERG-ULME (*f*)
Ulmus montana	127	Mountain elm ; wych elm ; Scotch elm (See *Ulmus scabra*)	BERGRÜSTER (*f*); BERGULME (*f*); HASELRÜSTER (*f*)
Ulmus montana vegeta	128	Huntingdon elm	HUNTINGDON-ULME (*f*)
Ulmus nitens	129	Smooth-leaved elm (See *Ulmus glabra*)	GLATTBLÄTTRIGE ULME (*f*)
Ulmus parvifolia	130	Small-leaved elm	KLEINBLÄTTRIGE ULME (*f*)
Ulmus pedunculata	131	(See *Ulmus effusa*)	FLATTERULME (*f*)
Ulmus polygana	132	(See *Zelkowa crenata*)	SIBERISCHE ULME (*f*)
Ulmus pumila	133	Dwarf elm (See *Ulmus microphylla*)	ZWERGULME (*f*)
Ulmus racemosa	134	Racemose elm	TRAUBENULME (*f*)
Ulmus scabra	135	(See *Ulmus montana*)	BERGRÜSTER (*f*)
Ulmus scabra atropurpurea	136	Purple-leaved elm ; purple-leaved mountain elm	ROTBLÄTTRIGE ULME (*f*); ROT--BLÄTTRIGE BERGRÜSTER (*f*)
Ulmus stricta	137	Cornish elm (See *Ulmus campestris*)	AUFRECHTE ULME (*f*)
Ulmus suberosa	138	Elm ; cork-barked elm (See *Ulmus campestris suberosa*)	FELDRÜSTER (*f*); ULME (*f*); ROTRÜSTER (*f*); ROTULME (*f*); KORKULME (*f*)
Ulva lactuca	139	Sea lettuce	MEERLATTICH (*m*)

Botanical Term.	No.	English Term.	German Term.
Ulvina aceti	140	(See *Mycoderma aceti*)	
Umbelliferæ	141	Umbelliferæ (umbelliferous plants)	DOLDENBLUMEN (*f.pl*)
Uncaria gambir	142	Gambir (gambier) shrub	GAMBIRPFLANZE (*f*)
Uncinia	143	Hooked sedge	HAKEN-SEGGE (*f*)
Uncinula	144	blight	HAKENMELTAU (*m*)
Uncinula adunca	145	Willow blight	WEIDEN-HAKENMELTAU (*m*)
Uncinula bicornis	146	Maple blight	AHORN-HAKENMELTAU (*m*)
Uncinula wallrothii	147	Sloe blight	SCHLEHEN-HAKENMELTAU (*m*)
Uniola	148	Sea oat	UNIOLA (*f*)
Uragoga ipecacuanha	149	(See *Psychotria ipecacuanha*)	
Urceolina aurea	150	Golden urn-flower (See *Urceolina pendula*)	GOLDENE KRUGBLUME (*f*)
Urceolina pendula	151	Drooping urn-flower (See *Urceolina aurea*)	HÄNGENDE KRUGBLUME (*f*)
Uredo	152	Uredo, rust	ROSTPILZ (*m*)
Uredo betulina	153	(See *Lecythea caprearum*)	
Uredo filicum	154	Fern rust	FARNROST (*m*)
Uredo palmarum	155	Palm rust	PALMENROST (*m*)
Uredo quercus	156	Oak-leaf rust	EICHENBLATTROST (*m*)
Urginea maritima	157	Squill; sea-onion (See also *Scilla maritima*)	MEERZWIEBEL (*f*)
Urocystis	158	Smut	BRAND (*m*)
Urocystis occulta	159	(See *Ustilago secalis*)	ROGGENSTENGELBRAND (*m*)
Urocystis parallela	160	(See *Ustilago secalis*)	
Urocystis violæ	161	Violet smut	VEILCHENSTENGELBRAND (*m*)
Uromyces	162	Rust	ROST (*m*); SCHWANZROST (*m*)
Uromyces alliorum	163	Garlic rust	LAUCH-SCHWANZROST (*m*)
Uromyces apiculata	164	Clover rust (See *Uromyces trifolii*)	
Uromyces appendiculatus	165	Bean rust; long-stemmed rust	BOHNEN-SCHWANZROST (*m*)
Uromyces betæ	166	Beet rust	ZUCKERRÜBEN-SCHWANZROST (*m*)
Uromyces concentrica	167	Hyacinth rust	HYAZINTHEN-SCHWANZROST (*m*)
Uromyces fabæ	168	Horse-bean rust	SAUBOHNEN-SCHWANZROST (*m*)
Uromyces geranii	169	Geranium rust	GERANIUM-SCHWANZROST (*m*)
Uromyces graminum	170	Cock's-foot rust	HAHNENFUSS-SCHWANZROST (*m*)
Uromyces pisi	171	Pea rust	ERBSEN-SCHWANZROST (*m*)
Uromyces polygoni	172	Corn-spurrey rust; polygonum rust (See *Trichobasis polygonorum*)	KNÖTERICH-SCHWANZROST (*m*)
Uromyces scrophulariæ	173	Fig-wort rust	KNOTENWURZ-SCHWANZROST (*m*)
Uromyces striatus	174	Saintfoin rust	ESPARSETTEN-SCHWANZROST (*m*)
Uromyces trifolii	175	Clover rust (See *Trichobasis fallens*)	KLEE-SCHWANZROST (*m*)
Urostigma	176	(See *Ficus urostigma*)	
Urtica	177	Nettle; stinging nettle	NESSEL (*f*); BRENNESSEL (*f*)
Urtica bœhmeria	178	(See *Urtica nivea*)	
Urticaceæ	179	Urticaceæ (See *Urtica* and *Parietaria*)	NESSELPFLANZEN (*f.pl*)
Urtica dioica	180	Great nettle; common nettle	NESSEL (*f*); GROSSE BRENNESSEL (*f*)
Urtica nivea	181	Chinese nettle; ramie (See *Bœhmeria nivea*)	CHINAGRAS (*n*); RAMIE (*f*)
Urtica pilulifera	182	Roman nettle	PILLENTRAGENDE BRENNESSEL (*f*); PILLENNESSEL (*f*); KUGEL-NESSEL (*f*)
Urtica urens	183	Small nettle	KLEINE BRENNESSEL (*f*)
Usnea barbata	184	Beard moss	BARTFLECHTE (*f*); GREISENBART-FLECHTE (*f*)
Ustilago	185	Smut, dust-brand	BRAND (*m*); BRANDPILZ (*m*)
Ustilago antherarum	186	Anther smut	ANTHERENSTAUBBRAND (*m*)
Ustilago avenæ	187	Oat smut	HAFERBRAND (*m*); HAFERSTAUB-BRAND (*m*); HAFERRUSS-BRAND (*m*); ECHTER HAFER-BRANDPILZ (*m*)
Ustilago carbo	188	Smut; burnt ear	RUSSBRAND (*m*)
Ustilago cardui	189	Thistle brand	DISTEL-STAUBBRAND (*m*); DISTEL-BRAND (*m*)
Ustilago grammica	190	Branded smut (See *Ustilago segetum*)	

Botanical Term.	No.	English Term.	German Term.
Ustilago hordei	191	Barley brand ; corn smut (See *Ustilago segetum*)	GERSTENBRAND (*m*); GERSTEN-STAUBBRAND (*m*)
Ustilago hypodytes	192	Grass smut	GRAS-STAUBBRAND (*m*)
Ustilago maydis	193	Maize brand ; Indian corn smut	BEULENBRAND (*m*); MAIS-BEU-LENBRAND (*m*)
Ustilago receptaculorum	194	Goatsbeard smut	BOCKSBART-STAUBBRAND (*m*)
Ustilago sacchari	195	Sugar-cane brand	ZUCKERROHR-STAUBBRAND (*m*)
Ustilago scabiosæ	196	Scabious brand	KNOPFKRAUT-STAUBBRAND (*m*)
Ustilago secalis	197	Rye smut, ustilago (See *Urocystis parallela* and *Urocystis occulta*)	ROGGENKORNBRAND (*m*); ROGGEN-BRAND (*m*)
Ustilago segetum	198	Barley smut, blackball, bunt-ear, dust-brand, corn smut (See *Ustilago hordei*)	RUSSBRAND (*m*); STAUBBRAND (*m*)
Ustilago sorghi	199	Millet brand	SORGHUMHIRSEN-STAUBBRAND (*m*)
Ustilago tragopogonis pratensis	200	Goat's-beard brand (See *Puccinia sparsa*)	BOCKSBART-STAUBBRAND (*m*)
Ustilago tritici	201	Wheat smut (See also *Tilletia caries*)	WEIZENBRAND (*m*)
Utricularia intermedia	202	Intermediate bladderwort	ZWISCHENFÖRMIGER WASSER-SCHLAUCH (*m*)
Utricularia minor	203	Smaller bladderwort; lesser bladderwort	KLEINER WASSERHELM (*m*)
Utricularia vulgaris	204	Greater bladderwort; common bladderwort	WASSERSCHLAUCH (*m*); WASSER-SCHLANGE (*f*); GEMEINER WASSERHELM (*m*)
Uvularia	205	Bell-wort	ZÄPFCHENKRAUT (*n*)
Vaccaria parviflora	206	cow-herb	KUHNELKE (*f*); KUHKRAUT (*n*)
Vaccinium	207	Whortleberry; bilberry; cranberry (See also *Empetrum*)	HEIDELBEERE (*f*); PREISSEL-BEERE (*f*); PICKELBEERE (*f*); RAUSCHBEERE (*f*)
Vaccinium macrocarpus	208	American cranberry (See *Oxycoccus macrocarpus*)	NORDAMERIKANISCHE MOOSBEERE (*f*)
Vaccinium myrtillus	209	Bilberry; whortleberry; blaeberry; blueberry	KRANBEERE (*f*); KRONBEERE (*f*); HEIDELBEERE (*f*); JUGEL-BEERE (*f*); BLAUBEERE (*f*); GROSSEL (*f*); MOSTJÖCKE (*f*)
Vaccinium oxycoccos	210	Cranberry; bog-berry, crowberry (See *Oxycoccus palustris*)	MOOSBEERE (*f*); KRANBEERE (*f*); KRANICHBEERE (*f*); TORF-SÜMPFE (*f*); RAUSCHBEERE (*f*); KRONSBEERE (*f*)
Vaccinium resinosum	211	Black whortleberry; huckleberry	SCHWARZE RAUSCHBEERE (*f*); SCHWARZE HEIDELBEERE (*f*)
Vaccinium uliginosum	212	Bog whortleberry	JUGELBEERE (*f*); KANNLBEERE (*f*); RAUSCHBEERE (*f*); TRUN-KELBEERE (*f*)
Vaccinium vitus-idæa	213	Cowberry; red wortleberry; low-bush cranberry, flowering box	PREISSELBEERE (*f*); PREISEL-BEERE (*f*)
Valeriana celtica	214	Nard, alpine valerian	NARDE (*f*); KELTISCHE NARDE (*f*); NARDENBALDRIAN (*m*); ALPENBALDRIAN (*m*); SPEIK-NARDE (*f*); SPEIK (*m*)
Valeriana dioica	215	Marsh valerian	SUMPFBALDRIAN (*m*)
Valeriana montana	216	Red valerian (See *Centranthus ruber* and *Valeriana ruber*)	ROTER BALDRIAN (*m*); ROTE SPORNBLUME (*f*)
Valeriana officinalis	217	Valerian; all-heal; common valerian; great wild valerian	BALDRIAN (*m*); KATZENWURZ (*f*); THERIAKWURZ (*f*)
Valeriana Phu aurea	218	Cretan spikenard	GROSSER BALDRIAN (*m*)
Valeriana pyrenaica	219	Pyrenean valerian	PYRENAISCHER BALDRIAN (*m*)
Valeriana ruber	220	(See *Valeriana montana*)	
Valerianella auricula	221	Sharp-fruited corn-salad	ACKERSALAT (*m*); FELDSALAT (*m*); SCHARFFRÜCHTIGES RAPÜNZ-CHEN (*n*)
Valerianella carinata	222	Lambs' lettuce; keeled corn-salad	RAPUNZEL (*f*); RAPÜNSCHEN (*n*) KIELFÖRMIGES RAPÜNZCHEN (*n*)

Botanical Term.	No.	English Term.	German Term.
Valerianella dentata	223	Narrow-fruited corn-salad	SCHMALFRÜCHTIGES RAPÜNZCHEN (*n*)
Valerianella olitoria	224	Corn salad, common corn-salad; lamb's-lettuce	RAPÜNZELCHEN (*n*); RABINZCHEN (*n*); RAPUNZEL (*f*); RAPÜNZCHEN (*n*); RAPÜNSCHEN (*n*); ACKERSALAT (*m*); FELDSALAT (*m*)
Vallisneria spiralis	225	Tape grass; eel grass	ROSETTBLÄTTRIGER SCHRAUBEN-STENGEL (*m*)
Vallota purpurea	226	Scarborough lily	VALLOTE (*f*); KAP-VALLOTE (*f*)
Vanda	227	Cowslip-scented orchid	SCHLÜSSELBLUMENDUFTENDE ORCHE (*f*)
Vanilla phalænopsis	228	Madagascar vanilla	MADAGASKAR-VANILLE (*f*)
Vanilla planifolia	229	Vanilla-(plant); vanilla orchid (See also *Heliotropium*)	VANILLENPFLANZE (*f*)
Variolaria lecanora	230	Litmus lichen, archil (See *Variolaria rocella*)	LACKMUSFLECHTE (*f*)
Variolaria rocella	231	Litmus lichen, archil (See *Variolaria lecanora* and *Roccella tinctoria*)	LACKMUSFLECHTE (*f*)
Vateria indica	232	Copal tree	KOPALBAUM (*m*)
Veratrum album	233	White hellebore (See *Veratrum viride*)	WEISSE NIESWURZ (*f*)
Veratrum nigrum	234	Black hellebore	SCHWARZE NIESWURZ (*f*)
Veratrum viride	235	White hellebore; green hellebore; American hellebore; Indian poke; American veratrum (See also *Helleborus viridis*)	GERMER (*m*); NIESWURZ (*f*); NIESWURZEL (*f*); GRÜNE NIES-WURZ (*f*); SCHAMPANIER-WURZEL (*f*)
Verbascum	236	(Common)-mullen, mullein	KÖNIGSKERZE (*f*); WOLLKRAUT (*n*); FACKELKRAUT (*n*)
Verbascum blattaria	237	Moth mullein	MOTTEN-WOLLKRAUT (*n*); SCHABENKRAUT (*n*)
Verbascum chaixii	238	Nettle-leaved mullein	NESSELBLÄTTRIGES WOLLKRAUT (*n*)
Verbascum floccosum	239	Hoary mullein	WOLLBLUME (*f*); WOLLKRAUT (*n*)
Verbascum lychnitis	240	White mullein	WEISSES WOLLKRAUT (*n*)
Verbascum nigrum	241	Dark mullein; black mullein	SCHWARZES WOLLKRAUT (*n*)
Verbascum olympicum	242	Olympian mullein	
Verbascum phlomoides	243	(See *Verbascum thapsus*)	
Verbascum phœniceum	244	Purple mullein	VIOLETTES WOLLKRAUT (*n*)
Verbascum pulverulentum	245	Hoary mullein	GRAUES WOLLKRAUT (*n*)
Verbascum thapsus	246	Great mullein, Aaron's rod	KÖNIGSKERZE (*f*); GROSSES WOLLKRAUT (*n*)
Verbascum virgatum	247	Twiggy mullein	GEZWEIGTES WOLLKRAUT (*n*)
Verbena aubletia	248	Rose vervain	EISENKRAUT (*n*)
Verbena officinalis	249	Vervain, verbena, vervein (See *Aloysia citriodora* and *Lippia citriodora*)	EISENKRAUT (*n*); EISENHART (*n*); VERBENE (*f*)
Vernonia	250	Ironweed	EISENKRAUT (*n*)
Veronica agrestis	251	Green field speedwell; procumbent speedwell	FELD-EHRENPREIS (*m*)
Veronica alpina	252	Alpine speedwell	ALPEN-EHRENPREIS (*m*)
Veronica anagallis	253	Water speedwell	GROSSE BACHBUNGE (*f*)
Veronica arvensis	254	Wall speedwell	ACKER-EHRENPREIS (*m*)
Veronica beccabunga	255	Brooklime	BACHBUNGE (*f*)
Veronica buxbaumii	256	Buxbaum's speedwell	BUXBAUMS-EHRENPREIS (*m*)
Veronica chamædrys	257	Germander speedwell; bird's eye	GAMANDER-EHRENPREIS (*m*); FRAUENBISS (*m*)
Veronica hederifolia	258	Ivy-leaved speedwell	EPHEUBLÄTTRIGER EHRENPREIS (*m*)
Veronica montana	259	Mountain speedwell	BERG-EHRENPREIS (*m*)
Veronica officinalis	260	Common speedwell veronica-(herb); decumbent speedwell	EHRENPREIS (*m*); MUNDKRAUT (*n*); MÄNNERTREU (*f*)
Veronica polita	261	Grey field speedwell	GLÄNZENDER EHRENPREIS (*m*)
Veronica saxatilis	262	Rock speedwell	FELSEN-ENRENPREIS (*m*)
Veronica scutellata	263	Marsh speedwell	SUMPF-EHRENPREIS (*m*)
Veronica serpyllifolia	264	Thyme-leaved speedwell	THYMIANBLÄTTRIGER EHREN-PREIS (*m*)
Veronica spicata	265	Spiked speedwell	

Botanical Term.	No.	English Term.	German Term.
Veronica triphyllos	266	Fingered speedwell	GEFINGERTER EHRENPREIS (*m*)
Veronica verna	267	Vernal (veronica) speedwell	FRÜHLINGS-EHRENPREIS (*m*)
Viborgia purpurea	268	Red broom (See *Cytisus purpureus*)	ROTER BOHNENSTRAUCH (*m*)
Viburnum lantana	269	Mealy Guelder-rose; wayfaring tree (See *Viburnum lantanoides*)	KOTSCHLINGE (*f*); WOLLIGER SCHNEEBALL (*m*); KANDEL--BEERE (*f*) TÜRKISCHE WEIDE (*f*)
Viburnum lantanoides	270	Hobble bush; American wayfaring tree (See *Viburnum lantana*)	SCHWINDELBEERBAUM (*m*); SCHIESSBEERSTRAUCH (*m*)
Viburnum opulus	271	Common Guelder-rose; snowball tree; dog rowan tree; wild guelder rose, water elder, cramp bark tree, high cranberry bush, cranberry tree squaw bush	GEMEINER SCHNEEBALL (*m*); HIRSCHHOLUNDER (*m*); ROSEN--HOLDER (*m*); WASSERAHORN (*m*); WASSERHOLDER (*m*); TÜRKISCHES PFEIFENHOLZ--BAUM (*m*); KALINKENHOLZ--BAUM (*m*)
Viburnum opulus sterilis	272	Double guelder rose; garden guelder rose; garden snowball tree	DOPPELBLÜTIGER SCHNEEBALL (*m*)
Viburnum prunifolium	273	American black haw, sheep-berry, sloe-leaved viburnum, sweet viburnum, stag bush	PFLAUMENBLÄTTRIGER SCHNEE--BALL (*m*)
Viburnum tinus	274	Laurustinus-(shrub); Laurestinus	SCHLINGE (*f*)
Vicia	275	Tares, wild vetch	KORNWICKE (*f*); WICKE (*f*)
Vicia angustifolia	276	Small-leaved vetch	SCHMALBLÄTTRIGE WICKE (*f*); FELDWICKE (*f*)
Vicia bithynica	277	Bithynian vetch	BITHYNISCHE WICKE (*f*)
Vicia cracca	278	Tufted vetch	VOGELWICKE (*f*)
Vicia dumetorum	279	Hedge vetch	HECKEN-WICKE (*f*)
Vicia ervilia	280	Ervilia vetch; lentil vetch	LINSENWICKE (*f*); KNOTEN--FRÜCHTIGE WICKE (*f*)
Vicia faba	281	Horse bean; broad bean (See *Faba vulgaris*)	FEIGENBOHNE (*f*); SAUBOHNE (*f*); PFERDEBOHNE (*f*); PUFF--BOHNE (*f*)
Vicia gracilis	282	Slender vetch	SCHLANKE WICKE (*f*)
Vicia hirsuta	283	Hairy tare; hairy vetch	HAARIGE WICKE (*f*); ZITTER--LINSE (*f*)
Vicia lathyroides	284	Spring vetch	PLATTERBSENARTIGE WICKE (*f*)
Vicia lutea	285	Yellow vetch	GELBE WICKE (*f*)
Vicia monanthos	286	Single-flowered vetch	EINBLÜTIGE WICKE (*f*)
Vicia narbonensis	287	Narbonne vetch	SCHEERERBSE (*f*); MANS-WICKE (*f*)
Vicia orobioides	288	(See *Vicia orobus* and *Orobus lathyroides*)	
Vicia orobus	289	Bitter vetch, upright vetch (See *Orobus*)	ERVE (*f*)
Vicia sativa	290	Common vetch; tares; wild vetch	FUTTERWICKE (*f*); KORNWICKE (*f*); HONIGWICKE (*f*); WICKE (*f*)
Vicia sepium	291	Bush vetch	ZAUN-WICKE (*f*)
Vicia sylvatica	292	Wood vetch	WALD-WICKE (*f*)
Vicia tetrasperma	293	Smooth tare; slender vetch	VIERSAMIGE WICKE (*f*)
Vicia villosa	294	Sand vetch	SAND-WICKE (*f*); PELLUSCHKE (*f*)
Victoria regia	295	Queen Victoria water lily; royal water lily; water platter (See *Euryale amazonica*)	KÖNIGLICHE WASSERLILIE (*f*)
Villarsia nymphæoides	296	(See *Limnanthemum peltatum*)	ICACINACEE (*f*)
Vinca	297	Periwinkle, evergreen	IMMERGRÜN (*n*); WINTERGRÜN (*n*)
Vinca major	298	Greater periwinkle	GROSSES SINGRÜN (*n*)
Vinca major variegata	299	Variegated greater periwinkle	VERSCHIEDENBLÄTTRIGES GROSSES SINGRÜN (*n*)
Vinca minor	300	Lesser periwinkle	IMMERGRÜN (*n*); KLEINES SIN--GRÜN (*n*)
Vinca rosea	301	Madagascar periwinkle; old maid	ROSENROTES SINGRÜN (*n*)

Botanical Term.	No.	English Term.	German Term.
Vincetoxicum officinale	302	White swallowwort, contrayerva (See *Cynanchum vincetoxicum*)	GIFTWENDE (*f*); GIFTWURZEL (*f*)
Viola alba	303	White-flowered violet; white violet	WEISSBLÜTIGES VEILCHEN (*n*)
Viola altaica	304	Altaian violet	ALTAI-VEILCHEN (*n*)
Viola ambigua	305	Ambiguous violet	ZWEIFELHAFTES VEILCHEN (*n*)
Viola arenaria	306	Sand violet	SAND-VEILCHEN (*n*)
Viola biflora	307	Twin-flowered violet	ZWEIBLÜTIGES VEILCHEN (*n*)
Viola calcarata	308	Spurred violet	LANGSPORNIGES VEILCHEN (*n*); BERGVIOLA (*f*)
Viola canina	309	Dog violet; wood violet	HUNDSVEILCHEN (*n*)
Viola collina	310	Hill violet	HÜGEL-VEILCHEN (*n*)
Viola cornuta	311	Horned violet	HORN-VEILCHEN (*n*)
Viola cucullata	312	Hollow-leaved violet	LÖFFELBLÄTTERIGES VEILCHEN (*n*)
Viola epipsila	313	Peat violet	TORF-VEILCHEN (*n*)
Viola gracilis	314	Olympian violet	SCHLANKES VEILCHEN (*n*)
Viola hirta	315	Hairy violet	RAUHHAARIGES VEILCHEN (*n*)
Viola lutea	316	Mountain violet	GELBES VEILCHEN (*n*)
Viola munbyana	317	Munby's violet	MUNBYS-VEILCHEN (*n*)
Viola odorata	318	Sweet violet; violet	MÄRZVEILCHEN (*n*); WOHLRIE--CHENDES VEILCHEN (*n*)
Viola odorata pallida plena	319	Neapolitan violet	BLASSES WOHLRIECHENDES VEIL--CHEN (*n*)
Viola palustris	320	Marsh violet	SUMPF-VEILCHEN (*n*)
Viola pedata	321	Bird's-foot violet	FUSSBLÄTTERIGES VEILCHEN (*n*)
Viola porphyrea	322	Porphyry violet	ROTBRAUNES VEILCHEN (*n*)
Viola rothamagensis	323	Rouen violet	ROUEN-VEILCHEN (*n*)
Viola sylvestris	324	Wood violet	WALDVEILCHEN (*n*)
Viola tricolor	325	Pansy; heartsease; Heart's-ease; field pansy; garden pansy	DREIFARBIGES VEILCHEN (*n*)
Viola uliginosa	326	Swamp violet	MOOR-VEILCHEN (*n*)
Virgilia lutea	327	Cape yellow-wood tree (See *Cladrastis tinctoria*)	KAP-GELBHOLZBAUM (*m*)
Viscaria alba	328	(See *Lychnis viscaria*)	PECHNELKE (*f*); MÜCKENFANG (*m*)
Viscum album	329	Mistletoe; common mistletoe; white-berried mistletoe (See also *Loranthus*)	KREUZHOLZ (*n*); MISTEL (*f*); WEISSE MISTEL (*f*)
Viscum quercinum	330	Oak mistletoe	EICHENMISTEL (*f*)
Vismia guianensis	331	Wax-tree; American gutta-gum tree; American gamboge	GUAYANA-GUMMIGUTTBAUM (*m*)
Visnea mocanera	332	Mocan tree; Canary Islands mocan tree	MOKANBAUM (*m*); VISNEA (*f*)
Vitex Agnus castus	333	Monk's pepper tree; tree of chastity; chaste tree	KEUSCHBAUM (*m*); SCHAFMÜLLEN (*m*); MÖNCHSPFEFFER (*m*)
Vitis æstivalis	334	North American vine; American summer vine	AMERIKANISCHE SOMMERREBE (*f*); NORD-AMERIKANISCHE REBE (*f*)
Vitis antarctica	335	Kangaroo vine, wild grape (See *Cissus antarctica*)	KLIMME (*f*)
Vitis apiifolia	336	Parsley-leaved vine (See *Vitis laciniosa*)	
Vitis arborea	337	Tree vine	BAUMREBE (*f*)
Vitis californica	338	Californian vine	KALIFORNISCHE REBE (*f*)
Vitis coignetiæ	339	Large-leaved Japanese vine	COIGNET-REBE (*f*)
Vitis cordifolia	340	Winter vine	WINTERREBE (*f*)
Vitis heterophylla humulifolia	341	Turquoise berry vine (See *Ampelopsis heterophylla*)	VERSCHIEDENBLÄTTRIGE DOLDEN--REBE (*f*)
Vitis inconstans	342	(See *Ampelopsis Veitchii*)	DREISPITZIGE DOLDENREBE (*f*)
Vitis labrusca	343	American fox grape	FUCHSREBE (*f*)
Vitis laciniosa	344	Parsley-leaved Vine (See *Vitis apiifolia*)	PETERSILIENWEIN (*m*)
Vitis purpurea	345	purple-leaved vine	PURPURBLÄTTRIGE DOLDENREBE (*f*)
Vitis quinquefolia	346	(See *Ampelopsis quinquefolia*) and also *Quinaria* and *Parthenocissus quinquefolia*)	GEMEINE DOLDENREBE (*f*); WIL--DER WEIN (*m*)
Vitis vinifera	347	Vine; grape vine	REBE (*f*); WEINREBE (*f*)

Botanical Term.	No.	English Term.	German Term.
Volkameria fragrans	348	(See *Clerodendron fragrans* and *Clerodendron siphonanthus*)	VOLKAMERIE (*f*); VOLKMANNIE (*f*)
Volvocineæ	349	(See *Volvox*)	WIMPERALGEN (*f.pl*)
Volvox globator	350	Volvox alga	WIMPERALGE (*f*)
Vossianthus	351	(See *Sparmannia africana*)	
Vriesia	352	(See *Tillandsia usneoides*)	
Wahlenbergia hederacea	353	Ivy-leaved bell-flower; creeping harebell (See *Campanula hederacea*)	EFEUBLÄTTRIGE WAHLENBERGIE (*f*)
Wahlenbergia saxicola	354	New Zealand bell-flower	WAHLENBERGIE (*f*); NEUSEE--LÄNDISCHE WAHLENBERGIE (*f*)
Waitzia aurea	355	Yellow-(Australian)-everlasting flower (See *Helipterum*)	GELBE WAITZIA (*f*)
Waldsteinia fragarioides	356	Barren strawberry	FÜNFLAPPIGE WALDSTEINIE (*f*)
Washingtonia	357	(See *Sequoia gigantea*)	
Watsonia rosea	358	(Rose)-bugle lily (See *Gladiolus pyramidatus*)	ROSENROTE WATSONIE (*f*)
Weingærtneria	359	(See *Aira canescens*)	
Wellingtonia gigantea	360	Wellingtonia, giant redwood (See *Sequoia gigantea*)	WELLINGTONIE (*f*); MAMMUT--BAUM (*m*)
Widdringtonia Whytei	361	Milanji cypress, African cypress	WIDDRINGTONIE (*f*); AFRIKA--NISCHE ZYPRESSE (*f*)
Winterana canella	362	(See *Canella alba*)	
Wistaria	363	Wistaria	WISTARIE (*f*)
Wistaria chinensis	364	Chinese kidney bean tree	CHINESISCHE WISTARIE (*f*)
Wistaria frutescens	365	Bush wistaria, American kidney bean tree	STRAUCHWISTARIE (*f*)
Wistaria polystachya	366	(See *Wistaria chinensis*)	
Wistaria sinensis	367	(See *Wistaria chinensis*)	
Wolfia arrhiza	368	(See *Lemna arrhiza*)	
Woodwardia radicans	369	Chain fern	(WURZELNDER)-GRÜBCHENFARN (*m*)
Xanthium strumarium	370	Burweed; button-bur	KNOPFKLETTE (*f*)
Xanthoria parietina	371	Wall moss (See *Theloschistes*)	WANDFLECHTE (*f*)
Xanthorrhœa australis	372	Grass-gum tree; grass tree	NUTTHARZBAUM (*m*)
Xanthoxylum americanum	373	Prickly ash, yellow-wood, angelica tree, tooth-ache tree (See *Zanthoxylum* and *Fagara clavaherculis*)	GELBHOLZ (*n*)
Xenodochus carbonarius	374	Burnet-chain brand	BRAND (*m*)
Xeranthemum annuum	375	Annual everlasting, immortelle	IMMORTELLE (*f*); PAPIERBLUME (*f*)
Xerophyllum asphodeloides	376	Turkey's beard	DÜRRPFLANZE (*f*)
Xylopia æthiopica	377	Ethiopian xylopia; negro pepper (See *Xylopia aromatica*)	NEGERPFEFFER (*m*); KUMBA--PFEFFER (*m*); MOHRENPFEFFER (*m*)
Xylopia aromatica	378	(See *Xylopia æthiopica*)	
Yucca	379	Yucca, dagger-plant	JUKKA (*f*); PALMLILIE (*f*); YUKKA (*f*)
Yucca aloifolia	380	Eve's thread	ALOENBLÄTTRIGE PALMLILIE (*f*)
Yucca filamentosa	381	Adam's needle	FADENTRAGENDE PALMLILIE (*f*)
Yucca gloriosa	382	Yucca	MONDBLUME (*f*); PRACHTALOE (*f*); SCHÖNE PALMLILIE (*f*)
Zamia purpuracea	383	Jamaica sago tree	STUTZSCHUPPEN-PALMFARN (*m*)
Zannichellia palustris	384	Horned pondweed	SUMPF-FADENBLATT (*n*)
Zantedeschia æthiopica	385	(See *Richardia africana*)	
Zanthoxylum	386	Prickly ash (See *Xanthoxylum americanum*)	ZAHNWEHBAUM (*m*)
Zanthoxylum pterota	387	(See *Fagara pterota* and *Xanthoxylum*)	
Zauschneria californica	388	Californian fuchsia	KALIFORNISCHE FUCHSIA (*f*); ZAUSCHNERIA (*f*)

Botanical Term.	No.	English Term.	German Term.
Zea mays	389	Indian corn ; maize	TÜRKISCHER WEIZEN (*m*) ; TÜR--KISCHES KORN (*n*) ; MAIS (*m*)
Zelkowa crenata	390	Siberian elm, water elm (See *Ulmus polygana*)	ZELKOWE (*f*)
Zephyranthes atamasco	391	Atamasco lily	ATAMASCO-ZEPHIRBLUME (*f*)
Zephyranthes candida	392	Peruvian swamp lily, zephyr flower, flower of the west wind	ZEPHIRBLUME (*f*)
Zingiber officinale	393	Ginger	INGWER (*m*)
Zinnia	394	Youth-and-old-age	ZINNIE (*f*)
Zizania aquatica	395	Water oats, water rice, Indian rice, Canadian rice	HAFERREIS (*m*) ; WASSERHAFER (*m*) ; WASSERREIS (*m*)
Zizyphus jujuba vulgaris	396	Jujube	JUDENDORN (*m*)
Zizyphus lotus	397	Lotus tree	LOTUSBAUM (*m*) ; LOTUSWEGE--DORN (*m*)
Zizyphus spina Christi	398	Christ's-thorn Zizyphus	LOTUSWEGEDORN (*m*)
Zostera marina	399	Grass-wrack ; common zostera	SEEGRAS (*n*) ; MEERGRAS (*n*)
Zostera nana	400	Dwarf grass-wrack ; dwarf zostera	KLEINE ZOSTERA (*f*) ; ZWERG--SEEGRAS (*n*)
Zygomeris flava	401	Double-jointed-podded amicia (See *Amicia zygomeris*)	JOCHBLATT-AMICIE (*f*)
Zygophyllum album	402	White bean-caper	WEICHHAARIGES BOHNENJOCH--BLATT (*n*)
Zygophyllum cornutum	403	Horned bean-caper	HORN-BOHNENJOCHBLATT (*n*)
Zygophyllum fabago	404	Syrian bean-caper	GEMEINES JOCHBLATT (*n*) ; BOHNEN-JOCHBLATT (*n*)

APPENDIX

CODE INDICATOR 70

A

ABBAUBELEUCHTUNG (*f*) : (coal)-face lighting (Mining) 405
ABBAULEUCHTE (*f*) : coal-face lighting (unit) fitting (Mining) 406
ABBAUSTOSS (*m*) : working face (Mining) 407
ABBRENNKONTAKT (*m*) : sparking (arcing, auxiliary break, secondary) contact (Mining) 408
ABBRENNSTÜCK (*n*) : arcing strip (Elect.) 409
ABDICHTUNG (*f*) : seal-(ing) ; packing ; stuffing 410
ABDROSSELUNG (*f*) : choking (Elect.) ; throttling 411
ABDRUCK (*m*) : print-(ing) ; copy 412
ABERREGEN : to de-energize 413
ABFALLSEIDEN (*f.pl*) : cut filaments (Art silk) 414
ABFLACHUNGSDROSSEL (*f*) : smoothing choke (Radio) 415
ABFRAGEAPPARAT (*m*) : exchange apparatus (Tel.) ; transmitter ; answering equipment ; operator's 'phone set 416
ABFRAGEGEHÄUSE (*n*) : (telephone)-exchange switchboard 417
ABFRAGEKLINKE (*f*) : jack (Tel.) 418
ABFRAGESTECKER (*m*) : plug (Tel.) ; answering plug 419
ABFRAGESTÖPSEL (*m*) : plug (Tel.) 420
ABFRAGEUMSCHALTER (*m*) : Kellogg (listening) key (Tel.) 421
ABFUHRROLLGANG (*m*) : run-out table (Rolling mills) 422
ABGANGSVERKEHR (*m*) : outgoing traffic 423
ABGEHEN : to depart, go, leave 424
ABGEHEND : outgoing 425
ABGESTIMMT : tuned, syntonized ; harmonic ; synchronized 426
ABGREIFPUNKT (*m*) : tapping (Radio) 427
ABGRIFF (*m*) : tap-(ping) (Radio) 428
ABHÄNGIGKEIT (*f*), **IN . . . VON** : as a function of ; depending 429
ABISOLIEREN : to insulate ; lag ; isolate 430
ABKLINGEN : to cool-(off) ; (*n*) : normal cooling 431
ABKLINGKURVE (*f*) : cooling curve 432
ABKLINGUNGSKONSTANTE (*f*) : cooling constant (the time in which a temperature drop of 50% is recorded) 433
ABLAUFKURVE (*f*) : time-lag curve (Elect.) 434
ABLAUFZEIT (*f*) : time lag (Elect.) ; (see also **ABLAUF**) 435
ABLÄUTEN : to ring off (Tel.) 436
ABLEITER (*m*) : lightning conductor 437
ABLEITUNGSELEKTRODE (*f*) : negative electrode ; cathode 438
ABLEITUNGSSTANGE (*f*) : lightning conductor 439
ABLEITUNGSSTROM (*m*) : leakage (fault) current ; earth current (Elect.) ; stray current 440
ABLEITUNGSWIDERSTAND (*m*) : leakance (Elect.) 441
ABLEITUNG (*f*) **ZUR ERDE** : fault (leak) to earth 442
ABLENKUNGSMESSER (*m*) : deflectometer 443
ABLENKUNGSREGEL (*f*) : deflection rule 444
ABLEUCHTLAMPE (*f*) : inspection (hand) lamp 445
ABMANTELN : to strip (Cable) ; (*n*) : stripping 446
ABNABELUNG (*f*) : cutting the umbilical cord (Med.) 447
ABNAHME (*f*) : decrease, diminution, decrement, drop, fall, reduction ; waning ; taking (off) over ; acceptance ; removal ; amputation 448
ABNEIGUNG (*f*) : antipathy ; aversion 449
ABONNENTENANSCHLUSS (*m*) : subscriber's connection (Tel.) 450
ABPFÄHLEN : to stake ; mark out plots of ground by means of stakes 451
ABREISSFUNKENSTRECKE (*f*) : quenching (breaking) spark gap (Elect.) 452
ABRÜSTUNG (*f*) : disarmament 453
ABSCHALTBAR : capable of being cut (out) off ; capable of being switched (turned) off 454
ABSCHALTUNG (*f*) : switching off ; disconnection ; breaking a circuit 455
ABSCHALTVERMÖGEN (*n*) : breaking (rupturing) capacity (Elect.) 456
ABSCHIEBVORRICHTUNG (*f*) : slab pusher (Rolling mills) 457
ABSCHIRMUNG (*f*) : shading (of Light) ; screen-(ing) ; cover-(ing) ; protection 458
ABSCHLUSSGLAS (*n*) : cover (diffusing) glass (Lighting) 459
ABSCHLUSSKONDENSATOR (*m*) : terminal condenser 460
ABSCHLUSSRINGÜBERTRAGER (*m*) : terminal ring transformer (Elect.) 461
ABSCHMELZDRAHT (*m*) : fuse-wire 462
ABSCHMELZVORRICHTUNG (*f*) : fuse (Elect.) 463
ABSCHWEMMUNG (*f*) : denudation 464
ABSPANNER (*m*) : step-down transformer (Elect.) 465
ABSPANNISOLATOR (*m*) : overhead line insulator (Elect.) ; terminal insulator 466
ABSPANNKLEMME (*f*) : straining clamp 467
ABSPANNSICHERUNG (*f*) : fuse for branch circuit (Elect.) 468
ABSTANDSCHELLE (*f*) : spacing saddle (Elect.) 469
ABSTELLVORRICHTUNG (*f*), **ELEKTRISCHE -** ; electrically operated stop motion (Textile) 470
ABSTIMMBAR : capable of being tuned (in-)(-to) (Radio) 471
ABSTIMMKONDENSATOR (*m*) : tuning condenser (Radio) 472
ABSZESS (*m*) : abscess (Med.) 473
ABTRAGUNG (*f*) : denudation 474
ABWÄRTSTRANSFORMATOR (*m*) : step-down transformer 475
ABWÄRTSTRANSFORMIERUNG (*f*) : step-down transformation 476
ABWASSERREINIGUNG (*f*) : water purification 477

ABZEHRUNG (*f*) : (see TUBERKULOSE) 478

ABZEICHEN (*n*) : emblem ; mark ; badge 479

ABZEICHNEN : to initial (documents) 480

ABZIEHER (*m*) : banksman (Mining) 481

ABZWEIGBRUNNEN (*m*) : cable branch-box ; cable pit 482

ABZWEIGDOSE (*f*) : branch box ; junction box 483

ABZWEIGKASTEN (*m*) : junction (branch) box (Elect.) ; distribution box 484

ABZWEIGMUFFE (*f*) : branch box ; coupling box ; sleeve 485

ABZWEIGSPULE (*f*) : differential coil 486

ABZWEIGSTATION (*f*) : branch exchange (Tel.) 487

ACCOLADE (*f*) : accolade ; brace ; bracket (connecting staves in music) 488

ACHILLESSEHNE (*f*) : Achilles' heel (Med.) 489

ACHROMASIE (*f*) : achromatism ; colourlessness 490

ACHROMATOPSIE (*f*) : achromatopsia ; achromatopsy ; colour-blindness 491

ACHSDYNAMO (*f*) : axle-driven dynamo 492

ACHTERSCHUTZ (*m*)*: figure eight protection (Elect.) 493

ACIDOMETER (*n*) : acidimeter 494

ACKERSCHACHTELHALM (*m*) : field horsetail ; paddock-pipes ; shave grass (*Equisetum arvense*) (Bot.) 495

ACUPUNKTUR (*f*) : acupuncture 496

ADER (*f*), GOLDENE- : haemorrhoid (Med.) 497

ADJUSTIERUNG (*f*) : adjustment ; setting ; alignment ; calibration 498

ADMITTANZ (*f*) : admittance (Elect.) 499

ADNEXEN (*pl.*): adnexa (structures lying adjacent to some other organ, as the ovaries and Fallopian tubes to the womb) (Med.) 500

ADNEXITIS (*f*) : inflammation of the adnexa (Med.) 501

AËROB : aerobic (inability of bacteria to live without air) 502

AËROBEN (*pl*), FAKULTATIVE- ; facultative aerobes 503

AËROBEN (*pl*), OBLIGATE- : obligate aerobes 504

AËROBIONTEN (*pl*), FAKULTATIVE- : facultative aerobes 505

AËROBIONTEN (*pl*), OBLIGATE- : obligate aerobes 506

AFTERGEGEND (*f*) : anal region (Med.) 507

AFTERMIETE (*f*) : sub-letting (of houses) 508

AHORNBLÄTTRIG : maple-leaved ; having leaves like maple (*acerifolius*) (Bot.) 509

AJAN- : from Ajan Bay in Asia (*ajanensis*) (Bot.) 510

AKKLIMATISIERUNG (*f*) : acclimatization 511

AKKORD (*m*), DOMINANTSEPTIMEN- : chord of the dominant seventh (Music) 512

AKKORD (*m*), DUR- : major chord (Music) 513

AKKORD (*m*), GEBROCHENER- : broken chord 514

AKKORD (*m*), MOLL- : minor chord 515

AKKORD (*m*), VERMINDERTER SEPTIMEN- : chord of the diminished seventh 516

AKQUISITEUR (*m*) : traveller ; salesman 517

AKROMEGALIE (*f*) : acromegaly (enlargement of the bones) (Med.) 518

AKTINO- : actino- (to do with radiation of heat or light) 519

AKTINOELEKTRISCH : actino-electric 520

AKTINOELEKTRIZITÄT (*f*) : actino-electricity (produced by direct heat radiation) 521

AKTINOMETER (*n*) : actinometer (for measuring actinic power of sun's rays and sensitivity of photographic plates and papers) 522

AKTINOMYKOSE (*f*) : actinomycosis ; streptotrichosis (a cattle disease due to ray fungus) 523

AKTIONSSTROM (*m*) : minute electric currents produced by the muscles of the heart 524

AKUSTISCH : acoustic-(al-)(-ly) ; audible 525

AKUSTISCHES MESSGERÄT (*n*) : sound-measuring apparatus 526

ALARM (*m*) : alarm 527

ALARMIEREN : to alarm 528

ALARMIERUNG (*f*) : alarm 529

ALBINISMUS (*m*) : albinism (Med.) 530

ALBINO (*m*) : albino (Med.) 531

ALIQUOTTON (*m*) : overtone ; upper partial ; harmonic, (of Sound) 532

ALKOHOLISMUS (*m*) : alcoholism (Med.) 533

ALLEGRO (*n*), ERSTES- : first movement ; sonata form in music 534

ALLERGIE (*f*) : allergy (Med.) ; hypersensitivity ; supersensitivity (from birth onwards) 535

ALLERGISCH : allergic ; hypersensitive ; supersensitive 536

ALLIGIEREN : to add, mix ; (*n*) : admixture (see VERMISCHEN) 537

ALLWÖCHENTLICH : every week ; weekly 538

ALPEN- : found in the lower alps (*alpestris ; alpigenus ; alpigalus*) ; belonging to the alps (*alpinus*) (Bot.) 539

ALT : old, late, antique, ancient, stale, pristine, not modern, archaic, obsolete, by-gone, primitive, out-of-date 540

ALTER (*n*) : age, period, epoch, era, old age, antiquity 541

ALTERNATIVLEITER (*m*) : neutral wire (Elect.) 542

ALTERSBLÖDSINN (*m*) : dementia (Med.) 543

ALTKARBONISCH : old carboniferous (Geol.) 544

AMALGAMIEREN : to amalgamate 545

AMORALISCH : unmoral ; non-moral 546

AMPERESTUNDENMESSER (*m*) : ampere/hour meter 547

AMPEREWINDUNGSZAHL (*f*) : (number of) ampere turns 548

AMPEREZAHL (*f*) : amperage (Elect.) 549

AMPHOTERISCHE SUBSTANZ (*f*) ; amphoteric substance (Electrochemistry) 550

AMPLITUDENVERZERRUNG (*f*) : amplitude distortion (Tel.) 551

AMPUTIEREN : to amputate (Med.) 552

AMTSENDIGENDE LEITUNG (*f*) : incoming line (Tel.) 553

AMTSGERICHT (*n*) : district court 554

AMTSLEITUNG (*f*) : exchange line (Tel.) 555

AMURSCH : from Amur (*amurensis*) (Bot.) 556

ANAËROB : anaerobic (Bacteriology) (living without free oxygen) 557

ANAËROBEN (*pl*) : anaerobes 558

ANAËROBIONTEN (*pl*) : anaerobes 559

ANAËROBIONTISCH : anaerobic 560

ANAËROBIOSE (*f*) : anaerobiosis 561

ANALGESIE (*f*) : analgesia (loss of sense of pain) (Med.) 562

ANAMNESE (*f*) : anamnesis ; recollection 563

ANAPHYLAKTISCH : anaphylactic ; highly susceptible ; highly sensitive ; supersensitive ; hypersensitive 564

ANAPHYLAXIE (*f*) : anaphylaxis ; supersensitivity ; hypersensitivity (reverse of immunity), high susceptibility (developed after birth) (Med.) 565

ANBEQUEMEN : to accommodate 566

ANDEUTUNG (*f*) : indication 567

ANFAHRMOMENT (*n*) : starting torque 568

ANFALL (*m*), EPILEPTISCHER- : epilepsy (Med.) 569

ANFÄLLIGKEIT (*f*) : liability (susceptibility) to (Med.) 570

ANFALL (*m*), URÄMISCHER- : uraemia (Med.) 571

ANFANGSSPANNUNG (*f*) : initial voltage 572

ANFANGSWIDERSTAND (*m*) : initial (starting) resistance 573

ANFRAGE (*f*), **AUF-** : on request ; on enquiry . 574
ANGELEGENHEIT (*f*) : affair ; matter 575
ANGESTELLTENKAMMER (*f*) : crew's quarters (in ships) 576
ANGESTELLTENVERSICHERUNG (*f*) : worker's insurance 577
ANGEZEIGTE (*m*) : accused ; defendant 578
ANGINA (*f*) : angina ; choking (Med.) 579
ANGIOM (*n*) : angioma (tumour of blood vessels) (Med.) 580
ANGREIFEND : deleterious (of chemicals) (see also **ANGREIFEN**) 581
ANGSTRÖMEINHEIT (*f*) : Angström unit (A.U.) (wave length is 100,000,000 A.U. per cm.) 582
ANHÄNGER (*m*) : trailer-(car) 583
ANISOTROPISCH : anisotropic ; unequal ; uneven ; unsymmetrical 584
ANKERFELD (*n*) : magnetic field (Elect.) 585
ANKERPLATTE (*f*) : armature end-plate 586
ANKERRÜCKWIRKUNG (*f*) : armature reaction (Elect.) ; back (E.M.F.) electro-motive force 587
ANKOCHEN : to boil up ; bring to the boil ; (*n*) : boiling-(up) 588
ANLASSAPPARAT (*m*) : starting (switch) device ; starter 589
ANLASSCHALTER (*m*) : starting switch (Elect.) 590
ANLASSCHALTUNG (*f*) : starting connection (Elect.) 591
ANLASSER (*m*), **ANKER-** : secondary starter 592
ANLASSER (*m*), **GEHÄUSE-** : primary starter 593
ANLASSER (*m*), **HAUPTSTROM-** : series starter (Elect.) 594
ANLASSER (*m*), **SELBST-** : self-starter (Motors) ; automatic starter (Elect.) 595
ANLASSER (*m*), **UMKEHR-** : starting and reversing switch 596
ANLASSREGULIERWIDERSTAND (*m*) : combined starter and regulator 597
ANLASSTELLUNG (*f*) : starting position 598
ANLASS- UND REGULIER-WIDERSTAND (*m*) : starting and regulating resistance 599
ANLASSWALZE (*f*) : (drum-type)-starter (Elect.) 600
ANLASSWIDERSTAND (*m*), **UMKEHR-** : reversing starting resistance 601
ANLAUFBEDINGUNGEN (*f.pl*) : starting conditions 602
ANLAUFSTROM (*m*) : starting current 603
ANLAUFSTROMSTÄRKE (*f*) : starting current 604
ANLAUFZEIT (*f*) : starting time ; time required for (duration of) starting up 605
ANLEGEMASCHINE (*f*) : spreading machine (Textile) 606
ANNÄHERUNG (*f*) : approximation 607
ANNEHMLICHKEIT (*f*) : advantage 608
ANODENKREIS (*m*) : anode circuit 609
ANODENSPANNUNG (*f*) : anode voltage ; plate (tension) voltage (Radio) 610
ANODENSTRAHLEN (*m.pl*) ; anodal rays 611
ANODENSTROMKREIS (*m*) : anode circuit (Radio) 612
ANODISCH : anodic 613
ANPEILEN : to obtain a fix ; (*n*) : direction finding (Radio) 614
ANREGERELAIS (*n*) : impulse relay ; exciting relay (Elect.) 615
ANRUFKLAPPE (*f*) : drop indicator (Tel.) 616
ANRUFRELAIS (*n*) : call relay (Tel.) 617
ANRUFSUCHER (*m*) : line finder (Tel.) 618
ANRUFTASTE (*f*) : call key 619
ANSCHLAG (*m*), **END-** : stop (for limiting a movement) 620
ANSCHLUSSBOLZEN (*m*) : terminal 621
ANSCHLUSSDOSE (*f*) : plug socket ; junction box 622
ANSCHLUSSDOSE (*f*) **UND STECKER** (*m*) : dip (Stage lighting) 623
ANSCHLUSSKASTEN (*m*) : terminal box (Elect.) 624
ANSCHLUSSKLEMME (*f*) : terminal 625
ANSCHLUSSLEISTE (*f*) : terminal (bar) strip (Elect.) 626
ANSCHLUSSMUFFE (*f*) : jointing sleeve 627
ANSCHLUSSROSETTE (*f*) : ceiling rose (Elect.) 628
ANSCHLUSS-SCHNUR (*f*) : connecting cord (Elect.) 629
ANSCHLUSS-SPANNUNG (*f*) : supply voltage (Elect.) 630
ANSCHLUSSTRANSFORMATOR (*m*) : terminal transformer (Tel.) 631
ANSCHLUSSÜBERTRAGER (*m*) : terminal transformer (Tel.) 632
ANSPRECHEN : to speak ; sound ; be regarded as ; operate ; function ; start up (Elect.) ; respond ; (*n*) : operating ; operation ; function-(ing) ; actuation ; response 633
ANSPRECHSPANNUNG (*f*) : operating voltage (Tel.) 634
ANSPRECHSTROM (*m*) : initial current (Elect.) 635
ANSPRECHZEIT (*f*) : transit time (of electric current to operate instruments) 636
ANSPRINGEN : to start up (of motors) 637
ANSTECKEN : to light ; infect 638
ANSTECKUNG (*f*) : infection (Med.) 639
ANSTEUERUNGSFEUER (*n*) : location beacon (for aircraft) 640
ANTENNE (*f*), **EMPFANGS-** : receiving aerial 641
ANTENNE (*f*), **GERICHTETE-** : directional aerial 642
ANTENNE (*f*), **MEHRDRÄHTIGE-** : multiple wire aerial 643
ANTENNENABSTIMMSPULE (*f*) : aerial tuning coil 644
ANTENNENBUCHSE (*f*) : aerial plug-socket (Radio) 645
ANTENNENERDUNG (*f*) : aerial earthing 646
ANTENNENKOPPELUNG (*f*) : aerial coupling 647
ANTENNENKREIS (*m*) : aerial circuit 648
ANTENNENLÄNGE (*f*) : length of aerial (Radio) 649
ANTENNENSPULE (*f*) : aerial tuning coil 650
ANTENNENSTROM (*m*) : aerial current 651
ANTHRAKOSIS (*f*) : anthracosis ; miner's phthisis (lung trouble due to coal dust) 652
ANTHRAZITISCH : anthracitic 653
ANTHRAZITSTADIUM (*n*) : anthracite (anthracitic) stage or state 654
ANTHROPOLOGIE (*f*) : anthropology 655
ANTHROPOMORPHOSIEREN : anthropomorphize 656
ANTHROPOPHAG : anthropophagous 657
ANTHROPOSOPHIE (*f*) : anthroposophy 658
ANTIK : antique 659
ANTIKATHODE (*f*) : anticathode ; target (of X-ray tubes) 660
ANTIKLINAL : anticlinal (sloping upwards) 661
ANTIKLINALE (*f*) : anticline ; saddleback (Geol.) 662
ANTIMAGNETISCH : antimagnetic 663
ANTIMONIERUNG (*f*) : electrolytic or galvanic precipitation of antimony 664
ANTLITZARTERIE (*f*) : facial artery (Med.) 665
ANWACHSEN : to grow ; increase ; mount up 666
ANZAPFDRUCK (*m*) : leak-off pressure ; pass-out pressure 667
ANZAPFTURBINE (*f*) : bleeder turbine ; interstage leak-off turbine 668
ANZEIGEN : to indicate (of instruments) ; inform ; report 669
ANZUGSMOMENT (*n*) : starting torque 670
AORTA (*f*) : aorta (Med.) 671
AORTA (*f*), **BAUCH-** : abdominal aorta (Med.) 672
AORTABOGEN (*m*) : arch of the aorta (Med.) 673

AORTA (*f*), **BRUST-** : thoracic (descending) aorta (Med.) 674
AORTENBOGEN (*m*) : arch of aorta (Med.) 675
APEPSIE (*f*) : faulty digestion ; indigestion ; dyspepsia ; apepsy (Med.) 676
APERIODISCH : aperiodic 677
APERIODIZITÄT (*f*) : aperiodicity 678
APHTHEN (*pl*) : apthae ; thrush (Med.) 679
APOCHROMATISCH : apochromatic 680
APOGAMIE (*f*) : (see **ZEUGUNGSVERLUST**) 681
APOGÄUM (*m*) : (see **ERDFERNE**) 682
APOGÄUMISCH : apogean 683
APOSPORIE (*f*) : apospory 684
APOSPORISCH : aposporous 685
APPENDIZITIS (*f*) : appendicitis (Med.) ; (see **BLINDDARMENTZÜNDUNG**) 686
APPOSITION : apposition (Grammar) 687
APPRETUR (*f*) : finish-(ing) ; dressing (of fabrics, etc.) 688
APTHENSEUCHE (*f*) : foot-and-mouth disease 689
APTIEREN (*n*) : conversion ; adaptation (of gas fittings for electric lighting) 690
ÄQUIPOTENTIALPUNKT (*m*) : point of equal potential 691
ÄQUIVALENTE LEITFÄHIGKEIT (*f*) : equivalent conductivity (Electrochemistry) 692
ARABISCH : arabian, from Arabia (*arabicus* ; *arabius*) (Bot.) 693
ARBEITEN, BLANK- : to burnish 694
ARBEITERSCHAFT (*f*) : operatives ; workers 695
ARBEITSDRAHT (*m*) : trolley wire (Elect.) 696
ARBEITSKONTAKT (*m*) : closing contact (Elect.) 697
ARBEITSKRAFT (*f*) : hand ; employee ; (*pl*) labour 698
ARBEITSKREIS (*m*) : external circuit 699
ARBEITSLEITUNG (*f*) : supply (wire) main (Elect.) 700
ARBEITSMESSER (*m*), **ELEKTRISCHER-** : Watt/hour meter 701
ARBEITSMESSER (*m*), **MECHANISCHER-** : dynamometer 702
ARBEITSMOTOR (*m*) : driving motor 703
ARBEITSSTROMKREIS (*m*) : open circuit (Tel.) 704
ARBEITSSTRÖMUNG (*f*) : active flow 705
ARBEITSWEISE : principle (method) of operation or working 706
ARCHITEKTONISCH : architectonic 707
ARHYTHMISCH : arhythmic ; irregular 708
ARKTIS (*f*) : arctic region 709
ARMIERUNG (*f*) : armouring (of cable) 710
ÄROMETER (*n*) : (see **SENKWAGE**) 711
ARRETIERLEISTE (*f*) : locking-bar ; stop (for limiting a movement) 712
ARSENIERUNG (*f*) : electrolytic or galvanic precipitation of arsenic 713
ARTERIE (*f*) : artery (Med.) 714
ARTERIENERWEITERUNG (*f*) : aneurysm ; aneurism ; (dilation or swelling of an artery) (Med.) 715
ARTERIENVERKALKUNG (*f*) : arteriosclerosis : hardening of the arteries (Med.) 716
ASBESTZEMENT (*m*) : asbestos cement 717
ASIATISCH : Asiatic ; of Asiatic origin ; from Asia (*asianus* ; *asius* ; *asiaticus*) (Bot.) 718
ASTASIE (*f*) : not influenced by the magnetism of the earth 719
ASTATISCH : astatic 720
ASTHENIE (*f*) : asthenia ; debility (Med.) 721
ASYNCHRONISMUS (*m*) : asynchronism 722
ATEM (*m*) : breath 723
ATEMNOT (*f*) : dyspnoea ; breathlessness (Med.) 724
ATEMSTOCKUNG (*f*) : apnoea ; stoppage of breathing (due to too great a supply of oxygen to the blood) (Med.) 725
ÄTHERSTRAHLUNG (*f*) : ether radiation 726
ATLAS- : from the Atlas mountains (*atlanticus*) (Bot.) 727
ATOMION (*n*) : (see **ANION**) 728
ATONIE (*f*) : atony (want of vitality in various organs and muscles) (Med.) 729
ATTRAPPE (*f*) : dummy (of goods exhibited in shop windows) 730
ÄTZUNG (*f*) : etching 731
AUDION (*n*) : (see **AUDIONRÖHRE**) 732
AUDIONRÖHRE (*f*) : audion valve (Radio) 733
AUFBAUEN : to assemble ; build up ; erect ; construct ; form 734
AUFBLINKEN (*n*) : flash-(ing) (of lighting) 735
AUFBLITZEN (*n*) : flash-(ing) (of lighting) 736
AUFDRÜCKEN : to impress upon ; force upon ; superimpose (of oscillations) 737
AUFFÄLLIG : noticeable 738
AUFFANGANTENNE (*f*) : receiving aerial 739
AUFFANGEN : to catch ; receive 740
AUFFANGSTANGE (*f*) : lightning conductor 741
AUFGABESTATION (*f*) : transmitting station ; transmitter 742
AUFGELEGT : laid up (of shipping) 743
AUFGESCHRUMPFT : shrunk on (Engineering) 744
AUFLEUCHTEN : to light ; illuminate ; (*n*) : lighting ; illumination 745
AUFMONTIEREN : to mount on ; attach to 746
AUFNAHMEFERTIG : ready for exposure (Phot.) 747
AUFNAHMELEITER (*m*) : producer (Radio) 748
AUFNAHMERAUM (*m*) : studio (Radio) 749
AUFNAHMEVERMÖGEN (*n*), **MAGNETISCHES-** : susceptibility 750
AUFRECHTERHALTUNG (*f*) : maintenance ; renewal (of patents) 751
AUFROLLVORRICHTUNG (*f*) : reeling apparatus ; reels (Paper) 752
AUFSAUGUNG (*f*) : adsorption ; absorption 753
AUFSCHLAGER (*m*) : onsetter (Mining) 754
AUFSICHTSSUCHER (*m*) : reflecting view-finder (Phot.) 755
AUFSPANNER (*m*) : step-up transformer 756
AUFSTOCKUNG (*f*) : addition (of a floor to a building) 757
AUFTAKT (*m*) : up beat (Music) 758
AUFTRAGS-OXYDKATHODE (*f*) : oxide-coated cathode 759
AUFWÄRTSTRANSFORMATOR (*m*) : step-up transformer 760
AUFWICKELSPULE (*f*) : winding-on bobbin (Textile) 761
AUFZIEHKURBEL (*f*) : winding (crank) handle 762
AUGENHEILKUNDE (*f*) : ophthalmology 763
AUGENKAMMER (*f*), **VORDERE-** : anterior chamber (of the eye) 764
AUGENKRANKHEIT (*f*) : (see **OPHTHALMIA**) 765
AUGENKRANKHEIT (*f*), **ÄGYPTISCHE-** : trachoma (an eye disease) (Med.) 766
AUGENPROTHESE (*f*) : artificial (glass) eye 767
AUGENSPIEGEL (*m*) : ophthalmoscope (an instrument for examination of the eye) (Med.) 768
AUGENTRIPPER (*m*) : blenorrhoea (Med.) 769
AUKUBENBLÄTTRIG : with aucuba-like leaves (*aucubifolius*) (Bot.) 770
AUSATMUNG (*f*) : breathing out ; expiration ; exhalation ; expulsion of air from the lungs (Med.) 771
AUSBAU (*m*) : development 772
AUSBESSERUNGSWERKSTATT (*f*) : repair shop 773
AUSBLÄSER (*m*) : blow-out ; blown-out shot (Mining) 774
AUSDAUERND : perennial 775

AUSEINANDERBAUEN : (see AUSEINANDERNEHMEN) 776
AUSFRANSEN : to fray (of rope) 777
AUSGANGSKREIS (*m*) : output circuit 778
AUSGANGSSTELLUNG (*f*) : initial (first, primary) position 779
AUSGANGSTRANSFORMATOR (*m*) : output transformer 780
AUSGANGSÜBERTRAGER (*m*) : output transformer 781
AUSGLEICHLEITUNG (*f*) : equalizing cable (Elect.) 782
AUSGLEICHSBECKEN (*n*) : equalizing basin (Water power) 783
AUSGLEICHSFEDER (*f*) : compensating (compensation) spring 784
AUSGLEICHSKAPAZITÄT (*f*) : balancing capacity 785
AUSGLEICHWIDERSTAND (*m*) : equalizing resistance 786
AUSGUSSMASSE (*f*) : (sealing)-compound (for cable boxes) 787
AUSHAUMASCHINE (*f*) : blanking machine (Metal.) 788
AUSHÖHLEN : to hollow out ; dish (Eng.) 789
AUSKLINKMASCHINE (*f*) : notching machine (Metal.) 790
AUSLADUNG (*f*) : projection ; effective radius (of cranes) 791
AUSLEGERLAUFKRAN (*m*) : travelling jib-crane 792
AUSLEGUNG (*f*) : interpretation 793
AUSLÖSEBÜRDE (*f*) : tripping load 794
AUSLÖSEGESTÄNGE (*n*) : tripping rods (Elect.) 795
AUSLÖSEKNOPF (*m*) : release (knob) button 796
AUSLÖSEMAGNET (*m*) : releasing magnet (Elect.) 797
AUSLÖSENOCKEN (*m*) : release cam (Mech.) 798
AUSLÖSER (*m*) : cut-out (Elect.) ; tripping relay (Elect.) 799
AUSLÖSERNENNSTROM (*m*) : (current)-rating of tripping relay 800
AUSLÖSESPULE (*f*) : tripping coil (Elect.) 801
AUSLÖSESTANGE (*f*) : tripping rod (Elect.) 802
AUSLÖSESTROM (*m*) : tripping current 803
AUSLÖSEVORRICHTUNG (*f*) : tripping device 804
AUSLÖSEWELLE (*f*) : release shaft 805
AUSLÖSEZEIT (*f*) : tripping time ; time lag (Elect.) 806
AUSLÖSUNG (*f*), **NULL-** : no-load release (Elect.) 807
AUSLÖSUNG (*f*), **NULLSPANNUNGS-** : no-voltage release 808
AUSLÖSUNG (*f*), **ÜBERLASTUNGS-** : overload release 809
AUSLÖSUNG (*f*), **UNVERZÖGERTE-** : instantaneous tripping 810
AUSLÖSUNG (*f*), **VERZÖGERTE-** : delayed tripping ; tripping lag (Elect.) 811
AUSNÜTZUNGSFAKTOR (*m*) : utilization factor (Lighting) ; load factor (Elect.) 812
AUSRUFSZEICHEN (*n*) : note of exclamation (!) 813
AUSSATZPILZ (*m*) : bacillus of leprosy 814
AUSSCHALTSTELLUNG (*f*) : " off " position (Elect.) 815
AUSSCHALTUHR (*f*) : time switch 816
AUSSCHLAGWINKEL (*m*) : angle of deflection 817
AUSSCHMÜCKEN : to embellish ; ornament ; (*n*) : embellishment ; ornamentation 818
AUSSCHNAPPEN : to snap out ; kick over (of a switch) 819
AUSSCHNEIDEN : to cut out ; excise 820
AUSSCHUSSWARE (*f*) : unsaleable goods ; throw-outs 821
AUSSCHWITZUNG (*f*) : exudation ; perspiration 822
AUSSENANTENNE (*f*) : outdoor aerial 823
AUSSENDEN : to transmit (Radio) ; send out ; emit 824
AUSSENDUNG (*f*) : sending ; transmission (Radio) ; emission (of rays) 825
AUSSENLEITER (*m* and *pl*) : outer wire-(s) ; external conductor-(s) 826
AUSSENLIEGEND : external 827
AUSSENPOLMASCHINE (*f*) : external pole dynamo 828
AUSSERTRITTFALLEN : to get out of step (of movements) 829
AUSSPEIEN : to spit out ; expectorate ; eject 830
AUSSTRAHLUNG (*f*) : radiation ; emission 831
AUSSTRAHLUNGSANTENNE (*f*) : transmitting aerial 832
AUSSTRAHLUNGSVERMÖGEN (*n*) : radiating capacity 833
AUSTAUSCHFÄHIGKEIT (*f*) : interchangeability 834
AUSTRALISCH : from Australia ; Australian ; of Australian origin (*australasicus*) (Bot.) 835
AUSWAHLFÄHIGKEIT (*f*) : selectivity 836
AUSWITTERUNG (*f*) : efflorescence 837
AUSWRINGEN : to wring ; (*n*) : wringing 838
AUSZIEHMASCHINE (*f*) ; drawing-out machine 839
AUSZIEHSCHACHT (*m*) : upcast shaft (Mining) 840
AVERTIN (*n*) : (see TRIBROMÄTHYLALKOHOL) 841
AXELARTERIE (*f*) : axillary artery (Med.) 842
AZETAL (*n*) : (see ACETAL) 843

B

BABYLONISCH : Babylonian ; from Babylon ; of Babylonian origin (*babylonicus* ; *babylonius*) (Bot.) 844
BAD (*n*), **ELEKTROMEDIZINISCHES-** : electromedical bath 845
BÄDERBESCHREIBUNG (*f*) : balneography 846
BAD (*n*), **GALVANISCHES-** : galvanic bath 847
BADWIDERSTAND (*m*) : bath resistance 848
BAHN (*f*), **ELEKTRISCHE-** : electric railway 849
BAJONETTFASSUNG (*f*) : bayonet (socket) lampholder 850
BAJONETTVERSCHLUSS (*m*) : bayonet adaptor (Elect.) 851
BAKTERIENTÖTEND : bactericidal 852
BAKTERIUM (*n*) : bacterium 853
BAKTERIZID : bactericidal 854
BAKTERIZIDIE (*f*) : bactericidal action 855
BALDACHIN (*m*) : canopy ; ceiling-rose (of lighting fittings) 856
BALDSCHUANISCH : from Baldschuan in central Asia (*baldschuanicus*) (Bot.) 857
BALEARISCH : from the Balearic Isles (*balearicus*) (Bot.) 858
BALGGESCHWULST (*f*) : atherome ; atheroma (tumour full of porridge-like matter) (Med.) 859
BALGGESCHWULSTARTIG : atheromatous (Med.) 860
BALLASTWIDERSTAND (*m*) : loading resistance 861
BALLENBRECHER (*m*) : bale (opener) breaker (Textile) 862
BALLENÖFFNER (*m*) : bale opener (Textile) 863
BALLISTIK (*f*) : ballistics 864
BALLISTISCH : ballistic 865
BALNEOLOGIE (*f*) : balneology (the science of treating disease by means of medicinal baths) 866

BALSAM-: having the soothing quality of balm, (*balsameus*; *balsamicus*); producing balsam (*balsamiferus*) (Bot.) 867
BALTISCH: from the Baltic coast; Baltic (*balticus*) (Bot.) 868
BANANENSTECKER (*m*): banana plug (Radio) 869
BANDBREITE (*f*): channel width (Tel.); (see also BAND) 870
BANDFILTER (*n*): band pass filter (Radio) 871
BANDFÖRMIG: band-shaped; ribbon-(tape-)like 872
BANDWEBSTUHL (*m*): ribbon loom (Textile) 873
BARRETTER (*m*): barretter; detector; barretter tube 874
BARTFLECHTE (*f*): *Tinea tonsurans* (Med.) 875
BARYUM-PLATINCYANÜRSCHIRM (*m*): barium platinocyanide screen (for X-rays) 876
BASTARD-ODERMENNIG-ÄHNLICH: like (resembling) floss-flower (*ageratoides*) (Bot.) 877
BASTFASERSTRECKE (*f*): drawing frame for bast fibre (Textile) 878
BATTERIE (*f*), **MIKROFON-**; microphone battery 879
BATTERIESPANNUNG (*f*): battery (voltage) tension 880
BAUCHHÖHLENERÖFFNUNG (*f*): laparotomy (Med.) 881
BAUGRUBE (*f*): (building)-excavation 882
BAUKÜNSTLER (*m*): architect 883
BÄUMMASCHINE (*f*): beaming machine (Textile) 884
BAUMWOLLGARN (*n*): cotton yarn; twist (Textile) 885
BAUMWOLLGEWEBE (*n*): cotton cloth (Textile) 886
BAUMWOLLKETTENSCHLICHTMASCHINE (*f*): cotton warp sizing machine (Textile) 887
BAUMWOLLSELFAKTOR (*m*): self-acting mule for cotton (Textile) 888
BAUMWOLLSPINNEREI (*f*): cotton mill (Textile) 889
BAUSTELLENBELEUCHTUNG (*f*): illumination of building sites 890
BAZILLUS (*m*): bacillus 891
BE (*n*): flat; ♭ (Music) 892
BEBUNG (*f*): vibrato (Music); oscillation; vibration 893
BECKENBINDEGEWEBSENTZÜNDUNG (*f*): parametritis (Med.) 894
BEDARFSMELDUNG (*f*): requisition 895
BEDECKTSAMIG: with seeds enclosed in ovaries fertilized through stigmas (*angiospermus*) (Bot.) 896
BEDIENUNGSLOS: automatic; not needing attention 897
BEERENFÖRMIG: berry-like; berry-shaped; in the form of a berry (*baccatus*) (Bot.) 898
BEEREN (*f.pl*), **MIT GELBEN-**: with yellow berries; yellow-berried (*baccis flavis*) (Bot.) 899
BEEREN (*f.pl*), **MIT GRÜNEN-**: with green berries; green-berried (*baccis viridis*) (Bot.) 900
BEEREN (*f.pl*), **MIT ROTEN-**: with red berries; red-berried (*baccis rubris*) (Bot.) 901
BEEREN (*f.pl*), **MIT SCHWARZEN-**: with black berries, black-berried (*baccis nigris*) (Bot.) 902
BEEREN (*f.pl*), **MIT WEISSEN-**: with white berries; white-berried (*baccis albis*) (Bot.) 903
BEERENTANG (*m*): floating gulf-weed (*Sargassum bacciferum*) (Bot.) 904
BEFEHLSFORM (*f*): imperative-(mood) (Grammar) 905
BEFLECHTUNG (*f*): braiding (of cables) 906
BEGICHTUNGSKÜBEL (*m*): furnace charging bucket (of cranes) 907
BEGRENZUNGSWIDERSTAND (*m*): limiting resistance 908
BEHELFSMÄSSIG: make-shift; substitute; expedient 909
BEIFÜGUNG (*f*): apposition (Grammar) 910
BEILAUF, MIT . . . RUND VERSEILT: served (of electric cables) 911
BEILSTEIN (*m*): nephrite (see NEPHRIT) 912
BEIN (*n*): bone (Med.); leg 913
BEISTRICH (*m*): comma 914
BEIWAGEN (*m*): trailer-(car) (of trams); trailer coach (of trains) 915
BEIZNARBE (*f*): etching mark (Metal.) 916
BEKÄMPFEN: to combat 917
BELASTBAR: capable of being (loaded) rated 918
BELASTUNG (*f*), **INDUKTIVE-**: inductive load 919
BELASTUNG (*f*), **REAKTIVE-**: reactive load 920
BELASTUNGSDIAGRAMM (*n*): load diagram 921
BELASTUNGSKURVE (*f*): load diagram 922
BELASTUNGSSPULE (*f*): loading coil 923
BELASTUNGSSTROMSTÄRKE (*f*), **ZULÄSSIGE-**: permissible current-(carrying capacity) (of cables) 924
BELEBUNGSMITTEL (*n*): excitant; stimulant 925
BELEUCHTERBRÜCKE (*f*): bridge (of stage lighting) 926
BELEUCHTUNGSANLAGE (*f*): lighting (installation) plant 927
BELEUCHTUNGSKÖRPER (*m*): lighting (unit) fitting 928
BELEUCHTUNGSSTÄRKE (*f*): illumination; intensity of lighting 929
BELEUCHTUNGSSTROMKREIS (*m*): lighting circuit (Elect.) 930
BELEUCHTUNGSTECHNIK (*f*): illumination engineering 931
BELEUCHTUNGSTECHNIKER (*m*): lighting engineer 932
BELÜFTUNG (*f*): ventilation 933
BEMUSTERUNG (*f*): sampling 934
BE-QUADRAT (*n*): natural; ♮ (Music) 935
BERGMITTEL (*n*): band -(of seam) (Mining) 936
BERGWERKDIREKTOR (*m*): manager (Mining) 937
BERUHIGUNGSWIDERSTAND (*m*): steadying resistance (Elect.) 938
BERÜHRUNGSSPANNUNG (*f*): voltage (the voltage of an electric current, with which it is possible to come into contact without serious consequences; 42 volts for human beings; 20 volts for animals) 939
BESATZMATERIAL (*n*): tamping material (Mining) 940
BESÄUMEN: to edge; trim (Paper-making): (*n*): edging (of boards) 941
BESCHLEUNIGUNGSMOMENT (*n*): moment of acceleration 942
BESCHWERUNGSVERFAHREN (*n*): weighting process (Textile) 943
BESENSTRAUCH (*m*): broom (*Sarothamnus scoparius*) (Bot.) 944
BESETZTZEICHEN (*n*): engaged signal (Tel.); engaged sign 945
BESETZUNG (*f*): tamping (Mining) 946
BESPINNUNG (*f*): braiding; covering (of cables) 947
BEST-: best; optimum 948
BETONKUPPEL (*f*): concrete dome 949
BETREUEN: to look after; care for; attend to; be responsible for (any kind of work) 950
BETRIEBSBEREIT: ready for (use) service; ready to (put into commission) start up 951
BETRIEBSFAKTOR (*m*): load factor (Elect.) 952
BETRIEBSFÜHRER (*m*): agent (Mining) 953
BETRIEBSKAPAZITÄT (*f*): wire-to-wire capacity (Telephony); mutual capacity 954
BETRIEBSSPANNUNG (*f*): line voltage; operating voltage; working pressure 955
BETRIEBSSTELLUNG (*f*): position when in (action) operation; working (running) position; "on" position (Elect.) 956
BETRIEBSVERLUST (*m*): operating losses 957

BEUTELTIER (*n*) : marsupial 958
BEWÄHREN, BESTENS- : to give the most satisfactory results 959
BEWEGGRUND (*m*) : motive 960
BEWEGUNGSNERV (*m*) : motor (efferent) nerve (Med.) 961
BEWEGUNGSZUSTAND (*m*) : (state of)-motion ; action ; operation ; working ; running 962
BEWICKLUNG (*f*) : winding ; covering 963
BEZUGSTEMPERATUR (*f*) : temperature of surrounding atmosphere (for which a measuring instrument is supplied to give a correct reading ; usually taken as 20° C.) (Elect.) 964
BIERDRUCKAPPARAT (*m*) : beer engine 965
BIFILAR : bifilar 966
BILDFUNK (*m*) : picture telegraphy 967
BILDFUNKÜBERTRAGUNG (*f*) : picture telegraphy ; picture transmission 968
BILDTELEGRAPHIE (*f*) : picture telegraphy 969
BILDTROMMEL (*f*) : transmitter (picture) drum (of picture telegraphy) 970
BILDÜBERTRAGUNG (*f*) : picture telegraphy ; picture transmission 971
BILDÜBERTRAGUNGSANLAGE (*f*) : picture telegraphy (installation, apparatus) equipment 972
BILDVERSCHLEIERND : fogging (of photographic plates) 973
BILDWERFER (*m*) : projector ; projection apparatus (Cinematography) 974
BILDWERFERRAUM (*m*) : projection room ; operating-box 975
BIMETALL (*n*) : bi-metal 976
BIMETALL-TINTENSCHREIBER (*m*) : bi-metallic ink recorder (Elect.) 977
BINDEHAUTKATARRH (*m*) : conjunctivitis (Med.) 978
BINDGEWEBE (*n*) : fibrous tissue (Med.) 979
BINDGEWEBSGESCHWULST (*f*) : fibroma (tumour of the fibrous tissue) (Med.) 980
BIOMAGNETISMUS (*m*) : animal magnetism 981
BIPOLAR : bipolar ; two-pole ; double pole 982
BIQUADRAT (*n*) : fourth power of a number 983
BIQUADRATISCH : biquadrate 984
BIRKENBLÄTTRIG : birch-leaved ; having leaves like the Birch tree (*betulifolius*) (Bot.) 985
BITTER : bitter (*amarus*) (Bot.) 986
BITUMINIT (*m*) : bog-head coal ; bitumenite coal 987
BLASENSPIEGEL (*m*) : cystoscope (Med.) 988
BLASENSPIEGELUNG (*f*) : cystoscopy (Med.) 989
BLASKERZE (*f*) : sparking-plug 990
BLATTLOS : leafless ; bare ; devoid of leaves (*aphyllus*) (Bot.) 991
BLAUFÄRBUNG (*f*) : cyanosis (Med.) 992
BLAUGRÜN : greenish (grey) blue (*caesius*) (Bot.) 993
BLAUPAUSMASCHINE (*f*) : blue print-(ing) machine 994
BLAUPROZESS (*m*) : (see **ZYANOTYPIE**) 995
BLEIKABEL (*n*) : lead-covered cable 996
BLEIMANTEL (*m*) : lead-sheathing (of cables) 997
BLEIMANTELLEITUNG (*f*) : lead-sheathed (lead-covered) cable 998
BLENDUNGSFREI : free from glare (Lighting) 999
BLINDDARMENTZÜNDUNG (*f*) : typhlitis (inflammation of the caecum) (Med.) 000
BLINDDARMREIZUNG (*f*), **CHRONISCHE-** : chronic irritation of the appendix (Med.) 001
BLINDELEMENT (*n*) : dummy element ; dummy (plug-socket) fuse-base (Elect.) 002
BLINDER FLECK (*m*) : blind spot (of the eye) 003
BLINDKOMPONENT (*m*) : wattless component (Elect.) ; reactive component (of current) 004
BLINDLEISTUNG (*f*) : wattless energy ; wattless component of energy (load factor = 0) (Elect.) ; reactive load 005
BLINDLEISTUNGSMASCHINE (*f*) : phase advancer ; rotary power factor corrector ; machine for improving the power factor (Elect.) 006
BLINDLEISTUNGSMESSER (*m*) : wattless current meter ; reactive volt-ampere meter 007
BLINDLEISTUNGSVERBRAUCH (*m*) : (see **BLINDVERBRAUCH**) 008
BLINDLEITWERT (*m*) : susceptance (Elect.) 009
BLINDSICHERUNG (*f*) : dummy fuse (Elect.) 010
BLINDSTROM (*m*) : wattless current ; wattless component of current ; magnetizing current ; reactive current 011
BLINDSTROMERZEUGUNG (*f*) : generation of wattless current 012
BLINDVERBRAUCH (*m*) : wattless current consumption (Elect.) 013
BLINDWIDERSTAND (*m*) : wattless resistance ; reactance (Elect.) 014
BLINDWIDERSTAND (*m*), **INDUKTIVER-** : inductance (Elect.) 015
BLINKAPPARAT (*m*) : flasher 016
BLINKFEUER (*n*) : flasher (Aerodrome lighting) 017
BLINKLICHTREKLAME (*f*) : flashing (advertising) signs (Lighting) 018
BLITZAUFFANGSPITZE (*f*) : lightning arrestor 019
BLITZAUFFANGSTANGE (*f*) : lightning conductor-(rod) 020
BLOCKDREH-ROLLGANG (*m*) : turning table (Rolling mills) 021
BLOCKDRÜCKER (*m*) : ingot feed device (Rolling mills) 022
BLOCKIERBARE STECKDOSE (*f*) : interlocking plug and socket (Elect.) 023
BLOCKIERUNGSSPULE (*f*) : choke ; choking coil 024
BLOCKKLEMME (*f*) : terminal block (Elect.) 025
BLOCKKONDENSATOR (*m*) : block condenser (Elect.) 026
BLOCKLAGER (*n*) : ingot store 027
BLOCKSYSTEM (*n*) : block signalling system 028
BLUME (*f*) : bloom ; flower ; bouquet (of Wines) ; brush (of Foxes) ; tail (of Rabbits and Bears) 029
BLUMENKRONE (*f*) : corolla (Bot.) 030
BLUMENKRONENLOS : without a corolla ; having flowers without petals (*apetalus*) (Bot.) 031
BLUTANHÄUFUNG (*f*) : hyperaemia ; hyperemia (excessive accumulation of blood) (Med.) 032
BLUTARMUT (*f*), **BÖSARTIGE-** : pernicious anaemia (Med.) 033
BLÜTENTRAGEND : flowering ; phanerogamous (Bot.) 034
BLUTERGUSS (*m*), **OBERFLÄCHLICHER-** : ecchymosis (collection of blood under the skin without swelling) ; haematoma (collection of blood under the skin with swelling) (Med.) 035
BLUTGERINNUNG (*f*) : thrombosis ; coagulation (Med.) 036
BLUTKÖRPERCHEN (*n*), **ROTES-** : red corpuscle (Med.) 037
BLUTKÖRPERCHEN (*n*), **WEISSES-** : white corpuscle ; leucocyte (Med.) 038
BLUTKREISLAUF (*m*) : blood-circulation (Med.) 039
BLUTPLASMA (*n*) : (see **BLUTFLÜSSIGKEIT**) 040
BLUTSCHWAMM (*m*) : (see **ANGIOM**) 041
BLUTSPEIEN (*n*) : spitting blood (see **BLUTHUSTEN** and **BLUTBRECHEN**) 042
BLUTSTOCKUNG (*f*) : (see **HÄMOSTASE**) 043
BLUTSTROM (*m*) : blood-stream 044
BLUTTRANSFUSION (*f*) : blood transfusion 045
BLUTÜBERFÜLLUNG (*f*) : hyperaemia (Med.) 046
BLUTÜBERTRAGUNG (*f*) : blood transfusion 047
BLUTUNG (*f*) : bleeding **048**

BLUTVERGIFTUNG (*f*) : septicaemia ; blood poisoning ; sepsis (Med.) 049
BOBINENSPULE (*f*) ; bobbin (Textile) 050
BOCKKRAN (*m*) : gantry crane 051
BODENENTWÄSSERUNG (*f*) : (soil)-drainage 052
BODENFRÄSE (*f*) : tiller ; tilling machine 053
BODENGRUBE (*f*) : pit 054
BOGENFÜHRUNG (*f*) : bowing (of stringed instruments) 055
BOGENGÄNGE (*m.pl*) : semi-circular canals (of the ear) (Med.) 056
BOGENLICHTKOHLE (*f*) ; arc-lamp carbon 057
BOLOMETER (*n*) : bolometer ; actinic (thermic) balance (for measuring radiant heat) 058
BOLOMETERDETEKTOR (*m*) : bolometric detector 059
BOLOMETERWELLENEMPFÄNGER (*m*) : bolometric detector 060
BOMBARDEMENT (*n*) : bombardment 061
BOMBARDIEREN : to bombard 062
BOMBARDIERUNG (*f*) : bombardment 063
BÖRDELSTOSS (*m*) : lip joint (Welding) 064
BORDFUNKSTELLE (*f*) : ship's radio station 065
BORDSCHALTER (*m*) : marine type switch 066
BRAUNKOHLENSTADIUM (*n*) : lignite (brown coal, lignitic) stage or state 067
BRAUNROT, DUNKEL- : dark purple (*atro-purpureus*) (Bot.) 068
BRECHERHAUS (*n*) : crusher house 069
BRECHMASCHINE (*f*) : breaker (Textile) 070
BRECHPRISMA (*n*) : refracting prism 071
BREITENGRAD (*m*) : degree of latitude 072
BREITHALTER (*m*) : temple (Textile) 073
BREITHALTER (*m*) **MIT ELF STACHELSCHEIBEN** : eleven-ring temple (Textile) 074
BREITSTRAHLSCHEINWERFER (*m*) : wide-angle floodlight 075
BREMSBELAG (*m*) : brake lining 076
BRENNFLECK (*m*) : focal spot 077
BRENNSTÜCK (*n*) : arcing piece (Elect.) 078
BRILLIEREN : to sparkle ; be brilliant 079
BRISANT : breaking-up ; destructive 080
BRISANZWIRKUNG (*f*) : breaking-up (destructive) action or effect 081
BRONCHIOLEN (*pl*) ; bronchioles (capillary bronchi) (Med.) 082
BRÜCKENVERZWEIGUNG (*f*) : Wheatstone bridge 083
BRUMMEN : to hum ; (*n*) : humming (of telegraph wires) 084
BRUNNENWASSERMESSER (*m*) : deep-well water meter 085
BRUSTDRÜSENENTZÜNDUNG (*f*) : mastitis (Med.) 086
BRUSTFELLEITERUNG (*f*) : empyema (Med.) 087
BRUSTMILCHDRÜSE (*f*) : mammary gland (see **BRUSTDRÜSE**) 088
BRUSTMILCHDRÜSENENTZÜNDUNG (*f*) : mastitis ; inflammation of the breast (Med.) 089
BRUSTWIRBEL (*m*) : dorsal vertebra (Med.) 090
BRUTAPPARAT (*m*) : incubator 091
BRUTTEMPERATUR (*f*) : incubation temperature (Bacteriology) 092
BUCHSBAUMBLÄTTRIG : box-leaved (*buxifolius*) (Bot.) 093
BUCHSKINSTUHL (*m*) : buckskin loom (Textile) 094
BÜGELMASCHINE (*f*) : ironing machine 095
BÜGELSTROMABNEHMER (*m*) : bow collector (of electric trains) 096
BÜHNENBELEUCHTUNG (*f*) : stage lighting 097
BÜHNENBELEUCHTUNGSKÖRPER (*m*) : stage lighting fitting 098
BÜHNENBILD (*n*) : stage scene 099
BÜHNENFUSSBODEN (*m*) : stage floor 100
BÜHNENÖFFNUNG (*f*) : proscenium opening 101
BÜHNENREGLER (*m*) : dimmer controller (Stage lighting) 102
BÜHNENSTELLWERK (*n*) : dimmer controller (Stage lighting) 103
BÜHNENTECHNISCHE EINRICHTUNG (*f*) : stage equipment 104
BÜHNENWIDERSTAND (*m*) : dimmer ; dimming resistance (Stage lighting) 105
BUNTGEWEBT : colour woven (Textile) 106
BÜRSTENABHEBER (m) : brush lifter (Elect.) 107
BÜRSTENABHEBEVORRICHTUNG (*f*) : brush-lifting (gear)-device (Elect.) 108
BÜRSTENBRÜCKE (*f*) : brush rocker (Elect.) 109
BÜRSTENFEUER (*n*) : brush sparking (Elect.) 110
BÜRSTENVERSTELLUNG (*f*) : brush shift 111
BÜSCHELENTLADUNG (*f*) : brush discharge (Elect.) 112
BUSSOLE (*f*), **SINUS-TANGENTEN-** : sine-tangent galvanometer 113

C

CAMPAGNE (*f*) : campaign (Sugar) 114
CEREISENSTEIN (*m*) : flint (for petrol lighters) 115
CERIUMOXYD (*n*) : Cerium oxide ; CeO_2 116
CHEMIKALIE (*f*) : chemical 117
CHEMISCHE EINWIRKUNG (*f*) : chemical action 118
CHLORALKALIANLAGE (*f*) : plant for the manufacture of caustic lye 119
CHOLERABAZILLUS (*m*) ; cholera bacillus ; comma bacillus (*Vibrio cholerae ; Spirillum cholerae ; Microspira comma*) 120
CHOLERASPALTPILZ (*m*) : (see **CHOLERABAZILLUS**) 121
CHOLERASPALTPILZ (*m*), **FALSCHER-** : Finkler-Prior bacillus (*Microspira Finkleri ; Vibrio proteus*) 122
CHROMBAD (*n*) : chromium (vat) bath 123
CHROMNIEDERSCHLAG (*m*) : chromium (coating) deposit 124
CHROMOGEN : chromogenic 125
CHROMÜBERZUG (*m*) : chromium (deposit) coating or layer 126
COMPOUNDMASCHINE (*f*) : compound-wound dynamo 127
COMPURVERSCHLUSS (*m*) : compur shutter (Phot.) 128
COTTONMASCHINE (*f*) : Cotton's-(Patents)-machine (Textile) 129
COTTONSTRUMPF (*m*) : full-fashioned hose (Textile) 130
CYSTITIS (*f*) : (see **HARNBLASENKATARRH**) 131
CYSTOSKOP (*n*) : cystoscope (Med.) 132
CYSTOSKOPIE (*f*) : cystoscopy (Med.) 133

D

DACHANTENNE (*f*) : roof aerial 134
DACHSTÄNDER (*m*) : roof standard 135
DALTONISMUS (*m*) : Daltonism (colour-blindness to red and blue) 136
DÄMMERSCHLAF (*m*) : twilight sleep (Med.) 137
DAMPFDYNAMO (*f*) : steam dynamo 138
DAMPFKOCHTOPF (*m*) : pressure boiler 139
DÄMPFUNG (*f*) : damping ; attenuation ; transmission loss (Tel.) 140
DÄMPFUNGSLEISTUNG (*f*) : damping effect 141
DÄMPFUNGSMESSER (*m*) : decremeter 142
DÄMPFUNGSMOMENT (*n*) : damping moment 143
DÄMPFUNGSVORRICHTUNG (*f*) : damping (arrangement) device 144
DÄMPFUNGSWERT (*m*) : attenuation 145
DÄMPFVORRICHTUNG (*f*) : damping (arrangement) device 146
DÄNGEL (*m*) : edge (of a scythe) 147
DÄNGELMASCHINE (*f*) : scythe-sharpening machine 148
DARBIETUNG (*f*) : presentation ; production ; (broadcast)-programme 149
DARLEGUNG (*f*) : exposition ; explanation ; demonstration 150
DARMAUSSCHEIDUNG (*f*) : excrement ; faeces (Med.) 151
DAUERERDSCHLUSS (*m*) : continuous fault to earth (Elect.) 152
DAUERKURZSCHLUSSTROM (*m*) : continuous (permanent value of)-short circuit current (Elect.) 153
DAUERSTROM (*m*) : permanent (continuous) current (Elect.) 154
DECIBEL (*n*) : decibel (see PHON) 155
DECIME (*f*) : tenth (Music) ; (see under TERZ) 156
DECKENLICHT (*n*) : ceiling light ; top light 157
DECKENROSETTE (*f*) : ceiling rose (Elect.) 158
DECKWALZE (*f*) : covering vortex-(of tail race) (Water power Eng.) 159
DEKADE (*f*) : decade 160
DEKLINATION (*f*) : declination 161
DEKORATIONSBELEUCHTUNG (*f*) : decorative (scenery) lighting (of stages) 162
DEKORATIONSSTOFF (*m*) : furnishing fabric (Textile) 163
DEKREMETER (*n*) : decremeter (for measuring equivalent logarithmic decrement) 164
DELNOTIZ (*f*) : copper market quotation (of the Vereinigung fuer die deutsche *E*lektrolyt-kupfer-Notiz) 165
DEMENZ (*f*) : dementia (Med.) 166
DEMENZ (*f*), SENILE- : senile decay (Med.) 167
DENDROMETER (*n*) : dendrometer (for measuring diameter and height of trees) 168
DENKSCHRIFT (*f*) : aide-memoire 169
DENUDATION (*f*) : denudation ; stripping 170
DEPOT (*n*) : depot ; storehouse ; magazine ; deposit 171
DETEKTORAPPARAT (*m*) : crystal set (Radio) 172
DETEKTOR (*m*), EINFACHER- : single (point) contact detector (Radio) 173
DETEKTORKOPPLUNG (*f*) : detector coupling (Radio) 174
DETEKTORKREIS (*m*) : detector circuit 175
DETEKTORKRISTALL (*m*) : (detector)-crystal (Radio) 176
DETEKTORRÖHRE (*f*) : detector valve 177
DEVISENGESCHÄFT (*n*) : foreign exchange business 178
DEVONISCH : Devonian (Geol.) 179
DEVONZEIT (*f*) : Devonian period (Geol.) 180
DEZIMALE (*f*) : decimal 181
DEZIMIEREN : to decimate 182
DIATHERMIE (*f*) : diathermy (Med.) 183
DICHTE (*f*), ELEKTRISCHE- : electrostatic flux density 184
DICKDARM (*m*), ABSTEIGENDER- : descending colon (Med.) 185
DICKDARM (*m*), AUFSTEIGENDER- : ascending colon (Med.) 186
DICKDARMENTZÜNDUNG (*f*) : colitis (inflammation of the colon) (Med.) 187
DICKDARM (*m*), QUERER- : transverse colon (Med.) 188
DIELEKTRISIERUNGSZAHL (*f*) : dielectric constant ; dielectric coefficient 189
DIENSTLEITUNG (*f*) : order wire (Tel.) 190
DIESELHORST-MARTIN-VERSEILUNG (*f*) : multiple twin formation (Tel.) 191
DIFFERENTIALAUSSCHALTER (*m*) : differential switch (Elect.) 192
DIFFERENTIALDROSSEL (*f*) : differential coil (Tel.) 193
DIFFERENTIALGETRIEBE (*n*) : differential gear (Mech.) 194
DINORM (*f*) : (*D*EUTSCHE *IN*DUSTRIE NORM) ; German Industrial Standard 195
DIOPTERGLAS (*n*) : dioptric glass 196
DIOPTRIK (*f*) : dioptrics 197
DIOPTRISCH : dioptric 198
DIOPTRISCHE GÜRTELLINSE (*f*) : central belt lens with refractors ; dispersive (dioptric) lens 199
DIPLOPIE (*f*) : diplopia ; diplopy ; double sight (Med.) 200
DISKANTSCHLÜSSEL (*m*) : soprano clef (C clef on the bottom line of the stave, making that line c') (Music) 201
DISLOKATION (*f*) : dislocation (Geol.) 202
DISLOZIEREN : to dislocate 203
DISRUPTIVE ENTLADUNG (*f*) : spark discharge (Elect.) 204
DISSONANZ (*f*), ELEKTRISCHE- : electrical dissonance (non-agreement between two alternating currents of different periodicity) 205
DISTANZ (*f*) : distance 206
DISTANZBLECH (*n*) : distance piece (Mech.) 207
DISTANZIEREN : to separate ; fix at a distance from 208
DOMINANTE (*f*) : dominant ; fifth (Music) 209
DOMINANTE EIGENSCHAFTEN (*f.pl*) : predominant characteristics (Med.) 210
DOMINANTTONART (*f*) : dominant key (Music) 211
DOMINIEREN : to dominate 212
DOPPEL-BE (*n*) : double flat (♭♭) (Music) 213
DOPPELBELICHTUNG (*f*) : double exposure (Phot.) 214
DOPPELERDSCHLUSS (*m*) : double earth circuit (Elect.) 215
DOPPELFOKUSRÖHRE (*f*) : double focus tube (X-rays) 216
DOPPELGEGENSPRECHEN (*n*) : quadruplex telegraphy 217
DOPPELGESTALTIG : biform 218
DOPPELGITTERRÖHRE (*f*) : double grid valve (Radio) 219
DOPPELKÄFIGMOTOR (*m*) : double (cage) bar motor (Elect.) 220
DOPPELNUTLÄUFER (*m*) : double (cage) bar rotor (Elect.) 221

DOPPELNUTMOTOR (*m*) : double (cage) bar motor (Elect.) 222
DOPPELREIHIG : in two rows 223
DOPPELSCHLAG (*m*) : turn (Music) 224
DOPPELSCHLUSSMASCHINE (*f*) : compound-wound dynamo 225
DOPPELSCHLUSSMOTOR (*m*) : compound-wound motor (Elect.) 226
DOPPELSPRACHIG : bilingual 227
DOPPELSPRECHBETRIEB (*m*) : phantom operation (Long distance Telephony) 228
DOPPELSTABLÄUFER (*m*) : double (cage) bar rotor (Elect.) 229
DOPPELSTABMOTOR (*m*) : double (cage) bar motor 230
DOPPEL-T-ANKER (*m*) : shuttle armature (Elect.) 231
DOPPELTDEKAPIERT : double pickled (Metal.) 232
DOPPELTELEGRAPHIE (*f*) : duplex telegraphy 233
DOPPELTELEMOTOR (*m*) : twin telemotor 234
DOPPELTRILLER (*m*) : trill in double (notes) stopping (Music) 235
DOPPELTSEHEN (*n*) : diplopia ; diplopy ; double sight (Med.) 236
DOSENWECKER (*m*) : under-dome bell-(alarm) 237
DRAHTDURCHHANG (*m*) : sag-(of wire) 238
DRAHTFÖRMIG : in the form of wire 239
DRAHTFUNK (*m*) : relayed wireless ; wireless transmitted by land-line (Radio) 240
DRAHTLAMPE (*f*) : metal filament lamp 241
DRAHTLITZE (*f*) : wire heald ; stranded wire 242
DRAHTSPANNUNG (*f*) : tension-(of wire) 243
DRAHTSPULE (*f*) : solenoid (Elect.) 244
DRAHTWINDE (*f*) : wire tensioner 245
DRAHTWURM (*m*) : wire worm (*Agriotes lineatus*) 246
DRAINAGE (*f*) : drainage (Med.) 247
DRAINSCHLAUCH (*m*) : drainage tube (Med.) 248
DREHEISENRELAIS (*n*) : moving iron relay 249
DREHEREI (*f*) : turning shop (Mech.) 250
DREHFELDMOTOR (*m*) : (see DREHSTROMMOTOR) 251
DREHFELDPRINZIP (*n*) : Ferraris principle 252
DREHKNOPF (*m*) : (rotary)-knob (Radio) 253
DREHKONDENSATOR (*m*) : variable condenser (Radio) 254
DREHSCHALTER (*m*) : rotary switch (Elect.) 255
DREHSPULENINSTRUMENT (*n*) : moving coil instrument 256
DREHSTROMANTRIEB (*m*) : (3-phase)-alternating-current drive 257
DREHSTROMDYNAMO (*f*) : three-phase alternator 258
DREHSTROMERREGUNG (*f*) : alternating current (A.C.) excitation (Elect.) 259
DREHSTROMERZEUGER (*m*) : three-phase alternating current generator 260
DREHSTROMGENERATOR (*m*) : turbo-generator ; three-phase alternating current generator 261
DREHSTROMKOLLEKTORMOTOR (*m*) : 3-phase commutator motor 262
DREHSTROMZÄHLER (*m*) : three-phase (alternating current) meter (Elect.) 263
DREHZAHLMESSER (*m*) : tachometer ; revolution counter ; speed indicator 264
DREHZAHLREGELUNG (*f*) : speed regulation 265
DREHZAPFEN (*m*) : pivot 266
DREIECKSTERNSCHALTER (*m*): star-delta switch (Elect.) 267
DREIELEKTRODENRÖHRE (*f*) : triode (Radio) 268
DREIETAGEN-ZWIRNMASCHINE (*f*) : twisting frame (having three tiers of twisting spindles) (Textile) 269
DREIGESTALTIG : trimorphous ; trimorphic 270
DREILEITERKABEL (*n*) : three-core cable 271
DREIPHASENGENERATOR (*m*) : three-phase current generator 272
DREIPHASENMOTOR (*m*) : three-phase motor 273
DREIPHASIGER WECHSELSTROM (*m*) : three-phase alternating current 274
DREIRÖHRENVERSTÄRKER (*m*) : three-valve amplifier (Radio) 275
DREISPÄNNERHAMMER (*m*) : three-man hammer 276
DREISTUFIG : three-stage 277
DREI-TAGE-FIEBER (*n*) : tertian fever (Med.) 278
DRITTE SCHIENE (*f*) : third rail (of electric traction) 279
DROSSELKLAPPENHAUS (*n*) : gate house (Water power) 280
DROSSELKONDENSATORKOPPLUNG (*f*) : choke coupling 281
DROSSELSPULENKUPPLUNGSELEMENT (*n*) : choke coupling unit 282
DROSSELWIDERSTAND (*m*) : choking resistance (Elect.) 283
DRUCKABFALL (*m*) : pressure (reduction) drop ; loss of head (Water) 284
DRUCKFORM (*f*) : forme (Printing) 285
DRUCKKESSEL (*m*) : pressure (boiler) tank 286
DRUCKKNOPF-LINIENWÄHLER (*m*) : push-button line selector (Tel.) 287
DRUCKKNOPFSCHALTER (*m*) : push-button switch 288
DRUCKKNOPFSTEUERUNG (*f*) : push-button control (Elect.) 289
DRUCKKONTAKT (*m*) : push-button contact 290
DRUCKMESSGERÄT (*n*) : pressure measuring (instrument) apparatus 291
DRUCKSCHEIBE (*f*) : (thrust)-washer 292
DRUCKSTELLE (*f*) : compression point (of arteries) (Med.) 293
DRUCKSTOCK (*m*) : block (Printing) 294
DRUCKSTOLLEN (*m*) : penstock 295
DRUCKTELEGRAPH (*m*) : type-printing telegraphic apparatus 296
DRUCKUNTERSCHIEDSPRINZIP (*n*) : differential pressure principle 297
DRÜSENBLÄTTRIG : with glandular leaves (*adenophyllus*) (Bot.) 298
DRÜSENENTZÜNDUNG (*f*) : adenitis (inflammation of a gland) (Med.) 299
DRÜSENSEKRET (*n*) : glandular secretion (Med.) 300
DUNKELRAUM (*m*), CROOKESCHER- : (Crooke's)-dark (space) spot, (between cathode and cathode glow) 301
DUNKELRAUM (*m*), FARADAYSCHER- : (Faraday's)-dark (space) spot (before cathode glow) (of X-ray tubes) 302
DUNKELRAUM (*m*), HITTORFSCHER- : (Hittorf's)-dark (space) spot (see DUNKELRAUM, CROOKESCHER-) 303
DUNKELSCHALTER (*m*) : dimmer ; dimming switch ; dimming resistance (Elect.) 304
DUODECIME (*f*) : twelfth (Music) (see under TERZ) 305
DUOLATERALSPULE (*f*) : honeycomb coil (Radio) 306
DUPLEXTELEGRAPHIE (*f*) : duplex telegraphy 307
DUR : major (Music) 308
DURCHBRENNEN : to burn out ; fuse ; melt ; blow-(out) 309
DURCHBRUCHSPANNUNG (*f*) : break-down voltage 310
DURCHDRINGUNGSFÄHIGKEIT (*f*) : penetrability ; permeability ; penetrative power ; degree of hardness of X-rays ; filtration 311
DURCHFLUTUNG (*f*) : magneto-motive force 312
DURCHFÜHRUNG (*f*) : lead-in (opening, pipe or tube) (Radio) 313
DURCHGANGSAMT (*n*) : through-line section (Tel.) 314
DURCHGANGSLEISTUNG (*f*) : continuous (load) rating 315

DURCHGEHEND : through-connected (Tel.) ; going through 316
DURCHHANG (*m*) : sag (of a rope or cable) 317
DURCHHÄNGEN : to sag ; (*n*) : sagging ; sag 318
DURCHHANGSKURVE (*f*) : sag curve (of overhead wires) 319
DURCHLASSBEREICH (*m*) : transmission range (Tel.) 320
DURCHLÄSSIGKEIT (*f*) : permeability ; dielectric constant ; penetrability 321
DURCHLÄSSIGKEIT (*f*), ELEKTRISCHE- : dielectric (constant) coefficient ; inductive capacity ; inductivity ; permittivity (Elect.) 322
DURCHLAUFERHITZER (*m*) : circulating water heater 323
DURCHLEUCHTUNG (*f*) : illumination ; irradiation 324
DURCHLEUCHTUNGSAPPARAT (*m*) : diaphanoscope 325
DURCHRATE (*f*) : through rate (Shipping) 326
DURCHSCHLAGSFESTIGKEIT (*f*) : dielectric strength ; electric limit ; breaking-down voltage (Elect.) ; disruptive (strength) power 327
DURCHSCHLAGSSPANNUNG (*f*) : break-down voltage 328
DURCHSCHMELZEN : to burn out ; fuse ; go (of fuses) ; blow 329
DURCHSICHTSUCHER (*m*) : vertical (direct) view-finder (Phot.) 330
DURDREIKLANG (*m*) : major triad (Music) 331
DYADISCH : Dyassic (Geol.) 332
DYAS (*f*) : Dyas ; Permian system (Geol.) 333
DYAS (*f*), OBERE- : upper Dyas (Geol.) 334
DYSPHAGIE (*f*) : difficulty in swallowing (Med.) 335
DYSPROSIUMOXYD (*n*) : dysprosium oxide ; Dy_2O_3 336

E

EBERRAUTENÄHNLICH : resembling southernwood (*abrotanoides*) (Bot.) 337
ECHODÄMPFUNG (*f*) : return loss (Tel.) ; damping (of sound) 338
ECHOSPERRER (*m*) : echo suppressor (Tel.) 339
EDISONFASSUNG (*f*) MIT E.27 GEWINDE : Edison screw lampholder (Elect.) 340
EDISONFASSUNG (*f*) MIT E.40 GEWINDE : Goliath lampholder (Elect.) 341
EFFEKT (*m*), ELEKTRISCHER- : wattage (Elect.) 342
EFFEKTGARN (*n*) : fancy yarn (Textile) 343
EFFEKTIVER WIDERSTAND (*m*) : effective resistance 344
EFFEKTIVE STROMSTÄRKE (*f*) : effective (mean) value of current intensity 345
EFFEKTIVSPANNUNG (*f*) : effective (mean) voltage 346
EFFLORESZENZ (*f*) : efflorescence 347
EFFLUVIOGRAPHIE (*f*) : galvanography 348
EICHBRETT (*n*) : test bench 349
EICHHORNKÄFIG (*m*) : squirrel-cage (Elect.) 350
EICHSPANNUNG (*f*) : calibration voltage (Elect.) 351
EICHSTROM (*m*) : calibrating current (Elect.) 352
EICHZÄHLER (*m*) : calibrating meter ; test meter 353
EIERISOLATOR (*m*) : egg insulator (Elect.) 354
EIFÖRMIG, BREIT- : elliptical ; oval ; ovate (Bot.) 355
EIFÖRMIG, UMGEKEHRT- : obovate ; inversely egg-shaped (Bot.) 356
EIGENABFALL (*m*) : spontaneous (drop) discharge 357
EIGENERREGUNG (*f*) : self-excitation (Elect.) 358
EIGENINDUKTIVITÄT (*f*) : self-inductivity ; self-inductance (Elect.) 359
EIGENSCHWINGUNGSZAHL (*f*) : natural vibration frequency 360
EIGENSTRAHLUNG (*f*) : fluorescence 361
EIGENTON (*m*) : individual (note) tone 362
EIGENWELLENLÄNGE (*f*) : natural wave-length 363
EIGENWORT (*n*) : adjective (Grammar) 364
EIGENWORT (*n*), SUBSTANTIVIERTES- : adjectival noun (Grammar) 365
EILÄNGLICH : long egg-shaped (Bot.) 366
EILANZETTLICH : ovata-lanceolate (between egg and lance shaped) (Bot.) 367
EILREGULIERUNG (*f*) : speed regulation 368
EINATMUNG (*f*) : breathing in ; inspiration ; inhalation ; drawing air into the lungs (Med.) 369
EINBAUMÖGLICHKEIT (*f*) : method (possibility) of installation 370
EINBRAND (*m*) : penetration (Welding) 371
EINBRUCHMELDEANLAGE (*f*) : burglar alarm 372
EINDEICHEN : to dike (see DEICHEN) 373
EINDEICHUNG (*f*) : construction of dikes ; dike construction 374
EINDRÄHTIG : unifilar ; single-wired ; having one wire 375
EINFACHVERKEHR (*m*) : simplex (operation) traffic (Tel.) 376
EINFÄDIG : unifilar ; single-threaded ; having one thread 377
EINFALLSTRECK (*f*) : dip roadway (Mining) 378
EINFARBENVERSATZ (*m*) : single colour dip (Stage lighting) 379
EINFÜHRUNGSDRAHT (*m*) : lead-in wire (Radio) 380
EINFÜHRUNGSISOLATOR (*m*) : lead-in insulator (Radio) 381
EINFÜHRUNGSPFEIFE (*f*) : lead-in tube or pipe (Radio) 382
EINFÜHRUNGSROHR (*n*) : lead-in tube or pipe (Radio) ; inlet tube or pipe 383
EINFÜHRUNGSSCHACHT (*m*) : cable pit 384
EINGANGSKREIS (*m*) : input circuit (Radio) 385
EINGANGSSPANNUNG (*f*) : input voltage ; initial voltage (Elect.) 386
EINGANGSTRANSFORMATOR (*m*) : input transformer 387
EINGANGSÜBERTRAGER (*m*) : input transformer (Elect.) 388
EINGESTRICHENE OKTAVE (*f*) : one-lined octave (Music) (the octave commencing on middle C) 389
EINGEWEIDENWÜRMER (*m.pl.*) : entozoa 390
EINGIPSEN : to cement in ; (*n*) : cementing in 391
EINGITTERRÖHRE (*f*) : single grid valve 392
EINHEIT (*f*), PHOTOMETRISCHE- : photometric unit 393
EINHEIT (*f*), PRAKTISCHE- : practical unit 394
EINHEITSFREQUENZ (*f*) : standard frequency 395
EINHEITSKERZE (*f*) : unit candle (lighting intensity of a paraffin candle 5 cm. high by 2 cm. dia.) 396
EINHEITSKRAFTLINIE (*f*) : unit magnetic line of force 397
EINHEITSLADUNG (*f*) : unit charge (Elect.) 398
EINHEITSLICHTQUELLE (*f*) : photometric unit ; light unit 399
EINHEITSPOL (*m*) : unit magnetic pole 400
EINKERBUNG (*f*) : notch-(ing) ; indenting ; indentation 401
EINKREISEMPFÄNGER (*m*) : single circuit (receiver ; receiving apparatus) set (Radio) 402

EINKURBEL- : single crank 403
EINLAUFBAUWERK (*n*) : head works ; intake-(works) (of Water power stations) 404
EINLAUFWERK (*n*) : head works ; intake-(works) (of Water power stations) 405
EINLEITERKABEL (*n*) : single-core cable 406
EINPHASEN-KOLLEKTORMOTOR (*m*) : single-phase commutator motor 407
EINPHASEN-KOMMUTATORMOTOR (*m*) : single-phase commutator motor 408
EINPHASENMASCHINE (*f*) : single-phase machine (if generating current " generator ", if using current " motor ") 409
EINREGULIEREN : to regulate (set ; adjust) correctly 410
EINRENKUNG (*f*) : setting ; re-setting (of bones) (Med.) 411
EINRÖHRENVERSTÄRKEP (*m*) : one valve amplifier (Radio) 412
EINRÜCKSTANGE (*f*) : striking rod (Mech.) 413
EINSATZROST (*m*) : grid-(shelf) (of cooking ranges) 414
EINSCHLAGWECKER (*m*) : single-stroke-(alarm) bell 415
EINSCHNAPPEN : to snap in (of a switch) 416
EINSCHNEIDEN : to cut into ; incise ; indent ; excavate ; notch ; engrave 417
EINSCHRAUBSICHERUNG (*f*) : screw-in fuse 418
EINSCHUBLEISTE (*f*) : slide (for lighting fittings) 419
EINSCHUSS (*m*) : weft (Textile) 420
EINSCHWINGUNG (*f*) : building-up (Tel.) 421
EINSCHWINGVORGANG (*m*) : building-up transient (Tel.) 422
EINSCHWINGZEIT (*f*) : building-up time (Tel.) 423
EINSPÄNNERHAMMER (*m*) : one-man hammer 424
EINSPRECHTRICHTER (*m*) : mouth-piece (of telephone) 425
EINSPRECHVORRICHTUNG (*f*) : mouth-piece (of telephone) 426
EINSTABLÄUFER (*m*) : single bar rotor (Elect.) 427
EINSTECKDOSE (*f*) : plug socket (Elect.) 428
EINSTECKEN : to plug (place ; put ; stick) in ; insert 429
EINSTECKKONTAKT (*m*) : plug-in contact 430
EINSTELLBEREICH (*m*) : (adjustable)-range (Elect.) 431
EINSTELLHEBEL (*m*) : setting (adjusting) lever 432
EINSTELLKNOPF (*m*) : control knob 433
EINSTELLREGLER (*m*) : adjusting regulator 434
EINSTELLUNGSPERIODE (*f*) : adjusting (setting) time 435
EINSTICHNADEL (*f*) : acupuncture needle 436
EINSTRAHLUNG (*f*) ; (see INSOLATION) 437
EINSYSTEMIG : (of a ; having a)-single (system) unit 438
EIN-TAG-FIEBER (*n*) : quotidian ague (Med.) 439
EINTÄGIG : ephemeral ; of one day 440
EINTRITTSDAMPFSPANNUNG (*f*) : initial steam pressure ; steam pressure at inlet 441
EINWATTLAMPE (*f*) : one Watt lamp 442
EINWEGIG : one way ; single way 443
EINWEGSCHALTER (*m*) : one-way switch 444
EINWELLIG ; sinusoidal 445
EINZELACHSANTRIEB (*m*) : individual axle drive (Railways) 446
EINZELANTRIEB (*m*) : individual drive (Mech.) 447
EINZELAUFNAHME (*f*) : single (picture) exposure (Phot.) 448
EINZELSCHLÄGER (*m*) : single stroke (bell) alarm 449
EINZIEHMASCHINE (*f*) : drawing-in machine 450
EINZIEHSCHACHT (*m*) ; downcast shaft (Mining) 451
EINZIEHWEG (*m*) : luffing distance (of cranes) 452
EISENBAHNTELEGRAPHIE (*f*) : railway telegraphy 453
EISENBAHNWAGENBELEUCHTUNG (*f*) : (railway)-carriage lighting 454
EISENBLECHPRÜFUNG (*f*) : sheet iron testing 455
EISENKERN (*m*), UNTERTEILTER- : laminated iron core 456
EISENKONSTANTAN-ELEMENT (*n*) : iron constantan couple 457
EISENKURZWAREN (*f.pl.*) : ironmongery 458
EISENOCKER (*m*) : (see OCKER) 459
EISENVERLUST (*m*) : iron loss ; hysteresis loss ; eddy-current loss (of transformers) 460
EISENWASSERSTOFF-WIDERSTAND (*m*) : iron-hydrogen resistance lamp 461
EISENWIDERSTAND (*m*) : iron resistance lamp 462
EISZEIT (*f*) : ice age ; glacial period (Geol.) 463
EITERKNÖTCHEN (*n*) : pimple ; pustule (Med.) 464
EITRIG : purulent ; mattery ; suppurative 465
ELASTANZ (*f*) : elastance (Elect.) 466
ELEFANTENKRANKHEIT (*f*) : Boucnemia ; Barbados leg ; elephantiasis (Med.) 467
ELEKTRIFIZIERUNG (*f*) : electrification ; electrifying 468
ELEKTRISATION (*f*) : electrical treatment 469
ELEKTRISCHE ARBEIT (*f*) : electric (power) energy (the unit is the Watt) 470
ELEKTRISCHE BEHANDLUNG (*f*) : electrical treatment 471
ELEKTRISCHE FELDSTÄRKE (*f*) : intensity (strength) of electric field ; electric force 472
ELEKTRISCHE LEISTUNG (*f*) : (see ELEKTRISCHE ARBEIT) 473
ELEKTRISCHES AUGE (*n*) : electric eye (selenium cell) 474
ELEKTRISCHES BAD (*n*) : galvanic bath ; electro-therapeutic bath 475
ELEKTRIZITÄTSMENGE (*f*) : quantity of electricity (product of current intensity × time) (Elect.) 476
ELEKTRIZITÄTSMESSUNG (*f*) : electrometry 477
ELEKTROCHEMISCHES ÄQUIVALENT (*n*) : electrochemical equivalent (the quantity of an element liberated by 1 amp. in 1 sec.) (Elect.) 478
ELEKTRODE (*f*), BLANKE- : bare wire electrode (Welding) 479
ELEKTRODE (*f*), NEGATIVE- : negative electrode ; cathode 480
ELEKTRODE (*f*), POSITIVE- : positive electrode ; anode 481
ELEKTRODYNAMIK (*f*) : electrodynamics 482
ELEKTRODYNAMISCHES MESSWERK (*n*) : dynamometer movement (Elect.) 483
ELEKTROENDOSMOSE (*f*) : electric endosmose 484
ELEKTROGRAPH (*m*) : apparatus for automatically recording atmospheric electricity 485
ELEKTROGRAPHIE (*f*) : galvanography 486
ELEKTROION (*n*) : (see KATION) 487
ELEKTRO-KARDIOGRAPH (*m*) : electro-cardiograph (for registering the electric currents produced by the muscles of the heart) 488
ELEKTROLYSEUR (*m*) : electrolyser 489
ELEKTROLYTCHLOR (*n*) : electrolytic bleaching lye (sodium hypochlorite NaOCl) 490
ELEKTROMAGNETBREMSE (*f*) : electro-magnetic brake 491
ELEKTROMAGNETIK (*f*) : electro-magnetism 492
ELEKTRONENBOMBARDEMENT (*n*) : electronic bombardment 493
ELEKTRONENEMISSION (*f*) : electronic emission ; emission of electrons (from a cathode) 494
ELEKTRONENRÖHRE (*f*) : electron-(ic) tube or valve 495
ELEKTRONENSTRAHLUNG (*f*) : electronic radiation 496

ELEKTRONENSTROM (*m*) : stream of electrons ; emission current 497
ELEKTRONIK (*f*) : electronics 498
ELEKTRON (*n*), NEGATIVES- : (see KATION) 499
ELEKTRON (*n*), POSITIVES- : (see ANION) 500
ELEKTROPHORESE (*f*) : cataphoresis ; electric (osmose) osmosis (Colloidal chemistry) 501
ELEKTROSTATIK (*f*) : electrostatics 502
ELEKTROTHERAPIE (*f*) : electrotherapy (healing of disease by electricity) 503
ELEKTROTONISCH : electro-tonic 504
ELEMENT (*n*), SEKUNDÄRES- : secondary (battery ; accumulator) cell (Elect:) 505
ELLIPSOIDISCH : elliptical 506
EMAILDRAHT (*m*) : enamelled wire (Elect.) 507
EMISSION (*f*) : emission 508
EMISSIONSSTROM (*m*) : emission current (varies with varying cathode temperatures) 509
EMITTIEREN : to emit ; give out ; exude 510
EMPFÄNGERRELAISKASTEN (*m*) : receiving relay (box) case 511
EMPFÄNGERRÖHRE (*f*) : receiving valve (Radio) 512
EMPFANGSANTENNE (*f*) : receiving aerial (Radio) 513
EMPFANGSAPPARATUR (*f*) : receiving (apparatus) set ; receiver (Radio) 514
EMPFANGSKREIS (*m*) : receiving (receiver) circuit 515
EMPFANGSSTÖRUNG (*f*) : interference (Radio) 516
EMPFANGSVERSTÄRKER (*m*) : reception amplifier (Picture telegraphy) 517
EMPFINDLICHKEIT (*f*) : susceptibility ; sensitiveness 518
EMPFINDSAM : perceptive ; sensitive ; irritable ; sensible ; susceptible (Med.) 519
EMPFINDUNGSNERV (*m*) : sensory (afferent) nerve (Med.) 520
EMULSEUR (*m*) : ejector 521
ENDANSCHLAG (*m*) : stop-(for limiting a movement) 522
ENDOMETRITIS (*f*) : endometritis (inflammation of the membrane lining the womb) (Med.) 523
ENDSTELLUNG (*f*) : terminal (end ; final) position 524
ENDVERSCHLUSS (*m*) : (see KABELENDVERSCHLUSS) 525
ENERGIEBEDARF (*m*) : (power)-input ; power consumption 526
ENERGIE (*f*), ELEKTRISCHE- : electrical energy (electric current doing useful work) 527
ENERGIE (*f*), KINETISCHE- : kinetic energy 528
ENERGIEKOMPONENTE (*f*) : energy component 529
ENERGIE (*f*), LATENTE- : latent energy 530
ENERGIE (*f*), MECHANISCHE- : mechanical energy 531
ENERGIE (*f*), POTENTIELLE- : potential energy 532
ENERGIESTOSS (*m*) : pulsation 533
ENERGIEÜBERTRAGUNG (*f*) : energy transmission ; electric power transmission 534
ENERGIEVERBRAUCH (*m*) : (power)-input ; power consumption 535
ENGLAGIG : in fine (close) layers 536
ENTARTUNG (*f*), FETTIGE- : fatty degeneration (Med.) 537
ENTLADESPANNUNG (*f*) : discharge voltage 538
ENTLADESTROMSTÄRKE (*f*) : discharging current (Elect.) 539
ENTLADEZEIT (*f*) : discharging time 540
ENTLADUNG (*f*), ALLMÄHLICHE- : gradual discharge 541
ENTLADUNGSSPANNUNG (*f*) : discharge (voltage)-potential 542
ENTLADUNGSSTROM (*m*) : discharging current (Elect.) 543
ENTNAHMESCHACHT (*m*) : discharge (tunnel) shaft 544
ENTROPIE (*f*), ELEKTRISCHE- : electrical entropy (electric current not doing useful work) 545
ENTSCHLAMMEN : to remove (mud ; slime ; scum) sediment 546
ENTWICKLUNGSARBEIT (*f*) : development work ; research work 547
ENTZERRER (*m*) : equalizer 548
EPILATION (*f*) : depilation 549
EPIPHYSIS (*f*) : epiphysis (growth or process formed by ossification) (Med.) 550
ERDBEBENLEHRE (*f*) : seismology (the science of earthquakes) 551
ERDBESCHLEUNIGUNG (*f*) : gravitation 552
ERDDRAHT (*m*) : earthing wire ; earth wire (Elect.) 553
ERDE (*f*), AN . . . LEGEN : to earth (Elect.) 554
ERDFERNE (*f*) : apogee 555
ERDKAPAZITÄT (*f*) : wire-to-earth capacity (Tel.) 556
ERDKRÜMMUNG (*f*) : curvature of the earth 557
ERDMAGNETISMUS (*m*) : terrestrial magnetism 558
ERDNÄHE (*f*) : perigee 559
ERDPLATTE (*f*) : earth-(ing) plate (Elect.) 560
ERDRÜCKLEITUNG (*f*) : earth return (Tel.) ; (see also ERDLEITUNG) 561
ERDSCHLUSSANZEIGER (*m*) ; fault indicator 562
ERDSCHLUSSRELAIS (*n*) : earth relay (Elect.) 563
ERDUNGSDRAHT (*m*) : earth wire (Elect.) 564
ERDUNGSKLEMME (*f*) : earthing terminal 565
ERDUNGSSCHALTER (*m*) : earthing switch (Tel.) 566
ERDUNGSSCHRAUBE (*f*) : earthing (screw) terminal 567
ERDWIDERSTAND (*m*) : earth resistance 568
ERLENBLÄTTRIG : alder-leaved ; having leaves like the Alder (*alnifolius*) (Bot.) 569
ERREGERGENERATOR (*m*) : exciter (Elect.) 570
ERREGERKREIS (*m*) : primary (exciting) circuit ; oscillatory circuit (Radio) 571
ERREGERMASCHINE (*f*) : exciter (Elect.) 572
ERREGERSTROMKREIS (*m*) : exciting circuit 573
ERSCHLAFFUNG (*f*), GEISTIGE- : lethargy 574
ERSCHÜTTERUNGSFREI : free from (shock) vibration 575
ERSCHÜTTERUNGSMASSAGE (*f*) : vibro-massage 576
ERSCHWERENDE BEDINGUNGEN (*f.pl.*) : restrictions 577
ERWÄGUNG (*f*) : consideration 578
ERWÄHNEN : to mention 579
ERWERBEN : to acquire ; obtain 580
ERYTHROZYT (*m*) : erythrocyte ; red blood corpuscle (Med.) 581
ETIOLIEREN : to etiolate ; blanch ; (*n*) : etiolation (disease of plants due to lack of light) 582
EXOSMOSE (*f*) : exosmosis 583

F

FADENBALLON (*m*) : balloon-(ing) thread (Textile) 584

FADENELEKTROMETER (*n*) : string electrometer (for measuring differences of potential) 585

FADENFÜHRER (*m*) : thread guide (Textile) 586

FADENGEFLECHT (*n*) : mycelium-(of fungi) (Bot.) 587

FADENKRAFT (*f*) : yarn tension (Textile) 588

FADENSPANNUNG (*f*) : filament (tension ; potential) voltage (Radio) 589

FADINGEFFEKT (*m*) : fading (Radio) 590

FAHNENABZUG (*m*) : rough proof (Printing) 591

FAHRBAHN (*f*) : track 592

FAHRDRAHT (*m*) : trolley wire (Electric traction) 593

FAHRDRAHTBUS (*m*) : trolley bus (Electric traction) 594

FAHRGAST (*m*) : passenger 595

FAHRGASTKAMMER (*f*) : stateroom (on ships) 596

FAHRGASTSCHIFF (*n*) : passenger (vessel) boat ; liner 597

FAHRLÄSSIG : careless ; negligent 598

FAHRLEITUNG (*f*) : trolley wire (Electric traction) 599

FAHRMOTOR (*m*) : portable motor (Elect.) 600

FAHRSCHIENE (*f*) : conductor rail 601

FALLBEIL (*n*) : guillotine 602

FALLBÜGEL (*m*) : chopper bar (Elect.) 603

FALLBÜGELINSTRUMENT (*n*) : chopper-bar instrument (Elect.) 604

FALLSCHEIBE (*f*) : drop indicator (Elect.) ; annunciator (indicator) disc (Elect.) 605

FALL (*m*), VON FALL ZU- : as occasion (serves) arises : taking each case on its own merits 606

FÄRBDISTEL (*f*) : (see SCHARTE) 607

FÄRBEHASPEL (*m*) : dye jigger 608

FARBENABWEICHUNG (*f*) : difference in colour ; chromatic aberration 609

FARBENBLINDHEIT (*f*) : achromatopsia ; achromatopsy ; colour-blindness 610

FARBENWIRKUNG (*f*) : (colour)-effect 611

FARBLOSIGKEIT (*f*) : achromatism ; colourlessness 612

FARBROLLE (*f*) : ink-(ed) roller 613

FARBSCHEIBE (*f*) : colour screen (Lighting) 614

FARBSCHREIBER (*m*) : ink recorder (Tel.) 615

FASERKOHLE (*f*) : fibrous coal (from fossilized conifers) 616

FASSLAMPE (*f*) : handlamp (for casks) 617

FÄULNISPFLANZE (*f*) : saprophyte (which lives on decaying organic matter) 618

FÄULNISVERHÜTEND : anti-fouling ; aseptic 619

FAULSCHLAMMKOHLE (*f*) : matt (dull) coal (coarsely granular ; low carbon content, high hydrogen content) 620

FEDERARBEIT (*f*) : pen work ; pen (and ink) drawing 621

FEDERKLAMMER (*f*) : spring clip 622

FEDERNDER SITZ (*m*) : spring (sprung) seat-(ing) 623

FEDERWERK (*n*) : clockwork 624

FEDERWICKELAPPARAT (*m*) : spring coiling apparatus 625

FEHLERORT (*m*) : location of fault ; faulty place 626

FEHLERORTBESTIMMUNG (*f*) : fault location 627

FEHLERTOLERANZ (*f*) : limit (percentage)- of error ; tolerance 628

FEHLWINKEL (*m*) : (angle of)- phase (difference) displacement -(between primary and secondary of transformers) 629

FEINFLYER (*m*) : roving frame (Textile) 630

FEINGARN (*n*) : finished yarn (Textile) 631

FEINGEWICHT (*n*) : (see PROBIERGEWICHT) 632

FEINKARDE (*f*) : finisher card (Textile) 633

FEINSTUFIG : fine-(ly) stepped ; finely graded 634

FELD (*n*), ELEKTRISCHES- : electric field 635

FELDINTENSITÄT (*f*) : field (density) strength ; (magnetic)-field intensity (Elect.) 636

FELDREGLER (*m*) : field regulator (Elect.) 637

-FELDRIG : having-(fields) panels 638

FERMATE (*f*) : pause ; 𝄐 (Music) 639

FERNAMT (*n*) : trunk exchange (Tel.) 640

FERNANRUF (*m*) : trunk call (Tel.) 641

FERNANTRIEB (*m*) : remote control-(gear) 642

FERNANZEIGE (*f*) : remote (reading)-indication (of instruments) 643

FERNDRUCKER (*m*) : teleprinter 644

FERNGEBER (*m*) : tele-transmitter (Tel.) 645

FERNGESPRÄCH (*n*) : long distance call ; trunk call (Tel.) 646

FERNKABEL (*n*) : trunk line (Tel.) ; long distance line 647

FERNKABELANLAGE (*f*) : long distance cable-(installation) 648

FERNMELDEANLAGE (*f*) : tele-indicator (Elect.) 649

FERNMELDETECHNIK (*f*) : communication engineering 650

FERNMESSANLAGE (*f*) : tele-metering equipment (Elect.) 651

FERNSCHREIBMASCHINE (*f*) : tele-typewriter ; teleprinter 652

FERNSPRECHAMT (*n*), AUTOMATISCHES- : automatic telephone exchange 653

FERNSPRECHANSCHLUSS (*m*) : telephone (station) connection 654

FERNSPRECHAPPARAT (*m*) : telephone apparatus 655

FERNSPRECHAUTOMAT (*m*) : automatic telephone 656

FERNSPRECHBETRIEB (*m*) : telephone service 657

FERNSPRECHEINRICHTUNG (*f*) : telephone apparatus 658

FERNSPRECHERANLAGE (*f*) : telephone installation 659

FERNSPRECHERANSCHLUSS (*m*) : telephonic connection 660

FERNSPRECHGEBÜHREN (*f.pl*) : telephone (charges) rates 661

FERNSPRECHKABEL (*n*) : telephone cable 662

FERNSPRECHÜBERTRAGER (*m*) : repeater (Tel.) 663

FERNSPRECHVERKEHR (*m*) : telephone (traffic)-service 664

FERNSPRECH-VERMITTLUNGSAMT (*n*) : telephone exchange 665

FERNSTEUEREINRICHTUNG (*f*) ; remote control (system ; apparatus) equipment 666

FERNSTEUERUNG (*f*) : distance (distant ; remote)- control 667

FERNSUMMIERUNGSANLAGE (*f*) : remote summation equipment (Elect.) 668

FERNÜBERTRAGUNG (*f*) : tele-transmission 669

FERRARISSCHEIBE (*f*) : Ferraris disc 670

FERSENBEIN (*n*) : tarsus ; ankle-bone (Med.) 671

FERTIGGEWEBE (*n*) : finished fabric (Textile) 672

FESTIGKEIT (*f*), DYNAMISCHE- : dynamic (mechanical) stability 673

FESTIGKEIT (*f*), ELEKTRISCHE- : breakdown voltage (voltage at which an insulating material of 1 cm. thickness will break down) ; electrical stability 674

FESTIGKEITSLEHRE (*f*) : theory of strength of materials 675

FESTIGKEIT (*f*), THERMISCHE- : thermal stability 676
FETTANSATZ (*m*) : adipose tissue (Med.) 677
FETTANSATZ (*m*), ÜBERFLÜSSIGER- : superfluous fat 678
FETTGESCHWULST (*f*) : lipoma (a fatty tumour) (Med.) 679
FETTPRESSE (*f*) : grease gun 680
FEUCHTGLÄTTE (*f*) : smoothing rolls (Paper) 681
FEUCHTGLÄTTWERK (*n*) : smoothing rolls (Paper) 682
FEUCHTIGKEIT (*f*), ABSOLUTE- : absolute (moisture) humidity (amount of water in grams, in 1 cubic metre of air) 683
FEUCHTIGKEIT (*f*), RELATIVE- : degree (percentage) of saturation 684
FEUCHTIGKEITSBESCHAFFENHEIT (*f*) : humidity 685
FEUCHTIGKEITSMESSER (*m*) : psychrometer (wet and dry bulb hydrometer) 686
FEUERALARM (*m*) : fire alarm 687
FEUERLÖSCHER (*m*) : fire extinguisher 688
FEUERVERZINKT : hot galvanized 689
FEUERVERZINNT : hot tinned 690
FIBRINÖS : fibrous (Med.) 691
FIEBERHARN (*m*) : the copious flow of thick urine during a fever (Med.) 692
FIEBER (*n*), HEKTISCHES- : hectic fever (remittent in its course) 693
FIEBER (*n*), INTERMITTIERENDES- : intermittent fever (periodic subsidence of symptoms) 694
FIEBER (*n*), KONTINUIERLICHES- : febricula (continued fever : temperature persists at about same level till crisis) 695
FIEBERKRIESE (*f*) : crisis (sudden change) 696
FIEBERKURVE (*f*) : the curve on a temperature chart 697
FIEBERLYSIS (*f*) : lysis (gradual change) 698
FIEBER (*n*), REKURRIERENDES- : recurrent fever (Med.) 699
FIEBER (*n*), REMITTIERENDES- : remittent fever (short daily diminution of symptoms) 700
FIEBERTABELLE (*f*) : temperature chart (Med.) 701
FIEBERTEMPERATUR (*f*) : febrile temperature 702
FIEBERVERLAUF (*m*) : course of the fever 703
FIEBER (*n*), WECHSEL- : intermittent fever ; ague 704
FIEBER (*n*), WUND- : traumatic fever (fever from wounds) 705
FIEBERZUSTAND (*m*) : febrile condition (Med.) 706
FILMZÄHLER (*m*) : footage (counter) indicator (of cinematograph films) 707
FILTERGESCHWINDIGKEIT (*f*) : filtering speed ; speed of flow-(through a filter) 708
FILTERKERZE (*f*) : filter candle 709
FILTERKREIS (*m*) : filter circuit ; wave trap (Radio) 710
FILTER (*n*), SCHNELL- : high-speed filter 711
FILTRIERVORGANG (*m*) : filtering process 712
FILZ (*m*), OBER- : upper felt (Papermaking) 713
FILZTROCKNER (*m*) : felt drying roll (Paper) 714
FILZTUCHWEBSTUHL (*m*) : felt loom (Textile) 715
FILZ (*m*), UNTER- : lower felt (Papermaking) 716
FILZWASCHPRESSE (*f*) : felt washing apparatus (Paper) 717
FINNENAUSSCHLAG (*m*) : Acne vulgaris (Med.) 718
FIRST (*m*) : roof (Mining) 719
FISSUR (*f*) : fissure (Med.) 720
FISTELSTIMME (*f*) : falsetto (of the voice) 721
FLÄCHENHELLIGKEIT (*f*) : lighting intensity (per unit of surface) 722
FLACHSSPINNMASCHINE (*f*) : flax spinning machine (Textile) 723
FLACHSTRAHLER (*m*) : distributing (lighting)-fitting 724
FLACHSTRICKMASCHINE (*f*) : flat-frame knitting machine (Textile) 725
FLAMMENBLUME (*f*) : phlox (Bot.) 726
FLAMMENTON (*m*) : coloured (variegated) clay 727
FLANKENDURCHMESSER (*m*) : pitch diameter-(of screw threads) 728
FLANKENKEHLNAHT (*f*) : (lateral)-strap fillet weld 729
FLANKENWINKEL (*m*) : angle of thread (of screw threads) 730
FLASCHENKREIS (*m*) : Leyden jar circuit (Elect.) (acts as a condenser) 731
FLATTEREFFEKT (*m*) : flutter (Tel.) (of Pupin coils) 732
FLECHTE (*f*), SCHERENDE- : Herpes tonsurans (Med.) 733
FLECHTE (*f*), ZEHRENDE- : (see LUPUS) 734
FLEIER (*m*) : flyer frame (Textile) : (see also FLÜGELSPINNMASCHINE) 735
FLIEDERTEE (*m*) : elder flower tea (Med.) 736
FLIEGERAUFNAHME (*f*) ; aerial photograph ; bird's-eye view 737
FLIEHKRAFTLÜFTER (*m*) : centrifugal fan 738
FLIEHKRAFTREIBUNGSKUPPLUNG (*f*) : centrifugal friction coupling 739
FLIESSBETRIEB (*m*) : mass production-(on the conveyor belt system) 740
FLORENZIUM (*n*) : Florenzium (a rare earth) 741
FLÖTZKLUFT (*f*) : fault (Mining) 742
FLUGBRAND (*m*) : smut ; dust-brand (*Ustilago*) (Bot.) 743
FLÜGELEINZELANTRIEB (*m*) : individually driven flyer spindle (Textile) 744
FLÜGELRADWASSERMESSER (*m*) : vane wheel water meter 745
FLÜGELSPINNMASCHINE (*f*) : flyer spinning frame (Textile) 746
FLUGHALLE (*f*) : hangar 747
FLUGMOTOR (*m*) : aero-engine 748
FLUGPLATZBELEUCHTUNG (*f*) : aerodrome lighting 749
FLUGPLATZSCHEINWERFER (*m*) : aerodrome searchlight 750
FLUGPOST (*f*) : air mail 751
FLUGSTRECKENFEUER (*n*) : airway beacon 752
FLUGWETTERWARTE (*f*) : aerodrome weather control station 753
FLUGZEUG (*n*) : aeroplane 754
FLUGZEUGZELLE (*f*) : fuselage 755
FLUSSABWÄRTS : down (river) stream 756
FLUSSAUFWÄRTS : up (river) stream 757
FLÜSSIGKEITSANLASSER (*m*) : liquid starter (Elect.) 758
FLÜSSIGKEITSWIDERSTAND (*m*) : liquid resistance (Elect.) 759
FLUSSLAUF (*m*) : flow ; stretch -(of a river) 760
FLUSSOHLE (*f*) : river (bed) bottom 761
FLUSSTAHLBLECH (*n*) : mild steel plate 762
FLUTLICHT (*n*) : flood light 763
FLUTLICHTLEUCHTE (*f*) : floodlight (unit) fitting 764
FLUX (*m*) : magnetic flux (see FLUSS) 765
FLYER (*m*) : (see FLEIER) 766
FOKUS (*m*) : focus ; focal spot (of X-ray tubes) 767
FOKUSPLATTENABSTAND (*m*) : focus-film distance (of X-ray photography) 768
FOKUSRÖHRE (*f*) : focus tube (of X-rays) 769
FOKUSRÖHRE (*f*), DOPPEL- : double focus tube 770
FOKUSRÖHRE (*f*), PUNKT- : spot focus tube 771
FOKUSRÖHRE (*f*), STRICH- : line focus tube 772
FÖRDERGERÜST (*n*) : headgear (Mining) 773
FÖRDERKORB (*m*) : cage (Mining) 774
FÖRDERKÜBEL (*m*) : sinking kibble (Mining) 775
FÖRDERMITTEL (*n*) : (coal)-conveyor 776
FÖRDERTURM (*m*) : headgear (Mining) 777

FORMERANLAGE (*f*), HANDBEDIENTE- : hand-(operated)-trimming plant (Textile) 778
FORMIERTE OXYDKATHODE (*f*) : combined oxide-(d) cathode 779
FORMWORT (*n*) : preposition (Grammar) 780
FORTFALLEN : to be (left out) omitted 781
FORTKOCHEN : to simmer ; (*n*) : simmering 782
FORTPFLANZUNGSGESCHWINDIGKEIT (*f*) : speed of propagation (Elect.) 783
FORTSCHELLKONTAKT (*m*) : continuous ringing contact (Tel.) 784
FOTOZELLE (*f*) : photo-electric cell 785
FRACHTER (*m*) : freighter ; tramp-(steamer) 786
FRAUENVEILCHEN (*n*) : (see NACHTVIOLE) 787
FRAUNHOFERLINIE (*f*) : Fraunhofer line ; fixed line -(in the spectrum) 788
FREIANTENNE (*f*) : outdoor aerial 789
FREILAUFKUPPLUNG (*f*) : free-wheel (coupling) clutch 790
FREILEITUNG (*f*) : overhead-(transmission) line ; overhead wire ; aerial (wire) cable 791
FREILEITUNGSNETZ (*n*) : overhead line system (Elect.) 792
FREILEITUNGSSICHERUNG (*f*) : aerial fuse (Elect.) 793
FREILUFTSCHALTANLAGE (*f*) : outdoor (open-air) transformer station (Elect.) 794
FREISPRECHER (*m*) : loud-speaking telephone 795
FREISTRAHLER (*m*) : dispersive (lighting)-fitting 796
FREISTRAHLTURBINE (*f*) : reaction turbine 797
FREIZEICHEN (*n*) : disengaged sign (Tel.) ; dialling tone 798
FREMDERREGT : separately excited (Elect.) 799
FREMDERREGUNG (*f*) : separate excitation 800
FREMDFELD (*n*) : stray field (Elect.) 801
FREMDFELDSCHUTZ (*m*) : protection against stray magnetic fields 802
FREQUENZBEREICH (*m*) : frequency band (Radio) 803
FREQUENZGEMISCH (*n*) : frequency mixture (Tel.) ; frequency complex 804
FREQUENZMESSER (*m*) : frequency meter (Elect.) 805
FREQUENZTRANSFORMATOR (*m*) : frequency changer 806
FREQUENZUMFORMER (*m*) : frequency changer (Elect.) 807
FREQUENZUMFORMUNG (*f*) : frequency conversion (Elect.) 808
FRIKTIONSANTRIEB (*m*) : friction drive 809
FRUCHTHYPHE (*f*) : spore-bearing hypha (Bacteriology) 810
FRÜCHTIG, ROT- : red-fruited ; red-fruiting ; bearing red fruit (Bot.) 811
FRÜCHTIG, SCHWARZ- : black-fruited ; black-fruiting ; bearing black fruit (Bot.) 812
FÜHRERHAUS (*n*) : control house ; engine house 813
FÜHRERSTAND (*m*) : driver's (stand ; cab) platform 814
FÜLLORT (*m*) : pit bottom (Mining) 815
FÜLLSÄURE (*f*) : accumulator-(charging) acid ; sulphuric acid 816
FUNDAMENTSCHIENE (*f*) : slide rail (Elect.) 817
FÜNFELEKTRODENRÖHRE (*f*) : pentode (Radio) 818
FÜNFERALPHABET (*n*) : five letter code 819
FÜNFLEITERSYSTEM (*n*) : five wire system (Elect.) 820
FUNKENFREI : free from sparking 821
FUNKENLÖSCHER (*m*) : spark (extinguisher) quencher 822
FUNKENLÖSCHUNG (*f*), MAGNETISCHE- : magnetic blow-out (Elect.) 823
FUNKENREISSER (*m*) : arcing (sparking ; secondary ; auxiliary break) contact (Elect.) 824
FUNKENSENDER (*m*) : spark transmitter (Radio) 825
FUNKEN (*m.pl*), TÖNENDE- : singing sparks (Radio) 826
FUNKER (*m*) : wireless (telegraphist) operator 827
FUNKPEILER (*m*) : radio direction finder 828
FUNKTURM (*m*) : radio tower 829
FUNKZWISCHENSTELLE (*f*) : (wireless) relay-(ing) station 830
FÜRWORT (*n*), BESITZANZEIGENDES- : possessive pronoun (Grammar) 831
FÜRWORT (*n*), PERSÖNLICHES- : personal pronoun (Grammar) 832
FUSSBODENKONTAKT (*m*) : floor contact (Elect.) 833
FUSSHEBEL (*m*) : pedal ; treddle ; treadle 834
-FÜSSIG : -footed (of animals) ; -foot (of organ pipes) 835
FUSSPFLEGE (*f*) : pedicure 836
FUSSRAMPE (*f*) : float ; footlights (of stage lighting) 837
FUSSTASTATUR (*f*) : pedals ; pedal board-(of an organ) 838
FUTTERKAMMER (*f*) : hayloft ; any place for storing fodder 839

G

GABELUNG (*f*) : bifurcation ; forking 840
GALLENBLASENENTZÜNDUNG (*f*) : cholecystitis (inflammation of the gall bladder) (Med.) 841
GALLENGANGSENTZÜNDUNG (*f*) : cholangitis (inflammation of the bile duct) (Med.) 842
GALLENKOLIK (*f*) : hepatic colic (Med.) 843
GALLENSTEINANFALL (*m*) : hepatic colic (Med.) 844
GALLENWEG (*m*) : bile duct ; biliary (ducts) vessels (Med.) 845
GALVANISEUR (*m*) : galvanizer 846
GANZTON (*m*) : whole tone (Music) 847
GARDINENSTUHL (*m*) : curtain loom (Textile) 848
GARNNUMMER (*f*) : count ; (number of)-counts (Textile) 849
GARNNUMMERZEIGER (*m*) : yarn counter (Textile) 850
GARNSPANNUNG (*f*) : yarn tension (Textile) 851
GÄRUNGSERZEUGER (*m*) : ferment 852
GASANSCHLUSS (*m*) : gas connection 853
GASFÜLLUNG (*f*) : gas content 854
GASGEHALT (*m*) : gas content 855
GASSCHMELZSCHWEISSVERFAHREN (*n*) : gas welding-(process) 856
GASTROTOMIE (*f*) : gastrotomy ; cutting open the stomach (Med.) 857
GASVERBRAUCH (*m*) : gas consumption 858
GAUTSCHPRESSE (*f*) : (wet)-press rolls (Paper) 859
GEARTET : of such a nature ; of this (that) nature, kind or sort 860
GEBÄRMUTTERHALS (*m*) : cervix (Med.) 861
GEBÄRMUTTERMUSKELENTZÜNDUNG (*f*) : metritis (Med.) 862
GEBÄRMUTTERSCHLEIMHAUTENTZÜNDUNG (*f*) ; endometritis (Med.) 863
GEBERRELAISKASTEN (*m*) : transmitting relay (box) case 864

GEBIRGSBESCHREIBUNG (*f*) : orography (geography of mountains) 865
GEBIRGSMESSUNG (*f*) : orometry (the measurement of mountain heights) 866
GEFÄLLE (*n*), NUTZBARES- : effective head-(of water) 867
GEGENDREHMOMENT (*n*) : resisting torque 868
GEGENINDUKTIVITÄT (*f*) : mutual inductance (Elect.) 869
GEGENMITSPRECHEN (*n*) : far-end overhearing (Tel.) 870
GEGENNEBENSPRECHEN (*n*) : far-end cross-talk (Tel.) 871
GEGENSPANNUNG (*f*) : back (counter) electromotive force (E.M.F.) 872
GEGENSPRECHEN (*n*) : duplex telegraphy (Tel.) 873
GEGENSTROMROLLE (*f*) : induction coil 874
GEGENTAKTSCHALTUNG (*f*) : push-pull (circuit) connection (Elect.) 875
GEGENTAKT-TRANSFORMATOR (*m*) : push-pull transformer (Elect.) 876
GEGENÜBERSPRECHEN (*n*) : far-end cross-talk (Tel.) 877
GEGENZEICHNEN : to countersign ; supply one of a number of signatures-(to documents) ; (see also UNTERZEICHNEN) 878
GEGENZEIGERSTROM (*m*) : anti-clockwise (current) circuit 879
GEHÄUSEOBERTEIL (*m*) : upper part of yoke (Elect. motors) ; upper part of (housing ; case) casing 880
GEHIRNKNÖCHELCHEN (*n.pl*) : the three small bones of the middle ear (*Malleus* : hammer) (*Incus* : anvil) (*Stapes* : stirrup) (Med.) 881
GEHIRNSCHLAG (*m*) : apoplexy (Med.) 882
GEHIRNVERKALKUNG (*f*) : arterio-sclerosis ; senile decay (Med.) 883
GEHÖLZKUNDE (*f*) ; dendrology (science of trees) 884
GEHÖRGANG (*m*) : meatus ; auditory canal (Med.) 885
GEHÖRGANG (*m*), ÄUSSERER- : external auditory (canal) meatus (Med.) 886
GEHÖRSINN (*m*) : sense of hearing 887
GEIGENSPIEL (*n*) : fiddle (violin) playing 888
GEISSELBAKTERIUM (*n*) : fluorescent bacillus (*Bacillus fluorescens* ; *Pseudomonas fluorescens*) 889
GEISSLERRÖHRE (*f*) : Geissler (vacuum) tube 890
GEKAPSELT : enclosed (Mech.) 891
GEKOPPELTER KREIS (*m*) : coupled circuit 892
GEKLÄGER (*n*) : deposit-(s) ; dregs ; foots ; bottoms 893
GELBUMRANDET : yellow-bordered (*aureo-marginatus*) (Bot.) 894
GELDKASSETTE (*f*) : till (for cash) 895
GELENKENTZÜNDUNG (*f*) : synovitis (Med.) 896
GELÖSCHT : with earth current choke coil (Elect.) : (see also LÖSCHEN) 897
GEMISCHT-WICKLUNG (*f*) : compound winding (Elect.) 898
GERÄUSCHMESSER (*m*) : audiometer ; noise meter ; phonometer (an instrument for measuring the intensity of sound) 899
GERÄUSCHSTÖRUNG (*f*) : interference ; interfering noises 900
GEREGELT : uniform ; regimented ; regular 901
GERICHTET : directed ; directional 902
GERINNSELBILDUNG (*f*) : thrombosis (Med.) 903
GERINNUNGSKÖRPERCHEN (*n*) : (see THROMBOZYT) 904
GERUCH (*m*), ELEKTRISCHER- : ozone smell 905
GERUNDIUM (*n*) : gerund (Grammar) 906
GESCHLOSSEN : closed ; enclosed ; locked ; shut 907
GESCHOSSGESCHWINDIGKEIT (*f*) : rate of fire (Mil.) 908
GESCHWINDIGKEITSABNAHME (*f*) : speed reduction ; deceleration ; slowing-up 909
GESCHWINDIGKEITSABSTUFUNG (*f*) : velocity staging 910
GESENKSCHMIEDEN (*n*) : drop forging 911
GESETZESGLEICHHEIT (*f*) : isonomy ; of equal legal or political right 912
GESICHTSGICHT (*f*) : facial neuralgia ; tic douloureux (Med.) 913
GESICHTSNERV (*m*) : facial nerve (Med.) 914
GESICHTSSCHUTZ (*m*) : shield (Welding) ; face (mask) guard 915
GESTELL (*n*), . . . TEILIGES- : . . . bay rack (Tel.) 916
GEWÄHRLEISTUNGSMANGEL (*m*) : breach of warranty 917
GEWEBE (*n*), BINDE- : connective tissue (Med.) 918
GEWEBE (*n*), GEMUSTERTES- : patterned (figured ; flowered) fabric-(s) (Textile) 919
GEWEBE (*n*), KUNSTSEIDE- : artificial silk fabric (Textile) 920
GEWEBESCHWÄCHE (*f*), ANGEBORENE- ; (see ASTHENIE) 921
GEWICHTSVERLEGUNG (*f*) : weight transfer-(ence) 922
GEWINDEAUTOMAT (*m*) : automatic (thread or screw) tapping machine 923
GEZEITENMASCHINE (*f*) : tide calculating (predicting) machine ; harmonic analyser 924
GIESSBETT (*n*) : pig bed (Blast furnaces) 925
GILLSPINNMASCHINE (*f*) : Gill spinning frame (Textile) 926
GITTERABLEITUNG (*f*) : grid leak (Radio) 927
GITTERABLEITUNGSWIDERSTAND (*m*) : grid leak resistance 928
GITTERELEKTRODE (*f*) : grid electrode 929
GITTERKREIS (*m*) : grid circuit 930
GITTERKREISIMPEDANZ (*f*) : input impedance 931
GITTERKREISKAPAZITÄT (*f*) : input capacity (Radio) 932
GITTERMASCHE (*f*) ; grid mesh 933
GITTERPLATTE (*f*) : grid plate 934
GITTERSTROM (*m*) : grid current (Radio) 935
GITTERSTROMKREIS (*m*) : grid circuit 936
GITTERVORSPANNUNG (*f*) : grid bias (Radio) 937
GLANZFÄDEN (*m.pl*) : bright picks (Artificial silk) 938
GLASBLÄSERSTAR (*m*) : glass-blower's cataract (Med.) 939
GLASGLEICHRICHTER (*m*) : glass-bulb rectifier 940
GLASSPINNEREI (*f*) : glass spinning 941
GLÄTTWERK (*n*) : smoothing rolls (Paper) 942
GLEICHDRUCKWÄRMESPEICHER (*m*) : constant pressure heat accumulator 943
GLEICHLIEGEND : homologous ; of a similar relation ; corresponding 944
GLEICHRICHTERKOLBEN (*m*) : (rectifier)-bulb 945
GLEICHRICHTERRÖHRE (*f*) : rectifier valve (Radio) 946
GLEICHRICHTERUNTERWERK (*n*) : rectifier sub-station (Elect.) 947
GLEICHRICHTERWIRKUNG (*f*) : rectification ; rectifying (action) effect 948
GLEICHSCHALTEN : to co-ordinate ; align ; bring into line with 949
GLEICHSTROMANTRIEB (*m*) : direct current (D.C.) drive (Elect.) 950
GLEICHSTROMERREGUNG (*f*) : direct current (D.C.) excitation (Elect.) 951
GLEICHSTROMKLINGELTRANSFORMATOR (*m*) : direct current (D.C.) bell transformer 952
GLEICHSTROMSTOSS (*m*) : direct current (D.C.) pulsation (Radio) 953
GLEICHUNG (*f*), BESTIMMTE- : determinate equation 954

GLEICHUNG (*f*), **DIOPHANTISCHE-** : indeterminate equation 955
GLEITDRAHTWIDERSTAND (*m*) : regulable resistance; sliding resistance 956
GLEITKONTAKTWIDERSTAND (*m*) : sliding resistance 957
GLEITSCHUH (*m*) : collector shoe (Electric traction) 958
GLETSCHERBAZILLUS (*m*) : glacier bacillus (*Bacillus nivalis*) 959
GLIMMENTLADUNG (*f*) : electrostatic discharge; static discharge; glow (corona) discharge 960
GLIMMLICHT (*n*) : glow 961
GLIMMVERLUST (*m*) : corona loss (Elect.) 962
GLOCKENSCHALE (*f*) : dome (gong) of a bell 963
GLOCKENSPINNMASCHINE (*f*) : cap spinning frame (Textile) 964
GLOSSAR (*n*) : glossary 965
GLÖSSE (*f*) : comment; gloss 966
GLOXYNIE (*f*) : gloxinia (Bot.) 967
GLUCKE (*f*) : incubator (Poultry farming) 968
GLUCKE (*f*), **KÜNSTLICHE-** : foster mother (Poultry farming) 969
GLÜHELEKTRONENEMISSION (*f*) : electronic emission caused by heating 970
GLÜHKÄFER (*m*) : firefly (*Photinus pyralis*) 971
GLÜHKATHODENGLEICHRICHTER (*m*) : mercury vapour arc rectifier; hot cathode rectifier 972
GLÜHKATHODENRÖHRE (*f*) : hot cathode tube (of X-rays); Coolidge tube 973
GLÜHLAMPENWIDERSTAND (*m*) : (glow)-lamp resistance (Elect.) 974
GLÜHLICHTÖLBRENNER (*m*) : incandescent oil burner 975
GLÜHLICHTSPANNUNG (*f*) : lamp voltage (Elect.) 976
GLÜHÖLBRENNER (*m*) : incandescent oil burner 977
GLÜHZUNDER (*m*) : forge scale (Metal) 978
GRABENBAGGER (*m*) : trenching machine 979
GRADIENT (*m*) : gradient; slope; incline; decline 980
-GRAD KELVIN : -degrees Kelvin (-degrees centigrade above absolute zero) (Determination of temperature of stars from their colour) 981
GRANNEN- : provided with bracts (*bracteatus*) (Bot.) 982
GRAPHISCHE KUNST (*f*) : graphic art 983
GRAPHITWIDERSTAND (*m*) : graphite (resistance) rheostat (Elect.) 984
GRATINIEREN : to grate 985
GREIFER-LAUFKRAN (*m*) : travelling crane with grab 986
GREISENHAFT : senile (Med.) 987
GRENZFREQUENZ (*f*) : cut-off frequency; limiting frequency; natural frequency 988
GRENZKREIS (*m*) : horizon; field of (view) vision 989
GROBFLYER (*m*) : stubbing frame (Textile) 990
GROBSPANNUNGSABLEITER (*m*) : coarse spark gap (Tel.) 991
GROSSGLEICHRICHTER (*m*) : heavy duty-(ironclad) rectifier 992
GRUBENKABEL (*n*) : mining cable 993
GRUNDEINRICHTUNG (*f*) : main (chief) equipment 994
GRUNDSCHWINGUNG (*f*) : natural-(aerial) oscillation 995
GRUNDTYPE (*f*) : fundamental (type) sort; prototype 996
GRUNDWALZE (*f*) : bottom vortex-(of tail race) (Water power Eng.) 997
GRUNDZEIT (*f*) : minimum (shortest) time 998
GRUPPENSCHALTER (*m*) : group (series-parallel) switch (Elect.) 999
GRUPPENWÄHLER (*m*) : group selector (Tel.) 000
GUMMIADERLEITUNG (*f*) : vulcanized indiarubber (V.I.R.) cable 001
GUMMIADERSCHNUR (*f*) : flex; flexible cord (Elect.) 002
GUMMIBLEIKABEL (*n*) : lead-covered (V.I.R.) cable 003
GUMMIFLUSS (*m*) : gumming (exudation of gum from the bark of stone fruit) 004
GUMMIKABEL (*n*) : rubber-covered cable 005
GUMMIMANTEL (*m*) : rubber sheath-(ing); rubber cover-(ing) 006
GUMMIRING (*m*) : indiarubber (ring) gasket 007
GUMMISCHLAUCHLEITUNG (*f*) : cabtyre-sheathed (C.T.S.) cable; tough rubber-sheathed (T.R.S.) cable 008
GUMMISCHLAUCHPENDEL (*n* and *m*) : cabtyre-sheathed (C.T.S.) flex-(ible cable) 009
GÜRTELLINSE (*f*), **DIOPTRISCHE-** ; central belt lens with refractors; dispersive (dioptric) lens 010
GUSSKRUSTE (*f*) : rough-cast surface (Metal.) 011
GÜTEZAHL (*f*) : quality factor 012
GUTTAPERCHADRAHT (*m*) : guttapercha covered wire 013
GYNÄKOLOGIE (*f*) : gynecology; gynaecology (Med) 014

H

HAARAUSFALL (*m*) : depilation 015
HAARAUSFALL (*m*), **KREISRUNDER-** : Alopecia areata; baldness (Med.) 016
HAARNADELKURVE (*f*) : hairpin bend (of roads) 017
HÄCKSELSCHNEIDER (*m*) : chaff cutting machine; chaff cutter 018
HACKSTROM (*m*) : pulsating direct current (used in electrotherapeutics) 019
HAFENVERWALTUNG (*f*) : harbour (commissioners) authorities 020
HAFERÄHNLICH : resembling oats (*avenaceus*) (Bot.) 021
HAFNIUMOXYD (*n*) : Hafnium oxide; HfO_2 022
HAHNFASSUNG (*f*) : switch-type lampholder (Elect.) 023
HAKENUMSCHALTER (*m*) : hook-switch (Tel.) 024
HALBAUTOMATISCH : semi-automatic 025
HALBINDIREKTE BELEUCHTUNG (*f*) : semi-indirect lighting 026
HALBLEITER (*m*) : bad conductor (Elect.) 027
HALBSEHEN (*n*) : (see **HEMIOPIE**) 028
HALBSELBSTTÄTIG : semi-automatic (Tel.) 029
HALBTONDRUCKSTÖCK (*m*) : half-tone block (Printing) 030
HALBWELLE (*f*), **ERSTE-** : first half cycle (Elect.) 031
HALSWIRBELKNOCHEN (*m*) : cervical vertebra (Med.) 032
HALTEZEICHEN (*n*) : (see **FERMATE**) 033
HÄMATOGLOBIN (*n*) : (see **HÄMOGLOBIN**) 034
HAMMER (*m*), **DREISPÄNNER-** : three-man hammer 035

HAMMER (*m*), KONTAKT- : contact finger (Elect.) 036
HÄMOPHILIE (*f*) : (see BLUTERKRANKHEIT) 037
HÄMOPTÖE (*f*) : haemoptysis (spitting or coughing blood from lower air passage); haematemesis (vomiting blood from stomach) (Med.) 038
HÄMOSTASE (*f*) : haemostasia ; stoppage of flow of blood ; congestion of blood (Med.) 039
HANDAMT (*n*) : manual exchange (Tel.) 040
HANDANTRIEB (*m*) : hand (drive) operation 041
HANDKAPAZITÄT (*f*) : hand capacity (Radio) 042
HANDREGELUNG (*f*) : hand (operation) regulation 043
HANDSCHALTER (*m*) : hand-operated switch 044
HANDWIDERSTAND (*m*) : hand-operated resistance (Elect.) 045
HANDWURZEL (*f*) : carpus ; wrist bones (Med.) 046
HÄNGEBANK (*f*) : bank (Mining) 047
HÄNGEDEKORATION (*f*) : drop scene (of stages) 048
HÄNGELAMPE (*f*) : pendant-(lamp) 049
HÄNGENDES (*n*) : roof (Mining) 050
HANKZÄHLER (*m*) : hank counter (Textile) 051
HARNBLASENKATARRH (*m*) : cystitis ; inflammation of the bladder (Med.) 052
HARNVERHALTUNG (*f*) : anury ; anuria ; anuresis (absence of micturition) (Med.) 053
HARTFASER (*f*) : hard fibre (Textile) 054
HARTMETALL (*n*) : antimony-tin alloy 055
HASELNUSSBRAUN : drab ; the colour of the hazel-nut shell (*avellana*) (Bot.) 056
HASENSCHARTE (*f*) : hare-lip 057
HASPELMASCHINE (*f*) : reeling (frame) machine (Textile) 058
HAUPTAPPARAT (*m*) : exchange telephone apparatus 059
HAUPTKABEL (*n*) : main cable (Elect.) 060
HAUPTKONTAKT (*m*) : main contact (Elect.) 061
HAUPTLEITUNG (*f*) : (supply)-main (Elect.) 062
HAUPTPATENT (*n*) : original patent 063
HAUPTSCHLAG (*m*) : down-beat (Music) 064
HAUPTSCHLUSSDYNAMO (*f*) : series excited dynamo 065
HAUPTSCHLUSSMOTOR (*m*) : series wound motor 066
HAUPTSEPTIMENAKKORD (*m*) : chord of the dominant seventh (Music) 067
HAUPTSTROMMASCHINE (*f*) : series dynamo (Elect.) 068
HAUPTSTROMREGLER (*m*) : series resistance (Elect.) 069
HAUPTSTROM (*m*), WATTVERBRAUCH IM- : series loss in watts (of electric meters) 070
HAUPTUHR (*f*) : master clock 071
HAUSBELEUCHTUNG (*f*) : auditorium (house) lighting (of theatres) 072
HAUSHALTUNGSMOTOR (*m*) : kitchen motor (for driving mechanical kitchen utensils) 073
HAUSWIRTSCHAFTSARTIKEL (*m.pl*) : household utensils 074
HAUTEFFEKT (*m*) : skin effect (Elect.) 075
HAUTKRUSTE (*f*) : scab (Med.) 076
HAUTPUSTEL (*f*) : (see QUADDEL) 077
HAUT (*f*), SERÖSE- : serous membrane (Med.) 078
HAUTTRANSPLANTATION (*f*) : skin grafting 079
HAUTTUBERKULOSE (*f*) : (see LUPUS) 080
HAUTÜBERPFLANZUNG (*f*) : skin grafting 081
HAUTWIRKUNG (*f*) : skin effect (Elect.) 082
HAVARIEKOMMISSAR (*m*) : average adjuster (Marine insurance) 083
HEAVISIDESCHICHT (*f*) : Heaviside layer (layer of ionized air about 50 miles above the surface of the earth) 084
HEBDREHWÄHLER (*m*) : Strowger type selector (Tel.) 085
HEBELÜBERSETZUNG (*f*) : leverage 086
HEBERRÖHRCHEN (*n*) : syphon (pipe) tube 087
HEBERSCHREIBER (*m*) : syphon recorder (Tel.) 088
HECHELMASCHINE (*f*) : hackling machine ; comber (Textile) 089
HEISSLAUFEN : to run hot (of water) 090
HEISSWASSERSPEICHER (*m*) : hot water (storage tank) accumulator ; calorifier 091
HEIZBATTERIE (*f*) : low-tension battery (Radio) 092
HEIZDRAHT (*m*) : heating (wire) filament 093
HEIZELEMENT (*n*) : heating element (Elect.) 094
HEIZFADEN (*m*) : heating filament 095
HEIZKISSEN (*m*) : heating pad (Elect. and Med.) 096
HEIZKREIS (*m*) : heating circuit 097
HEIZREGISTER (*m*) : heating resistance (for electric room heating) 098
HEIZSPANNUNG (*f*) : filament voltage (Radio) 099
HEIZSTROM (*m*) : heating current ; filament current (Radio) 100
HEIZSTROMKREIS (*m*) : filament circuit (Radio) 101
HEIZSTROMREGLER (*m*) : filament resistance (Elect.) 102
HEIZWIDERSTAND (*m*) : heating resistance (Elect.) 103
HELLIGKEITSMESSUNG (*f*) : astro-photometry (measurement of brightness of stars) 104
HEMIOPIE (*f*) : hemiopia ; hemiopsia ; hemianopsia ; half-blindness (in which vision partly obscured) (Med.) 105
HERAUFSETZEN : to step up ; increase ; boost ; augment ; put up ; raise 106
HERAUFTRANSFORMIEREN : to step up 107
HERAUFTRANSFORMIERUNG (*f*) : step-up (transforming) transformation (Elect.) 108
HERDPLATTE (*f*) : hob ; top-plate (of electric kitchen ranges) 109
HERTZ (*n*) : Hertz ; period ; cycle ; oscillation frequency 110
HERUNTERTRANSFORMIERUNG (*f*) : step-down (transforming) transformation (Elect.) 111
HERZBEUTELENTZÜNDUNG (*f*) : pericarditis (inflammation of the pericardium) (Med.) 112
HERZGERÄUSCH (*n*) : murmur (of the heart) 113
HERZINNENHAUT (*f*) : endocardium (Med.) 114
HERZJAGEN (*n*) : palpitation-(of the heart) 115
HERZKLAPPENFEHLER (*m*) : valvular disease of the heart (Med.) 116
HERZKONTRAKTION (*f*) : systole (contraction of the heart) (Med.) 117
HERZKRAMPF (*m*) : *Angina pectoris* (Med.) 118
HERZMUSKEL (*m*) : myocardium ; heart muscle (Med.) 119
HERZMUSKELENTZÜNDUNG (*f*) : myocarditis ; inflammation of the myocardium (Med.) 120
HERZMUSKELSCHWÄCHE (*f*) : myocarditis (Med.) 121
HERZNEUROSE (*f*) : neurasthenia ; neurosis of the heart (Med.) 122
HERZSCHLAG (*m*) : heart beat 123
HERZVERGRÖSSERUNG (*f*) : hypertrophy (enlargement) of the heart (Med.) 124
HERZZUSAMMENZIEHUNG (*f*) : systole (contraction of the heart) (Med.) 125
HETERODYNEEMPFANG (*m*) : heterodyne reception (Radio) 126
HEUBAZILLUS (*m*) : hay bacillus (*Bacillus subtilis*) 127
HILFSKONTAKT (*m*) : auxiliary break (arcing ; sparking ; secondary) contact (Elect.) 128
HILFSLINIE (*f*) : leger line (Music) 129
HILFSSTROMKREIS (*m*) : auxiliary circuit (Elect.) 130
HILFSWICKLUNG (*f*) : shunt (compensating) winding (Elect.) 131

HILFSWIDERSTAND (*m*) : auxiliary resistance 132
HILFSZWISCHENRAUM (*m*) : space between leger lines (Music) 133
HILUS (*m*) : (see LUNGENPFORTE) 134
HINDERNISFEUER (*n*) ; obstruction light (on aerodromes) 135
HINDURCHDIFFUNDIEREN : to diffuse through 136
HINDURCHDRINGEN : to penetrate ; go (pass) through ; permeate 137
HINTERANSICHT (*f*) : rear (back) elevation or view 138
HINTERHAUPTLAPPEN (*m*) : occipital lobe (of brain) (Med.) 139
HINTERHAUPTSARTERIE (*f*) : occipital artery (Med.) 140
HISTOLOGIE (*f*) : histology ; the science of organic tissue (Med.) 141
HITZDRAHT-MESSINSTRUMENT (*n*) : electro-calorimeter (an apparatus for measuring heat quantities) 142
HOCHANTENNE (*f*) : outdoor aerial (Radio) 143
HOCHBAU (*m*) : superstructure 144
HOCHEMPFINDLICH : highly (sensitive ; susceptible) selective 145
HOCHFREQUENZDROSSELSPULE (*f*) : high frequency (choke) choking coil (Elect.) 146
HOCHFREQUENZOFEN (*m*) : high frequency furnace (Elect.) 147
HOCHFREQUENZVERSTÄRKER (*m*) : high frequency (H.F.) amplifier (Radio) 148
HOCHGLANZFARBE (*f*) : high-gloss paint 149
HOCHKAPAZITIV : of high capacity (Elect.) 150
HOCHLEISTUNGSSICHERUNG (*f*) : fuse of high rupturing capacity (Elect.) 151
HOCHPOL (*m*) : anode ; positive (plus) pole (Elect.) 152
HOCHRAMPE (*f*) : elevated (loading ; unloading) platform 153
HOCHSPANNUNGSKABEL (*n*) : high-tension cable 154
HOCHSPANNUNGSLEITUNG (*f*) : high-tension (wire ; main) line (Elect.) 155
HOCHSPANNUNGSMAST (*m*) : mast for high-tension overhead wires 156
HOCHSPANNUNGSSEITE (*f*) : high-tension side (Elect.) 157
HÖCHSTGRAD (*m*) : superlative-(degree) (Grammar) 158
HÖCHSTSTROMZEITRELAIS (*n*) : maximum current time relay (Elect.) 159
HÖCHSTVERBRAUCHSZÄHLER (*m*) : maximum demand meter (Elect.) 160
HOCHVAKUUM (*n*) : high vacuum 161
HOCHVAKUUMRÖHRE (*f*) : thermionic valve (Radio) 162
HOCHZAPFENGESTELL (*n*) : bogie (Railways) 163
HODOMETER (*n*) : (see WEGMESSER) 164
HÖFLICHKEITSLIEFERUNG (*f*) : (see KULANZLIEFERUNG) 165
HÖHENFLUGMOTOR (*m*) : high-altitude aero-engine 166
HÖHENFÖRDERER (*m*) : elevator 167
HÖHENMESSUNG (*f*) : hypsometry (measurement of heights on earth's surface) 168
HÖHENSONNE (*f*) : mountain sun ; quartz mercury arc-(lamp) (Med.) 169
HOHLVENE (*f*), **OBERE-** : superior vena cava (Med.) 170
HOHLVENE (*f*), **UNTERE-** : inferior vena cava (Med.) 171
HOLUNDERBLÜTENTEE (*m*) : elder-flower tea (Med.) 172
HOLUNDERMARKKÜGELCHEN (*n*) : elder-pith ball 173
HOLZ (*n*), **NUTZ-** : timber 174
HOLZ (*n*), **TROCKENES-** : dry (seasoned) wood 175
HOLZWANNE (*f*) : wooden (tank) trough 176
HOMODYNEMPFÄNGER (*m*) : homodyne receiver 177
HONIGWABENSPULE (*f*) : honeycomb coil (Radio) 178
HÖRAPPARAT (*m*) : hearing apparatus : receiver (Tel.) 179
HÖRER (*m*) : receiver ; telephone ; loudspeaker 180
HÖRFREQUENZ (*f*) : tone (audible) frequency 181
HÖRGEMEINDE (*f*) : listener's (listening) circle (Radio) 182
HORIZONTBELEUCHTUNG (*f*) : horizon lighting (of stages) 183
HORIZONTLEUCHTE (*f*) : horizon lamp (Stage lighting) 184
HORIZONTWAND (*f*) : back cloth (of stages) 185
HÖRMUSCHEL (*f*) : ear-piece (Tel.) 186
HORNKNÖTCHEN (*n* and *pl*) : *Lichen pilaris* (Med.) 187
HÖRTRICHTER (*m*) : ear piece (Telephones) 188
HÖRUMSCHALTER (*m*) : listening (Kellogg) key (Telephones) 189
HUBWERK (*n*) : lifting mechanism 190
HÜFTARTERIE (*f*) : common iliac artery (Med.) 191
HÜFTSCHLAGADER (*f*) : common iliac artery (Med.) 192
HUMUSKOHLE (*f*) : (see GLANZKOHLE) 193
HUPE (*f*) : hooter ; siren 194
HYDROLYSIERT, NICHT- : unhydrolysed ; not hydrolysed (Electro-chemistry) 195
HYPERTRICHOSIS (*f*) : excessive hairiness (Med.) 196
HYPOZYKLOIDE (*f*) : hypocycloid 197
HYSTERESIS (*f*), **DIELEKTRISCHE-** : dielectric hysteresis 198
HYSTERESISVERLUST (*m*) : hysteresis loss 199
HYSTERETISCH : hysteresial 200

I

IDEENASSOZIATION (*f*) : association of ideas 201
IKTERISCHER ZUSTAND (*m*) : icteric ; jaundiced condition (Med.) 202
ILLATION (*f*) : (see SCHLUSSFOLGERUNG) 203
IMMUN : immune 204
IMPARTIBILITÄT (*f*) : indivisibility 205
IMPEDANZGLIED (*n*) : impedance element (Elect.) 206
IMPEDANZRELAIS (*n*) : impedance relay (Elect.) 207
IMPEDANZSCHUTZ (*m*) : impedance protective relay (Elect.) ; impedance protection 208
IMPULSTELEGRAPHIE (*f*) : impulse telegraphy 209
IMPULSÜBERTRAGER (*m*) : impulse repeater (Tel.) 210
INAPPELLABEL : final ; not subject to appeal (of decisions) 211
INDIFFERENTGLAS (*n*) : heat-resisting glass 212
INDIZIERTE PFERDEKRAFT (*f*) : I.H.P. : indicated horse-power 213
INDUKTANZ (*f*) : inductance (Elect.) 214
INDUKTION (*f*), **ELEKTROSTATISCHE-** : electrostatic induction 215

INDUKTIONSFREI : non-inductive (Elect.) 216
INDUKTIONSKAPAZITÄT (*f*) : inductive capacity 217
INDUKTIONSMOTOR (*m*) : induction motor (Elect.) 218
INDUKTIONSWIRKUNG (*f*) : induction ; inductive effect 219
INDUKTIONSZÄHLER (*m*) : induction meter (Elect.) 220
INDUKTIVE BELASTUNG (*f*) : inductive load 221
INDUKTIVER WIDERSTAND (*m*) : inductive resistance ; inductance 222
INDUKTORBETRIEB (*m*) : magneto operation (of bells) 223
INDUKTORIUM (*n*) : induction coil ; spark coil (Elect.) 224
INDUKTORRÜCKWIRKUNG (*f*) : (see ANKERRÜCKWIRKUNG) 225
INDUKTRISCH : inductive-(ly) 226
INDUZIEREND : inductive-(ly) 227
INFLUIEREN : to influence ; affect 228
INHIBIEREN : to inhibit ; (*n*) : inhibition 229
INKLINATION (*f*) : inclination 230
INKONSTANT : inconstant 231
INKONSTANZ (*f*) : inconstancy 232
INKREMENT (*n*) : increment ; addition ; increase 233
INKULPANT (*m*) : accuser ; prosecutor 234
INKULPAT (*m*) : accused 235
INNENANTENNE (*f*) : indoor (room) aerial (Radio) 236
INNENDEKORATION (*f*) : interior decorations ; soft furnishings (Textile) 237
INNENKABEL (*n*) : indoor cable (only for use in dry rooms) 238
INNENLIEGEND : internal 239
INNENWIDERSTAND (*m*) : internal resistance 240
INQUIRENT (*m*) : accused 241
INQUIRIST (*m*) : accuser 242
INSEKTENFRESSER (*m*) : insectivore (an insect-eating mammal) ; (*pl*) insectivora 243
INSOLATION (*f*) : insolation (treatment by exposure to sun's rays) (Med.) 244
INSTALLIEREN : to instal ; fit up ; equip ; erect ; put in 245
INSTANDSETZUNGSARBEIT (*f*) : repair-(work) ; re-conditioning 246
INSTAURIEREN : to instaurate ; renew ; repair ; restore 247
INSZENIERUNG (*f*) : scenery (of stages) 248
INTERFERENZ (*f*) : interference 249
INTERFERENZEMPFANG (*m*) : heterodyne reception (Radio) 250
INTERIMISTISCH : in the (meanwhile) interim 251
INTERMITTENZ (*f*) : intermittence ; periodic interruption 252
INTERPUNGIEREN : to punctuate (Grammar) 253
INTRAKARBON : middle of the Carboniferous period (Geol.) 254
INTRAVENÖS : intravenous 255
INZIDENZ (*f*) : incidence (see EINFALL) 256
INZIDENZFALL (*m*) : incidental-(case) 257
IONAL : ionic 258
IONISIERT : ionised ; split up into ions 259
IONISIERT, NICHT- : un-ionized ; not ionized 260
IONTOPHORESE (*f*) : (see ELEKTROPHORESE) 261
IRIDEKTOMIE (*f*) : iridectomy (cutting out a part of the iris of the eye) (Med.) 262
IRISIEREND : iridescent (coloured like a rainbow) 263
ISCHIASNERV (*m*) : sciatic nerve (Med.) 264
ISOCHIMENEN (*f.pl*) : lines of equal winter temperature ; isocheims 265
ISOCHROMATISCH : isochromatic ; of the same colour 266
ISOHYPSEN (*pl*) : (see SCHICHTLINIEN) 267
ISOKLIN : isoclinal ; isoclinic ; having the same magnetic inclination 268
ISOLATIONSDURCHSCHLAG (*m*) : break-down of insulation (Elect.) 269
ISOLATIONSFEHLER (*m*) : insulation (fault) defect 270
ISOLATORSTÜTZE (*f*) : insulator bracket 271
ISOLIERBÜCHSE (*f*) : insulating bush 272
ISOLIEREI (*n*) : egg insulator 273
ISOLIERROHR (*n*) : insulating (pipe) tube 274
ISOLIERSTÜTZE (*f*) : insulator bracket 275
ISOLIERZANGE (*f*) : insulated pliers 276
ISOMETRISCH : isometric ; of equal measure 277
ISONOMIE (*f*) : isonomy ; of equal legal or political right 278
ISOSTATISCH : isostatic (equilibrium due to equal pressure from all sides) 279
ISTBESTAND (*m*) : actual (amount ; stock) balance 280
ISTMENGE (*f*) : actual (amount) quantity 281

J

JACQUARDMASCHINE (*f*) : Jacquard knitting machine (Textile) 282
JALOUSIE (*f*) : venetian blind ; louvre 283
JAUCHEPUMPE (*f*) : sud (sump ; sewage) pump 284
JOCHBEIN (*n*) : cheek bone (Med.) 285
JODMETALLE (*f.pl*) : traces of metals found in the ashes of seaweed (potassium ; magnesium ; sodium) 286
JONISATION (*f*) : ionization 287
JONTOPHORESE (*f*) : (see ELEKTROPHORESE) 288
JUDENBLEI (*n*) : very pure lead 289
JUDENFOLIE (*f*) : (see STANNIOL) 290
JUDENGOLD (*n*) : (see MUSIVGOLD) 291
JUGENDIRRESEIN (*n*) : (see SCHIZOPHRENIE) 292
JUNKAZEEN (*pl*) : *Junceae* ; rushes (Bot.) 293
JURA (*f*) : Jurassic period (Geol.) 294
JUSTIERVORSCHRIFT (*f*) : instructions for the adjustment-(of) 295
JUSTIERWIDERSTAND (*m*) : regulating (adjusting) resistance 296
JUTEFEINKÖPER (*m*) : fine twilled sacking (Textile) 297
JUTEFEINLEINEN (*n*) : fine hessian (Textile) 298
JUTE-FLÜGELSPINNMASCHINE (*f*) : jute flyer frame (Textile) 299
JUTEKÖPER (*m*) : twilled sacking (Textile) 300
JUTELEINEN (*n*) : hessian (Textile) 301
JUTEÖFFNER (*m*) : jute opener (Textile) 302
JUTEQUETSCHMASCHINE (*f*) : jute softener (Textile) 303
JUTEREIBE (*f*) : jute softener (Textile) 304
JUTESACKLEINEN (*n*), DOPPEL- : double-warp bagging (Textile) 305
JUTESACKLEINEN (*n*), EINFACH- : single-warp bagging (Textile) 306
JUTESOFTENER (*m*) : jute softener (Textile) 307

K

KABELABZWEIGKASTEN (*m*) : cable (joint ; junction ; distribution) branch box 308
KABELABZWEIGMUFFE (*f*) : (see KABELMUFFE) 309
KABELANSCHLUSSKASTEN (*m*) : terminal box (Elect.) 310
KABELARMATUREN (*f.pl*) : cable (mountings ; fittings) accessories 311
KABEL (*n*), ARMIERTES- : armoured cable 312
KABELBEWEHRUNG (*f*) : armour-(ing) of a cable 313
KABELBRUNNEN (*m*) : cable pit 314
KABELDAMPFER (*m*) : cable laying and repair ship 315
KABEL (*n*), DOPPELADRIGES- : twin cable 316
KABEL (*n*) : EINADRIGES- : single-(core)-cable 317
KABEL (*n*), EINFACHES- : single-(core)-cable 318
KABELENDVERSCHLUSS (*m*) : cable (lug) socket 319
KABELGARNITUREN (*f.pl*) : cable (mountings ; fittings) accessories 320
KABELKREUZMUFFE (*f*) : four-way cable branch box 321
KABEL (*n*), KÜNSTLICHES- : artificial line (Tel.) 322
KABELLÖTWACHS (*n*) : sealing compound ; Chatterton compound 323
KABEL (*n*), MEHRADRIGES- : multicore cable 324
KABELNETZ (*n*) : cable (network) system 325
KABELROHR (*n*) : cable pipe 326
KABELSATTEL (*m*) : cable rack 327
KABELSCHUH (*m*) : cable (socket ; shoe ; eye) thimble 328
KABELTANK (*m*) : cable tank 329
KABELTROMMEL (*f*) : cable drum 330
KABELVERGUSSMASSE (*f*) : (sealing)-compound 331
KABELVERTEILER (*m*) : distribution board 332
KABELVERTEILUNGSKASTEN (*m*) : (see KABELABZWEIG-KASTEN) 333
KADENZ (*f*) : cadence (Music) 334
KADMIUMELEMENT (*n*) : cadmium cell (Elect.) 335
KADMIUMZELLE (*f*) : cadmium cell 336
KÄFIGANKER (*m*) : squirrel cage rotor (Elect.) 337
KAISERKRONE (*f*) : crown imperial (*Fritillaria imperialis*) (Bot.) 338
KALANDERHAUS (*n*) : calendering house 339
KALEBASSE (*f*) : calabash (see also KÜRBIS) 340
KALIFORNISCH : Californian ; from California (*californicus*) (Bot.) 341
KALIUM-NATRIUM (*n*), WEINSAURES- : (see KALIUM-NATRIUMTARTRAT) 342
KALIUMPERRHENAT (*n*) : potassium perrhenate 343
KALIUMRHENIUMCHLORID (*n*) : potassium rhenichloride ; K_2ReCl_6 344
KALIUMZELLE (*f*) : potassium cell (filled with helium, having as cathode potassium metal) 345
KALKPFLANZE (*f*) : a plant needing lime 346
KALKTUFF (*m*) : (see TUFFKALK) 347
KALKULOGRAPH (*m*) : calculograph (to measure and record time of telephone conversations) 348
KALKUTTAHANF (*m*) : jute 349
KALK (*m*), WIENER- : French chalk 350
KALOMELELEKTRODE (*f*) : calomel electrode 351
KALOMELELEMENT (*n*) : calomel cell 352
KÄLTEPERIODE (*f*) : Ice (age) period 353
KALTSCHERE (*f*) : cold shear (Rolling mills) 354
KAMMERTON (*m*) : the-(tuning)-note a' (440 vibrations per second) (Music) 355
KAMMGARNSELFAKTOR (*m*) : worsted mule (Textile) 356
KAMMGARNSPINNEREI (*f*) : worsted spinning-(mill) (Textile) 357
KÄMM-MASCHINE (*f*) : combing machine ; carding engine (Textile) 358
KANARIENGLAS (*n*) : yellow-coloured glass containing uranium oxide 359
KANONISCH : canonic (Music) 360
KANTENRECHEN (*m*) : Picot edge (Textile) 361
KANTENRECHENWIDERSTAND (*m*) : Picot edging resistance (Elect.) 362
KANÜLE (*f*) : cannula ; drainage tube (for drawing off fluid from the body) (Med.) 363
KAPAZITÄTSARM : (of)-low-capacity 364
KAPAZITÄTSAUSGLEICH (*m*) : capacity balance (Elect.) 365
KAPAZITÄTSKOPPLUNG (*f*) : capacity coupling (Elect.) 366
KAPAZITÄTSREAKTANZ (*f*) : capacity reactance (Elect.) 367
KAPAZITÄTSUNGLEICHHEIT (*f*) : capacity unbalance (Elect.) 368
KAPAZITÄTSWIRKUNG (*f*) : capacity effect (Elect.) 369
KAPAZITÄT (*f*), VERÄNDERLICHE- : variable capacity (Elect.) 370
KAPAZITIV : capacitive (Elect.) 371
KAPAZITIVER BLINDWIDERSTAND (*m*) : condensance ; capacity resistance 372
KAPAZITIVE REAKTANZ (*f*) : condensance ; capacity resistance (Elect.) 373
KAPILLARELEKTROMETER (*n*) : capillary electrometer 374
KARAFFE (*f*) : carafe ; decanter ; water-bottle 375
KARBON : carboniferous ; (*n*) : Carboniferous period (Geol.) 376
KARBONATHÄRTE (*f*) : temporary hardness-(of water, due to carbonate salts of lime and magnesia, which are precipitated to a considerable degree by merely heating the water) 377
KARBONFORMATION (*f*) : Carboniferous formation (Geol.) 378
KARBONISCH : carboniferous 379
KARBONZEIT (*f*) : Carboniferous period (Geol.) 380
KARBORUNDDETEKTOR (*m*) : crystal detector (Radio) 381
KARBORUNDKRISTALL (*m*) : carborundum crystal (see SILIZIUMKARBID) 382
KARBORUNDUMWIDERSTAND (*m*) : carborundum resistance 383
KARBOSILWIDERSTAND (*m*) : carborundum resistance 384
KARDIALGIE (*f*) : cardialgy ; cardialgia (Med.) 385
KARDIEREN : to card ; (*n*) : carding (Textile) 386
KARDIOGRAPH (*m*) : cardiograph (see ELEKTRO-KARDIOGRAPH) 387
KARDIOIDE (*f*) : cardioid (a heart-shaped curve) (Maths.) 388
KAROLUSZELLE (*f*) : Karolus (Kerr) cell (Picture telegraphy) 389
KAROOPTISCHER SPIEGEL (*m*) : sectioned mirror (carooptic mirror) reflector (in which the reflector is covered with diamond-shaped hollows in order to promote uniformity of illumination with floodlights) 390
KARTENSCHLÄGER (*m*) : card cutter (Textile) 391
KARTOFFELBAZILLUS (*m*) : potato bacillus (*Bacillus mesentericus fuscus* ; *B.m. ruber* ; *B.m. vulgatus* ; *Bacillus vulgatus*) 392
KASKADENANLASSUNG (*f*) : cascade starting (Elect.) 393
KASKADENBATTERIE (*f*) : cascade battery (battery of Leyden jars in series) 394
KASKADENSTROMWANDLER (*m*) : cascade converter (Elect.) 395

KASKADENUMFORMER (*m*) : cascade converter (Elect.) 396
KASKADENVERSTARKER (*m*) : multi-valve amplifier (Radio) 397
KASTENSPEISER (*m*) : hopper feeder (Textile) 398
KATADYNISIEREN : to sterilize water by means of the Katadyn process 399
KATADYN-SILBER (*n*) : silver as used in the Katadyn process of water sterilization 400
KATAPHORESE (*f*) : (see **ELEKTROPHORESE**) 401
KATHODENFALL (*m*) : cathode drop 402
KATHODENGEFÄLLE (*n*) : cathode drop 403
KATHODENGLIMMLICHT (*n*) : (cathode)-blue glow 404
KATHODENHOHLSPIEGEL (*m*) : cathode of (Roentgen) X-ray tube 405
KATHODENRAUM (*m*) : cathode dark (space) spot (the dark space between the cathode and the negative glow) 406
KATHODENSTRAHLENRÖHRE (*f*) : cathode ray tube ; Coolidge (Lenard ; Wehnelt) tube 407
KATHODENZERSTÄUBUNG (*f*) : cathode disintegration 408
KATOPTRIK (*f*) : catoptrics (the science of light reflection) 409
KAUSTISCHE SODA (*f*) : (see **ÄTZNATRON**) 410
KAUTSCHUKBAND (*n*) : rubber tape 411
KAUTSCHUKBAUM (*m*) : gum tree ; rubber tree (*Ficus elastica*) (Bot.) 412
KAUTSCHUKDRAHT (*m*) : caoutchouc-covered wire 413
KAUTSCHUKEMAIL (*n*) : caoutchouc enamel (for covering metal objects) 414
KAUTSCHUKREGENERAT (*n*) : reclaimed rubber 415
KAUTSCHUK (*m*), **REGENERIERTER-** : reclaimed rubber 416
KAUTSCHUKÜBERZUG (*m*) : rubber (coating ; covering) proofing 417
KEHLKOPFKUNDE (*f*) : laryngology (the science of the wind-pipe and its diseases) (Med.) 418
KEILRIEMENANTRIEB (*m*) : V-belt drive 419
KELCHGLOCKE (*f*) : church gong (of electric bells) 420
KELCHIG : cup-shaped ; in the nature or form of a calyx (*calycinus*) (Bot.) 421
KELLERISOLATOR (*m*) : drip-proof (porcelain ; cellar) insulator 422
KELLERSCHALTER (*m*) : totally enclosed waterproof switch-(for cellars and damp rooms) 423
KENNFADEN (*m*) : distinguishing thread (a coloured thread in cables indicating the maker) 424
KENNLAMPE (*f*) : pilot (light) lamp (of stage lighting) 425
KENNMARKE (*f*) : distinguishing mark 426
KERNTRANSFORMATOR (*m*) : core transformer (Elect.) 427
KERNTYPE (*f*) : core type (of transformers) 428
KERNVIERER (*m*) : central quad (of telephone cables) 429
KERREFFEKT (*m*) : Kerr effect (Picture telegraphy) 430
KERRZELLE (*f*) : Kerr cell (Picture telegraphy) 431
KERZENFASSUNG (*f*) : miniature lampholder for candle lamps (Elect.) 432
KERZENLAMPE (*f*) : candle lamp (for decorative lighting) (Elect.) 433
KESSELTAFEL (*f*) : boiler panel 434
KETTBEDRUCKT : warp printed (Textile) 435
KETTE (*f*), **GALVANISCHE-** : galvanic circuit 436
KETTELMASCHINE (*f*) : warping machine (Textile) 437
KETTENANTRIEB (*m*) : chain drive 438
KETTENSCHLICHTMASCHINE (*f*) : warp sizing machine (Textile) 439
KETTENSTUHL (*m*) : warp loom (Textile) 440
KETTGARNSPULMASCHINE (*f*) : warp winding frame (Textile) 441
KETTGEWEBE (*n*) : warp fabric (Textile) 442
KEUPER (*m*) : Upper New Red Sandstone (Geol.) 443
KEUPERFORMATION (*f*) : (see **KEUPER**) 444
KILOWATTSTUNDENZÄHLER (*m*) : kilowatt/hour meter (Elect.) 445
KINOLAMPE (*f*) : projection lamp (for cinemas) 446
KIPPKLAPPE (*f*) : drop indicator (in which the indicator falls sideways) 447
KIPPSCHALTER (*m*) : Kellogg (listening) key (Tel.) ; tumbler switch (Elect.) ; mercury (tipping switch) tipper 448
KIRCHENTON (*m*) : church mode (Music) ; ecclesiastical mode 449
KLANGFARBE (*f*) : timbre ; (sound ; tonal)-quality or colour (Music) 450
KLAPPENSCHRANK (*m*) : (telephone)-switchboard ; drop-indicator board 451
KLAPPENSCHRANK (*m*), **RÜCKSTELL-** : switchboard with self-restoring indicator (Tel.) 452
KLAPPENSCHRANK (*m*), **STAND-** : floor pattern switchboard (Tel.) 453
KLARLEGEN : to clear up ; make (clear ; plain ; evident) comprehensible ; (*n*) : clearing up 454
KLARLEGUNG (*f*) : clearing up ; making (clear ; plain ; evident) comprehensible 455
KLAVIATUR (*f*) : keyboard (as of an organ) 456
KLEBEMARKE (*f*) : stick-on label 457
KLEIENFLECHTE (*f*) : (see **KLEIENGRIND**) 458
KLEINGRANNIG : bearing a small awn (*aristulatus*) (Bot.) 459
KLEINMOTOR (*m*) : small motor ; fractional horsepower (H.P.) motor (Elect.) 460
KLEINSPANNUNG (*f*) : low voltage (up to 42 volts) (Elect.) 461
KLEINSTELLAMPE (*f*) : dimming lamp-(bulb) (Elect.) 462
KLEMMEN (*n*) : sticking ; jamming ; seizing (Mech.) ; clamping ; clipping ; binding ; locking 463
KLEMMENLEISTE (*f*) : terminal (block) strip (Elect.) 464
KLEMMENSTÜCK (*n*) : terminal block (Elect.) 465
KLEMMISOLATOR (*m*) : connector (Elect.) 466
KLEMMVORRICHTUNG (*f*) : clamping (locking) device 467
KLETTERTROMPETENARTIG : like a trumpet flower (*bignonioides*) (Bot.) 468
KLINGELANLAGE (*f*) : bell installation 469
KLINGELBATTERIE (*f*) : bell battery 470
KLINGELDRÜCKER (*m*) : push-button ; bell-push 471
KLINGELDRUCKKONTAKT (*m*) : push-button contact 472
KLINKENUMSCHALTER (*m*) : jack (Tel.) 473
KLINOMETER (*n*) : clinometer (for measuring angles of inclination) 474
KLISTIER (*n*) : clyster ; enema (Med.) 475
KLONISCH : clonic ; spasmodic movement (Med.) 476
KLYDONOGRAPH (*m*) : clydonograph ; klydonograph ; excess voltage meter (Elect.) 477
KLYDONOGRAPHISCH : klydonographic ; clydonographic 478
KNACKEN (*n*) : crackling (of a telephone membrane) 479
KNALLGERÄUSCH (*n*) : crackling 480
KNALLPATRONE (*f*) : (fog-signal)-detonator 481
KNALLSIGNALAUSLEGER (*m*) : fog-signal apparatus 482
KNETEN (*n*) : petrissage (of massage) ; kneading 483
KNETMASSAGE (*f*) : petrissage 484
KNITTERFÄHIGKEIT (*f*) : creasing property ; capacity for creasing (Textile) 485

KNITTERFESTIGKEIT (*f*) : resistance to (creasing) crushing ; non-creasing (capacity) property (Textile) 486
KNOCHENGEWEBE (*n*) : bone substance (Med.) 487
KNOCHENGEWEBEENTZÜNDUNG (*f*) : osteitis (inflammation of bone substance) ; osteomyelitis (inflammation of the bone marrow) (Med.) 488
KNOCHENLEHRE (*f*) : osteology (Med.) 489
KNOTENAMT (*n*) : central (main) exchange or station (Tel.) 490
KNOTENFREI : free from (knots) nodes 491
KNOTENROSE (*f*) : erythema nodosum (Med.) 492
KNÜPFUNG (*f*) : join ; fastening ; knotting (Textile) 493
KOBALTBAD (*n*) : cobalt bath 494
KOBALTNIEDERSCHLAG (*m*) : cobalt (precipitate) deposit 495
KOCHPLATTE (*f*) : hot-plate (of stoves) 496
KOHLEBÜRSTE (*f*) : carbon brush (Elect.) 497
KOHLEFADENLAMPE (*f*) : carbon filament lamp (Elect.) 498
KOHLEGRIES (*m*) : carbon granules 499
KOHLEKÖRNER (*n.pl*) : carbon granules 500
KOHLEKÖRNERMIKROPHON (*n*) : carbon granule microphone 501
KOHLEMIKROFON (*n*) : carbon microphone (Radio) 502
KOHLENBLOCK (*m*) : carbon block 503
KOHLENFORMATION (*f*) : coal measure (Geol.) 504
KOHLENGEBIRGE (*n*) : coal measure 505
KOHLENKÖRNER (*n.pl*) : carbon granules (Tel.) 506
KOHLENSTAB (*m*) : carbon (of arc lamps) 507
KOHLENSTOSS (*m*) : coal face (Mining) 508
KOHLEPLATTE (*f*) : carbon plate 509
KOHLEPULVER (*n*) : carbon (granules) powder 510
KOHLEWIDERSTAND (*m*) : carbon (resistance) rheostat (Elect.) 511
KOINZIDENZ (*f*) : coincidence 512
KOLIBAKTERIUM (*n*) : cholic bacterium (*Bacterium coli*) 513
KOLLEKTORFEUER (*n*) : arcing ; commutator sparking ; brush sparking 514
KOLLEKTORLAMELLE (*f*) : commutator (segment) bar 515
KOLLEKTORSEGMENT (*n*) : commutator (bar) segment 516
KOLLIMATION (*f*) : collimation 517
KOLLIMATOR (*m*) : collimator 518
KOLLIMIEREN : to collimate ; render parallel (of rays) 519
KOMBINATIONSSCHALTER (*m*) : series-parallel switch 520
KOMBINATIONSSCHALTUNG (*f*) : series-parallel connection 521
KOMEDONEN (*pl*) : black-heads (Med.) (see MITESSER) 522
KOMMANDOANLAGE (*f*) : signalling apparatus (Elect.) 523
KOMMUTATION (*f*) : commutation (Elect.) 524
KOMMUTATORMASCHINE (*f*) : commutating machine (Elect.) 525
KOMMUTATORMOTOR (*m*) : commutator motor (Elect.) 526
KOMPENDIÖS : compact ; compendious ; abridged 527
KOMPENSATION (*f*) : compensation 528
KOMPENSATIONSMASCHINE (*f*) : shunt dynamo (Elect.) 529
KOMPENSATIONSSPULE (*f*) : choke coil (Elect.) 530
KOMPENSATIONSWICKLUNG (*f*) : shunt (compensating) winding (Elect.) 531
KOMPOUNDDYNAMO (*f*) : compound wound (generator) dynamo (Elect.) 532
KOMPOUNDIERUNG (*f*) : compound winding (Elect.) 533
KOMPOUNDMOTOR (*m*) : compound wound motor (Elect.) 534
KOMPOUNDWICKLUNG (*f*) : compound winding (Elect.) 535
KOMPROMISS (*n*) : compromise 536
KONDENSANZ (*f*) : condensance (Elect.) ; capacity reactance 537
KONDENSATORENTLADUNG (*f*) : condenser discharge (Elect.) 538
KONDENSATORKETTE (*f*) : high pass filter (Elect.) 539
KONDENSATORKREIS (*m*) : condenser circuit (Elect.) 540
KONDENSATORPLATTE (*f*) : condenser plate (Elect.) 541
KONDENSATORRÜCKSTAND (*m*) : residual magnetism 542
KONDENSATORTELEPHON (*n*) : condenser telephone 543
KONDENSATOR (*m*), VERÄNDERLICHER- : variable condenser (Elect.) 544
KONDENSATORVERLUST (*m*) : condenser (loss) leakage 545
KONDENSATOR (*m*), VERSTELLBARER- : adjustable condenser (Elect.) 546
KONDENSATORWIRKUNG (*f*) : capacity (Radio) 547
KONDUKTANZ (*f*) : conductance (symbol Mho) (Elect.) 548
KONDUKTION (*f*) : conduction 549
KONDUKTIV : conductive 550
KONDUKTIVE ENTLADUNG (*f*) ; discharge through a conductor (Elect.) 551
KONDUKTIVITÄT (*f*) : conductivity 552
KONDUKTOR (*m*) : conductor (Elect.) 553
KONFERENZ (*f*) : conference ; meeting ; discussion ; congress 554
KONFERENZSAAL (*m*) : assembly room ; meeting room ; conference chamber 555
KONFERENZTEILNEHMER (*m*) : delegate ; one who takes part in a (meeting ; discussion) conference 556
KONGRUENZ (*f*) : congruence 557
KONJUGATION (*f*) : conjugation ; inflection (Grammar) 558
KONNEKTOR (*m*) : connector ; condenser with a capacity of 0·5–2 microfarads (Elect.) 559
KONSONANZ (*f*), ELEKTRISCHE- : electrical consonance (agreement between two alternating currents having the same periodicity) 560
KONSTANTER STROM (*m*) : direct current (Elect.) 561
KONSTANTES ELEMENT (*n*) : cell with constant E.M.F. 562
KONSTITUTION (*f*) : constitution ; diathesis (Med.) 563
KONSUMVERFALL (*m*) : reduction (falling off ; drop ; decrease) in consumption 564
KONTAKT (*m*) : contact (Elect.) 565
KONTAKTARM (*m*) : wiper (Tel.) ; trolley-arm (Electric traction) 566
KONTAKTBANK (*f*) : contact bank (Tel.) 567
KONTAKTBÜGEL (*m*) : bow collector (Electric traction) 568
KONTAKTDETEKTOR (*m*) : crystal detector (Radio) 569
KONTAKTDRAHT (*m*) : trolley wire ; overhead wire (Electric traction) 570
KONTAKTELEKTRIZITÄT (*f*) : contact E.M.F. ; galvanism 571
KONTAKTFINGER (*m*) : contact finger (Elect.) 572
KONTAKT (*m*), FLÜSSIGKEITS- : liquid contact 573
KONTAKT (*m*), GLEIT- : sliding contact 574
KONTAKTHAMMER (*m*) : contact finger (Elect.) 575

KONTAKTKRONE (*f*) : contact head (Elect.) 576
KONTAKTLEITUNG (*f*) : trolley (overhead) wire (Electric traction) 577
KONTAKTROLLE (*f*) : trolley (Electric traction) 578
KONTAKTRUTE (*f*) : contact arm ; collector (Electric traction) 579
KONTAKTSCHALTER (*m*) : contactor (for making and breaking a circuit) (Elect.) 580
KONTAKTSCHIENE (*f*) : conductor (third) rail ; live rail (of electric traction) 581
KONTAKT (*m*), SCHLEIF- : sliding (scraping) contact ; rubbing contact 582
KONTAKT (*m*), SCHRAUB- : screw contact 583
KONTAKTSCHUH (*m*) : collector shoe (Electric traction) 584
KONTAKTSTANGE (*f*) : trolley (arm) pole (Electric traction) 585
KONTAKTSTELLE (*f*) : point of contact 586
KONTAKT (*m*), UMSCHALT- : change-over contact (Elect.) 587
KONTAKTUNTERBRECHER (*m*) : (circuit-) breaker ; interrupter (Elect.) 588
KONTAKTWEG (*m*) : length of contact ; distance over which contact is made 589
KONTINUIERLICHER STROM (*m*) : continuous current (Elect.) 590
KONTRAHENT (*m*) : contracting party ; party to an agreement 591
KONTRAHIEREN : to contact ; agree 592
KONTRAKTSISTIERUNG (*f*) : cancellation of contract 593
KONTRAPUNKT (*m*) : counterpoint (Music) 594
KONTRAPUNKTIK (*f*) : counterpoint 595
KONTRAPUNKTISCH : contrapuntal-(ly) 596
KONTROLLAUTOMAT (*m*) : automatic controller ; mercury (tipper) tipping switch 597
KONTROLLGANG (*m*) : inspection (passage) gallery (Engineering) 598
KONTROLLNORMALE (*f*) : calibration standard 599
KONTROLLZÄHLER (*m*) : check meter (Elect.) 600
KONVEKTIONSSTROM (*m*) : convection current 601
KÖPERGEWEBE (*n*) : (see KÖPER) 602
KOPFARTERIE (*f*), ÄUSSERE- : external carotid artery (Med.) 603
KOPFARTERIE (*f*), INNERE- : internal carotid artery (Med.) 604
KOPFARTERIE (*f*), LINKE- : left common carotid artery (Med.) 605
KOPFARTERIE (*f*), RECHTE- : right common carotid artery (Med.) 606
KOPFDREHBANK (*f*) : chuck lathe (Mech.) 607
KOPFGESTELL (*n*) : head gear (Mining) 608
KOPFHÖRER (*m*) : head-phones ; ear-phone-(s) 609
KOPFHÖRER (*m* and *pl*), DOPPEL- : pair-(s) of ear (head) phones (Radio) 610
KOPF (*m*), MASCHINEN- : headstock (Textile) 611
KOPIERLAMPE (*f*) : printing arc lamp (for copying work) 612
KOPLIKSCHE FLECKEN (*n.pl*) : Koplik's spots (in measles) (Med.) 613
KOPPELUNGSSPULE (*f*) : coupling coil (Radio) 614
KOPPLUNG (*f*) : coupling 615
KOPPLUNG (*f*), ELEKTROSTATISCHE- : electro-static (capacitive ; condenser) coupling 616
KOPPLUNG (*f*), FESTE- : fixed coupling 617
KOPPLUNG (*f*), GALVANISCHE- : galvanic (direct) coupling 618
KOPPLUNG (*f*), INDUKTIVE- : inductive (indirect ; magnetic) coupling 619
KOPPLUNG (*f*), KAPAZITIVE- : capacitive (electro-static ; condenser) coupling 620
KOPPLUNG (*f*), KONDENSATOR- : condenser (capacitive ; electro-static) coupling 621
KOPPLUNG (*f*), LOSE- : loose coupling 622
KOPPLUNG (*f*), MAGNETISCHE- : magnetic (inductive ; indirect) coupling 623
KOPPLUNGSKOEFFIZIENT (*m*) : coupling coefficient (Radio) 624
KOPULIEREN (*n*) : (a method of)-grafting (Bot.) 625
KORKZIEHERREGEL (*f*) : cork-screw rule (Elect.) 626
KÖRNERMIKROPHON (*n*) : carbon granule microphone 627
KÖRNMASCHINE (*f*) : disintegrator ; granulating machine 628
KORONAENTLADUNG (*f*) : corona discharge (Elect.) (see GLIMMENTLADUNG) 629
KORONAERSCHEINUNG (*f*) : corona (phenomenon) discharge 630
KORONASCHUTZ (*m*) : corona protection (Elect.) 631
KORONAVERLUST (*m*) : corona (discharge ; effect) loss ; glow discharge 632
KÖRPERSCHLAGADER (*f*), GROSSE- : aorta (Med.) 633
KÖRPERSCHLUSS (*m*) : short-(circuit) (Elect.) 634
KORPUSKEL (*n*) : corpuscle 635
KORPUSKULARSTRAHL (*m*) : cathode ray 636
KORPUSKULARSTRAHLUNG (*f*) : corpuscular radiation 637
KORPUSKULARTHEORIE (*f*) : corpuscular theory 638
KORREKTURKOEFFIZIENT (*m*) : correction factor 639
KORRESPONDENZLEITUNG (*f*) : two-way wiring circuit 640
KORRESPONDENZSCHALTER (*m*) : two-way switch (Elect.) staircase (landing) switch 641
KOTFLÜGEL (*m*) : mudguard (of motor cars, etc.) 642
KÖTZER (*m*) : cop (Textile) 643
KÖTZERFÜLLE (*f*) : degree of fullness of the cop 644
KRAFTABSTUFUNG (*f*) : dynamics (Music) 645
KRAFTENDRÖHRE (*f*) : power valve (Radio) 646
KRAFTFLUSS (*m*), MAGNETISCHER- : magnetic flux 647
KRAFTKABEL (*m*) : power cable (Elect.) 648
KRAFTLINIEN (*f.pl*) : lines of magnetic force 649
KRAFTLINIENBILD (*n*) : magnetic figures 650
KRAFTLINIENBÜNDEL (*m*) : (see KRAFTRÖHRE) 651
KRAFTLINIENFELD (*n*) : magnetic field 652
KRAFTLINIENRICHTUNG (*f*) : direction of lines of magnetic force ; direction of magnetic flux 653
KRAFTLINIENSTREUUNG (*f*) : magnetic leakage ; dispersion of magnetic flux 654
KRAFTLINIENSTROM (*m*) : magnetic flux 655
KRAFTRÖHRE (*f*) : magnetic tube of force 656
KRAFTSTATION (*f*) : power station (Elect.) 657
KRAFTSTECKER (*m*) : power plug (Elect.) 658
KRAFTSTOFF (*m*) : fuel oil ; liquid fuel 659
KRAFTSTROM (*m*) : power current (not for lighting) 660
KRAFTTARIF (*m*) : power tariff ; tariff for power current 661
KRAFTÜBERTRAGUNG (*f*), ELEKTRISCHE- : electric power transmission 662
KRAFT (*f*), VERBORGENE- : latent power 663
KRAFTVERSTÄRKER (*m*) : power amplifier 664
KRAFTVERTEILUNG (*f*) : power distribution 665
KRAFTZÄHLER (*m*) : electricity meter for power current 666
KRAKEHLSUCHT (*f*) : rowdyism (Med.) 667
KRANKHEITSANLAGE (*f*) : diathesis ; predisposition to disease 668
KRANKHEITSENTSTEHUNG (*f*) : pathogeny 669
KRANKHEITSERKENNUNG (*f*) : diagnosis (Med.) 670
KRANKHEITSERREGEND : pathogenic (see PATHOGEN) ; morbific ; disease-producing 671
KRANKHEITSERSCHEINUNG (*f*) : symptomatology (science of the signs of disease) (Med.) 672

KRANKHEITSLEHRE (*f*) : pathology (the science of diseases) (Med.) 673
KRANKHEITSTRÄGER (*m*) : (disease)-carrier (Med.) 674
KRANKHEITSURSACHE (*f*) : aetiology (the science of the cause of disease) 675
KRANKHEITSVERHÜTUNG (*f*) : prophylaxis (Med.) 676
KRANKHEITSZUSTAND (*m*) : morbid (state) condition (Med.) 677
KRATERBILDUNG (*f*) : formation of a crater ; crater formation 678
KRATERFÖRMIG : crater-shaped 679
KRATZENWALZE (*f*) : card roller (Textile) 680
KREIDE (*f*), **MITTLERE-** : Middle Cretaceous (Geol.) 681
KREIDE (*f*), **UNTERE-** : lower chalk ; Lower Cretaceous (Geol.) 682
KREIS (*m*), **ANGEREGTER-** : secondary (excited) circuit (Radio) 683
KREIS (*m*), **HILFS-** : auxiliary circuit (Radio) 684
KREISSKALA (*f*) : circular scale 685
KREISSTROM (*m*) : molecular current (surrounding the molecules of Fe, Ni, and Co) 686
KREMPELEI (*f*) : carding (Textile) 687
KRETAZISCH : cretaceous (Geol.) 688
KRETAZISCH, UNTER- : Lower Cretaceous (Geol.) 689
KRETINISMUS (*m*) : cretinism (a form of idiocy) (Med.) 690
KREUZKÖTZERSPULMASCHINE (*f*) : cross pirn winder (Textile) 691
KREUZSCHALTER (*m*) : intermediate switch (used with two-way switches) 692
KREUZSPULMESSWERK (*n*) : cross-coil measuring apparatus (Elect.) 693
KREUZSTOSS (*m*) : double T weld 694
KREUZUNGSSCHALTER (*m*) : (see **KREUZSCHALTER**) 695
KREUZWIRBEL (*m*) : sacral vertebra (Med.) 696
KRIECHSTROM (*m*) : leakage current (Elect.) 697
KRIECHWEG (*m*) : creepage path (Elect.) 698
KRISTALLDETEKTOR (*m*) : crystal detector (Radio) 699
KRISTALLELEKTRIZITÄT (*f*) : (see **PIËZOELEKTRIZITÄT**) 700
KRONENSCHALTER (*m*) : lighting switch for the consecutive closing of three circuits, the fourth position being " off " (Elect.) 701
KRÜGERELEMENT (*n*) : Krueger element (Elect.) 702
KRUPPÖS : croupous (Med.) 703
KÜCHENMASCHINE (*f*) : kitchen (machine) appliance 704
KÜCHENMOTOR (*m*) : kitchen motor (for driving electric kitchen appliances) 705
KUGELBREMSVORRICHTUNG (*f*) : ball-drag (Textile) 706
KUGELFUNKENSTRECKE (*f*) : sphere (spark)-gap (Elect.) 707
KUGELMESSER (*m*) : spherometer (for measuring radii and curvature of spherical surfaces) 708
KÜHLBETT (*n*) : cooling bed (Rolling mills) 709
KUHSTALL (*m*) : byre ; cow house 710
KULANZLIEFERUNG (*f*) : gift (of goods in order to influence a later order) 711
KÜMPELPRESSE (*f*) : flanging press (Mech.) 712
KUNSTERZEUGNIS (*n*) : artefact ; artifact ; artificial product 713
KUNSTSEIDESPINNEREI (*f*) : rayon spinning mill (Textile) 714
KUPFERFINNE (*f*) : *Acne rosacea* (Med.) 715
KUPFERFLUORID (*n*) : cupric fluoride ; CuF_2 716
KUPFERHAUTSTAHLDRAHT (*m*) : steel-covered copper wire 717
KUPFERKUNSTSEIDE (*f*) : cuprammonium rayon (Textile) 718
KUPPELHORIZONT (*m*) : cyclorama (of stages) 719
KUPPLUNG (*f*), **DIREKTE-** : direct coupling 720
KUPPLUNGSBELAG (*m*) : clutch lining 721
KURBELINDUKTOR (*m*) : magneto 722
KURVENSCHEIBE (*f*) : cam (Mech.) 723
KURZARBEIT (*f*) : short time (in factories) 724
KURZBLÄTTRIG : short-leaved (*brachyphyllus*) (Bot.) 725
KURZSCHLIESSEFÄHIGKEIT (*f*) : short-circuiting capacity (Elect.) 726
KURZSCHLUSSART (*f*) : sort (kind) of short-circuit 727
KURZSCHLUSSCHLEIFE (*f*) : short-circuit loop 728
KURZSCHLUSSFEST : short-circuit proof (Elect.) 729
KURZSCHLUSSKÄFIG (*m*) : squirrel-cage (Elect.) 730
KURZSCHLUSSLÄUFER (*m*) : squirrel-cage rotor 731
KURZSCHLUSSMOTOR (*m*) : squirrel-cage motor 732
KURZSCHLUSSTROM (*m*) : short-circuit current 733
KURZSCHLUSSVORRICHTUNG (*f*) : short-circuiting device 734
KURZSCHLUSSWINKEL (*m*) : phase angle of short-circuit 735
KURZSCHLUSSZIFFER (*f*) : short-circuit factor (Elect.) 736
KURZSICHTIGKEIT (*f*) : myopia ; short sight (Med.) (see **MYOPIE**) 737
KURZWELLENEMPFÄNGER (*m*) : short-wave (receiver) set (Radio) 738
KURZWELLENTELEPHONIE (*f*) : short-wave telephony 739
KURZZÄHNIG : short-spiked ; bearing short spikes (*brachystachys*) (Bot.) 740
KÜSTENSTATION (*f*) : coastal station 741
KYMOGRAPHION (*n*) : kymograph ; cymograph (see **MYOGRAPHION**) 742
KYMOMETER (*n*) : wave meter 743
KYSTOSKOP (*n*) : cystoscope (for examination of disease by insertion in the human body) (Med). 744

L

LACKADERDRAHT (*m*) : lacquered and impregnated cotton covered (cable) wire (Elect.) 745
LADEHEBEL (*m*) : charging switch (Elect.) 746
LADENTHEKE (*f*) : counter 747
LADESCHALTTAFEL (*f*) : charging board (Elect.) 748
LADESPANNUNG (*f*) : charging voltage (Elect.) 749
LADESTADIUM (*n*) : state of charge (Elect.) 750
LADESTATION (*f*) : charging station (Elect.) 751
LADESTROMSTÄRKE (*f*) : charging current (Elect.) 752
LADEZEIT (*f*) : charging time (Elect.) 753
LADUNGSANZEIGER (*m*) : charge indicator (Elect.) 754
LADUNGSKAPAZITÄT (*f*) : charging capacity 755
LADUNGSRÜCKSTAND (*m*) : residual electricity 756
LAGERELEMENT (*n*) : uncharged (dry) storage battery 757
LAGERSCHILD (*n*) : end shield (of electric motors) 758
LAGERSTEIN (*m*) : jewel (for bearings of instruments) 759
LÄHMUNG (*f*), **UNVOLLSTÄNDIGE-** : slight or temporary paralysis ; paresis (Med.) 760

LAMELLENSICHERUNG (*f*) : laminated (strip) fuse (Elect.) 761
LAMELLIERT : laminated 762
LAMPENBATTERIE (*f*) : battery (bank) of lamps 763
LAMPENKOHLE (*f*) : arc lamp carbon 764
LAMPENREDUKTOR (*m*) : lamp resistance (Elect.) 765
LAMPENRHEOSTAT (*m*) : lamp resistance 766
LANDELEUCHTE (*f*) : landing light (Aerodromes) 767
LANDKABEL (*n*) : land cable (Elect.) 768
LANGFLACHSSPINNEREI (*f*) ; flax line spinning mill 769
LANGSIEBMASCHINE (*f*) : long wire machine (Paper making) 770
LANGSPITZIG : long pointed (*acuminatus*) (Bot.) 771
LANGSTAPELKUNSTWOLLE (*f*) : shoddy (Textile) 772
LÄNGSWELLE (*f*) : longitudinal wave-(of light) 773
LANZINIERENDE SCHMERZEN (*m.pl*) : shooting pains (Med.) 774
LÄPPEN (*n*) : buffing (Textile) 775
LÄRMZAHL (*f*) : unit of audibility (see also **PHON**) 776
LARYNGOSKOP (*n*) : laryngoscope (an instrument for examining the wind-pipe) (Med.) 777
LASTHEBEMAGNET (*m*) : lifting magnet 778
LAST (*f*), **TOTE-** : dead (weight) load 779
LATENZZEIT (*f*) : latent period (Med.) 780
LAUBMOOSE (*n.pl*) : *Musci* (the true mosses) (Bot.) 781
LÄUFERWICKLUNG (*f*) : rotor winding (Elect.) 782
LAUF (*m*), **GERÄUSCHLOSER-** : noiseless (operation ; working) running 783
LAUSCHAPPARAT (*m*) : listening apparatus ; hidden microphone 784
LAUTFERNSPRECHER (*m*) : loudspeaking telephone 785
LAUTSPRECHANLAGE (*f*) : loudspeaker (loudspeaking telephone) installation 786
LAUTSPRECHER (*m*) : loud-speaker 787
LAUTSTÄRKE (*f*) : signal strength ; volume (Radio) 788
LAUTSTÄRKEREGLER (*m*) : volume control (Radio) 789
LAUTVERSTÄRKER (*m*) : sound amplifier ; low frequency (L.F.) amplifier 790
L.C.-AUSGANG (*m*) : L.C. output (Radio) (L. = electrical symbol for inductance : L.C. = inductance capacity) 791
LEBENSMAGNETISMUS (*m*) : animal magnetism 792
LEBERATROPHIE (*f*) : atrophy of the liver 793
LEBERMOOSE (*n.pl*) : *Hepaticae* (the Liverworts) (Bot.) 794
LEBERSCHRUMPFUNG (*f*) : cirrhosis of the liver 795
LEBERSCHWUND (*m*) : atrophy of the liver 796
LEBERZIRRHOSE (*f*) : cirrhosis of the liver 797
LECITHIN (*n*), **SOJA-** : soya-bean lecithin 798
LECKSTROMMELDER (*m*) : leakage-(current) indicator 799
LEDERTREIBER (*m*) : leather picker (Textile) 800
LEERKONTAKT (*m*) : no-load contact (Elect.) 801
LEERLAUFARBEIT (*f*) : unnecessary work-(in business) ; frictional resistance (Mech.) 802
LEERLAUFSTROM (*m*) : no-load current (Elect.) 803
LEICHTPUPINISIERT : light loaded (Tel.) 804
LEINWANDHORIZONT (*m*) : cloth horizon ; back cloth (of stages) 805
LEISTUNGSABGABE (*f*) : output (Elect.) 806
LEISTUNGSAUFNAHME (*f*) : (power)-input (Elect.) 807
LEISTUNGSEFFEKT (*m*) : work ; power ; output 808
LEISTUNGSFAKTORMESSER (*m*) : power factor meter (Elect.) 809
LEISTUNGSSCHILD (*m*) : name-plate (Elect.) 810
LEISTUNGSVERBRAUCH (*m*) : energy consumption (Elect.) 811
LEITERGRUPPE (*f*) : group of conductors (Elect.) 812
LEITFÄHIGKEIT (*f*), **ÄQUIVALENTE-** : equivalent conductivity (Electro-chemistry) 813
LEITFÄHIGKEIT (*f*), **MOLEKULAR-** : molecular conductivity 814
LEITFÄHIGKEIT (*f*), **SPEZIFISCHE-** : specific conductivity 815
LEITUNG (*f*), **BIEGSAME-** : flexible cord (Elect.) 816
LEITUNG (*f*), **BLANKE-** : bare wire 817
LEITUNG (*f*), **ISOLIERTE-** : insulated (wire) cable ; lagged pipe-line 818
LEITUNG (*f*), **KRANKE-** : faulty line 819
LEITUNG (*f*), **KÜNSTLICHE-** : artificial line (Tel.) 820
LEITUNGSADER (*f*) : core (strand) of cable 821
LEITUNGSBELASTUNG (*f*) : carrying capacity (of cables and wires) 822
LEITUNGSBEMESSUNG (*f*) : size (gauge ; section) of wiring (Elect.) 823
LEITUNGSEINFÜHRUNG (*f*) : cable inlet (Elect.) 824
LEITUNGSFÜHRUNG (*f*) : feeder (Elect.) 825
LEITUNGSGESTÄNGE (*n*) : poles for overhead wiring ; telegraph poles 826
LEITUNGSIMPEDANZ (*f*) : line impedance 827
LEITUNGSKUPFER (*n*) : standard copper 828
LEITUNGSLÄNGE (*f*) : circuit length (Tel.) 829
LEITUNGSLITZE (*f*) : stranded wire ; flex-(ible cord) (Elect.) 830
LEITUNGSNETZ (*n*) : network (Elect.) 831
LEITUNGSNORMALIEN (*n.pl*) : wiring (regulations) standards (Elect.) 832
LEITUNGSPRÜFER (*m*) : line tester (Elect.) 833
LEITUNGSSCHNUR (*f*), **BIEGSAME-** : flex-(ible cord) (Elect.) 834
LEITUNGSTRAVERSE (*f*) : (cross)-arm on telegraph poles 835
LEITUNGSTROSSE (*f*) : stranded cable (for cranes, etc.) 836
LEITUNGSVERSTÄRKER (*m*) : repeater (Radio) 837
LEITUNGSWÄHLER (*m*) : line selector (Tel.) 838
LEITUNG (*f*), **UMHÜLLTE-** : covered (cable) wire 839
LEITVERMÖGEN (*n*) : carrying capacity ; conductivity ; conductance 840
LEITWERT (*m*), **SPEZIFISCHER-** : (see **LEITVERMÖGEN**) 841
LEONARDSCHALTUNG (*f*) : Ward-Leonard control 842
LEPRA (*f*) : leprosy (caused by *Bacterium leprae*) 843
LEUCHTDICHTE (*f*) : (intrinsic)-brilliance or brilliancy ; (surface)-brightness 844
LEUCHTFADEN (*m*) : filament (of electric lamps) 845
LEUCHTSALZ (*n*) : thorium nitrate (see **THORIUMNITRAT**) 846
LEUCHTSCHALTBILD (*n*) : illuminated connection diagram (Elect.) 847
LEUCHTSCHIRM (*m*) : fluorescent screen (of X-rays) ; luminous screen 848
LEUCHTSTÄRKE (*f*) : intensity (of a light source in standard candles) (Elect.) 849
LEUKODERM (*n*) : leucoderma ; leucodermia (white patches on the skin) (Med.) 850
LEUKOZYT (*m*) : leucocyte ; white-(blood)-corpuscle (Med.) 851
LIAS (*f*) : Lias (Geol.) 852
LIAS (*f*), **OBERE-** : Upper Lias 853
LIAS (*f*), **UNTERE-** : Lower Lias 854
LICHTABGABE (*f*) : (see **LICHTMENGE**) 855
LICHTAUSSTRAHLUNG (*f*) : dispersion (of light) 856
LICHTAUSSTRAHLUNGSWINKEL (*m*) : dispersive angle (of light) 857
LICHTBAND-INSTRUMENT (*n*) : light column instrument (which indicates by means of a column of light) 858

LICHTBOGENKRATER (*m*) : crater (in arc lamp carbon) 859
LICHTBOGENLÄNGE (*f*) : length of arc (Welding) 860
LICHTBOGENLÖSCHEND : arc quenching 861
LICHTBOGENLÖSCHUNG (*f*) : arc quenching (Elect.) 862
LICHTBOGEN-SCHWEISSUMFORMER (*m*) : arc welding set 863
LICHTBRECHUNGSLEHRE (*f*) : dioptrics 864
LICHTEINFALL (*m*) : incidence (of light) 865
LICHTENMASS (*n*) : clear opening ; inside diameter 866
LICHTER ABSTAND (*m*) : clear distance apart 867
LICHTHOFFREI : non-halation (Phot.) 868
LICHTHÜLLE (*f*) : glow 869
LICHTKOHLE (*f*) : arc lamp carbon 870
LICHTKRATER (*m*) : crater of the positive (upper) carbon (of arc lamps) 871
LICHTLEISTUNG (*f*) : (see LICHTMENGE) 872
LICHTMENGE (*f*) : quantity of light in lumen/seconds (Elect.) 873
LICHTNETZ (*n*) : lighting main (Elect.) 874
LICHTPOSITION (*f*) : position (source) of the light 875
LICHT (*n*), POSITIONS- : navigation light (of aeroplanes) 876
LICHTRELAIS (*n*) : light relay (Elect.) 877
LICHTSIGNALANLAGE (*f*) : light signalling apparatus or installation 878
LICHTSPANNUNG (*f*) : lamp voltage ; voltage of a lighting circuit 879
LICHTSPIELTHEATER (*n*) : cinematograph (picture) theatre ; cinema 880
LICHTSTECKER (*m*) : (5 amp.)-lighting plug 881
LICHTSTREUEND : diffusing (of glass) 882
LICHTSTREUUNG (*f*) : spreading (diffusion) of light 883
LICHTSTROM (*m*) : luminous flux ; lumen (Light unit) ; current for lighting purposes 884
LICHTTABLO (*n*) : lamp indicator-(board) ; illuminated indicator 885
LICHTTELEPHON (*n*) : (see PHOTOPHON) 886
LICHTTHERAPIE (*f*) : light therapy ; phototherapy (healing of disease by light) 887
LICHTVERTEILUNG (*f*) : distribution of light 888
LICHTVERTEILUNGSKURVE (*f*) : photometric (polar) curve (Lighting) 889
LICHTZÄHLER (*m*) : lighting (house-service) meter (Elect.) 890
LINDENBLÜTENTEE (*m*) : lime-flower tea (Med.) 891
LINIENFERNSPRECHER (*m*) : line telephone 892
LINIENRELAIS (*n*) : line relay (Tel.) 893
LINIENUMSCHALTER (*m*) : change-over switch (Tel.) 894
LINKSLAUF (*m*) : anti-clockwise (counter-clockwise) rotation 895
LINSENOPTIK (*f*) : system of lenses 896
LITZENDRAHT (*m*) : stranded wire, Litzen wire (Elect.) 897
LITZENLEITER (*m*) : stranded conductor (Elect.) 898
LITZENMASCHINE (*f*) : stranding machine (Textile) 899
LÖCHERPILZ (*m*) : *Polyporus* (Bot.) 900
LOCHSTREIFEN (*m*) : perforated tape (Tel.) 901
LOKALBATTERIE (*f*) : local battery (Tel.) 902
LOKSCHUPPEN (*m*) : (see LOKOMOTIVSCHUPPEN) 903
LONGITUDINALWELLE (*f*) : longitudinal wave (of light) 904
LOOG (*n*) : (see LOG) 905
LORE (*f*) : lorry ; truck (see LOWRY) 906
LÖSCHEINRICHTUNG (*f*), MIT- : with earth current choke coil (Elect.) 907
LÖSCHFUNKENSTRECKE (*f*) : quenched spark gap (Elect.) 908
LÖSCHKONDENSATOR (*m*) : extinguishing condenser (Elect.) 909
LÖSCHMEDIUM (*n*) : quenching (medium) agent 910
LÖSCHPULVER (*n*) : quenching powder (of electric fuses) 911
LÖTANSCHLUSS (*m*) : soldering (soldered) connection 912
LÖTFETT (*n*) : flux (for soft soldering) 913
LÖTVERBINDUNG (*f*) : soldered joint (see LÖTSTELLE) 914
LOWRY (*m*) : lorry ; truck (see LORE) 915
LUES (*f*) : syphilis (Med.) 916
LUFTABSCHEIDER (*m*) : air extractor (of pumps) 917
LUFTBEFEUCHTUNGSANLAGE (*f*) : (see LUFTBEFEUCHTIGUNGSANLAGE) 918
LUFTBLÄSCHEN (*n*) : air sac (of the lungs) 919
LUFTFAHNE (*f*) : air vane 920
LUFTISOLATION (*f*) : air insulation 921
LUFTPOST (*f*) : air mail 922
LUFTPOSTSTRECKE (*f*) : air mail route 923
LUFTSCHIEBER (*m*) : air vane 924
LUFTSCHRAUBE (*f*) : propeller (Aeroplanes) ; airscrew 925
LUFTSCHRAUBENNABE (*f*) : propeller (boss) hub ; airscrew (boss) hub 926
LUFTSCHÜTZ (*n*) : air break (Elect.) 927
LUMBALPUNKTION (*f*) : lumbar puncture (Med.) 928
LUMEN (*n*) : lumen (unit of lighting) 929
LUNGENARTERIE (*f*) : pulmonary artery (Med.) 930
LUNGENBASIS (*f*) : base of the lung 931
LUNGENBLÄHUNG (*f*) : distension (dilation ; dilatation) of the lungs (see LUNGENERWEITERUNG) 932
LUNGENBLÄSCHEN (*n* and *pl*) : air sac-(s) of the lungs 933
LUNGENERWEITERUNG (*f*) : (see LUNGENEMPHYSEM) 934
LUNGENGERÄUSCH (*n*) : crepitation 935
LUNGENGEWEBE (*n*) : lung tissue 936
LUNGENKAPILLAREN (*n.pl*) : capillaries of the lungs 937
LUNGENLAPPEN (*m*) : lobe of the lung 938
LUNGENOBERFLÄCHE (*f*) : surface of the lung 939
LUNGENPFORTE (*f*) : hilum (root of the lung) (Med.) 940
LUNGENSCHLAGADER (*f*) : pulmonary artery (Med.) 941
LUNGENSPITZE (*f*) : apex of the lung 942
LUNGENTÄTIGKEIT (*f*) : activity (functioning) of the lungs 943
LUPUS (*m*) : lupus (*Lupus vulgaris*) (tuberculosis of the face and nose) (Med.) 944
LÜSTERKUPPLUNG (*f*) : Edison-screw adaptor (Elect.) 945
LUX (*m*) : (see METERKERZE) 946
LUXATION (*f*) : sprain ; dislocation (Med.) 947
LYMPHANGIOM (*n*) : lymphatic angioma (Med.) 948
LYMPHDRÜSENENTZÜNDUNG (*f*) : lymphadenitis (Med.) 949
LYMPHGEFÄSSENTZÜNDUNG (*f*) : lymphangitis (Med.) 950

M

MACOSPINNEREI (*f*) : Egyptian cotton spinning-(industry) (Textile) 951
MADENSCHRAUBE (*f*) : grub screw (Engineering) 952
MAGENAUSGANG (*m*) : pylorus (Med.) ; pyloric end of stomach 953
MAGENBLUTUNG (*f*) : naematemesis (vomiting blood from the stomach) (Med.) 954
MAGENEINGANG (*m*) : cardiac end of stomach 955
MAGENKRAMPF (*m*) : cardialgy ; cardialgia (Med.) 956
MAGENPUMPE (*f*) : stomach pump 957
MAGENSCHLEIMHAUT (*f*) : mucous membrane of the stomach 958
MAGENSCHLEIMHAUTENTZÜNDUNG (*f*) : gastritis (Med.) 959
MAGENSTEIFUNG (*f*) : peristalsis (Med.) 960
MAGERSUCHT (*f*) : wasting away (Med.) 961
MAGNESIALICHT (*n*) : magnesium (light) flare 962
MAGNETBOHRUNG (*f*) : space occupied by the armature (Elect.) 963
MAGNETEXTRAKTION (*f*) : extraction of splinters from the eye by magnet 964
MAGNETFELD (*n*) : magnetic field (Elect.) 965
MAGNETGESTELL (*n*) : magnet frame (Elect.) 966
MAGNETISCHE DURCHLÄSSIGKEIT (*f*) : permeability (Elect.) 967
MAGNETISCHE LEGIERUNG (*f*) : any alloy with magnetic properties 968
MAGNETISCHER ERZSCHEIDER (*m*) : magnetic separator 969
MAGNETISCHE SÄTTIGUNG (*f*) : magnetic saturation 970
MAGNETISCHES AUFNAHMEVERMÖGEN (*n*) : susceptibility (Elect.) 971
MAGNETISCHE TRÄGHEIT (*f*) : magnetic hysteresis 972
MAGNETISIERUNGSSPULE (*f*) : magnetic (magnet) coil 973
MAGNETISMUS (*m*), KONSTANTER- : permanent magnetism 974
MAGNETISMUS (*m*), NATÜRLICHER- : earth magnetism 975
MAGNETLAUFKRAN (*m*) : travelling crane with lifting magnet 976
MAGNETOMOTORISCHE KRAFT (*f*) : magneto-motive force 977
MAGNETOPERATION (*f*) : (see MAGNETEXTRAKTION) 978
MAGNETRAD (*n*) : pole-wheel ; rotor (Elect.) 979
MAGNETREGULIERUNG (*f*) : field regulation (Elect.) 980
MAGNETTRAVERSE (*f*) : lifting magnet traverse (of cranes) 981
MAGNETZÜNDER (*m*) : magneto 982
MÄHBINDER (*m*) : cutter and binder (Agricultural) 983
MAKOGARN (*n*) : Egyptian cotton yarn (Textile) 984
MAKROSKOPISCH : macroscopic-(al)-(ly) (visible to the naked eye) 985
MANDELGESCHWÜR (*n*) : swollen (enlarged) tonsil-(s) (Med.) 986
MANGANINWIDERSTAND (*m*) : manganin resistance (Elect.) 987
MÄNGELHAFTUNG (*f*) : warranty ; guarantee (usually in respect of faulty material or workmanship) 988
MANIPULATOR (*m*) : morse (key) tapper 989
MANTELDRAHT (*m*) : metal-covered wire (Elect.) 990
MANTELGEKÜHLT : frame-cooled (of electric motors) 991
MANTELTRANSFORMATOR (*m*) : shell transformer (Elect.) 992
MARASMUS (*m*) : marasmus ; wasting away (Med.) 993
MARGE (*f*) : margin 994
MARINESCHALTER (*m*) : marine type switch (Elect.) ; waterproof, enclosed, rotary switch 995
MASCHINENAGGREGAT (*n*) : unit ; plant ; set ; installation (of machinery) 996
MASCHINENHALLE (*f*) : engine (house) room ; turbine house 997
MASCHINENSTELLER (*m*) : erector ; setter 998
MASSENEINHEIT (*f*) : unit of (weight) mass (equal to 1 gram) 999
MASSENTEILCHEN (*n*) : corpuscle 000
MASSGLEICH : isometric ; having the same measure 001
MASTAUSSCHALTER (*m*) : (see MASTSCHALTER) 002
MASTDARMABSCHNITT (*m*) : section of the intestines 003
MASTDARMKANAL (*m*) : intestinal canal 004
MASTDARMSCHRUNDE (*f*) : (rectal)-fissure 005
MASTITIS (*f*) : (see BRUSTMILCHDRÜSENENTZÜNDUNG) 006
MASTSCHALTER (*m*) : pole (mast) switch (for attaching to a mast carrying overhead high-tension wires) (Elect.) 007
MAST (*f*), VOLLE- : full crop of Beech mast (every 5–10 years from *Fagus sylvatica*) ; (see also BUCHECKER) 008
MATERFÄHIG : suitable for making a (stereotype) block from a cardboard matrix 009
MATTIERSALZ (*n*) : frosting mixture (for frosting glass) 010
MATTIERUNGSMITTEL (*n*) : delustrant (Textile) 011
MATTKOHLE (*f*) : matt (dull) coal (see FAULSCHLAMMKOHLE) 012
MATTVERCHROMUNG (*f*) : matt (dull) chromium plating ; matt-finished chromium plating 013
MAXIMALAUSSCHALTER (*m*) : overload release (Elect.) 014
MAXIMALAUTOMAT (*m*) : overload release (Elect.) 015
MEERESSCHWAMM (*m*) : sea (natural) sponge 016
MEGAMPERE (*n*) : megampere (1,000,000 amps.) 017
MEHLBIRNENBLÄTTRIG : with leaves like the Beam tree (*ariifolius*) (Bot.) 018
MEHRDRÄHTIG : multifilar ; having (many) a number of wires 019
MEHRFACHANTENNE (*f*) : multiple wire aerial 020
MEHRFACHTELEGRAPHIE (*f*) : multiplex telegraphy 021
MEHRFACHTELEPHONIE (*f*) : multiple way system of telephony 022
MEHRFACH-TELEPHONIEGERÄT (*n*) : multiplex telephone apparatus 023
MEHRFÄDIG : multifilar ; having (many) a number of threads 024
MEHRLEITERKABEL (*n*) : multiple (multi)-core cable 025
MEHRLEITERSYSTEM (*n*) : multi-wire system 026
MEHRPHASENKABEL (*n*) : polyphase cable 027
MELANOSE (*f*) : melanosis ; black cancer (Med.) 028
MENORRHAGIE (*f*) : menorrhagia (Med.) 029
MERKURIALISMUS (*m*) : (see QUECKSILBERVERGIFTUNG) 030
MESOZOIKUM (*n*) : Mesozoic-(period) (Geol.) 031

MESOZOISCH : mesozoic 032
MESSBRÜCKE (*f*) : measuring bridge (Elect.) 033
MESSCHRANK (*m*) : test board 034
MESSDRAHT (*m*) : test wire (Tel.) ; calibrated resistance wire 035
MESSDRUCK (*m*) : differential pressure 036
MESSERSCHALTER (*m*) : knife switch (Elect.) 037
MESSFLANSCH (*m*), **RUNDKAMMER-** : orifice flange 038
MESSGENAUIGKEIT (*f*) : accuracy (correctness) of measurement 039
MESSGENAUIGKEITSANGABE (*f*) : correct reading (of measuring instruments) 040
MESSKAMMER (*f*) : metering (measuring) chamber 041
MESSWANDLER (*m*) : instrument transformer (Elect.) 042
MESSWERK (*n*) : movement ; gearing (of measuring instruments) 043
METALLANLASSER (*m*) : metal starting resistance (Elect.) 044
METALLION (*n*) : metal ion (Electrolysis) 045
METALLVEREDLUNG (*f*), **GALVANISCHE-** : galvanic electro-plating 046
METEORISMUS (*m*) : meteorism ; flatulence (distension of abdomen by gas in the intestines) 047
METERKERZE (*f*) : lux (unit of lighting) ; meter candle (Elect.) 048
MHO (*n*) : mho ; reciprocal ohm ; Siemens (unit of conductivity in Electrochemistry) 049
MIGNONFASSUNG (*f*) : small Edison-screw lampholder 050
MIKROAMPERE (*n*) : micro-ampere ($\frac{1}{1,000,000}$ amp.) 051
MIKROFON (*n*) : (see **MIKROPHON**) 052
MIKROION (*n*) : (see **KATION**) 053
MIKROMILLIMETER (*n*) : micromillimeter ; micron, (1 μ = 0·001 mm.) (see **MIKRON**) 054
MIKRON (*n*) : micron ; μ ; micromillimeter (1 μ = 10,000 Angström units) 055
MILLIMIKRON (*n*) : millimicron ; $\mu\mu$ (1 $\mu\mu$ = 0·000001 mm.) 056
MILLIMOL (*n*) : millimole (see **MOL**) 057
MILZBRANDSTÄBCHENBAKTERIUM (*n*) : anthrax bacillus (*Bacterium anthracis ; Bacillus anthracis*) 058
MILZFARNBLÄTTRIG : with leaves resembling Spleenwort (*aspleniifolius*) (Bot.) 059
MINDERGESCHWINDIGKEIT (*f*) : narrowing speed (Textile) 060
MINDERSCHALTER (*m*) : narrowing-(speed) switch (Textile) 061
MINENSÄUBERUNGSARBEIT (*f*) : mine sweeping-(operations) 062
MINENSPRENGDRAHT (*m*) : shot-firing wire (Elect.) 063
MINENSUCHGERÄT (*n*) : mine-sweeping apparatus 064
MINENZÜNDER (*m*) : mine (exploder) detonator 065
MINERALMAGNETISMUS (*m*) : mineral magnetism 066
MINIMALAUSSCHALTER (*m*) : no-load release 067
MINUSELEKTRODE (*f*) : negative electrode ; cathode 068
MINUSPOL (*m*) : negative (minus) pole (Elect.) 069
MIOCÄN : Miocene (Geol.) 070
MISCHRAUM (*m*) : mixing chamber 071
MISCHRÖHRE (*f*) : modulator (Radio) 072
MITBETREUEN : to look after ; care for ; attend to ; be responsible for-(any sort of work) 073
MITGETEILTE AKTIVITÄT (*f*) : induced activity-(of radio-active substances) 074
MITHÖREN (*n*) : overhearing (Tel.) 075
MITHÖRKLINKE (*f*) : listening jack 076
MITHÖRSTÖPSEL (*m*) : listening jack (Tel.) 077
MITHÖRTASTE (*f*) : listening key (Tel.) 078
MITSCHWINGEN : to vibrate in sympathy with ; (*n*) : sympathetic vibration 079
MITSPRECHDÄMPFUNG (*f*) : overhearing attenuation (Tel.) 080
MITSPRECHEN (*n*) : (near-end)-overhearing (cross-talk between components of one cable) (Tel.) 081
MITTEL (*n*), **FÄULNISWIDRIGES-** : antiseptic 082
MITTELFREQUENZ (*f*) : medium frequency-(of sound waves) 083
MITTELLEITER (*m*) : neutral wire-(of electric cables) 084
MITTELOHR (*n*) : middle ear (Med.) 085
MITTELOHRENTZÜNDUNG (*f*) : otitis (inflammation of the middle ear) (Med.) 086
MITTELOHRKATARRH (*m*), **CHRONISCHER-** : chronic inflammation of the middle ear 087
MITTELSCHLÄCHTIGES WASSERRAD (*n*) : breast wheel 088
MITTELSTARK : medium heavy ; of medium strength ; medium strong 089
MITTELWORT (*n*) **DER GEGENWART** (*f*) : present participle (Grammar) 090
MITTELWORT (*n*) **DER VERGANGENHEIT** (*f*) : past participle (Grammar) 091
MÖBELSTOFF (*m*) : upholstery 092
MODULATOR (*m*) : modulator (Radio) 093
MOLEKULARKRAFT (*f*) : molecular force (such as cohesion or capillary attraction) 094
MOLEKULAR LEITFÄHIGKEIT (*f*) : molecular conductivity (Electro-chemistry) 095
MOLION (*n*) : molecule having a free electric charge 096
MOLLDREIKLANG (*m*) : minor triad (Music) 097
MÖLLERKÜBEL (*m*) : ore skip (see **MÖLLER**) 098
MOLTEBEERE (*f*) : cloudberry ; mountain bramble ; knotberry ; ground mulberry (*Rubus chamaemorus*) (Bot.) 099
MOMENTANSTROM (*m*) : surge (Elect.) 100
MOMENTAUSLÖSUNG (*f*) : quick make-and-break (Elect.) ; instantaneous (cut-out) release 101
MOMENTSCHALTER (*m*) : quick-make-and-break switch (Elect.) 102
MOMENTSCHALTUNG (*f*) : quick make-and-break (Electric switches) 103
MONOKARPISCH : monocarpic ; monocarpous (plants which die after once producing flowers and fruit) 104
MONTAGEHALLE (*f*) : fitting (erection) shop 105
MOOSE (*n.pl*) : (see **BRYOPHYTEN**) 106
MORBILITÄT (*f*) : ratio of the sick to the living (Med.) 107
MORPHIUMEINSPRITZUNG (*f*) : morphium injection (Med.) 108
MORTIFIZIEREN : to declare null and void ; mortify ; hurt (feelings) 109
MOSAISCHES GOLD (*n*) : (see **MUSIVGOLD**) 110
MOTIV (*n*), **UNBEWEGTES-** : still-(picture) (Cinematography) 111
MOTORDREHMOMENT (*n*) : motor torque 112
MOTORSCHIFF (*n*) : motor-(driven)-ship 113
MOTORWECKER (*m*) : motor-driven bell (Elect.) 114
MOULINAGE (*f*) : (see **ZWIRNEN**) (of silk) 115
MÜLLOFEN (*m*) : destructor ; incinerator 116
MÜLLVERWERTUNGSANLAGE (*f*) : dust destructor-(plant) 117
MÜLLVERWERTUNGSWERK (*n*) : dust destructor-(plant) 118
MUSKELENTZÜNDUNG (*f*) : myositis (see **MUSKELKRANKHEIT**) 119
MUSKELFLEISCH (*n*) : muscular tissue (Med.) 120
MUSKEL (*m*), **GLATTER-** : involuntary muscle 121

MUSKELÖS : muscular ; strong 122
MUSKEL (*m*), **QUERGESTREIFTER-** : striated muscle (Med.) 123
MUSKELSCHMERZ (*m*) : myalgia (Med.) 124
MUSKELSCHWAND (*m*) : atrophy ; wasting away of the muscles (Med.) 125
MUSKELSCHWUND (*m*) : (see **MUSKELATROPHIE**) 126
MUSKEL (*m*), **UNWILLKÜRLICHER-** : involuntary muscle (Med.) 127
MUSKEL (*m*), **WILLKÜRLICHER-** : voluntary muscle (Med.) 128
MUSKULATUR (*f*) : muscles ; muscular system (Med.) 129
MUSTERBEISPIEL (*n*) : paradigm ; pattern ; example 130
MUSTERKETTE (*f*) : pattern chain (Textile) 131
MUSTERUNG (*f*) : design (Textile) 132
MUSTERVORLAGE (*f*) : pattern (see **SCHABLONE**) 133
MUTTERPFLANZE (*f*) : prothallium (Bot.) 134
MUTTERSCHALTWERK (*n*) : master mechanism 135
MUTTERUHR (*f*) : master clock (Elect.) 136
MYKOSE (*f*) : mycosis (a disease caused by fission fungi) (Med.) 137
MYOGRAPHION (*n*) : myograph (for measuring amount and duration of muscular contraction) 138
MYOPIE (*f*) : myopia ; short-sight (Med.) 139
MYOPISCH : myopic 140
MYOSIS (*f*) : myosis ; permanent contraction of the pupil of the eye (Med.) 141
MYXÖDEM (*n*) : myxoedema (a disease of the sub-cutaneous and connective tissue) (Med.) 142
MYXOMYZETEN (*pl*) : myxomycetes (slime moulds or fungi) (Bot.) 143
MYZELIUM (*n*) : mycelium-(of fungi) (Bot.) 144

N

NACHARBEITEN (*n*) : checking ; overhauling 145
NACHBILDUNG (*f*) ; simulation (see **NACHBILDEN**) 146
NACHDREHSTEUERAPPARAT (*m*) : controller for " follow-up " type steering engine (Marine) 147
NACHEICHUNG (*f*) : recalibration 148
NACHJUSTIERUNG (*f*) : recalibration ; readjustment ; re-setting 149
NACHKNÜPFEN (*n*) : after-knotting (Textile) 150
NACHRICHTENTECHNIK (*f*) : communication engineering 151
NACHSCHABEN ; to scrape-(out)-(again) 152
NACHSCHWINDEN (*n*) : subsequent (shrinking) shrinkage 153
NACHSPANNEN : to re-tighten ; tighten-(up)-(again) 154
NACHSPANNSPINDEL (*f*) : tightening bolt 155
NACHTSTROM (*m*) : night current (usually at reduced rates) (Elect.) 156
NACHTWANDELN : to walk in one's sleep ; (*n*) : somnambulism ; sleep-walking (Med.) 157
NACHZISELIEREN : to re-chisel ; re-engrave 158
NACKENLEUKODERM (*n*) : leucoderma (leucodermia) of the neck (see **LEUKODERM**) 159
NADELAUSLESE (*f*) : needle selection (Textile) 160
NADELBLÄTTRIG : needle-leaved (*aciphyllus*) (Bot.) 161
NADELHALSLAGER (*n*) : pintle bearing 162
NADELSTAB (*m*) : needle bar (Textile) 163
NADELSTABSTRECKE (*f*) : spiral drawing frame ; Gill box (Textile) 164
NAHANZEIGE (*f*) : close-(range)-reading (of instruments) 165
NÄHERUNGSANGABE (*f*) : approximation 166
NÄHERUNGSFORMEL (*f*) : approximate formula 167
NÄHLICHT (*n*) : feeder-plate light (of sewing machines) 168
NAHT (*f*), **DURCHLAUFENDE-** : continuous weld 169
NAHT (*f*), **EINSEITIGE-** : weld on one side only 170
NAHT (*f*), **I-** : square butt weld 171
NAHT (*f*) **IM ZICKZACK** : staggered weld 172
NAHT (*f*), **LANGLOCH-** : elongated slot weld 173
NAHT (*f*), **RUNDLOCH-** : round slot weld 174
NAHT (*f*), **UNTERBROCHENE-** : intermittent weld 175
NAHT (*f*), **V-** : single V butt weld 176
NAHT (*f*), **X-** : double V butt weld 177
NAHT (*f*), **ZICKZACK-** : staggered weld 178
NAHT (*f*), **ZWEISEITIGE-** : weld on both sides 179
NARBENGEWEBE (*n*), **JUNGES-** : granulation (Med.) 180
NASENENTZÜNDUNG (*f*) : rhinitis (inflammation of the nose) (Med.) 181
NASENSCHLEIMHAUT (*f*) : mucous membrane of the nose (Med.) 182
NASENSCHLEIMHAUTENTZÜNDUNG (*f*) : coryza (inflammation of the mucous membrane of the nose) ; nasal catarrh (Med.) 183
NÄSSEDICHT : moisture-proof ; damp-proof 184
NASSPARTIE (*f*) : wet end (Paper) 185
NASSPRESSE (*f*) : (wet)-press rolls (Paper) 186
NASS-SPINNMASCHINE (*f*) : wet spinning frame (Textile) 187
NASSTÄUBER (*m*) ; wet sprayer 188
NATIONAL : national 189
NATIONALE (*n*) : personal (description) particulars (of individuals) ; general (description) particulars (of Firms, etc.) (such as Age, Business, etc.) 190
NATRIUM-ANTIMON-FLUORID (*n*) : sodium antimony fluoride ; fluoremetic ; $SbF_3.NaF$ 191
NATRON (*n*), **UNTERCHLORIGSAURES-** : (see **NATRIUM-HYPOCHLORIT**) 192
NATTERKOPF (*m*) : viper's bugloss (*Echium vulgare*) (Bot.) 193
NATURGLAS (*n*), **GEFÄRBTES-** : natural colour-(ed) glass (glass coloured in the molten state) 194
NATURHEILMETHODE (*f*) : nature cure (Med.) 195
NEBENANSCHLUSS (*m*) : side station (Tel.) 196
NEBENAPPARAT (*m*) : branch exchange (Tel.) 197
NEBENARM (*m*) : tributary (of rivers) ; bifurcation 198
NEBENGERÄUSCH (*n*) : interference (Radio) 199
NEBENSACHE (*f*) : secondary consideration 200
NEBENSÄCHLICH : accessory 201
NEBENSCHLAG (*m*) : up beat (Music) 202
NEBENSCHLUSSMASCHINE (*f*) : shunt dynamo (Elect.) 203
NEBENSCHLUSSREGLER (*m*) : shunt regulator (Elect.) 204
NEBENSCHLUSS (*m*), **WATTVERBRAUCH IM-** : shunt loss in watts (Elect.) 205
NEBENSPRECHABGLEICH (*m*) : cross-talk balance (Tel.) 206
NEBENSPRECHDÄMPFUNG (*f*) : cross-talk attenuation ; cross-talk transmission equivalent (Tel.) 207
NEBENSPRECHEN (*n*) ; (near end)-cross-talk 208

NEBENSPRECHEN (*n*), GEGEN- : far end cross-talk (Tel.) 209
NEBENSPRECHFREIHEIT (*f*) : freedom from cross-talk 210
NEBENSPRECHKOPPLUNG (*f*) : capacity unbalance (Tel.) 211
NEBENSPRECHWERT (*m*) : cross-talk value (Tel.) 212
NEBENSTELLE (*f*), FERNSPRECH- : (telephone)-side station 213
NEBENSTRAHLUNG (*f*) : secondary radiation (of X-rays) 214
NEBENUHR (*f*) : secondary clock (Elect.) 215
NEBENWIDERSTAND (*m*) : shunt resistance (Elect.) 216
NEEFSCHER HAMMER (*m*) : Neef's hammer (Elect.) 217
NEGERHIRSE (*f*) : *Pennisetum* (Bot.) 218
NEKTARIUM (*n*) : nectary (the part of a flower secreting or containing honey) (Bot.) 219
NENNAUFNAHME (*f*) : nominal consumption (Elect.) 220
NENNBÜRDE (*f*) ; nominal rating (maximum impedance at the secondary side of a current transformer without affecting its accuracy) (Elect.) 221
NENNFORM (*f*) : infinitive (Grammar) 222
NENNFREQUENZ (*f*) : nominal (rated) frequency (Elect.) 223
NENNLAST (*f*) : normal full-load-(speed) 224
NENNLAST (*f*), DREHMOMENT BEI- : full-load torque 225
NENNSPANNUNG (*f*) : nominal (rated) voltage 226
NENNSTROM (*m*) : nominal (rated) current 227
NENNSTROMSTÄRKE (*f*) : nominal (current) load (Elect.) 228
NEPER (*n*) : Neper ; Napier ; transmission unit ; T.U. (unit of attenuation) (Telephony) ; [1 decibel = 0·11513 nepers = 0·1 cross-talk units = 0·1 millionths of received current = 1·0846 M.S.C. (miles of standard cable)] 229
NERVENGESCHWULST (*f*) : neuroma (a nerve tumour) (Med.) 230
NERVENSCHEIDE (*f*) : perineurium (nerve sheath) (Med.) 231
NERV (*m*), MOTORISCHER- : motor (efferent) nerve 232
NERV (*m*), SENSIBILER- : sensory nerve ; afferent nerve (Med.) 233
NESTWURZ (*f*) : bird's-nest (orchid) orchis (*Neottia nidus-avis*) (Bot.) 234
NETZANSCHLUSS (*m*) : mains connection (Radio) 235
NETZANSCHLUSSVERSTÄRKER (*m*) : mains amplifier (Radio) 236
NETZDROSSEL (*f*) : mains choke (Elect.) 237
NETZEMPFÄNGER (*m*) : all-mains (receiver) set (Radio) 238
NETZENDSTUFE (*f*) : output power amplifier ; mains-operated power output stage (Radio) 239
NETZHEIZANODE (*f*) : battery eliminator (Radio) 240
NETZKNÜPFSTUHL (*m*) : net knitting frame (Textile) 241
NETZKONDENSATOR (*m*) : mains condenser (Elect.) 242
NETZSCHALTER (*m*) : main switch (Elect.) 243
NETZTRANSFORMATOR (*m*) : mains transformer (Elect.) 244
NETZTUCH (*n*) : biscuit bagging (Textile) 245
NEUROLOGIE (*f*) : neurology (the science of the nerves) (Med.) 246
NEUROM (*n*) : neuroma (a nerve tumour) (Med.) 247
NEXUS (*m*) : nexus ; link (see ZUSAMMENHANG) 248
NICHTFORMIERTE OXYDKATHODE (*f*) ; uncombined oxide-(d) cathode 249
NICHTGEBRAUCH (*m*), BEI- : when not (employed ; used ; wanted) in use 250
NIEDERFREQUENZOFEN (*m*) : low-frequency furnace (Elect.) 251
NIEDERFREQUENZVERSTÄRKER (*m*) : low-frequency (L.F.) amplifier (Radio) 252
NIEDLICH : agreeable ; charming ; enchanting ; pleasant (*blandus*) (Bot.) 253
NIERENBLUTUNG (*f*) : haematuria (blood in the urine) (Med.) 254
NIERENENTZÜNDUNG (*f*), CHRONISCHE- : Bright's disease ; chronic disease of the kidneys (Med.) 255
NIVEAUFLÄCHEN (*f.pl*) : surfaces of equal potential 256
NIVEAULINIEN (*f.pl*) : lines of equal level or height 257
NIVELLIERAPPARAT (*m*) : levelling (instrument) apparatus 258
NOCKENKLEMME (*f*) : lug-type terminal (Elect.) 259
NONE (*f*) : ninth (Music) (see under TERZ) 260
NORDHEMISPHÄRE (*f*) : northern hemisphere 261
NORMALELEKTRODE (*f*) : normal (auxiliary ; standard) electrode 262
NORMALKUPFER (*n*) : standard copper (must not exceed resistance of 17·84 ohms for 1 km. length and 1 sq. mm. section at 20° C.) 263
NORMALNULL (*f*) : standard zero (for all measurements of heights) 264
NORMALSOCKEL (*m*) : standard base 265
NORMALTON (*m*) : (see KAMMERTON) 266
NORMALUHR (*f*) : standard (time) clock 267
NOTBELEUCHTUNG (*f*) : emergency (stand-by) lighting 268
NOTE (*f*), DURCHGEHENDE- : passing note (Music) 269
NOTENBALKEN (*m*) : tail (connecting a group of notes) (Music) 270
NOTENFAHNE (*f*) : tail-(of a note) (Music) 271
NOTENPLAN (*m*) : staff ; stave (Music) 272
NOTENSCHWANZ (*m*) : tail-(of notes) (Music) 273
NOTLANDUNG (*f*) : forced landing-(of aeroplanes) 274
NOTWENDIG : necessary 275
NOTWENDIGKEIT (*f*) : necessity 276
NUANCIERUNG (*f*) : tinting ; shading 277
NULLEITER (*m*) : neutral wire (Elect.) 278
NULLEN (*n*) : earthing (of electrical apparatus) 279
NULLPOTENTIAL (*n*) : earth potential (Elect.) 280
NULLSCHALTER (*m*) : no-load release (Elect.) 281
NULLSPANNUNGSAUSLÖSUNG (*f*) : no-volt release (Elect.) 282
NULLSPANNUNGSRELAIS (*n*) : no-volt relay (Elect.) 283
NULLSTELLUNG (*f*) : " zero " position (Elect.) 284
NULLSTROMAUSLÖSUNG (*f*) : no-load release (Elect.) 285
NULLUNG (*f*) : earthing (of electrical apparatus) 286
NUMMERNSCHALTER (*m*) : dial-(of a telephone) 287
NUMMERNSCHEIBE (*f*) : switchboard (number) indicator (Tel.) 288
NUTATION (*f*) : bending ; oscillation ; fluctuation ; nodding (usually of oscillation of earth's axis) 289
NUTZBRENNDAUER (*f*) : life (of electric light bulbs until they have reached a drop of 20% in efficiency) 290
NUTZLAST (*f*) : actual (effective) load or carrying capacity 291
NUTZNIESER (*m*) : beneficiary 292
NUTZWIDERSTAND (*m*) : effective resistance 293

O

OBERARMARTERIE (*f*) : brachial artery (Med.) 294
OBERARMBEIN (*n*) : humerus ; bone of the upper arm (Med.) 295
OBERDOMINANTE (*f*) : dominant (Music) 296
OBERFLÄCHENBESTRAHLUNG (*f*) : surface irradiation (by X-rays) 297
OBERFLÄCHENKÜHLUNG (*f*) : surface cooling 298
OBERFLÄCHENLEITUNG (*f*) : surface conduction ; skin effect (Elect.) 299
OBERFLÄCHENSCHICHTWIDERSTAND (*m*) : impedance, apparent resistance 300
OBERKARBON (*n*) : Upper Carboniferous (Geol.) 301
OBERKARBONISCH : Upper Carboniferous 302
OBERKIEFER (*m*) : superior maxilla ; upper jaw (Med.) 303
OBERKOHLE (*f*) : upper (positive) carbon (Elect.) 304
OBERMEDIANTE (*f*) : mediant (Music) 305
OBERSCHENKELKNOCHEN (*m*) : femur (Med.) 306
OBERSCHWINGUNG (*f*) : upper partial (of oscillations) 307
OBERSCHWINGUNGSFREI : free from harmonics 308
OBERSTEIGER (*m*) : agent (Mining) 309
OBJEKTIVVORRICHTUNG (*f*) : focusing device-(of Stage lighting fittings) 310
OBSIGNIEREN : to seal officially-(by law) 311
OBSTIPATION (*f*) : constipation (Med.) 312
OBSTRUKTION (*f*) : obstruction ; constipation (Med.) 313
OBTURATOR (*m*) : obturator (a device for closing an opening) (Med. and Mil.) 314
ÖDLAND (*n*) : waste land 315
OHR (*n*), ÄUSSERES- : external ear 316
OHRBAND (*n*) : auricular ligament (Med.) 317
OHRBLUTADER (*f*) : auricular vein (Med.) 318
OHRBOCK (*m*) : tragus (a process on the front of the opening of the external ear) (Med.) 319
OHRBOLZEN (*m*) : ring-bolt 320
OHREISEN (*n*) : ring-bolt 321
OHRENSAUSEN (*n*) : rushing in the ear ; ear noises 322
OHRENTZÜNDUNG (*f*) : otitis (inflammation of the ear) (Med.) 323
OHRFLUSS (*m*) : otorrhoea (discharge from the ear) (Med.) 324
OHR (*n*), INNERES- : inner ear 325
OHRKNÖCHELCHEN (*n.pl*) : chain of bones of the middle ear 326
OHR (*n*), MITTLERES- : middle ear 327
OHRMUSCHEL (*f*) : auricle ; pinna ; ear lobe ; ear lobule ; ear flap ; external ear (Med.) 328
OHRPFROPF (*m*) : ear plug (for protecting the ear against shock) ; plug of wax in the ear 329
OHRSCHMALZ (*n*) : ear-wax (Med.) 330
OKTAVE (*f*), GROSSE- : great octave (octave shown by a capital letter) (Music) 331
OKTAVE (*f*), KLEINE- : small octave (octave denoted by a small letter) (Music) 332
ÖLGRUBE (*f*) : oil sump 333
OLIGOCÄN : Oligocene (Geol.) 334
OLIGOCÄN-MIOCÄN, OBER- : Upper Oligocene-Miocene 335
OLIGODYNAMIE (*f*) : latent power 336
ÖLSCHALTERNENNSTROM (*m*) : (current)-rating of oil switch 337
ÖLSCHÜTZ (*n*) : oil break (Elect.) 338
ÖLSPRITZPISTOLE (*f*) : grease gun 339
ÖLSUMPF (*f*) : oil sump 340
ÖLTRANSFORMATOR (*m*) : oil immersed transformer (Elect.) 341
ÖLWIDERSTAND (*m*) : oil-(cooled)-resistance (Elect.) 342
ÖNOLOGIE (*f*) : oenology, the science and study of wines 343
ÖNOMETER (*n*) : oenometer (a type of hydrometer for measuring the alcohol content of wines) 344
OPALÜBERFANGEN : opal-flashed ; opalescent (Glass) 345
OPHTHALMIATRIK (*f*) : (see OPHTHALMOLOGIE) 346
OPHTHALMOLOGIE (*f*) : ophthalmology 347
OPHTHALMOSKOP (*n*) : (see AUGENSPIEGEL) 348
OPTIMAL : optimum ; best 349
OPTIMUM (*n*) : optimum ; best 350
OPTOGRAMM (*n*) : optogram (an image of a luminous object) 351
OPTOMETER (*n*) : optometer (instrument for measuring distance of vision) 352
OPTOPHON (*n*) : optophone (instrument by which the blind are enabled to hear the printed word) 353
ORANGEAT (*n*) : candied orange peel 354
ORDNER (*m*) : file (Commercial) 355
ORGANOGEN : of organic origin 356
ORGASMUS (*m*) : orgasm, violent excitement of an organ (Med.) 357
ORGASTISCH : orgastic (Med.) 358
ORGELBAU (*m*) : organ (construction) building 359
ORGELTASTATUR (*f*) : organ key-board ; manual 360
ORTHOPÄDISCH : orthopaedic 361
ORTHOPNÖE (*f*) : orthopnoea (severe difficulty in breathing) (Med.) 362
ORTSSTROMKREIS (*m*) : local circuit (Tel.) 363
ORTSVERÄNDERLICH : portable ; movable 364
ORTSVERKEHR (*m*) : local traffic (Tel.) 365
OSMOSE (*f*), ELEKTRISCHE- : electric osmose 366
ÖSTERREICHISCH : from Austria ; Austrian ; of Austrian origin (*austriacus*) (Bot.) 367
OSZILLOGRAMM (*n*) : oscillogram (the graph made by an oscillograph)) 368
OSZILLOGRAPH (*m*) : oscillograph (an instrument for recording wave forms) 369
OTOSKOP (*n*) : otoscope (for viewing the interior of the ear) 370
OUTRIEREN : (see ÜBERTREIBEN) 371
OXYDKATHODE (*f*) : oxide-(d) cathode ; oxide-coated filament (of cathode-ray tubes) 372
OZÄNA (*f*) : (see STINKNASE) 373
OZEANISCH : oceanic 374
OZONANLAGE (*f*) : ozone plant 375
OZONBATTERIE (*f*) : battery of ozone generators 376
OZONGEMISCH (*n*) : ozone mixture 377
OZONISIERUNG (*f*) : ozonising ; ozonisation 378
OZONLUFT (*f*) : ozone air 379
OZONOMETER (*n*) : ozonometer ; ozometer (for determining the quantity of ozone in the atmosphere) 380

P

PAARVERSEILUNG (*f*) : paired (of cables) 381
PACINISCHE KÖRPERCHEN (*n*) : Pacinian corpuscles (of nerves in fingers) (Med.) 382
PÄDAGOGIK (*f*) : pedagogics ; pedagogism ; pedagogy 383
PÄDAGOGISCH : pedagogic-(al) ; pedantic 384
PAKTIEREN : to agree ; contract ; make (conclude) a pact 385
PALÄOZOIKUM (*n*) : Palaeozoic-(period) 386
PALÄOZOISCH : Palaeozoic 387
PAPAGEIENKRANKHEIT (*f*) : psittacosis ; parrot disease (Med.) 388
PAPIERBLEIKABEL (*n*) : paper-insulated lead (covered) sheathed cable 389
PAPIERKABEL (*n*) : paper-insulated cable 390
PAPPEBECHERKONDENSATOR (*m*) : paper (covered) cased condenser (Radio) 391
PARABOLSPIEGEL (*m*) : parabolic mirror reflector (Lighting) 392
PARALLELISMUS (*m*) : parallelism ; resemblance ; similarity 393
PARAMETRITIS (*f*) : parametritis (Med.) 394
PARASITISCH : parasitic 395
PARASITISCHER STROM (*m*) : eddy current (Elect.) 396
PAROXYSMUS (*m*) : (see **ANFALL**) 397
PARTNERANSCHLUSS (*m*) : shared service (Tel.) 398
PARZELLE (*f*) : parcel (piece) of land 399
PASSAGENWERK (*n*) : passage work (Music) 400
PASSCHRAUBE (*f*) : gauge screw-(of cartridge fuses) 401
PASSRING (*m*) : gauge ring-(of cartridge fuses) 402
PATENTVERLETZUNG (*f*) : infringement of patent 403
PATERNOSTERAUFZUG (*m*) : lift (passenger and goods) 404
PATERNOSTERWERK (*n*) : endless chain and bucket 405
PATHOGEN : pathogenic ; pathogenous (producing disease) (Med.) 406
PATHOGENIE (*f*) : pathogeny ; pathogenesis (the mode of the production of disease) (Med.) 407
PATHOGENISCH : (see **PATHOGEN**) 408
PATRONENSICHERUNG (*f*) : cartridge fuse (Elect.) 409
PAUKENHÖHLE (*f*) : chamber of the middle ear 410
PEGELMESSCHRANK (*m*) : level test board (Elect.) 411
PEGELWERT (*m*) : attenuation (factor) value (Tel.) 412
PEILANTENNE (*f*) : directional aerial (Radio) 413
PEILVERKEHR (*m*) : direction-finding service (Radio) 414
PEITSCHENWURM (*m*) : whip worm (*Trichocephalus dispar*) 415
PELAGISCHE TIERE (*n.pl*) : pelagian (pelagic) animals (inhabit the upper layer of the sea) 416
PELTIEREFFEKT (*m*) : Peltier effect (Elect.) 417
PELTONRAD (*n*) : Pelton wheel (Mech.) 418
PENDELKLAPPE (*f*) : pendulum type drop-indicator 419
PENDELMYOGRAPHION (*n*) : pendulum myograph (see **MYOGRAPHION**) 420
PENDELSTÜTZRAMPE (*f*) : tilting float (of Stage lighting) 421
PENDÜLE (*f*) : pendulum clock 422
PENTANLAMPE (*f*) : Pentane lamp ; Vernon-Harcourt Pentane lamp (a lighting standard = 10 candles) 423
PERFORIERMASCHINE (*f*) ; (paper)-punch 424
PERIGÄUM (*m*) : (see **ERDNÄHE**) 425
PERIGÄUMISCH : perigean 426
PERIGON (*n*) : perigon ; calyx (the outer, green covering of a flower) (Bot.) 427
PERIMETRITIS (*f*) : perimetritis (Med.) 428
PERIODENUMFORMER (*m*) : frequency changer (Elect.) 429
PERISPERM (*n*) : perisperm (outer albuminous covering of seed) (Bot.) 430
PERISTALTISCHE BEWEGUNG (*f*) : peristalsis (Med.) 431
PERKUSSIONSKRAFT (*f*) : percussive force 432
PERKUSSIONSZÜNDER (*m*) : percussion fuse 433
PERMEAMETER (*n*) : permeameter ; permeability tester 434
PERNIZIÖS : pernicious 435
PERRHENAT (*n*) : perrhenate 436
PERSONALEINSCHRÄNKUNG (*f*) : reduction in staff 437
PESTBAKTERIUM (*n*) : plague bacterium (*Bacterium pestis*) 438
PETRISSAGE (*f*) : petrissage (of massage) 439
PFEIFGRENZE (*f*) : singing limit 440
PFEIFPUNKT (*m*) : singing point 441
PFEIFPUNKTMETHODE (*f*) : singing point method 442
PFLANZENFRESSER (*m* and *pl*) : herbivore ; herbivora (herb- or plant-eating animals) 443
PFLEGEVORSCHRIFT (*f*) : instructions for the care of 444
PHÄNOMENOLOGISCH : as a phenomenon 445
PHANTOMKREIS (*m*) : phantom circuit (Tel.) 446
PHASENBELASTUNG (*f*), **GLEICHMÄSSIGE-** : balanced load (Elect.) 447
PHASENDIFFERENZ (*f*) : phase difference 448
PHASENFEHLER (*m*) : fault of phase 449
PHASENFOLGE (*f*) : order (sequence) of phases 450
PHASENGLEICHHEIT (*f*) : coincidence of phase 451
PHASENMESSER (*m*) : phase meter 452
PHASENNACHEILUNG (*f*) : phase lag 453
PHASENREGLER (*m*) : phase regulator 454
PHASENREGLUNG (*f*) : phase regulation 455
PHASENSCHIEBER (*m*) : phase shifter 456
PHASENSPANNUNG (*f*) : phase voltage ; voltage between phases 457
PHASENSPRUNG (*m*) : phase shift 458
PHASENSTROM (*m*) : phase current 459
PHASENTEILUNG (*f*) : phase splitting 460
PHASENUNTERSCHIED (*m*) : phase difference 461
PHASENVERSCHIEBUNG (*f*) : phase displacement 462
PHASENVERZÖGERUNG (*f*) : phase lag 463
PHASENVOREILUNG (*f*) : phase lead 464
PHASENWICKLUNG (*f*) : phase winding 465
PHASENWINKEL (*m*) : phase angle 466
PHASENZAHL-UMFORMUNG (*f*) : phase conversion 467
PHASOMETER (*n*) : (see **PHASENMESSER**) 468
PHON (*n*) : phone unit ; transmission unit ; decibel (db.) (unit of audibility or sound density) 469
PHONISCHES RAD (*n*) : phonic wheel ; Rayleigh synchronous motor (for determining frequency) 470
PHOTOGEN : photogenic 471
PHOTOPHON (*n*) : photophone ; optical telephone (for transmitting sounds by means of a beam of light) 472
PHOTOTHERAPIE (*f*) : phototherapy (healing by means of light) (Med.) 473
PHOTOZELLE (*f*) : photo-electric cell 474
PHYSIKUS (*m*) : medical officer of health for a district 475
PLÄDIEREN : to plead 476

PLANIMETRIERBAR : integrable by means of a planimeter 477
PLANIMETRIEREN : to integrate by means of a planimeter 478
PLANKTOLOG (*m*) : planktologist (a student of the animalcule in the sea) 479
PLANSACKLEINEN (*n*) : plain sacking (Textile) 480
PLATINENRADMASCHINE (*f*) : sinker wheel machine (Textile) 481
PLATTENSCHNEIDER (*m*) : disc (record) cutter (of gramophones) 482
PLATTENTELLER (*m*) : turn-table (of gramophone) 483
PLATTIERGESCHWINDIGKEIT (*f*) : dipping speed (Textile) 484
PLIOCÄN : pliocene (Geol.) 485
PLISSEEMASCHINE (*f*) : pleating machine (Textile) 486
PLÜCKERRÖHRE (*f*) : spectral (spectrum) tube (vacuum tube for studying the spectra of glowing gases) 487
PLUSPOL (*m*) : positive (plus) pole (Elect.) 488
POLRAD (*n*) : pole wheel ; rotor (Elect.) 489
POLSCHUH (*m*) : pole (shoe) piece (Elect.) 490
POLUMSCHALTBAR : pole-changing 491
POLUMSCHALTBARER MOTOR (*m*) : pole-changing motor 492
POLUMSCHALTUNG (*f*) : pole-changing-(gear) 493
PONIEREN : to pose ; place ; set 494
PORLING (*m*) : *Polyporus* (Bot.) 495
PORTAL-ABDECKUNG (*f*) : fly (of Stages) 496
PORTAL-SEITENRAHMEN (*m*) : wings (of Stages) 497
POSITION (*f*) : position ; place ; spot ; location 498
POSITIONSLICHT (*n*) : navigation light (of aeroplanes) 499
POSTHUM : posthumous ; after death 500
POSTKARBONISCH : Post-Carboniferous (Geol.) 501
POTENTIALDIFFERENZ (*f*) : potential difference ; voltage (Elect.) 502
POTENTIOMETER (*n*) : potentiometer (for measuring potential between two points) 503
PRÄFOLIATION (*f*) : prefoliation (in a bud, the folding of the leaves over the apex) (Bot.) 504
PRALLTRILLER (*m*) OHNE NACHSCHLAG (*m*) : mordant (Music) 505
PRATZEN-LAUFKRAN (*m*) : claw (grab) travelling crane 506
PRÄVENTIVIMPFUNG (*f*) : preventive inoculation (Med.) 507
PRELLBACKEN (*f.pl*) : buffers (Textile) 508
PRELLUNG (*f*) : chattering ; chatter (Tel.) (see also PRELLEN) 509
PRELLUNGSFREI : non-chattering (Tel.) 510
PRESSISOLIERSTOFF (*m*) : moulded insulating material 511
PRIMÄRAUSLÖSER (*m*) : primary tripping relay (Elect.) 512
PRIMÄRKRAFTMASCHINE (*f*) : prime mover (Mech.) 513
PRIMÄRREGLER (*m*) : primary regulator (Elect.) 514
PRINZIPSCHALTBILD (*n*) : theoretical diagram of connections (Elect.) 515
PROFILINSTRUMENT (*n*) : edgewise pattern instrument (Elect.) 516
PROFILRAHMENINSTRUMENT (*n*) : edgewise pattern instrument (Elect.) 517
PROJEKTIONSEBENE (*f*) : plane of projection 518
PROJEKTIONSLAMPE (*f*) : projector ; projection lamp (Stage lighting) 519
PROPELLER (*m*), BACKBORD- : port propeller 520
PROPELLER (*m*), STEUERBORD- : starboard propeller 521
PROSTATAHYPERTROPHIE (*f*) : increase in size of the prostate gland (Med.) 522
PROTHALLIUM (*n*) : prothallium (Bot.) 523
PROTHESE (*f*) : artificial limb 524
PROTOPHYTEN (*m.pl*) : protophyta (Bot.) 525
PROVENIENZ (*f*) : origin of wares or goods 526
PRÜFEINRICHTUNG (*f*) : testing (arrangement ; set) equipment 527
PSYCHOPATH : psychopathic (not quite normal) (Med.) 528
PSYCHROMETER (*n*) : (see FEUCHTIGKEITSMESSER) 529
PUFFERN : to act as a buffer-(for) 530
PULSATION (*f*) : pulsation ; beat ; throb ; impulse 531
PULSBESCHLEUNIGUNG (*f*) : tachycardia (Med.) 532
PUMPE (*f*), EINSPRITZ- : injection pump ; injector 533
PUMPE (*f*), ELEKTROWASSER- ; electrically driven water pump 534
PUMPE (*f*), JAUCHE- : sud (sump ; sewage) pump 535
PUMPE (*f*), KRAFTSTOFF- : petrol pump 536
PUMPE (*f*), KÜHLMITTEL- : sud pump 537
PUMPE (*f*), STOFF- : pulp pump (Papermaking) 538
PUMPE (*f*), UMWÄLZ- : rotary pump ; circulating (cooling) water pump 539
PUNKTFÖRMIG : punctiform 540
PUNKTFREQUENZ (*f*) : dot frequency (Tel.) 541
PUNKTIERMANIER (*f*) : shading by means of dots 542
PUNKTSCHWEISSMASCHINE (*f*) : spot welder ; spot welding machine 543
PUNZIERUNG (*f*) : punching ; stamping 544
PUPINISIEREN : to load (of telephone cables) 545
PUPINISIERUNG (*f*) : loading (Tel.) 546
PUPINSPULE (*f*) : Pupin (loading) coil (Tel.) 547
PUPINSPULENFELD (*n*) : loading section (Tel.) 548
PURKINJE PHÄNOMEN (*n*) : Purkinje effect (the effect on the appearance of colours due to intensity) 549
PUTZWALZENTUCH (*n*) : clearer cloth (Textile) 550

Q

QUADRUPLEX-TELEGRAPHIE (*f*) : quadruplex telegraphy 551
QUALMSICHER : smoke proof 552
QUARTANA (*f*) : quartan ague (Med.) 553
QUARTENZIRKEL (*m*) : circle of fourths (Music) 554
QUARZLAMPE (*f*) : (see QUECKSILBERDAMPFLAMPE) 555
QUECKSILBERCHLORÜRELEMENT (*n*) : calomel element (Elect.) 556
QUECKSILBEREINHEIT (*f*) : Siemens (unit of resistance of a column of mercury 1 m. long and 1 sq. mm. section, equivalent to 0·9407 ohms) 557
QUECKSILBERKONTAKT (*m*) : mercury contact 558
QUECKSILBERWIPPE (*f*) : mercury (tipper) tipping switch 559
QUERFELD (*n*) : opposing (cross) field (Elect.) 560
QUINTEN (*f.pl*), GLEICHLAUFENDE- : consecutive fifths (Music) 561
QUINTENZIRKEL (*m*) : circle of fifths (Music) 562
QUINTE (*f*), REINE- : perfect fifth (Music) 563
QUINTE (*f*), ÜBERMÄSSIGE- : augmented fifth (Music) 564
QUINTE (*f*), VERMINDERTE- : diminished fifth (Music) 565

R

RACHENBRÄUNE (*f*) : diphtheria (Med.) 566
RACHITIS (*f*) : rickets (Med.) 567
RÄDERKASTEN (*m*) : gear-box (see RÄDERGEHÄUSE) 568
RADIOGRAPHIE (*f*) : radiography ; X-ray photography 569
RADIOLOGIE (*f*) : radiology (the science of visible and invisible rays) 570
RADIOMIKROMETER (*n*) : radiomicrometer (for measuring minute variations in radiant heat) 571
RADIOPEILER (*m*) : radio direction finder 572
RADIOPHON (*n*) : radiophone (for producing sound by action of light and heat rays) 573
RADIOSPRECHMASCHINE (*f*) : radiogram ; radiogramophone 574
RADIOTHERAPIE (*f*) : radiotherapy (healing of disease by X-rays) (Med.) 575
RADIUM F (*n*) : (see POLONIUM) 576
RADON (*n*) : Radon (Rn) (atomic weight 222) 577
RADSATZANORDNUNG (*f*) : wheel arrangement (of engines) 578
RAHMENANTENNE (*f*) : frame aerial (Radio) 579
RÄNDELMASCHINE (*f*) : milling machine (Mech.) 580
RÄNDELMUTTER (*f*) : milled nut (Mech.) 581
RÄNDELSCHRAUBE (*f*) : milled screw 582
RÄNDERMASCHINE (*f*) : rib-(knitting)-machine (Textile) 583
RANGIERGELÄNDE (*n*) : sidings (of railways) 584
RASENHÄNGEBANK (*f*) : bank (Mining) 585
RASSELWECKER (*m*) : trembler bell (Elect.) 586
RASTERUNG (*f*) : screening (Picture telegraphy) 587
RASTERWEITE (*f*) : screen-(of blocks) 588
RAUCHGLAS (*n*) : smoked glass 589
RAUMKUNST (*f*) : (interior)-decorative art 590
RAUMLADUNG (*f*) : space charge (Elect.) 591
RAUMSKIZZE (*f*) : sketch of space available 592
RAUMWINKEL (*m*) : solid angle 593
RAUPENANTRIEB (*m*) : caterpillar drive 594
RAUPENKETTE (*f*) : caterpillar track 595
REAKTANZ (*f*) : reactance (Elect.) 596
REAKTANZSCHUTZ (*m*) : reactance protection (Elect.) 597
REAKTIONSSPULE (*f*) : reaction coil 598
REAKTIVIEREN : to re-start ; set going again ; reactivate 599
REALLASTEN (*f.pl*) : rates, taxes or tithes on land 600
RECEPTOR (*m*) : receiver ; receiving apparatus (Tel.) 601
RECHTSLAUF (*m*) : clockwise rotation ; dextro-rotation 602
REDHIBITION (*f*) : return of goods purchased on account of a fault hidden on purchase 603
REFAKTIE (*f*) : refund (compensation) for faulty parts purchased ; also refund of transport charges 604
REFLEXBEWEGUNG (*f*) : reflex action (Med.) 605
REFLEXVORGANG (*m*) : reflex action (Med.) 606
REGELSCHALTER (*m*) : dimmer ; dimming resistance (Elect.) 607
REGELSCHICHT (*f*) : normal (day) shift (of work) 608
REGELWALZE (*f*) : drum-(type) regulator (Elect.) 609
REGENAPPARAT (*m*) : sprinkler ; sprayer 610
REGENMESSER (*m*) : rain gauge ; pluviometer 611
REGENROHR (*n*) : rain-(down)-pipe 612
REGISTERWALZE (*f*) : brass tube roll (under the wire cloth) ; wet roll (Papermaking) 613
REIBUNGSANTRIEB (*m*) : friction drive 614
REIBUNGSGEWICHT (*n*) : adhesive weight 615
REIBUNGSWIDERSTANDSMOMENT (*n*) : moment of frictional resistance 616
REICHLICH BEMESSEN : of ample (dimensions ; capacity) size 617
REIFEVERFAHREN (*n*) : ripening process 618
REIHE (*f*), IN -SCHALTEN : to connect-(up)-in series ; to couple in series (Elect.) 619
REIHENABSCHALTUNG (*f*) : disconnection in series 620
REIHENANLAGE (*f*), FERNSPRECH- : intercommunication telephone equipment 621
REIHEN-PARALLELSCHALTUNG (*f*) : series-parallel connection (Elect.) 622
REIHENSCHLUSSMASCHINE (*f*) : series dynamo 623
REIHENSCHLUSSMOTOR (*m*) : series wound motor 624
REINRASSIG : pure bred ; thoroughbred 625
REINWASSERBEHÄLTER (*m*) : clean (pure) water (container ; reservoir) tank 626
REISSWOLF (*m*) : breaker ; willow (Textile) 627
REKLAMEBELEUCHTUNG (*f*) : sign-lighting ; illuminated advertising signs 628
REKTIFIKAT (*n*) : corrected copy (to replace a previous incorrect one) 629
RELAISSATZ (*m*) : set of relays (Elect.) 630
RELUKTANZ (*f*) : reluctance (ratio of magnetomotive force to resulting magnetic flux) (symbol S) 631
REPARATURWERKSTATT (*f*) : repair (mechanics) shop 632
REPRISE (*f*) : repeat (Music) 633
RESEKTION (*f*) : resection (cutting off part of an organ whilst preserving the remainder) (Med.) 634
RESERVAGEDRUCKEN (*n*) : resist printing (Textile) 635
RESERVAGEDRUCKEREI (*f*) : resist printing (Textile) 636
RESTDÄMPFUNG (*f*) : transmission equivalent ; overall loss (Tel.) 637
RETABLIEREN : (see WIEDERHERSTELLEN) 638
RETENTIONSKRAFT (*f*) : coercive force 639
REVERSIERANLASSER (*m*) : reversing switch (Elect.) 640
REZESSIV : recessive ; receding ; retiring ; withdrawing 641
REZIDIV : relapse (of a disease) 642
REZIPROKE (*f*) : reciprocal 643
REZIPROKES OHM (*n*) : reciprocal ohm ; mho (Electrochemistry) 644
RHENIUM (*n*) : Rhenium (Re) (a metal ; rare element of the manganese group. Position = Tantalum–Wolfram–Rhenium–Osmium–Iridium. Ordinal No. 75 ; Atomic weight 188·7 ; S.g. about 20 ; Specific heat 0·035 ; M.p. below that of Tantalum) 645
RHENIUMMETALL (*n*) : metallic Rhenium 646
RHENIUMSALZ (*n*) : Rhenium salt 647
RHENIUMTETRACHLORID (*n*) ; Rhenium tetrachloride ; $ReCl_4$ 648
RHEOTAN (*n*) : rheotan (resistance material ; specific resistance about 0·47–0·52) 649
RHEOTANDRAHT (*m*) : rheotan wire 650
RICHTANTENNE (*f*) : directional aerial 651
RICHTKRAFT (*f*), OHNE- : indifferent (Elect.) 652
RICHTUNGSRELAIS (*n*) : direction relay (Elect.) 653
RICHTUNGSSYSTEM (*n*) : directional system 654
RICHTVERSTÄRKER (*m*) : rectifying amplifier 655
RIECHENDE (*n*) : ozone (see OZON) 656
RIEMENANREGER (*m*) : belt shifting device ; striking gear (Mech.) 657
RIEMENDYNAMOMETER (*n*) : belt dynamometer 658

RIEMENROLLE (*f*) : jockey pulley (Mech.) 659
RIEMENSPANNROLLE (*f*) : belt tensioning device ; jockey pulley (Mech.) 660
RIESENWUCHS (*m*), **TEILWEISER** : (see **AKROMEGALIE**) 661
RIETDICHTENSYSTEM (*n*) : reed counts system (Textile) 662
RINGANKER (*m*) : ring armature (Elect.) 663
RING (*m*), **GRAMME'SCHER** : Gramme ring-(armature) 664
RINGKAMMER MESSFLANSCH (*m*) : orifice flange (of Water meters) 665
RINGMUTTER (*f*) : ring nut 666
RINGSPULE (*f*) : toroidal coil 667
RINGSTROMWANDLER (*m*) : ring-type current transformer (Elect.) 668
RINGTRANSFORMATOR (*m*) : ring transformer (Elect.) 669
RINGÜBERTRAGER (*m*) : repeating coil ; toroidal repeater (Tel.) ; ring transformer (Elect.) 670
RINGWICKLUNG (*f*) : ring winding (Elect.) 671
RIPPENKÜHLER (*m*) : radiator 672
RISTORNIEREN : to write back ; cancel (Accountancy) 673
ROHBAUMWOLLE (*f*) : raw cotton (Textile) 674
RÖHRE (*f*), **EUSTACHISCHE-** : Eustachian tube (Med.) 675
RÖHRE (*f*), **HARTE-** : hard tube (of thermionic valves) 676
RÖHRENEMPFÄNGER (*m*) : valve (receiver) set (Radio) 677
RÖHRENHALTER (*m*) : tube holder (of X-rays) ; valve holder (of Radio) 678
RÖHRENKENNLINIE (*f*) : valve characteristic (Radio) 679
RÖHRENSTATIV (*n*) : tube stand (of X-rays) 680
RÖHRENSUMMER (*m*) : (valve)-oscillator 681
RÖHRENVERSTÄRKER (*m*) : (valve)-amplifier 682
RÖHRE (*f*), **WEICHE-** : soft tube (of thermionic valves) 683
ROHRRÜCKLAUF (*m*) : shock absorber ; recoil (of a gun) 684
ROHRSCHELLE (*f*) : fixing clip (for tubes) 685
ROHRSTOSS (*m*) : pipe joint 686
ROHRSTRANG (*m*) : main-(pipe-line) 687
ROHRSTUTZEN (*m*) : pipe flange ; stand pipe 688
ROHRVERBINDUNGSMUFFE (*f*) : sleeve (for joining pipes) 689
ROLLAPPARAT (*m*) : reeling apparatus ; reels (Paper) 690
ROLLBINDE (*f*) : roller bandage (Med.) 691
ROLLENRICHTMASCHINE (*f*) : roller straightener (of Rolling mills) 692
ROLLENSTROMABNEHMER (*m*) : trolley pole (collector) (Electric traction) 693
ROLLENZÄHLWERK (*n*) : cyclometer-(dial) 694
ROLLFELD (*n*) : landing (surface) ground (of aeroplanes) 695
ROLLGANG (*m*) : continuation table (Rolling mills) 696
ROLLJALOUSIE (*f*) : roller shutter ; roll-top ; roll-front 697
ROLLKONTAKT (*m*) : roller contact (Elect.) 698
ROLLMOTOR (*m*) : roller (portable) motor 699
ROLLOFEN (*m*) : continuous furnace-(of the roller conveyor type) 700
RÖNTGENSPEKTROGRAPH (*m*) : Röntgen spectrograph 701
RÖNTGENTHERAPIE (*f*) : radiotherapy (healing of disease by X-rays) 702
ROSSKARTOFFEL (*f*) : (see **TOPINAMBURPFLANZE**) 703
ROSTPILZ (*m*) : rust (Bot.) 704
ROSTSICHER : rust proof 705
ROTGLUTSTRAHLER (*m*) : dull emitter-(valve) (Radio) 706
ROTLIEGENDES (*n*), **MITTEL-** : Middle Red Sandstone (Geol.) 707
ROTLIEGENDES (*n*), **OBER-** : Upper Red Sandstone (Geol.) 708
ROTLIEGENDES (*n*), **UNTER-** : Lower Red Sandstone (Geol.) 709
ROTORWICKLUNG (*f*) : rotor winding (Elect.) 710
ROTZKRANKHEITBAKTERIUM (*n*) : bacterium of glanders (*Bacterium mallei*) 711
RÜBENSCHNEIDER (*m*) : turnip cutter 712
RÜCKBEZÜGLICH : reflexive 713
RÜCKENMARK (*n*), **HÖHLENBILDUNG IM-** : syringomyelia (a disease of the spinal cord) (Med.) 714
RÜCKENMARKSSCHWINDSUCHT (*f*) : Tabes dorsalis (a wasting disease of the spine) ; locomotor ataxia ; posterior spinal sclerosis (Med.) 715
RÜCKFALLFIEBER (*n*) : recurrent fever (Med.) 716
RÜCKKOPPELN : to couple (connect) back ; (*n*) : back connection 717
RÜCKKOPPLUNG (*f*) : reactive coupling ; (feed) back coupling ; reaction (Radio) 718
RÜCKKOPPLUNG (*f*), **KAPAZITIVE-** : electrostatic reaction coupling 719
RÜCKKOPPLUNGSAUDION (*n*) : regenerative audion (Radio) 720
RÜCKKOPPLUNGSEMPFANG (*m*) : regenerative reception 721
RÜCKKOPPLUNGSEMPFÄNGER (*m*) : regenerative receiver 722
RÜCKKOPPLUNGSFREI : non-regenerative 723
RÜCKKOPPLUNGSKOEFFIZIENT (*m*) : coefficient of reaction 724
RÜCKKOPPLUNGSKONDENSATOR (*m*) : reaction condenser 725
RÜCKKOPPLUNGSSPULE (*f*) : reaction coil 726
RÜCKKOPPLUNGSVERSTÄRKUNG (*f*) : regenerative amplification 727
RÜCKLÖTVORRICHTUNG (*f*) : re-soldering apparatus (for fuses) 728
RÜCKSEITIGER ANSCHLUSS (*m*) : back-(of board) mounting or connection 729
RÜCKSTAND (*m*), **ELEKTRISCHER-** : residual magnetism 730
RÜCKSTELLBAR : capable of being re-set (of pointer or hands of a dial) 731
RÜCKSTELLEN : to make a (place to) reserve (Commercial) ; to re-set (of mechanism) 732
RÜCKSTELLKLAPPENSCHRANK (*m*) : switchboard with self-restoring indicator (Tel.) 733
RÜCKSTELLKNOPF (*m*) : re-setting knob (of Electric instruments) 734
RÜCKWÄRTIGER ANSCHLUSS (*m*) : back connection 735
RÜCKWATTRELAIS (*n*) : reverse power relay (Elect.) 736
RÜCKWOLLE (*f*) : shoddy (Textile) 737
RUDERA (*pl*) : (see **SCHUTT**) 738
RUDERAPFLANZE (*f*) : plants which prefer any waste such as refuse heaps 739
RUFMASCHINE (*f*) : ringing machine (Tel.) 740
RUFSATZ (*m*) : ringing (calling) set (Tel.) 741
RUFSTROM (*m*) : ringing current (Tel.) 742
RUFUMSETZER (*m*) : ring converter (Tel.) 743
RUHESTELLUNG (*f*) : position of (stoppage ; repose) rest ; off (zero) position 744
RUHEZEICHEN (*n*) : pause ; rest (Music) 745
RUHMKORFFSPULE (*f*) : Ruhmkorff coil ; induction coil ; spark coil 746
RÜHRBÜTTE (*f*) : stuff chest (Paper) 747

RUNDFUNK (*m*) : wireless ; radio ; broadcasting 748
RUNDFUNKEMPFÄNGER (*m*) : (wireless)-receiver or (receiving)-set 749
RUNDFUNKGERÄT (*n*) : wireless apparatus 750
RUNDFUNKLEITUNGSVERSTÄRKER (*m*) : broadcast relaying repeater 751
RUNDFUNKRÖHRE (*f*) : (wireless)-valve 752
RUNDFUNKSENDER (*m*) : (wireless)-transmitter ; transmitting station ; broadcasting station 753
RUNDINSTRUMENT (*n*) : round pattern instrument (Elect.) 754
RUND-JACQUARDMASCHINE (*f*) : circular Jacquard knitting machine (Textile) 755
RUNDSIEBMASCHINE (*f*) : drum type cardboard machine (Papermaking) 756
RUNDSTRICKMASCHINE (*f*) : circular knitting machine (Textile) 757
RUNDWIRKMASCHINE (*f*) : circular knitting machine (Textile) 758
RÜSCHSCHLAUCH (*m*) : (pure silk)-insulating sleeving (around resistances) (Radio) 759
RUSSABBLASVENTIL (*n*) : (see RUSSABBLASVORRICHTUNG) 760
RUSSBRAND (*m*) : (see STAUBBRAND) 761
RUSSKOHLE (*f*) : sooty coal 762
RUSSTAU (*m*) : *Fumago* (Bot.) 763

S

SAALVERDUNKLER (*m*) : dimmer ; dimming resistance (Elect.) 764
SAATGUTREINIGUNGSMASCHINE (*f*) : seed separator 765
SACHAROSE (*f*) : (see ROHRZUCKER) 766
SACKAUFZUG (*m*) : sack hoist 767
SAITENELEKTROMETER (*n*) : (see FADENELEKTROMETER) 768
SALPETERHAFEN (*m*) : nitrate port 769
SAMMELBAND (*n*) : collector belt 770
SANDFÄNGER (*m*) : strainer (Paper) 771
SANDSTRAHLGEBLÄSE (*n*) : sand-blast gun 772
SANIEREN, SICH- : to borrow money (in order to make oneself financially sound) (Commercial) 773
SAPRÄMIE (*f*) : sapraemia ; septic poisoning 774
SAPROGEN : saprogenic (producing or resulting from putrefaction) 775
SAPROPHYT : saprophytic ; (*n*) : saprophyte (growing on decaying organic matter) 776
SARKOM (*n*) : sarcoma (tumour of fleshy tissue) (Med.) 777
SÄTTIGUNGSSTROM (*m*) : saturation (voltage) current (Radio) 778
SATZ (*m*), DER ERSTE- : the first (allegro) movement (sonata form in Music) 779
SÄUBERUNG (*f*) : cleaning ; cleansing ; clearing ; sweeping-(mines at sea) ; mopping-up (Military) 780
SAUERSTOFF (*m*), AKTIVER- : ozone (O_3) 781
SÄUFER (*m*), STILLER- : secret drinker (Med.) 782
SÄUFERWAHNSINN (*m*) : delirium tremens (Med.) 783
SAUGBAGGER (*m*) : suction dredger 784
SAUGKASTEN (*m*) : suction box (Paper) 785
SAUGLEISTUNG (*f*) : suction (of vacuum cleaners) 786
SAUGÖFFNER (*m*) : exhaust opener (Textile) 787
SAUGÖFFNUNG (*f*) : suction (opening) inlet 788
SAUGSTRIEGEL (*m*) : suction curry comb (for use with vacuum cleaners) 789
SAUMPFAD (*m*) : bridle path 790
SCHACHT (*m*), BLINDER- : staple pit (Mining) 791
SCHÄDLINGSBEKÄMPFUNGSMITTEL (*n*) : pest destroyer ; insecticide 792
SCHAFPOCKEN (*f.pl*) : chicken-pox (Med.) 793
SCHÄLBLATTERN (*f.pl*) : pemphigus (Med.) 794
SCHALENGLOCKE (*f*) : dome gong (of bells) 795
SCHALLBECHER (*m*) : bell (of wind instrument) (Music) ; mouthpiece (Tel.) 796
SCHALLDICHT : sound-proof 797
SCHALLDOSE (*f*) : sound-box (of gramophone) 798
SCHALLMEMBRAN (*f*) : diaphragm (Tel.) 799
SCHALLMUSCHEL (*f*) : earpiece (of head-phones) 800
SCHALLPLATTE (*f*) : (gramophone)-disc or record 801
SCHALLPLATTENAPPARAT (*m*) : gramophone 802
SCHALLREFLEXION (*f*) : echo ; sound reflection 803
SCHALMEIGLOCKE (*f*) : sheep gong (of bells) 804
SCHALTBILD (*n*) : diagram of connections ; connection diagram (Elect.) 805
SCHALTERACHSE (*f*) : switch (spindle) rod 806
SCHALTERGRIFF (*m*) : switch (knob ; dolly) handle 807
SCHALTGERÄT (*n*) : switchgear 808
SCHALTGERÄT (*n*), WASSERDICHTES- : water-tight switchgear 809
SCHALTKONTAKT (*m*) : tripping contact (Elect.) 810
SCHALTLEISTUNG (*f*) : tripping (load) capacity (Elect.) 811
SCHALTSÄULE (*f*) : control (switch) pillar 812
SCHALTSTANGE (*f*) : switch operating rod 813
SCHALTSTÜCK (*n*) : contact piece 814
SCHALTSYSTEM (*n*) : system of (control) connections (Elect.) 815
SCHALTUHR (*f*) : time switch (Elect.) 816
SCHALTUNG (*f*), GEMISCHTE- : series-parallel connection (Elect.) 817
SCHALTWALZE (*f*) : switch drum (Elect.) 818
SCHALTWELLE (*f*) : switch (rod) shaft 819
SCHALTWERK (*n*) : (switch)-mechanism 820
SCHÄLWALD (*m*) : wood (forest) of oaks from which is obtained the bark for tanning 821
SCHÄRMASCHINE (*f*) : warping machine (Textile) 822
SCHARTE (*f*) : (dyer's)-saw wort (*Serratula tinctoria*) (Bot.) 823
SCHAUMKRAUT (*n*) : common bittercress ; cuckoo flower ; lady's smock ; meadow bittercress (*Cardamine pratensis*) (Bot.) 824
SCHEIBENZÄHLER (*m*) : disc meter 825
SCHEIBENZUGLEUCHTE (*f*) : tracker-wire operated lamps ; lighting fittings with colour screens (Stage lighting) 826
SCHEINLEITWERT (*m*) : admittance (Elect.) (reciprocal of impedance) 827
SCHEINSCHMAROTZER (*m*) : epiphyte (a plant growing upon another) 828
SCHEINTOD (*m*) : trance (Med.) 829
SCHEINWIDERSTAND (*m*) : impedance ; apparent resistance (Elect.) ; admittance (Elect.) 830
SCHEINWIDERSTANDSKURVE (*f*) : impedance curve 831
SCHEITELLAPPEN (*m*) : parietal lobe-(of brain) (Med.) 832
SCHERENARM (*m*) : (telescopic)-telephone arm 833
SCHICHTLINIEN (*f.pl*) : lines of equal sea level 834
SCHIEBESPULE (*f*) : sliding coil 835

SCHIEBEWIDERSTAND (*m*) : sliding (rheostat) resistance (Elect.) 836
SCHIEFHALS (*m*) : wry neck ; torticollis (Med.) 837
SCHIEFWUCHS (*m*) : scoliosis (Med.) 838
SCHIENBEINARTERIE (*f*) : anterior tibial artery (Med.) 839
SCHIENE (*f*), DRITTE- : conductor (third) rail (Electric traction) 840
SCHIFFAHRTSGESELLSCHAFT (*f*) : shipping (navigation) company 841
SCHIFFSEIGENGEWICHT (*n*) : displacement 842
SCHIFFSKREISEL (*m*) : ship's gyroscope 843
SCHIFFSLOOG (*n*) : (see LOG) 844
SCHIRMGITTERRÖHRE (*f*) : screened grid valve (Radio) 845
SCHIRMGLUCKE (*f*) : canopy hover (Poultry farming) 846
SCHIRMLEUCHTE (*f*) : intensive-(lighting)-fitting 847
SCHIRMPALME (*f*) : fan palm (*Corypha*) (Bot.) 848
SCHIZOPHRENIE (*f*) : dementia praecox (Med.) 849
SCHLAFKRANKHEIT (*f*) : sleeping sickness (Med.) (caused by *Trypanosoma gambiense*) 850
SCHLAGANFALL (*m*) : stroke ; seizure ; apoplexy 851
SCHLAGMASCHINE (*f*) : scutching machine ; scutcher (Textile) 852
SCHLAGRIEMEN (*m.pl*) : picking bands (Textile) 853
SCHLAGWORT (*n*) : caption ; slogan 854
SCHLAUCHLEITUNG (*f*) : hose-(pipe) 855
SCHLEICHSTROM (*m*) : stray current (Tel.) 856
SCHLEIFRINGMOTOR (*m*) : slip-ring motor (Elect.) 857
SCHLEIM ABSONDERN : to secrete mucus (Med.) 858
SCHLEIMBEUTELENTZÜNDUNG (*f*) : bursitis (Med.) 859
SCHLEIMHAUTENTZÜNDUNG (*f*), EITERIGE- : Blennorrhoea (Med.) 860
SCHLEIMPILZ (*m*) : slime mould ; slime fungus (see MYXOMYZETEN) 861
SCHLICHTEN (*n*) : sleeking (Leather) ; sizing (Textile) ; planishing (Metal) ; mediating ; smoothing ; planing ; dressing ; settling ; adjusting ; composing 862
SCHLINGBESCHWERDE (*f*) : difficulty in swallowing (Med.) 863
SCHLIPF (*m*) : slip (of electric motors) 864
SCHLUCKBESCHWERDE (*f*) : difficulty in swallowing (Med.) 865
SCHLUCKSEN : to hiccough ; (*n*) : hiccough (Med.) 866
SCHLÜSSEL (*m*), ALT- : alto clef (Music) 867
SCHLÜSSEL (*m*), BASS- : bass clef ; F clef (Music) 868
SCHLÜSSEL (*m*), C- : C clef (Music) 869
SCHLÜSSEL (*m*), TENOR- : tenor clef (Music) 870
SCHLÜSSEL (*m*), VIOLIN- : treble clef ; G clef (Music) 871
SCHLUSSFALL (*m*) : cadence (Music) 872
SCHLUSSFOLGERUNG (*f*) : illation ; inference ; conclusion 873
SCHLUSS (*m*), PLAGAL- : plagal cadence (Music) 874
SCHLUSSZEICHEN (*n*), DOPPELT- : double positive supervision (Tel.) 875
SCHLUSSZEICHEN (*n*), EINSEITIGES- : single positive supervision (Tel.) 876
SCHMALBLÄTTRIG : narrow-leaved (*angustifolius*) (Bot.) 877
SCHMAROTZERSTROM (*m*) : eddy current (Elect.) 878
SCHMELZDRAHT (*m*) : fuse wire (Elect.) 879
SCHMELZEINSATZ (*m*) : fusible plug ; fuse-(cartridge) 880
SCHMELZLEITER (*m*), HAUPT- : main fuse (Elect.) 881
SCHMELZLEITER (*m*), NEBEN- : auxiliary fuse (Elect.) 882
SCHMELZPATRONE (*f*) : fuse cartridge (Elect.) 883
SCHMELZSICHERUNG (*f*) : fuse ; cut-out ; fusible cut-out (Elect.) 884
SCHMELZVORRICHTUNG (*f*) : lubricator (Textile) 885
SCHMIEDEFEUERGEBLÄSE (*n*) : forge blower 886
SCHMIERFILZ (*m*) : lubricating felt 887
SCHMUCKWAREN (*f.pl*) : jewellery ; ornaments (for personal adornment) 888
SCHNARRE (*f*), WECHSELSTROM- : alternating current buzzer (Elect.) 889
SCHNARRWECKER (*m*) : buzzer 890
SCHNECKENFRÄSMASCHINE (*f*) : hobbing machine 891
SCHNEEBLINDHEIT (*f*) : snow blindness (Med.) 892
SCHNELLTELEGRAPHIE (*f*) : high-speed telegraphy 893
SCHNITZEL-SORTIERER (*m*) : chip screen (Papermaking) 894
SCHNÜRMASCHINE (*f*) : cording machine (Textile) 895
SCHNÜRRIEMEN (*m*) : (boot or shoe)-lace 896
SCHOFFÖR (*m*) : chauffeur 897
SCHONER (*m*) : protector ; protecting device 898
SCHONZEIT (*f*) : close season (of fish and game) 899
SCHOTENKLEE (*m*) : (common)-bird's-foot trefoil ; fingers and toes ; common lotus (*Lotus corniculatus*) (Bot.) 900
SCHRÄGAUFZUG (*m*) : inclined hoist 901
SCHRÄGSTRAHLER (*m*) : concentrating-(lighting)-fitting 902
SCHRANKENLEUCHTE (*f*) : track-crossing (reflector) lighting fitting (Elect.) 903
SCHRÄNKUNG (*f*) : setting-(of teeth of a saw) 904
SCHRAUBENLÜFTER (*m*) : stream-line fan 905
SCHRAUBSTÖPSEL (*m*) : screw plug (of fuses) 906
SCHREIBAPPARAT (*m*) : morse apparatus (Tel.) 907
SCHREIBKRAMPF (*m*) : writer's cramp (Med.) 908
SCHREIBTROMMEL (*f*) : recording drum (of meters) 909
SCHRITTLÄNGE (*f*) : length of (pace) step 910
SCHRITTZÄHLER (*m*) : pedometer 911
SCHROTTEFFEKT (*m*) : crackling-(noises) (Radio) 912
SCHROTTURM (*m*) : shot tower (for making lead shot) 913
SCHULFUNK (*m*) : wireless transmission to schools 914
SCHUPPENKELCH (*m*) : pappus (Bot.) 915
SCHUPPENKRANKHEIT (*f*) : Ichthyosis (a scaly skin disease) (Med.) 916
SCHUSSDICHTE (*f*) : density of picks (Textile) 917
SCHÜSSELFLECHTE (*f*) : Parmelia (Bot.) 918
SCHUSSFADEN (*m*) : weft thread (Textile) 919
SCHUSSPANNUNG (*f*) : weft tension (Textile) 920
SCHUSSPULE (*f*) : pirn (Artificial silk) 921
SCHUSSPULEN (*n*) : pirning (Textile) 922
SCHUSSPULENAUFWICKELN (*n*) : pirn winding (Textile) 923
SCHUSSPULMASCHINE (*f*) : pirn winder ; weft winding frame (Textile) 924
SCHUSSZÄHLER (*m*) : pick counter (Textile) 925
SCHÜTTELBOCK (*m*) : shake (Papermaking) 926
SCHÜTZ (*n*) : break (Elect.) 927
SCHUTZDRAHT (*m*) : earth wire (for use with SCHUTZSCHALTER, which see) (Elect.) 928
SCHÜTZENSTEUERUNG (*f*) : contactor control (Elect.) 929
SCHUTZFÄRBUNG (*f*) : protective colouring 930
SCHUTZHAFT (*f*) : protective custody 931
SCHUTZHAUBE (*f*) : protecting (cap) cover ; end cover (of electric motors) 932
SCHÜTZ (*n*), ÖLGEKÜHLTES- : oil-cooled break (Elect.) 933
SCHUTZRING (*m*), PORZELLAN- : porcelain skirt (for lampholders) (Elect.) 934

SCHUTZSCHALTER (*m*) : earth leakage-current protection switch (Elect.) 935
SCHUTZTRENNSCHALTER (*m*) : earth leakage (tripping device) release (Elect.) 936
SCHWACHSINN (*m*), MORALISCHER- : moral insanity (Med.) 937
SCHWÄCHUNGSWIDERSTAND (*m*) : fader ; volume control (Radio) 938
SCHWANENHALSSTÜTZE (*f*) : swan-neck bracket (for overhead wiring) 939
SCHWARZERDE (*f*) : black earth 940
SCHWEBEBÜHNE (*f*) : sinking scaffold (Mining) 941
SCHWEBUNGSEMPFANG (*m*) : heterodyne reception (Radio) 942
SCHWEFELREGEN (*m*) : clouds of pollen carried by the wind (Bot.) 943
SCHWEISSDRÜSE (*f*) : sweat gland (Med.) 944
SCHWEISSERKAPPE (*f*) : welding hood 945
SCHWEISSGENERATOR (*m*) : welding generator 946
SCHWEISSKOLBEN (*m*) : electrode holder (Welding) 947
SCHWEISSTAB (*m*) : electrode (welding) 948
SCHWEISSWULST (*m*) : reinforcement-(of welded seam) 949
SCHWENKBAR : swivelling 950
SCHWENKVORRICHTUNG (*f*) : swivelling device 951
SCHWERINDUSTRIE (*f*) : heavy industry 952
SCHWINDFLECHTE (*f*) : Lichen (a skin eruption) (Med.) 953
SCHWINGSCHIFFCHEN (*n*) : vibrating shuttle 954
SCHWINGUNGSKREIS (*m*) : oscillatory (oscillation) circuit 955
SCHWITZEN (*n*) : diaphoresis ; sweating ; perspiration 956
SCHWUNGMOMENT (*n*) : flywheel effect 957
SCHWURGERICHT (*n*) : jury 958
SECCION (*f*) : section ; part ; portion 959
SECHSKANTMUTTER (*f*) : hexagon nut 960
SECHSPOLIG : six prong-(ed) (of Wireless valves) 961
SEEHÖHE (*f*) : height above sea level 962
SEEKLAR : cleared, ready to put to sea 963
SEELENDURCHMESSER (*m*) : calibre (Mil.) 964
SEELENSTÖRUNG (*f*) : mental derangement ; insanity (Med.) 965
SEELENWANDERUNG (*f*) : metempsychosis 966
SEEMEILE (*f*) : nautical mile (6,080 feet) 967
SEEZEICHEN (*n*) : sea mark (buoys ; lighthouses ; lightships, etc.) 968
SEGGE (*f*) : (*see* RIEDGRAS) 969
SEHLINIE (*f*) : line of vision 970
SEHLOCH (*n*) : pupil (of the eye) (Med.) 971
SEHPURPUR (*m*) : visual purple ; rhodopsin (a substance in the retina of the eye) (Med.) 972
SEHROT (*n*) : (see SEHPURPUR) 973
SEHSCHÄRFE (*f*) : visual acuity ; clarity of vision 974
SEHWINKEL (*m*) : angle of (view) sight ; angle of vision 975
SEIDENMATT : frosted (of glass) 976
SEIDENZWIRNEREI (*f*) : silk twisting-(works) (Textile) 977
SEIDENZWIRNMASCHINE (*f*) : silk doubling frame (Textile) 978
SEIFENECHT : fast to soap-(liquor) 979
SEIFENSPIERE (*f*) : (see QUILLAJA) 980
SEILDURCHHANG (*m*) : sag (of a rope or cable) 981
SEILERBAHN (*f*) : rope walk 982
SEILEREI (*f*) : rope-(making)-works 983
SEILSCHEIBENGERÜST (*n*) : headgear (Mining) 984
SEILSCHLAGEN (*n*) : rope laying 985
SEILSCHLAGMASCHINE (*f*) : rope laying machine (Textile) 986
SEILWINDE (*f*) : winch 987
SEILWINDENHAUS (*n*) : winch house (Water power) 988
SEILZUG (*m*) : tracker wire (Stage lighting) 989
SEITENBAND (*n*) : fringing band (Tel.) 990
SEKTORFÖRMIGER LEITER (*m*) : shaped conductor (Elect.) 991
SEKUNDÄREMISSION (*f*) : secondary emission blocking effect (due to emission of metallic particles which settle on the grid) (in Valves) 992
SEKUNDÄRE STRAHLEN (*m.pl*) : secondary (scattered) rays (of X-rays) 993
SEKUNDÄRFILTER (*n*) : output filter (Radio) 994
SEKUNDÄRREGLER (*m*) : secondary regulator (Elect.) 995
SEKUNDÄRSEITE (*f*) : secondary side (Elect.) 996
SEKUNDE (*f*), GROSSE- : major second (Music) (see under TERZ) 997
SEKUNDIEREN : to second 998
SELBSTANLASSER (*m*) : automatic starter ; remote control starter (Elect.) 999
SELBSTANLAUFEND : self-starting (of machines) 000
SELBSTANSAUGEND : self-priming (of pumps) 001
SELBSTANSCHLUSS (*m*) : automatic connection 002
SELBSTANSCHLUSSAMT (*n*) : automatic-(telephone)-exchange 003
SELBSTBESTIMMUNGSRECHT (*n*) : right of self-determination 004
SELBSTENTLADUNG (*f*) : spontaneous discharge (Elect.) 005
SELBSTERREGUNG (*f*) : self-excitation (Elect.) 006
SELBSTINDUKTIONSSPULE (*f*) : self-induction coil ; variometer 007
SELBSTINDUKTIVITÄT (*f*) : self-inductance (Elect.) 008
SELBSTPOTENTIAL (*n*) : self-potential (Radio) 009
SELBSTSCHALTER (*m*) : automatic switch 010
SELBSTSCHLUSSAMT (*n*) : automatic exchange (Tel.) 011
SELBSTSTEUERUNGSAPPARAT (*m*) : automatic control apparatus 012
SELBSTUNTERBRECHUNG (*f*) : automatic (cut-out ; break) interruption 013
SELBSTVERWALTUNG (*f*) : self-government 014
SELENOGRAPHIE (*f*) : selenography (description of the moon) 015
SELFAKTOR (*m*), DIFFERENTIAL- : differential mule (Textile) 016
SELLERIEBLÄTTRIG : having celery-like leaves (*apiifolius*) (Bot.) 017
SEMIOTIK (*f*) : symptomatology (study of the signs of disease) (Med.) 018
SEMMELPILZ (*m*) : wheat fungus (*Polyporus confluens*) (Bot.) 019
SENDEANTENNE (*f*) : transmitting aerial (Radio) 020
SENDEVERSTÄRKER (*m*) : transmission amplifier (Picture telegraphy) 021
SENFSPIRITUS (*m*) : mustard oil dissolved in spirit 022
SENGMASCHINE (*f*) : gassing (singeing) machine (Textile) 023
SENIL : senile (Med.) 024
SENSORIUM (*n*) : (see BEWUSSTSEIN) 025
SEPSIS (*f*) : sepsis ; blood-poisoning (Med.) 026
SEPTICHÄMIE (*f*) : septaemia ; septicaemia (general illness due to wound infection of the blood) (Med.) 027
SEPTIME (*f*), GROSSE- : major seventh (Music) (see under TERZ) 028
SERIENMASCHINE (*f*) : series dynamo (Elect.) 029
SERIEN-PARALLELSCHALTER (*m*) : series-parallel switch (Elect.) 030

SERPENTINE (*f*) : serpentine (wavy) line 031
SETZHEBEL (*m*) : backstay (Mining) 032
SETZMANIER (*f*) : mannerism in (composing) writing 033
SEXTE (*f*), GROSSE- : major sixth (Music) (see under TERZ) 034
SEZIERMESSER (*n*) : scalpel (Med.) 035
SHEDDACH (*n*) : north-light roof-(for factories) 036
SICHERHEIT (*f*), DYNAMISCHE- : dynamic (mechanical) safety-(factor) 037
SICHERHEIT (*f*), THERMISCHE- : thermal safety-(factor) 038
SICHERUNGSELEMENT (*n*) : fuse base (of cartridge fuses) 039
SICHERUNGSTAFEL (*f*) : fuse board (Elect.) 040
SIDERISCHES JAHR (*n*) : sidereal year 041
SIEBBEINHÖHLE (*f*) : ethmoidal sinus (Med.) 042
SIEBKETTE (*f*) : band pass filter (Radio) 043
SIEB (*n*), OBER- : upper wire (Papermaking) 044
SIEBSCHALTUNG (*f*) : filter circuit (Radio) 045
SIEB (*n*), UNTER- : under wire (Papermaking) 046
SIEGELKUNDE (*f*) : (see SPHRAGISTIK) 047
SIGNALFEUER (*n*) : signal-(ling) light or fire 048
SIGNALTABLO (*n*) : indicator board (Elect.) 049
SILBERDEPOT (*n*) : silver deposit 050
SILURISCH : Silurian (Geol.) 051
SINNESNERV (*m*) : sensory nerve ; afferent nerve (Med.) 052
SINNESTÄUSCHUNG (*f*) : hallucination (Med.) 053
SINNESWAHRNEHMUNG (*f*) : perception ; sensation 054
SISTIEREN : to cancel ; abolish ; annul ; render void ; revoke ; inhibit ; stop ; hold up ; suspend 055
SISTIERUNG (*f*) : cancellation ; annulment ; abolition ; revocation ; inhibition ; suspension 056
SKABIOSE (*f*) : scabious (*Scabiosa*) (Bot.) 057
SKEPSIS (*f*) : scepticism 058
SKOLIOSE (*f*) : scoliosis ; curvature of the spine with bend to one side (Med.) 059
SKROFULOSE (*f*) : (see SKROPHULOSE) 060
SOCKELAUTOMAT (*m*) : circuit breaker (Elect.) 061
SOFFITTENLAMPE (*f*) : tubular lamp 062
SOFTENER (*m*) : softener (Textile) 063
SOG (*n*) : sucking ; suction ; wake ; wash (of a ship) 064
SÖHLIG : flat ; horizontal ; level (Mining) 065
SOLAR : solar 066
SOLBAD (*n*) : salt bath ; brine bath 067
SOLENOIDBREMSE (*f*) : magnetic brake 068
SOLLBESTAND (*m*) : theoretical (supposed ; expected) balance 069
SOLLMENGE (*f*) : theoretical (amount) quantity 070
SOLMISATION (*f*) : solmization ; solfaing (Music) 071
SOLÖZISMUS (*m*) : solecism 072
SOLSTITIUM (*n*) : solstice 073
SOLSTIZ (*f*) : (see SOLSTITIUM) 074
SOMATISCH : somatic ; corporeal ; bodily ; physical 075
SOMATOLOGIE (*f*) : somatology (the science of bodies, human or organic) 076
SOMNAMBULISMUS (*m*) : somnambulism ; sleep walking (Med.) 077
SONATENFORM (*f*) : sonata (first-movement) form (Music) 078
SONNENBESTRAHLUNG (*f*) : (see INSOLATION) 079
SONNENWEITE (*f*) : distance from the sun to the earth 080
SOORPILZ (*m*) : Oidium albicans (parasite causing thrush) (Med.) (see SCHWÄMMCHEN) 081
SORDO : muted ; damped 082
SORTIERGESTELL (*n*) : (sorting)-frame (Textile) 083
SORTIERRAUM (*m*) : sorting room 084
SOUFFLEURKASTEN (*m*) : prompter's box (of Stages) 085
SPALTUNGSIRRESEIN (*n*) : (see SCHIZOPHRENIE) 086
SPANNDRAHT (*m*) : strain-(ing) wire 087
SPANNUNGFÜHREND : live ; under tension (Elect.) 088
SPANNUNGSABNAHME (*f*) : voltage drop ; difference in potential 089
SPANNUNGSAUSGLEICHVORRICHTUNG (*f*) : (tension)-compensator (Textile) 090
SPANNUNGSGEFÄLLE (*n*) : voltage drop ; difference in potential 091
SPANNUNGSRÜCKGANGSAUSLÖSER (*m*) : no-volt release (Elect.) 092
SPANNUNGSSPULE (*f*) : shunt coil (Electric meters) 093
SPANNUNGSTEILER (*m*) : potentiometer ; potential (voltage) divider (Elect.) 094
SPANNUNGSUMFORMUNG (*f*) : voltage transformation 095
SPANNUNGSWANDLER (*m*) : potential (pressure ; voltage) transformer (Elect.) 096
SPANNUNGSZUSTAND (*m*) : state of strain (of Textile threads) 097
SPANNWEITE (*f*) : span ; amount (distance) of span 098
SPARTERIE (*f*) : wattle ; hurdle-work 099
SPARTOGRAS (*n*) : (see SPARTGRAS) 100
SPASMA (*f*) : spasm ; convulsion (Med.) 101
SPASMATISCH : spasmodic 102
SPASMODISCH : spasmodic 103
SPÄTKARBONISCH : Late Carboniferous (Geol.) 104
SPEKTROGRAPH (*m*) : spectrograph 105
SPEKULUM (*n*) : (see SPECULUM) 106
SPERRER (*m*) : suppressor 107
SPERRHEBEL (*m*) : locking lever (Mech.) 108
SPERRKREIS (*m*) : wave trap (Radio) 109
SPEZIFISCHE LEITFÄHIGKEIT (*f*) : specific conductivity (Electro-chemistry) 110
SPHÄROMETER (*n*) : (see KUGELMESSER) 111
SPHÄROMETRIE (*f*) : spherometry 112
SPHRAGISTIK (*f*) : sphragistics (study of engraved seals) 113
SPIEGELOPTIK (*f*) : system of mirrors 114
SPIEGELREFLEKTOR (*m*) : mirror reflector (Lighting) 115
SPIELDAUER (*f*) : cycle of operations ; duration (of a mechanical movement, game or performance) 116
SPIELFLÄCHE (*f*) : acting area (of stages) 117
SPIELFLÄCHENLEUCHTE (*f*) : acting area flood-(light) (Stage lighting) 118
SPIELFLÄCHENSCHEINWERFER (*m*) : acting area spot light (Stage lighting) 119
SPIELTISCH (*m*) : gramophone turntable 120
SPIERSTAUDE (*f*) : dropwort (*Spiraea filipendula*) (Bot.) 121
SPINDELBAUART (*f*) : construction (type) of spindle 122
SPINDELDREHZAHL (*f*) : spindle speed (Textile) 123
SPINDELSCHNUR (*f*) : spindle band (Textile) 124
SPINNAUTOMAT (*m*) : automatic spinning machine (Textile) 125
SPINNBETRIEB (*m*) : spinning (ring-frame) operation (Textile) 126
SPINNDÜSE (*f*) : spinning nozzle 127
SPINNFLÜGELMOTOR (*m*) : flyer spindle motor (Textile) 128

SPINNGEWEBENENTZÜNDUNG (*f*) : arachnoiditis (inflammation of the arachnoid membrane) (Med.) 129
SPINNMARGE (*f*) : spinner's margin (Textile) 130
SPINNREGLER (*m*) : spinning regulator 131
SPINNREGLERBETRIEB (*m*) : spinning regulator (drive) operation 132
SPINNTOPFMOTOR (*m*) : spinning pot motor (Textile) 133
SPINNWEBENHAUT (*f*), MITTLERE- : arachnid ; transparent membrane of dura mater (Med.) 134
SPINÖS : spinous ; thorny 135
SPITZBLÄTTRIG : with pointed leaves (*acutifolius*) (Bot.) 136
SPITZBLUMIG : with pointed flowers (*acutiflorus*) (Bot.) 137
SPITZBOGENSTIL (*m*) : Gothic-(style) 138
SPITZENLEISTUNG (*f*) : peak load (Mech.) 139
SPITZENSTUHL (*m*) : lace loom (Textile) 140
SPITZPOCKEN (*f.pl*) : chicken-pox (Med.) 141
SPOLIIEREN : to spoliate ; (*n*) : spoliation ; pillage ; plundering ; mutilation ; destruction ; alteration ; robbery 142
SPOLIIERUNG (*f*) : (see SPOLIIEREN) 143
SPONGIÖS : spongy ; sponge-like 144
SPRACHFREQUENZ (*f*) : voice (speech) frequency (Tel.) 145
SPRACHSCHWINGUNG (*f*) : voice (oscillation) vibration 146
SPRACHÜBERTRAGUNG (*f*) : speech transmission (Tel.) 147
SPRACHVERSTÄRKER (*m*) : speech (voice) amplifier (Radio) 148
SPRECHAPPARAT (*m*) : transmitter (Tel.) 149
SPRECHFÄHIGKEIT (*f*) : carrying capacity (of telephone cables) 150
SPRECHFREQUENZ (*f*) : voice frequency (Tel.) 151
SPRECHKREIS (*m*) : speaking circuit (Tel.) 152
SPRECHLÄNGE (*f*) : speaking (range) distance (Tel.) 153
SPRECHRICHTUNG (*f*) : speech direction (Tel.) 154
SPRECHSCHLÜSSEL (*m*) : Kellogg (listening) key (Tel.) 155
SPRECHTRICHTER (*m*) : mouthpiece (Tel.) 156
SPRECHUMSCHALTER (*m*) : Kellogg (speaking ; listening) key (Tel.) 157
SPRECHWEITE (*f*) : (see SPRECHLÄNGE) 158
SPRENGDRAHT (*m*) : shot-firing wire (Elect.) 159
SPRINGRING (*m*) : locking ring 160
SPRITZBLECH (*n*) : drip plate ; splash plate (of kitchen ranges) 161
SPRITZVERZINKT : cold galvanised 162
SPRITZWASSER (*n*) : spray ; splash 163
SPRITZWASSERDICHT : spray (shower ; drip ; splash) proof 164
SPRITZWASSERGESCHÜTZT : splash proof 165
SPRUNGBEIN (*n*) : talus ; ankle bone (Med.) 166
SPRUNGWEISE : in leaps and bounds ; in (jumps) jerks ; jerkily ; jumpy 167
SPULENABSTAND (*m*) : coil spacing ; spacing of coils ; distance between coils (Tel.) 168
SPULENFELD (*n*) : loading section (see PUPINSPULENFELD) 169
SPULENFÜLLE (*f*) : (degree of)-fullness of the cop (Textile) 170
SPULENHALTER (*m*) : coil holder (Radio) 171
SPULENKASTEN (*m*) : coil pot (Telephone cables) 172
SPULENSOCKEL (*m*) : coil holder (Radio) 173
SPULMASCHINE (*f*) : winding machine (Textile) 174
SPULSPINNMASCHINE (*f*) : bobbin spinning machine (Textile) 175
SPURKRANZAUFSCHWEISSMASCHINE (*f*) ; welding machine (for building up or revising worn tyres, wheel treads and flanges) 176
STACHELBLÄTTRIG : with leaves pointed like thorns (*aquifolius*) (Bot.) 177
STAFFELUNG (*f*) : grading 178
STAHLROHR (*n*), GESCHLITZTES- : split (tube) tubing (for electric wiring) 179
STAHLSTICHEL (*m*) : punch (for metal) 180
STAMMKREIS (*m*) : side (physical) circuit (Tel.) 181
STAMMLEITUNG (*f*) : side (physical) circuit (Tel.) 182
STANDBAHN (*f*) : portable railway 183
STÄNDERANSCHLUSS (*m*) : stator (lead) connection (Elect.) 184
STÄNDERPAKET (*n*) : stator core (Elect.) 185
STANDKLAPPENSCHRANK (*m*) : floor pattern switchboard (Tel.) 186
STANDROHRWASSERMESSER (*m*) : hydrant water meter 187
STANDSIGNALANLAGE (*f*) : pedestal-(type) signal system (Elect.) 188
STAPELFASER (*f*) : staple fibre (Textile) 189
STAPELFASERGARN (*n*) : staple fibre yarn (Textile) 190
STAPELFASERGEWEBE (*n*) : staple fibre fabric (Textile) 191
STAR (*m*), GRÜNER- : glaucoma (Med.) 192
STARKE KUPPLUNG (*f*) : rigid coupling (Mech.) 193
STAUBABSCHEIDER (*m*) : dust (extractor) separator 194
STAUBBEUTELTRAGEND : antheriferous ; having anthers (Bot.) 195
STAUBBLÜTE (*f*) : male flower (Bot.) 196
STAUBBRAND (*m*) : blight ; mildew ; rust (*Puccinia*) (Bot.) 197
STAUBFADEN (*m*) : filament (the stalk supporting the anther) (Bot.) 198
STAUBFADENLOS : sessile (Bot.) 199
STAUBFREI : free from dust 200
STAUBHAUT (*f*) : membrane of an anther (Bot.) 201
STAUBSAME (*m*) : pollen (the dust contained in the anther) (Bot.) 202
STAUBWEG (*m*) : pistil (Bot.) 203
STAUMAUER (*f*) : storage dam 204
STECHPALMENBLÄTTRIG : with leaves pointed like thorns ; holly-leaved (*aquifolius*) (Bot.) 205
STECHWINDE (*f*) : (see SARSAPARILLA) 206
STECKERSCHALTER (*m*) : switch-plug (Elect.) 207
STECKKONTAKT (*m*) : plug-in contact (Elect.) 208
STECKSCHLÜSSEL (*m*) : key ; box spanner 209
STEHLSUCHT (*f*) : kleptomania (Med.) 210
STEIGELEITUNG (*f*) : rising main ; house supply main (Elect.) 211
STEIGER (*m*) : surveying official (Mining) 212
STEIGERUNGSGRAD (*m*) : degree of comparison (Grammar) 213
STEILHEIT (*f*) : steepness ; precipitousness ; slope ; mutual conductance (of Valves) 214
STEINALTER (*n*) : Stone Age (Geol.) 215
STEINKOHLENFORMATION (*f*) : fossil-coal formation (of Upper Carboniferous Period) (Geol.) 216
STEINSTRECKE (*f*) ; cross-measure drift (Mining) 217
STEISSWIRBEL (*m*) : coccygeal vertebra (Med.) 218
STELLKNOPF (*m*) : setting (regulating) knob 219
STEPPSTICH (*m*) : lock-stitch (Textile) 220
STEREOCHROMIE (*f*) : stereochromy 221
STEREOSKOP (*n*) : stereoscope (an instrument which gives to pictures the appearance of solid bodies) 222
STERNDEUTEKUNST (*f*) : astrology 223
STERNDREIECKSCHALTER (*m*) : star-delta switch (Elect.) 224

STERNDREIECKSCHALTUNG (*f*) : star-delta connection (Elect.) 225
STERNFÜNFECK (*n*) : pentagram ; pentacle 226
STERNSCHALTUNG (*f*) : star connection (Elect.) 227
STERNVIERER (*m*) : (spiral)-quad (of cables) 228
STERNVIERERVERSEILUNG (*f*) : star quadded (of cables) 229
STEUERELEKTRODE (*f*) : grid (Elect.) 230
STEUERLEITUNG (*f*) : pilot wire (Stage lighting) 231
STEUERTAFEL (*f*) : control panel 232
STEUERVERSTÄRKER (*m*) : input amplifier (Radio) 233
STICHART (*f*) : type of stitch (Textile) 234
STICKMASCHINENANTRIEB (*m*) : embroidery machine drive (Textile) 235
STILB (*n*) : 1 Hefner candle/cm.^{2} = 1 Sb. (Lighting) (see LEUCHTDICHTE) 236
STIMME (*f*), LIEGENDE- : pedal point (in any part other than bass) (Music) 237
STIMMIG, EIN- : one (part) voice (Music) 238
STIMMIG, MEHR- : many (parts) voices (Music) 239
STIMMIG, VIEL- : many (parts) voices (Music) 240
STIMMLEHRE (*f*) : phonetics 241
STINKNASE (*f*) : ozaena (a foetid disease of the nose) (Med.) 242
STIRNBEIN (*n*) : frontal bone (Med.) 243
STIRNKEHLNAHT (*f*) : (longitudinal)-strap fillet weld (Welding) 244
STIRNLAPPEN (*m*) : frontal lobe (of brain) (Med.) 245
STOFFPUMPE (*f*) : stuff pump (Paper) 246
STÖPSELKOPF (*m*) : screw cap (of cartridge fuses) 247
STÖRSCHUTZGERÄT (*n*) : interference eliminator (Elect.) 248
STOSSFÖRDERUNG (*f*) : face transport (Mining) 249
STOSSFREI : quiet ; free from shock ; not in (jumps ; jolts) jerks ; smoothly 250
STOSSKURZSCHLUSSTROM (*m*) : instantaneous (initial value of) short circuit current (Elect.) 251
STOSSTEMPEL (*m*) : face timber (Mining) 252
STOSSZIMMERUNG (*f*) : face timber (Mining) 253
STRAHL (*m*), KURZWELLIGER- : hard (deeply penetrating) ray ; X-ray of short wave-length 254
STRAHL (*m*), LANGWELLIGER- : soft ray ; X-ray of long wave-length 255
STRAHLPILZKRANKHEIT (*f*) : Actinomycosis (disease due to ray fungus) (Med.) 256
STRAHLUNGSINTENSITÄT (*f*) : intensity (strength) of radiation 257
STRAMM SITZEN : to fit (sit) tightly 258
STRANGGARNSCHLICHTMASCHINE (*f*) : hank sizing machine (Textile) 259
STRANGLÄNGE (*f*) : hank length (Textile) 260
STRANGMESSVORRICHTUNG (*f*) : hank length measuring motion (Textile) 261
STRANGTROCKENMASCHINE (*f*) : hank drying machine (Textile) 262
STRATIGRAPHIE (*f*) : stratigraphy (Geol.) 263
STRATIGRAPHISCH : stratigraphic-(ally) (Geol.) 264
STREBBAU (*m*) : longwall (Mining) 265
STRECKBARKEIT (*f*) : ductility (Metal.) 266
STRECKENBELEUCHTUNG (*f*) : haulage road lighting (Mining) 267
STRECKMUSKEL (*m*) : extensor muscle (Med.) 268
STREICHGARN (*n*) : condenser yarn (Textile) 269
STREICHGARNKREMPEL (*f*) : condenser card (Textile) 270
STREICHGARNSELFAKTOR (*m*) : condenser mule (Textile) 271
STREICHGARNSPINNEREI (*f*) : condenser yarn spinning-(mill) (Textile) 272
STREICHMASSAGE (*f*) : effleurage ; stroking-(massage) 273
STREIFEN (*n*) : scraping ; rubbing ; stripping 274
STREIFENKOHLE (*f*) : (see STEINKOHLE) 275
STREUERGLAS (*n*) : spreading glass ; diffusing glass (Lighting) 276
STREUGLAS (*n*) : (see STREUERGLAS) 277
STREUSPIEGEL (*m*) : diffusing mirror (Lighting) 278
STREUSTRAHLEN (*m.pl*) : scatter rays 279
STREUUNG (*f*) : spreading ; spread ; diffusion ; dispersion ; dissipation ; scatter (of Light, etc.) 280
STREUUNGSGRAD (*m*) : degree of spread (Lighting) 281
STREUUNGSVERÄNDERUNG (*f*) : variation in spread (of Light rays) 282
STREUUNGSVERMÖGEN (*n*) : spreading (power) capacity (Lighting) 283
STRICKWAREN (*f.pl*) : knitted goods (Textile) 284
STROBOSKOP (*n*) : stroboscope (instrument for observing phases of a motion by means of a light which is periodically interrupted) 285
STROBOSKOPISCHE EINTEILUNG (*f*) : stroboscopic division ; synchroscopic division (as on turntable of a gramophone) 286
STROBOSKOPLAMPE (*f*) : stroboscope lamp 287
STROMABGEBER (*m*) : commutator (Elect.) 288
STROMABNAHMEKLEMME (*f*) : overhead line (terminal) clamp (Elect.) 289
STROMÄNDERUNG (*f*) : current variation (Elect.) 290
STROMANREGUNG (*f*) : operating current-(element) (Elect.) 291
STROMBEGRENZER (*m*) : current limiter (Elect.) 292
STROMFÜHREND : current-carrying ; live (Elect.) 293
STROMIMPULS (*m*) : current impulse (Elect.) 294
STROMLEISTUNG (*f*) : wattage (Elect.) 295
STROMPFAD (*m*) : current (path) circuit (Elect.) 296
STRÖMUNGSMESSGERÄT (*n*) : flow indicator 297
STROMVERSORGUNG (*f*) : electricity supply ; supply of current 298
STROMVOREILEN (*n*) : phase displacement (Elect.) 299
STROMWANDLER (*m*) : current (series) transformer (Elect.) 300
STROMWENDUNG (*f*) : commutation (Elect.) 301
STROMZUFUHR (*f*) : (current)-supply (Elect.) 302
STRUKTURELL : structural 303
STUFENLOS : direct ; not by steps or stages 304
STUHLUNG (*f*) : seating ; bed-plate ; support 305
STUHLVERSTOPFUNG (*f*) : constipation (Med.) 306
STUHLWIRKUNGSGRAD (*m*) : loom efficiency (Textile) 307
STUR : pertinacious ; persistent ; stubborn-(ly) 308
SUBSTANTIVIEREN : to make substantive ; substantiate 309
SULFATIERUNG (*f*) : sulphation 310
SULFIDIEREN : to sulphidize 311
SUMMENMESSUNG (*f*) : integration (Elect.) 312
SUMMERGERÄUSCH (*n*) : buzzing (Tel.) 313
SUMMIERUNG (*f*) : summation 314
SUMPFDOTTERBLUME (*f*) : marsh marigold (*Caltha palustris*) (Bot.) 315
SUMPFFIEBER (*n*) : malaria (Med.) 316
SÜSSKIRSCHE (*f*) : wild cherry ; bird cherry ; gean (heart) cherry (*Prunus avium*) (Bot.) 317
SUSZEPTANZ (*f*) : susceptance (Elect.) 318
SWANFASSUNG (*f*) : bayonet lamp-holder (Elect.) 319
SYKOMORE (*f*) : Egyptian fig tree (*Ficus sycomorus*) (Bot.) 320
SYMBIOTISCHER STICKSTOFF (*m*) : symbiotic nitrogen 321
SYMPATHISCHES NERVENSYSTEM (*n*) : sympathetic system (of nerves) 322

SYMPETAL : sympetalous , gamopetålous (Bot.) 323
SYNCHRONOSKOP (n) : synchronoscope (see STROBOSKOP) 324
SYNKLINAL : synclinal (sloping downward) 325
SYNKLINALE (f) : syncline ; trough (Geol.) 326
SYNKOPIERT : syncopated (Music) 327
SYNTONIE (f) : syntonism ; tuned or in harmony with each other (of oscillations or electric circuits) 328
SYNTONISCH : syntonic ; tuned 329
SYNTONISMUS (m) : (see SYNTONIE) 330

T

TABLO (n) : (fuse- ; meter-) board 331
TACHYMETER (n) : theodolite (for measuring horizontal distances) 332
TAGESDIENST (m) : day (work ; duty) shift 333
TAGESLICHTAPPARAT (m) : daylight apparatus (Lighting unit for providing artificial daylight for colour grading) (Textile) 334
TAGUNDNACHTGLEICHE (f) : equinox 335
TAKTE (m), IM- : in synchronism ; in time ; on the beat 336
TAKTMESSER (m) : metronome 337
TAMPON (m) : plug (Med.) 338
TAMPONADE (f) : plugging (Med.) 339
TANKSTELLE (f) : filling station (for motor-cars) 340
TAPOTEMENT (m) : tapping (of massage) 341
TARNEN : to camouflage 342
TARNKLEID (n) : camouflage 343
TARTSCHENFLECHTE (f) : Iceland moss (*Cetraria islandica*) (Bot.) 344
TAUCHANKER (m) : plunger armature (Elect.) 345
TAUCHSIEDER (m) : immersion heater (Elect.) 346
TAUMELBEWEGUNG (f) : tumbling movement ; rolling motion 347
TAUMELSCHEIBE (f) : tumbling disc (Mech.) 348
TAUSENDSCHÖNBLÄTTRIG : daisy-leaved ; having leaves like the daisy (*bellidifolius*) (Bot.) 349
TAXATOR (m) : valuer ; appraiser ; taxer ; estimator 350
TAXPROBE (f) : assay 351
TEEMASCHINE (f) : samovar (Elect.) 352
TEICHKOLBEN (m) : (see ROHRKOLBEN) 353
TEILEND : partitive 354
TEILNAHMLOSIGKEIT (f) : antipathy ; lack of interest 355
TEILNEHMERLEITUNG (f) : subscriber's line (Tel.) 356
TEKTONIK (f) : tectonics (the art of building or assembling) 357
TEKTONISCH : tectonic ; structural 358
TELEGRAPHENAMT (n) : telegraph office 359
TELEKRYPTOGRAPHIE (f) : telegram coding 360
TELEPHONABONNENT (m) : (telephone)-subscriber 361
TELEPHONBESITZER (m) : (telephone)-subscriber 362
TELEPHONMEMBRAN (f) : (telephone receiver)-diaphragm 363
TELE-RADIOPHON (n) : (see PHOTOPHON) 364
TELLURISCH : tellurian ; tellural (having reference to the earth) 365
TEMPERATURSTIEG (m) : temperature rise ; increase in temperature 366
TEMPORÄRER MAGNETISMUS (m) : temporary magnetism 367
TEPPICHKNÜPFSTUHL (m) : pile carpet loom (Textile) 368
TERPENTINBAUM (m) : turpentine tree (*Pistacia terebinthus*) (Bot.) 369
TERTIÄRFORMATION (f) : Tertiary formation (Geol.) 370
TERZ (f) : third (Music) 371
TERZ (f), GROSSE- : major third (Music) 372
TERZ (f), KLEINE- : minor third 373
TERZ (f), ÜBERMÄSSIGE- : augmented third 374
TERZ (f), VERMINDERTE- : diminished third 375
TESTLÖSUNG (f) : test solution 376
TEXTUR (f) : texture ; structure 377
TEXTVERFASSER (m) : librettist 378
THALLUSPFLANZEN (f.pl) : thallophyta (Bot.) (see THALLOPHYTEN) 379
THERMISCH WIDERSTANDSFÄHIG : refractive 380
THERMOPHOR (n) : an apparatus for (conserving) maintaining heat 381
THORIUMOXYDRÖHRE (f) : dull emitter-(valve) (Radio) 382
THROMBOZYT (m) : thrombus (clot of blood in thrombosis) (Med.) 383
TIDENRAD (n) : tide wheel (of a tide-calculating machine) 384
TIEFENWIRKUNG (f) : penetration 385
TIEFPASSFILTER (n) : low-pass filter (Elect.) 386
TIEFPOL (m) : cathode ; negative (minus) pole (Elect.) 387
TISCHTASTER (m) : desk (table) push-(button) (Elect.) 388
TONABNEHMER (m) : pick-up (of gramophones) 389
TONABSTAND (m) : interval (Music) 390
TONART (f), GROSSE- : major key (Music) 391
TONART (f), HARTE- : major key (Music) 392
TONART (f), HAUPT- : main key ; natural scale 393
TONART (f), KLEINE- : minor key (Music) 394
TONART (f), MÄNNLICHE- : major key (Music) 395
TONART (f), NEBEN- : any key except C major or A minor (Music) 396
TONART (f), NORMAL- : natural scale (C major or A minor) 397
TONART (f), PARALLEL- : parallel key 398
TONART (f), TRANSPONIERTE- : transposed key ; any key except C major or A minor (Music) 399
TONART (f), WEIBLICHE- : minor key (Music) 400
TONART (f), WEICHE- : minor key (Music) 401
TON (m), BRUMMENDER- : buzzing noise ; hum ; rhythmic hum (when carbons of arc lamps burning on A.C. current) (Elect.) 402
TONDICHTER (m) : tone poet ; composer 403
TONDICHTUNG (f) : tone poem ; composition 404
TONFALL (m) : cadence (Music) 405
TONFOLGE (f) : sequence (succession) of tones or sounds ; scale (Music) 406
TONFREQUENZ (f) : voice frequency ; audio-frequency (Tel.) 407
TONFREQUENZTELEGRAPHIE (f), 12-FACH- : twelve-channel (audio-frequency) voice frequency telegraphy 408
TONFÜHRUNG (f) : modulation 409
TONFUNKEN (m) : singing spark 410
TON (m), HALBER- : half-tone ; semitone (Music) 411
TON (m), HERRSCHENDER- : dominant (Music) 412
TONIKA (f) : key-note ; tonic (Music) 413

TONLEHRE (*f*) : acoustics ; phonetics 414
TONLEITER (*f*), CHROMATISCHE- : chromatic scale (Music) 415
TONLEITER (*f*), DIATONISCHE- : diatonic scale (Music) 416
TONLEITER (*f*), DUR- : major scale (Music) 417
TONLEITER (*f*), MOLL- : minor scale (Music) 418
TONMASS (*n*) : measure (Music) 419
TONMESSER (*m*) : monochord ; sonometer ; tonometer (an instrument for counting the vibrations of a sound) 420
TONMESSKUNST (*f*) : tonometry 421
TONNENLEGER (*m*) : buoy-laying vessel (Marine) 422
TONPRÜFER (*m*) : sound tester ; audiometer 423
TONREICHTUM (*m*) : tonal range ; richness of tone 424
TONREIHE (*f*) : scale ; series (succession ; sequence) of tones or sounds ; compass 425
TON (*m*), REINER- ; pure (sound) tone 426
TONSCHLUSS (*m*) : cadence 427
TONSCHLÜSSEL (*m*) : key ; clef (Music) 428
TONSCHWINGUNG (*f*) : sound (vibration) oscillation 429
TONSETZER (*m*) : composer 430
TONSETZUNG (*f*) : (written)-composition (Music) 431
TONSTÄRKE (*f*) : intensity-(of sound) ; loudness 432
TONSTÄRKEEINHEIT (*f*) : unit (of audibility) sound density (see PHON) 433
TONSTÜCK (*n*) : composition ; piece of music 434
TONSTUFE (*f*) : pitch ; interval (Music) 435
TONSTUFE (*f*), ERSTE- : unison (Music) 436
TONVERSTÄRKER (*m*) : (sound)-amplifier 437
TONWEITE (*f*) : interval (Music) 438
TONWERK (*n*) : composition (Music) 439
TON (*m*), ZISCH- : hiss-(ing sound) 440
TOPFSTRECKE (*f*) : can gill (Textile) 441
TOPINAMBURPFLANZE (*f*) : Jerusalem artichoke (*Helianthus tuberosus*) (see ROSSKARTOFFEL) (Bot.) 442
TORFBEERE (*f*) : (see MOLTEBEERE) 443
TOTER PUNKT (*m*) : dead-centre (of crankshafts) 444
TRACHOM (*n*) : trachoma (an eye disease) (Med.) 445
TRAFO (*m*) : abbreviation for TRANSFORMATOR, which see 446
TRAGKLEMME (*f*) : carrier clip 447
TRAGWALZE (*f*) : brass tube roll (supporting the wire-cloth of paper machines) 448
TRÄNKMASSE (*f*) : impregnating compound 449
TRANSFORMATORENHAUS (*n*) : transformer house (Elect.) 450
TRANSFORMATORENSTATION (*f*) : transformer sub-station (Elect.) 451
TRANSFUSION (*f*) : (blood)-transfusion 452
TRANSPORTBEREIT : ready for (moving) transport 453
TRANSPORTFÄHIGKEIT (*f*) : portability ; transportability 454
TRANSVERSALWELLE (*f*) : transverse wave (of light) 455
TRAVERTIN (*m*) : (see TUFFKALK) 456
TREIBRAD (*n*) : driving wheel 457
TREKKER (*m*) : tractor 458
TRENNBÜGEL (*m*) : disconnecting link (Elect.) 459
TRENNENDVERSCHLUSS (*m*) : cable head (Elect.) 460
TRENNSCHALTER (*m*) : contact breaker ; disconnector ; disconnecting switch (Telephony) ; isolating (link) switch (Elect.) 461
TRENNSCHÄRFE (*f*) : selectivity (Radio) 462
TRENNSICHERUNG (*f*) : (see ABSPANNSICHERUNG) 463
TREPANATION (*f*) : trepanation ; trephining ; trepanning (Med.) 464
TRESORANLAGE (*f*) : strong room 465
TREUHANDGESCHÄFT (*n*) : trustee business 466
TRIAS (*f*) : Trias ; Triassic period (Geol.) 467
TRIAS (*f*), UNTERE- : Lower Triassic (Geol.) 468
TRIBROMÄTHYLALKOHOL (*m*) : tri-brom-ethyl alcohol ; avertin (85% bromine) 469
TRICHINE (*f*) : trichina (*Trichina spiralis*) (an intestinal worm of humans and animals) (Med.) 470
TRICHINENKRANKHEIT (*f*) : trichinosis (Med.) 471
TRICHINOSE (*f*) : trichinosis (Med.) 472
TRICHOPHYTIE (*f*) : herpes tonsurans (Med.) 473
TRIEBKERN (*m*) : magnet (Elect.) 474
TRIEBWAGEN (*m*) : tractor car (of Tramways) ; tractor coach ; motor (carriage) coach (of Trains) 475
TRIGEMINUSNEURALGIE (*f*) : neuralgia of the trigeminal nerve (Med.) 476
TRIKOTAGE (*f*) : hosiery (Textile) 477
TRIMORPH : trimorphous ; trimorphic 478
TRIMORPHISMUS (*m*) : trimorphism 479
TRIOLE (*f*) : triole ; triplet (Music) 480
TRITT (*m*), AUS DEM -FALLEN : to get out of step (of movements) 481
TROCKENGLEICHRICHTER (*m*) : metal (copper film) rectifier ; dry rectifier 482
TROCKENPARTIE (*f*) : dry end (Paper) 483
TROCKENSTÄUBER (*m*) : dry sprayer (for spraying powders) 484
TROLLBLUME (*f*) : globe flower (*Trollius europaeus*) (Bot.) 485
TROMPETENBAUM (*m*) : catalpa, Indian bean ; bean tree (*Catalpa bignonioides*) (Bot.) 486
TROPFRAND (*m*) : drip-rim 487
TROPFWASSER (*n*) : drips ; drops 488
TROPFWASSERDICHT : drip proof 489
TROPFWASSERGESCHÜTZT : drip proof 490
TRUNKSUCHT (*f*) : drunkenness (Med.) 491
TÜBBINGS : tubbing (Mining) 492
TUBERKULOSE-BAKTERIUM (*n*) : bacterium of tuberculosis (*Bacterium tuberculosis*) 493
TUBERKULOSE-SPALTPILZ (*m*) : bacterium of tuberculosis 494
TUBEROSE (*f*) : tuberose (*Polianthes tuberosa*) (Bot.) 495
TÜLLSTUHL (*m*) : (bobbin)-net loom (Textile) 496
TURBOSATZ (*m*) : turbo-(generator)-set 497
TUSSAH (*f*) : Tussore (Shantung) silk (Textile) 498
TÜTTEL (*n*) : iota ; fraction ; point 499
TYPENSCHNELLSCHREIBER (*m*) : rapid printing telegraph 500
TYPHLITIS (*f*) : (see BLINDDARMENTZÜNDUNG) 501
TYPHUS (*m*), UNTERLEIBS- : typhoid (enteric) fever (*Typhus abdominalis*) (Med.) 502

U

ÜBERBLENDEN : to fade (in or out) (Radio) 503
ÜBERDRUCKVENTIL (*n*) : excess pressure valve 504
ÜBEREMPFINDLICH : super-sensitive ; allergic (Med.) 505
ÜBEREMPFINDLICHKEIT (*f*) : idiosyncrasy ; hypersensitiveness ; super-sensitiveness ; hypersensitivity ; super-sensitivity ; allergy ; hyperaesthesia ; hyperesthesia (excessive sensibility or sensitiveness) 506

ÜBERFLUTUNGSSICHER : watertight ; protected against overflowing 507
ÜBERGANGSKASTEN (*m*) : cable end box ; service box ; junction box (Elect.) 508
ÜBERHOLUNG (*f*) : overhaul-(ing) ; servicing ; maintenance 509
ÜBERLAGERUNGSEMPFANG (*m*) : heterodyne reception (Radio) 510
ÜBERLANDLINIE (*f*) : land-line (Tel.) 511
ÜBERLANDZENTRALE (*f*) : power station supplying electric current by overhead transmission ; grid power station 512
ÜBERLAPPTSTOSS (*m*) : lap joint 513
ÜBERLASTUNG (*f*) : overload-(ing) ; overcharge ; overcharging 514
ÜBERLASTUNGSSCHUTZ (*m*) : overload protection 515
ÜBERLEITUNG (*f*) : change-over 516
ÜBERPFLANZE (*f*) : epiphyte (a plant growing upon another) 517
ÜBERPUTZSCHALTER (*m*) : surface-(mounted) switch (Elect.) 518
ÜBERSCHIEBUNG (*f*) : fault (Mining) ; dislocation (Geol.) 519
ÜBERSCHREIEN (*n*) : howling (of Radio valves) 520
ÜBERSETZERANLAGE (*f*) : multi-language interpretation (installation) equipment 521
ÜBERSPANNUNGSMESSER (*m*) : excess voltage meter ; klydonograph (Elect.) 522
ÜBERSPRECHDÄMPFUNG (*f*) : cross-talk transmission equivalent (Tel.) 523
ÜBERSPRECHEN (*n*) : near-end cross-talk (cross-talk between different cables) (Tel.) 524
ÜBERSTOCKUNG (*f*) : overstocking (Commercial) 525
ÜBERSTROM (*m*) : overcurrent ; excess current ; overload ; rush (surge) of current (Elect.) 526
ÜBERSTROMAUSLÖSER (*m*) : overload (overcurrent ; excess current) release or relay (Elect.) 527
ÜBERSTROMAUSLÖSUNG (*f*) : overload release 528
ÜBERSTROMAUSSCHALTER (*m*) : overload release (Elect.) 529
ÜBERTEMPERATUR (*f*), **LEITER-** : temperature rise in wires (Elect.) 530
ÜBERTRAGENEN SINN (*m*), **IM-** : figuratively 531
ÜBERTRAGERAMT (*n*) : repeater station 532
ÜBERTRAGERAUSGANG (*m*) : transformer output (Radio) 533
ÜBERTRAGERTRANSFORMATOR (*m*) : transformer (Tel.) 534
ÜBERTRAGUNGSFÄHIGKEIT (*f*) : carrying capacity (of telephone cables) 535
ÜBERTRAGUNGSRELAIS (*n*) : (telephone)-relay 536
ÜBERTRAGUNGSSTATION (*f*) : repeater station 537
ÜBERWACHSTUM (*n*) : hypertrophy (excessive growth) (Med.) 538
UHR (*f*), **HAUPT-** : master clock 539
UHR (*f*), **NEBEN-** : secondary clock 540
UHRZEIGERSINNE (*m*), **IM-** : clockwise ; in a clockwise direction ; dextrorotary (rotating to the right) 541
UMBRUCH (*m*) : (type-) setting (Printing) 542
UMFLOREN : to (wind round with ; bind with) cover with gauze (of electric wires) 543
UMGEKEHRT EIFÖRMIG : inversely egg-shaped ; obovate (Bot.) 544
UMGEKEHRT HERZFÖRMIG : inversely heart-shaped (Bot.) 545
UMLAUFSTOLLEN (*m*) : by-pass tunnel 546
UMLEITEN : to deviate ; divert 547
UMMANTELN : to coat ; cover ; jacket ; case ; enclose 548
UMPOLEN : to change polarity ; (*n*) : change of polarity (Elect.) 549
UMRANDUNGSFEUER (*n*) : boundary light (of aerodromes) 550
UMRANGIERUNG (*f*) : rearrangement 551
UMRÜHRUNG (*f*) : agitation ; stirring ; mixing (paper pulp) 552
UMSCHALTKASTEN (*m*) : terminal box (Elect.) 553
UMSCHALTRELAIS (*n*) : change-over relay (Elect.) 554
UMSCHALTWALZE (*f*) : reversing switch drum (Elect.) 555
UMSETZER (*m*) : converter 556
UMSPANNER (*m*) : transformer (Elect.) 557
UMSPANNWERK (*n*) : transformer station (Elect.) 558
UMSTEUERUNGSWELLE (*f*) : reversing shaft 559
UMSTEUERWELLE (*f*) : (see **UMSTEUERUNGSWELLE**) 560
UMWICKLUNG (*f*) : (see **UMWICKELUNG**) 561
UNBEHINDERT : unrestricted ; uninterrupted 562
UNBERECHENBAR : incalculable ; imponderable 563
UNBESCHADET : without prejudice to ; save-(for) except-(for) 564
UNBEWEGLICHKEIT (*f*) : immobility ; immotility (Bot.) (incapable of motion) 565
UNBEWEHRT : unprotected ; not armoured (of cables) 566
UNDECIME (*f*) : eleventh (Music) (see under **TERZ**) 567
UNENDLICHE (*n*) : infinity 568
UNFALLGEFAHR (*f*) : risk of accident 569
UNGEBÜHR (*f*) **VOR GERICHT** : contempt of court (Law) 570
UNIPOLAR : single pole ; unipolar 571
UNMERKBAR : not (visible) noticeable ; unnoticeable ; imperceptible 572
UNPASSIERBAR : impassable 573
UNPUPINISIERT : not loaded (see **PUPINISIEREN**) 574
UNREGELMÄSSIGKEIT (*f*) : irregularity 575
UNSELBSTSTÄNDIG : heteronomous ; not autonomous ; divergent (from a type) ; dependent 576
UNSELBSTSTÄNDIGKEIT (*f*) : heteronomy 577
UNTERBEWUSSTSEIN (*n*) : subconscious-(ness) 578
UNTERDOMINANTE (*f*) : subdominant (Music) 579
UNTEREINANDER VERBUNDEN : interconnected 580
UNTERHAUTSZELLGEWEBEENTZÜNDUNG (*f*) : phlegmon (inflammation of the cellular tissue) (Med.) 581
UNTERKARBON (*n*) : Lower Carboniferous (Geol.) 582
UNTERKIEFERARTERIE (*f*) : facial artery (Med.) 583
UNTERKOHLE (*f*) : lower (negative) carbon (Elect.) 584
UNTERKÖRPER (*m*) : lower limb ; lower part of the body (Med.) 585
UNTERLAGERUNGSTELEGRAPHIE (*f*) : infra-acoustic telegraphy ; sub-audio telegraphy 586
UNTERLAGERUNGSTELEGRAPHIEGESTELL (*n*) : infra-acoustic telegraphy rack 587
UNTERMEDIANTE (*f*) : submediant (Music) 588
UNTERPUTZ : sunk ; concealed ; flush type (of electrical switches) 589
UNTERPUTZANLAGE (*f*) : flush installation (Elect.) 590
UNTERPUTZSCHALTER (*m*) : flush mounted switch (Elect.) 591
UNTERSCHRÄMMUNG (*f*) : undercutting (Mining) 592
UNTERSTATION (*f*) : (electricity-) sub-station 593
UNTERWASSER (*n*) : tail race (Hydro-electric) 594
UNTERWASSERGRABEN (*m*) : tail race (Hydro-electric) 595
UNTERWERK (*n*) : sub-station (Elect.) 596
UNTUNLICH : impracticable ; impossible 597
UNVERLIERBARE SCHRAUBE (*f*) : captive screw (Elect.) 598
UNVERWECHSELBAR : non-interchangeable 599
UNVERZERRT : undistorted ; clear 600
UNVERZÖGERT : instantaneous ; immediate ; without (lag) delay 601
URPFLANZEN (*f.pl*) : protophyta (Bot.) 602
UVIOLLICHT (*n*) : ultra-violet light 603

V

VAGABUNDIERENDE ERDSTRÖME (*m.pl*) : stray earth currents (Elect.) 604
VAKUUMSPEKTROGRAPH (*m*) : (see RÖNTGENSPEKTROGRAPH) 605
VALENZELEKTRON (*n*) : valency electron 606
VELLITÄT (*f*) : weak will ; disposition ; tendency ; inclination ; propensity 607
VENENENTZÜNDUNG (*f*) : phlebitis ; inflammation of the veins (Med.) 608
VENTILIERT GEKAPSELT : enclosed ventilated (of electric motors) 609
VENTILSPIELLEHRE (*f*) : valve (play) clearance gauge 610
VENTURI-EINSATZ (*m*) : venturi throat (Meters) 611
VERÄNDERBAR : variable ; adjustable 612
VERANLAGUNG (*f*) : disposition (Med.) 613
VERÄSTELN : to ramify ; branch 614
VERBALSUBSTANTIV (*n*) : verbal noun (Grammar) 615
VERBINDUNGSLEITUNG (*f*) : connecting main ; cord circuit (Telephones) 616
VERBINDUNGSSTRECKE (*f*) : stenton (Mining) 617
VERBLÖDUNG (*f*), FORTSCHREITENDE- : (see SCHIZOPHRENIE) 618
VERBUNDWASSERMESSER (*m*) : compound water meter 619
VERBUNDWICKLUNG (*f*) : compound winding (Elect.) 620
VERCHROMEN : to chromium-plate 621
VERCHROMT, HOCHGLANZ- : highly polished chromium plated 622
VERCHROMUNG (*f*) : chromium-plating 623
VERCHROMUNGSANLAGE (*f*) : chromium-plating plant 624
VERDAUUNGSKANAL (*m*) : alimentary canal ; alimentary (digestive) tract (Med.) 625
VERDRAHTUNG (*f*) : wiring 626
VERDUNKLER (*m*) : dimmer ; dimming resistance (Elect.) 627
VERDUNKLUNG (*f*) : black-out (of Stage lighting) 628
VERDUNKLUNG (*f*), ELEKTRISCHE- : electric (dimmers) dimming 629
VERDUNKLUNGSWIDERSTAND (*m*) : dimming resistance (Elect.) 630
VEREDLUNG (*f*) : enrichment ; improvement ; finish-(ing) 631
VEREDLUNGSMASCHINE (*f*) : finishing machine (Textile) 632
VERFLÜSSIGER (*m*) : condenser 633
VERFOLGUNGSSCHEINWERFER (*m*) : traversing projector (Stage lighting) 634
VERGASERMOTOR (*m*) : petrol engine 635
VERGEILEN : (see ETIOLIEREN) 636
VERGLEICHSWIDERSTAND (*m*) : standard resistance (for purposes of comparison) (Elect.) 637
VERGÜTBAR : capable of being improved (of metal) 638
VERHÄLTNISANTEIL (*m*) : quota ; proportion 639
VERHOLSPILL (*n*) : warping capstan (Marine) 640
VERHOLWINDE (*f*) : warping (winch) windlass (Marine) 641
VERKEHRSAMPEL (*f*) : traffic light signal 642
VERKEHRSSTOCKUNG (*f*) : traffic congestion 643
VERKETTETE SPANNUNG (*f*) : line voltage ; voltage between star point and phase (Elect.) 644
VERKLEMMUNG (*f*) : jamming ; sticking ; seizing 645
VERKLINKUNG (*f*) : lock-(ing device) 646
VERKRUSTUNG (*f*) : incrustation ; scabbing (Med.) 647
VERKÜRZUNGSKONDENSATOR (*m*) : (aerial)-shortening condenser (Radio) 648
VERLÄNGERUNGSSPULE (*f*) : aerial lengthening coil 649
VERLEGEN : to lay (cable) ; run (wires) 650
VERLEGEN, ZUM . . . IN RÖHREN : for drawing in tubes (of electric wires) 651
VERLEGUNGSTIEFE (*f*) : depth at which laid (of cables) 652
VERLETZUNG (*f*) : wound ; trauma (Med.) 653
VERMASCHT : meshed ; interlinked (of electric networks) 654
VERMESSUNGSINSTRUMENT (*n*) : surveying instrument 655
VERMITTLUNGSAMT (*n*) : (telephone)-exchange 656
VERMITTLUNGSZENTRALE (*f*) : (telephone)-exchange 657
VERRENKUNG (*f*) : dislocation (Med.) 658
VERRÜCKTHEIT (*f*) : paranoia ; paranoea (mental derangement) (Med.) 659
VERSATZLEUCHTE (*f*) : portable lighting fitting (for stages) 660
VERSCHEIMUNG (*f*) : stopping ; isolation-(of part of a mine) 661
VERSCHIEBEDIENST (*m*) : shunting-(operations) (of Railways) 662
VERSCHIEBUNGSWINKEL (*m*) : angle of displacement 663
VERSCHROTTEN : to throw away as (old iron) waste ; throw on the scrap heap ; scrap 664
VERSEILEN : to twist-(together) (of electric flexible wires) ; (*n*) : rope laying 665
VERSEILMASCHINE (*f*) : rope laying machine (Textile) 666
VERSEILT, MIT BEILAUF RUND- : served (of electric cables) 667
VERSEUCHEN : to foul ; become infected ; infect ; be fouled ; sow-(with mines) (Mil.) 668
VERSPANNUNG (*f*) : stiffening ; staying-(device or arrangement) ; stays ; stay-bolts 669
VERSTÄRKERABSCHNITT (*m*) : repeater section (Tel.) 670
VERSTÄRKERAMT (*n*) : repeater (station) exchange (Tel.) 671
VERSTÄRKEREINRICHTUNG (*f*) : repeater station equipment (Tel.) 672
VERSTÄRKERSCHRANK (*m*) : amplifier cabinet (Wireless) 673
VERSTÄRKUNGSFAKTOR (*m*) : amplification factor (Radio) 674
VERSTEINERUNGSKUNDE (*f*) : Palaeontology 675
VERSTELLVORRICHTUNG (*f*) : focusing device (of lighting fittings) 676
VERTEILEREINRICHTUNG (*f*) : distributor arrangement (Tel.) 677
VERTEILUNGSKASTEN (*m*) : distribution box (Elect.) 678
VERTEILUNGSSTROMKREIS (*m*) : distribution circuit (Elect.) 679
VERTEILUNGSTAFEL (*f*) : distribution board (Elect.) 680
VERTONER (*m*) : composer (of music) 681
VERTONUNG (*f*) : (musical)-setting ; composition 682
VERWACHSUNG (*f*) : adhesion (Med.) 683
VERWANDTE TONARTEN IN DUR : related major scales (Music) 684
VERWANDTE TONARTEN IN MOLL : related minor scales (Music) 685

VERWECHSLUNG (*f*), **ENHARMONISCHE-** : enharmonic change (Music) 686
VERWECHSLUNGSZEICHEN (*n*) : accidental (Music) 687
VERWURF (*m*) : fault (Mining) (see **VERWERFUNG**) 688
VERZERRUNG (*f*) : distortion 689
VERZÖGERTE AUSLÖSUNG (*f*) : tripping lag ; delayed tripping action (Elect.) 690
VERZÖGERUNGSWIDERSTAND (*m*) : impedance ; apparent resistance 691
VIBRATIONSMASSAGE (*f*) : vibro-massage ; vibratory massage 692
VIBRATOR (*m*) : vibrator (Tel.) 693
VIELADRIG : multicore (of cables) 694
VIELFACHWATTMETER (*n*) : integrating wattmeter (Elect.) 695
VIERDRAHTSCHALTUNG (*f*) : four-wire system (Tel.) 696
VIERDRAHTSPRECHKREIS (*m*) : four-wire speaking circuit (Tel.) 697
VIERELEKTRODENRÖHRE (*f*) : tetrode ; four electrode valve (Radio) 698
VIERER (*m*) : quad (of cables) 699
VIERERKREIS (*m*) : phantom circuit (Tel.) 700
VIERERPUPINISIERUNG (*f*) : phantom loading (Tel.) 701
VIERERSCHALTUNG (*f*) : phantom (connection) circuit (Tel.) 702
VIERERSEIL (*n*) : quad (of telephone cables) 703
VIERERSPULE (*f*) : phantom coil (Tel.) 704
VIERERVERSEILUNG (*f*) : quad (Tel.) 705
VIERERZAHL (*f*) : number of quads (Tel.) 706
VIERFARBENVERSATZ (*n*) : four-colour dip (Stage lighting) 707
VIERLEITERKABEL (*n*) : four-core cable 708
VIERSÄTZIG : having (in) four movements (Music) 709
VIER-TAGE-FIEBER (*n*) : quartan ague (Med.) 710
VIERTELPAUSE (*f*) : crotchet rest (Music) 711
VIERVIERTELTAKT (*m*) : common time ; $\frac{4}{4}$ time (Music) 712
VISIERSTAB (*m*) : dip-stick (for measuring the contents of casks) 713
VISKOSEFADEN (*m*) : viscose thread (Textile) 714
VISKOSEKESSEL (*m*) : viscose tank (Textile) 715
VISKOSELEITUNG (*f*) : viscose pipe (Textile) 716
VÖLKERKUNDE (*f*) : ethnography ; ethnology 717
VÖLKERWIRTSCHAFTSLEHRE (*f*) : national economy 718
VOLLNETZANSCHLUSSBETRIEB (*m*) : all-mains operation (Radio) 719
VOLTREGLER (*m*) : potentiometer (Elect.) 720
VOLUMENWASSERMESSER (*m*) : volumetric water meter 721
VORAUSNAHME (*f*) : anticipation (Music) 722
VORAUSZAHLUNG (*f*) : prepayment 723
VORBÜHNENBELEUCHTUNG (*f*) : front of stage lighting 724
VORDERANSCHLUSS (*m*) : front-(of board)-mounting or connection (Elect.) 725
VORDROSSEL (*f*) : series choke (choking) coil (Elect.) 726
VORGARNMASCHINE (*f*) : roving frame (Textile) 727
VORGARNSCHEIBE (*f*) : roving bobbin (Textile) 728
VORGELEGE (*n*), **DOPPELTES-** : double reduction gearing 729
VORGELEGEMOTOR (*m*) : geared motor 730
VORHALT (*m*) **NACH OBEN** : suspension resolving upwards (Music) 731
VORHALT (*m*) **NACH UNTEN** : suspension resolving downwards (Music) 732
VORKAMMER (*f*), **LINKE-** : left auricle (Med.) 733
VORKAMMER (*f*), **RECHTE-** : right auricle (Med.) 734
VORKARDE (*f*) : breaker card (Textile) 735
VORKEIM (*m*) : prothallium (Bot.) 736
VORMACHTSTELLUNG (*f*) : pre-eminence ; precedence ; leading position 737
VORMAGNETISIERUNG (*f*) : primary (magnetizing) magnetization (Elect.) 738
VORÖFFNER (*m*) : porcupine opener (Textile) 739
VORPOLIERUNG (*f*) : preliminary polish-(ing) 740
VORSCHALTGERÄT (*n*) : apparatus (device) for series connection (Elect.) 741
VORSCHLAG (*m*), **KURZER-** : acciaccatura (Music) 742
VORSCHLAG (*m*), **LANGER-** : appogiatura (Music) 743
VORSPANNUNG (*f*) : bias (Radio) ; grid (plus or minus) voltage or potential 744
VORSPINNMASCHINE (*f*) : roving frame (Textile) 745
VORSTEHHUND (*m*) : setter-(dog) 746
VORSTRASSE (*f*) : roughing mill (Rolling mills) 747
VORVERHANDLUNG (*f*) : preliminaries ; preliminary talk 748
VORWÄHLER (*m*) : pre-selector (Tel.) 749
VORWEGNAHME (*f*) : prolepsis ; anticipation 750
VORWIDERSTAND (*m*) : series resistance (Elect.) 751
VOUTE (*f*) : arch ; vault ; cornice 752
VOUTENBELEUCHTUNG (*f*) : cornice lighting 753
VOUTENBELEUCHTUNGSKÖRPER (*m*) : cornice lighting unit 754
VOUTENFORM (*f*) : shape of cornice 755

W

WACHSPFLAUME (*f*) : Mirabelle-(Plum) (Bot.) 756
WACKELKONTAKT (*m*) : loose contact (Elect.) 757
WADENBEINARTERIE (*f*) : posterior tibial artery (Med.) 758
WAHLANRUF (*m*) : selective call (Tel.) 759
WÄHLEREINRICHTUNG (*f*) : selector arrangement (Tel.) 760
WÄHLERSTÖPSEL (*m*) : selector plug (Tel.) 761
WAHLFÄHIGKEIT (*f*) : selectivity (Elect.) 762
WAHLSCHALTER (*m*) : selector switch (Elect.) 763
WAHLWEISE : alternatively 764
WALKEN (*n*) : petrissage (of massage) 765
WALKMASSAGE (*f*) : petrissage 766
WALZENGLÄTTWERK (*n*) : smoothing rolls (Paper) 767
WALZENKREMPEL (*f*) : roller card-(er) (Textile) 768
WALZENSCHALTER (*m*) : drum-type switch (Elect.) 769
WALZENTUCH (*n*) : roller cloth (Textile) 770
WALZENWRINGER (*m*) : roller (wringer) wringing machine 771
WALZENZÄHLWERK (*n*) : cyclometer (registering device) counter (of meters) 772
WANDBEKLEIDUNGEN (*f.pl*) : hangings (Textile) 773
WANDDURCHFÜHRUNG (*f*) : lead-in (Radio) 774
WANDERNIERE (*f*) : floating kidney (Med.) 775
WANDERROSE (*f*) : erysipelas (Med.) 776
WANDERUNGSZAHL (*f*) : migration (transport) number (Electro-chemistry) 777
WANDLADUNG (*f*) : anode (potential) charge 778
WANDLER (*m*) : transformer (Elect.) 779

WANDSCHELLE (*f*) : wall saddle (Elect.) 780
WANDVERKLEIDUNG (*f*) : panelling-(for walls) 781
WANGENBRAND (*m*) : noma ; cancrum oris ; water canker (ulcer of the mouth) (Med.) 782
WANGENHÖCKER (*m.pl*) : cheek-bones (Med.) 783
WARM AUFZIEHEN : to shrink on (Engineering) 784
WÄRMEDURCHLASSEND : diathermanous ; permeable to heat 785
WÄRMEFEST : heat-resisting 786
WARMSÄGE (*f*) : hot saw (Rolling mills) 787
WARMSCHERE (*f*) : hot shear (Rolling mills) 788
WARMWERDEN (*n*) : heating-(up) 789
WARNSIGNAL (*n*) : warning-(signal) 790
WASSER (*n*), AKTIVIERTES- : treated water ; water treated by Katadyn process of water sterilization 791
WASSERANSCHLUSS (*m*) : water connection 792
WASSERFLUGZEUG (*n*) : hydroplane 793
WASSERSCHLOSS (*n*) : penstock 794
WASSERSEIGE (*f*) : water-ring-(shaft) (Mining) 795
WASSERSTRAHLLUFTSAUGER (*m*) : ejector (Engineering) 796
WASSERWIDERSTAND (*m*) : water resistance (Elect.) 797
WATTABGABE (*f*) : output (Elect.) 798
WATTAUFNAHME (*f*) : input (Elect.) 799
WATTKOMPONENTE (*f*) : Watt component ; effective component 800
WATTLOSE KOMPONENTE (*f*) : Wattless component of current (Elect.) 801
WATTSEIDE (*f*) : floss (Textile) 802
WATTSTROM (*m*) : Watt component ; effective component of current (Elect.) 803
WATTVERBRAUCH (*m*) : consumption in watts (Elect.) 804
WEBBLATT (*n*) : loom reed (Textile) 805
WEBSTUHLANTRIEB (*m*) : loom drive (Textile) 806
WEBSTUHLGESTELL (*n*) : loom frame (Textile) 807
WEBSTUHLWIRKUNGSGRAD (*m*) : loom efficiency 808
WECHSELPOLIG : heteropolar 809
WECHSELSTROMMASCHINE (*f*) : alternator (Elect.) 810
WECHSELSTROMRÖHRE (*f*) : alternating current-(heated)-valve 811
WECHSELZAHL (*f*) : frequency (Elect.) 812
WEGERECHT (*n*) : way-leave 813
WEGMESSER (*m*) : hodometer (for measuring the distance of travel of a vehicle) 814
WEHNELTKATHODE (*f*) : oxide coated cathode 815
WEICHEISENINSTRUMENT (*n*) : soft iron meter (Elect.) 816
WEICHEISEN-STROMZEIGER (*m*) : moving iron ammeter (Elect.) 817
WEINKUNDE (*f*) : (see ÖNOLOGIE) 818
WEIN (*m*), WILDER- : Virginia creeper (*Ampelopsis quinquefolia*) (Bot.) 819
WEISSARMORIERT : white variegated (*argenteo-variegatis*) (Bot.) 820
WEISSBLÜTIG : white-flowering ; having white bloom (*albiflorus* : *albiflos*) (Bot.) 821
WEISSGEFÄRBT : white-coloured (*argenteo-pictis*) (Bot.) 822
WEISSGEFLECKT : white-spotted ; variegated white (*argenteo-variegatis*) (Bot.) 823
WEISSGEMALT : (see WEISSGEFÄRBT) 824
WEISSGERANDET : white-rimmed (*argenteo-marginatis*) (Bot.) 825
WELLENBEREICH (*m*) : wave band (Radio) 826
WELLENBUND (*m*) : shoulder (of a shaft) 827
WELLENENTDECKER (*m*) : detector (Radio) 828
WELLENFALLE (*f*) : wave trap (Radio) : filter circuit 829
WELLENSTUMPF (*m*) : shaft (spindle) extension ; shaft end 830
WELLENWIDERSTAND (*m*) : inductance (Elect.) ; impedance (Tel.) 831
WELLENZAHL (*f*) : frequency (see FREQUENZ) 832
WEMFALL (*m*) : dative case (Grammar) 833
WENDEGETRIEBE (*n*) : reversing gear (Mech.) 834
WENDEPOL (*m*) : interpole (Elect.) 835
WENFALL (*m*) : accusative case (Grammar) 836
WERKSTATTGERECHT : in line with workshop (practice) requirements 837
WERKZEUGTASCHE (*f*) : kit-bag ; tool bag 838
WERTDAUER (*f*) : value-(of a note) (Music) 839
WESFALL (*m*) : genitive (possessive) case (Grammar) 840
WETTERDISTEL (*f*) : Carline thistle (*Carlina acaulis*) (Bot.) 841
WETTERLUTTE (*f*) : air tube (Mining) 842
WETTERSCHEIDER (*m*) : brattice (Mining) 843
WETTERSTRECKE (*f*) : airway (Mining) 844
WETTERTÜR (*f*) : ventilation door (Mining) 845
WETTERZUG (*m*) : ventilator (Mining) 846
WICKLUNG (*f*) : (see WICKELUNG) 847
WIDERSTAND (*m*), ABGESTUFTER- : stepped resistance (Elect.) 848
WIDERSTAND (*m*), MAGNETISCHER- : reactance (Elect.) ; reluctance (Elect.) 849
WIDERSTANDSÄULE (*f*) : resistance column (Elect.) 850
WIDERSTANDSCHWEISSVERFAHREN (*n*) : electric resistance welding 851
WIDERSTANDSETALON (*n*) : standard resistance ; resistance standard (Elect.) 852
WIDERSTANDSFÄHIG, THERMISCH- : refractive 853
WIDERSTANDSKOEFFIZIENT (*m*) : co-efficient of resistance ; specific resistance (Elect.) 854
WIDERSTANDSLAUFZEIT (*f*) : time lag 855
WIDERSTANDSMESSWERK (*n*) : resistance measuring device 856
WIDERSTAND (*m*), SPEZIFISCHER- : resistivity ; specific resistance (Elect.) 857
WIDERSTANDSPIRALE (*f*) : resistance (choke ; choking) coil (Elect.) 858
WIDERSTANDSTUFE (*f*) : resistance (stage) step (Elect.) 859
WIDERSTAND (*m*), THERMISCHER- : thermal resistance 860
WIEDERAUFBAU (*m*) : re-assembly (see AUFBAUEN) ; rebuilding ; reconstruction 861
WIEDERAUFLADUNG (*f*) : re-charging (of accumulators) 862
WIEDERAUFSAUGEN : to resorb ; re-absorb 863
WIEDERAUFSAUGUNG (*f*) : resorption 864
WIEDEREINBAU (*m*) : replacement (see EINBAUEN) 865
WIEDEREINSTELLUNG (*f*) : to re-start ; set going again ; re-activate 866
WIEDERGEWINNBAR : recoverable 867
WIEDEROXYDATION (*f*) : re-oxydation 868
WIMPERIG GESÄGT : with bristly serratures (Bot.) 869
WIMPERIG GEZÄHNT : with bristly (serrations) serratures (Bot.) 870
WINDANZEIGER (*m*) : wind indicator 871
WINDER (*m*) : faller (Textile) 872
WINDFEGE (*f*) : winnowing machine 873
WINKELMESSINSTRUMENT (*n*) : theodolite 874
WINKELSCHNELLE (*f*) : angular velocity 875
WIPPTISCH (*m*) : lifting table (Rolling mills) 876
WIRBELBORSTE (*f*) : basil (*Satureia clinopodium*) (Bot.) 877
WIRBELDOSTEN (*m*) : basil (*Satureia clinopodium*) (Bot.) 878
WIRBELFÖRMIG : verticillate ; in a whorl (Bot.) 879

WIRBELGELENK (*n*) : vertebral (ligament) joint 880
WIRBELHAFT : whirling ; eddying ; rotatory 881
WIRBELKRAUT (*n*) : milk vetch (Bot.) 882
WIRBELLOSES TIER (*n*) : invertebrate ; (*pl*) : invertebrata 883
WIRBELSTROMBREMSE (*f*) : eddy current brake (Elect.) 884
WIRBELSTROMLÄUFER (*m*) : eddy current rotor (Elect.) 885
WIRBELSTROMVERLUST (*m*) : eddy current loss 886
WIRBELSUCHT (*f*) : staggers (of animals) 887
WIRBELUNG (*f*) : eddying ; whirling ; rotating 888
WIRKANTRIEB (*m*) : loom drive (Textile) 889
WIRKKOMPONENTE (*f*) : energy component (Elect.) ; actual component 890
WIRKLEISTUNG (*f*) : effective output 891
WIRKLEISTUNGSMESSER (*m*) : wattmeter (Elect.) 892
WIRKLEITWERT (*m*) : conductance (Elect.) 893
WIRKMASCHINE (*f*) : weaving machine ; knitting frame ; full-fashioned knitting frame (Textile) 894
WIRKMASCHINENANTRIEB (*m*) : knitting-frame drive (Textile) 895
WIRKSTROM (*m*) : power component of current (Elect.) 896
WIRKWAREN (*f.pl*) : woven goods (Textile) 897
WIRKWIDERSTAND (*m*) : effective resistance (Elect.) 898
WISCHERRELAIS (*n*) : leakage relay (Elect.) (see also **WISCHER**) 899
WOLKENAPPARAT (*m*) : cloud (machine) apparatus (Stage lighting) 900
WOLKENBILD (*n*) : cloud effect (Stage lighting) 901
WOLLBAUM (*m*) : Kapok tree (*Bombax ceiba* ; *Eriodendron anfractuosum*) (Bot.) 902
WOLLBLUME (*f*) : common kidney vetch ; Lady's finger (*Anthyllis vulneraria*) (Bot.) 903
WOLLDISTEL (*f*) : woolly thistle ; spear thistle (*Cirsium lanceolatum*) (Bot.) 904
WOLLERZEUGER (*m*) : wool producer 905
WOLLFARBEN : dyed in the wool 906
WOLLGRAS (*n*) : cotton grass (*Eriophorum vaginatum*) (Bot.) 907
WOLLKAMM (*m*) : card-(er) ; wool comb (Textile) 908
WOLLKÄMMEN (*n*) : wool combing 909
WOLLKRATZE (*f*) : (see **WOLLKREMPEL**) 910
WOLLKRATZMASCHINE (*f*) : (see **WOLLKREMPEL**) 911
WOLLKREMPEL (*m*) : (wool)-card ; carder ; carding machine 912
WOLLKREMPELMASCHINE (*f*) : (see **WOLLKREMPEL**) 913
WOLLMESSER (*m*) : eriometer ; erinometer (for measuring diameter of wool fibres) 914
WOLLPELZ (*m*) : sheep's skin 915
WOLLPRODUZENT (*m*) : wool producer 916
WOLLREISSER (*m*) : wool card-(er) 917
WOLLSCHERE (*f*) : wool shears 918
WOLLSEIDE (*f*) : wool-silk (Textile) 919
WOLLSORTIERER (*m*) : wool sorter 920
WOLLSPINNEREI (*f*) : wool spinning-(mill) (Textile) 921
WOLLSTREICHER (*m*) : wool comber ; carder 922
WOLLTRAGEND : bearing wool ; laniferous 923
WOLLWÄSCHEREIMASCHINE (*f*) : wool scouring machine (Textile) 924
WRINGER (*m*) : wringer ; wringing machine 925
WUCHERBLUME (*f*) : fever-few (*Pyrethrum*) (Bot.) 926
WUNDINFEKTION (*f*) : septicaemia (Med.) 927
WUNDMAL (*n*) : stigma (Med.) 928
WUNDSEKRET (*n*) : pus (Med.) 929
WUNDSTARRKRAMPF (*m*) : tetanus (Med.) 930
WURMFRÄSMASCHINE (*f*) : hobbing machine 931
WURZELBAZILLUS (*m*) : root bacillus (*Bacillus ramosus*) 932

X

XANTHOZYT (*m*) : xanthocyte ; erythrocyte (see **ERYTHROZYT**) 933
XEROPHYT (*m*) : xerophyte (a plant which flourishes in dry places) 934
XYLOLOGIE (*f*) : the science and study of wood 935
XYLOMETER (*n*) : an instrument for measuring the cubic contents of wood by displacement of water 936

Z

ZÄHLEREICHANLAGE (*f*) : meter-calibrating equipment 937
ZÄHLERPRÜFEINRICHTUNG (*f*) : meter testing (equipment) set 938
ZÄHLERTAFEL (*f*) : meter board (Elect.) 939
ZÄHLKETTE (*f*) : counting chain (Textile) 940
ZAHLUNGSFÄHIGKEIT (*f*) : solvency 941
ZAHNFLEISCHWUCHERUNG (*f*) : epulis ; a tumour of the gums (Med.) 942
ZAHNPROTHESE (*f*) : false (artificial) tooth or teeth ; denture 943
ZAHNRADANTRIEB (*m*) : gear drive (Mech.) 944
ZAHNTECHNIKER (*m*) : dentist 945
ZAUM (*m*), **PRONYSCHER-** : prony brake 946
ZEHENBEIN (*n*) : phalange ; toe-bone (Med.) 947
ZEIGERSTROM (*m*) : clockwise (current) circuit 948
ZEIGERZÄHLWERK (*n*) : multi-pointer dial 949
ZEITABSTAND (*m*) : period ; interval (of time) 950
ZEITABSTÄNDEN (*m.pl*), **IN ANGEMESSENEN-** : periodically ; at intervals 951
ZEITEINSTELLUNG (*f*) : timing ; time setting ; timing (mechanism) gear 952
ZEIT (*f*), **LEICHTE-** : up-beat ; weak beat ; unaccented period (Music) 953
ZEITLUPENAUFNAHME (*f*) : slow-motion picture (Cinematography) 954
ZEITRAFFERAUFNAHME (*f*) : rapid-motion picture (Cinematography) 955
ZEITRELAIS (*n*) : time relay (Elect.) 956

ZEIT (*f*), **SCHWERE-** : down beat ; strong beat ; accented period (Music) 957
ZEITWERK (*n*) : works ; clockwork ; timing (device) mechanism 958
ZEITZEICHEN (*n*) : time signal (Radio) 959
ZELLENKONTAKT (*m*) : cell switch contact (Elect.) 960
ZELLULOSETRIACETAT (*n*) : cellulose triacetate ; $C_6H_7O_5(CH_3CO)_3 + 3CH_3COOH$ 961
ZENTRALBATTERIEAMT (*n*) : central battery exchange (Tel.) 962
ZENTRIERBANK (*f*) : self-centring lathe 963
ZERFASERER (*m*) : disintegrator (Textile) 964
ZERLEGUNGSPRODUKT (*n*) : product of (dissociation) decomposition 965
ZERSPRATZEN : to decrepitate (of crystal) 966
ZERSTÄUBER (*m*) : (see **ZERSTÄUBUNGSAPPARAT**) 967
ZERSTREUTBLÄTTRIG : alternate-(leaved) (*alternifolius*) (Bot.) 968
ZERSTREUUNGSGITTER (*n*) : scatter grid (of X-rays) 969
ZETTELMASCHINE (*f*) : warping machine (Textile) 970
ZEUGDRUCKMASCHINE (*f*) : printing machine (Textile) 971
ZEUGUNGSVERLUST (*m*) : apogamy ; absence of sexual reproductive power (Med.) 972
ZIEGELSTEINSTÄRKE (*f*) : brick thickness 973
ZIMMERANTENNE (*f*) : room (indoor) aerial (Radio) 974
ZIMMEREI (*f*) : carpenter's shop 975
ZIMMERLINDE (*f*) : African hemp (*Sparmannia africana*) (Bot.) 976
ZINNKRAUT (*n*) : (see **ACKERSCHACHTELHALM**) 977
ZIRBE (*f*) : (see **ZIRBELFICHTE**) 978
ZISCHFUNKENSTRECKE (*f*) : quenched spark-gap (Elect.) 979
ZISELIEREN : to chisel ; engrave ; chase 980
ZOOPATHOLOGIE (*f*) : zoopathology (the science of animal diseases) 981
ZOOPHYT (*m*) : zoophyte (see **PFLANZENTIER**) 982
ZOOPHYTISCH : zoophytic 983
ZOTTENGESCHWULST (*f*) : papilloma (a tumour composed of papillae) (Med.) 984
ZUCKERSACK (*m*) : sugar-sack 985
ZUCKERSACKLEINEN (*n*) : hessian bagging (Textile) 986
ZUFUHR-ROLLGANG (*m*) : run-in table (Rolling mills) 987
ZUGBELEUCHTUNG (*f*) : train lighting 988
ZUGKRANKHEIT (*f*) : train sickness (Med.) 989
ZUGLEINE (*f*) : communication cord (in trains) 990
ZUGMESSGERÄT (*n*) : draught measuring (instrument) apparatus ; draught gauge 991
ZUGSCHALTER (*m*) : cord (pull) switch (Elect.) 992
ZUGSTANGENHEBEL (*m*) : draw-bar lever 993
ZULEITUNGSKABEL (*n*) : cable lead ; feeder (Elect.) 994
ZUNDERSCHICHT (*f*) : oxide scale (Welding) 995
ZUNDERSTELLE (*f*) : fused (places) spots-(in metal) 996
ZUNGENARTERIE (*f*) : lingual artery (Med.) 997
ZUNGEN-FREQUENZMESSER (*m*) : reed type of frequency meter (Elect.) 998
ZUSAMMENBALLUNG (*f*) : agglutination (Med.) 999
ZUSAMMENGESCHWEMMT : washed up ; washed together by floods 000
ZUSAMMENGESETZTHEIT (*f*) : composition 001
ZUSAMMENKLANG (*m*) : consonance ; harmony ; symphony ; sounding together 002
ZUSAMMENKLINGEN : to sound simultaneously ; harmonize ; synchronize 003
ZUSAMMENSCHNÜRUNG (*f*) : stricture (Med.) 004
ZUSATZMOTOR (*m*) : auxiliary motor 005
ZUSATZSPEISEWASSER (*n*) : make-up feed-(water) (of steam boilers) 006
ZUSCHAUERRAUM (*m*) : auditorium 007
ZUSCHLAGSTOFF (*m*) : aggregate material ; element-(of a compound) ; ingredient 008
ZWEIBANDTELEPHONIE (*f*) : two-channel telephony 009
ZWEIBLÄTTRIG : two-leaved (*bifolius* ; *bifoliatus*) (Bot.) 010
ZWEIBLÜTIG : two-flowered ; having two flowers (*biflorus*) (Bot.) 011
ZWEIDRÄHTIG : bifilar ; having two wires 012
ZWEIDRAHTSCHALTUNG (*f*) : two-wire system (Tel.) 013
ZWEIDRAHTSPRECHKREIS (*m*) : two-wire speaking circuit (Tel.) 014
ZWEIELEKTRODENRÖHRE (*f*) : diode (Radio) 015
ZWEIFÄDIG : bifilar ; having two threads 016
ZWEIGESTRICHENE OKTAVE (*f*) : two-lined octave (Music) 017
ZWEIG-FADENBAKTERIUM (*n*) : brown mould (*Cladothrix dichotoma* ; *Sphaerotilus dichotomus*) 018
ZWEIHÜFTIGER KOPF (*m*) : double-stepped barrel (of capstans) 019
ZWEILEITERKABEL (*n*) : two-core cable ; twin cable 020
ZWEINERVIG : having two (veins) nerves ; two (nerved) veined (*binervatus* ; *binervis*) (Bot.) 021
ZWEIRÖHRENVERSTÄRKER (*m*) : two-valve amplifier (Radio) 022
ZWERGPALME (*f*) : dwarf palm ; fan palm ; African hair palm (*Chamaerops humilis*) (Bot.) 023
ZWILLINGSLEITER (*m*) : twin conductor (Elect.) 024
ZWIRNEREI (*f*) : doubling (Textile) 025
ZWIRNMASCHINENANTRIEB (*m*) : doubling frame drive (Textile) 026
ZWIRNSCHEIBE (*f*) : twist wheel (Textile) 027
ZWIRNUNG (*f*) : twist (Textile) 028
ZWISCHENDOSE (*f*) : straight-through box (Elect.) 029
ZWISCHENRING (*m*) : adapting ring 030
ZWISCHENSCHICHT (*f*) : intermediate (coating) layer 031
ZWISCHENSTAATLICH : interstate 032
ZWISCHENTITEL (*m*) : sub-title (Cinematography) 033
ZWISCHENWIDERSTAND (*m*) : auxiliary (intermediate) resistance (Elect.) 034
ZWISCHENWIRT (*m*) : host ; (disease)-carrier (Med.) 035
ZWITTERION (*n*) : zwitter ion ; electrically neutral ion (Electro-chemistry) 036
ZWÖLFFINGERDARMGESCHWÜR (*n*) : duodenal ulcer (Med.) 037
ZYANEISENKALIUM (*n*) : (see **BLUTLAUGENSALZ**) 038
ZYANKALIUM (*n*) : (see **CYANKALIUM**) 039
ZYANOSE (*f*) : cyanosis (Med.) 040
ZYANOTYPIE (*f*) : blue-printing process 041
ZYMASE (*f*) : zymase (an enzyme) 042
ZYMOGEN : zymogenic 043
ZYSTITIS (*f*) : cystitis (Med.) 044
ZYTISUS (*m*) : cytisus (see **BOHNENBAUM**) 045
ZYTOLOGIE (*f*) : cytology (science of construction and life of cells) 046